Textbook of

MEDICAL MYCOLOGY

Textbook of MEDICAL MYCOLOGY

Dr. Jagdish Chander MD DNB MAMS
Professor & Head
Department of Microbiology
Government Medical College Hospital
Chandigarh, India

JAYPEE BROTHERS MEDICAL PUBLISHERS
The Health Sciences Publisher
New Delhi | London

Jaypee Brothers Medical Publishers (P) Ltd

Headquarters
EMCA House
23/23-B, Ansari Road, Daryaganj
New Delhi 110 002, India
Landline: +91-11-23272143, +91-11-23272703
+91-11-23282021, +91-11-23245672
E-mail: jaypee@jaypeebrothers.com

Overseas Office
JP Medical Ltd.
83, Victoria Street, London
SW1H 0HW (UK)
Phone: +44-20 3170 8910
E-mail: info@jpmedpub.com

Corporate Office
Jaypee Brothers Medical Publishers (P) Ltd.
4838/24, Ansari Road, Daryaganj
New Delhi 110 002, India
Phone: +91-11-43574357
Fax: +91-11-43574314
E-mail: jaypee@jaypeebrothers.com

EU GPSR Authorised Representative
Logos Europe, 9 rue Nicolas Poussin
17000, La Rochelle, France
Phone: +33 (0) 6 67 93 73 78
E-mail: Contact@logoseurope.eu

Website: www.jaypeebrothers.com
Website: www.jaypeedigital.com

Inquiries for bulk sales may be solicited at: jaypee@jaypeebrothers.com

Textbook of Medical Mycology

First Edition Published by: Interprint in July 1995

Second Edition Published by: Mehta Publishers in April 2002

Third Edition Published by: Mehta Publishers in October 2008

Fourth Edition: **2018**

Reprint: 2023, Reprint: October 2023, 2025, **2026**

ISBN: 978-93-86261-83-0

Printed at: Samrat Offset Pvt. Ltd.

Dedication

Dedicated to patients who died of fungal infections, as final diagnosis could not be timely ascertained during their lifetime and also to those who could not afford the cost of antifungal therapy thereby died despite establishment of fungal etiology while they were alive.

Foreword

This is great pleasure on my part, both personally as well as the President of Society for Indian Human and Animal Mycologists (SIHAM) to write Foreword to the fourth edition of the *Textbook of Medical Mycology*.

When my friend, Dr. Jagdish Chander, requested me to write a Foreword for the new edition, I pointed out that the Textbook is already a popular one among the medical students as well as the doctors and it does not require the lacing of a Foreword. But he insisted. Now, after going through the manuscript of the Textbook, I am happy that I am writing the Foreword for this excellent piece of work, which I am sure, will have a far better impact than before among the students of medicine and the faculty of Medical Microbiology.

The fungal infections are becoming increasingly important in the recent years for a number of reasons. Ample opportunities for world travel for various reasons have increased the risks of acquisition of exotic fungal infections among the general population. Introduction of modern life-saving therapies, such as antibiotics, immunosuppressive drugs, irradiation, along with the ominous AIDS pandemic, have contributed to the enormous increase in the incidence of opportunistic fungal infections among the susceptible hosts.

As a consequence, lots of new developments are taking place on different aspects of fungal infections. There are a number of novel sophisticated molecular techniques, which are being evolved for rapid and accurate diagnosis of various syndromes and identification of the fungal pathogens. Non-invasive imaging techniques are proving to be useful tools for an early diagnosis and assessment of the invasive fungal diseases. A number of effective topical and systemic antifungal agents are being added to the armamentarium of the clinicians. Therefore, to cope up with the changing scenario, medical professionals involved in the patient-care need updates on diagnostics and therapeutics on fungal diseases from time to time.

The author has done a commendable job of scanning the voluminous wealth of information and presenting it in a concise, lucid and palatable manner.

This fourth edition of the Textbook consists of Eight Sections, divided into 39 Chapters, which cover the entire spectrum of fungal infections, spanning from superficial to opportunistic infection along with the latest information on therapeutic modalities. The pseudofungal infections, mimicking typical fungal diseases, are also given in a separate Section. The Textbook encompasses interesting historical aspects pertaining to Medical Mycology and has valuable references for further reading by the students. It contains a very useful introductory chapter on the basics of medical mycology. The last Section has seven Appendices, pertaining to various procedural details mentioned in the text of the Book. One of them is exclusively devoted to the antifungal susceptibility testing.

This Textbook is ideally suited for medical students and junior doctors who are preparing themselves for the examinations and various entrance tests.

I strongly feel the present edition of this Textbook will also be appreciated by the students and teaching faculty in all the disciplines of medicine.

Prof. Niranjan Nayak
President
Society for Indian Human and Animal Mycologists
Professor, Department of Microbiology
Manipal College of Medical Sciences, Pokhara, Nepal
Formerly Professor, Department of Ocular Microbiology
All India Institute of Medical Sciences, New Delhi, India

Date: March 19, 2016
Place: Shimla (India)

Preface to the Fourth Edition

The primary objective of *Textbook of Medical Mycology* was to provide concise account of this subject to the students and health personnel having little or no background in this emerging field of medical sciences. The main emphasis was focused on epidemiological, clinical, immunological, pathological, diagnostic and therapeutic aspects of fungal diseases. This has been a Textbook of prime interest to microbiologists and allied specialists working in the field of infectious diseases. The first three editions of my concerted efforts were quite successful in achieving these goals during this period of last 22 years.

There are lot of developments during this period of eight and half years since publication of its third edition. There is tremendous increase in fungal infections in the recent past entailing explosion of published information on all the aspects of this upcoming field. This time it posed a challenging task in reality to keep pace with the latest developments, which has obligated me to revise the book as its fourth edition. This is extensively reorganized and updated edition. The entire text is backed by a up-to-date list of references for readers wanting further in-depth knowledge. At completion of every chapter, relevant literature is cited as Further Reading in the form of primary scientific papers and review articles published in biomedical journals. The data has been incorporated, based on literature available on fungal infections since the publication of its last edition in October 2008, commencing from January 2009 upto June 2017.

More number of colored clinical photographs including histopathological sections have been added to give a clearer picture of the lesions manifested in fungal diseases. At places, the line-drawings have also been made colored and some of the old photographs are as such retained. The non-invasive radiological techniques have advanced in localizing the underlying fungal infective process. Therefore, figures of relevant CT and MRI are inserted to delineate an in-depth understanding on the related issues.

Like other microbes, fungi also involve almost every organ of the body entailing invariably fatal outcome. Their importance has increased to a great extent as compared to the previous times when they used to be considered as merely contaminants. Therefore, keeping this fact in view, traditional fungal contaminants are being described in one of chapters on Mycology Laboratory Contaminants, which was removed in previous two editions.

The nomenclature pertaining to medical mycology is very versatile. Hence new names are being designated to some of the fungal diseases like Malasseziosis, Talaromycosis and Emergomycosis. For the convenience of discussion, Pseudofungal Infections have been kept in separate section. One Appendix is fully devoted to antifungal susceptibility testing (AFST) to meet out the difficulties faced by the clinicians to choose appropriate drug in the management. Due to the addition of newer and separate chapters on Microsporidiosis, Entomophthoramycosis and Mycology Laboratory Contaminants, the total number of chapters becomes thirty-nine and there are seven Appendices.

The medical mycology is based on conventional methods of identification as the 'gold standard'. However, the recently described molecular techniques have taken the lead in the diagnostic and taxonomical aspects. In this edition, more emphasis is although on conventional methods but at places, molecular techniques are referred briefly to highlight the issue. The clinicians will have a high index of suspicion about fungal infections after being acquainted to the book. The microbiologists as well as the pathologists will be able to substantiate their empirical approach in establishing the diagnosis by demonstrating either the causative fungus or its effects produced so that the specific therapeutic modalities are adopted well in time among the patients thereby their lives are saved from fatal fungal infections.

In nutshell, *Textbook of Medical Mycology* is now thoroughly revised and updated to include recently identified and emerging fungi and/or infections, innovative research and development of applications of new technology. However, there is always a scope of improvement hence I would like to welcome any constructive suggestions to make this textbook better in subsequent editions.

Date: June 30, 2017
Place: Chandigarh

Jagdish Chander

Preface to the First Edition

The incidence of fungal infections has increased largely. This may be because of the underlying predisposing factors like immunocompromised situations, such as the use of antimicrobial, immunosuppressive and anticancer agents, transplantations, endocrinal disorders, etc. Moreover, awareness about fungi among the medical personnel has increased, which has also contributed to an early diagnosis and prompt treatment of the life-threatening fungal infections. The expansion of medical mycology has been so rapid and extensive that it has now attained the status of an independent and a full-fledged specialty. In the coming years too, an increase in the fungal infections is anticipated. It makes the task of writing a Textbook on the subject quite onerous. The aim is to provide a concise account of medical mycology for students and personnel working in microbiology laboratory with a little or no background in medical mycology. Therefore, I have undertaken this endeavor, which is based on my experience of teaching medical undergraduates and postgraduates.

Already many books on this subject are available, which are excellent but they are mainly for reference. Therefore, a comprehensive book is required for the medical and paramedical students. It is observed that even the postgraduates rely on photocopies of the old notes in medical mycology. This Textbook attempts to present the essential information on fungi and their medical importance. The emphasis is essentially on the epidemiological, clinical, pathological, immunological, diagnostic and therapeutic aspects of fungal diseases. The Book is of prime interest to the medical microbiologists, especially medical mycologists and specialists working in the field of infectious diseases. It should prove useful for the microbiology students experiencing their first encounter with fungal agents of medical importance. It would also be of value to the clinicians concerned with these infections, particularly dermatologists.

All the chapters are organized in seven sections, as listed in the Table of Contents, including one section on Appendices. Their sequence is such that the readers will find it easier to trace the desired chapter. The application of mycology-related terms was essential in the text. Therefore, a Glossary of terms is given in one of the Appendices. The obsolete and rejected names of fungi/diseases have been excluded and only the latest and authentic ones are taken into consideration while writing the text.

Simple and self-explanatory illustrations, tables and flowcharts are inserted into the text. Each chapter, at its end, has a list of comprehensive up-to-date references as Further Reading. A list of reference books, which deals with fungal and related diseases, is given in Bibliography, as these books are not mentioned at the end of the chapters. The list of commonly used abbreviations is also given in the beginning of the Textbook.

The actinomycetes like *Nocardia*, *Streptomyces* and *Actinomadura* species are aerobic Gram-positive filamentous bacteria and not fungi but they cause manifestations, clinically resembling the fungal infections. Therefore, these are traditionally described along with the mycology text. In this Textbook, these organisms are described in Mycetoma and Allergic Diseases and not as full-fledged chapters. *Pneumocystis carinii*, which has been taxonomically a controversial organism and today knocking at the door of Medical Mycology, is described in a separate Chapter.

Basically, I have tried to focus on the fundamentals of fungi and diseases caused by them. It would be of interest to the broadest audience and I presume that my aim will be accomplished. I do not claim any perfection rather contrarily, I request my teachers, colleagues and students to write about its shortcomings and give their valuable suggestions as how to improve this Textbook in the subsequent editions.

Date: May 12, 1995
Place: Chandigarh

Jagdish Chander

Acknowledgments

I am grateful to Prof. Niranjan Nayak, formerly Professor, Department of Ocular Microbiology, AIIMS, New Delhi and President, Society for Indian Human and Animal Mycologists (SIHAM) and presently Professor, Department of Microbiology, Manipal College of Medical Sciences, Pokhara, Nepal for his constant support while composing this Textbook and for writing the Foreword of this edition.

I wish to express my gratitude to Dr Uma Handa, Professor & Head and Dr Rajpal Singh Punia, Professor, Department of Pathology, Government Medical College Hospital (GMCH), Chandigarh for their encouragement during compilation of present work and providing histopathological photomicrographs. I am thankful to Dr Rajiv Kumar, Professor & Head, Department of Pharmacology, Dr Deepak Aggarwal, Pulmonary Medicine and Dr Vaibhav Saini, ENT for going through some of the chapters. From our Department thanks are due to our faculty *i.e.* Dr Nidhi Singla, Dr Lipika Gautam and Neelam Gulati for critically checking the manuscript and providing photomicrographs. I am thankful to all resident doctors of our Department *i.e.* Drs Hena Rani, Shailpreet Sidhu, Kiran Bala, Mandeep Kaur, Ruby Jain, Neha Bansal, Neha Jain, Preety Thakur, Gursimran Mohi, Shivani Sharma, Prapti Bora, Arpandeep Tuli, Nidhi Tejan, Yashik Bansal, Pooja Singh, Dimpi Bhankhur, Ranu Soni, Vibha Mehta, Swati Sharma, Pooja Mishra, Alisha Bhagat, Anku Goel, Annamalai Anushya, Pallavi Dhawan, Manharpreet Kaur and Mani Bhushan Kumar for going through the manuscript at various stages. I am thankful to all our technical staff, particularly, Ms. Ruby Suria Dharia (Judy) and Mr. Sheetal Kumar for their sustained efforts and wholehearted support at every step of the preparation of this textbook.

I am also grateful to Prof. Kusum Joshi, former Professor & Head, Prof. BD Radotra, Professor and Dr Ritambhra Nada, Professor, Department of Histopathology, Postgraduate Institute of Medical Education and Research (PGIMER), Chandigarh, for providing the histopathological photomicrographs. I am thankful to Prof. Meera Sharma, former Head, Department of Medical Microbiology, PGIMER for her constant inspiration while preparing the manuscript. I am also thankful to Prof. Pushpa Talwar, former Head, Department of Medical Microbiology, PGIMER, with whom I had the privilege to work with in the field of medical mycology and whom we have lost recently. I would like to pay sincere tribute to her through this book. Thanks are due to Dr Tarun Narang, Assistant Professor, Department of Dermatology, PGIMER; Dr Gagandeep Singh, Assistant Professor, Department of Microbiology, AIIMS, New Delhi for critically reviewing some of the chapters.

A thorough reading of the manuscript by Dr Antarikshdeep, Professor, Department of Microbiology, Pandit Bhagwat Dayal Sharma Postgraduate Institute of Medical Sciences (PGIMS), Rohtak (Haryana) is duly acknowledged as it was valuable and timely and gave a positive shape to this endeavor. I am also thankful to Prof. SM Singh, former Head, Department of Biological Sciences, Rani Durgavati University, Jabalpur, Madhya Pradesh; Dr Mahendra Pal, former Head, Department of Veterinary Public Health, College of Veterinary and Animal Husbandry, Anand, Gujarat; Dr PS Nirwan, former Professor & Head, Dr Vijaylatha Rastogi, Associate Professor, Jawaharlal Nehru Medical College, Ajmer, Rajasthan; Dr Khuraijam Ranjana Devi, Professor, Regional Institute of Medical Sciences, Imphal (Manipur) for going through some of the chapters and providing clinical photographs and photomicrographs. I am thankful to Dr DP Lochan, Director General Health Services (DGHS), Haryana for his constant inspiration, support and encouragement.

I am indebted to Prof. K R Joshi, former Professor & Head, Department of Microbiology, Dr Sampurnanand Medical College, Jodhpur and Principal, Government Medical College, Kota, Rajasthan; Prof. HS Randhawa, Director and Professor of Medical Mycology, Vallabhbhai Patel Chest Institute (VPCI), New Delhi; Dr BM Hemashettar, former President, Society for Indian Human and Animal Mycologists; Prof. L N Mohapatra, former Professor & Head, Department of Microbiology, AIIMS, New Delhi for encouraging me to bring out the new edition. He had written the Foreword of the very first edition of this Textbook in May 1995. During this period we lost him also and this book is a sincere tribute to him as well.

I extend my sincere thanks to Dr Deepinder Chhina, Professor & Head, Dr Rama Gupta, Professor, Department of Microbiology, Dayanand Medical College Hospital, Ludhiana (Punjab); Dr Santwana Verma, Associate Professor,

Department of Microbiology, Indira Gandhi Medical College, Shimla (Himachal Pradesh); Dr. Shukla Das, Professor of Microbiology, University College of Medical Sciences, New Delhi; Dr Valinderjeet Singh Randhawa, Professor, Dr Charu Jain, Senior Resident, Lady Hardinge Medical College, New Delhi; Dr Geetanjali Sharma, Microbiologist, Central Food Laboratory (CFL-FSSAI), Kolkata; Dr K Umamaheswari, Dr ALM PGIBMS, University of Madras, Chennai for critically going through the manuscript and providing electron microscopy photographs of fungi.

I owe my gratefulness to Professor David W Denning, Professor of Medicine and Medical Mycology and Head, Regional Mycology Laboratory, Education and Research Centre, National Aspergillosis Centre, Wythenshawe Hospital, Manchester, UK for his encouraging support while writing this edition. My sincere thanks to Dr Elizabeth Johnson, Director, PHLS Mycology Reference Laboratory, Bristol, UK for encouragements in bringing out this book. Thanks to Dr Michael Petrou, Imperial College, London and Dr Vipul Rana Singh, Infectious Diseases specialist, Phoenix, Arizona (USA) for going through the manuscript and providing some of the clinical photographs. I am indebted to Prof. Carlos Pelleschi Taborda, University of Sao Paulo and Dr Renata Buccheri, Consultant, Infectious Diseases, Instituto de Infectologia Emilio Ribas, Sao Paulo (Brazil) for helping me in procuring some of the photomicrographs.

I am also thankful to Prof. Josep Guarro, Prof. Alberto Miguel Stchigel Glikman and Prof. Jose Francisco Cano Lira, Mycology Unit, Medical School, Universitat Rovira i Virgili, Reus (Spain) for encouraging me to bring out the new edition and providing some of the photomicrographs.

After submitting the entire manuscript to the press, while I was writing this Acknowledgment, I received the news that Dr Deanna Sutton has passed away on July 4, 2017. It is a big setback to the mycology community. Her teaching of basic mycology which we have incorporated in this book also and learning which we have grown in the subject are extraordinary. In this month of July, she jointly published a paper with us in *Fungal Planet* highlighting new species of *Saksenaea (S.loutrophoriformis)* simultaneously isolated from India and USA. I sincerely pay my tribute to the departed soul.

Thanks to the colleagues and students, who have critically gone through this edition of Textbook, for the encouragement they have given me and for the numerous probing questions, which they asked during this period, on the basis of which I was able to expand the material contained in the previous editions and to bring out the new one.

I would also like to thank the faculty and staff of our institute, whose support and help at various stages of the preparation of this edition have been valuable. I am also grateful to all those whose suggestions, comments and healthy criticisms were helpful in making this textbook precise and up-to-date.

The latest information on the subject in the form of textbooks, journals, both hard and electronic copies and other related material were provided from time to time by our Library Staff as well as PGIMER and I am indebted to all of them. I am also very thankful to the staff of our IT Centre for tackling the computer-related problems which cropped up while writing the manuscript.

My long-suffering family also deserves a special mention, particularly my wife, Anuradha, daughters Anjuman and Aarzoo as well as son Avijit for providing constant support, encouragement and whose forbearance allowed me to continue writing of this edition. I am very thankful to Mr R S Bains, Senior Advocate, Punjab and Haryana High Court, who stood by me at each and every step and I was able to accomplish writing of this edition.

I especially appreciate the constant support and encouragement of Shri Jitendar P Vij (Group Chairman) and Mr Ankit Vij (Group President) of Jaypee Brothers Medical Publishers (P) Ltd., New Delhi in publishing this Textbook and also Ms Chetna Malhotra Vohra (Associate Director - Content Strategy), Mr Vipin Kaushik (Team Leader, Typesetting Department) and Mr Rajeev Joshi (Team Leader, Graphic Department), who have been prompt, efficient and most helpful throughout in the completion of this endeavor.

There are many other contributors for this edition, whose names might have been inadvertently missed by me. I request that they do not feel offended in this regard. The love and affection of the students in microbiology and allied clinical areas have been a constant source of inspiration and encouragement. It is expected that the fourth edition of this textbook will also be able to fulfill its mission to pursue, promote and advance the cause of emerging biomedical science of mycology.

Date: July 7, 2017
Place: Chandigarh

Jagdish Chander

Contents

Section VI. Miscellaneous Mycoses

Section VII. Pseudofungal Infections

Section VIII. Appendices

Acronyms and Abbreviations

ABCD	Amphotericin B colloidal dispersion
ABLC	Amphotericin B lipid complex
ABPA	Allergic bronchopulmonary aspergillosis
ABPM	Allergic bronchopulmonary mycoses
AFAD	Allergic fungal airway disease
AFLP	Amplified Fragment Length Polymorphism
AFRS	Allergic fungal rhinosinusitis
AFST	Antifungal Susceptibility Testing
AmB	Amphotericin B
APACHE	Acute Physiology and Chronic Health Evaluation (scores)
API	Analytical Profile Index
APMIS	Acta Pathologica Microbiologica Immunologica Scandinavica
AP-PCR	Arbitrarily primed PCR
ATCC	American Type Culture Collection
AUC	Area under curve
BAL	Bronchoalveolar lavage (fluid)
BCE	Brown Color Effect
BDG	Beta-D-Glucan
BHI	Brain-heart infusion (agar/broth)
BMBL	Biosafety in Microbiological and Biomedical Laboratories
BMD	Broth microdilution
BLAST	Basic Local Alignment Search
BMT	Bone marrow transplant
BOD	Biochemical/Biological oxygen demand
BSA	Bird seed agar
BSL	Biohazard safety levels
CAB	Centre for Agriculture and Biosciences
CAPD	Continuous ambulatory peritoneal dialysis
CARD	Caspase-associated recruitment domain 9
CBS	Centraalbureau voor Schimmelcultures (Now, Westerdijk Fungal Biodiversity Institute)
CCSG	Coccidioidomycosis Cooperative Study Group
CDC	Centers for Disease Control and Prevention
Cf	Compare from
cfc	Continuous flow culture
CFPA	Chronic fibrosing pulmonary aspergillosis
CFU	Colony-forming unit
CFW	Calcofluor white (stain)
CGB	Canavanine glycine bromothymol blue
CHEF	Contour-clamped homogeneous electric field gel electrophoresis
CHHA	Cysteine-heart and hemoglobin agar
CIE	Counterimmunoelectrophoresis
CIOMS	Council for International Organization of Medical Sciences
CLSI	Clinical and Laboratory Standards Institute (since January 1, 2005, formerly NCCLS)
CMA	Cornmeal agar
CMCC	Chronic mucocutaneous candidiasis
CMI	Cell-mediated immunity
CMI	Commonwealth Mycological Institute
CNPA	Chronic Necrotizing Pulmonary Aspergillosis
CPA	Chronic pulmonary aspergillosis
CPA	Cyclopiazonic acid
CT	Computed tomography
CYP	Cytochrome P450 (enzymes)
D-AmB	Deoxycholate amphotericin B
DBB	Diazonium Blue B (dye)
DD	Double diffusion/disk diffusion
DHPS	Dihydropteroate synthase
DIM	Dermatophyte identification medium
DIN	Deutsches Institut für Normung, Germany
DMSO	Dimethyl sulfoxide
DPX	Dibutylphthalate xylene
DRBC	Dichloran Rose Bengal Chloramphenicol (agar)
DRIP	Dermocystidium, rosette agent, Ichthyophonus and Psorospermium (clade)
DTH	Delayed-type hypersensitivity
DTM	Dermatophyte test medium
e	Electronic (edition)
EAPCRI	European *Aspergillus* PCR Initiative
e.g.	*exempli gratia* - for example
ECIL	European Conference on Infections in Leukemia
ECMM	European Confederation of Med. Mycology
ed/edn	Editor(s)/Edition(s)
EDTA	Ethylene diamine tetra acetic acid
EIA	Enzyme immunoassay
ELISA	Enzyme-linked immunosorbent assay
EMBO	European Molecular Biology Organization
EORTC	European Organization for Research and Treatment of Cancer

et al	*et alii* 'and others' (more than one author)
EUCAST	European Committee for Antimicrobial Susceptibility Testing
f. sp.	*formae speciales*/special form
FA	Fluorescent antibodies
FEMS	Fed. of European Microbiological Societies
FESS	Functional Endoscopic Sinus Surgery
FFPE	Formalin fixed, paraffin-embedded
FIRE	Fungal Infection Risk Evaluation
FISH	Fluorescence *in situ* hybridization
FNA	Fine needle aspiration (biopsy/cytology)
FSSAI	Food Safety and Standards Authority of India
GAFFI	Global Action Fund for Fungal Infections
GalXM	Galactoxylomannan
GF	Gridley fungal (stain)
GLC	Gas liquid chromatography
GlcNAc	N-acetylglucosamine
GMS	Grocott-Gomori methenamine silver (stain)
Gr.	Greek
GTT	Germ tube test
GXM	Glucuronoxylomannan
GYE	Glucose-yeast extract (agar)
H&E	Hematoxylin and eosin (stain)
HA	Hemagglutination
HAI	Healthcare-associated infections
HACCP	Hazard Analysis Critical Control Point
HEPA	High efficiency particulate air (filter)
HPLC	High performance liquid chromatography
HRL	Histoplasmosis Reference Laboratory
HSCT	Hematopoietic Stem Cell Transplantation
Hsp	Heat shock proteins
i.e.	*id est* 'that is'
IA	Invasive aspergillosis
IARC	International Agency for Research on Cancer
ICBN	Internat. Code of Botanical Nomenclature
ICN	International Code of Nomenclature for Algae, Fungi and Plants
ICTF	Internat. Committee on Taxonomy of Fungi
ICZN	Internat. Code of Zoological Nomenclature
ID	Immunodiffusion
IFA	Indirect fluorescent antibodies
IFD	Invasive fungal disease
IFI	Invasive fungal infection
IFN	Interferon
IFT	Invasive fungal tracheobronchitis
IGS	Intergenic Spacer Region
IHRB	Industrial Health Research Board
IL	Interleukin
IMA	International Mycological Association
IMC	International Mycological Congress
IMI	International Mycological Institute
IPA	Invasive pulmonary aspergillosis
IRIS	Immune Reconstitution Inflamm. Syndrome
ISHAM	International Society for Human and Animal Mycology (Founded in July 1954, Paris)
ITS	Internal Transcribed Spacer
ITIS	Integrated Taxonomic Information System
kDa	Kilo Dalton
KOH	Potassium hydroxide
LA	Latex agglutination
La.	Latin
L-AmB	Liposomal amphotericin B
LAMP	Loop-Mediated Isothermal Amplification
LCB/LPCB	Lactophenol cotton blue (mount)
Lf	Lactoferrin
LFA/D	Lateral Flow Assay/Device
LIFE	Leading International Fungal Education
LPA	Latex particle agglutination
LSD	Lysergic acid diethylamide
LSU	Large-subunit
M	Mycelial (Saprotrophic phase of fungus)
mAb	Monoclonal antibody
MALDI	Matrix Assisted Laser Desorption Ionization
MEA	Malt-extract agar
MEC	Minimum effective concentration
MFC	Minimum fungicidal concentration
MIC	Minimum inhibitory concentration
MFSS	Masson-Fontana silver stain
MGG	May-Grunwald Giemsa (stain)
MLEE	Multi-locus enzyme electrophoresis
MLSA	Multi-locus sequence analysis
MLST	Multi-locus sequence typing
MOPS	Morpholinopropane Sulfonic acid (buffer)
MRI	Magnetic resonance imaging
MRL	Mycology Reference Laboratory
MS	Mass spectrometry (also see MALDI TOF)
MSG	Mycoses Study Group
MTCC	Microbial Type Culture Collection
MTT	Methylthiazyl tetrazolium
NCBI	National Center for Biotech. Information
NCCLS	National Committee for Clinical Laboratory Standards (now CLSI since January 1, 2005)
NCPF	National Collection of Pathogenic Fungi
NCTC	National Collection of Type Cultures
NCYC	National Collection of Yeasts Cultures
NDM	Non-dermatophytic molds

NTDs	Neglected Tropical Diseases
NSM	Non-sporulating molds
NIH	National Institute of Health (USA)
OSHA	Occupational Safety & Health Administration
OTA	Ochratoxin A
PAGE	Polyacrylamide gel electrophoresis
PAP	Papanicolaou (stain)
PATH	Prospective Antifungal Therapy
PAS	Periodic acid-Schiff (stain)
PBS	Phosphate-buffered saline
PCM	Paracoccidioidomycosis
PCP	*Pneumocystis* pneumonia
PJP	*Pneumocystis jirovecii* pneumonia
PCR	Polymerase chain reaction
PDA	Potato dextrose agar
PDT	Photodynamic therapy
PFBK	Primary fetal bovine kidney
PFGE	Pulsed field gel electrophoresis
PFGGE	Pulsed field gradient gel electrophoresis
PHLS	Public Health Laboratory Service (UK)
PHOL	Pal, Hasegawa, Ono, Lee (Stain)
PIM	Prototheca isolation medium
PIV	Pythium insidiosum vaccine
PMS	Pyrolysis-mass spectrometry
PNA	Peptide nucleic acid (PNA-FISH)
PNS	Paranasal sinuses
POCT	Point-of-Care Testing
ppb/ppm	Parts-per-billion (1×10^{-9})/million (1×10^{-6})
PRM	Peptidorhamnomannan (antigen)
PYG	Peptone-yeast extract glucose
QMAP	Quantamatrix Multiplexed Assay Platform
qPCR	Quantitative polymerase chain reaction
RAPD	Randomly amplified polymorphic DNA
RFLP	Restriction fragment length polymorphism
RIA	Radioimmunoassay
RPLA	Reverse passive latex agglutination
RPMI	Roswell Park Memorial Institute (Medium)
RSA	Rice starch agar
RT-PCR	Reverse transcriptase-PCR
SAFS	Severe asthma associated fungal sensitivity
SAM	Sinobronchial Allergic Mycosis
S/Suppl	Supplement (Issue of journal)
SD	Seborrheic dermatitis
SDA,SGA	Sabouraud dextrose (glucose) agar
SDD	Susceptible dose-dependent
SDS	Sodium dodecyl sulfate
SIHAM	Society for Indian Human and Animal Mycologists (Founded in 1996, Jabalpur, MP)
SMX/SMZ	Sulfamethoxazole
SOD	Superoxide dismutase
SOT	Solid organ transplant
Sp/Spp.	Species (Singular/Pleural)
SSA	Sunflower seed agar
SSCP	Single-stranded conformation polymorphism
SSKI	Saturated solution of potassium iodide
SSU	Small-subunit
Syn.	Synonymous (name)
TDM	Therapeutic drug monitoring
TD-PCR	Touchdown polymerase chain reaction
TEM	Transmission electron microscopy
TLC	Thin Layer Chromatography
TLR	Toll-like receptors
TMP	Trimethoprim
TOF	Time of flight (MALDI TOF MS)
TRANSNET	Transplant-Associated Infection Surveillance Network
TRM	Tetrazolium reduction medium
TSB	Tryptic soy broth
var.	Variety (of a fungal strain)
VFCE	Valley Fever Center for Excellence
vs.	Versus
WGST	Whole-Genome Sequence Typing
Y	Yeast or Parasitic phase of dimorphic fungi
YCB	Yeast carbon base
YEA	Yeast extract agar
YNB	Yeast nitrogen base
ZEN	Zearalenone (Mycotoxin)

SECTION I

General Topics in Medical Mycology

Chapters

Medical Mycology is a newly established discipline of medical sciences, which has attained immense importance during the terminal two decades of 20th century, particularly after the onset of AIDS pandemic. In the past, fungi were believed to be merely non-pathogenic, commensals or contaminants, with exception of a very few pathogenic ones. But now these are recognized as medically significant organisms causing potentially life-threatening diseases invariably with fatal outcome. Therefore, this medical speciality is gradually gaining more and more importance in the present context, especially among the immunocompromised and debilitated patients with one or the other underlying risk factors.

This Section covers some of the fundamental issues pertaining to fungi, which are generally applied to the field of Medical Mycology. Their morphological features are entirely different from other causative agents like viruses, bacteria and parasites because they exist as yeast, mycelial or even both morphological forms. The understanding of these features is also essential to have high index of suspicion thereby timely detection of fungi as the causative agent in clinical setting of a particular disease.

The nomenclature, classification and taxonomy of many medically significant fungi are settled as and when teleomorphic state of a fungus is discovered, however, many of these agents are still included in phylum Deuteromycota *i.e. 'fungi imperfecti'* as no teleomorphic state is yet discovered. Other agents looking similar to fungi are now regarded as pseudofungal organisms like *Prototheca, Rhinosporidium* and *Pythium* species. Taxonomically such organisms are either classified as protistan parasites or oomycetes. These taxonomical intricacies are pertinent to understand the basic disease process adopted by a causative fungal agent.

How fungi overcome defense mechanisms of human body among immunocompetent as well as immunocompromised hosts is separately dealt in one chapter of this Section. By the time fungal infection is recognized, patient is found to be already immunocompromised and remedial modalities become quite cumbersome thereby such infections invariably prove to be fatal. From prophylactic point of view, role of vaccination in Medical Mycology is still in its rudimentary stage. Hence general hygienic measures are recommended for prevention of fungal diseases.

A high index of suspicion is essential in any clinical setting for diagnosing fungal diseases, which is often lacking among medical specialists. The diagnostic techniques are utilized to prove or disprove the suspicion thereby therapeutic modalities are resorted, accordingly. Therefore, an appropriate diagnostic approach to fungal diseases is also described briefly in separate Chapter of this Section.

After establishing the final diagnosis, appropriate antifungals are to be instituted the therapeutic modality. However, some of the fungi, yeasts as well as mycelia, are developing resistance to the commonly used antifungals. It may be because of the indiscriminate use of such drugs merely on an empirical basis, without specific indication and/or appropriate antifungal susceptibility testing. The understanding of antifungals is of paramount importance thus commonly used antifungals have been described.

The diagnostic modalities and antifungal susceptibility testing are important components of the management of fungal diseases. Hence all the technical details pertaining to such issues are separately given in the Section as Appendices.

General Topics in Medical Mycology

CHAPTER 1

Introduction

The **fungi** are achlorophyllous eukaryotic organisms, which multiply sexually and asexually by production of spores. Their somatic structure is composed of either yeast or filament bordered by cell wall, which is mainly composed of chitin. The fungi, unlike plants, cannot produce their own food thus are called **heterotrophs**. It was generally believed in the past that fungi were plants and classified under kingdom Plantae. However, now these are classified under independent biological kingdom *i.e.* **Fungi** since 1969.

The **fungus** is essentially a Latin word that means **mushroom**. The branch of biological science, which deals with study of fungi, is known as **Mycology** (Myco+logy=-Fungus+study, Gr. *Mykes*, mushroom/fungus, *logos*, discourse/study). The term is basically derived from ***mykes***, Greek word, which is also used for mushroom. Hence **Medical Mycology** is the study of epidemiology, ecology, pathogenesis, diagnosis and therapeutic modalities of fungal diseases found prevalent among human beings. This discipline of Medical Microbiology is emerging rapidly and has now attained full-fledged and independent status in biomedical fields. The subject of mycology is essentially contemplation of medical sciences in newer perspectives.

Analogous to the field of Medical Mycology, Veterinary Mycology is study of fungal infections among animals, which often presents in similar way as those found in human beings. Sometimes, these are transmitted from animal-to-human beings and such infections are called zoonotic diseases. In recent reports cats are recognised as source of human sporotrichosis in Brazil establishing animal-to-man transmission. Hence reference of animal mycoses in the Textbook of Medical Mycology is essential to understand the epidemiology of a fungal disease.

The diseases of warm-blooded animals caused by fungi are known as **mycoses** (singular = mycosis). The prefix ***'myco'*** has been given to few bacterial organisms like ***Myco****bacterium* and ***Myco****plasma* or diseases like **Myco**sis **fungoides** *i.e.* cutaneous T-cell lymphoma, etc. that were speculated in the past to have similarities with either fungi or diseases caused by them. But later on it was realized that these organisms are not at all related to fungi. However, their nomenclature is still popular as misnomer because these are now well-accepted terms of the organisms and/or diseases.

Historical Perspective

The clinical manifestations of some of fungal infections are known since antiquity, however, systematic study of fungi is hardly one and half century old. In 1835, **Agostino Bassi** (1773-1856) in Italy established that fungus *(Beauveria bassiana)* was the cause of disease **(muscardine)** in silkworms (*Bombyx mori*), which could be transferred from one silkworm to the other. This fungus has shown to be widespread geographically and is currently used as a biological agent for the control of important insect pests in agriculture. Bassi is rightly acknowledged to be the first to refer the etiology of an animal disease to a microbial infection and is universally regarded as "Father of Mycology". On the basis of these findings he predicted that fungi could also cause infections in man.

In 1842, **David Gruby** demonstrated for the first time that infection of scalp **(favus)** was caused by fungus through its inoculation to the healthy skin. Gruby thereby fulfilled what later on became famous as Koch's postulates, comprising the criteria necessary for the acceptance of a microbe as the cause of a specific infectious disease. Since then knowledge of fungal infections, which was fragmentary to begin with, has increased exponentially over the years as an independent infectious disease speciality as **Medical Mycology**.

The discipline of Medical Mycology attained substantial recognition in the field of bioscience in 1910 when

French dermatologist and microbiologist, **Raymond Jacques Adrien Sabouraud** (1864-1938), published his seminal treatise, ***'Les Teignes'***. This monumental work was a comprehensive account of most of the known dermatophytes, which is still being referred by the mycologists. He laid down solid foundation of the field of Medical Mycology, which gained remarkable momentum. As a result official journal of ISHAM was initially named as *Sabouraudia* in his honour in 1961. The medium for cultivation of fungi still bears his name, which is universally used for primary isolation of pathogenic and non-pathogenic fungi from clinical as well as non-clinical specimens in the mycology laboratory. It is suitable for the growth of both yeasts as well as mycelial fungi. Therefore, **Raymond Sabouraud** has rightly been called "Father of Medical Mycology".

Similarly, P A Saccardo has played significant role in the establishment of field of Medical Mycology in earlier days. The other scientists who have played an active role in development of mycology are J I Schoenlein, Norman Conant, Chester Emmons, David Gruby, H P R Seeliger, J W Rippon, L Ajello, K J Kwon-Chung and Arvind A Padhye.

The introduction of fungi to man came into existence through the findings of Agostino Bassi in 1835 and the field of Medical Mycology has now completed more than 1½ century (mid-19th and 20th) as it has now entered into the 21st century. There have been lot of developments during this period, specially after advent of modern molecular techniques, which are dealing with diagnostic as well as taxonomical matters. But many issues are still lying unresolved before the mycologists, which include classification of pseudofungal organism and successful cultivation of *Lacazia loboi* and *Pneumocystis jirovecii* on artificial culture media. It is expected that molecular techniques will be able to resolve most of the intricacies of pending issues in the field of medical mycology in the times to come.

There are significant developments in treatment modalities of fungal infections and we have now achieved new prospects. However, till 1950s there was no specific antifungal agent available. Nystatin was discovered in 1951 and subsequently amphotericin B was introduced in 1957 and was sanctioned for use for human beings. In 1970s field was dominated by azole derivatives. Now, this is the most active field of interest, where potential drugs are being developed to treat fungal infections. By the end of 20th century, fungi have been reported to be developing drug resistance, among yeasts and mycelia in hospital as well as community acquired infections.

Milestones in Medical Mycology

The discovery of relationship of certain fungi to disease precedes work of even Louis Pasteur and Robert Koch. Some of the important milestones in Medical Mycology are given below:

1835 Bassi described fungal etiology of 'muscardine of silkworms.'

1839 Schoenlein studied fungal infection of scalp *i.e.* tinea capitis (favus). Remak, succeeded in growing the fungus on apples and reproduced the disease in animals and on his own forearm.

1839 Lagenbeck described yeast-like organism of thrush.

1842 Gruby independently isolated fungus on artificial media responsible for tinea capitis and produced disease by inoculation of healthy scalp.

1890 Sabouraud began publishing large numbers of articles on fungal disorders of skin, which eventually culminated in an enormous contribution to the field of Medical Mycology.

1928 Penicillin was discovered by Alexander Fleming as product from *Penicillium notatum (P.chrysogenum)*.

1934 According to Botanical Nomenclature, species concepts of dermatophytes was redefined by Chester Emmons.

1969 Separate Kingdom - **Fungi,** was created by Whittaker, in the five-kingdom classification.

1976 Organism causing human PCP *i.e. Pneumocystis carinii* was proposed as *P.jirovecii* by J K Frenkel, in honour of Czech Parasitologist **Otto Jirovec**, who described this microbe in humans but he is not widely recognised.

1999 *Pneumocystis carinii* strain infecting human was re-designated as *P.jirovecii.*

2002 *Loboa loboi* was re-designated as *Lacazia loboi* and *C.posadasii* was established as new species of genus *Coccidioides.*

2017 Penicilliosis Marneffei was renamed as Talaromycosis and Emmonsiosis as Emergomycosis.

The observations of both Bassi, in 1835 and Gruby in 1842, demonstrating that fungi could be cause of diseases, was long before Pasteur's Germ Theory came into existence fulfilling Koch's Postulates, about four decades before Koch actually formulated them in 1884.

The start of the modern age of mycology begins with Pier Antonio Micheli's (1679-1737) publication of **Nova Plantarum Genera**. Published in Florence, this seminal work laid the foundation of the systematic classification of grasses, mosses and fungi. He is considered as the founding 'Father of Scientific Mycology".

Invasive fungal infections are becoming an increasingly important cause of morbidity and mortality, particularly for immunocompromised human populations. The fungal pathogens belonging to genera *Candida, Cryptococcus* and *Aspergillus* collectively contribute to over 1 million human deaths annually.

Epidemiology

The fungi are widely found in environment and most of them are harmless commensals, contaminants or non-pathogenic agents. Some of the fungi are even useful to mankind in several ways. However, small number of these organisms are causing disease among men, animals and plants. This has been estimated by Hawksworth that only 5% of total fungal species in world have been identified which constitute about 70,000 species out of an estimated 1,500,000 but hardly 600 are usually recognized as primary pathogens of man and other mammals. Out of these fungal species, less than 100 are frequently encountered in routine clinical practice. It is estimated that about 1,500 new species of fungal genera are being described every year.

By 1995, approximately 70,000 fungal species have been accepted as compared to 5,000 viruses and 3,100 known bacteria. Yet remarkably, very few of these fungi cause disease in humans. Moreover, most of fungi exist as molds but there are number of pathogenic yeasts and some of them are dimorphic as well. The dimorphic fungi exist either as yeast or spherule in their parasitic form when causing infection to host and assume mold form when growing as saprotrophic in nature. Out of 1.5 million species of fungi estimated to exist all over the world and one-third of global fungal diversity found in India. According to Maheshwari there are 27,000 species of fungi, which have been recorded from India.

As fungal infections are not notifiable like viral, bacterial or parasitic diseases hence these are not given much attention and usually diagnosis is established very late when the patient is terminally sick. The approach to identify fungi in developing countries is on gross morphological features whereas in developed countries it is molecular based approach that is important due to availability of requisite infrastructure. Moreover, most of serious fungal infections are more common in developed countries due to underlying immunocompromised situations as compared to developing countries where majority of disease are due to low hygienic standards and various environmental factors.

The overall incidence and prevalence of mycotic infections is increasing, particularly during the last three decades. A major contributor to this emergence is growing number of immunocompromised and more susceptible individuals. Previously the epidemiological features of these diseases were not well-documented but the understanding of epidemiology of fungal infections has increased considerably in recent years, largely because of studies that have mainly focused on specific patient groups, hospital series and autopsy surveys.

Candida species is the fourth most common organism recovered from blood cultures in hospitalized patients. Aspergillosis is common in selected populations, such as bone marrow transplant recipients and lung is the most common clinical site. *Aspergillus* species is isolated in substantial number of solid organ and bone marrow transplant recipients with clinical findings of pneumonia. In such patients, high mortality rates are observed in relation to *Aspergillus* species.

The diseases, which were not prevalent in a particular area, are now being reported very commonly due to frequent travel across the globe. In the times to come, diseases like paracoccidioidomycosis and coccidioidomycosis which are currently found in restricted zones of endemicity in the New World, may be reported from other parts of world. Some of fungal infections are only limited to African continent like *Histoplasma capsulatum* var. *duboisii*. Blastomycosis is also found in Africa in addition to its native prevalence in the southeastern parts of North America.

The liberalization and globalization of economy has increased the gap between rich and poor as well as urban and rural population. Consequently privatization of health services has targeted poorest of poor due to migration of laborers thereby affecting epidemiology of fungal diseases.

The fungi are now recognized as significant cause of morbidity and mortality among man and animal. They have emerged as important etiological agents of opportunistic infections as well as full-fledged diseases as true pathogens. The invasive fungal infections were regarded as very rare till half century ago. Since that time there has been steady increase in number of patients suffering from life-threatening fungal infections, specially in prevailing immunocompromised circumstances. In recent years opportunistic fungal infections have emerged as one of interesting areas due to number of contributory factors. This calls for increased awareness about fungal diseases and their definitive diagnosis. Care must be taken to exclude such type of infection in cases where fungus is grown in culture but not seen in lesion by smear preparations and histopathological examination of tissue sections.

Like other organisms, Koch's postulates are applicable to fungal infections as well despite the fact that some of agents like *Lacazia loboi* are yet to be successfully grown on artificial culture media. The present challenges in Medical Mycology are diagnosis, treatment, taxonomy, mycology training and changes in diseases patterns. However, new and more effective tumour chemotherapeutics, methods of surgery and irradiation, complex treatment procedures such as autologous and allogenic bone marrow and blood stem cell therapy as well as organ transplantation, have enabled treatment of various solid tumors and systemic hematologic diseases. The use of such chemotherapeutics and immunosuppressants as therapeutic modalities result in general increase in opportunistic bacterial, viral parasitic as well as fungal infections. The epidemiological features of important fungi arc dcalt with subsequent chapters of this Textbook.

As such, the arthropods are not acting as biological vectors of the fungal diseases however; sometimes these may act as mechanical carriers of fungi. In this regard, cockroaches have been reported to the source of transmission of fungi in hospital set-up.

As per the statement of G C Ainsworth in 1966, that in reality it is the distribution of the medical mycologists or active investigators in this field and not the true distribution of fungal disease, which is reflected as its epidemiology.

Ecology of Fungi

The fungi are capable of existing and flourishing in a wide variety of environment as parasites, saprotrophic or symbionts. In the past they were known to be cave dwellers but presently they are found existing on innumerable places. They may be parasites of organic substrate, such as wood and other decaying plant parts, paper, leather, cloth, keratinous and chitinous substrate, oils and fats, resins and even petroleum and draw their nutrition.

The definite ecological groups of fungi known to exist as soil fungi, aquatic fungi, coprophilous fungi (grow on animal excreta), entomogenous fungi (parasites of insects) predacious fungi (capture small animals and protozoa), marine fungi, osmophilic fungi (from saline soil) and thermophilic fungi in which organism show remarkable adaptations not only to exploit nutritionally rich substrate but also to reproduce themselves. As decomposers, fungi are essential because along with bacteria, they recycle vital elements, such as nitrogen and phosphorus, back to the ecosystem.

Although beneficial effects of fungi far outweigh their harmful aspects but there are some disease-causing and destructive species, which are clinically significant. The number of saprotrophic fungal species far outnumbers those that are parasitic and impact on the health of man, animal and plants. This is well-known that nearly one hundred saprotrophic fungi, which may have their existence in any of common habitats, mentioned above can adapt to infect man and animals and cause diseases. It is also essential to acquire sound knowledge of biology of these fungi and their role in ecological niche. The limited data is available about survival of fungi that commonly cause health-care associated infections in immunocompromised patients on typical hospital materials. This indicates that many of fungi like *Candida*, *Aspergillus* and *Fusarium* species, which are responsible for health-care associated infections, survive for at least one day and often longer on fabrics and plastics routinely used in hospitals. These survival results indicate fabrics and plastics to serve as potential reservoirs or vectors for fungi.

There have been three ecological categories of infectious mycological agents based on their natural habitat. Therefore, traditionally fungi have been described as geophilic, zoophilic and anthropophilic. Now, a **fourth** category has also come into existence that is called **hydrophilic**, which encompasses free living organisms that live in aquatic habitats on non-living organic matters or on aquatic plants. The diseases, caused by such fungal (lacaziosis) and pseudofungal organisms (rhinosporidiosis and pythiosis) are currently categorized as hydrophilic infections.

Magnitude of the Problem

Mycology is a subject matter of immense interest to botanists as most of fungi are plant pathogens. The veterinarians also take lot of interest as fungi are significantly affecting health of domestic animals as well. Now, those fungi, which were supposed to be non-pathogenic, are reported significantly infective agents. Hence microbiologists and histopathologists are taking more interest in this branch to render diagnostic services in hospitals. The field of Medical Mycology that was virtually limited to case reports in the 19th century had attained status of full-fledged subject, as an important branch of medical science by the turn of the 20th century. It is estimated that during the 21st century, magnitude of problem is going to be increased because of the immunocompromised nature of hosts entailing in secondary as well as primary fungal infections.

Another way to estimate that this branch has expanded enormously in recent past, is by increased number of publications on issues related to Medical Mycology. Now, not only case reports but detailed prospective and retrospective studies are being published regularly in national and international journals, more in number and quality to previous times and as such subject of Medical Mycology has entirely changed over the years. Therefore, the field of Medical Mycology has evolved to the present independent status as significant branch of medical sciences.

Although fungi were recognized as disease causing agents much earlier but their significance was overshadowed by bacteriology and even virology. The attitude towards Medical Mycology was 'step-child' of our doctrine and research. It was least bothered subject in medical institutions as compared to other branches of medical sciences. But now there is 'obligatory' attention to this subject because clinicians face challenges due to underlying immunocompromised situations leading to life-threatening secondary fungal infections taking lives of innumerable patients. The fungal infections had such an impact that now-a-days no health personnel can ignore mycology, no matter whatsoever may be their speciality.

In the past, Mycology Sections of medical institutions, used to get only few skin scrapings from the Department of Dermatology and Venereology. In due course of time this field has evolved to an extent that Mycology Sections are now receiving clinical specimens from almost all specialties whether medical or surgical. Moreover, their number and quality has enormously increased in recent past due to greater awareness about fungi as disease causing agents and microbiology laboratories are now receiving more number of clinical specimens for mycological investigations. Therefore, due to the circumstantial compulsions, the microbiologists and pathologists have left their myopic attitude towards the fungal infections.

As fungal diseases are not notifiable diseases hence data on morbidity and mortality attributable to these diseases undoubtedly is understated. But incidence and prevalence of fungal infections has increased enormously. It may be due to underlying predisposing factors such as immunocompromised situations like use of corticosteroids, antimicrobials, immunosuppressive and anticancer drugs, bone marrow or solid organ transplants, HIV-positivity, metabolic disorders *i.e.* diabetes mellitus, etc as narrated in **Table 1.1**. Out of all these factors two are most important and prominent namely therapy among cancer patients and pandemic of the AIDS. Moreover, elderly patients, whose life span has been extended by treatment of cancer or other debilitating diseases, are more susceptible to secondary fungal infections than younger individuals. In addition to that there are changing parameters of mycoses like host population, emerging drug resistance and reporting of newer fungal pathogens.

Table 1.1. Predisposing Factors of Fungal Diseases.

- Widespread use of broad-spectrum antibiotics
- Prolonged use of steroids for more than three weeks
- Increased number of intravenous drug abusers
- Greater number of immunocompromised patients
- Increased number of AIDS cases with longer survival
- More aggressive therapeutic modalities of cancers
- More intervention in intensive care medicine in adults
- Catheter-borne infections among patients
- New and more widely used prosthetic devices
- Increased number of BMT and SOT procedures

The rapid growth of fungi such as *Candida albicans* in debilitated patient with non-mycotic disease may be minor factor as cause of death or in contrast the other fungal infections. Similarly, lack of experience in diagnostic techniques, inadequate laboratory infrastructure and lack of pathological autopsy examination, may limit recognition and reporting of many deaths due to fungal infections. These days, it is feasible to detect about 30% of diagnosis of fungal infections are made antemortem and rest 70% as postmortem. This paradigm is going to be shifted depending on the awareness among the medical personnel.

The basic need of hour is to create awareness among clinicians so that they are motivated to an extent as to send samples to Mycology Section of Microbiology Department. In most of advanced medical institutions, it is estimated that about five percent autopsied patients turn out to have died of fungal infections. At present in most of the teaching hospitals every third clinicopathological conference (CPC) is reporting deep-seated fungal infection as secondary invader responsible for terminal events of deceased patients. The autopsy findings also substantiate an increasing prevalence of invasive fungal infections among hospitalized patients along with significant changes in aetiology and underlying disease process. Therefore, pathological autopsy is an essential feedback for clinicians in particular setting of severely immunocompromised population and emphasizes constant and urgent need for more efforts towards prevention, diagnosis and treatment of fungal infections. The autopsy besides being diagnostic method, is also a means of medical quality control because it acts as deterrent on the diagnostic procedures.

The prognosis among patients of fungal infections as such is very poor. Therefore application of early diagnostic and therapeutic modalities is essential. A fruitful outcome depends on a very close association between Medical Mycology laboratory and clinician incharge of concerned specialty. It has also been observed that increased awareness among medical personnel has contributed to an early diagnosis and commencement of prompt and timely treatment in life-threatening fungal infections. Despite all the available modalities, Medical Mycology is still labeled as tip of the iceberg as little is known about it to the medical fraternity.

The radiological investigations, like ultrasonography, computed tomography, magnetic resonance imaging, have contributed significantly, in addition to routine x-rays examinations. Now, there are effective chemotherapeutic agents available and further modalities are being developed for treatment of fungal infections, therefore, their application requires proper techniques. The opportunistic fungal infections vary greatly in their clinical and pathological manifestations due to underlying illness of patients at risk as well as treatment modalities undertaken.

The newly isolated fungi may be morphologically atypical because of partial treatment given to patient. Therefore, if fungus does not fit into any description provided, serial transfers on Sabouraud dextrose agar may be necessary to restore typical morphology. When fungal growth is observed from clinical specimen, three principal characteristics to be evaluated are texture, growth rate and pigmentation. Therefore, using these criteria, conventional fungal identity can be narrowed down quickly to a limited group of fungi having similar characteristics.

Burden of Fungal Diseases

In a large number of countries two organizations *i.e.* Leading International Fungal Education (LIFE) and Global Action Fund for Fungal Infections (GAFFI) have conducted extensive campaign to document the burden of fungal diseases in respect to that country. This data has been published in various medical journals and cited in Further Reading of this Chapter as well.

Fungal Taxonomy

During the 20th Century, until around 1950s, botanists used the term fungi to include all members of 'plant kingdom' that did not have stems, roots, leaves and chlorophyll. By this definition, even some bacteria were included with fungal category. As fungi were initially classified with plants hence lot of botanical influence is seen on this subject. This may probably be the reason that Medical Mycology has been mainly subject of interest to the botanists. These organisms were transferred to an independent kingdom, after five-kingdom classification by Whittaker in 1969, based on their cell morphology that subsequently substantiated by molecular studies. Now, after recent increase in prevalence of fungal infections, especially in immunocompromised patients, it has become subject matter of great interest to microbiologists, pathologists, internists and the allied medical specialists, who regularly encounter such type of infections.

The Medical Mycologists encounter many new fungal genera and species that are known to have clinical relevance and every year many species are being added in the list of significant fungal pathogens. Many fungi, formerly considered as insignificant commensals or merely laboratory contaminants, have now taken new roles as agents of infectious diseases in patients with AIDS and other immunodeficiency conditions, in drug abusers and in debilitated individuals with various underlying diseases. The subject has reached a level of understanding where any fungus recovered as pure culture from any of the body sites must be considered as potential human pathogen.

In recent times, the human genome has been fully described and this knowledge, based on molecular techniques, is applicable to Medical Mycology as well so that diagnostic dilemma and taxonomic confusions are now feasible to be resolved. In this very sequence a long-standing taxonomical controversy of *Pneumocystis jirovecii* has been substantially resolved. *Microsporidium* species had a similar fate. These atypical fungi are shifted from kingdom

Protista to Fungi; thereby Parasitology to Mycology; hence being described in the Textbook of Medical Mycology.

The **pseudofungal** organisms like *Pythium* and *Prototheca* species are mycelial or unicellular microorganisms that are **not** classified in the kingdom Fungi and that produce infections with clinical and histopathological features resembling those caused by eumycetes. Taxonomically, on the basis of molecular studies, *Rhinosporidium seeberi* is also now considered as a protist, classified in Mesomycetozoea, recognized as DRIPs clade of protistan parasites.

The Medical Mycology is totally new and unfamiliar subject to medical personnel and presents difficulties, which are apparent while identifying newly isolated fungal agent. Even though procedures for direct examination and isolation have been mastered, mycologist is often left to confront fungal culture that may be totally unfamiliar both macroscopically and microscopically. This happens most frequently when person is new to this field or knows least what to do. In such a situation, even if the species identification is difficult, at least the isolate should be identified to the genus level and immediately conveyed to the clinician so that timely and definitive treatment is instituted then and there. In due course of time, further taxonomical studies can be performed for final identification of the clinical fungal strain.

Endemic Mycoses

The endemicity of some of the fungal diseases like histoplasmosis, blastomycosis, coccidioidomycosis, paracoccidioidomycosis as well as talaromycosis are restricted to limited geographical areas. This may be due to environmental and other factors, which favor growth of fungi in soil of these areas. Histoplasmosis and blastomycosis mostly afflict people in Mississippi and Ohio River Valleys, while coccidioidomycosis is found primarily in southwest desert regions of USA and paracoccidioidomycosis in certain parts of Latin America. However, talaromycosis is found prevalent in entirely different geographical zone away from Americas *i.e.* Southeast Asia. All these diseases are acquired through inhalation of air contaminated with conidia of mycelial phase and cause localized to diffuse pulmonary involvement entailing dissemination to distant body sites. Their clinical features are different but all of these fungi are dimorphic in nature.

There are now case reports of endemic mycoses from non-endemic areas also. In case of coccidioidomycosis, new area in northeast Brazil has been recognized. In this very context talaromycosis has been reported from Africa without patient's history of travel to its endemic zone. In India, it has been now significantly reported from Manipur, Mizoram, Nagaland and upper Assam. It may be because of its higher prevalence in adjoining countries like Myanmar and northern Thailand.

Imported Mycoses

Those fungal infections which are caused by endemically prevalent fungi, encountered in non-native locations are called imported mycoses. In this era of modern technology, the entire world has become a ***'global village'*** and borders of diseases are being demolished day-by-day. Now, due to frequent traveling all over the world in this era of jet age and underlying immunocompromised situations, zones of endemicity are not limited to their original native areas of prevalence. Such types of endemic fungal infections are also on the rise.

Emerging and Re-emerging Fungal Diseases

In the past fungi were considered to be merely non-pathogenic or simply laboratory contaminants with very few exceptions with cutaneous manifestations. But due to circumstantial immunocompromised background among the patients, these very non-pathogenic and contaminants have now proved to be significant pathogens and are encountered as emerging agents of life-threatening fungal diseases. In addition, some of the clinical entities which were not of great magnitude have recently re-emerged as important diseases. Most of the dimorphic fungi have re-emerged after the advent of AIDS and other immunocompromised situations.

In Medical Mycology as such there is an emergence and re-emergence of different types of fungal diseases of paramount significance, which are caused by *Saccharomyces cerevisiae, Candida auris, Scedosporium apiospermum, Talaromyces marneffei* and *Fusarium* species, which have recently been reported in literature. Outbreaks due to fungal infections do occur but only in sporadic form and epidemics have not been reported. For example, incidence of cryptococcal meningitis has recently increased to 1000-fold in the New York City alone.

In addition, during late September 2012, a multistate outbreak of CNS fungal infection and septic arthritis was caused due to an emerging fungus, *Exserohilum rostratum* in the USA, which is otherwise a plant pathogen. There were over 700 patients with 64 deaths, who received epidural injections of methylprednisolone produced at a Massachusetts compounding center and subsequently developed meningitis with or without posterior circulation stroke and/or spinal or para-spinal infection and more than 30 patients who received intra-articular injections of the same drug developed osteoarticular infections.

Some of the fungi, which have been reported as merely environmental contaminants, have now emerged as significant human pathogens. Some of mucormycetes like *Apophysomyces variabilis, Saksenaea erythrospora* and *Rhizopus homothallicus* have now emerged as significant cause of various forms of mucormycosis and if not timely treated may lead to fatal consequences even in immunocompetent individuals.

Medical Mycology and AIDS

Mycology as such was an unknown field of medical sciences except in some of clinical specialties like Dermatology and allied subjects dealing predominantly with dermatophytosis and candidiasis. The real augment came three and half decades ago when new virus was brought to the human civilization probably a spillover of biological warfare laboratories, which created havoc by abrogating immune system of its targeted victims. Consequently the index case was recognized in the USA in 1981. The novel agent was subsequently designated as Human Immunodeficiency Virus (HIV) in 1986, the labeled cause of Acquired Immunodeficiency Syndrome (AIDS).

This type of clinical setting gave an open invitation to all the secondary invaders of viral, bacterial, parasitic and even fungal origin. The patients virtually became microbiological culture plates and all these organisms thrived upon them. The fungi were already well-known agents invading individuals with weaker immune system, made their abode among AIDS patients thereby leading to fatal consequences. This duo of HIV and fungi attained medical acceptability to an extent that some of the diseases like pneumocystosis, cryptococcosis, talaromycosis, etc. became AIDS-defining illnesses. Therefore, the fungal disease/parameters started defining the viral infections for the first time in the history of medical sciences.

Now, every year fungal pathogens cause more than two million infections leading to high morbidity as well as mortality. Out of this two fungal genera *i.e. Candida* and *Cryptococcus* are together accounting for about 1.4 million infections with higher mortality rates. The patients with disturbed immune system like HIV/AIDS, are mainly vulnerable to such infections. Although there is substantial control on this disease due to introduction of ART but fungal infections are still on the higher rates.

Hence advent of AIDS has altered the incidence as well as prevalence of fungal infections thereby the course of action of the field of medical mycology. During this period since 1981, epidemiological scenario about fungal infections has entirely changed. The fungal isolates, which used to be discarded considering as merely laboratory contaminants are now found to be playing significant role in pathogenicity of the variety of infections. This is true with fungi affecting not only immunocompromised patients but about those capable of infecting even healthy individuals as well. Therefore, the impact of syndrome is such that even true fungal pathogen like *Histoplasma, Blastomyces, Coccidioides* and *Paracoccidioides* species, infect immunocompetent individuals as well as now immunocompromised patients also and that too in a very large number.

This was probably warfare experiment to begin with but in due course of time it has attained the level of a full-fledged pandemic as it is now reality and not fiction-based story of a syndrome of 'sealed' origin. In the times to come, fungal infections will be encountered more in number, killing thousands of patients with underlying abrogated immune system.

Biological Warfare and Fungi

During the events of nine-eleven (September 11, 2001), so-called greatest imperialist power of the world was 'shaken' to the hilt. This was followed by incidents of spreading white powdery material pertaining to *Bacillus anthracis*. All these developments led to apprehension that fungi also might be used on similar pattern. Most of the governments secretly resort on to biological warfare acts for their survival in power through state organized terrorism thereby giving rise to origin of counterproductive terrorist groups, who may retaliate by indulging in such activities.

As such till date no fungal agent has been apparently used in biological warfare but reports are available about their secondary metabolic products. In the past, ***Yellow Rain*** in Laos is one of examples, where mycotoxins (Trichothecene) were used as agent of biological warfare. Similarly, ***Agent Orange***, a herbicide and defoliant was also used during Vietnam War. The degradation of such material release dioxins, which cause serious health related maladies.

The danger of working with cultures of some of fungi like *Coccidioides* species, were even recognized as early as in 20th century. Now, in the changed international scenario, there is an apprehension that these fungi may be abused as biological warfare agents. Therefore, their interstate transport is restricted under provisions of US Anti-terrorism and Effective Death Penalty Act, 1996. These highly infectious agents require safe laboratory handling under Biosafety Level 3 facility.

Classification of Fungal Diseases

Most of the fungi are saprotrophic in nature as they use non-living organic material as source of their nutrition. They are significant scavengers in the ecosystems. Along with bacteria, they are important in recycling carbon, nitrogen and essential mineral nutrients. As far as their interaction with living world is concerned, they have various modes of existence like other microorganisms. The host-parasite relationship in fungi can be divided into three potential modes namely mutualism, commensalism and parasitism depending upon the basic interaction between fungus involved and the host. These are as follows:

(a) Commensalism: The fungus neither gets benefit nor harmed in the host-parasite relationship. There is usually no physiological interaction between two organisms. One organism uses other to get a better position in the environment.

(b) Mutualism: The fungus and host take mutual benefits from host-parasite relationship. For example in Mycorrhizae there is association of fungi with plant roots and in Lichens there is association with algae and cyanobacteria.

(c) Parasitism: The fungus gets benefits and host is harmed by host-parasite relationship. Most of the fungi causing disease in humans, animals and plants are placed in this category.

The basic mechanism of fungal pathogenicity is its ability to adapt to tissue environment, physiological barriers and temperature (37°C) thereby withstand lytic activity of hosts' cellular defense mechanism. The same fungal agent can produce different manifestations depending upon underlying situations of the host. Based on their wide-spectrum of adaptability, various fungi causing human mycoses can be categorized into following groups:

(a) Pathogenic Fungi: These fungi have ability to adapt to the tissue environment, which is quite marked. This is expressed as thermal dimorphism and fungi are able to cause infection even in the immunocompetent individuals.

(b) Opportunist Fungi: These fungi cause infections only in patients with immunodeficiency or other debilitating conditions, which carry high morbidity and mortality. Some sort of defect in immune system of the host is prerequisite for establishment of such fungal infection and these do not exhibit thermal dimorphism.

(c) Toxigenic Fungi: These fungal species cause illness or even death of patients and/or animals after ingestion of contaminated food by fungi (mycetismus) or their mycotoxin metabolites (mycotoxicosis).

(d) Allergenic Fungi: These fungal species act as allergens and cause various allergic manifestations in human beings.

The disease process is established on the basis of net outcome of host-parasite relationship. In addition, the fungal diseases in man can be classified according to the anatomical site of primary involvement.

(a) Superficial Mycoses: The fungal infection is limited to outermost layers of skin and its appendages. The immune response in such infections is rarely induced or it is very mild.

(b) Cutaneous Mycoses: In this type the infection extends deeper into epidermis and it also invades hair and nails. The disease process may evoke high inflammatory response in the host.

(c) Subcutaneous Mycoses: The infection is due to organisms of low pathogenicity usually following traumatic injury. It involves dermis, subcutaneous tissues and sometimes muscles and fasciae. The causal agents of subcutaneous mycoses require mechanical introduction of organisms into tissues through traumatic implantation.

(d) Systemic Mycoses: The infection primarily involve a site like lungs and later on disseminates systemically to distant body sites. Systemic mycoses along with the opportunistic fungal infections, is called **deep mycoses**.

(e) Opportunistic Mycoses: Besides the above-mentioned four categories of fungal infections, fifth group has come into prominent focus because of increasing use of immunosuppressive therapy and pandemic of AIDS and is called opportunistic group. The infectious agents normally are of very low pathogenic potential, which produce full-fledged disease only under changed circumstances, mostly involving host debilitation.

The fungal infections are not transmitted sexually as commonly seen in viral, bacterial and even protozoal diseases. However, balanoposthitis caused by *Candida* species is supposed to be transmitted by the sexual contact. Moreover, piedra is also taken into this category as higher rates are reported among homosexuals in Denmark. In some of the fungal infections like histoplasmosis and mucormycosis, genital lesions have definitely been reported but their mode of infections is not considered to be through sexual route.

Significance of Fungi

The fungi have recently gained more importance due to increased incidence of various fungal diseases. Normally, most of these organisms are found as soil saprotrophic but whenever there is some underlying cause, the very same organism becomes significantly pathogenic. In addition to pathogenic potential, significance of fungi lies in the economy of nature is described below:

(a) Useful Properties of Fungi

The fungi are pathogenic to man, animals and plants causing various types of diseases. On the other hand it is also known that most of the fungi are beneficial to mankind. Their ability to break down complex organic substrate of almost every type thus contributing to recycling of carbon and other elements making fungi as an important component of our terrestrial ecosystem. The industrial uses of fungi as food, fermenter in food industry, producers of antibiotics, drugs, etc. are innumerable and may surpass damages, which these agents do as pathogens of man and animals. However, to date penicillin, lovastatin, cyclosporin, griseofulvin, cephalosporin and ergometrine are the most important pharmaceutical products that have been extracted from different fungal species.

The importance of fungi has been more emphasized towards beneficial effects, which are given below:

- The edible fungi are widely used in industry as source of food, particularly wild or domesticated mushrooms of all shapes, sizes and colors, which belong to basidiomycetes. They are also put in salads, pizza and burgers.
- The fungi are used to alter texture, improve flavor and increase palatability and digestibility of natural and processed foods. The yeast, *Saccharomyces cerevisiae,* is used in leavening of bread and other baked products. Hence it is known as **Baker's yeast**. The blue mold, *Penicillium* species, is used in ripening process to prepare speciality cheeses such as blue cheese.
- Some of yeasts, *Candida fukuyamaensis,* are used to prepare special type of tea, known as Russian or Manchurian tea, which is presumed to be having some therapeutic potential in certain medical ailments. Hence it is consumed by many people under this impression. *Candida fukuyamaensis* is believed to be asexual state of *Pichia guilliermondii.*
- At industrial level fungi are used to produce alcohol (*Saccharomyces* species), fat (*Endomyces* species) and proteins (*Torulopsis* species) through fermentation. *Saccharomyces cerevisiae,* also called **Brewer's yeast**, is used in the fermentation processes that result in production of beers, wines and spirits. All citric acid used in soft drinks, candies, artificial lemon juice, baked goods, etc. are produced industrially by fermentation by certain fungi like *Aspergillus niger.*
- The fungi are now therapeutically being used as probiotics along with bacterial species like *Lactobacillus* and *Bifidobacterium.* In peculiar circumstances when the bacterial probiotic agents do not withstand onslaught of antimicrobials in antibiotic-associated diarrhoea, yeasts such as *Saccharomyces boulardii* can bear this assault thereby prove to be very effective in ameliorating such common alimentary ailments, while acting as probiotics. Hence *S.boulardii* is shown to be useful in the treatment (biotherapeutic agent) of acute infectious diarrhoea, the prevention or treatment of diarrhoea associated with antibiotic use and as an adjunctive therapy for *Helicobacter pylori* infection.
- The discovery of popular antibiotic penicillin from *Penicillium notatum* (now called *P.chrysogenum*) by Alexander Fleming in 1928 was real breakthrough in history of medical sciences. In 1941, it revolutionized the basic approach in treating infectious diseases. *P.chrysogenum* is mold that is widely distributed in nature and is often found living on foods and indoor environments. Many fungi like *P.chrysogenum* are used as source of several β-lactam antibiotics, most significantly penicillin. Another fungus, *Acremonium chrysogenum* (also called *Cephalosporium acremonium*) is used in manufacturing cephalosporin. *Cephalosporium chrysogenum* was isolated for the first time by Thirumalachar and Sukapure from Pimpri, Pune, India.
- The antifungal agent produced from fungi *(Penicillium griseofulvin)* is griseofulvin, is narrow-spectrum fungistatic antibiotic used for treating dermatophytes.

- *Aspergillus nidulans* is used in making anidulafungin.
- Fumagillin is produced from *Aspergillus fumigatus* and used in the treatment of microsporidiosis.
- In addition to β-lactam antibiotics (penicillins and cephalosporins), fungi are used to produce immunosuppressant like cyclosporin A, which is a primary metabolite of several fungi, including *Trichoderma polysporum* and *Cylindrocarpon lucidum*.
- The statin family of cholesterol-lowering drugs are derived from fungi. Simvastatin is prepared synthetically from fermentation product of *Aspergillus terreus*.
- The ergot produced by *Claviceps purpurea*, a fungal pathogen of rye, is used for inducing uterine contractions, controlling bleeding and alleviating localized vascular disorders *e.g.* migraine. The ergot alkaloids belong to two groups: (a) amine alkaloid - ergometrine and (b) aminoacid alkaloids - ergotamine. Ergometrine or its semisynthetic analogue methylergometrine is clinically used for oxytocic effects. As ergotamine has α-adrenergic blocking and vasoconstriction properties, it is clinically used in treating migraine.
- The fungi are now playing a vital role for producing natural products, most productive source of lead compounds in far reaching endeavor of new drug discovery. *Epicoccum* is known for its potential to produce diverse classes of biologically active secondary metabolites. Most of these fungal metabolites have cytotoxic, anticancer, antimicrobial and anti-diabetic activities.
- A few fungal species (*Saccharomyces cerevisiae, Neurospora crassa* and *Ustilago maydis*) are studied as model organisms. Such studies are useful to acquire more knowledge of the basic processes involved in biochemistry, physiology, genetics and molecular biology and these results are applicable to many organisms. The dung fungus, *Sordaria fimicola* has recently been exploited in genetics studies.
- The yeasts like *Saccharomyces cerevisiae, Hansenula polymorpha* and *Pichia* species have been used as heterogeneous host for preparation of recombinant vaccines for hepatitis B virus infection.
- Fungi like *Culicinomyces clavosporus* have been used as traps and pathogen to control mosquitoes in malaria eradication. The entomopathogenic genus, *Coelomomyces* is also known to be pathogenic to larva and develop within larval hemocoel of mosquitoes. The *Coelomomyces* species are obligate parasitic fungi that alternate their saprotrophic and gametophytic stages between mosquitoes and copepods, respectively.
- The pesticides of fungal origin have also been found to decrease malaria transmission in animal studies. After feeding on an infectious blood meal, mosquitoes have been exposed to surfaces coated with fungal biopesticides or entomopathogens, which have dramatically reduced their ability to transmit malaria.
- Some of biotechnological applications of fungi especially in the field of food, agriculture, etc. are very significant. The biological measures have been applied in agriculture in Havana where some of fungi are used *e.g. Beauveria bassiana*, to control banana root borer, sweet potato weevil, rice water weevil and sugarcane borer. *Verticillium lecanii* specifically is bred to control sweet potato white fly, one of the most damaging pests in Cuba. *Trichoderma* species is used to control soil-borne diseases that attack tobacco, tomatoes and pepper.
- Recently, one of the fungi (*Gliocladium roseum*), found prevalent in Ulmo tree in rain forests of Patagonia (South America), is tried to make biofuel. This digests the cellulose turning it into biofuel *i.e.* myco-diesel.

As an overall point of view, fungi have been used to cause man more satisfaction rather than trouble. They raised his bread, his wine and beer, carbonate his champagne, flavored his cheeses, furnished him edible mushrooms, produced drugs to stop bleeding, to reduce cholesterol, worked diligently to form antibiotics to help him fight bacterial as well as fungal infections and helped industry to produce many organics acids and other compounds. However, to a general mycologist they are even beautiful to behold and their unending variation is very fascinating, whereas to a medical mycologist they are now the dangerous and life-threatening organisms.

(b) Harmful Properties of Fungi

In addition to the useful properties, fungi are also important agents producing harmful effects, which are given below:

- The pathogenic fungi are responsible for superficial, subcutaneous, systemic and opportunistic infections in man and animals.
- Mycotoxicosis and mycetismus are becoming serious public health issues in most of the developing tropical and sub-tropical countries.
- The fungi can spoil improperly stored agricultural produce like grain, foodstuff, vegetables and fruits. These are commonly seen in eatables as ***'bread molds'***. The fungal pathogens are responsible for damages amounting about 10% of world's crops. It is estimated that as agents of plant disease, fungi cause annual crop losses of billions of dollars.

- The fungi, especially molds, can cause decay of fabrics, timber, leather, electrical insulation and other synthetic materials. There may be extensive loss following failure to protect material from ravages of fungi in warm humid climates. It is also speculated that mutated fungi, which were breeding in space station Mir were responsible for the deterioration of its equipment.
- The fungal growth on plastic products, especially on plasticize components, is known since 1940s. Each plastic product must be examined for its suitability to colonizing fungi. Saprotrophic fungi, especially in humid tropical areas, can colonize diskettes, audiotapes, videotapes and computer disks. The fungi belonging to genera like *Alternaria, Aspergillus, Epicoccum, Paecilomyces, Penicillium* and *Trichoderma* are observed on diskettes in the tropical countries.

Aeromycology

As fungal agents are part and parcel of soil and the environment, especially atmospheric air, which acts as the most common source of opportunistic as well as true fungal infections. The fungi in atmospheric air may be regularly monitored to know the magnitude of problem under which the population is living in an area. The fungal species constitute major component of air-spores. Various studies have revealed that fungal spores in air mostly come from fungi on plants or large fungal fruiting bodies that rise above soil surface and not from the ground. The soil itself is sinking into which most airborne spores finally disappear rather than their being a source. Many organisms found outside atmosphere will also be found inside buildings where they are introduced by air currents but main sources of contamination arise from within buildings from animals, humans or plants.

The aeromycology is study of airborne fungi, which is usually done with following purposes: (a) to know variety of fungi prevalent in particular region and also (b) to know seasonal variations of allergenic fungal spores. There are two methods for studying airborne fungi that may be used individually or concurrently; (i) slide exposure method and (ii) petri-dish exposure method. The combination of both these methods is ideal to minimize drawback of each other.

The value of aerobiology lies in the study of fungi that cause diseases to man and animals. Generally three types of diseases based on fungal origin are recognized: allergies, mycotoxicosis and mycoses. Hence inhalation of fungal spores and perhaps their metabolites may cause respiratory system disease such as bronchopulmonary aspergillosis, pulmonary mycotoxicosis and hypersensitivity reactions.

Surveillance of Fungal Infections

A regular monitoring of fungal infections is also required as it is done in case of viral, bacterial and parasitic infections. This is more important when large number of infections are healthcare-associated in nature and found prevalent in hospital environment. Moreover, these are found to be resistant to the most commonly used antifungal agents.

Outside hospital setup, surveillance of fungal infections is mandatory as certain outbreaks are reported in particular geographical areas. Some of diseases are prevalent in certain geographical pockets but in addition to that diseases are found prevalent in certain communities as well, which may be detected in time with watch of surveillance system.

Now molecular techniques are playing an important role not only in fungal diagnosis but in taxonomy and epidemiological aspects as well. These techniques have resolved many long-standing controversies on these aspects. Therefore, in any hospital set-up, regular surveillance of fungal infections is the most important, especially in epidemiological aspects of the healthcare-associated ones.

Safety in Medical Mycology Laboratory

There are certain fungal pathogens, which cause systemic infections associated with work related exposure to the staff working in diagnostic as well as research mycology laboratory. Four of these genera are established pathogenic fungi in which greatest potential of exposure is involved in such laboratories. These are: *Coccidioides immitis, C.posadasii, Histoplasma capsulatum, Sporothrix schenckii* and *Cryptococcus neoformans.* Additional species are being documented as etiological agents of mycotic morbidity and may pose potential threat for such type of infections. These agents usually take entry through respiratory route and may establish infection due to inhalation.

There are general safety measures like never smell or examine fungal culture in the open petri dish. The handling of clinical material as well as the fungal isolates should be done under biosafety hood. It is essential for both medical as well as classical mycologists to adhere standard safety measures during their laboratory work. The basic priority is to recognize potential risk of infection by these fungi. Secondly, acknowledgments that investigators have primary responsibility for containment of any organism with which personnel are working in laboratory to avoid any occupational health hazards. In diagnostic laboratory every clinical sample should be taken as highly infectious unless it is proven otherwise. Biosafety issues are not only important from medical point of view but because of legal angle also. Therefore, essential safety precautions and techniques should be adhered to in all research and diagnostic laboratories handling infective microorganisms including fungi. The general guidelines on laboratory safety practices are briefly described in **Chapter 36**.

The specifically recommended safety procedures and equipment are set out according to four **Biohazard Safety Levels** or **BioSafety Levels** (BSL) of laboratory facilities, dealing with potentially hazardous pathogens, moving through basic laboratories handling routine sample of low-risk microorganisms (BSL 1) to maximum containment laboratories working with highly infectious pathogens (BSL 4). These are also called corresponding levels of **Containment** (CL1 to 4), **Category** (Cat1 to 4) or **Protection** (P1 to 4) and the counterpart microorganisms as **Hazardous Group** (HG1 to 4).

A classification of fungi into biosafety categories is proposed and various criteria for attribution to Biosafety Levels have been laid down. The list of fungi stressing on ecological criteria is derived from these agents rather than host factors and include all species accepted in recent medical literature. Each fungal species is assigned a defined biohazard categories *i.e.* Biosafety Levels.

There are three primary goals for biological safety in medical mycology laboratory, which are as follows: (a) to reduce potential level of exposure of laboratory personnel; (b) to protect diagnostic and research work from exogenous contamination and (c) to prevent inadvertent contamination of environment outside laboratory.

The intent of dealing with this issue is to cover some of important aspects of biosafety; (i) standards in handling dimorphic fungal pathogens; (ii) principles and criteria of biosafety levels and classification of known medically important as well as environmental fungi according to their biosafety levels; (iii) medically important fungal waste and its safe disposal and (iv) biosafety and regulatory considerations in handling and mailing medically important fungi to a Reference Laboratory or Culture Collection.

The following safety procedures are adopted while working with potential fungal pathogens. These should be mandatory for all personnel posted in clinical or diagnostic laboratory. There are two most common means of getting infections by fungi in laboratory *i.e.* inhalation of conidia that are aerosolized and accidental inoculation through sharp object *e.g.* needles, scalpel blades and broken glasses while handling specimens.

The Biosafety Levels include laboratory practices and techniques, safety equipment and appropriate laboratory facilities and have been defined to guide those working with infectious agents. Keeping all types of microorganism in view, Centers for Disease Control and Prevention and National Institutes of Health, USA have divided Biosafety Levels into following categories, which are described below:

(a) Generally Regarded As Safe (GRAS): The harmless industrial strains like domesticated mutants widely applied in food production. GRAS refers to individual strains or clones whereas BSL refers to species in taxonomic sense.

(b) Biosafety Level 1: This level requires basic laboratory facilities and use of standard microbiological practices. It applies to use of specific microorganisms not known to cause disease in healthy human adults. These saprotrophic agents or plant pathogens are occupying non-vertebrate ecological niches or these are fungi utilizing dead animal products. The infections in this category are coincidental, superficial and non-invasive or mild in nature. *Saccharomyces cerevisiae*, which was categorized under GRAS, is now considered under this category because of substantial number of case reports by this organism.

(c) Biosafety Level 2: This level requires basic BSL 1 practices plus aprons, decontamination of all infectious wastes, limited access to laboratory, protective gloves and pasting of biohazard warning signs. This level is used when handling microorganisms of moderate-risk, which are associated with human diseases. These species are principally occupying non-vertebrate ecological niche but with relatively low ability to survive in vertebrate tissue. In severely immunocompromised patients they may cause deep or opportunistic mycoses. The pathogens causing superficial infections also fall into this category. Therefore, majority of fungi infecting man and animals are categorized under BSL 2.

Moreover, BSL 2 practices and facilities are recommended for handling and processing clinical specimens and identifying isolates of *Coccidioides* species. BSL 2 practices differ from BSL 1 practices in that (i) laboratory personnel have specific training in handling pathogenic agents and are directed by qualified scientists; (ii) access to laboratory is limited when work is being conducted; (iii) extreme precautions are taken with contaminated sharp items and (iv) procedures in which infectious aerosols or splashes may be created, are conducted in Class II biological safety cabinets.

(d) Biosafety Level 3: This level requires BSL 2 facilities and practices supplemented by controlled access to laboratory and use of special laboratory clothing and containment equipment *e.g.* biological safety cabinet. Biosafety level 3 facilities require a separate negative pressure room. This level is used when potentiality of infection from aerosols, autoinoculation or ingestion exists and when concerned personnel are handling microorganisms which may cause severe or lethal infections. These pathogens are potentially able to cause severe or deep mycoses in otherwise healthy patients. Earlier list of fungi included in this category was mycelial forms of dimorphic fungi but now it also includes additional other potentially pathogenic fungi like *Cladophialophora bantiana.*

(e) Biosafety Level 4: This is maximum containment level consisting of work with dangerous and exotic agents, which have high potential for causing life-threatening diseases. It does not include any of fungal pathogens as most of the agents covered under this level are mainly viral in origin.

All Mycology Laboratories are covered under purview of two levels *i.e.* BSL 2 and at the most BSL 3. The clinical specimens that may contain fungi pathogenic to man should be handled by using BSL 2 practices, containment equipment and other facilities. There is need for clinical laboratories to adhere to safety standards promulgated by CDC Office of Health and Safety in 5th edition of *'Biosafety in Microbiological and Biomedical Laboratories'* published in 2007. This document summarizes four recommended biosafety levels for infectious agents and addresses which types of biological safety cabinets are appropriate for each biosafety level and which are available online (***See* Appendix F**).

The standard precautions include use of Biological Safety Cabinet or hood when working with clinical material suspected of having fungi. These cabinets are most commonly used primary containment devices in laboratories working with infectious agents. The staff must be properly trained in use of these cabinets. Therefore, biological safety cabinets are essential protective element for investigators working with infectious or potentially infectious fungal pathogens. The Class II biological safety cabinets have vertical laminar airflow with high efficiency particulate air (HEPA) filtered supply and exhaust air, which are 99.99% efficient in removing particles of size 0.3 µm or larger. An additional advantage of using biological safety cabinets is that they also protect clinical specimens from extraneous and airborne contamination. Hence they protect worker, product as well as the environment. The types of biosafety cabinets and their proper use have been described in details in 3rd edition of CDC-NIH publication entitled *'Primary Containment for Biohazard: Selection, Installation and Use of Biological Safety Cabinets'* issued in September 2007.

BSL 2 procedures are specifically recommended for personnel working with clinical specimens, which may contain *Blastomyces dermatitidis, Coccidioides immitis, C.posadasii, Histoplasma capsulatum, Cryptococcus neoformans, Sporothrix schenckii* and pathogenic members of genera like *Trichophyton, Microsporum* and *Epidermophyton.* If culture of dimorphic fungus like *Blastomyces, Coccidioides* or *Histoplasma* is grown in its mycelial form, BSL 3 containment procedures should be adopted since conidia of these fungi have greater potential to cause infection through inhalation. In other words yeast (tissue) forms of some of dimorphic fungi are covered under BSL 2 and their counterpart mycelial forms strictly under provision of BSL 3 due to their potentially higher infectious nature.

Risk Groups

The biological agents are also categorized in Risk Groups (RG) based on their relative risk. Therefore, depending on the country or organization, this classification system may take the following factors into consideration: (i) Pathogenicity of the organism; (ii) Mode of transmission and host range; (iii) Availability of effective preventive measures like vaccines; (iv) Availability of effective treatment like antimicrobials. It is important to understand that biological agents are classified in a graded manner such that the level of hazard associated with RG1 being the lowest and RG4 being the highest. EHS Biosafety follows the NIH Guidelines categorization of Risk Groups as follows:

RG1: Are not associated with disease in healthy adult humans or animals.

RG2: Are associated with disease, which is rarely serious and for which preventative or therapeutic is often available.

RG3: Are associated with serious or lethal human disease for which preventative or therapeutics may be available.

RG4: Are associated with lethal human disease for which preventative or therapeutics are not readily available.

Outbreaks of Fungal Infections

Outbreaks of fungal infections following natural disasters and accidental/hospital acquisition are increasingly recognised. They have important public health and infection control consequences. Infections after natural disasters typically result from inhalation or cutaneous inoculation of environmental fungi. Clusters result from shared exposures. There have been >13 major global outbreaks in the past decade from air-borne, near-drowning or near-burial events encompassing fungi from *Coccidioides immitis* (*n*=3) to mucormycetes. Hurricane Katrina in 2008, the Boxing Day tsunami in 2004 and Joplin Tornado in 2011 'stand out' with horrendous soft tissue injuries and disseminated infection with a mortality of 30-80%. Healthcare infections after disasters can follow *e.g. Aspergillus* meningitis from contaminated syringes (Sri Lankan tsunami). Case clusters also occur through use of contaminated pharmaceutical products, fomites and environmental hospital exposure. Mucormycosis has resulted from contaminated bed linen and tongue depressors. Fungal endophthalmitis (*Fusarium, Bipolaris*) followed minor eye surgery after irrigation with contaminated triamcinolone and Brilliant Blue Green. Moreover, there were 752 cases (64 deaths) of *Exserohilum* CNS and joint infections from contaminated methylprednisolone vials used to treat back pain in the USA with a national crisis. Improving physician and public health awareness of mycoses is essential with an urgent need for infrastructure for case reporting and a network of alerts.

Index of Suspicion

In routine medical practice, the infectious diseases are usually considered to be caused by bacteria, viruses and seldom by the parasites. Fungus, as such the cause of infectious disease is rarely considered. Therefore, to tackle such infections, often antibacterial antibiotic are prescribed and underlying pathogenesis of fungal infection goes on increasing. By the time realization is there about the fungal infection, it is too late and situation becomes almost irreversible. Therefore, fungal infections should be considered as the differential diagnosis, wherever, a slightest doubt of fungal involvement is there. If the index of suspicion is kept very high then the additional diagnostic methodology is adopted along with the routine protocol and final diagnosis is reached without wastage of valuable time thereby proper antifungal treatment is instituted and life of the patient is saved. Hence a high index of suspicion is the key to an early diagnosis of fungal diseases, which is must to tackle such life-threatening infections.

Medical Mycology in India

In India, fungal infections are known since ancient civilization and have been mentioned in Aryan documents such as Atharva Veda, wherein mycetoma is described as Padavalmikam *i.e.* anthill foot. This finding was observed by John Gill in 1842 in Madurai district of a south Indian state, Tamil Nadu, which was subsequently designated as ***'madura foot'*** by Henry Carter in 1860.

In the beginning of 20th century, School of Tropical Medicine and Hygiene was established during the British regimen in 1920s in Kolkata (then Calcutta), the eastern zone of country. There was separate Department of Medical Mycology to cater services in this field of medical sciences, where Prof. Anisetti Thammayya did his innovative work in this field. In the western zone, Prof. K R Joshi consolidated different aspects of medical mycology, especially in the field of mycetoma and opportunistic fungal infections. Similarly, in southern zone, Prof. A S Thambiah, Prof. A Kamalam, Prof. Pankajalakshmi V Venugopal and Prof. P A Thomas in Tamil Nadu and Prof. B M Hemashettar in Karnataka, have brought Medical Mycology in the country to the level of substantial recognition. In the central India, mycological services were consolidated by Prof. S M Singh and his colleagues.

In northern side, Department of Medical Mycology was established at Vallabhbhai Patel Chest Institute, Delhi in 1959 to provide research, diagnostic, therapeutic mycological services in the national capital and its surrounding areas. Prof. H S Randhawa and Prof. H C Gugnani further developed this Department and currently being looked after by Prof. Anuradha Chowdhary. During the same period Mycology Sections were developed in All India Institute of Medical Sciences (AIIMS) and National Institute of Communicable Diseases (NICD), both in New Delhi. Prof. L N Mohapatra did pioneering work in this field at AIIMS, which was continued by Prof. Uma Banerjee and her colleagues.

The Division of Mycology at Postgraduate Institute of Medical Sciences and Research, Chandigarh was started under enthusiastic leadership of Prof. Pushpa Talwar in 1964 and is at present being looked after by Prof. Arunaloke Chakrabarti. This is one of the leading centers of Medical Mycology and providing diagnostic services as well as conducting research activities. Some of Indian institutions, where active work is currently going on, have been mentioned in **Appendix E** of this Textbook.

A curriculum was designed by Prof. R. Sambasiva and colleagues in 1999, wherein it is emphasized that postgraduates in Microbiology must have basic knowledge of theory and practical aspects in Medical Mycology, along with its other branches. In India, this branch of Microbiology has now attained substantially sound footing. The following salient features are also relevant to describe on this issue:

- Previously there were only a few case reports from India about fungal infections but now many prospective studies are being conducted mainly based on epidemiology and diagnostic modalities in the field of Medical Mycology.
- The international leading organizations in this field like ISHAM are helping medical institutions in India especially for training of staff, who handle difficulties related to management of fungal infections.
- The first edition of Textbook of Medical Mycology was published on July 1, 1995 which has modified trends in understanding of Medical Mycology at the grass root level in India and its neighboring countries followed by second edition in April 2002 and third edition in October 2008. It is expected that this fourth edition (2017) will also be able to further consolidate its endeavor in the times to come.
- The national body of Indian scientists interested in field of medical mycology came into existence. The Society of Indian Human and Animal Mycologists (SIHAM) was founded in February 1996 in its First Conference held at Jabalpur, Madhya Pradesh. Actually idea of making national body of mycologists was conceived in 1994 at time of 12th ISHAM Congress held at Adelaide (Australia), which was materialized after two years under the dynamic leadership of Prof. S M Singh, in the form of SIHAM, which is actively working for enhancing the cause of Medical Mycology in India. The National Conferences of SIHAM are being regularly held after a gap of two years. The very 1st Conference was held in Jabalpur (1996), 2nd in Jodhpur (1998), 3rd in Chennai (2000), 4th in Annamalai Nagar (2002), 5th in Chandigarh (2004), 6th in Hyderabad (2006), 7th in Mumbai (2008), 8th in New Delhi (2010), 9th in Siliguri (2012), 10th in Coimbatore (2014) and 11th in Shimla (2016). Now, 12th National SIHAM Conference will be held under the leadership of Dr. Jayanthi Savio in February 2018 at Bangalore (Karnataka).
- Many Seminars, Symposia, Workshops, Continuing Medical Education (CME) and Professional Development Programs (PDP) are being held in this field from time to time in different parts of the country. The relevant documents of these events are referred in **Appendix F** as Bibliography. A number of monographs and proceedings of these meetings are now available dealing with various aspects of Medical Mycology to provide firm footing in India.
- During this course of development, Mycology Reference Laboratories (MRL) are supposed to be designated to provide basic facilities for identification of fungi, teaching of graduates and postgraduates, other academic and service-oriented infrastructure in this unique field of medical sciences. The Indian Council of Medical Research (ICMR) has recognized the Division of Mycology as the 'Centre for Advanced Research in Medical Mycology' in the year 2005. It is providing two ICMR-WHO-PGI Training Courses in Diagnostic Medical Mycology in a calendar year: (i) summer course for the young faculty and medical scientists (May-June) and (ii) winter course for technical staff (December). In addition, it is the national reference centre for the identification, national culture collection as well as antifungal susceptibility testing of medically significant pathogenic fungi.
- The quality control in diagnostic mycology is essential to maintain standard in imparting the services. This may be internal or external quality control. In India periodic exercise was started by VPCI, New Delhi in the form of Proficiency Testing as external quality control. This service is now being provided as EQAS by the PGIMER, Chandigarh and dealt briefly in **Appendix E**. Therefore, auditing of Medical Mycology services is mandatory to impart good quality laboratory practices.

Fungal Repositories

The Budapest Treaty on international recognition of deposit of microorganisms for purposes of patent procedure was done at Budapest on April 28, 1977 and was amended on September 26, 1980. In India, based on these guidelines, the Microbial Type Culture Collection Center and Gene Bank (MTCC), IMTECH, Chandigarh was established and is functioning since October 4, 2002. In addition to other microorganisms, it is repository for the fungi, mainly isolated from the environment. The National Culture Collection of Pathogenic Fungi (NCCPF) was set-up at the PGIMER,

Chandigarh by the ICMR in 2010, where the fungal isolates can be deposited. The details of such issues are dealt in **Appendix E**.

Fungal Registries

In the recent times, lot of websites have been created which register data pertaining to any of the fungal issues, analyze it and periodically publish it from time to time. According to the feedback received, corrective measures are also taken. Some of the registries are: Mucormycetes *(www.zygomyco.net)*, Emerging Fungal Infections *(www.fungiscope.net)*, *Trichosporon (www.trireg.com)* and research collaboration focused on pediatric invasive fungal infections *i.e.* International Pediatric Fungal Network *(www.ipfn.org)*, etc.

Further Reading

1. Abdel-Azeem AM. The history, fungal biodiversity, conservation and future perspectives for mycology in Egypt. *IMA Fungus.* 2010; **1:** 123-42.
2. Adams RI, Miletto M, Taylor JW, *et al.* The diversity and distribution of fungi on residential surfaces. *PLoS One.* 2013; **8:** e78866. PMID: 24223861.
3. Akritidis N. Parasitic, fungal and prion zoonoses: An expanding universe of candidates for human disease. *Clin Microbiol Infect.* 2011; **17:** 331-5.
4. Alvarez Duarte E, Denning DW. Serious fungal infections in Chile. *Eur J Clin Microbiol Infect Dis.* 2017; **36:** 983-6.
5. Antinori S, Nebuloni M, Magni C, *et al.* Trends in the postmortem diagnosis of opportunistic invasive fungal infections in patients with AIDS: A retrospective study of 1,630 autopsies performed between 1984 and 2002. *Am J Clin Pathol.* 2009; **132:** 221-7.
6. Antinori S, Peri AM, Milazzo L. Fungal meningitis in England and Wales. *Lancet Infect Dis.* 2014; **14:** 921.
7. Aridogan IA, Izol V, Ilkit M. Superficial fungal infections of the male genitalia: A review. *Crit Rev Microbiol.* 2011; **37:** 237-44.
8. Armstrong-James D, Bicanic T, Brown GD, *et al.* AIDS-related mycoses: Current progress in the field and future priorities. *Trends Microbiol.* 2017; **25:** 428-30.
9. Armstrong-James D, Meintjes G, Brown GD. A neglected epidemic: Fungal infections in HIV/AIDS. *Trends Microbiol.* 2014; **22:** 120-7.
10. Arnold C. Fungal meningitis outbreak affects over 7000. *Lancet Neurol.* 2013; **12:** 429-30.
11. Badiane AS, Ndiaye D, Denning DW. Burden of fungal infections in Senegal. *Mycoses.* 2015; **58 (Suppl. 5):** 63-9.
12. Bassetti M, Righi E. Overview of fungal infections - The Italian experience. *Semin Respir Crit Care Med.* 2015; **36:** 796-805.
13. Batac MC, Denning D. Serious fungal infections in the Philippines. *Eur J Clin Microbiol Infect Dis.* 2017; **36:** 937-41.
14. Beardsley J, Denning DW, Chau NV, *et al.* Estimating the burden of fungal disease in Vietnam. *Mycoses.* 2015; **58 (Suppl. 5):** 101-6.
15. Ben R, Denning DW. Estimating the burden of fungal diseases in Israel. *Isr Med Assoc J.* 2015; **17:** 374-9.
16. Bidartondo MI, Read DJ, Trappe JM, *et al.* The dawn of symbiosis between plants and fungi. *Biol Lett.* 2011; **7:** 574-7.
17. Binder U, Lass-Florl C. Epidemiology of invasive fungal infections in the Mediterranean area. *Mediterr J Hematol Infect Dis.* 2011; **3:** e20110016.
18. Blackwell M. The fungi: 1, 2, 3 ...... 5.1 million species? *Am J Bot.* 2011; **98:** 426-38.
19. Brandt ME, Park BJ. Think fungus—Prevention and control of fungal infections. *Emerg Infect Dis.* 2013; **19:** 1688-9.
20. Brown GD, Denning DW, Levitz SM. Tackling human fungal infections. *Science.* 2012; **336:** 647.
21. Brown GD, Denning DW, Gow NA, *et al.* Hidden killers: Human fungal infections. *Sci Transl Med.* 2012; **4:** 165rv13.
22. Bustamante B, Denning DW, Campos PE. Serious fungal infections in Peru. *Eur J Clin Microbiol Infect Dis.* 2017; **36:** 943-8.
23. Casadevall A, Pirofski LA. *Exserohilum rostratum* fungal meningitis associated with methylprednisolone injections. *Future Microbiol.* 2013; **8:** 135-7.
24. Casadevall A, Pirofski LA. The weapon potential of human pathogenic fungi. *Med Mycol.* 2006; **44:** 689-96.
25. Chan Y, Selvaratnam V, Garg N. A fungating spica. *BMJ Case Rep.* 2015; pii: bcr2014206901. PMID: 25608979.
26. Chayakulkeeree M, Denning DW. Serious fungal infections in Thailand. *Eur J Clin Microbiol Infect Dis.* 2017; **36:** 931-5.
27. Chekiri-Talbi M, Denning DW. Burden of fungal infections in Algeria. *Eur J Clin Microbiol Infect Dis.* 2017; **36:** 999-1004.
28. Chowdhary A, Kathuria S, Agarwal K, *et al.* Recognizing filamentous basidiomycetes as agents of human disease: A review. *Med Mycol.* 2014; **52:** 782-97.
29. Chrdle A, Mallatova N, Vasakova M, *et al.* Burden of serious fungal infections in the Czech Republic. *Mycoses.* 2015; **58 (Suppl. 5):** 6-14.
30. Clark TA, Hajjeh RA. Recent trends in the epidemiology of invasive mycoses. *Curr Opin Infect Dis.* 2002; **15:** 569-74.
31. Colombo AL, Tobon A, Restrepo A, *et al.* Epidemiology of endemic systemic fungal infections in Latin America. *Med Mycol.* 2011; **49:** 785-98.
32. Colombo TE, Soares MM, D'Avilla SC, *et al.* Identification of fungal diseases at necropsy. *Pathol Res Pract.* 2012; **208:** 549-52.
33. Corzo-Leon DE, Armstrong-James D, Denning DW. Burden of serious fungal infections in Mexico. *Mycoses.* 2015; **58 (Suppl. 5):** 34-44.
34. Cox R, Sanchez J, Revie CW. Multi-criteria decision analysis tools for prioritising emerging or re-emerging infectious diseases associated with climate change in Canada. *PLoS One.* 2013; **8:** e68338. PMID: 23950868.

35. Czerucka D, Piche T, Rampal P. Yeast as probiotics - *Saccharomyces boulardii. Aliment Pharmacol Ther.* 2007; **26:** 767-78.
36. d'Enfert C. Hidden killers: Persistence of opportunistic fungal pathogens in the human host. *Curr Opin Microbiol.* 2009; **12:** 358-64.
37. Denning DW, Gugnani HC. Burden of serious fungal infections in Trinidad and Tobago. *Mycoses.* 2015; **58 (Suppl. 5):** 80-4.
38. Denning DW, Pleuvry A, Cole DC. Global burden of chronic pulmonary aspergillosis complicating sarcoidosis. *Eur Respir J.* 2013; **41:** 621-6.
39. Denning DW, Pleuvry A, Cole DC. Global burden of chronic pulmonary aspergillosis as a sequel to pulmonary tuberculosis. *Bull World Health Organ.* 2011; **89:** 864-72.
40. Denning DW. Minimizing fungal disease deaths will allow the UNAIDS target of reducing annual AIDS deaths below 500000 by 2020 to be realized. *Philos Trans R Soc Lond B Biol Sci.* 2016; **371 (1709).** pii: 20150468.
41. Desoubeaux G, Simon EG, Perrotin D, *et al.* The mobile team of Parasitology-Mycology, a medical entity for educational purposes to serve sick patients. *J Mycol Med.* 2014; **24:** 144-51.
42. Dignani MC. Epidemiology of invasive fungal diseases on the basis of autopsy reports. *F1000Prime Rep.* 2014; **6:** 81.
43. Dorgan E, Denning DW, McMullan R. Burden of fungal disease in Ireland. *J Med Microbiol.* 2015; **64:** 423-6.
44. Dufresne SF, Cole DC, Denning DW, *et al.* Serious fungal infections in Canada. *Eur J Clin Microbiol Infect Dis.* 2017; **36:** 987-92.
45. Espinel-Ingroff A. History of medical mycology in the United States. *Clin Microbial Rev.* 1996; **9:** 235-72.
46. Evans EGV, Bulmer GS, Aly R, *et al.* Training medical mycologists in developing countries. *Med Mycol.* 2000; **38:** S33-40.
47. Faini D, Maokola W, Furrer H, *et al.* Burden of serious fungal infections in Tanzania. *Mycoses.* 2015; **58 (Suppl. 5):** 70-9.
48. Fatima N, Ismail T, Muhammad SA, *et al. Epicoccum* sp., an emerging source of unique bioactive metabolites. *Acta Pol Pharm.* 2016; **73:** 13-21.
49. Fisher MC, Gow NA, Gurr SJ. Tackling emerging fungal threats to animal health, food security and ecosystem resilience. *Philos Trans R Soc Lond B Biol Sci.* 2016; **371 (1709).** pii: 20160332.
50. Fones HN, Fisher MC, Gurr SJ. Emerging fungal threats to plants and animals challenge agriculture and ecosystem resilience. *Microbiol Spectr.* 2017; 5(2). PMID: 28361733.
51. Frazer J. Fungi on the march. *Sci Am.* 2013; **309:** 50-7.
52. Gamaletsou MN, Drogari-Apiranthitou M, Denning DW, *et al.* An estimate of the burden of serious fungal diseases in Greece. *Eur J Clin Microbiol Infect Dis.* 2016; **35:** 1115-20.
53. Gandham NR, Jadhav SV, Sardar M, *et al.* The spectrum and aetiology of mycotic infections from a tertiary care hospital from Western part of India. *J Clin Diagn Res.* 2013; **7:** 2157-9.
54. Gangneux JP, Bougnoux ME, Hennequin C, *et al.* An estimation of burden of serious fungal infections in France. *J Mycol Med.* 2016; **26:** 385-90.
55. Giacomazzi J, Baethgen L, Carneiro LC, *et al.* The burden of serious human fungal infections in Brazil. *Mycoses.* 2016; **59:** 145-50.
56. Goldani LZ, Zavascki AP, Maia AL. Fungal thyroiditis: an overview. *Mycopathologia.* 2006; **161:** 129-39.
57. Goralska K, Blaszkowska J. Parasites and fungi as risk factors for human and animal health. *Ann Parasitol.* 2015; **61:** 207-20.
58. Govender NP, Chiller TM, Poonsamy B, *et al.* Neglected fungal disease in sub-Saharan Africa: A call to action. *Curr Fungal Infect Rep.* 2011; **5:** 224-32.
59. Gugnani HC, Denning DW. Burden of serious fungal infections in the Dominican Republic. *J Infect Public Health.* 2016; **9:** 7-12.
60. Gugnani HC, Denning DW. Estimated burden of serious fungal infections in Jamaica by literature review and modelling. *West Indian Med J.* 2015; **64:** 145-9.
61. Gugnani HC, Denning DW, Rahim R, *et al.* Burden of serious fungal infections in Bangladesh. *Eur J Clin Microbiol Infect Dis.* 2017; **36:** 993-7.
62. Guto JA, Bii CC, Denning DW. Estimated burden of fungal infections in Kenya. *J Infect Dev Ctries.* 2016; **10:** 777-84.
63. Hassan IA, Critten P, Isalska B, *et al.* Audit of laboratory mycology services for the management of patients with fungal infections in the northwest of England. *J Clin Pathol.* 2006; **59:** 759-63.
64. Hawksworth DL. Global species numbers of fungi: Are tropical studies and molecular approaches contributing to a more robust estimate? *Biodivers Conserv.* 2012; **21:** 2425-33.
65. Hawksworth DL. Mycological Research: Instructions and guidelines for authors. *Mycol Res.* 2007; **111:** 117-26.
66. Hay RJ. Fungal infections. *Clin Dermatol.* 2006; **24:** 201-12.
67. Hombalkar NN, Vaze D, Guha P, *et al.* Devastating penile mycosis leading to penile gangrene. *Urology.* 2013; **82:** 704-6.
68. Homei A. Specialization and medical mycology in the US, Britain and Japan. *Stud Hist Philos Biol Biomed Sci.* 2008; **39:** 80-92.
69. Horn DL, Fishman JA, Steinbach WJ, *et al.* Presentation of the PATH Alliance registry for prospective data collection and analysis of the epidemiology, therapy and outcomes of invasive fungal infections. *Diagn Microbiol Infect Dis.* 2007; **59:** 407-14.
70. Hsu LY, Wijaya L, Shu-Ting Ng E, Gotuzzo E. Tropical fungal infections. *Infect Dis Clin North Am.* 2012; **26:** 497-512.
71. Huh K, Ha YE, Denning DW, *et al.* Serious fungal infections in Korea. *Eur J Clin Microbiol Infect Dis.* 2017; **36:** 957-63.
72. Jabeen K, Farooqi J, Mirza S, Denning D, Zafar A. Serious fungal infections in Pakistan. *Eur J Clin Microbiol Infect Dis.* 2017; **36:** 949-56.
73. Jacobs CS, Etherton MR, Lyons JL. Fungal infections of the central nervous system. *Curr Infect Dis Rep.* 2014; **16:** 449.

74. Khandelwal N, Gupta V, Singh P. Central nervous system fungal infections in tropics. *Neuroimaging Clin N Am.* 2011; **21:** 859-66.
75. Khwakhali US, Denning DW. Burden of serious fungal infections in Nepal. *Mycoses.* 2015; 58 **(Suppl. 5):** 45-50.
76. Klassen-Fischer MK. Fungi as bioweapons. 2006; *Clin Lab Med.* **26:** 387-95.
77. Klimko N, Kozlova Y, Khostelidi S, *et al.* The burden of serious fungal diseases in Russia. *Mycoses.* 2015; **58 (Suppl. 5):** 58-62.
78. Knoke M, Bernhardt H, Schwesinger G. Is there a need for autopsies in the management of fungal disease? *Mycoses.* 2008; **51:** 291-300.
79. Kohler JR, Casadevall A, Perfect J. The spectrum of fungi that infects humans. *Cold Spring Harb Perspect Med.* 2014; **5:** a019273.
80. Kourbeti IS, Mylonakis E. Fungal central nervous system infections: Prevalence and diagnosis. *Expert Rev Anti Infect Ther.* 2014; **12:** 265-73.
81. Kugblenu RK, Reeves WK. Mortality from fungal diseases in the US Air Force from 1970 to 2013. *US Army Med Dep J.* 2016; **3-16:** 38-41.
82. Kume H, Yamazaki T, Togano T, *et al.* Epidemiology of visceral mycoses in autopsy cases in Japan: Comparison of the data from 1989, 1993, 1997, 2001, 2005 and 2007 in annual of pathological autopsy cases in Japan. *Med Mycol J.* 2011; **52:** 117-27.
83. Kutlubay Z, Yardımcı G, Kantarcıoglu AS, *et al.* Acral manifestations of fungal infections. *Clin Dermatol.* 2017; **35:** 28-39.
84. Lagrou K, Maertens J, Van Even E, *et al.* Burden of serious fungal infections in Belgium. *Mycoses.* 2015; **58 (Suppl. 5):** 1-5.
85. Larbcharoensub N, Srisuma S, Ngernprasertsri T, *et al.* Invasive fungal infection in Ramathibodi Hospital: A ten-year autopsy review. *J Med Assoc Thai.* 2007; **90:** 2630-7.
86. Lehrnbecher T, Frank C, Engels K, *et al.* Trends in the postmortem epidemiology of invasive fungal infections at a university hospital. *J Infect.* 2010; **61:** 259-65.
87. Lemos AA, Lemos JA, Prado MA, *et al.* Cockroaches as carriers of fungi of medical importance. *Mycoses.* 2006; **49:** 23-5.
88. Lewis RE, Cahyame-Zuniga L, Leventakos K, *et al.* Epidemiology and sites of involvement of invasive fungal infections in patients with haematological malignancies: a 20-year autopsy study. *Mycoses.* 2013; **56:** 638-45.
89. Ligon BL. Penicillin: its discovery and early development. *Semin Pediatr Infect Dis.* 2004; **15:** 52-7.
90. Lollis TR, Bradshaw WT. Fungal prophylaxis in neonates: A review article. *Adv Neonatal Care.* 2014; **14:** 17-23.
91. Lupi O, Tyring SK, McGinnis MR. Tropical dermatology: fungal tropical diseases. *J Am Acad Dermatol.* 2005; **53:** 931-54.
92. Mahajan VA, Mahajan AL, Harte BH, *et al.* Optimal environment for free flaps can also be a haven for fungi! *J Plast Reconstr Aesthet Surg.* 2009; **62:** e661-2.
93. Maheshwari R. Fungal biology in the 21st century. *Current Science.* 2005; **88:** 1406-17.
94. Malani AN, Kauffman CA. Changing epidemiology of rare mould infections: implications for therapy. *Drugs.* 2007; **67:** 1803-12.
95. Medina N, Samayoa B, Lau-Bonilla D, *et al.* Burden of serious fungal infections in Guatemala. *Eur J Clin Microbiol Infect Dis.* 2017; **36:** 965-9.
96. Mohapatra LN. Study of medical mycology in India: An overview. *Indian J Med Res.* 1989; **89:** 351-61.
97. Monga DP, Mohapatra LN. A compilation of published reports of mycoses in animals in India. *Mycopathologia.* 1980; **72:** 3-11.
98. Mora C, Tittensor DP, Adl S, *et al.* How many species are there on earth and in the ocean? *PLoS Biology.* 2011; **9:** e1001127.
99. Morris-Jones R. Fungal infections: An essential guide. *Practitioner.* 2006; **250:** 28-9, 31-2.
100. Mortensen KL, Denning DW, Arendrup MC. The burden of fungal disease in Denmark. *Mycoses.* 2015; **58 (Suppl. 5):** 15-21.
101. Moyad MA. Brewer's/baker's yeast *(Saccharomyces cerevisiae)* and preventive medicine: Part I. *Urol Nurs.* 2007; **27:** 560-1 and Part II. *Urol Nurs.* 2008; **28:** 73-5.
102. Muller J, Evans EG, Negroni R, *et al.* The role of ISHAM: How developing and developed countries can benefit from it. *Med Mycol.* 1998; **36**: S266-70.
103. Murthy JM, Sundaram C. Fungal infections of the central nervous system. *Handb Clin Neurol.* 2014; **121:** 1383-401.
104. Murthy JM. Fungal infections of the central nervous system: The clinical syndromes. *Neurol India.* 2007; **55:** 221-5.
105. Naggie S, Perfect JR. Molds: hyalohyphomycosis, phaeohyphomycosis and zygomycosis. *Clin Chest Med.* 2009; **30:** 337-53.
106. Naik V, Ahmed FU, Gupta A, *et al.* Intracranial fungal granulomas: A single institutional clinicopathologic study of 66 patients and review of the literature. *World Neurosurg.* 2015; **83:** 1166-72.
107. Naik V, Ahmed FU, Gupta A, *et al.* Intracranial fungal granulomas: A single institutional clinicopathologic study of 66 patients and review of the literature. *World Neurosurg.* 2015; **83:** 1166-72.
108. Negroni R, Ellis D, Bulmer G, *et al.* Teaching medical mycology in the year 2000. *Med Mycol.* 1998; **36:** S106-8.
109. Nyirjesy P, Sobel JD. Genital mycotic infections in patients with diabetes. *Postgrad Med.* 2013; **125:** 33-46.
110. Odds FC. Pathogenic fungi in the 21st century. *Trends Microbiol.* 2000; **8:** 200-201.
111. Oladele RO, Denning DW. Burden of serious fungal infection in Nigeria. *West Afr J Med.* 2014; **33:** 107-14.
112. Osmanov A, Denning DW. Burden of serious fungal infections in Ukraine. *Mycoses.* 2015; **58 (Suppl. 5):** 94-100.
113. Ostrosky-Zeichner L. 40 years of medical mycology at JAC. *J Antimicrob Chemother.* 2016; **71:** 3327-9.
114. Padhye AA, Bennett JE, McGinnis MR, *et al.* Biosafety considerations in handling medically important fungi. *Med Mycol.* 1998; **36:** S258-65.

115. Panagopoulou P, Filioti J, Farmaki E, *et al.* Filamentous fungi in a tertiary care hospital: Environmental surveillance and susceptibility to antifungal drugs. *Infect Control Hosp Epidemiol.* 2007; **28:** 60-7.
116. Park JS, Cho SH, Youn SK, *et al.* Epidemiological characterization of opportunistic mycoses between the years 2006 and 2010 in Korea. *J Microbiol Biotechnol.* 2016; **26:** 145-50.
117. Parkes-Ratanshi R, Achan B, Kwizera R, *et al.* Cryptococcal disease and the burden of other fungal diseases in Uganda; Where are the knowledge gaps and how can we fill them? *Mycoses.* 2015; **58 (Suppl. 5):** 85-93.
118. Pegorie M, Denning DW, Welfare W. Estimating the burden of invasive and serious fungal disease in the United Kingdom. *J Infect.* 2017; **74:** 60-71.
119. Perfect JR. Iatrogenic fungal meningitis: tragedy repeated. *Ann Intern Med.* 2012; **157:** 825-6.
120. Perfect JR. Invasive mycoses: Evolving challenges and opportunities in antifungal therapy - Introduction. *Am J Med.* 2012; **125:** S1-2.
121. Pfaller MA, Diekema DJ. Epidemiology of invasive mycoses in North America. Crit Rev Microbiol 2010; **36:** 1-53.
122. Pillai P, Lakhe M, Shetty A, *et al.* Pioneers in medical mycology and mycobacteriology. *J Assoc Physicians India.* 2016; **64:** 94-5.
123. Randhawa HS, Chowdhary A. Medical Mycology in India (1957-2007): Contributions by the VPCI Mycoses Group. *Indian J Chest Dis Allied Sci.* 2008; **50:** 19-32.
124. Richardson M, Lass-Florl C. Changing epidemiology of systemic fungal infections. *Clin Microbiol Infect.* 2008; **14 (Suppl. 4):** 5-24.
125. Rinaldi MG. Controversies in medical mycology. *Dermatology.* 1997; **194 (Suppl. 1):** 45-7.
126. Rodriguez-Tudela JL, Alastruey-Izquierdo A, Gago S, *et al.* Burden of serious fungal infections in Spain. *Clin Microbiol Infect.* 2015; **21:** 183-9.
127. Ruangritchankul K, Chindamporn A, Worasilchai N, *et al.* Invasive fungal disease in university hospital: A PCR-based study of autopsy cases. *Int J Clin Exp Pathol.* 2015; **8:** 14840-52.
128. Ruhnke M, Groll AH, Mayser P, *et al.* Estimated burden of fungal infections in Germany. *Mycoses.* 2015; **58 (Suppl. 5):** 22-8.
129. Sabino R, Verissimo C, Brandao J, *et al.* Serious fungal infections in Portugal. *Eur J Clin Microbiol Infect Dis.* 2017; **36:** 1345-52.
130. Sambasiva Rao R, Lalitha MK, Narang P. Curriculum designing for postgraduates in medical microbiology. *Indian J Med Microbiol.* 1999; **17:** 116-24.
131. Sekkides O. Bringing fungal infections in from the cold. *Lancet Infect Dis.* 2015; **15:** 884-5.
132. Sethi PK, Khanna L, Batra A, *et al.* Central nervous system fungal infections: Observations from a large tertiary hospital in northern India. *Clin Neurol Neurosurg.* 2012; **114:** 1232-7.
133. Sewell DL. Laboratory associated infections and biosafety. *Clin Microbiol Rev.* 1995; **8:** 389-405.
134. Seyedmousavi S, Guillot J, Tolooe A, *et al.* Neglected fungal zoonoses: Hidden threats to man and animals. *Clin Microbiol Infect.* 2015; **21:** 416-425.
135. Shimodaira K, Okubo Y, Nakayama H, *et al.* Trends in the prevalence of invasive fungal infections from an analysis of annual records of autopsy cases of Toho University. *Mycoses.* 2012; **55:** 435-43.
136. Shukla A, Bansal M, Husain M, Chhabra DK. Central nervous system mycosis: Analysis of 10 cases. *Indian J Pathol Microbiol.* 2014; **57:** 591-4.
137. Sinko J, Sulyok M, Denning DW. Burden of serious fungal diseases in Hungary. *Mycoses.* 2015; **58 (Suppl. 5):** 29-33.
138. Slavin MA, Chakrabarti A. Opportunistic fungal infections in the Asia-Pacific region. *Med Mycol.* 2012; **50:** 18-25.
139. Starkey J, Moritani T, Kirby P. MRI of CNS fungal infections: Review of aspergillosis to histoplasmosis and everything in between. *Clin Neuroradiol.* 2014; **24:** 217-30.
140. Steinbach WJ, Mitchell TG, Schell WA, *et al.* Status of medical mycology education. *Med Mycol.* 2003; **41:** 457-67.
141. Sukapure, RS, Thirumalachar, MJ. Studies on *Cephalosporium* species from India - I. *Mycologia.* 1963; **55:** 563-9.
142. Suzuki Y, Kume H, Togano T, *et al.* Epidemiology of visceral mycoses in autopsy cases in Japan: The data from 1989 to 2009 in the annual of pathological autopsy cases in Japan. *Med Mycol.* 2013; **51:** 522-6.
143. Taj-Aldeen SJ, Chandra P, Denning DW. Burden of fungal infections in Qatar. *Mycoses.* 2015; 58 **(Suppl. 5):** 51-7.
144. Tauber SC, Eiffert H, Kellner S, *et al.* Fungal encephalitis in human autopsy cases is associated with extensive neuronal damage but only minimal repair. *Neuropathol Appl Neurobiol.* 2014; **40:** 610-27.
145. Tilavberdiev SA, Denning DW, Klimko NN. Serious fungal diseases in the Republic of Uzbekistan. *Eur J Clin Microbiol Infect Dis.* 2017; **36:** 925-9.
146. US Department of Health and Human Services. Biosafety in Microbiological and Biomedical Laboratories, 5th edn. US Government Printing Office. Washington DC: 2007.
147. Vallabhaneni S, Mody RK, Walker T, *et al.* The global burden of fungal diseases. *Infect Dis Clin North Am.* 2016; **30:** 1-11.
148. Viegas C, Alves C, Carolino E, *et al.* Prevalence of fungi in indoor air with reference to gymnasiums with swimming pools. *Indoor and Built Environment.* 2010; **19:** 555-61.
149. Walsh TJ, Groll A, Hiemenz J, *et al.* Infections due to emerging and uncommon medically important fungal pathogens. *Clin Microbiol Infect.* 2004; **10:** S48-66.
150. Walsh TJ, Groll AH. Emerging fungal pathogens: Evolving challenges to immunocompromised patients for the twenty-first century. *Transpl Infect Dis.* 1999; **1:** 247-61.
151. Warnock DW, Campbell CK. Centenary review: Medical mycology. *Mycol Res.* 1996; **100:** 1153-62.
152. Warnock DW. Fungal diseases: An evolving public health challenge. *Med Mycol.* 2006; **44:** 697-705.

153. Wennergren G, Lagercrantz H. 'One sometimes finds what one is not looking for' (Sir Alexander Fleming): the most important medical discovery of the 20th century. *Acta Paediatr.* 2007; **96:** 141-4.
154. WHO. *Laboratory Biosafety Manual.* 3rd edn, Geneva: 2004.
155. Yao Z, Liao W. Fungal respiratory disease. *Curr Opin Pulm Med.* 2006; **12:** 222-7.
156. Yasmin AA. Seminal or male genital mycoses: A reason for hard recurrent vaginal mycoses. *J Mens Health.* 2009; **3:** 261.
157. Yoon HJ, Choi HY, Kim YK, *et al.* Prevalence of fungal infections using National Health Insurance data from 2009-2013, South Korea. *Epidemiol Health.* 2014; **36:** e2014017.
158. Zaki SM, Denning DW. Serious fungal infections in Egypt. *Eur J Clin Microbiol Infect Dis.* 2017; **36:** 971-4.
159. Zarrin M, Zarei Mahmoudabadi A. Central nervous system fungal infections: A review article. *Jundishapur J Microbiol.* 2010; **3:** 41-7.
160. Zurita J, Denning DW, Paz-Y-Mino A, *et al.* Serious fungal infections in Ecuador. *Eur J Clin Microbiol Infect Dis.* 2017; **36:** 975-81.

CHAPTER 2

Fungal Morphology

The fungi are diverse group of heterotrophic organisms that exist as saprotrophic, commensal or pathogens. Most of these are found over decaying vegetative material as well as in the soil. At the ambient temperature, fungi grow either as yeasts or spore-bearing molds, which are the infectious form to man and animals. Their morphological characteristics and functions are quite different from other infectious organisms like viruses, bacteria or parasites.

The science of Medical Mycology, has been termed as 'exercise in contemplative observation'. The characteristics of any fungal isolate are very significant, like its hyphal/conidial color and morphology on the obverse, pigmentation on reverse, process of conidium or spore formation and hyphal structures on microscopic examination. These morphological details, if examined together and correctly interpreted, lead to an accurate phenotypic identification of a particular fungus. Therefore, morphology is an important tool in understanding medical mycology as it constitutes the basis of conventional as well as molecular diagnostic techniques.

The Fungi are saprophytically living organisms, which are distantly related to plants and more closely related to animals but different from either of these groups. These can be recognized by following five characteristics:

- The fungal cells contain nuclei with chromosomes like plants and animals but unlike bacteria.
- The fungi cannot photosynthesize as they lack chlorophyll and are heterotrophic like animals.
- Fungi absorb their food as they are osmotrophic.
- They develop very diffuse bodies made of spreading network of very narrow, tubular, branching filaments called hyphae. These filaments exude enzymes and absorb food, at their growing tips. Although filaments are very narrow, they are very long and can explore and exploit food substrates very efficiently.
- They usually reproduce by means of spores, which develop on range of unique structures such as mushrooms, cup fungi and many other kinds of fruiting bodies are released from there.

The organisms that have all the above-mentioned features, can be found in two of the seven living Kingdoms- they make up an entire kingdom **Eumycota** or true fungi, their cell walls are made largely of chitin and those which are a part of kingdom Chromista, their cell walls are made of cellulose, which also include brown algae - wracks and kelps (seaweed). The Chromista may look very different but have similar swimming cells with two flagella at a stage in their lives.

The important morphological features, cell structure, reproduction, nutritional requirement and thermal dimorphism in medically significant fungi are described in this Chapter.

A. Fungal Morphology

The fungi are **eukaryotic**, unicellular or multicellular organisms, with range of internal membrane system, membrane bound organelles and well-defined cell wall, which is largely composed of polysaccharides and chitin. These show substantial morphological variation in size and shape but broadly divided into two main groups:

(a) Yeasts

(b) Molds

(a) Yeasts

The yeasts are unicellular fungi, which reproduce by an asexual process known as **budding**. The blastospore formed, further propagate yeast cells to its next generations. The budding with narrow-base is seen in *Candida*, *Cryptococcus* and many more yeasts **(Fig. 2.1A)**. The broad-based budding is seen in *Blastomyces dermatitidis*

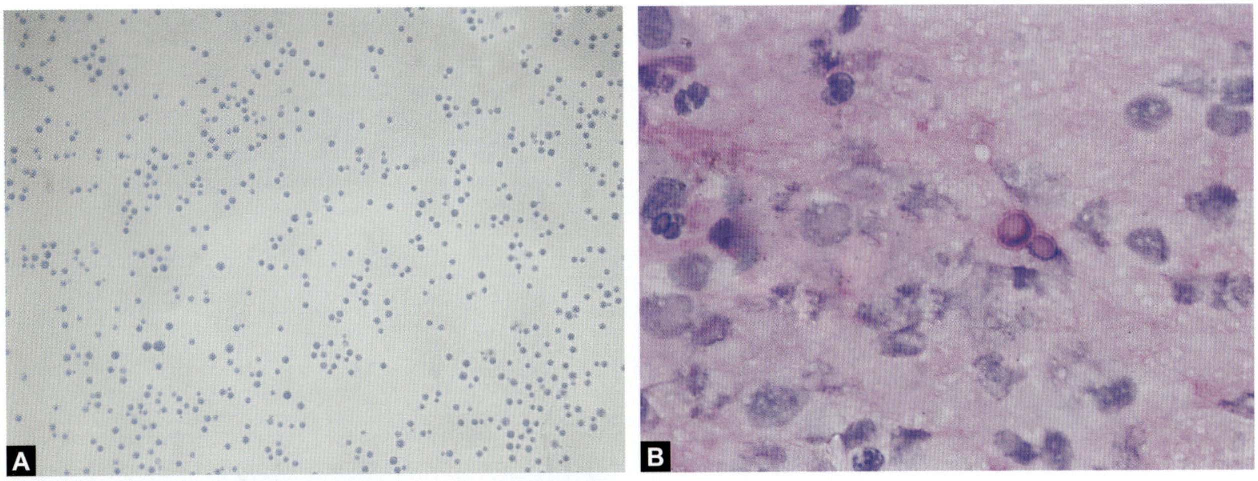

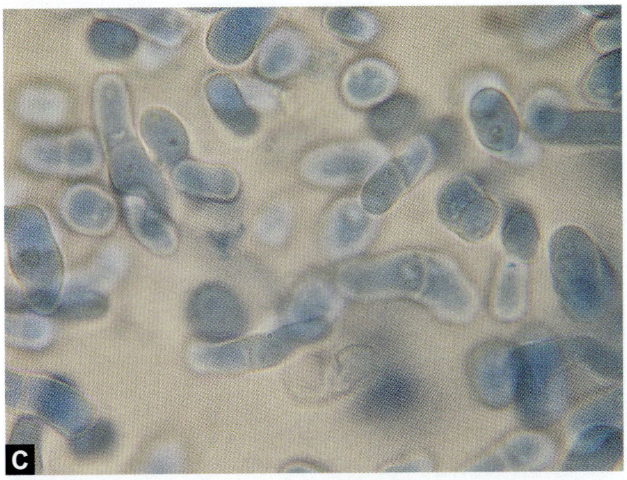

Figs. 2.1A to C. Yeast cells showing division by (A) budding with narrow-base (B) broad-base and (C) transverse septae (LCB × 200, PAS × 400, LCB × 400).

(Fig. 2.1B). Another way of multiplication in yeasts is called **fission** *i.e.* transverse septum formation (planate division) seen in non-pathogenic fungus such as *Schizosaccharomyces* species **(Fig. 2.1C)**. The cell develops a protuberance that enlarges and eventually separates from parent cell. Similar type of fission division is seen in dimorphic fungus *Talaromyces marneffei,* which is the only example from the pathogenic fungi. Therefore, except *Blastomyces dermatitidis* and *Talaromyces marneffei,* rest all medically important yeasts divide by narrow-based budding.

The yeasts may produce chains of elongated cells known as **pseudohyphae (Fig. 2.2)**, which resemble mycelia of molds and these may also produce true mycelia. Some yeasts also reproduce by sexual process and show teleomorph state as seen in *Cr.neoformans i.e. Filobasidiella neoformans.*

The yeasts are neither natural nor formal taxonomic group but are growth forms shown by a range of unrelated fungi. In some cases, they are merely a phase of growth in the life cycle of filamentous fungi, which takes place only under specific environmental conditions. Therefore, the term ***'yeast'*** is of no taxonomic significance rather it is only used to describe morphological form of a fungus. However, most of the yeasts are categorized under Ascomycota and a small percentage is placed in Basidiomycota.

The yeasts are ubiquitous in environment being found on fruits, vegetables and other plant materials (exogenous). However, some of them live as normal inhabitants in and/or on human body (endogenous). Therefore, yeasts sometimes may be found in clinical specimens as commensals without much clinical significance.

(b) Molds

The fungal spores germinate and grow into slender tubular thread-like structures called hyphae which may be

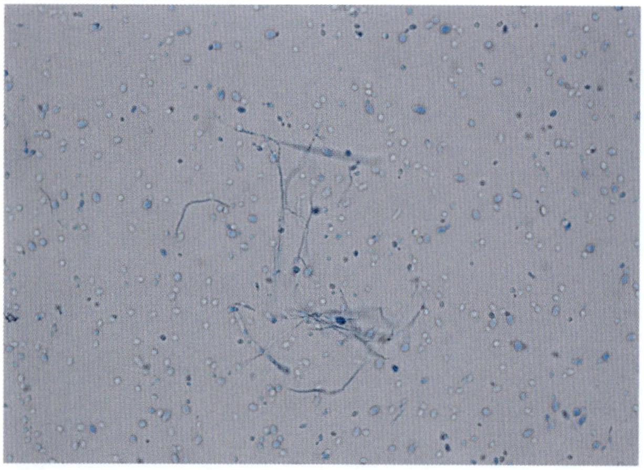

Fig. 2.2. Budding yeast cells with pseudohyphae seen in *Candida* species (LCB × 400).

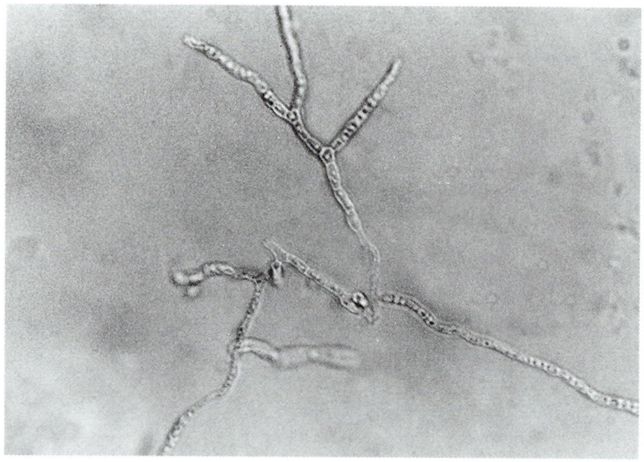

Fig. 2.3. Septate hyphae of fungi as seen in *Aspergillus* species (KOH × 400).

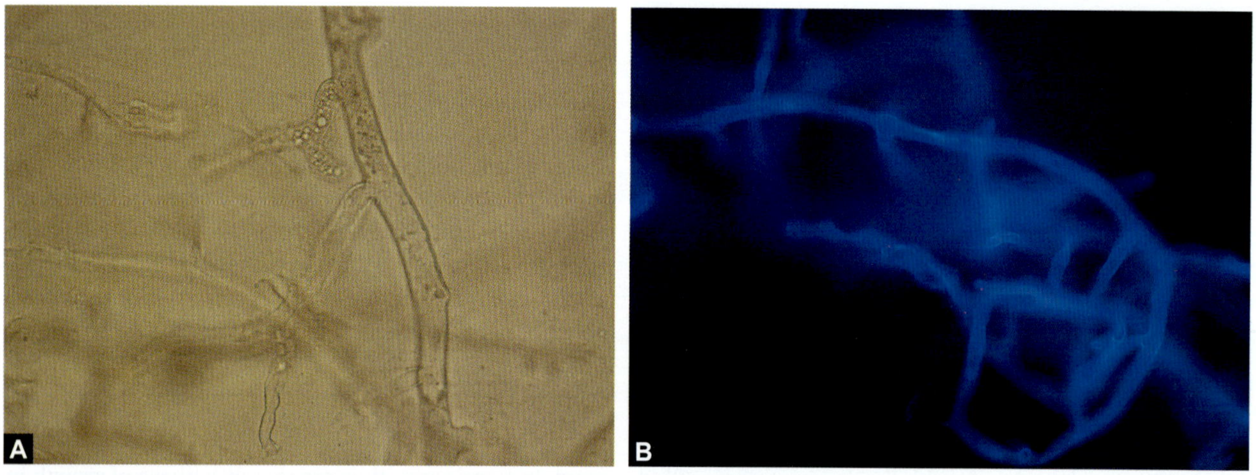

Figs. 2.4A and B. Non-septate broad ribbon-like hyphae with wide-angle branching seen in mucormycetes (KOH and CFW × 400).

septate or non-septate. All molds are composed of branching **hyphae**. They grow by apical extension, forming an interwoven mass called **mycelium**. The hypha is the structural unit of the mycelium. In most of fungi, the hyphae have regular cross-walls *i.e.* septate and the breadth of the cell walls remain almost parallel **(Fig. 2.3)** as seen in *Aspergillus, Fusarium, Scedosporium, Penicillium* and many other fungal genera, which is very important feature to differentiate from the phaeoid fungi, where this uniformity is not there rather many hyphal swellings are present.

However, in lower fungi the hyphae are usually sparsely septate, ribbon-like with wide-angle branching popularly known as non-septate as seen in mucormycetes **(Fig. 2.4A)**. Some of stainings like calcofluor white, are very useful in visualization of this feature whether septae are present or not **(Fig. 2.4B)**. Actually there is no existence of such a cell structure where no septae are present. But in mucormycetes these are distantly placed thereby are not visible in single field of microscopy hence called non-septate, which is the most familiar terminology among the microbiologists as well as pathologists. Somewhat similar type of sparsely septate (non-septate) hyphae are seen in *Pythium insidiosum* also, which is a hydrophilic pseudofungus and morphologically mimics mucormycetes thereby invariably misdiagnosed.

The septae may be complete, partial or perforated. The mucormycetes although have sparse septae, however, these are complete, whereas in other fungi, there are perforated septae.

The hyphae that grow as submerged or on the surface of culture medium are called **vegetative hyphae** because they are responsible for absorption of the nutrients. However, those which project above the surface of a medium are called **aerial hyphae** and produce specialized structures called **conidia**.

The **homothallic** fungi are self-fertile in which sexual reproduction takes place within thallus, where single mycelium is capable of reproducing sexually, whereas **heterothallic** are self-sterile and require two compatible thalli for sexual reproduction.

Although fungi exist either as yeast or mold but the traditional way of classification is empirical as it depends on the cell morphology and not on taxonomy, dividing it into four types, which are as follows:

(i) Yeasts: The yeasts are unicellular organisms and divide by budding. Most of these are considered as non-pathogenic like *Saccharomyces cerevisiae* but few are pathogenic as well *i.e. Cryptococcus* species.

(ii) Yeast-like: These fungi also reproduce by budding and exist as yeasts for a part of their life cycle but buds fail to get separated hence elongation takes place forming pseudohyphae as seen in *Candida* species. This group should not be misunderstood as dimorphic because both structures (yeasts and pseudohyphae) are not temperature dependent and show same morphology at 25°C as well as 37°C.

(iii) Molds: These fungi develop from spores, which germinate to form vegetative hyphae. The hyphae may or may not have septation. Dermatophytes, *Aspergillus*, *Penicillium* and *Mucor* are a few examples of molds.

(iv) Dimorphic Fungi: These fungi have two types of morphology at depending on different temperatures such as yeast forms or spherule at body temperature *i.e.* 37°C and filamentous form at room temperature *i.e.* 25°C. The true pathogenic fungi are usually dimorphic and endemic in nature.

Moreover, in some higher fungi the hyphae may be compacted together in such a way to form fungal tissue giving rise to macroscopic structures such as mushrooms and toadstools.

Most of the fungi are found as hyaline yeast or hyphal form but some of them are darkly colored known as **phaeoid** fungi. The term phaeoid is now being used replacing the older one *i.e.* **dematiaceous** fungi. Some of the phaeoid fungi, which are yeast-like in the primary culture but develop mycelial in due course of incubation like *Aureobasidium pullulans, Wangiella dermatitidis, Exophiala jeanselmei, Exophiala spinifera* and *Hortaea werneckii.* These are popularly called **black yeasts** (*See* **Figs. 8.1 and 8.2**).

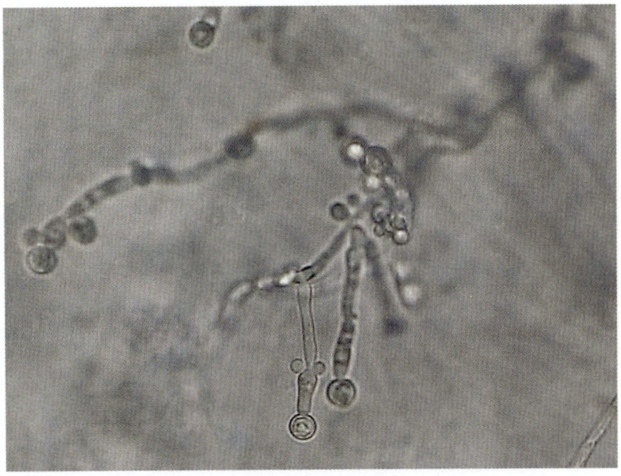

Fig. 2.5. Terminal chlamydospores seen in yeasts *i.e. Candida albicans* and *Candida dubliniensis* (CMA × 200).

(c) Vegetative Structures

There are several structures, which are formed by the vegetative mycelia that have no reproductive value but are of considerable importance in differentiating most of the clinically significant fungi. These arise following modifications of single vegetative cells or hyphae, for instance **chlamydospores** develop from cells that become enriched with nutritive materials and form a thick wall, which is resistant to adverse conditions. The chlamydospores are commonly seen in yeast-like fungi *i.e. Candida albicans* and *C.dubliniensis* (**Fig. 2.5**). They are usually larger than other cells and may be formed singly or in groups and in an intercalary, terminal or sessile position.

Another unicellular vegetative structure is **arthrospore** where many septae in the hyphae are formed to give chains of small cuboidal or rectangular spores with a slightly thickened wall that disarticulate at maturity. The arthrospore production is characteristic of dermatophyte and the mycelial phase of *Coccidioides* species (**Fig. 2.6**).

A brief description of some morphological structures is described here with examples of fungi where these are regularly encountered and is very helpful in identification of the clinical isolate. These are also described in subsequent Chapters *e.g.* **Chapter 10** on dermatophytosis, wherein the causative fungi have most of such morphological forms

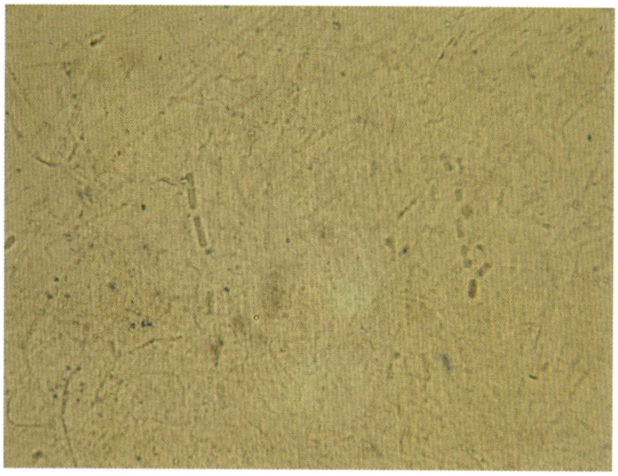

Fig. 2.6.. Alternate arthroconidia during mycelial phase of *Coccidioides* species grown at 25°C (LCB × 400).

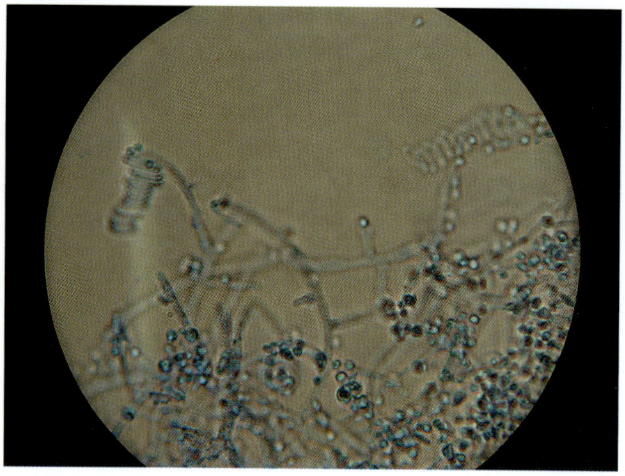

Fig. 2.7. Spiral hyphae usually seen in *Trichophyton mentagrophytes* (LCB × 400).

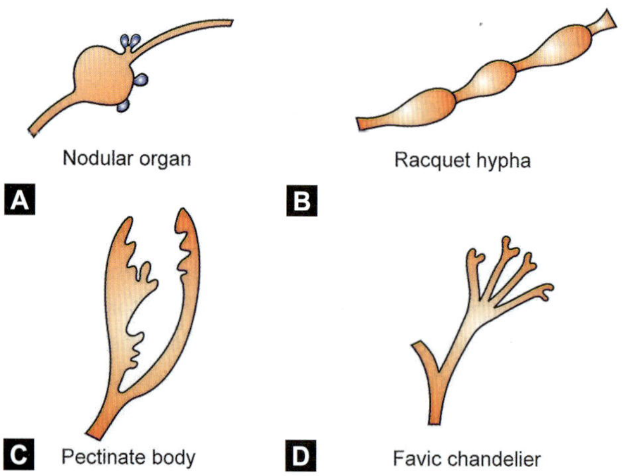

Figs. 2.8A to D. Diagrammatic sketches of various vegetative structures commonly seen in dermatophytes.

encountered during routine diagnostic services. These fungal morphological descriptions are:

(i) Spiral hyphae are corkscrew-like turns of mycelium, which are similar to the coiled filaments of actinomycetes *i.e. Streptomyces* species. These are observed in several pathogenic fungi, especially in dermatophytes *e.g. Trichophyton mentagrophytes* and rarely in *T.tonsurans* **(Fig. 2.7)**. The coils of *T.mentagrophytes* are uniform as compared to *T.tonsurans*.

(ii) Nodular organ or **knot body** is an enlargement in mycelium that consists of closely twisted hyphae. It is seen in the older fluffy portion of *Microsporum canis* and *T.mentagrophytes* **(Fig. 2.8A)**. They are more commonly demonstrated on the cornmeal agar as compared to Sabouraud dextrose agar.

(iii) Racquet mycelium is the hypha, which shows regular enlargement of one end of each segment, large and small ends being in opposition **(Fig. 2.8B)**. They are usually seen in some species of *Microsporum*, *Epidermophyton floccosum* and *T.mentagrophytes*.

(iv) Pectinate body is a unilateral, short, irregular projection or protuberance formed on one side of the hypha that gives it an appearance of a broken comb. They are also commonly seen in dermatophytes, especially *Microsporum audouinii* **(Fig. 2.8C)**.

(v) Favic chandeliers are many short, multiple branches appearing at the end of hypha. They resemble reindeer horn or chandelier. They are seen in *T.schoenleinii* and *T.violaceum* **(Fig. 2.8D)**.

(vi) Peridial hypha is a wide, indented, multi-septate hypha that may terminate in spiral structure as seen in *T.mentagrophytes*.

(vii) Pycnidium is a mycelial structure of mitosporic fungi resembling the fruiting body (perithecium) of some ascomycetes. This is a spherical to flask-shaped enclosed structure with an apical opening *i.e.* ostiole, inside which conidia are formed as seen in Coelomycetes. This is filled with asexually produced conidia. The body is large, up to several millimeters in diameter and may have a hard wall surrounded by peridial hyphae **(Fig. 2.9)**.

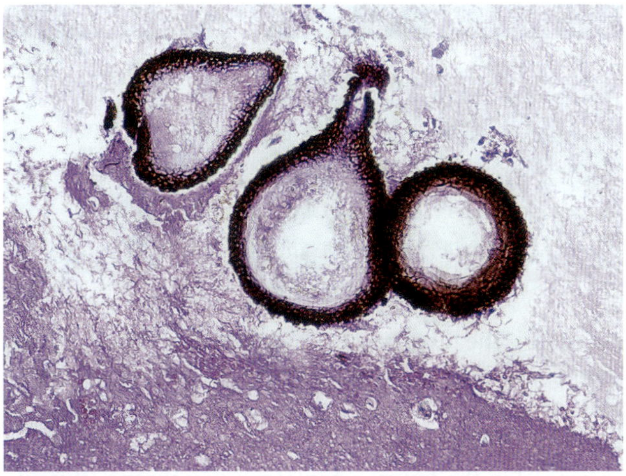

Fig. 2.9. Pycnidia of coelomycetes seen in the debris of nasal crust (H&E × 200)

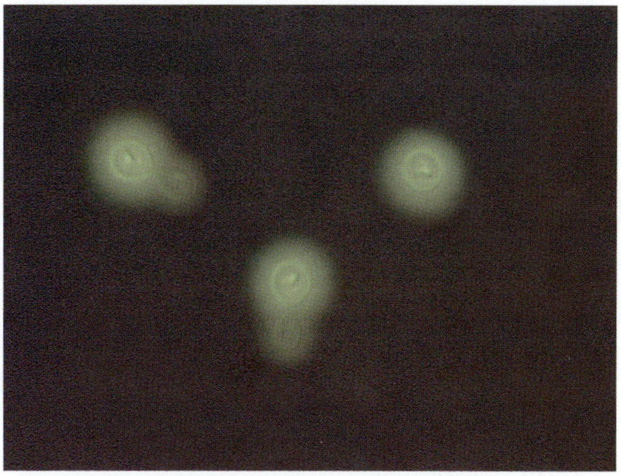

Fig. 2.10. Encapsulated budding yeasts of basidiomycetous fungus *Cryptococcus neoformans* (Nigrosin Staining × 400).

B. Cell Structure

The nuclei of all fungi contain nucleolus and chromosomes that are bound by a nuclear membrane like that of other eukaryotic organisms. The hyphal cells in septate hyphae may be uninucleate, binucleate or multinucleate. For the most part of life cycle, cellular and nuclear division are independent events, especially with respect to vegetative growth. The fungi have cellular organelles such as mitochondria, 80S ribosomes, vacuoles, lipid bodies and other storage inclusions. However, centriole, a barrel-shaped organelle found in most of the animal cells, is absent in fungi as in higher plants.

The cell wall of fungi consists of chitin, chitosan, glucan, mannan and other components in various combinations. It is unlike cell wall of plants, which is mainly composed of cellulose. The fungi are heterotrophs, therefore, require preformed organic compounds as carbon source. They do not contain chlorophyll in their cells. The lack of chlorophyll profoundly affects the functioning of fungi as they are not dependent of light and can occupy dark habitats. They can grow in any direction and invade the interior of the substrate with absorptive filaments.

The cell morphology of fungi is almost similar to any member of the Superkingdom **Eukaryota** and few of the salient features are described below:

(a) Capsule

Some fungi produce an extracellular polysaccharide in the form of **capsule** *e.g. Cryptococcus* species, which is demonstrated as a hallo surrounding the yeasts by negative stainings like India ink or nigrosin stain **(Fig. 2.10)**. The capsule determines virulence and plays an important role in eliciting host immune response and provides a basis for diagnostic tests like latex agglutination and lateral flow assay used in cryptococcosis.

(b) Cell Wall

The fungi possess a dynamic and plastic, multilayered rigid cell wall located external to the plasmalemma that is unlike the mammalian cells. The presence of cell wall is essential to several aspects of biology and pathogenicity of the fungi. Thus, cell wall is the structure responsible for maintaining shape that characterizes each growth form like yeast and hyphae and it acts as a permeability barrier that protects the protoplast against physical and osmotic injuries. In addition, cell wall plays nutritional roles and is the structure that mediates the initial interaction between fungus and the environment.

This cell wall is structurally and biochemically complex, containing **chitin**, water insoluble homopolymer of N-acetylglucosamine (GlcNAC) and polymer of N-acetyl-D-glucosamine in β 1-4 glycosidic linkage, as its structural foundation. Mannans, glucans and other complex polysaccharides in association with polypeptides are layered on chitin.

In filamentous fungi, biosynthesis of chitin occurs at the growing tip and is controlled by the activity of **chitin synthase**. This enzyme exists in cytosol in discrete membrane-bound packets called **chitosomes**. The latent state

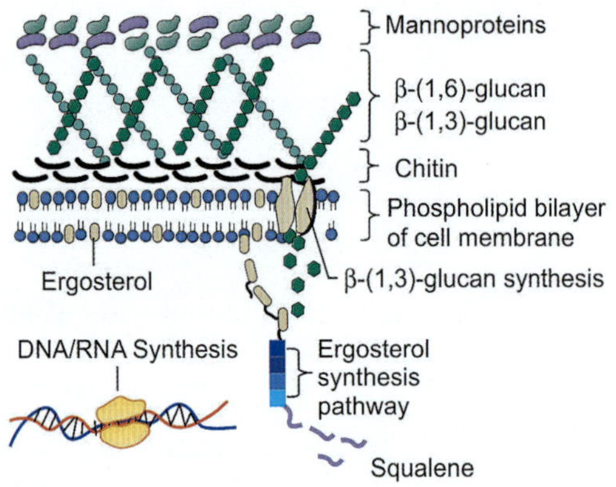

Fig. 2.11. Diagrammatic sketch of cell wall and cell membrane (plasmalemma) of fungi showing various layer-wise components.

(zymogen) of chitin synthase is the inactive form and active form is found in plasmalemma. Polymerization of chitin microfibrils occurs outside this membrane. It is also believed that many fungi have more than one chitin synthases.

There are different types of fungal derived materials, which may be used as crude or purified antigens. These antigens are prepared by a variety of physicochemical and immunochemical techniques. Hence the cell wall of pathogenic fungi is important for variety of reasons:

- This contains specific adhesive molecules that are contact point of organism for attachment and subsequent invasion.
- It acts as a protective barrier, which must be considered while choosing antifungal drugs that need to enter cell wall for their potentially effective action.
- The chitin and β-glucans found in pathogenic fungi are not present in the host, therefore, such compounds particularly the enzymes involved in their biosynthesis and degradation, are potentially safe target sites of the antifungal drugs **(Figs. 2.11 and *See* Fig. 6.1)**.

The association of certain polysaccharides, with broad fungal taxonomic groups, is shown at glance in **Table 2.1**.

Table 2.1. Major Fungal Cell Wall Polysaccharides.

Fungal Group	*Polysaccharides*
Lower aquatic fungi	Cellulose
Mucormycetes	Chitin, chitosan
Ascomycetous yeasts	β-glucans, mannans
Basidiomycetous yeasts	Chitin, mannans
Fungi with septate hyphae	Chitin, β-glucans

(c) Plasmalemma

The plasma membrane or plasmalemma is the outer boundary of protoplast enclosing complex cytosol, next to the cell wall and also called **cell membrane** or **cytoplasmic membrane**. The term plasma membrane is used more frequently while discussing prokaryotes. It consists of a single membrane composed of glycoproteins, lipid and ergosterol. The fungi possess ergosterol in contrast to cholesterol, which is the major sterol found in tissue of mammals. This fact is clinically very significant because most of the antifungal strategies are currently based on targeting ergosterol in the fungal cytoplasmic membrane. This site is targeted by azoles, allylamines and polyene macrolide antibiotics like amphotericin B and nystatin **(Fig. 2.11 and *See* Fig. 6.1)**. The recently classified atypical fungus, *Pneumocystis jirovecii,* is not having ergosterol in its cell membrane hence these antifungal drugs are not effective for treating pneumocystosis rather routine antimicrobials like co-trimoxazole are very effective.

(d) Cytosol

In comparison to bacteria, fungal cells possess complex **cytosol**. As fungi are eukaryotic organisms, the cells have nucleus, nuclear membrane, mitochondria, microvesicles, microtubules, ribosomes, golgi apparatus, double membrane endoplasmic reticulum and other cytoplasmic structures. The fungi have flat mitochondrial cristae. The fungal cell may be uninucleate or multinucleate. The nuclei of the fungi are enclosed by membrane and contain most of the cellular DNA. They have a true nucleolus rich in RNA. A unique property of the nuclear membrane is that it persists throughout the metaphase of mitotic cycle. This is in contrast to the nuclear membrane of plant and animal cells that dissolves and then re-forms after chromosomes segregate to their centromeres.

The fungi differ clearly from higher plants in structure, nutrition and reproduction. The fungal cytoplasm is enclosed by rigid cell wall that may contain cellulose or chitin-like substance. The fungi and bacteria are alike in lacking chlorophyll thus unable to photosynthesize but differ from each other in many ways. The points of differentiation are as follows:

- Having greater size and more complex morphological development. These can be unicellular or multicellular.

Table 2.2. Salient Differentiating Features between Fungi and Bacteria.

Features	*Fungi*	*Bacteria*
Taxonomic status	Eukaryotic	Prokaryotic
Cell wall composition	Multilayered; chitin, glucans, mannan, polysaccharides and peptides, capsule found in some members	Muramic and teichoic acids, peptidoglycan; lipopolysaccharides, capsule present in some members
Plasmalemma	Contain ergosterol except in *Pneumocystis*	Lack ergosterol except in *Mycoplasma*
Cytoplasmic contents	Include mitochondria, endoplasmic reticulum	Lack mitochondria and endoplasmic reticulum
Nucleus	True nucleus with nuclear membrane; paired chromosomes	Nuclear body equivalent to a single chromosome without nuclear membrane
Diameter	Average 5-10 μm; may be >100 μm in length	< 2 μm
Morphology	Yeast: ovoid, round Mold: filamentous branching hyphae, septate, non-septate	Coccal, bacillary; spirochaetal, some branching filamentous bacteria (*Actinomyces, Nocardia* species)
Spores	Asexual and sexual reproductive spores	Non-reproductive endospores
Staining attributes	Gram-positive; non-acid fast, specially stained with PAS and GMS	Gram-positive, Gram-negative; acid fast bacilli (*Mycobacterium, Nocardia* species)
Reproduction	Sexual (meiosis) and asexual (mitosis), both stages result in spore formation; yeasts forms of *Talaromyces marneffei* divide by binary fission	Binary fission

- Having rigid cell wall containing chitin, mannans and other polysaccharides.
- Presence of sterol in the cytoplasmic membrane.
- Presence of true nuclei with nuclear membrane and paired chromosomes.
- Modes of reproduction may be sexual or asexual and/ or both.

The fungi are eukaryotic and have properties that differ from the prokaryotic bacteria and important differentiating features between fungi and bacteria are given in **Table 2.2**.

(e) Woronin Bodies

The Woronin bodies are electron-dense, highly refractive, globose to oval microbodies found only in filamentous fungi, mainly of phylum Ascomycota and some Deuteromycota. These membrane-bound proteinaceous structures are located near septa. Their function is plugging of septal pores after hyphal wounding, which restricts the loss of cytoplasm to the sites of injury. In true yeasts and thalli transforming to yeast phase, these are not seen.

C. Reproduction in Fungi

The reproduction in fungi is by means of spores that are often produced in large numbers. There may be **asexual** or **sexual** cell divisions or reproduction. In the laboratory cultures, fungi mainly produce asexual spores. Therefore, almost all the fungi when they are described as primary cultural isolate, the description is usually about their asexual (imperfect) state because the sexual (perfect) state is rarely manifested, particularly after mating of the opposite types. *Pseudallescheria boydii*, an ascomycete mycelial fungus is exceptional, which is encountered in its sexual state in the clinical specimens, however, its asexual states may also be encountered *i.e. Scedosporium apiospermum* and *Graphium eumorphum*.

(a) Asexual Reproduction

The asexual reproduction is an outcome of mitosis, which involves budding or fission where total number of chromosomes remains same. Depending upon the growth conditions, many fungi can produce more than one type of spores. The laboratory molds, mainly produce asexual spores. As distinct from vegetative spores *i.e.* chlamydospores, arthrospores **(Fig. 2.12)**, asexual spores are usually formed on or in specialized structures called sporophores and are usually formed in large number as result of mitotic nuclear division, which may occur rapidly and frequently. Asexual spores vary greatly in size, shape, color and complexity but they are constant in these characteristics for a particular species. A fungus may produce more than one

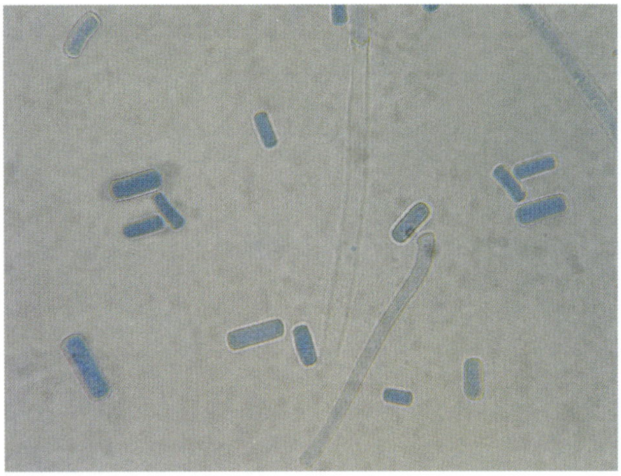

Fig. 2.12. Rectangular arthrospores of fungi seen in *Trichosporon* species (LCB × 400).

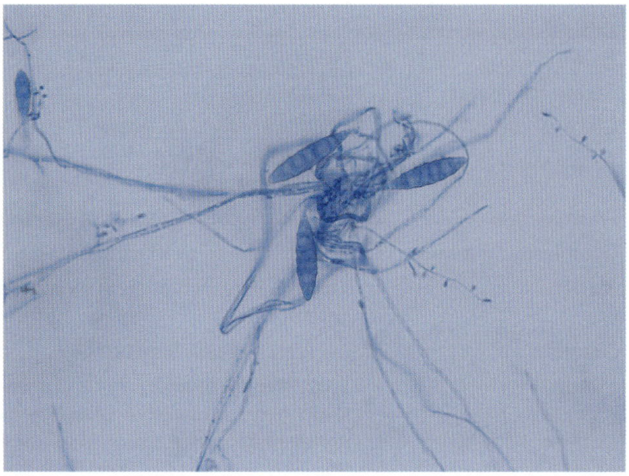

Fig. 2.13. Macroconidia and microconidia seen in dermatophytes species *Microsporum gypseum* (LCB × 400).

type of asexual spores and they are usually designated as microspores (**microconidia**) and macrospores (**macroconidia**) depending on the relative sizes as produced by *Microsporum gypseum* and shown in **Figure 2.13**. These spores may be borne endogenously within **sporangium** known as **sporangiospores** or they may be borne exogenously and called **conidiospores** or simply **conidia**. The arrangements of the conidia may be the following:

(i) Acropetal: Youngest conidium is at the top of the chain as seen in *Cladosporium* and *Alternaria* species **(Fig. 2.14A)**.

(ii) Basipetal: Youngest conidium is at the bottom of chain as in the phialidic *Aspergillus* **(Fig. 2.14B)** or annellidic *Scopulariopsis* species.

(iii) Sympodial: Characterized by continued growth after the main axis has produced terminal spore by the development of succession of apex, each of which originates below and to one side of the previous apex. Many of the fungi with sympodial conidiation have multicellular conidia, either with transverse septa - phragmoconidia or with both transverse and vertical septa - dictyoconidia. The sympodial arrangement of conidia in *Curvularia* species is shown in **Figure 2.14C**.

(iv) Microcycle Conidiation: As such in normal circumstances, conidia are produced from vegetative hyphae *i.e.* mycelia. However, when fungal species are subjected to stressful conditions, they exhibit an extremely simplified asexual life cycle wherein conidia that germinate directly generate further conidia, without forming mycelia. This phenomenon is called microcycle conidiation and has already been reported in more than a hundred fungal species.

(b) Sexual Reproduction

In sexual reproduction compatible strains mate and haploid strains fuse to form a diploid. It consists of **plasmogamy** (cytoplasmic fusion), **karyogamy** (union of two nuclei) and **meiosis** (haploid formation). The number of chromosomes is reduced to half; preceded by fusion of the protoplasm as well as nuclei of two cells. The fungi are taxonomically classified on the basis of sexual reproduction; however, these stages are rarely observed and very difficult to induce in routine diagnostic laboratory. This can be feasible with appropriate strains and special media are used for induction of sexual spores. The sexual reproduction showing three types of fungal spores (zygospore, ascospore and basidiospore) are also depicted in **Chapter 3**.

In addition, prior to mating in sexual reproduction, fungi communicate with other individuals chemically via pheromones. In every phylum at least one pheromone has been characterized and they range from sesquiterpenes and derivatives of the carotinoid pathway in chytridiomycetes and mucormycetes to oligopeptides in ascomycetes and basidiomycetes.

The paired nuclei in a dikaryotic hyphal cell do not fuse but undergo mitosis simultaneously. When a binucleate cell is ready to divide, hook-like outgrowth called **clamp-connection** arises between the two nuclei and one of these migrates into this branch as seen in **basidiomycetes** and is shown in **Figure 2.15**.

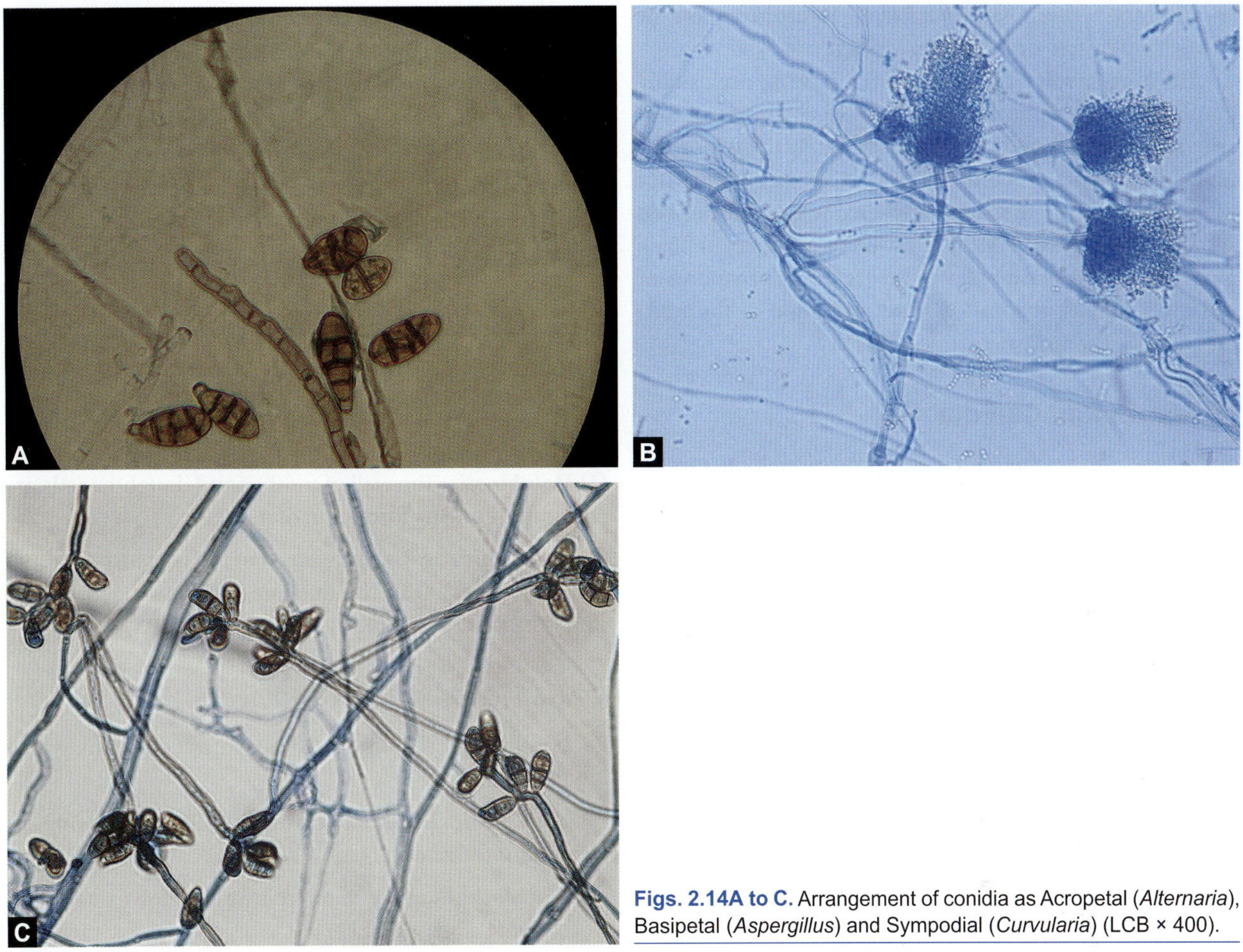

Figs. 2.14A to C. Arrangement of conidia as Acropetal (*Alternaria*), Basipetal (*Aspergillus*) and Sympodial (*Curvularia*) (LCB × 400).

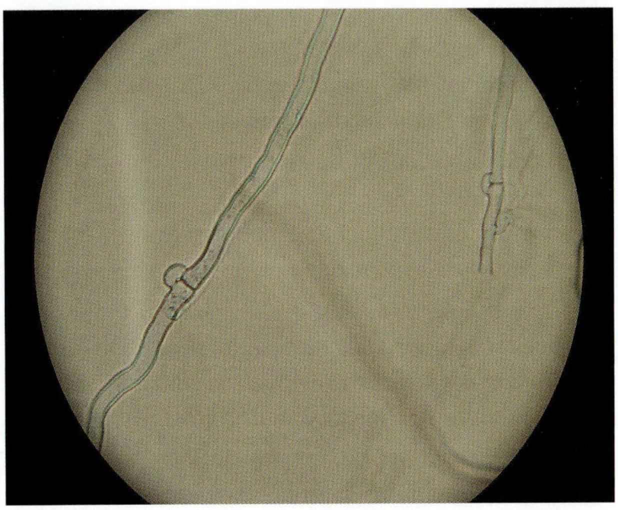

Fig. 2.15. Clamp connection seen in basidiomycetous fungus *Schizophyllum commune* (LCB × 400).

The spore bearing hyphae or sporophores usually differ from vegetative hyphae in one or more features *e.g.* limited growth, vertical position, nature of cell wall, shape of cells and characteristic branching pattern. These features are used as identification mark also.

In contrast to animals and higher plants, fungi are often haploid and meiosis takes place following fusion and not before the formation of sex cells or gametes. Except in certain yeasts, the haploid state is relatively short-lived. The sexual stage of life cycle is known as the **teleomorphic** or **perfect** state, which is the basis of fungal taxonomy thereby the nomenclature also.

There are three fundamental methods of sexual reproduction, which form the basis for dividing fungi into three **phyla**, namely **Mucormycetes**, **Ascomycetes** and **Basidiomycetes**.

All those fungi where sexual reproduction is unknown and the perfect state is yet to be recognized, are allocated an arbitrary phylum, which is called **Deuteromycota** or ***'Fungi Imperfecti'*** or ***'Mitosporic fungi'***.

The fungi involved in the human infections are usually found in their anamorphic state as vegetative form and asexual spores are demonstrated and rarely the teleomorphic state is encountered in the clinical specimens. But there are certain circumstances where the sexually derived spores appear to be very significant. For instance, in black piedra, fungal disease of hair, the asci and ascospores of causative fungus, *Piedraia hortae,* are found in the nodules, which develop on the affected hair.

(c) Spores and Conidia

The terms such as **spore** and **conidium** are no longer interchangeable. The spores are propagules that arise either from meiosis *i.e.* sexual reproduction (zygospores, ascospores or basidiospores) or by mitosis within sporangium *i.e.* asexual reproduction as seen in mucormycetes (sporangiospores). All other asexual, non-motile propagules are called conidia. They are usually borne-naked on specialized hyphae or hyphal branches called conidiophores. The distinction between various types of conidiogenous cell is significant for the identification of fungi.

(d) Conidiogenesis and Conidial Ontogeny

In the fungal asexual reproduction and classification, in 1979 Kendrick refined the concept of conidiogenesis in the *'Taxonomy of Fungi Imperfecti.'* As per the **Ontogenic Classification**, conidia of Fungi Imperfecti form adopt **two** general developmental methods of asexual conidiogenesis. These are called **blastic** and **thallic** process of conidia formation. The blastic or budded conidiation involves 'blowing out' and *de novo* growth of a part of the hyphal element, whereas the thallic conidiation involves conversion of a part of the preformed hyphae into conidium, which may require some enlargement of the cell and secondary wall growth. A conceptual understanding of conidiogenesis is essential for the identification of fungal isolates. The main points of conidial development include the origin of conidium, its wall, type of conidiogenous cell, arrangement of various conidia and site giving rise to conidium.

The conidia may arise from conidiogenous cells in several ways. In **blastic** ontogeny, the conidium originates from a narrow portion of the region, which swells before being cut off from parent cell by septum as seen in *Candida* species. In **thalloblastic** ontogeny, the conidium originates from broad region of the hypha and as with blastic ontogeny, the cell swells before delimitation by the septum. In **thallic** ontogeny, the conidia arise from the entire length of hypha; septa are laid down before each conidium swells as seen in *Geotrichum candidum.* **Arthric** conidia are specialized thallic conidia formed in chains that disarticulate readily. A brief narration of Ontogenic classification showing both these methods is given as below:

(a) Blastic Conidial Development

The blastic conidia are formed by 'blowing out' of the conidiogenous cell, which occurs in both yeasts as well as molds.

(i) Holoblastic: The wall of mother and daughter cells are continuous during formation of new cell.

(ii) Enteroblastic: The outer layer of cell wall of the mother cell is not present in the daughter cell *e.g. Acremonium, Bipolaris* and *Penicillium* species.

(b) Thallic Conidial Development

Each individual cell of hyphae becomes conidium, which is given below:

(i) Holothallic: The inner and outer walls are the same *e.g. Geotrichum* species.

(ii) Enterothallic: The outer cell wall dissolves away *e.g. Coccidioides* species.

(e) Types of Conidiogenesis

(i) Blastoconidium: The conidium is formed by budding directly from hyphae, pseudohyphae or single cell as seen in the yeasts *e.g. Candida albicans.*

(ii) Aleurioconidium: The conidiophore gives rise to truncated conidium formed when conidium expands directly from hyphae or conidiophore and fracture away from the base *e.g. Microsporum* species.

(iii) Annelloconidium: The conidiogenous cell (annelid) produces succession of conidia and bears multiple ring-like scars at the tip of the elongating annelid left by the released spores *e.g. Exophiala* and *Scopulariopsis* species.

(iv) Phialoconidium: A conidium produced by conidiogenous cell (phialide) that produces succession of blastic conidia without increasing in length *e.g. Aspergillus* species, *Penicillium* species and *Phialophora* species.

(v) Poroconidium: This is also called tretoconidia and sympodioconidia; wherein the conidiogenous structure continues to increase in length giving off (poro)conidia through pores often resulting in a bent (geniculate) appearance *e.g. Curvularia, Drechslera, Alternaria, Sporothrix schenckii.*

(vi) Arthroconidium: The pieces of septate hyphae fragment and become thallic conidia *e.g. Coccidioides* species, *Trichosporon* species.

(vii) Chlamydospore: This is a thick-walled enlarged hyphal swelling that forms during adverse environmental conditions. This may be terminal or intercalary and single or in chains. It is capable of germinating when growth conditions are not favorable. These are usually seen in *Candida albicans* and *C.dubliniensis.*

(viii) Sporangiospore: An asexual spore is formed by cleavage of the sac-like structure *i.e.* sporangium as seen in members of phylum Glomeromycota.

***(ix) Adiaconidium* (formerly *Adiaspore*):** Large, globose, thick-walled conidium released by fracture or dissolution of supporting hyphal cells, when introduced into host *in vivo* or incubated at an elevated temperature of 37-40°C *in vitro,* increases enormously in size without eventual reproduction or replication thereby developing as spherule, which is seen in adiaspiromycosis caused by two species of genus *Emmonsia i.e. Ea.crescens* and *Ea.parva.*

(x) Microcycle Conidiation: The usual life cycle of filamentous fungi commonly involves asexual sporulation after vegetative growth in response to environmental factors. The production of asexual spores is critical in the life cycle of most filamentous fungi. Normally, conidia are produced from vegetative hyphae *i.e.* mycelia. However, fungal species subjected to stressful conditions exhibit an extremely simplified asexual life cycle, in which the conidia that germinate, directly generate further conidia, without forming mycelia.

(f) Mycelia Sterilia

The mycelia sterilia are fast growing molds. They do not produce spores or conidia. The sterile isolates represent group of medically significant fungi, which are difficult to identify. Microscopically, hyphae appear hyaline or pigmented, septate and branching. They are commonly referred to as members of form-order **Mycelia Sterilia**. Either this group of fungi has lost the ability to produce spores therefore these are sterile or the laboratory conditions under which they grow are metabolically unfavorable for the production of spores. In either case, these fungi are classified on the basis of their lack of spore production. They are designated by the color of the colony as white, blue or brown mycelia sterilia.

These fungi have been shown to cause phaeohyphomycosis and mycetoma. When sterile phaeoid fungi are isolated, induction of spores or fruiting bodies should be tried on appropriate media. The cultures should be incubated at 25°C for a period of several weeks before being discarded. With the passage of time some of these fungi may develop structures that produce spores or conidia, such as ascocarps, pycnidia or synnemata, either in the agar or on the colony surface.

Some fungi do not sporulate on routine fungal culture media and may be confused with mycelia sterilia. In such circumstance, putting a slide culture of the fungal isolate is a must, as some of them may produce spores thereby can be identified to an extent. However, some of these fungi may be identified on the basis of molecular techniques up to their genus and species levels.

D. Fungal Dimorphism

The fungi grow either as yeasts or molds but some of them are capable of growing as both forms but under different circumstantial factors, mainly temperature. They grow as yeasts or spherules in living tissue or enriched media incubated at 37°C, while grow as sporulating hyphal structures at room temperature *i.e.* 25°C. The temperature of incubation primarily governs the change in somatic structure of such fungi. This feature is exhibited by organisms that cause endemic true systemic or subcutaneous fungal infections. The diseases caused by these dimorphic fungi and their corresponding causative agents are listed in **Table 2.3**.

They have yeast or spherule form *in vivo* and hyphal form in the natural environment or *in vitro*. Such fungi having both types of morphology in different circumstances are called the **dimorphic fungi**. There may be thermal, nutritional or tissue dimorphism. The yeast form can also be produced *in vitro* at 37°C on enriched media.

Both the phases of a particular fungus differ from one another in morphological, physiological, biochemical and antigenic properties, resistance to inhibitory substances and other allied features. These two morphological forms are also called yeast phase **(parasitic)** and mycelial phase **(saprotrophic)**.

Table 2.3. List of Mycotic Diseases caused by their Corresponding Dimorphic Fungi.

Sr. Nos.	*Mycotic Diseases*	*Dimorphic Fungi*
1.	Histoplasmosis	*Histoplasma capsulatum* var. *capsulatum, H.c.* var. *duboisii* and *H.c.* var. *farciminosum*
2.	Blastomycosis	*Blastomyces dermatitidis, B.gilchristii* and *B.percursus*
3.	Coccidioidomycosis	*Coccidioides immitis* and *C.posadasii*
4.	Paracoccidioidomycosis	*Paracoccidioides brasiliensis* and *P.lutzii*
5.	Sporotrichosis	*Sporothrix schenckii, S.brasiliensis, S.globosa* and *S.luriei*
6.	Talaromycosis	*Talaromyces marneffei*
7.	Emergomycosis	*Emergomyces pasteurianus, Es.africanus* and *Es.orientalis*
8.	Adiaspiromycosis	*Emmonsia crescens* and *Ea.parva*
9.	Mucormycosis	*Mucor circinelloides* and *Cokeromyces recurvatus*

This morphological change of these fungi is a response or adaptation for survival by the organism, finding itself in an unfavorable environment. Although temperature is the single most important factor affecting this transformation from mycelial to yeast and *vice versa* (M↔Y) but other factors individually or in combination also exert an influence like oxidation-reduction potential, availability of sulfhydryl groups, CO_2 concentration, cell density and age of the culture, etc.

The yeast cell morphology of each of these species is distinct. Mostly, several subcultures are needed before a smooth yeast colony is obtained. In many isolates, like *Histoplasma capsulatum, in vitro* conversion may be difficult even after several subcultures. All the dimorphic fungal pathogens do not have yeast form *in vivo i.e. Coccidioides* species grow in the form of a **spherule** with endosporulation in the host tissue. The yeast phase of *Talaromyces marneffei* is unique in the sense that it divides by **fission** (schizogony) and **not by budding** as is usually seen in other dimorphic fungi.

The thick-walled yeast cells of *Blastomyces dermatitidis* divide by broad-based budding producing daughter cells. However, in *Histoplasma capsulatum* it is by narrow-based budding. The mechanisms of transformation from mycelia to yeast (M→Y) is a result of heat-shock proteins induced uncoupling of oxidative phosphorylation.

The probability of finding these causative fungi in lesions of a disease, mandatorily compels the laboratory to incubate culture media, inoculated with clinical specimens, at least at two different temperatures simulating room (25°C) and body temperatures (37°C) in the diagnostic mycology services leading to their growth as mycelial and tissue phases, respectively. This is also pertinent to note that the mycelial phase of dimorphic fungi is highly infectious and requires appropriate biosafety precautions while handling them. It is a step higher than the precautions required for yeast phase.

Taxonomically one of the species of *Emmonsia i.e. Ea.crescens* is placed in family Onygenaceae and is considered to be a dimorphic fungus. The other species *i.e. Ea.parva* requires 40°C temperature to convert into adiaconidia. Morphologically, some of the mucormycetes like *Cokeromyces recurvatus* and *Mucor circinelloides* are converted to yeast form in the body being dimorphic fungi, may be confused with *Paracoccidioides brasiliensis* due to their multiple buddings.

The term 'dimorphism' is sometimes loosely applied to encompass morphological transformation of other fungi, such as *Candida* species, *Malassezia* species, *Wangiella dermatitidis* and *Phialophora verrucosa* as well as others causing chromoblastomycosis. This may be considered due to the production of pseudohyphae, hyphae or sclerotic bodies. But their morphological changes are not entirely dependent on temperature thereby in the strictest sense they cannot be termed as dimorphic. Hence the only genera, as mentioned above in **Table 2.3** are authentically designated as dimorphic fungi.

As such, only the yeast or spherule forms are seen in tissues but recently cases have been reported where hyphal forms of dimorphic fungi have been reported. These are unusual findings hence due care must be taken while making the final diagnosis after examining the tissue sections of lesions caused by such fungi.

E. Growth and Nutrition

Most of the fungi found in nature can grow readily on a simple source of nitrogen and carbohydrates. The medically significant fungi are mesophilic and have an optimal growth range considerably below the body temperature. The optimum temperature *in vitro* for a majority of the pathogenic fungi is between 25°C and 37°C. The fungi prefer acidic pH therefore most of the fungal culture media have pH range towards acidic side. These do not require light for their growth however it affects both sexual and asexual reproduction of many fungi.

The fungi are heterotrophic as they are unable to photosynthesize hence feed on preformed organic material such as cellulose and proteins from the host or substrate. They secrete digestive enzymes (exoenzymes), which break down live or dead materials into sugars and amino acids that can be absorbed through the fungal cell walls. Therefore, fungi digest and then ingest with the help of exoenzymes unlike animals which ingest and then digest the food material.

Most of the medically significant fungi are facultative parasites, capable of causing disease or living on dead organic matter. A few fungal and pseudofungal agents such as *Lacazia loboi* and *Rhinosporidium seeberi*, respectively, may be obligate parasites as they are not successfully grown outside host tissue on artificial culture media. Among the pathogenic fungi, several species of anthropophilic dermatophytes require thiamine or biotin for their growth. The nutritional requirements of parasitic forms of dimorphic fungi *in vitro* are more complex than those of their saprotrophic counterpart forms. The fungi store food like glycogen unlike plants, which store it as starch.

The fungi are not definitively known to produce important endotoxins, although intravenous injections of several fungi have been pyrogenic in animals. Certain fungi are known to produce exotoxins *in vitro* such as aflatoxin but none is yet known to produce toxin *in vivo*.

Thermophilic Fungi

The thermophilic organisms can be classified as those with an optimal growth temperature between 45°C and 80°C, hyperthermophiles are those with an optimum growth temperature above 80°C and mesophiles are those that grow optimally below 45°C. Thermophily is common in bacteria and Archaea, whereas hyperthermophiles are mainly confined to the Archaea. Only a small fraction of the estimated fungi is considered to be thermophilic and no fungus has been described as hyperthermophilic. However, Suryanarayanan *et al*, have reported from Chennai that some fungi may tolerate temperature above 100°C like *Chaetomella raphigera* and a few *Phoma* species. This finding will have significant implications in the field of medical mycology in future, if tenable on scientific grounds, in due course of time.

The known thermophilic fungi are limited to Sordariales, Eurotiales and Onygenales in Ascomycota and Mucorales with possibly an additional order harbouring *Calcarisporiella thermophila* as the basal fungi. However, there is no supporting evidence that thermophilic species belong to Basidiomycota.

Further Reading

1. Adams DJ. Fungal cell wall chitinases and glucanases. *Microbiology*. 2004; **150:** 2029-35.
2. Baumgardner DJ. Environmental determinants of dimorphic systemic mycoses - The macro and the micro. *Int J Epidemiol.* 2009; **38:** 1649-50.
3. Bowman B, Taylor JW, White TJ. Molecular evolution of the fungi: Human pathogens. *Mol Biol Evol.* 1992; **9:** 893-904.
4. Boyce KJ, Andrianopoulos A. Fungal dimorphism: The switch from hyphae to yeast is a specialized morphogenetic adaptation allowing colonization of a host. *FEMS Microbiol Rev.* 2015; **39:** 797-811.
5. Campbell CK. Conidiogenesis of fungi pathogenic for man. *Crit Rev Microbiol.* 1986; **12:** 321-41.
6. Cao C, Li R, Wan Z, *et al.* The effects of temperature, pH and salinity on the growth and dimorphism of *Penicillium marneffei*. *Med Mycol.* 2007; **45:** 401-7.
7. Clutterbuck AJ. Sexual and parasexual genetics of *Aspergillus* species. *Biotechnology*. 1992; **23:** 3-18.
8. Damialis A, Mohammad AB, Halley JM, *et al.* Fungi in a changing world: Growth rates will be elevated but spore production may decrease in future climates. *Int J Biometeorol.* 2015; **59:** 1157-67.
9. de Hoog GS, Yurlova NA. Conidiogenesis, nutritional physiology and taxonomy of *Aureobasidium* and *Hormonema*. *Antonie Van Leeuwenhoek.* 1994; **65:** 41-54.
10. Drouhet E, Huerre M. Yeast tissue phase of *Emmonsia pasteuriana* inoculated in golden hamster by intratesticular way. *Mycoses*. 1999; **42:** S11-8.
11. Duo-Chuan L. Review of fungal chitinases. *Mycopathologia*. 2006; **161:** 345-60.
12. Fernandez-Flores A, Saeb-Lima M, Arenas-Guzman R. Morphological findings of deep cutaneous fungal infections. *Am J Dermatopathol.* 2014; **36:** 531-53; Quiz 554-6.
13. Gauthier GM. Dimorphism in fungal pathogens of mammals, plants and insects. *PLoS Pathog.* 2015; **11:** e1004608.
14. Jain N, Hasan F, Fries BC. Phenotypic switching in fungi. *Curr Fungal Infect Rep.* 2008; **2:** 180-8.
15. Jung B, Kim S, Lee J. Microcycle conidiation in filamentous fungi. *Mycobiology*. 2014; **42:** 1-5.
16. Kirkland TN. A few shared up-regulated genes may influence conidia to yeast transformation in dimorphic fungal pathogens. *Med Mycol.* 2016; **54:** 648-53.

17. Levetin E. An atlas of fungal spores. *J Allergy Clin Immunol.* 2004; **113:** 366-8.
18. Magee PT. Which came first: The hypha or the yeast? *Science.* 1997; **277:** 52-3.
19. Manisha K, Panwar N. Morphopathological effects of isolated fungal species on human population. *Open Access Scientific Reports.* 2012; **1:** 521.
20. Marin-Felix Y, Stchigel AM, Cano-Lira JF, *et al. Emmonsiellopsis,* A new genus related to the thermally dimorphic fungi of the family Ajellomycetaceae. *Mycoses.* 2015; **58:** 451-60.
21. McGinnis MR, Sigler L, Bowman BH, *et al.* Impact of conidiogenesis, teleomorphic connections, pleomorphism and molecular genetics on evolving hyphomycetes systematics. *J Med Vet Mycol.* 1992; **30:** S261-9.
22. Morgenstern I, Powlowski J, Ishmael N, *et al.* A molecular phylogeny of thermophilic fungi. *Fungal Biol.* 2012; **116:** 489-502.
23. Odds FC. Biotyping of medically important fungi. *Curr Top Med Mycol.* 1985; **1:** 155-71.
24. Odds FC. Fungal dimorphism. *J Mycol Med.* 1998; **8:** 55-6.
25. Orlowski M. *Mucor* dimorphism. *Microbiol Rev.* 1991; **9:** 234-58.
26. Reiss E, Hearn VM, Poulain D, *et al.* Structure and function of the fungal cell wall. *J Med Vet Mycol.* 1992; **30:** S143-56.
27. Richards A, Gow NA, Veses V. Identification of vacuole defects in fungi. *J Microbiol Methods.* 2012; **91:** 155-63.
28. Riquelme M, Mourino-Perez RR. A comprehensive journey into the fungal cell: Editorial. *Mycologia.* 2016; **108:** 473-4.
29. Rippon JW. Dimorphism in pathogenic fungi. *CRC Crit Rev Microbiol.* 1980; **8:** 49-97.
30. Ruiz-Herrera J, Sentandreu R. Fungal cell wall synthesis and assembly. *Curr Top Med Mycol.* 1989; **3:** 168-217.
31. San-Blas G. The cell wall of fungal human pathogens: Its possible role in host-parasite relationships. *Mycopathologia.* 1982; **79:** 159-84.
32. Sigler L. Problems in application of the terms 'blastic' and 'thallic' to modes of conidiogenesis in some onygenalean fungi. *Mycopathologia.* 1989; **106:** 155-61.
33. Sil A, Andrianopoulos A. Thermally dimorphic human fungal pathogens - Polyphyletic pathogens with a convergent pathogenicity trait. *Cold Spring Harb Perspect Med.* 2014; **5:** a019794. PMID: 25384771.
34. Sohn K, Schwenk J, Urban C, *et al.* Getting in touch with *Candida albicans:* The cell wall of a fungal pathogen. *Curr Drug Targets.* 2006; **7:** 505-12.
35. Suryanarayanan TS, Govindarajulu MB, Thirumalai E, *et al.* Agni's fungi: Heat-resistant spores from the Western Ghats, southern India. *Fungal Biol.* 2011; **115:** 833-8.
36. Suryanarayanan TS. Heat-resistant fungal spores found in Western Ghats. *The Hindu,* Chennai, September 22, 2011.
37. Suryanarayanan TS. In Western Ghats, a new find: Fungi which can survive 100 degrees C. *Indian Express,* Bangalore, July 17, 2011.

CHAPTER 3

Fungal Taxonomy

Fungal Taxonomy is the science of naming, describing and classifying the fungi. This is actually science of grouping organisms according to their relationship by process of identification and classification. There is formation of system that breaks groups of things down into smaller and smaller categories that describe ways in which they are similar. It is theory and practice of naming and classifying organisms. It is ultimately matter of scientific evaluation and general agreement. It starts with broad categories and narrows down into more and more specific ones. It is basically placement of an organism into group based on its apparent natural relationship to other organisms. Since fungal taxonomy and disease nomenclature is constantly changing hence there are variations in names of fungi and their corresponding infective diseases caused by them.

Nomenclature is system under which proper name is applied to a particular organism. It is basically naming of an organism according to set of certain defined rules. Hence it is an essential adjunct to any system of classification. As fungi were earlier classified under kingdom Plantae, therefore, their nomenclature used to be governed by rules and recommendations of the International Code of Botanical Nomenclature under the auspices of periodically held International Botanical Congress. This was empowered to make necessary changes and is universally accepted as a governing body for application of scientific names.

Classification in biology is defined as arranging living things into groups in an orderly manner on the basis of their similarities and relationships. It has been mainly traditional rather than numerical and is usually based on readily observable morphological features. The classification systems of various organisms are historically based on phenotypic approach of observable characteristics. The identification and classification of fungi, unlike bacteria or viruses, relies mainly on morphological criteria. However, modern techniques have really paved the way of molecular basis of fungal taxonomy thereby consequently its classification also.

Most of the fungi are known by their anamorphic names as they have been in frequent use despite their parallel teleomorphic states/names are also known. However, in some of the fungi it is other way round that teleomorphic name is more popular than the anamorphic one. For instance, *Pseudallescheria boydii*, the name is based on teleomorphic state, which is more popular whereas it's other name, *Scedosporium apiospermum,* is based on anamorphic state, which is relatively less popular. As far as the issue of established names *vis-a-vis* frequently encountered sexual or asexual form is concerned, while dispatching the report, the new name may be conveyed along with the old one given in brackets so that clinician may get familiarized in due course of time.

The taxonomy being classification of organisms according to similarity hence it is highly influenced by phylogenetics but remains methodologically and logically distinct. It must precede nomenclature because after an organism has been placed in certain taxon only then a proper name can be assigned to it. There are different kinds of taxonomies related to various fields, however, in biology, taxonomy is classification of all living organisms. The current method of biological taxonomy was started by Carolus Linnaeus, wherein the organisms are arranged into groups within further groups on and on until an organism is finally defined within its own species or individual group.

The recently developed molecular techniques have further helped in advancing knowledge of taxonomy, entailing solution of long-standing taxonomic puzzles. Some of the isolates, which were misidentified on the basis of morphological features leading to misdiagnosis of underlying disease, have now been correctively identified on the basis of such techniques. On that very basis a few organisms are incorporated into kingdom Fungi (*Pneumocystis* and *Microsporidium*) and some of them are excluded from it (*Prototheca, Rhinosporidium* and *Pythium*), which were previously thought to be fungi. This advanced branch of biological sciences is known as **Molecular Taxonomy**.

In Medical Mycology, as practiced in other fields of biological sciences, **binomial system** of nomenclature is being followed. In binomial **nomenclature** there are genus and species - just two names, which replace long string of words previously used in **polynomial** system and it is certainly a better system over polynomial system. But there is lack of uniform approach towards recognition and description of new species of fungi, which has led to accumulation of, perplex number and variety of names in the literature. Names are usually descriptive but may indicate place of origin or may be given to honor scientists working in their field.

When a mycologist recognizes variety or other category below species level, nomenclature becomes **trinomial**. The third word in name (second epithet) might be subspecies (subsp.), variety (var.), forma (f.) or forma speciales (f. sp.), which are also italicized or underlined like the species. The trinomial nomenclature was proposed for *Pneumocystis carinii*, which included Latin name of host species. The rat prototype was called as *P.carinii* f. sp. *carinii* and human prototype was called as *P.carinii* f. sp. *hominis*, which is now called as *Pneumocystis jirovecii*.

In case of many fungi where only one or more species of particular genus was clinically or taxonomically significant, it is proposed to label as group of that species, on the basis of morphological as well as molecular based techniques. For example: *Aspergillus* species, *Sporothrix schenckii* and *Apophysomyces elegans*.

The taxonomy and nomenclature are entirely two different things. For example taxonomy of hydrophilic protista, *Rhinosporidium seeberi*, remained disputed throughout the 20th century (Fungus *vs.* Protist) but nomenclature of its genus as well as species remained stable during this entire one hundred years. However, its taxonomy has also now settled to a great extent as it is classified as Mesomycetozoea and not a fungus.

The general principle in fungal disease nomenclature promoted by the ISHAM is to provide description of specific pathology and causative organism (if known) in format 'pathology A caused by fungus X' or 'fungus X pathology A.' Adding suffix '-mycosis,' '-osis,' ' -iasis' or '-asis' to name of fungus should be avoided unless this gives an unambiguous identity to specific clinico-pathological process and also causal agent. However, certain names are retained for traditional reasons *e.g.* dermatophytosis, cryptococcosis, candidiasis.

Binomial Nomenclature

The modern system of classification in biosciences is binomial nomenclature, which has two words to its scientific name, the first one is genus and second is species regardless of rank. This was originally proposed by 18th century Swedish biologist, **Carolus Linnaeus** (1707-1778), in 1758 in the 10th edition of *'Systema Naturae.'* He devised this concise and precise system for plants and animals, using one Latin name to represent genus and second to distinguish species. He proposed that each type of organism should be given unique binomial terminology, consisting of two terms, to set it apart from every other type of organisms and it is called as Linnaean System of Nomenclature. Based on his contribution to this newly evolving field, he is popularly known as *'Father of Modern Taxonomy'* and 2007 was the year in honor of the tercentenary of his birth.

The first word in this binomial nomenclature represents **genus** to which various species belong. The second word denotes **species** itself within that genus. There is no plural for 'species' as singular and plural is the same word but while writing abbreviations these are used as **sp.** and **spp.** denoting singular and pleural, respectively. Originally, when Linnaeus founded this taxonomy, organisms were divided on the basis of visible physical characteristics. Now, they are separated on the basis of different kinds of unique features. Primarily these features are physical but secondarily can be behavioral as well. The **generic** names are generally nouns, which are capitalized and define genus whereas species names are generally adjectives, which are not capitalized and denote the species. Both names of genus as well as species must be either italicized *(Candida albicans)* or underlined (Candida albicans). As there are computer-generated writings these days hence scientific names are easily italicized thereby underlining has become almost obsolete and infructuous but it was the only option available during the period of manual typewriting.

The species names are descriptive, however, sometimes they are taken from scientists' name or geographical locality. When it is named after a person, in case of male scientist, nomenclature generally terminates as ***'ii'*** and in female it is, ***'eae'***. Just as species is one of the group of organisms that share certain characteristic(s) that has been decided by taxonomists to be sufficiently significant for that group to be considered genus, so is genus, which is one of the group that shares certain characteristics and makes up family. Species, genus and family are three most significant taxa, wherein most of the people are interested and seek to know as well as try to remember.

Since 1977, three competing nomenclatures for mycotic diseases have been proposed; one by *Council for International Organization of Medical Sciences* (*CIOMS*) in conjunction with WHO. In 1980, CIOMS published mycoses section of International Nomenclature of Diseases. Another nomenclature was by *British Society of Mycopathology* on behalf of Medical Research Council of Great Britain. The third was by *International Society for Human and Animal Mycology* (*ISHAM*). These divergent lists of proposed terms are acceptable to varying degrees by those, who are actively involved in human and animal mycoses.

The species is a group of phenotypically similar organisms, living in certain ways. It is the lowest category of any system of grouping used in taxonomy. It is group of actually or potentially interbreeding natural populations that are reproductively isolated from other such groups. However, in many cases such stringent criteria cannot be applied so in practice identification and description of species is largely based on morphological and physiological features. All the Latin names of biological groups, genera, species as well as subspecies are written in italics.

Some names of fungal diseases have been designated on the basis of names of **genus** and **species** by putting *'i'* as suffix like, histoplasmosis duboisii.

There are no hard and fast rules or regulations governing nomenclature of fungal infections and adding suffix -sis or -mycosis to generic name of causative fungus has designated most of disease names *i.e.* fungal name + sis. This system is not entirely satisfactory because same fungus may cause several distinct diseases *e.g. Aspergillus* may cause corneal ulcer, which is not labeled as aspergillosis rather it is called keratomycosis or mycotic keratitis. On the contrary, there are wide variety of several fungi (and bacteria), which may cause single clinical entity *e.g.* mycetoma.

Sometimes, diseases are named on the basis of geographical areas as in case of North American blastomycosis for Blastomycosis and similarly South American blastomycosis for Paracoccidioidomycosis. Now, diseases are not limited to their localized geographical areas rather reported from far away distances of their designated names hence such type of nomenclatures are not tenable. Therefore, these misleading names have been abandoned and alternative names are being given to such diseases.

Several fungal diseases have many names or synonyms. For example, *Candida* infections are known by name of Moniliasis, Torulopsis, Candidosis and Candidiasis. Therefore, it becomes very difficult to comprehend for the beginners, which disease is being referred to and that is the same or some different clinical entities.

Some of the names of fungi like *Talaromyces* ***marneffei***, *Candida* ***dublin****iensis* and *Paracoccidioides* ***brasil****iensis*, are based on name of scientist, city and country, respectively, analogous to various other branches of biological sciences.

The names have also been designated on the basis of cellular characteristics and/or clinical manifestations produced by fungus. For example, *Blastomyces dermatitidis* was named in 1898 on the basis of its asexual characteristics: fungus (***myces***) that reproduced by budding (***blasto***) recovered from skin of patient of dermatitis (***dermatitidis***). Subsequently in due course of time sexual states (ascospores) were identified completing life cycle and new name was given as *Ajellomyces dermatitidis* based on its perfect state but the old anamorphic name is widely acceptable hence still continuing as such.

Some names of fungi carry diametrically opposite meaning hence in their strictest sense supposed to be **misnomer**. But they are very popular among scientific community hence it is very difficult to change them at this stage. For example, *Histoplasma capsulatum*, nomenclature of this endemically prevalent dimorphic fungus gives apparent meaning as if it is protozoa and that too with a capsule. But the fact is that it is neither protozoan nor has any capsule. The acceptance to this name is such that even passage of a century could neither change the name of the genus nor its species. Similarly, chromoblastomycosis, where there is no budding yeast cells seen in the disease process caused by various dematiaceous fungi, which produce sclerotic bodies. Therefore, it was changed to simply chromomycosis to rectify this anomaly. But this new name created more confusion than to resolve it hence its old name, chromoblastomycosis, was again universally accepted as its official name and still going on as such till date.

However, recently many efforts have been made to list out internationally acceptable names of clinically significant fungi. There has been considerable measure of agreement at least for some of the most important diseases. These recommendations have been followed while writing this Textbook and where there is disagreement, choice of name has been made on the basis of latest accepted terminology. In such situations synonyms are also given to have link of particular disease under question. However, malassezıosis, talaromycosis and emergomycosis are being used for the first time in this Textbook for the diseases caused by *Malassezia*, *Talaromyces* and *Emergomyces* species, respectively.

Sometimes, names of fungi, disease caused by them or even scientists have different spellings. For example, Raimond Jacques Adrien Sabouraud is commonly written as Raymond Sabouraud. Both these names are correct. In this textbook such issues have been dealt with using commonly used spellings as per American English.

The understanding in nomenclature was put in order by systematic work of outstanding medical mycologists, Norman Conant, Chester Wilson Emmons, Libero Ajello, M R McGinnis, K J Kwon-Chung, J W Rippon, G S de Hoog, J Guarro and Arvind A Padhye.

Sensu lato: This is a Latin phrase, used after a binomial species name, often abbreviated as *s.l.*, to indicate a species complex represented by that species. Its literal meaning is '***in the broad sense***'.

Sensu stricto: This Latin phrase is added after a binomial species name to mean the taxon is being used in the sense of the original author or without taxa, which may otherwise be associated with it. However, its literal meaning is '***in the narrow sense***' and is abbreviated as *s.s.*

There is another unique feature in fungi that they are known by two names; one name is based on its asexual nature of reproduction and other one is based on sexual reproduction. This is not common with other biological kingdoms. The fungal taxonomists should address this issue also so that any sort of confusion is avoided.

Amsterdam Declaration

This declaration was on fungal nomenclature agreed at the international symposium held on April 19-20, 2011 under the auspices of the International Commission on the Taxonomy of Fungi at Amsterdam. Its purpose was to address the issue of whether or how the current system of naming pleomorphic fungi should be maintained or changed now that molecular data are routinely available. The issue was urgent as mycologists were following different practices. Moreover, no consensus was achieved by a Special Committee appointed in 2005 by the International Botanical Congress (Vienna Code) to advise on the intricacies. It recognized the need for an orderly transition to a single-name nomenclatural system for all fungi and to provide mechanisms to protect names that otherwise then become endangered. Priority to be given to the first described name, except where that is a younger name in general use when the first author to select a name of a pleomorphic monophyletic genus was to be followed and suggested controversial cases were referred to a body, such as the ICTF, which would report to the Committee for Fungi. If it was appropriate, the ICTF could be mandated to promote the implementation of the Declaration. This also becomes more important because the next venue of the 20th ISHAM Congress is again Amsterdam in June-July 2018.

Moreover, despite not forming part of the Declaration, were reports of discussions held during the symposium on the governance of the nomenclature of fungi and the naming of fungi known only from an environmental nucleic acid sequence in particular. There were possible amendments to the Draft BioCode of 2011 to allow for the needs of mycologists and suggested for further consideration to how fungus only known from the environment be described in future.

This should be noted that consequently taxonomic updates have taken effect as on **January 1, 2013** and dual nomenclature for pleomorphic fungi henceforth got discontinued. The ending of separate names for anamorphs and teleomorphs ***(One Fungus = One Name)*** has been a long-awaited change in mycology. The majority of fungi are likely to keep their more widely used anamorph name, however, a personal synopsis of the decisions made at the Nomenclature Section meeting of the IBC held at **Melbourne** (Australia) in **July 2011** is now provided with an emphasis on those which will affect the working practices or will otherwise be of interest to the mycologists.

The topics covered include the re-naming of the Code, the acceptance of English as an alternative to Latin for validating diagnoses, conditions for permitting electronic publication of names, mandatory deposit of key nomenclatural information in a recognized repository for the valid publication of fungal names, the discontinuance of dual nomenclature for pleomorphic fungi, clarification of the typification of sanctioned names and acceptability of names originally published under the zoological code. Collectively, these changes are the most fundamental to have been enacted at a single Congress since 1950s and herald the dawn of a new era in the practice of fungal nomenclature. The IBC Committee for Fungi during

Table 3.1. Decisions of Nomenclature Session during IMC9.

- The transference of the governance of the nomenclature of fungi from the International Botanical to International Mycological Congresses.
- The mandatory pre-publication deposit of nomenclatural information in a recognized depository for the valid publication of new fungal names.
- The acceptability of English as an alternative to Latin in the valid publication of fungal names.
- Requested the Committee for Fungi, the Special Committee on the Names of Pleomorphic Fungi and the International Commission on the Taxonomy of Fungi, to take note of the results of the questionnaire completed by delegates of IMC9 held at Edinburgh in 2010.

General Assembly of the IMA endorsed the decisions of the Nomenclature Session convened during IMC9 held on August 3-5, 2010 at Edinburgh **(Table 3.1)**.

International Code of Botanical Nomenclature

The application of names of fungi was governed by set of principles outlined by International Code of Botanical Nomenclature (ICBN). The ICBN was set of rules and recommendations dealing with formal botanical names that were given to plants. Its intent was that each taxonomic group of plants had only one correct name, accepted worldwide. The value of scientific name was that it was a label and may not necessarily be of descriptive value or even accurate.

The ICBN was applied not only to plants but also to other organisms traditionally studied by botanists. This included blue-green algae (Cyanobacteria); **fungi** including chytrids, oomycetes and slime molds; photosynthetic protists and taxonomically related non-photosynthetic groups. The ICBN consisted of 6 major Principles, 76 Articles and 4 Appendices. Several Articles were incorporated into main body of Code for Fungi from time to time as per the needs.

There is another way of analyzing this issue of nomenclature through informal names or Code systems. The ICBN **Vienna Code**, was adopted during the 17th International Botanical Congress (IBC) held at Vienna in July 2005. The IBC is being held after every six years and Code is designated after name of city where IBC is held like St. Louis Code or the Black Code (1999), Tokyo Code (1993), Berlin Code (1987), Sydney Code (1981), Leningrad Code (1975) and so on and so forth. The names of fungi were subject to change from time to time. The ICBN could only be changed by an International Botanical Congress, with International Association for Plant Taxonomy (IAPT) providing supporting infrastructure. Therefore, each new edition supersedes earlier editions and is retroactive back to 1753 as ICBN sets formal starting date of plant nomenclature on May 1, 1753, day of publication of 'Species Plantarum' by Linnaeus himself. The last one *i.e.* 18th IBC was held at Melbourne in year 2011, which is described separately in this very chapter, being one of the important milestones in fungal taxonomy. The 19th IBC-2017 will be held from July 23-29, 2017 at Shenzhen (China). Therefore, any sort of suggestions can be sent to secretariat of forthcoming IBC for incorporations as amendments, etc.

The ICBN was reviewed and revised every 6 years at nomenclatural sessions held in conjunction with International Botanical Congress (IBC). Normally proposals to change ICBN were published in journal *Taxon* and reviewed by committees and membership at large prior to the Congress. However, at IBC in Vienna, July 2005, in response to rapidly increasing problem of invalidation of names of parasites first described as protozoans but later shown to be fungi, small but significant as yet unpublished change was proposed at Congress for Art. 45.4 making an exception for fungi. This move to amend ICBN was formulated by attending mycological delegates and members of 'Committee for Fungi' present at Congress and was inspired by research on *Pneumocystis* names and anticipated major negative effects of reclassifying microsporidia as fungi. Previously only algal names had been protected when their types were originally classified as protozoans regulated by International Code of Zoological Nomenclature (ICZN) and afterwards reclassified as algae regulated by ICBN. The addition twice of phrase 'or fungi' after each mention of 'algae' in existing Art. 45.4 provided protection for validity of all effectively published microsporidian names meeting requirements of either ICZN or ICBN should they be reclassified as fungi. Additionally these changes re-validated name *Pneumocystis jirovecii* from 1976 under revised ICBN *i.e.* Vienna Code and changes also re-established 1912 validation and priority for names *Pneumocystis* and *P.carinii* as is explained further below.

The validity of names for fungi described under zoological Code, became disputed which did not require Latin description. Now sequence analyses imply *Pneumocystis* within kingdom Fungi, rendering previous protozoan name invalid under botanical Code. Therefore, change in Article 45.4 made exceptions for fungi described under other Codes, basically validating under botanical Code names previously available under ICZN. As there is shifting of their nomenclature from zoological to botanical Code hence sometimes, in loose terms such fungi were labeled as *'Zoological Fungi'*.

Moreover, ICBN did not recognize online publication of nomenclatural novelties, unless hard copies were simultaneously deposited. New text for electronically published periodicals was added that stipulated print and electronic copy must be identical in content and pagination, in 'platform-independent and printable' format like pdf and publicly available through Internet. Both publications must bore dates, included references and mention nomenclatural novelties in abstracts.

Independent of concerns about higher classification as fungi or protozoans, which affected choice of Codes, had been controversy over recognition of sub-generic taxa, taxonomic levels for these taxa (*i.e.* species *versus* formae speciales or other levels) and suitability of names. Therefore, entire field of systematics pertaining to *Pneumocystis,* from kingdom to *formae speciales* levels, had been under recent intensive debate. To research basic background, Redhead *et al,* investigated nomenclatural status of all *Pneumocystis* names, rigorously applying St. Louis Code of ICBN.

Hence under ICBN Art. 45.4 (St. Louis Code) as revised by Vienna Congress, name *P.jirovecii* was interpreted as being valid. Following recognition of *Pneumocystis* as fungus rather than protozoan since at that time, Latin description or diagnosis was required for all fungal names, Frenkel (1999) sought to validate name *P.jirovecii* by publishing Latin description. The necessity for Latin for *P.jirovecii* has now been removed by change in ICBN Art. 45.4.

The description of *P.jirovecii* clearly included cysts as well as trophozoites. Article 59 (St. Louis Code - ICBN) governed application of names with pleomorphic life cycles and it stated that correct name for nonlichen-forming fungus is earliest legitimate name typified by teleomorph.

The disease names that have been used throughout this Textbook are based on lists published independently by *CIOMS* and *ISHAM*. However, to help applied mycologists and microbiologists, *International Commission on the Taxonomy of Fungi (ICTF)* of *International Union of Microbiological Societies (IUMS)* had launched program of *Name changes in fungi of microbiological, industrial and medical importance* in year 1986. The objective of series was to select cases where name changes have been proposed in medical literature, to explain why these changes were made, more importantly to advise about whether it would be prudent for them to be generally adopted. The recommended and rejected names of general applicability of fungal diseases are given in an alphabetical order in **Table 3.2**.

Table 3.2. List of Old and New Names of Fungal Species and/or Diseases in Alphabetical Order.

Old Names	New Names
1. *Absidia corymbifera* or *Mycocladus corymbiferus*	*Lichtheimia corymbifera*
2. *Acremonium kiliense.*	*Sarocladium kiliense.*
3. *Blastoschizomyces capitatus (Geotrichum capitatum* or *Trichosporon capitatum)*	*Magnusiomyces capitatus*
4. *Emmonsia pasteuriana*	*Emergomyces pasteurianus*
5. Emmonsiosis	Emergomycosis
6. *Geosmithia argillacea*	*Rasamsonia*
7. *Hendersonula toruloidea*	*Neoscytalidium dimidiatum*
8. *Loboa loboi*	*Lacazia loboi*
9. Lobomycosis	Lacaziosis
10. *Madurella grisea*	*Trematosphaeria grisea*
11. *Paecilomyces lilacinus*	*Purpureocillium lilacinum*
12. Penicilliosis marneffei	Talaromycosis
13. *Penicillium marneffei*	*Talaromyces marneffei*
14. *Phialophora richardsiae*	*Pleurostomophora richardsiae*
15. *Pyrenochaeta romeroi*	*Medicopsis romeroi*
16. *Ramichloridium mackenziei*	*Rhinocladiella mackenziei*
17. *Scedosporium prolificans*	*Lomentospora prolificans*
18. *Trichophyton mentagrophytes*	*Arthroderma vanbreuseghemii*
19. *Wangiella dermatitidis*	*Exophiala dermatitidis*

ICBN *versus* ICN

The International Code of Nomenclature for algae, fungi, and plants **(ICN)** is the set of rules and recommendations dealing with the formal botanical names that are given to plants, fungi and a few other groups of organisms, all those 'traditionally treated as algae, fungi or plants'. This was formerly called the International Code of Botanical Nomenclature **(ICBN)**; the name was changed at the International Botanical Congress in Melbourne in July 2011 as part of the **Melbourne Code**, which eventually replaced the Vienna Code of 2005.

Melbourne Code

The International Mycological Association, founded in 1971, represents the interests of over 30,000 mycologists worldwide. The International Mycological Association (IMA) holds a regular meeting every three years. The proceedings of the 3–5 August 2010, IMC9 Edinburgh Nomenclature Sessions are already given in **Table 3.1**. Subsequently one more meeting of IMC10 was held on August 3-8, 2014 in Bangkok. The 11th International Mycological Congress (IMC11) will be held on July 16-21, 2018 in San Juan, Puerto Rico.

Whatever had already been discussed in IMC9 at Edinburgh as well as Amsterdam Declaration was further taken up during Nomenclature Section meeting held during 18th International Botanical Congress (IBC) in Melbourne (Australia) in July 2011. This meeting the saw the ratification of sweeping changes to the way scientists name new plants, algae and fungi. These followings are its salient features:

- For the first time in history the Code now permits electronic-only publication of names of new taxa; no longer will it be a requirement to deposit some paper copies in libraries.
- The title of the Code was broadened to make explicit that it applies not only to plants but also to algae and fungi.
- The requirement for a Latin validating diagnosis or description was changed to allow either English or Latin for these essential components of the publication of a new name (Article 39).
- "One fungus, one name" and "one fossil, one name" are important changes for fungi and for fossils; the concepts of anamorph and teleomorph (for fungi) as well as morphotaxa (for fossils) have been eliminated.
- As an experiment with 'registration of names', new fungal descriptions will require the use of an identifier from 'a recognized repository'; there are two recognized repositories so far (Index Fungorum and MycoBank).

Recognized Repositories

As per the Melbourne Code an experiment with registration of names, new fungal descriptions require the use of an identifier from a recognized repository. There are two recognized repositories so far *i.e.* MycoBank and Index Fungorum.

(a) MycoBank

The MycoBank, a registration system for fungi established in 2004 to capture all taxonomic novelties, acts as a coordination hub between repositories such as Index Fungorum and Fungal Names. Since January 2013, registration of fungal names is a mandatory requirement for valid publication under the International Code of Nomenclature for algae, fungi and plants (ICN). MycoBank is the online repository and nomenclatural registry of the International Mycological Association.

It provides a free service to the mycological and scientific society by databasing mycological nomenclatural novelties (new names and combinations) and associated data, such as descriptions, illustrations and DNA barcoding. Nomenclatural novelties are each allocated a unique MycoBank number to be cited in the publication where the nomenclatural novelty is introduced, to conform with the requirements of the International Code of Nomenclature for algae, fungi and plants.

(b) Index Fungorum

The Index Fungorum, the global fungal nomenclator coordinated and supported by the Index Fungorum Partnership, contains names of fungi (including yeasts, lichens, chromistan fungal analogues, protozoan fungal analogues and fossil forms) at all ranks. As a result of changes to the ICN (formerly ICBN) relating to registration of names and following the lead taken by MycoBank, Index Fungorum now provides a mechanism to register names of new taxa, new names, new combinations and new typifications and no login is required. Names registered at Index Fungorum can be published immediately through the Index Fungorum e-Publication facility - an authorized login is required for this.

ISHAM ITS Database

The currently used identification methods of agents causing human mycoses have serious limitations, which are time consuming and require special trained personnel. However to enable an informed choice for proper antifungal therapy an adequate identification upto species or at least genus level is essential. The DNA sequencing is an alternative to classical fungal identification. The Internal Transcribed Spacer (ITS) regions of the ribosomal DNA gene cluster is now widely used in clinical laboratories for fungal species identification. The quality controlled ITS sequence data has been generated by ISHAM representing the actual sequence variation found in a species.

There are eleven contributing research groups from all over the world. The database currently contains more than 3600 sequences representing 535 human/animal pathogenic fungal species. The users are encouraged to submit their full dataset to the curators of the database, to enable the build up of comprehensive global ITS database of clinically important fungal pathogens. Individuals interested in the sequences available from this website can download the latest version along with the taxonomy file (https://its.mycologylab.org).

Species Complexes

In many pathogenic fungi, use of different molecular markers has demonstrated that they are genetically more complex than was initially thought. These are - *Sporothrix schenckii, Pneumocystis jirovecii, Coccidioides immitis, Aspergillus fumigatus, Candida parapsilosis, Pseudallescheria boydii. Apophysomyces elegans* and *Saksenaea vasiformis*, etc. These organisms constitute species complexes which are often difficult to differentiate phenotypically but sometimes with different clinical manifestations and infecting different body sites or even different hosts.

Classification of Fungi

In 1969, Robert H Whittaker, an American biologist while categorizing living things gave independent status to Fungi, in a separate kingdom, in his descriptions of five-Kingdom system *i.e.* Monera, Protista, Fungi, Plantae and Animalia. This classification put forward status of fungi to the level of an independent Kingdom. The current classification scheme proposed by Whittaker and modified by Margulis and Schwartz is presently being followed. Out of five fundamental Kingdoms, first Kingdom includes only **Prokaryotes** and rest four are **Eukaryotes**.

The present system of classification has evolved from two-Kingdom classification, followed by three- and then four-Kingdom classification. Ernst Haeckel in 1866 proposed setting up of third Kingdom under name Protoctista, meaning very first to include all thalophytes and protozoans. This Kingdom also included bacteria and cyanobacterium *i.e.* blue-green algae. But three-Kingdom classification failed to properly accommodate prokaryotic forms.

In 1955, after an ambiguity for about a century, G W Martin reviewed history of the taxonomy of fungi and began to speculate separate evolution for these organisms. A noted biologist Herbert Copeland in 1956 re-designated Protoctista to Protista and envisioned **four** Kingdoms of Mychota (later referred to as Monera), Protista, Plantae and Animalia. The Protista included protozoa, fungi, slime molds and algae, except blue-green algae, which were grouped with bacteria as cyanobacteria.

The limitation of four-Kingdom classification was also realized because despite creation of kingdom Monera, it failed to account for inter-gradations between Protista and Animalia. From evolutionary point of view kingdom Monera is regarded as most primitive. The three- and four-Kingdom classifications were able to remove some of anomalies of two-Kingdom classification system but these systems, however, were not able to place Fungi appropriately *i.e.* the group of organisms, which lack chlorophyll. They were classified inappropriately under Protista in four-Kingdom classification despite the fact that they differed from Protista totally in form, function and their behavior. Moreover, they could neither be considered plants nor animals, therefore, in 1969, Robert Whittaker proposed a separate kingdom **'Fungi'**, for these organisms, thus proposing ***'Five-Kingdom System'*** of classification of the living world. Whittaker's scheme divided living organisms into five kingdoms, which retained basic prokaryote-eukaryote distinction.

In the five Kingdom classification, Monera contains prokaryote, whereas the eukaryotes are classified into four remaining Kingdoms. The kingdom Protista contains unicellular eukaryotic organisms *i.e.* Protozoa and unicellular eukaryotic algae. The multicellular organisms are split into three Kingdoms on the basis of mode of nutrition and other fundamental differences in organization. The kingdom Plantae includes multicellular, photosynthesizing organisms, higher plants and multicellular algae. Kingdom Fungi includes terrestrial organisms like molds, yeasts and mushrooms, which do not have chlorophyll but obtain their food by absorption. Fungi were regarded as plants in the past, which have lost their chlorophyll thereby adopted heterotrophic mode of nutrition. The non-chordates and chordates make up kingdom Animalia. Most of these forms ingest their food and digest it internally, although some parasitic forms are absorptive. The Protista are believed to have given rise to entire multicellular organisms, which have evolved independently.

The present five-Kingdom classification system is also not final solution to the intricacies of taxonomy, as it does not include viruses. They should probably have their own Kingdom since they are unlike any other group of organisms. The five-Kingdom classification is summarized below:

(i) Monera: This is regarded as most primitive kingdom from evolutionary point of view as organisms are prokaryotes, which are single-celled and have no nucleus or cellular organelles. It includes Bacteria, Actinomycetes and Cyanobacteria.

(ii) Protista: This kingdom includes simplest eukaryotes *i.e.* single-celled organisms that do possess nucleus *e.g.* Protozoa, Chytrids, Water Molds (Oomycota), Slime Molds (Myxomycota) and nucleated Algae.

(iii) Fungi: The Fungi is an independent kingdom, which has non-motile eukaryotes that lack flagella and develop from spores *e.g.* Yeasts, Molds and Mushrooms.

(iv) Plantae: These multicellular, autotrophic eukaryotes develop from embryos and also use chlorophyll and cell wall contains cellulose like all Plants, Liverworts, Mosses and higher Algae.

(v) Animalia: Multicellular eukaryotic heterotrophs develop from blastula like Mammals, Arthropods, Worms, Birds, Reptiles, Mollusks and Coelenterates.

The progress in classification did not stop at the five-kingdom level but has gone beyond it. In the six-kingdoms classification, living world is divided into two major groups, **Prokaryotes** and **Eukaryotes**. The Prokaryote group has further two-kingdoms (Archaebacteria and Eubacteria) and Eukaryote has four kingdoms (Protista, Plantae, Eumycota and Animalia). On the basis of recent developments in taxonomy now there are seven recognized Kingdoms, which are Archaebacteria, Eubacteria, Protista, Chromista, Plantae, Eumycota and Animalia. All members of kingdom Eumycota are fungi and two of phyla in kingdom Chromista are also treated as fungi.

The classification of fungi is primarily based on type of sexual reproduction; however, other phenotypic features like morphology, etc. are also taken into consideration. The morphological characteristics of fungi are further supported by fatty acids analysis data, zymogram pattern, DNA base ratio studies and DNA hybridization analysis. Restriction fragment length polymorphism (RFLP) is also very useful tool employed for the taxonomical studies.

The kingdom **Fungi** consists of terrestrial eukaryotic heterotrophs in which nutrition is by absorption. They are found as either yeasts or mycelial as well as dimorphic forms. The cell walls are mainly composed of chitin. The reproduction is sexual as well as asexual. There are about 1,50,000 species of fungi, making this kingdom the highest taxonomic category. They are organized into phyla, then into classes and orders, however, some authors avoid particular categories above orders. Mycologists previously used term Division instead of Phylum because it is more inherent to botany but during recent times term Phylum has replaced Division. Taxa higher than genus, of which family, order and class (in ascending order) are most commonly used in arranging fungi, are uninominal (one word) and have an author, who is seldom cited. The name of family ends in ***'-aceae'***, of an order in ***'-ales'*** and of class in ***'-mycetes.'*** The hierarchical manner of Kingdom Fungi by giving example of *Histoplasma capsulatum* is as given in **Table 3.3**. However, in scientific publication, name of author who had introduced that name is added *e.g. Histoplasma capsulatum* Darling. In this regard mnemonics helps to remember hierarchy of Linnaeus' Taxonomy as - **K**ing **P**hilip **C**an **O**nly **F**ind **G**reen **S**ocks, being used for the seven mandatory groupings *i.e.* **K**ingdom, **P**hylum, **C**lass, **O**rder, **F**amily, **G**enus and **S**pecies, respectively.

Table 3.3. Classification Scheme of the Kingdom Fungi.

Taxon	*Suffix*	*Teleomorph*	*Anamorph*
Phylum	-mycota	Ascomycota	Deuteromycetes
Class	-mycetes	Plectomycetes	Hyphomycetes
Order	-ales	Onygenales	Moniliales
Family	-aceae	Onygenaceae	Moniliaceae
Genus	-	*Ajellomyces*	*Histoplasma*
Species	-	*capsulatus*	*capsulatum*
Variety	-	*capsulatus*	*capsulatum*

The yeasts and molds are organized into various sub-groups on the basis of perfect (teleomorph) state and one of which is an artificial category, which contains fungi without known sexual (anamorph) state. The fungi those reproduce sexually are classified under three phyla—**Glomeromycota, Ascomycota** and **Basidiomycota** depending on morphology of their fruiting bodies/structures. The function of fruiting body is to produce sexual spores for dissemination. Although sexual states of many fungi are still not known, however, they are placed in appropriate taxonomical positions on the basis of other morphological features as well as molecular studies. The molecular techniques are useful tools in taxonomy to resolve long standing issues and puzzles in Medical Mycology. The anamorph fungi, which lack known sexual state, have been placed in fourth separate major high-level taxon called **Deuteromycetes** or ***'Fungi Imperfecti'*** or ***'Mitosporic Fungi'***.

The 'Fungi Imperfecti' denotes an artificial classification system for fungi that reproduce asexually. This is based on morphological and physiological characteristics selected by mycologists for their convenience. Unlike classification of fungi based on teleomorphs, 'Fungi Imperfecti' is not intended to reflect phylogenetic relationships. Though it might be speculated whether black yeast is an anamorph of an ascomycete or basidiomycete and this information should be used as the basis of classification. The Article 59 of ICBN clearly stipulates that anamorphs and teleomorphs are based upon different criteria and concepts. These fungi might have lost their ability to reproduce sexually or their sexual state may not have been discovered.

However, after use of molecular techniques in the fungal taxonomy, many previously 'unrelated' fungal species were found to be sexual and asexual stages of same fungus and many fungi that belonged to deuteromycetes were re-classified into other major phyla because of discovery of their sexual states. As and when sexual sporulation in any of these fungi is discovered, fungus is removed from Deuteromycetes to a suitable position in the above mentioned three phyla *i.e.* Glomeromycota, Ascomycota or Basidiomycota.

In the past ten years, with use of molecular strategies phylogenetic affinities of two more resilient pathogens studied in medical mycology, *Lacazia loboi* and *Rhinosporidium seeberi* were finally deciphered. These studies found that *L.loboi* was sister taxon to *Paracoccidioides brasiliensis* and *R.seeberi* was closely related to protistan spherical aquatic fish pathogens, located at the point where animals diverged from fungi, in the class Mesomycetozoea.

All fungi reproduce asexually however majority of them can reproduce sexually if the sexual state is observed for a fungal isolate. The terms **anamorph** (asexual) and **teleomorph** (sexual) are often used to describe taxonomic status of organisms. This is mentioned mostly in nomenclature used to refer a particular organism. For example anamorphic designation for organism causing histoplasmosis is *Histoplasma capsulatum*; if specific isolate exists in teleomorphic state, then it is labeled as *Ajellomyces capsulatus*.

This convention is used for all fungal organisms to maintain taxonomic uniformity. However, in clinical settings it is common to refer organisms by their asexual designations, since anamorph is only state usually isolated from clinical specimens and sexual state exists only under extremely controlled conditions of the fungal culture. However, in black piedra, causative fungus, *Piedraia hortae* is found as its teleomorphic state in clinical material, which is one of the exceptions.

The dual modality of fungal propagation *i.e.* sexual and asexual, has meant since last century there has been dual nomenclature. The fungus as a whole, comprises teleomorph and one or more anamorphs. The term **holomorph** has been reserved for fungi with teleomorphic sporulation together with all their sporulating or vegetative anamorphs. The anamorph and teleomorph generally develop at different times and on different substrates, although in mucormycetes they often occur together. The Article 59 of ICBN deals with this issue of one or more names of fungi.

Some names are given on the basis of their perfect state like *Pseudallescheria boydii* whereas in most of fungi names are based on anamorphic state of fungus as teleomorphic states are not usually seen in clinical lesions and require special conditions and cumbersome procedures to induce their sexual spores.

Another controversy resulted from replacement of terms anamorph and teleomorph with 'mitosporic fungi' and 'meiosporic fungi,' respectively in the 8th edition of Ainsworth and Bisby's Dictionary of Fungi published in 1995. This source is considered fundamental framework for fungal terminology and taxonomy. These changes have not been accepted by many workers who consider that anamorph and teleomorph states of fungus are determined not simply by type of cellular processes (meiosis and mitosis), which precede sporulation but also by morphological features. On the basis of teleomorphic states, following phyla are described:

(a) Glomeromycota

The previous phylum Zygomycota underwent drastic taxonomic changes in 2007. Hibbett *et al,* published a comprehensive phylogenetic classification of the kingdom Fungi that was based on well-supported monophyletic groups consistent across multiple phylogenetic studies. They used the available data generated by recent molecular phylogenetic studies, as well as the data contributed by diverse members of the fungal taxonomy community. They proposed to eliminate Zygomycota because the phylum was found to be polyphyletic and the taxa conventionally placed in Zygomycota were distributed among the phylum Glomeromycota and four subphyla. The Mucorales and Entomophthorales, which contain zoopathogenic fungi and two other orders including Kickxellales and Zoopagales were raised to the rank of subphyla: Mucoromycotina, Entomophthoromycotina, Kickxellomycotina and Zoopagomycotina.

In Mucoromycotina, the mycelia are broad, ribbon-like and non-septate or sparsely septate. The sexual fusion of hyphae leads directly to formation of resting, thick-walled cell, which is termed as **zygospore (Fig. 3.1)**, which divides by meiosis when it germinates. Usually the zygospore has suspensor cells on its both sides, however, in some of the species (*Rhizopus homothallicus*), the suspensor cells may be on one side only **(Figs. 3.2A to C)**. The asexual spores known as **sporangiospores** are borne **internally** within structure called **sporangium (Fig. 3.3)**.

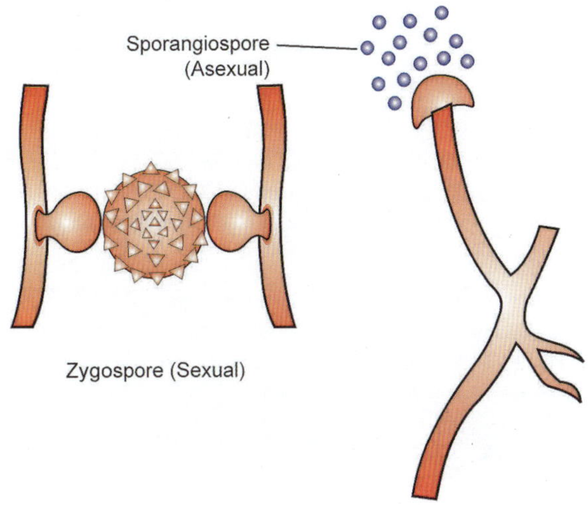

Fig. 3.1. Diagrammatic sketch of zygospore and burst sporangium releasing sporangiospores as seen in *Rhizopus* species.

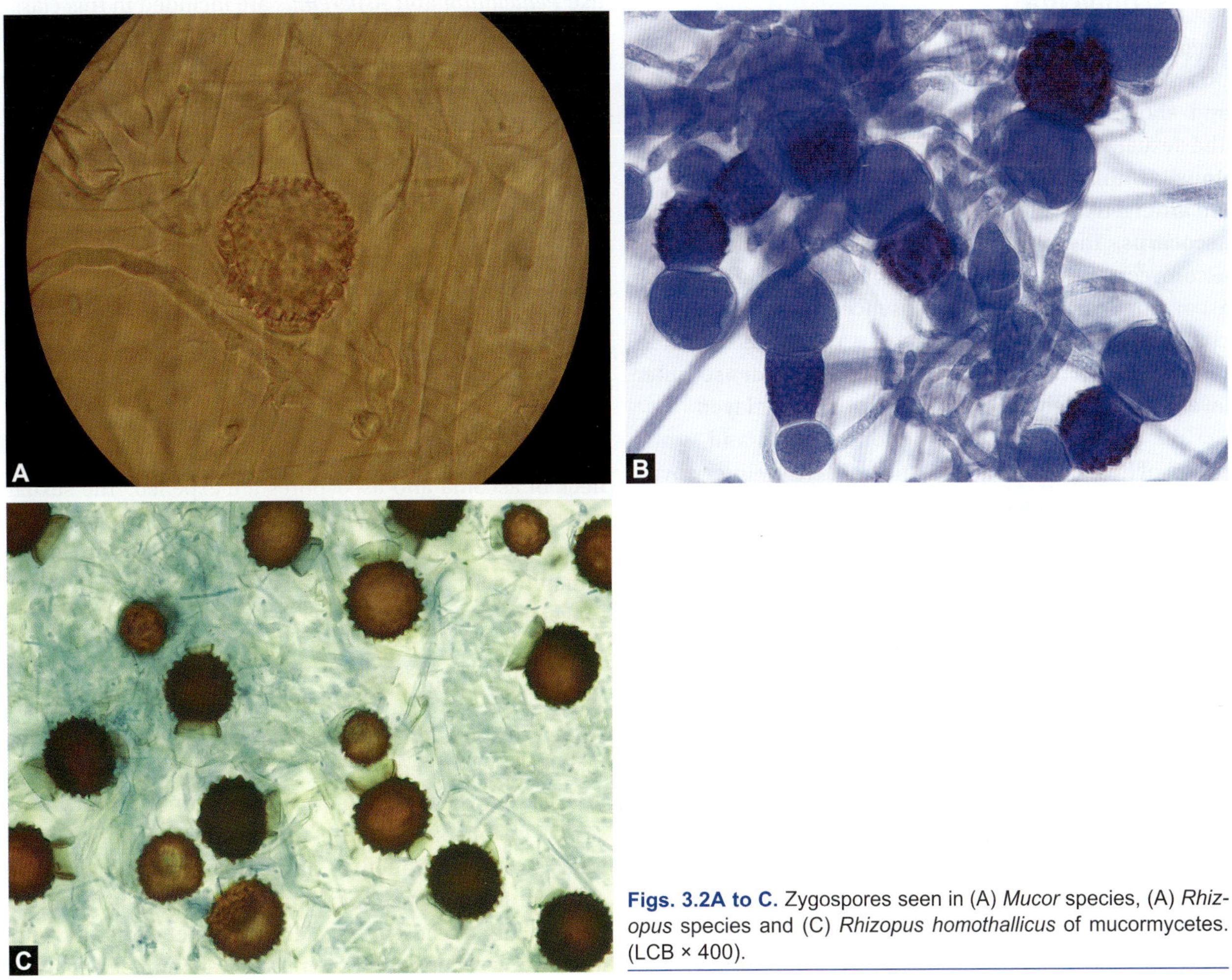

Figs. 3.2A to C. Zygospores seen in (A) *Mucor* species, (A) *Rhizopus* species and (C) *Rhizopus homothallicus* of mucormycetes. (LCB × 400).

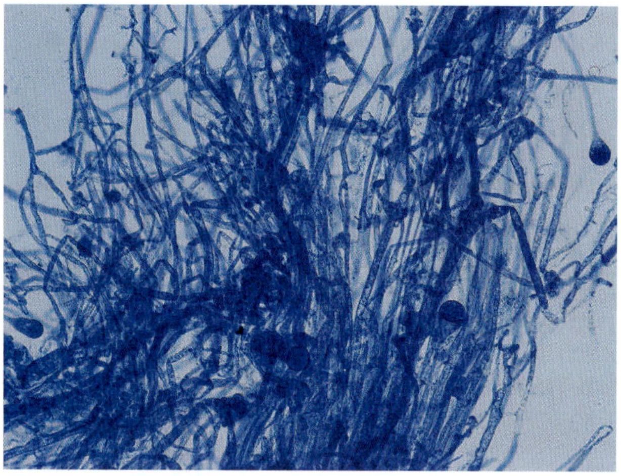

Fig. 3.3. Sporangia seen in *Apophysomyces variabilis* of mucormycetes (LCB × 200).

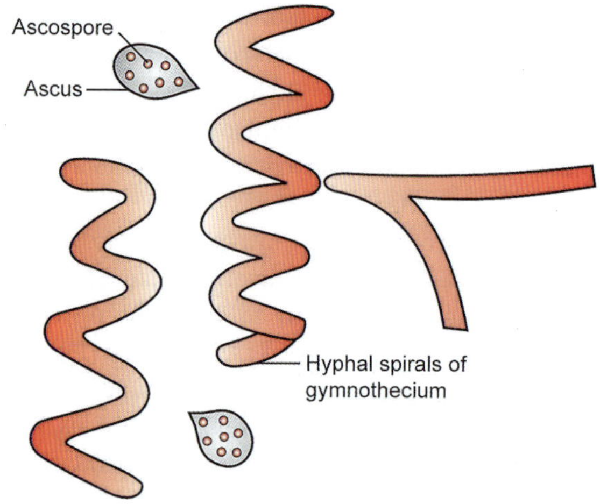

Fig. 3.4. Diagrammatic sketch of ascospores seen in *Ajellomyces dermatitidis* of Ascomycetes.

(b) Ascomycota

Ascomycota is largest phylum of kingdom Fungi. It comprises of approximately 75% described fungal species and about 80% of pathogenic and opportunistic fungi. The sexual spores *i.e.* **ascospores**, are produced **endogenously** within microscopic 'pods' called **asci**. These asci are arranged in different types of fruiting bodies called as **ascocarps**. The basic characteristic which differentiates ascomycetes from other fungi is presence of asci inside ascomata. However, even in absence of these important characteristics, ascomycetes can be recognized by their bi-layered hyphal walls with thin electron-dense outer layer and relatively electron-transparent inner layer.

The hyphae usually have perforated septa and sexual fusion of hyphae leads to formation of densely interwoven mass *i.e.* **sac** or **ascus** that contains characteristic meiotic reproductive products called asci as four to eight spores called as **ascospores (Fig. 3.4)**. The asci may be grouped together to form cup-shaped structure known as **perithecium**. They also produce asexual spores which are borne at tips of hyphae. On the basis of arrangement of asci and type of fruiting bodies produced, ascomycetes have been further grouped into six classes: Hemiascomycetes, Plectomycetes, Pyrenomycetes, Discomycetes, Laboulbeniomycetes and Loculoascomycetes.

The class Plectomycetes is characterized by presence of closed, more or less spherical fruiting bodies called as **cleistothecia (Figs. 3.5A and B)** with an irregular distribution of asci in cavity and ascospores are released after disintegration of thin walls of asci. The cosmopolitan genera like *Penicillium* and *Aspergillus* are included in this class. The order Onygenales has clearly defined groups of true dimorphic fungi like *Histoplasma* in family Onygenaceae and dermatophytes in Arthrodermataceae.

The Pyrenomycetes have pyriform (flask-shaped) fruiting bodies (perithecia) with usually saccate or cylindrical asci. The ascospores are forcibly extruded from ascus. The teleomorphs of several opportunistic fungi are included in this class. The class Bitunicate Pyrenomycetes or Loculoascomycetes comprises species with **bi-layered** asci, sometimes enclosed in stromatic ascomata. This sort of ascus has distinctly bi-layered wall with outer layer being rigid and inner layer being expansile. As it matures, thin outer layer splits and thick inner layer absorbs water and expands upwards. The ascus stretches up into narrow neck of ascoma and ascospores are expelled. The sexual states of numerous pathogenic phaeoid fungi are included in this class. The Hemiascomycetes comprises yeasts. In remaining two classes there is no medically significant fungal agent.

(c) Basidiomycota

There are about 25,000 species consisting of mushrooms, toadstools, shelf fungi, rusts and smut fungi. The hyphae in basidiomycetes usually have incomplete septa and sexual fusion of hyphae leads to formation of densely interwoven reproductive structure (mushroom) with characteristic microscopic, club-shaped organs called as **basidia**. The basidia are usually aseptate structures with four tiny projections, which are called sterigmata and each

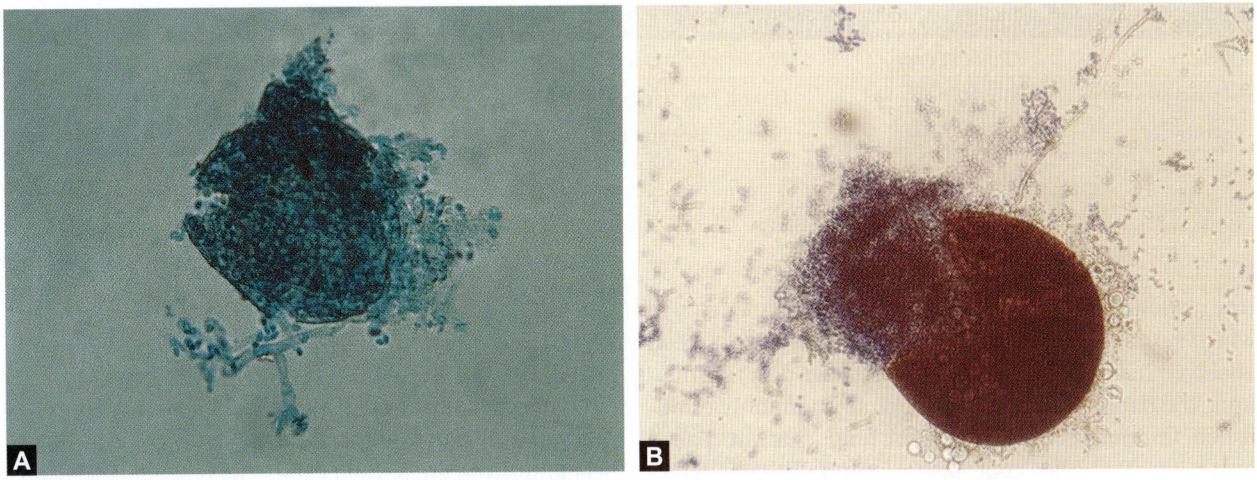

Figs. 3.5A and B. Bursting cleistothecium with ascospores of genus *Aspergillus* (LCB × 200).

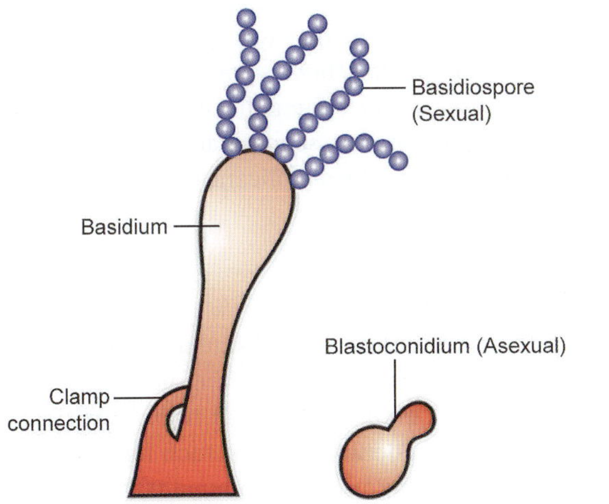

Fig. 3.6. Diagrammatic sketch of basidiospores seen in *Filobasidiella neoformans* of Basidiomycetes.

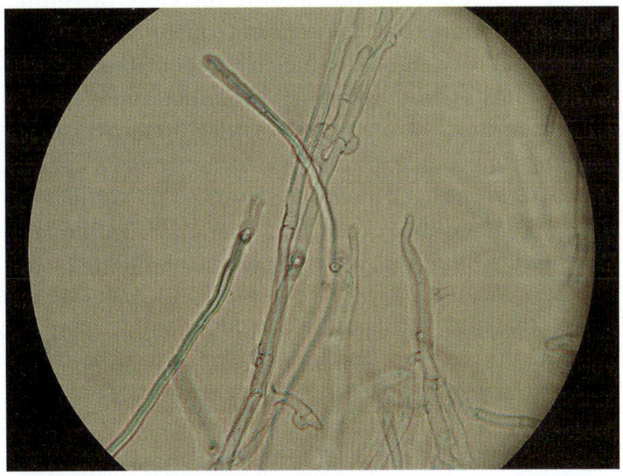

Fig. 3.7. Basidiomycetous fungus *Schizophyllum commune* showing clamp connection (LCB × 400).

sterigma bears haploid meiospore *i.e.* **basidiospore**. The basidiospores are **exogenously** produced on the basidium. The fruiting bodies, seen in Basidiomycota where basidia arranged are called as **basidiocarps**. The asexual spores are generally not found but vegetative reproduction by fragmentation is common. The asexual spores, if found (**conidia**), are borne externally at the tips of hyphae. There is formation of clamp-connection before making fruiting body **(Fig. 3.6)**.

The most significant yeasts/yeastlike pathogenic basidiomycetes to humans are *Cryptococcus*, *Malassezia* and *Trichosporon* species, whereas filamentous one is *Schizophyllum commune*.

The basidiomycetes of medical importance are placed in Classes -Basidiomycetes and Ustomycetes and in Orders - Agaricales, Poriales, Schizophyllales, Stereales, Tremellales and Ustilaginales. The emerging basidiomycetes pathogens are: *Schizophyllum commune* **(Fig. 3.7)**, *Hormographiella aspergillata* including its sexual state *i.e. Coprinopsis cinerea* (former *Coprinus cinereus)* and *Tilletiopsis minor*, which are found associated with various clinical presentations of hyalohyphomycosis.

(d) Deuteromycota (Fungi Imperfect)

The deuteromycota (Gr. for ***'second fungi'***) is considered formal phylum of kingdom Fungi. However, this is not

Table 3.4. Classification of Fungi Showing Different Phyla.

Phylum	*Growth Form*	*Hyphae (if present)*	*Sexual Propagules*	*Asexual Propagules*
Glomeromycota	Molds	Broad, Few Septa	Zygospores	Sporangiospores
Ascomycota	Molds, Yeasts	Narrow, Regular Septa	Ascospores	Conidia
Basidiomycota	Molds, Yeasts	Narrow, Regular Septa	Basidiospores	Conidia
Deuteromycetes	Molds, Yeasts	Narrow, Regular Septa	None	Conidia

true phylogenetic group rather it is an artificial category in which sexual process is either absent or has not yet been observed hence fungi are temporarily placed in this phylum. The deuteromycota do not fit into commonly established taxonomic classification of fungi that are based on biological species concepts or morphological characteristics of asexual structures because their sexual forms of reproduction are never observed and only the asexual mode of reproduction is only known. Therefore, this group of fungus produces only the asexual spores.

These are actually the ***'left-over'*** species that do not fit well into any of other groups of fungi and around 25,000 species are grouped in this phylum. Due to absence of sexual reproduction in their life cycle these are also called as ***'Fungi Imperfecti', 'Mitosporic Fungi'*** or ***'Anamorphic fungi'***. However, most of the members of this group morphologically resemble Ascomycetes. At present, deuteromycetes is second largest group of fungi after Ascomycota. In Holomorph Conference held in 1993, it was agreed to maintain the term **deuteromycetes** with lowercase ***'d'*** and not to formally recognize this group of organisms at particular rank.

The deuteromycetes are classified on the basis of morphology and ontogeny of conidia and conidiogenous cells. Conidia may develop by either blastic process (budding) or thallic process (fragmentation). The deuteromycetes are divided into **three form-classes**:

(i) Blastomycetes include asexual yeasts that produce spores following reproduction by budding.

(ii) Hyphomycetes, conidia are borne on hyphae, on specialized conidiophore and in many genera blastic process produces conidia.

(iii) Coelomycetes, spores are borne in pycnidium or acervulus. The genera belonging to this form-class are asexual fungi producing conidia within fruiting bodies, named conidiomata. These structures are spherical **(pycnidia)**, with conidiogenous cells lining inner cavity wall or cup-shaped **(acervuli)**, in which conidiogenous cells form palisade on surface of conidiomata. These two types of structure characterize two artificial *i.e.* form-orders-**Sphaeropsidales** and **Melanconiales**, respectively. *Lasiodiplodia* and *Pyrenochaeta* species are important coelomycetes causing eumycetoma.

When grown in culture, pycnidia can be confused with fruiting bodies of ascomycetes. The distinction between pycnidium and an ascoma should be made by squashing structure under coverslip. In case of ascomycetes, ascospores inside asci are visible whereas in pycnidia numerous free conidia are present usually in mucous masses.

The taxonomy of medically important **yeasts** is based not only on morphology but also on extensive physiological characterization. The systematic classification of fungi as simplified taxonomic scheme along with representative genera of clinically significant pathogens is given in **Table 3.4 and 3.5**.

Cryptic Fungal Species

There are morphologically indistinguishable fungal species in the main genus *e.g. Aspergillus* sections. Sometimes the clue comes from the elevated triazole MICs among such strains. The identification of such fungi is based on targeted sequencing using (i) internal transcriber spacer region (ii) beta-tubulin (iii) calmodulin or (iv) actin genes. In transplant patients -10% of isolates causing disease may belong to the these cryptic species.

Emmonsiellopsis

Recently a new genus related to the thermally dimorphic fungi of the family Ajellomycetaceae was established by Marin-Felix *et al* , who isolated two interesting fungi from fluvial sediments, which was collected in the North of Spain. They were morphologically related to the thermally dimorphic fungi of the family Ajellomycetaceae, however, the analysis of the internal transcribed spacer region of the rDNA and the domains D1 and D2 of the 28S rRNA gene sequences confirmed that they were different from all the species described in that family. They were accommodated in the new genus *Emmonsiellopsis* as *E.coralliformis*

Table 3.5. Classification of Medically Significant Fungi - Superkingdom Eukaryota, Kingdom Fungi.

Class/Order/Family	*Fungal Genera/Species*
Phylum Glomeromycota	
Mucorales	
Mucoraceae	*Lichtheimia , Mucor, Rhizopus, Rhizomucor,*
Cunninghamellaceae	*Cunninghamella*
Mortierellaceae	*Mortierella*
Saksenaeaceae	*Saksenaea, Apophysomyces*
Syncephalastraceae	*Syncephalastrum*
Thamnidiaceae	*Cokeromyces*
Entomophthorales	
Ancylistaceae	*Conidiobolus*
Basidiobolaceae	*Basidiobolus*
Trichomycetes	No Significant Human Pathogen
Phylum Ascomycota	
Basal Ascomycetes	
Pneumocystidales	*Pneumocystis jirovecii*
Pyrenomycetes (Unitunicate Asci)	
Microascales	
Microascaceae	*Pseudallescheria, Microascus, Scopulariopsis*
Ophiostomatales	*Ophiostoma, Sporothrix*
Sordariales	*Chaetomium, Neurospora*
Hypocreales	*Fusarium, Acremonium, Trichoderma*
Pyrenomycetes (Bitunicate Asci - Loculoascomycetes)	
Dothideales	
Mycosphaerellaceae	*Cladosporium,*
Herpotrichiellaceae	*Cladophialophora, Exophiala, Phialophora, Fonsecaea, Capronia, Rhinocladiella, Ochroconis*
Dothideaceae	*Hortaea werneckii Aureobasidium pullulans*
Piedraiaceae	*Piedraia hortae*
Pleosporaceae	*Bipolaris, Exserohilum, Curvularia, Alternaria, Botryomyces*
Plectomycetes (Cleistothecial Ascomata)	
Eurotiales	
Trichocomaceae	*Aspergillus, Penicillium,*
Thermoascaceae	*Paecilomyces*
Onygenales	
Onygenaceae	*Ajellomyces, Chrysosporium, Malbranchea*
Arthrodermataceae	*Arthroderma*

Class/Order/Family	*Fungal Genera/Species*
Phylum Basidiomycota	
Basidiomycetes	
Agaricales	*Coprinus, Hormographiella*
Poriales	*Bjerkandera*
Schizophyllales	*Schizophyllum commune*
Stereales	*Sporotrichum*
Tremellales	*Filobasidiella, Trichosporon, Malassezia, Cerinosterus*
Ustomycetes	
Ustilaginales	*Ustilago, Rhodotorula, Sporobolomyces, Tilletiopsis*
Phylum Deuteromycota	
Blastomycetes	
Ascogenous	*Saccharomyces, Dipodascus, Endomyces*
Basidiogenous	
Holobasidiomycetes	*Filobasidiella*
Teliomycetes	*Rhodosporidium, Leucosporidium*
Cryptococcaceae	*Candida, Malassezia, Trichosporon*
Hyphomycetes	
Moniliaceae	*Acremonium, Arthrographis, Aspergillus, Beauveria, Blastomyces, Chrysosporium, Coccidioides, Epidermophyton, Fusarium, Histoplasma, Madurella, Microsporum, Paecilomyces, Paracoccidioides, Penicillium, Scopulariopsis, Sporothrix, Trichophyton*
Dematiaceae	*Alternaria, Aureobasidium, Bipolaris, Cladosporium, Curvularia, Exophiala, Hortaea, Fonsecaea, Phialophora, Rhinocladiella, Sarcinomyces, Torula, Wangiella*
Coelomycetes	*Phoma, Macrophoma, Nattrassia, Pyrenochaeta\ Colletotrichum dematium*
Mycelia Sterilia	No Sporulation

Note: *Rhinosporidium seeberi*, *Pythium insidiosum* and *Prototheca* species are not classified in the Kingdom Fungi.

sp. nov. and *E.terrestris* sp. nov. The two species are distinguished mainly by the maximum temperature of growth (up to 33°C for *E.coralliformis* and to 42°C for *E.terrestris*), the dendritic mycelium of *E.coralliformis* and the conidial ornamentation (verrucose in *E.coralliformis* and spinulose in *E.terrestris*). In addition, the phylogenetic data demonstrated that *Ajellomyces griseus* also represents a new genus within the Ajellomycetaceae, namely *Helicocarpus*. This new genus is easily distinguished by the lack of asexual morph, the production of brownish gymnothecial ascomata and oblate to lenticular, sparingly pitted ascospores. The proposal of both new genera was confirmed by the analysis of actin gene sequences.

Organisms of Controversy

Since long it was disputed whether genus *Pneumocystis* is protozoa or fungus. Now, based on molecular techniques, it could be feasible to ascertain proper taxonomic status of this organism and it is now categorized as atypical fungus. Similarly, *Microsporidium*, has also recently been included in the Fungi.

Some of organisms are presently **not** classified in the kingdom **Fungi** but they are producing clinically and histopathologically very similar diseases, to that of fungal infections, are discussed in the differential diagnosis. They are clubbed under a common group called as **pseudofungi** or **parafungi**. Therefore, these are also significant organisms to be referred in this Chapter. However, details about these organisms are given subsequently in their respective chapters under separate section of this Textbook as Sector VII. These organisms are species of *Prototheca, Rhinosporidium* and *Pythium* genera.

Therefore, two organisms (*Pneumocystis* and *Microsporidium*) are now included, whereas three organisms (*Prototheca, Rhinosporidium* and *Pythium*), which were one or the other time confused with fungi, are now excluded from kingdom **Fungi**. *Pneumocystis jirovecii* is now settled taxonomy as well as nomenclature and rest four are briefly described below:

(i) Phylum Microspora

The phylum Microspora consists of obligate intracellular organisms. In humans *Microsporidium* is considered to be etiologic agent of opportunistic infections mostly in individuals with AIDS. The phylogenetic analysis of this organism indicates that it is related to Fungi rather than to Protozoa. Moreover, the structural features such as the presence of chitin in the spore wall, diplokaryotic nuclei and electron-dense spindle plaques associated with the nuclear envelope also suggest its possible relationship with Fungi.

(ii) Algae-like Organisms

Prototheca species are algae-like aerobic, unicellular organisms and lack chlorophyll thus being unable to produce energy by photosynthesis. The taxonomic status of *Prototheca* is considered to be possible mutation of green algae of genus *Chlorella* belonging to kingdom **Protista** as they sporulate in similar manner. Initially, they were considered as yeasts because colonies are yeast-like and therefore were included in kingdom Fungi. The genus *Prototheca* belongs to family **Chlorellaceae** which includes unicellular organisms that reproduce asexually by internal septation and cleavage. Currently re-classification of this genus defines three valid species, namely *P.wickerhamii, P.zopfii* and *P.stagnora* , first two are involved in human and bovine infections.

(iii) Mesomycetozoea

Rhinosporidium seeberi was previously classified as fungus on the basis of morphological and histochemical characteristics. But analysis of aligned sequence and inference of phylogenetic relationships showed that *R.seeberi* is protistan from novel clade of parasites that infect fish and amphibian host. The fluorescence *in situ* hybridization and *R.seeberi*-specific PCR showed that unique 18S rRNA sequence was also present in other tissues infected with this organism. The same is suggested by another group of workers that *R.seeberi* is not classic fungus rather first known human pathogen from group of fish parasites known as **DRIPs**, novel clade of aquatic protistan species. Therefore, at present *R.seeberi* is eukaryotic organism classified within protoctistan Mesomycetozoea.

(iv) Phylum Oomycota

These are heterotrophic, unicellular parasites or decomposers; cell walls composed of cellulose, not chitin as in fungi; asexual and sexual reproduction *e.g.* water molds, white rusts, downy mildews (*Phytophthora, Plasmopara*). There are about 580 species of oomycetes.

The taxonomy of genus *Pythium* is mainly based on morphological descriptions, however, these observations can now be supplemented with molecular characteristics on the basis of 18s rRNA and others sequences. The significant differences between oomycetes and fungi are depicted in **Table 39.1**. *Pythium insidiosum* currently belongs to kingdom **Stramenopila** under phylum **Oomycota** in class **Oomycetes** order **Pythiales** and family **Pythiaceae**. The members of this genus although have coenocytic mycelia but they are not considered to be true fungi. The molecular and biochemical analyses suggest that these are closer to algae and higher plants and classified amongst Stramenopiles, one of five major eukaryotic Kingdoms. Unlike most of eumycetes, members of this genus remain diploid throughout their life cycle with meiosis occurring in gametangia before fertilization. The sexual reproduction is an oogamy **(*See* Fig. 39.1)** while asexual reproduction takes place by formation of zoospores in filamentous or spherical types of sporangia.

Kingdom Chromista

The name Chromista was introduced by Cavalier-Smith in 1981. The Chromista are a paraphyletic eukaryotic supergroup, which may be treated as a separate kingdom or included among the Protista. They include all algae whose chloroplasts contain chlorophylls a and c, as well as various colorless forms that are closely related to them. These are surrounded by four membranes and are believed to have been acquired from some red alga. There are three different groups: (i) Heterokonts or stramenopiles - brown algae, diatoms, water molds, etc (ii) Haptophytes (iii) Cryptomonads.

Phylum Chytridiomycota

The Chytrids are heterotrophic decomposers and parasites with cell walls of chitin; live in water and soil. Most of them are unicellular and are characterized by their asexual state; zoospore (motile) with single whiplash flagellum oriented and located posteriorly. The zoospore is formed in sporangium and is released through an opening in wall and its release usually indicates death of thallus. It is to be noted that zoospores with single tinsel flagellum are found in diverse Protists, though morphology of some of these microbes is otherwise similar to some chytrids *e.g. Hyphochytrium*.

Assembling Fungal Tree of Life

Fungi are eukaryotes that form their own kingdom. Around 100000 fungal species have been named, although there are estimated to be from 1.5 to 3 million species worldwide . According to the Assembling the Fungal Tree of Life **(AFTOL)** project, the fungal kingdom can be divided into as many as 8-10 phyla. Among these, the Ascomycota and Basidiomycota groups together form the Dikarya, which represent approximately 98% of described fungal species. These ubiquitous organisms have adapted to a variety of ecological habitats. They are involved in the degradation of decomposing organic matter in nature but are also used in industry such as in the production of food, antibiotics and enzymes.

Further Reading

1. Adl SM, Simpson AG, Lane CE, *et al.* The revised classification of eukaryotes. *J Eukaryot Microbiol.* 2012; **59:** 429-93.
2. Arun Kumar TK, Blackwell M, Letcher PM, *et al.* Research and teaching with the AFTOL SBD: An informatics resource for fungal subcellular and biochemical data. *IMA Fungus.* 2013; **4:** 259-63.
3. Barr DJS. Evolution and kingdoms of organisms from the perspective of a mycologist. *Mycologia.* 1992; **84:** 1-11.
4. Bernhard M, Zautner AE, Steinmann J, *et al.* Towards proteomic species barcoding of fungi - An example using *Scedosporium/Pseudallescheria* complex isolates. *Fungal Biol.* 2016; **120:** 162-5.
5. Boekhout T, Gueidan C, de Hoog S, *et al.* Fungal taxonomy: New developments in medically important fungi. *Curr Fungal Infect Rep.* 2009; **3:** 170-78.
6. Braun U. The impacts of the discontinuation of dual nomenclature of pleomorphic fungi: The trivial facts, problems and strategies. *IMA Fungus.* 2012; **3:** 81-6.
7. Bruns T. Evolutionary biology: A kingdom revised. *Nature.* 2006; **443:** 758-61.
8. Bruns TD, White TJ, Taylor JW. Fungal molecular systematics. *Annu Rev Ecol Syst.* 1991; **22:** 525-64.
9. Calisher CH. Taxonomy: What's in a name? Doesn't a rose by any other name smell as sweet? *Croat Med J.* 2007; **48:** 268-70.
10. Cannon PF, Damm U, Johnston PR, *et al. Colletotrichum* - current status and future directions. *Stud Mycol.* 2012; **73:** 181-213.
11. Cannon PF. International Commission on the Taxonomy of Fungi (ICTF): Name changes in fungi of microbiological, industrial and medical importance. Part 1. *Microbiol Sci.* 1986; **3:** 168-71.
12. Cannon PF. International Commission on the Taxonomy of Fungi (ICTF): Name changes in fungi of microbiological, industrial and medical importance. Part 2. *Microbiol Sci.* 1986; **3:** 285-7.
13. Cannon PF. Name changes in fungi of microbiological, industrial and medical importance. Part 4. *Mycopathologia.* 1990; **111:** 75-83.
14. Cavalier-Smith T. 'Eukaryote kingdoms: seven or nine?' *Biosystems.* 1981; **14:** 461-81.

15. Cavalier-Smith T. A revised six-kingdom system of life. *Biol Rev Camb Philos Soc.* 1998; **73:** 203-66.
16. Cole GT. Models of cell differentiation in conidial fungi. *Microbiol Rev.* 1986; **50:** 95-132.
17. *Council for International Organization of Medical Sciences.* International Nomenclature of Diseases. 1st edn. Vol. II. Infectious Diseases. Part 2. *Mycoses.* CIOMS, Geneva; 1982.
18. Crous PW, Giraldo A, Hawksworth DL, *et al.* The genera of Fungi: Fixing the application of type species of generic names. *IMA Fungus.* 2014; **5:** 141-60.
19. Currah RS. Taxonomy of Onygenales: Arthrodermataceae, Gymnoascaceae, Myxotrichaceae and Onygenaceae. *Mycotaxon.* 1985; **24:** 1-216.
20. Daniel HM, Lachance MA, Kurtzman CP. On the reclassification of species assigned to *Candida* and other anamorphic ascomycetous yeast genera based on phylogenetic circumscription. *Antonie Van Leeuwenhoek.* 2014; **106:** 67-84.
21. de Hoog GS, Bowman B, Graser Y, *et al.* Molecular phylogeny and taxonomy of medically important fungi. *Med Mycol.* 1998; **36:** S52-6.
22. de Hoog GS, Chaturvedi V, Denning DW, *et al.* Name changes in medically important fungi and their implications for clinical practice. *J Clin Microbiol.* 2015; **53:** 1056-62.
23. de Hoog GS, Haase G, Chaturvedi V, *et al.* Taxonomy of medically important fungi in the molecular era. *Lancet Infect Dis.* 2013; **13:** 385-6.
24. de Hoog GS, Sigler L, Untereiner WA, *et al.* Changing taxonomic concepts and their impact on nomenclatural stability. *J Med Vet Mycol.* 1994; **32:** S113-22.
25. de Oliveira TB, Gomes E, Rodrigues A. Thermophilic fungi in the new age of fungal taxonomy. *Extremophiles.* 2015; **19:** 31-7.
26. Deshpande V, Wang Q, Greenfield P, *et al.* Fungal identification using a Bayesian classifier and the Warcup training set of internal transcribed spacer sequences. *Mycologia.* 2016; **108:** 1-5.
27. Dukik K, Munoz JF, Jiang Y, *et al.* Novel taxa of thermally dimorphic systemic pathogens in the Ajellomycetaceae (Onygenales). *Mycoses.* 2017; **60:** 296-309.
28. Farr DF, Rossman AY. Fungal databases, systematic mycology and microbiology laboratory, ARS, USDA. Retrieved December 31, 2016, from http://nt.ars-grin.gov/fungaldatabases.
29. Geiser DM, Aoki T, Bacon CW, *et al.* One fungus, one name: Defining the genus *Fusarium* in a scientifically robust way that preserves longstanding use. *Phytopathology.* 2013; **103:** 400-8.
30. Guarro J, Gene J, Stchigel AM. Developments in fungal taxonomy. *Clin Microbiol Rev.* 1999; **12:** 454-500.
31. Hawksworth DL, Crous PW, Redhead SA, *et al.* The Amsterdam Declaration on fungal nomenclature. *IMA Fungus.* 2011; **2:** 105-12.
32. Hawksworth DL, Lagreca S. New bottles for old wine: Fruit body types, phylogeny and classification. *Mycol Res.* 2007; **111:** 999-1000.
33. Hawksworth DL, McNeill J, de Beer ZW, *et al.* Names of fungal species with the same epithet applied to different morphs: How to treat them. *IMA Fungus.* 2013; **4:** 53-6.
34. Hawksworth DL. A new dawn for the naming of fungi: Impacts of decisions made in Melbourne in July 2011 on the future publication and regulation of fungal names. *IMA Fungus.* 2011; **2:** 155-62.
35. Hawksworth DL. Managing and coping with names of pleomorphic fungi in a period of transition. *IMA Fungus.* 2012; **3:** 15-24.
36. Hawksworth DL. Naming *Aspergillus* species: Progress towards one name for each species. *Med Mycol.* 2011; **49:** S70-6.
37. Hawksworth DL. Pandora's mycological box: Molecular sequences *vs.* morphology in understanding fungal relationships and biodiversity. *Rev Iberoam Micol.* 2006; **23:** 127-33.
38. Hawksworth DL. Possible house-keeping and other draft proposals to clarify or enhance the naming of fungi within the International Code of Nomenclature for algae, fungi and plants (ICN). *IMA Fungus.* 2014; **5:** 31-7.
39. Hawksworth DL. Proposals to clarify and enhance the naming of fungi under the International Code of Nomenclature for algae, fungi and plants. *IMA Fungus.* 2015; **6:** 199-205.
40. Heitman J. Microbial pathogens in the fungal kingdom. *Fungal Biol Rev.* 2011; **25:** 48-60.
41. Hibbett D, Abarenkov K, Koljalg U, *et al.* Sequence-based classification and identification of fungi. *Mycologia.* 2016; **108:** 1049-68.
42. Hibbett DS, Binder M, Bischoff JF, *et al.* A higher-level phylogenetic classification of the Fungi. *Mycol Res.* 2007; **111:** 509-47.
43. Hibbett DS, Taylor JW. Fungal systematics: Is a new age of enlightenment at hand? *Nat Rev Microbiol.* 2013; **11:** 129-33.
44. Hoffmann K, Discher S, Voigt K. Revision of the genus *Absidia* (Mucorales, Zygomycetes) based on physiological, phylogenetic, and morphological characters; thermotolerant *Absidia* spp. form a coherent group, Mycocladiaceae fam. nov. *Mycol Res.* 2007; **111:** 1169-83.
45. Hoffmann K, Walther G, Voigt K. *Mycocladus* vs. *Lichtheimia*: A correction (Lichtheimiaceae fam. nov., Mucorales, Mucoromycotina). *Mycol Res.* 2009; **113:** 277-8.
46. Houbraken J, de Vries RP, Samson RA. Modern taxonomy of biotechnologically important *Aspergillus* and *Penicillium* species. *Adv Appl Microbiol.* 2014; **86:** 199-249.
47. Howard DH. An introduction to the taxonomy and nomenclature of zoo-pathogenic fungi. *In:* Howard DH (ed). *Fungi Pathogenic for Human and Animals. Classification and Nomenclature.* Part A. Marcel Dekker, New York: 1983.
48. http://www.bgbm.org/iapt/nomenclature/code/default.htm
49. Irinyi L, Lackner M, de Hoog GS, *et al.* DNA barcoding of fungi causing infections in humans and animals. *Fungal Biol.* 2016; **120:** 125-36.
50. Irinyi L, Serena C, Garcia-Hermoso D, *et al.* International Society of Human and Animal Mycology (ISHAM)-ITS reference DNA barcoding database-the quality controlled standard tool for routine identification of human and animal pathogenic fungi. *Med Mycol.* 2015; **53:** 313-37.

51. Ishizaki H. Fungal taxonomy based on mitochondria DNA analysis. *Jpn J Med Mycol.* 1993; **34:** 243-51.
52. James TY, Kauff F, Schoch CL, *et al.* Reconstructing the early evolution of Fungi using a six-gene phylogeny. *Nature.* 2006; **443:** 818-22.
53. Jensen AB, Dromph KM. The causal agents of 'entomophthoramycosis' belong to two different orders: A suggestion for modification of the clinical nomenclature. *Clin Microbiol Infect.* 2005; **11:** 249-50.
54. Kirk PM, Stalpers JA, Braun U, *et al.* A without-prejudice list of generic names of fungi for protection under the International Code of Nomenclature for algae, fungi, and plants. *IMA Fungus.* 2013; **4:** 381-443.
55. Koljalg U, Nilsson RH, Abarenkov K, *et al.* Towards a unified paradigm for sequence-based identification of fungi. *Mol Ecol.* 2013; **22:** 5271-7.
56. Kraft CS, McAdam AJ, Carroll KC. A rose by any other name: Practical updates on microbial nomenclature for clinical microbiology. *J Clin Microbiol.* 2017; **55:** 3-4.
57. Kwon-Chung KJ, Bennett JE, Wickes BL, *et al.* The case for adopting the "Species Complex" nomenclature for the etiologic agents of cryptococcosis. *mSphere.* 2017; 2. pii: e00357-16. PMID: 28101535.
58. Kwon-Chung KJ. Phylogenetic spectrum of fungi that are pathogenic to humans. *Clin Infect Dis.* 1994; **19:** S1-7.
59. Kwon-Chung KJ. Taxonomy of fungi causing mucormycosis and entomophthoramycosis (zygomycosis) and nomenclature of the disease: Molecular mycologic perspectives. *Clin Infect Dis.* 2012; **54:** S8-15.
60. Levetin E, Horner WE, Scott JA, *et al.* Taxonomy of allergenic fungi. *J Allergy Clin Immunol Pract.* 2016; **4:** 375-85.e1.
61. Lutzoni F, Kauff F, Cox CJ, *et al.* Assembling the fungal tree of life: Progress, classification and evolution of subcellular traits. *Am J Bot.* 2004; **91:** 1446-80.
62. Ma LJ, Geiser DM, Proctor RH, *et al. Fusarium* pathogenomics. *Annu Rev Microbiol.* 2013; **67:** 399-416.
63. Manamgoda DS, Cai L, McKenzie EHC, *et al.* A phylogenetic and taxonomic re-evaluation of the *Bipolaris - Cochliobolus - Curvularia* complex. *Fungal Divers.* 2012; **56:** 131-44.
64. Manamgoda DS, Rossman AY, Castlebury LA, *et al.* A taxonomic and phylogenetic re-appraisal of the genus *Curvularia* (Pleosporaceae): Human and plant pathogens. *Phytotaxa.* 2015; **212:** 175-98.
65. Marin-Felix Y, Stchigel AM, Cano-Lira JF, *et al. Emmonsiellopsis*: A new genus related to the thermally dimorphic fungi of the family Ajellomycetaceae. *Mycoses.* 2015; **58:** 451-60.
66. Masatomo K, McGinnis MR. Nomenclature for fungus infections. *Int J Dermatol.* 1998; **37:** 825-6.
67. McGinnis MR, Ajello L, Schell WA. Mycotic diseases: A proposed nomenclature. *Int J Dermatol.* 1985; **24:** 9-15.
68. McGinnis MR, Rinaldi MG. Selected medically important fungi and some common synonyms and obsolete names. *Clin Infect Dis.* 1995; **21:** 277-8.
69. McGinnis MR, Rinaldi MG. Some medically important fungi and their common synonyms and obsolete names. *Clin Infect Dis.* 1997; **25:** 15-7.
70. McGinnis MR, Sigler L, Rinaldi MG. Some medically important fungi and their common synonyms and names of uncertain application. *Clin Infect Dis.* 1999; **29:** 728-30.
71. McGinnis MR. Recent taxonomic developments and changes in Medical Mycology. *Annu Rev Microbiol.* 1980; **34:** 109-35.
72. McLaughlin DJ, Hibbett DS, Lutzoni F, *et al.* The search for the fungal tree of life. *Trends Microbiol.* 2009; **17:** 488-97.
73. McNeill J, Stuessy TF, Turland NJ, *et al.* XVII International Botanical Congress: Preliminary mail vote and report of Congress action on nomenclature proposals. *Taxon.* 2005; **54:** 1057-64.
74. Mendoza L, Ajello L, Taylor JW. The taxonomic status of *Lacazia loboi* and *Rhinosporidium seeberi* has been finally resolved with the use of molecular tools. *Rev Iberoam Micol.* 2001; **18:** 95-8.
75. Mendoza L, Taylor JW, Ajello L. The class Mesomycetozoea: A heterogeneous group of microorganisms at the animal-fungal boundary. *Annu Rev Microbiol.* 2002; **56:** 315-44.
76. Morton JB, Msiska Z. Phylogenies from genetic and morphological characters do not support a revision of Gigasporaceae (Glomeromycota) into four families and five genera. *Mycorrhiza.* 2010; **20:** 483-96.
77. Moussa TA, Al-Zahrani HS, Kadasa NM, *et al.* Nomenclatural notes on *Nadsoniella* and the human opportunist black yeast genus *Exophiala. Mycoses.* 2017; **60:** 358-65.
78. Nagy E, Kredics L, Antal Z, *et al.* Molecular diagnosis, epidemiology and taxonomy of emerging medically important filamentous fungi. *Rev Med Microbiol.* 2004; **15:** 153-62.
79. No authors listed. International Commission on the Taxonomy of Fungi (ICTF): Name changes in fungi of microbiological, industrial and medical importance. Part 3. *Microbiol Sci.* 1988; **5:** 23-6.
80. Norvell LL, Hawksworth DL, Petersen RH, *et al.* IMC9 Edinburgh nomenclature sessions. *IMA Fungus.* 2010; **1:** 143-7.
81. Odds FC, Arai T, DiSalvo AF, *et al.* Nomenclature of fungal diseases: A report and recommendations from a Sub-Committee of the International Society for Human and Animal Mycology. *J Med Vet Mycol.* 1992; **30:** 1-10.
82. Odds FC, Rinaldi MG. Nomenclature of fungal diseases. *Curr Top Med Mycol.* 1995; **6:** 33-46.
83. Oehl F, Sieverding E, Palenzuela J, *et al.* Advances in Glomeromycota taxonomy and classification. *IMA Fungus.* 2011; **2:** 191-9.
84. Pitt JI, Taylor JW. *Aspergillus*, its sexual states and the new International Code of Nomenclature. *Mycologia.* 2014; **106:** 1051-62.
85. Prakash PY, Irinyi L, Halliday C, *et al.* Online databases for the taxonomy and identification of pathogenic fungi and proposal for a cloud-based dynamic data network platform. *J Clin Microbiol.* 2017; **55:** 1011-24.
86. Prillinger H, Lopandic K, Schweigkofler W, *et al.* Phylogeny and systematics of the fungi with special reference to the ascomycota and basidiomycota. *Chem Immunol.* 2002; **81:** 207-95.

87. Redhead SA, Demoulin V, Hawksworth DL, *et al.* Fungal Nomenclature at IMC10: Report of the Nomenclature Sessions. *IMA Fungus.* 2014; **5:** 449-62.
88. Rinaldi MG. Selected medically important fungi with common synonyms and other obsolete names. *Clin Infect Dis.* 1993: **16:** 610-1.
89. Robert V, Vu D, Amor ABH, *et al.* MycoBank gearing up for new horizons. *IMA Fungus.* 2013; **4:** 371-9.
90. Rossman AY. Lessons learned from moving to one scientific name for fungi. *IMA Fungus.* 2014; **5:** 81-9.
91. Ruggiero MA, Gordon DP, Orrell TM, *et al.* A higher level classification of all living organisms. *PLoS One.* 2015; **10:** e0119248.
92. Samson RA, Varga J. What is a species in *Aspergillus*? *Med Mycol.* 2009; **47:** S13-20.
93. Samson RA, Visagie CM, Houbraken J, *et al.* Phylogeny, identification and nomenclature of the genus *Aspergillus. Stud Mycol.* 2014; **78:** 141-73.
94. Samson RA. Problems caused by new approaches in fungal taxonomy. *Mycopathologia.* 1991; **116:** 149-50.
95. Sandoval-Denis M, Guarro J, Cano-Lira JF, *et al.* Phylogeny and taxonomic revision of Microascaceae with emphasis on synnematous fungi. *Stud Mycol.* 2016; **83:** 193-233.
96. Savitha SA, Sacchidanand SA, Gowda SK. Misnomers in dermatology: An update. *Indian J Dermatol.* 2013; **58:** 467-74.
97. Schoch CL, Robbertse B, Robert V, *et al.* Finding needles in haystacks: Linking scientific names, reference specimens and molecular data for Fungi. *Database (Oxford).* 2014; 1-21. pii: bau061. PMID: 24980130.
98. Schoch CL, Seifert KA, Huhndorf S, *et al.* Nuclear ribosomal internal transcribed spacer (ITS) region as a universal DNA barcode marker for Fungi. *Proc Natl Acad Sci USA.* 2012; **109:** 6241-6.
99. Schussler A, Schwartzott D, Walker C. A new phylum, the Glomeromycota: Phylogeny and evolution. *Mycol Res.* 2001 **105:** 1413-21.
100. Sigler L, Hawksworth DL. International Commission on the Taxonomy of Fungi (ICTF). Code of Practice for Systematic Mycologists. *Mycopathologia.* 1987; **99:** 3-7.
101. Sigler L, Hawksworth DL. International Commission on the Taxonomy of Fungi: Code of Practice for Systematic Mycologists. *Microbiol Sci.* 1987; **4:** 83-6.
102. Taylor J, Jacobson D, Fisher M. The evolution of asexual fungi: Reproduction, speciation and classification. *Annu Rev Phytopathol.* 1999; **37:** 197-246.
103. Taylor JW, Geiser DM, Burt A, *et al.* The evolutionary biology and population genetics underlying fungal strain typing. *Clin Microbiol Rev.* 1999; **12:** 126-46.
104. Taylor JW, Hibbett DS. Toward sequence-based classification of fungal species. *IMA Fungus.* 2013; **4:** 33-4.
105. Taylor JW, Jacobson DJ, Kroken S, *et al.* Phylogenetic species recognition and species concepts in fungi. *Fungal Genet Biol.* 2000; **31:** 21-32.
106. Taylor JW. One Fungus = One Name: DNA and fungal nomenclature twenty years after PCR. *IMA Fungus.* 2011; **2:** 113-20.
107. Untereiner WA, Scott JA, Naveau FA, *et al.* The Ajellomycetaceae, a new family of vertebrate-associated Onygenales. *Mycologia.* 2004; **96:** 812-21.
108. Valenzuela-Lopez N, Sutton DA, Cano-Lira JF, *et al.* Coelomycetous fungi in the clinical setting: Morphological convergence and cryptic diversity. *J Clin Microbiol.* 2017; **55:** 552-67.
109. Visagie CM, Houbraken J, Frisvad JC, *et al.* Identification and nomenclature of the genus *Penicillium. Stud Mycol.* 2014; **78:** 343-71.
110. Voigt K, Kirk PM. Recent developments in the taxonomic affiliation and phylogenetic positioning of fungi: Impact in applied microbiology and environmental biotechnology. *Appl Microbiol Biotechnol.* 2011; **90:** 41-57.
111. Walther G, Pawlowska J, Alastruey-Izquierdo A, *et al.* DNA barcoding in Mucorales: An inventory of biodiversity. *Persoonia.* 2013; **30:** 11-47.
112. Wang P, Kenyon C, de Hoog S, *et al.* A novel dimorphic pathogen, *Emergomyces orientalis* (Onygenales), agent of disseminated infection. *Mycoses.* 2017; **60:** 310-9.
113. Warnock DW. Name changes for fungi of medical importance, 2012 to 2015. *J Clin Microbiol.* 2017; **55:** 53-9.
114. Warren NG. Taxonomy and introduction. *Dermatol Clin.* 1996; **14:** 1-7.
115. Whittaker RH. New concepts of kingdoms of organisms. *Science.* 1969; **163:** 150-60.
116. WHO. International Statistical Classification of Diseases and Related Health Problems (ICD-10), 10th Revision. Geneva. World Health Organization; 2011.
117. Woese CR, Kandler O, Wheelis ML. Towards a natural system of organisms: Proposal for the domains Archaea, Bacteria and Eucarya. *Proc Natl Acad Sci.* 1990; **87:** 4576-9.
118. Yamanishi C, Alshahni MM, Sano A, *et al.* A new marker sequence for systematics of medically important fungi based on aminoacid sequence of the largest subunit of RNA polymerase I. *Med Mycol.* 2017; pii: myw098. PMID: 27811180.

CHAPTER 4

Immunity to Fungal Diseases

The experience about immunity to fungal diseases is limited as compared to the bacterial, viral or parasitic infections. The natural immunity to these infections is quite high, yet exact pathogenic mechanism of fungi, their virulence factors, mode of attachment to cell surfaces and predilection for invasion, are yet to be clearly elucidated. Therefore, infection depends on the exposure to sufficient inoculum size of organism and general resistance of host. Even then an asymptomatic or subclinical infection with particular fungal pathogen may provide substantial immunity. The immune deficiency and genetic differences predispose to infection with either established or opportunistic fungal pathogens.

During the past century, antibodies have been used clinically for therapy of various infectious diseases. A property of antibodies, not duplicated by antibacterial chemotherapy, is ability of specific antibodies to neutralize toxin and clear the antigen.

Some infections are accompanied by accumulation of microbial products in the host tissues *e.g.* capsular polysaccharide accumulates in tissues where infections are due to *Streptococcus pneumoniae, Klebsiella pneumoniae, Haemophilus influenzae* and *Cryptococcus neoformans.* The chronic nature of infection results in accumulation of very high levels of capsular polysaccharide in the serum as well as tissues.

The capsular polysaccharide of *Cr.neoformans* has been associated with variety of deleterious effects on the host immune system, including inhibition of leukocyte migration, cytokine dysregulation, L-selectin shedding, CD18 binding, brain cell edema and enhancement of HIV infection. Cryptococcal infections are generally incurable in HIV-infected persons because antifungal therapy does not eradicate infection in setting of severe immune suppression. Hence there has been considerable interest in developing adjunctive immune therapy for this infection and murine model monoclonal antibody (mAb) is currently in advanced pre-clinical development.

The term **pathogenicity** is taken as general ability of a species to induce diseases whereas **virulence** refers to quantitative ability for pathogenesis of a fungal species. These terms should not be used ambiguously and interchangeably, as they convey specific meaning in defense mechanism.

Fungal virulence factors operative in human infection, which are defined for major pathogens include, thermotolerance, melanin and toxin production and ability to either evade or suppress host defenses. Some of salient putative virulence factors for selected fungal pathogens have been listed in **Table 4.1**. Their individual or collective role in overall tissue invasion is tempered by host susceptibility. Various fungal immunogens, which are significant in immunity to fungal agents, are listed in **Table 4.2**.

There are important differences in host defenses that are primarily operative against different groups of mycoses, as listed in **Table 4.3**. The T cell-mediated immune mechanism predominates in containment of most of fungi, which have sufficient pathogenic potential to produce some degree of local or systemic infection in normal host. In comparison to many other microorganisms that infect humans, in fungi parasitism is an accidental phenomenon rather than an obligatory requirement for survival.

Mechanisms of Immunity

The defense to fungal infection involves both **innate** and **acquired immunity**. These can be further sub-divided into specific and non-specific immune responses. These two categories of responses to significant extent are intertwined and interdependent. The phagocytic cells and lymphocytes (both T and B cells) are believed to function together in protecting host against fungal pathogens but exact degree to which each cell involved is not yet fully known. Moreover, this differs with individual as well as type of infection.

Table 4.1. Virulence Factors of Common Fungal Pathogens.

Virulence Factors	*Functions*	*Fungal Pathogens*
A. Surface Components		
i. Cell wall glycoproteins	Adherence to epithelial surfaces	*Candida* species
ii. Melanin pigment	Shield against immunologically active cells, hydrolyses	*Cr.neoformans* *E.dermatitidis*
iii. Capsules, glucans	Anti-phagocytic	*Cr.neoformans*
B. Thermotolerance	Survive and replicate at 37°C	Human pathogens
C. Resistance to microbiocidal products of neutrophils *e.g.* H_2O_2, by dimorphic primary pathogen	Evasion of host defense mechanisms by tissue phase (yeast, spherule) of virulent dimorphic fungi	Primary pathogens *Blastomyces, Coccidioides, Histoplasma, Paracoccidioides, Sporothrix schenckii*
D. Epithelial cell and monocyte cytocidal activity	Evasion of host defense	*Candida albicans, Candida tropicalis*
E. Exoenzymes		
i. Elastase	Degrades elastin, scleroproteins, enhances invasion of elastin containing tissue (lung, skin, blood vessels)	*Aspergillus flavus, A.fumigatus,* Dermatophytes
ii. Alkaline protease	Degrades collagen, elastin, enhances invasion of lung tissue	*Aspergillus flavus, A.fumigatus, Rhizopus* species Dermatophytes
iii. Keratinase, collagenase	Degrades scleroproteins in skin	
iv. Acid protease	Cleavage of $IgA_{1,2}$	*Candida* species, *A.fumigatus*
F. Toxins		
i. Aflatoxin	Hepatotoxicity	*Aspergillus flavus*
ii. Endotoxin	Tissue necrosis	*A.flavus, A.fumigatus*
G. Dimorphism	Evasion of host defenses Environmental and tissue forms present different antigenic and surface features, requiring different host response or mechanisms to contain each form	True pathogens Opportunistic pathogens

Table 4.2. Medically Significant Fungal Immunogens.

Fungal Agents	*Immunogens*
A. Yeast-like and Opportunistic Fungi	
C.albicans	Mannan, Mannoprotein, Enolase
Cr.neoformans	Capsule, Cell wall and cell membrane
P.jirovecii	MSG, p55
B. Dimorphic Fungi	
H.capsulatum	Cell wall and cell membrane, Ribosomal-protein complex
B.dermatitidis	BAD (WI-1)
C.immitis	C-ASWS, Complement-fixing antigen, Chitinase, Antigen 2, 33-kDa antigen
P.brasiliensis	gp43

Table 4.3. Host Defense Mechanism in Fungal Infections.

A. Acquired T-cell-mediated response
Chronic mucocutaneous candidiasis
Dermatophytosis
Cryptococcosis
Coccidioidomycosis
Blastomycosis
Paracoccidioidomycosis
B. Anti-neutrophilic response
Disseminated candidiasis
Invasive aspergillosis
Mucormycosis

Non-specific Host Factors

In fungal infections, non-specific factors play an important role in primary defense. The passive protection is provided by intact skin and mucosal surfaces, which are generally considered resistant to the infection. The fatty acids like sebum also provide protection because of their antifungal activity. The serum has similar inhibitory effect on many of fungal pathogens. However, within living tissues, it is active and non-specific immune response, involving phagocytic cells that are important and capable of effectively dealing with invasion of small number of fungal cells.

The alveolar macrophages are important in engulfing cells in lungs, which are removed by ciliary action and coughing. Similar macrophages in liver and spleen can also remove fungi from blood stream.

It is clear from number of studies that T-lymphocytes together with monocytes play an important role in protection against fungal pathogens. It is also known that most of individuals with serious fungal infections have T-cell defects. The role of humoral immunity in protecting host against fungi is less clear and accordingly, it is not considered as significant as the cell-mediated immunity.

A variety of innate defense factors in saliva such as lysozymes and lactoferrin contribute to mucosal protection and modulate *Candida* populations in oral cavity. It is also known that in HIV-infected individuals, significant variations in concentrations of lysozyme and lactoferrin in saliva occur during the disease progression.

The use of corticosteroids is generally avoided in infectious diseases, particularly fungal for fear of suppressing immune reaction and promoting microbial dissemination.

Cellular Immunity

The cellular immunity is mainstay of the host defenses for most of the fungi. However, importance of phagocytic cells cannot be understated. It is in setting of granulocytopenia induced by cytotoxic chemotherapy for malignancies that most rapidly progressive and potentially overwhelming opportunistic fungal infections may occur, especially disseminated candidiasis, aspergillosis and mucormycosis. It has been shown that vegetative hyphal structures of *Aspergillus* and *Candida* species are ingested and killed by neutrophils. Activated neutrophils also release hydrogen peroxide and other cellular products, which are toxic to fungi. Thus, patients with chronic granulomatous disease or myeloperoxidase deficiency have more frequent and more severe infections with each of these two agents. Neutrophil effect seems less important for defense against *Cr.neoformans*, perhaps because of its large capsule.

Monocyte-macrophage activity, especially in respiratory tract, appears to be major defense mechanism against most of the fungi. Binding of fungi to macrophage cell surfaces can be mediated by cellular receptors for mannose, complement component C_3, IgG and β-glucans, with exact mechanism varying with causative fungal species.

The cell-mediated immunity has clear role in protection against fungal infections. The patients with Hodgkin's disease or sarcoidosis have peculiar predilection for cryptococcal infection, as do renal transplant recipients, whose immunosuppressive regimens are designed specifically to inhibit cell-mediated immunity to avoid graft rejection. For other fungi, role of this mechanism is probably related to antigen processing, macrophage activation and secretion of lymphokines such as interleukin-2 and interferon-γ.

Humoral Immunity

The precise role of humoral immunity in host defense against fungal infection is difficult to determine. Common wisdom assumes that opsonization by specific antibodies helps in phagocytosis and clearance mechanisms and there is experimental support primarily in *Candida* and *Cryptococcus* infections.

The serum opsonins may increase rate of phagocytosis of *Candida* by phagocytes but other studies show that specific antibody is not necessary for killing of ingested organisms.

In case of cryptococcal infection, specific anticapsular IgG facilitates ingestion of yeast by macrophages and protects passively immunized mice from an intravenous challenge and cell-deficient mice seem to have no particular predisposition to cryptococcal infection.

Complement, other arm of humoral immunity, seems to have role in host defense against fungal infection. Fungi cause activation of complement system and components of C_3 are deposited on fungal cell surface. Specific receptors for these C_3 fragments have been found on monocyte-macrophages, neutrophils and some lymphocytes thus may be significant in development of cellular immunity. In contrast to some bacterial cells, complement with specific antifungal antibodies is not lytic for fungi. By itself, antifungal IgG and perhaps other subclasses (even without formation of soluble immune complexes) have potential to inhibit cellular host responses.

The immunoglobulins of all classes are produced in fungal infection, with precise antibody response depending on pathogen involved, host and type of infection. The antibody responses are very useful in diagnosis of infections but evidence of protective effect of antibody is incomplete and presently is under evaluation. Transfer of specific antibodies observes partial protection. These antibodies, on the surface of fungal cells, help in phagocytosis by facilitating attachment of neutrophils and macrophages, which have Fc receptor sites on their surface. The direct antibody-mediated killing of fungi has not been fully substantiated. However, it has been shown that antibody enables monocytes to kill *Cr.neoformans* by non-phagocytic mechanism. The definite magnitude to which complement is involved as an opsonin in phagocytic killing of fungi is also vague. With *Candida* C_3 component of complement, activated by alternate pathway, has been directly implicated in phagocytic killing but it is not known if this commonly occurs with other fungal pathogens or not.

The microbial factors that help infections by damaging neutralizing host responses, which are found in bacterial infections, have been little studied in fungal infections. It is thought that many fungal infections, particularly opportunistic pathogens, such as *Candida*, exploit deficiencies in immune system rather than create susceptibility by regulating immunological response. However, importance of various fungal antigens in a disease process has yet to be explained. It is possible that some antigens, perhaps depending on their mode of presentation may cause anergy.

CARD9 Deficiency

This is Caspase Recruitment Domain family member 9 and is encoded by CARD9 gene, which is located on the long (q) arm of chromosome 9 at position 34.3. It is an important part of innate immune system. Human CARD9 deficiency is an autosomal recessive primary immunodeficiency disorder caused by biallelic mutations in gene CARD9. It is associated with spontaneous development of persistent and severe fungal infections that primarily localize to skin and subcutaneous tissue, mucosal surfaces and/or central nervous system.

When immune system recognizes fungi, it generates Th17 cells which in turn produce IL-17. The IL-17 pathway creates inflammation, sending other cytokines and white blood cells that fight fungal invaders and promote tissue repair. The CARD9-dependent production of TNF-α also enhances the candidacidal capacity of neutrophils, limiting fungal diseases. It is considered that mutations in CARD9 gene block (inhibit) the activity of IL-17 thus impair multiple signaling pathways that normally help to recognize fungi. Therefore, patients with autosomal recessive CARD9 mutations are predisposed to recurrent mucocutaneous and invasive fungal infections with *Candida* species, dermatophytes (*Trichophyton* species), phaeohyphomycetes (*Phialophora verrucosa*, *Exophiala* species) and *Aspergillus* species, particularity causing extrapulmonary aspergillosis. There are at least 15 CARD9 gene mutations, which are identified among individuals with familial candidiasis.

Immune Reconstitution Inflammatory Syndrome

IRIS is characterized by exaggerated atypical inflammation resulting in immunopathology. This is usually seen in response to the infective pathogens. Although reported in conditions like malignancy, chemotherapy recipients of SOT and HSCT, patients with primary immunodeficiencies, persons on steroid therapy and TNF-α inhibitors, IRIS is most commonly seen among AIDS patients. During IRIS, there is restoration of immune response rather than loss of fungal specific immune response, which leads to immunopathology thereby various manifestations.

This is largely driven by T-cells and/or innate immune responses, thus, infections which predominantly provoke B-cells and antibody response are less commonly a cause of IRIS. While infections which invoke cell-mediated immunity show IRIS. Easily treatable infections which result in complete pathogen and microbiological clearance with no latent reservoirs and no clinical latency, such as mucocutaneous candidiasis also do not seem to be associated with IRIS. However, pneumocystosis and cryptococcosis are two main invasive fungal infections, which result in IRIS among AIDS patients but common IFDs in SOT and HSCT recipients include invasive aspergillosis, mucormycosis, candidiasis along with cryptococcosis. In rheumatological patients on biologic therapy such as TNF-α blockers, common IFDs are histoplasmosis and cryptococcosis. During IRIS, all these IFDs have one factor in common and that is long latency thereby difficult to achieve mycological cure. The systemic manifestations are severer and there is flaring up of the pathogen-specific symptoms. Consequently there is increased morbidity, increased hospitalization, more medications and increased treatment cost.

Table 4.4. Iatrogenic Factors Predisposing to Fungal Infections.

Intervention	*Underlying Defects*	*Fungal Agents*
Antibacterial therapy	Decreased competing flora	*Candida, Aspergillus* species
Corticosteroids	Depressed CMI and chemotaxis by polymorphonuclear leukocytes	*Candida, Cryptococcus, Histoplasma, Aspergillus* species
Cytotoxic chemotherapy	Granulocytopenia, depressed CMI	*Candida, Aspergillus* species
Intravenous and arterial catheters	Integumentary breach	*Candida* species
Surgical intervention	Integumentary breach inoculation of fungal elements into sterile sites	*Candida, Aspergillus,* Mucormycetes
Transplantation	Depressed cell-mediated immunity	*Candida, Aspergillus, Cryptococcus* spp.

Hence there is an urgent need to identify such condition by the clinicians. Mycological culture positivity and quantification of fungal burden, can help in distinguishing the relapse from the inflammation caused due to the IRIS.

Iatrogenic Factors

The antibacterial therapy, if prolonged for a long time, may predispose to fungal infections even in immunocompetent host. Superficial *Candida* infections involving mucous membranes, especially vaginal epithelium, are very common. *Aspergillus* infection, particularly invasive forms, is also more common in patients subjected to antibiotic therapy. Interestingly, illnesses in which more superficial forms of aspergillosis are commonly seen, namely chronic sinusitis, chronic bronchitis, bronchiectasis and asthma, frequent antibacterial administration is rule, practice that probably promotes respiratory tract colonization by organism.

The use of corticosteroid promotes fungal infection via multiple mechanisms. Leukotaxis is depressed, transient T-cell lymphopenia and monocytopenia occurs and alterations in blastogenesis and monocyte functions also decreased response to lymphokines have been reported. As might be expected from such broad immunosuppressive effect, infections with *Candida, Cryptococcus, Histoplasma* and *Aspergillus* species have been reported in steroid administered patients. The use of nasal and lower respiratory corticosteroid aerosols for various indications has not totally reduced risk of fungal infections in respiratory tract. The incidence of thrush has been very high in persons who use inhaled steroids for treating asthma.

Cytotoxic chemotherapy, as is given for leukaemia and bone marrow transplantation, results in granulocytopenia and defects of cell-mediated immunity. TNF-α blockers pose higher risk of potential for opportunistic fungal infections.

Integumentary breaches, as seen in placement of intravenous or intra-arterial catheters, predispose patient primarily to infections with *Candida*, source usually being host's own flora. Similarly, in intravenous drug abusers who directly bypasses intact skin barrier, disseminated infections with *Candida*, *Aspergillus* and mucormycetes have occurred.

Surgical procedures can promote fungal infections in several ways but mechanisms are similar in that normally sterile site is exposed to exogenous or endogenous fungi and infection follows. Sternal wound and prosthetic valve infections due to *Candida* or *Aspergillus*, intra-abdominal candidal infections and superficial wound infections due to dressings contaminated with fungal spores have been seen. The complicated surgical patients may have multiple risk factors for fungal infections including prolonged course of antimicrobial, poor nutritional status and metabolic derangement, all contribute to fungal infections.

The immunosuppressive therapy used in transplantation primarily depresses cell-mediated immunity. Even with newer immunosuppressive agents such as cyclosporin A, superficial and systemic candidal infections occur. Cryptococcal infection correlates well with degree of immunosuppression, particularly when anti-lymphocyte serum is used to suppress rejection and aspergillosis in lungs and heart-lung transplants, is not uncommon. The iatrogenic predisposing conditions are listed in **Table 4.4**.

Non-immune Host Factors

The mucociliary action of mucous membranes is probably prime clearance mechanism active against inhaled fungal spores. Long chain fatty acids present on intact skin also probably serve to inhibit fungal colonization. Non-immune serum factors such as transferrin, which bind iron, an essential growth factor for *Candida* and β-globulin that promote clumping and phagocytosis of *Candida* species, are responsible for inhibitory effect of serum on some of the fungi.

Role of Melanins

In the nature melanin, protects fungal cell against ultraviolet radiations and also against host defenses in case of human or animal infections. Melanins represent virulence factors for several pathogenic fungi and number of fungi is increasing. Thus, albino mutants of several genera exhibit decreased virulence in mice. It is considered that phenomenon in relation to known chemical properties of melanin, beginning with biosynthesis from orthohydroquinone precursors which, when oxidized enzymatically to quinones, polymerize spontaneously to melanin.

The presence of melanin in cell wall of phaeoid (dematiaceous) fungi is one of the likely virulence factors. There are putative mechanisms by which melanin act as a virulence factor. It may confer a protective advantage by scavenging free radicals and hypochlorite that are produced by phagocytic cells in their oxidative burst and that would normally kill most organisms. In addition, melanin may bind to the hydrolytic enzymes, thereby preventing their action on the plasma membrane. These multiple functions may help explain the pathogenic potential of some phaeoid fungi, even in immunocompetent hosts.

Defects in Immune Response

The opportunistic fungal pathogens exploit impaired defenses in those patients where specific and non-specific mechanisms of defense are decreased. The true fungal pathogens also cause infections more frequently in such individuals and they are more likely to develop disseminated diseases.

Sometimes, immunodeficiencies that predispose to fungal infections are congenital and are associated with thymus defects *e.g.* chronic mucocutaneous candidiasis (CMCC). Most of the times fungal infections develop in patients with immune defects resulting from an underlying factor such as malignancy, immunosuppressive drugs, corticosteroid therapy, transplantation, etc.

The principal defect that arises due to above-mentioned predisposing factors is gross reduction in the number and functioning of neutrophils. The defects in B and T-cells function are also important and these may occur in certain types of malignancies. The treatment of these conditions greatly affects lymphocytic response particularly that of T-cells which may be eliminated for long periods depending upon drug used, doses and duration of therapy.

There is abundant scope for further studies concerning immunity to fungal infections. It is likely that proper understanding of mechanisms involved, will await further work on pathogens of fungal diseases and further knowledge of immunology in general.

Granulomatous Mycoses

The following fungal diseases present as granulomatous reactions. These are: coccidioidomycosis, cryptococcosis, candidiasis, sporotrichosis, histoplasmosis, aspergillosis, paracoccidioidomycosis, blastomycosis and infections by *Phialophora* and *Madurella* species.

Fungal Vaccines

The cellular and immunological defenses have long been recognized as critical in proper resolution of fungal infections. Immunosuppressive diseases and therapies allow overwhelming infection by fungi, such as *C.albicans, Cr. neoformans* and *Aspergillus fumigatus*, which are rarely encountered as systemic pathogens in immunocompetent host. The other fungi like *C.immitis, B.dermatitidis* and *H.capsulatum* cause a broad range of illnesses in immunocompetent individuals and other mammals, who do not have general immunological abnormalities. These organisms usually produce more severe disease in immunocompromised host.

The understanding of determinants of diseases caused by fungi is not yet complete; however, study of fungal vaccines might help to explain some mechanisms involved. Furthermore, as with many other types of infections, vaccination offers hope of preventing diseases. Vaccines for *C.immitis* have been studied most extensively. The other fungi of medical interest for which vaccines have been studied include *H.capsulatum, Cr.neoformans* and *B.dermatitidis.*

In case of an experimental guinea pig model, immunogenicities of live spore vaccine, killed hyphal cell wall vaccine and soluble cytoplasmic extract vaccine of *Trichophyton mentagrophytes* var. *erinacei* have been studied. Of these three vaccines, live spore vaccine has been found to be the most effective and closely simulated the type of immunity that develops following natural infection.

Ribosomal Vaccines

Histoplasma capsulatum ribosomal vaccine is effective in inducing protection against experimental histoplasmosis, at a rate similar to that induced by sublethal infection, with advantage of being subcellular safe antigen. The protection inducing capacity is attributed to γ-proteins and nature of immune response involved in protection is primarily cell-mediated immunity.

The *Candida* ribosomal vaccine induces protection against systemic candidiasis, elicits humoral and CMI responses, latter appearing to be correlated with protection. The *Microsporum canis* crude ribosomal fraction can stimulate cellular immune system.

Autophagy

Autophagy is a highly conserved eukaryotic mechanism whereby cells recycle cellular elements to survive under adverse conditions. Surprisingly, of the three fungal pathogens of greatest relevance to human health, only *Cr.neoformans* has been shown to require this process during infection. In contrast, autophagy is dispensable for the virulence of both *Candida albicans* and *Aspergillus fumigatus*. The divergent roles for autophagy in these opportunistic fungal species underscore the uniqueness of the host infection niche occupied by each fungus and provide insights into the evolutionary pressures that may have influenced the need for autophagy during infection. Further study of fungal autophagy may reveal the host signals, which induce this protective response and determine if these signals differ between host cells or tissues. In addition, a comprehensive understanding of the autophagy machinery in fungal pathogens may provide a rational basis for the design of future therapeutic interventions to improve outcome in patients who are at risk for these infections.

Cancer and Fungal Immunity

There are many infectious agents, which are classical risk factors for cancer like bacteria, viruses and parasites. However, there is very less evidence concerning the implication of mycotic diseases in carcinogenesis. The role of chronic candidiasis in the development of squamous cell carcinoma has been suspected since long as *Candida* species are found to be more prevalent in potentially malignant disorder and cancer of the oral mucosa. The other epidemiological evidence of a link between candidiasis and cancer is that is observed among patients with Autoimmune Polyendocrinopathy Candidiasis Ectodermal Dystrophy (APECED). The oral and esophageal carcinoma are frequent among such patients with chronic mucocutaneous candidiasis.

It is thought that production of nitrosamine and metabolism of procarcinogen are the causative mechanisms in which *Candida* species may be found to be involved in the development of oral cancer. Moreover, in chromoblastomycosis as well as lacaziosis chronic type of lesions may have a risk of undergoing malignant transformation. Even a diagnosis of paracoccidioidomycosis also appears to increase the risk of lung cancer among human beings.

Further Reading

1. Ackerman AL, Underhill DM. The mycobiome of the human urinary tract: Potential roles for fungi in urology. *Ann Transl Med.* 2017; **5:** 31.
2. Alonso-Monge R, Roman E, Arana DM, *et al.* Fungi sensing environmental stress. *Clin Microbiol Infect.* 2009; **15 (Suppl. 1):** 17-9.
3. Alspaugh JA. Virulence mechanisms and *Cryptococcus neoformans* pathogenesis. *Fungal Genet Biol.* 2015; **78:** 55-8.
4. Alves de Medeiros AK, Lodewick E, Bogaert DJ, *et al.* Chronic and invasive fungal infections in a family with CARD9 deficiency. *J Clin Immunol.* 2016; **36:** 204-9.
5. Amarsaikhan N, O'Dea EM, Tsoggerel A, *et al.* Isolate-dependent growth, virulence and cell wall composition in the human pathogen *Aspergillus fumigatus. PLoS One*; **9:** e100430.
6. Antachopoulos C, Katragkou A, Roilides E. Immunotherapy against invasive mold infections. *Immunotherapy.* 2012; **4:** 107-20.
7. Armstrong-James D, Harrison TS. Immunotherapy for fungal infections. *Curr Opi Microbiol.* 2012; **15:** 434-9.
8. Arzanlou M, Samadi R, Afsarian S, Badali H. An overview of the evolution of pathogenicity in human pathogenic fungi. *JBUMS.* 2015; **17:** 71-80.
9. Bairwa G, Hee Jung W, Kronstad JW. Iron acquisition in fungal pathogens of humans. *Metallomics.* 2017; **9:** 215-27.
10. Balloy V, Chignard M. The innate immune response to *Aspergillus fumigatus. Microbes Infect.* 2009; **11:** 919-27.
11. Ben-Ami R. Angiogenesis at the mold-host interface: A potential key to understanding and treating invasive aspergillosis. *Future Microbiol.* 2013; **8:** 1453-62.
12. Bergman PW, Bjorkhem-Bergman L. Is there a role for statins in fungal infections? *Expert Rev Anti Infect Ther.* 2013; **11:** 1391-400.
13. Binder U, Maurer E, Lass-Florl C. *Galleria mellonella:* An invertebrate model to study pathogenicity in correctly defined fungal species. *Fungal Biol.* 2016; **120:** 288-95.

14. Boddy L, Hiscox J. Fungal Ecology: Principles and mechanisms of colonization and competition by saprotrophic fungi. *Microbiol Spectr.* 2016; **4:** PMID: 28087930.
15. Bordon Y. Mucosal immunology: Fungal monopoly promotes allergy. *Nat Rev Immunol.* 2014; **14:** 138-9.
16. Borghi E, Borgo F, Morace G. Fungal biofilms: Update on resistance. *Adv Exp Med Biol.* 2016; **931:** 37-47.
17. Bozena DK, Iwona D, Ilona K, *et al.* The mycobiome - A friendly cross-talk between fungal colonizers and their host. *Ann Parasitol.* 2016; **62:** 175-84.
18. Brakhage AA, Bruns S, Thywissen A, *et al.* Interaction of phagocytes with filamentous fungi. *Curr Opin Microbiol.* 2010; **13:** 409-15.
19. Brand A. Hyphal growth in human fungal pathogens and its role in virulence. *Int J Microbiol.* 2012; 517529. PMID: 22121367.
20. Brown GD, Denning DW, Gow NA, *et al.* Hidden killers: Human fungal infections. *Sci Transl Med.* 2012; **4:** 165rv13.
21. Brown GD, Netea MG. Exciting developments in the immunology of fungal infections. *Cell Host Microbe.* 2012; **11:** 422-4.
22. Butler G. Fungal sex and pathogenesis. *Clin Microbiol Rev.* 2010; **23:** 140-59.
23. Calderone RA, Fonzi WA. Virulence factors of *Candida albicans. Trends Microbiol.* 2001; **9:** 327-35.
24. Capilla J, Clemons KV, Liu M, *et al. Saccharomyces cerevisiae* as a vaccine against coccidioidomycosis. *Vaccine.* 2009; **27:** 3662-8.
25. Carvalho A, Cunha C, Iannitti RG, *et al.* Host defense pathways against fungi: The basis for vaccines and immunotherapy. *Front Microbiol.* 2012; **3:** 176. PMID: 22590466.
26. Cassone A, Casadevall A. Recent progress in vaccines against fungal diseases. *Curr Opin Microbiol.* 2012; **15:** 427-33.
27. Cassone A. Fungal vaccines: Real progress from real challenges. *Lancet Infect Dis.* 2008; **8:** 114-24.
28. Chai LY, Vonk AG, Kullberg BJ, *et al.* Immune response to *Aspergillus fumigatus* in compromised hosts: From bedside to bench. *Future Microbiol.* 2011; **6:** 73-83.
29. Chandra J, Pearlman E, Ghannoum MA. Animal models to investigate fungal biofilm formation. *Methods Mol Biol.* 2014; **1147:** 141-57.
30. Chang CC, French MA. Immune reconstitution inflammatory syndrome in invasive fungal infections: What we know and what we need to know?. *Curr Clin Microbiol Rep.* 2016; **3:** 63-70.
31. Chaudhary N, Staab JF, Marr KA. Healthy human T-cell responses to *Aspergillus fumigatus* antigens. *PLoS One.* 2010; **5:** e9036.
32. Chotirmall SH, Al-Alawi M, Mirkovic B, *et al. Aspergillus*-associated airway disease, inflammation and the innate immune response. *Biomed Res Int.* 2013; 723129. PMID: 23971044.
33. Coutinho HD. Factors influencing the virulence of *Candida* spp. *West Indian Med J.* 2009; **58:** 160-3.
34. Cramer RA, Rivera A, Hohl TM. Immune responses against *Aspergillus fumigatus:* What have we learned? *Curr Opin Infect Dis.* 2011; **24:** 315-22.
35. Cramer RA. Secretion stress and fungal pathogenesis: A new, exploitable chink in fungal armor? *Virulence.* 2011; **2:** 1-3.
36. Cruz CR, Lam S, Hanley PJ, *et al.* Robust T cell responses to aspergillosis in chronic granulomatous disease: Implications for immunotherapy. *Clin Exp Immunol.* 2013; **174:** 89-96.
37. d'Enfert C, Vecchiarelli A, Brown AJ. Bioluminescent fungi for real-time monitoring of fungal infections. *Virulence.* 2010; **1:** 174-6.
38. de Mello TP, de Souza Ramos L, Braga-Silva LA, *et al.* Fungal biofilm - A real obstacle against an efficient therapy: Lessons from *Candida. Curr Top Med Chem.* 2017; PMID: 28056742.
39. Deepe GS Jr, Gibbons RS. Interleukins 17 and 23 influence the host response to *Histoplasma capsulatum. J Infect Dis.* 2009; **200:** 142-51.
40. Deepe GS Jr. Prospects for the development of fungal vaccines. *Clin Microbiol Rev.* 1997; **10:** 585-96.
41. Delsing CE, Gresnigt MS, Leentjens J, *et al.* Interferon-gamma as adjunctive immunotherapy for invasive fungal infections: A case series. *BMC Infect Dis.* 2014; **14:** 166.
42. Develoux M. Cancer and mycoses and literature review. *Bull Soc Pathol Exot.* 2017; **110:** 80-84.
43. Dowd SE, Delton Hanson J, *et al.* Survey of fungi and yeast in polymicrobial infections in chronic wounds. *J Wound Care.* 2011; **20:** 40-7.
44. Drewniak A, Gazendam RP, Tool AT, *et al.* Invasive fungal infection and impaired neutrophil killing in human CARD9 deficiency. *Blood.* 2013; **121:** 2385-92.
45. Drummond RA, Collar AL, Swamydas M, *et al.* CARD9-dependent neutrophil recruitment protects against fungal invasion of the central nervous system. *PLoS Pathog.* 2015; **11:** e1005293. PMID: 26679537.
46. Edwards JE Jr. Fungal cell wall vaccines: An update. *J Med Microbiol.* 2012; **61:** 895-903.
47. Ene IV, Brunke S, Brown AJ, Hube B. Metabolism in fungal pathogenesis. *Cold Spring Harb Perspect Med.* 2014; **4:** a019695. PMID: 25190251.
48. Etzioni A. Fungal infections: Blame the TH-17 cells. *Isr Med Assoc J.* 2011; **13:** 561-3.
49. Fallon J, Kelly J, Kavanagh K. *Galleria mellonella* as a model for fungal pathogenicity testing. *Methods Mol Biol.* 2012; **845:** 469-85.
50. Ferre EM, Rose SR, Rosenzweig SD, *et al.* Re-defined clinical features and diagnostic criteria in autoimmune polyendocrinopathy candidiasis ectodermal dystrophy. *JCI Insight.* 2016; **1:** pii: e88782. PMID: 27588307.
51. Fidel PL Jr. *Candida*-host interactions in HIV disease: relationships in oropharyngeal candidiasis. *Adv Dent Res.* 2006; **19:** 80-4.
52. Filler SG. Host-microbe interactions: Fungi - Editorial overview. *Curr Opin Microbiol.* 2011; **14:** 373-4.
53. Gafter-Gvili A. G6PD deficiency and fungal infections in patients with acute myeloid leukemia: Less enzyme more fungus. *Leuk Lymphoma.* 2017; 1-2. doi: 10.1080/10428194.2017.1330478. PMID: 28562136.

54. Garcia-Sherman MC, Lundberg T, Sobonya RE, *et al.* A unique biofilm in human deep mycoses: Fungal amyloid is bound by host serum amyloid P component. *NPJ Biofilms Microbiomes.* 2015; **1.** pii: 15009. PMID: 26366292.
55. Garcia-Solache MA, Casadevall A. Global warming will bring new fungal diseases for mammals. *mBio.* 2010; **1.** pii: e00061-10. PMID: 20689745.
56. Garcia-Vidal C, Viasus D, Carratala J. Pathogenesis of invasive fungal infections. *Curr Opin Infect Dis.* 2013; **26:** 270-6.
57. Gavino C, Cotter A, Lichtenstein D, *et al.* CARD9 deficiency and spontaneous central nervous system candidiasis: Complete clinical remission with GM-CSF therapy. *Clin Infect Dis.* 2014; **59:** 81-4.
58. Ghannoum MA, Jurevic RJ, Mukherjee PK, *et al.* Characterization of the oral fungal microbiome (mycobiome) in healthy individuals. *PLoS Pathog.* 2010; **6:** e1000713. PMID: 20072605.
59. Gilbert AS, Wheeler RT, May RC. Fungal pathogens: Survival and replication within macrophages. *Cold Spring Harb Perspect Med.* 2014; **5:** a019661. PMID: 25384769.
60. Glocker EO, Hennigs A, Nabavi M, *et al.* A homozygous CARD9 mutation in a family with susceptibility to fungal infections. *N Engl J Med.* 2009; **361:** 1727-35.
61. Gow NA, Netea MG. Medical mycology and fungal immunology: New research perspectives addressing a major world health challenge. *Philos Trans R Soc Lond B Biol Sci.* 2016; **371 (1709):** pii: 20150462. PMID: 28080988.
62. Guimaraes AJ, Martinez LR, Nosanchuk JD. Passive administration of monoclonal antibodies against *H.capsulatum* and other fungal pathogens. *J Vis Exp.* 2011; **48.** pii: 2532. PMID: 21372781.
63. Gupta AO, Singh N. Immune reconstitution syndrome and fungal infections. *Curr Opin Infect Dis.* 2011; **24:** 527-33.
64. Hallen-Adams HE, Suhr MJ. Fungi in the healthy human gastrointestinal tract. *Virulence.* 2017; **8:** 352-8.
65. Hamad M. Universal fungal vaccines: Could there be light at the end of the tunnel? *Hum Vaccin Immunother.* 2012; **8:** 1758-63.
66. Han Y, Rhew KY. Comparison of two *Candida* mannan vaccines: The role of complement in protection against disseminated candidiasis. *Arch Pharm Res.* 2012; **35:** 2021-7.
67. Hofs S, Mogavero S, Hube B. Interaction of *Candida albicans* with host cells: Virulence factors, host defense, escape strategies and the microbiota. *J Microbiol.* 2016; **54:** 149-69.
68. Hogan LH, Klein BS, Levitz SM. Virulence factors of medically important fungi. *Clin Microbiol Rev.* 1996; **9:** 469-88.
69. Hole C, Wormley FL Jr. Innate host defenses against *Cryptococcus neoformans. J Microbiol.* 2016; **54:** 202-11.
70. Hospenthal DR, Rogers AL. Immunology of fungal infection. *Curr Top Med Mycol.* 1995; **6:** 127-88.
71. Howard DH. Acquisition, transport and storage of iron by pathogenic fungi. *Clin Microbiol Rev.* 1999; **12:** 394-404.
72. Huai Y, Dong S, Zhu Y, et al. Vaccine against fungal infections: Genetically engineered virus nanofibers as an efficient vaccine for preventing fungal infection (Adv. Healthcare Mater. 7/2016). *Adv Healthc Mater.* 2016; **5:** 746.
73. Huffnagle GB, Noverr MC. The emerging world of the fungal microbiome. *Trends Microbiol.* 2013; **21:** 334-41.
74. Huseyin CE, O'Toole PW, Cotter PD, *et al.* Forgotten fungi - The gut mycobiome in human health and disease. *FEMS Microbiol Rev.* 2017; doi: 10.1093/femsre/fuw047. PMID: 28430946.
75. Hussein MR. Mucocutaneous Splendore-Hoeppli phenomenon. *J Cutan Pathol.* 2008; **35:** 979-88.
76. Ibrahim AS, Edwards JE Jr, Bryant R, *et al.* Economic burden of mucormycosis in the United States: Can a vaccine be cost-effective? *Med Mycol.* 2009; **47:** 592-600.
77. Ibrahim AS. Host cell invasion in mucormycosis: Role of iron. *Curr Opin Microbiol.* 2011; **14:** 406-11.
78. Iliev ID, Underhill DM. Striking a balance: Fungal commensalism versus pathogenesis. *Curr Opin Microbiol.* 2013; **16:** 366-73.
79. Ito JI, Lyons JM, Diaz-Arevalo D, *et al.* Vaccine progress. *Med Mycol.* 2009; **47:** S394-400.
80. Ito JI. T cell immunity and vaccines against invasive fungal diseases. *Immunol Invest.* 2011; **40:** 825-38.
81. Jacobson ES. Pathogenic roles for fungal melanins. *Clin Microbiol Rev.* 2000; **13:** 708-17.
82. Jayalakshmi SS, Reddy RG, Borgohain R, *et al.* Predictors of mortality in rhinocerebral mycosis. *Neurol India.* 2007; **55:** 292-7.
83. Jeyaprakasam NK, Razak MF, Ahmad NA, *et al.* Determining the pathogenic potential of non-sporulating molds isolated from cutaneous specimens. *Mycopathologia.* 2016; **181:** 397-403.
84. Jhingran A, Kasahara S, Shepardson KM, *et al.* Compartment-specific and sequential role of MyD88 and CARD9 in chemokine induction and innate defense during respiratory fungal infection. *PLoS Pathog.* 2015; **11:** e1004589. PMID: 25621893.
85. Jiang S. Immunity against fungal infections. *Immunol Immunogenet Insights.* 2016; **8:** 3-6.
86. Kalan L, Gardner SE, Grice EA. Reply to "Understanding the role of fungi in chronic wounds". *mBio.* 2016; 7. pii: e02033-16. PMID: 27923927.
87. Kalan L, Loesche M, Hodkinson BP, *et al.* Redefining the chronic-wound microbiome: Fungal communities are prevalent, dynamic and associated with delayed healing. *mBio.* 2016; 7. pii: e01058-16. PMID: 27601572.
88. Karkowska-Kuleta J, Rapala-Kozik M, Kozik A. Fungi pathogenic to humans: Molecular bases of virulence of *Candida albicans*, *Cryptococcus neoformans* and *Aspergillus fumigatus. Acta Biochim Pol.* 2009; **56:** 211-24.
89. Katragkou A, Roilides E. Immunotherapy of infections caused by rare filamentous fungi. *Clin Microbiol Infect.* 2012; **18:** 134-9.
90. Kim JY. Human fungal pathogens: Why should we learn? *J Microbiol.* 2016; **54:** 145-8.
91. Kohler JR, Hube B, Puccia R, *et al.* Fungi that infect humans. *Microbiol Spectr.* 2017; **5.** PMID: 28597822.
92. Kong E, Jabra-Rizk MA. The great escape: Pathogen versus host. *PLoS Pathog.* 2015; **11:** e1004661. PMID: 25764299.

93. Kontoyiannis DP. Manipulation of host angionеogenesis: A critical link for understanding the pathogenesis of invasive mold infections? *Virulence.* 2010; **1:** 192-6.
94. Konstantinovas C, de Oliveira Mendes TA, Vannier-Santos MA, *et al.* Modulation of human immune response by fungal biocontrol agents. *Front Microbiol.* 2017; **8:** 39.
95. Koutsouras GW, Ramos RL, Martinez LR. Role of microglia in fungal infections of the central nervous system. *Virulence.* 2016; 1-14. PMID: 27858519.
96. Kozel TR. Activation of the complement system by pathogenic fungi. *Clin Microbiol Rev.* 1996; **9:** 34-46.
97. Kwon-Chung KJ, Kozel TR, Edman JC, *et al.* Recent advances in biology and immunology of *Cryptococcus neoformans. J Med Vet Mycol.* 1992; **30:** S133-42.
98. Lanternier F, Cypowyj S, Picard C, *et al.* Primary immunodeficiencies underlying fungal infections. *Curr Opin Pediatr.* 2013; **25:** 736-47.
99. Lara-Lemus R, Alvarado-Vasquez N, Zenteno E, *et al.* Effect of *Histoplasma capsulatum* glucans on host innate immunity. *Rev Iberoam Micol.* 2014; **31:** 76-80.
100. Lawn SD, Wood R. Immune reconstitution inflammatory syndrome. *Lancet Infect Dis.* 2010; **10:** 833-4.
101. Lee WJ, Kim JY, Song CH, *et al.* Disruption of barrier function in dermatophytosis and pityriasis versicolor. *J Dermatol.* 2011; **38:** 1049-53.
102. Lehrnbecher T, Kalkum M, Champer J, *et al.* Immunotherapy in invasive fungal infection - Focus on invasive aspergillosis. *Curr Pharm Des.* 2013; **19:** 3689-712.
103. Levitz SM. *Aspergillus* vaccines: Hardly worth studying or worthy of hard study? *Med Mycol.* 2017; **55:** 103-8.
104. Lilic D. Unravelling fungal immunity through primary immune deficiencies. *Curr Opin Microbiol.* 2012; **15:** 420-6.
105. Lionakis MS. Genetic susceptibility to fungal infections in humans. *Curr Fungal Infect Rep.* 2012; **6:** 11-22.
106. Lionakis MS, Holland SM. CARD9: At the intersection of mucosal and systemic antifungal immunity. *Blood.* 2013; **121:** 2377-8.
107. Lionakis MS, Holland SM. Human invasive mycoses: Immunogenetics on the rise. *J Infect Dis.* 2015; **211:** 1205-7.
108. Lionakis MS, Iliev ID, Hohl TM. Immunity against fungi. *JCI Insight.* 2017; 2. pii: 93156. PMID: 28570272.
109. Liu M, Capilla J, Johansen ME, *et al. Saccharomyces* as a vaccine against systemic aspergillosis: 'The friend of man' a friend again? *J Med Microbiol.* 2011; **60 (Pt 10):** 1423-32.
110. Liu M, Clemons KV, Bigos M, *et al.* Immune responses induced by heat killed *Saccharomyces cerevisiae:* A vaccine against fungal infection. *Vaccine.* 2011; **29:** 1745-53.
111. Liu M, Clemons KV, Johansen ME, *et al. Saccharomyces* as a vaccine against systemic candidiasis. *Immunol Invest.* 2012; **41:** 847-55.
112. Madrid IM, Xavier MO, Mattei AS, *et al.* Role of melanin in the pathogenesis of cutaneous sporotrichosis. *Microbes Infect.* 2010; **12:** 162-5.
113. Magee PT. Fungal pathogenicity and morphological switches. *Nat Genet.* 2010; **42:** 560-1.
114. Maiti PK, Das S, Ghosh T, Dey R. Effects of potassium iodide on low avid immunological reactions: Probable mechanism of action on selective fungal infections. *Ann Med Health Sci Res.* 2013; **3:** 397-401.
115. Majumder T, Liu M, Chen V, *et al.* Killed *Saccharomyces cerevisiae* protects against lethal challenge of *Cryptococcus grubii. Mycopathologia.* 2014; **178:** 189-95.
116. Malone M, Dickson HG. Understanding the role of fungi in chronic wounds. *mBio.* 2016; 7. pii: e01898-16. PMID: 27923920.
117. Marciano BE, Spalding C, Fitzgerald A, *et al.* Common severe infections in chronic granulomatous disease. *Clin Infect Dis.* 2015; **60:** 1176-83.
118. Margalit A, Kavanagh K. The innate immune response to *Aspergillus fumigatus* at the alveolar surface. *FEMS Microbiol Rev.* 2015; **39:** 670-87.
119. Marine M, Bom VL, de Castro PA, *et al.* The development of animal infection models and antifungal efficacy assays against clinical isolates of *Trichosporon asahii, T.asteroides* and *T.inkin. Virulence.* 2015; **6:** 476-86.
120. Marodi L, Erdos M. Dectin-1 deficiency and mucocutaneous fungal infections. *N Engl J Med.* 2010; **362:** 367-8.
121. Martinez LR, Mihu MR, Gacser A, *et al.* Methamphetamine enhances histoplasmosis by immunosuppression of the host. *J Infect Dis.* 2009; **200:** 131-41.
122. Martinez M, Clemons KV, Stevens DA. Heat-killed yeast as a pan-fungal vaccine. *Methods Mol Biol.* 2017; **1625:** 23-30.
123. Medeiros AI, Sa-Nunes A, Turato WM, *et al.* Leukotrienes are potent adjuvant during fungal infection: Effects on memory T cells. *J Immunol.* 2008; **181:** 8544-51.
124. Medici NP, Del Poeta M. New insights on the development of fungal vaccines: From immunity to recent challenges. *Mem Inst Oswaldo Cruz.* 2015; **110:** 966-73.
125. Mendez-Tovar LJ. Pathogenesis of dermatophytosis and tinea versicolor. *Clin Dermatol.* 2010; **28:** 185-9.
126. Michalski C, Kan B, Lavoie PM. Antifungal immunological defenses in newborns. *Front Immunol.* 2017; **8:** 281.
127. Mora-Montes HM, Gacser A. Editorial: Recent advances in the study of the host-fungus interaction. *Front Microbiol.* 2016; **7:** 1694. PMID: 27833592.
128. Moretto MM, Khan IA, Weiss LM. Gastrointestinal cell mediated immunity and the microsporidia. *PLoS Pathog.* 2012; **8:** e1002775. PMID: 22807673.
129. Muller M, Wandel S, Colebunders R, *et al.* Immune reconstitution inflammatory syndrome in patients starting antiretroviral therapy for HIV infection: A systematic review and meta-analysis. *Lancet Infect Dis.* 2010; **10:** 251-61.
130. Murphy JW. Mechanisms of natural resistance to human pathogenic fungi. *Annu Rev Microbiol.* 1991; **45:** 509-38.
131. Nanjappa SG, Klein BS. Vaccine immunity against fungal infections. *Curr Opi Immunol.* 2014; **28:** 27-33.
132. Nguyen LD, Viscogliosi E, Delhaes L. The lung mycobiome: An emerging field of the human respiratory microbiome. *Front Microbiol.* 2015; **6:** 89.

133. Netea MG, Joosten LA, van der Meer JW, *et al.* Immune defence against *Candida* fungal infections. *Nat Rev Immunol.* 2015; **15:** 630-42.
134. Nicola AM, Albuquerque P, Martinez LR, *et al.* Macrophage autophagy in immunity to *Cryptococcus neoformans* and *Candida albicans. Infect Immun.* 2012; **80:** 3065-76.
135. No authors listed. Baker's yeast raises hope for a vaccine against aspergillosis: Study shows that inactivated Baker's yeast is protective against *Aspergillus* in mice. *Expert Rev Vaccines.* 2011; **10:** 1367.
136. Nosanchuk JD, Stark RE, Casadevall A. Fungal melanin: What do we know about structure? *Front Microbiol.* 2015; **6:** 1463.
137. Nosanchuk JD, Taborda CP. Vaccines and immunotherapy against fungi: The new frontier. *Front Microbiol.* 2013; **4:** 1-2.
138. Ordonez ME, Farraye FA, Di Palma JA. Endemic fungal infections in inflammatory bowel disease associated with anti-TNF antibody therapy. *Inflamm Bowel Dis.* 2013; **19:** 2490-500.
139. Palmer GE, Askew DS, Williamson PR. The diverse roles of autophagy in medically important fungi. *Autophagy.* 2008; **4:** 982-8.
140. Pamer EG. TLR polymorphisms and the risk of invasive fungal infections. *N Engl J Med.* 2008; **359:** 1836-8.
141. Parameswaran IG, Segal BH. Immunotherapy for fungal infections with special emphasis on central nervous system infections. *Neurol India.* 2007; **55:** 260-6.
142. Paredes K, Capilla J, Sutton DA, *et al.* Virulence of *Curvularia* in a murine model. *Mycoses.* 2013; **56:** 512-5.
143. Park HR, Voigt K. Interaction of zygomycetes with innate immune cells reconsidered with respect to ecology, morphology, evolution and infection biology: A mini-review. *Mycoses.* 2014; **57 (Suppl. 3):** 31-9.
144. Park SJ, Mehrad B. Innate immunity to *Aspergillus* species. *Clin Microbiol Rev.* 2009; **22:** 535-51.
145. Perfect JR, Cox GM. Virulence mechanisms for fungi. Part I & II. *Clin Microbial Newslet.* 2000; **22:** 113-19 & 121-5.
146. Perfect JR. The impact of the host on fungal infections. *Am J Med.* 2012; **125:** S39-51.
147. Pilmis B, Puel A, Lortholary O, *et al.* New clinical phenotypes of fungal infections in special hosts. *Clin Microbiol Infect.* 2016; **22:** 681-7.
148. Portuondo DL, Ferreira LS, Urbaczek AC, *et al.* Adjuvants and delivery systems for antifungal vaccines: Current state and future developments. *Med Mycol.* 2015; **53:** 69-89.
149. Posch W, Steger M, Wilflingseder D, *et al.* Promising immunotherapy against fungal diseases. *Expert Opin Biol Ther.* 2017; **17:** 861-70.
150. Powers-Fletcher MV, Kendall BA, Griffin AT, *et al.* Filamentous fungi. *Microbiol Spectr.* 2016; 4. PMID: 27337469.
151. Prohic A. Distribution of *Malassezia* species in seborrhoeic dermatitis: Correlation with patients' cellular immune status. *Mycoses.* 2010; **53:** 344-9.
152. Ramage G, Rajendran R, Gutierrez-Correa M, *et al. Aspergillus* biofilms: Clinical and industrial significance. *FEMS Microbiol Lett.* 2011; **324:** 89-97.
153. Ramage G, Robertson SN, Williams C. Strength in numbers: Antifungal strategies against fungal biofilms. *Int J Antimicrob Agents.* 2014; **43:** 114-20.
154. Ramirez-Amador VA, Espinosa E, Gonzalez-Ramirez I, *et al.* Identification of oral candidosis, hairy leukoplakia and recurrent oral ulcers as distinct cases of immune reconstitution inflammatory syndrome. *Int J STD AIDS.* 2009; **20:** 259-61.
155. Ramirez-Ortiz ZG, Means TK. The role of dendritic cells in the innate recognition of pathogenic fungi *(A.fumigatus, C.neoformans* and *C.albicans). Virulence.* 2012; **3:** 635-46.
156. Ribeiro AM, Souza AC, Amaral AC, *et al.* Nanobiotechnological approaches to delivery of DNA vaccine against fungal infection. *J Biomed Nanotechnol.* 2013; **9:** 221-30.
157. Richardson JP, Moyes DL. Adaptive immune responses to *Candida albicans* infection. *Virulence.* 2015; **6:** 327-37.
158. Rieber N, Singh A, Oz H, *et al.* Pathogenic fungi regulate immunity by inducing neutrophilic myeloid-derived suppressor cells. *Cell Host Microbe.* 2015; **17:** 507-14.
159. Rivera A. Protective immune responses to fungal infections. *Parasite Immunol.* 2014; **36:** 453-62.
160. Romani L. Cell mediated immunity to fungi: A reassessment. *Med Mycol.* 2008; **46:** 515-29.
161. Romani L. Immunity to fungal infections. *Nat Rev Immunol.* 2011; **11:** 275-88.
162. Roth S, Ruland J. Caspase recruitment domain-containing protein 9 signaling in innate immunity and inflammation. *Trends Immunol.* 2013; **34:** 243-50.
163. Roussey JA, Olszewski MA, Osterholzer JJ. Immunoregulation in fungal diseases. *Microorganisms.* 2016; **4:** pii: e47. PMID: 27973396.
164. Rouzaud C, Hay R, Chosidow O, *et al.* Severe dermatophytosis and acquired or innate immunodeficiency: A review. *J Fungi.* 2016; **2:** 4.
165. Safdar A. Immunotherapy for invasive mold disease in severely immunosuppressed patients. *Clin Infect Dis.* 2013; **57:** 94-100.
166. Saluja R, Metz M, Maurer M. Role and relevance of mast cells in fungal infections. *Front Immunol.* 2012; **3:** 146.
167. Santos E, Levitz SM. Fungal vaccines and immunotherapeutics. *Cold Spring Harb Perspect Med.* 2014; **4:** a019711.
168. Sardi Jde C, Pitangui Nde S, Rodriguez-Arellanes G, *et al.* Highlights in pathogenic fungal biofilms. *Rev Iberoam Micol.* 2014; **31:** 22-9.
169. Schmidt S, Schneider A, Demir A, *et al.* Natural killer cell-mediated damage of clinical isolates of mucormycetes. *Mycoses.* 2016; **59:** 34-8.
170. Schweizer A, Schroppel K. Experimental infection of rodent mammals for fungal virulence testing. *Methods Mol Biol.* 2009; **470:** 141-9.
171. Seed PC. The human mycobiome. *Cold Spring Harb Perspect Med.* 2014; **5:** a019810. PMID: 25384764.
172. Segal E. Vaccines against fungal infections. *Crit Rev Microbiol.* 1987; **14:** 229-71.
173. Sekkides O. Bringing fungal infections in from the cold. *Lancet Infect Dis.* 2015; **15:** 884-5.

174. Shankar SK, Mahadevan A, Sundaram C, *et al.* Pathobiology of fungal infections of the central nervous system with special reference to the Indian scenario. *Neurol India.* 2007; **55:** 198-215.
175. Sharma S, Gupta S, Shrivastava JN. Presence of virus-like particles in human pathogenic fungi: *Chrysosporium* spp. and *Candida albicans. Indian J Virol.* 2011; **22:** 104-10.
176. Sheppard DC, Filler SG. Host cell invasion by medically important fungi. *Cold Spring Harb Perspect Med.* 2014; **5:** a019687. PMID: 25367974.
177. Sherry L, Ramage G, Kean R, *et al.* Biofilm-forming capability of highly virulent, multidrug-resistant *Candida auris. Emerg Infect Dis.* 2017; **23:** 328-31.
178. Sorgo AG, Heilmann CJ, Brul S, *et al.* Beyond the wall: *Candida albicans* secret(e)s to survive. *FEMS Microbiol Lett.* 2013; **338:** 10-7.
179. Spellberg B. Vaccines for invasive fungal infections. *F1000 Med Rep.* 2011; **3:** 13.
180. Steele C, Wormley FL Jr. Immunology of fungal infections: Lessons learned from animal models. *Curr Opin Microbiol.* 2012; **15:** 413-9.
181. Stevens DA, Clemons KV, Liu M. Developing a vaccine against aspergillosis. *Med Mycol.* 2011; **49:** S170-6.
182. Suhr MJ, Hallen-Adams HE. The human gut mycobiome: Pitfalls and potentials - A mycologist's perspective. *Mycologia.* 2015; **107:** 1057-73.
183. Sun HY, Singh N. Opportunistic infection-associated immune reconstitution syndrome in transplant recipients. *Clin Infect Dis.* 2011; **53:** 168-76.
184. Taborda CP, da Silva MB, Nosanchuk JD, *et al.* Melanin as a virulence factor of *Paracoccidioides brasiliensis* and other dimorphic pathogenic fungi: A mini-review. *Mycopathologia.* 2008; **165:** 331-9.
185. Tavanti A, Naglik JR, Osherov N. Host-fungal interactions: Pathogenicity *versus* immunity. *Int J Microbiol.* 2012; 562480. PMID: 22693515.
186. Taylor PR, Leal SM Jr, Sun Y, *et al. Aspergillus* and *Fusarium* corneal infections are regulated by Th17 cells and IL-17-producing neutrophils. *J Immunol.* 2014; **192:** 3319-27.
187. Thomaz L, Garcia-Rodas R, Guimaraes AJ, *et al. Galleria mellonella* as a model host to study *Paracoccidioides lutzii* and *Histoplasma capsulatum. Virulence.* 2013; **4:** 139-46.
188. Tierney L, Kuchler K, Rizzetto L, *et al.* Systems biology of host-fungus interactions: Turning complexity into simplicity. *Curr Opin Microbiol.* 2012; **15:** 440-6.
189. Tiwari U, Das S, Tandon M, *et al.* Vaccines for fungal infections. *Natl Med J India.* 2015; **28:** 14-9.
190. Ueno K, Urai M, Ohkouchi K, *et al.* Dendritic cell-based vaccine against fungal infection. *Methods Mol Biol.* 2016; **1403:** 537-49.
191. Vecchiarelli A, d'Enfert C. Shedding natural light on fungal infections. *Virulence.* 2012; **3:** 15-7.
192. Vecchiarelli A, Pericolini E, Gabrielli E, *et al.* New approaches in the development of a vaccine for mucosal candidiasis: Progress and challenges. *Front Microbiol.* 2012; **3:** 294.
193. Vera-Cabrera L, Salinas-Carmona MC, Waksman N, *et al.* Host defenses in subcutaneous mycoses. *Clin Dermatol.* 2012; **30:** 382-8.
194. Vinh DC. Insights into human antifungal immunity from primary immunodeficiencies. *Lancet Infect Dis.* 2011; **11:** 780-92.Vyas SP, Khatri K, Goyal AK. Functionalized nanocarrier(s) to image and target fungi infected immune cells. *Med Mycol.* 2009; **47:** S362-8.
195. Wang SJ. *Candida* vaccines development from point view of US patent application. *Hum Vaccin.* 2011; **7:** 1165-71.
196. Wang X, van de Veerdonk FL, Netea MG. Basic genetics and immunology of *Candida* infections. *Infect Dis Clin North Am.* 2016; **30:** 85-102.
197. Wang ZK, Yang YS, Stefka AT, *et al.* Fungal microbiota and digestive diseases. *Aliment Pharmacol Ther.* 2014; **39:** 751-66.
198. Wheeler ML, Limon JJ, Underhill DM. Immunity to commensal fungi: Detente and disease. *Annu Rev Pathol.* 2017; **12:** 359-85.
199. Wiesner DL, Boulware DR. *Cryptococcus*-related immune reconstitution inflammatory syndrome (IRIS): Pathogenesis and its clinical implications. *Curr Fungal Infect Rep.* 2011; **5:** 252-61.
200. Williams C, Ramage G. Fungal biofilms in human disease. *Adv Exp Med Biol.* 2015; **831:** 11-27.
201. Williams C, Ranjendran R, Ramage G. Pathogenesis of fungal infections in cystic fibrosis. *Curr Fungal Infect Rep.* 2016; **10:** 163-9.
202. Wu W, Zhang R, Wang X, *et al.* Impairment of immune response against dematiaceous fungi in CARD9 knockout mice. *Mycopathologia.* 2016; **181:** 631-42.
203. Xin H, Cartmell J, Bailey JJ, *et al.* Self-adjuvanting glycopeptide conjugate vaccine against disseminated candidiasis. *PLoS One.* 2012; **7:** e35106. PMID: 22563378.
204. Zahur M, Afroz A, Rashid U, *et al.* Dermatomycoses: Challenges and human immune responses. *Curr Protein Pept Sci.* 2014; **15:** 437-44.
205. Zelante T, De Luca A, D'Angelo C, *et al.* IL-17/Th17 in anti-fungal immunity: What's new? *Eur J Immunol.* 2009; **39:** 645-8.
206. Zelante T, Pieraccini G, Scaringi L, *et al.* Learning from other diseases: Protection and pathology in chronic fungal infections. *Semin Immunopathol.* 2016; **38:** 239-48.
207. Zhang H, Chen H, Niu J, *et al.* Role of adaptive immunity in the pathogenesis of *Candida albicans* keratitis. *Invest Ophthalmol Vis Sci.* 2009; **50:** 2653-9. & 4551.
208. Zhang H, Li H, Li Y, *et al.* IL-17 plays a central role in initiating experimental *Candida albicans* infection in mouse corneas. *Eur J Immunol.* 2013; **43:** 2671-82.
209. Zhang JZ. Autoimmune diseases and fungal infections: Immunological mechanisms and therapeutic approaches. *Chin Med J (Engl).* 2009; **122:** 483-5.
210. Zubair S, Azhar A, Khan N, *et al.* Nanoparticle-based mycosis vaccine. *Methods Mol Biol.* 2017; **1625:** 169-211.
211. Zwinkels RL, Dawson L. Immunoparalysis in sepsis. *Neth J Med.* 2013; **71:** 243-5.

CHAPTER 5

Diagnosis of Fungal Diseases

A proper diagnosis of fungal diseases is being faced as one of the challenging task in medial practice. The clinical presentation is hard to interpret and findings of non-invasive methods like X-rays and computed tomographic scanning are helpful to know the extent of lesions but can not delineate specifically the nature of underlying disease process. The fungal culture results are available at the earliest in two to three days and blood and deep-tissue cultures from infections with focal lesions are invariably negative. The direct microscopy and histopathological examinations are relatively fast but they do not always confirm identification of infecting agent to species level.

In contrast, even though the latest generation of monoclonal antibody-based enzyme-linked immunosorbent assays for circulating *Aspergillus* and *Candida* antigens are specific but they lack sensitivity. Thus, rapid methods that should be sensitive and specific are needed and PCR is being standardized to fulfill essential requirements. A correct diagnosis provides specific treatment of such disease and may prove life-saving or at least stave off the complications produced by a fungal infectious process.

The biggest challenge before the medical mycologists is an early, prompt and correct diagnosis of fungal diseases. Like other microbial infections, diagnosis is based on combination of clinical observations and various laboratory investigations. The procedures include, recognition of fungal elements in tissue by direct examination, culture and detection of specific humoral and/or cell-mediated immune responses using serological methods, skin and *in vitro* testings for lymphocyte sensitization. The precise value of each of these components varies with the type of infection and causative fungal pathogens. All technical details of culture media, stains and laboratory procedures referred in this Chapter are given in the **Appendices**. The comprehensive diagnostic approach to fungal infections can be divided under following broad headings:

A. Fungal Clinical Diagnosis
B. Based on Conventional Methods
C. Based on Molecular Methods
D. Based on Recently Developed Techniques
E. Based on Miscellaneous Methods

A. FUNGAL CLINICAL DIAGNOSIS

The clinical criteria may give presumptive diagnosis of a fungal disease. The superficial and subcutaneous mycoses often produce characteristic lesions that strongly suggest their fungal etiology but they closely resemble other diseases. Moreover, it is not unusual to find that appearance of lesions is considerably modified and rendered atypical by prior therapy with topical steroids or other medicines.

In case of systemic mycosis, there is no sign or symptom that specifically suggests a fungal disease. These are very much similar to any of the viral, bacterial or parasitic disease. The establishment of an early diagnosis essentially increases chances of successful treatment thereby likelihood of the survival of patients. It is important that possibility of fungal involvement should be considered from the very outset of the disease process. Hence an **index of clinical suspicion** of fungal disease should be kept high for an early, prompt and correct diagnosis, which can be subsequently confirmed by various laboratory parameters.

In the recent times, clinical significance of fungal infections has been better recognized mainly due to the increased awareness among medical personnel. In combination with the modern imaging techniques for patients' evaluation, there is an improvement in the accuracy and rapidity of establishing diagnosis. The echocardiography has been used to show fungal vegetation on heart valves and similarly the usefulness of ophthalmological examination has also increased for endophthalmitis, a common complication of systemic candidiasis.

B. BASED ON CONVENTIONAL METHODS

The laboratory procedures are used to confirm the clinical suspicion to establish fungal cause in disease of unknown etiology or exclude fungal involvement where there are many differential diagnoses. When diagnosis of fungal infection is finally established, laboratory may help in choosing specific antifungal therapeutic regimen, monitoring course of disease and in due course confirming mycological cure as well. Sometimes, with opportunistic mycoses, the laboratory probably can provide only subjective evidence that is to be considered along with other clinical findings. In such cases close liaison between the clinician and mycology laboratory is essential.

The diagnosis of invasive fungal diseases (IFDs) is generally divided into three categories as **proven**, **probable** and **possible**. These 2002 criteria of European Organization for Research and Treatment of (EORTC) and the National Institute of Allergy and Infectious Diseases Mycoses Study Group (MSG) for classification of cases into possible, probable or proven were revised in 2008. The proven type of disease is defined as demonstration of fungal elements in infected tissue for most conditions irrespective of host factors or clinical features. The cases of probable disease require a host factor, clinical features and mycological evidence. The possible disease include cases with appropriate host factors and sufficient clinical evidence but no mycological support. These definitions have already been adopted by most of the practice guidelines meant for invasive fungal diseases. The most commonly identified fungal species associated with such infections are *Candida*, *Aspergillus*, Mucormycetes, *Cryptococcus* and *Pneumocystis* species.

This is very pertinent that laboratory receives correct type of specimen along with adequate clinical data so that appropriate investigation can be carried out, accordingly. The information related to travel or residence in any particular geographical area, animal contacts and occupation of patient enables the laboratory to direct or modify its procedures towards particular suspected fungus or group of fungal agents.

1. Sites and Types of Specimens

The site as well as nature of specimen is most important to proceed for final diagnosis. The sites are usually taken into considerations and to collect specimen depending on corresponding disease in a particular patient, which are described below:

(a) Superficial Mycoses: The diagnosis of superficial mycoses may be strongly suspected on clinical grounds and it is usually prudent and sometimes essential to seek laboratory support. Both direct microscopic examination and culture of selective materials should be made whenever possible.

In all these cases, cleaning of local site with 70% alcohol, which is allowed to evaporate before taking specimen, may be helpful and should always be done if greasy ointments or powders have already been applied as medication. The material is best collected into folded squares of paper thereby permitting drying of specimen, reducing bacterial contamination and providing conditions under which specimens can be stored for long periods without appreciable loss to viability of the fungal agents.

The dermatophytic lesion usually spreads outward in concentric fashion with healing in central region. Therefore, material should be collected by scraping outward from edges of active lesions with scalpel blade held at an angle of 90° to the skin surface. When there is least scaling, as with lesions of glabrous skin, it is preferable and sometimes necessary to use Cellophane tape to take adequate material for laboratory examination.

The specimens from scalp are best obtained by scraping with blunt scalpel so that they include hair-stubs, contents of plugged follicles and scales. Hair, which have been cut rather than plucked, are seldom satisfactory for investigations. In kerion, the hair that resist pull of epilation forceps can be left as such and which come out easily, are preferably selected. The ringworm of scalp, especially caused by anthropophilic species may cause small lesions that are difficult to detect clinically. This type of infection may produce fluorescence of infected hair and may be detected by Wood's lamp examination, which is helpful to detect scalp ringworm or to select material for culture when infection is caused by species that produce fluorescence of infected hairs. Wood's lamp was invented by Robert Wood and is an invaluable tool in practice of cutaneous medicine and is recommended for detection of fungal infections of hairs **(*See* Fig. 10.12)**.

The hairbrush sampling technique may be utilized to obtain specimens for fungal cultures. This involves thoroughly brushing scalp and using hairbrush to inoculate on to culture media.

In case of onychomycosis, patient should be off antifungal agents one week prior to collection of specimen and it should be taken from an appropriate site depending upon type of onychomycosis and the exact area involved.

After disinfecting with 70% alcohol, nail is clipped from free edge, taken as far back as possible and should include its full thickness. The fungus in distal portion of nail is often non-viable and if culture does fail, success can sometimes be obtained by culturing **borings** taken from base of the nail.

For the infections of mucous membranes, scrapings are preferred than the swabs. However, when swabs are to be used for sampling, the best way to send to laboratory in culture medium since yeasts rapidly loose their viability in dry swabs.

(b) Subcutaneous Mycoses: The scrapings or crusts from superficial parts of subcutaneous lesions may be satisfactory for microscopy and culture. The contamination with bacteria and saprotrophic fungi is common and aspirated samples of pus and/or biopsy specimens are more valuable. However, the biopsy may be avoided in suspected sporotrichosis because this may spread infection and hinder healing of lesions.

(c) Systemic Mycoses: In patients with suspected deep-seated mycoses specimens may be taken from as many potential sites as possible. The source include tissue removed for biopsy, pus, feces, urine, sputum, spinal fluid, blood and scrapings or swabs from edge of lesions. It is important to ensure that specimens such as sputum and urine are freshly collected. Twenty-four hours' sputum or urine from catheter bags is unsatisfactory since commensal yeasts multiply quite rapidly in such materials. If catheter piece is received as a sample, it should be rolled over the agar plate and kept in liquid media. A catheter specimen of urine is ideal but failing this midstream urine sample is satisfactory. Care should be taken not to over interpret high counts of yeast often encountered in patients with an indwelling urinary catheter.

2. Collection and Transport of Specimens

The exact diagnosis of fungal disease totally depends on collection of proper and adequate clinical specimens thereby selection of appropriate microbiological, serological and histopathological investigations. Many fungal infections are apparently similar to mycobacterial infections hence same specimen may be cultured for both fungi and mycobacteria. Most of systemic fungal infections have primary focus in lungs hence respiratory secretions are usually included among specimens selected for fungal culture. It should be emphasized that dissemination of such infections to distant body sites often occurs and fungi may be commonly recovered from non-respiratory sites. A proper collection of specimens and their prompt transport to clinical laboratory is essential for recovery of fungi on culture. Sometimes, contaminating bacteria or even fungi may rapidly overgrow some of the slow growing pathogenic fungi. Therefore, it is important to transport specimens to the clinical laboratory at the earliest to avoid overgrowth of contaminant, in addition to keeping viability of the causative fungus.

(a) Respiratory Specimens: The specimens from respiratory system for fungal culture are sputum, tracheal secretions, bronchoalveolar lavage (BAL) fluid and lung biopsy. The sputum sample of early morning should be collected after vigorous mouthwash. The flakes are chosen for processing of specimens which may be ideal source of detecting the causative agent. The sputum may be induced if cough is non-productive. Sometimes, bronchoscopy may also be required in respiratory mycoses to examine the lesions and collection of specimen from the affected sites.

For concentrating the specimen, 0.5 gm of N-acetyl L-cysteine (NALC), in 100 ml of freshly prepared sodium citrate buffer, is added and vortexed for 10-30 seconds. Then M/15 phosphate buffer with pH 7.0 is added double the volume already present in the tube, centrifuge at 1000 rpm for 15 minutes. Discard the supernatant from the deposit and use sediment to prepare smears as well as media inoculations.

A sufficient amount of the processed specimen should be inoculated on each medium. The interpretation of results with opportunistic pathogens such as *Candida* is more difficult because in sputum specimens it is commonly present as an oropharyngeal contaminant. In such cases specimens of bronchial secretions reflect flora of lungs more accurately. It is often unclear whether recovery of opportunistic fungi from sputum or bronchial secretions suggests tissue invasion or lung colonization and it may be necessary to use techniques such as bronchial brushing or lung biopsy to obtain material suitable for the fungal diagnosis. Therefore, merely presence of *Candida* in any of the respiratory specimen is not clinically significant unless it is demonstrated in the tissue.

The clinician should timely alert the mycology laboratory regarding any possibility of dimorphic pathogen when specimen is submitted for examination so that concerned staff becomes aware about precautionary measures and vigilantly look for their parasitic forms like yeasts and spherules in clinical specimens and their respective mycelial forms in due course of the incubation of fungal culture.

(b) Cerebrospinal Fluid: The cerebrospinal fluid collected for culture should be centrifuged and sediment should be placed on several areas on agar surface. The media used for recovery of fungi from CSF should contain no antimicrobial agents because it is normally a sterile fluid hence there is no requirement of inhibitory agents. Once submitted to laboratory, CSF specimens should be processed immediately. If prompt processing is not feasible, samples should be kept at room temperature or placed in an incubator at 30°C, since most organisms will continue to replicate in this environment.

(c) Blood Culture: The disseminated fungal diseases are more prevalent than previously recognized. The blood cultures provide an accurate method for establishing diagnosis. Only few fungal blood culture systems are available over past few years and clinical microbiology laboratories are not able to use most of them due to their high cost. However, in smaller setup it may be adequate to use biphasic Brain-Heart Infusion Agar Broth. Most of the fungi are detected within first four days of incubation. However, an occasional isolate of *Histoplasma capsulatum* may require incubation of about 10 to 14 days for recovery. The fungal blood culture media should be incubated at two temperatures *i.e.* 25°C and 37°C keeping in view of possibility of any dimorphic fungal agent in clinical specimens. They are sub-cultured after two days and seven days, respectively. The preliminary report is sent after seven days and final report after 28 days of subculture *i.e.* after 9 days and 35 days from receipt of specimens, respectively.

(d) Tissue, Bone Marrow and Body Fluids: The tissue specimens should be taken preferably from the pyogenic or necrotic area of the wound so that chances of isolation of fungi are increased. Sometimes in surgical branches, antimicrobials like povidone iodine are applied before doing any procedure even if it may be collection of sample for microbiological examination. It should be avoided as chances of isolation of fungi will be bleak and antimicrobials may be applied after collection of sample.

All these specimens should be cultured immediately as and when received by laboratory personnel to ensure recovery of fungi from these important sites. The tissues should be processed before culture by mincing or grinding. After processing, at least 1 ml of specimen should be spread on surface of the appropriate culture media and incubation should be done for a period of four weeks at 37°C. The bone marrow specimen may be placed directly on the surface of medium and incubated. The body fluids from sterile sites should be concentrated by centrifugation before culturing.

(e) Semen Culture: This is sometimes very much revealing particularly in histoplasmosis and cryptococcosis, when the disease remain somewhat hidden and patient has already been wrongly taking repeated courses of antitubercular treatment without any improvement.

(f) Skin, Hair and Nail: These specimens are usually taken for dermatophyte infection and may be obtained by scraping skin. The nail clippings should be taken from discolored, dystrophic or brittle parts of nails. Fungus in distal part of nail is often non-viable hence it may be visible on microscopy but fails to grow in culture. The hairs are removed by plucking them with forceps. The specimen should be placed in sterile petri dish or paper envelope and should not be refrigerated as low temperature may be detrimental to some of the dermatophytes. The nail should be cut into pieces before inoculation on to the media. The Sabouraud dextrose agar with chloramphenicol and cycloheximide is used for culturing dermatophytes. The cultures should be incubated at 25°C, 30°C and 37°C for a minimum period of four weeks before being reporting them to be as sterile.

When obtaining hair from patients with tinea capitis for fungal culture, it is important to secure the infected follicular portion of the shaft. In patients with very little hair remaining in the infected patch it may be impossible to grasp the hair with the forceps or tweezers. Under those circumstances, the root ends can be teased out of the follicular orifices with the tip of a scalpel blade. At times, the hairs are long enough to be epilated with a hemostat on tweezers. In these instances the hairs should be firmly grasped and the follicular portions removed for placement on agar for culture. Cut hairs are not adequate for culture because one misses the infected portion of the hair, which is still embedded in the scalp.

(g) Urine Culture: The urine specimens for fungal culture should be processed as early as possible after collection. The twenty-four hours' urine is not useful for fungal cultures. If there is delay in processing, specimens may be kept at 4°C for up to 12 hours. All the urine samples are to be centrifuged and sediment is inoculated using loop to provide adequate isolation of the colonies. Since urine is often contaminated with bacteria, it is necessary to use media containing antibacterial antibiotics to ensure isolation of fungi in their pure form.

(h) Vaginal Secretions: The diagnosis of vaginal candidiasis is better established with the help of clinical features and positive direct smear of secretions. The culture may be misleading also, since yeasts are part of normal vaginal flora of up to 20% of healthy women. However, cultures may be helpful for management of chronic recurring diseases or for monitoring the therapy.

(i) Stool Culture: The diagnosis of many fungal infections of gastrointestinal tract is better established by biopsy for histopathological examination of tissue than by culture of stool specimens. The positive cultures may be misleading because about 40% of healthy individuals and up to 75% of compromised patients are colonized with yeasts as commensals.

(j) Eye: In case of keratomycosis corneal scrapings are taken from base and margins of ulcer aseptically using Kimura's spatula under 4% xylocaine as local anesthetic. Aspiration of hypopyon is done with sterile needle in keratomycosis. The posterior chamber may also be aspirated in case of fungal endophthalmitis.

3. Processing of Specimens

The specimens for fungal diagnosis are processed as per protocol, depending upon nature of material collected in a particular case. Sometimes, pre-treatment of specimens is necessary before microscopic examination and/or fungal culture. The samples of CSF and urine are best centrifuged and deposit is re-suspended in a small volume of saline to detect scanty numbers of fungi. Sputum specimen is divided into two portions. One part has to be digested for homogenization, concentration by centrifugation and decontamination of the most of bacteria. The other part is used for the direct examination by various techniques.

It is often helpful and sometimes necessary to culture measured volumes of specimens. By estimating number of viable fungi, expressed as number of colony forming units (CFU) per ml, severity of an infection can be assessed and results of therapy can be monitored through periodic collection of specimens.

(a) Direct Examination

The direct demonstration of fungal elements is essential in establishing diagnosis by detecting presence of causative agent in the clinical material, which is co-related with the suspected disease. If fungal culture is taken as **'gold standard'** in Medical Mycology, direct demonstration of fungi in clinical specimen is **'gold mine'** which should be performed meticulously by an experienced person. The following procedures are adopted to directly observe fungi in clinical specimens:

(i) Wet Mounts: In the diagnostic mycology laboratory, slide and tube **KOH** mounts are prepared to establish the diagnosis. As an alternative, sodium hydroxide may also be used. Most of the specimens can be examined in wet mounts after partial digestion with 10-20% KOH. The clinical specimens like skin, hair and nails should be mounted under cover slip in KOH on slide. This clears material within 5-20 minutes, depending on its thickness. A slight warming over a low flame hastens digestion of keratin. However, care must be taken to avoid overheating thereby crystallization of slide material. KOH can also be supplemented with DMSO to increase clearing of fungi especially in skin scrapings and nail clippings.

The yeast cells and pseudohyphae are seen in KOH wet mount in clinical specimen **(Fig. 5.1A)**. There may be septate or non-septate hyphae in the clinical material **(Figs. 5.1B and C)**. When there is adipose tissue, then lot of bubbles are seen in the KOH preparation. This is the most important informative part of diagnostic procedures to detect fungal elements thereby subsequently to co-relate with cultural findings.

The specimens of sputum, centrifuged CSF or urine deposits may be examined without prior preparation; however, it is better to mix samples with equal volumes of KOH to clear off host cellular debris. The knack to differentiate between fungal structures and routinely encountered artifacts is gradually acquired with experience. The small size of yeast cells of *Histoplasma capsulatum* and *Sporothrix schenckii* as 4-6 μm makes it tricky to be detected in KOH wet mount, which is proved to be unsatisfactory and such fungi are invariably missed on direct demonstration. Hence Giemsa stained smears of sputum, pus and biopsies are usually advised in suspected patients of histoplasmosis.

The optical brighteners like **Calcofluor** or **Blankophor** are also used to prepare wet mounts, which are sensitive procedures offering excellent visualization of fungal morphology, particularly when clinical material is scanty. In addition, mixed infection by various types of fungi (*Aspergillus* and mucormycetes) is also better visualized by this technique. These dyes bind to polysaccharides with β-glucose linkages such as cellulose, glucans and chitins, which are abundant in fungal cell wall. The non-specific

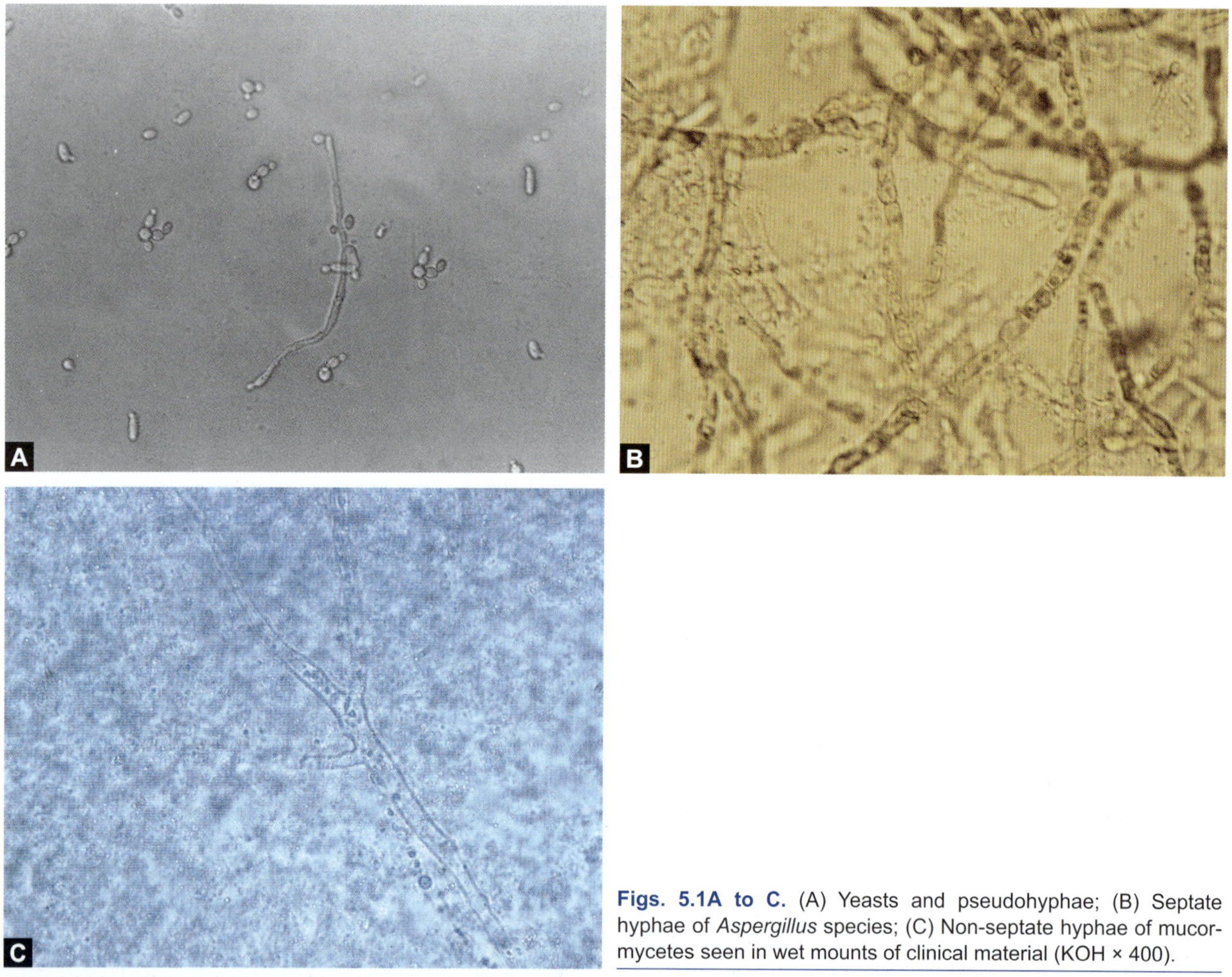

Figs. 5.1A to C. (A) Yeasts and pseudohyphae; (B) Septate hyphae of *Aspergillus* species; (C) Non-septate hyphae of mucormycetes seen in wet mounts of clinical material (KOH × 400).

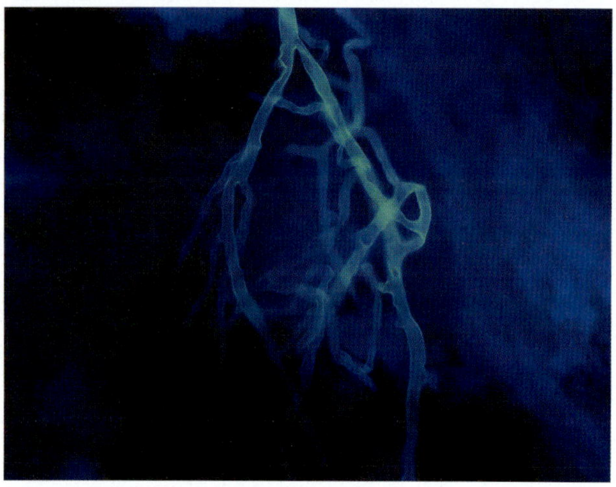

Fig. 5.2. Non-septate hyphae of mucormycetes seen in a fluorescent brightener wet mount (CFW × 400).

binding to fungal cell wall polysaccharides is in such a way that walls fluoresce brightly under ultraviolet illumination. Calcofluor supplemented with KOH, especially in scanty clinical material of corneal scrapings, can detect fungal elements. A fluorescence microscope with appropriate barrier filter (300 to 412 nm) is needed to view wet mounts of clinical specimens stained with Calcofluor or Blankophor. **Figure 5.2** shows fluorescing non-septate hyphae of mucormycetes in calcofluor white (CFW) stain under UV light of fluorescent microscope.

The wet mounts of **India ink** and **nigrosin** stainings are also very useful negative stainings procedures in making diagnosis of certain infections caused by encapsulated fungi like *Cryptococcus* species. In addition, neutral red staining is also simple and effective method for evaluation of viability of fungi.

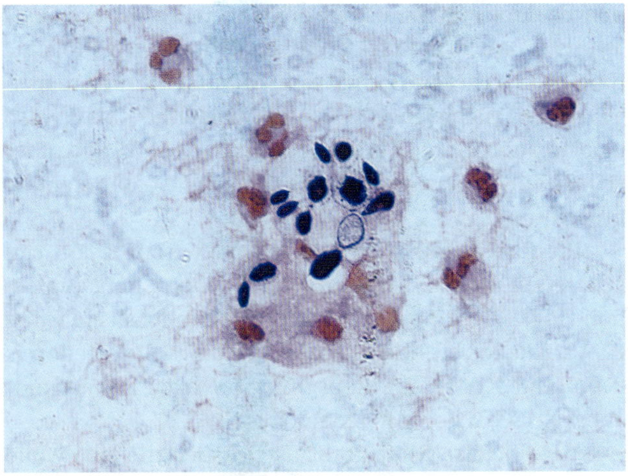

Fig. 5.3. Yeast cells with budding in *Cryptococcus* species seen in Gram-stained smear of pus aspirate (Gram's Stain × 400).

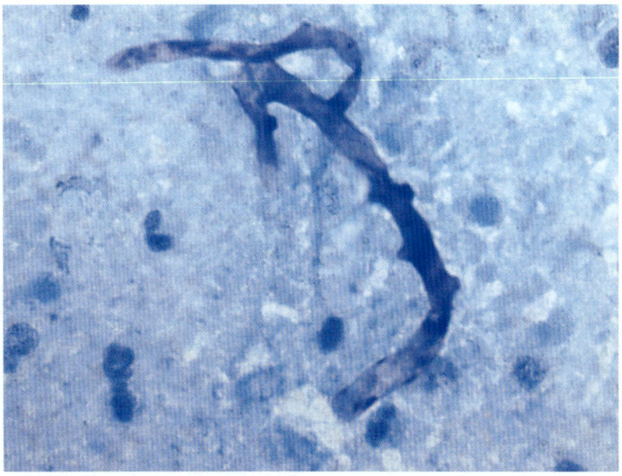

Fig. 5.4. Non-septate hyphae with right-angle branching seen in AFB smear as an incidental finding (ZN Stain × 400).

For direct microscopic diagnosis in Medical Mycology, **Vinyl Adhesive Tape (VAT)** preparation is also very useful tool among existing techniques for demonstration of certain fungi like *Malassezia* species, *Candida* species and dermatophytes. The collection and identification of fungal causative agents may also be performed easily by this technique, for setting of some deep-seated cutaneous mycoses, where infectious agents can be observed in horny layer of epidermis in transepidermal elimination (TEE) events. Therefore, it has also been recently tried successfully in chromoblastomycosis, lacaziosis and paracoccidioidomycosis. This is also called as **Scotch Technique**.

In wet mounts under low and high power microscopy, there are yeasts cells with or without pseudohyphae, septate hyphae, non-septate hyphae, spherules, arthrospores or any of fungal form. Sometimes aseptate hyphae of oomycetes, particularly, *Pythium insidiosum* may also be encountered confusing with mucormycetes. If none of the above mentioned fungal material is visible after thorough search in the wet mounts, report is prepared as ***'No fungal elements seen'***. The preliminary report is despatched on the basis of findings of direct demonstration in wet mounts with specific note to clinician incharge that ***'Fungal culture report to follow'***.

The direct microscopic examination of KOH wet mount is key test in diagnostic practice of Medical Mycology, which proves it to be a turning point in the management of fungal diseases. Moreover, these findings are also useful to establish **mycological cure** of clinical entity wherein KOH is negative in previous positive specimens followed by sterile fungal cultures. The clinical cure, followed by mycological cure, is essential in the management of various types of fungal diseases.

The yeast cells in the clinical material can be easily detected on Gram-stained smear **(Fig. 5.3)**. Sometimes, while examining other stainings, fungal elements may be incidentally visualised, as mucormycetes may be encountered in Ziehl Neelsen staining done for AFB **(Fig. 5.4)**.

(ii) Histopathology: A diagnostic mycology laboratory should have sufficient feedback from its Histopathology Division. This is essential because most of the times it is very difficult to decide whether fungal isolate in culture is significant or merely a contaminant. If fungus is demonstrated in tissue sections it provides ideal feedback and effective co-relation of significance of the particular cultural isolate.

The histopathological examination of biopsy and autopsy specimens is an excellent way to diagnose mycotic infections. The clinical autopsy is a significant diagnostic method and means of medical quality control. The fungi because of their size, morphological diversity and polysaccharide content, can be easily demonstrated and studied in tissue sections with special stains.

There is no specific tissue reaction, which is typical for fungal infection. In superficial infection histopathological picture is that of an acute, subacute or chronic dermatitis with folliculitis in inflammatory lesions of hairy skin.

In subcutaneous and systemic infections granulomatous reaction often with fibrosis or pyogenic inflammation is usually seen. The reaction is varying according to site and evolution of condition. With *Cryptococcus* species there is little tissue reaction. Therefore, diagnosis depends on demonstration of fungi within tissues.

The H&E stain is a routine procedure in histopathology laboratory and stains most of fungi sufficiently to be recognizable in tissue section. **Figure 5.5** shows fungi in H&E-stained section. The specific fungal stains, such as periodic acid-Schiff (PAS), **(Figs. 5.6A and B)** Grocott-Gomori's methenamine silver stain **(Fig. 5.7)** and Gridley stains are widely used for demonstrating fungi in tissues. In addition, Mayer's mucicarmine can also be used specifically to show capsular material of *Cryptococcus* and endospores and sporangia of *Rhinosporidium seeberi*. Alcian blue staining may be done to demonstrate acid mucin. However, causal agents of mucormycosis are unusual among pathogenic fungi in that they are faintly stained with PAS and are usually better visible in H&E-stained material. This is also ideal to put control slides for stainings like GMS and other stains with every batch to know the quality of the staining procedure.

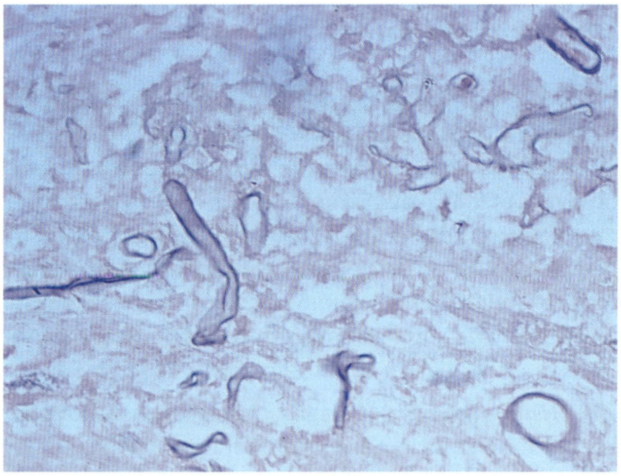

Fig. 5.5. Non-septate hyphae of mucormycetes species with wide angle branching in a tissue section (H&E × 400).

The fine-needle aspiration cytology from lesion can also reveal most of the information about a fungal infection **(Fig. 5.8)**. Biopsy may be taken from local site when FNAC fails to reveal any positive findings. If infection is deep-seated FNAC or biopsy may be taken under guidance of USG, CT or MRI techniques.

A major disadvantage of use of special stains while detecting etiological agents is that they mask natural color of fungal elements, making it almost impossible to decide whether an organism is hyaline or naturally pigmented. In case of phaeoid (dematiaceous) fungal infections, dark brown color of hyphae renders them relatively easy to be seen and special stains may not be strictly required and PAS staining of some sections is usually beneficial. The Masson-Fontana stain is specifically used for the demonstration of phaeoid fungi in tissues. In such a situation it is crucial to establish histopathological diagnosis of diseases caused by the phaeoid fungi. This is also pertinent to note that GMS stains all the fungal elements as black with greenish background, whereas Masson-Fontana stain only take up the phaeoid fungi as black in color with brownish background **(Fig. 5.9)**.

For a retrospective analysis, tissue sections previously stained by H&E, Giemsa and modified Gram procedures,

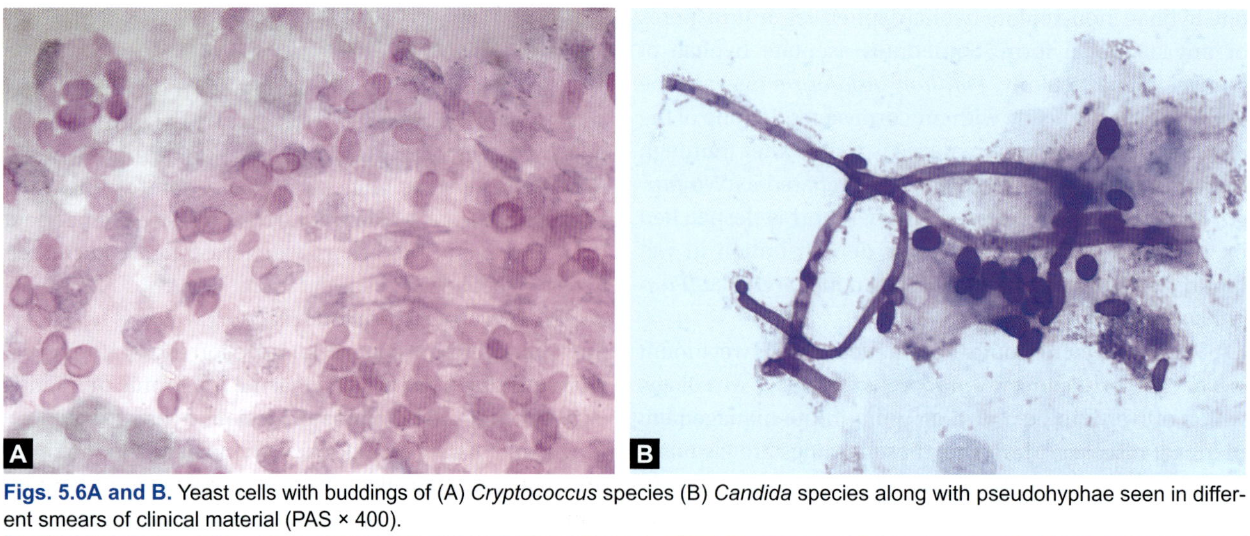

Figs. 5.6A and B. Yeast cells with buddings of (A) *Cryptococcus* species (B) *Candida* species along with pseudohyphae seen in different smears of clinical material (PAS × 400).

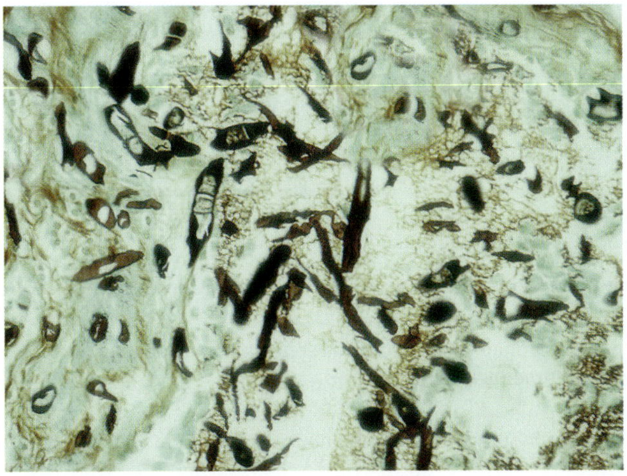

Fig. 5.7. Septate hyphae seen in *Trichosporon* species and may be confused with mucormycosis (GMS × 400).

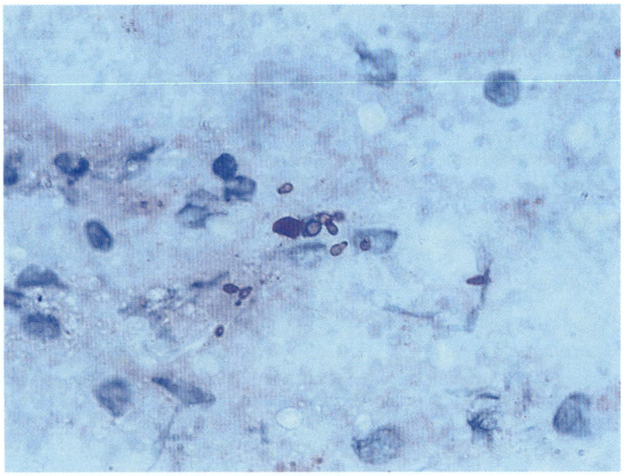

Fig. 5.8. Yeast cells with budding in *Cryptococcus* species seen in FNAC smear of clinical material (PAS × 400).

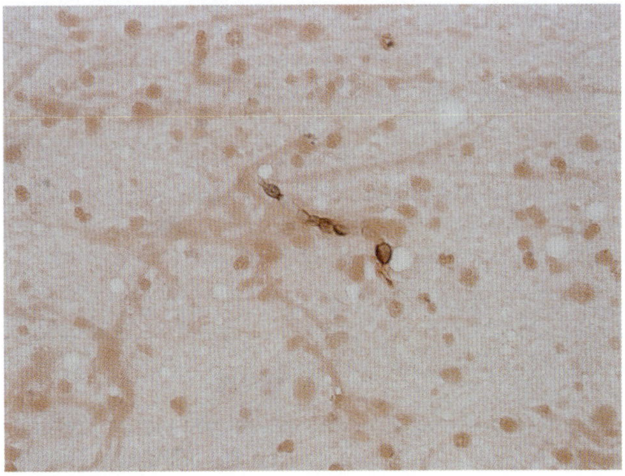

Fig. 5.9. Black colored septate hyphae with brown background in a patient of phaeohyphomycosis (MF Stain × 400).

can be decolonized in acid alcohol and then re-stained with specific reagents used for immunofluorescence, however, this is not possible with sections previously stained with GMS or PAS stains.

The significance of histopathological section is that it can detect the presence or absence of fungal elements along with its corresponding tissue reaction, if present. But its limitation is that further details of the involved fungus cannot be elucidated. For example, if there are septate hyphae with acute-angle branching, the descriptive report is always given as *'consistent with Aspergillus'* whereas similar findings may be seen in the infections caused by *Fusarium* species, *Scedosporium*/*Pseudallescheria* species and many other fungal agents. In such circumstances culture is the ideal modality to study in details about particular disease process and its causative fungus. Moreover, if histopathology shows neither fungal elements nor tissue reaction, there is likelihood of fungal isolate be merely a contaminant. Therefore, in combination both histopathology and fungal culture provide complete diagnostic lead to any mycotic infection. The typical appearance of different fungi in tissue is given in **Table 5.1**.

As such yeast or spherule forms of dimorphic fungi are seen in tissues but recently cases have encountered where hyphal forms have been reported. These are unusual findings hence care must be taken while making diagnosis in tissue sections of infections caused by such fungi.

(iii) Frozen-Section Biopsy: This has been observed that frozen section modality is adopted for making intra-operative diagnosis of a suspected malignancy. But if the same principal is applied to the suspected fungal infection, the exact etiology can be established, then and there, paving the way for specific antifungal therapy without wastage of time. There have been very good results seen in mucormycosis as well as fungal rhinosinusitis. The combined use of KOH wet mount, rapid Romanowsky and H&E stains may be diagnostically helpful. The frozen-section biopsy is a highly predictive tool for a rapid and effective diagnosis. It is very useful tool to guide the extent of surgical debridement and/or onset of antifungal therapy. However, permanent pathology section remains the gold standard for the fungal diagnosis.

Table 5.1. Summary of Histopathology of various Pathogenic Fungi and Spherule or Sporangia-Forming Organisms.

A. Yeast-like Appearance	
H.capsulatum	2-3 × 3-4 μm oval, narrow-neck budding, uni-nucleate yeast cells, intracellular with granulomatous tissue reaction.
B.dermatitidis	8-15 μm round multi-nucleate, broad-based yeast cells with pyogenic and granulomatous tissue reaction.
P.brasiliensis	2-30 μm multiple, budding, round cells with tiny pore between mother and daughter cell. The daughter cells are released when small in size.
S.schenckii	1-3 × 3-10 μm cigar-shaped cell or 2-10 μm round budding yeast cells with pyogenic and granulomatous tissue reaction.
C.albicans	3 × 5 μm oval, budding yeast cells with pseudohyphae, constrictions at septae and branching only at septations with neutrophilic and rarely granulomatous tissue reaction.
Cr.neoformans	4-6 μm round uni-nucleate cells with large capsule, narrow pore between mother and daughter cells which detach while small in size. The tissue reaction is variable often neutrophilic but may be granulomatous in immunocompetent patients.
Chromoblastomycosis	4-12 μm round or oval, brown, thick-walled sclerotic cells, often seen in clumps.
B. Mold Appearance	
Aspergillus species	2-5 μm wide hyphae, frequently septate, even diameter, Y-shaped acute-angle branching, with vascular invasion and necrosis. The other hyaline fungi like *Fusarium* and *Scedosporium/Pseudallescheria* species also have similar histopathological appearance.
Mucormycosis agents	4-5 μm wide hyphae, non-septate, uneven diameter, often branch at wide-angle, vascular invasion and necrosis. Differentiation should be made from pythiosis insidiosi.
C. Spherule-like Appearance	
Coccidioides species	5-60 μm thick walled, spherule with endospores with granulomatous tissue reaction.
Emmonsia species	50-500 μm and large thick-walled symmetrical, round to oval spherules, surrounded by fibrous granulomatous tissue reaction. Budding or endosporulation is never seen.
Prototheca species	8-20 μm sporangia containing 2-8 tightly packed sporangiospores giving 'spoke-wheel' appearance.
Rhinosporidium seeberi	200-300 μm sporangia having endospores of about 6-9 μm size with giant cells and lymphocytic reaction around the mature sporangia.

The use of intraoperative assessment of soft tissue disease for fungal etiology is a valuable tool in the management of fungal infections. The presence of fungal elements or necrosis/acute inflammation is used as criteria to make the diagnosis. The role of frozen sections is not to perfectly speciate the fungal pathogen but to describe the morphology and infectious process and provide a differential diagnosis of the candidate fungi. The importance of intraoperative culture in infectious cases cannot be understated and it is the responsibility of microbiologist/pathologists to inform surgeons that tissue is needed for culture. The frozen section evaluations are an important function of surgical pathology and can provide rapid and valuable information about a potential tissue diagnosis or other pathologic finding while patient is undergoing an operative procedure. This is a collaborative effort among the physicians keeping in view of the best interests of the patient. The tissue analysis under emergency circumstances may be difficult and the stakes are high hence requests for frozen section should be judiciously initiated and answered. This modality should be made compulsory in the hospital services.

(iv) Fluorescent-Antibody Staining: The fluorescent-antibody staining may be used to detect fungal antigen in clinical material such as pus, blood, CSF, tissue impression smears and in paraffin sections of formalin fixed tissue. But it is less satisfactory for sputum specimens. The main advantage of this technique lies in its ability to detect fungus when there are only few organisms present, as seen in pus from sporotrichosis. However, its use is limited by restricted availability of specific antisera.

(b) Fungal Culture

The isolation of fungi is not difficult as it apparently looks but with very few exceptions. The solid media are employed for fungal culture, as broths are not usually recommended except for fungal blood cultures where biphasic medium is used. The medium commonly employed is Emmons' modification of Sabouraud Dextrose Agar. The media may be supplemented with antibiotics, such as gentamicin and chloramphenicol to minimize bacterial contamination as well as cycloheximide to inhibit saprotrophic fungi. Cycloheximide should not be used for suspected specimen having *Cryptococcus*, *Talaromyces marneffei*, *Aspergillus* or *Scytalidium* species because these fungi are sensitive to this antibiotic. The special media may be used for isolation and to help rapid identification when identity of a particular pathogen is strongly suspected. For example, *Cr.neoformans* develops brown colonies on Bird Seed Agar. An extra set of media may also be inoculated without cycloheximide to rule out any inhibition of significant fungus, which may otherwise be taken as saprophyte.

The isolation of true pathogenic fungi is difficult and requires special media like BHIA, CHHA, with 10% CO_2 tension, inoculation in water of condensation, presence of sulfhydryl group in the medium and appropriate oxidation-reduction potential.

Many fungal pathogens have an optimum growth temperature below 37°C. Hence isolation from suspected superficial mycoses is usually attempted by incubating cultures at 30°C. For subcutaneous and systemic mycoses cultures are usually incubated at this lower temperature and at 37°C as well. With certain dimorphic pathogens such as *Histoplasma capsulatum* and *Blastomyces dermatitidis* mycelial form develops on the commonly used isolation media at 25°C and yeast at 37°C on enriched media, such as brain-heart infusion or blood agar. In the dimorphic fungi, for primary isolation from the clinical specimens, it is usually observed that mycelial form grows successfully at 25°C as compared to its counterpart yeast form at 37°C. However, yeast form is subsequently converted from its mycelial form and thereafter *vice versa* (M↔Y).

In addition to 25°C, 30°C and 37°C, fungal culture may be incubated at 45°C also because it helps in the growth of certain fungi as well as to differentiate them from others.

Fungi grow relatively slow and cultures should be kept incubated for at least four weeks and in some cases up to six weeks before being discarded as sterile. Usually, the positive results of culture are obtained within 7-10 days. In *Candida* and *Aspergillus* species, growth appear within 24-72 hours. Therefore, cultures should be examined for the expected growth, daily for the first week and twice a week for subsequent period. The subcultures are made from developing colonies of suspected pathogens if these are threatened by surrounding overgrowth or contamination by the saprotrophic fungi.

The specimens may be cultured on agar media in test tube slants or petri dish. The slants are preferred as compared to plates for isolation of dimorphic fungi as they are safer, require less space and are more resistant to desiccation during protracted incubation. All these fungi should always be handled in biosafety cabinet. If screw-cap containers are used then caps should not be fully tightened since lack of oxygen may inhibit fungal growth. The use of agar slants in screw-cap tubes sometimes may inhibit spore production by filamentous fungi. A sterile cotton stopper should replace screw cap once medium has been inoculated. The blood cultures should be inoculated into vented, biphasic blood culture bottles. The time frame for incubation of these cultures is 4-6 weeks. Sometimes, if smear is prepared from blood culture bottle, RBC may confuse with yeast cells therefore, LCB mount should be prepared from solid slant of blood culture bottle. The routine blood agar used in bacteriology sections should be incubated for longer period of time to provide coverage to the fungi as well *i.e.* discard after 48 hrs instead of 24 hrs.

The isolation rates from blood and bone marrow are higher when media are vented and aerated and by using biphasic medium. Moreover, chances of isolation are increased when multiple samples of blood are collected and larger volume is cultured. The lysis-centrifugation method is most dependable for isolation of fungi from blood but it is labor-intensive and not very cost-effective and hence usually not being used in routine diagnostic mycology laboratories. However, recent developments in blood culture media and automated systems have improved the isolation of fungi from blood as well as bone-marrow.

The biggest tragedy with fungal culture is that when any lesion is being suspected as malignancy by the clinician and biopsy material is sent to Pathology Department only for HPE in formalin and its Resident comes to Microbiology to know what has been grown. When the sample is not received by the Microbiology there is no question of fungal growth and the Pathology material cannot be utilized being dipped in formalin. Therefore, it is mandatory to send the sample for fungal culture in the saline to the Mycology/Microbiology even if primary suspicion is a malignancy. This has been observed that 5-10% of such samples turn out to be fungal diseases and not the malignancies.

The exoantigen tests are useful in early diagnosis of many fungi during there is course of incubation (*See* **Appendix C**).

(i) Interpretation of Fungal Cultures: The significance of fungal isolate depends on its source and identity. The isolation of an established pathogen such as *H.capsulatum* or *Coccidioides* species from any specimen is generally regarded as an evidence of infection. In case of commensal or opportunistic fungi that are otherwise considered as contaminants *e.g. Candida* and *Aspergillus*, following points should be considered:

- Growth of same strain in all culture tubes,
- Repeated isolation of same strain in multiple specimens,
- Immune status of patients and
- Serological evidence to confirm significance of isolate.

This has been observed that sometimes the patients are given antifungal agents, under the wrong impression about the diagnosis or a contaminant is grown in culture, which mislead the clinician to start the antifungal agent.

The isolation of these fungi from normally sterile sites like blood, CSF, pleural or peritoneal fluids usually provide reliable evidence of infection but when they are recovered from clinical material such as sputum, urine or pus, results must be interpreted with due care. Importance should be given to quantity of fungus isolated and further investigations are undertaken. Many species of fungi may cause disease in severely immunocompromised individuals and no isolate from this type of patient should be lightly discarded, presuming it to be a contaminant.

Most of fungal agents are not fastidious in their nutritional requirements and easily grow even on media used for bacterial isolation. However, failure to recover organism does not negate diagnosis as this may be due to insufficient specimen collection, failure to choose appropriate site, delayed transportation, incorrect isolation procedures or inadequate temperature and incubation periods.

In the diagnostic mycology, it important to report about the fungal isolate but equally important is what not to report as there are so many contaminants encountered daily. The over-reporting will mislead and confuse the treating clinician.

(ii) Identification of Fungal Isolates: It is pertinent to identify fungal isolate to the genus level and if possible to the species or variety level in diagnostic mycology laboratory for an ideal management of patient. For further identification or confirmation, help of Mycology Reference Laboratory may also be taken. The identification of mycelial fungus involves detailed study of its macroscopic and microscopic morphological features. Since the morphology can vary depending on medium and growth temperature, it may be necessary to study isolates on wide range of media at several temperatures.

After noting macroscopic features of colony, such as color and texture of growth, slide mounts should be made in LCB or PHOL stain to study morphological details in microscopic examination. It may be necessary to make several preparations, perhaps over a period, to see typical characteristic structures on which identification can be based. If delicate sporing structures are disrupted while making mounts, special techniques such as **Fungal Slide Cultures** should be used. It allows fungal growth to be examined in its original form without disturbing the spore arrangement. This is the simplest procedure and most revealing technique used in morphological identification of fungi rather it is the **Key Technique** in mycology.

The biochemical rather than morphological criteria are used for identification of yeasts. Important biochemical reaction are, ability to assimilate certain carbon and nitrogen compounds and to ferment sugars. The morphological features include cell shape, size, presence or absence of capsules, ability to produce hyphae or pseudohyphae, asexual and sexual modes of reproduction.

Identification of yeasts can involve use of wide range of carbon compounds and certain other investigative procedures. However, identification of medically important yeasts usually can be satisfactorily achieved with more limited range of tests. For most frequently encountered yeast pathogens several special tests have been evolved. In addition, two tests are routinely available for rapid identification of *C.albicans* and *C.dubliniensis* namely, production of:

- Germ tubes in serum after 1-2 hours incubation at 37°C (*See* **Fig. 20.10**)
- Chlamydospores under microaerophilic conditions on special media such as cornmeal agar or rice starch agar (*See* **Fig. 20.11**).

Media containing bird seed (*Guizotia abyssinica*) extract or caffeic acid/ferric citrate is also now widely used for rapid identification of *Cr.neoformans* that develops as brown colonies on these substrates.

The commercial kits for identification of medically important yeasts by their assimilation patterns are now commonly available. They are convenient to use and give results that are comparable in accuracy to those obtained by above-mentioned conventional procedures.

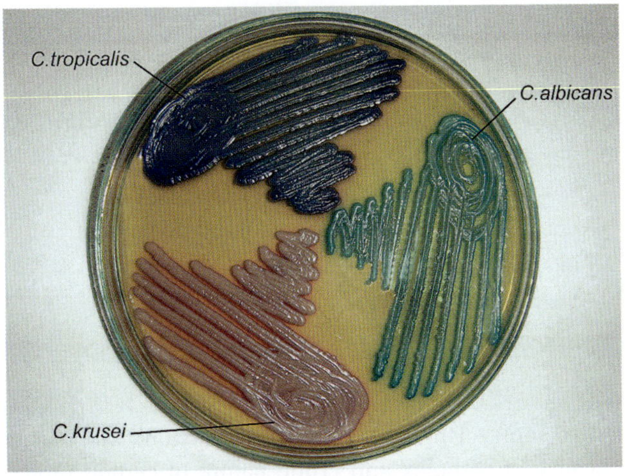

Fig. 5.10. CHROMagar Candida with colored growth of *Candida albicans* (Green), *C.krusei* (Pink) *C.tropicalis* (Blue).

The CHROMagar Candida is relatively new medium that distinguishes different *Candida* species by the color produced as a result of biochemical reactions. This is a chromogenic differential isolation medium that facilitates presumptive differentiation of clinically important yeast-like organisms. This is one of the media, which can be used for simultaneous isolation and presumptive identification of various *Candida* species like *C.albicans, C.krusei, C.tropicalis, C.glabrata, C.parapsilosis* and *C.dubliniensis* and other yeast-like colonies of *Trichosporon* and pseudofungal *Prototheca* species. This is based on direct detection of specific enzymatic activities by adding certain substrates of fluorochromes to media. It shortens time for presumptive identification of organisms on the basis of colony morphology and it also allows for easier detection of multiple yeasts present in the clinical specimens. This is pertinent to use a good quality medium otherwise results may be confusing and misleading also **(Fig. 5.10)**.

In addition, for identification of yeasts, API 20C AUX may also be used, which give results within 48-72-hours.

(c) Nonculture-based Methods

There is also essential requirement of nonculture-based methods for rapid and accurate diagnosis of life-threatening fungal infections. These include detection of circulating antigen, fungal constitutive macromolecules, fungus-specific metabolites and fungus-specific nucleic acid sequences. The last two methods are major advances in the diagnosis of fungal infections. Significant advancements have been made in the diagnosis and monitoring of invasive candidiasis. The detection of galactomannan polysaccharide antigen in serum and urine of patients with invasive aspergillosis can be done by ELISA.

The tests to detect circulating antigens of *Candida* species, in urine and serum as indicator of invasive disease are in various stages of development for use in clinical laboratory. The detectable *Candida*-specific products or antigens in clinical specimens includes D-arabinitol, cell-wall mannoprotein (CWMP), enolase, aspartyl proteinase and heat labile antigen. Unfortunately, none of the commercially available tests for detection of these antigens provide significant predictive value for definitive diagnosis of invasive candidiasis to recommend their routine use in clinical mycology laboratories.

Another potential antigen in diagnostic testing is (1, 3)-beta-D-glucan, which is major fungal cell wall constituent. It is observed that deep mycoses including aspergillosis and candidiasis are associated with high plasma levels of (1, 3)-beta-D-glucan, which is very low in normal individuals.

(i) G-test: This test detects circulating (1, 3)-beta-D-glucan with use of modification of limulus assay used for endotoxin and has sensitivity of ~20 pg/ml. Prospective studies of utility of G-test have shown that it can detect (1, 3)-beta-D-glucan in context of confirmed invasive mycoses, including invasive aspergillosis but it does not distinguish between species of *Candida, Aspergillus* or other medically significant fungi. It is also useful for the detection of *Pneumocystis* infections. It is not useful for detection of *Cryptococcus* species and mucormycetes.

(ii) Tests for Metabolites: An alternative approach to the diagnosis of disseminated fungal infections is detection of specific fungal metabolites in body fluids by gas liquid chromatography (GLC). This procedure is mainly evaluated in aspergillosis and candidiasis with good results.

Tests are now being introduced for detection of serious systemic infections in severely immunocompromised patients such as those patients who are undergoing bone marrow transplantation. Because of low levels of circulating antigen present in infected individuals, tests need to be extremely sensitive. Other tests like CFT, CCIEP, co-agglutination, ELISA, RIA and immunogold silver staining can be used for antigen detection.

(d) Immunology and Serology

The immunological and serological tests are also used in establishing the diagnosis of fungal diseases particularly to identify patients whose fungal cultures have been non-productive. Serology is helpful in interpreting clinical significance of positive cultures for *Aspergillus* because such fungi may be merely laboratory contaminants in a particular patient. Moreover, when new isolate is identified in a patient, serology may be deciding factor if antibodies are demonstrated against antigen of that particular isolate.

The specific immune response that results from exposure to cell wall, cytoplasmic or extracellular fungal antigen during infection can be used for diagnosis. By monitoring this response, prognosis of disease and outcome of therapy can be assessed. Recently galactomannan and glucan, two promising antigens, have been found to be useful not only for an early detection of fungal infections but therapeutic monitoring as well.

The skin tests and *in vitro* lymphocyte stimulation tests are used to detect cell-mediated changes and serological tests to detect humoral antibody production and/or presence of fungal antigens in body fluids. The type of immune response depends on fungus involved, form of disease and immunological status of the host.

Galactomannan gives cross-reactivity with *Exserohilum* species. The galactomannan index. If combined with lateral flow assay with any other test then it is the best one. False positive galactomannan Platelia due to piperacillin-tazobactam. Procalcitonin is not able to differentiate from bacterial and fungal sepsis.

1. Serological Tests

The serological tests are done either to demonstrate antigen or antibody in serum or body fluids of suspected fungal infections.

(i) Antibody Detection: Serological tests for antibody detection are of value for the diagnosis of systemic and subcutaneous mycoses. These may also be used to assess prognosis of fungal disease and response to antifungal therapy. For some of fungal diseases these tests are reliable but for others they provide only presumptive evidence of infection and results must be correlated with other clinical and laboratory findings. In certain cases there is need to improve specificity and some patients may have antibodies in absence of infection because they have been previously exposed to fungus either as saprotrophic or as commensal and hence test may lack sensitivity. Some of the patients may have little or no detectable antibody response because their infection is at an early stage or peculiarity of an underlying disease or due to chemotherapy.

Table 5.2. Serological Tests used in Medical Mycology.

• Agglutination:	Whole cell agglutination (WCA) Latex particle agglutination (LPA) Passive hemagglutination (PHA)
• Immunodiffusion (ID)	
• Counterimmunoelectrophoresis (CIE)	
• Complement fixation (CF)	
• Indirect fluorescent antibody (IFA)	
• Enzyme-linked immunosorbent assay (ELISA)	
• Radioimmunoassay: Solid phase, competitive RIA	

In some of the diseases, cross-reactions may occur because of complex and crude nature of certain fungal antigenic extract that cause difficulty with interpretation of results. Unfortunately, most unsatisfactory tests are those for opportunistic mycoses which are most difficult to diagnose by other means. An important feature in interpreting results of serology testing is to compare antibody titers between acute and convalescent serum specimens. For these infections, sequential tests with paired sera are more helpful than single investigation since in general, high or rapidly rising antibody or antigen titer suggests active fungal infection.

The techniques used for detection of antibodies and sometimes antigen, are essentially those used in routine medical microbiology to detect agglutinating, precipitating and complement-fixing antibodies of IgG, IgM and IgA subclasses of immunoglobulins **(Table 5.2)**. The predominant class of antibody produced depends on host, causative fungus and type of infection whereas technique decides type of antibody to be detected.

Until recently, most widely used test was immunodiffusion (ID) in agarose for detection of fungal precipitins and it was used as primary screening test even when more sensitive tests were available **(Fig. 5.11)**. However, CIE has now gained almost universal acceptance as more rapid and sensitive method of showing precipitins to fungal agents. It also offers more convenient way of determining precipitin titers. More sensitive tests such as ELISA and RIA are now being introduced for antibody detection but sometimes, as in candidiasis; increased sensitivity may not necessarily be of an advantage. Most of these tests are commercially available as serodiagnostic kits.

(ii) Antigen Detection: The serological tests for detection of antibodies are of limited value in early stages of infection, in patients with impaired immunity or immune response is not sufficient to raise significant level of antibodies. In such cases tests for detection of fungal antigens are of immense importance. The diagnosis of cryptococcosis using latex particle agglutination (LPA) test for detection of capsular antigen in cerebrospinal fluid and other body fluids, is well established. LPA can also be used in aspergillosis and candidiasis.

(iii) Immunohistochemistry: The technique of immunohistochemistry has proved to be important tool for accurate diagnosis of number of important mycoses in humans and animals, such as aspergillosis, candidiasis, cryptococcosis, blastomycosis, coccidioidomycosis, histoplasmosis, paracoccidioidomycosis, fusariosis, scedosporiosis/pseudallescheriasis, sporotrichosis, trichosporonosis, talaromycosis and mucormycosis. These techniques are also applicable to pneumocystosis and to non-mycotic infections caused by pseudofungal organisms *i.e.* algae such as protothecosis.

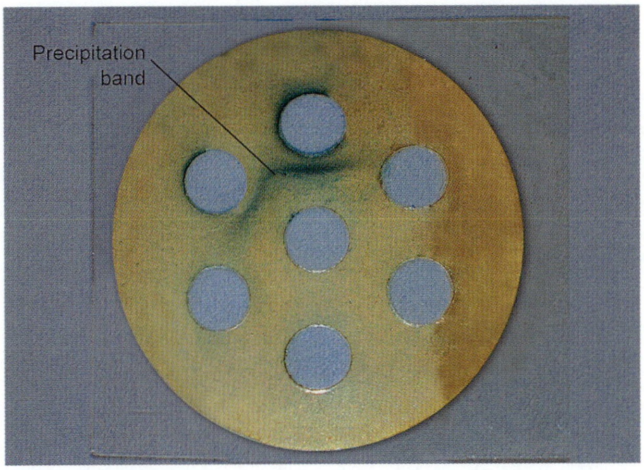

Fig. 5.11. Permanent mount of immunodiffusion with amido black solution showing the precipitation bands.

Apart from specificity of immunohistochemistry, application of fluorochromes is highly effective for localization of typical or atypical fungal elements in lesions when only few organisms are present. Occasionally, dual etiology of fungal infections may be suspected on the basis of morphological study and dual staining techniques have capacity for resolving this question by simultaneous and differential staining of two fungal species present in a tissue specimen. **Figure 5.12A** shows tissue section of aspergillosis and **Figure 5.12B** as mucormycosis.

2. Tests for Cell-mediated Immunity

Different tests for detection of cell-mediated immunity are performed *in vivo* and *in vitro*. These are given below:

(i) Skin Tests: Individuals who become infected with *Histoplasma* or *Coccidioides* develop delayed-type hypersensitivity (DTH) reaction within 1-14 days, which may persist for many years. This sensitization can be shown by skin tests in which induration and erythema occur 24-72 hours following intradermal inoculation of appropriate fungal antigen. In endemic areas most of the population may be skin test positive and consequently skin testing is more useful for epidemiological work than for diagnosis since positive reaction is more likely to suggest

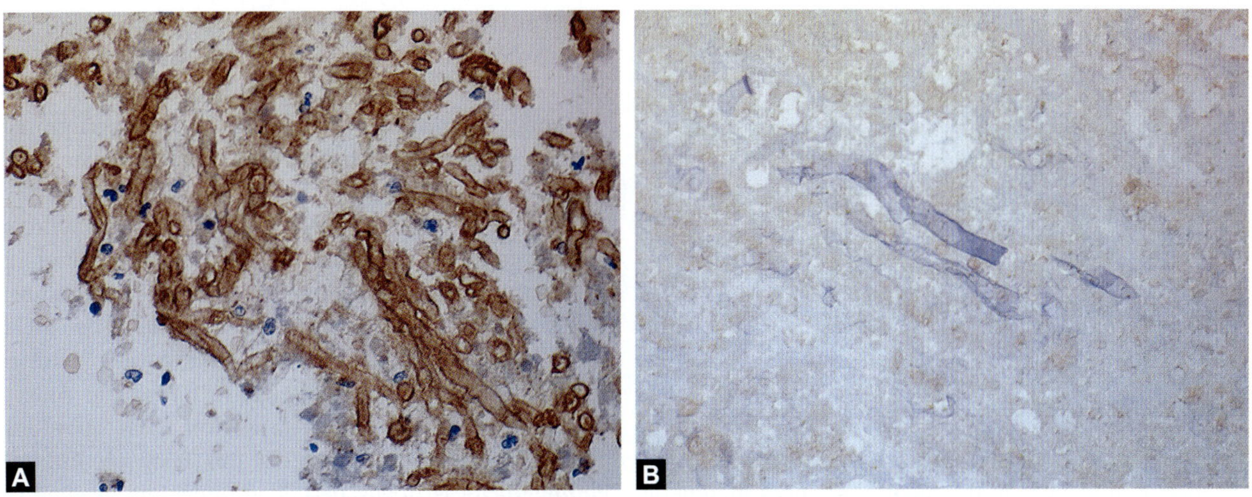

Figs. 5.12A and B. Immunohistochemistry of tissue sections showing (A) aspergillosis and (B) mucormycosis (IHC × 400).

Table 5.3. Various Skin Tests and Fungal Antigens Used.

Fungal Diseases	*Antigens*
• Histoplasmosis	Histoplasmin
• Coccidioidomycosis	Coccidioidin
• Blastomycosis	Blastomycin
• Dermatophytoses	Trichophytin
• Sporotrichosis	Sporotrichin
• Candidiasis	Candidin

previous infection rather than current disease. Their diagnostic usefulness is also limited by cross-reactions that may occur because of common antigens shared by major fungal pathogens. Therefore, it is preferable to test simultaneously with antigens from several fungi so that relative strength of reaction of each can be compared. The skin tests are used for following purposes:

- For establishing etiological diagnosis.
- For conducting epidemiological surveys.
- Immunological classification of subjects like atopic and non-atopic groups.
- To find out immunological status of patients as in immunodeficiency diseases.

The positive skin tests usually revert to negative in cases with severe or disseminated disease and therefore, they may be used as prognostic indicators. The negative skin test results may also be obtained when there is no infection, disease process is at an early stage and in those with defective cellular immunity. In allergic respiratory diseases, skin tests are very useful for establishing diagnosis. In these cases Type-I (immediate) and/or Type-III (arthus) reactions, associated with reaginic IgE antibodies occur.

The fungal infections where skin tests are commonly found useful for diagnosis as well as epidemiological surveys, are summarized in **Table 5.3**.

***(ii) In vitro* tests:** *In vitro* tests to show hypersensitivity, such as those for macrophage migration inhibition and blast formation of sensitized lymphocytes in presence of specific fungal antigens, are being increasingly used for diagnosis of fungal disease.

Radiodiagnosis

The radiological techniques have revolutionized the diagnostic modalities in the medical sciences in general and mycology in particular. A number of procedures, non-invasive as well as invasive, can clinch the fungal diagnosis at a very early stage of the diseases. The x-rays, ultrasonography, CT, MRI are some of such techniques. In addition USG or CT-guided procedures in combination prove to be very fruitful and revealing and material aspirated is further investigated to reach the diagnosis. However, radiology cannot prove the nature of fungal infection, which is supplemented with other laboratory techniques and parameters to clinch the final diagnosis.

Animals as Experimental Models

In medical mycology various animal models are occasionally used in establishing diagnosis, pathogenicity testing as well as study antifungals. The aim of using animal model is precisely to study pathogenicity of fungal strain and immune status, to evaluate therapeutic agents by screening methods and to determine *in vivo* potency of a compound under experimental conditions in order to obtain information about its antifungal value. In addition, invertebrate hosts are also used in some of the fungal species. The data obtained through these experiments is used to direct synthesis of new antifungal agents via substitution or modification of active structures, enabling new candidates to be selected for clinical trials. These models therefore must ensure that extrapolation of infection and results after prophylactic or therapeutic modality to natural and spontaneous infection in man and animals, is possible.

The animal models of fungal infections are a key tool in the advancement of human knowledge in the field of medical mycology and will remain as such in the future as well. There are different types of animal models of fungal infection, which have been developed, with murine models the most frequently used, for studies of pathogenesis, virulence, immunology, diagnosis, and therapy. The ability to control numerous variables in performing the model allows us to mimic human disease states and quantitatively monitor the course of the disease.

However, there is no single model which can address all questions and different animal species or different routes of infection can show somewhat different results. Therefore, the choice of which animal model to use must be made carefully, addressing issues of the type of human disease to mimic, the parameters to follow and collection of the appropriate data to answer those questions being asked. In every Chapter of this Textbook, **Animal Pathogenicity** is briefly described under the heading of Laboratory Diagnosis. The focus is on the most clinically important diseases affecting humans. Hence overall the animal models of fungal infection will continue to be valuable tools in addressing questions concerning fungal infections and contribute to deeper understanding of how these infections occur, progress and can be controlled and/or eliminated.

Moreover, the morbidity, mortality and economic burden associated with fungal infections, together with the emergence of fungal strains resistant to current antimicrobial agents, necessitate broadening the understanding of fungal pathogenesis and discovering new agents to treat these infections. Hence using **invertebrate hosts**, especially the nematode *Caenorhabditis elegans* and the model insects *Drosophila melanogaster* and *Galleria mellonella*, could help achieve these goals. The evolutionary conservation of several aspects of the innate immune response between invertebrates and mammals makes the use of these simple hosts an effective and fast screening method for identifying fungal virulence factors and testing potential antifungal compounds.

In India, the animal experimentation is looked after by the Committee for the Control and Supervision of Experiments on Animals (CPCSEA), under the Prevention of Cruelty to Animals Act, 1960. The Committee operates under the aegis of the Ministry of Environment and Forests and is rather duty bound to take all such measures as may be necessary to ensure that animals are not subjected to unnecessary pain or suffering before, during or after the performance of experiments on them. All establishments engaged in research and education involving animals, are required to comply with the various guidelines, norms and stipulations set out by the CPCSEA. The Establishment of Medical College Regulations, 2013 (Amendment) bans the use of vivisection in medical education thereby bans the use of live animal experiments in medical education.

Recently *Galleria mellonella* and murine infection models for the study of *Trichosporon* infections have been developed. The utility of these experimental models is demonstrated through the assessment of virulence and antifungal efficacy for clinical isolates of *Trichosporon asahii*, *T.asteroides* and *T.inkin*.

The uses of animal in medical mycology and various inoculation techniques have been dealt in all Chapters of the Textbook as well as in **Appendix C**.

C. BASED ON MOLECULAR METHODS

In recent times invasive fungal diseases (IFD) are being more frequently encountered in clinical practice. In order to effectively eliminate these infections, early diagnosis and species identification are of paramount importance. Traditional diagnostic methods, such as histopathology and culture, that are considered the gold standards, have low sensitivity and high turnaround time. The molecular techniques are now being developed for detection of fungal pathogens in clinical specimens including blood, serum, CSF and urine.

The molecular technique based diagnosis is helpful in two ways *i.e.* detection of fungi in the clinical material and secondly for the accurate identification of the pathogen obtained.

Recently polymerase chain reaction (PCR) has been used to detect segment of fungus specific DNA coding for cytochrome P450 L_1A_1 in clinical specimens, chitin synthase gene, 18S rRNA gene and aspartic protease gene for detection of *Candida* species.

This method can detect presence of as few as 10 organisms in 100 ml volume of variety of clinical specimens including urine, sputum and blood. The specificity of test is almost 100%, however, sensitivity in clinical specimen, is to be decided. This technique has been successfully used in candidiasis, cryptococcosis, invasive aspergillosis and pneumocystosis. The difficulty of PCR lies with false positivity. Panfungal PCR is also being tried involving large number of fungal agents. A multiplex protocol has been developed by Chakrabarti and his colleagues for early and rapid diagnosis of invasive candidiasis, aspergillosis and mucormycosis.

The most significant changes expected in coming years, to improve reliability of current PCR assays, is introduction of real-time PCR and automated DNA extraction. As detection of PCR products and data analysis are part of real-time instruments, there is no post-amplification handling. Thus, risk of contamination of environment with amplicons is sharply reduced. The likelihood of false positive results also is low and PCR assays can be routinely implemented. The future development of novel chemistries and improved real-time apparatus should promote use of multiplex real-time PCR assays. A combination with microarrays technology may also enhance rapidity of fungal identification.

In addition, DNA barcoding in fungi has emerged as a robust and standardized approach to species identification.

Molecular methods are often used to determine the relatedness among microorganisms (*i.e.* strain typing) which serves as an important tool for hospital microbiologists and public health epidemiologists. This may help to determine the point source during an outbreak. The phenotypic methods employed previously are less discriminatory as compared to genotypic methods. These Molecular methods can be non-amplification based typing methods such as Pulsed Field Gel Electrophoresis (PFGE). The amplification based typing methods are such as PCR-RFLP or rep-PCR. These methods are increasingly being used due to their given utility, ease of use and rapid turn around time.

The DNA sequencing is generally considered as 'gold standard' method of correct and accurate identification of microorganisms and is more reliable than hybridization techniques, which are based on single detection and discrimination of closely related species. The hybridization techniques are susceptible to cross-hybridization of probes. More interestingly, DNA sequencing provides nucleic acid information, which is core of every organism and it also assists to detect and analyze mutations and other characteristics of each species and its relation with other species.

An increasing number of molecular techniques for the diagnosis of fungal infections have been developed in the last few years, due to the growing prevalence of mycoses and the length of time required for diagnosis when classical microbiological methods are used. These methods are designed to resolve the following aspects of mycological diagnosis:

(a) Identification of fungi to species level by means of sequencing relevant taxonomic targets;
(b) early clinical diagnosis of invasive fungal diseases;
(c) detection of molecular mechanisms of resistance to antifungal agents; and
(d) molecular typing of fungi.

However, currently, these methods are restricted to highly developed laboratories. But some of these techniques will probably be available in daily clinical practice in near future. This emphasizes the need to use existing molecular techniques and to further develop newer means of detecting fungal infectious agents.

1. Hybridization Methods

The hybridization methods are based on detection of fungal pathogen without the use of nucleic acid amplification. They rely on the amplification of a signal generated, usually light or color as a result of successful hybridization of a nucleic acid probe with the target nucleic acid molecule *e.g.* Fluorescence *in situ* hybridization (FISH). It is a technique that uses fluorescent probes to identify target areas on the genomes of microbial pathogens in human samples, which can then be detected by fluorescent microscopy. This method has been used as an adjunct to culture and has high accuracy for rapid identification and differentiation of clinically important yeast in positive blood culture bottles.

2. Amplification Methods

Polymerase Chain Reaction (PCR) is one of the oldest and most widely used molecular methods in fungal diagnostics. PCR and its many modifications enable identification of fungal pathogens within human specimens, define the species and detect antimicrobial resistance markers. Their simplicity, ease of use, and short turnaround time are among their most important advantages over traditional techniques. Most PCR methods studied so far use primers to amplify sequences within the rRNA genes of fungal pathogens. Within the rRNA genes selected for amplification, there are multiple different options, such as the 18S ribosomal DNA (rDNA), the 28S rDNA, and the 5.8S rDNA, as well as internal transcriber regions between these genes. The less commonly used site for amplification are mitochondrial DNA of the fungus. Many different PCR assays have been developed over the years for invasive fungal infections. However, clinical reports of their sensitivities and specificities range considerably, from 43 to 100% and 64 to 100%, respectively.

Nevertheless, a major drawback of all traditional PCR techniques initially developed as potential fungal diagnostic tests is that they do not quantify the amount of amplified DNA. Therefore, there is no reliable way to identify the microbial burden within the human body. When it comes to IFDs, this becomes very significant, as fungi are frequent colonizers of human surfaces and this makes it impossible to determine if the identified fungal DNA is the result of the colonization or it represents an active infection.

A solution to this problem was given by the development of real-time PCR techniques. As the name suggests, real-time PCR is able to quantify the amount of amplified DNA in real time. As a result, real-time PCR techniques have largely replaced conventional PCR methods in clinical laboratories. Despite the great potential of PCR methods, several technical issues impede their use in routine. More specifically, fungal organisms and especially molds, have strong cell walls that are difficult to lyse and require complex methods for DNA isolation like enzymatic digestion processes that often rely on use of toxic chemicals, such as phenol-chloroform, mechanical disruption using glass beads and sonication. In an effort to overcome this barrier, automated extraction methods have been developed that are able to decrease the time for sample processing and lessen the possibility of errors. Another problem associated with fungal PCR is the potential for contamination. Fungi are ubiquitous in the environment and can easily contaminate surfaces and materials used in all steps of fungal PCR, including commercially available reagents and apparatus. Therefore, careful precautions and highly experienced personnel are necessary to avoid false-positive findings associated with contaminants. Another challenge is the choice of the best sample to evaluate the new tests. For example, the significance of *Aspergillus* sp. isolation from sputum samples is difficult to ascertain for critically ill patients, as it can be hard to differentiate between colonization and chronic infection. Additionally, without international standards, it is difficult to assess the agreement of quantitative data from different tests and thus to determine the clinical significance of various levels of fungal DNA. Also for many fungal infections, highly sensitive and specific serological assays are already available, thus lowering the need for molecular techniques. Due to the aforementioned issues, no single test has yet provided enough evidence of its accuracy to be incorporated into guidelines and thus PCR is not yet widely used in the diagnosis of fungal infections. However, for certain infections PCR has been studied and found useful as a potential method for diagnosis such as *Pneumocystis jirovecii* pneumonia (PCP), mucormycosis and even rarer fungal infections, such as coccidioidomycosis and scedosporiosis. Especially in the case of PCP, the PCR seems to be an excellent alternative to traditional methods. Some of the modifications of PCR employed for diagnosis of fungal infections are:

(a) Broad Range PCR: Primers designed to detect all organisms in group of interest while attempting to exclude as many organisms as possible that are not in this group. Detects organisms that are phylogenetically related. The main advantage is that any member of large group may be detected in a single reaction. However, the broad range can also detect same segment of DNA from many other G-C rich fungi that are closely related to the targeted group.

(b) Nested PCR: This assay is designed to increase the sensitivity and specificity of the assay. It involves the sequential use of two primer sets against same target and two rounds of PCR. The amplicon obtained in first amplification is then used as a target sequence for 2nd amplification using primers internal to those of 1st amplicon. This greatly increases the sensitivity. It also confers specificity as production of 2nd amplicon requires positive reaction of 1st amplicon, which automatically verifies the accuracy of 1st amplicon.

(c) Multiplex PCR: Use of primers specifically designed to amplify a region that is conserved among different fungal genera. The identification method is slightly more complex and is based either on sequencing of the amplified fragments of DNA or on the design of probes that bind to amplified fragments and have different melting temperatures, so they can be detected by melting curve analysis. This method can be combined with either standard, nested or real-time PCR and depending on the primer and probe design, it can detect either some or all fungal species. Multiplex PCR has also been tested as a method to detect fungal species in whole blood, serum or BAL fluid samples from patients at high risk for IFD. The results are variable but most studies report superior sensitivities and specificities of ~80 percent.

(d) Nucleic acid Sequence-based Amplification: The nucleic acid sequence-based amplification (NASBA) is a method very similar to PCR but differs in the sense that it amplifies mRNA by using an RNA polymerase instead of DNA, and it is isothermal. Arguably, its ability to detect mRNA gives it the advantage of detecting active disease instead of latent or previous infection and its isothermal nature, coupled with the fact that RNA is less stable than DNA, could decrease the chance of contamination.

(e) Fluorescence Resonance Energy Transfer: Two fluorophores, donor and acceptor are necessary to achieve fluorescence resonance energy transfer (FRET). The absorption spectrum of acceptor fluorophore should lie within the emission spectrum of donor fluorophore. The distance between the two fluorophores is a critical value. In most of the donor and acceptor fluorophore groups distance should be ≤ 10Å. Since the absorption spectrum of acceptor fluorophore falls within the emission spectrum of donor fluorophore, energy is transferred from donor to acceptor molecules and no fluorescence is generated. When the distance between the two fluorophores increases, energy cannot be transferred from donor to acceptor molecule and instead is released in solution in form of fluorescence which can be detected by various real time PCR machines. This technique can be used to identify various fungi.

(f) TaqMan: A fluorescent dye, typically FAM, is attached to 5′ end of the probe and a quencher, historically TAMRA, is attached at the 3′ end. As long as the two molecules (reporter and quencher) are maintained in close proximity, the fluorescence form the reporter is quenched and no fluorescence is detected at the reporter's dyes emission wavelength. TaqMan probes use a FRET quenching mechanism where quenching can occur over a fairly long distance so that as long as the quencher is on the same oligonucleotide as the fluorophore, quenching will occur.

(g) Molecular Beacons: They don't fluoresce when they are free in solution. However, when they hybridize to a nucleic acid strand containing a target sequence they undergo a conformational change that enables them to fluoresce brightly.

3. Sequencing-based Methods

The following methods are described under this heading:

(a) Sanger's Sequencing

The Sanger's sequencing is a technique for DNA sequencing, based upon the selective incorporation of chain-terminating dideoxynucleotides (ddNTPs) by DNA polymerase during *in vitro* DNA replication. It is also known as the chain termination method. It was developed by Frederick Sanger *et al* in 1977. The Sanger's method of DNA sequencing and more recently, pyrosequencing are becoming incorporated as routine assays in many modern molecular laboratories. These techniques are commonly used to identify wide-ranging genetic variations in human DNA and to identify bacteria, mycobacteria and fungi, particularly infectious agents that are difficult to identify using phenotypic methods or that are slow growing. Direct sequencing of DNA has the advantage of identifying the genetic variation in the context of flanking sequences, which offers the most reliable confirmation. When PCR-based assays are not available or inconclusive and need confirmation, DNA sequencing is often necessary.

(b) Pyrosequencing

This is a method of DNA sequencing *i.e.* determining the order of nucleotides in DNA, based on the "sequencing by synthesis" principle. It differs from Sanger sequencing, relying on the detection of pyrophosphate release on nucleotide incorporation, rather than chain termination with dideoxynucleotides. The technique was developed by Mostafa Ronaghi at the Royal Institute of Technology in Stockholm in 1996.

(c) Next-Generation Sequencing

The next-generation sequencing (NGS) is also known as high throughput sequencing. It is faster than the Sanger sequencing to analyze large genomes. The possibility that some of the patients diagnosed with Alzheimer's disease with disseminated fungal infection have been recently advanced by the demonstration of fungal proteins and DNA in nervous tissue from them. The NGS is used to identify fungal species present in the central nervous system of Alzheimer's disease patients. Initially, DNA was extracted from frozen tissue from different CNS regions of AD patient and the fungi in each region were identified by next-generation sequencing.

(d) Ultra Deep Sequencing

The ultra deep sequencing (UDS) appears to be a good method to improve laboratory identification of involved pathogens and subsequently adapt their therapeutic treatment. The UDS-based diagnostic approaches are ready to integrate conventional diagnostic testing to improve documentation of infectious disease thereby therapeutic management of patients. This approach has capacity to detect fungal pathogen in polymicrobial samples as well. The species of genus *Basidiobolus* was identified on the basis of UDS as *B.meristosporus*, new causative agent of basidiobolomycosis.

(e) DNA Bar Coding

The DNA barcoding first came to the attention of the scientific community in 2003 when Paul Hebert's research group at the University of Guelph, Ontario (Canada) published a paper titled "Biological identifications through DNA barcodes". The authors proposed a new system of species identification and discovery using a short section of DNA from a standardized region of the genome. That DNA sequence can be used to identify different species, in the same way a supermarket scanner uses the familiar black stripes of the UPC barcode to identify the purchases made there.

D. RECENTLY DEVELOPED TECHNIQUES

(a) AccuProbe

The AccuProbe culture identification products are tools for the identification of fungal, mycobacterial and bacterial pathogens, with sensitivities and specificities approaching hundred percent in most of the cases. These products allow for the detection of target organisms from primary cultures, eliminating the additional labor of purifying secondary cultures.

There are many commercially available research probes (DNA) for rapid identification of fungi like *Candida albicans, Cryptococcus neoformans, H.capsulatum, Blastomyces dermatitidis* and *Coccidioides immitis.* (AccuProbe System, Gen-Probe Inc., San Diego, California).

(b) PNA FISH

In the PNA FISH (Fluorescence In Situ Hybridization), FISH is cytogenetic technique to detect and localize the presence or absence of specific DNA sequences on chromosomes. Because of its high affinity and specificity to target DNA, PNA (Peptide Nucleic Acid) probes are ideal tools for FISH. The benefits are specific target binding, short hybridization time, low background, good reproducibility, and superior stability of the reagent. PNA FISH probes are also efficient at penetrating the tissues due to the small size and also no need of denaturation of the probe itself since it is a short single stranded oligonucleotide. The PNA FISH is used for identification of *Candida* species.

(c) MALDI-TOF Mass Spectrometry

There is another approach to identify fungi using matrix-assisted laser desorption/ionization time-of-flight mass spectrometry (MALDI-TOF MS). This is now being increasingly used by clinical microbiology laboratories to cope with the need for rapid, cost-effective and accurate identification of microorganisms. Several research teams have recently succeed in identifying moulds using MALDI-TOF MS, which was first adapted to bacteria, then to yeast identification. Since 2004, different commercial firms have released several ready-to-use MALDI-TOF MS platforms.

Although an old technique but in recent times, this has emerged as a widespread and reliable tool for fast identification and classification of fungi and other microorganisms. It is done by soft ionization technique used in mass spectrometry, using protein composition to compare patterns obtained from the yeast or mold being analyzed with a database of reference spectra. It allows analysis of biomolecules (biopolymers such as DNA, proteins, peptide and sugars) and organic molecules (such as polymers, dendrimers and other macromolecules), which are fragile and fragment when ionized by more conventional ionization methods. Pulsed laser is used to generate ions of high molecular weight samples, such as proteins and polymers. The detector used with MALDI is the time of flight mass detector (TOF-MS). The time of flight is a method where the ions are accelerated by an electric field, resulting in ions of the same strength to have the same kinetic energy. The time it takes for each ion to traverse the flight tube and arrive at the detector is based on its mass-to-charge ratio. Therefore the heavier ions have shorter arrival times compared to lighter ones.

The database has to be accurately curated so that the comparison organism is correctly identified by other techniques. Such databases are just being assembled and are now much smaller than DNA databases. However, MALDI-TOF mass spectrometry has the capacity to assign the species name to one of the newly created cryptic species, about which there may be little to find in the literature. It represents a strong challenge to microscopic and molecular biology methods.

Mass spectrometric peptide/protein profiles of fungi display peaks in the *m/z* region of 1000-20000, where a unique set of biomarker ions may appear facilitating a differentiation of samples at the level of genus, species or even strain. This is done with the help of a processing software and spectral database of reference strains, which should preferably be constructed under the same standardized experimental conditions.

Nowadays, commercial MALDI systems are accessible for biological research work as well as for diagnostic applications in clinical medicine, biotechnology and industry. They are employed namely in bacterial biotyping but numerous experimental strategies have also been developed for the analysis of fungi. MALDI-TOF is increasingly been applied to both mycelial fungi and yeasts. Members of many fungal genera such as *Aspergillus, Fusarium, Penicillium* or *Trichoderma* and also various yeasts from clinical samples *e.g. Candida* species, have been successfully identified by this technique.

However, there is no versatile method for fungi currently available even though the use of only a limited number of matrix compounds has been reported. Also a continuous extension of the database is necessary to further improve its reliability. Either intact cell/spore MALDI-TOF MS is chosen or an extraction of surface proteins is performed and then the resulting extract is measured. Furthermore, biotrophic fungi can be identified via a direct acquisition of MALDI-TOF mass spectra *e.g.* from infected tissue. More recently, but still at an infancy stage, MALDI-TOF MS is being used to perform strain typing as well as to determine antifungal drug susceptibility.

(d) Quantamatrix Multiplexed Assay Platform

The risk of death of patients increases by 7.6% each hour between the septic shock and the adjustment of antimicrobial therapy. Hence there is a need for an early detection of fungal blood stream infections. The Quantamatrix Multiplexed Assay Platform **(QMAP)** system plays important role in detection of candidemia, in addition to its utility in bacteremia. This is a molecular diagnostic tool with its capacity to analyze nucleic acids and proteins. It is a bead-based bioassay wherein magnetic microdisks are coated with silica. These disks contain carboxyl group that play major role in detection of pathogens.

It is basically a Multiplex PCR where 1,000 types of pathogen in 1-microwell for one sample can be detected. This assay is high throughput, fully automated, including the hybridization and washing steps as well as the scanning process for the interpretation of findings, rapid *i.c.* takes less than 3 hrs to test 96 samples including 15 min for DNA preparation and 1 hr for target DNA amplification. It can provide fast and accurate characterization of fungi at the genus, species and strain level. The yeasts like *Candida albicans, C.tropicalis, C.glabrata* and *C.parapsilosis* have been detected by using this newer technique, which is still under evaluation.

E. MISCELLANEOUS METHODS

The methods which are not categorized in the previous section are here clubbed can be described into following categories:

1. Point of Care Testing

The point of care testing (POCT) is defined as the medical testing at or near the site of patient care. It is a medical diagnostic testing performed outside the clinical laboratory in close proximity to where the patient is receiving care. It is typically performed by non-laboratory personnel and the results are used for clinical decision making. The driving notion behind this is to bring the test conveniently and immediately close to the patient. This increases the likelihood that the patient, physician and care team will receive the results quicker, which allows for immediate clinical management decisions to be made. The most promising POCT for the fungal diseases are *Aspergillus*-specific lateral-flow device test, *Aspergillus* proximity ligation antigen assay for invasive aspergillosis, which is antibody based test; cryptococcal lateral-flow assay for cryptococcosis based on antigen detection and loop-mediated isothermal amplification assay for histoplasmosis based on nucleic acid amplification.

2. T2Candida Panel

The need for an early detection of fungal blood stream infections is re-emphasized by the fact that risk of death increases by 7.6% each hour between septic shock and the adjustment of antimicrobial therapy. The T2Candida Panel has been introduced as a new class of infectious disease diagnostics that can rapidly detect and identify the causative pathogen of sepsis directly from a patient blood sample in a culture-independent manner. This test enables detection of *Candida* directly from the patient sample, a significant advance for the rapid and accurate diagnosis of invasive candidiasis.

3. Biosensors for Medical Mycology

Biosensor technology is significantly less developed for the detection and identification of fungi than for other microorganisms, not only in terms of commercially available, fully developed diagnostic devices but even at the prototype level. Hence there are possibilities of biosensor development and commercialization, raised by the persistence of immunocompromised fungal infections especially in developing countries. In parallel, by the need for their unambiguous, ultra-sensitive and early diagnosis – mainly for resource-limited settings and regions but also in developed countries by the growing importance of the point-of-care and end-user paradigms for healthcare. Some online reported biosensors for medical mycology are available and on the main nano-technology-based materials usually employed in biosensing are also in use.

4. Epidemiological Markers of Fungi

The source of infection caused by any fungus can be traced by applying various epidemiological markers. These are rather mandatory, particularly, in nosocomial fungal infections that are more difficult to manage. For epidemiological studies of fungal diseases the markers used, are given in **Table 5.4**. The details of these parameters are described in their respective Chapters.

Table 5.4. Epidemiological Markers Used in Fungal Infectons.
• Phage typing
• Secreted lethal factor typing
• Serotyping
• Morphotyping
• Mating typing
• Resistotyping
• Biotyping
• Protein electrophoresis (Immunoblot)
• Isoenzyme typing
• Restriction fragment length polymorphism (RFLP)
• Karyotyping
• Nucleic acid probes

5. Quality Control in Medical Mycology

The mycology reference laboratories identify fungal strains upto species level with conventional methods and where facilities are available, taking help of molecular techniques. In providing such services, maintenance of quality control is mandatory. This is defined as the aggregate of procedures and techniques so derived to detect, reduce and correct deficiencies in analytic process. It is rather an integral component of a laboratory assurance programme, focusing mainly on procedural and technical aspects of the test under question. It acts as sort of deterrent on the services being imparted by the laboratories. This may be **internal** or **external** quality control. Some of the centers are regularly running Mycology Proficiency Testing Program or EQAS, wherein unlabeled fungal strains are supplied to participating laboratories and results are compiled. They are graded according to their performance in this program.

The process of accreditation according to EN ISO 15189 now applies in all laboratories in Europe. This may be used by medical laboratories in developing their quality management systems and assessing their own competence. It can also be used for confirming or recognizing the competence of medical laboratories by laboratory customers, regulating authorities and accreditation bodies. The details of quality control program are given in **Appendix E**.

Further Reading

1. Affolter K, Tamm M, Jahn K, *et al.* Galactomannan in bronchoalveolar lavage for diagnosing invasive fungal disease. *Am J Respir Crit Care Med.* 2014; **190:** 309-17.
2. Agarwal T, Bandivadekar P, Satpathy G, *et al.* Detection of fungal hyphae using smartphone and pocket magnifier: Going cellular. *Cornea.* 2015; **34:** 355-7.
3. Agca H, Ener B, Yılmaz E, *et al.* Comparative evaluation of galactomannan optical density indices and culture results in bronchoscopic specimens obtained from neutropenic and non-neutropenic patients. *Mycoses.* 2014; **57:** 169-75.
4. Ahmad S, Khan Z. Invasive candidiasis: A review of non-culture-based laboratory diagnostic methods. *Indian J Med Microbiol.* 2012; **30:** 264-9.
5. Alanio A, Bretagne S. Difficulties with molecular diagnostic tests for mould and yeast infections: Where do we stand? *Clin Microbiol Infect.* 2014; **20 (Suppl. 6):** 36-41.
6. Alonso R, Pisa D, Aguado B, *et al.* Identification of fungal species in brain tissue from Alzheimer's disease by next-generation sequencing. *J Alzheimers Dis.* 2017; **58:** 55-67.
7. Angebault C, Lanternier F, Dalle F, *et al.* Prospective evaluation of serum β-glucan testing in patients with probable or proven fungal diseases. *Open Forum Infect Dis.* 2016; **3:** ofw128. PMID: 27419189.
8. Arendrup MC, Boekhout T, Akova M, *et al.* ESCMID and ECMM joint clinical guidelines for the diagnosis and management of rare invasive yeast infections. *Clin Microbiol Infect.* 2014; **20 (Suppl. 3):** 76-98.
9. Arvanitis M, Glavis-Bloom J, Mylonakis E. Invertebrate models of fungal infection. *Biochim Biophys Acta.* 2013; **1832:** 1378-83.
10. Arvanitis M, Anagnostou T, Fuchs BB, *et al.* Molecular and non-molecular diagnostic methods for invasive fungal infections. *Clin Microbiol Rev.* 2014; **27:** 490-526.
11. Avni T, Leibovici L, Paul M. PCR diagnosis of invasive candidiasis: Systematic review and meta-analysis. *J Clin Microbiol.* 2011; **49:** 665-70.
12. Axelson GK, Giorgadze T, Youngberg GA. Evaluation of the use of Congo red staining in the differential diagnosis of *Candida vs.* various other yeast-form fungal organisms. *J Cutan Pathol.* 2008; **35:** 27-30.
13. Babel DE. How to identify fungi. *J Am Acad Dermatol.* 1994; **31:** S108-11.
14. Babouee Flury B, Weisser M, Prince SS, *et al.* Performances of two different panfungal PCRs to detect mould DNA in formalin-fixed paraffin-embedded tissue: What are the limiting factors? *BMC Infect Dis.* 2014; **14:** 692.
15. Backx M, White PL, Barnes RA. New fungal diagnostics. *Br J Hosp Med (Lond).* 2014; **75:** 271-6.
16. Bader O. MALDI-TOF-MS-based species identification and typing approaches in medical mycology. *Proteomics.* 2013; **13:** 788-99.
17. Balajee SA, Borman AM, Brandt ME, et al. Sequence-based identification of *Aspergillus*, *Fusarium* and Mucorales species in the clinical mycology laboratory: Where are we and where should we go from here? *J Clin Microbiol.* 2009; **47:** 877-84.
18. Balajee SA, Sigler L, Brandt ME. DNA and the classical way: Identification of medically important molds in the 21st century. *Med Mycol.* 2007; **45:** 475-90.

19. Barnes RA, White PL, Bygrave C, *et al.* Clinical impact of enhanced diagnosis of invasive fungal disease in high-risk haematology and stem cell transplant patients. *J Clin Pathol.* 2009; **62:** 64-9.
20. Bazemore RA, Feng J, Cseke L, *et al.* Biomedically important pathogenic fungi detection with volatile biomarkers. *J Breath Res.* 2012; **6:** 016002. PMID: 22233561.
21. Beirao F, Araujo R. State of the art diagnostic of mold diseases: A practical guide for clinicians. *Eur J Clin Microbiol Infect Dis.* 2013; **32:** 3-9.
22. Bellanger AP, Grenouillet F, Henon T, *et al.* Retrospective assessment of β-D-(1,3)-glucan for presumptive diagnosis of fungal infections. *APMIS.* 2011; **119:** 280-6.
23. Benamore RE, Weisbrod GL, Hwang DM, *et al.* Reversed halo sign in lymphomatoid granulomatosis. *Br J Radiol.* 2007; **80:** e162-6.
24. Berkes CA, Chan LL, Wilkinson A, *et al.* Rapid quantification of pathogenic fungi by Cellometer image-based cytometry. *J Microbiol Methods.* 2012; **91:** 468-76.
25. Bille J. New nonculture-based methods for the diagnosis of invasive candidiasis. *Curr Opin Crit Care.* 2010; **16:** 460-4.
26. Bishop JA, Nelson AM, Merz WG, *et al.* Evaluation of the detection of melanin by the Fontana-Masson silver stain in tissue with a wide range of organisms including *Cryptococcus. Hum Pathol.* 2012; **43:** 898-903.
27. Borman AM, Linton CJ, Miles SJ, Johnson EM. Molecular identification of pathogenic fungi. *J Antimicrob Chemother.* 2008; **61 (Suppl. 1):** i7-12.
28. Borman AM, Szekely A, Palmer MD, *et al.* Assessment of accuracy of identification of pathogenic yeasts in microbiology laboratories in the United Kingdom. *J Clin Microbiol.* 2012; **50:** 2639-44.
29. Borras R, Rosello P, Chilet M, *et al.* Positive result of the *Aspergillus* galactomannan antigen assay using bronchoalveolar lavage fluid from a patient with an invasive infection due to *Lichtheimia ramosa. J Clin Microbiol.* 2010; **48:** 3035-6.
30. Brenier-Pinchart MP, Abaibou H, Berendsen T, *et al.* Usefulness of pan-fungal NASBA test for surveillance of environmental fungal contamination in a protected hematology unit. *Med Mycol.* 2014; **52:** 433-7.
31. Brownback KR, Pitts LR, Simpson SQ. Utility of galactomannan antigen detection in bronchoalveolar lavage fluid in immunocompromised patients. *Mycoses.* 2013; **56:** 552-8.
32. Brownback KR, Thomas LA, Simpson SQ. Role of bronchoalveolar lavage in the diagnosis of pulmonary infiltrates in immunocompromised patients. *Curr Opin Infect Dis.* 2014; **27:** 322-8.
33. Buchan BW, Ledeboer NA. Advances in identification of clinical yeast isolates by use of matrix-assisted laser desorption ionization-time of flight mass spectrometry. *J Clin Microbiol.* 2013; **51:** 1359-66.
34. Buchheidt D, Reinwald M, Hofmann WK, *et al.* Evaluating the use of PCR for diagnosing invasive aspergillosis. *Expert Rev Mol Diagn.* 2017; **17:** 603-10.
35. Buelow DR, Gu Z, Walsh TJ, Hayden RT. Evaluation of multiplexed PCR and liquid-phase array for identification of respiratory fungal pathogens. *Med Mycol.* 2012; **50:** 775-80.
36. Byard RW. Unusual patterned skin lesions caused by postmortem fungal activity. *Forensic Sci Med Pathol.* 2014; **10:** 651-3.
37. Cairns TC, Studholme DJ, Talbot NJ, *et al.* New and improved techniques for the study of pathogenic fungi. *Trends Microbiol.* 2016; **24:** 35-50.
38. Capilla J, Clemons KV, Stevens DA. Animal models: An important tool in mycology. *Med Mycol.* 2007; **45:** 657-84.
39. Carvalho A, Goldman GH. Editorial: An omics perspective on fungal infection: Toward next-generation diagnosis and therapy. *Front Microbiol.* 2017; **8:** 85. PMID: 28184220.
40. Cassagne C, Normand AC, L'Ollivier C, *et al.* Performance of MALDI-TOF MS platforms for fungal identification. *Mycoses.* 2016; **59:** 678-90.
41. Ceesay MM, Desai SR, Berry L, *et al.* A comprehensive diagnostic approach using galactomannan, targeted β-D-glucan, baseline computerized tomography and biopsy yields a significant burden of invasive fungal disease in at risk haematology patients. *Br J Haematol.* 2015; **168:** 219-29.
42. Challa S, Pamidi U, Uppin SG, *et al.* Diagnostic accuracy of morphologic identification of filamentous fungi in paraffin embedded tissue sections: Correlation of histological and culture diagnosis. *Indian J Pathol Microbiol.* 2014; **57:** 583-7.
43. Chalupova J, Raus M, Sedlarova M, *et al.* Identification of fungal microorganisms by MALDI-TOF mass spectrometry. *Biotechnol Adv.* 2014; **32:** 230-41.
44. Chander J, Chakrabarti A, Sharma A, *et al.* Evaluation of calcofluor staining in the diagnosis of fungal corneal ulcer. *Mycoses.* 1993; 36: 243-5.
45. Chen SC, Kontoyiannis DP. New molecular and surrogate biomarker-based tests in the diagnosis of bacterial and fungal infection in febrile neutropenic patients. *Curr Opin Infect Dis.* 2010; **23:** 567-77.
46. Cornely OA, Cuenca-Estrella M, Meis JF, *et al.* European Society of Clinical Microbiology and Infectious Diseases (ESCMID) Fungal Infection Study Group (EFISG) and European Confederation of Medical Mycology (ECMM) 2013 joint guidelines on diagnosis and management of rare and emerging fungal diseases. *Clin Microbiol Infect.* 2014; **20 (Suppl. 3):** 1-4.
47. Cornely OA. Galactomannan testing during mold-active prophylaxis. *Clin Infect Dis.* 2014; **59:** 1703-4.
48. Cornu M, Goudjil S, Kongolo G, *et al.* Evaluation of the (1,3)-β-D-glucan assay for the diagnosis of neonatal invasive yeast infections. *Med Mycol.* 2017; doi: 10.1093/mmy/myx021. PMID: 28371838.
49. Cuenca-Estrella M, Bassetti M, Lass-Florl C, *et al.* Detection and investigation of invasive mould disease. *J Antimicrob Chemother.* 2011; **66 (Suppl. 1):** i15-24.
50. Dadaci Z, Kılınc F, Ozer TT, *et al.* Periodic acid-Schiff staining demonstrates fungi in chronic anterior blepharitis. *Eye (Lond).* 2015; **29:** 1522-7.

51. Das P, Pandey P, Harishankar A, *et al.* A high yield DNA extraction method for medically important *Candida* species: A comparison of manual *versus* QIA cube-based automated system. *Indian J Med Microbiol.* 2016; **34:** 533-5.
52. De Pascale G, Posteraro B, Cutuli SL, *et al.* Why should we monitor (1-3)-β-D-glucan levels during invasive candidiasis? Just ask your ophthalmologist! *J Clin Microbiol.* 2013; **51:** 1645-6.
53. de Pauw BE, Patterson TF. Should the consensus guidelines' specific criteria for the diagnosis of invasive fungal infection be changed? *Clin Infect Dis.* 2005; **41:** S377-80.
54. Denning DW. The ambitious '95-95 by 2025' roadmap for the diagnosis and management of fungal diseases. *Thorax.* 2015; **70:** 613-4.
55. Deshpande P, Shetty A, Mehta A, *et al.* Standardization of fungal polymerase chain reaction for the early diagnosis of invasive fungal infection. *Indian J Med Microbiol.* 2011; **29:** 406-10.
56. Devi B, Jindal N, Mohan U. Identification of yeast using disk diffusion test. *Indian J Med Microbiol.* 1997; **15:** 157-8.
57. Dinand V, Anjan M, Oberoi JK, *et al.* Threshold of galactomannan antigenemia positivity for early diagnosis of invasive aspergillosis in neutropenic children. *J Microbiol Immunol Infect.* 2016; **49:** 66-73.
58. Dornbusch HJ, Groll A, Walsh TJ. Diagnosis of invasive fungal infections in immunocompromised children. *Clin Microbiol Infect.* 2010; **16:** 1328-34.
59. Dou YH, Du JK, Liu HL, Shong XD. The role of procalcitonin in the identification of invasive fungal infection - A systemic review and meta-analysis. *Diagn Microbiol Infect Dis.* 2013; **76:** 464-9.
60. Einsele H, Loeffler J. Contribution of new diagnostic approaches to antifungal treatment plans in high-risk haematology patients. *Clin Microbiol Infect.* 2008; **14 (Suppl. 4):** 37-45.
61. Falci DR, Stadnik CM, Pasqualotto AC. A review of diagnostic methods for invasive fungal diseases: Challenges and perspectives. *Infect Dis Ther.* 2017; **6:** 213-23.
62. Faria-Campos AC, Hanke LA, Batista PH, *et al.* An innovative electronic health records system for rare and complex diseases. *BMC Bioinformatics.* 2015; **16 (Suppl. 19):** S4. PMID: 26695733.
63. Fisher BT, Zaoutis TE, *et al.* Galactomannan antigen testing for diagnosis of invasive aspergillosis in pediatric hematology patients. *J Pediatric Infect Dis Soc.* 2012; **1:** 103-11.
64. Fontana C, Gaziano R, Favaro M, *et al.* (1-3)-β-D-glucan vs. galactomannan antigen in diagnosing invasive fungal infections (IFIs). *Open Microbiol J.* 2012; **6:** 70-3.
65. Fraczek MG, Kirwan MB, Moore CB, *et al.* Volume dependency for culture of fungi from respiratory secretions and increased sensitivity of *Aspergillus* quantitative PCR. *Mycoses.* 2014; **57:** 69-78.
66. Frickmann H, Loderstaedt U, Racz P, *et al.* Detection of tropical fungi in formalin-fixed, paraffin-embedded tissue: Still an indication for microscopy in times of sequence-based diagnosis? *Biomed Res Int.* 2015; 938721.
67. Fridlington E, Colome-Grimmer M, Kelly E, *et al.* Tzanck smear as a rapid diagnostic tool for disseminated cryptococcal infection. *Arch Dermatol.* 2006; **142:** 25-7.
68. Furfaro E, Mikulska M, Miletich F, *et al.* Galactomannan: Testing the same sample twice? *Transpl Infect Dis.* 2012; **14:** e38-9.
69. Gazzoni FF, Severo LC, Marchiori E, *et al.* Fungal diseases mimicking primary lung cancer: Radiologic-pathologic correlation. *Mycoses.* 2014; **57:** 197-208.
70. Ghelardi E, Pichierri G, Castagna B, *et al.* Efficacy of chromogenic *Candida* agar for isolation and presumptive identification of pathogenic yeast species. *Clin Microbiol Infect.* 2008; **14:** 141-7.
71. Gils S, Maertens J, Lagrou K. Pre-treatment of bronchoalveolar lavage fluid samples with SLsolution leads to false-negative *Aspergillus* galactomannan levels. *J Clin Microbiol.* 2016; **54:** 1171.
72. Gochhait D, Dey P, Rajwanshi A, *et al.* Spectrum of fungal and parasitic infections on fine needle aspiration cytology. *Diagn Cytopathol.* 2015; **43:** 450-5.
73. Gomez BL. Molecular diagnosis of endemic and invasive mycoses: Advances and challenges. *Rev Iberoam Micol.* 2014; **31:** 35-41.
74. Gompelmann D, Heussel CP, Schuhmann M, *et al.* The role of diagnostic imaging in the management of invasive fungal diseases-report from an interactive workshop. *Mycoses.* 2011; **54 (Suppl. 1):** 27-31.
75. Gorton RL, Ramnarain P, Barker K, *et al.* Comparative analysis of Gram's stain, PNA-FISH and Sepsityper with MALDI-TOF MS for the identification of yeast direct from positive blood cultures. *Mycoses.* 2014; **57:** 592-601.
76. Griffin AT, Hanson KE. Update on fungal diagnostics. *Curr Infect Dis Rep.* 2014; **16:** 415.
77. Gu Z, Buelow DR, Petraitiene R, *et al.* Quantitative multiplexed detection of common pulmonary fungal pathogens by labeled primer polymerase chain reaction. *Arch Pathol Lab Med.* 2014; **138:** 1474-80.
78. Guarner J, Brandt ME. Histopathologic diagnosis of fungal infections in the 21st century. *Clin Microbiol Rev.* 2011; **24:** 247-80.
79. Guimaraes MD, Marchiori E, de Souza Portes Meirelles G, *et al.* Fungal infection mimicking pulmonary malignancy: Clinical and radiological characteristics. *Lung.* 2013; **191:** 655-62.
80. Hage CA, Knox KS, Davis TE, *et al.* Antigen detection in bronchoalveolar lavage fluid for diagnosis of fungal pneumonia. *Curr Opin Pulm Med.* 2011; **17:** 167-71.
81. Halliday CL, Kidd SE, Sorrell TC, *et al.* Molecular diagnostic methods for invasive fungal disease: The horizon draws nearer? *Pathology.* 2015; **47:** 257-69.
82. Harris JL. Safe, low-distortion tape method for fungal slide mounts. *J Clin Microbiol.* 2000; **38:** 4683-4.
83. Hasseine L, Cassaing S, Robert-Gangneux F, *et al.* High negative predictive value diagnostic strategies for the reevaluation of early antifungal treatment: A multicenter prospective trial in patients at risk for invasive fungal infections. *J Infect.* 2015; **71:** 258-65.

84. Hay RJ, Jones RM. New molecular tools in the diagnosis of superficial fungal infections. *Clin Dermatol.* 2010; **28:** 190-6.
85. Hayes GE, Denning DW. Frequency, diagnosis and management of fungal respiratory infections. *Curr Opin Pulm Med.* 2013; **19:** 259-65.
86. Hazen KC. Respiratory fungal infections: Molecular diagnostic tests. *Clin Lab Med.* 2014; **34:** 351-64.
87. Heaton SM, Weintrob AC, Downing K, *et al.* Histopathological techniques for the diagnosis of combat-related invasive fungal wound infections. *BMC Clin Pathol.* 2016; **16:** 11. PMID: 27398067.
88. Herbrecht R, Berceanu A. Beta-D-glucan detection test: A step toward preemptive therapy for fungal infections in leukemic patients? *Clin Infect Dis.* 2008; **46:** 886-9.
89. Hohl TM. Overview of vertebrate animal models of fungal infection. *J Immunol Methods.* 2014; **410:** 100-12.
90. Hong G, Miller HB, Allgood S, *et al.* The use of selective-fungal culture media increases detection rates of fungi in the cystic fibrosis respiratory tract. *J Clin Microbiol.* 2017; **55:** 1122-30.
91. Hot A, Maunoury C, Poiree S, *et al.* Diagnostic contribution of positron emission tomography with [^{18}F] fluorodeoxyglucose for invasive fungal infections. *Clin Microbiol Infect.* 2011; **17:** 409-17.
92. Hussein MR. Mucocutaneous Splendore-Hoeppli phenomenon. *J Cutan Pathol.* 2008; **35:** 979-88.
93. Ibembe IN, Wiggin TR. An alternative to India ink stain. *Trop Doct.* 2015; **45:** 206-7.
94. Irinyi L, Lackner M, de Hoog GS, *et al.* DNA barcoding of fungi causing infections in humans and animals. *Fungal Biol.* 2016; **120:** 125-36.
95. Irinyi L, Serena C, Garcia-Hermoso D, *et al.* ISHAM-ITS reference DNA barcoding database - The quality controlled standard tool for routine identification of human and animal pathogenic fungi. *Med Mycol.* 2015; **53:** 313-37.
96. Jain G, Singh M, Singhla A *et al.* Unusual fungal bodies in conventional cervical smears: Report of nine cases. *Diagn Cytopathol.* 2015; **43:** 234-7.
97. Jain KK, Mittal SK, Kumar S, *et al.* Imaging features of central nervous system fungal infections. *Neurol India.* 2007; **55:** 241-50.
98. Jeong SJ, Lee JU, Song YG, *et al.* Delaying diagnostic procedure significantly increases mortality in patients with invasive mucormycosis. *Mycoses.* 2015; **58:** 746-52.
99. Johnson G, Ferrini A, Dolan SK, *et al.* Biomarkers for invasive aspergillosis: The challenges continue. *Biomark Med.* 2014; **8:** 429-51.
100. Johnson GL, Sarker SJ, Hill K, *et al.* Significant decline in galactomannan signal during storage of clinical serum samples. *Int J Mol Sci.* 2013; **14:** 12970-7.
101. Jung J, Park YS, Sung H, *et al.* Using immunohistochemistry to assess the accuracy of histomorphologic diagnosis of aspergillosis and mucormycosis. *Clin Infect Dis.* 2015; **61:** 1664-70.
102. Kain R. Histopathology. *Methods Mol Biol.* 2017; **1508:** 185-93.
103. Karageorgopoulos DE, Vouloumanou EK, Ntziora F, *et al.* β-D-glucan assay for the diagnosis of invasive fungal infections: A meta-analysis. *Clin Infect Dis.* 2011; **52:** 750-70.
104. Kato K, Onoda S, Asano J, *et al.* Evaluation of the clinical cutoff level of serum (1-3)-beta-D-glucan in patients with connective tissue diseases complicated by deep fungal infections. *Mod Rheumatol.* 2010; **20:** 366-9.
105. Kaufman L, Standard PG. Specific and rapid identification of medically important fungi by exoantigen detection. *Annu Rev Microbiol.* 1987; **41**: 209-25.
106. Kaufman L. Immunohistologic diagnosis of systemic mycoses: an update. *Eur J Epidemiol.* 1992; **8:** 377-82.
107. Kaufman L. Laboratory methods for the diagnosis and confirmation of systemic mycoses. *Clin Infect Dis.* 1992; **14:** S23-9.
108. Kedzierska A, Kochan P, Pietrzyk A, *et al.* Current status of fungal cell wall components in the immunodiagnostics of invasive fungal infections in humans: Galactomannan, mannan and (1-3)-beta-D-glucan antigens. *Eur J Clin Microbiol Infect Dis.* 2007; **26:** 755-66.
109. Khandelwal N, Sodhi KS, Sinha A, *et al.* Multidetector computed tomography and MR imaging findings in mycotic infections. *Radiol Clin North Am.* 2016; **54:** 503-18.
110. Khanna S, Oberoi JK, Datta S, *et al.* Variables affecting the performance of galactomannan assay in high-risk patients at a tertiary care centre in India. *Indian J Med Microbiol.* 2013; **31:** 34-9.
111. Khot PD, Fredricks DN. PCR-based diagnosis of human fungal infections. *Expert Rev Anti Infect Ther.* 2009; **7:** 1201-21.
112. Kimura M, McGinnis MR. Fontana-Masson-stained tissue from culture-proven mycoses. *Arch Pathol Lab Med.* 1998; **122:** 1107-11.
113. Kirani KR, Chandrika VS. Efficacy of in-house fluorescent stain for fungus. *Indian J Pathol Microbiol.* 2017; **60:** 57-60.
114. Ko JH, Peck KR, Lee JY, *et al.* Multiple myeloma as a major cause of false-positive galactomannan tests in adult patients with cancer. *J Infect.* 2016; **72:** 233-9.
115. Kono Y, Tsushima K, Yamaguchi K, *et al.* The utility of galactomannan antigen in the bronchial washing and serum for diagnosing pulmonary aspergillosis. *Respir Med.* 2013; **107:** 1094-100.
116. Kontoyiannis DP, Patterson TF. Diagnosis and treatment of invasive fungal infections in the cancer patient: recent progress and ongoing questions. *Clin Infect Dis.* 2014; **59:** S356-9.
117. Koo S, Thomas HR, Daniels SD, *et al.* A breath fungal secondary metabolite signature to diagnose invasive aspergillosis. *Clin Infect Dis.* 2014; **59:** 1733-40.
118. Kozel TR, Wickes B. Fungal diagnostics. *Cold Spring Harb Perspect Med.* 2014; **4:** a019299. PMID: 24692193.
119. Kumaraswamy Naik LR, Shetty P, Krishna Prasad MS, *et al.* Fluorescence of *Candida* in diagnosis of oral candidiasis. *Indian J Dent Res.* 2016; **27:** 618-22.
120. Lackner M, Lass-Florl C. Commercial molecular tests for fungal diagnosis from a practical point of view. *Methods Mol Biol.* 2017; **1508:** 85-105.

121. Lamoth F, Alexander BD. Nonmolecular methods for the diagnosis of respiratory fungal infections. *Clin Lab Med.* 2014; **34:** 315-36.
122. Lass-Florl C, Mayr A. Diagnosing invasive fungal diseases - limitations of microbiological diagnostic methods. *Expert Opin Med Diagn.* 2009; **3:** 461-70.
123. Lau A, Chen S, Sleiman S, *et al.* Current status and future perspectives on molecular and serological methods in diagnostic mycology. *Future Microbiol.* 2009; **4:** 1185-222.
124. Lau AF, Drake SK, Calhoun LB, *et al.* Development of a clinically comprehensive database and a simple procedure for identification of molds from solid media by matrix-assisted laser desorption ionization-time of flight mass spectrometry. *J Clin Microbiol.* 2013; **51:** 828-34.
125. Leake JL, Dowd SE, Wolcott RD, *et al.* Identification of yeast in chronic wounds using new pathogen-detection technologies. *J Wound Care.* 2009; **18:** 103-4, 106 & 108.
126. Leclercq A, Cinotti E, Labeille B, *et al. Ex vivo* confocal microscopy: A new diagnostic technique for mucormycosis. *Skin Res Technol.* 2016; **22:** 203-7.
127. Lehrnbecher T, Becker K, Groll AH. Current algorithms in fungal diagnosis in the immunocompromised host. *Methods Mol Biol.* 2017; **1508:** 67-84.
128. Lehrnbecher T, Robinson PD, Fisher BT, *et al.* Galactomannan, β-D-glucan and polymerase chain reaction-based assays for the diagnosis of invasive fungal disease in pediatric cancer and hematopoietic stem cell transplantation: A systematic review and meta-analysis. *Clin Infect Dis.* 2016; **63:** 1340-8.
129. Leroux S, Ullmann AJ. Management and diagnostic guidelines for fungal diseases in infectious diseases and clinical microbiology: Critical appraisal. *Clin Microbiol Infect.* 2013; **19:** 1115-21.
130. Levesque E, El Anbassi S, Sitterle E, *et al.* Contribution of (1,3)-beta-D-glucan to diagnosis of invasive candidiasis after liver transplantation. *J Clin Microbiol.* 2015; **53:** 771-6.
131. Lewis RE, Giannella M, Viale P. Serum galactomannan diagnosis of breakthrough invasive fungal disease. *Clin Infect Dis.* 2015; **60:** 1284.
132. Li S, Rong H, Guo Q, *et al.* Serum procalcitonin levels distinguish Gram-negative bacterial sepsis from Gram-positive bacterial and fungal sepsis. *J Res Med Sci.* 2016; **21:** 39.
133. Lim CS, Lim SL. New contrast stain for the rapid diagnosis of dermatophytic and candidal dermatomycoses. *Arch Dermatol.* 2008; **144:** 1228-9.
134. Limper AH. The changing spectrum of fungal infections in pulmonary and critical care practice: Clinical approach to diagnosis. *Proc Am Thorac Soc.* 2010; **7:** 163-8.
135. Litvintseva AP, Lindsley MD, Gade L, *et al.* Utility of (1-3)-β-D-glucan testing for diagnostics and monitoring response to treatment during the multistate outbreak of fungal meningitis and other infections. *Clin Infect Dis.* 2014; **58:** 622-30.
136. Liu JC, Modha DE, Gaillard EA. What is the clinical significance of filamentous fungi positive sputum cultures in patients with cystic fibrosis? *J Cyst Fibros.* 2013; **12:** 187-93.
137. Lu Y, Chen YQ, Guo YL, *et al.* Diagnosis of invasive fungal disease using serum (1→3)-β-D-glucan: A bivariate meta-analysis. *Intern Med.* 2011; **50:** 2783-91.
138. Lyons JL, Zhang SX. Current laboratory approaches to diagnosis of CNS fungal infections. *Future Microbiol.* 2016; **11:** 175-7.
139. Marchiori E, Zanetti G, Escuissato DL, *et al.* Reversed halo sign: High-resolution CT scan findings in 79 patients. *Chest.* 2012; **141:** 1260-6.
140. Marcos JY, Pincus DH. Fungal diagnostics: Review of commercially available methods. *Methods Mol Biol.* 2013; **968:** 25-54.
141. Marr KA. *Aspergillus* galactomannan index: A surrogate end point to assess outcome of therapy? *Clin Infect Dis.* 2008; **46:** 1423-5.
142. Martinez D, Ananda-Rajah MR, Suominen H, *et al.* Automatic detection of patients with invasive fungal disease from free-text computed tomography (CT) scans. *J Biomed Inform.* 2015; **53:** 251-60.
143. Maschmeyer G. Invasive fungal disease: Better survival through early diagnosis and therapeutic intervention. *Expert Rev Anti Infect Ther.* 2011; **9:** 279-81.
144. McCarthy MW, Petraitiene R, Walsh TJ. Nucleic acid amplification methodologies for the detection of pulmonary mold infections. *Expert Rev Mol Diagn.* 2017; **17:** 271-9.
145. McCarthy MW, Petraitiene R, Walsh TJ. Translational development and application of (1→3)-β-D-glucan for diagnosis and therapeutic monitoring of invasive mycoses. *Int J Mol Sci.* 2017; 18. pii: e1124. PMID: 28538702.
146. McCarthy MW, Walsh TJ. Molecular diagnosis of invasive mycoses of the central nervous system. *Expert Rev Mol Diagn.* 2017; **17:** 129-39.
147. McCarthy MW, Walsh TJ. PCR methodology and applications for the detection of human fungal pathogens. *Expert Rev Mol Diagn.* 2016; **16:** 1025-36
148. McClain CM, Van Horn GT, Chappell JD, *et al. Coccidioides, Cryptococcus* or *Blastomyces*? A diagnostic dilemma encountered during frozen section evaluation. *Pediatr Dev Pathol.* 2012; **15:** 71-5.
149. McMullan BJ, Desmarini D, Djordjevic JT, *et al.* Rapid microscopy and use of vital dyes: Potential to determine viability of *Cryptococcus neoformans* in the clinical laboratory. *PLoS One.* 2015; **10:** e0117186.
150. Mery A, Sendid B, Francois N, *et al.* Application of mass spectrometry technology to the early diagnosis of invasive fungal infections. *J Clin Microbiol.* 2016; **54:** 2786-97.
151. Miceli MH, Maertens J. Role of Nonculture-based tests, with an emphasis on galactomannan testing for the diagnosis of invasive aspergillosis. *Semin Respir Crit Care Med.* 2015; **36:** 650-61.
152. Mikulska M, Furfaro E, Viscoli C. Non-cultural methods for the diagnosis of invasive fungal disease. *Expert Rev Anti Infect Ther.* 2015; **13:** 103-17.
153. Miranda MF, Silva AJ. Vinyl adhesive tape also effective for direct microscopy diagnosis of chromomycosis, lobomycosis and paracoccidioidomycosis. *Diagn Microbiol Infect Dis.* 2005; **52:** 39-43.

154. Mohan H, Bal A, Aulakh R. Evaluation of skin biopsies for fungal infections: Role of routine fungal staining. *J Cutan Pathol.* 2008; **35:** 1097-9.
155. Monroe JR. The diagnostic value of a KOH. *JAAPA.* 2001; **14:** 50-1.
156. Montagna MT, Caggiano G, Borghi E, *et al.* The role of the laboratory in the diagnosis of invasive candidiasis. *Drugs.* 2009; **69 (Suppl. 1):** 59-63.
157. Montal S, Bousquet P, Rispail P, Tramini P. Evaluation of a new test for candidiasis diagnosis in elderly people. *Odontostomatol Trop.* 2012; **35:** 37-43.
158. Montone KT. Differentiation of *Fusarium* from *Aspergillus* species by colorimetric *in situ* hybridization in formalin-fixed, paraffin-embedded tissue sections using dual fluorogenic-labeled LNA probes. *Am J Clin Pathol.* 2009; **132:** 866-70.
159. Morrissey CO, Chen SC, Sorrell TC, *et al.* Galactomannan and PCR versus culture and histology for directing use of antifungal treatment for invasive aspergillosis in high-risk haematology patients: A randomised controlled trial. *Lancet Infect Dis.* 2013; **13:** 519-28.
160. Morrissey CO. Advancing the field: Evidence for new management strategies in invasive fungal infections. *Curr Fungal Infect Rep.* 2013; **7:** 51-58.
161. Mrazek C, Lass-Florl C. Biopsy procedures for molecular tissue diagnosis of invasive fungal infections. *Curr Infect Dis Rep.* 2011; **13:** 504-9.
162. Mukhopadhyay S. Role of histology in the diagnosis of infectious causes of granulomatous lung disease. *Curr Opin Pulm Med.* 2011; **17:** 189-96.
163. Mularoni A, Furfaro E, Faraci M, *et al.* High Levels of beta-D-glucan in immunocompromised children with proven invasive fungal disease. *Clin Vaccine Immunol.* 2010; **17:** 882-3.
164. Munson E, Endes T, Vaughan K, *et al.* Two cases of recovery of dimorphic pathogenic fungi via conventional BacT/ALERT microbial detection system media. *Mycopathologia.* 2009; **167:** 191-5.
165. Nawrot U, Kowalska-Krochmal B, Sulik-Tyszka B, *et al.* Evaluation of blood culture media for the detection of fungi. *Eur J Clin Microbiol Infect Dis.* 2015; **34:** 161-7.
166. Ng TY, Kang ML, Tan BH, *et al.* Case report: Enteral nutritional supplement as a likely cause of false-positive galactomannan testing. *Med Mycol Case Rep.* 2013; **3:** 11-3.
167. Norkin M, Wingard JR. Diagnostic strategies for invasive fungal infections in patients with hematologic malignancies and hematopoietic stem cell transplant recipients. *J Natl Compr Canc Netw.* 2013; **11:** 941-9.
168. Nye MB, Beard MA, Body BA. Diagnostic mycology: Controversies and consensus - what should laboratories do? Part I & II. *Clin Microbiol Newsl.* 2006; **28:** 121-7 & 129-34.
169. Onishi A, Sugiyama D, Kogata Y, *et al.* Diagnostic accuracy of serum 1,3-β-D-glucan for *Pneumocystis jirovecii* pneumonia, invasive candidiasis and invasive aspergillosis: systematic review and meta-analysis. *J Clin Microbiol.* 2012; **50:** 7-15.
170. Orlowski HLP, McWilliams S, Mellnick VM, *et al.* Imaging spectrum of invasive fungal and fungal-like infections. *Radiographics.* 2017; 160110. PMID: 28622118.
171. Ostrosky-Zeichner L. Invasive mycoses: Diagnostic challenges. *Am J Med.* 2012; **125:** S14-24.
172. Otto C, Buonocore D, Cohen N, *et al.* Comparison of fungal culture to surgical pathology exam in the detection of dimorphic fungi and the impact on treatment and outcomes: A 25-year retrospective review at a tertiary cancer center. *Am J Clin Pathol.* 2017; **147 (Suppl. 2):** S178.
173. Oz Y, Kiraz N. Diagnostic methods for fungal infections in pediatric patients: Microbiological, serological and molecular methods. *Expert Rev Anti Infect Ther.* 2011; **9:** 289-98.
174. Ozcan K, Ilkit M, Ates A, *et al.* Performance of chromogenic *Candida* agar and CHROMagar Candida in recovery and presumptive identification of monofungal and polyfungal vaginal isolates. *Med Mycol.* 2010; **48:** 29-34.
175. Palacios E, Rojas R, Rodulfa J, *et al.* Magnetic resonance imaging in fungal infections of the brain. *Top Magn Reson Imaging.* 2014; **23:** 199-212.
176. Pana ZD, Vikelouda K, Roilides E. Diagnosis of invasive fungal diseases in pediatric patients. *Expert Rev Anti Infect Ther.* 2016; **14:** 1203-13.
177. Panda A, Ghosh AK, Mirdha BR, *et al.* MALDI-TOF mass spectrometry for rapid identification of clinical fungal isolates based on ribosomal protein biomarkers. *J Microbiol Methods.* 2015; **109:** 93-105.
178. Panthagani AP, Tidman MJ. Diagnosis directs treatment in fungal infections of the skin. *Practitioner.* 2015; **259 (1786):** 25-9.
179. Pathak AK, Shukla VY. Signs of fungal infection in dead mimic the chronic torture. *J Forensic Sci.* 2017; doi: 10.1111/1556-4029.13350. PMID: 28168686.
180. Patterson TF. Clinical utility and development of biomarkers in invasive aspergillosis. *Trans Am Clin Climatol Assoc.* 2011; **122:** 174-83.
181. Perfect JR. Fungal diagnosis: How do we do it and can we do better? *Curr Med Res Opin.* 2013; 29 **(Suppl. 4):** 3-11.
182. Persat F, Ranque S, Derouin F, *et al.* Contribution of the (1,3)-beta-D-glucan assay for diagnosis of invasive fungal infections. *J Clin Microbiol.* 2008; **46:** 1009-13.
183. Pfaller MA, Wolk DM, Lowery TJ. T2MR and T2Candida: Novel technology for the rapid diagnosis of candidemia and invasive candidiasis. *Future Microbiol.* 2016; **11:** 103-17.
184. Pincus DH, Orenga S, Chatellier S. Yeast identification: past, present and future methods. *Med Mycol.* 2007; **45:** 97-121.
185. Pini P, Bettua C, Orsi CF, *et al.* Evaluation of serum (1 → 3)-β-D-glucan clinical performance: Kinetic assessment, comparison with galactomannan and evaluation of confounding factors. Infection. 2016; **44:** 223-33.
186. Posteraro B, De Carolis E, Vella A, *et al.* MALDI-TOF mass spectrometry in the clinical mycology laboratory: Identification of fungi and beyond. *Expert Rev Proteomics.* 2013; **10:** 151-64.

187. Pounder JI, Simmon KE, Barton CA, *et al.* Discovering potential pathogens among fungi identified as non-sporulating molds. *J Clin Microbiol.* 2007; **45:** 568-71.
188. Powers-Fletcher MV, Hanson KE. Nonculture diagnostics in fungal disease. *Infect Dis Clin North Am.* 2016; **30:** 37-49.
189. Prattes J, Heldt S, Eigl S, *et al.* Point of care testing for the diagnosis of fungal infections: Are we there yet? *Curr Fungal Infect Rep.* 2016; **10:** 43-50.
190. Presterl E, Parschalk B, Bauer E, *et al.* Invasive fungal infections and (1,3)-beta-D-glucan serum concentrations in long-term intensive care patients. *Int J Infect Dis.* 2009; **13:** 707-12.
191. Preuner S, Lion T. Towards molecular diagnostics of invasive fungal infections. *Expert Rev Mol Diagn.* 2009; **9:** 397-401.
192. Qin L, Zhao L, Tan C, *et al.* A novel method of combining periodic acid Schiff staining with Wright-Giemsa staining to identify the pathogens *Penicillium marneffei, Histoplasma capsulatum, Mucor* and *Leishmania donovani* in bone marrow smears. *Exp Ther Med.* 2015; **9:** 1950-4.
193. Raggam RB, Fischbach LM, Prattes J, *et al.* Detection of (1→3)-β-D-glucan in same-day urine and serum samples obtained from patients with haematological malignancies. *Mycoses.* 2015; **58:** 394-8.
194. Rajkumari N, Mathur P, Xess I, *et al.* Distribution of different yeasts isolates among trauma patients and comparison of accuracy in identification of yeasts by automated method versus conventional methods for better use in low resource countries. *Indian J Med Microbiol.* 2014; **32:** 391-7.
195. Rao S, Rajkumar A, Ehtesham M, *et al.* Autofluorescence: A screening test for mycotic infection in tissues. *Indian J Pathol Microbiol.* 2008; **51:** 215-7.
196. Reddy AK, Brahmaiah U, Narayen N, *et al.* Is blood agar an alternative to Sabouraud dextrose agar for the isolation of fungi in patients with mycotic keratitis. *Int Ophthalmol.* 2013; **33:** 251-4.
197. Reilly AA, Salkin IF, McGinnis MR, *et al.* Evaluation of mycology laboratory proficiency testing. *J Clin Microbiol.* 1999; **37:** 2297-305.
198. Rezusta A, de la Fuente S, Gilaberte Y, *et al.* Evaluation of incubation time for dermatophytes cultures. *Mycoses.* 2016; **59:** 416-8.
199. Ro JY, Lee SS, Ayala AG. Advantage of Fontana-Masson stain in capsule-deficient cryptococcal infection. *Arch Pathol Lab Med.* 1987; **111:** 53-7.
200. Rodriguez C, Weintrob AC, Dunne JR, *et al.* Clinical relevance of mold culture positivity with and without recurrent wound necrosis following combat-related injuries. *J Trauma Acute Care Surg.* 2014; **77:** 769-73.
201. Romanelli AM, Fu J, Herrera ML, *et al.* A universal DNA extraction and PCR amplification method for fungal rDNA sequence-based identification. *Mycoses.* 2014; **57:** 612-22.
202. Romanelli AM, Sutton DA, Thompson EH, *et al.* Sequence-based identification of filamentous basidiomycetous fungi from clinical specimens: A cautionary note. *J Clin Microbiol.* 2010; **48:** 741-52.
203. Rose SR, Vallabhajosyula S, Velez MG, *et al.* The utility of bronchoalveolar lavage beta-D-glucan testing for the diagnosis of invasive fungal infections. *J Infect.* 2014; **69:** 278-83.
204. Ruchel R, Schaffrinski M. Versatile fluorescent staining of fungi in clinical specimens by using the optical brightener Blankophor. *J Clin Microbiol.* 1999; **37:** 2694-6.
205. Sabour S. Interlaboratory and interstudy reproducibility of a novel lateral-flow device: A statistical issue. *J Clin Microbiol.* 2013; **51:** 1652.
206. San-Blas G, Burger E. Experimental medical mycological research in Latin America - a 2000-2009 overview. *Rev Iberoam Micol.* 2011; **28:** 1-25.
207. Sanguinetti M, Posteraro B. Identification of molds by MALDI-TOF Mass Spectrometry. *J Clin Microbiol.* 2017; **55:** 369-79.
208. Sartori A, Souza A, Zanon M, *et al.* Performance of magnetic resonance imaging in pulmonary fungal disease compared to high-resolution computed tomography. *Mycoses.* 2017 **60:** 266-72.
209. Schelenz S, Barnes RA, Barton RC, *et al.* British Society for Medical Mycology best practice recommendations for the diagnosis of serious fungal diseases. *Lancet Infect Dis.* 2015; **15:** 461-74.
210. Schroeder J, Schaffrinski M, Ruchel R. Optical brighteners in fungal diagnostics. *Mycoses.* 2006; **49 (Suppl. 2):** 14-7.
211. Schuetz AN. Invasive fungal infections: Biomarkers and molecular approaches to diagnosis. *Clin Lab Med.* 2013; **33:** 505-25.
212. Senn L, Robinson JO, Schmidt S, *et al.* 1, 3-Beta-D-glucan antigenemia for early diagnosis of invasive fungal infections in neutropenic patients with acute leukemia. *Clin Infect Dis.* 2008; **46:** 878-85.
213. Shah AA, Hazen KC. Diagnostic accuracy of histopathologic and cytopathologic examination of *Aspergillus* species. *Am J Clin Pathol.* 2013; **139:** 55-61.
214. Sharma P, Mukherjee A, Karunanithi S, *et al.* Potential role of 18F-FDG PET/CT in patients with fungal infections. *AJR Am J Roentgenol.* 2014; **203:** 180-9.
215. Shiogama K, Kitazawa K, Mizutani Y, *et al.* New Grocott stain without using chromic acid. *Acta Histochem Cytochem.* 2015; **48:** 9-14.
216. Singh L, Jain D, Madan K, *et al.* Pulmonary mycoses diagnosed using exfoliative cytology: Infection or colonization? *Acta Cytol.* 2013; **57:** 604-10.
217. Sinha K, Tendolkar U, Mathur M. Comparison of conventional broth blood culture technique and manual lysis centrifugation technique for detection of fungemia. *Indian J Med Microbiol.* 2009; **27:** 79-80.
218. Somogyvari F, Horvath A, Serly J, *et al.* Detection of invasive fungal pathogens by real-time PCR and high-resolution melting analysis. *In Vivo.* 2012; **26:** 979-83.
219. Sparagano O, Foggett S. Diagnosis of clinically relevant fungi in medicine and veterinary sciences. *Adv Appl Microbiol.* 2009; **66:** 29-52.
220. Springer J, Einsele H, Loeffler J. Molecular techniques in the diagnosis of deep and systemic mycosis. *Clin Dermatol.* 2012; **30:** 651-6.

221. Stacey KJ, Barbara RD. A critical appraisal of the role of the clinical microbiology laboratory in diagnosis of invasive fungal infections. *J Clin Microbiol.* 2011; **49:** S39-42.
222. Starkey J, Moritani T, Kirby P. MRI of CNS fungal infections: Review of aspergillosis to histoplasmosis and everything in between. *Clin Neuroradiol.* 2014; **24:** 217-30.
223. Stone NR, Gorton RL, Barker K, *et al.* Evaluation of PNA-FISH yeast traffic light for rapid identification of yeast directly from positive blood cultures and assessment of clinical impact. *J Clin Microbiol.* 2013; **51:** 1301-2.
224. Sundaram C, Shantveer GU, Umabala P, *et al.* Diagnostic utility of melanin production by fungi: Study on tissue sections and culture smears with Masson-Fontana stain. *Indian J Pathol Microbiol.* 2014; **57:** 217-22.
225. Sundaram C, Umabala P, Laxmi V, *et al.* Pathology of fungal infections of the central nervous system: 17 years' experience from Southern India. *Histopathology.* 2006; **49:** 396-405.
226. Talento AF, Dunne K, Joyce EA, *et al.* A prospective study of fungal biomarkers to improve management of invasive fungal diseases in a mixed specialty critical care unit. *J Crit Care.* 2017; **40:** 119-27.
227. Tang H, Liu C, Li Y, *et al.* Development of loop-mediated isothermal amplification (LAMP) assay for the rapid detection of *Alternaria alternata. J AOAC Int.* 2017; PMID: 27760588.
228. Tanriover MD, Ascioglu S, Altun B, *et al.* Galactomannan on the stage: Prospective evaluation of the applicability in routine practice and surveillance. *Mycoses.* 2010; **53:** 16-25.
229. Taremi M, Kleinberg ME, Wang EW, *et al.* Galactomannan antigen detection using bronchial wash and bronchoalveolar lavage in patients with hematologic malignancies. *Ann Clin Microbiol Antimicrob.* 2015; **14:** 50.
230. Teles F, Seixas J. The future of novel diagnostics in medical mycology. *J Med Microbiol.* 2015; **64:** 315-22.
231. Teles F. Biosensors for medical mycology. *In:* Biosensors and their Applications in Healthcare. Future Science Ltd. 2013; pp. 98-111.
232. Telles DR, Karki N, Marshall MW. Oral Fungal Infections: Diagnosis and management. *Dent Clin North Am.* 2017; **61:** 319-49.
233. Tepeoglu M, Ok Atılgan A, Ozdemir BH, *et al.* Role of bronchoalveolar lavage in diagnosis of fungal infections in liver transplant recipients. *Exp Clin Transplant.* 2015; **13 (Suppl. 1):** 331-4.
234. Theel ES, Doern CD. β-D-glucan testing is important for diagnosis of invasive fungal infections. *J Clin Microbiol.* 2013; **51:** 3478-83.
235. Thompson GR, Gomez BL. *Histoplasma, Blastomyces, Coccidioides* and other dimorphic fungi causing systemic mycoses. *In:* Manual of Clinical Microbiology, 11th edn. ASM. 2015. pp. 2109-27.
236. Thornton CR, Wills OE. Immunodetection of fungal and oomycete pathogens: Established and emerging threats to human health, animal welfare and global food security. *Crit Rev Microbiol.* 2015; **41:** 27-51.
237. Thornton CR. Lateral-flow device for diagnosis of fungal infection. *Curr Fungal Infect Rep.* 2013; **7:** 244-51.
238. Tortorano AM, Esposto MC, Prigitano A, *et al.* Cross-reactivity of *Fusarium* spp. in the *Aspergillus* galactomannan enzyme-linked immunosorbent assay. *J Clin Microbiol.* 2012; **50:** 1051-3.
239. Trubiano JA, Dennison AM, Morrissey CO, *et al.* Clinical utility of panfungal polymerase chain reaction for the diagnosis of invasive fungal disease: A single center experience. *Med Mycol.* 2016; **54:** 138-46.
240. Tsitsikas DA, Morin A, Araf S, *et al.* Impact of the revised (2008) EORTC/MSG definitions for invasive fungal disease on the rates of diagnosis of invasive aspergillosis. *Med Mycol.* 2012; **50:** 538-42.
241. Umabala P, Satheeshkumar T, Lakshmi V. Evaluation of fungichrom 1: A new yeast identification system. *Indian J Med Microbiol.* 2002; **20:** 160-2.
242. Valencia-Shelton F, Loeffelholz M. Non-culture techniques for the detection of bacteremia and fungemia. *Future Microbiol.* 2014; **9:** 543-59.
243. Valero C, de la Cruz-Villar L, Zaragoza O, *et al.* A new panfungal real-time PCR assay for the diagnosis of invasive fungal infections (IFIs). *J Clin Microbiol.* 2016; **54:** 2910-8.
244. Vehreschild JJ, Ruping MJ, Steinbach A, *et al.* Diagnosis and treatment of fungal infections in allogeneic stem cell and solid organ transplant recipients. *Expert Opin Pharmacother.* 2010; **11:** 95-113.
245. Vehreschild JJ. As galactomannan disappoints, our quest for a feasible diagnostic standard for invasive aspergillosis continues. *Am J Respir Crit Care Med.* 2014; **190:** 248-9.
246. Wang HY, Uh Y, Kim S, *et al.* Quantamatrix Multiplexed Assay Platform system for direct detection of bacteria and antibiotic resistance determinants in positive blood culture bottles. *Clin Microbiol Infect.* 2017; **23:** 333.e1-7.
247. Weinbergerova B, Kocmanova I, Racil Z, *et al.* Serological approaches. *Methods Mol Biol.* 2017; **1508:** 209-21.
248. West KL, Proia AD, Puri PK. Fontana-Masson stain in fungal infections. *J Am Acad Dermatol.* 2017; pii: S0190-9622(17)30304-3. PMID: 28392288.
249. Wheat LJ, Nguyen MH, Alexander BD, *et al.* Long-term stability at -20°C of *Aspergillus* galactomannan in serum and bronchoalveolar lavage specimens. *J Clin Microbiol.* 2014; **52:** 2108-11.
250. Wheat LJ. Approach to the diagnosis of the endemic mycoses. *Clin Chest Med.* 2009; **30:** 379-89.
251. White PL, Hibbitts SJ, Perry MD, *et al.* Evaluation of a commercially developed semiautomated PCR-surface-enhanced Raman scattering assay for diagnosis of invasive fungal disease. *J Clin Microbiol.* 2014; **52:** 3536-43.
252. White PL, Jones T, Whittle K, *et al.* Comparison of galactomannan enzyme immunoassay performance levels when testing serum and plasma samples. *Clin Vaccine Immunol.* 2013; **20:** 636-8.
253. Wiederhold NP, Thornton CR. Reply to "Interlaboratory and interstudy reproducibility of a novel lateral-flow device: A statistical issue". *J Clin Microbiol.* 2013; **51:** 1653.

254. Wiegand C, Bauer A, Brasch J, *et al.* Are the classic diagnostic methods in mycology still state of the art? *J Dtsch Dermatol Ges.* 2016; **14:** 490-4.
255. Willinger B. Culture-based techniques. *Methods Mol Biol.* 2017; **1508:** 195-207.
256. Wingard JR. Have novel serum markers supplanted tissue diagnosis for invasive fungal infections in acute leukemia and transplantation? *Best Pract Res Clin Haematol.* 2012; **25:** 487-91.
257. Yang M, Lee JH, Kim YK, *et al.* Identification of Mucorales from clinical specimens: A 4-year experience in a single institution. *Ann Lab Med.* 2016; **36:** 60-3.
258. Youngberg GA, Wallen E, Giorgadze G. Narrow-spectrum histochemical staining of fungi. *Arch Pathol Lab Med.* 2003; **127:** 1529-30.
259. Zalaudek I, Giacomel J, Cabo H, *et al.* Entodermoscopy: A new tool for diagnosing skin infections and infestations. *Dermatology.* 2008; **216:** 14-23.
260. Zhan X, Zhang L, Wang Z, *et al.* Reversed halo sign: Presents in different pulmonary diseases. *PLoS One.* 2015; **10:** e0128153. PMID: 26083865.
261. Zhang SX. Enhancing molecular approaches for diagnosis of fungal infections. *Future Microbiol.* 2013; **8:** 1599-611.
262. Zheng F, Zha H, Yang D, *et al.* Diagnostic values and limitations of (1,3)-β-D-glucans and galactomannan assays for invasive fungal infection in patients admitted to pediatric intensive care unit. *Mycopathologia.* 2017; **182:** 331-8.
263. Zhou X, Kong F, Sorrell TC, *et al.* Practical method for detection and identification of *Candida, Aspergillus* and *Scedosporium* spp. by use of rolling circle amplification. *J Clin Microbiol.* 2008; **46:** 2423-7.
264. Zimmermann N, Hagen MC, Schrager JJ, *et al.* Utility of frozen section analysis for fungal organisms in soft tissue wound debridement margin determination. *Diagn Pathol.* 2015; **10:** 188. PMID: 26470865.
265. Zoll J, Snelders E, Verweij PE, *et al.* Next-generation sequencing in the Mycology Lab. *Curr Fungal Infect Rep.* 2016; **10:** 37-42.

CHAPTER 6

Antifungal Therapy

The fungal diseases have become very common secondary infections to various predisposing factors as well as cause of substantial number of primary disorders. The treatment of these infections in immunocompromised patients is really a challenging task. Although many antifungal agents have been developed so far but only a few are clinically effective and safe to use. Most of these drugs are fungistatic except a few like amphotericin B, allylamine, benzylamine and morpholines, which are fungicidal in nature.

Historical Perspective

The field of antifungal chemotherapy is presently moving gradually, which began in 1903 with successful use of potassium iodide. There was little progress for the next half-century when nystatin was introduced in 1951, the first useful polyene antibiotic. Seven years later amphotericin B followed, which is still considered as the **gold standard** against which new systemic antifungals are compared. In 1959, M J Thirumalachar discovered hamycin from *Streptomyces pimprina* at Hindustan Antibiotics in India. Except for the development of flucytosine, there was again little progress until early 1970s when azoles came into existence.

The present era, which is characterized largely by modifications of azole drugs, began with ketoconazole and some other agents those can be given orally as well as have increased potency, decreased toxicity with broader spectrum of activity. Fluconazole is another broad-spectrum triazole, came in clinical practice in 1990, has been shown to be efficacious in various forms of candidiasis. Itraconazole, a broad-spectrum oral triazole came in clinical practice in 1993, has greatest advantage over imidazoles that it is active against aspergillosis and cryptococcosis, though it is efficacious against endemic mycoses as well.

The recent developments of resistance to the azole derivatives and increased prevalence of deep fungal infections have re-emphasized value of amphotericin B. Recent studies have shown ways to ameliorate well-known toxicities of this drug. A new approach is devised to make a complex of drug with lipids or entrap it into liposomes. The three lipid-based preparations were approved for clinical use (i) amphotericin B lipid complex in 1995; (ii) amphotericin B colloidal dispersion in 1996 and (iii) liposomal-encapsulated amphotericin B in 1997.

Therefore, keeping in view of increasing resistance and tolerability of drug in special circumstances like renal failure, lipid formulations and azoles are two mainstays for antifungal therapy. The latest azole approved was isavuconazole on March 6, 2015, mainly for invasive aspergillosis and mucormycosis. During the same period, the antifungal stewardship program was outlined in India under the leadership of Wattal *et al*, in 2015.

Current Scenario

There are many hindrances in the management of fungal diseases. The metabolism of human and fungal cells resembles substantially due to their **eukaryotic** nature. Therefore, any of the antifungals inhibiting a particular metabolic activity is bound to be toxic to human beings as well. There is poor penetration of drug in tissue because fungi infect relatively poorly vascularized areas. Furthermore, slow growth of fungi and granulomatous response of host tissue also decrease drug penetration into the target sites. In addition, poor absorption from gastrointestinal tract demands the use of parenteral route thereby entailing an increased toxicity of these drugs.

Three recent advancements have important impact upon the outcome of patients with mycotic infections. **Firstly**, standardization of *in vitro* antifungal susceptibility testing provides a consistent and reproducible data that may predict clinical response when used in conjunction

Table 6.1. Classification of Antifungal Drugs.

A. Antifungal Antibiotics

1. *Polyene Antibiotics*

Amphotericin B

(i) Conventional amphotericin B
  Amphotericin B deoxycholate

(ii) Lipid-based Formulations of AmB
  Amphotericin B lipid complex
  Amphotericin B colloidal dispersion
  Liposome-encapsulated AmB

Nystatin
Pimaricin
Hamycin

2. *Other Antibiotics*

Griseofulvin
Pradimicin

B. Synthetic Antifungals

1. *Thiocarbamates*

Tolnaftate

2. *Allylamines and Benzylamines*

Naftifine
Terbinafine
Butenafine

3. *Azoles*

(i) Imidazoles

Bifonazole	Butoconazole
Clotrimazole	Econazole
Fenticonazole	Ketoconazole
Miconazole	Omoconazole
Oxiconazole	Sulconazole

(ii) Triazoles

Fluconazole	Itraconazole
Voriconazole	Terconazole
Posaconazole	Ravuconazole

C. Miscellaneous Antifungals

Flucytosine
Ciclopiroxolamine
Amorolfine
Whitfield's ointment
Potassium iodide
Selenium sulfide
Undecylenic acid
Haloprogin
Triacetin
Echinocandin
Nikkomycin
Gentian violet paint

with individual patients' risk factors. **Secondly**, introduction of new antifungal agents like isavuconazole (triazole), provides more options for treating wide range of etiological agents in various clinical settings. **Thirdly**, molecular genetics have allowed scientists to begin understanding phylogenetic relationships among pathogenic fungi and their environmental and phytopathogenic relatives.

The efficacy of antifungal therapy is also dependent on several factors pertaining to drug, organism and host, which are briefly given below:

(a) **Drug factors** such as potency of drug, tissue penetration and distribution within body.
(b) **Organism factors** such as virulence, susceptibility to a given drug and development of resistance.
(c) **Host factors** such as underlying disease (diabetes mellitus), transplantation, immune status including alteration of normal mucosal flora, neutropenia and humoral as well as cell-mediated immunity.

An ideal antifungal drug should have broad-spectrum of its activity, it should be effective and safe *in vivo* and there should be no drug resistance.

CLASSIFICATION OF ANTIFUNGALS

The fungi have ergosterol in their cytoplasmic membrane as a major component. It is the most important site of action of many antifungals. There are four classes of antifungals, which inhibit synthesis of ergosterol *e.g.* polyenes, azoles, allylamines and morpholines. The rest of them act at various other sites of the fungal cell. They can also be classified as topical or systemic antifungal agents as well as on the basis of their route of administration.

Based on the source, these agents are divided in two broad categories and rest of the drugs are clubbed together in a miscellaneous group as shown in **Table 6.1**. There are some other related issues pertaining to the antifungal therapy and are dealt separately in the end of this Chapter.

A. Antifungal Antibiotics
B. Synthetic Antifungals
C. Miscellaneous Antifungals
D. Related Therapeutic Issues

A. ANTIFUNGAL ANTIBIOTICS

The antifungal antibiotics are produced by filamentous bacteria, like ***Streptomyces*** species or mycelial fungi, like ***Penicillium*** species. The following antibiotic groups are used as therapeutic antifungal agents in most of the commonly encountered fungal infections:

1. Polyene Antibiotics

The polyene macrolide antibiotics are produced from different species of filamentous bacteria belonging to genus *Streptomyces*. The chemical structure of these drugs has 4-7 conjugated double bonds hence is called as **polyene**

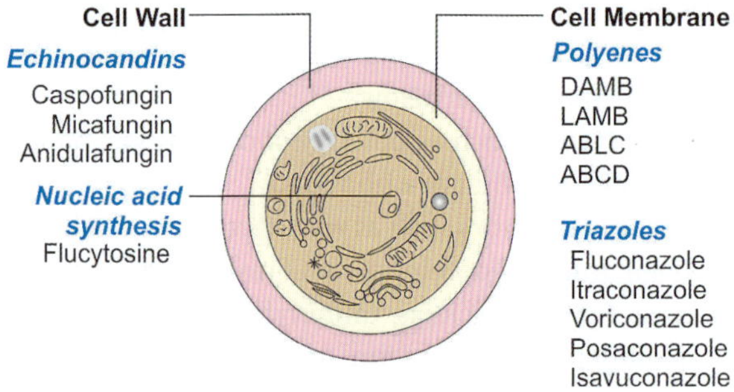

Fig. 6.1. Diagrammatic Representation Showing Target Sites of Fungal Cell with Corresponding Groups of Antifungals.

group. The mechanism of action is common in all these agents, which is essentially interference with sterol synthesis eventually entailing disruption in fungal cell as shown in **Figure 6.1 (also *See* Fig. 2.11)**. The following **four polyene antibiotics** are clinically important in the management of fungal infections:

(a) Amphotericin B
(b) Nystatin
(c) Pimaricin
(d) Hamycin

(a) Amphotericin B

This is broad-spectrum polyene antibiotic, which was obtained from the soil adjacent to the Orinoco River in Venezuela. This drug was discovered in 1957 and the following year in 1958, it was used for treating patients. Now, after almost sixty years, it is still considered as the **gold standard** for the treatment of most life-threatening mycosis. It is prepared from filamentous bacteria of actinomycetes group *i.e.* ***Streptomyces nodosus***.

It has greater affinity for ergosterol than cholesterol, later being a predominant sterol in mammalian cell wall. The pores formed by amphotericin B (AmB) increase permeability, so that essential molecules of fungal cell leak from the cytoplasm entailing arrest of fungal growth. Therefore, being a polyene, it binds to ergosterol in the fungal cell membrane thereby disrupting the membrane and killing the fungus. Moreover, induction of oxidative damage in the fungal cell also contributes to its fungicidal activity.

It has hydrophilic and hydrophobic regions. In its pure form it has very little solubility in aqueous solutions. It is **insoluble** in water hence given through infusions by complexing with bile salt deoxycholate. It is unstable at 37°C and potentially an effective fungicidal drug.

Amphotericin B is divided into two broad groups for its brief description:

(i) Conventional Amphotericin B
(ii) Lipid-based Formulations of Amphotericin B

(i) Conventional Amphotericin B: The conventional drug (Fungizone), powder for injection, is supplied in vials of 50 mg (50,000 IU) as deoxycholate amphotericin B. It is given through intravenous infusions in 5% dextrose over 2-4 hours, as it is not absorbed through oral intake. Intrathecal injections may also be given in fungal meningitis, particularly caused by *Coccidioides* species. It may be used for bladder irrigation in candiduria cases. Inhalation therapy by nebulization may be used in pulmonary aspergillosis.

A test dose before starting treatment may not be essential as it was earlier thought because initial intravenous infusion in itself is playing this role. The daily dose is 0.6 mg/kg of body weight (range 0.25-1.0 mg/kg) and is increased to level of 40-50 mg/day in due course of treatment and the total dose of entire antifungal course should not exceed 2 to 2½ gm. The course of therapy varies depending upon severity of disease, immune status of patient and type of fungal agent involved. Amphotericin B drops are also (0.15%) available for topical application like fungal keratitis.

Amphotericin B causes significant renal and infusion-related toxicities, which limits its therapeutic uses. The nephrotoxicity occurs frequently among patients undergoing antifungal therapy and it is dose related. The important manifestations are azotemia, reduced glomerular filtration rate, acidosis, hypocalcemia and inability to concentrate urine. The adverse effects reverse slowly and often completely after stoppage of therapy. It is generally recommended to perform renal function test as baseline parameters before starting treatment and thereafter periodically at frequent intervals during the course of therapy. Other side effects are headache, chills, fever, hemolytic anemia and hypokalemia. The nephrotoxic effects are increased when patients receive aminoglycosides, vancomycin and cyclosporin for simultaneous bacterial infections.

Amphotericin B has revolutionized antifungal treatment and improved prognosis among patients suffering from serious systemic fungal infections despite its toxicity. It still remains the drug of choice in almost all life-threatening situations arising due to fungi. In most of the developed countries due to the introduction of lipid-based formulations, manufacturing of the conventional amphotericin B has been stopped.

***(ii) Lipid-based Formulations of Amphotericin B*:** As conventional amphotericin B leads to severe renal toxicity and related adverse effects hence this salt is combined with lipid formulations to produce various preparations aiming to reduce toxicities, improve clinical tolerance and compliance of the drug. Therefore, it can be easily given in higher dosage as 3-5 mg/kg body weight. These preparations have potential efficacy in the treatment of systemic fungal infections due to their lesser toxicity to mammalian cells as compared to conventional amphotericin B.

There are three lipid formulations of amphotericin B with improved safety profiles, which are generally used for treatment of fungal infections. All these lipid-based AmB products differ in type of phospholipid and ratio of phospholipid : AmB, which are important determinants of fungicidal activity and toxicity.

The following three lipid-based preparations are approved for clinical use:

(i) Amphotericin B lipid complex
(ii) Amphotericin B colloidal dispersion
(iii) Liposomal-encapsulated amphotericin B

***(i) Amphotericin B lipid complex*:** Amphotericin B lipid complex (ABLC – Abelcet/Ampholip) is mixture of amphotericin B with dimyristoyl phosphatidyl choline (DMPC), dimyristoyl phosphatidyl glycerol (DMPG), sodium chloride and water for injection. It has molecular structure as **ribbon-like** sheet. It was the first lipid-based preparation to get approval by FDA in December 1995 in USA.

***(ii) Amphotericin B colloidal dispersion*:** Amphotericin B colloidal dispersion (ABCD - Amphocil, Amphotec) is a novel method of complexing amphotericin B with cholesterol sulfate in the lipid delivery system. It has a **disk-like** molecular structure and was approved by FDA in December 1996 in USA.

***(iii) Liposomal-encapsulated amphotericin B*:** The liposome-encapsulated amphotericin B (L-AmB - AmBisome/ Fungisome) is a true liposomal formulation. The liposomes are phospholipid vesicles used as target drug delivery system for AmB, in an attempt to mitigate its nephrotoxicity. These are biodegradable vesicles that consist of an aqueous environment surrounded by phospholipid bilayers. The various liposomes are prepared with different biochemical properties by altering various characteristics such as size, electrical charges, permeability and lipid composition for vesicles.

The phospholipids when dispersed in water, spontaneously develop closed **vesicles** composed of water surrounded by bilayered phospholipid membranes. Naturally occurring biodegradable phospholipids like lecithin, phosphatidyl serine are commonly used as liposomes. Multi-lamellar vesicles have also been designed, which look like onion layers. It differs from ABLC and ABCD in that it is made of uniformly sized unilamellar lipid vesicles. FDA approved this preparation in August 1997 in USA. AmBisome is available as 50 mg (50,000 IU) vial. L-AmB is also the only product with an approved indication for empirical treatment of febrile neutropenia in patients unresponsive to antibacterial therapy.

The main drawback of lipid preparations is their high cost, which is beyond the reach of majority of the patients in the developing countries. Therefore, more research is going on to have products with reasonable price, may be from this group or other groups of drugs.

The Indian version of liposomal amphotericin B (Fungisome) was originally prepared by Prof. B K Bachawat and colleagues in 1991. It was marketed in the year 2004 and is much cheaper with very good efficacy. It is available as parenteral preparations as 50, 25 and 10 mg percent. Its topical preparation (Fungisome gel - 0.1%) is also available, which is very effective in most of the systemic and cutaneous fungal infections like extensive lesions in cutaneous mucormycosis.

In addition to the indications in fungal diseases, amphotericin B is also used in the treatment of infections of non-fungal origin like Chagas' disease, visceral and cutaneous leishmaniasis, leprosy, schistosomiasis and infection caused by free-living amoebae.

(b) Nystatin

Nystatin was one of the first discovered antifungal antibiotics in 1951 and was abbreviated for **N**ew **Y**ork **Stat**e **In**stitute. This drug also belongs to group of polyene antibiotics and is obtained from ***Streptomyces noursei***. It combines with fungal cell membrane and interferes with vital cellular processes like respiration and glucose utilization. Nystatin exhibits fungistatic and fungicidal activity depending on its drug concentration, susceptibility of fungus, presence of blood, pus or tissue fluid, which reduce its activity. It is not absorbed after oral intake hence therapeutic use is limited to topical form in oral cavity and gastrointestinal tract. It is essentially insoluble in water and alcohol and is used topically for cutaneous and mucocutaneous infections, particularly caused by *Candida* species. It is relatively non-toxic but sometimes it may cause general side effects like nausea, vomiting and diarrhoea when given in higher doses. Nystatin (Mycostatin) is available as suspension, powder, cream, ointment and vaginal pessaries. The oral suspension has 1,00,000 units/ml.

(c) Pimaricin

This antibiotic is obtained from ***Streptomyces natalensis***. It shows efficacy in keratomycosis, particularly caused by resistant fungal strains like *Fusarium solani* and *Aspergillus* species. Pimaricin (Natamycin) is used as 5% ophthalmic ointment.

(d) Hamycin

This polyene antibiotic is isolated from ***Streptomyces pimprina***. It is similar to nystatin but relatively more water-soluble. Its use is restricted to topical application for oral thrush, cutaneous candidiasis and otomycosis. Insufflation of hamycin powder has been tried in pulmonary mycosis. Hamycin suspension contains 2,00,000 units/ml.

2. Other Antibiotics

In addition to antibiotics obtained from *Streptomyces* species, another groups of antibiotics are produced by *Penicillium* and *Actinomadura* species and used to treat superficial fungal infections. There are two drugs clubbed in this group:

(a) Griseofulvin

This is a narrow-spectrum oral antifungal antibiotic and was introduced in 1959. It was first produced from ***Penicillium griseofulvum*** however later on its preparation became feasible from other *Penicillium* species *(P.nigricans)* and even other fungal genera *(Khuskia oryzae)*. It inhibits fungal mitosis by interference with polymerized microtubules and spindle formation in dividing fungal cells hence it is rather a **fungistatic** and not fungicidal drug. It is very effective in treatment of almost all types of dermatophytoses. Hence it is the drug of choice for the most of dermatophyte infections, particularly chronic types, involving scalp and nail that do not respond to conventional topical antifungal therapy. A single dose of 2 gm of griseofulvin is also found to be effective in tinea capitis. Being a narrow-spectrum antibiotic, it is not having significant activity against other fungal infections like superficial candidiasis, pityriasis versicolor or deep mycoses.

Griseofulvin is used in a dose of 10 mg/kg body weight in two divided doses. The micronized forms of this drug are also available with increased ability of intestinal absorption. The drug absorption is also increased by intake of fatty acids but decreased by some of drugs like phenobarbitone. There is no major side effect, however, minor ones include headache, nausea, vomiting, bad taste, photosensitivity, skin rash, arthralgia, porphyria and exacerbation of pre-existing SLE. This drug is contraindicated in pregnancy, established porphyria and hepatic failure. It shows drug interaction with commonly used therapeutic agents hence due precautions should be taken, accordingly.

(b) Pradimicin

It is potent antifungal antibiotic produced from ***Actinomadura hibisca***. This drug acts through calcium dependent binding to mannans in cell wall. Its fungicidal activity has been shown against *Candida albicans* and *Aspergillus* species.

B. SYNTHETIC ANTIFUNGALS

The chemically synthetic preparations used in treatment of fungal infections can be broadly described in the following categories:

1. Thiocarbamates
2. Allylamines and Benzylamines
3. Azoles

1. Thiocarbamates

As a chemical family, the thiocarbamates are a group of pesticides with a wide range of uses. Some are used as fungicides, while others have herbicidal activity. However, one of them is commonly used for treating the fungal infections as well.

Tolnaftate

This topical drug was introduced in mid-1960s and is extensively being used for the treatment of dermatophytoses. It is given topically, simultaneously with systemic antifungal therapy and is indicated in tinea corporis and tinea cruris. It is available as 1% tolnaftate (Tinaderm) cream and lotion, which are applied over the affected site twice daily for a period of 1-3 weeks. Moreover, it is one of the important ingredients of the combination of topical agents used along with steroids, which is very popular among the general practitioners. However, it is less effective in tinea pedis because of poor penetrability and other hyper-keratinized lesions thereby not very effective in tinea capitis and tinea unguium also.

2. Allylamines and Benzylamines

Allylamine is a class of synthetic antifungal agents with activity against wide range of fungi. These agents selectively inhibit key enzyme, **squalene epoxidase**, which is required for fungal ergosterol biosynthesis. This inhibition is not mediated through Cytochrome P450 (CYP), as observed in azole derivatives and shown in **Figure 6.1**. Consequently, accumulation of squalene weakens the fungal cell membrane entailing its killing. The action is highly selective because these agents are inhibitory to fungal sterol synthesis more than that observed in mammalian cells. As allylamines and benzylamines act at an early step of this biosynthetic pathway leading to potential **fungicidal** activity instead of fungistatic activity, which is seen in other antifungals like azole derivatives.

They have broad-spectrum activity against dermatophytes, dimorphic fungi, *Aspergillus*, *Candida* and *Cryptococcus* species with both fungistatic as well as fungicidal activity. The following allylamines and benzylamines are clinically significant and are frequently used as therapeutic agents:

(a) Naftifine
(b) Terbinafine
(c) Butenafine

(a) Naftifine

Naftifine (Exoderil) is a topically active drug. Naftifine hydrochloride as 1% cream is prescribed for topical application in the treatment of tinea pedis, tinea cruris and tinea corporis.

(b) Terbinafine

Terbinafine (Lamisil or Sebifin) is an oral and topically active agent and it has been found to have high clinical efficacy against fungal infections of skin, nails, hair and some of the systemic mycoses. It inhibits biosynthesis of fungal ergosterol by inhibiting enzyme **squalene epoxidase**. Squalene accumulates in fungal cell and is believed to be responsible for or directly associated with, fungicidal properties of terbinafine. It is an effective fungicidal even at a very low concentration.

Terbinafine has high potency against dermatophytes as shown by its very low MIC values. There is fungicidal property, combined with its ability to penetrate rapidly into stratum corneum and nail plate hence a short course treatment is feasible with low potential for relapse. The fungicidal action of terbinafine is therefore clinically very significant.

In contrast to terbinafine, griseofulvin has fungistatic action and must be taken until all of the affected stratum corneum or nail is lost from body surface. Terbinafine works primarily as fungicidal hence every effective against dermatophytes. It has drug interaction with amitriptyline, caffeine, chlorpromazine and cimetidine.

Terbinafine is a broad-spectrum antifungal agent with fungicidal activity against molds and certain dimorphic fungi and is given as 250 mg tablets twice daily. The accepted dosage schedule for treatment with terbinafine is 6 weeks for onychomycosis of fingernail and 3 months in case of toenail. There is high efficacy and low relapse rates in toenail onychomycosis. For cutaneous infection, treatment regimen ranges 2-6 weeks. Terbinafine is also given as pulse therapy for dermatophyte infections. Terbinafine oral granules are used for tinea capitis in children. The activity against yeast is more species dependent, being fungicidal to *C.parapsilosis* but fungistatic for *C.albicans*.

(c) Butenafine

Butenafine hydrochloride is a new benzylamine derivative with a chemical structure and mode of action similar to allylamine. Butenafine is particularly active against dermatophytes, *Aspergillus* species and dimorphic fungi. The structure of butenafine resembles that of allylamine antifungals with benzyl group in place of allyl group. Like allylamines, butenafine is **fungicidal** owing to its inhibition of squalene epoxidation. This inhibition suppresses biosynthesis of ergosterol, an essential lipid component of fungal cell membranes and causes rapid accumulation of squalene. Butenafine, like allylamines, act at an early stage in ergosterol biosynthesis than azole antifungals resulting in potential for fungicidal rather than fungistatic activity. It also does not interfere with Cytochrome P450 dependent enzymes, which is inhibited by azole derivatives.

Butenafine has broad-spectrum activity and results in high clinical and mycological cure rates. It is used as topical therapy for common superficial fungal infections and effective in treatment of inter-digital tinea pedis and tinea cruris and is used as 1% cream (Mentax). It achieves and maintains high concentrations in epidermis, including

Flowchart 6.1. Sites of action of antifungals *i.e.* azoles, allylamines and benzylamines.

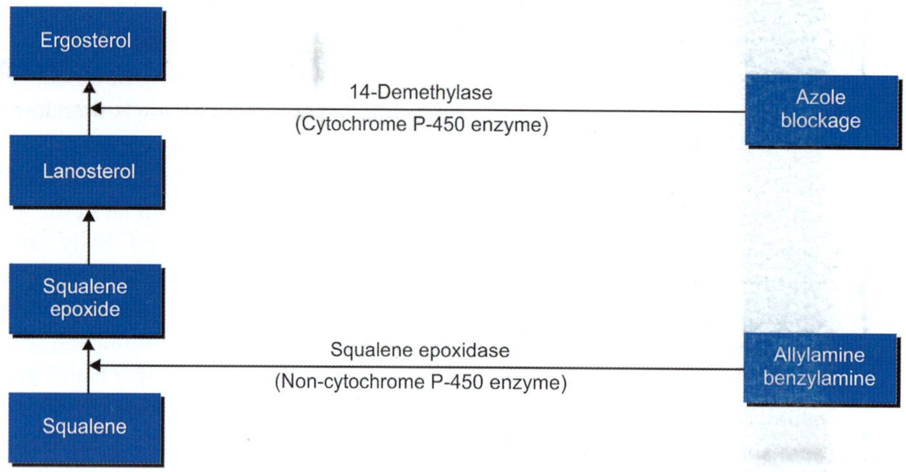

stratum corneum. The cure rates are high and drug remains effective several weeks after treatment is completed and relapse rates are low. The agent can be applied twice daily for as little as 1 week or once daily for 4 weeks for tinea pedis and once daily for 2 weeks for tinea cruris.

3. Azoles

In general azole derivatives are the most frequently used antimicrobial agents in routine clinical practice, not only for fungal but for protozoal, helminthic and even anaerobic bacterial infections. The basic structural unit of all azoles, used in medical mycology, is a five-member azole ring that is attached by carbon-nitrogen bonds to the other aromatic rings. The azoles are classified as **imidazoles** and **triazoles**, depending upon the presence of either **two** or **three** nitrogen in azole ring, respectively. Hence the imidazoles contain two nitrogen molecules in azole ring and triazoles have additional third nitrogen.

Mode of Action of Azoles

The azoles are fungistatic drugs and mode of action is inhibition of Cytochrome P450 (CYP) dependent C_{14} demethylation (sterol 14-α-demethylase enzyme) in biosynthesis of ergosterol present in fungal cell membrane. All azoles interact with Cytochrome P450 enzyme systems in fungal cells, resulting in impaired ergosterol biosynthesis, accumulation of 14 α-methylated sterols, a defective cell membrane and consequently fungal cell death as shown in the **Flowchart 6.1**. However, they also interact with mammalian Cytochrome P450 enzyme systems of host that are responsible for synthesis of many normal 'endobiotics' like cholesterol, retinoic acid, glucocorticoids, androgens, etc. Interference with synthesis of these key metabolic products accounts for the side effects of long-term azole therapy. Therefore, the baseline liver function tests are done initially and then after every 4-6 weeks during prolonged usage of azole to monitor effective course of therapy. These drugs also inhibit some of Gram-positive bacteria such as *Staphylococcus aureus*.

(a) Imidazoles

The imidazoles have two nitrogen molecules in azole ring. The following imidazoles are in common use - miconazole, clotrimazole, ketoconazole, econazole, tioconazole, sulconazole, isoconazole, bifonazole, oxiconazole and fenticonazole. The other imidazoles derivatives like thiabendazole, mebendazole and metronidazole have been used for treatment of parasitic and anaerobic bacterial infections. In the past, thiabendazole was also tried in a few fungal infections as well.

(i) Miconazole: This imidazole derivative has a broad-spectrum activity and has been used as a topical and oral agent. The topical preparation of miconazole is available as cream (1%) for superficial fungal infections like dermatophytoses, chronic mucocutaneous candidiasis, vulvovaginal candidiasis and pityriasis versicolor. However, systemic use is limited to *Pseudallescheria boydii* infection and certain forms of candidiasis where drug resistance to amphotericin B is inherently encountered. The daily dose of miconazole nitrate (Monistat) is 600 to 3600 mg in three divided doses.

***(ii) Clotrimazole*:** It is a tritylimidazole and is usually used topically. The drug is effective in dermatophytosis and *Candida* infections. One percent cream, solution and lotions of clotrimazole (Canesten/Lotrimin) are available for clinical use. This was the first oral azole to be used for fungal infections but not used presently through this route. It is most often used for vaginal candidiasis. It is effective in many systemic fungal diseases by giving through infusion. The use of clotrimazole is also limited by many side effects.

***(iii) Ketoconazole*:** This was the first successful oral azole antifungal agent used in 1980. Its oral bioavailability is critically dependent upon gastric pH and reduced in case of patients taking H_2 antagonists and antacids. Moreover, the protein binding is also very high. It used to be the drug of choice for chronic mucocutaneous candidiasis, blastomycosis, histoplasmosis, coccidioidomycosis, pityriasis versicolor and tinea corporis. It is not indicated in aspergillosis. The common side effects are anorexia, nausea, constipation, headache, hepatitis, reactions like pruritus and exanthema and inhibition of steroid hormone synthesis. When its oral intake came under question due to the side effects, it was extensively used topically in the shampoos, particularly for malasseziosis *i.e.* dandruff.

Ketoconazole is a potent inhibitor of adrenal androgen biosynthesis. Therefore, in a recent decision it has been ruled that it should not be used as a first-line treatment for any fungal infection because it may pose a risk for adrenal gland disturbances and potentially fatal liver injury. It is available as 200 mg tablets and dose is 200-400 mg/day. The liver function tests should be regularly monitored during therapy. If it exceeds two weeks duration and enzyme levels become triple during the course of therapy, its use should be discontinued.

The labels of ketoconazole now convey that the tablets are (a) no longer indicated for *Candida* and dermatophyte infections or (b) for skin and nail infections should not be used in patients with acute or chronic liver disease and (c) should only be used to treat endemic mycoses when patients do not respond to or can't tolerate other antifungal therapies.

(iv) Econazole: This is similar in activity to miconazole and clotrimazole when used topically. Sometimes, it may show side effects like irritation, burning sensation and itching. Econazole nitrate (Spectazole) cream (1%) is indicated for topical application in the treatment of tinea pedis, tinea cruris, tinea corporis, cutaneous candidiasis and pityriasis versicolor.

***(v) Bifonazole*:** This is a broad-spectrum antimycotic agent of imidazole group. It was introduced for therapy of superficial *Candida* infection and is found to be effective especially in vulvovaginal candidiasis, onychomycosis. Bifonazole shampoo (1%) has also been successfully tried in seborrheic dermatitis and pityriasis versicolor.

***(vi) Tioconazole*:** Tioconazole (Vagistat) is also a topical antifungal agent and is available as 6.5% cream.

***(vii) Oxiconazole*:** Oxiconazole nitrate 1% cream or lotion (Oxistat) is highly effective and well tolerated in a wide variety of tinea pedis infections.

(b) Triazoles

Triazole is a group of antifungal agents, which has three nitrogen molecules in the azole ring and includes fluconazole, itraconazole, voriconazole, terconazole, posaconazole and ravuconazole. These agents are used systemically and are safe as well as effective as compared to previously available imidazoles. They are preferably used for severe and disseminated fungal infections. Moreover, triazoles exert a reservoir effect hence used as intermittent pulse therapy also, particularly for superficial cutaneous infections.

***(i) Fluconazole*:** This triazole derivative was described in 1990 and has proper water solubility, oral absorption, extensive bioavailability independent of food or gastric pH, least protein binding and a sufficiently long half-life to allow once-a-day administration. It penetrates readily into CSF and is excreted unchanged in urine and feces, being metabolically stable with recovery over the most of administered dose.

It is effective in yeast-like fungi *e.g.* vaginal, oropharyngeal, esophageal and systemic candidiasis after oral administration. In culture proven acute cryptococcal meningitis, fluconazole (Diflucan/Forcan) is effective in dose of 100-200 mg/day even in AIDS patients. In histoplasmosis, blastomycosis, coccidioidomycosis and aspergillosis, the drug has shown mixed therapeutic response, after 1-12 months. Side effects are nausea, vomiting, abdominal pain, diarrhoea, headache and skin rash. The drug resistance to fluconazole has also been reported, particularly in *Candida* species and it is not effective in *C.krusei* and *C.glabrata*. Indiscriminate use of fluconazole, particularly as prophylactic agent in routine clinical practice, may be responsible for high drug resistance.

Fluconazole may be given as pulse therapy in dose of 150 mg once or twice weekly for a period of 2-3 months for treatment of superficial fungal infections like pityriasis versicolor, dermatophytoses and onychomycosis. A single dose therapy consisting of 400 mg fluconazole is also effective in pityriasis versicolor.

***(ii) Itraconazole*:** This triazole compound was first synthesized in 1980 and introduced for clinical use in 1993. It was found superior to other azoles in several respects. It has better distribution in tissues and its expanded half-life is 15-24 hours. It is a lipophilic compound characterized by good absorption after meals. It gains extensive distribution in all tissues except cerebrospinal fluid due to blood brain barrier. The drug is highly bound to plasma protein especially albumin and is degraded into several inactive metabolites and excreted primarily in bile and urine.

Itraconazole has less affinity for mammalian Cytochrome P450 than does ketoconazole hence there are less adverse effects due to very low hormonal suppression. No adjustment in dosage is required even with liver or kidney impairment. Due to longer expanded half-life, bioavailability of drug is augmented after daily dose of 100 mg/day and can achieve steady state in 10-14 days.

This drug is a broad-spectrum antifungal agent, active against most of the pathogenic yeasts as well as mycelial fungi except mucormycetes. It is very effective in superficial mycoses like dermatophytoses, oral and vaginal candidiasis, pityriasis versicolor and keratomycosis. It has also shown good response in systemic infections like disseminated aspergillosis, cryptococcosis, sporotrichosis, histoplasmosis, blastomycosis, coccidioidomycosis, paracoccidioidomycosis, phaeohyphomycosis, eumycetoma and chromoblastomycosis caused by *Cladosporium* species.

It may also be given in the immunocompromised patients. Moreover, it does not cause serious toxicity and there is no drug interaction with rifampicin, cyclosporin A and anticoagulants. However, itraconazole capsules are poorly absorbed especially in those on PPIs and H_2 blockers. Therefore, itraconazole solution is better absorbed on an empty stomach but is not as well tolerated as capsules. The side effects of itraconazole are headache, nausea and raised liver enzymes. It is available as 100 mg capsules (Sporanox/Canditral/Fungitrace) and daily oral dose is 100 to 400 mg. The drug may be given as pulse therapy in the dose of 200 mg twice daily for one week in a month for three consecutive months for the treatment of onychomycosis of both fingernails as well as toenails.

***(iii) Voriconazole*:** This is a novel triazole derivative with potent broad-spectrum activity against various fungi including those which are inherently supposed to be resistant to fluconazole, like non-*albicans Candida* species, *C.krusei.* Voriconazole is more effective than fluconazole in blocking candidial sterol biosynthesis, consistent with different antifungal potencies of these compounds. In addition to inhibition of sterol 14-α-demethylase enzyme in fungal cell biosynthesis, it is inhibitory to methylene dihydrolanosterol demethylation in some of yeasts and filamentous fungi, which might explain its increased activity against molds.

The strength of voriconazole lies in its strong activity against *Aspergillus* and *Cryptococcus,* its broad-spectrum activity against *Candida* species and low toxicity and good bioavailability. The drug can also act as fungicidal against *Aspergillus* species and is approved for treatment of invasive aspergillosis and for serious infections caused by *Scedosporium apiospermum* and *Fusarium* species in patients intolerant or refractory to other antifungals. The previous *in vitro* studies have shown that voriconazole has fungistatic activity against most yeast and many dematiaceous and hyaline filamentous fungi. It was licensed in the year 2002. The Indian version of voriconazole (Fungivor) is available as 200 mg tablets.

***(iv) Posaconazole*:** This is not only active *in vitro* against wide range of yeasts (*Candida* species) and molds (including *Aspergillus* species, mucormycetes and *Fusarium* species) but also appears to have an acceptable adverse-event profile and is well-tolerated. In *vitro,* posaconazole has potent and broad-spectrum activity against opportunistic, endemic and dermatophytic fungi. This activity extends to organisms that are often refractory to the existing triazoles, amphotericin B or echinocandins, such as *C.glabrata, C.krusei, Aspergillus terreus, Fusarium* species and mucormycetes.

This is available as an **oral** suspension and optimal exposure is achieved when the drug is administered in two to four divided doses along with food or nutritional supplement, particularly fat. Currently, posaconazole is only available as an oral suspension. Posaconazole was approved by the FDA and was licensed in 2006 and other regulatory bodies for the treatment of oropharyngeal candidiasis and the prophylaxis of invasive *Aspergillus* and *Candida* infections in severely immunocompromised patients. It is observed that posaconazole might be an attractive oral treatment alternative for mucormycosis. It is given as 800 mg/day, to begin with 200 QID as loading dose, followed by 400 BD.

Posaconazole is useful in the prophylaxis and treatment of fungal infections, particularly in neutropenic patients and bone marrow transplant recipients. However, it is administered orally and in seriously ill patients, absorption is often suboptimal. Therefore, keeping in view of this difficulty, its intravenous formulations are now available. Another issue with posaconazole is that it starts the effectivity after a fortnight or so. Hence it has to be started in advance of weaning other drug already going on so that antifungal cover can be kept maintained in the patient.

***(v) Isavuconazole*:** This is a new azole antifungal drug with a broad antifungal spectrum that includes yeasts, molds and dimorphic fungi. Its prodrug, isavuconazonium sulfate, was approved by FDA on March 6, 2015, mainly for treatment of two common and most challenging fungal infections in clinical practice, invasive aspergillosis and mucormycosis. It is available in both oral and intravenous formulations for once-a-day dosing and has favorable safety profile and drug interaction potential in comparison to voriconazole. Its role in the treatment of other fungal infections, besides aspergillosis and mucormycosis, remains to be determined. Similarly, its efficacy in prophylaxis against invasive fungal infections or its utility in patients with prior azole exposure is yet to be elucidated in the clinical studies.

Despite lack of standardization of antifungal susceptibility *in vitro* and poor correlations with efficacy *in vivo*, phaeoid fungi are usually more sensitive to triazoles and terbinafine then to amphotericin B. The newer azoles ravuconazole, voriconazole and posaconazole, all have *in vitro* activity comparable to itraconazole. They have potentiality to become drugs of choice in therapy of phaeoid fungal infections as well.

C. MISCELLANEOUS ANTIFUNGALS

Some of the drugs could not be placed under above-mentioned category of antibiotics and synthetic antifungal agents. Hence they are clubbed together and described under separate group as miscellaneous heading. Some of these agents like echinocandins may provide potent, broad-spectrum and versatile alternative to existing modalities for prevention of invasive mycoses in high-risk patients with neoplastic diseases. These antifungals are being described briefly as follows:

(a) Flucytosine

Flucytosine (5-Fluorocytosine) is a synthetic **fluoropyrimidine** and mainly used in treatment of infections caused by yeasts and phaeoid fungi. This is a **water-soluble** drug hence can be administered orally. It is converted by fungal **cytosine deaminase** to antimetabolite, 5-fluorouracil which inhibits **thymidylate synthetase** entailing inhibition of DNA synthesis.

This is a narrow-spectrum antifungal drug and is particularly used for treatment of cryptococcosis and chromoblastomycosis. The daily oral dose of flucytosine (5-FC/Ancoban/Ancotil) is 150 mg/kg six hourly. It can also serve as a companion drug to amphotericin B and in combination, it is having synergistic effects against most isolates of *Cryptococcus* and *Candida* species. However, primary and secondary resistance against flucytosine is frequently encountered.

Therefore, amphotericin B in combination with flucytosine prevents development of secondary resistance during the course of therapy. Antifungal drugs combination increases rate of microbial killing, shorten duration of therapy, avoid emergence of resistance, expand spectrum of activity and decrease related toxicities by allowing use of lower doses. The combination therapy is essentially useful in patients with chronic meningitis caused by *Cryptococcus* species.

Important side effects of this drug are gastrointestinal disturbances, neutropenia, thrombocytopenia, bone marrow depression and alopecia.

(b) Ciclopiroxolamine

This is analogue of pyridone and is unrelated to imidazole derivatives as apparently it looks like. However, unlike imidazoles and allylamines which inhibit synthesis of fungal cell walls, ciclopirox complexes polyvalent cations thus inhibiting metal-dependent enzymes including those responsible for peroxide degradation in fungal cell. The primary site of action is cell membrane where drug accumulates 200 times more than its surrounding medium. It inhibits membrane transfer of an amino acid, leucine. At higher concentrations, it begins to alter integrity of cell membrane and results in leakage of vital intracellular molecules. By inhibiting arachidonic acid cascade, the drug also has anti-inflammatory and anti-allergic activity.

It has a broad-spectrum activity against yeasts, molds including dermatophytes, actinomycetes and a variety of Gram-positive and Gram-negative bacteria. This makes it a useful agent for inflamed skin infections and dermatophytoses with secondary bacterial infection. It is available as 1% cream (Batrafen), commonly used as a topical agent against dermatophytosis and cutaneous candidiasis. The ciclopirox gel applied twice daily for 4 weeks is effective in the treatment of interdigital tinea pedis due to *T.rubrum*, *T.mentagrophytes* and *E.floccosum* and is associated with a low incidence of minor adverse effects. Its action is comparable to clotrimazole in its efficacy and side effects are also similar. As drug can penetrate into nails hence it is also indicated for treatment of onychomycosis. Ciclopirox nail lacquer is also used as a topical solution (8%) for onychomycosis. It has also been shown as very effective drug in the treatment of tinea nigra. Its topical application in the form of gel twice a day rapidly clears pigmented lesions in three days. Ciclopirox olamine cream 1% has been effective in treatment of tinea infections, cutaneous candidiasis, pityriasis versicolor and seborrheic dermatitis. The seborrheic dermatitis of scalp responds well to 1% ciclopirox shampoo once or twice weekly for 4 weeks.

(c) Morpholines

Amorolfine possesses excellent antifungal activity *in vitro* against several medically important fungi. The drug is highly efficacious against dermatophytes, phaeoid fungi and dimorphic pathogens and less active against molds like *Aspergillus* species.

Amorolfine is a unique, topically (0.25% cream) active fungicidal drug. It has lower minimum inhibitory concentration for dermatophytes than azole and naftifine. This has the same order of importance as that of terbinafine.

Amorolfine acts by inhibiting two separate enzymes, **reductase** and **isomerase** in pathway of ergosterol synthesis. The hyperfluidity of membrane not only lead to changes in its permeability adversely affecting fungal metabolic processes but also induce abnormal chitin deposition leading to growth abnormalities. It is found to be effective in inter-digital tinea pedis, tinea manuum, tinea cruris, tinea corporis and onychomycosis. The common side effects are local irritation, burning/stinging sensation and erythema.

(d) Whitfield's Ointment

This ointment is a mixture of benzoic acid (6%) and salicylic acid (3%) in a ratio of 2 : 1. The benzoic acid is mild fungistatic and salicylic acid is keratolytic and weak antifungal also. It is commonly used for treating pityriasis versicolor, tinea pedis and other types of dermatophytosis. It has been proved to be very effective and is considerably less expensive as it is used over a large surface area involved in such diseases.

(e) Castellani's Paint

This paint is carbol-fuchsin formulated by an Italian physician, Aldo Castellani in 1905, for the treatment of superficial fungal infections, when there were very few alternatives available. Active ingredient is phenol 1.5 percent. It was the first-aid antiseptic and drying agent and was very popularly used for the treatment of tinea imbricata (Tokelau). It is now used in otomycosis also.

Classical Castellani's solution contains 0.08 gm boric acid, 0.4 gm phenol, 0.04 gm fuchsine, 0.8 gm resorcinol, 0.4 ml acetone, 0.85 ml alcohol and 10 ml distilled water. Various modifications to the classical solution have been recommended. Fuchsine has antifungal activity and ethanol has antibacterial activity, whereas acetone makes the solution acidic. As there is toxicity of phenol, acetone and resorcinol, only topical application is recommended.

(f) Burow's Solution

Burow's solution was developed by Karl August von Burow, a German physician, as an ear drop. The solution is a colorless and acidic mixture containing aluminum acetate and can be prepared at different densities. In studies conducted using Burow's solution, 4%, 8% and 13% concentrations are used frequently. A ready-to-use preparation called Burow Galenik containing 4% aluminum acetate is available. In addition to the systemic toxic effects of aluminum acetate, its ototoxic effects have been reported. Burow's solution or 5% aluminum acetate solution should be used to reduce the swelling and remove the debris in otomycosis without perforation.

(g) Potassium Iodide

This is more than hundred years old drug and is being used for treating chronic subcutaneous fungal infections since the beginning of 20th century *i.e.* year 1903. The saturated solution of potassium iodide (SSKI) is the drug of choice for sporotrichosis, basidiobolomycosis and conidiobolomycosis.

The precise mechanism by which KI exerts its therapeutic effect against fungi is unknown, as it has no demonstrable inhibitory effect *in vitro*. Contrarily *Sporothrix schenckii, Basidiobolus* and *Conidiobolus* actually grow when plated in the presence of KI. But clinically, it is very effective drug against these very fungi. Therefore, it is still unclear whether KI acts against fungi by a fungicidal mechanism or by enhancing body's immunologic and non-immunologic defence mechanisms. It is contemplated that most probably it resolves the fungal granuloma so that body defence mechanism can attack the causative fungus.

This is used for cutaneous and lymphocutaneous sporotrichosis. In tropics, it is indicated for entomophthoramycosis, chronic, slowly progressive subcutaneous diseases caused by fungi *Basidiobolus* and *Conidiobolus*. It is also given in cutaneous cryptococcosis as well as pythiosis. Moreover, it is frequently used for inflammatory dermatoses of non-fungal origin like erythema nodosum, subacute nodular migratory panniculitis, nodular vasculitis, erythema multiforme and Sweet's syndrome.

For convenience, KI is usually administered in the form of a saturated solution (SSKI) as a dose of 50 mg/drop. SSKI is made by adding KI to hot purified water. This solution can be added to water, fruit juice or milk just before taking simply to protect gastrointestinal irritation. The initial dose is 5 drops thrice daily which can be increased to 5 drops per week amounting to 25-40 drops thrice daily with milk or juice and given for a period of 6-12 weeks.

Potassium iodide is not free from side effects. The excess quantity of iodine is counterproductive of thyroid hormone production. Hence thyroid gland stops producing hormone, which is called Wolff-Chaikoff effect. This is due to inhibition of organic binding of iodide in thyroid gland by circulating excess iodide thereby resulting in cessation of thyroid hormone synthesis. Therefore, periodic thyroid functions should be done while the patient is on SSKI therapy.

This is a very cheap drug hence unable to generate profits thereby it has never been an ideal choice for the pharmaceuticals. They propagate alternative drugs in place of SSKI. But in nutshell it is very effective and useful drug for the treatment of subcutaneous fungal infections.

(h) Selenium Sulfide

This is mainly used for treating pityriasis versicolor, dandruff and white piedra. It is mildly toxic and possesses unpleasant odor. Selenium sulfide (Selsun) is available in strength of 2.5% solution as shampoo for topical application.

(i) Undecylenic Acid

Undecylenic acid is available in numerous preparations including foams and soaps and in zinc, copper and calcium salts. The usual indication is tinea pedis (Athlete's foot). It is available as powder and ointment in strength of 10% for application to skin and 1% for mucous membranes.

(j) Haloprogin

This is a synthetic topical antifungal agent used in the treatment of superficial fungal infections of skin caused by dermatophytes, particularly athlete's foot. It also has some activity against cutaneous candidial infections. The cure rate is slightly higher and relapses are less frequent with haloprogin than with tolnaftate. It is available as 1% cream (Halotex). There were many reports of irritation and contact allergy entailing stoppage of its use.

(k) Triacetin

This is indicated for milder superficial fungal infection *e.g.* Athlete's foot. It is available as a cream, ointment and liquid (25% with cetylpyridinium and chloroxylenol) to be applied twice daily for one week even after symptoms have subsided.

(l) Echinocandins

The echinocandin analogues are such as pneumocandins and papulocandins. These are large cyclic peptides linked to long fatty acids and are shown as potentially effective antifungal agents. This novel class irreversibly inhibits (1, 3)-β-D-glucan synthase, the enzyme complex that forms glucan polymers in the fungal cell wall and absent in the mammalian cells. Hence they prevent cell wall synthesis of fungi by blocking this glucan synthase enzyme hence are cidal against yeasts and static against mycelial fungi.

The three candins *i.e.* caspofungin, micafungin and anidulafungin were licensed for use in fungal diseases in the years 2001, 2005 and 2006, respectively. These have efficient activity against *Candida* and *Aspergillus* species however moderately against *Histoplasma*, *Blastomyces* and *Scedosporium* species. But these are not effective against basidiomycetous yeasts such as *Cryptococcus*, *Trichosporon* and *Rhodotorula* species and upto some extent a few mycelial fungi. All these are available as intravenous preparations. Caspofungin is lipopeptide compound synthesied from a fermentation product of *Glarea lozoyensis* (1989), micafungin - *Coleophoma empetri* and anidulafungin - *Aspergillus nidulans.*

These are indicated for the treatment of invasive aspergillosis in patients who are either refractory or intolerant to other therapies with amphotericin B, its lipid formulation and/or itraconazole. In animal models it is found to have good efficacy against cell wall of *Pneumocystis jirovecii* also.

The echinocandins have mode of action that is different from all the other antifungals, consisting of inhibition of (1, 3)-β-D-glucan synthase, which produces an important component of the cell wall of many pathogenic fungi such as *Candida* and *Aspergillus* species. Inhibition of (1, 3)-β-D-glucan synthesis disrupts the structure of the growing cell wall thereby resulting in osmotic instability and death of susceptible yeast cells. The cell wall of *Cr.neoformans* consists mainly of α-(1, 3)- or α-(1, 6)-glucan therefore it is resistant to echinocandins. The target of echinocandins does not exist in mammalian cells so their toxicity is minimal. All three agents are water-soluble and available only for intravenous use. The candins are not indicated in endophthalmitis cases due to their problem with absorption in the eye.

The echinocandins, like penicillins show paradoxical effect or Eagle phenomenon, which refers to an observation of an increase in survivors, seen when testing the activity of an agent. Initially when an agent is added to a culture medium, the number of that survives drops, as you would expect. But after increasing the concentration beyond a certain point, the number of fungi that survive, paradoxically, increases.

(m) Sulfonamides

This group of drugs is used for the treatment of bacterial diseases in general. Actinomycetoma is also bacterial in origin hence it can be used for its treatment. But paracoccidioidomycosis is systemic and endemic fungal disease and sulfonamides are used for its treatment.

(n) Miltefosine

Miltefosine (MLT) is an alkyl-phospholipid analogue and was originally developed as an anti-tumor drug. It has also shown inhibitory activities against *Leishmania* species and *Trypanosoma cruzi*. In many countries of Latin America, India, Germany and others, it is frequently used for the treatment of leishmaniasis and breast cancer. Recently, it was approved by the FDA for cutaneous and visceral leishmaniasis.

In addition to its antiparasitic activity, investigators have demonstrated *in vitro* its antifungal activities against dimorphic fungi such as *Coccidioides posadasii*, *Histoplasma capsulatum* and *Sporothrix* species, clinically relevant moulds such as azole-susceptible and -resistant *Aspergillus*, *Fusarium solani* and *Fusarium oxysporum*, dermatophytes, *Scedosporium* species and yeasts including *Candida* species, *Cryptococcus neoformans* and *Cr.gattii*.

Moreover, it has displayed antibiofilm action against *Candida albicans* and non-*albicans Candida* species and *F.oxysporum*. *In vivo* studies have also demonstrated its efficacy in murine models of disseminated cryptococcosis and oral candidiasis. It has a fungicidal effect against both *Paracoccidioides* species and a sub-inhibitory concentration impacted melanogenesis.

(o) Probiotics as Antifungals

The recent clinical studies highlight that some of the probiotics may reduce *Candida* colonization on human mucosal surfaces, relieve signs and symptoms of fungal infection and enhance the antifungal effect of conventional therapy, implying that probiotics have a potential to sustain a healthy mucosal microbiota by acting both as prophylactic and adjunctive therapy against candidiasis. The *in vitro* studies suggest that the antifungal effect of probiotics is due to their interference with *Candida* biofilm development and hyphal differentiation.

However, it is premature to designate probiotics as an alternative to antifungals as yet, due to the paucity of available clinical trials. In particular, case-control clinical trials with adequate patient numbers are warranted, not only to ascertain the activity of the probiotic formulations but also to evaluate their dosage, administration schedules, side effects and biodynamics in humans. As with any formulations of live organisms, other concerns that need further investigation include the potential for selection of resistant strains, mutability and tolerability on prolonged use as well as pathogenic potential in immunocompromised patients. Therefore, probiotics may serve as worthy allies in the battle against chronic mucosal fungal infections.

(p) Nikkomycin

Nikkomycins are described as new antifungal agents which act through inhibitory action of chitin synthetase required for cell wall synthesis. Nikkomycins X and Z have been tried in *Candida albicans*, *Blastomyces dermatitidis* and *Coccidioides* species infections.

In addition to antifungal agents, cytokines are also tried in treatment of fungal infections like C-CSF, M-CSF,

IFN, IL-1 and TNF. The triphenylethylenes is also novel antifungal class, mainly used against cryptococcosis.

D. RELATED THERAPEUTIC ISSUES

Some of the issues which crop up during the course of therapeutic management and otherwise for an in-depth understanding on this topic are briefly described. The most important issue is antifungal susceptibility testing (AFST), which is given in detail as **Appendix-D**.

1. Antifungal Practice Guidelines

The practice guidelines are systematically developed statements issued by an organization to assist practitioners and patients in making decisions about appropriate health care for specific clinical circumstances. The attributes of good guidelines include validity, reliability, reproducibility, clinical applicability, clinical flexibility, clarity, multidisciplinary process, review of evidence and documentation. From time to time the guidelines on fungal diseases are released by various organizations like Infectious Disease Society of America, in general about infections or on a particular infection. On similar pattern Japanese *'Guidelines for the Diagnosis and Treatment of Deep-Seated Mycosis'* have also been released. After some time gap, their effectivity is reviewed thereby new guidelines are issued encompassing the latest developments in the field, overriding the already existing ones. Guidelines are general measures and are very helpful for medical personnel who are not experts in that area.

2. Antifungal Drug Resistance

Some of fungi do not respond to antifungal drug therapy due to development of resistance. In the beginning of antimicrobial therapy, this type of difficulty was being faced in bacterial or parasitic infections but now has cropped up in fungal diseases as well. The antifungal drug resistance can be divided into two broad categories:

(a) Clinical Resistance
(b) *In vitro* Resistance

(a) Clinical Resistance: The clinical resistance indicates lack of a clinical response to antifungal drugs used in a particular disease. More often clinical failure is due to low levels of drug in serum and/or tissue for several reasons, mainly noncompliance with medication regimen. A significant reason for clinical failure or resistance in AIDS patients is presence of severely immunosuppressive state of patient, where antifungal agents alone, including high-dose fungicidal agents, are unable to eradicate fungi.

(b) In vitro Resistance: *In vitro* resistance can be subdivided into **primary resistance** and **secondary (acquired) resistance**. The primary resistance is known as intrinsic or innate resistance and occurs when organism is naturally resistant to antifungal agent *e.g. C.krusei* and *C.auris* are known to be generally resistant to fluconazole. On the other hand, secondary or acquired resistance is described when isolate producing infection becomes resistant to antifungal agent during course of treatment. This form of resistance, which was rare in the past, is now most frequently reported in AIDS patients, who suffer from recurrent azole-resistant oropharyngeal or oesophageal candidiasis.

3. Farm Fungicide

The massive use of fungicides to protect European orchards, vineyards and grain fields may be contributing to resistance against drugs used to treat people with life-threatening infections of *Aspergillus fumigatus.* Although the overuse of antibiotics in animal husbandry is known to have caused resistance in the human population, this would be the first time a similar link is found between farm use of fungicides and human health. Invasive aspergillosis due to multi-azole-resistant *A.fumigatus* has emerged in the Netherlands since 1999, with 6.0-12.8% of patients harboring resistant isolates. The presence of a single resistance mechanism (denoted by TR/L98H), which consists of a substitution at codon 98 of cyp51A and a 34-bp tandem repeat in the gene-promoter region, was found in over 90% of clinical *A.fumigatus* isolates. This is consistent with a route of resistance development through exposure to azole compounds in the environment. Indeed, TR/L98H *A.fumigatus* isolates were cultured from soil and compost, were shown to be cross-resistant to azole fungicides and genetically related to clinical resistant isolates. Azoles are abundantly used in the environment and the presence of *A.fumigatus* resistant to medical triazoles is a major challenge because of the possibility of worldwide spread of resistant isolates. Reports of TR/L98H in other European countries indicate that resistance might already be spreading.

4. Antifungal Stewardship in India

Antifungal stewardship refers to a systematic program that promotes the optimal use of antifungals through the careful selection of agents based on patient profile, target organism, toxicity, costs and the likelihood of emergence and spread of resistance.

A vast number of pathogenic fungi, increasing susceptible population, invasive fungal infections, limited number of available drugs and increasing resistance, all necessitates the existence of antifungal stewardship. Key areas where antifungals are given empirically or presumptively are ICUs, hematology, transplant and oncology units. No doubt the task is not easy especially if we compare it to antibacterial stewardship programs where organisms grow overnight and getting an antimicrobial sensitivity testing report is much easier.

In antifungal stewardship program, there is need for hospitals to keep record of the antifungal prescription, consumption, response, therapeutic drug monitoring (TDM), therapeutic streamlining for escalation or de-escalation of drugs along with the patient outcome and total expenditure incurred during the course of treatment.

Since few guidelines/studies are available for stewardship, antifungals are often prescribed subjectively by the clinician, depending upon the patient's economical condition and comfort. Numerous reasons have been identified for improvement in the antifungal prescribing practices including failure to streamline, inappropriate agent selection, inappropriate duration, inappropriate route of administration, antifungal treatment unnecessary and inappropriate antifungal dose.

In India, on December 10, 2015, Delhi Chapter of Indian Association of Medical Microbiologists organized one-day workshop under the leadership of Wattal *et al*, in collaboration with British Society for Antimicrobial Chemotherapy, to design outline towards the development of antifungal stewardship program. This aimed at developing a road-map for optimizing better outcomes in patients with invasive fungal diseases while minimizing unintended consequences of antifungal use, ultimately leading to reduced healthcare costs and prevention development of resistance to the antifungals.

Moreover, this workshop was also a conclave of all stakeholders, eminent experts from India as well as United Kingdom *i.e.* clinical microbiologists, critical care specialists and infectious disease physicians. The issues discussed were like epidemiology, diagnostic and therapeutic algorithms in different healthcare settings in managing invasive fungal diseases. A consensus was formulated, outlining step-by-step approach to tackling the increasing incidence of invasive fungal diseases and antifungal drug resistance in Indian context.

5. Antifungal Therapeutic Drug Monitoring

The therapeutic drug monitoring **(TDM)** is measurement of drug levels in serum of patient during the course of treatment, who is on antifungals. The assay gives indication of whether suitable blood levels of the particular drug are being achieved or not. This is done to monitor levels of antifungal drugs during the course of treatment. It is indicated during the treatment with itraconazole and voriconazole. Excess concentration is demonstrated with flucytosine. The TDM is not required for amphotericin B, fluconazole, caspofungin and micafungin.

The burden of human disease related to medically important fungal pathogens is substantial. An improved understanding of antifungal pharmacology and antifungal pharmacokinetic-pharmacodynamics (PK/PD) has resulted in therapeutic drug monitoring becoming a valuable adjunct to the routine administration of some antifungal agents. TDM may increase the probability of a successful outcome, prevent drug-related toxicity and potentially prevent the emergence of antifungal drug resistance. However, much of the current evidence that supports TDM is circumstantial.

Therapeutic drug monitoring involves the measurement of plasma or serum drug concentration to adapt dosages to achieve predefined target concentrations that are associated with optimal clinical response while minimizing the chance of encountering toxicity. Most of the studies in the field of antifungal drugs have focused on the evidence that supports the use of TDM thereby emphasizing the breakpoints or target concentrations in general literature. Knowledge on the complete process of TDM including pharmacokinetics and relevant covariates, pharmacodynamic aspects, trials that are necessary to provide with evidence, translation of knowledge to other populations and pathogens and implications for the pre-analytical, analytical and post-analytical phases the process of TDM are pertinent to understand. For each individual step, recommendations are made for reference.

The sample collections are also very important as for oral therapy, random sample may be sufficient but for IV therapy pre-dose and post-dose specimens are taken between particular timings. About 200 μl blood is required per sample to perform these assays. All antiseptic precautions should be taken into account while performing these tests.

The drug levels can be measured using bioassay, mass spectrophotometry and high performance liquid chromatography. Out of these, the bioassay technique is the simplest as it does not require any expensive equipment.

However, it is very time consuming and has its limitation when the patient is on various combination of antifungal drugs. The mass spectrometry and high performance liquid chromatography are less time consuming but some drug extraction procedures are lengthy. The only limitation in these methods is requirement of the expensive instruments, which the routine diagnostic mycology laboratory may find difficult to afford.

(a) Bioassay Technique: In this technique, agar is seeded with an appropriate organism, which is sensitive to drug to be measured. This is poured into a large, square petri dish and once set, 36 wells are cut. These wells are loaded with standards of known concentration, quality control spiked samples and patients' sera. The plate is incubated for overnight *i.e.* approximately 18 hrs at 37°C. The zones of inhibition are observed around each well reflecting how much drug is present *i.e.* larger the zone size, greater the concentration of drug in sample. The diameters of zone of inhibition are measured with dial callipers and recorded on the result sheet and fed into the in Excel spreadsheet. The mean average of each sample is plotted on a semi-logarithmic and the line of best fit is drawn.

(b) Mass Spectrometry: Mass spectrometry is an analytical tool used for measuring the molecular mass of a sample.

(c) High Performance Liquid Chromatography: In certain cases like itraconazole is highly lipophilic drug. It is rarely detectable in body fluids such as urine and CSF hence the bioassays are not indicated rather if it is necessary for pharmacokinetic purposes, to measure these fluids by HPLC.

6. ARTEMIS Global Antifungal Surveillance

The ARTEMIS DISK global antifungal surveillance program is among the most comprehensive and long-running fungal surveillance programs. This is supported by grants from the Pfizer. It was designed to address many of the potential limitations of resistance surveillance studies namely:

(a) It is both longitudinal (1997 to present) and global (142 participating sites in 41 countries) in scope.
(b) It employs standardized Disk Diffusion and Broth Microdilution antifungal susceptibility test methods.
(c) Both internal quality control performed in each participating laboratory and external quality assurance measures are used to validate test results.
(d) Results are recorded electronically using the BIOMIC image analysis plate reader and are stored in a central database.
(e) Both *Candida* and non-*Candida* yeast isolates obtained from consecutive clinical samples from all body sites are tested locally thus avoiding misleading results based on biased selective testing.

Therefore, ARTEMIS program generates massive amounts of data that is externally validated and can be used to identify temporal and geographic trends in the species distribution of *Candida* and other opportunistic yeast as well as the resistance profiles of these organisms to fluconazole and voriconazole as determined by standardized CLSI Disk Diffusion methods.

7. Prospective Antifungal Therapy

A lot of progress was made in 1990-2000s with regard to new diagnostic tools in the field of invasive fungal infections, including the use of CT scans, the GM EIA, and the β-D-glucan test. The data suggests that traditional diagnostic modalities are used in the majority of invasive fungal infections, with culture remaining the most frequently used diagnostic test.

In the recent times lots of websites and/or databases have been created which register data pertaining to any of the fungal issues, analyze it and publish it periodically from time to time. The corrective measures are undertaken according to the feedback received and future course of action is contemplated. Keeping in view of this background, the prospective antifungal therapy **(PATH)** alliance registry was created in 2004 to undertake prospective surveillance of invasive fungal infections in patients hospitalized in tertiary care medical centres in North America. Its aim was to collect detailed descriptions of the diagnosis, fungal species, underlying conditions, treatment regimens and outcomes of such invasive fungal infections.

This is the biggest registry of invasive fungal infections from 25 institutions in North America providing a unique opportunity to study and further understand the epidemiology of even rare infections. Hence this registry provides a unique tool which can identify trends and outcomes pertaining to common invasive fungal infections, such as invasive candidiasis and invasive aspergillosis, those with limited data to date such as mucormycosis as well as invasive fungal infections due to rare molds and yeasts.

The PATH alliance is a comprehensive registry, which collects data and monitors trends in the epidemiology, diagnosis, treatment and outcomes of patients with invasive fungal infections. This long-term initiative was undertaken with the overarching goal of shaping improvements in care of patients with invasive fungal infections through the dedicated and detailed collection of data and analyses using advanced bioinformatic tools. The approaches and novel methods used in the creation of this database are documented.

The PATH alliance was established in order to perform prospective surveillance of all invasive fungal infections among patients hospitalized at selected tertiary care medical centers in North America. Its aim was to improve the understanding of the burden of invasive fungal infections in different patient groups, better define patients at risk and understand current approaches to the diagnosis and treatment of invasive fungal infections. The significant trends and treatment practices concerning yeasts and molds are observed. As enrollment continues, additional data is analyzed and published, which provides valuable information concerning the epidemiology, therapy and outcomes of invasive fungal infections.

Further Reading

1. Agrawal S, Barnes R, Bruggemann RJ, *et al.* The role of the multidisciplinary team in antifungal stewardship. *J Antimicrob Chemother.* 2016; **71 (Suppl. 2):** ii37-42.
2. Agrawal S, Hope W, Sink J, *et al.* Optimizing management of invasive mould diseases. *J Antimicrob Chemother.* 2011; **66 (Suppl. 1):** i45-53.
3. Agrawal S, Jones B, Barnes R, *et al.* A practical critique of antifungal treatment guidelines for haemato-oncologists. *Crit Rev Microbiol.* 2012; **38:** 203-16.
4. Aitken SL, Beyda ND, Shah DN, *et al.* Clinical practice patterns in hospitalized patients at risk for invasive candidiasis: Role of antifungal stewardship programs in an era of rapid diagnostics. *Ann Pharmacother.* 2014; **48:** 683-90.
5. Akers KS, Rowan MP, Niece KL, *et al.* Antifungal wound penetration of amphotericin and voriconazole in combat-related injuries: Case report. *BMC Infect Dis.* 2015; **15:** 184.
6. Akhtar N, Verma A, Pathak K. Topical delivery of drugs for the effective treatment of fungal infections of skin. *Curr Pharm Des.* 2015; **21:** 2892-913.
7. Ananda-Rajah MR, Kontoyiannis D. Isavuconazole: A new extended spectrum triazole for invasive mold diseases. *Future Microbiol.* 2015; **10:** 693-708.
8. Ananda-Rajah MR, Slavin MA, Thursky KT. The case for antifungal stewardship. *Curr Opin Infect Dis.* 2012; **25:** 107-15.
9. Arendrup MC, Cuenca-Estrella M, Lass-Florl C, *et al.* Breakpoints for antifungal agents: An update from EUCAST focussing on echinocandins against *Candida* spp. and triazoles against *Aspergillus* spp. *Drug Resist Updat.* 2013; **16:** 81-95.
10. Arendrup MC, Garcia-Effron G, Lass-Florl C, *et al.* Echinocandin susceptibility testing of *Candida* species: Comparison of EUCAST EDef 7.1, CLSI M27-A3, Etest, disk diffusion and agar dilution methods with RPMI and Iso-Sensitest media. *Antimicrob Agents Chemother.* 2010; **54:** 426-39.
11. Arendrup MC, Perlin DS. Echinocandin resistance: An emerging clinical problem? *Curr Opin Infect Dis.* 2014; **27:** 484-92.
12. Arendrup MC. Update on antifungal resistance in *Aspergillus* and *Candida. Clin Microbiol Infect.* 2014; **20 (Suppl. 6):** 42-8.
13. Arvanitis M, Mylonakis E. Characteristics, clinical relevance and the role of echinocandins in fungal-bacterial interactions. *Clin Infect Dis.* 2015; **61:** S630-4.
14. Ashbee HR, Barnes RA, Johnson EM, *et al.* Therapeutic drug monitoring (TDM) of antifungal agents: Guidelines from the British Society for Medical Mycology. *J Antimicrob Chemother.* 2014; **69:** 1162-76.
15. Austin N, Darlow BA, McGuire W. Prophylactic oral/topical non-absorbed antifungal agents to prevent invasive fungal infection in very low birth weight infants. *Cochrane Database Syst Rev.* 2009; **4:** CD003478.
16. Azie N, Neofytos D, Pfaller M, *et al.* The PATH (Prospective Antifungal Therapy) alliance registry and invasive fungal infections: Update 2012. *Diagn Microbiol Infect Dis.* 2012; **73:** 293-300.
17. Bailly S, Bouadma L, Azoulay E, *et al.* Failure of empirical systemic antifungal therapy in mechanically ventilated critically ill patients. *Am J Respir Crit Care Med.* 2015; **191:** 1139-46.
18. Becce F, Malghem J, Lecouvet FE, *et al.* Clinical images: voriconazole-induced periostitis deformans. *Arthritis Rheum.* 2012; **64:** 3490.
19. Benjamin Lash D, Jolliff J, Munoz A, *et al.* Cross-reactivity between voriconazole, fluconazole and itraconazole. *J Clin Pharm Ther.* 2016; **41:** 566-7.
20. Bhalaria MK, Naik S, Misra AN. Ethosomes: A novel delivery system for antifungal drugs in the treatment of topical fungal diseases. *Indian J Exp Biol.* 2009; **47:** 368-75.
21. Blyth CC, Hale K, Palasanthiran P, *et al.* Antifungal therapy in infants and children with proven, probable or suspected invasive fungal infections. *Cochrane Database Syst Rev.* 2010; **2:** CD006343.
22. Bonifaz A, Tirado-Sanchez A, Graniel MJ, *et al.* The efficacy and safety of sertaconazole cream (2%) in diaper dermatitis candidiasis. *Mycopathologia.* 2013; **175:** 249-54.
23. Botero Aguirre JP, Restrepo Hamid AM. Amphotericin B deoxycholate versus liposomal amphotericin B: Effects on kidney function. *Cochrane Database Syst Rev.* 2015; **11:** CD010481.
24. Brilhante RS, Caetano EP, Sidrim JJ, *et al.* Ciprofloxacin shows synergism with classical antifungals against *Histoplasma capsulatum* var. *capsulatum* and *Coccidioides posadasii. Mycoses.* 2013; **56:** 397-401.
25. Bruggemann RJ, Aarnoutse RE. Fundament and prerequisites for the application of an antifungal TDM service. *Curr Fungal Infect Rep.* 2015; **9:** 122-9.
26. Bulbake U, Doppalapudi S, Kommineni N, *et al.* Liposomal formulations in clinical use: An updated review. *Pharmaceutics.* 2017; 9. pii: e12. PMID: 28346375.

27. Bunyaratavej S, Leeyaphan C, Rujitharanawong C, *et al.* Efficacy of 5% amorolfine nail lacquer in *Neoscytalidium dimidiatum* onychomycosis. *J Dermatolog Treat.* 2016; **27:** 359-63.
28. Calderone R, Sun N, Gay-Andrieu F, *et al.* Antifungal drug discovery: The process and outcomes. *Future Microbiol.* 2014; **9:** 791-805.
29. Calo S, Shertz-Wall C, Lee SC, *et al.* Antifungal drug resistance evoked via RNAi-dependent epimutations. *Nature.* 2014; **513:** 555-8.
30. Campitelli M, Zeineddine N, Samaha G, *et al.* Combination antifungal therapy: A review of current data. *J Clin Med Res.* 2017; **9:** 451-6.
31. Campoy S, Adrio JL. Antifungals. *Biochem Pharmacol.* 2016. pii: S0006-2952(16)30422-1. PMID: 27884742.
32. Candoni A, Aversa F, Busca A, *et al.* Combination antifungal therapy for invasive mould diseases in haematologic patients. An update on clinical data. *J Chemother.* 2015; **27:** 1-12.
33. Carrillo-Munoz AJ, Tur-Tur C, Giusiano G, *et al.* Sertaconazole: An antifungal agent for the topical treatment of superficial candidiasis. *Expert Rev Anti Infect Ther.* 2013; **11:** 347-58.
34. Cavaleiro C, Salgueiro L, Goncalves MJ, *et al.* Antifungal activity of the essential oil of *Angelica major* against *Candida, Cryptococcus, Aspergillus* and dermatophyte species. *J Nat Med.* 2015; **69:** 241-8.
35. Chai LY, Kullberg BJ, Earnest A, *et al.* Voriconazole or amphotericin B as primary therapy yields distinct early serum galactomannan trends related to outcomes in invasive aspergillosis. *PLoS One.* 2014; **9:** e90176.
36. Chandrasekar P. Management of invasive fungal infections: A role for polyenes. *J Antimicrob Chemother.* 2011; **66:** 457-65.
37. Chen SC, Playford EG, Sorrell TC. Antifungal therapy in invasive fungal infections. *Curr Opin Pharmacol.* 2010; **10:** 522-30.
38. Chen SC, Slavin MA, Sorrell TC. Echinocandin antifungal drugs in fungal infections: A comparison. *Drugs.* 2011; **71:** 11-41.
39. Chitasombat MN, Kontoyiannis DP. The 'cephalosporin era' of triazole therapy: Isavuconazole, a welcomed newcomer for the treatment of invasive fungal infections. *Expert Opin Pharmacother.* 2015; **16:** 1543-58.
40. Chowdhary A, Hagen F, Curfs-Breuker I, *et al. In vitro* activities of eight antifungal drugs against a global collection of genotype *Exserohilum* isolates. *Antimicrob Agents Chemother.* 2015; **59:** 6642-5.
41. Chowdhary A, Sharma C, Hagen F, *et al.* Exploring azole antifungal drug resistance in *Aspergillus fumigatus* with special reference to resistance mechanisms. *Future Microbiol.* 2014; **9:** 697-711.
42. Clark NM, Grim SA, Lynch JP 3rd. Posaconazole: Use in the prophylaxis and treatment of fungal infections. *Semin Respir Crit Care Med.* 2015; **36:** 767-85.
43. Cortegiani A, Russotto V, Raineri SM, *et al.* The paradox of the evidence about invasive fungal infections prevention. *Critical Care.* 2016; **20:** 114.
44. Cortegiani A, Russotto V, Maggiore A, *et al.* Antifungal agents for preventing fungal infections in non-neutropenic critically ill patients. *Cochrane Database Syst Rev.* 2016; **1:** CD004920.
45. Cortegiani A, Russotto V, Raineri SM, *et al.* Antifungal prophylaxis: Update on an old strategy. *Eur J Clin Microbiol Infect Dis.* 2016; **35:** 1719-20.
46. Cortegiani A, Russotto V, Raineri SM, *et al.* Is it time to combine untargeted antifungal strategies to reach the goal of 'early' effective treatment? *Critical Care.* 2016; **20:** 241.
47. Cortegiani A, Russotto V, Giarratano A. Associations of antifungal treatments with prevention of fungal infection in critically ill patients without neutropenia. *JAMA.* 2017; **317:** 311-2.
48. Costa RO, Macedo PM, Carvalhal A, *et al.* Use of potassium iodide in dermatology: Updates on an old drug. *An Bras Dermatol.* 2013; **88:** 396-402.
49. Cowen LE, Sanglard D, Howard SJ, *et al.* Mechanisms of antifungal drug resistance. *Cold Spring Harb Perspect Med.* 2014; **5:** a019752.
50. Croxtall JD, Plosker GL. Sertaconazole: A review of its use in the management of superficial mycoses in dermatology and gynaecology. *Drugs.* 2009; **69:** 339-59.
51. Cuenca-Estrella M. Antifungal drug resistance mechanisms in pathogenic fungi: From bench to bedside. *Clin Microbiol Infect.* 2014; **20 (Suppl. 6):** 54-9.
52. Das S, Shivaprakash MR, Chakrabarti A. New antifungal agents in pediatric practice. *Indian Pediatr.* 2009; **46:** 225-31.
53. Dekkers BG, Bakker M, van der Elst KC, *et al.* Therapeutic drug monitoring of posaconazole: An update. *Curr Fungal Infect Rep.* 2016; **10:** 51-61.
54. Del Rosso JQ, Kircik LH. Optimizing topical antifungal therapy for superficial cutaneous fungal infections: focus on topical naftifine for cutaneous dermatophytosis. *J Drugs Dermatol.* 2013; **12:** S165-71.
55. Del Rosso JQ. The role of topical antifungal therapy for onychomycosis and the emergence of newer agents. *J Clin Aesthet Dermatol.* 2014; **7:** 10-8.
56. Delsing CE, Gresnigt MS, Leentjens J, *et al.* Interferon-gamma as adjunctive immunotherapy for invasive fungal infections: A case series. *BMC Infect Dis.* 2014; **14:** 166.
57. Denning DW, Bromley MJ. How to bolster the antifungal pipeline. *Science.* 2015; **347:** 1414-6.
58. Denning DW, Hope WW. Therapy for fungal diseases: Opportunities and priorities. *Trends Microbiol.* 2010; **18:** 195-204.
59. Denning DW, Perlin DS. Azole resistance in *Aspergillus*: A growing public health menace. *Future Microbiol.* 2011; **6:** 1229-32.
60. Denning DW, Perlin DS, Muldoon EG, *et al.* Delivering on antimicrobial resistance agenda not possible without improving fungal diagnostic capabilities. *Emerg Infect Dis.* 2017; **23:** 177-83.
61. Donnelly JP, Maertens J. The end of the road for empirical antifungal treatment? *Lancet Infect Dis.* 2013; **13:** 470-2.

62. Drew RH, Townsend ML, Pound MW, *et al.* Recent advances in the treatment of life-threatening, invasive fungal infections. *Expert Opin Pharmacother.* 2013; **14:** 2361-74.
63. Dreyfuss D, Ricard JD, Gaudry S. Amphotericin B deoxycholate for candidiasis in intensive care unit patients revisited: medical, ethical and financial implications. *Am J Respir Crit Care Med.* 2013; **187:** 661-3.
64. Eggimann P, Lamoth F, Marchetti O. On track to limit antifungal overuse! *Intensive Care Med.* 2009; **35:** 582-4.
65. Enserink M. Farm fungicides linked to resistance in a human pathogen. *Science.* 2009; **326:** 1173.
66. Falci DR, Pasqualotto AC. Profile of isavuconazole and its potential in the treatment of severe invasive fungal infections. *Infect Drug Resist.* 2013; **6:** 163-74.
67. Fernandez-Garcia R, de Pablo E, Ballesteros MP, *et al.* Unmet clinical needs in the treatment of systemic fungal infections: The role of amphotericin B and drug targeting. *Int J Pharm.* 2017; **525:** 139-48.
68. Fromtling RA. Overview of medically important antifungal drugs. *Clin Microbiol Rev.* 1988; **1:** 187-217.
69. George J, Reboli AC. Anidulafungin: When and how? The clinician's view. *Mycoses.* 2012; **55:** 36-44.
70. Ghannoum MA, Rice LB. Antifungal agents: Mode of action, mechanism of resistance and correlation of these mechanisms with bacterial resistance. *Clin Microbiol Rev.* 1999; **12:** 501-17.
71. Goncalves SS, Souza AC, Chowdhary A, *et al.* Epidemiology and molecular mechanisms of antifungal resistance in *Candida* and *Aspergillus. Mycoses,* 2016; **59:** 198-219.
72. Gotzsche PC, Johansen HK. Nystatin prophylaxis and treatment in severely immunodepressed patients. *Cochrane Database Syst Rev.* 2014; **9:** CD002033.
73. Goughenour KD, Rappleye CA. Antifungal therapeutics for dimorphic fungal pathogens. *Virulence.* 2017; **8:** 211-21.
74. Grover ND. Echinocandins: A ray of hope in antifungal drug therapy. *Indian J Pharmacol.* 2010; **42:** 9-11.
75. Guidelines for Prevention and Treatment of Opportunistic Infections in HIV-Infected Adults and Adolescents (http://aidsinfo.nih.gov/guidelines. Accessed on 31/03/2017).
76. Gupta AK, Lyons DC. The rise and fall of oral ketoconazole. *J Cutan Med Surg.* 2015; **19:** 352-7.
77. Gupta AK, Ryder JE, Cooper EA. Naftifine: A review. *J Cutan Med Surg.* 2008; **12:** 51-8.
78. Hamill RJ. Amphotericin B formulations: A comparative review of efficacy and toxicity. *Drugs.* 2013; **73:** 919-34.
79. Hassan I, Keen A. Potassium iodide in dermatology. *Indian J Dermatol Venereol Leprol.* 2012; **78:** 390-3.
80. Hatipoglu N, Hatipoglu H. Combination antifungal therapy for invasive fungal infections in children and adults. *Expert Rev Anti Infect Ther.* 2013; **11:** 523-35.
81. Heseltine P. Self-assembling plastic bottle technology could significantly reduce fungal infections. *Nanomedicine (Lond).* 2014; **9:** 185-6.
82. Hodge G, Cohen SH, Thompson GR 3rd. *In vitro* interactions between amphotericin B and hydrocortisone: Potential implications for intrathecal therapy. *Med Mycol.* 2015; **53:** 749-53.
83. Jadhav MP, Kshirsagar NA. Access to antifungal medicines in resource-poor countries. *Lancet Infect Dis.* 2014; **14:** 370-1.
84. Jebali A, Hajjar FH, Pourdanesh F, *et al.* Silver and gold nanostructures: Antifungal property of different shapes of these nanostructures on *Candida* species. *Med Mycol.* 2014; **52:** 65-72.
85. Johnson MD, Perfect JR. Use of antifungal combination therapy: Agents, order and timing. *Curr Fungal Infect Rep.* 2010; **4:** 87-95.
86. Jorgensen KJ, Gotzsche PC, Dalboge CS, *et al.* Voriconazole versus amphotericin B or fluconazole in cancer patients with neutropenia. *Cochrane Database Syst Rev.* 2014; **2:** CD004707.
87. Katragkou A, McCarthy M, Meletiadis J, *et al. In vitro* combination of isavuconazole with micafungin or amphotericin B deoxycholate against medically important molds. *Antimicrob Agents Chemother.* 2014; **58:** 6934-7.
88. Kauffman CA, Carver PL. Update on echinocandin antifungals. *Semin Respir Crit Care Med.* 2008; **29:** 211-9.
89. Kawakami H, Inuzuka H, Hori N, *et al.* Inhibitory effects of antimicrobial agents against *Fusarium* species. *Med Mycol.* 2015; **53:** 603-11.
90. Khanna D, Bharti S. Luliconazole for the treatment of fungal infections: An evidence-based review. *Core Evid.* 2014; **9:** 113-24.
91. Kidane YH, Lawrence C, Murali TM. Computational approaches for discovery of common immunomodulators in fungal infections: Towards broad-spectrum immunotherapeutic interventions. *BMC Microbiol.* 2013; **13:** 224.
92. Kidd SE, Goeman E, Meis JF, *et al.* Multi-triazole-resistant *Aspergillus fumigatus* infections in Australia. *Mycoses.* 2015; **58:** 350-5.
93. Kim JH, Williams K. Posaconazole salvage treatment for invasive fungal infection. *Mycopathologia.* 2014; **178:** 259-65.
94. Kiraz N, Dag I, Yamac M, *et al.* Antifungal activity of caspofungin in combination with amphotericin B against *Candida glabrata:* Comparison of disk diffusion, Etest and time-kill methods. *Antimicrob Agents Chemother.* 2009; **53:** 788-90.
95. Klepser M. The value of amphotericin B in the treatment of invasive fungal infections. *J Crit Care.* 2011; **26:** 225.e1-10.
96. Koehler P, Cornely OA. Contemporary strategies in the prevention and management of fungal infections. *Infect Dis Clin North Am.* 2016; **30:** 265-75.
97. Kohli V, Taneja V, Sachdev P, Joshi R. Voriconazole in newborns. *Indian Pediatr.* 2008; **45:** 236-8.
98. Kontoyiannis DP, Lewis RE. Treatment principles for the management of mold infections. *Cold Spring Harb Perspect Med.* 2014; **5:** pii: a019737.
99. Kontoyiannis DP. Invasive mycoses: Strategies for effective management. *Am J Med.* 2012; **125:** S25-38.
100. Kriengkauykiat J, Ito JI, Dadwal SS. Epidemiology and treatment approaches in management of invasive fungal infections. *Clin Epidemiol.* 2011; **3:** 175-91.

101. Krishnan-Natesan S. Terbinafine: A pharmacological and clinical review. *Expert Opin Pharmacother.* 2009; **10:** 2723-33.
102. Kullberg BJ, van de Veerdonk F, Netea MG. Immunotherapy: A potential adjunctive treatment for fungal infection. *Curr Opin Infect Dis.* 2014; **27:** 511-6.
103. Kuyucu N. Amphotericin B use in children: Conventional and lipid-based formulations. *Expert Rev Anti Infect Ther.* 2011; **9:** 357-67.
104. Kyriakidis I, Tragiannidis A, Munchen S, *et al.* Clinical hepatotoxicity associated with antifungal agents. *Expert Opin Drug Saf.* 2017; **16:** 149-65.
105. Lacerda JF, Oliveira CM. Diagnosis and treatment of invasive fungal infections: Focus on liposomal amphotericin B. *Clin Drug Investig.* 2013; **33:** S5-14.
106. Lepak A, Andes D. Fungal sepsis: Optimizing antifungal therapy in the critical care setting. *Crit Care Clin.* 2011; **27:** 123-47.
107. Lestner J, Hope WW. Itraconazole: An update on pharmacology and clinical use for treatment of invasive and allergic fungal infections. *Expert Opin Drug Metab Toxicol.* 2013; **9:** 911-26.
108. Lewis RE, Lortholary O, Spellberg B, *et al.* How does antifungal pharmacology differ for mucormycosis versus aspergillosis? *Clin Infect Dis.* 2012; **54:** S67-72.
109. Limper AH, Knox KS, Sarosi GA, *et al.* Treatment of fungal infections in adult pulmonary and critical care patients. *Am J Respir Crit Care Med.* 2011; **183:** 96-128.
110. Lollis TR, Bradshaw WT. Fungal prophylaxis in neonates: A review article. *Adv Neonatal Care.* 2014; **14:** 17-23.
111. Lortholary O, Fernandez-Ruiz M, Perfect JR. The current treatment landscape: Other fungal diseases (cryptococcosis, fusariosis and mucormycosis). *J Antimicrob Chemother.* 2016; **71(Suppl. 2):** ii31-6.
112. Lyon JP, Moreira LM, de Moraes PC, *et al.* Photodynamic therapy for pathogenic fungi. *Mycoses.* 2011; **54:** e265-71.
113. Lyon S. Antifungal prophylaxis: Why, what and how? *Future Microbiol.* 2016; **11:** 11-5.
114. Maertens J, Meersseman W, Van Bleyenbergh P. New therapies for fungal pneumonia. *Curr Opin Infect Dis.* 2009; **22:** 183-90.
115. Maertens JA, Raad II, Marr KA, *et al.* Isavuconazole versus voriconazole for primary treatment of invasive mould disease caused by *Aspergillus* and other filamentous fungi (SECURE): A phase 3, randomised-controlled, non-inferiority trial. *Lancet.* 2016; **387:** 760-9.
116. Maghu S, Desai VD, Sharma R. Comparison of efficacy of alternative medicine with allopathy in treatment of oral fungal infection. *J Tradit Complement Med.* 2015; **6:** 62-5.
117. Malani AN, Kerr LE, Kauffman CA. Voriconazole: How to use this antifungal agent and what to expect. *Semin Respir Crit Care Med.* 2015; **36:** 786-95.
118. Mathew BP, Nath M. Recent approaches to antifungal therapy for invasive mycoses. *Chem Med Chem.* 2009; **4:** 310-23.
119. Matsubara VH, Bandara HM, Mayer MP, *et al.* Probiotics as antifungals in mucosal candidiasis. *Clin Infect Dis.* 2016; **62:** 1143-53.
120. Mayr A, Aigner M, Lass-Florl C. Caspofungin: When and how? The microbiologist's view. *Mycoses.* 2012; **55:** 27-35.
121. Mayer FL, Kronstad JW. Discovery of a novel antifungal agent in the pathogen box. *mSphere.* 2017; 2. pii: e00120-17. PMID: 28435886.
122. McCarthy MW, Petraitis V, Walsh TJ. Combination therapy for the treatment of pulmonary mold infections. *Expert Rev Respir Med.* 2017; **11:** 481-9.
123. McCormack PL. Isavuconazonium: First global approval. *Drugs.* 2015; **75:** 817-22.
124. Miceli MH, Kauffman CA. Isavuconazole: A new broad-spectrum triazole antifungal agent. *Clin Infect Dis.* 2015; **61:** 1558-65.
125. Millikan LE. Current concepts in systemic and topical therapy for superficial mycoses. *Clin Dermatol.* 2010; **28:** 212-6.
126. Miyazaki T, Kohno S. Current recommendations and importance of antifungal stewardship for the management of invasive candidiasis. *Expert Rev Anti Infect Ther.* 2015; **13:** 1171-83.
127. Moore JN, Healy JR, Kraft WK. Pharmacologic and clinical evaluation of posaconazole. *Expert Rev Clin Pharmacol.* 2015; **8:** 321-34.
128. Mor V, Rella A, Farnoud AM, *et al.* Identification of a new class of antifungals targeting the synthesis of fungal sphingolipids. *MBio.* 2015; **6:** e00647.
129. Moran C, Benjamin D. Treatment of neonatal fungal infections. *Adv Exp Med Biol.* 2010; **659:** 129-38.
130. Moriyama B, Kadri S, Henning SA, *et al.* Therapeutic drug monitoring and genotypic screening in the clinical use of voriconazole. *Curr Fungal Infect Rep.* 2015; **9:** 74-87.
131. Morschhauser J. Regulation of multidrug resistance in pathogenic fungi. *Fungal Genet Biol.* 2010; **47:** 94-106.
132. Munoz P, Valerio M, Vena A, *et al.* Antifungal stewardship in daily practice and health economic implications. *Mycoses.* 2015; **58 (Suppl. 2):** 14-25.
133. Neoh CF, Snell G, Levvey B, *et al.* Antifungal prophylaxis in lung transplantation. *Int J Antimicrob Agents.* 2014; **44:** 194-202.
134. Nett JE, Andes DR. Antifungal agents: Spectrum of activity, pharmacology and clinical indications. *Infect Dis Clin North Am.* 2016; **30:** 51-83.
135. Niimi M, Firth NA, Cannon RD. Antifungal drug resistance of oral fungi. *Odontology.* 2010; **98:** 15-25.
136. Nucci M, Anaissie E. How we treat invasive fungal diseases in patients with acute leukemia: The importance of an individualized approach. *Blood.* 2014; **124:** 3858-69.
137. Odom AR. The triphenylethylenes, a novel class of antifungals. *mBio.* 2014 ; **5:** e01126-14.
138. Pappas PG. Antifungal clinical trials and guidelines: What we know and do not know. *Cold Spring Harb Perspect Med.* 2014; **4:** a019745.
139. Patel MS, Wright AJ, Kohn R, *et al.* Successful long-term management of invasive cerebral fungal infection following liver transplantation. *Mycoses.* 2015; **58:** 181-6.
140. Payne KD, Hall RG. Dosing of antifungal agents in obese people. *Expert Rev Anti Infect Ther.* 2016; **14:** 257-67.

141. Peman J, Canton E, Espinel-Ingroff A. Antifungal drug resistance mechanisms. *Expert Rev Anti Infect Ther.* 2009; **7:** 453-60.
142. Perfect JR. The antifungal pipeline: A reality check. *Nat Rev Drug Discov.* 2017; doi: 10.1038/nrd.2017.46. PMID: 28496146.
143. Perfect JR, Hachem R, Wingard JR. Update on epidemiology of and preventive strategies for invasive fungal infections in cancer patients. *Clin Infect Dis.* 2014; **59:** S352-5.
144. Perlin DS, Shor E, Zhao Y. Update on antifungal drug resistance. *Curr Clin Microbiol Rep.* 2015; **2:** 84-95.
145. Perlin DS. Echinocandin resistance in *Candida. Clin Infect Dis.* 2015; **61:** S612-7.
146. Perlin DS. Mechanisms of echinocandin antifungal drug resistance. *Ann N Y Acad Sci.* 2015; **1354:** 1-11.
147. Pettit NN, Carver PL. Isavuconazole: A new option for the management of invasive fungal infections. *Ann Pharmacother.* 2015; **49:** 825-42.
148. Pfaller MA, Diekema DJ, Gibbs DL, *et al.* Results from the ARTEMIS DISK Global Antifungal Surveillance Study, 1997 to 2007: A 10.5-year analysis of susceptibilities of *Candida* species to fluconazole and voriconazole as determined by CLSI standardized disk diffusion. *J Clin Microbiol.* 2010; **48:** 1366- 77.
149. Pfaller MA, Messer SA, Rhomberg PR, *et al. In vitro* activities of isavuconazole and comparator antifungal agents tested against a global collection of opportunistic yeasts and molds. *J Clin Microbiol.* 2013; **51:** 2608-16.
150. Pfaller MA. Antifungal drug resistance: Mechanisms, epidemiology and consequences for treatment. *Am J Med.* 2012; **125:** S3-13.
151. Pfaller MA, Castanheira M. Nosocomial candidiasis: Antifungal stewardship and the importance of rapid diagnosis. *Med Mycol.* 2016; **54:** 1-22.
152. Pinto E, Vale-Silva L, Cavaleiro C, *et al.* Antifungal activity of the clove essential oil from *Syzygium aromaticum* on *Candida, Aspergillus* and dermatophyte species. *J Med Microbiol.* 2009; **58 (Pt. 11):** 1454-62.
153. Prasad R, Shah AH, Rawal MK. Antifungals: Mechanism of action and drug resistance. *Adv Exp Med Biol.* 2016; **892:** 327-49.
154. Prasad R, Banerjee A, Shah AH. Resistance to antifungal therapies. *Essays Biochem.* 2017; **61:** 157-66.
155. Proesmans M, Vermeulen F, Vreys M, *et al.* Use of nebulized amphotericin B in the treatment of allergic bronchopulmonary aspergillosis in cystic fibrosis. *Int J Pediatr.* 2010; 376287.
156. Ramos A, Perez-Velilla C, Asensio A, *et al.* Antifungal stewardship in a tertiary hospital. *Rev Iberoam Micol.* 2015; **32:** 209-13.
157. Rex JH, Pfaller MA. Has antifungal susceptibility testing come of age? *Clin Infect Dis.* 2002; **35:** 982-9.
158. Robbins N, Wright GD, Cowen LE. Antifungal drugs: The current armamentarium and development of new agents. *Microbiol Spectr.* 2016; 4. PMID: 27763259.
159. Rossi DC, Spadari CC, Nosanchuk JD, *et al.* Miltefosine is fungicidal to *Paracoccidioides* spp. yeast cells but subinhibitory concentrations induce melanisation. *Int J Antimicrob Agents.* 2017; **49:** 465-71.
160. Russotto V, Cortegiani A, Raineri SM, *et al.* Antifungal stewardship in light of the updated evidence on untargeted antifungal treatment in critically ill patients. *Curr Clin Micro Rpt.* 2016; **3:** 171.
161. Russotto V, Cortegiani A, Raineri SM, *et al.* From bedside to bench: The missing brick for patients with fungal sepsis. *Critical Care.* 2016; **20:** 191.
162. Rybak JM, Marx KR, Nishimoto AT, *et al.* Isavuconazole: Pharmacology, pharmacodynamics and current clinical experience with a new triazole antifungal agent. *Pharmacotherapy.* 2015; **35:** 1037-51.
163. Sanglard D. Emerging threats in antifungal-resistant fungal pathogens. *Front Med (Lausanne).* 2016; **3:** 11.
164. Sanguinetti M, Posteraro B, Lass-Florl C. Antifungal drug resistance among *Candida* species: Mechanisms and clinical impact. *Mycoses.* 2015; **58 (Suppl. 2):** 2-13.
165. Schaenman JM. Is universal antifungal prophylaxis mandatory in lung transplant patients? *Curr Opin Infect Dis.* 2013; **26:** 317-25.
166. Scorzoni L, de Paula E Silva AC, Marcos CM, *et al.* Antifungal therapy: New advances in the understanding and treatment of mycosis. *Front Microbiol.* 2017; **8:** 36.
167. Scott LJ. Micafungin: A review in the prophylaxis and treatment of invasive *Candida* infections in paediatric patients. *Paediatr Drugs.* 2017; **19:** 81-90.
168. Sehgal IS, Agarwal R. Role of inhaled amphotericin in allergic bronchopulmonary aspergillosis. *J Postgrad Med.* 2014; **60:** 41-5.
169. Seyedmousavi S, Mouton JW, Melchers WJ, *et al.* The role of azoles in the management of azole-resistant aspergillosis: From the bench to the bedside. *Drug Resist Updat.* 2014; **17:** 37-50.
170. Seyedmousavi S, Verweij PE, Mouton JW. Isavuconazole, a broad-spectrum triazole for the treatment of systemic fungal diseases. *Expert Rev Anti Infect Ther.* 2015; **13:** 9-27.
171. Shah MK. Castellani's paint. *Indian J Dermatol Venereol Leprol.* 2003; **69:** 357-8.
172. Shahid SK. Newer patents in antimycotic therapy. *Pharm Pat Anal.* 2016; **5:** 115-34.
173. Sheehan DJ, Hitchcock CA, Sibley CM. Current and emerging azole antifungal agents. *Clin Microbiol Rev.* 1999; **12:** 40-79.
174. Shor E, Perlin DS. Coping with stress and the emergence of multidrug resistance in fungi. *PLoS Pathog.* 2015; **11:** e1004668.
175. Shrestha S, Jha AK, Pathak DT, *et al.* Ketoconazole or clotrimazole solution wash as a prophylaxis in management and prevention of fungal infection: A comparative study. *Nepal Med Coll J.* 2013; **15:** 31-3.
176. Sinko J, Bryan J. Latest trends in fungal epidemiology inform treatment choices and stewardship initiatives. *Future Microbiol.* 2012; **7:** 1141-6.
177. Slavin MA, Thursky KA, Worth LJ, *et al.* Introduction to the updated Australian and New Zealand consensus guidelines for the use of antifungal agents in the haematology/oncology setting, 2014. *Intern Med J.* 2014; **44:** 1267-76.

178. Souza AC, Amaral AC. Antifungal therapy for systemic mycosis and the nanobiotechnology era: Improving efficacy, biodistribution and toxicity. *Front Microbiol.* 2017; **8:** 336.
179. Srinivasan A, Lopez-Ribot JL, Ramasubramanian AK. Overcoming antifungal resistance. *Drug Discov Today Technol.* 2014; **11:** 65-71.
180. Steimbach LM, Tonin FS, Virtuoso S, *et al.* Efficacy and safety of amphotericin B lipid-based formulations - A systematic review and meta-analysis. *Mycoses.* 2017; **60:** 146-54.
181. Stergiopoulou T, Walsh TJ. Clinical pharmacology of antifungal agents to overcome drug resistance in pediatric patients. *Expert Opin Pharmacother.* 2015; **16:** 213-26.
182. Stevens DA. Advances in systemic antifungal therapy. *Clin Dermatol.* 2012; **30:** 657-61.
183. Stevens DA. Reflections on the approach to treatment of a mycologic disaster. *Antimicrob Agents Chemother.* 2013; **57:** 1567-72.
184. Sweileh WM, Sawalha AF, Al-Jabi S, *et al.* Bibliometric analysis of literature on antifungal triazole resistance: 1980-2015. *Germs.* 2017; **7:** 19-27.
185. Tada R, Latge JP, Aimanianda V. Undressing the fungal cell wall/cell membrane - The antifungal drug targets. *Curr Pharm Des.* 2013; **19:** 3738-47.
186. Tagliaferri E, Menichetti F. Treatment of invasive candidiasis: Between guidelines and daily clinical practice. *Expert Rev Anti Infect Ther.* 2015; **13:** 685-9.
187. Tang MM, Corti MA, Stirnimann R, *et al.* Severe cutaneous allergic reactions following topical antifungal therapy. *Contact Dermatitis.* 2013; **68:** 56-7.
188. Thakur M, Revankar SG. *In vitro* interaction of caspofungin and immunosuppressives against agents of mucormycosis. *J Antimicrob Chemother.* 2011; **66:** 2312-4.
189. Thompson GR 3rd, Rendon A, Ribeiro Dos Santos R, *et al.* Isavuconazole treatment of cryptococcosis and dimorphic mycoses. *Clin Infect Dis.* 2016; **63:** 356-62.
190. Thompson GR 3rd, Wiederhold NP. Isavuconazole: A comprehensive review of spectrum of activity of a new triazole. *Mycopathologia.* 2010; **170:** 291-313.
191. Tissot F, Agrawal S, Pagano L, *et al.* ECIL-6 Guidelines for the treatment of invasive candidiasis, aspergillosis and mucormycosis in leukemia and hematopoietic stem cell transplant patients. *Haematologica.* 2017; **102:** 433-44.
192. Trombino S, Mellace S, Cassano R. Solid lipid nanoparticles for antifungal drugs delivery for topical applications. *Ther Deliv.* 2016; **7:** 639-47.
193. Tseng HK, Perfect JR. Strategies to manage antifungal drug resistance. *Expert Opin Pharmacother.* 2011; **12:** 241-56.
194. Turel O. Newer antifungal agents. *Expert Rev Anti Infect Ther.* 2011; **9:** 325-38.
195. Vallabhaneni S, Chiller TM. Fungal infections and new biologic therapies. *Curr Rheumatol Rep.* 2016; **18:** 29.
196. van der Linden JW, Arendrup MC, Warris A, *et al.* Prospective multicenter international surveillance of azole resistance in *Aspergillus fumigatus. Emerg Infect Dis.* 2015; **21:** 1041-4.
197. van Hal SJ, Gilroy NM, Morrissey CO, *et al.* Survey of antifungal prophylaxis and fungal diagnostic tests employed in malignant haematology and haemopoietic stem cell transplantation (HSCT) in Australia and New Zealand. *Intern Med J.* 2014; **44:** 1277-82.
198. Verweij PE, Howard SJ, Melchers WJ, *et al.* Azole-resistance in *Aspergillus*: proposed nomenclature and breakpoints. *Drug Resist Updat.* 2009; **12:** 141-7.
199. Verweij PE, Snelders E, Kema GH, *et al.* Azole resistance in *Aspergillus fumigatus:* A side-effect of environmental fungicide use? *Lancet Infect Dis.* 2009; **9:** 789-95.
200. Walker RC, Zeuli JD, Temesgen Z. Isavuconazonium sulfate for the treatment of fungal infection. *Drugs Today (Barc).* 2016; **52:** 7-16.
201. Walsh TJ, Gamaletsou MN. Treatment of fungal disease in the setting of neutropenia. *Hematology.* 2013; **2013:** 423-7.
202. Wang T, Chen S, Sun J, *et al.* Identification of factors influencing the pharmacokinetics of voriconazole and the optimization of dosage regimens based on Monte Carlo simulation in patients with invasive fungal infections. *J Antimicrob Chemother.* 2014; **69:** 463-70.
203. Wattal C, Chakrabarti A, Oberoi JK, *et al.* Issues in antifungal stewardship: An opportunity that should not be lost. *J Antimicrob Chemother.* 2017; **72:** 969-74.
204. Wei X, Zhang Y, Lu L. The molecular mechanism of azole resistance in *Aspergillus fumigatus:* From bedside to bench and back. *J Microbiol.* 2015; **53:** 91-9.
205. Weiss E, Timsit JF. Management of invasive candidiasis in non-neutropenic ICU patients. *Ther Adv Infect Dis.* 2014; **2:** 105-15.
206. Wiederhold NP, Patterson TF. Emergence of azole resistance in *Aspergillus. Semin Respir Crit Care Med.* 2015; **36:** 673-80.
207. Wiederhold NP. Echinocandin resistance in *Candida* species: A review of recent developments. *Curr Infect Dis Rep.* 2016; **18:** 42.
208. Xie JL, Polvi EJ, Shekhar-Guturja T, *et al.* Elucidating drug resistance in human fungal pathogens. *Future Microbiol.* 2014; **9:** 523-42.
209. Xue SL, Li L. Oral potassium iodide for the treatment of sporotrichosis. *Mycopathologia.* 2009; **167:** 355-6.
210. Yamada K, Zaitz C, Framil VM, *et al.* Cutaneous sporotrichosis treatment with potassium iodide: A 24 year experience in Sao Paulo State, Brazil. *Rev Inst Med Trop Sao Paulo.* 2011; **53:** 89-93.
211. Zia Q, Azhar A, Kamal MA, *et al.* Super-aggregated form of amphotericin B: A novel way to increase its therapeutic index. *Curr Pharm Des.* 2016; **22:** 792-803.
212. Zumla A, Memish ZA, Maeurer M, *et al.* Emerging novel and antimicrobial-resistant respiratory tract infections: new drug development and therapeutic options. *Lancet Infect Dis.* 2014; **14:** 1136-49.

SECTION II

Superficial Cutaneous Mycoses

Chapters

The superficial cutaneous fungal infections involve the outer most covering of the skin and its appendages like hair and nails. The skin as such is a mechanically protective layer as well as cosmetically significant anatomical structure. The causative fungi colonize only the cornified layer of epidermis or supra-follicular portion of hair and usually do not penetrate into deeper anatomical sites. There is little tissue damage and immune response is also minimal. With the exception of dermatophytosis, the patients invariably try to neglect such infections and if at all seek medical advice, it is usually for cosmetic reasons that may not be because of the illness itself.

Malasseziosis is caused by various species of genus *Malassezia*, which used to be treated earlier as eukaryotic skin commensal. It is now posing serious threats of systemic infections and has undergone substantial taxonomical revisions. The group of lineages formally regarded as *M.furfur* in broad sense, has now fifteen valid species based on genetic and nutritional differences. The superficial infections also include changes in pigmentation of skin such as Tinea nigra. In other disease covered in this section *i.e.* there is formation of nodules along the hair shaft as seen in both White Piedra as well as Black Piedra. *Trichosporon* genus, causing white piedra, has also undergone extensive taxonomical changes defining now sixteen species relevant as human pathogens.

The dermatophytes are by far the most significant cutaneous fungi because of their widespread involvement of large section of population. The epidemiology of most of the clinical presentation of Dermatophytosis has substantially changed over the last few years. Now, *Trichophyton rubrum* is predominantly the prevalent species throughout the world. The infections of hair caused by the dermatophytes are dealt as dermatophytosis whereas those caused by other fungi are separately discussed as piedra. Similarly, infections of nails caused by dermatophytes are dealt as tinea unguium and others as onychomycosis. In addition, non-dermatophytic skin and nail involvement are likewise important groups of infections, dermatomycosis and onychomycosis, respectively, which were previously thought to be quite insignificant.

Although diagnosis and treatment of superficial mycoses are not very difficult but frequent recurrence and relapses are common particularly in pityriasis versicolor, dermatophytosis and onychomycosis. In addition, there is a rising trend of unresponsiveness to the commonly used antifungals for dermatophytosis as well as non-dermatophytic dermatomycosis. However, the resistance has to be thoroughly studied especially in cutaneous dermatophyte infection. Newer formulations are being tried for treating such refractory infections. The cutaneous manifestations observed in systemic fungal infections are deliberated up on in their respective Chapters.

CHAPTER 7

Malasseziosis

Malasseziosis is a fungal infection caused by various species of genus *Malassezia*. These species are usually considered as resident flora of man and animals. However, some species are responsible for variety of superficial as well as systemic infections, pityriasis versicolor being the most commonly presenting disease. The other conditions in which this lipophilic fungus plays important etiological role are seborrheic dermatitis, folliculitis and allied illnesses. Apart from superficial diseases, there are certain systemic infections as well because some species may induce catheter-associated sepsis in premature neonates and immunocompromised patients receiving parenteral lipid emulsions. The recent developments in molecular techniques in medical mycology have considerably helped to resolve most of the long-standing intricacies pertaining to the taxonomy of this genus.

Based on molecular biology of *Malassezia*, most of the controversies prevailing for the last so many years have resolved to a great extent entailing new interest in its clinical importance. This lipophilic yeast-like fungus with its natural habitat in stratum corneum of human skin as resident flora causes pityriasis versicolor and has been implicated in pathogenesis of many dermatoses. Now, there are fifteen well-characterized species of this genus involved in superficial as well as invasive, life-threatening systemic infections.

Historical Perspective

The lipophilic basidiomycetous yeasts *i.e. Malassezia* species have been recognized over a century as inhabitants of normal skin as well as the organisms involved with superficial cutaneous infections. Karl Ferdinand Eichstedt (1816-1892) reported the etiologic agent of pityriasis versicolor as early as in 1846 and Sluyter in 1847. Both these scientists gave the disease a name but could not designate its causative fungus. Their role in clinical conditions such as seborrheic dermatitis and scaly lesions of scalp was contemplated since long. Robin further described fungus in the scales of tinea versicolor in 1853. Because of the presence of filaments associated with the yeasts, he considered it to be a dermatophyte hence named *Microsporum furfur*. In 1873, Rivolta described double-contoured budding cells isolated from patient of psoriasis.

Malassez, in 1874, observed budding 'spores' of various shapes, which occurred in abnormal skin conditions. Bizzozero later on described both spherical and oval budding cells similar to 'Spores of Malassez'. These organisms were named *Saccharomyces sphaericus* and *S.ovalis*, respectively. The term seborrheic eczema was used for the first time by Unna in 1887. Subsequently in 1889, Baillon used the designation *Malassezia furfur* in his text, honoring Malassez, however, he also could not grow this organism. The genus *Pityrosporum* was proposed in 1904 by Sabouraud to describe budding yeast cells without hyphal elements and isolated from normal skin and fine scales of scalp (Gr. 'Pityron' = scale). He emphasized on variable morphology of the yeast cells and suggested the term *Pityrosporum malassezia*.

In 1913, Castellani and Chalmers for the first time isolated lipophilic oval-budding yeasts from normal skin as well as patients of seborrheic dermatitis and coined the name *Pityrosporum ovale* for the species based on previous descriptions. Acton and Panja from India considered *Pityrosporum* to be synonym of *Malassezia* in 1927 but no attention was given to this significant finding. During the contemporary period, Fred Weidman first isolated *Pityrosporum pachydermatis*, a zoophilic, non-obligatory lipophilic yeast, in 1925. It was named *pachydermatis* (Gr. = 'thick-skin') from an Indian rhinoceros with exfoliative dermatitis. Its detailed description was given in 1935 by C W Dodge. In 1939, Rhoda Benham discovered lipophilic nature of the genus *Malassezia*.

In 1951, Morris Gordon isolated and authenticated round, double-contoured budding yeast that produced spherical to oval buds in the scales of pityriasis versicolor as well as on normal skin. He renamed this organism *P.orbiculare* on the basis of morphology and presumed it to be distinct organism from *P.ovale*. By using a variety of culture conditions, induction of *P.ovale* and *P.orbiculare* to produce hyphae was demonstrated in the 1970s and these hyphae were indistinguishable from those seen in cases of pityriasis versicolor. The International Commission on Taxonomy of Fungi unified both genera in 1986 with the acceptance of the species name *Malassezia furfur*. In 1964, Dixon medium, described by Van Abbe, was used for the isolation of these species, which was subsequently modified to obtain better results. Another medium, described by Leeming and Notman, was introduced in 1987. Since early 1980s, *Malassezia* species have also been considered as agents responsible for causing opportunistic systemic infections, particularly in the premature neonates. In 1990, *M.sympodialis* was recognized as an independent species by Simmons and Gueho, based on the difference in the percentage of guanine and cytosine in the DNA (% GC) compared to *M.furfur*.

The taxonomy of genus *Malassezia* has always been a matter of controversy since its creation by Baillon in 1889 with *Malassezia furfur* as the generic type species. From earlier descriptions by Rivolta, Malassez and Bizzozero, more than a century ago, until successive contributions carried out by Sabouraud, Castellani, Chalmers and Gordon, the taxonomy of *Malassezia furfur* has been revised many times, based on its morphological criteria. However, difficulty of using micro-morphological descriptions as major method of taxonomic classification now seems to be clear. In 1996, extensive taxonomic revision established seven different species of genus *Malassezia*, namely—*M.furfur, M.globosa, M.obtusa, M.restricta, M.slooffiae, M.sympodialis* and *M.pachydermatis*. Due to advances in molecular taxonomy, subsequently after 2002, four more species were isolated from human skin namely - *M.dermatis, M.japonica, M.yamatoensis, M.arunalokei* and four new from animal skin *i.e. M.nana, M.caprae, M.equina* and *M.cuniculi*.

Mycology

The designation as *Malassezia*, previously used to describe mycelial phase of organism, whereas yeast phase was divided into two distinct species based on microscopic morphology: *(i) Pityrosporum orbiculare* - with round yeast cells and budding from narrow-neck; *(ii) Pityrosporum ovale*—with oval cells and budding from wide-neck. The former was considered to inhabit trunk and latter usually remained confined to scalp and face. *P.orbiculare* was regarded as an etiological agent of pityriasis versicolor as it was associated with trunk but both variants were designated as *M.furfur*, which is now formally accepted name for both phases of growth. Midgley retained two species *M.furfur* (*P.orbiculare*) and *M.ovalis*, which was further divided into three forms in 1989.

The physiological and serological studies showed that this organism possessed distinct cell-surface antigens. In 1990, Cunningham *et al*, defined Serovars A, B and C on the basis of these extracellular antigens. Microscopically, Serovars A and B had round, whereas Serovar C had oval blastospores. Historically, Serovars A and B were described as *P.orbiculare* and Serovar C as *P.ovale*. It was also shown that Serovar A was predominantly responsible for lesions over trunk, Serovar B over trunk and scalp while Serovar C was more prevalent on scalp of normal individuals. Cunningham's Serovars A, B and C are now identified with members of genus *Malassezia* as *M.sympodialis, M.globosa* and *M.restricta*, respectively.

The conidiogenesis in this fungus is percurrent type where conidia develop through a previous apex. The genus *Malassezia* was earlier classified to heterogeneous family **Cryptococcaceae** of form-class **Blastomycetes** in **Deuteromycetes** due to its ability to hydrolyze urea, positive reaction to staining with Diazonium Blue B (DBB) and laminar ultrastructure of cell wall. However, on the basis of its morphological features, taxonomically *Malassezia* is now classified in family **Malasseziaceae**, order **Malasseziales,** class **Malasseziomycetes** in subphylum **Ustilaginomycotina** of phylum **Basidiomycota** under the kingdom **Fungi**.

The taxonomy of this fungus and other lipophilic fungi has long been jumbled, including names such as *M.ovalis, P.malassezia, P.ovale*, etc., perhaps for some of the same fungi. For example, previously labeled species, *Pityrosporum orbiculare,* is now known as *Malassezia globosa*. The genus *Malassezia* now comprises of 15 species that differ from each other in cellular characteristics and morphology, molecular percentage G+C serotypes, RNA/DNA sequences, requirement of long-chain fatty acids from C12 to C24 series, catalase activity and temperature requirements. All the species are lipid-dependent except one *i.e. M.pachydermatis,* which is not a normal inhabitant of human skin but found on variety of animals, causing otitis externa in the dogs and is rarely isolated from humans patients as well. The invasive infections due to *M.pachydermatis* have been reported in high-risk infants and immunocompromised patients. The various species of genus *Malassezia*, with both new and old nomenclatures, are shown in **Table 7.1**.

Before 1990, only three *Malassezia* species were recognized namely *M.furfur*, *M.pachydermatis* and *M.sympodialis*. With developments in molecular techniques, new species were identified within *Malassezia* genus. In 1996, four new species added to genus *Malassezia* were *M.globosa*, *M.restricta*, *M.obtusa* and *M.slooffiae*. The epithet 'restricta' of *M.restricta* refers to its limited growth *in vitro* as it is only lipid-dependent species in which catalase reaction is negative besides *M.arunalokei*. Subsequently in 2005, four new species, infecting man and animals, were also added namely *M.dermatis*, *M.yamatoensis*, *M.japonica* and *M.nana*, thereby increasing the number of species to eleven. Tentatively one more species, infecting horses, has also been added as *Malassezia equi*, which has similar phenotypes and is genetically close to *M.sympodialis*. Some of salient physiological and biochemical characteristics of clinically significant *Malassezia* species are shown in **Table 7.2**.

Table 7.1. Nomenclature of various *Malassezia* Species.

New Nomenclature	*Old Nomenclature*
Malassezia furfur	*M.ovalis, P.malassezii, P.ovale*
M.pachydermatis	*Pityrosporum pachydermatis*
M.sympodialis	*M.furfur* - Serovar A, *M.ovalis*
M.globosa	*M.furfur* - Serovar B, *P.orbiculare*
M.restricta	*M.furfur* - Serovar C
M.obtusa	No previous nomenclature
M.slooffiae	No previous nomenclature

Although there are two distinct morphological forms of the genus *Malassezia*, it is not considered as dimorphic fungus in true sense as both phenotypic structures are simultaneously found irrespective of temperature variations. There is no thermal dimorphism observed in this fungus hence it cannot be treated as a dimorphic fungus in real sense.

Epidemiology

Malassezia species are lipophilic unipolar yeasts commonly recognized as commensals of normal skin over the sites rich in sebaceous glands of warm-blooded vertebrates that may be pathogenic under certain conditions, usually causing skin diseases like pityriasis versicolor and other infections.

Molecular methods are accurate tools in the identification and they lead to a better knowledge of the ecology and epidemiology of this genus. The molecular biological studies of *Malassezia* yeasts initially consisted of determining molecular percentage G+C content of chromosomal DNA. Other accurate methods of identification are comparison of ribosomal DNA sequences (SSU, D1/D2, ITS and IGS) and multilocus sequence analysis (mtrRNA, CHS2, RPB1, β-tubulin), PCR based methods (RFLP, RAPD, tFLP, DGGE, SSCP, real time PCR, AFLP). In molecular techniques, PCR and sequence based methods are widely used for routine identification purposes and for epidemiological surveys, however, all methods have their own pros and cons.

The pulsed-field gel electrophoresis (PFGE) studies have confirmed various aspects of the new taxonomic structure of genus *Malassezia* with all its species having been characterized by their individual karyotype. Besides karyotyping, molecular differentiation of *Malassezia* species

Table 7.2. Important Physiological and Biochemical Characteristics of Clinically Significant *Malassezia* Species.

Malassezia Species	*Lipid Dependence*	*Growth at >37°C*	*Catalase Reaction*	*Esculin Splitting*	*Cremophor (1-10%)*	*Tween 20 (High Conc.)*	*Tween 40 (0.1-10%)*	*Tween 80 (Low Conc.)*
Malassezia furfur	+	+	+	–	v	+	+	+
M.pachydermatis	–	+	v	v	v	+	+	+
M.sympodialis	+	+	+	+	–	–	+	+
M.globosa	+	–	+	–	–	–	–	–
M.obtusa	+	–	+	+	–	–	–	–
M.restricta	+	–	–	–	–	–	–	–
M.slooffiae	+	+	+	–	–	+	+	–

Note: + = Positive Reaction; – = Negative Reaction; v = Variable Reaction.

has also been successfully tried by PCR fingerprinting, restriction analysis and randomly amplified polymorphic DNA (RAPD) analysis. It has also been reported that although *Malassezia* species could be distinguished by RAPD typing but varying amounts of heterogeneity observed within species has rendered this method unreliable for species identification.

While karyotyping is very robust, its time consuming and labor intensive nature necessitates development of alternative molecular methods. A rapid and reliable molecular system is needed to facilitate epidemiological and related research studies, which may also be of potential utility in the reference laboratories. Therefore, pulsed-field gel electrophoresis is the only technique, which can reliably differentiate various strains of *Malassezia* species.

Immunity

Both humoral and cellular immunities have been investigated but with equivocal results. Serum antibodies against *M.furfur* are present in sera from healthy adults indicating that determining titer has no diagnostic or prognostic consequence. A cell-mediated immune response to *M.furfur* has also been demonstrated in patients and healthy individuals. Several exogenous and endogenous factors such as high temperature, humidity, greasy skin, use of corticosteroids and underlying immunodeficiency states can influence these yeasts to become pathogenic to man. The patients with atopic eczema (dermatitis) have been found to have specific IgE antibodies against *Malassezia* species.

Pathogenesis and Pathology

The genus *Malassezia* has undergone wide variety of changes in recent times and its taxonomical status and understanding of pathogenesis has been highlighted. The species grow readily on skin surface rich in sebum consisting of sterol ester, squalene, triglycerides and free fatty acids. Clinically, macular, erythematous, hyperpigmented (chromic) or hypopigmented (achromic) lesions with fine scaling are present in this disease.

These fungi interfere with melanin production. The hypopigmentation induced by this fungus can be explained on the basis of production of dicarboxylic acids, main component of which is **azelaic acid**. These acids act through competitive inhibition of DOPA tyrosinase and perhaps direct cytotoxic effect on hyperactive melanocytes. However, these acids have no effect on normal melanocytes in tissue culture. Establishment of re-pigmentation after mycological cure may take several months to even years.

Pityriasis versicolor is observed to cause invariably hypopigmented lesions in individuals with dark skin and hyperpigmented lesions in those with fair skin. The coexistence of hypo- and hyperpigmented lesions suggests multi-factor involvement other than pigmentary changes in pathogenesis of pityriasis versicolor. The pathogenesis of multiple pale-brownish hyperpigmentation in fair-skinned individuals is also not fully understood but two theories have been put forward: (i) increased thickness of keratin layer; (ii) an inflammatory cell infiltrate stimulating melanocytes to increase production of pigment. Electron microscopic studies have revealed abnormally large melanosomes in hyperpigmented lesions and smaller than normal melanosomes in hypopigmented ones. Melanosomes are tyrosinase-containing granules that synthesize melanin within melanocytes.

The biopsy of follicular lesions shows hyperkeratosis, plugging of follicular ostium by keratinaceous debris and numerous yeast-like cells within retained sebum. *Malassezia* yeasts require free fatty acids for survival. Usually, they are found in stratum corneum and in pilar folliculi over the sites with increased sebaceous gland activity such as chest and back. The yeasts hydrolyze triglycerides by enzyme lipase into free fatty acids and create long-chain and medium-chain fatty acids from free fatty acids thereby release of arachidonic acid, which is involved in inflammation of skin. Consequently, cell-mediated response is produced and alternative complement pathway is also activated leading to local inflammatory changes.

The exact mechanism as to how *Malassezia* causes inflammation is not known. It produces many enzymes (lipases and phospholipases) that can initiate an inflammatory response by releasing unsaturated free fatty acids from sebum lipids. Moreover, oleic acid produces irritant and desquamative effects on keratinocytes and arachidonic acid produces inflammatory cytokines and leads to inflammation and damage to stratum corneum.

Clinical Features

The fungal infections caused by species of *Malassezia* genus are termed as **malasseziosis** however this terminology is not much in use because its clinical entities are popular by their independent names. As these species are part of the normal flora of human skin and various other body sites rich in sebaceous glands hence most of the infections are endogenous in nature. Rather it is the only eukaryote found as commensal over the human skin. Therefore, the condition is not considered to be contagious in real sense. Once contemplated as only cosmetic botheration, *Malassezia* species have now joined ranks of opportunistic pathogens with potential to cause serious systemic disease also. Its implication in pathologic processes, including skin diseases to systemic infections, is the main issue of current investigations in order to determine the real pathogenic role of these basidiomycetous yeasts. The following clinical conditions are caused and/or associated with *Malassezia* species and described briefly under the following headings:

1. Pityriasis Versicolor
2. Seborrheic Dermatitis
3. Atopic Eczema/Dermatitis
4. Folliculitis
5. Systemic Malasseziosis

1. Pityriasis Versicolor

This is the most common skin disease caused by species of the genus *Malassezia*. This entity was known in the past as **tinea versicolor** under the misconception that it was caused by dermatophytes. Therefore, the term 'pityriasis' is preferred as nomenclature 'tinea' is usually reserved for various clinical types of dermatophytosis. As such pityriasis means a type of fine skin scaling. This is an asymptomatic mild, chronic recurrent superficial fungal infection of stratum corneum, characterized by patchy discoloration of skin ranging from hypo- to hyper-pigmentation with various shades. As implied in name 'versicolor' meaning variation in color as lesions may be hypopigmented, hyperpigmented, leukodermal, erythematous or dark brown (***'versi'*** means several). Recently, variants with different colors have been given various names like with red macule (pityriasis versicolor rubra) and another with black ones as (pityriasis versicolor nigra). There is also transformation from one variant to the other.

Most of the patients of pityriasis versicolor are healthy individuals and seek medical advice only on cosmetic grounds. Previously in certain geographical areas like Sri Lanka, it used to be considered as mark of beauty because rural women preferred to retain these lesions as so-called beauty spots rather than seeking medical advice for treatment. In vernacular language it is also called as ***'Gomara'*** literally meaning ***'tears of liquid gold'.***

In pityriasis versicolor although any part of body may be involved but neck, upper part of trunk, face and upper aspects of arms are most commonly affected. There are multiple well-defined, non-inflammatory, macular lesions with fine scaling which are discrete and vary in appearance *i.e.* hypo- and hyper-pigmented depending on degree of pigmentation of surrounding skin as seen in **Figures 7.1 to 7.5**. Occasionally, the lesions take papular appearance, especially in cases where hair follicles are involved and may result in folliculitis. The lesions are usually non-pruritic however in some patients slight to moderate itching have been observed. There may be presentation of atrophic lesions of pityriasis versicolor. It is thought to be caused by topical use of steroids and is considered as a new variant of the disease.

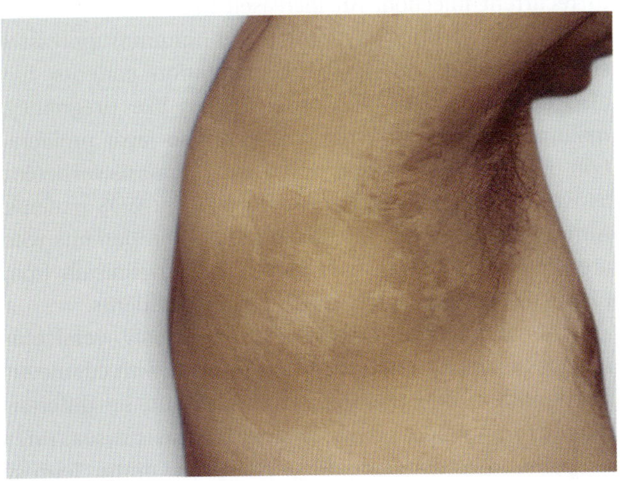

Fig. 7.1. Pityriasis versicolor showing well-demarcated, macular scaly lesions over the lateral wall of chest.

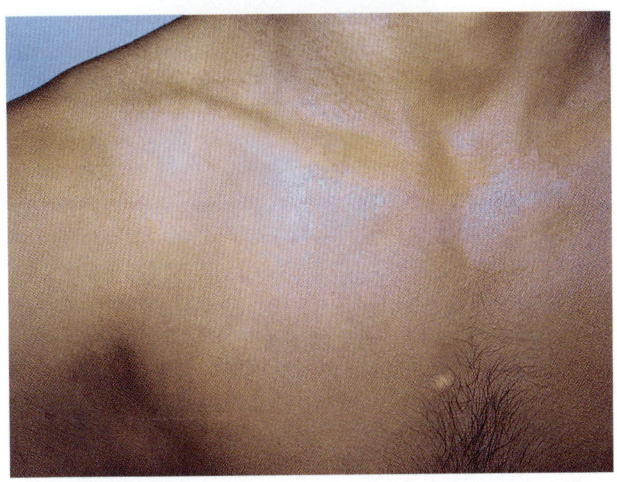

Fig. 7.2. Pityriasis versicolor showing hypopigmented scaly lesions over the neck and chest wall.

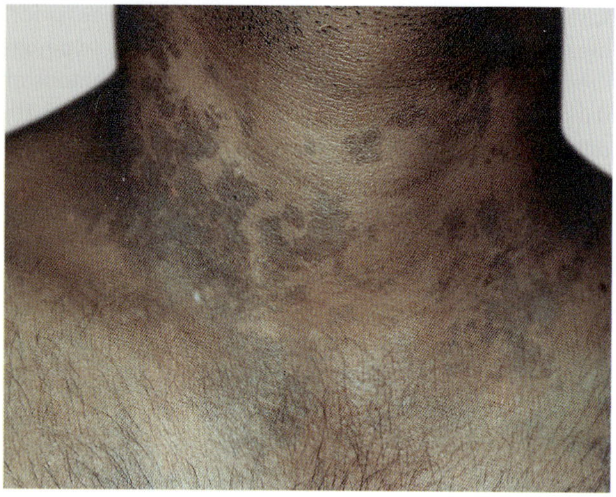

Fig. 7.3. Pityriasis versicolor showing hyperpigmented patches over the neck and chest wall.

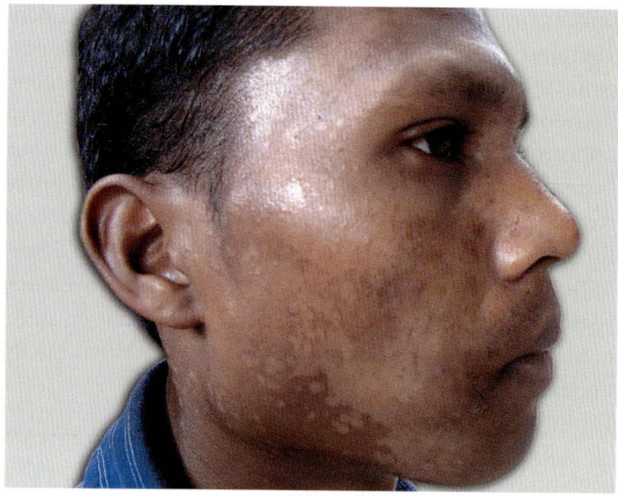

Fig. 7.4. Pityriasis versicolor showing extensive hypopigmented patches over the entire facial area including forehead.

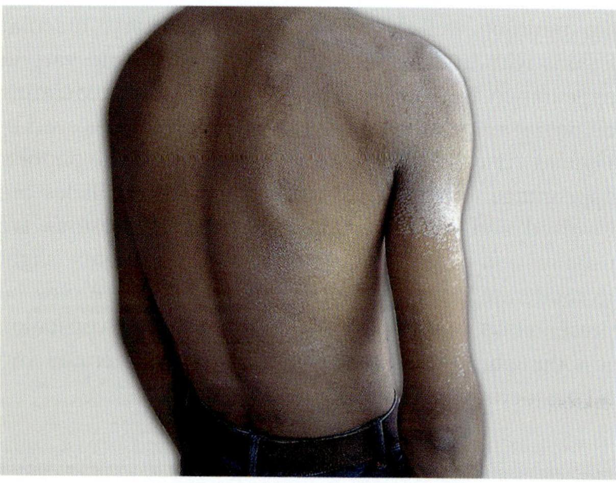

Fig. 7.5. Pityriasis versicolor showing extensive hypopigmented patches over the entire body including upper part of the arm.

The disease is related to various host and environmental factors. Both sexes are almost equally prone to have this condition but there are certain differences in susceptibility at different ages however, it is rare in childhood. The clinical features become common in late teenage and majority of cases occur in age range of 20 to 40 years with peak in early 20s. However, in India it is found to be very common between 10 to 30 years of age. There is no racial predilection in the incidence of disease. More cases are reported in warm moist climates than temperate zones. Incidence is also higher in individuals with illnesses causing fever, showing that excessive sweating is probably one of significant predisposing factors. *M.furfur* has long been presumed to be causative fungus of pityriasis versicolor, however, recent studies using cultural isolation and identification by morphological and physiological characteristics suggest that *M.globosa* is most commonly associated with this clinical entity. Presence of higher *Malassezia* load and mycelial form in the lesional area of pityriasis versicolor than non-lesional area proves its pathogenic role in the disease.

Pityriasis versicolor is neither contagious nor associated with poor hygiene. It has been observed to be more common in various conditions like malnutrition, heat, humidity and excessive sweating, which also predispose to recurrent infection. An increased incidence in various diseases like Cushing's syndrome and immunosuppressive states, especially in renal transplantation patients, has also been observed. The other factors like pregnancy, malnutrition and corticosteroids may also favor proliferation of this resident fungus of the skin. Pityriasis versicolor, as well as folliculitis, is known to occur in AIDS patients. Although disease may be more extensive in patients who are HIV-positive, however, it does not differ clinically from pityriasis versicolor seen in HIV-negative patients.

The differential diagnosis of pityriasis versicolor includes other common entities associated with cutaneous pigmentation/depigmentation. This should be differentiated from vitiligo, pityriasis alba, pityriasis rosea, tinea corporis, secondary syphilis, psoriasis, pinta, leprosy (Hansen's disease), chloasma, nevus anemicus, post-inflammatory hypopigmentation and post-kalaazar dermal leishmaniasis (PKDL). The distinct marginated macules without follicular prominence help to differentiate pityriasis versicolor from pityriasis alba found prevalent in children. Other possibilities include psoriasis, seborrheic dermatitis (both may coexist with pityriasis versicolor), confluent and reticulate papillomatosis (Gougerot-Carteaud), erythrasma and dermatophytosis. Wood's lamp examination is probably one of most useful tool for differentiating between last two diseases, as erythrasma fluoresce **coral-red** whereas tinea corporis does not fluoresce at all. The scaly lesions of pityriasis versicolor show **reddish** or **yellowish green** fluorescence. Moreover, pityriasis versicolor may present as circular dermatosis thereby mimicking pityriasis rotunda. The tuberous sclerosis shows hypopigmented macules, which are also known as 'ash-leaf spots'.

2. Seborrheic Dermatitis

Seborrheic dermatitis (SD) or seborrheic eczema is common inflammatory disorder of skin, characterized by erythema covered with white, greasy-looking, loose flakes (furfuraceous), itching and visible irritation. It is affecting approximately 3-5% of population and incidence is much higher in immunocompromised individuals particularly among those with AIDS, ranging from 30-80%. It is seen over those sites, which are rich in sebaceous glands, namely scalp, face, chest, back and flexural areas. There may be mild to marked erythema of nasolabial fold, often with scaling. The latter type may be precipitated by emotional stress. The scales are usually greasy, not dry, as commonly thought. Seborrhea is combination of Latin word sebum meaning *'grease'* and Greek word *'rhoea'* used for *'flow'*. The term refers to oily appearance of skin and not to the increased secretion of sebum as it literally means. Its mildest form is generally called as **dandruff**, wherein flaking and itching are restricted only to scalp and as such there is no visible inflammation.

This is chronic inflammatory dermatitis that predominately affects infants and young adults, often starting at adolescent age. It is characterized by fairly sharp, marginated, dull-red lesions covered with scales, which are often greasy in nature with variable pruritus. It is recognized as an early sign of AIDS, which may presumably be result of immunosuppression. An uncommon generalized form in infants may be linked to immunodeficiencies. A higher incidence and severity of disease has been observed in individuals with neutropenia and Parkinson's disease.

In recent times, Gupta *et al*, found that predominant species in seborrheic dermatitis patients was *M.globosa*, as opposed to *M.sympodialis* in normal skin. Nakabayashi *et al*, found both *M.globosa* and *M.restricta* on diseased skin but primarily *M.globosa* in the controls. Sandstrom *et al*, found *M.sympodialis* in both patients with seborrheic dermatitis and the controls. Therefore, this disease is mainly caused by *M.restricta, M.furfur, M.globosa, M.sympodialis* and *M.slooffiae*. It should be borne in mind that relative prevalence of these lipophilic species appears to vary with geographical region and the method used for diagnosis *i.e.* conventional culture *vs.* molecular. These studies strongly support the concept that *Malassezia* yeasts contribute to pathogenesis of seborrheic dermatitis. In a recent Indian study, Honnavar *et al*, have isolated a novel species *i.e. M.arunalokei* from the seborrheic dermatitis patients as well as healthy individuals.

Seborrhoeic dermatitis should be differentiated from acne, atopic eczema, candidiasis, dermatophytosis, Langerhans cell histiocytosis, pityriasis amiantacea, psoriasis, rosacea and SLE. Topical therapy primarily consists of antifungal agents and low-potency steroids. New topical calcineurin inhibitors *i.e.* immunomodulators such as tacrolimus ointment (0.03% for children and 0.1% for adults) and pimecrolimus cream (1%) may also be used for its treatment.

3. Atopic Eczema/Dermatitis

The yeasts of *Malassezia* species have also been implicated as an aggravating factor in **atopic eczema** or **dermatitis**. They have been proposed to act as allergens rather than infectious agents. The lesions are localized to scalp, face and neck of adult patients. It is defined as chronic intensely pruritic dermatosis, which affects genetically predisposed individuals with family history of atopy. The earliest clinical signs in development of atopic eczema are erythema, papules and pruritus. Subsequent scratching and rubbing may lead to secondary excoriations and lichenification. A vicious itch-scratch-itch cycle is established which allows penetration/interaction of *Malassezia* metabolites that could sensitize immune response and could be responsible for maintaining symptoms of atopic eczema for a long period. In first month of life, a yellowish desquamation on the scalp, known as 'cradle cap' may be one of the presentations of atopic eczema. Some of the species of another basidiomycete *i.e. Trichosporon*, particularly, *T.asahii*, have been reported from the skin of atopic eczema patients, however, their clinical significance is yet to be elucidated.

In atopic eczema, sensitization to *Malassezia* allergens can occur resulting in production of specific IgE. There are currently 14 known allergens mainly from its two species *i.e. M.furfur* and *M.sympodialis* and produce a variety of immunogenic proteins that elicit the production of specific IgE antibodies and may induce the release of pro-inflammatory cytokines. The *Malassezia*-specific IgE serum level is very significant in this disease. This may be divided in two types based on IgE level *i.e.* when increased it is extrinsic and when decreased it is intrinsic type of atopic eczema with corresponding severity *i.e.* high and low, respectively. In addition, *Malassezia* species induce autoreactive T-cells that cross-react between fungal proteins and their human counterparts. These mechanisms contribute to skin inflammation in atopic eczema and therefore influence the chronic course of this disease. Atopy patch test is useful modality to establish hypersensitivity to *Malassezia* species among patients of atopic eczema.

4. Folliculitis

The folliculitis caused by *Malassezia* species is a chronic inflammatory skin disorder, usually characterized by florid, acneiform, pruritic eruptions that rarely clear spontaneously. These are discrete, follicular pustules and erythematous papules, localized mainly to upper back, anterior portion of chest and to lesser extent to shoulder and arms. A typical patient is young woman complaining of itching, follicular papules and sparse pustules. The organism is present in ostium and central and deep segments of hair follicle.

Weary *et al,* first described this disease in a setting of antibiotic therapy in 1969. Later on Potter *et al,* in 1973 identified it as separate clinical entity with histological diagnosis. The direct microscopy of pustules shows presence of abundant *Malassezia* yeasts in the follicles in the absence of other microorganisms. It was known as *Pityrosporum* folliculitis at that time. *M.pachydermatis, M.globosa* and *M.furfur* are commonly involved species in this disease. An association of this infection with diabetes has often been observed. It is more prominent in the immunocompromised patients. Eosinophilic folliculitis is mainly seen in patients with advanced HIV infection and consists of pustule on the face and trunk. The erythematous papulofollicular lesions are seen in **Figure 7.6**. This should be differentiated from acne vulgaris and staphylococcal folliculitis, macronodular lesions of disseminated candidiasis as well as papulopustular lesions of candidiasis among heroin addicts. There are no comedones or cysts associated with *Malassezia* folliculitis and also no facial lesion as compared to acne.

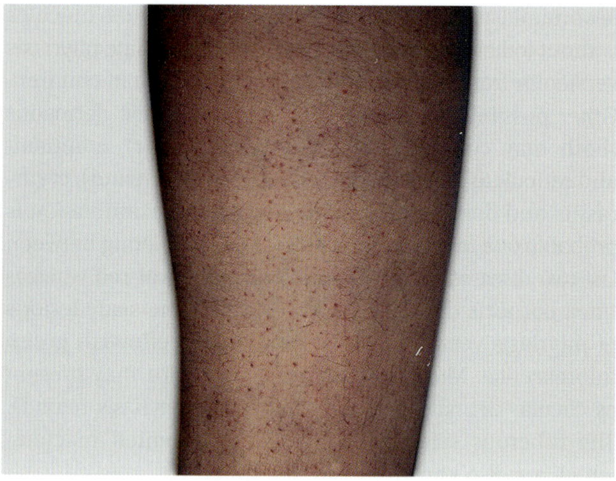

Fig. 7.6. *Malassezia* folliculitis seen over the upper part of arm.

5. Systemic *Malassezia* Infections

More recently, systemic diseases like septicemia and pneumonia have been reported among premature neonates and immunocompromised individuals due to lipid hyperalimentation. The portal of entry has been found to be intravenous lines and catheters. Two groups of patients are affected *i.e.* neonates on parenteral nutrition and immunosuppressed individuals.

The usual manifestation is lipid deposits in pulmonary arteries. The patient may have positive blood cultures and few develop multiple cutaneous pustules containing yeasts. The commonest species involved in such cases is - *M.pachydermatis.* There are reports of small outbreaks, particularly, in the nurseries. However, a study carried by Sriram *et al,* has contended that giving parenteral nutrition does not predispose to such types of fungal infections. Despite such caveats, outbreaks are still being documented. In a French University Hospital NICU, *M.pachydermatis* outbreak has recently been reported by Ilahi *et al,* in 2017.

In addition, *Malassezia* species have been found to be associated with variety of dermatoses like confluent and reticulate papillomatosis, otitis externa and onychomycosis. *M.furfur* may also cause obstructive dacryocystitis. Although psoriasis is considered as one of the differential diagnosis of *Malassezia* infections but at least three independent studies by Narang *et al*, Jagielski *et al* and Honnavar *et al*, have implicated species of this genus as its causative agent.

Laboratory Diagnosis

The laboratory diagnosis for the infections caused by the genus Malassezia is a challenging task. There are fifteen *Malassezia* species known and fourteen are lipophilic with very similar type of morphological, physiological and biochemical characteristics, therefore, conventional techniques are useful only to an extent to differentiate them. The molecular methods are accurate tools in their identification leading to better knowledge epidemiology of this genus.

In a routine diagnostic laboratory, the diagnosis of *Malassezia* infections is based on typical clinical picture and positive direct microscopy. Wood's lamp examination is also important, wherein UV light is emitted at an approximately 365 nm wavelength and scaly lesions usually show reddish or yellow green fluorescence. It also helps in revealing sub-clinical lesions, which are widely scattered as compared to the obvious clinical disease. Infection caused by non-fluorescent *Malassezia* species may be left undetected by Wood's lamp examination. The patient is asked to come without application of steroids and also without taking bath so that clinical picture remains unchanged. The fungal culture and histopathological examination may also be helpful in certain complex cases.

(a) Direct Examination

The skin from affected site is thoroughly sponged with 70% alcohol to remove surface contaminants. After drying, active edges of lesions are scrapped by using flame-sterilized no. 15 blade. Scales easily flake off in characteristic sheets by use of scalpel blade, edge of glass slide or even fingernail.

The scrapings are examined in KOH with DMSO, methylene blue or Albert's stain wet mounts. The Albert's stain is used as an alternative method in patients with pityriasis versicolor that it is less time consuming and as effective as KOH wet mount. It stains yeast cells and hyphae purple, clearly delineating the crisp details of fungi against a background of surrounding keratinocytes. The appearance can also be better delineated if Parker Quink Blue Black ink is incorporated in the KOH wet mount. Moreover, calcofluor white (CFW) staining is also used to avoid any confusion of the artifacts. Chlorazol Black E may be utilized for direct demonstration of fungal elements in the specimen. Kane's formulation may also be utilized, which consists of glycerol (10 ml), Tween 80 (10 ml), phenol (2.5 gm), methylene blue (1.0 gm), DW (480 ml). This formulation has an additional advantage of staining fungi as well as bacteria found in the differential diagnosis of *Malassezia* species like erythrasma.

Malassezia species are usually present in large quantity and clusters of round yeast cells, about 2 to 7 μm in size, with occasional budding. The hyphae are blunt, short, stout that may be curved with infrequent branching. Such characteristic forms are known as **'banana and grapes'** or **'spaghetti and meatballs'** which are diagnostic of *Malassezia* infections as seen in KOH wet mount in **Figure 7.7**. Almost similar findings are more clearer in the calcofluor white (CFW) smear examination as seen in **Figure 7.8**. Under scanning electron microscope, *Malassezia* species show liplike collarette around the point of bud initiation on the parent cell. The presence of collarette between mother and daughter cells is because the budding is phialidic and unipolar.

Another effective and reliable method of obtaining specimen, particularly in children is by applying piece of Scotch tape, with its adhesive side down, followed by pressing the tape firmly to recover scales. Then the tape is placed over 1-2 drops of KOH solution on the center of glass slide and examined under light microscope. This tape method has also been applied as reliable test for quantitative culture of *Malassezia* species in addition to scrub technique and contact plate method.

The histopathological examination shows typical hyphae and yeast cells in stratum corneum when stained with H&E, PAS or GMS stain. *Malassezia* is positive for many of the narrow-spectrum fungal stains it has staining pattern that differs from other yeasts. Therefore, it can be differentiated from *Candida* species by positive Alcian Blue and Masson-Fontana staining, from *Cryptococcus* species by positive Congo-Red staining and negative mucicarmine staining. The dermoscopy can also be used for diagnosis of pityriasis versicolor, which is an ancillary tool.

(b) Fungal Culture

The culture is essential to establish final diagnosis of *Malassezia* infections in routine clinical practice even though it is cumbersome to grow this organism. The fungal culture *vis-a-vis* clinical findings should be interpreted in the background that this fungus is also the normal flora of the human skin. As *Malassezia* is lipophilic fungus, therefore, lipids are

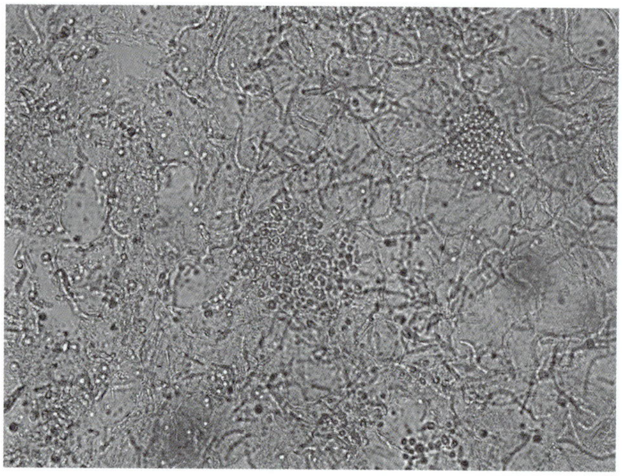

Fig. 7.7. *Malassezia* species showing short hyphae and clusters of yeast cells with occasional budding (KOH × 400).

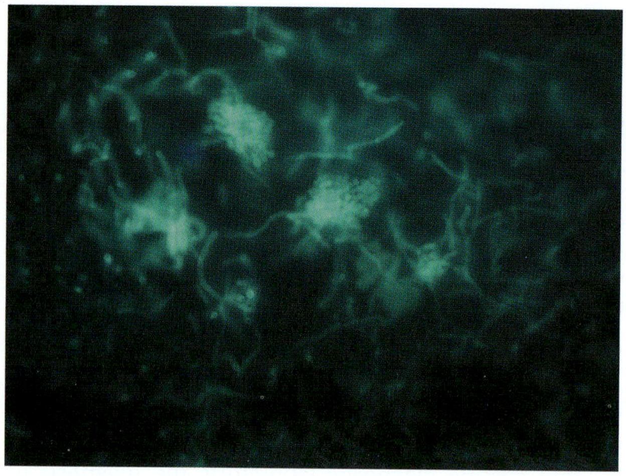

Fig. 7.8. *Malassezia* species showing more clear short hyphae and clusters of yeast cells with occasional budding (CFW × 400).

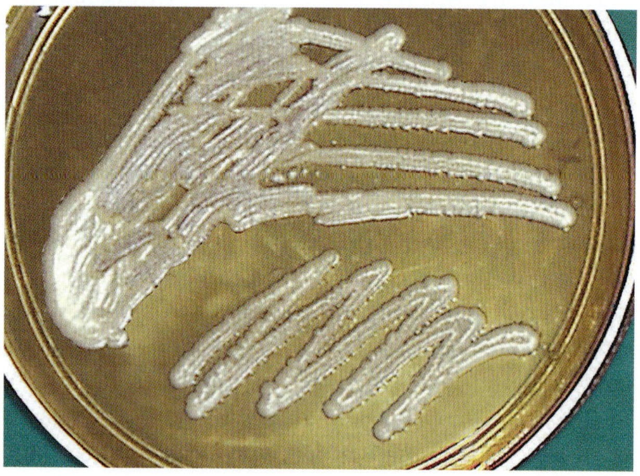

Fig. 7.9. Yeast-like colonies of *Malassezia* species grown on modified Dixon agar after a week of incubation.

incorporated into culture media, which include oleic acid, olive oil, glycerol monostearate and Tweens. Although SDA overlaid with olive oil has been frequently used in past but some of species like *M.globosa, M.restricta, M.obtusa* do not nurture properly on this medium hence more complex media are required. Dixon agar provides substantial growth and yields colonies with typical features hence helps in species identification **(Fig. 7.9)**. Moreover, Dixon's broth is used for growing *Malassezia* species. The modified version of Dixon agar is also used.

Leeming and Notman medium gives better results *i.e.* higher recovery rate, than variety of oil overlays and has longer shelf-life than poured plates. However, modified Dixon medium permits an easier colony counting and better appreciation of colony morphology hence is usually favored instead of Leeming and Notman formula. A new medium for rapid diagnosis and determination of antifungal susceptibility testing against *Malassezia* species has been devised and designated as Dutta and Dikshit Medium (DDM). Other alternative culture media like GYP-S agar can also be used which can provide growth and quantitation of colonies within 3-4 days. It contains glucose-yeast extract peptone agar plus olive oil, Tween 80 and glycerol monostearate. In the fungal blood culture bottle with lipid support, there are more chances of its isolation.

A new minimal medium for *M.furfur* consisting of only L-tryptophan and lipid source induces formation of brown pigmentation, which diffuses into agar. However, a few strains of *M.sympodialis* and *M.pachydermatis* fail to grow on this medium. Synthesis of pigments and fluorochromes by yeasts of genus *Malassezia* might help to explain several clinical characteristics of pityriasis versicolor.

Identification of the isolate to the genus level is performed by observation of lipid dependence, colony characteristics and cell morphology. Speciation is more difficult because carbon **assimilation** and **fermentation** techniques are **not** applicable to this genus. Their dependence on lipid precludes use of conventional carbon assimilation test, which are employed for identification of other yeasts. *Malassezia* species are basidiomycetes hence **urease positive**. Although *M.globosa* is most frequently isolated species, it is possible that many other species may have been wrongly identified as *M.furfur*.

The skin scrapings from infected sites are inoculated on SDA with chloramphenicol, actidione, Tween 80 and film of sterile olive oil in concentration of 10 ml per liter. Small, 3-6 mm cream-colored yellowish colonies, slightly raised with irregular edges, develop within 5-7 days at 32-35°C. *M.globosa* can be identified in most cases by its typical 'fried-egg' colony morphology. On microscopy, it shows typical globose cells with narrow budding base. On culture on modified Dixon medium, *M.globosa* usually produces only germ tubes but it can also develop short true mycelial branches. The LCB mount of colonies show 2-7 µm yeasts with many small 2-3 µm, bottle-shaped cells. Hyphae are occasionally seen in fungal cultures. *M.pachydermatis* is not an obligate lipophilic species hence does not require lipid as growth factor and can grow on conventional culture media like SDA.

The expansion of genus has made it slightly more difficult to identify its species. In routine diagnostic laboratory conventional approach is applied to recognize commonly implicated species by morphological characteristics, their ability to use several Tweens and catalase reaction. Some of these physiological and biochemical characteristics of *Malassezia* species are listed in **Table 7.2**.

Malassezia furfur utilizes Tween 20 (high concentration - 10%), Tween 40 (0.1-10%) and Tween 80 (low concentration - 0.1%) for its growth. A similar pattern is found in *M.sympodialis* except that ring of growth, rather solid zone with Tween 20, indicates that only low concentration of compound can be utilized by this species. *M.slooffiae* utilizes Tween 20 as well as Tween 40 but there is small zone with Tween 80, which indicates utilization of this compound only at higher concentration. *M.restricta* and *M.arunalokei* are catalase negative.

Other tests, which are found to be of diagnostic importance, are growth with 1-10% cremophor (castor oil) shown by most of isolates of *M.furfur* and splitting of esculin with strong reaction by *M.sympodialis* and *M.obtusa*. *M.globosa, M.restricta* and *M.obtusa* do grow with any of Tween compounds as source of lipid but they may cause some precipitation in medium and have more limited tolerance showing lack of growth above 37°C. The *Malassezia* species are resistant to benomyl.

Various preservation techniques have been studied for *Malassezia* species and the best recommended is lyophilization in liquid Dixon broth in 15% glycerol, immediately after first isolation to avoid loss of clinical isolates.

(c) Immunodiagnosis

The patients who have poor cellular immune response can be assessed by lymphocyte blastogenesis to specific fungal antigen. A transferable solid phase ELISA was developed for determination of antibody titers specific to *M.furfur* Serovars A, B and C in human sera. Other tests like fluorescent microscopy show green and orange fungal elements.

The molecular methods like restriction fragment length polymorphism (RFLP), polymerase chain reaction-fingerprinting and multilocus enzyme electrophoresis (MLEE) are also used to identify isolates besides epidemiological studies. Pulsed field gel electrophoresis (PFGE) may be useful to identify species of this medically important yeast and RAPD and AFLP analysis seems useful for screening large number of strains in epidemiological surveys. In recent times, MALDI-TOF-MS has also been used successfully in the identification of various *Malassezia* species.

As mentioned, the phenotypic identification based on microscopic and physiological method is difficult and time-consuming also. Hence molecular techniques have been useful in the identification and differentiation of *Malassezia* species, which provides their epidemiological information. The application of multiplex real-time PCR provides a sensitive and rapid identification system for *Malassezia* species, which may be applied in routine practice as well as epidemiological surveys. The use of multiplex real-time PCR for the specific detection of *Malassezia* species increases rates of detection rates thereby reduces time of result as compared to the fungal culture.

(d) Animal Pathogenicity

Malassezia species colonize a wide range of animals, however, may also cause disease to them. The common laboratory animals appear to be immune to this fungus when scales are applied. The trial studies conducted to infect guinea pigs and Swiss albino mice with *Malassezia* species, show hyperkeratosis, particularly at follicular ostia and pilous bulbs in experimental dermatitis. Rather *Malassezia* dermatitis is suspected among animals with inflammatory skin diseases characterized by erythematous or greasy lesions, especially in the intertriginous areas.

Treatment and Prophylaxis

The treatment of malasseziosis has gained favor with oral antifungal therapy since the introduction of ketoconazole in 1980s. Oral drugs are as effective as topical therapy and are preferred by the patients. The ideal regimen is short course of oral therapy yielding high clinical and mycological cure with low relapse rates.

The common therapeutic modalities in use are 1-2% imidazole creams, lotions and shampoos. Topical antifungal creams, including ketoconazole 2%, econazole nitrate 1%, clotrimazole 1% and miconazole nitrate 2%, are applied once or twice daily to the affected areas for 2 to 4 weeks. The main difficulty with the use of topical creams in pityriasis versicolor is its application over huge body surface area. Antifungal shampoos, including ketoconazole 2% and selenium sulfide 2.5%, are generally used as a body wash that is lathered on the scalp and affected skin for 5 to 10 minutes daily and then rinsed off. Seborrheic dermatitis of scalp also responds well to 1% ciclopirox shampoo applied once or twice weekly for 4 weeks.

Although *Malassezia* species are eliminated by the use of antifungals but it takes months to recover normal pigmentation even after adequate treatment. Hence patient should be counseled about this fact. However, if infection is left untreated, it waxes and wanes for months or even for years.

Short courses of itraconazole 200 mg daily for 7 days or fluconazole 400 mg for 3 days, are recommended. Alternatively, single dose consisting of 400 mg fluconazole is also effective for the treatment of pityriasis versicolor. In recurrent cases fluconazole is also given as **pulse therapy** in doses of 150 mg once or twice weekly for period of 2-3 months. Although a drug of choice for dermatophytes, oral terbinafine is not effective in tinea versicolor, as *Malassezia* species are not very sensitive to this drug and it does not reach an adequate concentration at the skin surface. Caution is advised when using oral antifungals as these drugs have potential hepatotoxicity and may interact with other medications.

Pityriasis versicolor generally responds to antifungal therapy but 60 to 80 percent of infections relapse or recur. The reasons assigned to recurrences are genetic, excessive sweating, hormonal factors and immunosuppression among these patients. Therefore, presence of positive cultures immediately after treatment with antifungal agents is not an unusual finding. The rapid reappearance of *Malassezia* species soon after antifungal treatment, is one of explanations for high relapse rates seen in this skin disorder. Intermittent application of 50% propylene glycol in water, twice daily for two weeks has been used to prevent such relapses.

The optimum antifungal susceptibility testing conditions for *Malassezia* species have been evaluated by modified Christensen's urea broth with 0.1% Tween 80 and 0.5% Tween 40 and modified broth microdilution method using fatty acid RPMI broth to ketoconazole, itraconazole and voriconazole. In the presence of various predisposing factors, recurrence is the major problem and to avoid it, prophylactic treatment schedule has been recommended. Gupta *et al,* have performed *in vitro* susceptibility of *Malassezia* species to ketoconazole, voriconazole, itraconazole and terbinafine thereby specific treatment can be advised to patients suffering from malasseziosis, accordingly. Danish workers Hald *et al,* have recently issued evidence-based guidelines for the treatment of *Malassezia* infections.

Recently Leong *et al,* have developed antifungal susceptibility testing of *Malassezia* species with an optimized colorimetric broth microdilution method. This is optimized broth microdilution assay compatible with the colorimetric indicator resazurin for the fast and efficient profiling of antifungal susceptibility in *Malassezia* species Determination of MIC values by visual reading of color change versus fluorescence reading is comparably reliable. Among the *Malassezia* strains tested, azoles such as voriconazole, itraconazole and posaconazole were the most effective antifungals. Terbinafine might be a treatment alternative for *Malassezia* infections.

In case of inflammatory skin conditions associated with *Malassezia* like seborrheic and atopic dermatitis, the addition of local anti-inflammatory therapy (corticosteroids or calcineurin inhibitors) is a prerequisite for rapid and effective control of exacerbations. One should always bear in mind that *Malassezia* species are integral components of the skin microbiota. Therefore therapeutic target should be controlling population of these yeasts with subsequent long-term antifungal treatment rather than eradicating it.

Further Reading

1. Affes M, Ben Salah S, Makni F, *et al.* Molecular identification of *Malassezia* species isolated from dermatitis affections. *Mycoses.* 2009; **52:** 251-6.
2. Agrawal V, Bhagwat AM, Sawant CS. Sesame oil incorporated medium for isolation and enumeration of lipophilic yeasts. *Int J Pharm Sci Res.* 2014; **5:** 2972-9.
3. Aguirre C, Euliarte C, Finquelievich J, *et al.* Fungemia and interstitial lung compromise caused by *Malassezia sympodialis* in a pediatric patient. *Rev Iberoam Micol.* 2015; **32:** 118-21.

4. Ahn JJ, Lim HK, Shin MK, *et al.* A case of pityriasis versicolor atrophicans. *Mycoses.* 2013; **56:** 358-60.
5. Akaza N, Akamatsu H, Sasaki Y, *et al. Malassezia* folliculitis is caused by cutaneous resident *Malassezia* species. *Med Mycol.* 2009; **47:** 618-24.
6. Akaza N, Akamatsu H, Takeoka S, *et al. Malassezia globosa* tends to grow actively in summer conditions more than other cutaneous *Malassezia* species. *J Dermatol.* 2012; **39:** 613-6.
7. Alvarez-Perez S, Blanco JL, Pelaez T, *et al. In vitro* amphotericin B susceptibility of *Malassezia pachydermatis* determined by the CLSI broth microdilution method and Etest using lipid-enriched media. *Antimicrob Agents Chemother.* 2014; **58:** 4203-6.
8. Ameen M. Epidemiology of superficial fungal infections. *Clin Dermatol.* 2010; **28:** 197-201.
9. Anane S, Chtourou O, Bodemer C, *et al. Malassezia* folliculitis in an infant. *Med Mycol Case Rep.* 2013; **2:** 72-4.
10. Andrade-FilhoJde S. Analogies in medicine: Spaghetti and meatballs. *Rev Inst Med Trop Sao Paulo.* 2013; **55.** 218.
11. Archana BR, Beena PM, Kumar S. Study of the distribution of *Malassezia* species in patients with pityriasis versicolor in Kolar region, Karnataka. *Indian J Dermatol.* 2015; **60:** 321.
12. Ashbee HR, Evans EGV. Immunology of diseases associated with *Malassezia* species. *Clin Microbiol Rev.* 2002; **15:** 21-57.
13. Ashbee HR. Update on the genus *Malassezia. Med Mycol.* 2007; **45:** 287-303.
14. Baker RM, Stegink RJ, Manaloor JJ, *et al. Malassezia* pneumonia: A rare complication of parenteral nutrition therapy. *J Parenter Enteral Nutr.* 2016; **40:** 1194-6.
15. Balakrishnan KP, Narayanswamy N, Mathews S, *et al.* Evaluation of some medicinal plants for their dandruff control properties. *Int J Pharm Biol Sci.* 2011; **2:** 38-45.
16. Barac A, Pekmezovic M, Milobratovic D, *et al.* Presence, species distribution and density of *Malassezia* yeast in patients with seborrhoeic dermatitis - A community-based case-control study and review of literature. *Mycoses.* 2015; **58:** 69-75.
17. Batra R, Boekhout T, Gueho E, *et al. Malassezia* Baillon, emerging clinical yeasts. *FEMS Yeast Res.* 2005; **5:** 1101-13.
18. Bentubo HD, Mantovani A, Yamashita JT, *et al.* Yeasts of the genital region of patients attending the dermatology service at Hospital Sao Paulo, Brazil. *Rev Iberoam Micol.* 2015; **32:** 229-34.
19. Bhargava P, Longhi LP. Peripheral smear with *Malassezia furfur. N Engl J Med.* 2007; **356:** e25.
20. Bikowski J. Facial seborrheic dermatitis: A report on current status and therapeutic horizons. *J Drugs Dermatol.* 2009; **8:** 125-33.
21. Blaes AH, Cavert WP, Morrison VA. *Malassezia:* Is it a pulmonary pathogen in the stem cell transplant population? *Transpl Infect Dis.* 2009; **11:** 313-7.
22. Borda LJ, Wikramanayake TC. Seborrheic dermatitis and dandruff: A comprehensive review. *J Clin Investig Dermatol.* 2015; **3.** PMID: 27148560.
23. Brockman J, Gibbs PM, Polenakovik H, *et al.* Acute appendicitis due to *Malassezia restricta. Mycoses.* 2012; **55:** e13-4.
24. BukvicMokos Z, Kralj M, Basta-Juzbasic A, *et al.* Seborrheic dermatitis: An update. *Acta Dermatovenerol Croat.* 2012; **20:** 98-104.
25. Cabanes FJ, Coutinho SD, Puig L, *et al.* New lipid-dependent *Malassezia* species from parrots. *Rev Iberoam Micol.* 2016; **33:** 92-9.
26. Cabanes FJ. *Malassezia* yeasts: How many species infect humans and animals? *PLoS Pathog.* 2014; **10:** e1003892.
27. Cafarchia C, Figueredo LA, Iatta R, *et al. In vitro* evaluation of *Malassezia pachydermatis* susceptibility to azole compounds using E-test and CLSI microdilution methods. *Med Mycol.* 2012; **50:** 795-801.
28. Cafarchia C, Gasser RB, Figueredo LA, *et al.* Advances in the identification of *Malassezia. Mol Cell Probes.* 2011; **25:** 1-7.
29. Cafarchia C, Iatta R, Immediato D, *et al.* Azole susceptibility of *Malassezia pachydermatis* and *Malassezia furfur* and tentative epidemiological cut-off values. *Med Mycol.* 2015; **53:** 743-8.
30. Cantrell WC, Elewksi BE. Can pityriasis versicolor be treated with 2% ketoconazole foam? *J Drugs Dermatol.* 2014; **13:** 855-9.
31. Carrillo-Munoz AJ, Rojas F, Tur-Tur C, *et al. In vitro* antifungal activity of topical and systemic antifungal drugs against *Malassezia* species. *Mycoses.* 2013; **56:** 571-5.
32. Chaudhary R, Singh S, Banerjee T, Tilak R. Prevalence of different *Malassezia* species in pityriasis versicolor in central India. *Indian J Dermatol Venereol Leprol.* 2010; **76:** 159-64.
33. Chowdhary A, Randhawa HS, Sharma S, *et al. Malassezia furfur* in a case of onychomycosis: Colonizer or etiologic agent? *Med Mycol.* 2005; **43:** 87-90.
34. Clark GW, Pope SM, Jaboori KA. Diagnosis and treatment of seborrheic dermatitis. *Am Fam Physician.* 2015; **91:** 185-90.
35. Crespo MJ, Abarca ML, Cabanos FJ. Evaluation of different preservative and storage methods for *Malassezia* species. *J Clin Microbiol.* 2000; **38:** 3872-75.
36. Crespo-Erchiga V, Gomez-Moyano E, Crespo M. Pityriasis versicolor and the yeasts of genus *Malassezia. Actas Dermosifiliogr.* 2008; **99:** 764-71.
37. Cullingham K, Hull PR. Atrophying pityriasis versicolor. *CMAJ.* 2014; **186:** 776.
38. Darabi K, Hostetler SG, Bechtel MA, *et al.* The role of *Malassezia* in atopic dermatitis affecting the head and neck of adults. *J Am Acad Dermatol.* 2009; **60:** 125-36.
39. Darlenski R, Kazandjieva J, Hristakieva E, *et al.* Atopic dermatitis as a systemic disease. *Clin Dermatol.* 2014; **32:** 409-13.
40. Das J, Majumdar M, Chakraborty U, *et al.* Oral itraconazole for the treatment of severe seborrhoeic dermatitis. *Indian J Dermatol.* 2011; **56:** 515-6.
41. Day T, Scurry J. Vulvar pityriasis versicolor in an immunocompetent woman. *J Low Genit Tract Dis.* 2014; **18:** e71-3.
42. Del Rosso JQ. Adult seborrheic dermatitis: A status report on practical topical management. *J Clin Aesthet Dermatol.* 2011; **4:** 32-8.

43. Denis J, Machouart M, Morio F, *et al.* Performance of matrix-assisted laser desorption ionisation-time of flight mass spectrometry for identifying clinical *Malassezia* isolates. *J Clin Microbiol.* 2017; **55:** 90-6.
44. de St Maurice A, Frangoul H, Coogan A, *et al.* Prolonged fever and splenic lesions caused by *Malassezia restricta* in an immunocompromised patient. *Pediatr Transplant.* 2014; **18:** e283-6.
45. Desai HB, Perkins PL, Procop GW. Granulomatous dermatitis due to *Malassezia sympodialis. Arch Pathol Lab Med.* 2011; **135:** 1085-7.
46. Dessinioti C, Katsambas A. Seborrheic dermatitis: Etiology, risk factors and treatments: Facts and controversies. *Clin Dermatol.* 2013; **31:** 343-51.
47. Difonzo EM, Faggi E, Bassi A, *et al. Malassezia* skin diseases in humans. *G Ital Dermatol Venereol.* 2013; **148:** 609-19.
48. Dikshit A, Tiwari AK, Mishra RK. New medium for rapid diagnosis and determination of antifungal testing against *Malassezia* spp: A potential candidate for industries. *Natl Acad Sci Lett.* 2013; **36:** 61-6.
49. Durdu M, Guran M, Ilkit M. Epidemiological characteristics of *Malassezia* folliculitis and use of the May-Grunwald-Giemsa stain to diagnose the infection. *Diagn Microbiol Infect Dis.* 2013; **76:** 450-7.
50. Dutta S, Bajaj AK, Basu S, *et al.* Pityriasis versicolor: Socioeconomic and clinico-mycologic study in India. *Int J Dermatol.* 2002; **41:** 823-4.
51. Elewski BE. Safe and effective treatment of seborrheic dermatitis. *Cutis.*2009; **83:** 333-8.
52. Ertam I, Aytimur D, Alper S. *Malassezia furfur* onychomycosis in an immunosuppressed liver transplant recipient. *Indian J Dermatol Venereol Leprol.* 2007; **73:** 425-6.
53. Faergemann J. Atopic dermatitis and fungi. *Clin Microbiol Rev.* 2002; **15:** 545-63.
54. Figueredo LA, Cafarchia C, Otranto D. Antifungal susceptibility of *Malassezia pachydermatis* biofilm. *Med Mycol.* 2013; **51:** 863-7.
55. Framil VM, Melhem MS, Szeszs MW, *et al.* New aspects in the clinical course of pityriasis versicolor. *An Bras Dermatol.* 2011; **86:** 1135-40.
56. Framil VM, Melhem MS, Szeszs MW, *et al.* Pityriasis versicolor circinata: Isolation of *Malassezia sympodialis* - Case report. *An Bras Dermatol.* 2010; **85:** 227-8.
57. Gaitanis G, Bassukas ID, Velegraki A. The range of molecular methods for typing *Malassezia. Curr Opin Infect Dis.* 2009; **22:** 119-25.
58. Gaitanis G, Magiatis P, Hantschke M, *et al.* The *Malassezia* genus in skin and systemic diseases. *Clin Microbiol Rev.* 2012; **25:** 106-41.
59. Gaitanis G, Velegraki A, Mayser P, *et al.* Skin diseases associated with *Malassezia* yeasts: Facts and controversies. *Clin Dermatol.* 2013; **31:** 455-63.
60. Gao Z, Perez-Perez GI, Chen Y, *et al.* Quantitation of major human cutaneous bacterial and fungal populations. *J Clin Microbiol.* 2010; **48:** 3575-81.
61. Gemmer CM, De Angelis YM, Theelen B, *et al.* Fast, noninvasive method for molecular detection and differentiation of *Malassezia* yeast species on human skin and application of the method to dandruff microbiology. *J Clin Microbiol.* 2002; **40:** 3350-7.
62. Glatz M, Bosshard PP, Hoetzenecker W, *et al.* The role of *Malassezia* spp. in atopic dermatitis. *J Clin Med.* 2015; **4:** 1217-28.
63. Glatz M, Bosshard P, Schmid-Grendelmeier P. The role of fungi in atopic dermatitis. *Immunol Allergy Clin North Am.* 2017; **37:** 63-74.
64. Gonzalez A, Sierra R, Cardenas ME, *et al.* Physiological and molecular characterization of atypical isolates of *Malassezia furfur. J Clin Microbiol.* 2009; **47:** 48-53.
65. Gupta A, Foley K. Antifungal treatment for pityriasis versicolor. *J Fungi.* 2015; **1:** 13-29.
66. Gupta AK, Batra R, Bluhm R, *et al.* Skin diseases associated with *Malassezia* species. *J Am Acad Dermatol.* 2004; **51:** 785-98.
67. Gupta AK, Kohli Y, Summerbell RC. Molecular differentiation of seven *Malassezia* species. *J Clin Microbiol.* 2000; **38:** 1869-75.
68. Gupta AK, Lane D, Paquet M. Systematic review of systemic treatments for tinea versicolor and evidence-based dosing regimen recommendations. *J Cutan Med Surg.* 2014; **18:** 79-90.
69. Gupta AK, Lyons DC. Pityriasis versicolor: An update on pharmacological treatment options. *Expert Opin Pharmacother.* 2014; **15:** 1707-13.
70. Gupta P, Chakrabarti A, Singhi S, *et al.* Skin colonization by *Malassezia* spp. in hospitalized neonates and infants in a tertiary care centre in north India. *Mycopathologia.* 2014; **178:** 267-72.
71. Haiduk J, Treudler R, Ziemer M. Atrophying tinea versicolor with epidermal atrophy. *J Dtsch Dermatol Ges.* 2016; **14:** 740-3.
72. Hald M, Arendrup MC, Svejgaard EL, *et al.* Evidence-based Danish guidelines for the treatment of *Malassezia*-related skin diseases. *Acta Derm Venereol.* 2015; **95:** 12-9.
73. Haninger DM, Snyder JW. Nausea and emesis in a complicated patient - *Malassezia pachydermatis. J Clin Microbiol.* 2013; **51:** 3475 & 3914.
74. Harada K, Saito M, Sugita T, *et al. Malassezia* species and their associated skin diseases. *J Dermatol.* 2015; **42:** 250-7.
75. Havlickova B, Czaika VA, Friedrich M. Epidemiological trends in skin mycoses worldwide. *Mycoses.* 2008; **51 (Suppl. 4):** 2-15.
76. Hay R. Superficial fungal infections. *Medicine.* 2013; **41:** 716-8.
77. Hay RJ. *Malassezia*, dandruff and seborrhoeic dermatitis: An overview. *Br J Dermatol.* 2011; **165 (Suppl. 2):** 2-8.
78. Heidrich D, Daboit TC, Stopiglia CD, *et al.* Sixteen years of pityriasis versicolor in metropolitan area of Porto Alegre, Southern Brazil. *Rev Inst Med Trop Sao Paulo.* 2015; **57:** 277-80.
79. Helou J, Obeid G, Moutran R, *et al.* Pityriasis versicolor: A case of resistance to treatment. *Int J Dermatol.* 2014; **53:** e114-6.

80. Hirai A, Kano R, Makimura K, *et al.* *Malassezia nana* sp. nov., a novel lipid-dependent yeast species isolated from animals. *Int J Syst Evol Microbiol.* 2004; **54:** 623-7.
81. Hiruma M, Cho O, Hiruma M, *et al.* Genotype analyses of human commensal scalp fungi, *Malassezia globosa* and *Malassezia restricta* on the scalps of patients with dandruff and healthy subjects. *Mycopathologia.* 2014; **177:** 263-9.
82. Holliday A, Grider D. Tinea versicolor. *N Engl J Med.* 2016; **374:** e11.
83. Honnavar P, Chakrabarti A, Dogra S, *et al.* Phenotypic and molecular characterization of *Malassezia japonica* isolated from psoriasis vulgaris patients. *J Med Microbiol.* 2015; **64:** 232-6.
84. Honnavar P, Prasad GS, Ghosh A, *et al.* *Malassezia arunalokei* sp. nov., A novel yeast species isolated from seborrhoeic dermatitis patients and healthy individuals from India. *J Clin Microbiol.* 2016; **54:** 1826-34.
85. Hort W, Mayser P. *Malassezia* virulence determinants. *Curr Opin Infect Dis.* 2011; **24:** 100-5.
86. Huang WW, Tharp MD.A case of tinea versicolor of the eyelids. *Pediatr Dermatol.* 2013; **30:** e242-3.
87. Hudson A, Carroll B, Kim SJ. Folliculocentric tinea versicolor. *Dermatol Online J.* 2017; 23. pii: 13030/qt5kj574bd. PMID: 28329492.
88. Iatta R, Cafarchia C, Cuna T, *et al.* Bloodstream infections by *Malassezia* and *Candida* species in critical care patients. *Med Mycol.* 2014; **52:** 264-9.
89. Iatta R, Figueredo LA, Montagna MT, *et al.* *In vitro* antifungal susceptibility of *Malassezia furfur* from bloodstream infections. *J Med Microbiol.* 2014; **63:** 1467-73.
90. Ibekwe PU, Ogunbiyi AO, Besch R, *et al.* The spectrum of *Malassezia* species isolated from students with pityriasis versicolor in Nigeria. *Mycoses.* 2015; **58:** 203-8.
91. Ilahi A, Hadrich I, Goudjil S, *et al.* Molecular epidemiology of a *Malassezia pachydermatis* neonatal unit outbreak. *Med Mycol.* 2017; doi: 10.1093/mmy/myx022. PMID: 28371911.
92. Ilahi A, Hadrich I, Neji S, *et al.* Real-time PCR identification of six *Malassezia* species. *Curr Microbiol.* 2017; **74:** 671-7.
93. Ishibashi Y, Kato H, Asahi Y, *et al.* Identification of the major allergen of *Malassezia globosa* relevant for atopic dermatitis. *J Dermatol Sci.* 2009; **55:** 185-92.
94. Jagielski T, Rup E, Ziolkowska A, *et al.* Distribution of *Malassezia* species on the skin of patients with atopic dermatitis, psoriasis and healthy volunteers assessed by conventional and molecular identification methods. *BMC Dermatol.* 2014; **14:** 3.
95. Janaki C, Sentamilselvi G, Janaki VR, *et al.* Unusual observations in the histology of pityriasis versicolor. *Mycopathologia.* 1997; **139:** 71-4.
96. Jena DK, Sengupta S, Dwari BC, *et al.* Pityriasis versicolor in the pediatric age group. *Indian J Dermatol Venereol Leprol.* 2005; **71:** 259-61.
97. Jubert E, Martin-Santiago A, Bernardino M, *et al.* Neonatal pityriasis versicolor. *Pediatr Infect Dis J.* 2015; **34:** 329-30.
98. Juntachai W, Kummasook A, Mekaprateep M, *et al.* Identification of the haemolytic activity of *Malassezia* species. *Mycoses.* 2014; **57:** 163-8.
99. Juntachai W, Oura T, Murayama SY, *et al.* The lipolytic enzymes activities of *Malassezia* species. *Med Mycol.* 2009; **47:** 477-84.
100. Kallini JR, Riaz F, Khachemoune A. Tinea versicolor in dark-skinned individuals. *Int J Dermatol.* 2014; **53:** 137-41.
101. Kaneko T, Makimura K, Abe M, *et al.* Revised culture-based system for identification of *Malassezia* species. *J Clin Microbiol.* 2007; **45:** 3737-42.
102. Kaneko T, Makimura K, Onozaki M, *et al.* Vital growth factors of *Malassezia* species on modified CHROMagar Candida. *Med Mycol.* 2005; **43:** 699-704.
103. Kaneko T, Murotani M, Ohkusu K, *et al.* Genetic and biological features of catheter-associated *Malassezia furfur* from hospitalized adults. *Med Mycol.* 2012; **50:** 74-80.
104. Kaneko T, Shiota R, Shibuya S, *et al.* Human external ear canal as the specific reservoir of *Malassezia slooffiae. Med Mycol.* 2010; **48:** 824-7.
105. Karakas M, Turac-Bicer A, Ilkit M, *et al.* Epidemiology of pityriasis versicolor in Adana, Turkey. *J Dermatol.* 2009; **36:** 377-82.
106. Karakatsanis G, Vakirlis E, Kastoridou C, *et al.* Co-existence of pityriasis versicolor and erythrasma. *Mycoses.* 2004; **47:** 343-5.
107. Karalezli A, Borazan M, Dursun R, *et al.* Impression cytology and ocular surface characteristics in patients with seborrhoeic dermatitis. *Acta Ophthalmol.* 2011; **89:** e137-41.
108. Kaujalgi R, Handa S, Jain A, Kanwar AJ. Ocular abnormalities in atopic dermatitis in Indian patients. *Indian J Dermatol Venereol Leprol.* 2009; **75:** 148-51.
109. Kaur M, Narang T, Bala M, *et al.* Study of the distribution of *Malassezia* species in patients with pityriasis versicolor and healthy individuals in tertiary care hospital, Punjab. *Indian J Med Microbiol.* 2013; **31:** 270-4.
110. Kaushik N, Pujalte GG, Reese ST. Superficial fungal infections. *Prim Care Clin Office Pract.* 2015; **42:** 501-16.
111. Kekki OM, Scheynius A, Poikonen S, *et al.* Sensitization to *Malassezia* in children with atopic dermatitis combined with food allergy. *Pediatr Allergy Immunol.* 2013; **24:** 244-9.
112. Kelly BP. Superficial fungal infections. *Pediatr Rev.* 2012; **33:** e22-37.
113. Kim GK. Seborrheic Dermatitis and *Malassezia* species: How are they related? *J Clin Aesthet Dermatol.* 2009; **2:** 14-7.
114. Kim JY, Hahn HJ, Choe YB, *et al.* Molecular biological identification of *Malassezia* yeasts using pyrosequencing. *Ann Dermatol.* 2013; **25:** 73-9.
115. Kim SY, Lee YW, Choe YB, *et al.* Progress in *Malassezia* research in Korea. *Ann Dermatol.* 2015; **27:** 647-57.
116. Kindo AJ, Sophia SK, Kalyani J, *et al.* Identification of *Mala_ ssezia* species. *Indian J Med Microbiol.* 2004; **22:** 179-81.
117. Kolecka A, Khayhan K, Arabatzis M, *et al.* Efficient identification of *Malassezia* yeasts by matrix-assisted laser desorption ionization-time of flight mass spectrometry. *Br J Dermatol.* 2014; **170:** 332-41.
118. Kucinskiene V, Sutkute A, Valiukeviciene S. Cutaneous fungal infection in a neonatal intensive care unit patient: a case report and literature review. *Pediatr Dermatol.* 2014; **31:** 267-70.

119. Kumar L, Verma S, Bhardwaj A, *et al.* Eradication of superficial fungal infections by conventional and novel approaches: A comprehensive review. *Artif Cells Nanomed Biotechnol.* 2014; **42:** 32-46.
120. Lally A, Casabonne D, Newton R, *et al.* Seborrheic dermatitis among Oxford renal transplant recipients. *J Eur Acad Dermatol Venereol.* 2010; **24:** 561-4.
121. Lange L, Alter N, Keller T, *et al.* Sensitization to *Malassezia* in infants and children with atopic dermatitis: Prevalence and clinical characteristics. *Allergy.* 2008; **63:** 486-7.
122. Lecerf P, Hay R.A new approach to the diagnosis and study of *Malassezia* infections. *Br J Dermatol.* 2014; **170:** 234.
123. Ledbetter EC, Starr JK. *Malassezia pachydermatis* keratomycosis in a dog. *Med Mycol Case Rep.* 2016; **10:** 24-6.
124. Lee JW, Kim BJ, Kim MN. Photodynamic therapy: New treatment for recalcitrant *Malassezia* folliculitis. *Lasers Surg Med.* 2010; **42:** 192-6.
125. Leeming JP, Notman FH. Improved methods for isolation and enumeration of *Malassezia furfur* from human skin. *J Clin Microbiol.* 1987; **25**: 2017-9.
126. Leong C, Buttafuoco A, Glatz M, *et al.* Antifungal susceptibility testing of *Malassezia* spp. with an optimized colorimetric broth microdilution method. *J Clin Microbiol.* 2017; **55:** 1883-93.
127. Levin NA. Beyond spaghetti and meatballs: Skin diseases associated with the *Malassezia* yeasts. *Dermatol Nurs.* 2009; **21:** 7-13, 51; Quiz 14.
128. Lim SL, Lim CS. New contrast stain for the rapid diagnosis of pityriasis versicolor. *Arch Dermatol.* 2008; **144:** 1058-9.
129. Lin X, Yuping R, Guangping Z. Advances in nomenclature and taxonomy of the genus *Malassezia. J Clin Dermatol.* 2000; **29:** 310-12.
130. Ljubojevic S, Lipozencic J, Basta-Juzbasic A. Contact allergy to corticosteroids and *Malassezia furfur* in seborrhoeic dermatitis patients. *J Eur Acad Dermatol Venereol.* 2011; **25:** 647-51.
131. Lyakhovitsky A, Shemer A, Amichai B. Molecular analysis of *Malassezia* species isolated from Israeli patients with pityriasis versicolor. *Int J Dermatol.* 2013; **52:** 231-3.
132. Manuel F, Ranganathan S. A new postulate on two stages of dandruff: A clinical perspective. *Int J Trichology.* 2011; **3:** 3-6.
133. Marcon MJ, Powell DA. Human infections due to *Malassezia* spp. *Clin Microbiol Rev.* 1992; **5:** 101-19.
134. Marques SA, Silva SB, Camargo RM, *et al.* Exuberant clinical presentation of probable *Malassezia* folliculitis in a young non-immunosuppressed patient. *An Bras Dermatol.* 2012; **87:** 459-62.
135. McNally B, McGraw T. Tinea versicolor. *J Spec Oper Med.* 2010; **10:** 107-10.
136. Messenger AG. Follicles, fungi and scalp conditions. *Br J Dermatol.* 2011; **165 (Suppl. 2):** 1. PMID: 21919895.
137. Metin A, Dilek N, Demirseven DD. Fungal infections of the folds (intertriginous areas). *Clin Dermatol.* 2015; **33:** 437-47.
138. Moon SY, Lee WJ, Lee SJ, *et al.* Pityriasis versicolor atrophicans: Is it true atrophy or pseudoatrophy? *J Cutan Pathol.* 2016; **43:** 187-9.
139. Morais PM, Cunha Mda G, Frota MZ. Clinical aspects of patients with pityriasis versicolor seen at a referral center for tropical dermatology in Manaus, Amazonas, Brazil. *An Bras Dermatol.* 2010; **85:** 797-803.
140. Moreno-Coutino G, Espinosa E, Garcia-Romero MT, *et al.* Novel presentation of immune reconstitution inflammatory syndrome: Folliculitis secondary to *Malassezia* spp. *Mycoses.* 2011; **54:** e252-4.
141. Morishita N, Sei Y. Microreview of pityriasis versicolor and *Malassezia* species. *Mycopathologia.* 2006; **162:** 373-6.
142. Mostafa WZ, Assaf MI, Ameen IA, *et al.* Hair loss in pityriasis versicolor lesions: A descriptive clinicopathological study. *J Am Acad Dermatol.* 2013; **69:** e19-23.
143. Nabatian AS, Millett CR, Heymann WR. What is your diagnosis? Folliculocentric tinea versicolor. *Cutis.* 2012; **90:** 113, 117-8.
144. Nagata R, Nagano H, Ogishima D, *et al.* Transmission of the major skin microbiota, *Malassezia*, from mother to neonate. *Pediatr Int.* 2012; **54:** 350-5.
145. Naim M, Varshney M, John VT, *et al. Malassezia* species infestation of the synovium: pictorial evidence in one case. *Histopathology.* 2011; **58:** 805-7.
146. Naldi L, Diphoorn J. Seborrhoeic dermatitis of the scalp. *BMJ Clin Evid.* 2015; pii: 1713.
147. Naldi L, Rebora A. Seborrheic dermatitis. *N Engl J Med.* 2009; **360:** 387-96.
148. Naldi L. Seborrhoeic dermatitis. *BMJ Clin Evid.* 2010; pii: 1713.
149. Narang T, Dogra S, Kaur I, *et al. Malassezia* and psoriasis: Koebner's phenomenon or direct causation? *J Eur Acad Dermatol Venereol.* 2007; **21:** 1111-2.
150. Narang T, Dogra S, Kaur I. Co-localization of pityriasis versicolor and BT Hansen's disease. *Int J Lepr Other Mycobact Dis.* 2005; **73:** 206-7.
151. Negroni R. Historical aspects of dermatomycoses. *Clin Dermatol.* 2010; **28:** 125-32.
152. Nenoff P, Kruger C, Ginter-Hanselmayer G, *et al.* Mycology - An update. Part 1: Dermatomycoses: causative agents, epidemiology and pathogenesis. *J Dtsch Dermatol Ges.* 2014; **12:** 188-209. Quiz 210; 188-211. Quiz 212.
153. Nenoff P, Kruger C, Schaller J, Ginter-Hanselmayer G, *et al.* Mycology - An update. Part 2: Dermatomycoses: Clinical picture and diagnostics. *J Dtsch Dermatol Ges.* 2014; **12:** 749-77.
154. Nenoff P, Kruger C, Paasch U, *et al.* Mycology - An update. Part 3: Dermatomycoses: topical and systemic therapy. *J Dtsch Dermatol Ges.* 2015; **13:** 387-410; Quiz 411.
155. Oh BH, Song YC, Lee YW, *et al.* Comparison of nested PCR and RFLP for identification and classification of *Malassezia* yeasts from healthy human skin. *Ann Dermatol.* 2009; **21:** 352-7.
156. Oliveri S, Trovato L, Betta P, *et al. Malassezia furfur* fungaemia in a neonatal patient detected by lysis-centrifugation blood culture method: First case reported in Italy. *Mycoses.* 2011; **54:** e638-40.
157. Park HK, Ha MH, Park SG, *et al.* Characterization of the fungal microbiota (mycobiome) in healthy and dandruff-afflicted human scalps. *PLoS One.* 2012; **7:** e32847.

158. Pasquetti M, Chiavassa E, Tizzani P, *et al.* Agar diffusion procedures for susceptibility testing of *Malassezia pachydermatis:* Evaluation of Mueller-Hinton agar plus 2% glucose and 0.5 μg/ml methylene blue as the test medium. *Mycopathologia.* 2015; **180:** 153-8.
159. Payle B, Serrano L, Bieley HC, *et al.* Albert's solution versus potassium hydroxide solution in the diagnosis of tinea versicolor. *Int J Dermatol.* 1994; **33:** 182-3.
160. Pedrosa AF, Lisboa C, Goncalves Rodrigues A. *Malassezia* infections: A medical conundrum. *J Am Acad Dermatol.* 2014; **71:** 170-6.
161. Petry V, Tanhausen F, Weiss L, *et al.* Identification of *Malassezia* yeast species isolated from patients with pityriasis versicolor. *An Bras Dermatol.* 2011; **86:** 803-6.
162. Prohic A, Jovovic Sadikovic T, Krupalija-Fazlic M, *et al.* *Malassezia* species in healthy skin and in dermatological conditions. *Int J Dermatol.* 2016; **55:** 494-504.
163. Prohic A, Kuskunovic-Vlahovljak S, Sadikovic TJ, *et al.* The prevalence and species composition of *Malassezia* yeasts in patients with clinically suspected onychomycosis. *Med Arch.* 2015; **69:** 81-4.
164. Prohic A, Simic D, Sadikovic TJ, *et al.* Distribution of *Malassezia* species on healthy human skin in Bosnia and Herzegovina: Correlation with body part, age and gender. *Iran J Microbiol.* 2014; **6:** 253-62.
165. Ramadan S, Sortino M, Bulacio L, *et al.* Prevalence of *Malassezia* species in patients with pityriasis versicolor in Rosario, Argentina. *Rev Iberoam Micol.* 2012; **29:** 14-9.
166. Ranganathan S, Mukhopadhyay T. Dandruff: The most commercially exploited skin disease. *Indian J Dermatol.* 2010; **55:** 130-4.
167. Rasi A, Naderi R, Behzadi AH, *et al.* *Malassezia* yeast species isolated from Iranian patients with pityriasis versicolor in a prospective study. *Mycoses.* 2010; **53:** 350-5.
168. Renati S, Cukras A, Bigby M. Pityriasis versicolor. *BMJ.* 2015; **350:** h1394.
169. Ricci G, Bellini F, Dondi A, *et al.* Atopic dermatitis in adolescence. *Dermatol Reports.* 2011; **4:** e1.
170. Rivard SC. Pityriasis versicolor: Avoiding pitfalls in disease diagnosis and therapy. *Mil Med.* 2013; **178:** 904-6.
171. Rodoplu G, Saracli MA, Gumral R, *et al.* Distribution of *Malassezia* species in patients with pityriasis versicolor in Turkey. *J Mycol Med.* 2014; **24:** 117-23.
172. Rojas FD, Cordoba SB, de Los Angeles Sosa M, *et al.* Antifungal susceptibility testing of *Malassezia* yeast: Comparison of two different methodologies. *Mycoses.* 2017; **60:** 104-111.
173. Rojas FD, Sosa Mde L, Fernandez MS, *et al.* Antifungal susceptibility of *Malassezia furfur, Malassezia sympodialis* and *Malassezia globosa* to azole drugs and amphotericin B evaluated using a broth microdilution method. *Med Mycol.* 2014; **52:** 641-6.
174. Roman J, Bagla P, Ren P, *et al.* *Malassezia pachydermatis* fungemia in an adult with multibacillary leprosy. *Med Mycol Case Rep.* 2016; **12:** 1-3.
175. Romano C, Feci L, Mancianti F, *et al.* Perineal and genital pityriasis versicolor due to *Malassezia globosa. J Eur Acad Dermatol Venereol.* 2015; **29:** 1857-8.
176. Romano C, Mancianti F, Nardoni S, *et al.* Identification of *Malassezia* species isolated from patients with extensive forms of pityriasis versicolor in Siena, Italy. *Rev Iberoam Micol.* 2013; **30:** 231-4.
177. Rosen T. Mycological considerations in the topical treatment of superficial fungal infections. *J Drugs Dermatol.* 2016; **15:** s49-55.
178. Rubenstein RM, Malerich SA. *Malassezia (Pityrosporum)* folliculitis. *J Clin Aesthet Dermatol.* 2014; **7:** 37-41.
179. Rudramurthy SM, Honnavar P, Chakrabarti A, *et al.* Association of *Malassezia* species with psoriatic lesions. *Mycoses.* 2014; **57:** 483-8.
180. Rudramurthy SM, Honnavar P, Dogra S, *et al.* Association of *Malassezia* species with dandruff. *Indian J Med Res.* 2014; **139:** 431-7.
181. Ryu HW, Cho JW, Lee KS. Pityriasis versicolor on penile shaft in a renal transplant recipient. *Ann Dermatol.* 2012; **24:** 345-7.
182. Saad M, Sugita T, Saeed H, *et al.* Molecular epidemiology of *Malassezia globosa* and *Malassezia restricta* in Sudanese patients with pityriasis versicolor. *Mycopathologia.* 2013; **175:** 69-74.
183. Salahi-Moghaddam A, Davoodian P, Jafari A, *et al.* Evaluation of pityriasis versicolor in prisoners: A cross-sectional study. *Indian J Dermatol Venereol Leprol.* 2009; **75:** 379-82.
184. Sampaio AL, Mameri AC, Vargas TJ, *et al.* Seborrheic dermatitis. *An Bras Dermatol.* 2011; **86:** 1061-71; Quiz 1072-4.
185. Santana JO, Azevedo FL, Campos Filho PC. Pityriasis versicolor: Clinical-epidemiological characterization of patients in the urban area of Buerarema-BA, Brazil. *An Bras Dermatol.* 2013; **88:** 216-21.
186. Saunders CW, Scheynius A, Heitman J. *Malassezia* fungi are specialized to live on skin and associated with dandruff, eczema and other skin diseases. *PLoS Pathog.* 2012; **8:** e1002701.
187. Schmidt JA. Seborrheic dermatitis: A clinical practice snapshot. *Nurse Pract.* 2011; **36:** 32-7.
188. Schwartz JR, Messenger AG, Tosti A, *et al.* A comprehensive pathophysiology of dandruff and seborrheic dermatitis - towards a more precise definition of scalp health. *Acta Derm Venereol.* 2013; **93:** 131-7.
189. Schwartz JR, Shah R, Krigbaum H, *et al.* New insights on dandruff/seborrhoeic dermatitis: The role of the scalp follicular infundibulum in effective treatment strategies. *Br J Dermatol.* 2011;**165 (Suppl. 2):** 18-23.
190. Schwartz RA. Superficial fungal infections. *Lancet.* 2004; **364:** 1173-82.
191. Shah A, Koticha A, Ubale M, *et al.* Identification and speciation of *Malassezia* in patients clinically suspected of having pityriasis versicolor. *Indian J Dermatol.* 2013; **58:** 239.
192. Sharma A, Rabha D, Ahmed G. *In vitro* antifungal susceptibility of *Malassezia* isolates from pityriasis versicolor lesions. *Indian J Dermatol Venereol Leprol.* 2017; **83:** 249-51.
193. Sharma A, Rabha D, Choraria S, *et al.* Clinicomycological profile of pityriasis versicolor in Assam. *Indian J Pathol Microbiol.* 2016; **59:** 159-65.

194. Sharma M, Kansal NK, Gautam RK. A case of Becker's nevus with pityriasis versicolor. *J Eur Acad Dermatol Venereol.* 2014; **28:** 1827-8.
195. Shokohi T, Afshar P, Barzgar A. Distribution of *Malassezia* species in patients with pityriasis versicolor in northern Iran. *Indian J Med Microbiol.* 2009; **27:** 321-4.
196. Siegfried E, Glenn E. Use of olive oil for the treatment of seborrheic dermatitis in children. *Arch Pediatr Adolesc Med.* 2012; **166:** 967.
197. Soares RC, Zani MB, Arruda AC, *et al. Malassezia* intra-specific diversity and potentially new species in the skin microbiota from Brazilian healthy subjects and seborrheic dermatitis patients. *PLoS One.* 2015; **10:** e0117921.
198. Song HS, Kim SK, Kim YC. Comparison between *Malassezia* folliculitis and non-*Malassezia* folliculitis. *Ann Dermatol.* 2014; **26:** 598-602.
199. Spence-Shishido A, Carr C, Bonner MY, *et al. In vivo* Gram staining of tinea versicolor. *JAMA Dermatol.* 2013; **149:** 991-2.
200. Sriram K, Meguid MM. Addition of lipids to parenteral nutrition does not cause fungal infections. *Nutrition.* 2015; **31:** 1443-6.
201. Sugita T, Takashima M, Kodama M, *et al.* Description of a new yeast species, *Malassezia japonica* and its detection in patients with atopic dermatitis and healthy subjects. *J Clin Microbiol.* 2003; **41:** 4695-9.
202. Sunenshine PJ, Schwartz RA, Janniger CK. Tinea versicolor. *Int J Dermatol.* 1998; **37:** 648-55.
203. Tan C, Zhu WY, Min ZS. Blaschkoid pityriasis versicolor. *Mycoses.* 2010; **53:** 366-8.
204. Tellechea O, Cravo M, Brinca A, *et al.* Pityriasis versicolor atrophicans. *Eur J Dermatol.* 2012; **22:** 287-8.
205. Thayikkannu AB, Kindo AJ, Veeraraghavan M. *Malassezia*-Can it be ignored? *Indian J Dermatol.* 2015; **60:** 332-9.
206. Tragiannidis A, Bisping G, Koehler G, *et al.* Mini-review: *Malassezia* infections in immunocompromised patients. *Mycoses.* 2010; **53:** 187-95.
207. Valentine MC. Follicular variant of seborrheic dermatitis: Is it identical to *Malassezia* folliculitis? *Skinmed.* 2011; **9:** 161-6.
208. Varada S, Dabade T, Loo DS. Uncommon presentations of tinea versicolor. *Dermatol Pract Concept.* 2014; **4:** 93-6.
209. Varade RS, Burkemper NM. Cutaneous fungal infections in the elderly. *Clin Geriatr Med.* 2013; **29:** 461-78.
210. Velegraki A, Cafarchia C, Gaitanis G, *et al. Malassezia* infections in humans and animals: Pathophysiology, detection and treatment. *PLoS Pathog.* 2015; **11:** e1004523.
211. Viana de Andrade AC, Pithon MM, *et al. Pityrosporum folliculitis* in an immunocompetent patient: Clinical case description. *Dermatol Online J.* 2013; **19(8):** 19273.
212. Viode C, Lejeune O, Turlier V, *et al.* Cathepsin S, a new pruritus biomarker in clinical dandruff/seborrhoeic dermatitis evaluation. *Exp Dermatol.* 2014; **23:** 274-5.
213. Vuran E, Karaarslan A, Karasartova D, *et al.* Identification of *Malassezia* species from pityriasis versicolor lesions with a new multiplex PCR method. *Mycopathologia.* 2014; **177:** 41-9.
214. Wahab MA, Ali ME, Rahman MH, *et al.* Single dose (400 mg) versus 7 day (200 mg) daily dose itraconazole in the treatment of tinea versicolor: A randomized clinical trial. *Mymensingh Med J.* 2010; **19:** 72-6.
215. Wang QM, Theelen B, Groenewald M, *et al.* Moniliellomycetes and Malasseziomycetes, two new classes in Ustilaginomycotina. *Persoonia.* 2014; **33:** 41-7.
216. White TC, Findley K, Dawson TL Jr, *et al.* Fungi on the skin: Dermatophytes and *Malassezia. Cold Spring Harb Perspect Med.* 2014; **4.** PMID: 25085959.
217. Wi HS, Na EY, Yun SJ, *et al.* The antifungal effect of light emitting diode on *Malassezia* yeasts. *J Dermatol Sci.* 2012; **67:** 3-8.
218. Yang YS, Shin MK, Haw CR. Atrophying pityriasis versicolor: Is this a new variant of pityriasis versicolor? *Ann Dermatol.* 2010; **22:** 456-9.
219. Youngchim S, Nosanchuk JD, Pornsuwan S, *et al.* The role of L-DOPA on melanization and mycelial production in *Malassezia furfur. PLoS One.* 2013; **8:** e63764.
220. Zeinali E, Sadeghi G, Yazdinia F, *et al.* Clinical and epidemiological features of the genus *Malassezia* in Iran. *Iran J Microbiol.* 2014; **6:** 354-60.
221. Zhang E, Tanaka T, Tajima M, *et al.* Characterization of the skin fungal microbiota in patients with atopic dermatitis and in healthy subjects. *Microbiol Immunol.* 2011; **55:** 625-32.
222. Zhang E, Tanaka T, Tsuboi R, *et al.* Characterization of *Malassezia* microbiota in the human external auditory canal and on the sole of the foot. *Microbiol Immunol.* 2012; **56:** 238-44.
223. Zhang H, Ran Y, Xie Z, *et al.* Identification of *Malassezia* species in patients with seborrheic dermatitis in China. *Mycopathologia.* 2013; **175:** 83-9.
224. Zhao Y, Li L, Wang JJ, *et al.* Cutaneous malasseziasis: Four case reports of atypical dermatitis and onychomycosis caused by *Malassezia. Int J Dermatol.* 2010; **49:** 141-5.
225. Zhou H, Tang XH, De Han J, *et al.* Dermoscopy as an ancillary tool for the diagnosis of pityriasis versicolor. *J Am Acad Dermatol.* 2015; **73:** e205-6.

CHAPTER

8

Tinea Nigra

Tinea nigra is an uncommon superficial fungal infection of horny layer of the epidermis. This disease is found among young adults and is characterized by presence of brown to black, pigmented, non-scaly, macular patches usually affecting palms, occasionally soles and very rarely other parts of body. It is caused by two phaeoid (dematiaceous) fungi *i.e.* ***Hortaea werneckii*** and ***Stenella araguata***. These fungi have predilection for palms and soles hence this disease is commonly termed as **tinea nigra palmaris** and **tinea nigra plantaris**, respectively. The predominantly involved fungus, ***H.werneckii***, is known by many synonyms like *Phaeoannellomyces werneckii*, *Exophiala werneckii* or *Cladosporium werneckii*. As the causative fungi are phaeoid in nature hence this clinical entity is also called as **superficial phaeohyphomycosis**.

Historical Perspective

In 1891, Alexandre Cerqueira from Bahia, Brazil, first gave brief description of this infection as **Keratomycosis nigricans palmaris** but he did not publish it. The earliest publications on tinea nigra and its pathological agent began in 1905, which are credited to Castellani. In 1916, Cerqueira's son, Castro Cerqueira-Pinto, published the case observed by his father, along with additional eight cases in his doctoral dissertation on the same topic *i.e.* Keratomycosis nigricans palmaris.

In 1921, **Paulo Parreiras Horta** (1884-1961), first isolated the fungus from patient and proposed a new taxon, *Cladosporium werneckii,* in honor of **Machado Werneck**, a Brazilian dermatologist, who was heading laboratory in which Horta had worked. Incidentally, during that very period, the nomenclature of this **genus** was designated as ***Hortaea***. Another fungus causing entirely different disease (Piedra), *Trichosporon hortai* (*Piedraia hortae),* was described during same period and its **species** name was assigned as ***'hortae'*** recognizing Horta's contribution in the field of Medical Mycology. Hence designating the genus name of one fungus *(Hortaea werneckii)* and species name of another *(Piedraia hortae)* was in honor of **Paulo Horta.**

Based on the fact that conidiogenous cells were annellides, von Arx transferred the fungus *Cladosporium werneckii* to genus *Exophiala*, by 1970, retaining the species name as such. In 1973, Dante Borelli (1920-1997), an Italian mycologist and his colleague Marcano, described new causative agent of tinea nigra palmaris that was endemic in Venezuela. This was named as *Cladosporium castellanii.* Subsequently in 1978, McGinnis classified this phaeoid fungal species as *Stenella araguata*. Therefore, *Cladosporium castellanii* is a synonym of *Stenella araguata*.

In 1984, Nishimura and Miyaji proposed that *Cladosporium werneckii* be moved to new genus, ***Hortaea***, as it exhibited both sympodial and annellidic conidiogenesis. Since then, agent of tinea nigra has usually been described as *H.werneckii*. However, some workers insisted that organism is phylogenetically related to members of the genus *Aureobasidium*, as it shared certain similarities with it. In 1985, McGinnis and Schell proposed placing the organism on the basis of conidiogenesis into a genus they called ***"Phaeoannellomyces"*** in the family, Phaeococcomycetaceae, a new phaeoid blastomycete taxon that included all black yeasts which replicate by **annellide** type of conidiation (***Phaeo = dark, annellomyces = ring***).

There was confusion prevailing among mycologists due to nature of conidiation in this fungus as to whether it was an **annellide** or **phialide** with collarette formation, entailing hurdles in discerning its taxonomy. Some workers consider using both designations, *Phaeoannellomyces* and *Hortaea,* as infructuous and prefer using the genus *Exophiala werneckii*. However, currently ***Hortaea werneckii*** is its accepted nomenclature. In this Textbook also, the organism shall continue to be referred to as *Hortaea werneckii*.

In the strictest sense, the term ***'tinea'*** should be restricted to disease processes caused by dermatophytes and ***'keratomycosis'*** for fungal infections of cornea. In 1992, Nomenclature Committee of ISHAM recommended its continued usage as 'tinea nigra' because of the long history of this nomenclature being in use in the literature. It took more than six decades, after Horta's isolation of the fungus, to place the **genus** in a taxonomically proper manner. However, the nomenclature of **species** as ***werneckii*** somehow remained stable throughout this period.

Mycology

The popular term '**dematiaceous**' used to describe dark colored fungi due to presence of melanin dihydroxynaphthalene pigment, was found to be epistemologically invalid. Hence it was subsequently replaced by a new term -**phaeoid** (Gk. ***'phaeios'***= ***'dusky'***). The phaeoid fungi have always attracted attention of botanists, medical mycologists and veterinarians because this group includes many significant pathogens infecting plants, humans and lower animals. The clinical and mycological spectrum of infections caused by phaeoid fungi is summarized in **Table 8.1**; however, their relevant details are described in subsequent Chapters of this Textbook.

The predominant agent of tinea nigra, *H.werneckii,* is a melanized yeast-like halophilic fungal species and a member of an arbitrary group of so-called **black yeasts.** This is group of rare extremophilic eukaryotes, which grow as solitary conidiogenous cells and collectively result in, at least initially, brown to black pasty colonies **(Figs. 8.1 and 8.2)**. The majority of black yeasts often occur as synanamorphs in association with polymorphic phaeoid hyphomycetes. The taxonomic status of this fungus was once transferred to a new genus, *Phaeoannellomyces,* parallel to genus *Hortaea* but that issue is now settled to a great extent.

Recently, Holker *et al,* have added another species in the genus *Hortaea i.e.* ***Hortaea acidophila***, an acid-tolerant black yeast from lignite, which grows in acidic substrates, as low as 0.6. It possesses at least two functional **laccases** that seem to be involved in melanin synthesis. Both these intra and extra-cellular laccases exhibit high stability at low pH. The role of new species in causation of tinea nigra has not yet been studied. Besides prevalence of *H.werneckii,* another fungal agent has also been reported as causative agent of tinea nigra *i.e. Stenella araguata* (formerly *Cladosporium castellanii*), which is an autochthonous fungus found in Venezuela.

The genus *Hortaea* was proposed to include all black yeasts, which replicate by annellide formation. The taxonomic position as mitosporic ascomycete, *H.werneckii,* was classified earlier in form-family **Dematiaceae** of form-class **Hyphomycetes** in phylum **Deuteromycetes**. But *H.werneckii,* is now classified in the family **Teratosphaeriaceae**, in the order **Capnodiales** of class **Dothideomycetes** of phylum **Ascomycota**.

Table 8.1. Clinical and Mycological Categories of Infections Caused by Phaeoid / Dematiaceous Fungi.

Infection Category	*Fungal Infection*	*Causative Phaeoid / Dematiaceous Fungi*
A. Superficial Mycoses	Tinea Nigra	*Hortaea werneckii*
	Black Piedra	*Piedraia hortae*
	Dermatomycoses	*Scytalidium dimidiatum*
	Onychomycosis	*Scytalidium dimidiatum, Onychocola, Alternaria* species
B. Subcutaneous Mycoses	Eumycetoma	*Exophiala jeanselmei, Curvularia lunata, C.geniculata*
	Chromoblastomycosis	*Fonsecaea pedrosoi, F.compacta, Phialophora verrucosa, Cladophialophora carrionii, Rhinocladiella aquaspersa*
	Phaeohyphomycosis	*Exophiala dermatitidis, E.jeanselmei, Cladophialophora bantiana, Chaetomium* species
C. Systemic Mycoses	Phaeohyphomycosis	Same as above in phaeohyphomycosis
D. Miscellaneous Mycoses	Keratomycosis	*Curvularia* species, *Alternaria* species, *Bipolaris* species, *Aureobasidium pullulans, Fonsecaea* species
	Allergic Rhinosinusitis	*Alternaria alternata, Bipolaris hawaiiensis, Curvularia lunata, Exserohilum rostratum*
	PNS Mycosis	*Bipolaris, Curvularia* species
	ABPM	*Bipolaris, Curvularia* species
	Brain abscess	*Cladophialophora bantiana, Ramichloridium mackenziei, Ochroconis gallopava*

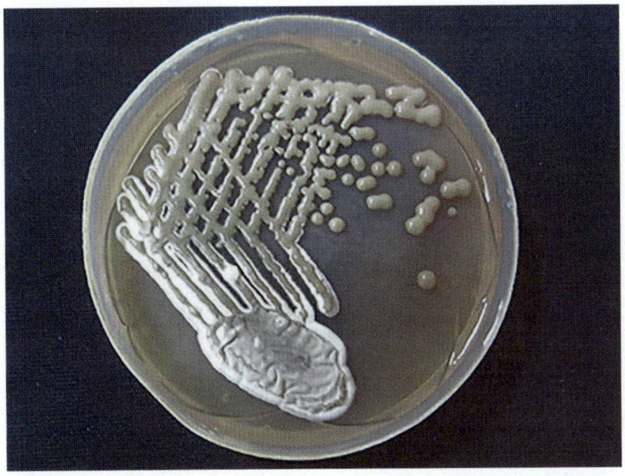

Fig. 8.1. White to blackish pasty colonies of black yeast on SDA in the initial phase of the growth.

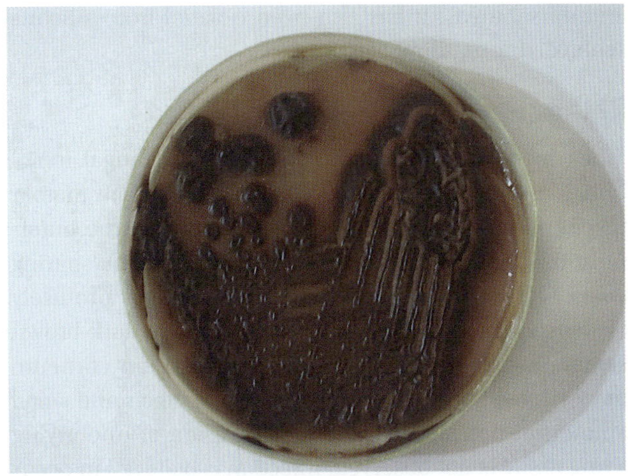

Fig. 8.2. Pasty colonies of black yeast after four weeks of incubation of the same plate of SDA as shown in Figure 8.1.

Epidemiology

Tinea nigra has a sporadic prevalence in many tropical and subtropical countries of the Central and South America, Caribbean countries, Africa, Europe, Southeast Asia, Australia, Japan and Far East Asia. An allochthonous case has been reported from Chile. In Southeast Asia, this has been reported from Sri Lanka, India, Burma, Southern China, Java, Sumatra and other adjoining countries. It is also no longer uncommon in USA where it has been recovered from coastal areas. It is relatively rare in Japan and till 2008 only a total of 28 cases were reported, most of them from subtropical Okinawa. Few cases have also been reported from temperate areas like Scotland and The Netherlands. From India, cases have been reported from different states as isolated case reports have appeared in the literature. Recently, five cases were observed in a single family from Goa, coastal state in southwestern India. There is no racial predilection and disease is more prevalent among children and young adults under the age of twenty years. Females are affected 3-5 times more frequently than males.

The fungus is ubiquitous and is found in abundance in soil, sewage, decaying vegetation, sticks of wood, humus, moldy salted fish and is present mainly at beaches. *H.werneckii* is a halophilic saprobe as it has been isolated from seashores and occasionally from sea fish. The natural ecological niche of this fungus is presumed to be marine habitat due to the salty environment of seashore. The species shows optimal reproduction in presence of 3-6% sodium chloride, which is the usual salt concentration of sea-water. However, it can tolerate higher concentrations of salinity and can grow in the presence of 10% sodium chloride. This may be one of the reasons of many patients acquiring infection at sea beaches during their holidays.

The lesions are mainly found on the hands of hyperhydrotic individuals. This may be due to the slightly raised salt concentration as a result of continuous evaporation. The skin becomes more hydrophobic than usual, thus is a suitable substrate to attract hydrophobic cells from environment. It has been suggested that primary adhesion of yeast cells of *H.werneckii* to solid substrates is based on hydrophobic interactions. In coastal areas, skin contamination can take place by contact with sand on the beach. In 1994, Uijthof *et al*, identified this fungus in areas with high saline concentration. Isolation of *H.werneckii* has also been reported from solar salterns of Cabo Rojo, Puerto Rico, These significant reports are considered due to hypersaline niche in Caribbean islands. *Stenella araguata* has not yet been isolated from the environment.

In tinea nigra, no significant predisposing factor has been identified and impairment of immune system does not appear to be relevant in pathogenesis. There is no association with genetic predisposition but most of the patients have hyperhidrosis as an associated condition. The reports of several family members with infection probably reflect common exposure. As a result of induced experimental infections, in man and guinea pigs, the incubation period is known to be about 10 to 15 days. The disease has not yet been reported in animals. In Japan, *H.werneckii* has recently been isolated from household guinea pig with dark superficial lesions on palm and dorsal areas. It has also been isolated from silicone scuba diving equipment in Spain.

Pathogenesis and Pathology

Tinea nigra being non-invasive superficial fungal infection, is associated with either no or minimal inflammatory reaction in dermis. The fungus is confined to the outermost dead layer of the skin *i.e.* stratum corneum, sparing deeper layer, stratum lucidum, where it grows profusely as densely packed and frequently branched, dark brown septate hyphae. There is thickening of stratum corneum in which hyphae are present with parakeratosis and small amount of perivascular infiltrate. Invasion of underlying living tissue is not observed. Infection is believed to occur as a result of inoculation from contaminating source such as soil, sewage, wood or compost subsequent to trauma to the affected part.

A pigmentary change in skin results in dark-colored macule due to accumulation of melanin-like substance present in the fungus. *H.werneckii* receives nourishment from utilization of decomposed lipids. Its tolerance to an environment with high salt concentration and low pH allows fungus to thrive on human skin. Higher incidence on palms and soles may be related to increased density of sweat pores at these sites by utilizing decomposed lipids. The disease is more common in females; however, any underlying hormonal influence has not yet been studied.

Clinical Features

The infection usually remains asymptomatic; therefore, patients may not seek medical advice for months or even years. The incubation period is variable but is usually 2 to 7 weeks and very rarely, long enough to be many years in some cases. In an experimental inoculation, it was found to be twenty years.

The clinical manifestations consist of circumscribed solitary, 1 to 5 cm well-defined, painless brown to black, macular, non-scaly, coin-sized flat patch enlarging by peripheral extension *i.e.* centrifugally and most often seen over palmar and plantar aspect of hands and feet, respectively. The lesions on palms tend to arise along sulci cutis of thenar prominence, without accompanying scales or inflammatory signs such as erythema. The other body sites such as neck, chest and dorsum of hands and phalanx may rarely be affected. The lesion is usually darkest at periphery with distinct margins and is often reported to resemble stain such as silver nitrate or India ink as shown in **Figure 8.3**. There is usually a solitary lesion as multiple ones are very rare. However, recently cases have been reported to have bilateral lesions also. Moreover, there is no scaling, erythema or induration, however, sometimes it may be associated with pruritus. Tinea nigra as such is considered a cosmetic rather than a true clinical condition.

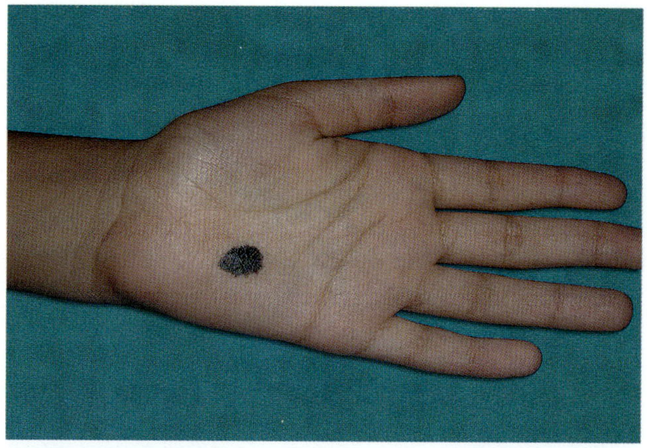

Fig. 8.3. Staining of palmer surface of hand due to chemical dye mimicking tinea nigra.

The reason for predilection for palmar surface is not yet known but it may be related to hyperhidrosis or excessive sweating as a risk factor. The patient may confuse it with melanoma thereby consult a dermatologist, which generally leads to early detection of tinea nigra. The recent increased public awareness regarding sun exposure and skin cancers has led to apprehension among many parents over potential malignant nature of their children's moles that ultimately paves the way for an early diagnosis.

There are many opportunistic fungi, which are isolated from patients with prolonged neutropenia and include phaeoid ones too. Now, *H.werneckii* is also increasingly recognized as a potential fungal pathogen at other anatomical sites. It has been reported to cause fungal endophthalmitis, non-skin infection in an immunocompetent individual following cataract surgery and isolated from blood as well as splenic abscess.

Dermoscopy

Tinea nigra is generally diagnosed on clinical grounds without much difficulty; however, use of dermoscopy for palmar or plantar pigmentation may enhance its recognition. There is ample evidence that dermoscopy facilitates the *in vivo* diagnosis of skin infections and infestations. The dermatoscope is a useful clinical adjunctive tool in

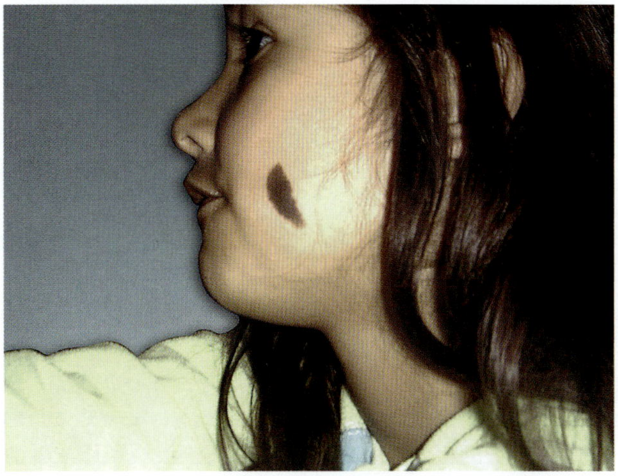

Fig. 8.4. Nevus on the face of a child mimicking tinea nigra.

differentiating tinea nigra from melanocytic lesions. On examinations, it shows a reticulate pattern, consisting of superficial fine, wispy, light-brown strands or pigmented spicules but no pigment network with a uniform brown color. The strands do not follow the furrows and ridges normally observed in the skin. There are no pigment networks, globules and stripes that would suggest a melanocytic neoplasm. Hence tinea nigra shows **parallel ridge pattern** on dermoscopy. As such, dermoscopy connects the dermatologists and entomologists, opening a new research field of Entodermoscopy.

Differential Diagnosis

Tinea nigra should be differentiated from other hyperpigmented lesions, especially when these are presenting as solitary lesions. The pigmented lesions of Addison's disease, melanocytic form of syphilis, pinta, yaws, lentigines, acral lentiginous form of malignant melanoma, palmar lichen planus and junctional melanocytic nevus of palm may be distinguished by microscopic examination of skin scrapings. A differential diagnosis (nevus) on the face of a child is shown in **Figure 8.4**. However, melanoma and nevi often appear slightly indurated and elevated. A reddish hue in black or brownish-black lesions, which can be seen in melanoma, is rarely seen in tinea nigra and the palmar lesions are uncommon in melanoma. Moreover, post-inflammatory melanosis, fixed drug eruptions, staining due to dyes, chemicals and pigments like silver nitrate or India ink should not be confused with tinea nigra. It is also differentiated from dermatophytoses and pityriasis versicolor by its absence of scaling, coloration and confinement of fungus to outer most zone of horny layer. The disease may mimic dermatomycosis caused by *Neoscytalidium dimidiatum* (former—*Scytalidium dimidiatum/Hendersonula toruloidea*), in which lesions are more furfuraceous and may affect epidermis, nails and other tissues. On the other hand, superficial lesions caused by another phaeoid fungus *Aureobasidium melanogenum* may mimic tinea nigra in an immunocompetent individuals. The blackish macular lesions should also be differentiated from new superficial fungal infection caused by another phaeoid fungus - *Coniosporium epidermidis*.

Laboratory Diagnosis

The laboratory diagnosis of tinea nigra is relatively easy and is established on the basis of demonstration of the fungus in skin scrapings followed by culture. Sometimes, skin biopsy is taken if malignant melanoma or junctional nevus is suspected. However, as such no investigative surgical intervention is usually required to establish the final diagnosis of tinea nigra.

(a) Direct Examination

The epidermal skin scrapings are taken with the help of scalpel blade from active border of macular lesions and examined in 20% KOH wet mount with DMSO, gentle heating for a minute may be performed for hastening clearing. There are brown to olivaceous-black pigmented, multi-branched septate hyphae seen on direct microscopy, practically representing the diagnostic finding for this disease. The hyphae are about 5-6 µm wide and yeast-like cells are 2-8 µm in diameter. The mycelia differ from dermatophytes, which produce hyaline, less branched hyphae along with absence of tapering contour of terminal branches. If there is any confusion, Masson-Fontana staining is done, which take up this phaeoid fungus as black in color with brownish background.

(b) Fungal Culture

The primary isolation of this fungus is done on SDA with actidione or on an equivalent medium. The cultures are incubated at 25-30°C as no growth is seen at 37°C. The colonies appear slowly and mature after about three weeks as moist, adherent, yeast-like growth. Initially the colonies are brown but rapidly become olive to shiny, greenish-black, tar-like, producing metallic sheen similar to ***'drop of oil'***. The growth shows budding yeast-like cells with occasional septa

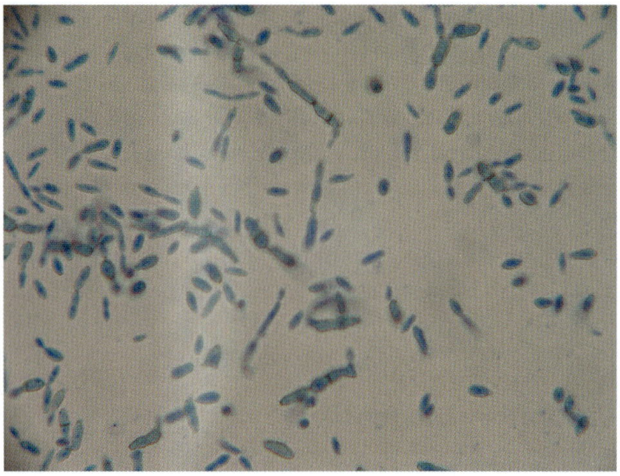

Fig. 8.5. Yeast-like cells of *H.werneckii* with transverse septa (LCB × 200).

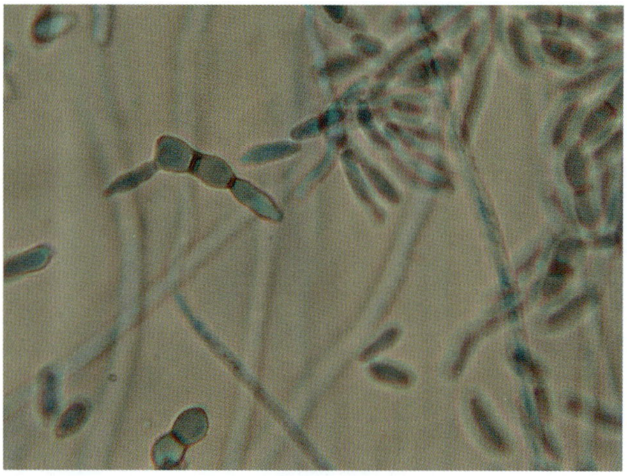

Fig. 8.6. Annelloconidia of *Hortae werneckii* (LCB × 400).

and many cells at different stages of cell division, producing characteristic two-celled (bi-cellular), oval structures with central darkly pigmented septa as seen in **Figure 8.5**. There are annelloconidia in addition to dematiaceous, septate hyphae with conidia on intercalary annellides **(Fig. 8.6)**.

In an older culture, mycelia and conidia predominate. The hyphae are dark, crooked laterally, producing conidia, which are single or two-celled. After liberation, conidia inflate and develop transverse and occasionally oblique septa; the liberated cells are converted into budding or chlamydospore-like cells. The hyphal elements are 5-6 µm in diameter and become densely septate brown and thick-walled on maturation. On slide culture, there are light brown, elliptic or peanut-shaped conidia comprised of one to two ampullaceous cells. The annellated zones are prominent, 1-2 µm wide with clearly visible annellations formed on the intercalary or lateral conidiogenous cells. The terminal portions of hyphae are usually seen as hyaline. There is black pigmentation on reverse of the culture plate. The oatmeal agar also shows slow growing colonies that are gray at first and later become oily, glistening and olivaceous-black. After several subcultures on oatmeal agar, colonies may become velvety due to production of aerial hyphae.

Morphologically, *H.werneckii* resembles *Aureobasidium* species and *Exophiala dermatitidis*, both produce black colored colonies on SDA. *E.dermatitidis* also grows at 42°C, with septate hyphae, spherical phialides and groups of small ellipsoidal phialoconidia. However, annellated zones of conidiogenous cell appear to be wider as compared to *H.werneckii*. *Aureobasidium* species produces hyaline hyphae. It shows irregular dichotomous branching and conidia arise in dense groups from small denticles arising from undifferentiated conidiogenous cells, which are intercalary in hyaline hyphae.

Hortaea werneckii can also be differentiated from species of genus *Exophiala (E.jeanselmei, E.spinifera* and *E.dermatitidis)* on the basis of biochemical characteristics. *H.werneckii* can hydrolyze casein but is unable to decompose tyrosine. Tolerance to 10% NaCl, lack of growth at 37°C and broad annellated zones differentiate it from various *Exophiala* species.

Exoantigen testing is also useful in differentiating *H.werneckii* from other phaeoid fungi pathogenic to man. *Stenella araguata,* another causative species endemic in Venezuela, differs from *H.werneckii* by formation of catenulate, cylindrical, septate and usually rough-walled conidia from simple or sympodial conidiogenous cells without annellation.

(c) Immunodiagnosis

There are no reliable immunological or serological tests reported so far to diagnose this disease. The molecular techniques like PCR and RAPD may be used to establish diagnosis. The PCR may be performed with primers specific to *H.werneckii* by direct sequencing with DNA segments from positive bands. The causative fungus is determined to establish various types of *H.werneckii* based on base sequences. The PCR-fingerprinting technique can also be applied to solve difficulties of genetic relatedness as it offers several advantages over other better-established molecular typing methods. It is more sensitive, faster and flexible while differentiating the isolates at various intraspecies levels.

(d) Animal Pathogenicity

The disease, tinea nigra, can be experimentally produced easily in guinea pigs by scarifying skin and applying pure culture of *H.werneckii* under a bandage. Its lesions appear in 10-15 days' time period.

Treatment and Prophylaxis

Although tinea nigra is a non-invasive fungal infection but it does not resolve spontaneously. There is good response to topical application of keratolytic agents such as Whitfield's ointment or salicylic acid (5-10%) ointment. The other agents useful are miconazole (1-2%), clotrimazole, econazole, isoconazole, thiabendazole (10%) or undecylenic acid, which are applied for a period of two weeks. Oral itraconazole has been successfully used in few cases in doses of 200 mg daily for three weeks, resulting in clinical cure. Similarly, topical butenafine, terbinafine and ciclopirox olamine gel have also been effectively used. Topical isoconazole cream with terbinafine has also been successfully tried. Ciclopirox olamine gel when applied twice a day, rapidly resolves lesions of tinea nigra within three days. Other antifungal agents like oral griseofulvin, topical tolnaftate or amphotericin B are not effective hence systemic antifungals as such are not recommended for its treatment. However, the systemic diseases caused by *H.werneckii* like endophthalmitis, bloodstream infections or abscesses, are to be treated with systemic antifungals like any other deep mycoses.

Most of the lesions in tinea nigra disappear within two to four weeks. However, prolonged therapy may occasionally be required for achieving cure as well as to prevent further relapse. After an effective therapy, relapses normally do not occur, however, recurrence may occur following re-exposure. There are no specific instructions or standard guidelines in the literature for prophylaxis hence prevention is dependent upon general hygienic measures undertaken by the public at large.

Further Reading

1. Bonifaz A, Badali H, de Hoog GS, *et al.* Tinea nigra by *Hortaea werneckii*, a report of 22 cases from Mexico. *Stud Mycol.* 2008; **61:** 77-82.
2. Bonifaz A, Gomez-Daza F, Paredes V, *et al.* Tinea versicolor, tinea nigra, white piedra and black piedra. *Clin Dermatol.* 2010; **28:** 140-5.
3. Burke WA. Tinea nigra: Treatment with topical ketoconazole. *Cutis.* 1993; **52:** 209-11.
4. Cabanes FJ, Bragulat MR, Castella G. *Hortaea werneckii* isolated from silicone scuba diving equipment in Spain. *Med Mycol.* 2012; **50:** 852-7.
5. Cabrera R, Sabatin N, Urrutia M, *et al.* Tinea nigra: A allochthonous case report in Chile. *Rev Chilena Infectol.* 2013; **30:** 90-3.
6. Castellani A. Tinea nigra. *Indian J Dermatol.* 1967; **12:** 45-50.
7. Cerqueira AG. Keratomycosis nigricans palmaris (Thesis). Salvador, Bahia, Brazil: *Faculdade de Medicina da Bahia (School of Medicine).* 1916.
8. Chadfield HW, Campbell CK. A case of tinea nigra in Britain. *Br J Dermatol.* 1972; **87:** 505-8.
9. Chen WT, Tu ME, Sun PL. Superficial phaeohyphomycosis caused by *Aureobasidium melanogenum* mimicking tinea nigra in an immunocompetent patient and review of published reports. *Mycopathologia.* 2016; **181:** 555-60.
10. Criado PR, Delgado L, Pereira GA. Dermoscopy revealing a case of tinea nigra. *An Bras Dermatol.* 2013; **88:** 128-9.
11. Dasgupta LR, Agarwal SC, Bedi BM. Tinea nigra palmaris from South India. *Sabouraudia.* 1975; **13:** 41-3.
12. de Hoog GS, Vicente VA, Gorbushina AA. The bright future of darkness - The rising power of black fungi: Black yeasts, microcolonial fungi and their relatives. *Mycopathologia.* 2013; **175:** 365-8.
13. Diniz LM. Study of nine observed cases of tinea nigra in Greater Vitoria (Espirito Santo state, Brazil) over a period of five years. *An Bras Dermatol.* 2004; **79:** 305-10.
14. Falcao EM, Trope BM, Martins NR, *et al.* Bilateral tinea nigra plantaris with good response to isoconazole cream: A case report. *Case Rep Dermatol.* 2015; **7:** 306-10.
15. Giraldi S, Abbage KT, Marinoni LP, *et al.* Tinea nigra: Six cases in Parana state. *An Bras Dermatol.* 2003; **78:** 593-600.
16. Gnanaguruvelan S, Janaki C, Sentamilselvi G, *et al.* Tinea nigra. *Indian J Dermatol Venereol Leprol.* 1998; **64:** 91-2.
17. Guarenti IM, Almeida HL Jr, Leitao AH, *et al.* Scanning electron microscopy of tinea nigra. *An Bras Dermatol.* 2014; **89:** 334-6.
18. Gunde-Cimerman N, Ramos J, Plemenitas A. Halotolerant and halophilic fungi. *Mycol Res.* 2009; **113:** 1231-41.
19. Gupta G, Burden AD, Shankland GS, *et al.* Tinea nigra secondary to *Exophiala werneckii* responding to itraconazole. *Br J Dermatol.* 1997; **137:** 483-4.
20. Hall J, Perry VE. Tinea nigra palmaris: Differentiation from malignant melanoma or junctional nevi. *Cutis.* 1998; **62:** 45-6.
21. Helm TN, Li A, Santalucia P. Tinea nigra. *Cutis.* 2011; **87:** 229 & 232.
22. Hemashettar BM, Patil CS, Siddaramappa B, *et al.* A case of tinea nigra from South India. *Indian J Dermatol Venereol Leprol.* 1985; **51:** 164-6.
23. Holker U, Bend J, Pracht R, *et al.* *Hortaea acidophila,* a new acid-tolerant black yeast from lignite. *Antonie van Leeuwenhoek.* 2004; **86:** 287-94.

24. Horta P. Sobre um caso de tinha preta e um novo cogumelo *(Cladosporium werneckii)*. *Ver Med Cir Brasil.* 1921; **29:** 269-74.
25. Huber CE, La Berge T, Schwiesow T, *et al. Exophiala werneckii* endophthalmitis following cataract surgery in an immunocompetent individual. *Ophthal Surg Lasers.* 2000; **31:** 417-22.
26. Hughes JR, Moore MK, Pembroke AC. Tinea nigra palmaris. *Clin Exp Dermatol.* 1993; **18:** 481-2.
27. Isaacs F, Reiss-Levy E. Tinea nigra plantaris: A case report. *Australas J Dermatol.* 1980; **21:** 13-5.
28. Kamalam A, Thambiah AS. Tinea nigra: First case report from Madras. *Mykosen.* 1982; **25:** 626-8.
29. Klokke AH, Durairaj P. The causal agents of superficial mycoses isolated in rural areas of South India. *Sabouraudia.* 1967; **5:** 153-8.
30. Larangeira de Almeida H Jr, Dallazem RN, Dossantos LS, *et al.* Bilateral tinea nigra in a temperate climate. *Dermatol Online J.* 2007; **13:** 25.
31. Li DM, Chen XR. A new superficial fungal infection caused by *Coniosporium epidermidis*. *J Am Acad Dermatol.* 2010; **63:** 725-7.
32. Madke B, Doshi B, Wankhede P, *et al.* Palmar lichen planus mimicking tinea nigra. *Indian J Dermatol.* 2013; **58:** 407.
33. Maia Abinader MV, Carvalho Maron SM, Araujo LO, *et al.* Tinea nigra dermoscopy: A useful assessment. *J Am Acad Dermatol.* 2016; **74:** e121-2.
34. Maria Flora MM. Tinea nigra in a family. *Indian J Dermatol.* 2003; **48:** 47-8.
35. McGinnis MR, Padhye AA. *Cladosporium castellanii* is a synonym of *Stenella araguata*. *Mycotaxon.* 1978; **7:** 415-8.
36. McGinnis MR, Schell WA, Carson J. *Phaeoannellomyces* and the *Phaeococcomycetaceae*, new dematiaceous blastomycete taxa. *J Med Vet Mycol.* 1985; **23:** 179-88.
37. McKinlay JR, Barrett TL, Ross EV. Tinea nigra. *Arch Pediatr Adolesc Med.* 1999; **153:** 305-6.
38. Mittag H. The fine structure of *Hortaea werneckii*. Mycoses. 1993; **36:** 343-50.
39. Mok WY. Nature and identification of *Exophiala werneckii*. *J Clin Microbiol.* 1982; **16:** 976-8.
40. Muellenhoff M, Cukrowski T, Morgan M, *et al.* Enlarging pigmented patches on the hand. *Int J Dermatol.* 2003; **42:** 810-11.
41. Muir J. Tinea nigra and dermoscopy. *Australas J Dermatol.* 2012; **53:** e14-5.
42. Nair SP. Brownish macule on the palm. *Indian Dermatol Online J.* 2015; **6:** 230-1.
43. Nazzaro G, Ponziani A, Cavicchini S. Tinea nigra: A diagnostic pitfall. *J Am Acad Dermatol.* 2016; **75:** e219-20.
44. Ng KP, Soo-Hoo TS, Na SL, *et al.* The mycological and molecular study of *Hortaea werneckii* isolated from blood and splenic abscess. *Mycopathologia.* 2005; **159:** 495-500.
45. Nishimura K, Miyaji M. Further studies on the phylogenesis of the genus *Exophiala* and *Hortaea*. *Mycopathologia.* 1985; **92:** 101-9.
46. Nishimura K, Miyaji M. *Hortaea*, a new genus to accommodate *Cladosporium werneckii*. *Jpn J Med Mycol.* 1984; **25:** 139-46.
47. Noguchi H, Hiruma M, Inoue Y, *et al.* Tinea nigra showing a parallel ridge pattern on dermoscopy. *J Dermatol.* 2015; **42:** 518-20.
48. Palmer SR, Bass JW, Mandojana R, *et al.* Tinea nigra palmaris and plantaris: A black fungus producing black spots on the palms and soles. *Pediatr Infect Dis J.* 1989; **8:** 48-50.
49. Paschoal FM, de Barros JA, de Barros DP, *et al.* Study of the dermatoscopic pattern of tinea nigra: report of 6 cases. *Skinmed.* 2010; **8:** 319-21.
50. Pegas JR, Criado PR, Lucena SK, *et al.* Tinea nigra: Report of two cases in infants. *Pediatr Dermatol.* 2003; **20:** 315-7.
51. Perez C, Colella MT, Olaizola C, *et al.* Tinea nigra: Report of twelve cases in Venezuela. *Mycopathologia.* 2005; **160:** 235-8.
52. Piliouras P, Allison S, Rosendahl C, *et al.* Dermoscopy improves diagnosis of tinea nigra: A study of 50 cases. *Australas J Dermatol.* 2011; **52:** 191-4.
53. Qazi AM, Sameem SF, Iffat SH. Tinea nigra. *Indian J Dermatol.* 2005; **50:** 40-1.
54. Reid BJ. *Exophiala werneckii* causing tinea nigra in Scotland. *Br J Dermatol.* 1998; **139:** 157-8.
55. Revankar SG, Sutton DA. Melanized fungi in human disease. *Clin Microbiol Rev.* 2010; **23:** 884-928.
56. Revankar SG. Epidemiology of black fungi. *Curr Fungal Infect Rep.* 2012; **6:** 283-7.
57. Rezusta A, Gilaberte Y, Betran A, *et al.* Tinea nigra: A rare imported infection. *J Eur Acad Dermatol Venereol.* 2010; **24:** 89-91.
58. Rosen T, Lingappan A. Rapid treatment of tinea nigra palmaris with ciclopirox olamine gel, 0.77%. *Skinmed.* 2006; **5:** 201-3.
59. Rossetto AL, Correa PR, Cruz RC, *et al.* A case of tinea nigra associated to a bite from a European rabbit *(Oryctolaguscuniculus, Leporidae):* The role of dermoscopy in diagnosis. *An Bras Dermatol.* 2014; **89:** 165-6.
60. Rossetto AL, Cruz RC, Haddad Junior V. Double-blind study with topical isoconazole and terbinafine for the treatment of one patient with bilateral tinea nigra plantaris and suggestions for new differential diagnosis. *Rev Inst Med Trop Sao Paulo.* 2013; **55:** 125-8.
61. Rossetto AL, Cruz RC, Haddad VJ. Tinea nigra presenting speckled or "salt and pepper" pattern. *Am J Trop Med Hyg.* 2014; **90:** 981.
62. Rossetto AL, Cruz RC. Spontaneous cure in a case of tinea nigra. *An Bras Dermatol.* 2012; **87:** 160-2.
63. Rossetto AL, Cruz RC. Tinea nigra in geographical forms of "heart" and "parrot beak". *An Bras Dermatol.* 2011; **86:** 389-90.
64. Rossetto AL, Cruz RC. Tinea nigra: Successful treatment with topical butenafine. *An Bras Dermatol.* 2012; **87:** 939-41.
65. Sarangi G, Dash D, Chayani N, *et al.* Bilateral tinea nigra of palm: A rare case report from eastern India. *Indian J Med Microbiol.* 2014; **32:** 86-8.

66. Saunte DM, Tarazooie B, Arendrup MC, *et al.* Black yeast-like fungi in skin and nail: It probably matters. *Mycoses.* 2012; **55:** 161-7.
67. Schneider J, La Casse A. Tinea nigra. *Cutis.* 2009; **84:** 292 & 299-300.
68. Schwartz RA. Superficial fungal infections. *Lancet.* 2004; **364:** 1173-82.
69. Seeliger HP, Seefried L. Aldo Castellani - An appraisal of his life and oeuvre. *Mycoses.* 1989; **32:** 391-7.
70. Severo LC, Bassanesi MC, Londero AT. Tinea nigra: Report of four cases observed in Rio Grande do Sul (Brazil) and a review of Brazilian literature. *Mycopathologia.* 1994; **126:** 157-62.
71. Sharmin S, Haritani K, Tanaka R, *et al.* The first isolation of *Hortaea werneckii* from a household guinea pig. *Jpn J Med Mycol.* 2002; **43:** 175-80.
72. Smith SB, Beals SL, Elston DM, *et al.* Dermoscopy in the diagnosis of tinea nigra plantaris. *Cutis.* 2001; **68:** 377-80.
73. Solak B, Unus Z. Tinea nigra on the fingers. *BMJ Case Rep.* 2015 Dec 7; pii: bcr2015211124.
74. Teixeira MM, Moreno LF, Stielow BJ, *et al.* Exploring the genomic diversity of black yeasts and relatives (Chaetothyriales, Ascomycota). *Stud Mycol.* 2017; **86:** 1-28.
75. Thomas CL, Samarasinghe V, Natkunarajah J, *et al.* Entodermoscopy: A spotlight on tinea nigra. *Int J Dermatol.* 2016; **55:** e117-8.
76. Tilak R, Singh S, Prakash P, *et al.* A case report of tinea nigra from north India. *Indian J Dermatol Venereol Leprol.* 2009; **75:** 538-9.
77. Tseng SS, Whittier S, Miller SR, *et al.* Bilateral tinea nigra plantaris mimicking melanoma. *Cutis.* 1999; **64:** 265-8.
78. Uezato H, Gushi M, Hagiwara K, *et al.* A case of tinea nigra palmaris in Okinawa, Japan. *J Dermatol.* 2006; **33:** 23-9.
79. Uijthof JM, de Cock AWAM, de Hoog GS, *et al.* Polymerase chain reaction-mediated genotyping of *Hortaea werneckii,* causative agent of tinea nigra. *Mycoses.* 1994; **37:** 307-12.
80. Xavier MH, Ribeiro LH, Duarte H, *et al.* Dermatoscopy in the diagnosis of tinea nigra. *Dermatol Online J.* 2008; **14:** 15.
81. Zalar P, de Hoog GS, Gunde-Cimerman N. Ecology of halotolerant Dothideaceous black yeasts. *Stud Mycol.* 1999; **43:** 38-48.

CHAPTER

9

Piedra

Piedra is the superficial infection of the hair shaft and characterized by the development of firm, irregular nodules, composed of fungal elements, cemented together to the hair. This condition is usually asymptomatic because as such there is no significant discomfort, thereby, patient often ignore leading to delayed medical attention. This also may be the reason for under-reporting of these cases. Piedra affects humans of both genders and individuals are equally affected irrespective of their age.

The literal meaning of 'piedra' is derived from Spanish word meaning 'stone' as nodules on the hair resemble small stones. Based on causal fungal genera and charac teristics of nodules, there are **two types** of piedra:

I. **White Piedra** caused by *Trichosporon* species
II. **Black Piedra** caused by *Piedraia hortae*

The clinical condition is categorized based on the color differentiation, although the etiological fungal agents responsible for each group are not related to each other. In white piedra, hyaline **basidiomycete** yeast-like fungi and in black one, phaeoid **ascomycetous** fungi are the causative agents. However, both types of piedra can be distinguished on the basis of shape, size and pigmentation of fungal cells constituting nodules, which are found around the hair cortex. As compared to black piedra, in white piedra, nodules do not have dark-brown to black coloration, the disruption of hair cuticle is relatively less extensive and there is lack of asci and ascospores on hair shaft. The hair shaft in piedra is normal on either side of cell mass, unlike in dermatophytosis *e.g.* tinea capitis where both, the base of hair shaft as well as hair follicles are involved. In this Chapter both the types of piedra are described for the convenience of the reader.

I. WHITE PIEDRA

White piedra is infection of the hair shaft characterized by soft, grayish-white to tan irregular nodules. This is caused by a group of yeast-like fungi belonging to the genus *Trichosporon*. The fungus is present as branched hyphae and arthrospores, both within and around the hair shaft. White piedra is also called as **trichosporonosis nodosa, tinea blanca, Beigel's disease**, **Chignon disease**. The systemic infections among immunocompromised patients, caused by various species of genus *Trichosporon* are termed as **trichosporonosis**, which are described in **Chapter 28** as Miscellaneous Opportunistic Mycoses.

Historical Perspective

White piedra was described by Beigel in 1865 in London as a fungal infection of both living as well as dead hair cut from hair chignon (buns) or wig, in his monograph titled, *'The Human Hair - Its Growth and Structure'*. He also isolated the causative fungus, which was erroneously classified in 1867 as the algae *"Pleurococcus beigelii"* by Friedrich Kuchenmeister and Rabenhorst. In 1890, Gustav Behrend, German dermatologist cultured a fungus from case of white piedra of beard and named it *Trichosporon ovoides*. There were reports of other *Trichosporon* species found to be involved in this disease but in 1902, French mycologist, Vuillemin considered all these fungi as variants of a single species and clubbed them as Trichosporon beigelii. In 1909, Beurmann cultured the cells collected from cutaneous lesion and denominated the fungus isolated as *Oidium cutaneum*, which was subsequently renamed as *Trichosporon cutaneum* by Ota in 1926. Diddens and Lodder in 1942 considered *T.beigelii* and *T.cutaneum* to be the same species.

The first case of white piedra in North America, was described by Scott in 1951. After two decades, Basu *et al*, reported the first such case from India in 1970. In 1981, Kamalam described four such cases from southern region of India. Gueho *et al,* carried out landmark study in 1992, wherein they performed extensive research and various species were carved out of the genus and **six species** were considered to be of clinical significance among human beings. However, at present the genus *Trichosporon* contains 51 accepted species, 16 of which are able to infect human hosts.

Mycology

In the literature, the designation *Trichosporon beigelii,* was used as an etiological agent of systemic trichosporonosis, whereas, this name is historically associated with white piedra of head. The word *Trichosporon* is derived from the Greek, which represents combination of 'Trichos' meaning hair and 'sporon' as spores. It has been observed that *Trichosporon* species involved in superficial infections are different from those responsible for invasive disease. Moreover, *Trichosporon beigelii* is considered to be identical to *Trichosporon cutaneum,* the fungal species frequently found on skin. *Trichosporon* should not be confused with *Trichophyton,* which is one of the dermatophyte genera.

Like *Malassezia,* genus *Trichosporon* also has undergone extensive review in recent times on the basis of morphological, biochemical, ultrastructural and biomolecular criteria, and currently it is classified in class **Basidiomycetes**. The reproduction in this class is predominantly by budding from yeast cells. In new classification of the genus *Trichosporon,* taxon *T.beigelii* was sub-divided into six newly defined distinct species mainly based on 26S rRNA sequences and DNA re-associations, which were primarily pathogenic to man. These six species belonged to class Basidiomycete and were: *T.ovoides, T.inkin, T.asahii, T.asteroides, T.cutaneum* and *T.mucoides*. These species were considered to be involved in mucosa-associated, superficial and systemic mycoses, in addition to being responsible for white piedra. However, a few other species also appeared in the literature as case reports making a total of 51 and out of these, 16 are found infecting the man.

The use of nomenclature as *Trichosporon beigelii* has now been abandoned in favor of new species. The genus *Trichosporon* earlier belonged to family **Cryptococcaceae,** sub-family **Trichosporoideae** of form-class **Blastomycetes** in **Deuteromycetes**. But at present, taxonomically the genus is classified in family **Trichosporonaceae** under the order **Tremellales**, class **Tremellomycetes**, subphylum **Agaricomycotina** in phylum **Basidiomycota**.

White piedra is not known to be caused by a species outside genus *Trichosporon* and is mainly caused by *T.ovoides* and *T.inkin*. Superficial infections, other than white piedra, are caused mainly by *T.asteroides* and *T.cutaneum*. On the other hand, *T.asahii* is frequently involved in deep-seated, disseminated type of infections. The common types of infections caused by *Trichosporon* species are shown in **Table 9.1**.

Table 9.1. Infections Caused by *Trichosporon* Species.
A. White Piedra
(i) Head-*Trichosporon ovoides*
(ii) Pubic-*Trichosporon inkin*
B. Other Cutaneous Lesions
Trichosporon cutaneum
Trichosporon asteroides
C. Systemic Infections
(i) Hematogenous dissemination-*Trichosporon asahii*
(ii) CNS disease-*Trichosporon mucoides*

Epidemiology

The causative agents of white piedra belong to the genus *Trichosporon,* which is yeast-like fungus that inhabits soil and occasionally constitutes substantial proportion of normal flora of human skin. Different species of the genus are known to cause white piedra, the superficial fungal infection characterized by nodules along the hair shaft. The species of this genus are also the etiological agents of other fungal diseases, such as onychomycosis. Furthermore, these fungi have emerged as opportunistic pathogens causing wide range of localized as well as disseminated invasive infections.

Trichosporon species may be present as normal flora of skin and also in the environment. The fungus can affect both humans as well as domestic animals such as horses. Usually, *Trichosporon ovoides* causes white piedra of head (capital), whereas, *T.asahii* and *T.inkin* involve the pubic area (crural). In some recent reports from India and Brazil, *T.inkin,* has been found to involve scalp too. In some of the reports, ethnic basis has been seen where it is found in Muslim young girls, may be because of the custom of wearing *hijab*. Rest of the species have been reported either to cause cutaneous or disseminated infection, which is termed as **trichosporonosis**, exclusively among immunocompromised as well as in neutropenic patients. Out of these, two species are frequently involved with deep fungal infections *i.e. T.asahii* and *T.mucoides*. Infections due to *T.asahii* are often associated with hematogenous dissemination, while *T.mucoides* is often recovered from patients with infection of central nervous system. The common manifestations produced by different species of *Trichosporon* are enumerated in **Table 9.1** and are also described in **Chapter 28**.

A significant number of species of *Trichosporon* are causative agents of white piedra. *T.asahii* is considered most commonly linked with white piedra, although some investigators believe that *T.ovoides* is the most common agent causing white piedra of scalp. White piedra has also been described in horses, monkeys, dogs and other animals.

Although white piedra is uncommon condition but it is widely distributed in temperate as well as tropical areas, including Eastern Europe, Asia, particularly Japan and South America. In the USA, occurrence of white piedra might be higher in black than in white people. This has now been reported from equatorial Africa also and its prevalence is high in Gabon. The young women are frequently affected but age and sex incidence varies from country to country, depending upon hairdressing fashions and social customs practiced in a particular society.

The exact source of infection is unknown but swimming in stagnant water has been suggested as a possibility since organism occurs in this environment in large number. The infection had never been reported previously in Africa but it is an autochthonous hair infection, which is widespread in central equatorial Africa. It affected mainly female population of urban area and was characterized by frequent occurrence in genito-pubic sites. In white piedra, familial outbreaks may also occur.

The mode of infection in man is not very clear. It is suggested that poor hygiene habits such as bathing in stagnant water may cause white piedra. The sexual transmission has been implicated as predisposing factor in cases of pubic white piedra. It has also been suggested that humidity acts as predisposing factor for white piedra of scalp. Rarely, *Trichosporon* species may be accompanied by *Candida parapsilosis* along the hair shafts, although it is unclear if this is really co-infection.

Pathogenesis and Pathology

White piedra is characterized by presence of softer nodules, which are white to light brown in color. The infection appears to start just beneath cuticle of hair shaft, possibly following damage. The organisms may grow inward and through shaft to form nodular swellings spaced irregularly along the axis. The hair is weakened at these points and hence may easily break. The growth occurs as collarette around hair shaft and consists of mycelia that rapidly fragment into arthrospores.

The nodules of white piedra are in the form of sheath that may extend around hair shaft. There may be extensive growth within hair giving rise to characteristic nodular swellings on hair shaft. The hyphae segment into individual cells, arthrospores, which are 2-4 μm in diameter. Synergistic effect has also been reported with some bacteria like *Brevibacterium* species.

Clinical Features

The disease is characterized by presence of soft, white, grayish or light brown nodules on hair shafts, which are composed of compact fungal elements. These are mainly seen on distal portion of facial and axillary hair, beard, moustache and pubic hair and less commonly affects the scalp, eyebrows and eyelashes. Sometimes, patient presents with pruritus or pain over the affected site. The disease may be accompanied by inflammation of skin folds.

As the organization texture of fungus is less in white piedra as compared to black piedra, fungal mass can easily be detached from the hair. The hair often breaks at the point of infection, leaving the ends clubbed or swollen. The nodule is greenish-brown mycelial mass, which consists of rectangular cells, usually 2-4 μm but may be up to 8 μm in diameter. The underlying skin is not affected. In white piedra, nodules are adhered less firmly and are softer as compared to black piedra. They may vary in color from white to light brown. The nodules are carried outwards, as hair grows, away from scalp.

The concretions of fungi found on hair are usually accompanied by bacteria, which usually belong to the genus *Corynebacterium*. In the individuals harboring infection with HIV, there is increased carriage rate of *Trichosporon* species in the perianal area. In severely neutropenic patients, systemic infections due to *Trichosporon* species have been described and these may affect different visceral organs including liver, spleen and heart. The cutaneous manifestations of disseminated infections include purpuric papules or papulovesicular lesions. There is evidence to suggest that some cases of white piedra may be sexually transmitted. In a case report by Zeller *et al*, *Cladosporium cladosporioides* has been considered as an unidentified cause of white piedra, which is one of the unusual findings because phaeoid or dematiaceous fungi have never been reported as cause of white piedra.

Differential Diagnosis

White piedra is to be differentiated from trichomycosis (trichobacteriosis) axillaris (caused by *Corynebacterium tenuis*), phthiriasis pubis and nits of pediculosis capitis. Pediculosis can be excluded easily as the nits do not completely encircle hair shaft. The trichomycosis axillaris shows cocco-bacilli, which are much smaller than rectangular or polygonal cells of *Trichosporon* species. The name could be misleading as trichomycosis is of bacterial origin rather than a fungal infection hence to avoid confusion it has been now re-designated as **trichobacteriosis axillaris**. On microscopy, the nodules in white piedra are circumferential around the hair shaft. As compared to black piedra, nodules of white piedra are more pale, soft and loosely adherent to hair shaft. Moreover, hair-shaft abnormalities such as monilethrix, trichorrhexis nodosa and trichoptilosis should also be kept as differential diagnosis.

The other conditions like peripilar casts or hair knots, often resulting in either delayed or inappropriate therapy may resemble white piedra. Under such circumstances, intrapilar invasion by hyphae may occur, leading to destruction of hair shaft and hair breakage. Tinea capitis affects base of hair shaft and follicle. Similarly, other conditions can be differentiated from white piedra, based on clinical appearance and by differences on routine KOH wet preparation. The organism causing trichomycosis axillaris gives pale-yellow fluorescence under Wood's lamp examination whereas white piedra does not fluoresce. *Trichosporon* species should be differentiated from *Geotrichum* species based on carbohydrate assimilation tests. Moreover, *Trichosporon* species being basidiomycete, produces urease and also blastoconidia.

Laboratory Diagnosis

The nodules are visible to the naked eye and can be white, tan or even reddish-green in color. They do not fluoresce on Wood's lamp examination. The affected hair can become brittle and break at points of cuticular invasion and swelling. The adjacent skin is not always affected. On direct microscopic examination of epilated hair, fungus is seen as sleeve-like concretions that are composed of hyphae and rectangular arthrospores within and around hair as shown diagrammatically in **Figure 9.1**. Blastoconidia are also produced, although difficult to appreciate. Some of the stains like Chlorazol Black E or Parker blue-black ink may be added to enhance visibility of fungal structures.

The fungal culture is done on SDA at 37°C with chloramphenicol added but without actidione, as the fungal growth is inhibited by cycloheximide. Moist yeast-like cream colored, soft and folded radiating colonies containing hyphae, blastospores and arthrospores can be seen after 2-3 days of incubation. The young culture is whitish and pasty in consistency. As fungal culture ages, colonies develop deep, radiating furrows and take yellowish color with creamy texture **(Fig. 9.2)**. Microscopic examination of the LCB mount reveals septate hyphae, rapidly fragmenting

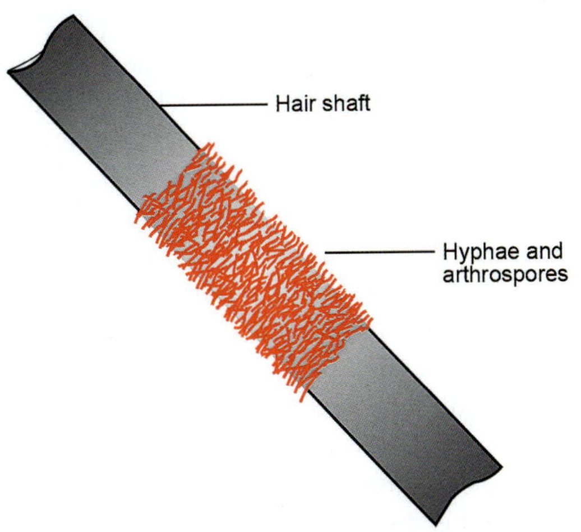

Fig. 9.1. White Piedra-*Trichosporon* species showing nodules consisting of hyphae and arthrospores on hair shaft.

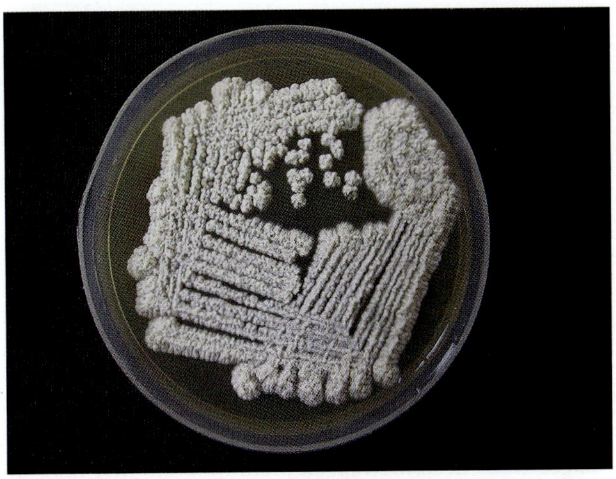

Fig. 9.2. Cream-colored, wrinkled, yeast-like colonies of *Trichosporon asahii* on SDA after one week of incubation.

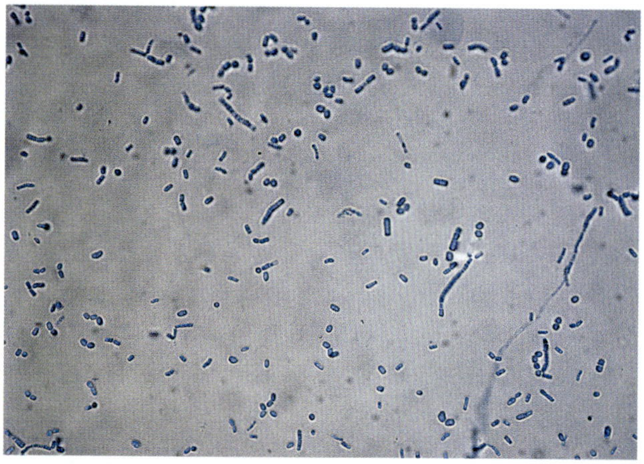

Fig. 9.3. Hyphae and rectangular arthrospores of *Trichosporon* species (LCB x 200).

to form rectangular arthrospores (2-4 × 3-9 μm), while the single cells round up to undergo budding **(Fig. 9.3)**.

Physiological studies are essential to establish species identification of genus *Trichosporon*. *T.beigelii* does not ferment carbohydrates but assimilates glucose, sucrose, galactose, maltose and lactose. It breaks down **urea** and also gives positive reaction with Diazonium Blue B (DBB) salts. These features help to distinguish *T.beigelii* from other species of the genus. The differentiation of species is also done on the basis of carbohydrate assimilation pattern of *Trichosporon* species as shown in **Table 28.1**. Three species, *T.cutaneum*, *T.inkin* and *T.ovoides* have been categorized as *T.cutaneum* or *T.beigelii* complex as they are difficult to be distinguished individually.

Slide agglutination can also be used to identify *Trichosporon* species, since these fungal species can be identified by specific serological procedure as they share serological relationship with species belonging to the genus *Cryptococcus*.

II. BLACK PIEDRA

Black piedra is also nodular type of infection of the hair shaft and is caused by phaeoid fungus, *Piedraia hortae*. This disease is also known by various names such as **trichomycosis nodularis**, **trichomycosis nodosa** and **tinea nodosa**.

Historical Perspective

Malgoi-Hoes described black piedra for the first time in 1901. **Paulo Parreiras Horta** (1884-1961), a Brazilian mycologist, systematically studied black piedra in 1911 thereby clearly differentiating it from the white variety. Hence in the year 1911, two varieties of piedra came into existence *i.e.* white and black, caused by two different fungi. He contemplated that the fungus causing this disease was also *Trichosporon* species, which was named as *Trichosporon hortai* by Brumpt in 1913 to highlight Horta's contributions in the field of mycology. In due course of time perfect state of this fungus was also discovered and ascospores were seen in nodules. It was then classified in phylum Ascomycota and re-designated as *Piedraia hortae* by Olimpo da Fonseca Filho and Antonio Eugenio de Area Leao in 1928. Subsequently, summarizing historical findings, Langeron in 1936 reviewed the available literature on both types of piedra. Unlike other fungi, nomenclature of the **genus** as well as **species** of black piedra remained more or less stable during this entire period of about a century.

Mycology

Piedraia hortae has a unique feature that it exists in the perfect state while colonizing hair shaft. Most of the fungi in other infections are found in their asexual states at the site of lesions. This was named *Piedraia hortae* after its teleomorphic state was established and its relation to ascomycetes became apparent. *Piedraia hortae* is thus placed in family **Piedraiaceae** under order **Capnodiales** in class **Dothideomycetes** in the phylum **Ascomycota** of kingdom Fungi.

Epidemiology

Black piedra is exclusively found in tropical countries, in the warm and humid climates of Central and South America, the Indies and Southeast Asia, in populations where hair care is usually done with oily substances. The causal fungus, *Piedraia hortae,* is believed to exist in soil and affects monkeys as well as man. In Africa, black piedra has been observed only in pelts of primates and never seen in humans. The disease is more common in males as compared to females but some of the recent studies have indicated that this difference in the prevalence of this disease between both sexes is insignificant.

The black piedra clearly differs from the white piedra with respect to the following three key points: (a) the causative agent is a dematiaceous filamentous fungus, *Piedraia hortae*; (b) the disease mainly involves the scalp and

(c) the geographical distribution is limited to tropical and subtropical areas i.e. South America and Southeast Asia.

Pathogenesis and Pathology

The infection starts under the cuticle of hair shaft with formation of stony hard, black nodules. The fungal mass may enlarge and grow outside hair and completely envelope the shaft. Black encrustations vary in size from microscopic structures to 1-2 millimeters and about 150 μm in depth. In the mature nodule, periphery is composed of aligned hyphal strands, whereas cells in central area are placed together to form pseudo-parenchymatous mass resembling the organized tissue. The fungus destroys cuticular layers of the hair and is able to penetrate deep into the cortex. Under such circumstances, intrapilar invasion by the hyphae may occur, leading to destruction of hair shaft and breakage of hair.

The nodules consist of either closely embedded spores in cement-like substance or tightly packed mass of branched hyphae with numerous one-celled asci containing ascospores. Some workers are of the opinion that *Piedraia hortae* is unable to penetrate cortex of hair shaft, whereas, other consider that it can penetrate but without extensive proliferation.

Clinical Features

Black piedra is characterized by formation of discrete gritty hard, brown-black nodules of about one millimeter diameter, firmly attached to hair shaft as diagrammatically shown in **Figure 9.4**. This disease affects mainly the hair of scalp. These nodules are composed of mass of fungal cells, dense ascostromata that are thicker, up to 150 μm at one end. They vary from microscopic size to 1-2 millimeters. Since the fungus grows into hair shaft, the hair may break easily at the site of infection.

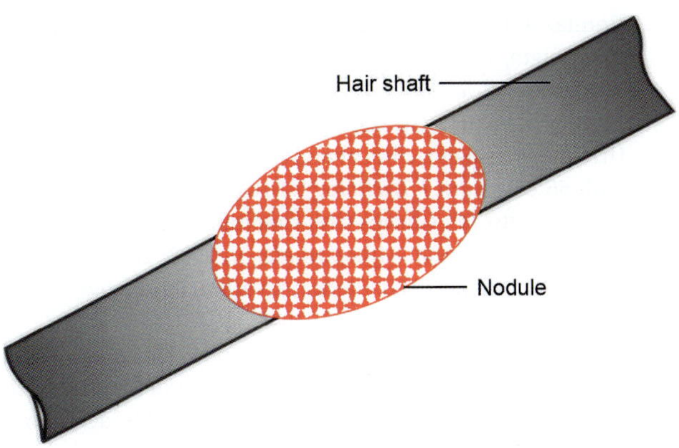

Fig. 9.4. Diagrammatic representation of Black Piedra *(Piedraia hortae)* showing nodules on the hair shaft.

The fungal mass may expand and in due course of disease process, grow outside the cuticle and completely encircle the hair shaft. As the nodules are gritty to feel, metallic sound may be heard when hair are combed. In most of the cases, **scalp hairs** are infected but moustache, beard and pubic hair may also be affected. Itching is usually absent in black piedra. If left untreated, disease may last for months or years. The medical advice is sometimes sought only for cosmetic reasons. Sometimes, there is mixed type of infection as black and white piedra may be found simultaneously in the same patient.

The mode of spread has not been defined, however, on the basis of available evidence; black piedra seems to be a contagious disease with predilection to affect certain communities. Pavithran described an outbreak of black piedra in women's hostel where 60% *i.e.* 15 inmates aged between 19 and 26 years, were found to have black gritty nodules on their long scalp hair. The probable mode of transmission of infection in this outbreak was traced to be use of a common comb for dressing scalp hair and sharing of pillows and bed sheets.

Differential Diagnosis

The clinical presentation of black piedra is so typical that diagnosis can be made easily; however, it should be differentiated from other similar types of diseases. These include nits of pediculosis and condition with abnormal hair growth *e.g.* trichorrhexis nodosa, etc. The presence of pruritus and distinctive shape of eggs, distinguish pediculosis from black piedra. The ovoid cells of black piedra may resemble arthrospores seen in tinea capitis (dermatophytosis) but hair shaft is normal on either side of cell mass in piedra whereas base of hair shaft and follicle are involved in tinea capitis. Therefore, alopecia does not occur in piedra, although hair shaft may break at the site of infection.

Laboratory Diagnosis

The diagnosis of black piedra is established by examining the crushed brittle nodule in 20% KOH and/or CFW wet mount. There are dark-colored thick-walled septate hyphae around surface of hair with many septations, which give appearance of arthrospores. There are round to oval asci containing 2-8 hyaline, aseptate, curved, banana-shaped (fusiform) ascospores (30 × 10 μm) that bear one or more whip-like appendages at both ends.

The hair sample is inoculated on SDA with chloramphenicol, glycerin and actidione, as *P.hortae* is not inhibited by cycloheximide. The culture shows very slow growing, adherent coal-black, cerebriform colonies, heaped up at the centre with flat periphery. Fungal cultures from clinical material usually yield only asexual form of fungus which consists of slow growing, brown to black mycelia at 25°C with reddish-brown diffusible pigment on agar. Microscopic examination reveals septate hyphae, chlamydospores and irregularly shaped hyphal elements. Such culture represents asexual phase of the fungus. Organisms in the sexual phase are difficult to grow in culture. The conidia and ascospores are rarely found on routine mycological cultures. Production of ascospores is enhanced if the SDA is diluted to one-tenth of normal strength with or without adding biotin. The microscopic examination of colonies in LCB mount reveals dark walled septate hyphae with chlamydospores and asci.

The teleomorphic state is found in older cultures and consists of specialized structures in which asci containing spindle-shaped ascospores develop. *Piedraia hortae* is unique among pathogenic fungi in producing sexual spores in its parasitic phase, when found infecting hair. The greenish-brown aerial mycelia eventually cover young glabrous colonies.

The micro-culture technique may also be employed using selective Dermatophyte Test Medium (DTM) for coating thin transparent plastic slide. The material to be studied is sampled with transparent adhesive tape, which is pressed on the surface of agar. This allows periodic microscopic examination of incubated culture which shows large irregular cells located around infected hair within two weeks. The cuticle of hair is raised, however, the cortex is spared and not infected.

Treatment and Prophylaxis

The ideal treatment in both types of piedra is shaving off hair of the affected part, which alone is usually sufficient for an effective and successful cure. The American Academy of Dermatology, Guidelines/Outcomes Committee also recommends complete removal of infected hair. Despite the fact that hair shaving of scalp is commonly used as an effective therapeutic procedure in men, it may not be feasible in women due to their habit of keeping long hair. An application of a topical azole derivative such as clotrimazole, econazole or ketoconazole on the affected hair may cure infection. White piedra can also be treated by using topical antifungals, including imidazoles, ciclopirox olamine, selenium sulfide (2.5%), precipitated sulfur in petroleum (6%), chlorhexidine solution, Castellani paint, zinc pyrithione and amphotericin B lotion.

All clinically significant *Trichosporon* species show high *in vitro* susceptibility to amphotericin B, clotrimazole, ketoconazole and itraconazole. Hence specific therapy for white piedra should be resorted, accordingly. Itraconazole and fluconazole, with their ability to bind keratin, persist in stratum corneum of scalp beyond initial infection, thus preventing and defending scalp from recurrences and re-infection. An oral azole antifungal medication for three to four weeks, in combination with topical azole antifungal shampoos for two to three months appears to be an effective treatment without the need for shaving. Terbinafine in a dose of 250 mg daily for a period of six weeks has also been used successfully in the treatment of black piedra. The treatment of white piedra remains difficult since scalp colonization serves as persistent reservoir for recurrence. However, relapses are common even after complete course of therapy. A good personal hygiene should be practiced to avoid infection with both types of piedra.

Further Reading

1. Adya KA, Inamadar AC, Palit A, *et al.* Light microscopy of the hair: A simple tool to "untangle" hair disorders. *Int J Trichology.* 2011; **3:** 46-56.
2. Basu N, Sanyal M, Banerjee AK, *et al.* White piedra in India: a case report. *Indian J Dermatol Venereol.* 1970; **36:** 154-5.
3. Bonifaz A, Gomez-Daza F, Paredes V, *et al.* Tinea versicolor, tinea nigra, white piedra and black piedra. *Clin Dermatol.* 2010; **28:** 140-5.
4. Bonifaz A, Vaquez-Gonzalez D, Fierro L, *et al.* Trichomycosis (trichobacteriosis): clinical and microbiological experience with 56 cases. *Int J Trichology.* 2013; **5:** 12-6.
5. Chagas-Neto TC, Chaves GM, Colombo AL. Update on the genus *Trichosporon*. *Mycopathologia.* 2008; **166:** 121-32.
6. Desai DH, Nadkarni NJ. Piedra: An ethnicity-related trichosis? *Int J Dermatol.* 2014; **53:** 1008-11.
7. Douchet C, Therizol-Ferly M, Kombila M, *et al.* White piedra and *Trichosporon* species in equatorial Africa. III. Identification of *Trichosporon* species by slide agglutination test. *Mycoses.* 1994; **37:** 261-4.
8. Finch J. Case of trichomycosis axillaris and erythrasma. *J Drugs Dermatol.* 2011; **10:** 1472-3.

9. Fischman O, Bezerra FC, Francisco EC, *et al. Trichosporon inkin*: An uncommon agent of scalp white piedra - report of four cases in Brazilian children. *Mycopathologia*. 2014; **178:** 85-9.
10. Ghorpade A, Ramanan C, Das M, *et al.* Black piedra: A case report. *Indian J Dermatol.* 2000; **45:** 20-1.
11. Ghorpade A, Ramanan C. White piedra. *J Eur Acad Dermatol Venereol.* 1994; **3**: 169-73.
12. Ghorpade A. Surrogate nits impregnated with white piedra - A case report. *J Eur Acad Dermatol Venereol.* 2004; **18:** 474-6.
13. Goldberg LJ, Wise EM, Miller NS. White piedra caused by *Trichosporon inkin*: A report of two cases in a northern climate. *Br J Dermatol.* 2015; **173:** 866-8.
14. Guidelines/Outcomes Committee, Drake LA, Dinehart SM, *et al.* Guidelines of care for superficial mycotic infections of the skin: Piedra. *J Am Acad Dermatol.* 1996; **34:** 122-4.
15. Gupta AK, Cooper EA, Ryder JE, *et al.* Optimal management of fungal infections of the skin, hair and nails. *Am J Clin Dermatol.* 2004; **5:** 225-37.
16. Huang CF, Liaw FY, Liu YC, *et al.* Can you identify this condition? Trichomycosis axillaris (TA). *Can Fam Physician.* 2013; **59:** 647-8.
17. Inacio CP, Rocha AP, Barbosa RD, *et al.* Experimental white piedra: A robust approach to ultrastructural analysis, scanning electron microscopy and etiological discoveries. *Exp Dermatol.* 2016: **25:** 79-81.
18. Kalter DC, Tschen JA, Cernoch PL, *et al.* Genital white piedra: Epidemiology, microbiology and therapy. *J Am Acad Dermatol.* 1986; **14:** 982-93.
19. Kamalam A, Thambiah S, Bagavandas M, *et al.* Mycoses in India - Study in Madras. *Trans R Soc Trop Med Hyg.* 1981; **75:** 92-100.
20. Khandpur S, Reddy BS. Itraconazole therapy for white piedra affecting scalp hair. *J Am Acad Dermatol.* 2002; **47:** 415-8.
21. Khatu SS, Poojary SA, Nagpur NG. Nodules on the hair: A rare case of mixed piedra. *Int J Trichology.* 2013; **5:** 220-3.
22. Kiken DA, Sekaran A, Antaya RJ, *et al.* White piedra in children. *J Am Acad Dermatol.* 2006; **55:** 956-61.
23. Kumaresan M, Deepa M. Trichonodosis. *Int J Trichology.* 2014; **6:** 31-3.
24. Ma DL, Vano-Galvan S. Trichomycosis axillaris. *N Engl J Med.* 2013; **369:** 1735.
25. Marine M, Brown NA, Riano-Pachon DM, *et al.* On and under the skin: Emerging basidiomycetous yeast infections caused by *Trichosporon* species. *PLoS Pathog.* 2015; **11:** e1004982.
26. Marques SA, Richini-Pereira VB, Camargo RM. White piedra and pediculosis capitis in the same patient. *An Bras Dermatol.* 2012; **87:** 786-7.
27. Pankajalaxmi VV, Taralaxmi VV, Paramasivan CN. *Trichosporon beigelii* infection in Tamilnadu. Indian *J Dermatol Venereol Leprol.* 1979; **45:** 136-8.
28. Pasricha JS, Nigam PK, Banerjee U. White piedra in Delhi. *Indian J Dermatol Venereol Leprol.* 1990; **56:** 56-7.
29. Pavithran K. An outbreak of black piedra in a women's hostel. *Indian J Dermatol Venereol Leprol.* 1988; **54:** 312-3.
30. Richini-Pereira VB, Camargo RM, Bagagli E, *et al.* White piedra: Molecular identification of *Trichosporon inkin* in members of the same family. *Rev Soc Bras Med Trop.* 2012; **45:** 402-4.
31. Roselino AM, Seixas AB, Thomazini JA, *et al.* An outbreak of scalp white piedra in a Brazilian children day care. *Rev Inst Med Trop Sao Paulo.* 2008; **50:** 307-9.
32. Roshan AS, Janaki C, Parveen B. White piedra in a mother and daughter. *Int J Trichology.* 2009; **1:** 140-1.
33. Sandoval-Tress C, Arenas-Guzman R, Guzman-Sanchez DA. Hair shaft yellow nodules in a pediatric female patient. *Skin Appendage Disord.* 2015; **1:** 62-4.
34. Saxena S, Uniyal V, Bhatt RP. Inhibitory effect of essential oils against *Trichosporon ovoides* causing piedra hair infection. *Braz J Microbiol.* 2012; **43:** 1347-54.
35. Sentamilselvi G, Janaki C, Murugusundram S. Trichomycoses. *Int J Trichology.* 2009; **1:** 100-7.
36. Shivaprakash MR, Singh G, Gupta P, *et al.* Extensive white piedra of the scalp caused by *Trichosporon inkin*: A case report and review of literature. *Mycopathologia.* 2011; **172:** 481-6.
37. Taj-Aldeen SJ, Al-Ansari HI, Boekhout T, *et al.* Co-isolation of *Trichosporon inkin* and *Candida parapsilosis* from a scalp white piedra case. *Med Mycol.* 2004; **42:** 87-92.
38. Tambe SA, Dhurat RS, Kumar CA, *et al.* Two cases of scalp white piedra caused by *Trichosporon ovoides. Indian J Dermatol Venereol Leprol.* 2009; **75:** 293-5.
39. Tendolkar U, Shinde A, Baveja S, *et al. Trichosporon inkin* and *Trichosporon mucoides* as unusual causes of white piedra of scalp hair. *Indian J Dermatol Venereol Leprol.* 2014; **80:** 324-7.
40. Thammayya A, Sanyal M. White piedra: Mycotic infection of the hair. *Indian J Med Res.* 1976; **64:** 554-6.
41. Therizol-Ferly M, Kombila M, Gomez de Diaz M, *et al.* White piedra and *Trichosporon* species in equatorial Africa. II. Clinical and mycological associations: An analysis of 449 superficial inguinal specimens. *Mycoses.* 1994; **37:** 255-60.
42. Therizol-Ferly M, Kombila M, Gomez de Diaz M, *et al.* White piedra and *Trichosporon* species in equatorial Africa. I. History and clinical aspects: An analysis of 449 superficial inguinal specimens. *Mycoses.* 1994; **37:** 249-53.
43. Uniyal V, Saxena S, Bhatt RP. Screening of some essential oils against *Trichosporon* species. *J Environ Biol.* 2013; **34:** 17-22.
44. Viswanath V, Kriplani D, Miskeen AK, *et al.* White piedra of scalp hair by *Trichosporon inkin. Indian J Dermatol Venereol Leprol.* 2011; **77:** 591-3.
45. Zawar V. Trichomycosis (Trichobacteriosis) axillaris. *J Dermatol Case Rep.* 2011; **5:** 36-7.
46. Zeller S, Lempert S, Goebeler M, *et al. Cladosporium cladosporioides:* A so far unidentified cause of white piedra. *Mycoses.* 2015; **58:** 315-7.
47. Zhuang K, Ran X, Dai Y, *et al.* An unusual case of white piedra due to *Trichosporon inkin* mimicking trichobacteriosis. *Mycopathologia.* 2016; **181:** 909-14.

CHAPTER 10

Dermatophytosis

Dermatophytoses are the most common types of superficial cutaneous fungal infections seen in men and animals. These are caused by group of closely related keratinophilic fungi, which are capable of invading keratinized tissues of skin and its appendages like hair and nail. They belong to three mycelial fungal genera *i.e. Trichophyton*, *Microsporum* and *Epidermophyton* and are collectively known as **dermatophytes**. The other frequently used terms like **tinea** and **ringworm** infections are synonym of dermatophytoses. In the recent times few cases of subcutaneous and deep-seated infections have been reported to be caused by dermatophytes.

In addition to that, there are fungi, which do not belong to these three genera but significantly involve the same anatomical sites *i.e.* skin, hair as well as nail. These have emerged as an important group of filamentous as well as yeast-like fungi causing various clinical entities. The diseases caused by non-dermatophytic fungi infecting **skin** are called **dermatomycoses**, whereas that of **hair** and **nail** are known as **piedra** and **onychomycosis**, respectively. Piedra is already described in **Chapter 9**, however, dermatomycoses and onychomycosis are being dealt in this Chapter along with dermatophytoses.

Historical Perspective

During 1830s various fungi were described as causative agents of beard and scalp infections. In 1837, Robert Remak first observed hyphae in scalp scraping. In 1839, Johann Lucas Schonlein described these filaments as molds and considered plants as the source of infection of scalp (tinea capitis). His pupil, Remak, succeeded in growing the fungus on apples and reproduced the disease on his own forearm. These studies were the first instance of a human disease being attributed to microorganisms and opened the new field of microbiology (mycology). In 1845, Remak gave an accurate description of the cultivated fungus from patients' lesions, supported by drawings of the microscopic fungal elements in culture and proposition of the valid name of the fungus *Achorion schoenleinii* (now, *Trichophyton schoenleinii*).

These observations were confirmed by series of works done by Hungarian physician, David Gruby, working in Paris between 1841 and 1844. He described clinical entities *i.e. favo*, the scalp disease caused by dermatophytes and demonstrated that fungi could also be cultured and transmitted to humans. He described ectothrix type of hair invasion of beard as well as scalp and named the causative agent as *Microsporum audouinii*. He also described subsequently endothrix type of hair invasion caused by *Herpes (Trichophyton) tonsurans.*

Prof. Domenico Majocchi (1849-1929), Italian dermatologist and physician, first described variant of tinea corporis popularly called **Majocchi's granuloma**, which is an uncommon infection of dermal and subcutaneous tissue by dermatophytes and he named this disorder as ***'granuloma tricofitico'*** in 1883. In 1892, Raymond Sabouraud began his studies of dermatophytes till 1938. He published monumental work, *Les Teignes* in 1910 wherein he classified dermatophytes into **four** genera, *Achorion, Epidermophyton, Microsporum* and *Trichophyton* primarily on the basis of clinical aspects of the disease, combined with cultural and microscopic observations. In the meantime, Wood's lamp was invented by a Baltimore physicist, Robert W. Wood that was used for the detection of infected hair in tinea capitis in 1925 and subsequently for other infections as well.

In 1934, Chester Emmons modified taxonomic scheme laid down by Sabouraud and other scientists and established new classification of dermatophytes on the basis of spore morphology and accessory structures. Consequently he eliminated genus *Achorion* and emphasized on rest of the **three** genera on the basis of mycological principles, which is being followed till date. In 1984,

Punsola and Guarro described *Keratinomyces,* as fourth genus of dermatophytes, psychrophilic species from soil, which grows at 20°C but its type species *Keratinomyces ceretanicus,* has not been found to be pathogenic to humans.

For the treatment of **tinea capitis**, very high dose ionizing radiations were given from 1948 to 1960 to more than 100,000 immigrating children into Israel, which led to 6000 deaths shortly and increased incidence of head and neck tumors in the long run. Baruch Modan *et al*, reported these findings in 1974, thereby Israeli government had to compensate the victims. **The Ringworm Children**, a documentary released in 2003, effectively highlighted the issue of irrational use of radiation therapy for treating an ordinary fungal infection.

The discovery of teleomorphs of *Trichophyton ajelloi* in 1959 by Dawson and Gentles using hair-bait technique of Vanbreuseghem, led to breakthrough research of teleomorphs of many dermatophytes and related keratinophilic fungi. Griffin and Stockdale in 1960s independently obtained teleomorphs of *Microsporum gypseum* complex, thereby vindicating Nannizzia's original observation. Now *Arthroderma* is the only genus recognized on the basis of teleomorphic states of dermatophytes encompassing many species of *Trichophyton* as well as *Microsporum* hence *Nannizzia* is presently an obsolete terminology of this genus.

The dermatophytosis remained a significant public health problem with many bizarre treatment modalities being prescribed. In 1958, Griseofulvin became available after breakthrough experimental works of Gentles in guinea pigs. This discovery revolutionized therapeutic approach to this disease. In 1980s, discovery of azole derivatives and allied group of antifungal drugs had significant impact in the management of dermatophytosis.

Mycopathologia publishes special issue on *"Dermatophytes and Dermatophytoses"* after every ten years. Now this time it has been published in February 2017 (Vol. 182 No. 1-2) covering most of the current aspects on this topic. In this issue there are drastic changes, particularly, related to taxonomy and allied areas. Hence readers are advised to go through the same for an update on this topic.

Mycology

The dermatophytes are hyaline septate molds with more than hundred species described. Forty-two species are considered valid and less than half of these are associated with human diseases. These are divided into **three** main anamorphic genera depending on their morphological characteristics, which are given below:

Trichophyton	24 species
Microsporum	16 species
Epidermophyton	2 species

The teleomorphic states of only 23 dermatophytes have been described which either belong to genus *Trichophyton* or *Microsporum* as sexual state of *Epidermophyton* has not yet been described. The other environmental species not infecting skin, hair or nail of men or animals are not called dermatophytes. *Trichophyton* should not be confused with *Trichosporon*, an emerging yeast-like basidiomycetous fungus.

Before 1986, teleomorphs of dermatophytes were classified into two genera, *Arthroderma* and *Nannizzia.* They were differentiated mainly on the basis of morphology of peridial hyphae. The genus *Arthroderma* was characterized by dumbbell-shaped peridial hyphal-cells with knobby protuberances at each end. Its anamorphs produce smooth-walled macroconidia and they belonged to genus *Trichophyton* or *Chrysosporium.* The genus *Nannizzia* was characterized by peridial hyphal cells with one or more interseptal constrictions but no knobby protuberance at the end of cells. Its anamorphs produce rough-walled macroconidia and belonged to only one genus, *Microsporum.*

The morphology and ontogeny of ascocarp, asci and ascospores were found to be identical in both these genera. Recently, mitochondrial DNA (mtDNA) analysis by restriction fragment length polymorphism (RFLP) has become useful tool for taxonomic studies. The phylogenetic analysis also shows that both genera, *Arthroderma* and *Nannizzia* were unified taxonomically in only one genus. Therefore, it was proposed by Weitzman *et al* in 1986 that genus *Nannizzia* be considered as synonym of genus *Arthroderma,* which was accepted unanimously.

This evaluation of descriptions defining these two genera concluded that all characters represent *continuum.* Consequently, these two taxa may be considered cogeneric and because of priority, therefore, correct generic name is *Arthroderma.* The teleomorphic states of dermatophytes, genus *Arthroderma* is placed in family **Arthrodermataceae** of order **Onygenales** in class **Plectomycetes** of phylum **Ascomycota**.

The *Trichophyton* species usually infect skin, hair and nails, *Microsporum* species infect skin and hair and **not** the **nails** and *Epidermophyton* species infect skin as well as nails but **not** the **hair**. Several specialized anthropophilic species like *T.concentricum* and *M.ferrugineum* consist of morphologically simplified asexual isolates with little or no ability to produce conidia. The taxonomic relationship among closely related *Trichophyton rubrum, T.tonsurans, T.violaceum* and *T.mentagrophytes* complex including teleomorphic species is of interest.

Non-dermatophytic dermatomycosis is infection of skin and nails by various species of fungi other than dermatophytes. Some of these are: *Candida* species, *Neoscytalidium dimidiatum, Fusarium oxysporum* and *Scopulariopsis brevicaulis.*

Epidemiology

The prevalence of dermatophytosis is governed by environmental conditions, personal hygiene and individual's susceptibility from place to place. The isolation of different species of dermatophytes also varies markedly from one ecological niche to another depending on their primary natural habitat. It is possible that dermatophytosis of some sites like genitalia is underestimated because of its common and self-healing nature.

Based on their ecological characteristics, dermatophytes are traditionally divided into three categories *i.e.* anthropophilic, zoophilic and geophilic species. The fungal species exclusively affecting humans are known as **anthropophilic** while those inhabitating with domestic and wild animals as well as birds are called **zoophilic** dermatophytes. A third group, frequently isolated from soil is known as **geophilic (Table 10.1)**. The distinction between geophilic and zoophilic species may not always be obvious and even may be controversial at times. For example, some of fungi like *T.simii* and *M.persicolor* are thought to be zoophilic by some workers but geophilic by others. The infections caused by anthropophilic species tend to be chronic and intractable and resultant inflammation is minimal. On the other hand, infections caused by geophilic and zoophilic species tend to be self-healing and resultant inflammation is more severe.

Many of saprotrophic soil fungi are closely related to dermatophytes, sharing ability to utilize keratin as growth substrate and it is believed that dermatophytes have evolved from these keratinophilic soil fungi. Therefore, all keratinophilic fungi may not be dermatophytes. In due course of time during this evolutionary process various dermatophyte species have become adapted to particular host and this has eventually led to development of epidemiological groups of anthropophilic, zoophilic and geophilic species. The anthropophilic agents have marked affinity for particular body sites. The infection of scalp is

Table 10.1. Ecological Classification of Dermatophytes.

Anthropophilic	*Zoophilic*	*Geophilic*
T.concentricum	*T.equinum*	*T.ajelloi*
T.megninii	*T.mentagrophytes*	*T.terrestre*
T.mentagrophytes	*var. mentagrophytes*	*M.amazonicum*
T.rubrum	*T.simii*	*M.cookei*
T.schoenleinii	*T.verrucosum*	*M.fulvum*
T.soudanense	*M.canis*	*M.gypseum*
T.tonsurans	*M.equinum*	*M.nanum*
T.violaceum	*M.gallinae*	*M.praecox*
M.audouinii	*M.persicolor*	*E.stockdaleae*
M.ferrugineum		
E.floccosum		

frequently caused by *T.tonsurans* and *M.canis* that of trunk and inguinal regions by *T.rubrum* whereas *T.interdigitale* affects the foot.

As dermatophytosis is prevalent throughout the world, it primarily depends on personal habits and living conditions of people. Maceration, occlusion and minor skin trauma are contributing factors in facilitating infection in addition to warm moist climate. Change in free fatty acids of scalp after puberty confers resistance to dermatophytosis. Certain traditional and cultural practices are correlated with greater occurrence like washing of extremities without drying, wearing tight clothings like saris tied around the waist, walking bare foot and mass shaving of heads with shared instruments. Other practices in certain communities like excessive use of hair oil may have a protective role. Moreover, certain occupations like agriculturists, farmers, athletes, wrestlers and those associated with animal-keeping are at a higher risk of acquiring these infections.

This is the only true contagious fungal infection in real sense. Infection is transmitted through fomites. The arthrospores are parasitic propagules and survive in environment for pretty long time. The persistence of *T.mentagrophytes* on fomites depends on the content of carbohydrate, which does not stimulate germination of conidia of this species. Carotenoid pigments found in arthrospores prolong their dormancy. Also, individuals with sub-clinical infections *e.g. T.tonsurans* carriers, are source of infection to susceptible hosts and may result in outbreaks.

Some species of dermatophytes are endemic in certain parts of world and have limited geographic distribution. Now, *Trichophyton soudanense, T.gourvilii* and *T.yaoundei* are geographically restricted to Central and West Africa. *Microsporum ferrugineum* predominates in Japan and surrounding areas. *Trichophyton concentricum* is confined to islands in South Pacific and small area in Central and South America. However, increasing mobility of world's population is disrupting several of these epidemiological patterns. In recent times, *T.tonsurans* has replaced *M.audouinii* as the principal agent of tinea capitis in USA, may be because of mass migration of individuals from Mexico and other Latin American countries where *T.tonsurans* usually predominates.

The most common etiologic agents of dermatophytosis in western countries are *Trichophyton rubrum* and *Microsporum canis*. The main changes in distribution of dermatophytes in Europe during 20th century have been near disappearance of certain anthropophilic species such as *T.violaceum*, *T.schoenleinii* and *T.tonsurans*. However, in certain European countries, where immigrants from former colonies in Africa, Asia and West Indies come to live in big cities, these species have been re-introduced but until now only in relatively small numbers. This might be compared to distribution of *T.tonsurans* in USA, where this pathogen is the most common cause of tinea capitis. It is presumed that Spanish-speaking immigrants from Central and South America have spread *T.tonsurans*. In southern and east European countries, anthropophilic fungi have been replaced by zoonotic species such as *M.canis* and *T.mentagrophytes*. A slight increase in infections with *M.canis* is also seen in some north-western countries but *T.rubrum* continues to predominate. *Microsporum distortum* is rare cause of tinea capitis in New Zealand and Australia and infection in lower animals such as dogs. It is similar to and possibly is a variant of *M.canis*. *T.rubrum* and *T.mentagrophytes* are dermatophytes commonly reported from the Middle East.

Hemashettar *et al,* isolated *T.yaoundei* from Karnataka, which has glabrous faint yellow-colored colonies. This is not prevalent beyond African countries, particularly Cameroon and Congo. Similarly *T.soudanense* is encountered in Africa and both have been reported for the first time in India. Other agents frequently reported from Africa include *M.audouinii*, *T.tonsurans* and *T.violaceum*.

Immunity

The host response to dermatophyte plays major role in the pathogenesis of dermatophytoses. The clinical manifestations are mostly due to immune response of host to the invading fungal species. Both humoral and cell-mediated responses, specific and non-specific defense mechanisms help to eliminate the fungus thereby preventing its invasion beyond the dead keratinized tissue. Some degree of acquired resistance in humans has been noted following the dermatophyte infection. An acute inflammatory reaction is correlated with mounting cell-mediated immunity to dermatophyte and infection generally resolves spontaneously or responds well to the treatment. This reaction is contrary to that seen frequently in individuals with chronic infections, who do not show cell-mediated response to these fungi. Moreover, apparently CMI, as expressed by delayed-type hypersensitivity response to dermatophyte antigen (trichophytin), accompanies acquired resistance to dermatophyte infections and is the main defense mechanism. Chronic dermatophytosis has been associated with CMI deficient conditions like primary immunodeficiency diseases, asthma, allergic rhinitis and connective tissue disorders or secondary immunosuppression as in HIV infection, cancer chemotherapy and prolonged use of steroids.

Keratinocyte proliferation and epidermal desquamation, lymphocytes, macrophages and mast cells are thought to be active against dermatophytes. Defective phagocytosis of peripheral blood leucocytes is associated with chronic trichophytosis. The prevalence of onychomycosis and tinea pedis is high in diabetes mellitus. Widespread tinea infections are encountered in patients with ichthyosis due to defective cutaneous barrier.

A broad-spectrum of exoenzymes or alternatively large quantity of a particular exoenzyme may be advantageous for fungal spread. The release of these fungal exoenzymes is also an important factor in invasion and utilization of host tissue by dermatophytes. However, same enzymes may also stimulate an inflammatory host response and thereby could be helpful in limiting fungal growth. There is some component of allergic response to dermatophyte in the disease process. However, some dermatophytes like *T.rubrum* may diminish the immune response. The fungal cell wall glycoprotein mannan may suppress CMI. In addition, it inhibits keratinocyte proliferation thereby slowing epidermal turnover resulting in chronic infection.

Dermatophytes of fingernails and toenails, in contrast to those at other body sites, are particularly difficult to eradicate with drug treatment. This is consequence of factors intrinsic to the nail *i.e.* hard, protective nail plate, sequestration of pathogens between nail bed and plate and slow growth of nail as well as relatively poor penetration of antifungal agents.

Pathogenesis and Pathology

The dermatophytes grow only within dead, keratinized tissue. The fungal cells produce keratinolytic proteases *in vivo* and *in vitro,* which provide means of entry into living cells. Fungal metabolic products diffuse through Malpighian layer of the epidermis to cause erythema, vesicles and even pustule formation along with pruritus. The hyphae become old, break into arthrospores, which are shed off in due course of time. This is partially responsible for central clearing of the **ringworm** types of lesions and owes its common name. Their *in vivo* activity is restricted to the zone of differentiation, newly differentiated keratin and for infection to persist, hyphal growth must keep pace with rate of keratin production. The hyphal tips growing down within hair shaft reach the edges of live keratinizing cells *i.e.* bulb but terminate above it in a fringe of hyphal ending called **Adamson's fringe**. The infective process ceases and healing occurs when the balance of fungus and host is tilted in favor of host thereby the upward movement of keratin carries active hyphae away from keratinous zone.

The great variation in clinical presentation is related to the involved species and strains of fungus, size of inoculum, involved sites and immune status of the host. As and when host's resistance is reduced, prevalence of dermatophytes can be extremely high and symptomatology can be very unusual. This strongly supports the fact that host's immunological factors determine the clinical course of the disease.

Some dermatophytic species, which are basically soil saprotrophs, with ability to digest keratinaceous debris in the soil, have now evolved to be capable of parasitizing keratinous tissues of the animals.

Clinical Features

The clinical manifestations of dermatophytosis are also called **tinea** or **ringworm** depending on anatomical site involved as shown in **Table 10.2**. The term *'tinea'* is derived from Latin word meaning, ***'worm'*** or ***'moth'***. This is used descriptively because of serpentine (snake-like) and circular or annular (ring-like) lesions that occur on skin, making it to appear as if worm is burrowing at the margins. The nomenclature of clinical entity is designated appending Latin word to an anatomical site to the term tinea. It should not be confused in any way with **taenia**, which is intestinal infestations caused by Cestodes.

The literal meaning of word ***'dermatophyte'*** is ***'skin plant.'*** The suffix ***'phyte'*** implies that these organisms are plants therefore in the present context, it is actually a misnomer because now fungi are phylogenetically not related to plants as were classified earlier and they are quite different and have their own kingdom since 1969, after Whitaker's five kingdom classification. The other synonyms were also suggested like trichophytosis or microsporosis, depending upon the causative genus involved *e.g. Trichophyton* or *Microsporum,* respectively but overall the term dermatophytosis is in common use.

Table 10.2. Clinical Types and Causative Dermatophytes.

Clinical Types	*Causative Dermatophyte Species*
Tinea capitis	*Trichophyton, Microsporum* species
Favus	*T.schoenleinii,* rarely *T.violaceum,M.gypseum*
Kerion	*T.verrucosum, T.mentagrophytes*
Tinea barbae	*T.rubrum, T.mentagrophytes, T.verrucosum*
Tinea corporis	*T.rubrum* and any other dermatophytes
Tinea imbricata	*T.concentricum*
Tinea cruris	*E.floccosum, T.rubrum, T.mentagrophytes*
Tinea unguium	*T.rubrum, T.mentagrophytes, E.floccosum*
Tinea manuum	*T.rubrum, E.floccosum, T.mentagrophytes*
Tinea pedis	*T.rubrum, T.mentagrophytes, E.floccosum*

The clinical importance of identifying species of dermatophyte is to find out probable source of infection. Moreover, there are some prognostic considerations as well. The anthropophilic group causes chronic infection and may be difficult to cure. The zoophilic and geophilic dermatophytes cause inflammatory lesions, which easily respond to therapy and occasionally heal spontaneously. The **chronic dermatophytosis** refers to persistent dermatophyte infection that runs a long course with episodes of remission and exacerbation for more than a year. The chronicity is considered in terms of duration and recurrences of dermatophyte infections. This is the most significant features encountered among the patients. The lesions in dermatophytosis are not symmetrical as seen in some of the non-infective disorders.

The dermatophytes have distinct clinical manifestations in different parts of the body. Each focus of infection is due to local inoculation. The inflammation is seen maximum at the advancing margins leaving central area with substantial clearing. The clinical features of dermatophyte infection result from combination of keratin destruction and inflammatory response generated in the host. The wide variations in clinical presentation depend upon species and probably strain of the fungus concerned, size

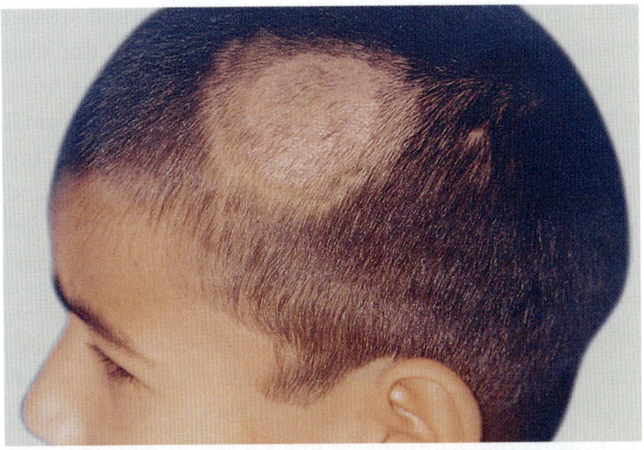

Fig. 10.1. Patch of alopecia with broken hair and ring formation at the periphery over the temporal area of the scalp.

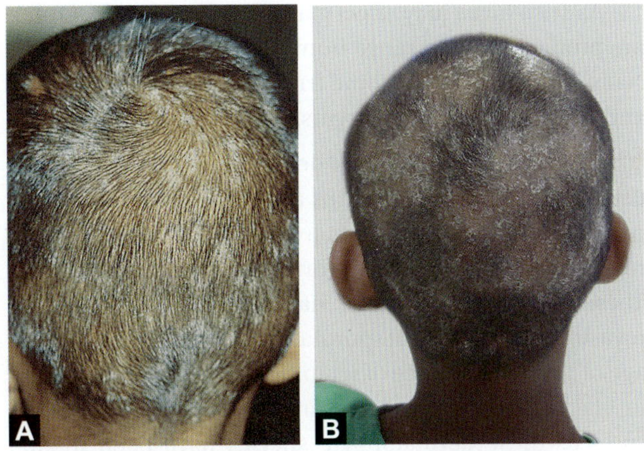

Figs. 10.2A and B. Patchy alopecia with crusting over the scalp in patients with kerion.

of inoculum, site of body infected and immune status of the host. The following common clinical conditions are produced by dermatophytes:

(a) Tinea Capitis

This is infection of shaft of scalp hairs and presents as the following clinical types:

(a) Inflammatory—Kerion, favus and agminate folliculitis.

(b) Non-inflammatory—Black dot, seborrheic dermatitis-like and grey patch.

The infected hair in tinea capitis appear dull and grey. The base of hair shaft as well as its follicle is involved in contrast to piedra where hair shaft is normal on either side of cell mass. There is breakage of hair at follicular orifice, which creates patches of alopecia with black dots of broken hair. A patch of alopecia with broken hair and ring formation at periphery of patient is shown in **Figure 10.1**. The predominant causative fungal species of tinea capitis belong to genus *Trichophyton*. The descriptions of various clinical types of tinea capitis are given below:

(i) Kerion: This is severely painful inflammatory reaction producing raised, circumscribed boggy mass on scalp, usually suppurating at multiple points (Gr.=***'honeycomb'***). It is severe form of dermatophytosis with deep, suppurative lesions on the scalp. The follicles may be seen with discharging pus. There may be sinus formation and rarely mycetoma-like grains may also be produced. Thick crusting with matting of adjacent hair is common. It may be followed by scarring and permanent alopecia in the area of inflammation and suppuration. Kerion is usually caused by zoophilic dermatophytes like *T.verrucosum* and *T.mentagrophytes*. Erythema nodosum has also been found in association with kerion infection. The patchy alopecia with crusting over scalp is shown in **Figures 10.2A and B**.

The **kerion celsi** is clinical form of tinea infection of scalp that usually involves children. The clinical features may develop as such but are more often result of an inflammatory reaction to tinea capitis. It presents clinically as sharply defined, painful edematous plaques, with pustules and abscesses that drain pus. There is an acute, intense inflammatory response of host, principally against antigens of dermatophyte involved and not to secondary bacterial infection. The etiological agents of kerion celsi are *T.verrucosum*, *T.mentagrophytes*, *M.canis* and *M.gypseum*, which sporadically infect the humans. Kerion celsi may progress to patchy or diffuse distribution and to severe hair loss with scarring alopecia.

(ii) Favus (Tinea Favosa): This condition is caused by *T.schoenleinii* and is seen sporadically which forms cup-like crusts (Latin scutulum = little shield) around infected follicles. The name favus has been derived from Latin word for honeycomb. The fungal growth within hair is minimal, which remain intact. The intense fungal growth within and around hair follicle produces waxy, honeycomb-like crust on scalp. It may lead to alopecia and scarring. Rarely *T.violaceum* and *M.gypseum* can also cause favus. It has been reported in Kashmir valley but it is becoming very rare due to an overall increased hygienic standard of the public at large. Favus among Bantus in South Africa is called witkop (whitehead). It is also prevalent in Middle East, South Eastern Europe and the Mediterranean region.

As such favus may affect glabrous skin where scutulae are produced which range from pin-head to 2 cm in size. Nails when affected become irregularly thickened, brittle and crusted under free margins.

(iii) Black-dot: The black-dot ringworm infection is usually caused by *T.tonsurans* and *T.violaceum*. The gray patch *M.audouinii* infections of 1950's have been now replaced by black-dot ringworm caused by *T.tonsurans.* These dermatophytes attack hair shaft by endothrix type invasion with abundant sporulation inside hair and breakage of hair near surface of scalp. This phenomenon results in black-dot appearance within an area of smooth scalp surface. These dermatophyte species have been identified by morphological and biochemical analysis. However, occasionally it is difficult to identify certain isolates since morphological characteristics of species may fail to form in a typical strain.

(iv) Ectothrix Infections: The arthrospores appear as mosaic sheath around hair or as chains on surface of hair shaft. The cuticle of hair remains intact. In small spored ectothrix infection, hyphae invade hair shafts at the level of mid-follicle. As hair grows out of follicle, hyphae burst out of shaft and cover hair surface with mass of small arthrospores. The small arthrospores are 2-5 µm and are present in *T.mentagrophytes, M.canis, M.audouinii* and *M.gypseum* infections. The large arthrospores are 8-12 µm in diameter in large dense chains around hair and is characteristic of *T.verrucosum* infection **(Table 10.3)**.

(v) Endothrix Infection: In case of endothrix infection, hyphae form arthrospores within hair shaft, which is severely weakened and the cuticle of hair is usually destroyed. The arthrospores are 3-4 µm in diameter and are observed in chains filling inside shortened hair stubs **(Figs. 10.3A to C and Table 10.3)**. T.*schoenleinii, T.tonsurans* and *T.violaceum* are usually responsible for endothrix type of hair invasion. *T.rubrum* although rarely infects hair but has been described to cause both ectothrix as well as endothrix infections.

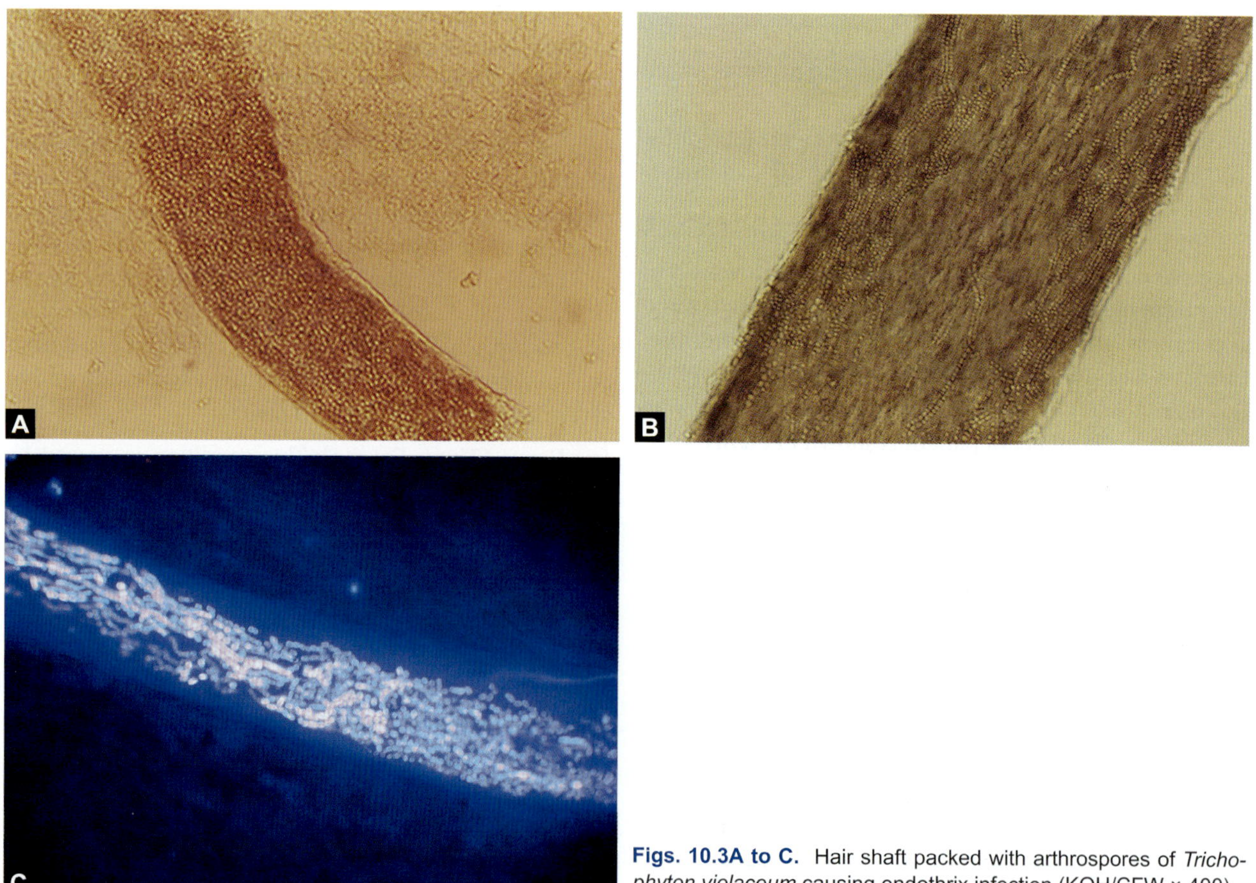

Figs. 10.3A to C. Hair shaft packed with arthrospores of *Trichophyton violaceum* causing endothrix infection (KOH/CFW × 400).

Table 10.3. Types of Hair Invasion by Dermatophytes.

Ectothrix	*Endothrix*
T.mentagrophytes	*T.schoenleinii*
M.canis	*T.tonsurans*
M.gypseum	*T.violaceum*
M.audouinii	*T.soudanense*
T.verrucosum	
T.rubrum	

Tinea corporis or fungal infection of other sites, secondary bacterial infection, *'id'* reaction and scarring alopecia may complicate tinea capitis. The differential diagnosis of tinea capitis includes seborrhoeic dermatitis, psoriasis, atopic dermatitis, alopecia areata, trichotillomania, syphilis, actinomycosis of skin, lupus erythematosus and Langhans cell histiocytosis. **Ear sign** may help in easy and rapid diagnosis, which is presence of erythematous papules, scaling or well to ill-defined plaques over helix, antihelix and retroauricular region (away from retroauricular fold) in tinea capitis while erythema, scaling or fissuring in retroauricular fold in seborrheic capitis. Inflammatory tinea capitis may have similarities with traction folliculitis or bacterial pyoderma *i.e.* furunculosis, folliculitis or impetigo.

Pityriasis amiantacea (PA) is a scalp disorder presenting with thick, silvery to yellowish, asbestos-like scales wrapping around and binding down tufts of hair. This is considered to be a reactive condition to several inflammatory diseases, which may affect the scalp, mainly psoriasis, atopic dermatitis and seborrheic dermatitis. However, although rarely, PA may be the presenting clinical pattern of dermatophyte infection (PA-like tinea capitis), thus emphasizing the importance of ruling out/confirming the possibility of tinea capitis when dealing with a case of PA. Dermoscopy as a supportive diagnostic tool is useful in diagnosing an instance of PA-like tinea capitis.

(b) Tinea Corporis

This entity is also called **tinea glabrosa** or **tinea glabrata circinata** due to the involvement of glabrous (non-hairy) skin of human body. It is a general term used for all dermatophyte infections of skin with the **exception** of scalp, face, hands, feet, groins and nails, which have their own independent nomenclatures. It may result from extension of infection from scalp, groin or beard but the term tinea corporis is properly applied to lesions originating on glabrous skin. It is characterized by erythematous-scaly lesions, annular, sharply marginated plaques with raised border, which may be single, multiple or confluent. There is partial central clearing of lesions with few pustules, which may be present along the border. It is the most common type of dermatophytosis in India and is usually caused by *T.rubrum* and other dermatophyte species. Annular erythematous scaly plaques of tinea corporis are shown in **Figures 10.4A to E**. Tinea corporis should be differentiated from pityriasis versicolor, pityriasis rosea, candidiasis and secondary syphilis.

Clinically atypical varieties of tinea corporis are increasingly being reported particularly in HIV-positive patients, although such presentations may rarely be seen in immunocompetent patients. The cases of bullous tinea corporis, mainly caused by *T.rubrum,* have also been reported.

Infection of hair follicles can lead to deep dermal inflammatory reaction very similar to kerion of scalp. This perifollicular pustular or well-circumscribed, elevated crusted granulomatous lesion is called **Majocchi's granuloma** or **Granuloma Trichophyticum** as shown in **Figure 10.5**. This is commonly seen in women with tinea pedis or onychomycosis (caused by *T.rubrum*) who shave their legs and is caused by infected hair penetrating wall of hair follicle. In a case report this deep infection by dermatophyte has also been reported in cardiac transplant patient.

A variant of tinea corporis, **Majocchi's granuloma** is localized cutaneous dermatophyte infection of dermal and subcutaneous tissue *i.e.* deep folliculitis. It is also known with its less popular synonym - **tinea profunda**. There are two types of clinical manifestations, which have been described *i.e.* perifollicular papules and nodules on the areas prone to trauma, usually in healthy individuals with chronic dermatophytosis and deep nodular plaque lesions or abscesses found in immunocompromised hosts. *T.rubrum* is the most common agent but recently other dermatophyte species like *T.tonsurans,* have also been reported.

This disease is a form of chronic nodular trichophytosis, usually affecting scalp, face, forearms, hands or legs. It is generally preceded or accompanied by acute trichophytosis like kerion, tinea of smooth skin and so on. Although incidence of Majocchi's granuloma is higher in childhood but it can be seen at any age.

Majocchi's granuloma appears to occur in two main groups of patients. One group is characterized by having limited immunologic defect, which predisposes them to chronic dermatophytosis and rarely to Majocchi's

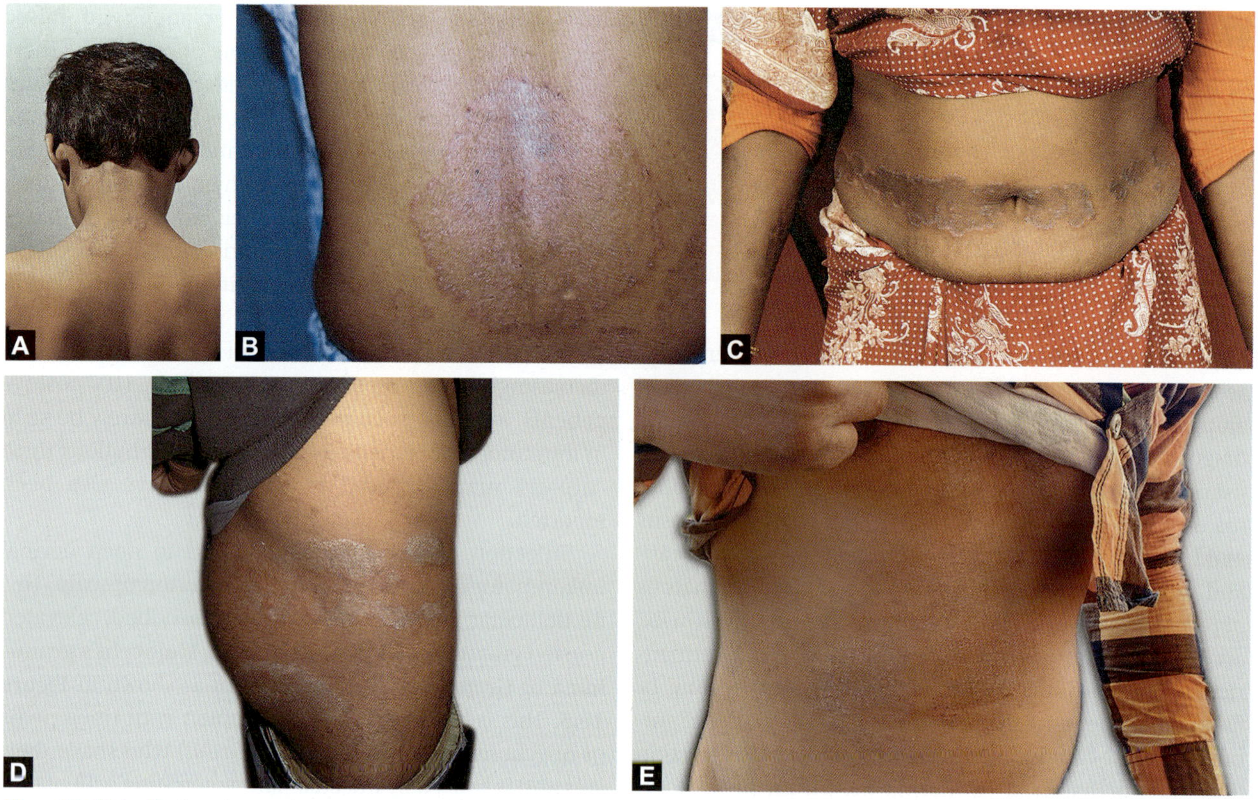

Figs. 10.4A to E. Annular erythematous scaly plaques with advancing margins of tinea corporis of glabrous skin over various parts.

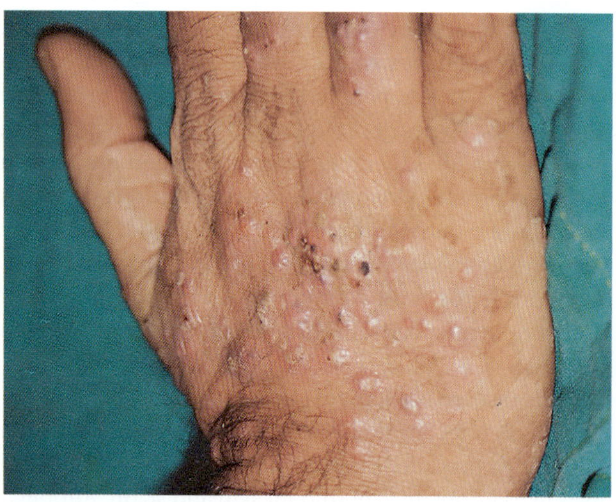

Fig. 10.5. Pustular, well-circumscribed, elevated and crusted lesions in Majocchi's dermatophytic granuloma of the hand.

granuloma. These patients show no tendency to systemic dissemination. The initiating factor is thought to be physical trauma *e.g.* shaving, which leads either directly or indirectly to disruption of follicle. Besides, potent anti-inflammatory action of topical steroids may cause local immunosuppression such that both antifungal activity and local cell-mediated immunity are overwhelmed. The fungi grow down into follicle thereby invade it. Topical steroid use can mask tinea infections leading to tinea incognito, which may result in Majocchi's granuloma. A long-term refillable prescription for antifungal and strong topical steroid combination should be avoided; instead, routine mycologic examination should be performed before the onset of treatment. When dermatophytosis is associated with pruritus severe enough requiring symptomatic relief, one can add low-potency steroids.

The other group is characterized by broader immunologic defects predisposing them to systemic dissemination of organism. This group of patients includes those with graft-versus-host reaction after bone marrow transplantation or patients on longstanding steroids, vincristine, cyclophosphamide or azathioprine or HIV-infected patients. These patients may also have physical trauma as initiating factor. Concerning etiology, *Phoma* species have been reported as secondary causative agents in addition

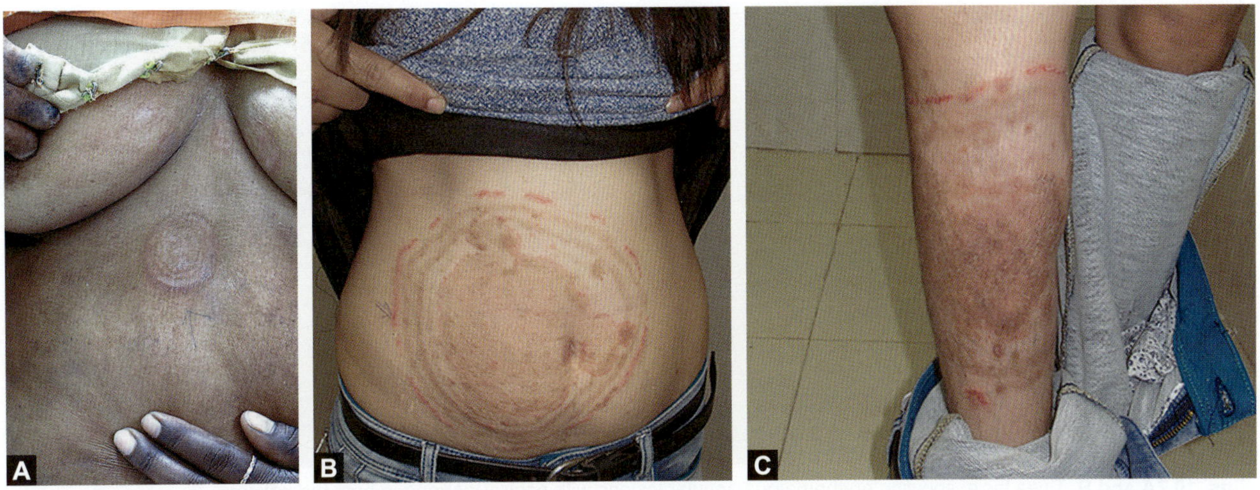

Figs. 10.6A to C. (A) Circular lesions of tinea imbricata over the upper part of abdomen. (B and C) Circular lesions of tinea pseudoimbricata seen as *"rings within the ring"* and *"double-edged tinea"* over the lower part of abdomen and right leg of a young girl.

to those previously mentioned. A recent review stated that Majocchi's granuloma occur mostly on buttocks and lower legs of patients with AIDS thereby may mimic Kaposi's sarcoma. Hence with an increase in organ transplantation and immunosuppression, physicians should consider diagnostic possibility of Majocchi's granuloma and therapeutic options, accordingly.

Dermatophytosis should be differentiated from **dermatophilosis** caused by bacterium *Dermatophilus congolensis*, which is an exudative, pustular dermatitis that mainly affects domestic cattle, sheep, horses and occasionally humans.

(c) Tinea Imbricata

Tinea imbricata or Tokelau is an unusual form of tinea corporis, caused only by *T.concentricum*. It is strictly anthropophilic dermatophyte, prevalent in limited rural geographical areas. It is found in China, India and particularly in Polynesian Islands (Tokelau, Fiji, Papua, Samoa and Tamana), as well as Central and South America including Mexico.

There are characteristic concentric and lamellar plaques of scales (imbricata - Latin, tiled), which spread out peripherally over many years, creating distinctive appearance from other tinea. These lesions cover almost all hairless skin but rarely involve nails, palm and soles **(Fig. 10.6A)**. There may be intense pruritus, which may lead to lichenification. The predisposing conditions include humidity, inheritance and immunologic factors. It is chronic and highly relapsing disease. There are good results with oral griseofulvin, terbinafine and topical combination with keratolytic medication such as Whitfield's ointments. This type of tinea these days is disappearing and hardly any case is encountered in Dermatology OPDs.

Tinea indecisiva or **tinea pseudoimbricata** is characterized by concentric scaly rings simulating tinea imbricata but caused by dermatophytes other than *Trichophyton concentricum.* It should be considered in a patient with tinea imbricata-like lesions with local immunosuppression caused by a non-*concentricum Trichophyton* or any other dermatophyte species. The typical concentric circles of tinea pseudoimbricata on the abdominal wall as well as over right leg of a young girl with ***"rings within the ring"*** and ***"double-edged tinea"*** appearance **(Figs. 10.6B and C)**. Such types of clinical presentation may be seen after the use of corticosteroids also as tinea incognito. Its diagnosis is confirmed by the appearance of ***"ring within the ring"*** lesions clinically and isolation of non-*concentricum* dermatophyte as the etiologic agent on mycological culture. The intradermal testing with *Trichophyton* extract shows fluctuating hypersensitivity responses. A daily oral terbinafine results in complete resolution in a month time.

(d) Tinea Gladiatorum

Tinea gladiatorum is an emerging infection in wrestlers and it affects their ability to compete. This is also found in other athletes depending on their playing habits. In wrestlers, it is found as a result of direct skin-to-skin

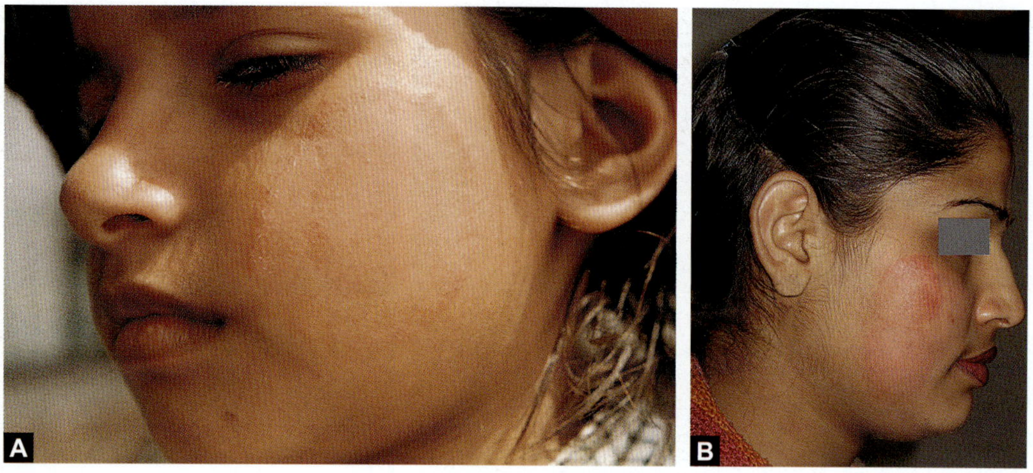

Figs. 10.7A and B. Tinea faciei showing erythematous annular lesions with central clearing in both the patients.

contact rather than via fomites such as wrestling mats. Most of the lesions are on arms, trunk, head and neck, corresponding to areas of maximum contact between wrestlers. Outbreaks of this infection have been reported. *T.tonsurans* is the commonest isolate. Fluconazole is effective and safe for primary prevention as well as treatment of tinea gladiatorum.

(e) Tinea Incognito

The term, **tinea incognito**, was first used by Ives and Marks, which refers to dermatophyte infection wherein clinical appearance was modified by prior application of topical corticosteroids. The use of these steroids apparently ameliorates symptoms and signs thereby offer clinical improvement during initial phase of treatment. However, consequently typical appearance is modified due to disturbance in classical clinical pattern of particular tinea leading to delay in diagnosis. Hence it is also called ***tinea atypica***. It is invariably mis-diagnosed as eczema. The pathogenesis is presumably linked to steroid-modified response of host to fungal infection rather than to direct pharmacological effect. In fact corticosteroids decrease resistance to infection and suppress inflammatory reaction. Therefore, such type of infection spreads and acquires a form quite different from the classical pattern of dermatophyte infections. Sometimes tinea incognito may present as tinea pseudoimbricata also. This is also reported with application of non-steroid topical immunomodulators *i.e.* calcineurin inhibitors like tacrolimus. It may resemble contact dermatitis, psoriasis, eczema, erythema migrans and herald patch of pityriasis rosea.

(f) Tinea Faciei

This is dermatophytic infection of skin that occurs on nonbearded regions of face. It is the most commonly misdiagnosed dermatophytic infection, essentially similar to tinea corporis. The causative agents of tinea faciei vary according to geographic region. The common clinical picture in tinea faciei is similar to other types of dermatophytosis. The patients usually present with history of photosensitivity and are likely to be treated with topical steroids with wrong diagnosis of inflammatory conditions hence it frequently takes form of tinea incognito. Topical antifungal drugs are less effective than in tinea corporis. **Figures 10.7A and B** show erythematous annular plaques with central clearing in a patient of tinea faciei.

(g) Tinea Barbae

This is ringworm infection of beard and moustache areas of face with invasion of coarse hair and is also called **barber's itch** or **tinea sycosis**. There are erythematous patches on face which show scaling, fragile lusterless hair and tendency to develop folliculitis as seen in tinea capitis. Two clinical types occur: (i) deep type produces nodular thickenings and kerion-like swellings and is usually caused by *T.verrucosum* and *T.mentagrophytes;* (ii) superficial crusted type shows less inflamed pustular folliculitis associated with *T. violaceum* and *T. rubrum*. Uncommonly, *E.floccosum* may cause widespread verrucous lesions known as verrucous epidermophytosis.

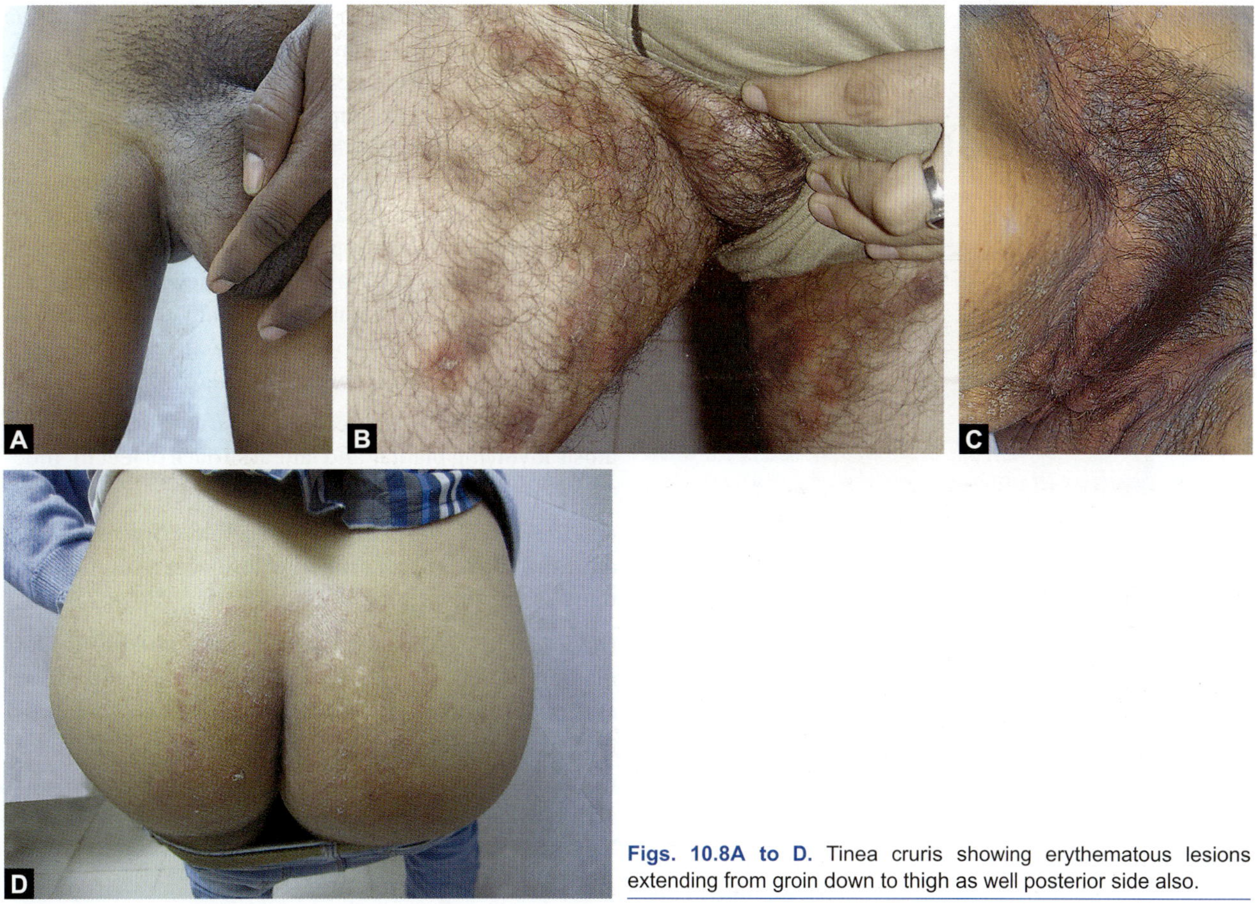

Figs. 10.8A to D. Tinea cruris showing erythematous lesions extending from groin down to thigh as well posterior side also.

(h) Tinea Cruris

This is dermatophytic infection of skin pertaining to groin area. However, Crus in Latin refers to the calf and not the groin hence tinea cruris, though commonly used term, is **tinea inguinalis** in the real sense. This occurs worldwide and is more prevalent in tropical countries. It is mostly seen in men with an underlying predisposing factor such as long-term use of tight-fitting garments. It involves perineum, scrotum and perianal area and may spread to inner third of buttock and occasionally to thigh. The appearance of tinea cruris can be seen in other intertriginous areas such as under pendulous breasts, axilla and around umbilicus of obese patients.

The borders of lesion are well delineated. In early stage there are erythematous plaques, arciform with sharp margins extending from groin down to thigh **(Figs. 10.8A to D)**. Scaling is variable and occasionally may mask inflammatory changes. Vesiculation is rare but dermal nodules forming beading along edge are commonly found in older lesions. It is also called ***'jock itch'*** as itching is predominant feature limited to inguinal area. *T.rubrum* and *E.floccosum* are common dermatophyte species responsible for this clinical entity.

(i) Tinea Manuum

This is dermatophytic infection of skin of palmar aspect of hands. The most common clinical manifestation is diffuse hyperkeratosis of palms and fingers **(Fig. 10.9)**. Infection of dorsal surface of hand presents no specific feature hence is considered as ringworm infection of glabrous skin under heading of tinea corporis **(*See* Fig. 10.11A)**. Mostly infection is caused by anthropophilic species like *T.rubrum*, *T.mentagrophytes* and *E.floccosum*. Tinea manuum should be differentiated from *'id'* reaction, psoriasis, contact dermatitis, secondary syphilis, dermatomycosis and pompholyx.

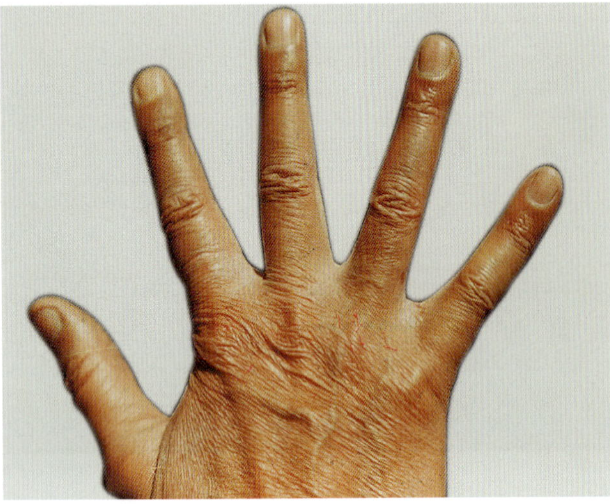

Fig. 10.9. Tinea manuum showing diffuse hyperkeratosis of the palm and fingers.

(j) Tinea Pedis

This is infection of plantar aspect of foot, toes and interdigital web spaces. The warmth and moisture produced by shoes are key factors in establishing and maintaining infection. It is frequently seen among individuals wearing shoes for long hours and popularly known as **Athlete's Foot**. During the initial period of the recognition of this disease, it was seen in athletes but as such it is found in the general population as well. In toe webs, scaling, fissuring, maceration and erythema may be associated with an itching or burning sensation. The small vesicles rupture and discharge thin fluid. Due to maceration and peeling, cracks appear which are prone to secondary bacterial infections. When secondary infection does occur, lymphangitis and lymphadenitis develop. The infection of sole may extend to sides of foot and therefore, it is also known as **Moccasin** or **Sandal ringworm**. It is also called one-hand-two-feet syndrome due to common involvement of one hand and both feet in a patient.

When infection becomes chronic, sole becomes hyperkeratotic and is often covered with fine scales. Anthropophilic species like *T.rubrum*, *T.mentagrophytes* and *E.floccosum* are common dermatophytes involved in tinea pedis throughout world. Tinea pedis may be divided into following four types on the basis of clinical presentation:

(a) Interdigitalis: There is maceration, fissuring, erythema and scaling in toe webs, especially between fourth and fifth toes. It may be subdivided into simplex and complex forms. The former presentation may consist of fine scaling and erythema; sometimes pruritus may be present. Tinea pedis interdigitalis can manifest with severity which is greater than seen in simplex variety. In complex variety, other presenting features include inflammation, fissuring, hyperkeratosis, maceration, erosions, burning/stinging and odor. In this presentation, both fungal and bacterial organisms are commonly present. The use of an antifungal agent as monotherapy may not be sufficient hence better results are observed when monotherapy or combination therapy that has broad-spectrum of activity against both fungal and bacterial organisms is used.

(b) Hyperkeratotic: This consists of scaling and hyperkeratosis involving plantar and lateral surface of feet in 'moccasin' distribution.

(c) Ulcerative: There is an acute ulcerative process usually affecting soles and associated with maceration, denudation of skin and oozing.

(d) Vesicular: This comprises of vesicles and blisters formation, usually near instep and adjacent plantar surface **(Figs. 10.10A to C)**. Sometimes pustules are also formed in this type but characteristically they are small and associated with clear vesicles. Large pustules, not associated with vesicles, are rare in tinea pedis.

(k) Tinea Unguium

Tinea unguium is dermatophyte infection of nail plates and is largely a disease of adults and is mainly caused by *T.rubrum*, *T.mentagrophytes* and *E.floccosum*. It begins under leading free edge of nail plate or along lateral nail fold and may continue until entire nail plate and nail bed are infected. **Figures 10.11A to F** show destruction of nail plates due to tinea unguium in patient of tinea corporis. There is accumulation of subungual debris in an opaque, chalky or yellowish thickened nail. The inflammatory changes of skin around nail *i.e.* nail fold which is called paronychia, is not included in either tinea unguium or onychomycosis. It should be differentiated from onychomycosis caused by *Candida* and other non-dermatophytic fungi, involving proximal portion of nail. A poor peripheral circulation is frequently implicated for resistance to treatment in tinea unguium.

(l) Onychomycosis

Onychomycosis is traditionally referred to non-dermatophytic infection of nail but now used as general term to denote all fungal infections of nail. Tinea unguium specifically describes dermatophytic invasion of nail. The term onychomycosis is derived from Greek word ***'onyx'***, nail and ***'mykes'***, fungus. It is used to describe fungal infection of nails caused, predominantly in 90% of cases by anthropophilic dermatophytes:-*T.rubrum*, *T.mentagrophytes* are usual pathogens. Yeast and non-dermatophyte mold infections are much less common. Onychomycosis affects

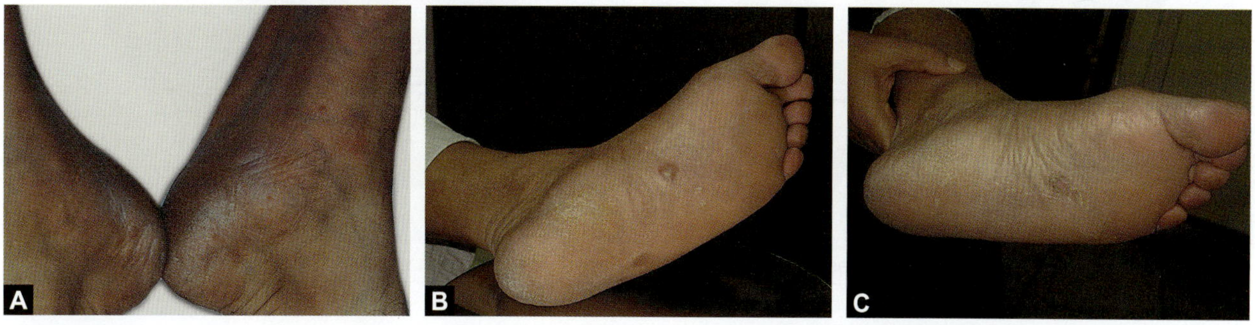

Figs. 10.10A to C. A. Tiny vesicles with erythematous annular borders over instep and heel. (B) Intact vesicle and (C) after its rupture.

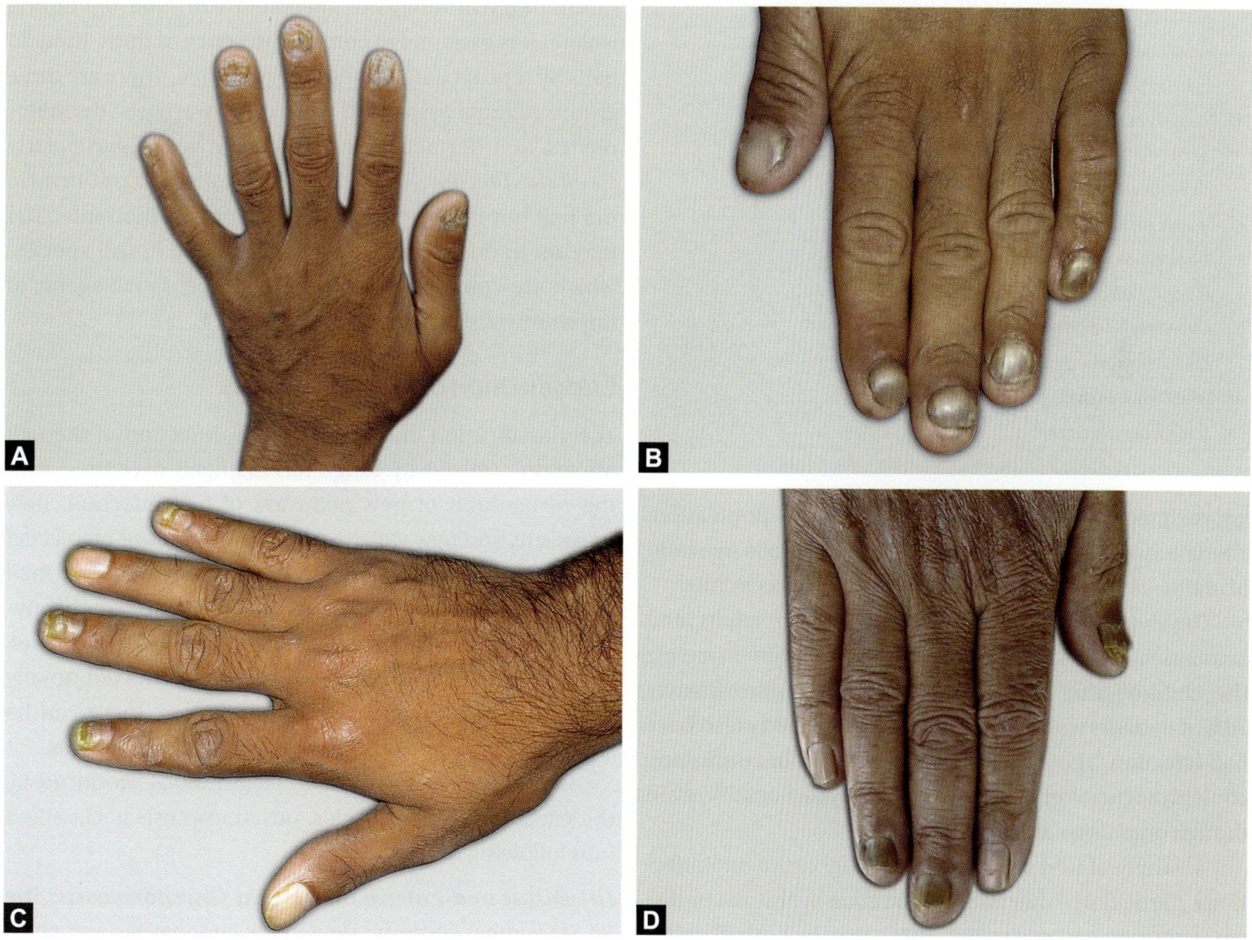

Figs. 10.11A to D

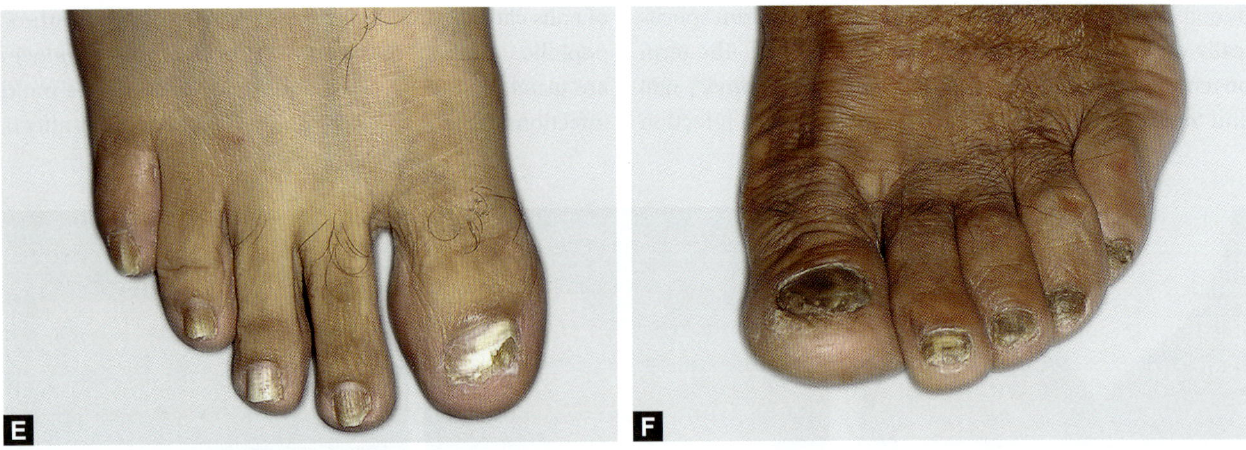

Figs. 10.11A and F. Destruction of nail plates due to tinea unguium / onychomycosis with various clinical presentations.

Table 10.4. Various Fungi Involved in Onychomycosis.
A. Dermatophytes
Trichophyton rubrum
Trichophyton mentagrophytes
Epidermophyton floccosum
B. Non-dermatophyte Fungi
Scopulariopsis brevicaulis
Acremonium species
Aspergillus flavus and *A.fumigatus*
Fusarium oxysporum
Neoscytalidium dimidiatum
Onychocola canadensis
Geotrichum candidum
C. Yeast-like Fungi
Candida albicans

approximately 6-13% of North American population. Toenails are more often affected than fingernails in a ratio of about 4:1 and it is difficult to treat them successfully.

Onychomycosis is common infection in adults and accounts for 20% of all nail diseases. It is less common among the children due to fast growing nails. Approximately 30% of patients with tinea elsewhere in the body also have nail infection. The use of KOH preparation for confirming clinical diagnosis of onychomycosis may be negative, even though organisms are present.

A variety of non-dermatophyte fungi cause nail infections, particularly after damage to tissue by trauma or diseases causing nail dystrophy. These may be both filamentous as well as yeast-like fungi. Most of them are present in environment and care must be taken in interpreting culture of infected material. It may be possible to correlate morphology of fungus in specimen with that of cultural isolate. It is essential to confirm presence of these fungi in nails by repeat cultures whenever possible. There are three groups of fungi associated with onychomycosis: dermatophytes, non-dermatophytic molds and yeasts particularly *Candida albicans*. The Mucorales as such do not involve the nail but if it is mechanically damaged, then infection may set in due to this group of fungi. The fungal species, which are commonly encountered as causative agents of onychomycosis, are listed in **Table 10.4**.

Classification of Onychomycosis

The classification of onychomycosis, infection of the nail apparatus caused by fungi, has changed over time with the recognition of new pathways of nail infection, new organisms and new variations in the appearance of infected nail. Taking into account published descriptions of nail morphology in fungal infection, the following forms of onychomycosis are recognized: distal and lateral subungual, superficial, endonyx, proximal subungual, mixed, totally dystrophic and secondary onychomycosis. These can be sub-divided by color and pattern of nail plate changes.

On the basis of various types of lesions produced by the causative fungal species, onychomycosis is classified into following four types:

(a) Distal and Lateral Subungual Onychomycosis: The DLSO is most common type of onychomycosis and characterized by invasion of nail bed and underside of nail plate. In this type dermatophytes predominate with occasional involvement of non-dermatophytic fungi. The commonest species is *T.rubrum* followed by *T.mentagrophytes, T.tonsurans* and *E.floccosum*. The ideal site of collection of specimen in DLSO is **nail bed** underside of nail plate from advancing edge, most proximal to cuticle.

(b) Proximal Subungual Onychomycosis (PSO): This usually affects fingernails, primarily as a result of *Candida* species. Fungal invasion of proximal nail fold is often visible through cuticle as whitish-yellow discoloration. Periungual inflammation may be quite marked and painful and in some cases associated with purulent discharge. These patients are frequently misdiagnosed as having bacterial infection. In immunocompromised patients, it may be caused by dermatophytes as well. The specimen is taken from **nail plate**, proximal nail bed as close to lunula.

(c) White Superficial Onychomycosis (WSO): This is surface infection of nail primarily when fungi invade superficial layers of nail plate directly and is caused by *T.mentagrophytes* and sometimes by hyaline mold like *Acremonium, Aspergillus* and *Fusarium* species. It may also be found in patients with AIDS. The surface scrapings of nail plate are ideal specimen to demonstrate the fungi.

(d) Total Dystrophic Onychomycosis (TDO): In this type there is total destruction of nail plate, which usually occurs as a result of infection by dermatophytes.

Sometimes combined lesion of DLSO and PSO may be present. Nail infection in patients with chronic mucocutaneous candidiasis is caused by *Candida albicans*. The organisms invade entire nail plate. There may be another syndromes like **onycholysis** and **paronychia** and found more commonly in females as compared to males and invariably affect middle finger.

In case of *Candida* infection, there is hard, thick, brownish-discolored nail plate that is striated or grooved. The nail does not become friable as found in tinea unguium. The nail plate is destroyed in untreated patients. Moreover, onychomycosis involves **proximal** portion of nail, whereas tinea unguium involves **distal** part. It is useful to take sample from both nail plates as well as subungual debris of infected nail. In case of heavy invasion by *Scopulariopsis brevicaulis*, brown spores may discolor nail producing cinnamon color, which is best seen in area of subungual hyperkeratosis.

A video-rate (real time) confocal microscope for imaging living human tissue has recently been developed for non-invasive imaging, which requires no specimen preparation. Confocal microscopy is technique for non-invasive optical sectioning through intact tissue and may be used to examine infected nails.

The differential diagnosis of onychomycosis is lichen planus, psoriasis particularly when it is bilateral, eczema, contact dermatitis, hyperkeratotic scabies, onychodystrophy and nail trauma. Onychomadesis is the total loss of the nail starting from proximal end of the nail, without discomfort or swelling, following temporary nail matrix arrest. It is a periodic idiopathic shedding of the nails beginning at the proximal end, possibly caused by the temporary arrest of the function of the nail matrix.

(m) Deep Dermatophytosis

In some of the case reports, dermatophytosis has been presented as deeper infections like abscess formation. It is caused by *Trichophyton* species, leading to superficial skin infection, affecting the outermost layer of epidermis *i.e.* stratum corneum. However, in immunocompromised patients, deeper invasion into dermis and even severe systemic infection with distant organ involvement can occur. Most cases of deeper dermal dermatophytosis described in the literature so far involved pre-existing superficial dermatophytosis. But now cases of dermatophyte abscesses caused by *T.rubrum* are being reported without pre-existing superficial dermatophytosis. A lack of experience with dermatophytosis could make clinicians underestimate the significance of positive dermatophyte fungal cultures obtained from deep soft tissue. Even without any superficial dermatophytosis lesion, fungi should be considered as a possible cause of deep soft tissue abscesses in immunocompromised patients and a culture for fungi as well as bacteria should be performed in these patients.

These dermatophyte fungi typically attack the skin, hair and nails because they require keratin for growth. In immunocompetent hosts, they cannot penetrate deeper than the stratum granulosum and their colonies are not usually observed in the layer over the epidermis. However, deeper dermis and subcutaneous dermatophyte infections can occur in patients with compromised immune systems, caused by solid organ transplantation, hematological malignancy, immunosuppressive therapy or congenital immunodeficiency. In these cases, the fungi can enter the bloodstream and disseminate to distant major organs, including the lymph nodes, liver, brain and bone. This often causes systemic infections, which can be fatal.

(n) 'Id' Reaction

Dermatophytid or **'id' reaction** is secondary eruption occurring in sensitized tinea patients because of circulation of allergenic products from primary site of infection. This is frequently found in patients with absence of delayed reaction to dermatophytic antigen *i.e.* **trichophytin**. An 'id' reaction may also develop following commencement of oral antifungal therapy and can be confused with an allergic drug reaction. This reaction may be result of fungal products being absorbed from skin and itching is the only symptom. The onset is sometimes accompanied by fever, anorexia, generalized lymphadenopathy, splenomegaly and leukocytosis. The morphological features and site of lesions are variable. Two main types of 'id' reactions are well established:

(a) Lichen scrofulosorum-like
(b) Pompholyx-like

(a) Lichen scrofulosorum-like: This is commonly associated with tinea capitis in children. Small grouped or diffusely scattered follicular lesions are found on the body. They are symmetrical and central in distribution but may extend to involve limbs and face. Horny spines are sometimes observed on top of involved follicles. The common cause of this type of reaction is kerion. This condition resembles lichen scrofulosorum hence is named, accordingly.

(b) Pompholyx-like: The lesions are frequently found on sides and flexor aspects of fingers and palms in patients with tinea pedis. Generally they are papular but may be vesicular, pompholyx-like, bullous or rarely pustular. They appear on sides of fingers and wrist or grouped on all over body. This is considered to be Type-III hypersensitivity reaction. The dermatophytid reaction responds to desensitization and resolves spontaneously with treatment of primary disease. The condition is sometimes aggravated during griseofulvin therapy or trichophytin injection.

Rarely the eruptions may be morbilliform, scarlatiniform or erysipelas-like. Erysipelas-like lesions are most commonly seen on the shin, appear elevated, sharply defined plaque like with toe-web tinea on the same side. This form of 'id' reaction responds to systemic steroids and treatment of the tinea.

Dermatophytids should be differentiated from pompholyx or other vesicular eruptions of hands and photosensitivity drug eruptions. Contact dermatitis may be difficult to exclude as cause of lesions. If cultured, lesions are found to be sterile. The 'id' reaction can also be produced by *Candida* species and known as candid reaction.

Chronic dermatophytosis syndrome is also called *Trichophyton rubrum* syndrome and is a rare condition caused by *T.rubrum*. It consists of skin lesions at four sites *i.e.* feet, usually soles; hands often palms; nails and a fourth site other than these, except for the groin. The KOH preparation should be positive from all the four sites however *T.rubrum* should be isolated in culture from at least three locations.

(o) Non-dermatophytic Dermatomycosis

This is infection of skin and nails caused by various fungi other than dermatophytes. These may be *Candida* species, *Neoscytalidium dimidiatum, Fusarium oxysporum, Onychocola canadensis, Trichosporon* species and many more. *Lasiodiplodia* species may cause onychomycosis. The details of these fungi are given in their respective **Chapters** along with cutaneous manifestations, produced by them. Hence the detailed description of these fungi are omitted from here just to avoid repetition.

Wood's Lamp Examination

The Wood's lamp was invented by Baltimore physicist, Robert W. Wood (1868-1955) in 1903. This was first used in dermatology practice for the detection of fungal infection of hair in 1925. Since then this device is useful in diagnosis and management of superficial cutaneous fungal infections. It is an important advancement in Medical Mycology, especially of scalp ringworm infections.

Wood's glass/filter consists of barium silicate with 9% nickel oxide. This Wood's filter is opaque to all light except for a band between 320 and 400 nm with a peak at 365 nm. Fluorescence of tissue occurs when light of shorter wavelengths, in this case 340-400 nm, initially emitted by Wood's light, is absorbed and radiation of longer wavelengths, usually visible light, is emitted. Hence it transmits long-wave ultraviolet light with peak of 365 nm that shows characteristic fluorescence produced by some microbial agents. This florescence has been used to demonstrate hair infection mainly by *Microsporum* species and one of *Trichophyton* species **(Fig. 10.12)**.

Surveys of large populations can be done rapidly and easily with the help of this lamp as infected hair show greenish fluorescence. It was an important tool in the management of individual patient consequently for control of epidemics. However, now it is becoming relatively less useful because of increase in prevalence of non-fluorescing fungal species like *T.tonsurans, T.verrucosum* as the causative agents of tinea capitis.

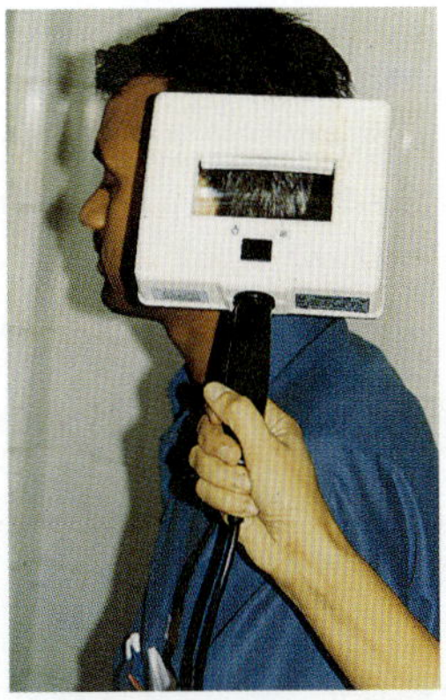

Fig. 10.12. Wood's Lamp Examination in a patient of tinea capitis.

The Wood's Lamp shows golden yellow fluorescence in pityriasis versicolor and *Malassezia* folliculitis to differentiate from other types of folliculitis. It shows bright green fluorescence in toe web lesions and wounds particularly burn wounds, infected by *Pseudomonas* and pink to coral-red in erythrasma. The patterns of fluorescence of different species of superficial and cutaneous mycotic and bacterial agents under Wood's Lamp are as follows:

- Bright green: *M.audouinii, M.canis* and *M.ferrugineum*
- Dull green: *Trichophyton schoenleinii*
- No fluorescence: All other dermatophytes
- Golden yellow: Pityriasis versicolor (*M.furfur*)
- Coral-red: Erythrasma (*Corynebacterium minutissimum*)
- Pale-yellow: Trichomycosis (trichobacteriosis) axillaris (*Corynebacterium tenuis*)

The chemical substance responsible for positive fluorescence is pteridine. Therefore, other materials containing this substance may also give false-positive fluorescence. The commonest source of error is bluish or purplish fluorescence produced by ointments containing petroleum jelly, scales, serum exudate, lint and dried soap. In a dark room the skin under Wood's light fluoresces faintly blue and dandruff commonly is bright blue-white. An inadequately darkened room, light reflected from observer's white coat and failure to remember that all fungi do not show fluorescence may also be responsible for wrong diagnosis. The characteristic fluorescence of various microorganisms under Wood's Lamp is summarized in **Table 10.5**.

Table 10.5. Fluorescence seen under Wood's Lamp.

Microorganisms	*Fluorescence Color*
Microsporum audouinii	*Bright-green*
Microsporum canis	*Bright-green*
Microsporum ferrugineum	*Blue-green*
Microsporum distortum	*Blue-green*
Microsporum gypseum	*Dull-yellow*
Trichophyton schoenleinii	*Dull-green*
Malassezia furfur	*Golden-yellow*
Corynebacterium minutissimum	Coral-red

Differential Diagnosis

The differential diagnosis of various clinical conditions caused by dermatophytes should be distinguished from other superficial cutaneous fungal infections, which have already been discussed with individual clinical types. However, the partially treated cases, particularly with steroids, pose a real challenge as the typical findings of dermatophytosis are changed. Sometimes bacterial infections may also be confused with tinea corporis like erythrasma **(Fig. 10.13)**.

Laboratory Diagnosis

The diagnosis of dermatophytosis is based on combination of clinical observations supplemented by laboratory investigations. The history of patient is essential regarding age, occupation, hobbies, living conditions with onset, duration and course of disease as well as intake of previous treatment. The clinical examination is done in well-lit room, may be with magnifying lens. The physician should observe distribution and type of lesions, concurrent disease and constitutional symptoms of patient.

In the laboratory, diagnosis depends on demonstration of causative pathogen in skin scrapings by microscopy, isolation of fungus in culture and serological tests. The scrapings should be taken from active margin of cutaneous lesion. If small vesicles or blisters are present, epithelium forming roof is an ideal source of clinical material. In case of onychomycosis, confocal microscopy is also useful to locate as well as to know the extent of lesions. When scrapings are to be sent through post, they should be folded in thick black paper.

(a) Direct Examination

The microscopic examination of KOH wet mounts of keratinous material is very simple and reliable. The clinical materials like skin scales, nail clippings or curettings from dystrophic subungual debris and hair stubs are usually

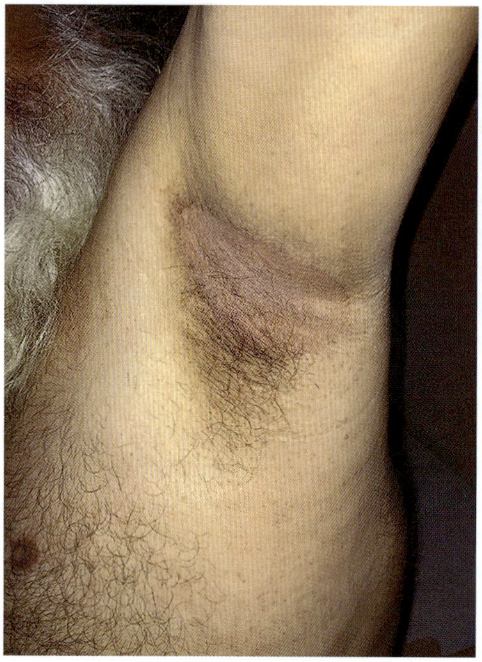

Fig. 10.13. Erythrasma of left axillary area in an adult patient.

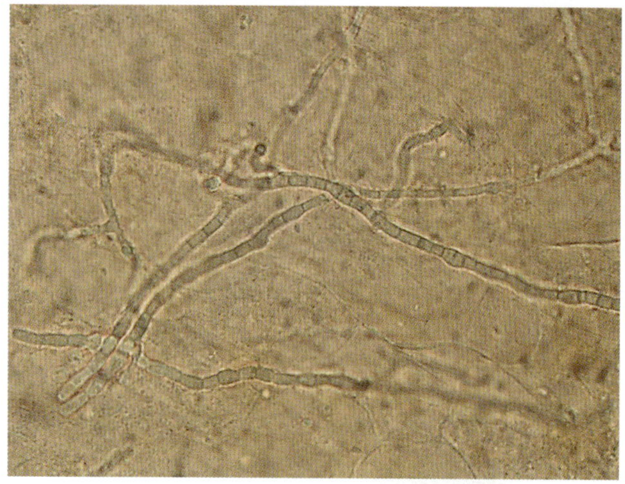

Fig. 10.14. Wet mount showing septate hyphae (KOH × 400).

examined. Disinfect affected site with alcohol before collecting clinical specimen preferably from advancing edge of lesions. Scrape from center to edge crossing margins. Vesicles and bullae are also clipped and examined thoroughly. In case of hair involvement, take basal root portion of hair by plucking and not by clipping, scrape scales and excavate hair for direct examination as well as culture. From nails, clippings are taken from an appropriate site depending on the type of nail infection.

In the KOH wet mount **(Fig. 10.14)** of collected material, fungus is seen as branching septate hyaline mycelia, which frequently show arthrospores production **(Figs. 10.15A and B)** particularly in the infected skin appendages like hair. Ectothrix and endothrix types of infections are distinguished in wet mount and arrangement of arthrospores can be made out **(*See* Figs. 10.3A to C)**. The demonstration of fungus in nails may be difficult and may be possible only after keeping clippings overnight in higher concentration of KOH. Moreover, nail clippings stained with PAS or GMS stain is more rewarding as compared to the KOH wet mount. Calcofluor fluorescent stain is often more useful in visualizing fungal hyphae or arthrospores **(Fig. 10.15C)**. The histopathological examination (HPE) of the skin biopsy may be stained with PAS to delineate the hyphae **(Fig. 10.16)**. For all types of clinical specimens, fungal hyphae must be differentiated from other artifacts **(*See* Appendix B)**.

(b) Fungal Culture

All three genera of dermatophytes are recognized namely, *Trichophyton*, *Microsporum* and *Epidermophyton*, based on morphology of macroconidia and microconidia **(Figs. 10.17A to E and Table 10.6)**. The macroconidia have typical shapes in different species and are important for their identification. In addition to dermatophytes, non-dermatophytic dermatomycotic agents are also demonstrated. The dermatophytes are conspicuously variable in their morphology and many isolates even fail to produce these characteristic spores. The unicellular microconidia and multicellular macroconidia, their shape and arrangement spore bearing hyphae and other associated features such as **spiral** hypha, **racquet** hypha, **nodular organ**, **pectinate body** and **favic chandelier** are also helpful in identification of dermatophyte species and are described in details in **Chapter 2** on morphology of the fungi.

The clinical specimens should be inoculated on fungal culture media irrespective of the findings of direct examination. Dermatophytes can grow easily on Sabouraud dextrose agar with antibiotics and cycloheximide and media are incubated at three temperatures *i.e.* 25°C, 30°C and 37°C. The growth is relatively slow and usually takes about 10 days to three weeks' time. The dermatophyte isolate can usually be distinguished from contaminant by compact growth around the inoculum and by the color of

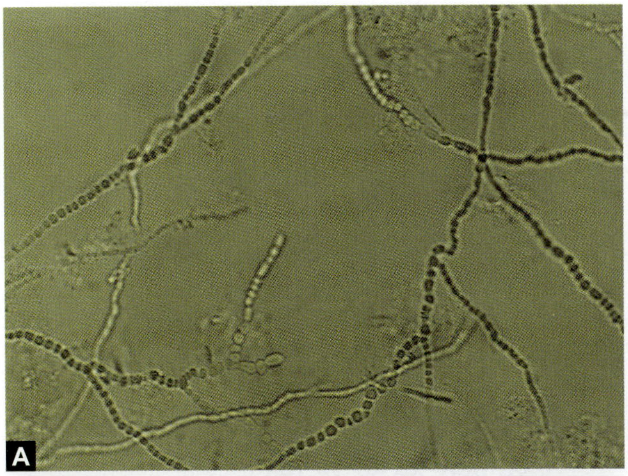

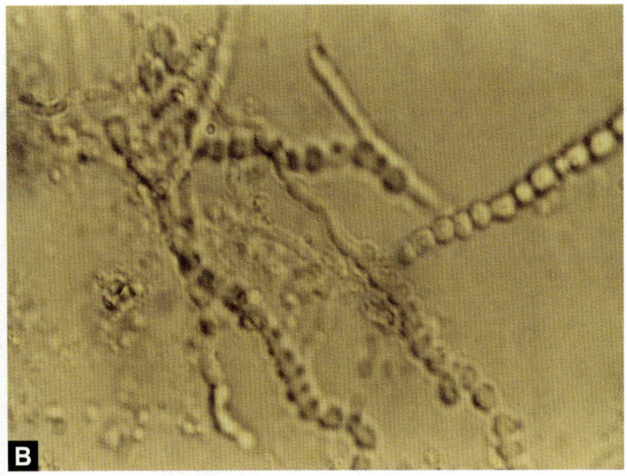

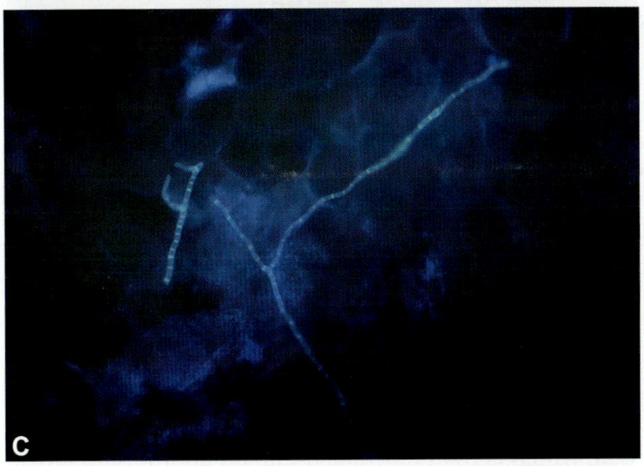

Figs. 10.15A to C. Septate hyaline hyphae and arthrospores of dermatophytes in wet mounts (KOH × 400/1000, CFW × 400).

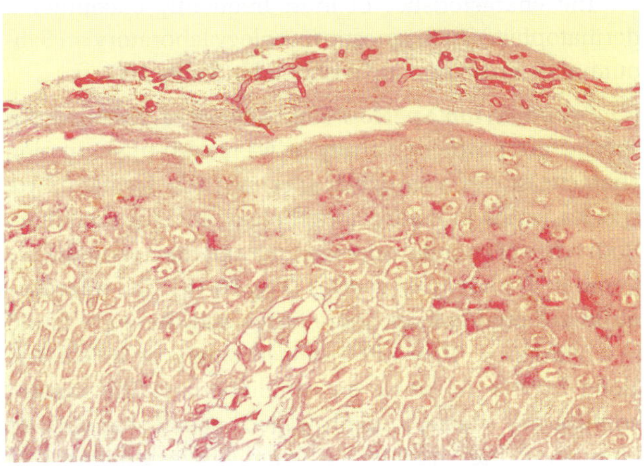

Fig. 10.16. HPE shows septate hyphae of dermatophytes involving cutaneous tissue (PAS × 400).

Table 10.6. Distribution of Conidia of Dermatophytes.

Dermatophytes	*Macroconidia*	*Microconidia*
Trichophyton	Rare, thin-walled, smooth	Abundant*
Microsporum	Numerous, thick-walled, rough**	Rare
Epidermophyton	Numerous smooth-walled	Absent

**T.schoenleinii* and ***M.audouinii* are exceptions

colony because the dermatophytes are never green, blue or black. It is said that dermatophytes have various pleasant colors but never dark colored. The rice grain medium may be used to differentiate *M.audouinii* from *M.canis*.

The **Dermatophyte Test Medium** (DTM) is used for presumptive identification of dermatophytes from fungal or bacterial contaminants found prevalent in cutaneous lesions. On incubation at 25°C, the dermatophytes turn medium red due to change in color of indicator phenol red by increased pH through their metabolic activity while

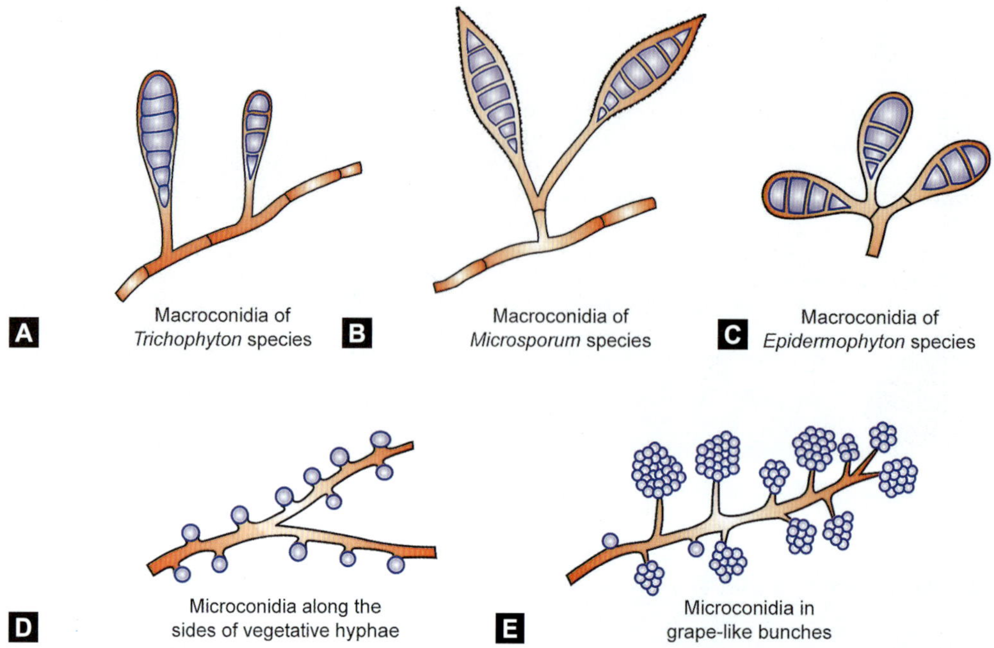

Figs. 10.17A to E. Various types of macroconidia and microconidia formation observed in dermatophytes.

Fig. 10.18. Diagrammatic representation of hair showing wedge-shaped perforation due to invasion by dermatophytes.

other fungi and bacteria do not. Similarly, dermatophyte identification medium (DIM) is also used for presumptive identification of dermatophytes and to avoid false-positive results associated with use of DTM (***See*** **Appendix A**). The other media like SDA with thiamine (1 mg%) can be used for sporulation and SDA with yeast extract can be used as enhancement medium for dermatophytes. If *T.verrucosum* is suspected, two tubes of media with thiamine are inoculated and incubated at 37°C.

Hair Perforation Test is performed to differentiate between *T.mentagrophytes* and *T.rubrum* as well as *M.canis* and *M.equinum*, respectively. The test is taken as positive when dermatophyte species show wedge-shaped perforations in hair **(Fig. 10.18)**. It is positive in *T.mentagrophytes* and *M.canis* and negative in *T.rubrum* and *M.equinum* (***See*** **Appendix B**).

Urease Test is done on Christensen's medium. *T. mentagrophytes* strain, hydrolyse urea thereby medium turns deep red while *T.rubrum* shows negative results. Urea broth may also be used which is more sensitive.

In case of **Hairbrush Sampling Technique**, adequate material from minimal lesions may be obtained by brushing scalp with sterile plastic hairbrush or massage pad, which is then inoculated into an appropriate fungal culture medium by pressing brush or pad spines into agar.

The characteristics of most frequently encountered dermatophytes in diagnostic mycology laboratory on Sabouraud dextrose agar are described in **Table 10.7**.

***(i) Trichophyton* Species:** The colonies of *Trichophyton* species are powdery, velvety or waxy with pigmentation characteristic of a particular species. Each dermatophyte species varies in colony morphology and pigmentation. Both types of conidia, microconidia and macroconidia are found in *Trichophyton* species **(Fig. 10.19)**. The **microconidia** are predominant spore forms and these are found in abundance. They are arranged singly or in clusters along hyphae or borne on conidiophores. The **macroconidia** are sparse, smooth-walled, pencil-shaped with blunt ends. They are 8-86 × 4-14 μm in size and have 1-12 septa. The conidia formation may also vary according to species under observation. The medium, on which the dermatophytes grow, greatly influences the characteristics of conidiation. Sometimes, the use of various nutritional substrates is essential to differentiate particular species.

Table 10.7. Morphological Characteristics of Commonly Encountered Dermatophytes.

Dermatophytes	Colony Morphology	Microscopic Morphology
T.rubrum	Velvety, red pigment on reverse	Tear-drop microconidia, few, long, pencil-shaped macroconidia
T.mentagrophytes	White to tan, cottony or powdery, pigment variable	Clusters of microconidia cigar-shaped macroconidia with terminal rat-tail filaments
T.tonsurans	Powdery to cream or yellow with central furrows	Abundant microconidia, thick-walled, irregular macroconidia
T.schoenleinii	Smooth, waxy, brownish	Hyphal swellings, chlamydospores, favic chandelier
T.violaceum	Very slow growing, waxy, violet to purple pigment	Distorted hyphae, conidia rare
M.audouinii	Velvety, brownish, slow growing	Thick-walled chlamydospores, conidia rare and irregular
M.canis	Cottony, orange pigment on reverse	Abundant, thick-walled spindle-shaped macroconidia with up to 15 septa
M.gypseum	Powdery, buff-colored	Abundant, thin-walled macroconidia with 4-6 septa
E.floccosum	Yellowish-green, powdery	Club-shaped macroconidia in clusters

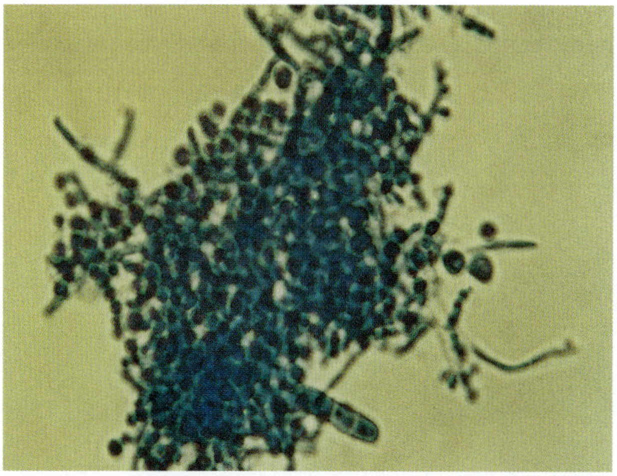

Fig. 10.19. *Trichophyton* species showing microconidia and macroconidia (LCB × 400).

The colonies of *T.mentagrophytes* range from granular to powdery and usually exhibit abundant **grape-like** clusters of subspherical microconidia on terminal branches. The granular form of *T.mentagrophytes* is world-wide in distribution. Sometimes, *T.mentagrophytes* may be confused with *T.simii*, which is morphologically very similar. Some cottony strains rarely develop tear-drop shaped microconidia along the sides of hyphae. The **spiral hyphae** are frequently seen in *T.mentagrophytes* **(Fig. 10.20)**. *T.rubrum* is the commonest species infecting man with its typical red pigment on the reverse **(Fig. 10.21)** and usually it also has clavate **teardrop-shaped** microconidia along sides of hyphae and in some of strains, these may be abundant **(Figs. 10.22A to C)**. The colonies of *T.rubrum* often develop port-wine colored pigmentation on the reverse. The African variety of *T.rubrum* shows chlamydospores within hyphae. The colonies of *T.tonsurans* are usually powdery and larger microconidia are usually numerous and clavate and may be borne on short branches **(Figs. 10.23A and B)**. The colonies of *T.tonsurans* sometimes visible as volcano crater like appearance. In *T.schoenleinii* microconidia are not found in abundance. At 37°C, *T.verrucosum* grows as chains of chlamydospores **(Fig. 10.24)**.

The nutritional tests for differentiation of *Trichophyton* species are done on casein agar base and its modification with inositol, thiamine and nicotinic acid and on ammonium nitrate agar base with and without histidine **(Fig. 10.25)**.

***(ii) Microsporum* Species:** The colonies are cottony, velvety or powdery, with white to brown pigmentation. The macroconidia are predominant conidial form in *Microsporum* species. They are large, rough-walled, multicellular, spindle-shaped and are formed at the ends of hyphae. A macroconidium is 6-160 × 6-25 µm in size with thin to thick walls 1-15 septae. The microconidia are scanty hence are usually not taken into consideration for differentiating species. *M.canis* forms numerous thick-walled, 8-15-celled macroconidia that frequently have curved or hooked spiny tips. A yellow orange pigment usually develops on reverse side of the colony.

In *M.gypseum,* colonies grow rapidly producing flat, spreading, powdery surface that is cinnamon-buff to brown, which consists of mass of macroconidia, which

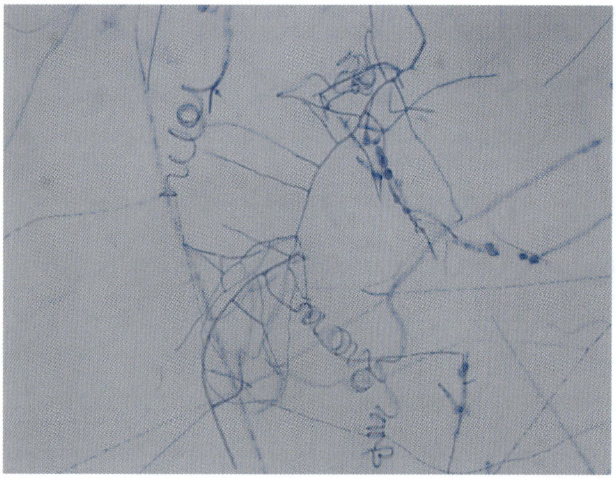

Fig. 10.20. Spiral hyphae of dermatophytes as seen in *Trichophyton mentagrophytes* (LCB × 200).

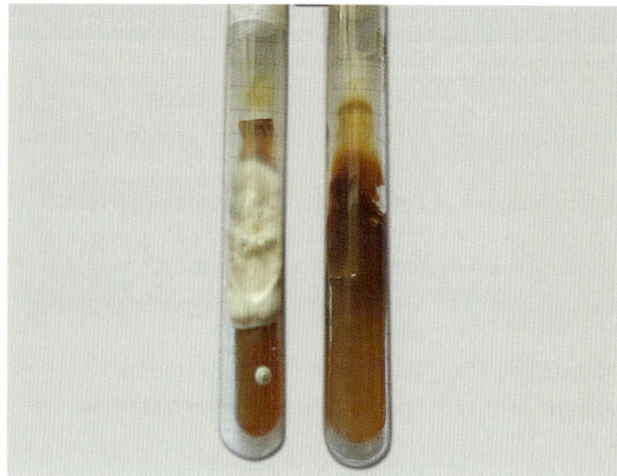

Fig. 10.21. White mycelial colonies of *Trichophyton rubrum* with red pigment on the reverse side.

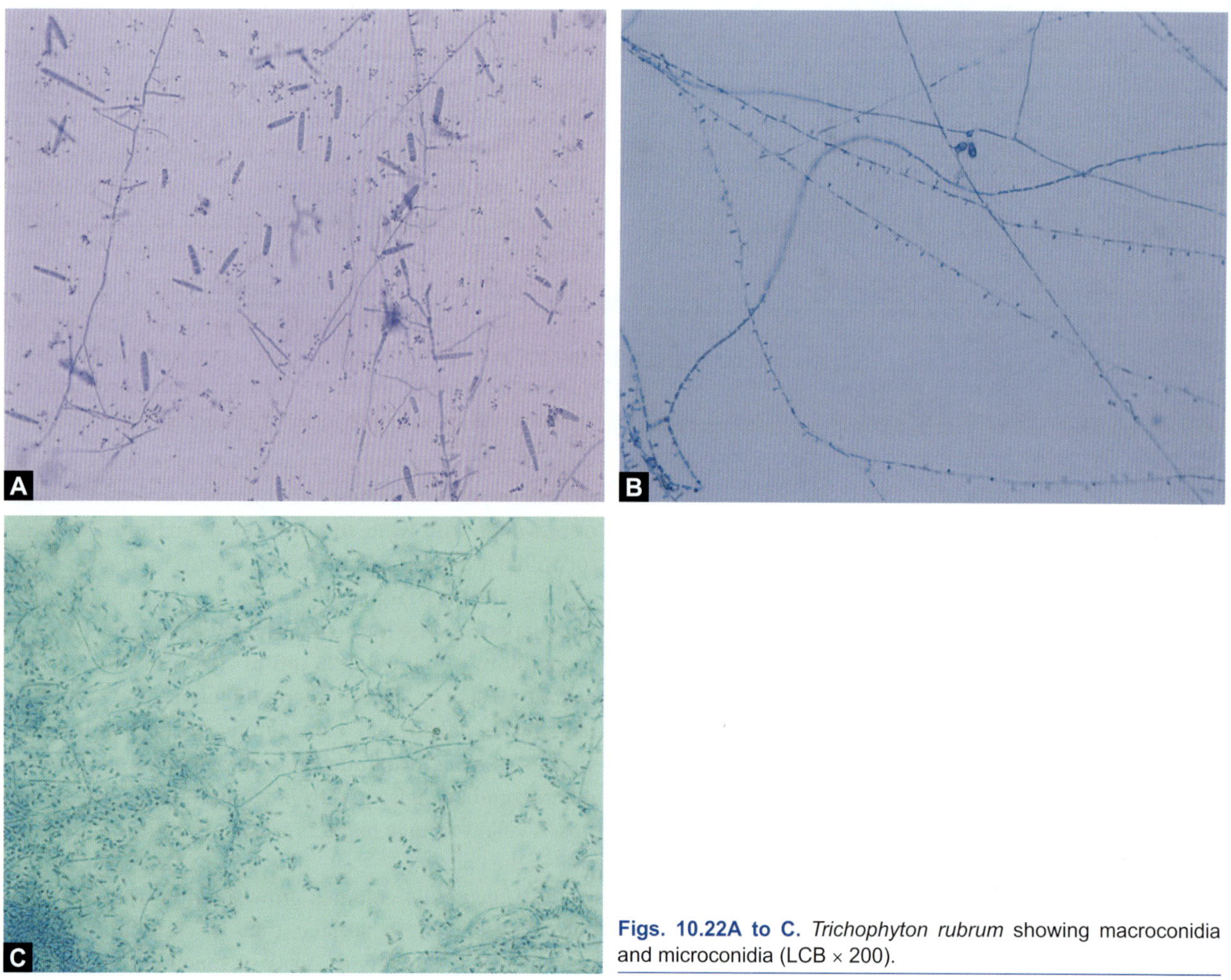

Figs. 10.22A to C. *Trichophyton rubrum* showing macroconidia and microconidia (LCB × 200).

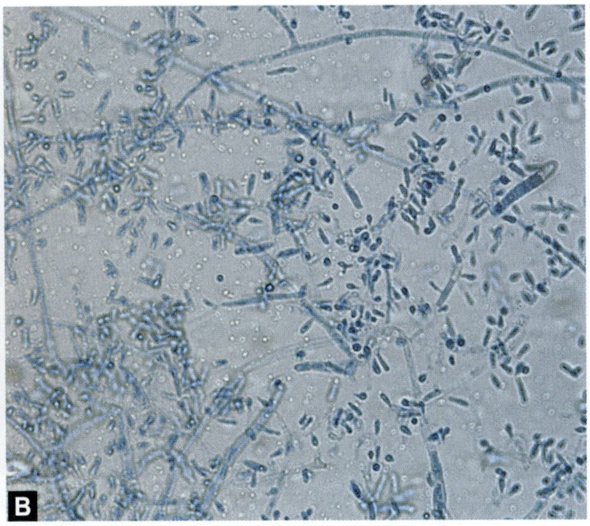

Figs. 10.23A and B. White mycelial colonies and microscopy of *T.tonsurans* showing microconidia and macroconidia (LCB × 200).

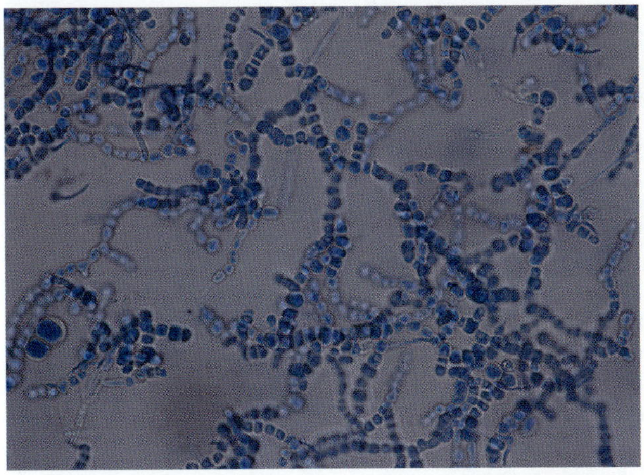

Fig. 10.24. *T.verrucosum* showing arthrospores (LCB × 200).

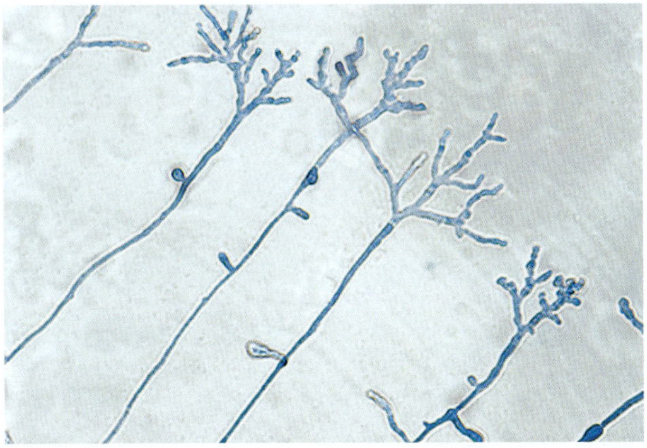

Fig. 10.25. *Trichophyton schoenleinii* showing favic chandelier (LCB × 200).

are produced in abundance. They are boat-shaped, rough-walled, 20-60 × 8-10 μm and have 4-6 septa **(Figs. 10.26A and B)**. Microconidia are very less in number. *M.audouinii* rarely forms conidia in colony but many thick-walled chlamydoconidia may be present. This fungus grows poorly on sterile rice grains whereas other *Microsporum* species show rapid growth **(Figs. 10.27A and B)**.

***(iii) Epidermophyton* Species:** The genus of *Epidermophyton* has only one species *i.e. E.floccosum*, which is the type species of this genus and **no teleomorphic state** has been observed. The relation of *E.floccosum* to other dermatophyte is of taxonomic interest and is identified by morphological features and biochemical analysis. This genus **does not** produce **microconidia** as compared to *Trichophyton* and *Microsporum,* which usually produce microconidia as well as macroconidia.

The colonies of *E.floccosum* are slow growing, powdery, greenish-brown ***(khaki-colored)*** with suede-like surface, often raised and folded in centre **(Fig. 10.28A)**. They mutate readily to form sterile white overgrowth and hyphae often undergo transformation into large chlamydoconidia. One to nine-celled, smooth thin walled, pear or club-shaped macroconidia with size of 20-60 × 4-16 μm are formed in abundance and are arranged in clusters **(Fig. 10.28B)**.

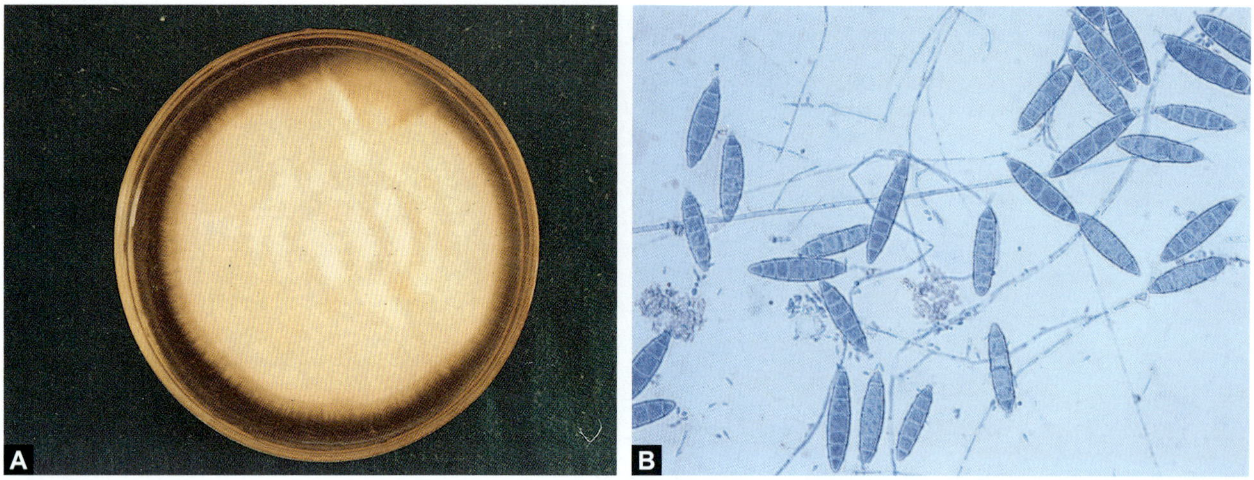

Figs. 10.26A and B. Colonial morphology and microscopic appearance of *Microsporum gypseum* showing macroconidia and microconidia (LCB × 400).

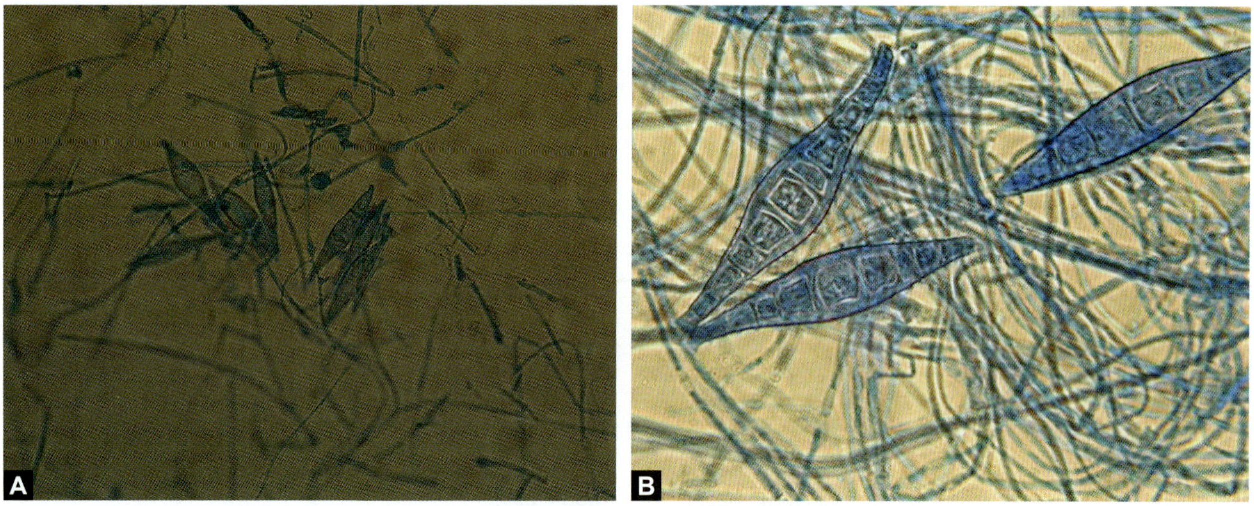

Figs. 10.27A and B. Spindle-shaped macroconidia of *Microsporum canis* (LCB × 200/400).

The other species of *Epidermophyton* like *E.inguinale* and *E.plicarum,* were eventually found to be synonyms of *E.floccosum* and merged with it. However, *E.stockdaleae,* remained quite some time in the literature due to its production of longer conidia with nine septations. But this is now considered as synonym of *Trichophyton ajelloi* and **not** separate species of genus *Epidermophyton.*

(iv)* Non-dermatophytic Fungal Species:** A large number of non-dermatophytic fungi are involved in skin and nail infections and some of them are listed in **Table 10.4**. These are described in subsequent **Chapters** wherein these are referred as the main cause of a particular disease along with onychomycosis but *Neoscytalidium* species, an emerging clinically significant agent is being described here. *Scytalidium* genus had three important species namely *Scytalidium lignicola, S.dimidiatum* and *S.hyalinum.** Although *S.lignicola* is type species but it is not a human pathogen except rare case reports. Moreover, the name genus *Scytalidium* is now changed to *Neoscytalidium,* which has the commonest infective species of this group. As their names indicate, *N.dimidiatum* is phaeoid and *N.hyalinum* is hyaline to an extent. Rather *N.hyalinum* is now considered to be melanin-less hyaline mutant of *N.dimidiatum.*

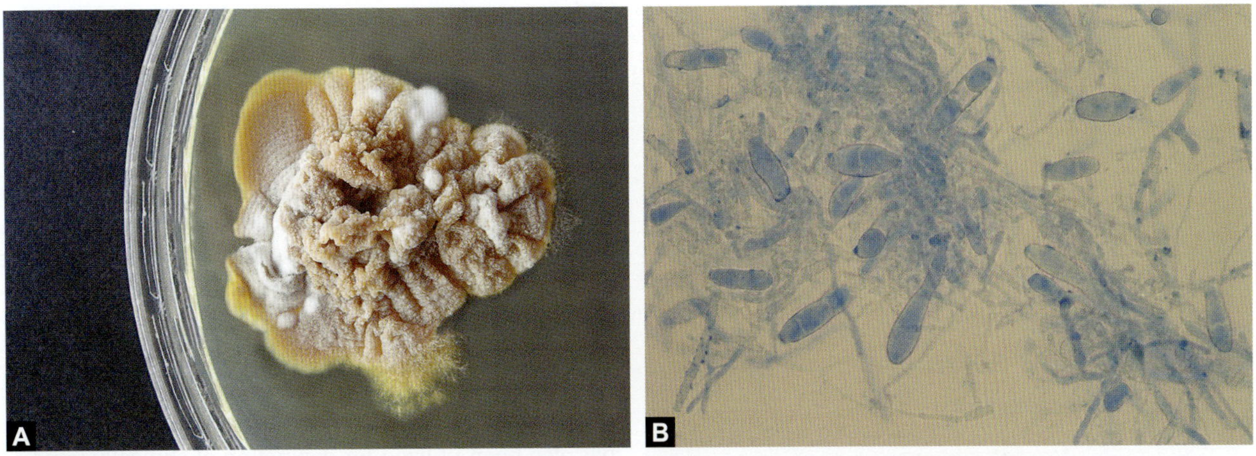

Figs. 10.28A and B. Khaki-colored mycelial growth and clubbed-shaped macroconidia of *Epidermophyton floccosum* (LCB × 400).

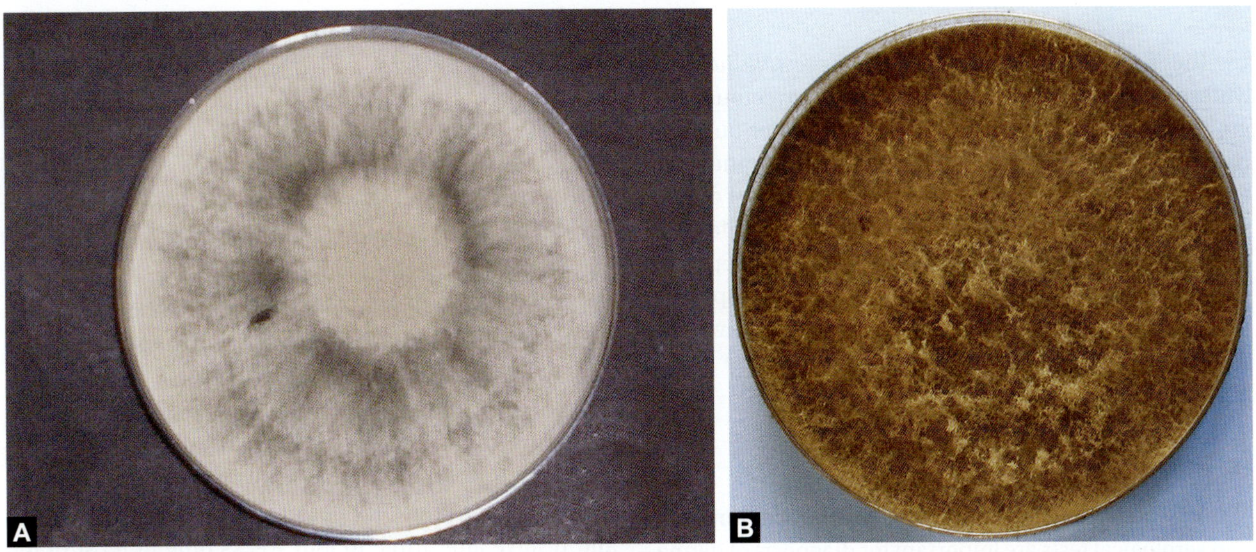

Figs. 10.29A and B. Greyish white mycelial colonies and brownish mycelial colonies of *Neoscytalidium dimidiatum* on SDA.

Neoscytalidium dimidiatum was previously known as *Hendersonula toruloidea* and it is synanamorph of *Nattrassia mangiferae*, a plant pathogen, which has pycnidial form of organism. It has two types of colony morphology and **actidione** is inhibitory to this genus. One is fast growing with dark-grey, floccose, aerial mycelia filling petri dish within 2-4 days and turns to brownish-black within a week time **(Figs. 10.29A and B)**. Microscopically, it shows wide-brown hyphae with long chains of fragmented arthrospores **(Fig. 10.30)**. The other form is slow growing olive-grey and within a week or so it reaches diameter of about two centimeters. Microscopically, it shows scanty brown arthrospores.

The colonies of *Nattrassia mangiferae* on Sabouraud dextrose agar, **without actidione**, are effuse, hairy and dark grey to blackish-brown on oatmeal agar. Microscopically these are seen as chains of arthrospores from undifferentiated broad, 4-7 µm, brown hyphae, often with slimy exudate.

Neoscytalidium hyalinum shows expanding angry-rat colonies, which are woolly, white to greyish on meat-extract agar and often deep ochraceus yellow pigment is produced. Microscopically hyaline hyphae, soon disarticulate into separate cells. The arthrospores are hyaline, cylindrical, 5-12 × 2.5-3.5 µm, locally swollen to subspherical, 4-6 µm wide and then becoming subdivided.

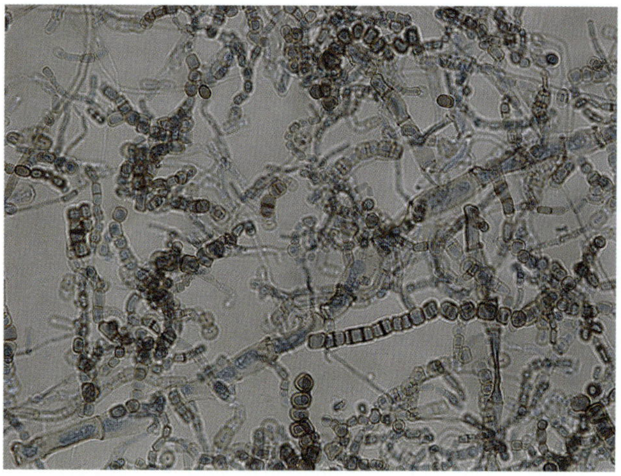

Fig. 10.30. Broad hyphae and chains of arthrospores seen in *Neoscytalidium dimidiatum* (LCB × 400).

The fungi belonging to genus *Neoscytalidium* are widespread around the world. These ascomycetous fungi are endemic in tropical or subtropical countries but these have species have different geographical distribution. The infections caused represent approximately 40% of dermatomycoses in these areas. A few cases of invasive infections due to *Neoscytalidium* species have also been reported, establishing ability of such fungi to behave as opportunists. Even if *Neoscytalidium* cases are rare in temperate countries, imported cases may increase in future due to frequent immigration and international travel.

(c) Immunodiagnosis

The skin tests with dermatophytic antigen, **trichophytin** and serological tests are important for diagnosis of dermatophytosis. The trichophytin is crude extract from dermatophytes, producing positive delayed-type hypersensitivity, tuberculin-like response in most of the adults. A galactomannan peptide is reactive component of the antigen. The carbohydrate portion is related to an immediate response whereas peptide moiety is associated with immunity. The patients without delayed-type or with an immediate-type reaction are more susceptible to chronic dermatophytosis. Various serological tests like immunodiffusion are done to establish diagnosis of dermatophytosis.

(d) Molecular Tools

The molecular methods offer the advantages of being rapid, reproducible and detection directly from the infected tissue also overcoming the observer's bias. The PCR is a highly sensitive technique being used as nested or multiplex PCR both in conventional and real-time formats. The most recent technique of MALDI-TOF is also being used in reference laboratories for diagnosis of dermatomycosis. Molecular analyses have recently been applied to dermatophytes in order to assess their relatedness and classify taxonomic boundaries. The PCR-finger printing for identification of species and varieties of common dermatophytes and related fungi utilizing as single primer simple repetitive oligonucleotide $(GACA)_4$ that has been previously used for *Cryptococcus* and *Candida* species. Chitin synthase 1 (CHS1) gene analysis of dermatophytes has also been utilized for distinction of clinically important species.

(e) Animal Pathogenicity

This modality is useful for studying nature of lesions and immunity produced by the organism. The mouse, guinea pig and rat are the commonly used animal models in dermatophytosis. The study areas are mainly focused on pathogenesis, host response or efficacy of antifungal therapy. These are not used for routine identification of dermatophytes but for experimental studies only.

Out of these commonly the animal pathogenicity testing is done on guinea pigs. *Microsporum canis, M.gypseum* and *T.mentagrophytes* may be established more readily in laboratory animals as compared to other species. The hair of area to be infected (usually dorsal part) are shaved and skin is scarified before applying the conidial and hyphal suspensions. The lesions develop within a week and resolve after three to four weeks time in most of the cases. Recently Cambier *et al* have reviewed this issue of models in dermatophytosis by emphasizing continuing need for the experimental animal studies. They have narrated approaches adopted so far and a possible solution to establishing a workable model in nude mice as well. In case of an experimental guinea pig model, immunogenicities of various vaccine have been studied.

Treatment and Prophylaxis

This is pertinent to establish the exact diagnosis of the infection before starting treatment for dermatomycosis so that specific therapeutic modalities can be adopted and monitored during its course. Various topical preparations like Whitfield's ointment, tolnaftate or azole derivatives, are used for all types of tinea. Topical management of onychomycosis with ciclopirox and amorolfine nail lacquers have modest effectivity.

Oral terbinafine and itraconazole are the coomonly used drugs for nail and scalp infections. Griseofulvin is no longer the drug of choice. However, its micronized and ultramicronized preparations are quite effective. It is given 500 mg daily in two divided doses. A single dose therapy of 2 gm griseofulvin may be given in tinea capitis. The chronic infections require prolonged oral and topical therapy for three to six months. Resistance to griseofulvin is being reported, particularly in *Trichophyton* species. In addition to increasing hygienic standards of population at large, probably widespread use of griseofulvin might have controlled these infections.

The reasonable regimens for treatment of tinea corporis in adults include:

- Terbinafine 250 mg per day for one to two weeks
- Itraconazole 200 mg per day for one week
- Fluconazole 150 to 200 mg once weekly for two to four weeks
- Griseofulvin microsize 500 to 1000 mg per day or griseofulvin ultramicrosize 375 to 500 mg per day for two to four weeks

Onychomycosis presents diagnostic dilemma to clinician, frequently leading to empirical treatment of disease with recently developed broad-spectrum antifungal agents. The clinician can now have trust that dermatophytic as well as non-dermatophytic pathogens involving nails can be eradicated effectively if timely and proper diagnosis is made.

The introduction of two antifungals *i.e.* itraconazole and terbinafine, has represented major advances in therapy for onychomycosis. Both drugs are highly effective and have in large parts replaced use of griseofulvin in treating onychomycosis, mainly because of relatively low efficacy of griseofulvin. Itraconazole, fluconazole and terbinafine have been used extensively for treatment of onychomycosis and safety profile for all these drugs appears excellent. The course of therapy may be 2–4 weeks in skin, 6–12 weeks in hair and 6-12 months in nail involvement. Itraconazole and terbinafine may be given as **pulse therapy** in patients of onychomycosis for better outcome.

The ***'clinical cure'*** is defined as disappearance of all lesions on each nail or residual disease of no more than 10% of original disease surface and ***'mycological cure'*** is defined as negative culture as well as negative microscopy. For nail treatment lasers have also been tried.

Non-dermatophytic molds are difficult to eradicate. Itraconazole is reported to be effective in nail infections caused by *Aspergillus, Fusarium* and *Scopulariopsis brevicaulis* with mycological and clinical cure rate upto about 88 percent. The resistance to infection may be acquired after primary infection. It varies in duration and degree depending on host, site and species of fungus causing infection. The drugs are ideally given on the basis of *in vitro* antifungal susceptibility testing of the fungal isolate. Prior to 1990s, no clinical resistance was reported to antifungals but now there is gradual increase in the number of recurrent cases and relapses. Resistance to terbinafine was reported in *T.rubrum* in the year 2003, which has now increased enormously.

The dermatophytosis and onychomycosis are presenting as refractory diseases in recent times, which is very frustrating for patients as well as the clinicians. It has also been observed that fresh crops of lesions start appearing even the patient is already on antifungal treatment. In such circumstances, it is ideal to start with the specific treatment based on the MIC of AFST for leading the clinician to a logical conclusion. The partially treated patients or those who take steroid combinations are bound to turn up as refractory cases. The latest topical preparations of azoles like sertaconazole or luliconazole may also be tried for the treatment of such dermatophytosis patients.

There are efforts made to induce protective immunity to infections caused by dermatophytes. In case of an experimental guinea pig model, immunogenicities of live spore vaccine, killed hyphal cell wall vaccine and soluble cytoplasmic extract vaccine of *T.mentagrophytes* var. *erinacei* have been studied. Of these three vaccines, live spore vaccine is the most effective and closely simulated type of immunity that develops following natural infection. The research directed towards management of dermatophyte infections by vaccination has yielded promising vaccine for prevention of *T.verrucosum* infections among the cattle.

Further Reading

1. Abd Elmegeed AS, Ouf SA, Moussa TA, *et al.* Dermatophytes and other associated fungi in patients attending to some hospitals in Egypt. Braz J Microbiol. 2015; **46:** 799-805.
2. Abdel-Rahman SM, Sugita T, Gonzalez GM, *et al.* Divergence among an international population of *Trichophyton tonsurans* isolates. *Mycopathologia.* 2010; **169:** 1-13.
3. Abraham PS. Treatment of ringworm by x-rays. 1910 [Classical Article]. Practitioner. 2010; **254 (1732):** 32.

4. Adams C, Athanasoula E, Lee W, *et al.* Environmental and genetic factors on the development of onychomycosis. *J Fungi.* 2015: **1:** 211-6.
5. Adamski Z, Kowalczyk MJ, Adamska K, *et al.* The first non-African case of *Trichophyton rubrum* var. *raubitschekii* or a urease-positive *Trichophyton rubrum* in Central Europe? *Mycopathologia.* 2014; **178:** 91-6.
6. Adhikari L, Das Gupta A, Pal R, *et al.* Clinico-etiologic correlates of onychomycosis in Sikkim. *Indian J Pathol Microbiol.* 2009; **52:** 194-7.
7. Afshar P, Vahedi L, Ghasemi M, *et al.* Epidemiology of tinea capitis in northeast Iran: A retrospective analysis from 1998 to 2012. *Int J Dermatol.* 2016; **55:** 640-4.
8. Agarwal U, Sitaraman S, Panse GG, *et al.* Useful sign to diagnose tinea capitis- 'ear sign'. *Indian J Pediatr.* 2012; **79:** 679-80.
9. Agarwal US, Saran J, Agarwal P. Clinico-mycological study of dermatophytes in a tertiary care centre in northwest India. *Indian J Dermatol Venereol Leprol.* 2014; **80:** 194.
10. Ahmad SM, Wani GM, Khursheed B. Kerion mimicking bacterial infection in an elderly patient. *Indian Dermatol Online J.* 2014; **5:** 494-6.
11. Ahmed R, Kharal SA, Durrani MA, *et al.* Frequency of *Candida* in onychomycosis. *J Pak Med Assoc.* 2013; **63:** 350-3.
12. Alejandro ML, Carlos RC. Tinea faciei: An old friend revisited. *J Pediatr Nurs.* 2014; **29:** 195-6.
13. Allevato MA. Diseases mimicking onychomycosis. *Clin Dermatol.* 2010; **28:** 164-77.
14. Almeida SR. Immunology of dermatophytosis. *Mycopathologia.* 2008; **166:** 277-83.
15. Ameen M, Lear JT, Madan V, *et al.* British Association of Dermatologists' guidelines for the management of onychomycosis 2014. *Br J Dermatol.* 2014; **171:** 937-58.
16. Ameen M. Epidemiology of superficial fungal infections. *Clin Dermatol.* 2010; **28:** 197-201.
17. Amir I, Foering KP, Lee JB. Revisiting office-based direct microscopy for the diagnosis of onychomycosis. *J Am Acad Dermatol.* 2015; **72:** 909-10.
18. Anane S, Chtourou O. Tinea capitis favosa misdiagnosed as tinea amiantacea. *Med Mycol Case Rep.* 2012; **2:** 29-31.
19. Ansari S, Hedayati MT, Zomorodian K, *et al.* Molecular characterization and *in vitro* antifungal susceptibility of 316 clinical isolates of dermatophytes in Iran. *Mycopathologia.* 2016; **181:** 89-95.
20. Arenas R, Moreno-Coutino G, Vera L, *et al.* Tinea incognito. *Clin Dermatol.* 2010; **28:** 137-9.
21. Aridogan IA, Izol V, Ilkit M. Superficial fungal infections of the male genitalia: A review. *Crit Rev Microbiol.* 2011; **37:** 237-44.
22. Asz-Sigall D, Tosti A, Arenas R. Tinea unguium: Diagnosis and treatment in practice. *Mycopathologia.* 2017; **182:** 95-100.
23. Atzori L, Pau M, Aste N, *et al.* Dermatophyte infections mimicking other skin diseases: A 154-person case survey of tinea atypica in the district of Cagliari (Italy). *Int J Dermatol.* 2012; **51:** 410-5.
24. Atzori L, Pau M, Aste N. Tinea atypica. *G Ital Dermatol Venereol.* 2013; **148:** 593-601.
25. Ayanlowo O, Oladele RO. Onychomycosis: Updates and management challenges. A review. *Niger Postgrad Med J.* 2014; **21:** 185-91.
26. Azambuja CV, Pimmel LA, Klafke GB, *et al.* Onychomycosis: Clinical, mycological and *in vitro* susceptibility testing of isolates of *Trichophyton rubrum. An Bras Dermatol.* 2014; **89:** 581-6.
27. Badali H, Mohammadi R, Mashedi O, *et al. In vitro* susceptibility patterns of clinically important *Trichophyton* and *Epidermophyton* species against nine antifungal drugs. *Mycoses.* 2015; **58:** 303-7.
28. Bakardzhiev I, Chokoeva A, Tchernev G, *et al.* Tinea profunda of the genital area. Successful treatment of a rare skin disease. *Dermatol Ther.* 2016; **29:** 181-3.
29. Baltazar LM, Santos DA. Perspective on animal models of dermatophytosis caused by *Trichophyton rubrum. Virulence.* 2015; **6:** 372-5.
30. Banerjee M, Ghosh AK, Basak S, *et al.* Comparative evaluation of effectivity and safety of topical amorolfine and clotrimazole in the treatment of tinea corporis. *Indian J Dermatol.* 2011; **56:** 657-62.
31. Barber K, Barber J. Onychomycosis: therapy directed by morphology and mycology. *Skin Therapy Lett.* 2009; **14:** 1-2.
32. Barua P, Barua S, Borkakoty B, *et al.* Onychomycosis by *Scytalidium dimidiatum* in green tea leaf pluckers: Report of two cases. *Mycopathologia.* 2007; **164:** 193-5.
33. Bassiri-Jahromi S, Sadeghi G, Paskiaee FA. Evaluation of the association of superficial dermatophytosis and athletic activities with special reference to its prevention and control. *Int J Dermatol.* 2010; **49:** 1159-64.
34. Beguin H, Goens K, Hendrickx M, *et al.* Is *Trichophyton simii* endemic to the Indian subcontinent? *Med Mycol.* 2013; **51:** 444-8.
35. Bell-Syer SE, Khan SM, Torgerson DJ. Oral treatments for fungal infections of the skin of the foot. *Cochrane Database Syst Rev.* 2012; **10:** CD003584.
36. Bennett D, Rubin AI. Dermatophytoma: A clinicopathologic entity important for dermatologists and dermatopathologists to identify. *Int J Dermatol.* 2013; **52:** 1285-7.
37. Bernhardt M. Onychomycosis. *Skinmed.* 2015; **13:** 240.
38. Berry A, Abramovici G, Chamlin SL. A 21-day-old boy with an annular eruption. Tinea faciei/Tinea capitis. *Pediatr Ann.* 2014; **43:** e16-8.
39. Bertanha L, Chiacchio ND. Nail clipping in onychomycosis. *An Bras Dermatol.* 2016; **91:** 688-90.
40. Bhagra S, Ganju SA, Kanga A, *et al.* Mycological pattern of dermatophytosis in and around Shimla hills. *Indian J Dermatol.* 2014; **59:** 268-70.
41. Bhagra S, Ganju SA, Sood A, *et al. Microsporum gypseum* dermatophytosis in a patient of acquired immunodeficiency syndrome: A rare case report. *Indian J Med Microbiol.* 2013; **31:** 295-8.
42. Bhatia VK, Sharma PC. Determination of minimum inhibitory concentrations of itraconazole, terbinafine and ketoconazole against dermatophyte species by broth microdilution method. *Indian J Med Microbiol.* 2015; **33:** 533-7.
43. Bhatia VK, Sharma PC. Epidemiological studies on dermatophytosis in human patients in Himachal Pradesh, India. *Springerplus.* 2014; **3:** 134.
44. Bhat YJ, Zeerak S, Kanth F, *et al.* Clinicoepidemiological and mycological study of tinea capitis in the pediatric

population of Kashmir Valley: A study from a tertiary care centre. *Indian Dermatol Online J.* 2017; **8:** 100-3.
45. Bhatta AK, Huang X, Keyal U, *et al.* Laser treatment for onychomycosis: A review. *Mycoses.* 2014; **57:** 734-40.
46. Blutfield M, Lohre J, Pawich D, *et al.* The immunologic response to *Trichophyton rubrum* in lower extremity fungal infections. *J Fungi.* 2015; **1:** 130-7.
47. Boaventura P, Soares P, Pereira D, *et al.* Head and neck lesions in a cohort irradiated in childhood for tinea capitis treatment. *Lancet Infect Dis.* 2011; **11:** 163-4 & 343.
48. Bond R. Superficial veterinary mycoses. *Clin Dermatol.* 2010; **28:** 226-36.
49. Bonifaz A, Rios-Yuil JM, Arenas R, *et al.* Comparison of direct microscopy, culture and calcofluor white for the diagnosis of onychomycosis. *Rev Iberoam Micol.* 2013; **30:** 109-11.
50. Bonifaz A, Rojas R, Tirado-Sanchez A, *et al.* Superficial mycoses associated with diaper dermatitis. *Mycopathologia.*2016; **181:** 671-9.
51. Bonifaz A, Vazquez-Gonzalez D. Tinea imbricata in the Americas. *Curr Opin Infect Dis.* 2011; **24:** 106-11.
52. Bouchara JP, Mignon B, Chaturvedi V. Dermatophytes and dermatophytoses: A thematic overview of state of the art and the directions for future research and developments. *Mycopathologia.* 2017; **182:** 1-4.
53. Brasch J, Muller S, Graser Y. Unusual strains of *Microsporum audouinii* causing tinea in Europe. *Mycoses.* 2015; **58:** 573-7.
54. Brasch J, Shimanovich I. Persistent fingernail onychomycosis caused by *Fusarium proliferatum* in a healthy woman. *Mycoses.* 2012; **55:** 86-9.
55. Brasch J, Varga J, Jensen JM, *et al.* Nail infection by *Aspergillus ochraceopetaliformis. Med Mycol.* 2009; **47:** 658-62.
56. Brasch J. Current knowledge of host response in human tinea. *Mycoses.* 2009; **52:** 304-12.
57. Brillowska-Dabrowska A, Swierkowska A, Lindhardt Saunte DM, *et al.* Diagnostic PCR tests for *Microsporum audouinii, M. canis* and *Trichophyton* infections. *Med Mycol.* 2010; **48:** 486-90.
58. Brissos J, Gouveia C, Neves C, *et al.* Remember kerion celsi. *BMJ Case Rep.* 2013; pii: bcr2013200594. PMID: 24005974.
59. Bunyaratavej S, Prasertworonun N, Leeyaphan C, *et al.* Distinct characteristics of *Scytalidium dimidiatum* and non-dermatophyte onychomycosis as compared with dermatophyte onychomycosis. *J Dermatol.* 2015; **42:** 258-62.
60. Burns C, Valentine J. Tinea imbricata. *N Engl J Med.* 2016; **375:** 2272.
61. Cafarchia C, Iatta R, Latrofa MS, *et al.* Molecular epidemiology, phylogeny and evolution of dermatophytes. *Infect Genet Evol.* 2013; **20:** 336-51.
62. Calista D, Ghetti E, Morri M. Squamous cell carcinomas of the scalp induced by ionizing radiation for tinea capitis. *G Ital Dermatol Venereol.* 2016; **151:** 574-6.
63. Cambier L, Heinen MP, Mignon B. Relevant animal models in dermatophyte research. *Mycopathologia.* 2017; **182:** 229-40.
64. Capoor MR, Agarwal S, Yadav S, *et al. Trichosporon mucoides* causing onychomycosis in an immunocompetent patient. *Int J Dermatol.* 2015; **54:** 704-7.
65. Chacon A, Franca K, Fernandez A, *et al.* Psychosocial impact of onychomycosis: A review. *Int J Dermatol.* 2013; **52:** 1300-7.
66. Chadeganipour M, Mohammadi R. Causative agents of onychomycosis: A 7-year study. *J Clin Lab Anal.* 2016; **30:** 1013-20.
67. Chen X, Jiang X, Yang M, *et al.* Systemic antifungal therapy for tinea capitis in children: An abridged Cochrane Review. *J Am Acad Dermatol.* 2017; **76:** 368-74.
68. Chen X, Jiang X, Yang M, *et al.* Systemic antifungal therapy for tinea capitis in children. *Cochrane Database Syst Rev.* 2016; **5:** CD004685. PMID: 27169520.
69. Cheng N, Rucker Wright D, *et al.* Dermatophytid in tinea capitis: Rarely reported common phenomenon with clinical implications. *Pediatrics.* 2011; **128:** e453-7.
70. Chiriac A, Brzezinski P, Chiriac AE, *et al.* Why is tinea an annular lesion with centrifugal growth? *Int J Surg Pathol.* 2015; **23:** 652-3.
71. Cho E, Lee YB, Park HJ, *et al.* Fungal melanonychia due to *Candida albicans. Int J Dermatol.* 2013; **52:** 1598-600.
72. Chollet A, Cattin V, Fratti M, *et al.* Which fungus originally was *Trichophyton mentagrophytes?* Historical review and illustration by a clinical case. *Mycopathologia.* 2015; **180:** 1-5.
73. Choudhary SV, Koley S, Mallick S, *et al.* Proximal subungual onychomycosis caused by *Aspergillus flavus* in a HIV-positive patient. *Indian J Dermatol Venereol Leprol.* 2009; **75:** 410-2.
74. Cinotti E, Fouilloux B, Perrot JL, *et al.* Confocal microscopy for healthy and pathological nail. *J Eur Acad Dermatol Venereol.* 2014; **28:** 853-8.
75. Coleman NW, Fleckman P, Huang JI. Fungal nail infections. *J Hand Surg Am.* 2014; **39:** 985-8.
76. Concha M, Nicklas C, Balcells E, *et al.* The first case of tinea faciei caused by *Trichophyton mentagrophytes* var. *erinacei* isolated in Chile. *Int J Dermatol.* 2012; **51:** 283-5.
77. Costa JE, Neves RP, Delgado MM, *et al.* Dermatophytosis in patients with human immunodeficiency virus infection: Clinical aspects and etiologic agents. *Acta Trop.* 2015; **150:** 111-5.
78. da Silva BC, Paula CR, Auler ME, *et al.* Dermatophytosis and immunovirological status of HIV-infected and AIDS patients from Sao Paulo city, Brazil. *Mycoses.* 2014; **57:** 371-6.
79. Dan P, Rawi R, Hanna S, *et al.* Invasive cutaneous *Trichophyton shoenleinii* infection in an immunosuppressed patient. *Int J Dermatol.* 2011; **50:** 1266-9.
80. Daniel RC. Onychomycosis: Burden of disease and the role of topical antifungal treatment. *J Drugs Dermatol.* 2013; **12:** 1263-6.
81. Das D, Das A, Das NK. Kerion. *Indian Pediatr.* 2014; **51:** 419-20.
82. Das NK, Ghosh P, Das S, *et al.* A study on the etiological and clinic-mycological correlation of fingernail onychomycosis in eastern India. *Indian J Dermatol Venerol Leprol.* 2008; **53:** 75-9.

83. de Almeida H Jr, Gotze F, Heckler G, *et al.* Trichomycosis capitis: First report of this localization and ultrastructural aspects. *Eur J Dermatol.* 2011; **21:** 823-4.
84. de Berker D. Fungal nail disease. *N Engl J Med.* 2009; **360:** 2108-16.
85. de Crignis G, Valgas N, Rezende P, *et al.* Dermatoscopy of onychomycosis. *Int J Dermatol.* 2014; **53:** e97-9.
86. de Hoog GS, Dukik K, Monod M, *et al.* Toward a novel multilocus phylogenetic taxonomy for the dermatophytes. *Mycopathologia.* 2017; **182:** 5-31.
87. De Respinis S, Monnin V, Girard V, *et al.* Matrix-assisted laser desorption ionization-time of flight (MALDI-TOF) mass spectrometry using the Vitek MS system for rapid and accurate identification of dermatophytes on solid cultures. *J Clin Microbiol.* 2014; **52:** 4286-92.
88. de Sa DC, Lamas AP, Tosti A. Oral therapy for onychomycosis: An evidence-based review. *Am J Clin Dermatol.* 2014; **15:** 17-36.
89. de Sousa Mda G, Santana GB, Criado PR, *et al.* Chronic widespread dermatophytosis due to *Trichophyton rubrum*: A syndrome associated with a *Trichophyton*-specific functional defect of phagocytes. *Front Microbiol.* 2015; **6:** 801.
90. Del Boz J, Padilla-Espana L, Crespo-Erchiga V. Sample taking and direct examination in dermatomycoses. *Actas Dermosifiliogr.* 2016; **107:** 65-67.
91. Deng S, de Hoog GS, Verweij PE, *et al.* *In vitro* antifungal susceptibility of *Trichophyton violaceum* isolated from tinea capitis patients. *J Antimicrob Chemother.* 2015; **70:** 1072-5.
92. Deng S, Hu H, Abliz P, *et al.* A random comparative study of terbinafine versus griseofulvin in patients with tinea capitis in Western China. *Mycopathologia.* 2011; **172:** 365-72.
93. Dev T, Saginatham H, Sethuraman G. Tinea faciei: Challenges in the diagnosis. *J Pediatr.* 2017. pii: S0022-3476(17)30456-0. PMID: 28433202.
94. Di Chiacchio N, Noriega LF, Gioia Di Chiacchio N, *et al.* Superficial black onychomycosis due to *Neoscytalidium dimidiatum. J Eur Acad Dermatol Venereol.* 2017. doi: 10.1111/jdv.14273. PMID: 28403528.
95. Dias MF, Bernardes-Filho F, Quaresma-Santos MV, *et al.* Treatment of superficial mycoses: Review. Part II. *An Bras Dermatol.* 2013; **88:** 937-44.
96. Dias MF, Quaresma-Santos MV, Bernardes-Filho F, *et al.* Update on therapy for superficial mycoses: Review article. Part I. *An Bras Dermatol.* 2013; **88:** 764-74.
97. Dogra S, Uprety S. The menace of chronic and recurrent dermatophytosis in India: Is the problem deeper than we perceive? *Indian Dermatol Online J.* 2016; **7:** 73-6.
98. Dogra S, Yadav S. What's new in nail disorders? *Indian J Dermatol Venereol Leprol.* 2011; **77:** 631-9.
99. Drakensjo IT, Chryssanthou E. Epidemiology of dermatophyte infections in Stockholm, Sweden: A retrospective study from 2005-2009. *Med Mycol.* 2011; **49:** 484-8.
100. Drira I, Neji S, Hadrich I, *et al.* Tinea manuum due to *Trichophyton erinacei* from Tunisia. *J Mycol Med.* 2015; **25:** 200-3.
101. Durdu M, Guran M, Kandemir H, *et al.* Clinical and laboratory features of six cases of *Candida* and dermatophyte folliculitis and a review of published studies. *Mycopathologia.* 2016; **181:** 97-105.
102. Durdu M, Ilkit M, Tamadon Y, *et al.* Topical and systemic antifungals in dermatology practice. *Expert Rev Clin Pharmacol.* 2017; **10:** 225-37.
103. Dutta B, Rasul ES, Boro B. Clinico-epidemiological study of tinea incognito with microbiological correlation. *Indian J Dermatol Venereol Leprol.* 2017; **83:** 326-31.
104. Eisman S, Sinclair R. Fungal nail infection: Diagnosis and management. *BMJ.* 2014; **348:** g1800. PMID: 24661991.
105. El-Gohary M, van Zuuren EJ, Fedorowicz Z, *et al.* Topical antifungal treatments for tinea cruris and tinea corporis. *Cochrane Database Syst Rev.* 2014; **8:** CD009992.
106. Eldridge ML, Chambers CJ, Sharon VR, *et al.* Fungal infections of the skin and nail: New treatment options. *Expert Rev Anti Infect Ther.* 2014; **12:** 1389-405.
107. Elewski B, Pariser D, Rich P, *et al.* Current and emerging options in the treatment of onychomycosis. *Semin Cutan Med Surg.* 2013; **32:** S9-12.
108. Elewski BE, Rich P, Tosti A, *et al.* Onychomycosis: An overview. *J Drugs Dermatol.* 2013; **12:** s96-103.
109. Elkeeb R, Hui X, Murthy N, *et al.* Emerging topical onychomycosis therapies - *Quo vadis*? *Expert Opin Emerg Drugs.* 2014; **19:** 489-95.
110. Ely JW, Rosenfeld S, Seabury Stone M. Diagnosis and management of tinea infections. *Am Fam Physician.* 2014; **90:** 702-10.
111. Errichetti E, Stinco G. Dermoscopy as a useful supportive tool for the diagnosis of pityriasis amiantacea-like tinea capitis. *Dermatol Pract Concept.*2016; **6:** 63-5.
112. Estela Cubells JR, Victoria Martinez AM, Martinez Leborans L, *et al.* Fluorescence microscopy as a diagnostic tool for dermatophytosis. *Am J Dermatopathol.* 2016; **38:** 208-10.
113. Evans JM, Wang AL, Elewski BE. Successful treatment of *Paecilomyces lilacinus* onychomycosis with efinaconazole and tavaborole. *Skin Appendage Disord.* 2016; **1:** 169-71.
114. Farwa U, Abbasi SA, Mirza IA, *et al.* Non-dermatophyte moulds as pathogens of onychomycosis. *J Coll Physicians Surg Pak.* 2011; **21:** 597-600 & 717.
115. Faway E, Cambier L, Mignon B, *et al.* Modeling dermatophytosis in reconstructed human epidermis: A new tool to study infection mechanisms and to test antifungal agents. *Med Mycol.* 2016. pii: myw111. PMID: 27760830.
116. Feih J, Ledeboer NA, Peppard WJ. A 5-year-old male with multiple pustules covering the entire scalp - Inflammatory tinea capitis. *J Clin Microbiol.* 2014; **52:** 1027 & 1312.
117. Fernandez MS, Rojas FD, Cattana ME, *et al.* *Aspergillus terreus* complex: An emergent opportunistic agent of onychomycosis. *Mycoses.* 2013; 56: 477-81.
118. Flint WW, Cain JD. Nail and skin disorders of the foot. *Med Clin North Am.* 2014; **98:** 213-25.
119. Friedman D, Friedman PC, Gill M. Reflectance confocal microscopy: An effective diagnostic tool for dermatophytic infections. *Cutis.* 2015; **95:** 93-7.

120. Fu M, Ge Y, Chen W, *et al.* Tinea faciei in a newborn due to *Trichophyton tonsurans. J Biomed Res.* 2013; **27:** 71-4.
121. Fuller LC, Barton RC, Mohd Mustapa MF, *et al.* British Association of Dermatologists' guidelines for the management of tinea capitis 2014. *Br J Dermatol.* 2014; **171:** 454-63.
122. Fuller LC. Changing face of tinea capitis in Europe. *Curr Opin Infect Dis.* 2009; **22:** 115-8.
123. Garcia C, Arenas R, Vasquez del Mercado E. Subungual black onychomycosis and melanonychia striata caused by *Aspergillus niger. Skinmed.* 2015; **13:** 154-6.
124. Garg J, Tilak R, Garg A, *et al.* Rapid detection of dermatophytes from skin and hair. *BMC Res Notes.* 2009; **2:** 60.
125. Ghadgepatil SS, Sharma YK, Misra R, *et al.* An unusual case of tinea capitis caused by *Trichophyton schoenleinii* in an elderly female. *Indian Dermatol Online J.* 2015; **6:** 49-50.
126. Ghannoum M, Isham N, Catalano V. A second look at efficacy criteria for onychomycosis: Clinical and mycological cure. *Br J Dermatol.* 2014; **170:** 182-7.
127. Ghannoum M, Isham N. Fungal nail infections (onychomycosis): A never-ending story? *PLoS Pathog.* 2014; 10:e1004105.
128. Ghannoum M. Azole resistance in dermatophytes prevalence and mechanism of action. *J Am Podiatr Med Assoc.* 2016; **106:** 79-86.
129. Gill M, Sachdeva B, Gill PS, Arora B, Deep A, Karan J. Majocchi's granuloma of the face in an immunocompetent patient. *J Dermatol.* 2007; **34:** 702-4.
130. Gits-Muselli M, Benderdouche M, Hamane S, *et al.* Continuous increase of *Trichophyton tonsurans* as a cause of tinea capitis in the urban area of Paris, France: A 5-year-long study. *Med Mycol.* 2016. pii: myw107. PMID: 27744309.
131. Gong J, Ran M, Wang X, *et al.* Development and evaluation of a novel real-time PCR for pan-dermatophyte detection in nail specimens. *Mycopathologia.* 2016; **181:** 51-7.
132. Grills N, Grills C, Spelman T, *et al.* Prevalence survey of dermatological conditions in mountainous north India. *Int J Dermatol.* 2012; **51:** 579-87.
133. Grover C, Arora P, Manchanda V. Comparative evaluation of griseofulvin, terbinafine and fluconazole in the treatment of tinea capitis. *Int J Dermatol.* 2012; **51:** 455-8.
134. Grover C, Arora P, Manchanda V. Tinea capitis in the pediatric population:A study from north India. *Indian J Dermatol Venereol Leprol.* 2010; **76:** 527-32.
135. Grover C, Khurana A. An update on treatment of onychomycosis. *Mycoses.* 2012; **55:** 541-51.
136. Grover C, Khurana A. Onychomycosis: Newer insights in pathogenesis and diagnosis. *Indian J Dermatol Venereol Leprol.* 2012; **78:** 263-70.
137. Grzybowski A, Pietrzak K. Robert Remak (1815-1865): Discoverer of the fungal character of dermatophytoses. *Clin Dermatol.* 2013; **31:** 802-5.
138. Gupta AK, Daigle D, Foley KA. Topical therapy for toenail onychomycosis: Anevidence-based review. *Am J Clin Dermatol.* 2014; **15:** 489-502.
139. Gupta AK, Drummond-Main C, Cooper EA, *et al.* Systematic review of non-dermatophyte mold onychomycosis: Diagnosis, clinical types, epidemiology and treatment. *J Am Acad Dermatol.* 2012; **66:** 494-502.
140. Gupta AK, Foley KA, Daigle D. Clinical trials of lasers for toenail onychomycosis: The implications of new regulatory guidance. *J Dermatolog Treat.* 2017; **28:** 264-70.
141. Gupta AK, Foley KA, Versteeg SG. Lasers for onychomycosis: Current status. *J Cutan Med Surg.* 2017; **21:** 114-6.
142. Gupta AK, MacLeod MA, Foley KA, *et al.* Fungal skin infections. *Pediatr Rev.* 2017; **38:** 8-22.
143. Gupta AK, Paquet M, Simpson FC. Therapies for the treatment of onychomycosis. *Clin Dermatol.* 2013; **31:** 544-54.
144. Gupta AK, Paquet M. Improved efficacy in onychomycosis therapy. *Clin Dermatol.* 2013; **31:** 555-63.
145. Gupta AK, Paquet M. Management of Onychomycosis in Canada in 2014. *J Cutan Med Surg.* 2015; **19:** 260-73.
146. Gupta AK, Paquet M. Systemic antifungals to treat onychomycosis in children: A systematic review. *Pediatr Dermatol.* 2013; **30:** 294-302.
147. Gupta AK, Simpson FC, Heller DF. The future of lasers in onychomycosis. *J Dermatolog Treat.* 2016; **27:** 167-72.
148. Gupta AK, Simpson FC. Diagnosing onychomycosis. *Clin Dermatol.* 2013; **31:** 540-3.
149. Gupta AK, Simpson FC. Laser therapy for onychomycosis. *J Cutan Med Surg.* 2013; **17:** 301-7.
150. Gupta AK, Uro M, Cooper EA. Onychomycosis therapy: Past, present, future. *J Drugs Dermatol.* 2010; **9:** 1109-13.
151. Gupta C, Das S, Ramachandran VG, *et al.* Possible role of trichophytin antigen in inducing impaired immunological clearance of fungus in onychomycosis. *Mycopathologia.* 2016; **181:** 247-51.
152. Haghani I, Shokohi T, Hajheidari Z, *et al.* Comparison of diagnostic methods in the evaluation of onychomycosis. *Mycopathologia.* 2013; **175:** 315-21.
153. Hanumanthappa H, Sarojini K, Shilpashree P, *et al.* Clinicomycological study of 150 cases of dermatophytosis in a tertiary care hospital in South India. *Indian J Dermatol.* 2012; **57:** 322-3.
154. Hawkins DM, Smidt AC. Superficial fungal infections in children. *Pediatr Clin North Am.* 2014; **61:** 443-55.
155. Hay RJ. Tinea capitis: Current status. *Mycopathologia.* 2017; **182:** 87-93.
156. Hayette MP, Sacheli R. Unusual species of dermatophytes: Rarely identified or new? *Mycopathologia.* 2017; **182:** 203-13.
157. Hayette MP, Sacheli S. Dermatophytosis, trends in epidemiology and diagnostic approach. *Curr Fungal Infect Rep.* 2015; **9:** 164-79.
158. Hemashettar BM, Nadig VS. First isolation of *Trichophyton soudanense* in India. *Indian J Pathol Microbiol.* 1980; **23:** 53-54.
159. Hiruma J, Ogawa Y, Hiruma M. *Trichophyton tonsurans* infection in Japan: Epidemiology, clinical features, diagnosis and infection control. *J Dermatol.* 2015; **42:** 245-9.
160. Hiruma M, Kano R, Sugita T, *et al.* Epidemiological aspects of *Trichophyton rubrum* var. *raubitschekii* in Japan. *J Dermatol.* 2012; **39:** 1000-1.

161. Hollander C, Visser J, de Haas E, *et al.* Effective single photodynamic treatment of *ex vivo* onychomycosis using a multifunctional porphyrin photosensitizer and green light. *J Fungi.* 2015; **1:** 138-53.
162. Hubka V, Cmokova A, Skorepova M, *et al. Trichophyton onychocola* sp. nov. isolated from human nail. *Med Mycol.* 2014; **52:** 285-92.
163. Hubka V, Dobiasova S, Dobias R, *et al. Microsporum aenigmaticum* sp. nov. from *M.gypseum* complex, isolated as a cause of tinea corporis. *Med Mycol.* 2014; **52:** 387-96.
164. Hwang SM, Suh MK, Ha GY. Onychomycosis due to non-dermatophytic molds. *Ann Dermatol.* 2012; **24:** 175-80.
165. Idriss MH, Khalil A, Elston D. The diagnostic value of fungal fluorescence in onychomycosis. *J Cutan Pathol.* 2013; **40:** 385-90.
166. Ilkit M, Durdu M, Karakas M. Cutaneous id reactions: A comprehensive review of clinical manifestations, epidemiology, etiology and management. *Crit Rev Microbiol.* 2012; **38:** 191-202.
167. Ilkit M, Durdu M. Tinea pedis: The etiology and global epidemiology of a common fungal infection. *Crit Rev Microbiol.* 2015; **41:** 374-88.
168. Ilkit M. Favus of the scalp: An overview and update. *Mycopathologia.* 2010; **170:** 143-54.
169. Inaoki M, Nishijima C, Miyake M, *et al.* Case of dermatophyte abscess caused by *Trichophyton rubrum:* A case report and review of the literature. *Mycoses.* 2015; **58:** 318-23.
170. Innocenti P, Pagani E, Vigl D, *et al.* Persisting *Paecilomyces lilacinus* nail infection following pregnancy. *Mycoses.* 2011; **54:** e880-2.
171. Isa-Isa R, Arenas R, Isa M. Inflammatory tinea capitis: Kerion, dermatophytic granuloma and mycetoma. *Clin Dermatol.* 2010; **28:** 133-6.
172. Ishizaki S, Sawada M, Suzaki R, *et al.* Tinea faciei by *Microsporum gypseum* mimicking allergic reaction following cosmetic tattooing of the eyebrows. *Med Mycol J.* 2012; **53:** 263-6.
173. Jain N, Sharma M. Distribution of dermatophytes and other related fungi in Jaipur city, with particular reference to soil pH. *Mycoses.* 2011; **54:** 52-8.
174. Jandial S, Sumbali G. Fusarial onychomycosis among gardeners: A report of two cases. *Indian J Dermatol Venereol Leprol.* 2012; **78:** 229.
175. Jarv H. Onychomycosis caused by *Onychocola canadensis:* The first report in Estonia and lessons to learn. *Mycoses.* 2015; **58:** 113-7.
176. Jayatilake JA, Tilakaratne WM, Panagoda GJ. Candidal onychomycosis: A mini-review. *Mycopathologia.* 2009; **168:** 165-73.
177. Jeelani S, Ahmed QM, Lanker AM, *et al.* Histopathological examination of nail clippings using PAS staining (HPE-PAS): Gold standard in diagnosis of onychomycosis. *Mycoses.* 2015; **58:** 27-32.
178. Jensen RH, Arendrup MC. Molecular diagnosis of dermatophyte infections. *Curr Opin Infect Dis.* 2012; **25:** 126-34.
179. Jerajani H, Janaki C, Kumar S, *et al.* Comparative assessment of the efficacy and safety of sertaconazole (2%) cream versus terbinafine cream (1%) versus luliconazole (1%) cream in patients with dermatophytoses: A pilot study. *Indian J Dermatol.* 2013; **58:** 34-8.
180. Jin KW, Jeon HS, Hyon JY, *et al.* A case of fungal keratitis and onychomycosis simultaneously infected by *Trichophyton* species. *BMC Ophthalmol.* 2014; **14:** 90.
181. John AM, Schwartz RA, Janniger CK. The kerion: An angry tinea capitis. *Int J Dermatol.* 2016. doi: 10.1111/ijd.13423. PMID: 27696388.
182. Joshi R. Adamson's fringe, Horatio George Adamson and Kligman's experiments and observations on tinea capitis. *Int J Trichology.* 2011; **3:** 14-9.
183. Jung MY, Shim JH, Lee JH, *et al.* Comparison of diagnostic methods for onychomycosis and proposal of a diagnostic algorithm. *Clin Exp Dermatol.* 2015; **40:** 479-84.
184. Kakourou T, Uksal U, European Society for Pediatric Dermatology. Guidelines for the management of tinea capitis in children. *Pediatr Dermatol.* 2010; **27:** 226-8.
185. Kanaan IC, Santos TB, Kac BK, *et al.* Majocchi's granuloma - case report. *An Bras Dermatol.* 2015; **90:** 251-3.
186. Kaur R, Panda PS, Sardana K, *et al.* Mycological pattern of dermatomycoses in a tertiary care hospital. *J Trop Med.* 2015; 157828.
187. Keisham C, Sarkar R, Khurana N, *et al.* Black dot tinea capitis caused by *Trichophyton rubrum* in an adult female presenting with cicatricial alopecia. *Indian J Dermatol Venereol Leprol.* 2015; **81:** 224.
188. Kershenovich R, Sherman S, Reiter O, *et al.* A unique clinicopathological manifestation of fungal infection: A case series of deep dermatophytosis in immunosuppressed patients. *Am J Clin Dermatol.* 2017. doi: 10.1007/s40257-017-0276-y. PMID: 28389891.
189. Khadka S, Sherchand JB, Pokharel DB, et al. Clinicomycological characterization of superficial mycoses from a tertiary care hospital in Nepal. *Dermatol Res Pract.* 2016; 9509705. PMID: 28003819.
190. Khanna U, Kumar Dhali T, D'Souza P, *et al.* Majocchi granuloma of the breast: A rare clinical entity. *Actas Dermosifiliogr.* 2016; **107:** 610-2.
191. Khare AK, Gupta LK, Mittal A, *et al.* Neonatal tinea corporis. *Indian J Dermatol.* 2010; **55:** 201.
192. Khosravi AR, Shokri H, Mansouri P. Immediate hypersensitivity and serum IgE antibody responses in patients with dermatophytosis. *Asian Pac J Allergy Immunol.* 2012; **30:** 40-7.
193. Khurana VK, Gupta RK, Pant L, *et al. Trichophyton rubrum* onychomycosis in an 8-week-old infant. *Indian J Dermatol Venereol Leprol.* 2011; **77:** 625.
194. Kim DM, Lee MH, Suh MK, *et al.* Onychomycosis caused by *Chaetomium globosum. Ann Dermatol.* 2013; **25:** 232-6.
195. Kim SH, Jo IH, Kang J, *et al.* Dermatophyte abscesses caused by *Trichophyton rubrum* in a patient without pre-existing superficial dermatophytosis: A case report. *BMC Infect Dis.* 2016; **16:** 298.

196. Kimura U, Yokoyama K, Hiruma M, *et al.* Tinea faciei caused by *Trichophyton mentagrophytes* (molecular type *Arthroderma benhamiae*) mimics impetigo: A case report and literature review of cases in Japan. *Med Mycol J.* 2015; **56:** E1-5.
197. Kizny Gordon A, McIver C, Kim M, *et al.* Clinical application of a molecular assay for the detection of dermatophytosis and a novel non-invasive sampling technique. *Pathology.* 2016; **48:** 720-6.
198. Klaassen KM, Dulak MG, van de Kerkhof PC, *et al.* The prevalence of onychomycosis in psoriatic patients: A systematic review. *J Eur Acad Dermatol Venereol.* 2014; **28:** 533-41.
199. Kondori N, Tehrani PA, Strombeck L, *et al.* Comparison of dermatophyte PCR kit with conventional methods for detection of dermatophytes in skin specimens. *Mycopathologia.* 2013; **176:** 237-41.
200. Krzysciak P, Al-Hatmi AM, Ahmed SA, *et al.* Rare zoonotic infection with *Microsporum persicolor* with literature review. Mycoses. 2015; **58:** 511-5.
201. Kundu D, Mandal L, Sen G. Prevalence of tinea capitis in school going children in Kolkata, West Bengal. *J Nat Sci Biol Med.* 2012; **3:** 152-5.
202. Lacarrubba F, Micali G, Tosti A. Scalp dermoscopy or trichoscopy. *Curr Probl Dermatol.* 2015; **47:** 21-32.
203. Lacarrubba F, Verzi AE, Micali G. Newly described features resulting from high-magnification dermoscopy of tinea capitis. *JAMA Dermatol.* 2015; **151:** 308-10.
204. Lai-Cheong J, McGrath J. Tinea. *N Engl J Med.* 2010; 363: e39.
205. Lakshmanan A, Ganeshkumar P, Mohan SR, *et al.* Epidemiological and clinical pattern of dermatomycoses in rural India. *Indian J Med Microbiol.* 2015; **33 (Suppl):** 134-6.
206. Lange M, Jasiel-Walikowska E, Nowicki R, *et al.* Tinea incognito due to *Trichophyton mentagrophytes. Mycoses.* 2010; **53:** 455-7.
207. Laniosz V, Wetter DA. What's new in the treatment and diagnosis of dermatophytosis? *Semin Cutan Med Surg.* 2014; **33:** 136-9.
208. Lanternier F, Pathan S, Vincent QB, *et al.* Deep dermatophytosis and inherited CARD9 deficiency. *N Engl J Med.* 2013; **369:** 1704-14.
209. Latha R, Sasikala R, Muruganandam N, *et al.* Onychomycosis due to ascomycete *Chaetomium globosum:* A case report. *Indian J Pathol Microbiol.* 2010; **53:** 566-7.
210. Latka C, Dey SS, Mahajan S, *et al.* Genome sequence of a clinical isolate of dermatophyte, *Trichophyton rubrum* from India. *FEMS Microbiol Lett.* 2015; **362:** fnv039.
211. Ledon JA, Savas J, Franca K, *et al.* Laser and light therapy for onychomycosis: A systematic review. *Lasers Med Sci.* 2014; **29:** 823-9.
212. Lee MH, Hwang SM, Suh MK, *et al.* Onychomycosis caused by *Scopulariopsis brevicaulis*: Report of two cases. *Ann Dermatol.* 2012; **24:** 209-13.
213. Lee WJ, Kim JY, Song CH, *et al.* Disruption of barrier function in dermatophytosis and pityriasis versicolor. *J Dermatol.* 2011; **38:** 1049-53.
214. Lee WJ, Kim SL, Jang YH, *et al.* Increasing prevalence of *Trichophyton rubrum* identified through an analysis of 115,846 cases over the last 37 years. *J Korean Med Sci.* 2015; **30:** 639-43.
215. Leite DP Jr, Amadio JV, Simoes Sde A, *et al.* Dermatophytosis in military in the central-west region of Brazil: Literature review. *Mycopathologia.* 2014; **177:** 65-74.
216. Li XF, Shen YN, Chen W, *et al.* A new medium for diagnosis ofdermatophyte infection. *Eur J Dermatol.* 2009; **19:** 34-7.
217. Liansheng Z, Xin J, Cheng Q, *et al.* Diagnostic applicability of confocal laser scanning microscopy in tinea corporis. *Int J Dermatol.* 2013; **52:** 1281-2.
218. Liddell L, Rosen T. Laser therapy for onychomycosis: Fact or fiction? *J Fungi.* 2015; **1:** 44-54.
219. Lillis JV, Dawson ES, Chang R, *et al.* Disseminated dermal *Trichophyton rubrum* infection - An expression of dermatophyte dimorphism? *J Cutan Pathol.* 2010; **37:** 1168-9.
220. Lim CS, Lim SL. New contrast stain for the rapid diagnosis of onychomycosis. *Arch Dermatol.* 2011; **147:** 981-2.
221. Lipner S, Scher RK. Onychomycosis: Current and future therapies. *Cutis.* 2014; **93:** 60-3.
222. Lipner SR, Scher RK. Management of onychomycosis and co-existing tinea pedis. *J Drugs Dermatol.* 2015; **14:** 492-4.
223. Luchsinger I, Bosshard PP, Kasper RS, *et al.* Tinea genitalis: A new entity of sexually transmitted infection? Case series and review of the literature. *Sex Transm Infect.* 2015; **91:** 493-6.
224. Lurati M, Baudraz-Rosselet F, Vernez M, *et al.* Efficacious treatment of non-dermatophyte mould onychomycosis with topical amphotericin B. *Dermatology.* 2011; **223:** 289-92.
225. Lyngdoh CJ, Lyngdoh WV, Choudhury B, *et al.* Clinicomycological profile of dermatophytosis in Meghalaya. *Int J Med Public Health.* 2013; **3:** 254-6.
226. Machouart M, Menir P, Helenon R, *et al.* *Scytalidium* and scytalidiosis: What's new in 2012? *J Mycol Med.* 2013; **23:** 40-6.
227. Macura AB, Krzysciak P, Skora M, *et al.* Case report: Onychomycosis dueto *Trichophyton schoenleinii. Mycoses.* 2012; **55:** e18-9.
228. Magalhaes AR, Nishikawa MM, Mondino SS, *et al.* *Trichosporon* isolation from human ungueal infections: Is there a pathogenic role? *An Bras Dermatol.* 2016; **91:** 173-9.
229. Majid I, Sheikh G, Kanth F, *et al.* Relapse after oral terbinafine therapy in dermatophytosis: A clinical and mycological study. *Indian J Dermatol.* 2016; **61:** 529-33.
230. Mantovani L, Ricci M, Ruina G, *et al.* A verrucous and ulcerated lesion of the leg: Challenge. Deep mycosis by *Trichophyton* species. *Am J Dermatopathol.* 2014; **36:** 243 & 263.
231. Mares M, Nastasa V, Apetrei IC, *et al.* Tinea corporis bullosa due to *Trichophyton schoenleinii:* Case report. *Mycopathologia.* 2012; **174:** 319-22.

232. Martínez E, Ameen M, Tejada D, *et al.* *Microsporum* spp. onychomycosis: Disease presentation, risk factors and treatment responses in an urban population. *Braz J Infect Dis.* 2014; **18:** 181-6.
233. Martinez-Herrera E, Moreno-Coutino G, Fernandez-Martinez RF, *et al.* Dermatophytoma: Description of 7 cases. *J Am Acad Dermatol.* 2012; **66:** 1014-6.
234. Martinez-Rossi NM, Peres NT, Rossi A. Pathogenesis of dermatophytosis: Sensing the host tissue. *Mycopathologia.* 2017; **182:** 215-27.
235. Martinez-Rossi NM, Persinoti GF, Peres NT, *et al.* Role of pH in the pathogenesis of dermatophytoses. *Mycoses.* 2012; **55:** 381-7.
236. Mask-Bull L, Msiii RP, Tarbox MB. America's first case of tinea pseudoimbricata. *Am J Dermat Venereol.* 2015; **4:** 15-7.
237. Mason D, Marks M. Bakua: Tinea imbricata in the Solomon islands. *Am J Trop Med Hyg.* 2015; **92:** 883.
238. Mathur M, Kedia SK, Ghimire RB. "Epizoonosis of dermatophytosis": A clinico-mycological study of dermatophytic infections in central Nepal. *Kathmandu Univ Med J.* 2012; **10:** 30-3.
239. Maulingkar SV, Pinto MJ, Rodrigues S. A clinico-mycological study of dermatophytoses in Goa, India. *Mycopathologia.* 2014; **178:** 297-301.
240. Mayer E, Izhak OB, Bergman R. Histopathological periodic acid-Schiff stains of nail clippings as a second-line diagnostic tool in onychomycosis. *Am J Dermatopathol.* 2012; **34:** 270-3.
241. McClanahan C, Wanat K. Tinea corporis in a wrestling team cheerleader. *Int J Women's Dermatol.* 2016; **2:** 143-4.
242. Mendez-Tovar LJ.Pathogenesis of dermatophytosis and tinea versicolor. *Clin Dermatol.* 2010; **28:** 185-9.
243. Meykadeh N, Waltermann K, Schaller M, *et al.* Bullous ulcerating tinea. *J Eur Acad Dermatol Venereol.* 2009; **23:** 846-7.
244. Miletta NR, Schwartz C, Sperling L. Tinea capitis mimicking dissecting cellulitis of the scalp: A histopathologic pitfall when evaluating alopecia in the post-pubertal patient. *J Cutan Pathol.* 2014; **41:** 2-4.
245. Mirmirani P, Tucker LY. Epidemiologic trends in pediatric tinea capitis: A population-based study from Kaiser Permanente Northern California. *J Am Acad Dermatol.* 2013; **69:** 916-21.
246. Miyajima Y, Satoh K, Uchida T, *et al.* Rapid real-time diagnostic PCR for *Trichophyton rubrum* and *Trichophyton mentagrophytes* in patients with tinea unguium and tinea pedis using specific fluorescent probes. *J Dermatol Sci.* 2013; **69:** 229-35.
247. Miyasato H, Yamaguchi S, Taira K, *et al.* Tinea corporis caused by *Microsporum gallinae:* First clinical case in Japan. *J Dermatol.* 2011; **38:** 473-8.
248. Mochizuki T, Anzawa K, Sakata Y, *et al.* Simple identification of *Trichophyton tonsurans* by chlamydospore-like structures produced in culture media. *J Dermatol.* 2013; **40:** 1027-32.
249. Mochizuki T, Takeda K, Anzawa K. Molecular markers useful for epidemiology of dermatophytoses. *J Dermatol.* 2015; **42:** 232-5.
250. Modan B, Baidatz D, Mart H, *et al.* Radiation-induced head and neck tumours. *Lancet.* 1974; **1 (7852):** 277-9.
251. Moraes RN, Ribeiro MC, Nogueira MC, *et al.* First report of *Tritirachium oryzae* infection of human scalp. *Mycopathologia.* 2010; **169:** 257-9.
252. Morales-Cardona CA, Valbuena-Mesa MC, Alvarado Z, *et al.* Non-dermatophyte mould onychomycosis: A clinical and epidemiological study at a dermatology referral centre in Bogota, Colombia. *Mycoses.* 2014; **57:** 284-93.
253. Moreno G, Arenas R. Other fungi causing onychomycosis. *Clin Dermatol.* 2010; **28:** 160-3.
254. Moriarty B, Hay R, Morris-Jones R. The diagnosis and management of tinea. *BMJ.* 2012; **345:** e4380.
255. Nagar R, Nayak CS, Deshpande S, *et al.* Subungual hyperkeratosis nail biopsy: A better diagnostic tool for onychomycosis. *Indian J Dermatol Venereol Leprol.* 2012; **78:** 620-4.
256. Naidu J. Growing incidence of cutaneous and ungual infections by non-dermatophyte fungi at Jabalpur (MP). *Indian J Pathol Microbiol.*1993; **36:** 113-8.
257. Narang K, Pahwa M, Ramesh V. Tinea capitis in the form of concentric rings in an HIV positive adult on antiretroviral treatment. *Indian J Dermatol.* 2012; **57:** 288-90.
258. Naseri A, Fata A, Khosravi AR. Tinea capitis due to *Microsporum vanbreuseghemii:* Report of two cases. *Mycopathologia.* 2012; **174:** 77-80.
259. Naseri A, Fata A, Najafzadeh MJ, *et al.* Surveillance of dermatophytosis in northeast of Iran (Mashhad) and review of published studies. *Mycopathologia.* 2013; **176:** 247-53.
260. Navarrete-Dechent C, Bajaj S, Marghoob AA, *et al.* Rapid diagnosis of tinea incognito using hand held reflectance confocal microscopy: A paradigm shift in dermatology? *Mycoses.* 2015; **58:** 383-6.
261. Negroni R. Historical aspects of dermatomycoses. *Clin Dermatol.* 2010; **28:** 125-32.
262. Nenoff P, Schetschorke C. Tinea faciei. *N Engl J Med.*2014; **370:** e31.
263. Neri I, Starace M, Patrizi A, *et al.* Corkscrew hair: a trichoscopy marker of tinea capitis in an adult white patient. *JAMA Dermatol.* 2013; **149:** 990-1.
264. Neupane S, Pokhrel DB, Pokhrel BM. Onychomycosis: Clinical pattern and prevailing fungi in Kathmandu. *Nepal Med Coll J.* 2011; **13:** 193-6.
265. Ngwogu AC, Ngwogu KO, Mba IEK, *et al.* Pattern of presentation of dermatomycosis in diabetic patients in Aba, South-eastern, Nigeria. *J Med Investig Pract.* 2014; **9:** 63-6.
266. Noguchi H, Jinnin M, Miyata K, *et al.* Clinical features of 80 cases of tinea faciei treated at a rural clinic in Japan. *Drug Discov Ther.* 2014; **8:** 245-8.
267. Noronha TM, Tophakhane RS, Nadiger S. Clinico-microbiological study of dermatophytosis in a tertiary-care hospital in north Karnataka. *Indian Dermatol Online J.* 2016; **7:** 264-71.

268. Nweze EI, Eke I. Dermatophytosis in northern Africa. *Mycoses.* 2016; **59:** 137-44.
269. Nweze EI. Dermatophytosis in western Africa: A review. *Pak J Biol Sci.* 2010; **13:** 649-56.
270. Nweze EI, Eke IE. Dermatophytes and dermatophytosis in the eastern and southern parts of Africa. *Med Mycol.* 2017. doi: 10.1093/mmy/myx025. PMID: 28419352.
271. Oanta A, Irimie M. Tinea on a Tattoo. *Acta Dermatovenerol Croat.* 2016; **24:** 223-4.
272. Oranje AP, de Waard-van der Spek FB. Recent developments in the management of common childhood skin infections. *J Infect.* 2015; **71:** S76-9.
273. Ortega-Springall MF, Arroyo-Escalante S, Arenas R. Onycholysis and chromonychia: A case caused by *Trichosporon inkin. Skin Appendage Disord.* 2016; **1:** 144-6.
274. Ortiz AE, Avram MM, Wanner MA. A review of lasers and light for the treatment of onychomycosis. *Lasers Surg Med.* 2014; **46:** 117-24.
275. Packeu A, Hendrickx M, Beguin H, *et al.* Identification of the *Trichophyton mentagrophytes* complex species using MALDI-TOF mass spectrometry. *Med Mycol.* 2013; **51:** 580-5.
276. Pai VV, Hanumanthayya K, Tophakhane RS, *et al.* Clinical study of tinea capitis in northern Karnataka: A three-year experience at a single institute. *Indian Dermatol Online J.* 2013; **4:** 22-6.
277. Paloni G, Valerio E, Berti I, *et al.* Tinea incognito. *J Pediatr.* 2015; **167:** 1450.e2.
278. Panda S, Verma S. The menace of dermatophytosis in India: The evidence that we need. *Indian J Dermatol Venereol Leprol.* 2017; **83:** 281-4.
279. Pandhi D, Verma P. Nail avulsion: Indications and methods (surgical nail avulsion). *Indian J Dermatol Venereol Leprol.* 2012; **78:** 299-308.
280. Papini M, Cicoletti M, Fabrizi V, *et al.* Skin and nail mycoses in patients with diabetic foot. *G Ital Dermatol Venereol.* 2013; **148:** 603-8.
281. Pariser D. The rationale for renewed attention to onychomycosis. *Semin Cutan Med Surg.* 2013; **32:** S1.
282. Park KY, Kim HK, Suh MK, *et al.* Unusual presentation of onychomycosis caused by *Exophiala (Wangiella) dermatitidis. Clin Exp Dermatol.* 2011; **36:** 418-9.
283. Patel GA, Schwartz RA. Tinea capitis: Still an unsolved problem? *Mycoses.* 2011; **54:** 183-8.
284. Patel GA, Wiederkehr M, Schwartz RA. Tinea cruris in children. *Cutis.* 2009; **84:** 133-7.
285. Patel NC, Silverman RA. Neonatal onychomadesis with candidiasis limited toaffected nails. *Pediatr Dermatol.* 2008; **25:** 641-2.
286. Pavlovic MD, Bulajic N. Great toenail onychomycosis caused by *Syncephalastrum racemosum. Dermatol Online J.* 2006; **12:** 7.
287. Perrier P, Monod M. Tinea manuum caused by *Trichophyton erinacei:* First report in Switzerland. *Int J Dermatol.* 2015; **54:** 959-60.
288. Petinataud D, Berger S, Contet-Audonneau N, *et al.* Molecular diagnosis of onychomycosis. *J Mycol Med.* 2014; **24:** 287-95.
289. Pharaon M, Gari-Toussaint M, Khemis A, *et al.* Diagnosis and treatment monitoring of toenail onychomycosis by reflectance confocal microscopy: Prospective cohort study in 58 patients. *J Am Acad Dermatol.* 2014; **71:** 56-61.
290. Pihet M, Govic YL. Reappraisal of conventional diagnosis for dermatophytes. *Mycopathologia.* 2017; **182:** 169-80.
291. Piraccini B, Alessandrini A. Onychomycosis: A review. *J Fungi.* 2015; **1:** 30-43.
292. Piraccini BM, Rech G, Tosti A. Photodynamic therapy of onychomycosis caused by *Trichophyton rubrum. J Am Acad Dermatol.* 2008; **59:** S75-6.
293. Pires CA, Cruz NF, Lobato AM, *et al.* Clinical, epidemiological and therapeutic profile of dermatophytosis. *An Bras Dermatol.* 2014; **89:** 259-64.
294. Pollak R, Siu W, Tatsumi Y, *et al.* Efinaconazole topical solution, 10%: Factors contributing to onychomycosis success. *J Fungi.* 2015; **1:** 107-14.
295. Poluri LV, Indugula JP, Kondapaneni SL. Clinicomycological study of dermatophytosis in south India. *J Lab Physicians.* 2015; **7:** 84-9.
296. Pontini P, Gorani A, Veraldi S. Onychomycosis by *Paecilomyces lilacinus. G Ital Dermatol Venereol.* 2016; **151:** 706-9.
297. Poziomczyk CS, Koche B, Becker FL, *et al.* Tinea pseudoimbricata caused by *M.gypseum* associated to crusted scabies. *An Bras Dermatol.* 2010; **85:** 558-9.
298. Prakash PY, Shrijana G, Indira B. Burden of onychomycosis in rural India: Clinical and mycological assessment of the disease severity using Naildex score. *Rev Iberoam Micol.* 2010; **27:** 152-3.
299. Proudfoot LE, Morris-Jones R. Kerion celsi. *N Engl J Med.* 2012; **366:** 1142.
300. Qadim HH, Golforoushan F, Azimi H, *et al.* Factors leading to dermatophytosis. *Ann Parasitol.* 2013; **59:** 99-102.
301. Queller J, Bhatia N. The dermatologist's approach to onychomycosis. *J Fungi.* 2015; **1:** 173-84.
302. Raghavendra KR, Yadav D, Kumar A, *et al.* The non-dermatophyte molds: Emerging as leading cause of onychomycosis in south-east Rajasthan. *Indian Dermatol Online J.* 2015; **6:** 92-7.
303. Raina D, Gupta P, Khanduri A. A first case of *Microsporum ferrugineum* causing tinea corporis in Uttarakhand. *Ann Trop Med Public Health.* 2016; **9:** 351-3.
304. Ranawaka RR, de Silva N, Ragunathan RW. Non-dermatophyte mold onychomycosis in Sri Lanka. *Dermatol Online J.* 2012; **18:** 7.
305. Ranawaka RR, de Silva N, Ragunathan RW. Onychomycosis caused by *Fusarium* sp. in Sri Lanka: Prevalence, clinical features and response to itraconazole pulse therapy in six cases. *J Dermatolog Treat.* 2008; **19:** 308-12.
306. Rao AG, Datta N. Tinea corporis due to *Trichophyton mentagrophytes* and *Trichophyton tonsurans* mimicking tinea imbricata. *Indian J Dermatol Venereol Leprol.* 2013; **79:** 554.
307. Rath S, Panda M, Sahu MC, *et al.* Bayesian analysis of two diagnostic methods for paediatric ringworm infections in a teaching hospital. *J Mycol Med.* 2015; **25:** 191-9.

308. Reddy KN, Srikanth BA, Sharan TR, *et al.* Epidemiological, clinical and cultural study of onychomycosis. *Am J Dermatol Venerol.* 2012; **1:** 35-40.
309. Rezaei-Matehkolaei A, Makimura K, Graser Y, *et al.* Dermatophytosis due to *Microsporum incurvatum:* Notification and identification of a neglected pathogenic species. *Mycopathologia.* 2016; **181:** 107-13.
310. Rich P, Elewski B, Scher RK, *et al.* Diagnosis, clinical implications and complications of onychomycosis. *Semin Cutan Med Surg.* 2013; **32:** S5-8.
311. Rippon JW. The changing epidemiology and emerging patterns of dermatophytes species. *Curr Top Med Mycol.* 1985; **1:** 208-34.
312. Rizzitelli G, Guanziroli E, Moschin A, *et al.* Onychomycosis caused by *Trichosporon mucoides. Int J Infect Dis.* 2016; **42:** 61-3.
313. Rodwell GE, Bayles CL, Towersey L, *et al.* The prevalence of dermatophyte infection in patients infected with human immunodeficiency virus. *Int J Dermatol.* 2008; **47:** 339-43.
314. Romero FA, Deziel PJ, Razonable RR. Majocchi's granuloma in solid organ transplant recipients. *Transpl Infect Dis.* 2011; **13:** 424-32.
315. Rose AE. Therapeutic update: Onychomycosis. *J Drugs Dermatol.* 2014; **13:** 1173-5.
316. Rosen T, Friedlander SF, Kircik L, *et al.* Onychomycosis: Epidemiology, diagnosis and treatment in a changing landscape. *J Drugs Dermatol.* 2015; **14:** 223-33.
317. Rosen T. Tinea and onychomycosis. *Semin Cutan Med Surg.* 2016; **35:** S110-3.
318. Rothmund G, Sattler EC, Kaestle R, *et al.* Confocal laser scanning microscopy as a new valuable tool in the diagnosis of onychomycosis - comparison of six diagnostic methods. *Mycoses.* 2013; **56:** 47-55.
319. Rotta I, Otuki MF, Sanches AC, *et al.* Efficacy of topical antifungal drugs in different dermatomycoses: A systematic review with meta-analysis. *Rev Assoc Med Bras.* 2012; **58:** 308-18.
320. Rotta I, Sanchez A, Goncalves PR, *et al.* Efficacy and safety of topical antifungals in the treatment of dermatomycosis: A systematic review. *Br J Dermatol.* 2012; **166:** 927-33.
321. Rotta I, Ziegelmann PK, Otuki MF, *et al.* Efficacy of topical antifungals in the treatment of dermatophytosis: A mixed-treatment comparison meta-analysis involving 14 treatments. *JAMA Dermatol.* 2013; **149:** 341-9.
322. Roy P, Bhatt P. *Nattrassia mangiferae:* An uncommon agent of onychomycosis. *Med J Armed Forces India.* 2015; **71:** 297-9.
323. Sageerabanoo, Malini A, Oudeacoumar P, *et al.* Onychomycosis due to *Trichosporon mucoides. Indian J Dermatol Venereol Leprol.* 2011; **77:** 76-7.
324. Sahai S, Mishra D, Tripathi P. Rare isolation of *Trichophyton soudanense* from three cases of superficial mycoses in Lucknow, India. *Indian J Pathol Microbiol.* 2010; **53:** 896-7.
325. Sahin GO, Dadaci Z, Ozer TT. Two cases of tinea ciliaris with blepharitis due to *Microsporum audouinii* and *Trichophyton verrucosum* and review of th literature. *Mycoses.* 2014; **57:** 577-80.
326. Sahoo AK, Mahajan R. Management of tinea corporis, tinea cruris, and tinea pedis: A comprehensive review. *Indian Dermatol Online J.* 2016; **7:** 77-86.
327. Sariguzel FM, Koc AN, Yagmur G, *et al.* Interdigital foot infections: *Corynebacterium minutissimum* and agents of superficial mycoses. *Braz J Microbiol.* 2014; **45:** 781-4.
328. Sarma S, Capoor MR, Deb M, *et al.* Epidemiologic and clinicomycologic profile of onychomycosis from north India. *Int J Dermatol.* 2008; **47:** 584-7.
329. Satter EK. Tinea imbricata. *Cutis.* 2009; **83:** 188-91.
330. Saunte DM, Tarazooie B, Arendrup MC, *et al.* Black yeast-like fungi in skin and nail: It probably matters. *Mycoses.* 2012; **55:** 161-7.
331. Schechtman RC. Non-dermatophytic filamentous fungi infection in South America - reality or misdiagnosis? *Dermatol Clin.* 2008; **26:** 271-83.
332. Scher RK, Rich P, Pariser D, *et al.* The epidemiology, etiology and pathophysiology of onychomycosis. *Semin Cutan Med Surg.* 2013; **32:** S2-4.
333. Seebacher C, Bouchara JP, Mignon B. Updates on the epidemiology of dermatophyte infections. *Mycopathologia.* 2008; **166:** 335-52.
334. Segal D, Wells MM, Rahalkar A, *et al.* A case of tinea incognito. *Dermatol Online J.* 2013; **19:** 18175.
335. Sehgal VN, Srivastava G, Dogra S, *et al.* Onychomycosis: An Asian perspective. *Skinmed.* 2010; **8:** 37-45.
336. Shah SR, Dalal BD, Modak MS. Non-dermatophytic onychomycosis by *Fusarium oxysporum* in an immunocompetent host. *J Mycol Med.* 2016; **26:** e18-21.
337. Sharma D, Capoor MR, Ramesh V, *et al.* A rare case of onychomycosis caused by *Emericella quadrilineata (Aspergillus tetrazonus). Indian J Med Microbiol.* 2015; **33:** 314-6.
338. Sharma M, Sharma R. Profile of dermatophytic and other fungal infections in Jaipur. *Indian J Microbiol.* 2012; **52:** 270-4.
339. Shemer A, Davidovici B, Grunwald MH, *et al.* New criteria for the laboratory diagnosis of non-dermatophyte moulds in onychomycosis. *Br J Dermatol.* 2009; **160:** 37-9.
340. Shemer A. Update: medical treatment of onychomycosis. *Dermatol Ther.* 2012; **25:** 582-93.
341. Shi D, Lu G, Mei H, *et al.* Onychomycosis due to *Chaetomium globosum* with yellowish black discoloration and periungual inflammation. *Med Mycol Case Rep.* 2016; **13:** 12-6.
342. Shimamura T, Kubota N, Shibuya K. Animal model of dermatophytosis. *J Biomed Biotechnol.* 2012; 125384.
343. Shivakumar V, Okade R, Rajkumar V, *et al.* Intermittent pulse-dosed terbinafine in the treatment of tinea corporis and/or tinea cruris. *Indian J Dermatol.* 2011; **56:** 121-2.
344. Shvarts S, Sevo G, Tasic M, *et al.* The tinea capitis campaign in Serbia in the 1950s. *Lancet Infect Dis.* 2010; **10:** 571-6.
345. Siegel-Itzkovich J. Israel compensates for ringworm treatment. *BMJ.* 1995; **310:** 350-1.
346. Sigurgeirsson B, Baran R. The prevalence of onychomycosis in the global population: A literature study. *J Eur Acad Dermatol Venereol.* 2014; **28:** 1480-91.

347. Simonnet C, Berger F, Gantier JC. Epidemiology of superficial fungal diseases in French Guiana: A three-year retrospective analysis. *Med Mycol.* 2011; **49:** 608-11.
348. Singal A, Khanna D. Onychomycosis: Diagnosis and management. *Indian J Dermatol Venereol Leprol.* 2011; **77:** 659-72.
349. Singh G, Kumar P, Joshi SC. Treatment of dermatophytosis by a new antifungal agent 'apigenin'. *Mycoses.* 2014; **57:** 497-506.
350. Skerlev M, Miklic P. The changing face of *Microsporum* spp. infections. *Clin Dermatol.* 2010; **28:** 146-50.
351. Smriti C, Anuradha S, Kamlesh T, *et al.* Tinea corporis due to *Trichophyton violaceum*: A report of two cases. *Indian J Med Microbiol.* 2015; **33:** 596-8.
352. Solis-Arias MP, Garcia-Romero MT. Onychomycosis in children. A review. *Int J Dermatol.* 2017; **56:** 123-30.
353. Sonthalia S, Singal A, Das S. Tinea cruris and tinea corporis masquerading as tinea indecisiva: Case report and review of the literature. *J Cutan Med Surg.* 2015; **19:** 171-6.
354. Steffen C. Dermatopathology in historical perspective: The man behind the eponym: Horatio George Adamson and Adamson's fringe. *Am J Dermatopathol.* 2001; **23:** 485-8.
355. Stewart CL, Rubin AI. Update: Nail unit dermatopathology. *Dermatol Ther.* 2012; **25:** 551-68.
356. Su H, Li L, Cheng B, *et al. Trichophyton rubrum* infection characterized by Majocchi's granuloma and deeper dermatophytosis: Case report and review of published literature. *Mycopathologia.* 2017; **182:** 549-54.
357. Summerbell RC, Moore MK, Starink-Willemse M, *et al.* ITS bar codes for *Trichophyton tonsurans* and *T.equinum. Med Mycol.* 2007; **45:** 193-200.
358. Sun PL, Ju YM. Onychomycosis caused by *Phaeoacremonium parasiticum:* First case report. *Mycoses.* 2011; **54:** 172-4.
359. Surendran K, Bhat RM, Boloor R, *et al.* A clinical and mycological study of dermatophytic infections. *Indian J Dermatol.* 2014; **59:** 262-7.
360. Tadepalli K, Gupta PK, Asati DP, *et al.* Onychomycosis due to *Cunninghamella bertholletiae* in an immunocompetent male from central India. *Case Rep Infect Dis.* 2015; 703240.
361. Tainwala R, Sharma Y. Pathogenesis of dermatophytoses. *Indian J Dermatol.* 2011; **56:** 259-61.
362. Tambosis E, Lim C. A comparison of the contrast stains, Chicago blue, chlorazole black, and Parker ink, for the rapid diagnosis of skin and nail infections. *Int J Dermatol.* 2012; **51:** 935-8.
363. Tan Y, Lin L, Feng P, *et al.* Dermatophytosis caused by *Trichophyton rubrum* mimicking syphilid: A case report and review of literature. *Mycoses.* 2014; **57:** 312-5.
364. Thakur R. Spectrum of dermatophyte infections in Botswana. *Clin Cosmet Investig Dermatol.* 2015; **8:** 127-33.
365. Thakur R. Tinea capitis in Botswana. *Clin Cosmet Investig Dermatol.* 2013; **6:** 37-41.
366. Thangaraju P, Giri V, Singh H, *et al.* Tinea barbae: In released from treatment (RFT) Hansen's disease patient. *J Clin Diagn Res.* 2014; **8:** YD01-2.
367. Tirado-Gonzalez M, Ball E, Ruiz A, *et al.* Disseminated dermatophytic pseudomycetoma caused by *Microsporum* species. *Int J Dermatol.* 2012; **51:** 1478-82.
368. Topaloglu Demir F, Karadag AS. Are dermatophytid reactions in patients with kerion Celsi much more common than previously thought? A prospective study. *Pediatr Dermatol.* 2015; **32:** 635-40.
369. Tosti A, Elewski BE. Onychomycosis: Practical approaches to minimize relapse and recurrence. *Skin Appendage Disord.* 2016; **2:** 83-7.
370. Tsunemi Y, Takehara K, Miura Y, *et al.* Diagnosis of tinea pedis by the dermatophyte test strip. *Br J Dermatol.* 2015; **173:** 1323-4.
371. Tsunemi Y, Takehara K, Miura Y, *et al.* Screening for tinea unguium by dermatophyte test strip. *Br J Dermatol.* 2014; **170:** 328-31.
372. Turan E, Erdemir AT, Gurel MS, *et al.* A new diagnostic technique for tinea incognito: *In vivo* reflectance confocal microscopy. Report of five cases. *Skin Res Technol.* 2013; **19:** e103-7.
373. Uprety S, Sharma R. Kerion - A boggy lump. *N Engl J Med.* 2016; **375:** 980.
374. van Zuuren EJ, Fedorowicz Z, El-Gohary M. Evidence-based topical treatments for tinea cruris and tinea corporis: A summary of a Cochrane systematic review. *Br J Dermatol.* 2015; **172:** 616-41.
375. Veeranna S. Wood's lamp: A modified method of examination. *Indian J Dermatol Venereol Leprol.* 2005; 71: 364-5.
376. Vena GA, Chieco P, Posa F, *et al.* Epidemiology of dermatophytoses: Retrospective analysis from 2005 to 2010 and comparison with previous data from 1975. *New Microbiol.* 2012; **35:** 207-13.
377. Veraldi S, Chiaratti A, Harak H. Onychomycosis caused by *Aspergillus versicolor. Mycoses.* 2010; **53:** 363-5.
378. Veraldi S, Giorgi R, Pontini P, *et al.* Tinea imbricata in an Italian child and review of the literature. *Mycopathologia.* 2015; **180:** 353-7.
379. Veraldi S, Pontini P, Nazzaro G. A case of tinea imbricata in an Italian woman. *Acta Derm Venereol.* 2015; **95:** 235-7.
380. Verma S, Hay RJ. Topical steroid-induced tinea pseudo-imbricata: A striking form of tinea incognito. *Int J Dermatol.* 2015; **54:** e192-3.
381. Verma S. Tinea pseudoimbricata. *Indian J Dermatol Venereol Leprol.* 2017; **83:** 344-5.
382. Verma SB. A closer look at the term "Tinea Incognito:" A factual as Well as grammatical inaccuracy. *Indian J Dermatol.* 2017; **62:** 219-20.
383. Verma SB, Vasani R. Male genital dermatophytosis - Clinical features and the effects of the misuse of topical steroids and steroid combinations - An alarming problem in India. *Mycoses.* 2016; **59:** 606-14.
384. Vinay K, Mahajan R, Sawatkar GU, *et al.* An unusual presentation of tinea cruris with bullous lesions. *J Cutan Med Surg.* 2013; **17:** 224-5.
385. Vyas A, Pathan N, Sharma R, *et al.* A clinicomycological study of cutaneous mycoses in Sawai Man Singh Hospital of Jaipur, North India. *Ann Med Health Sci Res.* 2013; **3:** 593-7.

386. Walker-Smith PK, Wlodek C, Sansom J. Tinea faciei caused by *Arthroderma benhamiae*. *BMJ*. 2017; **357:** doi: 10.1136/bmj.j2007. PMID: 28522542.
387. Wang HH, Lin YT. Bar code-like hair: Dermoscopic marker of tinea capitis and tinea of the eyebrow. *J Am Acad Dermatol*. 2015; **72:** S41-2.
388. Wang R, Hu Y, Tang H, *et al.* Majocchi granuloma in a pregnant woman. *Obstet Gynecol*. 2014; **124 (Suppl. 1):** 423-5.
389. Warycha MA, Leger M, Tzu J, *et al.* Deep dermatophytosis caused by *Trichophyton rubrum*. *Dermatol Online J*. 2011; **17:** 21.
390. Weaver G, Newmyer R, Yeo KT, *et al.* An unusual periorbital rash in a child. *Clin Infect Dis*. 2012; **55:** 844 & 885-6.
391. Weitzman I, Summerbell RC. The dermatophytes. *Clin Microbiol Rev*. 1995; **8:** 240-59.
392. Welsh O, Vera-Cabrera L, Welsh E. Onychomycosis. *Clin Dermatol*. 2010; **28:** 151-9.
393. Welsh O, Vera-Cabrera L. Red face and fungi infection. *Clin Dermatol*. 2014; **32:** 734-8.
394. Westerberg DP, Voyack MJ. Onychomycosis: Current trends in diagnosis and treatment. *Am Fam Physician*. 2013; **88:** 762-70.
395. White TC, Findley K, Dawson TL Jr, *et al.* Fungi on the skin: Dermatophytes and *Malassezia*. *Cold Spring Harb Perspect Med*. 2014; 4. pii: a019802.
396. Xavier AP, Oliveira JC, Ribeiro VL, *et al.* Epidemiological aspects of patients with ungual and cutaneous lesions caused by *Scytalidium* spp. *An Bras Dermatol*. 2010; **85:** 805-10.
397. Yadav P, Singal A, Pandhi D, *et al.* Clinico-mycological study of dermatophyte toenail onychomycosis in New Delhi, India. *Indian J Dermatol*. 2015; **60:** 153-8.
398. Yadav S, Saxena AK, Capoor MR, *et al.* Comparison of direct microscopic methods using potassium hydroxide, periodic acid Schiff and calcofluor white with culture in the diagnosis of onychomycosis. *Indian J Dermatol Venereol Leprol*. 2013; **79:** 242-3.
399. Yin B, Xiao Y, Ran Y, *et al.* *Microsporum canis* infection in three familial cases with tinea capitis and tinea corporis. *Mycopathologia*. 2013; **176:** 259-65.
400. Yu G, Jianhua W. Superficial white onychomycosis due to *Trichophyton rubrum* in a two-year-old child. *Indian J Dermatol Venereol Leprol*. 2013; **79:** 269.
401. Zaki SM, Ibrahim N, Aoyama K, *et al.* Dermatophyte infections in Cairo, Egypt. *Mycopathologia*. 2009; **167:** 133-7.
402. Zaraa I, Hawilo A, Aounallah A, *et al.* Inflammatory tinea capitis: A 12-year study and a review of the literature. *Mycoses*. 2013; **56:** 110-6.
403. Zhan P, Li D, Wang C, *et al.* Epidemiological changes in tinea capitis over the sixty years of economic growth in China. *Med Mycol*. 2015; **53:** 691-8.
404. Zhan P, Li ZH, Geng C, *et al.* A chronic disseminated dermatophytosis due to *Trichophyton violaceum*. *Mycopathologia*. 2015; **179:** 159-61.
405. Zhan P, Liu W. The changing face of dermatophytic infections worldwide. *Mycopathologia*. 2017; **182:** 77-86.
406. Zhou J, Chen M, Chen H, *et al.* *Rhodotorula minuta* as onychomycosis agent in a Chinese patient: First report and literature review. *Mycoses*. 2014; **57:** 191-5.
407. Ziegler W, Lempert S, Goebeler M, *et al.* Tinea capitis: Temporal shift in pathogens and epidemiology. *J Dtsch Dermatol Ges*. 2016; **14:** 818-25.
408. Zink A, Papanagiotou V, Todorova A, *et al.* Outbreak of *Microsporum audouinii* in Munich - The return of infectious fungi in Germany. *Mycoses*. 2014; **57:** 765-70.
409. Zisova LG, Dobrev HP, Tchernev G, *et al.* Tinea atypica: Report of nine cases. *Wien Med Wochenschr*. 2013; **163:** 549-55.
410. Zotti M, Machetti M, Perotti M, *et al.* A new species, *Aspergillus persii*, as an agent of onychomycosis. *Med Mycol*. 2010; **48:** 656-60.

SECTION III

Subcutaneous Mycoses

Chapters

This section comprises of heterogeneous group of infections, which are characterized by development of clinical manifestations in the subcutaneous tissues at the site of inoculation of etiological fungal agents. These are also called implantation mycoses, as the disease process usually starts by inoculation of fungi following a trivial trauma, which is the sole source of infection. Mycetoma is unique clinical entity of tropical and subtropical countries caused by various organisms, both fungal as well as bacterial in origin. Therefore, to complete description of such a disease in Textbook of Medical Mycology, both these groups are being described. It is the first fungal infection now declared as neglected tropical disease by the WHO. Another disease, botryomycosis, basically is a bacterial infection but clinically mimics mycetoma hence is briefly described in the same Chapter.

Sporotrichosis is caused by various species of dimorphic genus *Sporothrix.* It is prevalent worldwide and in India mainly in hilly areas of the entire sub-Himalayan belt. Out of all dimorphic fungi, this agent is commonly encountered in Indian patients as compared to the others. It used to be caused by singular agent but now a number of species are found to be involved and in India it is predominantly *S.globosa*. The phaeoid (dematiaceous) fungi now constitute significant group of organisms responsible for causing number of diseases. In this Section, two clinical entities are described based on their histopathological findings *i.e.* chromoblastomycosis and phaeohyphomycosis. The phaeohyphomycosis is an emerging fungal infection as day by day new agents are being added, which used to be considered as non-pathogenic to humans in the past.

Another recently re-named fungal infection - Lacaziosis, caused by *Lacazia loboi*, is now considered as hydrophilic in nature due to its affinity to water, however, it is yet to be successfully grown on artificial culture media. The description of other subcutaneous mycosis *i.e.* Entomophthoramycosis, caused by two genera *(Conidiobolus* and *Basidiobolus),* has been described separately in **Chapter 27** due to taxonomical reasons as well as their involvement now among the immunocompromised patients.

However, other subcutaneous infections are caused by many fungal species. As and when the diagnosis is established, therapeutic modality is mainly medical, however, occasionally surgical intervention may also be required. The conventional drug *i.e.* potassium iodide, which is already in use for more than a century, is still quite effective in case of sporotrichosis and other diseases.

There are many cutaneous and subcutaneous manifestations of systemic fungal infections due to their hematogenous dissemination and not because of direct inoculation to the skin. These are caused by dimorphic as well as opportunistic fungi hence are dealt in their corresponding Chapters and/or Sections of this Textbook.

CHAPTER 11

Mycetoma

Mycetoma is a slowly progressive, chronic granulomatous infection of skin and subcutaneous tissues with involvement of underlying fasciae and bones, usually affecting extremities. The **triad** of (i) tumefaction of affected tissue, (ii) formation of multiple draining sinuses and (iii) presence of granules in discharge, clearly defines this disease. The granules are tightly knit clusters of organisms within the infected tissues and are also called grains.

Mycetoma is popularly known as **Madura Foot** or **Maduramycosis**. These terms were used to describe when there was only involvement of foot and information about their prevalence in limited geographical areas. However, other anatomical sites and wider areas of tropical and subtropical countries are involved entailing application of such terms infructuous. Hence its terminology, as mycetoma is considered proper for this disease.

Further, the classification and nomenclature of mycetoma is actually based on the **nature** of infection *i.e.* fungal or bacterial, color of grains and cement-like matrix produced during the disease process. Broadly there are two categories of this entity, which are recognized namely, **eumycetoma** caused by fungi and **actinomycetoma** caused by higher bacteria of the class Actinomycetes. In addition, **botryomycosis**, exclusively a bacterial infection, which clinically mimics mycetoma hence it is also described in this Chapter.

Historical Perspective

The mycetoma has been referred in the ancient **Sanskrit** writings like the 4th collection (Samhita) of the Hindu scriptures, **Atharvaveda**, where it is mentioned as **Padavalmikam**, meaning anthill foot. In 1842, John Gill described this disease for the first time in India in a dispensary at **Madurai**, the capital of Pandyan Kingdom and now district of southern state, Tamil Nadu (India). Colebrook in 1846 confirmed Gill's observations and the term ***'Madura Foot'*** was used to designate the disease, incorporating name of this town in its nomenclature.

In 1860, Henry Vandyke Carter, who was Professor of Morbid Anatomy and Physiology at the Grant Medical College, Mumbai, India, established its fungal etiology classifying this infection according to color of the grains present in discharging sinuses and coined the term ***'Mycetoma'***. In 1874, he published a Monograph on this entity describing it as *'the fungal disease of India'* in general belief that this disease was only limited to Indian subcontinent.

Brumpt, in 1906, first isolated and described a fungus from black grains and gave binomial name *Madurella mycetomi*. Pinoy in 1913, taking consideration of the origin of causative organisms, categorized the disease in two groups, one caused by eumycetes *i.e.* true fungi, **Eumycetoma** and other by aerobic Gram-positive filamentous bacteria belonging to class Actinomycetes named **Actinomycetoma**. This etiological classification is followed till date as such despite passage of more than a century. The additional term, **Maduramycosis**, was given by Chalmer, Archibald and Christopherson, in 1916, to this entity, which was already labeled as mycetoma.

Periodic updates from time to time are made by the Mycetoma Research Center under the aegis of University of Khartoum in Sudan. This was started in 1991, is dedicated to continuing discovery and development of scientific knowledge and clinical skills applied to the care of mycetoma patients. Moreover, development of researches and creation of an international database about the disease, starting from local and international partnerships, are its objectives. An international conference was held on May 24-26, 2007 at Monterrey in Mexico dealing with various aspects of this disease.

The ISHAM Mycetoma Working Group is very active on the academic as well as side of this disease. On the basis of its efforts, the 69th World Health Assembly of WHO approved a resolution recognizing mycetoma as a

Table 11.1. Some of the Common Causative Agents of Eumycetoma with Characteristics of Grains.

Causative Fungal Agents	*Texture*	*Size (mm)*	*Shape*	*Cement-like Matrix*
A. Black Grain Eumycetoma				
Madurella mycetomatis	Hard	0.5-5.0	Oval to lobed	Present, Homogenous
Trematosphaeria grisea	Soft	0.3-0.6	Oval to lobed	Present, Peripheral
Exophiala jeanselmei	Soft	0.2-0.3	Irregular	Absent
Curvularia geniculata	Hard	0.5-1.0	Oval	Present, Peripheral
B. White Grain Eumycetoma				
Pseudallescheria boydii	Soft	0.5-1.0	Oval to lobed	Absent
Aspergillus nidulans	Soft	1.0-2.0	Oval	Absent
Acremonium falciforme	Soft	0.2-0.5	Oval	Absent
Fusarium species	Soft	0.2-0.6	Oval	Absent

Table 11.2. Causative Agents of Actinomycetoma.

Bacterial Agents	*Grains - Color*	*Size (mm)*
Actinomadura madurae	White, yellow	2.0
Actinomadura pelletieri	Pink to Red	0.5
Nocardia brasiliensis	Yellowish-White	< 0.5
Nocardia caviae	White to yellow	< 0.5
Nocardia asteroides	White to yellow	< 0.5
Nocardiopsis dassonvillei	Cream-colored	< 0.5
Streptomyces somaliensis	Yellowish-white	1.0

neglected tropical disease on May 28, 2016. This is the first fungal disease to be listed is this category. The new resolution will help to raise awareness of the disease. A wider recognition of the burden of mycetoma is expected to boost the development of control strategies and tools suitable for implementation in poor and remote areas where many cases are found prevalent.

Causative Organisms

The mycetoma is caused by two broad groups of organisms *i.e.* fungal and bacterial, which are briefly described below:

(a) Fungal Agents: Most of the agents causing eumycetoma are saprotrophic environmental fungi. Taxonomically these fungi are members of either **Deuteromycetes** or **Ascomycetes**. There are about 25 fungal species causing eumycetoma and some of them, frequently encountered are listed in **Table 11.1**. Out of these, ten species belonging to Deuteromycetes and three from Ascomycota are known to cause white-grain mycetoma. Rest of them are responsible for black grain mycetoma. The identification of these agents is mainly based on morphology of their fruiting bodies. Rarely coelomycetes species like *Medicopsis (Pyrenochaeta) romeroi* and *Leptosphaeria senegalensis* may also be encountered as causative agents of eumycetoma.

As such genus *Madurella* was originally based on tissue morphology and formation of sterile cultures on mycological media. There were two species initially namely, *M.mycetomatis* and *M.grisea*. Now, based on recent molecular studies have recognised five species: *Madurella mycetomatis, Trematosphaeria grisea (M.grisea), M.fahalii, M.pseudomycetomatis* and *M.tropicana*. These species are solated from soil and are major causative agents of mycetoma.

(b) Bacterial Agents: Actinomycetoma is caused by traumatic inoculation of exogenous Actinomycetes that normally dwell in soil and environment. In a *Textbook of Medical Mycology*, although description of bacterial agents is generally not required but due to peculiarity of this unique clinical entity, mycetoma, being caused by multiple agents, the description will not be complete without describing the bacterial causative agents. These are Gram-positive branching filamentous aerobic bacteria of class Actinomycetes and are listed in **Table 11.2**.

Epidemiology

The mycetoma is prevalent in almost all parts of the world; however, it is more common in tropical and subtropical countries of Asia, Africa and Central and South America, mainly between the tropics of Cancer and Capricorn, 23½° North and South of Equator. The area lying between 15 S and 30 N is said to be the 'Mycetoma belt'. Three countries *i.e.* Sudan, Mexico and India are predominantly affected due to this disease. The climatic conditions of the geographical areas also determine the distribution of this disease. The mycetoma is said to be more prevalent in places where rainfall is 50 to 500 mm with temperature of 15 to 25°C. But in India, mycetoma is distributed in widely different geographical areas *e.g.* large numbers of mycetoma cases are seen in Tamil Nadu, the moist southern part and Rajasthan, the dry western part of India.

The mycetoma is more prevalent in developing as compared to the developed countries and the incidence is more in rural than urban areas. The disease is commonly seen in young adults of 20 to 40 years. This may be explained by the fact that during this period of life the people are more active hence are more prone to trauma and thorn prick injury. Mycetoma is seen more commonly in men than women, in a ratio of 3.5 : 1 and the higher incidence in male may be due to higher risk of their exposure to injury during fieldwork as compared to women. Furthermore, the women are said to have some protective mechanism that may not allow the multiplication of the organism in the body and the female hormones may play role in this regard.

The incidence of mycetoma is mainly dependent on occupation as farmers, herdsman and field workers are more likely to come in contact with agents of mycetoma, present in the soil and thorny vegetation because of their habit of outdoor movement in the field being barefooted. In addition, the disease is also seen in other occupational groups like carpenters, drivers, builders and land workers who are at higher risk of minor penetrative trauma that facilitates easy entry of the causative elements.

The prevalence of etiological agents of mycetoma varies from place to place. The eumycetes accounts for about 40% and actinomycetes for 60% of mycetoma in the world. In Southeast Asia, India, Pakistan and neighboring countries, the ratio of eumycotic and actinomycotic mycetoma is about 35% and 65%, respectively except in Rajasthan. While in Europe actinomycetes account for about 30% and eumycetes for 70% of the reported cases of mycetoma. In Mexico, 98% of mycetoma is caused by actinomycetes and only 2% by eumycetes. 86% of the actinomycotic cases here are due to *Nocardia brasiliensis*, followed by *Actinomadura madurae, Streptomyces somaliensis, Nocardia asteroides, Actinomadura pelletieri* and *Nocardia caviae*. In Venezuela, most frequent etiological agents of eumycetoma are *Pyrenochaeta mackinnonii, Medicopsis romeroi* and *Trematosphaeria grisea*; actinomycetoma are *Actinomadura madurae* and *Nocardia brasiliensis*. In Africa, eumycetoma are more common and extensive. There is occurrence of sporadic cases of mycetoma, which may also be found scattered in other parts of the world.

Actinomycetoma is more common in India especially in southern (with a marginal difference) and eastern India, with predominance of eumycetoma mainly in the north-western part especially the dry desert regions of Rajasthan. However, sporadic cases of either type are reported from various parts of the country and the epidemiology is fast changing due to changes in the climate and human activities. From the eastern India, *Streptomyces viridis* has also been found among the patients of actinomycetoma, which is the first report of its kind. A few cases of mycetoma have also been described in animals like dogs, horses and goats.

Immunity

The infection in mycetoma appears to depend more upon exposure rather than factors intrinsic to the host. The general health of patients with mycetoma is unaffected unless decreased activity has brought about economic deprivation and malnutrition. The studies of delayed hypersensitivity reactions have been too scanty to establish their significance in understanding of the disease process, diagnosis or in epidemiological surveys.

Pathogenesis and Pathology

The mycetoma usually occurs by introduction of its causative agent present in saprotrophic soil source into subcutaneous tissue, probably by accidental trauma by thorns (thorn-prick) or splinter injury. This is evident by the fact that mycetoma is endemic in places where vegetation is abundant, occurring mainly in field workers and farmers, who work without protection of exposed parts. There are minor trauma and abrasions of skin, which also contribute in inoculation of the causative agents. In rural areas, habitual use of wicks (piece of straw from field soil) for removal of earwax may be responsible for mycetoma of the ear.

Mycetoma of back has been reported in people who carry goods like wood, grain bags, stone, etc. contaminated with soil saprotrophs. Similarly, mycetoma of head and neck is observed among people who usually carry bundles of wood on head and neck. This is due to minor abrasions caused, which favor entry of agents, thus reflecting particular occupation.

After the causative agent is introduced into skin, disease evolves slowly. The organisms are usually found in the center of microabscess formed by polymorphonuclear cells. The main characteristic of mycetoma infection is the presence of large aggregates of filaments of causative organisms, which are center of its inflammatory activity.

Table 11.3. Differentiating Clinical Features of Actinomycetoma and Eumycetoma.

Clinical Features	*Actinomycetoma*	*Eumycetoma*
Causative Organisms	Aerobic actinomycetes	Hyaline and phaeoid hyphomycetes
Tumor Mass	Multiple, diffuse with ill-defined margins	Usually single, with well-defined margins
Sinuses	Appear early and more in number	Appear late and relatively less in number
Opening of sinuses	Raised, inflamed and flared up	Flat opening and not flared up
Flap of Opening	Easily removed	Not easily removed
Discharge	Usually purulent	Serous or sero-sanguineous
Grains	White except *A.pelletieri* which is red	Black or white
Extent of Involvement	More extensive and obliterative with	Less extensive, only osteosclerotic lesions
		hypertrophic, punched out osteolytic lesions of bone

There are two important considerations, which are implicated in development of mycetoma as with other conditions; firstly, the effect of alteration of host resistance in emergence of infection and secondly ability of organisms to resist or evade host defense mechanism.

Classification of Mycetoma

Mycetoma is classified on the basis of etiological agent involved in the disease process. The hyaline and phaeoid hyphomycetes cause **eumycetoma** whereas aerobic actinomycetes cause **actinomycetoma** **(Table 11.3)**. There is another way of classifying mycetoma, which is based on color of grains produced, like black grain mycetoma and white or pale grain mycetoma. All bacterial agents produce white grain mycetoma with an exception of *A.pelletieri*, which produces red or pink grains, whereas fungal agents produce both black as well as white grain mycetoma. There is one more way to classify mycetoma, which is based on geographical areas (continents) and color of grains like European black, African black or Asian black mycetoma *vis-à-vis* European white, African white and Asian white mycetoma, respectively.

Clinical Features

The clinical presentation of mycetoma falls within a single clinical entity with many features common in most of cases. However, they may differ depending upon underlying etiological agent, affected anatomical site and stage of lesions as well as to size, shape and color of grains. The painless localized swollen lesions with multiple discharging sinuses and progressive destruction are common to all types of mycetomata. There is serous, sero-sanguineous or purulent discharge from the sinuses particularly following secondary bacterial infections. There are variations according to the causative organism, duration of illness and site of lesion.

A **triad** of clinical features, irrespective of underlying etiological agent, basically constitutes mycetoma. This consists of the followings:

- Tumor-like swelling *i.e.* tumefaction.
- Formation of multiple discharging/draining sinuses.
- Grains/granules discharging from sinuses.

Unless these three cardinal features are observed in any clinical entity, it cannot be labeled as mycetoma. In very rare instances, it is also observed that disease process may not yield grains at all and one has to rely on direct and cultural findings to substantiate the clinical diagnosis.

Within host tissue, the organisms lead to formation of compact mass of colonies *i.e.* grains of about 0.5-2 mm in diameter and their color depends on the underlying causative organism involved. *Actinomadura madurae* produces large-grain whereas *Nocardia* species produce small-grain mycetoma.

Although mycetoma mainly affects feet **(Figs. 11.1A to J)** but infections of other parts have been reported like hands **(Figs. 11.2A and B)**, shoulder **(Figs. 11.3A to C)**, abdomen **(Fig. 11.4)** buttocks and scalp. It may involve any site subjected to traumatic implantation of soil saprotrophs into subcutaneous tissue. The site of lesion often reflects particular occupation e.g. mycetoma of head and neck occurring more frequently in those individuals who carry bundles of woods on head or shoulders. Sometimes, mycetoma is also seen as recurrence even if amputation of the affected part is done **(Fig. 11.5)**.

The disease progresses slowly and usually takes long time, often years between infection and formation of

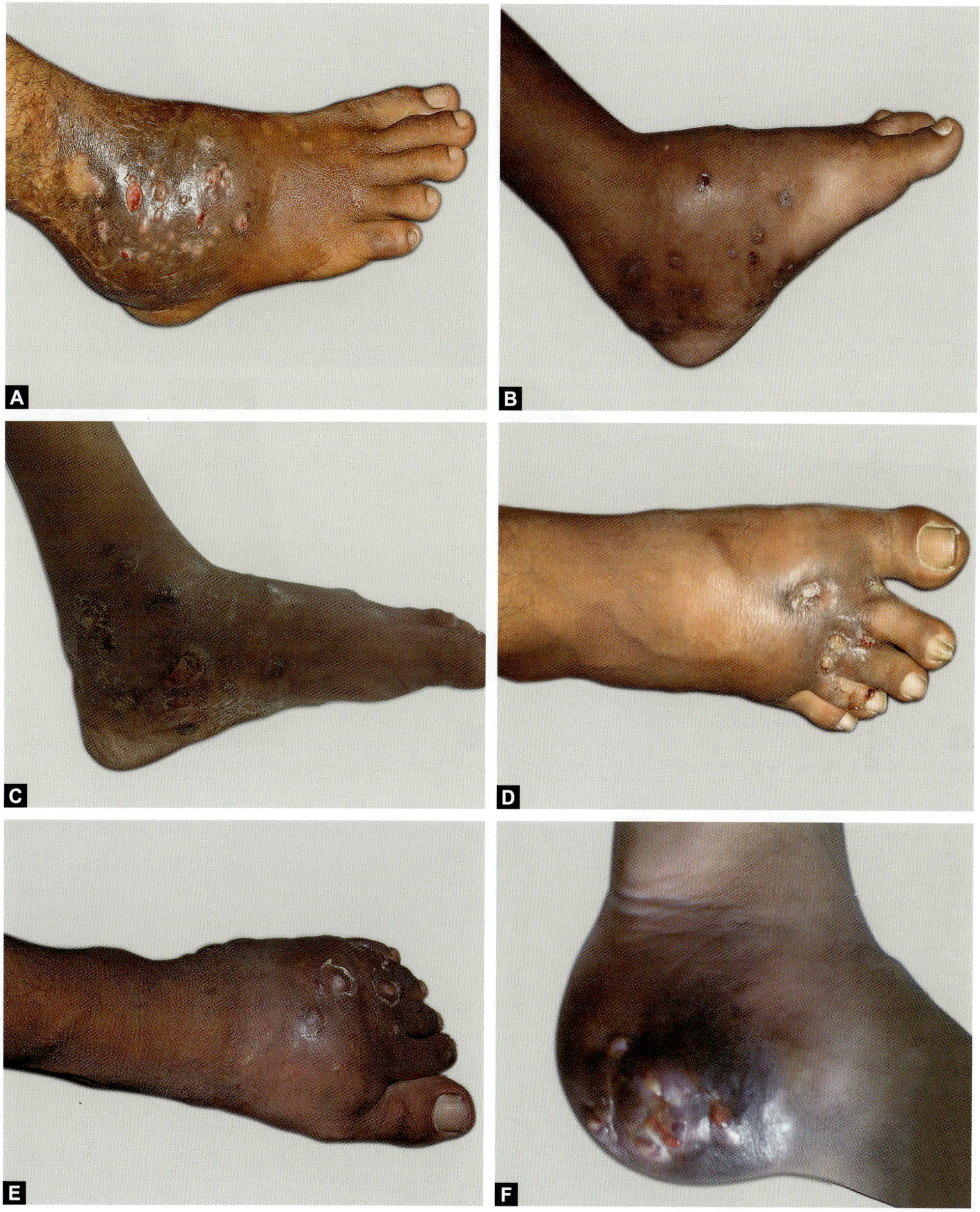

Figs. 11.1A to F. Mycetoma of foot presenting with various clinical manifestations with oozing discharge mixed with grains.

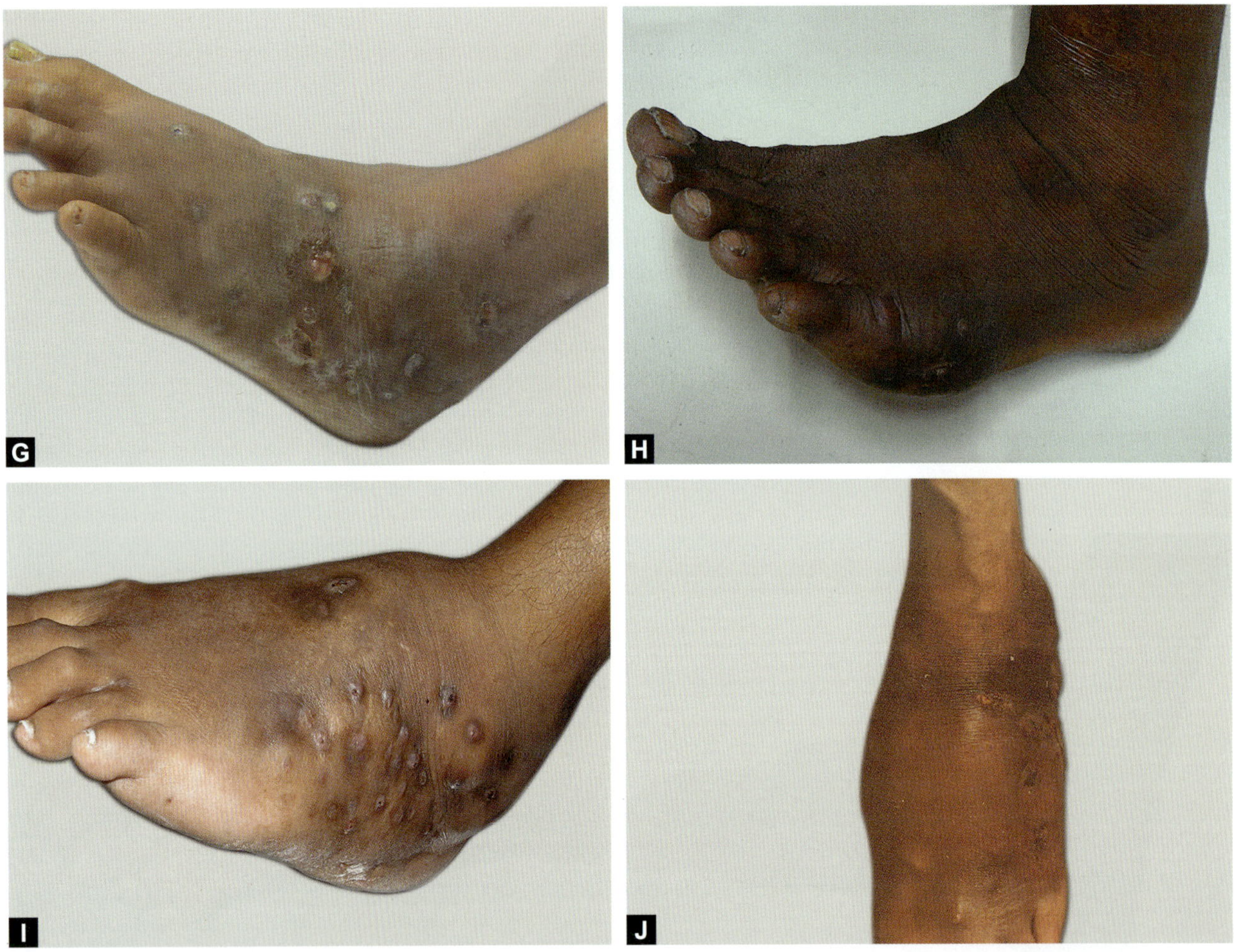

Figs. 11.1G to J. Mycetoma of foot presenting as various clinical manifestations with oozing discharge mixed with grains.

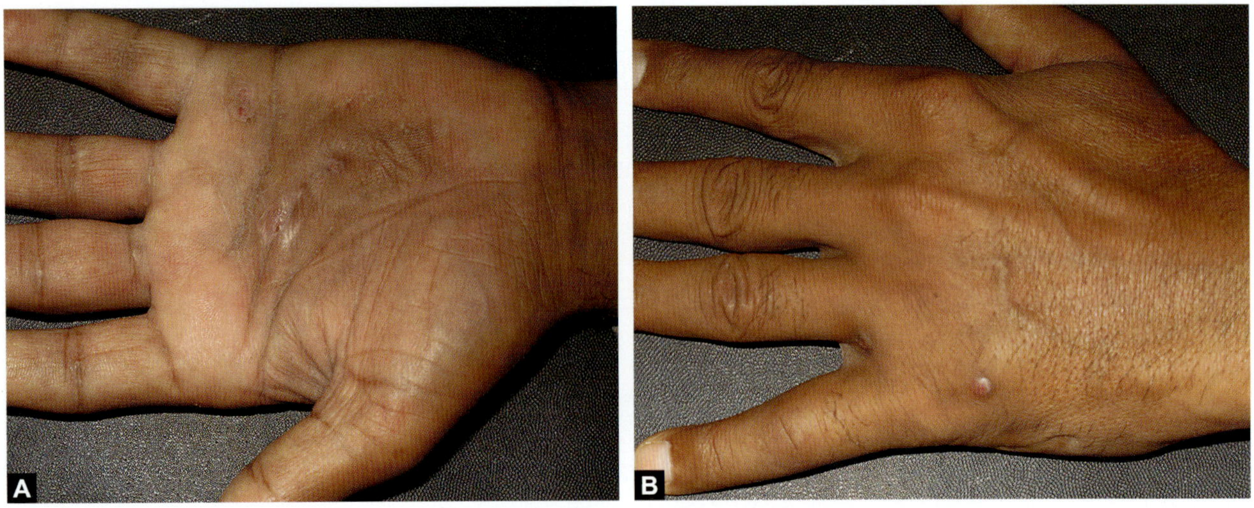

Figs. 11.2A and B. Mycetoma of hand showing lesions in the palm and dorsal surface of the same patient.

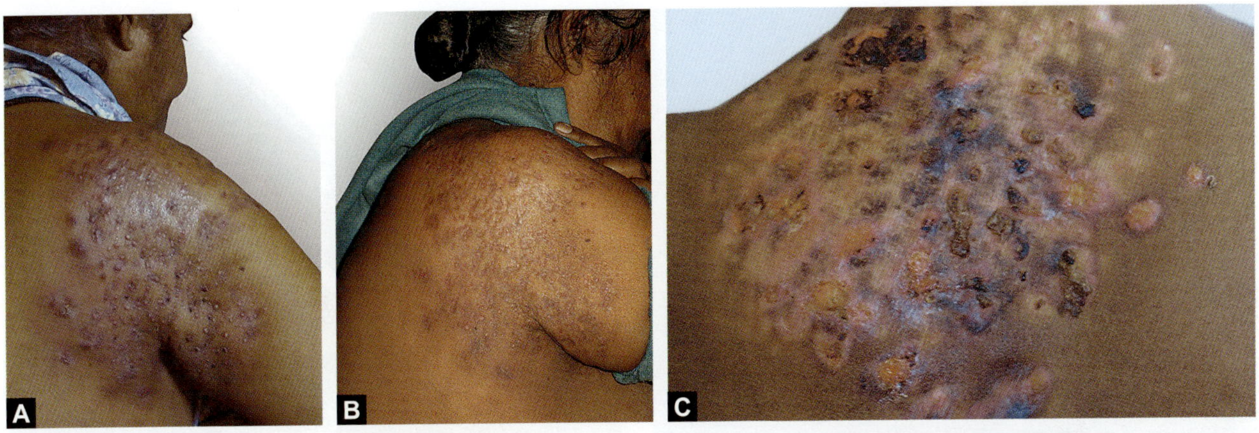

Figs. 11.3A to C. Actinomycetoma of shoulder caused by *Nocardia* species. A and B are from same patient before and after treatment.

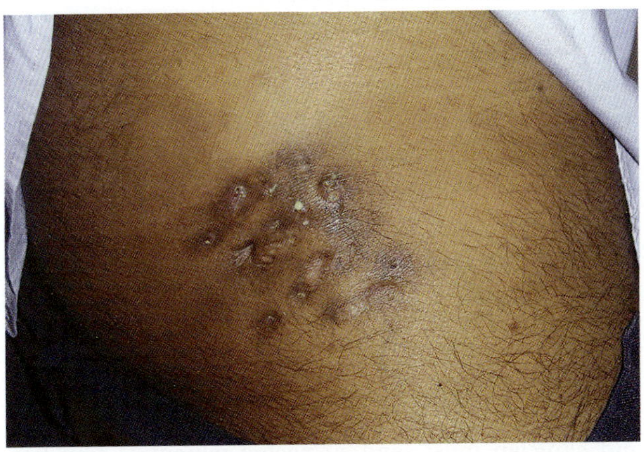

Fig. 11.4. Actinomycetoma involving the of abdominal wall of a patient caused by *Nocardia* species.

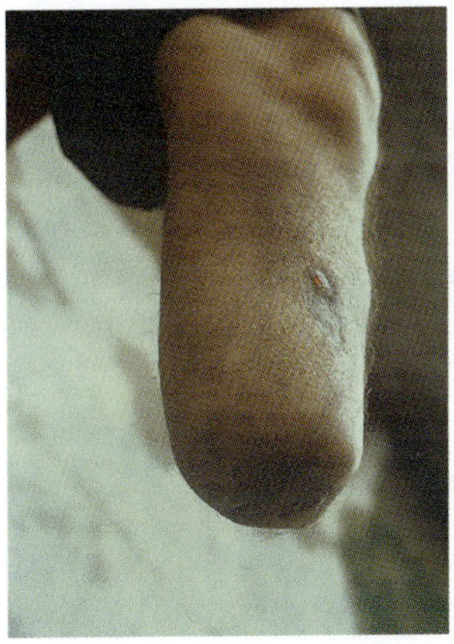

Fig. 11.5. Recurrence of lesions over amputated stump of patient of actinomycetoma caused by *Actinomadura madurae*.

characteristic lesions, which are firm, painless, localized subcutaneous nodules. It usually spreads by contiguity destroying surrounding structures, but spares tendons and nerves. In mycetoma of scalp, bones may be involved at an early stage. Hematogenous spread is also seen particularly in the infections caused by *Nocardia* and *Streptomyces* species. Recently possibility of spreading via lymphatics has also been observed.

When buttock and trunk are affected, spreading may be relatively rapid and extensive. In the infections of scalp, bones may take a shorter duration. Mycetoma is rarely a threat to life, but it causes severe incapacity. Actinomycetoma has been regarded to follow more extensive and obliterative involvement of bone with both lytic and hypertrophic changes as compared to eumycetoma since actinomycetes invade muscle more readily than fungi. The openings of actinomycetoma are inflamed, flared up and angry-looking as compared to eumycetoma (**See Table 11.3**).

The involved organisms of mycetoma may also cause other clinically significant diseases, such as nocardiosis, mycotic granuloma, phaeohyphomycosis, etc. Sometimes, fungus balls, caused by *Aspergillus* species, *Coccidioides* species, *Scedosporium apiospermum* or *Pseudallescheria boydii* in the preformed cavities, are erroneously labeled as mycetoma. These fungus balls should be referred to as aspergilloma or coccidioidoma depending upon causative organism because they lack in well-organized grain formation hence terminology like mycetoma should be avoided in such cases.

Similarly, mycelial aggregates formed by dermatophytes into deeper tissues, differ in many respects from grains of mycetoma. Such infection caused by dermatophytes

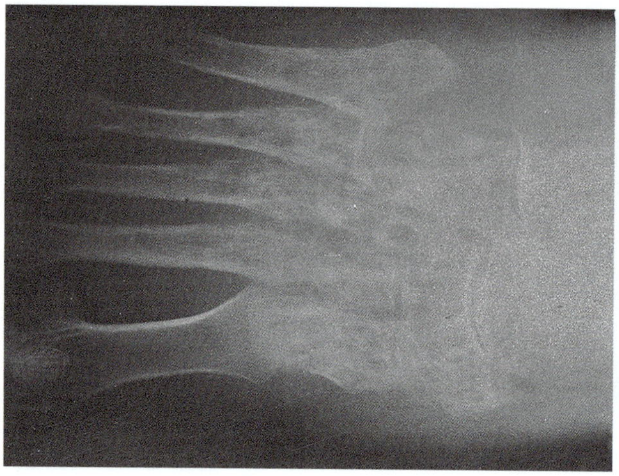

Fig. 11.6. Actinomycetoma of foot caused by *Actinomadura madurae* and X-rays showing the osteolytic lesions.

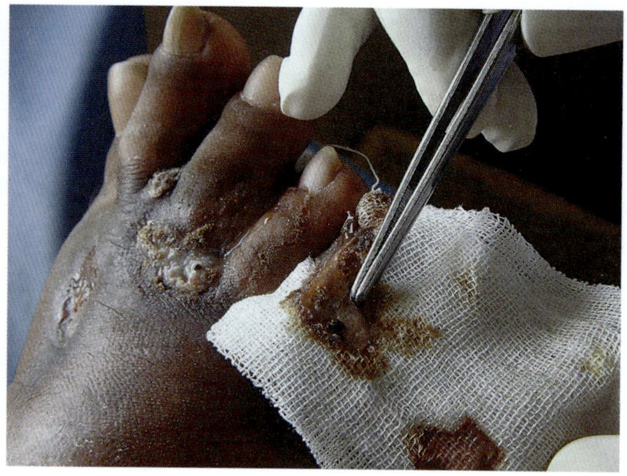

Fig. 11.7. Mycetoma oozing grains from foot seen while collection the clinical material on a sterile gauze.

is called **'pseudo-mycetoma'** rather than mycetoma, as there is formation of yellowish-white pseudograins without any cementing material. This is also known as **Majocchi's dermatophytic granuloma** as disease mycetoma is only labeled when three cardinal features are present in a clinical entity.

Radiodiagnosis

The radiological modalities like X-rays, USG, CT scans and/or MRI are used to determine the extent of the disease and involvement of the underlying tissue. Radiologically, mycetoma shows necrosis, generalized osteoporosis and fusion of smaller bones. The osteolytic lesions of actinomycetoma of foot **(Fig. 11.6)**. USG can differentiate eumycetoma, actinomycetoma and non-mycetomatous lesions. The ***"dot-in-circle"*** sign on MRI is easy to recognize and highly specific for mycetoma.

Differential Diagnosis

Mycetoma should be differentiated from pyogenic granuloma, osteomyelitis, atypical mycobacteriosis, leprosy, benign tumors, botryomycosis, actinomycosis and dermatophytosis. The dermatophyte infection of subcutaneous tissue of scalp, with formation of whitish-yellow grains, which is produced by *Microsporum* species, is loosely called dermatophytic mycetoma. There are several other diseases, which may mimic mycetoma at one or the other stage of their development. They are yaws, elephantiasis of foot, cellulitis, coccidioidomycosis, sporotrichosis and ulcerated hypertrophic lesions in chromoblastomycosis. It should not be confused with actinomycosis caused by anaerobic actinomycetes, which is usually **endogenous** in origin, instead of traumatic implantation.

Laboratory Diagnosis

A detailed history about occupation, trauma and geographical area of the patient is always valuable while establishing diagnosis of mycetoma. The need of gross examination of lesions by microbiologist is always underestimated and clinician must send suspected patient to him/her. As such Medical Mycology requires teamwork but in this disease, it is must along with lot of patience to recover the etiological agent.

A high index of clinical suspicion is required to reach eventually the final diagnosis. This may lead to detection of mycetoma cases that otherwise may be missed as examination of lesion often give clue to the type of mycetoma. Moreover, it facilitates collection of as many grains as needed for proper processing and recovery of the causative agent, which might be scanty in number.

The clinical sample in mycetoma is usually grains or granules but pus, exudates or biopsy tissue may also be taken. The lesions are thoroughly cleaned with antiseptics and grains are collected by pressing sinus from periphery thereby enhancing discharge mixed with granules, which are collected on sterile gauze **(Fig. 11.7)**. Alternatively, they can be collected with the help of a wire loop. If more grains are needed, flap of orifice of sinuses are opened and grains are collected in sterile petri dish.

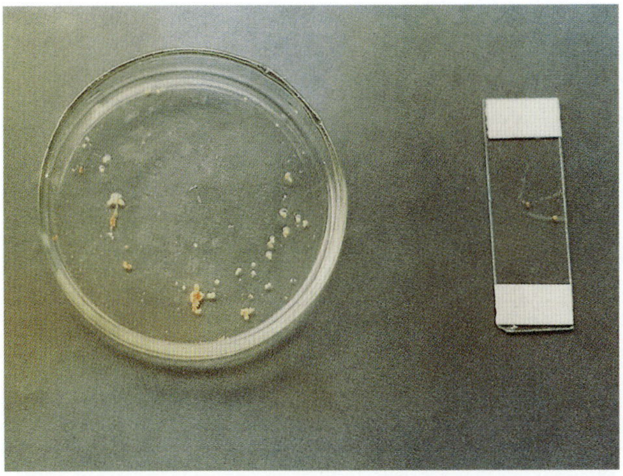

Fig. 11.8. Large-sized pale grains produced in actinomycetoma caused by *A.madurae* are washed with saline in petri dish.

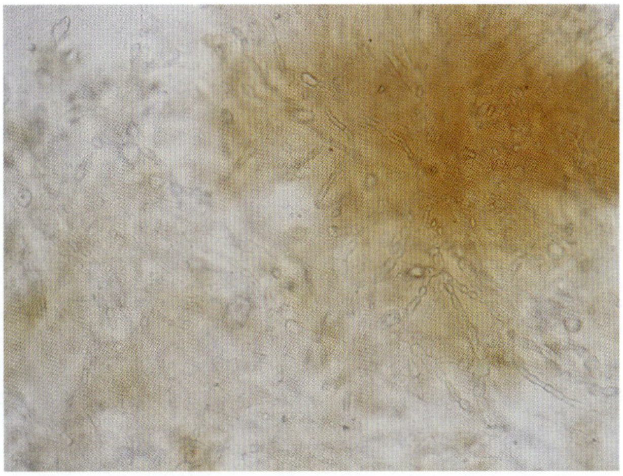

Fig. 11.9. Septate hyphae in wet mount of clinical material taken from eumycetoma of foot (KOH × 400).

The gross appearance of grains is of paramount value as they characterize certain etiological agents of this disease. The observation of characteristic size, shape, texture, color of grains and prevalence of agents in that geographical area may clinch the etiology by an experienced microbiologist. But final diagnosis always depends on direct demonstration followed by culture of the organism involved in the disease process.

The history and clinical examination of patient is also very much suggestive of mycetoma. It is extremely important to begin effective therapy based on specific microbiological diagnosis. The following procedures are helpful for establishing exact diagnosis of suspected case of mycetoma:

(a) Direct Examination

The laboratory examination of pus, exudate or biopsy material reveals presence of grains, which are primary diagnostic indicators. Eumycetoma grains can be seen with 2-6 µm wide interwoven hyphae with large, globose swollen cells of 10-15 µm or more at margin *i.e.* chlamydospores.

Actinomycotic grains have Gram-positive filamentous bacteria with diameter of 0.5 to 1 µm as well as their coccoid to bacillary forms. The clinical entity produced by these agents is indistinguishable except through observation of color, size, shape and consistency of the grains.

The characteristics of grains, color, texture, size, presence of cement-like matrix and hyaline or phaeoid hyphae are determining features for establishment of exact etiology of mycetoma (*See* **Tables 11.1 and 11.2**). The grains should be washed several times in sterile saline, crushed with sterile glass rod and cultured on different media (**Fig. 11.8**). These contain either actinomycetes or fungi, which is visualized in KOH wet mount and on histopathological examination.

Add a few drops of KOH and grains on glass slide, put coverslip and crush them gently by applying pressure over coverslip with glass rod or handle of loop. Leave this wet mount for few hours in petri dish with a moist filter paper.

The KOH wet mount is examined under microscope for looking characteristic hyphae, if any, for their dimension, septae or pigmentation, presence or absence of interhyphal cement substance and chlamydospores (**Fig. 11.9**). Actinomycotic grains show thin filaments with diameter of 0.5-1 µm with coccoid or bacillary forms while eumycotic grains show thick 2-6 µm wide hyphae with large swollen cells upto 15 µm at margin with or without chlamydospores.

For the diagnosis of actinomycetoma, as causative organisms are from bacterial origin, routine bacterial procedures are adopted. Gram's stain smear shows Gram-positive branching filamentous bacteria embedded in grain material (**Figs. 11.10A and B**). The modified Ziehl-Neelsen/acid-fast staining (Kinyoun's method) with 1% sulfuric acid, shows reddish-pink colored filamentous bacteria i.e. *Nocardia* species, whereas other actinomycetes are not acid-fast. Therefore, Kinyoun's staining should be performed routinely as *Nocardia* species show reddish-pink, acid-fast long filamentous bacteria (**Fig. 11.11**).

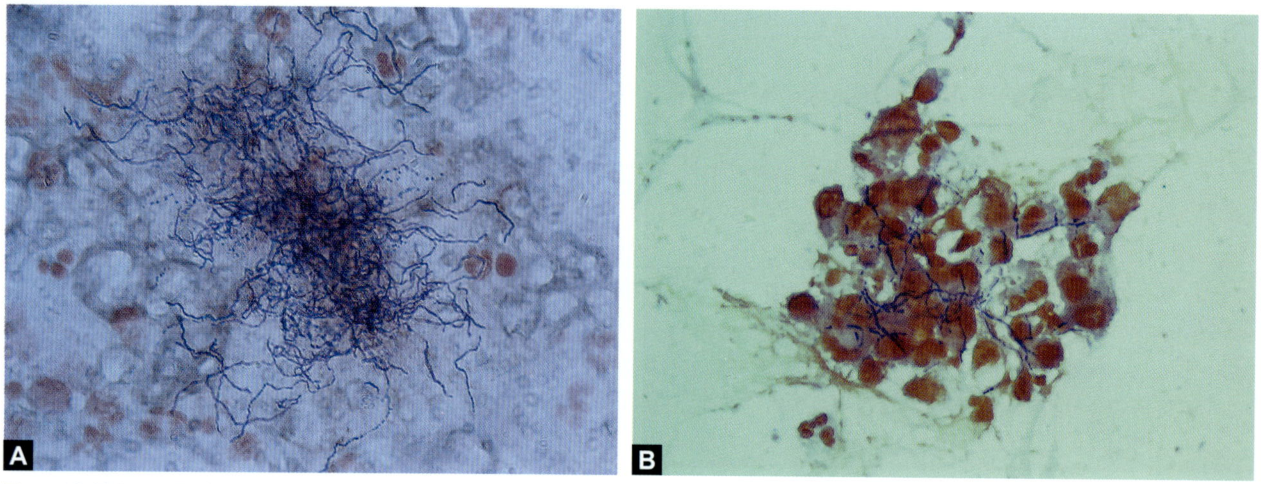

Figs. 11.10A and B. Gram-positive filamentous bacteria in grains of actinomycetoma (Gram's Staining × 1000).

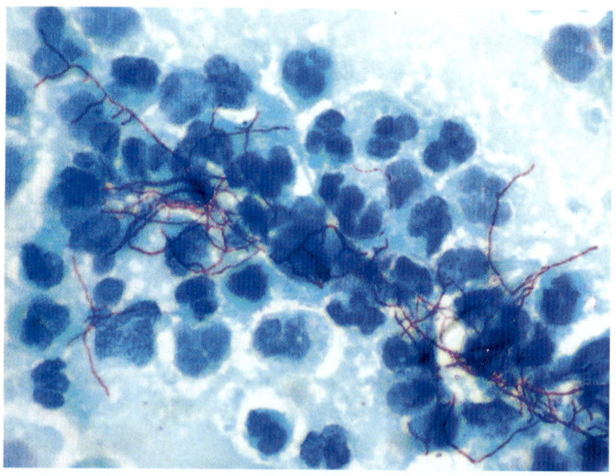

Fig. 11.11. Modified acid-fast staining showing pink-colored mass of *Nocardia* species (Ziehl-Neelsen Staining × 1000).

If wet mount smear shows septate hyphae in grain material, then special fungal stains are done to confirm such findings. The granulomatous reaction and palisade arrangement of organisms in mycetoma is seen in hematoxylin and eosin (H&E) staining **(Figs. 11.12A and B)**, PAS staining **(Figs. 11.13A to E)** and GMS staining **(Figs. 11.14A and B)** embedded in the cement substance.

(b) Culture

The culture of grains is done on different sets of media keeping in view possibilities of both fungi and bacteria. The sample of grains with deep biopsy is considered best as it contains viable organism.

When actinomycetoma, suspected by observation of Gram-positive filaments of crushed grains, these are washed thoroughly several times with normal saline without antibiotics and then inoculated on to the culture media *i.e.* Sabouraud dextrose agar without antibiotics, blood agar, Lowenstein-Jensen media and brain-heart infusion agar.

When eumycetoma is suspected by direct examination of grains then these are washed several times in normal saline with antibiotics - streptopenicillin, penicillin and inoculated on Sabouraud dextrose agar with antibiotic like chloramphenicol and gentamicin. Modified Sabouraud dextrose agar *i.e.* Emmons' modification is usually preferred. Cycloheximide (Actidione) should not be added in the culture media for isolation of eumycetoma agents because most of the causative fungi are sensitive to it **(Figs. 11.15A and B)**.

This is recommended to inoculate as many grains as possible on several plates because it enhances chances of isolation of the organism in duplicate sets. The cultures are incubated at 25°C, 37°C and 44°C because different levels of optimum temperatures are required for various organisms.

The specimens suspected of having bacterial agent are inoculated on Blood agar, LJ and SDA tubes/plates **(Figs. 11.16A to D)**.

The characteristics of frequently isolated agents of mycetoma are shown in **Tables 11.4** and **11.5** and brief description of both fungal and bacterial agents is given below:

(i) Madurella mycetomatis: The grains are <2 mm, firm to brittle, oval to lobed, black, hard, cemented together

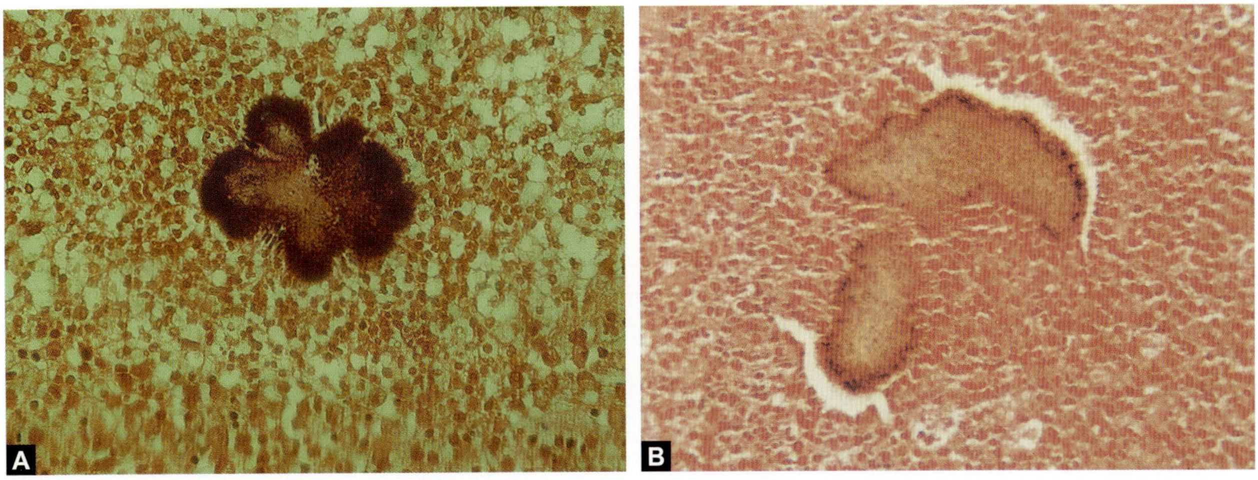

Figs. 11.12A and B. Histopathology showing mass of organisms in cement substance at margins of grains of mycetoma (H&E × 400).

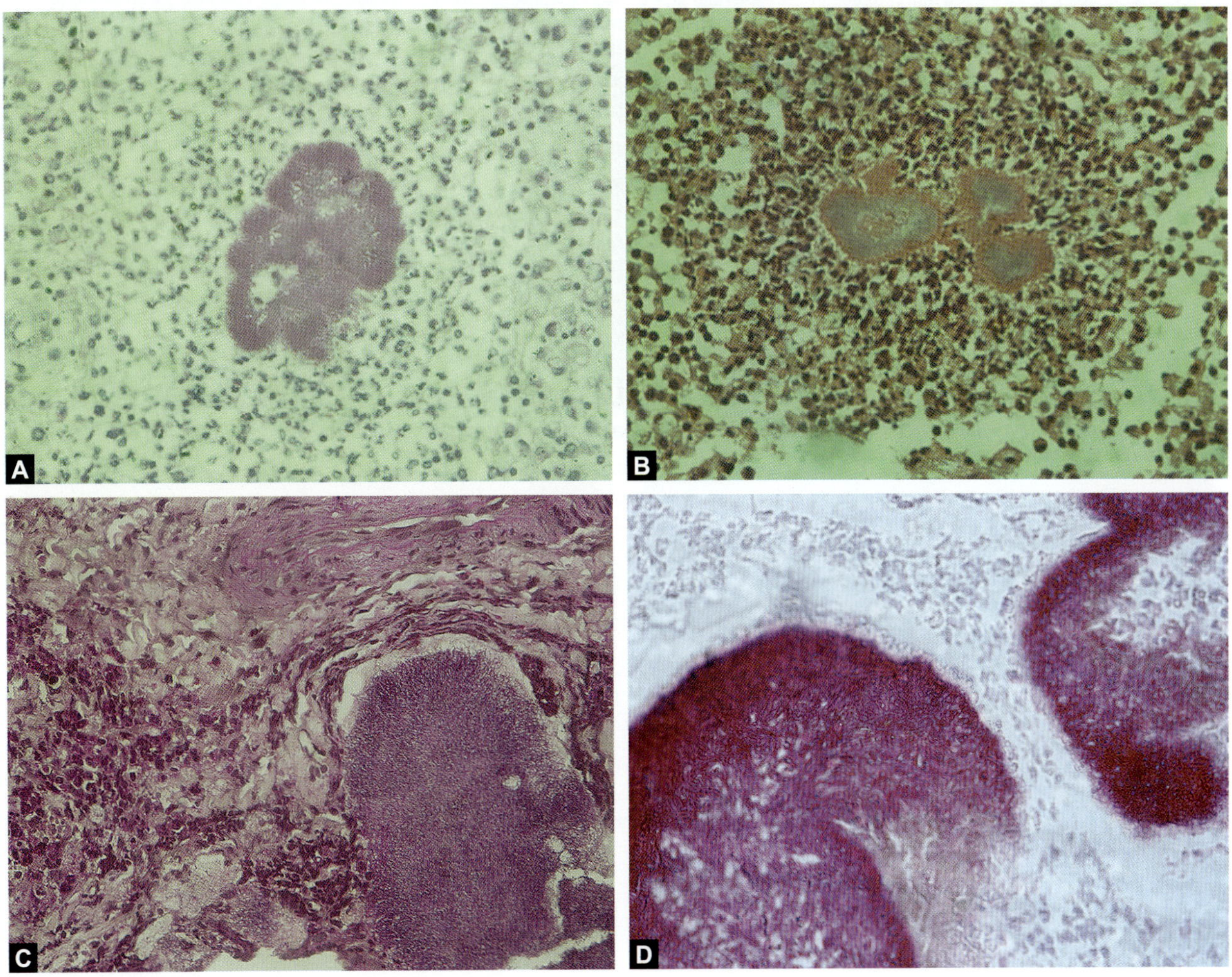

Figs. 11.13A to D

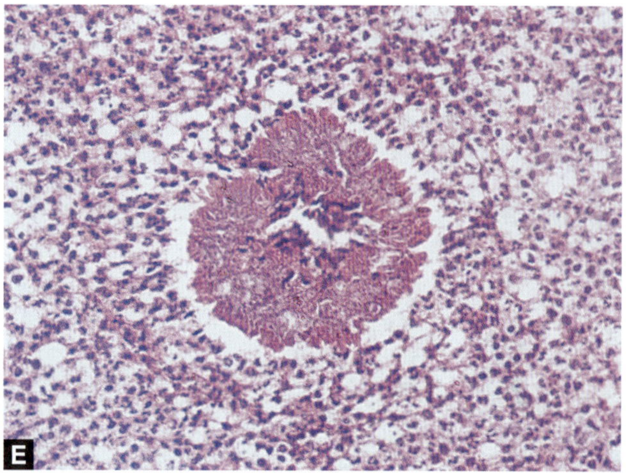

Figs. 11.13A to E. Eumycetoma with granulomatous reaction and palisade arrangement of hyphae in cement substance (PAS × 400).

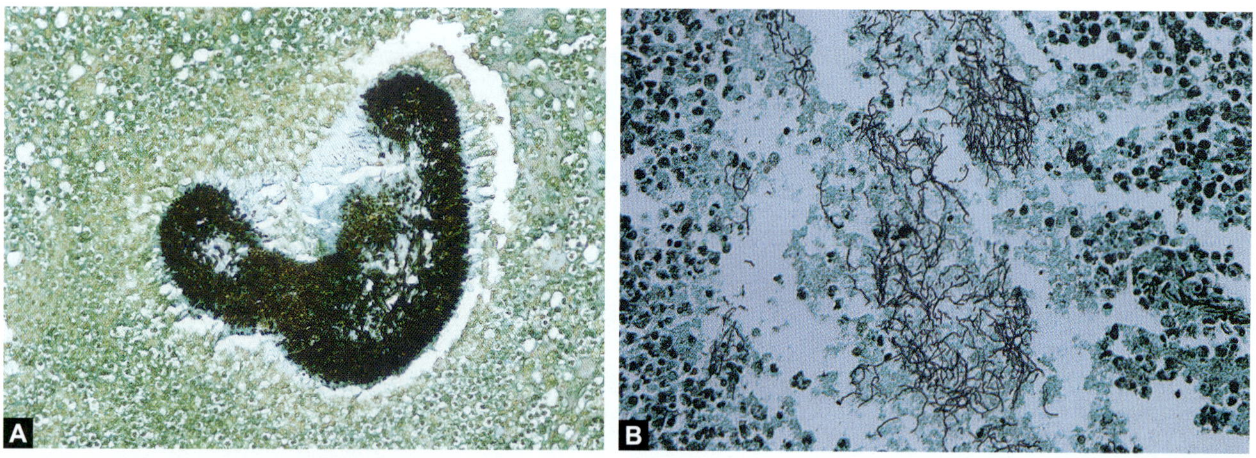

Figs. 11.14A and B. Mycetoma with granulomatous reaction and palisade arrangement of hyphae in cement substance of (A) eumycetoma and (B) actinomycetoma (GMS × 400).

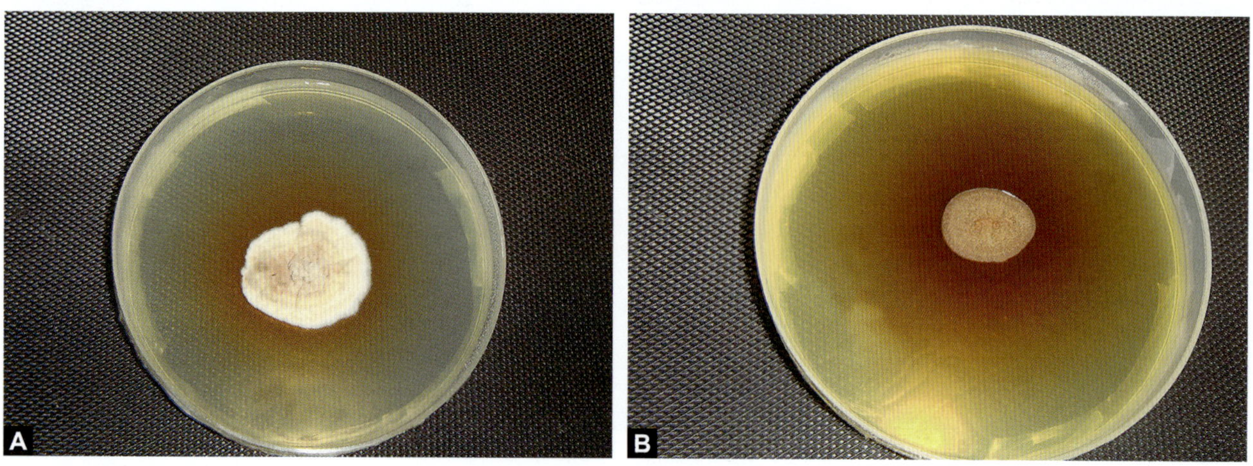

Figs. 11.15A and B. Mycelial growth of *Madurella mycetomatis* with diffusible yellowish-brown pigmentation on SDA.

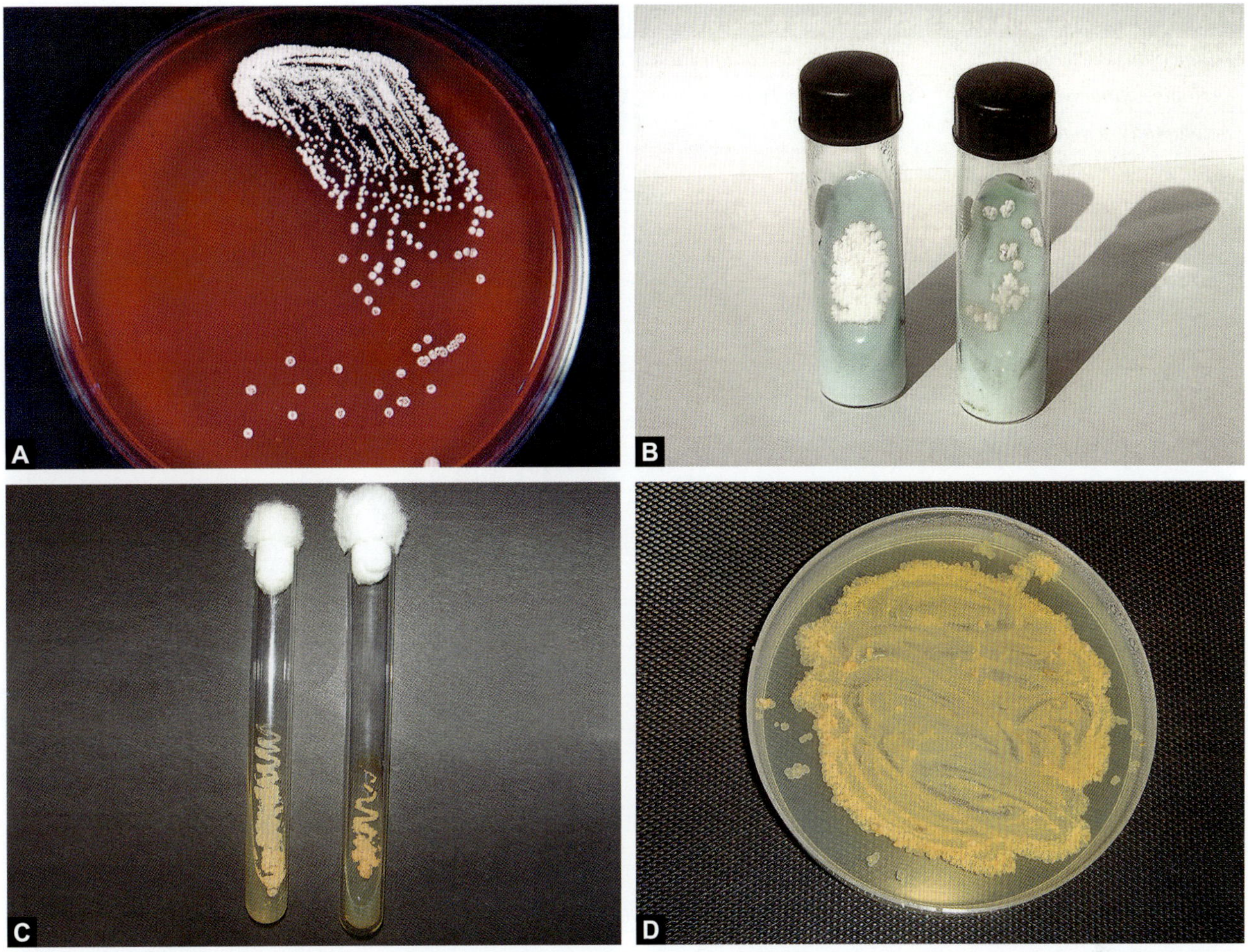

Figs. 11.16A to D. Culture of *Nocardia* species grown from actinomycetoma on blood agar, Lowenstein Jensen, SDA tubes and plate.

Table 11.4. Physiological Characteristics of Causative Agents of Eumycetoma.

	Utilization of					
Fungal Agents	*Glucose*	*Lactose*	*Maltose*	*Sucrose*	*Starch Hydrolysis*	*Protease Activity*
Madurella mycetomatis	+	+	+	–	+	+
Madurella grisea	+	–	+	+	+	+
Exophiala jeanselmei	+	+	+	+	+	–
Pseudallescheria boydii	+	–	+	+	+	–
Acremonium falciforme	+	–	–	+	+	–

and have consistency like that of coal. A second variety of grains is also seen like vesicular, 'open' grain, often with many large cyst cells. The colonies are smooth or folded, glabrous or powdery, leathery and range in color from white to yellowish-brown. There is usually brown diffusible pigment in agar making reverse of plate as dark brown **(*See* Figs. 11.15A and B)**. It is moderately fast growing at 37°C and mature in seven days but grows slowly at 25°C. The mycelia are on an average 1-5 µm, with moniliform hyphae from 2-6 µm and conidia are small, ovoid to globose. On Sabouraud dextrose agar, colonies show septate hyphae with many chlamydospores. On cornmeal

Table 11.5. Physiological Characteristics of the Causative Agents of Actinomycetoma.

Filamentous Bacteria	*Decomposition of Media*					
(Aerobic Actinomycetes)	*Casein*	*Tyrosine*	*Xanthine*	*Urease*	*AFB*	*Growth in Gelatin*
Actinomadura madurae	+	+	–	–	–	–
Actinomadura pelletieri	+	+	–	–	–	–
Nocardia brasiliensis	+	+	–	+	+	+
Nocardia caviae	–	–	+	+	+	–
Nocardia asteroides	–	–	–	+	+	–
Nocardiopsis dassonvillei	+	+	+	±	+	–
Streptomyces somaliensis	+	+	–	–	–	–

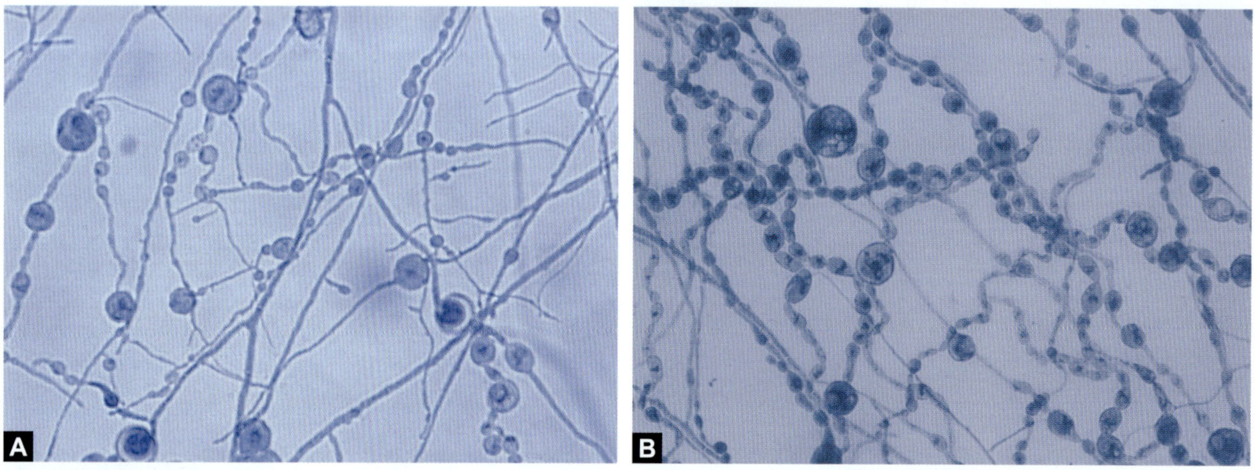

Figs. 11.17A and B. Phialides and conidia seen in culture of *Madurella mycetomatis* on SDA (LCB × 400).

agar (CMA), some strains produce rather short, pointed conidiophores that bear conidia at their tips of flask-shaped phialides **(Figs. 11.17A and B)**. It may also form large-black masses of modified hyphae. The species name of this genus is re-designated from ***'mycetomi'*** to ***'mycetomatis'***.

(ii) Trematosphaeria (Madurella) grisea: The grains are <1 mm, soft to firm, oval to folded, irregular with hyaline center and black periphery. In the section, the grain consists of two distinct zones - a weakly or non-pigmented central zone and a deeply pigmented periphery. The center contains a hyaline hyphal network. Similar type of morphology of grains is seen in mycetoma caused by *Medicopsis (Pyrenochaeta) romeroi*. The colonies are rapidly growing, measure 3-5 cm and gray or olive in color. They are raised, leathery and folded with short, gray aerial hyphae. The older colonies become reddish-brown. The hyphae are brown-colored, septate, 1-3 µm in diameter. There are mycelia of two widths along with chains of arthroconidia.

(iii) Exophiala jeanselmei: The black grains produced by *E.jeanselmei* are smallest among eumycetes measuring about 0.2-0.3 mm in size. The colonies on Sabouraud dextrose agar are restricted, initially moist and soon becoming velvety with olivaceous-green aerial mycelia. The yeast-like growth consists of sub-spherical cells with annellidic butt producing conidia often cohering long chains and changing over into hyphae. The conidiogenous cells of hyphae are intercalary or rocket-shaped, with conspicuous annellated zones, latter being slightly tapering, nearly smooth and produce narrow ellipsoidal conidia **(*See* Fig. 14.4)**.

(iv) Actinomadura madurae: The grains produced by *A.madurae* are white, soft, oval to lobed, rarely pinkish, largest in size measuring 2-5 mm **(*See* Fig. 11.8)**. The colonies rapidly grow on Lowenstein-Jensen medium, whereas slowly on Sabouraud dextrose agar. The growth appears as white to tan, pink or orange, glabrous, wrinkled, hard and adherent. The organism is Gram-positive with non-

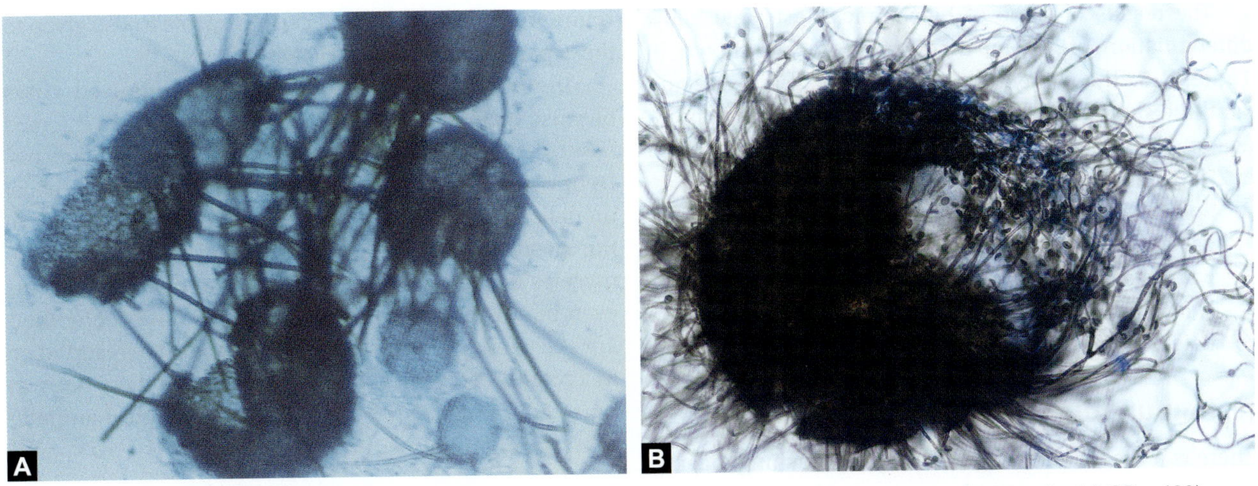

Figs. 11.18A and B. Pycnidia with setae of *Medicopsis (Pyrenochaeta) romeroi* isolated from eumycetoma patient (LCB × 400).

fragmentary delicate filaments. Short chains of spherical spores may also be formed in some isolates and these are negative for acid-fast bacilli stain.

(v) Actinomadura pelletieri: The grains are red, hard, small, oval to lobed and about 1 mm in size. The colonies are slow growing, either folded or wrinkled small dry and granular or glabrous and adherent. The color is at first pink to peach, developing into deep cranberry red with waxy surface. The branching filaments are delicate, 0.5 to 1 µm in diameter, which are not acid-fast and spores are rarely produced.

(vi) Nocardia brasiliensis: The grains are white to yellow, lobed, soft and small in size measuring less than 0.5 mm. The colonies are slow growing wrinkled or folded, yellow to orange in color, dry and chalky or glabrous with an earthy odor. All *Nocardia* species grow moderately rapid; mature in 7-9 days when inoculated on blood agar, Lowenstein-Jensen medium and SDA without antibiotics . The smears show Gram-positive, branching filaments that fragment into bacillary forms, acid-fast bacilli and chains of spores.

(vii) Nocardia caviae: It is also called *N.otitidiscaviarum*. The grains are similar to *N.brasiliensis*. The growth is moderately rapid, resembling both *N.brasiliensis* and *N.asteroides*. On special media, cream to peach colored colonies are formed. There are tufts of short aerial growth and organism is also acid-fast.

(viii) Nocardia asteroides: The grains are rare, white, soft, irregular, small in size measuring less than 0.5 mm. The colonies are glabrous, chalky, folded or wrinkled, white to orange pink in color and have aerial growth. Microscopically, morphology is very similar to *Nocardia brasiliensis*.

(ix) Streptomyces somaliensis: The grains are yellow, hard, round to oval, large and about 2 mm in size. The colonies are creamy, wrinkled with flaky growth, better seen on Lowenstein-Jensen medium. Sectoring of colonies is common, giving varying consistency and mottled colonizing areas of brown color may develop and underside often shows yellow color. Tufts of brownish aerial filaments develop which may become gray with passage of time. A diffusible brown pigment has been reported. The optimum growth is at 30°C. The organism is non-fragmenting and may form spores in chains. Gram-stained smears show delicate Gram-positive branching filaments 0.5 to 1 µm in diameter. They are non-acid-fast bacilli.

Some of the fungi pose difficulty in identification because these do not frequently sporulate *e.g. Medicopsis (Pyrenochaeta) romeroi*, which is very rarely isolated fungal species in eumycetoma cases. In such circumstances the only way out is identification by molecular based techniques. However, occasionally these may sporulate by putting slide cultures. The microscopic appearance of from culture *Medicopsis romeroi* is shown in **Figures 11.18A and B**.

The colony morphology of other agents of mycetoma like *Curvularia geniculata*, *Scedosporium apiospermum*, *Pseudallescheria boydii*, *Acremonium falciforme* and *Fusarium* species, have been described in their respective Chapters.

The results of some of biochemical tests that allow differentiation of aerobic Actinomycetes are shown in **Table 11.5**. The ability to hydrolyze casein, tyrosine and xanthine should be done with heavy streak inoculum on agar media in petri dish. The biochemical tests are carried at 25-30°C.

Special media for physiological tests like:

- gelatin medium for liquefaction test to detect proteolytic activity,
- starch medium for hydrolysis test to indicate amylolytic activity may be used.

(c) Immunodiagnosis

There is no regular serological study for reaching diagnosis or assessing prognosis of mycetoma. The immunological response of individuals with mycetoma is assessed by several serological methods such as complement fixation test, immunodiffusion, counterimmunoelectrophoresis, enzyme-linked immunosorbent assays and Western blot. Immunochemical stainings are also done in these cases. The skin tests are not useful in diagnosis but serodiagnostic techniques have been of some clinical significance. In most of tests, cell extracts rather than exoantigens are used and cross-reactivity with other diseases is a hindrance.

(d) Animal Pathogenicity

The grains containing lesions of mycetoma are not readily produced with causative organisms in experimental animals. The animal inoculation is done in guinea pig, mice and goat. The suspected organism is injected subcutaneously and scattered, delicate, Gram-positive and acid-fast filaments can be seen in inflammatory lesions and exudate.

Treatment and Prophylaxis

The management of mycetoma is highly challenging for the treating clinicians. Many drugs are tried with variable results. For **eumycetoma**, oral ketoconazole 200 mg twice daily and itraconazole 100 mg twice daily are given for 8-24 months. Conventional and liposomal formulations of amphotericin B are recommended for eumycetoma caused by *Madurella* and *Fusarium* species.

Many antibiotics and antimicrobial agents like sulfonamides (cotrimoxazole), tetracyclines, streptomycin, amoxycillin, clavulanate and amikacin have been used in the treatment of **actinomycetoma**. Combination therapy like sulfonamides and rifampicin may be given. **Welsh regimen** with amikacin and cotrimoxazole is also given to these patients for a period of five weeks.

The surgery may be required in protracted cases because most of the drugs do not penetrate into infected tissues adequately to eradicate the causative organisms. Debridement of infected tissue with split graft is usual modality. The prognosis differs markedly accordingly to the involved causative agents. The infections with actinomycetes respond well to therapy with antibacterial agents even in advanced forms but in eumycetoma chemotherapy is less effective and radical surgery is often resorted on as a radical measure. It is also observed that there is recurrence after medical treatment and even surgical amputation (*See* **Fig. 11.5**).

Botryomycosis

This is chronic, purulent, granule-producing, granulomatous lesion of skin, subcutaneous tissues and visceral organs, caused by several bacterial species. This is not a fungal disease as its name apparently sounds. The names **schizomycosis** and **granular bacteriosis** have also been proposed for this disease due to production of '**sulfur**' granules. In such cases neutrophilic reaction is most important whereas in mycetoma it is cement substance, which is present in between causative organisms. Bollinger described botryomycosis for the first time in 1870, while observing numerous granulomatous lesions that appeared after castration of horses.

This disease occurs more frequently in patients with deficient cell-mediated immunity. Clinically and histopathologically, it resembles to mycetoma as well as actinomycosis. The bacterial species involved are: *Staphylococcus aureus, Escherichia coli, Actinobacillus lignieresii, Pseudomonas aeruginosa*, non-hemolytic streptococci, *Proteus, Bacteriodes* and *Rhodococcus* species. Some of the important species are shown in **Table 11.6**.

Table 11.6. Causative Bacteria of Botryomycosis.

Bacterial Agents	*Color of Granules*
Staphylococcus aureus	White
Pseudomonas aeruginosa	White
Coagulase-negative Staphylococci	White
Escherichia coli	White
Proteus vulgaris	White
Propionibacterium acnes	White
Actinobacillus lignieresi	Yellow
Streptococcus species	White
Peptostreptococcus species	White

The lesions may develop anywhere on skin, head, pinnae, hands and feet. Disseminated lesions are usually seen in the debilitated patients. Several cases have been reported in children with cystic fibrosis as well as following surgery, skin abrasions, lacerations and piercing of ear. On the basis of clinical manifestations and distribution of lesions, botryomycosis is divided into two types:

(a) Cutaneous Type: In this type, the cutaneous lesions may range from simple inflammatory nodules to discharging ulcers or fistulae and are usually localized, affecting exposed surfaces such as hands, feet, head as well as intertriginous areas. Among the AIDS patients, the clinical presentation is usually non-specific type.

(b) Visceral Type: This type occurs in chronically ill and debilitated patients. The lungs are most frequently involved but dissemination to other visceral organs such as liver, kidney and lymph nodes has also been reported.

The differential diagnosis of botryomycosis, in addition to mycetoma includes tuberculosis, atypical mycobacterial infections, actinomycosis and sporotrichosis. Sometimes non-infectious conditions such as lymphoma and Kaposi's sarcoma may be confused with this clinical entity.

The clinical material is examined directly in wet preparations and routine differential stains. On histopathological examination, tissue reveals neutrophils, lymphocytes, eosinophils, plasma cells and fibroblasts as well as histiocytes and scattered foreign body giant cells. The '**sulfur**' granules are found in the center of granuloma and are covered by neutrophils. The eosinophilic shell around granules is PAS-positive, which is absent in case of staphylococcal abscesses. The culture is put up on routine media used in bacteriology like blood agar, chocolate agar, Lowenstein-Jensen medium as well as fungal culture medium like Sabouraud dextrose agar.

The treatment of botryomycosis depends on the nature of causative bacterium and its susceptibility pattern to various antimicrobial drugs.

Further Reading

1. Abd El-Bagi ME, Fahal AH. Mycetoma revisited. Incidence of various radiographic signs. *Saudi Med J.* 2009; **30:** 529-33.
2. Adhikari L, Dey S, Pal R. Mycetoma due to *Nocardia farcinica. J Glob Infect Dis.* 2010; **2:** 194-5.
3. Afroz N, Khan N, Siddiqui FA, *et al.* Eumycetoma versus actinomycetoma: Diagnosis on cytology. *J Cytol.*2010; **27:** 133-5.
4. Agarwal US, Besarwal RK, Gupta R, *et al.* Treatment of actinomycetoma foot - Our experience with ten patients. *J Eur Acad Dermatol Venereol.* 2013; **27:** 1505-13.
5. Aguilar-Donis A, Torres-Guerrero E, Arenas-Guzman R, *et al.* Mycetoma caused by *Phaeoacremonium parasiticum* - A case confirmed with β-tubulin sequence analysis. *Mycoses.* 2011; **54:** e615-8.
6. Ahmadzai H, Raley DA, Masters L, *et al.* An unusual case of a pituitary fossa aspergilloma in an immunocompetent patient mimicking infiltrative tumour. *J Surg Case Rep.* 2013; pii: rjt018. PMID: 24964429.
7. Ahmed AA, van de Sande WW, Fahal A, *et al.* Management of mycetoma: major challenge in tropical mycoses with limited international recognition. *Curr Opin Infect Dis.* 2007; **20:** 146-51.
8. Ahmed SA, Abbas MA, Jouvion G, *et al.* Seventeen years of subcutaneous infection by *Aspergillus flavus*; Eumycetoma confirmed by immunohistochemistry. *Mycoses.* 2015; **58:** 728-34.
9. Ahmed SA, de Hoog GS, Stevens DA, *et al. In vitro* antifungal susceptibility of coelomycete agents of black grain eumycetoma to eight antifungals. *Med Mycol.* 2015; **53:** 295-301.
10. Ahmed SA, Kloezen W, Duncanson F, *et al. Madurella mycetomatis* is highly susceptible to ravuconazole. *PLoS Negl Trop Dis.* 2014; **8:** e2942.
11. Ahmed SA, Kloezen W, Fahal AH, *et al. In vitro* interaction of currently used azoles with terbinafine against *Madurella mycetomatis. Antimicrob Agents Chemother.* 2015; **59:** 1373-4.
12. Ahmed SA, van de Sande WW, Desnos-Ollivier M, *et al.* Application of isothermal amplification techniques for identification of *Madurella mycetomatis*, the prevalent agent of human mycetoma. *J Clin Microbiol.* 2015; **53:** 3280-5.
13. Ahmed SA, van de Sande WW, Stevens DA, *et al.* Revision of agents of black-grain eumycetoma in the order Pleosporales. *Persoonia.* 2014; **33:** 141-54.
14. Ahmed SA, van den Ende BH, Fahal AH, *et al.* Rapid identification of black grain eumycetoma causative agents using rolling circle amplification. *PLoS Negl Trop Dis.* 2014; **8:** e3368.
15. Ajello L, Brown J, Macdonald E, *et al.* Actinomycetoma caused by *Nocardiopsis dassonvillei. Arch Dermatol.* 1987; **123:** 426.
16. Al-Hatmi AM, Bonifaz A, Tirado-Sanchez A, *et al. Fusarium* species causing eumycetoma: Report of two cases and comprehensive review of the literature. *Mycoses.* 2017; **60:** 204-12.
17. Al-Tawfiq JA, Amr SS. Madura leg due to *Exophiala jeanselmei* successfully treated with surgery and itraconazole therapy. *Med Mycol.* 2009; **47:** 648-52.
18. Al-Tawfiq JA. Chronic foot swelling with purulent discharge. *Int J Dermatol.* 2013; **52:** 1595-6.
19. Alam K, Maheshwari V, Bhargava S, *et al.* Histological diagnosis of madura foot (mycetoma): A must for definitive treatment. *J Glob Infect Dis.* 2009; **1:** 64-7.

20. Altoe A, Mendonca S, Taboada G, *et al.* Actinomycetoma of the foot. *Arthritis Rheumatol.* 2014; **66:** 2280.
21. Amatya R, Khanal B, Rijal A. *Syncephalastrum* species producing mycetoma-like lesions. *Indian J Dermatol Venereol Leprol.* 2010; **76:** 284-6.
22. Ameen M, Arenas R, Vasquez del Mercado E, *et al.* Efficacy of imipenem therapy for *Nocardia* actinomycetomas refractory to sulfonamides. *J Am Acad Dermatol.* 2010; **62:** 239-46.
23. Ameen M, Arenas R. Developments in the management of mycetomas. *Clin Exp Dermatol.* 2009; **34:** 1-7.
24. Ameen M, Arenas R. Emerging therapeutic regimes for the management of mycetomas. *Expert Opin Pharmacother.* 2008; **9:** 2077-85.
25. Ameen M. Managing mycetomas. *Trop Doct.* 2009; **39:** 66-8.
26. Anandi V, Jeya M, Subramaniyan CS, *et al.* Actinomycotic mycetoma of thumb. *Indian J Med Microbiol.* 1997; **15:** 43-44.
27. Antrobus R, Wong G, Jones J, *et al.* Mycetoma caused by *Acremonium* species in a patient with chronic granulomatous disease. *Case Reports Immunol.* 2016; 3209493.
28. Antunes J, Pacheco D, Travassos R, *et al.* Actinomycetoma of the chest wall attributed to *Nocardia nova* after reconstructive surgery. *Dermatol Online J.* 2012; **18:** 4.
29. Arenas R, Ameen M. Giant grains of *Nocardia* actinomycetoma. *Lancet Infect Dis.* 2010; **10:** 66.
30. Arenas R, Fernandez Martinez RF, Torres-Guerrero E, *et al.* Actinomycetoma: An update on diagnosis and treatment. *Cutis.* 2017; **99:** e11-5. PMID: 28319638.
31. Arenas R, Moreno-Coutino G, Welsh O. Classification of subcutaneous and systemic mycoses. *Clin Dermatol.* 2012; **30:** 369-71.
32. Asly M, Rafaoui A, Bouyermane H, *et al.* Mycetoma (Madura foot): A case report. *Ann Phys Rehabil Med.* 2010; **53:** 650-4.
33. Babcock MJ, Diaz L, Reddy S, *et al.* Painful swollen foot. Mycetoma (Madura foot). *Arch Dermatol.* 2011; **147:** 609-14.
34. Bachh AA, Haq I, Gupta R, *et al.* Pulmonary candidiasis presenting as mycetoma. *Lung India.* 2008; **25:** 165-7.
35. Bafghi MF, Heidarieh P, Soori T, *et al.* *Nocardia* isolation from clinical samples with the paraffin baiting technique. *Germs.* 2015; **5:** 12-6.
36. Bakshi R, Mathur DR. Incidence and changing pattern of mycetoma in western Rajasthan. *Indian J Pathol Microbiol.* 2008; **51:** 154-5.
37. Banashankari GS, Rudresh HK, Sreeja S, *et al.* Nocardial mycetoma: Re-infection with a different species at an unusual site - A rare case. *Indian J Pathol Microbiol.* 2013; **56:** 333-4.
38. Bashline B, Morrison M, Ramirez J, *et al.* Disseminated botryomycosis: A rare presentation. *J Drugs Dermatol.* 2014; **13:** 976-8.
39. Bassiri-Jahromi S. Mycetoma in Iran: Causative agents and geographic distribution. *Indian J Dermatol.* 2014; **59:** 529.
40. Basyigit I, Yazici A, Ucar EK, *et al.* Treatment for mycetoma with oral itraconazole in ankylosing spondylitis. *Rheumatol Int.* 2013; **33:** 1375-6.
41. Batran SE. Efficient staining of actinomycetoma and eumycetoma grains using henna extract. *Biotech Histochem.* 2015; **90:** 568-72.
42. Bedi TR, Kaur S, Kumar B. Red grain mycetoma of the scalp *(Actinomadura pelletieri):* a case report from India. *Mycopathologia.* 1978; **63:** 127-28.
43. Bekci T, Kesli R, Maden E, *et al.* Ankylosing spondylitis with mycetoma case treated with oral itraconazole. *Rheumatol Int.* 2010; **30:** 1397-9.
44. Beraud G, Desbois N, Coyo C, *et al.* Paradoxical response preceding control of *Scedosporium apiospermum* mycetoma with posaconazole treatment. *Infect Dis (Lond).* 2015; **47:** 830-3.
45. Besson FL, Blanc-Durand P, Meyer C, *et al.* PET/CT and MRI imaging of a eumycetoma of the right thigh. *Clin Nucl Med.* 2017; **42:** 61-3.
46. Bhat RM, Monteiro RC, Bala N, *et al.* Subcutaneous mycoses in coastal Karnataka in south India. *Int J Dermatol.* 2016; **55:** 70-8.
47. Bonifaz A, de Hoog S, McGinnis MR, *et al.* Eumycetoma caused by *Cladophialophora bantiana* successfully treated with itraconazole. *Med Mycol.* 2009; **47:** 111-4.
48. Bonifaz A, Tirado-Sanchez A, Calderon L, *et al.* Mycetoma: Experience of 482 cases in a single center in Mexico. *PLoS Negl Trop Dis.* 2014; **8:** e3102.
49. Bordoloi P, Nath R, Borgohain M, *et al.* Subcutaneous mycoses: An aetiological study of 15 cases in a tertiary care hospital at Dibrugarh, Assam, northeast India. *Mycopathologia.* 2015; **179:** 425-35.
50. Borman AM, Desnos-Ollivier M, Campbell CK, *et al.* Novel taxa associated with human fungal black-grain mycetomas: *Emarellia grisea* gen. nov., sp. nov., and *Emarellia paragrisea* sp. nov. *J Clin Microbiol.* 2016; **54:** 1738-45.
51. Bouziane M, Amriss O, Kadiri R, *et al.* The role of computed tomography in the exploration of Madura foot (pedal mycetoma). *Diagn Interv Imaging.* 2012; **93:** 884-6.
52. Brufman T, Ben-Ami R, Mizrahi M, *et al.* Mycetoma of the foot caused by *Madurella mycetomatis* in immigrants from Sudan. *Isr Med Assoc J.* 2015; **17:** 418-20.
53. Buonfrate D, Gobbi F, Angheben A, *et al.* Autochthonous cases of mycetoma in Europe: Report of two cases and review of literature. *PLoS One.* 2014; **9:** e100590.
54. Burki TK. Starting from scratch the unique neglect of mycetoma. *Lancet Infect Dis.* 2016; **16:** 1011-2.
55. Bustamante B, Ahmed SA, De Hoog GS, *et al.* *Phaeoacremonium sphinctrophorum* as a novel agent of eumycetoma. *JAMA Dermatol.* 2016; **152:** 1063-5.
56. Cameron K, Bannon K, Mittal V. Abdominal wall mass. Fungal mycetoma. *JAMA Surg.* 2014; **149:** 743-4.
57. Campos-Macias P, Arenas-Guzman R, Hernandez-Hernandez F. *Fusarium subglutinans:* A new eumycetoma agent. *Med Mycol Case Rep.* 2013; **2:** 128-31.
58. Capoor MR, Khanna G, Nair D, *et al.* Eumycetoma pedis due to *Exophiala jeanselmei. Indian J Med Microbiol.* 2007; **25:** 155-7.

59. Carison H. Mycetoma (Madura foot). *JAAPA*. 2011; **24:** 79.
60. Cartwright KE, Clark TW, Hussain AM, *et al.* Eumycetoma of the hand caused by *Leptosphaeria tompkinsii* and refractory to medical therapy with voriconazole. *Mycopathologia*. 2011; **172:** 311-5.
61. Cascio A, Mandraffino G, Cinquegrani M, *et al. Actinomadura pelletieri* mycetoma - An atypical case with spine and abdominal wall involvement. *J Med Microbiol.* 2011; **60:** 673-6.
62. Castro LG, Piquero-Casals J. Clinical and mycologic findings and therapeutic outcome of 27 mycetoma patients from Sao Paulo, Brazil. *Int J Dermatol.*2008; **47:** 160-3.
63. Castro-Matteotti B, Vera-Cabrera L, Ocampo-Candiani J, *et al.* Immune response to *Nocardia brasiliensis* extracellular antigens in patients with mycetoma. *Mycopathologia*. 2008; **165:** 127-34.
64. Cerar D, Malallah YM, Howard SJ, *et al.* Isolation, identification and susceptibility of *Pyrenochaeta romeroi* in a case of eumycetoma of the foot in the UK. *Int J Antimicrob Agents.* 2009; **34:** 617-8.
65. Chakrabarti A, Singh K. Mycetoma in Chandigarh and surrounding areas. *Indian J Med Microbiol.*1998; **16:** 64-5.
66. Chande C, Menon S, Gohil A, *et al.* Bilateral mycetoma - A case report. *Mycoses.* 2011; **54:** e606-8.
67. Chander J, Kanwar AJ, Lal P, *et al.* Recurrent actinomycetoma of amputated stump of the lower limb caused by *Actinomadura madurae. Mycoses.* 1999; **42:** 171.
68. Chander J, Singla N, Handa U. Human cervicofacial mycetoma caused by *Streptomyces griseus:* First case report. *J Microbiol Immunol Infect.* 2015; **48:** 703-5.
69. Chazan B, Colodner R, Polacheck I, *et al.* Mycetoma of the foot caused by *Cylindrocarpon lichenicola* in an immunocompetent traveler. *J Travel Med.* 2004; **11:** 331-2.
70. Chen B, Zhu LY, Xuan X, *et al.* Isolation of both *Pseudozyma aphidis* and *Nocardia otitidiscaviarum* from a mycetoma on the leg. *Int J Dermatol.* 2011; **50:** 714-9.
71. Chen HC, Wu SH, Hsiao YC. Unexpected disappearance of a mycetoma. *Am J Med Sci.* 2012; **343:** 243.
72. Cherian RS, Betty M, Manipadam MT, *et al.* The "dot-in-circle" sign - A characteristic MRI finding in mycetoma foot: A report of three cases. *Br J Radiol.* 2009; **82:** 662-5.
73. Chiapello LS, Dib MD, Nuncira CT, *et al.* Mycetoma of the scalp due to *Microsporum canis:* Histopathologic, mycologic and immunogenetic features in a 6-year-old girl. *Diagn Microbiol Infect Dis.* 2011; **70:** 145-9.
74. Chintaginjala A, Harshavardhan K, Kumar AS. Cutaneous botryomycosis: A rare case report. *Indian J Dermatol.* 2016; **61:** 126.
75. Chufal SS, Thapliyal NC, Gupta MK. An approach to histology-based diagnosis and treatment of Madura foot. *J Infect Dev Ctries.*2012; **6:** 684-8.
76. Chugh TD, Arora DR. Mycetoma caused by *Streptomyces somaliensis. Indian J Pathol Bacteriol.* 1975; **18:** 49-53.
77. Cordeiro F, Bruno C, Reis C. Mycetoma. *Am J Trop Med Hyg.* 2011; **85:** 791.
78. Crabol Y, Poiree S, Bougnoux ME, *et al.* Last generation triazoles for imported eumycetoma in eleven consecutive adults. *PLoS Negl Trop Dis.* 2014; **8:** e3232.
79. Damle DK, Mahajan PM, Pradhan SN, *et al.* Modified Welsh regimen: A promising therapy for actinomycetoma. *J Drugs Dermatol.* 2008; **7:** 853-6.
80. de Hoog GS, Ahmed SA, Najafzadeh MJ, *et al.* Phylogenetic findings suggest possible new habitat and routes of infection of human eumycetoma. *PLoS Negl Trop Dis.* 2013; **7:** e2229.
81. de Hoog GS, van Diepeningen AD, Mahgoub el-S, *et al.* New species of *Madurella*, causative agents of black-grain mycetoma.*J Clin Microbiol.*2012; **50:** 988-94.
82. de Palma L, Marinelli M, Pavan M, *et al.* A rare European case of madura foot due to actinomycetes. *Joint Bone Spine.* 2006; **73:** 321-4.
83. Desnos-Ollivier M, Bretagne S, Dromer F, *et al.* Molecular identification of black-grain mycetoma agents. *J Clin Microbiol.* 2006; **44:** 3517-23.
84. Desoubeaux G, Millon A, Freychet B, *et al.* Eumycetoma of the foot caused by *Exophiala jeanselmei* in a Guinean woman. *J Mycol Med.* 2013; **23:** 168-75.
85. Devi B, Behera B, Dash M, *et al.* Botryomycosis. *Indian J Dermatol.*2013; **58:** 406.
86. Dhingra M, Kaistha N, Bansal N, Solanki LS, Chander J, *et al. Nocardia nova* mycetoma over forehead in a lepromatous leprosy patient. *Dermatol Online J.* 2012; **18:** 3.
87. Di Benedetto G, Citro R, Longobardi A, *et al.* Giant *Candida* mycetoma in an ascending aorta tubular graft. *J Card Surg.* 2013; **28:** 557-60.
88. Difonzo EM, Massi D, Vanzi L, *et al. Madurella mycetomatis* mycetoma treated successfully with oral posaconazole. *J Chemother.* 2011; **23:** 243-4.
89. Dillard TA, Ortega I. Multiple endobronchial mycetomas with varied appearances and mixed fungal flora. *J Bronchology Interv Pulmonol.* 2013; **20:** 147-9.
90. Dlova NC, Mosam A. Actinomycetoma. *Skinmed.* 2012; **10:** 98.
91. Dlova NC, Mosam A. Mycetoma. Skinmed. 2012; **10:** 181.
92. Dresch TF, de Magalhaes TC, Pineiro-Maceira J, *et al.* Combined therapy for mycetoma: Medical and surgical. *Dermatol Surg.* 2010; **36:** 952-4.
93. Dupont B, Datry A, Poiree S, *et al.* Role of a NSAID in the apparent cure of a fungal mycetoma. *J Mycol Med.* 2016; **26:** 86-93.
94. Eadie K, Fahal A, van de Sande WW. *In vitro* activity of antiseptic solutions against *Madurella mycetomatis:* Implications for eumycetoma management. *Br J Dermatol.* 2015; **172:** 1657-9.
95. Efared B, Tahiri L, Boubacar MS, *et al.* Mycetoma in a non-endemic area: A diagnostic challenge. *BMC Clin Pathol.* 2017; **17:** 1. PMID: 28167862.
96. El Muttardi N, Kulendren D, Jemec B. Madura foot - mind the soil. *J Plast Reconstr Aesthet Surg.* 2010; **63:** e576-8.
97. El Shamy ME, Fahal AH, Shakir MY, *et al.* New MRI grading system for the diagnosis and management of mycetoma. *Trans R Soc Trop Med Hyg.* 2012; **106:** 738-42.
98. El-Sobky TA, Haleem JF, Samir S. Eumycetoma osteomyelitis of the calcaneus in a child: A radiologic-pathologic correlation following total calcanectomy. *Case Rep Pathol.* 2015; 129020. PMID: 26483983.

99. ELbadawi HS, Mahgoub E, Mahmoud N, *et al.* Use of immunoblotting in testing *Madurella mycetomatis* specific antigen. *Trans R Soc Trop Med Hyg.* 2016; **110:** 312-6.
100. Elfadil H, Fahal A, Kloezen W, *et al.* The *in vitro* antifungal activity of Sudanese medicinal plants against *Madurella mycetomatis*, the eumycetoma major causative agent. *PLoS Negl Trop Dis.* 2015; **9:** e0003488.
101. Estrada R, Chavez-Lopez G, Estrada-Chavez G, *et al.* Eumycetoma. *Clin Dermatol.* 2012; **30:** 389-96.
102. Estrada-Chavez GE, Vega-Memije ME, Arenas R, *et al.* Eumycotic mycetoma caused by *Madurella mycetomatis* successfully treated with antifungals, surgery and topical negative pressure therapy. *Int J Dermatol.* 2009; **48:** 401-3.
103. Ezaldeen EA, Fahal AH, Osman A. Mycetoma herbal treatment: The Mycetoma Research Centre, Sudan experience. *PLoS Negl Trop Dis.* 2013; **7:** e2400.
104. Fahal A, Mahgoub el S, El Hassan AM, *et al.* Head and neck mycetoma: The Mycetoma Research Centre experience. *PLoS Negl Trop Dis.* 2015; **9:** e0003587.
105. Fahal A, Mahgoub el S, El Hassan AM, *et al.* Mycetoma in the Sudan: An update from the Mycetoma Research Centre, University of Khartoum, Sudan. *PLoS Negl Trop Dis.* 2015; **9:** e0003679.
106. Fahal A, Mahgoub el S, El Hassan AM, *et al.* A new model for management of mycetoma in the Sudan. *PLoS Negl Trop Dis.* 2014; **8:** e3271.
107. Fahal AH, Finkelman MA, Zhang Y, *et al.* Detection of (1→3) β-D-glucan in eumycetoma patients. *J Clin Microbiol.* 2016; **54:** 2614-7.
108. Fahal AH. Mycetoma: A global medical and socio-economic dilemma. *PLoS Negl Trop Dis.* 2017; **11:** e0005509. PMID: 28426654.
109. Fahal AH, Rahman IA, El-Hassan AM, *et al.* The safety and efficacy of itraconazole for the treatment of patients with eumycetoma due to *Madurella mycetomatis. Trans R Soc Trop Med Hyg.* 2011; **105:** 127-32.
110. Fahal AH, Sabaa AH. Mycetoma in children in Sudan. *Trans R Soc Trop Med Hyg.* 2010; **104:** 117-21.
111. Fahal AH, Shaheen S, Jones DH. The orthopaedic aspects of mycetoma. *Bone Joint J.* 2014; **96-B:** 420-5.
112. Fahal AH. Mycetoma: A thorn in the flesh. *Trans R Soc Trop Med Hyg.* 2004; **98:** 3-11.
113. Fanucci E, Nezzo M, Neroni L, *et al.* Diagnosis and treatment of paranasal sinus fungus ball of odontogenic origin: Case report. *Oral Implantol (Rome).* 2014; **6:** 63-6.
114. Fernandes H, D'souza CR, Shekar JC, *et al.* Cytodiagnosis of actinomycetoma. *Diagn Cytopathol.* 2009; **37:** 506-8.
115. Fernandez-Flores A, Saeb-Lima M, Arenas-Guzman R. Morphological findings of deep cutaneous fungal infections. *Am J Dermatopathol.* 2014; **36:** 531-53; Quiz 554-6.
116. Fraser M, Borman AM, Johnson EM. Rapid and robust identification of the agents of black-grain mycetoma by MALDI-ToF mass spectrometry. *J Clin Microbiol.* 2017. pii: JCM.00417-17. PMID: 28592546.
117. Fuller LC. Revisiting antiseptic use in reducing post-surgical recurrence in mycetoma. *Br J Dermatol.* 2015; **172:** 1484-5.
118. Gabhane SK, Gangane N, Anshu. Cytodiagnosis of eumycotic mycetoma: A case report. *Acta Cytol.* 2008; **52:** 354-6.
119. Galor A, Karp CL, Forster RK, *et al.* Subconjunctival mycetoma after sub-Tenon's corticosteroid injection. *Cornea.* 2009; **28:** 933-5.
120. Garg A, Sujatha S, Garg J, *et al.* Eumycetoma due to *Curvularia lunata. Indian J Dermatol Venereol Leprol.* 2008; **74:** 515-6.
121. Garnica M, Nucci M, Queiroz-Telles F. Difficult mycoses of the skin: advances in the epidemiology and management of eumycetoma, phaeohyphomycosis and chromoblastomycosis. *Curr Opin Infect Dis.* 2009; **22:** 559-63.
122. Gilquin JM, Riviere B, Jurado V, *et al.* First case of actinomycetoma in France due to a novel *Nocardia* species, *Nocardia boironii* sp. nov. *mSphere.* 2016 Nov 23;1(6). pii: e00309-16. PMID: 27904882.
123. Goel RS, Kataria R, Sinha VD, *et al.* Craniocerebral maduromycosis. *J Neurosurg Pediatr.* 2012; **10:** 67-70.
124. Goel RS, Kataria R, Sinha VD. Clinical profile and management of craniocerebral *Madurella* mycetoma. Author reply. *J Neurosurg Pediatr.* 2013; **12:** 531.
125. Gooptu S, Ali I, Singh G, *et al.* Mycetoma foot. *J Family Community Med.* 2013; **20:** 136-8.
126. Gosselink C, Thomas J, Brahmbhatt S, *et al.* Nocardiosis causing pedal actinomycetoma: A case report and review of the literature. *J Foot Ankle Surg.* 2008; **47:** 457-62.
127. Gouveia C, Fraga A, Filipe P, *et al.* Mycetoma of the chest wall due to *Nocardia brasiliensis. Acta Derm Venereol.*2008; **88:** 508-10.
128. Grueber HL, Kumar TM. Mycetoma caused by *Streptomyces somaliensis* in north India. *Sabouraudia.* 1970; **8:** 108-11.
129. Guevara D, Wongkittiroch K, Goodman M, *et al. Acremonium* mycetoma: A case report and discussion. *Cutis.* 2011; **88:** 293-5, 299.
130. Gugnani HC, Ojukwu JO, Suseelan AV. Mycetoma of thumb caused by *Nocardia transvalensis. Mycopathologia.* 1982; **80:** 55-60.
131. Gugnani HC, Sehgal VN, Singh VK, *et al. 'Nocardia asteroides'* mycetoma of the foot. *J Eur Acad Dermatol Venereol.* 2002; **16:** 640-2.
132. Gulati V, Bakare S, Tibrewal S, *et al.* A rare presentation of concurrent *Scedosporium apiospermum* and *Madurella grisea* eumycetoma in an immunocompetent host. *Case Rep Pathol.* 2012; 154201. PMID: 23133772.
133. Gunathilake R, Perera P, Sirimanna G. *Curvularia lunata:* A rare cause of black-grain eumycetoma. *J Mycol Med.* 2014; **24:** 158-60.
134. Gupta MK, Banerjee T, Kumar D, *et al.* White grain mycetoma caused by *Scedosporium apiospermum* in north India: A case report. *Int J Low Extrem Wounds.* 2013; **12:** 286-8.
135. Gupta S, Jain K, Parmar C, *et al.* Mycetoma: Non-venerealperineal lesions. *Indian J Sex Transm Dis.* 2010; **31:** 39-41.
136. Gupta SK, Shukla VK, Khanna S. Primary actinomycotic mycetoma of the anterior abdominal wall: A case report. *J Postgrad Med.* 1990; **36:** 175-7.

137. Hamid HK, Mohamed MA, Babikir SM. Clinical profile and management of craniocerebral *Madurella* mycetoma. *J Neurosurg Pediatr.*2013; **12:** 529-30.
138. Hamid ME. Variable antibiotic susceptibility patterns among *Streptomyces* species causing actinomycetoma in man and animals. *Ann Clin Microbiol Antimicrob.* 2011; **10:** 24.
139. Hassan T, Nicholson S, Fahy R. Pneumothorax and empyema complicating *Scedosporium apiospermum* mycetoma: Not just a problem in the immunocompromised patients. *Ir J Med Sci.* 2011; **180:** 931-2.
140. Hay RJ, Fahal AH. Mycetoma: An old and still neglected tropical disease. *Trans R Soc Trop Med Hyg.* 2015; **109:** 169-70.
141. Hay RJ. Mycetoma, mycoses and pregnancy. *Acta Derm Venereol.*2015; **95:** 259-60.
142. Hazra B, Bandyopadhyay S, Saha SK, *et al.* A study of mycetoma in eastern India.*J Commun Dis.* 1998; **30:** 7-11.
143. Hemalata M, Prasad S, Venkatesh K, *et al.* Cytological diagnosis of actinomycosis and eumycetoma: A report of two cases. *Diagn Cytopathol.*2010; **38:** 918-20.
144. Hemashettar BM, Patil CS, Nagalotimath SJ, *et al.* Mycetoma due to *Exophiala jeanselmei:* A case report with description of the fungus. *Indian J Pathol Microbiol.* 1986; **29:** 75-8.
145. Hemashettar BM, Siddaramappa B, Munjunathaswamy BS, *et al. Phaeoacremonium krajdenii,* a cause of white grain eumycetoma. *J Clin Microbiol.* 2006; **44:** 4619-22.
146. Hemashettar BM, Siddaramappa B, Padhye AA, *et al.* White grain mycetoma caused by a *Cylindrocarpon* species in India. *J Clin Microbiol.* 2000; **38:** 4288-91.
147. Higgins HW, Vaidya DC, Telang GH. Pustules and sinus tracts involving the right arm of an elderly patient - Quiz case. Diagnosis: Eumycetoma. *Arch Dermatol.* 2012; **148:** 1199-200.
148. Hogade S, Metgud SC, Swoorooparani. Actinomycetes mycetoma. *J Lab Physicians.*2011; **3:** 43-5.
149. Ibrahim AI, El Hassan AM, Fahal A, *et al.* A histopathological exploration of the *Madurella mycetomatis* grain. *PLoS One.* 2013; **8:** e57774. PMID: 23483927.
150. Ichinomiya A, Nishimura K, Takenaka M, *et al.* Mycetoma caused by *Nocardia transvalensis* with repeated local recurrences for 25 years without dissemination to viscera. *J Dermatol.* 2014; **41:** 556-7.
151. Iriart X, Binois R, Fior A, *et al.* Eumycetoma caused by *Diaporthe phaseolorum* (*Phomopsisphaseoli*): A case report and a mini-review of *Diaporthe/Phomopsis* spp. invasive infections in humans. *Clin Microbiol Infect.* 2011; **17:** 1492-4.
152. Jain V, Makwana GE, Bahri N, *et al.* The "dot in circle" sign on MRI in maduramycosis: A characteristic finding. *J Clin Imaging Sci.* 2012; **2:** 66.
153. Janaki C, Sentamilselvi G, Janaki VR, *et al.* Case report: Eumycetoma due to *Curvularia lunata. Mycoses.* 1999; **42:** 345-6.
154. Javed F, Nazir R, Sharma M, *et al.* Dot-in-circle sign - A diagnostic MRI sign for "Madura Foot". *J Coll Physicians Surg Pak.* 2017; **27:** S8-S10.
155. Jimenez AL, Salvo NL. Mycetoma or synovial sarcoma? A case report with review of the literature. *J Foot Ankle Surg.* 2011; **50:** 569-76.
156. Joseph NM, Harish BN, Sistla S, *et al. Streptomyces* bacteremia in a patient with actinomycotic mycetoma. *J Infect Dev Ctries.* 2010; **4:** 249-52.
157. Joshi A, Acharya S, Anehosur VS, *et al.* Oral eumycetoma of infancy: A rare presentation and a brief review. *J Craniomaxillofac Surg.* 2014; **42:** 35-40.
158. Joshi KR, Mathur DR, Sharma JC, *et al.* Mycetoma caused by *Aspergillus nidulans* in India. *J Trop Med Hyg.* 1985; **88:** 41-4.
159. Joshi KR, Singhvi S. Serodiagnosis of mycetoma. *Trans R Soc Trop Med Hyg.* 1988; **82:** 334.
160. Joshi KR, Solanki A, Joshi YR. Mycetoma. 1st edn. Agrobios Publishers, Jodhpur, India: 2000; 1-107.
161. Joshi R. Treatment of actinomycetoma with combination of rifampicin and co-trimoxazole. *Indian J Dermatol Venereol Leprol.* 2008; **74:** 166-8.
162. Kalendar AM, Baykan H, Ozkan F, *et al.* Negative pressure wound therapy and skin graft in madura foot treatment. *Balkan Med J.* 2012; **29:** 214-7.
163. Kamalam A, Thambiah AS. A clinicopathological study of actinomycotic mycetomas caused by *Actinomadura madurae* and *Actinomadura pelletieri. Mycopathologia* 1987; **97:** 151-63.
164. Kanakis M, Chatzis A, Kytra V, *et al.* Multiple cardiac mycetomas in an immunosuppressed child. *Eur J Cardiothorac Surg.* 2014; **45:** 200.
165. Kang JW, Ku JS, Kim CH, *et al.* Intranasal mycetoma-induced Splendore-Hoeppli phenomenon. *Otolaryngol Head Neck Surg.* 2010; **142:** 456-7.
166. Kano R, Hattori Y, Murakami N, *et al.* The first isolation of *Nocardia veterana* from a human mycetoma. *Microbiol Immunol.* 2002; **46:** 409-12.
167. Kapoor S, Khunger N, Jain RK, *et al.* Iatrogenic actinomycetoma of neck and back successfully treated with sulphonamides. *Clin Exp Dermatol.* 2007; **32:** 95-6.
168. Kashima M, Kano R, Mikami Y, *et al.* A successfully treated case of mycetoma due to *Nocardia veterana.Br J Dermatol.* 2005; **152:** 1349-52.
169. Katkar VJ, Tankhiwale SS, Kurhade A. *Fusarium soloni* mycetoma. *Indian J Dermatol.* 2011; **56:** 315-7.
170. Kaur R, Maheshwari M. Hyperkeratotic warty skin lesion of foot caused by *Fusarium oxysporum. Indian J Dermatol.*2013; **58:** 159.
171. Kemper CA. Eumycetoma. *Curr Treat Option Infect Dis.* 2000; **2:** 533-8.
172. Khan FA, Hashmi S, Sarwari AR. Multiple subcutaneous mycetomas caused by *Pseudallescheria boydii:* Response to therapy with oral potassium iodide solution. *J Infect.* 2010; **60:** 178-81.
173. Khatri ML, Al-Halali HM, Fouad Khalid M, *et al.* Mycetoma in Yemen: clinicoepidemiologic and histopathologic study. *Int J Dermatol.* 2002; **41:** 586-93.
174. Kloezen W, Meis JF, Curfs-Breuker I, *et al. In vitro* antifungal activity of isavuconazole against *Madurella mycetomatis. Antimicrob Agents Chemother.* 2012; **56:** 6054-6.

175. Kloezen W, van Helvert-van Poppel M, Fahal AH, *et al.* A *Madurella mycetomatis* grain model in *Galleria mellonella* larvae. *PLoS Negl Trop Dis.* 2015;**9:** e0003926.
176. Koshi G, Padhye AA, Ajello L, Chandler FW. *Acremonium recifei* as an agent of mycetoma in India. *Am J Trop Med Hyg.*1979; **28:** 692-6.
177. Kothiwala SK, Purohit S, Meena M, *et al.* Black grain eumycetoma of the breast misdiagnosed as fibroadenoma. *Indian J Dermatol Venereol Leprol.* 2015; **81:** 521-3.
178. Kotwal N, Yanamandra U, Badwal S, *et al.* Mycetoma foot caused by *Aspergillus* in a diabetic patient. *Intern Med.* 2012; **51:** 517-8.
179. Kramer SC, Ryan M, Bourbeau P, *et al.* Fontana-positive grains in mycetoma caused by *Microsporum canis. Pediatr Dermatol.* 2006; **23:** 473-5.
180. Kresch-Tronik NS, Carrillo-Casas EM, Arenas R, *et al.* First case of mycetoma associated with *Nocardia takedensis. J Dermatol.* 2013; **40:** 135-6.
181. Kresch-Tronik NS, Carrillo-Casas EM, Arenas R, *et al. Nocardia harenae,* an uncommon causative organism of mycetoma: Report on two patients. *J Med Microbiol.* 2012; **61:** 1153-5.
182. Kriegsmann M, Harms A, Rickerts V, *et al.* Gunpowder or mycetoma? *Int J Surg Pathol.*2015; **23:** 373-4.
183. Kudur MH, Prakash P, Savitha M. *Fusarium solani* causing quasi-invasive infection of the foot in an immunocompetent middle-aged man from South India. *Indian J Dermatol.* 2013; **58:** 241.
184. Kulkarni RB, Patil RT, Praveena S. Actinomycotic mycetoma due to *Nocardia brasiliensis* in a case of leprosy. *Indian J Lepr.* 2001; **73:** 263-5.
185. Kumar J, Kumar A, Sethy P, *et al.* The dot-in-circle sign of mycetoma on MRI. *Diagn Interv Radiol.* 2007; **13:** 193-5.
186. Laohawiriyakamol T, Tanutit P, Kanjanapradit K, *et al.* The "dot-in-circle" sign in musculoskeletal mycetoma on magnetic resonance imaging and ultrasonography. *Springerplus.* 2014; **3:** 671.
187. Lee JK, Lee YJ, Park SS, *et al.* Clinical course and prognostic factors of pulmonary aspergilloma. *Respirology.* 2014; **19:** 1066-72.
188. Li EY, Yuen HK, Lung DC. Subconjunctival mycetoma as an unusual cause of tears with black deposits. *Arch Ophthalmol.* 2010; **128:** 1371-2.
189. Lichon V, Khachemoune A. Mycetoma: A review. *Am J Clin Dermatol.* 2006; **7:** 315-21.
190. Liu A, Maender JL, Coleman N, *et al.* Actinomycetoma with negative culture: A therapeutic challenge. *Dermatol Online J.* 2008; **14:** 5.
191. Ma KC, Pino A, Narula N, *et al. Scedosporium apiospermum* mycetoma in an immunocompetent patient without prior lung disease. *Ann Am Thorac Soc.* 2017; **14:** 145-7.
192. Machmachi H, Godineau N, Develoux M, *et al.* Black grain mycetoma caused by *Leptosphaeria tompkinsii. Med Mycol.* 2011; **49:** 186-9.
193. Magalhaes GM, Oliveira SC, Soares AC, *et al.* Mycetoma caused by *Nocardia caviae* in the first Brazilian patient. *Int J Dermatol.* 2010; **49:** 56-8.
194. Maheshwari S, Figueiredo A, Narurkar S, *et al. Madurella* mycetoma - A rare case with cranial extension. *World Neurosurg.* 2010; **73:** 69-71.
195. Mahgoub ES. Mycetoma. *Int J Dermatol.* 1985; **24:** 230-9.
196. Maina AM, Macharia JT. Alleviating a Nomad's anguish: Successful treatment of a case of leg mycetoma - A case report. *Case Rep Orthop.* 2012; 753174.
197. Maiti PK, Bandyopadhyay D, Dey JB, *et al.* Mycetoma caused by a new red grain mycetoma agent in two members of a family. *J Postgrad Med.* 2003; **49:** 322-4.
198. Maiti PK, Chakraborty B, Ghosh S, De A. Does the benefit of salvage amputation always outweigh disability in drug-failure mycetoma? A tale of two cases. *Indian J Dermatol.* 2015; **60:** 74-6.
199. Maiti PK, Ray A, Bandyopadhyay S. Epidemiological aspects of mycetoma from a retrospective study of 264 cases in West Bengal. *Trop Med Int Health.* 2002; **7:** 788-92.
200. Malone M, Gannass A, Bowling F. A chronic, destructive mycetoma infection in a diabetic foot in Saudi Arabia. *Int J Low Extrem Wounds.* 2011; **10:** 12-5.
201. Masoodi Z, Mansoor T, Ali WM, *et al.* Actinomycetoma leg ulcers in north India.*Wounds.*2014; **26:** 147-55.
202. Mathews S, Jadhav R, Reza A, *et al.* Actinomycetoma - The Welsh regimen in a rural Indian scenario. *Indian J Surg.* 2012; **74:** 480-2.
203. Mathur DR, Bharadwaj V, Vaishnav K, *et al.* Red grain mycetoma caused by *Actinomadura pelletieri* in western Rajasthan: report of two cases. *Indian J Pathol Microbiol.* 1993; **36:** 486-8.
204. Mattioni S, Develoux M, Brun S, *et al.* Management of mycetomas in France. *Med Mal Infect.*2013; **43:** 286-94.
205. McGinnis MR. Mycetoma. *Dermatol Clin.* 1996; **14:** 97-104.
206. McGrogan D, David MD, Roberts C, *et al.* Pseudotumoral presentation of fungating mycetoma caused by *Phaeoacremonium fuscum* in a renal transplant patient. *Transpl Infect Dis.* 2015; **17:** 897-903.
207. Mechow N, Goppner D, Franke I, *et al.* Cutaneous botryomycosis diagnosed long after an arm injury. *J Am Acad Dermatol.* 2014; **71:** e155-6.
208. Mencarini J, Antonelli A, Scoccianti G, *et al.* Madura foot in Europe: Diagnosis of an autochthonous case by molecular approach and review of the literature. *New Microbiol.* 2016; **39:** 156-9.
209. Mestre T, Vieira R, Coutinho J. Mycetoma of the foot - Diagnosis of the etiologic agent and surgical treatment. *Am J Trop Med Hyg.* 2015; **93:** 1-2.
210. Mhmoud NA, Ahmed SA, Fahal AH, *et al. Pleurostomophora ochracea,* a novel agent of human eumycetoma with yellow grains. *J Clin Microbiol.* 2012; **50:** 2987-94.
211. Mhmoud NA, Fahal AH, Mahgoub el S, *et al.* The combination of amoxicillin-clavulanic acid and ketoconazole in the treatment of *Madurella mycetomatis* eumycetoma and *Staphylococcus aureus* co-infection. *PLoS Negl Trop Dis.* 2014; **8:** e2959.
212. Mir F, Shakoor S, Khan MJ, *et al. Madurella mycetomatis* as an agent of brain abscess: Case report and review of literature. *Mycopathologia.* 2013; **176:** 429-34.

213. Mirza SH, Campbell C. Mycetoma caused by *Nocardia transvalensis*. *J Clin Pathol.* 1994; **47:** 85-6.
214. Mirza SH, Gardezi AH, Khan Y, *et al.* Subcutaneous facial mycosis in a child due to *Madurella mycetomatis*. *J Pak Med Assoc.* 2007; **57:** 466-8.
215. Mitja O, Hays R, Van Straten C, *et al.* Mycetoma caused by *Nocardia yamanashiensis*, Papua New Guinea. *Am J Trop Med Hyg.* 2012; **86:** 1043-5.
216. Mitja O, Marks M, Bertran L, *et al.* Integrated control and management of neglected tropical skin diseases. *PLoS Negl Trop Dis.* 2017; **11:** e0005136. PMID: 28103250.
217. Mohamed el SW, Mohamed el NA, YousifBel D, *et al.* Tongue actinomycetoma due to *Actinomadura madurae:* A rare clinical presentation. *J Oral Maxillofac Surg.* 2012; **70:** e622-4.
218. Mohamed el SW, Seif El Din N, Fahal AH. Multiple mycetoma lung secondaries from knee eumycetoma: An unusual complication. *PLoS Negl Trop Dis.* 2016; **10:** e0004735.
219. Mohamed NA, Fahal AH. Mycetoma pulmonary secondaries from a gluteal eumycetoma: An unusual presentation. *PLoS Negl Trop Dis.* 2016; **10:** e0004945.
220. Momin SB, Richardson BS, Bryan MG, *et al.* Mycetoma clinically masquerading as squamous cell carcinoma: Case report and literature review. *J Clin Aesthet Dermatol.* 2009; **2:** 26-31.
221. Morino G, Giaccardi GC, Brundisini R, *et al.* A unique case of bladder mycetoma. *Trop Doct.* 2002; **32:** 43-4, 124.
222. Motswaledi HM, Mathekga K, Sein PP, *et al. Paecilomyces lilacinus* eumycetoma. *Int J Dermatol.* 2009; **48:** 858-61.
223. Mowat E, Williams C, Jones B, *et al.* The characteristics of *Aspergillus fumigatus* mycetoma development: Is this a biofilm? *Med Mycol.* 2009; **47:** S120-6.
224. Munoz-Hernandez B, Noyola MC, Palma-Cortes G, *et al.* Actinomycetoma in arm disseminated to lung with grains of *Nocardia brasiliensis* with peripheral filaments. *Mycopathologia.* 2009; **168:** 37-40.
225. Murthy R, Swain JP. Concurrent mycetoma and chromomycosis. *Indian J Med Microbiol.* 2011; **29:** 437-9.
226. Nai GA, Stuani ML, Stuani LA. Oral cavity eumycetoma. *Rev Inst Med Trop Sao Paulo.* 2011; **53:** 165-8.
227. Nair PA, Kota RS. Black grain eumycetoma of the breast-misdiagnosed, incidental or iatrogenic? *Indian J Dermatol Venereol Leprol.* 2016; **82:** 180.
228. Nasr A, Abushouk A, Hamza A, *et al.* Th-1, Th-2 cytokines profile among *Madurella mycetomatis* eumycetoma patients. *PLoS Negl Trop Dis.* 2016; **10:** e0004862.
229. Nenoff P, van de Sande WW, Fahal AH, *et al.* Eumycetoma and actinomycetoma - An update on causative agents, epidemiology, pathogenesis, diagnostics and therapy. *J Eur Acad Dermatol Venereol.* 2015; **29:** 1873-83.
230. Oddy MJ, Blundell CM, Suvarna SK, Greig JM. Mycetoma of the foot. *Br J Hosp Med (Lond).*2009; **70:** 658.
231. Oliveira Fde M, Unis G, Hochhegger B, *et al. Scedosporium apiospermum* eumycetoma successfully treated with oral voriconazole: Report of a case and review of the Brazilian reports on scedosporiosis. *Rev Inst Med Trop Sao Paulo.* 2013; **55:** 121-3.
232. Omer RF, Seif El Din N, Abdel Rahim FA, *et al.* Hand mycetoma: The Mycetoma Research Centre experience and literature review. *PLoS Negl Trop Dis.* 2016; **10:** e0004886.
233. Ong DC, Khan R, Golledge C, *et al.* Case report: Eumycetoma and mycotic arthritis of the knee caused by *Arthrographis kalrae. J Orthop.* 2014; **12:** S140-4.
234. Padhi S, Uppin SG, Uppin MS, *et al.* Mycetoma in south India: Retrospective analysis of 13 cases and description of two cases caused by unusual pathogens: *Neoscytalidium dimidiatum* and *Aspergillus flavus. Int J Dermatol.*2010; **49:** 1289-96.
235. Padhye AA, Salkin IF. *Madurella pseudomycetomatis:* An invalidly published name. *J Clin Microbiol.* 2011; **49:** 1703.
236. Padilla-Desgarennes C, Vazquez-Gonzalez D, Bonifaz A. Botryomycosis. *Clin Dermatol.*2012; **30:** 397-402.
237. Palacios E, Jones W, Alvernia J. Sinonasal mycetoma. *Ear Nose Throat J.* 2008; **87:** 606-8.
238. Palit A, Ragunatha S, Inamadar AC. Actinomycetoma: Dramatic response to modified two-step regimen. *Int J Dermatol.* 2011; **50:** 446-9.
239. Pardo M, Bonifaz A, Valencia A, *et al.* Actinomycetoma by *Nocardia brasiliensis* in a girl with Down syndrome. *Dermatol Online J.* 2008; **14:** 9.
240. Parker L, Singh D, Biz C. The dot-in-circle sign in Madura foot. *J Foot Ankle Surg.* 2009; **48:** 690.e1-5.
241. Pattanaprichakul P, Bunyaratavej S, Leeyaphan C, *et al.* An unusual case of eumycetoma caused by *Exophiala jeanselmei* after a sea urchin injury. *Mycoses.* 2013; **56:** 491-4.
242. Pickert AJ, Nguyen X. Madura foot. *N Engl J Med.* 2012; **366:** e2.
243. Prasad PV, George RV, Paul EK, *et al.* Eumycotic mycetoma in the leg with a bone cyst in fibula. *Indian J Dermatol Venereol Leprol.* 2002; **68:** 174-5.
244. Prasanna S, Grover N, Bhatt P, *et al.* A case of *Aspergillus nidulans* causing white granule mycetoma. *Med J Armed Forces India.* 2016; **72:** 88-90.
245. Prinja A, Roberts C, Doherty T, *et al.* An unusual cause of an ankle mass. *BMJ Case Rep.* 2014; pii: bcr2014204253.
246. Pua BB, Mir DI, Gulati CM, *et al.* Transbronchial treatment of a bleeding pulmonary mycetoma refractory to arterial embolization. *J Vasc Interv Radiol.* 2013; **24:** 1583-5.
247. Pulikot AM, Bapat SS, Tolat S. Mycetoma of the sole. *Ann Trop Paediatr.* 2002; **22:** 187-90.
248. Queiroz-Telles F, McGinnis MR, Salkin I, *et al.* Subcutaneous mycoses. *Infect Dis Clin N Am.* 2003; **17:** 59-85.
249. Quintana ET, Wierzbicka K, Mackiewicz P, *et al. Streptomyces sudanensis* sp. nov., a new pathogen isolated from patients with actinomycetoma. *Antonie van Leeuwenhoek.* 2008; **93:** 305-13.
250. Rai VM, Balachandran C. Actinomycotic mycetoma of the hand: A rare occurrence. *Indian J Dermatol Venereol Leprol.* 2006; **72:** 178.
251. Rajendran C, Kandhari S. Mycetoma due to *Nocardia brasiliensis*: First case report from the Union Territory of Delhi. *Mycoses.* 1989; **32:** 476-81.

252. Ramam M, Bhat R, Garg T, *et al.* A modified two-step treatment for actinomycetoma. *Indian J Dermatol Venereol Leprol.* 2007; **73:** 235-9.
253. Ramasamy A, Kadom Z, Dubrey SW. Pulmonary mycetoma. *Br J Hosp Med (Lond).* 2014; **75:** 656.
254. Rao AG. Mycetoma. *Indian Dermatol Online J.* 2016; **7:** 353-5.
255. Rattanavong S, Vongthongchit S, Bounphamala K, *et al.* Actinomycetoma in SE Asia: The first case from Laos and a review of the literature. *BMC Infect Dis.* 2012; **12:** 349.
256. Reis LM, Lima BZ, ZilloFda C, *et al.* Dermoscopy assisting the diagnosis of mycetoma: Case report and literature review. *An Bras Dermatol.* 2014; **89:** 832-3.
257. Rimawi BH, Rimawi RH, Mirdamadi M, *et al.* A case of *Exophiala oligosperma* successfully treated with voriconazole. *Med Mycol Case Rep.* 2013; **2:** 144-7.
258. Rode M, Podboj J, Kogoj-Rode M. Sinus maxillaris mycetoma of odontogenic origin: case report. *Braz Dent J.* 2004; **15:** 248-50.
259. Rodriguez-Nava V, Couble A, Molinard C, *et al. Nocardia mexicana* sp. nov., a new pathogen isolated from human mycetomas. *J Clin Microbiol.* 2004; **42:** 4530-5.
260. Rohde LE, Torres FS, Foppa M, *et al.* Tumor like cardiac fungal mycetoma caused by *Scedosporium apiospermum* presenting as symptomatic ventricular tachycardia. *Circulation.* 2014; **129:** e488-9.
261. Rojas OC, Leon-Cachon RB, Moreno-Trevino M, *et al.* Molecular identification of unusual mycetoma agents isolated from patients in Venezuela. *Mycoses.* 2017; **60:** 129-35.
262. Rojas-Tula DG, Gomez-Fernandez M, Garcia-Lopez JJ, *et al.* Endobronchial cryotherapy for a mycetoma. *J Bronchology Interv Pulmonol.* 2013; **20:** 330-2.
263. Romano C. Subcutaneous mycoses. Part 1: Subcutaneous mycoses due to non-dermatophytes. *G Ital Dermatol Venereol.*2013; **148:** 621-31.
264. Rouphael NG, Talati NJ, Franco-Paredes C. A painful thorn in the foot: A case of eumycetoma. *Am J Med Sci.* 2007; **334:** 142-4.
265. Saha S, Dhar A, Karak AK. Mycetoma sans sinuses. *Indian J Dermatol Venereol Leprol.* 2006; **72:** 143-4.
266. Salipante SJ, Sengupta DJ, Hoogestraat DR, *et al.* Molecular diagnosis of *Actinomadura madurae* infection by 16S rRNA deep sequencing. *J Clin Microbiol.* 2013; **51:** 4262-5.
267. Sampaio FM, Galhardo MC, De Farias Cardoso R, *et al.* Eumycetoma on the foot caused by *Madurella mycetomatis:* Amputation after significant worsening during pregnancy. *Acta Derm Venereol.* 2015; **95:** 374-5.
268. Sampaio FM, Galhardo MC, Quintella LP, *et al.* Eumycetoma by *Madurella mycetomatis* with 30 years of evolution: Therapeutic challenge. *An Bras Dermatol.* 2013; **88 (6 Suppl. 1):** 82-4.
269. Sampaio FM, Wanke B, Freitas DF, *et al.* Review of 21 cases of mycetoma from 1991 to 2014 in Rio de Janeiro, Brazil. *PLoS Negl Trop Dis.* 2017; 11: e0005301. PMID: 28192433.
270. Sawatkar GU, Narang T, Shiva Prakash MR, *et al. Aspergillus:* An uncommon pathogen of eumycetoma. *Dermatol Ther.* 2017; **30:** PMID: 27612290.
271. Sayyad FE, Karp CL, Wong JR, *et al.* Marsupialized fungal mycetoma masquerading as conjunctival melanoma. *Cornea.* 2014; **33:** 744-6.
272. Schwartz E, Shpiro A. Madura foot or Philoctetes foot? *Isr Med Assoc J.* 2015; **17:** 442-4.
273. Scolding PS, Abbas MA, Omer RF, *et al. Madurella mycetomatis*-induced massive shoulder joint destruction: A management challenge. *PLoS Negl Trop Dis.* 2016; **10:** e0004849.
274. Sedlacek P, Vavra V, Masova I, *et al.* Successful therapy with ABLC, surgery and posaconazole for *Rhizopus microsporus* var. *rhizopodiformis* liver eumycetoma in a child with acute leukaemia. *Mycoses.*2009; **52:** 276-9.
275. Sen A, Pillay RS. Case report: Dot-in-circle sign - An MRI and USG sign for "madura foot". *Indian J Radiol Imaging.* 2011; **21:** 264-6.
276. Seol CA, Sung H, Kim DH, *et al.* The first Korean case of disseminated mycetoma caused by *Nocardia pseudobrasiliensis* in a patient on long-term corticosteroid therapy for the treatment of microscopic polyangiitis. *Ann Lab Med.* 2013; **33:** 203-7.
277. Shanbhag NU, Karandikar S, Deshmukkh PA. Disseminated orbital actinomycetoma: A case report. *Indian J Ophthalmol.* 2010; **58:** 60-3.
278. Sharma AM, Sharma N, Nat A, *et al.* Case report: Non-invasive management of Madura foot with oral posaconazole and ciprofloxacin. *Am J Trop Med Hyg.* 2014; **91:** 1259-62.
279. Sharma J, Nagraj A, Allapathi D, *et al.* Fungal endocarditis in a premature infant complicated by a right atrial mycetoma and inferior vena cava thrombosis. *Images Paediatr Cardiol.* 2009; **11:** 6-11.
280. Sharma NL, Mahajan VK, Agarwal S, *et al.* Nocardial mycetoma: Diverse clinical presentations. *Indian J Dermatol Venereol Leprol.* 2008; **74:** 635-40.
281. Shinde RS, Hanumantha S, Mantur BG, *et al.* A rare case of mycetoma due to *Curvularia. J Lab Physicians.* 2015; **7:** 55-7.
282. Singhal S. Pulmonary mycetoma. *Lung India.* 2012; **29:** 81-2.
283. Singhvi A, Joshi KR, Sharma JC, *et al.* Immunoglobulin profile in mycetoma. *Indian J Pathol Microbiol.* 1996; **39:** 203-6.
284. Sood A, Khanna N, Gandhi D, *et al.* Mycetoma involving the anterior mediastinum and chest wall. *J Eur Acad Dermatol Venereol.* 2002; **16:** 294-5.
285. Sousa JM, Wachholz PA, Sette CS, *et al.* Mycetoma caused by *Nocardia brasiliensis* in an immunocompetent patient. *J Dtsch Dermatol Ges.* 2014; **12:** 903-5.
286. Suleiman AM, Fahal AH. Oral cavity eumycetoma: A rare and unusual condition. *Oral Surg Oral Med Oral Pathol Oral Radiol.* 2013; **115:** e23-5.
287. Suleiman SH, Wadaella el S, Fahal AH. The surgical treatment of mycetoma. *PLoS Negl Trop Dis.* 2016; **10:** e0004690.
288. Sun PL, Peng PC, Wu PH, *et al.* Canine eumycetoma caused by *Cladophialophora bantiana* in a Maltese: Case report and literature review. *Mycoses.* 2013; **56:** 376-81.

289. Suodbakhsh A, Soleimani A, Emadi H. Photo clinic. Actinomycetoma. *Arch Iran Med.* 2011; **14:** 299-300.
290. Swaroopa D, Raju Y, Prasad AK, *et al.* A rare case of actinomycetoma of a leg. *J Assoc Physicians India.* 2016; **64:** 84. PMID: 27728085.
291. Szarf G, Obac AR, Puchnick A, *et al.* Mycetoma involving the heart. *Circulation.* 2013; **128:** e159-61.
292. Talwar P, Sehgal SC. Mycetomas in North India. *Sabouraudia.* 1979; **17:** 287-91.
293. Tellez I, Franco-Paredes C. A woman with chronic subcutaneous swelling of the right foot associated with sinus tracts discharging yellow grains. *PLoS Negl Trop Dis.* 2010; **4:** e772.
294. Tendolkar UM, Deodhar LP, Jerajani H, *et al.* Microbiological study of mycetoma patients from Bombay with special reference to actinomyces immunofluorescence. *Indian J Pathol Microbiol.* 1993; **36:** 245-52.
295. Thammayya A, Sanyal M, Basu N. *Pyrenochaeta romeroi* causing mycetoma pedis in India. *J Indian Med Assoc.* 1979; **73:** 66-7.
296. Thammayya A, Sanyal M. *Exophiala jeanselmei* causing mycetoma pedis in India. *Sabouraudia.* 1980; **18:** 91-5.
297. Tilak R, Singh S, Garg A, *et al.* A case of actinomycotic mycetoma involving the right foot. *J Infect Dev Ctries.* 2009; **3:** 71-3.
298. Tomczyk S, Deribe K, Brooker SJ, *et al.* Association between footwear use and neglected tropical diseases: A systematic review and meta-analysis. *PLoS Negl Trop Dis.* 2014; **8:** e3285.
299. Toplu Y, Toplu SA, Can S, *et al.* Fungus ball in concha bullosa: An unusual cause of retro-orbital pain. *J Craniofac Surg.* 2014; **25:** e138-40.
300. Toth EJ, Nagy GR, Homa M, *et al.* Recurrent *Scedosporium apiospermum* mycetoma successfully treated by surgical excision and terbinafine treatment: A case report and review of the literature. *Ann Clin Microbiol Antimicrob.* 2017; **16:** 31. PMID: 28410611.
301. Tsai WC, Lee CH, Wu WM, *et al.* Cutaneous manifestations of subcutaneous and systemic fungal infections in tropical regions: A retrospective study from a referral center in southern Taiwan. *Int J Dermatol.* 2017; **56:** 623-9.
302. van Belkum A, Fahal A, van de Sande WW. Mycetoma caused by *Madurella mycetomatis:* A completely neglected medico-social dilemma. *Adv Exp Med Biol.* 2013; **764:** 179-89.
303. van Belkum A, Fahal AH, van de Sande WW. *In vitro* susceptibility of *Madurella mycetomatis* to posaconazole and terbinafine. *Antimicrob Agents Chemother.*2011; **55:** 1771-3.
304. van de Sande WW, Fahal AH, Bakker-Woudenberg IA, *et al. Madurella mycetomatis* is not susceptible to the echinocandin class of antifungal agents. *Antimicrob Agents Chemother.* 2010; **54:** 2738-40.
305. van de Sande WW, Fahal AH, Goodfellow M, *et al.* Merits and pitfalls of currently used diagnostic tools in mycetoma. *PLoS Negl Trop Dis.* 2014; **8:** e2918.
306. van de Sande WW, Maghoub el S, *et al.* The mycetoma knowledge gap: Identification of research priorities. *PLoS Negl Trop Dis.* 2014;**8:** e2667.
307. van de Sande WW, van Vianen W, Ten Kate M, *et al.* Amphotericin B but not itraconazole is able to prevent grain formation in experimental *Madurella mycetomatis* mycetoma in mice. *Br J Dermatol.* 2015; **173:** 1561-2.
308. van de Sande WW. Global burden of human mycetoma: A systematic review and meta-analysis. *PLoS Negl Trop Dis.* 2013; **7:** e2550.
309. van Hellemond JJ, Vonk AG, de Vogel C, *et al.* Association of eumycetoma and schistosomiasis. *PLoS Negl Trop Dis.* 2013; **7:** e2241.
310. Venkatswami S, Sankarasubramanian A, Subramanyam S. The madura foot: Looking deep. *Int J Low Extrem Wounds.* 2012; **11:** 31-42.
311. Venugopal PV, Venugopal TV, Pandurangan CN. *Leptosphaeria senegalensis* causing mycetoma pedis in Madras. *Indian J Dermatol Venereol Leprol.* 1980; **46:** 364-7.
312. Venugopal PV, Venugopal TV, Pandurangan CN. *Leptosphaeria tompkinsii* mycetoma. *Int J Dermatol.* 1990; **29:** 432-3.
313. Venugopal PV, Venugopal TV. Pale grain eumycetomas in Madras. *Australas J Dermatol.* 1995; **36:** 149-51.
314. Venugopal PV, Venugopal TV. Red grain mycetoma of the scalp due to *Actinomadura pelletieri* in Madurai. *Indian J Pathol Microbiol.* 1990; **33:** 384-6.
315. Veraldi S, Grancini A, Venegoni L, *et al.* Mycetoma caused by *Aspergillus nidulans. Acta Derm Venereol.* 2016; **96:** 118-9.
316. Verma R, Vasudevan B, Sahni AK, *et al.* First reported case of *Aspergillus nidulans* eumycetoma in a sporotrichoid distribution. *Int J Dermatol.* 2015; **54:** 74-7.
317. Verma SB, Nayak S, Pasale RK, *etal.* Late complication after tropic storm accident: Subcutaneous and intracranial actinomycetoma. *Int Wound J.* 2008; **5:** 655-9.
318. Viguier M, Lafaurie M. Actinomycetoma. *N Engl J Med.* 2015; **372:** 264.
319. Vilela R, Duarte OM, Rosa CA, *etal.* A case of eumycetoma due to *Madurella grisea* in northern Brazil. *Mycopathologia.* 2004; **158:** 415-8.
320. Vinay D, Ramasubramanian V, Gopalakrishnan R, *et al.* Botryomycosis in a lung cavity. *Lung India.* 2016; **33:** 540-2.
321. Vongphoumy I, Dance DA, Dittrich S, *et al.* Case report: Actinomycetoma caused by *Nocardia aobensis* from Lao PDR with favourable outcome after short-term antibiotic treatment. *PLoS Negl Trop Dis.* 2015; **9:** e0003729.
322. Wadal A, Elhassan TA, Zein HA, *et al.* Predictors of post-operative mycetoma recurrence using machine-learning algorithms: The Mycetoma Research Center experience. *PLoS Negl Trop Dis.* 2016; **10:** e0005007.
323. Welsh O, Al-Abdely HM, Salinas-Carmona MC, *et al.* Mycetoma medical therapy. *PLoS Negl Trop Dis.* 2014; **8:** e3218.
324. Welsh O, Arenas R. Subcutaneous mycoses. *Clin Dermatol.*2012; **30:** 367-8.
325. Welsh O, Morales-Toquero A, Vera-Cabrera L, *et al.* Actinomycetoma of the scalp after a car accident. *Int J Dermatol.*2011; **50:** 854-7 & 1178.
326. Welsh O, Salinas MC, Rodriguez MA. Treatment of eumycetoma and actinomycetoma. *Curr Top Med Mycol.* 1995; **6:** 47-71.

327. Welsh O, Vera-Cabrera L, Salinas-Carmona MC. Mycetoma. *Clin Dermatol.* 2007; **25:** 195-202.
328. Welsh O, Vera-Cabrera L, Welsh E, *et al.* Actinomycetoma and advances in its treatment. *Clin Dermatol.* 2012; **30:** 372-81.
329. Werlinger KD, Yen Moore A. Eumycotic mycetoma caused by *Cladophialophora bantiana* in a patient with systemic lupus erythematosus. *J Am Acad Dermatol.* 2005; **52:** S114-7.
330. White EA, Patel DB, Forrester DM, *et al.* Madura foot: Two case reports, review of the literature and new developments with clinical correlation. *Skeletal Radiol.* 2014; **43:** 547-53.
331. Xiujiao X, Hong S, Ai-e X. Eumycetoma due to *Acremonium falciforme* acquired in China. *Mycoses.* 2012; **55:** e4-7.
332. Yan J, Deng J, Zhou CJ, *et al.* Phenotypic and molecular characterization of *Madurella pseudomycetomatis* sp. nov., A novel opportunistic fungus possibly causing black-grain mycetoma. *J Clin Microbiol.* 2010; **48**: 251-7.
333. Yeh I, Dhanireddy S. Madura foot caused by *Actinomadura madurae* in a pregnant woman. *Arch Dermatol.* 2010; **146:** 1189-90.
334. Yera H, Bougnoux ME, Jeanrot C, *et al.* Mycetoma of the foot caused by *Fusarium solani*: Identification of the etiologic agent by DNA sequencing. *J Clin Microbiol.* 2003; **41:** 1805-8.
335. Yousif BM, Fahal AH, Shakir MY. A new technique for the diagnosis of mycetoma using fixed blocks of aspirated material. *Trans R Soc Trop Med Hyg.* 2010; **104:** 6-9.
336. Zampella JG, Kwatra SG, Kazi N, *et al.* Madura foot caused by *Gordonia terrae* misdiagnosed as *Nocardia. Australas J Dermatol.* 2016. doi: 10.1111/ajd.12516. PMID: 27270783.
337. Zarei Mahmoudabadi A, Zarrin M. Mycetomas in Iran: A review article. *Mycopathologia.* 2008; **165:** 135-41.
338. Zijlstra EE, van de Sande WW, Fahal AH. Mycetoma: A long journey from neglect. *PLoS Negl Trop Dis.* 2016; **10:** e0004244.
339. Zijlstra EE, van de Sande WW, Welsh O, *et al.* Mycetoma: A unique neglected tropical disease. *Lancet Infect Dis.* 2016; **16:** 100-12.
340. Zoroquiain P, Alghamdi S, Arthurs B, *et al.* Eyelid mycetoma masquerading as sebaceous carcinoma. *Ophthal Plast Reconstr Surg.* 2017; **33:** S101-4.

CHAPTER 12

Sporotrichosis

Sporotrichosis is an acute or chronic, pyogranulomatous, polymorphic fungal infection of subcutaneous tissues of humans and animals. This usually remains localized to skin or may show lymphatic spread with occasional dissemination to distant parts of body. It is characterized by development of noduloulcerative lesions in skin, subcutaneous tissue and along the lymphatic tracts. These are usually found over the exposed parts of body like hands, arms or legs. The commonest clinical presentation is lymphocutaneous form but fixed cutaneous, systemic types may also present. Very rarely a disseminated form may be seen involving skeletal, respiratory or central nervous systems. In the recent years, a large number of zoonotic cases of this disease have also been reported.

For the last more than a century, it was presumed that the disease is caused by sole dimorphic fungus *i.e.* ***Sporothrix schenckii.*** But on the basis of molecular techniques, this is realized that it is caused by ***S.schenckii*** **species complex** with many **cryptic** species. These are all thermally dimorphic saprotrophic fungi found prevalent in the environment and have worldwide distribution. The clinical condition is traditionally known by various synonym like **Schenck's** disease, **Rose Gardener's** disease, **Rose Handler's** disease as well as **Rose Thorn** disease.

This is a very unique dimorphic fungal disease in many aspects. Its endemicity is widespread and not like other ones confined to limited geographical areas. In comparison to other dimorphic fungal diseases, it is the one wherein respiratory route is not very significant. The causative agent, *Sporothrix* species, are isolated more commonly from the environment as compared to other dimorphic fungi, which are very rarely isolated. This is the only fungal disease where zoonotic spread is considered very significant, particularly from cats, changing the paradigm of acquisition and transmission of fungi.

Historical Perspective

Sporotrichosis was originally described in 1898 by a medical student, **Benjamin Schenck,** at Johns Hopkins Hospital in Baltimore, Maryland (USA). Two years later, in 1900, Hektoen and Perkins in Chicago reported its second case. They gave detailed morphological description of the pathogen along with results of their studies involving laboratory animals. In 1903, De Beurmann and Ramond used potassium iodide for the first time as therapeutic agent of this disease, which is successfully being used as such for the last more than a century.

In 1907, Lutz and Splendore observed **asteroid body** for the first time in pus discharging from cutaneous lesions associated with this disease. Almost simultaneously Greco described its formation in humans and rats. Subsequently several authors confirmed these findings. In 1912, De Beurmann and Gougerot published a complete treatise on this disease. In 1914, sporotrichosis was diagnosed for the first time in Guateng gold mines of South Africa. In the Western Europe, most cases originated from France, where about two hundreds were reported by 1920. In 1932, L M Ghosh reported the first case of sporotrichosis from Kolkata, which represented endemic focus in northeast belt of undivided India.

In 1963, **H I Lurie** observed that five out of about three thousand patients in South Africa had disease that deviated from classic lymphangitic form by having extra-cutaneous involvement. In 1969, new variety of pathogen was designated by Ajello and Kaplan in his honor as *S.schenckii* var. *luriei* (now elevated to species level) based on cases having significantly different tissue morphology as compared to typical ones. Only four cases of this variety have been reported till date, one each from South Africa, Italy and India. In the fourth case, diagnosis was retrospectively amended from blastomycosis. Padhye *et al*, reported

Indian case from Chandigarh in 1992. In 1970, Lynch recognized the role of *S.schenckii* as an opportunist infection in immunocompromised individuals.

The outbreaks of disease have also been reported in South Africa between 1941-44; USA in 1988, which affected 84 people who handled sphagnum moss and Peru during 1995-97. A cluster of cases has also been reported from Western Australia between 2000-03. From 1998 onwards, outbreak of this disease is going on in southeastern Brazil involving thousands of human as well as pets like cats and dogs.

In 2007, Marimon *et al*, in their landmark study proposed three new clinically significant species of genus *Sporothrix* *i.e.* *S.brasiliensis*, *S.globosa* and *S.mexicana* depending on chitin synthase, β-tubulin and calmodulin gene analysis as well as phenotypic characters like morphology of the sessile pigmented conidia, growth at 30, 35 and 37°C as well as assimilation of sucrose, raffinose and ribitol. Moreover, in another study status of *S.schenckii* var. *luriei* was elevated to the species level as *S.luriei* in the year 2008, with the same description as that of its variety.

Sporotrichosis is a potentially re-emerging disease in areas of South Africa where traditional, rather than heavily mechanized, gold mining techniques are used and surveillance should be instituted for this disease at sentinel mine locations. *S.schenckii sensu stricto* was identified as the causative pathogen in a group of miners with lymphocutaneous disease at a South African gold mines.

In the recent times, maximum number of cases of sporotrichosis were reported from South American subcontinent. Therefore, realizing gravity of the situation, the very 1st international meeting on this disease was organised by the Working Group on *Sporothrix* and Sporotrichosis (WGSS) under the aegis of ISHAM on October 2-4, 2013 at Rio de Janeiro (Brazil). The 2nd one will be held on September 11-12, 2017 in Guanajuato (Mexico) under the leadership of Prof. Hector M. Mora Montes.

Mycology

The earlier binomial nomenclature of causative dimorphic fungus of sporotrichosis was *Sporothrix schenckii,* with two varieties *i.e.* *S.schenckii* var. *schenckii* and *S.schenckii* var. *luriei*. *S.schenckii* is found as saprotroph in nature whereas environmental niche of *S.schenckii* var. *luriei* is not yet known. Now, *S.schenckii* var. *luriei* has been allotted an independent genus status as *S.luriei*. There is absence of sessile dark-pigmented conidia, which is typical of the *S.schenckii* species complex and presence of very long conidia are the two key morphological features to distinguish *S.schenckii* and *S.luriei*. However, based on molecular studies, it is demonstrated that *S.schenckii* is genetically a complex species. Hence three new clinically important species were proposed by Marimon *et al*, which are: *S.brasiliensis*, *S.globosa* and *S.mexicana*.

There are 51 species included in the genus *Sporothrix* which are non-pathogenic environmental fungi that are closely related to decaying wood, plants, and soil. However, members of the ***Sporothrix schenckii* complex** are pathogenic, including *S.brasiliensis, S.schenckii sensu stricto (s.str.), S.globosa* and *S.luriei,* the causative agents of human as well as animal disease.

The causative fungi of the sporotrichosis are now named as ***S.schenckii* species complex**. Its gene sequencing has revealed following species, mentioned in an alphabetical order: *S.albicans, S.brasiliensis, S.globosa, S.luriei, S.mexicana, S.pallida* and *S.schenckii (*both *sensu stricto* and *sensu lato).* There are five different clades also depending upon the geographical distribution of a particular species. The genus *Sporothrix* is divided into five Clades *i.e.* Clade I to V depending upon geographical distribution of a particular species. All clinical isolates belong to Clade I *(S.brasiliensis),* II *(S.schenckii)* and III *(S.globosa)* and environmental isolates belong to Clade IV *(S.mexicana)* and V *(S.albicans)*. The common agents of human or animal sporotrichosis are *S.brasiliensis, S.schenckii* and *S.globosa.* A few cases, however are now caused by environmental species also like *S.mexicana* and *S.pallida* and by rare one-*S.luriei.*

As this organism has dark-brown conidia due to production of melanin, some workers propose that taxonomically it may be placed in the category of phaeoid fungi. The teleomorphic state of *S.schenckii* has not been conclusively demonstrated despite many attempts at mating various clinical and environmental isolates. Until molecular techniques were developed, *S.schenckii* remained included in form-class **Hyphomycetes** of phylum mitosporic fungi *i.e.* **Deuteromycetes**. On the basis of sequences of SSU rDNA, it is shown that *S.schenckii* phylogenetically lies within sexual genus *Ophiostoma stenoceras*.

Therefore, at present causative species of genus *Sporothrix* are taxonomically placed in Family **Ophiostomataceae** under Order **Ophiostomatales** of Class **Pezizomycotina** of Phylum **Ascomycota**. This is a unique dimorphic genus and is phylogenetically distinct from other dimorphic fungi, which are placed in Family Onygenaceae of Order

Onygenales under Class Plectomycetes. Furthermore, 18s rDNA sequences reveal significant differences between *S.schenckii* and *Ophiostoma stenoceras*. The latter one was previously considered as the possible teleomorph of *S.schenckii*.

Earlier, this organism was also placed in genus *Sporotrichum* as *Sporotrichum schenckii* but members of that genus are classified as **Basidiomycetes** in the Order **Steerages** and are characterized by formation of large hyphae with clamp-connection and large, one-celled, thick-walled golden conidia. These are neither dimorphic nor pathogenic to man or animals.

A new species, *Sporothrix cyanescens* was added to genus *Sporothrix* but it was found to be non-pathogenic in experimental animals. Subsequently it was transferred to genus *Cerinosterus cyanescens* as it is found to possess septal structure that showed affinity to basidiomycetes and placed in **Tremellales** of phylum **Basidiomycota**.

Now, various species of the genus *Sporothrix* are geographically confined to particular areas, however, in this textbook *S.schenckii* is being treated as the representative species for conventional description of this subject matter.

Epidemiology

Sporotrichosis is globally prevalent more frequently in tropical and subtropical climatic condition of high humidity (65%) and moderate temperature of 25-28ºC. Although this is endemic mycoses but endemicity is quite different due to its worldwide prevalence. It has no typical endemic geographical region as seen in other dimorphic fungi like histoplasmosis, blastomycosis, coccidioidomycosis and even *Talaromyces* infection, however; hyperendemic foci are seen in many countries. It has now been reported from more than 14,000 humans and animals throughout the world. In addition to isolated case reports it has been observed in many **outbreaks** also.

The exact environmental niche of *S.schenckii* is not yet known but it is found as saprotroph associated with plant debris and soil. It is isolated from various natural sources and dead vegetation such as wood, bark, leaves, straw, sphagnum moss and soil. It is the only dimorphic fungus, which has been isolated most frequently from various environmental sources as compared to others. The fungus is commonly found on decaying vegetation hence is easily introduced into skin following minor trauma. The **sphagnum moss** is applied to roots of tree seedling mainly in USA in order to humidify and protect them during shipment from central nurseries. It is also a common constituent of the soil mixture for **bonsai trees**. As there is frequent association between sphagnum moss with this disease, cultivating bonsai trees in USA is taken as its risk factor. In Mexico, hay grass, used as packing material for pottery, fruit, other items and contaminated farming tools, act as the source of infection.

Sporotrichosis is predominantly, a disease of young population due to their frequent contact with natural sources of infection. The outdoor rural workers, gardeners, florists, landscapers, mine-workers and carpenters are usually prone to get the infection. They are exposed to soil and plant debris where fungus is saprotrophic and has opportunity to penetrate into skin through minor wound. Therefore, it is considered as an occupational hazard to persons who have somehow direct contact with plants or soil such as nursery and forestry workers. Males are affected three times more commonly as compared to females because of their outdoor occupation thereby chances of increased exposure. However, during the recent outbreaks it is observed that the number of females is also substantial thereby reaching almost equal to the males. The disease was earlier not known to be transmitted from man-to-man but now the cases of person-to-person transmission are being reported, in addition to its zoonotic spread, which is one of the unique feature.

Sporotrichosis is encountered in Central and South America, predominantly in Brazil, Mexico Peru, Guatemala, Columbia and Uruguay, parts of USA, South Africa, India, Australia, Japan. It is rare in Europe but now cases are being reported from Spain, Italy, France, Portugal, Greece and UK. Recently 7 cases have been reported/reviewed from Turkey spanning over a period of nine decades. It is reported in epidemic proportions as well. The largest outbreak occurred during 1941 to 1944 when 2825 cases were seen among gold miners in Witwatersrand, Johannesburg, South Africa, where source of infection was found to be timber contaminated with *S.schenckii*. In 1988, another large outbreak was reported from USA that stretched over 15 states and affected 84 individuals, spreading due to sphagnum moss obtained from single Wisconsin distributor.

A cluster of cases was observed between the years 2000 to 2003 in Busselton-Margaret River region of Western Australia. The epidemiological investigation and mycological studies implicated moldy hay distributed through a commercial supplier as the source of this outbreak. Outbreaks have been reported in Northern Territory, Darling

Downs region of Queensland as well as northern New South Wales in Australia.

In India, sporotrichosis is prevalent in the northern states in the **sub-Himalayan** hilly areas ranging from Himachal Pradesh to Assam. Chakrabarti *et al*, in a review summarized that 117 patients of sporotrichosis have been reported from India; 75 from northern states, 39 from north-eastern states and 3 from southern states. In 1999, Ghosh *et al*, have identified hyperendemic area in **Kangra** district of Himachal Pradesh, the sub-Himalayan state. Hemashettar *et al*, have reported new endemic focus of this disease in two adjoining south-western states of India *i.e.* northern Karnataka and southern Maharashtra, particularly in alcoholics and immunocompromised patients. Therefore, in India there are at least **three** recognized endemic zones representing (a) north-eastern (b) northern and (b) southern states of the country. In recent times, even Manipur is also recognized as an endemic focus of this disease in India. Three autochthonous cases of lymphocutaneous disease from east and south districts of Sikkim have also been reported.

The increasing number of reports of *Sporothrix* infection in immunocompromised patients, mainly the HIV infected population, suggests it as an emerging global health problem concomitant with the AIDS pandemic. *S.globosa* requires slightly lower (~35°C) as compared to the body temperature of 37°C and this may be a possible reason of its being not involved with systemic or disseminated disease among the population of the Asia Pacific region.

The remote rural south central highlands, an area of hyperendemicity has been identified in Peru, where highest incidence is found among children aged 7-14 years. This epidemic is ongoing for the last three decades *i.e.* 1988 onwards, in the province of Abancay (Peru), which is affecting primarily the pediatric population, predominantly as lymphocutaneous form with presence of lesions over the facial region.

There is an ongoing outbreak in Brazil also since 1998 involving mostly housewives and children who care for cats, which are infected by this fungus. The metropolitan regions of Rio Grande do Sul, Rio de Janeiro and Sao Paulo in the South and Southeastern of Brazil are hyperendemic for the last two decades for cat-associated disease and *S.brasiliensis* is implicated in majority of the cases in this region. *S.brasiliensis* is considered the most virulent among the complex species and with a high prevalence in Brazil being geographically restricted to this country only.

The Brazilian outbreak is a huge zoonosis, whereas contemporary outbreak of almost similar dimensions ongoing in Jilin Province of northeast China is a sapronosis *i.e.* from soil. Moreover, there is another area adjacent to Rio de Janeiro *i.e.* Espirito Santo, where there is classical transmission of this disease, which is not through the cat and species involved is *S.schenckii* instead of *S.brasiliensis*, even in one case as *S.globosa*. Therefore, the epidemiology of sporotrichosis is very unique having multifaceted features.

Immunity

As *S.schenckii* is ubiquitously present in the environment, natural immunity to disease is fairly high, however, antibodies do not provide protection against its infection. The cell-mediated immunity is probably more important in limiting the extent of infection as more severe and disseminated forms of disease are observed in patients with AIDS. In a study on mice it has been observed that antigen-specific T-cells release macrophage-activating cytokine IFN-γ, which activates macrophages thereby exerting protective immunity against *S.schenckii* via enhanced fungicidal activity. It has been demonstrated that melanin protects this fungus from oxidation by certain antimicrobial compounds and from attacks by macrophages. Melanin also protects from ultraviolet solar irradiation and alters host defense mechanisms by reducing phagocytosis through scavenging of reactive oxygen and nitrogen.

Pathogenesis and Pathology

Infective form of this disease is usually spore but now yeast form may also be taken to this effect because in case of zoonotic transmission, it is the yeast, which infects the human beings. Its pathogenicity depends on various virulence factors like thermotolerance, melanin production, ergosterol peroxide, virulence related proteins and adhesion. *S.schenckii* produces two extracellular proteinases, namely proteinase I and II. Proteinase I is serine proteinase, inhibited by chymostatin, while proteinase II is an aspartic proteinase, inhibited by pepstatin. These enzymes have important role in fungal invasion and growth. Moreover, detection of antibodies against these proteinases also contributes to serodiagnosis of this disease. The fixed cutaneous lesions may rarely present secondary sporotrichids formation also.

The **asteroid bodies** of sporotrichosis consist of central yeast of *S.schenckii*, surrounded by eosinophilic spicules probably derived from neutrophils of abscess at center of which extracellular sporotrichotic asteroid body is located. It can be demonstrated immunohistochemically. These are different from the asteroid bodies seen in other fungal infections like lobomycosis, which are found in giant cells.

The histopathologic features of primary cutaneous sporotrichosis are combination of granulomatous and pyogenic reactions. In 1963, Lurie described three granulomatous patterns, which are as follows:-

(i) Sporotrichoid Type: There are masses of epithelioid histiocytes, which have tendency to form concentric zones. The central area of lesion consists of neutrophils or necrotic material surrounded by an infiltrate of neutrophils, plasma cells and lymphocytes.

(ii) Tuberculoid Type: In this reaction, the involved area merges into zone of epithelioid cells mixed with fibroblasts, lymphocytes and Langhans' giant cells.

(iii) Foreign Body Type: There is foreign body type of granulomatous reaction, without any evidence of pyogenic component.

Clinical Features

Sporotrichosis is a polymorphic disease caused by many species of genus *Sporothrix*. These are important thermodimorphic fungi of medical importance and cause lymphatic and subcutaneous infection. This is usually neither invasive nor highly virulent organism and tends to produce insidious and chronic infection in humans as well as domestic animals. The index of clinical suspicion should be kept high about this disease so that diagnosis can be promptly established or ruled out for further management at the earliest. The disease is broadly categorized into three clinical types. Most of these are outcome of traumatic implantation and pulmonary type is due to inhalation of conidia. These are briefly described as follows:-

1. Lymphocutaneous Sporotrichosis
2. Fixed cutaneous Sporotrichosis
3. Extra-cutaneous Sporotrichosis

1. Lymphocutaneous Sporotrichosis

This classical variety is the commonest type of sporotrichosis constituting more than 75% cases with proximal advancement of fungus through lymphatics. It follows subcutaneous implantation of spores in a penetrating wound caused by thorn or splinter; through cat scratches or bites. It is generally an infection of upper and lower extremities, which is generally subjected to trauma. There are rare instances of needle-prick or laboratory acquired infection. The incubation period varies from 8 to 30 days with an average of **three** weeks' duration.

The initial manifestation is appearance of a small, firm, non-tender, mobile subcutaneous nodule, which eventually becomes violaceous and may ulcerate to expose ragged necrotic-base forming sporotrichotic chancre. The nodules develop proximal to the primary lesion following the lymphatic distribution and often ulcerate. A small-ulcerated lesion develops at the site of injury within one to two weeks time and remains localized but fails to heal with routine topical therapy.

The clinical photograph in **Figure 12.1** shows typical ascending lymphocutaneous lesions at the inoculation site of ring finger and over left hand as well as the wrist. In **Figure 12.2** the nodulo-ulcerative lesions are seen advanced to right forearm and arm along the lymphatic tract in another patient. All the **Figures 12.3 to 12.7** are having almost same finding as explained in the legends. Similarly, in **Figure 12.8** the nodulo-ulcerative lesions are seen over the dorsal surface of right great toe and in **Figure 12.9** these are further extended through lymphatics to the lower limb and even up to the thigh in the same patient.

The typical course of disease is characterized by involvement of lymphatics that drain primary site of inoculation and development of characteristic linear nodulo-ulcerative secondary lesions along lymphatics. The development of

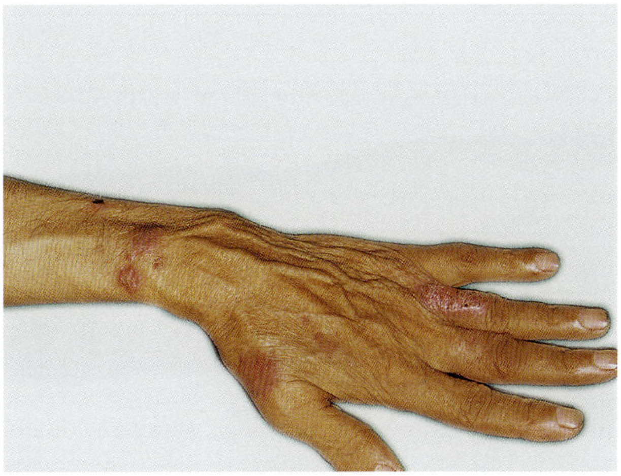

Fig. 12.1. Nodulo-ulcerative lesions of lymphocutaneous sporotrichosis arranged in ascending order over ring finger, dorsum of hand and wrist.

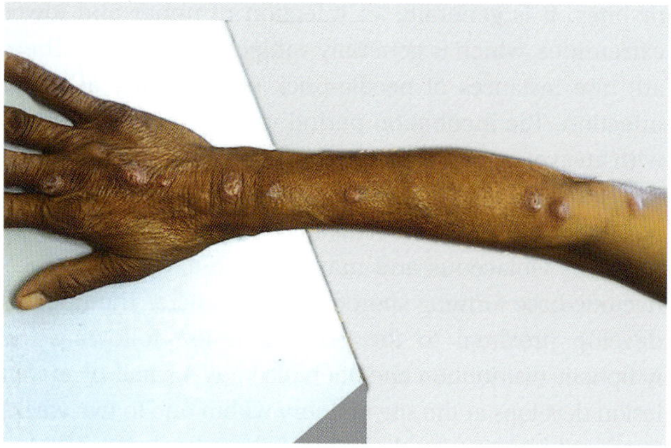

Fig. 12.2. Nodulo-ulcerative lesions of lymphocutaneous sporotrichosis arranged in ascending order on the right wrist and forearm.

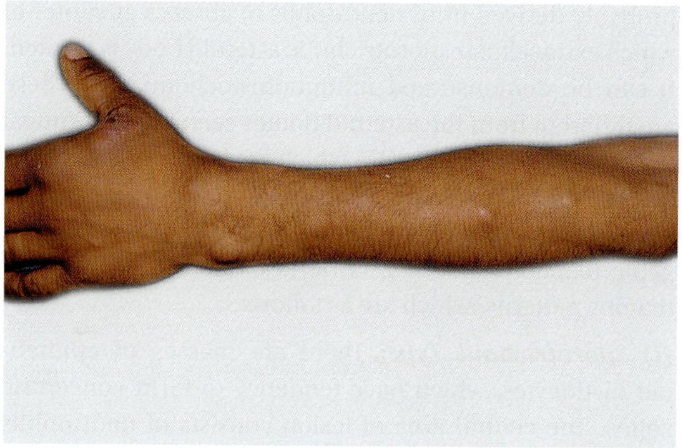

Fig. 12.3. Nodular lesions of lymphocutaneous sporotrichosis arranged in ascending order over thumb, dorsum of hand, wrist and forearm.

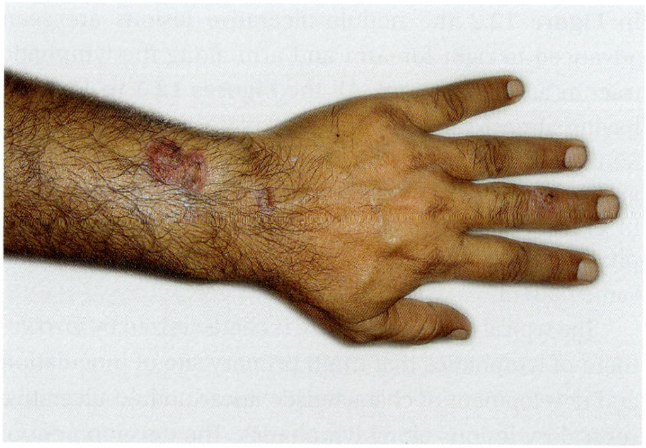

Fig. 12.4. Nodulo-ulcerative lesions of lymphocutaneous sporotrichosis arranged in ascending order over dorsum of hand and wrist.

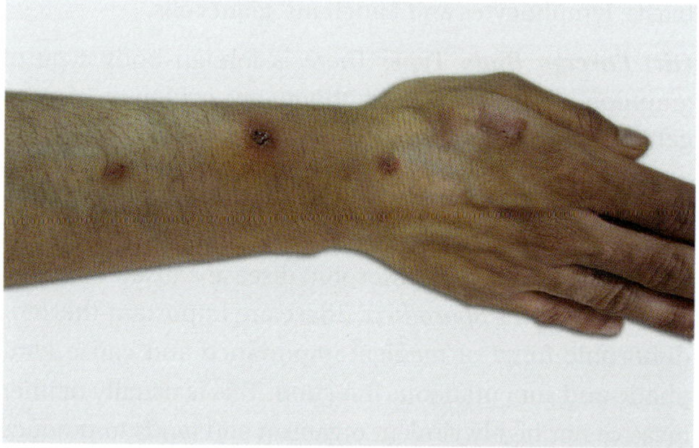

Fig. 12.5. Nodulo-ulcerative lesions of lymphocutaneous sporotrichosis arranged in ascending order over thumb, dorsum of hand, wrist and forearm.

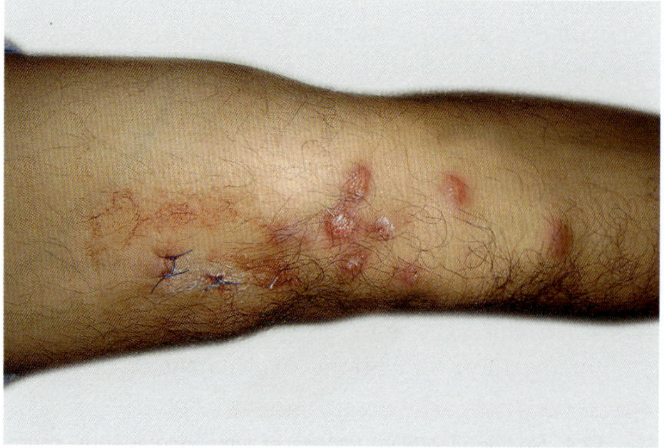

Fig. 12.6. Nodulo-ulcerative lesions of lymphocutaneous sporotrichosis arranged in ascending order over upper limb.

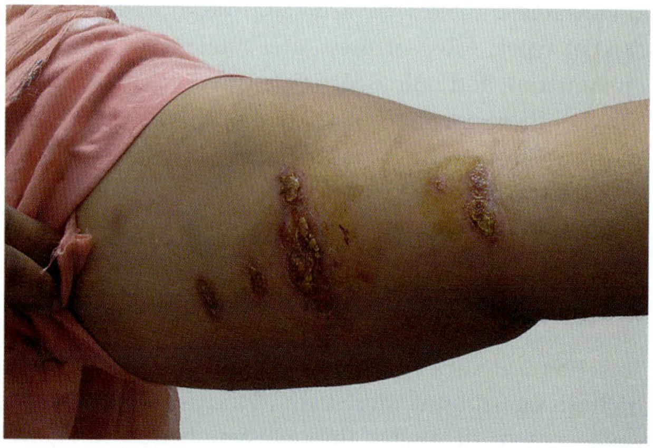

Fig. 12.7. Nodulo-ulcerative lesions of lymphocutaneous sporotrichosis arranged in ascending order over left arm.

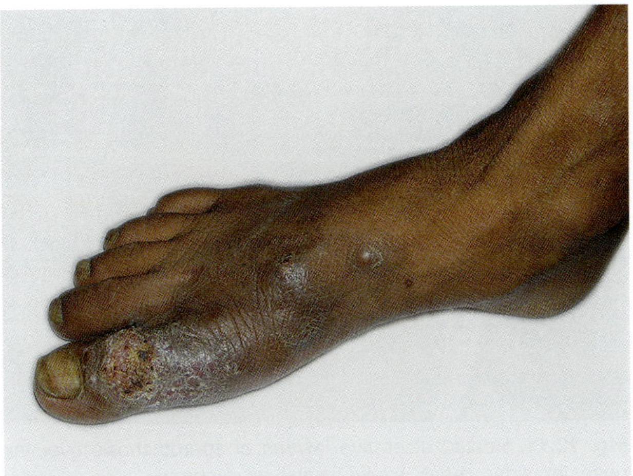

Fig. 12.8. Nodulo-ulcerative lesions of lymphocutaneous sporotrichosis arranged in ascending order on the right foot.

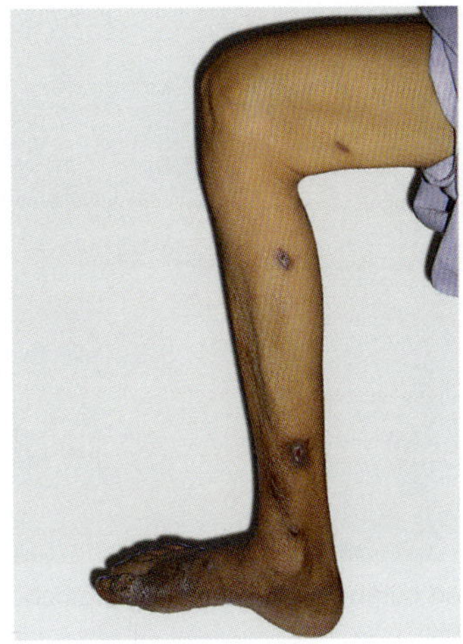

Fig. 12.9. Nodulo-ulcerative lesions of lymphocutaneous sporotrichosis arranged in ascending order on the right leg.

lymphangitic nodules secondary to open skin lesions is typical of this disease. This type of lesions are its pathognomonic and give an adjective, **sporotrichoid** to such sort of description in medical literature. The lymph nodes become swollen and inevitably suppurate and connecting lymphatic channel becomes indurated and **cord-like**. Sometimes psoriasiform cutaneous lesions are also rarely reported. Rarely solitary tumoral, vegetating and friable lesion in the periungual region on the finger of the hand and patient present with lymphadenopathy in ascending distribution along the affected limb.

2. Fixed Cutaneous Sporotrichosis

There are cases of cutaneous sporotrichosis, wherein lesions remain localized to inoculation site and there is no tendency towards lymphatic or systemic spread. This type is commonly seen in individuals with higher degree of immunity because of subclinical infection due to prior exposure of the organism. The lesions commonly appear on face, neck, trunk or arm and vary widely in their appearance *i.e.* ulcerative, verrucose, papular, acneform, gummatous or erythematoid plaques. This type is also known as **plaque** sporotrichosis due to its frequent clinical presentation. The disease is confused with many other similar looking clinical settings, particularly cutaneous tuberculosis wherein there is no improvement despite full course of ATT or even MDR therapy.

In the clinical photograph of the fixed cutaneous type of disease in **Figure 12.10**, there is single papular lesion on the right cheek and ulcerative lesion over the nose. On the left side of face of this patient, there are extensive ulcerative lesions over nose, cheek, upper eyelid as well as lateral side of eyebrow **(Fig. 12.11)**. After one year of completion of treatment, all the lesions disappeared **(Fig. 12.12)**. Moreover, there was no recurrence after eight years of treatment with no medication in between period **(Fig. 12.13)**. Sometimes, papular lesions of this type may be seen over uncommon sites like chest wall (bilateral) as seen in **Figure 12.14** along with the close up view in **Figure 12.15**. The bilateral anterior chest wall lesions got healed after a fortnight treatment with potassium iodide **(Fig. 12.16)**.

3. Extracutaneous Sporotrichosis

The involvement of anatomical sites away form the skin are clubbed under this heading. They may be ranging mild form to serious and life threatening disseminated forms also. These are as follows:

(a) Pulmonary Type: The pulmonary manifestations present as two forms: (a) **primary** pulmonary infection, which usually shows cavitary pattern among immunocompetent individuals and (b) as part of **multifocal** infection, which presents as non-cavitary disease among immunocompromised patients. The disease may present as primary pulmonary infection without involvement of cutaneous or subcutaneous tissue. It is in contrast to the other systemic dimorphic fungi, which are mainly routed through the respiratory tract. This extracutaneous infection results from inhalation of conidia and it is analogous to primary

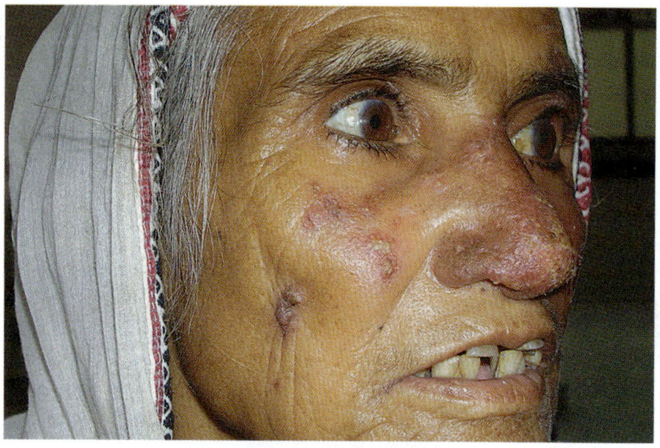

Fig. 12.10. Fixed cutaneous lesions of sporotrichosis on the right side of the cheek.

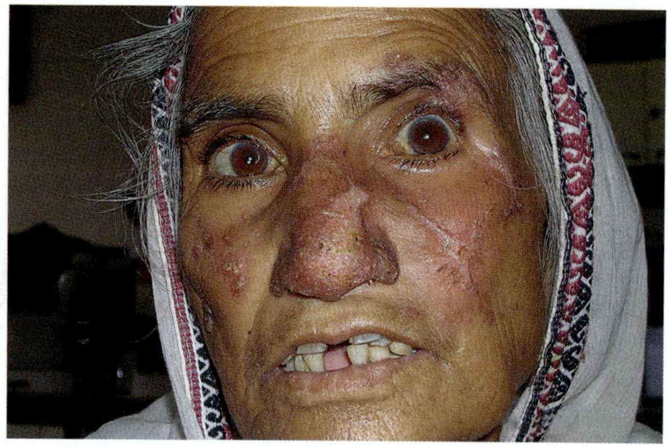

Fig. 12.11. Healed ulcerative lesions of sporotrichosis over the nose, left-sided cheek, eyelid as well as eyebrow.

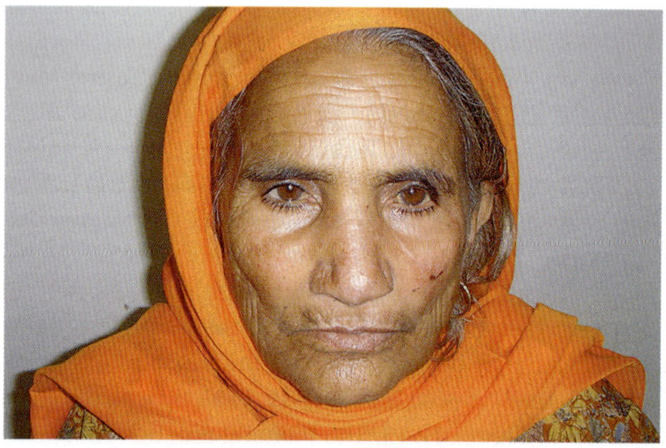

Fig. 12.12. All the lesions disappeared after taking treatment with KI for a period of six weeks. Photograph taken after one year.

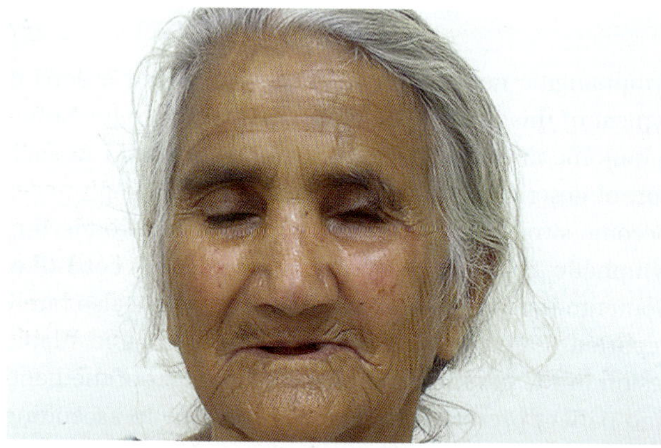

Fig. 12.13. Photograph taken after eight years and there was no lesion left and patient became normal.

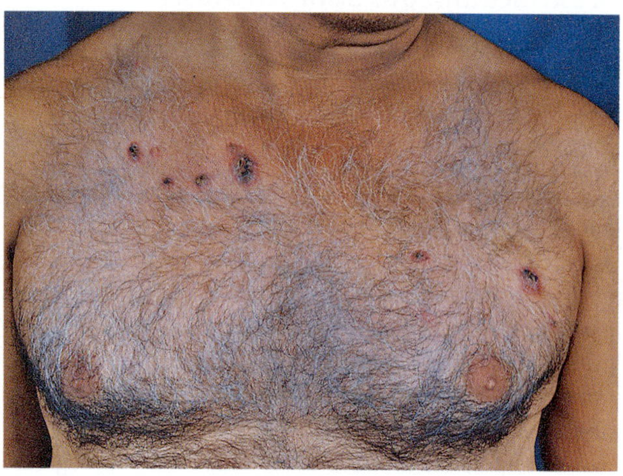

Fig. 12.14. Linear pyogenic bilateral lymphocutaneous lesion of sporotrichosis over anterior chest wall.

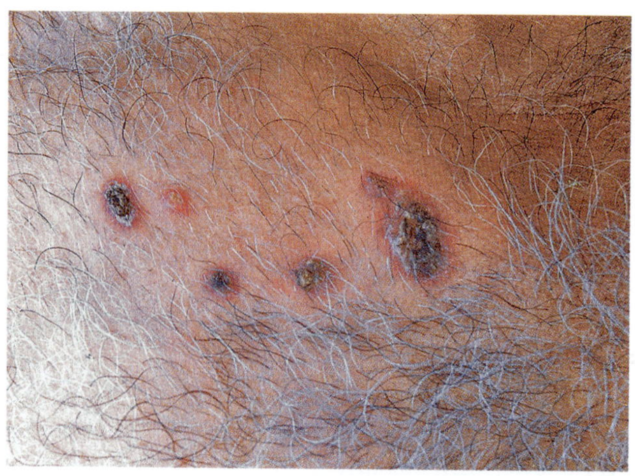

Fig. 12.15. The close-up view of linear pyogenic bilateral lymphocutaneous lesion of sporotrichosis shown in **Figure 12.14**.

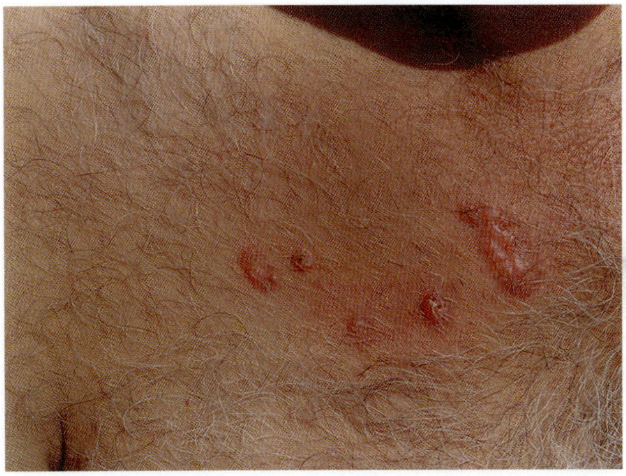

Fig. 12.16. Linear pyogenic lymphocutaneous lesions of sporotrichosis over anterior check wall after two weeks' treatment with KI.

pulmonary tuberculosis, histoplasmosis or coccidioidomycosis. Once conidia of *S.schenckii* enter lungs, they may form nodules, cavities or diffuse reticulonodular infiltrates and thus can simulate variety of other infectious diseases as well as neoplasia. A single, cavitary lesion of upper lobe is the most distinctive feature. Gradually chronic pneumonitis sets in with nodular masses and development of thin-walled cavities with fibrosis and pleural effusion. The nodules correspond to solid granulomas, many of which have central suppuration or necrosis. The solitary residual fibrocaseous nodule **(sporotrichoma)**, an infrequent pathogenic manifestation of pulmonary sporotrichosis, closely resembles solitary fibrocaseous pulmonary nodules, produced more commonly by *Histoplasma capsulatum* and *Coccidioides* species. A second type of pulmonary infection primarily involving lymph nodes is also described. The disease is acute and rapidly progressive but complete resolution of lesions is observed.

(b) Mucocutaneous Type: In this form, lesions develop in mouth, pharynx, vocal cords or nose. The lesions are erythematous, ulcerative and suppurative at first and eventually become granulomatous, vegetative or papillomatous. The infection is usually accompanied by pain that is unusual in cutaneous type. The regional lymph nodes become enlarged and firm. A case of penile sporotrichosis has also been reported by Kimyai-Asadi *et al,* which was initially diagnosed to be as pyoderma.

(c) Osteoarticular Type: Osteoarticular sporotrichosis has also been reported, which occurs more frequently among alcoholics. The infection is secondary to contiguous spread but is more likely caused by hematogenous spread to one or more joints. The joints most commonly affected are the knee, elbow, wrist and ankle. Tenosynovitis is another manifestation and may present as carpal tunnel syndrome. Rare case of urinary involvement in a renal transplant recipient has been reported.

(d) Ocular Type: In ocular manifestations, it is multifocal choroiditis due to disseminated sporotrichosis can occur in profoundly immunosuppressed patients with HIV/AIDS. Ocular involvement is being reported from Jilin Province of northeast China and Peru. The cases of ocular adnexa, particularly eyelid and acute dacryocystitis caused by *Sporothrix* species are now being reported. The involvement of upper eyelid as well as lateral side of left eyebrow is seen in **Figure 12.11**.

(e) Disseminated Type: The disseminated form of the disease is very rare and may occur by hematogenous spread from primary lesion or from suppurating lymph nodes. Dissemination may be manifested by many widespread skin lesions, which begin as subcutaneous nodules and become papules, pustules, gummata or confluent areas of folliculitis. Dissemination to visceral organs is rarely observed but pyelonephritis, orchitis and mastitis may develop at times. It is mostly seen in patients suffering from AIDS. The disease is also shown involving meninges hence cerebrospinal fluid should be examined for *S.schenckii*-antibody, in addition to other fungal agents, in a patient with chronic meningitis where no other cause is identified by conventional diagnostic tests. Fungemia may also be caused particularly among the immunocompromised patient like that of AIDS.

Sporotrichosis in Animals

The zoonotic transmission of sporotrichosis is observed particularly in pets like domestic cats as well as dogs. It is also documented in number of other animals including armadillos, camel, cattle, pig, donkey, horse and birds. The domestic pets may be an important carrier for agents of sporotrichosis to the humans. It is a enzootic disease and in Brazil it is wildly found prevalent in the cats and dogs as feline and canine sporotrichosis, respectively. Brazil was not only in the news due to Zika virus hoax or Olympic games but because of increased incidence of this disease also and that too found rampantly in the pets and their almost corresponding number among the humans. Since 1998 its outbreak is going on in Rio Grande do Sul, Rio de Janeiro and Sao Paulo in the South and Southeastern areas of Brazil. It continued by the turn of 21st century and now there are reports revealing ~4000 culture-proven human cases, who were identified and treated, mainly among middle-aged women, with domiciliary or professional contact with infected cats, almost same number *i.e.* 4120. It is suggested that feline transmission to humans is facilitated by unusually large number of organisms typically found in cutaneous lesions of infected cats. In Brazil kissing of cats is a very common practice. Therefore, now cat-to-cat and cat-to-man is one of the feasible modes of infection in this disease.

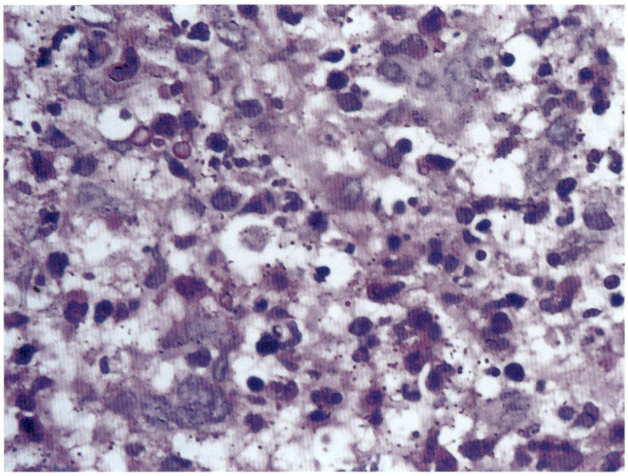

Fig. 12.17. Budding yeast cells seen in histopathological tissue sections of a patient of sporotrichosis (PAS x 400).

Radiodiagnosis

The radiological is mainly useful in pulmonary type of sporotrichosis. The findings are patches of pneumonitis with widespread areas of miliary infiltration mimicking tuberculosis. Mediastinal widening and hilar lymphadenopathy may also be observed due to involvement of tracheobronchial lymph nodes.

Differential Diagnosis

The disease should be differentiated from cutaneous leishmaniasis, nocardiosis, tuberculosis verrucosa cutis, chromoblastomycosis, paracoccidioidomycosis, blastomycosis, cutaneous anthrax, granuloma annulare, foreign body granuloma, Majocchi's granuloma, subcutaneous abscesses of tularemia, primary syphilis, cat-scratch disease and chronic staphylococcal ecthyma. Swimming pool granuloma caused by atypical mycobacteria such as *Mycobacterium marinum* may also simulate sporotrichosis. Differentiation of the fixed form of sporotrichosis from pyoderma gangrenosum can be very difficult. The differential diagnosis of sporotrichal arthritis is often delayed because it can be confused with tuberculous arthritis, sarcoidosis, rheumatoid arthritis and gout. The pulmonary disease must be distinguished from other chronic cavitary infections like tuberculosis, histoplasmosis, coccidioidomycosis and sarcoidosis.

Laboratory Diagnosis

In laboratory, diagnosis of sporotrichosis can be established by direct microscopy and culture of pus aspirated from series of proximal subcutaneous nodules and moist, ulcerated lesions. The KOH wet mount is of little or no value because of small size as well as scanty number of fungi in pus, exudates or aspirates hence their presence may easily be overlooked.

(a) Direct Examination

The clinical material is collected like pus, exudate and aspirate from nodules, curettage or swabbing from open lesions. The KOH wet mount preparation of material shows small, elongated yeast cells. The sensitivity of KOH wet mount is very low due to paucity of organism in the clinical material as well as its size which is very small. The Gram-stained smear shows Gram-positive irregularly stained yeast cells that are very few in number. Usually the inexperienced observer may miss the yeasts due to the same reasons of their smaller size and less number in the clinical material. In such a situation, calcofluor white stain proves to be very useful. In case of respiratory involvement, BAL fluid is examined for establishing the diagnosis.

The fine-needle aspiration cytology or biopsy from nodule may also be performed, which shows round, oval or cigar-shaped yeast cells in H&E, PAS or GMS stains without any capsule. In the photomicrograph **Figure 12.17**, the rounded budding yeast cells are seen in PAS staining. In tissue sections, the organism appears as **cigar-shaped bodies** of size 3-5 µm or wider in diameter. In another PAS stained smear the rounded and elongated yeast cells are seen as pink colored **(Fig. 12.18)**. Asteroid bodies may also be seen in direct smear examination of clinical specimens. Being dimorphic fungus, the yeast cells are demonstrated in direct smear from clinical material

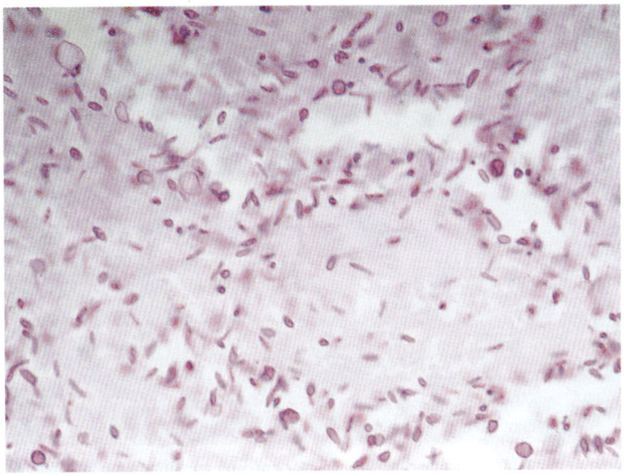

Fig. 12.18. Budding yeast cells of *Sporothrix schenckii* seen along with cigar-shaped cells (PAS × 400).

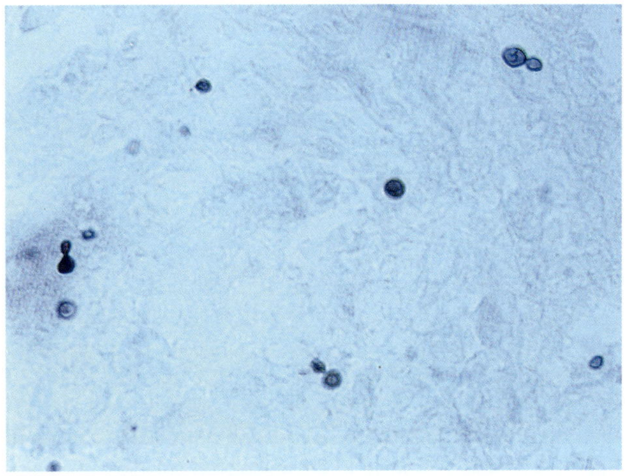

Fig. 12.19. Budding yeast cells of *Sporothrix schenckii* seen in the GMS staining (GMS × 400).

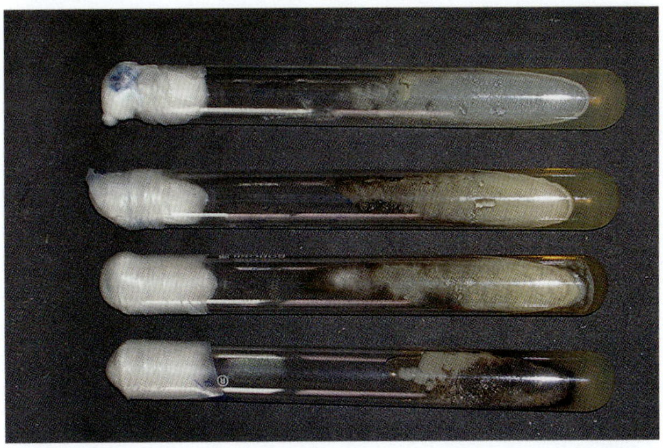

Fig. 12.20. Moist, off-white cream-colored colonies of *Sporothrix schenckii* converted to grey, brown or black color after two weeks.

and very rarely pseudohyphal or filamentous forms of *S.schenckii* are reported. In the **Figure 12.19**, the small budding yeast cells are seen as black colored in the GMS staining. The smaller size of budding yeast cells in this disease may confuse with *Histoplasma capsulatum* or even *Candida glabrata*.

The immunohistochemical avidini-biotin complex immunoperoxidase is even better than PAS or GMS stains for identification of *S.schenckii* in the tissue. The fungi are surrounded by refractile, eosinophilic halo, called **Splendore-Hoeppli** phenomenon, which is believed to be due to immune complex deposition around the organisms. The basis of asteroid body is this phenomenon and is best seen on histopathological examination with H&E.

In the presence of thick-walled horizontal and longitudinal septate and classical **eyeglass**-shaped cells, there is a possibility that tissue form of *S.luriei* may be there. It is often difficult to differentiate this form of sporotrichosis in the tissue from adiaspiromycosis, blastomycosis and coccidioidomycosis.

(b) Fungal Culture

The fungal culture remains **gold standard** for establishing definitive diagnosis of sporotrichosis, however, it takes slightly longer time of around 3-4 weeks' duration. On conventional media, *S.schenckii* grows well over a wide temperature range. The clinical material like pus, aspirated from lesions, should be inoculated on two sets of SDA, BHIA with actidione, blood agar, chocolate agar as well as Pine's medium and incubated at 25°C and 37°C. The colonies at 25°C are initially moist, off-white to cream-colored which appear within 3-5 days and turn gray, brown or black leathery in about 10-14 days **(Fig. 12.20)**. The colonies may remain moist, membranous or fine grayish velvety and vary from cream to black in color. In **Figure 12.21**, the SDA plate is showing the yeast-like cream colored colonies. **Figure 12.22** shows off-white mycelial growth being converted into the grayish-back from the center.

S.schenckii being dimorphic fungus shows delicate mycelia/hyphae at room temperature. The hyphae are considerably more slender, than those of most molds, the diameter being ~2 μm. The mycelial strands are intertwined with each other to give appearance of loose **twisted**

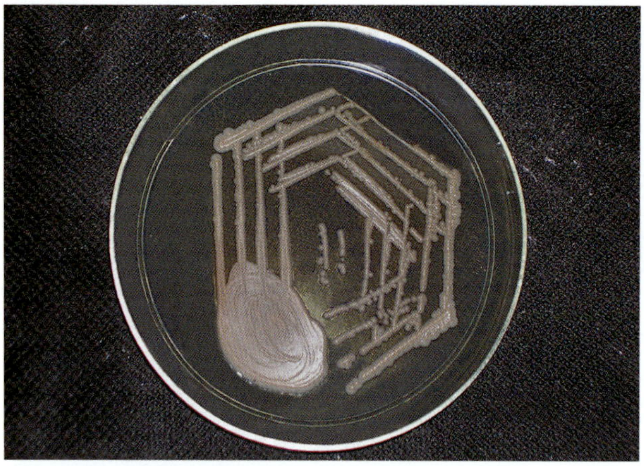

Fig. 12.21. Moist, off-white to cream-colored colonies of *Sporothrix schenckii* grown on SDA plate at 37°C.

Fig. 12.22. Mycelial colonies of *Sporothrix schenckii* converted to grey, brown or black color on SDA after two weeks at 25°C.

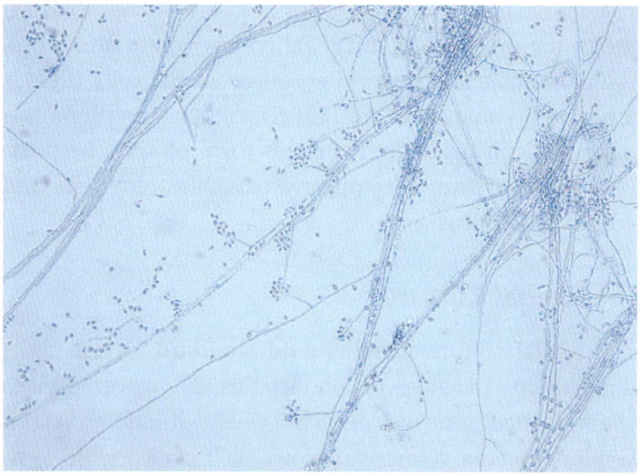

Fig. 12.23. *Sporothrix schenckii* showing two types of sporulation, along hyphae and flower-like pattern (LCB × 200).

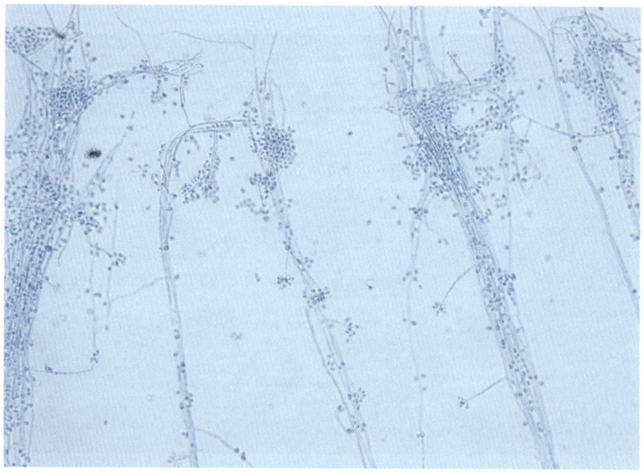

Fig. 12.24. *Sporothrix schenckii* showing sporulation giving a palm tree-like appearance (LCB × 200).

ropes. There are two types of sporulation; spore borne individually on delicate sterigma along hyphae and pyriform spores borne in small groups; each spore is attached by its own fragile sterigma to a common conidiophore. Moreover, slender conidiophores of varying length arise perpendicular from hyphae and taper at tip thus forming a **flower-like pattern** on delicate stalk **(Fig. 12.23)**. The conidia arise both laterally and at tips of delicate sterigma on conidiophore and are described to give a **palm tree-like** appearance **(Fig. 12.24)**. They arise directly from mycelium or small stems called denticles **(Fig. 12.25)**.

Sporulation increases as colony ages and conidia are formed not only along the length of conidiophore but may form sleeves of spores along hyphae. The LCB mount of colony grown at 37°C shows round, oval, ellipsoidal or fusiform, budding yeast cells.

The demonstration of dimorphism is essential for definitive identification of *S.schenckii* and mold (25°C) to yeast (37°C) conversion should be demonstrated in all the suspected isolates and *vice versa* (M↔Y) thereby its differentiation from non-pathogenic saprotrophic species of genus *Sporothrix*. There are three important factors essential for conversion of mycelial to yeast phase which are: (a) temperature at 37°C (b) medium rich in amino acids, vitamins particularly thiamine and biotin and (c) humid atmosphere with higher concentration of 5-10% carbon dioxide.

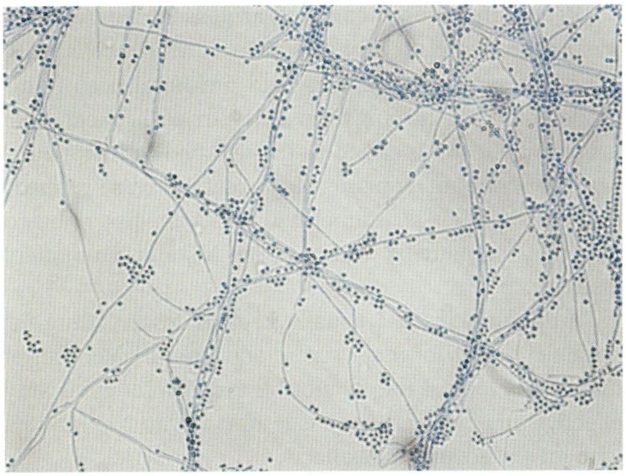

Fig. 12.25. *Sporothrix schenckii* showing conidia arise directly from denticles (LCB × 400).

Sporothrix luriei produces light tan colony that becomes white during maintenance in laboratory. The conidia are slightly longer and in tissues it forms large thick-walled cells measuring 15-20 μm in size. Some of the cells multiply by budding, while others by fission. Sometimes, it may acquire similar tissue forms as that of *S.schenckii.*

The slide culture of mycelial isolate is also put up to examine typical conidial arrangement of *S.schenckii*. It is also useful to differentiate from look-alike fungi *i.e. Acrodontium* species. The exoantigen test is done by micro-immunodiffusion procedure, used by Kaufman and Standard, taking control antigen from yeast phase culture of the clinical isolate.

(c) Immunodiagnosis

The immunodiagnosis is done with *in vivo* skin tests and *in vitro* serological tests in the laboratory. However, staining by direct fluorescent antibody (DFA) technique to demonstrate antigen is highly specific method and reveals asteroid bodies in almost 90% cases. The immunodiagnostic techniques used in sporotrichosis are given below:-

(i) *Skin Tests*: The mycelial and yeast phase antigens of *S.schenckii* are used in skin tests but results are variable with poor specificity. **Sporotrichin** preparations may have varying antigenic composition depending on fungal growth conditions. They are not routinely used for diagnostic purpose. The sporotrichin is prepared from filtrate of 2-3 months mycelial growth in glucose-yeast-extract-casamino acid liquid medium at room temperature.

The standard sporotrichin skin test consists of intradermal injection of 0.1 ml of 1:1000 dilution of heat-killed, packed yeast cells. A reaction of more than 5 mm diameter of induration and erythema after 24-48 hours is taken as positive. This test is useful in determining exposure to fungus and carrying out epidemiological surveys.

Skin tests with yeast phase antigen *i.e.* peptido-rhamnomannan (PRM) antigen test is also done in similar way.

(ii) *Serological Tests*: The latex particle or whole cell agglutination tests are done for detecting extra-cutaneous form of disease. These tests have some prognostic value since titers decline on successful therapy. Other tests that may be helpful are complement fixation test and immunodiffusion. The agglutination and immunodiffusion tests are useful for both diagnosis and assessing prognosis of this disease.

(iii) *Molecular Methods*: The DNA typing methods have provided useful epidemiological information on infections due to variety of pathogenic fungi including *Aspergillus fumigatus, Candida albicans* and *Histoplasma capsulatum*. Cooper *et al*, have demonstrated that restriction fragment length profiles of whole-cell DNA, which vary widely among clinical isolates of *S.schenckii* and hence were helpful in identifying the source of 1988 epidemic of sporotrichosis in North America.

(d) Animal Pathogenicity

This is essential done to differentiate saprotrophic species of *Sporothrix* from *S.schenckii*, which look otherwise morphologically very similar. Only *S.schenckii* gives positive animal pathogenicity test. Mice and ratsare susceptible to *S.schenckii*. Saline suspension (0.5-1 ml) of cells from yeast phase or mycelial fragments and conidia from filamentous cultures should be inoculated intraperitoneally into male Swiss albino mice. On autopsy findings, approximately after 7-10 days, show features of orchitis. The cigar-shaped cells and large spherical bodies up to 8 μm in diameter are found in abundance in testes of mice. Inoculation of clinical material directly into experimental animal is not useful for diagnosis since generally very few organisms are present to initiate infection.

Treatment and Prophylaxis

The Infectious Diseases Society of America releases clinical practice guidelines for the management of sporotrichosis from time to time. The last one was released about ten years ago in November 2007 replacing the previous ones laid down in 2000, describing therapeutic modalities in various clinical settings of this disease.

Historically, saturated solution of potassium iodide (SSKI) is the classical drug of choice traditionally given for cutaneous and lymphocutaneous types of sporotrichosis for the last one-century and till date it remains the best one being a wonder drug. It is being regularly put in use as first line drug because of its effectiveness and low cost. The initial dose is 5 drops thrice daily, which can be increased by 5 drops per week amounting to 25-40 drops thrice daily for 6-12 weeks. It is associated with a few side effects like metallic taste, salivary gland swelling, rash and fever. It is taken with milk or fruit juices to mitigate gastrointestinal side effects. The recommended dose in children is 3 drops 3 times a day, increasing by 1 drop/kg/dose to a maximum of 25 drops thrice a day. Therefore, the dose is 0.2-0.3 gm/day for children and 0.5-1.5 gm/day for adults. The dosage is measured using standard eye-dropper in which 1 drop = 50 mg of SSKI.

The precise mechanism by which KI exerts its therapeutic effect is not known. The use of iodides is empirical and probably therapeutic effect is mediated by enhancement of body's immunological and non-immunological defense mechanisms rather than any fungistatic or fungicidal activity of KI. It is presumed that granuloma is resolved by KI, followed by elimination of fungus through natural defense system. Most of the disorders for which KI is indicated display neutrophil in early stages of disease and modulation of neutrophils' chemotaxis may also be affected by administration of KI. This agent will probably remain drug of choice in this disease for times to come because of its high clinical efficacy and low cost, particularly among the developing countries. However, it is not effective in pulmonary or systemic disease, where others salts like azole derivatives (itraconazole) are indicated.

Oral ketoconazole (10 mg/kg/day) or itraconazole (100-200 mg/day) are also used for long-term treatment. Itraconazole is better than ketoconazole and has now become the drug of choice for treatment of sporotrichosis. Itraconazole in the form of liquid is more effective than the capsule, however, the liquid form is not available in India. Terbinafine is also found to be effective in a dose of 250 mg twice daily for treatment of cutaneous sporotrichosis administered for an average period of 18 weeks. However, its dose may be increased to 500 mg twice daily in severe clinical types of the disease when other drugs are either not feasible or contraindicated.

Before the introduction of azoles, amphotericin B was the drug of choice for treatment of lymphocutaneous, pulmonary and other unifocal and multifocal deep forms of sporotrichosis showing relapse. The triazole, itraconazole is currently recommended as the first choice therapeutic agent for lymphocutaneous and visceral disease whereas amphotericin B is recommended for severe disseminated infection, in patients with AIDS, CNS and pulmonary disease.

Amphotericin B is usually kept as reserve drug for systemic therapy of more severe and disseminated fungal disease, its topical preparation in the form of gel is very effective and cures cutaneous lesions within two months without any adverse side effects.

As *S.schenckii* does not tolerate >39°C therefore hyperthermia has been tried successfully as an adjuvant therapy to systemic antifungals. A heat-evolving pad is attached to lesions for about two hours per day.

The surgical techniques like knee arthroplasty, used in active tuberculous arthritis, is also tried for treatment of chronic sporotrichal arthritis. When fixed-type sporotrichosis is small and well localized, surgical excision can be the treatment of choice. The disease usually takes favorable course when it is timely managed.

There are attempts to develop a vaccine, still under way. **Sporotrichin** in the form of subcutaneous injection once a week has been prescribed for 50 Japanese patients and results were found to be encouraging. This vaccine may be used as supplementary treatment for patients, who do not respond properly to potassium iodide or to local heat therapy. However, as such no vaccine is currently available for use in this disease. Immunization with monoclonal antibody against *S.schenckii* adhesion has recently been found to be effective in murine model.

For the prevention of the disease, certain common measures have been found to be useful like wearing of gloves while handling of moldy hay. However, to control the ongoing epidemic and for public health purposes, it is necessary that an effective and sustainable therapeutic regimen applied to cats under field conditions. In addition, public awareness programs on prophylaxis are required, encouraging the following: responsible ownership, castration, cremation of dead cats, confinement of cats inside the home, limitation of the number of cats per household, regular cleaning of the dwellings and proper healthcare for the animals, as well as general public health measures such as basic sanitation, regular garbage collection and cleaning of empty lots. Further epidemiological and environmental studies are needed to better understand this ongoing epidemic in Brazil and Peru.

Further Reading

1. Agarwal S, Gopal K, Umesh, *et al.* Sporotrichosis in Uttarakhand (India): A report of nine cases. *Int J Dermatol.* 2008; **47:** 367-71.
2. Akhtar N, Begum M, Saleh AA, *et al.* Cutaneous sporotrichosis. *Mymensingh Med J.* 2010; **19:** 458-61.
3. Alba-Fierro CA, Perez-Torres A, Toriello C, *et al.* Molecular components of *Sporothrix schenckii* complex that induce immune response. *Curr Microbiol.* 2016; **73:** 292-300.
4. Alberici F, Paties CT, Lombardi G, *et al. Sporothrix schenckii* var. *luriei* as the cause of sporotrichosis in Italy. *Eur J Epidemiol.* 1989; **5:** 173-7.
5. Almeida-Paes R, de Oliveira LC, Oliveira MM, *et al.* Phenotypic characteristics associated with virulence of clinical isolates from the *Sporothrix* complex. *Biomed Res Int.* 2015; 212308. PMID: 25961005.
6. Almeida-Paes R, de Oliveira MM, Freitas DF, *et al.* Sporotrichosis in Rio de Janeiro, Brazil: *Sporothrix brasiliensis* is associated with atypical clinical presentations. *PLoS Negl Trop Dis.* 2014; **8:** e3094. PMID: 25233227.
7. Aronowitz PB, Gilroy M, Christiansen KN. Disseminated sporotrichosis with osteolytic bone involvement. *J Gen Intern Med.* 2017; doi: 10.1007/s11606-017-4048-4. PMID: 28349410.
8. Arora U, Aggarwal A, Arora RK. Sporotrichosis in Amritsar - a case report. *Indian J Pathol Microbiol.* 2003; **46:** 442-3.
9. Assi M, Lakkis IE, Wheat LJ. Cross-reactivity in the *Histoplasma* antigen enzyme immunoassay caused by sporotrichosis. *Clin Vaccine Immunol.* 2011; **18:** 1781-2.
10. Aung AK, Spelman DW, Thompson PJ. Pulmonary sporotrichosis: An evolving clinical paradigm. *Semin Respir Crit Care Med.* 2015; **36:** 756-66.
11. Aung AK, Teh BM, McGrath C, *et al.* Pulmonary sporotrichosis: Case series and systematic analysis of literature on clinico-radiological patterns and management outcomes. *Med Mycol.* 2013; **51:** 534-44.
12. Bachmeyer C, Buot G, Binet O, *et al.* Fixed cutaneous sporotrichosis: An unusual diagnosis in west Europe. *Clin Exp Dermatol.* 2006; **31:** 479-81.
13. Bahr NC, Janssen K, Billings J, *et al.* Respiratory failure due to possible donor-derived *Sporothrix schenckii* infection in a lung transplant recipient. *Case Rep Infect Dis.* 2015; 925718. PMID: 26697244.
14. Banerjee AK, Basu N, Roychoudhury DS, *et al.* Sporotrichosis. *Bull Calcutta Sch Trop Med.* 1971; **19:** 56-7.
15. Banerjee S, Jaiswal AK. Lymphocutaneous sporotrichosis treated with three different modalities. *Indian J Dermatol Venereol Leprol.* 2003; **69:** 186-7.
16. Barros MB, de Almeida Paes R, Schubach AO. *Sporothrix schenckii* and sporotrichosis. *Clin Microbiol Rev.* 2011; **24:** 633-54.
17. Barros MB, Schubach AO, Schubach TM, *et al.* An epidemic of sporotrichosis in Rio de Janeiro, Brazil: Epidemiological aspects of a series of cases. *Epidemiol Infect.* 2008; **136:** 1192-6.
18. Baruah BD, Saikia TC, Bhuyan RN. Sporotrichosis in Assam: A clinical and mycological study. *J Indian Med Assoc.* 1976; **67:** 223-9.
19. Bhattacharjee P, Brodell RT. A lesion with lymphangitic spread in a gardener. Exposure to soil increases risk of sporotrichosis. *Postgrad Med.* 2004; **115:** 21-3.
20. Bhutia PY, Gurung S, Yegneswaran PP, *et al.* A case series and review of sporotrichosis in Sikkim. *J Infect Dev Ctries.* 2011; **5:** 603-8.
21. Biancardi AL, Freitas DF, Valviesse VR, *et al.* Multifocal choroiditis in disseminated sporotrichosis in patients with HIV/AIDS. *Retin Cases Brief Rep.* 2017; **11:** 67-70.
22. Bonifaz A, Fierro L, Saul A, *et al.* Cutaneous sporotrichosis. Intermittent treatment (pulses) with itraconazole. *Eur J Dermatol.* 2008; **18:** 61-4.
23. Bonifaz A, Saul A, Paredes-Solis V, *et al.* Sporotrichosis in childhood: Clinical and therapeutic experience in 25 patients. *Pediatr Dermatol.* 2007; **24:** 369-72.
24. Bonifaz A, Tirado-Sanchez A. Cutaneous disseminated and extracutaneous sporotrichosis: Current status of a complex disease. *J Fungi.* 2017; **3:** 6.
25. Bonifaz A, Vazquez-Gonzalez D. Diagnosis and treatment of lymphocutaneous sporotrichosis: What are the options? *Curr Fungal Infect Rep.* 2013; **7:** 252-9.
26. Bonifaz A, Vazquez-Gonzalez D. Sporotrichosis: An update. *G Ital Dermatol Venereol.* 2010; **145:** 659-73.
27. Borba-Santos LP, Gagini T, Ishida K, *et al.* Miltefosine is active against *Sporothrix brasiliensis* isolates with *in vitro* low susceptibility to amphotericin B or itraconazole. *J Med Microbiol.* 2015; **64:** 415-22.
28. Borba-Santos LP, Rodrigues AM, Gagini TB, *et al.* Susceptibility of *Sporothrix brasiliensis* isolates to amphotericin B, azoles and terbinafine. *Med Mycol.* 2015; **53:** 178-88.
29. Brilhante RS, Rodrigues AM, Sidrim JJ, *et al. In vitro* susceptibility of antifungal drugs against *Sporothrix brasiliensis* recovered from cats with sporotrichosis in Brazil. *Med Mycol.* 2016; **54:** 275-9.
30. Bunce PE, Yang L, Chun S, *et al.* Disseminated sporotrichosis in a patient with hairy cell leukemia treated with amphotericin B and posaconazole. *Med Mycol.* 2012; **50:** 197-201.
31. Burns MJ, Kapadia NN, Silman EF. Sporotrichosis. *West J Emerg Med.* 2009; **10:** 204.
32. Bustamante B, Campos PE. Sporotrichosis treatment: Overview and update. *Curr Fungal Infect Rep.* 2011; **5:** 42-8.
33. Bustamante B, Lama JR, Mosquera C, *et al.* Sporotrichosis in human immunodeficiency virus infected Peruvian patients: Two case reports and literature review. *Infect Dis Clin Pract.* 2009; **17:** 78-83.
34. Camacho E, Leon-Navarro I, Rodriguez-Brito S, *et al.* Molecular epidemiology of human sporotrichosis in Venezuela reveals high frequency of *Sporothrix globosa*. *BMC Infect Dis.* 2015; **15:** 94. PMID: 25880588.

35. Capoor MR, Ramesh V, Khanna G, *et al.* Sporotrichosis in Delhi among the migrant population from Uttarakhand, India. *Trop Doct.* 2011; **41:** 46-8.
36. Carlos IZ, Sassa MF, da Graça Sgarbi DB, *et al.* Current research on the immune response to experimental sporotrichosis. *Mycopathologia.* 2009; **168:** 1-10.
37. Chakrabarti A, Bonifaz A, Gutierrez-Galhardo MC, *et al.* Global epidemiology of sporotrichosis. *Med Mycol.* 2015; **53:** 3-14.
38. Chakrabarti A, Roy SK, Dhar S, *et al.* Sporotrichosis in northwest India. *Indian J Med Res.* 1994; **100:** 62-5.
39. Chakrabarti A, Slavin MA. Endemic fungal infections in the Asia-Pacific region. *Med Mycol.* 2011; **49:** 337-44.
40. Chang S, Hersh AM, Naughton G, *et al.* Disseminated cutaneous sporotrichosis. *Dermatol Online J.* 2013; **19:** 20401.
41. Chen F, Yan BL, An L, *et al.* Is there a relation between percentages of T helper cells and prognosis for sporotrichosis? *Br J Dermatol.* 2016; **174:** 1398-400.
42. Civila ES, Bonasse J, Conti-Diaz IA, *et al.* Importance of the direct fresh examination in the diagnosis of cutaneous sporotrichosis. *Int J Dermatol.* 2004; **43:** 808-10.
43. Costa RO, Bernardes-Engemann AR, Azulay-Abulafia L, *et al.* Sporotrichosis in pregnancy: Case reports of 5 patients in a zoonotic epidemic in Rio de Janeiro, Brazil. *An Bras Dermatol.* 2011; **86:** 995-8.
44. Costa RO, de Mesquita KC, Damasco PS, *et al.* Infectious arthritis as the single manifestation of sporotrichosis: Serology from serum and synovial fluid samples as an aid to diagnosis. *Rev Iberoam Micol.* 2008; **25:** 54-6.
45. Cotino Sanchez A, Torres-Alvarez B, Gurrola Morales T, *et al.* Mycosis fungoides-like lesions in a patient with diffuse cutaneous sporotrichosis. *Rev Iberoam Micol.* 2015; **32:** 200-3.
46. Criseo G, Malara G, Romeo O, *et al.* Lymphocutaneous sporotrichosis in an immunocompetent patient: A case report from extreme southern Italy. *Mycopathologia.* 2008; **166:** 159-62.
47. Curi AL, Felix S, Azevedo KM, *et al.* Retinal granuloma caused by *Sporothrix schenckii. Am J Ophthalmol.* 2003; **136:** 205-7.
48. da Rosa AC, Scroferneker ML, Vettorato R, *et al.* Epidemiology of sporotrichosis: A study of 304 cases in Brazil. *J Am Acad Dermatol.* 2005; **52:** 451-9.
49. Daniel Da Rosa W, Gezuele E, Calegari L, *et al.* Asteroid body in sporotrichosis. Yeast viability and biological significance within the host immune response. *Med Mycol.* 2008; **46:** 443-8.
50. Das S, Banerjee G, Biswas I. Fixed variant of cutaneous sporotrichosis: A rare entity in non-endemic belt. *Indian J Dermatol.* 2006: **51:** 209-10.
51. de Araujo ML, Rodrigues AM, Fernandes GF, *et al.* Human sporotrichosis beyond the epidemic front reveals classical transmission types in Espirito Santo, Brazil. *Mycoses.* 2015; **58:** 485-90.
52. de Araujo T, Marques AC, Kerdel F. Sporotrichosis. *Int J Dermatol.* 2001; **40:** 737-42.
53. de Beer ZW, Duong TA, Wingfield MJ. The divorce of *Sporothrix* and *Ophiostoma*: Solution to a problematic relationship. *Stud Mycol.* 2016; **83:** 165-91.
54. de Camargo Ferreira E Vasconcellos E, Valete-Rosalino CM, Bastos de Lima Barros M, *et al.* Breast sporotrichosis. *Breast J.* 2010; **16:** 658-9.
55. de Lima Barros MB, Alves Costa DL, Schubach TM, *et al.* Endemic of zoonotic sporotrichosis: Profile of cases in children. *Pediatr Infect Dis J.* 2008; **27:** 246-50.
56. de Lima Barros MB, Schubach AO, de Vasconcellos Carvalhaes de Oliveira R, *et al.* Treatment of cutaneous sporotrichosis with itraconazole - study of 645 patients. *Clin Infect Dis.* 2011; **52:** e200-6.
57. de Macedo PM, Sztajnbok DC, Camargo ZP, *et al.* Dacryocystitis due to *Sporothrix brasiliensis:* A case report of a successful clinical and serological outcome with low-dose potassium iodide treatment and oculoplastic surgery. *Br J Dermatol.* 2015; **172:** 1116-9.
58. de Meyer EM, de Beer ZW, Summerbell RC, *et al.* Taxonomy and phylogeny of new wood- and soil-inhabiting *Sporothrix* species in the *Ophiostoma stenoceras-Sporothrix schenckii* complex. *Mycologia.* 2008; **100:** 647-61.
59. de Oliveira MM, de Almeida-Paes R, de Medeiros Muniz M, *et al.* Sporotrichosis caused by *Sporothrix globosa* in Rio de Janeiro, Brazil: Case report. *Mycopathologia.* 2010; **169:** 359-63.
60. de Oliveira MM, Verissimo C, Sabino R, *et al.* First autochthone case of sporotrichosis by *Sporothrix globosa* in Portugal. *Diagn Microbiol Infect Dis.* 2014; **78:** 388-90.
61. Devi KR, Devi MU, Singh TN, *et al.* Emergence of sporotrichosis in Manipur. *Indian J Med Microbiol.* 2006; **24:** 216-9.
62. Dey NC. Sporotrichosis (a rare disease in India). *Indian J Dermatol.* 1969; **15:** 21-3.
63. Dhingra D, Durrheim D, Porigneaux P. Sporotrichosis outbreak and mouldy hay in NSW. *Aust Fam Physician.* 2015; **44:** 217-21.
64. Dias NM, Oliveira MM, Santos C, *et al.* Sporotrichosis caused by *Sporothrix mexicana*, Portugal. *Emerg Infect Dis.* 2011; **17:** 1975-6.
65. Dixon D, Salkin IF, Duncan RA, *et al.* Isolation and characterization of *Sporothrix schenckii* from clinical and environmental sources associated with the largest US epidemic of sporotrichosis. *J Clin Microbiol.* 1991; **29:** 1106-13.
66. do Socorro Costa da Silva M, Ferreira ER, Campos LR, *et al.* Painful ulcers in a 5-year-old girl. *Eur J Pediatr.* 2011; **170:** 949-50.
67. Engle J, Desir J, Bernstein JM. A rose by any other name. *Skinmed.* 2007; **6:** 139-41.
68. Eustace KE, Sampaio FM, Lyra MR, *et al.* Cutaneous disseminated sporotrichosis complicated by osteomyelitis. *Acta Derm Venereol.* 2013; **93:** 192-3.
69. Evans SJ, Kenyon NJ, Lew EY, *et al.* Cavitary lung disease in a dry-rot worker. *Chest.* 2003; **124:** S319-20.
70. Falqueto A, Bravim Maifrede S, Araujo Ribeiro M. Unusual clinical presentation of sporotrichosis in three members of one family. *Int J Dermatol.* 2012; **51:** 434-8.

71. Fan B, Wang JF, Zheng B, *et al.* Clinical features of 10 cases of eyelid sporotrichosis in Jilin Province (Northeast China). *Can J Ophthalmol.* 2016; **51:** 297-301.
72. Feeney KT, Arthur IH, Whittle AJ, *et al.* Outbreak of sporotrichosis, Western Australia. *Emerg Infect Dis.* 2007; **13:** 1228-31.
73. Fernandez-Silva F, Capilla J, Mayayo E, *et al.* Virulence of *Sporothrix luriei* in a murine model of disseminated infection. *Mycopathologia.* 2012; **173:** 245-9.
74. Ferreira CP, do Valle AC, Freitas DF, *et al.* Pregnancy during a sporotrichosis epidemic in Rio de Janeiro, Brazil. *Int J Gynaecol Obstet.* 2012; **117:** 294-5.
75. Ferreira CP, Galhardo MC, Valle AC. Cryosurgery as adjuvant therapy in cutaneous sporotrichosis. *Braz J Infect Dis.* 2011; **15:** 181-3.
76. Ferreira LC, Barroso PF, Tonomura E, *et al.* Osteomyelitis caused by *Sporothrix schenckii* in an immunocompetent patient. *Rev Soc Bras Med Trop.* 2016; **49:** 527-9.
77. Firinci F, Erdemir Y, Cetin B, *et al.* A 6-year-old boy with a nodule on his hand. *Pediatr Ann.* 2012; **41:** 360-1.
78. Fischman Gompertz O, Rodrigues AM, *et al.* Atypical clinical presentation of sporotrichosis caused by *Sporothrix globosa* resistant to itraconazole. *Am J Trop Med Hyg.* 2016; **94:** 1218-22.
79. Freitas DF, de Siqueira Hoagland B, do Valle AC, *et al.* Sporotrichosis in HIV-infected patients: Report of 21 cases of endemic sporotrichosis in Rio de Janeiro, Brazil. *Med Mycol.* 2012; **50:** 170-8.
80. Freitas DF, do Valle AC, de Almeida Paes R, *et al.* Zoonotic sporotrichosis in Rio de Janeiro, Brazil: A protracted epidemic yet to be curbed. *Clin Infect Dis.* 2010; **50:** 453.
81. Freitas DF, Lima IA, Curi CL, *et al.* Acute dacryocystitis: Another clinical manifestation of sporotrichosis. *Memorias do Instituto Oswaldo Cruz.* 2014; **109:** 262-4.
82. Freitas DF, Lima MA, de Almeida-Paes R, *et al.* Sporotrichosis in the central nervous system caused by *Sporothrix brasiliensis. Clin Infect Dis.* 2015; **61:** 663-4.
83. Freitas DF, Santos SS, Almeida-Paes R, *et al.* Increase in virulence of *Sporothrix brasiliensis* over five years in a patient with chronic disseminated sporotrichosis. *Virulence.* 2015; **6:** 112-20.
84. Freitas DF, Valle AC, Cuzzi T, *et al.* Sweet syndrome associated with sporotrichosis. *Br J Dermatol.* 2012; **166:** 212-3.
85. Freitas DF, Valle AC, da Silva MB, *et al.* Sporotrichosis: An emerging neglected opportunistic infection in HIV-infected patients in Rio de Janeiro, Brazil. *PLoS Negl Trop Dis.* 2014; **8:** e3110.
86. Fujii H, Tanioka M, Yonezawa M, *et al.* A case of atypical sporotrichosis with multifocal cutaneous ulcers. *Clin Exp Dermatol.* 2008; **33:** 135-8.
87. Fulghum K, Mollet T, Nguyen J. Fixed cutaneous sporotrichosis. *Cutis.* 2015; **96:** 218 & 227-8.
88. Galhardo MC, De Oliveira RM, Valle AC, *et al.* Molecular epidemiology and antifungal susceptibility patterns of *Sporothrix schenckii* isolates from a cat-transmitted epidemic of sporotrichosis in Rio de Janeiro, Brazil. *Med Mycol.* 2008; **46:** 141-51.
89. Galhardo MC, Silva MT, Lima MA, *et al. Sporothrix schenckii* meningitis in AIDS during immune reconstitution syndrome. *J Neurol Neurosurg Psychiatry.* 2010; **81:** 696-9.
90. Gandhi N, Chander R, Jain A, *et al.* Atypical cutaneous sporotrichosis in an immunocompetent adult: Response to potassium iodide. *Indian J Dermatol.* 2016; **61:** 236.
91. Gerhard R, de Moscoso PC, Gabbi TV, *et al.* Fine-needle aspiration biopsy of disseminated sporotrichosis: A case report. *Diagn Cytopathol.* 2008; **36:** 174-7.
92. Gewehr P, Jung B, Aquino V, *et al.* Sporotrichosis in renal transplant patients. *Can J Infect Dis Med Microbiol.* 2013; **24:** e47-9.
93. Ghosh A, Chakrabarti A, Hemashettar BM, Maiti PK. *In vitro* susceptibility pattern of *Sporothrix schenckii* strains isolated from three centers in India. *Indian J Med Res.* 2001; **113:** 214-20.
94. Ghosh A, Maity PK, Hemashettar BM, *et al.* Physiological characters of *Sporothrix schenckii* isolates. *Mycoses.* 2002; **45:** 449-54.
95. Gilaberte Y, Aspiroz C, Alejandre MC, *et al.* Cutaneous sporotrichosis treated with photodynamic therapy: An *in vitro* and *in vivo* study. *Photomed Laser Surg.* 2014; **32:** 54-7.
96. Gold JA, Derado G, Mody RK, *et al.* Sporotrichosis-associated hospitalizations, United States, 2000-2013. *Emerg Infect Dis.* 2016; **22:** 1817-20.
97. Gordhan A, Ramdial PK, Morar N, *et al.* Disseminated cutaneous sporotrichosis: A marker of osteoarticular sporotrichosis masquerading as gout. *Int J Dermatol.* 2001; **40:** 717-9.
98. Gori S, Lupetti A, Moscato G, *et al.* Pulmonary sporotrichosis with hyphae in a human immunodeficiency virus-infected patient: A case report. *Acta Cytol.* 1997; **41:** 519-21.
99. Govender NP, Maphanga TG, Zulu TG, *et al.* An outbreak of lymphocutaneous sporotrichosis among mine-workers in South Africa. *PLoS Negl Trop Dis.* 2015; **9:** e0004096.
100. Gremiao ID, Miranda LH, Reis EG, *et al.* Zoonotic epidemic of sporotrichosis: Cat to human transmission. *PLoS Pathog.* 2017; **13:** e1006077. PMID: 28103311.
101. Gremiao ID, Menezes RC, Schubach TM, *et al.* Feline sporotrichosis: Epidemiological and clinical aspects. *Med Mycol.* 2015; **53:** 15-21.
102. Guo H, Xu XG, Guan XH, *et al.* Psoriasiform cutaneous sporotrichosis. *Int J Dermatol.* 2015; **54:** e364-5.
103. Gurcan S, Konuk E, Kiliç H, *et al.* Sporotrichosis, a disease rarely reported from Turkey and an overview of Turkish literature. *Mycoses.* 2007; **50:** 426-9.
104. Gutierrez-Galhardo MC, do Valle AC, Fraga BL, *et al.* Disseminated sporotrichosis as a manifestation of immune reconstitution inflammatory syndrome. *Mycoses.* 2010; **53:** 78-80.
105. Haldar N, Sharma MK, Gugnani HC. Sporotrichosis in north-east India. *Mycoses.* 2007; **50:** 201-4.
106. Hampton DE, Adesina A, Chodosh J. Conjunctival sporotrichosis in the absence of antecedent trauma. *Cornea.* 2002; **21:** 831-3.
107. Hardman S, Stephenson I, Jenkins DR, *et al.* Disseminated *Sporothrix schenckii* in a patient with AIDS. *J Infect.* 2005; **51:** e73-7.

108. Haruna K, Shiraki Y, Hiruma M, *et al.* A case of lymphangitic sporotrichosis occurring on both forearms with a published work review of cases of bilateral sporotrichosis in Japan. *J Dermatol.* 2006; **33:** 364-7.
109. Hassan K, Turker T, Zangeneh T. Disseminated sporotrichosis in an immunocompetent patient. *Case Reports Plast Surg Hand Surg.* 2016; **3:** 44-7.
110. Hattori M, Yoshiike T, Sonoda T, *et al.* A case of lymphocutaneous sporotrichosis occurring at the nail bed. *Mycoses.* 2011; **54:** e663-5.
111. Hay RJ, Morris-Jones R. Outbreaks of sporotrichosis. *Curr Opin Infect Dis.* 2008; **21:** 119-21.
112. Henry LR, Danaher PJ, Boseley ME. Laryngeal sporotrichosis mimicking Merkel cell carcinoma recurrence. *Otolaryngol Head Neck Surg.* 2005; **132:** 336-8.
113. Hogan BK, Hospenthal DR. Update on the therapy for sporotrichosis. *Drug Benefit Trends.* 2010; **22:** 49-52.
114. Hu S, Chung WH, Hung SI, *et al.* Detection of *Sporothrix schenckii* in clinical samples by a nested PCR assay. *J Clin Microbiol.* 2003; **41:** 1414-8.
115. Ishizaki H, Kawasaki M, Anzawa K, *et al.* Mitochondrial DNA analysis of *Sporothrix schenckii* in India, Thailand, Brazil, Colombia, Guatemala and Mexico. *Jpn J Med Mycol.* 2009; **50:** 19-26.
116. Iyengar SS, Khan JA, Brusco M, *et al.* Cutaneous *Sporothrix schenckii* of the human eyelid. *Ophthal Plast Reconstr Surg.* 2010; **26:** 305-6.
117. Kamal A, Orenstein R. Disseminated sporotrichosis. *J Hosp Med.* 2010; **5:** E29-30.
118. Kamalam A, Thambiah AS. Sporotrichosis: First case report from Madras. *Mykosen.* 1982; **25:** 576-8.
119. Kashima T, Honma R, Kishi S, *et al.* Bulbar conjunctival sporotrichosis presenting as a salmon-pink tumor. *Cornea.* 2010; **29:** 573-6.
120. Kauffman CA, Bustamante B, Chapman SW, *et al.* IDSA Clinical practice guidelines for the management of sporotrichosis. *Clin Infect Dis.* 2007; **45:** 1255-65.
121. Kazemi A, Razi A. Sporotrichosis in Iran. *Rev Iberoam Micol.* 2007; **24:** 38-40.
122. Khabie N, Boyce TG, Roberts GD, *et al.* Laryngeal sporotrichosis causing stridor in a young child. *Int J Pediatr Otorhinolaryngol.* 2003; **67:** 819-23.
123. Khan ZU, Randhawa HS, Kowshik T, *et al.* The pathogenic potential of *Sporotrichum pruinosum* isolated from the human respiratory tract. *J Med Vet Mycol.* 1988; **26:** 145-51.
124. Kini S, Pal D, Kowshik T, *et al.* Sporotrichosis in India: First authentic case report from the north-western region and a critical literature review. *J Med Vet Mycol.* 1986; **24:** 289-95.
125. Kohler LM, Hamdan JS, Ferrari TC. Successful treatment of a disseminated *Sporothrix schenckii* infection and *in vitro* analysis for antifungal susceptibility testing. *Diagn Microbiol Infect Dis.* 2007; **58:** 117-20.
126. Kumar R, Kaushal V, Chopra H, *et al.* Pansinusitis due to *Sporothrix schenckii. Mycoses.* 2005; **48:** 85-8.
127. La Hoz RM, Baddley JW. Subcutaneous fungal infections. *Curr Infect Dis Rep.* 2012; **14:** 530-9.
128. Lacerda CM, Martins EM, de Resende MA, *et al.* Gamma radiation effects on *Sporothrix schenckii* yeast cells. *Mycopathologia.* 2011; **171:** 395-401.
129. Lauermann F, Lyra M, Gaudio R. Sporotrichosis mimicking keratoacanthoma. *Am J Trop Med Hyg.* 2012; **86:** 741.
130. Lederer HT, Sullivan E, Crum-Cianflone NF. Sporotrichosis as an unusual case of osteomyelitis: A case report and review of the literature. *Med Mycol Case Rep.* 2016; **11:** 31-5.
131. Lembo S, Cirillo T, Marasca C, *et al.* Sporotrichosis: Long-term treatment and follow-up in a diabetic patient from Southern Italy. *G Ital Dermatol Venereol.* 2016; **151:** 576-8.
132. Liu X, Zhang Z, Hou B, *et al.* Rapid identification of *Sporothrix schenckii* in biopsy tissue by PCR. *J Eur Acad Dermatol Venereol.* 2013; **27:** 1491-7.
133. Lloret A, Hartmann K, Pennisi MG, *et al.* Sporotrichosis in cats: ABCD guidelines on prevention and management. *J Feline Med Surg.* 2013; **15:** 619-23.
134. Lopes JO, Alves SH, Benevenga JP, *et al.* Filamentous forms of *Sporothrix schenckii* in material from human lesions. *J Med Vet Mycol.* 1992; **30:** 403-6.
135. Lopes-Bezerra LM, Mora-Montes HM. *Sporothrix* and sporotrichosis: Contributions from the first international meeting sponsored by the Working Group on *Sporothrix* and sporotrichosis of the ISHAM, Rio de Janeiro, Brazil, October, 2013. *Med Mycol.* 2015; **53:** 1-2.
136. Lopes-Bezerra LM, Schubach A, Costa RO. *Sporothrix schenckii* and sporotrichosis. *An Acad Bras Cienc.* 2006; **78:** 293-308.
137. Lopez-Romero E, Reyes-Montes Mdel R, Perez-Torres A, *et al. Sporothrix schenckii* complex and sporotrichosis, an emerging health problem. *Future Microbiol.* 2011; **6:** 85-102.
138. Lurie HI. Five unusual cases of sporotrichosis from South Africa showing lesions in muscles, bones and viscera. *Br J Surg.* 1963; **50:** 585-91.
139. Lyra MR, Nascimento ML, Varon AG, *et al.* Immune reconstitution inflammatory syndrome in HIV and sporotrichosis coinfection: Report of two cases and review of the literature. *Rev Soc Bras Med Trop.* 2014; **47:** 806-9.
140. Lyra MR, Valle AC, Pimentel MI, *et al.* Scar examination in sporotrichosis: An additional tool for clinical diagnosis. *Indian J Dermatol Venereol Leprol.* 2015; **81:** 290-2.
141. Macedo PM, Lopes-Bezerra LM, Bernardes-Engemann AR, *et al.* New posology of potassium iodide for the treatment of cutaneous sporotrichosis: Study of efficacy and safety in 102 patients. *J Eur Acad Dermatol Venereol.* 2015; **29:** 719-24.
142. Madrid H, Cano J, Gene J, *et al. Sporothrix globosa*, a pathogenic fungus with widespread geographical distribution. *Rev Iberoam Micol.* 2009; **26:** 218-22.
143. Madrid IM, Mattei AS, Fernandes CG, *et al.* Epidemiological findings and laboratory evaluation of sporotrichosis: A description of 103 cases in cats and dogs in southern Brazil. *Mycopathologia.* 2012; **173:** 265-73.

144. Mahajan VK, Mehta KS, Chauhan PS, *et al.* Fixed cutaneous sporotrichosis treated with topical amphotericin B in an immune suppressed patient. *Med Mycol Case Rep.* 2015; **7:** 23-5.
145. Mahajan VK, Sharma NL, Shanker V, *et al.* Cutaneous sporotrichosis: Unusual clinical presentations. *Indian J Dermatol Venereol Leprol.* 2010; **76:** 276-80.
146. Mahajan VK, Sharma NL, Sharma RC, *et al.* Cutaneous sporotrichosis in Himachal Pradesh, India. *Mycoses.* 2005; **48:** 25-31.
147. Mahajan VK. Sporotrichosis: An overview and therapeutic options. *Dermatol Res Pract.* 2014; 272376. PMID: 25614735.
148. Mahlberg MJ, Patel R, Rosenman K, Cheung W, *et al.* Fixed cutaneous sporotrichosis. *Dermatol Online J.* 2009; **15:** 5. PMID: 19891913.
149. Mahmoudi S, Zaini F, Kordbacheh P, *et al. Sporothrix schenckii* complex in Iran: Molecular identification and antifungal susceptibility. *Med Mycol.* 2016; **54:** 593-9.
150. Maia DC, Goncalves AC, Ferreira LS, *et al.* Response of cytokines and hydrogen peroxide to *Sporothrix schenckii* exoantigen in systemic experimental infection. *Mycopathologia.* 2016; **181:** 207-15.
151. Manohar A, Nizlan MN. Chronic nonhealing ulcer of the right thumb with multiple subcutaneous nodules. *Orthopedics.* 2008; **31:** 710.
152. Marimon R, Cano J, Gene J, *et al. Sporothrix brasiliensis, S.globosa* and *S.mexicana*, three new *Sporothrix* species of clinical interest. *J Clin Microbiol.* 2007; **45:** 3198-206.
153. Marimon R, Gene J, Cano J, *et al.* Molecular phylogeny of *Sporothrix schenckii. J Clin Microbiol.* 2006; **44:** 3251-6.
154. Marimon R, Gene J, Cano J, *et al. Sporothrix luriei:* A rare fungus from clinical origin. *Med Mycol.* 2008; **46:** 621-5.
155. Mario DN, Guarro J, Santurio JM, *et al. In vitro* and *in vivo* efficacy of amphotericin B combined with posaconazole against experimental disseminated sporotrichosis. *Antimicrob Agents Chemother.* 2015; **59:** 5018-21.
156. Marques GF, Martins AL, Sousa JM, *et al.* Characterization of sporotrichosis cases treated in a dermatologic teaching unit in the state of Sao Paulo, Brazil, 2003-2013. *An Bras Dermatol.* 2015; **90:** 273-5.
157. Martinez-Alvarez JA, Perez-Garcia LA, Flores-Carreon A, *et al.* The immune response against *Candida spp.* and *Sporothrix schenckii. Rev Iberoam Micol.* 2014; **31:** 62-6.
158. Martins CP, Nakamura R, Schechtman RC, *et al.* Periungual sporotrichosis: A diagnostic challenge. *Int J Dermatol.* 2017; **56:** 209-11.
159. Martins-Costa GM, Bonamigo RR. Facial verrucous sporotrichosis in an adult during treatment for rosacea. *Int J Dermatol.* 2014; **53:** e124-6.
160. Martis J, Kamath GH, Nandakishore B, *et al.* Linear nodules along the arm. *Indian J Dermatol Venereol Leprol.* 2009; **75:** 435-6.
161. Mata-Essayag S, Delgado A, Colella MT, *et al.* Epidemiology of sporotrichosis in Venezuela. *Int J Dermatol.* 2013; **52:** 974-80.
162. Mauermann ML, Klein CJ, Orenstein R, *et al.* Disseminated sporotrichosis presenting with granulomatous inflammatory multiple mononeuropathies. *Muscle Nerve.* 2007; **36:** 866-72.
163. McGuinness SL, Boyd R, Kidd S, *et al.* Epidemiological investigation of an outbreak of cutaneous sporotrichosis, Northern Territory, Australia. *BMC Infect Dis.* 2016; **16:** 16. PMID: 26758905.
164. Medeiros KB, Landeiro LG, Diniz LM, *et al.* Disseminated cutaneous sporotrichosis associated with ocular lesion in an immunocompetent patient. *An Bras Dermatol.* 2016; **91:** 537-9.
165. Mehta A, Marie MB, Justin R, *et al.* Primary pulmonary sporotrichosis: A case report. *Chest.* **126:** 2004; S945-6.
166. Mehta KI, Sharma NL, Kanga AK, *et al.* Isolation of *Sporothrix schenckii* from the environmental sources of cutaneous sporotrichosis patients in Himachal Pradesh, India: results of a pilot study. *Mycoses.* 2007; **50:** 496-501.
167. Mello CX, Schubach Ade O, Madeira Mde F. Can yeast-like form of *Sporothrix schenckii* confuse the direct parasitological diagnosis of American cutaneous leishmaniasis? *Rev Soc Bras Med Trop.* 2011; **44:** 127.
168. Mesa-Arango AC, Del Rocio Reyes-Montes M, Perez-Mejia A, *et al.* Phenotyping and genotyping of *Sporothrix schenckii* isolates according to geographic origin and clinical form of sporotrichosis. J Clin Microbiol. 2002; **40:** 3004-11.
169. Milby AH, Pappas ND, O'Donnell J, *et al.* Sporotrichosis of the upper extremity. *Orthopedics.* 2010; **33.** PMID: 20415315.
170. Mitra AN, Das S, Sinha R, *et al.* Sporotrichosis of maxillary sinuses in a middle aged female patient from rural area of Eastern India. *J Clin Diagn Res.* 2016; **10:** DD01-2.
171. Mohan N, Jayaseelan E, Abraham A, *et al.* Cutaneous sporotrichosis in Bangalore, southern India. *Int J Dermatol.* 2004; **43:** 269-72.
172. Montenegro BL, Arnold JC. North American dimorphic fungal infections in children. *Pediatr Rev.* 2010; **31:** e40-8.
173. Mora-Montes HM, Dantas Ada S, Trujillo-Esquivel E, *et al.* Current progress in the biology of members of the *Sporothrix schenckii* complex following the genomic era. *FEMS Yeast Res.* 2015; 15. pii: fov065. PMID: 26260509.
174. Moreira JA, Freitas DF, Lamas CC. The impact of sporotrichosis in HIV-infected patients: A systematic review. *Infection.* 2015; **43:** 267-76.
175. Morgado FN, Schubach AO, Barros MB, *et al.* The *in situ* inflammatory profile of lymphocutaneous and fixed forms of human sporotrichosis. *Med Mycol.* 2011; **49:** 612-20.
176. Morris-Jones R. Sporotrichosis. *Clin Exp Dermatol.* 2002; **27:** 427-31.
177. Moussa TAA, Kadasa NMS, Al Zahrani HS, *et al.* Origin and distribution of *Sporothrix globosa* causing sapronoses in Asia. *J Med Microbiol.* 2017; **66:** 560-9.
178. Nayyar E, Taimur S. Sporotrichoid lesions: An age-old conundrum. *Am J Med.* 2015; **128:** e11-3.
179. Newton PN, Chung WH, Phetsouvanh R, *et al.* Sporotrichosis, Plain of Jars, Lao People's Democratic Republic. *Emerg Infect Dis.* 2005; **11:** 1496-7.

180. Neyra E, Fonteyne PA, Swinne D, *et al.* Epidemiology of human sporotrichosis investigated by amplified fragment length polymorphism. *J Clin Microbiol.* 2005; **43:** 1348-52.
181. Nunes Mario DA, Denardi LB, Brayer Pereira DI, *et al. In vitro* photodynamic inactivation of *Sporothrix schenckii* complex species. *Med Mycol.* 2014; **52:** 770-3.
182. Oliveira DC, Lopes PG, Spader TB, *et al.* Antifungal susceptibilities of *Sporothrix albicans, S.brasiliensis* and *S.luriei* of the *S.schenckii* complex identified in Brazil. *J Clin Microbiol.* 2011; **49:** 3047-9.
183. Oliveira MM, Almeida-Paes R, Gutierrez-Galhardo MC, *et al.* Molecular identification of the *Sporothrix schenckii* complex. *Rev Iberoam Micol.* 2014; **31:** 2-6.
184. Oliveira MM, Almeida-Paes R, Muniz MM, *et al.* Phenotypic and molecular identification of *Sporothrix* isolates from an epidemic area of sporotrichosis in Brazil. *Mycopathologia.* 2011; **172:** 257-67.
185. Oliveira MM, Santos C, Sampaio P, *et al.* Development and optimization of a new MALDI-TOF protocol for identification of the *Sporothrix* species complex. *Res Microbiol.* 2015; **166:** 102-10.
186. Orofino-Costa R, Unterstell N, Carlos Gripp A, *et al.* Pulmonary cavitation and skin lesions mimicking tuberculosis in a HIV negative patient caused by *Sporothrix brasiliensis. Med Mycol Case Rep.* 2013; **2:** 65-71.
187. Ottonelli Stopiglia CD, Magagnin CM, Castrillon MR, *et al.* Antifungal susceptibilities and identification of species of the *Sporothrix schenckii* complex isolated in Brazil. *Med Mycol.* 2014; **52:** 56-64.
188. Padhye AA, Kaufman L, Durry E, *et al.* Fatal pulmonary sporotrichosis caused by *Sporothrix schenckii* var. *luriei* in India. *J Clin Microbiol.* 1992; **30:** 2492-4.
189. Paixao AG, Galhardo MC, Almeida-Paes R, *et al.* The difficult management of disseminated *Sporothrix brasiliensis* in a patient with advanced AIDS. *AIDS Res Ther.* 2015; **12:** 16.
190. Papaiordanou F, da Silveira BR, Abulafia LA. Hypersensitivity reaction to *Sporothrix schenckii:* Erythema nodosum associated with sporotrichosis. *Rev Soc Bras Med Trop.* 2015; **48:** 504.
191. Pappas PG, Tellez I, Deep AE, *et al.* Sporotrichosis in Peru: Description of an area of hyperendemicity. *Clin Infect Dis.* 2000; **30:** 65-70.
192. Parekh PK, Butler DF. What is your diagnosis? Periorbital granulomatous plaque. *Pediatr Dermatol.* 2011; **28:** 457-8.
193. Patel A, Mudenda V, Lakhi S, *et al.* A 27-year-old severely immunosuppressed female with misleading clinical features of disseminated cutaneous sporotrichosis. *Case Rep Dermatol Med.* 2016; 9403690. PMID: 26881148.
194. Patel AB, Shaikh S. Fixed cutaneous sporotrichosis. *Indian Pediatr.* 2006; **43:** 73-4.
195. Queiroz-Telles F, Nucci M, Colombo AL, *et al.* Mycoses of implantation in Latin America: An overview of epidemiology, clinical manifestations, diagnosis and treatment. *Med Mycol.* 2011; **49:** 225-36.
196. Quintella LP, Passos SR, do Vale AC, *et al.* Histopathology of cutaneous sporotrichosis in Rio de Janeiro: A series of 119 consecutive cases. *J Cutan Pathol.* 2011; **38:** 25-32.
197. Rajendran C, Ramesh V, Misra RS, *et al.* Sporotrichosis in Nepal. *Int J Dermatol.* 1990; **29:** 716-8.
198. Ramesh V, Singh A, Pahwa M, Capoor M. A recalcitrant facial plaque: Fixed cutaneous sporotrichosis. *Int J Dermatol.* 2011; **50:** 367-8.
199. Ramirez Soto MC. Is epidemic of sporotrichosis in Abancay, Peru, caused by zoonotic transmission of *Sporothrix*? *Rev Iberoam Micol.* 2016; **33:** 256-8.
200. Ramirez Soto MC, Andagua-Castro J, Lizarraga-Trujillo J. Palpebral sporotrichosis in a 6-year-old child. *Int J Dermatol.* 2016; **55:** e625-6.
201. Ramirez Soto MC. Effects of potassium iodide in concentrations of TSH, tT3 and tT4 in serum of subjects with sporotrichosis. *Mycoses.* 2014; **57:** 460-5.
202. Ramirez Soto MC. Sporotrichosis in the ocular adnexa: 21 cases in an endemic area in Peru and review of the literature. *Am J Ophthalmol.* 2016; **162:** 173-9.
203. Ramirez Soto MC. Sporotrichosis: The story of an endemic region in Peru over 28 years (1985 to 2012). *PLoS One.* 2015; **10:** e0127924. PMID: 26030742.
204. Ramirez Soto MC. Sporotrichosis among children of a hyperendemic area in Peru: An 8-year retrospective study. *Int J Dermatol.* 2017; doi: 10.1111/ijd.13643. PMID: 28488313.
205. Ramli R, Abd Rashid AH, Phang KS, *et al.* Sporotrichosis atypical presentation as a soft tissue tumour. *Malays J Pathol.* 2009; **31:** 143-5.
206. Rammaert B, Gamaletsou MN, Zeller V, *et al.* Dimorphic fungal osteoarticular infections. *Eur J Clin Microbiol Infect Dis.* 2014; **33:** 2131-40.
207. Ramos-e-Silva M, Vasconcelos C, Carneiro S, *et al.* Sporotrichosis. *Clin Dermatol.* 2007; **25:** 181-7.
208. Randhawa HS, Chand R, Mussa AY, *et al.* Sporotrichosis in India: First case in a Delhi resident and an update. *Indian J Med Microbiol.* 2003; **21:** 12-6.
209. Ranjana KH, Chakrabarti A, Kulachandra M, *et al.* Sporotrichosis in Manipur: Report of two cases. *Indian J Dermatol Venereol Leprol.* 2001; **67:** 86-8.
210. Rathi SK, Ramam M, Rajendran C. Localized cutaneous sporotrichosis lasting for 10 years. *Indian J Dermatol Venereol Leprol.* 2003; **69:** 239-40.
211. Reis BD, Cobucci FO, Zacaron LH, *et al.* Sporotrichosis in an unusual location - Case report. *An Bras Dermatol.* 2015; **90:** S84-7.
212. Ribeiro BN, Ribeiro RN, Penna CR, *et al.* Bone involvement by *Sporothrix schenckii* in an immunocompetent child. *Pediatr Radiol.* 2015; **45:** 1427-30.
213. Rodrigues AM, Bagagli E, de Camargo ZP, *et al. Sporothrix schenckii sensu stricto* isolated from soil in an armadillo's burrow. *Mycopathologia.* 2014; **177:** 199-206.
214. Rodrigues AM, Choappa RC, Fernandes GF, *et al. Sporothrix chilensis* sp. nov. (Ascomycota: Ophiostomatales), A soil-borne agent of human sporotrichosis with mild-pathogenic potential to mammals. *Fungal Biol.* 2016; **120:** 246-64.
215. Rodrigues AM, de Hoog G, Zhang Y, *et al.* Emerging sporotrichosis is driven by clonal and recombinant *Sporothrix* species. *Emerg Microbes Infect.* 2014; **3:** e32.

216. Rodrigues AM, de Hoog GS, de Camargo ZP. Molecular diagnosis of pathogenic *Sporothrix* species. *PLoS Negl Trop Dis.* 2015; **9:** e0004190.
217. Rodrigues AM, de Hoog S, de Camargo ZP. Emergence of pathogenicity in the *Sporothrix schenckii* complex. *Med Mycol.* 2013; **51:** 405-12.
218. Rodrigues AM, de Melo Teixeira M, de Hoog GS, *et al.* Phylogenetic analysis reveals a high prevalence of *Sporothrix brasiliensis* in feline sporotrichosis outbreaks. *PLoS Negl Trop Dis.* 2013; **7:** e2281. PMID: 23818999.
219. Rodrigues AM, Najafzadeh MJ, de Hoog GS, *et al.* Rapid identification of emerging human-pathogenic *Sporothrix* species with rolling circle amplification. *Front Microbiol.* 2015; **6:** 1385. PMID: 26696992.
220. Rodriguez G, Sarmiento L. The asteroid bodies of sporotrichosis. *Am J Dermatopathol.* 1998; **20:** 246-9.
221. Roldan-Marin R, Contreras-Ruiz J, Arenas R, *et al.* Fixed sporotrichosis as a cause of a chronic ulcer on the knee. *Int Wound J.* 2009; **6:** 63-6.
222. Romero-Cabello R, Bonifaz A, Romero-Feregrino R, *et al.* Disseminated sporotrichosis. *BMJ Case Rep.* 2011; pii: bcr1020103404. PMID: 22700076.
223. Rufino R, Marques BL, Costa CH. Rose Gardener's disease. *Thorax.* 2013; **68:** 602.
224. Ruiz-Villaverde R, Galan-Gutierrez M. Skin nodules with a linear distribution - A case study. *Aust Fam Physician.* 2011; **40:** 801-2.
225. Saleh M, Grauman E, Saccio M, *et al.* Sporotrichosis. *Ann Emerg Med.* 2008; **52:** 616 & 625.
226. Sanchotene KO, Brandolt TM, Klafke GB, *et al. In vitro* susceptibility of *Sporothrix brasiliensis:* Comparison of yeast and mycelial phases. *Med Mycol.* 2017. doi: 10.1093/mmy/myw143. PMID: 28472490.
227. Sanchotene KO, Madrid IM, Klafke GB, *et al. Sporothrix brasiliensis* outbreaks and the rapid emergence of feline sporotrichosis. *Mycoses.* 2015; **58:** 652-8.
228. Sandhu K, Gupta S. Potassium iodide remains the most effective therapy for cutaneous sporotrichosis. *J Dermatolog Treat.* 2003; **14:** 200-2.
229. Santos RCV, de Almeida Vaucher R, Alves SH. Current trends in sporotrichosis. *Fungal Genom Biol.* 2012; **2:** e110.
230. Sassa MF, Ferreira LS, Ribeiro LC, *et al.* Immune response against *Sporothrix schenckii* in TLR-4-deficient mice. *Mycopathologia.* 2012; **174:** 21-30.
231. Schechtman RC, Crignis GS, Pockstaller MP, *et al.* Molluscum-like lesions in a patient with sporotrichosis. *An Bras Dermatol.* 2011; **86:** 1217-9.
232. Schechtman RC. Sporotrichosis: Part I & II. *Skinmed.* 2010; **8:** 216-210 & 275-80.
233. Schubach A, Barros MB, Wanke B. Epidemic sporotrichosis. *Curr Opin Infect Dis.* 2008; **21:** 129-33.
234. Shah SK, Hicks C, De Los Santos M, *et al.* Muddying the water for a clear diagnosis. *J Bronchology Interv Pulmonol.* 2015; **22:** 178-9.
235. Sharma NL, Mahajan VK, Verma N, *et al.* Cutaneous sporotrichosis: an unusual clinico-pathologic and therapeutic presentation. *Mycoses.* 2003; **46:** 515-8.
236. Sharma NL, Mehta KI, Mahajan VK, *et al.* Cutaneous sporotrichosis of face: Polymorphism and reactivation after intralesional triamcinolone. *Indian J Dermatol Venereol Leprol.* 2007; **73:** 188-90.
237. Sharma VK, Kaur S, Kumar B, Talwar P, *et al.* Sporotrichosis in northwestern India: report of 5 cases. *Indian J Dermatol Venereol Leprol.* 1988; **54:** 142-7.
238. Sharon VR, Kim J, Sudhakar S, *et al.* Disseminated cutaneous sporotrichosis. *Lancet Infect Dis.* 2013; **13:** 95.
239. Sigler L, Harris JL, Dixon DM, *et al.* Microbiology and potential virulence of *Sporothrix cyanescens,* a fungus rarely isolated from blood and skin. *J Clin Microbiol.* 1990; **28:** 1009-15.
240. Silva-Vergara ML, de Camargo ZP, Silva PF, *et al.* Disseminated *Sporothrix brasiliensis* infection with endocardial and ocular involvement in an HIV-infected patient. *Am J Trop Med Hyg.* 2012; **86:** 477-80.
241. Singh MF, Fernandes SR, Samara AM. *Sporothrix schenckii* infection mimicking sarcoidosis. *Rheumatology.* 2004; **43:** 248-9.
242. Singh P, Sharma RC, Gupta ML, *et al.* Sporotrichosis in Himachal Pradesh (India). *Indian J Med Sci.* 1983; **37:** 101-3.
243. Singhai M, Rawat V, Verma P, Jha PK, Shree D, Goyal R, Umesh. Primary pulmonary sporotrichosis in a sub-Himalayan patient. *J Lab Physicians.* 2012; **4:** 48-9.
244. Sivagnanam S, Bannan AM, Chen SC, et al. Sporotrichosis (*Sporothrix schenckii* infection) in the New South Wales mid-north coast, 2000-2010. *Med J Aust.* 2012; **196:** 588-90.
245. Smego RA Jr, Castiglia M, Asperilla MO. Lymphocutaneous syndrome. A review of non-*Sporothrix* causes. *Medicine (Baltimore).* 1999; **78:** 38-63.
246. Song Y, Li SS, Zhong SX, *et al.* Report of 457 sporotrichosis cases from Jilin province, northeast China, a serious endemic region. *J Eur Acad Dermatol Venereol.* 2013; **27:** 313-8.
247. Song Y, Yao L, Zhong SX, *et al.* Infant sporotrichosis in northeast China: A report of 15 cases. *Int J Dermatol.* 2011; **50:** 522-9.
248. Sood N, Gugnani HC, Joshi A, *et al.* Sporotrichosis: Report of first case from the State of Jharkhand, India. *J Medical Mycol.* 2004; **14:** 52-3.
249. Subedi S, Kidd SE, Baird RW, *et al.* Case report: Sporotrichosis from the Northern Territory of Australia. *Am J Trop Med Hyg.* 2014; **91:** 1263-8.
250. Suchitha S, Vijaya B, Sunila R, *et al.* Sporotrichosis in Mysore: A case report to emphasise the role of histopathology. *Indian J Pathol Microbiol.* 2008; **51:** 45-6.
251. Suzuki R, Yikelamu A, Tanaka R, *et al.* Studies in phylogeny, development of rapid identification methods, antifungal susceptibility and growth rates of clinical strains of *Sporothrix schenckii* complex in Japan. *Med Mycol J.* 2016; **57:** E47-57.
252. Takenaka M, Yoshizaki A, Utani A, *et al.* A survey of 165 sporotrichosis cases examined in Nagasaki prefecture from 1951 to 2012. *Mycoses.* 2014; **57:** 294-8.
253. Talwar P. Sporotrichosis: a case report. *Indian J Dermatol Venereol Leprol.* 1977; **43:** 106-9.

254. Tambini R, Farina C, Fiocchi R, *et al.* Possible pathogenic role for *Sporothrix cyanescens* isolated from a lung lesion in a heart transplant patient. *J Med Vet Mycol.* 1996; **34:** 195-8.
255. Tang MM, Tang JJ, Gill P, *et al.* Cutaneous sporotrichosis: A six-year review of 19 cases in a tertiary referral center in Malaysia. *Int J Dermatol.* 2012; **51:** 702-8.
256. Tellez MD, Batista-Duharte A, Portuondo D, *et al. Sporothrix schenckii* complex biology: Environment and fungal pathogenicity. *Microbiology.* 2014; **160:** 2352-65.
257. Tilak R, Kumari V, Bansal M, *et al.* Lymphocutaneous sporotrichosis in an adolescent girl presenting as mycetoma. *Int J Low Extrem Wounds.* 2012; **11:** 184-6.
258. Tlougan BE, Podjasek JO, Patel SP, *et al.* Neonatal sporotrichosis. *Pediatr Dermatol.* 2009; **26:** 563-5.
259. Trilles L, Fernandez-Torres B, Dos Santos Lazera M, *et al. In vitro* antifungal susceptibilities of *Sporothrix schenckii* in two growth phases. *Antimicrob Agents Chemother.* 2005; **49:** 3952-4.
260. Tsuji G, Takei K, Takahara M, *et al.* Simultaneous onset of sporotrichosis in mother and daughter. *J Dermatol.* 2016; doi: 10.1111/1346-8138.13645. PMID: 27774634.
261. Vaishampayan S, Borde P. An unusual presentation of sporotrichosis. *Indian J Dermatol.* 2013; **58:** 409.
262. Vasquez-del-Mercado E, Arenas R, Padilla-Desgarenes C. Sporotrichosis. *Clin Dermatol.* 2012; **30:** 437-43.
263. Verma S, Verma GK, Singh G, *et al.* Sporotrichosis in sub himalayan India. *PLoS Negl Trop Dis.* 2012; **6:** e1673.
264. Watanabe M, Hayama K, Fujita H, *et al.* A case of sporotrichosis caused by *Sporothrix globosa* in Japan. *Ann Dermatol.* 2016; **28:** 251-2.
265. Williams BA, Jennings TA, Rushing EC, *et al.* Sporotrichosis on the face of a 7-year-old boy following a bicycle accident. *Pediatr Dermatol.* 2013; **30:** e246-7.
266. Wong SM, Tang JJ. Disseminated sporotrichosis in a patient with a past history of lepromatous leprosy: A case report. *Med Mycol.* 2012; **50:** 404-6.
267. Xavier MO, Bittencourt LR, Silva CM, *et al.* Atypical presentation of sporotrichosis: Report of three cases. *Rev Soc Bras Med Trop.* 2013; **46:** 116-8.
268. Xirotagaros G, Drogari-Apiranthitou M, Panayiotides IG, *et al.* Imported lymphocutaneous sporotrichosis in Greece. *Br J Dermatol.* 2015; **173:** 291-3.
269. Xue S, Gu R, Wu T, *et al.* Oral potassium iodide for the treatment of sporotrichosis. *Cochrane Database Syst Rev.* 2009; **4:** CD006136. PMID: 19821356.
270. Xue SL, Li L. Oral potassium iodide for the treatment of sporotrichosis. *Mycopathologia.* 2009; **167:** 355-6.
271. 000. Xu, TH, Lin JP, Gao XH, *et al.* Identification of *Sporothrix schenckii* of various mtDNA types by nested PCR assay. *Med Mycol.* **48:** 161-5.
272. Yamada K, Zaitz C, Framil VM, *et al.* Cutaneous sporotrichosis treatment with potassium iodide: A 24 year experience in Sao Paulo State, Brazil. *Rev Inst Med Trop Sao Paulo.* 2011; **53:** 89-93.
273. Yamaguchi T, Ito S, Takano Y, *et al.* A case of disseminated sporotrichosis treated with prednisolone, immunosuppressants and tocilizumab under the diagnosis of rheumatoid arthritis. *Intern Med.* 2012; **51:** 2035-9.
274. Yap FB. Disseminated cutaneous sporotrichosis in an immunocompetent individual. *Int J Infect Dis.* 2011; **15:** e727-9.
275. Yegneswaran PP, Sripathi H, Bairy I, *et al.* Zoonotic sporotrichosis of lymphocutaneous type in a man acquired from a domesticated feline source: report of a first case in southern Karnataka, India. *Int J Dermatol.* 2009; **48:** 1198-200.
276. Yelverton CB, Stetson CL, Bang RH, *et al.* Fatal sporotrichosis. *Cutis.* 2006; **78:** 253-6.
277. Zhang X, Huang H, Feng P, *et al. In vitro* activity of itraconazole in combination with terbinafine against clinical strains of itraconazole-insensitive *Sporothrix schenckii. Eur J Dermatol.* 2011; **21:** 573-6.
278. Zhang Y, Hagen F, Stielow B, *et al.* Phylogeography and evolutionary patterns in *Sporothrix* spanning more than 14000 human and animal case reports. *Persoonia.* 2015; **35:** 1-20.
279. Zhang Y, Hagen F, Wan Z, *et al.* Two cases of sporotrichosis of the right upper extremity in right-handed patients with diabetes mellitus. *Rev Iberoam Micol.* 2016; **33:** 38-42.
280. Zhang Y, Pyla V. Cancer-like lesions in a patient with sporotrichosis. *Int J Dermatol.* 2014; **53:** e311-2.
281. Zhang Y, Pyla V. Nasal sporotrichosis in children. *Int J Dermatol.* 2014; **53:** e342-3.
282. Zhang Y, Pyla V. Palpebral sporotrichosis. *Int J Dermatol.* 2014; **53:** e356-7.
283. Zhang Y, Pyla V. Sweet's syndrome-like sporotrichosis. *Int J Dermatol.* 2014; **53:** e324-5.
284. Zhang Y, Wang Y, Cong L, *et al.* Eyelid sporotrichosis: Unique clinical findings in 72 patients. *Australas J Dermatol.* 2016; **57:** 44-7.
285. Zhang YQ, Xu XG, Zhang M, *et al.* Sporotrichosis clinical and histopathological manifestations. *Am J Dermatopathol.* 2011; **33:** 296-302.
286. Zhang Z, Liu X, Lv X, *et al.* Variation in genotype and higher virulence of a strain of *Sporothrix schenckii* causing disseminated cutaneous sporotrichosis. *Mycopathologia.* 2011; **172:** 439-46.
287. Zhao MD, Zhou X, Liu TT, *et al.* Morphological and physiological comparison of taxa comprising the *Sporothrix schenckii* complex. *J Zhejiang Univ Sci B.* 2015; **16:** 940-7.

CHAPTER 13

Chromoblastomycosis

Chromoblastomycosis is a slowly progressing localized fungal infection of skin and subcutaneous tissue mostly involving exposed parts of body without tendency to disseminate. This is caused by **phaeoid (dematiaceous) fungi** and is characterized by polymorphic, verrucoid, crusted or ulcerated lesions. The causative fungi have a unique morphology, as these are present as **sclerotic bodies** in the lesions, which is pathognomonic feature of this disease.

Chromoblastomycosis (CBM) is often regarded as synonymous to **chromomycosis**, the term is rejected because chromoblastomycosis was the name used originally for the specific pathology. Moreover, chromomycosis is an all-embracing term that amounts to synonym of **phaeohyphomycosis** (PHM) thus creating more confusions in nomenclature of diseases caused by various phaeoid fungi.

Historical Perspective

In 1911, **Alexandrino Pedroso** described chromoblastomycosis when he observed the presence of **sclerotic** or **muriform bodies** in a patient at Sao Paulo, Brazil. The organism, isolated from nodular and ulcerated lesions of foot and leg, was identified as a phaeoid fungus, *Fonsecaea pedrosoi*. The work of Pedroso was not published until 1920, however, in the meantime H Rudolph, a German physician, working in Brazil, who was unaware of Pedroso's work, published the very first clinical description of this disease in 1914.

Although chromoblastomycosis is a disease of tropical and subtropical countries but Lane and Medlar reported the first case in 1915 from Boston, New England having verrucous lesions on the foot of an Italian immigrant. They used the term verrucous dermatitis, which is still followed as one of the synonyms of this disease. In this case fungus was isolated as *Phialophora verrucosa* by an eminent mycologist, Roland Thaxter, who was not given due credit to his contribution and the sclerotic bodies were named as **Medlar bodies**, in honor of Medlar.

The term chromoblastomycosis (***'chromo'*** = colored, ***'blasto'*** = budding, ***'mycosis'*** = fungal infection) was first proposed by Terra *et al*, in 1922 to define polymorphic fungal disease located on the lower limbs, consisting of nodules or verrucous plaques which could possibly ulcerate and develop into hyperkeratosis and acanthosis of the affected epithelial tissues.

A large number of cases began to be recorded where phaeoid fungi were found in tissues as hyphal elements and not as sclerotic bodies or planate-dividing yeast cells. These findings led to the creation of more confusion over the nomenclature of chromoblastomycosis. Hence in 1935, Moore and Almeida proposed the term ***'chromomycosis'*** to include even those cases showing hyphal form of the fungus in tissue. The term chromomycosis instead of resolving the matter of unsettled nomenclature of this disease rather created more confusion because it was to be used not only for chromoblastomycosis but also for a variety of other unrelated diseases caused by the phaeoid fungi. There was a period during which number of cases were reported under this broader heading of chromomycosis.

Fonseca was later honored by Pablo Negroni, who in 1936 created the genus *Fonsecaea* to accommodate the most frequent agent of chromo(blasto)mycosis, *Fonsecaea pedrosoi*. This agent was isolated in 1913 by Alexandrino Pedroso in Brazil and described as *Hormodendrum pedrosoi* by Brumpt in 1922. Pablo Negroni's nomenclature survives to this date, paying tribute to two important Brazilian mycologists of the early 20th century. The second most important agent of chromo(blasto)mycosis is *Cladosporium* (now *Cladophialophora*) *carrionii*, described for the first time in 1954 by A. Trejos, a Costa Rican mycologist, who in this way acknowledged the contributions made in the field by his Puerto Rican colleague Arturo Carrion.

There was another ground of prevailing confusion between chromoblastomycosis and phaeohyphomycosis as some of fungi like *Phialophora verrucosa* or *Fonsecaea pedrosoi* could produce both these diseases but with distinct histopathological findings.

During the mid-seventies, in 1974, a consensus emerged that the disease should be reverted to its old terminology as **chromoblastomycosis** for cases having verrucous lesions and showing rounded sclerotic forms of the fungus in tissues despite absence of budding and to use the term **phaeohyphomycosis** for those clinical entities showing hyphal forms.

Ajello *et al*, who proposed the term phaeohyphomycosis, addressed to the prevailing intricacies of nomenclature. This term included infections where causative agents produced any combination of phaeoid yeast-like cells, hyphae and pseudohyphae in tissue. The characteristic muriform cells, seen in tissue of patients with chromoblastomycosis, were **absent** in phaeohyphomycosis. The term **chromoblastomycosis** was exclusively reserved to encompass infections caused by phaeoid fungi in which **muriform cells** were present in the tissues.

Concurrently, Ajello *et al*, advocated for discarding the term chromomycosis considering it to be superfluous. The concept of phaeohyphomycosis has now been defined so that it reflects a combined clinical, pathological and mycological relationship of the disease process caused by phaeoid fungi without sclerotic bodies. The clinical component of this concept was expanded to reflect on the major clinical categories of phaeohyphomycosis: superficial, cutaneous, subcutaneous and systemic. These categories were characterized on the basis of anatomical site involved, extent of tissue injury and host response to causative phaeoid fungal pathogen.

In medical sciences as such if a misnomer becomes widely acceptable, it is not easy to replace it even if an appropriate alternative is there. In the present context, the terminology as chromoblastomycosis was designated in 1922, which was changed to chromomycosis in 1935 due to above-mentioned reasons but prevailing confusions compelled the investigators to re-designate it again as chromoblastomycosis in 1974 because no suitable alternative term was available despite the fact that there are no budding yeast cells. Therefore, the official name of the disease caused by phaeoid fungi with sclerotic bodies is now **chromoblastomycosis**. This is a unique fungal infection where the nomenclature is histopathology based rather than other features.

Table 13.1. Five Common Agents of Chromoblastomycosis.

- *Fonsecaea pedrosoi*
- *Phialophora verrucosa*
- *Cladophialophora carrionii*
- *Fonsecaea compacta*
- *Rhinocladiella aquaspersa*

Mycology

The disease is caused by several darkly pigmented (dark brown, olivaceous or black) exogenous fungi of form-family **Dematiaceae** of **Hyphomycetes** under phylum **Deuteromycetes**. The recent DNA sequence data indicate that nearly all etiological agents of phaeohyphomycosis, chromoblastomycosis and phaeoid taxa of eumycetoma belong to a single clade of **loculoascomycetes** fungi. Therefore, taxonomical position of the causative agents of chromoblastomycosis lies in the family **Herpotrichiellaceae** of order **Dothideales** of class **Bitunicate Pyrenomycetes** *i.e.* **Loculoascomycetes**.

The genera involved in chromoblastomycosis are *Fonsecaea, Phialophora, Cladophialophora* and *Rhinocladiella*. The five principal phaeoid fungal species recognized as agents of chromoblastomycosis are listed in **Table 13.1**.

Occasionally other fungi like *Exophiala dermatitidis, Exophiala spinifera* and *Chaetomium* species are also involved in chromoblastomycosis. However, infections resulting from *Exophiala dermatitidis* are seen in tissue as short chains of spherical cells that have septa and only rarely have sclerotic bodies. On this basis, it is mainly considered as an agent of phaeohyphomycosis rather than chromoblastomycosis.

Epidemiology

Chromoblastomycosis is a common disease among rural workers in tropical and subtropical countries of Central and South America as well as Africa. This is found prevalent in Mexico, Cuba and other parts of Latin America, Africa and Madagascar. Rather Madagascar is considered as a storehouse of this disease. This is quite common in China, especially in Shandong and Henan provinces. The first case was reported in 1952 and since then over 500 cases from 20 provinces have been reported. It may occasionally be seen in temperate zones but it is most frequently encountered in warmer climates where people go barefoot and wear minimal clothing. These fungi are widely distributed as saprotrophic organisms in soil and decaying vegetation in all types of climates. The mode of infection is by inoculation of phaeoid fungi into cutaneous tissues through trauma. The disease most frequently involves lower extremities. This is in contrast to Venezuela where trunk is most frequently involved.

Like mycetoma, chromoblastomycosis is also mostly seen in males residing in rural areas. In Japan, however, incidence is found to be equal in both sexes. This disease has also been reported in domestic animals like dogs and horses. The infection is non-contagious as it is not transmitted from animal-to-humans or man-to-man as the infective form of the causative fungi is the mycelial form, whereas it is found as sclerotic cells in man.

The clinical entity is caused by several phaeoid molds identical in their tissue form *i.e.* sclerotic cells. In humid tropical areas, *Fonsecaea pedrosoi* is the most commonly encountered etiological agent whereas in dry and desert areas incidence of *Cladophialophora carrionii* predominates. In the United States, *Phialophora verrucosa* is the most commonly isolated agent of chromoblastomycosis. It is reported as an imported infection in the United Kingdom.

Chromoblastomycosis was first reported from India in 1957. The peculiar distribution of disease in sub-Himalayan belt and coastal areas appears to be due to hot and humid climates, which are essential for the growth of these fungi, however, the north-west arid zone of the country is free from disease.

Immunity

The immunological aspects of chromoblastomycosis are not fully known but evidence so far demonstrates that host response to fungal antigens occurs at both cellular and humoral levels. Antibodies have been reported with different procedures like double diffusion in agar gel. The skin test antigens are prepared in research settings and limited testing is found reactive among patients diagnosed with chromoblastomycosis. Antibodies to phaeoid fungi have been raised in rabbits to examine antigenic relationships with various fungal species and to identify phaeoid isolates by exoantigen testing. Although natural immunity to fungal infections in general and to chromoblastomycosis in particular, is not well understood, it is known to have no protective effect, as the disease has very chronic and debilitating course and prognosis.

Pathogenesis and Pathology

The tissue form of dematiaceous fungal infections depends on resistance of host, tissue in which fungus is growing and relative virulence of the infecting strain. In phaeohyphomycosis, there are dematiaceous hyphae in the tissues. If yeasts-like cells are present, they differ from sclerotic bodies as they divide in a single-plane, have thinner walls and tend to occur in chains. Histopathologically, disease is characterized by pseudoepitheliomatous hyperplasia (PEH), fibrosis and microabscess formation in epidermis along with presence of pigmented sclerotic bodies. The histopathological appearance of these fungi is distinct and consists of dark brown, subglobose, multicellular, thick-walled round cells, 5-12 μm in size that divide by splitting along septal planes by transverse septa and these internal septae are in more than one plane. These are called **muriform cells, Medlar bodies**, **copper-penny** or **sclerotic bodies** and are a pathognomonic feature of chromoblastomycosis as these are not seen in any other fungal infection. These bodies are considered to be in intermediate vegetative state between yeasts and hyphae. Medlar bodies may either be present singly or in clusters and both within giant cells or free in the dermis, as well as in epidermal microabscesses and in the stratum corneum. Ultrastructural studies demonstrate that Medlar bodies can be characterized as pathogens with either active metabolic and synthetic organelles or as dormant spore-like structures. The term sclerotic bodies should be avoided as muriform is the ideal terminology.

The muriform cells are planate-dividing yeast-like cells with formation of internal septa and are light yellow-brown in color giving **chestnut** appearance. By definition, mature muriform cells have intersecting cross-walls. Many cells are found in their younger stage of development and have single or no septa. Medlar bodies appear to develop an adaptive mechanism that facilitates survival of the organism in host tissues despite attack of the defensive processes of the body.

Hyphal strands are also observed particularly when the organism metastasizes to other parts. The septate cells may be seen singly, in pairs or in clusters. In most cases, the pigment is melanin and specifically, dihydroxynaphthalene melanin. Melanin is produced in cell walls of muriform bodies which results in dark-brown coloration that can easily be seen in histopathological section stained with H&E. Melanin pigmentation of phaeoid fungi can also be demonstrated by Fontana-Masson stain.

Over a period of many years there is development of highly organized granulomatous reaction associated with extensive fibrosis in dermis and subcutaneous tissue. Most of the patients exhibit organized mixed mycotic granuloma (OMMG), granulomatous reaction modified by presence of polymorphonuclear neutrophils. There is a characteristic tissue response termed pseudoepitheliomatous hyperplasia (PEH), with epithelium playing an important role in transepidermal elimination (TEE) of fungus. The infected tissue exhibits an epithelial overgrowth caused by an abnormal multiplication in number of cells in normal tissue. There may be ***'black dots'*** visible on the surface of lesions, which represent sites where blood, inflammatory debris and fungal cells *i.e.* sclerotic bodies are expelled from skin through transepidermal elimination, which may be sampled for skin scrapings and culture.

Clinical Features

This is a chronic disease and its onset is insidious. After traumatic inoculation, the initial process can evolve into pleomorphic lesions leading to dense dermal fibrosis and edema. The lesions of chromoblastomycosis are characterized by slow developing polymorphic nodular, verrucose, tumoral, plaques and cicatricial skin lesions **(Figs. 13.1A to G)**. The verrucose lesions frequently ulcerate and may be raised about 1-3 cm above the skin level with rough irregular surfaces, giving cauliflower-like appearance hence it is also called verrucose dermatitis. The black scabs over lesions are a very significant clue for making diagnosis. The infection is confined to skin and subcutaneous tissues and not disseminated to deeper organs of body. Initially the lesions are solitary but slow and localized spread occurs. Satellite lesions may also develop by autoinoculation.

The lower legs are the most frequently affected parts of the body and rarely shoulders, arms, hands, buttocks, ears, chest, face and abdomen may also be involved. The hematogenous and lymphatic dissemination, as seen in sporotrichosis, is rarely observed. Sometimes, secondary bacterial infection may result in lymphatic obstruction and consequently results in elephantiasis of legs. The lesions may persist for decades if neglected or unsuccessfully treated.

An attempt was made by Carrion to classify chromoblastomycosis into five different clinical forms; nodular, tumoral, verrucous, plaque-like and scarring. These possibly represent different stages of the disease or vary with anatomical site involved. This diversity might have corresponded to various evolutionary phases of the disease process hence such classification is no longer in use.

Differential Diagnosis

Chromoblastomycosis may be differentiated from similar types of skin lesions observed in blastomycosis, lacaziosis, protothecosis, sporotrichosis, keratoacanthoma, tuberculosis verrucosa cutis, Hansen's disease, leishmaniasis, mycetoma, candidiasis, yaws, tertiary syphilis, paracoccidioidomycosis and phaeohyphomycosis. Unlike mycetoma in which morphologic appearance of granules yields a clue to identity, whereas causative fungus in chromoblastomycosis cannot be identified by appearance of sclerotic bodies alone. Therefore, identification is made by supplementing cultural isolation thereby colony morphology and microscopic appearance of sporulation pattern of the isolate grown from the clinical lesions.

Laboratory Diagnosis

The term, chromoblastomycosis, is now restricted to phaeoid fungal species capable of forming muriform cells in host tissue, cause local acanthosis and have chaetothyrialean etiology. As the disease is usually confused with many related illnesses, therefore, exact diagnosis should be established before resorting to specific antifungal therapy. The demonstration of sclerotic bodies followed by culture of causative phaeoid fungus are two essential modalities to establish the final diagnosis. Serology although specific is usually not available due to scarcity of requisite infrastructure.

(a) Direct Examination

The skin scrapings ('black dots') or biopsy specimens from dry, crusty material recovered from the surface of lesions and examined microscopically in potassium hydroxide wet mount **(Fig. 13.2)**. The **muriform bodies** (sclerotic/medlar bodies) are the characteristic tissue form of all causative agents of chromoblastomycosis irrespective of the species involved, as *in vivo* morphology of *Fonsecaea pedrosoi* is identical to that of *Phialophora verrucosa*. A vinyl adhesive tape (VAT) preparation (Scotch Test) has been proved to be useful for direct microscopic examination. The biopsy specimen is stained with Hematoxylin and Eosin **(Figs. 13.3A to G)**, Giemsa stain and Masson-Fontana stain. The sclerotic bodies are so characteristic in wet

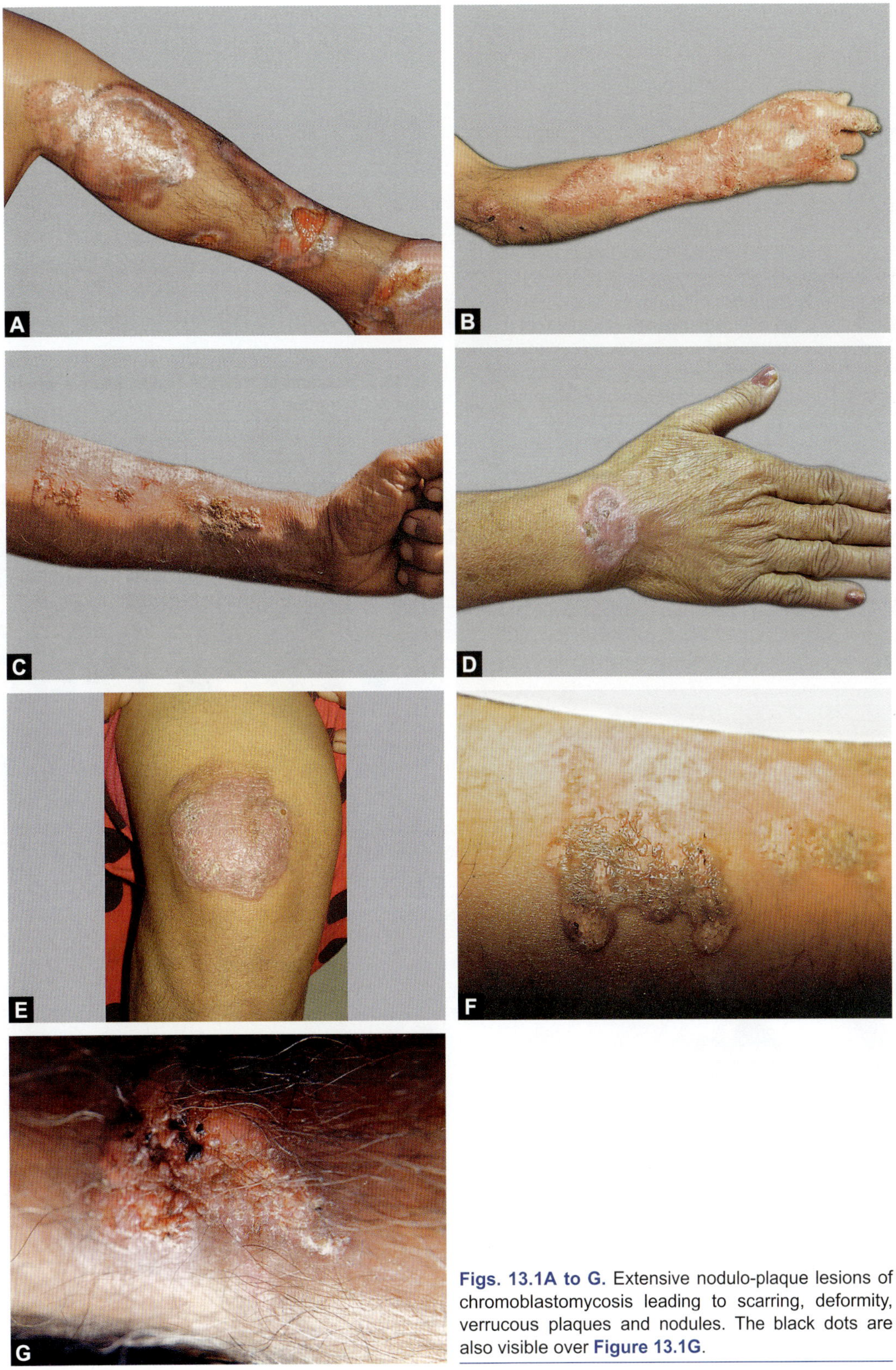

Figs. 13.1A to G. Extensive nodulo-plaque lesions of chromoblastomycosis leading to scarring, deformity, verrucous plaques and nodules. The black dots are also visible over **Figure 13.1G**.

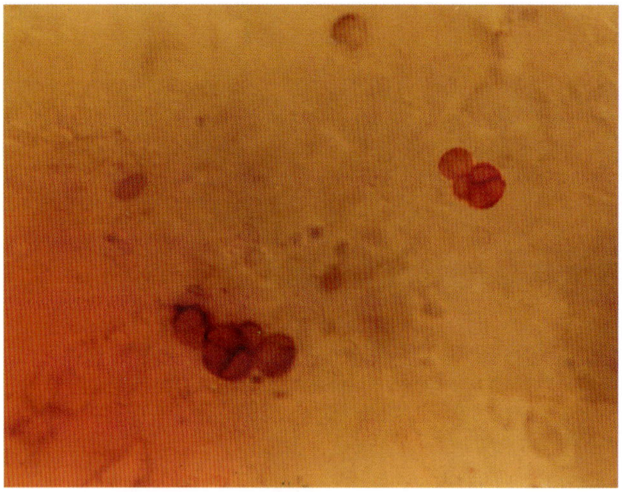

Fig. 13.2. Muriform or sclerotic bodies seen in wet mount from a lesion (KOH × 400).

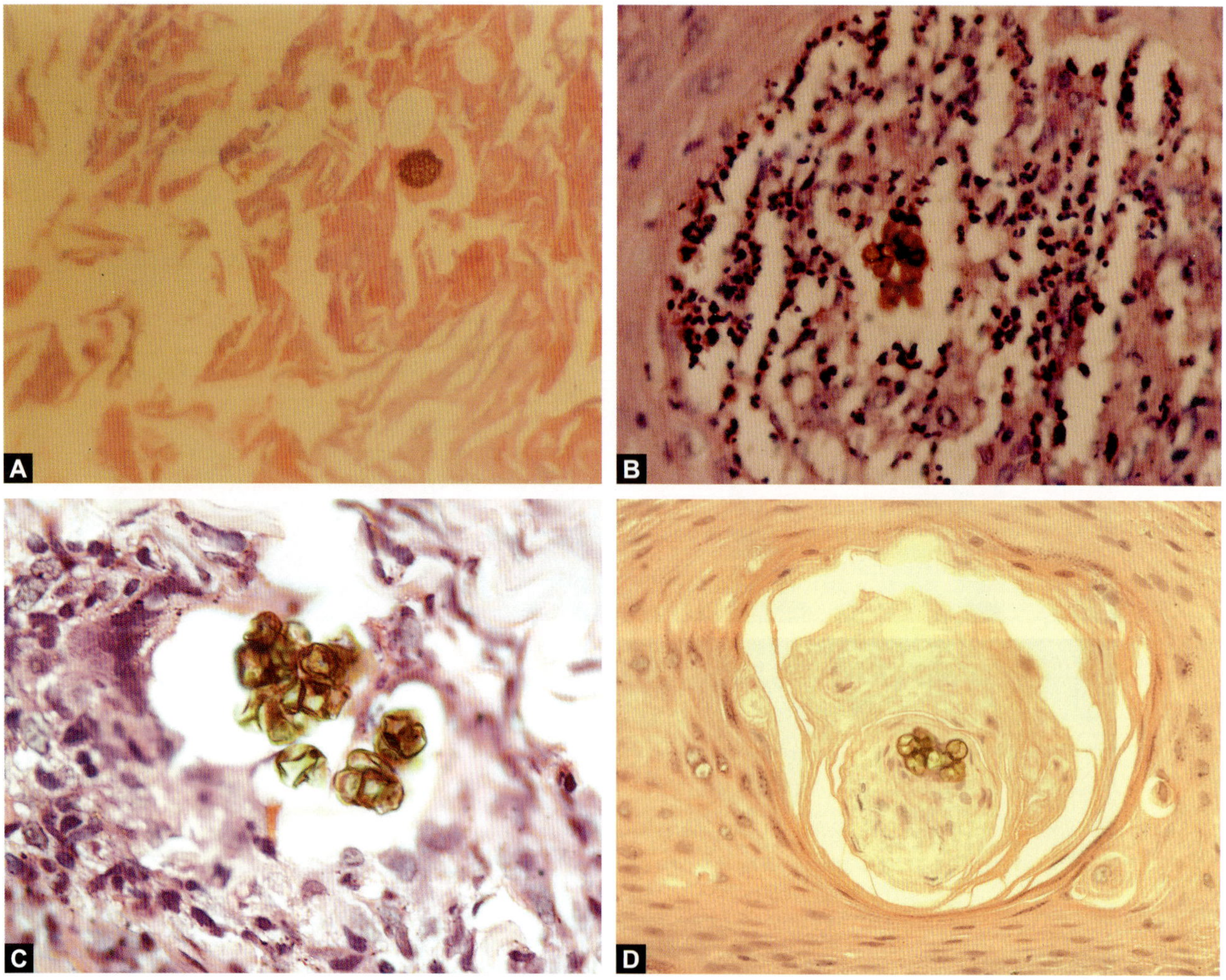

Figs. 13.3A to D. Muriform or sclerotic bodies seen in histopathological tissue sections (H&E × 400).

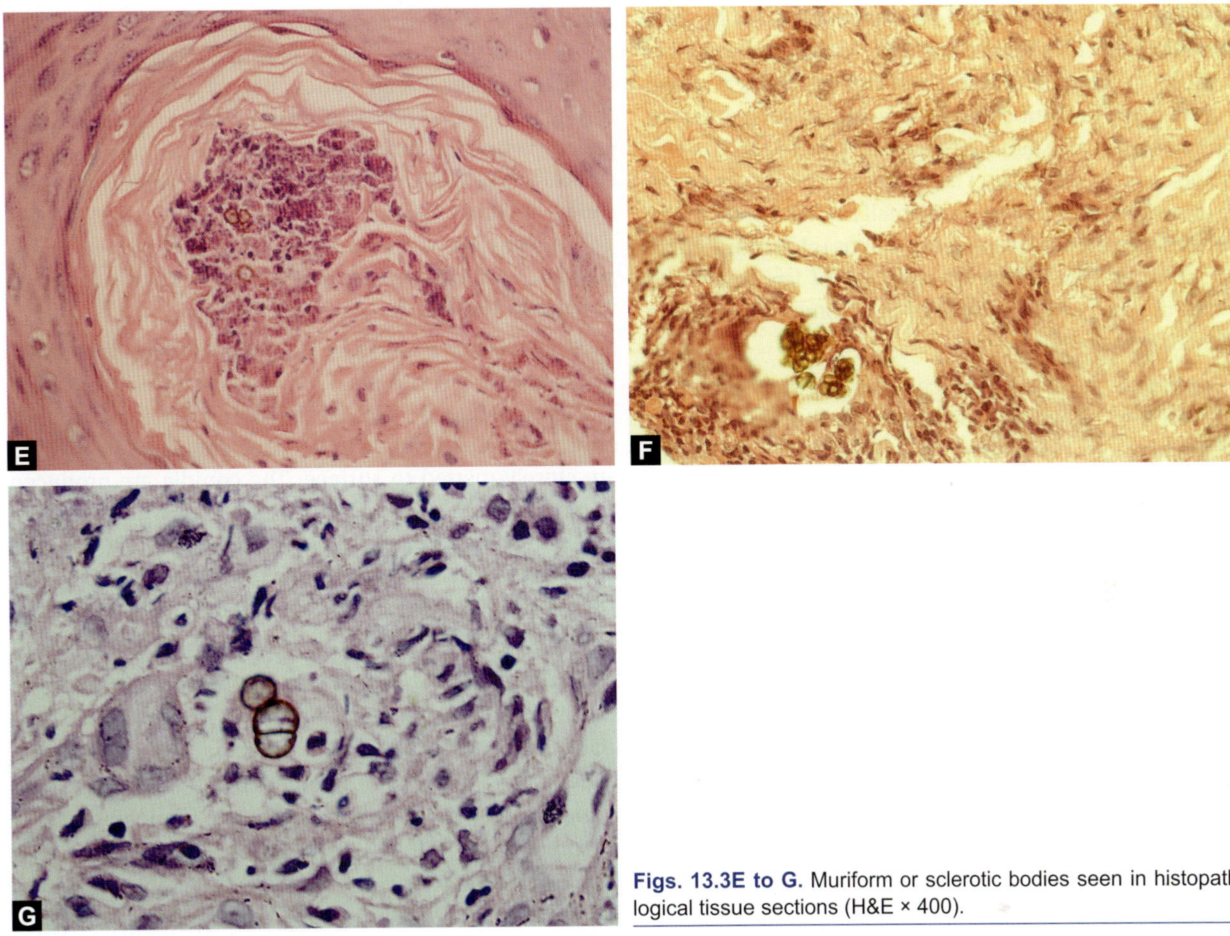

Figs. 13.3E to G. Muriform or sclerotic bodies seen in histopathological tissue sections (H&E × 400).

mounts as well as in stained smears that other stainings like Gomori methenamine silver stain and calcofluor white stainings are seldom required. In addition, Ziehl-Neelsen and Wade-Fite stainings are also useful to demonstrate the Medlar bodies.

(b) Fungal Culture

The specimen is collected from lesions and is plated on Sabouraud dextrose agar with cycloheximide. The cultures are incubated at 25°C and 37°C for a period of four weeks. Most of these fungi grow very slowly however in some cases growth may be visible by one to two weeks also. The identification of causative phaeoid fungus is established only after isolation of strain on Sabouraud dextrose agar with actidione and other antibiotics because these organisms are not inhibited by cycloheximide. These are identified according to the pattern and type of sporulation exhibited by a particular isolate as listed in **Table 13.2**. They may exhibit more than one pattern of sporulation due to **pleomorphism** and consequently confusion may arise concerning the correct taxonomic placement of the fungal isolate.

Table 13.2. Conidiogenesis in Phaeoid Fungal Genera.

Fungal Genera	*Type of Conidiogenous Cells*
• *Cladosporium*	Blastoconidia in chains
• *Phialophora*	Phialides with collarettes
• *Exophiala*	Annellides
• *Rhinocladiella*	Annellides and Phialides
• *Fonsecaea*	Sympodulae, also phialides and blastoconidia

This is slightly difficult to differentiate pathogenic phaeoid fungi on the basis of their macroscopic appearance. Their colonies are floccose, grey to dark brown or black in color with black reverse. *Exophiala* species, however, seen as so-called **black yeasts** are easily differentiated from

Cladosporium, *Fonsecaea* and *Phialophora* species by their yeast-like black glistening growth in primary culture. This yeast-like appearance progressively changes to the mycelial form with age and subcultures. The fungal isolates grown in liquid media also allow differentiation between *Exophiala* species and other fungi as the former produces yeast-like growth while others grow as small mycelial pellets. The final identification of the phaeoid fungus is best studied on slide cultures. The colony morphology of few important species of phaeoid fungi involved in chromoblastomycosis is described as follows:

1. Fonsecaea pedrosoi: The colonies are slow growing, dark green to brown or even black in color. The characteristics of colony morphology may vary between individual isolate. Three distinct types of sporulation may be seen in an isolated strain.

(a) Cladosporium Type: There is a simple stalk that serves as conidiophore and is slightly enlarged at distal end. The conidia are 3-6 μm in length and 1.5-3 μm in diameter. Separate conidia are brown to green in color with projections at the ends that represent previous junction of one conidium with another which are called as **disjunctor**. The **shield cells** are usually seen at the confluence of multiple chains. This type of conidia formation predominates in *Fonsecaea pedrosoi* as shown in **Figure 13.4A**.

(b) Rhinocladiella Type: The conidiophores are simple and sometimes not differentiated from vegetative hyphae. The conidia are formed like bristles on **bottle-brush** along the sides of swollen, knotted, club-shaped, linear central conidiospores that have developed terminally or laterally on hyphae as shown in **Figure 13.4B**.

(c) Phialophora Type: There is a distinct conidiogenous cell called phialide, which occurs terminally or along mycelium. The structure is generally **flask-shaped** and has rounded, oval or elongated base, constricted neck and an opening that may have a flaring collarette or lip. The conidia are developed endogenously and are extruded through the neck. They may accumulate around neck area, giving picture of ***'flowers in vase.'*** The conidia are oval, smooth-walled and hyaline and have no attachment scars as shown in **Figure 13.4C**.

The dark brown to black colored colonies of *Fonsecaea pedrosoi* with *Cladosporium* type of sporulation is shown in **Figures 13.5A to D**.

2. Phialophora verrucosa: The colonies develop slowly as greenish-brown to black on Sabouraud dextrose agar. The dome-shaped colonies becomes flattened as they grow and spread rapidly. The olive-gray colored mycelia are closely matted to surface. The conidia are produced from characteristic flask-shaped phialides with prominent dark cup-shaped collarettes. These phialoconidia are hyaline and held in clusters within a drop of slimy substance. The phialides are thin-walled oval cells about 1½-4 μm in size as shown in **Figures 13.6A to G**.

3. Cladophialophora carrionii: This organism grows slowly, producing small, smooth or folded, dark olive-black, compact colonies. The edges are entire and are marginated by black submerged hyphae. The elongate conidiophores produce long, flexuose, branching chains of conidia 1.5 × 3.0 by 2.0 × 7.5 μm in size as shown in **Figure 13.7**. Only *Cladosporium* type of conidiation is regularly found. Shield cells and disjunctors are also seen. On the basis of large subunit RNA analysis studies, previously designated *Cladosporium carrionii* has been re-classified as *Cladophialophora carrionii*.

4. Fonsecaea compacta: This fungus grows slowly on Sabouraud dextrose agar forming a heaped, brittle colony, dark green to black in color with short aerial mycelia.

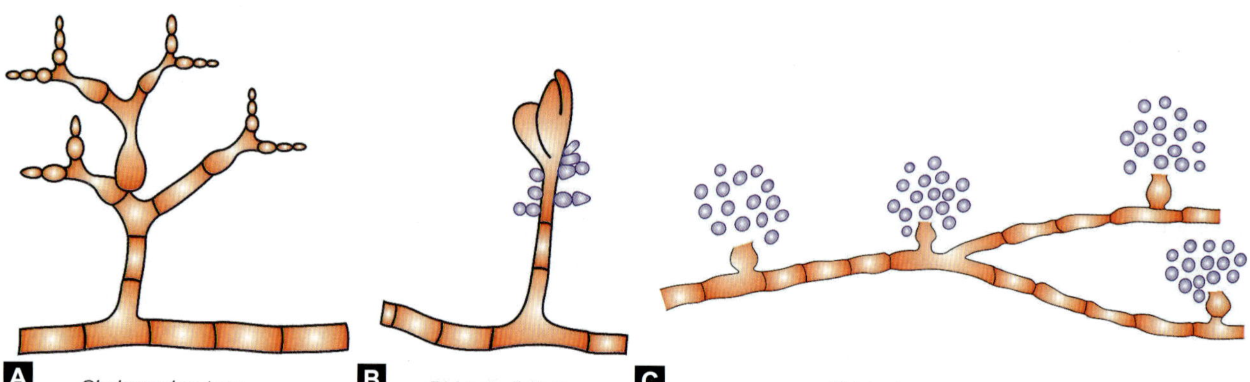

A *Cladosporium* type **B** *Rhinocladiella* type **C** *Phialophora* type

Figs. 13.4A to C. Various types of sporulation patterns seen in phaeoid fungi causing chromoblastomycosis.

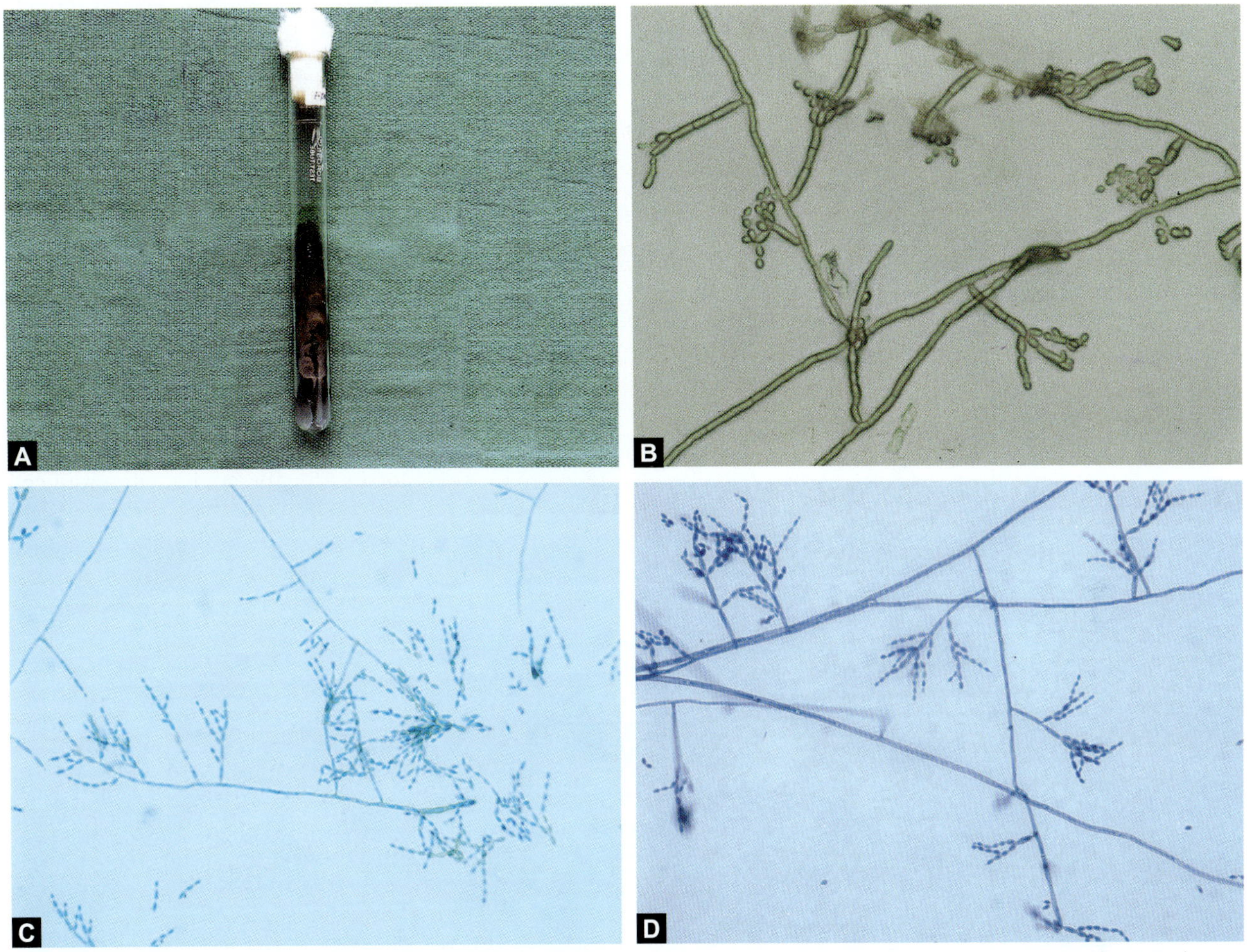

Figs. 13.5A to D. Dark brown to black mycelial colonies of *Fonsecaea pedrosoi* with *Cladosporium* type of sporulation (LCB × 400).

There are terminal and lateral conidiophores with chains of nearly spherical conidia, close together as compact heads. The conidia are 2-3 μm in size. The growth on cornmeal agar may show *Phialophora* type of conidiospores as shown in **Figure 13.8**.

5. Rhinocladiella aquaspersa: The colonies are moderately rapid growing, floccose olivaceous-black with dark reverse. The hyphae are pale olivaceous, smooth or slightly rough-walled. The conidiophores are straight, unbranched, thick-walled, dark brown. Conidiogenous cells are usually terminal, cylindrical with crowded slightly pigmented, unpigmented or faintly pigmented conidial scars. The conidia are subhyaline, smooth and thin walled, one or occasionally two-cells, ellipsoidal to clavate 2 × 5 μm arranged like a bottle brush **(Figs. 13.9A and B)**. Occasionally annellides may be present. The other species, *Rhinocladiella similis* is considered as resistant to commonly used antifungals.

(c) Immunodiagnosis

The demonstration of circulating antibodies in this disease was first made by using double immunodiffusion or indirect immunofluorescence techniques, however, with inconsistent results. More recently, the antigen profile of *Fonsecaea pedrosoi* was characterized by immunoprecipitation and iodination of surface proteins. ELISA has also been developed and its efficiency for diagnostic purposes has been evaluated on reference patients. The serological tests are usually not employed for diagnostic purposes due to lack of infrastructure in diagnostic laboratories. Moreover, diagnosis by smear or biopsy is quite reliable making such procedures infructuous.

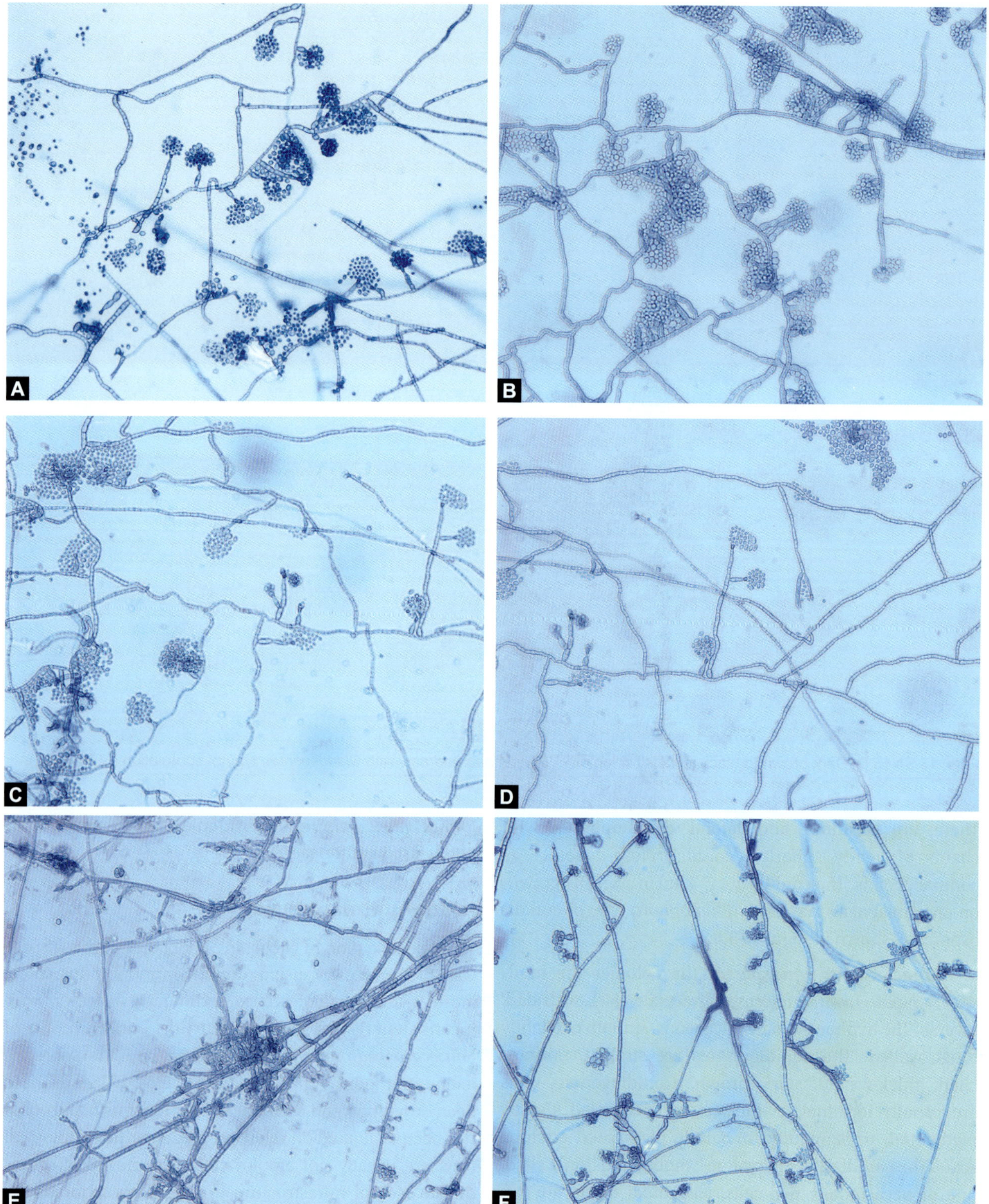

Figs. 13.6A to F

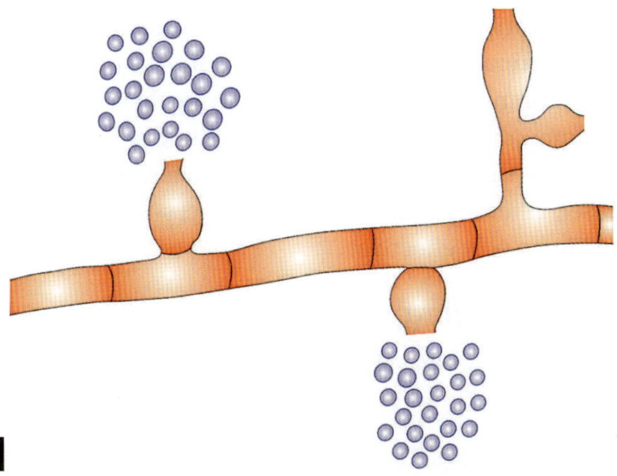
G

Figs. 13.6A to G. Microscopic appearance of *Phialophora verrucosa* from cultures and diagrammatic representation (LCB × 400).

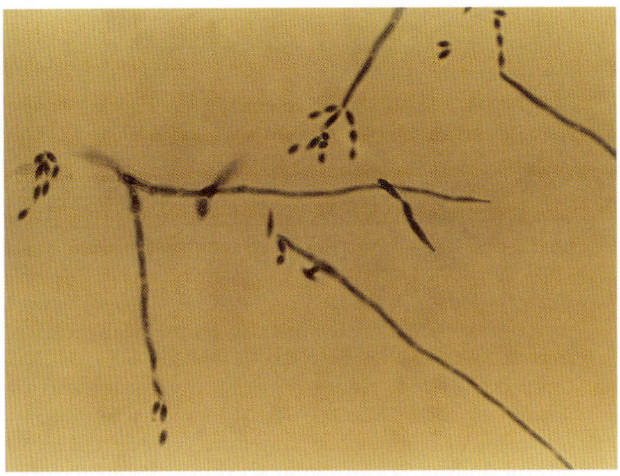

Fig. 13.7. Microscopic appearance of *Cladophialophora carrionii* (LCB × 400).

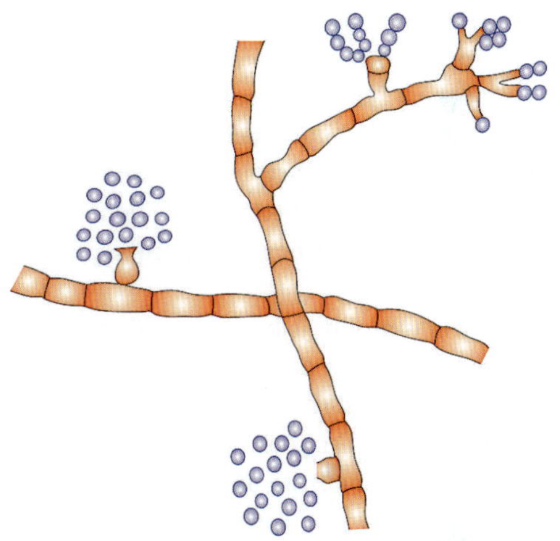

Fig. 13.8. Diagrammatic representation of *Fonsecaea compacta*.

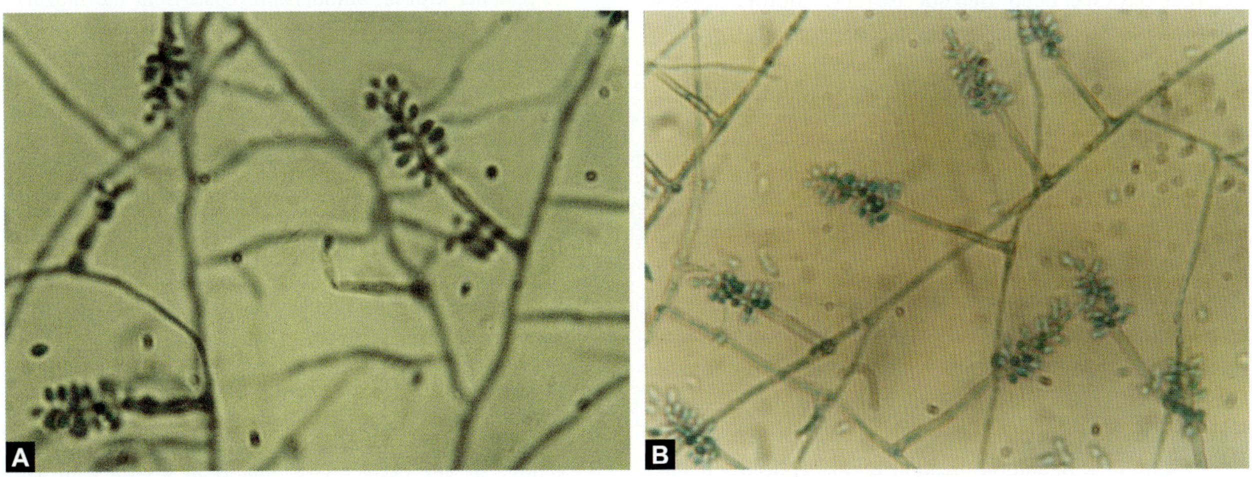

Figs. 13.9A and B. On microscopy bottle-brush appearance of *Rhinocladiella aquaspersa* (LCB × 400).

(d) Animal Pathogenicity

The experimental infections have been successfully produced in rats and mice by intraperitoneal inoculation. The cutaneous lesions of chromoblastomycosis in mice on histopathology show granulomas, muriform cells and necrotic foci. The murine model can be used for studying other aspects of infection, such as humoral and cellular immunological mechanisms during evolution of disease, host resistance and efficacy of antifungal drugs used in this clinical entity. The fungal culture is preferred over animal inoculation for routine identification of etiological agent of chromoblastomycosis.

Treatment and Prophylaxis

The therapeutic modalities for chromoblastomycosis may be cryotherapy (topical liquid nitrogen), thermotherapy, laser therapy, chemotherapy and surgery. Surgical excision can be carried out when small areas are affected as seen in an early, solitary lesion. Antifungal therapy is not satisfactory but it is often the only option for massive lesions. Chromoblastomycosis responds poorly to the available therapies.

The most commonly used drug is flucytosine which acts by inhibiting nucleic acid synthesis and is given orally as 50-150 mg/kg per day, in four divided doses. Secondary drug resistance to flucytosine has been reported. Therefore, azoles like itraconazole and fluconazole have been used effectively. The absence of toxicity in itraconazole therapy is of great advantage and prolonged therapy for 3-6 months can be continued.

A combination of flucytosine and amphotericin B on alternate days has been tried successfully. Another possible combination is flucytosine and itraconazole and/or local application of liquid nitrogen as cryotherapy. Both methods may reduce duration of itraconazole treatment. Another option for this subcutaneous mycosis is terbinafine as daily dose of 500 mg.

Hyperthermic treatment with a disposable chemical pocket warmer has also been successfully tried in Japan as it is done in case of sporotrichosis. The disease caused by *Exophiala dermatitidis* is considered as the most resistant form. Relapses are frequently reported in chromoblastomycosis. At present two antifungal agents, itraconazole and terbinafine have significantly improved the treatment and prognosis of the disease.

Despite lack of standardization in susceptibility testing and poor co-relations with efficacy *in vitro*, phaeoid fungi are usually more sensitive to triazoles and terbinafine then to amphotericin B. Among new azoles, including ravuconazole, voriconazole and posaconazole, all have *in vitro* activity comparable to itraconazole. They have a potential to become drugs of choice in therapy of phaeoid fungal infections. The use of physically ablative modalities such as laser, photodynamic therapy and cryosurgery is also found to be very effective.

Further Reading

1. Agarwalla A, Khanal B, Garg VK, *et al.* Chromoblastomycosis: Report of two cases from Nepal. *J Dermatol.* 2002; **29:** 315-9.
2. Agarwal R, Singh G, Ghosh A, *et al.* Chromoblastomycosis in India: Review of 169 cases. *PLoS Negl Trop Dis.* 2017; [in press].
3. Ameen M. Chromoblastomycosis: Clinical presentation and management. *Clin Exp Dermatol.* 2009; **34:** 849-54.
4. Ameen M. Managing chromoblastomycosis. *Trop Doct.* 2010; **40:** 65-7.
5. Anjaneyan G, Jagadeesan S, Thomas J. Cytodiagnostic copper pennies in chromoblastomycosis. *Indian Dermatol Online J.* 2016; **7:** 145-6.
6. Antonello VS, Appel da Silva MC, Cambruzzi E, *et al.* Treatment of severe chromoblastomycosis with itraconazole and 5-flucytosine association. *Rev Inst Med Trop Sao Paulo.* 2010; **52:** 329-31.
7. Apte G, Gedam JR, Poojary S, *et al.* Chromoblastomycosis in a case of borderline lepromatous leprosy with recurrent type II lepra reaction. *Lepr Rev.* 2011; **82:** 310-5.
8. Arguello-Guerra L, Gatica-Torres M, Dominguez-Cherit J. Chromomycosis. *BMJ Case Rep.* 2016; pii: bcr2016215391.
9. Attapattu MC. Chromoblastomycosis: A clinical and mycological study of 71 cases from Sri Lanka. *Mycopathologia.* 1997; **137:** 145-51.
10. Avelar-Pires C, Simoes-Quaresma JA, Moraes-de Macedo GM, *et al.* Revisiting the clinical and histopathological aspects of patients with chromoblastomycosis from the Brazilian Amazon region. *Arch Med Res.* 2013; **44:** 302-6.
11. Azad K, Khanna G, Capoor MR, *et al.* *Cladophialophora carrionii*: An aetiological agent of cutaneous chromoblastomycosis from a non-endemic area, north India. *Mycoses.* 2011; **54:** e217-9.
12. Azevedo CM, Marques SG, Santos DW, *et al.* Squamous cell carcinoma derived from chronic chromoblastomycosis in Brazil. *Clin Infect Dis.* 2015; **60:** 1500-4.
13. Badali H, Bonifaz A, Barron-Tapia T, *et al.* *Rhinocladiella aquaspersa*, proven agent of verrucous skin infection and a novel type of chromoblastomycosis. *Med Mycol.* 2010; **48:** 696-703.
14. Badali H, Fernandez-Gonzalez M, Mousavi B, *et al.* Chromoblastomycosis due to *Fonsecaea pedrosoi* and *F. monophora* in Cuba. *Mycopathologia.* 2013; **175:** 439-44.
15. Bahk WJ, Chang ED, Chun KA, *et al.* Chromomycosis presenting as soft-tissue mass: Report of a case with MRI features. *Skeletal Radiol.* 2009; **38:** 177-80.

16. Bandyopadhyay A, Majumdar K, Gangopadhyay M, *et al.* Cutaneous chromoblastomycosis mimicking tuberculosis verrucosa cutis: Look for copper pennies! *Turk Patoloji Derg.* 2015; **31:** 223-5.
17. Bansal AS, Prabhakar P. Chromomycosis: A twenty-year analysis of histologically confirmed cases in Jamaica. *Trop Geogr Med.* 1989; **41:** 222-26.
18. Basílio FM, Hammerschmidt M, Mukai MM, *et al.* Mucormycosis and chromoblastomycosis occurring in a patient with leprosy Type 2 reaction under prolonged corticosteroid and thalidomide therapy. *An Bras Dermatol.* 2012; **87:** 767-71.
19. Bassas-Vila J, Fuente MJ, Guinovart R, *et al.* Chromoblastomycosis: Response to combination therapy with cryotherapy and terbinafine. *Actas Dermosifiliogr.* 2014; **105:** 196-8.
20. Battu V, Raman M, Pasricha JS, *et al.* Chromomycosis with some unusual features. *Indian J Dermatol Venereol Leprol.* 1986; **52:** 181-3.
21. Bayles MA. Chromomycosis. *Curr Top Med Mycol.* 1995; **6:** 221-43.
22. Bhagwat PV, Tophakhane RS, Kudligi C, *et al.* Multiple asymptomatic verrucous plaques over the legs. *Indian J Dermatol Venereol Leprol.* 2010; **76:** 86.
23. Bharti R, Malhotra SK, Bal MS, *et al.* Chromoblastomycosis. *Indian J Dermatol Venereol Leprol.* 1995; **61:** 54-5.
24. Bhat RM, Monteiro RC, Bala N, *et al.* Subcutaneous mycoses in coastal Karnataka in south India. *Int J Dermatol.* 2016; **55:** 70-8.
25. Bonifaz A, Carrasco-Gerard E, Saul A. Chromoblastomycosis: Clinical and mycologic experience of 51 cases. *Mycoses.* 2001; **44:** 1-7.
26. Bonifaz A, Vazquez-Gonzalez D, Perusquia-Ortiz AM. Subcutaneous mycoses: Chromoblastomycosis, sporotrichosis and mycetoma. *J Dtsch Dermatol Ges.* 2010; **8:** 619-28.
27. Brischetto A, Kidd S, Baird R. First reported Australian case of *Cladophialophora arxii*: Features consistent with possible primary pulmonary chromoblastomycosis. *Am J Trop Med Hyg.* 2015; **92:** 791-3.
28. Brown M, Pasvol G. Chromoblastomycosis. *N Engl J Med.* 2005; **352:** e19.
29. Calvo E, Pastor FJ, Mayayo E, *et al.* Antifungal therapy in an athymic murine model of chromoblastomycosis by *Fonsecaea pedrosoi. Antimicrob Agents Chemother.* 2011; **55:** 3709-13.
30. Calvo E, Pastor FJ, Salas V, *et al.* Histopathology and antifungal treatment of experimental murine chromoblastomycosis caused by *Cladophialophora carrionii. J Antimicrob Chemother.* 2012; **67:** 666-70.
31. Camara-Lemarroy CR, Soto-Garcia AJ, Preciado-Yepez CI, *et al.* Case of chromoblastomycosis with pulmonary involvement. *J Dermatol.* 2013; **40:** 746-8.
32. Carolina Rojas O, Leon-Cachon RB, Perez-Maya AA, *et al.* Phenotypic and molecular identification of *Fonsecaea pedrosoi* strains isolated from chromoblastomycosis patients in Mexico and Venezuela. *Mycoses.* 2015; **58:** 267-72.
33. Carrion AL. Chromoblastomycosis. *Ann N Y Acad Sci.* 1950; **50:** 1255-82.
34. Chandran V, Sadanandan SM, Sobhanakumari K. Chromoblastomycosis in Kerala, India. *Indian J Dermatol Venereol Leprol.* 2012; **78:** 728-33.
35. Chavan SS, Kulkarni MH, Makannavar JH. 'Unstained' and 'de stained' sections in the diagnosis of chromoblastomycosis: A clinico-pathological study. *Indian J Pathol Microbiol.* 2010; **53:** 666-71.
36. Chavan SS, Reddy P. Cytological diagnosis of chromoblastomycosis. *J Cytol.* 2013; **30:** 276-7.
37. Chen Y, Yin S, Li M, *et al.* A case of chromoblastomycosis by *Fonsecaea nubica* indicating a possible insect route of transmission. *Mycoses.* 2016; **59:** 662-7.
38. Chowdhary A, Guarro J, Randhawa HS, *et al.* A rare case of chromoblastomycosis in a renal transplant recipient caused by a non-sporulating species of *Rhytidhysteron. Med Mycol.* 2008; **46:** 163-6.
39. Correia RT, Valente NY, Criado PR, *et al.* Chromoblastomycosis: Study of 27 cases and review of medical literature. *An Bras Dermatol.* 2010; **85:** 448-54.
40. Criado PR, Careta MF, Valente NY, *et al.* Extensive long-standing chromomycosis due to *Fonsecaea pedrosoi:* Three cases with relevant improvement under voriconazole therapy. *J Dermatolog Treat.* 2011; **22:** 167-74.
41. da Silva MB, da Silva JP, Sirleide Pereira Yamano S, *et al.* Development of natural culture media for rapid induction of *Fonsecaea pedrosoi* sclerotic cells *in vitro. J Clin Microbiol.* 2008; **46:** 3839-41.
42. Daboit TC, Magagnin CM, Heidrich D, *et al.* A case of relapsed chromoblastomycosis due to *Fonsecaea monophora:* Antifungal susceptibility and phylogenetic analysis. *Mycopathologia.* 2013; **176:** 139-44.
43. Daboit TC, Massotti Magagnin C, Heidrich D, *et al. In vitro* susceptibility of chromoblastomycosis agents to five antifungal drugs and to the combination of terbinafine and amphotericin B. *Mycoses.* 2014; **57:** 116-20.
44. Dashatwar D, Kar S, Gangane N, *et al.* Chromoblastomycosis in a resident of a leprosarium. *Lepr Rev.* 2015; **86:** 102-7.
45. de A, Gharami RC, Datta PK. Verrucous plaque on the face: What is your diagnosis? *Dermatol Online J.* 2010; **16:** 6.
46. de Guzman L, Perlman DC, Hubbard CE. Septic arthritis and osteomyelitis due to the chromoblastomycosis agent *Fonsecaea pedrosoi. Am J Orthop.* 2012; **41:** 328-31.
47. de Hoog GS, Attili-Angelis D, Vicente VA, *et al.* Molecular ecology and pathogenic potential of *Fonsecaea* species. *Med Mycol.* 2004; **42:** 405-16.
48. de Hoog GS, Nishikaku AS, Fernandez-Zeppenfeldt G, *et al.* Molecular analysis and pathogenicity of the *Cladophialophora carrionii* complex, with the description of a novel species. *Stud Mycol.* 2007; **58:** 219-34.
49. de Sousa Mda G, Belda W Jr, Spina R, *et al.* Topical application of imiquimod as a treatment for chromoblastomycosis. *Clin Infect Dis.* 2014; **58:** 1734-7.
50. de Azevedo CM, Gomes RR, Vicente VA, *et al. Fonsecaea pugnacius,* a novel agent of disseminated chromoblastomycosis. *J Clin Microbiol.* 2015; **53:** 2674-85.

51. Deng S, de Hoog GS, Badali H, *et al. In vitro* antifungal susceptibility of *Cladophialophora carrionii*, an agent of human chromoblastomycosis. *Antimicrob Agents Chemother.* 2013; **57:** 1974-7.
52. Deng S, de Hoog GS, Pan W, *et al.* Three isothermal amplification techniques for rapid identification of *Cladophialophora carrionii*, an agent of human chromoblastomycosis. *J Clin Microbiol.* 2014; **52:** 3531-5.
53. Deng S, Tsui CK, Gerrits van den Ende AH, *et al.* Global spread of human chromoblastomycosis is driven by recombinant *Cladophialophora carrionii* and predominantly *clonal Fonsecaea species. PLoS Negl Trop Dis.* 2015; **9:** e0004004.
54. Dube B, Dube R. Chromoblastomycosis in India: A review and report of two new cases. *Indian J Med Sci.* 1966; **20:** 12-6.
55. Dupont C, Duong TA, Mallet S, *et al.* Unusual presentation of chromoblastomycosis due to *Cladophialophora carrionii* in a renal and pancreas transplant recipient patient successfully treated with posaconazole and surgical excision. *Transpl Infect Dis.* 2010; **12:** 180-3.
56. Elgart GW. Chromoblastomycosis. *Dermatol Clin.* 1996; **14:** 77-83.
57. Erstad D, Sell N, Cauley R, *et al.* Autografted electrical burn complicated by cutaneous chromoblastomycosis. *Eplasty.* 2015; **15:** ic59.
58. Esterre P, Queiroz-Telles F. Management of chromoblastomycosis: Novel perspectives. *Curr Opin Infect Dis.* 2006; **19:** 148-52.
59. Fatemi MJ, Bateni H. Oral chromoblastomycosis: A case report. *Iran J Microbiol.* 2012; **4:** 40-3 & 101.
60. Franca K, Villa RT, Bastos VR, *et al.* Auricular chromoblastomycosis: A case report and review of published literature. *Mycopathologia.* 2011; **172:** 69-72.
61. Fransisca C, He Y, Chen Z, *et al.* Molecular identification of chromoblastomycosis clinical isolates in Guangdong. *Med Mycol.* 2017. pii: myw140. PMID: 28053146.
62. Fukushiro R. Chromomycosis in Japan. *Int J Dermatol.*1983; **22:** 221-9.
63. Gimenes VM, Criado PR, Martins JE, *et al.* Cellular immune response of patients with chromoblastomycosis undergoing antifungal therapy. *Mycopathologia.* 2006; **162:** 97-101.
64. Gomes NM, Bastos TC, Cruz KS, *et al.* Chromoblastomycosis: An exuberant case. *An Bras Dermatol.* 2014; **89:** 351-2.
65. Gomes RR, Vicente VA, Azevedo CM, *et al.* Molecular epidemiology of agents of human chromoblastomycosis in Brazil with the description of two novel species. *PLoS Negl Trop Dis.* 2016; **10:** e0005102. PMID: 27893750.
66. Gonzalez GM, Rojas OC, Bocanegra-Garcia V, *et al.* Molecular diversity of *Cladophialophora carrionii* in patients with chromoblastomycosis in Venezuela. *Med Mycol.* 2013; **51:** 170-7.
67. Gonzalez GM, Rojas OC, Gonzalez JG, *et al.* Chromoblastomycosis caused by *Rhinocladiella aquaspersa. Med Mycol Case Rep.* 2013; **2:** 148-51.
68. Gugnani HC, Egere JU, Suseelan AV, *et al.* Chromomycosis caused by *Phialophora pedrosoi* in Eastern Nigeria. *J Trop Med Hyg.*1978; **81:** 208-10.
69. Gupta AK, Taborda PR, Sanzovo AD. Alternate week and combination itraconazole and terbinafine therapy for chromoblastomycosis caused by *Fonsecaea pedrosoi* in Brazil. *Med Mycol.* 2002; **40:** 529-34.
70. Harshan V. Chromoblastomycosis due to *Phialophora verrucosa. Indian J Dermatol Venereol Leprol.* 1994; **60:** 95-6.
71. Heidrich D, Gonzalez GM, Pagani DM, *et al.* Chromoblastomycosis caused by *Rhinocladiella similis:* Case report. *Med Mycol Case Rep.* 2017; **16:** 25-27.
72. Hernandez-Hernandez F, Vargas-Arzola J, Rios-Cruz OP, *et al.* First case of chromoblastomycosis due to *Phoma insulana. Enferm Infecc Microbiol Clin.* 2016. pii: S0213-005X(16) 30259-2. PMID: 27726899.
73. Hu Y, Huang X, Lu S, *et al.* Photodynamic therapy combined with terbinafine against chromoblastomycosis and the effect of PDT on *Fonsecaea monophora in vitro. Mycopathologia.* 2015; **179:** 103-9.
74. Jakopp B, Stamm B, Eyer D, *et al.* Hidden under a cauliflower-like skin tumor: Chromoblastomycosis. *Case Rep Infect Dis.* 2013; 450153. PMID: 23864969.
75. Jaleel A, Celine MI, Sobhanakumari K, *et al.* Mycetoma-like chromoblastomycosis: A diagnostic dilemma. *Int J Dermatol.* 2017; **56:** 563-6.
76. Jamil A, Lee YY, Thevarajah S. Invasive squamous cell carcinoma arising from chromoblastomycosis. *Med Mycol.* 2012; **50:** 99-102.
77. Jayalakshmi P, Looi LM, Soo-Hoo TS. Chromoblastomycosis in Malaysia. *Mycopathologia.* 1990; **109:** 27-31.
78. Kalabhavi AS. Chromoblastomycosis - Review article. *Int J Cur Res.* 2013; **5:** 1691-5.
79. Kampirapap K, Reangchainam S, Ornpaew P, *et al.* Chromoblastomycosis masquerading as dermatophytosis, with the description of a new opportunistic species. *Southeast Asian J Trop Med Public Health.* 2015; **46:** 105-9.
80. Kapoor S. Chromoblastomycosis in a Solomon islander. *Med J Aust.* 2013; **198:** 140.
81. Kawtar I, Salim G, Mariame M, *et al.* Sporotrichoid chromomycosis. *Dermatol Online J.* 2013; **19:** 20394.
82. Khan K, Mondal K, Dutta R, *et al.* Unusual presentation of cutaneous chromoblastomycosis. *Am J Dermatopathol.* 2017; **39:** 159-61.
83. Khan S, Kumar A, Vinod V, *et al.* Chromoblastomycosis due to *Fonsecaea pedrosoi:* An old wine in a rare bottle. *J Infect Dev Ctries.* 2015; **9:** 325-9.
84. Kim DM, Hwang SM, Suh MK, *et al.* Chromoblastomycosis caused by *Fonsecaea pedrosoi. Ann Dermatol.* 2011; **23:** 369-74.
85. Knox J, Marshall C. Chromoblastomycosis in a Solomon islander. *Med J Aust.* 2012; **197:** 350.
86. Krishna S, Shenoy MM, Pinto M, *et al.* Two cases of axillary chromoblastomycosis. *Indian J Dermatol Venereol Leprol.* 2016; **82:** 455-6.
87. Krzysciak PM, Pindycka-Piaszczynska M, Piaszczynski M. Chromoblastomycosis. *Postepy Dermatol Alergol.* 2014; **31:** 310-21.
88. Kullavanijaya P, Rojanavanich V. Successful treatment of chromoblastomycosis due to *Fonsecaea pedrosoi* by

the combination of itraconazole and cryotherapy. *Int J Dermatol.*1995; **34:** 804-7.
89. Kumar B, Kaur I, Chakrabarti A, *et al.* Treatment of deep mycoses with itraconazole. *Mycopathologia.* 1991; **115:** 169-74.
90. Kumar B. Chromoblastomycosis in India: two more cases. *Int J Dermatol.* 2000; **39:** 800.
91. Kutty MK, Majumder M. Chromoblastomycosis. *Southeast Asian J Trop Med Public Health.*1971; **2:** 86-7.
92. La Hoz RM, Baddley JW. Subcutaneous fungal infections. *Curr Infect Dis Rep.* 2012; **14:** 530-9.
93. Leeyaphan C, Hau C, Takeoka S, *et al.* Immune response in human chromoblastomycosis and eumycetoma - Focusing on human interleukin-17A, interferon-gamma, tumour necrosis factor-alpha, interleukin-1 beta and human beta-defensin-2. *Mycoses.* 2017; doi: 10.1111/myc.12526. PMID: 27402133.
94. Li Y, Wan Z, Li R. *In vitro* activities of nine antifungal drugs and their combinations against *Phialophora verrucosa. Antimicrob Agents Chemother.* 2014; **58:** 5609-12.
95. Liu ZH, Xia XJ. Successful sequential treatment with itraconazole and ALA-PDT for chromoblastomycosis because of *Alternaria alternata. Dermatol Ther.* 2014; **27:** 357-60.
96. Lokuhetty MD, Alahakoon VS, Kularatne BD, *et al.* Zeil Neelson and Wade-Fite stains to demonstrate medlar bodies of chromoblastomycosis. *J Cutan Pathol.* 2007; **34:** 71-2.
97. Long H, Zhang G, Lu Q. A persistent red crusted plaque on the back. *JAMA.* 2013; **310:** 1730-1.
98. Lopez Martinez R, Mendez Tovar LJ. Chromoblastomycosis. *Clin Dermatol.* 2007; **25:** 188-94.
99. Lu S, Lu C, Zhang J, *et al.* Chromoblastomycosis in mainland China: A systematic review on clinical characteristics. *Mycopathologia.* 2013; **175:** 489-95.
100. Lyon JP, Moreira LM, de Carvalho VS, *et al. In vitro* photodynamic therapy against *Fonsecaea pedrosoi* and *Cladophialophora carrionii. Mycoses.* 2013; **56:** 157-61.
101. Lyon JP, Pedroso e Silva Azevedo Cde M, Moreira LM, *et al.* Photodynamic antifungal therapy against chromoblastomycosis. *Mycopathologia.* 2011; **172:** 293-7.
102. Machado AP, Silva MR, Fischman O. Local phagocytic responses after murine infection with different forms of *Fonsecaea pedrosoi* and sclerotic bodies originating from an inoculum of conidiogenous cells. *Mycoses.* 2011; **54:** 202-11.
103. Machado AP, Silva MR, Fischman O. Prolonged infection by *Fonsecaea pedrosoi* after antigenic co-stimulation at different sites in experimental murine chromoblastomycosis. *Virulence.* 2010; **1:** 29-36.
104. Mandal RK, Banerjee S, Kumar P, *et al.* Non-healing verrucous plaque over upper limb for 1 year in a tea garden worker. *Dermatol Online J.* 2013; **19:** 12.
105. Marques GF, Masuda PY, Sousa JM, *et al.* Clinical and demographic profile of chromoblastomycosis in a referral service in the Midwest of Sao Paulo state (Brazil). *An Bras Dermatol.* 2015; **90:** 140-2.
106. Matsumoto T, Matsuda T. Chromoblastomycosis and phaeohyphomycosis. *Semin Dermatol.*1985; **4:** 240-51.
107. Matsumoto T, Padhye AA, Ajello L. Medical significance of so-called black yeasts. *Eur J Epidemiol.*1987; **3:** 87-95.
108. McDaniel P, Walsh DS. Chromoblastomycosis in western Thailand. *Am J Trop Med Hyg.* 2010; **83:** 448.
109. McGinnis MR. Chromoblastomycosis and phaeohyphomycosis: new concepts, diagnosis and mycology, *J Am Acad Dermatol.*1983; **8:** 1-16.
110. Menezes N, Varela P, Furtado A, *et al.* Chromoblastomycosis associated with *Fonsecaea pedrosoi* in a carpenter handling exotic woods. *Dermatol Online J.* 2008; **14:** 9.
111. Milam CP, Fenske NA. Chromoblastomycosis. *Dermatol Clin.* 1989; **7:** 219-25.
112. Minotto R, Bernardi CD, Mallmann LF, *et al.* Chromoblastomycosis: A review of 100 cases in the state of Rio Grande do Sul, Brazil. *J Am Acad Dermatol.* 2001; **44:** 585-92.
113. Mishra A, Tripathi K, Biswal P, *et al.* Chromoblastomycosis of chin masquerading as facial wart. *Indian J Pathol Microbiol.* 2011; **54:** 221-2.
114. Mittal A, Agarwal N, Gupta LK, et al. Chromoblastomycosis from a non-endemic area and response to itraconazole. *Indian J Dermatol.* 2014; **59:** 606-8.
115. Miyagi H, Yamamoto Y, Kanamori S, *et al.* Case of chromoblastomycosis appearing in an Okinawa patient with a medical history of Hansen's disease. *J Dermatol.* 2008; **35:** 354-61.
116. Mohanty L, Mohanty P, Padhi T, *et al.* Verrucous growth on leg. *Indian J Dermatol Venereol Leprol.* 2006; **72:** 399-400.
117. Mohanty P, Vivekanandh K, Mohapatra L, *et al.* Squamous cell carcinoma in long-standing chromoblastomycosis. *Indian J Dermatol Venereol Leprol.* 2017; **83:** 363-5.
118. Mohapatra LN, Sood VK, Grueber HLE. Chromoblastomycosis in India and Nepal. *Mykosen.* 1967; **10:** 309-14.
119. Morio F, Barbarot S, Pineau S, *et al.* A Cutaneous lesion in a 66-year-old traveler returning from Thailand. *J Clin Microbiol.* 2016; **54:** 1179. & 1409.
120. Mouchalouat Mde F, Galhardo MC, Fialho PC, *et al. Cladophialophora carrionii:* A rare agent of chromoblastomycosis in Rio de Janeiro State, Brazil. *Rev Inst Med Trop Sao Paulo.* 2008; **50:** 351-3.
121. Mouchalouat Mde F, Gutierrez Galhardo MC, Zancope-Oliveira RM, *et al.* Chromoblastomycosis: A clinical and molecular study of 18 cases in Rio de Janeiro, Brazil. *Int J Dermatol.* 2011; **50:** 981-6.
122. Mugleston BJ, Usatine RP, Rosen T. Wide morphologic variability of chromoblastomycosis in the Western Hemisphere. *Skinmed.* 2016; **14:** 423-7.
123. Muhammed K, Nandakumar G, Asokan KK, *et al.* Lymphangitic chromoblastomycosis. *Indian J Dermatol Venereol Leprol.* 2006; **72:** 443-5.
124. Mukherjee B, Sengupta D, Bhattacharyya KK. Chromoblastomycosis. *J Indian Med Assoc.* 1975; **64:** 338-40.
125. Murthy R, Swain JP. Concurrent mycetoma and chromomycosis. *Indian J Med Microbiol.* 2011; **29:** 437-9.

126. Najafzadeh MJ, Badali H, Illnait-Zaragozi MT, *et al. In vitro* activities of eight antifungal drugs against 55 clinical isolates of *Fonsecaea* spp. *Antimicrob Agents Chemother.* 2010; **54:** 1636-8.
127. Najafzadeh MJ, Gueidan C, Badali H, *et al.* Genetic diversity and species delimitation in the opportunistic genus *Fonsecaea. Med Mycol.* 2009; **47:** 17-25.
128. Najafzadeh MJ, Rezusta A, Cameo MI, *et al.* Successful treatment of chromoblastomycosis of 36 years duration caused by *Fonsecaea monophora. Med Mycol.* 2010; **48:** 390-3 & 560.
129. Najafzadeh MJ, Sun J, Vicente V, *et al. Fonsecaea nubica* sp. nov. A new agent of human chromoblastomycosis revealed using molecular data. *Med Mycol.* 2010; **48:** 800-6.
130. Najafzadeh MJ, Sun J, Vicente VA, *et al.* Molecular epidemiology of *Fonsecaea* species. *Emerg Infect Dis.* 2011; **17:** 464-9.
131. Namratha N, Nadgir S, Kale M, *et al.* Chromoblastomycosis due to *Cladosporium carrionii. J Lab Physicians.* 2010; **2:** 47-8.
132. Narendranath S, Sudhakar GK, Pai MR, *et al.* Safety and efficacy of oral potassium iodide in chromoblastomycosis. *Int J Dermatol.* 2010; **49:** 341-3.
133. Naveen KN, Shetty PC, Naik AS, *et al.* Chromoblastomycosis presenting as a phagedenic ulcer on the face. *Int J Dermatol.* 2012; **51:** 576-8.
134. Ohashi T, Ohtsuka M, Kikuchi N, *et al.* A case of chromoblastomycosis caused by *Fonsecaea pedrosoi* in a patient with rheumatoid arthritis. *Int J Rheum Dis.* 2015; **18:** 580-1.
135. Padhye AA, Hampton AA, Hampton MT, *et al.* Chromoblastomycosis caused by *Exophiala spinifera. Clin Infect Dis.* 1996; **22:** 331-5.
136. Panicker NK, Chandanwale SS, Sharma YK, *et al.* Chromoblastomycosis: Report of two cases on face from urban industrial area. *Indian Dermatol Online J.* 2013; **4:** 371-3.
137. Paniz-Mondolfi AE, Colella MT, Negrin DC, *et al.* Extensive chromoblastomycosis caused by *Fonsecaea pedrosoi* successfully treated with a combination of amphotericin B and itraconazole. *Med Mycol.* 2008; **46:** 179-84.
138. Patel U, Chu J, Patel R, *et al.* Subcutaneous dematiaceous fungal infection. *Dermatol Online J.* 2011; **17:** 19.
139. Pavithran K. Chromoblastomycosis in a residual patch of leprosy. *Indian J Lepr.*1988; **60:** 444-7.
140. Pavithran K. Disseminated chromoblastomycosis. *Indian J Dermatol Venereol Leprol.* 1991; **57:** 155-6.
141. Penjor D, Khizuan AK, Chong AW, *et al.* A case of nasal chromoblastomycosis causing epistaxis. *J Laryngol Otol.* 2014; **128:** 1117-9.
142. Phadke SA, Mhatre HC. Primary chromoblastomycosis of brain: A case report. *Indian J Pathol Microbiol.* 1981; **24:** 61-6.
143. Pho MT, Burgin S. Right buttock rash for thirty years in a patient from China. *Clin Infect Dis.* 2009; **49:** 432 & 477-8.
144. Piepenbring M, Caceres Mendez OA, *et al.* Chromoblastomycosis caused by *Chaetomium funicola*: A case report from western Panama. *Br J Dermatol.* 2007; **157:** 1025-9.
145. Pindycka-Piaszczynska M, Krzysciak P, Piaszczynski M, *et al.* Chromoblastomycosis as an endemic disease in temperate Europe: First confirmed case and review of the literature. *Eur J Clin Microbiol Infect Dis.* 2014; **33:** 391-8.
146. Pires CA, Xavier MB, Quaresma JA, *et al.* Clinical, epidemiological and mycological report on 65 patients from the Eastern Amazon region with chromoblastomycosis. *An Bras Dermatol.* 2012; **87:** 555-60.
147. Pradeepkumar NS, Joseph NM. Chromoblastomycosis caused by *Cladophialophora carrionii* in a child from India. *J Infect Dev Ctries.* 2011; **5:** 556-60.
148. Pradhan SV, Talwar OP, Ghosh A, *et al.* Chromoblastomycosis in Nepal: A study of 13 cases. *Indian J Dermatol Venereol Leprol.* 2007; **73:** 176-8.
149. Queiroz-Telles F, Esterre P, Perez-Blanco M, *et al.* Chromoblastomycosis: An overview of clinical manifestations, diagnosis and treatment. *Med Mycol.* 2009; **47:** 3-15.
150. Queiroz-Telles F, Santos DW. Challenges in the therapy of chromoblastomycosis. *Mycopathologia.* 2013; **175:** 477-88.
151. Queiroz-Telles F. Chromoblastomycosis: A neglected tropical disease. *Rev Inst Med Trop Sao Paulo.* 2015; **57 (Suppl. 19):** 46-50.
152. Raj HJ, Majumdar B, Jain A, Maiti PK, Chatterjee G. A clinico-mycological study on suspected cases of chromoblastomycosis: Challenges in diagnosis and management. *J Clin Diagn Res.* 2015; **9:** WC01-4.
153. Rajendran C, Ramesh V, Misra RS, *et al.* Chromoblastomycosis in India. *Int J Dermatol.* 1997; **36:** 29-33.
154. Rajendran C. *Ascosubramania* gen. nov., and its *Fonsecaea*-like anamorph causing chromoblastomycosis in India. *J Med Vet Mycol.* 1997; **35:** 335-9.
155. Ranawaka RR, Amarasinghe N, Hewage D. Chromoblastomycosis: Combined treatment with pulsed itraconazole therapy and liquid nitrogen cryotherapy. *Int J Dermatol.* 2009; **48:** 397-400.
156. Randhawa HS, Budimulja U, Bazaz-Malik G, *et al.* Recent developments in the diagnosis and treatment of subcutaneous mycoses. *J Med Vet Mycol.* 1994; **32 (Suppl. 1):** 299-307.
157. Rasul ES, Hazarika NK, Sharma A, *et al.* Chromoblastomycosis. *J Assoc Physicians India.* 2007; **55:** 149-51.
158. Revankar SG, Sutton DA. Melanized fungi in human disease. *Clin Microbiol Rev.* 2010; **23:** 884-928.
159. Riddel CE, Surovik JG, Chon SY, *et al.* Fungal foes: Presentations of chromoblastomycosis post-hurricane Ike. *Cutis.* 2011; **87:** 269-72.
160. Rodriguez B, Karnes J. Plaques on dorsal hands & arms. *J Fam Pract.* 2015; **64:** e11-3.
161. Rojas OC, Gonzalez GM, Moreno-Trevino M,*et al.* Chromoblastomycosis by *Cladophialophora carrionii* associated with squamous cell carcinoma and review of published reports. *Mycopathologia.* 2015; **179:** 153-7.
162. Rosen T, Overholt M. Persistent viability of the Medlar body. *Int J Dermatol.* 1996; **35:** 96-8.
163. Roy AD, Das D, Deka M. Chromoblastomycosis - A clinical mimic of squamous carcinoma. *Australas Med J.* 2013; **6:** 458-60.

164. Roy P, Prasanna S, Laxmikant DV, *et al.* Chromoblastomycosis caused by *Cladophialophora carrionii* in a skin graft recipient. *Med J Armed Forces India.* 2016; **72:** 389-92.
165. Salgado CG, da Silva MB, Yamano SS, *et al.* Cutaneous localized annular chromoblastomycosis. *J Cutan Pathol.* 2009; **36:** 257-61.
166. Santmyire A. The effectiveness of a multifocal training to improve the treatment of chromoblastomycosis in rural Madagascar. *J Health Care Poor Underserved.* 2016; **27:** 993-1010.
167. Santos AL, Palmeira VF, Rozental S, *et al.* Biology and pathogenesis of *Fonsecaea pedrosoi,* the major etiologic agent of chromoblastomycosis. *FEMS Microbiol Rev.* 2007; **31:** 570-91.
168. Saxena AK, Jain S, Ramesh V, *et al.* Chromoblastomycosis: Demonstration of abundant microorganisms on microscopy of a scaly crust following intralesional corticosteroids. *J Eur Acad Dermatol Venereol.* 2015; **29:** 189-90.
169. Shanthala GB, Rudresh SM, Nagarathnamma T. Chromoblastomycosis: A case report. *J Clin Diagn Res.* 2011; **5:** 865-6.
170. Sharma A, Hazarika NK, Gupta D. Chromoblastomycosis in sub-tropical regions of India. *Mycopathologia.* 2010; **169:** 381-6.
171. Sharma N, Marfatia YS. Genital elephantiasis as a complication of chromoblastomycosis: A diagnosis overlooked. *Indian J Sex Transm Dis.* 2009; **30:** 43-5.
172. Sharma NL, Sharma RC, Grover PS, *et al.* Chromoblastomycosis in India. *Int J Dermatol.* 1999; **38:** 846-51.
173. Sharma NL, Sharma VC, Mahajan V, *et al.* Chromoblastomycosis with underlying osteolytic lesion. *Mycoses.* 2007; **50:** 517-9.
174. Sharma VK, Kumar B, Kaur I, *et al.* Side lab diagnosis of chromoblastomycosis. *Indian J Dermatol Venereol Leprol.* 1985; **51:** 157-9.
175. Shi D, Zhang W, Lu G, *et al.* Chromoblastomycosis due to *Fonsecaea monophora* misdiagnosed as sporotrichosis and cutaneous tuberculosis in a pulmonary tuberculosis patient. *Med Mycol Case Rep.* 2016; **11:** 57-60.
176. Shivananda PG, Ganesan V, Pai KR, *et al.* Chromomycosis in man in an unusual site: Case report. *Indian J Dermatol Venereol Leprol.* 1978; **44:** 369-71.
177. Shresta D, Kumar R, Durgapal P, *et al.* Isolated nasal chromoblastomycosis. *Indian J Pathol Microbiol.* 2014; **57:** 519-21.
178. Silva-Rocha WP, Cardoso FJ, Colalto W, *et al.* Clinical improvement of chromoblastomycosis refractory to itraconazole successfully treated with high dose of terbinafine. *J Dermatol.* 2013; **40:** 775-6.
179. Siqueira IM, de Castro RJA, Leonhardt LCM, *et al.* Modulation of the immune response by *Fonsecaea pedrosoi* morphotypes in the course of experimental chromoblastomycosis and their role on inflammatory response chronicity. *PLoS Negl Trop Dis.* 2017; **11:** e0005461. PMID: 28355277.
180. Skupsky H, Junkins-Hopkins J. Counterfeit Pennies: Distinguishing chromoblastomycosis from phaeohyphomycotic infections. *Am J Dermatopathol.* 2017; **39:** 485-7.
181. Slesak G, Inthalad S, Strobel M, *et al.* Chromoblastomycosis after a leech bite complicated by myiasis: A case report. *BMC Infect Dis.* 2011; **11:** 14.
182. Smith LC, Fisher KR, Granese J, *et al.* Verrucous plaques on the arm. *Arch Dermatol.* 2009; **145:** 195-200.
183. Son YM, Kang HK, Na SY, *et al.* Chromoblastomycosis caused by *Phialophora richardsiae. Ann Dermatol.* 2010; **22:** 362-6.
184. Sooraj YS, Nainan GK, Eapen M, *et al.* Chromoblastomycosis in a renal allograft recipient. *Indian J Nephrol.* 2013; **23:** 235-6.
185. Brun S, Zumelzu C, Hoanganh MB, *et al.* First case of chromoblastomycosis from Bangladesh. *Med Mycol Case Rep.* 2015; **10:** 1-3.
186. Spiker A, Ferringer T. Chromoblastomycosis. *Cutis.* 2015; **96:** 224 & 267-8.
187. Srinivas SM, Gowda VK, Mahantesh S, *et al.* Chromoblastomycosis associated with bone and central nervous involvement system in an immunocompetent child caused by *Exophiala spinifera. Indian J Dermatol.* 2016; **61:** 324-8.
188. Takeuchi A, Anzawa K, Mochizuki T, *et al.* Chromoblastomycosis caused by *Phialophora verrucosa* on the hand. *Eur J Dermatol.* 2015; **25:** 274-5.
189. Tan H, Xu Y, Lan XM, *et al.* Chromoblastomycosis due to *Fonsecaea monophora* in a man with nephritic syndrome. *Mycopathologia.* 2015; **179:** 447-52.
190. Tang J, Zhuang K, Ran X, *et al.* Chromoblastomycosis caused by *Cladophialophora carrionii.* *Indian J Dermatol Venereol Leprol.* 2017; **83:** 482-5.
191. Thuile T, Hoepfl R, Puelzl P, *et al.* Exophytic and ulcerated nodule on the hand. *J Dtsch Dermatol Ges.* 2015; **13:** 342-5.
192. Torres E, Beristain JG, Lievanos Z, Arenas R. Chromoblastomycosis associated with a lethal squamous cell carcinoma. *An Bras Dermatol.* 2010; **85:** 267-70.
193. Torres-Guerrero E, Isa-Isa R, Isa M, *et al.* Chromoblastomycosis. *Clin Dermatol.* 2012; **30:** 403-8.
194. Troncoso A, Bava J. Chromoblastomycosis. *N Engl J Med.* 2009; **361:** 2165.
195. Veasey JV, Machado Bde A, Lellis RF, *et al.* Tumoral chromoblastomycosis: A rare manifestation with typical complementary exams. *An Bras Dermatol.* 2015; **90:** 907-8.
196. Veasey JV, Machado Bde A, Lellis RF, *et al.* Tumoral chromoblastomycosis: A rare manifestation with typical complementary exams. *An Bras Dermatol.* 2015; **90:** 907-8.
197. Verma GK, Verma S, Singh G, *et al.* A case of extensive chromoblastomycosis from north India. *Braz J Microbiol.* 2014; **45:** 275-7.
198. Verma KC, Chaudhary SD, Chugh TD, *et al.* Chromoblastomycosis. *Indian J Dermatol Venereol Leprol.* 1977; **43:** 35-7.
199. Verma P, Karmakar S, Pandhi D, *et al.* Chromoblastomycosis caused by *Cladophialophora bantiana* in a renal transplant recipient from Delhi, India. *Skinmed.* 2015; **13:** 251-4.
200. Verma S, Verma GK, Singh G, *et al.* Facial chromoblastomycosis in sub-Himalayan region misdiagnosed as cutaneous leishmaniasis: Brief report and review of Indian literature. *Dermatol Online J.* 2012; **18:** 3.

201. Vijaya D, Kumar BH. Chromoblastomycosis. *Mycoses.* 2005; **48:** 82-4.
202. Vitale RG, de Queiroz Telles F, de Hoog S. The dark world of black fungi—A major area of concern. Preface. *Med Mycol.* 2009; **47:** 1.
203. Vitale RG, Perez-Blanco M, De Hoog GS. *In vitro* activity of antifungal drugs against *Cladophialophora* species associated with human chromoblastomycosis. *Med Mycol.* 2009; **47:** 35-40.
204. Weedon D, van Deurse M, Allison S, *et al.* Chromoblastomycosis in Australia: An historical perspective. *Pathology.* 2013; **45:** 489-91.
205. Wu PA, Turner ML, Cowen EW, *et al.* Sixty-year-old man with slowly expanding nodular plaque on the thigh. *J Am Acad Dermatol.* 2010; **63:** 1083-7.
206. Xi L, Lu C, Sun J, *et al.* Chromoblastomycosis caused by a meristematic mutant of *Fonsecaea monophora. Med Mycol.* 2009; **47:** 77-80.
207. Xi L, Sun J, Lu C, *et al.* Molecular diversity of *Fonsecaea (Chaetothyriales)* causing chromoblastomycosis in southern China. *Med Mycol.* 2009; **47:** 27-33.
208. Yanagihara S, Kobayashi H, Kamo R, *et al.* Chromoblastomycosis caused by *Fonsecaea nubica:* First report from Japan. *J Dermatol.* 2015; **42:** 833-4.
209. Yang Y, Hu Y, Zhang J, *et al.* A refractory case of chromoblastomycosis due to *Fonsecaea monophora* with improvement by photodynamic therapy. *Med Mycol.* 2012; **50:** 649-53.
210. Yang YP, Li W, Huang WM, *et al.* Chromoblastomycosis caused by *Fonsecaea*: Clinicopathology, susceptibility and molecular identification of seven consecutive cases in Southern China. *Clin Microbiol Infect.* 2013; **19:** 1023-8.
211. Yap FB. Chromoblastomycosis in Sarawak, East Malaysian Borneo. *Trans R Soc Trop Med Hyg.* 2010; **104:** 168-9.
212. Yap FB. Chromoblastomycosis. *Int J Infect Dis.* 2010; **14:** e543-4.
213. Zarei M, Schneider S, Villasante A, *et al.* Chromoblastomycosis in a diabetic patient without a history of trauma. *Am J Dermatopathol.* 2015; **37:** e112-5.
214. Zhang J, Xi L, Lu C, *et al.* Successful treatment for chromoblastomycosis caused by *Fonsecaea monophora*: A report of three cases in Guangdong, China. *Mycoses.* 2009. **52:** 176-81.
215. Zhu CY, Yang YP, Sheng P, *et al.* Cutaneous chromoblastomycosis caused by *Veronaea botryosa* in a patient with pemphigus vulgaris and review of published reports. *Mycopathologia.* 2015; **180:** 123-9.

CHAPTER 14

Phaeohyphomycosis

Phaeohypho**mycosis** is subcutaneous and systemic infection, caused by various heterogeneous groups of **phaeoid** (dematiaceous) fungi. These fungi are found in hyphal form in tissues and not as thick-walled muriform cells as seen in chromoblastomycosis. The disease has been defined on histopathological grounds rather than clinical entity. Some of the phaeoid fungi are involved in both, phaeohyphomycosis and chromoblastomycosis however clinical and histopathological findings are distinct which is mainstay for final diagnosis. The description of disease is based on case reports of patients, who are infected by various phaeoid fungal agents.

The term phaeohyphomycosis is derived from Greek word ***'phaios'*** meaning dark and refers to brownish-black color of fungi *in vivo* as well as *in vitro*. It comprises group of infections ranging from superficial, cutaneous or subcutaneous to disseminated invasive disease whose etiological agents produce yeast-like cells, pseudohyphae and/or septate hyphae in tissues but certainly not the sclerotic bodies.

Historical Perspective

The phaeoid fungi are heterogeneous groups of organisms unified by their production of melanin pigments. **Ajello** *et al*, in 1974 originally coined this term **phaeohyphomycosis** to describe superficial, cutaneous and subcutaneous or systemic infections caused by specific group of phaeoid fungi in which tissue invasion forms were both filamentous and pigmented. Now, the term phaeohyphomycosis is well accepted and widely used in clinical practice of Medical Mycology. It combines clinical features and histopathological findings where these fungi are seen in tissue sections as brownish hyphal elements, pseudohyphae, yeast cells or combination of these forms, irrespective of their taxonomical identity.

The first case of cutaneous infection due to *Alternaria alternata* was described in 1933 by Barsok, however, it was without histological documentation. The first case of this disease as phaeohyphomycosis was reported by Binford *et al*, in 1952. There was confusion prevailing among the scientific community about nomenclature of this clinical entity. Therefore, in 1974, Ajello defined phaeohyphomycosis, limiting it to subcutaneous and systemic infection caused by dark-walled hyphae in the tissue as well as in culture. Most of the cases reported as cutaneous phaeohyphomycosis are included under subcutaneous phaeohyphomycosis because causative fungi are seen deep to dermis. In contrast to phaeohyphomycosis, **hyalohyphomycosis** is fungal infection caused by **hyaline** fungi exhibiting hyphal tissue form and described in details in **Chapter 31**.

In 1981, the term phaeohyphomycosis was expanded by Ajello to include subcutaneous and systemic infections caused by all those members of **Hyphomycetes** and **Coelomycetes** of phylum **Deuteromycetes** whose tissue forms are filamentous and phaeoid. The recent DNA sequence data indicate that nearly all etiological agents of phaeohyphomycosis, chromoblastomycosis and phaeoid taxa of eumycetoma belong to single clade of **loculoascomycetes** fungi. The list of etiologic agents is increasing as number of new case reports are being added almost every day.

The nomenclature of *Ochroconis gallopavum* has been controversial. This is formerly *Dactylariaconstricta* var. *gallopava*. The fungus *O.gallopavum* was initially termed *Diplorhinotrichum gallopavum* and was later re-classified as *Dactylaria gallopava* in 1968. In 1983, organism was further re-classified and assigned to genus *Ochroconis*. However, this classification has not been entirely accepted. Salkin and Dixon used morphologic features and proposed *D.gallopava* be regarded as *D.constricta* var. *gallopava*. Therefore, some investigators still prefer to use genus *Dactylaria*.

Some of the phaeoid fungi remained confined to certain geographical areas *e.g. Rhinocladiella mackenziei* (formerly *Ramichloridium mackenziei)* used to be found in Middle-East countries but now it has been reported outside also. Badali *et al*, reported it from India in 2010 and subsequently reports came from Afghanistan by Cristini *et al*, Pakistan by Jableen *et al* and other countries as well.

In September 2012, the CDC began investigating an outbreak of fungal meningitis among patients who had received contaminated preservative-free methyl prednisolone acetate injections from the New England Compounding Center in Framingham, Massachusetts (USA). In this outbreak, thousands of patients were potentially exposed to the tainted corticosteroids and 752 cases (64 deaths) of *Exserohilum* CNS and joint infections were reported while treating back pain with a national crisis. However, establishing diagnosis of fungal meningitis during this nationwide outbreak was slightly difficult because little was known about the natural history of such an unexpected disease in a developed country.

Mycology

The fungi involved are phaeoid in some part of their life cycle and are characterized by brown melanin pigment in the cell walls. The dark pigmentation is caused by deposition of dihydroxynaphthalene melanin. The classification of phaeoid fungi is complex because of increasing number of pathogenic species and frequent changes in nomenclature. There are about 100 different organisms that have been commonly implicated in phaeohyphomycosis. The frequently changing taxonomy of agents of phaeohyphomycosis is difficult, especially for non-specialist in the field of Medical Mycology. They vary taxonomically and include members in form-family **Dematiaceae** of **Hyphomycetes**, form-order **Sphaeropsidales** of **Coelomycetes** and various unitunicate **Ascomycetes** and other groups. The taxonomical status of the majority of causative agents of phaeohyphomycosis is placed in family **Herpotrichiellaceae** of order **Dothideales** of class **Bitunicate Pyrenomycetes** *i.e.* **Loculoascomycetes**. Some of species like *Chaetomium* are placed in **Unitunicate Pyrenomycetes** under order **Sordariales**.

Phaeohyphomycosis comprise vast array of opportunistic fungal infections characterized by presence of different types of melanized fungal elements in tissue. These can be clearly detected with Masson-Fontana stain because they become darkly colored in brown background in contrast to fungi present in hyalohyphomycosis which remain colorless.

Table 14.1. Morphological Grouping of Dematiaceous Fungi.

Type of Conidia	*Dematiaceous Fungi*
Muriform macroconidia	*Alternaria, Ulocladium, Epicoccum, Stemphylium* species
Macroconidia with transverse septae	*Curvularia, Drechslera, Ochroconis gallopava*
Microconidia	*Cladosporium, Phialophora verrucosa, Fonsecaea pedrosoi, Exophiala dermatitidis*
Miscellaneous	*Exophiala dermatitidis, Chaetomium* species

Over the years, the list of fungi causing phaeohyphomycosis has grown considerably and will likely continue to expand as number of agents are being added regularly on the basis of cases reported in the literature. At present more than 100 species of diverse phaeoid fungi belonging to 55 unrelated genera have been reported as causative agents of phaeohyphomycosis. The most common genera involved in phaeohyphomycosis are *Bipolaris, Curvularia, Exserohilum, Alternaria, Exophiala, Phialophora, Cladophialophora, Cladosporium, Chaetomium* species and *Ochroconis (Dactylaria)* species. In addition, *Phialophora parasitica, Curvularia pallescens, Nattrassia mangiferae* have also been rarely reported **(Table 14.1)**.

The genus *Phialemonium*, proposed by Gams and McGinnis in 1983 for fungal intermediate between genera *Acremonium* and *Phialophora* and *Phialemonium curvatum* is involved in phaeohyphomycotic cyst. The genus *Phaeoacremonium* is newly established pathogen of phaeohyphomycosis that is intermediate between genus *Acremonium* and *Phialophora*. *Rhinocladiella mackenziei* (formerly *Ramichloridium mackenziei)* used to be reported mainly from the Middle East countries and being encountered outside this zone.

All species of phaeoid fungi do not belong to Hyphomycetes as genus *Phoma* belong to Coelomycetes. *Phoma* is common plant pathogen but very rarely reported as human pathogen causing subcutaneous phaeohyphomycosis. Some of the species of this genus like *Phoma hibernica, Phoma minutella* and *Phoma sorghina* have been reported in human cases.

The agents that cause this infection are primarily recognized as soil saprotrophs, plant pathogens and contaminants living in the environment. With growing number of patients who are immunocompromised, phaeohyphomycotic agents are increasingly being reported as cause of human disease. More than 100 species belonging to at least 60 genera are known to be agents of phaeohyphomycosis. Of these species, many have been reported only occasionally, while others such as *Bipolaris, Curvularia, Exserohilum* and *Alternaria* species, have frequently been involved in human infections. The phaeoid fungi can grow/tolerate higher temperature.

Some scientists think that *Scedosporium* and *Scopulariopsis* species should be included in phaeohyphomycosis rather than hyalohyphomycosis. Their argument is that some element of melanin is produced by these fungi during their life cycle. There are case reports in the literature to this effect also. Patel *et al,* have described a case of phaeohyphomycosis caused by *Scopulariopsis brumptii* in a liver transplant recipient in 1994. However, others are contesting it and traditionally these fungi are being described as hyalohyphomycosis and not phaeohyphomycosis. In this Textbook such fungi of gray-area are described in **Chapter 31** on **Hyalohyphomycosis.**

Epidemiology

The phaeoid fungi are widely distributed in nature as contaminant and are not common human pathogens. These fungi are found in soil, decaying vegetation and rotten woods. These saprotrophic fungi are ubiquitous in nature and are being recognized with increasing frequency as cause of human diseases. The observation may, in part, be attracted to improvements in diagnostic microbiological techniques for identification of these organisms. However, expanding population of immunocompromised patients is likely to be responsible for an increase in number of infections due to phaeoid fungi in future.

In 1983, McGinnis redefined boundary of phaeohyphomycosis and included many well-established and distinct clinical entities such as tinea nigra, black piedra, eumycetoma and mycotic keratitis but the original concept of Ajello is still being followed.

Some of the phaeoid fungi are having affinity to the CNS, particularly *Cladophialophora bantiana.* Moreover, *Ochroconis gallopavum* is also known for its neurotropic potential and predilection to cause severe necrotizing encephalitis in humans, cats and birds, including poultry, grey-winged trumpeters (*Psophiacreptans*) and Japanese quail (*Coturnixcoturnix japonica*). Infections in fish, amphibians and reptiles have also been reported.

A few phaeoid fungi remained confined to certain geographical pockets *e.g. Rhinocladiella mackenziei* used to be found prevalent in Middle-East countries mainly causing cerebral phaeohyphomycosis however now it is being reported outside this belt also. Badali *et al,* have reported this fungus from India in 2010 and subsequently reports came from Afghanistan, Pakistan and other countries. This shows how the epidemiology of phaeoid fungi is rapidly changing by involving larger spectrum of population.

In late September 2012, a multistate outbreak of CNS fungal infection and septic arthritis was caused due to *Exserohilum rostratum* in the USA. There were over 700 patients who received epidural injections of methylprednisolone produced at a single compounding center and subsequently developed meningitis with or without posterior circulation stroke and/or spinal or para-spinal infection, and more than 30 patients who received intra-articular injections of the same drug developed osteoarticular infections.

This disastrous outbreak caused by the injection of corticosteroid contamination primarily by *Exserohilum rostratum* occurred in a Massachusetts compounding pharmacy. This mold, a plant pathogen, has rarely been associated with human infection and might be considered as non-virulent. Virulence, however, is not really a property of an organism *per se* but rather a combination of factors occurring at the pathogen/potential host interface. Clearly, by introducing this organism into the body with corticosteroids (which can affect the local immune response), it can produce quite serious and potentially lethal infections, as occurred in this large outbreak of USA.

This is surprising, with as many positive blood or catheter cultures that occurred and the cluster was not recognized soon. It is vitally important that compounding pharmacies practice extreme infection control to prevent such potentially lethal complications. It is not clear how often the compounding pharmacy was inspected in regard to its infection control practices.

In recent times many species of phaeoid nature have been implicated in phaeohyphomycosis. Some of them are described here but rest are appended in the Further Reading for a ready reference. Chander *et al,* in 2017 have reported *Rhytidhysteron rufulum* as an emerging phaeoid agent and there are more cases of this fungus from India and other parts of the world. Similarly, *Thielavia subthermophila,* is also encountered as one of the emerging phaeoid fungus as reported by Badali and his colleagues.

Immunity

The host factors, predisposing to this infection, have not been properly studied. However, disease is found more in immunocompromised patients. Due to the lack of immunological studies, serological tests are also not available for establishing diagnosis of phaeohyphomycosis. The role of melanin is considered as the most significant in this group of fungi.

Pathogenesis and Pathology

Multiple stellate abscesses progress to single circumscribed lesion with central cavity filled with pus and surrounded by fibrous wall. The margins of these abscesses and granulomas are composed of giant cells, epithelioid cells, histiocytes, plasma cells and lymphocytes. The fungi are found in or adjacent to purulent areas. Despite phaeoid nature of causative fungi, brown pigment may not always be apparent and hyphae may also appear hyaline in lesions in H&E stain. The GMS staining masks natural brown color of phaeoid hyphae. Therefore, confirmation of presence of these hyphae can be achieved by using melanin-specific technique, such as Masson-Fontana staining.

In *Cladophialophora bantiana* besides formation of the most common phaeomycotic cyst there may be pseudoepitheliomatous hyperplasia, mixed inflammation with granulomatous component and intraepidermal neutrophilic microabscess formation.

Clinical Features

The fungal agents causing phaeohyphomycosis are mostly plant pathogens found in soil, infecting subcutaneous tissue by producing solitary lesions. There are four clinical types, which have been described on the basis of involved anatomical site and degree of tissue invasion by causative fungi. The lesions are superficial - confined to stratum corneum; cutaneous/corneal - invasion and destruction of keratinised tissue; subcutaneous and disseminated - generally occurring in immunocompromised hosts and are associated with high mortality.

Superficial phaeohyphomycosis is caused by *Hortaea werneckii* and known as tinea nigra, is described in **Chapter 8** and other types have been dealt in their respective Section.

According to phaeoid fungus involved and anatomical site affected phaeohyphomycosis can be classified as follows:

1. Cutaneous Phaeohyphomycosis
2. Subcutaneous Phaeohyphomycosis
3. Invasive and cerebral Phaeohyphomycosis
4. Paranasal Sinus Phaeohyphomycosis

1. Cutaneous Phaeohyphomycosis

The colonization of skin by phaeoid fungi may be seen over cracked, fissured areas of sole of feet, ulceration and lichenified or eczematized skin. The lesions are reported from face, upper part of chest including neck and leg of a patient as shown in **Figures 14.1A to C**.

2. Subcutaneous Phaeohyphomycosis

This is localized cutaneous infection following deep inoculation of fungus into subcutaneous tissues. The lesions are found on feet, legs, hands, arms, back or other body sites. The nodules develop as indolent, usually small subcutaneous masses. The subcutaneous masses may be over the face involving maxillary area as seen in **Figures 14.2A and B**. The lesions are often palpable while skin and tissue surrounding cyst are relatively normal (phaeomycotic cyst). There may be cystic lesions of solitary nature and present at various anatomical sites **(Figs. 14.3A to E)**. Infection has also occurred following intramuscular injections.

3. Invasive/Cerebral Phaeohyphomycosis

The invasive and systemic phaeohyphomycosis are rare but most serious forms of disease. Cerebral phaeohyphomycosis often presents with brain abscess that results from hematogenous spread of neurotropic phaeoid fungus *Cladophialophora bantiana*. There are few reports of extracranial infections.

The patients are usually healthy before infection sets in but may also be immunocompromised due to any underlying condition. The symptoms and signs of infection are largely non-specific and are reflection of pressure effect of lesions like headache and papilledema followed by hemiparesis. The localizing signs vary greatly and are as a result of tissue destruction.

The duration of symptoms varies, ranging from weeks to as long as five years. The patients have variety of underlying conditions and specific symptoms that reflect size and location of lesion. The frontal lobe is mostly affected and encapsulated abscesses filled with brown hyphae are

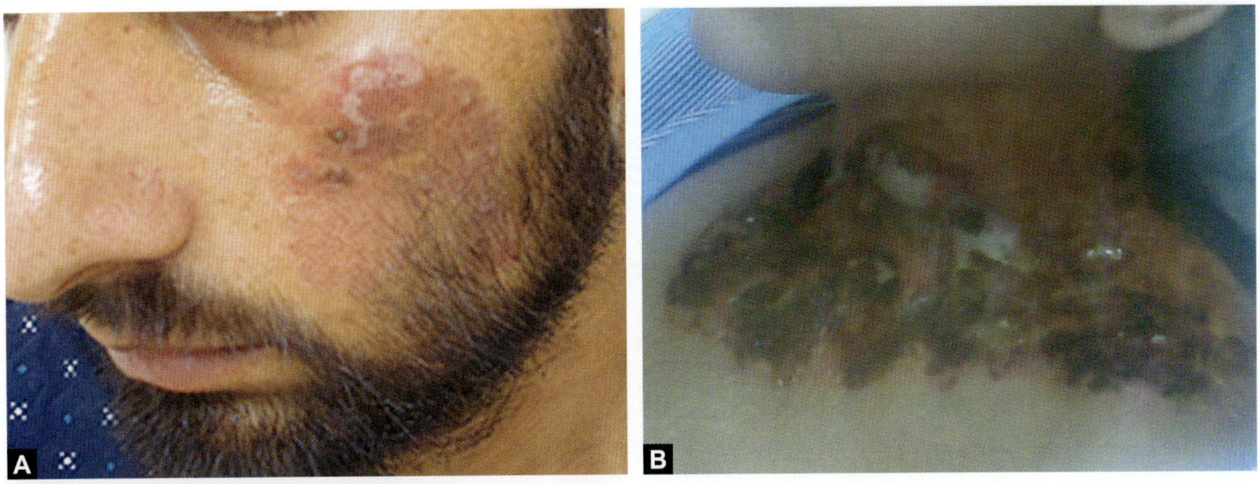

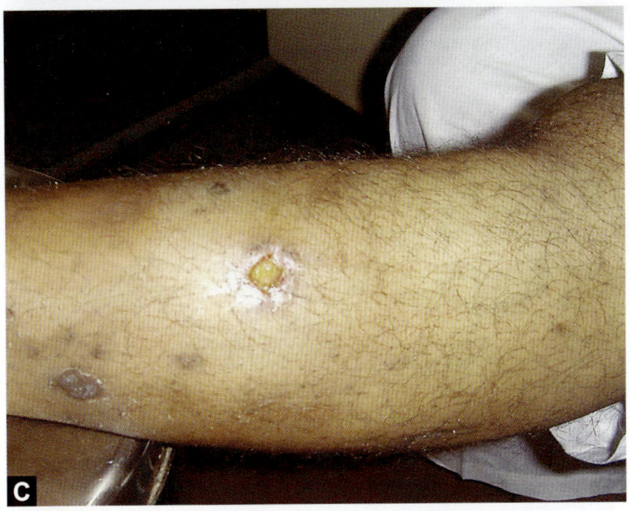

Figs. 14.1A to C. Extensive cutaneous phaeohyphomycosis over (A) Face (B) neck and upper part of chest in a child and (C) leg.

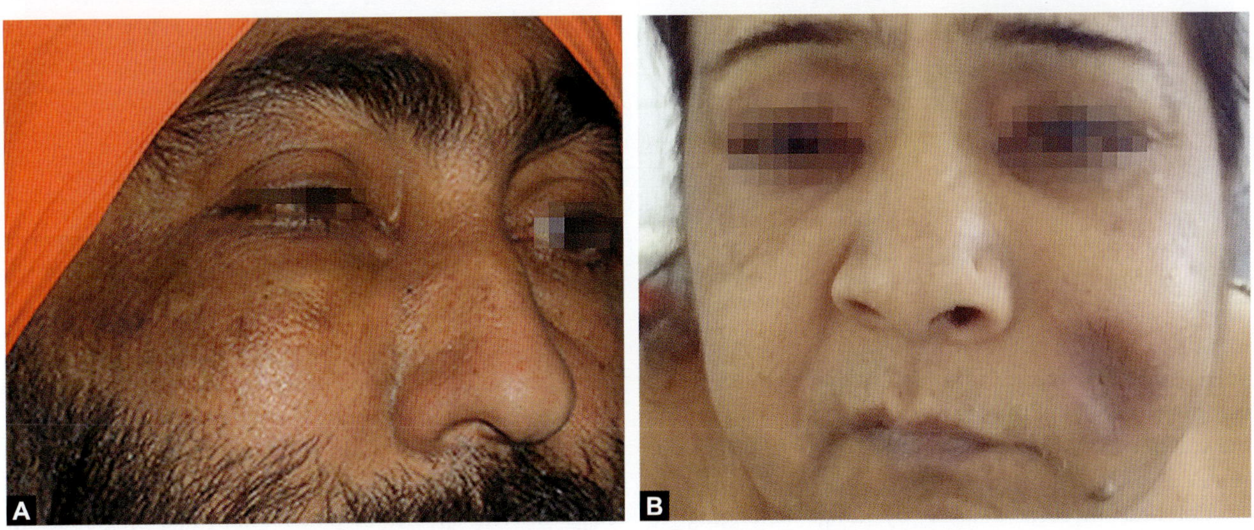

Figs. 14.2A and B. Subcutaneous phaeohyphomycosis of face on the right and left sids of the patients.

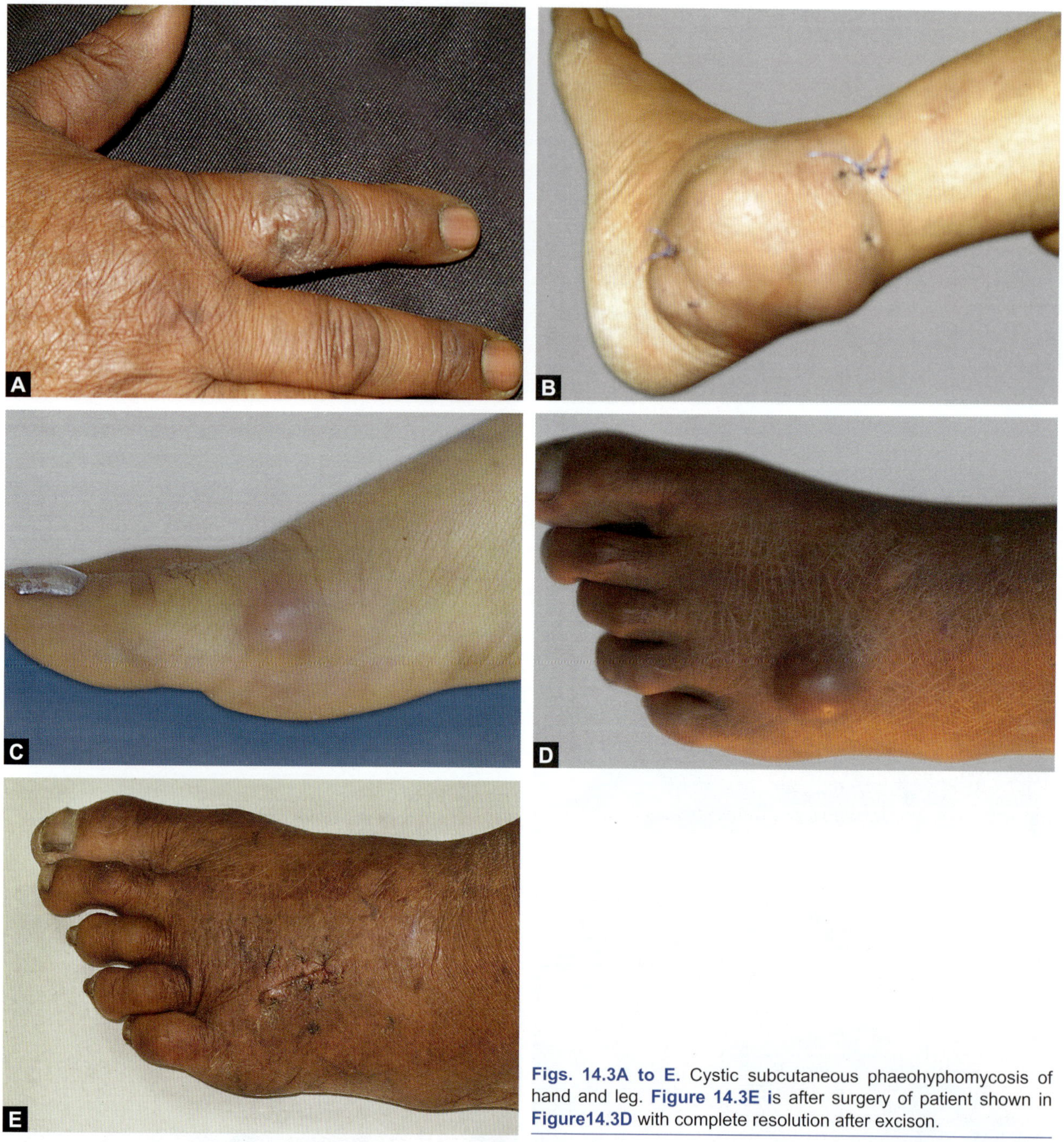

Figs. 14.3A to E. Cystic subcutaneous phaeohyphomycosis of hand and leg. **Figure 14.3E** is after surgery of patient shown in **Figure14.3D** with complete resolution after excison.

present in nervous tissue. In cerebral phaeohyphomycosis, computed tomography is significantly helpful in assessing the extent of involvement before and after surgical excision.

Cerebral mycosis caused by fungi belonging to genus *Chaetomium* is an emerging group of opportunistic pathogens in intravenous drug abusers and renal transplant recipients. *Chaetomium* species are traditionally found as soil fungi and are more often implicated in cutaneous infections and occasionally in systemic phaeohyphomycosis. Several species of *Chaetomium* like *Ch.globosum, Ch.funicolum, Ch.strumarium* and *Ch.atrobrunneum* have

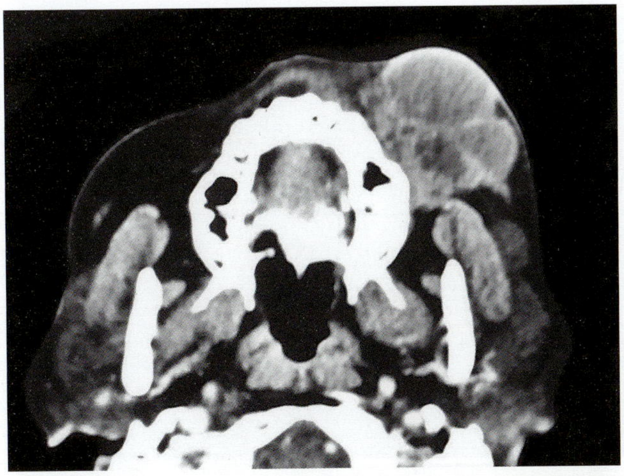

Fig. 14.4. CT scan of subcutaneous phaeohyphomycosis of face of the patient shown in Figure 14.2B.

been reported as causative agents of opportunistic infections in humans. *Chaetomium atrobrunneum* has also been isolated from brain abscess.

The fungal infections of CNS are often misdiagnosed as pyogenic meningitis, cerebral malaria, intracranial hemorrhage and space-occupying lesions.

4. Paranasal Sinus Phaeohyphomycosis

This is an indolent disease that may remain confined to the sinus cavity or spread to adjoining structures taking over months to years. The most common species involved are from two fungal genera *i.e. Bipolaris* and *Curvularia,* in addition to other phaeoid organisms. Computed tomography is helpful in establishing the diagnosis of sinusitis, particularly delineating the extent of involvement.

Radiodiagnosis

Deep-seated phaeohyphomycotic lesions are localized by imaging techniques like computed tomography, magnetic resonance imaging that can be subsequently confirmed by fine needle aspiration cytology or biopsy of lesion and fungal cultures. Both T1 and T2-weighted images with and without contrast enhancement are suggested and in central nervous system there are ***'double ring sign'***. The cerebral phaeohyphomycosis show typically an irregular, variably contrast enhancing mass with surrounding hypodense edema, pattern mimicking metastatic lesions or high-grade gliomas. The CT Scan of subcutaneous phaeohyphomycosis of face of the patient shown in **Figure 14.2B** is depicting soft tissue swelling over the left facial area **(Fig. 14.4)**.

Differential Diagnosis

Clinically phaeomycotic cyst is mistaken for ganglion, fibroma and lipoma, which can be elucidated on direct microscopy and histopathological examination of material taken from lesion. In CNS involvement, these presentations are clinically labeled as bacterial brain abscess before final diagnosis is made. The diagnosis of other fungal diseases like cryptococcosis, histoplasmosis, coccidioidomycosis and sporotrichosis may also be considered as differential of phaeohyphomycosis.

On clinical grounds, cutaneous *Alternaria* infection may simulate different granulomatous diseases, such as foreign body granuloma, sporotrichosis, leishmaniasis, subcutaneous sarcoidosis, infundibular cysts, xanthoma and other deep mycoses.

Laboratory Diagnosis

The diagnosis of phaeohyphomycosis is established in the laboratory by direct demonstration and culture of phaeoid fungus involved in various types of lesions of this disease. The serological tests or experimental studies in animals are really not available. The following protocol is followed for reaching the final diagnosis:

(a) Direct Examination

The aspirates from cysts and curettings from plaques, nodules and drained abscesses are examined in KOH wet mounts **(Figs. 14.5A to E)**. These fungi are usually pigmented and dark brown in color. This feature is quite variable and sometimes even hyaline mycelial strands may be visible therefore, special stains like Masson-Fontana are used to elicit presence of melanin in the hyphae so that diagnosis of phaeohyphomycosis is confirmed.

The fungal hyphae are 3-4 µm in diameter in variable shapes and sizes, with cyst cells up to 25 µm in diameter. The fungi are seen as short or elongated hyphal forms, irregularly swollen, septate, branched or unbranched, yeast-like cells, single or in short chains and occasionally as spherical elements that may resemble chlamydospores. The sclerotic bodies, as seen in chromoblastomycosis and grains as seen in eumycetoma, are absent in phaeohyphomycosis.

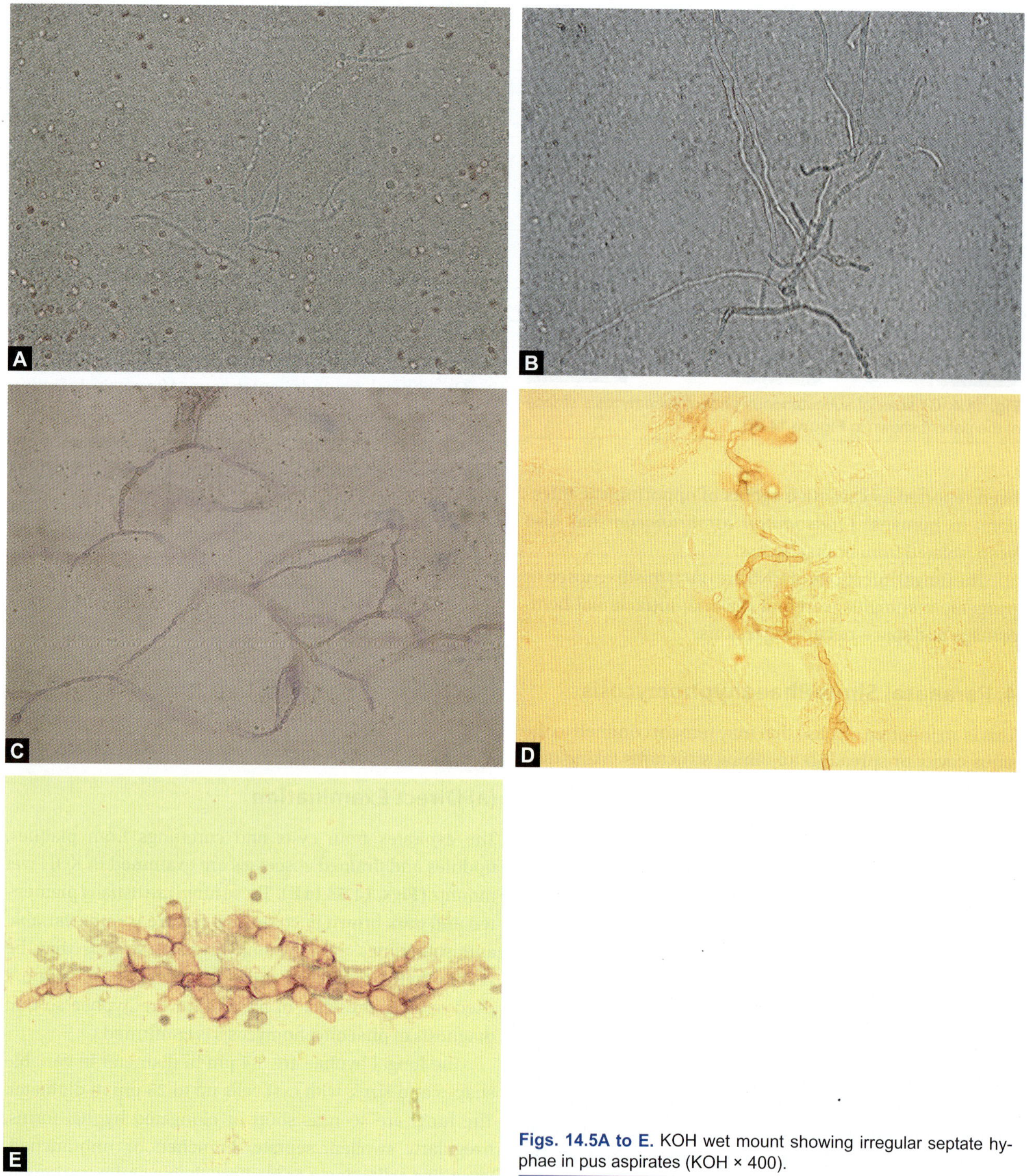

Figs. 14.5A to E. KOH wet mount showing irregular septate hyphae in pus aspirates (KOH × 400).

All phaeoid fungi are strongly stained with Giemsa-positive in cytoplasm and walls as non-phaeoid fungi. Hence Giemsa stained (MGG) smear is prepared from the FNAC, showing the phaeoid fungi **(Figs. 14.6A to D)**. In the routine hematoxylin and eosin (H&E) these fungi are visualised in the tissue sections **(Figs. 14.7A and B)**. The

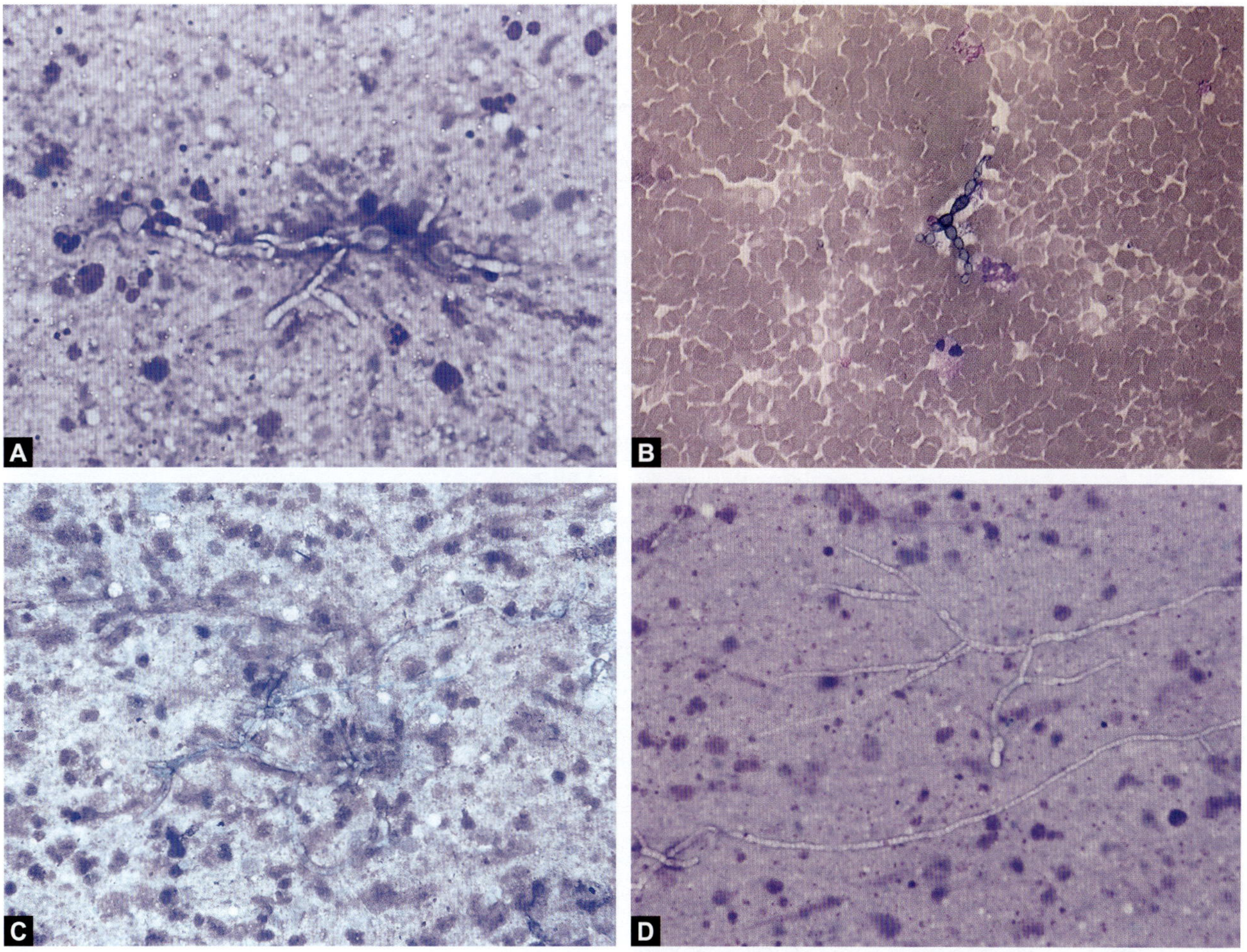

Figs. 14.6A to D. Giemsa staining showing septate hyphae with different morphological features, swollen and narrow, yeast-like swellings and there is no consistency in the breadth of the hyphae (MGG × 400).

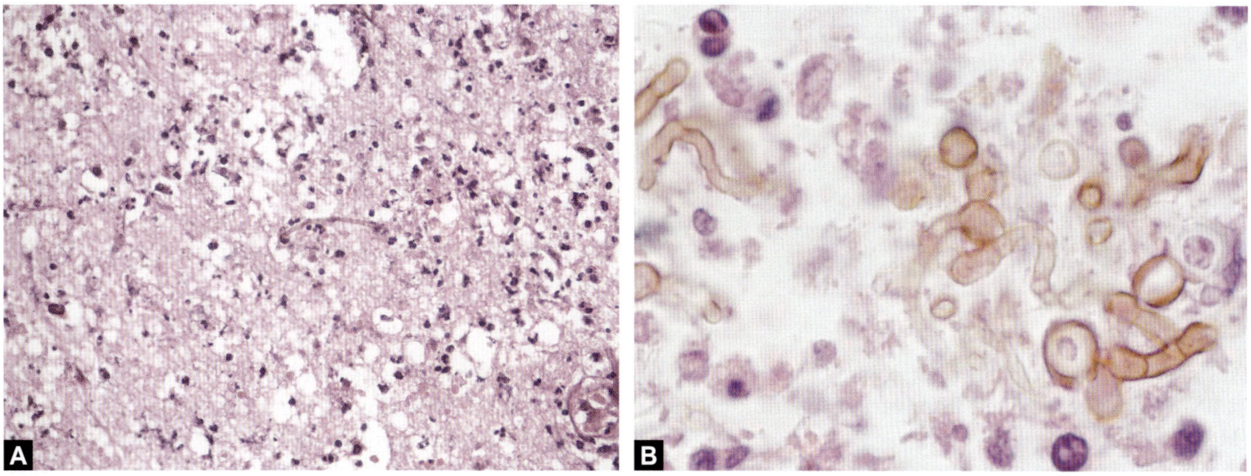

Figs. 14.7A and B. Histopathological sections showing irregular septate hyphae (H&E × 200 / 400)..

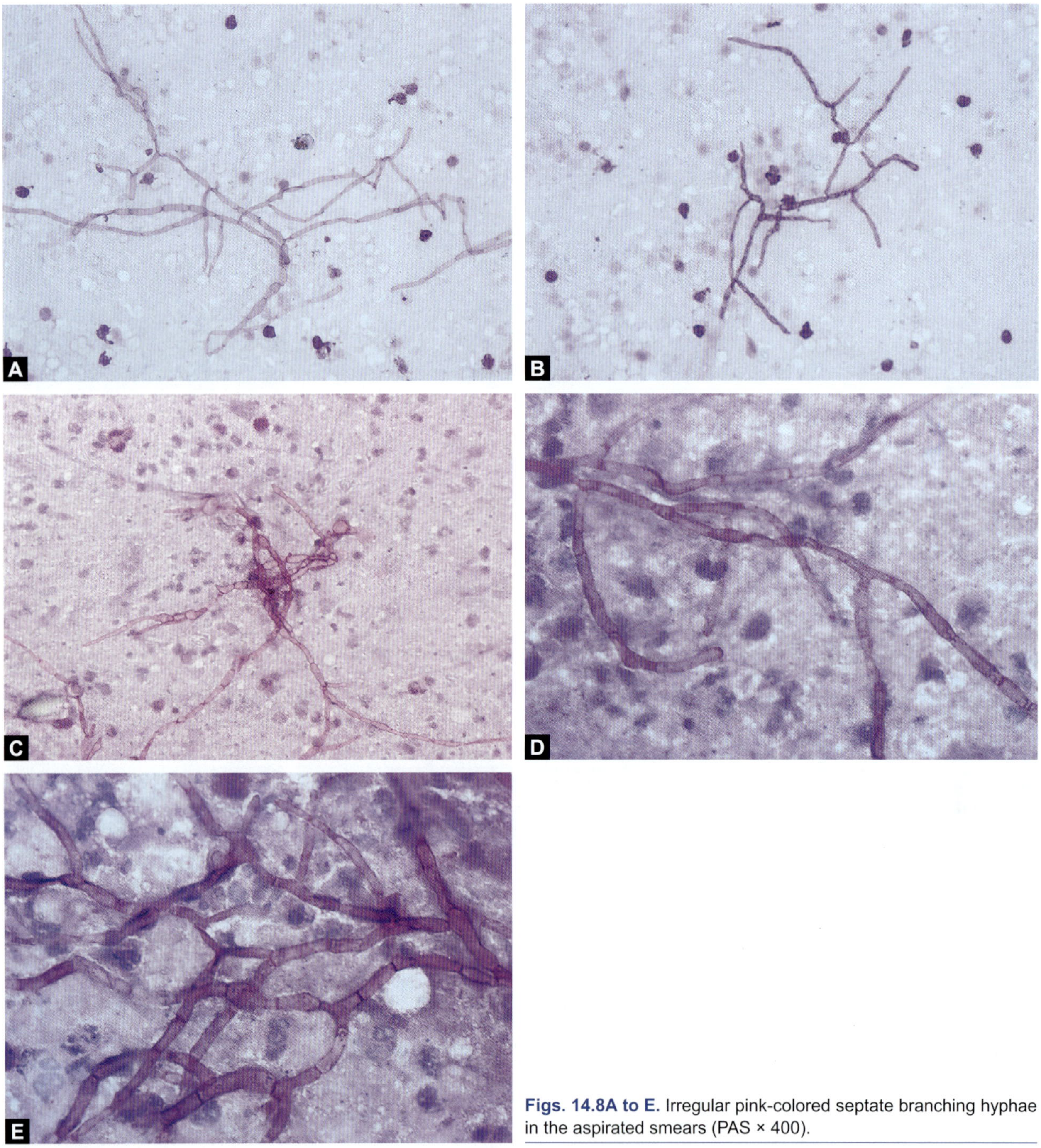

Figs. 14.8A to E. Irregular pink-colored septate branching hyphae in the aspirated smears (PAS × 400).

PAS staining showing the pink colored septate hyphae in FNAC smears prepared from various anatomical sites **(Figs. 14.8A to E)**.

Masson-Fontana stain specifically shows presence of melanin in cell walls and septa of hyphae of phaeoid fungi with lowered microscope condenser. This is also pertinent to note that GMS stains all the fungal elements as black with greenish background, whereas MF stain only take up the phaeoid fungi as black in color with brownish background. Masson-Fontana Staining showing the septate

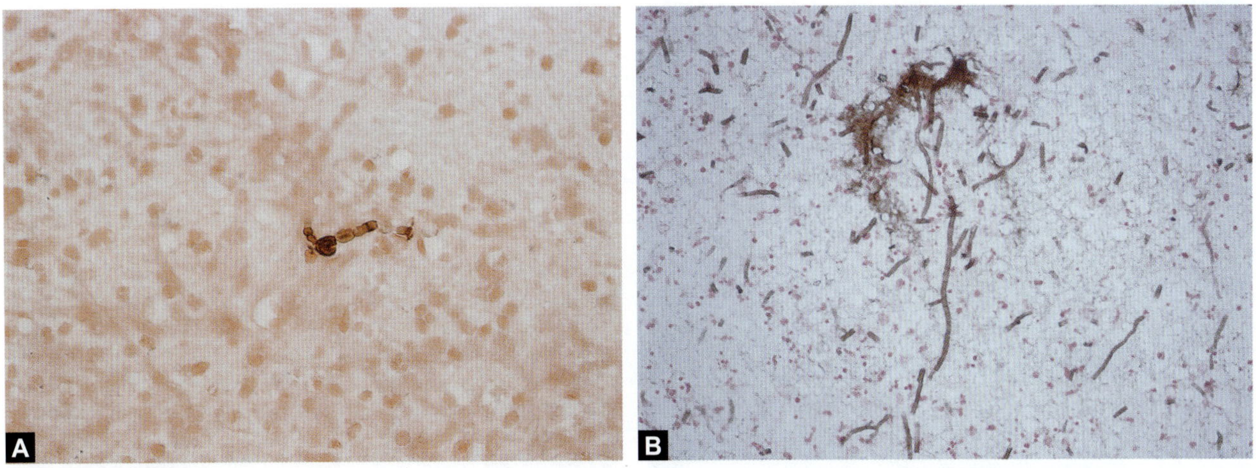

Figs. 14.9A and B. Irregular blackish septate branching hyphae in the aspirated smears with brown background (MF Stain × 400).

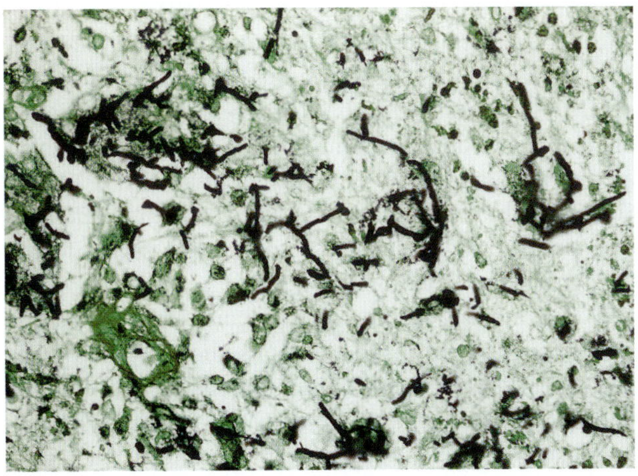

Fig. 14.10. Irregular septate branching hyphae in tissue section (GMS × 400).

hyphae in the tissues **(Figs. 14.9A to B)**. The GMS staining showing black colored hyphae in the brain tissue of a patient **(Fig. 14.10)**.

(b) Fungal Culture

The specimen from clinical lesion is plated on SDA with cycloheximide. They are incubated at 25°C and 37°C for a period of four weeks. Most of these fungi grow slowly but in some cases growth may be visible by 1-2 weeks.

In the beginning, the growth sometimes show white mycelia and not suspected to be as phaeoid type but after 7-10 days it starts changing its color as brownish-black. Similarly, some of these phaeoid agents grow initially as whitish yeasts and in due course of time turn as ***'black yeasts'***. On prolonged incubation they become mycelial in nature. These are: *Aureobasidium pullulans*, *Exophiala dermatitidis*, *Exophiala jeanselmei* and *Exophiala spinifera*. The black yeast as such is not any formal taxonomic group but this term is applied to a wide range of unrelated ascomycetous as well as basidiomycetous fungi that are able to produce budding cells at some stage in their life cycle **(*See* Figs. 8.1 and 8.2)** On the pattern of black yeasts, the terminology as black molds is used. This term is not scientific but is widely used to refer *Stachybotrys* molds (also "toxic black mold").

These phaeoid fungi are also described in other **Chapters (Nos. 8, 9, 13, 29, 33, 34 and 36)**. However, ultural characteristics of some of the commonly encountered ones involved in phaeohyphomycosis are given below in an alphabetical order:

1. *Alternaria* Species: From more than 80 species of *Alternaria* identified, only 8 have been involved in human or animal infections. The most common species implicated in human disease are *A.alternata, A.dianthicola, A.chartarum* and *A.tenuissima*. *Alternaria* has been implicated in several cases of keratomycosis, osteomyelitis, rhinosinusitis, peritonitis or granulomatous pulmonary nodules. *Alternaria infectoria*, on the basis of its very small 1 or 2-celled conidia often arranged in long chains and presenting with very long secondary conidiophores. Its morphological and microscopical descriptions are also given in **Chapters 2, 33 and 36 (*See* Figs. 2.14A, 33.1, 33.2 and 36.7)**.

2. *Bipolaris* Species: *Bipolaris* species, previously classified in genus *Drechslera*, are group of pigmented, phaeoid,

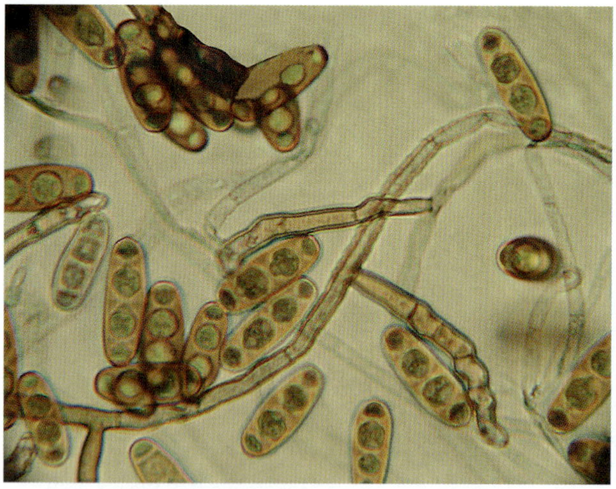

Fig. 14.11. *Bipolaris* species showing sympodially-arranged conidia with distoseptae (LCB × 400).

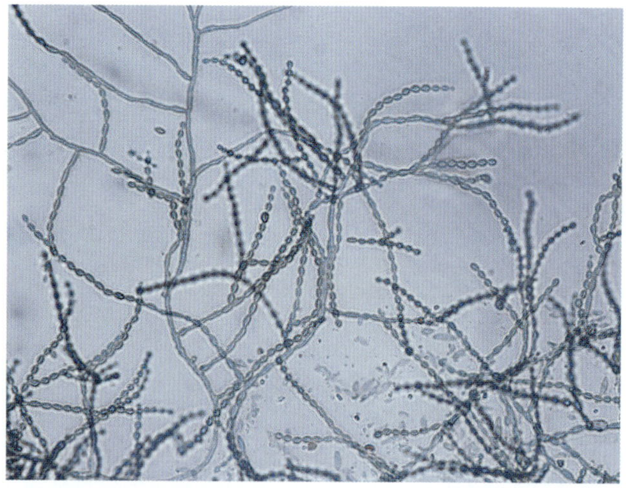

Fig. 14.12. Branching long conidial chains of *Cladophialophora bantiana* (LCB × 400).

filamentous fungi ubiquitous in environment. They are recognized pathogens of plants but also cause infections, *albeit* rarely, in animals and humans. The characters of hilum and germ tube production are important features. *B.spinifera* has less number of false septa as compared to the *B.hawaiiensis*, which has more septa and can tolerate higher temperature of 43°C **(Fig. 14.11)**.

3. *Chaetomium* Species: These are widespread worldwide in soil and plant debris and belong to broad genus of saprotrophic ascomycetes. These are characterized by densely hairy, egg-shaped fruiting bodies (perithecia) containing asci, which in turn enclose 4-8 brown spores (ascospores). The perithecial hairs can take variety of forms, depending upon the species. The ascospores collect in dense mass outside perithecium. Human cases of *Chaetomium* infection are relatively rare but case reports are increasing in the literature and new species such as *C.atrobrunneum, C.strumarium, C.globosum* and *C.perlucidum* are implicated especially in immunosuppressed patients. The species are involved in systemic mycosis like subcutaneous fistulous nodules, peritonitis, chromoblastomycosis and onychomycosis and some of allergic conditions and mycotoxicosis.

4. *Cladophialophora bantiana:* This fungus is highly neurotropic and invariably prove fatal hence frequently isolated from brain abscess. The older names *Cladosporium trichoides, Cladosporium bantianum, Xylohypha bantiana* and *Xylohypha emmonsia* are synonyms of this species.

This was transferred to genus *Xylohypha* in 1986 because it produces conidiophores that are indistinguishable from vegetative hyphae. There are also one-celled blastoconidia without evident darkly pigmented hila, which are borne in long, sparsely branching chains **(Fig. 14.12)**. Subsequently de Hoog *et al,* transferred it to genus *Cladophialophora* in 1995. The colonies are spreading, slightly folded olive grey to brown in color.

There are now some of the reports implicating this fungus as an agent of subcutaneous phaeohyphomycosis and other systemic sites like lungs but mainly involves the brain. Chakrabarti *et al,* have reviewed 124 cases of cerebral phaeohyphomycosis caused by it in the recent past.

Cladophialophora bantiana and *Cladosporium carrionii* are morphologically alike, however, *C.carrionii* has slower growth rate and shorter conidia (2-3 × 4-5 µm *vs.* 2-2.5 × 4-7 µm, with some being 3 × 15-20 µm), is dermatotropic rather than neurotropic and has maximum growth temperature of 35-38°C in comparison to 42-43°C for *Cladophialophora bantiana.*

5. *Curvularia* Species: The colonies of *Curvularia* species are woolly, initially white becoming grayish-black with black reverse. The conidiophores are simple or branched, bent at points where conidia originate hence called as sympodial geniculate growth. The conidia (poroconidia) are 8-14 × 21-35 µm in size and have three to four septae, almost always curved at **third** cell from base, which is larger and darker than the others. The conidia are borne in acropetal pattern and usually have curved appearance, having broad points at either end. The cell in middle of

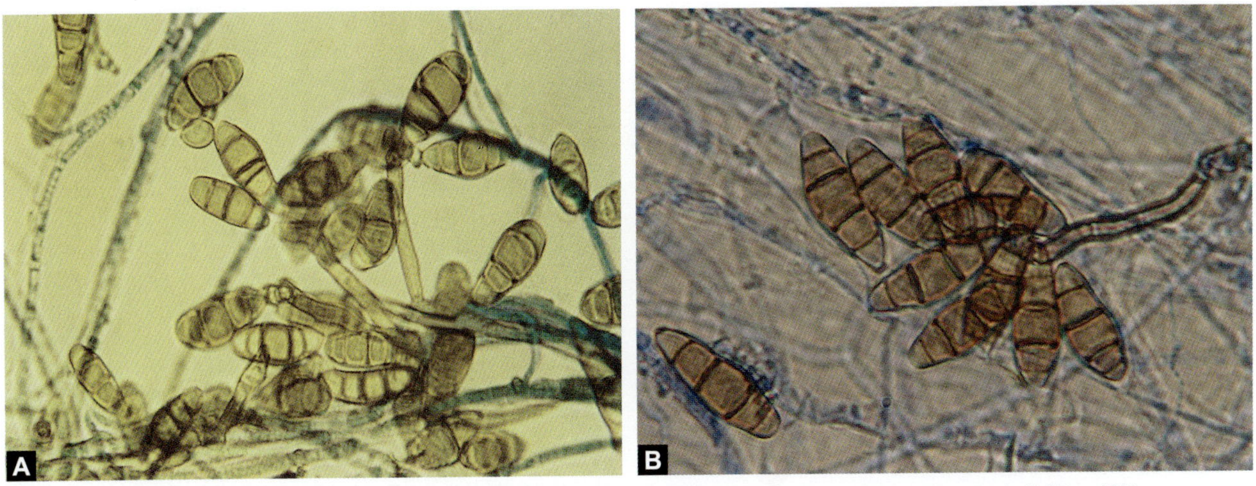

Figs. 14.13A and B. *Curvularia lunata* and *C.genticulalta* showing 3 and 4 septae in the conidia, respectively (LCB × 400).

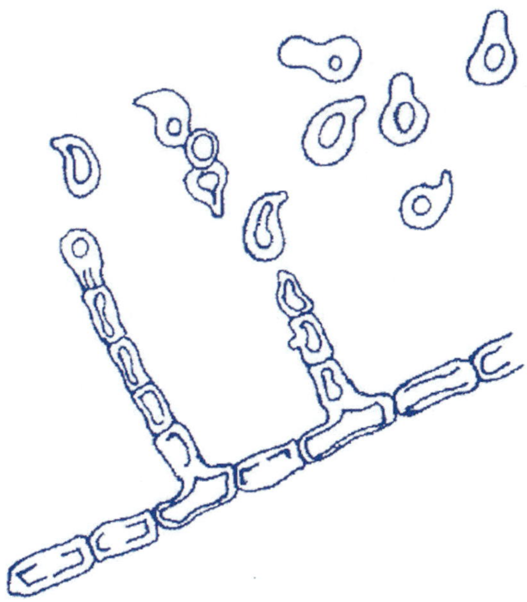

Fig. 14.14. Diagrammatic representation of *Exophiala dermatitidis*.

curve is often larger than those toward end hence name of genus. The conidia of *C.lunata* have 3 septae *i.e.* 4 cells, while those of *C.geniculata* mostly have 4 septae *i.e.* 5 cells. *Cochliobolus lunatus* is the teleomorph of *Curvularia lunata* **(Figs. 14.13A and B)**. Its morphological and microscopical descriptions are also given in **Chapters 2, 29 and 36 (*See* Figs. 2.14C and 29.14)**.

This genus is an environmental saprotroph hence found in water, soil and vegetables. *Curvularia* infections in humans are relatively uncommon despite ubiquitous presence of this soil dwelling dematiaceous fungus in the environment. Originally thought to be sole pathogen of plants, *Curvularia* has been described as pathogen of humans and animals during the last half-century. In addition to phaeohyphomycosis, it can cause onychomycosis, keratitis, rhinosinusitis, eumycetoma, pneumonia, ABPM, endocarditis, dialysis-associated peritonitis, cerebral abscess and disseminated fungal infections. Since majority of cases are immunocompromised hence it has also emerged as significant opportunistic pathogen in the recent past.

6. *Exophiala dermatitidis:* The young colonies of *Exophiala dermatitidis* are black, soft, moist, shiny and yeast-like, popularly known as **black yeasts**. Approximately after four weeks' time, colonies develop as mycelial growth that changes organism to filamentous form. This species exhibits yeast-like phase in its early growth and consists of unicellular elements that bud at one, two or three points. After few days, through germination of yeast-like cells, hyphal structure develops which spread out to form filamentous layer around central growth. Along the course of hyphae, numerous clusters of blastospores are produced by successive budding of cells from minute protuberances on hyphal wall **(Fig. 14.14)**. Previously it used to be known as *Wangiella dermatitidis*.

7. *Exophiala jeanselmei:* The colonies on SDA agar are black and slimy with many yeast-like cells. They become effuse, mycelial and covered with aerial hyphae. With prolonged maintenance, colonies become velvety and

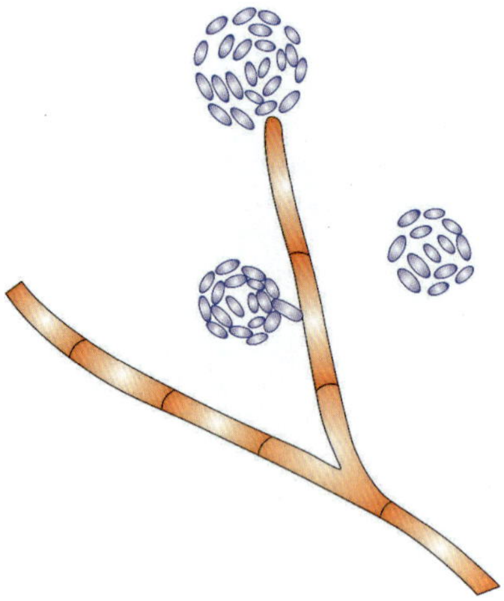

Fig. 14.15. Diagrammatic representation of *Exophiala jeanselmei.*

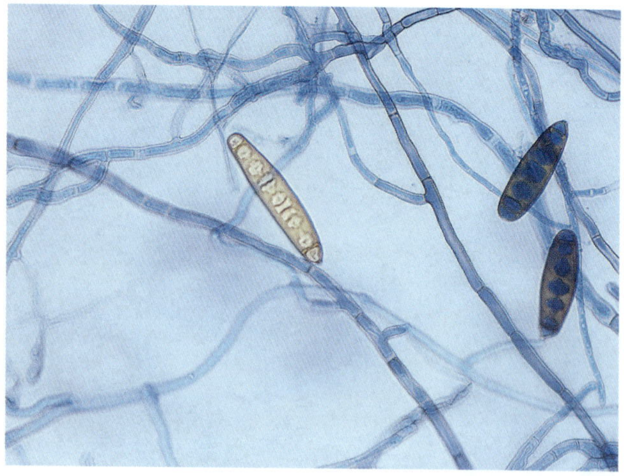

Fig. 14.16. Conidia of *Exserohilum rostratum* having distoseptae with truncated hilum and clear-cut septum near hilum (LCB × 400).

lose its slimy appearance. The hyphae are brown, septate and smooth producing branched or unbranched conidiophores laterally or on apex. The conidiogenous cells are brown, cylindrical annellides, which are either integrated with hyphae or produced at tip of conidiophore or hyphae. The conidia are one-celled, subglobose, elliptical to cylindrical, smooth, hyaline and aggregated in masses that tend to slide down conidiophore or along hyphae on microscopic mount **(Fig. 14.15)**.

8. *Exserohilum* Species: These fungi are ubiquitous and cosmopolitan pathogens, usually associated with soil and plants. The main portals of entry in humans are the airways by inhalation. Taxonomically, *Exserohilum*, together with *Bipolaris* and *Alternaria*, belongs to family Dematiaceae of order Moniliales. This is characterized by woolly, gray to blackish-brown, rapidly growing colonies with an olivaceous black reverse. The conidia are straight, curved or slightly bent, ellipsoidal to fusiform and are formed apically through a pore being poroconidia on a sympodially elongating geniculate conidiophore. They have distoseptate with a protruding and truncated hilum. There is clear-cut septum near the hilum, which helps in differentiation from *Bipolaris*. The common one is *E.rostratum*. In case of *E.mcginnis*, there is rough surface **(Fig. 14.16)**.

9. *Nattrassia mangiferae:* The colonies of *Nattrassia mangiferae* are effuse, hairy and dark grey to blackish-brown on oatmeal agar. Microscopically they are seen as chains of arthroconidia from undifferentiated broad, 4-7 µm, brown hyphae, often with slimy exudate and pycnidia. It is synanamorph of *Neoscytalidium dimidiatum* **(*See* Figs. 10.29A and B; 10.30)**.

10. *Ochroconis gallopavum:* The colonies of *Ochroconis gallopavum* on SDA, without cycloheximide, are velvety, greyish brown, with distinctive reddish brown pigment that diffuses into medium resulting into halo around colony. On microscopic examination, there are septate, sub-hyaline to brown hyphae bearing lateral short, slender tapering conidiophores sympodially bearing conidia on cylindrical denticles. The conidia are predominately two-celled, clavate to cylindrical in shape, smooth-walled, pale-brown and measure 11-18 × 2.5-4.5 µm **(Figs. 14.17A to D)**.

11. *Phaeoacremonium parasiticum:* The colonies are initially velvety and white-gray then develop as gray-brown pigment and clusters of brown aerial hyphae over a period of one week. LCB mount shows pigmented septate hyphae with brown, cylindrical, slightly tapered conidiophores with clusters of oblong, hyaline conidia **(Fig. 14.18)**.

12. *Ulocladium* species: *Ulocladium* species are usually recovered from soil. However, when encountered in clinical specimens, they are usually considered as contaminants; however, there are few reports of their involvement in cases of human phaeohyphomycosis. It is described also in **Chapter 36** and shown in **Figure 36.15**.

In addition to the above-mentioned phaeoid species, there are large number of fungi, which are encountered through case reports and now being regularly reported from lots of clinical presentations. Therefore, identification

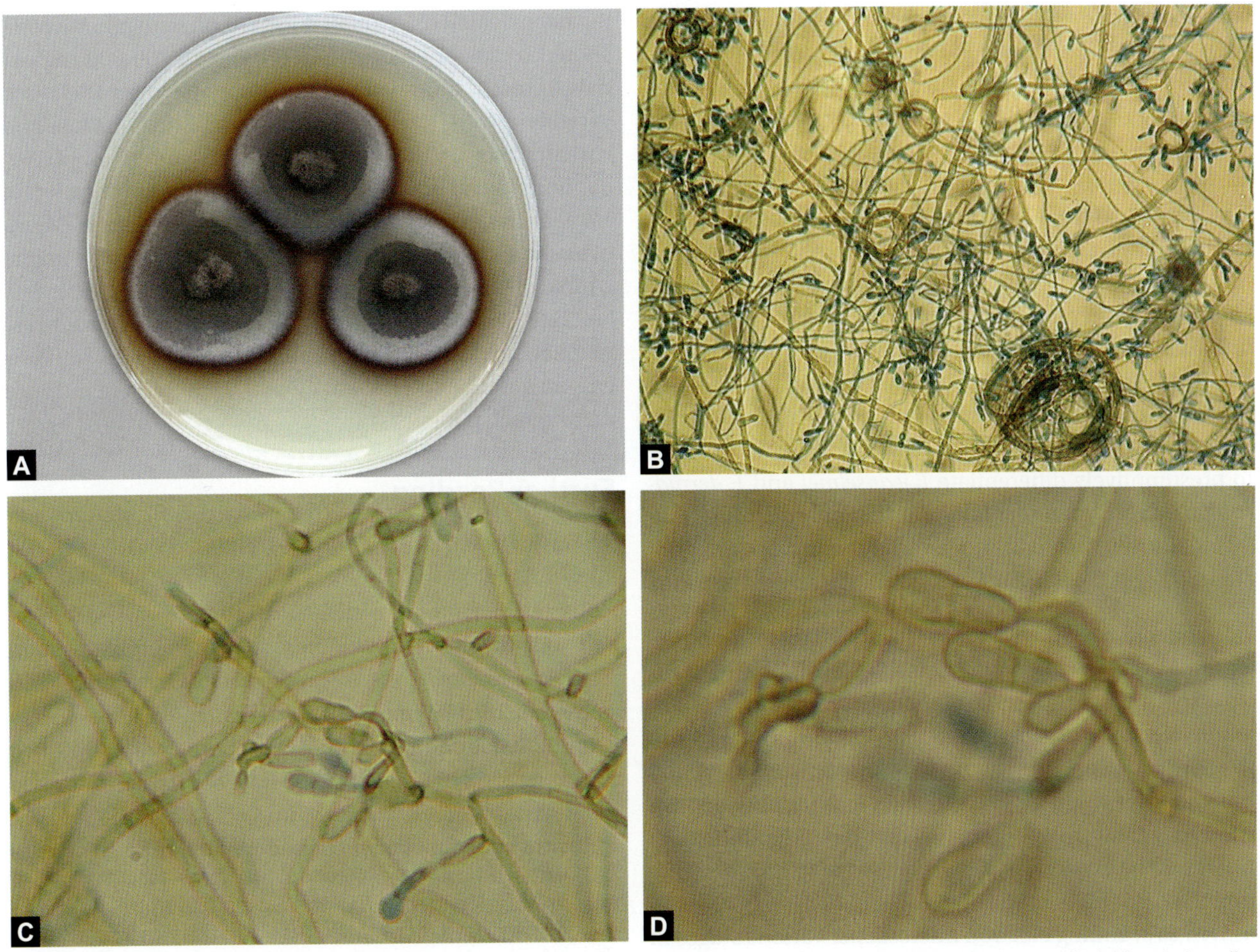

Figs. 14.17A to D. Velvety, grayish brown with distinctive reddish brown pigment diffusing into medium and two-celled, clavate to cylindrical conidia of *Ochroconis gallopavum* (LCB × 200 / 400).

Fig. 14.18. Cylindrical, tapered conidiophores with clusters of oblong conidia *Phaeoacremonium parasiticum* (LCB × 400).

should be made up to the species level for better management. If there is no sporulation, slide culture may be put up. If still there is no spore induction, then identification may be done on the bases of molecular techniques involving sequencing and other modalities. This is emerging as group as non-sporulating mold (NSM) being faced regularly by the microbiologists. Hence, the only option left is sequence based identification in such circumstances.

(c) Immunodiagnosis

The phaeoid fungus like *Exophiala werneckii* (now *Hortaea werneckii)* is routinely identified on the basis of its annellidic conidiogenous cells and the production of two-celled conidia. But without the proper recognition of its annellidic mode of conidiogenesis, isolates of *H.werneckii* can be confused with other species of *Exophiala* genus. In such a situation, exoantigen based tests have been developed, which are capable of specifically identifying *H.werneckii*. The exoantigen tests permit the rapid and specific identification of *H.werneckii* isolates.

The exoantigen test also helps differentiation of *E.jeanselmei* and *E.dermatitidis* from one another as well as from other *Exophiala* species, *Fonsecaea* species, *Phialophora* species, *Cladosporium* species, *Rhinocladiella* species and *Sporothrix schenckii* by presence or absence of lines of identity or of partial identity or lines of non-identity. Therefore, using adsorbed serum eliminates the difficulty with cross-reactivity seen with non-adsorbed serum. Thus, with an adsorbed serum as the reagent, it is feasible to presumptively differentiate *E.jeanselmei* and *E.dermatitidis* from one another and from other phaeoid fungi also.

(d) Animal Pathogenicity

The phaeoid fungi cause disease among large number of warm- as well as cold-blooded animals. In one of the studies, experimental infection of the rabbit eye with *Exophiala dermatitidis* was performed, which was isolated from the corneal ulcer of a patient. This fungus was inoculated into the front chamber and the vitreous body. The disease showed a trend to spontaneous recovery. The individual phases of the experimental infection were followed by histopathology and electron microscopy, both transmission as well as scanning electron microscope. Various stages of development of the polymorphous fungus, mycelial filaments and yeast-like cells, are demonstrated. The fungal culture is always preferred than inoculation into the experimental animal for routine identification of causative etiological agent of phaeohyphomycosis.

Treatment and Prophylaxis

The subcutaneous forms of phaeohyphomycosis are usually treated by local excision but invasive infections require combination therapy with intravenous amphotericin B and oral flucytosine. The antifungal drugs such as amphotericin B, flucytosine, ketoconazole, fluconazole, itraconazole and terbinafine have been used in the past with variable success. Due to non-availability of effective medical treatment, early recognition and excision of lesion is indicated to avoid further complications especially in immunocompromised patients. The prognosis of cerebral phaeohyphomycosis is poor and rarely survival of patients may not exceed one year. The patients with non-life-threatening deep-seated or systemic forms of phaeohyphomycosis should be given itraconazole in dose of 200-600 mg per day. In recent times voriconazole has also been tried with variable results. From the azole group, itraconazole and voriconazole are the mainstay of the treatment of phaeohyphomycosis. Itraconazole may be given for duration of 1½ years otherwise there are chances of recurrence. Moreover, there may be development of drug resistance in due course of time of treatment.

This has also been observed that if the localized cystic phaeohyphomycotic lesion is excised *in toto*, antifungals may not be required at all and there is no recurrence among such patients.

Further Reading

1. Abdolrasouli A, Gonzalo X, Jatan A, *et al.* Subcutaneous phaeohyphomycosis cyst associated with *Medicopsis romeroi* in an immunocompromised host. *Mycopathologia.* 2016; **181:** 717-21.
2. Abraham LK, Joseph E, Thomas S, *et al.* Subcutaneous phaeohyphomycosis: A clinicopathological study. *Int Surg J.* 2014; **1:** 140-3.
3. Adler A, Yaniv I, Samra Z, *et al. Exserohilum:* An emerging human pathogen. *Eur J Clin Microbiol Infect Dis.* 2006; **25:** 247-56.
4. Agarwal A, Singh SM. A case of cutaneous phaeohyphomycosis caused by *Exserohilum rostratum*, its *in vitro* sensitivity and review of literature. *Mycopathologia.* 1995; **131:** 9-12.
5. Agarwal R, Kalita J, Marak RS, *et al.* Spectrum of fungal infection in a neurology tertiary care center in India. *Neurol Sci.* 2012; **33:** 1305-10.
6. Agarwal S, Chouhan SS. Subcutaneous abscess caused by dematiaceous fungus. *Indian J Pathol Bacteriol.* 1969; **12:** 42-5.
7. Aggarwal A, Salunke P, Mohindra S, *et al.* A case of cerebral phaeohyphomycosis in an immunocompetent patient: Emphasis on intraoperative findings. *Neurol India.* 2014; **62:** 551-2.
8. Ahmed SA, Desbois N, Quist D, *et al.* Phaeohyphomycosis caused by a novel species, *Pseudochaetosphaeronema martinelli. J Clin Microbiol.* 2015; **53:** 2927-34.
9. Ahmed SA, Hofmuller W, Seibold M, *et al. Tintelnotia*, a new genus in Phaeosphaeriaceae harbouring agents of cornea and nail infections in humans. *Mycoses.* 2017; **60:** 244-53.
10. Ahmed SA, Stevens DA, van de Sande WW, *et al. Roussoella percutanea*, a novel opportunistic pathogen causing subcutaneous mycoses. *Med Mycol.* 2014; **52:** 689-98.
11. Ajello L. Phaeohyphomycosis: Definition and etiology. *Pan Am Health Org Sci Publ.* 1975; **304:** 126-133.
12. Al-Tawfiq JA, Boukhamseen A. Cerebral phaeohyphomycosis due to *Rhinocladiella mackenziei* (formerly

Ramichloridium mackenziei): Case presentation and literature review. *J Infect Public Health.* 2011; **4:** 96-102.
13. Alabaz D, Kibar F, Arikan S, *et al.* Systemic phaeohyphomycosis due to *Exophiala (Wangiella)* in an immunocompetent child. *Med Mycol.* 2009; **47:** 653-7.
14. Alastruey-Izquierdo A, Cuesta I, Ros L, *et al.* Antifungal susceptibility profile of clinical *Alternaria* spp. identified by molecular methods. *J Antimicrob Chemother.* 2011; **66:** 2585-7.
15. Almagro-Molto M, Haas A, Melcher C, *et al.* First case of *Roussoella percutanea* bursitis. *Diagn Microbiol Infect Dis.* 2017; **87:** 172-4.
16. Amr SS, Al-Tawfiq JA. Aspiration cytology of brain abscess from a fatal case of cerebral phaeohyphomycosis due to *Ramichloridium mackenziei. Diagn Cytopathol.* 2007; **35:** 695-9.
17. Anandan V, Nayak V, Sundaram S, *et al.* An association of *Alternaria alternata* and *Scopulariopsis brevicaulis* in cutaneous phaeohyphomycosis. *Indian J Dermatol Venereol Leprol.* 2008; **74:** 244-7.
18. Anandi V, John TJ, Walter A, *et al.* Cerebral phaeohyphomycosis caused by *Chaetomium globosum* in a renal transplant recipient. *J Clin Microbiol.* 1989; **27:** 2226-9.
19. Angurana SK, Suthar R, Mehta A, *et al.* Cerebral phaeohyphomycosis: Fulminant encephalitic presentation. *Indian J Pediatr.* 2017. doi: 10.1007/s12098-017-2352-y. PMID: 28462506.
20. Antonucci A, Ghetti P, Iozzo I. Recurrent subcutaneous phaeohyphomycosis caused by *Exophiala* sp. associated with squamocellular carcinoma. *Int J Dermatol.* 2008; **47:** 1323-4.
21. Aoyama Y, Nomura M, Yamanaka S, *et al.* Subcutaneous phaeohyphomycosis caused by *Exophiala xenobiotica* in a non-Hodgkin lymphoma patient. *Med Mycol.* 2009; **47:** 95-9.
22. Arakaki O, Asato Y, Yagi N, *et al.* Phaeohyphomycosis caused by *Exophiala jeanselmei* in a patient with polymyalgia rheumatica. *J Dermatol.* 2010; **37:** 367-73.
23. Aranegui B, Feal C, Garcia CP, *et al.* Subcutaneous phaeohyphomycosis caused by *Exophiala jeanselmei* treated with wide surgical excision and posaconazole: Case report. *Int J Dermatol.* 2013; **52:** 255-6.
24. Arnold C. Fungal meningitis outbreak affects over 700. *Lancet Neurol.* 2013; **12:** 429-30.
25. Aragon-Miguel R, Calleja-Algarra A, Morales-Raya C, *et al. Alternaria infectoria* skin infection in a renal transplant recipient: An emerging phaeohyphomycosis of occidental countries? *Int J Dermatol.* 2017; **56:** e153-5.
26. Babic MN, Zupancic J, Gunde-Cimerman N, *et al.* Ecology of the human opportunistic black yeast *Exophiala dermatitidis* indicates preference for human-made habitats. *Mycopathologia.* 2017. doi: 10.1007/s11046-017-0134-8. PMID: 28447292.
27. Badali H, Carvalho VO, Vicente V, *et al. Cladophialophora saturnica* sp. nov., a new opportunistic species of Chaetothyriales revealed using molecular data. *Med Mycol.* 2009; **47:** 51-62.
28. Badali H, Chander J, Bansal S, Aher A, Borkar SS, Meis JF, De Hoog GS. First autochthonous case of *Rhinocladiella mackenziei* cerebral abscess outside the Middle East. *J Clin Microbiol.* 2010; **48:** 646-9.
29. Badali H, Chander J, Bayat M, Seyedmousavi S, Sidhu S, Rani H, Attri A, Handa U, Meis JF, de Hoog GS. Multiple subcutaneous cysts due to *Exophiala spinifera* in an immunocompetent patient. *Med Mycol.* 2012; **50:** 207-13.
30. Badali H, Chander J, Gulati N, Attri A, Chopra R, Najafzadeh MJ, Chhabra S, Meis JF, De Hoog GS. Subcutaneous phaeohyphomycotic cyst caused by *Pyrenochaeta romeroi. Med Mycol.* 2010; **48:** 763-8.
31. Badali H, Chander J, Gupta A, Rani H, Punia RS, de Hoog GS, Meis JF. Fatal cerebral phaeohyphomycosis in an immunocompetent individual due to *Thielavia subthermophila*: Case report and review of the literature. *J Clin Microbiol.* 2011; **49:** 2336-41.
32. Badali H, de Hoog GS, Curfs-Breuker I, *et al.* Use of amplified fragment length polymorphism to identify 42 *Cladophialophora* strains related to cerebral phaeohyphomycosis with *in vitro* antifungal susceptibility. *J Clin Microbiol.* 2010; **48:** 2350-6.
33. Badali H, de Hoog GS, Sudhadham M, *et al.* Microdilution *in vitro* antifungal susceptibility of *Exophiala dermatitidis,* a systemic opportunist. *Med Mycol.* 2011; **49:** 819-24.
34. Badali H, Gueidan C, Najafzadeh MJ, Bonifaz A, *et al.* Biodiversity of the genus *Cladophialophora. Stud Mycol.* 2008; **61:** 175-91.
35. Badali H, Najafzadeh MJ, van Esbroeck M, *et al.* The clinical spectrum of *Exophiala jeanselmei,* with a case report and in vitro antifungal susceptibility of the species. *Med Mycol.* 2010; **48:** 318-27.
36. Bagla P, Loeffelholz M, Blanton LS. Cerebral phaeohyphomycosis by *Fonsecaea monophora:* Report in a patient with AIDS and a ring enhancing lesion. *Med Mycol Case Rep.* 2016; **12:** 4-7.
37. Berger JS, Cusumano LR, Derose JJ, *et al. Exophiala (Wangiella) dermatitidis* prosthetic aortic valve endocarditis and prosthetic graft infection in an immune competent patient. *Case Rep Infect Dis.* 2017; 4839314. PMID: 28484654.
38. Baradkar VP, Kumar S. Subcutaneous granulomatous infection caused by *Phaeoacremonium infalitipes* on foot. *Indian J Dermatol.* 2011; **56:** 244-5.
39. Baradkar VP, Mathur M, Kumar S. Phaeohyphomycosis of subcutaneous tissue caused by *Phaeoacremonium parasiticum. Indian J Med Microbiol.* 2009; **27:** 66-9.
40. Barua P, Rudramurthy SM, Devi U, *et al. Exophiala mesophila* from respiratory tract of an immunocompromised patient. *Int J Health Sci Res.* 2015; **5:** 449-53.
41. Ben-Ami R, Lasala PR, Lewis RE, *et al.* Lack of galactomannan reactivity in dematiaceous molds recovered from cancer patients with phaeohyphomycosis. *Diagn Microbiol Infect Dis.* 2010; **66:** 200-3.
42. Ben-Ami R, Lewis RE, Raad II, *et al.* Phaeohyphomycosis in a tertiary care cancer center. *Clin Infect Dis.* 2009; **48:** 1033-41.

43. Bhardwaj S, Capoor MR, Kolte S, *et al.* Phaeohyphomycosis due to *Exophiala jeanselmei:* An emerging pathogen in India - Case report and review. *Mycopathologia.* 2016; **181:** 279-84.
44. Blasi B, Tafer H, Tesei D, *et al.* From glacier to Sauna: RNA-Seq of the human pathogen black fungus *Exophiala dermatitidis* under varying temperature conditions exhibits common and novel fungal response. *PLoS One.* 2015; **10:** e0127103.
45. Bohelay G, Robert S, Bouges-Michel C, *et al.* Subcutaneous phaeohyphomycosis caused by *Exophiala spinifera* in a European patient with lymphoma: A rare occurrence case report and literature review. *Mycoses.* 2016; **59:** 691-6.
46. Bonifaz A, Davoudi MM, de Hoog GS, *et al.* Severe disseminated phaeohyphomycosis in an immunocompetent patient caused by *Veronaea botryosa. Mycopathologia.* 2013; **175:** 497-503.
47. Borkar SA, Sharma MS, Rajpal G, *et al.* Brain abscess caused by *Cladophialophora bantiana* in an immunocompetent host: Need for a novel cost-effective antifungal agent. *Indian J Med Microbiol.* 2008; **26:** 271-4.
48. Borman AM, Fraser M, Szekely A, *et al.* Rapid identification of clinically relevant members of the genus *Exophiala* by matrix-assisted laser desorption ionization-time of flight mass spectrometry and description of two novel species, *Exophiala campbellii* and *Exophiala lavatrina. J Clin Microbiol.* 2017; **55:** 1162-76.
49. Boyce RD, Deziel PJ, Otley CC, *et al.* Phaeohyphomycosis due to *Alternaria* species in transplant recipients. *Transpl Infect Dis.* 2010; **12:** 242-50.
50. Bras S, Sabino R, Laureano A, *et al.* Cutaneous infection by different *Alternaria* species in a liver transplant recipient. *Med Mycol Case Rep.* 2015; **8:** 1-4.
51. Brasch J, Busch JO, de Hoog GS. Cutaneous phaeohyphomycosis caused by *Alternaria infectoria. Acta Derm Venereol.* 2008; **88:** 160-1.
52. Brokalaki EI, Sommerwerck U, von Heinegg EH, *et al. Ochroconis gallopavum* infection in a lung transplant recipient: Report of a case. *Transplant Proc.* 2012; **44:** 2778-80.
53. Cai Q, Lv GX, Jiang YQ, *et al.* The first case of phaeohyphomycosis caused by *Rhinocladiella basitona* in an immunocompetent child in China. *Mycopathologia.* 2013; **176:** 101-5.
54. Calabro G, Nino M, Gallo L, *et al.* Cutaneous alternariosis in a kidney transplantation recipient: Report of a case. *J Dermatolog Treat.* 2008; **19:** 246-8.
55. Campolina SS, Caligiorne RB, Rezende-Silva S, *et al.* A skin infection mimicking chromoblastomycosis by a Capnodialean fungus. *Med Mycol.* 2009; **47:** 81-5.
56. Capoor MR, Agarwal P, Goel M, *et al.* Invasive pulmonary mycosis due to *Chaetomium globosum* with false-positive galactomannan test: A case report and literature review. *Mycoses.* 2016; **59:** 186-93.
57. Carrasco-Zuber JE, Navarrete-Dechent C, Bonifaz A, *et al.* Cutaneous involvement in the deep mycoses: A literature review. Part I - Subcutaneous Mycoses. *Actas Dermosifiliogr.* 2016; **107:** 806-15.
58. Castro AS, Oliveira A, Lopes V. Pulmonary phaeohyphomycosis: A challenge to the clinician. *Eur Respir Rev.* 2013; **22:** 187-8.
59. Castro LG, da Silva Lacaz C, Guarro J, *et al.* Phaeohyphomycotic cyst caused by *Colletotrichum crassipes. J Clin Microbiol.* 2001; **39:** 2321-4.
60. Caviedes MP, Torre AC, Eliceche ML, *et al.* Cutaneous phaeohyphomycosis. *Int J Dermatol.* 2017; **56:** 415-20.
61. Ch K, Thejaswids P, Kini H, *et al.* Phaeohyphomycotic cyst in the foot by *Exophiala. J Clin Diagn Res.* 2014; **8:** ND10-1.
62. Chahal J, Dhotar HS, Anastakis DJ. Phaeohyphomycosis infection leading to flexor tendon rupture: A case report. *Hand (NY).* 2009; **4:** 335-8.
63. Chakrabarti A, Kaur H, Rudramurthy SM, *et al.* Brain abscess due to *Cladophialophora bantiana:* A review of 124 cases. *Med Mycol.* 2016; **54:** 111-9.
64. Chalkias S, Alonso CD, Levine JD, *et al.* Emerging pathogen in immunocompromised hosts: *Exophiala dermatitidis* mycosis in graft-versus-host disease. *Transpl Infect Dis.* 2014; **16:** 616-20.
65. Chander J, Singla N, Kundu R, Handa U, Chowdhary A. Phaeohyphomycosis caused by *Rhytidhysteron rufulum* and review of literature. *Mycopathologia.* 2017; **182:** 403-7.
66. Chang X, Li R, Yu J, *et al.* Phaeohyphomycosis of the central nervous system caused by *Exophiala dermatitidis* in a 3-year-old immunocompetent host. *J Child Neurol.* 2009; **24:** 342-5.
67. Chen CY, Lu PL, Lee KM, *et al.* Acute meningitis caused by *Cladosporium sphaerospermum. Am J Med Sci.* 2013; **346:** 523-5.
68. Chen M, Zhang J, Dong Z, *et al.* Cutaneous phaeohyphomycosis caused by *Exophiala dermatitidis:* A case report and literature review. *Indian J Dermatol Venereol Leprol.* 2016: **82:** 173-7.
69. Chen Q, Jiang JR, Zhang GZ, *et al.* Resolving the *Phoma* enigma. *Stud Mycol.* 2015; **82:** 137-217.
70. Chen YC, Su YC, Tsai CC, *et al.* Subcutaneous phaeohyphomycosis caused by *Exophiala jeanselmei. J Microbiol Immunol Infect.* 2014; **47:** 546-9.
71. Chhabra V, Rastogi S, Barua M, *et al. Alternaria alternata* infection associated osteomyelitis of maxilla: A rare disease entity. *Indian J Dent Res.* 2013; **24:** 639-41.
72. Chhonkar A, Kataria D, Tambe S, *et al.* Three rare cases of cutaneous phaeohyphomycosis. *Indian J Plast Surg.* 2016; **49:** 271-4.
73. Chiller TM, Roy M, Nguyen D, *et al.* Clinical findings for fungal infections caused by methylprednisolone injections. *N Engl J Med.* 2013; **369:** 1610-9.
74. Chintagunta S, Arakkal G, Damarla SV, *et al.* Subcutaneous phaeohyphomycosis in an immunocompetent individual: A case report. *Indian Dermatol Online J.* 2017; **8:** 29-31.
75. Choi J, Lee Y, Chung HS, *et al.* Subcutaneous phaeohyphomycosis caused by *Phaeoacremonium* species in a kidney transplant patient: The first case in Korea. *Korean J Lab Med.* 2011; **31:** 201-4.
76. Chowdhary A, Guarro J, Randhawa HS, *et al.* A rare case of chromoblastomycosis in a renal transplant recipient

caused by a non-sporulating species of *Rhytidhysteron*. *Med Mycol.* 2008; **46:** 163-6.
77. Chowdhary A, Meis JF, Guarro J, *et al.* ESCMID and ECMM joint clinical guidelines for the diagnosis and management of systemic phaeohyphomycosis: Diseases caused by black fungi. *Clin Microbiol Infect.* 2014; **20:** 47-75.
78. Chowdhary A, Perfect J, de Hoog GS. Black molds and melanized yeasts pathogenic to humans. *Cold Spring Harb Perspect Med.* 2014; **5:** a019570. PMID: 25384772.
79. Colombier MA, Alanio A, Denis B, *et al.* Dual Invasive infection with *Phaeoacremonium parasiticum* and *Paraconiothyrium cyclothyrioides* in a renal transplant recipient: Case report and comprehensive review of the literature of *Phaeoacremonium* phaeohyphomycosis. *J Clin Microbiol.* 2015; **53:** 2084-94.
80. Costa RO. Subcutaneous phaeohyphomycosis. *An Bras Dermatol.* 2010; **85:** 727-8.
81. Coussens E, Rogge S, Haspeslagh M, *et al.* Cutaneous infection by *Alternaria infectoria* in a liver transplant recipient: A case report. *Acta Gastroenterol Belg.* 2014; **77:** 256-8.
82. Coutinho I, Teixeira V, Gameiro A, *et al.* Cutaneous alternariosis - A case series of an increasing phaeohyphomycosis. *J Eur Acad Dermatol Venereol.* 2015; **29:** 2053-4.
83. Crawford SJ, Chen SC, Halliday C, *et al. Microsphaeropsis arundinis* skin and soft tissue infection in renal transplant recipients: Three case reports and a review of the literature. *Transpl Infect Dis.* 2015; **17:** 915-20.
84. Cristini A, Garcia-Hermoso D, Celard M, *et al.* Cerebral phaeohyphomycosis caused by *Rhinocladiella mackenziei* in a woman native to Afghanistan. *J Clin Microbiol.* 2010; **48:** 3451-4.
85. Cunha D, Amaro C, Vieira MR, *et al.* Phaeohyphomycosis caused by *Alternaria infectoria* presenting as multiple vegetating lesions in a renal transplant patient. *Rev Iberoam Micol.* 2012; **29:** 44-6.
86. Daboit TC, Duquia RP, Magagnin CM, *et al.* A case of *Exophiala spinifera* infection in southern Brazil: Molecular identification and antifungal susceptibility. *Med Mycol Case Rep.* 2012; **1:** 72-5.
87. da Cunha KC, Sutton DA, Fothergill AW, *et al. In vitro* antifungal susceptibility and molecular identity of 99 clinical isolates of the opportunistic fungal genus *Curvularia*. *Diagn Microbiol Infect Dis.* 2013; **76:** 168-74.
88. Daglar D, Akman-Karakas A, Ozhak-Baysan B, *et al.* Cutaneous *Alternaria infectoria* infection diagnosed by molecular techniques in a renal transplant patient. *Clin Lab.* 2014; **60:** 1569-72.
89. de Gruyter J, Woudenberg JH, Aveskamp MM, *et al.* Redisposition of *Phoma*-like anamorphs in Pleosporales. *Stud Mycol.* 2013; **75:** 1-36.
90. de Hoog GS, Attili-Angelis D, Vicente VA, *et al.* Molecular ecology and pathogenic potential of *Fonsecaea* species. *Med Mycol.* 2004; **42:** 405-16.
91. de Hoog GS, Gueho E, Masclaux F, *et al.* Nutritional physiology and taxonomy of human-pathogenic *Cladosporium-Xylohypha* species. *J Med Vet Mycol.* 1995; **33:** 339-47.
92. de Hoog GS, Horre R. Molecular taxonomy of the *Alternaria* and *Ulocladium* species from humans and their identification in the routine laboratory. *Mycoses.* 2002; **45:** 259-76.
93. de Hoog GS, Queiroz-Telles F, Haase G, *et al.* Black fungi: Clinical and pathogenic approaches. *Med Mycol.* 2000; **38:** S243-50.
94. de Hoog GS, Yurlova NA. Conidiogenesis, nutritional physiology and taxonomy of *Aureobasidium* and *Hormonema*. *Antonie Van Leeuwenhoek.* 1994; **65:** 41-54.
95. de Hoog GS. Significance of fungal evolution for the understanding of their pathogenicity, illustrated with agents of phaeohyphomycosis. *Mycoses.* 1997; **40:** S5-8.
96. Deng S, Pan W, Liao W, *et al.* Combination of amphotericin B and flucytosine against neurotropic species of melanized fungi causing primary cerebral phaeohyphomycosis. *Antimicrob Agents Chemother.* 2016; **60:** 2346-51.
97. Desmet S, Smets L, Lagrou K, *et al. Cladophialophora bantiana* osteomyelitis in a renal transplant patient. *Med Mycol Case Rep.* 2016; **12:** 17-20.
98. Desoubeaux G, Garcia D, Bailly E, *et al.* Subcutaneous phaeohyphomycosis due to *Phialemoniopsis ocularis* successfully treated by voriconazole. *Med Mycol Case Rep.* 2014; **5:** 4-8.
99. Dessinioti C, Soumalevris A, Papadogeorgaki E, *et al.* Cutaneous alternariosis in an immunocompetent patient successfully treated with oral fluconazole. *Eur J Dermatol.* 2013; **23:** 545-7.
100. Didehdar M, Gokanian A, Sofian M, *et al.* First fatal cerebral phaeohyphomycosis due to *Rhinocladiella mackenziei* in Iran, based on ITS rDNA. *J Mycol Med.* 2015; **25:** 81-6.
101. Diernaes JE, Hjuler KF, Kristensen L, *et al.* Subcutaneous phaeohyphomycosis due to *Alternaria dennisii* in a 85-year-old immunocompromised patient. *Acta Derm Venereol.* 2016; **96:** 701-2.
102. Dinh A, Levy B, Bouchand F, *et al.* Subcutaneous phaeohyphomycosis due to *Pyrenochaeta romeroi* mimicking a synovial cyst. *Front Microbiol.* 2016; **7:** 1405.
103. Dionne B, Neff L, Lee SA, *et al.* Pulmonary fungal infection caused by *Neoscytalidium dimidiatum*. *J Clin Microbiol.* 2015; **53:** 2381-4.
104. Doymaz MZ, Seyithanoglu MF, Hakyemez I, *et al.* A case of cerebral phaeohyphomycosis caused by *Fonsecaea monophora*, A neurotropic dematiaceous fungus and a review of the literature. *Mycoses.* 2015; **58:** 187-92.
105. Duquia RP, de Almeida HL Jr, Vettorato G, *et al.* Ecthyma-like phaeohyphomycosis caused by *Cladosporium cladosporioides*. *Mycoses.* 2010; **53:** 541-3.
106. Elinav H, Izhar U, Benenson S, *et al.* Invasive *Scytalidium dimidiatum* infection in an immunocompetent adult. *J Clin Microbiol.* 2009; **47:** 1259-63.
107. Essabbah N, Gorsane I, Youssef M, *et al.* Cutaneous alternariosis in a renal transplant recipient. *Cutis.* 2014; **93:** 237-40.
108. Eswarappa M, Varma PV, Madhyastha R, *et al.* Unusual fungal infections in renal transplant recipients. *Case Rep Transplant.* 2015; 292307. PMID: 25815239.
109. Evans N, Gunew M, Marshall R, *et al.* Focal pulmonary granuloma caused by *Cladophialophora bantiana* in a domestic short haired cat. *Med Mycol.* 2011; **49:** 194-7.

110. Fan YM, Huang WM, Li SF, *et al.* Cutaneous phaeohyphomycosis of foot caused by *Curvularia clavata*. *Mycoses.* 2009; **52:** 544-6.
111. Fathy H, Abdel-Razek MM, Abdelgaber S, *et al.* Subcutaneous phaeohyphomycosis in immunocompetent child caused by *Exophiala jeanselmei*. *Int J Dermatol.* 2012; **51:** 1267-70.
112. Ferreira Ide S, Teixeira G, Abecasis M. *Alternaria alternata* invasive fungal infection in a patient with Fanconi's anemia after an unrelated bone marrow transplant. *Clin Drug Investig.* 2013; **33:** S33-6.
113. Ferringer T. A 79-year-old man presented with a 1.2-cm nodule on the left arm. The best diagnosis is: Phaeohyphomycosis. *Cutis.* 2012; **89:** 111 & 114-5.
114. Frank T, Esquenazi Y, Nigo M, *et al.* Disseminated phaeohyphomycosis with brain abscess and biliary invasion due to *Bipolaris* spp. in an immunocompetent patient. *Ann Clin Lab Sci.* 2016; **46:** 439-42.
115. Friedman BJ, Mehta D, Chaffins ML. Friable nodules on the elbow of a transplant patient. *JAMA Dermatol.* 2016; **152:** 567-8.
116. Fukai T, Hiruma M, Ogawa Y, *et al.* A case of phaeohyphomycosis caused by *Exophiala oligosperma* successfully treated with local hyperthermia. *Med Mycol J.* 2013; **54:** 297-301.
117. Furudate S, Sasai S, Numata Y, *et al.* Phaeohyphomycosis caused by *Phaeoacremonium rubrigenum* in an immunosuppressive patient: A case report and review of the liter ature. *Case Rep Dermatol.* 2012; **4:** 119-24.
118. Galipothu S, Kalawat U, Ram R, *et al.* Cutaneous fungal infection in a renal transplantation patient due to a rare fungus belonging to order Pleosporales. *Indian J Med Microbiol.* 2015; **33:** 165-7.
119. Gao LJ, Yu J, Wang DL, *et al.* Recalcitrant primary subcutaneous phaeohyphomycosis due to *Phialophora verrucosa*. *Mycopathologia.* 2013; **175:** 165-70.
120. Garg N, Devi IB, Vajramani GV, *et al.* Central nervous system cladosporiosis: An account of ten culture-proven cases. *Neurol India.* 2007; **55:** 282-8.
121. Garinet S, Tourret J, Barete S, *et al.* Invasive cutaneous *Neoscytalidium* infections in renal transplant recipients: A series of five cases. *BMC Infect Dis.* 2015; **15:** 535.
122. Ge YP, Lv GX, Shen YN, *et al.* First report of subcutaneous phaeohyphomycosis caused by *Ochroconis tshawytschae* in an immunocompetent patient. *Med Mycol.* 2012; **50:** 637-40.
123. George IA, Mathews MS, Karthik R, *et al.* Fatal cerebral abscess caused by *Cladophialophora bantiana*. *J Assoc Physicians India.* 2008; **56:** 470-2.
124. Giraldo A, Sutton DA, Samerpitak K, *et al.* Occurrence of *Ochroconis* and *Verruconis* species in clinical specimens from the United States. *J Clin Microbiol.* 2014; **52:** 4189-201.
125. Gomes J, Vilarinho C, Duarte Mda L, *et al.* Cutaneous phaeohyphomycosis caused by *Alternaria alternata* unresponsive to itraconazole treatment. *Case Rep Dermatol Med.* 2011; 385803. PMID: 23198173.
126. Gordon RA, Sutton DA, Thompson EH, *et al.* Cutaneous phaeohyphomycosis caused by *Paraconiothyrium cyclothyrioides*. *J Clin Microbiol.* 2012; **50:** 3795-8.
127. Gramaje D, Mostert L, Groenewald JZ, *et al.* *Phaeoacremonium*: from esca disease to phaeohyphomycosis. *Fungal Biol.* 2015; **119:** 759-83.
128. Guarro J, Alves SH, Gene J, *et al.* Two cases of subcutaneous infection due to *Phaeoacremonium* spp. *J Clin Microbiol.* 2003; **41:** 1332-6.
129. Guarro J, Gugnani HC, Sood N, *et al.* Subcutaneous phaeohyphomycosis caused by *Wallemia sebi* in an immunocompetent host. *J Clin Microbiol.* 2008; **46:** 1129-31.
130. Guarro J, Nucci M, Akiti Tiyomi, *et al.* *Phialemonium* fungemia: Two documented nosocomial cases. *J Clin Microbiol.* 1999; **37:** 2493-7.
131. Guegan S, Garcia-Hermoso D, Sitbon K, *et al.* Ten-year experience of cutaneous and/or subcutaneous infections due to *Coelomycetes* in France. *Open Forum Infect Dis.* 2016; **3:** ofw106.
132. Gugnani HC, Ramesh V, Sood N, *et al.* Cutaneous phaeohyphomycosis caused by *Cladosporium oxysporum* and its treatment with potassium iodide. *Med Mycol.* 2006; **44:** 285-8.
133. Gugnani HC, Sidiqqui N, Magulike MO. Cerebral phaeohyphomycosis: report of a case from Nigeria. *Mycoses.* 1998; **41:** 431-2.
134. Gupta AJ, Singh M, Yadav S, *et al.* Phaeohyphomycosis breast masquerading as fibroadenoma in a young teenage girl. *Diagn Cytopathol.* 2017. doi: 10.1002/dc.23755. PMID: 28631431.
135. Harris JE, Sutton DA, Rubin A, *et al.* *Exophiala spinifera* as a cause of cutaneous phaeohyphomycosis: case study and review of the literature. *Med Mycol.* 2009; **47:** 87-93.
136. Hasei M, Takeda K, Anzawa K, *et al.* Case of phaeohyphomycosis producing sporotrichoid lesions. *J Dermatol.* 2013; **40:** 638-40.
137. Hauck EF, McGinnis M, Nauta HJ. Cerebral phaeohyphomycosis mimics high-grade astrocytoma. *J Clin Neurosci.* 2008; **15:** 1061-6.
138. Hoffmann CC, Danucalov IP, Purim KS, *et al.* Infections caused by dematiaceous fungi and their anatomoclinical correlations. *An Bras Dermatol.* 2011; **86:** 138-41.
139. Hong KH, Kim JW, Jang SJ, Yu E, Kim EC. Liver cirrhosis caused by *Exophiala dermatitidis*. *J Med Microbiol.* 2009; **58 (Pt. 5):** 674-7.
140. Hsiao YW, Chia JH, Lu CF, *et al.* Molecular diagnosis and therapeutic experience of subcutaneous *Pyrenochaeta romeroi* infection: A case report and review of the literature. *Int J Dermatol.* 2013; **52:** 1237-40.
141. Hsu CC, Chang SS, Lee PC, *et al.* Cutaneous alternariosis in a renal transplant recipient: A case report and literature review. *Asian J Surg.* 2015; **38:** 47-57.
142. Huang HK, Liu CE, Liou JH, *et al.* Subcutaneous infection caused by *Corynespora cassiicola*, A plant pathogen. *J Infect.* 2010; **60:** 188-90.
143. Hubka V, Mencl K, Skorepova M, *et al.* Phaeohyphomycosis and onychomycosis due to *Chaetomium* spp., including

the first report of *Chaetomium brasiliense* infection. *Med Mycol.* 2011; **49:** 724-33.
144. Hughart R, Merrick M, Adelaja OT, Bleasdale SC, Harrington A, Tsoukas M. Cutaneous phaeohyphomycosis caused by *Biatriospora mackinnonii* in a renal transplant recipient. *JAAD Case Rep.* 2016; **2:** 230-2.
145. Ikram A, Hussain W, Satti ML, Wiqar MA. Invasive infection in a young immunocompetent soldier caused by *Scytalidium dimidiatum. J Coll Physicians Surg Pak.* 2009; **19:** 64-6.
146. Isa-Isa R, Garcia C, Isa M, *et al.* Subcutaneous phaeohyphomycosis (mycotic cyst). *Clin Dermatol.* 2012; **30:** 425-31.
147. Jabeen K, Farooqi J, Zafar A, *et al. Rhinocladiella mackenziei* as an emerging cause of cerebral phaeohyphomycosis in Pakistan: A case series. *Clin Infect Dis.* 2011; **52:** 213-7.
148. Jahan-Tigh RR, Shelton M, Rapini R. Extensive *Exserohilum* infection in a burn patient. *JAAD Case Rep.* 2015; **1:** 188-90.
149. Jaiswal S, Vij M, Prasad N, *et al.* Diagnostic pitfalls in cytological diagnosis of subcutaneous fungal infection in renal transplant recipients. *Diagn Cytopathol.* 2012; **40:** 255-61.
150. Jelinek AG, Fuller C, Baykal A, *et al.* Headaches and hemiparesis in an immunocompetent inmate. *Neuropathology.* 2014; **34:** 314-7.
151. Joshi A, Singh R, Shah MS, *et al.* Subcutaneous mycosis and fungemia by *Aureobasidium pullulans:* A rare pathogenic fungus in a post allogeneic BM transplant patient. *Bone Marrow Transplant.* 2010; **45:** 203-4.
152. Joshi P, Agarwal S, Singh G, *et al.* "A fine needle aspiration cytology in time saves nine" - Cutaneous phaeohyphomycosis caused by *Exophiala jeanselmei* in a renal transplant patient: Diagnosis by fine needle aspiration cytology. *J Cytol.* 2016; **33:** 55-7.
153. Juhas E, Reyes-Mugica M, Michaels MG, *et al. Exserohilum* infection in an immunocompromised neonate. *Pediatr Dermatol.* 2013; **30:** e232-3.
154. Jung NY, Kim E. Cerebral phaeohyphomycosis: A rare cause of brain abscess. *J Korean Neurosurg Soc.* 2014; **56:** 444-7.
155. Kamalapur MG, Patil PB, Joshi SK, *et al.* The role of magnetic resonance imaging in distinguishing the fungal from the non-fungal multiple brain abscesses. *Egypt J Radiol Nucl Med.* 2011; **42:** 397-404.
156. Kan T, Takahagi S, Kamegashira A, *et al.* Disseminated subcutaneous phaeohyphomycosis caused by *Exophiala oligosperma* in a patient with Wegener's granulomatosis. *Acta Derm Venereol.* 2013; **93:** 356-7.
157. Kantarcioglu AS, Guarro J, de Hoog GS, *et al.* A case of central nervous system infection due to *Cladophialophora bantiana. Rev Iberoam Micol.* 2016; **33:** 237-41.
158. Kantarcioglu AS, Guarro J, De Hoog S, *et al.* An updated comprehensive systematic review of *Cladophialophora bantiana* and analysis of epidemiology, clinical characteristics and outcome of cerebral cases. *Med Mycol.* 2016. pii: myw124. doi: 10.1093/mmy/myw124. PMID: 28007938.
159. Karuppal R, Kumaran CM, Marthya A, *et al.* Tibial osteomyelitis due to *Fonsecaea pedrosoi* in an immunocompetent patient: Case report. *J Foot Ankle Surg.* 2009; **48:** 569-72.
160. Kashgari TQ, Al-Miniawi H, Moawad Hanna MK. Cerebral phaeohyphomycosis caused by *Ramichloridium mackenziei* in the eastern province of Saudi Arabia. *Ann Saudi Med.* 2000; **20:** 457-60.
161. Katragkou A, Pana ZD, Perlin DS, *et al. Exserohilum* infections: Review of 48 cases before the 2012 United States outbreak. *Med Mycol.* 2014; **52:** 376-86.
162. Kaur R, Bala K. Unilateral renal phaeohyphomycosis due to *Bipolaris spicifera* in an immunocompetent child - rare case presentation and review of literature. *Mycoses.* 2015; **58:** 437-44.
163. Kerchner KR, Swing DC Jr, Williford PR. A verrucous plaque of the eyebrow. *Am J Dermatopathol.* 2009; **31:** 506-8.
164. Khader A, Ambooken B, Binitha MP, *et al.* Disseminated cutaneous phaeohyphomycosis due to *Cladophialophora bantiana. Indian J Dermatol Venereol Leprol.* 2015; **81:** 491-4.
165. Khan SA, Hasan AS, Capoor MR, *et al.* Calcaneal osteomyelitis caused by *Exophiala jeanselmei* in an immunocompetent child. A case report. *J Bone Joint Surg Am.* 2007; **89:** 859-62.
166. Khan Z, Ahmad S, Kapila K, *et al. Pyrenochaeta romeroi:* A causative agent of phaeohyphomycotic cyst. *J Med Microbiol.* 2011; **60:** 842-6.
167. Khan ZU, Lamdhade SJ, Johny M, *et al.* Additional case of *Ramichloridium mackenziei* cerebral phaeohyphomycosis from the Middle East. *Med Mycol.* 2002; **40:** 429-33 & 470.
168. Khetan SP, Agrawal VA, QaziMS. Cerebral phaeohyphomycosis - Could early diagnosis have saved the patient? *Indian J Med Microbiol.* 2014; **32:** 440-2.
169. Kindo AJ, Ramalakshmi S, Giri S, *et al.* A fatal case of prostatic abscess in a post-renal transplant recipient caused by *Cladophialophora carrionii. Saudi J Kidney Dis Transpl.* 2013; **24:** 76-9.
170. Kollipara R, Peranteau AJ, Nawas ZY, *et al.* Emerging infectious diseases with cutaneous manifestations: Fungal, helminthic, protozoan and ectoparasitic infections. *J Am Acad Dermatol.* 2016; **75:** 19-30.
171. Kondo RN, Pontello R Jr, Pontello R, *et al.* Subcutaneous phaeohyphomycosis. *An Bras Dermatol.* 2014; **89:** 169-70.
172. Kondori N, Gilljam M, Lindblad A, *et al.* High rate of *Exophiala dermatitidis* recovery in the airways of patients with cystic fibrosis is associated with pancreatic insufficiency. *J Clin Microbiol.* 2011; **49:** 1004-9.
173. Kondo Y, Hiruma M, Matsushita A, *et al.* Cutaneous phaeohyphomycosis caused by *Veronaea botryosa* observed as sclerotic cells in tissue. *Int J Dermatol.* 2007; **46:** 625-7.
174. Koo S, Klompas M, Marty FM. *Fonsecaea monophora* cerebral phaeohyphomycosis: case report of successful surgical excision and voriconazole treatment and review. *Med Mycol.* 2010; **48:** 769-74.
175. Kuan CS, Cham CY, Singh G, *et al.* Genomic analyses of *Cladophialophora bantiana,* a major cause of cerebral

phaeohyphomycosis provides insight into its lifestyle, virulence and adaption in host. *PLoS One.* 2016; **11:** e0161008.

176. Kulkarni M, Jamale T, Hase N, *et al.* Subcutaneous phaeohyphomycosis caused by *Pyrenochaeta romeroi* in a kidney transplant recipient: A case report. *Exp Clin Transplant.* 2017; **15:** 226-7.
177. Kumar GN, Nair SP. Phaeohyphomycosis presenting as a solitary nodulocystic lesion in a renal transplant patient. *Indian Dermatol Online J.* 2015; **6:** 359-61.
178. Kumar KK, Hallikeri K. Phaeohyphomycosis. *Indian J Pathol Microbiol.* 2008; **51:** 556-8.
179. Kumarguru B, Srinivas T, Jagadish MH. A case of subcutaneous phaeohyphomycosis in a diabetic patient: A cryptic entity. *J Glob Infect Dis.* 2014; **6:** 45-6.
180. Lahiri Mukhopadhyay S, Mahadevan A, Bahubali VH, *et al.* A rare case of multiple brain abscess and probably disseminated phaeohyphomycosis due to *Cladophialophora bantiana* in an immunosuppressed individual from India. *J Mycol Med.* 2017. pii: S1156-5233(16)30233-5. doi: 10.1016/j.mycmed.2017.04.002. PMID: 28478966.
181. Lakshmi V, Padmasri C, Umabala P, *et al.* Cerebral phaeohyphomycosis due to *Cladophialophora bantiana. Indian J Med Microbiol.* 2008; **26:** 392-5.
182. Lanternier F, Barbati E, Meinzer U, *et al.* Inherited CARD9 deficiency in 2 unrelated patients with invasive *Exophiala* infection. *J Infect Dis.* 2015; **211:** 1241-50.
183. Larbcharoensub N, Chongtrakool P, Wirojtananugoon C, *et al.* Treatment of a brain abscess caused by *Scedosporium apiospermum* and *Phaeoacremonium parasiticum* in a renal transplant recipient. *Southeast Asian J Trop Med Public Health.* 2013; **44:** 484-9.
184. LaRocco A Jr, Netzer RC. Central venous catheters as a risk factor for disseminated phaeohyphomycosis? *Clin Infect Dis.* 2002; **35:** 1022-4.
185. Larsen CG, Arendrup MC, Krarup E, *et al.* Subcutaneous phaeohyphomycosis in a renal transplant recipient successfully treated with voriconazole. *Acta Derm Venereol.* 2009; **89:** 657-8.
186. Leahy TR, Punnett AS, Richardson SE, *et al.* Molecular identification of phaeohyphomycosis due to *Alternaria infectoria* in a patient with acute myeloid leukemia—A case report. *Diagn Microbiol Infect Dis.* 2010; **66:** 318-21.
187. Lee KC, Kim MJ, Chae SY, *et al.* A case of phaeohyphomycosis caused by *Exophiala lecanii-corni. Ann Dermatol.* 2016; **28:** 385-7.
188. Levenstadt JS, Poutanen SM, Mohan S, *et al. Pleurostomophora richardsiae* - An insidious fungus presenting in a man 44 years after initial inoculation: A case report and review of the literature. *Can J Infect Dis Med Microbiol.* 2012; **23:** 110-3.
189. Li DM, de Hoog GS, Saunte DM, *et al. Coniosporium epidermidis* sp. nov. A new species from human skin. *Stud Mycol.* 2008; **61:** 131-6.
190. Li DM, de Hoog GS. Cerebral phaeohyphomycosis - A cure at what lengths? *Lancet Infect Dis.* 2009; **9:** 376-83.
191. Lin SC, Sun PL, Ju YM, *et al.* Cutaneous phaeohyphomycosis caused by *Exserohilum rostratum* in a patient with cutaneous T-cell lymphoma. *Int J Dermatol.* 2009; **48:** 295-8.
192. Lin YP, Li W, Yang YP, *et al.* Cutaneous phaeohyphomycosis caused by *Exophiala spinifera* in a patient with systemic lupus erythematosus. *Lupus.* 2012; **21:** 548-51.
193. Litchevski V, Goldschmidt A, Nass D, *et al.* Cerebral phaeohyphomycosis in an immunocompetent patient: A case report and literature summary. *Clin Neurol Neurosurg.* 2014; **124:** 179-81.
194. Liu AW, Bateman AC, Greenbaum A, *et al.* Cutaneous phaeohyphomycosis in a hematopoietic stem cell transplant patient caused by *Alternaria rosae:* First case report. *Transpl Infect Dis.* 2017; 19 (3). doi: 10.1111/tid.12698. PMID: 28295973.
195. Lv GX, Ge YP, Shen YN, *et al.* Phaeohyphomycosis caused by a plant pathogen, *Corynespora cassiicola. Med Mycol.* 2011; **49:** 657-61.
196. Ma X, Gu Y, Liu X, *et al.* Phaeohyphomycotic dermatitis in a giant panda (*Ailuropodame lanoleuca*) caused by *Cladosporium cladosporioides. Med Mycol Case Rep.* 2013; **2:** 119-21.
197. Machouart M, Menir P, Helenon R, *et al. Scytalidium* and scytalidiosis: What's new in 2012? *J Mycol Med.* 2013; **23:** 40-6.
198. Madhugiri VS, Bhagavatula ID, Mahadevan A, *et al.* An unusual infection, an unusual outcome—*Fonsecaea pedrosoi* cerebral granuloma. *J Neurosurg Pediatr.* 2011; **8:** 229-32.
199. Madke B, Khopkar U. Phaeohyphomycotic cyst. *Indian Dermatol Online J.* 2015; **6:** 223–5.
200. Madrid H, da Cunha KC, Gene J, *et al.* Novel *Curvularia* species from clinical specimens. *Persoonia.* 2014; **33:** 48-60.
201. Mahajan VK, Chauhan PS, Mehta KS, *et al.* Subcutaneous phaeohyphomycosis of the face presenting as rhinoentomophthoramycosis. *Int J Dermatol.* 2013; **52:** 1105-8.
202. Mahajan VK, Sharma V, Prabha N, *et al.* A rare case of subcutaneous phaeohyphomycosis caused by a *Rhytidhysteron* species: A clinico-therapeutic experience. *Int J Dermatol.* 2014; **53:** 1485-9.
203. Malakzai MO, Sahak JG, Campbell R, *et al.* Multifocal but non-disseminated phaeohyphomycosis in a healthy man via a unique mechanism: Ejection from motor vehicle accident into a vegetable field in Afghanistan resulting in multiple contaminated skin wounds. *J Cutan Pathol.* 2017; doi: 10.1111/cup.12942. PMID: 28349548.
204. Manamgoda DS, Cai L, Bahkali AH, *et al. Cochliobolus:* An overview and current status of species. *Fungal Divers.* 2011; **51:** 3-42.
205. Manamgoda DS, Rossman AY, Castlebury LA, *et al.* The genus *Bipolaris. Stud Mycol.* 2014; **79:** 221-88.
206. Mani RS, Chickabasaviah YT, Nagarathna S, *et al.* Cerebral phaeohyphomycosis caused by *Scytalidium dimidiatum*: A case report from India. *Med Mycol.* 2008; **46:** 705-11.

207. Manoharan M, Shanmugam N, Veeriyan S. A rare case of a subcutaneous phaeomycotic cyst with a brief review of literature. *Malays J Med Sci.* 2011; **18:** 78-81.
208. Mansour A, Jordan K. Disseminated *Cladophialophora bantiana* disease in a patient with prediabetes. *BMJ Case Rep.* 2014; pii: bcr2014206426.
209. Marques SA, Camargo RM, Summerbell RC, *et al.* Subcutaneous phaeohyphomycosis caused by *Phaeoacremonium parasiticum* in a renal transplant patient. *Med Mycol.* 2006; **44:** 671-6.
210. Martanto W, Tee SI, Pan JY. *Exophiala* deep fungal infection complicating dermatitis artefacta of the arms responding to itraconazole. *J Eur Acad Dermatol Venereol.* 2014; **28:** 1262-3.
211. Martinez-Gonzalez MC, Verea MM, Velasco D, *et al.* Three cases of cutaneous phaeohyphomycosis by *Exophiala jeanselmei. Eur J Dermatol.* 2008; **18:** 313-6.
212. Martinez-Lamas L, Alvarez M, Llovo J, *et al.* Phaeohyphomycosis caused by *Cladophialophora bantiana. Rev Iberoam Micol.* 2014; **31:** 203-6.
213. Mason SE, Dueker ND, Stratton CW, Whetsell WO Jr. An 80-year-old man with a ring-enhancing right basal ganglia lesion. *Brain Pathol.* 2008; **18:** 288-91.
214. Mattei AS, Severo CB, Guazzelli LS, *et al.* Cutaneous infection by *Diaporthe phaseolorum* in Brazil. *Med Mycol Case Rep.* 2013; **2:** 85-7.
215. Maudgil A, Johnson Z, Rogers N, Mudhar HS. Unusual ocular presentations of ocular phaeohyphomycosis. *Eye (Lond).* 2016; **30:** 1517-9.
216. Mazzurco JD, Ramirez J, Fivenson DP. Phaeohyphomycosis caused by *Phaeoacremonium* species in a patient taking infliximab. *J Am Acad Dermatol.* 2012; **66:** 333-5.
217. McCarty TP, Baddley JW, Walsh TJ, *et al.* Phaeohyphomycosis in transplant recipients: Results from the Transplant Associated Infection Surveillance Network (TRANSNET). *Med Mycol.* 2015; **53:** 440-6.
218. McGinnis MR, Borelli D, Padhye AA, *et al.* Re-classification of *Cladosporium bantianum* in the genus *Xylohypha. J Clin Microbiol.* 1986; **23:** 1148-51.
219. McGinnis MR, Rinaldi MG, Winn RE. Emerging agents of phaeohyphomycosis: pathogenic species of *Bipolaris* and *Exserohilum. J Clin Microbiol.* 1986; **24:** 250-9.
220. McGinnis MR. Taxonomy of *Exophiala jeanselmei* (Langeron) McGinnis and Padhye. *Mycopathologia.* 1978; **65:** 79-87.
221. McNeil CJ, Luo RF, Vogel H, *et al.* Brain abscess caused by *Phaeoacremonium parasiticum* in an immunocompromised patient. *J Clin Microbiol.* 2011; **49:** 1171-4.
222. Michelon M, Greenlaw S, O'Donnell P, *et al.* Multifocal cutaneous alternariosis in a 70-year-old Kenyan renal transplant patient. *Dermatol Online J.* 2014; **20.** pii: 13030/qt45s7m4mk. PMID: 25046464.
223. Minotto R, Corte LD, Silva MV, *et al.* Phaeohyphomycosis of the ungual apparatus - case report. *An Bras Dermatol.* 2014; **89:** 649-51.
224. Mirhendi H, Fatemi MJ, Bateni H, *et al.* First case of disseminated phaeohyphomycosis in an immunocompetent individual due to *Alternaria malorum. Med Mycol.* 2013; **51:** 196-202.
225. Mishra D, Singal M, Rodha MS, *et al.* Subcutaneous phaeohyphomycosis of foot in an immunocompetent host. *J Lab Physicians.* 2011; **3:** 122-4.
226. Mishra K, Das S, Goyal S, *et al.* Subcutaneous mycoses caused by *Rhytidhysteron* species in an immunocompetent patient. *Med Mycol Case Rep.* 2014; **5:** 32-4.
227. Miyagawa F, Shobatake C, Fukumoto T, *et al.* Cutaneous phaeohyphomycosis caused by *Exophiala jeanselmei* in a healthy individual. *J Dermatol.* 2017 Feb 13. doi: 10.1111/1346-8138.13766. PMID: 28191648.
228. Mohapatra M, Satyanarayana S. Cytohistomorphology of subcutaneous phaeohypomycosis. *J Cytol.* 2013; **30:** 211-2.
229. Monaganti S, Santos CA, Markwardt A, *et al.* Pulmonary phaeohyphomycosis caused by *Phaeoacremonium* in a kidney transplant recipient: Successful treatment with posaconazole. *Case Rep Med.* 2014; 902818.
230. Moody MN, Tschen J, Mesko M. Cutaneous *Curvularia* infection of the forearm. *Cutis.* 2012; **89:** 65-8.
231. Morais OO, Porto C, Coutinho AS, *et al.* Infection of the lymphatic system by *Aureobasidium pullulans* in a patient with erythema nodosum leprosum. *Braz J Infect Dis.* 2011; **15:** 288-92.
232. Morio F, Berre JY, Garcia-Hermoso D, *et al.* Phaeohyphomycosis due to *Exophiala xenobiotica* as a cause of fungal arthritis in an HIV-infected patient. *Med Mycol.* 2012; **50:** 513-7.
233. Moss HB, Fowler WC, Schell WA, *et al.* Dematiaceous fungi mimicking a conjunctival melanoma. *Ophthal Plast Reconstr Surg.* 2013; **29:** e65-6.
234. Mudholkar VG, Acharya AS, Ramteke RV. Disseminated cutaneous phaeohyphomycosis in immunocompetent child. *Pediatr Dermatol.* 2011; **28:** 30-1.
235. Mukai Y, Nureki S, Hata M, *et al. Exophiala dermatitidis* pneumonia successfully treated with long-term itraconazole therapy. *J Infect Chemother.* 2014; **20:** 446-9.
236. Naggie S, Perfect JR. Molds: Hyalohyphomycosis, phaeohyphomycosis and zygomycosis. *Clin Chest Med.* 2009; **30:** 337-53.
237. Najafzadeh MJ, Fata A, Naseri A, *et al.* Implantation phaeohyphomycosis caused by a non-sporulating *Chaetomium* species. *J Mycol Med.* 2014; **24:** 161-5.
238. Najafzadeh MJ, Suh MK, Lee MH, *et al.* Subcutaneous phaeohyphomycosis caused by *Exophiala equina*, with susceptibility to eight antifungal drugs. *J Med Microbiol.* 2013; **62:** 797-800.
239. Najafzadeh MJ, Sutton DA, Keisari MS, *et al.* In vitro activities of eight antifungal drugs against 104 environmental and clinical isolates of *Aureobasidium pullulans. Antimicrob Agents Chemother.* 2014; **58:** 5629-31.
240. Nandedkar S, Bajpai T, Bhatambare GS, *et al.* Cerebral phaeohyphomycosis: A rare case from central India. *Asian J Neurosurg.* 2015; **10:** 142-4.
241. Nath R, Barua S, Barman J, *et al.* Subcutaneous mycosis due to *Cladosporium cladosporioides* and *Bipolaris cynodontis* from Assam, North-East India and review of published literature. *Mycopathologia.* 2015; **180:** 379-87.

242. Nayyar C, Gulati N, Sherwal BL. A rare case of phaeohyphomycosis caused by *Fonsecaea pedrosoi* in a child with nephrotic syndrome. *Indian J Nephrol.* 2016; **26:** 220-2.
243. Nelson G, Fermo O, Thakur K, *et al.* Resolution of a fungal mycotic aneurysm after a contaminated steroid injection: A case report. *BMC Res Notes.* 2014; **7:** 327.
244. Neoh CY, Tan SH, Perera P. Cutaneous phaeohyphomycosis due to *Cladophialophora bantiana* in an immunocompetent patient. *Clin Exp Dermatol.* 2007; **32:** 539-40.
245. Ng CY, de Hoog S, Li HE, *et al.* Cutaneous *Exophiala oligosperma* infection in a patient with bullous pemphigoid with a review of the literature. *Mycopathologia.* 2017; **182:** 539-47.
246. Nobrega JP, Rosemberg S, Adami AM, *et al. Fonsecaea pedrosoi* cerebral phaeohyphomycosis ('chromoblastomycosis'): First human culture-proven case reported in Brazil. *Rev Inst Med Trop Sao Paulo.* 2003; **45:** 217-20.
247. Noguchi H, Hiruma M, Matsumoto T, *et al.* Subcutaneous cystic phaeohyphomycosis due to *Pleurostomophora richardsiae. J Dermatol.* 2017; **44:** e62-3.
248. Nomura M, Maeda M, Seishima M. Subcutaneous phaeohyphomycosis caused by *Exophiala jeanselmei* in collagen disease patient. *J Dermatol.* 2010; **37:** 1046-50.
249. Oberlin KE, Nichols AJ, Rosa R, *et al.* Phaeohyphomycosis due to *Exophiala* infections in solid organ transplant recipients: Case report and literature review. *Transpl Infect Dis.* 2017. doi: 10.1111/tid.12723. PMID: 28500778.
250. Ocampo MA, Kanitakis J, Bienvenu AL, *et al.* Phaeohyphomycosis caused by *Pyrenochaeta romeroi* mimicking a plantar wart in a kidney transplant recipient. *Transpl Infect Dis.* 2012; **14:** e173-4.
251. Ochiai H, Kawano H, Minato S, *et al.* Cerebral phaeohyphomycosis: Case report. *Neuropathology.* 2012; **32:** 202-6.
252. Ogawa MM, Galante NZ, Godoy P, *et al.* Treatment of subcutaneous phaeohyphomycosis and prospective follow-up of 17 kidney transplant recipients. *J Am Acad Dermatol.* 2009; **61:** 977-85.
253. Ogawa MM, Peternelli MP, Enokihara MM, *et al.* Spectral manifestation of melanized fungal infections in kidney transplant recipients: Report of six cases. *Mycopathologia.* 2016; **181:** 379-85.
254. Oliveira WR, Borsato MF, Dabronzo ML, *et al.* Phaeohyphomycosis in renal transplantation: Report of two cases. *An Bras Dermatol.* 2016; **91:** 89-92.
255. Osmond GW, Walters RW, Puri PK. Cutaneous alternariosis microscopically mimicking blastomycosis. *J Cutan Pathol.* 2011; **38:** 923-5.
256. Oztas E, Odemis B, Kekilli M, *et al.* Systemic phaeohyphomycosis resembling primary sclerosing cholangitis caused by *Exophiala dermatitidis. J Med Microbiol.* 2009; **58:** 1243-6.
257. Packeu A, Lebecque P, Rodriguez-Villalobos H, *et al.* Molecular typing and antifungal susceptibility of *Exophiala* isolates from patients with cystic fibrosis. *J Med Microbiol.* 2012; **61:** 1226-33.
258. Padhye AA, Davis MS, Baer D, *et al.* Phaeohyphomycosis caused by *Phaeoacremonium inflatipes. J Clin Microbiol.* 1998; **36:** 2763-5.
259. Padhye AA, Davis MS, Reddick A, *et al. Mycoleptodiscus indicus*: A new etiologic agent of phaeohyphomycosis. *J Clin Microbiol.* 1995; **33:** 2796-7.
260. Padhye AA, Karpati A, Rosenthal SA, *et al.* Subcutaneous phaeohyphomycotic abscess caused by *Pleurophomopsis lignicola. Med Mycol.* 2004; **42:** 129-34.
261. Pai VV, Naveen KN, Hanumanthayya K, *et al.* Subcutaneous phaeohyphomycosis presenting as a innocuous pustule. *Indian J Dermatol.* 2013; **58:** 159.
262. Palavutitotai N, Chongtrakoo P, Ngamskulrungroj P, *et al. Nocardia beijingensis* psoas abscess and subcutaneous phaeohyphomycosis caused by *Phaeoacremonium parasiticum* in a renal transplant recipient: The first case report in Thailand. *Southeast Asian J Trop Med Public Health.* 2015; **46:** 1049-54.
263. Palmisano A, Morio F, Le Pape P, *et al.* Multifocal phaeohyphomycosis caused by *Exophiala xenobiotica* in a kidney transplant recipient. *Transpl Infect Dis.* 2015; **17:** 297-302.
264. Papacostas LJ, Henderson A, Choong K, *et al.* An unusual skin lesion caused by *Lasiodiplodia theobromae. Med Mycol Case Rep.* 2015; **8:** 44-6.
265. Parente JN, Talhari C, Ginter-Hanselmayer G, *et al.* Subcutaneous phaeohyphomycosis in immunocompetent patients: Two new cases caused by *Exophiala jeanselmei* and *Cladophialophora carrionii. Mycoses.* 2011; **54:** 265-9.
266. Patel AK, Patel KK, Darji P, *et al. Exophiala dermatitidis* endocarditis on native aortic valve in a postrenal transplant patient and review of literature on *E.dermatitidis* infections. *Mycoses.* 2013; **56:** 365-72.
267. Patel R, Gustaferro CA, Krom RA, *et al.* Phaeohyphomycosis due to *Scopulariopsis brumptii* in a liver transplant recipient. *Clin Infect Dis.* 1994; **19:** 198-200.
268. Patra S, Vij M, Kancharla R, *et al.* Systemic *Wangiella dermatitidis* infection presenting as liver mass and obstructive cholangiopathy. *Trop Gastroenterol.* 2013; **34:** 277-9.
269. Pec J, Minarikova E, Zaborska D, *et al.* Treatment of dermal and subcutaneous phaeohyphomycosis of 55 years' duration. *Int J Dermatol.* 2008; **47:** 526-9.
270. Pereira RR, Nayak CS, Deshpande SD, *et al.* Subcutaneous phaeohyphomycosis caused by *Cladophialophora boppii. Indian J Dermatol Venereol Leprol.* 2010; **76:** 695-8.
271. Pincus LB, Schwartz BS, Cunningham G, *et al.* Cutaneous phaeohyphomycosis caused by *Cladophialophora bantiana* in a scar after treatment with intralesional corticosteroid injections. *J Am Acad Dermatol.* 2009; **61:** 537-8.
272. Ponnuswamy K, Muthureddy Y, Sigamani K. Two cases of multiple subcutaneous cystic phaeohyphomycosis in immunocompromised patients with a rare causative organism. *Indian J Dermatol.* 2014; **59:** 421.
273. Porter JG, Fine JD, Zwerner JP. Crateriform papule on the left knee in a 7-year-old boy. *Clin Exp Dermatol.* 2014; **39:** 97-8.
274. Pospisil L, Bucek J, Moster M, *et al.* Experimental phaeohyphomycosis. *Mycoses.* 1990; **33:** 469-75.
275. Qiu-Xia C, Chang-Xing L, Wen-Ming H, *et al.* Subcutaneous phaeohyphomycosis caused by *Cladosporium sphaerospermum. Mycoses.* 2008; **51:** 79-80.

276. Queiroz-Telles F, Nucci M, Colombo AL, *et al.* Mycoses of implantation in Latin America: An overview of epidemiology, clinical manifestations, diagnosis and treatment. *Med Mycol.* 2011; **49:** 225-36.
277. Raclavsky V, Novotny R. *Burkholderia cepacia* selective agar can be useful for recovery of *Exophiala dermatitidis* from sputum samples of cystic fibrosis patients. *J Cyst Fibros.* 2016; **15:** e19.
278. Radhakrishnan D, Jayalakshmi G, Madhumathy A, *et al.* Subcutaneous phaeohyphomycosis due to *Exophiala spinifera* in an immunocompromised host. *Indian J Med Microbiol.* 2010; **28:** 396-9.
279. Rai MK. *Phoma sorghina* infection in human being. *Mycopathologia.* 1989; **105**: 167-70.
280. Rajendran C, Khaitan BK, Mittal R, *et al.* Phaeohyphomycosis caused by *Exophiala spinifera* in India. *Med Mycol.* 2003; **41:** 437-41.
281. Rajput DK, Mehrotra A, Srivastav AK, *et al.* Cerebral phaeohypomycosis mimicking glioma. *Pan Arab J Neurosurg.* 2011; **15:** 87-90
282. Rammaert B, Aguilar C, Bougnoux ME, *et al.* Success of posaconazole therapy in a heart transplanted patient with *Alternaria infectoria* cutaneous infection. *Med Mycol.* 2012; **50:** 518-21.
283. Rantala M, Attia S, Koukila-Kahkola P, *et al. Cladophialophora bantiana* as an emerging pathogen in animals: Case report of equine endometritis and review of the literature. *J Clin Microbiol.* 2015; **53:** 3047-53.
284. Ravisankar S, Chander RV. Cerebral phaeohyphomycosis: Report of a rare case with review of literature. *Neurol India.* 2013; **61:** 526-8.
285. Rawal YB, Kalmar JR. Intraoral phaeohyphomycosis. *Head Neck Pathol.* 2012; **6:** 481-5.
286. Raza H, Khan RU, Anwar K, *et al.* Visceral phaeohyphomycosis caused by *Alternaria alternata* offering a diagnostic as well as a therapeutic challenge. *Saudi J Kidney Dis Transpl.* 2015; **26:** 339-43.
287. Revankar S. Phaeohyphomycosis in transplant patients. *J Fungi.* 2016; **2:** 2. doi: 10.3390/jof2010002.
288. Revankar SG, Sutton DA. Melanized fungi in human disease. *Clin Microbiol Rev.* 2010; **23:** 884-928.
289. Revankar SG. *Cladophialophora bantiana* brain abscess in an immunocompetent patient. *Can J Infect Dis Med Microbiol.* 2011; **22:**149-50.
290. Revankar SG. Dematiaceous fungi. *Mycoses.* 2007; **50:** 91-101.
291. Richey PM, Radfar A, Damavandy AA. Ulcerative pretibial lesions in the setting of multifactorial immunosuppression. *JAMA Dermatol.* 2016; **152:** 85-6.
292. Rimawi BH, Rimawi RH, Mirdamadi M, *et al.* A case of *Exophiala oligosperma* successfully treated with voriconazole. *Med Mycol Case Rep.* 2013; **2** :144-7.
293. Rinaldi MG. Phaeohyphomycosis. *Dermatol Clin.* 1996; **14:** 147-53.
294. Robert T, Talarmin JP, Leterrier M, *et al.* Phaeohyphomycosis due to *Alternaria infectoria:* A single-center experience with utility of PCR for diagnosis and species identification. *Med Mycol.* 2012; **50:** 594-600.
295. Romano C. Subcutaneous mycoses. Part 1: Subcutaneous mycoses due to non-dermatophytes. *G Ital Dermatol Venereol.* 2013; **148:** 621-31.
296. Rosow L, Jiang JX, Deuel T, *et al.* Cerebral phaeohyphomycosis caused by *Bipolaris spicifera* after heart transplantation. *Transpl Infect Dis.* 2011; **13:** 419-23.
297. Rossetto AL, Dellatorre G, Persio RA, *et al.* Subcutaneous phaeohyphomycosis on the scrotum caused by *Exophiala jeanselmei:* case report. *An Bras Dermatol.* 2010; **85:** 517-20.
298. Rudnicka L, Lukomska M. *Alternaria* scalp infection in a patient with alopecia areata. Coexistence or causative relationship? *J Dermatol Case Rep.* 2012; **6:** 120-4.
299. Russo JP, Raffaeli R, Ingratta SM, *et al.* Cutaneous and subcutaneous phaeohyphomycosis. *Skinmed.* 2010; **8:** 366-9.
300. Sadigursky D, Nogueira E Ferreira L, *et al.* Phaeohyphomycosis infection in the knee. *Rev Bras Ortop.* 2016; **51:** 231-4.
301. Saegeman V, Bammens B, De Munter P, *et al.* How to isolate a patient with *Cladophialophora bantiana* infection? An opinion. *J Hosp Infect.* 2015; **91:** 89-90.
302. Saha R, Rudra S. Phaeomycotic cyst - A case report. *J Indian Med Assoc.* 2005; **103:** 555-6.
303. Saint-Jean M, St-Germain G, Laferriere C, *et al.* Hospital-acquired phaeohyphomycosis due to *Exserohilum rostratum* in a child with leukemia. *Can J Infect Dis Med Microbiol.* 2007; **18:** 200-2.
304. Sakata Y, Kitayama A, Yoshimura R, *et al.* Case of cutaneous phaeohyphomycosis caused by *Phaeoacremonium* sp. in a renal transplant recipient. *J Dermatol.* 2015; **42:** 263-6.
305. Salido-Vallejo R, Linares-Sicilia MJ, Garnacho-Saucedo G, *et al.* Subcutaneous phaeohyphomycosis due to *Alternaria infectoria* in a renal transplant patient: Surgical treatment with no long-term relapse. *Rev Iberoam Micol.* 2014; **31:** 149-51.
306. Samerpitak K, Gerrits van den Ende AH, Menken SB, *et al.* Three new species of the genus *Ochroconis. Mycopathologia.* 2015; **180:** 7-17.
307. Sandoval-Denis M, Gene J, Sutton DA, *et al. Acrophialophora*, a poorly known fungus with clinical significance. *J Clin Microbiol.* 2015; **53:** 1549-55.
308. Sandoval-Denis M, Sutton DA, Martin-Vicente A, *et al. Cladosporium* species recovered from clinical samples in the United States. *J Clin Microbiol.* 2015; **53:** 2990-3000.
309. Sang H, Zheng XE, Kong QT, *et al.* A rare complication of ear piercing: A case of subcutaneous phaeohyphomycosis caused by *Veronaea botryosa* in China. *Med Mycol.* 2011; **49:** 296-302.
310. Sang H, Zheng XE, Zhou WQ, *et al.* A case of subcutaneous phaeohyphomycosis caused by *Cladosporium cladosporioides* and its treatment. *Mycoses.* 2012; **55:** 195-7.
311. Santos DW, Camargo LF, Gonçalves SS, *et al.* Melanised fungal infections in kidney transplant recipients: Contributions to optimize the clinical management. *Clin Microbiol Infect.* 2017; **23:** 333.e9-14.
312. Santos DW, Padovan AC, Melo AS, *et al.* Molecular identification of melanized non-sporulating moulds: A useful

tool for studying the epidemiology of phaeohyphomycosis. *Mycopathologia.* 2013; **175:** 445-54.
313. Sato T, Yaguchi T. A case of phaeohyphomycosis of the face caused by *Exophiala oligosperma* in an immunocompromised host. *J Dtsch Dermatol Ges.* 2013; **11:** 1087-9.
314. Schieffelin JS, Garcia-Diaz JB, Loss GE Jr, *et al.* Phaeohyphomycosis fungal infections in solid organ transplant recipients: Clinical presentation, pathology and treatment. *Transpl Infect Dis.* 2014; **16:** 270-8.
315. Secnikova Z, Juzlova K, Vojackova N, *et al.* The rare case of *Alternaria alternata* cutaneous and pulmonary infection in a heart transplant recipient treated by azole antifungals. *Dermatol Ther.* 2014; **27:** 140-3.
316. Segner S, Jouret F, Durant JF, *et al.* Cutaneous infection by *Alternaria infectoria* in a renal transplant patient. *Transpl Infect Dis.* 2009; **11:** 330-2.
317. Severo CB, Oliveira Fde M, Pilar EF, *et al.* Phaeohyphomycosis: A clinical-epidemiological and diagnostic study of eighteen cases in Rio Grande do Sul, Brazil. *Mem Inst Oswaldo Cruz.* 2012; **107:** 854-8.
318. Seyedmousavi S, Guillot J, de Hoog GS. Phaeohyphomycoses, emerging opportunistic diseases in animals. *Clin Microbiol Rev.* 2013; **26:** 19-35.
319. Seyedmousavi S, Netea MG, Mouton JW, *et al.* Black yeasts and their filamentous relatives: Principles of pathogenesis and host defense. *Clin Microbiol Rev.* 2014; **27:** 527-42.
320. Seyfarth F, Goetze S, Graser Y, Kaatz M, *et al.* Successful treatment of cutaneous alternariosis with caspofungin in a renal transplant recipient. *Mycoses.* 2012; **55:** 457-62.
321. Sharma NL, Mahajan V, Sharma RC, *et al.* Subcutaneous phaeohyphomycosis in India - A case report and review. *Int J Dermatol.* 2002; **41:** 16-20.
322. Sharma S, Capoor MR, Singh M, *et al.* Subcutaneous phaeohyphomycosis caused by *Pyrenochaeta romeroi* in a rheumatoid arthritis patient: A case report with review of the literature. *Mycopathologia.* 2016; **181:** 735-43.
323. Shashikala RP, Saikumar A, Ramesh K, *et al.* *Alternaria alternata* in a case of mass in the lung. *Indian J Med Microbiol.* 2013; **31:** 308-10.
324. Shirbur S, Telkar S, Goudar B, *et al.* Recurrent phaeohyphomycosis: A case report. *J Clin Diagn Res.* 2013; **7:** 2015-6.
325. Shrivastava A, Tadepalli K, Goel G, *et al.* Melanized fungus as an Epidural abscess: A diagnostic and therapeutic challenge. *Med Mycol Case Rep.* 2017; **16:** 20-4.
326. Shukla A, Bansal M, Husain M, *et al.* Central nervous system mycosis: Analysis of 10 cases. *Indian J Pathol Microbiol.* 2014; **57:** 591-4.
327. Shukla M, Vidyarani, Mathews A, *et al.* Phaeohyphomycosis masquerading as a palatal neoplasm in a patient who is immunocompetent. *BMJ Case Rep.* 2009; pii: bcr03. 2009.1721.
328. Silva WC, Goncalves SS, Santos DW, *et al.* Species diversity, antifungal susceptibility and phenotypic and genotypic characterisation of *Exophiala* spp. infecting patients in different medical centres in Brazil. *Mycoses.* 2017; **60:** 328-37.
329. Silveira CJ, Amaral J, Gorayeb RP, *et al.* Fungal meningoencephalitis caused by *Alternaria:* A clinical case. *Clin Drug Investig.* 2013; **33:** S27-31.
330. Singal A, Pandhi D, Bhattacharya SN, *et al.* Phaeohyphomycosis caused by *Exophiala spinifera:* A rare occurrence. *Int J Dermatol.* 2008; **47:** 44-7.
331. Singh A, Singh PK, Kumar A, Chander J, *et al.* Molecular and matrix-assisted laser desorption ionization-time of flight mass spectrometry characterization of clinically significant melanized fungi in India. *J Clin Microbiol.* 2017; **55:** 1090-1103.
332. Singh G, Shivaprakash MR, De D, *et al.* Chronic disfiguring facial lesions in an immunocompetent patient due to *Exophiala spinifera:* A case report and review of literature. *Mycopathologia.* 2012; **174:** 293-9.
333. Singh N, Agarwal R, Gupta D. Improvement with long-term itraconazole therapy for *Fonsecaea pedrosoi*-related mediastinal phaeohyphomycosis. *Eur Respir J.* 2008; **32:** 528-9.
334. Singh SM, Agarwal A, Naidu J, *et al.* Cutaneous phaeohyphomycosis caused by *Phialophora richardsiae* and the effect of topical clotrimazole in its treatment. *Antonie van Leeuwenhoek.* 1992; **61:** 51-5.
335. Singh SM, Naidu J, Pouranik M. Ungual and cutaneous phaeohyphomycosis caused by *Alternaria alternata* and *Alternaria chlamydospora. J Med Vet Mycol.* 1990; **28:** 275-8.
336. Skovrlj B, Haghighi M, Smethurst ME, *et al. Curvularia* abscess of the brain stem. *World Neurosurg.* 2014; **82:** 241.
337. Smith RM, Schaefer MK, Kainer MA, *et al.* Fungal infections associated with contaminated methylprednisolone injections. *N Engl J Med.* 2013; **369:** 1598-609.
338. Sood N, Gugnani HC, Guarro J, *et al.* Subcutaneous phaeohyphomycosis caused by *Alternaria alternata* in an immunocompetent patient. *Int J Dermatol.* 2007; **46:** 412-3.
339. Sood S, Vaid VK, Sharma M, *et al.* Cerebral phaeohyphomycosis by *Exophiala dermatitidis. Indian J Med Microbiol.* 2014; **32:** 188-90.
340. Sorkin M, Kung TA, Chung KC. Attritional extensor tendon rupture in a patient with *Phialophora verrucosa* tenosynovitis: Case report. *Hand (NY).* 2015; **10:** 342-5.
341. Standard PG, Padhye AA, Kaufman L. Exoantigen test for the rapid identification of *Exophiala spinifera. J Med Vet Mycol.* 1991; **29:** 273-7.
342. Sudhadham M, Gerrits van den Ende AH, Sihanonth P, *et al.* Elucidation of distribution patterns and possible infection routes of the neurotropic black yeast *Exophiala dermatitidis* using AFLP. *Fungal Biol.* 2011; **115:** 1051-65.
343. Suh MK, Lee HC, Kim DM, *et al.* Molecular phylogenetics of *Exophiala* species isolated from Korea. *Ann Dermatol.* 2012; **24:** 287-94.
344. Sujit Kumar GS, Dugar M, Chacko G. Cerebral phaeohyphomycosis presenting as an intraventricular mass. *Neurol India.* 2006; **54:** 102-3.
345. Suri P, Chhina DK, Kaushal V, *et al.* Cerebral phaeohyphomycosis due to *Cladophialophora bantiana* - A case report and review of literature from India. *J Clin Diagn Res.* 2014; **8:** DD01-5.

346. Sutton DA, Rinaldi MG, Kielhofner M. First US report of subcutaneous phaeohyphomycosis caused by *Veronaea botryosa* in a heart transplant recipient and review of the literature. *J Clin Microbiol.* 2004; **42:** 2843-6.
347. Suzuki K, Nakamura A, Fujieda A, *et al.* Pulmonary infection caused by *Exophiala dermatitidis* in a patient with multiple myeloma: A case report and a review of the literature. *Med Mycol Case Rep.* 2012; **1:** 95-8.
348. Taj-Aldeen SJ, Almaslamani M, Alkhalf A, *et al.* Cerebral phaeohyphomycosis due to *Rhinocladiella mackenziei* (formerly *Ramichloridium mackenziei*): A taxonomic update and review of the literature. *Med Mycol.* 2010; **48:** 546-56.
349. Taj-Aldeen SJ, El Shafie S, Alsoub H, *et al.* Isolation of *Exophiala dermatitidis* from endotracheal aspirate of a cancer patient. *Mycoses.* 2006; **49:** 504-9.
350. Tambasco D, D'Ettorre M, Bracaglia R, *et al.* A suspected squamous cell carcinoma in a renal transplant recipient revealing a rare cutaneous phaeohyphomycosis by Alternaria infectoria. *J Cutan Med Surg.* 2012; **16:** 131-4.
351. Teixeira MR, Assuncao C, Gomes R, *et al.* A case of subcutaneous phaeohyphomycosis associated with leprosy. *Infect Disord Drug Targets.* 2017; doi: 10.2174/187152651 7666170530124905. PMID: 28558644.
352. Tendolkar UM, Kerkar P, Jerajani H, *et al.* Phaeohyphomycotic ulcer caused by *Phialophora verrucosa*: successful treatment with itraconazole. *J Infect.* 1998; **36:** 122-5.
353. Teran CG, Downes K, Medows M. Fatal *Bipolaris spicifera* infection in an immunosuppressed child. *BMJ Case Rep.* 2014; pii: bcr2013009703. PMID: 24493105.
354. Tokuhisa Y, Hagiya Y, Hiruma M, *et al.* Phaeohyphomycosis of the face caused by *Exophiala oligosperma. Mycoses.* 2011; **54:** e240-3.
355. Tong Z, Chen SC, Chen L, *et al.* Generalized subcutaneous phaeohyphomycosis caused by *Phialophora verrucosa:* Report of a case and review of literature. *Mycopathologia.* 2013; **175:** 301-6.
356. Tran AN, Boyer JD. Phaeohyphomycosis of the hand treated using Mohs micrographic surgery. *Dermatol Surg.* 2013; **39:** 1276-9.
357. Tsang CC, Chan JF, Ip PP, *et al.* Subcutaneous phaeohyphomycotic nodule due to *Phialemoniopsis hongkongensis* sp. nov. *J Clin Microbiol.* 2014; **52:** 3280-9.
358. Tsang CC, Chan JF, Trendell-Smith NJ, *et al.* Subcutaneous phaeohyphomycosis in a patient with IgG4-related sclerosing disease caused by a novel ascomycete, *Hongkongmyces pedis* gen. et sp. nov.: First report of human infection associated with the family Lindgomycetaceae. *Med Mycol.* 2014; **52:** 736-47.
359. Tsujioka K, Tanaka R, Anzawa K, *et al.* A case of cutaneous phaeohyphomycosis caused by *Exophiala lecanii-corni* showing a seasonal fluctuation of skin lesions. *Med Mycol J.* 2015; **56:** J117-21.
360. Umabala P, Lakshmi V, Murthy AR, *et al.* Isolation of *Nodulisporium* species from a case of cerebral phaeohyphomycosis. *J Clin Microbiol.* 2001; **39:** 4213-8.
361. Umemoto N, Demitsu T, Kakurai M, *et al.* Two cases of cutaneous phaeohyphomycosis due to *Exophiala jeanselmei:* diagnostic significance of direct microscopical examination of the purulent discharge. *Clin Exp Dermatol.* 2009; **34:** e351-3.
362. Vallabhaneni S, Purfield AE, Benedict K, *et al.* Cardiothoracic surgical site phaeohyphomycosis caused by *Bipolaris* mould, multiple US states, 2008-2013: A clinical description. *Med Mycol.* 2016; **54:** 318-21.
363. Vasant JA, Maggiani F, Borman AM. Subcutaneous mycotic cyst caused by *Roussoella percutanea* in a UK renal transplant patient. *Mycopathologia.* 2017 Feb 11. doi: 10.1007/s11046-017-0121-0. PMID: 28190203.
364. Vasquez AM, Lake J, Ngai S, *et al.* Fungal bloodstream infections associated with a compounded intravenous medication at an outpatient oncology clinic - New York City. *MMWR.* 2016; **65:** 1274-5.
365. Vasquez-del-Mercado E, Lammoglia L, Arenas R. Subcutaneous phaeohyphomycosis due to *Curvularia lunata* in a renal transplant patient. *Rev Iberoam Micol.* 2013; **30:** 116-8.
366. Venkateshwar S, Ambroise MM, Asir GJ, *et al.* A rare case report of subcutaneous phaeohyphomycotic cyst caused by *Exophiala oligosperma* in an immunocompetent host with literature review. *Mycopathologia.* 2014; **178:** 117-21.
367. Verma P, Karmakar S, Pandhi D, *et al.* Chromoblastomycosis caused by *Cladophialophora bantiana* in a renal transplant recipient from Delhi, India. *Skinmed.* 2015; **13:** 251-4.
368. Verma R, Roy P, Vasudevan B, *et al.* Subcutaneous phaeohyphomycosis caused by *Bipolaris hawaiiensis* in an immunocompetent patient. *Indian J Dermatol Venereol Leprol.* 2014; **80:** 554-6.
369. Verma R, Vasudevan B, Badwal S, *et al.* Rare case of subcutaneous mycosis with intrathoracic extension due to *Chaetomium strumarium. Clin Exp Dermatol.* 2015; **40:** 622-5.
370. Vermeire SE, de Jonge H, Lagrou K, *et al.* Cutaneous phaeohyphomycosis in renal allograft recipients: Report of 2 cases and review of the literature. *Diagn Microbiol Infect Dis.* 2010; **68:** 177-80.
371. Vicente VA, Attili-Angelis D, Pie MR, *et al.* Environmental isolation of black yeast-like fungi involved in human infection. *Stud Mycol.* 2008; **61:** 137-44.
372. Wakamatsu K, Takahata Y, Tokuhisa Y, *et al.* Two cases of phaeohyphomycosis due to *Exophiala jeanselmei. J Dermatol.* 2008; **35:**178-80.
373. Wang L, She X, Lv G, *et al.* Cutaneous and mucosal phaeohyphomycosis caused by *Exophiala spinifera* in a pregnant patient: Case report and literature review. *Mycopathologia.* 2013; **175:** 331-8.
374. Wang L, Wang C, Shen Y, *et al.* Phaeohyphomycosis caused by *Exophiala spinifera:* An increasing disease in young females in mainland China? Two case reports and review of five cases reported from mainland China. *Mycoses.* 2015; **58:** 193-6.
375. Wang X, Wang W, Lin Z, *et al.* CARD9 mutations linked to subcutaneous phaeohyphomycosis and TH17 cell deficiencies. *J Allergy Clin Immunol.* 2014; **133:** 905-8.

376. Welfringer A, Vuong V, Argy N, *et al.* A rare fungal infection: Phaeohyphomycosis due to *Veronaea botryosa* and review of literature. *Med Mycol Case Rep.* 2017; **15:** 21-4.
377. Wong EH, Revankar SG. Dematiaceous molds. *Infect Dis Clin North Am.* 2016; **30:** 165-78.
378. Woo PC, Ngan AH, Tsang CC, *et al.* Clinical spectrum of *Exophiala* infections and a novel *Exophiala* species, *Exophiala hongkongensis. J Clin Microbiol.* 2013; **51:** 260-7.
379. Xue Y, Chen H, Hu S, *et al.* Cutaneous phaeohyphomycosis on the auricle due to *Veronaea botryosa. Eur J Dermatol.* 2011; **21:** 418-9.
380. Yadav S, Agarwal R, Singh S, *et al. Pyrenochaeta romeroi* causing subcutaneous phaeohyphomycotic cyst in a diabetic female. *Med Mycol Case Rep.* 2015; **8:** 47-9.
381. Yan XX, Yu CP, Fu XA, *et al.* CARD9 mutation linked to *Corynespora cassiicola* infection in a Chinese patient. *Br J Dermatol.* 2016; **174:** 176-9.
382. Yarita K, Sano A, Murata Y, *et al.* Pathogenicity of *Ochroconis gallopava* isolated from hot springs in Japan and a review of published reports. *Mycopathologia.* 2007; **164:** 135-47.
383. Yasui M, Hattori M, Ikegami N, *et al.* A case of alternariosis successfully treated with local hyperthermia. *Mycoses.* 2011; **54:** e623-6.
384. Yew SM, Chan CL, Lee KW, *et al.* A five-year survey of dematiaceous fungi in a tropical hospital reveals potential opportunistic species. *PLoS One.* 2014; **9:** e104352.
385. Yong LK, Wiederhold NP, Sutton DA, *et al.* Morphological and molecular characterization of *Exophiala polymorpha* sp. nov. isolated from sporotrichoid lymphocutaneous lesions in a patient with myasthenia gravis. *J Clin Microbiol.* 2015; **53:** 2816-22.
386. Yoon YA, Park KS, Lee JH, *et al.* Subcutaneous phaeohyphomycosis caused by *Exophiala salmonis. Ann Lab Med.* 2012; **32:** 438-41.
387. Yu J, Yang S, Zhao Y, Li R. A case of subcutaneous phaeohyphomycosis caused by *Chaetomium globosum* and the sequences analysis of *C.globosum. Med Mycol.* 2006; **44:** 541-5.
388. Yusupov N, Merve A, Warrell CE, *et al.* Multiple brain abscesses caused by *Rhinocladiella mackenziei* in an immunocompetent patient: A case report and literature review. *Acta Neurochir (Wien).* 2017. doi: 10.1007/s00701-017-3141-0. PMID: 28365816.
389. Zeng JS, Sutton DA, Fothergill AW, *et al.* Spectrum of clinically relevant *Exophiala* species in the United States. *J Clin Microbiol.* 2007; **45:** 3713-20.
390. Zhou YB, Chen P, Sun TT, *et al.* Acne-like subcutaneous phaeohyphomycosis caused by *Cladosporium cladosporioides*: A rare case report and review of published literatures. *Mycopathologia.* 2016; **181:** 567-73.

CHAPTER 15

Lacaziosis

Lacaziosis is a slowly progressive, chronic granulomatous fungal infection of skin and subcutaneous tissue. The disease is characterized by typical keloidal, nodular and verrucous lesions but usually without affecting either mucosal surfaces or internal organs of the body. This rare clinical entity is only prevalent in the New World tropics, particularly Brazilian Amazon region, implicating only two mammals *i.e.* man and dolphin. It used to be known as **Lobomycosis** in the past. It is also known with many synonyms such as **keloidal blastomycosis** or **Lobo's disease**.

The recently re-designated fungus, ***Lacazia loboi***, is the causative organism for this disease, which is yet to be successfully grown on artificial culture media. There is a similarly manifesting illness, reported in dolphins but without substantiation of a tissue diagnosis of fungal etiology, is called **lacaziosis-like disease**. As *L.loboi* has affinity towards water and moist areas hence it is broadly considered within **hydrophilic** group of organisms where two pseudofungi *i.e. Rhinosporidium seeberi* and *Pythium insidiosum*, are also placed.

Historical Perspective

Lacaziosis was first observed in 1931 by Brazilian dermatologist, Jorge Lobo working in the city of Recife, from a 52-year-old man who was working as a rubber collector in the Amazon basin of Brazil. The patient presented with gradually progressing confluent, multinodular keloidal lesions in the lumbosacral area. This was thought to be a new type of cutaneous mycosis, which he termed as **blastomycosis queloidiforme**. He observed that the causative organism resembled *Paracoccidioides brasiliensis*, another endemic mycosis found prevalent in the same overlapping geographical areas of South American subcontinent. Therefore, he thought it to be a milder form of paracoccidioidomycosis. However, based on clinical, histopathological and mycological findings, it turned out to be an entirely distinct and a novel clinical entity.

More cases were reported from Brazil and other Latin American countries and in 1938 it was referred to as Jorge Lobo's blastomycosis. Its causative organism has undergone an extensive classification and synonymy phase, with more frequently used terms, *Paracoccidioides loboi*, being introduced by Almeida and Lacaz In 1949. Ciferri *et al*, described the etiological fungus of this endemic disease in 1956. Two years later in 1958, Borelli proposed name of disease as **lobomycosis** and its causative agent as *Loboa loboi*.

In 1971, Fonseca and Lacaz again proposed the previous name, *Paracoccidioides loboi*, for the etiological agent because of its antigenic similarity; however, the term *Loboa loboi* was preferred and retained. Moreover, other binomial designations like *Glenosporella loboi, Glenosporopsis amazonica, Lobomyces loboi* and *Blastomyces loboi* were also proposed during initial days of its nomenclature but subsequently they were merged with either *P.brasiliensis* or correspondingly similar fungi of their clade.

This fungus was earlier presumed to exist only in soil but in 1971, Migaki *et al*, demonstrated the disease in bottlenose dolphins (*Tursiops truncatus*) thus providing a clue that perhaps it could also be a **hydrophilic** organism with an aquatic habitat. Thus the first report of non-human infection was documented in an Atlantic bottlenose female dolphin from Sarasota Bay, Florida, USA. Later on in 1973, dolphins with lacaziosis were identified in Europe as well, along the Atlantic coast of France and Spain. In addition, the dolphin's caretaker, a resident of The Netherlands presented with this disease in 1983. He was supposed to be the first human case of lacaziosis reported from outside Latin America as well as acquiring infection from animal. In due course of time, more cases of lacaziosis among dolphins were confirmed, including the one from the Texas coast.

The species epithet ***loboi*** was proposed by Fonseca and Leao and endorsed by Lacaz in 1996 to honor Brazilian dermatologist Jorge Lobo because of his comprehensive description of this disease along with detailed findings of fungal etiology. In 1999, Taborda *et al*, proposed re-designation of genus name of the causative fungus of lobomycosis from ***Loboa*** to ***Lacazia***. This nomenclature was adopted to acknowledge the major contributions made by **Carlos da Silva Lacaz** in understanding this enigmatic disease. The intention to change the name was to honor not only Jorge Lobo, discoverer of organism but also to acknowledge the most important personality of Brazilian mycology throughout second half of the 20th century, Carlos da Silva Lacaz (19.09.1915-23.04.2002). This re-designation was done during his lifetime, however, after his passing away on April 23, 2002, it was real tribute to him for his contribution as well as memories and subsequently acceptance of the changes in nomenclature were testimony, with passage of time. The mycology world celebrated his **Centennial Anniversary** in 2015 with great enthusiasm and fervor by publishing special issue of *Rev Inst Med Trop Sao Paulo* {2015; **57 (Suppl. 19):** 1-64} devoted to him.

This is only fungus in Medical Mycology where the designations of both the genus ***(Loboa,*** till 1999), its species ***(loboi)***, the name of disease **(Lobomycosis)** as well as its synonym **(Lobo's disease)**, all were assigned in honor of a single clinician, Jorge Lobo. At present, therefore, *Lacazia loboi* is regarded as the valid binomial nomenclature of fungus, previously known as *Loboa loboi*. In 2005, Vilela *et al*, introduced the new name of this clinical entity as **Lacaziosis**. It is tradition in Medical Mycology that designation of diseases are assigned on genus names of etiologic agents. On the same analogy, this new name (lacaziosis) was floated and it has now replaced the older name (lobomycosis), which was given about six decades ago in 1958. Now, the newly proposed nomenclature of the disease has already been accepted to a great extent. The genus name, *Loboa*, ceased to exist and was replaced by *Lacazia* in 1999, thereby, corresponding name of disease also got replaced from lobomycosis to lacaziosis in 2005.

The author paid tribute to Prof. Carlos da Silva Lacaz by visiting his Placard, Statue **(Fig. 15.1)** as well as Museum and delivered the Memorial Oration to the students and faculty at the Institute of Tropical Medicine, University of Sao Paulo, SP, (Brazil) on March 31, 2017.

Mycology

The genus *Lacazia* has a single member species, which is the causative agent of lacaziosis. In 1999, Taborda *et al*, proposed its binomial name as *Lacazia loboi*, replacing extant one *i.e. Loboa loboi*. This organism exists as lemon-shaped, globose yeast cells with hyaline double refractile wall, which is consistent in their size ranging between 6 to 12 μm. There are unbranched or linear chains of 4 to 7 cells due to sequential budding linked to one another by a tubular connection or isthmus. This typical bead-like arrangement of yeasts resembles ripened seedpods of prickly **acacia** (*Acacia nilotica* sub. sp. *indica*), which is also constricted between seeds **(Fig. 15.2)**. However, some of these yeast cells may have secondary buddings entailing branched or radiating chains of cells. Such type of

Fig. 15.1. Statue of Prof. Carlos da Silva Lacaz on the front gate of Institute of Tropical Medicine, University of Sao Paulo, SP, Brazil.

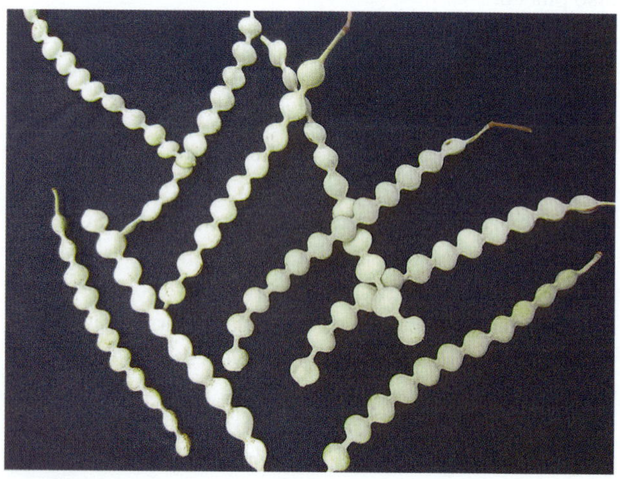

Fig. 15.2. Ripened seedpods of prickly *Acacia nilotica* sub. sp. *indica* wherein individual seedpods joined by tube-like bridges.

morphological arrangement of yeast cells in tissue may sometimes resemble ***'mariner's wheel'*** form of *P.brasiliensis.*

This fungus has not been successfully cultivated *in vitro* till date thus precluding use of conventional methods for classifying it to a member of kingdom Fungi. Moreover, antifungal drugs have also no significant effect on this organism. Therefore, due to such unusual findings, this was one of debatable issues whether to classify it as a fungus or not. However, this is now settled as molecular techniques have analysed it to be a taxonomically ascomycetous fungus.

It is also speculated that this uncultivable organism might be a dimorphic fungus where yeast form is detectable in the tissues but its saprobic phase exists in the environmental niche may be soil, vegetation or aquatic. It might be because this fungus requires a slightly lower temperature for its growth. This could also be a plausible explanation for its cutaneous and subcutaneous manifestations, as involvement of deeper tissues is not usually seen.

Vilela *et al*, have recently introduced a new molecular strategy to investigate unknown proteins of interest in immunology and pathogenesis of infections caused by *L.loboi.* Based on the observation of Herr *et al*, wherein they indicated that *P.brasiliensis* could be the phylogenetic sister group of *L.loboi*, Vilela *et al*, suggested that both these pathogens, endemic in South America, might be sharing DNA sequences.

However, despite several decades of traditional taxonomic investigations, relationship of this fungus to other organisms remained elusive. But when Herr *et al*, amplified 18S small-subunit rDNA gene and 600-bp chitin synthase-2 gene of *L.loboi* by using universal primers, this ambiguity came to an end. It was found that this unique organism was phylogenetic sister taxon of dimorphic onygenal pathogen, *P.brasiliensis*, which is also found endemic in Latin America. Although *L.loboi* and *P.brasiliensis* share a close genomic sequence but there are significant differences between them. Further, in-depth studies after successful *in vitro* isolation of fungus will highlight unexplored phylogenetic areas. Therefore, based on recent molecular analysis, *Lacazia loboi* is currently classified within order **Onygenales** under phylum **Ascomycota** where most of the **dimorphic fungi** are placed.

Epidemiology

Lacaziosis is a chronic, cutaneous and subcutaneous mycosis prevalent in the New World tropics. This is found in humid, tropical and subtropical forests where mean temperature is above 24°C and annual rainfall is over 2000 mm. These areas are located about 200 meters above the sea level and belong to similar biographical complexes. The infection is more prevalent particularly among those individuals who work in close proximity of thick vegetation, with high temperature and humidity as found in the Amazon rain forests. Therefore, disease occurs among individuals residing in rural areas and involved in occupation, such as farming, hunting and rubber tapping hence it frequently affects farmers, rubber, plant and mineral workers. All age groups are affected, with the mean age of onset among young males being between 20 and 40 years.

The areas of endemicity and geographical distribution of lacaziosis is more or less overlapping with that of paracoccidioidomycosis. In endemic areas, most of the patients are inhabitants of Amazon basin of central and western Brazil, the country which constitutes almost one-fourth of the total cases of the world. Approximately, 550 cases have been reported mainly from South and Central America and only five cases from Europe and three from North America.

The **Brazilian** and **Colombian** regions have presented the largest number of cases. In 1999, Opromolla *et al*, reported 40 new cases of lacaziosis in Brazil. One of the most significant aspects of disease is its high incidence among an indigenous tribe, **Cayabi Indians** living in Brazilian state of **Mato Grosso** in the Amazon region. These account for about 60 percent of cases, out of total reported in Brazil and due to involvement of this ethnic group, the clinical condition used to be once known as Cayabi leprosy. Moreover, natives of Brazilian rain forest recognize this disease as ***'piraip'*** or ***'miraip'***, meaning in their local vernacular (Tupi language) as ***'burning'***. There is case report of an old man where the disease is reported along with lepromatous leprosy, in the Amazon region of Brazil, which is first such an instance in English literature.

In addition to Brazil and Colombia, the human lacaziosis has also been sporadically reported from seven other South American countries *i.e.* Suriname, Venezuela, Peru, Guyana, French Guiana, Ecuador and Bolivia and three Central American countries *i.e.* Costa Rica, Panama and Mexico and only five cases have been reported from Europe, one of them resulted from handling of the dolphin. The African American and American Indians are commonly affected with lacaziosis. There are no reports of this disease from some of adjoining countries to the endemic ones in Latin America like Chile, El Salvador, Guatemala and Nicaragua. The geographical distribution of lacaziosis in the endemic areas is shown over the map of Central and South America **(Fig.15.3)**.

Fig. 15.3. Map showing geographical distribution of endemic areas of lacaziosis in Central as well as South America.

A few **imported** cases of lacaziosis have also been reported from non-endemic areas outside Latin America. In the European continent, case reports are available from France, Germany and Greece. However, the very first case in Europe was from The Netherlands, who was presumed to have been got infected from dolphins, while handling them. From North America (Atlanta, Georgia), Burns, *et al*, reported a case in the year 2000 involving chest wall of a person who had previously traveled to Venezuela and was exposed to extremely high water pressure while walking under Angel Falls in Canaima Amazon forest. Elsayed *et al*, reported another case from North America (Calgary, Canada) in 2004. One more case was reported from New York by Arju *et al*, who had traveled to Brazil, making a total of three cases from the North America. The occurrence of lacaziosis was supposed to be confined to New World tropical countries but its recent report in African patients has broken this belief. In 2008, Al-Daraji *et al*, have reported two South African patients, who were found to be suffering from this disease but never visited South America. However, both patients had traveled to Mexico and Palestine and might have acquired the infection as 'cause and effect' relationship, remains circumstantial, due to lack of confirmatory history.

All attempts to isolate this uncultivable fungus from the lesions of affected individuals till date have been unsuccessful. It is contemplated that the source of this infection is exogenous. Therefore, *L.loboi* is believed to be saprophyte of soil or vegetation whereas no reservoir has been identified till date. The natural habitat of organism has been hypothesized as aquatic environment proposing it to be a **hydrophilic fungus**. As it is thought that this fungus might be associated with water thereby gain entry through trauma-inflicted wounds. The trauma to skin seems to be the mode of acquiring infection and development of lesions.

Therefore, due to extensive outdoor exposure, rubber workers, farmers, miners, fishermen and hunters are at higher risk. The site of involvement may be quite extensive since spread by autoinoculation occurs over the years leading to new crops of satellite nodular lesions. The fungus apparently proliferates in skin because it prefers cool body sites. Therefore, Rippon once suggested that inability to culture the organism and its clinical restriction to cooler extremities of body indicate that it might be a parasite of lower animals.

In addition to man, infection has also been reported in **two** genera of dolphins from **three** different areas: (i) marine bottle-nose dolphins (*Tursiops truncatus*) found off the coast of Florida, (ii) Indo-Pacific bottlenose dolphins (*Tursiops aduncus*) found in Indo-Pacific sea (iii) fresh-water/Guiana dolphins (*Sotalia guianensis*) found in Suriname river estuaries. Therefore, with emergence of this disease among dolphins, which inhabit marine as well as fresh water, there is strong possibility of an aquatic natural habitat of *L.loboi*. However, as such there is no evidence of natural transmission of this disease from man-to-man or animal-to-man with exception of the very first case from Europe, which was probably due to an occupational exposure. In another case it was reported to be transmitted from experimental mice-to-man in Sao Paulo, Brazil.

Immunity

Very little information is available about immunity in lacaziosis because *L.loboi* has not yet been successfully cultured *in vitro*. Therefore, the inability to cultivate the etiological fungus has hampered all endeavors of gaining insight into its immunology. However, studies about its immunogenic components are feasible by use of heterologous fungal antigens. *Lacazia loboi* seems to share antigens with *P.brasiliensis*. The rabbit serum raised against *P.brasiliensis* reacts positively with histological material from biopsies of patients, indicating an antigenic relationship between both these fungi. In addition, sera from patients with lacaziosis react with *P.brasiliensis* in the indirect immunofluorescence assays, however, these are not capable of defining specifically that which antigens are involved in cross-reaction. It is thought that *P.brasiliensis* gp43 might represent main cross-reacting antigen and suggest that this antigen is shared between these two fungi.

Pathogenesis and Pathology

The pathogenesis of lacaziosis still remains obscure to a great extent. The granulomatous lesions consist of a mass of globose yeasts, uniform in size with average being 9 µm, usually form single chain of about 4 to 7 cells, which are joined by bridges, however, branching may also be seen. The most distinguishing pathological feature lies in dermis, which shows granulomatous inflammatory reaction consisting of packed histiocytes and Langhans giant cells containing numerous yeasts. The inflammatory cells include many foamy macrophages, multinucleated foreign body giant cells and scattered lymphocytes. Many of these phagocytes contain intact or fragmented fungal cells. The lymphocytes and plasma cells may be seen in scattered clusters. The involvement of lymph nodes in some of the cases substantiates lymphatic spread of the organism.

The appearance of epidermis is variable showing atrophy, hyperkeratosis, hyperpigmentation or parakeratosis. There is no evidence of necrosis or suppuration in the lesions of lacaziosis. The atrophied epidermis neither shows pseudo-epitheliomatous hyperplasia nor intra-epidermal abscesses that are common in other fungal infections like blastomycosis or coccidioidomycosis.

Asteroid bodies may be observed in giant cells in lacaziosis, which are also seen in other clinical entities like sarcoidosis, leprosy, Miescher's granuloma and foreign body tissue reactions. However, as such there is no morphological or pathological relationship with asteroid bodies of sporotrichosis.

Clinical Features

The disease entity, lacaziosis, is a progressive, chronic, localized, subcutaneous fungal infection of an insidious onset characterized by presence of macular, papular, keloidal, verrucous, nodular lesions or vegetating crusty plaques and tumors. The incubation period of disease varies from months to years. The initial infection takes place at the site of trauma into the skin. Most of the times, trauma is too trivial to be recollected by the patient. The disease process takes an extremely slow course and years may elapse before patient seeks medical advice. The lesions often begin as small macules or papules that may occur at sites of minor trauma. Occasionally, single lesion may regress to form a scar. However, the disease never disappears, as organisms are seen in scar tissue. The patients usually lack other systemic symptoms in this disease.

The infection is confined only to dermis as neither mucosal surface nor deeper tissues are involved. The lesions remain fairly well-localized and apparently do not affect general condition of the patient. The size of the lesions depends upon duration of infection. There may be single, small, hard nodule at an early stage, which becomes elevated and subsequently as keloidal type. As lesions become older, verrucous and ulcerative forms of disease may also occur. The colour of nodule may be slightly brownish with telangiectasis, however, erythema is absent as surrounding cutaneous tissue is normal.

The lesions may be single or multiple and found anywhere over the body but invariably the exposed parts are often affected like the lower limbs, ears, upper limbs and face in order of frequency. The other sites such as chest, scapula, lumbosacral area, buttocks and scrotum may also be involved. The lesions may be hypo- or hyperpigmented. The mucous membrane or internal organs of body are not affected, however, it has been found to affect regional lymph nodes in around 10 to 25% of cases. The older lesions typically become verrucoid and ulcerative with satellite lesions resulting from autoinoculation.

The very early stages of disease are commonly described as ***'wart-like'*** infiltrative lesions. The typical keloid-like skin lesions appear only after several months. They increase in size by contiguity but can also spread from one site to another by autoinoculation. The lesions have well-defined lobulated margins, which are not attached to deeper structures. The epidermis may be shiny, atrophic and discolored. The disease may also spread proximally from extremities, suggesting the lymphatic spread.

The lesions are invariably painless with slight pruritus. The older ones become verrucoid and may even ulcerate. The organism remains dormant, as disease is present for a long period of many years, may be for 40 to 50 years. The male patients outnumber females by a ratio of more than 20 : 1 since majority of them are engaged in outdoor agricultural and forest activities in rural areas. A case of this disease from Brazil has been reported to be associated with HIV infection. As there is no systemic spread of organism hence usually deeper organs are spared and visceral lesions do not occur. However, there are recent reports of disseminated lesions in this disease, which may be attributed to lymphatic spread.

The monomorphic concept of lesions was supposed to be obsolete hence Silva in 1972 described other forms of disease besides **keloidal type**, where many types of lesions are present simultaneously. The dermal lesions are polymorphic, ranging from macules, papules, keloidal nodules and plaques to verrucous and ulcerated, all of them may be present in single patient. These may be **infiltrative type** and usually associated with initial stage of infection, **gummatous type** in which organisms are lodged in deeper tissues, **ulcerative type** where epidermis is lost and secondary bacterial infections crop up and **verrucous type** observed in old and chronic lesions. Therefore, these **five types** of lesions may exist in this disease and more than one type may be present in a single patient. In this disease, there is no tendency for the lesions to heal spontaneously. Sometimes squamous cell carcinoma may also develop due to chronic nature of the lesions.

Lacaziosis in Dolphins

The natural infection in animals has so far been observed in **two** genera of **three** species of dolphins namely, marine bottlenose dolphins (*Tursiops truncatus*), Indo-Pacific bottlenose dolphins (*Tursiops aduncus*) and freshwater/ Guiana dolphins (*Sotalia guianensis*). The disease in dolphins presents with whitish-gray to slightly pinkish verrucous lesions predominately on their **dorsal aspect**. The lesions are histopathologically somewhat similar to human infections but the size of the yeasts is **smaller** one suggesting a possibility that organisms may not be identical in both hosts. The lesions may ulcerate and form large plaques also. The presence of lesions in dolphins solely over dorsal aspect suggests that development of lacaziosis is secondary to a traumatic event, probably during fishing or riding pleasure boats. The dolphins with lacaziosis have been found frequently along the coasts of Florida and Texas, Spanish-French coast, south Brazilian coast as well as Suriname river estuary.

Lacaziosis-like Disease

The causative organism of *L.loboi* does not grow on artificial culture media creating hindrance in its diagnostic, therapeutic and research activities. Now, there are reports of another apparently similar looking infection called **Lacaziosis-like Disease** (LLD) because its histologic diagnosis is still lacking. As this is mainly involving the dolphins, it is also called cetacean lacaziosis-like disease. It is found in coastal bottlenose dolphins from Colombia, Ecuador, Peru and Brazil; Guiana dolphins in Brazil and Indo-Pacific bottlenose dolphins (*T.aduncus*) in the tropical lagoon of Mayotte in the Indian Ocean between Mozambique and Madagascar. This is highly reminiscent of lobomycosis but its exact aetiology is still unknown. Moreover, an Indo-Pacific bottlenose dolphin (*Tursiops aduncus*) stranded in Kagoshima, Japan had severe skin lesions characterized by granulomatous reactions and hyperkeratosis that were similar to those of lacaziosis but no fungal organism was observed in the skin lesion. In the absence of a tissue diagnosis, the disease is called lacaziosis-like disease. Complete identification and phylogenetic classification of its etiological agent still needs to be elucidated but preliminary phylogenetic analyses have shown a closer relationship of the LLD agent with *Paracoccidioides* spp. rather than to *L.loboi.*

Differential Diagnosis

Clinically, lacaziosis should be differentiated from vegetating type of chromoblastomycosis, sporotrichosis, anergic cutaneous leishmaniasis, paracoccidioidomycosis, psoriasis, nodular lepromatous or tuberculoid type of leprosy, cutaneous tuberculosis, tropical pyodermatitis, eczematoid dermatitis, eumycetoma, squamous cell carcinoma, melanoma, Kaposi's angiosarcoma and true keloids caused by either burns or trauma.

Histopathologically, the budding yeasts, closely resemble those of *P.brasiliensis*; however, its larger mother cell with smaller multiple daughter cells help to distinguish it from *L.loboi*. Other fungal disease like blastomycosis and African histoplasmosis may also mimic lacaziosis due to similarity of yeast cells in tissues. The refractile globose yeasts producing blastoconidia with all cells being of almost equal size and some of them found in chains are consistently seen with *L.loboi* and not with other fungal infections. In addition, cell wall of *L.loboi* contains constitutive melanin that is readily detectable by Masson-Fontana stain, which is not seen in *P.brasiliensis*. The other dimorphic fungi like *Blastomyces dermatitidis* and *Histoplasma capsulatum* var. *duboisii* should be differentiated, as there are no chains of yeast cells. Moreover, both *S.schenckii* and *H.capsulatum* var. *capsulatum* have smaller yeast cells of 2 to 8 μm whereas *L.loboi* has 6 to 12 μm.

The regressive lesions as observed in leprosy and filariasis may mimic lacaziosis. There is no simultaneous development of lesions of mucous membrane in lacaziosis as it is observed in leishmaniasis and paracoccidioidomycosis. The affected pinna in lacaziosis may present similar to be seen in chondritis, sebaceous cysts and gout. In case of wrestlers, chondritis due to repeated blunt trauma to pinna look similar to as that of lacaziosis.

Laboratory Diagnosis

In the laboratory, final diagnosis of lacaziosis is established on the basis of demonstration of characteristic yeast cells in direct examination of wet mounts and stained preparation of clinical specimen because culture is not feasible due to inability of this fungus to grow in artificial media.

(a) Direct Examination

The clinical material is obtained by curettage or biopsy by surgical excision and is homogenized to examine in KOH and/or CFW wet mounts. A vinyl adhesive tape (VAT) preparation (Scotch Test) has been proved to be useful for the direct microscopic examination. *Lacazia loboi* is usually intracellular although extracellular forms may be seen. The narrow-based budding yeast cells are globose to elliptical or lemon-shaped, doubly-refractile, thick-walled and multinucleate but fairly uniform in size ranging from 6-12 µm, with an average size of 10 µm as seen in GMS staining **(Fig. 15.4)**. There may be chains of 20 or even more hourglass-like yeast cells with a large connecting isthmus to each other by a narrow-neck as grossly seen in the fruits of acacia. The organisms usually reproduce by budding at one or rarely more points similar to as that of *P.brasiliensis*. However, consistency in its size and characteristic chain-like arrangement of yeast cells in *L.loboi,* clearly distinguish it from *P.brasiliensis*.

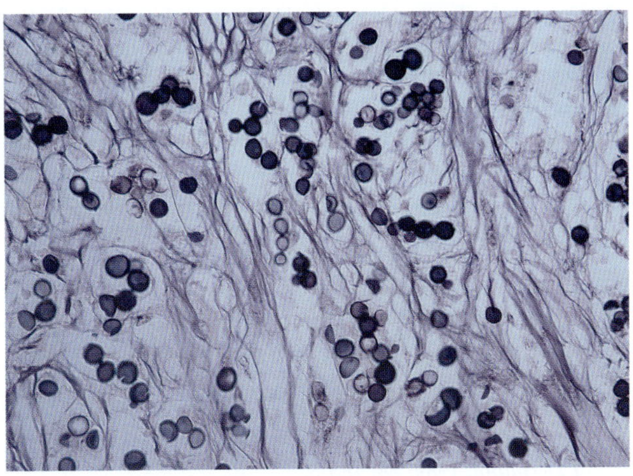

Fig. 15.4. Budding yeasts of *Lacazia loboi* in chains wherein cells joined by tube-like bridges in dermis of a patient (GMS × 400).

On H&E staining, there is a diffuse, superficial and deep granulomatous dermatitis with multinucleated giant cells. There are faintly stained or unstained fungal walls with or without a stained protoplasmic mass, appearing like vacuole. Sometimes, halo-like structures are seen surrounding the fungal cells, which are mucicarmine negative. The intracellular and extracellular unstained cells with thick doubly refractile walls give a ***'sieve-like'*** pattern to the granulomatous inflammation. There is dispersed granulomatous infiltrate with giant cells, macrophages, epithelioid cells and numerous rounded yeasts in chains of 4 to 7. The epidermis is usually atrophic but may show pseudoepitheliomatous hyperplasia or ulceration. The yeasts are mostly intracellular within giant cells and macrophages. Fibrosis is infrequent and there is no necrosis.

The yeast cells of *L.loboi* are deeply stained with special stains such as PAS, GMS and Gridley fungal staining. In Gridley stain, walls of the organism appear red and protoplasmic mass as yellowish-brown. In certain cells, granular material may appear more sharply defined with PAS than Gridley stain. The connecting tubes between mother and daughter cells are best demonstrated with GMS stain as compared to other fungal stains. For determination of viability of *L.loboi,* fluorescein diacetate-ethidium bromide (FD-EB) vital staining has been found to be specific and sensitive method that can reliably be used.

There is formation of dermal granulomas in which giant cells phagocytose an overwhelming number of fungi. Intracytoplasmic asteroid bodies have been observed in giant cells in about 25% of cases, which occurs in other clinical entities as mentioned above, however, with significant morphological differences. An inflammatory infiltrate of histiocytes is present and dermis may be completely replaced by histiocytic granuloma.

(b) Fungal Culture

Like *Rhinosporidium seeberi* and *Mycobacterium leprae*, *L.loboi* has not been successfully grown till date *in vitro* in artificial culture media. Therefore, this fungus does not fulfill conventional four postulates laid down by Robert Koch by the turn of 19th century. The humans have been inoculated both voluntarily and involuntarily however without much success. There is also one well-documented case of experimental inoculation in human volunteer who developed single lesion 3 months after intradermal injection of infected tissue.

(c) Immunodiagnosis

There is no standardized serological procedure for diagnosis of lacaziosis due to absence of evidence regarding development of specific antibodies. A skin test, with antigen **lobin**, is described but its specificity or utility has not been established. The skin tests done with paracoccidioidin are found to be negative in patients of lacaziosis. The fungal rDNA has also been isolated from dolphin tissue infected with *L.loboi*. A roboprobe derived from one of the isolated fragments was subsequently hybridized to *L.loboi* in paraffin sections providing first molecular clue supporting hypothesis that this organism is a true fungus.

(d) Animal Pathogenicity

The experimental inoculation of laboratory animals with suspensions of *L.loboi* cells from either human or dolphin lesions rarely results in transmission of infection. This disease has been successfully transmitted to armadillos, tortoises, mice and hamsters. The mice have been successfully infected with injections of *L.loboi* cells into footpads through nine generations. Typical chronic nodules appear slowly over a period of many months. Other routes of inoculation like intraperitoneal, intravenous or intratesticular were not found to be successful. It can also be propagated by inoculation in hamsters (*Mesocricetus auratus*) and armadillos (*Euphractus sexcintus*) resulting in nodular lesions, which are similar to human lesions. Opromolla *et al*, have tried to produce lesions by inoculation of *L.loboi* into subcutaneous tissue of hamster cheek pouch. There was experimental reproduction of the disease in BALB/c mice inoculated with *L.loboi* yeast cells obtained from a previously infected mouse.

Treatment and Prophylaxis

Lacaziosis neither resolves spontaneously nor it has any effective medical treatment. The antifungals reported to be effective, in the empirical management of other cutaneous and subcutaneous fungal infections, have proven unsuccessful and there is no satisfactory therapeutic approach for tackling this disease. The drugs like 5-fluorocytosine, itraconazole and amphotericin B as well as other antimicrobials such as sulfonamides, have been tried with unsatisfactory clinical response, which may be due to the slow generation time. An anti-leprotic drug, clofazimine in the dose of 100-200 mg/day, is used with good results in some of the patients. Posaconazole is stated to have quite successful results. Moreover, combination of clofazimine and itraconazole is also claimed to be effective in the treatment.

Hence keeping in view of the mixed therapeutic results, management by complete surgical excision with wide margins is the only ideal treatment of choice. There is frequent recurrence after excision, therefore, care must be taken to avoid contamination of healthy tissue in the surrounding vicinity while performing surgery. Cryosurgery is also employed in its treatment as performed in case of chromoblastomycosis. A combination of itraconazole, cryosurgery with liquid nitrogen has been quite successful. Moreover, electrodessication is useful in early stages of disease. As such it is not life-threatening disease because it does not affect overall the general health of infected individuals and as such there is no death reported till date. The disease can be prevented by taking routine measures by avoiding traumatic injuries while performing outdoor activities.

Further Reading

1. Al-Daraji WI, Husain E, Robson A. Lobomycosis in African patients. *Br J Dermatol.* 2008; **159:** 234-6.
2. Al-Daraji WI. Cutaneous lobomycosis: A delayed diagnosis. *Am J Dermatopathol.* 2008; **30:** 575-7.
3. Arju R, Kothadia JP, Kaminski M, *et al.* Jorge Lobo's disease: A case of keloidal blastomycosis (lobomycosis) in a non-endemic area. *Ther Adv Infect Dis.* 2014; **2:** 91-6.
4. Azulay RD, Carneiro JA, Da Graca M, *et al.* Keloidal blastomycosis (Lobo's disease) with lymphatic involvement: A case report. *Int J Dermatol.* 1976; **15:** 40-2.
5. Barbosa AS, Diorio SM, Pedrini SC, *et al.* Nutritional status and immune response in murine experimental Jorge Lobo's disease. *Mycoses.* 2015; **58:** 522-30.
6. Barboza TC, Quaresma JA, de Brito AC, *et al.* Jorge Lobo's disease: Immunohistochemical characterization of dendritic cells in cutaneous lesions. *Mycopathologia.* 2015; **179:** 269-74.
7. Baruzzi RG, Rodrigues DA, Michalany NS, *et al.* Squamous-cell carcinoma and lobomycosis (Jorge Lobo's disease). *Int J Dermatol.* 1989; **28:** 183-5.
8. Belone AF, Madeira S, Rosa PS, *et al.* Experimental reproduction of the Jorge Lobo's disease in BALB/c mice

inoculated with *Lacazia loboi* obtained from a previously infected mouse. *Mycopathologia.* 2002; **155:** 191-4.

9. Bermudez L, Van Bressem MF, Reyes-Jaimes O, *et al.* Lobomycosis in man and lobomycosis-like disease in bottlenose dolphin, Venezuela. *Emerg Infect Dis.* 2009; **15:** 1301-3.
10. Bessesen BL, Oviedo L, Burdett Hart L, *et al.* Lacaziosis-like disease among bottlenose dolphins *Tursiops truncatus* photographed in Golfo Dulce, Costa Rica. *Dis Aquat Organ.* 2014; **107:** 173-80.
11. Brun AM. Lobomycosis in three Venezuelan patients. *Int J Dermatol.* 1999; **38:** 302-5.
12. Burdett Hart L, Rotstein DS, Wells RS, *et al.* Lacaziosis and lacaziosis-like prevalence among wild, common bottlenose dolphins *Tursiops truncatus* from the west coast of Florida, USA. *Dis Aquat Organ.* 2011; **95:** 49-56.
13. Burns RA, Roy JS, Woods C, *et al.* Report of the first human case of lobomycosis in the United States. *J Clin Microbiol.* 2000; **38:** 1283-5.
14. Bustamante B, Seas C, Salomon M, *et al.* Lobomycosis successfully treated with posaconazole. *Am J Trop Med Hyg.* 2013; **88:** 1207-8.
15. Caldwell DK, Caldwell MC, Woodard JC, *et al.* Lobomycosis as a disease of the Atlantic bottle-nosed dolphin (*Tursiops truncatus* Montagu, 1821). *Am J Trop Med Hyg.* 1975; **24:** 105-14.
16. Camargo ZP, Baruzzi RG, Maeda SM, *et al.* Antigenic relationship between *Loboa loboi* and *Paracoccidioides brasiliensis* as shown by serological methods. *Med Mycol.* 1998; **36:** 413-7.
17. Carneiro FP, Maia LB, Moraes MA, *et al.* Lobomycosis: Diagnosis and management of relapsed and multifocal lesions. *Diagn Microbiol Infect Dis.* 2009; **65:** 62-4.
18. Carvalho KA, Floriano MC, Enokihara MM, *et al.* Jorge Lobo's disease. *An Bras Dermatol.* 2015; **90:** 586-8.
19. Cheuret M, Miossec C, Milley S, *et al.* A 43-year-old Brazilian man with a chronic ulcerated lesion. *Clin Infect Dis.* 2014; **59:** 264, 314-5.
20. Ciferri R, Carneiro LS, Campos S, *et al.* Advances in the knowledge of the fungus of Jorge Lobo's disease. *J Trop Med Hyg.* 1956; **59:** 214-5.
21. Cowan, D. F. Lobo's disease in a bottlenose dolphin (*Tursiops truncatus*) from Matagorda Bay, Texas. *J Wildl Dis.* 1993; **29:** 488-9.
22. da Costa SH, Da Silva M, Almeida HJR, *et al.* An unusual association between chromoblastomycosis and Jorge Lobo's disease in the same patient from the Amazon region. *Am J Agric Biol Sci.* 2008; **3:** 575-8.
23. de Souza MN, Schlosser AR, da Silva-Nunes M. Lobomycosis of the lower limb in an Amazonian patient. *Am J Trop Med Hyg.* 2015; **93:** 675-6.
24. Elsayed S, Kuhn SM, Barber D, *et al.* Human case of lobomycosis. *Emerg Infect Dis.* 2004; **10:** 715-8.
25. Elston DM, Sangueza OP. Fungal foes: *Lacazia loboi. Cutis.* 2009; **83:** 67-8.
26. Esperon F, Garcia-Parraga D, Belliere EN, *et al.* Molecular diagnosis of lobomycosis-like disease in a bottlenose dolphin in captivity. *Med Mycol.* 2012; **50:** 106-9.
27. Fischer M, Chrusciak Talhari A, Reinel D, *et al.* Successful treatment with clofazimine and itraconazole in a 46 year old patient after 32 years duration of disease. *Hautarzt.* 2002; **53:** 677-81.
28. Fonseca JJ. Lobomycosis. *Int J Surg Pathol.* 2007; **15:** 62-3.
29. Francesconi F, Francesconi V. Lobomycosis. *N Engl J Med.* 2011; **364:** e2.
30. Francesconi VA, Klein AP, Santos AP, *et al.* Lobomycosis: Epidemiology, clinical presentation and management options. *Ther Clin Risk Manag.* 2014; **10:** 851-60.
31. Fuchs J, Milbradt R, Pecher SA. Lobomycosis (keloidal blastomycosis): Case reports and overview. *Cutis.* 1990; **46:** 227-34.
32. Goihman-Yahr M. Update on Jorge Lobo's disease. *Int J Dermatol.* 2008; **47:** 630.
33. Haubold EM, Cooper CR Jr, Wen JW, *et al.* Comparative morphology of *Lacazia loboi* (syn. *Loboa loboi*) in dolphins and humans. *Med Mycol.* 2000; **38:** 9-14.
34. Herr RA, Tarcha EJ, Taborda PR, *et al.* Phylogenetic analysis of *Lacazia loboi* places this previously uncharacterised pathogen within the dimorphic Onygenales. *J Clin Microbiol.* 2001; **39:** 309-14.
35. Ihara GM, Massone C, Schettini AP, *et al.* Leprosy and lobomycosis: First report from the Amazon region. *Lepr Rev.* 2015; **86:** 195-201.
36. Kanashiro-Galo L, Pagliari C, Barboza TC, *et al.* Th17 and regulatory T cells contribute to the *in situ* immune response in skin lesions of Jorge Lobo's disease. *Med Mycol.* 2016; **54:** 23-8.
37. Kiszka J, Van Bressem MF, Pusineri C. Lobomycosis-like disease and other skin conditions in Indo-Pacific bottlenose dolphins *Tursiops aduncus* from the Indian Ocean. *Dis Aquat Organ.* 2009; **84:** 151-7.
38. Lacaz CS. *Paracoccidioides loboi* (Fonseca Filho *et* Area Leao, 1940) Almeida *et* Lacaz, 1948-1949. Description of the fungus in Latin. *Rev Inst Med Trop Sao Paulo.* 1996; **38:** 229-31.
39. Lupi O, Tyring SK, McGinnis MR. Tropical dermatology: Fungal tropical diseases. *J Am Acad Dermatol.* 2005; **53:** 931-54.
40. Mendoza L, Vilela R, Rosa PS, *et al.* *Lacazia loboi* and *Rhinosporidium seeberi*: A genomic perspective. *Rev Iberoam Micol.* 2005; **22:** 213-6.
41. Migaki G, Valerio MG, Irvine B, *et al.* Lobo's disease in an Atlantic bottlenosed dolphin. *J Am Vet Assoc.* 1971; **159:** 578-82.
42. Miranda MF, Unger DA, Brito AC, *et al.* Jorge Lobo's disease with restricted labial presentation. *An Bras Dermatol.* 2011; **86:** 373-4.
43. Murdoch ME, Mazzoil M, McCulloch S, *et al.* Lacaziosis in bottlenose dolphins *Tursiops truncatus* along the coastal Atlantic Ocean, Florida, USA. *Dis Aquat Organ.* 2010; **92:** 69-73.

44. Nalovic K, Pradinaud RA. Anergic cutaneous leishmaniasis versus lobomycosis. *Int J Dermatol.* 1996; **35:** 438-40.
45. Negroni R. Centennial Anniversary of Prof. Carlos da Silva Lacaz (1915-2015). *Rev Inst Med Trop Sao Paulo.* 2015; **57 (Suppl. 19):** 3-4.
46. Nogueira L, Mendes L, Rodrigues CA, *et al.* Lobomycosis and squamous cell carcinoma. *An Bras Dermatol.* 2013; **88:** 293-5.
47. Norton SA. Dolphin-to-human transmission of lobomycosis? *J Am Acad Dermatol.* 2006; **55:** 723-4.
48. Okay TS. Tribute to Prof. Carlos da Silva Lacaz. *Rev Inst Med Trop Sao Paulo.* 2015; **57 (Suppl. 19):** 1.
49. Oliveira Carneiro FR, da Cunha Fischer TR, Brandao CM, *et al.* Disseminated infection with *Lacazia loboi* and immunopathology of the lesional spectrum. *Hum Pathol.* 2015; **46:** 334-8.
50. Opromolla DV, Nogueira ME. Inoculation of *Lacazia loboi* into the subcutaneous tissue of the hamster cheek pouch. *Rev Inst Med Trop Sao Paulo.* 2000; **42:** 119-23.
51. Opromolla, DV, Belone AF, Taborda PR, *et al.* Lymph node involvement in Jorge Lobo's disease: Report of two cases. *Int J Dermatol.* 2003; **42:** 938-41.
52. Pagliari C, Kanashiro-Galo L, Silva AA, *et al.* Plasmacytoid dendritic cells in cutaneous lesions of patients with chromoblastomycosis, lacaziosis and paracoccidioidomycosis: A comparative analysis. *Med Mycol.* 2014; **52:** 397-402.
53. Pang KR, Wu JJ, Huang DB, *et al.* Subcutaneous fungal infections. *Dermatol Ther.* 2004; **17:** 523-31.
54. Paniz-Mondolfi A, Talhari C, Sander Hoffmann L, *et al.* Lobomycosis: An emerging disease in humans and delphinidae. *Mycoses.* 2012; **55:** 298-309.
55. Paniz-Mondolfi AE, Reyes Jaimes O, Davila Jones L. Lobomycosis in Venezuela. *Int J Dermatol.* 2007; **46:** 180-5.
56. Paniz-Mondolfi AE, Sander-Hoffmann L. Lobomycosis in inshore and estuarine dolphins. *Emerg Infect Dis.* 2009;**15:** 672-3.
57. Papadavid E, Dalamaga M, Kapniari I, *et al.* Lobomycosis: A case from Southeastern Europe and review of the literature. *J Dermatol Case Rep.* 2012; **6:** 65-9.
58. Quaresma JA, Brito MV, Sousa JR, *et al.* Analysis of microvasculature phenotype and endothelial activation markers in skin lesions of lacaziosis (lobomycosis). *Microb Pathog.* 2015; **78:** 29-36.
59. Quaresma JA, Unger D, Pagliari C, *et al.* Immunohistochemical study of Langerhans cells in cutaneous lesions of the Jorge Lobo's disease. *Acta Trop.* 2010; **114:** 59-62.
60. Ramos-E-Silva M, Aguiar-Santos-Vilela F, Cardoso-de-Brito A, *et al.* Lobomycosis: Literature review and future perspectives. *Actas Dermosifiliogr.* 2009; **100 (Suppl. 1):** 92-100.
61. Reif JS, Mazzoil MS, McCulloch SD, *et al.* Lobomycosis in Atlantic bottlenose dolphins from the Indian River Lagoon, Florida. *J Am Vet Med Assoc.* 2006; **228:** 104-8.
62. Reif JS, Peden-Adams MM, Romano TA, *et al.* Immune dysfunction in Atlantic bottlenose dolphins *(Tursiops truncatus)* with lobomycosis. *Med Mycol.* 2008: **46:** 1-11.
63. Reif JS, Schaefer AM, Bossart GD. Lobomycosis: Risk of zoonotic transmission from dolphins to humans. *Vector Borne Zoonotic Dis.* 2013; **13:** 689-93.
64. Rodriguez G, Barrera GP. The asteroid body of lobomycosis. *Mycopathologia.* 1996-1997; **136:** 71-4.
65. Rodriguez-Toro G, Tellez N. Lobomycosis in Colombian Amer Indian patients. *Mycopathologia.* 1992; **120:** 5-9.
66. Rodriguez-Toro G. Lobomycosis. *Int J Dermatol.* 1993; **32:** 324-32.
67. Rosa PS, Soares CT, Belone Ade F, *et al.* Accidental Jorge Lobo's disease in a worker dealing with *Lacazia loboi* infected mice: A case report. *J Med Case Reports.* 2009; **3:** 67-72.
68. Rotstein DS, Burdett LG, McLellan W, *et al.* Lobomycosis in offshore bottlenose dolphins (*Tursiops truncatus*), North Carolina. *Emerg Infect Dis.* 2009; **15:** 588-90.
69. Sacristan C, Ressio RA, Castilho P, *et al.* Lacaziosis-like disease in *Tursiops truncatus* from Brazil: A histopathological and immunohistochemical approach. *Dis Aquat Organ.* 2016; **117:** 229-35.
70. Saint-Blancard P, Maccari F, Le Guyadec T, *et al.* Lobomycosis: A mycosis seldom observed in metropolitan France. *Ann Pathol.* 2000; **20:** 241-4.
71. Salgado CG, Tavares LS, Plautz HL, *et al.* Enzymatic isolation of *Lacazia loboi* cells from skin lesions of lobomycosis. *Med Mycol.* 2009; **47:** 119-23.
72. Schaefer AM, Reif JS, Guzman EA, *et al.* Toward the identification, characterization and experimental culture of *Lacazia loboi* from Atlantic bottlenose dolphin *(Tursiops truncatus). Med Mycol.* 2016; **54:** 659-65.
73. Silva D. Micose de Jorge Lobo. *Rev Soc Bras Med Trop.* 1972; **6:** 85-98.
74. Sousa PP, Schettini AP, Rodrigues CA, *et al.* Lacaziosis - Unusual clinical presentation. *An Bras Dermatol.* 2015; **90:** 268-9.
75. Symmers WS. A possible case of Lobo's disease acquired in Europe from a bottlenosed dolphin (*Tursiops truncatus*). *Bull Soc Pathol Exot Filiales.* 1983; **76:** 777-84.
76. Taborda PR, Taborda VA, McGinnis MR. *Lacazia loboi* gen. nov., comb. nov., the etiologic agent of lobomycosis. *J Clin Microbiol.* 1999; **37:** 2031-3.
77. Tajima Y, Sasaki K, Kashiwagi N, *et al.* A case of stranded Indo-Pacific bottlenose dolphin *(Tursiops aduncus)* with lobomycosis-like skin lesions in Kinko-wan, Kagoshima, Japan. *J Vet Med Sci.* 2015; **77:** 989-92.
78. Talhari C, Brunasso AM, Talhari S, *et al.* Nodules on the right ear. *Dermatol Pract Concept.* 2012; **2:** 31-3.
79. Talhari C, Chrusciak-Talhari A, de Souza JV, *et al.* Exfoliative cytology as a rapid diagnostic tool for lobomycosis. *Mycoses.* 2009; **52:** 187-9.
80. Talhari C, Oliveira CB, de Souza Santos MN, *et al.* Disseminated lobomycosis. *Int J Dermatol.* 2008; **47:** 582-3.

81. Talhari C, Rabelo R, Nogueira L, *et al.* Lobomycosis. *An Bras Dermatol.* 2010; **85:** 239-40.
82. Talhari S, Talhari C. Lobomycosis. *Clin Dermatol.* 2012; **30:** 420-4.
83. Van Bressem MF, Simoes-Lopes PC, Felix F, *et al.* Epidemiology of lobomycosis-like disease in bottlenose dolphins *Tursiops* spp. from South America and southern Africa. *Dis Aquat Organ.* 2015; **117:** 59-75.
84. Vilani-Moreno FR, Belone Ade F, Rosa PS, *et al.* Evaluation of the vital staining method for *Lacazia loboi* through the experimental inoculation of BALB/c mice. *Med Mycol.* 2003; **41**: 211-6.
85. Vilani-Moreno FR, Belone AF, Soares CT, *et al.* Immunohistochemical characterization of the cellular infiltrate in Jorge Lobo's disease. *Rev Iberoam Micol.* 2005; **22:** 44-9.
86. Vilani-Moreno FR, Lauris JR, Opromolla DV. Cytokine quantification in the supernatant of mononuclear cell cultures and in blood serum from patients with Jorge Lobo's disease. *Mycopathologia.* 2004; **158:** 17-24.
87. Vilela R, Martins JE, Pereira CN, *et al.* Molecular study of archival fungal strains isolated from cases of lacaziosis (Jorge Lobo's disease). *Mycoses.* 2007; **50:** 470-4.
88. Vilela R, Mendoza L, Rosa PS, *et al.* Molecular model for studying the uncultivated fungal pathogen *Lacazia loboi. J Clin Microbiol.* 2005; **43:** 3657-61.
89. Xavier MB, Ferreira MM, Quaresma JA, *et al.* HIV and lacaziosis, Brazil. *Emerg Infect Dis.* 2006; **12:** 526-7.
90. Xavier MB, Libonati RMF, Unger D, *et al.* Macrophage and TGF-β immunohistochemical expression in Jorge Lobo's disease. *Human Pathol.* 2008; **39:** 269-74.

SECTION IV

Systemic Mycoses

Chapters

All the fungal infections included in this Section are caused by dimorphic fungi, which are primarily endemic in certain parts of American continents *i.e.* Western Hemisphere and rarely encountered as case reports in rest of the world. These were described more than a century ago and are mainly restricted to particular geographical areas. The cases of coccidioidomycosis and paracoccidioidomycosis, which are being reported outside the endemic areas, are usually considered as 'imported mycoses' found among travellers and immigrants.

These fungal infections are generally acquired by inhalation of propagules of the saprotrophic phase found at ambient temperature *i.e.* air-borne conidia. The respiratory system is the primary route of entry and infections are commonly self-limiting, being asymptomatic in >95% of patients. In rest of them these are symptomatic presenting as full-fledged disease and may disseminate to other anatomical sites of body through hematogenous route. Cutaneous manifestations are secondary to hematogenous and lymphatic dissemination. These infections are caused by true fungal pathogens since healthy, immunocompetent individuals are also infected.

There are four diseases described in this Section, which are traditionally grouped together due to their similarities in epidemiological, clinical and morphological features being dimorphic in nature. Three of these found prevalent in North America, were thought to be protozoal in initial period of their recognition, however, subsequently in due course of time these were identified correctly as fungal in nature. The systemic fungal infections found exclusively in immunocompromised hosts are described in a separate Section on Opportunistic Mycoses. Both Systemic and Opportunistic fungal infections, in combination, are also known as Deep Mycoses.

All the agents causing systemic mycoses have a mycelial phase when grown at an ambient temperature but have an entirely different morphology in tissues *i.e.* spherules filled with endospores in case of *Coccidioides* species whereas rest others form budding yeast cells. The narrow-based budding is seen in *Histoplasma* species, broad-based in *Blastomyces* species and multiple buddings in *Paracoccidioides* species. One of the emerging genus of dimorphic fungi, *Emergomyces (Es.pasteurianus, Es.africanus* and *Es.orientalis)* has been described under Miscellaneous Mycoses, in the Chapter on Adiaspiromycosis and Emergomycosis due to the taxonomic reasons.

All these fungi are highly infectious in nature hence due care should be taken while processing clinical specimens or handling cultural isolates. *Coccidioides* species is the most infectious fungal agent and requires BSL-3 precautions. In the laboratory, conversion of isolates from mycelial-to-yeast and *vice versa* (M↔Y), is necessary for establishing diagnosis. However, molecular techniques are now commonly employed for establishing the final diagnosis.

In this Section epidemiology, clinical features, diagnosis and treatment of four major systemic mycoses are described *i.e.* histoplasmosis, blastomycosis and coccidioidomycosis, found prevalent in North America and paracoccidioidomycosis, in South America. All these diseases were previously named on the basis of geographical regions but this feature was not consistent hence such nomenclature is not in much use. These infections are usually not found prevalent in India except few indigenous or imported reports, however, paracoccidioidomycosis has never been reported. In most of the cases, amphotericin B is initial drug of choice, followed by one of azoles or any other drug for maintenance therapy.

CHAPTER 16

Histoplasmosis

Histoplasmosis is systemic granulomatous disease caused by three varieties of **dimorphic** fungus, ***Histoplasma capsulatum*** *i.e. H.c.* var. *capsulatum*, *H.c.* var. *duboisii* and *H.c.* var. *farciminosum*. The most important variety of histoplasmosis is caused by *H.c.* var. *capsulatum*, which was once thought to be an *'obscure'* disease but it is now being recognized with an increasing occurrence. The scattered cases are being reported throughout the world, mainly from areas with temperate climate along the river valleys. However, most of them are seen in endemic areas in central and southeastern United States bordering Ohio and Mississippi rivers. It generally occurs as an asymptomatic or relatively mild, self-limiting, pulmonary infection. An acute or chronic disseminated disease may also be there with a poor prognosis. The causative fungus is exceptional among pathogenic fungi that it is an **intracellular** uninucleate organism involving cells of reticuloendothelial system *i.e.* macrophages and giant cells.

This disease is also called as classic histoplasmosis, Caver's and Miner's disease, Cave disease, Bat disease, Ohio Valley disease, Tingo Maria fever, reticuloendotheliosis and cytomycosis

Historical Perspective

The first case of histoplasmosis was reported on December 7, 1905 by **Samuel Taylor Darling** while performing an autopsy on a patient in Panama, believing it to be an encapsulated *Plasmodium* (***'plasmo'***-dium). Subsequently, he encountered two more cases and realized that despite striking similarities to causative organism of kalaazar (*Leishmania donovani*), he was confronted with a different organism. In the tissue section, cells were recognizable by hyaline halos around stained cytoplasm, which led to impression that it was an encapsulated organism. However, he noted that no kinetoplast was visible and believed that he was observing new organism that he named *Histoplasma capsulata* (now *capsulatum*). The literal meaning of this nomenclature was that intracellular organisms he saw in ***'histio'***-cytes resembled ***'plasmo'***-dium and seemed to have ***capsule*** also. On retrospective analysis, only one out of these three assessments of Darling was found to be accurate *i.e. H.capsulatum* is an intracellular pathogen of macrophages (histiocytes) but is a fungus without a capsule.

The name of the organism as *Histoplasma capsulatum* was suggested by appearance of yeast cells within macrophages, which was erroneously considered to be a protozoan. Despite its name, which was based on early histopathological descriptions of lesions, *Histoplasma capsulatum* is neither a protozoan nor an encapsulated organism; rather it is a dimorphic non-capsulated fungus. In honor of discoverer of the causative agent, infection was designated as **Darling's disease**.

In 1913, Henrique da Rocha-Lima, was first to suspect that organism isolated by Darling was rather yeast than a protozoa while he was comparing sections of epizootic lymphangitis of horses. In 1929, Katherine Dodd diagnosed first case of histoplasmosis ante-mortem in a child. In 1934, de Monbreun discovered dimorphism of *Histoplasma capsulatum* and concluded that saprotrophic form of *H.capsulatum* probably existed in nature. In 1944, Amos Christie established that it was not rare disease rather frequent pulmonary infection. Emmons established environmental reservoir when he isolated *H.capsulatum* from soil in 1948. The first case of histoplasmosis in India was described by Panja and Sen in 1954 from Kolkata.

In 1952, Vanbreuseghem described a new species of *Histoplasma capsulatum*. This organism was later on recognized by Drouhet in 1957 as variant of *Histoplasma capsulatum* and was thus designated *Histoplasma capsulatum* var. *duboisii*. Mating studies have confirmed that *H.c.* var. *capsulatum* and *H.c.* var. *duboisii* are variants of same species. In addition, Duncan JT in 1958 further highlighted the mycological features of this variety in details.

The third variety although observed in 1873 by Rivolta, was placed in genus *Histoplasma* by Ciferri and Redaelli in 1934 and finally proposed as *H.c.* var. *farciminosum* by Weeks and colleagues in 1985. In 1959, ocular *Histoplasma* syndrome was defined by Woods, which is popularly known as Presumed Ocular *Histoplasma* Syndrome (POHS).

Mycology

The causative organism of the disease, *Histoplasma capsulatum,* is thermally dimorphic fungus and takes longer time for primary isolation, may be upto 6 weeks to grow on SDA at 25-30°C, as a white to buff mycelial colonies. When the mycelial growth is incubated at 37°C on enriched media such as BHI agar with blood and cysteine, conidia germinate and get converted to yeast cells. The conversion from mycelial-to-yeast (M→Y) phase is also feasible *in vivo* by inoculating in an appropriate experimental animal.

There are **three varieties** of this dimorphic endemic fungus, *Histoplasma capsulatum.* These are: *H.c.* var. *capsulatum, H.c.* var. *duboisii* and *H.c.* var. *farciminosum,* causing histoplasmosis capsulati (classic histoplasmosis), histoplasmosis duboisii (African histoplasmosis) and histoplasmosis farciminosi (epizootic histoplasmosis), respectively. The first two are primarily pathogenic to human beings whereas third one is pathogenic to animals. The human pathogens are also called as small-celled and large-celled species for var. *capsulatum* and var. *duboisii,* respectively, due to two different sizes of yeast cells.

On SDA at 25°C, colonies are initially white or buff brown. These two morphological types are popularly termed as A **(albino)** and B **(brown)**. Both types may be isolated from same patient and on subculture, brown type shows sectoring and eventual conversion to albino type.

The albino type growth consists of white, fine aerial hyphae **(Fig. 16.1)** and brown type consists of flat colonies with light tan or dark brown in color within seven days **(Fig. 16.2)**. Later on brown pigment diffuses into the medium. The albino type may become buff colored with passage of time. At 37°C, colonies grow as granular to rough, mucoid, cream-colored turning tan to brown in couple of weeks **(Fig. 16.3)**.

The mycelial phase of *H.capsulatum* has septate hyphae which produce two types of unicellular asexual conidia. The tuberculate **macroconidia** are most prominent, thick-walled, larger in size *i.e.* 8-16 μm, emerge from short, tubular conidiophore. The **microconidia** are smooth-walled broadly elliptical, small in size measuring 2-4 μm, arise on short stalks from undifferentiated hyphae and are abundant in primary cultures **(Fig. 16.4)**. Incidentally, size of macroconidia is almost equivalent to yeast cells of *H.c.* var. *duboisii* and that of microconidia is equivalent to yeast cells of *H.c.* var. *capsulatum*.

The slide culture of fungal isolate is put up to examine typical conidiation of *Histoplasma* along with their arrangement taking due biosafety precautions. In mycelial phase, differentiation is not feasible between *H.c.* var. *capsulatum* and *H.c.* var. *duboisii,* as both have same morphological structures, however, it can be established on conversion to yeast phase or by DNA probes.

Fig. 16.1. Mycelial colonies of *Histoplasma capsulatum* showing white aerial hyphae at 25°C (albino type).

Fig. 16.2. Mycelial colonies of *Histoplasma capsulatum* showing white aerial hyphae at 25°C (brown type).

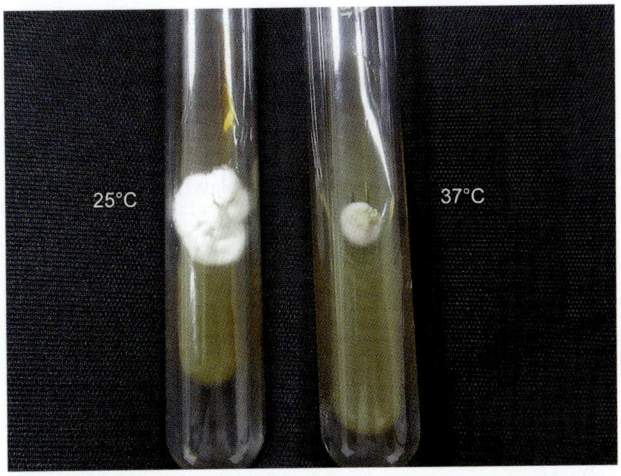

Fig. 16.3. Mycelial colonies of *H.capsulatum* showing white aerial hyphae at 25°C (albino type) and yeast like colonies 37°C.

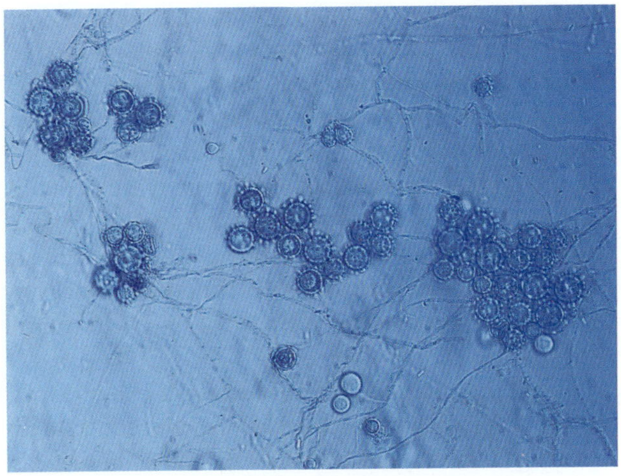

Fig. 16.4. Macroconidia (tuberculate) and microconidia from mycelial growth of *Histoplasma capsulatum* (LCB × 400),

The sexual state of fungus was discovered by Kwon-Chung in 1973, who proposed its name as *Emmonsiella capsulata*. Subsequently in 1979, a new term, *Ajellomyces capsulatus* was proposed by McGinnis and Katz and *Emmonsiella* was considered to be synonym of *Ajellomyces* on taxonomic grounds, which belong to family **Onygenaceae**, in order **Onygenales** under class **Plectomycetes** in phylum **Ascomycota**. The teleomorphic states of *Histoplasma capsulatum, Blastomyces dermatitidis* and *Emmonsia crescens* are clubbed together and belong to same genus *i.e. Ajellomyces*. The co-culture of isolates with opposite mating types can produce perfect state of this fungus and is designated *Ajellomyces capsulatus*. Hence both varieties are heterothallic and compatible (+) and (-) mating types unite to form ascomycetous sexual state. Their respective three counterpart teleomorphic states are *Ajellomyces capsulatus*. var. *capsulatus*, *Ajellomyces capsulatus* var. *duboisii* and *Ajellomyces capsulatus* var. *farciminosus*.

Currently, *H.capsulatum* is classified into the kingdom Fungi, subkingdom Dikarya, phylum Ascomycota, class Eurotiomycetes, order Onygenales and family Onygenaceae and/or Ajellomycetaceae according to the analyses of six loci: 18S, 5.8S and 28S of the rRNA genes; EF1 (Elongation factor-1); RPB1and RPB2 (RNA polymerase II subunits 1 and 2). It is considered that *Histoplasma* has now six cryptic species.

There are at least eight clades within *H.capsulatum sensu lato* as per the phylogenetic studies, which are geographically distributed as follows: North American classes 1 and 2 (NAm 1 and NAm 2), Latin American groups A and B (LAm A and LAm B), Australian, Netherlands (Indonesian), Eurasian and African clades. However, with the exception of the Eurasian cluster, these clades are considered phylogenetic species.

Epidemiology

The initial cases of the histoplasmosis were encountered in relatively non-endemic areas like Panama and Minnesota but subsequently, it was realized that it is endemic to the South eastern part of USA. This is prevalent in various geographical areas throughout world but mostly afflicts people residing along the Ohio, Mississippi and St. Lawrence Rivers of USA and also in Latin America. The disease predominates in central and southeastern parts of United States and in certain Latin American countries particularly Argentina, Brazil, Venezuela and Uruguay. Less frequently, it has also been found in northern Chile, Peru, Colombia, Guiana, Bolivia and Paraguay.

In USA alone, infection rate is estimated to be around 2,00,000 persons per year. The infection may also be found prevalent outside these endemic areas. Sporadic cases and minor endemic foci have been reported from Africa, Asia and Europe. During outbreaks in Indianapolis in 1980s, histoplasmosis was AIDS-defining illness and subsequent complication in nearly one-third of AIDS patients.

The causative agent, *H.capsulatum*, has been isolated from soil, which appears to be source of infection for human beings. The soil with high nitrogen content is especially related to droppings of starling, chickens and bats. Areas with highest incidence of histoplasmosis correspond to areas, where starlings, other birds or bats commonly roost. *H.capsulatum* flourishes in soil fertilized by bird droppings or bat guano. The birds themselves are not infected probably because of their high body temperature. The fungus is present as yeast in tissues and as filamentous form in environment. The areas of endemicity are defined in epidemiological surveys conducted usually by histoplasmin skin tests that are positive to specific fungal antigen.

The precise reasons for this endemic distribution pattern are unknown but are thought to include moderate climate, high humidity and characteristic soil of these areas. The soil is reservoir of this fungus where organism exists as multicellular, filamentous form. In the infected tissue, only unicellular yeast form is found. Infection from man-to-man or from animal-to-man is unknown. The infection takes place by inhalation of microconidia, which are the **infective forms** due to their smaller size. Accidental infections have occurred among laboratory workers while handling culture of *H.capsulatum*.

Adult men are affected more commonly than women but in children no such sex difference in its prevalence. The infection tends to occur in rural areas and is usually associated with sources such as chicken coops, pigeon roosts, bat caves and old decaying wooden structures. High humidity provides nidus for infection.

The chronic granulomatous disease found in tropical Africa is due to *H.c.* var. *duboisii*. Therefore, in African continent all the three varieties of *Histoplasma capsulatum* are found prevalent.

In Asia including India, case reports of histoplasmosis have been thoroughly reviewed by Randhawa and colleagues in 1970 and 1994, respectively. Another review of literature by Padhye *et al*, from India described epidemiology of 25 authentic cases from 1968 to 1992. Moreover, the histoplasmin skin test surveys could detect only small percentage of positive reactors in various studies. There is no specific endemic focus yet recognized in India and further epidemiological study is essential.

Histoplasmosis is the most common endemic mycosis in the United States. The causative organism, *H.capsulatum*, is a soil-based dimorphic fungus. It is endemic to the Ohio and Mississippi River Valleys and certain parts of Central and South America; however, it also exists in microfoci elsewhere in the world and it has been reported from every continent other than Antarctica. The growth of the fungus is promoted by specific environmental conditions such as temperature, precipitation, humidity, soil acidity and nitrogen concentration.

In May 2017, eighteen firefighters were found to be infected with histoplasmosis in Brazil, who attended a rescue course in cave of Brazlandia.

There are three types of histoplasmosis depending on variety of organism, which are described below:

I. Histoplasmosis capsulati
II. Histoplasmosis duboisii
III. Histoplasmosis farciminosi

I. HISTOPLASMOSIS CAPSULATI

This is also called as classic histoplasmosis or small form histoplasmosis. It is highly infectious mycosis and is caused by *H.c.* var. *capsulatum*. It primarily affects lungs and is generally asymptomatic. The disease may rarely become chronic, progressive and even fatal.

Immunity

As this is an intracellular mycosis of reticuloendothelial system involving spleen, liver, adrenals, kidneys, skin, CNS and other body organs, cell-mediated immunity plays essential role in defense against *H.capsulatum*. With the development of specific cellular immunity, cytokines, macrophages and other cellular defense mechanisms come into play to inhibit the organism entailing halt in the progression of disease. The patients with AIDS are at higher risk for progressive infection with *H.capsulatum* because their cellular immunity is severely impaired. However, in mild cases, it resolves within 2-3 weeks, leaving lymphocyte-rich infiltrate around lesion.

Histopathologically, acute phase of disease results in lesions in which yeast cells are found within histiocytes but few other cells like neutrophils and lymphocytes are also present. The chronic disease often results in epithelioid cell granuloma. The molecular genetics has shown that crucial virulence gene encoding calcium binding protein (CBP1) is required for *H.capsulatum* parasitism of macrophages and virulence in mice. The disruption of *CBP1* gene results in decreased survival of *H.capsulatum* yeast in cultured macrophages. In HIV-infected patients, highly active antiretroviral therapy (HAART) produces partial recovery of cellular immunity. However, this may also induce unexpected clinical manifestations classified as immune reconstitution inflammatory syndrome (IRIS).

Pathogenesis and Pathology

The infection by *H.capsulatum* develops when microconidia are inhaled and subsequently get converted to yeasts in alveolar macrophages in the lungs. This may also occur when old foci of infection get reactivated following immunosuppression, which decreases cellular immunity. The oval yeast cells parasitize macrophages and modulated by T-lymphocytes, results in localized granulomatous inflammation, which may assist dissemination to reticuloendothelial system. If disease establishes, there is proliferation of histiocytes or development of epithelioid and giant cells, forming granuloma with or without central caseation. All phagocytic cells of reticuloendothelial system are involved, including those in liver, spleen, lymph nodes and bone marrow, so that cytoplasm is studded with masses of fungal cells. Histoplasmosis may be considered fungal homologue of bacterial infection like tuberculosis since both *H.capsulatum* and *Mycobacterium tuberculosis* exploit macrophages as their host entailing acute or persistent pulmonary, disseminated as well as reactivation disease.

Clinical Features

The infection remains asymptomatic in about 90-95% of cases. The asymptomatic form of histoplasmosis is indicated by presence of positive histoplasmin skin test without any focus of infection, which is very common in endemic areas. The development of asymptomatic or symptomatic disease appears to depend on intensity of exposure to conidia and status of cellular immune response of host, which is manifested solely by positive histoplasmin skin test.

The illness may appear years after leaving endemic areas where infection was acquired. This type of infection is called reactivation histoplasmosis and occurs in patients with serious underlying disorders, such as organ transplantation or AIDS.

The disease may be classified according to duration of infection as acute, subacute or chronic; site of involvement as pulmonary, extrapulmonary or disseminated and by the pattern of infection as primary or reactivation. The various clinical types of classic histoplasmosis capsulati are as follows:-

1. Acute Pulmonary Histoplasmosis
2. Chronic Pulmonary Histoplasmosis
3. Cutaneous, Mucocutaneous Histoplasmosis
4. Disseminated Histoplasmosis

1. Acute Pulmonary Histoplasmosis

The acute type of pulmonary histoplasmosis covers entire range of infection in normal host, including rare type of early and late complications. More than 95% of immunocompetent person remain asymptomatic following low-inoculum infection with *H.capsulatum*. In few symptomatic cases, clinical manifestations of the disease are non-specific and incubation period is about 10-16 days. The onset resembles influenza-like illness (**Cave Disease**) *i.e.* generalized malaise with fever, headache, chills, profuse sweating, sore throat, dry cough, chest pain and dyspnoea. There may be erythema multiforme or nodosum at later stages. The radiological examination of lungs shows small, scattered pulmonary infiltrate and hilar lymphadenopathy. The dense nodules disseminated in lung field may reveal localized parenchymatous infiltrate. There may be calcification of lungs at later stage. The course of the disease varies from case to case that may be benign or severe.

2. Chronic Pulmonary Histoplasmosis

This is a variant of acute pulmonary histoplasmosis. It remains latent for long time and then gradually produces similar symptoms as in acute disease but in more pronounced form. The disease presents with hemoptysis and apical or subapical cavities. The patients lose weight and may have ulcerative lesions over the lips, mouth, nose, larynx and intestine. There are typical pulmonary lesions with central necrotic area. The chronic pulmonary disease is found predominantly among adult Caucasian males, whereas no significant difference in racial susceptibility has been documented for acute pulmonary histoplasmosis.

The lesion becomes uniformly calcified and is surrounded by peripheral fibrous zone with calcification as more or less unbroken concentric rings *i.e.* **histoplasmoma**. It is an enlarging fibrous mass that develops around healed primary focus of infection in the lungs. There is fibrosis and formation of concentric layers of collagen tissue and subsequently it undergoes calcification. In the beginning it is 2-4 mm and over a period of 10-15 years attains a size of 3-4 cm. Radiologically, it resembles primary or secondary pulmonary neoplasia. The calcification is more common in histoplasmosis than in tuberculosis and coccidioidomycosis. It is also unusual in cryptococcosis and blastomycosis.

3. Cutaneous/Mucocutaneous Histoplasmosis

The skin and mucous membrane lesions are secondary to one of the above described forms. There is petechial or ecchymotic purpura, usually on the abdominal wall and thorax. As far as mucous lesions are concerned, these are predominantly found in oral cavity. In the a study of 25 authentic cases in India, 19 had lesions confined to oral cavity, confirming earlier observation that histoplasmosis in Indian patients tends to occur primarily in extrapulmonary sites, particularly involving the oral cavity.

4. Disseminated Histoplasmosis

This form may occur at any age but it is seen especially in children under two years and in adolescent. It is manifested as fever, anorexia, weight-loss, deterioration of general condition, anemia, leukopenia, constant hepatosplenomegaly and lymphadenopathy. It has been seen that individuals infected with HIV are at greater risk of developing disseminated histoplasmosis.

In 1985, patients suffering from disseminated histoplasmosis and positive serology for HIV were defined as having AIDS according to criteria of the CDC, USA. This disease is most commonly diagnosed endemic mycosis in AIDS with poor prognosis. Three factors correlated with severe manifestations of hosts are respiratory failure, shock and death in patients with AIDS during an outbreak.

The disseminated disease has also been reported in patients having undergone organ transplantation. Infections of thyroid gland, sinuses, prostate and epididymis with *H.capsulatum* have also been rarely reported as a part of disseminated disease, in geographical areas where histoplasmosis is endemic. The CNS histoplasmosis manifesting as an intramedullary spinal cord abscess is extremely rare.

The gastrointestinal histoplasmosis has no specific symptomatology. It can present with bloody diarrhea, which may mimic ulcerative colitis or Crohn's disease and can cause intestinal obstruction, perforation or gastrointestinal bleeding. The distal ileum is most common site involved, followed by colon and stomach. The infection has increasingly been reported in HIV infected individuals in areas where histoplasmosis is not endemic. Therefore, it is imperative that histoplasmosis be kept as differential diagnosis when dealing with cases of acute abdomen in AIDS patients.

The ocular manifestations of histoplasmosis are rarely described at the time of initial infection. The late sequelae of subclinical infection is called Presumed Ocular Histoplasmosis Syndrome **(POHS),** which is characterized by distinct features such as atrophic ***'histo spots'***, peripapillary atrophy, choroid neovascular membrane and linear streaks of choroidal atrophy. When macula is involved, metamorphosis and other symptoms develop. Fluorescein angiography and laser photocoagulation are recommended for diagnosis and management, respectively.

Radiodiagnosis

The radiological findings of histoplasmosis are significant to establish the diagnosis. The lungs show small, scattered pulmonary infiltrates with hilar lymphadenopathy. The dense nodules disseminated in lung fields may suggest localized parenchymatous infiltrates. There may also be calcification in lungs at later stage.

Differential Diagnosis

The clinical and pathological features of histoplasmosis can closely mimic tuberculosis. Localized pneumonitis is easily mistaken for *Mycoplasma pneumoniae, Legionella, Coxiella burnetii* or *Chlamydia pneumoniae* infections. Sometimes, lesions of histoplasmosis may be confused with syphilis, carcinoma and even paracoccidioidomycosis. Talaromycosis is usually found among AIDS patients in Southeast Asia, resembles *Histoplasma* in tissues however on careful examination can be differentiated by the dividing unique transverse fission cells. The cutaneous lesions of histoplasmosis may be confused with emergomycosis, which is also having circumstantial background of HIV infections.

On histopathology, the uninucleate yeasts of *H.capsulatum* should be distinguished from other intracellular organisms like *Leishmania donovani, Toxoplasma gondii, Talaromyces marneffei* as well as *Emergomyces pasteurianus* and *Es.africanus* The yeast size in case of *Es.orientalis* is slightly bigger *i.e.* 4-10 μm hence may confuse with African histoplasmosis. The first two are protozoa and rest all are dimorphic fungi in their parasitic phase. *Leishmania donovani* contains kinetoplast, which is not present in yeasts of *H.capsulatum*. The tachyzoites of *T.gondii,* being protozoa are not stained with GMS stain. The broad-based yeasts of *B.dermatitidis*, spherules of various sizes in *Coccidioides* species and capsule of *Cr.neoformans* are also differentiating features. It may sometimes be confused with yeasts of *Candida glabrata* due to their smaller size.

Laboratory Diagnosis

The diagnosis of histoplasmosis requires high index of clinical suspicion and awareness about advantages

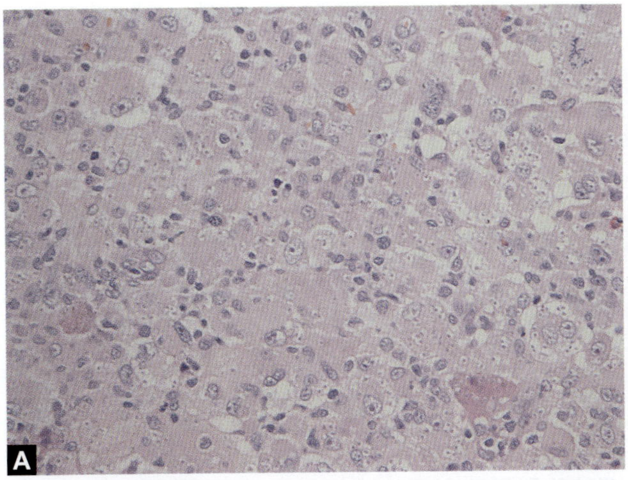
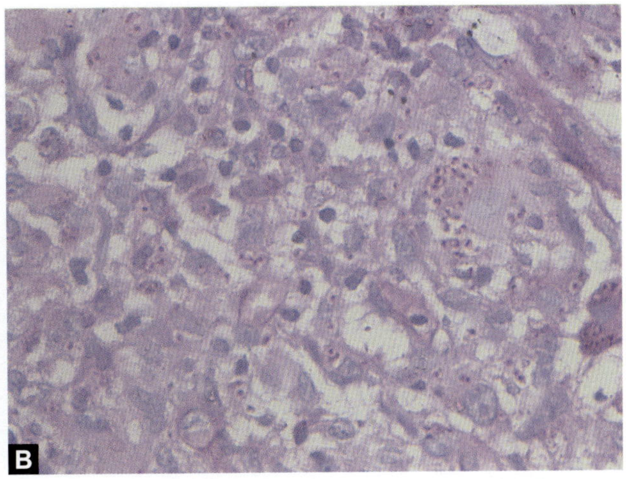
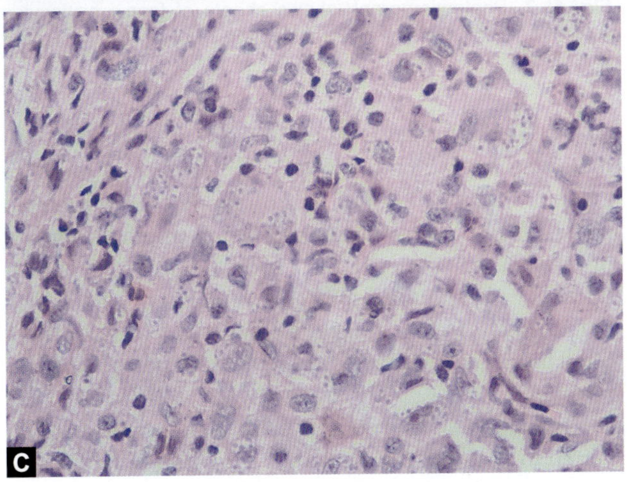

Figs. 16.5A to C. Multiple yeast cells of *Histoplasma capsulatum* within histiocytes of alveoli in renal transplant recipient (H&E x 400).

and limitations of tests commonly used to establish this disease. The specimens helpful in establishing diagnosis are sputum, bone marrow and lymph nodes aspirate/ biopsy and peripheral blood film. The blood smears can be made from buffy coat of centrifuged citrated blood or from bottom of tube, where sedimented, heavily infested cells are found. The biopsy of lesions from skin, mucous membranes, lymph nodes or tissues from autopsy material should be examined. Smears can also be made directly from moist specimens. Direct examination for demonstration of *H.capsulatum* is best accomplished with special stains. Detection of antigen as well as antibodies is also done from serum or urine. Definitive identification requires conversion from mycelial-to-yeast form and *vice versa* (M↔Y), demonstration of specific reactivity with anti-*H.capsulatum* antisera *i.e.* exoantigen test or reactivity with molecular probes.

(a) Direct Examination

The conventional KOH wet mount, prepared from clinical sample, shows tiny yeast cells. The size of yeast cells is too small to an extent that it is invariably missed because it is not stained in this preparation. Therefore, thick and thin smears are prepared from peripheral blood, bone marrow or other suitable clinical material like buffy coat and stained with calcofluor white, Giemsa or Wright stains. The fungus appears as small, oval yeast cells, 2-4 μm, within mononuclear or polymorphonuclear cells and occasionally in giant cells **(Figs. 16.5A to C)**. The Giemsa stain shows cell wall as light blue with clear space between wall and dark blue protoplasm. In PAS stain **(Fig. 16.6)**, wall is stained violet, red to pink with pallor protoplasm filling cell. The fluorescent antibody technique has been found useful in diagnosis of histoplasmosis in tissues. The yeast cells are too small and may be confused with *Candida glabrata,*

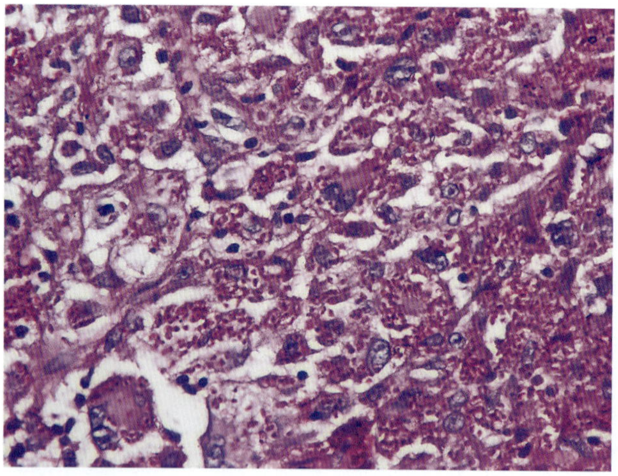

Fig. 16.6. Photomicrograph shows sheets of foamy histiocytes with intracellular *Histoplasma capsulatum* (PAS x 400).

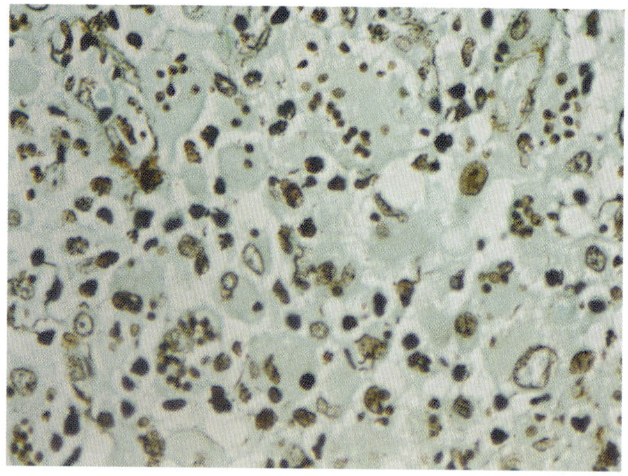

Fig. 16.7. Photomicrograph shows sheets of foamy histiocytes with intracellular *Histoplasma capsulatum* (GMS x 400).

which often colonizes oropharynx but not found intracellularly like *H.capsulatum*. The tissue often shows granulomatous response in immunocompetent patients, whereas reaction is variable in immunocompromised patients and may be predominately histiocytic, with phagocytosis of small yeast cells with narrow-neck budding. In the GMS Stain the yeast cells of *H.capsulatum* are seen as black colored **(Fig. 16.7)**.

(b) Fungal Culture

The clinical material is inoculated on SDA with antibacterial antibiotics and actidione and is incubated at 25°C and 37°C. Another set of culture on BHI agar with same antimicrobials is also inoculated and incubated at 25°C. The BHI agar without antibiotics is supplemented with 6% blood and is incubated at 37°C. Actidione may be inhibitory to *H.capsulatum* hence it should not be used in the primary isolation media. If at all it is to be used, the growth is very slow. When there is no growth on primary isolation, clinical material can be inoculated on **yeast-extract phosphate agar** with ammonium hydroxide. All cultures should be examined daily during first week and twice weekly for further three weeks period before being declared as sterile.

The fungal culture at 37°C for demonstration of yeast phase is not much beneficial for primary isolation but conversion from mycelial-to-yeast (M→Y) phase is useful to confirm identity of isolates. Special media like BHI agar, BHIB, Kelley's agar and Pine's medium are used for conversion of fungal isolate. After subculture, blastoconidia develop which are oval to spherical and measure 2-4 µm. Yeasts with similar morphology are present in tissues. Although macroconidia give presumptive diagnosis of *Histoplasma* species but resemble those of saprotrophic fungi like *Sepedonium* species and *Renispora flavissima* hence likely to create confusion in its identification. Similarly, microconidia resemble to that of genus *Chrysosporium*.

Therefore, a suspected mycelial isolate must be converted to yeast form or be shown to exhibit **H** and **M** exoantigens to identify it as genus *Histoplasma*. These soluble antigens are unique for *Histoplasma* varieties and are reliable modalities for identification of organism. Sporulation can be induced on potato dextrose agar. These days conventional methods like mycelial-to-yeast (M→Y) conversion may not be essential as there are DNA probes available for final identification.

For conversion to yeast form, blood-glucose-cysteine (BGC) agar or BHI agar with blood **(*See* Appendix A)** are commonly used. *H.capsulatum* requires presence of reducing conditions such as free thiol group provided from cysteine, which is critical factor for conversion and maintenance of yeast form at 37°C. For definitive identification of culture, nucleic acid hybridization test, exoantigen test or conversion to yeast form are essential. Among these tests, nucleic acid hybridization offers the most rapid method of diagnosis.

Isolation of *H.capsulatum* directly from the environment by culture of soil suspensions on agar media has rarely been successful. When cultures are handled for identification or sent to reference laboratories, sealing or taping primary tube or plate is undertaken to prevent aerosolization of organisms. The handling of culture of *Histoplasma* species like that of *Coccidioides* requires Biohazard Safety Level 3 infrastructure.

(c) Immunodiagnosis

In histoplasmosis, detection of antibodies is very pertinent to establish the diagnosis. These appear 4-6 weeks after onset of primary acute disease and decline after 2-5 years. The antigens used in immunological tests are derived from two sources: (a) sterile culture filtrate (**histoplasmin**) of asparagine-glucose broth in which **mycelial** form of *H.capsulatum* is grown for a period of 3 to 6 months (b) merthiolated suspension of **whole yeast cells** in saline. Both antigens are used because neither type detects antibodies in all the cases.

Histoplasmin is the most important antigen in assessing antibody response. It contains two antigens: **H** antigen, against which antibodies are formed during **active** histoplasmosis and **M** antigen, which is abundant in mycelial form and against which antibodies may be produced in both **active** and **chronic** histoplasmosis. Both **H** and **M** antigens are regarded as being species-specific proteins, in contrast to other components of histoplasmin, such as carbohydrate antigen, which may cross-react with *Coccidioides* and *Blastomyces* species. Histoplasmin is also found to play useful role as skin test reagent and also as reagent in **immunodiffusion (ID)** and **complement fixation tests (CFT)** for detection of antibody responses.

(i) *Histoplasmin Skin Test:* The intradermal test with histoplasmin is useful from an epidemiological point of view but it has no utility for making diagnosis in the endemic areas. Hence histoplasmin test remains useful in epidemiological studies to identify endemic areas of histoplasmosis and evaluation of immunity prevailing in certain population. It is performed by intradermal injection of 0.1 ml of antigen in the forearm. The test is based on delayed-type hypersensitivity (DTH) and is analogous to tuberculin test but in this case, the antigen, protein-polysaccharide, is culture-filtrate and not purified protein derivative. The test is positive after 48 hours with an induration of 5 mm. A positive histoplasmin test shows past or present infection but cannot differentiate between them and negative test does not exclude the disease. This test is of little diagnostic or prognostic value and may be misleading because in most of hypersensitive patients, serological titer may become

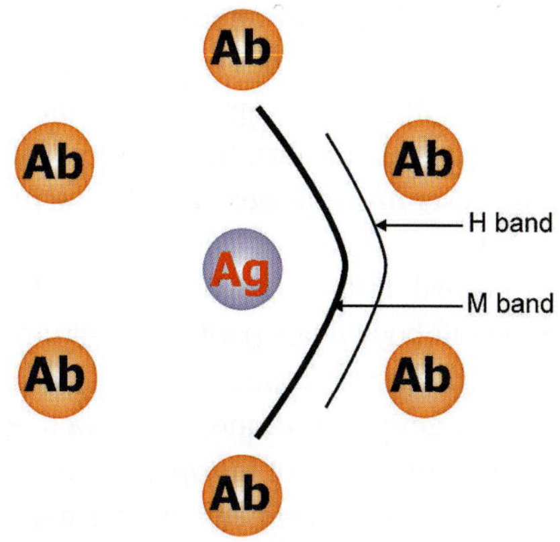

Fig. 16.8. Immunodiffusion H and M bands of precipitation seen among patients of histoplasmosis.

elevated because of skin testing with reagent (anamnestic response). Moreover, skin tests should be avoided in individuals from an endemic area who have pulmonary disease and fever of unknown origin or hepatosplenomegaly. If they are tested, resultant positive skin test will stimulate an increase in antibodies thereby misleading information.

(ii) *Serological Tests:* Immunodiffusion, latex agglutination and complement fixation tests are commonly used, often in combination, to detect serum antibodies. A titer of 1 : 32 or higher is suggestive of histoplasmosis. However, four-fold rise in titer of antibodies is also taken as the diagnostic criteria. Individuals from endemic areas may show cross-reactions mainly with other systemic fungal diseases like blastomycosis and coccidioidomycosis.

In immunodiffusion test, histoplasmin antigen is used which gives ~80% positivity among the infected individuals. There are two major precipitin bands of diagnostic significance, namely **H** and **M** bands corresponding the antibodies against **H** and **M** antigens. The **H** band is finer line near the well containing serum antibodies and **M** band is thicker line near antigen containing well **(Fig. 16.8)**. The **H** band shows **active** infection and **M** band suggests either acute or **chronic** infection or previous skin test with histoplasmin. The presence of an **M** band is less specific for active infection and may persist after recovery or may be demonstrated by histoplasmin skin testing. The **H** band has higher diagnostic value during the first six months of infection since it disappears after active stage of disease is over. Therefore, presence of both **H** and **M** bands is highly specific for active histoplasmosis.

Use of an exoantigen test permits confirmation when sufficient growth occurs. A monoclonal antibody (mAb) of IgG_1 subclass, with greater affinity to yeast than mycelial phase of *H.capsulatum*, has been raised. It has been predominantly found to recognize molecule of 80-kDa by immunoblotting. Both immunodiffusion and complement fixation tests must be performed to achieve maximum sensitivity.

The serological tests for detection of *Histoplasma*-derived **antigen** in body fluids particularly serum and urine may be used especially in immunosuppressed individuals. The *Histoplasma* antigen immunoassay is well-established method for rapid diagnosis of histoplasmosis. The specificity of antigen detection has been greater than 98%. The cross-reactions with detection of urine antigen is seen in paracoccidioidomycosis, blastomycosis, talaromycosis and African histoplasmosis. The serum antigen detection cross-reacts with rheumatoid factor and antithymocyte globulin. These organisms share similar antigens, thus cross-reactivity can be explained. However, cross-reactions are of limited clinical importance since specific diagnosis usually can be made on the basis of epidemiological features or positive cultures. Any type of queries regarding diagnostic problems can be sent to **Histoplasmosis Reference Laboratory (HRL)**, Indianapolis, USA.

Although mycelial-to-yeast form conversion and exoantigen testing may be used to confirm mold as *H.capsulatum*. These days, *H.capsulatum* AccuProbe (GEN-PROBE, Inc. San Diego, California), a probe that recognizes specific rDNA target sequence, is commonly used for more rapid and accurate results. A comparative evaluation of chemiluminescent DNA probe and an exoantigen test for rapid identification of *H.capsulatum* has also been done by Padhye *et al* in 1992. The lysis centrifugation technique is most effective method of recognizing *H.capsulatum* from blood specimens.

(d) Animal Pathogenicity

A number of animals are susceptible to infection with mycelial and yeast phase of organism such as dogs, guinea pigs, hamsters and rabbits, all being used as models of histoplasmosis. The mouse is an ideal laboratory experimental animal for isolation of *H.capsulatum* from pathological material as well as from air, soil or bat and bird guano. Despite the fact that natural mode of acquiring infection is by inhalation of air-borne conidia but only few models are developed adopting this route of inoculation using mycelial form of *H.capsulatum*. However, inoculation of yeasts forms suspended in saline by intravenous route has proven to be an ideal reproducible method of establishing systemic infection. This method has also been used to evaluate therapeutic trials in both normal and immunosuppressed animals.

Animal pathogenicity was one of important methods for primary isolation of organism from clinical specimens or soil survey samples and confirmation of an unidentified strain by conversion but after discovery of exoantigen tests and currently molecular probes, its frequency of usage has decreased.

Treatment and Prophylaxis

The management of histoplasmosis includes an induction phase of treatment to achieve clinical remission and maintenance phase to prevent relapse of disease. Amphotericin B remains mainstay for treatment of disseminated and other severe forms of histoplasmosis. This is highly effective for inducing remission in about 80% of patients, usually within first week of therapy. The disease usually takes favorable course when timely treatment is instituted. However, treatment failures are also seen in severely ill patients or in those who are suffering from CNS infection.

Oral itraconazole is an effective alternative to amphotericin B for all types of histoplasmosis in immunocompetent host with exception of meningitis. The recommended dose is 400 mg daily for 6 to 12 months until maximal clinical and radiological benefits have been achieved. The introduction of azoles, such as itraconazole, has shifted therapeutic modalities of histoplasmosis easier from indoor to outdoor patients' department. However, in AIDS patients relapse following completion of therapy is serious issue and life-long suppressive therapy must be considered. An untreated disseminated histoplasmosis is progressively fatal infection in AIDS patients.

For prevention of histoplasmosis, regular cleaning of farm buildings and chicken houses, is required; thus markedly reducing risks of exposure to disease. This fungus does not infect birds and is not found in their droppings. But when these droppings are allowed to compost, they create conditions in soil, which promote growth of *H.capsulatum*.

Invariably, exposure cannot be prevented since people are unaware that they are being exposed. There are recommended work practices and guidelines on selection of respirators and other items of personal protective equipment for protecting workers as described in booklet published by National Institute for Occupational Safety and Health

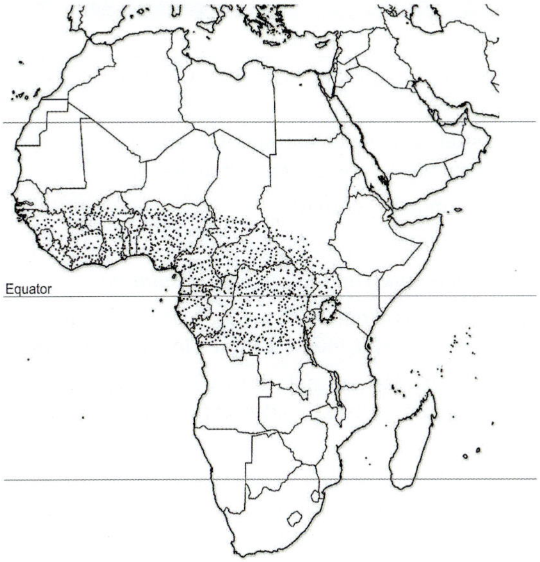

Fig. 16.9. Map showing geographical distribution of African histoplasmosis in central-western Africa around the equator.

and Centers for Diseases Control and Prevention, Atlanta from time to time.

A ribosomal vaccine is effective in inducing protection against experimental histoplasmosis, at a rate similar to that induced by sublethal infection, with advantage of being subcellular safe antigen. The protection inducing capacity is attributed to γ-proteins and nature of immune response involved and is primarily cell-mediated immunity.

II. HISTOPLASMOSIS DUBOISII

This is also called as African histoplasmosis and is caused by *H.c.* var. *duboisii,* which are pear-shaped yeast cells and found as groups in giant cells. It is primarily disease of cutaneous, subcutaneous and osseous tissues with little evidence of lesions in pulmonary or internal organs as there is very less involvement of reticuloendothelial system. The route of infection is presumed to be respiratory *i.e.* inhalation.

Its endemicity is essentially restricted solely to sub-Saharan Africa involving equatorial central and West African countries falling within vicinity of both sides of equator, between latitudes of 20°N and 10°S as well as island of Madagascar. The disease has been reported from 19 African countries mainly Nigeria, Niger, Senegal, Congo, Zaire, Kenya, Gabon and Uganda **(Fig. 16.9)**. In 1994, Gugnani *et al,* isolated this variety from soil collected

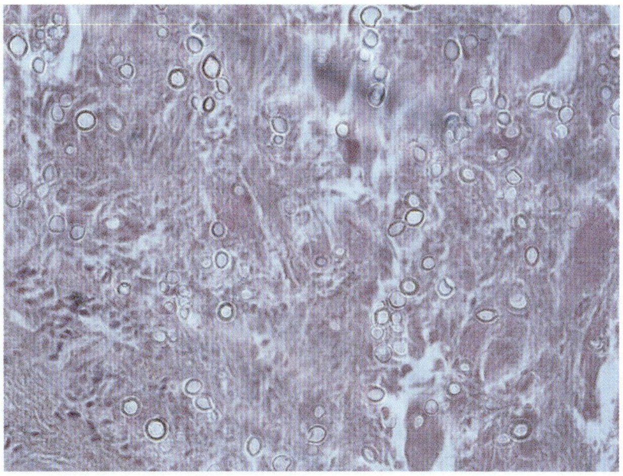

Fig. 16.10. Thick-walled oval yeast cells of *Histoplasma capsulatum* var. *duboisii* (H&E × 400).

from southeastern Nigerian bat cave in the state of Anambra. There is a report from Kerala state of south India by Ravindran *et al,* in 2015, which is being claimed as first indigenous case of this disease, as patient has never gone abroad.

This disease is not only geographically distinct but also differs from classic histoplasmosis in distribution of lesions as well as morphology of causative fungus. The common sites of involvement are skin and bones. The skin lesions present as small papules that often develop an umbilicated center, nodules, abscesses or ulcers. Subcutaneous manifestations include granulomata and abscesses which discharge through sinuses to skin. The bony lesions are well circumscribed and osteolytic. The long bones and flat bones of skull are typical sites of invasion. Respiratory tract is presumed to be portal of entry despite very less manifestations of pulmonary disease. The involvement of other sites like lymph nodes and gastrointestinal tract is also less common. This infection is very rarely associated with AIDS. The cutaneous lesions may mimic skin carcinoma.

In its mycelial form, this fungus is virtually identical to *H.c.* var. *capsulatum* but differs in tissue phase *in vivo* as well as *in vitro*. It has thick-walled oval yeast cells, 7-15 µm about three to four times larger than that of *H.capsulatum* **(Fig. 16.10)**. Therefore, the diagnosis of this variety is mainly histopathological with corroborative clinical findings. It is one of the varieties of *H.capsulatum* because organisms isolated from both varieties are almost identical for the mycelial growth in culture at 25°C hence are regarded as variants of single species. However, differentiating features

Table 16.1. Differentiating Features between African and Classic Histoplasmosis.

Differentiating Features	*African Histoplasmosis*	*Classic Histoplasmosis*
1. Areas of Endemicity	Confined to African Subcontinent	Mainly Central and Eastern USA, isolated cases worldwide
2. Clinical Presentations	Papules, nodules, ulcerative lesions over skin, subcutaneous tissue and bones	Systemic infection of respiratory, cutaneous, mucocutaneous and disseminated disease
3. Incidence of Diseases	Rarely reported	Commonly reported
4. Association with AIDS	Rare	Common
5. Causative Organisms	*Histoplasma capsulatum* var. *duboisii*	*Histoplasma capsulatum* var. *capsulatum*
6. Yeast Morphology on HPE	Narrow-based, yeast cells of 7-15 μm found clustered in giant cells & histiocytes	Yeast cells of one-third to fourth in size (2-4 μm) of var. *duboisii*, found scattered in the histiocytes
7. Differential Diagnosis on HPE	Blastomycosis, lacaziosis and emergomycosis *(Es.orientalis)*	Leishmaniasis, toxoplasmosis, talaromycosis and emergomycosis *(Es.pasteurianus/Es.africanus)*

of *H.c.* var. *duboisii* from *H.c.* var. *capsulatum* are as follows:

- Lesser frequency of pulmonary or internal organ involvement or calcification.
- Higher tendency to involve lymph nodes, skin and bones with granulomatous lesions.
- Larger, thick-walled, ovoid yeast cells of 7-15 μm with more prominent bud scar or isthmus.
- There is prominent giant cells formation.

The differentiating features between these two varieties are also summed up at a glance in **Table 16.1**.

When mold form is incubated at 37°C, fungus gets converted to yeast cells that are similar in size and shape to those of *H.c.* var. *capsulatum* in early period of their growth, however, subsequently they increase in size. Due to its larger size *in vivo,* it may be confused with *B.dermatitidis* but can be distinguished by the fact that budding cells of *Blastomyces* are multinucleate with broad-base, which do not grow intracellularly. In lacaziosis although the yeast cells are found in chains but if cells are scanty and not in chains, it can be confused with it. Therefore, histopathological examination of tissue demonstrating *H.c.* var. *duboisii* is necessary for establishing laboratory diagnosis.

However, recently in 2017 Valero *et al,* established that conventional methods for diagnosing this disease are not reliable and only molecular based methods *i.e.* multilocus sequence analysis (MLSA) can differentiate between these two fungi and/or diseases. They have categorically stated that (i) historical statements about clinical characteristics of the infection caused by *H.capsulatum* var. *duboisii* are useless to distinguish between varieties in their clinical setting; (ii) classical microbiological methods are not conclusive either and (iii) MLSA allowed defining an African cluster that included *H.capsulatum* var. *capsulatum* and *H.capsulatum* var. *duboisii* strains from international collections and all the strains isolated from patients, supporting previous studies and suggesting that classical classification should further be revised.

The treatment of this entity is same as that of histoplasmosis capsulati and it is treated with amphotericin B and azole derivatives, including posaconazole. The isolated lesions respond to surgical excision also.

III. HISTOPLASMOSIS FARCIMINOSI

This type of histoplasmosis is also called as epizootic lymphangitis (EZL) and is caused by *H.capsulatum* var. *farciminosum*. The organism is relatively host specific as disease is commonly seen in horses, mules as well as dogs and very rarely in human beings. Among the horses, clinical presentation varies with four forms *i.e.* asymptomatic, ocular, cutaneous, and respiratory. However, mixed forms may also be sometimes observed. The cutaneous form is characterized by multifocal pyogranulomatous subcutaneous nodules, which progress along the lymphatic system, with the coalescence of nodules producing a corded appearance. If clinical lesions are located on the limbs, progression of the disease can result in severe lameness. The respiratory form is characterized by pyogranulomatous lesions within the nasal mucosa and lung parenchyma, with potential for multisystem pathology.

It is commonly found prevalent in countries of Africa, Middle East, Japan, Philippines and Far East. In Ethiopia there is large population of horses, it is commonly reported. However, it has been eradicated from Europe.

In tissue, yeast cells resemble *H.c.* var. *capsulatum*. This is an extremely slow growing fungus usually taking about three months for the growth to appear and tends to have less aerial hyphae. The colony are gray, moist and with glabrous surface. On SDA, it produces comparatively short irregularly curved and branched aerial hyphae and bears small, irregular, round to oval bodies that are described as 'rudimentary aleuriospores.' Soil-extract agar may be used to induce sporulation. The molecular methods like PCR based techniques have recently been established in studies carried out in Ethiopia for the rapid detection of this disease in the clinical samples of horses.

Oral itraconazole in dose of 200 to 600 mg per day is useful for treating most of the cases. In widespread infections amphotericin B may be necessary. The solitary cutaneous lesions may be removed surgically. In this disease there may be relapse even after full course of therapy.

Further Reading

1. Abou Zahr A, Saad Aldin E, Yunyongying P. *Histoplasma* epiglottitis in a patient with Crohn's disease maintained on infliximab, prednisone, and azathioprine. *Int J Infect Dis.* 2013; **17:** e650-2.
2. Addidle M, Pynn J, Skellern P, *et al.* Chronic disseminated histoplasmosis with prolonged latency. *J Infect.* 2010; **61:** 272-4.
3. Adenis A, Nacher M, Hanf M, *et al.* HIV-associated histoplasmosis early mortality and incidence trends: From neglect to priority. *PLoS Negl Trop Dis.* 2014; **8:** e3100.
4. Adenis A, Nacher M, Hanf M, *et al.* Tuberculosis and histoplasmosis among human immunodeficiency virus-infected patients: A comparative study. *Am J Trop Med Hyg.* 2014; **90:** 216-23.
5. Adenis AA, Aznar C, Couppie P. Histoplasmosis in HIV-infected patients: A review of new developments and remaining gaps. *Curr Trop Med Rep.* 2014; **1:** 119-28.
6. Adhi M, Jusufbegovic D, Schaal S. Sequential chorioretinal changes in presumed ocular histoplasmosis syndrome analyzed using spectral domain optical coherence tomography. *Ocul Immunol Inflamm.* 2016: 1-9. PMID: 27082105.
7. Agarwal P, Capoor MR, Singh M, *et al.* An unusual presentation of disseminated histoplasmosis: Case report and review of pediatric immunocompetent patients from India. *Mycopathologia.* 2015; **180:** 359-64.
8. Agarwal S, Moodley J, Ajani Goel G, *et al.* A rare trigger for macrophage activation syndrome. *Rheumatol Int.* 2011; **31:** 405-7.
9. Aggarwal A, Garg S. Isolated pancreatic histoplasmosis: An unusual suspect of pancreatic head mass in an immunocompetent host. *Perm J.* 2015; **19:** e145-7.
10. Agnihotri M, Naik L, Fernandes G, *et al.* Diagnosis of histoplasmosis on lymph node fine needle aspiration cytology utilizing Giemsa stain: A report of three cases. *Anal Quant Cytopathol Histpathol.* 2014; **36:** 299-302.
11. Agudelo CA, Restrepo CA, Molina DA, *et al.* Tuberculosis and histoplasmosis co-infection in AIDS patients. *Am J Trop Med Hyg.* 2012; **87:** 1094-8.
12. Ahuja A, Mathur SR, Iyer VK, *et al.* Histoplasmosis presenting as bilateral adrenal masses: Cytomorphological diagnosis of three cases. *Diagn Cytopathol.* 2012; **40:** 729-31.
13. Ahumada F, Perez D, de Gorgolas M, *et al.* Subacute histoplasmosis with focal involvement of the epiglottis: importance of differential diagnosis. *Case Rep Otolaryngol.* 2014; 235975.
14. Aide MA. Histoplasmosis. *J Bras Pneumol.* 2009; **35:** 1145-51.
15. Aisenberg G, Marcos LA, Ogbaa I. Recurrent histoplasmosis in AIDS mimicking a colonic carcinoma. *Int J STD AIDS.* 2009; **20:** 429-30.
16. Akin L, Herford AS, Cicciu M. Oral presentation of disseminated histoplasmosis: A case report and literature review. *J Oral Maxillofac Surg.* 2011; **69:** 535-41.
17. Ala-Kauhaluoma M, Aho I, Ristola M, *et al.* Involvement of intraocular structures in disseminated histoplasmosis. *Acta Ophthalmol.* 2010; **88:** 493-6.
18. Alave J, Bustamante B, Soto L, *et al.* Acalculous cholecystitis caused by *Histoplasma capsulatum* in a severely immunosuppressed HIV-infected patient. *J Infect Dev Ctries.* 2011; **5:** 235-8.
19. Almeida Mde A, Pizzini CV, Damasceno LS, *et al.* Validation of western blot for *Histoplasma capsulatum* antibody detection assay. *BMC Infect Dis.* 2016; **16:** 87.
20. Alverson B, Alexander N, LeGolvan MP, *et al.* A human immunodeficiency virus-positive infant with probable congenital histoplasmosis in a nonendemic area. *Pediatr Infect Dis J.* 2010; **29:** 1055-7.
21. Andrade AI, Donato M, Previgliano C, *et al.* Histoplasmosis brain abscesses in an immunocompetent adult. A case report and literature review. *Neuroradiol J.* 2014; **27:** 334-8.
22. Ansari HA, Saeed N, Khan N, *et al.* Laryngeal histoplasmosis. *BMJ Case Rep.* 2016; pii: bcr2016216423.
23. Antinori S. *Histoplasma capsulatum:* More widespread than previously thought. *Am J Trop Med Hyg.* 2014; **90:** 982-3.
24. Antonello VS, Zaltron VF, Vial M, *et al.* Oropharyngeal histoplasmosis: Report of eleven cases and review of the literature. *Rev Soc Bras Med Trop.* 2011; **44:** 26-9.
25. Anza K, Mamatha G, Sandhya G, *et al.* Syphilis of fungal world: Novel skin manifestations of histoplasmosis in an immunocompetent host. *Indian J Dermatol.* 2012; **57:** 504.
26. Aparna V. Histoplasmosis mimicking bronchogenic carci noma. *J Assoc Physicians India.* 2016; **64:** 54-5.
27. Arango M, Castaneda E, Agudelo CI, *et al.* Histoplasmosis: Results of the Colombian national survey, 1992-2008. *Biomedica.* 2011; **31:** 344-56.
28. Arango-Bustamante K, Restrepo A, Cano LE, *et al.* Diagnostic value of culture and serological tests in the diagnosis of histoplasmosis in HIV and non-HIV Colombian patients. *Am J Trop Med Hyg.* 2013; **89:** 937-42.

29. Ardizzoni A, Baschieri MC, Manca L, *et al.* The mycoarray as an aid for the diagnosis of an imported case of histoplasmosis in an Italian traveler returning from Brazil. *J Travel Med.* 2013; **20:** 336-9.
30. Arghya B, Kaushik M, Mimi G, *et al.* Cytodiagnosis of cutaneous histoplasmosis in HIV positive patient initially presenting with multiple umbilicated disseminated skin nodules. *Diagn Cytopathol.* 2013; **41:** 459-62.
31. Armin BB, Nabili V, Wang MB. Disseminated histoplasmosis caused by *Histoplasma capsulatum. Arch Otolaryngol Head Neck Surg.* 2011; **137:** 307 & 309-10.
32. Aronis ML, dos Santos RP, Goldani LZ. Disseminated *Histoplasma capsulatum* and *Cryptococcus neoformans* co-infection in patients with AIDS. *Mycopathologia.* 2011; **172:** 233-6.
33. Arora N, Mishra A, Santra P, *et al.* Hairy cell leukemia with ascites and coexistent histoplasmosis. *Indian J Pathol Microbiol.* 2015; **58:** 125-6.
34. Ashbee HR, Evans EG, Viviani MA, *et al.* Histoplasmosis in Europe: Report on an epidemiological survey from the ECMM Working Group. *Med Mycol.* 2008; **46:** 57-65.
35. Assi M, Martin S, Wheat LJ, *et al.* Histoplasmosis after solid organ transplant. *Clin Infect Dis.* 2013; **57:** 1542-9.
36. Assi MA, Sandid MS, Baddour LM, *et al.* Risk factor analysis of *Histoplasma capsulatum* fungemia. *Med Mycol.* 2010; **48:** 85-9.
37. Azar MM, Hage CA. Laboratory diagnostics for histoplasmosis. *J Clin Microbiol.* 2017; **55:** 1612-20.
38. Azar MM, Malinis MF. Disseminated histoplasmosis with skin lesions and osteomyelitis in a patient from the Philippines. *Am J Trop Med Hyg.* 2016; **95:** 70-4.
39. Azhar JK, Jacqueline HS, Tony LK, *et al.* Bilateral adrenal histoplasmosis: Endoscopic ultrasound-guided fine needle aspiration as a method of diagnosis and assessment. *Med J Malaysia.* 2011; **66:** 504-6.
40. Baddley JW, Sankara IR, Rodriquez JM, *et al.* Histoplasmosis in HIV-infected patients in a southern regional medical center: Poor prognosis in the era of highly active antiretroviral therapy. *Diagn Microbiol Infect Dis.* 2008; **62:** 151-6.
41. Badyal RK, Gupta R, Vaiphei K. Diffuse perforated necrotising amoebic colitis with histoplasmosis in an immunocompetent individual presenting as an acute abdomen. *BMJ Case Rep.* 2013; pii: bcr2013009059.
42. Bagla P, Sarria JC. Disseminated histoplasmosis in early human immunodeficiency virus infection. *Am J Med Sci.* 2017; **353:** 293-5.
43. Bahr NC, Antinori S, Wheat LJ, *et al.* Histoplasmosis infections worldwide: Thinking outside of the Ohio River valley. *Curr Trop Med Rep.* 2015; **2:** 70-80.
44. Bahr NC, Lee D, Stauffer WM, *et al.* Seroprevalence of histoplasmosis in Somali, Burmese and Hmong refugees residing in Thailand and Kenya. *J Immigr Minor Health.* 2017; doi: 10.1007/s10903-017-0595-3. PMID: 28474198.
45. Bahr NC, Sarosi GA, Meya DB, *et al.* Seroprevalence of histoplasmosis in Kampala, Uganda. *Med Mycol.* 2016; **54:** 295-300.
46. Baig WW, Attur RP, Chawla A, *et al.* Epididymal and prostatic histoplasmosis in a renal transplant recipient from southern India. *Transpl Infect Dis.* 2011; **13:** 489-91.
47. Banerjee AK, Thammayya A, Basu N, *et al.* Localised histoplasmosis treated with amphotericin. *Bull Calcutta Sch Trop Med.* 1971; **19:** 91-3.
48. Bansal A, Drewek R. Sarcoidosis and histoplasmosis: Is one a consequence of the other? A case report and review of the literature. *Case Rep Rheumatol.* 2015; 108459.
49. Baradkar VP, Tendolkar U, Baveja S, *et al.* A rare case of *Histoplasma* fungemia in an AIDS patient. *Indian J Med Microbiol.* 2011; **29:** 188-91.
50. Barket S, Collins B, Halusic E, *et al.* A chronic nonhealing gingival mass: Histoplasmosis. *J Am Dent Assoc.* 2013; **144:** 491-4.
51. Bates R, Wan SH, Sohail MR. 44-year-old man with abdominal pain, fever and bloody diarrhea. *Mayo Clin Proc.* 2015; **90:** e59-62.
52. Bava J, Troncoso A. Giemsa and Grocott in the recognition of *Histoplasma capsulatum* in blood smears. *Asian Pac J Trop Biomed.* 2013; **3:** 418-20.
53. Benedict K, Derado G, Mody RK. Histoplasmosis-associated hospitalizations in the United States, 2001-2012. *Open Forum Infect Dis.* 2016; **3:** ofv219. PMID: 26894201.
54. Benedict K, Mody RK. Epidemiology of histoplasmosis outbreaks, United States, 1938-2013. *Emerg Infect Dis.* 2016; **22:** 370-8.
55. Benedict K, Thompson GR 3rd, Deresinski S, *et al.* Mycotic infections acquired outside areas of known endemicity, United States. *Emerg Infect Dis.* 2015; **21:** 1935-41.
56. Benmassaoud A, Maedler C, Soulellis CA. Histoplasmosis of the small bowel acquired while camping. *Am J Gastroenterol.* 2015; **110:** 1653.
57. Berger JR, Greenberg RN. Isolated central nervous system histoplasmosis in an immunocompetent patient: 53-month hiatus to diagnosis and treatment. *J Neurovirol.* 2010; **16:** 472-4.
58. Berkowitz R, Chiang WK. Update on emerging infections: News from the Centers for Disease Control and Prevention. *Ann Emerg Med.* 2014; **63:** 353-6.
59. Bhagwat PV, Hanumanthayya K, Tophakhane RS, *et al.* Two unusual cases of histoplasmosis in human immunodeficiency virus-infected individuals. *Indian J Dermatol Venereol Leprol.* 2009; **75:** 173-6.
60. Bhansali A, Das S, Dutta P, *et al.* Adrenal histoplasmosis: Unusual presentations. *J Assoc Physicians India.* 2012; **60:** 54-8.
61. Bhari N, Pahadiya P, Arava S, *et al.* Histoplasmosis mimicking non-Hodgkin lymphoma in a 40-year-old man with AIDS. *Int J STD AIDS.* 2017; **28:** 312-4.
62. Bharti P, Bala K, Gupta N. Cutaneous manifestation of underlying disseminated histoplasmosis in an immunocompetent host of nonendemic area with reversible CD4 cell depletion and its recovery on antifungal therapy. *Mycopathologia.* 2015; **180:** 223-7.
63. Bhattacharya JB, Rani P, Aggarwal R, *et al.* Primary cutaneous histoplasmosis masquerading as lepromatous leprosy. *J Clin Diagn Res.* 2017; **11:** ED01-2.

64. Bhattacharyya I. Diagnostic discussion. Histoplasmosis. Todays FDA. 2012; **24:** 32-3, & 35-7.
65. Bhojwani N, Hartman JB, Taylor DC, *et al.* Non-disseminated histoplasmosis of the trachea. *Clin Respir J.* 2016; **10:** 255-8.
66. Bitterman R, Oren I, Geffen Y, *et al.* Prolonged fever and splinter hemorrhages in an immunocompetent traveler with disseminated histoplasmosis. *J Travel Med.* 2013; **20:** 57-9.
67. Bolen RD, Bushnell CD, Reynolds PS. Acute ischemic strokes from small vessel vasculitis due to disseminated histoplasmosis infection. *J Neurol Sci.* 2015; **352:** 125-6.
68. Bonifaz A, Chang P, Moreno K, *et al.* Disseminated cutaneous histoplasmosis in acquired immunodeficiency syndrome: Report of 23 cases. *Clin Exp Dermatol.* 2009; **34:** 481-6.
69. Bonsignore A, Orcioni GF, Barranco R, *et al.* Fatal disseminated histoplasmosis presenting as FUO in an immunocompetent Italian host. *Leg Med.* 2017; **25:** 66-70.
70. Borges-Costa J, Marques T, Soares-Almeida L, *et al.* Progressive disseminated histoplasmosis as a presentation of AIDS in a patient from the Congo: The role of skin biopsy. *Trop Doct.* 2011; **41:** 251-2.
71. Braun-Falco M, Meiss F. Cutaneous histoplasmosis. *Lancet Infect Dis.* 2008; **8:** 734.
72. Bravo EA, Zegarra AJ, Piscoya A, *et al.* Dimorphic fungal coinfection as a cause of chronic diarrhea and pancolitis. *Case Rep Med.* 2011; 960638. PMID: 21837244.
73. Brazao-Silva MT, Mancusi GW, Bazzoun FV, *et al.* A gingival manifestation of histoplasmosis leading diagnosis. *Contemp Clin Dent.* 2013; **4:** 97-101.
74. Brilhante RS, Fechine MA, Mesquita JR, *et al.* Histoplasmosis in HIV-positive patients in Ceara, Brazil: Clinical-laboratory aspects and *in vitro* antifungal susceptibility of *Histoplasma capsulatum* isolates. *Trans R Soc Trop Med Hyg.* 2012; **106:** 484-8.
75. Brilhante RS, Ribeiro JF, Lima RA, *et al.* Evaluation of the genetic diversity of *Histoplasma capsulatum* var. *capsulatum* isolates from north-eastern Brazil. *J Med Microbiol.* 2012; **61 (Pt. 12):** 1688-95.
76. Brummer E, Katsuhiko K. *Histoplasma capsulatum*: Master evader of innate immunity. *Med Mycol J.* 2014; **55:** e57-62.
77. Bruno MA, Parodi M, Olmedo I, *et al.* Massive gastrointestinal bleeding in AIDS patients secondary to *Histoplasma* and cytomegalovirus infection. *Int J Surg Case Rep.* 2016; **25:** 128-31.
78. Buitrago MJ, Canteros CE, Frias De Leon G, *et al.* Comparison of PCR protocols for detecting *Histoplasma capsulatum* DNA through a multicenter study. *Rev Iberoam Micol.* 2013; **30:** 256-60.
79. Buitrago MJ, Gonzalo-Jimenez N, Navarro M, *et al.* A case of primary cutaneous histoplasmosis acquired in the laboratory. *Mycoses.* 2011; **54:** e859-61.
80. Caceres DH, Scheel CM, Tobon AM, *et al.* Validation of an enzyme-linked immunosorbent assay that detects *Histoplasma capsulatum* antigenuria in Colombian patients with AIDS for diagnosis and follow-up during therapy. *Clin Vaccine Immunol.* 2014; **21:** 1364-8.
81. Caceres DH, Tobon AM, Cleveland AA, *et al.* Clinical and laboratory profile of persons living with human immunodeficiency virus/acquired immune deficiency syndrome and histoplasmosis from a Colombian hospital. *Am J Trop Med Hyg.* 2016; **95:** 918-24.
82. Caceres DH, Zuluaga A, Arango-Bustamante K, *et al.* Implementation of a training course increased the diagnosis of histoplasmosis in Colombia. *Am J Trop Med Hyg.* 2015; **93:** 662-7.
83. Cairoli E, Tafuri J, Olivari D. Laryngeal histoplasmosis in systemic lupus erythematosus: First reported case. *Lupus.* 2010; **19:** 1354-5.
84. Camara LA, Miraglia K, Leclair SJ. *Histoplasma capsulatum* in New England: A case study. *Clin Lab Sci.* 2014; **27:** 6-12.
85. Canelo-Aybar C, Cuadra-Urteaga J, Atencia F, *et al.* Human T Lymphotropic virus-1 associated gastrointestinal histoplasmosis in Peru. *J Infect Dev Ctries.* 2011; **5:** 484-8.
86. Cantor AG, John J, Brown B. Metrorrhagia in a human immunodeficiency virus positive woman caused by cervical histoplasmosis. *Am J Med Sci.* 2010; **339:** 285-7.
87. Cao C, Bulmer G, Li J, *et al.* Indigenous case of disseminated histoplasmosis from the *Penicillium marneffei* endemic area of China. *Mycopathologia.* 2010; **170:** 47-50.
88. Carlucci JG, Halasa N, Creech CB, *et al.* Vertical transmission of histoplasmosis associated with anti-tumor necrosis factor therapy. *J Pediatric Infect Dis Soc.* 2016; **5:** e9-12.
89. Carrasco-Zuber JE, Navarrete-Dechent C, Bonifaz A, *et al.* Cutaneous involvement in the deep mycoses: A review. Part II - Systemic mycoses. *Actas Dermosifiliogr.* 2016; **107:** 816-22.
90. Castelli AA, Rosenthal DG, Bender Ignacio R, *et al.* Hemophagocytic lymphohistiocytosis secondary to human immunodeficiency virus-associated histoplasmosis. *Open Forum Infect Dis.* 2015; **2:** ofv140. PMID: 26566535.
91. Castro Alegria LA. Laryngeal histoplasmosis: Report of the first case in Colombia. *Colomb Med (Cali).* 2015; **46:** 202-3.
92. Catania J, Martin SS, Corey GR, *et al.* Diagnostic dilemma in a returning traveler with fever. *Diagn Microbiol Infect Dis.* 2013; **77:** 85-6.
93. Chakrabarti A, Slavin MA. Endemic fungal infections in the Asia-Pacific region. *Med Mycol.* 2011; **49:** 337-44.
94. Chande C, Menon S, Gohil A, *et al.* Cutaneous histoplasmosis in AIDS. *Indian J Med Microbiol.* 2010; **28:** 404-6.
95. Chandra H, Chandra S, Sharma A. Histoplasmosis on bone marrow aspirate cytological examination associated with hemophagocytosis and pancytopenia in an AIDS patient. *Korean J Hematol.* 2012; **47:** 77-9.
96. Chang P, Rodas C. Skin lesions in histoplasmosis. *Clin Dermatol.* 2012; **30:** 592-8.
97. Chang YT, Huang SC, Hu SY, *et al.* Disseminated histoplasmosis presenting as haemolytic anaemia. *Postgrad Med J.* 2010; **86:** 443-4.
98. Chatterjee D, Chatterjee A, Agarwal M, *et al.* Disseminated histoplasmosis with oral manifestation in an immunocompetent patient. *Case Rep Dent.* 2017; 1323514. PMID: 28255468.

99. Chaudhari D, Mckinney J, Hubbs D, *et al.* Mediastinal histoplasmosis presenting as dysphagia: A case report with literature review. *Clin J Gastroenterol.* 2013; **6:** 315-8.
100. Chawla A, Chawla K, Thomas J. Genitourinary histoplasmosis in post-renal transplant patient: Diagnostic dilemma. *Indian J Urol.* 2012; **28:** 359-61.
101. Chejara RS, Nawal CL, Agrawal MK, *et al.* Progressive disseminated histoplasmosis with Coomb's positive hemolytic anemia in an immunocompetent host. *J Assoc Physicians India.* 2016; **64:** 78-9.
102. Chheda LV, Ferketich AK, Carroll CP, *et al.* Smoking as a risk factor for choroidal neovascularization secondary to presumed ocular histoplasmosis syndrome. *Ophthalmology.* 2012; **119:** 333-8.
103. Choi SJ, Choi HS, Chun JY, *et al.* Subacute progressive disseminated histoplasmosis in immunocompetent patient. *Korean J Intern Med.* 2016; **31:** 999-1002.
104. Choudhary N, Aggarwal I, Dutta D, *et al.* Acquired perforating dermatosis and Addison's disease due to disseminated histoplasmosis: Presentation and clinical outcomes. *Dermatoendocrinol.* 2013; **5:** 305-8.
105. Choudhary NS, Puri R, Kotecha H, *et al.* An uncommon etiology of adrenal enlargement and fever in a patient of cirrhosis diagnosed with EUS guided FNAC. *Trop Gastroenterol.* 2014; **35:** 258-60.
106. Choudhary NS, Puri R, Paliwal M, *et al.* Histoplasmosis presenting as pancreatic head mass lesion and gastric outlet obstruction in an immunocompetent patient. *Endoscopy.* 2014; **46 (Suppl. 1):** e411-2.
107. Cionni DA, Lewis SA, Petersen MR, *et al.* Analysis of outcomes for intravitreal bevacizumab in the treatment of choroidal neovascularization secondary to ocular histoplasmosis. *Ophthalmology.* 2012; **119:** 327-32.
108. Clement CG, Loeffelholz MJ, Eltorky MA, *et al.* *Mycobacterium haemophilum* and *Histoplasma capsulatum* coinfection in a renal transplant patient. *J Clin Microbiol.* 2011; **49:** 3425-8.
109. Colaiacovo R, de Castro AC, Shiang C, *et al.* Disseminated histoplasmosis: A rare cause of multiple ulcers in the gastrointestinal tract. *Endoscopy.* 2011; **43 (Suppl. 2):** e216.
110. Cormier H, Perez N, Blanchet D, *et al.* Fatal histoplasmosis in a non-HIV patient in French Guiana. *J Mycol Med.* 2012; **22:** 189-91.
111. Correia FG, Alencar CH, Damasceno LS, *et al.* Spatial distribution of disseminated histoplasmosis and AIDS co-infection in an endemic area of Northeastern Brazil. *Rev Soc Bras Med Trop.* 2016; **49:** 227-31.
112. Corti M, Villafane MF, Palmieri O, *et al.* Rupioid histoplasmosis: First case reported in an AIDS patient in Argentina. *Rev Inst Med Trop Sao Paulo.* 2010; **52:** 279-80.
113. Cottle LE, Gkrania-Klotsas E, Williams HJ, *et al.* A multinational outbreak of histoplasmosis following a biology field trip in the Ugandan rainforest. *J Travel Med.* 2013; **20:** 83-7.
114. Couturier MR, Graf EH, Griffin AT. Urine antigen tests for the diagnosis of respiratory infections: Legionellosis, histoplasmosis, pneumococcal pneumonia. *Clin Lab Med.* 2014; **34:** 219-36.
115. Cuellar-Rodriguez J, Avery RK, Lard M, *et al.* Histoplasmosis in solid organ transplant recipients: 10 years of experience at a large transplant center in an endemic area. *Clin Infect Dis.* 2009; **49:** 710-6.
116. Cunha VS, Pereira PR, Sprinz E. Medical imagery. Cutaneous histoplasmosis as initial presentation of AIDS. *Int J Infect Dis.* 2010; **14:** e271-2.
117. da Silva RM Jr, da Silva Neto JR, Santos CS, *et al.* Fluorescent *in situ* hybridization of pre-incubated blood culture material for the rapid diagnosis of histoplasmosis. *Med Mycol.* 2015; **53:** 160-4.
118. Dalal P, Chernin L, Swender D, *et al.* Histoplasmosis in the olecranon bursa of a patient with idiopathic CD4 lymphocytopenia. *Ann Allergy Asthma Immunol.* 2011; **107:** 543-4.
119. Dall Bello AG, Severo CB, Guazzelli LS, *et al.* Histoplasmosis mimicking primary lung cancer or pulmonary metastases. *J Bras Pneumol.* 2013; **39:** 63-8.
120. Dall Bello AG, Severo CB, Oliveira Fde M, *et al.* Histoplasmosis presenting with multiple pulmonary nodules. A case mimicking radiological features of pulmonary metastasis. *Rev Inst Med Trop Sao Paulo.* 2013; 55. pii: S0036-46652013000300209.
121. Dalvin LA, Thom SB, Garcia JJ, *et al.* Scleritis due to histoplasmosis. *Cornea.* 2016; **35:** 402-4.
122. Damasceno LS, Leitao TM, Taylor ML, *et al.* The use of genetic markers in the molecular epidemiology of histoplasmosis: A systematic review. *Eur J Clin Microbiol Infect Dis.* 2016; **35:** 19-27.
123. Damasceno LS, Ramos AN Jr, Alencar CH, *et al.* Disseminated histoplasmosis in HIV-infected patients: Determinants of relapse and mortality in a north-eastern area of Brazil. *Mycoses.* 2014; **57:** 406-13.
124. Damm DD. Chronic ulceration of tongue. Histoplasmosis. *Gen Dent.* 2014; **62:** 79-80.
125. Dantas KC, Freitas RS, Moreira AP, *et al.* The use of nested Polymerase Chain Reaction (nested PCR) for the early diagnosis of *Histoplasma capsulatum* infection in serum and whole blood of HIV-positive patients. *An Bras Dermatol.* 2013; **88:** 141-3.
126. Dasmasceno LS, Novaes AR Jr, Alencar CH, *et al.* Disseminated histoplasmosis and AIDS: Relapse and late mortality in endemic area in North-Eastern Brazil. *Mycoses.* 2013; **56:** 520-6.
127. Dawood H. A case of immune reconstitution syndrome to disseminated histoplasmosis. *J Int Assoc Physicians AIDS Care (Chic).* 2011; **10:** 277-9.
128. de Aguiar Cordeiro R, de Farias Marques FJ, de Aguiar Cordeiro R, *et al.* Synthesis and antifungal activity in vitro of isoniazid derivatives against *Histoplasma capsulatum* var. *capsulatum*. *Antimicrob Agents Chemother.* 2014; **58:** 2504-11.
129. De D, Nath UK. Disseminated histoplasmosis in immunocompetent individuals - not a so rare entity in India. *Mediterr J Hematol Infect Dis.* 2015; **7:** e2015028.
130. De Lavaissiere M, Manceron V, Bouree P, *et al.* Reconstitution inflammatory syndrome related to histoplasmosis, with a hemophagocytic syndrome in HIV infection. *J Infect.* 2009; **58:** 245-7.

131. de Paulo LF, Rosa RR, Durighetto Júnior AF. Primary localized histoplasmosis: Oral manifestations in immunocompetent patients. *Int J Infect Dis.* 2013; **17:** e139-40.
132. Deepe GS Jr, Buesing WR. Deciphering the pathways of death of *Histoplasma capsulatum*-infected macrophages: Implications for the immunopathogenesis of early infection. *J Immunol.* 2012; **188:** 334-44.
133. Delfino E, Di Biagio A, Chandrapatham K, Viscoli C, Prinapori R. Disseminated histoplasmosis with mucocutaneous immune reconstitution inflammatory syndrome in an HIV-infected patient. *AIDS Res Hum Retroviruses.* 2015; **31:** 274-5.
134. Deodhar D, Frenzen F, Rupali P, *et al.* Disseminated histoplasmosis: A comparative study of the clinical features and outcome among immunocompromised and immunocompetent patients. *Natl Med J India.* 2013; **26:** 214-5.
135. Dhawan J, Verma P, Sharma A, *et al.* Disseminated cutaneous histoplasmosis in an immunocompetent child, relapsed with itraconazole, successfully treated with voriconazole. *Pediatr Dermatol.* 2010; **27:** 549-51.
136. Diaz RI, Sigler EJ, Rafieetary MR, *et al.* Ocular histoplasmosis syndrome. *Surv Ophthalmol.* 2015; **60:** 279-95.
137. Dlova NC. Disseminated cutaneous histoplasmosis. *Skinmed.* 2010; **8:** 110.
138. Doleschal B, Rodhammer T, Tsybrovskyy O, *et al.* Disseminated histoplasmosis: A challenging differential diagnostic consideration for suspected malignant lesions in the digestive tract. *Case Rep Gastroenterol.* 2016; **10:** 653-60.
139. Dubois A, Janssens PG, Brutsaert P, *et al.* A case of African histoplasmosis; with a mycological note on *Histoplasma duboisii* n. sp. *Ann Soc Belg Med Trop* (1920). 1952; **32:** 569-84.
140. Dufresne SF, Le Blanc RE, Zhang SX, *et al.* Histoplasmosis and subcutaneous nodules in a kidney transplant recipient: Erythema nodosum versus fungal panniculitis. *Transpl Infect Dis.* 2013; **15:** e58-63.
141. Duncan JT. Tropical African histoplasmosis. *Trans R Soc Trop Med Hyg.* 1958; **52:** 468-74.
142. Duncan SJ, Kadaria D. A woman with a lung mass and multiple pulmonary nodules. *Am J Case Rep.* 2015; **16:** 367-70.
143. Durand ML, Lin DT, Juliano AF, *et al.* A 78-year-old woman with chronic sore throat and a tonsillar mass. *N Engl J Med.* 2014; **371:** 1535-43.
144. Dussouil AS, Allardet-Servent J, Dunogeant L, *et al.* Disseminated histoplasmosis partially mimicking a dermatomyositis in a patient with rheumatoid arthritis. *Br J Dermatol.* 2015; **173:** 797-800.
145. Dylewski J. Acute pulmonary histoplasmosis. *CMAJ.* 2011; **183:** e1090.
146. Ebenye CM. A case of disseminated histoplasmosis detected in peripheral blood smear staining revealing AIDS at terminal phase in a female patient from Cameroon. *Case Rep Med.* 2012; 215207. PMID: 23251167.
147. Ecka RS, Bhatia P, Varma N, *et al.* Disseminated histoplasmosis with peripheral blood spill over. *Indian J Pediatr.* 2014; **81:** 313-4.
148. Ecka RS, Sharma M, Tomar V. Disseminated histoplasmosis in an immunocompetent haweli dweller: A diagnosis and follow-up by endoscopic ultrasound-guided fine-needle aspiration. *J Cytol.* 2015; **32:** 142-4.
149. Edwards JA, Rappleye CA. *Histoplasma* mechanisms of pathogenesis - one portfolio doesn't fit all. *FEMS Microbiol Lett.* 2011; **324:** 1-9.
150. Edwards PC, Hess S, Saini T, *et al.* Twenty-eight-year-old woman with recent onset of painful oral erosions. *J Oral Maxillofac Surg.* 2010; **68:** 1890-5.
151. Egressy Kv, Mohammed M, Ferguson JS. The use of endobronchial ultrasound in the diagnosis of subacute pulmonary histoplasmosis. *Diagn Ther Endosc.* 2015; 510863.
152. Ehrlich R, Ciulla TA, Maturi R, *et al.* Intravitreal bevacizumab for choroidal neovascularization secondary to presumed ocular histoplasmosis syndrome. *Retina.* 2009; **29:** 1418-23.
153. Eid AJ, Leever JD, Husmann K. Compartmentalized *Histoplasma capsulatum* infection of the central nervous system. *Case Rep Infect Dis.* 2015; 581415.
154. Eissenberg LP, Goldman WE. *Histoplasma* variation and adaptive strategies for parasitism: New perspective on histoplasmosis. *Clin Microbiol Rev.* 1991; **4:** 411-21.
155. Elansari R, Abada R, Rouadi S, *et al.* *Histoplasma capsulatum* sinusitis: Possible way of revelation to the disseminated form of histoplasmosis in HIV patients: Case report and literature review. *Int J Surg Case Rep.* 2016; **24:** 97-100.
156. Elias NA, Cuestas ML, Sandoval M, *et al.* Rapid identification of *Histoplasma capsulatum* directly from cultures by multiplex PCR. *Mycopathologia.* 2012; **174:** 451-6.
157. Ellickalputhenpura BE, Torres HA. Histoplasmosis: An asymptomatic to life-threatening infection. *JAAPA.* 2012; **25:** 18, 25-6.
158. Eloubeidi MA, Luz LP, Crowe DR, *et al.* Bilateral adrenal gland enlargement secondary to histoplasmosis mimicking adrenal metastases: Diagnosis with EUS-guided FNA. *Diagn Cytopathol.* 2010; **38:** 357-9.
159. Emani M, Malhotra S, Rana F. Bone marrow aspirate showing *Histoplasma capsulatum. Blood.* 2014; **123:** 957.
160. Escobar B, Maldonado VN, Ansari S, *et al.* Antigen negative gastrointestinal histoplasmosis in an AIDS patient. *Am J Case Rep.* 2014; **15:** 90-3.
161. Espinoza JV, Lasave AF, Guzman-Blanco M, *et al.* Disseminated histoplasmic multifocal choroiditis in AIDS. *Ophthalmic Surg Lasers Imaging Retina.* 2014; **45:** 179-80.
162. Estrada-Bellmann I, Camara-Lemarroy CR, Flores-Cantu H, *et al.* Hemichorea in a patient with HIV-associated central nervous system histoplasmosis. *Int J STD AIDS.* 2016; **27:** 75-7.
163. Faiolla RC, Coelho MC, Santana Rde C, *et al.* Histoplasmosis in immunocompetent individuals living in an endemic area in the Brazilian Southeast. *Rev Soc Bras Med Trop.* 2013; **46:** 461-5.
164. Falci DR, Hoffmann ER, Paskulin DD, *et al.* Progressive disseminated histoplasmosis: A systematic review on the performance of non-culture-based diagnostic tests. *Braz J Infect Dis.* 2017; **21:** 7-11.

165. Falster L, Marin MB, Gomes JL. Histoplasmosis diagnosed after arthroscopy of the knee: Case report. *Braz J Infect Dis.* 2015; **19:** 546-8.
166. Fandino-Devia E, Rodriguez-Echeverri C, Cardona-Arias J, *et al.* Antigen detection in the diagnosis of histoplasmosis: A meta-analysis of diagnostic performance. *Mycopathologia.* 2016; **181:** 197-205.
167. Farrell SK, Hiremath GS, Subramanian PS, *et al.* The peaks and valleys of ocular histoplasmosis syndrome. *Lancet Infect Dis.* 2010; **10:** 212.
168. Federico C, Sergio P. Histoplasmosis mimicking felty syndrome. *J Clin Rheumatol.* 2012; **18:** 278.
169. Ferguson-Paul K, Mangum S, Porter A, *et al.* Hemophagocytic lymphohistiocytosis and progressive disseminated histoplasmosis. *Emerg Infect Dis.* 2016; **22:** 1119-21.
170. Ferrari TC, Soares JM, Salles JM, *et al.* Laryngeal histoplasmosis in an immunocompetent patient from a non-endemic region: Case report. *Mycoses.* 2009; **52:** 539-40.
171. Figueira JA, Camilo Junior D, Biasoli ER, *er al.* Oral ulcers associated with bone destruction as the primary manifestation of histoplasmosis in an immunocompetent patient. *J Eur Acad Dermatol Venereol.* 2017; doi: 10.1111/jdv.14257. PMID: 28380270.
172. Fischer GB, Mocelin H, Severo CB, *et al.* Histoplasmosis in children. *Paediatr Respir Rev.* 2009; **10:** 172-7.
173. Fitzhugh VA, Maniar KP, Kim MK, *et al.* Adrenal histoplasmosis. *Diagn Cytopathol.* 2010; **38:** 188-9.
174. Flores-Franco RA, Gomez-Diaz A, de Jesus Fernandez-Alonso A. Chronic progressive disseminated histoplasmosis in a Mexican cockfighter. *Am J Trop Med Hyg.* 2015; **92:** 4-5.
175. Fortuna G, Mignogna MD. Oral histoplasmosis of a healthy man in a non-endemic area. *Infection.* 2011; **39:** 497-8.
176. Fowler B, Shen C, Mastellone J, *et al.* Acute systemic histoplasmosis associated with chorioretinitis in an immunocompetent adolescent. *Digit J Ophthalmol.* 2011; **17:** 31-5.
177. França CM, Cavalcante EG, Ribeiro AS, *et al.* Disseminated histoplasmosis in a juvenile lupus erythematosus patient. *Acta Reumatol Port.* 2012; **37:** 276-9.
178. Frank PN, Palma Diaz MF, Uslan DZ, *et al.* Histoplasmosis mimicking malignant adenopathy during diagnostic thyroid lobectomy. *JAMA Otolaryngol Head Neck Surg.* 2014; **140:** 664-6.
179. Freifeld AG, Wheat LJ, Kaul DR. Histoplasmosis in solid organ transplant recipients: early diagnosis and treatment. *Curr Opin Organ Transplant.* 2009; **14:** 601-5.
180. Frias De Leon MG, Arenas Lopez G, Taylor ML, *et al.* Development of specific sequence-characterized amplified region markers for detecting *Histoplasma capsulatum* in clinical and environmental samples. *J Clin Microbiol.* 2012; **50:** 673-9.
181. Frias-De-Leon MG, Ramirez-Barcenas JA, Rodriguez-Arellanes G, *et al.* Usefulness of molecular markers in the diagnosis of occupational and recreational histoplasmosis outbreaks. *Folia Microbiol (Praha).* 2017; **62:** 111-6.
182. Friedman A, Solomon G, Segal-Maurer S, *et al.* Sudden onset of verrucous plaques to the face and trunk: A case of cutaneous histoplasmosis in the setting of HIV. *Dermatol Online J.* 2008; **14:** 19.
183. Fritzsche C, Loebermann M, Aepinus C, *et al.* Vaginal ulceration and local lymphadenopathy in an African immigrant. *Clin Infect Dis.* 2009; **48:** 441-2 & 493-4.
184. Furuie JL, Sun J, do Nascimento MM, *et al.* Molecular identification of *Histoplasma capsulatum* using rolling circle amplification. *Mycoses.* 2016; **59:** 12-9.
185. Gailey MP, Keeney ME, Jensen CS. A cytomorphometric analysis of pulmonary and mediastinal granulomas: Differentiating histoplasmosis from sarcoidosis by fine-needle aspiration. *Cancer Cytopathol.* 2015; **123:** 51-8.
186. Gailey MP, Klutts JS, Jensen CS. Fine-needle aspiration of histoplasmosis in the era of endoscopic ultrasound and endobronchial ultrasound: Cytomorphologic features and correlation with clinical laboratory testing. *Cancer Cytopathol.* 2013; **121:** 508-17.
187. Gajendra S, Jha B, Sahni T, *et al.* Disseminated histoplasmosis in an immunocompetent host presenting as pancytopenia with bilateral adrenal masses. *Turk J Haematol.* 2015; **32:** 191-2.
188. Gajendra S, Sharma R, Goel S, *et al.* Adrenal histoplasmosis in immunocompetent patients presenting as adrenal insufficiency. *Turk Patoloji Derg.* 2016; **32:** 105-11.
189. Gandhi V, Ulyanovskiy P, Epelbaum O. Update on the spectrum of histoplasmosis among Hispanic patients presenting to a New York City municipal hospital: A contemporary case series. *Respir Med Case Rep.* 2015; **16:** 60-4.
190. Ganesan N, Sharma R, Phansalkar MD, *et al.* Disseminated histoplasmosis in an immunocompetent patient diagnosed on bone marrow aspirate - A rare presentation from a non-endemic area. *J Clin Diagn Res.* 2015; **9:** ED07-8.
191. Garcia-Guinon A, Torres-Rodriguez JM, Ndidongarte DT, *et al.* Disseminated histoplasmosis by *Histoplasma capsulatum* var. *duboisii* in a paediatric patient from the Chad Republic, Africa. *Eur J Clin Microbiol Infect Dis.* 2009; **28:** 697-9.
192. Garg A, Wadhera R, Gulati SP, *et al.* Tongue histoplasmosis mimicking malignancy. *J Assoc Physicians India.* 2011; **59:** 591-2.
193. Ge L, Zhou C, Song Z, *et al.* Primary localized histoplasmosis with lesions restricted to the mouth in a Chinese HIV-negative patient. *Int J Infect Dis.* 2010; **14 (Suppl. 3):** e325-8.
194. George AA, Bindra M, Mohanraj P. Rare presentation of disseminated histoplasmosis in an immunocompetent host. *Am J Trop Med Hyg.* 2015; **93:** 1125-7.
195. George P, Nayak N, Anoop TM, *et al.* Pulmonary histoplasmosis mimicking carcinoma lung. *Lung India.* 2015; **32:** 663-5.
196. Ghosh R, Mishra P, Sen S, Maiti PK, Chatterjee G. Experience of varied presentation of chronic progressive disseminated histoplasmosis in immunocompetent patients: A diagnostic conundrum. *Indian J Dermatol.* 2016; **61:** 580.
197. Ghosh S, Das T, Kundu AK, *et al.* Disseminated cutaneous histoplasmosis in HIV infection. *J Assoc Physicians India.* 2012; **60:** 43.

198. Gill D, Dean R, Virk J, *et al.* Unusual presentation of disseminated histoplasmosis. *Am J Emerg Med.* 2017; **35:** 668.e3-4.
199. Golda N, Feldman M. Histoplasmosis clinically imitating cutaneous malignancy. *J Cutan Pathol.* 2008; **35 (Suppl. 1):** 26-8.
200. Goldani LZ, Aquino VR, Lunardi LW, *et al.* Two specific strains of *Histoplasma capsulatum* causing mucocutaneous manifestations of histoplasmosis: Preliminary analysis of a frequent manifestation of histoplasmosis in southern Brazil. *Mycopathologia.* 2009; **167:** 181-6.
201. Gomez BL. Histoplasmosis: Epidemiology in Latin America. *Curr Fungal Infect Rep.* 2011; **5:** 199-205.
202. Goncalves D, Ferraz C, Vaz L. Posaconazole as rescue therapy in African histoplasmosis. *Braz J Infect Dis.* 2013; **17:** 102-5.
203. Gopal K, Singh S, Gupta R. Fine needle aspiration cytology in bilateral adrenal masses in a 58-year-old man. *Acta Cytol.* 2010; **54:** 234-6.
204. Gopalakrishnan R, Nambi PS, Ramasubramanian V, *et al.* Histoplasmosis in India: Truly uncommon or uncommonly recognised? *J Assoc Physicians India.* 2012; **60:** 25-8.
205. Goswami RP, Pramanik N, Banerjee D, Raza MM, Guha SK, Maiti PK. Histoplasmosis in eastern India: The tip of the iceberg? *Trans R Soc Trop Med Hyg.* 1999; **93:** 540-2.
206. Gotoff R, Franceschelli J, Prichard J, *et al.* A 78-year-old male with a 4-month history of fever, fatigue, and weight loss. *J Clin Microbiol.* 2011; **49:** 771 & 1194.
207. Gravdahl DJ, Gardetto JS, Hurley JR, *et al.* Pulmonary histoplasmosis producing a spindle cell "pseudotumor". *Am J Clin Pathol.* 2011; **136:** 410-5.
208. Gregory G, Linda R. Disseminated histoplasmosis complicating HIV infection. *Blood.* 2012; **120:** 5101.
209. Grover K, Zainah H, Bhatnagar S, *et al.* A rare case of diffuse alveolar hemorrhage secondary to acute pulmonary histoplasmosis. *Case Rep Infect Dis.* 2015; 821749.
210. Grover S, Selhi PK, Sood N, *et al.* "Polka Dot Macrophages" on cytology of bilateral adrenal masses - Nailing disseminated histoplasmosis. *Diagn Cytopathol.* 2017; doi: 10.1002/dc.23758. PMID: 28548709.
211. Gugnani HC, Egere JU, Larsh H. Skin sensitivity to *capsulatum* and *duboisii* histoplasmins in Nigeria. *J Trop Med Hyg.* 1991; **94:** 24-6.
212. Gugnani HC, Muotoe-Okafor FA, Kaufman L, *et al.* A natural focus of *Histoplasma capsulatum* var. *duboisii* in a bat cave. *Mycopathologia.* 1994; **127:** 151-7.
213. Gugnani HC. Histoplasmosis in Africa: A review. *Indian J Chest Dis Allied Sci.* 2000; **42:** 271-7.
214. Gulati SP, Gupta A, Wadhera R, Deep A, Kalra R. Histoplasmosis of larynx in immunocompetent patients mimicking carcinoma: Report of two cases. *J Infect Dis Antimicrob Agents.* 2008; **25:** 145-9.
215. Gupta H, Tankhiwale SS. A case of bilateral eyelid histoplasmosis mistaken as basal cell carcinoma. *Can J Ophthalmol.* 2017; **52:** e45-6.
216. Gupta N, Aragaki A, Wikenheiser-Brokamp KA, *et al.* Occam's razor or Hickam's dictum? *J Bronchology Interv Pulmonol.* 2012; **19:** 216-9.
217. Gupta N, Arora SK, Rajwanshi A, *et al.* Histoplasmosis: Cytodiagnosis and review of literature with special emphasis on differential diagnosis on cytomorphology. *Cytopathology.* 2010; **21:** 240-4.
218. Gupta N, Fox CM, Grisolano SW. Disseminated histoplasmosis with colonic ulcers in a patient receiving infliximab. *Gastrointest Endosc.* 2009; **70:** 597-8.
219. Gupta P, Bhardwaj M. Cytodiagnosis of disseminated histoplasmosis in an immunocompetent individual with molluscum contagiosum-like skin lesions and lymphadenopathy. *J Cytol.* 2016; **33:** 163-5.
220. Gupta V, Singhal V, Singh MK, *et al.* Disseminated histoplasmosis with hypercalcemia. *J Am Acad Dermatol.* 2013; **69:** e250-1.
221. Hage C, Kleiman MB, Wheat LJ. Histoplasmosis in solid organ transplant recipients. *Clin Infect Dis.* 2010; **50:** 122-4.
222. Hage CA, Azar MM, Bahr N, *et al.* Histoplasmosis: Up-to-date evidence-based approach to diagnosis and management. *Semin Respir Crit Care Med.* 2015; **36:** 729-45.
223. Hage CA, Bowyer S, Tarvin SE, *et al.* Recognition, diagnosis and treatment of histoplasmosis complicating tumor necrosis factor blocker therapy. *Clin Infect Dis.* 2010; **50:** 85-92.
224. Hage CA, Davis TE, Fuller D, *et al.* Diagnosis of histoplasmosis by antigen detection in BAL fluid. *Chest.* 2010; **137:** 623-8.
225. Hage CA, Horan DJ, Durkin M, *et al.* *Histoplasma capsulatum* preferentially induces IDO in the lung. *Med Mycol.* 2013; **51:** 270-9.
226. Hage CA, Ribes JA, Wengenack NL, *et al.* A multicenter evaluation of tests for diagnosis of histoplasmosis. *Clin Infect Dis.* 2011; **53:** 448-54.
227. Hage CA, Wheat LJ. Diagnosis of pulmonary histoplasmosis using antigen detection in the bronchoalveolar lavage. *Expert Rev Respir Med.* 2010; **4:** 427-9.
228. Hanf M, Adenis A, Couppie P, *et al.* HIV-associated histoplasmosis in French Guiana: recent infection or reactivation? *AIDS.* 2010; **24:** 1777-8.
229. Hariri OR, Minasian T, Quadri SA, *et al.* Histoplasmosis with deep CNS involvement: Case presentation with discussion and literature review. *J Neurol Surg Rep.* 2015; **76:** e167-72.
230. Harnalikar M, Kharkar V, Khopkar U. Disseminated cutaneous histoplasmosis in an immunocompetent adult. *Indian J Dermatol.* 2012; **57:** 206-9.
231. Hasmoni MH, Shah AS, Ayoub S, *et al.* Disseminated histoplasmosis presenting with chronic ulcerative tongue lesions in a patient with diabetes. *BMJ Case Rep.* 2010; pii: bcr.06.2010.3120. PMID: 22791841.
232. Haydoura S, Wallentine J, Lopansri B, *et al.* Disseminated histoplasmosis in allogeneic bone marrow transplant: A diagnosis not to be missed. *Transpl Infect Dis.* 2014; **16:** 822-6.
233. Hegerova LT, Lin Y. Disseminated histoplasmosis: A cause of hemophagocytic syndrome. *Mayo Clin Proc.* 2013; **88:** e123.

234. Heilbrunn BR, Weinstein E, Walthall JD. Erythema nodosum associated with pulmonary histoplasmosis. *Ann Emerg Med.* 2008; **52:** 606 & 615.
235. Heller HM, Wu CC, Pierce VM, *et al.* A 29-year-old man with abdominal pain, fever and weight loss. *N Engl J Med.* 2013; **369:** 1453-61.
236. Hernandez JM, Munoz-Cadavid CO, Hernandez DL, *et al.* Detection of *Histoplasma capsulatum* DNA in peripheral blood from a patient with ocular histoplasmosis syndrome. *Med Mycol.* 2012; **50:** 202-6.
237. Hess J, Fondell A, Fustino N, *et al.* Presentation and treatment of histoplasmosis in pediatric oncology patients: Case series and review of the literature. *J Pediatr Hematol Oncol.* 2017; **39:** 137-40.
238. Hoffmann ER, Daboit TC, Paskulin DD, et al. Disseminated histoplasmosis and AIDS: A prospective and multicentre study to evaluate the performance of different diagnostic tests. *Mycoses.* 2017; **60:** 20-4.
239. Hong JH, Stetsenko GY, Pottinger PS, *et al.* Cutaneous presentation of disseminated histoplasmosis as a solitary peri-anal ulcer. Case report and discussion. *J Cutan Pathol.* 2016; **43:** 438-43.
240. Horwath MC, Fecher RA, Deepe GS Jr. *Histoplasma capsulatum*, lung infection and immunity. *Future Microbiol.* 2015; **10:** 967-75.
241. Huang CC, Wu CJ, Chang CM, *et al.* A man with fever, gingival plaque and diffuse lymphadenopathy. *Clin Infect Dis.* 2011; **53:** 1118-9 & 1164-5.
242. Ibrahim KY, Carvalho NB, Mimicos EV, *et al.* Cutaneous and bone marrow histoplasmosis after 18 years of renal allograft transplant. *Mycopathologia.* 2014; **178:** 273-8.
243. Ijaz A, Choudhury D. A case of rare, fungal peritonitis caused by *Histoplasma capsulatum* in a patient on CAPD. *Nat Rev Nephrol.* 2010; **6:** 435-9.
244. Iqbal F, Schifter M, Coleman HG. Oral presentation of histoplasmosis in an immunocompetent patient: A diagnostic challenge. *Aust Dent J.* 2014; **59:** 386-8.
245. Iriart X, Blanchet D, Menard S, *et al.* A complementary tool for management of disseminated *Histoplasma capsulatum* var. *capsulatum* infections in AIDS patients. *Int J Med Microbiol.* 2014; **304:** 1062-5.
246. Jaimes A, Muvdi S, Alvarado Z, *et al.* Perforation of the nasal septum as the first sign of histoplasmosis associated with AIDS and review of published literature. *Mycopathologia.* 2013; **176:** 145-50.
247. Jain M, Revankar SG. A case of peritoneal histoplasmosis in a patient receiving chronic ambulatory peritoneal dialysis. *Mycoses.* 2012; **55:** 99-100.
248. Jain P, Hu S, Jabbour E, *et al.* Disseminated histoplasmosis as pseudo Richter's transformation in a patient with chronic lymphocytic leukemia. *Am J Hematol.* 2015; **90:** 752-3.
249. Jaiswal S, Vij M, Chand G, *et al.* Diagnosis of adrenal histoplasmosis by fine needle aspiration cytology: An analysis based on five cases. *Cytopathology.* 2011; **22:** 323-8.
250. Jana M, Hari S, Arava SK. Disseminated histoplasmosis manifested by laryngopharyngeal and adrenal lesions in an HIV-negative individual. *J Indian Med Assoc.* 2010; **108:** 615.
251. Jazwinski A, Naggie S, Perfect J. Immune reconstitution syndrome in a patient with disseminated histoplasmosis and steroid taper: maintaining the perfect balance. *Mycoses.* 2011; **54:** 270-2.
252. Jinno S, Gripshover BM, Lemonovich TL, *et al. Histoplasma capsulatum* prosthetic valve endocarditis with negative fungal blood cultures and negative *Histoplasma* antigen assay in an immunocompetent patient. *J Clin Microbiol.* 2010; **48:** 4664-6.
253. John AM, Vasanthi N, Khurana A, *et al.* An unusual cause for rings in the brain. *Hong Kong Med J.* 2012; **18:** 346-7.
254. John M, Koshy JM, Mohan S, *et al.* Histoplasmosis presenting as a laryngeal ulcer in an immunocompetent host. *J Assoc Physicians India.* 2015; **63:** 69-71.
255. Johnson JA, Loyd JE, Wheat LJ, *et al.* A case series and review of histoplasmosis infection in the neck. *Arch Otolaryngol Head Neck Surg.* 2010; **136:** 916-9.
256. Johnston KT, Dang PA, Specht MC, et al. A 53-year-old woman with pain and a mass in the breast. *N Engl J Med.* 2016; **375:** 1172-80.
257. Johnston RB Jr, Thareja S, Shenefelt PD. Disseminated histoplasmosis in a renal transplant patient. *Cutis.* 2013; **91:** 295-9.
258. Jorns JM, Knoepp SM. Asteroid bodies in lymph node cytology: Infrequently seen and still mysterious. *Diagn Cytopathol.* 2011; **39:** 35-6.
259. Joshi KR, Chhangani DL, Ramdeo IN, *et al.* Oral histoplasmosis in India. *J Laryngol Otol.* 1973; **87:** 165-70.
260. Kabangila R, Semvua K, Rambau P, *et al.* Pulmonary histoplasmosis presenting as chronic productive cough, fever, and massive unilateral consolidation in a 15-year-old immune-competent boy: A case report. *J Med Case Rep.* 2011; **5:** 374.
261. Kadaria D, Archie DS, Sultan Ali I, *et al.* Dual time point positron emission tomography/computed tomography scan in evaluation of intrathoracic lesions in an area endemic for histoplasmosis and with high prevalence of sarcoidosis. *Am J Med Sci.* 2013; **346:** 358-62.
262. Kadaria D, Muthiah MP, Sinclair SE. Unusual presentations of disseminated histoplasmosis. *Tenn Med.* 2012; **105:** 39-40.
263. Kairalla R, Kawano-Dourado L, Baldi B, *et al.* Diffuse micronodules with spontaneous resolution. *Am J Respir Crit Care Med.* 2011; **184:** 859.
264. Kajfasz P, Basiak W. Outbreak of pulmonary histoplasmosis involving a group of four Polish travellers returning from Ecuador. *Int Marit Health.* 2012; **63:** 59-62.
265. Kalongolera L, Kamiza S, Bates J, *et al.* Histoplasmosis in a Malawian patient on ART. *Malawi Med J.* 2013; **25:** 93.
266. Kamili Qu, Menter A. Atypical presentation of histoplasmosis in a patient with psoriasis and psoriatic arthritis on infliximab therapy. *J Drugs Dermatol.* 2010; **9:** 57-60.
267. Kandi V, Vaish R, Palange P, *et al.* Chronic pulmonary histoplasmosis and its clinical significance: An under-reported systemic fungal disease. *Cureus.* 2016; **8:** e751.

268. Kapelusznik L, Arumugam V, Caplivski D, *et al.* Disseminated histoplasmosis presenting as AIDS cholangiopathy. *Mycoses.* 2011; **54:** 262-4.
269. Karunanithi S, Kumar G, Sharma SK, *et al.* Staging and response of sternal histoplasmosis by 18F-FDG PET/CT. *Clin Nucl Med.* 2015; **40:** 231-3.
270. Kasaliwal R, Malhotra G, Bukan A, *et al.* 18F-FDG PET as a monitoring tool to assess treatment response in bilateral adrenal histoplasmosis. *Clin Nucl Med.* 2014; **39:** 576-8.
271. Kash N, Jahan-Tigh RR, Efron-Everett M, *et al.* Primary mucocutaneous histoplasmosis in an immunocompetent patient. *Dermatol Online J.* 2014; **21.** pii: 13030/qt87614080.
272. Kashif M, Tariq H, Ijaz M, *et al.* Disseminated histoplasmosis and secondary hemophagocytic syndrome in a non-HIV patient. *Case Rep Crit Care.* 2015; 295735.
273. Kathuria S, Capoor MR, Yadav S, *et al.* Disseminated histoplasmosis in an apparently immunocompetent individual from north India: A case report and review. *Med Mycol.* 2013; **51:** 774-8.
274. Kathuria S, Singh PK, Meis JF, Chowdhary A. *In vitro* antifungal susceptibility profile and correlation of mycelial and yeast forms of molecularly characterized *Histoplasma capsulatum* strains from India. *Antimicrob Agents Chemother.* 2014; **58:** 5613-6.
275. Kauffman CA, Miceli MH. Histoplasmosis and blastomycosis in solid organ transplant recipients. *J Fungi.* 2015; **1:** 94-106.
276. Kauffman CA. Histoplasmosis: A clinical and laboratory update. *Clin Microbiol Rev.* 2007; **20:** 115-32.
277. Kauffman CA. Histoplasmosis. *Clin Chest Med.* 2009; **30:** 217-25.
278. Kikuchi K, Sugita T, Makimura K, *et al.* Is *Histoplasma capsulatum* a native inhabitant of Japan? *Microbiol Immunol.* 2008; **52:** 455-9.
279. Klassen-Fischer M, McEvoy P, Neafie RC, *et al.* Accurate diagnosis of infection with *Histoplasma capsulatum* var. *duboisii. Clin Infect Dis.* 2004; **38:** 595-6.
280. Klein IP, Martins MA, Martins MD, *et al.* Diagnosis of HIV infection on the basis of histoplasmosis-related oral ulceration. Spec Care Dentist. 2016; **36:** 99-103.
281. Knox KS, Hage CA. Histoplasmosis. *Proc Am Thorac Soc.* 2010; **7:** 169-72.
282. Koene RJ, Catanese J, Sarosi GA. Adrenal hypofunction from histoplasmosis: A literature review from 1971 to 2012. *Infection.* 2013; **41:** 757-9.
283. Koepsell SA, Hinrichs SH, Iwen PC. Applying a real-time PCR assay for *Histoplasma capsulatum* to clinically relevant formalin-fixed paraffin-embedded human tissue. *J Clin Microbiol.* 2012; **50:** 3395-7.
284. Koh PS, Roslani AC, Vimal KV, *et al.* Concurrent amoebic and histoplasma colitis: A rare cause of massive lower gastrointestinal bleeding. *World J Gastroenterol.* 2010; **16:** 1296-8.
285. Koley S, Mandal RK, Khan K, *et al.* Disseminated cutaneous histoplasmosis, an initial manifestation of HIV, diagnosed with fine needle aspiration cytology. *Indian J Dermatol.* 2014; **59:** 182-5.
286. Kollipara R, Hans A, Hall J, *et al.* A case report of primary cutaneous histoplasmosis requiring deep tissue sampling for diagnosis. *Dermatol Online J.* 2014; 20. pii: 13030/qt11b53041.
287. Konik EA, Bremer M, Lin PT, *et al.* Severe pulmonic valve regurgitation due to *Histoplasma* endocarditis. *Echo Res Pract.* 2015; **2:** K21-4.
288. Kooblall M, Keane B, Murray G, *et al.* Histoplasmosis mimicking primary lung neoplasm. *BMJ Case Rep.* 2014; pii: bcr2013203335. PMID: 24728897.
289. Kothari D, Chopra S, Bhardwaj M, *et al.* Persistence of *Histoplasma* in adrenals 7 years after antifungal therapy. *Indian J Endocrinol Metab.* 2013; **17:** 529-31.
290. Krakauer M, Prendes MA, Wilkes B, *et al.* A case of orbital histoplasmosis. *Ophthal Plast Reconstr Surg.* 2016; **32:** e71-2.
291. Kroetz DN, Deepe GS. The role of cytokines and chemokines in *Histoplasma capsulatum* infection. *Cytokine.* 2012; **58:** 112-7.
292. Kumari M, Udayakumar M, Kaushal M, *et al.* Unusual presentation of disseminated histoplasmosis in an immunocompetent patient. *Diagn Cytopathol.* 2017; doi: 10.1002/dc.23742. PMID: 28474430.
293. Labhardt N, Rickerts V, Popescu S, *et al.* TB or not TB - Persistent cough fever and night sweats in a 46-year-old traveler returning from South America. *Travel Med Infect Dis.* 2015; **13:** 346-8.
294. Lad D, Malhotra P, Maskey D, *et al.* Pyrexia, lung nodules, granulomas: Pulmonary lymphomatoid granulomatosis. *Indian J Hematol Blood Transfus.* 2014; **30 (Suppl. 1):** 418-21.
295. Lakshman A, Dhir V, Kumar N, *et al.* Miliary tuberculosis in an Indian lady: Looking beyond miliary tuberculosis. *Lung India.* 2015; **32:** 492-4.
296. Larbcharoensub N, Boonsakan P, Aroonroch R, *et al.* Adrenal histoplasmosis: A case series and review of the literature. *Southeast Asian J Trop Med Public Health.* 2011; **42:** 920-5.
297. Ledtke C, Rehm SJ, Fraser TG, *et al.* Endovascular infections caused by *Histoplasma capsulatum:* A case series and review of the literature. *Arch Pathol Lab Med.* 2012; **136:** 640-5.
298. Ledtke C, Tomford JW, Jain A, *et al.* Clinical presentation and management of histoplasmosis in older adults. *J Am Geriatr Soc.* 2012; **60:** 265-70.
299. Lee AI, Koo S, Vaidya A, *et al.* A bird's-eye view of fever. *N Engl J Med.* 2011; **365:** 1727-32.
300. Lee T, Woods C, O'Hagan A. Broncholithiasis from histoplasmosis in a pediatric patient: Case reports and review of literature. *J Pediatric Infect Dis Soc.* 2013; **2:** 76-9.
301. Lehur AC, Zielinski M, Pluvy J, *et al.* Case of disseminated histoplasmosis in a HIV-infected patient revealed by nasal involvement with maxillary osteolysis. *BMC Infect Dis.* 2017; **17:** 328. PMID: 28476105.
302. Leitao TD, Jorge IF, Freitas AE. Enlarged hilar lymph node due to *Histoplasma. Rev Soc Bras Med Trop.* 2016; **49:** 663.

303. Letranchant L, Debourgogne A, Doco-Lecompte T, *et al.* Fatal *Histoplasma capsulatum* mitral endocarditis in a French patient treated for rheumatoid arthritis. *Mycopathologia.* 2012; **173:** 183-6.
304. Lim LT, Ruzmetova N, Ballinger SH, *et al.* Acute pulmonary histoplasmosis in a patient with uveitis after infliximab therapy. *Int Ophthalmol.* 2011; **31:** 349-51.
305. Lim SY, Kijsirichareanchai K, Winn R. Progressive disseminated histoplasmosis in systemic lupus erythematosus-an unusual presentation of acute tenosynovitis and a literature review. *Clin Rheumatol.* 2013; **32:** 135-9.
306. Lise ML, Godinho RN, Brollo FM, *et al.* Cutaneous ulcer in an immunosuppressed patient with adult onset Still's disease: Primary cutaneous histoplasmosis? *An Bras Dermatol.* 2014; **89:** 532-4.
307. Listernick R. A 6-year-old male with daily fever accompanied by nausea and abdominal pain. *Pediatr Ann.* 2014; **43:** 210-3.
308. Lo MM, Mo JQ, Dixon BP, Czech KA. Disseminated histoplasmosis associated with hemophagocytic lymphohistiocytosis in kidney transplant recipients. *Am J Transplant.* 2010; **10:** 687-91.
309. Lobo V, Joshi A, Khatavkar P, *et al.* Pulmonary histoplasmosis in a renal allograft recipient. *Indian J Nephrol.* 2014; **24:** 120-3.
310. Lofgren SM, Kirsch EJ, Maro VP, *et al.* Histoplasmosis among hospitalized febrile patients in northern Tanzania. *Trans R Soc Trop Med Hyg.* 2012; **106:** 504-7.
311. Lopez LF, Valencia Y, Tobon AM, *et al.* Childhood histoplasmosis in Colombia: Clinical and laboratory observations of 45 patients. *Med Mycol.* 2016; **54:** 677-83.
312. Lorchirachonkul N, Foongladda S, Ruangchira-Urai R, *et al.* Prosthetic valve endocarditis caused by *Histoplasma capsulatum:* The first case report in Thailand. *J Med Assoc Thai.* 2013; **96:** S262-5.
313. Loughan AR, Perna R, Hertza J. Cognitive impairment and memory loss associated with histoplasmosis: A case study. *Clin Neuropsychol.* 2014; **28:** 514-24.
314. Loulergue P, Bastides F, Baudouin V, *et al.* Literature review and case histories of *Histoplasma capsulatum* var. *duboisii* infections in HIV-infected patients. *Emerg Infect Dis.* 2007; **13:** 1647-52.
315. Luckett K, Dummer JS, Miller G, *et al.* Histoplasmosis in patients with cell-mediated immunodeficiency: Human immunodeficiency virus infection, organ transplantation, and tumor necrosis factor-α inhibition. *Open Forum Infect Dis.* 2015; **2:** ofu116. PMID: 26034750.
316. Lunardi LW, Wagner R, Dos Santos CC, *et al.* Complete blood count alterations in disseminated histoplasmosis. *Rev Bras Hematol Hemoter.* 2015; **37:** 263-5.
317. Mackowiak PA. A well-preserved culprit. *Clin Infect Dis.* 2013; **56:** 704 & 747-8.
318. Maharaja K, Gupta V, Khandpur S. Nonhealing tongue ulcer in an Indian man. JAMA Dermatol. 2015; **151:** 333-4.
319. Mahy S, Bel B, Chavanet P, *et al.* Disseminated histoplasmosis with cutaneous lesions in an HIV patient. *Eur J Dermatol.* 2011; **21:** 128-9.
320. Maiti PK. Disseminated histoplasmosis - comment 1. *Indian J Med Microbiol.* 2007; **25:** 427.
321. Maltos AL, Maia E, Oliveira CD. Disseminated histoplasmosis diagnosed in a peripheral blood smear. *Rev Soc Bras Med Trop.* 2017; **50:** 148.
322. Makol A, Wieland CN, Ytterberg SR. Articular involvement in disseminated histoplasmosis in a kidney transplant patient taking azathioprine. *J Rheumatol.* 2011; **38:** 2692-3.
323. Mambie A, Pasquet A, Melliez H, *et al.* A case of immune reconstitution inflammatory syndrome related to a disseminated histoplasmosis in an HIV-1 infected patient. *AIDS.* 2013; **27:** 2170-2.
324. Mandal R, Mondal K, Datta S, *et al.* A clinicopathological study of peripheral lymph nodes in HIV-infected patients with special reference to CD4+ T-cell counts: Experience from a tertiary care institution in Darjeeling (India). *Diagn Cytopathol.* 2015; **43:** 971-7.
325. Mandal SK, Mani S, Sau D, *et al.* Bilateral adrenal histoplasmosis in an immunocompetent patient. *J Assoc Physicians India.* 2012; **60:** 47.
326. Mandengue CE, Ngandjio A, Atangana PJ. Histoplasmosis in HIV-infected persons, Yaoundé, Cameroon. *Emerg Infect Dis.* 2015; **21:** 2094-6.
327. Mansoor CA, Bhargavan PV, Rajanish R, *et al.* Disseminated histoplasmosis. *Indian J Orthop.* 2013; 47:639-42.
328. Manzini M, Lavinsky-Wolff M. Nasal histoplasmosis without lung involvement in an immunocompromised patient. *Braz J Otorhinolaryngol.* 2012; **78:** 136.
329. Marchiori E, Melo SM, Vianna FG, *et al.* Pulmonary histoplasmosis presenting with the reversed halo sign on high-resolution CT scan. *Chest.* 2011; **140:** 789-91.
330. Marchiori E, Zanetti G, Hochhegger B, *et al.* Reversed halo sign on computed tomography: State-of-the-art review. *Lung.* 2012; **190:** 389-94.
331. Marchiori E, Zanetti G, Hochhegger B. Small interstitial nodules. *J Bras Pneumol.* 2015; **41:** 250.
332. Marianelli LG, Frassone N, Marino M, *et al.* Immune reconstitution inflammatory syndrome as histoplasmosis osteomyelitis in South America. *AIDS.* 2014; **28:** 1848-50.
333. Marques N, Lebre A, Marques F, *et al.* Isolated oral histoplasmosis presenting as fever of unknown origin in a Portuguese hemodialysis patient. *Mycopathologia.* 2013; **176:** 89-93.
334. Marques SA, Hortense J, Requena CB, *et al.* Disseminated cutaneous histoplasmosis in elderly patients. An uncommon presentation. *Rev Iberoam Micol.* 2015; **32:** 131-2.
335. Marques SA, Silvares MR, Camargo RM, *et al.* Cutaneous histoplasmosis disclosing an HIV-infection. *An Bras Dermatol.* 2013; **88:** 420-3.
336. Martin-Iguacel R, Kurtzhals J, Jouvion G, *et al.* Progressive disseminated histoplasmosis in the HIV population in Europe in the HAART era. Case report and literature review. *Infection.* 2014; **42:** 611-20.
337. Mata-Essayag S, Colella MT, Rosello A, *et al.* Histoplasmosis: A study of 158 cases in Venezuela, 2000-2005. *Medicine.* 2008; **87:** 193-202.

338. Mathews DM, John R, Verghese V, *et al.* *Histoplasma capsulatum* infection with extensive lytic bone lesions mimicking LCH. *J Trop Pediatr.* 2016; **62:** 496-9.
339. Mathews MS. Disseminated histoplasmosis - comment 2. *Indian J Med Microbiol.* 2007; **25:** 427-8.
340. Mathur P, Zurlo JJ, Crook TJ. The intricate relationship of histoplasmosis and sarcoidosis: A case report. *J Med Case Rep.* 2014; **8:** 235.
341. Matos Figueroa JR, Vazquez Torres OL, *et al.* PET-CT scan positive pulmonary nodule revealing histoplasmosis: A case report. *Bol Asoc Med P R.* 2010; **102:** 47-50.
342. McCabe MP, Heck RK. *Histoplasma* osteomyelitis simulating giant-cell tumor of the distal part of the radius: A case report. *J Bone Joint Surg Am.* 2010; **92:** 708-14.
343. McKinsey DS, McKinsey JP. Pulmonary histoplasmosis. *Semin Respir Crit Care Med.* 2011; **32:** 735-44.
344. McLeod DS, Mortimer RH, Perry-Keene DA, *et al.* Histoplasmosis in Australia: Report of 16 cases and literature review. *Medicine (Baltimore).* 2011; **90:** 61-8.
345. McNeeley MF, Chung JH, Bhalla S, *et al.* Imaging of granulomatous fibrosing mediastinitis. *AJR Am J Roentgenol.* 2012; **199:** 319-27.
346. Mehta A, Desai S, Desai N, *et al.* Disseminated histoplasmosis: Missed opportunistic infection in a HIV-infected patient. *BMJ Case Rep.* 2013; pii: bcr2013201062.
347. Mehta BM, Hashkes PJ, Avery R, *et al.* A 21-year-old man with Still's disease with fever, rash and pancytopenia. *Arthritis Care Res (Hoboken).* 2010; **62:** 575-9.
348. Mehta V, De A, Balachandran C, *et al.* Mucocutaneous histoplasmosis in HIV with an atypical ecthyma like presentation. *Dermatol Online J.* 2009; **15:** 10. PMID: 19450403.
349. Merin MR, Fung MA, Eisen DB, *et al.* Histoplasmosis presenting as a cutaneous malignancy of the eyelid. *Ophthal Plast Reconstr Surg.* 2011; **27:** e41-2.
350. Min Z, Baddley JW, Rodriguez JM, *et al.* Cross-reactivity of *Aspergillus* galactomannan in an HIV-infected patient with histoplasmosis. *Med Mycol Case Rep.* 2012; **1:** 119-22.
351. Mishra DP, Ramamurthy S, Behera SK. Histoplasmosis presenting as isolated cervical lymphadenopathy: A rare presentation. *J Cytol.* 2015; **32:** 188-90.
352. Mohammed S, Sinha M, Chavan P, *et al.* Oral histoplasmosis masquerading as oral cancer in HIV-infected patient: A case report. *Med Mycol Case Rep.* 2012; **1:** 85-7.
353. Mohan M, Fogel B, Eluvathingal T, *et al.* Gastrointestinal histoplasmosis in a patient after autologous stem cell transplant for multiple myeloma. *Transpl Infect Dis.* 2016; **18:** 939-41.
354. Moreno-Coutino G, Hernandez-Castro R, Toussaint-Caire S, *et al.* Histoplasmosis and skin lesions in HIV: A safe and accurate diagnosis. *Mycoses.* 2015; **58:** 413-5.
355. Moriones Robayo CA, Guerra Ortiz CP. Histoplasmosis laryngeal: Report first case in Colombia. *Colomb Med (Cali).* 2014; **45:** 186-9.
356. Mudawi HM, Elamin EM, Baraka OZ, *et al.* Addison's disease due to *Histoplasma duboisii* infection of the adrenal glands. *Saudi Med J.* 2008; **29:** 904-6.
357. Mukherjee A, Talukdar P, Karak A, *et al.* Disseminated histoplasmosis in an immunocompetent individual: A myth or reality? *Trop Doct.* 2016; **46:** 117-8.
358. Mukherjee A, Tangri R, Verma N, *et al.* Chronic disseminated histoplasmosis bone marrow involvement in an immunocompetent patient. *Indian J Hematol Blood Transfus.* 2010; **26:** 65-7.
359. Mukherjee AK, Mukherjee D, Mukhopadhyay M. Histoplasmosis in India: a clinicopathological review with report of a case in a child. *Indian J Pathol Microbiol.* 1986; **29:** 263-70.
360. Mukherjee T, Basu A. Disseminated histoplasmosis presenting as a case of erythema nodosum and hemophagocytic lymphohistiocytosis. *Med J Armed Forces India.* 2015; **71:** S598-600.
361. Mukhopadhyay S, Doxtader EE. Visibility of *Histoplasma* within histiocytes on Hematoxylin and Eosin distinguishes disseminated histoplasmosis from other forms of pulmonary histoplasmosis. *Hum Pathol.* 2013; **44:** 2346-52.
362. Mukhopadhyay S, Katzenstein AL. Biopsy findings in acute pulmonary histoplasmosis: Unusual histologic features in 4 cases mimicking lymphomatoid granulomatosis. *Am J Surg Pathol.* 2010; **34:** 541-6.
363. Munante-Cardenas JL, de Assis AF, Olate S, *et al.* Treating oral histoplasmosis in an immunocompetent patient. *J Am Dent Assoc.* 2009; **140:** 1373-6.
364. Muniz LB, Franco T, Ramos LM, *et al.* Oral exfoliative cytology in the diagnosis of histoplasmosis. *Cytopathology.* 2012; **23:** 204-5.
365. Muniz Mde M, Morais E Silva Tavares P, *et al.* Comparison of different DNA-based methods for molecular typing of *Histoplasma capsulatum. Appl Environ Microbiol.* 2010; **76:** 4438-47.
366. Munoz C, Gomez BL, Tobon A, *et al.* Validation and clinical application of a molecular method for identification of *Histoplasma capsulatum* in human specimens in Colombia, South America. *Clin Vaccine Immunol.* 2010; **17:** 62-7.
367. Muotoe-Okafor FA, Gugnani HC. *In vitro* interactions between *Histoplasma capsulatum* var. *duboisii* and other fungi. *Mycoses.* 1997; **40:** 309-12.
368. Muraosa Y, Toyotome T, Yahiro M, *et al.* Detection of *Histoplasma capsulatum* from clinical specimens by cycling probe-based real-time PCR and nested real-time PCR. *Med Mycol.* 2016; **54:** 433-8.
369. Murphy RA, Gounder L, Manzini TC, *et al.* Challenges in the management of disseminated progressive histoplasmosis in human immunodeficiency virus-infected patients in resource-limited settings. *Open Forum Infect Dis.* 2015; **2:** ofv025. PMID: 26034774.
370. Muzyka BC, Epifanio RN. Update on oral fungal infections. *Dent Clin North Am.* 2013; **57:** 561-81.
371. Myint T, Al-Hasan MN, Ribes JA, *et al.* Temporal trends, clinical characteristics, and outcomes of histoplasmosis in a tertiary care center in Kentucky, 2000 to 2009. *J Int Assoc Provid AIDS Care.* 2014; **13:** 100-5.

372. Myint T, Anderson AM, Sanchez A, *et al.* Histoplasmosis in patients with human immunodeficiency virus/acquired immunodeficiency syndrome (HIV/AIDS): multicenter study of outcomes and factors associated with relapse. *Medicine (Baltimore).* 2014; **93:** 11-8.
373. N'Golet A, N'Gouoni BG, Moukassa D, *et al.* Maxillary African histoplasmosis: Unusual diagnostic problems of an unusual presentation. *Pathol Res Pract.* 2005; **200:** 841-4.
374. Nacher M, Adenis A, Adriouch L, *et al.* What is AIDS in the Amazon and the Guianas? Establishing the burden of disseminated histoplasmosis. *Am J Trop Med Hyg.* 2011; 84: 239-40.
375. Nacher M, Adenis A, Aznar C, *et al.* How many have died from undiagnosed human immunodeficiency virus-associated histoplasmosis, a treatable disease? Time to act. *Am J Trop Med Hyg.* 2014; **90:** 193-4.
376. Nacher M, Adenis A, Basurko C, *et al.* Primary prophylaxis of disseminated histoplasmosis in HIV patients in French Guiana: Arguments for cost effectiveness. *Am J Trop Med Hyg.* 2013; **89:** 1195-8.
377. Nacher M, Adenis A, Blanchet D, *et al.* Risk factors for disseminated histoplasmosis in a cohort of HIV-infected patients in French Guiana. *PLoS Negl Trop Dis.* 2014; **8:** e2638.
378. Nacher M, Adenis A, Mc Donald S, *et al.* Disseminated histoplasmosis in HIV-infected patients in South America: A neglected killer continues on its rampage. *PLoS Negl Trop Dis.* 2013; **7:** e2319.
379. Nacher M, Adenis A, Sambourg E, *et al.* Histoplasmosis or tuberculosis in HIV-infected patients in the Amazon: What should be treated first? *PLoS Negl Trop Dis.* 2014; **8:** e3290.
380. Naeem F, Metzger ML, Arnold SR, Adderson EE. Distinguishing benign mediastinal masses from malignancy in a histoplasmosis-endemic region. *J Pediatr.* 2015; **167:** 409-15.
381. Nagoda M, Uloko AE, Babashani M, *et al.* Histoplasmosis: An elusive re-emerging chest infection. *Niger J Clin Pract.* 2012; **15:** 235-7.
382. Naina HV, Thomas CF Jr, Harris S. Histoplasmosis and asplenia. Thorax. 2010; **65:** 372.
383. Nakshabendi R, Berry AC, Torres-Miranda D, *et al.* Primary *Histoplasma capsulatum* enterocolitis mimicking peptic and inflammatory bowel disease. *Case Rep Gastrointest Med.* 2016; 7139573.
384. Naniwadekar A, Malhotra A. Gastrointestinal histoplasmosis. *J Gastroenterol Hepatol.* 2008; **23:** 668.
385. Narang V, Sinha T, Sandhu AS, *et al.* Clinically inapparent bilateral adrenal masses due to histoplasmosis. *Eur Urol.* 2009; **55:** 518-21.
386. Narayana N, Gifford R, Giannini P, *et al.* Oral histoplasmosis: An unusual presentation. *Head Neck.* 2009; **31:** 274-7.
387. Nasa M, Patel N, Lipi L, *et al.* Gastrointestinal histoplasmosis and CMV co-Infection in an immunocompetent host. *J Assoc Physicians India.* 2017; **65:** 94-5.
388. Nascimento CR, Soares RR, Martelli AC, *et al.* Borderline tuberculoid leprosy associated with primary cutaneous histoplasmosis. *An Bras Dermatol.* 2012; **87:** 146-7.
389. Natarajan M, Swierzbinski MJ, Maxwell S, *et al.* Pulmonary *Histoplasma* infection after allogeneic hematopoietic stem cell transplantation: Case report and review of the literature. *Open Forum Infect Dis.* 2017; **4:** ofx041. PMID: 28470019.
390. Natarajan SK, Choo ZW, Rao AD, *et al.* Gastrointestinal manifestation of disseminated histoplasmosis in a non-endemic region. *ANZ J Surg.* 2017; **87:** 207.
391. Neglected Histoplasmosis in Latin America Group. Disseminated histoplasmosis in Central and South America, the invisible elephant: The lethal blind spot of international health organizations. *AIDS.* 2016; **30:** 167-70.
392. Nett RJ, Skillman D, Riek L, *et al.* Histoplasmosis in Idaho and Montana, USA, 2012-2013. *Emerg Infect Dis.* 2015; **21:** 1071-2.
393. Newman SL, Smulian AG. Iron uptake and virulence in *Histoplasma capsulatum. Curr Opin Microbiol.* 2013; **16:** 700-7.
394. Nguyen FN, Kar JK, Zakaria A, *et al.* Isolated central nervous system histoplasmosis presenting with ischemic pontine stroke and meningitis in an immune-competent patient. *JAMA Neurol.* 2013; **70:** 638-41.
395. Nico MM, Higashino HR, Yeh Li H, *et al.* Exuberant mucocutaneous lesions of disseminated histoplasmosis in a human immunodeficiency virus (HIV) positive patient. *Int J Dermatol.* 2009; **48:** 157-9.
396. Nielsen JS, Fick TA, Saggau DD, *et al.* Intravitreal anti-vascular endothelial growth factor therapy for choroidal neovascularization secondary to ocular histoplasmosis syndrome. *Retina.* 2012; **32:** 468-72.
397. Niemann N, Hochman FL, Huang RS. Histoplasmosis as a possible cause of retroperitoneal fibrosis and median arcuate ligament syndrome: A case report. *Int J Surg Case Rep.* 2014; **5:** 473-5.
398. Nieto-Ríos JF, Aristizabal-Alzate A, Ocampo C, *et al.* Disseminated histoplasmosis and haemophagocytic syndrome in two kidney transplant patients. *Nefrologia.* 2012; **32:** 683-4.
399. Nieto-Rios JF, Serna-Higuita LM, Guzman-Luna CE, *et al.* Histoplasmosis in renal transplant patients in an endemic area at a reference hospital in Medellin, Colombia. *Transplant Proc.* 2014; **46:** 3004-9.
400. Nimitvilai S, Thammaprasert W, Vinyuvat S. Histoplasmosis myositis: A case report and literature review. *Southeast Asian J Trop Med Public Health.* 2015; **46:** 738-42.
401. Norgan AP, Berbari EF, Roberts GD, *et al.* A 79-year-old man with swelling and crusted cutaneous ulceration of both hands. *Clin Infect Dis.* 2010; **50:** 871 & 933-4.
402. Norkaew T, Ohno H, Sriburee P, *et al.* Detection of environmental sources of *Histoplasma capsulatum* in Chiang Mai, Thailand, by nested PCR. *Mycopathologia.* 2013; **176:** 395-402.
403. Norman FF, Martín-Davila P, Fortun J, *et al.* Imported histoplasmosis: Two distinct profiles in travelers and immigrants. *J Travel Med.* 2009; **16:** 258-62.
404. Nosanchuk JD, Gacser A. *Histoplasma capsulatum* at the host-pathogen interface. *Microbes Infect.* 2008; **10:** 973-7.

405. Nosanchuk JD, Zancope-Oliveira RM, Hamilton AJ, *et al.* Antibody therapy for histoplasmosis. *Front Microbiol.* 2012; **3:** 21.
406. Nunes JO, Pillon KR, Bizerra PL, *et al.* The simultaneous occurrence of histoplasmosis and cryptococcal fungemia: A case report and review of the literature. *Mycopathologia.* 2016; **181:** 891-7.
407. Nyalakonda H, Albuerne M, Suazo Hernandez LP, *et al.* Central nervous system histoplasmosis in acquired immunodeficiency syndrome. *Am J Med Sci.* 2016; **351:** 177-186.
408. O'Donovan E, Abdalla S. Histoplasmosis capsulatum in an immunosuppressed patient. *Br J Haematol.* 2013; **162:** 571.-Ohji G, Kikuchi K, Inoue K, *et al.* Progressive disseminated histoplasmosis in an immunocompetent patient as an under-recognized imported mycosis in Japan. *J Infect Chemother.* 2010; **16:** 443-5.
409. O'Shaughnessy MA, Tande AJ, Vasoo S, *et al.* A rare diagnosis: Recognizing and managing fungal tenosynovitis of the hand and upper extremity. *J Hand Surg Am.* 2017; **42:** e77-89.
410. Ohno H, Ogata Y, Suguro H, *et al.* An outbreak of histoplasmosis among healthy young Japanese women after traveling to Southeast Asia. *Intern Med.* 2010; **49:** 491-5.
411. Ohno H, Tanabe K, Umeyama T, *et al.* Application of nested PCR for diagnosis of histoplasmosis. *J Infect Chemother.* 2013; **19:** 999-1003.
412. Oikawa F, Carvalho D, Matsuda NM, *et al.* Histoplasmosis in the nasal septum without pulmonary involvement in a patient with acquired immunodeficiency syndrome: Case report and literature review. *Sao Paulo Med J.* 2010; **128:** 236-8.
413. Oliveira PB, Abreu MA, Mattos AL. Importance of multidisciplinary approach in diagnosis of histoplasmosis. *An Bras Dermatol.* 2016; **91:** 362-4.
414. Ollague Sierra JE, Ollague Torres JM. New clinical and histological patterns of acute disseminated histoplasmosis in human immunodeficiency virus-positive patients with acquired immunodeficiency syndrome. *Am J Dermatopathol.* 2013; **35:** 205-12.
415. Olson TC, Bongartz T, Crowson CS, *et al.* Histoplasmosis infection in patients with rheumatoid arthritis, 1998-2009. *BMC Infect Dis.* 2011; **11:** 145.
416. Orenstein LA, Klein R, Crawford G, *et al.* A healthy patient with iron deficiency anemia and a perianal papule. *Clin Infect Dis.* 2016; **62:** 361 & 397-8.
417. Orsi AT, Nogueira L, Chrusciak-Talhari A, *et al.* Histoplasmosis and AIDS co-infection. *An Bras Dermatol.* 2011; **86:** 1025-6.
418. Ozdalga E, Gratzinger D. Histoplasmosis presenting with ulcers on the soft palate. *J Gen Intern Med.* 2012; **27:** 1219.
419. Pa J, Ray AN, Sherp P, *et al.* Primary cutaneous histoplasmosis simulating molluscum contagiosum. *J Assoc Physicians India.* 2013; **61:** 498-500.
420. Padhye AA, Pathak AA, Katkar VJ, *et al.* Oral histoplasmosis in India: a case report and overview of cases reported during 1968-1992. *J Med Vet Mycol.* 1994; **32:** 93-103.
421. Padhye AA, Smith G, McLaughlin D, *et al.* Comparative evaluation of a chemiluminescent DNA probe and an exoantigen test for rapid identification of *Histoplasma capsulatum*. *J Clin Microbiol.* 1992; **30:** 3108-11.
422. Padma S, Sreehar S. 18F FDG PET/CT identifies unsuspected bilateral adrenal histoplasmosis in an elderly immunocompromised patient. *Indian J Med Res.* 2014; **139:** 786-7.
423. Paixao M, Miot HA, Avancini J, *et al.* Primary cutaneous histoplasmosis developed in the penis of an immunocompetent patient. *An Bras Dermatol.* 2015; **90:** 255-7.
424. Pal N, Adhikary M. Cutaneous histoplasmosis in a HIV seronegative patient. *J Nat Sci Biol Med.* 2013; **4:** 477-9.
425. Pamnani R, Rajab JA, Githang'a J, *et al.* Disseminated histoplasmosis diagnosed on bone marrow aspirate cytology: Report of four cases. *East Afr Med J.* 2009; **86:** S102-5.
426. Pan B, Chen M, Pan W, *et al.* Histoplasmosis: A new endemic fungal infection in China? Review and analysis of cases. *Mycoses.* 2013; **56:** 212-21.
427. Panchabhai TS, Bais RK, Pyle RC, *et al.* An apple-core lesion in the colon: An infectious etiology. *J Glob Infect Dis.* 2011; **3:** 195-8.
428. Panda PK, Jain S, Sood R, *et al.* Typical facial lesions: A window of suspicion for progressive disseminated histoplasmosis - A case of Asian prototype. *Case Rep Infect Dis.* 2016; 2865241. PMID: 27752372.
429. Panja G, Sen S. A unique case of histoplasmosis. *J Indian Med Assoc.* 1954; **23:** 257-8.
430. Parihar A, Tomar V, Ojha BK, *et al.* Magnetic resonance imaging findings in a patient with isolated *Histoplasma* brain abscess. *Arch Neurol.* 2011; **68:** 534-5.
431. Parikh F, Londhe VA, Khude S, *et al.* Gastrointestinal histoplasmosis mimicking abdominal tuberculosis. *J Assoc Physicians India.* 2009; **57:** 76-8.
432. Park H, Shafer D. Disseminated histoplasmosis in peripheral blood smear. *Blood.* 2014; **123:** 1445.
433. Pasqualotto AC, Oliveira FM, Severo LC. *Histoplasma capsulatum* recovery from the urine and a short review of genitourinary histoplasmosis. *Mycopathologia.* 2009; **167:** 315-23.
434. Passos AN, Kohara VS, de Freitas RS, *et al.* Immunological assays employed for the elucidation of an histoplasmosis outbreak in Sao Paulo, SP. *Braz J Microbiol.* 2015; **45:** 1357-61.
435. Passos L, Talhari C, Santos M, *et al.* Histoplasmosis-associated immune reconstitution inflammatory syndrome. *An Bras Dermatol.* 2011; **86:** S168-72.
436. Pastor TA, Holcomb MJ, Motaparthi K, *et al.* Disseminated histoplasmosis mimicking secondary syphilis. *Dermatol Online J.* 2011; **17:** 10.
437. Patel DM, Heath J. A 58-year-old man with a cough and fever. *Am J Med Sci.* 2013; **345:** 150-3.
438. Patel M, Lu F, Hannaway M, *et al.* Fibrosing mediastinitis: A rare complication of histoplasmosis. *BMJ Case Rep.* 2015; pii: bcr2015212774. PMID: 26538131.
439. Patel N, Bronze MS. *Histoplasma* infection of aortofemoral bypass graft. *Am J Med Sci.* 2014; **347:** 421-4.
440. Patel NM, Schwartz DJ. Lower GI bleeding from ileocolonic histoplasmosis. *Gastrointest Endosc.* 2011; **74:** 1404-6.

441. Patil K, Mahima VG, Prathibha Rani RM. Oral histoplasmosis. *J Indian Soc Periodontol.* 2009; **13:** 157-9.
442. Peigne V, Dromer F, Elie C, *et al.* Imported acquired immunodeficiency syndrome-related histoplasmosis in metropolitan France: A comparison of pre-highly active anti-retroviral therapy and highly active anti-retroviral therapy eras. *Am J Trop Med Hyg.* 2011; **85:** 934-41.
443. Pellaton C, Cavassini M, Jaton-Ogay K, *et al. Histoplasma capsulatum* var. *duboisii* infection in a patient with AIDS: Rapid diagnosis using polymerase chain reaction-sequencing. *Diagn Microbiol Infect Dis.* 2009; **64:** 85-9.
444. Peppa D, Plumb I, du Parcq J, *et al.* Gastrointestinal histoplasmosis in an HIV-infected patient living in a non-endemic area. *Int J STD AIDS.* 2008; **19:** 864-5.
445. Pereira GH, Almeida LY, Okubo RS, *et al.* Pulmonary histoplasmosis presenting with a halo sign on CT in an immunocompetent patient. *J Bras Pneumol.* 2013; **39:** 523-4.
446. Pereira GH, Padua SS, Park MV, *et al.* Chronic meningitis by histoplasmosis: Report of a child with acute myeloid leukemia. *Braz J Infect Dis.* 2008; **12:** 555-7.
447. Perko R, Messinger Y, Moertel C. Pseudometastasis secondary to histoplasmosis infection: False-positive PET/CT findings. *Pediatr Blood Cancer.* 2010; **54:** 621-3.
448. Pervez MM, Cobb B, Matin N, *et al.* Disseminated histoplasmosis in a patient with advanced HIV disease - lessons learnt from Bangladesh. *J Health Popul Nutr.* 2010; **28:** 305-7.
449. Pettit AC, Raynor MB, Schwartz HS, *et al.* Histoplasmosis masquerading as a rheumatoid nodule in an immunocompromised host: A case report. *JBJS Case Connect.* 2014; **4:** pii: e75.
450. Pettit NN, Pisano J, Weber S, *et al.* Hepatic failure in a patient receiving itraconazole for pulmonary histoplasmosis - Case report and literature review. *Am J Ther.* 2016; **23:** e1215-21.
451. Phillips J, Staszewski H, Garrison M. Successful treatment of secondary hemophagocytic lymphohistiocytosis in a patient with disseminated histoplasmosis. *Hematology.* 2008; **13:** 282-5.
452. Pourfarziani V, Taheri S. Is pulmonary histoplasmosis a risk factor for acute renal failure in renal transplant recipients? *Saudi J Kidney Dis Transpl.* 2009; **20:** 643-5.
453. Putot A, Perrin S, Jolivet A, *et al.* HIV-associated disseminated histoplasmosis in western French Guiana, 2002-2012. *Mycoses.* 2015; **58:** 160-6.
454. Qian Q, Humayun H, Humayun Y, *et al.* Granulomatous interstitial nephritis associated with disseminated histoplasmosis in an immunocompetent patient. *Am J Kidney Dis.* 2011; **58:** 1018-21.
455. Qualtieri J, Stratton CW, Head DR, *et al.* PCR detection of *Histoplasma capsulatum* var. *capsulatum* in whole blood of a renal transplant patient with disseminated histoplasmosis. *Ann Clin Lab Sci.* 2009; **39:** 409-12.
456. Quilter LA, Kleiman MB, Kirsch E, *et al.* Disseminated histoplasmosis of infancy in one of the twins. *Pediatr Infect Dis J.* 2012; **31:** 990-1.
457. Quinter S, Cheng CL, Prakash N, *et al.* Disseminated histoplasmosis presenting as panniculitis in two immunosuppressed patients. *Dermatol Online J.* 2012; **18:** 3.
458. Qureshi A. A case of histoplasmosis mimicking tuberculosis. *J Pak Med Assoc.* 2008; **58:** 457-8.
459. Radhakrishnan S, Adulkar NG, Kim U. Primary cutaneous histoplasmosis mimicking basal cell carcinoma of the eyelid: A case report and review of literature. *Indian J Pathol Microbiol.* 2016; **59:** 227-8.
460. Rafee Y, Xavier A, Antoine M, *et al.* An infant with fever, hepatosplenomegaly and pancytopenia. *Pediatr Infect Dis J.* 2008; **27:** 571.
461. Raina RK, Mahajan V, Sood A, *et al.* Primary cutaneous histoplasmosis in an immunocompetent host from a non-endemic area. *Indian J Dermatol.* 2016; **61:** 467.
462. Rajput A, Bence-Bruckler I, Huebsch L, *et al.* Disseminated histoplasmosis associated with acquired hemophagocytic lymphohistiocytosis. *Clin Case Rep.* 2015; **3:** 195-6.
463. Ramireddy S, Wanger A, Ostrosky L. An instructive case of CNS histoplasmosis in an immunocompetent host. *Med Mycol Case Rep.* 2012; **1:** 69-71.
464. Rana C, Krishnani N, Kumari N, *et al.* Rectal histoplasmosis in Job's syndrome. *Indian J Gastroenterol.* 2013; **32:** 64-5.
465. Rana C, Krishnani N, Kumari N. Bilateral adrenal histoplasmosis in immunocompetent patients. *Diagn Cytopathol.* 2011; **39:** 294-6.
466. Rana C, Kumari N, Krishnani N. Adrenal histoplasmosis: A diagnosis on fine needle aspiration cytology. *Diagn Cytopathol.* 2011; **39:** 438-42.
467. Randhawa HS, Chaturvedi S, Khan ZU, *et al.* Epididymal histoplasmosis diagnosed by isolation of *Histoplasma capsulatum* from semen. *Mycopathologia.* 1995; **131:** 173-7.
468. Randhawa HS, Khan ZU. Histoplasmosis in India: current status. *Indian J Chest Dis Allied Sci.* 1994; **36:** 193-213.
469. Randhawa HS. Occurrence of histoplasmosis in Asia. *Mycopathol Mycol Appl.* 1970; **41:** 75-89.
470. Ranganath S, McGhie TA, LaRosa S, *et al.* Clinical and laboratory diagnosis of central nervous system histoplasmosis. *Am J Med.* 2012; **125:** e1-2.
471. Rangel-Castilla L, Hwang SW, White AC, *et al.* Neuroendoscopic diagnosis of central nervous system histoplasmosis with basilar arachnoiditis. *World Neurosurg.* 2012; **77:** 399.e9-13.
472. Rangwala F, Putcharoen O, Bowonwatanuwong C, *et al.* Histoplasmosis and penicilliosis among HIV-infected Thai patients: A retrospective review. *Southeast Asian J Trop Med Public Health.* 2012; **43:** 436-41.
473. Ranjan R, Jain D, Singh L, *et al.* Differentiation of *Histoplasma* and *Cryptococcus* in cytology smears: A diagnostic dilemma in severely necrotic cases. *Cytopathology.* 2015; **26:** 244-9.
474. Rappo U, Beitler JR, Faulhaber JR, *et al.* Expanding the horizons of histoplasmosis: Disseminated histoplasmosis in a renal transplant patient after a trip to Bangladesh. *Transpl Infect Dis.* 2010; **12:** 155-60.

475. Rastogi P, Sharma P, Kumar N, *et al.* A case of histoplasmosis in a patient with MDS/MPN-U. *Blood Res.* 2016; **51:** 206-7.

476. Ravindran S, Sobhanakumari K, Celine M, *et al.* African histoplasmosis: The first report of an indigenous case in India. *Int J Dermatol.* 2015; **54:** 451-5.

477. Ray A, Sanyal D. A rare case of Addison's disease due to bilateral adrenal histoplasmosis presenting with hypoglycaemia. *J Assoc Physicians India.* 2016; **64:** 45-46.

478. Regnier-Rosencher E, Dupont B, Jacobelli S, *et al.* Late occurrence of *Histoplasma duboisii* cutaneous and pulmonary infection 18 years after exposure. *J Mycol Med.* 2014; **24:** 229-33.

479. Reis F, Franca MC Junior, Nucci A, *et al.* Chronic meningitis, hydrocephalus and spinal paraplegia in non-systemic histoplasmosis. *Arq Neuropsiquiatr.* 2016; **74:** 514-5.

480. Rekha A, Kindo AJ, Sounderarajan P, *et al.* Infection with *Histoplasma capsulatum* in a renal transplant recipient. *Saudi J Kidney Dis Transpl.* 2010; **21:** 1115-7.

481. Reynolds DJ, Andersen CA, Hoskote SS, *et al.* Lobar pulmonary consolidation in an immunocompromised host. *Clin Med (Lond).* 2016; **16:** 595-8.

482. Rezende RE, Brunaldi MO, Girao MS, *et al.* Case report: Esophageal histoplasmosis associated with disseminated tuberculosis in acquired immunodeficiency syndrome. *Am J Trop Med Hyg.* 2009; **80:** 347-50.

483. Richaud C, Chandesris MO, Lanternier F, *et al.* Imported African histoplasmosis in an immunocompetent patient 40 years after staying in a disease-endemic area. *Am J Trop Med Hyg.* 2014; **91:** 1011-4.

484. Richer SM, Smedema ML, Durkin MM, *et al.* Improved diagnosis of acute pulmonary histoplasmosis by combining antigen and antibody detection. *Clin Infect Dis.* 2016; **62:** 896-902.

485. Richmond BW, Worrell JA, Bastarache JA, *et al.* Histoplasmomas of uncommon size. *Chest.* 2013; **143:** 1795-8.

486. Riddell J 4th, Kauffman CA, Smith JA, *et al.* *Histoplasma capsulatum* endocarditis: multicenter case series with review of current diagnostic techniques and treatment. *Medicine (Baltimore).* 2014; **93:** 186-93.

487. Righi E, Lugano M, Assi M, *et al.* Histoplasmosis in a lung transplant recipient from a nonendemic area. *Transpl Int.* 2014; **27:** e99-e101.

488. Rihana NA, Kandula M, Velez A, *et al.* Histoplasmosis presenting as granulomatous hepatitis: Case report and review of the literature. *Case Rep Med.* 2014; 879535.

489. Rimtepathip PP, Erickson A, Katner H, *et al.* Disseminated histoplasmosis presenting as pelvic inflammatory disease. *Clin Case Rep.* 2015; **4:** 76-80.

490. Riviere S, Denis B, Bougnoux ME, *et al.* Serum *Aspergillus* galactomannan for the management of disseminated histoplasmosis in AIDS. *Am J Trop Med Hyg.* 2012; **87:** 303-5.

491. Robinson WS, Arnold SR, Michael CF, *et al.* Case report of a young child with disseminated histoplasmosis and review of hyper-immunoglobulin E syndrome (HIES). *Clin Mol Allergy.* 2011; **9:** 14.

492. Rocha-Silva F, Figueiredo SM, Silveira TT, *et al.* Histoplasmosis outbreak in Tamboril cave-Minas Gerais state, Brazil. *Med Mycol Case Rep.* 2013; **4:** 1-4.

493. Rodriguez-Cerdeira C, Arenas R, Moreno-Coutino G, *et al.* Systemic fungal infections in patients with human immunodeficiency virus. *Actas Dermosifiliogr.* 2014; **105:** 5-17.

494. Rodriguez-Waitkus PM, Bayat V, George E, *et al.* Gastrointestinal histoplasmosis in a hepatitis C-infected individual. *Mycopathologia.* 2013; **176:** 161-4.

495. Rog CJ, Rosen DG, Gannon FH. Bilateral adrenal histoplasmosis in an immunocompetent man from Texas. *Med Mycol Case Rep.* 2016; **14:** 4-7.

496. Romano RC, Soape MM, Thirumala S, *et al.* Disseminated histoplasmosis mimicking metastatic disease of the colon and omentum: Report of a case and literature review. *Arab J Gastroenterol.* 2015; **16:** 66-8.

497. Rosado-Odom VM, Daoud J, Johnson R, *et al.* Cutaneous presentation of progressive disseminated histoplasmosis nine years after renal transplantation. *Transpl Infect Dis.* 2013; **15:** e64-9.

498. Rose AS, Knox KS, Hage CA. A 24-year-old man with mediastinal mass, dyspnea and a pleural effusion. *Chest.* 2008; **134:** 200-3.

499. Ross JJ, Koo S, Lee AI, *et al.* A bird's-eye view of fever. *N Engl J Med.* 2011; **365:** e33. PMID: 21995411.

500. Sahaza JH, Perez-Torres A, Zenteno E, *et al.* Usefulness of the murine model to study the immune response against *Histoplasma capsulatum* infection. *Comp Immunol Microbiol Infect Dis.* 2014; **37:** 143-52.

501. Salimnia H, Brown P, Lephart P, *et al.* Hyphal and yeast forms of *Histoplasma capsulatum* growing within 5 days in an automated bacterial blood culture system. *J Clin Microbiol.* 2012; **50:** 2833-4.

502. Samaila MO, Abdullahi K. Cutaneous manifestations of deep mycosis: An experience in a tropical pathology laboratory. *Indian J Dermatol.* 2011; **56:** 282-6.

503. Sampaio Ide L, Freire AK, Ogusko MM, *et al.* Selection and optimization of PCR-based methods for the detection of *Histoplasma capsulatum* var. *capsulatum. Rev Iberoam Micol.* 2012; **29:** 34-9.

504. Sanmani L, Randall CJ, Palfrey J, *et al.* Hoarseness of voice in an AIDS patient: A rare presenting feature of disseminated histoplasmosis. *Int J STD AIDS.* 2011; **22:** 115-6.

505. Sansom S, Shah A, Hussain S, *et al.* *Histoplasma capsulatum:* An unusual case of pericardial effusion and coarctation of the aorta. *J Clin Med Res.* 2016; **8:** 254-6.

506. Santos JW, Michel GT, Lazzarotto M, *et al.* Chronic cavitary pulmonary histoplasmosis. *J Bras Pneumol.* 2009; **35:** 1161-4.

507. Santos L, Santos-Martinez G, Magana-Ortiz JE, *et al.* Acute histoplasmosis in three Mexican sewer workers. *Occup Med (Lond).* 2013; **63:** 77-9.

508. Sanyal M, Basu N, Roy A, Thammayya A. Primary oral histoplasmosis in India. *Sabouraudia.* 1973; **11:** 115-8.

509. Sanyal M, Thammayya A, Khan KP, *et al.* Disseminated histoplasmosis: report of a case. *Indian J Pathol Microbiol.* 1976; **19:** 179-84.

510. Sanyal M, Thammayya A. *Histoplasma capsulatum* in the soil of Gangetic Plain in India. *Indian J Med Res.* 1975; **63:** 1020-8.
511. Saremi N, Ameli F, Rahman WF, *et al.* Bilateral adrenal masses in a diabetic patient: An unusual presentation of histoplasmosis. *Pathology.* 2016; **48:** S154.
512. Sarkar P, Basu K, Mallick Sinha MG. A rare case of juvenile systemic lupus erythematosus with disseminated histoplasmosis. *Indian J Dermatol.* 2016; **61:** 700.
513. Sarkar S, Saha K, Maji A, *et al.* Non-resolving pneumonia: A rare presentation of progressive disseminated histoplasmosis. *Lung India.* 2014; **31:** 73-5.
514. Scantlebury CE, Pinchbeck GL, Loughnane P, *et al.* Development and evaluation of a molecular diagnostic method for rapid detection of *Histoplasma capsulatum* var. *farciminosum*, the causative agent of epizootic lymphangitis, in equine clinical samples. *J Clin Microbiol.* 2016; **54:** 2990-9.
515. Scheel CM, Samayoa B, Herrera A, *et al.* Development and evaluation of an enzyme-linked immunosorbent assay to detect *Histoplasma capsulatum* antigenuria in immunocompromised patients. *Clin Vaccine Immunol.* 2009; **16:** 852-8.
516. Scheel CM, Zhou Y, Theodoro RC, *et al.* Development of a loop-mediated isothermal amplification method for detection of *Histoplasma capsulatum* DNA in clinical samples. *J Clin Microbiol.* 2014; **52:** 483-8.
517. Schiavo DN, Bauer PR, Iyer VN, *et al.* Fatal fulminant necrotizing pneumonia: A case report. *J Med Case Rep.* 2014; **8:** 37.
518. Schulze AB, Heptner B, Kessler T, *et al.* Progressive histoplasmosis with hemophagocytic lymphohistiocytosis and epithelioid cell granulomatosis: A case report and review of the literature. *Eur J Haematol.* 2017; **99:** 91-100.
519. Schuster JE, Wushensky CA, Di Pentima MC. Chronic primary central nervous system histoplasmosis in a healthy child with intermittent neurological manifestations. *Pediatr Infect Dis J.* 2013; **32:** 794-6.
520. Schwartz IS, Kenyon C, Thompson III GR. Endemic mycoses: What's new about old diseases? *Curr Clin Micro Rpt.* 2016; **3:** 71-80.
521. Schwenk HT, Vo P, Moffitt K, *et al.* Allograft-transmitted *Histoplasma capsulatum* infection in a solid organ transplant recipient. *J Pediatric Infect Dis Soc.* 2013; **2:** 270-3.
522. Segel MJ, Rozenman J, Lindsley MD, *et al.* Histoplasmosis in Israeli travelers. *Am J Trop Med Hyg.* 2015; **92:** 1168-72.
523. Sehgal S, Chawla R, Loomba PS, *et al.* Gastrointestinal histoplasmosis presenting as colonic pseudotumour. *Indian J Med Microbiol.* 2008; **26:** 187-9.
524. Senechal A, Ferry T, Boibieux A, *et al.* Imported pulmonary histoplasmosis in three French cavers after a trip to Cuba. *J Travel Med.* 2012; **19:** 64-5.
525. Sethi SK, Wadhwani N, Jha P, *et al.* Uncommon cause of fever in a pediatric kidney transplant recipient: Answers. *Pediatr Nephrol.* 2016 Nov 22. PMID: 27878380.
526. Shah BK, Serban K, Tandon M. Diagnosis of progressive disseminated histoplasmosis on bone marrow biopsy. *Int J Hematol.* 2009; **90:** 433-4.
527. Shahani L. Disseminated histoplasma and CMV infection presenting as subacute intestinal obstruction in an immunocompromised patient. *BMJ Case Rep.* 2012; pii: bcr0820114732.
528. Sharma A, Gupta R, Ahuja A, *et al.* Renal allograft recipient with co-existing BK virus nephropathy and pulmonary histoplasmosis: Report of a case. *Clin Exp Nephrol.* 2010; **14:** 641-4.
529. Sharma A, Nawal CL, Agarwal MK, *et al.* Interesting case of disseminated histoplasmosis. *J Assoc Physicians India.* 2016; **64:** 116.
530. Sharma D, McKendry A, Nageshwaran S, *et al.* A case of oral ulceration and disseminated histoplasmosis in HIV infection. *Int J STD AIDS.* 2012; **23:** 522-3.
531. Sharma LC, Falodia J, Kalla K, *et al.* Esophageal histoplasmosis in a renal allograft recipient. *Saudi J Kidney Dis Transpl.* 2013; **24:** 764-7.
532. Sharma R, Lipi L, Gajendra S, *et al.* Gastrointestinal histoplasmosis: A case series. *Int J Surg Pathol.* 2017; doi: 10.1177/1066896917709945. PMID: 28530163.
533. Sharma S, Gajendra S, Lipi L, *et al.* Tumiform laryngeal histoplasmosis. *Int J Surg Pathol.* 2017; **25:** 321-2.
534. Sharma S, Gupta P. *Histoplasma capsulatum* in the peripheral-blood smear in a patient with AIDS. *Indian J Pathol Microbiol.* 2011; **54:** 212-3.
535. Sharma S, Sehgal S. Disseminated histoplasmosis diagnosed on bone marrow aspiration in an immunocompetent patient. *Blood Res.* 2015; **50:** 183-4.
536. Sharma S, Singh P, Sahu KK, *et al.* Histoplasmosis in pleural effusion in a 23-year-old man with mixed-phenotype acute leukemia. *Lab Med.* 2017; doi: 10.1093/labmed/lmx021. PMID: 28379418.
537. Sharma SK, Gupta S, Durgapal P, *et al.* Disseminated histoplasmosis in a patient with aplastic anemia. *Indian J Hematol Blood Transfus.* 2014; **30:** 143-4.
538. Shersher DD, Hong E, Breard J, *et al.* Anterior mediastinal mass secondary to histoplasmosis. *Ann Thorac Surg.* 2012; **93:** e9-10.
539. Shine MM, Shah MM, Hanna CA, *et al.* Vulvar histoplasmosis as a rare cause of genital ulceration. *Obstet Gynecol.* 2013; **122 (2 Pt 2):** 449-52.
540. Shinha T, Zabarsky G. Acalculous Cholecystitis Due to *Histoplasma capsulatum* in a patient with HIV infection. *ACG Case Rep J.* 2015; **2:** 245-6.
541. Shirali A, Kini J, Vupputuri A, *et al.* Disseminated histoplasmosis with conjunctival involvement in an immunocompromised patient. *Indian J Sex Transm Dis.* 2010; **31:** 35-8.
542. Shukla A, Shah C, Hardik P, *et al.* A probable case of histoplasmosis presenting as portal hypertension and bone lesion in a case of common variable immunodeficiency syndrome. *J Postgrad Med.* 2015; **61:** 49-50.
543. Shweihat YR, Samant RS, Joshi M, *et al.* Endobronchial ultrasound-guided transbronchial needle aspiration in a geographical region with endemic histoplasmosis infection. *J Bronchology Interv Pulmonol.* 2010; **17:** 295-300.
544. Silva TC, Tremea CM, Zara AL, *et al.* Prevalence and lethality among patients with histoplasmosis and AIDS in the midwest region of Brazil. *Mycoses.* 2017; **60:** 59-65.

545. Simon S, Veron V, Boukhari R, *et al.* Detection of *Histoplasma capsulatum* DNA in human samples by real-time polymerase chain reaction. *Diagn Microbiol Infect Dis.* 2010; **66:** 268-73.
546. Singh N, Pizanis C, Davis J. A difficult diagnosis: Acute histoplasmosis. *Clin Kidney J.* 2012; **5:** 244-6.
547. Singh S, Chhabra S, Goyal R, *et al.* Cytodiagnosis of histoplasmosis: Case reports from two patients with variable clinical presentation. *Diagn Cytopathol.* 2012; **40:** 1090-2.
548. Singh S, Kalra R, Chhabra S, *et al.* Variable clinical presentations of histoplasmosis: A report of six cases. *Trop Doct.* 2012; **42:** 32-4.
549. Singh V, Gupta P, Khatana S, *et al.* A nonhealing ulcer of mandibular alveolar ridge. *Oral Surg Oral Med Oral Pathol Oral Radiol.* 2014; **117:** 272-6.
550. Sinha S, Sardana K, Garg VK. Disseminated histoplasmosis with initial oral manifestations. *J Dermatol Case Rep.* 2013; **7:** 25-6.
551. Siraj F, Manucha V. Pharyngeal histoplasmosis presenting as a tumor mass in an immunocompetent host. *J Glob Infect Dis.* 2010; **2:** 70-1.
552. Sizemore TC. Rheumatologic manifestations of histoplasmosis: A review. *Rheumatol Int.* 2013; **33:** 2963-5.
553. Smulian AG. Invasive models of histoplasmosis. *Methods Mol Biol.* 2012; **845:** 519-25.
554. Sonavane AD, Sonawane PB, Chandak SV, *et al.* Disseminated histoplasmosis with haemophagocytic lymphohistiocytosis in an immunocompetent host. *J Clin Diagn Res.* 2016; **10:** OD03-5.
555. Soza GM, Patel M, Readinger A, *et al.* Disseminated cutaneous histoplasmosis in newly diagnosed HIV. *Proc (Bayl Univ Med Cent).* 2016; **29:** 50-1.
556. Spec A, Barrios CR, Ahmad U, *et al.* AST to ALT Ratio is elevated in disseminated histoplasmosis as compared to localized pulmonary disease and other endemic mycoses. *Med Mycol.* 2016. pii: myw106. PMID: 27744308.
557. Spinner MA, Paulin HN, Wester CW. Duodenal histoplasmosis presenting with upper gastrointestinal bleeding in an AIDS patient. *Case Rep Gastrointest Med.* 2012; 515872.
558. Srinivasan A, Kleiman MB, Debelenko L, *et al.* False-negative *Histoplasma* antigen in acute pulmonary histoplasmosis: the value of urinary concentration by ultrafiltration and heat denaturation of serum proteins in detection of *Histoplasma* antigen. *Pediatr Infect Dis J.* 2009; **28:** 447-9.
559. Srinivasan J, Ooi WW. Successful treatment of histoplasmosis brain abscess with voriconazole. *Arch Neurol.* 2008; **65:** 666-7.
560. Srirangalingam U, Selvaratnam R, Monson JP, *et al.* A case of hiccoughs. *Clin Med (Lond).* 2011; **11:** 366-8 & 475.
561. Srivastava B, King B, Galan A. An unusual clinical and histologic presentation of disseminated cutaneous histoplasmosis. *J Am Acad Dermatol.* 2011; **65:** e146-8.
562. St-Pierre J, Rioux-Masse B, Hou H, *et al.* A 58-year-old renal transplant recipient with fever and progressive dyspnea. *Clin Infect Dis.* 2014; **58:** 1738 & 1778-9.
563. Stander S, Molatjana RE, Dreyer WP, *et al.* Oral histoplasmosis. *SADJ.* 2012; **67:** 230-3.
564. Steiner SJ, Kleiman MB, Corkins MR, *et al.* Ileocecal histoplasmosis simulating Crohn disease in a patient with hyperimmunoglobulin E syndrome. *Pediatr Infect Dis J.* 2009; **28:** 744-6.
565. Stevenson J, Taylor C. Posterior reversible encephalopathy syndrome in disseminated histoplasmosis and advanced HIV infection. *Int J STD AIDS.* 2014; **25:** 611-3.
566. Subbalaxmi MV, Umabala P, Paul R, *et al.* A rare presentation of progressive disseminated histoplasmosis in an immunocompetent patient from a non-endemic region. *Med Mycol Case Rep.* 2013; **2:** 103-7.
567. Subedee A, Van Sickels N. Hemophagocytic syndrome in the setting of AIDS and disseminated histoplasmosis: Case report and a review of literature. *J Int Assoc Provid AIDS Care.* 2015; **14:** 391-7.
568. Sun NZ, Augustine JJ, Gerstenblith MR. Cutaneous histoplasmosis in renal transplant recipients. *Clin Transplant.* 2014; **28:** 1069-74.
569. Swaika A, Ailawadhi S, Menke DM. Disseminated histoplasmosis in a renal transplant patient. *Int J Hematol.* 2014; **100:** 315-6.
570. Swartzentruber S, LeMonte A, Witt J, *et al.* Improved detection of *Histoplasma* antigenemia following dissociation of immune complexes. *Clin Vaccine Immunol.* 2009; **16:** 320-2.
571. Swartzentruber S, Rhodes L, Kurkjian K, *et al.* Diagnosis of acute pulmonary histoplasmosis by antigen detection. *Clin Infect Dis.* 2009; **49:** 1878-82.
572. Tai YF, Kullmann DM, Howard RS, *et al.* Central nervous system histoplasmosis in an immunocompetent patient. *J Neurol.* 2010; **257:** 1931-3.
573. Tajan J, Espasa M, Sala M, *et al.* Disseminated infection by *Mycobacterium sherrisii* and *Histoplasma capsulatum* in an African HIV-infected patient. *Am J Trop Med Hyg.* 2013; **88:** 914-7.
574. Takahashi K, Sasaki T, Nabaa B, *et al.* Pulmonary lymphatic drainage to the mediastinum based on computed tomographic observations of the primary complex of pulmonary histoplasmosis. *Acta Radiol.* 2012; **53:** 161-7.
575. Talhari C, Talhari S, Massone C. Widespread crusted lesions in a patient with HIV - Quiz Case. *Arch Dermatol.* 2009; **145:** 1447-52.
576. Talwar P, Mishra SS, Dash RJ. Disseminated histoplasmosis. *Indian J Pathol Microbiol.* 1978; **21:** 357-60.
577. Teixeira Mde M, Patane JS, Taylor ML, *et al.* Worldwide phylogenetic distributions and population dynamics of the genus *Histoplasma. PLoS Negl Trop Dis.* 2016; **10:** e0004732.
578. Tella SH, Abuzaid A. Disseminated histoplasmosis in an HIV patient with CD4 count of 1 cell/μL. *BMJ Case Rep.* 2014; pii: bcr2013202906.
579. Teoh JW, Hassan F, Mohamad Yunus MR. Laryngeal histoplasmosis: An occupational hazard. *Singapore Med J.* 2013; **54:** e208-10.
580. Thammayya A, Sanyal M. Histoplasmosis with cutaneous nodules and oral granuloma caused by *Histoplasma capsulatum* in a Bengalee. *Indian J Med Res.* 1979; **69:** 428-35.

581. Thapa S, Jha SC, Trotter AB. Persistent fever and skin lesions due to histoplasmosis in a boy from rural Nepal. *Am J Trop Med Hyg.* 2016; **94:** 249-50.
582. Theel ES, Harring JA, Dababneh AS, *et al.* Reevaluation of commercial reagents for detection of *Histoplasma capsulatum* antigen in urine. *J Clin Microbiol.* 2015; **53:** 1198-203.
583. Theel ES, Jespersen DJ, Harring J, *et al.* Evaluation of an enzyme immunoassay for detection of *Histoplasma capsulatum* antigen from urine specimens. *J Clin Microbiol.* 2013; **51:** 3555-9.
584. Theel ES, Ramanan P. Clinical significance of low-positive *Histoplasma* urine antigen results. *J Clin Microbiol.* 2014; **52:** 3444-6.
585. Theel ES. Reply to "Low-positive *Histoplasma* antigen results in the MVista assay should not be assumed to be false positive". *J Clin Microbiol.* 2014; **52:** 4446.
586. Threlkeld ZD, Broughton R, Khan GQ, *et al.* Isolated *Histoplasma capsulatum* meningoencephalitis in an immunocompetent child. *J Child Neurol.* 2012; **27:** 532-5.
587. Thuruthumaly C, Yee DC, Rao PK. Presumed ocular histoplasmosis. *Curr Opin Ophthalmol.* 2014; **25:** 508-12.
588. Tichindelean C, East JW, Sarria JC. Disseminated histoplasmosis presenting as granulomatous epididymoorchitis. *Am J Med Sci.* 2009; **338:** 238-40.
589. Timothy AD, Jegalian A, Lichtin AE. A 54-year-old woman with pancytopenia. *Cleve Clin J Med.* 2011; **78:** 815-20.
590. Townsend JL, Shanbhag S, Hancock J, *et al.* Histoplasmosis-induced hemophagocytic syndrome: A case series and review of the literature. *Open Forum Infect Dis.* 2015; **2:** ofv055.
591. Tsai YJ, Lin YH, Hsu CH, *et al.* ^{18}F-fluorodeoxyglucose positron emission tomography for the initial evaluation and monitoring of therapeutic response in bilateral adrenal histoplasmosis. *Clin Imaging.* 2013; **37:** 791-3.
592. Tsiodras S, Drogari-Apiranthitou M, Pilichos K, *et al.* An unusual cutaneous tumor: African histoplasmosis following mudbaths: Case report and review. *Am J Trop Med Hyg.* 2012; **86:** 261-3.
593. Turashvili G, Cunningham KS. Bilateral adrenal histoplasmosis in a man with chronic alcoholism. *J Microbiol Immunol Infect.* 2016; **49:** 797-8.
594. Tyagi R, Kaur A, Selhi PK, *et al.* Disseminated histoplasmosis: A fatal opportunistic infection in disguise. *J Lab Physicians.* 2016; **8:** 129-31.
595. Tyler KL, Johnson EC, Cantu DS, *et al.* A 20-year-old woman with headache and transient numbness. Neurohospitalist. 2013; **3:** 101-10.
596. Ubesie AC, Okafo OC, Ibeziako NS, *et al.* Disseminated histoplasmosis in a 13-year-old girl: A case report. *Afr Health Sci.* 2013; **13:** 518-21.
597. Untanu RV, Akbar S, Graziano S, *et al.* Histoplasmosis-induced hemophagocytic lymphohistiocytosis in an adult patient: A case report and review of the literature. *Case Rep Infect Dis.* 2016; 1358742. PMID: 27830097.
598. Updesh M, Sachdeva S, Das R. Peripheral smear discloses histoplasmosis. *Blood.* 2010; **115:** 3653.
599. Vaid N, Patel P. A case of haemophagocytic syndrome in HIV-associated disseminated histoplasmosis. *Acute Med.* 2011; **10:** 142-4.
600. Valdez GD, Raval A, Myers J, *et al.* Gastrointestinal bleeding one complaint, two culprits. *Am J Med.* 2009; **122:** e3-4.
601. Valero C, Gago S, Monteiro MC, *et al.* African histoplasmosis: New clinical and microbiological insights. *Med Mycol.* 2017; doi: 10.1093/mmy/myx020. PMID: 28431110.
602. van Doorn-Schepens ML, Peters EJ, van Vugt RM, *et al.* Progressive disseminated histoplasmosis mimicking a flare of systemic lupus erythematosus: A European case report. *JMM Case Rep.* 2016; **3:** e005035.
603. van Koeveringe MP, Brouwer RE. *Histoplasma capsulatum* reactivation with haemophagocytic syndrome in a patient with chronic lymphocytic leukaemia. *Neth J Med.* 2010; **68:** 418-21.
604. Vantilcke V, Boukhari R, Jolivet A, *et al.* Fever in hospitalized HIV-infected patients in western French Guiana: First think histoplasmosis. *Int J STD AIDS.* 2014; **25:** 656-61.
605. Vasudevan B, Ashish B, Amitabh S, *et al.* Primary cutaneous histoplasmosis in a HIV-positive individual. *J Glob Infect Dis.* 2010; **2:** 112-5.
606. Veeravagu A, Ludwig C, Camara-Quintana JQ, *et al.* Fungal infection of a ventriculoperitoneal shunt: Histoplasmosis diagnosis and treatment. *World Neurosurg.* 2013; **80:** 222. e5-13.
607. Vergidis P, Avery RK, Wheat LJ, *et al.* Histoplasmosis complicating tumor necrosis factor-α blocker therapy: A retrospective analysis of 98 cases. *Clin Infect Dis.* 2015; **61:** 409-17.
608. Vergidis P, Walker RC, Kaul DR, *et al.* False-positive *Aspergillus* galactomannan assay in solid organ transplant recipients with histoplasmosis. *Transpl Infect Dis.* 2012; **14:** 213-7.
609. Vidyanath S, Shameena P, Sudha S, *et al.* Disseminated histoplasmosis with oral and cutaneous manifestations. *J Oral Maxillofac Pathol.* 2013; **17:** 139-42.
610. Vinicki JP, Tiraboschi IN, Fernandez D, *et al.* Necrotizing vasculitis secondary to disseminated histoplasmosis simulating pyoderma gangrenosum. *Rheumatology (Oxford).* 2013; **52:** 2304-5.
611. Vishwanath S, Girish N, Mukhyaprana P, *et al.* Disseminated histoplasmosis in an AIDS patient: A rare case from southern India. *Trop Doct.* 2009; **39:** 247-8.
612. Vitale MA, Roden AC, Rizzo M. Tenosynovitis of the wrist and thumb and carpal tunnel syndrome caused by *Histoplasma capsulatum:* Case report and review of the literature. *Hand (NY).* 2015; **10:** 54-9.
613. Vite-Garin T, Estrada-Barcenas DA, Cifuentes J, *et al.* The importance of molecular analyses for understanding the genetic diversity of *Histoplasma capsulatum*: An overview. *Rev Iberoam Micol.* 2014; **31:** 11-5.
614. Vivan RH, Leonello PC, Nagashima LA, *et al.* Soluble components of *Histoplasma capsulatum* var. *capsulatum* have hemagglutinin activity and induce syngeneic hemophagocytosis *in vitro. Mycopathologia.* 2010; **169:** 151-7.

615. voss Gonzalez T, Mattos e Dinato SL, Sementilli A, *et al.* Atypical presentation of histoplasmosis in an immunocompromised patient. *An Bras Dermatol.* 2015; **90 (3 Suppl. 1):** 32-5.
616. Vyas S, Kalra N, Das PJ, *et al.* Adrenal histoplasmosis: An unusual cause of adrenomegaly. *Indian J Nephrol.* 2011; **21:** 283-5.
617. Wang Y, Pan B, Wu J, *et al.* Detection and phylogenetic characterization of a case of *Histoplasma capsulatum* infection in mainland China. *Am J Trop Med Hyg.* 2014; **90:** 1180-3.
618. Warren WA, Markert RJ, Stewart ED. Pulmonary nodule tracking using chest computed tomography in a histoplasmosis endemic area. *Clin Imaging.* 2015; **39:** 417-20.
619. Wee EW, Lim SG, Wee A, *et al.* Disseminated histoplasmosis presenting as fever and jaundice. *Ann Acad Med Singapore.* 2009; **38:** 739-40.
620. Weeks RJ, Padhye AA, Ajello L. *Histoplasma capsulatum* variety *farciminosum*: a new combination for *Histoplasma farciminosum. Mycologia.* 1985; **77:** 964-70.
621. Wheat LJ, Azar MM, Bahr NC, *et al.* Histoplasmosis. *Infect Dis Clin North Am.* 2016; **30:** 207-27.
622. Wilhelm L, Martins N, Loreto ES. Histoplasmosis associated with Addison's disease. *Rev Soc Bras Med Trop.* 2013; **46:** 123.
623. Wilkes MF, Miller DM, Mitchell MD, *et al.* Investigation of choroidal neovascularization risk alleles in ocular histoplasmosis. *Ophthalmology.* 2014; **121:** 1487-8.e1.
624. Winn JN, Akbar S, Theivanayagam S, *et al.* Disseminated histoplasmosis presenting as colonic pseudotumor. *Endoscopy.* 2013; **45 (Suppl. 2):** e155-6.
625. Wisanuyotin S, Jiravuttipong A, Siritunyaporn S. Disseminated histoplasmosis in a renal transplant child. *Southeast Asian J Trop Med Public Health.* 2012; **43:** 992-6.
626. Wong KF, Cheng NH. Fever and productive cough in a patient with AIDS. *Clin Infect Dis.* 2011; **52:** 646-7 & 688-9.
627. Wongprommek P, Chayakulkeeree M. Clinical characteristics of histoplasmosis in Siriraj Hospital. *J Med Assoc Thai.* 2016; **99:** 257-61.
628. Woods JP. Revisiting old friends: Developments in understanding *Histoplasma capsulatum* pathogenesis. *J Microbiol.* 2016; **54:** 265-76.
629. Xiong XF, Fan LL, Kang M, *et al.* Disseminated histoplasmosis: A rare clinical phenotype with difficult diagnosis. *Respirol Case Rep.* 2017; **5:** e00220. PMID: 28250930.
630. Xu Z, German G, Jessamine P, *et al.* Disseminated histoplasmosis diagnosed by peripheral blood film in a patient with chronic lymphocytic leukaemia. *Br J Haematol.* 2013; **162:** 572.
631. Yang B, Lu L, Li D, *et al.* Colonic involvement in disseminated histoplasmosis of an immunocompetent adult: Case report and literature review. *BMC Infect Dis.* 2013; **13:** 143.
632. Yang JD, Jumani MI, Coulter KS, *et al.* Disseminated histoplasmosis presenting with intestinal ulcers and adrenal insufficiency. *Am J Med.* 2014; **127:** e19-20.
633. Yap ES, Chee YL. Disseminated *Histoplasma capsulatum* diagnosed on peripheral blood film. *Eur J Haematol.* 2013; **90:** 440.
634. Ye C, Zhang G, Wang J, *et al.* Histoplasmosis presenting with solitary pulmonary nodule: Two cases mimicking pulmonary metastases. *Niger J Clin Pract.* 2015; **18:** 304-6.
635. Youness H, Michel RG, Pitha JV, *et al.* Tracheal and endobronchial involvement in disseminated histoplasmosis: A case report. *Chest.* 2009; **136:** 1650-3.
636. Young H, Tomsick S, Storfa A, *et al.* Arm pain and fistulous tract in a 45-year-old Liberian woman. *Clin Orthop Relat Res.* 2011; **469:** 1800-3.
637. Zafar N. Histoplasmosis in fine-needle aspiration cytology of axillary lymph node in an HIV-positive woman with clinical suspicion of lymphoma. *Diagn Cytopathol.* 2009; **37:** 584-5.
638. Zahn M, Hesson M, Morton R, *et al.* Granulomatous pleuritis caused by histoplasmosis in a healthy child. *Pediatr Pulmonol.* 2011; **46:** 729-31.
639. Zarrinpar A, Lin GY, Lonergan JT. A rare cause of an ileocecal fistula in an AIDS patient. Gastrointestinal infection by *Histoplasma capsulatum* infection identified with internal transcribed spacer primer sets. *Gastroenterology.* 2013; **144:** 697 & 857-8.
640. Zhang B, Martin VK, Motamedi N, *et al.* Isolated ileocolonic histoplasmosis in an HIV-negative patient. *Dig Dis Sci.* 2017; doi: 10.1007/s10620-017-4617-8. PMID: 28547647.
641. Zhang X, Gibson B Jr, Daly TM. Evaluation of commercially available reagents for diagnosis of histoplasmosis infection in immunocompromised patients. *J Clin Microbiol.* 2013; **51:** 4095-101.
642. Zhu C, Wang G, Chen Q, *et al.* Pulmonary histoplasmosis in a immunocompetent patient: A case report and literature review. *Exp Ther Med.* 2016; **12:** 3256-60.
643. Zhu LL, Wang J, Wang ZJ, *et al.* Intestinal histoplasmosis in immunocompetent adults. *World J Gastroenterol.* 2016; **22:** 4027-33.
644. Zida A, Niamba P, Barro-Traore F, *et al.* Disseminated histoplasmosis caused by *Histoplasma capsulatum* var. *duboisii* in a non-HIV patient in Burkina Faso: Case report. *J Mycol Med.* 2015; **25:** 159-62.

CHAPTER 17

Blastomycosis

Blastomycosis is an acute or chronic pyogranulomatous systemic fungal infection of lungs disseminating to skin, bones and genitourinary system. This disease is caused by three species of genus ***Blastomyces*** *i.e. B.dermatitidis, B.gilchristii* and *B.percursus*, the soil-dwelling thermally **dimorphic** fungi. It belongs to group of deep-seated mycoses and is found endemic in North America. In addition, it is also described in circumscribed African regions and other parts of world as well. In USA, it is one of the three major endemic mycoses and is found prevalent in north, central and mid-eastern provinces as well as adjoining areas of Canada including Great Lakes. This is also known by the name of its discoverers *i.e.* **Gilchrist's disease** and on the basis of its geographical area of endemicity as **North American Blastomycosis.**

In contrast to many fungal infections, blastomycosis is a true fungal infection and typically occurs in healthy host. Hence it has potential to infect physiologically and immunologically normal individuals and if left untreated may lead to poor prognosis.

Historical Perspective

Casper Gilchrist first reported this disease in May 1894 from Baltimore (USA), in a patient with localized skin lesions, clinically diagnosed as scrofuloderma. He saw 'numerous curious bodies in tissue, which he thought were protozoa. Gilchrist and Stokes, in their preliminary report of another case in 1896, tentatively identified causative organism as an *Oidium*. In the follow-up of same patient in 1898, they definitively identified causative organism and named it as blastomycete, refuting earlier description as protozoa, based on extensive cultural and experimental animal inoculations thereby named it *Blastomyces dermatitidis* and disease as **blastomycetic dermatitis**. The clinical entity was also designated subsequently to honor its discoverer as **Gilchrist's disease**.

In 1902, Walker and Montgomery first time described systemic infection caused by this fungus. In 1906, Primrose reported first Canadian case of blastomycosis. In early 1900s, illness became known as **'Chicago's disease'** because of number of cases reported in that area. Later on it was known as **'North American Blastomycosis'** due to its wide prevalence in northern USA and adjoining Canada.

The thermal basis of its dimorphism was first described by Hamburger in 1907 by maintaining incubator at different temperatures. This infection showed substantial number of clinical manifestations in skin and hence name of species was originally described as ***'dermatitidis'***. Subsequently, involvement of deeper structures and more disseminated nature of infection were recognized in due course of time. Blastomycosis was first reported in head and neck region by New in 1917.

In the beginning, it was thought to be a protozoal infection, which was subsequently identified correctly as fungus. This was not until 1951, following detailed pathologic description of blastomycosis by Schwarz and Baum discovering that most cases of the disorder were acquired through respiratory, not cutaneous, route. In 1952, Broc and Haddad reported this disease from Tunisia in Africa. Subsequently additional cases were also reported involving this entirely different continent as that of North America. Despite differences in prevalent strains of *B.dermatitidis*, it is not clear whether natural history of this disease is significantly different in these two continents. As such none of the manifestations described in Africans falls outside known clinical spectrum of blastomycosis in North America. Hence geography based terminology as **'North American'** became infructuous.

The literal meaning of blastomycosis is 'yeast disease' but it is specifically applied to disease caused by *B.dermatitidis*. At present existing terminology is well-established as, **blastomycosis**, due to its long term usage despite the fact that it conveys a vague meaning of the disease.

In 2013, another species was described as a significant pathogen in the genus *Blastomyces i.e.* ***B.gilchristii***, which subsumes certain strains previously assigned to *B.dermatitidis*. In 2017, Dukik *et al*, reported another novel dimorphic pathogen, producing broad-based budding cells at 37°C and occurring outside North America, proved to belong genus *Blastomyces* and is described as ***B.percursus***, making a total three causative species of this genus.

Mycology

Blastomycosis is caused by three species of genus *Blastomyces* namely, *B.dermatitidis, B.gilchristii* and *B.percursus*, however, *Blastomyces dermatitidis* is considered as its representative species. It is a thermally dimorphic fungus, existing in its saprotrophic as mycelial phase at ambient temperature and as **broad-based** budding yeast cells with size of 8-15 µm in tissue phase. The yeasts have characteristic thick, doubly-refractile cell wall and single wide-base budding with multinucleate (8-12 nuclei) cells. There is slight variation in size, shape, staining and cell wall thickness of yeast forms in different fields of the same specimen in immunocompetent hosts. However, greatest variation is seen in the yeast forms in immunosuppressed patients to the extent that sometimes even hyphal forms may be present. The transformation from mycelial to yeast form is as a result of heat shock protein induced uncoupling of oxidative phosphorylation.

Several laboratories have used exoantigen techniques to identify and differentiate isolates of this fungus. On the basis of immunological methods, it appears that two serotypes exist, **AK** and **K** variants. The former is found worldwide, whereas latter is prevalent in Africa. The mycelial form has no morphologically unique characteristic that allow definitive identification.

The co-culture of isolates having opposite mating types results in formation of distinctive structures like spiral hyphae and ascocarps which contain ascospores. Soil-extract agar may be used for mating of *B.dermatitidis* strains. These sexual structures classify perfect state of this fungus as an ascomycete. The teleomorphic state is identified as ***Ajellomyces dermatitidis*** placed in family **Onygenaceae** under order **Onygenales** of class **Plectomycetes** in phylum **Ascomycota**. The genus of this organism is same as that of teleomorphic states of *Histoplasma capsulatum, Emmonsia crescens* and *Emergomyces* species.

Epidemiology

The infection of blastomycosis results from inhalation of spores from soil but natural habitat remains poorly defined and its exact reservoir is not yet known. *B.dermatitidis,* unlike *H.capsulatum*, is rarely cultured from the soil even in endemic areas. It is believed that organism is present in moist soil but flourishes only in narrow, yet undefined, ecological niche.

Blastomycosis is co-endemic with histoplasmosis over much of the central USA, east of the Mississippi River. The endemic areas of are eastern, central and mid-western states of USA and eastern Canada, particularly Quebec province. It is endemic in Ohio and Mississippi River valleys, Great Lakes and to lesser extent in Missouri and Arkansas River basins. Additional endemic areas have been found in Minnesota, extending northward to Canadian provinces bordering United States like southern Manitoba and southwest Ontario, including St. Lawrence River basin.

Some of the outbreaks of this disease have been reported in north-central Wisconsin, Minnesota, Illinois and in eastern states of Virginia and North Carolina. The most notable outbreak occurred in June 1984 involving 95 students near Eagle River in Vilas County of northern Wisconsin who camped along a beaver pond and explored a beaver lodge. *B.dermatitidis* was isolated from wood on the pond bank as well as from the lodge. It is thought that this fungus resides in moist, organically rich soils typically near waterways. The epidemiological studies suggest that within these endemic areas, there are scattered foci of persistent or recurrent hyperendemicity. It is speculated that epidemiology of this infection can now be extended to include the area west of Mississippi River valley as cases have been reported from Colorado also.

The geographic spread of blastomycosis now appears to be considerably wider than it was previously estimated. Infection has been identified throughout Africa from north coast to South Africa and extending as far west as Nigeria. In the African endemic zone, most of the patients have been described in Zimbabwe. The cases have been reported from India, Middle East particularly Lebanon, Israel and Saudi Arabia. The isolated cases have also been reported from wide geographical areas of South America, Central America and Europe.

This disease is not frequently reported from India except few cases, which are thirteen in number including both human and animals. In 1982, Khan *et al*, while striving for *H.capsulatum*, reported isolation of *B.dermatitidis* from lungs of rat-tailed bats (*Rhinopoma hardwickei hardwickei*). The first authentic report of autochthonous human case from India was by Randhawa *et al*, in 1983. The isolation was done from bronchial aspirate of asthmatic female patient who had never traveled abroad.

She was resident of Rudain, small town in Badaun district of Uttar Pradesh, situated about 250 km southeast of Delhi. Subsequently Jambhekar reported a case of disseminated blastomycosis from Madhya Pradesh in 1988. Cases have also been reported from Mumbai, Vellore, Bangalore and few other medical institutions of India.

The infection in blastomycosis, if remain either undiagnosed or untreated, may advance as life-threatening even in immunocompetent hosts. Among all endemic mycoses, blastomycosis has been the least associated with immune dysfunction but now there are reports establishing that it may be more common in immunocompromised patients than previously contemplated. The patients with AIDS and other immunosuppressive conditions are more prone to dissemination with an aggressive clinical course thereby leading to fatal consequences.

Immunity

The immune response to *B.dermatitidis* in normal host is not fully understood, however, intact cellular immunity is significant in prevention of progression of disease and lymphocyte reactivity is marker of specific cellular immunity. The innate and adaptive mechanisms limit infection of *B.dermatitidis* and promote clearance of fungus which have been characterized, include polymorphonuclear leukocytes, mononuclear phagocytes and antigen specific T-lymphocytes. The macrophages and/or H_2O_2 are capable of preventing transformation of conidia to yeast phase, without killing the organism. Moreover, macrophages activated by cytokines are capable of *in vitro* killing yeasts of *B.dermatitidis*.

The macrophages lose ability to kill organisms by blocking activation with inhibitors of protein synthesis, which implies that macrophages may be of much significance in defense against this fungus. However, antigens of *B.dermatitidis* that stimulate clearance of fungus have not been identified. A 120 kDa glycoprotein Ag on the surface of *B.dermatitidis* yeast cell wall *i.e. B.dermatitidis* adhesin (BAD - former WI-1), which being a surface adhesin, binds this fungus to complement and CD14 receptors on host cells and is an immunodominant antigen. It is also postulated that virulence is lost during storage of isolates of *B.dermatitidis* due to spontaneous mutations.

Pathogenesis and Pathology

The infection caused by *B.dermatitidis* begins with inhalation of spores into lungs but the disease is often detected only after dissemination. If organism evades non-specific host defense mechanisms, it undergoes phase transition to the yeast cells. The yeasts increase in number in parenchyma of lung and spread to other organs via bloodstream. With the development of immunity, inflammatory pyogranulomatous reactions occur at initial pulmonary site and at widespread foci of infection.

The initial response to fungus is suppurative and is followed by formation of granulomas. This mixed neutrophilic and mononuclear cell response is distinctive of blastomycosis, although necrosis or fibrosis may also occur. The granulomas of blastomycosis do not caseate, as seen in tuberculosis. Despite spontaneous resolution of pneumonia in some of the cases, endogenous reactivation may occur at either pulmonary or extrapulmonary sites, with or without previous therapy. The histopathological studies show pseudoepitheliomatous hyperplasia, intraepidermal abscesses and polymorphous dermal infiltrate.

Blastomycosis, like histoplasmosis, is a T-cell opportunistic infection with highly aggressive course in patients who are suffering from AIDS and most of them have diffuse infiltrates. They have widely disseminated disease and mortality is also high. However, it is not considered as one of the AIDS-defining illnesses.

Clinical Features

The initial infection in blastomycosis results from inhalation of conidia into lungs. Although *B.dermatitidis* is highly infectious but symptomatic disease seems to develop among less than half of those infected. This is great masquerader hence it can present in several ways, either as acute or chronic. It involves one or several body sites, most common being lungs, followed by skin and bone. The clinical spectrum is highly variable including asymptomatic infection, acute or chronic pneumonia and disseminated disease involving skin, bones, CNS and other anatomical sites. The clinical course tends to be more aggressive in immunocompromised hosts than in those who are immunocompetent and most deaths occur during first few weeks of disease. The disseminated disease in immunocompetent hosts is uncommon.

The symptomatic infection is seen in ~50% of cases, who usually present as flu-like illness with fever, chills, productive cough, myalgia, arthralgia and pleuritic chest pain. There is weight-loss, malaise, fatigue and other non-specific complaints which are fairly common and offer little diagnostic help. Its mortality rate is ~8-10%. There are alveolar infiltrates, mass lesions and fibronodular infiltrates, which may be seen radiologically.

The males are affected six times more as compared to females. The adults are more frequently affected than children, who constitute a small percentage of cases and it is very rare among infants younger than one year. This disease may be suspected when there are isolated skin manifestations, focal lesions of bone, CNS and other visceral organs. The male patients also have involvement of genitourinary system.

These days opportunistic infection of blastomycosis is not as rare as it used to be reported previously. Moreover, disease may be suspected if patients have recently traveled to endemic eastern areas of United States, mid-west in South Dakota and Minnesota, southern Canada. It should also be taken into consideration if patients have AIDS, leukemia/lymphoma, on corticosteroid therapy and immunosuppressed in any other way. The disease can manifest in following clinical forms:

1. Pulmonary Blastomycosis

The acute pulmonary phase of illness may be asymptomatic or self-limiting. The clinical spectrum may range from an asymptomatic, subclinical pulmonary infection to severe, life-threatening disease. The incubation period on an average is ~45 days. Primary pulmonary blastomycosis is usually asymptomatic or may present with relatively mild, flu-like illness. However, within short period of time following infection, disease becomes chronic and may disseminate to distant body organs. The acute disease mimics bacterial pneumonia and presents with fever, chills and productive cough. Hence pulmonary blastomycosis should be suspected, in case of non-resolving pneumonia, despite standard antibiotic therapy specifically in the patients who had been to endemic areas. The chronic cases resemble pulmonary tuberculosis or bronchogenic carcinoma and radiological findings show diffuse infiltration in lung fields.

2. Cutaneous Blastomycosis

This is the commonest manifestation of disseminated blastomycosis and name of species as ***'dermatitidis'*** is derived from this very fact. It is mainly seen over the exposed parts of body like face, neck and hands. The incubation period varies from one to two weeks. The lesions begin with one or more papules or nodules which breakdown and may lead to fistula formation. A purulent foul smelling fluid is discharged from fistula, which is sometimes mixed with blood. There may be non-healing ulcerative lesions.

Primary cutaneous blastomycosis is extremely rare and occurs exclusively due to accidental inoculation while working in laboratory, autopsy rooms, in veterinarians. In isolated cutaneous cases there is no visceral involvement. The cutaneous blastomycosis of perianal skin is rare hence their diagnosis is frequently overlooked. They present with friable, violaceous, papulopustular lesion, with heaped-up edges along anal verge. The skin lesions of blastomycosis can be mistaken for squamous cell carcinoma, atypical pyoderma gangrenosum, non-tuberculous mycobacterial infections and other fungal infections, especially coccidioidomycosis or paracoccidioidomycosis.

3. Osseous Blastomycosis

Osteomyelitis develops in about one-third of patients with disseminated blastomycosis and most commonly involves bones are vertebra, pelvis, sacrum, skull, ribs and long bones in their descending order. Vertebral blastomycosis should be a diagnostic consideration for any patient with destructive vertebral lesion and paravertebral abscess, even if pulmonary and cutaneous manifestations, typical of blastomycosis, are lacking. Whenever vertebral tuberculosis is included in the differential diagnosis, blastomycosis should also be suspected. Most of the patients with vertebral blastomycosis respond to therapy with amphotericin B or as has been noted more recently, with itraconazole. Blastomycosis arthritis is usually uncommon and is monoarticular but few cases involving multiple joints have been reported. Overall, the most frequently affected one is knee joint.

4. Disseminated Blastomycosis

In the disseminated infection, chronic pulmonary disease persists and spreads to most of the organs, body tissues including bones. This spreads mainly by hematogenous route. There is formation of abscesses and granulomatous lesions. Bone involvement, breast abscess and sinuses in testis may be formed. Genitourinary infection is fourth most common manifestation of blastomycosis after involvement of lungs, skin and bones. It has been reported in males almost exclusively as prostatitis or epididymo-orchitis. Unusual occurrences of pericardial effusion and erythema nodosum have also been reported. The infection with *B.dermatitidis* may infrequently be an opportunistic, notably in patients who are in late stages of AIDS, transplant recipients and patients being treated with immunosuppressive or cytotoxic chemotherapy.

5. Miscellaneous Types of Blastomycosis

In the absence of an identifiable pulmonary or cutaneous disease, laryngeal blastomycosis can present a diagnostic challenge to both clinician and mycologist. Typically, patients with laryngeal disease complain of insidious onset of hoarseness and occasionally pain and dysphasia for several months. On examination, these lesions may be erythematous, ulcerated and fungating, thus mimicking carcinoma. The microscopic examination of mucosal lesions often shows pseudoepitheliomatous hyperplasia that may be florid and difficult to distinguish from laryngeal squamous cell carcinoma.

Blastomycosis of central nervous system occurs in ~4% of patients and recurrent blastomycosis is very rare. In endemic areas, this disease may cause serious illness among pregnant patients, which may prove fatal to fetus. However, in some of them there is spontaneous remission after delivery even without institution of antifungal therapy. The patients may also have retropharyngeal abscess.

Blastomycosis in Animals

In addition to human beings, blastomycosis has also been seen in dogs, cats and cows. The involvement of human and dogs share similar geographical distribution of disease and are exposed to same natural nidus of organism. However, there is no evidence of animal-to-human transmission nor is infection ordinarily spread from person-to-person. Therefore, history of infection to dogs still becomes relevant. This observation merely reflects the fact that dog has been in the same environment and acts as a sentinel event for human illness.

Radiodiagnosis

The chest X-rays findings in pulmonary blastomycosis are variable and there is no characteristic radiographic pattern. The lesions range from single or multiple round densities of variable sizes in any lung field to segmental or lobar consolidation. There are patchy alveolar infiltrates in one or more lobes, more commonly in lower lobes. The hilar enlargement is common but significant pleural effusion and cavitation is unusual. These findings may mimic those of pulmonary tuberculosis or bronchogenic carcinoma.

Similarly, radiographic appearance of bone involvement in blastomycosis is non-specific hence cannot be differentiated from other forms of osteomyelitis. However, lesions of long and short bones are classified as either focal or diffuse. The focal pattern may have sclerotic margin and may expand slowly. Periosteal reaction is generally present in long bones but absent in short bones. Diffuse osteomyelitis is characterized by rapid bone destruction (**moth-eaten pattern**) with vigorous periosteal reaction and infection may invade adjacent joints or produce draining sinuses. Pulmonary blastomycosis lesions mimicking respiratory malignancy have been reported on positron emission tomographic (PET) scanning.

Differential Diagnosis

The patients of blastomycosis are usually diagnosed as tuberculosis on clinical grounds being similar type of respiratory diseases. When full-fledged patients of blastomycosis present to physicians, most of them are already on antitubercular treatment. Pulmonary blastomycosis mimics malignancy and differentiation may prove to be a challenging task. Some of features in blastomycosis are same as that of atypical pneumonia. Pseudoepitheliomatous hyperplasia in blastomycosis can be mistaken for keratoacanthoma. Out of all endemic mycoses, mucous lesions of blastomycosis are confused with paracoccidioidomycosis.

In tissues, *B.dermatitidis* may resemble *H.capsulatum* var. *duboisii* and immature spherules of *C.immitis* and its smaller form resembles to *H.capsulatum* var. *capsulatum*. The other yeast infections that may be considered in differential diagnosis are, cryptococcosis showing budding by narrow base with capsule, paracoccidioidomycosis with multiple budding or 'spoke-wheel' appearance and histoplasmosis with uninucleate yeast cells and narrow-base budding.

The blastomycosis-like pyoderma vegetans or gangrenosum, unusual exaggerated vegetating reaction to prolonged primary or secondary bacterial infections may be confused with cutaneous blastomycosis. It should also to be differentiated from emergomycosis, particularly caused by *Emergomyces orientalis,* wherein the yeast size is much bigger *i.e.* 4-10 μm as compared to other species of this genus. The clinical entities which mimic blastomycosis are shown in **Table 17.1**.

Laboratory Diagnosis

The demonstration of organism in tissues or exudates, growth on culture media and detection of antigen or antibody in serum specimen, are required to establish laboratory diagnosis of the disease. Blastomycin, culture filtrate antigen from mycelial phase is used in skin test and yeast phase antigen for complement fixation and precipitation tests.

Table 17.1. Diseases Clinically Mimicking Blastomycosis.

Blastomycosis	*Mimicking Diseases*
Acute Pulmonary	Bacterial pneumonia
Chronic Pulmonary	Tuberculosis, malignancy like bronchogenic carcinoma
Cutaneous	Basal cell or squamous cell carcinoma, pyoderma vegetans or gangrenosum keratoacanthoma, emergomycosis
Nasal Lesions	Leishmaniasis, paracoccidioidomycosis
CNS	Malignancy, meningioma
Lymphadenopathy	Lymphoma, amyloidosis
Miliary	Tuberculosis, carcinomatosis
Hypercalcemia	Sarcoidosis

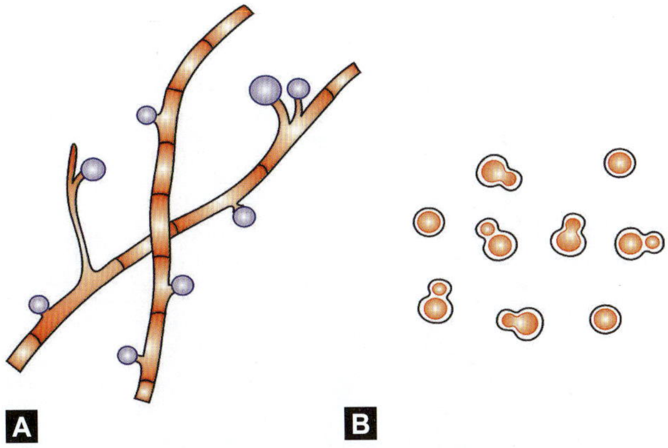

Figs. 17.1A and B. Diagrammatic representation of *Blastomyces dermatitidis* showing mycelia with spherical conidia at 25°C and broad-based yeast cells at 37°C.

(a) Direct Examination

The clinical specimens like sputum, bronchoalveolar lavage fluid, biopsy or pus from abscess are taken and examined for presence of yeast cells in KOH wet mount and calcofluor white staining. Urine collected for culture after prostate massage may also reveal yeast cells. The characteristic double-contoured, thick-walled, multi-nucleated yeast forms with single **broad-based** budding daughter cells are observed in microscopic examination of unstained wet mount of clinical specimen, cytology preparation or also in histopathology section. The yeast cells of small form blastomycosis found in Africa are usually mistaken for histoplasmosis. Multiple nuclei may be seen in well-fixed tissue. The organisms are often difficult to identify with H&E staining and may require special staining like PAS and GMS for its identification. The most sensitive method for detecting less number of organisms is histopathological section is GMS staining, which shows broad-based yeasts with granulomatous inflammatory response.

The differentiation between *B.dermatitidis* and *Cryptococcus* species may also be difficult in some cases. The size of organism is similar and it may be difficult to determine whether broad-based or narrow-based budding is present, especially when very few organisms are found. Histopathological examination with PAS staining of microabscesses shows presence of round yeast cells. **Splendore-Hoeppli phenomenon** may also be observed. The unusual filamentous forms have been rarely described in tissue specimens.

Papanicolaou staining of sputum and other respiratory specimens is another quick and reliable means of making an early diagnosis. The fungi are seen as large pink to red structures bearing their typical morphology. The main difficulty with Papanicolaou smears is that they are usually performed to screen for malignancy and even characteristic appearance of *B.dermatitidis* may be overlooked unless trained personnel look for them.

(b) Fungal Culture

The culture of sputum or biopsy specimens like skin scrapings, small pieces from tissues, sputum or pus from multiple minute lesions can be done on blood agar, Sabouraud dextrose agar, brain-heart infusion agar and blood-glucose-cysteine agar at 25°C and 37°C. The growth is very slow and takes about two to four weeks.

On SDA at 25-30°C, colonies look yeast-like at an early stage and later on hyphal projections develop on surface and finally entire surface becomes downy or fluffy white within 14 days and some strains may grow slowly. The old culture may become tan colored.

The microscopic examination of mount from colony shows mycelium consisting of fine branched and septate hyphae. The mold form of *B.dermatitidis* produces conidia measuring 2-10 μm, which are located on short terminal or lateral branches. These conidia are typically spherical, ovoid or pyriform in shape **(Figs. 17.1A and 17.2A)**. They resemble microconidia of *H.capsulatum* but are smooth. Thick-walled chlamydospores of 7-18 μm may also be observed.

The colony and conidia of *B.dermatitidis* morphologically resemble those of *Chrysosporium* species and cannot be distinguished from an early culture of *H.capsulatum*

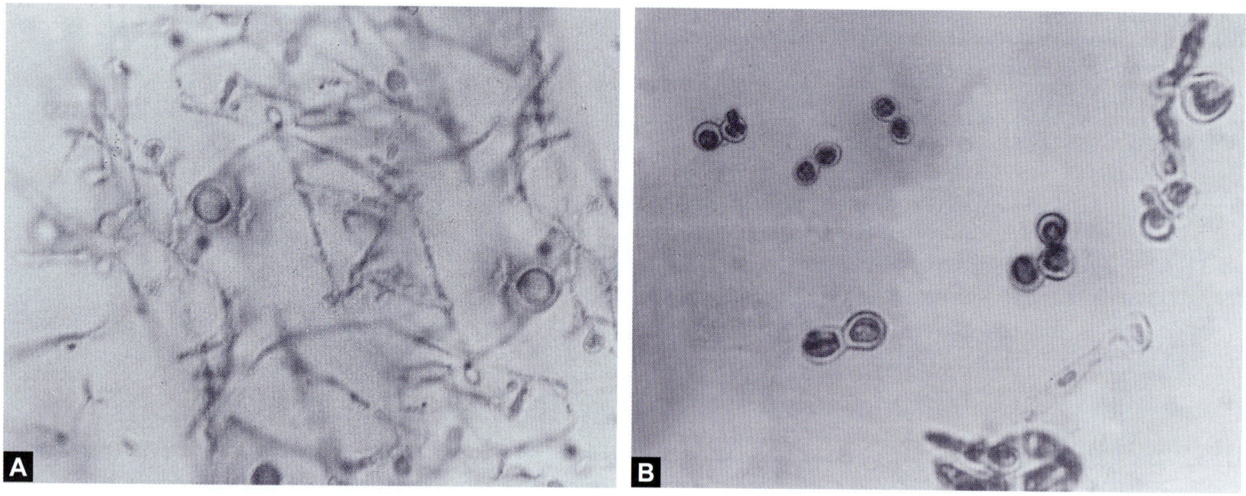

Figs. 17.2A and B. Mycelia with spherical conidia of *B.dermatitidis culture* at 25°C and broad-based yeast cells at 37°C (LCB × 400).

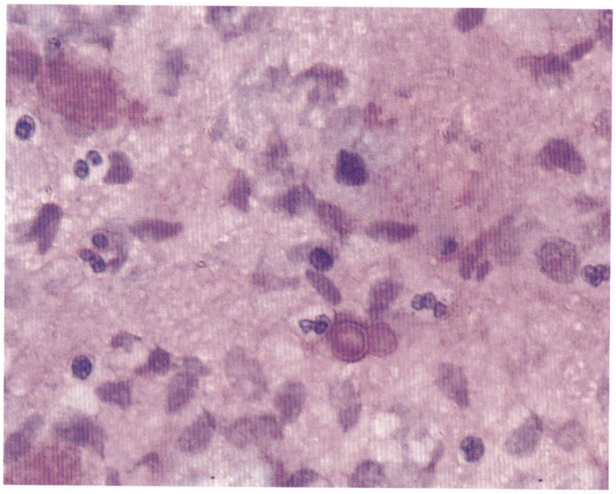

Fig. 17.3. *Blastomyces dermatitidis* showing broad-based yeast cells in smear of patient of blastomycosis (PAS × 400).

having only hyphae, conidiophores and microconidia. The slide culture of mycelial isolate is put up to examine typical conidial arrangement of *B.dermatitidis* and also for comparing similar looking fungi.

The isolate of *B.dermatitidis* is identified by several morphological characteristics. The classic means of identification is *in vitro* conversion of this dimorphic fungus from yeast to mycelial form and/or *vice versa* (M ↔ Y). This is done on brain-heart infusion agar, blood-glucose-cysteine agar and ordinary blood agar depending upon the availability in the laboratory. The seed-based media (cottonseed agar) are reported as more fruitful than BHIA with or without blood as well as Kelley's agar for mycelial-to-yeast conversion (M→Y) of this fungus. Cysteine-heart agar with rabbit blood or Cysteine-heart and hemoglobin agar (CHHA) for the same *in vitro* mycelial-to-yeast conversion, may also be used.

At 37°C, on brain-heart infusion agar and blood agar, colonies are white and fluffy at first but later on become smooth and cream or yellowish in color. The slow growing colonies develop cream to tan in color, soft wrinkled and waxy in appearance. On enriched media at 37°C, it grows as yeasts with colonies that are folded, pasty and moist.

The yeast cells are thick-walled and spherical, produce single bud with characteristic **broad-base** attachment between bud and parent cell. The ease of diagnosis should be further augmented by large size of yeast phase organism (8 to 15 μm) and easily recognizable *'figure of eight morphology'* with characteristic thick, doubly refractile cell wall and single broad-based budding **(*See* Fig. 17.1B)**. The same is shown in the LCB mount **(Fig. 17.2B)** as well as in PAS-stained smear from a patient **(Fig. 17.3)**. However, the daughter cell is often as large as the mother cell before detachment. The giant yeast forms ranging between 20-40 μm have also been reported in some of cases of blastomycosis.

One of the studies showed that African isolates of *B.dermatitidis* belong to a distinct serotype and share a unique antigen, *i.e.* **K** antigen whereas typical antigenic determinant *i.e.* **A** antigen as seen in most strains of this fungus, including North American forms and is absent from African cultures. These isolates differ from North American isolates in colonial and microscopic morphology.

The exoantigen test can also be performed for an early establishment of diagnosis. The serotypes of *B.dermatitidis* have been identified by exoantigen analysis of yeast phase (antigen **A**) of the organisms. The exoantigen test is a rapid and reliable alternative for conversion and produces specific precipitate, designated **A**. The exoantigen technique has proven dependable for identifying mycelial anamorph of the small tissue form of *B.dermatitidis*. Exoantigen test does not detect antigen **K**. The soil extract agar is the best medium to isolate this fungus.

(c) Immunodiagnosis

The serological tests include those for complement-fixating antibodies, immunodiffusion precipitin bands, enzyme immunoassays and radioimmunoassays, all using antigen **A**. The delayed hypersensitivity of skin to blastomycin may also be done. The serological and immunological findings of blastomycosis are vague and these are unreliable for diagnosis. At present two antigenic preparations are used to detect immune response to infection by *B.dermatitidis i.e.* cell free culture-filtrate of mycelial phase (**blastomycin**) and inactivated **whole yeast phase cells**. The results have been reported better with enzyme-linked immunosorbent assay techniques using yeast-phase antigen. Therefore, serodiagnosis of blastomycosis is difficult owing to potentially low rates of sensitivity and specificity.

The skin testing with blastomycin is, unfortunately, no better diagnostic procedure. This mycelial-phase antigen does not provide sensitivity or specificity for reliable assessment of patients' condition and no longer obtainable clinically. The skin test and serological investigations are only of limited diagnostic value because of cross-reactions with histoplasmosis and coccidioidomycosis. There is a relatively high proportion of non-reactors also.

There are newer and more purified *B.dermatitidis* antigens, such as antigen **A**, BAD and ASWS (alkali-soluble water-soluble) antigen which have stimulated renewed interest in skin testing. These preparations have been used in various *in vitro* assays but large-scale human skin surveys have not been reported. The blastomycin skin test antigen is not available commercially and its current use is limited to investigational purposes.

An immunodiffusion test that is specific for blastomycosis has been developed. It is based on availability of yeast phase culture filtrate, which is designated as antigen **A**. The serological tests for diagnosis of blastomycosis have had inadequate sensitivity but results can be improved with enzyme immunoassay (EIA) with antigen **A** of *B.dermatitidis*. The Premier Blastomyces EIA kit (Meridian Diagnostics, Inc, Cincinnati, Ohio) can detect antibody against the *B.dermatitidis* antigen **A**. The newer tests like immunofluorescence, Western blotting and DNA probes are also used in establishing the diagnosis of blastomycosis. The exact identification of *Blastomyces* is done on the basis of molecular based techniques to reach the final diagnosis upto species level.

(d) Animal Pathogenicity

Intraperitoneal injections of conidia and mycelial fragments produce infection in mice, guinea pig, dogs, rats and hamsters. The rabbits are not susceptible to *B.dermatitidis*. The yeast form is equally effective in causing extensive lesions and even death of mice. The murine model has successfully allowed quantitative assessments of disease initiation and progression. Therefore, virulence, immune response and study of new antifungal agents are also feasible on the experimental animals.

Treatment and Prophylaxis

The evidence-based guidelines for the management of patients with blastomycosis were released by the IDSA in June 2008 replacing previous ones published in 2000. The majority of patients require antifungal therapy although spontaneous cure may be seen in a few immunocompetent individuals with acute pulmonary blastomycosis. Treatment options include amphotericin B, itraconazole and fluconazole. The immunocompetent individuals are given itraconazole if disease is disseminated to other parts of body and immunosuppressed individuals are given amphotericin B.

The azole antifungal like itraconazole is the drug of choice in initial phase of treatment in immunocompetent patients with no life-threatening and non-meningeal type of infection. Its dose is 200 to 400 mg/day for 6 months. Itraconazole has replaced ketoconazole as first-line therapy due to less toxicity despite its higher cost. The practice of treating with fluconazole has been restricted because it is moderately effective.

Amphotericin B remains drug of choice in life-threatening manifestations of infection or in cases of CNS involvement. It is given intravenously as 0.7 mg/kg daily for a period 8 weeks but cavitary lesions in lungs and severe multi-organ disease may require 10-12 weeks of therapy to prevent relapse. Itraconazole has even replaced amphotericin B as drug of choice to treat life-threatening blastomycosis. Blastomycotic meningitis should always be treated with intravenous amphotericin B because oral agents poorly penetrate blood brain barrier. *B.dermatitidis* is also susceptible to newer antifungal agents like chitin synthetase inhibitors - nikkomycin as seen in *Coccidioides* species.

The relapse in blastomycosis following primary response to therapy is well-documented. Most of the patients who relapse do so within a year after completion of treatment but some patients may have relapse many years later. Thus, a minimum of one-year follow up after therapy is essential for an accurate identification of all relapsed cases.

The only way to prevent infection is by not visiting endemic areas. The exposure occurs when fungus is inhaled and there is no way to avoid that. But it is important that public should be aware of possible risk, especially if individual is immunosuppressed. As such the normal individuals need not worry too much about the same.

At present there is no vaccination available to prevent blastomycosis. But recent studies suggest excellent potential for live recombinant vaccine strain of *B.dermatitidis*, which lacks the surface antigen, BAD, an important virulence factor for the organism which mediates binding to the target cells. This vaccine has been highly immunogenic in experimental animals and is being evaluated for clinical trials in human volunteers as well.

Further Reading

1. Abdalla M, Sax PE, Mostaghimi A, *et al.* On the nose. *N Engl J Med.* 2015; **373:** 955-61.
2. Abdwani R, Houghton K, Dobson S, *et al.* An adolescent with both Wegener's granulomatosis and chronic blastomycosis. *Pediatr Rheumatol Online J.* 2008; **6:** 13.
3. Abrams E, Dean HJ, Bunge MK, *et al.* Blastomycosis of the central nervous system in a child: A rare cause of hypopituitarism. *J Pediatr Endocrinol Metab.* 2011; **24:** 775-7.
4. Akture E, Salamat S, Baskaya MK. Intracranial blastomycosis presenting as an enhancing cerebellopontine mass. *Turk Neurosurg.* 2013; **23:** 252-5.
5. Al-Nassar S, Macnair T, Lipschitz J, *et al.* Blastomycosis presenting as multiple splenic abscesses: Case report and review of the literature. *Can J Infect Dis Med Microbiol.* 2010; **21:** 53-6.
6. Alhaji M, Sadikot RT. Respiratory failure due to blastomycosis infection in a patient with hypertension, cirrhosis and chronic pancreatitis. *Expert Rev Respir Med.* 2013; **7:** 583-5.
7. Almony A, Kraus CL, Apte RS. Successful treatment of choroidal blastomycosis with oral administration of voriconazole. *Can J Ophthalmol.* 2009; **44:** 334-5.
8. Alpern JD, Bahr NC, Vazquez-Benitez G, *et al.* Diagnostic delay and antibiotic overuse in acute pulmonary blastomycosis. *Open Forum Infect Dis.* 2016; **3:** ofw078.
9. Anderson EJ, Ahn PB, Yogev R, *et al.* Blastomycosis in children: A study of 14 cases. *J Pediatric Infect Dis Soc.* 2013; **2:** 386-90.
10. Anderson JL, Dieckman JL, Reed KD, *et al.* Canine blastomycosis in Wisconsin: A survey of small-animal veterinary practices. *Med Mycol.* 2014; **52:** 774-9.
11. Anderson JL, Meece JK, Hall MC, *et al.* Evidence of delayed dissemination or re-infection with *Blastomyces* in two immunocompetent hosts. *Med Mycol Case Rep.* 2016; **13:** 9-11.
12. Arnett MV, Fraser SL, Grbach VX. Pulmonary blastomycosis diagnosed in Hawaii. *Southeast Asian J Trop Med Public Health.* 2008; **39:** 701-5.
13. Arora K, Dawkins RL, Bauer DF, *et al.* Intracranial blastomycotic abscess mimicking malignant brain neoplasm: Successful treatment with voriconazole and surgery. *Surg Neurol Int.* 2015; **6:** 174.
14. Avninder S, Malhotra P, Chand K. Massive cerebral blastomycoma from a nonendemic region diagnosed on squash smear cytology. *Acta Cytol.* 2010; **54:** 364-6.
15. Azar MM, Assi R, Relich RF, *et al.* Blastomycosis in Indiana: Clinical and epidemiologic patterns of disease gleaned from a multicenter retrospective study. *Chest.* 2015; **148:** 1276-84.
16. Azar MM, Relich RF, Schmitt BH, *et al.* Cutaneous blastomycosis masquerading as pyoderma gangrenosum. *J Clin Microbiol.* 2014; **52:** 1298-300.
17. Babady NE, Buckwalter SP, Hall L, *et al.* Detection of *Blastomyces dermatitidis* and *Histoplasma capsulatum* from culture isolates and clinical specimens by use of real-time PCR. *J Clin Microbiol.* 2011; **49:** 3204-8.
18. Baddley JW, Winthrop KL, Patkar NM, *et al.* Geographic distribution of endemic fungal infections among older persons, United States. *Emerg Infect Dis.* 2011; **17:** 1664-9.
19. Baker T, Patel A, Halteh P, *et al.* Blastomycosis during pregnancy: A case report and review of the literature. *Diagn Microbiol Infect Dis.* 2017; **88:** 145-51.
20. Bariola JR, Hage CA, Durkin M, *et al.* Detection of *Blastomyces dermatitidis* antigen in patients with newly diagnosed blastomycosis. *Diagn Microbiol Infect Dis.* 2011; **69:** 187-91.
21. Bariola JR, Perry P, Pappas PG, *et al.* Blastomycosis of the central nervous system: A multicenter review of diagnosis and treatment in the modern era. *Clin Infect Dis.* 2010; **50:** 797-804.
22. Bariola JR, Vyas KS. Pulmonary blastomycosis. *Semin Respir Crit Care Med.* 2011; **32:** 745-53.
23. Barocas JA, Gauthier GM. Peritonitis caused by *Blastomyces dermatitidis* in a kidney transplant recipient: case report and literature review. *Transpl Infect Dis.* 2014; **16:** 634-41.
24. Baumgardner DJ, Temte JL, Gutowski E, *et al.* The differential diagnosis of pulmonary blastomycosis using case vignettes: A Wisconsin Network for Health Research (WiNHR) study. *WMJ.* 2011; **110:** 68-73.
25. Baumgardner DJ. Microecology of *Blastomyces dermatitidis:* The ammonia hypothesis. *Med Mycol.* 2009; **47:** 745-52.

26. Bednarczyk JM, Kethireddy S, White CW, *et al.* Extracorporeal membrane oxygenation for blastomycosis-related acute respiratory distress syndrome: A case series. *Can J Anaesth.* 2015; **62:** 807-15.
27. Bibbo C, Spellman J. Successful salvage of complicated calcaneal blastomycosis in disseminated disease with staged surgical reconstruction and local-systemic antifungal therapy. *J Foot Ankle Surg.* 2014; **53:** 472-9.
28. Bonifaz A, Morales D, Morales N, *et al.* Cutaneous blastomycosis: An imported case with good response to itraconazole. *Rev Iberoam Micol.* 2016; **33:** 51-4.
29. Bradsher RW Jr. Pulmonary blastomycosis. *Semin Respir Crit Care Med.* 2008; **29:** 174-81.
30. Bradsher RW Jr. The endemic mimic: Blastomycosis an illness often misdiagnosed. *Trans Am Clin Climatol Assoc.* 2014; **125:** 188-202; Discussion 202-3.
31. Brady RC, Crawford AH. Clavicular pain of 2 months' duration in a 9-year-old girl. *Pediatr Rev.* 2015; **36:** 545-7.
32. Brick KE, Agger WA. Successful treatment of brainstem blastomycosis with fluconazole. *Clin Med Res.* 2012; **10:** 72-4.
33. Brick KE, Drolet BA, Lyon VB, *et al.* Cutaneous and disseminated blastomycosis: A pediatric case series. *Pediatr Dermatol.* 2013; **30:** 23-8.
34. Bridger N, Fanella S. The basketball player with back pain. *Pediatr Infect Dis J.* 2011; 30: 915-6.
35. Brown DA, Whealy MA, Van Gompel JJ, *et al.* Diagnostic dilemma in primary *Blastomyces dermatitidis* meningitis: Role of neurosurgical biopsy. *Case Rep Neurol.* 2015; **7:** 63-70.
36. Brown EM, McTaggart LR, Zhang SX, *et al.* Phylogenetic analysis reveals a cryptic species *Blastomyces gilchristii*, sp. nov. within the human pathogenic fungus *Blastomyces dermatitidis. PLoS One.* 2013; **8:** e59237.
37. Brown JE, Blumin JH. Disseminated blastomycosis. *Arch Otolaryngol Head Neck Surg.* 2012; **138:** 605-7.
38. Bruce Light R, Kralt D, Embil JM, *et al.* Seasonal variations in the clinical presentation of pulmonary and extrapulmonary blastomycosis. *Med Mycol.* 2008; **46:** 835-41.
39. Bush JW, Wuerz T, Embil JM, *et al.* Outcomes of persons with blastomycosis involving the central nervous system. *Diagn Microbiol Infect Dis.* 2013; **76:** 175-81.
40. Carlos WG, Rose AS, Wheat LJ, *et al.* Blastomycosis in Indiana: Digging up more cases. *Chest.* 2010; **138:** 1377-82.
41. Carman WF, Frean JA, Crewe-Brown HH, *et al.* Blastomycosis in Africa: A review of known cases diagnosed between 1951 and 1987. *Mycopathologia.* 1989; **107:** 25-32.
42. Castillo CG, Kauffman CA, Miceli MH. Blastomycosis. *Infect Dis Clin North Am.* 2016; **30:** 247-64.
43. Cerullo L, Zussman J, Young L. An unusual presentation of blastomycosis-like pyoderma (pyoderma vegetans) and a review of the literature. *Cutis.* 2009; **84:** 201-4.
44. Chakrabarti A, Slavin MA. Endemic fungal infections in the Asia-Pacific region. *Med Mycol.* 2011; **49:** 337-44.
45. Chander B, Deb P, Sarkar C, *et al.* Cerebral blastomycosis: A case report. *Indian J Pathol Microbiol.* 2007; **50:** 821-4.
46. Chapman SW, Dismukes WE, Proia LA, *et al.* Clinical practice guidelines for the management of blastomycosis: 2008 update by the IDSA. *Clin Infect Dis.* 2008; **46:** 1801-12.
47. Cheikh Rouhou S, Racil H, Ismail O, *et al.* Pulmonary blastomycosis: A case from Africa. *Scientific World Journal.* 2008; **8:** 1098-103.
48. Choptiany M, Wiebe L, Limerick B, *et al.* Risk factors for acquisition of endemic blastomycosis. *Can J Infect Dis Med Microbiol.* 2009; **20:** 117-21.
49. Codifava M, Guerra A, Rossi G, *et al.* Unusual osseous presentation of blastomycosis in an immigrant child: A challenge for European pediatricians. *Ital J Pediatr.* 2012; **38:** 69.
50. Connolly P, Hage CA, Bariola JR, *et al. Blastomyces dermatitidis* antigen detection by quantitative enzyme immunoassay. *Clin Vaccine Immunol.* 2012; **19:** 53-6.
51. Connolly SL, Frank C, Thompson CA, *et al.* Dual infection with *Pythium insidiosum* and *Blastomyces dermatitidis* in a dog. *Vet Clin Pathol.* 2012; **41:** 419-23.
52. Craig J, Al Habeeb A, Leis JA. Budding Issues: Blastomycosis. *Am J Med.* 2015; **128:** 1182-5.
53. Curry S, Morris SK, Richardson SE, *et al.* Blastomycosis infection in an adolescent patient with Hodgkin lymphoma. *Pediatr Blood Cancer.* 2017; **64:** PMID: 27663466.
54. Dalcin D, Ahmed SZ. Blastomycosis in northwestern Ontario, 2004 to 2014. *Can J Infect Dis Med Microbiol.* 2015; **26:** 259-62.
55. Dalcin D, Rothstein A, Spinato J, *et al. Blastomyces gilchristii* as cause of fatal acute respiratory distress syndrome. *Emerg Infect Dis.* 2016; **22:** 306-8.
56. Day SR, Weiss DB, Hazen KC, *et al.* Successful treatment of osseous blastomycosis without pulmonary or disseminated disease and review of the literature. *Diagn Microbiol Infect Dis.* 2014; **79:** 242-4.
57. Derkach Y, Brown CW Jr. Cutaneous blastomycosis. *Dermatol Nurs.* 2008; **20:** 305-6.
58. Dhamija A, D'Souza P, Salgia P, *et al.* Blastomycosis presenting as solitary nodule: A rare presentation. *Indian J Dermatol.* 2012; **57:** 133-5.
59. Dimar JR 2nd, Puno RM, Nowacki MR, *et al.* Surgery for blastomycosis of the spine. *Am J Orthop (Belle Mead NJ).* 2014; **43:** e266-71.
60. Dobre MC, Smoker WR, Kirby P. A case of solitary *Blastomyces dermatitidis* meningitis. *Clin Neurol Neurosurg.* 2011; **113:** 665-7.
61. Dukik K, Munoz JF, Jiang Y, *et al.* Novel taxa of thermally dimorphic systemic pathogens in the Ajellomycetaceae (Onygenales). *Mycoses.* 2017; **60:** 296-309.
62. Emer JJ, Spear JB. Primary cutaneous blastomycosis as a cause of acute respiratory distress syndrome: Case report and literature review. *J Clin Aesthet Dermatol.* 2009; **2:** 22-30.
63. Fanella S, Skinner S, Trepman E, *et al.* Blastomycosis in children and adolescents: A 30-year experience from Manitoba. *Med Mycol.* 2011; **49:** 627-32.
64. Fanella S, Walkty A, Bridger N, *et al.* Gastric lavage for the diagnosis of pulmonary blastomycosis in pediatric patients. *Pediatr Infect Dis J.* 2010; **29:** 1146-8.

65. Federer AE, Haughom BD, Levy DM, *et al. Blastomyces* tenosynovitis of the foot and ankle: A case report and review of the literature. *J Foot Ankle Surg.* 2015; **54:** 1183-7.
66. Fisher KR, Baselski V, Beard G, *et al.* Pustular blastomycosis. *J Am Acad Dermatol.* 2009; **61:** 355-8.
67. Frisch CD, Gross BC, Link MJ, *et al.* Blastomycosis of the temporal bone mimicking glomus jugulare tumor. *Otol Neurotol.* 2016; **37:** e173-5.
68. Frost HM, Anderson J, Ivacic L, *et al.* Blastomycosis in children: An analysis of clinical, epidemiologic and genetic features. *J Pediatric Infect Dis Soc.* 2017; **6:** 49-56.
69. Frost HM, Novicki TJ. *Blastomyces* antigen detection for diagnosis and management of blastomycosis. *J Clin Microbiol.* 2015; **53:** 3660-2.
70. Gandhi V, Singh A, Woods GL, *et al.* A 66-year-old woman with fever, cough and a tongue lesion. *Chest.* 2015; **147:** e140-7.
71. Grim SA, Proia L, Miller R, *et al.* A multicenter study of histoplasmosis and blastomycosis after solid organ transplantation. *Transpl Infect Dis.* 2012; **14:** 17-23.
72. Guidry JA, Downing C, Tyring SK. Deep fungal infections, blastomycosis-like pyoderma and granulomatous sexually transmitted infections. *Dermatol Clin.* 2015; **33:** 595-607.
73. Guimaraes MD, Marchiori E, de Souza Portes Meirelles G, *et al.* Fungal infection mimicking pulmonary malignancy: Clinical and radiological characteristics. *Lung.* 2013; **191:** 655-62.
74. Gundelly P, Young M. Pustulonodular lesion on the nose. *JAMA.* 2014; **312:** 2564-5.
75. Gupta J, Patel G, Epelbaum O. Reversal of fortune: Central nervous system blastomycosis. *Am J Med.* 2016; **129:** e109-12.
76. Hage CA, Knox KS, Wheat LJ. Endemic mycoses: Overlooked causes of community acquired pneumonia. *Respir Med.* 2012; **106:** 769-76.
77. Hakim H, Apiwattanakul N, Miyairi I, *et al.* Blastomycosis: The Great Mimicker. *Clin Pediatr (Phila).* 2011; **50:** 1156-60.
78. Hamann JC, Marberry K. Sentinel presentation of disseminated *Blastomyces dermatitidis* infection as hip pain in a young adult: a case report. *J Bone Joint Surg Am.* 2010; **92:** 469-72.
79. Hankins CL. Blastomycotic hand infections. *Scand J Plast Reconstr Surg Hand Surg.* 2009; **43:** 166-70.
80. Harris JR, Blaney DD, Lindsley MD, *et al.* Blastomycosis in man after kinkajou bite. *Emerg Infect Dis.* 2011; **17:** 268-70.
81. Harvey AM, Mody DR, Amrikachi M. Disseminated blastomycosis diagnosed by fine-needle aspiration of the thyroid. *Diagn Cytopathol.* 2011; **39:** 446-50.
82. Haselow D, Saccente M, Vyas K, *et al.* Geographic distribution of endemic fungal infections among older persons, United States. *Emerg Infect Dis.* 2012; **18:** 360-1.
83. Hatch WO, Scalarone GM. Development of a slide agglutination assay for detection of blastomycosis. *Microbiol Immunol.* 2013; **57:** 756-61.
84. Hussein R, Khan S, Levy F, *et al.* Blastomycosis in the mountainous region of northeast Tennessee. *Chest.* 2009; **135:** 1019-23.
85. Jahangir AA, Heck RK. Blastomycosis: Case report of an isolated lesion in the distal fibula. *Am J Orthop (Belle Mead NJ).* 2010; **39:** e22-4.
86. Jain R, Singh K, Lamzabi I, *et al.* Blastomycosis of bone: A clinicopathologic study. *Am J Clin Pathol.* 2014; **142:** 609-16.
87. Jambhekar NA, Shrikhande SS, Advani SH, *et al.* Disseminated blastomycosis: a case report. *Indian J Pathol Microbiol.* 1988; **31:** 330-3.
88. Jetmore TM, Phan J, Yekinni AO. Blastomycosis of the nose: A case report. *Ear Nose Throat J.* 2016; **95:** e28-30.
89. Johnson RR, Vora SS. Blastomycotic multifocal arthritis and osteomyelitis in the urban setting. *J Rheumatol.* 2013; **40:** 1627-9.
90. Kaka AS, Sarosi GA. Disseminated blastomycosis. *N Engl J Med.* 2017; **376:** e9.
91. Kapila A, Motiani R, Chhabra L, *et al.* Peritoneal blastomycosis: A hidden mystery unfolds itself. *Conn Med.* 2014; **78:** 537-9.
92. Kar PM, Montalvo FA, Salazar DM, *et al.* Blastomycosis: Case report investigating a persistent pulmonary lesion in an immunocompromised patient. *J Ky Med Assoc.* 2009; 107: 259-63.
93. Khakoo R, Roidad N. Vesico-pustular blastomycosis. *Int J Infect Dis.* 2012; **16:** e763-4.
94. Khan ZU, Randhawa HS, Lulla M. Isolation of *Blastomyces dermatitidis* from the lungs of a bat, *Rhinopoma hardwickei hardwickei* Gray, in Delhi. *Sabouraudia.* 1982; **20:** 137-44.
95. Khuu D, Shafir S, Bristow B, *et al.* Blastomycosis mortality rates, United States, 1990-2010. *Emerg Infect Dis.* 2014; **20:** 1789-94.
96. Kobraei KB, Wesson SK. Blastomycosis-like pyoderma: response to systemic retinoid therapy. *Int J Dermatol.* 2010; **49:** 1336-8.
97. Kochar PS, Lath CO, Klein AP, *et al.* Multimodality imaging in cranial blastomycosis, a great mimicker: Case-based illustration with review of clinical and imaging findings. *Indian J Radiol Imaging.* 2016; **26:** 120-5.
98. Koh H, Demos TC, Lomasney LM, *et al.* Radiologic case study. Blastomycosis of bone. *Orthopedics.* 2011; **34:** 409 & 486.
99. Kralt D, Light B, Cheang M, *et al.* Clinical characteristics and outcomes in patients with pulmonary blastomycosis. *Mycopathologia.* 2009; **167:** 115-24.
100. Kroll RR, Grossman RF. Pulmonary blastomycosis in a professional diver: An occupational risk. *Can Respir J.* 2013; **20:** 340-2.
101. Kruse AL, Zwahlen RA, Bredell MG, *et al.* Primary blastomycosis of oral cavity. *J Craniofac Surg.* 2010; **21:** 121-3.
102. Kumar A, Sreehari S, Velayudhan K, *et al.* Autochthonous blastomycosis of the adrenal: First case report from Asia. *Am J Trop Med Hyg.* 2014; **90:** 735-9.
103. Kyaw TH, Sullivan L, Klingsberg RC. A 45-year-old woman with 3 weeks of cough and night sweats. *Chest.* 2016; **149:** e87-90.

104. Lahm T, Neese S, Thornburg AT, *et al.* Corticosteroids for blastomycosis-induced ARDS: A report of two patients and review of the literature. *Chest.* 2008; **133:** 1478-80.
105. Lahmiti S, Baki S. Thoracic blastomycosis from Morocco. *Scientific World Journal.* 2009; **9:** 200.
106. Lemke MA, Baumgardner DJ, Brummitt CF, *et al.* Blastomycosis in urban southeastern Wisconsin. *WMJ.* 2009; **108:** 407-10.
107. Lin MY, Chihara S, Smith KY, *et al.* Classic pyomyositis of the extremities as an unusual manifestation of *Blastomyces dermatitidis:* A report of two cases. *Mycoses.* 2010; **53:** 356-9.
108. Listernick R. A 13-year-old boy with elevated BUN and creatinine. *Pediatr Ann.* 2013; **42:** 96-8.
109. Litvinov IV, St-Germain G, Pelletier R, *et al.* Endemic human blastomycosis in Quebec, Canada, 1988-2011. *Epidemiol Infect.* 2013; **141:** 1143-7.
110. Liu A, Shear N, Wong J, *et al.* An atypical case of mucocutaneous blastomycosis in the Greater Toronto Area. *J Cutan Med Surg.* 2017; doi: 10.1177/1203475416686928. PMID: 28300442.
111. Lopez-Martinez R, Mendez-Tovar LJ. Blastomycosis. *Clin Dermatol.* 2012; **30:** 565-72.
112. MacDonald K, Lagace-Wiens P, Embil J. Hemorrhagic pustules in an aboriginal man. *CMAJ.* 2014; **186:** 607-10.
113. MacFadden DR, Gold WL, Richardson SE, *et al.* A breast mass in a 56-year-old man. *CMAJ.* 2011; **183:** 1875-8.
114. MacLean IS, Day SR, Moore CC, *et al.* Blastomycosis infection of the knee treated with staged total knee arthroplasty. *Knee.* 2015; **22:** 669-71.
115. Madigan T, Fatemi Y, Theel ES, *et al.* Central nervous system blastomycosis in children - A case report and review of the literature. *Pediatr Infect Dis J.* 2017; **36:** 679-84.
116. Mahiquez M, Bunton KL, Carney G, *et al.* Non-surgical treatment of lumbosacral blastomycosis involving L2-S1: A case report. *Spine (Phila Pa 1976).* 2008; **33:** e442-6.
117. Mansour MK, Ackman JB, Branda JA, *et al.* Case 32-2015. A 57-year-old man with severe pneumonia and hypoxemic respiratory failure. *N Engl J Med.* 2015; **373:** 1554-64.
118. Mardini J, Nguyen B, Ghannoum M, *et al.* Treatment of chronic pulmonary blastomycosis with caspofungin. *J Med Microbiol.* 2011; **60 (Pt. 12):** 1875-8.
119. Mason AR, Cortes GY, Cook J, *et al.* Cutaneous blastomycosis: A diagnostic challenge. *Int J Dermatol.* 2008; **47:** 824-30.
120. McCann DA, Smith HL. Infliximab-associated *Blastomyces dermatitidis* in treatment of ulcerative colitis. *Colorectal Dis.* 2013; **15:** e102-3.
121. McKinnell JA, Pappas PG. Blastomycosis: New insights into diagnosis, prevention and treatment. *Clin Chest Med.* 2009; **30:** 227-39.
122. Meece JK, Anderson JL, Gruszka S, *et al.* Variation in clinical phenotype of human infection among genetic groups of *Blastomyces dermatitidis. J Infect Dis.* 2013; **207:** 814-22.
123. Meece JK, Anderson JL, Klein BS, *et al.* Genetic diversity in *Blastomyces dermatitidis:* implications for PCR detection in clinical and environmental samples. *Med Mycol.* 2010; **48:** 285-90.
124. Moinuddin S, Barazi H, Moinuddin M. Acute blastomycosis thyroiditis. *Thyroid.* 2008; **18:** 659-61.
125. Monaco WE, Batsis JA. A case of disseminated blastomycosis in Vermont. *Diagn Microbiol Infect Dis.* 2013; **75:** 423-5.
126. Montenegro BL, Arnold JC. North American dimorphic fungal infections in children. *Pediatr Rev.* 2010; **31:** e40-8.
127. Morjaria S, Otto C, Moreira A, *et al.* Ribosomal RNA gene sequencing for early diagnosis of *Blastomyces dermatitidis* infection. *Int J Infect Dis.* 2015; **37:** 122-4.
128. Motswaledi HM, Monyemangene FM, Maloba BR, *et al.* Blastomycosis: A case report and review of the literature. *Int J Dermatol.* 2012; **51:** 1090-3.
129. Mouser H, Miller FH, Berggruen SM. Blastomycosis. *Radiology.* 2016; **280:** 972-7.
130. Mukherji BB, Sengupta TN. North American blastomycosis. *J Indian Med Assoc.* 1968; **51:** 80-1.
131. Munich SA, Johnson AK, Ahuja SK, Venizelos A, Byrne RW. Large cerebellar mass lesion: A rare intracranial manifestation of blastomycosis. *Surg Neurol Int.* 2013; **4:** 141.
132. Muzyka BC, Epifanio RN. Update on oral fungal infections. *Dent Clin North Am.* 2013; **57:** 561-81.
133. Nagalotimath SJ, Malur PR, Mastiholimath RD, *et al.* Blastomycosis: letter to editor. *Indian J Pathol Microbiol.* 1990; **33:** 398.
134. Nagrath SP, Atal PR, Kishore B, *et al.* Pulmonary blastomycosis. *J Indian Med Assoc.* 1971; **57:** 144-5.
135. Neal PM, Nikolai A. Systemic blastomycosis diagnosed by prostate needle biopsy. *Clin Med Res.* 2008; **6:** 24-8.
136. Nemeth NM, Campbell GD, Oesterle PT, *et al.* Red fox as sentinel for *Blastomyces dermatitidis*, Ontario, Canada. *Emerg Infect Dis.* 2016; **22:** 1275-7.
137. O'Guinn DJ, Serletis D, Kazemi N. Fungal osteomyelitis with vertebral re-ossification. *Int J Surg Case Rep.* 2016; **19:** 1-3.
138. Ortega-Loayza AG, McCall CO. Cutaneous blastomycosis. *N Engl J Med.* 2013; **368:** e13.
139. Ortega-Loayza AG, Nguyen T. Cutaneous blastomycosis: A clue to a systemic disease. *An Bras Dermatol.* 2013; **88:** 287-9.
140. Owen WR, Thurs K, Kim DS. Blastomycosis presenting as a nonhealing wound. *Adv Skin Wound Care.* 2012; **25:** 321-3.
141. Padhye AA, Smith G, Standard PG, *et al.* Comparative evaluation of chemiluminescent DNA probe assays and exoantigen tests for rapid identification of *Blastomyces dermatitidis* and *Coccidioides immitis. J Clin Microbiol.* 1994; **32:** 867-70.
142. Paganini CB, Ferreira AB, Minanni CA, *et al.* Blastomycosis: A differential diagnosis of periampullary tumors. *Pancreas.* 2010; **39:** 1120-2.
143. Pappas PG. Blastomycosis. *Semin Respir Crit Care Med.* 2004; **25:** 113-21.
144. Patel AJ, Gattuso P, Reddy VB. Diagnosis of blastomycosis in surgical pathology and cytopathology: Correlation with microbiologic culture. *Am J Surg Pathol.* 2010; **34:** 256-61.
145. Patel HV, Kute VB, Vanikar AV, *et al. Blastomyces dermatitidis* in a renal transplant recipient. *Saudi J Kidney Dis Transpl.* 2014; **25:** 1042-5.

146. Patel KR, Szczodry M, Neckrysh S, *et al.* Anterior cervical corpectomy and fusion for blastomycosis causing destruction of C6 vertebra: A case report. *J Med Case Rep.* 2015; **9:** 271.
147. Patel PD, Trabue CH, Myers JW, *et al.* Lingering lesions. *Am J Med.* 2008; **121:** 691-2.
148. Pelly L, Al Juaid A, Fanella S. Severe blastomycosis in infants. *Pediatr Infect Dis J.* 2014; **33:** 1189-91.
149. Pemberton JD, Vidor I, Sivak-Callcott JA, *et al.* North American blastomycosis of the eyelid. *Ophthal Plast Reconstr Surg.* 2009; **25:** 230-2.
150. Permpalung N, Kaewpoowat Q, Prasidthrathsint K, *et al.* Pulmonary blastomycosis: A new endemic area in New York state. *Mycoses.* 2013; **56:** 592-5.
151. Pfaff BL, Agger WA, Volk TJ. Blastomycosis diagnosed in a nonhyperendemic area. *WMJ.* 2014; **113:** 11-9.
152. Pfister JR, Archer JR, Hersil S, *et al.* Non-rural point source blastomycosis outbreak near a yard waste collection site. *Clin Med Res.* 2011; **9:** 57-65.
153. Phelps PO, Seagrave Z, Williams KM. Blastomycosis in the eyelid of a native Chicagoan. *Ophthalmology.* 2015; **122:** 1015.
154. Plamondon M, Lamontagne F, Allard C, *et al.* Corticosteroids as adjunctive therapy in severe blastomycosis-induced acute respiratory distress syndrome in an immunosuppressed patient. *Clin Infect Dis.* 2010; **51:** e1-3.
155. Polcari IC, Reindel R, Chadwick E, *et al.* Skin ulcers and bone pain in a healthy 6-year-old female. *Pediatr Dermatol.* 2013; **30:** 749-50.
156. Qian Q, Urman G, Cederroth T, *et al.* A man with fever and cough. *J Clin Microbiol.* 2016; **54:** 825 & 1177.
157. Randhawa HS, Chaturvedi VP, Kini S, *et al.* *Blastomyces dermatitidis* in bats: first report of its isolation from the liver of *Rhinopoma hardwickei hardwickei* Gray. *Sabouraudia.* 1985; **23:** 69-76.
158. Randhawa HS, Chowdhary A, Kathuria S, Roy P, Misra DS, Jain S, Chugh TD. Blastomycosis in India: Report of an imported case and current status. *Med Mycol.* 2013; **51:** 185-92.
159. Randhawa HS, Khan ZU, Gaur SN. *Blastomyces dermatitidis* in India: first report of its isolation from clinical material. *Sabouraudia.* 1983; **21:** 215-21.
160. Randhawa HS. Respiratory and systemic mycoses: an overview. *Indian J Chest Dis Allied Sci.* 2000; **42:** 207-19.
161. Rao GR, Narayan BL, Durga Prasad BK, *et al.* Disseminated blastomycosis in a child with a brief review of the Indian literature. *Indian J Dermatol Venereol Leprol.* 2013; **79:** 92-6.
162. Ray D, Jairaj PS, Date A. North American blastomycosis in a south Indian girl: Report of a case. *Indian J Tub.* 1995; **42:** 43-5.
163. Reed KD, Meece JK, Archer JR, *et al.* Ecologic niche modeling of *Blastomyces dermatitidis* in Wisconsin. *PLoS One.* 2008; **3:** e2034.
164. Resch M, Kurz K, Schneider-Brachert W, *et al.* Extracorporeal membrane oxygenation (ECMO) for severe acute respiratory distress syndrome (ARDS) in fulminant blastomycosis in Germany. *BMJ Case Rep.* 2009; pii: bcr07.2008.0392.
165. Richer SM, Smedema ML, Durkin MM, *et al.* Development of a highly sensitive and specific blastomycosis antibody enzyme immunoassay using *Blastomyces dermatitidis* surface protein BAD-1. *Clin Vaccine Immunol.* 2014; **21:** 143-6.
166. Rodriguez RC, Cornock E, White VA, Dolman PJ. Eyelid blastomycosis in British Columbia. *Can J Ophthalmol.* 2012; **47:** e1-2.
167. Ronald S, Strzelczyk J, Moore S, *et al.* Computed tomographic scan evaluation of pulmonary blastomycosis. *Can J Infect Dis Med Microbiol.* 2009; **20:** 112-6.
168. Roos KL, Gerdes LA, Kurz K, *et al.* An enhancing brainstem lesion in a patient with a history of worldwide travel. *Neurology.* 2011; **77:** 1756-60.
169. Roy M, Benedict K, Deak E, *et al.* A large community outbreak of blastomycosis in Wisconsin with geographic and ethnic clustering. *Clin Infect Dis.* 2013; **57:** 655-62.
170. Rucci J, Eisinger G, Miranda-Gomez G, *et al.* Blastomycosis of the head and neck. *Am J Otolaryngol.* 2014; **35:** 390-5.
171. Saccente M, Woods GL. Clinical and laboratory update on blastomycosis. *Clin Microbiol Rev.* 2010; **23:** 367-81.
172. Saeed AI, Williams AL, Bradley RF, *et al.* Bronchoscopic view of pulmonary blastomycosis. *J Bronchology Interv Pulmonol.* 2009; **16:** 266-9.
173. Salas-Alanis JC, Martinez MF, Garcia-Melendez M, *et al.* Blastomycosis imported to Monterrey, Mexico: Fifth case reported in Mexico. *Mycoses.* 2013; **56:** 495-7.
174. Sarkar PK, Malhotra P, Sriram PS. Rapid progression of pulmonary blastomycosis in an untreated patient of chronic lymphocytic leukemia. *Case Rep Med.* 2014; 514382.
175. Saucier J, Gauthier G. Verrucous lesions and ectropion in an immunocompetent individual. *Clin Infect Dis.* 2012; **55:** 1390-1 & 1426-8.
176. Savio J, Muralidharan S, Macaden RS, *et al.* Blastomycosis in a South Indian patient after visiting an endemic area in USA. *Med Mycol.* 2006; **44:** 523-9.
177. Schmutz JS, Ramey NA, Gauthier GM, *et al.* Severe oculofacial sequelae of cutaneous *Blastomyces dermatitidis.* *Ophthal Plast Reconstr Surg.* 2016; **32:** e61-2.
178. Schwartz IS, Embil JM, Sharma A, *et al.* Management and outcomes of acute respiratory distress syndrome caused by blastomycosis: A Retrospective case series. *Medicine (Baltimore).* 2016; **95:** e3538.
179. Scott AD, Dhir A. Intracranial blastomycosis. *Can Assoc Radiol J.* 2012; **63:** 73-6.
180. Scuderi S, O'Brien B, Robertson I, *et al.* Heterogeneity of blastomycosis-like pyoderma: A selection of cases from the last 35 years. *Australas J Dermatol.* 2017; **58:** 139-41.
181. Seitz AE, Adjemian J, Steiner CA, *et al.* Spatial epidemiology of blastomycosis hospitalizations: Detecting clusters and identifying environmental risk factors. *Med Mycol.* 2015; **53:** 447-54.
182. Seitz AE, Younes N, Steiner CA, *et al.* Incidence and trends of blastomycosis-associated hospitalizations in the United States. *PLoS One.* 2014; **9:** e105466.
183. Sharma AK, Narang P. Primary cutaneous inoculation blastomycosis with some unusual features. *Australas J Dermatol.* 1997; **38:** 101-2.

184. Shukla S, Singh S, Jain M, *et al.* Paediatric cutaneous blastomycosis: A rare case diagnosed on FNAC. *Diagn Cytopathol.* 2009; **37:** 119-21.
185. Sidamonidze K, Peck MK, Perez M, *et al.* Real-time PCR assay for identification of *Blastomyces dermatitidis* in culture and in tissue. *J Clin Microbiol.* 2012; **50:** 1783-6.
186. Siderits RH, Ouattara O, Marcus A, *et al.* Case study documenting the diagnosis of idiopathic CD4+ lymphocytopenia in a patient with atypical fungal infection (disseminated blastomycosis) by FNA of adrenal mass. *Cytojournal.* 2010; **7:** 13.
187. Siemieniuk RA, MacFadden DR, Matukas LM. Disseminated blastomycosis with cutaneous involvement in a Canadian railroad worker. *Lancet.* 2015; **385:** 883.
188. Silhan L, Reed RM. A blast from the past. *Am J Med.* 2011; **124:** 1132-5.
189. Smith JA, Gauthier G. New developments in blastomycosis. *Semin Respir Crit Care Med.* 2015; **36:** 715-28.
190. Smith JA, Kauffman CA. Blastomycosis. *Proc Am Thorac Soc.* 2010; **7:** 173-80.
191. Smith JA, Kauffman CA. Endemic fungal infections in patients receiving tumour necrosis factor-alpha inhibitor therapy. *Drugs.* 2009; **69:** 1403-15.
192. Smith JA, Kauffman CA. Pulmonary fungal infections. *Respirology.* 2012; **17:** 913-26.
193. Smith JA, Riddell J 4th, Kauffman CA. Cutaneous manifestations of endemic mycoses. *Curr Infect Dis Rep.* 2013; **15:** 440-9.
194. Smith RJ, Boos MD, Burnham JM, *et al.* Atypical cutaneous blastomycosis in a child with juvenile idiopathic arthritis on infliximab. *Pediatrics.* 2015; **136:** e1386-9.
195. Stavrakis C, Narayan A, Voronel O. Cerebral blastomycosis: Radiologic-pathologic correlation of solitary CNS blastomycosis mass-like infection. *J Clin Imaging Sci.* 2015; **5:** 30.
196. Steinberg BE, Goldenberg NM, Laffey JG. Extracorporeal membrane oxygenation for blastomycosis-related severe ARDS: A new indication as a rescue therapy? *Can J Anaesth.* 2015; **62:** 731-5.
197. Sterkel AK, Mettelman R, Wuthrich M, *et al.* The unappreciated intracellular lifestyle of *Blastomyces dermatitidis. J Immunol.* 2015; **194:** 1796-805.
198. Sung VW. Blastomycosis. *Headache.* 2009; **49:** 461-2.
199. Surprenant D, Kaniszewska M, Hutchens K, *et al.* Blastomycosis and pregnancy: An unusual postpartum disease course. *Case Rep Dermatol.* 2015; **7:** 107-12.
200. Ta M, Flowers SA, Rogers PD. The role of voriconazole in the treatment of central nervous system blastomycosis. *Ann Pharmacother.* 2009; **43:** 1696-700.
201. Taghizadeh F, Raghuwanshi M RS. Uncontrolled pneumonia in a 49-year-old white man. *Lab Med.* 2014; **45:** e123-7.
202. Taub AL, Nelsen DD, Nasser R, *et al.* Perianal North American blastomycosis. *Cutis.* 2015; **96:** e23-6.
203. Tsigrelis C, Baddour LM. Skeletal muscle abscess caused by *Blastomyces dermatitidis. Am J Med Sci.* 2009; **337:** 293-4.
204. Vachhani JA, Lee WC, Desanto JR, *et al.* Magnetic resonance spectroscopy imaging characteristics of cerebral blastomycosis. *Surg Neurol Int.* 2014; **5:** 80.
205. Van Der Veer J, Lewis RJ, Emtiazjoo AM, *et al.* Cross-reactivity in the Platelia™ *Aspergillus* enzyme immunoassay caused by blastomycosis. *Med Mycol.* 2012; **50:** 396-8.
206. Vindenes T, Gardiner BJ, Nierenberg N, *et al.* Pneumonia, septic arthritis and brain abscesses in a construction worker. *Clin Infect Dis.* 2015; **60:** 1722-3 & 1684-5.
207. Walsh CM, Morris SK, Brophy JC, *et al.* Disseminated blastomycosis in an infant. *Pediatr Infect Dis J.* 2006; **25:** 656-8.
208. Wang S, Martini MC, Groth JV, *et al.* Dermatoscopic and clinicopathologic findings of cutaneous blastomycosis. *J Am Acad Dermatol.* 2015; **73:** e169-70.
209. Werner A, Norton F. Blastomycosis. *Compend Contin Educ Vet.* 2011; **33:** e1-5.
210. Wheat LJ. Approach to the diagnosis of the endemic mycoses. *Clin Chest Med.* 2009; **30:** 379-89.
211. Wineland A, Siegel E, Francis C, *et al.* Fine-needle aspiration diagnosis of thyroid blastomycosis. *Endocr Pract.* 2008; **14:** 224-8.
212. Wuthrich M, Krajaejun T, Shearn-Bochsler V, *et al.* Safety, tolerability and immunogenicity of a recombinant, genetically engineered, live-attenuated vaccine against canine blastomycosis. *Clin Vaccine Immunol.* 2011; **18:** 783-9.
213. Youssef D, Raval B, El-Abbassi A, *et al.* Pulmonary blastomycosis during pregnancy: Case report and review of the literature. *Tenn Med.* 2013; **106:** 37-9.
214. Youssef D, Shams W, Ganote CE, *et al.* Negative image of *Blastomyces* on diff-quik stain. *Acta Cytol.* 2011; **55:** 377-81.
215. Yousuf T, Kramer J, Kopiec A, *et al.* In Search for equilibrium: Immunosuppression *versus* opportunistic infection. *J Clin Med Res.* 2016; **8:** 175-7.
216. Zhao TM, Gao J, She DY, *et al.* Blastomycosis in China: A case report and literature review. *Chin Med J (Engl).* 2011; **124:** 4368-71.
217. Zhu D, Chen H, Yu L. Probable pulmonary blastomycosis in an immunocompetent person. *Int J Infect Dis.* 2017; **59:** 86-9.

CHAPTER 18

Coccidioidomycosis

Coccidioidomycosis primarily is a fungal infection of the respiratory system of humans and wide variety of animals. This is caused by inhaling airborne spores of two species of dimorphic genus ***Coccidioides*** *i.e.* ***C.immitis*** native to California and ***C.posadasii,*** prevalent outside California. These fungi inhabit semiarid desert soils exclusively of the Western Hemisphere. The disease is endemic in the desert regions of southwestern United States, northwestern Mexico and some parts of Central and South America. This is also known as San Joaquin Valley Fever after name of the famous valley of California. The other synonyms are also frequently used like Valley fever, desert rheumatism, desert fever, desert bumps or Posadas-Wernicke disease or simply ***'kok-see'***.

Most of coccidioidal infections are usually sub-acute and self-limiting and take place in otherwise healthy individuals. However, in a few cases it produces protracted illness, such as chronic pneumonia or extra-pulmonary dissemination to skin, bones, meninges and other body sites. It has also emerged as significant systemic mycoses in patients with poor cell-mediated immunity such as cases of lymphoma, solid organ transplantation and co-infection with HIV, which often prove to be life-threatening. Out of all pathogenic fungi, *Coccidioides* is considered to be the most potential agent of bioterrorism.

Historical Perspective

Coccidioidomycosis is one of the oldest human mycoses. The first case of this disease was a 36 years old Argentinean soldier, named Domingo Ezcurra, reported in 1892 by a medical student, Alejandro Posadas, working under supervision of Roberto Wernicke in Buenos Aires, while he was performing histopathological examination. The patient was diagnosed erroneously as neoplastic skin disease *i.e.* ***mycosis fungoides*** but true clinical picture became clear during subsequent eleven years' regular follow up of this case. Posadas noted microscopically an organism that resembled protozoan of chickens, belonging to group referred to as ***'Coccidia'*** that are characterized by cyst containing endocystic cells (from Greek ***kokkis***, **"little berry"**). Hence suffix *'-oides'* was given to genus designating as *Coccidioides,* which means *'similar to'* as that of protozoan parasite called *Coccidia.*

The second known case of this disease was a Portuguese farm laborer who had immigrated to San Joaquin Valley from Azores in 1886. This case had skin lesions and was thoroughly studied in California by Emmett Rixford and Thomas Caspar Gilchrist in 1896. Due to similarity to protozoan coccidia and severity of disease caused by it, the organism was named *Coccidioides immitis* (coccidia-like; *im* - not; *mitis* - mild). In Latin, ***"immitis"*** is also used for "harsh" describing the clinical course.

The fungal nature of organism remained unrecognized until 1900, when filamentous mold, believed for several years to be contaminant of cultures, was shown by William Ophuls and Herbert Moffitt by inoculating laboratory animals to change its morphological form in tissue resembling protozoa. Ophuls conclusively demonstrated transition between two fungal forms and conditions under which they occur. This provided substantiation of dimorphic nature of causative fungus. Consequently disease was classified to be fungal in nature and the organism was identified as *Coccidioides immitis.* The designation, *C.immitis,* was originally assigned to this organism presuming it to be protozoan and subsequently same name was endorsed when its fungal nature was established, which is continuing as such for the last more than a century.

In 1932, Stewart and Meyer isolated *C.immitis* for the first time from California soil. In late 1930s, it was recovered from sputum of many patients suffering from well-known, self-limiting pulmonary infection, now popularly called Valley fever. In 1938, Dickson and Gifford elucidated pathogenesis of this disease and clarified the relationship between primary and the secondary infection. The first Canadian case of coccidioidomycosis in human was reported in 1952. Subsequently specific endemic areas were recognized and it was established that disease was found prevalent only in Western Hemisphere. Another milestone was established as discovery of amphotericin B in late 1950s, which was beginning of therapeutic era of fungal diseases. The addition of oral azole group of drugs in 1970s to armamentarium followed, thereby opening another landmark in coccidioidomycosis therapy.

The DNA evidence based on molecular techniques shown by Fisher *et al*, in 2002 established that there were in fact two species of the genus *Coccidioides*. The one, still called *C.immitis*, restricted to California. The strains found **outside** California seemed to be distinct species, which was named *C.posadasii*, in honour of Alejandro Posadas. Therefore, the second significantly pathogenic species of this genus came into existence after a gap of more than a century.

The decennial meeting of Coccidioidomycosis Study Group is being regularly held to discuss various aspects of this disease and last one *i.e.* Sixth International Symposium, was held at Stanford, California, August 23-26, 2006 and its proceedings were published by the New York Academy of Sciences (*Ann N Y Acad Sci.* 2007; **1111:** 1-462). This Symposium marked fiftieth year of existence of the Study Group and progress made over half century. Its objective was to comprehensively address biology, pathology, therapy and preventive measures of coccidioidomycosis. In addition, other topics in molecular biology and genetics, immunology, epidemiology, occupational health were also widely covered. Many medical, surgical and clinical laboratory specialists involved in human, veterinary medicine, antifungal therapeutics as well as drug development attended this decennial academic event. The 7th International Coccidioidomycosis Symposium will be held on August 10-13, 2017 again at Stanford, California, this time after a gap of eleven years.

Mycology

The causative organisms of coccidioidomycosis are dimorphic fungi existing in two forms, mycelial at ambient temperature in the environment and as a spherule in the tissue. The term ***'Coccidia'*** is actually derived from their tissue (parasitic) phase resembling Protozoa belonging to the genus ***Coccidia*** that cause coccidiosis in domestic fowls and rabbits. The species epithet ***'immitis'*** is of Latin derivation and signifies ***'not mild'*** depicting serious nature, as it is probably most virulent of all fungal pathogens.

For the last one century, only single species was thought to be the causative agent of coccidioidomycosis, However now it is established that the disease is caused by two species of *Coccidioides*. The distinction between these species is based on genetic analysis and epidemiological features-*C.immitis* from California and *C.posadasii* from outside California, generally referred to as 'Californian' and 'non-Californian' species, respectively. However, morphology of these two species, clinical manifestations of disease and therapeutic options are essentially identical. Both are also listed as 'Select Agents' by the Department of Health and Human Services thereby requiring strict regulation for laboratory testing. The studies in Arizona have documented increasing number of cases in humans and animals.

The endospores are released *in vivo*; enlarge in size producing endosporulating spherules that in turn release second generation of endospores that is again able to yield an average of 200 to 300 endospores entailing pathogenic events to continue. The efficiency of endosporulation as reproductive process accounts for rapid growth of organisms. The spherules rapidly increase in size and undergo segmentation to form peripheral compartments surrounding central cavity. The endospores, approximately 4-10 µm, develop in packets within these peripheral compartments.

In this fungus, the process of conidiogenesis is **enteroarthric** as arthroconidia develop within fertile hyphae. The conidial release is **rhexolytic** type, wherein an intermediate cell is disrupted and lysed leaving behind wall remnants. The mycelial-to-spherule conversion is performed *in vivo* in experimental animals as well as *in vitro*. This fungus is not that temperature dependent as seen in other dimorphic fungi. The parasitic cycle is also reproduced *in vitro* by growing the organism in defined glucose-salts medium at 37°C in the presence of 20% CO_2. When *Coccidioides* species are grown on glucose-yeast extract (GYE) agar, these give rise to endosporulating parasitic cells *i.e.* spherule, similar to when arthroconidia infect mammalian tissues.

As *Coccidioides* species are soil-dwelling dimorphic fungi, they grow as white fluffy filamentous mold in soil and as spherule at body temperature. The hyphal form may rarely be found in human tissue, which is an unusual feature. The presence of CNS plastic devices appears to be associated with morphological reversion to saprotrophic form. This reversion has implications for diagnosis and therapy and may increase risk of obstruction of such devices.

A proline rich antigen (PRA) has been analyzed for differential expression and variation among isolates of *Coccidioides* species. It is from coccidioidal cell wall, primarily of parasitic spherule phase.

The life cycle of *Coccidioides* species involves development of hyphae, arthroconidia and spherule. The sexual state of fungus has not yet been observed. Hence, taxonomic placement of *Coccidioides* species is in the form-family **Moniliaceae** of the form-class **Hyphomycetes** in **Deuteromycetes**. Several morphological features indicate that organism may be placed in family **Onygenaceae** of order **Onygenales** of class **Eurotiomycetes** in phylum **Ascomycota** of kingdom **Fungi**.

Epidemiology

Coccidioidomycosis is disease acquired by people of all ethnic groups, whether healthy individuals or immunocompromised who reside or travel in certain areas of Western Hemisphere including newly recognized area in Brazil **(Fig. 18.1)**. It is **'endemic mycosis'** because of its persistent prevalence in these geographical areas. *Coccidioides* species are normal inhabitant of sandy and somewhat alkaline soil of Lower Sonoran Desert Life Zone, where it grows in mycelial form. This area is ecologically classified as semiarid, with high summer temperatures and mild winters and plants such as creosote brush, cacti and yuccas. It specifically inhabits alkaline soil though it is strangely absent from deserts of Africa and Asia. The imported (non-indigenous) cases of this disease may be observed following travel to endemic areas. Therefore, physicians outside endemic regions should consider coccidioidomycosis as differential diagnosis of respiratory infection if patient has ever traveled through desert of southwestern USA or lived in an endemic area.

Fig. 18.1. Map showing geographical distribution of endemic areas of coccidioidomycosis in the New World.

The cases of coccidioidomycosis are sporadic but there are occasional outbreaks among people who, as a group, are involved in disturbance of soil. More cases occur in hot, dry summer and fall succeeding rainy winters and springs than in months following lighter rainy season. Despite being highly infectious organism, it is not ***'contagious'*** disease because person-to-person transmission is not feasible as infective stage is arthroconidium, which is found in environment and not in human beings, who have spherules. However, there are now reports that it may be transmitted while undergoing transplantation surgeries when donor is having the underlying infection.

The causative fungal species are geophilic dimorphic pathogens with distribution restricted to southwestern areas of United States. They cause disease of major public health importance in endemic area and it is estimated that annually about 150,000 new cases occur in United States, most of them resolve spontaneously of their own. The patients with primary infection are asymptomatic 60% of the time and 40% have a wide spectrum of clinical manifestations, most commonly a community-acquired pneumonia. However, overall < 1% develop extrapulmonary manifestations. The infection with this organism has reached epidemic proportions in endemic areas with greater frequency of international travel.

The pathogen is also being increasingly recognized in parts of the world where it is not endemic in people and who have history of travel to the zone of endemicity. There are ongoing large-scale land development for renewable energy projects in the Antelope Valley, located in the Western Mojave Desert, has been blamed for increased fugitive dust emissions and coccidioidomycosis incidence among the general public in recent years.

In South America, coccidioidomycosis is prevalent in Argentina, Paraguay, Bolivia, Colombia and western Venezuela particularly in Falcon and Lara states. It is less frequent in Ecuador, Peru, Chile and Uruguay. Recently disease has been found endemic in semiarid zone of northeastern Brazil, in contiguous states of Bahia, Ceara, Maranhao and newly reported cases in north-eastern state of Piaui. It is thought that there was some missing link that existed in catenulate geographic distribution of *Coccidioides* species, which has been put in place from north-central California to Argentina's south-central province.

The Valley Fever is primarily prevalent in six states of the USA *i.e.* San Joaquin and Central Valleys of **California** and in hot, desert regions of southern **Arizona** (specially in Phoenix and Tucson areas), southern **Nevada**, southern **Utah**, southern **New Mexico**, western **Texas** (especially around El Paso) and Mexico (in states of Sonora and Chihuahua).

In United States, there are more than 4 millions people residing in areas where this fungus is found prevalent in the soil. The climatic conditions that are important in establishing boundaries of these endemic areas consist of season of adequate rainfall to allow fungus to grow. It is followed by long hot dry period during which fungus produces large number of infectious arthroconidia, which are easily dispersed in dust by recurring winds. It is endemic in south-western United States, adjacent areas of northern Mexico, Central and South America. Majority of inhabitants of these areas are known to have infection of this disease. As coccidioidomycosis is persistently confined to geographically limited countries of North, Central and South America hence it is also described as disease of **New World**.

The areas of highest endemicity have semiarid climate and include central San Joaquin Valley and Kern County in **California**, Manicopa and Pima Counties in **Arizona** and Western and Southern Counties in **Texas**. In Arizona, highest prevalence of infections occurs from June to July and from October to November. In California, risk of infection is highest from June to November, without late summer break. There is virtually no overlap with endemic areas of histoplasmosis, which include great River Valleys of central United States, although two endemic areas connect just west to San Antonio, Texas. Recently there was marked increase in number of cases in western United States as reported to Arizona Department of Health Services. Similarly, San Luis Obispo County in California has reported 47 cases of valley fever during the first four months of 2017 with 5 deaths and 11 additional cases are under investigation. It is very significant findings as 18 cases were reported in 2015 and 49 in 2016 from this very County.

An epidemic of coccidioidomycosis occurred in San Joaquin valley in 1978 as a result of air-borne dissemination of organism during severe dust storm. A similar increase of cases was reported during 1991-92, related to drought, dust storms, construction work and an influx of new susceptible persons and abundant rain during March 1991 and February-March 1992. Following the marked increase in cases, isolates were studied by RFLP and two major genotypic groups were identified.

An outbreak has been described in Ventura County following an earthquake on January 17, 1994, which was centered in Northridge, California by identifying factors that increased risk for acquiring acute coccidioidomycosis infection. Both location and timing of cases suggest that outbreak was caused when arthroconidia spread in dust clouds generated by earthquake.

During last 1½ decade, coccidioidomycosis has become an emerging fungal disease particularly among immunosuppressed patients. Actually emerging infectious diseases have been defined as *'infections that have newly existed in a population or have existed but are rapidly increasing in incidence or geographic range'*. This is not a new disease in its endemic area being more than 100 years old. However, recently number of cases has increased dramatically, especially in South-western USA and clinical symptoms of this illness have changed in patients with AIDS. Hence it is being considered as re-emerging disease in certain pockets of endemic areas.

This disease is not perceived to have substantial impact outside areas classically considered as endemic because of its apparent regional confinement. However, some of cases have been reported outside endemic areas in other continents. Most of them have significant history of travel or residence in endemic areas. But in some of them, there was no such history and still they had full-blown disease. This can be explained by their exposure to material or items like cotton bales imported from endemic countries. For example, in Japan 16 cases of coccidioidomycosis have been reported by 1998 and most of them had visited endemic areas. Therefore, this disease was designated as notifiable fungal disease in Japan in 1999. This number grew to >40 cases by the year 2007 and 3-4 new cases are found every year. There is no original case report of this disease from India till date however; imported cases have been described from Mumbai, Chennai, Bangalore and Delhi.

Among various outcomes possible when body is exposed to live fungi, infection represents most potential life-threatening disease. While major risk to laboratory personnel is from handling of live cultures, however, there are a few associated with collection, transport and processing of specimens of body fluids, biopsy and autopsy tissues.

Occupational hazards are in those occupations where there is exposure to soil dust, including agricultural workers, construction workers, telephone pole diggers and archaeological students. Like *Histoplasma, Coccidioides* species are also subject of containment practices in microbiology laboratories, owing to their potential danger of laboratory-acquired infection. The dangers of working with cultures of *Coccidioides* species were recognized early in 20th century. Therefore, safe laboratory handling of such highly infectious fungi require Biosafety Level 3 facilities.

Moreover, both *C.immitis* as well as *C.posadasii,* are potential agents of bioterrorism with aerosol delivery. These are on US Government Occupational Safety and Health Administration (OSHA) list of biological agents and toxins that have potential to pose threat to public health and safety and that can be utilized by various governments in their state-organized activities or even by terrorist organizations. As such, in USA and elsewhere, these organisms fall under stringent regulations that govern their possession, use and transfer. Hence interstate transport of infectious material is restricted under provisions of US Anti-terrorism and Effective Death Penalty Act, 1996 (Department of Health and Human Services). As such there is no fungal organism currently on the CDC list of category A or B agents although it could be argued that some should be considered as potential category C agents. It is not a notifiable disease in Europe.

Immunity

Coccidioides species are probably the most virulent of fungal pathogens. The majority of individuals infected with this fungus recover without significant sequelae. However, ~5% of patients develop severe and life-threatening disease. There are several factors that contribute to its virulence. Cell wall of infectious arthroconidia is anti-phagocytic as it inhibits phagocytosis. Hence after evading host's innate defenses, arthroconidia are converted to spherules, which in turn release endospores. The spherule and endospore forms of *Coccidioides* species also provide major barriers to host defense as spherules are too large to be engulfed by phagocytes. Moreover, endospores are covered by remnants of anti-phagocytic inner cell wall of spherules. Each spherule releases hundreds of endospores, overwhelming defense mechanisms of non-immune host.

There are several host factors that also contribute to pathogenesis. Protection against this disease and related fungal respiratory infections, like blastomycosis and paracoccidioidomycosis, is correlated with strong response of cell-mediated immunity, which is believed to play defensive role in infections caused by *Coccidioides* species. The impairment of cell-mediated immune system is associated with a more severe disease. Therefore, general lowering of cell-mediated immunity is responsible for the onset of this disease. In addition, elevated levels of estradiol and progesterone in pregnancy can enhance growth of *Coccidioides* species. The infection stimulates protective immune response that limits initial illness and confers life-long immunity to subsequent infections.

Pathogenesis and Pathology

The lung is the usual portal of entry of this fungus and is the most commonly infected organ. The infection occurs following inhalation of arthroconidia of mycelial form of *Coccidioides* species or reactivation of latent infection in immunocompromised patients. In the host, arthroconidia swell to form thick-walled spherules, which enlarge in size and develop endospores. The mature spherules mechanically rupture, releasing endospores that spread locally and disseminate to extra-pulmonary sites.

The inflammatory response usually contains both (i) pyogenic response with participation of neutrophils in early stage and (ii) predominantly granulomatous tissue reaction with lymphocytes, plasma, epithelioid and giant cells. In the lungs, calcification is also common, which is unusual elsewhere. The cellular immunity is key defense mechanism in coccidioidomycosis, serving to arm macrophages to halt progression of infection. As AIDS patients lack cellular immunity so they suffer from severe progressive forms of coccidioidomycosis.

Clinical Features

The infection in man and animals results from inhalation of dust containing infective units, **arthroconidia**. Approximately 60% of infections are asymptomatic and 90% of symptomatic infections resolve without specific treatment. The individuals contracting coccidioidomycosis have self-limiting primary pulmonary illness that is often asymptomatic. The usual course of disease in otherwise healthy people is complete recovery within a span of six months. In most cases, body's immune response is effective and no specific course of treatment is necessary. However, few cases may develop progressive and sometimes even fatal disease. Dissemination also occurs weeks, months or years after primary respiratory infection. Unlike tuberculosis, coccidioidomycosis primarily affects bones and not joints, however, joints may secondarily be involved by extension from adjacent sites having osteomyelitis. The recovery from infection gives life-long immunity to re-infection. The clinical features of disease can be described under following headings for convenience of discussion:

1. Pulmonary Coccidioidomycosis
2. Disseminated Coccidioidomycosis
3. Chronic Coccidioidal Meningitis
4. Coccidioidomycosis in AIDS Patients

1. Pulmonary Coccidioidomycosis

The infection is acquired through inhalation of air-borne arthroconidia and an average incubation period is about two weeks. Spores are 2-5 μm in diameter and are able to reach terminal bronchioles and alveoli. This form of disease develops after 1-4 weeks of infection and is usually asymptomatic to begin with and may be marked only by conversion of negative to positive skin reaction. However, symptomatic pulmonary infection presents with acute flu-like symptoms such as headache, fever, cough, rash, malaise and myalgia, which are usually self-limiting. A few individuals develop hypersensitivity reaction after 1-2 weeks as skin rash like erythema nodosum or erythema multiforme in about 20% of those with primary disease, most often in white adult females. In endemic areas, atypical pneumonia with arthralgia (desert rheumatism) or erythema nodosum (bumps) is suggestive of primary coccidioidomycosis. This symptom complex is called San Joaquin Valley fever and is self-limiting.

The infection tends to be indolent, progressing with infiltrates and fibrosis taking over weeks and months' time. Sometimes, pulmonary complications like bronchopneumonia may also be seen. The radiological changes can be observed as coin-like lesions and thin-walled cavity or **coccidioidoma** as chronic pulmonary residual lesions, which may heal or become chronic in nature.

About 5% of patients develop **lung cavities (Fig. 18.2)** after their initial infection of Valley Fever. These cavities occur most often in older adults, usually without symptoms and about 50% of them disappear within two years. Occasionally, these cavities rupture, causing chest pain and difficulty in breathing hence may require surgical repair. The MRI and CT scans are very useful in defining the affected anatomical sites, which may also be helpful as guide to surgical procedures.

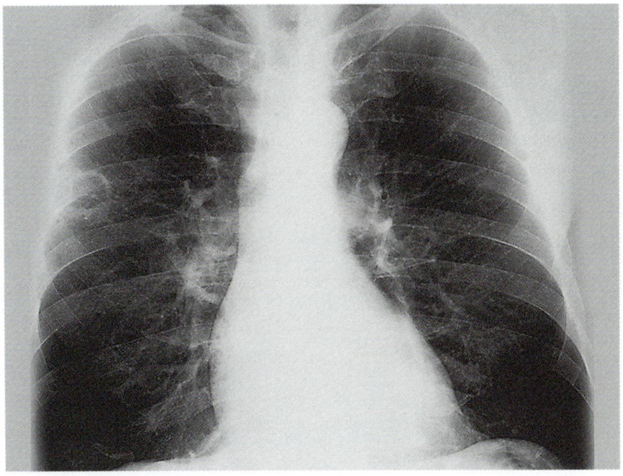

Fig. 18.2. Chest X-rays PA view showing classic thin-walled cavity in the right lung of a coccidioidomycosis patient.

2. Disseminated Coccidioidomycosis

The disseminated variety is defined as coccidioidal disease that spreads beyond pulmonary parenchyma or hilar node which occurs in ~1% of individuals infected with *Coccidioides* species. The disease may be found in virtually every organ of body except gastrointestinal tract. The dissemination is through hematogenous route starting from an active, latent or residual pulmonary focus. The most common site of disease spread is skin, where granulomatous lesions develop. There are varieties of histopathological appearances from clusters of neutrophils to classical granulomas. The additional subcutaneous involvement may present as cold abscess. Moreover, pericardial involvement is also there but very rare with only 23 cases reported in the medical literature till date.

The second most frequent types of disseminated infections are osteomyelitis and synovitis. A wide variety of bones are involved, including vertebrae, skull, ribs and long bones. There are skeletal lesions with punched-out lytic or permeative patterns of destruction and sclerotic areas. As far as joint involvement is concerned, disease shows some predilection for knee, ankle, wrist and elbow. The multiple osteoarticular lesions are osteomyelitic type. Disseminated coccidioidomycosis is comparable to tuberculosis in which lesions develop in many organs, bones and CNS. Less than 1% of patients develop disseminated disease, which involves CNS in ~33 to 50% of cases. The clinical course of disease often includes remissions and exacerbations and may even prove to be fatal. This holds true in some races like Filipinos, Blacks and Mexicans as

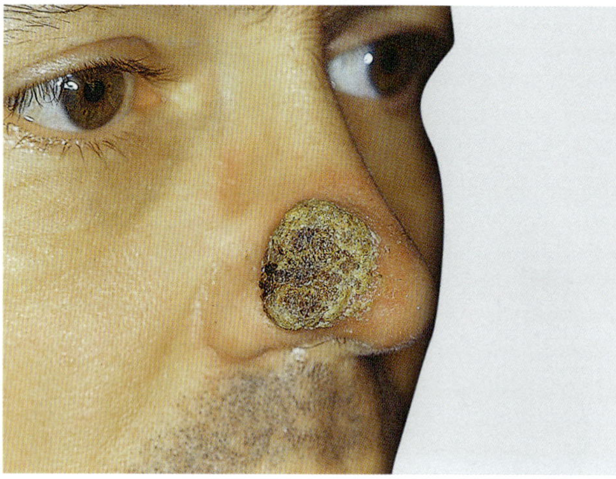

Fig. 18.3. Primary cutaneous ulcerated plaque over nose in healthy adult male from endemic area of coccidioidomycosis.

well as in pregnant women. The additional loci of dissemination include eye, larynx, epididymis and peritoneum. There may be involvement of CNS along with other organs.

The miliary coccidioidomycosis is characterized by development of numerous small 'millet seed' granulomas throughout lungs and other organs. This pattern indicates hematogenous or lymphatic spread and has been associated with high mortality rate. The previous reports have suggested that this disorder occurs almost exclusively in immunocompromised patients, with most patients succumbing to progressive respiratory failure. Now, it has been reported in immunocompetent patients as well with low mortality rate and better outcome as compared to earlier reports.

This disease may take serious turn and may lead to frequently fatal complication in liver transplant patients who have lived in an area of endemicity. The incidence of coccidioidomycosis may be increasing in solid organ transplantation (SOT) patients as a result of local environmental conditions, which are conducive to acquisition of infection. A high index of clinical suspicion followed by an aggressive diagnostic procedure is necessary for establishing proper diagnosis of coccidioidomycosis in liver transplant recipients. The other individuals at increased risk of disseminated disease are those with immune system deficiencies. In areas of southwestern USA where Valley Fever is endemic, it is one of the most frequent opportunistic infections among HIV-infected patients. The individuals who are immunocompromised due to organ transplantation, Hodgkin's disease, diabetes, pregnancy or chronic corticosteroid therapy also have an increased risk of developing disseminated disease.

The cervical lymph nodes may be inflamed and coalesce and rupture through skin via proliferative and verrucous fistulae thus mimicking classical pathology of scrofuloderma. The patient may present with primary cutaneous coccidioidomycosis wherein initial infection of disease is a nodule or plaque ulcerated at the center resembling syphilitic, tubercular or sporotrichotic chancre **(Fig. 18.3)**. The skin lesion may be accompanied by lymphangitis and corresponding regional lymphadenopathy. Although symptoms of disease are quite variable but often patient has an allergic reaction to circulating fungus, producing reddening of skin known as 'desert bumps'. Isolated cases of thyroid involvement have been described in setting of disseminated coccidioidomycosis.

3. Chronic Coccidioidal Meningitis

The disease can disseminate widely into meninges as chronic meningitis, which is frequent site of extrapulmonary involvement and occurs in one-third to one-half of all patients with dissemination. Meningitis usually involves basilar portions of brain where thick opaque membrane is formed. This membrane frequently obstructs flow of cerebrospinal fluid or blocks re-absorption resulting in fatal hydrocephalus.

4. Coccidioidomycosis in AIDS Patients

Coccidioidomycosis is one of the most frequent opportunistic infection among HIV-infected patients in southwestern USA. The disease in AIDS patients is nearly always severe and was included in surveillance case definition for AIDS in 1987. There is a high incidence of disseminated infection in AIDS patients living in endemic regions as compared to general population, which may be explained in part by reactivation of latent infection. The patients, who are immunocompromised due to underlying circumstances like solid organ transplantation, Hodgkin's disease, diabetes mellitus, third trimester of pregnancy or even long-term corticosteroid therapy, have an increased risk of developing disseminated disease. The clinical presentation in these patients is seen in same manner as other immunodeficient individuals.

Radiodiagnosis

The radiological changes in coccidioidomycosis can be seen as small or large and single or multiple infiltrates in

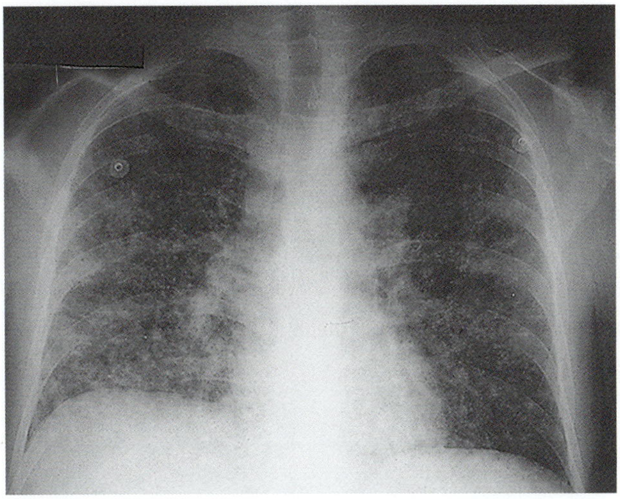

Fig. 18.4. Chest X-rays PA view showing diffuse infiltrate of pulmonary coccidioidomycosis in a patient of AIDS.

primary pulmonary disease on chest X-rays PA view **(Fig. 18.4)**. There may be hilar or mediastinal lymphadenopathy and cavitary or coin-like lesions in the lungs. The patient may show pleural or pleuro-pericardial effusion. The osteolytic lesions in bones are also seen over the affected sites. The MRI and CT scans are useful in localizing affected areas and serve as effective tool for guidance to surgical procedures.

The lung nodules and masses may mimic cancer on cross-sectional or PET imaging, especially in an older or immunocompromised patient population, with high risk for malignancy.

Differential Diagnosis

The primary disease is often confused with any other acute pulmonary infections such as influenza, primary atypical pneumonia, bronchitis, broncho-pneumonia or simply common cold. Mostly primary form of disease resolves uneventfully and without requiring any sort of treatment. Secondary coccidioidomycosis requires therapy so that it is imperative to make specific diagnosis. The disease must be differentiated from tuberculosis, neoplasia, other mycotic infections, syphilis, tularemia, glanders and osteomyelitis of bacterial origin. The AIDS patients with diffuse pulmonary infiltrates present special diagnostic enigma because coccidioidomycosis may mimic *Pneumocystis* pneumonia or may be a co-infection. The spherule should also be differentiated from sporangium of *Rhinosporidium seeberi* and that of adiaspiromycosis, which is much bigger in size. The young spherule mimics sporangium of *Prototheca* species. The differences in morphological features of spherule and similar structures producing fungal and pseudofungal organisms are shown in **Table 18.1**.

Laboratory Diagnosis

The presentation of coccidioidomycosis is very similar to other pulmonary or extra-pulmonary illnesses therefore identifying exact cause of disease requires specific laboratory investigations. The diagnosis depends on presentation of symptomatic clinical syndrome. Information about an accurate travel history or residence in area of endemicity is significant among such patients. The diagnosis is confirmed by fungal culture of specimens on appropriate culture media, serology and histopathological examination of tissue biopsy. The serological and skin tests either use coccidioidin, culture filtrate antigen from mycelial phase of *Coccidioides* or spherulin, an extract of spherules. As this is highly infectious organism hence laboratory personnel should practice Biohazard Safety Level 3 containment, which is mandatory for handling of the *Coccidioides* species.

(a) Direct Examination

The clinical specimens like sputum, gastric contents, cerebrospinal fluid, exudate or pus are examined microscopically for the presence of doubly refractile thick-walled globular spherules of about 20-80 μm. The immature spherules are smaller and without endospores. At maturity when spherule wall ruptures endospores are released in surrounding tissue and subsequently enlarge to form new spherules, which can be visualized in direct examination. The endospores released into tissues from ruptured spherules may be mistaken for non-budding yeasts.

The spherules are visualized properly in KOH, CFW or Papanicolaou preparations from respiratory secretions, BAL fluid or pus material and may take up routine Gram's stain. The histopathological stainings like H&E, PAS and GMS stains are performed to demonstrate spherules **(Figs. 18.6 and 18.7)**. The intact spherules are surrounded by granulomatous inflammatory infiltrate by lymphocytes, plasma, epithelioid and giant cells, whereas ruptured spherules are surrounded by neutrophils. These features distinguish *Coccidioides* from *R.seeberi*. The endospores and/or young spherules may be mistaken for non-budding yeast cells of *B.dermatitidis* or *Cr.neoformans*. In tissues, *Coccidioides* typically appears as spherule containing endospores but sometimes organisms may produce atypical

Table 18.1. Morphological Features of Spherules and Similar Structures Produced by Fungal and Pseudofungal Organisms.

Morphological Features	*C.immitis*	*Ea.crescens*	*P.wickerhamii*	*R.seeberi*
Taxonomic Status	Fungus	Fungus	Protista	Protista
Size of Spherules/Sporangia	60-100 µm	50-500 µm	7-13 µm	10-450 µm
Thickness of Spherule Walls	1-2 µm	10-70 µm	-	3-5 µm
Size of Endospores	2-5 µm	Not Applicable	2-11 µm	6-10 µm
Max. Number of Endospores	200-300	Not Applicable	50	16,000
Arthroconidia/Hyphae	Present	Not Applicable	Not Applicable	Not Applicable
Pathological Reaction of Host	Necrotic granuloma	Fibrotic granuloma	Necrotizing epithelioid granuloma	Acute/chronic inflammatory
Growth on Artificial Media	Mycelial	Mycelial	Yeast-like	Non-cultivable
Special Fungal Stain - GMS	Positive	Positive	Positive	Positive
PAS	Positive	Positive	Positive	Positive
Mucicarmine	Negative	Negative	Negative	Positive

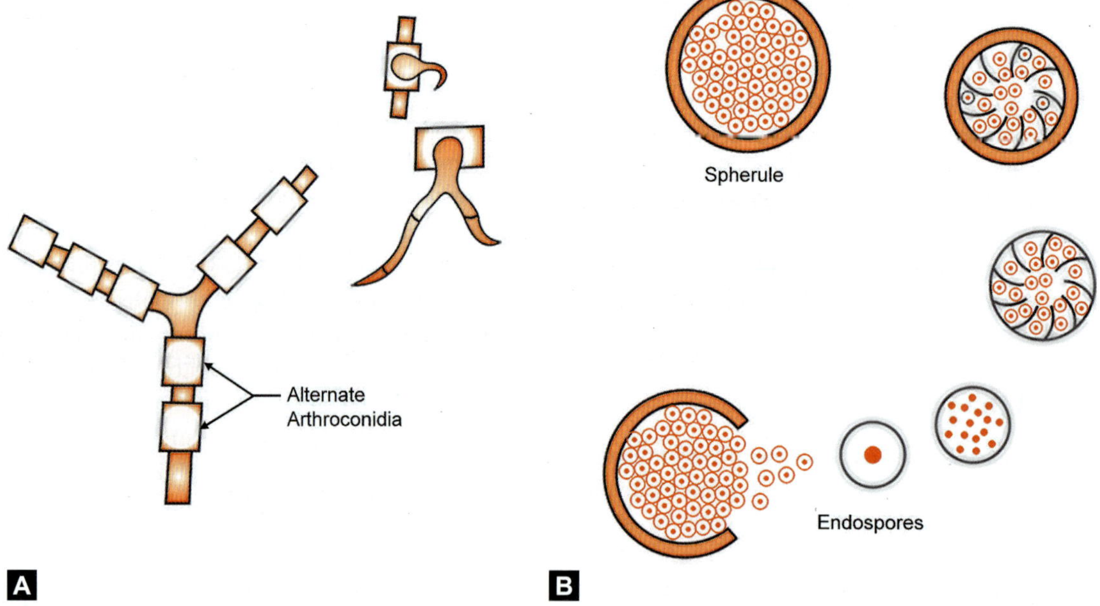

Figs. 18.5A and B. *Coccidioides* species showing arthroconidia in mycelial phase at 25°C and spherules in tissue phase 37°C.

forms that may be confused with other fungal pathogens. The other organisms, which give spherule-like appearance, are *Emmonsia* species, *R.seeberi* and *Prototheca* species.

The hyphal form has also been occasionally found in a few case reports in direct examination. A report by Kaufman *et al,* brings attention to potentially diverse morphological forms of this organism, which may be demonstrated in tissues. Within a single patient, they have observed forms resembling *Prototheca* and *B.dermatitidis* as well as cells forming germ tubes, branching septate and moniliform hyphae. None of later structure was pathognomonic for any particular fungal pathogen. Ultimately diagnosis was established by immunofluorescence staining with *C.immitis*-specific conjugate. It points out the value of using multiple methods for establishing diagnosis of specific fungal infection. The importance of conserving tissue

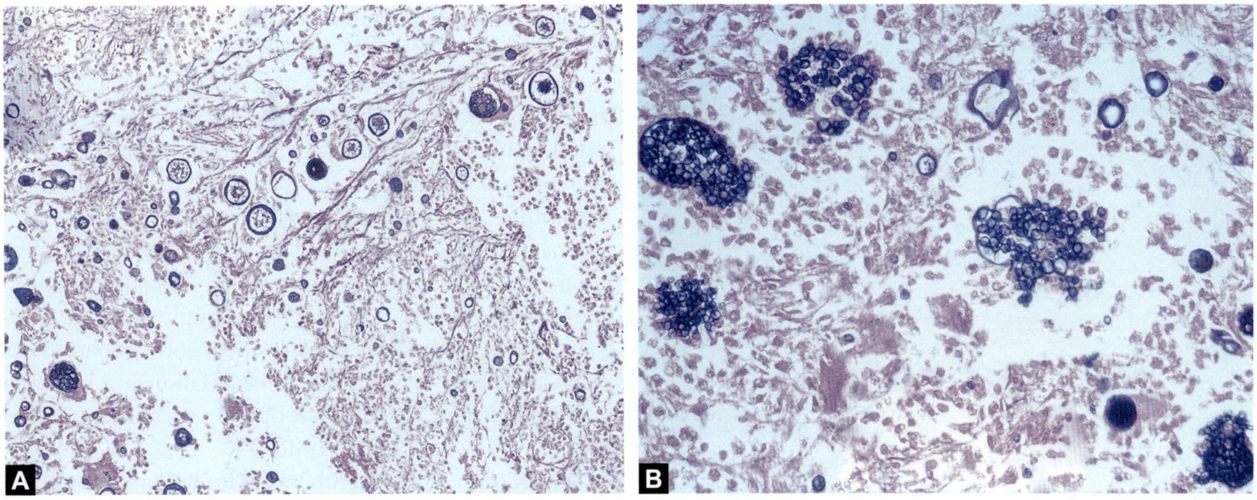

Figs. 18.6A and B. *Coccidioides* species showing spherules in tissue phase at body temperature 37°C (H&E × 200 / 400).

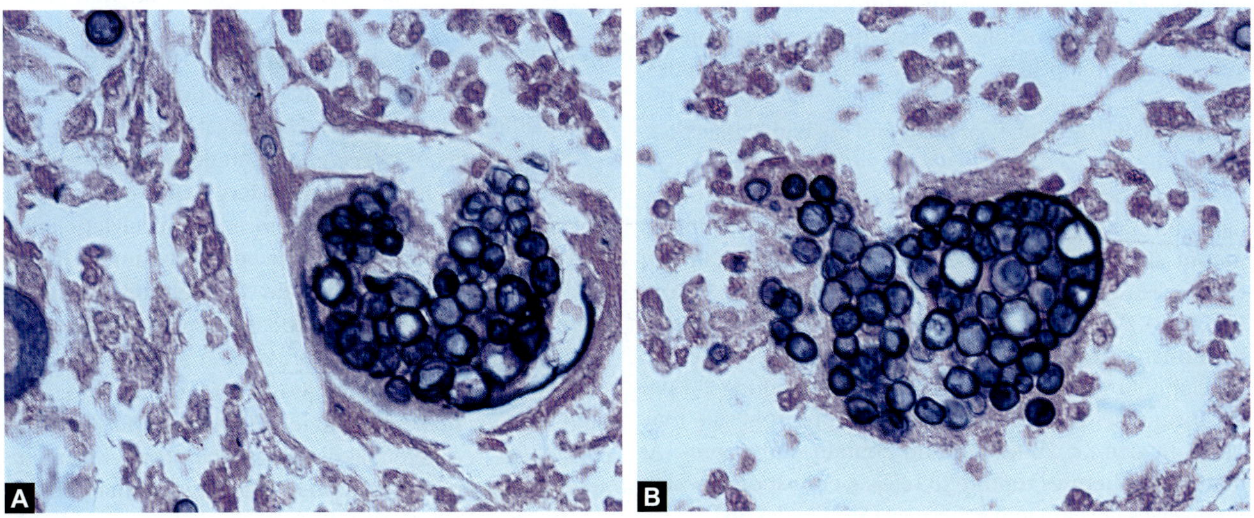

Figs. 18.7A and B. *Coccidioides* spherules in tissue phase at body temperature 37°C just before and after rupture (H&E × 1000).

for culture cannot be over-emphasized. The laboratory investigators must be aware of diverse manifestations of coccidioidomycosis in tissues.

(b) Fungal Culture

The recovery of *C.immitis* or *C.posadasii* on culture remains the 'gold standard' for a definitive diagnosis. The infected clinical material should be collected and cultivated in well-stoppered narrow-neck bottles or culture tube slants containing Sabouraud dextrose agar, blood agar and brain-heart infusion agar with actidione. The colonies at 25°C appear within 3-5 days, which are initially moist and smooth but rapidly become suede-like to downy greyish white with tan to brown underside. The tan cottony colonies develop moderately rapid as moist membranous culture at first and later on have abundant aerial mycelia but arthroconidia may not be apparent by ten days or so. The culture of *Coccidioides* species differs from other dimorphic fungi because under standard laboratory conditions, it grows as mold at 25°C as well as 37°C.

They form alternate, thick walled, characteristically barrel-shaped arthroconidia of 2 × 4 μm size and empty cells. The alternate cells between arthroconidia undergo

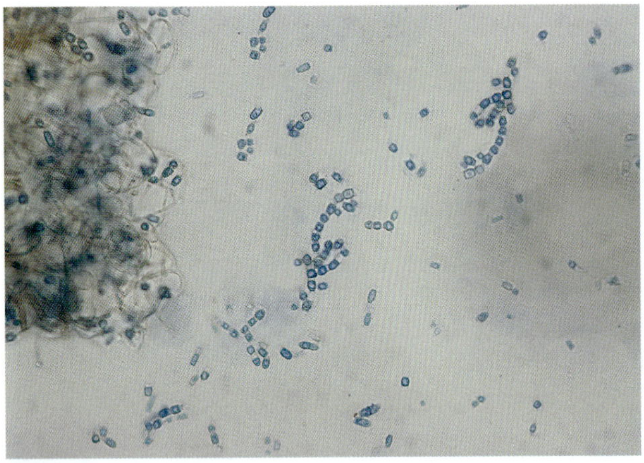

Fig. 18.8. *Coccidioides* species showing alternate arthroconidia during mycelial phase of fungus grown at 25°C (LCB x 400).

autolysis to release intervening arthroconidia **(rhexolytic type)**. The arthroconidia are mature infectious propagules that develop from alternate cells on hyphal stalks. The hyphae fragment easily and release the conidia. The arthroconidia are light, float in air and are highly infectious in nature.

For laboratory demonstration, use killed culture by using formalin before making any preparations. The LCB mount shows branching septate hyphae and chains of thick-walled rectangular arthroconidia **(*See* Figs. 18.5A and Fig. 18.8)**. When they are inoculated into animals or inhaled by human, these conidia become spherical, enlarge and develop into spherules. The endospores measuring 2-4 µm are formed within spherule and fill it. The spherule may range in size *i.e.* 20-80 µm and contain endospores. At maturity, spherules rupture to release endospores, which themselves develop into spherules **(*See* Fig. 18.5B)**.

Out of four dimorphic true pathogenic fungi causing systemic mycosis, *Coccidioides* species grow most quickly, taking about 4-5 days, whereas *H.capsulatum*, *B.dermatitidis* and *P.brasiliensis* grow slowly taking about 2-3 weeks. Since the arthroconidia of *Coccidioides* species are highly infectious hence all requisite precautions, like use of biosafety hood in the laboratory, should be taken seriously.

In both mycelial and parasitic forms, the dimorphic fungal species of *Coccidioides* show wide morphological variations. Some of these forms may be even mistaken for other organisms. The variable manifestations of this fungus have potential to establish misdiagnosis.

The exoantigen test, although old, may be employed as confirmatory test for coccidioidomycosis by using heat stable (**HS**) or heat labile (**HL** or **F**) antigens. The antibodies to specific heat-labile antigens (HL or F) may be detected. The exoantigen test has proven to be highly sensitive and specific for identifying *Coccidioides* species in its typical or atypical mold forms. The slant extracts when concentrated to 25 times, are excellent source of diagnostic exoantigen. The newer molecular based techniques, particularly DNA probes, are now replacing exoantigen tests.

Invariably *Coccidioides* is confused with look-alike fungi, which also present with similar mycelial morphological features. The commonest example is non-pathogenic fungus, *Malbranchea* species. Sequences of partial 18S rDNA gene can demonstrate close relationship between *Coccidioides* and *Malbranchea* due to production of alternate, barrel-shaped arthroconidia.

In addition, other members of family Gymnoascaceae in order Onygenales of class Plectomycetes in phylum Ascomycota like *Auxarthron*, *Uncinocarpus* and *Arachniotus*, may also confuse with *Coccidioides*, which are non-pathogenic. Similarly arthroconidia of *Geotrichum candidum* may be confused with that of *Coccidioides*. The conversion of mycelial-to-spherule form in experimental animal can finally differentiate between two fungal species.

The cultures of *Coccidioides* should **never** be **sniffed** while examining isolates in laboratory as it is one of highly infectious fungal pathogens. All procedures involving clinical specimens and cultures must be performed under Biohazard Safety Level 3 precautions. The cultures should be done in screw-capped tubes and not in petri dishes.

The mycelial forms of *Coccidioides* are not limited to visceral lesions, bronchial washings and cerebrospinal fluid but may present in cutaneous lesions and should be considered in differential diagnosis when hyphal elements are seen in skin tissues upon histopathologic analysis.

The two genetically distinct species of *Coccidioides* cannot be distinguished phenotypically or clinically hence molecular techniques should be adopted to make species level diagnosis. The nucleic acid amplification tests (NAAT) are also introduced for detection of *Coccidioides* species in specimens.

(c) Immunodiagnosis

At present there are two sources of antigen, both cell free culture filtrates and are used in serological reactions. These are mycelial phase (**coccidioidin**) and spherule phase (**spherulin**) of *Coccidioides* species. Coccidioidin was used until 1980s, when it was replaced by spherulin. For epidemiological purposes, difference between these two reagents is negligible.

The coccidioidin is sterile filtrate of broth in which mycelia are grown, whereas spherulin is sterile filtrate of broth in which spherules are grown. These materials give positive skin tests (in dilution up to 1:10,000) in infected individuals. Same antigens are used in ID, LPA, CFT and other tests. However, in low dilution of 1:10, these antigens cross-react with other fungal antigens. Some antisera give highly specific immunofluorescence tests with spherules in tissues. The latest and perhaps best test for detection of IgM antibodies is ELISA, which is both sensitive and specific.

(i) Skin Test: The skin test is done by intradermal inoculation of coccidioidin. A test is considered as positive when zone size of reaction is >5 mm in 24-48 hours. However, it does not distinguish between present and past infection and the negative skin test does not actually rule out infection caused by *Coccidioides* species. The skin test is useful although cross-reaction may occur in patients with histoplasmosis and blastomycosis. Routine skin testing with coccidioidin in areas where coccidioidomycosis is endemic is not predictive of this disease and should not be performed.

(ii) Serological Tests: The serological tests are mainly based on detection of **antibodies** in suspected patients of coccidioidomycosis. All these tests play significant role in establishing an early diagnosis. The precipitin test is most useful for detection of early, primary infection or exacerbation of existing disease. The precipitins appear 1-3 weeks after infection but are seldom detectable after 2-6 weeks or in patients with disseminated coccidioidomycosis. The latex particle agglutination (LPA) test gives similar results to that of precipitin test but it is not very specific. The complement fixation test (CFT) is most useful for detecting disseminated disease. The complement-fixing antibodies appear 2 to 3 months after infection and persist until recovery or till survival of the patient.

The tests for **antigen** detection have not yet proven useful diagnostically and are not available for clinical use. The **DNA probes** for *Coccidioides* species like that of *B.dermatitidis* have also been evaluated as useful in diagnostic armamentarium.

An antigen of *Coccidioides* species that cross-reacts in *Histoplasma* antigen assay is present in urine of patients who have coccidioidomycosis. These findings support potential for use of *Histoplasma* antigen test for diagnosis of coccidioidomycosis in areas of endemicity for *Coccidioides* species.

The differentiation between species into *C.immitis* and *C.posadasii* is best achieved by specialized molecular techniques, which are normally available in advanced clinical diagnostic laboratories.

(d) Animal Pathogenicity

Many domestic and native animals are susceptible to both species of *Coccidioides* like mice, rats, guinea pigs, primates and dogs. However, the primary animal used for experimental infection for these fungi remains the mouse. The developmental cycle of parasitic phase is identical in animal and human tissues. There are five routes of infection used in these murine models. They include intranasal instillation, intraperitoneal, intravenous, intrathecal inoculation of arthroconidia and intracranial injection of endospores. In animal models of infection, inhalation of as few as 10 spores is sufficient to cause infection. Intraperitoneal injections of arthroconidia and hyphal fragments from virulent strains can produce fatal disease in three weeks of inoculation. Animal inoculation may also be used for primary isolation and culture confirmation of *Coccidioides* species. As there is development of life-long immunity after infection with these fungi hence many investigators have used murine models to address queries relating to immunology of host-resistance such as immunoprophylaxis and cytokine expression of disease. Therefore, the use of animal models has contributed greatly to an overall understanding of this disease with respect to its pathogenesis, how the host responds, prevention as well as treatment.

Treatment and Prophylaxis

The IDSA has released its revised treatment guidelines for coccidioidomycosis in 2016 replacing the previous ones of year 2000 and 2005. The decision, whether to treat disease or not and type of therapy, depends on stage of infection, recognized risk factors that influence disease and nature of lesions. Most of patients of coccidioidomycosis recover without treatment and have life-long immunity. The management involves defining extent of infection and assessing host factors that predispose to disease severity. The cases with relatively localized acute pulmonary infections and no circumstantial risk factors of complications usually require only periodic re-assessment to demonstrate resolution of the self-limiting process. On the other hand, patients with extensive spread of infection or at high risk of complications because of immunosuppression or other pre-existing factors, require variety of treatment modalities, which may include antifungal therapy, surgical debridement and/or even both. Surgical drainage of abscesses in bones or joints is commonly performed.

There are several antifungal drugs like, amphotericin B, indicated for rapidly progressive disease and azoles for chronic, mild to moderate disease. Miconazole and ketoconazole were used for treatment of coccidioidomycosis. Ketoconazole offered more convenient and cost-effective alternative for ambulatory patients. The dose of ketoconazole was 200-400 mg/day for 6 months. It is now not being used due to side effects. The triazoles like fluconazole, itraconazole, voriconazole and posaconazole have shown efficacy with minimal toxicity. The prognosis is poor in disseminated cases particularly in immunocompromised patients with meningeal manifestations. The meningeal involvement is particularly difficult to treat like joint infections, partly because of poor penetration of amphotericin B into the CSF.

When oral azole antifungals were introduced, they provided safe and effective alternative to intravenous therapy in patients with disseminated coccidioidomycosis. Initially with ketoconazole and later on with triazoles *i.e.* fluconazole and itraconazole, it was clear that with oral therapy some patients with meningitis could achieve both remission of symptoms and changes in CSF toward normal. The posaconazole is used successfully in cases of disseminated non-meningeal coccidioidomycosis that previously failed to respond to other azoles and Amphotericin B or even both. The syrup of posaconazole (400 mg BD) has been recommended as a possible treatment option in fluconazole failure and it is now available as tablets and intravenous injections in some countries.

The treatment for CNS involvement has been successfully accomplished with itraconazole or fluconazole, avoiding necessity of administering intrathecal amphotericin B. The adjunctive interferon-gamma (IFN-γ) is used in successful treatment of severe coccidioidomycosis. Combination therapy with caspofungin and fluconazole may be an alternative treatment for diffuse coccidioidal pneumonia with non-meningeal extrapulmonary involvement that is unresponsive to amphotericin B therapy. However, determining best course of therapy remains a challenge. Further clinical studies evaluating this combination are required.

The coccidioidomycosis has long been recognized as one of the most difficult endemic mycosis to treat. The rates of relapse after apparently successful treatment with amphotericin B have not been well-defined but relapse has been recognized as challenge since drug was first used in the treatment of coccidioidomycosis in late 1950s. Hence relapse remains one of the most significant problems in treatment of this disease. Serial complement fixation titers determination after therapy is indicated for patients with coccidioidomycosis and rising titer warrant administration of therapy, even in absence of evident disease, since relapse is likely.

Primary prophylaxis for HIV-infected persons who live in areas where coccidioidomycosis is endemic is not routinely recommended. The patients who complete initial therapy of coccidioidomycosis should be administered life-long suppressive therapy *i.e.* secondary prophylaxis or long-term maintenance therapy, either 400 mg of oral fluconazole per day or 200 mg of itraconazole twice a day.

There is need to increase awareness about coccidioidomycosis among physicians and travellers to regions, where *Coccidioides* species are endemic, through publications in primary care and especially medical journals, presentation at national and regional professional meetings, posting information on Internet and adding disease to list of other known travel-related health risks in bulletins, magazines and books. The measures recommended for protection of workers against histoplasmosis and coccidioidomycosis may be available from their respective centers.

The detailed information about disease is available at **Valley Fever Center for Excellence** (VFCE) through electronic or print media, which was established in 1995 at Tucson, Arizona to promote education, research and patient-care for coccidioidomycosis. It provides information to public, physician consultations with VFCE physicians, referrals for patients and also promotes research into various aspects of disease.

The risk of exposure to coccidioidomycosis is eminent particularly where disease is endemic. Therefore, vaccine for coccidioidomycosis may be useful particularly in these areas. Vaccination of persons at risk of contracting coccidioidomycosis is a feasible approach to control this insidious fungal disease. The rationale for immunoprotection is based on observation that natural infection by *Coccidioides* species almost always confers life-long immunity to this disease.

Further Reading

1. Adam RD, Elliott SP, Taljanovic MS. The spectrum and presentation of disseminated coccidioidomycosis. *Am J Med.* 2009; **122:** 770-7.

2. Al-Daraji WI, Al-Mahmoud RM, Ali MA. Disseminated coccidioidomycosis: A case report from the United kingdom. *Int J Clin Exp Pathol.* 2009; **2:** 494-7.
3. Alavi K, Atla PR, Haq T, *et al.* Coccidioidomycosis masquerading as eosinophilic ascites. *Case Rep Gastrointest Med.* 2015; 891910. PMID: 26266062.
4. Alfreijat M, Wilhelmi B. A case of a positive *Coccidioides* stool culture in an immunocompetent patient with disseminated coccidioidomycosis. *IDCases.* 2017; **8:** 89-91.
5. Allen JE, Belafsky PC. Laryngeal coccidioidomycosis with vocal fold paralysis. *Ear Nose Throat J.* 2011; **90:** e1-5.
6. Aly FZ, Millius R, Sobonya R, *et al.* Cytologic diagnosis of coccidioidomycosis: Spectrum of findings in southern Arizona patients over a 10 year period. *Diagn Cytopathol.* 2016; **44:** 195-200.
7. Ampel NM, Giblin A, Mourani JP, *et al.* Factors and outcomes associated with the decision to treat primary pulmonary coccidioidomycosis. *Clin Infect Dis.* 2009; **48:** 172-8.
8. Ampel NM, Hector RF. Measuring cellular immunity in coccidioidomycosis: The time is now. *Mycopathologia.* 2010; **169:** 425-6.
9. Ampel NM. Coccidioidomycosis in persons infected with HIV-1. *Ann N Y Acad Sci.* 2007; **1111:** 336-42.
10. Ampel NM. Coccidioidomycosis: A review of recent advances. *Clin Chest Med.* 2009; **30:** 241-51.
11. Ampel NM. New perspectives on coccidioidomycosis. *Proc Am Thorac Soc.* 2010; **7:** 181-5.
12. Ampel NM. The complex immunology of human coccidioidomycosis. *Ann N Y Acad Sci.* 2007; **1111:** 245-58.
13. Ampel NM. The diagnosis of coccidioidomycosis. *F1000 Med Rep.* 2010; **2:** doi: 10.3410/M2-2. PMID: 20948866.
14. Ampel NM. The treatment of coccidioidomycosis. *Rev Inst Med Trop Sao Paulo.* 2015; **57 (Suppl. 19):** 51-6.
15. Ampel NM. What's behind the increasing rates of coccidioidomycosis in Arizona and California? *Curr Infect Dis Rep.* 2010; **12:** 211-6.
16. Anderson N, Ledeboer NA. Things are not always what they appear to be. *J Clin Microbiol.* 2011; **49:** 3727 & 4030.
17. Angelo KM, Nnedu ON. Rare manifestations of coccidioidomycosis. *J La State Med Soc.* 2013; **165:** 137-9.
18. Ansari Pirsaraei P, Tabsh K, Lentz J. Rare dysuria: Prostatic abscess due to disseminated coccidioidomycosis. *Urol Case Rep.* 2016; **9:** 12-4.
19. Anstead GM, Graybill JR. Coccidioidomycosis. *Infect Dis Clin N Am.* 2006; **20:** 621-43.
20. Antony SJ, Parikh MS, Friedman G. Coccidioidomycosis involving the cranium: A case report and review of current literature. *Infect Disord Drug Targets.* 2015; **15:** 202-6.
21. Arbeloa-Gutierrez L, Kuberski T, Johnson SM, *et al.* Reactivation of coccidioidomycosis: A prosthetic joint infection in Spain. *Eur J Clin Microbiol Infect Dis.* 2016; **35:** 183-6.
22. Arora NP, Taneja V, ReyesSacin C, *et al.* Coccidioidomycosis masquerading as malignancy. *BMJ Case Rep.* 2012; pii: bcr1220115357. PMID: 22665569.
23. Austen S, van der Weegen W, Verduin CM, *et al.* Coccidioidomycosis infection of a total knee arthroplasty in a nonendemic region. *J Arthroplasty.* 2013; **28:** 375.e13-5.
24. Awasthi S. Dendritic cell-based vaccine against *Coccidioides* infection. *Ann N Y Acad Sci.* 2007; **1111:** 269-74.
25. Azadeh N, Chang YH, Kusne S, *et al.* The impact of early and brief corticosteroids on the clinical course of primary pulmonary coccidioidomycosis. *J Infect.* 2013; **67:** 148-55.
26. Azadeh N, Rank MA, Lewis JC, *et al.* Wheezes and desert breezes: When asthma and valley fever collide. *J Asthma.* 2016; **53:** 125-32.
27. Baden LR, Digumarthy SR, Guimaraes AS, *et al.* A 60-year-old male renal-transplant recipient with renal insufficiency, diabetic ketoacidosis and mental status changes. *N Engl J Med.* 2009; **361:** 1980-9.
28. Bajema KL, Dalesandro MF, Fredricks DN, *et al.* Disseminated coccidioidomycosis presenting with intramedullary spinal cord abscesses: Management challenges. *Med Mycol Case Rep.* 2017; **15:** 1-4.
29. Bamberger DM, Pepito BS, Proia LA, *et al.* Cerebrospinal fluid *Coccidioides* antigen testing in the diagnosis and management of central nervous system coccidioidomycosis. *Mycoses.* 2015; **58:** 598-602.
30. Baptista-Rosas RC, Hinojosa A, Riquelme M. Ecological niche modeling of *Coccidioides* spp. in western North American deserts. *Ann N Y Acad Sci.* 2007; **1111:** 35-46.
31. Benedict K, Purfield AE, Mohle-Boetani J, *et al.* Awareness and environmental exposures related to coccidioidomycosis among inmates at two California prisons, 2013. *J Correct Health Care.* 2016; **22:** 157-63.
32. Bercovitch RS, Catanzaro A, Schwartz BS, *et al.* Coccidioidomycosis during pregnancy: A review and recommendations for management. *Clin Infect Dis.* 2011; **53:** 363-8.
33. Berg N, Ryscavage P, Kulesza P. The utility of fine needle aspiration for diagnosis of extrapulmonary coccidioidomycosis: A case report and discussion. *Clin Med Res.* 2011; **9:** 130-3.
34. Berli JU, Campbell WN, Katz RD. Coccidioidomycosis causing osteomyelitis of the hand in an immunocompetent patient. *Hand (NY).* 2015; **10:** 562-4.
35. Berry CD, Stevens DA, Hassid EI, *et al.* A new method for the treatment of chronic fungal meningitis: Continuous infusion into the cerebrospinal fluid for coccidioidal meningitis. *Am J Med Sci.* 2009; **338:** 79-82.
36. Beshoy Y, Nneji J, Buxbaum J. A case of the hiccups in the setting of colonic ulcers. *Gastroenterology.* 2015; **148:** e8-9.
37. Bhardwaj H, Bhardwaj B, Levin DC, *et al.* An unusual case of pulmonary coccidioidomycosis presenting as pneumothorax. *J Okla State Med Assoc.* 2013; **106:** 315 & 317-8.
38. Binnicker MJ, Popa AS, Catania J, *et al.* Meningeal coccidioidomycosis diagnosed by real-time polymerase chain reaction analysis of cerebrospinal fluid. *Mycopathologia.* 2011; **171:** 285-9.
39. Bird GR, Libke RD, Billelo JF, *et al.* Disseminated coccidioidomycosis with clinically evident splenomegaly in an immunocompetent host, First case reported in the literature. *Mcgill J Med.* 2009; **12:** 14.

40. Blair JE, Chang YH, Cheng MR, *et al.* Characteristics of patients with mild to moderate primary pulmonary coccidioidomycosis. *Emerg Infect Dis.* 2014; **20:** 983-90.
41. Blair JE, Chang YH, Ruiz Y, *et al.* Distance from construction site and risk for coccidioidomycosis, Arizona, USA. *Emerg Infect Dis.* 2014; **20:** 1464-71.
42. Blair JE, Coakley B, Santelli AC, *et al.* Serologic testing for symptomatic coccidioidomycosis in immunocompetent and immunosuppressed hosts. *Mycopathologia.* 2006; **162:** 317-24.
43. Blair JE, Mayer AP, Currier J, *et al.* Coccidioidomycosis in elderly persons. *Clin Infect Dis.* 2008; **47:** 1513-8.
44. Blair JE. Coccidioidal meningitis: Update on epidemiology, clinical features, diagnosis, and management. *Curr Infect Dis Rep.* 2009; **11:** 289-95.
45. Blair JE. Coccidioidomycosis in patients who have undergone transplantation. *Ann N Y Acad Sci.* 2007; **1111:** 365-76.
46. Blair JE. State-of-the-art treatment of coccidioidomycosis skeletal infections. *Ann N Y Acad Sci.* 2007; **1111:** 422-33.
47. Blodget E, Geiseler PJ, Larsen RA, *et al.* Donor-derived *Coccidioides immitis* fungemia in solid organ transplant recipients. *Transpl Infect Dis.* 2012; **14:** 305-10.
48. Bonifaz A, Vázquez-González D, Perusquía-Ortiz AM. Endemic systemic mycoses: Coccidioidomycosis, histoplasmosis, paracoccidioidomycosis and blastomycosis. *J Dtsch Dermatol Ges.* 2011; **9:** 705-14; Quiz 715.
49. Borchers AT, Gershwin ME. The immune response in coccidioidomycosis. *Autoimmun Rev.* 2010; **10:** 94-102.
50. Bradley G, Adger H Jr, Splinter A, *et al.* Index of suspicion. *Pediatr Rev.* 2010; **31:** 117-23.
51. Bravo R, Pelayo-Katsanis LO, Shehab ZM, *et al.* Diagnostic and treatment challenges for the pediatric hematologist oncologist in endemic areas for coccidioidomycosis. *J Pediatr Hematol Oncol.* 2012; **34:** 389-94.
52. Breedlove B. Stillness, Light and distance. *Emerg Infect Dis.* 2017; **23:** 372-3.
53. Brown J, Benedict K, Park BJ, *et al.* Coccidioidomycosis: Epidemiology. *Clin Epidemiol.* 2013; **5:** 185-97.
54. Brugiere O, Forget E, Biondi G, *et al.* Coccidioidomycosis in a lung transplant recipient acquired from the donor graft in France. *Transplantation.* 2009; **88:** 1319-20.
55. Buhrow SM. Coccidioidomycosis: A differential diagnosis for visitors to the southwest. *Am J Nurs.* 2013; **113:** 52-5.
56. Buijze GA, Kok P, Botha-Scheepers SA, *et al.* Fungal disease of the Western hemisphere: A patient with coccidioidomycosis. *Eur J Intern Med.* 2009; **20:** 319-22.
57. Cadena J, Hartzler A, Hsue G, *et al.* Coccidioidomycosis and tuberculosis co-infection at a tuberculosis hospital: Clinical features and literature review. *Medicine (Baltimore).* 2009; **88:** 66-76.
58. Campbell M, Kusne S, Renfree KJ, *et al.* Coccidioidal tenosynovitis of the hand and wrist: Report of 9 cases and review of the literature. *Clin Infect Dis.* 2015; **61:** 1514-20.
59. Cannella AP, Vinetz JM. A young man evaluated for suspicion of lymphoma. *Am J Trop Med Hyg.* 2014; **91:** 440-1.
60. Canteros CE, Velez H A, Toranzo AI, *et al.* Molecular identification of *Coccidioides immitis* in formalin-fixed, paraffin-embedded (FFPE) tissues from a Colombian patient. *Med Mycol.* 2015; **53:** 520-7.
61. Capilla J, Clemons KV, Liu M, *et al.* *Saccharomyces cerevisiae* as a vaccine against coccidioidomycosis. *Vaccine.* 2009; **27:** 3662-8.
62. Capoor MR, Sen B, Varshney P, *et al.* Coccidioidomycosis masquerading as skeletal tuberculosis: An imported case and review of coccidioidomycosis in India. *Trop Doct.* 2014; **44:** 25-8.
63. Carmichael JK. Coccidioidomycosis in HIV-infected persons. *Clin Infect Dis.* 2006; **42:** 1059-60.
64. Carpenter JB, Feldman JS, Leyva WH, *et al.* Clinical and pathologic characteristics of disseminated cutaneous coccidioidomycosis. *J Am Acad Dermatol.* 2010; **62:** 831-7.
65. Carvalho C, Ferreira I, Gaiao S, *et al.* Cerebral coccidioidomycosis after renal transplantation in a non-endemic area. *Transpl Infect Dis.* 2010; **12:** 151-4.
66. Castanon-Olivares LR, Laniado-Laborin R, Concepcion T, *et al.* Clinical comparison of two Mexican coccidioidins. *Mycopathologia.* 2010; **169:** 427-30.
67. Castro S, Bernardes I. Coccidioidal cerebral abscess with peripheral restricted diffusion. *J Neuroradiol.* 2009; **36:** 162-4.
68. Castro-Lopez N, Hung CY. Immune response to coccidioidomycosis and the development of a vaccine. *Microorganisms.* 2017; 5. pii: e13. PMID: 28300772.
69. Catanzaro A. Coccidioidomycosis. *Semin Respir Crit Care Med.* 2004; **25:** 123-8.
70. Centers for Disease Control and Prevention (CDC). Increase in reported coccidioidomycosis - United States, 1998-2011. *MMWR.* 2013; **62:** 217-21.
71. Centers for Disease Control and Prevention (CDC). Increase in Coccidioidomycosis - California, 2000-2007. *MMWR.* 2009; **58:** 105-9.
72. Chan O, Low SW, Urcis R, *et al.* Coccidioidomycosis with pericardial involvement: Case report and literature review. *Am J Med.* 2016; **129:** e21-5.
73. Char DH, Crawford JB, Bertolucci G, *et al.* Intraocular coccidioidomycosis simulating a neoplasm. *Br J Ophthalmol.* 2012; **96:** 218-9.
74. Chastain CA, Cox GM, Powers MA. Dyspnea and hypoxia in a 64-year-old male traveler. *Chest.* 2012; **142:** 532-5.
75. Chaudhary S, Meinke L, Ateeli H, *et al.* Coccidioidomycosis among persons undergoing lung transplantation in the coccidioidal endemic region. *Transpl Infect Dis.* 2017; doi: 10.1111/tid.12713. PMID: 28452423.
76. Chen S, Erhart LM, Anderson S, *et al.* Coccidioidomycosis: Knowledge, attitudes and practices among healthcare providers - Arizona, 2007. *Med Mycol.* 2011; **49:** 649-56.
77. Chiller TM, Galgiani JN, Stevens DA. Coccidioidomycosis. *Infect Dis Clin N Am.* 2003; **17:** 41-57.
78. Christian MD. Biowarfare and bioterrorism. *Crit Care Clin.* 2013; **29:** 717-56.
79. Chung CR, Lee YC, Rhee YK, *et al.* Pulmonary coccidioidomycosis with peritoneal involvement mimicking lung cancer with peritoneal carcinomatosis. *Am J Respir Crit Care Med.* 2011; **183:** 135-6.

80. Clemons KV, Antonysamy MA, Danielson ME, *et al.* Whole glucan particles as a vaccine against systemic coccidioidomycosis. *J Med Microbiol.* 2015; **64:** 1237-43.
81. Clemons KV, Capilla J, Stevens DA. Experimental animal models of coccidioidomycosis. *Ann N Y Acad Sci.* 2007; **1111:** 208-24.
82. Cole GT, Hung CY, Sanderson SD, *et al.* Novel strategies to enhance vaccine immunity against coccidioidomycosis. *PLoS Pathog.* 2013; **9:** e1003768.
83. Cole GT, Hurtgen BJ, Hung CY. Progress toward a human vaccine against coccidioidomycosis. *Curr Fungal Infect Rep.* 2012; **6:** 235-44.
84. Collins J, Dy M, Clemens C, *et al.* Not just a simple pneumothorax. *Pediatr Infect Dis J.* 2011; **30:** 361 & 364.
85. Colson AJ, Vredenburgh L, Guevara RE, *et al.* Large-scale land development, fugitive dust and increased coccidioidomycosis incidence in the Antelope Valley of California, 1999-2014. *Mycopathologia.*2017; **182:** 439-58.
86. Compton F, Everett J, Wanger A, *et al.* Coccidioidomycosis presented as a pelvic mass in an otherwise healthy female. *Ann Clin Lab Sci.* 2015; **45:** 585-7.
87. Cordeiro Rde A, Brilhante RS, Rocha MF, *et al.* Twelve years of coccidioidomycosis in Ceara State, Northeast Brazil: Epidemiologic and diagnostic aspects. *Diagn Microbiol Infect Dis.* 2010; **66:** 65-72.
88. Cordeiro Rde A, e Silva KR, Brilhante RS, *et al. Coccidioides posadasii* infection in bats, Brazil. *Emerg Infect Dis.* 2012; **18:** 668-70.
89. Cordeiro Rde A, Fechine MA, Brilhante RS, *et al.* Serologic detection of coccidioidomycosis antibodies in northeast Brazil. *Mycopathologia.* 2009; **167:** 187-90.
90. Corpolongo A, Giancola ML, Ghirga P, *et al.* An imported case of acute pulmonary coccidioidomycosis in an Italian traveller. *Infection.* 2014; **42:** 921-4.
91. Cox RA, Magee DM. Coccidioidomycosis: Host response and vaccine development. *Clin Microbiol Rev.* 2004; **17:** 804-39.
92. Crawford P, Peterson T, Honeycutt D. Clinical inquiries: Which risk factors and signs and symptoms are associated with coccidioidomycosis? *J Fam Pract.* 2014; **63:** 747 & 756.
93. Crum NF, Ballon-Landa G. Coccidioidomycosis in pregnancy: case report and review of the literature. *Am J Med.* 2006; **119:** 993.e11-7.
94. Crum-Cianflone NF. Coccidioidomycosis in the US military: A review. *Ann N Y Acad Sci.* 2007; **1111:** 112-21.
95. Cummings KC, McDowell A, Wheeler C, *et al.* Point-source outbreak of coccidioidomycosis in construction workers. *Epidemiol Infect.* 2010; **138:** 507-11.
96. D'Avino A, Di Giambenedetto S, Fabbiani M, *et al.* Coccidioidomycosis of cervical lymph nodes in an HIV-infected patient with immunologic reconstitution on potent HAART: A rare observation in a nonendemic area. *Diagn Microbiol Infect Dis.* 2012; **72:** 185-7.
97. Das R, McNary J, Fitzsimmons K, *et al.* Occupational coccidioidomycosis in California: Outbreak investigation, respirator recommendations and surveillance findings. *J Occup Environ Med.* 2012; **54:** 564-71.
98. de Aguiar Cordeiro R, Patoilo KR, Praciano SB, *et al.* Antigens of *Coccidioides posadasii* as an important tool for the immunodiagnosis of coccidioidomycosis. *Mycopathologia.* 2013; **175:** 25-32.
99. de Macedo RC, Rosado AS, da Mota FF, *et al.* Molecular identification of *Coccidioides* spp. in soil samples from Brazil. *BMC Microbiol.* 2011; **11:** 108.
100. de Perio MA, Niemeier RT, Burr GA. *Coccidioides* exposure and coccidioidomycosis among prison employees, California, United States. *Emerg Infect Dis.* 2015; **21:** 1031-3.
101. Del Rocio Reyes-Montes M, Perez-Huitron MA, Ocana-Monroy JL, *et al.* The habitat of *Coccidioides* spp. and the role of animals as reservoirs and disseminators in nature. *BMC Infect Dis.* 2016; **16:** 550.
102. Deresinski S. *Coccidioides immitis* as a potential bioweapon. *Semin Respir Infect.* 2003; **18:** 216-9.
103. Desai NR, McGoey R, Troxclair D, *et al.* Coccidioidomycosis in nonendemic area: case series and review of literature. *J La State Med Soc.* 2010; **162:** 97-103.
104. Deus Filho Ad. Coccidioidomycosis. *J Bras Pneumol.* 2009; **35:** 920-30.
105. DiCaudo DJ. Coccidioidomycosis: A review and update. *J Am Acad Dermatol.* 2006; **55:** 929-45.
106. DiCaudo DJ. Coccidioidomycosis. *Semin Cutan Med Surg.* 2014; **33:** 140-5.
107. Dierberg KL, Marr KA, Subramanian A, *et al.* Donor-derived organ transplant transmission of coccidioidomycosis. *Transpl Infect Dis.* 2012; **14:** 300-4.
108. Dimitrova D, Mason WH, Shaham B. Monoarticular coccidioidal synovitis in a pediatric patient. *Pediatr Infect Dis J.* 2014; **33:** 100-2.
109. Dimitrova D, Ross L. Coccidioidomycosis: Experience from a children's hospital in an area of endemicity. *J Pediatric Infect Dis Soc.* 2016; **5:** 89-92.
110. Downie BA, Rosser G, Atkin CM, *et al.* An unusual cause of a pulmonary opacity in a visitor to the UK. *BMJ Case Rep.* 2012; pii: bcr2012006706. PMID: 22879005.
111. Drake KW, Adam RD. Coccidioidal meningitis and brain abscesses: Analysis of 71 cases at a referral center. *Neurology.* 2009; **73:** 1780-6.
112. Duarte-Escalante E, Frias-De-Leon MG, Zuniga G, *et al.* Molecular markers in the epidemiology and diagnosis of coccidioidomycosis. *Rev Iberoam Micol.* 2014; **31:** 49-53.
113. Duarte-Escalante E, Zuniga G, Frias-De-Leon MG, *et al.* AFLP analysis reveals high genetic diversity but low population structure in *Coccidioides posadasii* isolates from Mexico and Argentina. *BMC Infect Dis.* 2013; **13:** 411.
114. Dubey D, Narayan RN, Motiwala A, *et al.* Spherules in spine: Vertebral coccidioidomycosis. *Neurology.* 2014; **83:** e158.
115. Duggan PT, Deegan AP, McDonnell TJ. Case of coccidioidomycosis in Ireland. *BMJ Case Rep.* 2016; pii: bcr2016215898.
116. Duplessis CA, Tilley D, Bavaro M, *et al.* Two cases illustrating successful adjunctive interferon-γ immunotherapy in refractory disseminated coccidioidomycosis. *J Infect.* 2011; **63:** 223-8.

117. Durkin M, Connolly P, Kuberski T, *et al.* Diagnosis of coccidioidomycosis with use of the *Coccidioides* antigen enzyme immunoassay. *Clin Infect Dis.* 2008; **47:** e69-73.
118. Dworak DP, Kapustiak J, Ehklassi TA, *et al.* Primary cutaneous coccidioidomycosis of the eyelid: A case report. *Ophthal Plast Reconstr Surg.* 2016; **32:** e40-1.
119. El Abd OH, Fusco HN, Gomba L, *et al.* Coccidioidomycosis infection presenting with thoracic spinal pain. *PM&R.* 2012; **4:** 450-5.
120. El Dib NA, Eldessouky NM, El Sherbini SA, *et al.* Disseminated coccidioidomycosis in a 5-year-old Sudanese boy. *J Trop Pediatr.* 2014; **60:** 260-3.
121. Elgafy H, Miller J, Meyers S, *et al.* Disseminated coccidioidomycosis of the spine in an immunocompetent patient. *Am J Orthop (Belle Mead NJ).* 2014; **43:** e181-4.
122. Elgafy H, Miller J, Meyers S, *et al.* Disseminated coccidioidomycosis of the spine. *Spine J.* 2015; **15:** 1478-9.
123. Ellerbrook L, Laks S. Coccidioidomycosis osteomyelitis of the knee in a 23-year-old diabetic patient. *Radiol Case Rep.* 2015; **10:** 1034.
124. Engelthaler DM, Chiller T, Schupp JA, *et al.* Next-generation sequencing of *Coccidioides immitis* isolated during cluster investigation. *Emerg Infect Dis.* 2011; **17:** 227-32.
125. Escobedo EV, Ho CH, Homans JD. An 18-year-old man with bilateral violaceous cervical adenopathy. *JAMA Pediatr.* 2014; **168:** 773-4.
126. Evans K, Calhoun RF, Black H, *et al.* Disseminated primary coccidioidomycosis of the chest wall. *J Thorac Cardiovasc Surg.* 2010; **140:** e78-9.
127. Farber SA, Rubin AI, Micheletti RG. Disseminated coccidioidomycosis masquerading as a ruptured epidermal inclusion cyst in a healthy young adult from Philadelphia. *Int J Dermatol.* 2015; **54:** e441-2.
128. Fernandez A, Landaburu F, Lopez-Daneri G, *et al.* Phenotypic and molecular identification of *Coccidioides posadasii* in a patient evaluated for bilateral lung transplantation. *Rev Iberoam Micol.* 2012; **29:** 245-8.
129. Filho RE, Bandeira SP, Brillhante RS, *et al.* Biochemical characterization of an in-house *Coccidioides* antigen: Perspectives for the immunodiagnosis of coccidioidomycosis. *Molecules.* 2012; **17:** 7854-63.
130. Fisher BT, Chiller TM, Prasad PA, *et al.* Hospitalizations for coccidioidomycosis at forty-one children's hospitals in the United States. *Pediatr Infect Dis J.* 2010; **29:** 243-7.
131. Fisher MC, Koenig GL, White TJ, *et al.* Molecular and phenotypic description of *Coccidioides posadasii* sp. nov., previously recognized as the non-California population of *Coccidioides immitis. Mycologia.* 2002; **94:** 73-84.
132. Fitterer C, Berg Z, Halfdanarson TR, *et al.* Coccidioidomycosis in patients with selected solid organ cancers: A case series and review of medical literature. *Mycopathologia.* 2016; **181:** 787-98.
133. Flaherman VJ, Hector R, Rutherford GW. Estimating severe coccidioidomycosis in California. *Emerg Infect Dis.* 2007; **13:** 1087-90.
134. Flores-Franco RA, Aguayo-Yong R. Primary coccidioidal pleural effusions successfully treated with adjunctive corticosteroid therapy. *Respiration.* 2010; **79:** 85-6.
135. Foerter J, Sundell J, Vroman P. PET/CT: First-line examination to assess disease extent of disseminated coccidioidomycosis. *J Nucl Med Technol.* 2016; **44:** 212-3.
136. Forrester JD, Gomez CA, Forrester JA, *et al.* First case of mesh infection due to *Coccidioides* spp. and literature review of fungal mesh infections after hernia repair. *Mycoses.* 2015; **58:** 582-7.
137. Forrester JD, Guo HH, Weiser TG. Coccidioidomycosis: Surgical issues and implications. *Surg Infect (Larchmt).* 2016; **17:** 645-55.
138. Gabrielian A, Hariprasad SM. New onset of bilateral multifocal coccidioidomycosal choroiditis in a patient on oral fluconazole. *Can J Ophthalmol.* 2010; **45:** 419-20.
139. Gago S, Buitrago MJ, Clemons KV, *et al.* Development and validation of a quantitative real-time PCR assay for the early diagnosis of coccidioidomycosis. *Diagn Microbiol Infect Dis.* 2014; **79:** 214-21.
140. Galgiani JN, Ampel NM, Blair JE, *et al.* Executive Summary: 2016 Infectious Diseases Society of America (IDSA) Clinical Practice Guideline for the Treatment of Coccidioidomycosis. *Clin Infect Dis.* 2016; **63:** 717-22.
141. Galgiani JN, Ampel NM, Blair JE, *et el.* 2016 Infectious Diseases Society of America (IDSA) Clinical Practice Guideline for the Treatment of Coccidioidomycosis. *Clin Infect Dis.* 2016; **63:** e112-46.
142. Galgiani JN. Coccidioidomycosis: Changing perceptions and creating opportunities for its control. *Ann N Y Acad Sci.* 2007; **1111:** 1-18.
143. Galgiani JN. Elements of style in managing coccidioidomycosis. *Clin Infect Dis.* 2013; **56:** 1586-8.
144. Gallo ES, Pehoushek JF, Crowson AN. An exophytic nasal nodule. Coccidioidomycosis. *Arch Dermatol.* 2010; **146:** 789-94.
145. Garcia SC, Salas Alanis JC, Flores MG, *et al.* Coccidioidomycosis and the skin: A comprehensive review. *An Bras Dermatol.* 2015; **90:** 610-9.
146. Garrett AL, Chang YH, Ganley K, *et al.* Uphill both ways: Fatigue and quality of life in valley fever. *Med Mycol.* 2016; **54:** 310-7.
147. Garza-Chapa JI, Martinez-Cabriales SA, Ocampo-Garza J, *et al.* Cold subcutaneous abscesses as the first manifestation of disseminated coccidioidomycosis in an immunocompromised host. *Australas J Dermatol.* 2016; **57:** e49-52.
148. Gildardo JM, Leobardo VA, Nora MO, *et al.* Primary cutaneous coccidioidomycosis: case report and review of the literature. *Int J Dermatol.* 2006; **45:** 121-3.
149. Gobbi F, Angheben A, Farina C, *et al.* Coccidioidomycosis: First imported case in Italy. *J Travel Med.* 2012; **19:** 192-4.
150. Gokhale S, Punia H, Ohri VC, *et al. Malbranchea,* an imitator of *Coccidioides immitis*: A case report. *Indian J Med Microbiol.* 1995, **13:** 48-50.
151. Goldenholz DM, Mehra N, Dahlin B, *et al.* Fungus in the brain: Coccidioidomycosis meningoencephalitis. *Neurology.* 2013; **80:** e82.
152. Guevara RE, Motala T, Terashita D. The changing epidemiology of coccidioidomycosis in Los Angeles (LA) County, California, 1973-2011. *PLoS One.* 2015; **10:** e0136753.

153. Hagos G, Esayas R. Pulmonary coccidioidomycosis presenting as a mass, an uncommon disease entity in Ethiopia. *Ethiop Med J.* 2015; **53:** 35-7.
154. Hamer L, Castillo J. Coccidioidomycosis presenting as a massive pleural effusion in a postpartum woman. *Indian J Chest Dis Allied Sci.* 2006; **48:** 59-62.
155. Hammoudeh ZS, Lettieri S. Management of temporomandibular joint coccidioidomycosis. *Cranio.* 2016; **34:** 402-5.
156. Hardesty DA, Ramey W, Afrasiabi M, *et al.* Patient outcomes and surgical complications in coccidioidomycosis-related hydrocephalus: An institutional review. *J Neurosurg.* 2014; **121:** 785-9.
157. Hartmann CA, Aye WT, Blair JE. Treatment considerations in pulmonary coccidioidomycosis. *Expert Rev Respir Med.* 2016; **10:** 1079-91.
158. Hector R, Rutherford GW. The public health need and present status of a vaccine for the prevention of coccidioidomycosis. *Ann N Y Acad Sci.* 2007; **1111:** 259-68.
159. Hector RF, Rutherford GW, Tsang CA, *et al.* The public health impact of coccidioidomycosis in Arizona and California. *Int J Environ Res Public Health.* 2011; **8:** 1150-73.
160. Helig D, Giampoli EJ. Mycelial form of *Coccidioides immitis* diagnosed in bronchoalveolar lavage. *Diagn Cytopathol.* 2007; **35:** 535-6.
161. Himpe U, Verbeken E, Goegebuer T, *et al.* Persistent cough in a young smoker. Coccidioidomycosis infection, pulmonary nodular presentation. *Respiration.* 2009; **78:** 114-6.
162. Hirschmann JV. The early history of coccidioidomycosis: 1892-1945. *Clin Infect Dis.* 2007; **44:** 1202-7.
163. Ho AK, Shrader MW, Falk MN, *et al.* Diagnosis and initial management of musculoskeletal coccidioidomycosis in children. *J Pediatr Orthop.* 2014; **34:** 571-7.
164. Ho J, Fowler P, Heidari A, *et al.* Intrathecal amphotericin B: A 60 year experience in treating coccidioidal meningitis. *Clin Infect Dis.* 2017. pii: ciw794. PMID: 27927853.
165. Ho L, Schnall S, Schiller F, *et al.* Metacarpal coccidioidal osteomyelitis. *Am J Orthop (Belle Mead NJ).* 2011; **40:** 34-6.
166. Homans JD, Spencer L. Itraconazole treatment of non-meningeal coccidioidomycosis in children: Two case reports and review of the literature. *Pediatr Infect Dis J.* 2010; **29:** 65-7.
167. Huang JY, Bristow B, Shafir S, *et al.* Coccidioidomycosis-associated deaths, United States, 1990-2008. *Emerg Infect Dis.* 2012; **18:** 1723-8.
168. Humphrey PA. Fungal prostatitis caused by *Coccidioides. J Urol.* 2014; **191:** 215-6.
169. Hung CY, Xue J, Cole GT. Virulence mechanisms of *Coccidioides. Ann N Y Acad Sci.* 2007; **1111:** 225-35.
170. Indhirajanti S, Maartense E, Posthuma EF, *et al.* Pulmonary coccidioidomycosis: Import illness and the importance of travel history. *Neth J Med.* 2009; **67:** 353-5.
171. Islam S, Hsieh CJ, Lee BP. A teenage male with acute worsening of chronic back pain. Disseminated coccidioidomycosis. *Pediatr Infect Dis J.* 2012; **31:** 1209 & 1214-5.
172. Iyengar S, Chambers CJ, Chang S, *et al.* Subcorneal pustular dermatosis associated with *Coccidioides immitis. Dermatol Online J.* 2015; **21:** pii: 13030/qt35r6z6bx.
173. Jaroszewski DE, Halabi WJ, Blair JE, *et al.* Surgery for pulmonary coccidioidomycosis: A 10-year experience. *Ann Thorac Surg.* 2009; **88:** 1765-72.
174. Jehangir W, Tadepalli GS, Sen S, *et al.* Coccidioidomycosis and blastomycosis: Endemic mycotic co-infections in the HIV Patient. *J Clin Med Res.* 2015; **7:** 196-8.
175. Jinno S, Chang S, Jacobs MR. *Coccidioides* thyroiditis in an HIV-infected patient. *J Clin Microbiol.* 2012; **50:** 2535-7.
176. Johnson L, Gaab EM, Sanchez J, *et al.* Valley fever: Danger lurking in a dust cloud. *Microbes Infect.* 2014; **16:** 591-600.
177. Jones JM, Koski L, Khan M, *et al.* Coccidioidomycosis: An underreported cause of death - Arizona, 2008-2013. *Med Mycol.* 2017; doi: 10.1093/mmy/myx041. PMID: 28595294.
178. Johnson R, Kernerman SM, Sawtelle BG, *et al.* A reformulated spherule-derived coccidioidin (Spherusol) to detect delayed-type hypersensitivity in coccidioidomycosis. *Mycopathologia.* 2012; **174:** 353-8.
179. Johnson RH, Einstein HE. Amphotericin B and coccidioidomycosis. *Ann N Y Acad Sci.* 2007; **1111:** 434-41.
180. Johnson SM, Carlson EL, Fisher FS, *et al.* Demonstration of *Coccidioides immitis* and *Coccidioides posadasii* DNA in soil samples collected from Dinosaur National Monument, Utah. *Med Mycol.* 2014; **52:** 610-7.
181. Johnson SM, Carlson EL, Pappagianis D. *Coccidioides* species determination: Does sequence analysis agree with restriction fragment length polymorphism? *Mycopathologia.* 2015; **179:** 373-9.
182. Jude CM, Nayak NB, Patel MK, *et al.* Pulmonary coccidioidomycosis: Pictorial review of chest radiographic and CT findings. *Radiographics.* 2014; **34:** 912-25.
183. Kahn A, Carey EJ, Blair JE. Universal fungal prophylaxis and risk of coccidioidomycosis in liver transplant recipients living in an endemic area. *Liver Transpl.* 2015; **21:** 353-61.
184. Kakarla UK, Kalani MY, Sharma GK, *et al.* Surgical management of coccidioidomycosis of the spine: Clinical article. *J Neurosurg Spine.* 2011; **15:** 441-6.
185. Kantarcioglu AS, Sandoval-Denis M, Aygun G, *et al.* First imported coccidioidomycosis in Turkey: A potential health risk for laboratory workers outside endemic areas. *Med Mycol Case Rep.* 2014; **3:** 20-5.
186. Kassis C, Zaidi S, Kuberski T, *et al.* Role of *Coccidioides* antigen testing in the cerebrospinal fluid for the diagnosis of coccidioidal meningitis. *Clin Infect Dis.* 2015; **61:** 1521-6.
187. Kaufman L, Valero G, Padhye AA. Misleading manifestations of *Coccidioides immitis in vivo. J Clin Microbiol.* 1998; **36:** 3721-3.
188. Ke Y, Smith CW, Salaru G, *et al.* Unusual forms of immature sporulating *Coccidioides immitis* diagnosed by fine-needle aspiration biopsy. *Arch Pathol Lab Med.* 2006; **130:** 97-100.
189. Keckich DW, Blair JE, Vikram HR, *et al.* Reactivation of coccidioidomycosis despite antifungal prophylaxis in solid organ transplant recipients. *Transplantation.* 2011; **92:** 88-93.
190. Keckich DW, Blair JE, Vikram HR. *Coccidioides* fungemia in six patients, with a review of the literature. *Mycopathologia.* 2010; **170:** 107-15.

191. Kim MM, Blair JE, Carey EJ, *et al.* Coccidioidal pneumonia, Phoenix, Arizona, USA, 2000-2004. *Emerg Infect Dis.* 2009; **15:** 397-401.
192. Kim MM, Vikram HR, Kusne S, *et al.* Treatment of refractory coccidioidomycosis with voriconazole or posaconazole. *Clin Infect Dis.* 2011; **53:** 1060-6.
193. Kirkland TN. The quest for a vaccine against coccidioidomycosis: A neglected disease of the Americas. *J Fungi.* 2016; **2:** 34.
194. Kishi K, Fujii T, Takaya H, *et al.* Pulmonary coccidioidomycosis found in healthy Japanese individuals. *Respirology.* 2008; **13:** 252-6.
195. Knox KS. Perspective on coccidioidomycosis and histoplasmosis. *Am J Respir Crit Care Med.* 2014; **189:** 752-3.
196. Kokseng SL, Blair JE. Subclinical dissemination of coccidioidomycosis in a liver transplant recipient. *Mycopathologia.* 2011; **172:** 223-6.
197. Kokseng SL, Blair JE. Successful kidney transplantation after coccidioidal meningitis. *Transpl Infect Dis.* 2011; **13:** 285-9.
198. Kormos WA, Wu CC, Branda JA, *et al.* A 56-year-old man with cough, hypoxemia, and rash. *N Engl J Med.* 2010; **363:** 2046-54.
199. Kuberski T, Herrig J, Pappagianis D. False-positive IgM serology in coccidioidomycosis. *J Clin Microbiol.* 2010; **48:** 2047-9.
200. Kumar B, Singh S, Kaur S. Coccidioidin skin hypersensitivity in a section of north Indian population. *Indian J Chest Dis Allied Sci.* 1980; **22:** 236-8.
201. Kumar KS, Narasimhan A, Gopalakrishnan R, *et al.* Coccidioidomycosis in Chennai. *J Assoc Physicians India.* 2011; **59:** 122-4.
202. Kuprian M, Schofield C, Bennett S. Symptomatic hepatitis secondary to disseminated coccidioidomycosis in an immunocompetent patient. *BMJ Case Rep.* 2014; pii: bcr2013202144.
203. Kusne S, Taranto S, Covington S, *et al.* Coccidioidomycosis transmission through organ transplantation: A report of the OPTN *ad hoc* Disease Transmission Advisory Committee. *Am J Transplant.* 2016; **16:** 3562-7.
204. Kwok HK, Chan JW, Li IW, *et al.* Coccidioidomycosis as a rare cause of pneumonia in non-endemic areas: A short exposure history should not be ignored. *Respirology.* 2009; **14:** 617-20.
205. Lammering JC, Iv M, Gupta N, *et al.* Imaging spectrum of CNS coccidioidomycosis: Prevalence and significance of concurrent brain and spinal disease. *AJR Am J Roentgenol.* 2013; **200:** 1334-46.
206. Lan F, Tong YZ, Huang H, *et al.* Primary pulmonary coccidioidomycosis in China. *Respirology.* 2010; **15:** 722-5.
207. Langelier C, Baxi SM, Iribarne D, *et al.* Beyond the superficial: *Coccidioides immitis* fungaemia in a man with fever, fatigue and skin nodules: A case of an emerging and evolving pathogen. *BMJ Case Rep.* 2014; pii: bcr2014205333.
208. Laniado-Laborin R. Coccidioidomycosis and other endemic mycoses in Mexico. *Rev Iberoam Micol.* 2007; **24:** 249-58.
209. Laniado-Laborin R. Expanding understanding of epidemiology of coccidioidomycosis in the Western hemisphere. *Ann N Y Acad Sci.* 2007; **1111:** 19-34.
210. Lauer A, Baal JD, Baal JC, *et al.* Detection of *Coccidioides immitis* in Kern County, California, by multiplex PCR. *Mycologia.* 2012; **104:** 62-9.
211. Lee CH, Wilcox L, Chorneyko K, *et al. Coccidioides immitis*: two cases of misidentified mycosis. *Can Respir J.* 2008; **15:** 377-9.
212. Lee JM, Graciano AL, Dabrowski L, *et al.* Coccidioidomycosis in infants: A retrospective case series. *Pediatr Pulmonol.* 2016; **51:** 858-62.
213. Levy ER, McCarty JM, Shane AL, *et al.* Treatment of pediatric refractory coccidioidomycosis with combination voriconazole and caspofungin: A retrospective case series. *Clin Infect Dis.* 2013; **56:** 1573-8.
214. Li YC, Calvert G, Hanrahan CJ, *et al.* Coccidioidomycosis infection of the patella mimicking a neoplasm - Two case reports. *BMC Med Imaging.* 2014; **14:** 8.
215. Litvintseva AP, Marsden-Haug N, Hurst S, *et al.* Valley fever: Finding new places for an old disease: *Coccidioides immitis* found in Washington State soil associated with recent human infection. *Clin Infect Dis.* 2015; **60:** e1-3.
216. Loudin M, Clayburgh DR, Hakki M. *Coccidioides immitis* cervical lymphadenitis complicated by esophageal fistula. *Case Rep Infect Dis.* 2016; 8715405.
217. Luna-Isaac JA, Muniz-Salazar R, Baptista-Rosas RC, *et al.* Genetic analysis of the endemic fungal pathogens *Coccidioides posadasii* and *Coccidioides immitis* in Mexico. *Med Mycol.* 2014; **52:** 156-66.
218. Luo R, Greenberg A, Stone CD. Hospitalized burden and outcomes of coccidioidomycosis: A nationwide analysis, 2005-2012. *Med Mycol.* 2017; **55:** 368-74.
219. Majeed A, Ullah W, Hamadani AA, *et al.* First reported case of peroneal tenosynovitis caused by *Coccidioides immitis* successfully treated with fluconazole. *BMJ Case Rep.* 2016; pii: bcr2016216804. PMID: 27903576.
220. Malo J, Luraschi-Monjagatta C, Wolk DM, *et al.* Update on the diagnosis of pulmonary coccidioidomycosis. *Ann Am Thorac Soc.* 2014; **11:** 243-53.
221. Mangold AR, DiCaudo DJ, Blair JE, *et al.* Chronic interstitial granulomatous dermatitis in coccidioidomycosis. *Br J Dermatol.* 2016; **174:** 881-4.
222. Marsden-Haug N, Goldoft M, Ralston C, *et al.* Coccidioidomycosis acquired in Washington State. *Clin Infect Dis.* 2013; **56:** 847-50.
223. Marsden-Haug N, Hill H, Litvintseva AP, *et al. Coccidioides immitis* identified in soil outside of its known range - Washington, 2013. *MMWR.* 2014; **63:** 450.
224. Martirosyan NL, Skoch JM, Zaninovich O, *et al.* A paradigm for the evaluation and management of spinal coccidioidomycosis. *Surg Neurol Int.* 2015; **6:** 107.
225. Masannat FY, Ampel NM. Coccidioidomycosis in patients with HIV-1 infection in the era of potent antiretroviral therapy. *Clin Infect Dis.* 2010; **50:** 1-7.
226. Mathisen G, Shelub A, Truong J, *et al.* Coccidioidal meningitis: Clinical presentation and management in the fluconazole era. Medicine (Baltimore). 2010; **89:** 251-84.

227. Mayer AP, Files JA, Biria N. There is a fungus among us. *J Womens Health (Larchmt).* 2012; **21:** 798-9.
228. Mayer AP, Morris MF, Panse PM, *et al.* Does the presence of mediastinal adenopathy confer a risk for disseminated infection in immunocompetent persons with pulmonary coccidioidomycosis? *Mycoses.* 2013; **56:** 145-9.
229. McAninch EA, Xu C, Lagari VS, *et al.* Coccidioidomycosis thyroiditis in an immunocompromised host post-transplant: Case report and literature review. *J Clin Endocrinol Metab.* 2014; **99:** 1537-42.
230. McCarty JM, Demetral LC, Dabrowski L, *et al.* Pediatric coccidioidomycosis in central California: A retrospective case series. *Clin Infect Dis.* 2013; **56:** 1579-85.
231. McGwire B, Marr B, Zhou XP, *et al.* Iliopsoas coccidioidomycotic abscess with associated intra-abdominal extension in an immunocompetent patient. *BMJ Case Rep.* 2012; pii: bcr2012007436. PMID: 23188867.
232. Mease L. Pulmonary and extrapulmonary coccidioidomycosis, active component, U.S. Armed Forces, 1999-2011. *MSMR.* 2012; **19:** 2-4.
233. Mendez LA, Flores SA, Martinez R, *et al.* Ulcerated lesion of the tongue as manifestation of systemic coccidioidomycosis. *Case Rep Med.* 2017; 1489501. PMID: 28386282.
234. Mendoza N, Blair JE. The utility of diagnostic testing for active coccidioidomycosis in solid organ transplant recipients. *Am J Transplant.* 2013; **13:** 1034-9.
235. Mendoza N, Noel P, Blair JE. Diagnosis, treatment and outcomes of coccidioidomycosis in allogeneic stem cell transplantation. *Transpl Infect Dis.* 2015; **17:** 380-8.
236. Mirochnik BD, Lev S, Weingarten EP. Disseminated coccidioidal spondylodiskitis. *Radiology.* 2014; **272:** 914-8.
237. Moffat A, Lovecchio F. Presentations of coccidioidomycosis in the ED. *Am J Emerg Med.* 2016; **34:** 906-10.
238. Montgomery MW, Yawetz S, Levy BD, *et al.* Back to the history. *N Engl J Med.* 2017; **376:** 1783-8.
239. Montone KT, Litzky LA, Feldman MD, *et al. In situ* hybridization for *Coccidioides immitis* 5.8S ribosomal RNA sequences in formalin-fixed, paraffin-embedded pulmonary specimens using a locked nucleic acid probe: A rapid means for identification in tissue sections. *Diagn Mol Pathol.* 2010; **19:** 99-104.
240. Motley BD, Grabowski M, Prayson RA. *Coccidioides* parenchymal cerebral abscess in the setting of lymphoma. *J Clin Neurosci.* 2015; **22:** 40-1.
241. Munoz-Hernandez B, Palma-Cortes G, Cabello-Gutierrez C, *et al.* Parasitic polymorphism of *Coccidioides* spp. *BMC Infect Dis.* 2014; **14:** 213.
242. Nagler AR, Kim BS, Gormley RH, *et al.* Verrucous nodule on the upper lip. *Cutis.* 2014; **93:** e15-7.
243. Narang V, Garg B, Sood N, *et al.* Primary cutaneous coccidioidomycosis: First imported case in north India. *Indian J Dermatol.* 2014; **59:** 422.
244. Navalkar KA, Johnston SA, Woodbury N, *et al.* Application of immunosignatures for diagnosis of valley fever. *Clin Vaccine Immunol.* 2014; **21:** 1169-77.
245. Nazarullah A, Nilson R, Maselli DJ, *et al.* Incidence and aetiologies of pulmonary granulomatous inflammation: A decade of experience. *Respirology.* 2015; **20:** 115-21.
246. Nelson EC, Thompson GR 3rd, Vidovszky TJ. Disseminated coccidioidomycosis. *Arch Surg.* 2012; **147:** 95-6.
247. Nelson JK, Giraldeau G, Montoya JG, *et al.* Donor-derived *Coccidioides immitis* endocarditis and disseminated infection in the setting of solid organ transplantation. *Open Forum Infect Dis.* 2016; **3:** ofw086. PMID: 27413765.
248. Nesbit LA, Knox KS, Nguyen CT, *et al.* Immunological characterization of bronchoalveolar lavage fluid in patients with acute pulmonary coccidioidomycosis. *J Infect Dis.* 2013; **208:** 857-63.
249. Newlon HR, Lambiase MC. Disseminated cutaneous coccidioidomycosis masquerading as lupus pernio. *Cutis.* 2010; **86:** 25-8.
250. Nguyen C, Barker BM, Hoover S, *et al.* Recent advances in our understanding of the environmental, epidemiological, immunological and clinical dimensions of coccidioidomycosis. *Clin Microbiol Rev.* 2013; **26:** 505-25.
251. Nicolasora N, Cinti SK, Riddell Iv J, *et al.* A 53-year-old man with a large air bubble in his chest. *Clin Infect Dis.* 2008; **47:** 823-4 & 853-5.
252. Nimer N, Camp T. Coccidioidomycosis. *Pediatr Rev.* 2015; **36:** 181-2.
253. No Authors Listed. Coccidioidomycosis, active component, US Armed Forces, January 2000 - June 2012. *MSMR.* 2012; **19:** 10.
254. No authors listed. Etymologia: *Coccidioides. Emerg Infect Dis.* 2015; **21:** 1033.
255. No Authors Listed. Historical perspective: Coccidioidomycosis in the US military and military-associated populations. *MSMR.* 2012; **19:** 5-6.
256. Noble JA, Nelson RG, Fufaa GD, *et al.* Effect of geography on the analysis of coccidioidomycosis-associated deaths, United States. *Emerg Infect Dis.* 2016; **22:** 1821-3.
257. Oberoi JK, Wattal C, Aggarwal PK, *et al.* Pulmonary coccidiomycosis in New Delhi, India. *Infection.* 2012; **40:** 699-702.
258. Ocampo-Garza J, Castrejon-Perez AD, Gonzalez-Saldivar G, *et al.* Cutaneous coccidioidomycosis: A great mimicker. *BMJ Case Rep.* 2015; pii: bcr2015211680. PMID: 26216930.
259. Odio CD, Marciano BE, Galgiani JN, *et al.* Risk factors for disseminated coccidioidomycosis, United States. *Emerg Infect Dis.* 2017; **23:** 308-11 & 376.
260. Ondo AL, Zlotoff BJ, Mings SM, *et al.* Primary cutaneous coccidioidomycosis: An incidental finding. *Clin Exp Dermatol.* 2010; **35:** e42-3.
261. Ong AT, Mahajan H, Chen SC, *et al.* Coccidioidal septic arthritis: Lessons learned from a clinical and laboratory perspective. *Med J Aust.* 2012; **196:** 705-6.
262. Pappagianis D, Johnson SM. Revision and return of a coccidioidal skin test reagent. *Mycopathologia.* 2012; **174:** 351-2.
263. Pappagianis D, Zimmer BL. Serology of coccidioidomycosis. *Clin Microbiol Rev.* 1990; **3:** 247-68.
264. Patel S, Lee RH. The case of the sinister spores: The patient was hospitalized for a menacing infection in the second trimester of pregnancy. *Am J Obstet Gynecol.* 2013; **208:** 417.e1.
265. Patel S, Snyder L. A sore throat in the southwest. *Am J Med.* 2009; **122:** 233-5.

266. Patel SM, Patel SS, Myers R. A 37-year-old man with a chronic cough. *Cleve Clin J Med.* 2012; **79:** 141-50.
267. Patel SS, Cardile AP. Diffuse skeletal coccidioidomycosis in an AIDS patient. *Int J Infect Dis.* 2013; **17:** e928-9.
268. Ramsi M, Alvira C, Purohit P, *et al.* Haemophagocytic lymphohistiocytosis associated with coccidioidomycosis. *BMJ Case Rep.* 2014; pii: bcr2014205681. PMID: 25139924.
269. Reach P, Paugam A, Kahan A, *et al.* Coccidioidomycosis of the spine in an immunocompetent patient. *Joint Bone Spine.* 2010; **77:** 611-3.
270. Rempe S, Sachdev MS, Bhakta R, *et al. Coccidioides immitis* fungemia: clinical features and survival in 33 adult patients. *Heart Lung.* 2007; **36:** 64-71.
271. Rendon A, Rendon RA, Bauerle O. Empyema necessitatis: Unique presentation in a coccidioidomycosis case. *Am J Respir Crit Care Med.* 2015; **191:** 964-6.
272. Reyes N, Onadeko OO, Luraschi-Monjagatta Mdel C, *et al.* Positron emission tomography in the evaluation of pulmonary nodules among patients living in a coccidioidal endemic region. *Lung.* 2014; **192:** 589-93.
273. Ricciotti RW, Shekhel TA, Blair JE, *et al.* Surgical pathology of skeletal coccidioidomycosis: A clinical and histopathologic analysis of 25 cases. *Am J Surg Pathol.* 2014; **38:** 1672-80.
274. Rivard SC, Satter E. Primary cutaneous coccidioidomycosis presenting as a recurrent preauricular cyst. *Skinmed.* 2016; **14:** 142-4
275. Rixford E, Gilchrist TC. Two cases of protozoan (coccidioidal) infection of the skin and other organs. *Johns Hopkins Hospital Reports.* 1896; **1:** 209-68.
276. Rodriguez-Morales AJ, Jimenez-Canizales CE, Mondragon-Cardona A, *et al.* So, if I travel to "Venezuela", can I get coccidioidomycosis? *Infection.* 2014; **42:** 1067-8.
277. Ruddy BE, Mayer AP, Ko MG, *et al.* Coccidioidomycosis in African Americans. *Mayo Clin Proc.* 2011; **86:** 63-9.
278. Ryu JH, Tian X, Baqir M, *et al.* Diffuse cystic lung diseases. *Front Med.* 2013; **7:** 316-27.
279. Sachdev MS, Blair JE, Mulligan DC, *et al.* Coccidioidomycosis masked by symptoms of end-stage liver disease in transplant candidates. *Transpl Infect Dis.* 2007; **9:** 153-5.
280. Sachdeva M, Ronaghi R, Mills PK, *et al.* Complications and yield of computed tomography-guided transthoracic core needle biopsy of lung nodules at a high-volume academic center in an endemic coccidioidomycosis area. *Lung.* 2016; **194:** 379-85.
281. Saettele TM, Jimenez CA. Pleural coccidioidomycosis. *BMJ Case Rep.* 2014; pii: bcr2014206380. PMID: 25422353.
282. Sakata KK, Grys TE, Chang YH, *et al.* Serum procalcitonin levels in patients with primary pulmonary coccidioidomycosis. *Ann Am Thorac Soc.* 2014; **11:** 1239-43.
283. Satyanarayan A, Klotz S, Han L, *et al.* Coccidioidomycosis of the genitourinary tract: A case report and discussion. *Urology.* 2014; **84:** e30-1.
284. Saubolle MA. Laboratory aspects in the diagnosis of coccidioidomycosis. *Ann N Y Acad Sci.* 2007; **1111:** 301-14.
285. Savio J, Srinivasa H, Mathew J, Chakraborty A, McNamara DR, Hemashettar BM, Padhye AA. Coccidioidomycosis in India: Report of a third imported case. *J Mycol Med.* 2011; **21:** 33-6.
286. Schein R, Homans J, Larsen RA, *et al.* Posaconazole for chronic refractory coccidioidal meningitis. *Clin Infect Dis.* 2011; **53:** 1252-4.
287. Schieffelin JS, Torrellas M, Lartchenko S, *et al.* How natural disasters change natural patterns: Coccidioidomycosis imported to New Orleans. *J La State Med Soc.* 2013; **165:** 145-9.
288. Schuetz AN, Pisapia D, Yan J, *et al.* An atypical morphologic presentation of *Coccidioides* spp. in fine-needle aspiration of lung. *Diagn Cytopathol.* 2012; **40:** 163-7.
289. Schweigert M, Dubecz A, Beron M, *et al.* Pulmonary infections imitating lung cancer: Clinical presentation and therapeutical approach. *Ir J Med Sci.* 2013; **182:** 73-80.
290. Sebro R, Aparici CM, Hernandez-Pampaloni M. FDG PET/CT evaluation of pathologically proven pulmonary lesions in an area of high endemic granulomatous disease. *Ann Nucl Med.* 2013; **27:** 400-5.
291. Seitz AE, Prevots DR, Holland SM. Hospitalizations associated with disseminated coccidioidomycosis, Arizona and California, USA. *Emerg Infect Dis.* 2012; **18:** 1476-9.
292. Sheff KW, York ER, Driebe EM, *et al.* Development of a rapid, cost-effective TaqMan Real-Time PCR Assay for identification and differentiation of *Coccidioides immitis* and *Coccidioides posadasii. Med Mycol.* 2010; **48:** 466-9.
293. Shehab ZM. Coccidioidomycosis. *Adv Pediatr.* 2010; **57:** 269-86.
294. Shekhel TA, Ricciotti RW, Blair JE, *et al.* Surgical pathology of pleural coccidioidomycosis: A clinicopathological study of 36 cases. *Hum Pathol.* 2014; **45:** 961-9.
295. Sheppard JE, Switlick DN. *Coccidioides immitis* osteomyelitis of the radius presenting as Ewing's sarcoma. *Orthopedics.* 2008; **31:** 607.
296. Short JH, Bradley C, Blair JE, *et al.* Integrative medicine preferences among coccidioidomycosis (Valley Fever) patients. *J Altern Complement Med.* 2017; **23:** 135-9.
297. Shubitz LF. Comparative aspects of coccidioidomycosis in animals and humans. *Ann N Y Acad Sci.* 2007; **1111:** 395-403.
298. Simons CM, Stratton CW, Kim AS. Peripheral blood eosinophilia as a clue to the diagnosis of an occult *Coccidioides* infection. *Hum Pathol.* 2011; **42:** 449-53.
299. Singh G, Patel T, Hu S. Disseminated cutaneous coccidioidomycosis in a liver transplant patient. *JAAD Case Rep.* 2015; **1:** 225-6.
300. Singh VR, Smith DK, Lawrence J, *et al.* Coccidioidomycosis in patients infected with human immunodeficiency virus: Review of 91 cases at a single institution. *Clin Infect Dis.* 1996; **23:** 563-8.
301. Smith G, Hoover S, Sobonya R, *et al.* Abdominal and pelvic coccidioidomycosis. *Am J Med Sci.* 2011; **341:** 308-11.
302. Sobonya RE, Yanes J, Klotz SA. Cavitary pulmonary coccidioidomycosis: Pathologic and clinical correlates of disease. *Hum Pathol.* 2014; **45:** 153-9.

303. Sondermeyer G, Lee L, Gilliss D, *et al.* Coccidioidomycosis-associated hospitalizations, California, USA, 2000-2011. *Emerg Infect Dis.* 2013; **19:** 1590-7.
304. Sondermeyer GL, Lee LA, Gilliss D, *et al.* Coccidioidomycosis-associated deaths in California, 2000-2013. *Public Health Rep.* 2016; **131:** 531-5.
305. Sondermeyer GL, Lee LA, Gilliss D, *et al.* Epidemiology of pediatric coccidioidomycosis in California, 2000-2012. *Pediatr Infect Dis J.* 2016; **35:** 166-71.
306. Sotello D, Rivas M, Fuller A, *et al.* Coccidioidomycosis with diffuse miliary pneumonia. *Proc (Bayl Univ Med Cent).* 2016; **29:** 39-41.
307. Spinello IM, Johnson RH, Baqi S. Coccidioidomycosis and pregnancy: A review. *Ann N Y Acad Sci.* 2007; **1111:** 358-64.
308. Spinello IM, Munoz A, Johnson RH. Pulmonary coccidioidomycosis. *Semin Respir Crit Care Med.* 2008; **29:** 166-73.
309. Stagliano D, Epstein J, Hickey P. Fomite-transmitted coccidioidomycosis in an immunocompromised child. *Pediatr Infect Dis J.* 2007; **26:** 454-6.
310. Stern NG, Galgiani JN. Coccidioidomycosis among scholarship athletes and other college students, Arizona, USA. *Emerg Infect Dis.* 2010; **16:** 321-3.
311. Stevens DA, Clemons KV, Levine HB, *et al.* Expert opinion: What to do when there is *Coccidioides* exposure in a laboratory. *Clin Infect Dis.* 2009; **49:** 919-23.
312. Stevens DA, Zhang Y, Finkelman MA, *et al.* CSF (1,3)-beta-D-glucan testing is useful in diagnosis of coccidioidal meningitis. *J Clin Microbiol.* 2016; **54:** 2707-10.
313. Stevens DA. Coccidioidomycosis. *N Engl J Med.* 1995; **332:** 1077-82.
314. Stieglitz E, Hsiang MS, Simko JP, *et al.* Pulmonary coccidiomycosis masquerading as refractory metastatic Ewing sarcoma. *J Pediatr Hematol Oncol.* 2014; **36:** e57-60.
315. Stockamp NW, Thompson GR 3rd. Coccidioidomycosis. *Infect Dis Clin North Am.* 2016; **30:** 229-46.
316. Storage TR, Segal J, Brown J. Peritoneal coccidioidomycosis: A rare case report and review of the literature. *J Gastrointestin Liver Dis.* 2015; **24:** 527-30.
317. Studer HM. A rare cause of unintentional weight loss, anhedonia and cognitive decline. *JAAPA.* 2013; **26:** 36-9.
318. Subedi S, Broom J, Caffery M, *et al.* Coccidioidomycosis in returned Australian travellers. *Intern Med J.* 2012; **42:** 940-3.
319. Suh JD, Kirsch CF, Hirschowitz SL, *et al.* Disseminated coccidioidomycosis of the infratemporal fossa. *Arch Otolaryngol Head Neck Surg.* 2009; **135:** 616 & 618-9.
320. Sunenshine RH, Anderson S, Erhart L, *et al.* Public health surveillance for coccidioidomycosis in Arizona. *Ann N Y Acad Sci.* 2007; **1111:** 96-102.
321. Sutton DA. Diagnosis of coccidioidomycosis by culture: Safety considerations, traditional methods and susceptibility testing. *Ann N Y Acad Sci.* 2007; **1111:** 315-25.
322. Szeyko LA, Taljanovic MS, Dzioba RB, *et al.* Vertebral coccidioidomycosis: presentation and multidisciplinary management. *Am J Med.* 2012; **125:** 304-14.
323. Taljanovic MS, Adam RD. Musculoskeletal coccidioidomycosis. *Semin Musculoskelet Radiol.* 2011; **15:** 511-26.
324. Tamerius JD, Comrie AC. Coccidioidomycosis incidence in Arizona predicted by seasonal precipitation. *PLoS One.* 2011; **6:** e21009.
325. Tan LA, Kasliwal MK, Nag S, *et al.* Rapidly progressive quadriparesis heralding disseminated coccidioidomycosis in an immunocompetent patient. *J Clin Neurosci.* 2014; **21:** 1049-51.
326. Tang CG, Nuyen BA, Puligandla B, *et al.* The coccidioidomycosis conundrum: A rare parotid mass. *Perm J.* 2014; **18:** 86-8.
327. Tang TH, Tsang OT. Fungal infection from sweeping in the wrong place. *N Engl J Med.* 2011; **364:** e3.
328. Teixeira MM, Barker BM. Use of population genetics to assess the ecology, evolution, and population structure of *Coccidioides. Emerg Infect Dis.* 2016; **22:** 1022-30.
329. Tepeoglu M, Erinanc H, Ozdemir H, *et al.* Pulmonary coccidioidomycosis after a renal transplant in a nonendemic region. *Exp Clin Transplant.* 2014; **12:** 71-3.
330. Theodoropoulos N, Angarone M. Lip lesion in a solid organ transplant recipient. *Clin Infect Dis.* 2012; **54:** 1332 & 1373-4.
331. Thomas S, Basu S, Dutta R, *et al.* Coccidioidomycosis presenting as liver abscess. *Indian J Gastroenterol.* 2001; **20:** 113-4.
332. Thompson G 3rd, Wang S, Bercovitch R, *et al.* Routine CSF analysis in coccidioidomycosis is not required. *PLoS One.* 2013; **8:** e64249.
333. Thompson GR 3rd, Bays DJ, Johnson SM, *et al.* Serum (1-3)-β-D-glucan measurement in coccidioidomycosis. *J Clin Microbiol.* 2012; **50:** 3060-2.
334. Thompson GR 3rd, Stevens DA, Clemons KV, *et al.* Call for a California coccidioidomycosis consortium to face the top ten challenges posed by a recalcitrant regional disease. *Mycopathologia.* 2015; **179:** 1-9.
335. Thompson GR 3rd. Pulmonary coccidioidomycosis. *Semin Respir Crit Care Med.* 2011; **32:** 754-63.
336. Thompson GR, Sharma S, Bays DJ, *et al.* Coccidioidomycosis: Adenosine deaminase levels, serologic parameters, culture results and polymerase chain reaction testing in pleural fluid. *Chest.* 2013; **143:** 776-81.
337. Tintelnot K, de Hoog GS, Antweiler E, *et al.* Taxonomic and diagnostic markers for identification of *Coccidioides immitis* and *Coccidioides posadasii. Med Mycol.* 2007; **45:** 385-93.
338. Tiu CT, Cook J, Pineros DF, *et al.* Pneumothorax in a young man in Brooklyn, New York. *Clin Infect Dis.* 2011; **53:** 1255 & 1296-7.
339. Togashi RH, Aguiar FM, Ferreira DB, *et al.* Pulmonary and extrapulmonary coccidioidomycosis: Three cases in an endemic area in the state of Ceara, Brazil. *J Bras Pneumol.* 2009; **35:** 275-9.
340. Tonelli AR, Khalife WT, Cao M, *et al.* Spherules, hyphae and air-crescent sign. *Am J Med Sci.* 2008; **335:** 504-6.
341. Tortorano AM, Carminati G, Tosoni A, *et al.* Primary cutaneous coccidioidomycosis in an Italian nun working in South America and review of published literature. *Mycopathologia.* 2015; **180:** 229-35.

342. Trible R, Edgerton N, Hayek S, *et al.* Antiretroviral therapy-associated coccidioidal meningitis. *Emerg Infect Dis.* 2013; **19:** 163-5.
343. Tsang CA, Anderson SM, Imholte SB, *et al.* Enhanced surveillance of coccidioidomycosis, Arizona, USA, 2007-2008. *Emerg Infect Dis.* 2010; **16:** 1738-44.
344. Turabelidze G, Aggu-Sher RK, Jahanpour E, *et al.* Coccidioidomycosis in a state where it is not known to be endemic - Missouri, 2004-2013. *MMWR.* 2015; **64:** 636-9.
345. Turaka K, Gordon AJ, Patel S, *et al.* Diagnostic and therapeutic challenges. *Retina.* 2013; **33:** 661-5.
346. Twarog M, Thompson GR 3rd. Coccidioidomycosis: Recent updates. *Semin Respir Crit Care Med.* 2015; **36:** 746-55.
347. Verghese S, Arjundas D, Krishnakumar KC, *et al.* Coccidioidomycosis in India: Report of a second imported case. *Med Mycol.* 2002; **40:** 307-9.
348. Vikram HR, Blair JE. Coccidioidomycosis in transplant recipients: A primer for clinicians in nonendemic areas. *Curr Opin Organ Transplant.* 2009; **14:** 606-12.
349. Vikram HR, Dhaliwal G, Saint S, *et al.* Clinical problem-solving. A recurrent problem. *N Engl J Med.* 2011; **364:** 2148-54.
350. Vikram HR, Dosanjh A, Blair JE. Coccidioidomycosis and lung transplantation. *Transplantation.* 2011; **92:** 717-21.
351. Vinh DC. Coccidioidal meningitis: Disseminated disease in patients without HIV/AIDS. *Medicine (Baltimore).* 2011; **90:** 87.
352. Vogler AJ, Nottingham R, Parise KL, *et al.* Effective disinfectants for *Coccidioides immitis* and *C.posadasii. Appl Biosaf.* 2015; **20:** 154-8.
353. Vucicevic D, Blair JE, Binnicker MJ, *et al.* The utility of *Coccidioides* polymerase chain reaction testing in the clinical setting. *Mycopathologia.* 2010; **170:** 345-51.
354. Vucicevic D, Carey EJ, Blair JE. Coccidioidomycosis in liver transplant recipients in an endemic area. *Am J Transplant.* 2011; **11:** 111-9 & 639.
355. Wack EE, Ampel NM, Sunenshine RH, *et al.* The return of delayed-type hypersensitivity skin testing for coccidioidomycosis. *Clin Infect Dis.* 2015; **61:** 787-91.
356. Wang XL, Wang S, An CL. Mini-review of published reports on coccidioidomycosis in China. *Mycopathologia.* 2015; **180:** 299-303.
357. Wang ZY, Wen SL, Ying KJ. A case study of imported pulmonary coccidioidomycosis. *J Zhejiang Univ Sci B.* 2011; **12:** 298-302.
358. Warnock DW. *Coccidioides* species as potential agents of bioterrorism. *Future Microbiol.* 2007; **2:** 277-83.
359. Waterman BR, Waterman SM, McCoy AC, *et al.* Coccidioidal osteomyelitis of the patella. *Orthopedics.* 2010; **33.** doi: 10.3928/01477447-20100225-30. PMID: 20415313.
360. Weatherhead JE, Barrows BD, Stager CE, *et al.* Gastrointestinal coccidioidomycosis. *J Clin Gastroenterol.* 2015; **49:** 628-9.
361. Welsh O, Vera-Cabrera L, Rendon A, *et al.* Coccidioidomycosis. *Clin Dermatol.* 2012; **30:** 573-91.
362. Wheat LJ. Approach to the diagnosis of the endemic mycoses. *Clin Chest Med.* 2009; **30:** 379-89.
363. Wheeler C, Lucas KD, Mohle-Boetani JC. Rates and risk factors for coccidioidomycosis among prison inmates, California, USA, 2011. *Emerg Infect Dis.* 2015; **21:** 70-5.
364. Whiston E, Taylor JW. Genomics in *Coccidioides*: Insights into evolution, ecology and pathogenesis. *Med Mycol.* 2014; **52:** 149-55.
365. White FN, Zedek DC, Collins DL, *et al.* Granuloma annulare arising in association with pulmonary coccidioidomycosis. *Dermatol Online J.* 2012; **18:** 7.
366. Wilken JA, Marquez P, Terashita D, *et al.* Coccidioidomycosis among cast and crew members at an outdoor television filming event - California, 2012. *MMWR.* 2014; **63:** 321-4.
367. Wilken JA, Sondermeyer G, Shusterman D, *et al.* Coccidioidomycosis among workers constructing solar power farms, California, USA, 2011-2014. *Emerg Infect Dis.* 2015; **21:** 1997-2005.
368. Williams PL. Coccidioidal meningitis. *Ann N Y Acad Sci.* 2007; **1111:** 377-84.
369. Wise HZ, Hung CY, Whiston E, *et al.* Extracellular ammonia at sites of pulmonary infection with *Coccidioides posadasii* contributes to severity of the respiratory disease. *Microb Pathog.* 2013; **59-60:** 19-28.
370. Wiwanitkit V. TaqMan real-time PCR assay for *Coccidioides. Med Mycol.* 2010; **48:** 679-80.
371. Yau AA. Risk factors and epidemiology of coccidioidomycosis demonstrated by a case of spontaneous pulmonary rupture of cavitary coccidioidomycosis. *Case Rep Infect Dis.* 2016; 8165414.
372. Yoon HJ, Clemons KV. Vaccines against *Coccidioides. Korean J Intern Med.* 2013; **28:** 403-7.
373. Yu JJ, Kirkland TN, Hall LK, *et al.* Characterization of a serodiagnostic complement fixation antigen of *Coccidioides posadasii* expressed in the non-pathogenic fungus *Uncinocarpus reesii. J Clin Microbiol.* 2005; **43:** 5462-9.
374. Yurkanin JP, Ahmann F, Dalkin BL. Coccidioidomycosis of the prostate: A determination of incidence, report of 4 cases and treatment recommendations. *J Infect.* 2006; **52:** e19-25.
375. Zaid S, Mahmoud A, Barry M, *et al.* The head of the pancreas: Another unusual site of coccidioidomycosis. *Surg Infect (Larchmt).* 2013; **14:** 554-8.
376. Zangeneh TT, Malo J, Luraschi-Monjagatta C, *et al.* Positive (1-3) β-D-glucan and cross reactivity of fungal assays in coccidioidomycosis. *Med Mycol.* 2015; **53:** 171-3.
377. Zhou S, Ma Y, Chandrasoma P. Small bowel dissemination of coccidioidomycosis. *Case Rep Pathol.* 2015; 403671.

CHAPTER 19

Paracoccidioidomycosis

Paracoccidioidomycosis is subacute, acute or chronic granulomatous systemic fungal infection. This disease is caused by thermally dimorphic fungus, ***Paracoccidioides***, which has two species ***P.brasiliensis*** and ***P.lutzii***. It involves primarily lungs and subsequently disseminates to various mucosal surfaces, skin, lymph nodes and occasionally to internal organs in otherwise healthy individuals. The infection, however, remains subclinical or limited to lungs in initial stage but sometimes when progressive disease is established, it may lead to severe morbidity as well as mortality.

This disease is seen predominantly among adult males residing in rural belt of geographically confined endemic areas of Latin America; with majority of cases encountered in Brazil. Imported cases diagnosed in North America, Europe and Asia represent patients who had previously lived in endemic areas. In certain clinical settings like presentation of patient with oral lesions, it is being referred to as an emerging deep mycosis. As it is the most common type of endemic mycosis of the South American continent hence called as **South American blastomycosis.** The disease is also known by number of synonyms such as **Lutz's mycosis**, **Brazilian blastomycosis, paracoccidioidal granuloma** or **paracoccidioidal blastomycosis**.

Historical Perspective

In 1908, **Adolfo Lutz** (1855-1940) cultured and identified causative agent of paracoccidioidomycosis in Brazil from cervical lymph nodes of two patients. In further studies on this fungus, he noted its dimorphic nature although he did not name the organism. Between 1909 and 1912, **Alfonso Splendore** (1871–1953), an Italian bacteriologist working at Sao Paulo, observed new cases of this disease and isolated the etiological agent giving its detailed account along with clinical and histopathological characteristics. He considered it to be ascomycetous yeast and designated its name, *Zymonema brasiliense*. In 1930, **Floriano Paulo de Almeida** (1898–1977) comprehensively characterized the disease and its causative organism by creating new genus *Paracoccidioides,* thereby designating it *P.brasiliensis,* as the species name had already been given. Due to contribution of these three scientists in the early 20th century on this clinical entity in Brazil, it was also called **Lutz-Splendore-de Almeida** disease.

In early period of studies on this organism, it was speculated to be similar to another dimorphic fungus *Coccidioides*, which was earlier confused with Protozoa belonging to Coccidia. Therefore, prefix *'para-'* was assigned to name it as *Paracoccidioides,* which meant *'near/similar to'* thereby its name implied *'like-Coccidioides'*. Hence in descriptive language *Paracoccidioides* literally means *'an organism which is like another one that is like protozoa coccidia.* Both, suffix as *'-iodes'* mentioned in *Coccidioides* and prefix as *'para-'* have made the nomenclature of this fungal disease as lengthy as consisting of 22 letters, which is difficult to pronounce and even spell hence some medical mycologists, in this fast moving electronic era of SMS, may start writing it simply **'PCM'** *or* **'Pmycosis'**.

This is also commonly referred to as **Lutz's mycosis**, in honor of Adolfo Lutz, who identified causative fungus of this disease more than a century ago. The Bacteriological Institute in Sao Paulo was later renamed in his honor as Instituto Adolfo Lutz. Within endemic areas, it is the most prevalent systemic mycosis in Brazil, which account for almost 80% of cases hence species epithet of fungus was designated by Splendore as ***'brasiliensis'***. It is continuing as such for the last one century without any change. In general, designations based on geographical areas of endemicity in most of fungal diseases were subsequently abandoned due to deceptive inference, as there has been exponential increase in number of such cases even beyond their defined boundaries.

The first Pan American Symposium on this clinical entity was held in 1971 at Medellin in Colombia, which was sponsored by the Pan American Health Organization. The term **paracoccidioidomycosis (PCM)** was officially recognised in this first ever international forum focused on this disease. Therefore, the World Health Organization officially acknowledged it as paracoccidioidomycosis in 1971. Subsequently similar meetings were held periodically and IX International Meeting was held in 2005 in Brazil. As this disease was first described in 1908 by Brazilian researcher, Adolfo Lutz, *Mycopathologia* has published a special issue commemorating its centennial discovery (*Mycopathologia*. 2008; **165:** 179-352). In addition, X International Congress on this disease was held on August 7-10, 2008 again at Medellin, Colombia as a mark of its centennial celebrations. The next one, XI International Congress on paracoccidioidomycosis, was held on May 1-4, 2011 at Sao Paulo. The XII International Congress was held on September 11-24, 2014 at the Brazilian capital Brasilia. Now, its XIII International Congress will be held on October 22-25, 2017 at Foz do Iguacu (Brazil) under chairmanship of Prof. Carlos Pelleschi Taborda, University of Sao Paulo, Brazil.

A new species of the genus *Paracoccidioides* was identified as *P.lutzii* by Teixeira, *et al* using multilocus sequencing studies in 2014. As it was referred in Chapter 15 (Lacaziosis), the mycology world celebrated **Centennial Anniversary** of **Carlos da Silva Lacaz** in the year 2015 with great enthusiasm and fervor by publishing special issue of *Rev Inst Med Trop Sao Paulo* {2015; **57 (Suppl. 19):** 1-64} dealing with topics pertaining to paracoccidioidomycosis.

Mycology

Paracoccidioides brasiliensis, the etiologic agent, is a thermally dimorphic fungus that in patients and cultures at 37°C adopts a yeast form while at lower temperatures it behaves as a mold that bears the infectious conidia. The genus *Paracoccidioides* has only two medically significant species *i.e. P.brasiliensis* and *P.lutzii.* The other species previously described as *P.cerebriformis*, was found to be belonging to genus *Aspergillus*.

Paracoccidioides being thermally dimorphic fungus, its main feature is presence of multiple budding yeast cells, which exist *in vivo* among human host and experimental animals and *in vitro* when cultures are kept at 37°C. The yeast-like colonies grow at 35-37°C within week's time on suitable enriched media, which are soft, off-white to cream colored, wrinkled and rough to pasty in appearance.

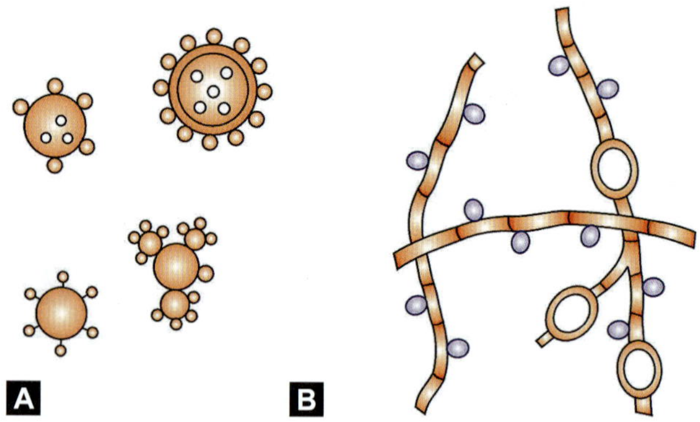

Figs. 19.1.A and B. *Paracoccidioides brasiliensis* showing multiple budding yeasts (37°C) and mycelial growth (25°C).

The yeast cells are characteristically round to oval, globose to pyriform that are somewhat variable in size (3-30 μm) and reproduce by multiple budding. The spherical mother cell is surrounded by multiple peripheral cells with synchronous thin-necked buds *i.e.* daughter cells all over its surface. They produce many small buds of 2-10 μm hence budding (daughter) yeast cells resemble ***'Mariner's*** or ***'Pilot's Wheel'*** **(Fig. 19.1A)**. The buds of various sizes, few larger ones may present on same mother yeast cell, giving typically ***'Mickey Mouse'*** appearance. Some of parent cells may appear empty also. Other yeast-like fungi such as *Candida tropicalis* may sometimes show multiple buddings thereby should not be confused with *P.brasiliensis*.

The mycelial colonies at 25°C are very slow growing and have propensity to form heaped craters, with agar dug out within depths of crater. Microscopically, mycelial form is characterized by hyphae measuring 1-2 μm, which bear single-celled conidia **(Fig. 19.1B)**. There are numerous intercalary chlamydospores, which may be seen as globose or pyriform conidia similar to those produced by *Blastomyces dermatitidis*.

The production of conidia by mycelial form has been considered to be rare event. However, when proper substrate and cultural conditions are provided, fungus sporulates, producing various types of conidia. These propagules are produced after relatively long incubation of about 3-4 weeks at room temperature but remain adherent to parent mycelium. The cultures for primary isolation from clinical specimens should be incubated for at least six weeks before being discarded as sterile. The conversion from mycelial-to-yeast (M→Y) phase is essential for definitive identification of an isolate. The heat shock proteins (70-kDa) may possibly play an important role in phase transition process.

Paracoccidioides exists in asexual state (anamorph) and its perfect state (teleomorph) has recently been explored by Teixeira *et al*, in 2014. The features related to sexual reproduction like coiled constricted hyphae and knob-like structures have been observed, indicating the formation of young ascocarps. Moreover, multiple nuclei were found in coiled constricted hyphae, which may possibly be as a consequence of nuclear migration during mating. However, as such no cleistothecium or ascus production has so far been detected. The presence and expression of sexual machinery as well the ability to produce sex-related structures indicates that mating may occur in the life cycle of the *Paracoccidioides* species.

Paracoccidioides lutzii: This species of the genus used to be formerly known as 'Pb01-like' strains in the *P.brasiliensis* complex. It was then proposed as a new species based on phylogenetic and comparative genomics data, recombination analysis and morphological characteristics. The conidia of *P.lutzii* are elongated, different from those of *P.brasiliensis.* Moreover, *P.lutzii* occurs in the central and northern regions of Brazil. Therefore, comparison of *P.brasiliensis* and *P.lutzii* is quite essential, which may have significant clinical implications for the diagnosis and treatment of this disease.

The absence of sexual reproduction, was actually impeding its classical genetic analysis. It is taxonomically placed into form-family **Moniliaceae**, in form-class **Hyphomycetes,** of Phylum **Deuteromycota**. In addition, phylogenetic studies based on large subunit (28S) ribosomal DNA sequence revealed close relationship with *Histoplasma capsulatum* and *Blastomyces dermatitidis.* Therefore, by using molecular techniques, *P.brasiliensis* has been classified as member of Family **Onygenaceae,** under Order **Onygenales** and Phylum **Ascomycota**, where other dimorphic fungi are placed.

Epidemiology

Paracoccidioidomycosis (PCM) is a systemic fungal disease with restricted endemicity to Latin America (Mexico to Argentina), Brazil accounts for the largest number of cases. The causative dimorphic fungus, *P.brasiliensis,* exists as saprotroph in nature. The infection is generally considered to follow inhalation of conidia from air, with lungs being usual site of primary infection. The disease, in its inception and during subsequent development, presents more or less similar to blastomycosis or coccidioidomycosis. The organism enters human host usually via respiratory route, followed by hematogenous dissemination to different body sites as is manifested in different clinical forms of paracoccidioidomycosis.

The infection is acquired at an early age and before puberty, the prevalence is equal in both sexes. The highest incidence of clinical disease is found in adults, predominantly in men to the extent of about 90%, aged between 20-50 years. PCM has a peculiar gender distribution with preference for adult males at a ratio of ≥ 11 to 1 and afflicts predominantly adult males engaged in agriculture. It has been suggested that this marked sex difference could be explained by inhibitory action displayed by estrogen on mycelial-to-yeast (M→Y) transition of this dimorphic fungus. The inhibition of yeast phase is at least partially dependent on binding of estrogen by fungus which may account for distinct gender difference in the incidence of this disease. The disease is not transmitted directly from person-to-person.

This clinical entity is geographically restricted to Central and South America extending from Mexico down to Argentina between latitudes 23° N - 35° S. It is commonly found in Brazil, Argentina, Venezuela, Colombia and Ecuador **(Fig. 19.2)**. In Brazil, it is most commonly distributed in south, central and southeast areas with high prevalence. In geographical areas where paracoccidioidomycosis is highly endemic, incidence of disease is estimated to be

Fig. 19.2. Map showing endemic geographical areas of paracoccidioidomycosis in Central and South American Countries.

approximately 1-3 clinical cases per 100,000 inhabitants per year. The endemic areas have been delineated by data taken from extensive skin test surveys and case reports. The Paracotos area of Venezuela, a coffee growing belt, is characterized by 'humid mountain forest' and shows high skin sensitivity rate to this infection.

The disease is rare in dry north-eastern states and tropical rain forest areas of Manaus and Amazons in Brazil. In the continents where disease is endemic, it has not been reported from Guyana, Suriname, French Guiana and Chile in South America; Nicaragua and Belize in Central America and Caribbean islands. The absence of cases in some of these Latin American countries is thought to be due to lack of favorable climatic conditions for the growth of *P.brasiliensis*.

The patients reported from outside endemic areas have either traveled or previously resided there. More than 60 such scattered cases have been reported from North America, Europe, Africa, Middle East and Japan. Out of these, 18 cases of this disease have been reported from Japan by the year 2011. The imported cases are expected to continue rising because of increase in frequent international travel and immigration. However, no case of this disease has yet been reported from India.

Similar to *Blastomyces dermatitidis*, *P.brasiliensis* is also difficult to be cultured from soil even in endemic areas. Its ecological niche has not been determined and has proved to be a difficult task for medical mycologists. Several factors have contributed to this intricacy such as rare isolation of fungus from environment and prolonged latent period of disease. Although micro-niche of *P.brasiliensis* in nature is still unknown, it has been speculated that fungus exists as saprotroph in soil where it grows in mycelial form. It is thought that fungus resides in environment that has high humidity and average temperature of about 23°C. In endemic areas, disease is also found to be associated with AIDS where mortality is as high as 30-45%. The HIV infection is rampant in urban areas of south-central Brazil where incidence of this disease is also very high.

The dormancy of fungus may be the reason for no outbreaks reported with this disease. *P.brasiliensis* was first isolated from armadillos found in Amazonian region where mycosis is uncommon. Now, it has been described to be frequently isolated from nine-banded armadillos (*Dasypus novemcinctus*) from **hyperendemic** region of disease. The fungus has been isolated from soil of coffee plantation farm, which suggests that this is one of its habitats and supports hypothesis of acquisition of disease during agricultural activity in these areas. It is still a serious health problem in endemic areas, where rural workers continue to be heavily affected. There are areas where up to 60% of population is skin test positive in these endemic countries.

The genotypic studies revealed variations of *P.brasiliensis*, which was assigned to the cryptic species PS1 (more common and isolated from different regions of South America), PS2 (isolates from Brazil and Venezuela), PS3 (isolates from Colombia) and PS4 (apparently limited to Venezuela). *P.lutzii*, a new species of the genus *Paracoccidioides*, was also identified by means of multilocus sequencing studies. The recent data contribute to a better understanding of the ecology of *P.lutzii*, suggesting that it is not restricted to specific regions of Brazil as previously contemplated.

Immunity

In this disease cell-mediated immunity is considered as the most important host defence mechanism. There is reactivation of this disease as a result of immune deficiency that takes place in patients of AIDS, those undergoing transplantation and cancer treatment.

The external layer of the cell wall is electron dense and in the yeast form it has predominantly α-1-3-glucan whereas in the mycelial form it has β-1-3-glucan fibrils. The internal layer is electron transparent, thick, composed of chitin and juxtaposed to the cytoplasmic membrane. The amount of α-1-3-glucan in the cell wall correlates directly with degree of virulence of the fungus. The main antigen, which elicits humoral and cellular immune responses, is 43-kilodalton glycoprotein (gp43). This antigen was first isolated by Puccia *et al*, from concentrated dialyzed supernatant fluids of fungal cultures in yeast phase.

However, several other glycoproteins have been described as being widely recognized in the sera of patients of this disease. These antigens may also be relevant in immune response of host, immunopathogenesis and diagnosis of this fungal infection. There is cross-reaction of gp43 with antigens of other fungi like *Lacazia loboi*. There is high production of γ-globulin and specific antibodies but severe depletion of cellular immunity facilitates growth and dissemination of this causative fungus.

Pathogenesis and Pathology

The organism causing this disease gets converted into yeast form after taking entry into the host. This process is regarded as primary step for establishment of infection. However, only small percentage of individuals infected by *P.brasiliensis* develops the clinical disease, possibly in part because of genetically determined inter-individual variability of host immunity. The fungus is actively phagocytosed by mononuclear cells and is frequently found in cytoplasm of the giant cells. The acute form is characterized by progressive systemic involvement with abundance of viable fungi, non-specific inflammatory infiltrate and lack of granulomatous pattern. It is associated with depressed cellular immune response and high levels of specific antibodies.

The chronic form is characterized by clinical evolution with localized or systemic involvement and lesions present granulomatous pattern with very few fungi. The morphogenesis of granulomatous reaction in this disease is not well-developed. However, this reaction has been associated with many factors, some of them related to fungus and others related to host response. The clinical and experimental evidence indicates relevance of humoral and/or cellular immune responses in pathogenesis and evolution of this disease.

CLINICAL FEATURES

The clinical manifestations of paracoccidioidomycosis range from an asymptomatic **infection** to **disease** process with either acute or chronic form. It is mostly a chronic disease with acute/subacute cases accounting for less than 15% of all reports. In 1986, an International Committee classified this disease into two main forms *i.e.* **Acute/ Subacute Form (Juvenile Type)** and unifocal or multifocal **Chronic Form (Adult Type)**, depending on evolution and localization of lesions. The acute or subacute form is characterized by involvement of reticuloendothelial system and equally affects both sexes, whereas chronic form shows predominantly pulmonary and/or mucocutaneous involvement and is prevalent in adult males. More than 90% of patients who progress to symptomatic disease are males with estrogen-mediated inhibition of mycelial-to-yeast transformation and majority of them are agriculturists. These clinical forms have been described on the basis of pathogenesis and immunological status of host and are given in **Table 19.1**.

Table 19.1. Clinical Forms of Paracoccidioidomycosis.

Clinical forms
A. Paracoccidioidomycosis: Infection
B (i) Paracoccidioidomycosis: Disease (Duration Based)
1. Acute or Subacute Forms (Juvenile Type)
Severe
2. Chronic Form (Adult Type)
Unifocal
Multifocal
3. Quiescent or Sequelae (Latent Form)
B (ii) Paracoccidioidomycosis: Disease (Sites Based)
1. Mucocutaneous Lesions
2. Lymphatic Lesions
3. Visceral Lesions

A. Paracoccidioidomycosis: Infection

The infection is of subclinical type and is detected only by the skin test positivity. The individuals who have specifically positive skin test reaction, with no clinical sign or symptoms, are categorized under the heading of infection. Some of these patients may show cross-reaction with individuals who have been exposed to other fungi such as *Histoplasma capsulatum*.

B. (i) Paracoccidioidomycosis: Disease (Based on Duration)

The disease often presents as **triad** of pulmonary, oral and skin lesions. It begins as indolent, relatively asymptomatic pneumonia but spreads via bloodstream and lymphatic system to mucocutaneous surfaces, gastrointestinal tract, lymph nodes and other organs of body thereby almost all parts of body. The disease has multiple manifestations but two progressive clinical forms are recognized *i.e.* acute and chronic. The various forms of disease are described as follows:

1. Acute/Subacute Form (Juvenile Type)

In juvenile type of the disease, patients present with subacute or acute onset of disease. It is called as juvenile because most of patients are in their first three decades of life. This is less common variety and accounts for about 5% of cases and usually affects young adults and has rapidly progressive course with high mortality rate. It is characterized by involvement of reticuloendothelial system and is found equally in both sexes. It can present as an acute infection, usually in children and adolescents, in which superficial and/or visceral lymph node enlargement is major manifestation.

It is divided in two subtypes-**grave**, with rapid onset and progressive course with poor general condition affecting lymph nodes, liver, spleen and bone marrow and **moderate**, with less rapid onset and progressive course than grave type, with better general condition affecting only one system or lymph channels. The signs and symptoms often are more prominent in one or other group of organs, resulting in variety of clinical pictures; (a) gastrointestinal symptoms like diarrhea, abdominal pain and/or masses sometimes masquerading as Crohn's disease; (b) signs and symptoms of osteoarticular involvement, simulating arthritis or osteomyelitis; (c) acute abdominal syndrome. Mucosal involvement is infrequent and pulmonary involvement is found to be rare, although disease may begin as respiratory infection or with cutaneous lesions.

2. Chronic Form (Adult Type)

This form of disease is long lasting with slow onset and gradual changes in general condition. It may be moderate or grave and divided in two sub-types—**unifocal** that affects only one organ or system and **multifocal** that affects more than one organ or system of body. The chronic progressive adult form accounts for majority of cases and affects predominantly males of age 30 years or more. The adult form can be restricted to lungs or disseminate even at onset of disease.

The chronic form results from reactivation of quiescent lung lesions. Usually disease begins in lungs and may then spread by bronchogenic, lymphatic or hematogenous route. Most cases present with lung involvement in addition to mucous membrane lesions of upper respiratory and digestive tracts.

The chest X-rays PA view reveals mixed alveolar and interstitial process, usually bilateral, symmetrical, parahilar and located in middle and lower lobes of lungs. Mandibular or cervical lymph node enlargement is frequently associated with oral lesions. Skin, adrenal glands, CNS and bones may also be involved.

3. Quiescent or Sequelae (Latent Form)

The paracoccidioidomycosis can either be eliminated or further progress once established, thereby, leading to formation of quiescent or latent foci in a manner dependent on both fungal and host factors. The primary infection occurs in the lungs, where it causes chronic granulomatous inflammation of the parenchyma. However, this disease can progress by fungal proliferation and dissemination through both hematogenous and lymphatic routes, with an outcome compromising several extrapulmonary tissues. This is rather residual form of paracoccidioidomycosis and shows signs or symptoms of old active wound with scars.

B. (ii) Paracoccidioidomycosis: Disease (Based on Anatomical Sites)

The paracoccidioidomycosis may also be described in relation to involvement of various anatomical sites and lesions produced thereby, which are briefly described below:

1. Mucocutaneous Lesions

The lesions of mucous membrane and skin are either due to direct inoculation or secondary manifestations. These lesions are usually painful and appear infiltrated or ulcerative. The ulcerative lesions are found in oral cavity, on lips, tongue, conjunctiva, eyelids, mucous membrane of body orifices and less commonly on nasopharynx or larynx and may cause dysphonia. The ulcers become granulomatous and spread over mucous membranes, so-called mulberry-like erosions. The skin lesions can be extremely polymorphic and often appear as warty, hyperkeratotic plaques or ulcerative and crusted. They are commonly found near mouth, anus or genitalia. The mucocutaneous lesions may be seen without apparent lung involvement. There is one case report about Parinaud's oculoglandular syndrome, which is a curious medical condition characterized by granulomatous conjunctivitis associated with homolateral cervical and either pre- or retro-auricular lymphadenopathy. However, it has also been associated with many other infectious diseases.

2. Lymphatic Lesions

In lymphatic type of lesions, lymphadenopathy commonly begins first in cervical lymph nodes on one side and then infection involves other side and further spreads on to submaxillary, preauricular and supraclavicular areas. When lymph nodes are massively enlarged there is **bull neck** appearance. The lymphadenopathy is occasionally complicated by draining sinuses and may involve cervical or mediastinal lymph nodes. Sometimes they are transformed into abscesses or caseous necrosis, opening on skin through sinuses or ulcerations and clinically resemble scrofuloderma.

3. Visceral Lesions

In visceral lesions, there is disseminated infection classically presenting with fever, weight loss and anorexia as well as triad of pulmonary, mucosal and skin lesions. The lung involvement is found in almost 90% of the cases. The lesions may be in CNS, bones, joints, eye and endocrine glands. The lung involvement usually presents non-specifically with progressive cough and dyspnea with diffuse inspiratory crepitations. The radiological examination of chest often reveals bilateral patchy pneumonia with nodular, miliary shadows and/or cavities superimposed on diffuse interstitial or fibrotic pattern, typically involving mid and lower lung zones sparing apices.

This disease may be associated with tuberculosis in 10-12% cases, which may occur before or after progressive form of this mycosis. The frequency of this association varies from region to region. Neuroparacoccidioidomycosis occurs more frequently than previously recognized and can be fatal if treatment is delayed. In CNS, this fungus produces basal meningitis or isolated granulomas (pseudotumor forms) in spinal cord, brain stem, cerebellum or cerebral hemispheres. **Paracoccidioidoma**, a large cavitary mass, designates an asymptomatic solitary pulmonary paracoccidioidal nodule enclosed by fibrous capsule and occasionally calcified. The common sequelae of treated disease are fibrotic scarring in affected organs, particularly lungs. The adrenal involvement may be permanent in the form of chronic adrenal insufficiency. The infection is difficult to eradicate and relapses are common. Spondylodiscitis caused by the fungus *P.brasiliensis* is uncommon. However, in patients with chronic low-back pain who live or used to live in endemic regions, this infection should be considered as a possible differential diagnosis. Carpal tunnel syndrome may also be observed in patients of paracoccidioidomycosis.

Differential Diagnosis

Paracoccidioidomycosis should be distinguished from other wasting diseases such as pulmonary tuberculosis, histoplasmosis, cryptococcosis, mucocutaneous leishmaniasis, leprosy, sporotrichosis, actinomycosis, rhinoscleroma and disseminated malignancy. The lymphangitic form simulates tuberculosis or Hodgkin's disease. The diffuse skin eruptions may assume multiple aspects resembling syphilis, psoriasis and lymphomas. When localized lesions are present they must be differentiated from leishmaniasis, sporotrichosis, tuberculosis and chromoblastomycosis. Lacaziosis has also been confused with paracoccidioidomycosis but upon isolation, the fungus turned out to be *P.brasiliensis*. If lesions have sarcoid-like appearance, they simulate systemic lupus erythematosus or sarcoidosis.

There is some resemblance to blastomycosis, due to many similarities with this disease but the endemic areas do not overlap, which is predominantly prevalent in North America. The differential diagnosis of neuro-paracoccidioidomycosis is cerebral abscess, gliomas, metastasis and neurocysticercosis. But histopathologically, the lesions are well-defined with central zone of necrosis containing many fungi and surrounded by cellular infiltrate, which includes epithelioid cells, giant cells, lymphocytes and plasma cells.

Laboratory Diagnosis

A detailed history of travel to endemic areas and clinical examination are relevant to be elicited in patients before proceeding for the laboratory diagnosis. This disease should be suspected in patients with an appropriate travel history who have had weight loss and have pulmonary, mucosal and cutaneous lesions. The diagnosis is based mainly on demonstration of large, budding yeast cells with distinctive morphology in clinical material and detection of gp43 antigen or humoral antibody to this antigen. The 'gold standard' for diagnosis is isolation of fungus in culture and detection of multiple budding yeast cells by direct examination of biological fluids or biopsy specimens.

(a) Direct Examination

The direct microscopic examination in 10% KOH or CFW wet mount of sputum, BAL fluid, CSF, pus, crusts from granulomatous lesions or biopsy material reveals rounded refractile yeast cells of *P.brasiliensis,* greatly varying in size from 2-10 to 30 µm or more **(Fig. 19.3)**. They are usually present as single cell or short chain of cells, as characteristic multipolar budding is seen infrequently in clinical material, often difficult to differentiate from other yeast-like fungal pathogens. A vinyl adhesive tape (VAT) preparation (Scotch Test) has been proved to be useful for direct microscopic examination. Immunofluorescence methods are also used to demonstrate these yeast cells.

The histopathological sections are stained with H&E, in which fungal cell wall remains indistinct but nuclei are

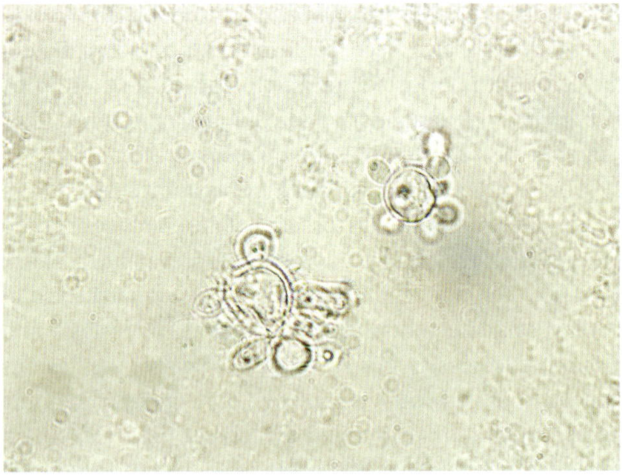

Fig. 19.3. *Paracoccidioides brasiliensis* showing multiple budding yeast cells in wet mount from a PCM patient (KOH × 400).

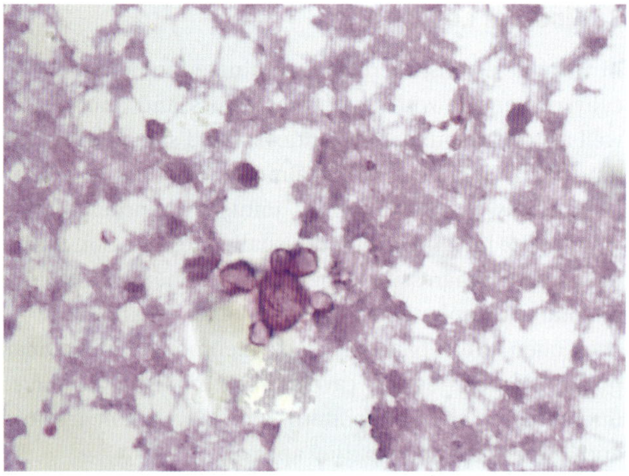

Fig. 19.4. *Paracoccidioides brasiliensis* showing multiple budding yeasts in a smear from PCM patient (PAS × 400).

often clearly delineated and are basophilic. The PAS smear may also be prepared from the clinical material **(Fig. 19.4)**. The GMS stain demonstrates the morphology best as multiple budding yeast cells **(Figs. 19.5A to C)**. The granulomata are also investigated by using immunohistochemical techniques and monoclonal antibodies against T-lymphocytes. In juvenile form histopathology reveals loose granulomata with necrosis and large number of characteristic round thick-walled multiple-budding yeast cell whereas in chronic form there are compact granulomata with epithelioid and giant cells containing characteristic forms of *P.brasiliensis*. In histological sections, yeasts may be missed or confused with other dimorphic fungi such as *Histoplasma* species, immature spherule of *Coccidioides* species and even budding yeasts of *Cryptococcus* species. Therefore, it is essential that multiple budding forms should be demonstrated for definitive evidence of *P.brasiliensis*. The Ziehl-Neelsen staining for demonstrating acid-fast bacilli is also performed to establish or rule out any co-existing tuberculosis.

Some of emerging mucormycetes, which are dimorphic in nature like *Cokeromyces recurvatus* and *Mucor circinelloides*, may be confused with *P.brasiliensis* in direct KOH wet mount or on histopathology due to multiple buddings of their mother yeasts cells.

(b) Fungal Culture

This fungus can grow on SDA with antibacterial antibiotics and actidione; however, the growth of both yeast and mycelial phases of *P.brasiliensis* are slow on primary isolation. The growth at 35-37°C appears within week or so, which is yeast-like soft, off-white to cream colored, wrinkled and rough to pasty in appearance **(Fig. 19.6)**. On BHI agar, the colonies are butyrous and cerebriform. The LCB mount shows oval or globose multiple budding yeast cells 3-30 μm in size with small buds attached by narrow neck, encircling the parent cell **(Fig. 19.7)**. The colonies at 25°C are flat to wrinkled, leathery at first and later on developing tufts of aerial mycelia, which eventually cover colonies. The color of mycelial colony varies from white to tan, with yellowish-brown reverse. *Chrysosporium* species is morphologically similar to *B.dermatitidis* and may also be confused with *P.brasiliensis*. The slide culture of mycelial isolate is useful to induce sporulation and thereby to study their arrangement. The asexual conidia have no characteristic appearance in *P.brasiliensis* and hence its identification depends on conversion from mycelial-to-yeast (M→Y) phase. The conversion can be achieved on BHI agar supplemented with 5% sheep blood or glutamine, which takes 14 days time or longer. The exoantigen test by using antigens 1, 2 and 3 is also useful for specific identification of fungus at an early stage of incubation **(*See* Appendix C)**.

(c) Immunodiagnosis

The immunodiagnosis of paracoccidioidomycosis is based on skin testing and serology. It may be of less importance among immunocompromised patients because of reduced antibody production.

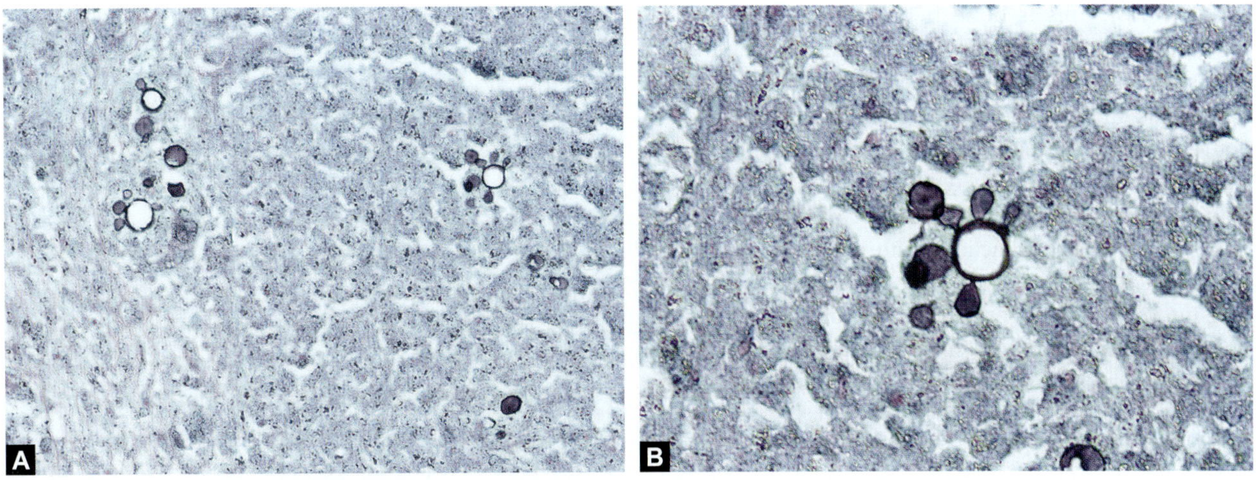

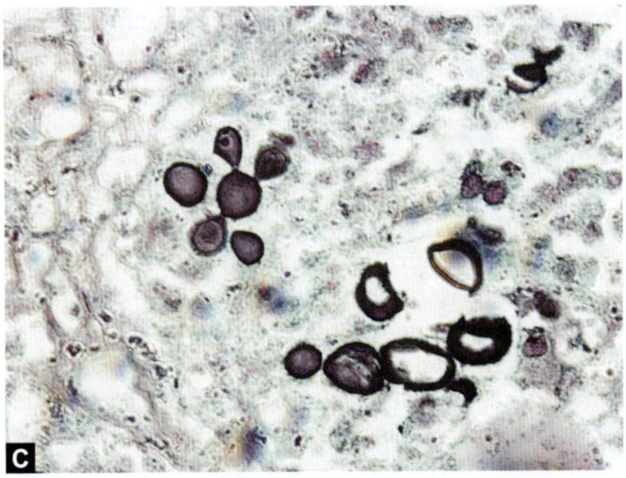

Figs. 19.5A to C. *P.brasiliensis* showing multiple budding yeasts in tissue sections from a PCM patient. (GMS × 200, 400 & 400).

Fig. 19.6. *Paracoccidioides brasiliensis* showing off-white, wrinkled, pasty yeast colonies on SDA after 1 & 2 weeks at 37°C.

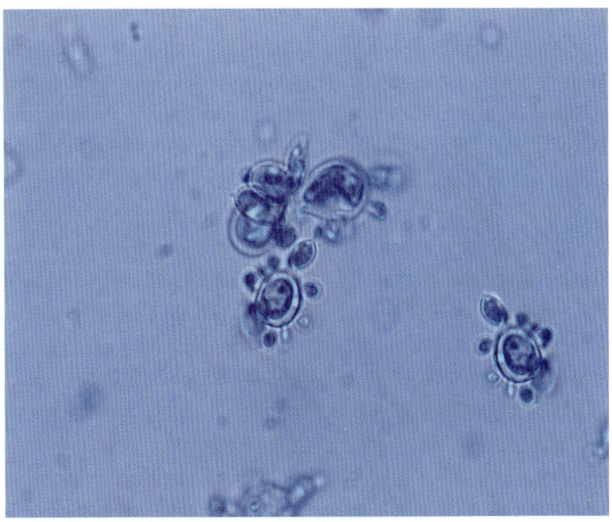

Fig. 19.7. Paracoccidioides *brasiliensis* showing multiple budding yeasts from the culture on SDA at 37°C (LCB × 400).

(i) Skin Test: This is done by intradermal injection of **paracoccidioidin** and local site is examined for induration. A positive skin test may represent previous exposure to the fungus and does not necessarily infer presence of an active disease. There is limited diagnostic value of skin tests because of negative reactions even in individuals with advanced disease or in immunocompromised condition. Moreover, cross-reactions with other fungal antigens in patients of histoplasmosis and sporotrichosis may confuse interpretation of results. The use of purified antigen gp43 is found to be more specific. The paracoccidioidin skin test reactivity has prognostic value during the course of therapy, as it indicates, at least, partial restoration of cellular immunity. The epidemiological surveys are conducted mainly by skin tests among population residing in endemic areas.

(ii) Serological Tests: The serological tests are useful for diagnosis as well as for follow-up of patients as prognostic indicator for monitoring response to therapy. The serological diagnosis is based on detection of specific antibodies and various methods for detecting circulating antigens. The main diagnostic antigen is gp43 and sera of about 90% of patients react with this antigen in immunodiffusion tests and virtually 100% are positive in immunoblot assays.

The tests for antibodies detection like immunodiffusion, complement fixation and ELISA are useful for diagnosis and prognosis. The double immunodiffusion test for circulating antibodies has sensitivity and specificity of 91% and 100%, respectively. A positive immunodiffusion test usually indicates presence of an active lesion even in absence of clinical manifestations. The number of precipitin bands correlates positivity with severity of disease. Counterimmunoelectrophoresis (CIE) is also very useful with about 95% sensitivity and 100% specificity among these patients.

The complement fixation test is of low specificity but is quantitative and can be used for follow up of response to therapy. The precipitin tests and complement fixation tests, when used together, detect about 98% of infections. The estimation of antibody response in paracoccidioidomycosis by CIE and ELISA shows better clinical correlation than complement fixation.

The ELISA has given new perception to diagnosis for detection and quantitation of IgG antibodies against gp43 antigen of *P.brasiliensis*. It is introduced because it allows processing of large number of samples simultaneously and uses crude preparation, which have at least 18 antigens, thus being more representative of whole fungus, whereas CIE employs culture filtrate antigen whose major component is 43-kDa fraction. The cross-reactions particularly in precipitin test with other fungi may rarely be found. The antibody levels decline on successful therapy.

The detection of *P.brasiliensis* circulating antigens in body fluids offers potential advantages over testing for antibody, both for initial diagnosis and for follow up of patients. Several techniques for detection of antigenemia in paracoccidioidomycosis have been utilized, with variable results.

New strategies for detection of circulating antigens by use of monoclonal antibodies have been studied. Antigen detection tests may be more effective than antibody tests for diagnosing this disease mainly in immunocompromised patients, who have decreased humoral antibody response. The newly developed immunoblot and EIA tests are capable of detecting antigens in urine specimens also. The detection of *P.brasiliensis* antigens in urine thus appears to be promising method for diagnosis and monitoring outcome of treatment thereby reducing chances of relapse in infection. The quantitation of 87-kDa *P.brasiliensis* antigen is useful in follow up of patients with acute, multifocal and unifocal forms of disease.

Recently, hybridization techniques as molecular diagnostic tools have been introduced for detection of this systemic fungal infection. *P.brasiliensis* PCR assay for diagnosis using 18S rRNA may be advantageous due to its rapidity, sensitivity and minimized health risk as compared to other methods of diagnosis in tissue and has been evaluated in experimental animals like mice.

Moreover, MALDI-TOF can also be used as the method of choice for species differentiation of genus *Paracoccidioides* benefiting clinical and diagnostic studies aiming to determine possible differences between paracoccidioidomycosis caused by *P.brasiliensis* and *P.lutzii* species.

(d) Animal Pathogenicity

As such the experimental animal models have contributed much to the knowledge of fungal infections, pathogenesis and their corresponding therapeutic modalities. This holds true for the animal models used in case of the primary pathogens, *P.brasiliensis.* This is experimentally shown that mice, guinea pigs, rabbits, rats and hamsters (*Mesocricetus auratus*) are susceptible to infection by *P.brasiliensis.* The lesions produced are histopathologically very similar to those observed in human beings. The disease has been demonstrated in internal organs of nine-banded armadillos (*Dasypus novemcinctus*), which is one of the natural reservoir/host and may provide valuable clues in search for natural habitat of this fungus. The armadillos, like humans are probably infected by airborne conidia of *P.brasiliensis*. It is unlikely that disease is transmitted to humans by animals since yeast form of fungus as found in armadillo tissues is of low infectivity. The geographical distribution of this animal overlaps areas of endemicity of paracoccidioidomycosis in various Latin American countries.

Treatment and Prophylaxis

The patients require long-term antifungal therapy of about 6-12 months or even longer duration and should be reviewed periodically as relapses are common with different therapeutic regimens. The standard antifungals are itraconazole and fluconazole with amphotericin B followed by sulfamethoxazole or trimethoprim-sulfamethoxazole. Allylamine like terbinafine has also been successfully used without significant side effects and no sign of relapse after completion of the treatment.

The amphotericin B was first used in late 1950s and is still prescribed in severe disseminated disease. A cumulative dose of 1.2–3.0 gm administered intravenously, is combined with sulfonamides as amphotericin B alone is not usually curative. A combination therapy for severe disseminated disease produces remission of symptoms in about 50-70% cases but as many as 20-30% successfully treated patients get relapse.

A triazole derivative, itraconazole is highly effective in its treatment. It has proved to be about ten times more active than ketoconazole because of negligible interference with liver, gonadal or adrenal functions or bone marrow integrity. There are fewer side effects and lower relapse rates hence it is the drug of choice in this disease. The patients with acute juvenile form require treatment for 6 months or more whereas with chronic form, especially those with unifocal lesions require treatment only for 3-6 months. On the other hand, some of patients with multifocal involvement need therapy for 12 months or more. The high dose of itraconazole may suppress adrenal gland function.

The sulfonamides were first introduced in 1940s even prior to the introduction of amphotericin B and are the mainstay in therapy of this disease. Like pneumocystosis this systemic mycosis is also amenable to sulfonamides, alone or in combination with trimethoprim. Therefore, either sulfadiazine or any of long-acting compounds like sulfadimethoxine may be given. Due to low cost and relative low toxicity, sulfonamides constitute an important part in the treatment especially in developing countries. The sulfonamide in doses of 4-6 gm/day is shown to be 70% effective in inducing remission. It is relatively inexpensive and non-toxic agent that has fallen out of favor because of requirement of prolonged treatment of about five years and high relapse rate ~25% following its withdrawal.

There is scarcity of reports of this disease among AIDS patients, which may be explained by widespread use of trimethoprim-sulfamethoxazole as prophylactic agent for *Pneumocystis* pneumonia thereby indirectly providing treatment for this disease. Despite specific therapy with different regimens, an overall mortality among AIDS patients is very high. The specific diagnosis in this disease is established late although available therapy is usually successful in controlling this disease however patients who survive usually develop residual fibrotic lesions that heavily impair their quality of life.

As there is no precise niche of this fungus therefore the preventive strategies cannot be contemplated. The DNA-based vaccine has been tried against murine paracoccidioidomycosis by using gp43 gene from *P.brasiliensis*.

Further Reading

1. Abreu e Silva MA, Salum FG, Figueiredo MA, *et al.* Important aspects of oral paracoccidioidomycosis - A literature review. *Mycoses.* 2013; **56:** 189-99.
2. Abreu MAMM, Lastoria JC, Mattos ALA, *et al.* Paracoccidioidomycosis with sarcoid-like lesions: A diagnostic challenge. *Rev Soc Bras Med Trop.* 2017; **50:** 273-6.
3. Abud LG, Queiroz RM, Abud TG. Concomitant pulmonary and central nervous system paracoccidioidomycosis with cerebellar abscess. *Rev Soc Bras Med Trop.* 2015; **48:** 789.
4. Agudelo CA, Munoz C, Ramirez A, *et al.* Identification of *Paracoccidioides brasiliensis* in adrenal glands biopsies of two patients with paracoccidioidomycosis and adrenal insufficiency. *Rev Inst Med Trop Sao Paulo.* 2009; **51:** 45-8.
5. Albano AP, Klafke GB, Brandolt TM, *et al.* Wild animals as sentinels of *Paracoccidioides brasiliensis* in the state of Rio Grande do Sul, Brazil. *Mycopathologia.* 2014; **177:** 207-15.
6. Albuquerque CF, da Silva SH, Camargo ZP. Improvement of specificity of an enzyme-linked immunosorbent assay for diagnosis of paracoccidioidomycosis. *J Clin Microbiol.* 2005; **43:** 1944-6.
7. Alegre-Maller AC, Mendonca FC, da Silva TA, *et al.* Therapeutic administration of recombinant Paracoccin confers protection against *Paracoccidioides brasiliensis* infection: Involvement of TLRs. *PLoS Negl Trop Dis.* 2014; **8:** e3317. PMID: 25474158.

8. Alencar Marques S, Petrechen L, Kakuda AC, *et al.* Paracoccidioidomycosis and HIV coinfection presenting severe disease even with CD4+ cells in the normal range. *Rev Iberoam Micol.* 2014; **31:** 154-5.
9. Almeida FA, Neves FF, Mora DJ, *et al.* Paracoccidioidomycosis in Brazilian patients with and without human immunodeficiency virus infection. *Am J Trop Med Hyg.* 2017; **96:** 368-72.
10. Alvarenga JA, Martins DE, Kanas M, *et al.* Paracoccidioidomycosis in the spine: Case report and review of the literature. *Sao Paulo Med J.* 2016; **134:** 263-7.
11. Alvarez M, Pina DR, de Oliveira M, *et al.* Objective CT-based quantification of lung sequelae in treated patients with paracoccidioidomycosis. *Medicine.* 2014; **93:** e167.
12. Alves FL, Ribeiro MA, Hahn RC, *et al.* Transposable elements and two other molecular markers as typing tools for the genus *Paracoccidioides. Med Mycol.* 2015; **53:** 165-70.
13. Alves GR, Lauda Filho JD, Haygert CJ. Neuroparacoccidioidomycosis. *Rev Soc Bras Med Trop.* 2011; **44:** 654.
14. Ameen M, Talhari C, Talhari S. Advances in paracoccidioidomycosis. *Clin Exp Dermatol.* 2010; **35:** 576-80.
15. Amorim Pellicioli AC, Neves-Silva R, Santos-Silva AR, *et al.* Synchronous oral paracoccidioidomycosis and pulmonary tuberculosis in an immunocompetent patient. *Mycopathologia.* 2015; **179:** 459-64.
16. Andrade MG, Medrado AP, de Brito IC, *et al.* Oral paracoccidioidomycosis: A case without lung manifestations. *J Contemp Dent Pract.* 2007; **8:** 92-8.
17. Arantes TD, Bagagli E, Nino-Vega G, *et al. Paracoccidioides brasiliensis* and *Paracoccidioides lutzii,* A secret love affair. *Rev Inst Med Trop Sao Paulo.* 2015; **57 (Suppl. 19):** 25-30.
18. Arantes TD, Theodoro RC, Da Graca Macoris SA, *et al.* Detection of *Paracoccidioides* spp. in environmental aerosol samples. *Med Mycol.* 2013; **51:** 83-92.
19. Araujo SA, Espindola BM, Pedroso ER. Cutaneous disseminated paracoccidioidomycosis. *Am J Trop Med Hyg.* 2012; **86:** 1.
20. Araujo Sde A, Prado Lde G, Veloso JM, *et al.* Case of recurrent paracoccidioidomycosis: 25 years after initial treatment. *Braz J Infect Dis.* 2009; **13:** 394.
21. Assis-Marques MA, Oliveira AF, Ruas LP, *et al. Saccharomyces cerevisiae* expressing Gp43 protects mice against *Paracoccidioides brasiliensis* infection. *PLoS One.* 2015; **10:** e0120201.
22. Auwaerter PG, Dove J, Mackowiak PA. Simon Bolivar's medical labyrinth: An infectious diseases conundrum. *Clin Infect Dis.* 2011; **52:** 78-85.
23. Azenha MR, Caliento R, Brentegani LG, *et al.* A retrospective study of oral manifestations in patients with paracoccidioidomycosis. *Braz Dent J.* 2012; **23:** 753-7.
24. Azevedo RS, Gouvea AF, Lopes MA, *et al.* Synchronous oral paracoccidioidomycosis and oral squamous cell carcinomas with submandibular enlargement. *Med Mycol.* 2011; **49:** 84-9.
25. Barbosa de Paulo LF, Silva de Faria L, Francisco Durighetto A Jr. Endemic oral paracoccidioidomycosis: Clinical presentation, management and outcomes. *Int J Infect Dis.* 2014; **19:** 109-10.
26. Barreto MM, Marchiori E, Amorim VB, *et al.* Thoracic paracoccidioidomycosis: Radiographic and CT findings. *Radiographics.* 2012; **32:** 71-84.
27. Barreto MM, Marchiori E, de Brito A, *et al.* CT morphological features of the reversed halo sign in pulmonary paracoccidioidomycosis. *Br J Radiol.* 2015; **88 (1055):** 20150246.
28. Barrozo LV, Benard G, Silva ME, *et al.* First description of a cluster of acute/subacute paracoccidioidomycosis cases and its association with a climatic anomaly. *PLoS Negl Trop Dis.* 2010; **4:** e643.
29. Barrozo LV, Mendes RP, Marques SA, *et al.* Climate and acute/subacute paracoccidioidomycosis in a hyper-endemic area in Brazil. *Int J Epidemiol.* 2009; **38:** 1642-9.
30. Batista J Jr, de Camargo ZP, Fernandes GF, *et al.* Is the geographical origin of a *Paracoccidioides brasiliensis* isolate important for antigen production for regional diagnosis of paracoccidioidomycosis? *Mycoses.* 2010; **53:** 176-80.
31. Batista MV, Sato PK, Pierrotti LC, *et al.* Recipient of kidney from donor with asymptomatic infection by *Paracoccidioides brasiliensis. Med Mycol.* 2012; **50:** 187-92.
32. Batista VG, Toledo MS, Straus AH, *et al.* Glycolipid sensing and innate immunity in paracoccidioidomycosis. *Mycopathologia.* 2014; **178:** 153-62.
33. Bellissimo-Rodrigues F, Bollela VR, Da Fonseca BA, *et al.* Endemic paracoccidioidomycosis: Relationship between clinical presentation and patients' demographic features. *Med Mycol.* 2013; **51:** 313-8.
34. Bellissimo-Rodrigues F, Machado AA, Martinez R. Paracoccidioidomycosis epidemiological features of a 1,000-cases series from a hyperendemic area on the southeast of Brazil. *Am J Trop Med Hyg.* 2011; **85:** 546-50.
35. Bellissimo-Rodrigues F, Vitali LH, Martinez R. Serological diagnosis of paracoccidioidomycosis in HIV-coinfected patients. *Mem Inst Oswaldo Cruz.* 2010; **105:** 904-7.
36. Benard G. An overview of the immunopathology of human paracoccidioidomycosis. *Mycopathologia.* 2008; **165:** 209-21.
37. Benard G, Campos AF, Netto LC, *et al.* Treatment of severe forms of paracoccidioidomycosis: Is there a role for corticosteroids? *Med Mycol.* 2012; **50:** 641-8.
38. Benard G, Costa AN, Leopercio AP, *et al.* Chronic paracoccidioidomycosis of the intestine as single organ involvement points to an alternative pathogenesis of the mycosis. *Mycopathologia.* 2013; **176:** 353-7.
39. Benard G, Costa AN, Ravanini J, *et al.* Fatal acute respiratory distress syndrome in a patient with paracoccidioidomycosis: First case report. *Med Mycol.* 2010; **48:** 542-5.
40. Benard G, Patzina RL, Schwab JB, *et al.* Fatal septic shock due to a disseminated chronic form of paracoccidioidomycosis in an aged woman. *Med Mycol.* 2012; **50:** 407-11.
41. Berni AD, Elias Junior E, Arao Filho A, *et al.* Protein-losing enteropathy in paracoccidioidomycosis identified by scintigraphy: Report of three cases. *Braz J Infect Dis.* 2010; **14:** 540-3.
42. Bertoni TA, Perenha-Viana MC, Patussi EV, *et al.* Western blotting is an efficient tool for differential diagnosis of

paracoccidioidomycosis and pulmonary tuberculosis. *Clin Vaccine Immunol.* 2012; **19:** 1887-8.
43. Bocca AL, Amaral AC, Teixeira MM, *et al.* Paracoccidioidomycosis: Eco-epidemiology, taxonomy and clinical and therapeutic issues. *Future Microbiol.* 2013; **8:** 1177-91.
44. Bonilla-Abadia F, Velez JD, Zarate-Correa LC, *et al.* Overinfection by *Paracoccidioides brasiliensis* in gouty crystal arthritis. *Case Rep Med.* 2012; 128103.
45. Borges SR, Silva GM, Chambela Mda C, *et al.* Itraconazole *vs.* trimethoprim-sulfamethoxazole: A comparative cohort study of 200 patients with paracoccidioidomycosis. *Med Mycol.* 2014; **52:** 303-10.
46. Braga Gde M, Hessel G, Pereira RM. Hepatic involvement in pediatric patients with paracoccidioidomycosis: A clinical and laboratory study. *Mycopathologia.* 2013; **176:** 279-86.
47. Bravo EA, Zegarra AJ, Piscoya A, *et al.* Chronic diarrhea and pancolitis caused by paracoccidioidomycosis: A case report. *Case Rep Med.* 2010; 140505.
48. Bravo EA, Zegarra AJ, Piscoya A, *et al.* Dimorphic fungal coinfection as a cause of chronic diarrhea and pancolitis. *Case Rep Med.* 2011; 960638.
49. Brazao-Silva MT, Andrade MF, Franco T, *et al.* Paracoccidioidomycosis: A series of 66 patients with oral lesions from an endemic area. *Mycoses.* 2011; **54:** e189-95.
50. Brunaldi MO, Rezende RE, Zucoloto S, *et al.* Co-infection with paracoccidioidomycosis and human immunodeficiency virus: Report of a case with esophageal involvement. *Am J Trop Med Hyg.* 2010; **82:** 1099-101.
51. Buccheri R, Khoury Z, Barata LC, Benard G. Incubation period and early natural history events of the acute form of paracoccidioidomycosis: Lessons from patient with a single *Paracoccidioides* spp. exposure. *Mycopathologia.* 2016; **181:** 435-9.
52. Buitrago MJ, Bernal-Martinez L, Castelli MV, *et al.* Histoplasmosis and paracoccidioidomycosis in a non-endemic area: A review of cases and diagnosis. *J Travel Med.* 2011; **18:** 26-33.
53. Cabral LA, Lima CF, Coutinho de Oliveira ML, *et al.* Oral exfoliative cytology for the diagnosis of paracoccidioidomycosis in a patient with human immunodeficiency virus: A case report. *Acta Cytol.* 2010; **54:** 1127-9.
54. Cabrera LG, Santos AF, Andrade UV, *et al.* Spontaneous pneumothorax in paracoccidioidomycosis patients from an endemic area in Midwestern Brazil. *Mycoses.* 2017; **60:** 124-8.
55. Caldeira AC, Cas Kd, Pinto TP, *et al.* Small vessel vasculitis as first manifestation of paracoccidioidomycosis - case report. *An Bras Dermatol.* 2011; **86:** 1208-12.
56. 000. Camacho E, Nino-Vega GA. *Paracoccidioides* spp.: Virulence factors and immune-evasion strategies. *Mediators Inflamm.* 2017; 5313691. PMID: 28553014.
57. Campos FF, Johann S, Cota BB, *et al.* Antifungal activity of trichothecenes from *Fusarium* sp. against clinical isolates of *Paracoccidioides brasiliensis. Mycoses.* 2011; **54:** e122-9.
58. Carlesse F, Santolaya ME. Important mycoses in children in South America. *Curr Fungal Infect Rep.* 2016; **10:** 10-23.
59. Carneiro Rda C, Miranda BG, Camilo Neto C, *et al.* Juvenile paracoccidioidomycosis in urban area: Report of two cases. *Braz J Infect Dis.* 2010; **14:** 77-80.
60. Catano JC, Aguirre HD. Disseminated paracoccidioidomycosis. *Am J Trop Med Hyg.* 2013; **88:** 407-8.
61. Catano JC, Morales M. Cutaneous paracoccidioidomycosis. *Am J Trop Med Hyg.* 2015; **93:** 433-4.
62. Cavalcante Rde S, Sylvestre TF, Levorato AD, *et al.* Comparison between itraconazole and cotrimoxazole in the treatment of paracoccidioidomycosis. *PLoS Negl Trop Dis.* 2014; **8:** e2793.
63. Cazzo E, Ferrer JA, Chaim EA. Obstructive jaundice secondary to paracoccidioidomycosis. *Trop Gastroenterol.* 2015; **36:** 46-7.
64. Cermeno J, Cermeno J, Godoy G, *et al.* Epidemiological study of paracoccidioidomycosis and histoplasmosis in a suburb of San Felix city, Bolivar state, Venezuela. *Invest Clin.* 2009; **50:** 213-20.
65. Coelho MG, Severo CB, de Mattos Oliveira F, *et al.* Paracoccidioidomycosis mimicking sarcoidosis: A review of 8 cases. *Mycopathologia.* 2016; **181:** 137-43.
66. Colaiacovo R, Ganc RL, Leone AC, *et al.* Diagnosis of left adrenal paracoccidioidomycosis by endoscopic ultrasound fine needle aspiration. *Endoscopy.* 2011; **43 (Suppl. 2):** e236-7.
67. Correa-de-Castro B, Pompilio MA, Odashiro DN, *et al.* Unifocal bone paracoccidioidomycosis, Brazil. *Am J Trop Med Hyg.* 2012; **86:** 470-3.
68. Corti M, Trione N, Risso D, *et al.* Disseminated paracoccidioidomycosis with a single brainstem lesion. A case report and literature review. *Neuroradiol J.* 2010; **23:** 454-8.
69. Costa AN, Benard G, Albuquerque AL, *et al.* The lung in paracoccidioidomycosis: New insights into old problems. *Clinics (Sao Paulo).* 2013; **68:** 441-8.
70. Costa AN, Marchiori E, Benard G, *et al.* Lung cysts in chronic paracoccidioidomycosis. *J Bras Pneumol.* 2013; **39:** 368-72.
71. Costa CC, Guimaraes Vde C, Rebouças MN, *et al.* Paracoccidioidomycosis of the larynx: Cases report. *Braz J Otorhinolaryngol.* 2013; **79:** 790.
72. Costa PF, Fernandes GF, dos Santos PO, *et al.* Characteristics of environmental *Paracoccidioides brasiliensis* isolates. *Mycopathologia.* 2010; **169:** 37-46.
73. Costa TA, Bazan SB, Feriotti C, *et al.* In pulmonary paracoccidioidomycosis IL-10 deficiency leads to increased immunity and regressive infection without enhancing tissue pathology. *PLoS Negl Trop Dis.* 2013; **7:** e2512.
74. Coutinho ZF, Wanke B, Travassos C, *et al.* Hospital morbidity due to paracoccidioidomycosis in Brazil (1998-2006). *Trop Med Int Health.* 2015; **20:** 673-80.
75. Cruz RC, Werneck SM, Oliveira CS, *et al.* Influence of different media, incubation times and temperatures for determining the MICs of seven antifungal agents against *Paracoccidioides brasiliensis* by microdilution. *J Clin Microbiol.* 2013; **51:** 436-43.
76. da Silva Jde F, de Oliveira HC, Marcos CM, *et al.* Advances and challenges in paracoccidioidomycosis serology caused

by *Paracoccidioides* species complex: An update. *Diagn Microbiol Infect Dis.* 2016; **84:** 87-94.

77. Dall Bello AG, Severo CB, Oliveira Fde M, *et al.* Disseminated paracoccidioidomycosis (simulating metastatic lung cancer) and *Strongyloides stercoralis* hyper-infestation in a steroid-treated patient. *J Clin Microbiol.* 2011; **49:** 2054-5.
78. de Abreu E Silva MA, Salum FG, Figueiredo MA, *et al.* Interrelationship of clinical, histomorphometric and immunohistochemical features of oral lesions in chronic paracoccidioidomycosis. *J Oral Pathol Med.* 2013; 42: 235-42.
79. de Almeida SM, Roza TH. HIV Immune recovery inflammatory syndrome and central nervous system paracoccidioidomycosis. *Mycopathologia.* 2017; **182:** 363-6.
80. de Almeida Soares CM, Mendes-Giannini MJ, Felipe MS, Chaturvedi V. A centennial: Discovery of *Paracoccidioides brasiliensis. Mycopathologia.* 2008; **165:** 179-81.
81. de Arruda PF, Gatti M, de Arruda JG, *et al.* Prostatic paracoccidioidomycosis with a fatal outcome: A case report. *Med Case Rep.* 2013; **7:** 126.
82. de Camargo ZP. Serology of paracoccidioidomycosis. *Mycopathologia.* 2008; **165:** 289-302.
83. de Carli ML, Cardoso BC, Malaquias LC, *et al.* Serum antibody levels correlate with oral fungal cell numbers and influence the patients' response to chronic paracoccidioidomycosis. *Mycoses.* 2015; **58:** 356-61.
84. de Castro LF, Ferreira MC, da Silva RM, *et al.* Characterization of the immune response in human paracoccidioidomycosis. *J Infect.* 2013; **67:** 470-85.
85. de Freitas RS, Dantas KC, Garcia RS, *et al. Paracoccidioides brasiliensis* causing a rib lesion in an adult AIDS patient. *Hum Pathol.* 2010; **41:** 1350-4.
86. de Jesus Carlos N, Pinto DA, Ono MA, *et al.* Immunoblotting of soluble antigens in *Paracoccidioides brasiliensis* culture. *Microbiol Immunol.* 2014; **58:** 212-4.
87. de Lacorte Singulani J, Scorzoni L, de Paula E Silva AC, *et al.* Evaluation of the efficacy of antifungal drugs against *Paracoccidioides brasiliensis* and *Paracoccidioides lutzii* in a *Galleria mellonella* model. *Int J Antimicrob Agents.* 2016; **48:** 292-7.
88. de Macedo PM, Almeida-Paes R, Freitas DF, Varon AG, *et al.* Acute juvenile paracoccidioidomycosis: A 9-year cohort study in the endemic area of Rio de Janeiro, Brazil. *PLoS Negl Trop Dis.* 2017; **11:** e0005500. PMID: 28355221.
89. de Melo Braga G, Hessel G, Pereira RM, *et al.* Hepatic involvement in paediatric patients with paracoccidioidomycosis: A histological study. *Histopathology.* 2014; **64:** 256-62.
90. de Oliveira Gondak R, Mariano FV, dos Santos Silva AR, *et al.* Single oral paracoccidioidomycosis mimicking other lesions: Report of eight cases. *Mycopathologia.* 2012; **173:** 47-52.
91. de Oliveira HC, Assato PA, Marcos CM, *et al. Paracoccidioides*-host Interaction: An overview on recent advances in the paracoccidioidomycosis. *Front Microbiol.* 2015; **6:** 1319.
92. de Oliveira HC, da Silva Jde F, Scorzoni L, *et al.* Importance of adhesins in virulence of *Paracoccidioides* spp. *Front Microbiol.* 2015; **6:** 303.
93. de Pinho DB, da Costa Neves T, Celem LR, *et al.* Enlarged, painful cervical and axillary lymph nodes in chronic paracoccidioidomycosis. *J Dermatol Case Rep.* 2014; **8:** 50-4.
94. de Souza SP, Jorge VM, Xavier MO. Paracoccidioidomycosis in southern Rio Grande do Sul: A retrospective study of histopathologically diagnosed cases. *Braz J Microbiol.* 2014; **45:** 243-7.
95. de Souza Vianna LM, Pirani Carneiro F, Calvalca Tavares A, *et al.* Cytological diagnosis of paracoccidioidomycosis: A report of four cases. *Diagn Cytopathol.* 2013; **41:** 374-6.
96. Di Martino Ortiz B, Moreno T, Galeano G, *et al.* Acute disseminated paracoccidioidomycosis with molluscoid lesions in a young woman. *Actas Dermosifiliogr.* 2015; **106:** 597-9.
97. Di MartinoOrtiz B, Rodriguez-Oviedo ML, Rodriguez-Masi M. Chronic multifocal paracoccidioidomycosis in an immunocompetent adult. *Actas Dermosifiliogr.* 2012; **103:** 645-6.
98. Dias L, de Carvalho LF, Romano CC. Application of PCR in serum samples for diagnosis of paracoccidioidomycosis in the southern Bahia-Brazil. *PLoS Negl Trop Dis.* 2012; **6:** e1909.
99. Dias OM, Baldi BG, Costa AN, *et al.* Acute respiratory distress syndrome in lung paracoccidioidomycosis. *Am J Respir Crit Care Med.* 2016; **193:** e55-6.
100. Dias PH, Zeni M, Taques GR, *et al.* Genitourinary paracoccidioidomycosis complicated with urinary outflow obstruction - A report of two cases and a review of the literature. *Clinics (Sao Paulo).* 2010; **65:** 1207-10.
101. dos Santos PO, Rodrigues AM, Fernandes GF, *et al.* Immunodiagnosis of paracoccidioidomycosis due to *Paracoccidioides brasiliensis* using a latex test: Detection of specific antibody anti-gp43 and specific antigen gp43. *PLoS Negl Trop Dis.* 2015; **9:** e0003516.
102. dos Santos VM, Xavier RM, Cortes JA, *et al.* Pseudotumoral neuroparacoccidioidomycosis: One case report. *Mycopathologia.* 2008; **166:** 155-8.
103. Duani H, Nunes VR, Assumpcao AB, *et al.* Bilateral paracoccidioidomycotic iliopsoas abscess associated with ileo-colonic lesion. *Rev Soc Bras Med Trop.* 2012; **45:** 649-51.
104. Fabris LR, Andrade UV, Ferreira Dos Santos A, *et al.* Decreasing prevalence of the acute/subacute clinical form of paracoccidioidomycosis in Mato Grosso do Sul State, Brazil. *Rev Inst Med Trop Sao Paulo.* 2014; **56:** 121-5.
105. Facanha MC, Souza Ade Q, Cafe VS, *et al.* Paracoccidioidomycosis: description of two cases autochthonous in Ceara. *Braz J Infect Dis.* 2010; **14:** 86-8.
106. Fernandes FF, Alves VO, Sánchez TE, *et al.* Chylothorax in paracoccidioidomycosis. *Rev Inst Med Trop Sao Paulo.* 2016; **58.** pii: S0036-46652016005000407.
107. Ferreira MS. Paracoccidioidomycosis. *Paediatr Respir Rev.* 2009; **10:** 161-5.
108. Fiol FS, Oliveira Sde J, Barberato-Filho S, *et al.* Paracoccidioidomycosis: Evaluation of treatment and patient profile. *Braz J Infect Dis.* 2013; **17:** 720-1.

109. Fortes MR, Miot HA, Kurokawa CS, *et al.* Immunology of paracoccidioidomycosis. *An Bras Dermatol.* 2011; **86:** 516-24.
110. Francesconi F, da Silva MT, Costa RL, *et al.* Long-term outcome of neuroparacoccidioidomycosis treatment. *Rev Soc Bras Med Trop.* 2011; **44:** 22-5.
111. Francesconi F, Francesconi do Valle AC, Silva MT, *et al.* International issues: Meningoencephalitis due to *Paracoccidioides brasiliensis. Neurology.* 2008; **71:** e65-7.
112. Francesconi F, Rocha MA, Bonilla LF, *et al.* A rapidly growing lesion on the lip - Quiz case. *Arch Dermatol.* 2009; **145:** 1325-30.
113. Freitas RM, Prado R, Prado FL, *et al.* Pulmonary paracoccidioidomycosis: Radiology and clinical-epidemiological evaluation. *Rev Soc Bras Med Trop.* 2010; **43:** 651-6.
114. Garcia NG, Oliveira DT, Pereira AA, *et al.* Extensive cutaneous lesions in paracoccidioidomycosis successfully treated with itraconazole and β-glucan. *Int J Dermatol.* 2014; **53:** e168-70.
115. Gava P, de Melo AS, Marchiori E, *et al.* Intestinal and appendiceal paracoccidioidomycosis. *Radiol Bras.* 2015; **48:** 126-7.
116. Gaviria M, Rivera V, Munoz-Cadavid C, *et al.* Validation and clinical application of a nested PCR for paracoccidioidomycosis diagnosis in clinical samples from Colombian patients. *Braz J Infect Dis.* 2015; **19:** 376-83.
117. Gegembauer G, Araujo LM, Pereira EF, *et al.* Serology of paracoccidioidomycosis due to *Paracoccidioides lutzii. PLoS Negl Trop Dis.* 2014; **8:** e2986.
118. Girardi FM, Scrofernecker ML. Oral paracoccidiodomycosis mimicking lip carcinoma. *Braz J Infect Dis.* 2016; **20:** 103-4.
119. Girardi FM, Scroferneker ML, Gava V, *et al.* Head and neck manifestations of paracoccidioidomycosis: An epidemiological study of 36 cases in Brazil. *Mycopathologia.* 2012; **173:** 139-44.
120. Godoy MC, Viswanathan C, Marchiori E, *et al.* The reversed halo sign: Update and differential diagnosis. *Br J Radiol.* 2012; **85:** 1226-35.
121. Goes HF, Duraes SM, Lima Cdos S, *et al.* Paracoccidioidomycosis in a renal transplant recipient. *Rev Inst Med Trop Sao Paulo.* 2016; **58:** 12.
122. Goldani LZ. Gastrointestinal paracoccidioidomycosis: An overview. *J Clin Gastroenterol.* 2011; **45:** 87-91.
123. Goldani LZ, Wirth F. Animal models and antifungal agents in paracoccidioidomycosis: An overview. *Mycopathologia.* 2017; doi: 10.1007/s11046-017-0130-z. PMID: 28324244.
124. Gonzalez A, Hernandez O. New insights into a complex fungal pathogen: The case of *Paracoccidioides* spp. *Yeast.* 2016; **33:** 113-28.
125. Grossklaus Dde A, Tadano T, Breder SA, *et al.* Acute disseminated paracoccidioidomycosis in a 3 year-old child. *Braz J Infect Dis.* 2009; **13:** 242-4.
126. Gryschek RC, Pereira RM, Kono A, *et al.* Paradoxical reaction to treatment in 2 patients with severe acute paracoccidioidomycosis: A previously unreported complication and its management with corticosteroids. *Clin Infect Dis.* 2010; **50:** e56-8.
127. Guery R, Merat C, Blanchet D, *et al.* Coinfection with tuberculosis and paracoccidioidomycosis in French Guiana: A common misdiagnosis. *Int J Dermatol.* 2015; **54:** e436-9.
128. Hahn RC, Rodrigues AM, Fontes CJ, *et al.* Fatal fungemia due to *Paracoccidioides lutzii. Am J Trop Med Hyg.* 2014; **91:** 394-8.
129. Igami TZ, Solano F, Preti RC, *et al.* Retinal involvement of paracoccidioidomycosis: A case report. *Trop Med Health.* 2012; **40:** 149-53.
130. Isolan GR, Vieira DM, Hehn F, *et al.* Paracoccidioidomycosis simulating brain tumor. *Surg Neurol Int.* 2014; **5:** 134.
131. Jham BC, Fernandes AM, Duraes GV, *et al.* The importance of intraoral examination in the differential diagnosis of paracoccidioidomycosis. *Braz J Otorhinolaryngol.* 2008; **74:** 946.
132. Kioshima ES, Aliperti F, Maricato JT, *et al.* A synthetic peptide selectively kills only virulent *Paracoccidioides brasiliensis* yeasts. *Microbes Infect.* 2011; **13:** 251-60.
133. Koishi AC, Vituri DF, Dionizio Filho PS, *et al.* A semi-nested PCR assay for molecular detection of *Paracoccidioides brasiliensis* in tissue samples. *Rev Soc Bras Med Trop.* 2010; **43:** 728-30.
134. Konno FT, Maricato J, Konno AY, *et al. Paracoccidioides brasiliensis* GP43-derived peptides are potent modulators of local and systemic inflammatory response. *Microbes Infect.* 2012; **14:** 517-27.
135. Laccourreye O, Mirghani H, Brasnu D, *et al.* Imported acute and isolated glottic paracoccidioidomycosis. *Ann Otol Rhinol Laryngol.* 2010; **119:** 89-92.
136. Lambertucci JR, Vale TC, Voieta I. Concomitant progressive multifocal leukoencephalopathy and disseminated paracoccidioidomycosis in an AIDS patient. *Rev Soc Bras Med Trop.* 2010; **43:** 758.
137. Lenhard-Vidal A, Assolini JP, Ono MA, *et al. Paracoccidioides brasiliensis* and *P.lutzii* antigens elicit different serum IgG responses in chronic paracoccidioidomycosis. *Mycopathologia.* 2013; **176:** 345-52.
138. Leon M, Alave J, Bustamante B, *et al.* Human T lymphotropic virus 1 and paracoccidioidomycosis: A probable association in Latin America. *Clin Infect Dis.* 2010; **51:** 250-1.
139. Lima Junior FV, Savarese LG, Monsignore LM, *et al.* Computed tomography findings of paracoccidioidomycosis in musculoskeletal system. *Radiol Bras.* 2015; **48:** 1-6
140. Lima Pde S, Chung D, Bailao AM, *et al.* Characterization of th *Paracoccidioides* hypoxia response reveals new insights into pathogenesis mechanisms of this important human pathogenic fungus. *PLoS Negl Trop Dis.* 2015; **9:** e0004282.
141. Lima TB, Domingues MA, Caramori CA, *et al.* Pancreatic paracoccidioidomycosis simulating malignant neoplasia: Case report. *World J Gastroenterol.* 2013; **19:** 5750-3.
142. Lima TC, Bezerra RO, Siqueira LT, *et al.* Para-coccidioidomycosis in a liver transplant recipient. *Rev Soc Bras Med Trop.* 2017; **50:** 138-40.
143. Lopez-Martinez R, Hernandez-Hernandez F, Mendez-Tovar LJ, *et al.* Paracoccidioidomycosis in Mexico: Clinical and epidemiological data from 93 new cases (1972-2012). *Mycoses.* 2014; **57:** 525-30.

144. Loth EA, Castro SV, Silva JR, *et al.* Occurrence of 102 cases of paracoccidioidomycosis in 18 months in the Itaipu Lake region, Western Parana. *Rev Soc Bras Med Trop.* 2011; **44:** 636-7.
145. Lucinda LR, Polanski JF. Unusual otolaryngologic manifestations of paracoccidioidomycosis: A case report and review of literature. *Am J Trop Med Hyg.* 2017; **96:** 1136-8.
146. Machado GC, Moris DV, Arantes TD, *et al.* Cryptic species of *Paracoccidioides brasiliensis:* Impact on paracoccidioidomycosis immunodiagnosis. *Mem Inst Oswaldo Cruz.* 2013; **108:** 637-43.
147. Marchiori E, Dalston M, Zanetti G, *et al.* Paracoccidioidomycosis: Another cause of sternal osteomyelitis. *Joint Bone Spine.* 2012; **79:** 323-4.
148. Marchiori E, Ferreira EC, Zanetti G, *et al.* Whole-body magnetic resonance imaging for the evaluation of thoracic involvement in disseminated paracoccidioidomycosis. *J Bras Pneumol.* 2013; **39:** 248-50.
149. Marchiori E, Valiante PM, Mano CM, *et al.* Paracoccidioidomycosis: High-resolution computed tomography-pathologic correlation. *Eur J Radiol.* 2011; **77:** 80-4.
150. Marchiori E, Zanetti G, Escuissato DL, *et al.* Reversed halo sign: High-resolution CT scan findings in 79 patients. *Chest.* 2012; **141:** 1260-6.
151. Marchiori E, Zanetti G, Hochhegger B, *et al.* Reversed halo sign on computed tomography: State-of-the-art review. *Lung.* 2012; **190:** 389-94.
152. Marques AP, Oliveira SM, Rezende GR, *et al.* Evaluation of *Paracoccidioides brasiliensis* infection by gp 43 intradermal test in rural settlements in Central-West Brazil. *Mycopathologia.* 2013; **176:** 41-7.
153. Marques S, Batalha P, Britto J, *et al.* Paracoccidioidomycosis simulating a soft tissue tumor. *Int J Dermatol.* 2014; **53:** e319-20.
154. Marques SA, Camargo RM, Abbade LP, *et al.* Paracoccidioidomycosis: An unusual presentation in a young girl disclosing an unnoted HIV-infection. *Med Mycol.* 2010; **48:** 182-7.
155. Marques SA, Lastoria JC, Camargo RM, *et al.* Paracoccidioidomycosis: Acute-subacute clinical form, juvenile type. *An Bras Dermatol.* 2016; **91:** 384-6.
156. Marques SA, Lastoria JC, Marques ME. Paracoccidioidomycosis in a patient with cervical cancer. *An Bras Dermatol.* 2011; **86:** 587-8.
157. Marques SA, Lastoria JC, Putinatti MS, *et al.* Paracoccidioidomycosis: Infiltrated, sarcoid-like cutaneous lesions misinterpreted as tuberculoid leprosy. *Rev Inst Med Trop Sao Paulo.* 2008; **50:** 47-50.
158. Marques SA, Tangoda LK, Camargo RM, *et al.* Paracoccidioidomycosis of external genitalia: Report of six new cases and review of the literature. *An Bras Dermatol.* 2012; **87:** 235-40.
159. Marques SA. Fungal infections of the mucous membrane. *Dermatol Ther.* 2010; **23:** 243-50.
160. Marques SA. Paracoccidioidomycosis: Epidemiological, clinical, diagnostic and treatment up-dating. *An Bras Dermatol.* 2013; **88:** 700-11.
161. Marques SA. Paracoccidioidomycosis. *Clin Dermatol.* 2012; **30:** 610-5.
162. Marques-da-Silva SH, Rodrigues AM, de Hoog GS, *et al.* Occurrence of *Paracoccidioides lutzii* in the Amazon region: Description of two cases. *Am J Trop Med Hyg.* 2012; **87:** 710-4.
163. Martinez R, Moya MJ. Primary complex of paracoccidioidomycosis and hypereosinophilia. *J Bras Pneumol.* 2009; **35:** 1259-62.
164. Martinez R. Epidemiology of paracoccidioidomycosis. *Rev Inst Med Trop Sao Paulo.* 2015; **57 (Suppl. 19):** 11-20.
165. Martinez R. New trends in paracoccidioidomycosis epidemiology. *J Fungi.* 2017; **3:** 1.
166. Martinez R. Paracoccidioidomycosis: The dimension of the problem of a neglected disease. *Rev Soc Bras Med Trop.* 2010; **43:** 480.
167. Martins JFS, Castilho ML, Cardoso MAG, *et al.* Identification of *Paracoccidioides brasiliensis* by gold nanoprobes. *Proc SPIE.* 2012; **8219:** 82190Z.
168. Matos WB, Dos Santos GM, Silva VE, *et al.* Paracoccidioidomycosis in the state of Maranhao, Brazil: Geographical and clinical aspects. *Rev Soc Bras Med Trop.* 2012; **45:** 385-9.
169. Medeiros VL, Arruda L. Sarcoid-like lesions in paracoccidioidomycosis: immunological factors. *An Bras Dermatol.* 2013; **88:** 113-6.
170. Mendes-Giannini MJ, Monteiro da Silva JL, de Fatima da Silva J, *et al.* Interactions of *Paracoccidioides brasiliensis* with host cells: recent advances. *Mycopathologia.* 2008; **165:** 237-48.
171. Mendonca JA, Peron Filho F, Schincariol NB, *et al.* Musculoskeletal ultrasound findings in paracoccidioidomycosis. *Rev Bras Reumatol Engl Ed.* 2016; **56:** 75-8.
172. Mendes JF, Klafke GB, Albano AP, *et al.* Para-coccidioidomycosis infection in domestic and wild mammals by *Paracoccidioides lutzii. Mycoses.* 2017; **60:** 402-6.
173. Menezes VM, Soares BG, Fontes CJ. Drugs for treating paracoccidioidomycosis. *Cochrane Database Syst Rev.* 2006; **2:** CD004967.
174. Michelan MS, Fernandes Ede A, Freitas LF, *et al.* Osteomyelitis and pyoarthritis resulting from local paracoccidioidomycosis in an immunocompetent patient: A case report. *J Med Case Rep.* 2012; **6:** 342.
175. Miranda EJ, Gonçalves LG, França FO. Chronic pulmonary paracoccidioidomycosis in an AIDS patient and *Mycobacterium gordonae* as confounding factor. *Braz J Infect Dis.* 2011; **15:** 92-3.
176. Monsignore LM, Martinez R, Simao MN, *et al.* Radiologic findings of osteoarticular infection in paracoccidioidomycosis. *Skeletal Radiol.* 2012; **41:** 203-8.
177. Morejon KM, Machado AA, Martinez R. Paracoccidioidomycosis in patients infected with and not infected with human immunodeficiency virus: A case-control study. *Am J Trop Med Hyg.* 2009; **80:** 359-66.
178. Moreno L, Figueredo M, Sadigursky M, *et al.* A man with newly diagnosed HIV/AIDS with unusual severe opportunistic infection and no AIDS-defining disease. *Clin Infect Dis.* 2014; **58:** 1430-1 & 1484-5.

179. Moreto TC, Marques ME, de Oliveira ML, *et al.* Accuracy of routine diagnostic tests used in paracoccidioidomycosis patients at a university hospital. *Trans R Soc Trop Med Hyg.* 2011; **105:** 473-8.
180. Munoz JF, Gallo JE, Misas E, *et al.* Genome update of the dimorphic human pathogenic fungi causing paracoccidioidomycosis. *PLoS Negl Trop Dis.* 2014; **8:** e3348.
181. Nakamura R, Valgas N, Bichara RM, *et al.* Paracoccidioidomycosis: Chronic adult unifocal form. *Int J Dermatol.* 2012; **51:** 195-6.
182. Nascimento CR, Delanina WF, Soares CT. Paracoccidioidomycosis: Sarcoid-like form in childhood. *An Bras Dermatol.* 2012; **87:** 486-7.
183. Nishikaku AS, Peraçoli MT, Bagagli E, *et al.* Experimental infections with *Paracoccidioides brasiliensis* obtained from armadillos: comparison to clinical isolates. *Braz J Infect Dis.* 2008; **12:** 57-62.
184. Nobrega de Almeida J Jr, Del Negro GM, Grenfell RC, *et al.* Matrix-assisted laser desorption ionization-time of flight mass spectrometry for differentiation of the dimorphic fungal species *Paracoccidioides brasiliensis* and *Paracoccidioides lutzii. J Clin Microbiol.* 2015; **53:** 1383-6.
185. Odashiro AN, Odashiro PR, Fernandes PI, *et al.* Eyelid and conjunctival paracoccidioidomycosis simulating carcinoma. *Int Ophthalmol.* 2011; **31:** 63-7.
186. Oliveira EV, Almeida MT, Turatti A, *et al.* Paracoccidioidomycosis and cryptococcosis with localized skin manifestations: Report of two cases in the elderly. *An Bras Dermatol.* 2016; **91:** 243-4.
187. Onda H, Komine M, Murata S, *et al.* Imported paracoccidioidomycosis in Japan. *Dermatol Online J.* 2011; **17:** 11.
188. Pecanha PM, de Souza S, Falqueto A, *et al.* Amphotericin B lipid complex in the treatment of severe paracoccidioidomycosis: A case series. *Int J Antimicrob Agents.* 2016; **48:** 428-30.
189. Pedreira Rdo P, Guimaraes EP, de Carli ML, *et al.* Paracoccidioidomycosis mimicking squamous cell carcinoma on the dorsum of the tongue and review of published literature. *Mycopathologia.* 2014; **177:** 325-9.
190. Pedroso VS, Lyon AC, Araujo SA, *et al.* Paracoccidioidomycosis case series with and without central nervous system involvement. *Rev Soc Bras Med Trop.* 2012; **45:** 586-90.
191. Pereira CM, Gasparetto PF, Alves VF, *et al.* Paracoccidioidomycosis in pregnancy: An atypical case report. *Braz J Infect Dis.* 2012; **16:** 489-90.
192. Pereira GH, Lanzoni VP, Beirao EM, *et al.* Disseminated fungal infection with adrenal involvement: Report of two HIV-negative Brazilian patients. *Rev Inst Med Trop Sao Paulo.* 2015; **57:** 527-30.
193. Pereira GH, Santos AQ, Park M, *et al.* Bone marrow involvement in a patient with paracoccidioidomycosis: A rare presentation of juvenile form. *Mycopathologia.* 2010; **170:** 259-61.
194. Pereira PM, Akel PB, Lima LL, *et al.* Multifocal paracoccidioidomycosis: A diagnostic challenge due to late cutaneous manifestation. *An Bras Dermatol.* 2011; **86:** 149-52.
195. Perenha-Viana MC, Gonzales IA, Brockelt SR, *et al.* Serological diagnosis of paracoccidioidomycosis through a Western blot technique. *Clin Vaccine Immunol.* 2012; **19:** 616-9.
196. Pitz Ade F, Koishi AC, Tavares ER, *et al.* An optimized one-tube, semi-nested PCR assay for *Paracoccidioides brasiliensis* detection. *Rev Soc Bras Med Trop.* 2013; **46:** 783-5.
197. Pontes AM, Borborema J, Correia CR, *et al.* A rare paracoccidioidomycosis diagnosis in a kidney transplant receptor: Case report. *Transplant Proc.* 2015; **47:** 1048-50.
198. Pontes HA, Guimaraes DM, Pontes FS, *et al.* Kaposi sarcoma and paracoccidioidomycosis in the same fragment of oral mucosa biopsy: A rare association in human immunodeficiency virus-positive patient. A case report. *Diagn Microbiol Infect Dis.* 2011; **69:** 196-9.
199. Porro AM, Rotta O. Cutaneous and pulmonary paracoccidioidomycosis in a patient with a malignant visceral tumor. *An Bras Dermatol.* 2011; **86:** 1220-1.
200. Puccia R, Vallejo MC, Matsuo AL, *et al.* The *Paracoccidioides* cell wall: Past and present layers toward understanding interaction with the host. *Front Microbiol.* 2011; **2:** 257.
201. Queiroz Junior Lde P, de Camargo ZP, Tadano T, *et al.* Serological and antigenic profiles of clinical isolates of *Paracoccidioides* spp. from Central Western Brazil. *Mycoses.* 2014; **57:** 466-72.
202. Queiroz RM, Gomes MP, Valentin MV. Pulmonary paracoccidioidomycosis showing reversed halo sign with nodular/coarse contour. *Radiol Bras.* 2016; **49:** 59-60.
203. Queiroz-Telles F, Escuissato DL. Pulmonary paracoccidioidomycosis. *Semin Respir Crit Care Med.* 2011; **32:** 764-74.
204. Radisic MV, Linares L, Afeltra J, *et al.* Acute pulmonary involvement by paracoccidioidomycosis disease immediately after kidney transplantation: Case report and literature review. *Transpl Infect Dis.* 2017; **19:** doi: 10.1111/tid.12655. PMID: 28039947.
205. Ramirez-Olivencia G, Ramirez-Rubio O, Gonzalez PR, *et al.* Paracoccidioidomycosis in a Spanish missionary. *J Travel Med.* 2010; **17:** 139-40.
206. Ramos-E-Silva M, Saraiva Ldo E. Paracoccidioidomycosis. *Dermatol Clin.* 2008; **26:** 257-69.
207. Reis F, Collier PP, Souza TF, *et al.* Neuroparacoccidioidomycosis (NPCM): Magnetic resonance imaging (MRI) findings. *Mycopathologia.* 2013; **175:** 181-6.
208. Resende LS, Mendes RP, Bacchi MM, *et al.* Bone marrow necrosis related to paracoccidioidomycosis: The first eight cases identified at autopsy. *Histopathology.* 2009; **54:** 486-9.
209. Restrepo A, Benard G, de Castro CC, *et al.* Pulmonary paracoccidioidomycosis. *Semin Respir Crit Care Med.* 2008; **29:** 182-97.
210. Restrepo A, Gomez BL, Tobon A. Paracoccidioidomycosis: Latin America's own fungal disorder. *Curr Fungal Infect Rep.* 2012; **6:** 303-11.
211. Ribeiro EB, Pochat VD, Magaldi EN, *et al.* Complex nasal reconstruction after paracoccidioidomycosis infection - Case report. *An Bras Dermatol.* 2015; **90 (3 Suppl. 1):** 203-5.

212. Rigobello FF, Marquez AS, Lopes JD, *et al.* Patients with chronic-form paracoccidioidomycosis present high serum levels of IgE anti-*Paracoccidioides brasiliensis* Gp70. *Mycopathologia.* 2013; **175:** 307-13.
213. Rittner GM, Munoz JE, Marques AF, *et al.* Therapeutic DNA vaccine encoding peptide P10 against experimental paracoccidioidomycosis. *PLoS Negl Trop Dis.* 2012; **6:** e1519.
214. Roberto TN, Rodrigues AM, Hahn RC, *et al.* Identifying *Paracoccidioides* phylogenetic species by PCR-RFLP of the alpha-tubulin gene. *Med Mycol.* 2016; **54:** 240-7.
215. Rodrigues DR, Fernandes RK, Balderramas Hde A, *et al.* Interferon-gamma production by human neutrophils upon stimulation by IL-12, IL-15 and IL-18 and challenge with *Paracoccidioides brasiliensis. Cytokine.* 2014; **69:** 102-9.
216. Rodrigues Gda S, Severo CB, Oliveira Fde M, *et al.* Association between paracoccidioidomycosis and cancer. *J Bras Pneumol.* 2010; **36:** 356-62.
217. Ruas LP, Pereira RM, Braga FG, *et al.* Severe paracoccidioidomycosis in a 14-year-old boy. *Mycopathologia.* 2016; **181:** 915-20.
218. Ruiz e Resende LS, Yasuda AG, Mendes RP, *et al.* Paracoccidioidomycosis in patients with lymphoma and review of published literature. *Mycopathologia.* 2015; **179:** 285-91.
219. Safe IP, Valle FF, Maia DC, *et al.* Extra-pulmonary manifestations of paracoccidioidomycosis associated with acquired immunodeficiency syndrome: A case report. *An Bras Dermatol.* 2014; **89:** 150-3.
220. Sargenti Neto S, Paulo LF, Rosa RR. Oral paracoccidioidomycosis as a differential diagnosis of oral cancer. *Rev Soc Bras Med Trop.* 2012; **45:** 777.
221. Sarti EC, de Oliveira SM, dos Santos LF, *et al.* Paracoccidioidal infection in HIV patients at an endemic area of paracoccidioidomycosis in Brazil. *Mycopathologia.* 2012; **173:** 145-9.
222. Sarvepalli S, Gorthi K, Many WJ Jr, *et al.* A 41-year-old Hispanic man with nonproductive cough. *Clin Infect Dis.* 2009; **48:** 1741-2 & 1794-5.
223. Savarese LG, Monsignore LM, de Andrade Hernandes M, *et al.* Magnetic resonance imaging findings of paracoccidioidomycosis in the musculoskeletal system. *Trop Med Int Health.* 2015; **20:** 1346-54.
224. Sbeghen MR, Zanata TB, Macagnan R, *et al. Paracoccidioides brasiliensis* infection in small wild mammals. *Mycopathologia.* 2015; **180:** 435-40.
225. Scorzoni L, de Paula E Silva AC, Singulani Jde L, *et al.* Comparison of virulence between *Paracoccidioides brasiliensis* and *Paracoccidioides lutzii* using *Galleria mellonella* as a host model. *Virulence.* 2015; **6:** 766-76.
226. Severo CB, Dall Bello AG, Oliveira Fde M, *et al.* Pleural effusion an unusual feature of paracoccidioidomycosis: Report of two new cases with a systematic review of the literature. *Mycopathologia.* 2013; **175:** 323-30.
227. Shankar J, Restrepo A, Clemons KV, *et al.* Hormones and the resistance of women to paracoccidioidomycosis. *Clin Microbiol Rev.* 2011; **24:** 296-313.
228. Shikanai-Yasuda MA, Conceicao YM, Kono A, *et al.* Neoplasia and paracoccidioidomycosis. *Mycopathologia.* 2008; **165:** 303-12.
229. Shikanai-Yasuda MA. Paracoccidioidomycosis treatment. *Rev Inst Med Trop Sao Paulo.* 2015; **57 (Suppl. 19):** 31-7.
230. Silva AA, Sotto MN, Dpuarte MI, *et al.* Regulatory T cells in cutaneous lesions of patients with paracoccidioidomycosis. *Microb Pathog.* 2013; **65:** 36-40.
231. Silva AR, Vargas PA, Ribeiro AC, *et al.* Fine needle aspiration cytology in the diagnosis of cervical tuberculosis and paracoccidioidomycosis. *Cytopathology.* 2010; **21:** 66-8.
232. Silva Ferreira C, de Castro Ribeiro EM, Miranda Goes Ad, *et al.* Current strategies for diagnosis of paracoccidioidomycosis and prospects of methods based on gold nanoparticles. *Future Microbiol.* 2016; **11:** 973-85.
233. Silva Neto GR, Ganc RL, Colaiacovo R, *et al.* Paracoccidioidomycosis simulating cancer diagnosed by endoscopic ultrasound-guided fine needle aspiration. *Endoscopy.* 2012; **44 (Suppl. 2):** e439-40.
234. Silva-Vergara ML, Rocha IH, Vasconcelos RR, *et al.* Central nervous system paracoccidioidomycosis in an AIDS patient: Case report. *Mycopathologia.* 2014; 177: 137-41.
235. Silveira-Gomes F, Marques-da-Silva SH. Effects of pretreating serum samples on the performance of a latex agglutination test for serodiagnosis of paracoccidioidomycosis. *Clin Vaccine Immunol.* 2012; **19:** 386-90.
236. Silveira-Gomes F, Sarmento DN, Pinto TM, *et al.* Development and evaluation of a latex agglutination test for the serodiagnosis of paracoccidioidomycosis. *Clin Vaccine Immunol.* 2011; **18:** 604-8.
237. Souza PV, Pinto WB, Matas SL. Paracoccidioidomycosis: A rare cause of infectious encephalomyelopathy. *Arq Neuropsiquiatr.* 2014; **72:** 904-5.
238. Sperandio FF, Giudice FS, Coelho AP, *et al.* Difficulty in diagnosing oral paracoccidioidomycosis after topical nystatin usage. *Gen Dent.* 2012; **60:** e44-6.
239. Steinbruck K, Fernandes R, Campos CF, *et al.* Intrahepatic biliary paracoccidioidomycosis mimicking cholangiocarcinoma. *J Gastrointestin Liver Dis.* 2014; **23:** 239.
240. Sylvestre TF, Franciscone Silva LR, Cavalcante Rde S, *et al.* Prevalence and serological diagnosis of relapse in paracoccidioidomycosis patients. *PLoS Negl Trop Dis.* 2014; **8:** e2834.
241. Taborda CP, Uran ME, Nosanchuk JD, *et al.* Paracoccidioidomycosis: Challenges in the development of a vaccine against an endemic mycosis in the Americas. *Rev Inst Med Trop Sao Paulo.* 2015; **57 (Suppl. 19):** 21-4.
242. Takayama A, Itano EN, Sano A, *et al.* An atypical *Paracoccidioides brasiliensis* clinical isolate based on multiple gene analysis. *Med Mycol.* 2010; **48:** 64-72.
243. Tavares Dos Santos H, Augusto Benevenuto de Andrade B, Fernandes D, *et al.* Chronic paracoccidioidomycosis in a woman with Crohn disease. *Dermatol Online J.* 2017; 23. pii: 13030/qt1gp8z9gx. PMID: 28541872.
244. Teixeira Mde M, Theodoro RC, Derengowski Lda S, *et al.* Molecular and morphological data support the existence of a sexual cycle in species of the genus *Paracoccidioides. Eukaryot Cell.* 2013; **12:** 380-9.

245. Teixeira MM, Theodoro RC, Nino-Vega G, *et al.* *Paracoccidioides* species complex: Ecology, phylogeny, sexual reproduction and virulence. *PLoS Pathog.* 2014; **10:** e1004397. PMID: 25357210.
246. Teixeira Mde M, Theodoro RC, Oliveira FF, *et al.* *Paracoccidioides lutzii* sp. nov.: Biological and clinical implications. *Med Mycol.* 2014; **52:** 19-28.
247. Teixeira MM, Theodoro RC, de Carvalho MJ, *et al.* Phylogenetic analysis reveals a high level of speciation in the *Paracoccidioides* genus. *Mol Phylogenet Evol.* 2009; **52:** 273-83.
248. Teles FR, Martins ML. Laboratorial diagnosis of paracoccidioidomycosis and new insights for the future of fungal diagnosis. *Talanta.* 2011; **85:** 2254-64.
249. Theodoro RC, Teixeira Mde M, Felipe MS, *et al.* Genus *Paracoccidioides:* Species recognition and biogeographic aspects. *PLoS One.* 2012; **7:** e37694.
250. Thind SK, Taborda CP, Nosanchuk JD. Dendritic cell interactions with *Histoplasma* and *Paracoccidioides. Virulence.* 2015; **6:** 424-32.
251. Torres Esteche V, Arteta Z, Torres G, *et al.* Concomitant pulmonary paracoccidioidomycosis and pulmonary histoplasmosis: A rare case. *J Bras Pneumol.* 2012; **38:** 264-8.
252. Torres JR, Murillo J. Simon Bolivar's remains revisited: Getting out or deeper into his medical Labyrinth? *Clin Infect Dis.* 2011; **53:** 305; Author reply 305-6.
253. Torres-Pereira C, Giovanini AF, Stramandinoli RT, *et al.* Oral paracoccidioidomycosis and pulmonary tuberculosis co-infection: Relevance of oral biopsy in establishing the diagnosis and therapeutic approach. *Int J Infect Dis.* 2009; **13:** 114-6.
254. Travassos LR, Taborda CP, Colombo AL. Treatment options for paracoccidioidomycosis and new strategies investigated. *Expert Rev Anti Infect Ther.* 2008; **6:** 251-62.
255. Travassos LR, Taborda CP. New advances in the development of a vaccine against paracoccidioidomycosis. *Front Microbiol.* 2012; **3:** 212.
256. Travassos LR, Taborda CP. Paracoccidioidomycosis vaccine. *Hum Vaccin Immunother.* 2012; **8:** 1450-3.
257. Travassos LR, Taborda CP. Linear epitopes of *Paracoccidioides brasiliensis* and other fungal agents of human systemic mycoses as vaccine candidates. *Front Immunol.* 2017; **8:** 224.
258. Trindade AH, Meira HC, Pereira IF, *et al.* Oral paracoccidioidomycosis: Retrospective analysis of 55 Brazilian patients. *Mycoses.* 2017; doi: 10.1111/myc.12622. PMID: 28422366.
259. Tubino PV, Sarmento BJ, dos Santos VM, *et al.* Synchronous oral paracoccidioidomycosis and esophageal carcinoma. *Mycopathologia.* 2012; **174:** 157-61.
260. Valera ET, Mori BM, Engel EE, *et al.* Fungal infection by *Paracoccidioides brasiliensis* mimicking bone tumor. *Pediatr Blood Cancer.* 2008; **50:** 1284-6.
261. Vermelho MB, Correia AS, Michailowsky TC, *et al.* Abdominal alterations in disseminated paracoccidioidomycosis: Computed tomography findings. *Radiol Bras.* 2015; **48:** 81-5.
262. Vidal MS, Del Negro GM, Vicentini AP, *et al.* Serological diagnosis of paracoccidioidomycosis: High rate of inter-laboratorial variability among medical mycology reference centers. *PLoS Negl Trop Dis.* 2014; **8:** e3174.
263. Vieira Gde D, Alves Tda C, Lima SM, *et al.* Paracoccidioidomycosis in a western Brazilian Amazon State: Clinical-epidemiologic profile and spatial distribution of the disease. *Rev Soc Bras Med Trop.* 2014; **47:** 63-8.
264. Vilela R, Bossart GD, St Leger JA, *et al.* Cutaneous granulomas in dolphins caused by novel uncultivated *Paracoccidioides brasiliensis. Emerg Infect Dis.* 2016; **22:** 2063-9.
265. von Glehn F, Damasceno A, Miotto N, *et al.* Carpal tunnel syndrome with paracoccidioidomycosis. *Emerg Infect Dis.* 2012; **18:** 1390-2.
266. Wagner G, Moertl D, Eckhardt A, *et al.* Chronic paracoccidioidomycosis with adrenal involvement mimicking tuberculosis - A case report from Austria. *Med Mycol Case Rep.* 2016; **14:** 12-6.
267. Webber LP, Martins MD, de Oliveira MG, *et al.* Disseminated paracoccidioidomycosis diagnosis based on oral lesions. *Contemp Clin Dent.* 2014; **5:** 213-6.
268. Wiwanitkit V. Paracoccidioidomycosis and cancer. *J Bras Pneumol.* 2010; **36:** 666.
269. Woyciechowsky TG, Dalcin DC, dos Santos JW, et al. Paracoccidioidomycosis induced by immunosuppressive drugs in a patient with rheumatoid arthritis and bone sarcoma: Case report and review of the literature. *Mycopathologia.* 2011; **172:** 77-81.
270. Xavier MO, Pasqualotto AC, Cardoso IC, *et al.* Cross-reactivity of *Paracoccidioides brasiliensis, Histoplasma capsulatum and Cryptococcus* species in the commercial Platelia *Aspergillus* enzyme immunoassay. *Clin Vaccine Immunol.* 2009; **16:** 132-3.
271. Zambuzzi-Carvalho PF, Fernandes AG, Valadares MC, *et al.* Transcriptional profile of the human pathogenic fungus *Paracoccidioides lutzii* in response to sulfamethoxazole. *Med Mycol.* 2015; **53:** 477-92.

SECTION V

Opportunistic Mycoses

Chapters

The opportunistic mycoses are those fungal infections which are found among patients with underlying predisposing conditions. Such clinical entities are essentially disease processes produced by fungi of lesser pathogenic potential in the host whose immunological defense mechanism is weakened by endogenous causes like cancer, leukemia or exogenous causes like immunosuppressive therapy, AIDS, etc. These opportunistic organisms are considered as vultures of the fungal world. In normal circumstances most of them are found to be non-pathogenic but wherever an opportunity arises, then an ordinary looking contaminant also becomes significant pathogen. These infections involve different body systems and invariably prove fatal if timely care is not provided to the patient.

In contrast to the endemic mycoses, they are found to occur in any geographical area and their clinical manifestations are variable depending on the underlying predisposition and immune status of the patient. These diseases are responsible for deep mycoses and are important cause of high morbidity and mortality among the hospitalized patients. Rather the patient becomes actually the victim of circumstances. In addition, the prevalence of nosocomial fungal infections particularly candidiasis has increased in the recent past to a great extent. There is an emergence of non-*albicans Candida* in the blood stream infections caused by many species of this genus particularly *C.tropicalis, C.krusei, C.glabrata, C.auris* and *C.haemulonii* since the introduction of fluconazole. Similarly, there is an emergence of non-*neoformans Cryptococcus* species also which were not taken as pathogenic till recent times. There emergence of *Cryptococcus gattii* still continues in Vancouver, the western part of Canada.

The change in nomenclature of human strains of atypical fungus, *Pneumocystis* from *P.carinii* to *P.jirovecii,* is highlighted in chapter on Pneumocystosis. The emergence of HIV pandemic has changed the entire scenario of opportunist fungal infections. The fungal species, which are endemic and supposed to be true pathogens in their prevalence, are now implicated even among AIDS patients, changing all epidemiological aspects. *Talaromyces marneffei,* another opportunist fungus, is found prevalent in Southeast Asia, analogous to the true fungi like *Histoplasma capsulatum, Blastomyces dermatitidis* and *Coccidioides immitis* in North America and *Paracoccidioides brasiliensis* in South America, respectively.

Microsporidiosis, recently shifted to medical mycology, is given in a separate Chapter. Mucormycosis, one of the emerging opportunistic mycoses in India, is emphasized substantially including its newer species of genus *Apophysomyces* and *Saksenaea*. There is emergence of opportunistic aspergillosis and is being dealt in a separate chapter. There is also an emergence of drug resistance among certain fungi causing opportunistic mycoses. The newly emerged opportunistic infections and nosocomial fungal infections, which have not been categorized anywhere in this Textbook, are collectively described in Chapter on Miscellaneous Opportunistic Mycoses at the end of this Section.

A large number of immunosuppressed patients, widening range of resistance to antifungal drugs mean that the common practice of identification or exclusion of these fungi alone is no longer adequate unless their presence is clinically correlated. Some of the infections, which were not actually considered as opportunistic in nature like Entomophthoramycosis, have been described in this section in a separate chapter due to many such reports.

CHAPTER 20

Candidiasis

Candidiasis is the commonest fungal disease found in humans affecting mucosa, skin, nails and internal organs. It is caused by various species of yeast-like fungi belonging to genus ***Candida*** with ***Candida albicans*** as the representative species. The infection may be acute or chronic, superficial or deep and its clinical spectrum is wide. It is found mainly as secondary infection in individuals with some underlying immunocompromised condition and very rarely as the primary disease.

Historical Perspective

The history of candidiasis is very old as the disease was described in ancient times and gradually its etiological agents, diagnostic procedures and therapeutic modalities came into existence. The first known description of candidal infections as oral thrush in patients with underlying diseases may be found in Hippocrates' *'Epidemics'* from the fourth century BC. Rosen von Rosenstein and Underwood identified candidal infections in pediatric patients and made the first description of thrush in modern medicine. In 1839, Lagenbeck described yeast-like organism of thrush. In 1842, David Gruby described the fungus responsible for the true thrush of children before the Academy of Science in Paris. In 1844, Bennett isolated the fungus from sputum of patient suffering from tuberculosis. Later on it was also isolated from other sites like vagina, blood and brain by various workers. In 1853, Robin named the fungus *Oidium albicans*. *Cryptococcus glabratus* was first isolated from grapes in 1894 by Berlese and was described in humans by Anderson for the first time in 1917. In 1938, Lodder and de Vries transferred this fungus to genus *Torulopsis glabratus.*

The nomenclature of fungus presently known as *Candida albicans* was previously designated as *Odium albicans* and *Monilia albicans*. The literal meaning of *Candida* is glowing white in Latin. Berkhout in 1923 structured genus *Candida* to encompass these yeasts bringing end to the prevailing confusion, which was officially accepted as name of the genus in the 8th Botanical Congress held in Paris in 1954. Due to its genus name as *Monilia*, the disease used to be called **moniliasis**. Subsequently, similar fungi, which were known by different names like *Torulopsis* were also brought into the domain by merging into this genus in 1978 and giving new name as *Candida glabrata*. Viswanathan and Randhawa isolated *Candida viswanathii* in 1959, which was designated in honor of Dr Raman Viswanathan, the first Director of the Vallabhbhai Patel Chest Institute (VPCI), Delhi. In 1995, *Candida dubliniensis*, was described as new species by Sullivan and colleagues from Dublin, Ireland.

There are two popular terms, **candidosis** and **candidiasis**, which are frequently used and both are correct, however, candidosis is often preferred in some of the European countries like United Kingdom, France and Italy as well as Canada giving disease name on the basis of pattern followed in other fungal diseases by putting ***'osis'*** as suffix. On the other hand candidiasis is used in the United States of America and rest of the world.

There are many common names for this disease because the fungus is so commonly associated with infections. The word thrush is derived from ancient Scandinavian or Anglo-Saxon words for the disease. The French word for the condition is *'le Muguet'*, which means *'lily-of-the-valley.'*

This is usually said that we lead inextricably mycotic lives: Yeasts leaven our bread, ferment our wine and beer and inhabit our skins, mouths, and gastrointestinal tracts; however, not all is harmony. Hippocrates reported aphthous ulcers consistent with thrush in patients with severe debilitation, but it was not until the 1840s that in the newly emerging field of clinical experimental medicine that thrush—as well as other mycotic conditions, including ringworm—was recognized as being caused by transmissible fungi. In 1923 Christine Marie Berkhout named what we now call *Candida albicans*, for the white robe, toga candida, worn by Roman senators and senatorial.

In 2009, *Candida auris* was first found in an ear infection in a patient in Japan. Now, nearly three dozen people in the USA have been diagnosed with this deadly and highly drug-resistant yeast infection. The federal health officials first warned clinicians in June 2016 to be on the lookout for the emerging pathogen that has been spreading around the world.

This is now evident that *C.auris* has emerged as an important nosocomial pathogen with clonal inter- and intra-hospital transmission and it has become widespread across several countries. A total of 77 patients from 7 states in the USA have been diagnosed with this deadly and highly drug-resistant yeast infection, maximum from New York. The federal health officials first warned clinicians in June 2016 to be on the lookout for the emerging pathogen that has been spreading around the world. It has been reported in a dozen countries *i.e.* Colombia, India, Israel, Kenya, Kuwait, Panama, Pakistan, South Africa, South Korea, Venezuela, UK and USA.

Mycology

There are 163 anamorphic species of genus *Candida* with teleomorphs in at least thirteen genera. Nearly twenty of these species are considered to be significant pathogens causing various infections in the human beings and out of these seven are well-known opportunistic pathogens. Some of these species are listed below:

Candida albicans
Candida tropicalis
Candida krusei
Candida glabrata
Candida auris
Candida haemulonii
Candida guilliermondii
Candida parapsilosis
Candida lusitaniae
Candida kefyr
Candida rugosa
Candida dubliniensis
Candida viswanathii

The commonest pathogenic species of this genus is *Candida albicans*. The genus *Torulopsis* has been merged within the genus *Candida* by general consensus among the medical mycologists. *Candida stellatoidea* is now included within *C.albicans* and *C.pseudotropicalis* is re-named as *C.kefyr.*

There are two serotypes of *Candida albicans* named as serotype **A** and **B**, which are based on their differences between mannan components of cell wall. Serotype **A** is found to be antigenically related to *C.tropicalis* and serotype **B** was related to *C.stellatoidea*, currently considered synonym of *C.albicans*. The genus *Candida* belongs to the phylum **Fungi Imperfecti**, the order **Moniliales** and family **Cryptococcaceae**.

It is now feasible to identify the teleomorphic states of *Candida* on the basis of newer molecular techniques, however, it is not feasible to keep a single name of the genus despite similar clinical conditions produced by the organisms.

Epidemiology

At present there are 163 acknowledged anamorphic species of the genus *Candida,* which are found in different habitats. The agents of human candidiasis have a comparatively restricted natural distribution and have been discovered primarily in association with man and animals. *Candida albicans* holds an exceptional position, constituting about 40%, opposed to about ten non-*albicans Candida* species of medical importance i.e. 60%. The increase in the incidence of *Candida* species over the past two decades is significant and non-*albicans Candida* species continue to replace *C.albicans* at most of the clinical sites like bloodstream infections. Candidiasis, aspergillosis and mucormycosis constitute 80% of infections caused by fungi.

Candida species are component of normal flora of human beings. They are commonly found on the skin, throughout gastrointestinal tract and female genital tract particularly higher in vagina during pregnancy. This has been observed that most of the times it is found as an endogenous infection due to its commensal nature.

A variety of factors are known to predispose both superficial and deep-seated candidiasis. All these factors act either by altering the balance of normal microbial flora of the body or by lowering the host resistance. Mostly, numerous factors operate collectively and full-fledged infection takes place. The yeasts of genus *Candida* have been the **fourth** most common primary blood stream organisms in the United States and the **seventh** most common pathogens to cause health-care associated infections during the last four decades. Similarly, rates of candidemia have substantially increased in Europe as well as in the developing countries. Now, *Candida* species constitute the third to fourth most common cause of health-care associated blood stream infections.

During the last century two important pre-disposing factors have been noticed to be important. One was the advent of antibacterial antibiotics and their indiscriminate use led to the increased incidence of candidiasis and secondly the emergence of pandemic of AIDS. Candidiasis is undoubtedly the most common fungal infection in HIV-infected individuals. It manifests most commonly as thrush, oropharyngeal candidiasis and may also present as esophagitis and even skin rash. Oral candidiasis is declining in certain countries like United Kingdom. The application of DNA-based typing methods has advanced the understanding of the epidemiology of oropharyngeal candidiasis in AIDS. Most studies thus have shown that in a given patient, the same DNA subtype of *C.albicans* is maintained throughout multiple infections.

Although *Candida albicans* remains the most common causative agent of both superficial and deep fungal infections, an increasing incidence of less common species of *Candida* has also been documented in the last few years. These species include *C.tropicalis, C.krusei, C.glabrata* and *C.parapsilosis* and tend to be less susceptible to azoles, particularly fluconazole, than *C.albicans.*

The epidemiology of yeast infections is evolving and species in the *Candida haemulonii* complex are now being recognized as an important cause of candidiasis. However, in 2012, *C.haemulonii* complex was re-classified as two species and one variety: *C.haemulonii* (former gp. I), *C.duobushaemulonii* (former gp. II) and *C.haemulonii* var. *vulnera.*

This is also observed that every year human fungal pathogens cause more than two million infections leading to high morbidity as well as mortality. Out of this two fungal genera *i.e. Candida* and *Cryptococcus* are together accounting for about 1.4 million infections with higher mortality rates. The patients with dysfunctional immune systems such as with HIV/AIDS, are mainly vulnerable to such infections.

Immunity

As such immune response to *Candida* species is basically shaped by its commensal character. There is a crucial role of finding out between colonization and invasion at mucosal surfaces. This is along with antifungal host defense mechanisms used during mucosal or systemic infection, with *Candida* species, differ substantially. Moreover, how the innate sensing of fungi by pattern recognition receptors and the interplay of both myeloid and lymphoid immune cells with non-immune cells, including platelets and epithelial cells, shapes host immunity to *Candida* species. The data based on recent studies suggest that both innate and adaptive immune systems display memory characteristics after encountering *Candida* species.

The classical T-cell mediated immunity is important in protection from persistent high-level mucocutaneous *Candida* colonization. A variety of systems are operative to augment the function of these effector cells. Serum factors, the complement system and antibody may serve this function in part. Additionally, augmentation of phagocytic cell function appears to occur by the activity of NK cells and immune T-cells, mediated primarily by the cytokines IFN-γ and TNF. The other cytokines including IL-3, GM-CSF, M-CSF and G-CSF may also be involved. Early in infection or in hosts with impaired T-cell function, NK cells may be the primary source of such cytokines. T-cells may then perform this role more effectively and become the primary source of cytokine, augmenting the function of phagocytic cells.

When yeast cells of *Candida* penetrate mucous membrane or skin barrier, ultimate effectors of destruction of organism are phagocytic cells, both polymorphonuclear *i.e.* neutrophils and mononuclear *i.e.* monocytic cells. These cells provide phagocytic and candidacidal activity following fungal penetration. These processes involve myeloperoxidase and superoxide or cationic proteins. The polymorphonuclear activity is of significance in protection from the deep-seated forms of candidiasis, as an increased rate of infections is associated with neutropenia.

1. *Candida* Antigens

The antigenic structure of *Candida* has been studied primarily to identify antigens useful for improved identification of *Candida* species or strain of *Candida albicans* and for diagnosis of infection. Antigens useful for serological applications are identified by agglutination reactions with antisera raised against surface components of yeast cells. The antigens considered for diagnosis of deep-seated infections have been associated with cytoplasmic components of the organism because they may become available in amounts sufficient to evoke humoral response only when fungal cells are destroyed by the host. The antigens of *Candida* species can be divided into two main broad groups:

(i) Cell Wall Antigens
(ii) Cytoplasmic Antigens

(i) Cell Wall Antigens: The serological studies to identify *Candida* species and serotypes of *C.albicans* have been based on agglutination reactions that use antisera to components of cell wall. As mannan is the major antigenic component of *Candida* cell wall, these antisera must recognize subtle variations in the linkage or number of mannose residues in side chains of molecule and primary structure of peptide component of mannan, all of which have yet to be fully defined. The glucan polymers, which are in greater abundance than mannans in the *C.albicans* cell wall, are immunologically less active. The cell wall of *C.albicans* is not antigenically consistent as its dynamic structure and antigenic characters vary in response to growth conditions *in vitro* and *in vivo*.

(ii) Cytoplasmic Antigens: The number of antigenic components in extracts of broken cells is also considerably significant. These antigens have not yet been associated with specific cytoplasmic components of the cell.

2. Virulence Factors of *Candida* Species

As the antigenic components have been identified immunologically, without emphasis on biological functions of associated species hence virulence factors have been characterized in terms of function without consideration of antigenicity. These virulence factors are as follows:

(i) Adhesin
(ii) Enzymes
(iii) Toxins
(iv) Complement Receptors
(v) Phenotypic Switching

(i) Adhesin: The ability of *Candida* species to adhere to host cells correlates with their virulence and clearly plays major role in pathogenesis of candidiasis. Binding of *Candida* to epithelial and endothelial cells is controlled by adhesin of surface of fungus that interact with receptors of host cells.

(ii) Enzymes: There are fourteen hydrolytic enzyme activities, which have been described in *Candida* species but two of these activities may have role in pathogenesis of the disease process. These are proteases, lipases, phospholipases, esterases, phosphatases, etc. Secretion of hydrolytic enzymes such as secreted aspartyl proteinase (Sap) serves to be an important determinant for pathogenicity. There are three types of aspartyl proteinases and two types of serine proteinases, involved as virulence factors in the pathogenicity of *Candida* species. These enzymes have received much attention in the past, as they are known to mediate pathogenesis, particularly by facilitating the hyphal invasion especially seen in disseminated candidiasis.

(iii) Toxins: The glycoprotein extracts of *Candida* cell walls, like bacterial endotoxins, are lethal and pyrogenic and induce anaphylactic shock in various animal models with variable potencies.

(iv) Complement Receptors: A quality recently described for *Candida albicans* that might influence its virulence is an ability to bind complement-derived opsonins. Receptors for C3d and iC3b, CR2 and CR3 were first described as surface molecules of phagocytic leukocytes that mediate endocytosis of complement-opsonized microorganisms.

(v) Phenotypic Switching: The phenotypic switching denotes the ability of organisms of single strain to switch reversibly at high frequencies among different colony phenotypes. Due to this ability, it can grow in variety of morphological forms, ranging from unicellular budding yeast (blastospore) to filamentous pseudohyphae and even true hyphae. Such switching could enable adaptation to different and/or changing conditions in the host and thereby assist the fungus in evading the host's defense system. It is considered analogous to the phase variation in bacteria. This ability to switch between morphologies is often considered contributing to virulence.

Pathogenesis and Pathology

The most studied *Candida* species, *C.albicans,* has several known virulence factors contributing to its pathogenicity that include adherence to epithelial and endothelial cells, proteinase production, pseudohyphae formation, phenotypic switching, phospholipase production and antigenic modulation as a result of pseudohyphae formation. After yeast cells of *Candida* encounter particular host tissue, colonization takes place at the local site or it invades deeper into the host tissue. Apart from its fate, both specific and non-specific adhesive properties are required to allow binding at these sites. *C.albicans* and *C.tropicalis* produce aspartyl proteinase *in vivo*, which is an important factor of pathogenicity.

The transformation into the hyphal form is observed during an active infection. It is believed that phospholipase concentrated at the hyphal tip may be related to the greater invasiveness of this form as compared to the yeast form. Moreover, the hyphae being larger than the yeast form are more resistant to phagocytosis and thus morphological change contributes to the increased pathogenic potential of the fungus.

Most of the manifestations are associated with biofilm formation with cells in biofilms displaying properties dramatically different from free-living cells grown under normal laboratory conditions. Biofilms formation in *Candida dubliniensis* may represent key factor for survival of this species, which seems to be particularly well-adapted to colonization of the oral cavity with important clinical repercussions.

The pathogenicity in *Candida* also depends on a complex array of microorganism-related putative virulence factors. These include the yeast-to-mycelium transition, antigenic variability, phenotypic switching, adhesion to host cells and tissues, cell surface hydrophobicity, molecular mimicry and production of extracellular enzymes. Most of the biological functions related to pathogenicity and virulence reside in the fungal cell wall.

The factors predisposing people to candidiasis include AIDS, burn patients, young individual, pregnancy, high fruit diets, steroids, antibiotic therapy, immunosuppressants, cancer treatments, heart surgery, genetic deficiency, endocrine deficiency diabetes, use of catheters and unsterile needles.

Clinical Features

The clinical manifestations of candidiasis are extremely varied, ranging from acute, subacute, chronic and episodic. Involvement may be localized to the mouth, throat, skin, scalp, vagina, fingers, toes, nails, bronchi, lungs or gastrointestinal tract. It may also be systemic as in septicemia (circulating in the blood and causing damage to blood vessels and sometimes blood cells), endocarditis and meningitis. The pathologic processes evoked are diverse and vary from irritation and inflammation to chronic and acute suppuration or granulomatous response.

The *Candida* species are found commensal on mucosal surfaces of the body but they can cause disease as and when conditions are favorable. These yeast-like fungi typically colonize mucocutaneous surfaces, which can be portals of entry into deeper tissues when the host defenses are compromised. They may cause simple lesion to even the life-threatening systemic infections. The clinical manifestations of candidiasis can be divided into two broad categories as given in the **Table 20.1**. Some of the important clinical features pertaining to an individual species have also been described along with the cultural characteristics of that particular species.

Table 20.1. Clinical Classification of Candidiasis.

Table 20.1. Clinical Classification of Candidiasis.
I. Infectious Diseases
A. Mucocutaneous Manifestations
Oral: thrush, stomatitis, glossitis, cheilitis
Alimentary: esophagitis, gastritis
Vulvovaginitis, balanitis, balanoposthitis
Chronic mucocutaneous candidiasis
Ocular candidiasis
B. Cutaneous Manifestations
Intertriginous and generalized
Paronychia and onychomycosis
Diaper dermatitis
Candidal granuloma
C. Systemic Manifestations
Urinary tract
Endocarditis
Pulmonary candidiasis
Meningitis
Candidemia
Dissemination
Arthritis
Osteomyelitis
Endophthalmitis
II. Allergic Diseases
Candidids
Eczema
Asthma
Gastritis

I. Infectious Diseases

Candida species produce various infectious manifestations among humans depending upon their immune status and underlying predisposing factors, which are described below:

A. Mucocutaneous Manifestations

There are various mucosal sites involved in candidiasis, which are briefly described below:

1. ***Oral Candidiasis:*** This is the most common form of disease produced by colonization of *Candida* species and it is also known as oral thrush. The oropharyngeal candidiasis occurs in infants, individuals with diabetes mellitus, those receiving antibacterial antibiotics and in patients infected with HIV. The infection may diffuse or may remain confined to buccal mucosa, gums, tongue or palate. The acute infection begins with congestive reddening of the mucous membrane that gives dry, smooth, shiny varnish-like appearance. It is accompanied by thirst, metallic taste and sensation of dryness and burning at the local site. *Candida albicans* is the main cause of oral candidiasis in patients with HIV infection and AIDS.

2. ***Angular Cheilitis:*** The patient presents with sore, erythematous, fissured lesions affecting the angles of the

mouth and is commonly associated with denture stomatitis. This may be associated with iron deficiency anemia or vitamin B_{12} deficiency. A significant number of patients with orofacial granulomatosis have angular cheilitis and it may also be seen among AIDS patients.

3. ***Alimentary:*** The esophagitis may be asymptomatic or may cause burning pain in the substernal area, epigastrium or throat. On esophagoscopy there are white mucosal plaques, more notable in the distal third of the esophagus which resemble oral thrush. It is also seen in acute cases as well as in advanced HIV infection. Almost 90% of AIDS patients suffer from oropharyngeal or esophageal candidiasis at some stage of their disease.

The term *Candida* esophagitis (CE) was introduced by B E Kodsi in 1976. Its severe form may lead to development of strictures, hemorrhage, esophagotracheal fistula and a consequent decrease in quality of life. The severity is classified on macroscopic findings on endoscopy into four degrees **(Kodsi's Classification)**: (a) Grade 1, a few raised white plaques up to 2 mm in size with mucosal hyperemia but without edema or ulcerations; (b) Grade 2, multiple raised white plaques greater than 2 mm in size, with mucosal hyperemia and edema but no ulceration; (c) Grade 3, confluent, linear, and nodular elevated plaques with mucosal hyperemia and ulcerations; (d) Grade 4, findings of grade 3 with increased friability of the mucous membranes and occasional narrowing of the lumen.

The stomach is the second to esophagus to be affected in gastrointestinal candidiasis. The symptoms are more readily attributable to the underlying gastritis than to *Candida*. Sometimes, uncommon complication like peritonitis may be there due to chronic ambulatory peritoneal dialysis. There is abdominal pain and tenderness with or without nausea, vomiting or low-grade fever.

4. ***Vulvovaginitis, Balanitis, Balanoposthitis:*** Vaginal candidiasis is one of the most common infections seen in general practice. This affects primarily young and middle-aged females, particularly during their active reproductive life. About three-quarter of all women suffer at least one episode of this condition during their lifetime and around half of them have recurrence. The candidal vulvovaginitis is mainly caused by *C.albicans* and *C.glabrata*. The capacity of the yeast to assume pathogenic character depends on general factors that can predispose to this infection. There is curd-like vaginal discharge, itching, burning sensation and dyspareunia among these patients. Some women complain of recurrent or increasing symptoms preceding each menstrual period. On examination, vulval erythema and edema may be observed.

In males, balanitis or balanoposthitis are also caused by different *Candida* species. It mainly affects non-circumcised and diabetic men. There is pruritus, associated with erythema, whitish lesions over the glans penis and sometimes erosions and edema of the prepuce that may lead to phimosis. It is also thought to be an allergic manifestation of *Candida* antigen. The consequence of fungal infections, such as colpitis and endometritis, caused by *Candida albicans*, may be infertility. This may lead to decreased spermatozoon motility also as per some of the reports.

The diagnosis of genital candidiasis should not be on clinical grounds alone as similar types of manifestations may be produced in other infective or allergic conditions. Therefore, evidence based laboratory support should be sought for confirmation of the diagnosis. As the transmission of the infection may be through sexual activity hence it is taken as one of the sexually transmitted diseases.

5. ***Chronic Mucocutaneous Candidiasis:*** Chronic mucocutaneous candidiasis (CMC) is common in the patients with deficient cell-mediated immunity, particularly beginning in infancy or later in the childhood. The lesions of oral mucosa may be the initial observation, which could be followed by endocrinopathy or secondary to some rare diseases, such as HIV infection, Di George syndrome, high IgE syndrome and severe combined immunodeficiency (SCID). There are lesions at multiple sites including mouth, skin and fingernails. Hyperplastic and nodule-like lesion with non-removable whitish patches and deep fissure are the most common oral manifestations of these patients. This condition is heterogeneous group of clinical syndromes characterized by chronic, resistant type of candidiasis of skin, nails and oropharynx. Despite extensive cutaneous involvement there is no propensity for occurrence of disseminated visceral candidiasis. Various underlying conditions such as hypoparathyroidism, hypoadrenalism, hypothyroidism and presence of circulating auto-antibodies have been associated with CMC. Sometimes, there is total failure of T-cell response to the test antigen. The delayed-type of hypersensitivity to *Candida* antigens is also absent but circulating antibodies against it are found in high titer. It is refractory to the currently available therapy and persists for years. Therefore, it is important to be familiar with the clinical appearances of CMC to make an accurate diagnosis and to prompt the physicians to look for systemic disorders. Furthermore, their potential systemic disorders should be concerned to avoid the recurrence of *Candida* infection.

6. ***Ocular Candidiasis:*** Candidal keratoconjunctivitis is one of the complications that occur following long-term use of topical corticosteroids. There is conjunctival edema, cheesy discharge in the conjunctival sac and progressive corneal ulceration leading to fungal keratitis. In addition, the most common cause of endogenous fungal endophthalmitis is *Candida albicans* that follows fungemia. These clinical entities are described in more details in **Chapter 29** on **Oculomycosis**.

Table 20.2. Predisposing Factors Associated with Candidiasis.

- Natural receptive states like infancy, old age, pregnancy
- Changes in local bacterial flora secondary to antibiotics
- Changes in epithelial surfaces due to moisture, trauma
- Primary or secondary T-lymphocyte defects to underlying diseases *e.g.* AIDS or immunosuppression
- Neutropenia primary or secondary to disease *e.g.* immunosuppression
- Endocrine disease like diabetes mellitus
- Patients on total parenteral nutrition (TPN)
- Miscellaneous conditions like zinc or iron deficiency

B. Cutaneous Manifestations

The cutaneous manifestations of candidiasis are also very significant clinical features, which are given below:

1. ***Intertrigo:*** This is an inflammatory lesion of the skin folds. The superimposed infection can be accompanied by lymphangitis and enlarged lymphnodes. It occurs predominantly in female patients.

2. ***Paronychia and Onychomycosis:*** Paronychia is an inflammation of the nail fold usually affecting the hand nails and at times nails of the feet. There is localization of infection to the distal part of the nail. It is frequently confused with tinea unguium hence should be carefully differentiated from it.

3. ***Diaper Dermatitis:*** Cutaneous candidiasis is common cause of diaper rash in infants. Maceration and wet diapers predispose infants to diaper dermatitis. It is also known as napkin candidiasis. There are maculopapules and vesicles coalescing into patches with satellite pustules. A moist, macular erythematous rash is most marked in the intertriginous areas of gluteal crease, perineum and inguinal fold, however, there is characteristic sparing of depth of the creases.

The perianal candidiasis commonly affects young infants and is characterized by patchy or confluent erythematous, papulovesicular lesions. Diaper dermatitis is ubiquitous problem with high preponderance rates during the infancy period. A combination of causes include mechanical friction, moisture, prolonged contact with urine and feces, rise in fecal enzyme activity, variation in skin pH and various topical applications. Secondary infection with bacteria and *C.albicans* originating from the gastrointestinal tract plays major pathogenic role. Other contributory factors are young age, frequency of episodes of diarrhea and infectious diseases, antibiotic therapy, immunosuppressive drugs, topical dermatitis and overall health status of the child. Therefore, an appropriate therapy should be directed to the underlying cause and local pathology. Diaper dermatitis commonly becomes secondarily infected with polymicrobial organisms.

4. ***Candidal Granuloma:*** This condition is commonly found in children. It is described as primary vascularized papules covered with thick, adherent, yellow-brown crust over face, scalp, fingernails, trunk and legs. These may develop into horns or protrusions up to two cm in length. On histopathological examination, the lesions appear as poorly organized granulomas, with giant cells and chronic inflammatory reaction.

C. Systemic Manifestations

The following systemic infections in candidiasis are most commonly observed. Various predisposing factors associated with the candidiasis have been shown in **Table 20.2** which can precipitate these infections.

1. ***Urinary Tract Candidiasis:*** It is commonly seen in females who get the infection from vulva or vagina. The infection may involve the bladder and kidneys. It is reported in association with disseminated candidiasis, diabetes mellitus, pregnancy and prolonged administration of antibiotics and use of unclean catheters.

2. ***Candiduria:*** Candiduria is presence of yeast cells in urine and it is an increasingly common finding in hospitalized patients. Most of patients with candiduria are asymptomatic as there are no associated signs and symptoms. However, quantitative cultures with colony count of $>10^5$/ml of urine is associated with infection in patients without indwelling urinary catheters. In contrast, clinically significant renal candidiasis has been reported even with low colony counts of 10^3/ml of urine. Moreover, pyuria usually supports diagnosis of *Candida* infection. It is marker for hematogenous seeding to the kidneys. It involves vast majority of patients, however, candiduria most likely reflects colonization or infection of lower urinary tract or collecting system of kidneys.

3. ***Endocarditis and Pericarditis:*** Candidemia can cause subacute endocarditis, particularly in patients with previously abnormal native or prosthetic valve. Pericarditis due to *Candida* species although is rare clinical entity, associated with thoracic surgery and immunosuppression.

4. ***Pulmonary Candidiasis:*** The pulmonary lesions arise from hematogenous seeding, causing diffuse reticulonodular streaking, which is very difficult to see on chest X-ray unless other pulmonary disorder is present. It is not very common condition and perhaps difficult to establish. *Candida* pneumonia in the immunocompromised host should be considered life-threatening infection and treated with systemic antifungals. It is more of an autopsy finding than an identifiable clinical entity during life.

5. ***Meningitis:*** Candidial meningitis is more common with extremes of age *i.e.* pediatric and elderly individuals instead of adults. This is seen in low-birth weight neonates with candidemia. It is also seen in patients with hematologic malignancies, complicated neurosurgical procedures and intracerebral prosthetic devices such as ventriculo-peritoneal shunts. Hydrocephalus may also result from chronic meningitis or due to shunt obstruction.

6. ***Candidemia and Septicemia:*** The candidal septicemia usually occurs in the immunocompromised patients. Invariably, it is fatal condition in neonates. It may be iatrogenic also. Common species involved are *C.albicans*, *C.tropicalis* and *C.krusei*. *C.tropicalis* has emerged as an important pathogen in neutropenic patients and *C.parapsilosis* infection is common in individuals receiving parenteral nutrition. This is one of the most important entities in the hospitalized patients receiving TPN solution. Hence the highest attention must be paid to prevent the external contamination of TPN solution since the infections are associated with high mortality. *Candida glabrata* may also cause fungemia.

7. ***Disseminated Candidiasis:*** The disseminated candidiasis occurs frequently in individuals with hematologic malignancies, patients undergoing broad-spectrum antibiotic therapies, catheterization, in bone marrow and organ transplant recipients or in patients undergoing immunosuppressive therapy. Severe neutropenia is considered to be more significant predisposing factor for life-threatening infections. Other factors include use of intravenous catheters, post-surgical conditions, severe burns, hyperalimentation and diabetes mellitus. The hematogenous infection is more likely to be fatal when caused by *C.tropicalis* than other species of genus *Candida*. The disseminated candidiasis continues to be major complication in immunocompromised hosts, which is not so in patients with AIDS. These patients develop serious infections of the oropharynx and upper gastrointestinal tract but rarely experience systemic disease. Various clinical forms of candidiasis in the HIV patients are given in the **Table 20.3**. *Candida krusei* is an emerging pathogen, especially in immunocompromised hosts. The adherence of this organism both to host epithelial surfaces and to catheter and prosthetic surfaces, appears to be important in the pathogenesis of superficial as well as systemic candidiasis.

Table 20.3. Clinical Forms of Candidiasis in HIV Patients.

- Asymptomatic oral carriage
- Oropharyngeal thrush
- Acute atrophic erythema
- Perleche (angular cheilitis)
- Leukoplakia
- Esophagitis
- Laryngitis
- Vulvovaginitis, balanitis
- Hematogenous dissemination

8. ***Arthritis:*** Prosthetic or rheumatoid joints are prone to infection by *Candida* species mainly through hematogenous route. It may be due to direct inoculation during joint surgery or intra-articular corticosteroid injections. A delayed diagnosis leads to destruction of cartilage or loosening of the prosthesis.

9. ***Osteomyelitis:*** It is also hematogenous infection. There is indolent onset of fever and back pain followed later by radiculopathy. A well-defined osteolytic lesion with paravertebral pus is a characteristic feature.

10. ***Endophthalmitis:*** This is the manifestation of disseminated infections *i.e.* candidemia rather provides the only physical findings that are pathognomonic of invasive candidiasis. However, it may occur as sequel to an otherwise harmless episode of benign fungemia. Initially, the patient complains of cloudy vision or scotoma. Deep bulbar pain or feeling of irritation in the eye may occur later as visual acuity worsens. The most characteristic finding is one or more focal, white retinal lesions that may be associated with haziness in the overlying vitreous. This may lead to blindness if an early effective therapy is not started. But sizeable minority has no symptoms. It is rare complication of ocular surgery such as in intraocular lens implantation.

In addition, Acute Zonal Outer Optic Retinitis (AZOOR) has been linked with some of the *Candida* species like

C.fomata, wherein patients usually see spots in the outer boundaries of the field of vision, which should not be confused with endophthalmitis.

11. ***Nosocomial Candidiasis:*** *Candida* species have been implicated in the nosocomial infections and are important cause of fatal consequences among the hospitalized patients. Candidemia is the most extensively studied nosocomial invasive fungal infection. The frequency of nosocomial bloodstream infections by *Candida* species has risen dramatically in the last two decades. The arrival of antifungal drugs with better tolerability than conventional amphotericin B has resulted in widespread use of systemic antifungal therapy. With the introduction of new systemic antifungals, however, there have been major shifts in the epidemiology of candidal bloodstream infection towards species with less susceptibility to the available antifungal agent.

12. ***Candida Pancreatitis:*** The first reports of pancreatic abscesses secondary to *Candida albicans* were published in the 1980s but its potential role in SAP was suspected later. Although bacterial infections are most commonly encountered, fungal infections are increasingly being recognized. Candida is the most common fungal infection. The occurrence of fungal infection in patients with acute pancreatitis adversely affects the clinical course, leading to a higher incidence of systemic complications, and possibly mortality as well. Important risk factors for fungal infection in patients with acute pancreatitis include broad-spectrum antibiotics, prolonged hospitalization and surgical/endoscopic interventions, use of total parenteral nutrition (TPN) and mechanical ventilation. Patients with higher severity of pancreatitis are at a greater risk. The pathogenesis of fungal infection in patients with acute pancreatitis is multifactorial. Translocation of microorganisms across the gut epithelium, lymphocyte dysfunction, and the virulence of the invading microorganisms play important roles. Histological demonstration of fungi remains the gold standard of diagnosis, but a positive biopsy is rarely obtained. The role of biomarkers in the diagnosis is being investigated. As early diagnosis and treatment can lead to improved outcome, a high index of suspicion is required for prompt diagnosis. Limiting the use of broad-spectrum antibiotics, early introduction of enteral nutrition, and timely change of vascular catheters are important preventive strategies. The role of antifungal prophylaxis remains controversial. Surgical necrosectomy with antifungal therapy is the most widely used treatment approach. Clinical trials on antifungal prophylaxis are needed, and indications for surgical intervention need to be clearly defined. Severe acute pancreatitis (SAP) is associated with significant morbidity and mortality. The majority of deaths related to SAP are the result of infectious complications.

II. Allergic Diseases

The following manifestations may present as allergic reactions to the candidal antigens:

Candidid: This is an allergic manifestation to the metabolites of *Candida*. This clinical form is less commonly seen and involves lesions mediated by the host's immune response to candidal infection. It is analogous to the dermatophytid reaction in appearance, morphology, distribution and genesis of clinical lesions. Both these conditions are called the **'id' reactions**. There is pompholyx type of vesicular lesions in interdigital spaces of hands or other parts of body. They disappear following resolution of infection by treatment or after desensitization.

Other Allergic Manifestations: The cutaneous allergy as urticaria and eczema due to *Candida* species have been documented. Similarly, bronchial asthma caused by *Candida* species has also been reported. Hypersensitivity resulting from the overgrowth of *Candida* species has also been implicated in gastritis and irritable bowel syndrome (IBS).

Radiodiagnosis

The radiological techniques have been very helpful to establish the diagnosis of the deep-seated candidiasis to localize the site and extent of the lesions. Computed tomography and magnetic resonance imaging scans are the most useful procedures.

Differential Diagnosis

Oral thrush can mimic bacterial infections like sore throat caused by *Staphylococcus aureus*, *Corynebacterium diphtheriae*, leukoplakia, lichen planus and tertiary syphilis. Vaginal candidiasis may be confused with trichomoniasis, bacterial vaginosis and herpes simplex infections. The differential diagnosis of systemic infection must include other mycoses, tuberculosis, neoplasia or chronic bacterial infections. In all these cases fungal culture is essential to confirm or rule out the diagnosis of candidiasis.

Laboratory Diagnosis

An early diagnosis of candidiasis should be established as the infections have high mortality. The difficulty for diagnosis lies due to the absence of specific symptoms and signs as well as opportunistic nature of the yeasts. Their presence in normally colonized body sites of immunocompromised patients does not substantiate infection and they are seldom isolated from deeper organs or tissues. The following protocol is instituted to validate the diagnosis of suspected cases of candidiasis in the laboratory:

(a) Direct Examination

The clinical specimens are collected depending upon the site of involvement *i.e.* from superficial lesions or deep-seated infections. Whitish patches from the mucous membrane of the mouth, vagina, skin or nails scrapping, sputum are collected with the help of sterile swabs. These are examined in KOH wet mount or normal saline preparation **(Fig. 20.1)**.

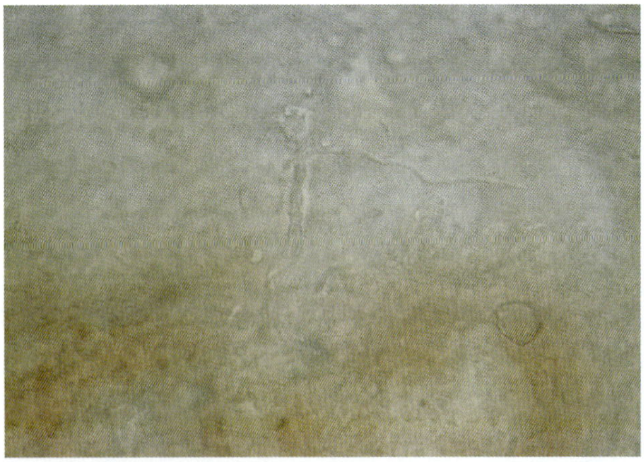

Fig. 20.1. Wet mount showing yeast cells and pseudohyphae, as well as one true hypha also, seen in candidiasis (KOH × 400).

Gram's staining is performed to see presence of yeast and pseudohyphae of *Candida* species. The yeast cells are approximately 4-8 μm with budding and pseudohyphae **(Figs. 20.2A and B)**. The Gram's smear may be visible as Gram-negative also due to over-decolorization otherwise all the fungi are Gram-positive in nature. The presence of pseudohyphae shows colonization and tissue invasion hence their demonstration in the direct smear of tissue is highly significant. Some species like *Candida glabrata, C.rugosa* and *C.lusitaniae* do not form pseudohyphae and this fact must be kept in mind while doing microscopy. Calcofluor white stain may also be used to highlight the fungal elements in the smear. In addition, PAP smear, a routine diagnostic procedure in obstetrics and gynecology, may show yeast cells of *Candida* species **(Fig. 20.3)**. Similarly, while examining other stainings, fungal elements may be incidentally visualized, as *Candida* species encountered in Ziehl-Neelsen staining routinely done for acid-fast bacilli **(Fig. 20.4)**.

The biopsy specimens are kept in tube containing KOH for an overnight period at 37°C and after mincing these are examined under the microscope for yeast cells and pseudohyphae. The other diagnostic stainings done for histopathological examination like H&E **(Fig. 20.5)**, periodic acid-Schiff, **(Figs. 20.6 and 20.7)** and Gomori's methenamine silver stain **(Fig. 20.8)** also done for the demonstration of *Candida* species in tissue sections.

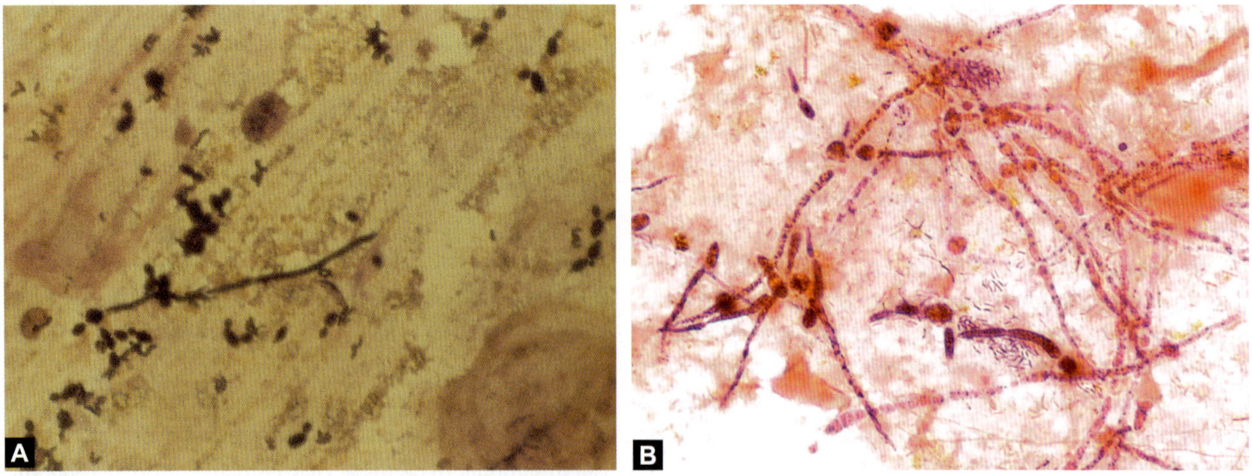

Figs. 20.2A and B. Gram-positive yeast cells and pseudohyphae of *Candida* species seen in clinical smear (Gram's Staining × 400).

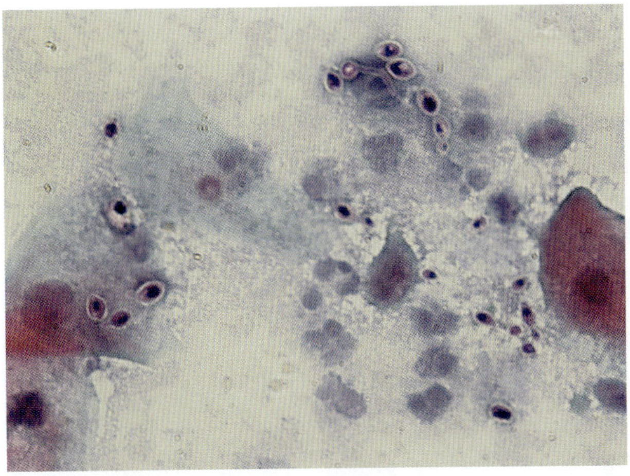

Fig. 20.3. Yeast cells of *Candida* species seen cervical smear prepared as gynecological investigations (PAP × 400).

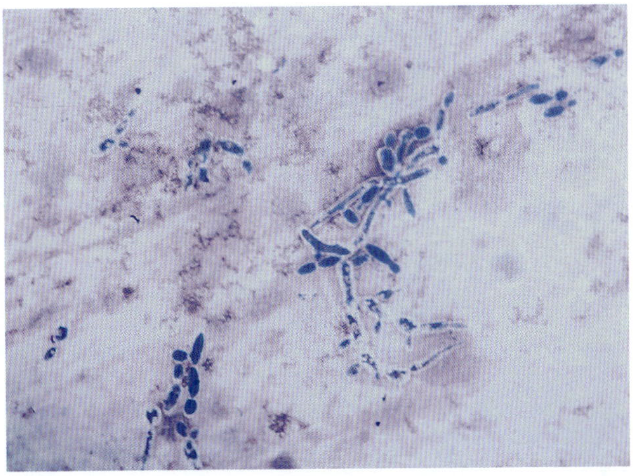

Fig. 20.4. Yeast cells and pseudohyphae of *Candida* species in AFB smear as an incidental finding (Ziehl-Neelsen Staining × 400).

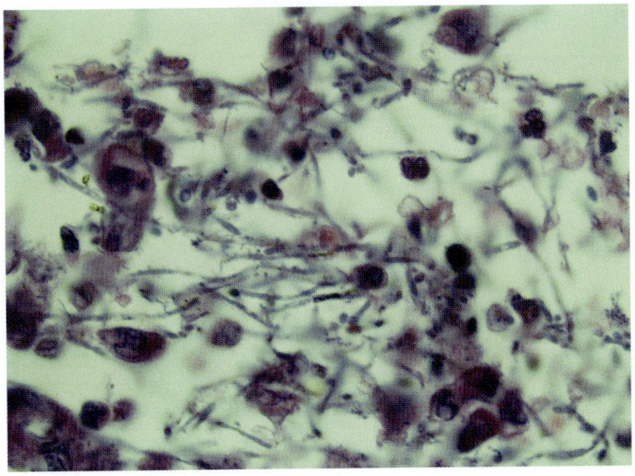

Fig. 20.5. Yeast cells and pseudohyphae of *Candida* species (H&E × 400).

(b) Fungal Culture

The clinical specimens can be cultured on Sabouraud dextrose agar with antibacterial antibiotics and incubated at 25°C and 37°C. The colonies appear in 2-3 days as cream colored, smooth and pasty **(Fig. 20.9)**. Sometimes growth may be observed after an overnight incubation as seen in bacteria. The bacterial culture media like blood agar may also be used for growing the *Candida* species. The LCB mount is prepared from the colonies to examine for the presence of yeast cells and pseudohyphae **(*See* Fig. 20.1)**. Very rarely true hyphae may be seen. Gram staining may also be performed from the culture isolates. For systemic infection like candidemia, blood culture is done in biphasic medium like brain-heart infusion agar-broth and incubated at both the temperatures. The detection of *Candida* species in blood culture is one of the most important advances in the recent diagnostic procedures because in earlier times, blood culture techniques often did not recover *Candida* species.

The growth of *Candida* species is also seen on Tetrazolium Reduction Medium (TRM) and compared to the standard colors to identify various species **(*See* Table 20.7)**. The *Candida* isolates are identified by standard protocols that include germ tube formation **(Fig. 20.10)**, chlamydospore production **(Fig. 20.11)** and sugar fermentation and assimilation tests. The salient cultural and morphological features pertaining to individual species of *Candida* are given herewith:

(i) Candida albicans: (Latin - candidus - glowing white) *Candida albicans* is an opportunistic fungal pathogen with host interaction abilities that range from commensal through life-threatening disseminated diseases. The interplay between *Candida* and host defense is essential in determining outcome of the infection. The colonies are cream-colored, pasty and smooth. The production of germ tube may be demonstrated **(Fig. 20.10)**. On cornmeal agar at 25°C, there are large thick-walled, terminal chlamydospores, characteristic of this species **(Fig. 20.11)**. On CHROMagar Candida, the colonies of *C.albicans* appear as light green to bluish-green in color **(Fig. 20.12)**. Due to the relatively high DNA homology between *C.albicans* and *C.stellatoidea*, the latter has been re-classified as sucrose negative variant of *C.albicans*.

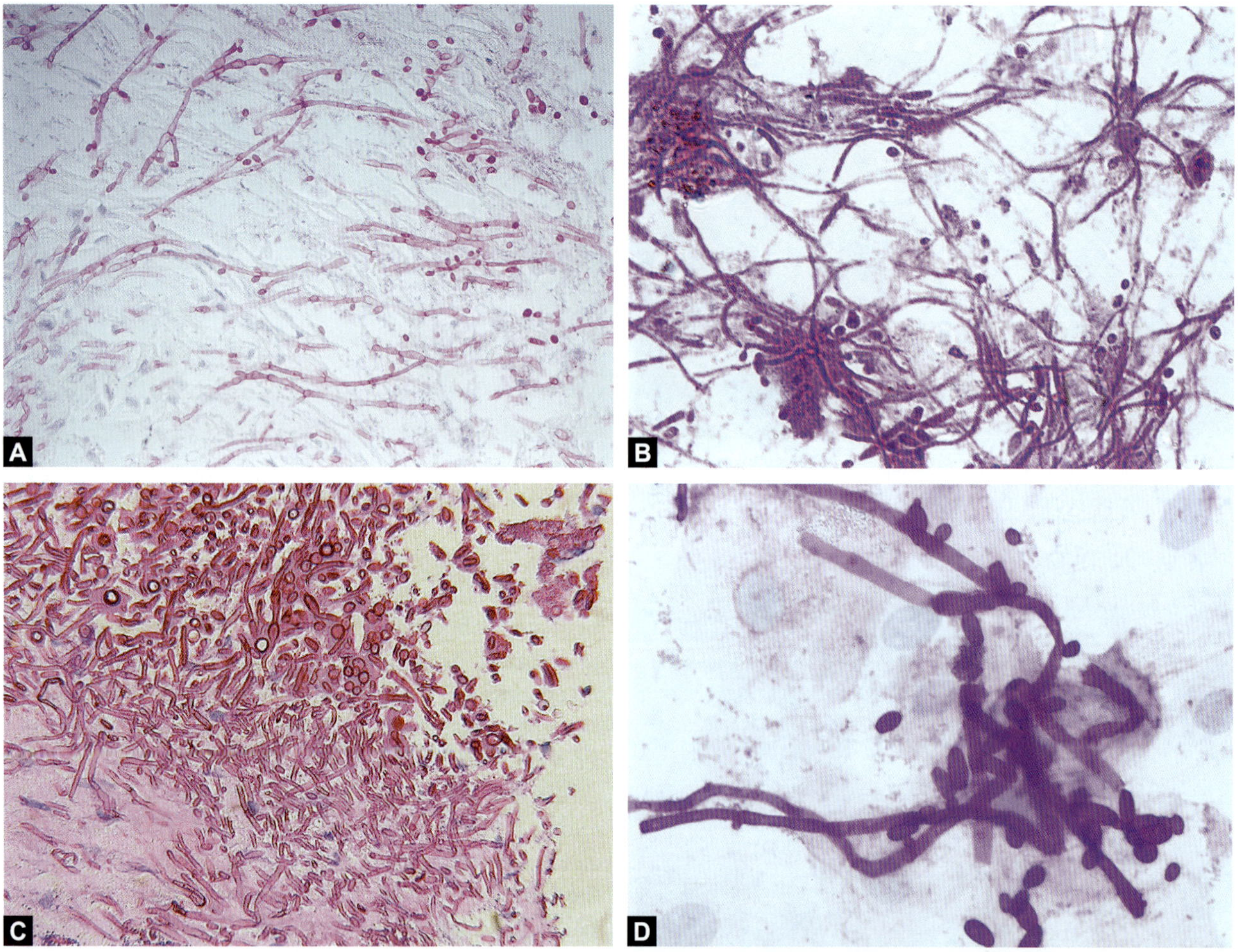

Figs. 20.6A to D. Smears / sections showing pink-colored yeast cells and pseudohyphae of *Candida* species (PAS × 400).

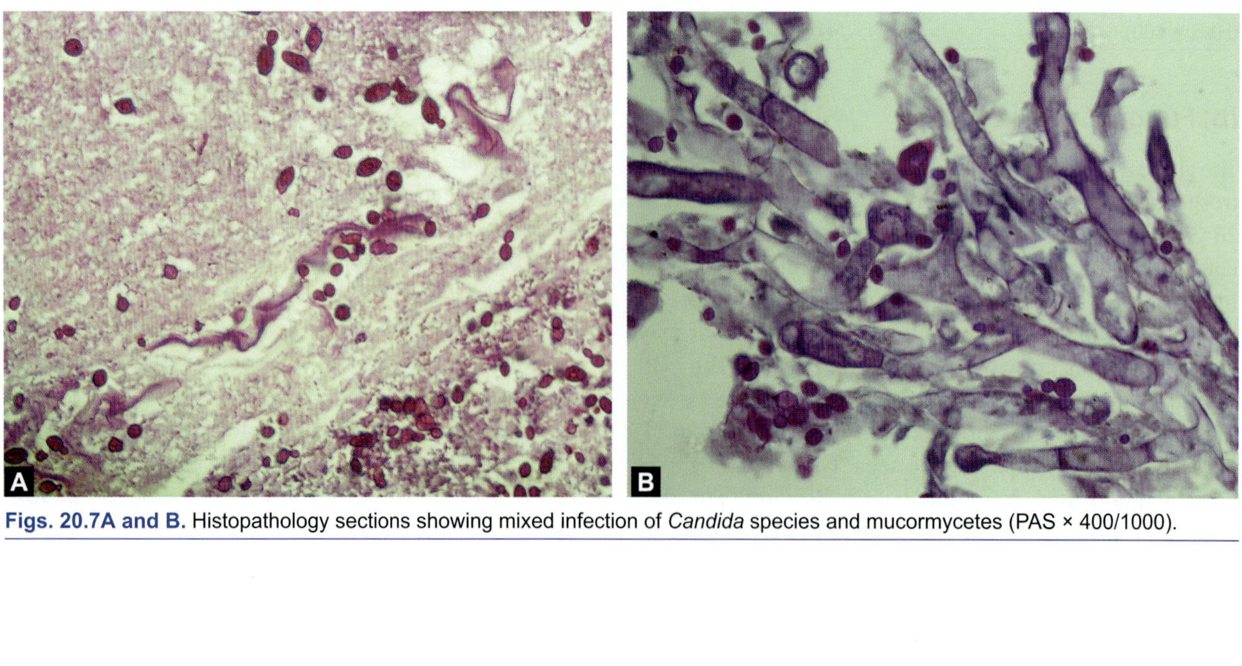

Figs. 20.7A and B. Histopathology sections showing mixed infection of *Candida* species and mucormycetes (PAS × 400/1000).

Fig. 20.8. Histopathology section showing yeast cells and pseudohyphae of *Candida* species (GMS × 400).

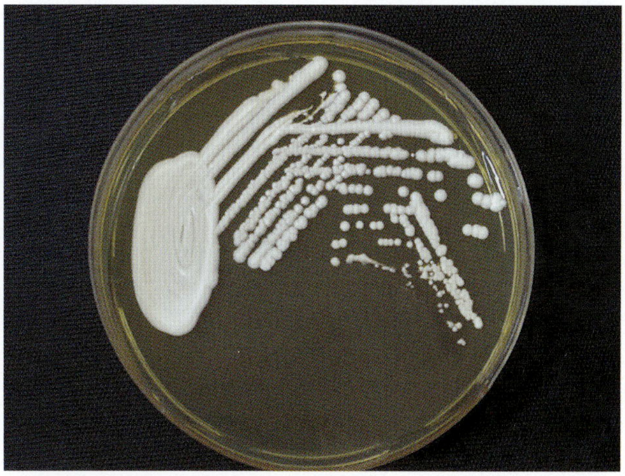

Fig. 20.9. Growth of *Candida* species seen on Sabouraud' dextrose agar after three days.

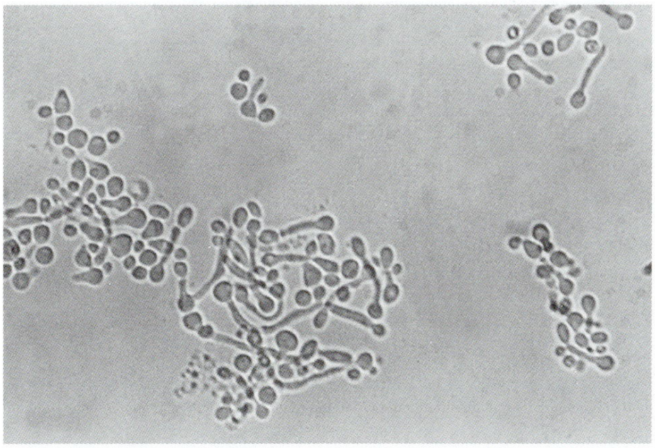

Fig. 20.10. Germ tube seen in *Candida albicans* (Serum × 400).

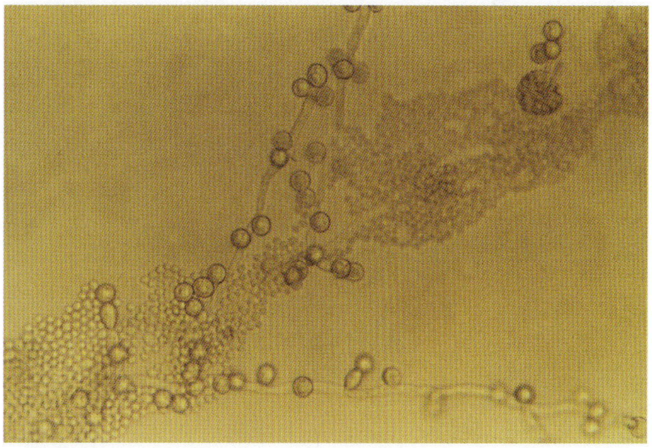

Fig. 20.11. Terminal chlamydospores seen in *Candida albicans* (CMA × 400).

(ii) Candida tropicalis: *C.tropicalis* grossly looks similar to *C.albicans* hence cannot be differentiated from it on colony morphology or growth rate. Its colonies are cream-colored to off white, glistening to dull, soft, smooth or wrinkled with mycelial fringe. On cornmeal agar, it forms blastospores singly or in small groups. It shows fir tree appearance on CMA **(Fig. 20.13)**. It has gained more importance in patients with neutropenia and hematologic malignancies. Moreover, it is common agent of hospital acquired infections. There is thin pellicle on the top of Sabouraud dextrose broth. It is not associated with oral thrush.

(iii) Candida krusei: The colonies are flat, dull and dry and after incubation for month, they are greenish-yellow, dull, soft and smooth or wrinkled. There is dense growth of mycelia extending as lateral fringe around the colonies. The elongated cells have tree-like arrangement or crossed match-stick appearance **(Fig. 20.14)**. In some of the strains, blastospores are also produced, which are elongate and grow in verticillate branches from pseudohyphae. Like *C.tropicalis*, it is important in patients of hematologic malignancies and blood and bone marrow transplant recipients. There is thick pellicle on the top of the Sabouraud dextrose broth. On CHROMagar Candida medium it gives pinkish to purplish-colored colonies, which may confuse with *C.glabrata* due to their similarity in production of pink colored colonies **(Fig. 20.12)**. This species is found to be inherently resistant to most of the antifungal agents, particularly the azoles derivatives.

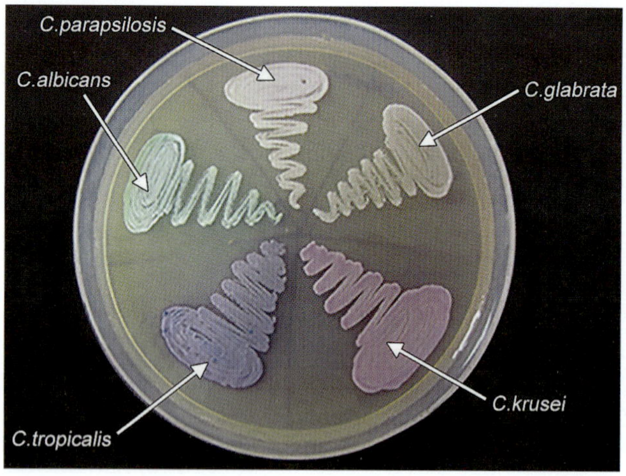

Fig. 20.12. CHROMagar Candida showing growth of different *Candida* species as indicated with their colors (CMA × 400).

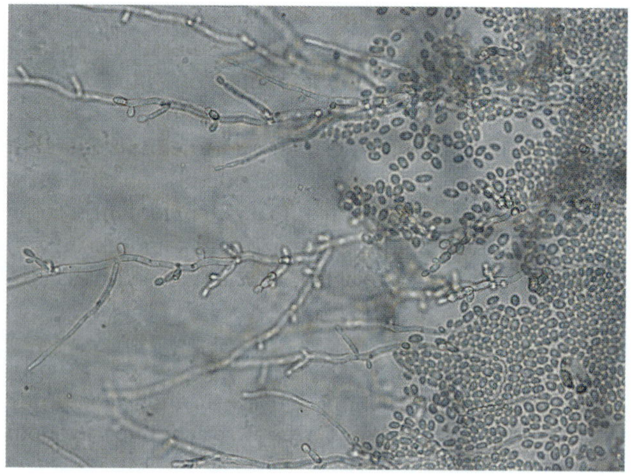

Fig. 20.13. Growth of *Candida tropicalis* on CMA (CMA × 400).

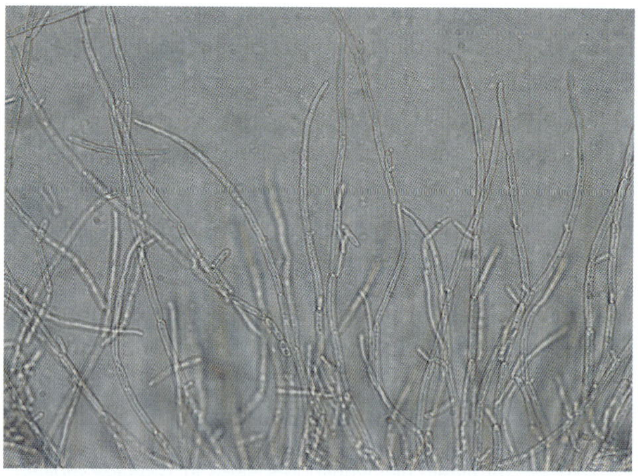

Fig. 20.14. Growth of *Candida krusei* on CMA (CMA × 400).

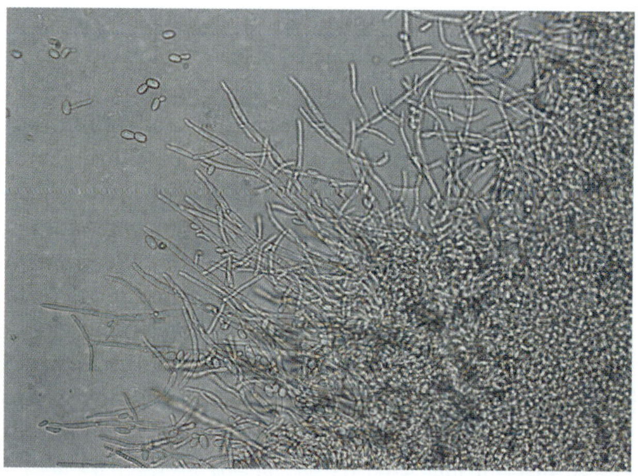

Fig. 20.15. Growth of *Candida parapsilosis* on CMA (CMA × 400).

(iv) Candida parapsilosis: The colony morphology is creamy, sometimes developing in lacy pattern. It produces blastospores singly or in small clusters that are seen along the pseudohyphae **(Fig. 20.15)**. On CHROMagar Candida medium it gives cream-colored colonies however, slightly larger than *Prototheca* species **(Fig. 20.12)**. The outstanding feature of this species is presence of large mycelial elements that are called giant cells.

This is having low pathogenic potential as compared to other *Candida* species and is associated with both localized and deep-seated infections, particularly in neonates. Fungemia in these cases is generally related to presence of central venous catheter as well as to the use of parenteral nutrition (TPN). It is hypothesized that biofilm formation in *C.parapsilosis* is responsible for such infections, which helps in sticking to external material of catheters, etc. It is also reported as the causative agent of endocarditis, endophthalmitis, septic arthritis and peritonitis. Superficial infections include vaginitis, which is clinically indistinguishable from more common infection due to *C.albicans*. In addition to being frequent isolate from several healthy cutaneous sites, *C.parapsilosis* is frequently isolated from pathological lesions of nails and skin. It is not associated with oral thrush.

(v) Candida glabrata: This species was considered as relatively non-pathogenic and normal flora of healthy individuals and rarely causing serious infection in humans. However, following the widespread and increased use of immunosuppressive therapy together with broad-spectrum antifungal therapy, the frequency of mucosal and systemic infections caused by *C.glabrata* has significantly increased particularly in older age groups.

Over the past decade incidence of bloodstream infection caused by *Candida* species has increased five to ten fold to become the **fourth** leading cause of health-care associated bloodstream infections in the hospitalized patients. The incidence of non-*albicans Candida* bloodstream infection has also been increasing and now surpasses *C.albicans*. The most common of these non-*albicans Candida* species is *C.glabrata* that accounts for 11-16% of all candidemia cases. *C.glabrata* fungemia is associated with severe complications more among than other *Candida* species. However, in certain centers of India, *C.tropicalis* is the commonest species which is isolated, among non-*albicans Candida*.

The health-care associated fungemia by *C.glabrata* is not just disease of debilitated and neutropenic patients but affect wide variety of individuals, entailing a very high mortality. This species exists as small blastospores under all environmental conditions. It is the only species of genus *Candida* that does not produce pseudohyphae and hence it was earlier described with other names as *Torulopsis glabrata*. However, it was determined that the ability to produce pseudohyphae was not reliable distinguishing feature for members of genus *Candida* and it was finally classified as *Candida glabrata*.

The colonies are glistening, smooth, cream-colored which are relatively indistinguishable from those of other *Candida* species except for their relative size, which is quite small. The colony size remains very tiny on blood agar medium hence there are chances of their missing during the initial period of incubation. There is neither mycelial fringes on agar media nor pseudohyphae formation. In a LCB mount they are seen as small round or ovoid yeast cells of 3×5 µm in size. On CHROMagar Candida medium, *C.glabrata* colonies appear pink to purple, in contrast to *C.albicans*, which are green to bluish-green. However, it may confuse with *C.krusei* due to its pinkish coloration.

It ferments and assimilates only glucose and trehalose as compared to other commonly encountered species, *C.albicans*, which ferments and/or assimilates large number of sugars. There are some commercially available rapid tests like Glabrata RTT, based on the ability of *C.glabrata* to hydrolyze trehalose but not maltose, where results are available within 20 minutes for rapid identification of this fungus. *C.glabrata* infections are difficult to treat as they are often resistant to many azole derivatives, especially fluconazole.

(vi) Candida kefyr: The colonies are creamy, smooth and after prolonged incubation become cream to yellowish, somewhat dull, soft and smooth. The pseudohyphae are in abundance in most of the strains. The cells are very elongate; they fall apart and lie parallel to each other like logs in stream. Blastospores are not abundant however; when present they are elongated and verticillate. Earlier this species was known as *C.pseudotropicalis*. This is used as standard strain in the Antifungal Susceptibility Testing of the yeasts as most of the strains are found to be sensitive to the commonly used antifungal drugs.

(vii) Candida guilliermondii: The colonies are thin flat, glossy, cream to pink, glistening, smooth or dull, wrinkled. Variable development of pseudohyphae that is very fine and short with small cells, sometimes bearing ramified chains of small ovoid blastospores.

(viii) Candida dubliniensis: This is newly recognized opportunistic pathogen that has been linked to oropharyngeal candidiasis in HIV-infected patients. It has also been observed in blood isolates from bone marrow transplant patients, denture wearers, cancer patients, infants and elderly, oral and vaginal isolates from non-HIV-infected patients. This was named ***'dubliniensis'*** because its first isolation was from AIDS patients from Dublin, Ireland in 1995 and is now isolated worldwide. Its importance is based on rapidly developing resistance to fluconazole, common antifungal drug used for the treatment of mycoses.

It was initially difficult to distinguish it from other *Candida* species in standard clinical laboratory procedures because of its closely shared phenotypic (germ tube, chlamydospore and true hypha formation) and genotypic characteristics with *C.albicans*. In case of *C.dubliniensis* and *Candida albicans,* genotypic characterization rather than phenotypic characters distinguish between these two species. However, more recently, phenotypic characteristics, including carbon assimilation, growth temperature (the former will grow at 45°C), immunofluorescence and DNA based molecular approaches have been used to distinguish *C.dubliniensis* from other species of *Candida* genus.

The use of CHROMagar Candida medium, temperature differential studies, xylose assimilation and β-glucosidase activity and nucleic acid hybridization studies provide means of differentiation. Odds and colleagues employed several of the tests noted above to study large collection of the *Candida* species. The detection and differentiation of *C.dubliniensis* from *C.albicans* in primary cultures requires application of newer strategies.

Since *C.dubliniensis* is normally found in mixed culture with *C.albicans*, therefore, color of colonies in primary culture on CHROMagar Candida is to be observed. This medium is usually recommended for identifying mixed yeast cultures. Dark-green colonies are indicative of *C.dubliniensis*, in contrast to light-green colonies of *C.albicans* **(Table 20.4)**.

Table 20.4. Features of *C.albicans* and *C.dubliniensis*.

Characteristics	*C.albicans*	*C.dubliniensis*
Germ Tube	+	++
Chlamydospores	+	+++
CHROMagar	Light Green	Dark Green
Growth at 42-45°C	Grow (+)	Does not grow (-)

(ix) Candida pelliculosa: It is grown in high-sugar containing media and colonies are cream-colored, glistening, moist and butyrous. The budding cells are spherical to elliptical, 2-4 × 2-6 µm. Pseudohyphae may be present. ***Pichia anomala*** is the teleomorphic state of the ascosporogenous yeast, which was previously known as *Hansenula anomala*. The species is recognized by fermentation of sucrose and the following growth tests like L-rhamnose -, soluble starch +, nitrate + and growth without vitamins is positive. *Pichia ciferrii* differs by presence of septate hyphae. It can cause fungemia in neonatal intensive care unit, interstitial lung disease and endocarditis. An outbreak of *Pichia anomala* has been reported in 1996-98 in pediatric ward from Chandigarh by Chakrabarti and colleagues, wherein a Junior Resident was found to be the source of infection.

(x) Candida lusitaniae: Although this is normal flora of human gastrointestinal tract but is now emerging as a significant pathogen causing fungemia particularly in patients with malignancies. The other types of infections may be peritonitis, UTI and meningitis. There may be secondary resistance to amphotericin B. *Clavispora lusitaniae* is its teleomorphic state.

(xi) Candida rugosa: It is an emerging fungus, which was first identified as a cause of catheter related fungemia in US in1985. Recently, *C.rugosa* has been cited as an emerging species causing candidemia. A study by Dube *et al* reported 15 cases of candidemia due to *C.rugosa* in burn patients in a US hospital.

(xii) Candida auris: This species was first described after isolation from external ear discharge of a patient in Japan. Subsequent studies confirmed an association with chronic otitis media in 15 patients from South Korea, with evidence of clonal transmission and resistance to certain triazoles. *C.auris* has since been reported from a wide spectrum of clinical manifestations, ranging from colonization through deep-seated infections and candidemia. The fungus was also isolated from patients' mattresses, windowsills, chairs, and countertops.

Candida auris fungemia is associated with a high mortality rate, therapeutic failure and widespread resistance to several classes of antifungals. Furthermore, correct identification of *C.auris* isolates is complicated by the fact that many commercially available biochemical-based tests can misidentify it as the phylogenetically related species *Candida haemulonii*, which presents an additional challenge for an appropriate patient management.

Now, it is evident that *C.auris* has emerged as an important nosocomial pathogen with clonal inter- and intra-hospital transmission and it has become widespread across several countries. A total of 77 patients from 7 states in the USA have been diagnosed with this deadly and highly drug-resistant yeast infection, maximum from New York. The federal health officials first warned clinicians in June 2016 to be on the lookout for the emerging pathogen that has been spreading around the world. It has been reported in a dozen countries *i.e.* Colombia, India, Israel, Kenya, Kuwait, Panama, Pakistan, South Africa, South Korea, Venezuela, UK and USA.

The identification of the *C.auris* isolates is undertaken by using CHROMagar Candida medium, Vitek-2 YST, API 20C, Auxacolor and sequencing of internal transcribed spacer (ITS) and D1/D2 domains of the ribosomal RNA gene, followed by microbroth dilution susceptibility testing. Some of the isolates are misidentified as *C.haemulonii* and *Rhodotorula glutinis* by Vitek-2 YST, API 20C and Auxacolor testing, respectively **(Figs. 20.16 and 20.17)**.

(xiii) Candida haemulonii Complex: The emerging fungal pathogens comprising the *Candida haemulonii* complex (*Candida haemulonii*, *Candida haemulonii* var. *vulnera* and *Candida duobushaemulonii*) are notable for their antifungal resistance. *Candida haemulonii*, one of the non-*albicans Candida* species, is an emerging yeast pathogen, which is resistant to amphotericin B and other antifungals such as azoles. Members of the *Candida haemulonii* species complex are uncommon yeasts that cause bloodstream and deep-seated infections and consist of two genotypically distinguishable species *i.e.* *C.haemulonii*

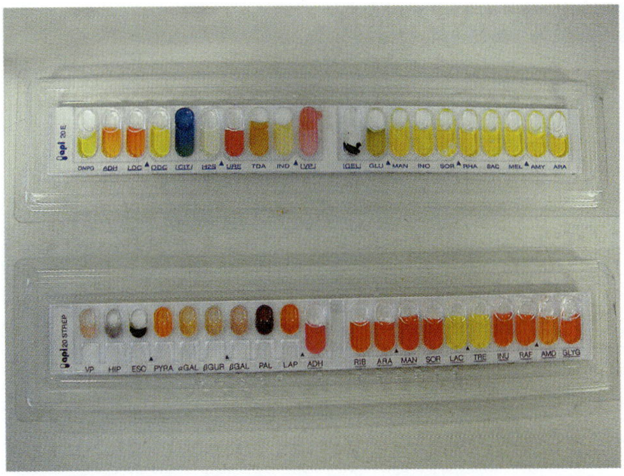

Fig. 20.16. API 20C testing for yeast identification

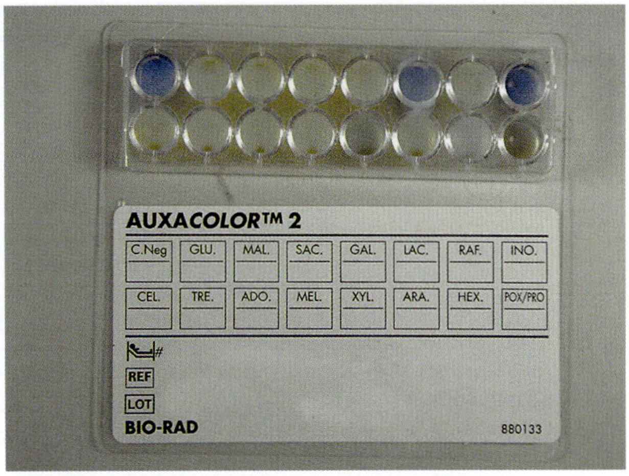

Fig. 20.17. Auxacolor testing for yeast identification.

Table 20.5. Differences between Hyphae and Pseudohyphae.

Features	*Hyphae*	*Pseudohyphae*
Growth Process	Apical elongation	Budding
Terminal Cell	Longer, cylindrical	Shorter, spherical
Cell Wall	Parallel	Constricted at septa
Septa	Straight, perpedicular	Curved or pinched
Origin of branches	No constriction, septum not required	Constricted and septate

and *C.duobushaemulonii* and a variety, *C.haemulonii* var. *vulnera*. These species and other relatives of *C.haemulonii, i.e. Candida auris* and *Candida pseudohaemulonii*, cannot be differentiated by the conventional yeast identification methods used in microbiology laboratories; thus, the true distributions of *C.haemulonii* and its sibling species remain unknown. During the last few years, clinical treatment failures for *Candida haemulonii* infections associated with resistance to amphotericin B and reduced susceptibility to azoles and echinocandins have been reported.

(c) Germ Tube Test

This procedure is used for presumptive identification of *Candida* species and is called germ tube test (GTT). The culture of *Candida* species is treated with bovine, sheep, horse or normal human serum and incubated at 37°C for 2-4 hours. A drop of suspension is examined on the slide under the microscope. The germ tubes are seen as long tube-like projections extending from the yeast cells. There is no constriction at the point of attachment to the yeast cell as seen in case of pseudohyphae **(*See* Fig. 20.10)**. The differences between hyphae and pseudohyphae are shown in **Table 20.5**. The germ tubes are formed within two hours of incubation in *C.albicans* and *C.dubliniensis* and not in other species of this genus. In addition, Mueller-Hinton agar is also used as medium for germ tube formation in both these *Candida* species. The demonstration of the germ tube is also known as **Reynolds-Braude phenomenon**. A new milk medium has been used for the germ tube and chlamydospores production in *C.albicans*. Similarly, amino acid synthetic medium also exhibits germ tubes and mycelial development at 25°C.

(d) Chlamydospore Formation

The suspected strain of the *Candida* isolates is grown on cornmeal agar (CMA) or rice starch agar (RSA) and incubated at 25ºC. It shows formation of large, highly refractile, thick-walled, terminal chlamydospores after 2 to 3 days of incubation **(*See* Fig. 20.11)**. These are seen in the clinical isolates of *C.albicans* and *C.dubliniensis*.

Chlamydospores are refractile, thick-walled cells that are produced under nutritionally poor, oxygen-limited conditions when incubated at room temperature. There are several media, such as corn meal agar and cream of rice agar, which have been used for the production of chlamydospores. Recently, media such as Staib's agar, Pal's agar, casein agar and tobacco agar have also been reported to selectively induce chlamydosporulation in *C.dubliniensis*, enabling its differentiation from *C.albicans*. Moreover,

Table 20.6. Fermentation Reactions of *Candida* Species.

Candida species	Glu	Mal	Suc	Lac
C.albicans	AG	AG	–	–
C.tropicalis	AG	AG	AG	–
C.kefyer	AG	AG	AG	–
C.guilliermondii	AG	–	AG	–
C.parapsilosis	AG	–	–	–
C.krusei	AG	–	–	–
C.glabrata	AG	–	–	–

Note: Glu = Glucose, Mal = Maltose, Suc = Sucrose, Lac = Lactose;+ = Positive Reaction, – = Negative Reaction, A = Acid Production G = Gas Production.

Table 20.7. Assimilation Reactions of Different *Candida* Species and their growth on TRM.

Candida Species	Glu	Mal	Suc	Lac	Cel	Gal	Tre	Raff	Mel	Xyl	Ino	Dul	TRM
Candida albicans	+	+	+	+	+	+	+	–	–	+	–	–	PP
C.tropicalis	+	+	+	–	+	+	+	–	–	+	–	–	M
C.kefyer	+	–	+	+	+	+	–	+	–	+	–	–	SP
C.parapsilosis	+	+	+	–	–	+	+	–	–	+	–	–	RP
C.guilliermondii	+	+	+	–	+	+	+	+	+	+	–	+	PsP
C.krusei	+	–	–	–	–	–	–	–	–	+	–	–	DP
C.glabrata	+	–	–	–	–	–	+	–	–	+	–	+	PP

Note: Glu = Glucose, Mal = Maltose, Suc = Sucrose, Lac = Lactose, Cel = Cellobiose, Gal = Galactose, Tre = Trehalose, Raf = Raffinose, Mel = Melibiose, Xyl = Xylose, Ino = Inositol, Dul = Dulcitol; + = Positive Reaction, – = Negative Reaction, V = Variation.
Tetrazolium Reduction Medium (TRM): PP = Pale Pink, OP = Orange Pink, M = Maroon, SP = Salmon Pink, RP = Rose Pink, PsP = Pink and Pasty, DP = Pink and Dry.

preliminary findings regarding chlamydosporulation by *C.albicans* and *C.dubliniensis* on new medium *i.e.* mustard agar, have also been reported.

(e) Biochemical Tests

As in bacteriology, the biochemical tests like sugar fermentation and assimilation are of immense importance for the identification of the yeast isolates. API 20C and Auxacolor testing for identification of the most of the clinically significant yeast species has also been evaluated. A set of these tests used in the routine mycology laboratory is given in **Figures 20.16 and 20.17** as well as in the **Tables 20.6 and 20.7**.

For the identification of unknown yeast, isolated in the diagnostic mycology laboratory, schematic approach is followed by doing series of tests, which is given in the **Flowchart 20.1**.

In Auxacolor, there are problems in the identification of *C.parapsilosis*.

(f) CHROMagar Candida

This is rapid, plate-based test for the simultaneous isolation and identification of various *Candida* species. This is relatively new medium that distinguishes different *Candida* species by color as a result of biochemical reactions. This can be used for simultaneous isolation and presumptive identification of various *Candida* species like *C.albicans, C.krusei, C.tropicalis, C.glabrata, C.parapsilosis* and *C.dubliniensis* and other yeast-like colonies **(Fig. 20.12)**. It is possible to identify the principal yeast species by their morphological characteristics within 24 hours and to isolate pathogenic fungi within 24-48 hours. This is based on the direct detection of specific enzymatic activities by adding certain substrates of fluorochromes to the media.

Flowchart 20.1. Identification scheme of commonly encountered yeast isolates.

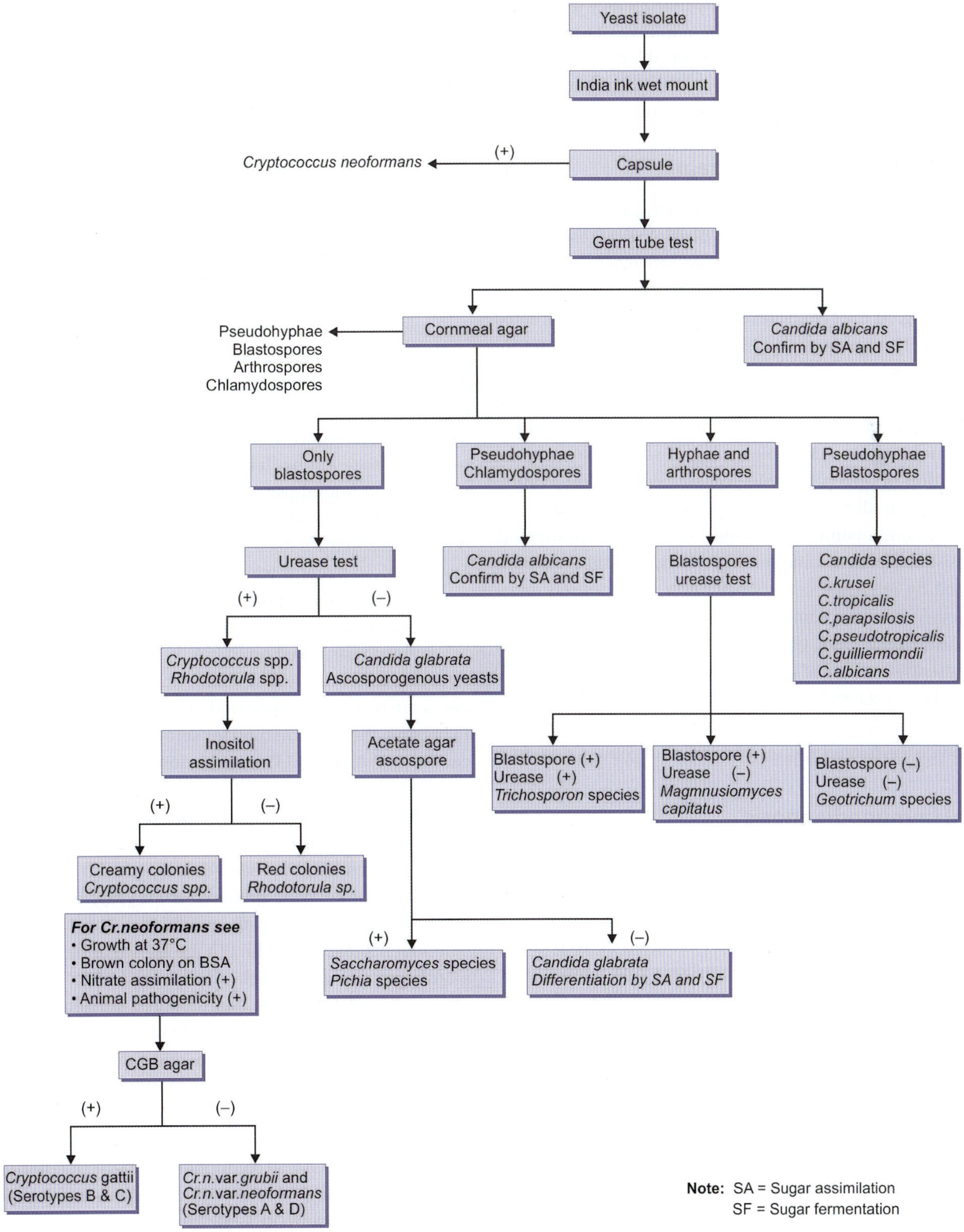

Note: SA = Sugar assimilation
SF = Sugar fermentation

It shortens the time for presumptive identification of the organisms on the basis of colony morphology and it also allows for easier detection of multiple yeasts present in clinical specimen **(*See* Appendix A)**. The medium should be procured from standard firm as the indigenous type of firm-made media do not yield good results.

(g) Typing of *Candida* Strains

The strains of various *Candida* species can be differentiated at the phenotypic level by physiological tests and by detection of phenotypically expressed macro-molecular structures with tests such as serotyping and isoenzyme profiling. At the genetic level, strains can be differentiated by the electrophoretic karyotype patterns of their chromosome by restriction fragment length polymorphism (RFLP) and Southern blot hybridization patterns generated with number of probes based on chitin synthase gene. Neither phenotypic nor genotypic typing of strains alone is the solution of the epidemiological studies therefore combination of both typing procedures is to be employed.

A combination of methods such as morphotyping, resistance patterns and serological agglutination reactions used by reference centers has never been entirely satisfactory. Of these combined methods, morphotyping plus resistotyping (M-R typing) is the most suitable which is presently available. Pyrolysis-mass spectrometry (PMS) has been successfully used in rapid inter-strain comparison for many bacterial species and has been used to differentiate the related species and strains of genus *Candida*.

(h) Immunodiagnosis

Numerous serological and molecular techniques have been developed for the diagnosis of *Candida* species. Efforts have been made to find either antibody against *Candida albicans* molecules or *Candida*-derived molecules whose presence in patients' sera could indicate tissue invasion. Tests have been developed to detect *Candida albicans* proteins, metabolites, DNA and polysaccharides. Various approaches for the serological diagnosis of candidiasis have been concentrated on the detection of *Candida albicans*-derived molecules. These molecules have been detected either on the basis of their antigenicity or through biochemical-enzymatic procedures. More recent progress has been made on the latter methods and kits are commercially available for the detection of arabinitol and glucans whereas PCR-based tests for *Candida*-DNA detection are routinely performed in some of the laboratories.

Mannan is major structural component of the cell wall of yeasts and the principal surface antigen that is available for immune interaction with colonized or infected hosts. It is large molecular weight protein-polysaccharide whose carbohydrate portion contains backbone chains of repeating mannose units in α (1,6) - linkage and numerous α (1,2) - and α (1,3) - oligomannoside side chains. The whole cell agglutination by specific antisera identifies two major serotypes of *C.albicans*, serotype A and B and mannan comprises the type specific antigen.

The serological tests available currently lack specificity and sensitivity. Therefore, the results must be interpreted with care. Various tests, like ID and CIE are used for the demonstration of precipitin to somatic extracts of *C.albicans* and *C.parapsilosis*. A positive test does not necessarily indicate infection since the antigens used are unable to differentiate antibodies formed during mucosal colonization and from those produced during deep infection. Similarly, negative antibody test does not necessarily rule out the possibility of deep-seated candidiasis in immunocompromised patients who are incapable of mounting an adequate antibody response. In such cases more sensitive assay system such as ELISA gives better results.

Antigen tests based on ELISA, RIA, CIE, PHA and LPA to detect either cell wall mannan or cytoplasmic components are being developed and antigen detection is likely to become the main method for serodiagnosis of systemic candidiasis. A rapid ELISA can be done to shorten the time to one-fourth as compared to CIE and standard ELISA procedure for the detection of *C.albicans* IgG antibodies. *C.albicans* cytoplasmic antigen enzyme immunoassay (CACP antigen EIA) has been developed with antibodies raised against antigens prepared from yeast cells grown under standardized growth conditions. The *C.albicans* components reactive in the EIA are shown to be predominantly protein with associated carbohydrates. The diagnostic value of anti-*Candida* antibodies, antigen cell wall mannan and characteristic agglutination antigens is clinically significant. Various cytoplasmic antigens may often be another marker for detection of invasive candidiasis. *Candida* enolase antigen, anti-*Candida* enolase antibody, HSP-90 related *Candida* antigen and different *Candida* metabolites and cell wall components have been reported. Some of the serological tests performed in *Candida* infections have been shown in **Table 20.8**.

Table 20.8. Serological Tests for Invasive Candidiasis.

A.	**Detection of Antibodies**
	Slide agglutination
	Immunodiffusion
	Phytohemagglutination
	Coelectosynersis
	Immunoprecipitation
	A and B immunofluorescence
B.	**Nonspecific Candida Antigens**
	Latex agglutination
	Immunoblotting
C.	**Cell Wall Components**
	Cell Wall Mannoprotein (CWMP)
	β-(1,3)-D-glucan
D.	**Candida Enolase Antigen Testing**

(i) Detection of Metabolites

There are certain species of *Candida* produce the metabolite D-arabinitol including *C.albicans, C.tropicalis, C.parapsilosis, C.guilliermondii* and *C.pseudotropicalis* but not *C.krusei* or *C.glabrata*. Serum arabinitol can be measured by gas liquid chromatography. Arabinitol is excreted by the kidneys and its clearance mirrors that of creatinine. So to adjust for its accumulation during renal insufficiency the arabinitol/creatinine ratio is calculated because this is independent of renal function. For many years the labor intensity of the GLC process required to measure arabinitol was major limitation to this approach. However, now an enzymatic fluorometric method for determining serum arabinitol provides an alternative to GLC.

There are certain species of *Candida* produce the metabolite D-arabinitol including *C.albicans, C.tropicalis, C.parapsilosis, C.guilliermondii* and *C.pseudotropicalis* but not *C.krusei* or *C.glabrata*. Serum arabinitol can be measured by gas liquid chromatography. Arabinitol is excreted by the kidneys and its clearance mirrors that of creatinine. So to adjust for its accumulation during renal insufficiency the arabinitol/creatinine ratio is calculated because this is independent of renal function. For many years the labor intensity of the GLC process required to measure arabinitol was major limitation to this approach. However, now an enzymatic fluorometric method for determining serum arabinitol provides an alternative to GLC.

D-mannose and D-arabinitol are metabolites of *Candida* species that can be detected in sera by gas liquid chromatography. By using suitable standard, serum concentrations of the metabolites can be found out. The detection of circulating candidal antigens as diagnostic marker of disseminated candidiasis has been the subject of intensive research for more than two decades. This reflects inherent problems with conventional diagnostic procedures, lack of specific clinical features and insensitivity of blood cultures and the continuing upward trend in incidence. Candidemia was the fourth commonest cause of nosocomial septicemia in recent national surveillance carried out in the USA and accounted for about 8% of hospital acquired blood stream infections. Mortality rates among patients with candidemia remain much higher than those of the commonly encountered bacterial causes of septicemia. Estimates vary from 38 to 80%. Early initiation of antifungal therapy is critical in reducing the mortality of disseminated candidiasis. Therefore, the challenge for diagnostic tests is to provide evidence of infection more quickly than occurs with blood cultures or to give positive result in patients with candidiasis in whom blood cultures remain consistently negative.

Moreover, G-test for detection of glucan may also be performed in diagnosing candidiasis.

(j) Enolase

Enolase is an abundant cytoplasmic enzyme typical of yeasts and involved in the glycolytic pathway. Detection of serum enolase as diagnostic marker was facilitated by the development of liposomal sandwich-type immunoassay. This was developed by Becton Dickinson (Philadelphia, PA) in collaboration with Walsh and colleagues.

(k) Proteinase

Extracellular proteinase activity has been demonstrated, by auxanographic methods, for most isolates of *C.albicans, C.tropicalis* and *C.parapsilosis* while various isolates of *C.glabrata, C.guilliermondii, C.pseudotropicalis* and *C.krusei* were negative. A secretory acid proteinase antigen and specific antibodies have been demonstrated in patients infected with *C.albicans*

(l) Cell Wall Mannoprotein

Of the antigens that circulate during systemic infection *C.albicans* cell wall mannoprotein was the first to be characterized. It exists as immune complexes which have to be dissociated before the antigen can be detected. It is stable antigen resistant to boiling, proteinase treatment and acidic pH. It is labile to periodate oxidation and binds to concanavalin A. Galactomannan is an analogous surface carbohydrate of the fungal cell wall of *Aspergillus fumigatus.* Immunoassay for galactomannan and mannoprotein are specific for invasive aspergillosis and disseminated candidiasis, respectively.

(m) Skin Tests

The skin tests may also be performed to evaluate cell-mediated immunity *in vivo* by testing delayed-type hypersensitivity to candidal antigens. The *in vitro* assays may also be done such as lymphocyte transformation to test cell-mediated immunity of the patients. These tests do not have significant role in the diagnosis of candidiasis.

(n) PNA FISH

The peptide nucleic acid fluorescent *in situ* hybridization (PNA FISH) method is useful for the identification of *Candida* species and is comparable with the conventional methods. This testing is a practical, reliable and rapid, results in 90 minutes, method for the identification of *Candida* strains at species level isolated from blood cultures.

(o) T2Candida Panel

T2Candida panel is a new diagnostic modality which runs on a fully automated instrument. The panel detects five clinically significant species of *Candida i.e. C.albicans, C.tropicalis, C.parapsilosis, C.krusei and C.glabrata* directly from the blood. This decreases time for identification, culture is not needed, thus, providing a faster result taking an average time as 4.4 hrs. It is based on T2 magnetic resonance technology which introduces super paramagnetic particles coated with Candida specific binding particles in blood. The binding results in disruption of magnetic fields in surrounding water molecules which is then detected. This system requires minimal skill and can detect as low as 1CFU/ml.

(p) Animal Pathogenicity

The understanding of both the host response as well as the mechanisms of fungal virulence is helped by the development of various animal models that recapitulate human disease with remarkable results. The mice and rabbits are most commonly employed experimental animals in the mycology laboratory. These are susceptible to various *Candida* species. These models allow researchers to probe the main host immune components by use of genetically modified mice *i.e.* knockouts and transgenics or mice treated with depleting antibodies. Moreover, requirements for microbial virulence can be probed using different strains of *Candida* either from clinical isolates or genetically modified laboratory strains. These provide a powerful tool to dissect the interplay between pathogen and host thereby increasing the understanding of disease process caused by various species of *Candida* genus.

Treatment and Prophylaxis

The recent clinical practice guidelines for the management of candidiasis were released by IDSA in 2016. As far as the treatment of oral and mucocutaneous lesions of candidiasis is concerned, 1% gentian violet is locally applied. Nystatin can be used to the resistant mucosal lesions as suspension containing 2,00,000 units/ml. The azole creams like clotrimazole, miconazole, ketoconazole and econazole are also topically applied. For the systemic lesions intravenous infusion of amphotericin B is indicated. The azoles like ketoconazole, fluconazole and itraconazole are given orally. Itraconazole solution is also available for topical application. Amphotericin B can be combined with flucytosine in systemic candidiasis. Flucytosine should not be given alone because of apprehension of emergence of drug resistance in the yeasts of this genus.

Drug resistance has also been reported in miconazole, particularly in *C.albicans* serotype B. In case of drug resistance, antifungal susceptibility testing can be done by disk diffusion and broth dilution and accordingly, specific antifungal regimen is instituted. However, delay in establishing proper diagnosis of candidiasis results delay in treatment eventually leading to poor prognosis among these patients.

As drug resistance in *Candida* species is now very common, therefore, antifungal susceptibility testing should be performed by determining MICs as per the CLSI guidelines of approved document M27-A or EUCAST guidelines (**See Appendix D**). This can play major role in the management of resistant candidal infections, particularly, caused by non-*albicans Candida* species. In addition, Etest may also provide the simplest, most reproducible method of antifungal susceptibility testing and allow hospitals to develop suitable antifungal biograms.

Most of the fluconazole resistant strains recovered from clinical specimens, particularly blood, are non-*albicans Candida* species, especially *C.tropicalis, C.krusei* and *C.glabrata*. *C.krusei* and *C.glabrata* are inherently resistant to azole and is an emerging frequent cause of fungemia in patients with hematologic malignancies. The treatment guidelines by IDSA recommend testing for azole sensitivity in all cases of blood stream infections. The patients should be treated for a period of two weeks after being declared as culture negative of *Candida* bloodstream infection.

The echinocandin group of drugs are indicated in case of drug resistance in *Candida* although it is emerging in this group also. However, there is no other option except these ones, particularly when the infection is due to *C.auris* or *C.haemulonii*. The details of antifungals and AFST used for managing candidiasis are given in **Chapter 6** and **Appendix D**, respectively. The other modalities are also in use like chemoprophylaxis and immunotherapy. In addition *Candida* Score is also helpful guide in deciding initiation of treatment in this disease.

(a) Chemoprophylaxis: There is no effective active or passive agent for immunization against deep-seated candidal infections, therefore, the mainstay of prophylaxis lies on appropriate antifungal chemotherapy. Newer antifungals like azole group of drugs are commonly given for prophylaxis. The basic idea is to decrease fungal load in the gastrointestinal tract so that risk of infection by *Candida* species is reduced. It is also observed that due to frequent prophylactic use of various antifungals, particularly azole derivatives, there is an increase in non-*albicans Candida* species, particularly, *C.krusei* and *C.glabrata,* which are invariably found resistant to the commonly used antifungals.

(b) Immunotherapy: As prolonged neutropenia is one of the most important risk factors for the development of serious candidial bloodstream infections hence **cytokines** may enhance the activity of neutrophils against *Candida* species. Immunotherapy has also been undertaken as an **adjunct** to antimicrobial treatment. It is given by the use of cytokines such as granulocyte colony stimulating factor (G-CSF) or granulocyte macrophage colony stimulating factor (GM-CSF). It is difficult to draw some conclusion on the clinical validity on this issue at this stage as these findings are based on experimental data.

(c) Candida Score: In a large, observational, multicentric study in Spain, a new *Candida* score was developed to help in deciding as to which patients need prophylactic therapy. A total of four independent risk factors were taken into consideration. These are multisite colonization (1 point), surgery upon ICU admission (1 point), total parenteral nutrition (1 point) and severe sepsis (2 points). A score of more than 2.5 is found to be associated with increased likelihood of invasive candidiasis.

Further Reading

1. Abe M, Kimura M, Araoka H, *et al.* Serum (1,3)-beta-D-glucan is an inefficient marker of breakthrough candidemia. *Med Mycol.* 2014; **52:** 835-40.
2. Achkar JM, Fries BC. *Candida* infections of the genitourinary tract. *Clin Microbiol Rev.* 2010; **23:** 253-73.
3. Adam H, Groenewald M, Mohan S, *et al.* Identification of a new species, *Candida subhashii,* as a cause of peritonitis. *Med Mycol.* 2009; **47:** 305-11.
4. Adhikary R, Joshi S. Species distribution and anti-fungal susceptibility of candidaemia at a multi-super-specialty center in Southern India. *Indian J Med Microbiol.* 2011; **29:** 309-11.
5. Afzal Azim, Armin Ahmed, Rungmei Marak. Invasive candidiasis. *N Eng J Med.* 2016. **374:** 793-5.
6. Ahmad A, Khan AU. Prevalence of *Candida* species and potential risk factors for vulvovaginal candidiasis in Aligarh, India. *Eur J Obstet Gynecol Reprod Biol.* 2009; **144:** 68-71.
7. Ahmad S, Khan Z, Asadzadeh M, *et al.* Performance comparison of phenotypic and molecular methods for detection and differentiation of *Candida albicans* and *Candida dubliniensis. BMC Infect Dis.* 2012; **12:** 230.
8. Ahmed A, Azim A, Baronia AK, *et al.* Invasive candidiasis in non-neutropenic critically ill - Need for region-specific management guidelines. *Indian J Crit Care Med.* 2015; **19:** 333-9.
9. Ahmed A, Azim A, Baronia AK, *et al.* Risk prediction for invasive candidiasis. *Indian J Crit Care Med.* 2014; **18:** 682-8.
10. Al-Sweih N, Khan ZU, Ahmad S, *et al. Kodamaea ohmeri* as an emerging pathogen: A case report and review of the literature. *Med Mycol.* 2011; **49:** 766-70.
11. Alam MZ, Alam Q, Jiman-Fatani A, *et al. Candida* identification: A journey from conventional to molecular methods in medical mycology. *World J Microbiol Biotechnol.* 2014; **30:** 1437-51.
12. Alastruey-Izquierdo A, Mandelblat M, Ben Ami R, *et al.* Multilocus sequence typing of *Candida albicans* isolates from candidemia and superficial candidiasis in Israel. *Med Mycol.* 2013; **51:** 755-8.
13. Alcoba-Florez J, Laich F, Perez-Roth E, *et al.* First reported case of catheter-related fungemia due to *Candida mengyuniae. J Clin Microbiol.* 2011; **49:** 3429-31.
14. Alothman AF, Al-Musawi T, Al-Abdely HM, *et al.* Clinical practice guidelines for the management of invasive *Candida* infections in adults in the Middle East region: Expert panel recommendations. *J Infect Public Health.* 2014; **7:** 6-19.
15. Altinmakas E, Guo M, Kundu UR, *et al.* Computed tomography and 18 F-fluorodeoxyglucose positron emission tomography/computed tomography findings in adrenal candidiasis and histoplasmosis: Two cases. *Clin Imaging.* 2015; **39:** 1115-8.
16. Andrew M, John S, Daniel K. Iatrogenic *Candida glabrata* fungal adrenalitis after endoscopic ultrasound cyst aspiration. *Surgical Infections Case Reports.* 2016; **1:** 132-4.
17. Andrew NH, Ruberu RP, Gabb G. The first documented case of *Candida dubliniensis* leptomeningeal disease in an immunocompetent host. *BMJ Case Rep.* 2011; pii: bcr0620114384. PMID: 22687683.

18. Anil Kumar V, Francis E, Sreehari S, *et al.* Tricuspid valve mycetoma in an infant successfully treated by excision and complex tricuspid valve repair followed by fluconazole therapy. *Surg Infect (Larchmt).* 2014; **15:** 134-40.
19. Anuradha S, Agarwal SK, Prakash A, Singh NP, Kaur R. *Candida* sake - A rare cause of fungal endocarditis. *Med J Malaysia.* 2008; **63:** 75-6.
20. Arendrup MC. *Candida* and candidaemia. Susceptibility and epidemiology. *Dan Med J.* 2013; **60:** B4698.
21. Arendrup MC. Epidemiology of invasive candidiasis. *Curr Opin Crit Care.* 2010; **16:** 445-52.
22. Ariff S, Saleem AF, Soofi SB, *et al.* Clinical spectrum and outcomes of neonatal candidiasis in a tertiary care hospital in Karachi, Pakistan. *J Infect Dev Ctries.* 2011; **5:** 216-23.
23. Arora BB, Maheshwari M, Devgan N, Arora DR. Prevalence of trichomoniasis, vaginal candidiasis, genital herpes, chlamydiasis and actinomycosis among urban and rural women of Haryana, India. *J Sex Transm Dis.* 2014; 963812.
24. Arora N, Parihar M, Mishra D, *et al. Candida parapsilosis* in peripheral blood. *Indian J Pathol Microbiol.* 2014; **57:** 645.
25. Arora U, Jagdev M, Jindal N. Immunosuppression level in HIV-positive patients with oropharyngeal candidiasis. *Indian J Med Microbiol.* 2009; **27:** 174-5.
26. Arsenault AB, Bliss JM. Neonatal candidiasis: New insights into an old problem at a unique host-pathogen interface. *Curr Fungal Infect Rep.* 2015; **9:** 246-52.
27. Aruna C, Seetharam K. Congenital candidiasis. *Indian Dermatol Online J.* 2014; **5:** S44-7.
28. Asayama N, Nagata N, Shimbo T, *et al.* Relationship between clinical factors and severity of esophageal candidiasis according to Kodsi's classification. *Dis Esophagus.* 2014; **27:** 214-9.
29. Asmundsdottir LR, Erlendsdottir H, Gottfredsson M. Nationwide study of candidemia, antifungal use and antifungal drug resistance in Iceland, 2000 to 2011. *J Clin Microbiol.* 2013; **51:** 841-8.
30. Avni T, Leibovici L, Paul M. PCR diagnosis of invasive candidiasis: Systematic review and meta-analysis. *J Clin Microbiol.* 2011; **49:** 665-70.
31. Awasthi AK, Jain A, Awasthi S, *et al.* Epidemiology and microbiology of nosocomial pediatric candidemia at a northern Indian tertiary care hospital. *Mycopathologia.* 2011; **172:** 269-77.
32. Aznar-Marin P, Galan-Sanchez F, Marin-Casanova P, *et al. Candida nivariensis* as a new emergent agent of vulvovaginal candidiasis: Description of cases and review of published studies. *Mycopathologia.* 2016; **181:** 445-9.
33. Bacconi A, Richmond GS, Baroldi MA, *et al.* Improved sensitivity for molecular detection of bacterial and *Candida* infections in blood. *J Clin Microbiol.* 2014; **52:** 3164-74.
34. Badiee P, Alborzi A, Farhoudi F. A case of *Candida* mediastinitis after dental extraction. *J Infect Dev Ctries.* 2011; **5:** 75-8.
35. Badiee P, Kordbacheh P, Alborzi A, *et al.* Early detection of systemic candidiasis in the whole blood of patients with hematologic malignancies. *Jpn J Infect Dis.* 2009; **62:** 1-5.
36. Bajwa SK, Bajwa SJ, Jindal R, *et al*, Jindal R. Candidiasis: An unusual cause of persistent high-grade fever in mid-pregnancy. *Int J Crit Illn Inj Sci.* 2013; **3:** 217-9.
37. Ballal M, Chakraborty R, Bhandary S, *et al. Candida tropicalis* in a case of cholangiocarcinoma with cholangitis at a tertiary care hospital in Manipal. *Med Mycol Case Rep.* 2013; **2:** 95-7.
38. Banerjee P, Kaur R, Uppal B. Study of fungal isolates in patients with chronic diarrhea at a tertiary care hospital in north India. *J Mycol Med.* 2013; **23:** 21-6.
39. Baradkar VP, Kumar S. Species identification of *Candida* isolates obtained from oral lesions of HIV infected patients. *Indian J Dermatol.* 2009; **54:** 385-6.
40. Baradkar VP, Mathur M, Kumar S. Cerebral *Candida* abscess in an infant. *Indian J Med Microbiol.* 2009; **27:** 70-2.
41. Baradkar VP, Mathur M, Kumar S. HiChrome Candida Agar for identification of *Candida* species. *Indian J Pathol Microbiol.* 2010; **53:** 93-5.
42. Baradkar VP, Mathur M, Kumar S. Neonatal septicaemia in a premature infant due to *Candida dubliniensis. Indian J Med Microbiol.* 2008; **26:** 382-5.
43. Baradkar VP, Mathur M, Sonavane A, *et al.* Candidal infections of ventriculoperitoneal shunts. *J Pediatr Neurosci.* 2009; **4:** 73-5.
44. Bartoletti M, Cervera C, Hoyo I, *et al.* Incidence and outcome of early *Candida* peritonitis after liver and pancreas transplantation. *Mycoses.* 2013; **56:** 162-7.
45. Bassetti M, Marchetti M, Chakrabarti A, *et al.* A research agenda on the management of intra-abdominal candidiasis: Results from a consensus of multinational experts. *Intensive Care Med.* 2013; **39:** 2092-106.
46. Bassetti M, Merelli M, Righi E, *et al.* Epidemiology, species distribution, antifungal susceptibility and outcome of candidemia across five sites in Italy and Spain. *J Clin Microbiol.* 2013; **51:** 4167-72.
47. Bassetti M, Mikulska M, Viscoli C. Bench-to-bedside review: Therapeutic management of invasive candidiasis in the intensive care unit. *Crit Care.* 2010; **14:** 244.
48. Bassetti M, Righi E, Ansaldi F, *et al.* A multicenter multinational study of abdominal candidiasis: Epidemiology, outcomes and predictors of mortality. *Intensive Care Med.* 2015; **41:** 1601-10.
49. Basu S, Suri S, Kumar A. Scrotal abscess owing to *Candida albicans* in a newborn. *Paediatr Int Child Health.* 2013; **33:** 53-5.
50. Behera B, Singh RI, Xess I, *et al. Candida rugosa:* A possible emerging cause of candidaemia in trauma patients. *Infection.* 2010; **38:** 387-93.
51. Ben-Ami R, Berman J, Novikov A, *et al.* Multidrug-resistant *Candida haemulonii* and *C. auris,* Tel Aviv, Israel. *Emerg Infect Dis.* 2017; **23:** 195-203 & 375.
52. Benjamin DK Jr, Stoll BJ, Gantz MG, *et al.* Neonatal candidiasis: epidemiology, risk factors and clinical judgment. *Pediatrics.* 2010; **126:** e865-73.
53. Beyda ND, Chuang SH, Alam MJ, *et al.* Treatment of *Candida famata* bloodstream infections: Case series and review of the literature. *J Antimicrob Chemother.* 2013; **68:** 438-43.
54. Bhai N, Tendolkar U, Baradkar V, *et al.* Paediatric oropharyngeal and cutaneous candidiasis with special reference to *Candida dubliniensis. J Med Microbiol.* 2014; **63:** 518-21.

55. Bhatt M, Sarangi G, Paty BP, *et al.* Biofilm as a virulence marker in *Candida* species in nosocomial blood stream infection and its correlation with antifungal resistance. *Indian J Med Microbiol.* 2015; **33 (Suppl):** 112-4.
56. Bhavasar RS, Goje SK, Takalkar AA, *et al.* Detection of *Candida* by calcofluor white. *Acta Cytol.* 2010; **54:** 679-84.
57. Bisht V, Voort JV. Obstructive renal candidiasis in infancy. *Eur J Pediatr.* 2011; **170:** 1227-35.
58. Bismay K, Mathew A, Rajesh R, *et al.* Disseminated candidiasis 18 years after renal transplantation. *Indian J Nephrol.* 2012; **22:** 462-5.
59. Bliss JM. *Candida parapsilosis:* An emerging pathogen developing its own identity. *Virulence.* 2015; **6:** 109-11.
60. Bolotin-Fukuhara M, Fairhead C. *Candida glabrata:* A deadly companion? *Yeast.* 2014; **31:** 279-88.
61. Bonfietti LX, Szeszs MW, Chang MR, *et al.* Ten-year study of species distribution and antifungal susceptibilities of *Candida* bloodstream isolates at a Brazilian tertiary hospital. *Mycopathologia.* 2012; **174:** 389-96.
62. Borman AM, Szekely A, Johnson EM. Comparative pathogenicity of United Kingdom isolates of the emerging pathogen *Candida auris* and other key pathogenic *Candida* species. *mSphere.* 2016; **1.** pii: e00189-16.
63. Borman AM, Szekely A, Linton CJ, *et al.* Epidemiology, antifungal susceptibility, and pathogenicity of *Candida na* isolates from the United Kingdom. *J Clin Microbiol.* 2013; **51:** 967-72.
64. Botero-Calderon L, Benjamin DK Jr, Cohen-Wolkowiez M. Advances in the treatment of invasive neonatal candidiasis. *Expert Opin Pharmacother.* 2015; **16:** 1035-48.
65. Bow EJ, Evans G, Fuller J, *et al.* Canadian clinical practice guidelines for invasive candidiasis in adults. *Can J Infect Dis Med Microbiol.* 2010; **21:** e122-50.
66. Brunke S, Hube B. Two unlike cousins: *Candida albicans* and *C.glabrata* infection strategies. *Cell Microbiol.* 2013; **15:** 701-8.
67. Buchan BW, Ledeboer NA. Advances in identification of clinical yeast isolates by use of matrix-assisted laser desorption ionization-time of flight mass spectrometry. *J Clin Microbiol.* 2013; **51:** 1359-66.
68. Buchanan PJ, Mast BA, Lottenberg L, *et al.* *Candida albicans* necrotizing soft tissue infection: A case report and literature review of fungal necrotizing soft tissue infections. *Ann Plast Surg.* 2013; **70:** 739-41.
69. Calvo B, Melo AS, Perozo-Mena A, *et al.* First report of *Candida auris* in America: Clinical and microbiological aspects of 18 episodes of candidemia. *J Infect.* 2016; **73:** 369-74.
70. Carneiro HA, Mavrakis A, Mylonakis E. *Candida* peritonitis: An update on the latest research and treatments. *World J Surg.* 2011; **35:** 2650-9.
71. Cascio A, Bartolotta M, Venneri A, *et al.* A case of *Candida krusei* peritonitis secondary to duodenal perforation due to *Candida duodenitis. Mycopathologia.* 2011; **171:** 51-5.
72. Cendejas-Bueno E, Kolecka A, Alastruey-Izquierdo A, *et al.* Reclassification of the *Candida haemulonii* complex as *Candida haemulonii* (*C.haemulonii* Group I), *C.duobushaemulonii* sp. nov. (*C.haemulonii* Group II) and *C.haemulonii* var. *vulnera* var. nov.: Three multi-resistant human pathogenic yeasts. *J Clin Microbiol.* 2012; **50:** 3641-51.
73. Cevik R, Tekin R, Gem M. *Candida* arthritis in a patient diagnosed with spondyloarthritis. *Rev Soc Bras Med Trop.* 2016; **49:** 793-5.
74. Chai LY, Denning DW, Warn P. *Candida tropicalis* in human disease. *Crit Rev Microbiol.* 2010; **36:** 282-98.
75. Chakrabarti A, Chander J, Kasturi P, *et al.* Candidaemia: A 10 years study in an Indian teaching hospital. *Mycoses.* 1992; **35:** 47-51.
76. Chakrabarti A, Chatterjee SS, Rao KL, *et al.* Recent experience with fungaemia: Change in species distribution and azole resistance. *Scand J Infect Dis.* 2009; **41:** 275-84.
77. Chakrabarti A, Rudramurthy SM, Kale P, *et al.* Epidemiological study of a large cluster of fungaemia cases due to *Kodamaea ohmeri* in an Indian tertiary care centre. *Clin Microbiol Infect.* 2014; **20:** O83-9.
78. Chakrabarti A, Sood P, Rudramurthy SM, *et al.* Incidence, characteristics and outcome of ICU-acquired candidemia in India. *Intensive Care Med.* 2015; **41:** 285-95.
79. Chakrabarti A. *Candida glabrata* candidemia. *Indian J Crit Care Med.* 2015; **19:** 138-9.
80. Chalmers C, Gaur S, Chew J, *et al.* Epidemiology and management of candidaemia - A retrospective, multicentre study in five hospitals in the UK. *Mycoses.* 2011; **54:** e795-800.
81. Chander J, Singla N, Sidhu SK, Gombar S. Epidemiology of *Candida* blood stream infections: Experience of a tertiary care centre in North India. *J Infect Dev Ctries.* 2013; **7:** 670-5.
82. Chapman B, Slavin M, Marriott D, *et al.* Changing epidemiology of candidaemia in Australia. *J Antimicrob Chemother.* 2017; **72:** 1103-8.
83. Chaves GM, Tercarioli GR, Padovan AC, *et al.* *Candida mesorugosa* sp. nov., a novel yeast species similar to *Candida rugosa*, isolated from a tertiary hospital in Brazil. *Med Mycol.* 2013; **51:** 231-42.
84. Chen CH, Chen WL, Yen HC. *Candida albicans* lumbar spondylodiscitis in an intravenous drug user: A case report. *BMC Res Notes.* 2013; **6:** 529.
85. Chopra T, Bhargava A, Kumar S, *et al.* *Candida kefyr* endocarditis in a patient with hypertrophic obstructive cardiomyopathy. *Am J Med Sci.* 2010; **339:** 188-9.
86. Chow BD, Linden JR, Bliss JM. *Candida parapsilosis* and the neonate: Epidemiology, virulence and host defense in a unique patient setting. *Expert Rev Anti Infect Ther.* 2012; **10:** 935-46.
87. Chow H, Hurley JC. (1→3) β-D-glucan as a prognostic marker. *Clin Infect Dis.* 2013; **56:** 466.
88. Chowdhary A, Anil Kumar V, Sharma C, *et al.* Multidrug-resistant endemic clonal strain of *Candida auris* in India. *Eur J Clin Microbiol Infect Dis.* 2014; **33:** 919-26.
89. Chowdhary A, Randhawa HS, Kowshik T, *et al.* Application of hypertonic Sabouraud glucose agar for differentiation of *Candida dubliniensis* from *Candida albicans. Diagn Microbiol Infect Dis.* 2011; **69:** 440-2.
90. Chowdhary A, Sharma C, Duggal S, *et al.* New clonal strain of *Candida auris*, Delhi, India. *Emerg Infect Dis.* 2013; **19:** 1670-3.

91. Chowdhary A, Voss A, Meis JF. Multidrug-resistant *Candida auris:* 'new kid on the block' in hospital-associated infections? *J Hosp Infect.* 2016; **94:** 209-212.
92. Chowdhary A, Sharma C, Meis JF. *Candida auris:* A rapidly emerging cause of hospital-acquired multidrug-resistant fungal infections globally. *PLoS Pathog.* 2017; **13:** e1006290.
93. Chumpitazi BF, Lebeau B, Faure-Cognet O, *et al.* Characteristic and clinical relevance of *Candida* mannan test in the diagnosis of probable invasive candidiasis. *Med Mycol.* 2014; **52:** 462-71.
94. Clancy C, Shields R, Nguyen, M. Invasive candidiasis in various patient populations: Incorporating non-culture diagnostic tests into rational management strategies. *J Fungi.* 2016; **2:** 10. doi:10.3390/jof2010010.
95. Clancy CJ, Nguyen MH. Diagnostic methods for detection of blood-borne candidiasis. *Methods Mol Biol.* 2016; **1356:** 215-38.
96. Clancy CJ, Nguyen MH. Emergence of *Candida auris:* An international call to arms. *Clin Infect Dis.* 2017; **64:** 141-3.
97. Clancy CJ, Nguyen MH. Finding the "missing 50%" of invasive candidiasis: How nonculture diagnostics will improve understanding of disease spectrum and transform patient care. *Clin Infect Dis.* 2013; **56:** 1284-92.
98. Cobo F, Rodriguez-Granger J, Sampedro A, *et al. Candida* prosthetic joint infection. A review of treatment methods. *J Bone Jt Infect.* 2017; **2:** 114-21.
99. Cohen-Wolkowiez M, Benjamin DK Jr. Fluconazole therapeutic drug monitoring in children with cancer: Not today. *Clin Infect Dis.* 2014; **59:** 1534-6.
100. Colomba C, Trizzino M, Imburgia C, *et al. Candida glabrata* meningitis and endocarditis: A late severe complication of candidemia. *Int J Infect Dis.* 2014; **29:** 174-5.
101. Colombo AL, Garnica M, Aranha Camargo LF, *et al. Candida glabrata:* An emerging pathogen in Brazilian tertiary care hospitals. *Med Mycol.* 2013; **51:** 38-44.
102. Concia E, Azzini AM, Conti M. Epidemiology, incidence and risk factors for invasive candidiasis in high-risk patients. *Drugs.* 2009; **69 (Suppl. 1):** 5-14.
103. Conti HR, Huppler AR, Whibley N, *et al.* Animal models for candidiasis. *Curr Protoc Immunol.* 2014; **105:** 19.6.1-17.
104. Corona A, Singer M. Early-onset candidemia: an increasing problem? *Crit Care Med.* 2009; **37:** 2652-3.
105. Coronado-Castellote L, Jimenez-Soriano Y. Clinical and microbiological diagnosis of oral candidiasis. *J Clin Exp Dent.* 2013; **5:** e279-86.
106. da Costa VG, Quesada RM, Abe AT, *et al.* Nosocomial bloodstream *Candida* infections in a tertiary-care hospital in South Brazil: A 4-year survey. *Mycopathologia.* 2014; **178:** 243-50.
107. da Silva CM, de Carvalho Parahym AM, *et al.* Fungemia by *Candida pelliculosa (Pichia anomala)* in a neonatal intensive care unit: A possible clonal origin. *Mycopathologia.* 2013; **175:** 175-9.
108. Dan M, Leshem Y, Yeshaya A. Performance of a rapid yeast test in detecting *Candida* spp. in the vagina. *Diagn Microbiol Infect Dis.* 2010; **67:** 52-5.
109. Dar MS, Sreedar G, Shukla A, *et al.* An *in vitro* study of antifungal drug susceptibility of *Candida* species isolated from human immunodeficiency virus seropositive and human immunodeficiency virus seronegative individuals in Lucknow population Uttar Pradesh. *J Oral Maxillofac Pathol.* 2015; **19:** 205-11.
110. Das I, Nightingale P, Patel M, *et al.* Epidemiology, clinical characteristics and outcome of candidemia: Experience in a tertiary referral center in the UK. *Int J Infect Dis.* 2011; **15:** e759-63.
111. De Luca C, Guglielminetti M, Ferrario A, *et al.* Candidemia: Species involved, virulence factors and antimycotic susceptibility. *New Microbiol.* 2012; **35:** 459-68.
112. De Pascale G, Antonelli M. *Candida* colonization of respiratory tract: To treat or not to treat, will we ever get an answer? *Intensive Care Med.* 2014; **40:** 1381-4.
113. De Rosa FG, Trecarichi EM, Montrucchio C, *et al.* Mortality in patients with early- or late-onset candidaemia. *J Antimicrob Chemother.* 2013; **68:** 927-35.
114. de Toro M, Torres MJ, Maite R, *et al.* Characterization of *Candida parapsilosis* complex isolates. *Clin Microbiol Infect.* 2011; **17:** 418-24.
115. Debska-Slizien A, Chrobak L, Bzoma B, *et al. Candida* arteritis in kidney transplant recipients: Case report and review of the literature. *Transpl Infect Dis.* 2015; **17:** 449-55.
116. Delaloye J, Calandra T. Invasive candidiasis as a cause of sepsis in the critically ill patient. *Virulence.* 2014; **5:** 161-9.
117. Deorukhkar SC, Saini S, Mathew S. Non-*albicans Candida* infection: An emerging threat. *Interdiscip Perspect Infect Dis.* 2014; 615958. PMID: 25404942.
118. Deshpande A, Gaur S, Bal AM. Candidaemia in the non-neutropenic patient: A critique of the guidelines. *Int J Antimicrob Agents.* 2013; **42:** 294-300.
119. Dhakal BP, Tribble CG, Bergin JD, *et al.* Recurrent *Candida* prosthetic endocarditis over fifteen years managed with medical therapy and four valvular surgeries: A case report and review of literature. *J Cardiothorac Surg.* 2015; **10:** 105.
120. Dineshshankar J, Sivakumar M, Karthikeyan M, *et al.* Immunology of oral candidiasis. *J Pharm Bioallied Sci.* 2014; **6:** S9-12.
121. Divi V, Gupta R, Sataloff RT, *et al.* Laryngeal candidiasis. *Ear Nose Throat J.* 2010; **89:** 526.
122. Donders GG, Sobel JD. *Candida* vulvovaginitis: A store with a buttery and a show window. *Mycoses.* 2017; **60:** 70-2.
123. Downey LC, Smith PB, Benjamin DK Jr, *et al.* Recent advances in the detection of neonatal candidiasis. *Curr Fungal Infect Rep.* 2010; **4:** 17-22.
124. Doymaz M, Yampierre C, Tarchini G, *et al.* Demonstration and utility of clustered pseudohyphae on Gram-stained smears from *Candida albicans*-positive blood cultures. *Diagn Microbiol Infect Dis.* 2010; **66:** 449-51.
125. Duchaussoy AC, Rose A, Talbot JJ, *et al.* Gastrointestinal granuloma due to *Candida albicans* in an immunocompetent cat. *Med Mycol Case Rep.* 2015; **10:** 14-7.
126. Dufresne SF, Marr KA, Sydnor E, *et al.* Epidemiology of *Candida kefyr* in patients with hematologic malignancies. *J Clin Microbiol.* 2014; **52:** 1830-7.

127. Duggan S, Leonhardt I, Hunniger K, *et al.* Host response to *Candida albicans* bloodstream infection and sepsis. *Virulence.* 2015; **6:** 316-26.
128. Dutta A, Palazzi DL. *Candida* non-*albicans* versus *Candida albicans* fungemia in the non-neonatal pediatric population. *Pediatr Infect Dis J.* 2011; **30:** 664-8.
129. Echeverria PM, Kett DH, Azoulay E. *Candida* prophylaxis and therapy in the ICU. *Semin Respir Crit Care Med.* 2011; **32:** 159-73.
130. Eddouzi J, Hofstetter V, Groenewald M, *et al.* Characterization of a new clinical yeast species, *Candida tunisiensis* sp. nov., isolated from a strain collection from Tunisian hospitals. *J Clin Microbiol.* 2013; **51:** 31-9.
131. Elhoufi A, Ahmadi A, Asnaashari AM, *et al.* Invasive candidiasis in critical care setting, updated recommendations from "Invasive Fungal Infections - Clinical Forum", Iran. *World J Crit Care Med.* 2014; **3:** 102-12.
132. Ells R, Kock JL, Pohl CH. *Candida albicans* or *Candida dubliniensis? Mycoses.* 2011; **54:** 1-16.
133. Emara M, Ahmad S, Khan Z, *et al. Candida auris* candidemia in Kuwait, 2014. *Emerg Infect Dis.* 2015; **21:** 1091-2.
134. Erami M, Afzali H, Heravi MM, *et al.* Recurrent arthritis by *Candida glabrata,* a diagnostic and therapeutic challenge. *Mycopathologia.* 2014; **177:** 291-8.
135. Escribano P, Guinea J, Marcos-Zambrano L, *et al.* Does identification to species level provide sufficient evidence to confirm catheter-related fungemia caused by *Candida albicans*? *Med Mycol.* 2013; **51:** 769-73.
136. Farmakiotis D, Kyvernitakis A, Tarrand JJ, *et al.* Early initiation of appropriate treatment is associated with increased survival in cancer patients with *Candida glabrata* fungaemia: A potential benefit from infectious disease consultation. *Clin Microbiol Infect.* 2015; **21:** 79-86.
137. Farmakiotis D, Tarrand JJ, Kontoyiannis DP. Drug-resistant *Candida glabrata* infection in cancer patients. *Emerg Infect Dis.* 2014; **20:** 1833-40.
138. Farooqi JQ, Jabeen K, Saeed N, *et al.* Invasive candidiasis in Pakistan: Clinical characteristics, species distribution and antifungal susceptibility. *J Med Microbiol.* 2013; **62(Pt 2):** 259-68.
139. Fichadia H, Layman C. *Candida parapsilosis* tenosynovitis in an immunocompetent patient: Case report and review of literature. *J Hand Surg Am.* 2015; **40:** 993-6.
140. Fisher BT, Ross RK, Localio AR, *et al.* Decreasing rates of invasive candidiasis in pediatric hospitals across the United States. *Clin Infect Dis.* 2014; **58:** 74-7.
141. Fleming L, Ng A, Paden M, *et al.* Fungal osteomyelitis of calcaneus due to *Candida albicans:* A case report. *J Foot Ankle Surg.* 2012; **51:** 212-4.
142. Gamaletsou MN, Kontoyiannis DP, Sipsas NV, *et al. Candida* osteomyelitis: Analysis of 207 pediatric and adult cases (1970-2011). *Clin Infect Dis.* 2012; **55:** 1338-51.
143. Gamaletsou MN, Rammaert B, Bueno MA, *et al. Candida* arthritis: Analysis of 112 pediatric and adult cases. *Open Forum Infect Dis.* 2015; **3:** ofv207. PMID: 26858961.
144. Ganesan V, Kumar GS, Ponnusamy SS. *Candida* endocarditis: The insidious killer. *J Glob Infect Dis.* 2015; **7:** 48-50.
145. Ganguly S, Mitchell AP. Mucosal biofilms of *Candida albicans. Curr Opin Microbiol.* 2011; **14:** 380-5.
146. Garcia J, Soch K, Matthew E, *et al.* Endocarditis caused by *Candida dubliniensis. Am J Med Sci.* 2013; **346:** 237-9.
147. Garcia-Cuesta C, Sarrion-Perez MG, Bagan JV. Current treatment of oral candidiasis: A literature review. *J Clin Exp Dent.* 2014; **6:** e576-82.
148. Gautam H, Kaur R, Goyal R, *et al.* Oral thrush to candidemia: A morbid outcome. *J Int Assoc Physicians AIDS Care (Chic).* 2010; **9:** 325-7.
149. Georgiadou SP, Tarrand J, Sipsas NV, *et al.* Candiduria in haematologic malignancy patients without a urinary catheter: Nothing more than a frailty marker? *Mycoses.* 2013; **56:** 311-4.
150. Geusau A, Antoniewicz L, Poitschek C, *et al. In vitro* susceptibility of *Candida* isolates from organ transplant recipients to newer antifungals. *Mycopathologia.* 2014; **177:** 143-56.
151. Giri S, Kindo AJ, Kalyani J. Candidemia in intensive care unit patients: A one year study from a tertiary care center in South India. *J Postgrad Med.* 2013; **59:** 190-5.
152. Giri S, Kindo AJ. A review of *Candida* species causing blood stream infection. *Indian J Med Microbiol.* 2012; **30:** 270-8.
153. Goncalves B, Ferreira C, Alves CT, *et al.* Vulvovaginal candidiasis: Epidemiology, microbiology and risk factors. *Crit Rev Microbiol.* 2016; **42:** 905-27.
154. Greenberg RG, Benjamin DK Jr. Neonatal candidiasis: Diagnosis, prevention and treatment. *J Infect.* 2014; **69:** S19-22.
155. Guducuoglu H, Gultepe B, Otlu B, *et al. Candida albicans* outbreak associated with total parenteral nutrition in the neonatal unit. *Indian J Med Microbiol.* 2016; **34:** 202-7.
156. Guinea J. Global trends in the distribution of *Candida* species causing candidemia. *Clin Microbiol Infect.* 2014; **20 (Suppl. 6):** 5-10.
157. Guitard J, Angoulvant A, Letscher-Bru V, *et al.* Invasive infections due to *Candida norvegensis* and *Candida inconspicua:* Report of 12 cases and review of the literature. *Med Mycol.* 2013; **51:** 795-9.
158. Hakamifard A, Khorvash F, Raisi A. Isolated pulmonary candidiasis in a patient with diabetes mellitus: A rare case report. *Adv Biomed Res.* 2016; **5:** 146.
159. Haltmeier T, Inaba K, Effron Z, *et al. Candida* score as a predictor of worse outcomes and mortality in severely injured trauma patients with positive *Candida* cultures. *Am Surg.* 2015; **81:** 1067-73.
160. Hani U, Shivakumar HG, Vaghela R, *et al.* Candidiasis: A fungal infection - current challenges and progress in prevention and treatment. *Infect Disord Drug Targets.* 2015; **15:** 42-52.
161. Hasibeder W, Halabi M. *Candida* peritonitis. *Minerva Anestesiol.* 2014; **80:** 470-81.
162. Hebecker B, Naglik JR, Hube B, *et al.* Pathogenicity mechanisms and host response during oral *Candida albicans* infections. *Expert Rev Anti Infect Ther.* 2014; **12:** 867-79.
163. Held J, Kohlberger I, Rappold E, *et al.* Comparison of (1→3)-β-D-glucan, mannan/anti-mannan antibodies and Cand-Tec *Candida* antigen as serum biomarkers for candidemia. *J Clin Microbiol.* 2013; **51:** 1158-64.

164. Hoarau G, Kerdraon O, Lagree M, *et al.* Detection of (1,3)-β-D-glucans in situ in a *Candida albicans* brain granuloma. *J Infect.* 2013; **67:** 622-4.
165. Hof H, Eigner U, Maier T, *et al.* Differentiation of *Candida dubliniensis* from *Candida albicans* by means of MALDI-TOF mass spectrometry. *Clin Lab.* 2012; **58:** 927-31.
166. Holland SM, Vinh DC. Yeast infections - Human genetics on the rise. *N Engl J Med.* 2009; **361:** 1798-801.
167. Hope WW, Castagnola E, Groll AH, *et al.* ESCMID guideline for the diagnosis and management of *Candida* diseases 2012: Prevention and management of invasive infections in neonates and children caused by *Candida* spp. *Clin Microbiol Infect.* 2012; **18 (Suppl. 7):** 38-52.
168. Horn D, Neofytos D, Fishman J, *et al.* Use of the PATH alliance database to measure adherence to IDSA Guidelines for the therapy of candidemia. *Eur J Clin Microbiol Infect Dis.* 2007; **26:** 907-14.
169. Horn DL, Neofytos D, Anaissie EJ, *et al.* Epidemiology and outcomes of candidemia in 2019 patients: Data from the Prospective Antifungal Therapy Alliance registry. *Clin Infect Dis.* 2009; **48:** 1695-703.
170. Idelevich EA, Grunastel B, Becker K. Rapid detection and identification of candidemia by direct blood culturing on solid medium by use of lysis-centrifugation method combined with matrix-assisted laser desorption ionization-time of flight mass spectrometry (MALDI-TOF MS). *J Clin Microbiol.* 2017; **55:** 97-100.
171. Ikegaya S, Tai K, Shigemi H, *et al.* Fulminant candidemia diagnosed by prompt detection of pseudohyphae in a peripheral blood smear. *Am J Med Sci.* 2012; **343:** 419-20.
172. Ilkit M, Guzel AB. The epidemiology, pathogenesis and diagnosis of vulvovaginal candidosis: A mycological perspective. *Crit Rev Microbiol.* 2011; **37:** 250-61.
173. Jackel D, Lai K. *Candida glabrata* sepsis associated with chorioamnionitis in an *in vitro* fertilization pregnancy: Case report and review. *Clin Infect Dis.* 2013; **56:** 555-8.
174. Jaijakul S, Vazquez JA, Swanson RN, *et al.* (1,3)-β-D-glucan as a prognostic marker of treatment response in invasive candidiasis. *Clin Infect Dis.* 2012; **55:** 521-6.
175. Jain M, Dogra V, Mishra B, Thakur A, Loomba PS, Bhargava A. Candiduria in catheterized intensive care unit patients: Emerging microbiological trends. *Indian J Pathol Microbiol.* 2011; **54:** 552-5.
176. Jalalat S, Hunter L, Yamazaki M, *et al.* An outbreak of *Candida albicans* folliculitis masquerading as *Malassezia* folliculitis in a prison population. *J Correct Health Care.* 2014; **20:** 154-62.
177. Jamil B, Mukhtar Bokhari MT, Saeed A, *et al.* Candidiasis: Prevalence and resistance profiling in a tertiary care hospital of Pakistan. *J Pak Med Assoc.* 2017; **67:** 688-97.
178. Janum S, Afshari A. Central venous catheter (CVC) removal for patients of all ages with candidaemia. *Cochrane Database Syst Rev.* 2016; **7:** CD011195.
179. Jorge VC, Cardoso C, Noronha C, *et al.* 'Fungal spondylodiscitis in a non-immunocompromised patient'. *BMJ Case Rep.* 2012; pii: bcr1220115337.
180. Kalaiarasan K, Singh R, Chaturvedula L. Fungal profile of vulvovaginal candidiasis in a tertiary care hospital. *J Clin Diagn Res.* 2017; **11:** DC06-9.
181. Karacaer Z, Oncul O, Turhan V, *et al.* A surveillance of nosocomial *Candida* infections: Epidemiology and influences on mortalty in intensive care units. *Pan Afr Med J.* 2014; **19:** 398.
182. Kauffman CA, Fisher JF, Sobel JD, *et al. Candida* urinary tract infections - Diagnosis. *Clin Infect Dis.* 2011; **52:** S452-6.
183. Kauffman CA. Complications of candidemia in ICU patients: Endophthalmitis, osteomyelitis, endocarditis. *Semin Respir Crit Care Med.* 2015; **36:** 641-9.
184. Kaufman DA. Challenging issues in neonatal candidiasis. *Curr Med Res Opin.* 2010; **26:** 1769-78.
185. Kaur R, Dhakad MS, Goyal R, Kumar R. Emergence of non-albicans Candida species and antifungal resistance in intensive care unit patients. *Asian Pac J Trop Biomed.* 2016; **6:** 451.
186. Kelly MS, Benjamin DK Jr, Smith PB. The epidemiology and diagnosis of invasive candidiasis among premature infants. *Clin Perinatol.* 2015; **42:** 105-17.
187. Khan EA, Choudhry S, Fatima M, *et al.* Clinical spectrum, management and outcome of neonatal candidiasis. *J Pak Med Assoc.* 2015; **65:** 1206-9.
188. Khan Z, Ahmad S, Joseph L, *et al. Candida dubliniensis:* an appraisal of its clinical significance as a bloodstream pathogen. *PLoS One.* 2012; **7:** e32952.
189. Kim GY, Jeon JS, Kim JK. Isolation frequency characteristics of *Candida* species from clinical specimens. *Mycobiology.* 2016; **44:** 99-104.
190. Kim J, Sudbery P. *Candida albicans*, a major human fungal pathogen. *J Microbiol.* 2011; **49:** 171-7.
191. Kim MN, Shin JH, Sung H, *et al. Candida haemulonii* and closely related species at 5 University Hospitals in Korea: Identification, antifungal susceptibility and clinical features. *Clin Infect Dis.* 2009; **48:** e57-61.
192. Kim TH, Park BR, Kim HR, *et al. Candida dubliniensis* screening using the germ tube test in clinical yeast isolates and prevalence of *C.dubliniensis* in Korea. *J Clin Lab Anal.* 2010; **24:** 145-8.
193. Kiraz N, Oz Y. Species distribution and *in vitro* antifungal susceptibility of clinical *Candida* isolates from a university hospital in Turkey over a 5-year period. *Med Mycol.* 2011; **49:** 126-31.
194. Kirby JE, Branch-Elliman W, LaSalvia MT, *et al.* Investigation of a *Candida guilliermondii* pseudo-outbreak reveals a novel source of laboratory contamination. *J Clin Microbiol.* 2017; **55:** 1080-9.
195. Kitajima S, Yasui A. Oral maxillary exostosis. *N Engl J Med.* 2015; **373:** 1457.
196. Klatte JM, Newland JG, Jackson MA. Incidence, classification and risk stratification for *Candida* central line-associated bloodstream infections in pediatric patients at a tertiary care children's hospital, 2000-2010. *Infect Control Hosp Epidemiol.* 2013; **34:** 1266-71.
197. Kluger N, Jokinen M, Krohn K, *et al.* Gastrointestinal manifestations in APECED syndrome. *J Clin Gastroenterol.* 2013; **47:** 112-20.
198. Kochhar R, Noor MT, Wig J. Fungal infections in severe acute pancreatitis. *J Gastroenterol Hepatol.* 2011; **26:** 952-9.

199. Kocyigit I, Unal A, Sipahioglu MH, *et al.* Peritonitis due to *Candida rugosa:* The first case report. *Perit Dial Int.* 2010; **30:** 576-7.
200. Koehler P, Tacke D, Cornely OA. Our 2014 approach to candidaemia. *Mycoses.* 2014; **57:** 581-3.
201. Kojic EM, Darouiche RO. *Candida* infections of medical devices. *Clin Microbiol Rev.* 2004; **17:** 255-67.
202. Kordalewska M, Zhao Y, Lockhart SR, *et al.* Rapid and accurate molecular identification of the emerging multi-drug resistant pathogen *Candida auris. J Clin Microbiol.* 2017; pii: JCM.00630-17. PMID: 28539346.
203. Kothari A, Sagar V. Epidemiology of *Candida* bloodstream infections in a tertiary care institute in India. *Indian J Med Microbiol.* 2009; **27:** 171-2.
204. Kratzer C, Graninger W, Lassnigg A, *et al.* Design and use of *Candida* scores at the intensive care unit. *Mycoses.* 2011; **54:** 467-74.
205. Kullberg BJ, Arendrup MC. Invasive Candidiasis. *N Engl J Med.* 2015; **373:** 1445-56.
206. Kumar A, Prakash A, Singh A, *et al. Candida haemulonii* species complex: An emerging species in India and its genetic diversity assessed with multilocus sequence and amplified fragment-length polymorphism analyses. *Emerg Microbes Infect.* 2016; **5:** e49.
207. Kumar D, Banerjee T, Pratap CB, Tilak R. Itraconazole-resistant *Candida auris* with phospholipase, proteinase and hemolysin activity from a case of vulvovaginitis. *J Infect Dev Ctries.* 2015; **9:** 435-7.
208. Kumar D, Kumar A, Singh S, *et al.* Candidemia-induced pediatric sepsis and its association with free radicals, nitric oxide and cytokine level in host. *J Crit Care.* 2015; **30:** 296-303.
209. Kumar S, Bandyopadhyay MK, Bhattacharyya K, *et al.* A Rare case of pancreatic abscess due to *Candida tropicalis. J Glob Infect Dis.* 2011; **3:** 396-8.
210. Kumar S, Kalam K, Ali S, *et al.* Frequency, clinical presentation and microbiological spectrum of candidemia in a tertiary care center in Karachi, Pakistan. *J Pak Med Assoc.* 2014; **64:** 281-5.
211. Kumar S, Singhi S, Chakrabarti A, *et al.* Probiotic use and prevalence of candidemia and candiduria in a PICU. *Pediatr Crit Care Med.* 2013; **14:** e409-15.
212. Kumar S, Singhi S. Role of probiotics in prevention of *Candida* infection in critically ill children. *Mycoses.* 2013; **56:** 204-11.
213. Kumari V, Banerjee T, Kumar P, *et al.* Emergence of non-*albicans Candida* among candidal vulvovaginitis cases and study of their potential virulence factors, from a tertiary care center, north India. *Indian J Pathol Microbiol.* 2013; **56:** 144-7.
214. Lacroix C, Gicquel A, Sendid B, *et al.* Evaluation of two matrix-assisted laser desorption ionization-time of flight mass spectrometry (MALDI-TOF MS) systems for the identification of *Candida* species. *Clin Microbiol Infect.* 2014; **20:** 153-8.
215. Lau AF, Kabir M, Chen SC, *et al. Candida* colonization as a risk marker for invasive candidiasis in mixed medical-surgical intensive care units: development and evaluation of a simple, standard protocol. *J Clin Microbiol.* 2015; **53:** 1324-30.
216. Lee WG, Shin JH, Uh Y, *et al.* First three reported cases of nosocomial fungemia caused by *Candida auris. J Clin Microbiol.* 2011; **49:** 3139-42.
217. Leon C, Ostrosky-Zeichner L, Schuster M. What's new in the clinical and diagnostic management of invasive candidiasis in critically ill patients. *Intensive Care Med.* 2014; **40:** 808-19.
218. Leon C, Ruiz-Santana S, Saavedra P, *et al.* Usefulness of the "Candida score" for discriminating between *Candida* colonization and invasive candidiasis in non-neutropenic critically ill patients: A prospective multicenter study. *Crit Care Med.* 2009; **37:** 1624-33.
219. Leroy G, Lambiotte F, Thevenin D, *et al.* Evaluation of "*Candida* score" in critically ill patients: A prospective, multicenter, observational, cohort study. *Ann Intensive Care.* 2011; **1:** 50.
220. Lewis RE. Overview of the changing epidemiology of candidemia. Curr Med Res Opin. 2009; **25:** 1732-40.
221. Li D, Zhang W, Zheng S, *et al.* Surveillance study of candidemia in cancer patients in North China. *Med Mycol.* 2013; **51:** 378-84.
222. Li J, Shan Y, Fan S, Liu X. Prevalence of *Candida nivariensis* and *Candida bracarensis* in vulvovaginal candidiasis. *Mycopathologia.* 2014; **178:** 279-83.
223. Liew YX, Teo J, Too IA, *et al. Candida* surveillance in surgical intensive care unit (SICU) in a tertiary institution. *BMC Infect Dis.* 2015; **15:** 256.
224. Lim CS, Rosli R, Seow HF, *et al. Candida* and invasive candidiasis: Back to basics. *Eur J Clin Microbiol Infect Dis.* 2012; **31:** 21-31.
225. Lionakis MS. New insights into innate immune control of systemic candidiasis. *Med Mycol.* 2014; **52:** 555-64.
226. Lisboa C, Santos A, Dias C, *et al. Candida* balanitis: Risk factors. *J Eur Acad Dermatol Venereol.* 2010; **24:** 820-6.
227. Liu M, Huang S, Guo L, *et al.* Clinical features and risk factors for blood stream infections of *Candida* in neonates. *Exp Ther Med.* 2015; **10:** 1139-44.
228. Lockhart SR, Etienne KA, Vallabhaneni S, *et al.* Simultaneous emergence of multidrug-resistant *Candida auris* on 3 continents confirmed by Whole-Genome Sequencing and epidemiological analyses. *Clin Infect Dis.* 2017; **64:** 134-40.
229. Lockhart SR, Iqbal N, Cleveland AA, *et al.* Species identification and antifungal susceptibility testing of *Candida* bloodstream isolates from population-based surveillance studies in two U.S. cities from 2008 to 2011. *J Clin Microbiol.* 2012; **50:** 3435-42.
230. Lockhart SR. Do hospital microbiology laboratories still need to distinguish *Candida albicans* from *Candida dubliniensis? J Clin Microbiol.* 2011; **49:** 4415.
231. Lortholary O, Petrikkos G, Akova M, *et al.* ESCMID guideline for the diagnosis and management of *Candida* diseases 2012: Patients with HIV infection or AIDS. *Clin Microbiol Infect.* 2012; **18 (Suppl. 7):** 68-77.
232. Madhavan P, Jamal F, Chong PP, *et al.* Identification of local clinical *Candida* isolates using CHROMagar Candida™ as a primary identification method for various *Candida* species. *Trop Biomed.* 2011; **28:** 269-74.

233. Magobo RE, Corcoran C, Seetharam S, *et al.* *Candida auris*-associated candidemia, South Africa. *Emerg Infect Dis.* 2014; **20:** 1250-1.
234. Martini A, Gottin L, Menestrina N, *et al.* Procalcitonin levels in surgical patients at risk of candidemia. *J Infect.* 2010; **60:** 425-30.
235. Martins N, Ferreira IC, Barros L, *et al.* Candidiasis: Predisposing factors, prevention, diagnosis and alternative treatment. *Mycopathologia.* 2014; **177:** 223-40.
236. Marty M, Bourrat E, Vaysse F, *et al.* Direct microscopy: A useful tool to diagnose oral candidiasis in children and adolescents. *Mycopathologia.* 2015; **180:** 373-7.
237. Mastrolorenzo A, Giomi B, Cipollini EM, *et al.* Mycetomatoid infection of the penis by *Candida albicans. Int J Dermatol.* 2012; **51:** 1082-5.
238. Maubon D, Garnaud C, Calandra T, *et al.* Resistance of *Candida* spp. to antifungal drugs in the ICU: Where are we now? *Intensive Care Med.* 2014; **40:** 1241-55.
239. Maurya V, Srivastava A, Mishra J, *et al.* Oropharyngeal candidiasis and *Candida* colonization in HIV positive patients in northern India. *J Infect Dev Ctries.* 2013; **7:** 608-13.
240. Mayer FL, Wilson D, Hube B. *Candida albicans* pathogenicity mechanisms. *Virulence.* 2013; **4:** 119-28.
241. McCarty TP, Pappas PG. Invasive Candidiasis. *Infect Dis Clin North Am.* 2016; **30:** 103-24.
242. Menon T, Herrera M, Periasamy S, *et al.* Oral candidiasis caused by *Kodamaea ohmeri* in a HIV patient in Chennai, India. *Mycoses.* 2010; **53:** 458-9.
243. Mikulska M, Del Bono V, Ratto S, *et al.* Occurrence, presentation and treatment of candidemia. *Expert Rev Clin Immunol.* 2012; **8:** 755-65.
244. Min YW, Kim E, Son HJ, *et al.* Antifungal treatment is not required for immunocompetent individuals with asymptomatic esophageal candidiasis. *Medicine (Baltimore).* 2015; **94:** e1969. PMID: 26559270.
245. Modrzewska B, Kurnatowski P. Selected pathogenic characteristics of fungi from the genus *Candida. Ann Parasitol.* 2013; **59:** 57-66.
246. Mokaddas E, Khan ZU, Ahmad S, *et al.* Value of (1-3)-β-D-glucan, *Candida* mannan and *Candida* DNA detection in the diagnosis of candidaemia. *Clin Microbiol Infect.* 2011; **17:** 1549-53.
247. Mokaddas E, Khan ZU, Ahmad S. Prevalence of *Candida dubliniensis* among cancer patients in Kuwait: a 5-year retrospective study. *Mycoses.* 2011; **54:** e29-34.
248. Montravers P, Boudinet S, Houissa H. *Candida* and severe acute pancreatitis: We won't be fooled again. *Crit Care.* 2013; **17:** 137.
249. Montravers P, Dupont H, Eggimann P. Intra-abdominal candidiasis: The guidelines-forgotten non-candidemic invasive candidiasis. *Intensive Care Med.* 2013; **39:** 2226-30.
250. Moragues MD, Rementeria A, Sevilla MJ, *et al.* *Candida* antigens and immune responses: Implications for a vaccine. *Expert Rev Vaccines.* 2014; **13:** 1001-12.
251. Morii D, Seki M, Binongo JN, *et al.* Distribution of *Candida* species isolated from blood cultures in hospitals in Osaka, Japan. *J Infect Chemother.* 2014; **20:** 558-62.
252. Moriyama B, Gordon LA, McCarthy M, *et al.* Emerging drugs and vaccines for candidemia. *Mycoses.* 2014; **57:** 718-33.
253. Motoa G, Munoz JS, Onate J, *et al.* Epidemiology of *Candida* isolates from intensive care units in Colombia from 2010 to 2013. *Rev Iberoam Micol.* 2017; **34:** 17-22.
254. Mtibaa L, Fakhfakh N, Kallel A, *et al.* Vulvovaginal candidiasis: Etiology, symptomatology and risk factors. *J Mycol Med.* 2017; **27:** 153-8.
255. Mylonakis E, Clancy CJ, Ostrosky-Zeichner L, *et al.* T2 magnetic resonance assay for the rapid diagnosis of candidemia in whole blood: A clinical trial. *Clin Infect Dis.* 2015; **60:** 892-9.
256. Naglik JR, Richardson JP, Moyes DL. *Candida albicans* pathogenicity and epithelial immunity. *PLoS Pathog.* 2014; **10:** e1004257. PMID: 25121985.
257. Nawange SR, Singh K, Naidu J, Singh SM. Naturally acquired systemic dual infection caused by *Candida famata (Debaryomyces hansenii)* and *Candida catenulate* in albino rats bred for sale in the market at Jabalpur (Madhya Pradesh), India. *Mycoses.* 2010; **53:** 173-5.
258. Negri M, Martins M, Henriques M, *et al.* Examination of potential virulence factors of *Candida tropicalis* clinical isolates from hospitalized patients. *Mycopathologia.* 2010; **169:** 175-82.
259. Nemcova E, Cernochova M, Ruzicka F, *et al.* Rapid identification of medically important *Candida* isolates using high resolution melting analysis. *PLoS One.* 2015; **10:** e0116940.
260. Ng KP, Kuan CS, Kaur H, *et al.* *Candida* species epidemiology 2000-2013: A laboratory-based report. *Trop Med Int Health.* 2015; **20:** 1447-53.
261. Nobile CJ, Johnson AD. Candida albicans biofilms and human disease. *Annu Rev Microbiol.* 2015; **69:** 71-92.
262. Nobrega de Almeida J Jr, Assy JG, Levin AS, *et al.* *Candida haemulonii* complex species, Brazil, January 2010-March 2015. *Emerg Infect Dis.* 2016; **22:** 561-3.
263. Nucci M, Queiroz-Telles F, Alvarado-Matute T, *et al.* Epidemiology of candidemia in Latin America: A laboratory-based survey. *PLoS One.* 2013; **8:** e59373.
264. Oberoi JK, Wattal C, Goel N, *et al.* Non-*albicans Candida* species in blood stream infections in a tertiary care hospital at New Delhi, India. *Indian J Med Res.* 2012; **136:** 997-1003.
265. Odds FC. Molecular phylogenetics and epidemiology of *Candida albicans. Future Microbiol.* 2010; **5:** 67-79.
266. Oeser C, Lamagni T, Heath PT, *et al.* The epidemiology of neonatal and pediatric candidemia in England and Wales, 2000-2009. *Pediatr Infect Dis J.* 2013; **32:** 23-6.
267. Pahwa N, Kumar R, Nirkhiwale S, *et al.* Species distribution and drug susceptibility of *Candida* in clinical isolates from a tertiary care centre at Indore. *Indian J Med Microbiol.* 2014; **32:** 44-8.
268. Pan N, Herzog R, Blanco JS, *et al.* *Candida albicans* osteomyelitis in an infant: A case report and literature review. *J Pediatr Orthop B.* 2013; **22:** 491-7.
269. Papon N, Courdavault V, Clastre M, *et al.* Emerging and emerged pathogenic *Candida* species: beyond the *Candida albicans* paradigm. *PLoS Pathog.* 2013; **9:** e1003550.

270. Pappas PG, Kauffman CA, Andes DR, *et al.* Clinical Practice Guideline for the Management of Candidiasis: 2016 Update by the Infectious Diseases Society of America. *Clin Infect Dis.* 2016; **62:** e1-50.
271. Pappas PG, Kauffman CA, Andes DR, *et al.* Executive Summary: Clinical Practice Guideline for the Management of Candidiasis: 2016 Update by the Infectious Diseases Society of America. *Clin Infect Dis.* 2016; **62:** 409-17.
272. Paredes K, Sutton DA, Cano J, *et al.* Molecular identification and antifungal susceptibility testing of clinical isolates of the *Candida rugosa* species complex and proposal of the new species *Candida neorugosa. J Clin Microbiol.* 2012; **50:** 2397-403.
273. Parmeland L, Gazon M, Guerin C, *et al. Candida albicans* and non-*Candida albicans* fungemia in an institutional hospital during a decade. *Med Mycol.* 2013; **51:** 33-7.
274. Pate A, Baltazar G, Jung YS, *et al.* Total parenteral nutrition changes species distribution in patients with fungemia. *Am Surg.* 2014; **80:** e278-80.
275. Peman J, Zaragoza R. Current diagnostic approaches to invasive candidiasis in critical care settings. *Mycoses.* 2010; **53:** 424-33.
276. Pendleton KM, Huffnagle GB, Dickson RP. The significance of *Candida* in the human respiratory tract: Our evolving understanding. *Pathog Dis.* 2017; **75:** PMID: 28423168.
277. Pereira GH, Muller PR, Szeszs MW, *et al.* Five-year evaluation of bloodstream yeast infections in a tertiary hospital: The predominance of non-*C.albicans Candida* species. *Med Mycol.* 2010; **48:** 839-42.
278. Peters BM, Yano J, Noverr MC, *et al. Candida* vaginitis: When opportunism knocks, the host responds. *PLoS Pathog.* 2014; **10:** e1003965.
279. Pfaller MA, Diekema DJ, Gibbs DL, *et al.* Geographic variation in the frequency of isolation and fluconazole and voriconazole susceptibilities of *Candida glabrata:* An assessment from the ARTEMIS Disk Global Antifungal Surveillance Program. *Diagn Microbiol Infect Dis.* 2010; **67:** 162-71.
280. Pfeiffer CD, Samsa GP, Schell WA, *et al.* Quantitation of *Candida* CFU in initial positive blood cultures. *J Clin Microbiol.* 2011; **49:** 2879-83.
281. Playford EG, Lipman J, Sorrell TC. Management of invasive candidiasis in the intensive care unit. *Drugs.* 2010; **70:** 823-39.
282. Playford EG, Lipman J, Sorrell TC. Prophylaxis, empirical and preemptive treatment of invasive candidiasis. *Curr Opin Crit Care.* 2010; **16:** 470-4.
283. Prevost N, English JC 3rd. *Candida* rib osteomyelitis: Erythema and nodule in midline scar. *J Am Acad Dermatol.* 2013; **69:** e205-7.
284. Prim N, Remacha A, Sanchez-Reus F, *et al.* Candidaemia detected on direct blood smears. *Eur J Haematol.* 2013; **90:** 536-7.
285. Pruller F, Wagner J, Raggam RB, *et al.* Automation of serum (1→3)-beta-D-glucan testing allows reliable and rapid discrimination of patients with and without candidemia. *Med Mycol.* 2014; **52:** 455-61.
286. Pulcrano G, Iula DV, Vollaro A, *et al.* Rapid and reliable MALDI-TOF mass spectrometry identification of *Candida* non-*albicans* isolates from bloodstream infections. *J Microbiol Methods.* 2013; **94:** 262-6.
287. Punithavathy PM, Nalina K, Menon T. Antifungal susceptibility testing of *Candida tropicalis* biofilms against fluconazole using calorimetric indicator resazurin. *Indian J Pathol Microbiol.* 2012; **55:** 72-4.
288. Quindos G. Epidemiology of candidaemia and invasive candidiasis. A changing face. *Rev Iberoam Micol.* 2014; **31:** 42-8.
289. Raghunath BV, Gowrishankar BC, Narendrababu M, *et al.* Successful management of a renal fungal ball in a pretermature neonate: A case report and review of literature. *J Indian Assoc Pediatr Surg.* 2013; **18:** 121-3.
290. Ramla S, Sharma V, Patel M. Influence of cancer treatment on the *Candida albicans* isolated from the oral cavities of cancer patients. *Support Care Cancer.* 2016; **24:** 2429-36.
291. Ramos LS, Figueiredo-Carvalho MH, Barbedo LS, *et al. Candida haemulonii* complex: Species identification and antifungal susceptibility profiles of clinical isolates from Brazil. *J Antimicrob Chemother.* 2015; **70:** 111-5.
292. Ratcliffe L, Davies J, Anson J, *et al. Candida pelliculosa* meningitis as an opportunistic infection in HIV: The first reported case. *Int J STD AIDS.* 2011; **22:** 54-6.
293. Rautemaa R, Ramage G. Oral candidosis - Clinical challenges of a biofilm disease. *Crit Rev Microbiol.* 2011; **37:** 328-36.
294. Reed EE, West JE, Keating EA, *et al.* Improving the management of candidemia through antimicrobial stewardship interventions. *Diagn Microbiol Infect Dis.* 2014; **78:** 157-61.
295. Rodrigues CF, Silva S, Henriques M. *Candida glabrata:* A review of its features and resistance. *Eur J Clin Microbiol Infect Dis.* 2014; **33:** 673-88.
296. Rodrigues ME, Silva S, Azeredo J, *et al.* Novel strategies to fight *Candida* species infection. *Crit Rev Microbiol.* 2016; **42:** 594-606.
297. Romeo O, Criseo G. *Candida africana* and its closest relatives. *Mycoses.* 2011; **54:** 475-86.
298. Romeo O, Delfino D, Costanzo B, *et al.* Molecular characterization of Italian *Candida parapsilosis* isolates reveals the cryptic presence of the newly described species *Candida orthopsilosis* in blood cultures from newborns. *Diagn Microbiol Infect Dis.* 2012; **72:** 234-8.
299. Ruan SY, Kuo YW, Huang CT, *et al.* Infections due to *Candida haemulonii:* Species identification, antifungal susceptibility and outcomes. *Int J Antimicrob Agents.* 2010; **35:** 85-8.
300. Rudramurthy SM, Chakrabarti A, Paul RA, *et al. Candida auris* candidaemia in Indian ICUs: Analysis of risk factors. *J Antimicrob Chemother.* 2017; **72:** 1794-801.
301. Ruhnke M. Antifungal stewardship in invasive Candida infections. *Clin Microbiol Infect.* 2014; **20 Suppl 6:** 11-8.
302. Sajith KG, Dutta AK, Sahni RD, *et al.* Is empiric therapy with fluconazole appropriate for esophageal candidiasis? *Indian J Gastroenterol.* 2014; **33:** 165-8.
303. Salim N, Moore C, Silikas N, *et al.* Chlorhexidine is a highly effective topical broad-spectrum agent against *Candida* spp. *Int J Antimicrob Agents.* 2013; **41:** 65-9.
304. Salvatore CM, Chen TK, Toussi SS, *et al.* (1→3)-β-D-glucan in cerebrospinal fluid as a biomarker for *Candida*

and *Aspergillus* infections of the central nervous system in pediatric patients. *J Pediatric Infect Dis Soc.* 2016; **5:** 277-86.
305. Saravu K, Vishwanath S, Shastry BA, *et al.* Candidal renal papillary necrosis conquered. *J Assoc Physicians India.* 2013; **61:** 573-4.
306. Saraydaroglu O, Coskun H, Elezoglu B. An interesting entity mimicking malignancy: Laryngeal candidiasis. *J Int Med Res.* 2010; **38:** 2146-52.
307. Sardana V, Pandey A, Madan M, *et al.* Neonatal candidemia: A changing trend. *Indian J Pathol Microbiol.* 2012; **55:** 132-3.
308. Sardi JC, Scorzoni L, Bernardi T, *et al. Candida* species: Current epidemiology, pathogenicity, biofilm formation, natural antifungal products and new therapeutic options. *J Med Microbiol.* 2013; **62 (Pt 1):** 10-24.
309. Sarika J, Bhawna S, Rajni G, *et al.* Late-onset prosthetic valve fungal endocarditis due to *Candida albicans:* A case report and review of literature from India. *J Immunol Tech Infect.* 2015; **4:** 1.
310. Sarkar R, Rathod GP. Clinicopathologic assessment of *Candida* colonization of oral leukoplakia. *Indian J Dermatol Venereol Leprol.* 2014; **80:** 413-8.
311. Sarma S, Kumar N, Sharma S, *et al.* Candidemia caused by amphotericin B and fluconazole resistant *Candida auris. Indian J Med Microbiol.* 2013; **31:** 90-1.
312. Sensoy G, Belet N. Invasive *Candida* infections in solid organ transplant recipient children. *Expert Rev Anti Infect Ther.* 2011; **9:** 317-24.
313. Shah SS, Parikh RS, Vaswani LP. Dangerously candid: Detection of yeast cells in peripheral blood smear in a case of candidaemia. *J Assoc Physicians India.* 2015; **63:** 74-5.
314. Shan Y, Fan S, Liu X, *et al.* Prevalence of *Candida albicans*-closely related yeasts, *Candida africana* and *Candida dubliniensis,* in vulvovaginal candidiasis. *Med Mycol.* 2014; **52:** 636-40.
315. Sharma C, Kumar N, Pandey R, *et al.* Whole genome sequencing of emerging multidrug resistant *Candida auris* isolates in India demonstrates low genetic variation. *New Microbes New Infect.* 2016; **13:** 77-82.
316. Sharma C, Muralidhar S, Xu J, *et al.* Multilocus sequence typing of *Candida africana* from patients with vulvovaginal candidiasis in New Delhi, India. *Mycoses.* 2014; **57:** 544-52.
317. Sharma C, Wankhede S, Muralidhar S, *et al. Candida nivariensis* as an etiologic agent of vulvovaginal candidiasis in a tertiary care hospital of New Delhi, India. *Diagn Microbiol Infect Dis.* 2013; **76:** 46-50.
318. Sharma D, Choudhary M, Shastri S, *et al.* Neonatal liver abscesses due to *Candida* infection in a preterm infant, secondary to malpositioned umbilical lines - A rare entity. *Pathog Glob Health.* 2015; **109:** 84-7.
319. Sharma S, Gangwal K. Neonatal *Candida* arthritis. *Indian J Orthop.* 2014; **48:** 339-42.
320. Sharon V, Fazel N. Oral candidiasis and angular cheilitis. *Dermatol Ther.* 2010; **23:** 230-42.
321. Shibasaki S, Aoki W, Nomura T, *et al.* An oral vaccine against candidiasis generated by a yeast molecular display system. *Pathog Dis.* 2013; **69:** 262-8.
322. Shields RK, Nguyen MH, Clancy CJ. Clinical perspectives on echinocandin resistance among *Candida* species. *Curr Opin Infect Dis.* 2015; **28:** 514-22.
323. Shivaswamy U, Neelambike SM. A study of candidiasis in HIV reactive patients in a tertiary care hospital, Mysore - South India. *Indian J Dermatol Venereol Leprol.* 2014; **80:** 278.
324. Silva CM, Carvalho-Parahym AM, Macedo DP, *et al.* Neonatal candidemia caused by *Candida haemulonii:* Case report and review of literature. *Mycopathologia.* 2015; **180:** 69-73.
325. Silveira-Gomes F, Sarmento DN, Espirito-Santo EP, *et al.* Differentiation between *Candida albicans* and *Candida dubliniensis* using hypertonic Sabouraud broth and tobacco agar. *Rev Soc Bras Med Trop.* 2011; **44:** 457-60.
326. Singh A, Verma R, Murari A, *et al.* Oral candidiasis: An overview. *J Oral Maxillofac Pathol.* 2014; **18:** S81-5.
327. Singh R, Parija SC. *Candida parapsilosis:* An emerging fungal pathogen. *Indian J Med Res.* 2012; **136:** 671-3.
328. Singhi S, Deep A. Invasive candidiasis in pediatric intensive care units. *Indian J Pediatr.* 2009; **76:** 1033-44.
329. Singla N, Gulati N, Kaistha N, Chander J. *Candida* colonization in urine samples of ICU patients: Determination of etiology, antifungal susceptibility testing and evaluation of associated risk factors. *Mycopathologia.* 2012; **174:** 149-55.
330. Sivabalan S, Mahadevan S, Srinath MV. Recurrent oral thrush. *Indian J Pediatr.* 2014; **81:** 394-6.
331. Skrobik Y, Laverdiere M. Why *Candida* sepsis should matter to ICU physicians. *Crit Care Clin.* 2013; **29:** 853-64.
332. Sobel JD. Changing trends in the epidemiology of *Candida* blood stream infections: A matter for concern? *Crit Care Med.* 2010; **38:** 990-2.
333. Sobel JD, Fisher JF, Kauffman CA, *et al. Candida* urinary tract infections - Epidemiology. *Clin Infect Dis.* 2011; **52:** S433-6.
334. Sobel JD. Recurrent vulvovaginal candidiasis. *Am J Obstet Gynecol.* 2016; **214:** 15-21.
335. Song YB, Suh MK, Ha GY, *et al.* Antifungal susceptibility testing with Etest for *Candida* species isolated from patients with oral candidiasis. *Ann Dermatol.* 2015; **27:** 715-20.
336. Spampinato C, Leonardi D. *Candida* infections, causes, targets and resistance mechanisms: Traditional and alternative antifungal agents. *Biomed Res Int.* 2013; 204237.
337. Spampinato C, Leonardi D. Molecular fingerprints to identify *Candida* species. *Biomed Res Int.* 2013; 923742.
338. Spanu T, Posteraro B, Fiori B, *et al.* Direct MALDI-TOF Mass Spectrometry assay of blood culture broths for rapid identification of *Candida* species causing bloodstream infections: An observational study in two large microbiology laboratories. *J Clin Microbiol.* 2012; **50:** 176-9.
339. Spiliopoulou A, Dimitriou G, Jelastopulu E, *et al.* Neonatal intensive care unit candidemia: Epidemiology, risk factors, outcome and critical review of published case series. *Mycopathologia.* 2012; **173:** 219-28.
340. Steinbach W. Pediatric invasive candidiasis: Epidemiology and diagnosis in children. *J Fungi.* 2016; **2:** 5.
341. Steinbach WJ, Roilides E, Berman D, *et al.* Results from a prospective, international, epidemiologic study of invasive candidiasis in children and neonates. *Pediatr Infect Dis J.* 2012; **31:** 1252-7.

342. Stone NR, Gorton RL, Barker K, *et al.* Evaluation of PNA-FISH yeast traffic light for rapid identification of yeast directly from positive blood cultures and assessment of clinical impact. *J Clin Microbiol.* 2013; **51:** 1301-2.
343. Su SY, Chao CM. Invasive *Candida* infections in critically ill patients. *Crit Care Med.* 2015; **43:** e322-3.
344. Swanson JR, Vergales J, Kaufman DA, *et al.* Cost analysis of fluconazole prophylaxis for prevention of neonatal invasive candidiasis. *Pediatr Infect Dis J.* 2016; **35:** 519-23.
345. Taff HT, Mitchell KF, Edward JA, *et al.* Mechanisms of *Candida* biofilm drug resistance. *Future Microbiol.* 2013; **8:** 1325-37.
346. Taj-Aldeen SJ, Kolecka A, Boesten R, *et al.* Epidemiology of candidemia in Qatar, the Middle East: Performance of MALDI-TOF MS for the identification of *Candida* species, species distribution, outcome and susceptibility pattern. *Infection.* 2014; **42:** 393-404.
347. Takahashi Y, Nagata N, Shimbo T, *et al.* Long-term trends in esophageal candidiasis prevalence and associated risk factors with or without HIV infection: Lessons from an endoscopic study of 80,219 patients. *PLoS One.* 2015; **10:** e0133589.
348. Tan BH, Chakrabarti A, Li RY, *et al.* Incidence and species distribution of candidaemia in Asia: A laboratory-based surveillance study. *Clin Microbiol Infect.* 2015; **21:** 946-53.
349. Taneja J, Mishra B, Thakur A, Dogra V, Loomba P. *Candida albicans* meningitis with ventriculitis in a four-year-old child. *Neurol India.* 2009; **57:** 682-4.
350. Tang HJ, Liu WL, Lin HL, *et al.* Clinical manifestations and prognostic factors of central line-associated candidemia. *Am J Infect Control.* 2014; **42:** 1238-40.
351. Tang HJ, Liu WL, Lin HL, *et al.* Epidemiology and prognostic factors of candidemia in cancer patients. *PLoS One.* 2014; **9:** e99103.
352. Tellapragada C, Eshwara VK, Johar R, *et al.* Antifungal susceptibility patterns, in vitro production of virulence factors and evaluation of diagnostic modalities for the speciation of pathogenic *Candida* from blood stream infections and vulvovaginal candidiasis. *J Pathog.* 2014; 142864.
353. Teoh F, Pavelka N. How chemotherapy increases the risk of systemic candidiasis in cancer patients: Current paradigm and future directions. *Pathogens.* 2016; 5. pii: e6. doi: 10.3390/pathogens5010006. PMID: 26784236.
354. Thulkar J, Kriplani A, Agarwal N. Utility of pH test & Whiff test in syndromic approach of abnormal vaginal discharge. *Indian J Med Res.* 2010; **131:** 445-8.
355. Tobudic S, Kratzer C, Lassnigg A, *et al.* Antifungal susceptibility of *Candida albicans* in biofilms. *Mycoses.* 2012; **55:** 199-204.
356. Todd B. Clinical Alert: *Candida auris. Am J Nurs.* 2017; **117:** 53-5.
357. Tragiannidis A, Tsoulas C, Groll AH. Invasive candidiasis and candidaemia in neonates and children: Update on current guidelines. *Mycoses.* 2015; **58:** 10-21.
358. Tragiannidis A, Tsoulas C, Kerl K, *et al.* Invasive candidiasis: Update on current pharmacotherapy options and future perspectives. *Expert Opin Pharmacother.* 2013; **14:** 1515-28.
359. Trikudanathan G, Navaneethan U, Vege SS. Intra-abdominal fungal infections complicating acute pancreatitis: a review. *Am J Gastroenterol.* 2011; **106:** 1188-92.
360. Trofa D, Gacser A, Nosanchuk JD. *Candida parapsilosis,* an emerging fungal pathogen. *Clin Microbiol Rev.* 2008; **21:** 606-25.
361. Turner SA, Butler G. The Candida pathogenic species complex. *Cold Spring Harb Perspect Med.* 2014; **4:** a019778.
362. Ukekwe FI, Nwajiobi C, Agbo MO, *et al.* Candidiasis, A rare cause of gastric perforation: A case report and review of literature. *Ann Med Health Sci Res.* 2015; **5:** 314-6.
363. Ulu Kilic A, Alp E, Cevahir F, *et al.* Epidemiology and cost implications of candidemia, a 6-year analysis from a developing country. *Mycoses.* 2017; **60:** 198-203.
364. Vanhee LM, Meersseman W, Lagrou K, *et al.* Rapid and direct quantification of viable *Candida* species in whole blood by use of immunomagnetic separation and solid-phase cytometry. *J Clin Microbiol.* 2010; **48:** 1126-31.
365. van Prehn J, Menke-van der Houven van Oordt CW, de Rooij ML, *et al.* Hepato-splenic candidiasis without prior documented candidemia: An under-recognized diagnosis? *Oncologist.* 2017; pii: theoncologist.2017-0019. PMID: 28469041.
366. Vela JI, Rosello N, Diaz-Cascajosa J, *et al.* Combined retinal detachment and *Candida* chorioretinitis. *BMJ Case Rep.* 2010; pii: bcr0120102648.
367. Verma SP, Dubashi B, Narayanan P, *et al.* A case of pediatric acute lymphoblastic leukemia with invasive candidiasis: Short review. *Indian J Hematol Blood Transfus.* 2014; **30 (Suppl. 1):** 101-4.
368. Vijaya D, Dhanalakshmi TA, Kulkarni S. Changing trends of vulvovaginal candidiasis. *J Lab Physicians.* 2014; **6:** 28-30.
369. Wang XJ, Sui X, Yan L, *et al.* Vaccines in the treatment of invasive candidiasis. *Virulence.* 2015; **6:** 309-15.
370. Wisplinghoff H, Ebbers J, Geurtz L, *et al.* Nosocomial bloodstream infections due to *Candida* spp. in the USA: Species distribution, clinical features and antifungal susceptibilities. *Int J Antimicrob Agents.* 2014; **43:** 78-81.
371. Xu WM, Shui W, Lin JC, *et al.* The serum glucan level and pathological changes of antifungal treatment for lower respiratory tract infection of *Candida albicans. Med Mycol.* 2015; **53:** 153-9.
372. Yaman G, Akyar I, Can S. Evaluation of the MALDI TOF-MS method for identification of *Candida* strains isolated from blood cultures. *Diagn Microbiol Infect Dis.* 2012; **73:** 65-7.
373. Yapar N. Epidemiology and risk factors for invasive candidiasis. *Ther Clin Risk Manag.* 2014; **10:** 95-105.
374. Zhang J, Silao FG, Bigol UG, *et al.* Calcineurin is required for pseudohyphal growth, virulence and drug resistance in *Candida lusitaniae. PLoS One.* 2012; **7:** e44192.

CHAPTER 21

Cryptococcosis

Cryptococcosis is an acute, subacute or chronic fungal disease caused by encapsulated basidiomycetous yeasts belonging to genus ***Cryptococcus***. There are many species of this genus prevalent ubiquitously in the environment, however, only a very few are usually pathogenic to men and animals. The clinical entities caused by these fungi are essentially opportunistic infections, which primarily involve lungs. Then infection disseminates to extrapulmonary sites through hematogenous route to various body organs particularly CNS, causing life-threatening meningoencephalitis among patients with and without impaired immune function. The disease is usually an opportunistic infection in patients with compromised immunity as a consequence of HIV infection, steroid administration, cancer chemotherapy, sarcoidosis, diabetes or inherited immune system defects. It is responsible for high morbidity as well as mortality particularly among immunocompromised patients unless it is diagnosed and treated at an early stage.

The predominant species causing cryptococcosis, ***Cryptococcus neoformans,*** was considered as *'sleeping giant'* among fungal diseases by Ajello in 1970 but it has now become an *'awakening giant'* since last three decades by the turn of century, particularly after emergence of AIDS pandemic. Now, there is an overall increase in number of cryptococcosis cases worldwide caused by various species/varieties of *Cryptococcus*. As the initial cases were reported from the European continent hence it was termed as **European blastomycosis** to differentiate it from the North and South American blastomycosis. It has also been known by other synonyms like **Torulopsis** or **Busse-Buschke disease**, etc.

Historical Perspective

Cryptococcus neoformans has been recognized as significant human pathogen for over a century. Sanfelice first reported this fungus in Italy from fermenting peach juice in 1894. In the same year, it was reported to be pathogenic organism among human beings by Otto Busse and Buschke in Germany from tibial lesion of female with sarcoma. The epithet ***'neoformans'*** for species was assigned presuming that this fungus caused tumor or cancer that was essentially found to be its underlying predisposing factor in due course of time. In 1950, Benham named it as *Cryptococcus neoformans.* In 1955, Emmons isolated this fungus from natural habitat, particularly pigeon droppings and nitrogenous soil contaminated with avian guanos.

In 1970s Kwon-Chung elucidated sexual reproduction in *Cr.neoformans* under specific laboratory conditions and taxonomical status of fungus was explained in detail. Consequently, in the same year, *Cryptococcus* was differentiated into **two varieties** *i.e.* var. *neoformans* and var. *gattii* followed by **five serotypes**. Moreover, their clinical and epidemiological significance was also described. In 1971, Staib *et al,* reported isolation of *Cr.neoformans* from ripe peach fruit and showed that wide variety of plant substrates supported *in vitro* colonization by this fungus.

The non-avian sources like fruits, vegetables and dairy products were found as natural habitats of this fungus. In 1990, Ellis *et al,* in Australia have shown association of *Cr. neoformans* var. *gattii* with various species of *Eucalyptus* trees. Moreover, in Australia, majority of human isolates are from koalas, a native animal that lives in close association with *Eucalyptus*. There are five hosts *Eucalyptus* identified *i.e. E.camaldulensis, E.tereticornis, E.rudis, E.gomphocephala* (tuart) and *E.blakelyi,* which exhibiting same genetic fingerprint when identified by RAPD and PCR. *Cr.neoformans* var. *gattii* was isolated for the first time from north India by Padhye *et al,* in 1993. In the year 1999, Franzot and colleagues have described **third variety** of this fungus, which is known as var. *grubii*. In 2002, Kwon-Chung *et al,* accepted var. *gatti* as variety, which was earlier proposed by themselves. In the year 2006, *Cr.neoformans* var. *gattii* was given full-fledged species level as ***Cryptococcus gattii.***

When the AIDS pandemic came to surface and fluconazole arrived in the market, the first International Conference on *Cryptococcus* and Cryptococcosis (ICCC) was held in Jerusalem in 1989. The object was to review almost all aspects of emerging disease during those years thereby to take necessary corrective efforts to tackle this opportunistic AIDS-defining illness. Subsequently this conference was held every three years, in Milan, Paris, London, Adelaide, Boston, Nagasaki, Charleston and Amsterdam. The last one *i.e.* 10th was held in Foz do Iguacu (Brazil) in March 2017 and 11th will be held at Kampala (Uganda) in 2020.

Mycology

Cryptococcus neoformans is a haploid yeast, which is distributed worldwide in association with pigeon droppings and is the most common cause of fungal meningoencephalitis in immunosuppressed hosts. The literal meaning of nomenclature is derived from Greek words ***'krypto'*** meaning ***'hidden'***, ***'kokkos'*** as ***'berry'***, ***'neos'*** for ***'new'*** and Latin word ***'forma'*** used for ***'form'*** making it as *Cryptococcus neoformans.* In contrast, relatively less common sibling species *Cryptococcus gattii* is restricted to tropical and subtropical climates, inhabiting trees and their hollows, apparently infecting immunocompetent individuals.

As different species of genus *Cryptococcus* are prevalent in environment but majority of them are not clinically significant. There is total of 37 species of this genus currently known and only **two** are medically significant. *Cryptococcus neoformans* is representative species which is pathogenic to man and animals. It is divided into three varieties and five serotypes.

There are 37 species of *Cryptococcus*, classically divided into two categories - *Cr.neoformans* and non-*neoformans* species, with former being pathogenic in humans and latter are generally considered to be non-pathogenic except a few case reports. As mentioned earlier, *gattii* variant has now been elevated to species level as *Cryptococcus gattii* since 2006. In this chapter, at places it is mentioned as variety just to make its description easier.

These species are the following:

1. *Cryptococcus neoformans*
2. Non-*neoformans Cryptococcus* species
   (a) Cryptococcus albidus
   (b) Cryptococcus laurentii

The clinical manifestations are mainly caused by different varieties of *Cr.neoformans.* Non-*neoformans Cryptococcus* species like *Cr.albidus* and *Cr.laurentii* may rarely be pathogenic to human beings. Recently, these species have also been isolated from HIV-positive patients. There are broadly two groups of *Cryptococcus i.e. Cryptococcus neoformans* complex and *Cryptococcus gattii* complex. However, due to the availability of MALDI-TOF and MLST, now seven species concept has also been put forward by Hagen *et al*, in 2015. This Chapter is mainly based on existing conventional classification.

1. *Cryptococcus neoformans*

The mycological features of *Cr.neoformans* are described under the following headings:

(a) Serotypes: The serotype classification of genus *Cryptococcus* is based on antigenic differences detected with adsorbed rabbit sera. *Cr.neoformans* exhibits antigenic variability of mucopolysaccharide capsule that permits distinction into five serotypes: **A**, **B**, **C**, **D** and **AD**. The five-serotype classification is based on immunologic reactivity of cryptococcal capsule with immune sera. These are basically distinguished according to their capsule, specifically differences in proportions of xylose and glucuronic residues, in degree of mannose substitution and in percentage of O-acetyl attachments.

Of five serotypes, **AD** type is probably least studied. The existence of serotype **AD** is reported in between serotype **A** and **D** but it is still controversial. Serotype **AD**, however was not recognized by other authors, who classified reference strains previously serotype **AD** as either **A** or **D** using immunofluorescence assay with monoclonal antibodies. Now, more sensitive molecular typing methods are employed as alternative tools to serotyping methods for epidemiological studies. The PCR-fingerprinting using $(GACA)_4$ as single primer proved *Cr.neoformans* serotypes **A**, **D** and **AD** and was able to distinguish among serotype **AD** strains. The chemical analysis of capsular polysaccharide of serotype **AD** strains shows that it is very similar to that of serotype **A** and **D** but differs primarily in ratio of substituted to un-substituted mannose residues. Restriction fragment length polymorphism, multilocus enzyme electrophoresis and random amplified polymorphic DNA analysis reveal that **AD** isolates comprise discrete group that can be distinguished from both serotypes **A** and **D**. However, there is no significant difference in causing human illness by **A** or **D**.

Some of the strains of this organism, which do not show agglutination with antisera to any of the above serotypes, are labeled as **untypeable** serotypes.

(b) Varieties: *Cr.neoformans* is further divided into three varieties as per biochemical, morphological and genetic characteristics with corresponding four-serotype classification based on antigenic differences resulting from structural variation of major capsular polysaccharides - glucuronoxylomannan. More than 90% of yeast cell envelope is carbohydrate and glucuronoxylomannan is main capsular polysaccharide produced *in vivo* and *in vitro*. The difference in glucuronoxylomannan structure among strains is responsible for antigenic differentiation of *Cr.neoformans* strains, resulting in separation of this fungus into five serotypes known as A, B, C, D and AD, which are given below:

i. *Cr.neoformans* var. *grubii* (Serotype **A**)
ii. *Cr.neoformans* var. *gattii* (Serotypes **B** and **C**)
iii. *Cr.neoformans* var. *neoformans* (Serotype **D**)

Each variety has its serotypes: as variety *grubii*—serotype **A**; variety *gattii*—serotypes are **B** and **C** and variety *neoformans*—serotype **D**. These varieties of *Cr.neoformans* were originally distinguished into four serotypes: var. *neoformans* serotypes **A** and **D** and var. *gattii* serotypes **B** and **C**. Some of isolates reacted to both **A** and **D** antisera which were designated as fifth serotype **AD** which was subsequently identified using panel of reciprocally adsorbed antisera. The structure of capsular polysaccharide responsible for antigenic reaction was investigated and found chemically different in these five serotypes **(Table 21.1)**.

Table 21.1. Classification of Pathogenic *Cryptococcus* spp.

Serotypes	*Species and Varieties*	*Molecular Types*
A	*Cr.n.* var. *grubii*	VN I, VN II
B	*Cr.gattii*	VG I, VG II, VG III, VG IV
C	*Cr.gattii*	VG I, VG II, VG III, VG IV
D	*Cr.n.* var. *neoformans*	VN IV
AD	*Cr.neoformans*	VN III

The concept of two varieties was widely accepted and emergence of AIDS has prompted recognition of differences among varieties, especially in their epidemiology and ecology. Since agent of cryptococcosis in AIDS patients invariably belonged to var. *neoformans* even in geographical areas where var. *gattii* is prevalent, it is important to study possible differences in pathogenesis of these two varieties.

In 1999, new variety was proposed on the basis of genetic differences as *Cr. n.* var. *grubii* comprising serotype **A**. Consequent to that *Cr.neoformans* strains are subdivided into varieties *Cr. n.* var. *grubii* (serotype **A**) and *Cr. n.* var. *neoformans* (serotype **D**). The specific epithet *grubii* was designated in honor of David Gruby, Hungarian physician working in Paris, who was real founder of dermatomycology to recognize and prove in mid-19th century (1841-44) that dermatophytic infections were caused by fungi.

(c) Mating Types: *Cr.neoformans* is anamorphic state of fungus and its perfect state is *Filobasidiella neoformans* (Serotypes **A** and **D**) and *Filobasidiella bacillispora* (Serotypes **B** and **C**) that taxonomically belongs to order **Tremellales** in phylum **Basidiomycota**. This has been observed in laboratory and not definitively in nature that yeast exists in **two mating types**, **α** and ***a*** and produces basidiomycetous teleomorphs when two types are crossed which is determined by single locus, two-allele system. The **α**-type is 30 to 40-fold more frequent than ***a*** type in environmental and clinical isolates and show greater virulence in mice. The teleomorph is described in genus *Filobasidiella* and discovery of heterothallism in *Cryptococcus* allowed classic genetic studies to be performed. The teleomorph is sexual state of fungus and is characterized by production of basidiospores (sexual propagule). Asexual reproduction in *Cryptococcus* is anamorph state that is characterized by production of budding yeast cells (asexual propagule). For *Cr.gattii* it is postulated that principal infectious propagule is basidiospore and exposure to host tree, *Eucalyptus camaldulensis*, is required to initiate infection in man and animals.

This pathogenic yeast has a defined sexual cycle, which allows classical genetic analysis. Molecular biology approaches, including transformation and gene disruption by homologous recombination and animal models for studies of virulence are both well-developed.

(d) Newer Classification: The analysis of cryptococcal strain variability, particularly at the molecular level, has resolved distinct lineages with the consequence of a taxonomic revision that divides *Cr.neoformans* and *Cr.gattii* into seven *Cryptococcus* species. Similarly, analysis of host factors in so called "immune-competent" individuals revealed previously unrecognized risk factors. Research on these species has established them as important model organisms to understand gene evolution and function in other fungi and eukaryotes. The stage is set for the refinement of research directions, leading ultimately to better treatment of this monophyletic clade of pathogens in the genus *Cryptococcus*.

Table 21.2. Proposed New Classification and Taxonomic Changes for *Cryptococcus neoformans* and *Cr.gattii* Complexes.

Sr. Nos.	*Current Species Names*	*Genotype by RFLP*	*Proposed Species Names*
1.	*Cr.neoformans* var. *grubii*	VN I VN II VN III	*Cr.neoformans*
2.	*Cr.neoformans* var. *neoformans*	VN IV	*Cr.deneoformans*
3.	*Cr.neoformans* inter-variety hybrid	VN III	*Cr.neoformans* *Cr.deneoformans* hybrid
4.	*Cryptococcus gattii (Cr.gattii)*	VG I VG III VG II VG IV VG IV/VG IIIc	VN IV
5.	*Cr.neoformans* var. *neoformans* *Cr.gattii* AFLP/VG I hybrid	-	*Cr.deneoformans* *Cr.gattii* hybrid
6.	*Cr.neoformans* var. *grubii* *Cr.gattii* AFLP/VG I hybrid	-	*Cr.neoformans* *Cr.gattii* hybrid
7.	*Cr.neoformans* var. *grubii* *Cr.gattii* AFLP6/VG II hybrid	-	*Cr.deneoformans* *Cr.deuterogattii* hybrid

As the genus *Cryptococcus* has been proposed to be reclassified into seven groups based on latest taxonomical changes, the description is given in tabulated form in **Table 21.2**. The basis of this classification is newer diagnostic modalities like MLST and MALDI-TOF MS rather than the conventional technical procedures. Due to this reason, some scientists have not agreed to the proposed classification with the argument that most of the mycology laboratories do not have this much of advanced molecular infrastructure hence already existing conventional classification should be continued as such. However, the other scientists are in favor of newly proposed classification and polemic is still going on.

2. Non-*neoformans Cryptococcus* species

The non-*neoformans Cryptococcus* species have been generally regarded as saprotrophic and rarely reported as human pathogens. However, the incidence of infection due to these organisms has increased over the past half-century, with *Cr.laurentii* and *Cr.albidus*, together, responsible for 80% of reported cases. Conditions associated with impaired cell-mediated immunity are important risks for non-*neoformans Cryptococcus* infections and prior azole prophylaxis has been associated with antifungal resistance. As the clinical manifestations of non-*neoformans Cryptococcus* infections are often indistinguishable from those caused by *Cr.neoformans* hence high index of suspicion remains essential to facilitate early diagnosis thereby prompt treatment for such infections.

Cryptococcus albidus is aerobic yeast, non-*neoformans* species of genus *Cryptococcus* and rare cause of human disease. The *Cryptococcus* species are classified as pathogenic or non-pathogenic based on their virulence. *Cr.neoformans* has been classically regarded as the only pathogenic cryptococcal species, while *Cr.terreus, Cr.laurentii, Cr.gastricus, Cr.luteolus, Cr.uniguttulatus* and *Cr.albidus* have been regarded as non-pathogenic species. Although *Cr.albidus* shares some capsular antigens with *Cr.neoformans,* as shown by cross-reactions by fluorescent antibody staining technique, latex agglutination test used for detection of cryptococcal capsular polysaccharide antigen appears to be specific for *Cr.neoformans.* Cross-reaction with other cryptococcal species does not seem to occur.

In the environment there are many non-*neoformans Cryptococcus* species prevalent but *Cr.laurentii* or *Cr.albidus* are also found associated with human infections. The number of non-*neoformans Cryptococcus* cases has increased as opportunistic fungal infections.

The spectrum of clinical infections due to these species ranges from skin lesions to fungemia. Most cases of non-*neoformans Cryptococcus* fungemia were nosocomially acquired and found to be associated with indwelling intravenous catheters and neutropenia. Potential sources for bloodstream infection include dissemination from pulmonary source or transmission along intravenous catheters.

Both these species are described as cause of pulmonary infection and thus acquisition of organism by inhalation may be possible. The natural habitat of *Cr.laurentii* and *Cr.albidus* is still unknown. The limited data about *in vitro* susceptibility testing of non-*neoformans* cryptococci show these species to be more resistant to fluconazole and flucytosine than most of *Cr.neoformans* species. Hence taxonomic research and taxonomy-based identification of non-*neoformans Cryptococcus* isolates is essential from management point of view.

(a) Cr.albidus: This species is commonly isolated from skin of healthy individuals as well as indoor or outdoor air as rare cause of disease. Only dozen cases have been reported till date. This yeast was isolated from CSF, lung abscess, pleural fluid and blood specimens from immunocompromised but non-HIV patients. There is no explanation that how human beings develop infection from *Cr.albidus*. It may be possible that patients are infected during implantation of chamber for perfusion and acquire classical nosocomial infection.

(b) Cr.adeliensis: This species has been described as a separate species within *Cr.albidus* clade and can be misidentified as *Cr.albidus* and it is phenotypically variable. It is cream-colored and smooth to dry on SDA, shows touch of brown-color-effect on *Guizotia abyssinica*-agar after two weeks of incubation and *in vitro* intolerance at 35°C. This xylanase producing basidiomycetous yeast was first isolated from Antarctica. In contrast to type strain from Antarctic, European isolates do not tolerate 10% NaCl and 0.01% cycloheximide. *Cr.adeliensis* has been isolated from clinical samples of respiratory tract as well as in bird droppings from Berlin and Hanover. A case of meningitis caused by *Cr.adeliensis* in patient with acute myeloid leukemia, has recently been reported.

(c) Cr.laurentii: This species is now responsible for both deep seated infections such as fungemia and meningitis and superficial infections such as keratitis. Generally *Cr. laurentii* infection clinically presents with febrile illness, hypotension, bone marrow recipients, cardiac valvular involvement and choroidal infection and endophthalmitis. Recently it has been isolated from the immunocompromised diarrhea patients. There is an emergence of drug-resistant strains of *Cr.laurentii* hence awareness about this infection should be emphasized especially among immunocompromised patients.

The other non-*neoformans Cryptococcus* species like *Cr.diffluens* and *Cr.liquefaciens* frequently colonize skin of patients with atopic dermatitis. The presence of specific IgE antibodies against *Cr.diffluens* and *Cr.liquefaciens* has been investigated in sera of atopic dermatitis patients by using an enzyme immunoassay. The results of various studies suggest that fungal components from *Cr.diffluens* and *Cr.liquefaciens* may act as allergens and play significant role in pathogenesis of atopic dermatitis.

Epidemiology

Cryptococcosis is a systemic mycosis caused by species of the *Cryptococcus neoformans* complex and *Cryptococcus gattii* complex, affecting human and animal hosts. There are 37 species of genus *Cryptococcus,* which are prevalent ubiquitously in the environment but only *Cr.grubii, Cr.gattii* and *Cr.neoformans* are clinically significant as they are generally considered to be potentially pathogenic species. These are encapsulated yeasts that cause meningitis, pneumonia and disseminated infections, mostly in immunocompromised hosts. Approximately, 5-13% of patients with AIDS suffer from cryptococcal meningoencephalitis.

Sporadic cases of disease are seen throughout the world but pandemic of AIDS is one of the significant triggering factors for increased incidence of two most common clinical entities among these patients *i.e.* esophageal candidiasis and cryptococcosis, besides other opportunistic mycoses. The infections with non-*neoformans* cryptococci like *Cr.laurentii* and *Cr.albidus* are also reported but very rarely.

A rapid upsurge in cryptococcosis worldwide has followed pandemic of acquired immunodeficiency syndrome. AIDS accounted for doubling incidence of cryptococcosis in 1980s in most of the countries. Substantial changes in epidemiology of disease have occurred with evolution of the pandemic. As the syndrome developed, there was marked increase initially but subsequent fall in frequency of AIDS-associated cryptococcosis as a result of widespread use of prophylactic antifungal agent like fluconazole for other fungal infections such as oral candidiasis in HIV-positive patients. Cryptococcosis is the second most common fungal infection after candidiasis in HIV-infected individuals and potentially the most serious one.

Cryptococcus being ubiquitous fungus, is isolated from many saprotrophic sources in nature such as house dust and soil. In general, number of patients with underlying immunocompromised conditions have increased and correspondingly incidence of cryptococcosis, is also gradually on the rise. Previously very few cases used to be reported from human beings but incidence has increased alarmingly throughout world during last three decades and it is now commonly encountered pathogenic fungal agent. This is also significant particularly when broad-spectrum antibiotics and corticosteroids are being used indiscriminately in unexplained maladies. There is frequent association of this fungus with increased use of corticosteroids, immunosuppressive therapy, renal transplantation thereby prolonged survival of patients with an underlying malignancy. Moreover, greater awareness about fungal infections among clinicians and laboratory personnel is also one of the important contributory factors for an early recognition of this disease.

The relationship between organism and pigeons (*Columba livia*) has been studied extensively. This fungus is isolated from pigeon nests, droppings and old buildings. This association is explained by the fact that they contain nitrogenous compounds particularly creatinine that favor growth of this yeast. The pigeons do not suffer from disease probably because of their high body temperature than humans, ranging from 41.5 to 43°C. In addition to pigeons, *Cryptococcus* species have been isolated from excrements of a variety of avian species like chickens, parrots, sparrows, starlings, turtledoves, canaries and skylarks. In this disease, transmission from animal-to-man or man-to-man, has not been reported.

On the basis of biological differences, serotype **A** is associated with var. *grubii*, serotype **D** with var. *neoformans* whereas serotypes **B** and **C** with *Cr.gattii.* The varieties have teleomorphs, *Filobasidiella neoformans* and *Filobasidiella bacillispora,* which can be distinguished genetically, serologically, biochemically and epidemiologically. *Cr.neoformans* var. *grubii* is prevalent throughout world and its major environmental source is avian droppings but more recently it is found in association with rotting wood debris of *Cassia grandis* (Pink shower tree) and other trees.

The natural habitat of *Cr.gattii* (*Filobasidiella bacillispora*) was unknown which is found to be prevalent mainly in tropical and subtropical areas. It was isolated in 1990 from its natural habitat in association with *Eucalyptus camaldulensis* (river red gum) and *E.tereticornis* (forest red gum), which are extensively exported overseas from Australia, followed by many related species of *Eucalyptus* trees. There is seasonal association between *Cr.gattii* and flowering of these trees because cultures of air samples taken from flowering trees yield, whereas samples taken when trees are not in flowering season do not yield cryptococci. *Cr.gattii* is also isolated from two other members of red gum group, *E.rudis* (flooded gum) and *E.blakelyi* (Blakely's red gum).

In India, *Cr.gattii* was also reported from northern India for the first time in 1993 followed by reports from other states also. It was reported from Chandigarh, Karnataka and Tamil Nadu. In some case reports it is found in immunocompromised patients as well, contrary to belief that it is always found in immunocompetent individuals. In rest of the reports, only *Cr. n.* var. *neoformans* has been found. Earlier serotyping was not usually done as most of the isolates reported from different centers in India have only studied upto either genus or species level. Now, the isolates are being sent to different referral centers for serotyping and it is observed that *Cr.gattii* is also frequently found prevalent in India.

There are differences in geographical distribution of serotypes from both clinical and environmental sources. Serotype **A** predominates among most areas of world with exception of certain northern European countries. Serotype **D** is common in countries of Europe like Denmark, Switzerland and Italy. It is also reported from France, Germany and United States. A study in Italy showed that incidence of serotype **D** was 71% and skin lesion were observed only in serotype **D** infections. In France prevalence of serotype **D** has markedly increased in recent times. It is also reported from New York City. The geographic differences in prevalence of serotype A and D of *Cr.neoformans* may reflect climatic tolerance and/or virulence.

Serotype **B** is found in tropical and sub-tropical areas whereas serotype **C** is found predominantly in North America, particularly southern California, Southeast Asia and Australia and bat guano in Brazil. Serotype **C** is relatively rare than other serotypes and its natural habitat is not yet known. Majority of clinical isolates of serotype **C** were found in southern California and were reported in 1998 in tropical almond tress (*Terminalia cattappa*) in Columbia. Because of institution of ART, rapid increase in AIDS spread is curtailed thereby cryptococcosis also. Despite a worldwide pandemic of this disease, UK has less number of patients, may be less than one hundred during entire span of a year.

Cryptococcus neoformans var. *neoformans* included isolates that were determined by immunological reactivity of their capsular polysaccharides to be serotype **A** and those that were determined to be serotype **D**. However, analyses of ***URA5*** sequences and DNA fingerprinting patterns suggest significant genetic differences between two serotypes. Therefore, it was proposed to recognize these

Table 21.3. Prevalence of Serotypes of *Cryptococcus* spp.

Serotypes	*Geographical Distribution*
A	Worldwide including North USA
B	Australia, Southeast Asia, Central Africa, Brazil, Venezuela, California, Italy, Canada
C	Southern California, Colombia
D	Europe: Denmark, Italy, Switzerland, France
AD	Worldwide

genotypic distinctions, as well as previously reported phenotypic differences; by restricting *Cr. n.* var. *neoformans* to isolates that are serotype **D** and described new variety as *Cr. n.* var. *grubii* for serotype **A** isolates. The geographical distribution of different serotypes of *Cr.neoformans* is shown in **Table 21.3**.

Cr. n. var. *neoformans* **A** and **D** are more likely to affect immunosuppressed urban dwellers and associated with high mortality. On the contrary, *Cr.gattii* affects mainly immunocompetent rural dweller hosts and causes severe illness and neurological damage but low mortality. It was identified in 1970 and natural habitat was established after two decades in 1991. A tree common in South Australia and southern California, *Eucalyptus camaldulensis* (river red gum), is identified as ecological niche of serotype **B**. *Cr.gattii* is found in Australia, Brazil, Cambodia, Hawaii, southern California, Mexico, Paraguay, Thailand, Vietnam, Nepal, Central Africa and India. In Papua New Guinea, in contrast to other countries, cryptococcal meningitis occurs predominantly among immunocompetent, previously healthy adults. *Cr.gattii* was implicated in 95% of cases and ocular complications were very common. Rarely, cases of infection with *Cr.gattii* in patients with AIDS have been reported. Infection due to *Cr.gattii* is remarkably similar to infection caused by *Blastomyces dermatitidis*. Both these infections occur rarely in patients with AIDS and tend to be associated with focal nodular lesions and appear to be controlled by immune system responses that are very different from classical cellular immune response among the immunocompromised hosts.

Cryptococcosis is more prevalent among male patients as compared to females. The disseminated form of disease is frequently found in immunocompromised patients. In India, pattern of cryptococcal infection mimics Americans' where nearly 50% of cases are among immunocompromised patients and rest are in normal, immunocompetent individuals, contrary to that in United Kingdom, where most of infections are found in the immunocompromised patients.

There is an unprecedented outbreak of cryptococcosis caused by *Cr.gattii* on Vancouver Island of British Columbia in Canada, which has affected both human and animal hosts with normal immunity. More than 160 human cases, including at least 6 fatalities, have been reported. This is the largest island on Pacific coast of North America, located about 50 km west of BC mainland. The cases have been exposed largely on the eastern coast of Vancouver Island. However, vast majority of rest of BC remains unaffected by this outbreak. Moreover, Vancouver Island is a major tourist destination, with about 7.5 million visits each year. There is also intercontinental transmission of *Cr.gattii* from this outbreak in a tourist from Denmark who visited this island. The case indicated a potential risk for tourism-related acquisition of this environmental fungus. The endemic acquisition of *Cr.gattii* has been occurring in province of BC since 1999. However, there is no recent increase or clustering of cases. All over the world cryptococcosis is coming down but in sub-Saharan Africa and India, it is still quite high.

This is also to note that every year human fungal pathogens cause more than two million infections leading to high morbidity as well as mortality. Out of this two fungal genera *i.e. Candida* and *Cryptococcus* are together accounting for about 1.4 million infections with higher mortality rates. The patients with dysfunctional immune systems such as with HIV/AIDS, are mainly vulnerable to such infections.

Immunity

Even after more than a century since discovery of genus *Cryptococcus*, its exact mechanism of virulence still remains unresolved. The immune response is primarily cell-mediated but humoral immunity also has role in determining response to the infective process. Although anti-cryptococcal antibodies can be detected in substantial proportion of patients with active or recent cryptococcosis, measurement of these antibodies has not proven to be useful in diagnosis because their substantial level has also been observed in proportion of healthy people who presumably suffered previously from self-limiting subclinical cryptococcal infection or mounted humoral response to some other cross-reacting polysaccharide antigen. The exact percentage of people with such antibody levels depends on their occupation, residential environment and other demographic factors. The macrophages are central to immune response to *Cryptococcus*, through antigen presentation and co-stimulation of T-cells, secretion of chemokines as well as cytokines and once activated as final effector cells. There is significant difference between inflammatory responses among AIDS and non-AIDS patients as later have granulomatous lesions supporting role of CMI in cryptococcal meningoencephalitis whereas HIV patients do not have granulomatous reaction.

Virulence Factors of *Cryptococcus* Species

The virulence factors of *Cryptococcus* that have been studied independently are:

- Sialic acids
- Capsular polysaccharide
- Melanin production
- Urease production
- Phospholipase
- Mating types
- Mannitol production
- Superoxide dismutase secretions

Out of the above-mentioned virulence factors, some of them are briefly described as follows:

(i) Antigens: The culture filtrates of *Cryptococcus* yield at least three well-characterized and distinct antigens, one major and two minor antigens; primarily of glucuronoxylomannan (GXM), predominant polysaccharide of *Cryptococcus* capsule, to lesser extent readily soluble galactoxylomannan (GalXM) complex and a mannoprotein (MP). The cryptococcal polysaccharide is composed of 80% GXM, its major antigenic component and 20% GalXM. The mannoprotein is an important protein of cell wall of *Cryptococcus*. GXM, GalXM and MP are serologically distinct and at least two of these components, GXM and MP have different effects on immune system. The antigen responsible for CMI appears to be a mannoprotein that is probably associated with cell wall. The difference in cryptococcal polysaccharide structure among strains result in antigenic differentiation that are used to classify cryptococcal strains into five serotypes known as **A**, **B**, **C**, **D** and **AD**.

(ii) Capsule: *Cryptococcus* is unique among pathogenic fungi due to production of a mucinous capsule. This is polysaccharide in nature and is composed of xylose, mannose and glucuronic acid. It is an important virulence determinant contributing about 26% of virulence of *Cryptococcus* and is considered to be one of classic virulence factors for any of the encapsulated organism. The capsule is essentially immunologically inert which accounts for paucity of inflammatory cellular response.

The binding of isolated capsular polysaccharides to a capsule-free mutant of *Cryptococcus* confers resistance to phagocytosis. The major capsular polysaccharide, GXM of *Cryptococcus* displays structural heterogeneity that can be demonstrated in single serotype. The relationship of this heterogeneity and presence or absence of individual structural chemical groups to virulence is not defined. The capsular material tends to be produced and shed from outer surface of capsule and this antigenic material can be detected in all body fluids especially serum and CSF. In AIDS patients, capsule may not be fully appreciated and likely to be missed on India ink preparation. Circulating cryptococcal polysaccharide can also be considered as virulence factor and earlier it goes into body fluids more it perpetuates progression of disease. Soluble intravascular capsular polysaccharide prevents leukocytes from migrating into sites of acute or chronic inflammation, causing neutrophils and lymphocytes to shed their surface L-selectin and induces immune responses that downregulate protective immunity.

Cryptococcus is unusual among fungal pathogens because it has polysaccharide capsule. The capsule is anti-phagocytic and functions as virulence factor in a manner similar to that of classical encapsulated bacterial pathogens (*Streptococcus pneumoniae* or *Klebsiella pneumoniae*). Phagocytosis is an important host defense mechanism because internalization in phagocytic cells results in enhanced killing of fungal cells. In *Cryptococcus*, immune effectors cells are unable to phagocytose encapsulated yeasts in the absence of opsonins. Unfortunately, *Cryptococcus* infection may be accompanied by lack of opsonins since polysaccharide capsule is poorly antigenic and often fails to elicit an antibody response and acquired complement deficiency may result in depletion of this component. Based on these findings, an anti-cryptococcal conjugate vaccine consisting of GXM and tetanus toxoid has been tried in animal models.

(iii) Melanin: Another significant virulence trait of *Cryptococcus*, is melanin, contributing about 14% of virulence. It possesses a unique enzyme, phenoloxidase, which catalyzes conversion of L-dihydroxyphenylalanine (DOPA) to melanin. Melanin serves as an antioxidant that protects organism from intracellular killing by phagocytic cells. In addition, melanin prevents macrophages from producing TNF-α that is needed for generation of protective *Cryptococcus* immune response. It produces copper-dependent enzyme called **laccase**, which catalyzes formation of melanin-like pigment where fungus is grown on substrates containing polyphenolic or polyaminobenzene compounds. The melanin synthetic pathway can use catecholamines that may provide an explanation for neurotropism in *Cryptococcus*. The melanin production is involved in pathogenic processes. In *Cryptococcus*, abilities to melanize include protection against oxidants, temperature extremes, iron reduction, ultraviolet light, amphotericin B and microbiocidal peptides and macrophages.

The basidiospore is infectious propagules of *Filobasidiella neoformans* that can cause natural cryptococcosis, particularly in immunocompromised patients. The portal of entry is probably inhalation and inoculum size required is not very big.

The mechanism by which mating types affect virulence has not yet been fully defined. The characteristic of yeast to produce mannitol has been implicated in virulence by its ability to scavenge free radicals, thereby inactivating intracellular killing mechanisms. Moreover, high concentration of mannitol in CNS may contribute to brain edema. The rats infected with *Cryptococcus* mount quick granulomatous response associated with increased inducible nitric oxide synthase expression and reduction in pulmonary burden.

Pathogenesis and Pathology

Cryptococcus is the only encapsulated fungus that infects man. The initial infection is initiated upon inhalation of spores or desiccated yeast. The inflammatory response to inhaled cryptococci produces primary lung lymph node complex that restrains spread of yeasts. In the lungs, fungus proliferates in alveolar space and in immunocompetent subjects the infection is normally contained in this organ. The fungus may remain dormant in lungs until immune system weakens and then can reactivate and disseminate to CNS and other body sites. In immunocompromised subjects, dissemination of the yeast cells from lung to brain can occur intracellularly and extracellularly, eventually leading to development of a life-threatening disease.

Cryptococcus has strong predilection for CNS, however, the reason for this unique feature is unclear but existence of receptor on nerve cells for legend on yeast has been proposed. In addition, it is shown to possess series of potential virulence factors such as capsule production, melanin synthesis, high temperature growth between 37 to 39°C, α-mating type locus, urease and production of extracellular phospholipase promoting the disease process. The non-capsulated cells are readily phagocytosed and killed, whereas encapsulated cells are relatively more resistant.

The host factors are also significant in prognosis of this infection. Host resistance depends primarily on cell-mediated immunity. The CD4+ cells, cytotoxic lymphocytes, natural killer cells, activated macrophages and various cytokines, Interleukin-12, granulocyte-macrophage colony-stimulating factor and interferon-γ, all are implicated in successful host response to *Cryptococcus* species. The loss of an efficient cell-mediated response is common for fungal meningitis. In addition, there are convincing data that humoral immunity has an important role in prevention of *Cryptococcus* infections. Both complement and antibodies are shown to be an integral part of an effective immune response. These observations have led to strategies in which monoclonal antibodies are also used.

Clinical Features

Cryptococcosis is caused by various species of encapsulated basidiomycetous yeast of genus *Cryptococcus*. The clinical manifestations of disease are influenced more strongly by host immune status than by the cryptococcal variety. The disease usually occurs between ages of 30 and 60 years and is uncommon in childhood. The respiratory tract is usual portal of entry for *Cryptococcus* and subsequently it disseminates into different body organs.

The disease is broadly categorized into two types depending on immune status of host *i.e.* infection in immunocompetent individuals and in immunocompromised patients. The severity of infection also depends upon cryptococcal species involved in addition to underlying predisposing factors. The lungs and brain are most commonly affected organs by this organism followed by other body sites.

The incubation period of cryptococcosis is usually two to four weeks. The disease is presumed to be resulting from inhalation of infectious air-borne organisms from environment. The fungal propagules are deposited in respiratory tract where they may be transient inhabitants or may colonize mucosa.

Cryptococcosis is now predominant cause of fatal fungal infection in patients with AIDS and ranks **fourth** among CNS infections. It may occur at any point of time during the course of AIDS but is most common when CD4+ count falls below 200 cells/mm^3. The extra-pulmonary cryptococcosis is one of AIDS-defining diseases. The frequent use of corticosteroids is second most important risk factor for cryptococcal meningitis.

The clinical features of cryptococcosis may be described depending upon anatomical sites involved, under the following headings:

1. Pulmonary Cryptococcosis
2. CNS Cryptococcosis
3. Visceral Cryptococcosis
4. Osseous Cryptococcosis
5. Cutaneous Cryptococcosis

1. Pulmonary Cryptococcosis

The respiratory route is portal of entry for propagules in pulmonary cryptococcosis that subsequently disseminate to extrapulmonary sites and present as cerebrospinal or generalized cryptococcosis. The clinical picture of disease does not differ from any other respiratory infections. Mostly, there is an asymptomatic colonization therefore it may not be feasible to establish diagnosis on the basis of clinical symptoms or sputum examination unless it is suspected in view of existing cryptococcosis of CNS in a particular patient or due to underlying immunocompromised situation.

The patients may present as self-limiting pneumonitis. The onset is indolent and symptoms in patients are dry cough, dull chest pain and mild or no fever. Pleural effusions and hilar adenopathy are uncommon and calcification is very rare. The chest radiographs of these patients show following patterns:

(a) There are discrete, solitary, moderately dense areas of infiltration appearing in upper and sometimes in lower lung fields.
(b) There is broader and more diffuse interstitial infiltrate involving upper or lower lung areas similar to those often observed in *Pneumocystis jirovecii* pneumonia.
(c) Extensive peribronchial infiltrates which resemble active tuberculosis.
(d) Small gelatinous granules present all over lung fields resembling miliary tuberculosis and usually found in association with lymphoma and leukaemia.

2. CNS Cryptococcosis

The infection of brain and meninges leading to meningoencephalitis is the most commonly encountered clinical presentation of cryptococcosis. The organisms probably come across less phagocytic response and it has been thought that selective nutritional factors for yeasts are present. Simple sources of nitrogen, such as asparagine and creatinine are found in CSF. Moreover, inhibitory factors, as found in serum, are not present in CSF. The neurological findings are those typical of basilar meningitis, such as cranial nerve palsies and those associated with increased intracranial pressure entailing hydrocephalus including visual loss.

Cryptococcal meningitis (CM) is the most common cause of meningitis in adults in Africa, accounting for over 500,000 deaths annually. In the USA, it is more common than all causes of bacterial meningitis combined, at an incidence of 1.1 per 100,000 *vs.* 0.728 per 100,000, respectively. As such the mortality due to AIDS-related cryptococcal meningitis is usually >50% in low-middle income countries.

The clinical features depend on pace of onset of disease. It can manifest as diffuse meningoencephalitis or well-circumscribed granuloma of brain or spinal cord. The features may be similar to those of non-specific meningoencephalitis or brain tumor. The patients usually present with manifestations similar to those of intracranial space occupying lesion **(ICSOL)**. There is fairly diffuse, bilateral dull headache. The fever is often low grade or absent until late during the infection. In addition, there is nausea, dizziness, irritability, impaired memory and subsequently ataxia supervenes. As infection progresses, lassitude, fits, blurred vision, photophobia, double vision or blind spot (scotoma) in visual fields may be noted.

Overt meningeal signs such as neck stiffness and photophobia are seen in about one-third of patients. **Kernig's** and **Brudzinski's** signs may be positive in some of the patients. The onset of coma is sudden, sometimes accompanied by respiratory arrest. It is most frequent cause of death among patients suffering from cryptococcosis. In such cases detection of organism and/or antigen from CSF is essential to establish final diagnosis.

The dissemination of disease may result in intracranial mass lesions known as **cryptococcoma**. Such enlarging granulomatous cerebral lesion has radiological appearance similar to a growing neoplasm. Its formation is seen in immunocompetent hosts involving lungs and brain and it is rare in patients with AIDS. The patients who develop cryptococcoma often have worse outcomes as compared to those without cryptococcoma. The CSF of such patients may be negative for antigen and culture may turn out be sterile hence organism have to be grown from brain biopsy specimens. In lungs it has been seen that these are usually associated with *Cr.gattii*.

Cryptococcal meningitis takes more protracted course as compared to tubercular meningitis, the feature helpful in differential diagnosis. Altered mental state is associated with poor prognosis hence diagnosis of cryptococcosis should be considered among patients presenting with progressive dementia. Disseminated cryptococcal meningoencephalitis is still fatal in approximately 30% cases despite antifungal therapy. *Cr.gattii* mainly involves parenchyma of CNS and therefore, in true sense this species may be considered as primary pathogen instead of an opportunistic fungal agent. The lesions spread diffusely and prominently throughout brain so disease would be more properly termed a meningoencephalitis rather than meningitis.

The patients of chronic meningitis, caused by *Cryptococcus* species are invariably confused as tubercular meningitis on clinical grounds. Therefore, they might have taken a therapeutic trial or even full course of anti-tubercular therapy without any improvement prior to establishment of the final diagnosis. The patients of *Cr.gattii* infection may present with intracranial hypertension that may be threat to both vision as well as the life. Therefore, shunting of ventricular fluid is required for mechanical drainage, using intraventricular or lumbar drain or daily lumber punctures in such situation. Formation of pseudocyst among these patients is also reported at terminal end of ventriculoperitoneal (VP) shunt in the abdominal cavity. The infection of CNS determines prognosis and sequelae of this disease in due course of time. In the CNS, there is Swiss-cheese type parenchyma of the brain tissue.

3. Visceral Cryptococcosis

The lungs represent common site of primary infection however, most often meningeal manifestations caused by this neurotropic fungus, are the presenting features. Moreover, any organ or tissue of body is subjected to invasion in the form of granulomatous lesions that sometimes resemble even malignancy. Cryptococcosis is a major cause of neuropathological complications in patients with AIDS. Cryptococcal infection usually spreads from primary focus to invade optic nerve, meninges or choroid. Visual loss in patients without endophthalmitis has been explained to result from optic neuritis or from effects of increased intracranial pressure. The histopathological examination of choroid can reveal granulomatous inflammation with microabscesses and giant cells centered around the capsule of organism.

There are two distinct patterns of visual loss: **rapid** or **slow**. The rapid one is characterized by onset of profound visual loss over a period as short as 12 hours before or early in course of therapy and clinical syndrome, which is strongly suggestive of optic neuritis.

The other pattern is characterized by slow but progressive visual loss that typically begins later during therapy and may be due to effects of increased intracranial pressure. While initial defect may be mild, patients with slow visual loss can progress to severe loss over weeks to months. The factors that appear to predict either pattern of visual loss is the presence of papilledema, elevated CSF opening pressure and encapsulated yeast cells in India ink preparation.

Cryptococcal prostatitis is also reported in patients with chronic lymphocytic leukemia. Primary abdominal lymphonodular cryptococcosis particularly in children has been reported. Primary adrenal insufficiency with bilateral adrenal masses and meningitis has also been reported due to disseminated cryptococcosis.

In the patients with **cryptococcaemia** there are three main predisposing conditions *i.e.* AIDS, immunosuppressive therapy and liver cirrhosis. Such patients have high risk of acute mortality hence require an early diagnosis and prompt antifungal therapy. The sepsis syndrome and meningeal involvement are common. Those with liver cirrhosis have particularly grave prognosis, even after adjusting for the effect of age, timing of antifungal therapy, initial APACHE II score and severity of sepsis.

4. Osseous Cryptococcosis

The bony lesions are found in very few cases of cryptococcosis. These are associated with pain and swelling of local area for a long period. The lesions spread slowly without periosteal proliferation. There are osteolytic lesion and often spread to skin by extension or following surgical exploration. The X-rays of lesions show one or more well-circumscribed areas of osteolysis, without periosteal elevation and rarely with invasion of adjacent joints. In the absence of disseminated cryptococcosis, isolated cryptococcal lesions may be clinically mistaken for any primary neoplastic disease.

5. Cutaneous Cryptococcosis

Although cryptococcosis mainly involves CNS but the most common extra-neural site of infection is skin. About 10-15% of systemic cryptococcal infections become manifested in skin. The isolated cutaneous cryptococcal lesions in otherwise healthy individuals are rare. These types of manifestations are caused by serotype **D**, mainly in old patients who are usually on corticosteroids.

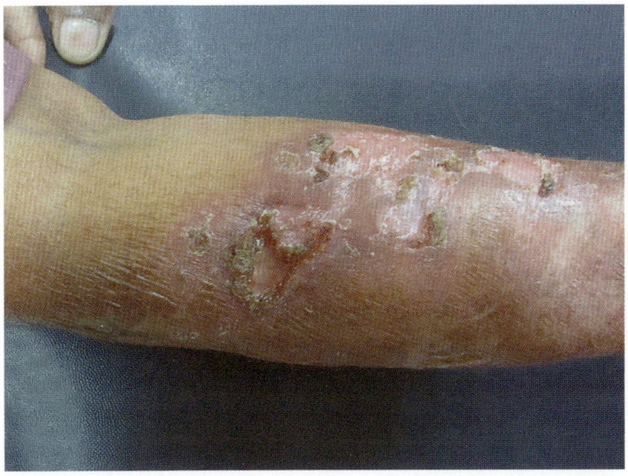

Fig. 21.1. Extensive ulcerative lesions of cryptococcosis over right forearm of a female patient.

The cutaneous cryptococcosis is of two types, primary and secondary. The primary cutaneous cryptococcosis occurs only in the skin without any evidence of systemic involvement. But it may result in systemic disease sooner or later during progressive course of infection. The localized lesion may be caused by traumatic inoculation. The scalp, face and neck are frequently involved. The lesions are variable and consist of papules, subcutaneous nodules, infiltrated plaques, abscesses ulcers, granulomata or cellulitis **(Fig. 21.1)**. The lesions resemble molluscum contagiosum with 3-6 mm in diameter, pink, smooth hemispherical and semi-translucent, with slight central depression **(Figs. 21.2A and B)**. There may be lobulated abscess formation in the subcutaneous tissue. This is not suspected to be cryptococcosis, however, when FNAC is done gelatinous frank gelatinous pus is aspirated, which contains numerous cryptococci **(Figs. 21.3A to C)**.

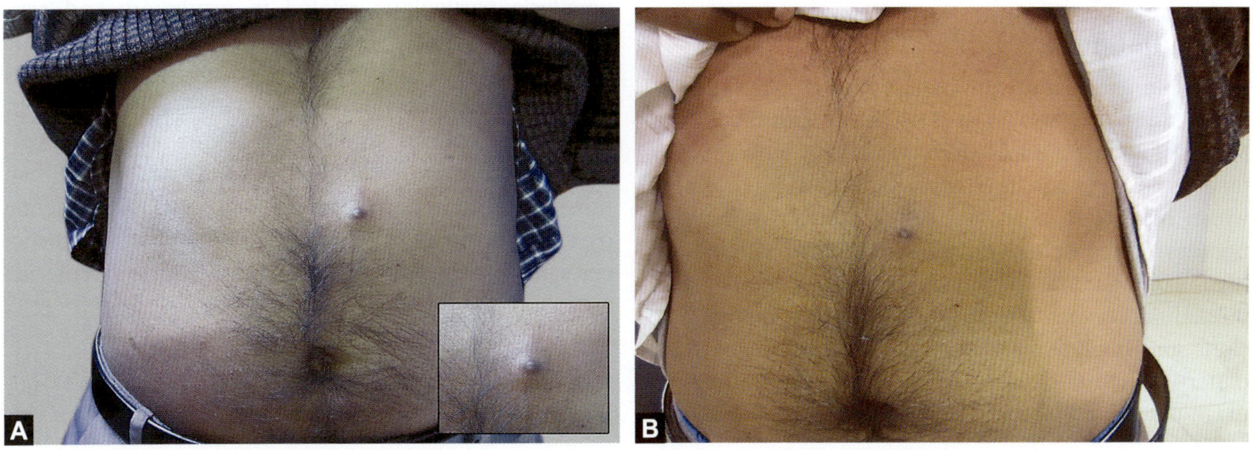

Figs. 21.2A and B. Solitary lesion of primary cutaneous cryptococcosis over the upper part of abdominal wall, its closer view (inset) and after healing of the lesion.

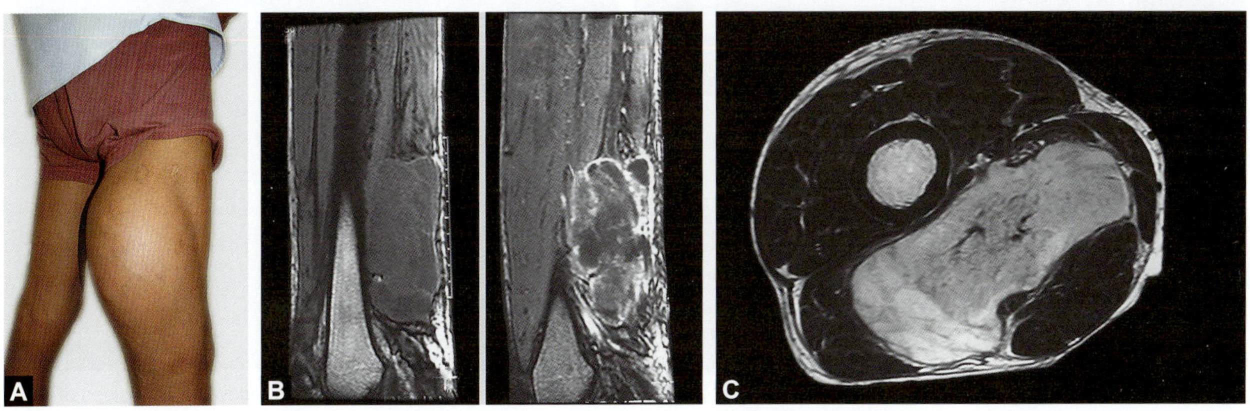

Figs. 21.3A to C. A. Large subcutaneous lobulated cystic swelling due to cryptococcosis over posterior side of right distal thigh; B. MRI done in initial period of disease without contrast and with contrast delineating swelling (C) along with its transverse section.

The secondary cutaneous cryptococcosis occurs following manifestations of disease at other sites. The hematogenous lesions to skin occur in very few cases of cryptococcosis, particularly in immunosuppressed patients. In the beginning there is painless papule and as lesion enlarges to several centimeters in diameter, center often becomes shiny and flat. With further progression, center becomes depressed and ulcerates, draining a thin exudate that contains several cryptococci that simulates Kaposi's sarcoma. The clinical significance of these lesions lies in an early diagnosis of disseminated cryptococcosis.

There are two types of histological reactions to infection with *Cryptococcus* that may occur in skin *i.e.* gelatinous and granulomatous. These lesions should be differentiated from histoplasmosis, pneumocystosis and molluscum contagiosum. The mucocutaneous lesions may also present in ~10-15% of cases consequent to disseminated disease.

Radiodiagnosis

In addition to clinical pulmonary findings, radiological modalities like X-rays chest **(Figs. 21.4A and B)**, CT and MRI are helpful for establishing an early diagnosis of cryptococcosis. There are three main patterns of involvement, which have been described, *i.e.* (i) solitary or multiple nodules, (ii) consolidation, often multifocal, patchy, segmental or lobar and (iii) interstitial micronodular or reticulonodular lesions. There may be cavitation, pleural effusions and hilar lymphadenopathy accompanying these patterns. Therefore, based on these findings, differential diagnosis should be taken into account like tuberculosis, histoplasmosis, aspergillosis, pneumocystosis, cytomegalovirus pneumonia, Kaposi's sarcoma and lung cancer.

In CNS, there are numerous clustered tiny foci that are hyperintense on T2-weighted images and hypodense non-enhancing on post contrast T1-weighted images, located relatively bilateral symmetrical in basal ganglia and in mid-brain. A viable cerebral cryptococcoma can present as a ring-pattern lesion with its central cavity showing hyperintensity on both T2-weighted images and apparent diffusion coefficient maps and hypointensity on diffusion-weighted imaging **(Figs. 21.5A and B)**. These features are different from that of a pyogenic abscess or a tuberculoma but may mimic those of a brain tumor and toxoplasmosis.

Differential Diagnosis

The cryptococcosis is most often mistaken for malignant neoplasia particularly related to respiratory and central nervous systems. This may be confused with chronic tuberculosis or any other bacterial disease like brucellosis or nocardiosis. It may mimic fungal diseases like candidiasis, histoplasmosis and coccidioidomycosis. Meningitis takes more protracted course in cryptococcosis than tuberculosis but may resemble encephalitis, dementia psychosis or dementia paralytica. Sometimes, cryptococcosis may also be found concomitantly with tuberculosis.

Cryptococcosis in Animals

Cryptococcosis has been reported among variety of animals. This is the most common systemic mycosis of cats

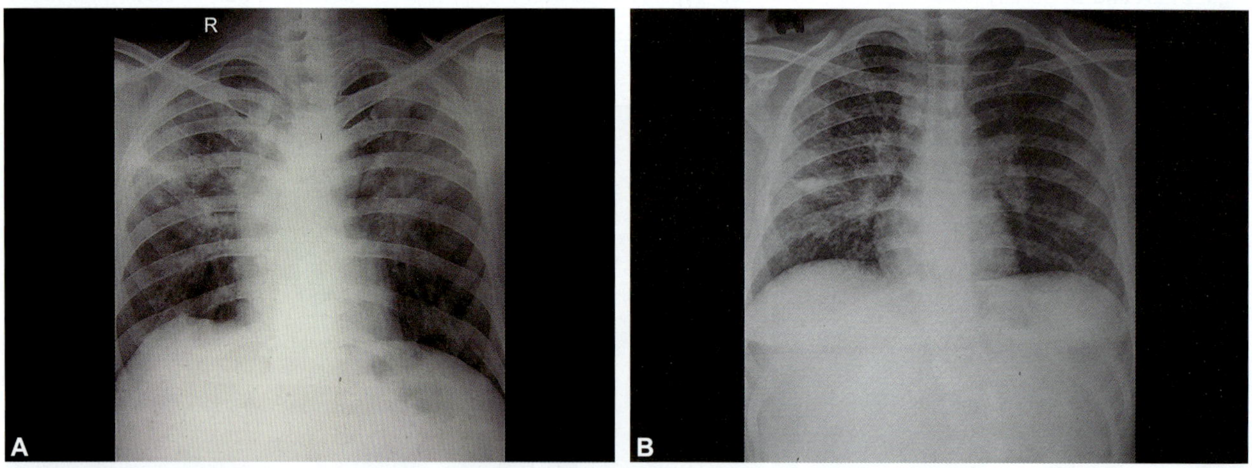

Figs. 21.4A and B. X-rays PA view of chest of two different patients of cryptococcosis showing infiltrating shadows of both sides of lungs predominately over the hilar areas.

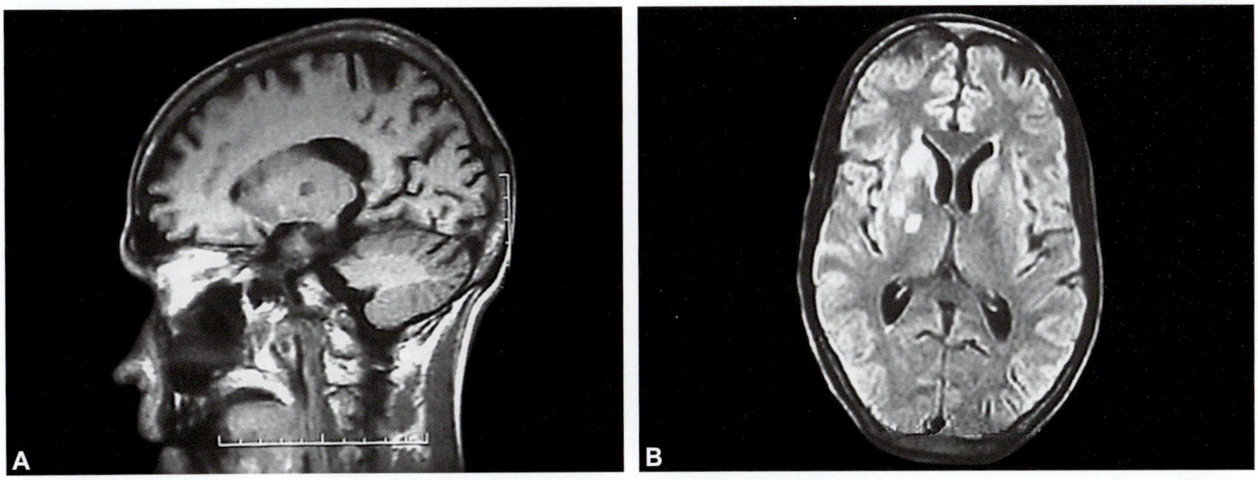

Figs. 21.5A and B. CNS cryptococcosis: T1 W-Images show hypointense lesions in head of caudate nucleus and lentiform nucleus on the right side and these lesions are hyperintense on T2 W and FLAIR sequences.

however, other domestic mammals may also be infected like dogs, horses, goats, sheep, squirrels, cows and other cattle. Moreover, other less frequently encountered species of genus *Cryptococcus*, such as *Cr.albidus* (horses, cats and dogs), *Cr.magnus* (cats), *Cr.laurentii* (dog) and *Cr.flavescens* (dog) are occasionally reported as etiological agents of fungal infections. Feline cryptococcosis occurs worldwide, being most frequently reported in Australia, Canada and USA. The natural route of infection is usually through inhalation. The most common sites of infection are the nasal cavity, skin, lymph nodes, CNS and eyes. It is not considered a contagious or anthropozoonotic disease, representing a fungal infection acquired from the environment, with animals serving as potential sentinel hosts for human exposure. However, some cases of zoonotic transmission of *Cr.neoformans* from birds, raised as pets, have been described.

This is also a devastating disease of koalas and difficult to treat. It is caused predominantly by the *Cr.gattii* species complex because of association between arboreal koalas and *Eucalyptus*. The free-living koalas with cryptococcosis and most captive koala cases have been caused by *Cr.gattii* VGI/AFLP4. Amongst captive koalas, there have also been a small number of cases caused by *Cr.gattii* VGII/AFLP6 in Australia. In the environment it has been recovered from bird excreta, dust, soil and arboreal material.

Laboratory Diagnosis

The laboratory diagnosis of cryptococcosis is established by demonstration of encapsulated budding yeast cells in direct smears of clinical specimens by various stainings, culture, demonstration of cryptococcal antigen in serum and/or CSF and finally by animal pathogenicity testing.

(a) Direct Examination

The clinical specimens are collected as per the anatomical site involved, mainly CSF, serum and other body fluids. Direct microscopy of specimen, India ink or 10% Nigrosin with formalin wet mount, shows round to globular budding yeast cells ranging from 5-20 μm in size with distinct halo. A wide refractile gelatinous capsule surrounds organisms that may be twice as thick as diameter of yeast cell itself as seen in **Figure 21.6**. Sometimes, number of organisms in the specimen is too less and may be missed out by the observer. In such circumstances, it is better to centrifuge the specimen before making the preparation. India ink staining is positive in about 75% of cryptococcosis cases. A modified India ink preparation can also be employed using 2% chromium mercury. This technique allows clear identification of external and internal structures of the organism. The demonstration of pathogen in an India ink preparation is pathognomonic for *Cryptococcus* infection but observer must distinguish yeast cells from artifacts such as erythrocytes, leukocytes and talc granules acquired from gloves. The analysis of CSF characteristically shows numerous organisms and few lymphocytes. Most of the patients have 10-100 lymphocytes/mm^3, reduced CSF glucose and elevated proteins levels.

As a routine procedure in mycology laboratory, KOH wet mount shows budding yeast cells with numerous pus cells **(Figs. 21.7A and B)**. If the capsule is not prominent,

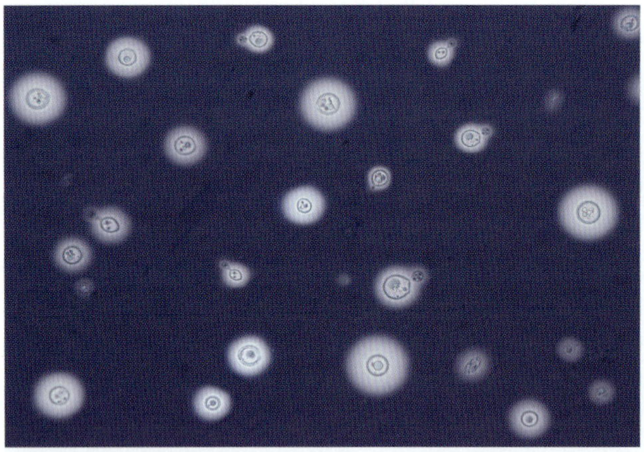

Fig. 21.6. Budding yeast cells with clear halo in black background of negative staining (India Ink × 400).

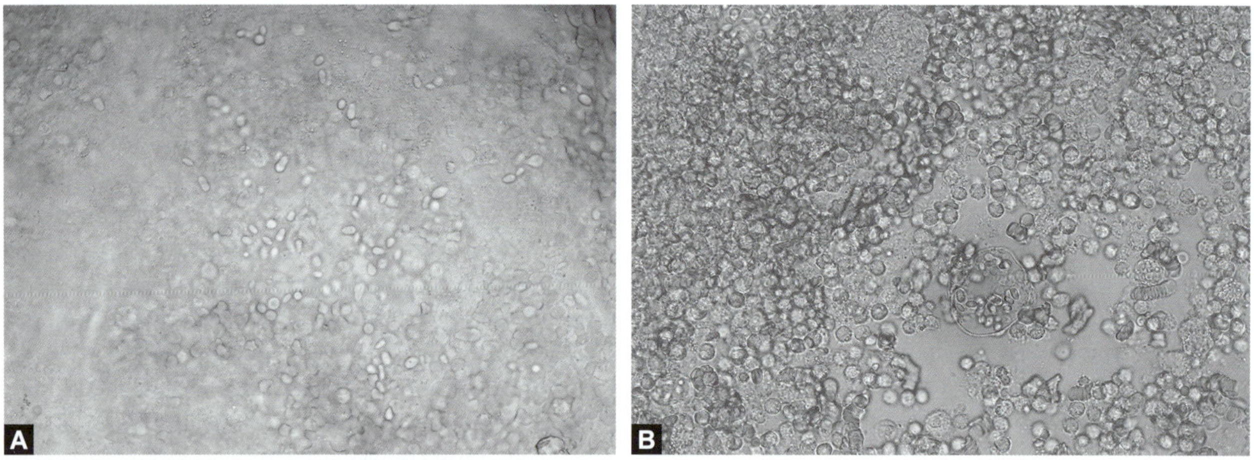

Figs. 21.7A and B. Budding yeast cells seen in pyogenic lesion along with numerous pus cells (KOH × 400).

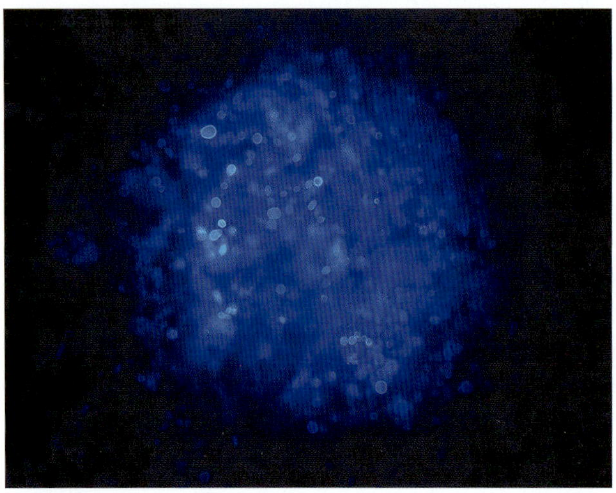

Fig. 21.8. Budding yeast cells of *Cryptococcus* species seen with fluorescent brightener (CFW × 400).

it may be confused with other yeast-like organisms. However, when the number of organisms is scanty, fluorescent brightener may be used to delineate the finding in calcofluor white staining **(Fig. 21.8)**.

The Gram's staining is usually done as routine processing of clinical specimens in every microbiology laboratory and these organisms are easily visible as Gram-positive budding yeast cells **(Fig. 21.9)**. The FNAC smear are stained with Giemsa and examined for budding yeast cells **(Fig. 21.10)**. The organism can also be seen in Papanicolaou preparations of CSF or other clinical material. In some primary isolates capsule may not develop until it is grown in 1% peptone solution or subjected to mouse pathogenicity test. The hyphae are not seen with yeast forms but pseudohyphal forms may be very rarely visualized when yeast cells are under stress by circumstantial conditions such as elevated temperature, etc.

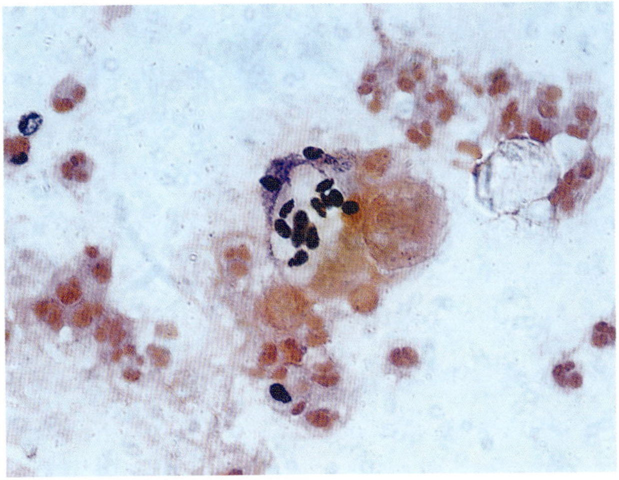

Fig. 21.9. Budding yeast cells seen in pyogenic lesion along with numerous pus cells (Gram's Staining × 400).

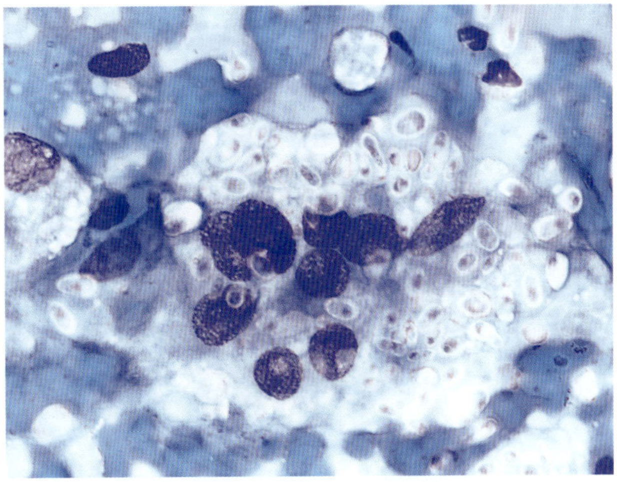

Fig. 21.10. Budding yeast cells of *Cryptococcus neoformans* seen in pyogenic lesion along with numerous pus cells and may be confused with *Histoplasma capsulatum* (MGG × 400).

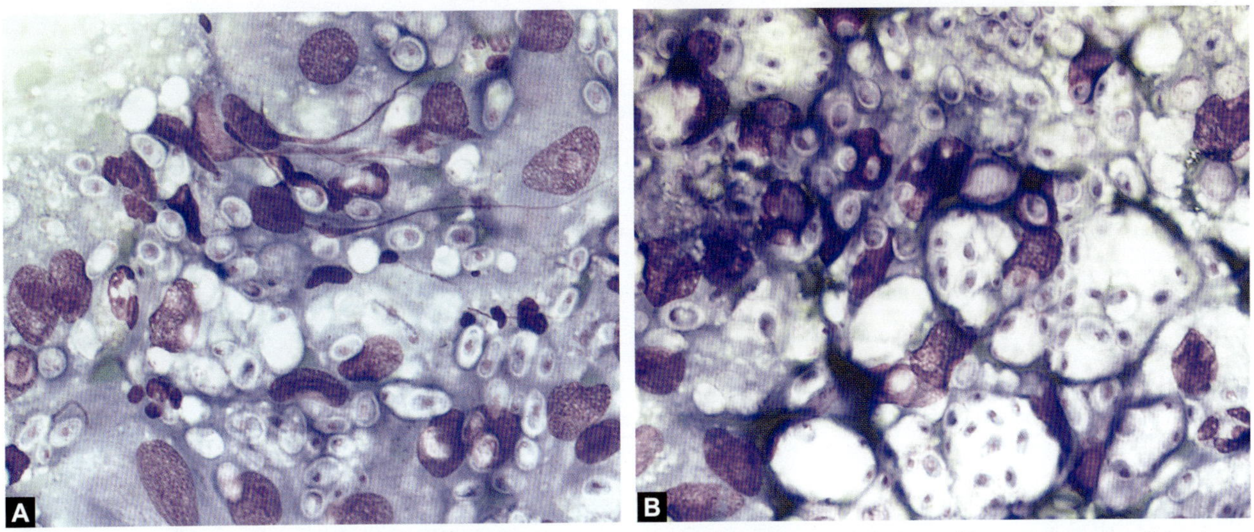

Figs. 21.11A and B. Budding yeast cells of *Cryptococcus neoformans* seen in pyogenic lesion along with numerous pus cells and may be confused with *Histoplasma capsulatum* (H&E × 400).

The smear may be examined with H&E **(Figs. 21.11A and B)**, periodic acid-Schiff **(Figs. 21.12A and B; 21.13A and B)**. Moreover, *Cryptococcus* is also carminophilic like *Rhinosporidium seeberi* and shows rose-colored capsule with Mayer's mucicarmine staining **(Figs. 21.14A and B)**. Alcian-blue stain is also done to demonstrate capsule of the organism **(Fig. 21.15)**. Masson-Fontana staining shows production of melanin by yeasts hence it is highly significant for detecting capsule-deficient cryptococci in tissues which are otherwise missed out. This stain may also be combined with capsular stainings like mucicarmine or Alcian blue.

(b) Fungal Culture

For fungal culture, two sets of Sabouraud dextrose agar with antibiotics are inoculated and incubated at 25°C and 37°C, separately, over a period of four weeks. The culture media should be **without actidione** as it is inhibitory to the growth of *Cryptococcus* species. For this reason diphenyl is

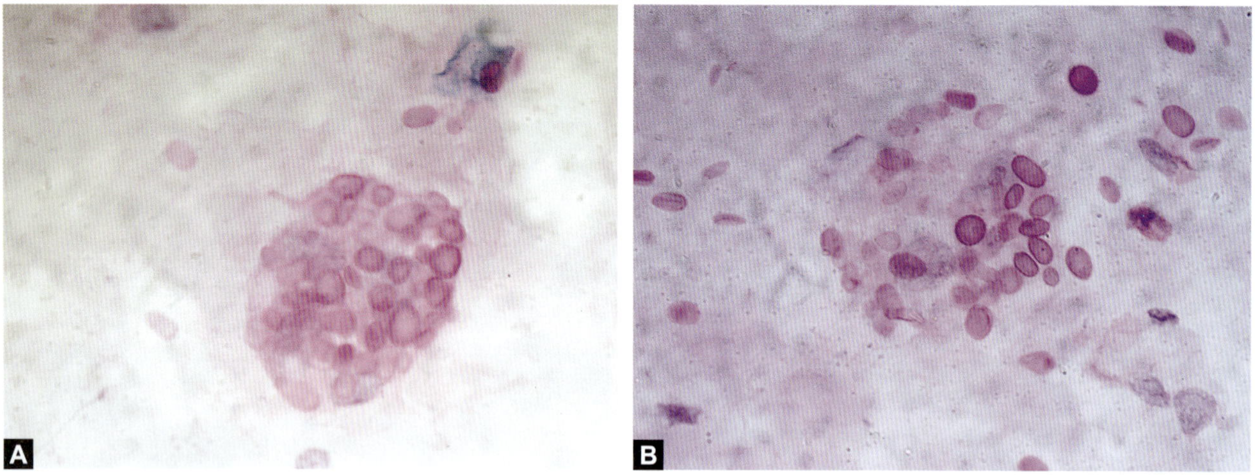

Figs. 21.12A and B. Budding yeast cells of *Cryptococcus* species seen in pyogenic lesion along with numerous pus cells (PAS × 400).

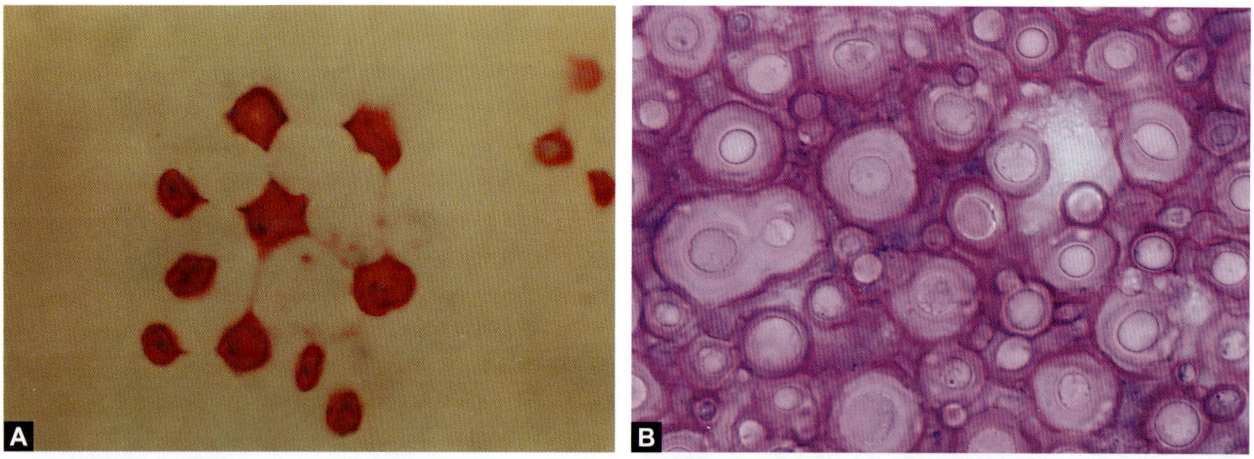

Figs. 21.13A and B. Budding yeast cells of *Cryptococcus* species with prominent capsule (PAS × 400 / 1000).

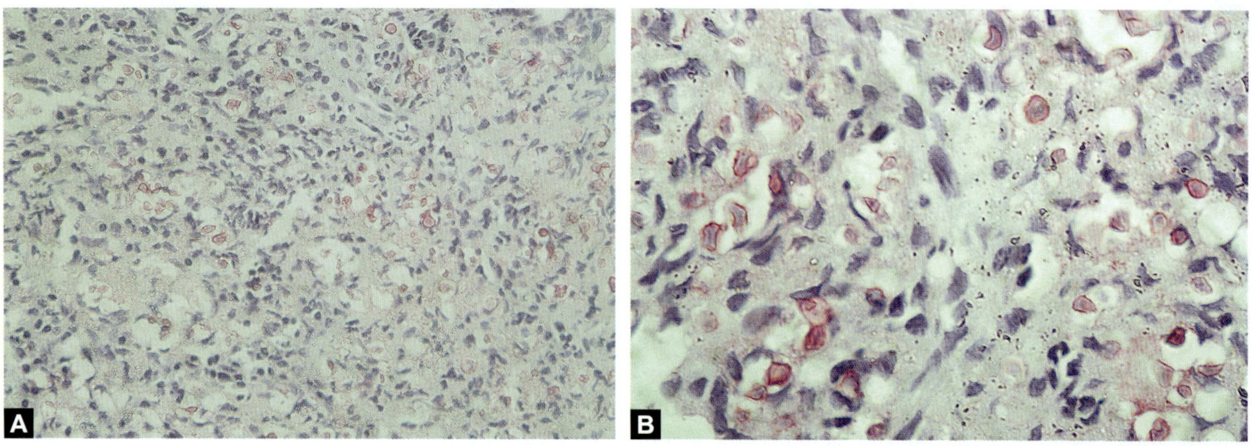

Figs. 21.14A and B. Budding yeast cells of *Cryptococcus* species along with numerous pus cells (Mayer's Mucicarmine × 200 / 400).

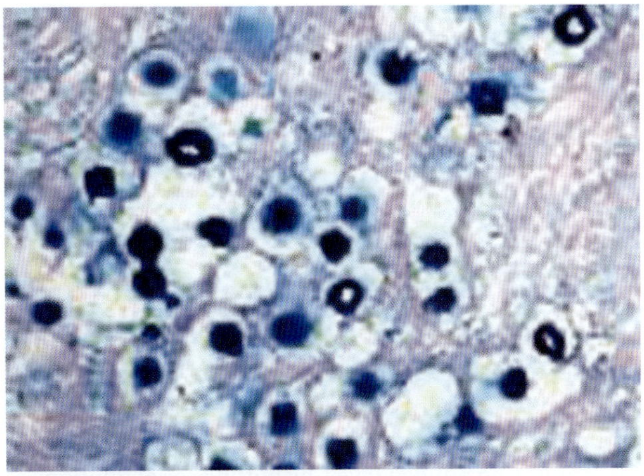

Fig. 21.15. Budding yeast cells of *Cryptococcus* species with prominent capsule (Alcian Blue × 400).

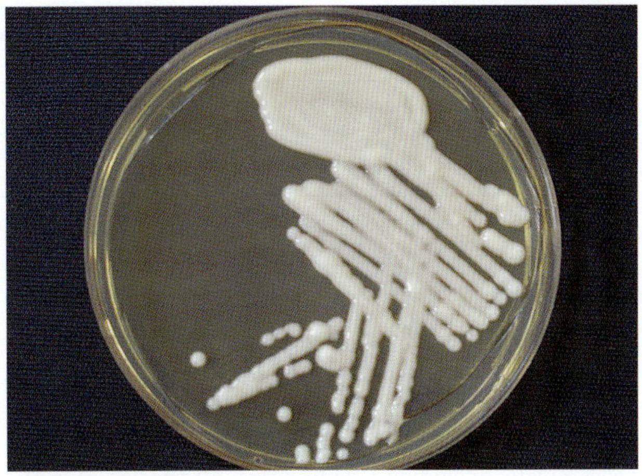

Fig. 21.16. Yeast-like mucoid growth of *Cryptococcus* species on SDA after three days of incubation.

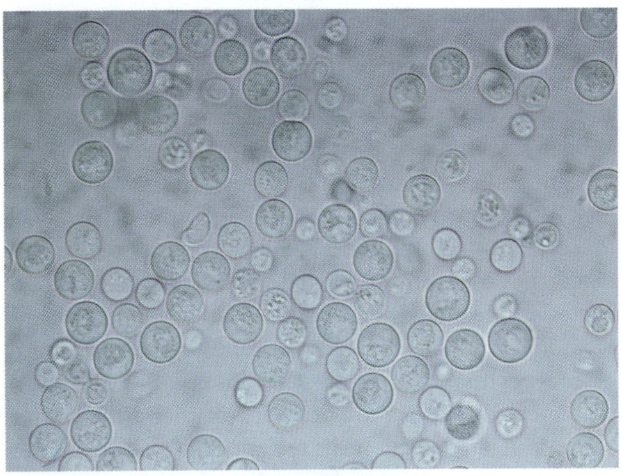

Fig. 21.17. Budding yeast cells of *Cryptococcus* species (Phase Contrast × 200).

used instead of actidione in birdseed agar (Staib's agar), which is inhibitory to saprotrophic fungi. For primary isolation, blood agar, Sabouraud dextrose agar, brain-heart infusion agar, cysteine-heart hemoglobin agar, birdseed agar and sunflower seed agar, are used. All cultures are examined every day during first week and twice a week during next three weeks. The birdseed agar if turns out to be sterile may be discarded after a fortnight or so.

The *Cryptococcus* species produce brown-color-effect when grown on medium containing extract of niger seed (*Guizotia abyssinica*) or sunflower seed (*Helianthus annuus*). The growth can be seen as early as 48 hours on Staib's agar or Pal's sunflower seed agar (SSA). These are also inoculated at 30°C and examined daily for fortnight before being discarded as sterile. There are brown-colored colonies after few days due to conversion of substrate to melanin by phenoloxidase.

A modification of Dichloran Rose-Bengal Chloramphenicol Medium (DRBCm) is also being used to detect colonies of *Cr.neoformans*, particularly from food or environment. Nandhakumar *et al*, have described use of mustard seed agar for the differentiation of *Cryptococcus neoformans*. In addition, Tendolkar *et al*, used tobacco agar, new medium for the pigment production in *Cryptococcus neoformans*. The details of all these media used for the isolation of *Cryptococcus* are described in **Appendix A**.

The growth of *Cryptococcus* is yeast-like and highly mucoid, cream to buff-colored on Sabouraud dextrose agar **(Fig. 21.16)**. On repeated sub-cultures, capsule may not be seen prominently as it is seen in a fresh specimen. **Figure 21.17** shows phase contrast microscopy of subcultured *Cryptococcus* strain, where the capsule is not prominent. The organisms are identified by colony characteristics and microscopic appearance in LCB or PHOL stained mounts. The mucoid appearance of colonies is because of production of polysaccharides. The capsule production can be improved by growing organisms on chocolate agar at 37°C in CO_2 incubator. The urine and sputum should also be cultured, even without any clinical or laboratory evidence suggesting involvement of genitourinary or respiratory system. Negative cultures do not rule out cryptococcosis because often very small number of organisms are present in the CSF.

Table 21.4. Assimilation and Other Biochemical Characteristics of Medically Significant *Cryptococcus* Species.

Species	*Mal*	*Suc*	*Tre*	*Gal*	*Cel*	*Xyl*	*Raf*	*Lac*	*Duc*	*Mel*	*Ino*	*Urease*	NO_3-NO_2
Cr.neoformans	+	+	+	+	+	+	+	–	+	–	+	+	–
Cr.albidus	+	+	+	+	+	+	+	+	+	+	+	+	+
Cr.laurentii	+	+	+	+	+	+	+	+	+	+	+	+	–

Note: Mal = Maltose, Suc = Sucrose, Tre = Trehalose, Gal = Galactose, Cel = Cellobiose, Xyl = Xylose, Raf = Raffinose, Lac = Lactose, Dul = Dulcitol, Mel = Melibiose, Ino = Inositol; + = Positive Reaction; - = Negative Reaction.

The suspected yeast isolates are finally identified by standard biochemical tests as shown in **Table 21.4**. A Flowchart is also given in the last Chapter as **Flowchart 20.1** depicting identification scheme of commonly encountered yeast isolates including *Cryptococcus*. All species of genus *Cryptococcus* are non-fermentative, assimilate inositol and produce urease. Therefore, confirmation of *Cryptococcus* species is carried out by browning of colony on BSA or SSA; growth at 37°C; hydrolysis of Christensen's urea agar, inositol and nitrate assimilation and positive mice pathogenicity.

The biotyping is essential for species/variety differentiation. *Cr.gattii* is differentiated from other two varieties on the basis of their growth on different media such as L-canavanine-glycine-bromothymol blue (CGB) agar where *Cr.gattii* turns it to characteristically cobalt-blue color and var. *grubii* and var. *neoformans* remain as such *i.e.* yellow to light-green colored; glycine cycloheximide phenol red agar (GCP) and same medium without cycloheximide *i.e.* glycine phenol red agar **(*See* Appendix A)**. D-proline agar is also an efficient medium for differentiation of both these varieties. Only *Cr.gattii* utilizes D-proline and D-tryptophan as sole source of nitrogen and detected by incubating assimilation agar at 25°C.

The creatinine-dextrose-bromothymol-blue-thymine (CDBT) medium for thymine assimilation is positive in serotype **D** and negative in serotype **A** and is also utilized for differentiation of *Cr. n.* var. *grubii* and *Cr. n.* var. *neoformans*. Rapid urease test may also be used to identify genus of this fungus and positive results can be seen within 10-30 min. The comparative salient epidemiological, clinical, mycological and therapeutic features of infection caused by species/varieties of *Cryptococcus* are shown in **Table 21.5**.

(c) Immunodiagnosis

The serological tests are used for demonstration of cryptococcal polysaccharide capsular antigen (CrAg) as well as antibodies in serum, CSF and urine. The antigen detection is done with latex agglutination - **Crypto-LA** test. This test is qualitative as well as quantitative hence it has diagnostic and prognostic values. The quantitative test can be done for monitoring antigen levels by putting serial dilutions of serum during follow up of the patients. The titer is highest in serum, intermediate in CSF and lowest in urine. Antigen titer in CSF stays for longer duration and is directly proportional to mortality of patient whereas in serum level such co-relation is not observed. A negative serum test for cryptococcal antigen excludes cryptococcal meningitis. The bronchoalveolar lavage (BAL) fluid for cryptococcal antigen screening may also be used.

Sometimes, the agglutination test may give false positive results in rheumatological disorder due to presence of serum antiglobulins (rheumatoid factor); disseminated trichosporonosis caused by *Trichosporon* species; bacterial infections by *Capnocytophaga canimorsus* and *Stomatococcus* species. This difficulty can be overcome by pre-treatment of serum with pronase, however, CSF specimens do not require pronase treatment. The cryptococcal antigen detection kits occasionally may cross-react strongly with unrelated fungi. Therefore, it is ideal to cross-check every positive reacting clinical specimen with control or test kit from second manufacturer. A cryptococcal antigen titer greater than or equal to 1 : 8 is taken as significant.

The **Eiken latex agglutination test** was introduced to detect cryptococcal antigen in serum, particularly in pulmonary cryptococcosis. Pre-treatment with pronase succeeded in eliminating false-positive reactions due to rheumatoid factor and false negative due to antigen-antibody complex formation. It is claimed to be more sensitive than Crypto-LA test. Pronase enhances sensitivity of Eiken test, which appears to be more useful in patients with pulmonary cryptococcal disease and its use may avoid needless lung biopsy. ELISA is also useful for the detection of antigen.

The **Lateral Flow Assay** (LFA) is easy to perform as point-of-care testing. It is accurate and efficacy is similar

Table 21.5. Comparative Features of Different Species / Varieties of *Cryptococcus neoformans*.

Features	*Cr. n. var. grubii*	*Cr. n. var. neoformans*	*Cryptococcus gattii*
Serotypes	A	D	B and C
Mating Types	α	α and ***a***	α and ***a***
Teleomorphs	*F. n. var. neoformans*	*F. n. var. neoformans*	*F. n. var. bacillispora*
Prevalence	Worldwide	Europe	Tropical, Subtropical
Habitat	Avian Excreta	Avian Excreta	Eucalyptus Trees
Immune Status of Patient	HIV - Positive	On corticosteroids	HIV - Negative
Virulence of Organism	High	High	Low
Onset of Disease	Acute	Acute	Insidious
Fever	++	++	+
Neurological Sequelae	+	+	++
Focal CNS Lesions	+	+	++
Focal Pulmonary Lesions	+	+	+++
Cryptococcaemia	++	++	+
Skin Involvement	Less Frequent	Frequent	Not known
Blastospores	Rounded	Rounded	Elliptical
CDBT medium for thymine assimilation	No growth	Growth with orange color change	Growth with blue green color change
CGB Agar	No change	No change	Cobalt Blue color
GCP Agar	Negative	Negative	Positive
D-proline	Not utilized	Not utilized	Utilized
D-tryptophan	Not utilized	Not utilized	Utilized
Malate Assimilation	Negative	Negative	Positive
Mortality	++	++	+
Treatment	Shorter	Shorter	Longer with relapse

to latex agglutination test and higher than blood and urine culture. There is low frequency of cross reactions.

The cryptococcal antigen in blood can be detected a median of 22 day before symptoms of meningitis develop, thus enabling the identification of patients who could receive pre-emptive therapy, which significantly increases survival times. In 2011, the WHO recommended routine screening for cryptococcal antigen in low CD4 count HIV patients and subsequently 22 countries have incorporated similar screening strategies into their national guidelines.

Antibodies detection may be performed by agglutination, indirect fluorescent antibody test and complement fixation test. These are not very reliable tests due to cross-reactions with other diseases. Although anti-cryptococcal antibody titer has not proved useful for the diagnosis but there is speculation that it is of prognostic value, with presence of antibody signaling an enhanced likelihood of cure in humans with cryptococcal meningoencephalitis.

The restriction fragment length polymorphism analysis is an effective tool in taxonomical and epidemiological studies. Pyrolysis-mass spectrometry (PMS), as typing method, has proved to be rapid, simple and inexpensive technique for this organism. Other methods like PCR and electrophoretic karyotyping analysis have been applied in taxonomic studies of *Cryptococcus* species. These may have some potential application in typing but none can compete combination of general applicability, speed, simplicity and economy of pyrolysis-mass spectrometry.

The serotyping of clinical isolates is performed using different commercial kits to know the prevalence of particular variety or serotype of these encapsulated fungi. These kits are costly, therefore, serotyping facilities are not available in most of routine diagnostic microbiology laboratories except a few Mycology Reference Centers where *Cryptococcus* isolates/strains can be sent by speed post/courier for confirmation and serotyping. The serotyping is carried out using slide agglutination technique against factor sera by using Crypto Check agglutination test kit. Rapid, accurate procedures for identifying varieties and serotypes of *Cryptococcus* have been established.

The molecular techniques are also utilized in diagnosis, particularly from epidemiological point of view. The PCR fingerprinting using primers $(GACA)_4$, $(CAC)_5$ and FM1 result in clear and reproducible assignment of *Cryptococcus* strains to *Cr.neoformans* and *Cr.gattii*, respectively and in addition, it can confirm serovar assignment. Species-specific DNA-probe based on ribosomal ITS sequences has also been developed.

(d) Animal Pathogenicity

In case of cryptococcosis, animal pathogenicity testing is commonly done as it is one of the criteria to differentiate between pathogenic and non-pathogenic strains of *Cryptococcus* genus. The suspected clinical specimen or saline suspension of two to four days old culture of *Cryptococcus* (1×10^6 cells) is injected into Swiss albino mice. The route of inoculation may be intracerebral (0.02 ml), intravenous into tail vein (0.1-0.2 ml) or intraperitoneal (0.5 ml). The inoculated mice die on an average within seven to ten days. If mice do not die, autopsy is performed after period of two weeks of inoculation. The autopsied brain tissue is grossly mucoid to touch. The yeast cells and capsule of *Cryptococcus* can be demonstrated from various body fluids and viscera by routine conventional methods. The intravenous route is the most standardized and quantitative one, however, intrathecal inoculation simulates the natural infection of disease in the best way.

However, the laboratory diagnosis of cryptococcosis caused by non-*neoformans Cryptococcus* species may be difficult for the following reasons:

- Microscopic and chemical analyses of the CSF may reveal slightly abnormal values for cell counts, protein and lactate concentrations.
- The cryptococcal antigen test may be negative, as strains are usually non-encapsulated.
- *Cryptococcus* species other than *Cr.neoformans*, including *Cr.adeliensis*, may fail to grow *in vitro* at 37°C. Therefore, it is important to incubate an additional medium inoculated with CSF specimens at a decreased temperature if cryptococcosis is suspected.
- Due to heterogeneity of non-*neoformans Cryptococcus* group, reliable species identification requires sequence analysis of ribosomal gene fragments. *Cryptococcus* species other than *Cr.neoformans* may be unable to grow *in vitro* at 37°C and they may be resistant to antifungal agents, including azoles.

Treatment and Prophylaxis

The choice of treatment of cryptococcosis depends on both site of involvement and host's immune status. The therapeutic modalities before 1980s were limited to amphotericin B, flucytosine and miconazole. But during last three decades, newer azoles like fluconazole are found to be very useful in its management. In addition to that extended spectrum triazoles like voriconazole, posaconazole and ravuconazole are found to be potential *in vitro* hence are tried in various treatment modalities. The IDSA released its guidelines on cryptococcosis in the year 2010. The following regimens are used in the treatment of cryptococcosis.

(a) Antifungal Drugs: Amphotericin B is still the '**gold standard**' because of rapid sterilization of CNS. Flucytosine is also useful antifungal agent, however, risk of development of resistance limits its use as monotherapy. It is mainly used in combination with amphotericin B to reduce toxicity of the later. There are two types of *modus operandi*, which are followed to treat cryptococcosis. These are **induction therapy** and **maintenance therapy**. In induction therapy amphotericin B is given as 0.7 mg/kg/day and flucytosine as 100 mg/kg/day for a period of two weeks. In non-AIDS patients with cryptococcal meningitis, flucytosine is given as 150 mg/kg/day. After two weeks of treatment, therapy is shifted to fluconazole 400 mg/kg/day to complete total of ten weeks period. Fluconazole is given 200 mg/day as life-long maintenance therapy to prevent relapses. Unlike other azoles - ketoconazole and itraconazole, fluconazole is relatively less protein bound and adequately penetrates blood brain barrier. This leads to high concentration of active and free drug in CSF. A number of lipid formulations of amphotericin B, primarily designated to reduce toxicity are used like ABLC, ABCD or liposomal preparations. It takes more than two weeks for drug treatments for cryptococcal meningitis to sterilize the CSF. *In vitro* and animal studies lend support to use of combinations of amphotericin B, flucytosine and fluconazole for treating cryptococcosis.

As with most of the other opportunist infections, management of cryptococcosis in patients with AIDS, further complicated by curative therapy, unusual and chronic suppressive therapy is likely to be necessary. Relapse probably represents failure to eradicate infection rather than new primary infection. *Cr.neoformans* infection with AIDS is usually incurable because antifungal therapy does not eradicate infection. As far as the role of steroids is concerned, it is not very helpful.

The drug resistance is also noted in *Cr.neoformans*. The isolates, clinically and *in vitro* resistant to amphotericin B and fluconazole, are recovered in patients with relapsed cryptococcal meningitis. Heteroresistance to fluconazole is an intrinsic property of *Cryptococcus*. Although resistant organisms are still rare, potential for an increasing number of cases exists because of frequent use of such antifungal agents. Antifungal susceptibility is done as per CLSI standard method (M27-A), Etest and urea broth dilution test.

Moreover, Advancing Cryptococcal meningitis Treatment for Africa (ACTA) has taken a very good initiative to tackle this disease in this continent.

In a recent study in infected mice by Ribeiro *et al*, as an animal model, treated with atorvastatin and fluconazole, showed increased survival, improved clinical condition as well as reduced fungal burden in the lungs and brain. This is the first study to perform *in vivo* tests with such combination for treating this disease. Although it is premature to state but results suggest that atorvastatin may be an important therapeutic adjuvant for cryptococcosis.

(b) Serum Therapy: Monoclonal antibodies, directed against capsular polysaccharide, are tried at experimental level. Immunotherapy is an attractive option if treatment strategies can be developed which might prove effective in immune deficiencies which occur in patients with AIDS.

In response to limitations of current treatment, several investigators have explored efficacy of passive therapy in the form of monoclonal antibodies (mAbs) directed against capsular polysaccharide of *Cr.neoformans*. The limited clinical experience suggests that passive immunotherapy for *Cr.neoformans* is well-tolerated and that serum antigen titers may be reduced after antibody administration. Anticapsular monoclonal antibodies prolong survival and decrease organ-fungal burden in animal models of *Cr.neoformans* infection and enhance therapeutic efficacy of amphotericin B, fluconazole and flucytosine. The mechanism of antibody-mediated protection is believed to include enhancement of macrophage and natural killer cell-mediated phagocytosis.

(c) Prognosis: The prognosis of disease depends upon various factors. At least there are five major factors, which determine outcome of cryptococcal meningitis. First, underlying disease is the most important determinant of outcome. The second factor is burden of yeasts determined by India ink preparation and high antigen titers and third is development of intracranial pressure. Fourth, impact of host response is major determinant of outcome of infection. The patients with poor inflammatory response such as 20 WBC/mm^3 or less in CSF do not respond to therapy as well as those with more brisk inflammatory response. Finally, signs and symptoms at the time of presentation, in particular decreased level of consciousness, suggest an advanced disease thereby poor prognosis.

(d) Prophylaxis: The epidemiological studies of this disease suggest that exposure to *Cr.neoformans* is unlikely to be avoided; therefore, prevention must be either on chemotherapy or immunization. The patient should avoid exposure to areas that may be heavily contaminated with *Cr.neoformans* such as sites contaminated with pigeon droppings, particularly old buildings. The prophylactic use of fluconazole has reduced rate of cryptococcal meningitis in HIV-infected patients.

(e) Vaccine: Based on rationale that antibodies can be protective, conjugate vaccine of glucuronoxylomannan, major capsule component, linked to tetanus toxoid, was introduced in 1991. This could protect mice and was shown to be immunogenic in normal human subjects but has not been tested in individuals who are at risk of cryptococcal infection.

Immune Reconstitution Inflammatory Syndrome

The paradoxical cryptococcosis-associated **Immune Reconstitution Inflammatory Syndrome** (IRIS) occurs in ~25% of HIV-infected patients with cryptococcal meningitis after they commence antiretroviral therapy (ART). This is also observed in 5-10% solid organ transplant recipients with cryptococcosis with increased risk of allograft failure.

As such pathogenesis of IRIS remains poorly understood, with few biomarkers to predict IRIS risk or to establish an early diagnosis. The risk factors for paradoxical IRIS are low inflammatory response and CD4 cell count at baseline, rapid immune restoration from this low baseline and high organism or antigen load at baseline at ART initiation. The detailed immune mechanisms are still unclear. The latest information provides insight into the pathogenesis of IRIS that could be applied to developing diagnostic tests or targeted immunomodulatory treatments.

A rapidly fungicidal induction therapy, allowing prompt initiation of ART from around three weeks in resource-limited settings in the context of amphotericin B induction at a time when organism and antigen loads are low, may reduce overall mortality without exacerbating paradoxical IRIS, compared with initiation of ART at later time points. The recent cohorts studies suggest early recognition and management can reduce the mortality associated with paradoxical IRIS. Unmasking IRIS is preventable through screening for cryptococcal antigen prior to the ART and preemptive antifungal therapy for those testing positive, although prospective studies are also needed. An optimal antifungal induction and the judicious ART timings, together with early recognition and management of developing cases, with thorough exclusion of alternative diagnoses, should help reduce paradoxical IRIS-related mortality due to *Cryptococcus* species. The unmasking IRIS cases should be preventable keeping in view of the latest available information.

Further Reading

1. Abid MB, De Mel S, Limei MP. Disseminated cryptococcal infection in an immunocompetent host mimicking plasma cell disorder: A case report and literature review. *Clin Case Rep.* 2015; **3:** 319-24.
2. Abuav R, McGirt LY, Kazin RA. Cryptococcal panniculitis in an immunocompromised patient: A case report and review of the literature. *Cutis.* 2010; **85:** 303-6.
3. Adachi M, Tsuruta D, Imanishi H, *et al.* Necrotizing fasciitis caused by *Cryptococcus neoformans* in a patient with pemphigus vegetans. *Clin Exp Dermatol.* 2009; **34:** e751-3.
4. Adams P. Cryptococcal meningitis: A blind spot in curbing AIDS. *Lancet.* 2016; **387:** 1605-6.
5. Agadi JB, Madni NA, Nanjappa V, *et al.* Cryptococcal osteomyelitis of the skull in a patient with transient lymphopenia. *Neurol India.* 2010; **58:** 300-2.
6. Aguiar PADF, Pedroso RDS, Borges AS, *et al.* The epidemiology of cryptococcosis and the characterization of *Cryptococcus neoformans* isolated in a Brazilian University Hospital. *Rev Inst Med Trop Sao Paulo.* 2017; **59:** e13.
7. Ajesh K, Sreejith K. *Cryptococcus laurentii* biofilms: Structure, development and antifungal drug resistance. *Mycopathologia.* 2012; **174:** 409-19.
8. Akinosoglou K, Melachrinou M, Siagris D, *et al.* Good's syndrome and pure white cell aplasia complicated by *Cryptococcus* infection: A case report and review of the literature. *J Clin Immunol.* 2014; **34:** 283-8.
9. Albuquerque PC, Rodrigues ML. Research trends on pathogenic *Cryptococcus* species in the last 20 years: A global analysis with focus on Brazil. *Future Microbiol.* 2012; **7:** 319-29.
10. Alhaji M, Sadikot RT. Cryptococcal endocarditis. *South Med J.* 2011; **104:** 363-4.
11. Andrade-Silva L, Ferreira-Paim K, *et al.* Susceptibility profile of clinical and environmental isolates of *Cryptococcus neoformans* and *Cryptococcus gattii* in Uberaba, Minas Gerais, Brazil. *Med Mycol.* 2013; **51:** 635-40.
12. Animalu C, Mazumder S, Cleveland KO, *et al. Cryptococcus uniguttulatus* meningitis. *Am J Med Sci.* 2015; **350:** 421-2.
13. Antachopoulos C, Walsh TJ. Immunotherapy of *Cryptococcus* infections. *Clin Microbiol Infect.* 2012; **18:** 126-33.
14. Antinori S, Corbellino M, Galimberti L, *et al.* Cryptococcal meningitis and systemic lupus erythematosus. *J Emerg Med.* 2014; **47:** 323-5.
15. Antinori S, Ridolfo A, Fasan M, *et al.* AIDS-associated cryptococcosis: A comparison of epidemiology, clinical features and outcome in the pre- and post-HAART eras. Experience of a single centre in Italy. *HIV Med.* 2009; **10:** 6-11.
16. Antinori S. New insights into HIV/AIDS-associated cryptococcosis. *ISRN AIDS.* 2013; 471363. PMID: 24052889.
17. Anuradha S, H AN, Dewan R, Kaur R, *et al.* Asymptomatic cryptococcal antigenemia in people living with HIV (PLHIV) with severe immunosuppression: Is routine CrAg screening indicated in India? *J Assoc Physicians India.* 2017; **65:** 14-7.
18. Araujo BS, Bay M, Reichert R, *et al.* Intra-abdominal cryptococcosis by *Cryptococcus gattii:* Case report and review. *Mycopathologia.* 2012; **174:** 81-5.
19. Arsic Arsenijevic V, Pekmezovic MG, Meis JF, *et al.* Molecular epidemiology and antifungal susceptibility of Serbian *Cryptococcus neoformans* isolates. *Mycoses.* 2014; **57:** 380-7.
20. Asadi Gharabaghi M, Allameh SF. Primary pulmonary cryptococcosis. *BMJ Case Rep.* 2014; pii: bcr2014203821.
21. Asanuma Y, Fujimoto H, Nakabayashi H, *et al.* Extradural cryptococcoma at the sacral spine without bone involvement in an immunocompetent patient. *J Orthop Sci.* 2014; **19:** 1040-5.
22. Ashwini BR, Raghupathi AR, Srinarthan A. Cryptococcosis of bone marrow: A case report with review of literature. *J Clin Diagn Res.* 2014; **8:** 158-9.
23. Babu AK, Gopalakrishnan R, Sundararajan L. Pulmonary cryptococcosis: An unusual presentation. *Lung India.* 2013; **30:** 347-50.
24. Badali H, Alian S, Fakhim H, *et al.* Cryptococcal meningitis due to *Cryptococcus neoformans* genotype AFLP1/VNI in Iran: A review of the literature. *Mycoses.* 2015; **58:** 689-93.
25. Baddley JW, Forrest GN *et al.* Cryptococcosis in solid organ transplantation. *Am J Transplant.* 2013; **13 (Suppl. 4):** 242-9.
26. Baddley JW, Schain DC, Gupte AA, *et al.* Transmission of *Cryptococcus neoformans* by organ transplantation. *Clin Infect Dis.* 2011; **52:** e94-8.
27. Badiye A, Patnaik M, Deshpande A, *et al.* Think fungus not just a crypto-meningitis in AIDS! *J Assoc Physicians India.* 2012; **60:** 21-4.
28. Baer S, Baddley JW, Gnann JW, *et al.* Cryptococcal disease presenting as necrotizing cellulitis in transplant recipients. *Transpl Infect Dis.* 2009; **11:** 353-8.
29. Baillif S, Delas J, Asrargis A, *et al.* Multimodal imaging of bilateral cryptococcal choroiditis. *Retina.* 2013; **33:** 249-51.
30. Bamba S, Lortholary O, Sawadogo A, *et al.* Decreasing incidence of cryptococcal meningitis in West Africa in the

era of highly active antiretroviral therapy. *AIDS.* 2012; **26:** 1039-41.

31. Banerjee P, Haider M, Trehan V, Mishra B, Thakur A, Dogra V, Loomba P. *Cryptococcus laurentii* fungemia. *Indian J Med Microbiol.* 2013; **31:** 75-7.
32. Bansal N, Shah R, Patel A, *et al.* Hypercalcemia as a primary manifestation of cryptococcal immune reconstitution syndrome - A rare presentation. *Am J Emerg Med.* 2015; **33:** 598.e3-4.
33. Baradkar V, Mathur M, De A, *et al.* Prevalence and clinical presentation of cryptococcal meningitis among HIV seropositive patients. *Indian J Sex Transm Dis.* 2009; **30:** 19-22.
34. Barnett JA. A history of research on yeasts 14: Medical Yeasts. Part 2, *Cryptococcus neoformans. Yeast.* 2010; **27:** 875-904.
35. Barragan NC, Sorvillo F, Kuo T. Cryptococcosis-related deaths and associated medical conditions in the United States, 2000-2010. *Mycoses.* 2014; **57:** 741-6.
36. Bartlett KH, Cheng PY, Duncan C, *et al.* A decade of experience: *Cryptococcus gattii* in British Columbia. *Mycopathologia.* 2012; **173:** 311-9.
37. Basu S, Jain P, Ansari S, *et al.* Disseminated cryptococcosis in an AIDS patient with unusual clinical presentation. *Rev Iberoam Micol.* 2008; **25:** 179-81.
38. Bava AJ, Troncoso A. Detection of *Cryptococcus neoformans* in faecal matter: A novel presentation of disseminated cryptococcosis. *J Infect Dev Ctries.* 2009; **3:** 572-4.
39. Bavishi AV, McGarry TM. A case of pulmonary cryptococcosis caused by capsule-deficient *Cryptococcus neoformans* in an immunocompetent patient. *Respir Care.* 2010; **55:** 937-41.
40. Bedi NG, Nawange SR, Singh SM, Naidu J, Kavishwar A. Seasonal prevalence of *Cryptococcus neoformans* var. *grubii* and *Cryptococcus gattii* inhabiting *Eucalyptus terreticornis* and *Eucalyptus camaldulensis* trees in Jabalpur City of Madhya Pradesh, Central India. *J Mycol Med.* 2012; **22:** 341-7.
41. Begon E, Bachmeyer C, Thibault M, *et al.* Necrotizing fasciitis due to *Cryptococcus neoformans* in a diabetic patient with chronic renal insufficiency. *Clin Exp Dermatol.* 2009; **34:** 935-6.
42. Bellissimo-Rodrigues F, Baciotti M, Zanatto MP, *et al.* Cutaneous cryptococcosis due to *Cryptococcus gattii* in a patient on chronic corticotherapy. *Rev Soc Bras Med Trop.* 2010; **43:** 211-2.
43. Bennett J. Companion drugs for amphotericin B in cryptococcal meningitis: Flucytosine, fluconazole or......nothing? *Clin Infect Dis.* 2012; **54:** 129-30.
44. Bernabe DG, Veronese LA, Miyahara GI, *et al.* Cryptococcosis parotid diagnosed by FNAC in a non-immunosuppressed patient. *Cytopathology.* 2010; **21:** 343-5.
45. Bernal-Martinez L, Gomez-Lopez A, Castelli MV, *et al.* Susceptibility profile of clinical isolates of non-*Cryptococcus neoformans*/non-*Cryptococcus gattii Cryptococcus* species and literature review. *Med Mycol.* 2010; **48:** 90-6.
46. Bestard J, Siddiqi ZA. Cryptococcal meningoencephalitis in immunocompetent patients: Changing trends in Canada. *Neurology.* 2010; **74:** 1233-5.
47. Bhat V, Vira H, Khattry N, *et al. Cryptococcus laurentii* diarrhea post hematopoietic stem cell transplant. *Transpl Infect Dis.* 2017; 19(2). PMID: 28083955.
48. Bhuyan P, Pattnaik K, Kar A, *et al.* Cryptococcal lymphadenitis in HIV: A chance diagnosis by FNAC. *Diagn Cytopathol.* 2013; **41:** 456-8.
49. Biancheri D, Kanitakis J, Bienvenu AL, *et al.* Cutaneous cryptococcosis in solid organ transplant recipients: Epidemiological, clinical, diagnostic and therapeutic features. *Eur J Dermatol.* 2012; **22:** 651-7.
50. Bielska E, May RC. What makes *Cryptococcus gattii* a pathogen? *FEMS Yeast Res.* 2016; 16. pii: fov106. PMID: 26614308.
51. Bisson GP, Molefi M, Bellamy S, *et al.* Early versus delayed antiretroviral therapy and cerebrospinal fluid fungal clearance in adults with HIV and cryptococcal meningitis. *Clin Infect Dis.* 2013; **56:** 1165-73.
52. Bothra M, Selvaperumal P, Kabra M, *et al.* Disseminated cryptococcosis. *Indian Pediatr.* 2014; **51:** 225-6.
53. Bouklas T, Fries BC. Aging: An emergent phenotypic trait that contributes to the virulence of *Cryptococcus neoformans. Future Microbiol.* 2015; **10:** 191-7.
54. Boulware DR, Meya DB, Muzoora C, *et al.* Timing of antiretroviral therapy after diagnosis of cryptococcal meningitis. *N Engl J Med.* 2014; **370:** 2487-98.
55. Boulware DR, Meya DB. Antiretroviral therapy after cryptococcal meningitis. *N Engl J Med.* 2014; **371:** 1166-7.
56. Brizendine KD, Baddley JW, Pappas PG. Pulmonary cryptococcosis. *Semin Respir Crit Care Med.* 2011; **32:** 727-34.
57. Brown SJ, George S, Braithwaite K. A puzzling case of cryptococcal meningitis. *S Afr Med J.* 2014; **104:** 720.
58. Byrnes EJ 3rd, Bartlett KH, Perfect JR, *et al. Cryptococcus gattii:* An emerging fungal pathogen infecting humans and animals. *Microbes Infect.* 2011; **13:** 895-907.
59. Byrnes EJ 3rd, Li W, Lewit Y, *et al.* Emergence and pathogenicity of highly virulent *Cryptococcus gattii* genotypes in the northwest United States. *PLoS Pathog.* 2010; **6:** e1000850.
60. Cai X, Liu K, Liang Y, *et al.* Isolated biliary cryptococcosis manifesting as obstructive jaundice in an immunocompetent adult. *Int J Med Sci.* 2012; **9:** 200-6.
61. Calista F, Tomei F, Assalone P, *et al. Cryptococcus laurentii* diarrhea in a neoplastic patient. *Case Rep Oncol Med.* 2015; 216458. PMID: 25692059.
62. Capoor MR, Mandal P, Deb M, *et al.* Current scenario of cryptococcosis and antifungal susceptibility pattern in India: A cause for reappraisal. *Mycoses.* 2008; **51:** 258-65.
63. Casadevall A. Cryptococci at the brain gate: Break and enter or use a Trojan horse? *J Clin Invest.* 2010; **120:** 1389-92.
64. Cerrati EW, Myssiorek D. External nasal lesion in a middle-aged man. *JAMA Otolaryngol Head Neck Surg.* 2014; **140:** 569-70.
65. Chai SM, Teoh SC. Ocular cryptococcosis as a presenting manifestation of cryptococcal meningitis in a patient with HIV. *Int J STD AIDS.* 2012; **23:** 377-8.
66. Chaiwarith R, Vongsanim S, Supparatpinyo K. Cryptococcal meningitis in HIV-infected patients at Chiang Mai

University Hospital: A retrospective study. *Southeast Asian J Trop Med Public Health.* 2014; **45:** 636-46.

67. Chakrabarti A, Varma SC, Roy P, Sakhuja V, Chander J, *et al.* Cryptococcosis in and around Chandigarh: An analysis of 65 cases. *Indian J Med Microbiol.* 1995; **13:** 65-9.
68. Chan M, Lye D, Win MK, *et al.* Clinical and microbiological characteristics of cryptococcosis in Singapore: Predominance of *Cryptococcus neoformans* compared with *Cryptococcus gattii. Int J Infect Dis.* 2014; **26:** 110-5.
69. Chander J, Sapra RK, Talwar P. Incidence of cryptococcosis in and around Chandigarh, India during the period 1982-91. *Mycoses.* 1994; **37:** 23-6.
70. Chandrasekar PH, Revankar SG. Cryptococcosis in a febrile renal transplant recipient. *JAMA.* 2013; **310:** 748.
71. Chandrashekar UK, Acharya V, Varghese GK, *et al.* An unusual presentation of pulmonary cryptococcosis with co-existing disseminated tuberculosis in an AIDS patient. *Trop Doct.* 2012; **42:** 60-2.
72. Chang C, Chen S. Colliding epidemics and the rise of cryptococcosis. *J Fungi.* 2016, **2:** 1. doi: 10.3390/jof2010001.
73. Chang CC, Dorasamy AA, Gosnell BI, *et al.* Clinical and mycological predictors of cryptococcosis-associated immune reconstitution inflammatory syndrome. *AIDS.* 2013; **27:** 2089-99.
74. Chang CC, Sorrell TC, Chen SC. Pulmonary cryptococcosis. *Semin Respir Crit Care Med.* 2015; **36:** 681-91.
75. Chang CM, Tsai CC, Tseng CE, *et al.* Donor-derived *Cryptococcus* infection in liver transplant: Case report and literature review. *Exp Clin Transplant.* 2014; **12:** 74-7.
76. Chang YL, Hung SH, Liu CH, *et al.* Cryptococcal infection of the vocal folds. *Southeast Asian J Trop Med Public Health.* 2013; **44:** 1043-6.
77. Chaturvedi AK, Wormley FL Jr. Cryptococcus antigens and immune responses: Implications for a vaccine. *Expert Rev Vaccines.* 2013; **12:** 1261-72.
78. Chaturvedi AK, Wormley FL Jr. Methodology for anti-cryptococcal vaccine development. *Methods Mol Biol.* 2017; **1625:** 129-40.
79. Chaturvedi V, Chaturvedi S. *Cryptococcus gattii:* A resurgent fungal pathogen. *Trends Microbiol.* 2011; **19:** 564-71.
80. Chau TT, Mai NH, Phu NH, *et al.* A prospective descriptive study of cryptococcal meningitis in HIV uninfected patients in Vietnam - high prevalence of *Cryptococcus neoformans* var. *grubii* in the absence of underlying disease. *BMC Infect Dis.* 2010; **10:** 199.
81. Chaya R, Padmanabhan S, Anandaswamy V, *et al.* Disseminated cryptococcosis presenting as cellulitis in a renal transplant recipient. *J Infect Dev Ctries.* 2013; **7:** 60-3.
82. Chayakulkeeree M, Wangchinda P. Clinical characteristics and outcomes of patients with cryptococcal meningoencephalitis in a resource-limited setting. *J Med Assoc Thai.* 2014; **97:** S26-34.
83. Chen J, Liu S, Xiong Z, *et al.* Cryptococcal infection of the femoral bone similar with pathologic features of vascular tumors: A case report and review of literature. *Int J Clin Exp Pathol.* 2015; **8:** 8551-4.
84. Chen M, Pan WH, Boekhout T. *Cryptococcus gattii* infections in China: Extent of the problem? *Chin Med J (Engl).* 2013; **126:** 203-5.
85. Chen M, Wang X, Yu X, *et al.* Pleural effusion as the initial clinical presentation in disseminated cryptococcosis and fungaemia: An unusual manifestation and a literature review. *BMC Infect Dis.* 2015; **15:** 385.
86. Chen SC, Korman TM, Slavin MA, *et al.* Antifungal therapy and management of complications of cryptococcosis due to *Cryptococcus gattii. Clin Infect Dis.* 2013; **57:** 543-51.
87. Chen SC, Liu JC, Tseng GC, *et al.* Rare presentation of pulmonary cryptococcosis as a calcified nodule. *Intern Med.* 2011; **50:** 169-70.
88. Chen SC, Meyer W, Sorrell TC. *Cryptococcus gattii* infections. *Clin Microbiol Rev.* 2014; **27:** 980-1024.
89. Chen SC, Slavin MA, Heath CH, *et al.* Clinical manifestations of *Cryptococcus gattii* infection: Determinants of neurological sequelae and death. *Clin Infect Dis.* 2012; **55:** 789-98.
90. Chen YC, Chang TY, Liu JW, *et al.* Increasing trend of fluconazole-non-susceptible *Cryptococcus neoformans* in patients with invasive cryptococcosis: A 12-year longitudinal study. *BMC Infect Dis.* 2015; **15:** 277.
91. Cheong JW, McCormack J. Fluconazole resistance in cryptococcal disease: Emerging or intrinsic? *Med Mycol.* 2013; **51:** 261-9.
92. Chimalizeni Y, Tickell D, Connell T. Evidence behind the WHO guidelines: Hospital care for children: What is the most appropriate anti-fungal treatment for acute cryptococcal meningitis in children with HIV? *J Trop Pediatr.* 2010; **56:** 4-12.
93. Chipungu C, Veltman JA, Jansen P, *et al.* Feasibility and acceptability of cryptococcal antigen screening and prevalence of cryptocococcemia in patients attending a resource-limited HIV/AIDS clinic in Malawi. *J Int Assoc Provid AIDS Care.* 2015; **14:** 387-90.
94. Chipungu GA, Christians SJ, Oliver SP. Cutaneous cryptococcosis erroneously diagnosed as *Histoplasma capsulatum* infection. *S Afr Med J.* 2008; **98:** 85-6.
95. Chopra S, Capoor MR, Mallik R, *et al.* Pulmonary cryptococcosis in HIV-sero-negative patients: Case series from India. *Mycoses.* 2015; **58:** 288-93.
96. Chowdhary A, Prakash A, Randhawa HS, *et al.* First environmental isolation of *Cryptococcus gattii*, genotype AFLP5, from India and a global review. *Mycoses.* 2013; **56:** 222-8.
97. Chowdhary A, Randhawa HS, Sundar G, *et al. In vitro* antifungal susceptibility profiles and genotypes of 308 clinical and environmental isolates of *Cryptococcus neoformans* var. *grubii* and *Cryptococcus gattii* serotype B from north-western India. *J Med Microbiol.* 2011; **60 (Pt. 7):** 961-7.
98. Chowdhary A, Randhawa HS, Prakash A, *et al.* Environmental prevalence of *Cryptococcus neoformans* and *Cryptococcus gattii* in India: An update. *Crit Rev Microbiol.* 2012; **38:** 1-16.
99. Chuang YM, Ku SC, Liaw SJ, *et al.* Disseminated *Cryptococcus neoformans* var. *grubii* infections in intensive care units. *Epidemiol Infect.* 2010; **138:** 1036-43.

100. Cicora F, Petroni J, Formosa P, *et al.* A rare case of *Cryptococcus gattii* pneumonia in a renal transplant patient. *Transpl Infect Dis.* 2015; **17:** 463-6.
101. Cicora F, Petroni J, Roberti J. Cryptococcosis presenting as a colonic ulcer in a kidney transplant recipient: A case report. *Transplant Proc.* 2015; **47:** 2786-7.
102. Cleveland KO, Gelfand MS, Rao V. Posaconazole as successful treatment for fungemia due to *Cryptococcus albidus* in a liver transplant recipient. *QJM.* 2013; **106:** 361-2.
103. Cocker R, McNair SA, Kahn L, *et al.* Isolated adrenal cryptococcosis, diagnosed by fine-needle aspiration biopsy: A case report. *Diagn Cytopathol.* 2014; **42:** 899-901.
104. Coelho C, Bocca AL, Casadevall A. The intracellular life of *Cryptococcus neoformans. Annu Rev Pathol.* 2014; **9:** 219-38.
105. Cogliati M, Chandrashekar N, Esposto MC, *et al. Cryptococcus gattii* serotype-C strains isolated in Bangalore, Karnataka, India. *Mycoses.* 2012; **55:** 262-8.
106. Cogliati M, Zamfirova RR, Tortorano AM, *et al.* Molecular epidemiology of Italian clinical *Cryptococcus neoformans* var. *grubii* isolates. *Med Mycol.* 2013; **51:** 499-506.
107. Cogliati M. Global molecular epidemiology of *Cryptococcus neoformans* and *Cryptococcus gattii:* An atlas of the molecular types. *Scientifica (Cairo).* 2013; 675213. PMID: 24278784.
108. Conti F, Spinelli FR, Colafrancesco S, *et al.* Acute longitudinal myelitis following *Cryptococcus laurentii* pneumonia in a patient with systemic lupus erythematosus. *Lupus.* 2015; **24:** 94-7.
109. Cordoba S, Vivot W, Szusz W, *et al.* Comparison of different *in vitro* tests to detect *Cryptococcus neoformans* not susceptible to amphotericin B. *Mycopathologia.* 2015; **179:** 359-71.
110. Corral JE, Lima S, Quezada J, *et al.* Cryptococcal osteomyelitis of the skull. *Med Mycol.* 2011; **49:** 667-71.
111. Corti M, Priarone M, Negroni R, *et al.* Ventriculoperitoneal shunts for treating increased intracranial pressure in cryptococcal meningitis with or without ventriculomegaly. *Rev Soc Bras Med Trop.* 2014; **47:** 524-7.
112. Corti M, Solari R, Cangelosi D, *et al.* Sudden blindness due to bilateral optic neuropathy associated with cryptococcal meningitis in an AIDS patient. *Rev Iberoam Micol.* 2010; **27:** 207-9.
113. Crabtree JN, Okagaki LH, Wiesner DL, *et al.* Titan cell production enhances the virulence of *Cryptococcus neoformans. Infect Immun.* 2012; **80:** 3776-85.
114. da Cunha Colombo ER, Mora DJ, Silva-Vergara ML. Immune reconstitution inflammatory syndrome (IRIS) associated with *Cryptococcus neoformans* infection in AIDS patients. *Mycoses.* 2011; **54:** e178-82.
115. Dall Bello AG, Severo CB, Schio S, *et al.* First reported case of cellulitis due to *Cryptococcus gattii* in lung transplantation recipient: A case report. *Dermatol Online J.* 2013; **19:** 20395. PMID: 24314772.
116. Das R, Muldrew KL, Posligua WE, *et al.* Cryptococcal retropharyngeal abscess. *Travel Med Infect Dis.* 2010; **8:** 322-5.
117. Das S, Datt S, Roy P, *et al.* Sporadic occurrence of cryptococcal meningitis in HIV-seronegative patients: Uncommon etiology? *Indian J Pathol Microbiol.* 2017; **60:** 236-8.
118. Datsis AC, Tsintoni A, Tasoula A, *et al.* Isolated cutaneous cryptococcosis in a immunocompromised patient cured without antifungals. *Int J Dermatol.* 2009; **48:** 440-1.
119. Datta K, Bartlett KH, Baer R, *et al.* Spread of *Cryptococcus gattii* into Pacific Northwest region of the United States. *Emerg Infect Dis.* 2009; **15:** 1185-91.
120. Day JN, Chau TT, Lalloo DG. Combination antifungal therapy for cryptococcal meningitis. *N Engl J Med.* 2013; **368:** 2522-3.
121. Day JN, Chau TT, Wolbers M, *et al.* Combination antifungal therapy for cryptococcal meningitis. *N Engl J Med.* 2013; **368:** 1291-302.
122. De Oliveira R, Barbosa L, Jeunon T, *et al.* Ulcerated plaque in the nasal dorsum in a patient with multiple myeloma. *Clin Exp Dermatol.* 2013; **38:** 678-80.
123. Debourgogne A, Iriart X, Blanchet D, *et al.* Characteristics and specificities of *Cryptococcus* infections in French Guiana, 1998-2008. *Med Mycol.* 2011; **49:** 864-71.
124. Del Poeta M, Casadevall A. Ten challenges on *Cryptococcus* and cryptococcosis. *Mycopathologia.* 2012; **173:** 303-10.
125. Del Poeta M, Chaturvedi V. *Cryptococcus* and cryptococcosis in the twenty-first century. *Mycopathologia.* 2012; **173:** 283-5.
126. Desalermos A, Kourkoumpetis TK, Mylonakis E. Update on the epidemiology and management of cryptococcal meningitis. *Expert Opin Pharmacother.* 2012; **13:** 783-9.
127. Dhana A. Diagnosis of cryptococcosis and prevention of cryptococcal meningitis using a novel point-of-care lateral flow assay. *Case Rep Med.* 2013; 640216. PMID: 24319464.
128. Dharwadkar A, Vimal S, Buch AC, *et al.* HIV infection presenting as bone marrow cryptococcosis. *Adv Biomed Res.* 2014; **3:** 144.
129. Dogbey P, Golden M, Ngo N. Cryptococcal lymphadenitis: An unusual initial presentation of HIV infection. *BMJ Case Rep.* 2013; pii: bcr2013010316. PMID: 24014328.
130. Drain PK, Moosa Y, Powderly WG. Antiretroviral therapy after cryptococcal meningitis. *N Engl J Med.* 2014; **371:** 1165.
131. Dromer F, Bernede-Bauduin C, Guillemot D, *et al.* Major role for amphotericin B-flucytosine combination in severe cryptococcosis. *PLoS One.* 2008; **3:** e2870. PMID: 18682846.
132. Dubbels M, Granger D, Theel ES. Low *Cryptococcus* antigen titers by a lateral flow assay should be interpreted cautiously in patients without a prior diagnosis of cryptococcal infection. *J Clin Microbiol.* 2017; pii: JCM.00751-17. PMID: 28566315.
133. Du L, Yang Y, Gu J, *et al.* Systemic review of published reports on primary cutaneous cryptococcosis in immunocompetent patients. *Mycopathologia.* 2015; **180:** 19-25.
134. Dzoyem JP, Kechia FA, Ngaba GP, *et al.* Prevalence of cryptococcosis among HIV-infected patients in Yaounde, Cameroon. *Afr Health Sci.* 2012; **12:** 129-33.
135. Edwards LJ, Price RN, Krause VL, *et al.* Detection of *Mycobacterium leprae* by PCR testing of sputa from a patient with pulmonary cryptococcus coinfection in northern Australia. *J Clin Microbiol.* 2014; **52:** 3811-2.
136. El-Kersh K, Chaddha U, Cavallazzi R, *et al.* The use of endobronchial ultrasound-guided fine-needle aspiration

(EBUS-FNA) in the diagnosis of lymphatic cryptococcosis. *BMJ Case Rep.* 2014; pii: bcr2014207005. PMID: 25253488.

137. Elhence P, Bansal R. Cryptococcosis presenting as anterior neck swelling in an immunocompetent man: A case report. *Acta Cytol.* 2010; **54:** 1130-2.
138. Endo JO, Klein SZ, Pirozzi M, *et al.* Generalized *Cryptococcus albidus* in an immunosuppressed patient with palmopustular psoriasis. *Cutis.* 2011; **88:** 129-32.
139. Enoki E, Maenishi O, Chikugo T, *et al.* Coinfection of *Aspergillus* and *Cryptococcus* in post-tuberculosis pulmonary cavity. *Pathol Int.* 2012; **62:** 574-6.
140. Escandon P, Lizarazo J, Agudelo CI, *et al.* Evaluation of a rapid lateral flow immunoassay for the detection of cryptococcal antigen for the early diagnosis of cryptococcosis in HIV patients in Colombia. *Med Mycol.* 2013; **51:** 765-8.
141. Espinel-Ingroff A, Chowdhary A, Gonzalez GM, *et al.* Multicenter study of isavuconazole MIC distributions and epidemiological cutoff values for the *Cryptococcus neoformans-Cryptococcus gattii* species complex using the CLSI M27-A3 broth microdilution method. *Antimicrob Agents Chemother.* 2015; **59:** 666-8.
142. Espino Barros Palau A, Morgan ML, Foroozan R, *et al.* Neuro-ophthalmic presentations and treatment of cryptococcal meningitis-related increased intracranial pressure. *Can J Ophthalmol.* 2014; **49:** 473-7.
143. Fabbri A, Grazzini M, Vannucci F, *et al.* Cryptococcosis: An unusual cause of tracheal obstruction. *Intern Med.* 2013; **52:** 1279.
144. Fallah H, Watson A, Henderson CJ, *et al.* Cryptococcosis presenting as upper limb cellulitis and ulceration: A case series. *Australas J Dermatol.* 2011; **52:** 288-91.
145. Fang W, Fa Z, Liao W. Epidemiology of *Cryptococcus* and cryptococcosis in China. *Fungal Genet Biol.* 2015; **78:** 7-15.
146. Fang W, Hong N, Li Y, *et al.* Cryptococcosis in patients with nephrotic syndrome: A pooled analysis of cases. *Mycopathologia.* 2017; **182:** 517-25.
147. Farber SA, Micheletti RG. Cryptococcal meningitis presenting with headache and a pustular eruption in a heart transplant patient. *Transpl Infect Dis.* 2015; **17:** 716-8.
148. Farlow AW. The challenges and possibilities of reducing deaths from cryptococcal meningitis in sub-Saharan Africa. *PLoS Med.* 2012; **9:** e1001318. PMID: 23049487.
149. Farrer RA, Voelz K, Henk DA, *et al.* Micro-evolutionary traits and comparative population genomics of the emerging pathogenic fungus *Cryptococcus gattii. Philos Trans R Soc Lond B Biol Sci.* 2016; **371(1709):** pii: 20160021.
150. Feng X, Fu X, Ling B, *et al.* Development of a singleplex PCR assay for rapid identification and differentiation of *Cryptococcus neoformans* var. *grubii, Cryptococcus neoformans* var. *neoformans, Cryptococcus gattii* and hybrids. *J Clin Microbiol.* 2013; **51:** 1920-3.
151. Ferry T, Moos D, Radenne S, *et al.* Primary cutaneous cryptococcosis in a liver transplant recipient. *BMJ Case Rep.* 2011; pii: bcr0220113814. PMID: 22692485.
152. Fishman JA. Editorial commentary: Immune reconstitution syndrome: How do we "tolerate" our microbiome? *Clin Infect Dis.* 2015; **60:** 45-7.
153. Franco-Paredes C, Womack T, Bohlmeyer T, *et al.* Management of *Cryptococcus gattii* meningoencephalitis. *Lancet Infect Dis.* 2015; **15:** 348-55.
154. Freij JB, Freij BJ. The earliest account of human cryptococcosis (Busse-Buschke Disease) in a woman with chronic osteomyelitis of the tibia. *Pediatr Infect Dis J.* 2015; **34:** 1278.
155. Gago S, Serrano C, Alastruey-Izquierdo A, *et al.* Molecular identification, antifungal resistance and virulence of *Cryptococcus neoformans* and *Cryptococcus deneoformans* isolated in Seville, Spain. *Mycoses.* 2017; **60:** 40-50.
156. Ganju SA, Bhagra S, Guleria RC, *et al.* Occupational exposure to Human Immunodeficiency Virus infection: A case missed is a life lost. *Indian J Med Microbiol.* 2013; **31:** 98-9.
157. Garcia-Santibanez RC, Gill V, Yancovitz S, *et al.* Neuroinvasive cryptococcosis in an immunocompetent patient with a negative spinal fluid cryptococcus antigen. *Case Rep Infect Dis.* 2015; 857539. PMID: 25954558.
158. Garrett L, Marr K, West S, *et al.* 74-year-old man from the pacific northwest with fever and a lung mass. *Chest.* 2011; **140:** 814-7.
159. Gates-Hollingsworth MA, Kozel TR. Serotype sensitivity of a lateral flow immunoassay for cryptococcal antigen. *Clin Vaccine Immunol.* 2013; **20:** 634-5.
160. Gazzoni AF, Severo CB, Salles EF, *et al.* Histopathology, serology and cultures in the diagnosis of cryptococcosis. *Rev Inst Med Trop Sao Paulo.* 2009; **51:** 255-9.
161. Georgi A, Schneemann M, Tintelnot K, *et al. Cryptococcus gattii* meningoencephalitis in an immunocompetent person 13 months after exposure. *Infection.* 2009; **37:** 370-3.
162. Ghatalia PA, Vick A, Vattoth S, *et al.* Reversible blindness in cryptococcal meningitis with normal intracranial pressure: Case report and review of the literature. *Clin Infect Dis.* 2014; **59:** 310-3.
163. Ghosh A, Tilak R, Bhushan R, *et al.* Lymphnodal co-infection of *Cryptococcus* and *Histoplasma* in a HIV-infected patient and review of published reports. *Mycopathologia.* 2015; **180:** 105-10.
164. Gibson JF, Johnston SA. Immunity to *Cryptococcus neoformans* and *C.gattii* during cryptococcosis. *Fungal Genet Biol.* 2015; **78:** 76-86.
165. Gill H, Ip AH, So JC, *et al.* Disseminated cryptococcosis mimicking a lymphoma. *Eur J Haematol.* 2012; **88:** 275-6.
166. Girardin M, Greloz V, Hadengue A. Cryptococcal gastroduodenitis: A rare location of the disease. *Clin Gastroenterol Hepatol.* 2010; **8:** e28-9.
167. Gogia A, Mehta P, Raina V. Cryptococcal meningitis in chronic lymphocytic leukemia. *Indian J Cancer.* 2013; **50:** 301.
168. Goldbach H, McMahon C, Boos MD. An exophytic mass on the mandible of an immunocompromised man. *Clin Infect Dis.* 2014; **58:** 540 & 596-7.
169. Goldman JD, Vollmer ME, Luks AM. Cryptococcosis in the immunocompetent patient. *Respir Care.* 2010; **55:** 1499-503.
170. Gopal M, McCrosson S, Edmonds P, *et al.* Cryptococcosis of the upper genital tract. *AIDS Patient Care STDS.* 2009; **23:** 71-3.

171. Gordon DH, Stow NW, Yapa HM, *et al.* Laryngeal cryptococcosis: Clinical presentation and treatment of a rare cause of hoarseness. *Otolaryngol Head Neck Surg.* 2010; **142:** S7-9.
172. Govender NP, Roy M, Mendes JF, *et al.* Evaluation of screening and treatment of cryptococcal antigenaemia among HIV-infected persons in Soweto, South Africa. *HIV Med.* 2015; **16:** 468-76.
173. Greenlee JE. Cryptococcal encephalopathy without persisting cerebrospinal fluid pleocytosis, a diagnostic challenge: Case report and review of the literature. *Clin Neurol Neurosurg.* 2013; **115:** 1897-9.
174. Guan C, Chen H, Shao C, *et al.* Intralobar pulmonary sequestration complicating with cryptococcal infection. *Clin Respir J.* 2015; **9:** 22-6.
175. Guinea J, Hagen F, Pelaez T, *et al.* Antifungal susceptibility, serotyping and genotyping of clinical *Cryptococcus neoformans* isolates collected during 18 years in a single institution in Madrid, Spain. *Med Mycol.* 2010; **48:** 942-8.
176. Gullo FP, Rossi SA, Sardi Jde C, *et al.* Cryptococcosis: Epidemiology, fungal resistance and new alternatives for treatment. *Eur J Clin Microbiol Infect Dis.* 2013; **32:** 1377-91.
177. Gunda DW, Bakshi FA, Rambau P, *et al.* Pulmonary cryptococcosis presenting as acute severe respiratory distress in a newly diagnosed HIV patient in Tanzania: A case report. *Clin Case Rep.* 2015; **3:** 749-52.
178. Gupta A, Capoor MR, Gupta S, *et al.* Concomitant infections of influenza A H1N1 and disseminated cryptococcosis in an HIV seropositive patient. *J Lab Physicians.* 2015; **7:** 134-6.
179. Gupta N, Sachdev A, Gupta D, *et al.* Disseminated cryptococcosis in an immunocompetent toddler. *Indian Pediatr.* 2017; **54:** 145-6.
180. Gupta P, Malik S, Khare V, *et al.* A fatal case of meningitis caused by *Cryptococcus neoformans* var. *grubii* in an immunocompetent male. *J Infect Dev Ctries.* 2011; **5:** 71-4.
181. Gupta R, Kushwaha S, Behera S, *et al.* Vertebro-cerebral cryptococcosis mimicking tuberculosis: A diagnostic dilemma in countries with high burden of tuberculosis. *Indian J Med Microbiol.* 2012; **30:** 245-8.
182. Gupta V, Karnik ND, Agrawal D, *et al.* A case of lung mass: A common association between uncommon diseases. *BMJ Case Rep.* 2014; pii: bcr2014205632. PMID: 25422330.
183. Gurung S, Sherpa NT, Yoden Bhutia P, *et al. Cryptococcus gatti* serotype B isolated in Sikkim (North-East India) - A new geographical niche. *Med Mycol Case Rep.* 2012; **1:** 27-8.
184. Gutch RS, Nawange SR, Singh SM, *et al.* Antifungal susceptibility of clinical and environmental *Cryptococcus neoformans* and *Cryptococcus gattii* isolates in Jabalpur, a city of Madhya Pradesh in Central India. *Braz J Microbiol.* 2015; **46:** 1125-33.
185. Hadano Y, Yoshii H, Hayashi M, *et al.* A rare case report of central line-associated bloodstream infection caused by *Cryptococcus arboriformis. Intern Med.* 2015; **54:** 1141-3.
186. Haddad N, Cavallaro MC, Lopes MP, *et al.* Pulmonary cryptococcoma: A rare and challenging diagnosis in immunocompetent patients. *Autops Case Rep.* 2015; **5:** 35-40.
187. Haddow LJ, Colebunders R, Meintjes G, *et al.* Cryptococcal immune reconstitution inflammatory syndrome in HIV-1-infected individuals: Proposed clinical case definitions. *Lancet Infect Dis.* 2010; **10:** 791-802.
188. Hagen F, Ceresini PC, Polacheck I, *et al.* Ancient dispersal of the human fungal pathogen *Cryptococcus gattii* from the Amazon rainforest. *PLoS One.* 2013; **8:** e71148.
189. Hagen F, Colom MF, Swinne D, *et al.* Autochthonous and dormant *Cryptococcus gattii* infections in Europe. *Emerg Infect Dis.* 2012; **18:** 1618-24.
190. Hagen F, Hare Jensen R, Meis JF, *et al.* Molecular epidemiology and *in vitro* antifungal susceptibility testing of 108 clinical *Cryptococcus neoformans sensu lato* and *Cryptococcus gattii sensu lato* isolates from Denmark. *Mycoses.* 2016; **59:** 576-84.
191. Hagen F, Khayhan K, Theelen B, *et al.* Recognition of seven species in the *Cryptococcus gattii/Cryptococcus neoformans* species complex. *Fungal Genet Biol.* 2015; **78:** 16-48.
192. Hagen F, van Assen S, Luijckx GJ, *et al.* Activated dormant *Cryptococcus gattii* infection in a Dutch tourist who visited Vancouver Island (Canada): A molecular epidemiological approach. *Med Mycol.* 2010; **48:** 528-31.
193. Haidar G, Singh N. *Cryptococcus*: Shedding new light on an inveterate yeast. *J Fungi.* 2015; **1:** 115-29.
194. Han G, Kwon SH, Song HJ, *et al.* Disseminated cryptococcosis presenting as cellulitis. *Indian J Dermatol Venereol Leprol.* 2017; **83:** 89-91.
195. Hansen J, Slechta ES, Gates-Hollingsworth MA, *et al.* Large-scale evaluation of the immuno-mycologics lateral flow and enzyme-linked immunoassays for detection of cryptococcal antigen in serum and cerebrospinal fluid. *Clin Vaccine Immunol.* 2013; **20:** 52-5.
196. Harris J, Lockhart S, Chiller T. *Cryptococcus gattii:* Where do we go from here? *Med Mycol.* 2012; **50:** 113-29.
197. Harris JR, Lindsley MD, Henchaichon S, *et al.* High prevalence of cryptococcal infection among HIV-infected patients hospitalized with pneumonia in Thailand. *Clin Infect Dis.* 2012; **54:** e43-50.
198. Harris JR, Lockhart SR, Debess E, *et al. Cryptococcus gattii* in the United States: Clinical aspects of infection with an emerging pathogen. *Clin Infect Dis.* 2011; **53:** 1188-95.
199. Harris JR, Lockhart SR, Sondermeyer G, *et al. Cryptococcus gattii* infections in multiple states outside the US Pacific Northwest. *Emerg Infect Dis.* 2013; **19:** 1620-6.
200. Harris RM, Stillman IE, Goldsmith JD, *et al.* Pathological rib fracture and soft tissue mass simulating malignancy - *Cryptococcus*, an unsuspected culprit. *Diagn Microbiol Infect Dis.* 2015; **81:** 189-91.
201. Hassan H, Cotton MF, Rabie H. Complicated and protracted cryptococcal disease in HIV-infected children. *Pediatr Infect Dis J.* 2015; **34:** 62-5.
202. Henao-Martínez AF, Beckham JD. Cryptococcosis in solid organ transplant recipients. *Curr Opin Infect Dis.* 2015; **28:** 300-7.

203. Hernandez YU, Monzon JF, Delgado-Hernandez R, *et al.* Cryptococcoma of the brain in an immunocompetent man. *Natl Med J India.* 2013; **26:** 216-7.
204. Hiraga A, Yatomi M, Ozaki D, *et al.* Cryptococcosis mimicking lung cancer with brain metastasis. *Clin Neurol Neurosurg.* 2015; **135:** 93-5.
205. Hirsch HH, Wehrle-Wieland E, Battegay M. Antiretroviral therapy after cryptococcal meningitis. *N Engl J Med.* 2014; **371:** 1166.
206. Ho SW, Ang CL, Ding CS, *et al.* Necrotizing fasciitis caused by *Cryptococcus gattii. Am J Orthop (Belle Mead NJ).* 2015; **44:** e517-22.
207. Hu WL, Hu HJ, Dai N, *et al.* Extrahepatic biliary cryptococcosis: A case report. *J Dig Dis.* 2011; **12:** 500-4.
208. Hu Z, Xu C, Wei H, *et al.* Solitary cavitary pulmonary nodule may be a common CT finding in AIDS-associated pulmonary cryptococcosis. *Scand J Infect Dis.* 2013; **45:** 378-89.
209. Huang CJ, You DL, Lee PI, *et al.* Characteristics of integrated ^{18}F-FDG PET/CT in pulmonary cryptococcosis. *Acta Radiol.* 2009; **50:** 374-8.
210. Huang HR, Fan LC, Rajbanshi B, *et al.* Evaluation of a new cryptococcal antigen lateral flow immunoassay in serum, cerebrospinal fluid and urine for the diagnosis of cryptococcosis: A meta-analysis and systematic review. *PLoS One.* 2015; **10:** e0127117.
211. Huston SM, Mody CH. Cryptococcosis: An emerging respiratory mycosis. *Clin Chest Med.* 2009; **30:** 253-64.
212. Hu Z, Chen J, Wang J, *et al.* Radiological characteristics of pulmonary cryptococcosis in HIV-infected patients. *PLoS One.* 2017; **12:** e0173858. PMID: 28301552.
213. Iatta R, Hagen F, Fico C, *et al. Cryptococcus gattii* infection in an immunocompetent patient from Southern Italy. *Mycopathologia.* 2012; **174:** 87-92.
214. Idnurm A, Lin X. Rising to the challenge of multiple *Cryptococcus* species and the diseases they cause. *Fungal Genet Biol.* 2015; **78:** 1-6.
215. Ikeda T, Kaminaka C, Yamamoto Y, *et al.* Disseminated cryptococcosis-induced skin ulcers in a patient with autoimmune hepatitis. *Case Rep Dermatol.* 2014; **6:** 98-102.
216. Illnait-Zaragozi MT, Martinez-Machin GF, Fernandez-Andreu CM, *et al. Cryptococcus* and cryptococcosis in Cuba. A minireview. *Mycoses.* 2014; **57:** 707-17.
217. Illnait-Zaragozí MT, Martinez-Machín GF, Fernandez-Andreu CM, *et al.* Environmental isolation and characterisation of *Cryptococcus* species from living trees in Havana city, Cuba. *Mycoses.* 2012; **55:** e138-44.
218. Im H, Chae JD, Yoo M, *et al.* First case of continuous ambulatory peritoneal dialysis-related peritonitis caused by *Cryptococcus arboriformis. Ann Lab Med.* 2014; **34:** 328-31.
219. Iqbal N, De Bess EE, Wohrle R, *et al.* Correlation of genotype and *in vitro* susceptibilities of *Cryptococcus gattii* strains from the Pacific Northwest of the United States. *J Clin Microbiol.* 2010; **48:** 539-44.
220. Ishikawa H, Kasahara K, Sato S, *et al.* Simple and rapid method for the detection of *Filobasidiella neoformans* in a probiotic dairy product by using loop-mediated isothermal amplification. *Int J Food Microbiol.* 2014; **178:** 107-12.
221. Islam S, Das A, Islam N. Cryptococcosis in organ transplantation. *Mymensingh Med J.* 2010; **19:** 142-3.
222. Iverson SA, Chiller T, Beekmann S, *et al.* Recognition and diagnosis of *Cryptococcus gattii* infections in the United States. *Emerg Infect Dis.* 2012; **18:** 1012-5.
223. Jackson A, Hosseinipour MC. Management of cryptococcal meningitis in sub-Saharan Africa. *Curr HIV/AIDS Rep.* 2010; **7:** 134-42.
224. Jackson A, van der Horst C. New insights in the prevention, diagnosis, and treatment of cryptococcal meningitis. *Curr HIV/AIDS Rep.* 2012; **9:** 267-77.
225. Jackson AT, van der Horst CM. Cryptococcosis in AIDS: New data but questions remain. *Clin Infect Dis.* 2016; **62:** 588-9.
226. Jackson NA, Herring DB. Primary capsule-deficient cutaneous cryptococcosis in a sporotrichoid pattern in an immunocompetent host. *Cutis.* 2015; **96:** e26-9.
227. Jacobson ME, Griesser MJ, Paloski MD, *et al.* Isolated *Cryptococcus neoformans* osteomyelitis of the proximal femur: A case report and review of literature. *Orthop Surg.* 2012; **4:** 190-3.
228. Jadhav MP, Bamba A, Shinde VM, *et al.* Liposomal amphotericin B (Fungisome) for the treatment of cryptococcal meningitis in HIV/AIDS patients in India: A multicentric, randomized controlled trial. *J Postgrad Med.* 2010; **56:** 71-5.
229. Jain BB, Bose D, Mondal R, *et al.* Disseminated cryptococcosis in an immunocompetent child. *Turk Patoloji Derg.* 2017; **33:** 77-80.
230. Jain D, Najjar M, Azher Q, *et al.* Cryptococcal sternal osteomyelitis in a healthy woman: A review of *Cryptococcus neoformans. BMJ Case Rep.* 2013; pii: bcr2013009129.
231. Jain K; Mruthyunjaya, Ravishankar R. Cryptococcal abscess and osteomyelitis of the proximal phalanx of the hand. *Indian J Pathol Microbiol.* 2011; **54:** 216-8.
232. Jain TK, Karunanithi S, Bal C, *et al.* 18F-FDG PET/CT Imaging in adrenal cryptococcosis. *Clin Nucl Med.* 2017; **42:** e194-5.
233. Jarvis J, Meintjes G. Cryptococcal meningitis - A neglected killer. *S Afr Med J.* 2011; **101:** 244-5.
234. Jarvis JN, Bicanic T, Loyse A, *et al.* Determinants of mortality in a combined cohort of 501 patients with HIV-associated cryptococcal meningitis: Implications for improving outcomes. *Clin Infect Dis.* 2014; **58:** 736-45.
235. Jarvis JN, Percival A, Bauman S, *et al.* Evaluation of a novel point-of-care cryptococcal antigen test on serum, plasma and urine from patients with HIV-associated cryptococcal meningitis. *Clin Infect Dis.* 2011; **53:** 1019-23.
236. Jarvis JN, Wainwright H, Harrison TS, *et al.* Pulmonary cryptococcosis misdiagnosed as smear-negative pulmonary tuberculosis with fatal consequences. *Int J Infect Dis.* 2010; **14 (Suppl. 3):** e310-2.
237. Jeng JY, Tomblinson CM, Ocal IT, *et al.* Laryngeal cryptococcosis: Literature review and guidelines for laser ablation of fungal lesions. *Laryngoscope.* 2016; **126:** 1625-9.
238. Jhamb R, Kashyap B, Das S, *et al.* Symptomatic relapse of HIV-associated cryptococcal meningitis: Recurrent cryptococcal meningitis or *Cryptococcus*-related immune

reconstitution inflammatory syndrome? *Int J STD AIDS.* 2014; **25:** 369-72.

239. Jitmuang A, Panackal AA, Williamson PR, *et al.* Performance of the cryptococcal antigen lateral flow assay in non-HIV-related cryptococcosis. *J Clin Microbiol.* 2016; **54:** 460-3.
240. Johannson KA, Huston SM, Mody CH, *et al. Cryptococcus gattii* pneumonia. *CMAJ.* 2012; **184:** 1387-90.
241. Johannsson B, Callaghan JJ. Prosthetic hip infection due to *Cryptococcus neoformans:* Case report. *Diagn Microbiol Infect Dis.* 2009; **64:** 76-9.
242. Joo HS, Ha JK, Hwang CJ, *et al.* Lumbar cryptococcal osteomyelitis mimicking metastatic tumor. *Asian Spine J.* 2015; **9:** 798-802.
243. Joshi NS, Fisher BT, Prasad PA, *et al.* Epidemiology of cryptococcal infection in hospitalized children. *Pediatr Infect Dis J.* 2010; **29:** e91-5.
244. Kabanda T, Siedner MJ, Klausner JD, *et al.* Point-of-care diagnosis and prognostication of cryptococcal meningitis with the cryptococcal antigen lateral flow assay on cerebrospinal fluid. *Clin Infect Dis.* 2014; **58:** 113-6.
245. Kadam D, Chandanwale A, Bharadwaj R, *et al.* High prevalence of cryptococcal antigenaemia amongst asymptomatic advanced HIV patients in Pune, India. *Indian J Med Microbiol.* 2017; **35:** 105-8.
246. Kannangai R, Sachithanandham J, Mahadevan A, *et al.* Association of neurotropic viruses in HIV-infected individuals who died of secondary complications of tuberculosis, cryptococcosis or toxoplasmosis in South India. *J Clin Microbiol.* 2013; **51:** 1022-5.
247. Kao C, Goldman DL. Cryptococcal disease in HIV-infected children. *Curr Infect Dis Rep.* 2016; **18:** 27.
248. Kao CC, Wu VC, Sun HY, *et al.* Paradoxical cryptococcal immune reconstitution inflammatory syndrome in advanced chronic kidney disease. *Int Urol Nephrol.* 2013; **45:** 1505-9.
249. Kaplan JE, Vallabhaneni S, Smith RM, *et al.* Cryptococcal antigen screening and early antifungal treatment to prevent cryptococcal meningitis: A review of the literature. *J Acquir Immune Defic Syndr.* 2015; **68:** S331-9.
250. Karmakar P, Jeyarajah R, Ramasubramanian V. Isolated cryptococcal osteomyelitis in immunocompetent patient. *J Indian Med Assoc.* 2011; **109:** 592 & 594.
251. Katchanov J, Jefferys L, Tominski D, *et al.* Cryptococcosis in HIV-infected hospitalized patients in Germany: Evidence for routine antigen testing. *J Infect.* 2015; **71:** 110-6.
252. Kaur G, Manucha V, Verma K. Cryptococcal osteomyelitis of the rib diagnosed on fine needle aspiration cytology. *Acta Cytol.* 2010; **54 (Suppl. 5):** 1056-7.
253. Kaur H, Zaman K, Thapa BR, *et al.* Fatal cryptococcosis involving multiple sites in an immunocompetent child. *Indian J Med Microbiol.* 2015; **33 (Suppl):** 148-50.
254. Kawamoto K, Miyoshi H, Suzuki T, *et al.* Clinicopathological features of cryptococcal lymphadenitis and a review of literature. *J Clin Exp Hematop.* 2017; PMID: 28592745.
255. Kelly S, Marriott D. Miliary pulmonary cryptococcosis. *Med Mycol Case Rep.* 2014; **6:** 22-4.
256. Kennedy E, Vanichanan J, Rajapreyar I, *et al.* A pseudo-outbreak of disseminated cryptococcal disease after orthotopic heart transplantation. *Mycoses.* 2016; **59:** 75-9.
257. Kessler B, Bally F, Hewer E, Sendi P. Delayed diagnosis of cryptococcal meningoencephalitis due to negative cryptococcal antigen test. *BMJ Case Rep.* 2013; pii: bcr2012007980.
258. Khan A, Jamil B, Ali R, *et al.* Tuberculous and cryptococcal meningitis in a setting with high TB and low HIV prevalence. *J Coll Physicians Surg Pak.* 2009; **19:** 487-91.
259. Khan ZU, Randhawa HS, Chehadeh W, *et al. Cryptococcus neoformans* serotype A and *Cryptococcus gattii* serotype B isolates differ in their susceptibilities to fluconazole and voriconazole. *Int J Antimicrob Agents.* 2009; **33:** 559-63.
260. Khayhan K, Hagen F, Pan W, *et al.* Geographically structured populations of *Cryptococcus neoformans* variety *grubii* in Asia correlate with HIV status and show a clonal population structure. *PLoS One.* 2013; **8:** e72222.
261. Khuraijam R, Lungran P, Yoihenba K, *et al.* Pancytopenia and cutaneous cryptococcosis as an indicator disease of acquired immune deficiency syndrome. *Indian J Med Microbiol.* 2015; **33:** 439-42.
262. Khurana RN, Javaheri M, Rao N. Ophthalmic manifestations of immune reconstitution inflammatory syndrome associated with *Cryptococcus neoformans. Ocul Immunol Inflamm.* 2008; **16:** 185-90.
263. Kiggundu R, Rhein J, Meya DB, *et al.* Unmasking cryptococcal meningitis immune reconstitution inflammatory syndrome in pregnancy induced by HIV antiretroviral therapy with postpartum paradoxical exacerbation. *Med Mycol Case Rep.* 2014; **5:** 16-9.
264. Kikuchi N, Hiraiwa T, Ishikawa M, *et al.* Cutaneous cryptococcosis mimicking pyoderma gangrenosum: A report of four cases. *Acta Derm Venereol.* 2016; **96:** 116-7.
265. Kim YS, Lee IH, Kim HS, *et al.* Pulmonary cryptococcosis mimicking primary lung cancer with multiple lung metastases. *Tuberc Respir Dis (Seoul).* 2012; **73:** 182-6.
266. King VS, Winder MJ. Multiple cerebral cryptococcomas in an immunocompetent man: An unlikely diagnosis. *ANZ J Surg.* 2014; **84:** 588-90.
267. Klausner JD, Govender N, Oladoyinbo S, *et al.* Preventing AIDS deaths: Cryptococcal antigen screening and treatment. *Lancet Infect Dis.* 2012; **12:** 431-2.
268. Klein KR, Hall L, Deml SM, *et al.* Identification of *Cryptococcus gattii* by use of L-canavanine glycine bromothymol blue medium and DNA sequencing. *J Clin Microbiol.* 2009; **47:** 3669-72.
269. Kohno S, Kakeya H, Izumikawa K, *et al.* Clinical features of pulmonary cryptococcosis in non-HIV patients in Japan. *J Infect Chemother.* 2015; **21:** 23-30.
270. Kong QT, Zhou WQ, Feng J, *et al.* Isolated skull cryptococcosis in an immunocompetent patient. *Mycopathologia.* 2013; **175:** 187-91.
271. Koshy JM, Mohan S, Deodhar D, *et al.* Clinical diversity of CNS cryptococcosis. *J Assoc Physicians India.* 2016; **64:**15-9.
272. Kothavade RJ, Oberai CM, Valand AG, *et al.* Disseminated cryptococcosis and fluconazole resistant oral candidiasis in a patient with acquired immunodeficiency syndrome (AIDS). *J Infect Dev Ctries.* 2010; **4:** 674-8.

273. Kothiwala SK, Prajapat M, Kuldeep CM, *et al.* Cryptococcal panniculitis in a renal transplant recipient: Case report and review of literature. *J Dermatol Case Rep.* 2015; **9:** 76-80.
274. Kozel TR, Bauman SK. CrAg lateral flow assay for cryptococcosis. *Expert Opin Med Diagn.* 2012; **6:** 245-51.
275. Kresch ZA, Espinosa-Heidmann D, Harper T, *et al.* Disseminated *Cryptococcus* with ocular cryptococcoma in a human immunodeficiency virus-negative patient. *Int Ophthalmol.* 2012; **32:** 281-4.
276. Kulkarni A, Sinha M, Anandh U. Primary cutaneous cryptococcosis due to *Cryptococcus laurentii* in a renal transplant recipient. *Saudi J Kidney Dis Transpl.* 2012; **23:** 102-5.
277. Kumar A, Gopinath S, Dinesh KR, *et al.* Infectious psychosis: Cryptococcal meningitis presenting as a neuropsychiatry disorder. *Neurol India.* 2011; **59:** 909-11.
278. Kumar S, Wanchu A, Chakrabarti A, *et al.* Cryptococcal meningitis in HIV infected: Experience from a North Indian tertiary center. *Neurol India.* 2008; **56:** 444-9.
279. Kumari R, Raval M, Dhun A. Cryptococcal choroid plexitis: Rare imaging findings of central nervous system cryptococcal infection in an immunocompetent individual. *Br J Radiol.* 2010; **83:** e14-7.
280. Kunadharaju R, Choe U, Harris JR, *et al. Cryptococcus gattii*, Florida, USA, 2011. *Emerg Infect Dis.* 2013; **19:** 519-21.
281. Kushner YB, Brimo F, Schwartzman K, *et al.* A rare case of pulmonary cryptococcal inflammatory myofibroblastic tumor diagnosed by fine needle aspiration cytology. *Diagn Cytopathol.* 2010; **38:** 447-51.
282. Kuttiatt V, Sreenivasa P, Garg I, *et al.* Cryptococcal lymphadenitis and immune reconstitution inflammatory syndrome: Current considerations. *Scand J Infect Dis.* 2011; **43:** 664-8.
283. Kwon-Chung KJ, Bennett JE, Wickes BL, *et al.* The case for adopting the "Species Complex" nomenclature for the etiologic agents of cryptococcosis. *mSphere.* 2017; 2. pii: e00357-16. PMID: 28101535.
284. Kwon-Chung KJ, Fraser JA, Doering TL, *et al. Cryptococcus neoformans* and *Cryptococcus gattii*, the etiologic agents of cryptococcosis. *Cold Spring Harb Perspect Med.* 2014; 4a019760. PMID: 24985132.
285. Kwon-Chung KJ, Perfect JR, Levitz SM. A chronological history of the International Conference on *Cryptococcus* and Cryptococcosis (ICCC), an invaluable forum for growth of the cryptococcal research field and clinical practice. *Mycopathologia.* 2012; **173:** 287-93.
286. Kwon-Chung KJ, Saijo T. Is *Cryptococcus gattii* a primary pathogen? *J. Fungi.* 2015; **1:** 154-67.
287. La Hoz RM, Pappas PG. Cryptococcal infections: Changing epidemiology and implications for therapy. *Drugs.* 2013; **73:** 495-504.
288. Laaks D, Smit DP, Meyer D. Cryptococcal IRIS in the anterior segment of the eye. *AIDS.* 2013; **27:** 489-90.
289. Lazzara M, Joshi A. Disseminated cryptococcosis involving the head and neck. *BMJ Case Rep.* 2014; pii: bcr2013202306.
290. Leao CA, Ferreira-Paim K, Andrade-Silva L, *et al.* Primary cutaneous cryptococcosis caused by *Cryptococcus gattii* in an immunocompetent host. *Med Mycol.* 2011; **49:** 352-5.
291. Lee MW, Kim EJ, Ko JY, *et al.* Cryptococcal panniculitis and myositis: A diverse manifestation of cryptococcal infection. *J Dermatol.* 2014; **41:** 362-3.
292. Lee Y, Kim JH, Seo H, *et al.* A case of disseminated cryptococcosis mimicking lung cancer and prostate cancer in an immunocompetent patient: A review of the literature. *Korean J Med.* 2015; **89:** 91-6.
293. Leechawengwongs M, Milindankura S, Sathirapongsasuti K, *et al.* Primary cutaneous cryptococcosis caused by *Cryptococcus gattii* VGII in a tsunami survivor from Thailand. *Med Mycol Case Rep.* 2014; **6:** 31-3.
294. Li F, Yang HM, Wang HW. Disseminated cryptococcosis with widespread necrotizing fasciitis and cryptococcemia occurring in an immunosuppressed patient. *Ann Dermatol.* 2014; **26:** 273-5.
295. Li J, Wang N, Hong Q, *et al.* Duodenal cryptococcus infection in an AIDS patient: Retrospective clinical analysis. *Eur J Gastroenterol Hepatol.* 2015; **27:** 226-9.
296. Li SS, Mody CH. *Cryptococcus. Proc Am Thorac Soc.* 2010; **7:** 186-96.
297. Liapis K, Taussig D, Cotter FE, *et al.* Cutaneous cryptococcosis in Hodgkin lymphoma. *Br J Haematol.* 2014; **164:** 467.
298. Liaw SJ, Wu HC, Hsueh PR. Microbiological characteristics of clinical isolates of *Cryptococcus neoformans* in Taiwan: Serotypes, mating types, molecular types, virulence factors and antifungal susceptibility. *Clin Microbiol Infect.* 2010; **16:** 696-703.
299. Lin GY, Lin TY, Lee JT, *et al.* An isolated cryptococcoma mimicking nasopharyngeal cancer. *Infection.* 2015; **43:** 129-30.
300. Lin YY, Shiau S, Fang CT. Risk factors for invasive *Cryptococcus neoformans* diseases: A case-control study. *PLoS One.* 2015; **10:** e0119090. PMID: 25747471.
301. Lindsley MD, Mekha N, Baggett HC, *et al.* Evaluation of a newly developed lateral flow immunoassay for the diagnosis of cryptococcosis. *Clin Infect Dis.* 2011; **53:** 321-5.
302. Liu K, Ding H, Xu B, *et al.* Clinical analysis of non-AIDS patients pathologically diagnosed with pulmonary cryptococcosis. *J Thorac Dis.* 2016; **8:** 2813-21.
303. Liu TB, Perlin DS, Xue C. Molecular mechanisms of cryptococcal meningitis. *Virulence.* 2012; **3:** 173-81.
304. Liu XZ, Wang QM, Goker M, *et al.* Towards an integrated phylogenetic classification of the Tremellomycetes. *Stud Mycol.* 2015; **81:** 85-147.
305. Liu Y, Qunpeng H, Shutian X, *et al.* Fatal primary cutaneous cryptococcosis: Case report and review of published literature. *Ir J Med Sci.* 2016; **185:** 959-63.
306. Liyanage DS, Pathberiya LP, Gooneratne IK, *et al.* Cryptococcal meningitis presenting with bilateral complete ophthalmoplegia: A case report. *BMC Res Notes.* 2014; **7:** 328.
307. Lizarazo J, Escandon P, Agudelo CI, *et al.* Retrospective study of the epidemiology and clinical manifestations of *Cryptococcus gattii* infections in Colombia from 1997-2011. *PLoS Negl Trop Dis.* 2014; **8:** e3272. PMID: 25411779.

308. Lomes NR, Melhem MS, Szeszs MW, Martins MA, Buccheri R. Cryptococcosis in non-HIV/non-transplant patients: A Brazilian case series. *Med Mycol.* 2016; **54:** 669-76.
309. Longley N, Harrison TS, Jarvis JN. Cryptococcal immune reconstitution inflammatory syndrome. *Curr Opin Infect Dis.* 2013; **26:** 26-34.
310. Lortholary O, Harrison TS. Prevention of AIDS-associated cryptococcosis in resource-poor areas. *Lancet Infect Dis.* 2011; **11:** 892-4.
311. Lourens A, Jarvis JN, Meintjes G, *et al.* Rapid diagnosis of cryptococcal meningitis by use of lateral flow assay on cerebrospinal fluid samples: Influence of the high-dose "hook" effect. *J Clin Microbiol.* 2014; **52:** 4172-5.
312. Louro R, Ferreira R, Pinheiro C, *et al.* Fungal meningitis in an immunocompetent patient. *Clin Drug Investig.* 2013; **33:** S47-50.
313. Loyse A, Bicanic T, Jarvis JN. Combination antifungal therapy for cryptococcal meningitis. *N Engl J Med.* 2013; **368:** 2522.
314. Loyse A, Moodley A, Rich P, *et al.* Neurological, visual and MRI brain scan findings in 87 South African patients with HIV-associated cryptococcal meningoencephalitis. *J Infect.* 2015; **70:** 668-75.
315. Loyse A, Thangaraj H, Easterbrook P, *et al.* Cryptococcal meningitis: Improving access to essential antifungal medicines in resource-poor countries. *Lancet Infect Dis.* 2013; **13:** 629-37.
316. Luo FL, Tao YH, Wang YM, *et al.* Clinical study of 23 pediatric patients with cryptococcosis. *Eur Rev Med Pharmacol Sci.* 2015; **19:** 3801-10.
317. Mackowick PA, Klein CN, Mohanraj BS, *et al.* Rapidly progressive skin lesions requiring admission in a young, HIV-infected man. *Clin Infect Dis.* 2013; **56:** 117 & 159-60.
318. Mah-Lee R, Barrow G. A case of systemic cryptococcal disease in HIV infection. *West Indian Med J.* 2013; **62:** 374-6.
319. Mahale K, Patil S, Ravikumar, *et al.* Prevalence of cryptococcal meningitis among immunocompetent and immunocompromised individuals in Bellary, South India: A prospective study. *J Clin Diag Res.* 2012; **6:** 388-92.
320. Majumder S, Mandal SK, Bandyopadhyay D. Prognostic markers in AIDS-related cryptococcal meningitis. *J Assoc Physicians India.* 2011; **59:** 152-4.
321. Makadzange AT, McHugh G. New approaches to the diagnosis and treatment of cryptococcal meningitis. *Semin Neurol.* 2014; **34:** 47-60.
322. Makino Y, Nishiyama O, Sano H, *et al.* Cavitary pulmonary cryptococcosis with an *Aspergillus* fungus ball. *Intern Med.* 2014; **53:** 2737-9.
323. Malheiro L, Lazzara D, Xerinda S, *et al.* Cryptococcal meningoencephalitis in a patient with hyper immunoglobulin M (IgM) syndrome: A case report. *BMC Res Notes.* 2014; **7:** 566. PMID: 25155248.
324. Manabe YC, Nonyane BA, Nakiyingi L, *et al.* Point-of-care lateral flow assays for tuberculosis and cryptococcal antigenuria predict death in HIV infected adults in Uganda. *PLoS One.* 2014; **9:** e101459. PMID: 25000489.
325. Marques SA, Bastazini I Jr, Martins AL, *et al.* Primary cutaneous cryptococcosis in Brazil: report of 11 cases in immunocompetent and immunosuppressed patients. *Int J Dermatol.* 2012; **51:** 780-4.
326. Marr KA, Datta K, Pirofski LA, *et al.* *Cryptococcus gattii* infection in healthy hosts: A sentinel for subclinical immunodeficiency? *Clin Infect Dis.* 2012; **54:** 153-4.
327. Marr KA. *Cryptococcus gattii* as an important fungal pathogen of western North America. *Expert Rev Anti Infect Ther.* 2012; **10:** 637-43.
328. Marr KA. *Cryptococcus gattii:* The tip of the iceberg. *Clin Infect Dis.* 2011; **53:** 1196-8.
329. Martin-Blondel G, Ysebaert L. Disseminated cryptococcosis. *N Engl J Med.* 2014; **370:** 1741.
330. Martinez-Giron R. Cryptococcal yeast cells on a sputum smear. *Diagn Cytopathol.* 2010; **38:** 270-1.
331. May RC, Stone NR, Wiesner DL, *et al.* *Cryptococcus:* From environmental saprophyte to global pathogen. *Nat Rev Microbiol.* 2016; **14:** 106-17.
332. Maziarz EK, Perfect JR. Cryptococcosis. *Infect Dis Clin North Am.* 2016; **30:** 179-206.
333. Mazumder SA, Cleveland KO. Cryptococcal meningitis after neurosurgery. *Am J Med Sci.* 2010; **339:** 582-3.
334. McCulloh RJ, Phillips R, Perfect JR, *et al.* *Cryptococcus gattii* genotype VGI infection in New England. *Pediatr Infect Dis J.* 2011; **30:** 1111-4.
335. McDonald T, Wiesner DL, Nielsen K. *Cryptococcus. Curr Biol.* 2012; **22:** R554-5.
336. McKenney J, Bauman S, Neary B, *et al.* Prevalence, correlates and outcomes of cryptococcal antigen positivity among patients with AIDS, United States, 1986-2012. *Clin Infect Dis.* 2015; **60:** 959-65.
337. McMullan BJ, Halliday C, Sorrell TC, *et al.* Clinical utility of the cryptococcal antigen lateral flow assay in a diagnostic mycology laboratory. *PLoS One.* 2012; **7:** e49541.
338. McMullan BJ, Sorrell TC, Chen SC. *Cryptococcus gattii* infections: Contemporary aspects of epidemiology, clinical manifestations and management of infection. *Future Microbiol.* 2013; **8:** 1613-31.
339. McTaggart L, Richardson SE, Seah C, *et al.* Rapid identification of *Cryptococcus neoformans* var. *grubii, C.neoformans* var. *neoformans* and *C.gattii* by use of rapid biochemical tests, differential media, and DNA sequencing. *J Clin Microbiol.* 2011; **49:** 2522-7.
340. Medaris LA, Ponce B, Hyde Z, *et al.* Cryptococcal osteomyelitis: A report of 5 cases and a review of the recent literature. *Mycoses.* 2016; **59:** 334-42.
341. Meya D, Rajasingham R, Nalintya E, *et al.* Preventing cryptococcosis-shifting the paradigm in the era of highly active antiretroviral therapy. *Curr Trop Med Rep.* 2015; **2:** 81-89.
342. Meyer AC, Jacobson M. Asymptomatic cryptococcemia in resource-limited settings. *Curr HIV/AIDS Rep.* 2013; **10:** 254-63.
343. Mfinanga S, Chanda D, Kivuyo SL, *et al.* Cryptococcal meningitis screening and community-based early adherence support in people with advanced HIV infection starting antiretroviral therapy in Tanzania and Zambia:

An open-label, randomized controlled trial. *Lancet.* 2015; **385:** 2173-82.

344. Miglia KJ, Govender NP, Rossouw J, *et al.* Analyses of pediatric isolates of *Cryptococcus neoformans* from South Africa. *J Clin Microbiol.* 2011; **49:** 307-14.
345. Mihara T, Izumikawa K, Kakeya H, *et al.* Multilocus sequence typing of *Cryptococcus neoformans* in non-HIV associated cryptococcosis in Nagasaki, Japan. *Med Mycol.* 2013; **51:** 252-60.
346. Miozzo I, Aquino VR, Duarte M, *et al. Cryptococcus neoformans* as a rare cause of hospital infection. *Infect Control Hosp Epidemiol.* 2010; **31:** 315-7.
347. Mistry N, Tan K, Shokravi M, *et al. Cryptococcus gattii* infections with cutaneous involvement. *J Cutan Med Surg.* 2011; **15:** 236-7.
348. Mitchell TG, Perfect JR. Cryptococcosis in the era of AIDS: 100 years after the discovery of *Cryptococcus neoformans. Clin Microbiol Rev.* 1995; **8:** 515-48.
349. Mittal N, Collignon P, Pham T, *et al.* Cryptococcal infection of the larynx: Case report. *J Laryngol Otol.* 2013; **127:** S54-6.
350. Mohamad I, Abdullah B, Salim R, *et al.* Cryptococcosis: A rare fungal infection of the tongue. *Southeast Asian J Trop Med Public Health.* 2010; **41:** 1188-91.
351. Molina-Leyva A, Ruiz-Carrascosa JC, Leyva-Garcia A, *et al.* Cutaneous *Cryptococcus laurentii* infection in an immunocompetent child. *Int J Infect Dis.* 2013; **17:** e1232-3.
352. Morita S, Shirai T, Asada K, *et al.* Pulmonary cryptococcosis presenting with a large cavity. *Respirol Case Rep.* 2014; **2:** 61-3.
353. Munson E, Olson R. A 38-year-old male with a 3-month history of abdominal pain, constipation and headache. *J Clin Microbiol.* 2012; **50:** 3415 & 3819.
354. Musubire AK, Meya DB, Lukande R, *et al.* Gastrointestinal cryptococcoma - Immune reconstitution inflammatory syndrome or cryptococcal relapse in a patient with AIDS? *Med Mycol Case Rep.* 2015; **8:** 40-3.
355. Mylonakis E, Muse VV, Mino-Kenudson M. A 74-year-old man with pemphigus vulgaris and lung nodules. *N Engl J Med.* 2011; **365:** 1043-50.
356. Nagarathna S, Kumari HB, Arvind N, *et al.* Prevalence of *Cryptococcus gattii* causing meningitis in a tertiary neurocare center from south India: A pilot study. *Indian J Pathol Microbiol.* 2010; **53:** 855-6.
357. Nagotkar L, Shanbag P, Mauskar A, *et al.* Fulminant intracranial hypertension due to cryptococcal meningitis in a child with nephrotic syndrome. *Indian J Crit Care Med.* 2011; **15:** 176-8.
358. Nair JP, Athavale AU, Gawande S, *et al.* Disseminated cryptococcosis with caverno-oesophageal fistula in a case of idiopathic CD4+ T-lymphocytopenia. *J Assoc Physicians India.* 2014; **62:** 66-9.
359. Nalintya E, Kiggundu R, Meya D. Evolution of cryptococcal antigen testing: What is new? *Curr Fungal Infect Rep.* 2016; 1-6. PMID: 27158322.
360. Nankeu S, Djaha JM, Saint Marcoux B, *et al.* Disseminated cryptococcosis revealed by vertebral osteomyelitis in an immunocompetent patient. *Joint Bone Spine.* 2012; **79:** 629-31.
361. Narvaez-Moreno B, Bernabeu-Wittel J, Zulueta-Dorado T, *et al.* Primary cutaneous cryptococcosis of the penis. *Sex Transm Dis.* 2012; **39:** 792-3.
362. Nascimento E, Bonifacio da Silva ME, Martinez R, *et al.* Primary cutaneous cryptococcosis in an immunocompetent patient due to *Cryptococcus gattii* molecular type VGI in Brazil: A case report and review of literature. *Mycoses.* 2014; **57:** 442-7.
363. Nayal B, Veena, Niveditha S, *et al.* Detection of cryptococcosis in peripheral blood smear: A case report. *Int J Appl Basic Med Res.* 2011; **1:** 116-7.
364. Negroni R. Cryptococcosis. *Clin Dermatol.* 2012; **30:** 599-609.
365. Nguyen HP, Gordon RA, Hamill RJ, *et al.* A polymorphic, mucocutaneous eruption in a patient with end-stage renal disease. *Dermatol Online J.* 2013; **19:** 20024.
366. Ni W, Huang Q, Cui J. Disseminated cryptococcosis initially presenting as cellulitis in a patient suffering from nephrotic syndrome. *BMC Nephrol.* 2013; **14:** 20. PMID: 23336386.
367. Nidhi A, Meena A, Sreekumar A, *et al.* Corticosteroid-induced cryptococcal meningitis in patient without HIV. *BMJ Case Rep.* 2017; pii: bcr2016216496. PMID: 28052943.
368. Nigam C, Gahlot R, Kumar V, *et al.* Central nervous system cryptococcosis among a cohort of HIV infected patients from a university hospital of north India. *J Clin Diagn Res.* 2012; **6:** 1385-7.
369. Nirwan PS, Dalal AS, Peters BP, Rastogi VL, Mehta K. Cryptococcal meningitis in HIV negative infant. *Indian J Med Microbiol.* 2003; **21:** 145.
370. Njei B, Kongnyuy EJ, Kumar S, *et al.* Optimal timing for antiretroviral therapy initiation in patients with HIV infection and concurrent cryptococcal meningitis. *Cochrane Database Syst Rev.* 2013; **2:** CD009012.
371. O'Meara TR, Alspaugh JA. The *Cryptococcus neoformans* capsule: A sword and a shield. *Clin Microbiol Rev.* 2012; **25:** 387-408.
372. Odashima K, Takayanagi N, Ishiguro T, *et al.* Pulmonary cryptococcosis with endobronchial lesions and meningitis. *Intern Med.* 2014; **53:** 2731-5.
373. Offiah CE, Naseer A. Spectrum of imaging appearances of intracranial cryptococcal infection in HIV/AIDS patients in the anti-retroviral therapy era. *Clin Radiol.* 2016; **71:** 9-17.
374. Ogundeji AO, Albertyn J, Pohl CH, *et al.* Method for identification of *Cryptococcus neoformans* and *Cryptococcus gattii* useful in resource-limited settings. *J Clin Pathol.* 2016; **69:** 352-7.
375. Oh DY, Madhusoodhan PP, Springer DJ, *et al.* Cryptococcal osteomyelitis in an adolescent survivor of T-cell acute lymphoblastic leukemia. *Pediatr Infect Dis J.* 2015; **34:** 662-6.
376. Okachi S, Wakahara K, Kato D, *et al.* Massive mediastinal cryptococcosis in a young immunocompetent patient. *Respirol Case Rep.* 2015; **3:** 95-8.
377. Oladele RO, Akanmu AS, Nwosu AO, *et al.* Cryptococcal antigenemia in Nigerian patients with advanced human

immunodeficiency virus: Influence of antiretroviral therapy adherence. *Open Forum Infect Dis.* 2016; **3:** ofw055.
378. Olszewski MA, Zhang Y, Huffnagle GB. Mechanisms of cryptococcal virulence and persistence. *Future Microbiol.* 2010; **5:** 1269-88.
379. Orsini J, Nowakowski J, Delaney V, *et al.* Cryptococcal infection presenting as cellulitis in a renal transplant recipient. *Transpl Infect Dis.* 2009; **11:** 68-71.
380. Osawa R, Alexander BD, Lortholary O, *et al.* Identifying predictors of central nervous system disease in solid organ transplant recipients with cryptococcosis. *Transplantation.* 2010; **89:** 69-74.
381. Padhye AA, Chakrabarti A, Chander J, *et al. Cryptococcus neoformans* var. *gattii* in India. *J Med Vet Mycol.* 1993; **31:** 165-8.
382. Pal M, Dave P. Cryptococcosis: An emerging airborne mycosis of global concern. *Air Water Borne Diseases.* 2016; **5:** 127.
383. Pal P, Ray S, Patra SK, *et al.* Disseminated cryptococcosis in an apparently immunocompetent patient presenting with primary intraventricular haemorrhage. *BMJ Case Rep.* 2015; pii: bcr2015210250. PMID: 26494714.
384. Paliwal VK, Gupta PK, Rai P, Verma R. Cryptococcal meningitis in a patient with hepatitis C virus related decompensated cirrhosis: Coincidental or immunologically related? *Trop Gastroenterol.* 2012; **33:** 146-8.
385. Pan B, Chen M, Jia H, Pan W, Liao W. Multiple subcutaneous abscesses: A rare presentation of cutaneous cryptococcosis. *Indian J Dermatol Venereol Leprol.* 2013; **79:** 118-9.
386. Pan W, Khayhan K, Hagen F, *et al.* Resistance of Asian *Cryptococcus neoformans* serotype A is confined to few microsatellite genotypes. *PLoS One.* 2012; **7:** e32868.
387. Pan W, Liao W, Hagen F, *et al.* Meningitis caused by *Filobasidium uniguttulatum:* Case report and overview of the literature. *Mycoses.* 2012; **55:** 105-9.
388. Panackal AA, Dekker JP, Proschan M, *et al.* Enzyme immunoassay *versus* latex agglutination cryptococcal antigen assays in adults with non-HIV-related cryptococcosis. *J Clin Microbiol.* 2014; **52:** 4356-8.
389. Panackal AA, Komori M, Kosa P, *et al.* Spinal arachnoiditis as a complication of cryptococcal meningoencephalitis in non-HIV previously healthy adults. *Clin Infect Dis.* 2017; **64:** 275-83.
390. Panda S, Kumar MV, Bagchi S, *et al.* Migratory skin lesions in a renal transplant recipient. *Nephrology (Carlton).* 2014; **19:** 661-2.
391. Panigrahi MK, Kumar NN, Jaganathan V, *et al.* Pulmonary cryptococcosis with cryptococcal meningitis in an immunocompetent host. *Lung India.* 2014; **31:** 152-4.
392. Pappas PG. Cryptococcal infections in non-HIV-infected patients. *Trans Am Clin Climatol Assoc.* 2013; **124:** 61-79.
393. Pappas PG. Cryptococcosis in the developing world: An elephant in the parlor. *Clin Infect Dis.* 2010; **50:** 345-6.
394. Pappas PG. Editorial commentary: An expanded role for therapeutic lumbar punctures in newly diagnosed AIDS-associated cryptococcal meningitis? *Clin Infect Dis.* 2014; **59:** 1615-7.
395. Park SH, Kim M, Joo SI, *et al.* Molecular epidemiology of clinical *Cryptococcus neoformans* isolates in Seoul, Korea. *Mycobiology.* 2014; **42:** 73-8.
396. Parkes-Ratanshi R, Wakeham K, Levin J, *et al.* Primary prophylaxis of cryptococcal disease with fluconazole in HIV-positive Ugandan adults: A double-blind, randomised, placebo-controlled trial. *Lancet Infect Dis.* 2011; **11:** 933-41.
397. Pasa CR, Chang MR, Hans-Filho G. Post-trauma primary cutaneous cryptococcosis in an immunocompetent host by *Cryptococcus gattii* VGII. *Mycoses.* 2012; **55:** e1-3.
398. Patel C, Rabindra R, Hashemi N, *et al.* Chronic meningitis: A diagnostic challenge highlighted in a case of cryptococcal meningoencephalitis in an apparently immunocompetent older woman. *BMJ Case Rep.* 2011; pii: bcr0820114724.
399. Patel M, Beckerman KP, Reznik S, *et al.* Transplacental transmission of *Cryptococcus neoformans* to an HIV-exposed premature neonate. *J Perinatol.* 2012; **32:** 235-7.
400. Patel NC, Ewanowski C, Kupiec-Banasikowska A, *et al.* Disseminated *Cryptococcus neoformans:* Case report and review of the literature. *Cutis.* 2009; **84:** 93-6.
401. Patel NC, Kupiec-Banasikowska A, Kauffman CL. Immediate diagnosis of *Cryptococcus* fungal infection using touch preparation analysis. *Arch Dermatol.* 2009; **145:** 501-2.
402. Patel S, Shin GY, Wijewardana I, *et al.* The prevalence of cryptococcal antigenemia in newly diagnosed HIV patients in a Southwest London cohort. *J Infect.* 2013; **66:** 75-9.
403. Patil SA, Katyayani S, Arvind N. Significance of antibody detection in the diagnosis of cryptococcal meningitis. *J Immunoassay Immunochem.* 2012; **33:** 140-8.
404. Pau M, Lallai C, Aste N, *et al.* Primary cutaneous cryptococcosis in an immunocompetent host. *Mycoses.* 2010; **53:** 256-8.
405. Pazare AR. Cryptococcosis: The ubiquitous yeast. *J Assoc Physicians India.* 2016; **64:** 11-12.
406. Perfect JR, Bicanic T. Cryptococcosis diagnosis and treatment: What do we know now. *Fungal Genet Biol.* 2015; **78:** 49-54.
407. Perfect JR, Dismukes WE, Dromer F, *et al.* Clinical Practice Guidelines for the Management of Cryptococcal Disease: 2010 Update by the Infectious Diseases Society of America. *Clin Infect Dis.* 2010; **50:** 291-322.
408. Perfect JR. Cryptococcosis: A model for the understanding of infectious diseases. *J Clin Invest.* 2014; **124:** 1893-5.
409. Perfect JR. Efficiently killing a sugar-coated yeast. *N Engl J Med.* 2013; **368:** 1354-6.
410. Perfect JR. The triple threat of cryptococcosis: It's the body site, the strain and/or the host. *mBio.* 2012; 3. pii: e00165-12. PMID: 22782526
411. Persad P, Patel R, Stephens J, *et al.* Explosive nasofacial cryptococcosis. *Dermatol Online J.* 2010; **16:** 5.
412. Philip KJ, Kaur R, Sangeetha M, *et al.* Disseminated cryptococcosis presenting with generalized lymphadenopathy. *J Cytol.* 2012; **29:** 200-2.
413. Phillips PH, Angtuaco E, VanHemert RL, *et al.* Curtains. *Surv Ophthalmol.* 2012; **57:** 284-91.
414. Poojary S, Khatu S. Disseminated cryptococcosis in a diabetic patient. *Cutis.* 2014; **94:** 91-5.

415. Posteraro B, Vella A, Cogliati M, *et al.* Matrix-assisted laser desorption ionization-time of flight mass spectrometry-based method for discrimination between molecular types of *Cryptococcus neoformans* and *Cryptococcus gattii. J Clin Microbiol.* 2012; **50:** 2472-6.
416. Powel MS, Alizadeh AA, Budvytiene I, *et al.* First isolation of *Cryptococcus uzbekistanensis* from an immunocompromised patient with lymphoma. *J Clin Microbiol.* 2012; **50:** 1125-7.
417. Prakash PY, Sugandhi RP. Neuropsychiatric manifestation of confusional psychosis due to *Cryptococcus neoformans* var. *grubii* in an apparently immunocompetent host: A case report. *Cases J.* 2009; **2:** 9084. PMID: 20062721.
418. Prasad KT, Sehgal IS, Shivaprakash MR, *et al.* Uncommon mycosis in a patient with diabetes. *BMJ Case Rep.* 2016; 2016. pii: bcr2016214453. PMID: 26917800.
419. Price MS, Perfect JR. Host defenses against cryptococcosis. *Immunol Invest.* 2011; **40:** 786-808.
420. Pyrgos V, Seitz AE, Steiner CA, *et al.* Epidemiology of cryptococcal meningitis in the US: 1997-2009. *PLoS One.* 2013; **8:** e56269. PMID: 23457543.
421. Qadir I, Ali F, Malik UZ, *et al.* Isolated cryptococcal osteomyelitis in an immunocompetent patient. *J Infect Dev Ctries.* 2011; **5:** 669-73.
422. Qureshi A, Wray D, Rhome R, *et al.* Detection of antibody against fungal glucosylceramide in immunocompromised patients: A potential new diagnostic approach for cryptococcosis. *Mycopathologia.* 2012; **173:** 419-25.
423. Rai AK, Synmon B, Basumatary LJ, *et al.* Cryptococcal myelitis: A rare manifestation in immunocompetent patients. *Neurol India.* 2014; **62:** 321-2.
424. Rai S, Marak RS, Jain S, *et al.* Posterior fossa midline cryptococcoma in a patient with idiopathic CD4 lymphocytopenia. *Indian J Med Microbiol.* 2012; **30:** 367-70.
425. Rajagopala S, Panjwani A, Agarwal R, *et al.* Unusual lung mass. *Intern Med J.* 2012; **42:** 468-9.
426. Rajasingham R, Boulware DR. HIV care: ART adherence support and cryptococcal screening. *Lancet.* 2015; **385 (9983):** 2128-9. PMID: 25765697.
427. Rajasingham R, Rolfes MA, Birkenkamp KE, *et al.* Cryptococcal meningitis treatment strategies in resource-limited settings: A cost-effectiveness analysis. *PLoS Med.* 2012; **9:** e1001316. PMID: 23055838.
428. Rajpal S, Dwivedi S, Chaudhary SC. Disseminated cryptococcosis in an HIV-negative patient. *Trop Doct.* 2009; **39:** 114-5.
429. Ramdial PK, Sing Y, Deonarain J, *et al.* Pediatric renal cryptococcosis: Novel manifestations in the acquired immunodeficiency syndrome era. *Int J Surg Pathol.* 2011; **19:** 386-92.
430. Ramkillawan Y, Dawood H, Ferreira N. Isolated cryptococcal osteomyelitis in an immune-competent host: A case report. *Int J Infect Dis.* 2013; **17:** e1229-31.
431. Randhawa H, Kowshik T, Khan ZU. Efficacy of swabbing versus conventional technique for isolation of *Cryptococcus neoformans* from decayed wood in tree trunk hollows. *Med Mycol.* 2005; **43:** 67-71.
432. Ranjan P, Jana M, Krishnan S, *et al.* Disseminated cryptococcosis with adrenal and lung involvement in an immunocompetent patient. *J Clin Diagn Res.* 2015; **9:** OD04-5.
433. Rawat D, Capoor MR, Nair D, *et al.* Concomitant TB and cryptococcosis in HIV-infected patients. *Trop Doct.* 2008; **38:** 251-2.
434. Reddy BY, Shaigany S, Schulman L, *et al.* Fatal cryptococcal panniculitis in a lung transplant recipient. *J Drugs Dermatol.* 2015; **14:** 519-22.
435. Ribeiro NQ, Costa MC, Magalhaes TFF, *et al.* Atorvastatin as a promising anticryptococcal agent. *Int J Antimicrob Agents.* 2017; **49:** 695-702.
436. Rigby AL, Glanville AR. Miliary pulmonary cryptococcosis in an HIV-positive patient. *Am J Respir Crit Care Med.* 2012; **186:** 200-1.
437. Rivera V, Gaviria M, Munoz-Cadavid C, *et al.* Validation and clinical application of a molecular method for the identification of *Cryptococcus neoformans/Cryptococcus gattii* complex DNA in human clinical specimens. *Braz J Infect Dis.* 2015; **19:** 563-70.
438. Rivet-Danon D, Guitard J, Grenouillet F, *et al.* Rapid diagnosis of cryptococcosis using an antigen detection immunochromatographic test. *J Infect.* 2015; **70:** 499-503.
439. Rodrigues ML. Funding and innovation in diseases of neglected populations: The paradox of cryptococcal meningitis. *PLoS Negl Trop Dis.* 2016; **10:** e0004429. PMID: 26964103
440. Rohatgi S, Pirofski LA. Host immunity to *Cryptococcus neoformans. Future Microbiol.* 2015; **10:** 565-81.
441. Rolfes MA, Hullsiek KH, Rhein J, *et al.* The effect of therapeutic lumbar punctures on acute mortality from cryptococcal meningitis. *Clin Infect Dis.* 2014; **59:** 1607-14.
442. Rolston KV. Cryptococcosis due to *Cryptococcus gattii. Clin Infect Dis.* 2013; **57:** 552-4.
443. Romeo O, Scordino F, Chillemi V, *et al. Cryptococcus neoformans/Cryptococcus gattii* species complex in southern Italy: An overview on the environmental diffusion of serotypes, genotypes and mating-types. *Mycopathologia.* 2012; **174:** 283-91.
444. Roy M, Chiller T. Preventing deaths from cryptococcal meningitis: From bench to bedside. *Expert Rev Anti Infect Ther.* 2011; **9:** 715-7.
445. Ruan Q, Zhu Y, Chen S, *et al.* Disseminated cryptococcosis with recurrent multiple abscesses in an immunocompetent patient: A case report and literature review. *BMC Infect Dis.* 2017; **17:** 369. PMID: 28558705.
446. Sabiiti W, May RC. Mechanisms of infection by the human fungal pathogen *Cryptococcus neoformans. Future Microbiol.* 2012; **7:** 1297-313.
447. Safdar N, Abad CL, Narayan S, *et al.* Keeping an open mind. *N Engl J Med.* 2009; **360:** 72-6.
448. Saha DC, Xess I, Biswas A, *et al.* Detection of *Cryptococcus* by conventional, serological and molecular methods. *J Med Microbiol.* 2009; **58 (Pt. 8):** 1098-105.
449. Sandhu JS, Makkar V, Sohal PM, *et al.* Isolated cavitary lung disease in a renal allograft recipient: A rare manifestation of cryptococcus. *Indian J Nephrol.* 2013; **23:** 381-3.

450. Sangarlangkarn A, Rughwani N. Hidden in a headache: Cryptococcal meningitis in a septuagenarian. *J Am Geriatr Soc.* 2015; **63:** 200-1.
451. Sarkis RA, Mays M, Isada C, *et al.* MRI findings in cryptococcal meningitis of the non-HIV population. *Neurologist.* 2015; **19:** 40-5.
452. Satish S, Rajesh R, Shashikala S, *et al.* Cryptococcal sepsis in small vessel vasculitis. *Indian J Nephrol.* 2010; **20:** 159-61.
453. Scriven JE, Graham LM, Schutz C, *et al.* Flow cytometry to assess cerebrospinal fluid fungal burden in cryptococcal meningitis. *J Clin Microbiol.* 2016; **54:** 802-4.
454. Sellers B, Hall P, Cine-Gowdie S, *et al. Cryptococcus gattii:* An emerging fungal pathogen in the Southeastern United States. *Am J Med Sci.* 2012; **343:** 510-1.
455. Sethupathi M, Yoganathan K. Late onset of cryptococcal cervical lymphadenitis following immune reconstitution inflammatory syndrome in a patient with AIDS. *BMJ Case Rep.* 2015; pii: bcr2014206601.
456. Severo CB, Bruno RM, Oliveira Fde M, *et al.* A case of miliary pulmonary cryptococcosis and review of literature. *Mycopathologia.* 2015; **179:** 313-5.
457. Severo CB, Pinto GL, Sotilli J, *et al.* Cryptococcuria as manifestation of disseminated cryptococcosis: Staib agar as a selective identification medium. *Mycoses.* 2011; **54:** e760-6.
458. Shah NB, Shoham S, Nayak S. *Cryptococcus neoformans* prosthetic joint infection: Case report and review of the literature. *Mycopathologia.* 2015; **179:** 275-8.
459. Shah VB, Patil PA, Agrawa V, *et al.* Primary cryptococcal prostatitis - rare occurrence. *J Assoc Physicians India.* 2012; **60:** 57-9.
460. Sharma A, Aggarwal S, Sharma V. Molluscum-like lesions in cryptococcal meningitis. *Intern Emerg Med.* 2011; **6:** 181-2.
461. Sharma D, Singh N, Kaushal S, *et al.* Isolated cutaneous cryptococcosis in clinically unsuspected idiopathic CD4 lymphocytopenia. *J Cytol.* 2014; **31:** 230-2.
462. Sharma S, Gupta N, Behera D, *et al.* Detection of cryptococcosis in liquid-based sputum cytology. *Cytopathology.* 2017; **28:** 177-8.
463. Shimoda M, Saraya T, Tsujimoto N, *et al.* Fatal disseminated cryptococcosis resembling miliary tuberculosis in a patient with HIV infection. *Intern Med.* 2014; **53:** 1641-4.
464. Shorman M, Evans D, Gibson C, *et al.* Cases of disseminated cryptococcosis in intravenous drug abusers without HIV infection: A new risk factor? *Med Mycol Case Rep.* 2016; **14:** 17-9.
465. Shukla BS, Giglio P, Couch JE, *et al.* Concomitant lymphoma and cryptococcosis in a patient with acquired immune deficiency syndrome. *Am J Med Sci.* 2010; **340:** 340-2.
466. Sidrim JJ, Costa AK, Cordeiro RA, *et al.* Molecular methods for the diagnosis and characterization of *Cryptococcus*: A review. *Can J Microbiol.* 2010; **56:** 445-58.
467. Silva DC, Martins MA, Szeszs MW, *et al.* Susceptibility to antifungal agents and genotypes of Brazilian clinical and environmental *Cryptococcus gattii* strains. *Diagn Microbiol Infect Dis.* 2012; **72:** 332-9.
468. Singh N, Sifri CD, Silveira FP, *et al.* Cryptococcosis in patients with cirrhosis of the liver and posttransplant outcomes. *Transplantation.* 2015; **99:** 2132-41.
469. Singh N. How I treat cryptococcosis in organ transplant recipients. *Transplantation.* 2012; **93:** 17-21.
470. Singh P, Joshi D, Gangane N. *Cryptococcus* peritonitis a rare manifestation in HIV positive host: A case report. *Diagn Cytopathol.* 2011; **39:** 365-7.
471. Singh R, Joshi D, Gupta A, *et al.* Large pulmonary cryptococcoma and cryptococcal meningitis in an immunocompetent patient: A case report. *Diagn Cytopathol.* 2010; **38:** 929-31.
472. Singh R, Xess I. Multiple osseous involvements in a case of disseminated cryptococcosis. *Indian J Orthop.* 2010; **44:** 336-8.
473. Singh U, Aditi, Aneja P, *et al.* Cryptococcal meningitis associated with tuberculosis in HIV infected patients. *Indian J Tuberc.* 2013; **60:** 180-3.
474. Sivalingam SK, Saligram P, Natanasabapathy S, *et al.* Covert cryptococcal meningitis in a patient with systemic lupus erythematous. *J Emerg Med.* 2012; **42:** e101-4.
475. Sivasangeetha K, Harish BN, Sujatha S, *et al.* Cryptococcal meningoencephalitis diagnosed by blood culture. *Indian J Med Microbiol.* 2007; **25:** 282-4.
476. Sloan DJ, Parris V. Cryptococcal meningitis: Epidemiology and therapeutic options. *Clin Epidemiol.* 2014; **6:** 169-82.
477. Smith FL, Mercurio MG, Poligone B. Primary cutaneous cryptococcosis presenting as an extensive eroded plaque. *Cutis.* 2017; **99:** e16-8.
478. Smith IM, Stephan C, Hogardt M, *et al.* Cryptococcosis due to *Cryptococcus gattii* in Germany from 2004-2013. *Int J Med Microbiol.* 2015; **305:** 719-23.
479. Smith RM, Mba-Jonas A, Tourdjman M, *et al.* Treatment and outcomes among patients with *Cryptococcus gattii* infections in the United States Pacific Northwest. *PLoS One.* 2014; **9:** e88875.
480. Soni ZJ, Reubenson G, De Maayer T, *et al.* Neonatal cryptococcosis: Beware of false-positive results. *J Pediatric Infect Dis Soc.* 2012; **1:** 250-3.
481. Sow D, Tine RC, Sylla K, *et al.* Cryptococcal meningitis in Senegal: Epidemiology, laboratory findings, therapeutic and outcome of cases diagnosed from 2004 to 2011. *Mycopathologia.* 2013; **176:** 443-9.
482. Spec A, Raval K, Powderly WG. End-stage liver disease is a strong predictor of early mortality in cryptococcosis. *Open Forum Infect Dis.* 2015; **3:** ofv197. PMID: 26835475.
483. Sperry BW, Howard EW, Gitterman S, *et al.* A case of cryptogenic dyspnea: Disseminated cryptococcosis. *Am J Med.* 2014; **127:** 707-10.
484. Spies FS, de Oliveira MB, Krug MS, *et al.* Cryptococcosis in patients living with hepatitis C and B viruses. *Mycopathologia.* 2015; **179:** 307-12.
485. Spiliopoulou A, Anastassiou ED, Christofidou M. Primary cutaneous cryptococcosis in immunocompetent hosts. *Mycoses.* 2012; **55:** e45-7.

486. Springer DJ, Chaturvedi V. Projecting global occurrence of *Cryptococcus gattii. Emerg Infect Dis.* 2010; **16:** 14-20.
487. Springer DJ, Phadke S, Billmyre B, *et al. Cryptococcus gattii,* no longer an accidental pathogen? *Curr Fungal Infect Rep.* 2012; **6:** 245-56.
488. Srichatrapimuk S, Sungkanuparph S. Integrated therapy for HIV and cryptococcosis. *AIDS Res Ther.* 2016; **13:** 42.
489. Srikanta D, Santiago-Tirado FH, Doering TL. *Cryptococcus neoformans:* Historical curiosity to modern pathogen. *Yeast.* 2014; **31:** 47-60.
490. Srinivasan R, Gupta N, Shifa R, *et al.* Cryptococcal lymphadenitis diagnosed by fine needle aspiration cytology: A review of 15 cases. *Acta Cytol.* 2010; **54:** 1-4.
491. Srivastava GN, Tilak R, Yadav J, *et al.* Cutaneous *Cryptococcus*: Marker for disseminated infection. *BMJ Case Rep.* 2015; pii: bcr2015210898. PMID: 26199299.
492. Steele KT, Thakur R, Nthobatsang R, *et al.* In-hospital mortality of HIV-infected cryptococcal meningitis patients with *C.gattii and C.neoformans* infection in Gaborone, Botswana. *Med Mycol.* 2010; **48:** 1112-5.
493. Suchitha S, Sheeladevi CS, Sunila R, *et al.* Disseminated cryptococcosis in an immunocompetent patient: A case report. *Case Rep Pathol.* 2012; 652351. PMID: 22953139.
494. Sugiura K, Sugiura N, Yagi T, *et al.* Cryptococcal cellulitis in a patient with bullous pemphigoid. *Acta Derm Venereol.* 2013; **93:** 187-8.
495. Sun HY, Alexander BD, Huprikar S, *et al.* Predictors of immune reconstitution syndrome in organ transplant recipients with cryptococcosis: implications for the management of immunosuppression. *Clin Infect Dis.* 2015; **60:** 36-44.
496. Sun HY, Alexander BD, Lortholary O, *et al.* Cutaneous cryptococcosis in solid organ transplant recipients. *Med Mycol.* 2010; **48:** 785-91.
497. Sun HY, Alexander BD, Lortholary O, *et al.* Unrecognized pretransplant and donor-derived cryptococcal disease in organ transplant recipients. *Clin Infect Dis.* 2010; **51:** 1062-9.
498. Sun L, Chen H, Shao C, *et al.* Pulmonary cryptococcosis with trachea wall invasion in an immunocompetent patient: A case report and literature review. *Respiration.* 2014; **87:** 324-8.
499. Sundar R, Rao L, Vasudevan G, *et al.* Gastric cryptococcal infection as an initial presentation of AIDS: A rare case report. *Asian Pac J Trop Med.* 2011; **4:** 79-80.
500. Suner L, Mathis S. Disseminated cryptococcosis in bone marrow. *Blood.* 2014; **123:** 3070. PMID: 24949507.
501. Suwantarat N, Dalton JB, Lee R, *et al.* Large-scale clinical validation of a lateral flow immunoassay for detection of cryptococcal antigen in serum and cerebrospinal fluid specimens. *Diagn Microbiol Infect Dis.* 2015; **82:** 54-6.
502. Tabassum S, Rahman A, Herekar F, *et al.* Cryptococcal meningitis with secondary cutaneous involvement in an immunocompetent host. *J Infect Dev Ctries.* 2013; **7:** 680-5.
503. Talwar P, Sharma M. Incidence and diagnostic aspects of cryptococcosis in Chandigarh (India) during the period 1970-1980. *Indian J Pathol Microbiol.* 1986; **29:** 45-52.
504. Tan WP, Tan SH, Tan AW. An extensive painful leg ulcer in a patient with rheumatoid arthritis. Primary cutaneous cryptococcosis. *Clin Exp Dermatol.* 2010; **35:** e46-7.
505. Taneja J, Bhargava A, Loomba P, Dogra V, Thakur A, Mishra B. Cryptococcal granulomas in an immunocompromised HIV-negative patient. *Indian J Pathol Microbiol.* 2008; **51:** 553-5.
506. Taneja J, Mishra B, Bhargava A, Loomba P, Dogra V, Thakur A. Cryptococcal meningitis in a tertiary care hospital. *Jpn J Med Mycol.* 2009; **50:** 95-9.
507. Tang MW, Clemons KV, Katzenstein DA, *et al.* The cryptococcal antigen lateral flow assay: A point-of-care diagnostic at an opportune time. *Crit Rev Microbiol.* 2016; **42:** 634-42.
508. Tarai B, Kher V, Kotru P, *et al.* Early onset primary pulmonary cryptococcosis in a renal transplant patient. *Indian J Med Microbiol.* 2010; **28:** 250-2.
509. Tay ST, Rohani MY, Hoo TS, *et al.* Epidemiology of cryptococcosis in Malaysia. *Mycoses.* 2010; **53:** 509-14.
510. Taylor-Smith LM, May RC. New weapons in the *Cryptococcus* infection toolkit. *Curr Opin Microbiol.* 2016; **34:** 67-74.
511. Tilak R, Prakash P, Nigam C, *et al.* Cryptococcal meningitis with an antecedent cutaneous cryptococcal lesion. *Dermatol Online J.* 2009; **15:** 12.
512. Tintelnot K, Hagen F, Han CO, *et al.* Pitfalls in serological diagnosis of *Cryptococcus gattii* infections. *Med Mycol.* 2015; **53:** 874-9.
513. Tore O, Akcaglar S, Kazak E, *et al.* Multiple intracranial abscesses due to *Cryptococcus neoformans:* An unusual clinical feature in an immunocompetent patient and a short review of reported cases. *Med Mycol.* 2010; **48:** 398-401.
514. Trilles L, Meyer W, Wanke B, Guarro J, Lazera M. Correlation of antifungal susceptibility and molecular type within the *Cryptococcus neoformans / C.gattii* species complex. *Med Mycol.* 2012; **50:** 328-32.
515. Trilles L, Wang B, Firacative C, Lazera MS. *et al.* Identification of the major molecular types of *Cryptococcus neoformans* and *C.gattii* by hyperbranched rolling circle amplification. *PLoS One.* 2014; **9:** e94648. PMID: 24736745.
516. Tripathi S, Patro I, Mahadevan A, *et al.* Glial alterations in tuberculous and cryptococcal meningitis and their relation to HIV co-infection - A study on human brains. *J Infect Dev Ctries.* 2014; **8:** 1421-43.
517. Tseng HK, Liu CP, Ho MW, *et al.* Microbiological, epidemiological and clinical characteristics and outcomes of patients with cryptococcosis in Taiwan, 1997-2010. *PLoS One.* 2013; **8:** e61921. PMID: 23613973.
518. Uejio CK, Mak S, Manangan A, *et al.* Climatic influences on *Cryptococcus gattii* populations, Vancouver Island, Canada, 2002-2004. *Emerg Infect Dis.* 2015; **21:** 1989-96.
519. Ulett KB, Cockburn JW, Jeffree R, *et al.* Cerebral cryptococcoma mimicking glioblastoma. *BMJ Case Rep.* 2017; pii: bcr2016218824. PMID: 28188169.
520. Valente ES, Lazzarin MC, Koech BL, *et al.* Disseminated cryptococcosis presenting as cutaneous cellulitis in an adolescent with systemic lupus erythematosus. *Infect Dis Rep.* 2015; **7:** 5743. PMID: 26294948.

521. Varma A, Kwon-Chung KJ. Heteroresistance of *Cryptococcus gattii* to fluconazole. *Antimicrob Agents Chemother.* 2010; **54:** 2303-11.
522. Velamakanni SS, Bahr NC, Musubire AK, *et al.* Central nervous system cryptococcoma in a Ugandan patient with Human Immunodeficiency Virus. *Med Mycol Case Rep.* 2014; **6:** 10-3.
523. Vellayappan BA, Bharwani L. Cryptococcemia in a patient with glioblastoma: Case report and literature review. *Can J Neurol Sci.* 2012; **39:** 852-4.
524. Venkatachala S, Naik DR, Shanthakumari S, *et al.* Acquired immunodeficiency syndrome presenting as marrow cryptococcosis. *Indian J Pathol Microbiol.* 2010; **53:** 904-6.
525. Vidal JE, Boulware DR. Lateral flow assay for cryptococcal antigen: An important advance to improve the continuum of HIV care and reduce cryptococcal meningitis-related mortality. *Rev Inst Med Trop Sao Paulo.* 2015; **57 (Suppl. 19):** 38-45.
526. Vidal JE, Penalva de Oliveira AC, Dauar RF, *et al.* Strategies to reduce mortality and morbidity due to AIDS-related cryptococcal meningitis in Latin America. *Braz J Infect Dis.* 2013; **17:** 353-62.
527. Vieira MA, Costa CH, Ribeiro JC, *et al.* Soap bubble appearance in brain magnetic resonance imaging: Cryptococcal meningoencephalitis. *Rev Soc Bras Med Trop.* 2013; **46:** 658-9.
528. Vijayan T, Chiller T, Klausner JD. Sensitivity and specificity of a new cryptococcal antigen lateral flow assay in serum and cerebrospinal fluid. *Med Lab Obs.* 2013; **45:** 16-20.
529. Vitale RG, Pascuccelli V, Afeltra J. Influence of capsule size on the *in vitro* activity of antifungal agents against clinical *Cryptococcus neoformans* var. *grubii* strains. *J Med Microbiol.* 2012; **61 (Pt. 3):** 384-8.
530. Wald-Dickler N, Blodget E. Cryptococcal disease in the solid organ transplant setting: review of clinical aspects with a discussion of asymptomatic cryptococcal antigenemia. *Curr Opin Organ Transplant.* 2017. doi: 10.1097/MOT.0000000000000426. PMID: 28562416.
531. Walraven CJ, Gerstein W, Hardison SE, *et al.* Fatal disseminated *Cryptococcus gattii* infection in New Mexico. *PLoS One.* 2011; **6:** e28625. PMID: 22194869.
532. Walsh TL, Bhanot N, Murillo MA, *et al.* Creeping skin lesions: Primary cutaneous cryptococcosis. *Am J Med.* 2017; **130:** 666-8.
533. Wang C, Jia N, Zhang L, *et al.* Imaging findings of cryptococcal infection of the thoracic spine. *Int J Infect Dis.* 2014; **29:** 162-5.
534. Wang H, Ling C, Chen C, *et al.* Evaluation of ventriculoperitoneal shunt in the treatment of intracranial hypertension in the patients with cryptococcal meningitis: A report of 12 cases. *Clin Neurol Neurosurg.* 2014; **124:** 156-60.
535. Wang H, Yuan X, Zhang L. Latex agglutination: Diagnose the early *Cryptococcus neoformans* test of capsular polysaccharide antigen. *Pak J Pharm Sci.* 2015; **28 (Suppl. 1):** 307-11.
536. Wang J, Bartelt L, Yu D, *et al.* Primary cutaneous cryptococcosis treated with debridement and fluconazole monotherapy in an immunosuppressed patient: A case report and review of the literature. *Case Rep Infect Dis.* 2015; 131356.
537. Wang J, Ju HZ, Yang MF. Pulmonary cryptococcosis and cryptococcal osteomyelitis mimicking primary and metastatic lung cancer in ^{18}F-FDG PET/CT. *Int J Infect Dis.* 2014; **18:** 101-3.
538. Wang J, Zeng Y, Luo W, *et al.* The role of *Cryptococcus* in the immune system of pulmonary cryptococcosis patients. *PLoS One.* 2015; **10:** e0144427. PMID: 26637129.
539. Wang X, Li W, Sun S, *et al.* Know your enemy: How to build and vanquish a global fungal scourge. *Mycopathologia.* 2012; **173:** 295-301.
540. Warkentien T, Crum-Cianflone NF. An update on *Cryptococcus* among HIV-infected patients. *Int J STD AIDS.* 2010; **21:** 679-84.
541. Williams DA, Kiiza T, Kwizera R, *et al.* Evaluation of fingerstick cryptococcal antigen lateral flow assay in HIV-infected persons: A diagnostic accuracy study. *Clin Infect Dis.* 2015; **61:** 464-7.
542. Williamson PR. Advancing translational immunology in HIV-associated cryptococcal meningitis. *J Infect Dis.* 2013; **207:** 1793-5.
543. Wysocki JD, Said SM, Papadakis KA. An uncommon cause of abdominal pain and fever in a patient with Crohn's disease. *Gastroenterology.* 2015; **148:** e12-3. PMID: 25824348.
544. Xie LX, Chen YS, Liu SY, *et al.* Pulmonary cryptococcosis: Comparison of CT findings in immunocompetent and immunocompromised patients. *Acta Radiol.* 2015; **56:** 447-53.
545. Xie X, Xu B, Yu C, *et al.* Clinical analysis of pulmonary cryptococcosis in non-HIV patients in south China. *Int J Clin Exp Med.* 2015; **8:** 3114-9.
546. Xue X, Wu H, Wang K, *et al.* Cryptococcosis by *Cryptococcus gattii* in China. *Lancet Infect Dis.* 2015; **15:** 1135-6.
547. Yamakawa H, Yoshida M, Yabe M, *et al.* Correlation between clinical characteristics and chest computed tomography findings of pulmonary cryptococcosis. *Pulm Med.* 2015; 703407. PMID: 25767722.
548. Yang Y, Sang J, Pan W, *et al.* Cryptococcal meningitis in patients with autoimmune hemolytic anemia. *Mycopathologia.* 2014; **178:** 63-70.
549. Yang Y, Shen YN, Zong WK, Cui PG. Disseminated cryptococcosis. *Indian J Dermatol Venereol Leprol.* 2016; **82:** 206-8.
550. Ye F, Xie JX, Zeng QS, *et al.* Retrospective analysis of 76 immunocompetent patients with primary pulmonary cryptococcosis. *Lung.* 2012; **190:** 339-46.
551. Yeh CH, Wang CS, Yeh TC, *et al.* Central nervous system infection caused by cryptococcus. *Intern Med.* 2013; **52:** 2387-8.
552. Yeung VA, Azzam R, Dendle C, *et al.* Cryptococcemia in primary HIV infection. *Int J STD AIDS.* 2016; **27:** 1231-3.
553. Yigit N, Wu WW, Covey S, *et al.* Cryptococcosis in bone marrow following treatment for Hodgkin lymphoma. *Int J Hematol.* 2015; **101:** 211-2.

554. Yoneda T, Itami Y, Hirayama A, *et al.* Cryptococcal necrotizing fasciitis in a patient after renal transplantation - A case report. *Transplant Proc.* 2014; **46:** 620-2.
555. Yong L. Cryptococcal immune reconstitution syndrome in HIV-negative patients. *Int J Infect Dis.* 2011; **15:** e884.
556. Yu JQ, Tang KJ, Xu BL, *et al.* Pulmonary cryptococcosis in non-AIDS patients. *Braz J Infect Dis.* 2012; **16:** 531-9.
557. Yuanjie Z, Jianghan C, Nan X, *et al.* Cryptococcal meningitis in immunocompetent children. *Mycoses.* 2012; **55:** 168-71.
558. Zaidi M, Qureshi S, Shakoor S, *et al.* Abdominal lymphonodular cryptococcosis in an immunocompetent child. *Case Rep Pediatr.* 2015; 347403.
559. Zhang C, Du L, Cai W, *et al.* Isolated hepatobiliary cryptococcosis manifesting as obstructive jaundice in an immunocompetent child: Case report and review of the literature. *Eur J Pediatr.* 2014; **173:** 1569-72.
560. Zhang Y, Yu YS, Tang ZH, *et al.* Cryptococcal osteomyelitis of the scapula and rib in an immunocompetent patient. *Med Mycol.* 2012; **50:** 751-5.
561. Zhou HX, Lu L, Chu T, *et al.* Skeletal cryptococcosis from 1977 to 2013. *Front Microbiol.* 2015; **5:** 740.
562. Zhou HX, Ning GZ, Feng SQ, *et al.* Cryptococcosis of lumbar vertebra in a patient with rheumatoid arthritis and scleroderma: Case report and literature review. *BMC Infect Dis.* 2013; **13:** 128. PMID: 23496879.
563. Zhou J, Lv J, Pan Y, *et al.* Unilateral adrenal cryptococcosis on FDG PET/CT. *Clin Nucl Med.* 2017; **42:** 565-6.

CHAPTER 22

Pneumocystosis

Pneumocystosis is an opportunistic fungal infection of respiratory system leading to atypical pneumonia. This is primarily a disease of alveoli and is caused by taxonomically unique fungus, ***Pneumocystis jirovecii***. The atypical pneumonia, previously thought to be caused by *Pneumocystis carinii,* a species remain confined to mice and to human beings. This is popularly known as *Pneumocystis* pneumonia (**PCP**), a serious opportunistic infection found among immunocompromised patients and is responsible for high morbidity as well as mortality. It is also one of the frequently encountered AIDS-defining diseases, which came to the international attention when the index case was reported in early 1980s in USA. It is also called as **interstitial plasma cell pneumonia** due to the intense plasma cell infiltrate in its initial description among malnourished infants.

Historical Perspective

Pneumocystis is unicellular eukaryotic organism with ubiquitous geographic distribution, which was discovered nearly a century ago. **Carlos Chagas**, while investigating etiology of new disease affecting Brazilian rail-road workers, initially discovered this organism in lungs of guinea pigs in 1909, which were co-infected with *Trypanosoma cruzi.* After one year in 1910, an Italian biologist **Antonio Carini** observed same organism in the lungs of rats but it was thought as variant of genus *Trypanosoma.* In 1912, French couple **Delanoe** and **Delanoe** at Pasteur Institute, Paris correctly identified in the material sent by Carini and presented their findings that it was **cystic** form of a unique organism hence named it as *Pneumocystis carinii.* The genus name of organism was derived from ***pneumo*** = lung, ***cyst*** = cyst-like structure and the species as ***carinii,*** in honor of Antonio Carini, a unique contribution by this husband-and-wife team.

In 1930s and 1940s, during Second World War as well as post war period, *P.carinii* was found to be associated with epidemic form of interstitial plasma cell pneumonia among malnourished orphaned infants, who were born prematurely in Central and Eastern Europe. In 1951, Czech Parasitologist, **Otto Jirovec** along with Joseph Vanek, demonstrated that *Pneumocystis* was the etiologic agent of such type of pneumonia. In mid-1950s, pentamidine isethionate was reported to be useful therapeutic agent and by 1975 trimethoprim-sulfamethoxazole (TMP-SMZ) was found to be an effective combination for pneumocystosis.

In **1976**, four years after Otto Jirovec's death, Jacob Frenkel proposed to use name, ***Pneumocystis carinii*** for species infecting rats and ***Pneumocystis jirovecii*** those infecting humans, thus separating on the basis of **host** and serological reactions, since interspecies infection was not experimentally possible. *Pneumocystis* derived from different hosts have very different DNA sequences, indicating multiple species. Due to the genetic and functional disparities, the organism that causes human PCP was named *Pneumocystis jirovecii.* However, new nomenclature did not draw much attention of the scientific community at that time.

In 1981, the disease was encountered sporadically among immunocompromised patients in USA and that was the beginning of a new saga of **AIDS-era**. Sandy Ford at CDC apprehended shortage of pentamidine isethionate thereby arranged and supplied the drug for treating this new and emerging PCP. As the infections caused by this organism came to the prominence, incidentally its taxonomy, whether it was fungus or protozoa, also became focal point of a major debate. The knowledge of the different species remained rudimentary until the mid-eighties when DNA analysis revealed its extensive diversity. The cumulative molecular genetic information in 1988 firmly established that *Pneumocystis* was more closely related to fungi than to protozoa. Hence it is now thought to be member of fungal lineage of eukaryotes. Based on demonstration of rRNA homology, between *Pneumocystis* and some other fungi like *Saccharomyces cerevisiae*, organism was classified as an atypical fungus under kingdom **Fungi**.

The nomenclature in mycology is so versatile that the name of *Pneumocystis carinii* strains, causing infections among **humans**, was officially changed to *Pneumocystis jirovecii*, on the basis of DNA sequence data in 1999. However, acronym of disease was retained as such with same abbreviation but different full form *i.e.* **PCP** - '*PneumoCystis* Pneumonia'. However, in medical literature at places its abbreviation **PJP** is also being referred *i.e.* '*Pneumocystis jirovecii* Pneumonia' but not in common use hence both abbreviations are interchangeably applied for this disease.

Therefore, *Pneumocystis* is in focus of the medical community since detection of index case of AIDS, firstly due to its role in human pathology, then due to its status whether it is protozoa or fungus and now because of its species' name, which has officially been re-designated in honor of Otto Jirovec. It is pronounced as ***'yee-row-vet-see'*** and is terminating with double ***'ii'*** and not with single ***'i'***. This is pertinent to mention because even in the current medical journals and books, *P.carinii* is found to be stated as the cause of PCP and if at all new name is referred, it is often written with single ***'i'*** and not the correct one with double ***'ii'***.

Taxonomy of *P.jirovecii*

For last many years, taxonomic status of this unicellular eukaryote, *Pneumocystis,* remained as matter of big controversy, which has settled now. The organism, while encountered in diagnostic mycology laboratory, is known to be quite different from other fungi. The molecular techniques have played a key role in advancing understanding of the taxonomy of this organism as well as epidemiology of pneumocystosis. The taxonomic position, re-grouped under taxon *Pneumocystis,* is based on morphological, biochemical, enzymatic, molecular and therapeutic studies. It shares biological features of the kingdom **Fungi** as well as **Protista**.

Prior to **1988**, it was not clear whether *Pneumocystis* was protozoa or fungus as classification was essentially based on its morphology, ultrastructure, growth requirements and susceptibility to various antimicrobial agents. Therefore, on the basis of these fairly weak criteria, *Pneumocystis* was generally considered to be a **protozoa** belonging to class **Sporozoa**, sub-class **Coccidia** or to an unidentified group of taxonomically uncertain status possibly related to **Microsporidia**. However, at the same time it was contested that *Pneumocystis* was more alike fungus than protozoa.

Now, phylogenetic studies have placed *Pneumocystis* in unique status in between ascomycetes and basidiomycetes. It is considered to be **fungus** and is phylogenetically classified as an **ascomycete**. Accordingly, it is related to fission yeast *Schizosaccharomyces pombe* and regarded as only member of order Pneumocystidales, although immunological and molecular data as well as host species significantly support existence of different species and variants. The biochemical analysis of cyst wall demonstrates β-(1,3)-D-glucan, compound found in the cell wall of fungi, which is used as one of the diagnostic biomarkers.

The inability to successfully culture this significant pulmonary pathogen on artificial media is major hindrance in understanding the cell biology and biochemistry. Consequently, little is known about gene regulation and expression in *Pneumocystis*. This organism was considered to be Protozoa on the basis following points to substantiate taxonomic status of this organism as Protozoa:

(a) *Pneumocystis* does not grow *in vitro* on fungal culture media but require tissue culture/cell lines for its growth and viability.
(b) There is absence of ergosterol in cytoplasmic membrane of *Pneumocystis* and abundant cholesterol is present as bulk sterol hence it is insensitive to antifungal drugs, which target mainly ergosterol synthesis.
(c) *Pneumocystis* is susceptible to anti-protozoan agents like pentamidine and TMP-SMZ and not to the broad-spectrum antifungal drugs.

Now, based on the observation of conventional as well as recent molecular studies, it is considered to be a fungus, which are as follows:

(a) *Pneumocystis* takes fungal stains like Gomori's methenamine silver stain.
(b) It possesses chitin at all stages of its life cycle *i.e.* cyst form, intracystic bodies and trophozoites. Moreover, several enzymes, including chitinase attack cell wall of *Pneumocystis* as well as true fungi.
(c) The protein synthesis elongation factor-3 (EF3) and thymidylate synthase of *Pneumocystis* are more homologous to those of ascomycetous fungi.
(d) *Pneumocystis* and fungi have similar cyst wall ultrastructures *i.e.* mitochondria with lamellar cristae and cyst forms containing intracystic bodies resembling those of ascospores whereas protozoal mitochondria have vesicular and tubular cristae.

(e) The ribosomal RNA studies reveal that **16S-like RNA** of *Pneumocystis* shares substantial sequence homology with various species of **Ascomycota**. The sequence study of **5S-rRNA** showed that its phylogenetic position was closely associated with **Rhizopoda, Myxomycota** and **Zygomycota** but not with ascomycete species. The use of this gene has now been replaced by 16S-like RNA gene.

Therefore, keeping these points in view, particularly small-subunit rRNA studies, it shows more similarity to fungi than protozoa hence phylogenetically *Pneumocystis* is now placed in kingdom **Fungi**. However, while deciding nomenclature to *Pneumocystis*, there is shifting from International Code of Zoological Nomenclature (ICZN) to International Code of Botanical Nomenclature (ICBN). The names *P.carinii* and *P.jirovecii* were introduced in a belief that they were protozoa hence covered under ICZN norms but as Fungus, now ICBN norms are to be applied.

Therefore, these names could not meet criteria for valid publication under ICBN until July 2005, when Vienna Code was changed to enable names, originally described as protozoan to be automatically treated as valid. Both names then became free of nomenclatural obstacles and could be used although are to be spelt with double ***'ii'*** as *P.carinii* as well as *P.jirovecii* and **not** with single ***'i'*** as *P.carini* and *P.jiroveci*. Under ICBN, application of names can be fixed by permanently attaching them to a collection different from original one to maintain current usage, subject to approval of the Committee for Fungi.

In this organism, **trinomial nomenclature** was also adopted in 1994 that included Latin name of the host species, which was *P.jirovecii* f. sp. *hominis* (human), *P.jirovecii* f. sp. *jirovecii* (rat prototype), *P.jirovecii* f. sp. *rattus* (rat variant), *P.jirovecii* f. sp. *mustelae* (ferret), *P.jirovecii* f. sp. *oryctolagi* (rabbit). There was genetic diversity, observed among isolates of *Pneumocystis* from same host species. Two distinct types of *Pneumocystis* named 'prototype' and 'variant' were identified in infected lungs of rats. The term *formae speciales* distinguished physiological variants of an organism by adaptation or restriction to particular host. Therefore, it was proposed that organisms found in each mammalian species be given **tripartite name** based on the host of origin. Following system of provisional trinomial nomenclature, eight *formae speciales* of *Pneumocystis* were identified in various hosts.

The phylogenetic classification of *P.jirovecii* is based on the gene comparisons with other fungi, which places it in the phylum Ascomycota; subphylum Taphrinomycotina; class Pneumocystidomycetes; order Pneumocystidales; genus *Pneumocystis*.

Life Cycle of *P.jirovecii*

The life cycle of *P.jirovecii* in lungs comprises of asexual phase of cell division by an apparent haploid trophic form and sexual phase leading to genesis of thick-walled reproductive cyst containing eight intracystic bodies. Although organism is now classified as fungus but description of sexual and asexual forms is still used as in protozoan terminology like **trophozoite**, **cyst** and **sporozoite**. The term sporozoite is synonymous to an **intracystic body**.

In the times to come, terminology of trophozoite may be changed to yeast cell and intracystic bodies/sporozoite as (endo)spore. However, substitution for term cyst for **ascus** should be treated with caution. In the kingdom Fungi ascus is not merely spore but specialized form of structure typically exhibited by class **Ascomycete**, generation of which is strictly associated with sexual reproduction. In this Chapter, more familiar and traditional protozoan terms are being used to describe this fungus.

Pneumocystis jirovecii is a unicellular **eukaryote** with proposed intrapulmonary life cycle consisting of trophozoites and cysts. Its life cycle is divided into three main stages: **trophozoite** (vegetative), **cystic** and **sporozoite** as shown in **Figure 22.1**. The transition phase between trophic and cystic stages is called as **precyst (sporocyst)**.

The **trophozoites** are principal components of foamy substance filling alveoli of involved lungs. They are pleomorphic, tiny bodies 2-5 μm in size and exist in clusters. They are covered with tubular projections, which facilitate attachment to epithelial cells and increase absorptive surface. Instead of rigid, thick cell wall, trophic forms of *P.jirovecii* have **thin** flexible, fragile external layer containing β-glucan and glycoprotein.

The **precyst,** also known as sporocyst is an intermediate stage of sexual phase of reproduction leading to cyst formation. It is presumed that mating event first occurs to produce zygote that initiates sporogenesis. The zygotic nucleus undergoes meiosis and subsequent mitosis takes place within sporocyst.

The **cystic form** is frequently observed stage of organism in clinical specimens. These are large, disc-like structures, 4-6 μm in size, oval-shaped, **thick-walled** and possessing up to eight **intracystic bodies** or **sporozoites**. The sporozoites are oval, amoeboid or peach-shaped and 1-2 μm

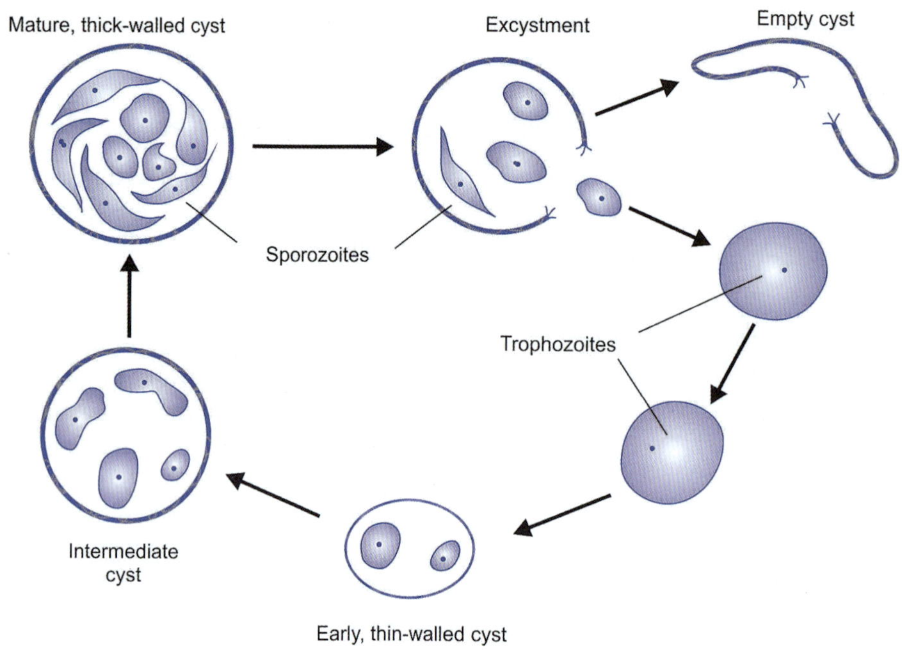

Fig. 22.1. Diagrammatic representation of the life cycle of *Pneumocystis jirovecii.*

in length. They are best demonstrated by Giemsa stain, showing basophilic cytoplasm and reddish-purple nuclei in eosinophilic mass. The sporozoites are extruded through cyst wall after rupture of mature cyst. They are subsequently converted into trophozoites. The empty cysts are seen as **navicular** structures, which are stained black by GMS stain or reddish-purple by Toluidine blue O. Some of trophozoites become encysted and produce eight **haploid** daughter trophozoites that are also known as intracystic bodies. In **asexual phase** of life cycle, trophic forms multiply by binary fission as seen in some fungi like *Schizosaccharomyces pombe* and not by budding.

The environmental form of this pathogen has never been identified. The rodents are implicated as reservoir of infection although these strains are now different from human strains. *P.jirovecii* may be having morphologically unidentified environmental phase of its life cycle.

Epidemiology

Pneumocystis jirovecii is an organism of low virulence found in the lungs of man and its counterpart species *P.carinii* in animals. The disease caused by this genus, pneumocystosis, is one of commonest opportunistic infections among AIDS patients. However, it was not until 1981 in the wake of pandemic of AIDS when *P.jirovecii* drew attention of scientists, physicians and other medical specialists. Actually, pneumocystosis has played key role in the natural history of AIDS hence the medical science is 'indebted' to this disease, as it has given new dimensions to the infections in the immunocompromised circumstances. The importance of *P.jirovecii* has increased over the half-century and last three decades in particular. The molecular techniques have played the key role in advancing understanding of epidemiology of pneumocystosis.

The clinical entity produced by this organism, PCP or **interstitial plasma cell pneumonia** gets its name from distinctive lung infiltrate. It is thought that disease is acquired through droplets inhalation. It is seen as an opportunistic infection predominantly in patients with impaired cell-mediated immunity. The prevalence and mortality due to PCP has essentially declined during last decade in the industrialized countries due to implementation of primary and secondary prophylaxis but it still exists in certain parts of developing world due to low socioeconomic status of patients and other associated factors. In India, although incidence of HIV infection is contemplated to be increasing but case reports of pneumocystosis are scarce in literature. Even premier medical institutions have reported very few cases thereby it is one of the underreported fungal infections.

As far as the prevalence of this disease in the developed and developing world is concerned, as such there is no difference and it is almost same. However, it was contemplated previously that developed countries have more number of cases. This difference was due to the infrastructure and facilities available for diagnosis hence developing world was underreporting. But now facilities are available at many a places.

The conditions predisposing to PCP include cytotoxic and corticosteroid therapy in hematologic malignancies like lymphoma, chronic myelogenous or lymphocytic leukemia, Hodgkin's disease and solid tumors; iatrogenically induced immunosuppression, congenital immunodeficiency, Cushing's syndrome due to hypercorticosteroidism, protein-calorie malnutrition as seen in marasmus and even old age. It is significant cause of morbidity as well as mortality in immunosuppressed patients, especially those with AIDS or following solid organ transplantation. Person-to-person transmission is now reported as occurrence of outbreaks among debilitated infants and in hospitals caring for immunocompromised patients. Therefore, the person-to-person mode is also there in this disease in contrast to the earlier reactivation and *de novo* concepts.

The natural reservoir and transmission routes of *P.jirovecii* remain unknown despite recognized existence of opportunistic pathogen causing PCP in immunocompromised patients since 1960s. It is thought to be widespread in environment. Although transmission of *Pneumocystis* among rodents occurs through airborne route, it is unlikely that infected rodents serve as zoonotic reservoir for human infection since rat derived *Pneumocystis* strains are different from those of humans. Despite this evidence suggesting that new acquisition of *P.jirovecii* might occur later in life, exact route of transmission remains unknown. There is still uncertainty about infective stage and environmental source of this organism.

Despite availability of effective treatments, *Pneumocystis* pneumonia remains significant cause of mortality and morbidity in immunocompromised patients. Although introduction of Highly Active Antiretroviral Therapy (HA-ART) as well as Facilitated Integrated ART (FI-ART) have substantially reduced the incidence of this infection in HIV population but it is still common amongst those unaware of their HIV status and who have no access to such treatments. The recent advances in transplant medicine and chemotherapy have also increased population of immunocompromised patients at risk of developing PCP.

Immunity

The host immune system plays an important role in suppressing *P.jirovecii* and maintaining presumably latent state of infection in immunosuppressed host. It is probably not the underlying disorder that predisposes to development of PCP in patients with cancer; rather it is the type and intensity of the cytotoxic or immunosuppressive therapy, used to treat the cancer patients responsible for development of PCP. During last decade, pulmonary carriage of *P.jirovecii* in healthy persons has been re-evaluated and PCP is now frequently considered to result from *de novo* infection rather from re-activation of latent infection. Mainly T-cells are involved in immune process, which govern suppression of latent infection. However, no virulence factor in *P.jirovecii* genome has been described.

There are **two** major groups of antigens of *P.jirovecii* that have been identified. The most widely studied is 95-140 kDa moiety, termed **major surface glycoprotein** (MSG or gpA) that is highly immunogenic, exhibits shared and species-specific antigenic determinants and contains protective B- and T-cells epitopes. *P.jirovecii* is coated with this abundant and highly immunogenic surface antigen. This appears to be involved in *P.jirovecii* attachment on alveolar epithelial cells and with mannose receptors on macrophages by interacting with fibronectin. MSG can induce cellular immune response mediated by T-cells and can elicit cytokine secretion. Thus, it may play pivotal role not only in host-organism interaction but also in provoking the immune response. The other antigen complex is a **glycoprotein** that migrates as a broad band of 35-45 kDa in human strains and of 45-55 kDa in rat strains of *Pneumocystis* and serves as marker of infection.

Pathogenesis and Pathology

Pneumocystis jirovecii is an extracellular pathogen that initiates infection within alveoli via preferential attachment of trophozoites to alveolar epithelium. Various mechanisms in pathogenesis of infection by this organism have been postulated. Once organism is inhaled, it escapes defense of upper respiratory tract and is deposited in the alveoli. The trophic forms preferentially attach to alveolar type I epithelial cells and proliferate in lungs of immunocompromised hosts, provoking severe pneumonitis. Alveolar type II cell hypertrophy, macrophages infiltrate and filling of alveolar spaces with foamy eosinophilic material, are most typical changes. Alveolar macrophages are first line of defense against *P.jirovecii* and principal effector in removing organism from lungs. The organisms produce foci of necrosis and cellular debris in extrapulmonary sites. *P.jirovecii* may occasionally elicit other host responses like granulomatous lesions and cavitary lung disease. The AIDS patients have deficient T4 cells and development of PCP correlates with depletion of CD4+ cells below 200/mm^3. It has been postulated that cell-mediated immunity is of paramount importance in protecting against this infection.

Pneumocystosis is host specific as atypical pneumonia in humans is unlikely to be acquired from reservoir of *Pneumocystis* species of other mammalian host. The molecular biological techniques now show that *Pneumocystis* is fungus and that different strains of organism infect different mammalian species. Impaired cellular immunity has long been considered to be an important predisposing factor in development of pneumocystosis. The foamy eosinophilic exudate and 'honeycomb' appearance of lung tissue are described as hallmarks of PCP.

The DNA sequences and genotypage have shown that variations exist among samples of *P.jirovecii*. But the molecular biology is helpful in the study of the mechanisms of transmission, which can only occur in the same host and the different resistance as well as providing a better understanding of the relationship between host and pathogen. The PCP among immunosuppressed patients was previously thought to result from the reactivation of a latent infection acquired in early childhood. However, it is now believed to result from a new infection from an exogenous source.

Clinical Features

Pneumocystis is a pathogen of both man and animals mainly affecting the lungs. In fact, it is no longer considered a zoonosis anymore. It causes fatal pneumonia in immunocompromised individuals, especially among AIDS patients. Most of the time infection remains asymptomatic. The clinical course of pneumocystosis is insidious and incubation period is about 4 to 8 weeks and patient usually complains of respiratory symptoms. PCP is exclusively seen in patients with defective cell-mediated immunity. It is associated with initial AIDS-related manifestation in about 25% of patients and occurs in more than 50% of them in their terminal stage. It was hypothesized that pneumocystosis in immunocompromised hosts may result from *de novo* acquisition of organism from an exogenous source as well as reactivation of an endogenous infection. The clinical manifestations caused by *P.jirovecii* can be divided into two broad categories:

1. Pulmonary Pneumocystosis
2. Extrapulmonary Pneumocystosis

1. Pulmonary Pneumocystosis

The pulmonary manifestations present as PCP, which progresses insidiously, with gradually increasing non-productive cough over a period of weeks or even months. It is accompanied by dyspnea, fever and occasionally sputum production or hemoptysis. There is an alveolar or interstitial infiltrate and invariably the course is complicated by secondary infections due to bacteria or other opportunistic fungi. Cyanosis is late sign of severe hypoxia among these patients.

The presentation of PCP in children generally involves cyanosis, flaring of nasal ala with mild cough usually without fever and intercostal retraction in severe cases. Nosocomial, epidemic forms of disease have also been noted in premature newborns and infants with malnutrition, congenital immunodeficiency or in those patients who are debilitated due to other underlying diseases. All these cases essentially had infections of prematurely born infants or 'sickly' neonates.

The PCP is associated with considerably high morbidity and mortality, predominantly among patients in whom number of CD4+ cells has fallen below 200/mm^3. This has also been reported in individuals with intact immune system. The chest X-rays show bilateral pulmonary infiltrates and biopsy may be indicated to confirm diagnosis. It is thought that fresh cases usually represent new infection with organism rather than reactivation of latent disease.

2. Extrapulmonary Pneumocystosis

The extrapulmonary manifestations caused by *P.jirovecii* have been described in 0.5-2.5% of persons with AIDS. These occur mainly in cases with advanced HIV infection who are either taking no prophylaxis or only on aerosolized pentamidine. The clinical manifestations are seen depending on the anatomical site involved, which may occur with or without involvement of lungs. The main extrapulmonary sites involved are lymph nodes, bone marrow, spleen, liver, stomach, small intestine, pancreas and eyes.

The central nervous system involvement with *P.jirovecii* usually occurs as a late complication of AIDS and probably represents hematogenous dissemination. The thyroid involvement is rare but should be suspected in HIV-positive individuals with CD4+ cells below 200/mm^3 on prophylactic inhalatory pentamidine who present with neck enlargement with or without pain and clinical and laboratory evidence of hypothyroidism. In a study, Zavascki *et al*, have reviewed 15 cases of thyroiditis caused by this fungus. In another report of 2008, fourteen cases of AIDS-associated otic pneumocystosis have been reported with external auditory canal masses and otorrhea.

Radiodiagnosis

The chest X-rays PA view of pneumocystosis patients reveal gradual spreading of perihilar haziness with granular components or formation of indistinct nodules (diffuse mottling). There are bilateral diffuse alveolar or interstitial pulmonary infiltrates with characteristic **'ground-glass'** appearance, which is classic finding of this disease. Thin section HRCT of chest may show classical **'crazy-paving'** pattern due to combination of alveolitis and inter-lobular septal involvement **(Fig. 22.2)**.

Differential Diagnosis

The differential diagnosis of pneumocystosis includes gamut of diseases caused by opportunistic pulmonary pathogens in patients with profoundly depressed cell-mediated immunity or AIDS. Among these are tuberculosis, histoplasmosis, cryptococcosis, toxoplasmosis, cytomegalovirus infection, bacterial pneumonia, lymphomas and Kaposi's sarcoma. Cardiogenic and non-cardiogenic pulmonary edema need to be considered as differential diagnosis in critically ill patients.

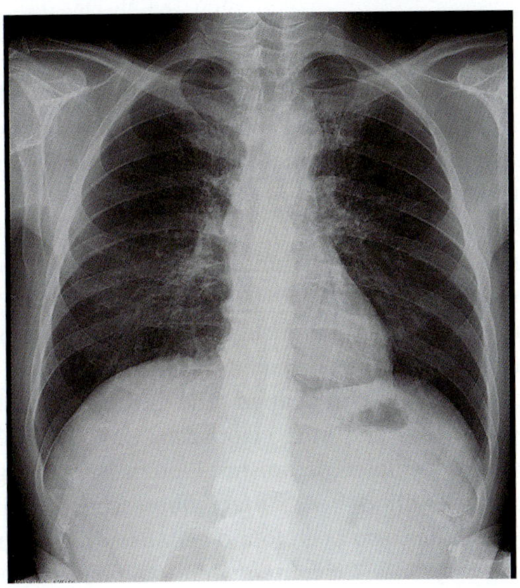

Fig. 22.2. X-rays PA view of pneumocystosis patient showing perihilar haziness with diffuse mottling.

Laboratory Diagnosis

Pneumocystosis being a fungal infection generally presenting as atypical pneumonia that can be investigated by techniques, which are now used routinely in most of hospitals in developed countries and to some extent in developing countries as well. The handling of clinical material is different from routine mycology samples, as organism has to be actively searched within very short time to establish diagnosis. Since it is a treatable condition, rapid and accurate diagnosis is essential to save the life of patient. The T-helper (CD4+) lymphocytes counts of <200/mm^3 or below 20% of total number of T-cells suggest that patient is severely immunocompromised and an underlying HIV-related infection is likely to be present.

The pulmonary function tests show reduction in vital capacity, total lung capacity and single breath diffusing capacity for carbon monoxide. The gallium lung scan is a sensitive, although non-specific test for establishing diagnosis of PCP. Currently, the 'gold standard' technique for diagnosis of *P.jirovecii* involves colorimetric and immunofluorescent staining of bronchoalveolar lavage (BAL) fluid with sensitivity and specificity values of about 95% each. However, PCR based assays have been shown to increase diagnostic sensitivity in an induced sputum specimen with even smaller number of *P.jirovecii.* The induced sputum may be taken by forceful coughing triggered by inhaled hypertonic saline. Extrapulmonary pneumocystosis is diagnosed by demonstration of *P.jirovecii* cysts or trophozoites in body fluids drawn from the affected anatomical site.

In patients with AIDS, diagnosis of PCP is usually made from induced sputum. In contrast, among immunocompromised patients with conditions other than AIDS, induced sputum specimen is rarely positive and diagnosis requires BAL fluid, bronchial washing or tissue biopsy. The improved diagnostic yield from induced sputum for patients with AIDS is attributed to higher organism burden. The measurement of plasma β-D-glucan may be the best non-invasive diagnostic and therapeutic indicator of PCP in non-AIDS individuals, following radiotherapy.

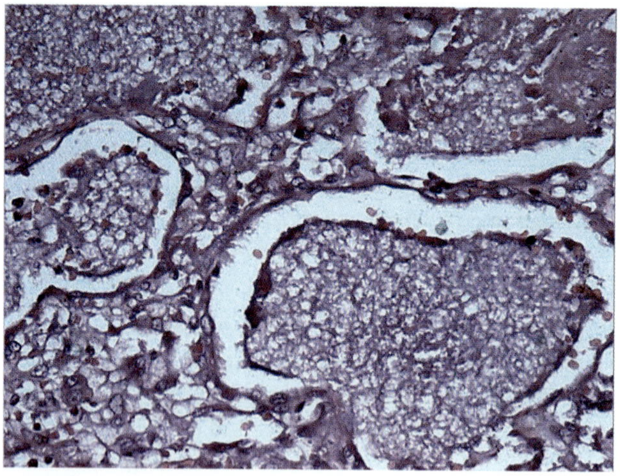

Fig. 22.3. Frothy edema fluid in alveoli of lung in a post-renal transplant recipient patient suggestive of PCP (H&E × 200).

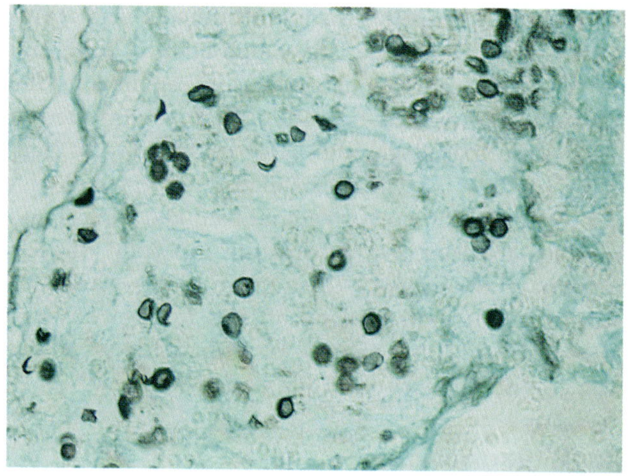

Fig. 22.4. Cluster of *Pneumocystis jirovecii* cysts in frothy edema fluid in alveolar spaces in a renal allograft recipient (GMS × 400).

(a) Direct Examination

The direct examination is the mainstay of diagnosis of pneumocystosis. This depends on demonstration of organism in various clinical specimens and tissues. It is established by visualizing characteristic organisms in expectorated sputum, BAL fluid or lung biopsy. Out of all BAL remains ideal sample in cytological staining of respiratory specimens as it also provides information about burden of organisms, host inflammatory response and presence of other opportunistic infections. The yield of *P.jirovecii* in induced sputum is higher ~50-90%. The lung biopsy in pneumocystosis shows frothy edema fluid in alveoli of patients **(Fig. 22.3)**.

The organism is routinely detected by different staining procedures. Selective staining methods, such as GMS, Toluidine Blue O/Cresyl echt violet and Gram-Weigert stains are widely used which stain wall of **cyst** and are popular because they provide easy identification of this form showing as clusters of organism as seen in **Figure 22.4**. They specifically stain cystic stage of *P.jirovecii* that under ideal conditions is easily identifiable. Gram-Weigert stain is easiest, convenient and most sensitive among all histochemical stain for diagnosis of pneumocystosis.

The trophozoites are more fragile than cysts and care must be taken not to damage them during sample processing. Other stains such as Wright-Giemsa or its variant *i.e.* Giemsa-like rapid stain *e.g.* Diff-Quik, polychrome methylene blue are rapid and easy to perform but are not selective stain for internal contents of cystic nuclei of developmental stages and trophic forms hence require experience for proper interpretation thereby making an accurate identification difficult. The phase-contrast microscopy provides rapid identification of *P.jirovecii* particularly when combined with vital dyes. The tinctorial stain like May-Grunwald Giemsa (MGG) and Papanicolaou (PAP), stain intracystic sporozoites as well as trophozoites, thereby providing significant information about various developmental stages of life cycle of organism.

Giemsa stains the trophozoites as deep blue with no staining of cyst walls. Its modifications such as Diff-Quick, which is a rapid Giemsa technique taking only 30 seconds and May-Grunwald have shown better sensitivity and specificity ranging from 50-84%.

Immunofluorescence assays have been developed with polyclonal and monoclonal antibodies to both trophozoites and cystic stages. These methods are most successful but false negative result may be a drawback.

The presence of *P.jirovecii* in tissue is suspected and confirmed by demonstration of typical, granular, foamy, honeycombed material by H&E stain. The alveolar walls are infiltrated with mononuclear cells with predominance of plasma cells. The cyst wall of organism may have folds and cysts may be **cup-shaped**. The Giemsa stain shows reddish nucleus and blue cytoplasm. GMS stain shows mass of black-colored organisms. Fluorescent conjugated monoclonal antibody *i.e.* direct fluorescence antibody (DFA) stain has also been found very useful in diagnosis of pneumocystosis. Calcofluor white stain has been used, particularly in BAL fluid with number of advantages including rapidity, ease of stain performance and low cost.

If specimens are repeatedly negative, lung biopsy should be performed. Transbronchial biopsy in failed BAL fluids, is very useful but skilled bronchoscopist must perform such procedure and multiple biopsy samples should be obtained. Sometimes, it is associated with complications like pneumothorax and bleeding hence procedure should be carried out taking due precautions. The percutaneous needle aspiration, relatively safe procedure, is also used to obtain specimens from children.

The methods involved in direct microscopic examination have two disadvantages because *P.jirovecii* is not stable outside human body hence delay in sample transport might result in false-negative results. Secondly, exact quantification, needed for therapy monitoring, is also not possible.

(b) Fungal Culture

The fungal culture is taken as **'gold standard'** in medical mycology, is not very useful for routine diagnosis of pneumocystosis as organism has not been successfully grown. Unlike other fungal pathogens, due to fastidious growth requirement of *P.jirovecii*, it is yet to be grown *in vitro*. Cushion *et al*, tried variety of cell lines with limited success using A549, a continuous cell line derived from human lung adenocarcinoma cells. Subsequently, other workers tried with inoculum derived from peripheral blood mononuclear cell from AIDS patients with PCP. The yield of *P.jirovecii* is low and host cell contamination remains an important impediment. Bartlett and Smith investigated ability of human embryonic lung fibroblasts (WI-38) under varying conditions to support *P.jirovecii* growth. A system has also been devised that allows this organism to be cultured axenically and continuously, thus providing starting point for optimization of culture conditions and improvement in its efficacy.

(c) Immunodiagnosis

The serum antibodies to *P.jirovecii* have been measured by complement fixation test, indirect fluorescent test and enzyme immunoassay with whole organism or through soluble extract of antigen. These are non-specific tests because of high prevalence of serum antibodies in general population. The antibody testing has been shown to be useful epidemiologically but it is inconclusive in diagnosing PCP, as majority of healthy individuals as well as patients with pneumocystosis have essentially significant titer.

Immunologic staining with monoclonal antibodies specific for *P.jirovecii* from human lung may allow more accurate diagnosis than histochemical diagnosis. Immunologic staining may be reserved for specimens that are negative by rapid histochemical stainings. β-(1, 3)-D-glucan estimation can be done, as its amount in cysts is more as compared to the trophozoites. A β-D-glucan level >31 pg/mL in serum is found to be useful marker for diagnosis of PCP.

The DNA probe has been used successfully to detect *P.jirovecii* in BAL fluid and appear to be promising as a diagnostic tool. The developments of DNA amplification techniques have given rise to highly sensitive molecular diagnostic assay. These molecular amplification techniques offer increased sensitivity but are often cumbersome and may give positive results in asymptomatic individuals, presumably reflecting colonization or subclinical infection. The PCR-mediated detection has been developed on the basis of thymidylate synthase sequences. The assay protocol can be varied and has been applied to detect specific *P.jirovecii* DNA. Single-step PCR and Southern blot hybridization have also been used successfully to detect this organism in BAL fluid and induced sputum samples. As compared to other fungal infections, PCR in this disease is highly sensitive and specific and can be adopted for routine use in clinical microbiology laboratory. *P.jirovecii*-DNA can now be detected by nested PCR in non-invasive oropharyngeal samples obtained by simple rinsing of mouth, with sensitivity of about 80%. A rapid and specific real time PCR has been developed that is applicable to clinical diagnosis of pneumocystosis.

(d) Animal Pathogenicity

The disease can be produced experimentally by giving methylprednisolone (4mg/kg/wk) and cyclophosphamide (33 mg/kg/wk) in rats, mice or rabbits. The experimental models may be of two types; in first case there is provocation of organisms' resident flora within lungs or acquired from ambient air by chronic immunosuppressive administration. In second type, organisms are instilled in trachea of lightly anesthetized, immunosuppressed *P.jirovecii* rats. The murine model has also been developed to investigate contribution of local inflammation to pathogenesis of PCP hence the role of host's response to this fungus.

Treatment and Prophylaxis

As such there is no utility of routine antifungal agents in treating PCP because causative organism is not susceptible to these drugs due to lack of ergosterol in its cytoplasmic membrane. A combination of **trimethoprim** (TMP) and **sulfamethoxazole** (SMZ or SMX) is the drug of choice for first-line prophylaxis/treatment of all forms of pneumocystosis, which is commonly used anti-protozoal drug. This combination inhibits two key enzymes in folate metabolism hence has been used in this disease for >30 years. This is usually given orally in doses of trimethoprim as 15-20 mg/kg/day and sulfamethoxazole 75-100 mg/kg/day. SMZ targets enzyme dihydropteroate synthase (DHPS) essential for folate biosynthesis and TMP acts on dihydrofolate reductase (DHFR). Despite proven efficacy of TMP-SMZ in both AIDS and non-AIDS patients, studies have shown that failure to respond clinically to this combination is observed in 10-40% of patients and factors leading to treatment failure are not well understood.

For the patients who cannot tolerate TMP-SMZ, the approved second-line therapy is **pentamidine isethionate** in doses of 4 mg/kg/day and administered intravenously. An optimum three weeks' course is required for treatment. Alternate therapies and combinations have been tried. However, combination therapy of these drugs is no more effective than either agent used alone and may even increase the risk of adverse effects.

The other drugs like **dapsone** may also be tried which is another commonly used prophylactic agent and like SMZ, it targets enzyme DHPS. Dapsone is widely used in patients intolerant to sulfonamides. Recovery from PCP is not accompanied by development of acquired immunity and patient is at risk of recurrence as long as predisposing factors exist. The prognosis of disease is poor particularly among immunodeficient patients with delayed treatment. Both regimens have approximately equal efficacy and carry substantial risk of toxicity.

The armamentarium of therapeutic agents has enlarged to include **alternative therapies**, which may be effective and less toxic in mild to moderate episodes with arterial PO_2 of breathing room air greater than 70 mm Hg. In these patients, intravenous medications may not be necessary. Regimens include oral TMP-SMZ, trimethoprim-dapsone and clindamycin-primaquine and trimetrexate (dihydrofolate reductase inhibitor). **Atovaquone**, hydroxynaphthoquinone, in dose of 750 mg orally thrice a day for three weeks, has also been approved as second-line treatment for mild to moderate PCP, with PaO_2 greater than 60 mm Hg in patients intolerant to TMP-SMZ. This drug has advantage of oral administration and has fewer adverse reactions than TMP-SMZ. It has also been recommended for patients with G-6-PD deficiency, as they are intolerant to TMP-SMZ or pentamidine.

As the routine antifungals are not effective in this disease, some of recent ones like caspofungin, an echinocandin, which is a 1,3-β-D-glucan synthase inhibitor, has been shown to have efficacy against cysts of *P.jirovecii* due to glucan synthase inhibition as tested in animal models.

The use of **adjunctive corticosteroids** in treatment of PCP may be indicated when arterial oxygen partial pressure is less than 70 mm Hg and should be started in early course of illness usually when antimicrobial drugs are started. The corticosteroids administered together with antimicrobial therapy, can lead to more rapid improvement in symptoms and gas exchange and increase survival in patients with moderate to severe PCP as compared to patients without corticosteroids.

Primary prophylaxis in pneumocystosis is given in those patients who never had PCP *(i)* if their CD4+ cell count is $<200/mm^3$ or *(ii)* if they have oral thrush or *(iii)* unexplained fever >100°F for more than two weeks' duration. **Secondary prophylaxis** is given in those HIV-infected persons who have already had an episode of PCP.

Until the time an environmental niche of *P.jirovecii* is identified and more is known about spread of organism among people and role of reservoir hosts, ways to avoid acquiring organisms cannot be determined with certainty. The primary and secondary prophylaxis have led to significant decline of PCP in developed countries particularly during the last decade. As PCP is a relapsing condition hence all patients should be given secondary prophylaxis, on recovery, with oral cotrimoxazole.

There are **three** prophylactic agents commonly used for both, primary and secondary prophylaxis of PCP among patients with AIDS: TMP-SMZ, dapsone and aerosolized pentamidine. However, none of these agents is fully effective and PCP is common illness with 'breakthrough' cases often encountered. The **aerosolized pentamidine** is given 300 mg/day for three weeks. It is generally well tolerated but proves less efficacious than TMP-SMZ for prevention of PCP. There is no role of nebulization with pentamidine. Moreover, it poses risk of possible dissemination of the air-borne pathogens during administration. In many circumstances, clinical presentation and approach to diagnosis have been altered by failed prophylactic regimen. The strategies for prevention against PCP in susceptible patients, have been developed with an expectation to decrease both incidence and mortality. The standard guidelines for prophylaxis of PCP are as follows:

(a) Patients with CD4+ cell count of < 200/mm^3 and those with higher counts who have esophageal candidiasis or unexplained fever of >100°F for more than two weeks should receive prophylaxis for PCP. A life-long prophylaxis is recommended for all HIV-positive patients with CD4+ cell count less than 200/mm^3.
(b) All patients with previous episode of PCP should receive prophylaxis.
(c) Patients without history of sensitivity to TMP-SMZ should receive this agent as prophylaxis.
(d) If TMP-SMZ is not tolerated, alternative regimens should be chosen on the basis of convenience, cost and tolerability of patients.

Various reports have also shown that immune reconstitution resulting from HAART may allow some patients to discontinue primary prophylaxis in PCP. Despite major advances in diagnosis and management, this disease remains significant health problem in non-HIV patients and respiratory failure is associated with high mortality rate for both HIV-positive as well as HIV-negative patients.

There are reports of resistant isolates of *P.jirovecii* emerging against first-line drugs, may be because of point mutations in its DHPS gene, which has now been identified and sequenced by PCR. The single-stranded conformation polymorphism (SSCP) assay has been developed which is simple and highly sensitive method for rapid identification of *P.jirovecii* DHPS mutations.

Some of preliminary clinical studies suggest that concomitant use of TMP-SMZ and caspofungin may provide synergistic activity against *P.jirovecii* by completely inhibiting its life cycle. This may be crucial in the treatment of severely hypoxemic patients with PCP.

Immunization

The experimental work on immunization for pneumocystosis is still in early stages, which may be taken as research challenge for further investigation into this disease. The possibility of immunization of individuals at high risk of PCP may be used in future as prophylaxis. This is supported by the fact that successful attempts have been made to elicit protective antibody responses to major surface glycoprotein immunization in an immunosuppressed rat model. The immunization with major surface glycoprotein may protect persons at high risk of PCP. The candidates for immunization include newly diagnosed malignancy, pre-organ transplant cases or patients with HIV infection having CD4 lymphocyte count still in normal range.

Further Reading

1. Aboualigalehdari E, Zarei Mahmoudabadi A, Fatahinia M, *et al.* The prevalence of *Pneumocystis jirovecii* among patients with different chronic pulmonary disorders in Ahvaz, Iran. *Iran J Microbiol.* 2015; **7:** 333-7.
2. Ainoda Y, Hirai Y, Fujita T, *et al.* Analysis of clinical features of non-HIV *Pneumocystis jirovecii* pneumonia. *J Infect Chemother.* 2012; **18:** 722-8.
3. Alanio A, Desoubeaux G, Sarfati C, *et al.* Real-time PCR assay-based strategy for differentiation between active *Pneumocystis jirovecii* pneumonia and colonization in immunocompromised patients. *Clin Microbiol Infect.* 2011; **17:** 1531-7.
4. Alanio A, Hauser PM, Lagrou K, *et al.* ECIL guidelines for the diagnosis of *Pneumocystis jirovecii* pneumonia in patients with haematological malignancies and stem cell transplant recipients. *J Antimicrob Chemother.* 2016; **71:** 2386-96.
5. Alli OA, Ogbolu DO, Ademola O, *et al.* Molecular detection of *Pneumocystis jirovecii* in patients with respiratory tract infections. *N Am J Med Sci.* 2012; **4:** 479-85.
6. Anand S, Samaniego M, Kaul DR. *Pneumocystis jirovecii* pneumonia is rare in renal transplant recipients receiving only one month of prophylaxis. *Transpl Infect Dis.* 2011; **13:** 570-4.
7. Andama AO, Cattamanchi A, Davis JL, *et al.* Modified Giemsa method for confirmation of *Pneumocystis pneumonia* in low-income countries. *BMJ Case Rep.* 2009; pii: bcr02.2009.1580. PMID: 21691388.
8. Antoine R, Caroline B, François V, *et al.* All patients with leukemia are not equally at risk of contracting *Pneumocystis jirovecii* pneumonia. *Am J Med.* 2015; **128:** e9.
9. Armstrong-James D, Copas AJ, Walzer PD, *et al.* A prognostic scoring tool for identification of patients at high and low risk of death from HIV-associated *Pneumocystis jirovecii* pneumonia. *Int J STD AIDS.* 2011; **22:** 628-34.
10. Asai N, Aoshima M, Ohkuni Y, *et al.* A successful diagnostic case of *Pneumocystis* pneumonia by the loop-mediated isothermal amplification method in a patient with dermatomyositis. *J Infect Chemother.* 2012; **18:** 965-9.
11. Avino LJ, Naylor SM, Roecker AM. *Pneumocystis jirovecii* pneumonia in the non-HIV-infected population. *Ann Pharmacother.* 2016; **50:** 673-9.
12. Azoulay E, Bergeron A, Chevret S, *et al.* Polymerase chain reaction for diagnosing *Pneumocystis* pneumonia in non-HIV immunocompromised patients with pulmonary infiltrates. *Chest.* 2009; **135:** 655-61.
13. Babic-Erceg A, Vilibic-Cavlek T, Erceg M, *et al.* Prevalence of *Pneumocystis jirovecii* pneumonia (2010-2013): The first Croatian report. *Acta Microbiol Immunol Hung.* 2014; **61:** 181-8.

14. Baker FN, Cushion MT, Porollo A. A quantitative model to estimate drug resistance in pathogens. *J Fungi (Basel)*. 2016; 2. pii: 30.
15. Bartlett MS, Lee CH. Airborne spread of *Pneumocystis jirovecii*. *Clin Infect Dis*. 2010; **51:** 266.
16. Bartlett MS, Smith JW. *Pneumocystis carinii*, an opportunist in immunocompromised patients. *Clin Microbiol Rev*. 1991; **4:** 137-49.
17. Bava AJ, Romero M, Prieto R, *et al*. A case report of pulmonary coinfection of *Strongyloides stercoralis* and *Pneumocystis jiroveci*. *Asian Pac J Trop Biomed*. 2011; **1:** 334-6.
18. Blount RJ, Djawe K, Daly KR, *et al*. Ambient air pollution associated with suppressed serologic responses to *Pneumocystis jirovecii* in a prospective cohort of HIV-infected patients with *Pneumocystis* pneumonia. *PLoS One*. 2013; **8:** e80795. PMID: 24236202.
19. Bonilla-Abadia F, Betancurt JF, Pineda JC, *et al*. *Pneumocystis jirovecii* pneumonia in two patients with systemic lupus erythematosus after rituximab therapy. *Clin Rheumatol*. 2014; **33:** 415-8.
20. Boondireke S, Mungthin M, Tan-ariya P, *et al*. Evaluation of sensitivity of multiplex PCR for detection of *Mycobacterium tuberculosis* and *Pneumocystis jirovecii* in clinical samples. *J Clin Microbiol*. 2010; **48:** 3165-8.
21. Borstnar S, Lindic J, Tomazic J, *et al*. *Pneumocystis jirovecii* pneumonia in renal transplant recipients: A national center experience. *Transplant Proc*. 2013; **45:** 1614-7.
22. Botterel F, Cabaret O, Foulet F, *et al*. Clinical significance of quantifying *Pneumocystis jirovecii* DNA by using real-time PCR in bronchoalveolar lavage fluid from immunocompromised patients. *J Clin Microbiol*. 2012; **50:** 227-31.
23. Brakemeier S, Durr M, Bachmann F, *et al*. Risk evaluation and outcome of *Pneumocystis jirovecii* pneumonia in kidney transplant patients. *Transplant Proc*. 2016; **48:** 2924-2930.
24. Brubaker R, Redhead SA, Stringer JR, *et al*. Misinformation about *Pneumocystis*. *Clin Exp Dermatol*. 2009; **34:** e426-7.
25. Calderon EJ, de Armas Y, Panizo MM, *et al*. *Pneumocystis jirovecii* pneumonia in Latin America. A public health problem? *Expert Rev Anti Infect Ther*. 2013; **11:** 565-70.
26. Calderon EJ, Dei-Cas E. *Pneumocystis* infection: Unraveling the colonization-to-disease shift. *Expert Rev Anti Infect Ther*. 2010; **8:** 259-62.
27. Calderon EJ, Friaza V, Dapena FJ, *et al*. *Pneumocystis jirovecii* and cystic fibrosis. *Med Mycol*. 2010; **48:** S17-21.
28. Calderon EJ, Gutierrez-Rivero S, Durand-Joly I, *et al*. *Pneumocystis* infection in humans: Diagnosis and treatment. *Expert Rev Anti Infect Ther*. 2010; **8:** 683-701.
29. Calderon EJ. *Pneumocystis* infection: Seeing beyond the tip of the iceberg. *Clin Infect Dis*. 2010; **50:** 354-6.
30. Calderon Sandubete E, de Armas Rodríguez Y, Capo de Paz V. *Pneumocystis jirovecii*: One hundred years of history. *Rev Cubana Med Trop*. 2011; **63:** 97-116.
31. Caselli D, Petris MG, Rondelli R, *et al*. Single-day trimethoprim/sulfamethoxazole prophylaxis for *Pneumocystis* pneumonia in children with cancer. *J Pediatr*. 2014; **164:** 389-92. e1.
32. Catherinot E, Lanternier F, Bougnoux ME, *et al*. *Pneumocystis jirovecii* pneumonia. *Infect Dis Clin North Am*. 2010; **24:** 107-38.
33. Chandola P, Lall M, Sen S, *et al*. Outbreak of *Pneumocystis jirovecii* pneumonia in renal transplant recipients on prophylaxis: Our observation and experience. *Indian J Med Microbiol*. 2014; **32:** 333-6.
34. Chang CH, Ruan SY, Li CC, *et al*. Non-human immunodeficiency virus *Pneumocystis jiroveci* pneumonia. *Respirology*. 2013; **18:** 191-2.
35. Chang H, Shih LY, Wang CW, *et al*. Granulomatous *Pneumocystis jiroveci* pneumonia in a patient with diffuse large B-cell lymphoma: Case report and review of the literature. *Acta Haematol*. 2010; **123:** 30-3.
36. Chapman JR, Marriott DJ, Chen SC, *et al*. Post-transplant *Pneumocystis jirovecii* pneumonia - A re-emerged public health problem? *Kidney Int*. 2013; **84:** 240-3.
37. Chatzikyrkou C, Clajus C, Haubitz M, *et al*. Hypercalcemia and *Pneumocystis* pneumonia after kidney transplantation: Report of an exceptional case and literature review. *Transpl Infect Dis*. 2011; **13:** 496-500.
38. Chawla K, Martena S, Gurung B, *et al*. Role of PCR for diagnosing *Pneumocystis jirovecii* pneumonia in HIV-infected individuals in a tertiary care hospital in India. *Indian J Pathol Microbiol*. 2011; **54:** 326-9.
39. Cheng WL, Ko WC, Lee NY, *et al*. Pneumomediastinum in patients with AIDS: A case report and literature review. *Int J Infect Dis*. 2014; **22:** 31-4.
40. Chew LC, Maceda-Galang LM, Tan YK, *et al*. *Pneumocystis jirovecii* pneumonia in patients with autoimmune disease on high-dose glucocorticoid. *J Clin Rheumatol*. 2015; **21:** 72-5.
41. Chien JY, Liu CJ, Chuang PC, *et al*. Evaluation of the automated Becton Dickinson MAX real-time PCR platform for detection of *Pneumocystis jirovecii*. *Future Microbiol*. 2017; **12:** 29-37.
42. Choe PG, Kang YM, Kim G, *et al*. Diagnostic value of direct fluorescence antibody staining for detecting *Pneumocystis jirovecii* in expectorated sputum from patients with HIV infection. *Med Mycol*. 2014; **52:** 326-30.
43. Chou CW, Chao HS, Lin FC, *et al*. Clinical usefulness of HRCT in assessing the severity of *Pneumocystis jirovecii* pneumonia: A cross-sectional study. *Medicine (Baltimore)*. 2015; **94:** e768.
44. Chou CW, Lin FC, Tsai HC, *et al*. The impact of concomitant pulmonary infection on immune dysregulation in *Pneumocystis jirovecii* pneumonia. *BMC Pulm Med*. 2014; **14:** 182.
45. Choukri F, Menotti J, Sarfati C, *et al*. Quantification and spread of *Pneumocystis jirovecii* in the surrounding air of patients with *Pneumocystis* pneumonia. *Clin Infect Dis*. 2010; **51:** 259-65.
46. Chua KY, Halliday CL, Grote D, *et al*. Colonisation with *Pneumocystis jirovecii* in Australian infants. *Pathology*. 2015; **47:** 489-90.
47. Church DL, Ambasta A, Wilmer A, *et al*. Development and validation of a *Pneumocystis jirovecii* real-time polymerase

chain reaction assay for diagnosis of *Pneumocystis* pneumonia. *Can J Infect Dis Med Microbiol.* 2015; **26:** 263-7.

48. Church JA, Fitzgerald F, Walker AS, *et al.* The expanding role of co-trimoxazole in developing countries. *Lancet Infect Dis.* 2015; **15:** 327-39.
49. Clark A, Hemmelgarn T, Danziger-Isakov L, *et al.* Intravenous pentamidine for *Pneumocystis carinii/jiroveci* pneumonia prophylaxis in pediatric transplant patients. *Pediatr Transplant.* 2015; **19:** 326-31.
50. Cooley L, Dendle C, Wolf J, *et al.* Consensus guidelines for diagnosis, prophylaxis and management of *Pneumocystis jirovecii* pneumonia in patients with haematological and solid malignancies, 2014. *Intern Med J.* 2014; **44:** 1350-63.
51. Cordonnier C, Cesaro S, Maschmeyer G, *et al. Pneumocystis jirovecii* pneumonia: Still a concern in patients with haematological malignancies and stem cell transplant recipients. *J Antimicrob Chemother.* 2016; **71:** 2379-85.
52. Costa JM, Botterel F, Cabaret O, *et al.* Association between circulating DNA, serum (1-3)-β-D-glucan, and pulmonary fungal burden in *Pneumocystis* pneumonia. *Clin Infect Dis.* 2012; **55:** e5-8.
53. Coutsoudis A, Coovadia HM, Kindra G. Time for new recommendations on co-trimoxazole prophylaxis for HIV-exposed infants in developing countries? *Bull World Health Organ.* 2010; **88:** 949-50.
54. Coyle PV, McCaughey C, Nager A, *et al.* Rising incidence of *Pneumocystis jirovecii* pneumonia suggests iatrogenic exposure of immune-compromised patients may be becoming a significant problem. *J Med Microbiol.* 2012; **61 (Pt 7):** 1009-15.
55. Craker LR. Late presentation of *Pneumocystis jiroveci* pneumonia after cardiac transplantation. *BMJ Case Rep.* 2010; pii: bcr0620103111. PMID: 22791580.
56. Creemers-Schild D, Kroon FP, Kuijper EJ, *et al.* Treatment of *Pneumocystis* pneumonia with intermediate-dose and step-down to low-dose trimethoprim-sulfamethoxazole: lessons from an observational cohort study. *Infection.* 2016; **44:** 291-9.
57. Cuetara MS, Alhambra A, Chaves F, *et al.* Use of a serum (1-3)-beta-D-glucan assay for diagnosis and follow-up of *Pneumocystis jiroveci* pneumonia. *Clin Infect Dis.* 2008; **47:** 1364-6.
58. Curbelo J, Galvan JM, Aspa J. Updates on *Aspergillus, Pneumocystis* and other opportunistic pulmonary mycoses. *Arch Bronconeumol.* 2015; **51:** 647-653.
59. D'Avignon LC, Schofield CM, Hospenthal DR. *Pneumocystis* pneumonia. *Semin Respir Crit Care Med.* 2008; **29:** 132-40.
60. Dalpke AH, Hofko M, Zimmermann S. Development and evaluation of a real-time PCR assay for detection of *Pneumocystis jirovecii* on the fully automated BD MAX platform. *J Clin Microbiol.* 2013; **51:** 2337-43.
61. Damiani C, Choukri F, Le Gal S, *et al.* Possible nosocomial transmission of *Pneumocystis jirovecii. Emerg Infect Dis.* 2012; **18:** 877-8.
62. Damiani C, Le Gal S, Da Costa C, *et al.* Combined quantification of pulmonary *Pneumocystis jirovecii* DNA and serum (1-3)-β-D-glucan for differential diagnosis of *Pneumocystis* pneumonia and *Pneumocystis* colonization. *J Clin Microbiol.* 2013; **51:** 3380-8.
63. Damiani C, Le Gal S, Goin N, *et al.* Usefulness of (1,3) β-D-glucan detection in bronchoalveolar lavage samples in *Pneumocystis* pneumonia and *Pneumocystis* pulmonary colonization. *J Mycol Med.* 2015; **25:** 36-43.
64. Damiani C, Le Gal S, Lejeune D, *et al.* Serum (1,3)-beta-D-glucan levels in primary infection and pulmonary colonization with *Pneumocystis jirovecii. J Clin Microbiol.* 2011; **49:** 2000-2.
65. Das CK, Mirdha BR, Singh S, *et al.* Use of induced sputum to determine the prevalence of *Pneumocystis jirovecii* in immunocompromised children with pneumonia. *J Trop Pediatr.* 2014; **60:** 216-22.
66. de Armas Rodriguez Y, Wissmann G, Muller AL, *et al. Pneumocystis jirovecii* pneumonia in developing countries. *Parasite.* 2011; **18:** 219-28.
67. de Boer MG, Kroon FP, le Cessie S, *et al.* Risk factors for *Pneumocystis jirovecii* pneumonia in kidney transplant recipients and appraisal of strategies for selective use of chemoprophylaxis. *Transpl Infect Dis.* 2011; **13:** 559-69.
68. De Castro N, Xu F, Porcher R, *et al. Pneumocystis jirovecii* pneumonia in renal transplant recipients occurring after discontinuation of prophylaxis: A case-control study. *Clin Microbiol Infect.* 2010; **16:** 1375-7.
69. de Leeuw BH, Voskuil WS, Maraha B, *et al.* Evaluation of different real time PCRs for the detection of *Pneumocystis jirovecii* DNA in formalin-fixed paraffin-embedded bronchoalveolar lavage samples. *Exp Mol Pathol.* 2015; **98:** 390-2.
70. Debourgogne A, Favreau S, Ladriere M, *et al.* Characteristics of *Pneumocystis* pneumonia in Nancy from January 2007 to April 2011 and focus on an outbreak in nephrology. *J Mycol Med.* 2014; **24:** 19-24.
71. Del Palacio A, Cuetara MS, Llenas-Garcia J, *et al.* Serum (1, 3)-{beta}-D-glucan assay for the diagnosis of *Pneumocystis jiroveci* pneumonia. *Clin Vaccine Immunol.* 2010; **17:** 202-3.
72. del Rio C, Barragan M, Franco-Paredes C. *Pneumocystis* pneumonia. *N Engl J Med.* 2004; **351:** 1262-3.
73. Deng X, Zhuo L, Lan Y, *et al.* Mutational analysis of *Pneumocystis jirovecii* dihydropteroate synthase and dihydrofolate reductase genes in HIV-infected patients in China. *J Clin Microbiol.* 2014; **52:** 4017-9.
74. Desmet S, Van Wijngaerden E, Maertens J, *et al.* Serum (1,3)-beta-D-glucan as a tool for diagnosis of *Pneumocystis jirovecii* pneumonia in patients with human immunodeficiency virus infection or hematological malignancy. *J Clin Microbiol.* 2009; **47:** 3871-4.
75. Dini L, du Plessis M, Frean J, *et al.* High prevalence of dihydropteroate synthase mutations in *Pneumocystis jirovecii* isolated from patients with *Pneumocystis* pneumonia in South Africa. *J Clin Microbiol.* 2010; **48:** 2016-21.
76. Diri R, Anwer F, Yeager A, *et al.* Retrospective review of intravenous pentamidine for *Pneumocystis* pneumonia prophylaxis in allogeneic hematopoietic stem cell transplantation. *Transpl Infect Dis.* 2016; **18:** 63-9.

77. Djawe K, Daly KR, Levin L, *et al.* Humoral immune responses to *Pneumocystis jirovecii* antigens in HIV-infected and uninfected young children with *Pneumocystis* pneumonia. *PLoS One.* 2013; **8:** e82783.
78. Djawe K, Daly KR, Vargas SL, *et al.* Seroepidemiological study of *Pneumocystis jirovecii* infection in healthy infants in Chile using recombinant fragments of the *P.jirovecii* major surface glycoprotein. *Int J Infect Dis.* 2010; **14:** e1060-6.
79. Ebner L, Walti LN, Rauch A, *et al.* Clinical course, radiological manifestations and outcome of *Pneumocystis jirovecii* pneumonia in HIV patients and renal transplant recipients. *PLoS One.* 2016; **11:** e0164320.
80. Eddens T, Kolls JK. Pathological and protective immunity to *Pneumocystis* infection. *Semin Immunopathol.* 2015; **37:** 153-62.
81. Elias S, Almogi-Hazan O, Aker M, *et al.* A diagnostic challenge: PCP in a non-HIV patient. QJM. 2011; **104:** 889-91.
82. Elnadi NA, Almasry AE, Abosdera MM. Prevalence and effect of a combined treatment on *Pneumocystitis* pneumonia. *J Egypt Soc Parasitol.* 2013; **43:** 457-62.
83. Enomoto T, Azuma A, Kohno A, *et al.* Differences in the clinical characteristics of *Pneumocystis jirovecii* pneumonia in immunocompromized patients with and without HIV infection. *Respirology.* 2010; **15:** 126-31.
84. Esteves F, Cale SS, Badura R, *et al.* Diagnosis of *Pneumocystis* pneumonia: Evaluation of four serologic biomarkers. *Clin Microbiol Infect.* 2015; **21:** 379. e1-10.
85. Esteves F, Gaspar J, De Sousa B, *et al.* Clinical relevance of multiple single-nucleotide polymorphisms in *Pneumocystis jirovecii* pneumonia: Development of a multiplex PCR-single-base-extension methodology. *J Clin Microbiol.* 2011; **49:** 1810-5.
86. Esteves F, Lee CH, de Sousa B, *et al.* (1,3)-beta-D-glucan in association with lactate dehydrogenase as biomarkers of *Pneumocystis* pneumonia (PCP) in HIV-infected patients. *Eur J Clin Microbiol Infect Dis.* 2014; **33:** 1173-80.
87. Esteves F, Medrano FJ, de Armas Y, *et al.* *Pneumocystis* and pneumocystosis: First meeting of experts from Latin-American and Portuguese-speaking countries - A mini-review. *Expert Rev Anti Infect Ther.* 2014; **12:** 545-8.
88. Evans RA, Clifford TM, Tang S, *et al.* Efficacy of once-weekly dapsone dosing for *Pneumocystis jirovecii* pneumonia prophylaxis post-transplantation. *Transpl Infect Dis.* 2015; **17:** 816-21.
89. Ewald H, Raatz H, Boscacci R, *et al.* Adjunctive corticosteroids for *Pneumocystis jiroveci* pneumonia in patients with HIV infection. *Cochrane Database Syst Rev.* 2015; **4:** CD006150.
90. Fan LC, Lu HW, Cheng KB, *et al.* Evaluation of PCR in bronchoalveolar lavage fluid for diagnosis of *Pneumocystis jirovecii* pneumonia: A bivariate meta-analysis and systematic review. *PLoS One.* 2013; **8:** e73099.
91. Fillatre P, Decaux O, Jouneau S, *et al.* Incidence of *Pneumocystis jiroveci* pneumonia among groups at risk in HIV-negative patients. *Am J Med.* 2014; **127:** 1242.e11-7.
92. Fillaux J, Berry A. Real-time PCR assay for the diagnosis of *Pneumocystis jirovecii* pneumonia. *Methods Mol Biol.* 2013; **943:** 159-70.
93. Fong S, Daly KR, Tipirneni R, *et al.* Antibody responses against *Pneumocystis jirovecii* in health care workers over time. *Emerg Infect Dis.* 2013; **19:** 1612-9.
94. Frenkel JK. *Pneumocystis jiroveci* n. sp. from man: Morphology, physiology and immunology in relation to pathology. *Natl Cancer Inst Monogr.* 1976; **43:** 13-30.
95. Friaza V, Morilla R, Respaldiza N, *et al.* *Pneumocystis jiroveci* dihydropteroate synthase gene mutations among colonized individuals and *Pneumocystis* pneumonia patients from Spain. *Postgrad Med.* 2010; **122:** 24-8.
96. Fritzsche C, Ghanem H, Koball S, *et al.* High *Pneumocystis jirovecii* colonization rate among haemodialysis patients. *Infect Dis (Lond).* 2017; **49:** 132-6. PMID: 27684384.
97. Gabardi S, Millen P, Hurwitz S, *et al.* Atovaquone versus trimethoprim-sulfamethoxazole as *Pneumocystis jirovecii* pneumonia prophylaxis following renal transplantation. *Clin Transplant.* 2012; **26:** E184-90.
98. Gago S, Esteban C, Valero C, *et al.* A multiplex real-time PCR assay for identification of *Pneumocystis jirovecii, Histoplasma capsulatum* and *Cryptococcus neoformans/ Cryptococcus gattii* in samples from AIDS patients with opportunistic pneumonia. *J Clin Microbiol.* 2014; **52:** 1168-76.
99. Gal SL, Hery-Arnaud G, Ramel S, *et al.* *Pneumocystis jirovecii* and cystic fibrosis in France. *Scand J Infect Dis.* 2010; **42:** 225-7.
100. Gianella S, Haeberli L, Joos B, *et al.* Molecular evidence of interhuman transmission in an outbreak of *Pneumocystis jirovecii* pneumonia among renal transplant recipients. *Transpl Infect Dis.* 2010; **12:** 1-10.
101. Gigliotti F, Limper AH, Wright T. *Pneumocystis. Cold Spring Harb Perspect Med.* 2014; **4:** a019828. PMID: 25367973.
102. Gilroy SA, Bennett NJ. Pneumocystis pneumonia. *Semin Respir Crit Care Med.* 2011; **32:** 775-82.
103. Goldman AS, Goldman LR, Goldman DA. What caused the epidemic of *Pneumocystis* pneumonia in European premature infants in the mid-20th century? *Pediatrics.* 2005; **115:** e725-36.
104. Gonzalez BE, Faverio LA, Marty FM, *et al.* Elevated serum beta-D-glucan levels in immunocompromised children with clinical suspicion for *Pneumocystis jirovecii* pneumonia. *Clin Vaccine Immunol.* 2011; **18:** 1202-3.
105. Gopal R, Rapaka RR, Kolls JK. Immune reconstitution inflammatory syndrome associated with pulmonary pathogens. *Eur Respir Rev.* 2017; 26. pii: 160042.
106. Goto N, Futamura K, Okada M, *et al.* Management of *Pneumocystis jirovecii* pneumonia in kidney transplantation to prevent further outbreak. *Clin Med Insights Circ Respir Pulm Med.* 2015; **9 (Suppl. 1):** 81-90.
107. Goto N, Oka S. *Pneumocystis jirovecii* pneumonia in kidney transplantation. *Transpl Infect Dis.* 2011; **13:** 551-8.
108. Green MR. A modicum of caution for blood (1, 3)-β-D-glucan testing for *Pneumocystis jirovecii* in HIV-infected patients. *Clin Infect Dis.* 2011; **53:** 1039-40.

109. Grewal P, Brassard A. Fact or fiction: Does the non-HIV/AIDS immunosuppressed patient need *Pneumocystis jiroveci* pneumonia prophylaxis? An updated literature review. *J Cutan Med Surg.* 2009; **13:** 308-12.
110. Grier DD, Lewis Z, Palavecino EL. Bone marrow involvement by *Pneumocystis jiroveci. Br J Haematol.* 2009; **145:** 149.
111. Grubbs JA, Baddley JW. *Pneumocystis jirovecii* pneumonia in patients receiving tumor-necrosis-factor-inhibitor therapy: Implications for chemoprophylaxis. *Curr Rheumatol Rep.* 2014; **16:** 445.
112. Guigue N, Alanio A, Menotti J, *et al.* Utility of adding *Pneumocystis jirovecii* DNA detection in nasopharyngeal aspirates in immunocompromised adult patients with febrile pneumonia. *Med Mycol.* 2015; **53:** 241-7.
113. Gupta R, Iyer VK, Mirdha BR, *et al.* Role of cytology and polymerase chain reaction based detection of *Pneumocystis jirovecii* infection in bronchoalveolar lavage fluid. *Acta Cytol.* 2010; **54:** 296-302.
114. Gupta R, Mirdha BR, Guleria R, *et al.* Genotypic variation of *Pneumocystis jirovecii* isolates in India based on sequence diversity at mitochondrial large subunit rRNA. *Int J Med Microbiol.* 2011; **301:** 267-72.
115. Gupta R, Mirdha BR, Guleria R, *et al.* Diagnostic significance of nested polymerase chain reaction for sensitive detection of *Pneumocystis jirovecii* in respiratory clinical specimens. *Diagn Microbiol Infect Dis.* 2009; **64:** 381-8.
116. Gutierrez S, Respaldiza N, Campano E, *et al. Pneumocystis jirovecii* colonization in chronic pulmonary disease. *Parasite.* 2011; **18:** 121-6.
117. Haddad TM, Vallabhajosyula S, Nawaz MS, *et al.* Fatal *Pneumocystis jirovecii* pneumonia in a HIV-negative adult. *BMJ Case Rep.* 2015; pii: bcr2015210117. PMID: 26311008.
118. Hagiya H, Miyake T, Kokumai Y, *et al.* Co-infection with invasive pulmonary aspergillosis and *Pneumocystis jirovecii* pneumonia after corticosteroid therapy. *J Infect Chemother.* 2013; **19:** 342-7.
119. Hardak E, Brook O, Yigla M. Radiological features of *Pneumocystis jirovecii* pneumonia in immunocompromised patients with and without AIDS. *Lung.* 2010; **188:** 159-63.
120. Hardak E, Neuberger A, Yigla M, *et al.* Outcome of *Pneumocystis jirovecii* pneumonia diagnosed by polymerase chain reaction in patients without human immunodeficiency virus infection. *Respirology.* 2012; **17:** 681-6.
121. Harris JR, Marston BJ, Sangrujee N, *et al.* Cost-effectiveness analysis of diagnostic options for *Pneumocystis* pneumonia (PCP). *PLoS One.* 2011; **6:** e23158. PMID: 21858013.
122. Harris K, Maroun R, Chalhoub M, *et al.* Unusual presentation of *Pneumocystis pneumonia* in an immunocompetent patient diagnosed by open lung biopsy. *Heart Lung Circ.* 2012; **21:** 221-4.
123. Hartel PH, Shilo K, Klassen-Fischer M, *et al.* Granulomatous reaction to *Pneumocystis jirovecii:* Clinicopathologic review of 20 cases. *Am J Surg Pathol.* 2010; **34:** 730-4.
124. Hauser P, Rabodonirina M, Nevez G. *Pneumocystis jirovecii* genotypes involved in *Pneumocystis* pneumonia outbreaks among renal transplant recipients. *Clin Infect Dis.* 2013; **56:** 165-6.
125. Hauser PM, Bille J, Lass-Florl C, *et al.* Multicenter, prospective clinical evaluation of respiratory samples from subjects at risk for *Pneumocystis jirovecii* infection by use of a commercial real-time PCR assay. *J Clin Microbiol.* 2011; **49:** 1872-8.
126. Hawksworth DL. Responsibility in naming pathogens: The case of *Pneumocystis jirovecii,* the causal agent of *Pneumocystis* pneumonia. *Lancet Infect Dis.* 2007; **7:** 3-5.
127. Hayes GE, Denning DW. Frequency, diagnosis and management of fungal respiratory infections. *Curr Opin Pulm Med.* 2013; **19:** 259-65.
128. Held J, Koch MS, Reischl U, *et al.* Serum (1→3)-β-D-glucan measurement as an early indicator of *Pneumocystis jirovecii* pneumonia and evaluation of its prognostic value. *Clin Microbiol Infect.* 2011; **17:** 595-602.
129. Held J, Wagner D. β-D-Glucan kinetics for the assessment of treatment response in *Pneumocystis jirovecii* pneumonia. *Clin Microbiol Infect.* 2011; **17:** 1118-22.
130. Hernandez-Hernandez F, Frealle E, Caneiro P, *et al.* Prospective multicenter study of *Pneumocystis jirovecii* colonization among cystic fibrosis patients in France. *J Clin Microbiol.* 2012; **50:** 4107-10.
131. Horwedel TA, Bowman LJ, Saab G, *et al.* Benefits of sulfamethoxazole-trimethoprim prophylaxis on rates of sepsis after kidney transplant. *Transpl Infect Dis.* 2014; **16:** 261-9.
132. Huaringa AJ, Francis WH. Pulmonary alveolar proteinosis: A case report and world literature review. *Respirol Case Rep.* 2016; **4:** e00201.
133. Hyun MH, Sim JK, Oh JY, *et al.* Centrilobular cysts of the lung. *Am J Med Sci.* 2016; **351:** e1.
134. Iriart X, Bouar ML, Kamar N, *et al. Pneumocystis* pneumonia in solid-organ transplant recipients. *J Fungi.* 2015; **1:** 293-331.
135. Iriart X, Challan Belval T, Fillaux J, *et al.* Risk factors of *Pneumocystis* pneumonia in solid organ recipients in the era of the common use of post-transplantation prophylaxis. *Am J Transplant.* 2015; **15:** 190-9.
136. Jain SB, Wig N, Nagpal SJ, *et al.* Evaluation of the current management protocols for prophylaxis against *Pneumocystis jiroveci* pneumonia and other opportunistic infections in patients living with HIV/AIDS. *AIDS Care.* 2011; **23:** 846-50.
137. Janmeja AK, Mohapatra PR, Shivaprakash MR, *et al.* Concurrent infection of *Pneumocystis* pneumonia and pulmonary tuberculosis in an HIV-seronegative patient. *Indian J Chest Dis Allied Sci.* 2008; **50:** 369-71.
138. Jarboui MA, Sellami A, Sellami H, *et al.* Molecular diagnosis of *Pneumocystis jiroveci* pneumonia in immunocompromised patients. *Mycoses.* 2010; **53:** 329-33.
139. Javier B, Susana L, Santiago G, *et al.* Pulmonary co-infection by *Pneumocystis jiroveci* and *Cryptococcus neoformans. Asian Pac J Trop Biomed.* 2012; **2:** 80-2.

140. Jiang X, Mei X, Feng D, *et al.* Prophylaxis and treatment of *Pneumocystis jiroveci* pneumonia in lymphoma patients subjected to rituximab-contained therapy: A systemic review and meta-analysis. *PLoS One.* 2015; **10:** e0122171.
141. Kaji Y, Ohara G, Kagohashi K, *et al.* Pneumomediastinum in a patient with *Pneumocystis jirovecii* pneumonia. *Intern Med.* 2012; **51:** 2251.
142. Kalin M, Kristinsson SY, Cherif H, *et al.* Fatal *Pneumocystis jiroveci* pneumonia in ABVD-treated Hodgkin lymphoma patients. *Ann Hematol.* 2010; **89:** 523-5.
143. Kalkanis A, Judson MA, Napier MB. *Pneumocystis jerovecii* pneumonia in a patient with untreated chronic lymphocytic leukaemia: A novel case and postulations concerning the mechanism. *BMJ Case Rep.* 2013; pii: bcr2013202124.
144. Kamada T, Furuta K, Tomioka H. *Pneumocystis* pneumonia associated with human immunodeficiency virus infection without elevated (1→3)-β-D glucan: A case report. *Respir Med Case Rep.* 2016; **18:** 73-5.
145. Kanne JP, Yandow DR, Meyer CA. *Pneumocystis jiroveci* pneumonia: High-resolution CT findings in patients with and without HIV infection. *AJR Am J Roentgenol.* 2012; **198:** W555-61.
146. Karageorgopoulos DE, Qu JM, Korbila IP, *et al.* Accuracy of β-D-glucan for the diagnosis of *Pneumocystis jirovecii* pneumonia: A meta-analysis. *Clin Microbiol Infect.* 2013; **19:** 39-49.
147. Karageorgopoulos DE, Vouloumanou EK, Ntziora F, *et al.* β-D-glucan assay for the diagnosis of invasive fungal infections: A meta-analysis. *Clin Infect Dis.* 2011; **52:** 750-70.
148. Karam MB, Mosadegh L. Extra-pulmonary *Pneumocystis jiroveci* infection: A case report. *Braz J Infect Dis.* 2014; **18:** 681-5.
149. Kaur R, Gautam H, Maheshwari M, *et al.* Concurrent *Cryptococcal* and *Pneumocystis* pneumonia along with pulmonary tuberculosis in an HIV-positive patient: Lessons learned for early management. *J Int Assoc Physicians AIDS Care (Chic).* 2011; **10:** 146-9.
150. Kaur R, Katariya P, Dhakad MS, *et al.* An unusual case of cystic fibrosis associated *Pneumocystis jiroveci* pneumonia in an infant. *Case Rep Infect Dis.* 2016; 9206707. PMID: 28070430.
151. Kaur R, Panda PS, Dewan R. Profile of *Pneumocystis* infection in a tertiary care institute in North India. *Indian J Sex Transm Dis.* 2016; **37:** 143-6.
152. Kaur R, Wadhwa A, Bhalla P, *et al.* *Pneumocystis* pneumonia in HIV patients: A diagnostic challenge till date. *Med Mycol.* 2015; **53:** 587-92.
153. Kawano S, Maeda T, Suzuki T, *et al.* Loop-mediated isothermal amplification with the procedure for ultra rapid extraction kit for the diagnosis of *Pneumocystis* pneumonia. *J Infect Chemother.* 2015; **21:** 224-6.
154. Kelly DM, Cronin S. PCP prophylaxis with use of corticosteroids by neurologists. *Pract Neurol.* 2014; **14:** 74-6.
155. Kelly MN, Shellito JE. Current understanding of *Pneumocystis* immunology. *Future Microbiol.* 2010; **5:** 43-65.
156. Khalife S, Aliouat EM, Aliouat-Denis CM, *et al.* First data on *Pneumocystis jirovecii* colonization in patients with respiratory diseases in North Lebanon. *New Microbes New Infect.* 2015; **6:** 11-4.
157. Khodadadi H, Mirhendi H, Mohebali M, *et al.* *Pneumocystis jirovecii* colonization in non-HIV-infected patients based on nested-PCR detection in bronchoalveolar lavage samples. *Iran J Public Health.* 2013; **42:** 298-305.
158. Kim HS, Shin KE, Lee JH. Single nodular opacity of granulomatous *Pneumocystis jirovecii* pneumonia in an asymptomatic lymphoma patient. *Korean J Radiol.* 2015; **16:** 440-3.
159. Kim SJ, Lee J, Cho YJ, *et al.* Prognostic factors of *Pneumocystis jirovecii* pneumonia in patients without HIV infection. *J Infect.* 2014; **69:** 88-95.
160. Kofteridis DP, Valachis A, Velegraki M, *et al.* Predisposing factors, clinical characteristics and outcome of *Pneumocystis jirovecii* pneumonia in HIV-negative patients. *J Infect Chemother.* 2014; **20:** 412-6.
161. Kono M, Yamashita H, Kubota K, *et al.* FDG PET Imaging in *Pneumocystis* pneumonia. *Clin Nucl Med.* 2015; **40:** 679-81.
162. Kostakis ID, Sotiropoulos GC, Kouraklis G. *Pneumocystis jirovecii* pneumonia in liver transplant recipients: A systematic review. *Transplant Proc.* 2014; **46:** 3206-8.
163. Kuik KT, Trubiano J, Worth LJ, *et al.* *Pneumocystis jirovecii* pneumonia following everolimus treatment of metastatic breast cancer. *Med Mycol Case Rep.* 2014; **6:** 34-6.
164. Kumar N, Bazari F, Rhodes A, *et al.* Chronic *Pneumocystis jiroveci* presenting as asymptomatic granulomatous pulmonary nodules in lymphoma. *J Infect.* 2011; **62:** 484-6.
165. Lanaspa M, O'Callaghan-Gordo C, Machevo S, *et al.* High prevalence of *Pneumocystis jirovecii* pneumonia among Mozambican children <5 years of age admitted to hospital with clinical severe pneumonia. *Clin Microbiol Infect.* 2015; **21:** 1018.e9-15.
166. Langlois ME, Lorillou M, Ferry T, *et al.* Cystic lung lesions revealing a *Pneumocystis jirovecii* and *Aspergillus flavus* co-infection in an HIV-infected patient. *Int J Infect Dis.* 2015; **37:** 143-4.
167. Le Gal S, Damiani C, Rouille A, *et al.* A cluster of *Pneumocystis* infections among renal transplant recipients: molecular evidence of colonized patients as potential infectious sources of *Pneumocystis jirovecii. Clin Infect Dis.* 2012; **54:** e62-71.
168. Le Gal S, Damiani C, Virmaux M, *et al.* A 39-year-old man with human immunodeficiency virus infection presenting with an alveolo-interstitial pulmonary syndrome. *J Clin Microbiol.* 2013; **51:** 2809 & 3165.
169. Le Gal S, Pougnet L, Damiani C, *et al.* *Pneumocystis jirovecii* in the air surrounding patients with *Pneumocystis* pulmonary colonization. *Diagn Microbiol Infect Dis.* 2015; **82:** 137-42.
170. Lee HY, Kang HS, Lee HY, *et al.* Clinical significance of positive *Pneumocystis jirovecii* polymerase chain reaction in non-human immunodeficiency virus immunocompromised patients in a real practice. *Korean J Intern Med.* 2017; **32:** 478-85.

171. Lee JY, Park HJ, Kim YK, *et al.* Cellular profiles of bronchoalveolar lavage fluid and their prognostic significance for non-HIV-infected patients with *Pneumocystis jirovecii* pneumonia. *J Clin Microbiol.* 2015; **53:** 1310-6.
172. Lee NY, Chang CM, Lee CC, *et al.* Improved outcome of *Pneumocystis* pneumonia by early treatment. *J Microbiol Immunol Infect.* 2012; **45:** 163-4.
173. Lee SC, Feenstra J, Georghiou PR. *Pneumocystis jiroveci* pneumonitis complicating ruxolitinib therapy. *BMJ Case Rep.* 2014; pii: bcr2014204950. PMID: 24891492.
174. Lee SH, Huh KH, Joo DJ, *et al.* Risk factors for *Pneumocystis jirovecii* pneumonia (PJP) in kidney transplantation recipients. *Sci Rep.* 2017; **7:** 1571. PMID: 28484270.
175. Lee SM, Cho YK, Sung YM, *et al.* A case of pneumonia caused by *Pneumocystis jirovecii* resistant to trimethoprim-sulfamethoxazole. *Korean J Parasitol.* 2015; **53:** 321-7.
176. Lemiale V, Debrumetz A, Delannoy A, *et al.* Adjunctive steroid in HIV-negative patients with severe *Pneumocystis* pneumonia. *Respir Res.* 2013; **14:** 87.
177. Leth S, Jensen-Fangel S, Ostergaard L, *et al.* *Pneumocystis jirovecii* pneumonia in patients with end-stage renal disease: A comparison with the general population. *Scand J Infect Dis.* 2014; **46:** 704-11.
178. Lerdlamyong K, Sukpanichnant S, Rattanaumpawan P. Bone marrow infection by *Pneumocystis jirovecii. Br J Haematol.* 2017; **176:** 849.
179. Liebling M, Rubio E, Ie S. Prophylaxis for *Pneumocystis jiroveci* pneumonia: Is it a necessity in pulmonary patients on high-dose, chronic corticosteroid therapy without AIDS? *Expert Rev Respir Med.* 2015; **9:** 171-81.
180. Llibre JM, Revollo B, Vanegas S, *et al.* *Pneumocystis jirovecii* pneumonia in HIV-1-infected patients in the late-HAART era in developed countries. *Scand J Infect Dis.* 2013; **45:** 635-44.
181. Lopez-Sanchez C, Falco V, Burgos J, *et al.* Epidemiology and long-term survival in HIV-infected patients with *Pneumocystis jirovecii* pneumonia in the HAART era: Experience in a university hospital and review of the literature. *Medicine (Baltimore).* 2015; **94:** e681.
182. Louis M, Guitard J, Jodar M, *et al.* Impact of HIV infection status on interpretation of quantitative PCR for detection of *Pneumocystis jirovecii. J Clin Microbiol.* 2015; **53:** 3870-5.
183. Lowe DM, Rangaka MX, Gordon F, *et al.* *Pneumocystis jirovecii* pneumonia in tropical and low and middle income countries: a systematic review and meta-regression. *PLoS One.* 2013; **8:** e69969.
184. Lundgren IS, Englund JA, Burroughs LM, *et al.* Outcomes and duration of *Pneumocystis jiroveci* pneumonia therapy in infants with severe combined immunodeficiency. *Pediatr Infect Dis J.* 2012; **31:** 95-7.
185. Maeda T, Babazono A, Nishi T, *et al.* Quantification of the effect of chemotherapy and steroids on risk of *Pneumocystis jiroveci* among hospitalized patients with adult T-cell leukaemia. *Br J Haematol.* 2015; **168:** 501-6.
186. Maertens J, Cesaro S, Maschmeyer G, *et al.* ECIL guidelines for preventing *Pneumocystis jirovecii* pneumonia in patients with haematological malignancies and stem cell transplant recipients. *J Antimicrob Chemother.* 2016; **71:** 2397-404.
187. Maillet M, Maubon D, Brion JP, *et al.* *Pneumocystis jirovecii (Pj)* quantitative PCR to differentiate *Pj* pneumonia from *Pj* colonization in immunocompromised patients. *Eur J Clin Microbiol Infect Dis.* 2014; **33:** 331-6.
188. Maini R, Henderson KL, Sheridan EA, *et al.* Increasing *Pneumocystis* pneumonia, England, UK, 2000-2010. *Emerg Infect Dis.* 2013; **19:** 386-92.
189. Maitte C, Leterrier M, Le Pape P, *et al.* Multilocus sequence typing of *Pneumocystis jirovecii* from clinical samples: How many and which loci should be used? *J Clin Microbiol.* 2013; **51:** 2843-9.
190. Mane A, Gujar P, Chandra J, *et al.* *Pneumocystis jirovecii* infection and the associated dihydropteroate synthase (DHPS) and dihydrofolate reductase (DHFR) mutations in HIV-positive individuals from Pune, India. *Mycopathologia.* 2015; **179:** 141-5.
191. Maschmeyer G, Helweg-Larsen J, Pagano L, *et al.* ECIL Guidelines for treatment of *Pneumocystis jirovecii* pneumonia in non-HIV-infected haematology patients. *J Antimicrob Chemother.* 2016; **71:** 2405-13.
192. Matos O, Esteves F. *Pneumocystis jirovecii* multilocus gene sequencing: Findings and implications. *Future Microbiol.* 2010; **5:** 1257-67.
193. Matos O. *Pneumocystis jirovecii* pneumonia in Africa: Impact and implications of highly sensitive diagnostic technologies. *N Am J Med Sci.* 2012; **4:** 486-7.
194. Matsumura Y, Ito Y, Iinuma Y, *et al.* Quantitative real-time PCR and the (1→3)-β-D-glucan assay for differentiation between *Pneumocystis jirovecii* pneumonia and colonization. *Clin Microbiol Infect.* 2012; **18:** 591-7.
195. Matsumura Y, Ito Y, Yamamoto M, *et al.* *Pneumocystis* polymerase chain reaction and blood (1→3)-β-D-glucan assays to predict survival with suspected *Pneumocystis jirovecii* pneumonia. *J Infect Chemother.* 2014; **20:** 109-14.
196. McCaughan JA, Courtney AE. *Pneumocystis jiroveci* pneumonia in renal transplantation: Time to review our practice? *Nephrol Dial Transplant.* 2012; **27:** 13-5.
197. Messiaen PE, Cuyx S, Dejagere T, *et al.* The role of CD4-cell count as discriminatory measure to guide chemoprophylaxis against *Pneumocystis jirovecii* pneumonia in human immunodeficiency virus-negative immunocompromised patients: A systematic review. *Transpl Infect Dis.* 2017; doi: 10.1111/tid.12651. PubMed PMID: 28035717.
198. Mok HP, Hart E, Venkatesan P. Early development of immune reconstitution inflammatory syndrome related to *Pneumocystis* pneumonia after antiretroviral therapy. *Int J STD AIDS.* 2014; **25:** 373-7.
199. Monroy-Vaca EX, de Armas Y, Illnait-Zaragozí MT, *et al.* Prevalence and genotype distribution of *Pneumocystis jirovecii* in Cuban infants and toddlers with whooping cough. *J Clin Microbiol.* 2014; **52:** 45-51.

200. Montes-Cano MA, Chabe M, Fontillon-Alberdi M, *et al.* Vertical transmission of *Pneumocystis jirovecii* in humans. *Emerg Infect Dis.* 2009; **15:** 125-7.
201. Montesinos I, Brancart F, Schepers K, *et al.* Comparison of 2 real-time PCR assays for diagnosis of *Pneumocystis jirovecii* pneumonia in human immunodeficiency virus (HIV) and non-HIV immunocompromised patients. *Diagn Microbiol Infect Dis.* 2015; **82:** 143-7.
202. Mori S, Sugimoto M. *Pneumocystis jirovecii* infection: An emerging threat to patients with rheumatoid arthritis. *Rheumatology (Oxford).* 2012; **51:** 2120-30.
203. Mori S, Sugimoto M. *Pneumocystis jirovecii* pneumonia in rheumatoid arthritis patients: Risks and prophylaxis recommendations. *Clin Med Insights Circ Respir Pulm Med.* 2015; **9 (Suppl. 1):** 29-40.
204. Morris A, Norris KA. Colonization by *Pneumocystis jirovecii* and its role in disease. *Clin Microbiol Rev.* 2012; **25:** 297-317.
205. Morris AM, Masur H. A serologic test to diagnose *Pneumocystis* pneumonia: Are we there yet? *Clin Infect Dis.* 2011; **53:** 203-4.
206. Msaad S, Yangui I, Bahloul N, *et al.* Do inhaled corticosteroids increase the risk of *Pneumocystis* pneumonia in people with lung cancer? *World J Clin Cases.* 2015; **3:** 843-7.
207. Mulhall AM, McCormack FX. A 25-year-old man with fatigue, dyspnea and pulmonary cysts. *Ann Am Thorac Soc.* 2015; **12:** 120-3.
208. Mulpuru S, Knoll G, Weir C, *et al. Pneumocystis* pneumonia outbreak among renal transplant recipients at a North American transplant center: Risk factors and implications for infection control. *Am J Infect Control.* 2016; **44:** 425-31.
209. Munson E, Paul M. A 30-year-old male with a 4-week history of shortness of breath, productive cough and weight loss. *J Clin Microbiol.* 2011; **49:** 3113 & 3447.
210. Musallam N, Bamberger E, Srugo I, *et al. Legionella pneumophila* and *Pneumocystis jirovecii* coinfection in an infant treated with adrenocorticotropic hormone for infantile spasm: Case report and literature review. *J Child Neurol.* 2014; **29:** 240-2.
211. Nakazato T, Mihara A, Sanada Y, *et al. Pneumocystis jiroveci* pneumonia detected by FDG-PET. *Ann Hematol.* 2010; **89:** 839-40.
212. Nankivell BJ, Firacative C, Kable K, *et al.* Molecular epidemiology linking multihospital clusters of opportunistic *Pneumocystis jirovecii* pneumonia. *Clin Infect Dis.* 2013; **57:** 1058-9.
213. Nasnas R, Matta MY, El Helou G, *et al.* Brain abscesses in HIV-positive patients due to *Pneumocystis jiroveci. J Int Assoc Physicians AIDS Care (Chic).* 2012; **11:** 169-71.
214. Neuwelt AJ, Nguyen TM, Fu R, *et al.* Incidence of *Pneumocystis jirovecii* pneumonia after temozolomide for CNS malignancies without prophylaxis. *CNS Oncol.* 2014; **3:** 267-73.
215. Nevez G, Chabe M, Rabodonirina M, *et al.* Nosocomial *Pneumocystis jirovecii* infections. *Parasite.* 2008; **15:** 359-65.
216. Ng B, Dipchand A, Naftel D, *et al.* Outcomes of *Pneumocystis jiroveci* pneumonia infections in pediatric heart transplant recipients. *Pediatr Transplant.* 2011; **15:** 844-8.
217. Ng VL, Yajko DM, Hadley WK. Extrapulmonary pneumocystosis. *Clin Microbiol Rev.* 1997; **10:** 401-18.
218. Nickel P, Schurmann M, Albrecht H, *et al.* Clindamycin-primaquine for *Pneumocystis jiroveci* pneumonia in renal transplant patients. *Infection.* 2014; **42:** 981-9.
219. Nitzan O, Peretz A, Pastukh N, *et al.* Should prophylaxis for *Pneumocystis jirovecii* pneumonia be considered in patients with a CD4+ count higher than 200 cells/µl? *AIDS.* 2015; **29:** 2533-4.
220. Nobile A, Valenti A, Aubert JD, *et al.* Granulomatous reaction to *Pneumocystis jirovecii* diagnosed in a bronchoalveolar lavage: A case report. *Acta Cytol.* 2015; **59:** 284-8.
221. Nou DD, Vaidya A, Dichtel L, *et al.* A woman with fever and dyspnea. *N Engl J Med.* 2014; **370:** e25. PMID: 24716703.
222. Nowaseb V, Gaeb E, Fraczek MG, *et al.* Frequency of *Pneumocystis jirovecii* in sputum from HIV and TB patients in Namibia. *J Infect Dev Ctries.* 2014; **8:** 349-57.
223. O'Neal CB, Ball SC. Splenic pneumocystosis: An atypical presentation of extrapulmonary *Pneumocystis* infection. *AIDS Read.* 2008; **18:** 503-8.
224. Okafor PN, Wasan SK, Farraye FA. *Pneumocystis jiroveci* pneumonia in patients with inflammatory bowel disease: A survey of prophylaxis patterns among gastroenterology providers. *Inflamm Bowel Dis.* 2013; **19:** 812-7.
225. Onishi A, Sugiyama D, Kogata Y, *et al.* Diagnostic accuracy of serum 1,3-β-D-glucan for *Pneumocystis jiroveci* pneumonia, invasive candidiasis and invasive aspergillosis: systematic review and meta-analysis. *J Clin Microbiol.* 2012; **50:** 7-15.
226. Oren I, Hardak E, Finkelstein R, *et al.* Polymerase chain reaction-based detection of *Pneumocystis jirovecii* in bronchoalveolar lavage fluid for the diagnosis of *Pneumocystis* pneumonia. *Am J Med Sci.* 2011; **342:** 182-5.
227. Orgel E, Rushing T. Efficacy and tolerability of intravenous pentamidine isethionate for *Pneumocystis jiroveci* prophylaxis in a pediatric oncology population. *Pediatr Infect Dis J.* 2014; **33:** 319-21.
228. Orsi CF, Bettua C, Pini P, *et al.* Detection of *Pneumocystis jirovecii* and *Aspergillus* spp. DNA in bronchoalveolar lavage fluids by commercial real-time PCR assays: Comparison with conventional diagnostic tests. *New Microbiol.* 2015; **38:** 75-84.
229. Parisaei M, Hemelaar J, Govind A. HIV in pregnancy: A case of *Pneumocystis (carinii) jiroveci* pneumonia. *Arch Gynecol Obstet.* 2010; **281:** 1-3.
230. Park YK, Jung HC, Kim SY, *et al.* Spontaneous pneumomediastinum, pneumopericardium, and pneumothorax with respiratory failure in a patient with *AIDS* and *Pneumocystis jirovecii* pneumonia. *Infect Chemother.* 2014; **46:** 204-8.
231. Parobek CM, Jiang LY, Patel JC, *et al.* Multilocus microsatellite genotyping array for investigation of genetic epidemiology of *Pneumocystis jirovecii. J Clin Microbiol.* 2014; **52:** 1391-9.

232. Patel KB, Gleason JB, Diacovo MJ, *et al.* *Pneumocystis* pneumonia presenting as an enlarging solitary pulmonary nodule. *Case Rep Infect Dis.* 2016; 1873237.
233. Patrat-Delon S, Gangneux JP, Lavoue S, *et al.* Correlation of parasite load determined by quantitative PCR to clinical outcome in a heart transplant patient with disseminated toxoplasmosis. *J Clin Microbiol.* 2010; **48:** 2541-5.
234. Pena ZG, Byers HR, Lehmer LM, *et al.* Mixed *Pneumocystis* and *Cryptococcus* cutaneous infection histologically mimicking xanthoma. *Am J Dermatopathol.* 2013; **35:** e6-10.
235. Pereira RM, Muller AL, Zimerman RA, *et al.* High prevalence of *Pneumocystis jirovecii* colonization among HIV-positive patients in southern Brazil. *Med Mycol.* 2014; **52:** 804-9.
236. Perez-Ordono L, Hoyo I, Sanclemente G, *et al.* Late-onset *Pneumocystis jirovecii* pneumonia in solid organ transplant recipients. *Transpl Infect Dis.* 2014; **16:** 324-8.
237. Pfeifer K, Kalra V, Adebowale A, *et al.* Apical *Pneumocystis jiroveci* as an AIDS defining illness: A case report illustrating a change in the paradigm. *J Radiol Case Rep.* 2014; **8:** 15-24.
238. Phipps LM, Chen SC, Kable K, *et al.* Nosocomial *Pneumocystis jirovecii* pneumonia: Lessons from a cluster in kidney transplant recipients. *Transplantation.* 2011; **92:** 1327-34.
239. Pierre-Louis N, Dlugos B. A rare cause of cavitary lung lesions. *JAAPA.* 2017; **30:** 28-31.
240. Pliquett RU, Asbe-Vollkopf A, Hauser PM, *et al.* A *Pneumocystis jirovecii* pneumonia outbreak in a single kidney-transplant center: Role of cytomegalovirus co-infection. *Eur J Clin Microbiol Infect Dis.* 2012; **31:** 2429-37.
241. Ponce CA, Gallo M, Bustamante R, *et al.* *Pneumocystis* colonization is highly prevalent in the autopsied lungs of the general population. *Clin Infect Dis.* 2010; **50:** 347-53.
242. Proudfoot R, Cox R, Phillips B, *et al.* UK survey of *Pneumocystis jirovecii* pneumonia (PJP) prophylaxis use in paediatric oncology patients. *Arch Dis Child.* 2015; **100:** 115. PMID: 25324566.
243. Queener SF, Cody V, Pace J, *et al.* Trimethoprim resistance of dihydrofolate reductase variants from clinical isolates of *Pneumocystis jirovecii. Antimicrob Agents Chemother.* 2013; **57:** 4990-8.
244. Ramalho J, Bacelar Marques ID, *et al.* *Pneumocystis jirovecii* pneumonia with an atypical granulomatous response after kidney transplantation. *Transpl Infect Dis.* 2014; **16:** 315-9.
245. Ray A, Khong B, Khong HT. A case report of *Pneumocystis jiroveci* pneumonia in a patient with metastatic breast cancer. *Anticancer Res.* 2016; **36:** 6673-6.
246. Reid AB, Chen SC, Worth LJ. *Pneumocystis jirovecii* pneumonia in non-HIV-infected patients: New risks and diagnostic tools. *Curr Opin Infect Dis.* 2011; **24:** 534-44.
247. Reinbolt RE, Alam S, Layman R, *et al.* *Pneumocystis jiroveci* pneumonia in an atypical host. *Clin Breast Cancer.* 2012; **12:** 138-41.
248. Revathy M, Therese KL, Bagyalakshmi R, *et al.* Application of real time polymerase chain reaction targeting kex 1 gene & its comparison with the conventional methods for rapid detection of *Pneumocystis jirovecii* in clinical specimens. *Indian J Med Res.* 2014; **140:** 406-13.
249. Ricks DM, Chen K, Zheng M, *et al.* Dectin immunoadhesins and *Pneumocystis* pneumonia. *Infect Immun.* 2013; **81:** 3451-62.
250. Riebold D, Enoh DO, Kinge TN, *et al.* *Pneumocystis jirovecii* colonisation in HIV-positive and HIV-negative subjects in Cameroon. *Trop Med Int Health.* 2014; **19:** 643-655.
251. Rispail P. Pneumocysta (Protista: Protomycota), a new phylum for *Pneumocystis*? *J Mycol Med.* 2000; **10:** 51-61.
252. Robert-Gangneux F, Belaz S, Revest M, *et al.* Diagnosis of *Pneumocystis jirovecii* pneumonia in immunocompromised patients by real-time PCR: A 4-year prospective study. *J Clin Microbiol.* 2014; **52:** 3370-6.
253. Roblot F, Le Moal G, Kauffmann-Lacroix C, *et al.* *Pneumocystis jirovecii* pneumonia in HIV-negative patients: A prospective study with focus on immunosuppressive drugs and markers of immune impairment. *Scand J Infect Dis.* 2014; **46:** 210-4.
254. Roembke F, Heinzow HS, Gosseling T, *et al.* Clinical outcome and predictors of survival in patients with *Pneumocystis jirovecii* pneumonia - Results of a tertiary referral centre. *Clin Respir J.* 2014; **8:** 86-92.
255. Rostved AA, Sassi M, Kurtzhals JA, *et al.* Outbreak of *Pneumocystis* pneumonia in renal and liver transplant patients caused by genotypically distinct strains of *Pneumocystis jirovecii. Transplantation.* 2013; **96:** 834-42.
256. Roux A, Canet E, Valade S, *et al.* *Pneumocystis jirovecii* pneumonia in patients with or without AIDS, France. *Emerg Infect Dis.* 2014; **20:** 1490-7.
257. Roux A, Gonzalez F, Roux M, *et al.* Update on pulmonary *Pneumocystis jirovecii* infection in non-HIV patients. *Med Mal Infect.* 2014; **44:** 185-98.
258. Sabur N, Kelly MM, Gill MJ, *et al.* Granulomatous *Pneumocystis jiroveci* pneumonia associated with immune reconstituted HIV. *Can Respir J.* 2011; **18:** e86-8.
259. Saito T, Shime N, Itoh K, *et al.* Disseminated aspergillosis following resolution of *Pneumocystis* pneumonia with sustained elevation of beta-glucan in an intensive care unit: A case report. *Infection.* 2009; **37:** 547-50.
260. Salerno D, Mushatt D, Myers L, *et al.* Serum and BAL beta-D-glucan for the diagnosis of *Pneumocystis* pneumonia in HIV positive patients. *Respir Med.* 2014; **108:** 1688-95.
261. Saltzman RW, Albin S, Russo P, *et al.* Clinical conditions associated with PCP in children. *Pediatr Pulmonol.* 2012; **47:** 510-6.
262. Samuel CM, Whitelaw A, Corcoran C, *et al.* Improved detection of *Pneumocystis jirovecii* in upper and lower respiratory tract specimens from children with suspected *Pneumocystis* pneumonia using real-time PCR: A prospective study. *BMC Infect Dis.* 2011; **11:** 329.
263. Sarwar S, Carey B, Hegarty JE, *et al.* Low incidence of *Pneumocystis jirovecii* pneumonia in an unprophylaxed liver transplant cohort. *Transpl Infect Dis.* 2013; **15:** 510-5.
264. Sarwar Zubairi AB, Shahzad H, Zafar A. Clinical outcomes of *Pneumocystis* pneumonia from a tertiary care centre in Pakistan. *J Pak Med Assoc.* 2016; **66:** 1367-71.

265. Sasso M, Chastang-Dumas E, Bastide S, *et al.* Performances of four real-time PCR assays for the diagnosis of *Pneumocystis jirovecii* pneumonia. *J Clin Microbiol.* 2016; **54:** 625-30.
266. Schultz MG, Bloch AB. In Memoriam: Sandy Ford (1950–2015). *Emerg Infect Dis.* 2016; **22:** 764-5.
267. Shah N, Owen L, Bhagani S. 'Occam's Scissors': opportunistic infections in advanced HIV infection. *BMJ Case Rep.* 2013; pii: bcr2013009544. PMID: 23833087.
268. Shteinberg M, Shaked-Mishan P, Kinarti A, *et al.* Asymptomatic carriage of *Pneumocystis jirovecii* and cytomegalovirus in lungs of immunocompetent patients. *Lung.* 2014; **192:** 875-9.
269. Sidhu VK, Foisy MM, Hughes CA. Discontinuing *Pneumocystis jirovecii* pneumonia prophylaxis in HIV-infected patients with a CD4 cell count <200 cells/mm^3. *Ann Pharmacother.* 2015; **49:** 1343-8.
270. Siegel M, Masur H, Kovacs J. *Pneumocystis jirovecii* pneumonia in human immunodeficiency virus infection. *Semin Respir Crit Care Med.* 2016; **37:** 243-56.
271. Singh P, Singh S, Mirdha BR, *et al.* Evaluation of loop-mediated isothermal amplification assay for the detection of *Pneumocystis jirovecii* in immunocompromised patients. *Mol Biol Int.* 2015; 819091. PMID: 26664746.
272. Singh Y, Mirdha BR, Guleria R, *et al.* Molecular detection of DHFR gene polymorphisms in *Pneumocystis jirovecii* isolates from Indian patients. *J Infect Dev Ctries.* 2015; **9:** 1250-6.
273. Singh Y, Mirdha BR, Guleria R, *et al.* Circulating genotypes of *Pneumocystis jirovecii* and its clinical correlation in patients from a single tertiary center in India. *Eur J Clin Microbiol Infect Dis.* 2017. doi: 10.1007/s10096-017-2977-9. PMID: 28401321.
274. Sivam S, Sciurba FC, Lucht LA, *et al.* Distribution of *Pneumocystis jirovecii* in lungs from colonized COPD patients. *Diagn Microbiol Infect Dis.* 2011; **71:** 24-8.
275. Sokulska M, Kicia M, Wesołowska M, *et al. Pneumocystis jirovecii* - From a commensal to pathogen: Clinical and diagnostic review. *Parasitol Res.* 2015; **114:** 3577-85.
276. Song Y, Ren Y, Wang X, *et al.* Recent advances in the diagnosis of *Pneumocystis* pneumonia. *Med Mycol J.* 2016; **57:** E111-6.
277. Stern A, Green H, Paul M, *et al.* Prophylaxis for *Pneumocystis* pneumonia (PCP) in non-HIV immunocompromised patients. *Cochrane Database Syst Rev.* 2014; **10:** CD005590.
278. Strich JR, Jerussi TD, Wiestner A, *et al. Pneumocystis jirovecii* pneumonia in a treatment-naive patient with chronic lymphocytic leukemia. *Infect Dis Clin Pract.* 2016; **24:** e86-7. PMID: 27917032.
279. Stringer JR, Beard CB, Miller RF, *et al.* A new name *(Pneumocystis jiroveci)* for *Pneumocystis* from humans. *Emerg Infect Dis.* 2002; **8:** 891-6.
280. Stringer JR, Beard CB, Miller RF. Spelling *Pneumocystis jirovecii. Emerg Infect Dis.* 2009; **15:** 506.
281. Stringer JR, Edman JC, Cushion MT, *et al.* The fungal nature of *Pneumocystis. J Med Vet Mycol.* 1992; **30:** S271-8.
282. Stringer JR. *Pneumocystis carinii* - What is it, exactly? *Clin Microbiol Rev.* 1996; **9:** 489-98.
283. Struijk GH, Gijsen AF, Yong SL, *et al.* Risk of *Pneumocystis jiroveci* pneumonia in patients long after renal transplantation. *Nephrol Dial Transplant.* 2011; **26:** 3391-8.
284. Su YS, Lu JJ, Perng CL, *et al. Pneumocystis jirovecii* pneumonia in patients with and without human immunodeficiency virus infection. *J Microbiol Immunol Infect.* 2008; **41:** 478-82.
285. Subedee A, Kimmel W, Donato AA. Never too early for the opportunists! *BMJ Case Rep.* 2013; pii: bcr2013009612.
286. Suraj KP, Narayan KV, Edakalavan J, *et al.* Diffuse parenchymal lung diseases with clinicoradiological discordance: Role of transbronchial lung biopsy as a diagnostic tool - An observational study. *J Clin Diagn Res.* 2016; **10:** OC01-4.
287. Tamai K, Tachikawa R, Tomii K, *et al.* Prognostic value of bronchoalveolar lavage in patients with non-HIV *Pneumocystis* pneumonia. *Intern Med.* 2014; **53:** 1113-7.
288. Tanaka Y, Saraya T, Kurai D, *et al.* Spontaneous resolution of *Pneumocystis jirovecii* pneumonia on high-resolution computed tomography in a patient with renal cell carcinoma. *Am J Case Rep.* 2014; **15:** 496-500.
289. Tasaka S, Kobayashi S, Yagi K, *et al.* Serum (1→3) β-D-glucan assay for discrimination between *Pneumocystis jirovecii* pneumonia and colonization. *J Infect Chemother.* 2014; **20:** 678-81.
290. Tasaka S, Tokuda H. Recent advances in the diagnosis of *Pneumocystis jirovecii* pneumonia in HIV-infected adults. *Expert Opin Med Diagn.* 2013; **7:** 85-97.
291. Tasaka S. *Pneumocystis* pneumonia in human immunodeficiency virus-infected adults and adolescents: Current concepts and future directions. *Clin Med Insights Circ Respir Pulm Med.* 2015; **9 (Suppl. 1):** 19-28.
292. Teh BW, Azzato FA, Lingaratnam SM, *et al.* Molecular diagnosis of *Pneumocystis jirovecii* in patients with malignancy: Clinical significance of quantitative polymerase chain reaction. *Med Mycol.* 2014; **52:** 427-32.
293. Thomas CF Jr, Limper AH. *Pneumocystis* pneumonia. *N Engl J Med.* 2004; **350:** 2487-98.
294. Tia T, Putaporntip C, Kosuwin R, *et al.* A highly sensitive novel PCR assay for detection of *Pneumocystis jirovecii* DNA in bronchoalveolar lavage specimens from immunocompromised patients. *Clin Microbiol Infect.* 2012; **18:** 598-603.
295. Tipirneni R, Daly KR, Jarlsberg LG, *et al.* Healthcare worker occupation and immune response to *Pneumocystis jirovecii. Emerg Infect Dis.* 2009; **15:** 1590-7.
296. To KK, Hung IF, Xu T, *et al.* Clinical significance of *Pneumocystis jiroveci* in patients with active tuberculosis. *Diagn Microbiol Infect Dis.* 2013; **75:** 260-5.
297. To KK, Wong SC, Xu T, *et al.* Use of nasopharyngeal aspirate for diagnosis of *Pneumocystis* pneumonia. *J Clin Microbiol.* 2013; **51:** 1570-4.
298. Travis J, Hart E, Helm J, *et al.* Retrospective review of *Pneumocystis jirovecii* pneumonia over two decades. *Int J STD AIDS.* 2009; **20:** 200-1.
299. Tschudy J, Michail S. Disseminated histoplasmosis and *Pneumocystis* pneumonia in a child with Crohn disease receiving infliximab. *J Pediatr Gastroenterol Nutr.* 2010; **51:** 221-2.

300. Tyagi AK, Mirdha BR, Guleria R, *et al.* Study of dihydropteroate synthase (DHPS) gene mutations among isolates of *Pneumocystis jiroveci. Indian J Med Res.* 2008; **128:** 734-9.
301. Tyagi AK, Mirdha BR, Luthra K, *et al.* Dihydropteroate synthase (DHPS) gene mutation study in HIV-Infected Indian patients with *Pneumocystis jirovecii* pneumonia. *J Infect Dev Ctries.* 2010; **4:** 761-6.
302. Tyagi AK, Mirdha BR, Luthra K, *et al. Pneumocystis jirovecii* dihydropteroate synthase (DHPS) genotypes in non-HIV-immunocompromised patients: A tertiary care reference health centre study. *Med Mycol.* 2011; **49:** 167-71.
303. Unnewehr M, Friederichs H, Bartsch P, *et al.* High diagnostic value of a new real-time *Pneumocystis* PCR from bronchoalveolar lavage in a real-life clinical setting. *Respiration.* 2016; **92:** 144-9.
304. Urabe N, Ishii Y, Hyodo Y, *et al.* Molecular epidemiologic analysis of a *Pneumocystis* pneumonia outbreak among renal transplant patients. *Clin Microbiol Infect.* 2016; **22:** 365-71.
305. Valade S, Azoulay E, Damiani C, *et al. Pneumocystis jirovecii* airborne transmission between critically ill patients and health care workers. *Intensive Care Med.* 2015; **41:** 1716-8.
306. van Groeningen I, Arnoldus J, Perenboom R, *et al.* Red fist and muscle weakness with a rare complication. *BMJ Case Rep.* 2014; pii: bcr2013202457. PMID: 24557476.
307. van Halsema C, Johnson L, Baxter J, *et al.* Diagnosis of *Pneumocystis jirovecii* pneumonia by detection of DNA in blood and oropharyngeal wash, compared with sputum. *AIDS Res Hum Retroviruses.* 2016; **32:** 463-6.
308. Vananuvat P, Suwannalai P, Sungkanuparph S, *et al.* Primary prophylaxis for *Pneumocystis jirovecii* pneumonia in patients with connective tissue diseases. *Semin Arthritis Rheum.* 2011; **41:** 497-502.
309. Vanspauwen MJ, Knops VE, Bruggeman CA, *et al.* Molecular epidemiology of *Pneumocystis jiroveci* in human immunodeficiency virus-positive and -negative immunocompromised patients in The Netherlands. *J Med Microbiol.* 2014; **63:** 1294-302.
310. Vargas SL, Pizarro P, Lopez-Vieyra M, *et al. Pneumocystis* colonization in older adults and diagnostic yield of single versus paired noninvasive respiratory sampling. *Clin Infect Dis.* 2010; **50:** e19-21.
311. Vas A, Laws P, Marsland A, *et al.* Acute generalised exanthematous pustulosis induced by *Pneumocystis jirovecii* pneumonia prophylaxis with dapsone. *Int J STD AIDS.* 2013; **24:** 745-7.
312. Waitt P. Ward round-sudden increase of breathlessness in a patient with *Pneumocystis* pneumonia (PCP) and haematemesis. *Malawi Med J.* 2009; **21:** 28 & 32.
313. Walzer PD. *Pneumocystis carinii:* A historical perspective. *Semin Respir Infect.* 1998; **13:** 279-82.
314. Wang DD, Zheng MQ, Zhang N, *et al.* Investigation of *Pneumocystis jirovecii* colonization in patients with chronic pulmonary diseases in the People's Republic of China. *Int J Chron Obstruct Pulmon Dis.* 2015; **10:** 2079-85.
315. Wang Y, Zhao Z, Lu H, *et al.* Fungal infection involvement in primary biliary cirrhosis: A review of 2 cases. *Exp Ther Med.* 2017; **13:** 489-94.
316. Wasserman S, Engel ME, Griesel R, *et al.* Burden of *Pneumocystis* pneumonia in HIV-infected adults in sub-Saharan Africa: A systematic review and meta-analysis. *BMC Infect Dis.* 2016; **16:** 482. PMID: 27612639.
317. Wasserman S, Engel ME, Mendelson M. Burden of *Pneumocystis* pneumonia in HIV-infected adults in sub-Saharan Africa: Protocol for a systematic review. *Syst Rev.* 2013; **2:** 112. PMID: 24330755.
318. Wickramasekaran RN, Jewell MP, Sorvillo F, *et al.* The changing trends and profile of pneumocystosis mortality in the United States, 1999-2014. *Mycoses.* 2017; doi: 10.1111/myc.12636. PMID: 28568970.
319. Wilson JW, Limper AH, Grys TE, *et al. Pneumocystis jirovecii* testing by real-time polymerase chain reaction and direct examination among immunocompetent and immunosuppressed patient groups and correlation to disease specificity. *Diagn Microbiol Infect Dis.* 2011; **69:** 145-52.
320. Wood BR, Komarow L, Zolopa AR, *et al.* Beta-glucan for *Pneumocystis* pneumonia diagnosis in persons with AIDS: authors' reply. *AIDS.* 2013; **27:** 2967-8.
321. Wood BR, Komarow L, Zolopa AR, *et al.* Test performance of blood beta-glucan for *Pneumocystis jirovecii* pneumonia in patients with AIDS and respiratory symptoms. AIDS. 2013; **27:** 967-72.
322. Wuerz TC, Bow EJ, Seftel MD. Potential consequences of essential drug shortages in Canada: Brain abscess due to *Nocardia farcinica* associated with dapsone prophylaxis for *Pneumocystis jirovecii* pneumonia. *Can J Infect Dis Med Microbiol.* 2013; **24:** 159-61.
323. Yang CY, Shih CJ, Yang WC, *et al.* Aggressive immunosuppressant reduction and long-term rejection risk in renal transplant recipients with *Pneumocystis jiroveci* pneumonia. *Exp Clin Transplant.* 2012; **10:** 344-9.
324. Yang CY, Yang AH, Yang WC, Lin CC. Risk factors for *Pneumocystis jirovecii* pneumonia in glomerulonephritis patients receiving immunosuppressants. *Intern Med.* 2012; **51:** 2869-75.
325. Yang JJ, Huang CH, Liu CE, *et al.* Multicenter study of trimethoprim/sulfamethoxazole-related hepatotoxicity: Incidence and associated factors among HIV-infected patients treated for *Pneumocystis jirovecii* pneumonia. *PLoS One.* 2014; **9:** e106141.
326. Zakout GA. Beware the "red herring". *Transpl Infect Dis.* 2016; **18:** 765-7.
327. Zhou Y, Shetty J, Pins MR. Small-intestine *Pneumocystis jiroveci* pseudotumor as an acquired immunodeficiency syndrome-presenting illness: Report of a case and review of the literature. *Arch Pathol Lab Med.* 2012; **136:** 1001-3.
328. Zmarlicka M, Martin ST, Cardwell SM, *et al.* Tolerability of low-dose sulfamethoxazole/trimethoprim for *Pneumocystis jirovecii* pneumonia prophylaxis in kidney transplant recipients. *Prog Transplant.* 2015; **25:** 210-6.

CHAPTER 23

Microsporidiosis

Microsporidiosis is an emerging opportunistic fungal infection caused by various microsporidial species. This has recently been transferred from Parasitology to Medical Mycology due to taxonomic reasons. This is found among individuals with immunocompromised backgrounds like AIDS patients, organ transplant recipients, travelers, contact lens wearers, children and the elderly. It includes a wide range of clinical manifestations involving various systems of body presenting as syndromes, collectively called **microsporidiosis**.

The microsporidia are obligate intracellular organism and have undergone recent taxonomic change and re-classified from **Protozoa** to the kingdom **Fungi**. They are widely distributed in the nature with ~1300 species characterized so far. However, microsporidia have only recently been documented to parasitize humans and more research is needed to understand this newly emerging infectious disease.

There are 15 species of Microsporidia currently known to infect humans. ***Enterocytozoon bieneusi*** and ***Encephalitozoon intestinalis***, are the most common causes of human infections and are associated mainly with diarrhea and systemic diseases. Species of microsporidia infecting humans have been identified in water sources as well as in wild, domestic and food-producing farm animals, raising concerns of their water-borne, food-borne and even zoonotic transmission.

The infection among humans, microsporidiosis, is primarily found as an opportunistic infection, especially among those infected with HIV or those who have undergone organ transplants. Some species, however, have also been known to parasitize individuals with healthy immune systems. Beyond the human realm, Microsporidia are primarily parasites in fisheries, veterinary medicines and pest management.

Microsporidiosis is an emerging disease in hosts from aquatic as well as terrestrial biomes. The human infections are often derived from contact with animals and the environment. The immune suppression allows this opportunistic infection to set in thereby onset of the disease. The animal-human food chain provides a portal for transmission and emergence of this disease.

Microsporidia are unicellular, eukaryotic obligate intracellular microorganisms with a size of ~1-2 µm. The phylum Microsporidia includes 170 genera and ~1300 species, of which at least 15 have been detected in human infections. Microsporidia have been isolated from mammals, reptiles, fish, birds and insects. Mature microsporidia spores have a thick three-layered wall and are resistant to the chlorine concentrations used in drinking water. These spores may pass through water filters due to their smaller size. The transmission is believed to occur via the fecal-oral route, contaminated food, water and infected animals or humans.

As stated earlier that the most common microsporidia species affecting humans are *Enterocytozoon bieneusi* and *Encephalitozoon intestinalis*. These species primarily settle into the gastrointestinal system and cause chronic diarrhea entailing weight loss. The species may give rise to a disseminated infection and can involve the urogenital and respiratory systems, as well as the eyes. In the era of molecular diagnostics, studies have increased aiming to understand natural sources, life cycles, transmission routes and virulence of the microsporidia.

Historical Perspective

The Phylum Microspora was discovered in 1800s but the first human case was described late only in 1959 in a Japanese child. The emergence of microsporidiosis is associated with the rise and spread of HIV; microsporidiosis is primarily found among patients with AIDS or who are otherwise immunocompromised like organ transplant patients. However, at least three species of *Nosema* and one of *Brachiola* have been documented in immunocompetent patients. Microsporidia are considered casual, accidental or opportunistic agents in humans.

Microsporidia, unicellular parasites of animals and protists with highly reduced mitochondria, are now included in phylum under the kingdom Fungi, based on analyses of various studies. The latter study concluded that Microsporidia are the sister group of the rest of the Fungi and should not be classified as true Fungi but that topology does not conflict with the delimitation of the monophyletic Fungi as proposed. The analysis of James *et al*, in 2006 suggested that *Rozella*, which was not sampled by Liu *et al*, is the sister group of Microsporidia. However, no subdivision of the Microsporidia is proposed, owing to a lack of well-sampled multilocus analyses of this group hence included under one of the opportunistic mycoses.

The nomenclatural status of Microsporidia has remained ambiguous. It was treated as a phylum under the zoological Code (International Commission on Zoological Nomenclature, 1999) but there is disagreement about the correct author citation and it is uncertain if the name would be valid under the botanical Code. This uncertainty arises as *Microsporidium,* Balbiani, 1884 appears to be a later synonym of *Nosema,* Naegeli, 1857. Before the status of the Microsporidia can be resolved, it will be necessary to decide whether the nomenclature of the group as a whole should be governed by the zoological or the botanical Code although the latter now allows names of fungi described under the zoological Code to be accepted. The final decision requires input from the community of scientists who study Microsporidia. This long going taxonomical controversies further makes it more interesting.

Mycology

Microsporidia are obligate, intracellular, opportunistic fungi infecting a wide range of invertebrate and vertebrate hosts, including human beings. Of 15 species belonging to 8 genera known to infect humans, *Enterocytozoon bieneusi, Encephalitozoon intestinalis, E.cuniculi,* and *E.hellem* are the most common ones. *Microsporidium* is basically a collective generic name for microsporidia. The microsporidia are intracellular sporulating eukaryotic, ubiquitous in the environment, unicellular organisms belonging to the phylum Microspora. All Microsporidia are obligate, spore-forming, intracellular organisms that invade vertebrates and invertebrates. Microsporidia are a group of obligate intracellular parasitizing fungi that have risen over the past two decades from obscure organisms to well-recognized human pathogens. There are 15 species reported to infect humans and causing more severe symptoms in immunocompromised individuals. Microsporidia of the genus *Encephalitozoon* and *Enterocytozoon bieneusi* are the most frequent causes of life-threatening chronic diarrhea and systemic disease among HIV patients and acute, self-limited diarrhea in immunocompetent persons.

A characteristic feature of microsporidia is the polar tube or polar filament found in the spore used to infiltrate host cells as the invasive property. The term microsporidia is also used as a general nomenclature for the obligate intracellular organisms belonging to the phylum Microsporidia. To date, ~1,300 species belonging to 170 genera have been described as infecting a wide range of vertebrate and invertebrate hosts.

Microsporidial spores are the only form found outside the host, which survive in environment at 4°C for more than a year. The mode of transmission is not yet demonstrable, presumed to be inhalation, ingestion or inoculation.

Microsporidia, are characterized by the production of resistant spores that vary in size, depending on the species. They possess a unique organelle, the polar tubule or polar filament, which is coiled inside the spore as demonstrated by its ultrastructure. The microsporidia spores of species associated with human infection measure from 1 to 4 μm and that is a useful diagnostic feature. There are at least 15 microsporidian species that have been identified as human pathogens: *Anncaliia* (formerly *Brachiola*) *algerae, A.connori, A.vesicularum, Encephalitozoon cuniculi, E.hellem, E.intestinalis, Enterocytozoon bieneusi, Microsporidium ceylonensis, M.africanum, Nosema ocularum, Pleistophora* species, *Trachipleistophora hominis, T.anthropophthera, Vittaforma corneae* and *Tubulinosema acridophagus*. However, *Encephalitozoon intestinalis* was previously named *Septata intestinalis* but it is now reclassified as *Encephalitozoon intestinalis* based on its similarity at the morphologic, antigenic and molecular levels to other species of this genus. A list of commonly encountered Microsporidia among humans is given at a glance in **Table 23.1**.

Based on recent data, it is now known that some domestic and wild animals may be naturally infected with the following microsporidian species: *Encephalitozoon cuniculi, Encephalitozoon intestinalis* and *Enterocytozoon bieneusi*. Birds, especially parrots (parakeets, love birds, budgies) are naturally infected with *Encephalito-*

Table 23.1. Pathogenic Microsporidial Species in Humans.
Anncaliia (Syn. *Nosema, Brachiola) algerae*
Anncaliia (Syn. *Nosema) connori*
Anncaliia (Syn. *Brachiola) vesicularum*
Encephalitozoon cuniculi
Encephalitozoon hellem
Encephalitozoon intestinalis
Encephalitozoon species (undetermined)
Endoreticulatus species
Enterocytozoon bieneusi
Microsporidium ceylonensis, M.africanum
Nosema ocularum
Pleistophora ronneafiei
Trachipleistophora anthropopthera
Trachipleistophora hominis
Tubulinosema acridophagus
Vittaforma corneae

zoon hellem. However, *Enterocytozoon bieneusi* and *V.corneae* have been identified in surface waters and spores of *Nosema* species (likely *A.algerae*) have been identified in ditch water. Recently *Tubulinosema acridophagus*, an insect parasite, is implicated in two cases of disseminated microsporidiosis.

Microsporidiosis can cause infection of the intestine, lung, kidney, brain, sinuses, muscles and eyes. Although there are over 1,300 species of microsporidia, the most prevalent pathogens in humans include *Enterocytozoon bieneusi, Encephalitozoon cuniculi* and *Encephalitozoon intestinalis*.

The scientific classification of Microsporidia has evolved through time with growing scientific research in the area and the specifics are still currently debated. Initially thought to be a protozoan (kingdom Protista), recent studies using DNA techniques indicate phylum **Microspora** should be classified under the kingdom **Fungi** or at least as a sister kingdom to Fungi. The class, order and family within the Microspora phylum are also frequently revised and debated. Traditionally, species were identified by observing the physical characteristics of the spore, life cycle and relationship with the host cell. However, recent scientific studies using genetic tools (namely ribosomal RNA sequencing) have challenged this approach and suggest genetic markers a more correct method for scientific classification. More research is still required to better understand the origins of Microspora and of individual species. Despite all this, there are now over 1300 species identified in 170 genera. However, currently, at least 15 species distributed in 8 genera are known to infect humans.

Microsporidia are primitive eukaryotes with well-defined nuclei and plasma membrane but lack some typical organelles found in more typical eukaryotes mainly mitochondria, stacked golgi and peroxisomes. Microsporidia spores are all round and oblong and those associated with human infection tend to be about 1-4 µm in size (an important feature for diagnosis as some species are often mistaken for bacteria). All have a characteristic coiled polar tube, tubule or filament, layered polarplast, a posterior vacuole (thought to function as golgi) and a protective exospore made of proteins and chitin. Chitin is responsible for the spores' high environmental resistance.

During infection, the cell membranes of the spore (also called the sporoplasm) is injected into a host cell though osmotic pressure. After the infection, *Microsporidium* relies on host cell machinery for energy and begins to multiply inside the host cell cytoplasm. Species differ in their relationships with host cell; some species alter host cell functioning to induce more nutrient absorption and cell growth to accommodate the agent. Microsporidia may reproduce sexually or asexually. In asexual reproduction, nuclear division takes place and form one or more pairs of nuclei and cellular division may isolate the nuclei or pair them in a diplokaryon arrangement (merogony- binary fission). Sexual reproduction is not well understood but it thought to involve autogamous fusion and reorganization of genetic material (schizogony - multiple fission).

Epidemiology

Microsporidia are now identified as the cause of opportunistic fungal infections predominantly in immunodeficient individuals such as among AIDS patients. However, the global epidemiology of human microsporidiosis is poorly understood and the ability of Microsporidia to survive and multiply in immunocompetent hosts still remains unsolved. Moreover, the diagnosis and clinical management of microsporidiosis cases have improved significantly in recent times, still epidemiology of human microsporidiosis is unclear.

Microsporidia have emerged as causative agents of opportunistic infections among AIDS and other immunodeficient patients. Several species of Microsporidia can cause disease in humans. Intestinal microsporidiosis due to *Encephalitozoon* (*Septata*) *intestinalis* and *Enterocyto-*

zoon bieneusi are most frequently reported among immunocompromised people including patients with AIDS and other immunocompromised patients such as transplant recipients, mainly renal transplant. *Encephalitozoon cuniculi* and *E.hellem* are less prevalent among immunodeficient patients. Infections with microsporidia in immunocompetent individuals such as travelers have also been described.

Since the studies examining prevalence of microsporidiosis have been limited to patients who are infected with HIV or who have AIDS, the global epidemiology of human microsporidiosis is poorly understood. Variation of spore shedding intensity of microsporidia was shown in both humans and animals. However, there are no reports on the spore shedding pattern of microsporidia in immunocompetent individuals.

As mentioned, microsporidia are extremely widespread. They infect nearly every organism on earth from honey-bees and silkworm to mammals and birds. Relatively little is known about the epidemiology of microsporidia, as transmission and infection pathways are still somewhat unclear. Though active microsporidia spores are found in water sources in developed and developing nations, microsporidiosis remains primarily a disease of HIV-positive and AIDS patients. Microsporidia are reported to infect 39% of AIDS patients with diarrhea and 30% of AIDS patients with *Cryptosporidium* species. Despite our relatively recent discovery of this pathogen, the infection among AIDS patients is remarkable and the organism will be of growing importance in the future as HIV continues to spread and more research is undertaken to understand the role microsporidia play in the human health. Other risk group chronic T-cell leukemia, idiopathic CD4 T-lymphocytopenia.

Microsporidiosis is an emerging and opportunistic infection in AIDS patients, organ transplant recipients, children, travelers, contact lens wearers and the elderly. It is associated with a wide range of clinical syndromes of microsporidiosis in humans. The disease is caused by microsporidia, obligate intracellular microorganisms that are recently reclassified from protozoa to fungi.

The 15 species of microsporidia currently known to infect humans, *Enterocytozoon bieneusi* and *Encephalitozoon intestinalis*, are the most common causes of human infections and are associated with diarrhea and systemic disease. Species of microsporidia infecting humans have been identified in water sources as well as in wild, domestic and food-producing farm animals, raising concerns of water-borne, food-borne and zoonotic transmission. It is primarily the disease of insects. Mysterious disease of silkworms threatened viability of the silk industry - *Nosema bombycis*.

Various molecules are tested for treating microsporidiosis in humans with variable success rates. Albendazole is effective against *Encephalitozoon* species such as *Encephalitozoon intestinalis* but not against *Enterocytozoon bieneusi*. This species has shown excellent clinical therapeutic response to direct action with fumagillin, but this drug is toxic when administered systematically to mammals. Its analog, TNP 470, could be promising alternative. Further work is necessary to identify other drugs, which are both effective and devoid of adverse effects. The introduction of highly active antiretroviral therapy (HAART) which, by suppressing replication of HIV, leads to maintenance or restoration of relative immunocompetence and consequent suppression or elimination of opportunistic infections, including microsporidiosis.

Immunity

This disease is implicated in causing morbidity in immunocompromised and immunocompetent individuals. The immunity to microsporidiosis depends upon functional T-lymphocytes and is based on the greater severity of disease in AIDS patients with declining $CD4^+$ T-cell levels and development of lethal experimental microsporidia infections in mice depleted of $CD4^+$ and $CD8^+$ T-cells.

As it is acquired via an oral route, gut-associated lymphoid tissues is found to be critical for the initial control of this infection. Studies have demonstrated that intraepithelial lymphocytes play a dominant role in this protection; however, the mechanisms involved in the trafficking of these cells and their role in long-term immunity remain to be elucidated.

The experimental murine models and *ex vivo* human studies demonstrated the importance of the Th1 cytokines such as interferon-gamma, tumor necrosis factor-alpha and interleukin-12 along with a role for nitric oxide, in resistance to *Encephalitozoon* species. $CD8^+$ intraepithelial lymphocytes are observed to increase rapidly after oral administration of *E.cuniculi* to mice. These cells appear to participate in pro-inflammatory responses via IFN-gamma production and cytotoxic activity and also contributed to immune regulation via IL-10 secretion.

Pathogenesis and Pathology

The infective form of microsporidia is the resistant spore and it can survive for an extended period of time in the

CHAPTER 25

Aspergillosis

Aspergillosis is a systemic fungal infection found among immunocompromised as well as immunocompetent individuals. This is primarily pulmonary infection with involvement of other body sites like paranasal sinuses and cutaneous tissues. It is caused by several species of genus ***Aspergillus***, which are ubiquitously found in the environment worldwide in the decaying organic matter. In most serious type of infections, invasion of lung tissue and dissemination to vital body organs may take place leading to fatal consequences among the immunocompromised patients. It is the commonest opportunistic fungal disease after candidiasis.

The hyphae of genus *Aspergillus* being hyaline in nature hence it is also a type of **hyalohyphomycosis** but is popularly known by its independent and well-established name, *i.e.* aspergillosis. The toxicity due to ingestion of foods contaminated with mycotoxins, produced by various species of genus *Aspergillus* among healthy individuals, is **not** termed as aspergillosis rather mycotoxicosis or known by the name of toxin produced, *e.g.* aflatoxicosis and is dealt in **Chapter 35**—Mycotoxicosis and Mycetismus. Similarly other infections like otomycosis, keratomycosis, endophthalmitis, onychomycosis, allergic fungal diseases and fungal rhinosinusitis are essentially caused by various *Aspergillus* species but known by their own names hence they are described separately in the respective Chapters of this Textbook.

Historical Perspective

Aspergillus infections in humans were initially described in the eighteenth century. The first attempt to define genus *Aspergillus* was made by Micheli in 1729. In same year he described, named and illustrated fungus nicely in his famous work, *Nova Geneva Plantarum*. He noted that pattern of conidial head of *Aspergillus* with its spore heads radiating from central structure, resembled to **aspergillum**, brush or perforated globe used for sprinkling holy water by the priests hence he named genus as *Aspergillus*. He described experiments of culturing fungi from spores also both in gardens and on pieces of melon. During that period his observations were ahead of his time and their significance was overlooked until more than a century, later it was Link who first named *Aspergillus flavus* in 1809.

John Hughes Bennett (1812-1875) was first to describe aspergillosis. In his publication of 1842, he made the very first description of *Aspergillus*, growing in lung tissue of humans. *Aspergillus fumigatus* was first found in the lungs of bird by George W Fresenius in 1863, who also applied the term aspergillosis to this respiratory disease. Mackenzie first reported paranasal sinus mycosis in 1893 and since then numerous cases have been reported from different parts of world. In 1926, Tom and Church first classified genus and accepted 69 *Aspergillus* species in 11 groups. By the year 1945, Thom and Raper had accepted 80 species of genus *Aspergillus*.

The first case of invasive disease was described in 1953 and since then there has been increasing recognition of invasive aspergillosis as an important factor in morbidity and mortality among patients with severe immunosuppression. By the year 1965 previous classification of *Aspergillus* was declared outdated and Raper and Fennel detailed 151 species in 18 different groups. By then, most species with teleomorph had been identified. Additional work led by Samson and Pitt further refined species designations with use of new technologies such as thin-layer chromatography of secondary metabolites and DNA hybridization.

Mycology

Aspergillus species are saprotrophic molds in nature and found in decaying organic matter worldwide. At present there are 339 species, which are now accepted and new species continue to be described and added to this list. Of the recognized species of this genus, only **twenty** have been verified to cause human infections and **three** of them are consistently and regularly encountered as etiological agents of over 95% diseases caused by members of genus which are: *Aspergillus fumigatus, A.flavus* and *A.niger*. The other common species of this genus are: *A.terreus, A.glaucus, A.nidulans, A.oryzae, A.versicolor* and *A.clavatus*. However, *A.fumigatus* is the most common species found in human infections all over the globe but in India and other southeast Asian countries, it is *Aspergillus flavus*.

The taxonomy of species within *Aspergillus* genus is gradually undergoing revision with the use of molecular techniques and is not yet complete. Most of the species in the genus *Aspergillus* reproduce asexually and are classified in the phylum **Deuteromycota**. However, some species, in addition to their asexual reproduction, characteristically reproduce sexually by formation of ascospores and taxonomically placed in order **Eurotiales** of family **Trichocomaceae** in class **Plectomycetes** of phylum **Ascomycota**.

The aspergilli produce conidia in **basipetal** manner, which results in chains of asexual conidia where youngest conidium is at base and the oldest is at the tip of chain. The conidiogenous cell is termed **phialide**. The conidiophore is hypha-like structure that enlarges at its apex to form swollen vesicle. The base of conidiophore, where it originates from parent vegetative hypha, is termed **foot-cell**. The phialides may arise directly from vesicle in **uniseriate** and from metulae as **biseriate**. These phialides, in some species, cover entire surface of vesicle, whereas in others they may cover upper half to three-quarters. They may vary in color in different species from hyaline to light or dark pigmentation however hyphae are always hyaline.

Epidemiology

All the species of *Aspergillus* are common saprotrophic in soil and on decaying plant material and their spores are ubiquitous in environment. The human beings are regularly exposed to these spores particularly people working with decaying vegetation, like moldy hay in agriculture, are maximally exposed. *Aspergillus* species most commonly invade the respiratory tract, followed by the gastrointestinal tract and brain, in addition to the liver, kidneys, thyroid, heart, diaphragm and sinuses. The disease process generally follows inhalation of spores but fungus may also gain direct entry to body tissues through wounds or during surgery. In addition to the infective process, these are also responsible for the toxic manifestations like aflatoxicosis.

Aspergillosis is considered as the second most common fungal infection requiring hospitalization in United States. The frequency and relative importance of these infections is on the rise in developed countries, which is possibly related to increased number of immunocompromised patients, owing to improved survival in AIDS, malignancies and more intensive cytotoxic therapy, transplantation for organ dysfunctions and better therapy and prophylaxis for other fungal infections like candidiasis. *Aspergillus* infections in patients with AIDS have been reported worldwide. The precise incidence is hard to estimate owing to difficulty in diagnosis of clinically unsuspected disease but on an average ~4% of patients develop aspergillosis.

There is widespread use of fluconazole, which has no clinically useful activity against invasive aspergillosis and it has not reduced the number of cases of this infection, which remains most common invasive mold infection worldwide. As a result of these factors, burden of infection-related hospitalizations in developed countries has also increased to great extent. The nosocomial outbreaks of invasive aspergillosis are attributed to the environmental sources.

Aspergillus fumigatus is commonly encountered as etiological agent of systemic aspergillosis in man and animals. The infections are mostly acquired through inhalation of light airborne conidia, which this fungus produces in abundance during its saprotrophic existence in the environment. The fungus is widespread in nature and it has been isolated from virtually all over the world and soil is considered to be the main source.

Immunity

There is little tendency for invasive disease unless phagocyte number or function in the body is reduced. Both skin and lungs have phagocytic defense mechanisms. Macrophages phagocytose and destroy *Aspergillus* conidia, whereas polymorphonuclear (PMNs) leukocytes and monocytes (MNCs) damage *Aspergillus* hyphae via oxidative and non-oxidative mechanisms. Macrophages and neutrophils are mainstay of pulmonary host defense against *Aspergillus* species. The keratin and epidermal barriers of skin serve as additional front line of mechanical host defense. The smaller, 3 to 5 μm conidia of *Aspergillus* species easily reach alveolar spaces, which have no equivalent mechanical barrier.

The host defense mechanism to invasive aspergillosis is primarily an innate mechanism as compared to allergic forms of aspergillosis, acquired responses appear to contribute more towards pathology than the control of disease.

Pathogenesis and Pathology

The pathological responses caused by species of genus *Aspergillus* vary in severity, clinical course and may occur as primary as well as secondary infections. The spores of this opportunistic fungus, *A.fumigatus* may bypass upper respiratory tract defenses and reach distal bronchial airway and pulmonary alveoli because of their smaller size (<5 μm) and aerodynamic properties. In this region host relies primarily upon phagocytic cells to remove spores efficiently. If however, phagocytic cells are unable to clear spores quickly, germination and fungal colonization may occur. The principal phagocytic cells responsible for maintaining sterility of lower respiratory tract are pulmonary alveolar macrophages, although other cells including polymorphonuclear leukocytes and lymphocytes are known to play an important role in removal of aspergilli.

Clinical Features

The genus *Aspergillus* consists of ubiquitous molds, which cause variety of clinical conditions in immunosuppressed as well as healthy individuals. The term aspergillosis indicates tissue invasion or allergic disease caused by any of the species of genus *Aspergillus*. The infection may be primary or secondary depending on mode of infection, which vary in severity as well as clinical course. Clinically, response produced by various species of this genus, include the followings:

- Superficial infection of skin, nail, external auditory canal, burn eschar, paranasal sinuses and other sites.
- Colonization in preformed cavities and debilitated tissues without extension within respiratory tract often called aspergilloma.
- Invasive, inflammatory, granulomatous and necrotizing infection of the lungs.
- Disseminated disease, commonly involving lungs, brain, kidneys, skin, heart, eyes and other organs.
- Allergy and sequelae to the presence of conidia or transient growth of organism in body orifices.
- Toxicity due to ingestion of foods contaminated with mycotoxins or other metabolites produced by aspergilli in healthy host.

Therefore, clinical manifestations of aspergillosis and allied diseases are largely determined by local or general immunological and physiological state of the host. However, it is observed that various forms of disease may integrate and overlap to each other. There are several clinical types of aspergillosis, which have been described so far. The frequently encountered clinical conditions produced by genus are given below:

1. Pulmonary Diseases
2. CNS Aspergillosis
3. PNS Aspergillosis
4. *Aspergillus* Endocarditis
5. Cutaneous Aspergillosis
6. Miscellaneous Forms

1. Pulmonary Diseases

The pulmonary disease caused by *Aspergillus* species may be classified under **three categories** depending upon whether host is atopic or immunocompromised. However, sometimes it may be difficult to categorize these cases under specific category, as more than one form of disease may be present simultaneously. These categories are as follows:

(a) Allergic Aspergillosis: Mostly hypersensitivity to *Aspergillus* species manifests as **bronchial asthma**, which may be found as such or with pulmonary eosinophilia or extrinsic allergic alveolitis. This follows inhalation of spores usually in atopic individuals who are sensitive to wide range of allergens and have raised IgE levels. Asthma with eosinophilia is the most chronic form of disease and results in progressive lung damage and decreased lung functions. It is characterized by episodes of lung consolidation and fleeting shadows on chest x-rays. The various types of hypersensitivity reactions observed in different clinical manifestations of aspergillosis are shown in the **Table 25.1**.

The second variety observed in atopic individuals is **allergic bronchopulmonary aspergillosis** (ABPA). The heavy and repeated exposure to spores of *Aspergillus* species may result in an allergic alveolitis in which individual suffers from breathlessness, fever and malaise after few hours of exposure. Repeated attacks of disease may lead to progressive lung damage. In majority of cases fungus grows in larger airways to produce plugs of entangled mycelia and mucous, which block segments of lung tissue and sometimes even entire lobe. These mucous plugs may be coughed out, which is the diagnostic feature of this disease. During clinical course of ABPA, total serum IgG antibodies declines by 50 to 75% after treatment with prednisolone. The diagnostic criteria of ABPA are shown in the **Table 25.2**.

The obstructive aspergillosis is characterized by progressive cough, which is sometimes productive of bronchial casts and composed entirely of *Aspergillus* hyphae, in contrast to mucous plugs of ABPA in which hyphae are scanty. There is no wheezing sounds or eosinophilia but

Table 25.1. Hypersensitivity Reactions in Aspergillosis.

Clinical Forms	Hypersensitivity Reaction
A. Allergic Forms	
Atopic Host	
Asthma	I only
ABPA	I and III
Non-atopic Host	
Extrinsic Allergic Alveolitis	III only or with I or IV
B. Non-invasive Colonization	
Aspergilloma	I, II, or None
Non-pulmonary, Local Infection	
C. Invasive Aspergillosis	
Pulmonary	None or any
Chronic Necrotizing Pulmonary	
Disseminated	

Table 25.2. Commonly used Diagnostic Criteria of ABPA.

Diagnostic Criteria	Positivity
• Bronchial Asthma	100
• Pulmonary Infiltrates	91-100
• Fleeting Shadows	+
• Central Bronchiectasis	+
• Eosinophilia in Blood	80-100
• Immune responses to *A.fumigatus* antigen	
Type I (Wheal and Flare)	100
Type III (Arthus)	100
Serum Precipitins	84-91
Total Serum IgE	> 1000 ng/ml
• Sputum	
Eosinophilia	44-100
History of Plugs	74
Culture of *A.fumigatus*	46-83

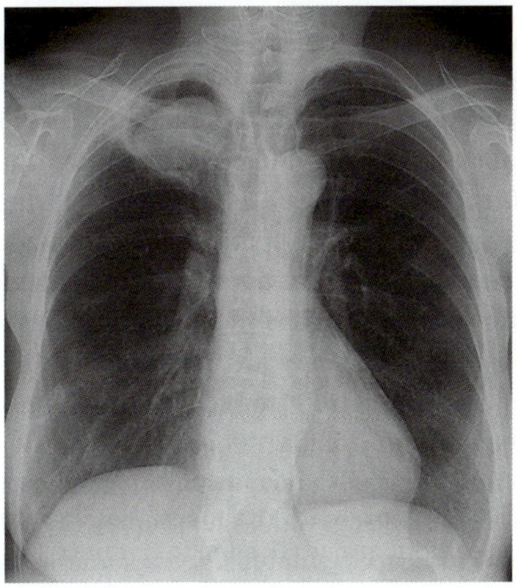

Fig. 25.1. X-ray PA view showing fungus ball in the right upper lobe of a patient of aspergilloma.

patient rapidly develops hypoxia. The chest X-rays show areas of collapse and at bronchoscopy some airways are found to be completely blocked by fungal casts. When these blockages are removed bronchial mucosa appears to be normal. It is probably precursor to locally invasive pseudomembranous *Aspergillus* tracheobronchitis.

(b) Aspergilloma: The fungus colonizes in pre-existing cavity that is often tubercular. In the lungs, there is compact mass of fungal mycelia that is often surrounded by dense fibrous walls. It is also called as **fungus ball**. Aspergilloma are usually solitary and vary in size, approximately 8-10 cm in diameter. Radiologically, they have characteristic appearance of well-defined mobile opacity within cavity, usually with crescent of air at the upper margin, popularly known as **air crescent** or **Monod's sign** (Fig. 25.1).

The patients of aspergilloma are either asymptomatic or have only moderate degree of cough and sputum production. However, occasional hemoptysis may occur particularly when fungus is actively growing. Sometimes, involvement of major blood vessel results into massive hemorrhage.

(c) Invasive Aspergillosis: This is an increasingly common presentation and an important cause of morbidity and mortality in immunocompromised hosts, especially in patients with acute leukemia, bone marrow, solid organ transplantation or even AIDS. The attributable mortality rate remains extremely high despite treatment with antifungals, such as amphotericin B and azole derivatives itraconazole and voriconazole. The outcome of about 85% patients in invasive aspergillosis is invariably fatal.

Invasive aspergillosis is well known to occur in various immunocompromised populations, such as neutropenic patients with hematologic malignancies, patients with advanced stage of HIV infection and children with chronic granulomatous diseases. The transplantation is an increasingly important group of immunocompromised patients with high susceptibility for invasive aspergillosis and has very high mortality.

This is common in severely immunocompromised patients who are on immunosuppressive drugs, corticosteroids or broad-spectrum antibiotics. There is widespread

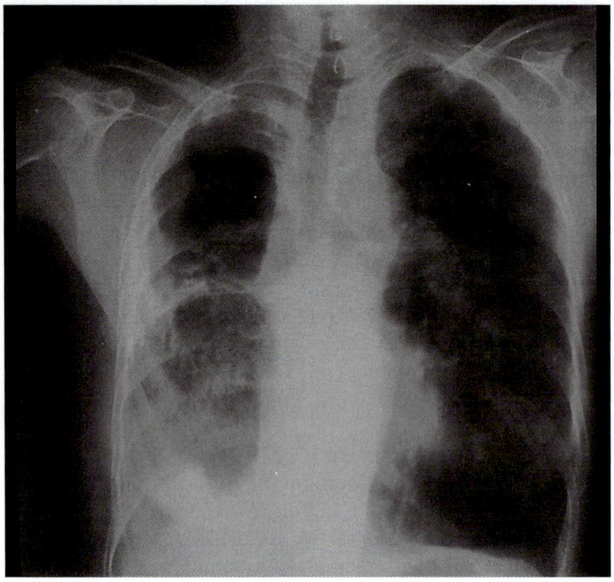

Fig. 25.2. X-ray PA view showing fibrosis in the right lower lobe of a patient of chronic pulmonary necrotizing aspergillosis.

growth of fungus in lung tissue that may disseminate to involve other organs particularly kidneys and brain. The disease has very poor prognosis and usually diagnosis is established on autopsy.

Various types of infections and graft rejections have been leading causes of morbidity and mortality among the transplant recipients for longer time. Although recent advances in immunosuppression, prophylaxis and treatment of infections such as cytomegalovirus have resulted in improved outcome in transplant recipients but invasive aspergillosis still remains scourge of transplantation medicine and surgery. In addition, *Aspergillus* species have emerged as most common infectious cause of mortality in bone marrow/stem cell transplant recipients.

(d) Chronic Necrotizing Pulmonary Aspergillosis: CNPA is also called semi-invasive pulmonary aspergillosis. It is an uncommon, indolent pulmonary infection commonly seen in patients with altered local defense due to any pre-existing respiratory disease such as chronic obstructive pulmonary disease, bronchiectasis, pneumoconiosis and mildly compromised systemic defense due to diabetes mellitus, alcoholism, poor nutrition or low dose corticosteroid therapy. This is usually found among middle-aged to older individuals who present with fever, productive cough and weight loss that occurs over a period of months and usually mimick tuberculosis often entailing delay in establishing exact diagnosis. In contrast to the patients with aspergilloma, CNPA nearly always presents with pulmonary or systemic symptoms. In addition, hemoptysis, the most common symptom in patients with aspergilloma, is reported in ~10% of patients with CNPA and is rarely an isolated symptom. The aspergilloma is a non-invasive form of aspergillosis, which may develop in a healthy host where fungus colonizes a pre-existing cavity whereas CNPA may create its own cavity among immunocompromised hosts and then grow as a relatively non-invasive organism. The radiological findings usually consist of consolidation areas in the upper lobe initially, which develops progressive cavitation over several weeks to months **(Fig. 25.2)**. The cavitation may be associated with intra-cavitary soft tissue opacity, which can be recognized radiographically with the appearance of an air crescent sign, which may be visualized on chest CT well before it is seen on the chest radiograph.

2. CNS Aspergillosis

The aspergillosis of central nervous system is a serious fungal infection and majority of cases occur by hematogenous dissemination from pulmonary or gastrointestinal focus, more often in immunocompromised individuals. *Aspergillus* infection accounts for ~5% of CNS fungal infections. Such life-threatening infections are still commonly diagnosed at autopsy and have very poor prognosis even with *ante mortem* diagnosis. Most commonly causative species is *Aspergillus fumigatus* and others like *A.flavus*, *A.terreus* and *A.versicolor* have also been found to cause CNS infection.

Aspergillosis of CNS has varied clinical manifestations that range from abscesses to granulomas **(Fig. 25.3)**, rhinocerebral form to meningitis. Intracranial mass lesions are most common, either solitary or multiple, followed by granuloma and meningitis and ventriculitis. True intracranial aneurysms caused by *Aspergillus* are now diagnosed more frequently. The patients may also present with *Aspergillus* epidural abscess.

The cranial and intracranial invasion by contiguous spread has been found following infection within sinuses. Clinical syndromes include encephalitis, meningoencephalitis, stroke like syndrome, intracranial space occupying lesion, skull base syndrome and intra-orbital space occupying lesion. *Aspergillus* meningoencephalitis is seen primarily in patients with prolonged neutropenia, as part of disseminated invasive aspergillosis or in patients undergoing intense immunosuppression for solid organ and bone marrow transplantation with graft rejection.

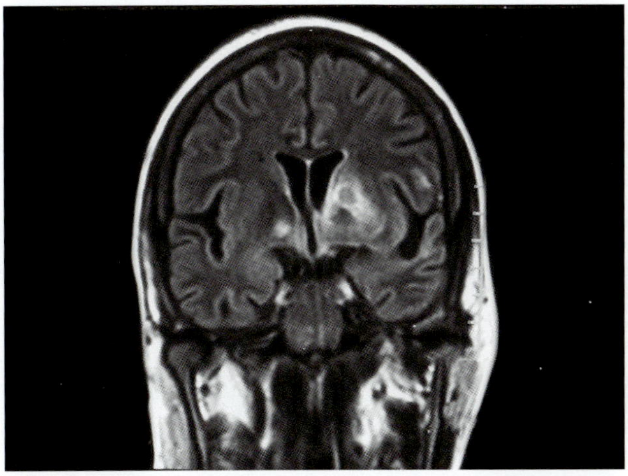

Fig. 25.3. CT scan showing intracranial *Aspergillus* granuloma in the left basal nuclei seen in coronal section of a patient.

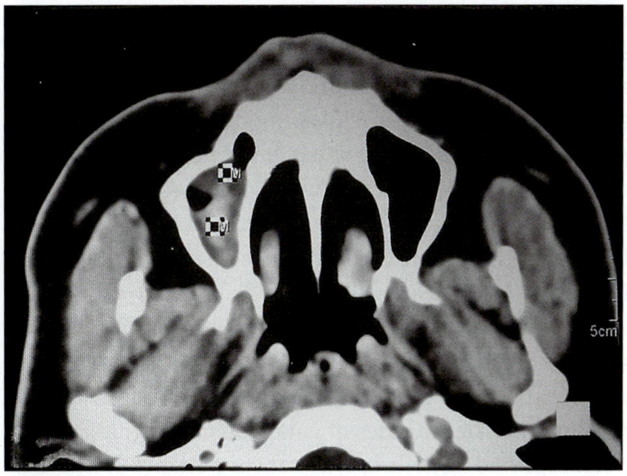

Fig. 25.4. CT scan of PNS showing fungal granuloma due to *Aspergillus* in the right maxillary sinus of a patient.

The CNS aspergillosis should be included in differential diagnosis of HIV-infected patients who present with non-specific neurologic signs and symptoms. The computed tomogram and magnetic resonance imaging of head show low-density lesions or hemorrhagic infarcts.

3. PNS Aspergillosis

The colonization and invasion of paranasal sinuses by various species of *Aspergillus* has been substantially observed as clinically significant. The nasal and paranasal sinus aspergillosis is classically described as having four types: allergic, non-invasive, invasive and fulminant despite the fact there are many controversial issues yet to be settled. The allergic form is similar to ABPA and is thought to be combination of Type I and Type III hypersensitivity to the *Aspergillus* antigens. Non-invasive form results from formation of an aspergilloma or fungus ball and behaves as chronic sinusitis. Both these forms are considered saprotrophic and extra-mucosal and are benign in nature **(Fig. 25.4)**.

The invasive form behaves like malignant neoplasia presenting with facial mass and proptosis. It is slowly progressive and locally destructive. Fulminant disease is angioinvasive, rapidly destructive and often fatal. The infection may spread through bone to adjoining areas like orbit of eye and even to the brain. In some countries like India and Sudan, *A.flavus* is the commonest causative agent of paranasal sinus mycosis. The symptomatology of rhinocerebral aspergillosis is non-specific and patients may present with proptosis and polypoid lesions in nose. PNS aspergillosis should be differentiated from similar disease caused by some phaeoid fungi. This topic is discussed separately in detail in **Chapter 34** as Fungal Rhinosinusitis of this Textbook.

4. *Aspergillus* Endocarditis

The *Aspergillus* may cause endocarditis in immunocompromised patients and particularly in those who have had prior cardiac surgery. Less commonly, infection occurs in patients who have not undergone cardiac surgery and these patients are often immunocompromised. *Aspergillus* species are most common fungal organism after *Candida* species implicated in endocarditis following cardiothoracic surgery.

The patients present with fever and multiple embolic strokes because of lesions characterized by large fungal vegetations on heart valves and high frequency of embolization. The other risk factors include hyperalimentation, antibiotic therapy, intravenous drug abuse, concomitant bacterial endocarditis and immunosuppression. Sometimes patients may also present as myocarditis also. Echocardiography reveals vegetations in majority of cases. Generally patients have poor prognosis and successful treatment depends on antifungal therapy and surgical removal of infected tissue.

5. Cutaneous Aspergillosis

This type of aspergillosis is apparent in skin and can present as primary cutaneous aspergillosis derived from direct external inoculation through contact with objects such as arm boards, external catheter sites, surgical or other traumas. The range of clinical findings of primary cutaneous aspergillosis includes nodules, molluscum-like papules, plaques and ulcers. Early detection of primary cutaneous aspergillosis in conjunction with combined medical and surgical intervention can successfully eradicate infection, helpful in treating locally destructive disease and preventing disseminated aspergillosis.

There may be secondary cutaneous aspergillosis usually on the chest wall in patients with pulmonary aspergillosis and caused by dissemination of invasive disease to skin. Most of the cases of cutaneous aspergillosis have been reported in neutropenic cancer patients. It has also emerged as significant infection among variety of immunocompromised patients, including neonates, patients with organ transplants or chronic granulomatous disease, those with autoimmune diseases who receive corticosteroids and patients with burn wounds. In patients with HIV, cutaneous aspergillosis occurs with high frequency.

6. Miscellaneous Forms

Different species of *Aspergillus* may also cause keratomycosis, endophthalmitis, otomycosis and onychomycosis that are described in their respective Chapters. Occasionally patients may also present with mastoiditis and osteomyelitis. Most of the patients of fungal mastoiditis are immunocompromised, particularly with deficient CMI and it is caused by *Aspergillus* species.

Radiodiagnosis

A detailed radiological evaluation of patients of aspergillosis is essential. This is done with computed tomography **(Fig. 25.5)** or magnetic resonance imaging. T1-weighted and T2-weighted images are taken to delineate exact extent of lesions. The radiodense shadows are due to calcium and magnesium salts inside the fungal granuloma. Ultrasonography of the affected internal organs may also be helpful. High resolution computed tomography scan of chest is one of most important procedures for diagnosing invasive aspergillosis. The physicians should be aware that CNS might be the only site of *Aspergillus* involvement and include CNS aspergillosis in differential diagnosis of HIV-infected patients presenting with focal neurological signs and symptoms especially when head CT scan reveals hypodense lesions.

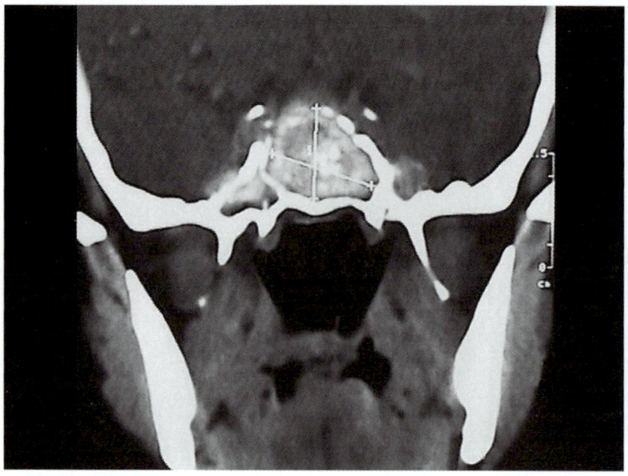

Fig. 25.5. CT scan of PNS showing fungal granuloma due to *Aspergillus* in the left sphenoid sinus of a patient.

Differential Diagnosis

The patients presenting with any of the clinical form of aspergillosis should be differentiated from deep mycotic infections, particularly with immunocompromised conditions. The differential diagnosis of cutaneous aspergillosis include ecthyma gangrenosum caused by *Pseudomonas* or *Candida* species, herpes simplex virus infection, mucormycosis, cryptococcosis and phaeohyphomycosis.

The *Aspergillus* granuloma should be differentiated from other granulomatous diseases as well as neoplasia. It is also observed that such patients are wrongly given radiotherapy under the impression of malignant swellings.

Laboratory Diagnosis

Aspergillosis in man and different animals is primarily confirmed using conventional means of culturing causative fungal organism from clinical material on standard fungal media such as Sabouraud dextrose agar. Microscopy and culture both in combination increase the diagnostic yield. During past few years, progress has been made in development and evaluation of diagnostic tests and procedures for invasive aspergillosis. In patients with positive fungus cultures, *Aspergillus* species are second most common isolates after *Candida* species but positive cultures alone may not indicate pathogenic process.

As these fungal species exist everywhere in environment, diagnosis of invasive aspergillosis is established by repeated isolation of organisms in culture. The collection of infected material is done in sterile container for

Table 25.3. Tests done in Clinical Types of Aspergillosis.

Disease	*Microscopy*	*Culture*	*Serology*	*Skin Test*
ABPA	+	+	Moderate +	++
Aspergilloma	+	+	Strong +	-
Invasive-aspergillosis	+	+	Weakly +/-	-

Table 25.4. Characteristics of *Aspergillus* Species.

Species	*Phialides*	*Color of Conidia*
A.fumigatus	Uniseriate	Grey, green or blue-green
A.flavus	Bi or Uniseriate	Yellow to green
A.niger	Biseriate	Black
A.terreus	Biseriate	Orange to brown
A.nidulans	Biseriate	Dark green

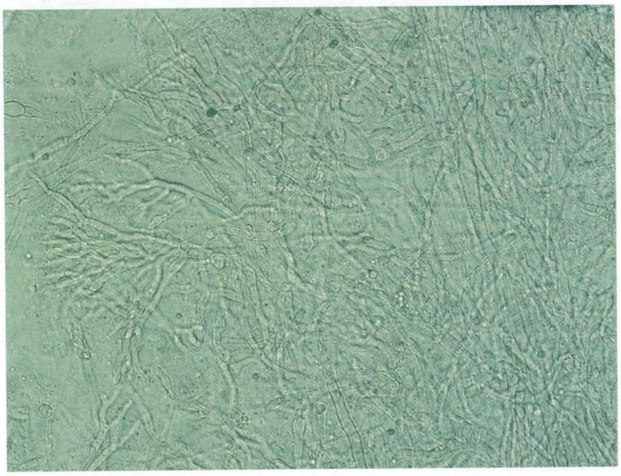

Fig. 25.6. Septate hyphae of *Aspergillus* species with acute-angle branching seen in wet mount of clinical material (KOH × 400).

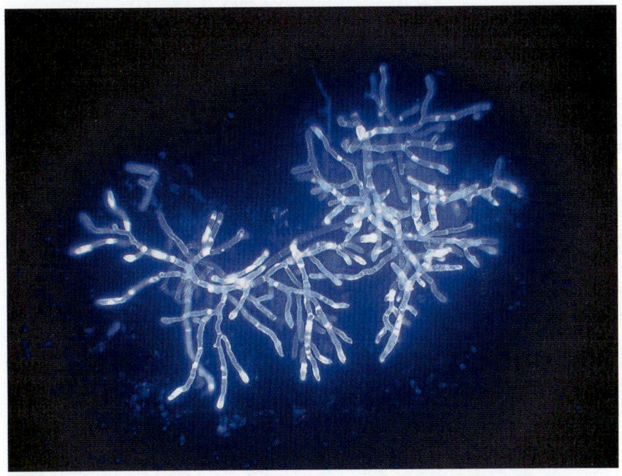

Fig. 25.7. Fluorescent septate hyphae of *Aspergillus* species with acute-angle branching seen in clinical material (CFW × 400).

laboratory examination. Various laboratory tests performed in different forms of aspergillosis are shown in **Table 25.3**. The important characteristics related to conidiation of commonly encountered species of genus *Aspergillus* are shown in the **Table 25.4**.

(a) Direct Examination

The specimens like sputum, bronchoalveolar lavage (BAL) fluid or transbronchial biopsy are taken depending on site of involvement. Biopsy is also taken in case of fungal granuloma. The direct examination of clinical specimens is done in 10% KOH for demonstration of hyaline septate hyphae of *Aspergillus* species. The septate hyphae are 3-6 μm in diameter with **dichotomous** branching **(Fig. 25.6)**. Calcofluor white stain **(Fig. 25.7)**, fluorescent-antibody techniques also demonstrate septate hyphae. Distorted and broader hyphae may be mistaken for those of mucormycetes but hyphae of latter are generally broader, upto 15 μm, have non-parallel walls which often appear collapsed and acutely twisted and are infrequently septate. Now, cases of mixed aspergillosis and mucormycosis infections are frequently reported, wherein it becomes imperative to differentiate both the diseases.

The biopsy material is kept in tube KOH for an overnight at 37°C and wet mount is prepared to see for septate hyphae. Histopathological examination of biopsy material stained with hematoxylin and eosin **(Figs. 25.8A to C)**, periodic acid-Schiff **(Figs. 25.9A and B)**, Grocott-Gomori's methenamine silver stain **(Fig. 25.10)** and other fungal stains show acute-angled (~45°) branching, hyaline septate hyphae with neutrophilic to granulomatous response. The hyphae exhibit characteristic dichotomous acute-angle branching, often giving finger-like appearance to the branching elements This is also angioinvasive fungus like mucormycetes hence vascular invasion and thrombosis can also be seen in aspergillosis.

In invasive aspergillosis, hyphae proliferate extensively throughout the tissue, often in parallel or radial arrays. Aspergilli colonizing pulmonary cavitary lesions grow as tangled masses of hyphae and in chronic infection may

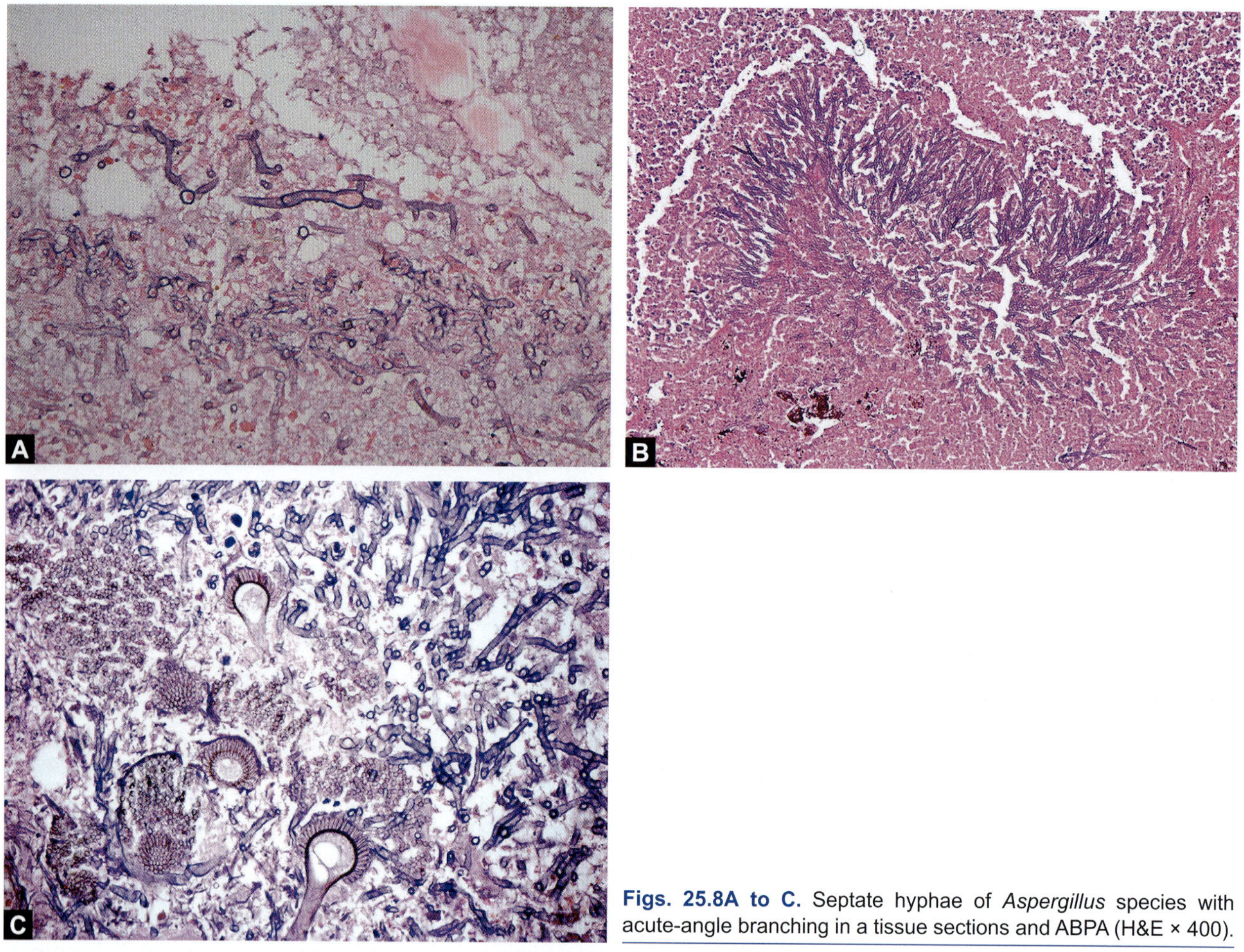

Figs. 25.8A to C. Septate hyphae of *Aspergillus* species with acute-angle branching in a tissue sections and ABPA (H&E × 400).

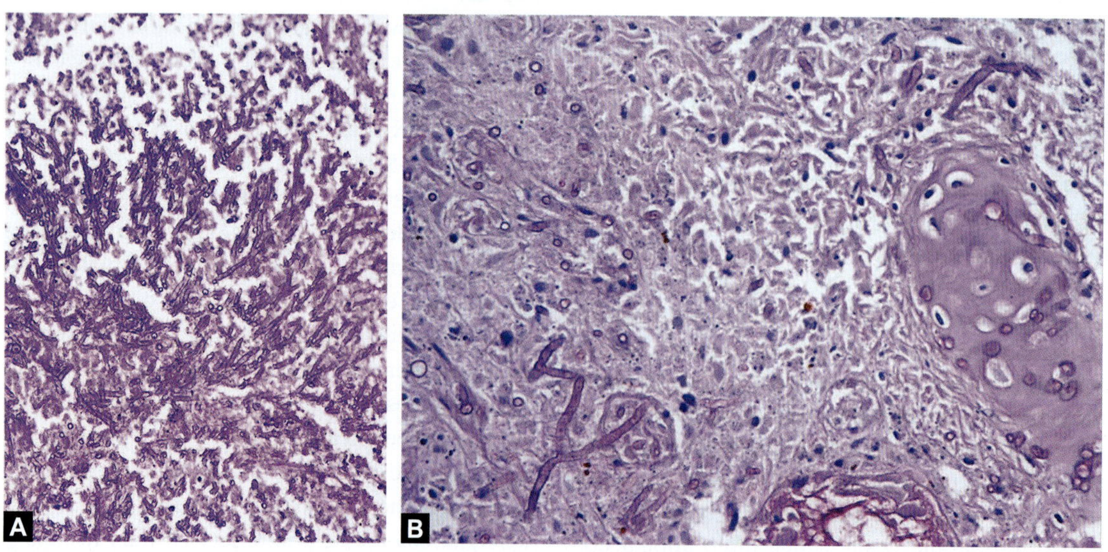

Figs. 25.9A and B. Septate hyphae of *Aspergillus* species with acute-angle branching in a tissue section (PAS × 400).

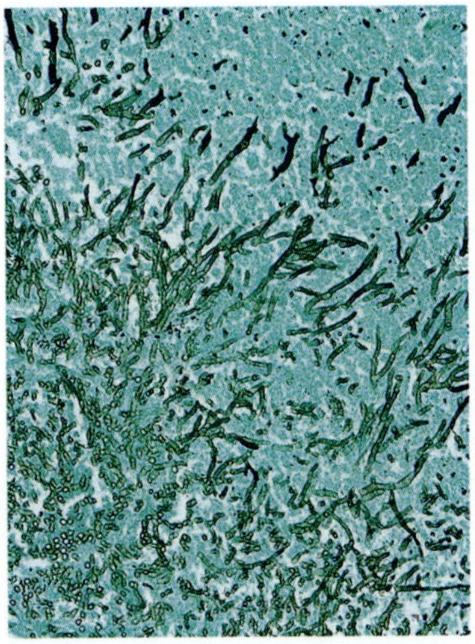

Fig. 25.10. Septate hyphae of *Aspergillus* species with acute-angle branching in a tissue section (GMS × 400).

exhibit atypical hyphal features such as swellings measuring up to 12 µm in diameter and/or absence of conspicuous septa as seen in *Aspergillus* fungal balls.

Direct demonstration of septate hyphae or other findings like vesicle in clinical specimens, is essential to substantiate significance of clinical isolates in culture. It can rule out any possibility of assigning importance to contaminating *Aspergillus* species grown in culture. Sometimes, repeated demonstration in direct smear as well as culture, establish diagnosis of aspergillosis beyond doubt. The direct findings do not speciate the organism as to whether it is *Aspergillus* or any other hyalohyphomycete like *Fusarium* species or *Scedosporium* species, etc. Hence, fungal culture is imperative in such a situation for establishing the specific diagnosis.

(b) Fungal Culture

The pathogenic *Aspergillus* species generally grow easily and relatively quickly on routine mycological as well as bacteriological media in diagnostic mycology laboratory. The material is inoculated on Sabouraud dextrose agar with antibiotics and without cycloheximide at 25°C and 37°C, respectively. The cultures should be examined daily during first week and twice a week for further four weeks before being considered the plate/tube as sterile.

Subculture of an isolate to Czapek Dox agar and 2% malt extract agar with incubation at 25°C allows identification of most aspergilli, using standard monographs and taxonomic keys. Potato dextrose agar is particularly useful for inducing sporulation so that identification of isolate becomes easier.

The positive urine cultures nearly always imply metastatic abscesses in kidney and are diagnostic of invasive aspergillosis. *Aspergillus* isolates that are regarded to be of clinical significance, should be identified to the species level. *Aspergillus* species are seldom recovered from blood, urine or CSF, although cultures of blood have been positive in occasional patients with endocarditis. Exoantigen tests are useful for identification to genus level and in case of *A.fumigatus* to species level.

The contamination of cultural media during incubation with different aspergilli is not uncommon, therefore, histopathological examination of tissue specimen from the patient is highly recommended as a backup for culture isolation and to correlate its clinicopathological significance.

The diagnosis of CNPA is usually suggested by its clinical course and isolation of fungus from pulmonary secretions; negative cultures for other pathogens and failure to respond to antibacterial or antimycobacterial therapy. Diagnostic confirmation requires histologic evidence of local lung tissue invasion by septate hyphae, consistent with *Aspergillus* species; however, this is often difficult to obtain. Both transbronchial and percutaneous biopsy have low diagnostic yields for locally invasive aspergillosis when compared with autopsy findings. Sputum is also unavailable for culture in many cases; even when present, the sensitivity of such culture is probably in the order of 50% to 60%. Similar values have been found for culture of respiratory tract secretions sampled by bronchoalveolar lavage or bronchial washing or brushing. A positive sputum culture may also reflect simple colonization. However, in the appropriate clinical setting, repeated positive sputum or BAL fluid cultures have been found to be a reliable method of diagnosis.

The description of some of the frequently encountered *Aspergillus* species is given below:

(i) Aspergillus flavus: The colonies are velvety, yellow to green or brown. Reverse is golden to red brown. The conidiophores are of variable length, rough, pitted and spiny; phialides are single and double cover entire vesicle, point out all directions **(Figs. 25.11A and 25.12A to D)**.

(ii) Aspergillus fumigatus: The colonies are velvety or powdery at first, turning to smoky-green. Reverse is white to tan. The conidiophore is smooth, phialides single (uniseriate), usually cover upper half vesicle, parallel to the axis of stalk **(Figs. 25.11B; 25.13A and B; 25.14A to D)**.

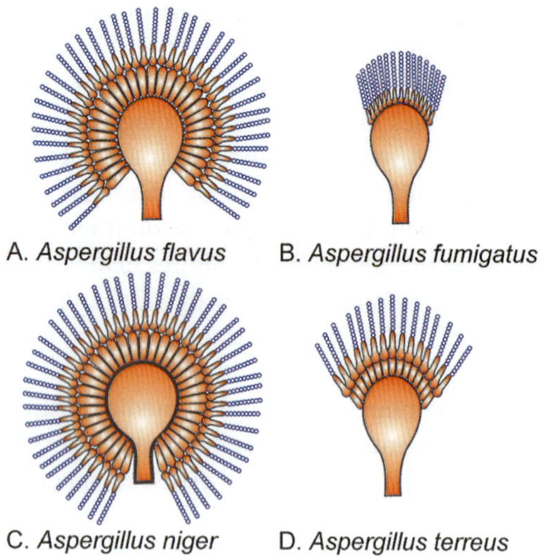

Fig. 25.11. Diagrammatic representation of commonly encountered *Aspergillus* species.

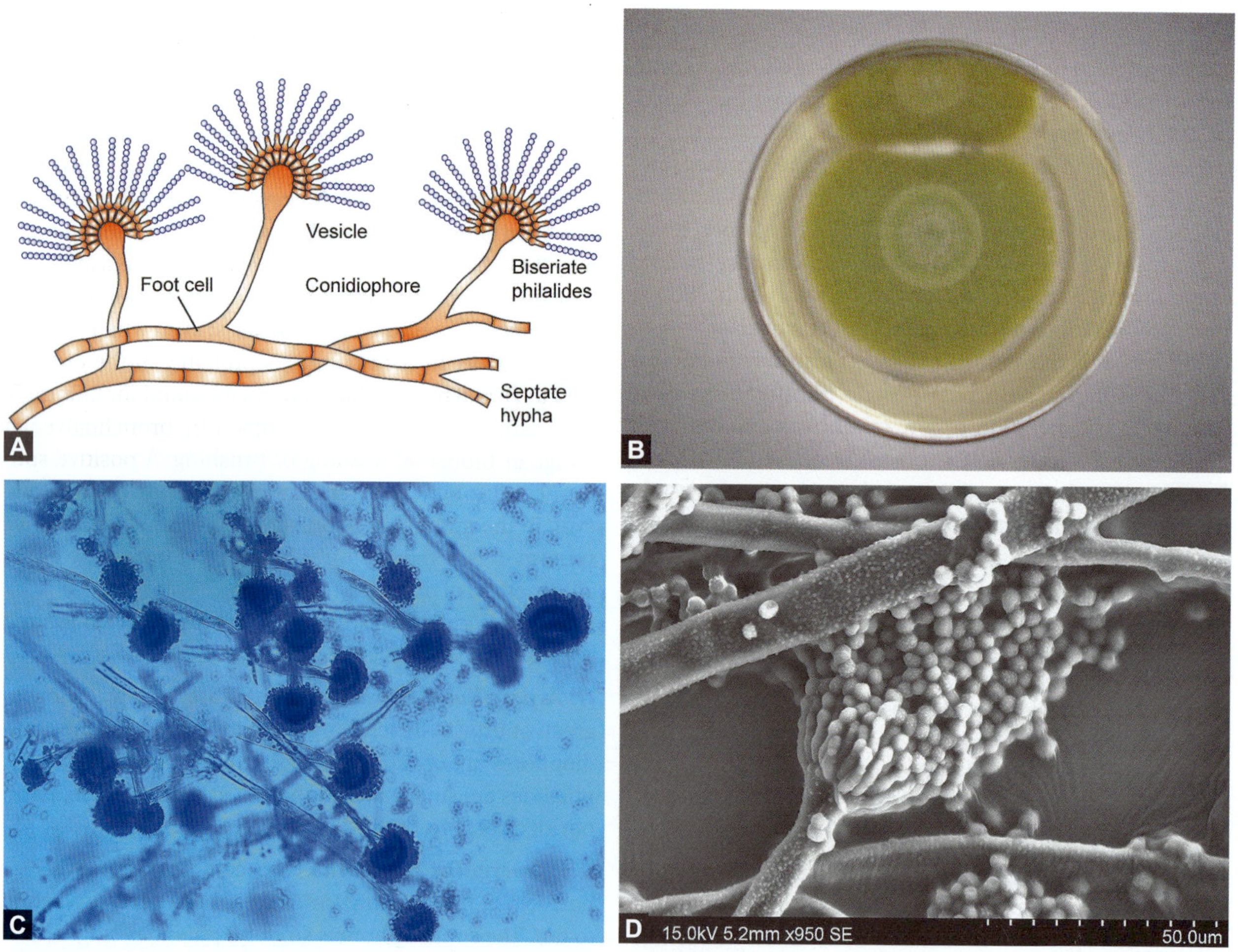

Figs. 25.12A to D. (A) Diagrammatic representation *Aspergillus flavus* (B) its yellowish-green mycelial colonies (C) conidia, vesicles, conidiophores and septate hyphae (LCB × 400) and (D) Electron microscopy appearance.

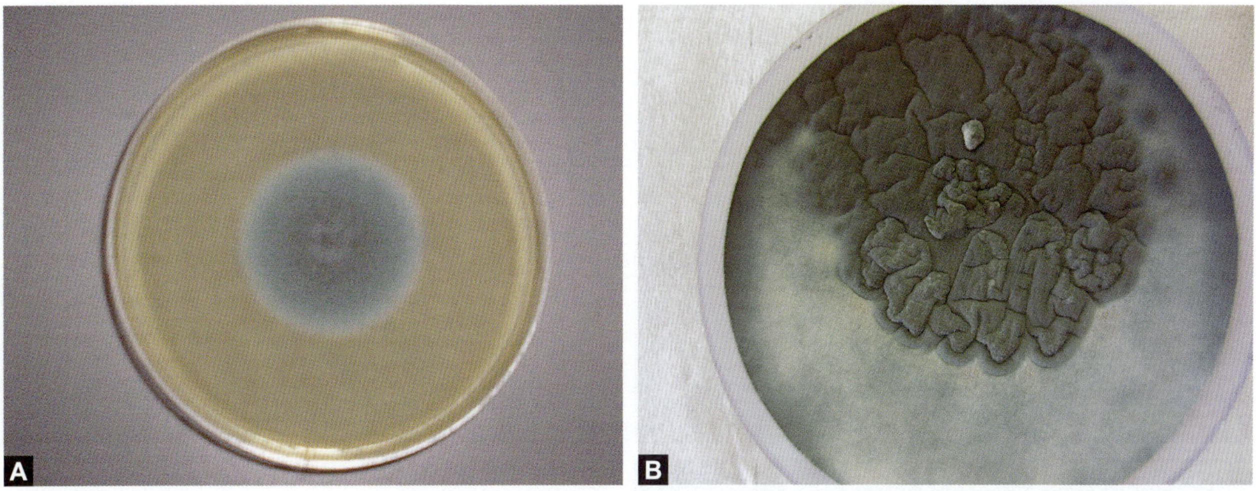

Figs. 25.13A and B. Smoky green-colored colonies of *Aspergillus fumigatus* on Sabouraud's dextrose agar plates.

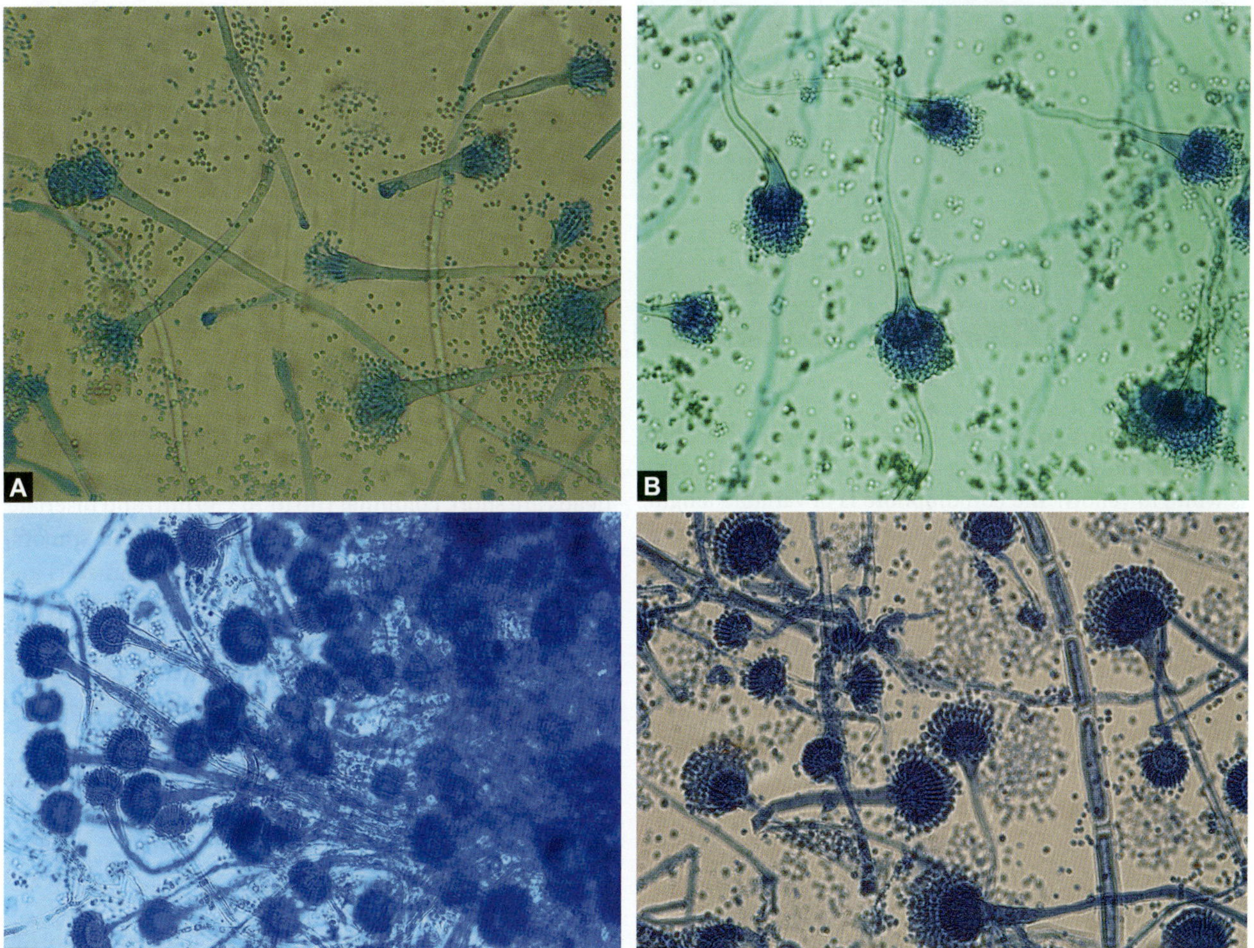

Figs. 25.14A to D. Conidia, uniseriate phialides, vesicles, conidiophores and septate hyphae of *Aspergillus fumigatus* (LCB × 400).

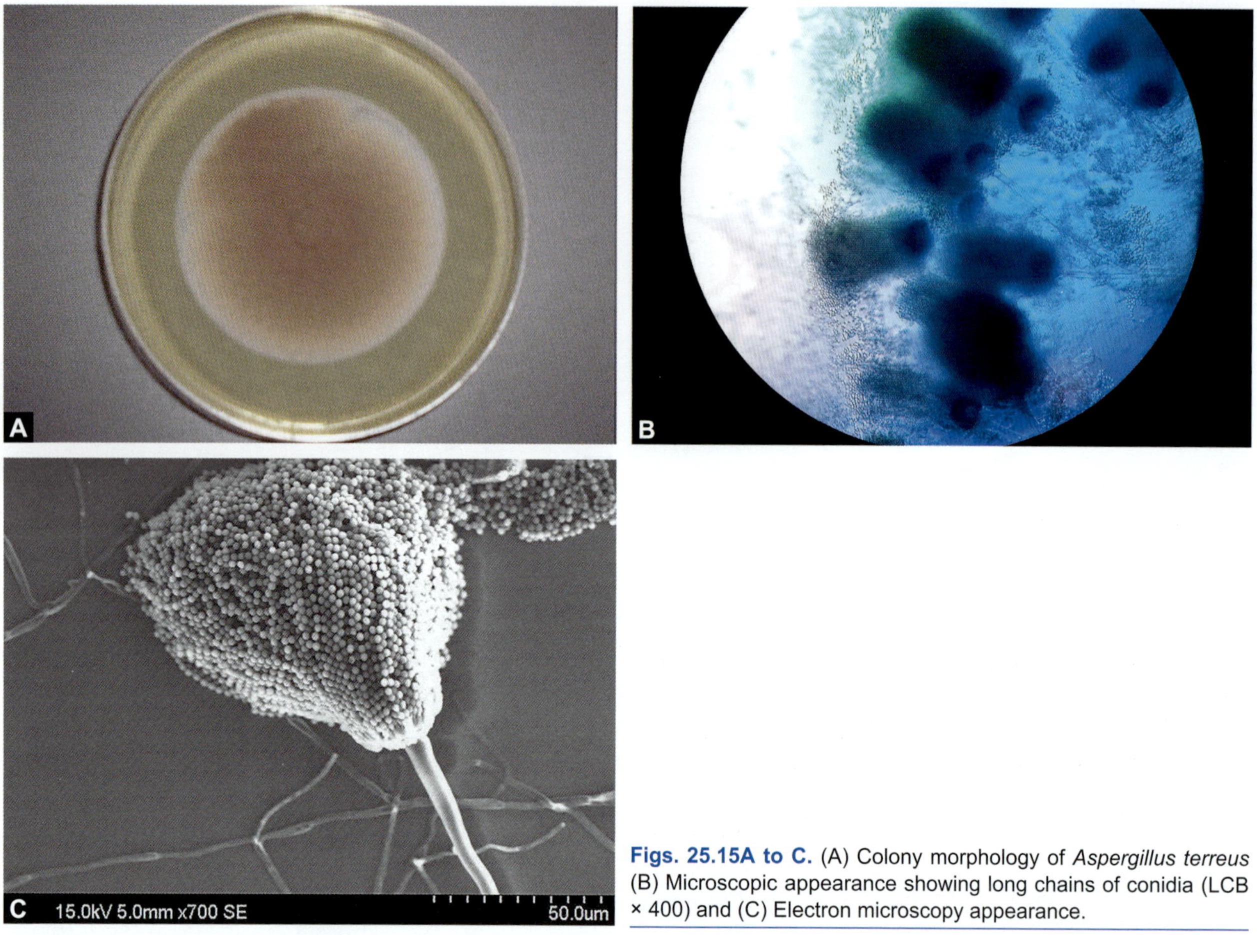

Figs. 25.15A to C. (A) Colony morphology of *Aspergillus terreus* (B) Microscopic appearance showing long chains of conidia (LCB × 400) and (C) Electron microscopy appearance.

(iii) Aspergillus niger: The colonies are woolly at first white to yellow, then turning dark brown to black. Reverse is white to yellow. The conidiophore is of variable length, phialides are biseriate *i.e.* arranged in two rows, covering entire vesicle, form radiate head **(Fig. 25.11C. Also *See* Figs. 30.5; 30.6A and B; 30.7A and B)**.

(iv) Aspergillus terreus: The colonies are usually velvety cinnamon brown. Reverse is white to brown. The conidiophores are short smooth, phialides are in two rows, compactly columnar and conidia are very small *i.e.* 2 μm in size **(Figs. 25.11D and 25.15A to C)**.

(v) Aspergillus glaucus: The colony morphology is green with yellow areas, occasionally, brown. Reverse is yellowish to maroon. The conidiophores are variable in length, smooth. The phialides are single, radiate to very loosely columnar.

(vi) Aspergillus nidulans: The colonies are typically plain green in color with dark red-brown cleistothecia developing within and upon the conidial layer. The reverse may be olive to drab-gray or purple-brown. Conidial heads are short, columnar (up to 70 x 30 μm) and biseriate. Conidiophores are usually short, brownish and smooth-walled. Conidia are globose (3.0-3.5 μm in diameter) and rough-walled. Cleistothecium of its teleomorph, *Emericella nidulans*, show numerous reddish-brown ascospores and are often surrounded by mass of Hulle cells which may be upto 25 μm in size **(Figs. 25.16A and B)**. *Aspergillus nidulans* is a typical soil fungus with worldwide distribution. It is used in making anidulafungin and reported as causative agent of aspergillosis in humans and animals.

(c) Immunodiagnosis

The immunological tests are used as important tools in establishing diagnosis of various clinical forms of aspergillosis. This is especially true for diagnosis of aspergilloma

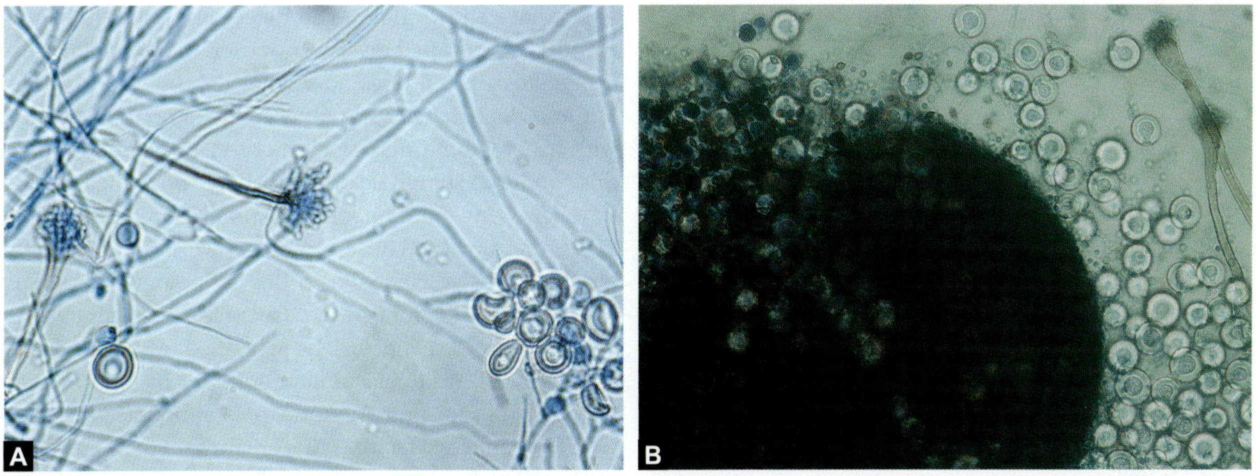

Figs. 25.16A and B. Microscopic appearance of *Aspergillus nidulans* along with bursting cleistothecia and Hulle cells (LCB × 400).

in which high percentage of patients demonstrate precipitating IgG antibodies. In case of allergic bronchopulmonary aspergillosis, diagnostic criteria include positive skin test reactions to *Aspergillus* antigens and elevated levels of IgE and specific IgE and IgG precipitating antibodies to *Aspergillus* species in serum. Detection of circulating antibodies or *Aspergillus* antigens also helps to define etiology of infection in patients yielding negative cultures, in patients whom certain forms of aspergillosis are masked by or mistaken for other diseases and in patients in whom two or more fungi are found to be co-existing.

The use of serology may be an answer to diagnosis and several serodiagnostic tools have been used as alternative or to substitute culture isolation. These include methods such as indirect immunofluorescent, immunoelectrophoresis, immunoelectrodiffusion, enzyme-linked immunosorbent assay, enzyme-linked immunofiltration assay and immunoblotting.

(i) Detection of Antibody: Immunodiffusion test by commercial or laboratory prepared antigen is done in all suspected patients to get clue or to establish final diagnosis. A sensitive enzyme-linked immunosorbent assay using biotin-avidin amplifications system (BALISA) is also employed in determination of antibody isotypes. The antigens currently employed in these assays are crude culture filtrate, cell wall or cytoplasmic extracts of *A.fumigatus*.

(ii) Detection of Antigen: Demonstrable antibodies against aspergilli may be absent in patients with immunosuppression and their detection may be limited by antifungal therapy or by acute and fulminant course of infection. Hence in patients with invasive infection, antigen detection may be helpful in establishing an early and accurate diagnosis. A variety of tests for detection of soluble antigens of *Aspergillus* species in serum, urine or other body fluids and even within host phagocytic cells, have been developed. Of these, latex agglutination, RIA, ELISA and BALISA, latex agglutination and immunoblotting have been most commonly used methods. Ray and colleagues have demonstrated Western blot analysis for *Aspergillus* antigens detection in CSF. An ELISA for detecting *Aspergillus* galactomannan is used to establish an early diagnosis. Low concentration of galactomannan, cell wall component of *Aspergillus*, appears to be an indicator of invasive disease. However, negative result does not rule out diagnosis of aspergillosis.

(iii) Molecular Techniques: The molecular techniques appear to be highly sensitive and specific, even when blood cultures, serological assays and other examinations detect no pathogens and result can be obtained on the same day. The DNA probes have been developed for an efficient, more reliable and rapid diagnostic approach to probably most fatal mycotic disease especially among immunocompromised patients.

In case of disseminated infection, when disease is associated with immunosuppression, serodiagnostic approach fails due to anergy. Recently, advanced laboratory diagnostic procedures, based on DNA techniques have been introduced. The usefulness of nested PCR assay for detection of *Aspergillus* species DNA is evaluated in bronchoalveolar lavage fluid specimen but it cannot differentiate between infection and colonization. In addition

Table 25.5. Epidemiological Markers of *Aspergillus* Species.
A. Phenotyping
Morphotyping
Biotyping
Chemotyping
Secondary Metabolites
Isoenzyme Pattern
Protein Profiles
Ubiquinone Typing
B. Genotyping
Restriction fragment length polymorphism (RFLP)
PCR-based Amplification
G + C Values
DNA Complementarity
RNA Sequence

to detection of invasive fungal infections, PCR can also be used for monitoring response to antifungal therapy. A quantitative PCR helps to monitor fungal burden in relation to the response to treatment.

(iv) Molecular Typing: The typing of isolates in clinical situation requires vital information such as source of infection, whether patient is infected with more than one isolate or there are geographically localized isolates. There are several molecular typing techniques, which have been applied to fulfill necessary criteria of reproducibility and discrimination to various *Aspergillus* species particularly *A.fumigatus* and *A.niger*. These typing techniques can be based on phenotypic characters, such as antigens, enzymes, morphology and biochemistry. Genotypic methods involve analysis of genomic DNA, mitochondrial DNA (mtDNA) and ribosomal DNA (rDNA). The restriction fragment length polymorphism (RFLP) analysis involves detection of variation in genome by using enzymes (restriction endonucleases) which cleave specific DNA sequences. The epidemiological markers related to infections caused by *Aspergillus* genus are shown in **Table 25.5**.

(v) Detection of Fungal Metabolites: Another way of diagnosing aspergillosis particularly in the absence of culture is to demonstrate distinctive metabolic product in fluids or tissues of infected patients. The **G-test** developed by Japanese workers detects circulating β-(1,3)-D-glucan with use of modification of limulus assay for endotoxin and has sensitivity of ~20 pg/ml. Factor G is horse-shoe crab coagulation factor. Prospective studies of utility of G-test have shown that it can detect β-(1,3)-D-glucan in context of confirmed invasive mycoses, including invasive aspergillosis but it does not distinguish between species of *Candida*, *Aspergillus* or other fungi. Sera from patients with aspergilloma may also be G-test positive, although this condition should be distinguishable on clinical basis.

High concentration of D-mannitol, fungal metabolite is found by GLC in serum of rats with experimentally induced invasive aspergillosis. This has potential for use as diagnostic marker for invasive aspergillosis in humans.

The AFMP1 gene has been cloned which encodes antigenic cell wall galactomannoprotein in *A.fumigatus*. AFMP1 codes for protein, Afmp1p, of aminoacid residues with few sequence features that are present in Mp1p, antigenic cell wall mannoprotein in *Talaromyces marneffei* as well as several other cell wall proteins of *Saccharomyces cerevisiae* and *Candida albicans*.

(vi) Skin Tests: Skin test reactivity to *Aspergillus* antigen extracts is used for patients with suspected allergic bronchopulmonary aspergillosis, atopic dermatitis or allergic asthma sensitized by *Aspergillus* species. Intradermal skin tests are performed using 0.1 ml of antigen (1000 PNU/ml Aspergillin) in all patients and controls. Type I hypersensitivity showing erythema and wheal is noted within 1 hour, Type III-Arthus reaction develops in 4 to 10 hours and Type IV showing induration of more than 5 mm diameter after 24 hours.

(d) Animal Pathogenicity

The laboratory animals can be experimentally infected with *Aspergillus* species. Rabbits and mice are model animals to demonstrate lesions of aspergillosis. Intravenous injection of conidia is followed by development of granulomatous lesions in various organs, particularly kidneys. The animals may die depending on concentration of conidia being given through the inoculum. Chakrabarti *et al*, have developed rabbit model for demonstration of paranasal sinus mycosis caused by *Aspergillus* species.

Treatment and Prophylaxis

The treatment depends on clinical type of aspergillosis in a particular patient. The IDSA published its Practice Guidelines for diagnosis and management in 2016 replacing previous documents of 2000 and 2008. Most of the hospital settings broadly follow these Guidelines for the management of their patients. Amphotericin B is the drug of choice for disseminated aspergillosis whereas granulomas has to be removed surgically. Oral itraconazole, a triazole antifungal with activity against *Aspergillus* species, is an alternative to amphotericin B for all types of aspergillosis. However, its use has been limited due to the fact that its oral absorption is erratic in critically ill patients. The outcome of therapy for invasive aspergillosis is dependent not only on intensive antifungal therapy but also on recovery from underlying host immune defects such as resolution of neutropenia.

Unlike *Aspergillus* fungal ball, in which medical therapy has very little benefit, successful treatment of CNPA has been reported. Treatment with itraconazole, followed by amphotericin B or surgical excision or intracavitary amphotericin is suggested. The dose and duration of therapy should be based on clinical response. Moreover, maintenance therapy with itraconazole can be considered in patients with residual parenchymal scarring. Voriconazole and micafungin are useful in refractory cases. In this disorder, diagnosis is usually delayed, patients are poor operative candidates and post-operative complications are common. Treatment outcome is likely to be influenced by the severity of co-morbid conditions, the extent of the underlying disease, delays in diagnosis and initiation of effective therapy. In any developing country, cavity in upper zone of chest radiograph is considered as a case of pulmonary tuberculosis even in the negative acid-fast bacilli staining. Diagnosis of chronic necrotizing pulmonary aspergillosis must be considered in those patient in which disease progress along with evolution of fungus ball after a trial of antitubercular treatment. The angioinvasion in aspergillosis and mucormycosis leads to thrombus formation which roam in circulation and it is not like bacterial infections.

The disease usually takes favorable course when timely treatment is initiated. The mortality in transplant recipients with invasive aspergillosis remains very high. Successful treatment depends on early diagnosis and aggressive antifungal therapy along with measures to reduce immunosuppression or augment immune system. *A.terreus* is less susceptible to amphotericin B than *A.fumigatus,* both *in vitro* and *in vivo*.

Newer treatment options include early surgical intervention in cases of localized disease; use of liposomal preparations of amphotericin B, itraconazole or voriconazole and immunotherapy with G-CSF or GM-CSF and interferon-γ. Voriconazole is given at most advanced stage of development and appears to be at least as effective as itraconazole and possibly amphotericin B in treatment of invasive aspergillosis. Similarly posaconazole is also being clinically tried. The patients with chronic granulomatous disease are treated with itraconazole and interferon-γ. In addition, echinocandin group of drugs are indicated in case of drug resistance in *Aspergillus* although it is emerging in this group also. However, there is no other option except these ones, particularly when the infection is due to *A.terreus* or *A.fumigatus.* The details of antifungals and AFST used for managing aspergillosis are given in **Chapter 6** and **Appendix D**, respectively.

Although number of prophylactic regimens are in widespread use but none has shown conclusively to reduce incidence of invasive aspergillosis. Reducing environmental exposure to *Aspergillus*, early diagnosis of infection by means of antigen detection in serum and use of high resolution computed tomography (HRCT) are alternatives to use of prophylaxis but have not been systematically evaluated.

Besides maintenance of ventilation systems and minimization of high risk factors, single most important infection control measure to prevent healthcare associated transmission of invasive aspergillosis is heightened awareness about pathogenesis and epidemiology of aspergillosis. The CDC has published revised recommendations for preventing healthcare-associated pulmonary aspergillosis. Periodic review of microbiology, histopathology and autopsy data is recommended as routine surveillance for healthcare-associated aspergillosis, especially in specialized units where high-risk of such patients are being treated.

Further Reading

1. Abdin MZ, Ahmad MM, Javed S. Advances in molecular detection of *Aspergillus*: An update. *Arch Microbiol.* 2010; **192:** 409-25.
2. Ader F. Invasive pulmonary aspergillosis in patients with chronic obstructive pulmonary disease: An emerging fungal disease. *Curr Infect Dis Rep.* 2010; **12:** 409-16.
3. Agarwal R, Vishwanath G, Aggarwal AN, *et al.* Itraconazole in chronic cavitary pulmonary aspergillosis: A randomised controlled trial and systematic review of literature. *Mycoses.* 2013; **56:** 559-70.
4. Agarwal S, Bohara S, Thakran A, *et al.* Pulmonary hydatid disease with co-existent aspergillosis: An incidental finding. *Indian J Med Microbiol.* 2013; **31:** 85-6.
5. Agarwal S, Das A, Gochhait D, *et al. Aspergillus* mediastinitis presenting as superior mediastinal syndrome in an immunocompetent child. *Pediatr Infect Dis J.* 2015; **34:** 221-2.
6. Aguado JM, Vazquez L, Fernandez-Ruiz M, *et al.* Serum galactomannan versus a combination of galactomannan and polymerase chain reaction-based *Aspergillus* DNA

detection for early therapy of invasive aspergillosis in high-risk hematological patients: a randomized controlled trial. *Clin Infect Dis.* 2015; **60:** 405-14.

7. Ahuja A, Aulakh BS, Cheena DK, *et al. Aspergillus* fungal balls causing ureteral obstruction. *Urol J.* 2009; **6:** 127-9.
8. Ajith C, Dogra S, Radotra BD, *et al.* Primary cutaneous aspergillosis in an immunocompetent individual. *J Eur Acad Dermatol Venereol.* 2006; **20:** 738-9.
9. Al-Abdely HM, Alothman AF, Salman JA, *et al.* Clinical practice guidelines for the treatment of invasive *Aspergillus* infections in adults in the Middle East region: Expert panel recommendations. *J Infect Public Health.* 2014; **7:** 20-31.
10. Al-Ogaili Z, Chapeikin G, Palmer D. Primary aspergillosis of bilateral laryngoceles. *Case Rep Med.* 2014; 384271. PMID: 24639876.
11. Al-Tawfiq JA, Al-Abdely HM. Vertebral osteomyelitis due to *Aspergillus fumigatus* in a patient with chronic granulomatous disease successfully treated with antifungal agents and interferon-gamma. *Med Mycol.* 2010; **48:** 537-41.
12. Al-Wathiqi F, Ahmad S, Khan Z. Molecular identification and antifungal susceptibility profile of *Aspergillus flavus* isolates recovered from clinical specimens in Kuwait. *BMC Infect Dis.* 2013; **13:** 126. PMID: 23496810.
13. Alastruey-Izquierdo A, Mellado E, Cuenca-Estrella M. Current section and species complex concepts in *Aspergillus*: Recommendations for routine daily practice. *Ann N Y Acad Sci.* 2012; **1273:** 18-24.
14. Alevizakos M, Farmakiotis D, Mylonakis E. Updated practice guidelines for the diagnosis and management of aspergillosis: Challenges and opportunities. *J Thorac Dis.* 2016; **8:** e1767-70.
15. Aliyali M, Badali H, Shokohi T, *et al.* Co-infection of pulmonary hydatid cyst and aspergilloma: Case report and systematic review. *Mycopathologia.* 2016; **181:** 255-65.
16. Aller-Garcia AI, Castro-Mendez C, Alastruey-Izquierdo A, *et al.* Case series study of invasive pulmonary aspergillosis. *Mycopathologia.* 2017; **182:** 505-15.
17. Amatya R, Khanal B, Mahato PK, *et al. Aspergillus* species in clinical specimen: A seven year prevalence study from eastern Nepal. *Nepal Med Coll J.* 2011; **13:** 175-7.
18. Ambasta A, Carson J, Church DL. The use of biomarkers and molecular methods for the earlier diagnosis of invasive aspergillosis in immunocompromised patients. *Med Mycol.* 2015; **53:** 531-57.
19. Anila KR, Somanathan T, Mathews A, *et al.* Fruiting bodies of *Aspergillus*: An unusual finding in histopathology. *Lung India.* 2013; **30:** 357-9.
20. Antinori S, Corbellino M, Meroni L, *et al. Aspergillus* meningitis: A rare clinical manifestation of central nervous system aspergillosis. Case report and review of 92 cases. *J Infect.* 2013; **66:** 218-38.
21. Aquino VR, Nagel F, Andreolla HF, *et al.* The performance of real-time PCR, galactomannan and fungal culture in the diagnosis of invasive aspergillosis in ventilated patients with chronic obstructive pulmonary disease (COPD). *Mycopathologia.* 2012; **174:** 163-9.
22. Arikan-Akdagli S. Azole resistance in *Aspergillus*: Global status in Europe and Asia. *Ann N Y Acad Sci.* 2012; **1272:** 9-14.
23. Armstead J, Morris J, Denning DW. Multi-country estimate of different manifestations of aspergillosis in cystic fibrosis. *PLoS One.* 2014; **9:** e98502. PMID: 24914809.
24. Arora V, Nagarkar NM, Dass A, Malhotra A. Invasive rhino-orbital aspergillosis. *Indian J Otolaryngol Head Neck Surg.* 2011; **63:** 325-9.
25. Arvanitis M, Mylonakis E. Diagnosis of invasive aspergillosis: Recent developments and ongoing challenges. *Eur J Clin Invest.* 2015; **45:** 646-52.
26. Asare KA, Jahng M, Pincus JL, *et al.* Sternal osteomyelitis caused by *Aspergillus fumigatus* following cardiac surgery: Case and review. *Med Mycol Case Rep.* 2012; **2:** 4-6.
27. Asimakopoulos P, Supriya M, Kealey S, *et al.* A case-based discussion on a patient with non-otogenic fungal skull base osteomyelitis: Pitfalls in diagnosis. *J Laryngol Otol.* 2013; **127:** 817-21.
28. Askew DS, Kontoyiannis DP, Clemons KV. Advances against aspergillosis: Biology, host response, diagnosis and treatment. *Mycopathologia.* 2014; **178:** 321-4.
29. Aslan M, Oz Y, Aksit F, *et al.* Potential of polymerase chain reaction and galactomannan for the diagnosis of invasive aspergillosis in patients with febrile neutropenia. *Mycoses.* 2015; **58:** 343-9.
30. Avni T, Levy I, Sprecher H, *et al.* Diagnostic accuracy of PCR alone compared to galactomannan in bronchoalveolar lavage fluid for diagnosis of invasive pulmonary aspergillosis: A systematic review. *J Clin Microbiol.* 2012; **50:** 3652-8.
31. Badali H, Fakhim H, Zarei F, *et al. In vitro* activities of five antifungal drugs against opportunistic agents of *Aspergillus* Nigri Complex. *Mycopathologia.* 2016; **181:** 235-40.
32. Baddley JW, Andes DR, Marr KA, *et al.* Factors associated with mortality in transplant patients with invasive aspergillosis. *Clin Infect Dis.* 2010; **50:** 1559-67.
33. Baddley JW. Clinical risk factors for invasive aspergillosis. *Med Mycol.* 2011; **49:** S7-12.
34. Badiee P, Alborzi A. Detection of *Aspergillus* species in bone marrow transplant patients. *J Infect Dev Ctries.* 2010; **4:** 511-6.
35. Bal A, Agarwal AN, Das A, *et al.* Chronic necrotising pulmonary aspergillosis in a marijuana addict: A new cause of amyloidosis. *Pathology.* 2010; **42:** 197-200.
36. Bal A, Bagai M, Mohan H, Dalal U. Aspergilloma in a pulmonary hydatid cyst: A case report. *Mycoses.* 2008; **51:** 357-9.
37. Balajee SA. *Aspergillus terreus* complex. *Med Mycol.* 2009; **47:** S42-6.
38. Baradkar VP, Mathur M, Kumar S. Uncommon presentation of pulmonary aspergilloma. *Indian J Med Microbiol.* 2009; **27:** 270-2.
39. Barchiesi F, Mazzocato S, Mazzanti S, *et al.* Invasive aspergillosis in liver transplant recipients: Epidemiology, clinical characteristics, treatment and outcomes in 116 cases. *Liver Transpl.* 2015; **21:** 204-12.

40. Barnes PD, Marr KA. Aspergillosis: Spectrum of disease, diagnosis and treatment. *Infect Dis Clin North Am.* 2006; **20:** 545-61.
41. Barnes RA. Directed therapy for fungal infections: Focus on aspergillosis. *J Antimicrob Chemother.* 2013; **68:** 2431-4.
42. Barrs VR, van Doorn TM, Houbraken J, *et al. Aspergillus felis* sp. nov., an emerging agent of invasive aspergillosis in humans, cats and dogs. *PLoS One.* 2013; **8:** e64871.
43. Batista MV, Costa SF, Shikanai-Yasuda MA, *et al.* Current treatment options for invasive aspergillosis. *Drugs Today (Barc).* 2013; **49:** 213-26.
44. Batra S, Arora S, Meshram H, *et al.* A rare etiology of cauda equina syndrome. *J Infect Dev Ctries.* 2011; **5:** 79-82.
45. Baxter CG, Dunn G, Jones AM, *et al.* Novel immunologic classification of aspergillosis in adult cystic fibrosis. *J Allergy Clin Immunol.* 2013; **132:** 560-6.e10.
46. Bergeron A, Porcher R, Menotti J, *et al.* Prospective evaluation of clinical and biological markers to predict the outcome of invasive pulmonary aspergillosis in hematological patients. *J Clin Microbiol.* 2012; **50:** 823-30.
47. Bernal-Martínez L, Alastruey-Izquierdo A, Cuenca-Estrella M. Diagnostics and susceptibility testing in *Aspergillus. Future Microbiol.* 2016; **11:** 315-28.
48. Bernardeschi C, Foulet F, Ingen-Housz-Oro S, *et al.* Cutaneous invasive aspergillosis: Retrospective multicenter study of the French invasive-aspergillosis registry and literature review. *Medicine (Baltimore).* 2015; **94:** e1018.
49. Bhagat M, Rapose A. Rapidly progressing dual infection with *Aspergillus* and *Rhizopus*: When soil inhabitants become deadly invaders. *BMJ Case Rep.* 2016; pii: bcr2016217535. PMID: 27932434.
50. Bhaskar N, Mohammad K, Kaur V. *Aspergillus* brain abscess. *J Ark Med Soc.* 2013; **110:** 41-2.
51. Bhaskaran A, Kabbani D, Singer LG, *et al.* (1,3) β-D-glucan in bronchoalveolar lavage of lung transplant recipients for the diagnosis of invasive pulmonary aspergillosis. *Med Mycol.* 2017; **55:** 164-72.
52. Bhatnagar T, Bhatnagar AK. Pleural aspergillosis in an otherwise healthy individual. *Lung India.* 2014; **31:** 155-7.
53. Bienvenu AL, Walter T, Boulez F, *et al.* Acute liver failure may lead to lethal pulmonary aspergillosis. *J Clin Gastroenterol.* 2010; **44:** 593-4.
54. Binder U, Lass-Florl C. New insights into invasive aspergillosis - From the pathogen to the disease. *Curr Pharm Des.* 2013; **19:** 3679-88.
55. Boch T, Buchheidt D, Spiess B, *et al.* Direct comparison of galactomannan performance in concurrent serum and bronchoalveolar lavage samples in immunocompromised patients at risk for invasive pulmonary aspergillosis. *Mycoses.* 2016; **59:** 80-5.
56. Bokhari R, Baeesa S, Al-Maghrabi J, *et al.* Isolated cerebral aspergillosis in immunocompetent patients. *World Neurosurg.* 2014; **82:** e325-33.
57. Boonsarngsuk V, Niyompattama A, Teosirimongkol C, *et al.* False-positive serum and bronchoalveolar lavage *Aspergillus* galactomannan assays caused by different antibiotics. *Scand J Infect Dis.* 2010; **42:** 461-8.
58. Bowyer P, Moore CB, Rautemaa R, *et al.* Azole antifungal resistance today: Focus on *Aspergillus. Curr Infect Dis Rep.* 2011; **13:** 485-91.
59. Bretagne S. Primary diagnostic approaches of invasive aspergillosis - Molecular testing. *Med Mycol.* 2011; **49:** S48-53.
60. Bua J, Marchetti F, Zennaro F, *et al.* A lung mass: When the history helps the diagnosis. *BMJ Case Rep.* 2011; pii: bcr0120113718. PMID: 22679144.
61. Butros SR, Tierce ML 4th, Marks A, *et al.* Retroperitoneal aspergilloma. *Pediatr Radiol.* 2013; **43:** 886-9.
62. Buzina W. *Aspergillus* - Classification and antifungal susceptibilities. *Curr Pharm Des.* 2013; **19:** 3615-28.
63. Cadena J, Thompson GR 3rd, Patterson TF. Invasive aspergillosis: Current strategies for diagnosis and management. *Infect Dis Clin North Am.* 2016; **30:** 125-42.
64. Cai X, Ni W, Wei C, *et al.* Diagnostic value of the serum galactomannan and (1,3)-β-D-glucan assays for invasive pulmonary aspergillosis in non-neutropenic patients. *Intern Med.* 2014; **53:** 2433-7.
65. Camargo JF, Husain S. Immune correlates of protection in human invasive aspergillosis. *Clin Infect Dis.* 2014; **59:** 569-77.
66. Capone A, Di Bella S, Chinello P, *et al.* Multiple bilateral pulmonary "fungus balls" in an immunocompetent patient unsuitable for surgical treatment: Efficacy of azoles treatment. *Mycopathologia.* 2013; **175:** 361-3.
67. Carlesse F, Marcos AC, Seber A, *et al.* Renal aspergillosis in a 6-year-old male with Burkitt's lymphoma. *Pediatr Infect Dis J.* 2015; **34:** 679-80.
68. Carvalho A, Cunha C, Bistoni F, *et al.* Immunotherapy of aspergillosis. *Clin Microbiol Infect.* 2012; **18:** 120-5.
69. Cateau E, Randriamalala E, Elsendoorn A, *et al.* Fatal-mixed cutaneous zygomycosis-aspergillosis: A case report. *Mycopathologia.* 2013; **176:** 423-7.
70. Cerqueira LB, de Francisco TM, Gasparetto JC, *et al.* Development and validation of an HPLC-MS/MS method for the early diagnosis of aspergillosis. *PLoS One.* 2014; **9:** e92851.
71. Chai LY, Hsu LY. Recent advances in invasive pulmonary aspergillosis. *Curr Opin Pulm Med.* 2011; **17:** 160-6.
72. Chakrabarti A, Chatterjee SS, Das A, *et al.* Invasive aspergillosis in developing countries. *Med Mycol.* 2011; **49:** S35-47.
73. Chambers ST, Bhandari S, Scott-Thomas A, *et al.* Novel diagnostics: Progress toward a breath test for invasive *Aspergillus fumigatus. Med Mycol.* 2011; **49:** S54-61.
74. Chatterjee D, Bal A, Singhal M, *et al.* Fibrosing mediastinitis due to *Aspergillus* with dominant cardiac involvement: Report of two autopsy cases with review of literature. *Cardiovasc Pathol.* 2014; **23:** 354-7.
75. Chawla K, Kosaraju K, Rayasam S, *et al.* Clinicomicrobiological profile of chronic pulmonary aspergillosis from a tertiary care centre in southern India. *J Clin Diagn Res.* 2013; **7:** 2712-5.
76. Chen L, Liu Y, Wang W, *et al.* Adrenal and hepatic aspergillosis in an immunocompetent patient. *Infect Dis (Lond).* 2015; **47:** 428-32.

77. Chen S, Pu JL, Yu J, *et al.* Multiple *Aspergillus* cerebellar abscesses in a middle-aged female: Case report and literature review. *Int J Med Sci.* 2011; **8:** 635-9.
78. Chen Y, Cameron A. Aspergillosis after liver transplantation in the context of common variable immunodeficiency: Case report. *Transpl Infect Dis.* 2013; **15:** 540-4.
79. Cheon S, Yang MK, Kim CJ, *et al.* Disseminated aspergillosis in the immunocompetent host: A case report and literature review. *Mycopathologia.* 2015; **180:** 217-22.
80. Chong GL, van de Sande WW, Dingemans GJ, *et al.* Validation of a new *Aspergillus* real-time PCR assay for direct detection of *Aspergillus* and azole resistance of *Aspergillus fumigatus* on bronchoalveolar lavage fluid. *J Clin Microbiol.* 2015; **53:** 868-74.
81. Chong GM, Maertens JA, Lagrou K, *et al.* Diagnostic performance of galactomannan antigen testing in cerebrospinal fluid. *J Clin Microbiol.* 2016; **54:** 428-31.
82. Chotirmall SH, Mirkovic B, Lavelle GM, *et al.* Immunoevasive *Aspergillus* virulence factors. *Mycopathologia.* 2014; **178:** 363-70.
83. Chou HW, Chuang MK, Chen YS. Rapid recurrence of right ventricular aspergilloma. *Eur J Cardiothorac Surg.* 2013; **44:** 965.
84. Choudhury N, Khan AB, Tzvetanov I, *et al.* Multiple *Aspergillus* brain abscesses after liver transplantation. *Transplantation.* 2014; **97:** e72-3.
85. Chowdhary A, Kathuria S, Randhawa HS, *et al.* Isolation of multiple-triazole-resistant *Aspergillus fumigatus* strains carrying the TR/L98H mutations in the cyp51A gene in India. *J Antimicrob Chemother* 2012; **67:** 362-6.
86. Chowdhary A, Kathuria S, Xu J, *et al.* Emergence of azole-resistant *Aspergillus fumigatus* strains due to agricultural azole use creates an increasing threat to human health. *PLoS Pathog.* 2013; **9:** e1003633.
87. Chowdhary A, Sharma C, Hagen F, *et al.* Exploring azole antifungal drug resistance in *Aspergillus fumigatus* with special reference to resistance mechanisms. *Future Microbiol.* 2014; **9:** 697-711.
88. Chowdhary A, Sharma C, Kathuria S, *et al.* Prevalence and mechanism of triazole resistance in *Aspergillus fumigatus* in a referral chest hospital in Delhi, India and an update of the situation in Asia. *Front Microbiol.* 2015; **6:** 428.
89. Chowdhury FH, Haque MR, Khan SK, *et al.* Cerebral aspergilloma in a SLE patient: A case report with short literature review. *Asian J Neurosurg.* 2014; **9:** 58-61.
90. Clemons KV, Perlin DS, Richardson M. Preface for Advances Against Aspergillosis. *Ann N Y Acad Sci.* 2012; 1273: vii-viii.
91. Colmenero I, Alonso-Sanz M, Casco F, *et al.* Cutaneous aspergillosis mimicking pancreatic and gouty panniculitis. *J Am Acad Dermatol.* 2012; **67:** 789-91.
92. Cristina ML, Sartini M, Spagnolo AM. Health care-acquired aspergillosis and air conditioning systems. *J Prev Med Hyg.* 2009; **50:** 3-8.
93. Cruciani M, Mengoli C, Loeffler J, *et al.* Polymerase chain reaction blood tests for the diagnosis of invasive aspergillosis in immunocompromised people. *Cochrane Database Syst Rev.* 2015; **10:** CD009551.
94. Crum-Cianflone NF. Invasive aspergillosis associated with severe influenza infections. *Open Forum Infect Dis.* 2016; **3:** ofw171. PMID: 27704024.
95. Cucchetto G, Cazzadori A, Conti M, *et al.* Treatment of chronic pulmonary aspergillosis with voriconazole: Review of a case series. *Infection.* 2015; **43:** 277-86.
96. Cumbo-Nacheli G, de Sanctis J. A near fatal case of invasive aspergillosis. *Am J Case Rep.* 2012; **13:** 169-72.
97. Cunha C, Aversa F, Lacerda JF, *et al.* Genetic PTX3 deficiency and aspergillosis in stem-cell transplantation. *N Engl J Med.* 2014; **370:** 421-32.
98. Cunha C, Aversa F, Romani L, *et al.* Human genetic susceptibility to invasive aspergillosis. *PLoS Pathog.* 2013; **9:** e1003434. PMID: 23950708.
99. Cunha C, Kurzai O, Carvalho A. PTX3 deficiency and aspergillosis. *N Engl J Med.* 2014; **370:** 1666-7.
100. Cunha C, Kurzai O, Loffler J, *et al.* Neutrophil responses to aspergillosis: New roles for old players. *Mycopathologia.* 2014; **178:** 387-93.
101. D'Haese J, Theunissen K, Vermeulen E, *et al.* Detection of galactomannan in bronchoalveolar lavage fluid samples of patients at risk for invasive pulmonary aspergillosis: Analytical and clinical validity. *J Clin Microbiol.* 2012; **50:** 1258-63.
102. D'sa SR, Singh S, Satyendra S, *et al.* Case report of *Aspergillus* osteomyelitis of the ribs in an immunocompetent patient. *J Glob Infect Dis.* 2013; **5:** 118-20.
103. da Silva TV, Carneiro LC, Ramos Fdos S, *et al.* PCR as a screening test for invasive aspergillosis in haematological patients: A pilot study. *Mycopathologia.* 2014; **177:** 111-4.
104. Dabkana TM, Pindiga UH, Mayun AA, *et al.* Aspergillosis of the bone. *Ann Afr Med.* 2015; **14:** 62-4.
105. Dagenais TR, Keller NP. Pathogenesis of *Aspergillus fumigatus* in invasive aspergillosis. *Clin Microbiol Rev.* 2009; **22:** 447-65.
106. Davies S, Guidry C, Politano A, *et al.* *Aspergillus* infections in transplant and non-transplant surgical patients. *Surg Infect (Larchmt).* 2014; **15:** 207-12.
107. Davoudi S, Anderlini P, Fuller GN, *et al.* A long-term survivor of disseminated *Aspergillus* and Mucorales infection: An instructive case. *Mycopathologia.* 2014; **178:** 465-70.
108. de Azevedo PZ, Sylvestre TF, Cavalcante Rde S, *et al.* Evaluation of the double agar gel immunodiffusion test and of the enzyme-linked immunosorbent assay in the diagnosis and follow-up of patients with chronic pulmonary aspergillosis. *PLoS One.* 2015; **10:** e0134841.
109. de Boer MG, Halkes CJ, van de Vosse E. PTX3 deficiency and aspergillosis. *N Engl J Med.* 2014; **370:** 1665-6.
110. de Mol M, de Jongste JC, van Westreenen M, *et al.* Diagnosis of invasive pulmonary aspergillosis in children with bronchoalveolar lavage galactomannan. *Pediatr Pulmonol.* 2013; **48:** 789-96.
111. Del Rosario RC, Halloush RA, Khasawneh FA. A 56-year-old man with undifferentiated fever in the ICU and endobronchial plaques. *Chest.* 2013; **143:** 566-9.
112. Denis B, Guiguet M, de Castro N, *et al.* Relevance of EORTC criteria for the diagnosis of invasive aspergillosis

in HIV-infected patients and survival trends over a 20-year period in France. *Clin Infect Dis.* 2015; **61:** 1273-80.

113. Denning DW, Bowyer P. Voriconazole resistance in *Aspergillus fumigatus:* Should we be concerned? *Clin Infect Dis.* 2013; **57:** 521-3.
114. Denning DW, Cadranel J, Beigelman-Aubry C, *et al.* Chronic pulmonary aspergillosis: Rationale and clinical guidelines for diagnosis and management. *Eur Respir J.* 2016; **47:** 45-68.
115. Denning DW, Pleuvry A, Cole DC. Global burden of chronic pulmonary aspergillosis complicating sarcoidosis. *Eur Respir J.* 2013; **41:** 621-6.
116. Desai R, Ross LA, Hoffman JA. The role of bronchoalveolar lavage galactomannan in the diagnosis of pediatric invasive aspergillosis. *Pediatr Infect Dis J.* 2009; **28:** 283-6.
117. Desai SR, Hedayati V, Patel K, *et al.* Chronic aspergillosis of the lungs: Unravelling the terminology and radiology. *Eur Radiol.* 2015; **25:** 3100-7.
118. Desgranges FP, Hernu R, Philit F, *et al.* Chronic necrotising pulmonary aspergillosis: An uncommon aetiology of pneumothorax. *BMJ Case Rep.* 2014; pii: bcr2014204137.
119. Desoubeaux G, Bailly E, Chandenier J. Diagnosis of invasive pulmonary aspergillosis: Updates and recommendations. *Med Mal Infect.* 2014; **44:** 89-101.
120. Devaraj U, Priya R, Savio J, *et al. Aspergillus* - The great masquerader. *Indian J Cancer.* 2011; **48:** 137-8.
121. Dimopoulos G, Frantzeskaki F, Poulakou G, *et al.* Invasive aspergillosis in the intensive care unit. *Ann N Y Acad Sci.* 2012; **1272:** 31-9.
122. Ding WY, Chan T, Yadavilli RK, *et al.* Aspergilloma and massive haemoptysis. *BMJ Case Rep.* 2014; pii: bcr2013200019.
123. Doloi PK, Baruah DK, Goswami SC, *et al.* Primary aspergillosis of the larynx: A case report. *Indian J Otolaryngol Head Neck Surg.* 2014; **66 (Suppl. 1):** 326-8.
124. Dos Santos VM, da Trindade MC, de Souza DW, *et al.* A 76-year-old man with a right lung adenocarcinoma and invasive aspergillosis. *Mycopathologia.* 2013; **176:** 113-8.
125. Dotis J, Roilides E. Osteomyelitis due to *Aspergillus* species in chronic granulomatous disease: An update of the literature. *Mycoses.* 2011; **54:** e686-96.
126. Duarte RF, Sanchez-Ortega I, Cuesta I, *et al.* Serum galactomannan-based early detection of invasive aspergillosis in hematology patients receiving effective antimold prophylaxis. *Clin Infect Dis.* 2014; **59:** 1696-702.
127. Dutta M, Jotdar A, Kundu S, *et al.* Primary laryngeal aspergillosis in the immunocompetent state: A clinical update. *Braz J Otorhinolaryngol.* 2015. pii: S1808-8694(15)00173-1.
128. Eigl S, Prattes J, Lackner M, *et al.* Multicenter evaluation of a lateral-flow device test for diagnosing invasive pulmonary aspergillosis in ICU patients. *Crit Care.* 2015; **19:** 178.
129. Er Dedekarginoglu B, Savas Bozbas S, Ulubay G, *et al.* Culture-positive pulmonary aspergillosis infection: Clinical and laboratory features of solid-organ transplant recipients. *Exp Clin Transplant.* 2017; **15 (Suppl. 1):** 214-8.
130. Erdem H, Uzunlar AK, Yildirim U, *et al.* Diffuse infiltration of *Aspergillus* hyphae in the thyroid gland with multinodular goiter. *Indian J Pathol Microbiol.* 2011; **54:** 814-6.
131. Ergene U, Akcali Z, Ozbalci D, *et al.* Disseminated aspergillosis due to *Aspergillus niger* in immunocompetent patient: A case report. *Case Rep Infect Dis.* 2013; 385190.
132. Escribano P, Marcos-Zambrano LJ, Pelaez T, *et al.* Sputum and bronchial secretion samples are equally useful as bronchoalveolar lavage samples for the diagnosis of invasive pulmonary aspergillosis in selected patients. *Med Mycol.* 2015; **53:** 235-40.
133. Faber J, Moritz N, Henninger N, *et al.* Rapid detection of common pathogenic *Aspergillus* species by a novel real-time PCR approach. *Mycoses.* 2009; **52:** 228-33.
134. Faruqi S, Hughes D, Chauhan S, *et al.* Aspergilloma. *Intern Med J.* 2009; **39:** 64.
135. Faruqi S, Thirumaran M, Smith G. A "mobile" crescent. *Am J Med Sci.* 2010; **339:** 98.
136. Felton IC, Simmonds NJ. *Aspergillus* and cystic fibrosis: Old disease - new classifications. *Curr Opin Pulm Med.* 2014; **20:** 632-8.
137. Fifer HM, Chiodini PL, Lucas SB, *et al.* Invasive aspergillosis associated with pulmonary hydatid disease. *Travel Med Infect Dis.* 2012; **10:** 272-4.
138. Fisher BT. The role of biomarkers for diagnosis of and therapeutic decisions related to invasive aspergillosis in children. *Curr Fungal Infect Rep.* 2013; **7:** 7-14.
139. Fisher CE, Stevens AM, Leisenring W, *et al.* The serum galactomannan index predicts mortality in hematopoietic stem cell transplant recipients with invasive aspergillosis. *Clin Infect Dis.* 2013; **57:** 1001-4.
140. Florescu DF, Iwen PC, Hill LA, *et al.* Cerebral aspergillosis caused by *Aspergillus ustus* following orthotopic heart transplantation: Case report and review of the literature. *Clin Transplant.* 2009; **23:** 116-20.
141. Fournier A, Pantet O, Guerid S, *et al.* Effective treatment of invasive *Aspergillus fumigatus* infection using combinations of topical and systemic antifungals in a severely burned patient. *J Burn Care Res.* 2015; **36:** e85-9.
142. Gabrielli E, Fothergill AW, Brescini L, *et al.* Osteomyelitis caused by *Aspergillus* species: A review of 310 reported cases. *Clin Microbiol Infect.* 2014; **20:** 559-65.
143. Gamaletsou MN, Rammaert B, Bueno MA, *et al. Aspergillus* arthritis: Analysis of clinical manifestations, diagnosis and treatment of 31 reported cases. *Med Mycol.* 2017; **55:** 246-54.
144. Gamaletsou MN, Rammaert B, Bueno MA, *et al. Aspergillus* osteomyelitis: Epidemiology, clinical manifestations, management and outcome. *J Infect.* 2014; **68:** 478-93.
145. Gamson K, Ding Y, Aronowitz P. Invasive aspergillosis. *J Gen Intern Med.* 2014; **29:** 686-7.
146. Gangopadhyay M, Majumdar K, Bandyopadhyay A, *et al.* Invasive primary aspergillosis of the larynx presenting as hoarseness and a chronic non-healing laryngeal ulcer in an immunocompetent host: A rare entity. *Ear Nose Throat J.* 2014; **93:** 265-8.
147. Garcia-Hermoso D, Alanio A, Cabaret O, *et al.* High diversity of non-sporulating moulds in respiratory specimens of immunocompromised patients: Should all the species be reported when diagnosing invasive aspergillosis? *Mycoses.* 2015; **58:** 557-64.

148. Gautier M, Normand AC, L'Ollivier C, *et al. Aspergillus tubingensis:* A major filamentous fungus found in the airways of patients with lung disease. *Med Mycol.* 2016; **54:** 459-70.
149. Gautier M, Normand AC, Ranque S. Previously unknown species of *Aspergillus. Clin Microbiol Infect.* 2016; **22:** 662-9.
150. Gazzoni FF, Hochhegger B, Severo LC, *et al. Aspergillus fumigatus* fungus ball in the native lung after single lung transplantation. *J Bras Pneumol.* 2013; **39:** 393-5.
151. Geltner C, Lass-Florl C. Invasive pulmonary aspergillosis in organ transplants - Focus on lung transplants. *Respir Investig.* 2016; **54:** 76-84.
152. Genzen JR, Kenney B. Central nervous system *Aspergillus* infection after epidural analgesia: Diagnosis, therapeutic challenges and literature review. *Diagn Microbiol Infect Dis.* 2009; **65:** 312-8.
153. George S, Al-Waheeb S. *Aspergillus* fungal arteritis causing vascular anastomotic rupture and loss of commercially transplanted kidney. *Indian J Pathol Microbiol.* 2014; **57:** 643-5.
154. Georgiadou SP, Kontoyiannis DP. The impact of azole resistance on aspergillosis guidelines. *Ann N Y Acad Sci.* 2012; **1272:** 15-22.
155. Georgiadou SP, Sipsas NV, Marom EM, *et al.* The diagnostic value of halo and reversed halo signs for invasive mold infections in compromised hosts. *Clin Infect Dis.* 2011; **52:** 1144-55.
156. Gheith S, Saghrouni F, Bannour W, *et al.* Characteristics of invasive aspergillosis in neutropenic haematology patients (Sousse, Tunisia). *Mycopathologia.* 2014; **177:** 281-9.
157. Ghosh GC, Chawla MP, Chatterjee K, *et al.* Chronic necrotising aspergillosis presenting as secondary spontaneous pneumothorax in an immunocompetent patient. *J Assoc Physicians India.* 2014; **62:** 44-5.
158. Ghosh I, Raina V, Kumar L, *et al.* Serum galactomannan assay for diagnosis of probable invasive aspergillosis in acute leukemia and hematopoietic stem cell transplantation. *Indian J Med Paediatr Oncol.* 2013; **34:** 74-9.
159. Gjeorgjievski M, Amin MB, Cappell MS. Characteristic clinical features of *Aspergillus* appendicitis: Case report and literature review. *World J Gastroenterol.* 2015; **21:** 12713-21.
160. Godet C, Goudet V, Laurent F, *et al.* Nebulised liposomal amphotericin B for *Aspergillus* lung diseases: Case series and literature review. *Mycoses.* 2015; **58:** 173-80.
161. Godet C, Philippe B, Laurent F, *et al.* Chronic pulmonary aspergillosis: An update on diagnosis and treatment. *Respiration.* 2014; **88:** 162-74.
162. Goel MK, Juneja D, Jain SK, *et al.* A rare presentation of *Aspergillus* infection as empyema thoracis. *Lung India.* 2010; **27:** 27-9.
163. Golmia R, Bello I, Marra A, *et al. Aspergillus fumigatus* joint infection: A review. *Semin Arthritis Rheum.* 2011; **40:** 580-4.
164. Gonçalves SS, Stchigel AM, Cano JF, *et al. Aspergillus novoparasiticus:* A new clinical species of the section Flavi. *Med Mycol.* 2012; **50:** 152-60.
165. Grech R. Aspergilloma. *N Engl J Med.* 2010; **362:** 1030.
166. Gregg KS, Kauffman CA. Invasive aspergillosis: Epidemiology, clinical aspects and treatment. *Semin Respir Crit Care Med.* 2015; **36:** 662-72.
167. Guazzelli LS, Severo CB, Hoff LS, *et al. Aspergillus fumigatus* fungus ball in the pleural cavity. *J Bras Pneumol.* 2012; **38:** 125-32.
168. Guazzelli LS, Unis G, Xavier MO, *et al.* Fungus ball in HIV-infected patients. *Rev Inst Med Trop Sao Paulo.* 2009; **51:** 345-8.
169. Guinea J, Bouza E. Current challenges in the microbiological diagnosis of invasive aspergillosis. *Mycopathologia.* 2014; **178:** 403-16.
170. Guinea J, Sandoval-Denis M, Escribano P, *et al. Aspergillus citrinoterreus*, a new species of section *Terrei* isolated from samples of patients with non-hematological predisposing conditions. *J Clin Microbiol.* 2015; **53:** 611-7.
171. Guinea J, Torres-Narbona M, Gijon P, *et al.* Pulmonary aspergillosis in patients with chronic obstructive pulmonary disease: Incidence, risk factors and outcome. *Clin Microbiol Infect.* 2010; **16:** 870-7.
172. Guitard J, Sendid B, Thorez S, *et al.* Evaluation of a recombinant antigen-based enzyme immunoassay for the diagnosis of non-invasive aspergillosis. *J Clin Microbiol.* 2012; **50:** 762-5.
173. Gupta A, Khaira A, Lata S, *et al.* Primary cutaneous aspergillosis in renal transplant recipient. *Saudi J Kidney Dis Transpl.* 2009; **20:** 848-9.
174. Gupta K, Gupta P, Mathew JL, *et al.* Fatal disseminated *Aspergillus penicillioides* infection in a three-month-old infant with suspected cystic fibrosis: Autopsy case with review of literature. *Pediatr Dev Pathol.* 2016; **19:** 506-11.
175. Gupta KL, Rajaram KG, Joshi K, *et al.* Progression of hepatic aspergillosis following second renal transplantation in a patient with recurrent glomerulonephritis. *Indian J Pathol Microbiol.* 2012; **55:** 580-2.
176. Hachem R, Gomes MZ, El Helou G, *et al.* Invasive aspergillosis caused by *Aspergillus terreus:* An emerging opportunistic infection with poor outcome independent of azole therapy. *J Antimicrob Chemother.* 2014; **69:** 3148-55.
177. Hadrich I, Makni F, Neji S, *et al.* Invasive aspergillosis: Resistance to antifungal drugs. *Mycopathologia.* 2012; **174:** 131-41.
178. Haiduven D. Nosocomial aspergillosis and building construction. *Med Mycol.* 2009; **47:** S210-6.
179. Hall WA. Improving outcomes in cerebral aspergillosis. *World Neurosurg.* 2014; **82:** e139-40.
180. Han SB, Kim SK, Lee JW, *et al.* Serum galactomannan index for early prediction of mortality in immunocompromised children with invasive pulmonary aspergillosis. *BMC Infect Dis.* 2015; **15:** 271.
181. Haranaga S, Hirai J, Karimata Y, *et al.* Haloed centrilobular sign: Early diagnosis of airway-invasive aspergillosis. *Intern Med.* 2014; **53:** 2259-60.
182. Heng SC, Morrissey O, Chen SC, *et al.* Utility of bronchoalveolar lavage fluid galactomannan alone or in combination with PCR for the diagnosis of invasive aspergillosis in adult hematology patients: a systematic review and meta-analysis. *Crit Rev Microbiol.* 2015; **41:** 124-34.

183. Henriet SS, Verweij PE, Warris A. *Aspergillus nidulans* and chronic granulomatous disease: A unique host-pathogen interaction. *J Infect Dis.* 2012; **206:** 1128-37.
184. Heylen L, Maertens J, Naesens M, *et al.* Invasive aspergillosis after kidney transplant: Case-control study. *Clin Infect Dis.* 2015; **60:** 1505-11.
185. Hoenigl M, Koidl C, Duettmann W, *et al.* Bronchoalveolar lavage lateral-flow device test for invasive pulmonary aspergillosis diagnosis in haematological malignancy and solid organ transplant patients. *J Infect.* 2012; **65:** 588-91.
186. Hoenigl M, Prattes J, Spiess B, *et al.* Performance of galactomannan, beta-D-glucan, *Aspergillus* lateral-flow device, conventional culture and PCR tests with bronchoalveolar lavage fluid for diagnosis of invasive pulmonary aspergillosis. *J Clin Microbiol.* 2014; **52:** 2039-45.
187. Hofman V, Dhouibi A, Butori C, *et al.* Usefulness of molecular biology performed with formaldehyde-fixed paraffin embedded tissue for the diagnosis of combined pulmonary invasive mucormycosis and aspergillosis in an immunocompromised patient. *Diagn Pathol.* 2010; **5:** 1.
188. Horn BW, Moore GG, Carbone I. Sexual reproduction in *Aspergillus flavus. Mycologia.* 2009; **101:** 423-9.
189. Horn D, Sae-Tia S, Neofytos D. *Aspergillus* osteomyelitis: Review of 12 cases identified by the Prospective Antifungal Therapy Alliance registry. *Diagn Microbiol Infect Dis.* 2009; **63:** 384-7.
190. Howard SJ. Multi-resistant aspergillosis due to cryptic species. *Mycopathologia.* 2014; **178:** 435-9.
191. Hoyo I, Sanclemente G, de la Bellacasa JP, *et al.* Epidemiology, clinical characteristics, and outcome of invasive aspergillosis in renal transplant patients. *Transpl Infect Dis.* 2014; **16:** 951-7.
192. Huang G, Liu Q, Ye X, *et al.* Invasive pulmonary aspergillosis: A rare complication after microwave ablation. *Int J Hyperthermia.* 2014; **30:** 412-7.
193. Hubka V, Kubatova A, Mallatova N, *et al.* Rare and new etiological agents revealed among 178 clinical *Aspergillus* strains obtained from Czech patients and characterized by molecular sequencing. *Med Mycol.* 2012; **50:** 601-10.
194. Hubka V, Lyskova P, Frisvad JC, *et al. Aspergillus pragensis* sp. nov. discovered during molecular re-identification of clinical isolates belonging to *Aspergillus* section *Candidi. Med Mycol.* 2014; **52:** 565-76.
195. Iatta R, Nuccio F, Immediato D, *et al.* Species distribution and *in vitro* azole susceptibility of *Aspergillus* section *Nigri* isolates from clinical and environmental settings. *J Clin Microbiol.* 2016; **54:** 2365-72.
196. Indramohan P, Rohit A, Kanchanamala M, *et al.* Culture-negative *Aspergillus* peritonitis diagnosed by peritoneal biopsy. *Perit Dial Int.* 2013; **33:** 464-5.
197. Iqbal N, Irfan M, Zubairi AB, *et al.* Clinical manifestations and outcomes of pulmonary aspergillosis: Experience from Pakistan. *BMJ Open Respir Res.* 2016; **3:** e000155.
198. Izumikawa K, Takazono T, Kohno S. Chronic *Aspergillus* infections of the respiratory tract: Diagnosis, management and antifungal resistance. *Curr Opin Infect Dis.* 2010; **23:** 584-9.
199. Izumikawa K, Tashiro M, Kohno S. Management of chronic pulmonary aspergillosis. *Ann N Y Acad Sci.* 2012; **1272:** 40-8.
200. Izumikawa K, Tashiro T, Tashiro M, *et al.* Pathogenesis and clinical features of chronic pulmonary aspergillosis - Is it possible to distinguish CNPA and CCPA clinically? *J Infect Chemother.* 2014; **20:** 208-12.
201. Izumikawa K, Yamamoto Y, Mihara T, *et al.* Bronchoalveolar lavage galactomannan for the diagnosis of chronic pulmonary aspergillosis. *Med Mycol.* 2012; **50:** 811-7.
202. Izumikawa K. Recent advances in chronic pulmonary aspergillosis. *Respir Investig.* 2016; **54:** 85-91.
203. Jain S, Kapoor G. Invasive aspergillosis in children with acute leukemia at a resource-limited oncology center. *J Pediatr Hematol Oncol.* 2015; **37:** e1-5.
204. Jenks JD, Preziosi M. A challenging case of invasive pulmonary aspergillosis after near-drowning: A case report and literature review. *Infect Dis Clin Pract.* 2015; **23:** 227-30.
205. Jensen J, Guinea J, Torres-Narbona M, *et al.* Post-surgical invasive aspergillosis: An uncommon and under-appreciated entity. *J Infect.* 2010; **60:** 162-7.
206. Jha AK, Bansal D, Chakrabarti A, *et al.* Serum galactomannan assay for the diagnosis of invasive aspergillosis in children with haematological malignancies. *Mycoses.* 2013; **56:** 442-8.
207. Jhun BW, Jeon K, Eom JS, *et al.* Clinical characteristics and treatment outcomes of chronic pulmonary aspergillosis. *Med Mycol.* 2013; **51:** 811-7.
208. Jin E, Wang LM, Li QY, *et al.* Chronic necrotizing pulmonary aspergillosis in an immunocompetent patient: Report of a rare case. *Infection.* 2014; **42:** 565-8.
209. Johnson GL, Sarker SJ, Nannini F, *et al. Aspergillus*-specific lateral-flow device and real-time PCR testing of bronchoalveolar lavage fluid: A combination biomarker approach for clinical diagnosis of invasive pulmonary aspergillosis. *J Clin Microbiol.* 2015; **53:** 2103-8.
210. Jones AM, Horsley A, Denning DW. What is the importance of classifying *Aspergillus* disease in cystic fibrosis patients? *Expert Rev Respir Med.* 2014; **8:** 389-92.
211. Jung J, Kim MY, Lee HJ, *et al.* Comparison of computed tomographic findings in pulmonary mucormycosis and invasive pulmonary aspergillosis. *Clin Microbiol Infect.* 2015; **21:** 684.e11-8.
212. Jung SW, Kim MW, Cho SK, *et al.* A case of endobronchial aspergilloma associated with foreign body in immunocompetent patient without underlying lung disease. *Tuberc Respir Dis (Seoul).* 2013; **74:** 231-4.
213. Kadota J. Update and recent advances on the management of invasive and chronic pulmonary aspergillosis. *Respir Investig.* 2016; **54:** 75.
214. Kalokhe AS, Rouphael N, El Chami MF, *et al. Aspergillus* endocarditis: A review of the literature. *Int J Infect Dis.* 2010; **14:** e1040-7.
215. Kane S, Pinto JM, Dadzie CK, *et al.* Aspergilloma caused by *Aspergillus versicolor. Pediatr Infect Dis J.* 2014; **33:** 891.
216. Kantartzis SN, Dacic S, Strollo DC. AIRP best cases in radiologic-pathologic correlation: Pulmonary sarcoidosis

complicated by aspergilloma formation. *Radiographics.* 2012; **32:** 469-73.

217. Karthik RK, Sudarsanam TD. An unusual cause of empyema thoracis. *Indian J Med Sci.* 2009; **63:** 30-2.
218. Kasliwal MK, Agrawal D, Sharma BS. Frontal intra-axial mass lesion in an adult. Invasive aspergillosis. *J Clin Neurosci.* 2011; **18:** 1230 & 1283.
219. Kathuria S, Sharma C, Singh PK, *et al.* Molecular epidemiology and *in vitro* antifungal susceptibility of *Aspergillus terreus* species complex isolates in Delhi, India: Evidence of genetic diversity by amplified fragment length polymorphism and microsatellite typing. *PLoS One.* 2015; **10:** e0118997.
220. Kaur R, Mehra B, Dhakad MS, *et al.* Pulmonary aspergillosis as opportunistic mycoses in a cohort of human immunodeficiency virus-infected patients: Report from a tertiary care hospital in North India. *Int J Health Sci (Qassim).* 2017; **11:** 45-50.
221. Kaur S, Singh S. Biofilm formation by *Aspergillus fumigatus. Med Mycol.* 2014; **52:** 2-9.
222. Kawakami Y, Tagami T, Kusakabe T, *et al.* Disseminated aspergillosis associated with tsunami lung. *Respir Care.* 2012; **57:** 1674-8.
223. Khan MA, Dar AM, Kawoosa NU, *et al.* Clinical profile and surgical outcome for pulmonary aspergilloma: Nine year retrospective observational study in a tertiary care hospital. *Int J Surg.* 2011; **9:** 267-71.
224. Khandelwal A, Basheer N, Mahapatra AK. Multiple intracranial fungal abscesses in an immunocompetent infant treated surgically. *J Pediatr Neurosci.* 2014; **9:** 159-61.
225. Khurana U, Raghubanshi G, Handa U, Chander J. "Fruiting bodies" of *Aspergillus flavus:* A rare finding in histopathology. *Int J Surg Pathol.* 2012; **20:** 60-1.
226. Kim da S, Jeong JS, Kim SR, *et al.* A case report of mass-forming *Aspergillus* tracheobronchitis successfully treated with voriconazole. *Medicine (Baltimore).* 2015; **94:** e1434.
227. Kim HW, Moon MH, Lee JW. Myocardial abscess, a rare form of cardiac aspergillosis. *Arch Cardiovasc Dis.* 2014; **107:** 415-7.
228. Kim JY, Yoo JW, Oh M, *et al.* ^{18}F-fluoro-2-deoxy-D-glucose positron emission tomography/computed tomography findings are different between invasive and non-invasive pulmonary aspergillosis. *J Comput Assist Tomogr.* 2013; **37:** 596-601.
229. Kim SH, Kim MY, Hong SI, *et al.* Invasive pulmonary aspergillosis-mimicking tuberculosis. *Clin Infect Dis.* 2015; **61:** 9-17.
230. Kimura S, Akahoshi Y, Nakano H, *et al.* False-positive *Aspergillus* galactomannan and its kinetics in allogeneic hematopoietic stem cell transplantation. *J Infect.* 2015; **70:** 520-40.
231. Koehler P, Tacke D, Cornely OA. Aspergillosis of bones and joints — A review from 2002 until today. *Mycoses.* 2014; **57:** 323-35.
232. Kolwijck E, van de Veerdonk FL. The potential impact of the pulmonary microbiome on immunopathogenesis of *Aspergillus*-related lung disease. *Eur J Immunol.* 2014; **44:** 3156-65.
233. Kontoyiannis DP, Mathur M, Chen YB, *et al.* A 41-year-old man with fever and abdominal pain after stem-cell transplantation. *N Engl J Med.* 2014; **370:** 1637-46.
234. Kose S, Cavdar G, Senger SS, *et al.* Central nervous system aspergillosis in an immunocompetent patient. *J Infect Dev Ctries.* 2011; **5:** 313-5.
235. Kosmidis C, Denning DW. The clinical spectrum of pulmonary aspergillosis. *Thorax.* 2015; **70:** 270-7.
236. Kosmidis C, Muldoon EG. Challenges in the management of chronic pulmonary aspergillosis. *Med Mycol.* 2017; **55:** 63-8.
237. Kothari A, Pillai BS, Bhan A. Pacing lead endocarditis due to *Aspergillus fumigatus. Indian J Med Microbiol.* 2010; **28:** 72-3.
238. Kotwal A, Biswas D, Kakati B, *et al. Aspergillus nidulans* causing primary cutaneous aspergillosis in an immunocompetent patient. *Cutis.* 2015; **95:** e1-3.
239. Koulenti D, Garnacho-Montero J, Blot S. Approach to invasive pulmonary aspergillosis in critically ill patients. *Curr Opin Infect Dis.* 2014; **27:** 174-83.
240. Koulenti D, Vogelaers D, Blot S. What's new in invasive pulmonary aspergillosis in the critically ill. *Intensive Care Med.* 2014; **40:** 723-6.
241. Kourkoumpetis TK, Desalermos A, Muhammed M, *et al.* Central nervous system aspergillosis: A series of 14 cases from a general hospital and review of 123 cases from the literature. *Medicine (Baltimore).* 2012; **91:** 328-36.
242. Kousha M, Tadi R, Soubani AO. Pulmonary aspergillosis: A clinical review. *Eur Respir Rev.* 2011; **20:** 156-74.
243. Krel M, Petraitis V, Petraitiene R, *et al.* Host biomarkers of invasive pulmonary aspergillosis to monitor therapeutic response. *Antimicrob Agents Chemother.* 2014; **58:** 3373-8.
244. Krishnan S, Manavathu EK, Chandrasekar PH. *Aspergillus flavus:* An emerging non-*fumigatus Aspergillus* species of significance. *Mycoses.* 2009; **52:** 206-22.
245. Kumar J, Seith A, Kumar A, *et al.* Chest wall and mediastinal nodal aspergillosis in an immunocompetent host. *Diagn Interv Radiol.* 2009; **15:** 176-8.
246. Kumar N, Mishra M, Singhal A, *et al.* Aspergilloma coexisting with idiopathic pulmonary fibrosis: A rare occurrence. *J Postgrad Med.* 2013; **59:** 145-8.
247. Kurosaki F, Bando M, Nakayama M, *et al.* Clinical features of pulmonary aspergillosis associated with interstitial pneumonia. *Intern Med.* 2014; **53:** 1299-306.
248. Kwon JW, Hong SH, Choi SH, *et al.* MRI findings of *Aspergillus* spondylitis. *AJR Am J Roentgenol.* 2011; **197:** W919-23.
249. Kwon-Chung KJ, Sugui JA. *Aspergillus fumigatus* - What makes the species a ubiquitous human fungal pathogen? *PLoS Pathog.* 2013; **9:** e1003743.
250. Kupeli E, Ulubay G, Bayram Akkurt S, *et al.* Invasive pulmonary aspergillosis in heart transplant recipients. *Exp Clin Transplant.* 2015; **13 (Suppl. 1):** 352-5.
251. Lackner M, Coassin S, Haun M, *et al.* Geographically predominant genotypes of *Aspergillus terreus* species complex in Austria: A microsatellite typing study. *Clin Microbiol Infect.* 2016; **22:** 270-6.

252. Lackner M, Lass-Florl C. Up-date on diagnostic strategies of invasive aspergillosis. *Curr Pharm Des.* 2013; **19:** 3595-614.
253. Lamoth F, Juvvadi PR, Steinbach WJ. Advances in *Aspergillus fumigatus* pathobiology. *Front Microbiol.* 2016; **7:** 43.
254. Lanzarin LD, Mariano LC, Macedo MC, *et al.* Conidial heads (fruiting bodies) as a hallmark for histopathological diagnosis of angioinvasive aspergillosis. *Autops Case Rep.* 2015; **5:** 9-18.
255. Lass-Florl C, Roilides E, Loffler J, *et al.* Mini-review: Host defence in invasive aspergillosis. *Mycoses.* 2013; **56:** 403-13.
256. Lass-Florl C. *Aspergillus terreus:* How inoculum size and host characteristics affect its virulence. *J Infect Dis.* 2012; **205:** 1192-4.
257. Lass-Florl C. Susceptibility testing in *Aspergillus* species complex. *Clin Microbiol Infect.* 2014; **20 (Suppl. 6):** 49-53.
258. Latge JP. *Aspergillus fumigatus* and aspergillosis. *Clin Microbiol Rev.* 1999; **12:** 310-50.
259. Lazarescu RE, Vinelli M. Dissemination of invasive aspergillosis: Diagnostic and management dilemmas. *BMJ Case Rep.* 2014; pii: bcr2014204642.
260. Lazaro M, Ramos A, Ussetti P, *et al. Aspergillus* endocarditis in lung transplant recipients: Case report and literature review. *Transpl Infect Dis.* 2011; **13:** 186-91.
261. Lazovic B, Stajic Z, Putnikovic B. Pulmonary aspergilloma. *Med Arch.* 2012; **66:** 420-2.
262. Le X, Jain P, O'Brien S. *Aspergillus* pseudomembranous necrotizing tracheitis. *Am J Hematol.* 2013; **88:** 242.
263. Leeflang MM, Debets-Ossenkopp YJ, Wang J, *et al.* Galactomannan detection for invasive aspergillosis in immunocompromised patients. *Cochrane Database Syst Rev.* 2015; **12:** CD007394.
264. Lestrade PP, Meis JF, Arends JP, *et al.* Diagnosis and management of aspergillosis in the Netherlands: A national survey. *Mycoses.* 2016; **59:** 101-7.
265. Lin YC, Chen SC. Cavity with ball-in-hole lesion in the lung. *Eur J Intern Med.* 2016; **31:** e5-6.
266. Liss B, Vehreschild JJ, Bangard C, *et al.* Our 2015 approach to invasive pulmonary aspergillosis. *Mycoses.* 2015; **58:** 375-82.
267. Loeffler J, Hafner J, Mengoli C, *et al.* Prospective biomarker screening for diagnosis of invasive aspergillosis in high-risk pediatric patients. *J Clin Microbiol.* 2016; **55:** 101-109.
268. Logotheti M, Kotsovili-Tseleni A, Arsenis G, *et al.* Multiplex PCR for the discrimination of *A.fumigatus, A.flavus, A.niger* and *A.terreus. J Microbiol Methods.* 2009; **76:** 209-11.
269. Lokuhetty MD, Wijesinghe HD, Weerasundera B, *et al.* Iatrogenic *Aspergillus* infection of the central nervous system in a pregnant woman. *Indian J Pathol Microbiol.* 2009; **52:** 427-9.
270. Lopez-Medrano F, Silva JT, Fernandez-Ruiz M, *et al.* Risk factors associated with early invasive pulmonary aspergillosis in kidney transplant recipients: Results from a multinational matched case-control study. *Am J Transplant.* 2016; **16:** 2148-57.
271. Lyon GM, Abdul-Ali D, Loeffler J, *et al.* Development and evaluation of a calibrator material for nucleic acid-based assays for diagnosing aspergillosis. *J Clin Microbiol.* 2013; **51:** 2403-5.
272. Macesic N, Morrissey CO, Liew D, *et al.* Is a biomarker-based diagnostic strategy for invasive aspergillosis cost effective in high-risk haematology patients? *Med Mycol.* 2017; pii: myw141. PMID: 28131991.
273. Maheshwari V, Varshney M, Alam K, *et al.* Aspergilloma lung mimicking tuberculosis. *BMJ Case Rep.* 2011; pii: bcr0420114051.
274. Mansh M, Binstock M, Williams K, *et al.* Voriconazole exposure and risk of cutaneous squamous cell carcinoma, *Aspergillus* colonization, invasive aspergillosis and death in lung transplant recipients. *Am J Transplant.* 2016; **16:** 262-70.
275. Marchiori E, Godoy MC, Zanetti G, *et al.* The reversed halo sign. Another CT finding useful for distinguish invasive pulmonary aspergillosis and pulmonary lymphoma. *Eur J Radiol.* 2011; **79:** e96-7.
276. Marchiori E, Guimaraes MD, Hochhegger B, *et al.* Importance of the reversed halo sign for the diagnosis of angioinvasive pulmonary aspergillosis. *Respir Med.* 2014; **108:** 1238-9.
277. Marchiori E, Zanetti G, Hochhegger B, *et al.* Reversed halo sign on computed tomography: State-of-the-art review. *Lung.* 2012; **190:** 389-94.
278. Marr KA, Schlamm HT, Herbrecht R, *et al.* Combination antifungal therapy for invasive aspergillosis: A randomized trial. *Ann Intern Med.* 2015; **162:** 81-9.
279. Martin-Pena A, Aguilar-Guisado M, Espigado I, *et al.* Antifungal combination therapy for invasive aspergillosis. *Clin Infect Dis.* 2014; **59:** 1437-45.
280. Martinez-Giron R, Martinez-Torre S, Mosquera-Martinez J. Bladder aspergillosis detected by urine cytology. *Diagn Cytopathol.* 2015; **43:** 403-6.
281. Martinez-Herrera EO, Frias De-Leon MG, Duarte-Escalante E, *et al.* Fungal diversity and *Aspergillus* species in hospital environments. *Ann Agric Environ Med.* 2016; **23:** 264-9.
282. Marui S, de Lima Pereira AC, *et al.* Suppurative thyroiditis due to aspergillosis: A case report. *J Med Case Rep.* 2014; **8:** 379.
283. Marzolf G, Sabou M, Lannes B, *et al.* Magnetic resonance imaging of cerebral aspergillosis: imaging and pathological correlations. *PLoS One.* 2016; **11:** e0152475.
284. Masih A, Singh PK, Kathuria S, *et al.* Identification by molecular methods and matrix-assisted laser desorption ionization-time of flight mass spectrometry and antifungal susceptibility profiles of clinically significant rare *Aspergillus* species in a referral chest hospital in Delhi, India. *J Clin Microbiol.* 2016; **54:** 2354-64.
285. Maturu VN, Agarwal R. Itraconazole in chronic pulmonary aspergillosis: In whom, for how long and at what dose? *Lung India.* 2015; **32:** 309-12.
286. McCormick A, Loeffler J, Ebel F. *Aspergillus fumigatus:* Contours of an opportunistic human pathogen. *Cell Microbiol.* 2010; **12:** 1535-43.
287. Meis JF, Chowdhary A, Rhodes JL, *et al.* Clinical implications of globally emerging azole resistance in *Aspergillus*

fumigatus. Philos Trans R Soc Lond B Biol Sci. 2016; **371 (1709)**. pii: 20150460. PMID: 28080986.

288. Meng XC, Jiang T, Yi SH, *et al.* Renal aspergillosis after liver transplantation: Clinical and imaging manifestations in two cases. *World J Gastroenterol.* 2014; **20:** 18495-502.
289. Miceli MH, Goggins MI, Chander P, *et al.* Performance of lateral flow device and galactomannan for the detection of *Aspergillus* species in bronchoalveolar fluid of patients at risk for invasive pulmonary aspergillosis. *Mycoses.* 2015; **58:** 368-74.
290. Misch EA, Safdar N. Updated guidelines for the diagnosis and management of aspergillosis. *J Thorac Dis.* 2016; **8:** e1771-6.
291. Mohammed AP, Dhunputh P, Chiluka R, *et al.* An unusual case of invasive aspergillosis in an immunocompetent individual. *BMJ Case Rep.* 2015; pii: bcr2015210381.
292. Mohapatra S, Xess I, Swetha JV, *et al.* Primary cutaneous aspergillosis due to *Aspergillus niger* in an immunocompetent patient. *Indian J Med Microbiol.* 2009; **27:** 367-70.
293. Mokaddas E, Burhamah MH, Ahmad S, *et al.* Invasive pulmonary aspergillosis due to *Aspergillus terreus:* Value of DNA, galactomannan and (1->3)-beta-D-glucan detection in serum samples as an adjunct to diagnosis. *J Med Microbiol.* 2010; **59:** 1519-23.
294. Mora D, Barba P, Ruiz I, *et al.* Primary gastrointestinal aspergillosis six months after allogeneic hematopoietic cell transplantation: A case report. *Transpl Infect Dis.* 2013; **15:** e107-10.
295. Munta K, Gopal PB, Vigg A. Invasive aspergillosis in near-drowning non-neutropenic patient. *Indian J Crit Care Med.* 2015; **19:** 739-42.
296. Muthu V, Agarwal R. A report of a successfully treated case of ABPA in a HIV-infected individual. *BMJ Case Rep.* 2014; pii: bcr2014206236.
297. Mutschlechner W, Grif K, Blum G, *et al.* Rep-PCR and RAPD-PCR fingerprinting of *Aspergillus terreus. Med Mycol.* 2013; **51:** 876-9.
298. Najafi N, Shokohi T, Basiri A, *et al. Aspergillus terreus*-related ureteral obstruction in a diabetic patient. *Iran J Kidney Dis.* 2013; **7:** 151-5.
299. Nakamoto K, Takayanagi N, Kanauchi T, *et al.* Prognostic factors in 194 patients with chronic necrotizing pulmonary aspergillosis. *Intern Med.* 2013; **52:** 727-34.
300. Nandi S, Santra A, Ghoshal L, *et al.* Interstitial lung disease in systemic scleroderma, complicated with bilateral pulmonary aspergilloma: An unusual association. *J Clin Diagn Res.* 2015; **9:** OD11-3.
301. Nasa P, Yadav R, Nagrani SK, *et al.* Invasive pulmonary aspergillosis in an immunocompetent patient with severe dengue fever. *Indian J Crit Care Med.* 2014; **18:** 323-5.
302. Nasim A, Baqi S, Zeeshan SM, *et al.* Chronic necrotizing pulmonary aspergillosis in a renal transplant recipient. *J Pak Med Assoc.* 2011; **61:** 1242-4.
303. Nath A, Agarwal R, Singh A, *et al.* An unusual cause of hoarseness in an elderly asthmatic. *Mycoses.* 2010; **53:** 544-6.
304. Nedel WL, Pasqualotto AC. Treatment of infections by cryptic *Aspergillus* species. *Mycopathologia.* 2014; **178:** 441-5.
305. Negri CE, Goncalves SS, Xafranski H, *et al.* Cryptic and rare *Aspergillus* species in Brazil: Prevalence in clinical samples and *in vitro* susceptibility to triazoles. *J Clin Microbiol.* 2014; **52:** 3633-40.
306. Neofytos D, Railkar R, Mullane KM, *et al.* Correlation between circulating fungal biomarkers and clinical outcome in invasive aspergillosis. *PLoS One.* 2015; **10:** e0129022.
307. Nguyen J, Manera R, Minutti C. *Aspergillus* thyroiditis: A review of the literature to highlight clinical challenges. *Eur J Clin Microbiol Infect Dis.* 2012; **31:** 3259-64.
308. Nicolle A, de la Blanchardiere A, Bonhomme J, *et al. Aspergillus* vertebral osteomyelitis in immunocompetent subjects: Case report and review of the literature. *Infection.* 2013; **41:** 833-40.
309. Nicolle MC, Benet T, Vanhems P. Aspergillosis: Nosocomial or community-acquired? *Med Mycol.* 2011; **49:** S24-9.
310. Nitschke A, Sachs P, Suby-Long T, *et al.* Monod sign. *J Thorac Imaging.* 2013; **28:** W120.
311. Nucci M, Anaissie E. Fungal infections in hematopoietic stem cell transplantation and solid-organ transplantation - Focus on aspergillosis. *Clin Chest Med.* 2009; **30:** 295-306.
312. O'Gorman CM, Fuller H, Dyer PS. Discovery of a sexual cycle in the opportunistic fungal pathogen *Aspergillus fumigatus. Nature.* 2009; **457:** 471-4.
313. Oda M, Saraya T, Wakayama M, *et al.* Calcium oxalate crystal deposition in a patient with aspergilloma due to *Aspergillus niger. J Thorac Dis.* 2013; **5:** e174-8.
314. Ohara S, Tazawa Y, Tanai C, *et al.* Clinical characteristics of patients with *Aspergillus* species isolation from respiratory samples: Comparison of chronic pulmonary aspergillosis and colonization. *Respir Investig.* 2016; **54:** 92-7.
315. Oliva A, Flori P, Hennequin C, *et al.* Evaluation of the *Aspergillus* Western blot IgG kit for diagnosis of chronic aspergillosis. *J Clin Microbiol.* 2015; **53:** 248-54.
316. Oz Y, Aslan M, Aksit F, *et al.* The effect of clinical characteristics on the performance of galactomannan and PCR for the diagnosis of invasive aspergillosis in febrile neutropenic patients. *Mycoses.* 2016; **59:** 86-92.
317. Page ID, Richardson M, Denning DW. Antibody testing in aspergillosis – *quo vadis*? *Med Mycol.* 2015; **53:** 417-39.
318. Page ID, Richardson MD, Denning DW. Comparison of six *Aspergillus*-specific IgG assays for the diagnosis of chronic pulmonary aspergillosis (CPA). *J Infect.* 2016; **72:** 240-9.
319. Palmero ML, Pope E, Brophy J. Sporotrichoid aspergillosis in an immunocompromised child: A case report and review of the literature. *Pediatr Dermatol.* 2009; **26:** 592-6.
320. Pan Z, Fu M, Zhang J, *et al.* Diagnostic accuracy of a novel lateral-flow device in invasive aspergillosis: A meta-analysis. *J Med Microbiol.* 2015; **64:** 702-7.
321. Panackal AA, Bennett JE, Williamson PR. Treatment options in invasive aspergillosis. *Curr Treat Options Infect Dis.* 2014; **6:** 309-25.
322. Panchabhai TS, Bandyopadhyay D, Alraiyes AH, *et al.* A 60-year-old woman with cough, dyspnea, and atelectasis 19 years after liver transplant. Chest. 2015; **148:** e122-5.
323. Panda PK, Mavidi SK, Wig N, *et al.* Intracranial aspergillosis in an immunocompetent young woman. *Mycopathologia.* 2017; **182:** 527-38.

324. Pandey P, Dixit AK, Tanwar A, *et al.* Pulmonary echinococcal cyst with a filamentous fungus co-infection. *Ghana Med J.* 2013; **47:** 148-52.
325. Park SH, Choi SM, Lee DG, *et al.* Serum galactomannan strongly correlates with outcome of invasive aspergillosis in acute leukaemia patients. *Mycoses.* 2011; **54:** 523-30.
326. Pasqualotto AC. Differences in pathogenicity and clinical syndromes due to *Aspergillus fumigatus* and *Aspergillus flavus. Med Mycol.* 2009: **47:** S261-70.
327. Passera E, Rizzi A, Robustellini M, *et al.* Pulmonary aspergilloma: Clinical aspects and surgical treatment outcome. *Thorac Surg Clin.* 2012; **22:** 345-61.
328. Pastor FJ, Guarro J. Treatment of *Aspergillus terreus* infections: A clinical problem not yet resolved. *Int J Antimicrob Agents.* 2014; **44:** 281-9.
329. Patel N, Saleh A, Raoof S, *et al.* Tracheobronchial manifestations of aspergillosis. *J Bronchology Interv Pulmonol.* 2010; **17:** 45-53.
330. Patiroglu T, Unal E, Karakukcu M, *et al.* Multiple fungal brain abscesses in a child with acute lymphoblastic leukemia. *Mycopathologia.* 2012; **174:** 505-9.
331. Patterson KC, Strek ME. Diagnosis and treatment of pulmonary aspergillosis syndromes. *Chest.* 2014; **146:** 1358-68.
332. Patterson TF, Thompson GR 3rd, Denning DW, *et al.* Executive Summary: Practice guidelines for the diagnosis and management of aspergillosis: 2016 Update by the IDSA. *Clin Infect Dis.* 2016; **63:** 433-42.
333. Patterson TF, Thompson GR 3rd, Denning DW, *et al.* Practice guidelines for the diagnosis and management of aspergillosis: 2016 Update by the IDSA. *Clin Infect Dis.* 2016; **63:** e1-60.
334. Paul S, Singh V, Sankhwar S, *et al.* Renal aspergillosis secondary to renal instrumentation in immunocompetent patient. *BMJ Case Rep.* 2013; pii: bcr2013200306.
335. Peghin M, Monforte V, Martin-Gomez MT, *et al.* 10 years of prophylaxis with nebulized liposomal amphotericin B and the changing epidemiology of *Aspergillus* spp. infection in lung transplantation. *Transpl Int.* 2016; **29:** 51-62.
336. Pendleton M, Denning DW. Multifocal pulmonary aspergillomas: Case series and review. *Ann N Y Acad Sci.* 2012; **1272:** 58-67.
337. Perlin DS, Zhao Y. Molecular diagnostic platforms for detecting *Aspergillus. Med Mycol.* 2009; **47:** S223-32.
338. Pettit AC, Kropski JA, Castilho JL, *et al.* The index case for the fungal meningitis outbreak in the United States. *N Engl J Med.* 2012; **367:** 2119-25.
339. Phute SU, Bhakre JB. *Aspergillus fumigatus* meningitis in a preterm. *Indian Pediatr.* 2010; **47:** 980.
340. Pillai J, Mubeen M, Chaudhari A, *et al. Aspergillus* pacemaker lead endocarditis. *Asian Cardiovasc Thorac Ann.* 2013; **21:** 211-4.
341. Pradhan B, Jindal A. Invasive pulmonary aspergillosis in an immunocompetent patient with severe dengue fever. *Indian J Crit Care Med.* 2014; **18:** 637-8.
342. Prakash G, Thulkar S, Arava SK, *et al.* Cerebral *Aspergillus* infection in pediatric acute lymphoblastic leukemia induction therapy. *Indian J Med Paediatr Oncol.* 2012; **33:** 236-8.
343. Prasad A, Agarwal K, Deepak D, *et al.* Pulmonary aspergillosis: What CT can offer before it is too late! *J Clin Diagn Res.* 2016; **10:** TE01-5.
344. Prasad R, Garg SR. Progressive increase in cavitation with the evolution of fungus ball: A clue to the diagnosis of chronic necrotizing pulmonary aspergillosis. *Lung India.* 2009; **26:** 95-7.
345. Prattes J, Flick H, Pruller F, *et al.* Novel tests for diagnosis of invasive aspergillosis in patients with underlying respiratory diseases. *Am J Respir Crit Care Med.* 2014; **190:** 922-9.
346. Prattes J, Lackner M, Eigl S, *et al.* Diagnostic accuracy of the *Aspergillus*-specific bronchoalveolar lavage lateral-flow assay in haematological malignancy patients. *Mycoses.* 2015; **58:** 461-9.
347. Puri V, Johnson SM, Munfakh NA. Fungal osteomyelitis masquerading as a chest wall tumor. *Ann Thorac Surg.* 2011; **91:** 304.
348. Radotra BD, Salunke P, Parthan G, *et al.* True mycotic aneurysm in a patient with gonadotropinoma after trans-sphenoidal surgery. *Surg Neurol Int.* 2015; **6:** 193.
349. Raj KA, Srinivasamurthy BC, Nagarajan K, *et al.* A rare case of spontaneous *Aspergillus* spondylodiscitis with epidural abscess in a 45-year-old immunocompetent female. *J Craniovertebr Junction Spine.* 2013; **4:** 82-4.
350. Rajalingham S, Anshar FM. Chronic necrotizing pulmonary aspergillosis presenting as bilateral pleural effusion: A case report. *J Med Case Rep.* 2012; **6:** 62.
351. Ramirez M, Castro C, Palomares JC, *et al.* Molecular detection and identification of *Aspergillus* spp. from clinical samples using real-time PCR. *Mycoses.* 2009; **52:** 129-34.
352. Ranjan R, Mishra S, Ranjan S. *Aspergillus* vertebral osteomyelitis in an immunocompetent person. *Neurol India.* 2010; **58:** 806-8.
353. Rao P. *Aspergillus* infection in urinary tract post-ureteric stenting. *Indian J Med Microbiol.* 2015; **33:** 316-8.
354. Ratermann KL, Ereshefsky BJ, Fleishaker EL, *et al.* Fulminant invasive pulmonary aspergillosis after a near-drowning accident in an immunocompetent patient. *Ann Pharmacother.* 2014; **48:** 1225-9.
355. Ray A, Suri JC, Sen MK, *et al.* Cavitating lung disease due to concomitant drug resistant tuberculosis and invasive pulmonary aspergillosis in a post-partum patient: A case report. *Indian J Tuberc.* 2015; **62:** 50-3.
356. Reischies F, Hoenigl M. The role of surgical debridement in different clinical manifestations of invasive aspergillosis. *Mycoses.* 2014; **57 (Suppl. 2):** 1-14.
357. Reischies FM, Raggam RB, Prattes J, *et al.* Urine galactomannan-creatinine ratio for detection of Invasive aspergillosis in patients with hematologic malignancies. *J Clin Microbiol.* 2016; **54:** 771-4.
358. Richardson MD, Page ID. *Aspergillus* serology: Have we arrived yet? *Med Mycol.* 2017; **55:** 48-55.
359. Risslegger B, Zoran T, Lackner M, *et al.* A prospective international *Aspergillus terreus* survey: An EFISG, ISHAM and ECMM joint study. *Clin Microbiol Infect.* 2017; pii: S1198-743X(17)30220-3. PMID: 28412383.

360. Riviere S, Lortholary O, Michon J, *et al.* *Aspergillus* endocarditis in the era of new antifungals: Major role for antigen detection. *J Infect.* 2013; **67:** 85-8.
361. Robinson A, Fien S, Grassi MA. Non-healing scalp wound infected with *Aspergillus niger* in an elderly patient. *Cutis.* 2011; **87:** 197-200.
362. Rofaiel R, Turkistani Y, McCarty D, *et al.* Fungal mobile mass on echocardiogram: Native mitral valve *Aspergillus fumigatus* endocarditis. *BMJ Case Rep.* 2016; pii: bcr2016217281.
363. Rogdo B, Kahlert C, Diener PA, *et al.* Primary cutaneous aspergillosis in a preterm neonate. *BMJ Case Rep.* 2014; pii: bcr2014204752.
364. Rogers TR, Morton CO, Springer J, *et al.* Combined real-time PCR and galactomannan surveillance improves diagnosis of invasive aspergillosis in high risk patients with haematological malignancies. *Br J Haematol.* 2013; **161:** 517-24.
365. Roilides E, Pana ZD. Application of diagnostic markers to invasive aspergillosis in children. *Ann N Y Acad Sci.* 2012; **1272:** 1-8.
366. Rudramurthy SM, Chakrabarti A, Geertsen E, *et al.* *In vitro* activity of isavuconazole against 208 *Aspergillus flavus* isolates in comparison with 7 other antifungal agents: assessment according to the methodology of the European Committee on Antimicrobial Susceptibility Testing. *Diagn Microbiol Infect Dis.* 2011; **71:** 370-7.
367. Russo A, Giuliano S, Vena A, *et al.* Predictors of mortality in non-neutropenic patients with invasive pulmonary aspergillosis: Does galactomannan have a role? *Diagn Microbiol Infect Dis.* 2014; **80:** 83-6.
368. Sadarangani M, Harvey M, McDonald A, *et al.* Brain abscesses due to *Aspergillus nidulans* infection during induction chemotherapy for acute lymphoblastic leukemia. *J Pediatr Hematol Oncol.* 2015; **37:** e384-6.
369. Safdar A. Immunomodulation therapy for invasive aspergillosis: Discussion on myeloid growth factors, recombinant cytokines and antifungal drug immune modulation. *Curr Fungal Infect Rep.* 2010; **4:** 1-7.
370. Sagan D, Gozdziuk K. Surgery for pulmonary aspergilloma in immunocompetent patients: No benefit from adjuvant antifungal pharmacotherapy. *Ann Thorac Surg.* 2010; **89:** 1603-10.
371. Salzer HJ, Heyckendorf J, Kalsdorf B, *et al.* Characterization of patients with chronic pulmonary aspergillosis according to the new ESCMID/ERS/ECMM and IDSA guidelines. *Mycoses.* 2017; **60:** 136-42.
372. Samson RA, Visagie CM, Houbraken J, *et al.* Phylogeny, identification and nomenclature of the genus *Aspergillus.* *Stud Mycol.* 2014; **78:** 141-73.
373. Sandherr M, Maschmeyer G. Pharmacology and metabolism of voriconazole and posaconazole in the treatment of invasive aspergillosis: Review of the literature. *Eur J Med Res.* 2011; **16:** 139-44.
374. Sato FR, Sawazaki R, Berretta D, *et al.* Aspergillosis of the maxillary sinus associated with a zygomatic implant. *J Am Dent Assoc.* 2010; **141:** 1231-5.
375. Savio J, Menon NR, Sudharma AR, *et al.* Galactomannan assay and invasive pulmonary aspergillosis - Comparison of the test performance at an in-house and the kit cut-off. *J Clin Diagn Res.* 2016; **10:** DC01-4.
376. Schweer KE, Bangard C, Hekmat K, *et al.* Chronic pulmonary aspergillosis. *Mycoses.* 2014; **57:** 257-70.
377. Segal BH, Cornely O, Bromley M. Proceedings from the 4th Advances Against Aspergillosis Conference, February 4-6, 2010, Rome, Italy. *Med Mycol.* 2011; **49:** S5-176.
378. Segal BH. Aspergillosis. *N Engl J Med.* 2009; **360:** 1870-84.
389. Segundo JB, da Silva MA, Filho WE, *et al.* Cerebral aspergillosis in a patient with leprosy and diabetes: A case report. *BMC Res Notes.* 2014; **7:** 689.
380. Sethi P, Saluja R, Jindal N, *et al.* Invasive aspergillosis in an immunocompetent host. *J Oral Maxillofac Pathol.* 2012; **16:** 297-300.
381. Seyedmousavi S, Guillot J, Arne P, *et al.* *Aspergillus* and aspergilloses in wild and domestic animals: A global health concern with parallels to human disease. *Med Mycol.* 2015; **53:** 765-97.
382. Shah AA, Hazen KC. Diagnostic accuracy of histopathologic and cytopathologic examination of *Aspergillus* species. *Am J Clin Pathol.* 2013; **139:** 55-61.
383. Shamim MS, Enam SA, Ali R, *et al.* Craniocerebral aspergillosis: A review of advances in diagnosis and management. *J Pak Med Assoc.* 2010; **60:** 573-9.
384. Sharma A, Singhal M, Thulkar S. Fungal infection - From bad to ugly. *J Assoc Physicians India.* 2010; **58:** 111.
385. Sharma D, Dorgan E, Douglas H, *et al.* *Aspergillus* infection in pulmonary cavitating lesions with right atrial myxoma. *Asian Cardiovasc Thorac Ann.* 2014; **22:** 1090-2.
386. Sharma S, Dubey SK, Kumar N, *et al.* 'Monod' and 'air crescent' sign in aspergilloma. *BMJ Case Rep.* 2013; pii: bcr2013200936.
387. Sharma S, Yenigalla BM, Naidu SK, *et al.* Primary cutaneous aspergillosis due to *Aspergillus tamarii* in an immunocompetent host. *BMJ Case Rep.* 2013; 010128.
388. Sharma SK, Gupta N, Arav S, *et al.* Pulseless right upper limb: An unusual manifestation of invasive pulmonary aspergillosis in acute myeloid leukemia. *J Assoc Physicians India.* 2012; **60:** 119-22.
389. Shashidhar N, Tripathy SK, Balasubramanian S, *et al.* *Aspergillus* spondylodiscitis in an immunocompetent patient following spinal anesthesia. *Orthop Surg.* 2014; **6:** 72-7.
390. Sherif R, Segal BH. Pulmonary aspergillosis: Clinical presentation, diagnostic tests, management and complications. *Curr Opin Pulm Med.* 2010; **16:** 242-50.
391. Shin B, Koh WJ, Jeong BH, *et al.* Serum galactomannan antigen test for the diagnosis of chronic pulmonary aspergillosis. *J Infect.* 2014; **68:** 494-9.
392. Shivaprakash MR, Geertsen E, Chakrabarti A, *et al.* *In vitro* susceptibility of 188 clinical and environmental isolates of *Aspergillus flavus* for the new triazole isavuconazole and seven other antifungal drugs. *Mycoses.* 2011; **54:** e583-9.
393. Sidani C, Freiser ME, Saigal G, *et al.* Unusual case of cerebral aspergillosis with clinical and imaging findings mimicking lymphoma. *Neuroradiol J.* 2013; **26:** 290-6.

394. Siddappa S, Mythri KM, Kowsalya R, *et al.* An unusual case of non-disseminated bladder aspergillosis in a setting of transitional cell carcinoma. *Indian J Med Microbiol.* 2012; **30:** 106-8.
395. Simmonds L, Mitchell S, White B, *et al. Aspergillus niger* infection in an immunosuppressed patient confined solely to the brain. *BMJ Case Rep.* 2017; pii: bcr2016218658. doi: 10.1136/bcr-2016-218658. PMID: 28331021.
396. Sindhwani G, Rawat J, Kesarwani V. Haemoptysis after four years of lobectomy for aspergilloma. *Indian J Chest Dis Allied Sci.* 2013; **55:** 43-4.
397. Singh H. Aspergilloma in a patient with a history of TB. *BMJ Case Rep.* 2014; pii: bcr2013203276.
398. Singh N, Husain S, *et al.* Invasive aspergillosis in solid organ transplant recipients. *Am J Transplant.* 2009; **9:** S180-91.
399. Singh UP, Aneja P, Aditi, *et al.* Co-existence of HIV, active tuberculosis and aspergilloma in a single individual - A case report. *Indian J Tuberc.* 2013; **60:** 55-8.
400. Smith NL, Denning DW. Underlying conditions in chronic pulmonary aspergillosis including simple aspergilloma. *Eur Respir J.* 2011; **37:** 865-72.
401. Snelders E, Melchers WJ, Verweij PE. Azole resistance in *Aspergillus fumigatus:* A new challenge in the management of invasive aspergillosis? *Future Microbiol.* 2011; **6:** 335-47.
402. Solak Y, Atalay H, Nar A, *et al. Aspergillus* thyroiditis in a renal transplant recipient mimicking subacute thyroiditis. *Transpl Infect Dis.* 2011; **13:** 178-81.
403. Soman SO, Vijayaraghavan G, Padmaja NP, *et al.* Aspergilloma of the heart. *Indian Heart J.* 2014; **66:** 238-40.
404. Srikumar T, Pabbathi S, Fernandez J, *et al. Aspergillus terreus* brain abscess complicated by tension pneumocephalus in a patient with angiosarcoma. *Am J Case Rep.* 2017; **18:** 33-37.
405. Steinbach WJ. Special Issue "*Aspergillus fumigatus:* From diagnosis to therapy" *J Fungi.* 2016; **2:** 31.
406. Steinbach WJ. Invasive aspergillosis in pediatric patients. *Curr Med Res Opin.* 2010; **26:** 1779-87.
407. Steinbach WJ. Pediatric invasive aspergillosis. *Pediatr Infect Dis J.* 2010; **29:** 964-5.
408. Stern JB, Wyplosz B, Validire P, *et al.* Bulky mediastinal aspergillosis mimicking cancer in an immunocompetent patient. *Ann Thorac Surg.* 2014; **98:** 1472-5.
409. Stevens DA, Melikian GL. Aspergillosis in the 'non-immunocompromised' host. *Immunol Invest.* 2011; **40:** 751-66.
410. Stevens DA. Clinical aspergillosis for basic scientists. *Med Mycol.* 2009; **47:** S 1-4.
411. Sugui JA, Kwon-Chung KJ, Juvvadi PR, *et al. Aspergillus fumigatus* and related species. *Cold Spring Harb Perspect Med.* 2014; **5:** a019786.
412. Sugui JA, Peterson SW, Clark LP, *et al. Aspergillus tanneri* sp. nov., a new pathogen that causes invasive disease refractory to antifungal therapy. *J Clin Microbiol.* 2012; **50:** 3309-17.
413. Sugui JA, Vinh DC, Nardone G, *et al. Neosartorya udagawae (Aspergillus udagawae)*, An emerging agent of aspergillosis: How different is it from *Aspergillus fumigatus? J Clin Microbiol.* 2010; **48:** 220-8.
414. Sun L, Zhang L, Wang K, *et al.* Fungal osteomyelitis after arthroscopic anterior cruciate ligament reconstruction: A case report with review of the literature. *Knee.* 2012; **19:** 728-31.
415. Sundaram C, Murthy JM. Intracranial *Aspergillus* granuloma. *Patholog Res Int.* 2011; 157320. PMID: 22191079.
416. Sundarray C, Panda S, Ray R. Primary vocal cord aspergillosis in a non-immunocompromised host. *J Indian Med Assoc.* 2011; **109:** 200.
417. Sutton DA, Wickes BL, Romanelli AM, *et al.* Cerebral aspergillosis caused by *Aspergillus granulosus. J Clin Microbiol.* 2009; **47:** 3386-90.
418. Swoboda-Kopec E, Sikora M, Piskorska K, *et al.* Diagnosis of invasive pulmonary aspergillosis. *Adv Exp Med Biol.* 2017; **944:** 27-33.
419. Tahir C, Garbati M, Nggada HA, *et al.* Primary cutaneous aspergillosis in an immunocompetent patient. *J Surg Tech Case Rep.* 2011; **3:** 94-6.
420. Tak V, Mathur P, Xess I, *et al.* A case of dual infection in a paediatric trauma victim of primary cutaneous aspergillosis caused by *Aspergillus flavus* and *Aspergillus terreus. Indian J Med Microbiol.* 2013; **31:** 193-6.
421. Takeda K, Imamura Y, Takazono T, *et al.* The risk factors for developing of chronic pulmonary aspergillosis in non-tuberculous mycobacteria patients and clinical characteristics and outcomes in chronic pulmonary aspergillosis patients coinfected with non-tuberculous mycobacteria. *Med Mycol.* 2016; **54:** 120-7.
422. Tang Q, Tian S, Yu N, *et al.* Development and evaluation of loop-mediated isothermal amplification method for rapid detection of *Aspergillus fumigatus. J Clin Microbiol.* 2016; **54:** 950-5.
423. Tashiro T, Izumikawa K, Tashiro M, *et al.* A case series of chronic necrotizing pulmonary aspergillosis and a new proposal. *Jpn J Infect Dis.* 2013; **66:** 312-6.
424. Thada ND, Prasad SC, Alva B, *et al.* A rare case of suppurative aspergillosis of the thyroid. *Case Rep Otolaryngol.* 2013; 956236. PMID: 24159398.
425. Thakar S, Chickabasaviah YT, Hegde AS. Spontaneous resolution of invasive cerebral aspergillosis following partial resection in a medically untreated infant. *J Neurosurg Pediatr.* 2012; **10:** 71-4.
426. Thompson GR 3rd, Patterson TF. Pulmonary aspergillosis: Recent advances. *Semin Respir Crit Care Med.* 2011; **32:** 673-81.
427. Thornton CR. Breaking the mould - Novel diagnostic and therapeutic strategies for invasive pulmonary aspergillosis in the immune deficient patient. *Expert Rev Clin Immunol.* 2014; **10:** 771-80.
428. Tilve SS, Shah AB, Prabhudesai PP, *et al.* An unusual case of invasive *Aspergillus* ulcerative tracheobronchitis without involvement of lung parenchyma in a post-renal transplant patient. *Indian J Chest Dis Allied Sci.* 2013; **55:** 221-3.
429. Tiwari V, Khatri K, Khan SA, *et al.* Disseminated *Aspergillus flavus* following septic arthritis in an immunocompetent patient: A case report. *BMC Res Notes.* 2014; **7:** 709.

430. Tochigi N, Ishiwatari T, Okubo Y, *et al.* Histological study of chronic pulmonary aspergillosis. *Diagn Pathol.* 2015; **10:** 153.
431. Tragiannidis A, Roilides E, Walsh TJ, *et al.* Invasive aspergillosis in children with acquired immunodeficiencies. *Clin Infect Dis.* 2012; **54:** 258-67.
432. Tresallet C, Seman M, Hoang C, *et al.* Gastric perforation from potential primary digestive aspergillosis. *Surgery.* 2010; **148:** 158-9.
433. Tripathy SR, Mishra SS, Deo RC, *et al.* Aqueductal stenosis has a new dimension - aspergillosis: A rare case report and review of the literature. *World Neurosurg.* 2015; **84:** 1494. e13-6.
434. Trotter SC, Sammons DL. Primary cutaneous aspergillosis. *Cutis.* 2013; **92:** 219 & 223-4.
435. Uchiyama M. Invasive pulmonary aspergillosis diagnosed by bronchoalveolar lavage. *BMJ Case Rep.* 2012; pii: bcr0320126144. PMID: 22707689.
436. Uhrin GB, Jensen RH, Korup E, *et al.* Recurrent prosthetic valve endocarditis caused by *Aspergillus delacroxii* (formerly *Aspergillus nidulans* var. *echinulatus*). *Med Mycol Case Rep.* 2015; **10:** 21-3.
437. Vaideeswar P, Vyas M, Katewa A, *et al.* Piggyback mycosis: Pulmonary hydatid cyst with a mycotic co-infection. *Mycoses.* 2010; **53:** 265-8.
438. Vaideeswar P. *Aspergillus* pancarditis manifesting as hospital-acquired infection: Report of two cases and review of literature. *Cardiovasc Pathol.* 2010; **19:** e253-7.
439. van der Linden JW, Arendrup MC, Melchers WJ, *et al.* Azole resistance of *Aspergillus fumigatus* in immunocompromised patients with invasive aspergillosis. *Emerg Infect Dis.* 2016; **22:** 158-9.
440. van der Linden JW, Arendrup MC, Warris A, *et al.* Prospective multicenter international surveillance of azole resistance in *Aspergillus fumigatus. Emerg Infect Dis.* 2015; **21:** 1041-4.
441. van Ingen J, van der Lee HA, Rijs AJ, *et al.* High-level pan-azole-resistant aspergillosis. *J Clin Microbiol.* 2015; **53:** 2343-5.
442. Vehreschild JJ, Vehreschild MJ, Cornely OA. PTX3 deficiency and aspergillosis. *N Engl J Med.* 2014; **370:** 1666.
443. Venugopal TV, Venugopal PV. Primary cutaneous aspergillosis from Tamilnadu: Diagnosed by fine needle aspiration cytology. *Med Mycol Case Rep.* 2012; **1:** 103-6.
444. Verghese S, Chellamma T, Cherian KM. Osteomyelitis of the rib caused by *Aspergillus flavus* following cardiac surgery. *Mycoses.* 2009; **52:** 91-3.
445. Vermeulen E, Lagrou K, Verweij PE. Azole resistance in *Aspergillus fumigatus:* A growing public health concern. *Curr Opin Infect Dis.* 2013; **26:** 493-500.
446. Vermeulen E, Maertens J, De Bel A, *et al.* Nationwide surveillance of azole resistance in *Aspergillus* diseases. *Antimicrob Agents Chemother.* 2015; **59:** 4569-76.
447. Vermeulen E, Maertens J, Meersseman P, *et al.* Invasive *Aspergillus niger* complex infections in a Belgian tertiary care hospital. *Clin Microbiol Infect.* 2014; **20:** O333-5.
448. Verweij PE, Ananda-Rajah M, Andes D, *et al.* International expert opinion on the management of infection caused by azole-resistant *Aspergillus fumigatus. Drug Resist Updat.* 2015; **21-22:** 30-40.
449. Verweij PE, Chowdhary A, Melchers WJ, *et al.* Azole resistance in *Aspergillus fumigatus:* Can we retain the clinical use of mold-active antifungal azoles? *Clin Infect Dis.* 2016; **62:** 362-8.
450. Verweij PE, Varga J, Houbraken J, *et al. Emericella quadrilineata* as cause of invasive aspergillosis. *Emerg Infect Dis.* 2008; **14:** 566-72.
451. Vijayvargiya P, Javed I, Moreno J, *et al.* Pituitary aspergillosis in a kidney transplant recipient and review of the literature. *Transpl Infect Dis.* 2013; **15:** e196-200.
452. Vinh DC, Shea YR, Sugui JA, *et al.* Invasive aspergillosis due to *Neosartorya udagawae. Clin Infect Dis.* 2009; **49:** 102-11.
453. Vohra S, Taylor R, Aronowitz P. The tell-tale heart: *Aspergillus fumigatus* endocarditis in an immunocompetent patient. *Hosp Pract.* 2013; **41:** 117-21.
454. Waller S, Raglow Z, Lemons S, *et al.* Microwave ablation of a large renal aspergilloma. *Transpl Infect Dis.* 2014; **16:** 496-500.
455. Walsh TJ, Wissel MC, Grantham KJ, *et al.* Molecular detection and species-specific identification of medically important *Aspergillus* species by real-time PCR in experimental invasive pulmonary aspergillosis. *J Clin Microbiol.* 2011; **49:** 4150-7.
456. Wang ES. Guideline recommendations for the prophylaxis of invasive aspergillosis in AML. *Clin Adv Hematol Oncol.* 2016; **14:** 876-9.
457. Wang HC, Chang K, Lu PL, *et al.* Fatal invasive aspergillosis: A rare co-infection with an unexpected image presentation in a patient with dengue shock syndrome. *Clin Respir J.* 2017; **11:** 248-53.
458. Warris A. Azole-resistant aspergillosis. *J Infect.* 2015; **71:** S121-5.
459. Warris A. The biology of pulmonary *Aspergillus* infections. *J Infect.* 2014; **69:** S36-41.
460. Wattier RL, Ramirez-Avila LJ. Pediatric invasive aspergillosis. *J Fungi.* 2016; **2:** 19.
461. Webb BJ, Blair JE, Kusne S, *et al.* Concurrent pulmonary *Aspergillus fumigatus* and *Mucor* infection in a cardiac transplant recipient: A case report. *Transplant Proc.* 2013; **45:** 792-7.
462. Wessolossky M, Welch VL, Sen A, *et al.* Invasive *Aspergillus* infections in hospitalized patients with chronic lung disease. *Infect Drug Resist.* 2013; **6:** 33-9.
463. Wheat LJ. Approach to the diagnosis of invasive aspergillosis and candidiasis. *Clin Chest Med.* 2009; **30:** 367-77.
464. White PL, Parr C, Thornton C, *et al.* Evaluation of real-time PCR, galactomannan enzyme-linked immunosorbent assay (ELISA) and a novel lateral-flow device for diagnosis of invasive aspergillosis. *J Clin Microbiol.* 2013; **51:** 1510-6.
465. White PL, Wingard JR, Bretagne S, *et al. Aspergillus* polymerase chain reaction: Systematic review of evidence for

clinical use in comparison with antigen testing. *Clin Infect Dis.* 2015; **61:** 1293-303.

466. Winterstein AR, Bohndorf K, Vollert K, *et al.* Invasive aspergillosis osteomyelitis in children - A case report and review of the literature. *Skeletal Radiol.* 2010; **39:** 827-31.
467. Yamakawa H, Takayanagi N, Ishiguro T, *et al.* An autopsy case of pulmonary aspergillosis with fungus ball formation in an artificial aortic graft. *Intern Med.* 2013; **52:** 2117-9.
468. Yan X, Zong F, Kong H, *et al.* Pulmonary fungal diseases in immunocompetent hosts: A single-center retrospective analysis of 35 subjects. *Mycopathologia.* 2016; **181:** 513-21.
469. Yasuda M, Nagashima A, Haro A, *et al.* Aspergilloma mimicking a lung cancer. *Int J Surg Case Rep.* 2013; **4:** 690-2.
470. Yii AC, Koh MS, Lapperre TS, *et al.* The emergence of *Aspergillus* species in chronic respiratory disease. *Front Biosci (Schol Ed).* 2017; **9:** 127-38.
471. Yoon YK, Kang EH, In KH, *et al.* Unilateral ureteral obstruction caused by *Aspergillus,* subgenus *Nidulantes* in a patient on steroid therapy: A case report and review of the literature. *Med Mycol.* 2010; **48:** 647-52.
472. Yu J, Mu X, Li R. Invasive pulmonary aspergillosis due to *Emericella nidulans* var. *echinulata,* successfully cured by voriconazole and micafungin. *J Clin Microbiol.* 2013; **51:** 1327-9.
473. Yuan L, Prayson RA. Optic nerve aspergillosis. *J Clin Neurosci.* 2015; **22:** 1191-3.
474. Yuchong C, Dingheng Z, Zhizhong Y, *et al.* Aspergillosis of biliary tract after liver transplantation: A case report. *Mycopathologia.* 2010; **170:** 117-21.
475. Zarrinfar H, Mirhendi H, Makimura K, *et al.* Use of mycological, nested PCR and real-time PCR methods on BAL fluids for detection of *Aspergillus fumigatus* and *A.flavus* in solid organ transplant recipients. *Mycopathologia.* 2013; **176:** 377-85.
476. Zhang RR, Wang SF, Lu HW, *et al.* Clinical investigation of misdiagnosis of invasive pulmonary aspergillosis in 26 immunocompetent patients. *Int J Clin Exp Med.* 2014; **7:** 4139-46.
477. Zhang XB, Chen GP, Lin QC, *et al.* Bronchoalveolar lavage fluid galactomannan detection for diagnosis of invasive pulmonary aspergillosis in chronic obstructive pulmonary disease. *Med Mycol.* 2013; **51:** 688-95.
478. Zhou XM, Li P, Zhao L, *et al.* Lung carcinosarcoma masked by tracheobronchial aspergillosis. *Intern Med.* 2015; **54:** 1905-7.
479. Zhu LP, Chen XS, Wu JQ, *et al. Aspergillus* vertebral osteomyelitis and ureteral obstruction after liver transplantation. *Transpl Infect Dis.* 2011; **13:** 192-9.
480. Zou M, Tang L, Zhao S, *et al.* Systematic review and meta-analysis of detecting galactomannan in bronchoalveolar lavage fluid for diagnosing invasive aspergillosis. *PLoS One.* 2012; **7:** e43347.

CHAPTER

26

Mucormycosis

Mucormycosis is an acute opportunistic infection caused by several fungi belonging to phylum **Glomeromycota**. These saprotrophic fungi are found ubiquitously in the soil and environment. In the past, these used to be traditionally considered as non-pathogenic to man and animals. They used to be rather treated as fungal contaminants in the diagnostic microbiology laboratory. But the understanding about mucormycetes has entirely changed as these are now **emerging** pathogenic organisms invariably entailing to fatal consequences especially when an obvious underlying predisposing factor exists in a particular clinical setting. The cases are being increasingly reported among healthy individuals also where no obvious underlying risk factor is present.

In the tissue, these fungi are seen as broad, **non-septate hyphae** but two of them as multiple budding yeast cells. They have predilection to invade blood vessels being angioinvasive in nature thereby leading to extensive necrosis of the surrounding area and forming embolism. Since morphology of all these organisms is indistinguishable in histopathological sections hence culture is essential for an exact identification of the causative fungal species. The course and outcome of the disease differs according to anatomical site affected as well as nature of fungal species involved in a particular patient.

Mucormycosis is an umbrella term used for diseases caused by many non-septate filamentous fungal species. Earlier classified under Orders **Mucorales** and **Entomophthorales**, they were previously considered subordinate members of phylum Zygomycota and now are elevated to the rank of sub-phylum **Mucoromycotina** and **Entomophthoromycotina** under phylum **Glomeromycota**. Some investigators prefer to call mycoses caused by members belonging to sub-phylum **Mucoromycotina** as **mucormycosis** and to sub-phylum **Entomophthoromycotina** as **entomophthoromycosis** or subcutaneous mucormycosis, which is further sub-divided into two types *i.e.* **conidiobolomycosis** and **basidiobolomycosis**.

As there are fungi in same mucormycetes class, which produce syndromes quite different from those by members of sub-phylum **Mucoromycotina** and thus some workers prefer more restrictive designation, mucormycosis, as an appropriate term. However, due to changes in taxonomy and expanded range of etiological agents, term mucormycosis should be preferred over old but still frequently used term, zygomycosis. The older synonym like **phycomycosis**, is now an obsolete term, which denoted that infection was once caused by Phycomycetes.

Unlike other fungal diseases, even the AIDS pandemic could not significantly affect the incidence and prevalence of mucormycosis during last 3½ decades. However, upsurge in the number of diabetics has really changed the entire scenario more or less like an epidemic in Southeast Asia posing a very serious health threat. Mucormycosis is going to destroy India shortly, in a couple of years.

Historical Perspective

In 1855, Kurchenmeister described case of mucormycosis in a patient of neoplastic lung on the basis of its histopathology which was probably first authentic human case. Furbringer in 1876, described pulmonary mucormycosis for the first time, which was caused by *Absidia (now, Lichtheimia)*. At that time, most of pathogenic mucormycetes were originally classified as members of genus *Mucor*. These organisms were later re-assigned different genera and families within sub-phylum **Mucoromycotina**. In 1884, Lichtheim established pathogenicity of sub-phylum **Mucoromycotina** in rabbits and described two species: *Mucor corymbifera (*later on *Absidia,* then *Lichtheimia)* and *Mucor rhizopodiformis* (now *Rhizopus).*

Platauf coined term ***'Mycosis Mucorina'*** and first described a well-documented case of systemic infection with gastric and rhinocerebral involvement more than a century ago in 1885, which was popularly called as **mucormycosis** during those days.

In 1943, Gregory and colleagues, in a series of three fatal cases associated with diabetic ketoacidosis, reported more typical findings of advanced rhinocerebral mucormycosis, proptosis and ophthalmoplegia.

Saksenaea vasiformis was first described by Dr SB Saksena in 1953 as only species of a new genus of mucormycete from forest soil in India, which is found associated with a number of human infections. In 1955, Harris reported cure of mucormycosis in a pediatric patient. Subsequently, with increasing recognition of typical features of mucormycosis and discovery of amphotericin B in 1957 as a breakthrough therapeutic agent, universal mortality of disease had been dramatically reversed during those days. Subsequently genus *Apophysomyces* was first described by Dr PC Misra and colleagues in 1979 as a soil isolate from mango orchard in India, which has now emerged as a significant cause of mucormycosis.

In the year 2008, the Henry Schueler 41 & 9 Foundation came into existence in his loving memory to advance research for pediatric leukemia and fungal infections *(http://www.henryschueler.org)*. The Foundation also sponsored the first **International Conference** on **Mucormycosis** in Chicago in 2010, which generated research and publications in *Clin Infect Dis (2012; 54 Suppl. 1: S1-85), "Advances Against Mucormycosis: A Tribute to the Memory and Courage of Hank Schueler"*.

Almost there were parallel developments in Europe. A meeting of **International Forum** on **Zygomycosis** was held in 2008 at Athens (Greece) under the leadership of Georgios Petrikkos, wherein most of the aspects of this emerging disease were discussed at length. Subsequently two more meetings of this Forum were organized in 2010 (Porto Heli) and 2014 (Marathon), in the same country.

Many newer species of mucorales have been reported in recent times. Tendolkar *et al*, have reported a case of *Saksenaea erythrospora* in a case of fungal rhinosinusitis in 2015 first time from India followed by five cases by Chander *et al*, in 2017. A new species of *Saksenaea* was also reported as *S.loutrophoriformis* by Sutton *et al*, in 2017 from India and USA in two different cases making a total of four species *i.e. S.vasiformis, S.erythrospora, S.oblongispora* and *S.loutrophoriformis.*

Apophysomyces has also now five clinically significant species *i.e. A.elegans, A.trapeziformis, A.ossiformis, A.variabilis* and *A.mexicanus*. In India, *A.variabilis*, is the most common and *A.trapeziformis* and *A.mexicanus* are found prevalent in USA and Mexico, respectively.

Mycology

The mucormycetes is a group of lower fungi and their hyphae are generally non-septate, sparsely septate or pauci-septate. However, when these septa occur they have **solid cross-walls** with **no pores** and there is no flow of cytoplasmic material between cells. These reproduce asexually by way of sporangiospores formed within a sac called a **sporangium** and/or by means of conidial development. These fungi also reproduce sexually by formation of a single, dark, thick-walled spore called **zygospore**. Based on zygospore formation, these fungi were classified under phylum Zygomycota, which has now been included in **Glomeromycota**.

The traditional phylum Zygomycota has been divided among the phylum Glomeromycota and four subphyla pending resolution of further taxonomic questions. In these groups of lower fungi, the thallus is pauci-septate and consists of wide, hyaline, branched hyphal elements. The asexual spores (termed sporangiospores) are nonmotile and are often produced inside a closed sac, termed a **sporangium**, the wall of which ruptures to release them, although in some genera the spores are formed around a vesicle at the tip of the sporangiophore. Sexual reproduction leads to the formation of a single large zygospore with a thickened wall. Most of the medically important species are heterothallic and do not form their sexual structures unless two compatible isolates come in contact.

The sub-phylum Mucoromycotina contains the order Mucorales, which is the most clinically important and includes the genera *Lichtheimia* (formerly *Absidia*), *Mucor*, *Rhizomucor* and *Rhizopus* species. The sub-phylum Entomophthoromycotina contains one order of medical importance, the Entomophthorales. The Entomophthorales includes the genera *Basidiobolus* and *Conidiobolus*, the agents of subcutaneous infections.

Keeping in view of recent taxonomic changes, currently none of the terms Zygomycetes, Zygomycotina or Zygomycota are valid for taxonomic denomination of this group of fungi because they do not reflect their polyphyletic origin. Consequently, the term zygomycosis has been replaced by mucormycosis. The nomenclature of disease *i.e.* mucormycosis comes from Mucorales and not from genus *Mucor*.

The nature of spore bearing structure, location of rhizoids, type of zygospores produced and physiological

Table 26.1. Common Causative Agents of Mucormycosis and Entomophthoramycosis - Phylum Glomeromycota.

A. Order: Mucorales-Mucormycosis
Rhizopus arrhizus (Old name R.oryzae)
R.microsporus var. *rhizopodiformis*
Mucor racemosus
Rhizomucor pusillus
Lichtheimia corymbifera (Mycocladus corymbiferus or Absidia corymbifera)
Apophysomyces Species
Cunninghamella bertholletiae
Saksenaea Species
Cokeromyces recurvatus
Syncephalastrum recemosum
B. Order: Entomophthorales-Entomophthoramycosis
1. Conidiobolomycosis-*Conidiobolus coronatus, Conidiobolus incongruus* and *Conidiobolus lamprauges*
2. Basidiobolomycosis-*Basidiobolus ranarum*

characteristics, are used to differentiate mucormycete genera and species. In addition, length and ramification of sporangiophores, diameter of sporangia, presence and form of apophysis and shape of columella are also used for identification. In some of species like *Cokeromyces recurvatus* and *Mucor circinelloides* dimorphic conversion to yeast form is observed. *Cunninghamella bertholletiae,* a member of family Cunninghamellaceae, has specific mycological characters such as formation of monosporic sporangiola and conspicuous terminal swelling of conidiophores.

The etiological agents in an approximate order of their frequency as pathogens are shown in **Table 26.1**. In this Chapter clinical description is given on the basis of broad division of Orders of mucormycetes under two headings *i.e.* Mucorales and Entomophthorales, as fungi belonging to other Orders of Glomeromycota are not clinically significant.

In 1978, the genus *Rhizomucor* was taken up for *Mucor*-like fungi forming stolons and rudimentary rhizoids and expressing thermophilism. Molecular phylogenetic studies based on nuclear small- and large-subunit rRNA gene sequences support the distinction of *R.pusillus* from *Mucor* species. *Mucor, Rhizopus, Cunninghamella* and *Lichtheimia* species are more closely related to each other than any is to *R.pusillus,* with the exception of *Lichtheimia corymbifera,* which also shares antigenic similarities.

Fig. 26.1. White mycelial growth of *Apophysomyces variabilis* after three days of incubation on Sabouraud dextrose agar.

A. Order: Mucorales

The fungal species belonging to order Mucorales are ubiquitously present in the environment. They are found in food items, soil, air and frequently encountered as laboratory contaminants also. The relevance of isolates in clinical material may be difficult to establish if coenocytic hyphal elements are not seen in direct examination of KOH wet mount or histopathological section. Hence repeated isolation of the organism from consecutive specimens provides strong evidence that it may be relevant.

The morphological features of some species of order Mucorales are collectively given below. The growth on culture in all the cases is rapid, cottony mycelial on SDA **(Fig. 26.1)**. The genus *Mucor* is characterized by rapidly growing light dull to grey colonies and **absence** of rhizoids and no distinct columella **(Fig. 26.2)**. On the other hand *Rhizopus* has **nodal** rhizoids directly beneath **unbranched** sporangiophores and ovoid columella **(Fig. 26.3)**. In genus *Lichtheimia*, rhizoids are internodal and not directly beneath sporangiophore with pear-shaped columella and spherical or ovoid sporangiospores **(Fig. 26.4)**. The genus *Rhizomucor* includes three known species: *R.pusillus, R.miehei* and *R.tauricus,* however, *R.pusillus* being commonest, which are clearly distinct from *Mucor* by morphological features and their thermophilic nature. There is development of sporangiophores from aerial hyphae or stolons with single and weakly branched rhizoids.

Only a few cases of infections caused by other groups of Mucorales have been reported. The genus *Syncephalastrum* has sporangiospores in a row within sporangial sac

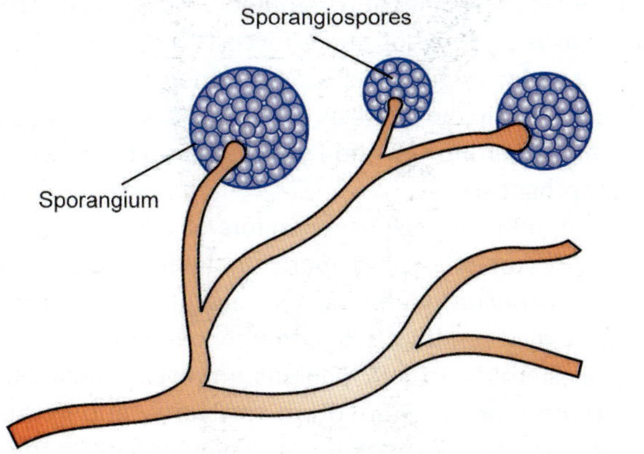

Fig. 26.2. Non-septate hyphae, branched sporangiophores with columellae and sporangia of *Mucor* species.

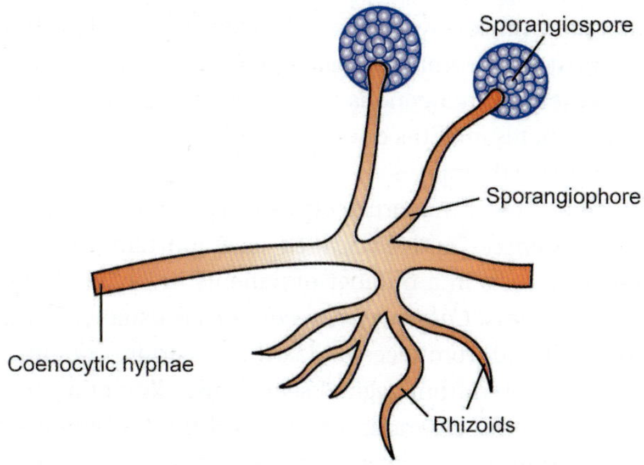

Fig. 26.3. Non-septate hyphae with rhizoids, sporangiophores and sporangia of *Rhizopus* species.

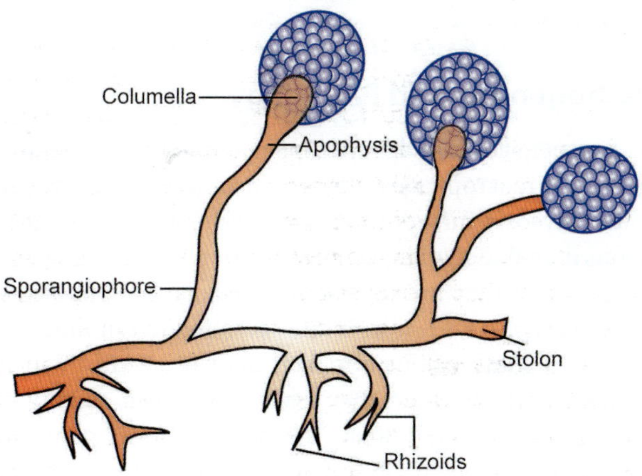

Fig. 26.4. Non-septate hyphae with rhizoids, stolon, sporangiophores and sporangia of *Lichtheimia (Absidia)* species.

known as **merosporangium**. In genus *Cunninghamella,* spores are globose sporangiola (reduced sporangium) at tips of a few sporangiophores and contain from one to as many as 30 spores. The genus *Saksenaea* has a flask-shaped sporangium with a vase-like extension.

Apophysomyces species closely resembles *L.corymbifera* but it is distinguished by its prominent, funnel or bell-shaped (campanulate) apophyses, which is popularly called as champagne glass appearance. The colonies of *Apophysomyces* genus are rapidly growing white to pale yellow. The sporangiophores are generally single, **unbranched**, taper towards apex, topped by pyriform sporangia. There is characteristic **darkening** of sporangiophore just below the apophysis, whereas in *L.corymbifera* there is septum at this site. The columellae are hemispherical and sporangiospores are cylindrical and smooth walled. *Apophysomyces* is implicated in many deep-seated infections.

Cokeromyces recurvatus is usually recovered from soil, rodent and lizard dung. It has been reported to be a colonizer of vagina and urinary bladder and seen as multiple **budding yeasts** in tissues as it is **dimorphic** in nature like *Mucor circinelloides.* Both these mucormycetes are not to be confused with *Paracoccidioides brasiliensis* due to multiple budding yeast cells in the tissue or at 37°C *in vitro.*

Epidemiology

All clinical forms of mucormycosis occur following infection by fungi, which are found as saprotrophic in the environment and on a wide range of substrates including soil. Their spores are widely distributed, growing on leaf litter and other decaying carbohydrate substrates. These are not as abundant in free air as *Cladosporium* and *Alternaria* but can be found in large number in damp interiors and around composing vegetation and even present as ***'bread molds'*** appearing as grayish, fluffy and rapidly spreading growth. Because of their rapid growth and prolific spore-forming capacity, inhalation of conidia is a routine occurrence. It is now third most common invasive mycoses after candidiasis and aspergillosis.

The virulence of these fungi is clearly of low order and sporadic infections occur throughout world particularly in severely debilitated patients. Being an opportunistic infection, mucormycosis is produced by causative fungi in a host whose immunological defense mechanisms are weakened by endogenous causes like malignancy, leukemia, diabetes mellitus or exogenous causes like immunosuppressive therapy.

Now, newer mucormycetes species are emerging like *Apophysomyces* and *Saksenaea* as important causes of mucormycosis in a number of patients involving various organ systems. *Cokeromyces recurvatus* is usually isolated from soil and from feces of lizard and certain rodents. It is now isolated from vaginal secretions, alimentary tract, bladder, urine, pleura and peritoneal fluid. Members of Entomophthorales like *Basidiobolus ranarum*, although are known for subcutaneous infection but are now causing systemic infections very similar to gastrointestinal mucormycosis.

The spectrum of diseases and etiological agents of mucormycosis in animals is somewhat different from that associated with human infections. There is no set of established predisposing factors and chronic granulomatous disease and cutaneous lesions are more commonly encountered. *Lichtheimia corymbifera* is the most frequently recorded agent from systemic infections among animals.

As such mucormycosis is acutely presenting disease but in recent times cases of chronic presentations have also been reported creating confusion in arriving the final diagnosis. *Mucor irregularis* is being reported from such cases.

The breakthrough mucormycosis cases have also been reported after voriconazole administration among patients with hematologic malignancies who receive hematopoietic stem-cell transplants or intensive chemotherapy.

Immunity

The neutrophils appear to play a major role in defense of host from infection caused by mucormycetes. In normal hosts, macrophages prevent initiation of infection by phagocytosis and oxidative killing of spores. On the other hand in hosts with diabetes mellitus and other underlying factor, monocytes/macrophages are dysfunctional and fail to suppress spore germination process. The hosts with neutrophil defects, either qualitative or quantitative, are predisposed to infection by fungal agents of this class. The clinical entities include rhinocerebral mucormycosis in diabetic ketoacidosis and invasive infection with *Rhizopus, Lichtheimia* and *Cunninghamella* species in case of treated malignancies and granulocytopenia. Like *Aspergillus* species, agents of mucormycosis have a predilection for **vascular invasion** causing thrombosis and infarction thereby leading to necrosis of affected tissue. In such type of clinical setting, infection with mucormycetes carry very poor prognosis.

The other predisposing factors of mucormycosis are granulocytopenia, particularly in hematological malignancies, uncontrolled diabetes mellitus and immunosuppressive conditions, in addition to long-term intake of corticosteroids. In HIV patients however, mucormycosis cases have been reported but very rarely and frequency is not common like candidiasis, cryptococcosis and other opportunistic fungal infections, which is one of exceptional feature of mucormycosis. It may be because HIV infection does not appear to increase risk of developing mucormycosis probably because neutrophils, as opposed to T-lymphocytes, play a major role in defense against Mucorales. If it is there, it occurs in context of intravenous drug abusers.

Pathogenesis and Pathology

As in aspergillosis, phagocytes, polymorphonuclear neutrophils and macrophages play a critical role in patients of mucormycosis. Infection can occur by inhalation, percutaneous inoculation or ingestion. The spores are inhaled into lungs where they are ingested by alveolar macrophages. These cells are known to inhibit germination of ingested spores to some extent but their activity to kill them is limited. Further, when they evade antifungal activity of macrophages and germinate into mycelial form, polymorphonuclear neutrophils and peripheral monocytes are expected to work against fungi. The former, which possess property of fungicidal activities against mucormycetes are known to play an important role in defense against mucormycosis and therefore, leukocytopenic patients are extremely susceptible. The increased risk of mucormycosis in patients with ketoacidosis may also be due to release of iron bound to proteins. Due to ketoacidosis, low serum pH diminishes the phagocytic effect of macrophages, chemotactic and oxidative burst of neutrophils.

The commonest encountered mucormycetes, *Rhizopus arrhizus*, has several virulence factors like angioinvasive nature, growth at or above body temperature, production of destructive enzymes, dormant spores which are resistant to destruction at extremes of temperature along with active ketone reductase system and hydroxamate siderophores. The diabetic patients with ketoacidosis are usually more affected by mucormycosis. *Rhizopus* species have an active ketone reductase system and thrive in high glucose and acidotic conditions. These patients also have decreased phagocytic activity because of an impaired glutathione pathway. The exact mechanism of increased susceptibility among these patients remains unknown, probably a combination of metabolic abnormalities present in the patients with diabetes. Hyperglycemia or acidosis alone does not permit fungal growth *in vivo* although acidosis without hyperglycemia has been reported with invasive mucormycosis. The normal serum inhibits *Rhizopus* whereas serum from patients of diabetic ketoacidosis stimulates its growth.

It has been observed that patients, on dialysis and iron overload, who are being treated with deferoxamine, an iron chelator, are more susceptible to mucormycosis. It is probably because Mucorales use this chelator as a siderophore to obtain more iron. The other risk factors include neutropenia, high-dose systemic steroids, protein-calorie malnutrition, solid organ and bone marrow transplants, immunodeficiency, leukemia and intravenous drug abusers who may inject spores of Mucorales with the drugs and then present with space occupying lesions of central nervous system.

Mucormycetes have a predilection for elastic lamina of large and small arteries causing thrombosis, hemorrhage and infarction. Hence angioinvasion, thrombosis, ischemia and tissue necrosis are primary histopathological features of mucormycosis. In AIDS patients, neutrophil function remains relatively intact and this may be probable reason for less frequency of mucormycosis among HIV positive individuals, therefore, as such it does not seem to be a significant risk factor.

Clinical Features

Mucormycosis is a very rapidly progressive disease thereby may prove fatal if timely diagnosis is not made entailing delay in institution of specific treatment. The main reason of its rapidity is that it involves blood vessels, being angioinvasive in nature. It presents as following **six** clinical types on the basis of anatomical site involved in a particular patient. Each one is associated with certain underlying disorders, relationship and prevalence of variable factors. However, recently cases have been reported among immunocompetent individuals without any obvious underlying risk factors. These clinical types are given below:

1. Rhinocerebral Mucormycosis
2. Pulmonary Mucormycosis
3. Cutaneous Mucormycosis
4. Gastrointestinal Mucormycosis
5. Isolated Renal Mucormycosis
6. Disseminated Mucormycosis

1. Rhinocerebral Mucormycosis

This is most common and fulminating type of mucormycosis, which may lead to fatal consequences within a week of onset of disease if left untreated. It is invariably associated either with acute onset of diabetes mellitus, hyperglycemia and ketoacidosis or with debilitating diseases such as leukemia and lymphoma. Most commonly, it spreads from nasal mucosa to turbinate bones, paranasal sinuses, orbit and palate with eventual extension into brain where massive invasion of blood vessels cause major infarct. The destruction of bones is often revealed by imaging techniques like CT or MRI. Several species of mucormycetes have been reported as etiological agents but majority of cases are caused by *Rhizopus arrhizus*.

The presenting symptoms include facial pain, headache, lethargy and in advanced cases, loss of vision. If there is no obvious reason for diplopia in diabetics, mucormycosis may be suspected. The physical examination reveal brownish, bloodstained nasal discharge on the affected side, black eschar on palate due to hemorrhage and tissue necrosis, fixed and dilated pupil and global proptosis and ptosis with dysfunction of cranial nerves, especially fifth and seventh nerves. There is an extensive and rapid destruction of surrounding tissues. Sometimes, spread may also occur to lungs, gastrointestinal tract, skin and occasionally to other organs. The rhinocerebral mucormycosis is usually fatal as patient dies within a week's time and invariably diagnosed on autopsy, if clinician could not reach tentative diagnosis during life.

The signs and symptoms of orbital mucormycosis include chemosis, periorbital cellulitis, ophthalmoplegia, proptosis, ptosis, abrupt visual loss, orbital pain and facial hypoesthesia **(Figs. 26.5 and 26.6)**. The infection may disseminate to lungs, other organ systems and may spread from orbit into brain, leading to frontal lobe necrosis and abscesses formation. These features result from invasion of fungus through cribriform plate. The fungi invade blood vessels and cause necrotizing vasculitis, resulting in thrombosis of vessel lumen. Unless diagnosed and treated early, this type of mucormycosis is often fatal due to cerebral involvement. An overall mortality among these patients is very high.

Figs. 26.5A to D. Rhino-orbital mucormycosis of various patients.

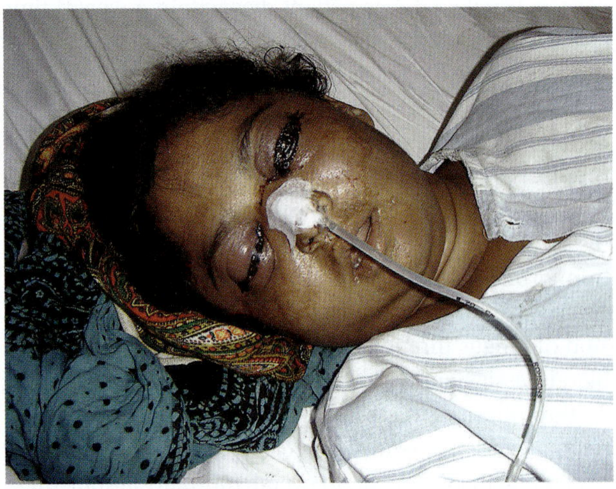

Fig. 26.6. Rhino-orbito-cerebral mucormycosis in a patient.

2. Pulmonary Mucormycosis

The mucormycetes may present as pulmonary disease through inhalation of sporangiospores. The patients are severely immunocompromised by virtue of an absolute lack of circulating neutrophils, secondary to hematologic malignancy like leukemia, lymphoma, profound immunosuppression or bone marrow transplantation. The lesions may be focal or diffuse and usually uncommon in patients having underlying diabetes mellitus in comparison to rhinocerebral type. The invasion of blood vessel may result to destruction of lung parenchyma. The usual course of disease takes about 1 to 4 weeks from onset to terminal stages of illness. The clinical manifestations are nonspecific and may include chest pain, dyspnea and hemoptysis. The chest X-rays show evidence of infiltration, usually

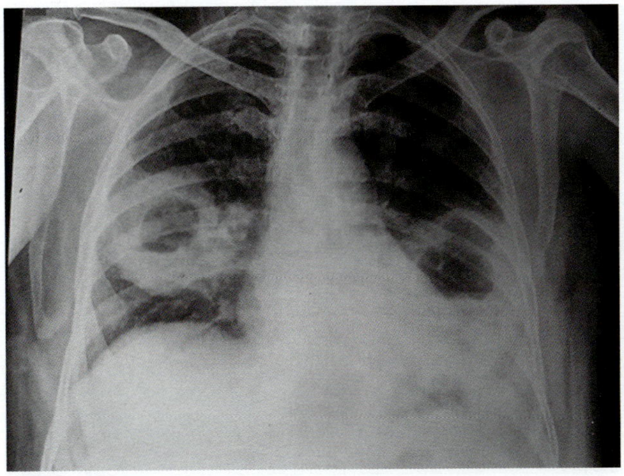

Fig. 26.7. Chest X-rays PA view of a patient of pulmonary mucormycosis.

involving an anatomical segment but typically progressing to involve multiple adjoining areas in the same lung **(Fig. 26.7)**. The chest high resolution computed tomography (HRCT) scan is the best method of determining the extent of pulmonary mucormycosis. This entity is suspected when patients have a reverse halo sign on CT of the chest, along with the right clinical findings. Prognosis is better in localized lesion as compared to the diffuse or bilateral involvement of the lungs.

Very rarely members of order Mucorales are found to involve preformed cavities as fungus balls with their radiological picture similar to that of aspergilloma. Hypersensitivity pneumonitis due to *Rhizopus* has been reported in Scandinavian Sawmill workers (so-called Wood trimmer's disease) and in farm workers.

3. Cutaneous Mucormycosis

The clinical manifestations of cutaneous mucormycosis are varied and range from pustules or vesicles to wounds with wider areas of necrotic zones. In their early stages, lesions resemble ecthyma gangrenosum; cotton-like growth may be seen over surface of tissues, a clinical sign known as ***'hairy pus'***. The cutaneous type of mucormycosis can be either primary infection or secondary to the disseminated form. It is relatively benign form of disease in comparison to the other clinical types. The primary cutaneous lesions have rarely been reported which is due to direct inoculation of fungus into skin. It usually occurs in patients with severe burns, superficial ulcers in diabetics, gangrenous cellulitis in trauma patients, infections associated with contaminated elasticized wound dressings. The lesions are painful; progressively enlarge with nodular ecchymotic areas of cutaneous infarction **(Figs. 26.8A to D)**. Necrotizing fasciitis may be secondary to cutaneous or subcutaneous mucormycosis and has a very high mortality rate.

The histopathological examination shows focal necrosis of parenchyma and adipose tissue. Biopsy and culture of skin lesions should be performed as early as possible, which is diagnostic procedure of choice in this disease. *Lichtheimia corymbifera* and *Saksenaea* have also been reported in cutaneous mucormycosis particularly among diabetic patients.

The primary cutaneous mucormycosis caused by *Saksenaea* and *Apophysomyces* should be suspected when blistering, necrotic lesions develop within a few days of traumatic injury due to potential contamination of wound with dust or soil. Cutaneous mucormycosis has a much more favorable prognosis than more deeply invasive forms but it is still associated with significant fatality and long-term morbidity. Cutaneous mucormycosis may also be observed as a manifestation of disseminated disease in patients with leukemia.

Apophysomyces causes cutaneous and soft tissue infections following trauma, burns and invasive procedures in previously healthy patients. It tends to invade and grow within vascular lumen like other species of mucormycetes. The vascular invasion frequently causes thrombosis leading to ischemia and thereby tissue necrosis. Histopathological examination of most cases infected with *Apophysomyces* reveals hyphae in these thrombi. The rapidly progressive necrosis usually occurs within days after inoculation and explained by fast growing fungi within vessels. The possibility of *Apophysomyces* infection must be considered in differential diagnosis if a patient is seen with progressive necrosis of wound, especially when there is ineffective response to routine antibacterial chemotherapy in an otherwise healthy patient.

4. Gastrointestinal Mucormycosis

The gastrointestinal mucormycosis occurs rarely accounting for ~7% of all cases of mucormycosis, most often involving stomach. It is primarily found among patients suffering from extreme malnutrition and is believed to be acquired by ingesting food contaminated with fungal spores. Ingestion of fermented milk, porridge and alcohol made from corn and herbal products has been implicated

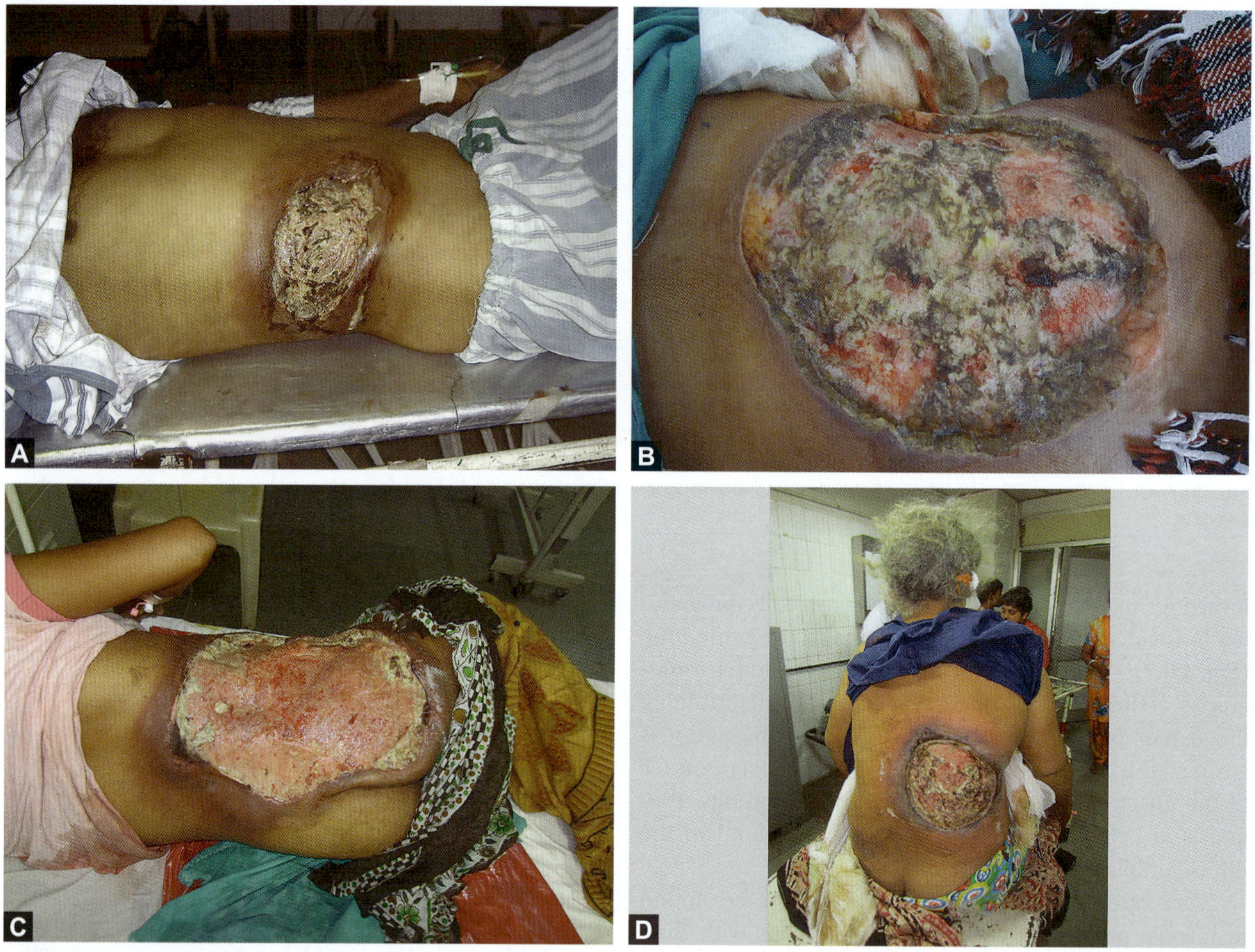

Figs. 26.8A to D. Extensive primary cutaneous mucormycosis of (A) right-sided abdominal wall (B) anterior abdominal wall (C) right gluteal region and (D) posterior side of chest wall.

in gastrointestinal mucormycosis. The lesions in stomach are followed by colon, ileum and esophagus. This may follow surgery from filthy trauma to abdomen or contaminated ileostomy. Ulceration of gastric mucosa with thrombosis of associated vessels has been observed. The disease is usually found in adult patients but has also been reported in neonates.

In this category, mucormycetes either **colonize** or **invade** gastric mucosa. In invasive variant, there is involvement of vessel walls producing thrombosis, hemorrhage, necrosis and ulceration of local tissue and usually has a fatal outcome. Intestinal tract involvement is rare with terminal ileum, cecum and colon being primarily affected sites.

The signs and symptoms of abdominal mucormycosis vary considerably and depend on site as well as extent of involvement. These are non-specific like abdominal pain, diarrhea, hematemesis and melena. The patients on dialysis are reported to develop mucormycosis as most of them are being treated with desferrioxamine (deferoxamine, desferal), which is used as chelator either for aluminum or iron and may be for both.

The agents of gastrointestinal mucormycosis are *Lichtheimia corymbifera* of Mucorales and *Basidiobolus ranarum* of Entomophthorales. The gastrointestinal mucormycosis caused by members of order Entomophthorales is very rare clinical entity. Khan and colleagues have reported basidiobolomycosis of rectum. In gastric mucormycosis brushing cytology may be helpful for establishing diagnosis.

Formerly, diagnosis of gastric mucormycosis used to be made on the basis of postmortem examination and rarely feasible during life time of patient. More recently, endoscopic biopsies have established diagnosis at an earlier stage, allowing attempt to a successful treatment. Gastric brushing is less invasive than biopsy and allows early identification of organisms as compared to histopathological processing.

5. Isolated Renal Mucormycosis

The isolated renal mucormycosis is one of the emerging clinical entity, which is an unusual cause of renal infarction and may be fatal if not timely detected. Any of the species of Mucorales may infect the kidneys. The patients usually present with flank pain, fever and pyuria. In due course of time there is infarct of renal tissue leading to hematuria. Most of these patients are presumed to be suffering from bacterial pyelonephritis hence do not respond to routinely prescribed antibacterial antibiotics. In the immunocompromised patients there are clinical features of acute pyelonephritis. There is renal enlargement with or without perinephric collection on ultrasonography. In such a situation, contrast enhanced CT or MRI is indicated to have more clear view. In case of infarction or multiple abscesses in parenchyma, renal biopsy is indicated to establish tissue diagnosis. A combination of aggressive surgical and medical treatment may improve the outcome of patients. The overall survival of isolated renal mucormycosis is around 65% and mortality is almost cent percent for those who have not undergone nephrectomy. Therefore, a high index of suspicion and awareness among clinicians about the protean manifestations of mucormycosis is significant to establish timely and proper diagnosis of such a fatal condition. In a recent review, a total of 42 cases have been reported in the literature by Rungmei *et al,* in 2010, which has now crossed the figure as fifty.

6. Disseminated Mucormycosis

The mucormycetes may become widely disseminated affecting lungs, kidney, gastrointestinal tract, heart and brain however lungs being the most commonly involved site. The clinical syndromes most frequently reported include pneumonia, stroke, subarachnoid hemorrhage, brain abscess, cellulitis or gangrene of a skin structure. In addition, gastrointestinal bleeding, peritonitis, acute myocardial infarction, hepatitis and superior *vena cava* syndrome have been related to mucormycosis. The vessel invasion with infarction of surrounding parenchyma is seen with a striking absence of inflammatory cells, however, antemortem diagnosis is very rare. In cardiac patients, ultrastructural studies have revealed presence of a thick-walled fungus that participates in formation of thrombus, which crosses coronary arteries in a 'corkscrew' pattern. Thin-walled branching structures are also observed occluding micro-circulation of preserved cardiac areas. These structures could be plasma membranes left after fungus loses its thick wall.

The portal of entry into human body is probably through traumatized skin followed by invasion of blood vessels and dissemination to body organs. This has been reported with infections caused by *Apophysomyces, Saksenaea* and *Rhizopus* species. Cerebral dissemination is caused by hematogenous and lymphogenous seeding by abscess formation, particularly in intravenous drug abusers. In mucormycotic aneurysms (cerebral), numerous non-septate hyphae are seen obstructing cerebral arteries. There is formation of perinephric abscess also that can be visualized on ultrasonography, CT or MRI. Other infections without rhinocerebral involvement may occur in severely immunocompromised patients.

Disseminated infections like endophthalmitis and prosthetic mitral valve mucormycosis have also been established due to *Mucor* species. In addition, *Cokeromyces recurvatus* is also found to be responsible for deep-seated fungal infections depending on underlying pathologic conditions. The cases of hepatic mucormycosis are also reported, wherein an acute and rapidly fatal infection is often diagnosed on postmortem. It has been found to be associated with contamination of herbal medications. Osteomyelitis is another form and bone lesions are usually adjunct to other forms of mucormycosis. It has been reported at base of cranium, in bones of feet and hands and in humerus, tibia, femur and vertebrae.

Radiodiagnosis

It has been established that CT and particularly MRI are most helpful in enabling an early detection of orbital **(Figs. 26.9A to C)**, sinus, meningeal, intraparenchymal, cerebral lesions as well as intracranial vascular occlusion, even before clinical signs develop. T1 and T2-weighted images are taken to know extent of lesions. These imaging techniques are helpful in defining extent of soft tissue and bone destruction and are more useful in planning surgical intervention in addition to establishing the diagnosis.

Differential Diagnosis

Rhinocerebral mucormycosis is an acute and distinctive condition but it can be confused with cavernous sinus

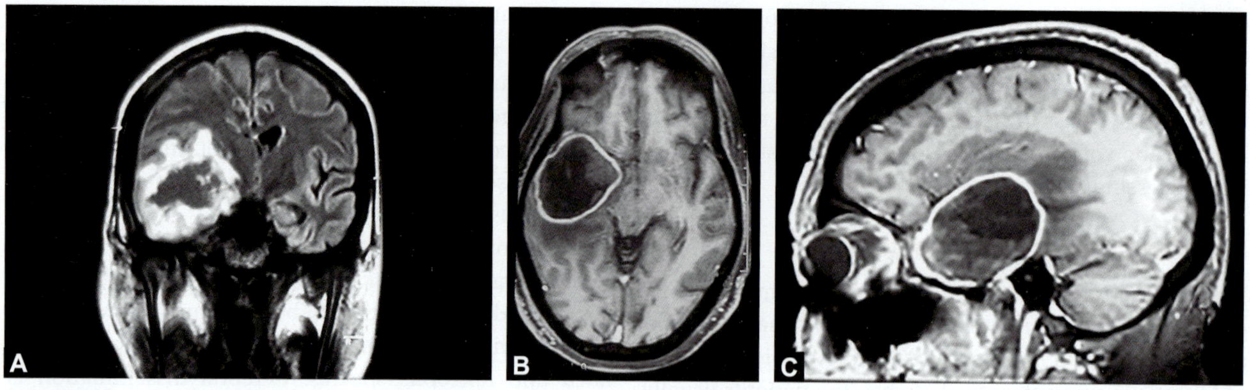

Figs. 26.9A to C. Intracranial mucormycosis of temporal area seen in different CT scan sections.

thrombosis, bacterial orbital cellulitis or rhinocerebral aspergillosis or scedosporiosis/pseudallescheriasis. Clinically, pulmonary mucormycosis cannot be distinguished from gram-negative bacterial pneumonia, aspergillosis or scedosporiosis. The histopathological examination is useful in revealing the correct diagnosis.

The cutaneous and subcutaneous lesions caused by Oomycetes like *Pythium insidiosum* are invariably confused with mucormycosis. However, hyphae are intensely stained with PAS than H&E in case of pythiosis insidiosi, which is other way round in mucormycosis and is helpful in differentiation of these similar looking clinical entities. Some of their morphological and clinical features are described comparatively in **Table 39.2**.

Laboratory Diagnosis

The laboratory diagnosis of mucormycosis is slightly difficult because of rapid and fulminating course of disease and doubtful significance of isolates, which are usually encountered as laboratory contaminants. Therefore, detection of fungus in tissues is supplemented to establish reliability of cultural isolate. The clinical materials from infected area, nasal discharge or sputum may rarely contain fungal elements and deeper tissue section may be required. Detection of β-glucan in patients' serum, which is often helpful in diagnosis of other systemic mycosis, is of little or no significance because mucormycetes produce it very little.

(a) Direct Examination

The microscopic examination of nasal discharge or biopsy material in KOH wet mount shows characteristic broad, non-septate, ribbon-like hyphae with wide-angle or right-angle branching at irregular intervals **(Figs. 26.10A and B)**. The calcofluor white stain delineate in a better way **(Fig. 26.11)**. These hyphae usually are very sparsely distributed in infected tissue and therefore special fungal stains are mandatory. The mucormycetes are very well seen in H&E **(Figs. 26.12A and B)** but are poorly stained with PAS **(Figs. 26.13A to C)**. The other staining done are Gridley, Gram's and GMS **(Figs. 26.14A and B)**. The inflammatory response is neutrophilic and associated with infarct, vascular and perineural invasion. Sometimes, distorted hyphae are also seen in the tissues. The disease may be established on the basis of FNAC from the affected site. It is also pertinent to note that *Cokeromyces recurvatus* and *Mucor circinelloides* are not seen as hyphae in tissues rather as yeast-like forms. The presence of large spherical yeast with multiple budding daughter cells can be observed in direct smear thereby it should not be confused with systemic dimorphic fungus, *Paracoccidioides brasiliensis*.

The hyphal elements of mucormycetes can be recognized as non-septate thick-walled (10-20 μm across) hyphae, with wide-angle branching. Due to absence of cross-walls, fluids from hyphae are free to escape and during handling of biopsy tissues, hyphae collapse and crinkle giving characteristic ribbon-like appearance. They do not radiate from a single point in tissue. These features distinguish from slender hyphae of *Aspergillus*, *Fusarium* and/or *Scedosporium* species, which have regular **dichotomous** branching and frequent septation.

Therefore, it is recommended **not** to **homogenize** tissue material and specimens must be directly inoculated on to culture media to keep viability of fungal cells. The invasive mucormycosis is correctly diagnosed with tissue

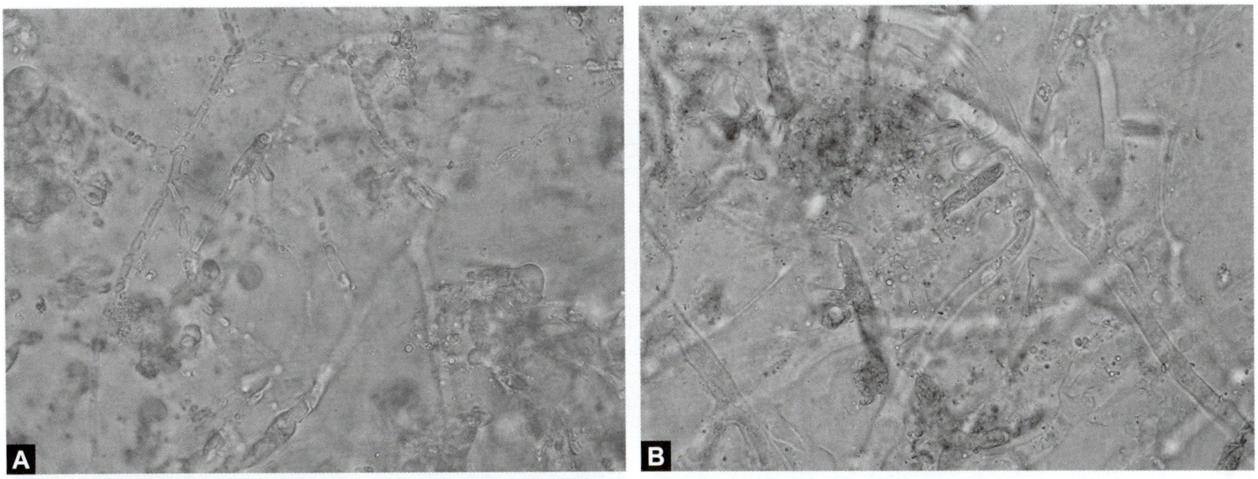

Figs. 26.10A and B. Broad, ribbon-like, non-septate hyphae with wide angle branching seen in wet mount (KOH × 400).

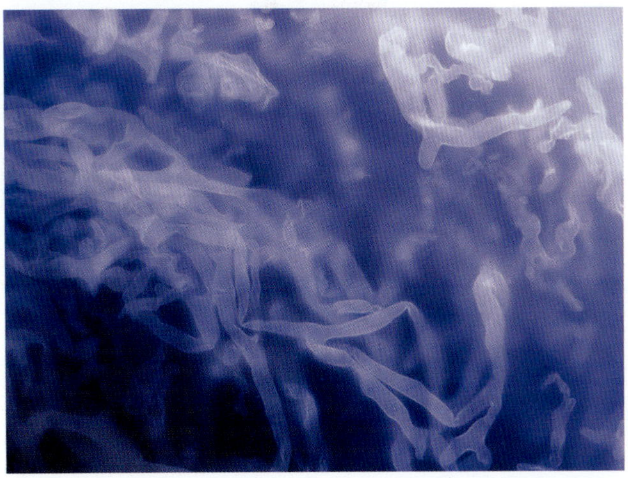

Fig. 26.11. Broad, ribbon-like, non-septate hyphae with wide angle branching stained with fluorescent brightener (CFW × 400).

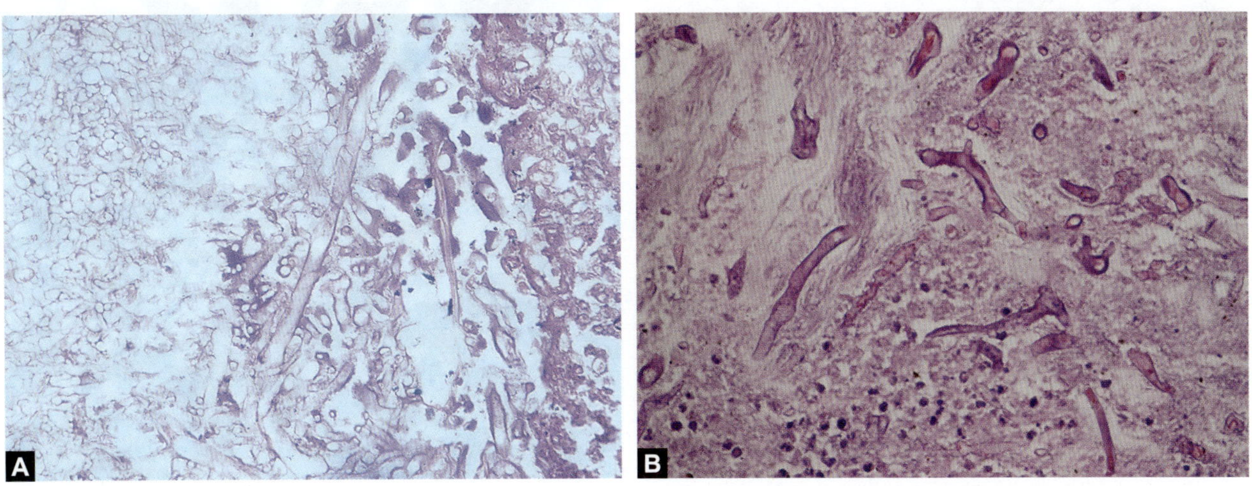

Figs. 26.12A and B. Non-septate hyphae of mucormycetes with wide-angle branching in a tissue sections (H&E × 400).

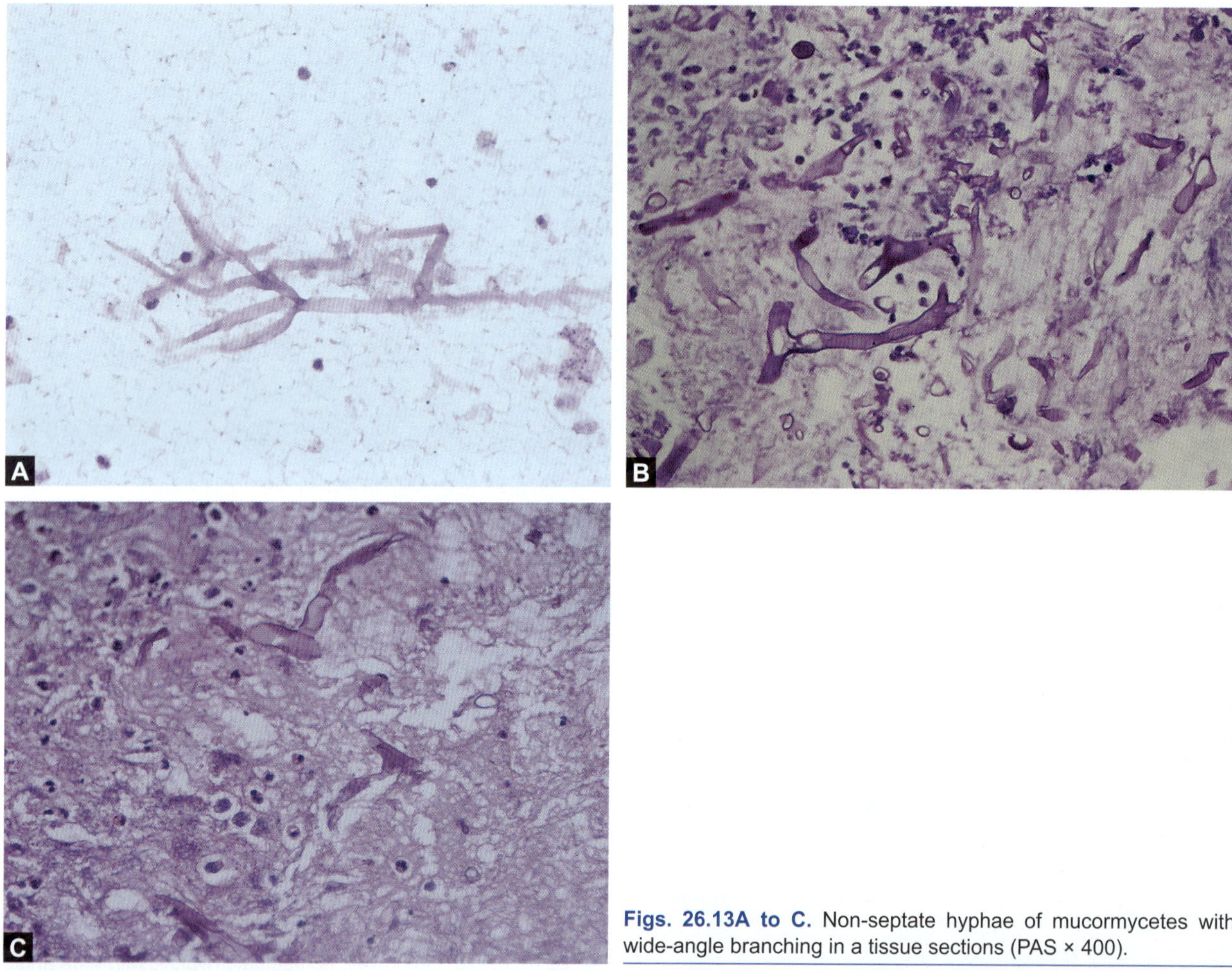

Figs. 26.13A to C. Non-septate hyphae of mucormycetes with wide-angle branching in a tissue sections (PAS × 400).

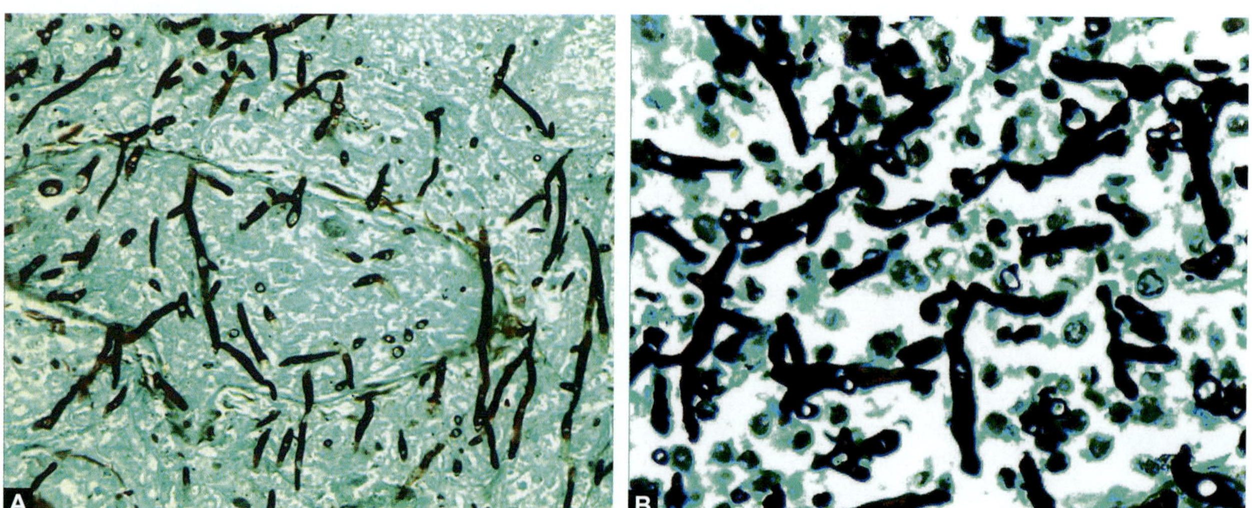

Figs. 26.14A and B. Non-septate hyphae of mucormycetes with wide-angle branching in a tissue sections (GMS × 200 / 400).

biopsy and microscopic identification of pathogen. Immunochemical staining methods can be useful in diagnosis of mucormycosis and their differentiation from other fungal infections. The direct examination is pertinent to differentiate from *Pythium* species also which give almost similar look but comparatively slightly thinner ones.

Moreover, keeping in view of the rapidity of disease process in mucormycosis, frozen section evaluation is an important modality, which can provide fast and valuable information about a potential tissue diagnosis or other pathologic finding while patient is undergoing an operative procedure.

(b) Fungal Culture

The hyphal elements of mucormycetes are prone to physical damage, therefore, specimens must be directly inoculated onto fungal culture media avoiding excessive grinding. The mucormycetes can be easily grown on conventional media like SDA with antibiotics at both temperatures *i.e.* 25°C and 37°C. The Mucorales grow on standard fungal culture media without cycloheximide, as it is inhibitory to most of them except *Apophysomyces* species.

Although the mucormycetes are not fastidious fungi still they fail to grow sometimes during primary isolation. In about fifty percent of cases there is no growth despite direct demonstration of the fungi. The reason is that these are sparsely septate fungi and while handling/processing of biopsy entire cytoplasm oozes out losing viability of the organism. Therefore, in addition to conventional solid culture media, BHI broth may also be used where biopsy material is inoculated as such. It has been observed that it gives an additional advantage when other media turn out to be sterile and growth is seen in BHI broth. In addition, to avoid growth failure, some portion of tissue may be kept in water added with a few drops of Malt yeast broth. In such circumstances, it is better to do molecular technique directly from the sample thereby an early diagnosis can be established.

The rapidly growing mycelial colonies are floccose, dense and have hairy appearance. The mycelia are described as fibrous or cotton-candy growth, which is very vigorous hence some of mucormycetes of order Mucorales are called as ***'lid-lifters'*** as they press upon lid of petri dish from below.

Most of members of *Mucoraceae* easily sporulate on standard fungal media such as SDA, CMA or PDA but *Apophysomyces* species, *Saksenaea* species and *Mortierella wolfii* may produce only sterile hyphae without any spores under these routine cultural conditions. Sporulation of these isolates may be stimulated using various spore induction techniques designed by various workers from time to time *e.g.* Ellis *et al*, Padhye *et al* and Mohapatra *et al*, with very good results.

Microscopic examination of growth confirms coenocytic nature of hyphae and sporulating species produce characteristic sporangia that contain from 50 to 100,000 sporangiospores. *Apophysomyces* and *Saksenaea* species are now being frequently encountered as the significant causative agent of mucormycosis. They differ from other pathogenic mucormycetes species in their cultural behavior that they fail to sporulate on routine fungal culture media used in mycology laboratory like SDA, CMA and PDA. Hence Czapek Dox agar and slide culture in humid atmosphere are tried for induction of spores thereby final identification of these isolates. The sporulation may also be achieved by Padhye and Ajello technique, wherein saline agar (1%) supplemented with grass and hay is used. Sporulation by **soil extract medium** designed by Mohapatra *et al* gives very good results. *C.recurvatus* and *Mucor circinelloides* are thermally dimorphic fungi that grow *in vitro* as yeast-like at 37°C.

Most of the fungi of the order Mucorales reproduce asexually as well as sexually and both types of reproductive structures are essential to speciate mucormycetes. As far as **asexual sporulation** is concerned, members of family Mucoraceae have characteristic sporangia of different size and shapes, which contain numerous sporangiospores. In addition, presence and location of rhizoids, morphology of apophysis and columellae, branching or non-branching pattern of conidiophores, are other essential morphological features to differentiate among various genera.

However, **three genera** of Mucorales do not produce sporangia rather they have vesicles, somewhat like *Aspergillus* species, which bears spores either through **sterigmata** (sporangioles - *Cunninghamella* and *Cokeromyces*) or finger-like structure (merosporangia - *Syncephalastrum*). Therefore, one should be vigilant enough not to get confused with *Aspergillus* species while examining the cultural isolate in LCB mount under microscope, particularly during initial period of growth of these three mucormycetes.

The **sexual sporulation** is also equally significant in establishing identity of mucormycetes. The sexual spores in some of species may be encountered in routine cultures more frequently and are produced without much difficulty

Fig. 26.15. White mycelial growth of *A.variabilis* growing in the container after 72 hours over the specimen without adding anything.

as compared to other fungi. However, in some of the species these are to be actively induced by providing different substrates.

The biochemical reactions and nutritive requirements are not very helpful to speciate these fungi, however, temperature requirements of growth do vary between species. In some of Mucorales, exoantigen test have been tried for establishing their identification.

The morphology of *Apophysomyces* in culture is somewhat similar to that of *Lichtheimia* species. However, *Apophysomyces* has dark brown campanulate or funnel-shaped apophysis and an exceptionally darkly pigmented, thick-walled zone of sporangiophore just below apophysis. In addition, *Apophysomyces* has foot cell that resemble those produced by *Aspergillus* or *Fusarium* species.

The morphological descriptions of commonly encountered mucormycetes in clinical material, are given below in an alphabetical order:

1. *Apophysomyces* Species*:* The colonies of *Apophysomyces* grow as floccose aerial mycelium, demonstrating confluent growth in 24 to 48 hours on conventional fungal culture media exceptionally irrespective of presence of cycloheximide. The growth is very aggressive and is similar to that of other ***'lid-lifting'*** Mucorales. The older colonies are cream or brownish gray **(Fig. 26.15 and *See* Fig. 26.1)**. This fungus produces stolons and rhizoids and short coenocytic sporangiophores that terminate in a columellate, apophysate, multispored sporangium with a deliquescent wall. Some of apophyses are constricted below sporangium. The sporangiospores are globose to ellipsoid, smooth-walled. Sporangiophores sometimes end in a **'foot-cell-like'** structure from which arises tuft of rhizoids. As observed in *S.vasiformis*, it is difficult to get sporulation on ordinary culture media. However methods have been devised to produce identifiable reproductive structures on modified water culture. The apophyses are thick-walled and light brown and have a prominent bell or flask shape or **'champagne glass'** shaped, feature, differentiating it from *Lichtheimia corymbifera*, which has less prominent flask-shaped apophyses **(Figs. 26.16A and B)**. Moreover, just below apophysis, *Apophysomyces* has a brown-pigmented thickening in sporangiophore, which is lacking in *Lichtheimia* species. Sexual reproduction with zygospore production has not been seen in this fungus. *A.elegans* and *S.vasiformis* produce rhizoidal complexes from **foot-cell-like** structures at base of sporangiophore. These rhizoids are submerged into culture medium, anchoring fungus onto substrate. Sporangiophores often have a hyperpigmented region of thickening just below apophysis. This finding is unique to *Apophysomyces* genus. At present *Apophysomyces* has five clinically significant species *i.e. A.elegans, A.trapeziformis, A.ossiformis, A.variabilis* and *A.mexicanus.* In India, *A.variabilis*, is the most common. *A.trapeziformis* and *A.mexicanus* are found prevalent in USA and Mexico, respectively.

2. *Cokeromyces recurvatus:* This fungus was described by Shanor *et al* in 1950 from Illinois and Gilman *et al* isolated it from Iowa (USA) in 1957. Subsequently Benny *et al,* described its yeast conversion at 37°C in the tissue section in 1985, which is very rare and unusual feature in mucormycetes. On BHI agar with 5% sheep blood,yeast-like colonies are 3-5 mm in two days and are pasty, gray and slightly wrinkled with large spherical, thick walled cells bearing multiple buds. The mold-like colonies of 15-30 mm in size appear in 3-5 days at room temperature, containing branching, broad, non-septate hyphae of 5-15 µm in diameter and thick-walled, dark brown-black zygospores borne on suspensors. After about ten days of incubation at 30°C, hyaline, smooth-walled, broad sporangiophores with terminating vesicles, elongated, recurved stalks or pedicels, globose sporangiola and sporangiospores as well as zygospores are visible. It produces relatively short sporangiophores that arise directly from the substrate and the apex terminates in a single vesicle that bears several recurved to twisted and contorted pedicels. Sporangia are formed

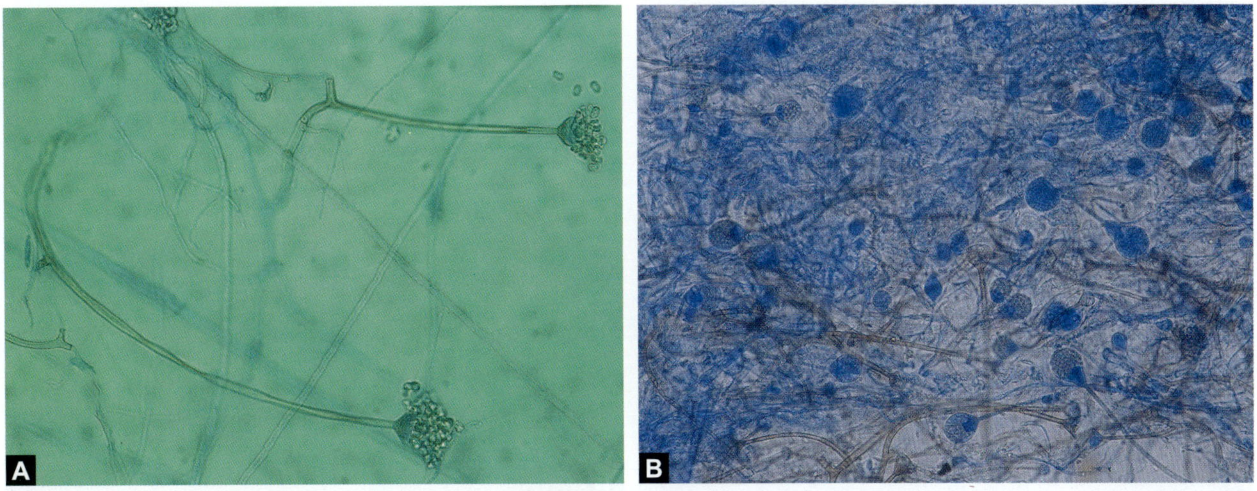

Figs. 26.16A and B. *Apophysomyces variabilis* showing champagne cup appearance (LCB × 400).

at the apex of each pedicel that are globose, columellate, multispored and they have a persistent wall. Zygospores with rough, dark walls and opposed suspensors are formed just above the surface of the substrate. The zygospores and sporangia are formed when grown under 24 hours of fluorescent light but under alternating fluorescent light and dark, the colony formed rings because sporangia are not produced without light. Direct sunlight inhibit sporulation and placing the cultures close to a window receiving incidental sunlight induce sporangial formation but inhibited zygospore production. Incubation temperature is thought to be important for reproduction of this fungus. About ten cases have been reported in the literature caused by this fungus.

3. *Cunninghamella bertholletiae:* The colonies of this genus are rapidly growing and gray in color. *Cunninghamella bertholletiae* is type species encountered in some of the infections. The sporangiophores are erect with a whorl of short lateral branches at apical region, each branch ending in a swollen **vesicle**. One-spored sporangioles, cover almost entire surface **(Figs. 26.17A and B)**. The sporangiospores are oval to tear-shaped. Zygospores are formed in all levels of aerial hyphae, which are having a reddish-brown zygosporangial wall and are formed between opposed suspensors of more or less equal in size.

4. *Lichtheimia* Species: The current name for thermotolerant *Absidia*-like fungi is *Lichtheimia* under family Lichtheimiaceae was described by Hoffmann *et al*, in 2009. There are only two infective species of this genus encountered among human beings constituting about 5% of mucormycosis cases *i.e. Lichtheimia corymbifera* and *L.ramosa* (Old names: *Absidia corymbifera/Mycocladus corymbiferus*). The colonies are fast growing, floccose, white at first becoming pale gray with age. The sporangiophores are hyaline to faintly pigmented, branched and long arising solitary from stolons. They are present in groups of three or in whorls of up to seven. The arrangement of whorls or clusters is called as ***'corymbs'*** hence name of *Lichtheimia* species is given as *corymbifera*. The sporangiophore has **septation**, a single septum is formed in subtending sporangiophore so that walls do not collapse, which helps in differentiation from other mucormycetes like *Apophysomyces* species.

The genus *Lichtheimia* is characterized by differentiation of hyphae into arched stolons bearing more or less verticillate sporangiophores at **internode** and rhizoids formed at point of contact with substrate (at node). Rhizoids, which are usually scarce, are situated between sporangiophores on stolons. This feature differentiates species of *Lichtheimia* from genus *Rhizopus*, where sporangiophore arise from nodes and are therefore found opposite to rhizoids. The sporangia are relatively small, globose, pyriform- or pear-shaped and have a prominent conical columella and are supported by characteristic flask or funnel-shaped apophysis **(*See* Fig. 26.4; Figs. 26.18A to C)**. This feature distinguishes *Lichtheimia* from genera *Mucor* and *Rhizomucor*, which have large, globose sporangia **without** an apophysis. Zygospores are round to somewhat flattened, measuring 50-80 µm in diameter. They are brown with rough walls, with several equatorial ridges.

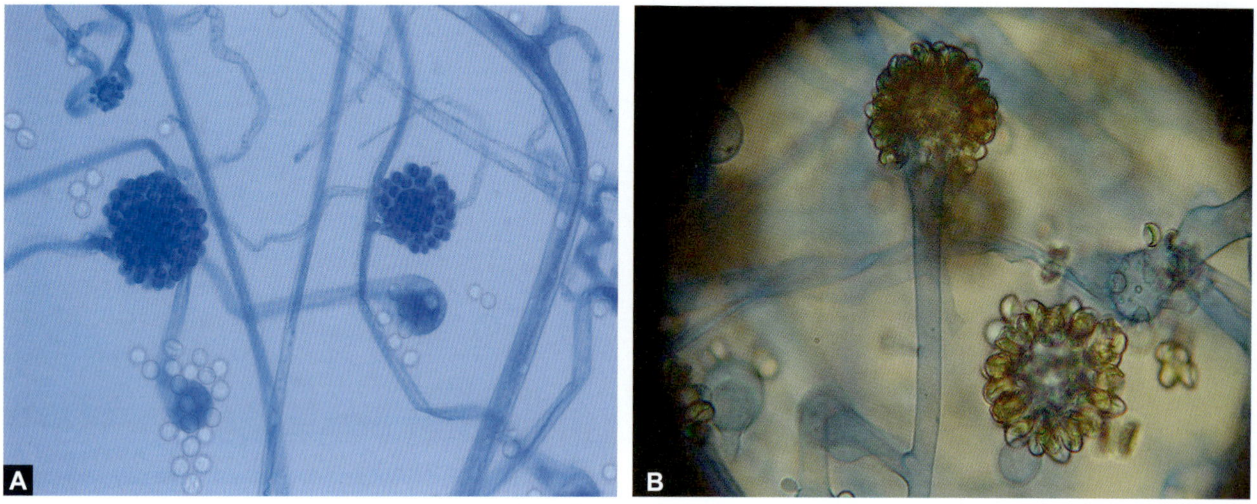

Figs. 26.17A and B. Microscopic appearance of *Cunninghamella bertholletiae* (LCB × 400).

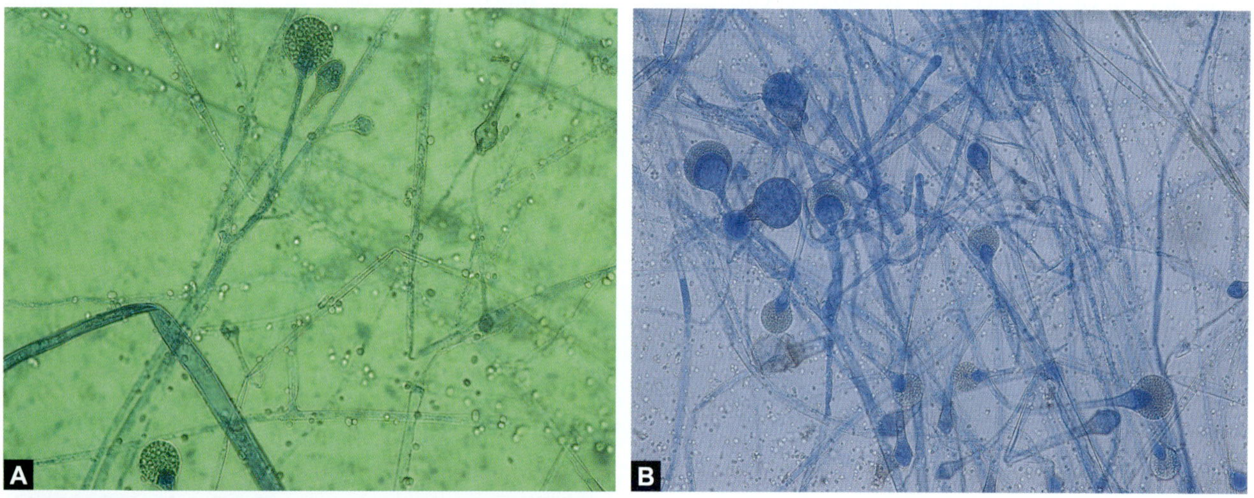

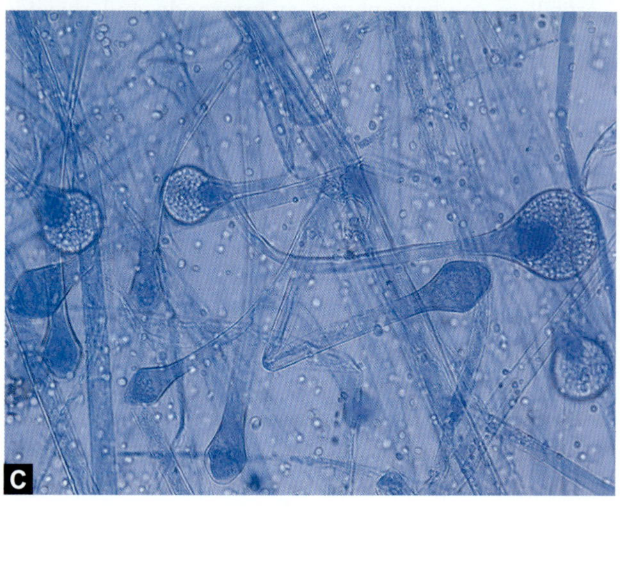

Figs. 26.18A to C. Microscopic appearance of *Lichtheimia corymbifera* (LCB × 400).

The revision of the genus *Absidia* was recently performed on the basis of phylogenetic, physiological and morphological characteristics. A new family (Mycocladiaceae) and the genus *Mycocladus* were proposed to accommodate the three species *Mycocladus corymbiferus* (formerly *Absidia corymbifera), M.blakesleeana* and *M.hyalospora.* More recently, it was suggested that additional nomenclatural changes were necessary and the names *Lichtheimiaceae* and *Lichtheimia* were proposed for the family and the genus, respectively.

L.corymbifera and *L.ramosa* are very similar both macroscopically and microscopically but some differences that delineate these two species. First, by culturing isolates on MEA plates at 30°C for 3 to 4 days, compact growth characterizes *L.corymbifera,* while more effuse growth is suggestive of *L.ramosa.* The sporangiospores of *L.corymbifera* are smooth, hyaline and ellipsoidal when they are mature while those of *L.ramosa* are smooth, lightly colored and more ellipsoidal.

5. *Mucor* Species: The colonies of *Mucor* species are rapidly growing, light to dull gray or brownish, with abundant aerial and substrate mycelia. The hyphae are broad with thin walls, mostly non-septate, sporangiophores are **branched**, bearing spherical sporangia at the tip of each branch. The sporangia are thin-walled with large columellae but **without apophysis** becoming dark as sporangiospores mature. The sporangia are more or less globose and all are of the same type. When sporangia rupture, fragments of wall remain at base of columellae. The sporangiospores are globose, oval or cylindrical with smooth walls. The hyphae are **without** rhizoids or stolons **(*See* Fig. 26.2)**. Zygospores, where known, have more or less equal, opposed, non-appendaged suspensors.

6. *Rhizopus* Species: The commonest species involved *Rhizopus arrhizus,* which used to be previously known as *R.oryzae.* The colonies of *Rhizopus* species are fast spreading white to light gray in the beginning but become darker as sporangia mature. The hyphae are broad non-septate with thin walls submerged or aerial stolon with brown root-like short branches *i.e.* **rhizoids**, under points where sporangiospore develops. The large rhizoids are on stolon immediately adjacent to sporangiophore known as **'nodal'** position. The sporangiophores are hyaline to brown, erect, single or in groups of 2-3, producing sporangia with columellae at tip. The sporangia are globose, hyaline at first, turning dark brown, grey or black as sporangiospores inside it mature. The sporangiospores have longitudinal **striations**, which is a distinctive feature of *R.arrhizus* **(*See* Fig. 26.3; Figs. 26.19A to C)**. The sporangiophores and the spores, both are smaller in size in case *R.microsporus* as its name indicates and may be seen in the same field while doing microscopy, in contrast to the *R.arrhizus* due to longer size of its sporangiophore. Taxonomically, all these species of *Rhizopus* genus have been clubbed together as *Rhizopus microsporus* group. Kokkayil P, *et al,* in 2017 reported three cases of *R.homothallicus* from Delhi and reviewed this emerging fungus in its clinical perspectives in India.

7. *Rhizomucor pusillus:* This is found in a variety of food items including grains, seeds, nuts and beans. It shows wider temperature growth range from 20°–60°C. The colonies are initially colorless to white, becoming gray with passage of time. They have floccose mycelia, sporangiophores are branched irregularly and extensively but sometimes are unbranched. The rhizoids are poorly formed and may be present but occur in between and not at base of sporangiophores. The sporangia are spherical, brown or gray, which rupture at maturity releasing enclosed sporangiospores, which are smooth-walled, hyaline measure 3-5 μm in diameter and may be globose to broadly ellipsoidal. The columellae are slightly pyriform to subglobose and there is **no apophysis**. Chlamydospores are not produced. *Rhizomucor pusillus* is the commonest species of this genus, which is heterothallic, require opposite-orientation mating strains for production of spherical to slightly compressed zygospores. The zygospore formation may be stimulated at higher temperatures of incubation *i.e.* 30–40°C.

8. *Saksenaea vasiformis* Complex: The colonies of *Saksenaea* grow rapidly producing hyaline white, cottony aerial mycelium with a white reverse. The colonies grown on conventional fungal media are generally composed of sterile hyphae. Sparse sporulation on direct culture is occasionally seen. Sterile hyphae may be stimulated to sporulate by stressing fungus under low-nutrient growth conditions. The sporangia are typically long-necked and flask-shaped, arising from dichotomously branched, darkly pigmented rhizoids. They become filled with oblong shaped sporangiospores that are eventually extruded out of apical opening of sporangium. Sporangiophores are distinctive, flask-shaped **(Figs. 26.20A to C)**. Rhizoids often arise near base of sporangiophore, which are dichotomous and get darkened as culture advances in its age.

The sporulation in *Saksenaea* species does not occur in routine culture but requires modified water culture.

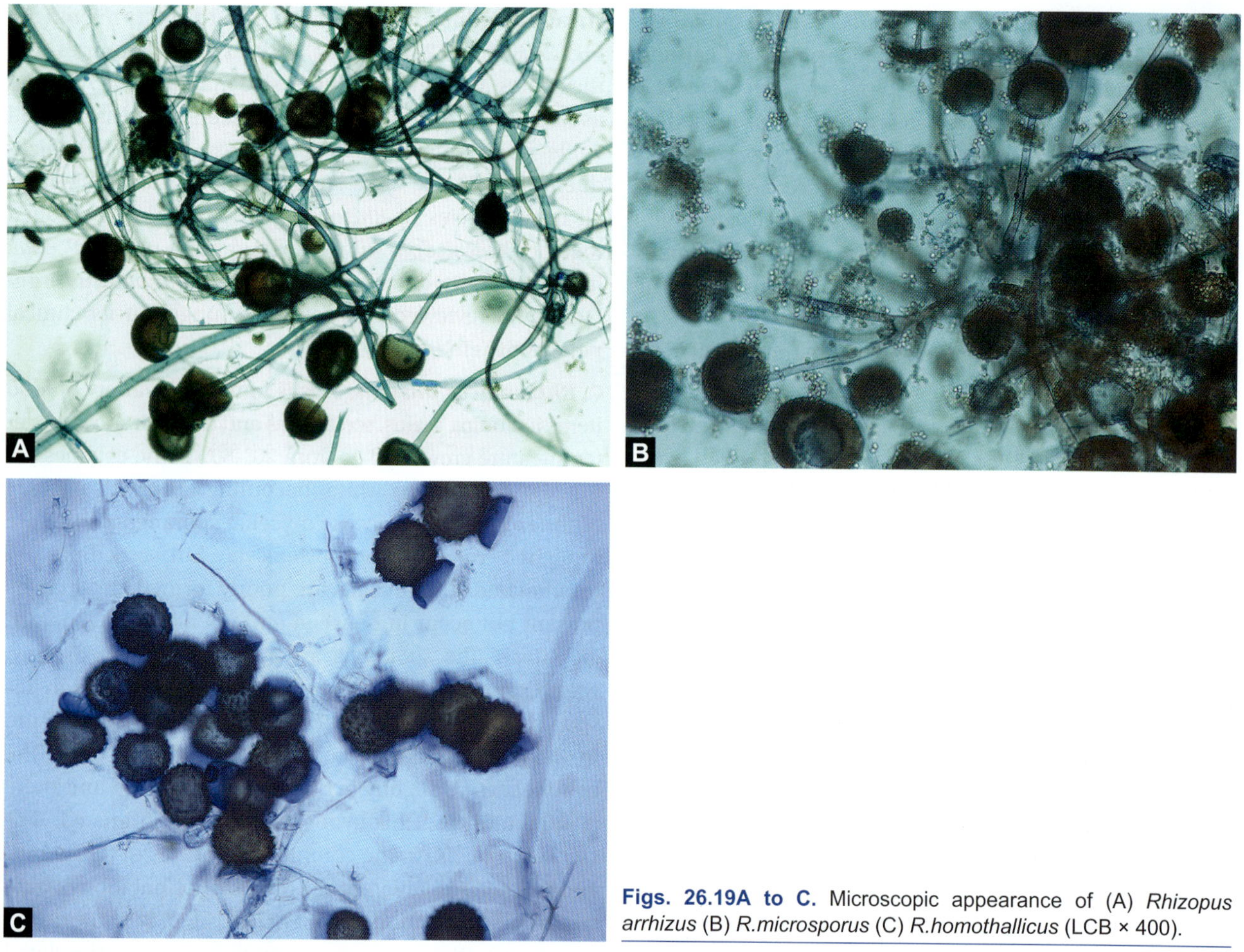

Figs. 26.19A to C. Microscopic appearance of (A) *Rhizopus arrhizus* (B) *R.microsporus* (C) *R.homothallicus* (LCB × 400).

This fungus has additional feature of having a gelatinous plug, occluding apical opening of its sporangium. Sporangiospore release is mediated by dissolution of gelatinous plug, allowing spores to flow out from sporangium. Once sporulation occurs, fungus can be easily identified by its characteristic flask-shaped sporangium and rhizoids complex, as originally described by Dr. S B Saksena. Flask-shaped sporangia form on short sporangiophores. The sporangiophore terminates in a dome-shaped columella. Sporangiophores are unbranched like *Apophysomyces* species. The sporangium forms above columella, with tapered base that extends into a thin neck portion, giving sporangium an overall ***flask- or 'vase'-like*** shape for which organism is named. At the base of sporangiospore, a compact tuft of rhizoids is produced. This rhizoidal complex is brownish color. Like *Apophysomyces,* sexual reproduction with zygospore production has not been seen in this fungus also.

Now, it is established that *Saksenaea vasiformis* is a complex of species. New species have been added *i.e. S.oblongispora,* which is characterized by oblong sporangiospores and unable to grow at 42°C. Another species, *S.erythrospora,* is characterized by large sporangiophores and sporangia with ellipsoid sporangiospores, which are seen biconcave in their lateral view. *S.loutrophoriformis* is new species described by Sutton *et al,* in 2017 from India and USA in two different cases. Out of total four species of *Saksenaea* (*S.vasiformis, S.erythrospora, S.oblongispora* and *S.loutrophoriformis*), all are being reported from India except *S.oblongispora.*

9. *Syncephalastrum recemosum:* The colonies of *Syncephalastrum* species are cottony with aerial mycelia that quickly cover the entire agar surface. The aerial mycelia are white at first that later turn grey to black against a white background. The reverse is white. There are chains

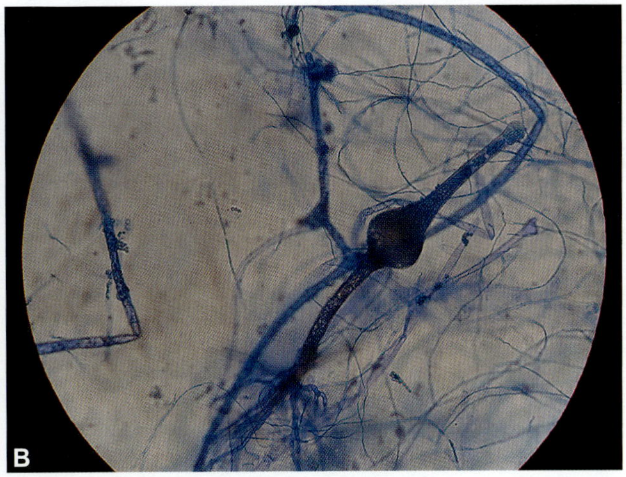

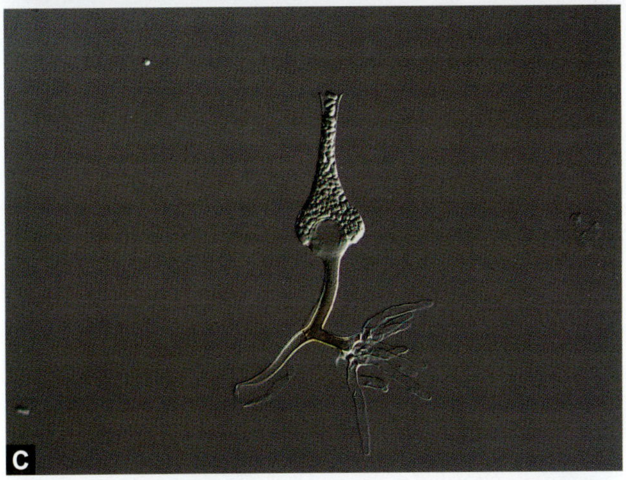

Figs. 26.20A to C. (A) White mycelial growth of *Saksenaea erythrospora* after three days of incubation on SDA. (B and C) Microscopic appearance of *Saksenaea vasiformis* and *Saksenaea loutrophoriformis* (LCB × 400).

of spores formed within long, finger-like tubular sporangia called as **merosporangia** that project from gently swollen tip of short branches of aerial hyphae called sporangiophores **(Figs. 26.21A to F)**. They lack phialides, the feature is helpful to differentiate this genus from microscopically somewhat similar structure of *Aspergillus* genus.

10. *Thamnostylum* Species: There are two clinically important species of this genus *i.e. T.lucknowense* and *T.piriforme.* The first one is already reported from Delhi by Xess *et al* and second is being reported from GMCH, Chandigarh from a patient of cutaneus mucormycosis. The colonies of *T.lucknowense* are fast-growing, initially white and cottony, reaching a diameter of approximately 10 cm after a week incubation. They gradually turn pale yellow to olive brown after a week at 25°C. Aerial sporangiophores are produced from the substrate mycelium and stolons; these usually produced a terminal sporangium and a subterminal cluster of obpyriform sporangiola. Dehisced sporangia bear a distinct columella and a collar. Whorls of obpyriform, recurved pedicillate sporangiola are typically formed on subterminal vesicles on the sporangiophores. Sporangiola mostly produced 8 to 16 sporangiospores that are oval, smooth and hyaline and measure 3 to 7 by 2 to 3 m. *T.lucknowense* was originally isolated from soil in Lucknow, India and designated *Helicostylum lucknowense.* Subsequently, Von Arx and Upadhyay transferred this species to *Thamnostylum* based on the shared production of stolons and rhizoids, apophysate sporangia and pyriforme sporangiola borne on circinate stalks. *T.piriforme* is fast growing, cottony and white to gray on SDA. It forms many fertile vesicles along the sporangiophores. Both possess sporangiola borne on recurved stalks. The rhizoids are occasionally seen **(Figs. 26.22A and B)**.

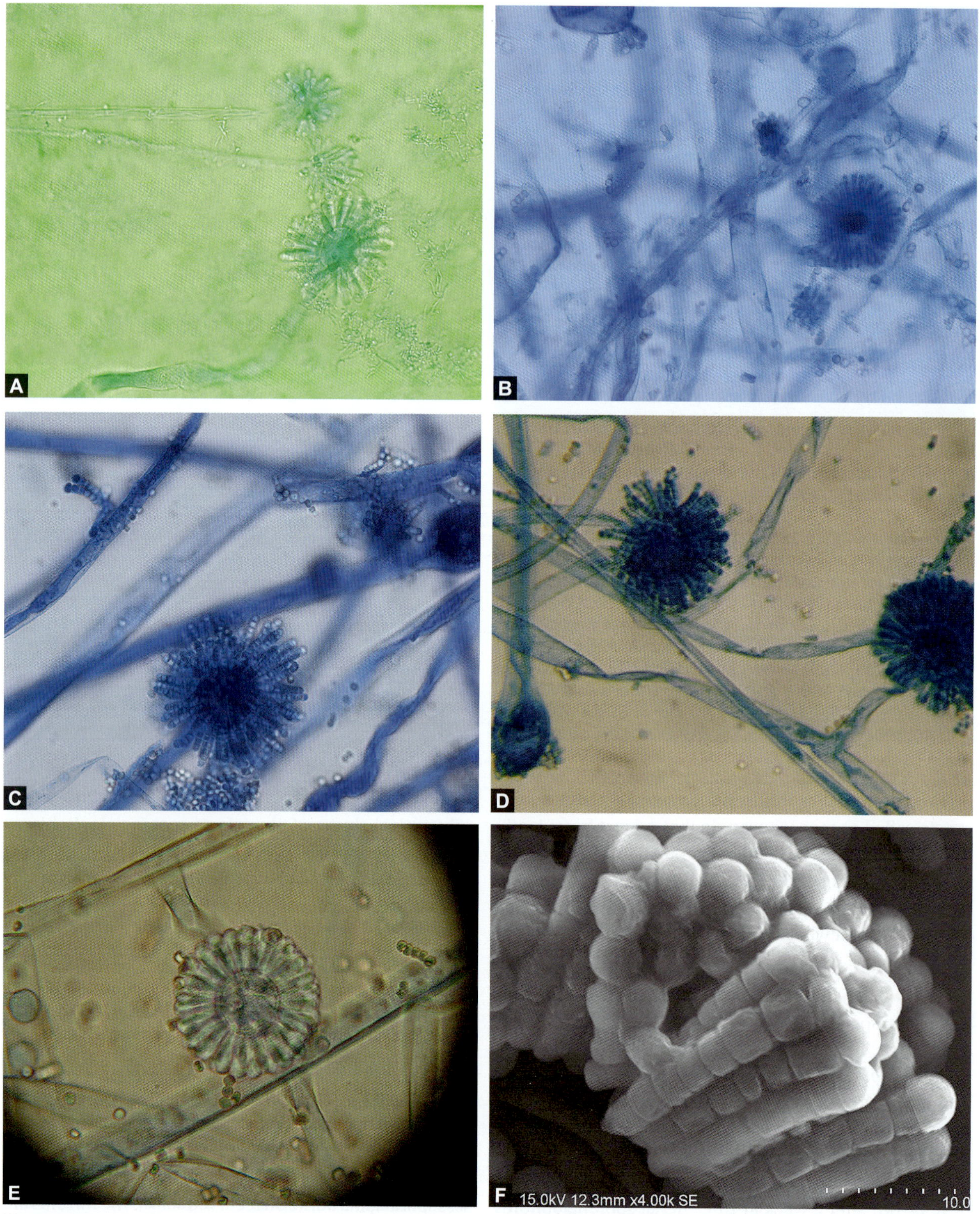

Figs. 26.21A to F. Microscopic appearance of *Syncephalastrum racemosum* (LCB × 400).

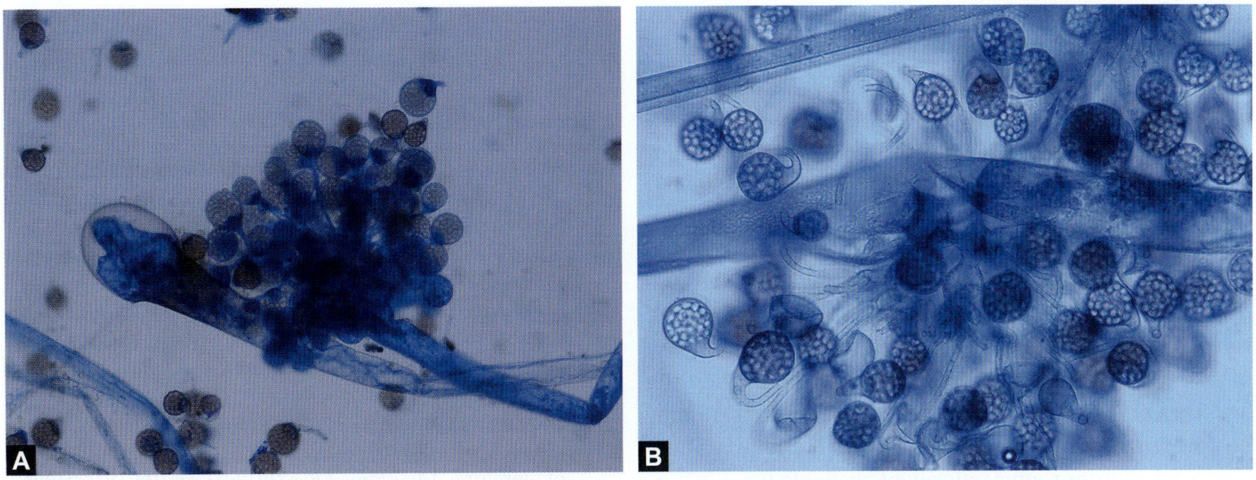

Figs. 26.22A and B. *Thamnostylum piriforme* showing several pedicellate sporangiola (LCB × 200 / 400).

(c) Immunodiagnosis

There is no satisfactory serological test, which can be relied upon to establish the final diagnosis of any of clinical forms of mucormycosis or their causative fungal agent. In a study serological diagnosis of antibody production is 73% sensitive and 100% specific when using homogenate antigens. Serological cross-reactions have been reported among other potentially pathogenic species. Ouchterlony technique of patients' serum demonstrate multiple precipitation bands against cytoplasmic antigen prepared from *Basidiobolus ranarum* isolate. ELISA may also be performed to detect antibodies produced in invasive types of infections. However, serodiagnosis of mucormycosis still remains investigational and cannot yet be recommended for routine use.

(d) Animal Pathogenicity

Mucormycosis has been tried in experimental animals to demonstrate lesions produced by various Mucorales species. Infection in laboratory has been induced in mice and rabbits to define pathogenicity and mechanism of sporangiospore germination *in vivo*. Among clinically significant mucormycetes causing human infection, animal models are available for *Lichtheimia corymbifera, Rhizopus arrhizus, Rhizopus microsporus* var. *rhizopodiformis, Rhizomucor pusillus* and *C.bertholletiae*. Mice are most commonly used animals but experimental models of guinea pigs and rabbits are also tried.

Pre-treatment of experimental animals with cyclophosphamide, corticosteroids, alloxan or streptozocin is done to create an immunocompromised state. Treatment with deferoxamine, an iron chelator, is also used to make animal models. In terms of route of infection, usually airborne route is used for pathophysiological studies in pulmonary infection models but sometimes intravenous injection is preferred, particularly for performing antifungal studies.

Treatment and Prophylaxis

Mucormycosis is life-threatening condition invariably prove to be fatal within very short span, particularly when specific diagnosis is not established well in time thereby delay in institution of proper antifungal therapy. The therapy can be divided in four concomitant approaches. These are (i) rapid correction of underlying predisposing factor of the host like diabetic ketoacidosis; (ii) surgical debridement of necrotizing tissue if this is feasible; (iii) antifungal therapy and (iv) consideration of adjunctive therapy such as hyperbaric oxygen.

The combination of surgical debridement and antifungal drugs is required for an ideal treatment of mucormycosis. The temptation for small daily piecemeal type bedside debridement should be avoided, which further aggravates morbidity and mortality among the patients. It should be preferably done under general anesthesia taking adequate healthy tissue for debridement. Simultaneously management of underlying risk factor, if any, is also essential. There are only few drugs, which are available for treating mucormycosis *i.e.* intravenous amphotericin B, posaconazole and isavuconazole. It is observed that start with higher doses of antifungal drugs instead of step-wise increment. Surgical debridement is the mainstay of therapy for cutaneous mucormycosis and topical amphotericin B is a useful adjunct in concentration of 5 mg/ml, which is applied with a gauge. Skin grafting to repair defects remaining after healing of surgical wound is usually successful. The liposomal amphotericin B is given successfully.

The azole antifungal agents as such are not useful in treating mucormycosis due to lack of both *in vivo* and *in vitro* susceptibility. However, only posaconazole and isavuconazole are found useful in mucormycosis. In case of posaconazole, it should be kept in mind that the drug becomes effective after a fortnight or so. This period may be covered by any other drug like amphotericin B. There is no oral chemoprophylactic agent available for this fungal disease. Cytokines such as interferon-γ and granulocyte-macrophage colony-stimulating factors (GM-CSF) have also been used to treat mucormycosis.

There were no standard guidelines, defined by any of the organizations, as it is available in the case of candidiasis, cryptococcosis or aspergillosis. However, now ESCMID and ECMM have issued joint clinical guidelines for the diagnosis and management of mucormycosis as well as simultaneously similar guidelines released by ECIL, both in the year 2013. The consensus has been derived from experiences of clinicians treating such cases, while framing these guidelines. However, ECIL has further amended its guidelines in 2016 for candidiasis and aspergillosis but no substantial change was done in case of mucormycosis.

This has also been observed that if the entire necrotic material is meticulously removed, antifungals may not be required. The granulation tissue start coming from the base of the wound and it heals only with proper and regular dressing.

As the guidelines sometimes prove to be toothless because there in nothing mandatory. Hence keeping in view of the rapid and devastating course of mucormycosis, every institution should resort on to its local mandatory protocol in Emergency Department, wherein KOH wet mount is to be done in half an hour of the arrival of suspected patient followed by surgical debridement within two hours. Then antifungal regimen is to be instituted as per clinical and financial conditions of patient. Mucormycosis is a surgical emergency to be taken at top priority as slight delay will kill the patient. Proper counseling of patient should also be done keeping in view of the cost of antifungal regimen *vis-a-vis* outcome of the patient, who otherwise tends to leave against medical advice (LAMA). If such measures are not followed in letter and spirit, one is bound to lose the patient of a fully treatable disease.

Further Reading

1. Abboud CS, Bergamasco MD, Baia CE, *et al.* Case report of hepatic mucormycosis after liver transplantation: Successful treatment with liposomal amphotericin B followed by posaconazole sequential therapy. *Transplant Proc.* 2012; **44:** 2501-2.
2. Abidi MZ, Coelho-Prabhu N, Hargreaves J, *et al.* Mucormycosis in patients with inflammatory bowel disease: Case series and review of the literature. *Case Rep Med.* 2014; 637492. PMID: 24872818.
3. Abtahi SM, Omrani M, Karimifar M, *et al.* Wegener's granulomatosis and mucoromycosis: A case study and review of literature. *Adv Biomed Res.* 2012; **1:** 23. PMID: 23210082.
4. Abu El-Naaj I, Leiser Y, Wolff A, *et al.* The surgical management of rhinocerebral mucormycosis. *J Craniomaxillofac Surg.* 2013; **41:** 291-5.
5. Abuzayed B, Al-Abadi H, Al-Otti S, *et al.* Neuronavigation-guided endoscopic endonasal resection of extensive skull base mucormycosis complicated with cerebral vasospasm. *J Craniofac Surg.* 2014; **25:** 1319-23.
6. Adegbola SO, Banerjee A, Mulcahy MP. Rhinocerebral mucormycosis. *Br J Hosp Med (Lond).* 2009; **70:** 298-9.
7. Afolayan O, Copeland H, Hargrove R, *et al.* Successful treatment of invasive pulmonary mucormycosis in an immunocompromised patient. *Ann Thorac Surg.* 2016; **101:** e117-9.
8. Agaronov M, Ratkiewicz I, Lawlor M, *et al. Cokeromyces recurvatus* identified in lung biopsy: Case report of a non-pathogenic fungus, highlighting its potential histologic mimics. *Ann Clin Lab Sci.* 2015; **45:** 209-14.
9. Aggarwal D, Chander J, Janmeja AK, Katyal R. Pulmonary tuberculosis and mucormycosis co-infection in a diabetic patient. *Lung India.* 2015; **32:** 53-5.
10. Agrawal P, Saikia U, Ramanaathan S, *et al.* Neonatal small intestinal zygomycosis misdiagnosed as intussusception in a two-day-old child with a review of the literature. *Fetal Pediatr Pathol.* 2013; **32:** 418-21.
11. Ahamed SK, Al Thobaiti Y. Mucormycosis: A challenge for diagnosis and treatment: Two case reports and review of literature. *Oral Health Dent Manag.* 2014; **13:** 703-6.
12. Ahmadinejad Z, Khazraiyan H, Ghanbari F, *et al.* Cutaneous mucormycosis in a diabetic patient following traditional dressing. *Case Rep Dermatol Med.* 2013; 894927. PMID: 23991340.
13. Air EL, Vagal AA, Kendler A, *et al.* Isolated cerebellar mucormycosis, slowly progressive over 1 year in an immunocompetent patient. *Surg Neurol Int.* 2010; **1:** 81.
14. Al Akhrass F, Debiane L, Abdallah L, *et al.* Palatal mucormycosis in patients with hematologic malignancy and stem cell transplantation. *Med Mycol.* 2011; **49:** 400-5.
15. Al Barbarawi MM, Allouh MZ. Successful management of a unique condition of isolated intracranial mucormycosis

in an immunocompetent child. *Pediatr Neurosurg.* 2015; **50:** 165-7.
16. Al-Jabri S, Al-Abri M, Al-Hinai A, *et al.* Bilateral ocular ischaemic syndrome - Rare complication of rhinocerebral mucormycosis in an Omani Patient: Case report and literature review. *Sultan Qaboos Univ Med J.* 2013; **13:** 137-42.
17. Al-Mendalawi MD. Irreversible fatal renal failure resulting from isolated renal mucormycosis. *Saudi J Kidney Dis Transpl.* 2015; **26:** 597-8.
18. Al-Otaibi F, Albloushi M, Alhindi H, *et al.* Carotid artery occlusion by rhinoorbitocerebral mucormycosis. *Case Rep Surg.* 2012; 812420. PMID: 23198252.
19. Al-Tarrah K, Abdelaty M, Behbahani A, *et al.* Cutaneous mucormycosis postcosmetic surgery: A case report and review of the literature. *Medicine (Baltimore).* 2016; **95:** e4185.
20. Al-Zaydani IA, Al-Hakami AM, Joseph MR, *et al.* Aggressive cutaneous zygomycosis caused by *Apophysomyces variabilis* in an immunocompetent child. *Med Mycol Case Rep.* 2015; **10:** 11-3.
21. Alanio A, Garcia-Hermoso D, Mercier-Delarue S, *et al.* Molecular identification of Mucorales in human tissues: Contribution of PCR electrospray-ionization mass spectrometry. *Clin Microbiol Infect.* 2015; **21:** 594.e1-5.
22. Alastruey-Izquierdo A, Hoffmann K, de Hoog GS, *et al.* Species recognition and clinical relevance of the zygomycetous genus *Lichtheimia* (Syn. *Absidia pro parte, Mycocladus*). *J Clin Microbiol.* 2010; **48:** 2154-70.
23. Alexander BD, Schell WA, Siston AM, *et al.* Fatal *Apophysomyces elegans* infection transmitted by deceased donor renal allografts. *Am J Transplant.* 2010; **10:** 2161-7.
24. Almannai M, Imran H, Estrada B, *et al.* Successful treatment of rhino-orbital mucormycosis with posaconazole and hyperbaric oxygen therapy. *Pediatr Hematol Oncol.* 2013; **30:** 184-6.
25. Alrefai AH, Berger JR, Saadeh RS. Against all odds: Surviving rhino-orbital-cerebral mucormycosis: A case report. *J Neurovirol.* 2012; **18:** 144-7.
26. Alsuhaibani AH, Al-Thubaiti G, Al Badr FB. Optic nerve thickening and infarction as the first evidence of orbital involvement with mucormycosis. *Middle East Afr J Ophthalmol.* 2012; **19:** 340-2.
27. Altini C, Niccoli Asabella A, Ferrari C, *et al.* (18)F-FDG PET/CT contribution to diagnosis and treatment response of rhino-orbital-cerebral mucormycosis. *Hell J Nucl Med.* 2015; **18:** 68-70.
28. Alvarez E, Cano J, Stchigel AM, *et al.* Two new species of *Mucor* from clinical samples. *Med Mycol.* 2011; **49:** 62-72.
29. Alvarez E, Garcia-Hermoso D, Sutton DA, *et al.* Molecular phylogeny and proposal of two new species of the emerging pathogenic fungus *Saksenaea*. *J Clin Microbiol.* 2010; **48:** 4410-6.
30. Alvarez E, Stchigel AM, Cano J, Sutton DA, Fothergill AW, Chander J, *et al.* Molecular phylogenetic diversity of the emerging mucoralean fungus *Apophysomyces*: Proposal of three new species. *Rev Iberoam Micol.* 2010; **27:** 80-9.
31. Alvi AR, Shamsi G. *Rhizopus* necrotizing fasciitis of caesarean wound - A rare life threatening condition. *J Coll Physicians Surg Pak.* 2009; **19:** 579-81.
32. Anand J, Ghazala K, Chong VH. Massive lower gastrointestinal bleeding secondary to colonic mucormycosis. *Med J Malaysia.* 2011; **66:** 266-7.
33. Ananthaneni AR, Undavalli SB, Velagapudi RP, *et al.* Mucormycosis: An atrocious mate of patients with diabetes. *BMJ Case Rep.* 2013; pii: bcr20130096000.
34. Anders UM, Taylor EJ, Martel JR, *et al.* Acute orbital apex syndrome and rhino-orbito-cerebral mucormycosis. *Int Med Case Rep J.* 2015; **8:** 93-6.
35. Angali RK, Jeshtadi A, Namala VA, *et al.* Fatal rhino-orbito-cerebral mucormycosis in a healthy individual. *J Oral Maxillofac Pathol.* 2014; **18:** 460-3.
36. Angeline, Hanifah M, Balachandran G, *et al.* Subcutaneous *Mucor* zygomycosis with potential life-threatening visceral complication. *Indian J Med Microbiol.* 2013; **31:** 182-4.
37. Antonov NK, Tang R, Grossman ME. Utility of touch preparation for rapid diagnosis of cutaneous mucormycosis. *JAAD Case Rep.* 2015; **1:** 175-7.
38. Antony SJ, Parikh MS, Ramirez R, *et al.* Gastrointestinal mucormycosis resulting in a catastrophic outcome in an immunocompetent patient. *Infect Dis Rep.* 2015; **7:** 6031.
39. Aras MH, Kara MI, Erkilic S, *et al.* Mandibular mucormycosis in immunocompromised patients: Report of two cases and review of the literature. *J Oral Maxillofac Surg.* 2012; **70:** 1362-8.
40. Arnaiz-Garcia ME, Alonso-Pena D, Gonzalez-Vela Mdel C, *et al.* Cutaneous mucormycosis: Report of five cases and review of the literature. *J Plast Reconstr Aesthet Surg.* 2009; **62:** e434-41.
41. Arndt S, Aschendorff A, Echternach M, *et al.* Rhino-orbital-cerebral mucormycosis and aspergillosis: Differential diagnosis and treatment. *Eur Arch Otorhinolaryngol.* 2009; **266:** 71-6.
42. Arockiaraj J, Balaji G, Ashok A, *et al.* Amphotericin B cement beads: A good adjunctive treatment for musculoskeletal mucormycosis. *Indian J Orthop.* 2012; **46:** 369-72.
43. Arroyo MA, Schmitt BH, Davis TE, *et al.* Detection of the dimorphic phases of *Mucor circinelloides* in blood cultures from an immunosuppressed female. *Case Rep Infect Dis.* 2016; 3720549. PMID: 27777804.
44. Austin CL, Finley PJ, Mikkelson DR, *et al.* Mucormycosis: A rare fungal infection in tornado victims. *J Burn Care Res.* 2014; **35:** e164-71.
45. Avcu G, Karapinar DY, Yazici P, *et al.* Difficult diagnosis of invasive fungal infection predominantly involving the lower gastrointestinal tract in acute lymphoblastic leukaemia. *Med Mycol Case Rep.* 2016; **11:** 1-4.
46. Ayaz M, Moein R. Myocutaneous mucormycosis in a diabetic burnt patient led to upper extremity amputation: A case report. *Bull Emerg Trauma.* 2017; **5:** 58-62.
47. Azarpira N, Ashraf MJ, Kazemi K, *et al.* Rhinomaxillary mucormycosis in a renal transplant recipient: Case report. *Exp Clin Transplant.* 2012; **10:** 605-8.

48. Badiee P, Arastefar A, Jafarian H. Comparison of histopathological analysis, culture and polymerase chain reaction assays to detect mucormycosis in biopsy and blood specimens. *Iran J Microbiol.* 2013; **5:** 406-10.
49. Badiee P, Jafarpour Z, Alborzi A, *et al.* Orbital mucormycosis in an immunocompetent individual. *Iran J Microbiol.* 2012; **4:** 210-4.
50. Badior M, Trigo F, Eloy C, *et al. Mucor* infection: Difficult diagnosis. *Clin Drug Investig.* 2013; **33:** S19-21.
51. Bae MS, Kim EJ, Lee KM, *et al.* Rapidly progressive rhino-orbito-cerebral mucormycosis complicated with unilateral internal carotid artery occlusion: A case report. *Neurointervention.* 2012; **7:** 45-9.
52. Bala K, Chander J, Handa U, Punia RS, Attri AK. A prospective study of mucormycosis in north India: Experience from a tertiary care hospital. *Med Mycol.* 2015; **53:** 248-57.
53. Bansal M, Martin SR, Rudnicki SA, *et al.* A rapidly progressing Pancoast syndrome due to pulmonary mucormycosis: A case report. *J Med Case Rep.* 2011; **5:** 388.
54. Baradkar VP, Kumar S. Cutaneous zygomycosis due to *Saksenaea vasiformis* in an immunocompetent host. *Indian J Dermatol.* 2009; **54:** 382-4.
55. Baradkar VP, Mathur M, Panda M, *et al.* Sino-orbital infection by *Syncephalastrum recemosum* in chronic hepatorenal disease. *J Oral Maxillo Facial Pathol.* 2008; **12:** 45-7.
56. Baradkar VP, Mathur M, Taklikar S, *et al.* Fatal rhino-orbito-cerebral infection caused by *Saksenaea vasiformis* in an immunocompetent individual: First case report from India. *Indian J Med Microbiol.* 2008; **26:** 385-7.
57. Barizzi J, Merlo E, Grassi P, *et al.* Vaginal colonisation by *Mucor circinelloides.* Case report with cytopathology, molecular sequencing and epidemiology. *Cytopathology.* 2016; **27:** 491-4.
58. Barnajian M, Gioia W, Iordache F, *et al.* Mucormycosis-induced colon perforation after renal transplantation. *Surg Infect (Larchmt).* 2014; **15:** 665-6.
59. Beatty N, Al Mohajer M. Primary cutaneous mucormycosis developing after incision and drainage of a subcutaneous abscess in an immunocompetent host. *BMJ Case Rep.* 2016. pii: bcr2015213700. PMID: 26729834.
60. Bellazreg F, Hattab Z, Meksi S, *et al.* Outcome of mucormycosis after treatment: Report of five cases. *New Microbes New Infect.* 2014; **6:** 49-52.
61. Benachinmardi KK, Rajalakshmi P, Veenakumari HB, *et al.* Successful treatment of primary cerebral mucormycosis: Role of microbiologist. *Indian J Med Microbiol.* 2016; **34:** 550-3.
62. Bergantim R, Rios E, Trigo F, *et al.* Invasive co-infection with *Aspergillus* and *Mucor* in a patient with acute myeloid leukemia. *Clin Drug Investig.* 2013; **33:** S51-5.
63. Bernal-Martinez L, Buitrago MJ, Castelli MV, *et al.* Development of a single tube multiplex real-time PCR to detect the most clinically relevant Mucormycetes species. *Clin Microbiol Infect.* 2013; **19:** e1-7.
64. Berne JD, Villarreal DH, McGovern TM, *et al.* A fatal case of post-traumatic gastric mucormycosis. *J Trauma.* 2009; **66:** 933-5.
65. Bhadani PP, Bhadani UK, Thapliyal N, *et al.* A rare presentation of invasive rhino-orbital mucormycosis in an immunocompetent young girl: a case report. *Indian J Pathol Microbiol.* 2007; **50:** 785-6.
66. Bharathi R, Arya AN. Mucormycosis in an immunocompetent patient. *J Oral Maxillofac Pathol.* 2012; **16:** 308-9.
67. Bhutada K, Borkar SS, Mendiratta DK, *et al.* Successful treatment of peritonitis by *C.bertholletiae* in a chronic kidney failure patient on continuous ambulatory peritoneal dialysis after kidney rejection. *Singapore Med J.* 2012; **53:** e106-9.
68. Bialek R, Konrad F, Kern J, *et al.* PCR based identification and discrimination of agents of mucormycosis and aspergillosis in paraffin wax embedded tissue. *J Clin Pathol.* 2005; **58:** 1180-4.
69. Bibashi E, de Hoog GS, Pavlidis TE, *et al.* Wound infection caused by *Lichtheimia ramosa* due to a car accident. *Med Mycol Case Rep.* 2012; **2:** 7-10.
70. Binder U, Maurer E, Lass-Florl C. Mucormycosis - From the pathogens to the disease. *Clin Microbiol Infect.* 2014; **20 (Suppl. 6):** 60-6.
71. Bini R, Addeo A, Maganuco L, *et al.* The role of surgery in a case of diffuse mucormycosis with haematemesis and gastric necrosis. *Ann R Coll Surg Engl.* 2014; **96:** e31-3.
72. Bird J, Telang G, Robinson-Bostom L. Two pink nodules in a patient with acute myeloid leukemia. *J Cutan Pathol.* 2014; **41:** 483-6.
73. Bist SS, Varshney S, Bisht M, *et al.* Isolated palate ulcer due to mucormycosis. *Indian J Otolaryngol Head Neck Surg.* 2008; **60:** 79-82.
74. Biswas D, Kotwal A, Kakati B, *et al.* Amphotericin B resistant *Apophysomyces elegans* causing rhino-oculo-cerebral mucormycosis in an immunocompetent host. *J Clin Diagn Res.* 2015; **9:** DD01-2.
75. Bitar D, Morizot G, Van Cauteren D, *et al.* Estimating the burden of mucormycosis infections in France (2005-2007) through a capture-recapture method on laboratory and administrative data. *Rev Epidemiol Sante Publique.* 2012; **60:** 383-7.
76. Bitar D, Van Cauteren D, Lanternier F, *et al.* Increasing incidence of zygomycosis (mucormycosis), France, 1997-2006. *Emerg Infect Dis.* 2009; **15:** 1395-401.
77. Blanchet D, Dannaoui E, Fior A, *et al. Saksenaea vasiformis* infection, French Guiana. *Emerg Infect Dis.* 2008; **14:** 342-4.
78. Bonifaz A, Macias B, Paredes-Farrera F, *et al.* Palatal zygomycosis: Experience of 21 cases. *Oral Dis.* 2008; **14:** 569-74.
79. Bonifaz A, Stchigel AM, Guarro J, *et al.* Primary cutaneous mucormycosis produced by the new species *Apophysomyces mexicanus. J Clin Microbiol.* 2014; **52:** 4428-31.
80. Bonifaz A, Tirado-Sanchez A, Calderon L, *et al.* Mucormycosis in children: A study of 22 cases in a Mexican hospital. *Mycoses.* 2014; **57 (Suppl. 3):** 79-84.
81. Bonifaz A, Vazquez-Gonzalez D, Tirado-Sanchez A, *et al.* Cutaneous zygomycosis. *Clin Dermatol.* 2012; **30:** 413-9.
82. Busca A, Limerutti G, Locatelli F, *et al.* The reversed halo sign as the initial radiographic sign of pulmonary zygomycosis. *Infection.* 2012; **40:** 77-80.

83. Camara-Lemarroy CR, Gonzalez-Moreno EI, Rodriguez-Gutierrez R, *et al.* Clinical features and outcome of mucormycosis. *Interdiscip Perspect Infect Dis.* 2014; 562610. PMID: 25210515.
84. Camargo JF, Yakoub D, Cho-Vega JH. Successful treatment of primary cutaneous mucormycosis complicating anti-TNF therapy with a combination of surgical debridement and oral posaconazole. *Mycopathologia.* 2015; **180:** 187-92.
85. Campbell A, Cooper C, Davis S. Disseminated mucormycosis in a paediatric patient: *Lichtheimia corymbifera* successfully treated with combination antifungal therapy. *Med Mycol Case Rep.* 2014; **6:** 18-21.
86. Canete-Gomez J, Segura-Sampedro JJ, Reguera-Rosal J. An unusual cause of pneumoperitoneum. *Gastroenterology.* 2014; **147:** e3-5.
87. Carceller F, Onoro G, Buitrago MJ, *et al. Cunninghamella bertholletiae* infection in children: Review and report of two cases with disseminated infection. *J Pediatr Hematol Oncol.* 2014; **36:** e109-14.
88. Casqueiro J, Casqueiro J, Alves C. Infections in patients with diabetes mellitus: A review of pathogenesis. *Indian J Endocrinol Metab.* 2012; **16:** S27-36.
89. Chahal HS, Abgaryan N, Lakshminarayanan R, *et al.* Orbital mucormycosis following periorbital cutaneous infection. *Ophthal Plast Reconstr Surg.* 2015; PMID: 25853503.
90. Chakrabarti A, Chatterjee SS, Das A, *et al.* Invasive zygomycosis in India: Experience in a tertiary care hospital. *Postgrad Med J.* 2009; **85:** 573-81.
91. Chakrabarti A, Dhaliwal M. Epidemiology of mucormycosis in India. *Curr Fungal Infect Rep.* 2013; **7:** 287-92.
92. Chakrabarti A, Shivaprakash MR, Curfs-Breuker I, *et al. Apophysomyces elegans:* epidemiology, amplified fragment length polymorphism typing and *in vitro* antifungal susceptibility pattern. *J Clin Microbiol.* 2010; **48:** 4580-5.
93. Chakrabarti A, Singh R. Mucormycosis in India: Unique features. *Mycoses.* 2014; **57 (Suppl. 3):** 85-90.
94. Chakravarti A, Bhargava R, Bhattacharya S. Cutaneous mucormycosis of nose and facial region in children: A case series. *Int J Pediatr Otorhinolaryngol.* 2013; **77:** 869-72.
95. Chambers CJ, Reyes Merin M, Fung MA, *et al.* Primary cutaneous mucormycosis at sites of insulin injection. *J Am Acad Dermatol.* 2011; **64:** e79-81.
96. Chamdine O, Gaur AH, Broniscer A. Effective treatment of cerebral mucormycosis associated with brain surgery. *Pediatr Infect Dis J.* 2015; **34:** 542-3.
97. Chander J, Kaur J, Attri A, Mohan H. Primary cutaneous zygomycosis from a tertiary care centre in north-west India. *Indian J Med Res.* 2010; **131**: 765-70.
98. Chander J, Kaur M, Bhalla M, Punia RS, Singla N, Bhola K, Alastruey-Izquierdo A, Stchigel AM, Guarro J. Changing epidemiology of Mucoralean fungi: Chronic cutaneous infection caused by *Mucor irregularis. Mycopathologia.* 2015; **180:** 181-6.
99. Chander J, Singla N, Kaur M, Punia RS, Attri A, Alastruey-Izquierdo A, Cano-Lira JF, Stchigel AM, Guarro J. *Saksenaea erythrospora*, an emerging mucoralean fungus causing severe necrotizing skin and soft tissue infections - A study from a tertiary care hospital in north India. *Infect Dis (Lond).* 2017; **49:** 170-7.
100. Chander J, Stchigel AM, Alastruey-Izquierdo A, Jayant M, Bala K, Rani H, Handa U, Punia RS, Dalal U, Attri AK, Monzon A, Cano-Lira JF, Guarro J. Fungal necrotizing fasciitis, an emerging infectious disease caused by *Apophysomyces* (Mucorales). *Rev Iberoam Micol.* 2015; **32:** 93-8.
101. Chandran R, Mishra P, Pawar DK, *et al.* Internal jugular vein thrombosis from rhino-cerebral mucormycosis: Be careful before cannulation. *Indian J Crit Care Med.* 2014; **18:** 546-7.
102. Charles P, Kahn JE, Ackermann F, *et al.* Renal mucormycosis complicating extracorporeal membrane oxygenation. *Med Mycol.* 2013; **51:** 193-5.
103. Chaudhry A, Hirano SA, Hayes TJ, *et al.* Fatal rhino-orbito-cerebral mucormycosis in a patient with liver disease. *J Am Acad Dermatol.* 2011; **65:** 241-3.
104. Chawla B, Sharma S, Kashyap S, *et al.* Primary orbital mycosis in immunocompetent infants. *J AAPOS.* 2011; **15:** 211-3.
105. Chawla N, Reddy SJ, Agrawal M. Ileocolic mucormycosis causing intestinal obstruction. *Indian J Med Microbiol.* 2012; **30:** 373-4.
106. Cheng VC, Chan JF, Ngan AH, *et al.* Outbreak of intestinal infection due to *Rhizopus microsporus. J Clin Microbiol.* 2009; **47:** 2834-43.
107. Chethan K, Prasad S, Ramachandra L. Ileocolic mucormycosis - An unusual cause of a mass in the right iliac fossa. *Can J Infect Dis Med Microbiol.* 2012; **23:** e65-6.
108. Chhaya V, Gupta S, Arnaout A. Mucormycosis causing giant gastric ulcers. *Endoscopy.* 2011; **43 (Suppl. 2):** e289-90.
109. Chi M, Kim HJ, Basham R, *et al.* Temporal artery calciphylaxis presenting as temporal arteritis in a case of rhinoorbitocerebral mucormycosis. *Ophthal Plast Reconstr Surg.* 2015; **31:** e132-5.
110. Chinen K, Matsumoto H, Fujioka Y. Cardiac mucormycosis presenting as a "fungus ball" in the left atrium. *Intern Med.* 2009; **48:** 1781-2.
111. Chitasombat MN, Kontoyiannis DP. Treatment of mucormycosis in transplant patients: Role of surgery and of old and new antifungal agents. *Curr Opin Infect Dis.* 2016; **29:** 340-5.
112. Cho E, Kim MH. What is your diagnosis? Rhino-orbital-cerebral mucormycosis. *Cutis.* 2014; **94:** 168 & 195-6.
113. Choi HL, Shin YM, Lee KM, *et al.* Bowel infarction due to intestinal mucormycosis in an immunocompetent patient. *J Korean Surg Soc.* 2012; **83:** 325-9.
114. Choi WT, Chang TT, Gill RM. Gastrointestinal zygomycosis masquerading as acute appendicitis. *Case Rep Gastroenterol.* 2016; **10:** 81-7.
115. Choo JY, Park CM, Lee HJ, *et al.* Sequential morphological changes in follow-up CT of pulmonary mucormycosis. *Diagn Interv Radiol.* 2014; **20:** 42-6.
116. Chow V, Khan S, Balogun A, *et al.* Invasive rhino-orbito-cerebral mucormycosis in a diabetic patient - The need for prompt treatment. *Med Mycol Case Rep.* 2014; **8:** 5-9.

117. Chowdhary A, Kathuria S, Singh PK, *et al.* Molecular characterization and *in vitro* antifungal susceptibility of 80 clinical isolates of mucormycetes in Delhi, India. *Mycoses.* 2014; **57 (Suppl. 3):** 97-107.
118. Christy MR, Shridharani SM. Appearance of a rapidly expanding facial eschar in a severely injured trauma patient. *JAMA.* 2012; **307:** 1080-1.
119. Chugh KS, Padhye AA, Chakrabarti A, *et al.* Renal zygomycosis in otherwise healthy hosts. *J Mycol Med.* 1996; **6:** 22-25.
120. Chugh KS, Sakhuja V, Gupta KL, *et al.* Renal mucormycosis: computerised tomography findings and their diagnostic significance. *Am J Kidney Dis.* 1993; **22:** 393-97.
121. Chung JH, Godwin JD, Chien JW, *et al.* Pulmonary mucormycosis. *Radiology.* 2010; **256:** 667-70.
122. Cloyd JM, Brown J, Sinclair T, *et al.* Gastrointestinal mucormycosis initially manifest as hematochezia from arterio-enteric fistula. *Dig Dis Sci.* 2014; **59:** 2905-8.
123. Colon M, Romaguera J, Mendez K, *et al.* Mucormycosis of the vulva in an immunocompromised pediatric patient. *Bol Asoc Med P R.* 2013; **105:** 65-7.
124. Corey KE, Gupta NK, Agarwal S, *et al.* A 55-year-old woman with autoimmune hepatitis, cirrhosis, anorexia and abdominal pain. *N Engl J Med.*; **369:** 1545-53.
125. Cornely OA, Arikan-Akdagli S, Dannaoui E, *et al.* ESCMID and ECMM joint clinical guidelines for the diagnosis and management of mucormycosis 2013. *Clin Microbiol Infect.* 2014; **20 (Suppl. 3):** 5-26.
126. Cornely OA. Current controversies in treating invasive zygomycosis. *J Pediatr Hematol Oncol.* 2010; **32:** 83-4.
127. Cortez J, Gomes BC, Speidel A, *et al.* Mind the gap: Management of an emergent and threatening invasive fungal infection - A case report of rhino-orbital-cerebral and pulmonary mucormycosis. *Med Mycol Case Rep.* 2013; **2:** 79-84.
128. Corzo-Leon DE, Chora-Hernandez LD, Rodriguez-Zulueta AP, *et al.* Diabetes mellitus as the major risk factor for mucormycosis in Mexico: Epidemiology, diagnosis and outcomes of reported cases. *Med Mycol.* 2017; doi: 10.1093/mmy/myx017. PMID: 28431008.
129. Dabritz J, Attarbaschi A, Tintelnot K, *et al.* Mucormycosis in paediatric patients: Demographics, risk factors and outcome of 12 contemporary cases. *Mycoses.* 2011; **54:** e785-8.
130. Dai Y, Walker JW, Halloush RA, Khasawneh FA. Mucormycosis in two community hospitals and the role of infectious disease consultation: A case series. *Int J Gen Med.* 2013; **6:** 833-8.
131. Dalal P. An unusual cause of renal failure: Disseminated mucormycosis. *J Nephropathol.* 2012; **1:** 188-9.
132. Dan M. Mucormycosis of the head and neck. *Curr Infect Dis Rep.* 2011; **13:** 123-31.
133. Dang CJ, Li YJ, Zhan FH, *et al.* The appearance of pulmonary mucormycosis on FDG PET/CT. *Clin Nucl Med.* 2012; **37:** 801-3.
134. Danion F, Aguilar C, Catherinot E, *et al.* Mucormycosis: New developments into a persistently devastating infection. *Semin Respir Crit Care Med.* 2015; **36:** 692-705.
135. Dave SP, Vivero RJ, Roy S. Facial cutaneous mucormycosis in a full-term infant. *Arch Otolaryngol Head Neck Surg.* 2008; **134:** 206-9.
136. Davies BW, Smith JM, Hink EM, *et al.* Increased Incidence of rhino-orbital-cerebral mucormycosis after Colorado flooding. *Ophthal Plast Reconstr Surg.* 2017; **33:** e1-3.
137. Davoudi S, Graviss LS, Kontoyiannis DP. Healthcare-associated outbreaks due to Mucorales and other uncommon fungi. *Eur J Clin Invest.* 2015; **45:** 767-73.
138. Davuodi S, Manshadi SA, Salehi MR, *et al.* Fatal cutaneous mucormycosis after kidney transplant. *Exp Clin Transplant.* 2015; **13:** 82-5.
139. de Chaumont A, Pierret C, Janvier F, *et al.* Mucormycosis: A rare complication of an amputation. *Ann Vasc Surg.* 2014; **28:** 1035. e15-9.
140. de Hoog S, Ibrahim AS, Voigt K. Zygomycetes: An emerging problem in the clinical laboratory. *Mycoses.* 2014; **57 (Suppl. 3):** 1. PMID: 25175704.
141. de Leonardis F, Perillo T, Giudice G, *et al.* Recurrent rhino-ocular-cerebral mucormycosis in a leukemic child: A case report and review of pediatric literature. *Pediatr Rep.* 2015; **7:** 5938.
142. Dehority W, Willert J, Pong A. Zygomycetes infections in pediatric hematology oncology patients: a case series and review of the literature. *J Pediatr Hematol Oncol.* 2009; **31:** 911-9.
143. dela Cruz WP, Calvano TP, Griffith ME, *et al.* Invasive *Apophysomyces variabilis* infection in a burn patient. *J Clin Microbiol.* 2012; **50:** 2814-7.
144. Devana SK, Bora GS, Mavuduru RS, *et al.* Successful management of renal mucormycosis with antifungal therapy and drainage. *Indian J Urol.* 2016; **32:** 154-5.
145. Devi SC, Kanungo R, Barreto E, *et al.* Favorable outcome of amphotericin B treatment of zygomycotic necrotizing fasciitis caused by *Apophysomyces elegans. Int J Dermatol.* 2008; **47:** 407-9.
146. Devidayal, Menon P, Vaiphei K, *et al.* Pleuropulmonary zygomycosis in a diabetic child. *Clin Pediatr Endocrinol.* 2009; **18:** 35-40.
147. Dhakar MB, Rayes M, Kupsky W, *et al.* A cryptic case: Isolated cerebral mucormycosis. *Am J Med.* 2015; **128:** 1296-9.
148. Dhaliwal HS, Singh A, Sinha SK, *et al.* Diagnosed only if considered: Isolated renal mucormycosis. *Lancet.* 2015; **385:** 2322.
149. Dhooria S, Agarwal R, Chakrabarti A. Mediastinitis and bronchial perforations due to mucormycosis. *J Bronchology Interv Pulmonol.* 2015; **22:** 338-42.
150. Dhua AK, Sinha S, Sarin YK, *et al.* Isolated mucormycosis in a post-pyeloplasty kidney in an immunocompetent child. *J Indian Assoc Pediatr Surg.* 2012; **17:** 132-4.
151. Di Carlo P, Pirrello R, Guadagnino G, *et al.* Multimodal surgical and medical treatment for extensive rhinocerebral mucormycosis in an elderly diabetic patient: A case report and literature review. *Case Rep Med.* 2014; 527062.
152. Di Pentima MC, Chan S, Powell J, *et al.* Topical amphotericin B in combination with standard therapy for severe

necrotizing skin and soft-tissue mucormycosis in an infant with bilineal leukemia: Case report and review. *J Pediatr Hematol Oncol.* 2014; **36:** e468-70.

153. Dien Bard J, Mangahis A, Hofstra TC, *et al.* First case report of bloodstream infection by *Rhizomucor pusillus* in a child with hemophagocytic lymphohistiocytosis. *Med Mycol Case Rep.* 2014; **5:** 20-3.
154. Dimaka K, Mallis A, Naxakis SS, *et al.* Chronic rhinocerebral mucormycosis: A rare case report and review of the literature. *Mycoses.* 2014; **57:** 699-702.
155. Dioverti MV, Cawcutt KA, Abidi M, *et al.* Gastrointestinal mucormycosis in immunocompromised hosts. *Mycoses.* 2015; **58:** 714-8.
156. Diwakar A, Dewan RK, Chowdhary A, *et al.* Zygomycosis - A case report and overview of the disease in India. *Mycoses.* 2007; **50:** 247-54.
157. Doan L, Al-Khafaji A. Zygomycosis-induced rash after liver transplant. *Am J Gastroenterol.* 2017; **112:** 533.
158. Do GW, Jung SW, Jun JB, *et al.* Colonic mucormycosis presented with ischemic colitis in a liver transplant recipient. *World J Gastroenterol.* 2013; **19:** 3508-11.
159. Dodemont M, Hites M, Bailly B, *et al.* When you can't see the wood for the trees. *Mucor circinelloides:* A rare case of primary cutaneous zygomycosis. *J Mycol Med.* 2015; **25:** 151-4.
160. Dogra V, Talwar D, Saxena R, *et al.* Trilogy of sequential infections in a diabetic male. *Respirol Case Rep.* 2015; **3:** 155-8.
161. Dolatabadi S, de Hoog GS, Meis JF, *et al.* Species boundaries and nomenclature of *Rhizopus arrhizus* (Syn. *R.oryzae*). *Mycoses.* 2014; **57 (Suppl. 3):** 108-27.
162. Dolatabadi S, Kolecka A, Versteeg M, *et al.* Differentiation of clinically relevant Mucorales *Rhizopus microsporus* and *R.arrhizus* by matrix-assisted laser desorption ionization time-of-flight mass spectrometry (MALDI-TOF MS). *J Med Microbiol.* 2015; **64:** 694-701.
163. Dolatabadi S, Najafzadeh MJ, de Hoog GS. Rapid screening for human-pathogenic Mucorales using rolling circle amplification. *Mycoses.* 2014; **57 (Suppl. 3):** 67-72.
164. Dolatabadi S, Scherlach K, Figge M, *et al.* Food preparation with mucoralean fungi: A potential biosafety issue? *Fungal Biol.* 2016; **120:** 393-401.
165. Donnelley MA, Zhu ES, Thompson GR 3rd. Isavuconazole in the treatment of invasive aspergillosis and mucormycosis infections. *Infect Drug Resist.* 2016; **9:** 79-86.
166. Donnelly JP, Lahav M. Deferasirox as adjunctive therapy for mucormycosis. *J Antimicrob Chemother.* 2012; **67:** 519-20.
167. Drogari-Apiranthitou M, Panayiotides I, Galani I, *et al.* Diagnostic value of a semi-nested PCR for the diagnosis of mucormycosis and aspergillosis from paraffin-embedded tissue: A single center experience. *Pathol Res Pract.* 2016; **212:** 393-7.
168. Dua R, Poate T, Abe F, *et al.* Lingual necrosis secondary to mucormycosis. *Br J Oral Maxillofac Surg.* 2012; **50:** e96-8.
169. Dube P, Saroa R, Palta S. Co-infections in Intensive Care Unit with pulmonary tuberculosis and mucormycosis: A clinical dilemma. *Indian J Crit Care Med.* 2016; **20:** 191-3.
170. Duffy J, Harris J, Gade L, *et al.* Mucormycosis outbreak associated with hospital linens. *Pediatr Infect Dis J.* 2014; **33:** 472-6.
171. Duggal S, Rongpharpi SR. *Mucor*-culture/molecular analysis necessary for identification. *Indian J Med Microbiol.* 2015; **33 (Suppl):** 163-4.
172. Dusart A, Duprez T, Van Snick S, *et al.* Fatal rhinocerebral mucormycosis with intra-cavernous carotid aneurysm and thrombosis: A late complication of trans-sphenoidal surgery? *Acta Neurol Belg.* 2013; **113:** 179-84.
173. Dussaule C, Nifle C, Therby A, *et al.* Teaching neuroimages: Brain MRI aspects of isolated cerebral mucormycosis. *Neurology.* 2012; **78:** e93.
174. Dwarakanath S, Kumar V, Blackburn J, *et al.* A rare case of multi-chambered fungal endocarditis from a virulent *Cunninghamella* infection. *Eur Heart J.* 2014; **35:** 343.
175. Ebadi M, Alavi S, Ghojevand N, *et al.* Infantile splenorenopancreatic mucormycosis complicating neuroblastoma. *Pediatr Int.* 2013; **55:** e152-5.
176. Echaiz JF, Burnham CA, Bailey TC. A case of *Apophysomyces trapeziformis* necrotizing soft tissue infection. *Int J Infect Dis.* 2013; **17:** e1240-2.
177. Eickhardt S, Braendstrup P, Clasen-Linde E, *et al.* A non-fatal case of invasive zygomycete *(Lichtheimia corymbifera)* infection in an allogeneic haematopoietic cell transplant recipient. *APMIS.* 2013; **121:** 456-9.
178. Einollahi B, Rostami Z. Mucormycosis after kidney transplantation. *Saudi J Kidney Dis Transpl.* 2013; **24:** 576-7.
179. El-Mahallawy HA, Khedr R, Taha H, *et al.* Investigation and management of a *Rhizomucor* outbreak in a pediatric cancer hospital in Egypt. *Pediatr Blood Cancer.* 2016; **63:** 171-3.
180. Elinav H, Zimhony O, Cohen MJ, *et al.* Rhinocerebral mucormycosis in patients without predisposing medical conditions: A review of the literature. *Clin Microbiol Infect.* 2009; **15:** 693-7.
181. Ellis JJ, Ajello L. An unusual source of *Apophysomyces elegans* and a method for stimulating sporulation of *Saksenaea vasiformis. Mycologia.* 1982; **74:** 144-5.
182. Elsiesy H, Saad M, Shorman M, *et al.* Invasive mucormycosis in a patient with liver cirrhosis: Case report and review of the literature. *Hepat Mon.* 2013; **13:** e10858.
183. Enani MA, Alharthi BN, Dewanjee N, *et al.* Spontaneous gastric ulcer perforation and acute spleen infarction caused by invasive gastric and splenic mucormycosis. *J Glob Infect Dis.* 2014; **6:** 122-4.
184. Erami M, Shams-Ghahfarokhi M, Jahanshiri Z, *et al.* Rhinocerebral mucormycosis due to *Rhizopus oryzae* in a diabetic patient: A case report. *J Mycol Med.* 2013; **23:** 123-9.
185. Ermak D, Kanekar S, Specht CS, *et al.* Looks like a stroke, acts like a stroke but it's more than a stroke: A case of cerebral mucormycosis. *J Stroke Cerebrovasc Dis.* 2014; **23:** e403-4.
186. Ervens J, Ghannoum M, Graf B, *et al.* Successful isavuconazole salvage therapy in a patient with invasive mucormycosis. *Infection.* 2014; **42:** 429-32.
187. Etienne KA, Gillece J, Hilsabeck R, *et al.* Whole genome sequence typing to investigate the *Apophysomyces* outbreak

following a tornado in Joplin, Missouri, 2011. *PLoS One.* 2012; **7:** e49989. PMID: 23209631.

188. Farmakiotis D, Kontoyiannis DP. Mucormycoses. *Infect Dis Clin North Am.* 2016; **30:** 143-63.
189. Fathi P, Vranis NM, Paryavi E. Diagnosis and treatment of upper extremity mucormycosis infections. *J Hand Surg Am.* 2015; **40:** 1032-4.
190. Fernandez JF, Maselli DJ, Simpson T, *et al.* Pulmonary mucormycosis: What is the best strategy for therapy? *Respir Care.* 2013; **58:** e60-3.
191. Ferrell BA, Tolle JJ. Invasive endobronchial mucormycosis. *Am J Respir Crit Care Med.* 2014; **190:** e28.
192. Forrester JD, Chandra V, Shelton AA, *et al.* Gastrointestinal mucormycosis requiring surgery in adults with hematologic malignant tumors: Literature review. *Surg Infect (Larchmt).* 2015; **16:** 194-202.
193. Galasso S, Chilelli NC, Burlina S, *et al.* Palpebral ptosis, fixed mydriasis and exophthalmos: An uncommon presentation of new diabetes mellitus. *Intern Emerg Med.* 2013; **8:** 89-91.
194. Gamaletsou MN, Sipsas NV, Roilides E, *et al.* Rhino-orbital-cerebral mucormycosis. *Curr Infect Dis Rep.* 2012; **14:** 423-34.
195. Garcia-Hermoso D, Hoinard D, Gantier JC, *et al.* Molecular and phenotypic evaluation of *Lichtheimia corymbifera (*formerly *Absidia corymbifera)* complex isolates associated with human mucormycosis: Rehabilitation of *L.ramosa. J Clin Microbiol.* 2009; **47:** 3862-70.
196. Garcia-Martinez J, Lopez-Medrano F, Alhambra A, *et al.* Rhinocerebral zygomycosis caused by *Saksenaea vasiformis* in a diabetic patient. *Mycoses.* 2008; **51:** 549-53.
197. Garcia-Pajares F, Sanchez-Antolin G, Almohalla Alvarez C, *et al.* Cutaneous mucormycosis infection by *Absidia* in two consecutive liver transplant patients. *Transplant Proc.* 2012; **44:** 1562-4.
198. Garcia-Rodriguez J, Quiles-Melero I, Humala-Barbier K, *et al.* Isolation of *Cunninghamella blakesleeana* in an immunodepressed patient. *Mycoses.* 2012; **55:** 463-5.
199. Garg J, Sujatha S, Garg A, *et al.* Nosocomial cutaneous zygomycosis in a patient with diabetic ketoacidosis. *Int J Infect Dis.* 2009; **13:** e508-10.
200. Garg PK, Gupta N, Gautam V, *et al.* Gastric zygomycosis: unusual cause of gastric perforation in an immunocompetent patient. *South Med J.* 2008; **101:** 449-50.
201. Garg R, Gupta VV, Ashok L. Rhinomaxillary mucormycosis: A palatal ulcer. *Contemp Clin Dent.* 2011; **2:** 119-23.
202. Garg R, Marak RS, Verma SK, *et al.* Pulmonary mucormycosis mimicking as pulmonary tuberculosis: A case report. *Lung India.* 2008; **25:** 129-31.
203. Garnacho-Montero J, Olaechea P, Alvarez-Lerma F, *et al.* Epidemiology, diagnosis and treatment of fungal respiratory infections in the critically ill patient. *Rev Esp Quimioter.* 2013; **26:** 173-88.
204. Gayathri Devi HJ, Mohan Rao KN, Prathima KM, *et al.* Pulmonary mucormycosis presenting with vocal cord paralysis. *Respir Med Case Rep.* 2013; **9:** 15-7.
205. Gelman A, Valdes-Rodriguez R, Bhattacharyya S, *et al.* A case of primary cutaneous mucormycosis caused by minor trauma. *Dermatol Online J.* 2015; 21. pii: 13030/qt8qn2g3zv.
206. Gen R, Horasan ES, Vaysoglu Y, Arpaci RB, *et al.* Rhino-orbito-cerebral mucormycosis in patients with diabetic ketoacidosis. *J Craniofac Surg.* 2013; **24:** e144-7.
207. Georgiadou SP, Sipsas NV, Marom EM, *et al.* The diagnostic value of halo and reversed halo signs for invasive mold infections in compromised hosts. *Clin Infect Dis.* 2011; **52:** 1144-55.
208. Geramizadeh B, Kazemi K, Shamsaifar AR, *et al.* Isolated renal mucormycosis after liver transplantation: An unusual case report. *Iran Red Crescent Med J.* 2012; **14:** 447-50.
209. Ghafur A, Shareek PS, Senthur NP, *et al.* Mucormycosis in patients without cancer: A case series from a tertiary care hospital in South India. *J Assoc Physicians India.* 2013; **61:** 305-8. Reply from author. 2014; **62:** 549.
210. Ghuman MS, Kaur S, Bhandal SK, *et al.* Bilateral optic nerve infarction in rhino-cerebral mucormycosis: A rare magnetic resonance imaging finding. *J Neurosci Rural Pract.* 2015; **6:** 403-4.
211. Godara SM, Kute VB, Goplani KR, *et al.* Mucormycosis in renal transplant recipients: predictors and outcome. *Saudi J Kidney Dis Transpl.* 2011; **22:** 751-6.
212. Goel A, Rao SV, Mathews MS, *et al.* Rhino-orbital zygomycosis in a patient with pregnancy related liver disease and multi-organ failure. *Indian J Gastroenterol.* 2011; **30:** 49-50.
213. Goel P, Jain V, Sengar M, *et al.* Gastrointestinal mucormycosis: A success story and appraisal of concepts. *J Infect Public Health.* 2013; **6:** 58-61.
214. Goel S, Palaskar S, Shetty VP, *et al.* Rhinomaxillary mucormycosis with cerebral extension. *J Oral Maxillofac Pathol.* 2009; **13:** 14-7.
215. Gomes MZ, Lewis RE, Kontoyiannis DP. Mucormycosis caused by unusual mucormycetes, non-*Rhizopus*, -*Mucor* and -*Lichtheimia* species. *Clin Microbiol Rev.* 2011; **24:** 411-45.
216. Gomez-Camarasa C, Rojo-Martin MD, Miranda-Casas C, *et al.* Disseminated infection due to *Saksenaea vasiformis* secondary to cutaneous mucormycosis. *Mycopathologia.* 2014; **177:** 97-101.
217. Gonzalez-Abad MJ, Alonso Sanz M. Zygomycosis in children: Disseminated infection caused by *Cunninghamella bertholletiae. Arch Bronconeumol.* 2013; **49:** 35.
218. Grannan BL, Yanamadala V, Venteicher AS, *et al.* Use of external ventriculostomy and intrathecal anti-fungal treatment in cerebral mucormycotic abscess. *J Clin Neurosci.* 2014; **21:** 1819-21.
219. Guarro J, Chander J, Alvarez E, et al. *Apophysomyces variabilis* infections in humans. *Emerg Infect Dis.* 2011; **17:** 134-5.
220. Gupta A, Jain S, Agrawal C, *et al.* Successful outcome of mucormycosis in two children on induction therapy for acute lymphoblastic leukemia. *Indian J Med Paediatr Oncol.* 2013; **34:** 313-6.
221. Gupta A, Lal C, Dogra PM, *et al.* Insulin site wound in a renal allograft recipient. *Saudi J Kidney Dis Transpl.* 2011; **22:** 134-5.

222. Gupta K, Nada R, Joshi K, *et al.* Can ascending infection from bladder serve as the portal of entry for primary renal zygomycosis? *Mycopathologia.* 2010; **170:** 357-60.
223. Gupta KL, Gupta A. Mucormycosis and acute kidney injury. *J Nephropathol.* 2012; **1:** 155-9.
224. Gupta KL, Joshi K, Bhat A, *et al.* Mucormycosis of the transplanted kidney with renal papillary necrosis. *Exp Clin Transplant.* 2013; **11:** 554-7.
225. Gupta KL, Joshi K, Kohli HS, *et al.* Mucormycosis (zygomycosis) of renal allograft. *Clin Kidney J.* 2012; **5:** 502-7.
226. Gupta KL, Joshi K, Sud K, *et al.* Renal zygomycosis: An under diagnosed cause of acute renal failure. *Nephrol Dial Transplant.* 1999; **14:** 2720-5.
227. Gupta N, Kumar R, Prasad R. Reversed halo sign. *Indian J Chest Dis Allied Sci.* 2014; **56:** 247-8.
228. Gupta R, Parelkar SV, Oak S, *et al.* Neonatal lingual and gastrointestinal mucormycosis in a case of low anorectal malformation - A rare presentation. *J Pediatr Surg.* 2011; **46:** 745-8.
229. Gupta V, Rajagopalan N, Patil M, *et al. Aspergillus* and mucormycosis presenting with normal chest X-ray in an immunocompromised host. *BMJ Case Rep.* 2014; pii: bcr2014204022. PMID: 24717585.
230. Gupta V, Singh SK, Kakkar N, *et al.* Splenic and renal mucormycosis in a healthy host: Successful management by aggressive treatment. *Trop Gastroenterol.* 2010; **31:** 57-8.
231. Gurevich M, Levi I, Steinberg R, *et al.* Mucormycosis in a liver allograft: Salvage re-transplantation and targeted immunosuppressive management. *Transpl Infect Dis.* 2012; **14:** e97-101.
232. Gutierrez-Acevedo M, Balleste-Frank CR, Fernandez-Gonzalez F, *et al.* Immunocompromised patients: What a challenge! *Bol Asoc Med P R.* 2012; **104:** 62-4.
233. Gutierrez-Delgado EM, Trevino-Gonzalez JL, Montemayor-Alatorre A, *et al.* Chronic rhino-orbito-cerebral mucormycosis: A case report and review of the literature. *Ann Med Surg (Lond).* 2016; **6:** 87-91.
234. Guymer C, Khurana S, Suppiah R, *et al.* Successful treatment of disseminated mucormycosis in a neutropenic patient with T-cell acute lymphoblastic leukaemia. *BMJ Case Rep.* 2013; pii: bcr2013009577. PMID: 23904418.
235. Hadaschik E, Koschny R, Willinger B, *et al.* Pulmonary, rhino-orbital and cutaneous mucormycosis caused by *Rhizomucor pusillus* in an immunocompromised patient. *Clin Exp Dermatol.* 2012; **37:** 355-7.
236. Hallur V, Singh G, Rudramurthy SM, *et al.* Demodex mite infestation of unknown significance in a patient with rhinocerebral mucormycosis due to *Apophysomyces elegans* species complex. *J Med Microbiol.* 2013; **62:** 926-8.
237. Hamdi A, Mulanovich VE, Matin SF, *et al.* Isolated renal mucormycosis in a transplantation recipient. *J Clin Oncol.* 2015; **33:** e50-1.
238. Hamdi T, Karthikeyan V, Alangaden GJ. Mucormycosis in a renal transplant recipient: Case report and comprehensive review of literature. *Int J Nephrol.* 2014; 950643. PMID: 24688793.
239. Hamilos G, Samonis G, Kontoyiannis DP. Pulmonary mucormycosis. *Semin Respir Crit Care Med.* 2011; **32:** 693-702.
240. Hammond SP, Bialek R, Milner DA, *et al.* Molecular methods to improve diagnosis and identification of mucormycosis. *J Clin Microbiol.* 2011; **49:** 2151-3.
241. Harman M, Uçmak D, Dal T. A rare case of mucormycosis in the scalp. *Acta Med Port.* 2013; **26:** 754-7.
242. Harrasser N, Banke IJ, Hauschild M, *et al.* Clinical challenge: Fatal mucormycotic osteomyelitis caused by *Rhizopus microsporus* despite aggressive multimodal treatment. *BMC Infect Dis.* 2014; **14:** 488.
243. Hatahet MH, Narayanan M, Cleaves C, *et al.* Disseminated mucormycosis in a patient with recent kidney transplantation: A case report and review of the literature. *Case Rep Nephrol Urol.* 2013; **3:** 58-63.
244. Hawkes JE, Florell SR, Wada DA. Chronic, painful, non-healing ulcer on the right arm following minor trauma. *JAMA Dermatol.* 2015; **151:** 787-8.
245. Hayat M, Mushtaq S, Saba S, *et al.* Rhino-orbital-mucormycosis as a presenting manifestation of gestational diabetes mellitus. *Indian J Endocrinol Metab.* 2011; **15:** S65-6.
246. Hemashettar BM, Kagal AE, Patil CS, *et al.* Rhinoorbital mucormycosis treated with amphotericin B with successful outcome. *J Assoc Physicians India.* 1985; **33:** 302-4.
247. Hemashettar BM, Patil RN, O'Donnell K, *et al.* Chronic rhinofacial mucormycosis caused by *Mucor irregularis (Rhizomucor variabilis)* in India. *J Clin Microbiol.* 2011; **49:** 2372-5.
248. Heydari AA, Fata A, Mojtabavi M. Chronic cutaneus mucormycosis in an immunocompetent female. *Iran Red Crescent Med J.* 2013; **15:** 254-5.
249. Hirano T, Yamada M, Sato K, *et al.* Invasive pulmonary mucormycosis: Rare presentation with pulmonary eosinophilia. *BMC Pulm Med.* 2017; **17:** 76. PMID: 28454572.
250. Hirayama Y, Yajima N, Kaimori M, *et al.* Disseminated infection and pulmonary embolization of *Cunninghamella bertholletiae* complicated with hemophagocytic lymphohistiocytosis. *Intern Med.* 2013; **52:** 2275-9.
251. Hoffmann K, Discher S, Voigt K. Revision of the genus *Absidia* (Mucorales, Zygomycetes) based on physiological, phylogenetic and morphological characters; thermotolerant *Absidia* spp. form a coherent group, Mycocladiaceae fam. nov. *Mycol Res.* 2007; **111:** 1169-83.
252. Hoffmann K, Walther G, Voigt K. *Mycocladus vs. Lichtheimia:* A correction (Lichtheimiaceae fam. nov., Mucorales, Mucoromycotina). *Mycol. Res.* 2009; **113:** 277-8.
253. Hofman V, Dhouibi A, Butori C, *et al.* Usefulness of molecular biology performed with formaldehyde-fixed paraffin embedded tissue for the diagnosis of combined pulmonary invasive mucormycosis and aspergillosis in an immunocompromised patient. *Diagn Pathol.* 2010; **5:** 1.
254. Honeybul S, Morrison DA. Skull vault destruction after rhinocerebral mucormycosis. *World Neurosurg.* 2012; **78:** 553.e1-4.

255. Hong HL, Lee YM, Kim T, *et al.* Risk factors for mortality in patients with invasive mucormycosis. *Infect Chemother.* 2013; **45:** 292-8.
256. Hospenthal DR, Chung KK, Lairet K, *et al.* *Saksenaea erythrospora* infection following combat trauma. *J Clin Microbiol.* 2011; **49:** 3707-9.
257. Hrncirova K, Lengerova M, Kocmanova I, *et al.* Rapid detection and identification of mucormycetes from culture and tissue samples by use of high-resolution melt analysis. *J Clin Microbiol.* 2010; **48:** 3392-4.
258. Hung HC, Shen GY, Chen SC, *et al.* Pulmonary mucormycosis in a patient with systemic lupus erythematosus: A diagnostic and treatment challenge. *Case Rep Infect Dis.* 2015; 478789. PMID: 26185693.
259. Hyvernat H, Dunais B, Burel-Vandenbos F, *et al.* Fatal peritonitis caused by *Rhizopus microsporus. Med Mycol.* 2010; **48:** 1096-8.
260. Ibrahim AS, Edwards JE Jr, Bryant R, *et al.* Economic burden of mucormycosis in the United States: Can a vaccine be cost-effective? *Med Mycol.* 2009; **47:** 592-600.
261. Ibrahim AS, Kontoyiannis DP. Update on mucormycosis pathogenesis. *Curr Opin Infect Dis.* 2013; **26:** 508-15.
262. Ibrahim AS, Spellberg B, Walsh TJ, *et al.* Pathogenesis of mucormycosis. *Clin Infect Dis.* 2012; **54:** S16-22.
263. Ibrahim AS. Host-iron assimilation: Pathogenesis and novel therapies of mucormycosis. *Mycoses.* 2014; **57 (Suppl. 3):** 13-7.
264. Ilyas S, Al-Abbadi MA, Raval B, *et al.* *Mucor* causing non-healing skin ulcer diagnosed by scrape cytology: Description of unusual presentation. *Diagn Cytopathol.* 2011; **39:** 714-5.
265. Imai Y, Adachi Y, Kimura T, *et al.* An autopsy case of pulmonary fissure induced by zygomycosis. *Int J Gen Med.* 2013; **6:** 575-9.
266. Imbernon A, Agud JL, Cuetara MS, *et al.* Successful therapy of progressive rhino-orbital mucormycosis caused by *Rhizopus arrhizus* with combined and sequential antifungal therapy, surgery and hyperbaric therapy. *Med Mycol Case Rep.* 2014; **6:** 51-4.
267. Ingram PR, Suthananthan AE, Rajan R, *et al.* Cutaneous mucormycosis and motor vehicle accidents: Findings from an Australian case series. *Med Mycol.* 2014; **52:** 819-25.
268. Iqbal N, Irfan M, Jabeen K, *et al.* Chronic pulmonary mucormycosis: An emerging fungal infection in diabetes mellitus. *J Thorac Dis.* 2017; **9:** e121-5.
269. Irga N, Kosiak W, Jaworski R, *et al.* Hyperthyroidism secondary to disseminated mucormycosis in a child with acute lymphoblastic leukemia: Case report and a review of published reports. *Mycopathologia.* 2013; **175:** 123-7.
270. Irtan S, Lamerain M, Lesage F, *et al.* Mucormycosis as a rare cause of severe gastrointestinal bleeding after multivisceral transplantation. *Transpl Infect Dis.* 2013; **15:** e235-8.
271. Ishak RS, Kurban M, Abbas O. Erythematous purpuric indurated plaque over right forearm. *Pediatr Dermatol.* 2014; **31:** 249-50.
272. Islam N, Bhattacharyya I. Diagnostic discussion. Zygomycosis (mucormycosis). *Todays FDA.* 2014; **26:** 45-7.
273. Iwen PC, Sigler L, Noel RK, *et al.* *Mucor circinelloides* was identified by molecular methods as a cause of primary cutaneous zygomycosis. *J Clin Microbiol.* 2007; **45:** 636-40.
274. Jain D, Kohli K. Neonatal gastrointestinal mucormycosis clinically mimicking necrotizing enterocolitis. *Eur J Pediatr Surg.* 2009; **19:** 405-7.
275. Jain S, Kumar S, Kaushal A. Rhinocerebral mucormycosis with isolated sixth nerve palsy in an immunocompetent patient. *Med J Malaysia.* 2011; **66:** 376-8.
276. Jain SK, Kaza RC, Tanwar R. Mucormycosis of the anterior chest wall presenting as a soft tissue tumour. *J Wound Care.* 2011; **20:** 176-8.
277. Jayakrishnan B, Al Aghbari J, Rizavi D, *et al.* Chronic renal failure presenting for the first time as pulmonary mucormycosis with a fatal outcome. *Case Rep Nephrol.* 2015; 589537. PMID: 25664192.
278. Jeong SJ, Lee JU, Song YG, *et al.* Delaying diagnostic procedure significantly increases mortality in patients with invasive mucormycosis. *Mycoses.* 2015; **58:** 746-52.
279. Jha R, Gude D, Chennamsetty S, *et al.* Intracranial hypertension: An unusual presentation of mucormycosis in a kidney transplant recipient. *Indian J Nephrol.* 2013; **23:** 130-2.
280. Jha VK, Borpujari PJ, Shenoy G, *et al.* Empyema with pleuropulmonary mucormycosis. *J Assoc Physicians India.* 2013; **61:** 665-7.
281. Jiang N, Zhao G, Yang S, *et al.* A retrospective analysis of eleven cases of invasive rhino-orbito-cerebral mucormycosis presented with orbital apex syndrome initially. *BMC Ophthalmol.* 2016; **16:** 10. PMID: 26758904.
282. Jimenez Caballero PE, Falcon Garcia AM, *et al.* Multiple subcortical strokes caused by mucormycosis in a patient with lymphoma. *Arq Neuropsiquiatr.* 2012; **70:** 69-70.
283. Johnson CB, Ahmeti M, Tyroch AH, *et al.* Gastric mucormycosis as a cause of life-threatening upper gastrointestinal bleeding in a trauma patient. *Am Surg.* 2010; **76:** e76-7.
284. Johnson JB, Affolter KE, Samadder NJ. A rare cause of hematochezia: Colon mucormycosis. *Clin Gastroenterol Hepatol.* 2013; **11:** A22.
285. Joichi Y, Chijimatsu I, Yarita K, *et al.* Detection of *Mucor velutinosus* in a blood culture after autologous peripheral blood stem cell transplantation: A pediatric case report. *Med Mycol J.* 2014; **55:** e43-8.
286. Joos ZP, Patel BC. Intraorbital irrigation of amphotericin B in the treatment of rhino-orbital mucormycosis. *Ophthal Plast Reconstr Surg.* 2017; **33:** e13-6.
287. Juan YH, Saboo SS, Lin YC, *et al.* Reverse halo sign in pulmonary mucormycosis. *QJM.* 2014; **107:** 777-8.
288. Kaide CG, Khandelwal S. Hyperbaric oxygen: Applications in infectious disease. *Emerg Med Clin North Am.* 2008; **26:** 571-95.
289. Kaiser P, Maggio EM, Pfammatter T, *et al.* Histopathological evidence of invasive gastric mucormycosis after transarterial chemoembolization and liver transplantation. *Infection.* 2014; **42:** 779-83.
290. Kalva N, Somaraju V, Puli S. A fatal case of gastrointestinal mucormycosis in immunosuppressed host. *Med J Armed Forces India.* 2013; **69:** 285-7.

291. Kanemaru M, Tashima S, Yamazaki A, *et al.* Disseminated mucormycosis due to *Rhizopus oryzae* diagnosed by skin biopsy. *J Dermatol.* 2015; **42:** 100-1.
292. Kang D, Jiang X, Wan H, *et al. Mucor irregularis* infection around the inner canthus cured by amphotericin B: A case report and review of published literatures. *Mycopathologia.* 2014; **178:** 129-33.
293. Kang SH, Kim HS, Bae MN, *et al.* Fatal breakthrough mucormycosis in an acute myelogenous leukemia patient while on posaconazole prophylaxis. *Infect Chemother.* 2015; **47:** 49-54.
294. Kataria SP, Sharma J, Singh G, *et al.* Primary breast mucormycosis: FNAC diagnosis of a rare entity. *Diagn Cytopathol.* 2016; **44:** 761-3.
295. Katragkou A, Walsh TJ, Roilides E. Why is mucormycosis more difficult to cure than more common mycoses? *Clin Microbiol Infect.* 2014; **20 (Suppl. 6):** 74-81.
296. Katsantonis NG, Hunter JB, O'Connell BP, *et al.* Temporal bone mucormycosis. *Ann Otol Rhinol Laryngol.* 2016; **125:** 850-3.
297. Kaur R, Bala K, Ahuja RB, *et al.* Primary cutaneous mucormycosis in a patient with burn wounds due to *Lichtheimia ramosa. Mycopathologia.* 2014; **178:** 291-5.
298. Kaushik R, Chander J, Gupta S, et al. Fatal primary cutaneous zygomycosis caused by *Saksenaea vasiformis*: Case report and review of literature. *Surg Infect (Larchmt).* 2012; **13:** 125-9.
299. Kaushik R. Primary cutaneous zygomycosis in India. *Indian J Surg.* 2012; **74:** 468-75.
300. Kawasaki M, Kawakami N, Kawai K, *et al.* Cutaneous mucormycosis in bone marrow transplantation recipients. *Eur J Dermatol.* 2012; **22:** 578-9.
301. Kazak E, Aslan E, Akalın H, *et al.* A mucormycosis case treated with a combination of caspofungin and amphotericin B. *J Mycol Med.* 2013; **23:** 179-84.
302. Kennedy KJ, Daveson K, Slavin MA, *et al.* Mucormycosis in Australia: Contemporary epidemiology and outcomes. *Clin Microbiol Infect.* 2016; **22:** 775-81.
303. Kermani W, Bouttay R, Belcadhi M, *et al.* ENT mucormycosis. Report of 4 cases. *Eur Ann Otorhinolaryngol Head Neck Dis.* 2016; **133:** 83-6.
304. Khan AA, Garg A, Dhawan S, *et al.* T-cell non-Hodgkin's lymphoma with colesional mucormycosis presenting as palatal perforation: A case report. *J Indian Med Assoc.* 2012; **110:** 499-500.
305. Khan AA, Kumaran V, Jain D, *et al.* Hepatic mucormycosis in a patient of acute lymphoblastic leukemia: A case report with literature review. *Indian J Hematol Blood Transfus.* 2013; **29:** 96-8.
306. Khan ZU, Ahmad S, Brazda A, *et al. Mucor circinelloides* as a cause of invasive maxillofacial zygomycosis: An emerging dimorphic pathogen with reduced susceptibility to posaconazole. *J Clin Microbiol.* 2009; **47:** 1244-8.
307. Khan ZU, Ahmad S, Mokaddas E, *et al. Actinomucor elegans* var. *kuwaitiensis* isolated from the wound of a diabetic patient. *Antonie van Leeuwenhoek.* 2008; **94:** 343-52.
308. Khandelwal A, Gupta P, Gupta A, *et al.* Renal mucormycosis in aplastic anemia: A novel presentation. *Int Urol Nephrol.* 2013; **45:** 285-8.
309. Khandelwal D, Gadodia A, Sood R, *et al.* Disseminated zygomycosis with renal involvement simulating malignancy in a diabetic patient. *Indian J Urol.* 2012; **28:** 347-9.
310. Khandelwal N, Gupta V, Singh P. Central nervous system fungal infections in tropics. *Neuroimaging Clin N Am.* 2011; **21:** 859-66.
311. Kim MJ, Park PW, Ahn JY, *et al.* Fatal pulmonary mucormycosis caused by *Rhizopus microsporus* in a patient with diabetes. *Ann Lab Med.* 2014; **34:** 76-9.
312. Kim ST, Kim WS, Lee HH, *et al.* Successful treatment of invasive rhinopulmonary mucormycosis with an indolent presentation by combined medical and surgical therapy. *J Craniofac Surg.* 2013; **24:** e182-4.
313. Kim YI, Kang HC, Lee HS, *et al.* Invasive pulmonary mucormycosis with concomitant lung cancer presented with massive hemoptysis by huge pseudoaneurysm of pulmonary artery. *Ann Thorac Surg.* 2014; **98:** 1832-5.
314. Kimura M, Araoka H, Uchida N, *et al. Cunninghamella bertholletiae* pneumonia showing a reversed halo sign on chest computed tomography scan following cord blood transplantation. *Med Mycol.* 2012; **50:** 412-6.
315. Kimura M, Nishimura K, Enoki E, *et al.* Chlamydospores of *Rhizopus microsporus* var. *rhizopodiformis* in tissue of pulmonary mucormycosis. *Mycopathologia.* 2012; **174:** 441-50.
316. Klimko NN, Khostelidi SN, Volkova AG, *et al.* Mucormycosis in haematological patients: Case report and results of prospective study in Saint Petersburg, Russia. *Mycoses.* 2014; **57 (Suppl. 3):** 91-6.
317. Koehler P, Tacke D, Cornely OA. Bone and joint infections by Mucorales, *Scedosporium, Fusarium* and even rarer fungi. *Crit Rev Microbiol.* 2016; **42:** 158-71.
318. Kogure Y, Nakamura F, Shinozaki-Ushiku A, *et al.* Pulmonary mucormycosis with embolism: Two autopsied cases of acute myeloid leukemia. *Int J Clin Exp Pathol.* 2014; **7:** 3449-53.
319. Kokkayil P, Pandey M, Agarwal R, *et al. Rhizopus homothallicus* causing invasive infections: Series of three cases from a single centre in north India. *Mycopathologia.* 2017; doi: 10.1007/s11046-017-0153-5. PMID: 28623532.
320. Kolekar JS. Rhinocerebral mucormycosis: A retrospective study. *Indian J Otolaryngol Head Neck Surg.* 2015; **67:** 93-6.
321. Kompoti M, Michalia M, Kallitsi G, *et al.* Fatal cutaneous *Saksenaea vasiformis* infection in a critically ill trauma patient. *Mycoses.* 2011; **54:** e599-601.
322. Kontoyiannis DP, Azie N, Franks B, *et al.* Prospective Antifungal Therapy (PATH) Alliance: Focus on mucormycosis. *Mycoses.* 2014; **57:** 240-6.
323. Kontoyiannis DP, Lewis RE, Lortholary O, *et al.* Future directions in mucormycosis research. *Clin Infect Dis.* 2012; **54:** S79-85.
324. Kontoyiannis DP, Lewis RE. How I treat mucormycosis. *Blood.* 2011; **118:** 1216-24.

325. Kontoyiannis DP. Invasive mycoses: Strategies for effective management. *Am J Med.* 2012; **125:** S25-38.
326. Koshy CG, Shah S, Mammen T. Subcutaneous emphysema of the chest: Could it be pulmonary mucormycosis? *Thorax.* 2010; **65:** 280.
327. Kulkarni RV, Thakur SS. Invasive gastric mucormycosis - A case report. *Indian J Surg.* 2015; **77 (Suppl. 1):** 87-9.
328. Kumar AR, Hunt P, Ritter EM, *et al.* Successful knee extensor mechanism reconstruction in a warfare-related open lower extremity injury complicated by mucormycosis infection: A case report. *J Orthop Trauma.* 2012; **26:** e7-10.
329. Kumar Debata P, Keshari Panda S, Dash A, *et al.* "An unusual presentation of colonic mucormycosis mimicking carcinoma colon - A surgeon's perspective". *Int J Surg Case Rep.* 2015; **10:** 248-51.
330. Kumar JA, Babu P, Prabu K, Kumar P. Mucormycosis in maxilla: Rehabilitation of facial defects using interim removable prostheses: A clinical case report. *J Pharm Bioallied Sci.* 2013; **5:** S163-5.
331. Kumar M, Tanwar P, Radhika S, *et al.* Fine needle aspiration cytology of primary mucormycosis of the breast in a young immunocompetent pregnant woman. *Cytopathology.* 2013; **24:** 411-2.
332. Kumar MM, Poovazhagi V, Anbalagan S, *et al.* Rhino-orbito-cerebral mucormycosis in a child with diabetic ketoacidosis. *Indian J Crit Care Med.* 2014; **18:** 334-5.
333. Kumar N, Hanks ME, Chandrasekaran P, *et al.* Gain-of-function signal transducer and activator of transcription 1 (STAT1) mutation-related primary immunodeficiency is associated with disseminated mucormycosis. *J Allergy Clin Immunol.* 2014; **134:** 236-9.
334. Kumar NS, Padala RK, Tirupati S, *et al.* Rhinocerebral mucormycosis with top of basilar artery syndrome. *J Stroke Cerebrovasc Dis.* 2016; **25:** 378-82.
335. Kumar P, Begum CZ, Thirumaran P, *et al.* Rhinocerebral mucormycosis in systemic lupus erythematosus. *Indian J Dermatol.* 2013; **58:** 158.
336. Kumar S, Pushkarna A, Sharma V, Ganesamoni R, Nada R. Fournier's gangrene with testicular infarction caused by mucormycosis. *Indian J Pathol Microbiol.* 2011; **54:** 847-8.
337. Kursun E, Turunc T, Demiroglu YZ, *et al.* Evaluation of 28 cases of mucormycosis. *Mycoses.* 2015; **58:** 82-7.
338. Kutlu M, Ergin C, Bir F, *et al.* Pulmonary mucormycosis due to *Lichtheimia ramosa* in a patient with HIV infection. *Mycopathologia.* 2014; **178:** 111-5.
339. Kuy S, He C, Cronin DC 2nd. Renal mucormycosis: A rare and potentially lethal complication of kidney transplantation. *Case Rep Transplant.* 2013; 915423. PMID: 24251062.
340. Kwan LP, Choy CB, Chan TM, Suen WS, Yap DY. Successful treatment of pulmonary *Rhizopus* infection with surgical resection and posaconazole in a renal transplant recipient. *Nephrology (Carlton).* 2013; **18:** 74-5.
341. Kwok M, Maurice A, Carroll J, *et al.* Gastrointestinal mucormycosis in an immunocompromised host. *ANZ J Surg.* 2017; doi: 10.1111/ans.13955. PMID: 28419686.
342. Lacarriere E, Lacaze L, Schwarz L, *et al.* First case of gastrointestinal mucormycosis in an immunocompromised patient with gallbladder and duodenum involvement. *Infection.* 2011; **39:** 595-8.
343. Lackner M, Caramalho R, Lass-Florl C. Laboratory diagnosis of mucormycosis: Current status and future perspectives. *Future Microbiol.* 2014; **9:** 683-95.
344. Lai MC, Zhang W, Yang Z, *et al.* First case report of isolated penile mucormycosis in a liver transplantation recipient. *Int J Infect Dis.* 2014; **29:** 208-10.
345. Lalayanni C, Baliakas P, Xochelli A, *et al.* Outbreak of cutaneous zygomycosis associated with the use of adhesive tape in haematology patients. *J Hosp Infect.* 2012; **81:** 213-5.
346. Lalwani S, Govindasamy M, Gupta M, *et al.* Gastrointestinal mucormycosis - Four cases with different risk factors, involving different anatomical sites. *Indian J Gastroenterol.* 2012; **31:** 139-43.
347. Langford S, Trubiano JA, Saxon S, *et al.* Mucormycete infection or colonisation: Experience of an Australian tertiary referral centre. *Mycoses.* 2016; **59:** 291-5.
348. Lanternier F, Dannaoui E, Morizot G, *et al.* A global analysis of mucormycosis in France: the RetroZygo Study (2005-2007). *Clin Infect Dis.* 2012; **54:** S35-43.
349. Lanternier F, Sun HY, Ribaud P, *et al.* Mucormycosis in organ and stem cell transplant recipients. *Clin Infect Dis.* 2012; **54:** 1629-36.
350. Larcher G, Dias M, Razafimandimby B, *et al.* Siderophore production by pathogenic mucorales and uptake of deferoxamine B. *Mycopathologia.* 2013; **176:** 319-28.
351. Lass-Florl C. Zygomycosis: Conventional laboratory diagnosis. *Clin Microbiol Infect.* 2009; **15 (Suppl. 5):** 60-5.
352. Lawhon SD, Corapi WV, Hoffmann AR, *et al.* *In utero* infection of a calf by *Saksenaea erythrospora* resulting in neonatal abomasitis and dermatitis. *J Vet Diagn Invest.* 2012; **24:** 990-3.
353. Layios N, Canivet JL, Baron F, *et al.* *Mortierella wolfii*-associated invasive disease. *Emerg Infect Dis.* 2014; **20:** 1591-2. PMID: 25153198.
354. Lazar SP, Lukaszewicz JM, Persad KA, *et al.* Rhinocerebral *Mucor circinelloides* infection in immunocompromised patient following yogurt ingestion. *Del Med J.* 2014; **86:** 245-8.
355. LeBlanc RE, Meriden Z, Sutton DA, *et al.* *Cunninghamella echinulata* causing fatally invasive fungal sinusitis. *Diagn Microbiol Infect Dis.* 2013; **76:** 506-9.
356. Lechevalier P, Hermoso DG, Carol A, *et al.* Molecular diagnosis of *Saksenaea vasiformis* cutaneous infection after scorpion sting in an immunocompetent adolescent. *J Clin Microbiol.* 2008; **46:** 3169-72.
357. Lee JS, Kim HC, Park SW, *et al.* A case of isolated pulmonary mucormycosis in an immunocompetent host. *Tuberc Respir Dis (Seoul).* 2013; **74:** 269-73.
358. Lee SC, Billmyre RB, Li A, *et al.* Analysis of a food-borne fungal pathogen outbreak: Virulence and genome of a *Mucor circinelloides* isolate from yogurt. *mBio.* 2014; **5:** e01390-14.
359. Lee SC, Li A, Calo S, *et al.* Calcineurin plays key roles in the dimorphic transition and virulence of the human pathogenic zygomycete *Mucor circinelloides. PLoS Pathog.* 2013; **9:** e1003625. PMID: 24039585.

360. Lee SH, Son YG, Sohn SS, *et al.* Successful treatment of invasive gastric mucormycosis in a patient with alcoholic liver cirrhosis: A case report. *Exp Ther Med.* 2014; **8:** 401-404.
361. Legouge C, Caillot D, Chretien ML, *et al.* The reversed halo sign: Pathognomonic pattern of pulmonary mucormycosis in leukemic patients with neutropenia? *Clin Infect Dis.* 2014; **58:** 672-8.
362. Legrand M, Gits-Muselli M, Boutin L, *et al.* Detection of circulating Mucorales DNA in critically ill burn patients: Preliminary report of a screening strategy for early diagnosis and treatment. *Clin Infect Dis.* 2016; **63:** 1312-7.
363. Leithauser M, Kahl C, Aepinus C, *et al.* Invasive zygomycosis in patients with graft-versus-host disease after allogeneic stem cell transplantation. *Transpl Infect Dis.* 2010; **12:** 251-7.
364. Lekakis LJ, Lawson A, Prante J, *et al.* Fatal *Rhizopus* pneumonia in allogeneic stem cell transplant patients despite posaconazole prophylaxis: Two cases and review of the literature. *Biol Blood Marrow Transplant.* 2009; **15:** 991-5.
365. Lelievre L, Garcia-Hermoso D, Abdoul H, *et al.* Posttraumatic mucormycosis: A nationwide study in France and review of the literature. *Medicine (Baltimore).* 2014; **93:** 395-404.
366. Lengerova M, Racil Z, Hrncirova K, *et al.* Rapid detection and identification of mucormycetes in bronchoalveolar lavage samples from immunocompromised patients with pulmonary infiltrates by use of high-resolution melt analysis. *J Clin Microbiol.* 2014; **52:** 2824-8.
367. Lewis RE, Georgiadou SP, Sampsonas F, *et al.* Risk factors for early mortality in haematological malignancy patients with pulmonary mucormycosis. *Mycoses.* 2014; **5 7:** 49-55.
368. Lewis RE, Kontoyiannis DP. Epidemiology and treatment of mucormycosis. *Future Microbiol.* 2013; **8:** 1163-75.
369. Lewis RE, Lortholary O, Spellberg B, *et al.* How does antifungal pharmacology differ for mucormycosis versus aspergillosis? *Clin Infect Dis.* 2012; **54:** S67-72.
370. Li DM, Lun LD. *Mucor irregularis* infection and lethal midline granuloma: A case report and review of published literature. *Mycopathologia.* 2012; **174:** 429-39.
371. Li H, Hwang SK, Zhou C, *et al.* Gangrenous cutaneous mucormycosis caused by *Rhizopus oryzae:* A case report and review of primary cutaneous mucormycosis in China over Past 20 years. *Mycopathologia.* 2013; **176:** 123-8.
372. Li KW, Wen TF, Li GD. Hepatic mucormycosis mimicking hilar cholangiocarcinoma: A case report and literature review. *World J Gastroenterol.* 2010; **16:** 1039-42.
373. Li WF, He C, Liu XF, *et al.* A diagnosis neglected for 6 years: Report of a misdiagnosed case of pulmonary mucormycosis and review of the literature. *Chin Med J (Engl).* 2010; **123:** 2480-2.
374. Lim MY, Alker AP, Califano S, *et al.* Concurrent disseminated nocardiosis and GI mucormycosis in a stem-cell transplantation recipient. *J Clin Oncol.* 2016; **34:** e84-6.
375. Lin E, Moua T, Limper AH. Pulmonary mucormycosis: Clinical features and outcomes. *Infection.* 2017; doi: 10.1007/s15010-017-0991-6. PMID: 28220379.
376. Lineberry KD, Boettcher AK, Blount AL, *et al.* Cutaneous mucormycosis of the upper extremity in an immunocompetent host: Case report. *J Hand Surg Am.* 2012; **37:** 787-91.
377. Liu Y, Wu H, Huang F, *et al.* Utility of ^{18}F-FDG PET/CT in diagnosis and management of mucormycosis. *Clin Nucl Med.* 2013; **38:** e370-1.
378. Lo OS, Law WL. Ileocolonic mucormycosis in adult immunocompromised patients: A surgeon's perspective. *World J Gastroenterol.* 2010; **16:** 1165-70.
379. London V, Chong S. Cutaneous mucormycosis: A case report. *Cutis.* 2012; **89:** 167-8.
380. Long B, Koyfman A. Mucormycosis: What emergency physicians need to know? *Am J Emerg Med.* 2015; **33:** 1823-5.
381. Lu XL, Najafzadeh MJ, Dolatabadi S, *et al.* Taxonomy and epidemiology of *Mucor irregularis,* agent of chronic cutaneous mucormycosis. *Persoonia.* 2013; **30:** 48-56.
382. Lumbang WA, Caufield BA. Vesicular eruption on the arm of an infant. *Dermato Online J.* 2010; **16:** 13.
383. Lundy JB, Driscoll IR. Experience with proctectomy to manage combat casualties sustaining catastrophic perineal blast injury complicated by invasive *Mucor* soft-tissue infections. *Mil Med.* 2014; **179:** e347-50.
384. Lunge SB, Sajjan V, Pandit AM, *et al.* Rhinocerebrocutaneous mucormycosis caused by *Mucor* species: A rare causation. *Indian Dermatol Online J.* 2015; **6:** 189-92.
385. Luo Y, Zeng F, Huang X, *et al.* Successful treatment of a necrotizing fasciitis patient caused by *Mucor indicus* with amphotericin B and skin grafting. *Mycopathologia.* 2014; **177:** 187-92.
386. Ma J, Jia R, Li J, *et al.* Retrospective clinical study of eighty-one cases of intracranial mucormycosis. *J Glob Infect Dis.* 2015; **7:** 143-50.
387. Machicado JD, Younes M, Wolf DS. A rare cause of gastrointestinal bleeding in the intensive care unit. Healthcare-associated mucormycosis. *Gastroenterology.* 2014; 146: 911 & 1136-7.
388. Mahadevaiah AH, Rajagopalan N, Patil M, *et al.* Coinfection of pulmonary mucormycosis and aspergillosis presenting as bilateral vocal cord palsy. *BMJ Case Rep.* 2013; pii: bcr2013009615. PMID: 23964036.
389. Mahajan R, Paul G, Chopra P, *et el.* Mucormycosis masquerading as an endobronchial tumor. *Lung India.* 2014; **31:** 308-10.
390. Maheshwari R. Review of orbital exenteration from an eye care centre in Western India. *Orbit.* 2010; **29:** 35-8.
391. Mahmood A, Chaump M, Knoll B, *et al.* Pulmonary zygomycosis in a diabetic patient: Treated with pneumonectomy and antifungal agents. *R I Med J (2013).* 2014; **97:** 44-5.
392. Mahmud A, Lee R, Munfus-McCray D, *et al. Actinomucor elegans* as an emerging cause of mucormycosis. *J Clin Microbiol.* 2012; **50:** 1092-5.
393. Malhotra S, Duggal S, Bhatia NK, *et al.* Rhinocerebral zygomycosis with pulmonary aspergillosis in a non-HIV-infected patient: An unusual case report from India. *J Med Microbiol.* 2009; **58 (Pt 1):** 146-50.

394. Malik AN, Bi WL, McCray B, *et al.* Isolated cerebral mucormycosis of the basal ganglia. *Clin Neurol Neurosurg.* 2014; **124:** 102-5.
395. Malkan AD, Wahid FN, Rao BN, *et al.* Aggressive *Cunninghamella* pneumonia in an adolescent. *J Pediatr Hematol Oncol.* 2014; **36:** 581-2.
396. Mallis A, Mastronikolis SN, Naxakis SS, Papadas AT. Rhinocerebral mucormycosis: An update. *Eur Rev Med Pharmacol Sci.* 2010; **14:** 987-92.
397. Mandal PK, Bhattacharyya NK, Mookerjee SK, *et al.* Primary squamous cell carcinoma with mucormycosis in a diabetic foot ulcer. *J Indian Med Assoc.* 2013; **111:** 125-6.
398. Mandegari E, Fu L, Arambu C, *et al.* Mucormycosis rhinosinusitis at diagnosis of acute lymphoblastic leukemia: Diagnostics and management challenges in a low-middle-income country. *J Pediatr Hematol Oncol.* 2015; **37:** e173-7.
399. Mangaraj S, Sethy G, Patro MK, *et al.* A rare case of subcutaneous mucormycosis due to *Syncephalastrum racemosum:* Case report and review of literature. *Indian J Med Microbiol.* 2014; **32:** 448-51.
400. Mani RK, Mishra V, Sharma M, *et al.* Isolated pulmonary mucormycosis. *BMJ Case Rep.* 2017; 2017. pii: bcr2017219342. doi: 10.1136/bcr-2017-219342.
401. Manjunatha BS, Das N, Sutariya RV, *et al.* Mucormycosis of the hard palate masquerading as carcinoma. *Clin Pract.* 2012; **2:** e28.
402. Mantadakis E, Samonis G. Clinical presentation of zygomycosis. *Clin Microbiol Infect.* 2009; **15 (Suppl. 5):** 15-20.
403. Martinello M, Nelson A, Bignold L, *et al.* "We are what we eat!" Invasive intestinal mucormycosis: A case report and review of the literature. *Med Mycol Case Rep.* 2012; **1:** 52-5.
404. Marty FM, Ostrosky-Zeichner L, Cornely OA, *et al.* Isavuconazole treatment for mucormycosis: A single-arm open-label trial and case-control analysis. *Lancet Infect Dis.* 2016; **16:** 828-37.
405. Mathur NB, Gupta A. Neonatal zygomycosis with gastric perforation. *Indian Pediatr.* 2013; **50:** 699-701.
406. Matsudate Y, Murao K, Urano Y, *et al.* Primary cutaneous mucormycosis caused by *Mucor irregularis* in an immunocompetent patient. *J Dermatol.* 2015; **42:** 267-8.
407. Matsumoto K, Yamamoto W, Ohgusa E, *et al.* Disseminated *Cunninghamella bertholletiae* infection with septic pulmonary embolism after allogeneic bone marrow transplantation. *Transpl Infect Dis.* 2014; **16:** 304-6.
408. Mattioni J, Portnoy JE, Moore JE, *et al.* Laryngotracheal mucormycosis: Report of a case. *Ear Nose Throat J.* 2016; **95:** 29-39.
409. Maturu VN, Agarwal R. Reversed halo sign: A systematic review. *Respir Care.* 2014; **59:** 1440-9.
410. McCrory MC, Moore BA, Nakagawa TA, *et al.* Disseminated mucormycosis in an adolescent with newly diagnosed diabetes mellitus. *Pediatr Infect Dis J.* 2014; **33:** 1094-6.
411. Mehmood M, Halloush RA, Khasawneh FA. A 29-year-old man with hospital-acquired cavitary pneumonia. *Can Respir J.* 2014; **21:** 208-10.
412. Metzen D, Bohm H, Zimmermann M, *et al.* Mucormycosis of the head and neck. *J Craniomaxillofac Surg.* 2012; **40:** e321-7.
413. Millon L, Larosa F, Lepiller Q, *et al.* Quantitative polymerase chain reaction detection of circulating DNA in serum for early diagnosis of mucormycosis in immunocompromised patients. *Clin Infect Dis.* 2013; **56:** e95-101.
414. Misra P, Roy P, Dutta V, *et al.* Fatal massive hemoptysis: An autopsy study. *Med J Armed Forces India.* 2015; **71:** S234-6.
415. Misra PC, Srivastava KJ, Lata K. *Apophysomyces,* a new genus of the Mucorales. *Mycotaxon* 1979; **8:** 377-82.
416. Mitchell ME, McManus M, Dietz J, *et al.* *Absidia corymbifera* endocarditis: Survival after treatment of disseminated mucormycosis with radical resection of tricuspid valve and right ventricular free wall. *J Thorac Cardiovasc Surg.* 2010; **139:** e71-2.
417. Mitchell TA, Hardin MO, Murray CK, *et al.* Mucormycosis attributed mortality: A seven-year review of surgical and medical management. *Burns.* 2014; **40:** 1689-95.
418. Mittal T, Pulle MV, Dey A, *et al.* Gastric mucormycosis presenting as gastrocolic fistula: A rare entity. *Indian J Surg.* 2016; **78:** 511-2.
419. Mohamed MS, Abdel-Motaleb HY, Mobarak FA. Management of rhino-orbital mucormycosis. *Saudi Med J.* 2015; **36:** 865-8.
420. Mohammadi R, Meidani M, Mostafavizadeh K, *et al.* Case series of rhinocerebral mucormycosis occurring in diabetic patients. *Caspian J Intern Med.* 2015; **6:** 243-6.
421. Mohammadi R, Nazeri M, Sayedayn SM, *et al.* A successful treatment of rhinocerebral mucormycosis due to *Rhizopus oryzae. J Res Med Sci.* 2014; **19:** 72-4.
422. Mohanty D, Dhar M, Dwivedi S. Mucormycosis. *Trop Doct.* 2010; **40:** 127-8.
423. Mohapatra S, Xess I, Shwetha JV, *et al.* Soil extract media for sporulation of *Apophysomyces elegans. Indian J Pathol Microbiol.* 2010; **53:** 897-8.
424. Mohindra S, Gupta B, Gupta K, Bal A. Tracheal mucormycosis pneumonia: a rare clinical presentation. *Respir Care.* 2014; **59:** e178-81.
425. Mohta A, Neogi S, Das S. Gastrointestinal mucormycosis in an infant. *Indian J Pathol Microbiol.* 2011; **54:** 664-5.
426. Mondal PK, Mondal SK, Mondal TK, *et al.* Mucormycosis of pouch of Douglas in a diabetic woman. *J Glob Infect Dis.* 2012; **4:** 172-4.
427. Moosavi Movahed M, Hosamirudsari H, Mansouri F, *et al.* Spontaneous pneumothorax followed by reversed halo sign in immunocompromised patient with pulmonary mucormycosis. *Med Mycol Case Rep.* 2015; **9:** 22-5.
428. Moreira J, Varon A, Galhardo MC, *et al.* The burden of mucormycosis in HIV-infected patients: A systematic review. *J Infect.* 2016; **73:** 181-8.
429. Morton J, Nguyen V, Ali T. Mucormycosis of the intestine: A rare complication in Crohn's disease. *Gastroenterol Hepatol (NY).* 2012; **8:** 137-40.
430. Motaleb HY, Mohamed MS, Mobarak FA. A fatal outcome of rhino-orbito-cerebral mucormycosis following tooth extraction: A case report. *J Int Oral Health.* 2015; **7 (Suppl. 1):** 68-71.

431. Mundra RK, Gupta Y. Mucormycosis of nose and paranasal sinuses with orbital complication in young diabetic. *Indian J Otolaryngol Head Neck Surg.* 2008; **60:** 360-4.
432. Munoz J, Hughes A, Guo Y. Mucormycosis-associated intracranial hemorrhage. *Blood Coagul Fibrinolysis.* 2013; **24:** 100-1.
433. Muqeetadnan M, Rahman A, Amer S, *et al.* Pulmonary mucormycosis: An emerging infection. *Case Rep Pulmonol.* 2012; 120809. PMID: 23304605.
434. Murphy AD, Williamson PA, Vesely M. Reconstruction of an extensive peri-orbital defect secondary to mucormycosis in a patient with myelodysplasia. *J Plast Reconstr Aesthet Surg.* 2013; **66:** e69-71.
435. Mutchnick S, Soares D, Shkoukani M. To exenterate or not? An unusual case of pediatric rhinocerebral mucormycosis. *Int J Pediatr Otorhinolaryngol.* 2015; **79:** 267-70.
436. Mysorekar VV, Rao SG. Cytomegalovirus pneumonia with pulmonary mucormycosis. *Indian J Pathol Microbiol.* 2008; **51:** 294-5.
437. Nagarkar NM, Verma H, Punia RPS. Co-existing mucormycosis with aspergillosis in a patient with diabetes mellitus - First case report. *Otolaryngology Online Journal.* 2014; **4:** 257.
438. Nain PS, Matta H, Singh K, Chhina D, Trehan M, Batta N. Post-operative abdominal wall mucormycosis - A case series. *Indian J Surg.* 2015; **77 (Suppl. 2):** 253-6.
439. Nair V, Sharma RK, Khanna A, *et al.* Pulmonary mucormycosis diagnosed by convex probe endobronchial ultrasound-guided fine needle aspiration of cavity wall. *Lung India.* 2017; **34:** 179-81.
440. Nam Y, Jung J, Park SS, *et al.* Disseminated mucormycosis with myocardial involvement in a renal transplant recipient. *Transpl Infect Dis.* 2015; **17:** 890-6.
441. Nandigam K. SWAN MRI revealing multiple microhemorrhages secondary to septic emboli from mucormycosis. *Neurology.* 2013; **81:** 199-200.
442. Nandu V, Nagral A, Khubchandani S, *et al.* A rare cause of upper GI bleeding. *Gut.* 2009; **58:** 1039 & 1048.
443. Nandwani A, Jha PK, Duggal R, *et al.* Invasive gastric mucormycosis and cytomegalovirus infection in an ABO incompatible renal transplant recipient. *Indian J Nephrol.* 2015; **25:** 373-6.
444. Narayanan MLS, Narayanan CD, Kindo AJ, *et al.* Fatal fungal infection: The living dead. *J Surg Case Rep.* 2014; pii: rju104. PMID: 25352577.
445. Narayanan S, Panarkandy G, Subramaniam G, *et al.* The "black evil" affecting patients with diabetes: A case of rhino orbito cerebral mucormycosis causing Garcin syndrome. *Infect Drug Resist.* 2017; **10:** 103-8.
446. Navanukroh O, Jitmuang A, Chayakulkeeree M, *et al.* Disseminated *Cunninghamella bertholletiae* infection with spinal epidural abscess in a kidney transplant patient: Case report and literature review. *Transpl Infect Dis.* 2014; **16:** 658-65.
447. Nawange SR, Singh SM, Naidu J, *et al.* Zygomycosis caused by *Rhizopus microsporus* and *Rhizopus oryzae* in Madhya Pradesh (MP), Central India: A report of two cases. *Mycopathologia.* 2012; **174:** 171-6.
448. Nayagam LS, Vijayanand B, Balasubramanian S. Isolated renal mucormycosis in an immunocompetent child. *Indian J Nephrol.* 2014; **24:** 321-3.
449. Nayak S, Jindal A. Rhino-orbito-cerebral mucormycosis in a child with diabetic ketoacidosis. *Indian J Crit Care Med.* 2014; **18:** 638-9.
450. Neblett Fanfair R, Benedict K, Bos J, *et al.* Necrotizing cutaneous mucormycosis after a tornado in Joplin, Missouri, in 2011. *N Engl J Med.* 2012; **367:** 2214-25.
451. Neto FM, Camargo PC, Costa AN, *et al.* Fungal infection by Mucorales order in lung transplantation: 4 case reports. *Transplant Proc.* 2014; **46:** 1849-51.
452. Nichols L, Ocque RZ, Daly I. Zygomycosis associated with HIV infection and liver transplantation. *Patholog Res Int.* 2011; 545981. PMID: 21331168.
453. Nilesh K, Malik NA, Belgaumi U. Mucormycosis in a healthy elderly patient presenting as oro-antral fistula: Report of a rare incidence. *J Clin Exp Dent.* 2015; **7:** e333-5.
454. Nirmala SV, Lalitha V, Sivakumar N, *et al.* Mucormycosis associated with juvenile diabetes. *J Indian Soc Pedod Prev Dent.* 2011; **29:** S87-91.
455. Nithyanandam S, Correa MA. Rhino-orbital mucormycosis and aspergillosis: Differences in outcome, clinical and imaging characteristics. *Eur Arch Otorhinolaryngol.* 2010; **267:** 161-2.
456. No autors listed. Cluster of rare fungal infections identified after tornado. *BMJ.* 2012; **345:** e8382.
457. Non L, Sta Cruz JP, Tuazon S. Sudden death in a patient with bone marrow transplant by a fungus among us. *BMJ Case Rep.* 2014; pii: bcr2014207403. PMID: 25391828.
458. Nunes DM, Rocha AJ, Rosa Júnior M, Silva CJ. "Black turbinate sign" : A potential predictor of mucormycosis in cavernous sinus thrombophlebitis. *Arq Neuropsiquiatr.* 2012; **70:** 78. PMID: 22218482.
459. Odronic SI, Scheidemantel T, Tuohy MJ, *et al.* Two cases of *Cokeromyces recurvatus* in liquid-based Papanicolaou tests and a review of the literature. *Arch Pathol Lab Med.* 2012; **136:** 1593-6.
460. Okubo Y, Ishiwatari T, Izumi H, *et al.* Pathophysiological implication of reversed CT halo sign in invasive pulmonary mucormycosis: A rare case report. *Diagn Pathol.* 2013; **8:** 82.
461. Oladeji S, Amusa Y, Olabanji J, *et al.* Rhinocerebral mucormycosis in a diabetic case report. *J West Afr Coll Surg.* 2013; **3:** 93-102.
462. Oswal NP, Gadre PK, Sathe P, *et al.* Mucormycosis of mandible with unfavorable outcome. *Case Rep Dent.* 2012; 257940. PMID: 22779014.
463. Pagano L, Cornely OA, Busca A, *et al.* Combined antifungal approach for the treatment of invasive mucormycosis in patients with hematologic diseases: A report from the SEIFEM and FungiScope registries. *Haematologica.* 2013; **98:** e127-30.
464. Page AV, Evans AJ, Snell L, *et al.* Primary cutaneous mucormycosis in a lung transplant recipient: case report and concise review of the literature. *Transpl Infect Dis.* 2008; **10:** 419-25.

465. Pahwa M, Pahwa AR, Girotra M, *et al.* Isolated renal mucormycosis in a healthy immunocompetent patient: Atypical presentation and course. *Korean J Urol.* 2013; **54:** 641-3.
466. Pajpani M, Webb R. Lingual necrosis caused by mucormycosis in a patient with aplastic anaemia: Case report. *Br J Oral Maxillofac Surg.* 2014; **52:** e144-6.
467. Pak J, Tucci VT, Vincent AL, *et al.* Mucormycosis in immunochallenged patients. *J Emerg Trauma Shock.* 2008; **1:** 106-13.
468. Palacio-Bedoya F, Cadena JA, Thompson GR, *et al.* A non-invasive renal fungus ball caused by *Rhizopus* - A previously unreported manifestation of zygomycosis. *Med Mycol.* 2010; **48:** 866-9.
469. Pan AS, Srinath L. Mucormycosis in a patient with AIDS receiving systemic steroids. *J Am Osteopath Assoc.* 2013; **113:** 708-11.
470. Pana ZD, Seidel D, Skiada A, *et al.* Invasive mucormycosis in children: An epidemiologic study in European and non-European countries based on two registries. *BMC Infect Dis.* 2016; **16:** 667. PMID: 27832748.
471. Panigrahi MK, Manju R, Kumar SV, *et al.* Pulmonary mucormycosis presenting as non-resolving pneumonia in a patient with diabetes mellitus. *Respir Care.* 2014; **59:** e201-5.
472. Paonam S, Bag S, Mavuduru RS, *et al.* Isolated bilateral renal mucormycosis masquerading as renal abscess in an immunocompetent individual: A lesson learnt. *Case Rep Urol.* 2014; 304380. PMID: 25587482.
473. Paquette C, Slater SE, McMahon MD, *et al. Cokeromyces recurvatus* in a cervical papanicolaou test: A case report of a rare fungus with a brief review of the literature. *Diagn Cytopathol.* 2016; **44:** 419-21.
474. Park W, Jang M, Hwang E, *et al.* Allograft mucormycosis due to *Rhizopus microsporus* in a kidney transplant recipient. *Transplant Proc.* 2014; **46:** 623-5.
475. Parsi K, Itgampalli RK, Vittal R, *et al.* Perineural spread of rhino-orbitocerebral mucormycosis caused by *Apophysomyces elegans. Ann Indian Acad Neurol.* 2013; **16:** 414-7.
476. Pastor FJ, Ruiz-Cendoya M, Pujol I, *et al. In vitro* and *in vivo* antifungal susceptibilities of the Mucoralean fungus *Cunninghamella. Antimicrob Agents Chemother.* 2010; **54:** 4550-5.
477. Patel A, Bishburg E, Nagarakanti S. Mucormycosis in an HIV-infected renal transplant patient: A case report and review of the literature. *Am J Case Rep.* 2014; **15:** 74-8.
478. Patel AK, Patel KK, Patel K, *et al.* Mucormycosis at a tertiary care centre in Gujarat, India. *Mycoses.* 2017; **60:** 407-11.
479. Patel AK. Mucormycosis in patients without cancer: A case series from a tertiary care hospital in South India. *J Assoc Physicians India.* 2014; **62:** 548.
480. Patel AK, Vora HJ, Patel KK, *et al.* Elderly diabetic patient with surgical site mucormycosis extending to bowel. *J Glob Infect Dis.* 2010; **2:** 186-8.
481. Patra S, Vij M, Chirla DK, *et al.* Unsuspected invasive neonatal gastrointestinal mucormycosis: A clinicopathological study of six cases from a tertiary care hospital. *J Indian Assoc Pediatr Surg.* 2012; **17:** 153-6.
482. Peixoto D, Gagne LS, Hammond SP, *et al.* Isavuconazole treatment of a patient with disseminated mucormycosis. *J Clin Microbiol.* 2014; **52:** 1016-9.
483. Peixoto D, Hammond SP, Issa NC, *et al.* Green herring syndrome: Bacterial infection in patients with mucormycosis cavitary lung disease. *Open Forum Infect Dis.* 2014; **1:** ofu014. PMID: 25734087.
484. Pellicelli AM, D'Ambrosio C, Villani R, *et al.* Liver cirrhosis and rhino-orbital mucormycosis, a possible but rare association: Description of a clinical case and literature review. *Braz J Infect Dis.* 2009; **13:** 314-6.
485. Pepeler MS, Acar K, Güzel Tunçcan O, *et al.* A proven case of cutaneous *Rhizopus* infection presenting with severe limb pain very soon after induction treatment in a patient with acute lymphoblastic leukemia. *Case Rep Hematol.* 2015; 285360. PMID: 25785211.
486. Petraitis V, Petraitiene R, Antachopoulos C, *et al.* Increased virulence of *Cunninghamella bertholletiae* in experimental pulmonary mucormycosis: Correlation with circulating molecular biomarkers, sporangiospore germination and hyphal metabolism. *Med Mycol.* 2013; **51:** 72-82.
487. Petrikkos G, Drogari-Apiranthitou M. Zygomycosis in immunocompromised non-haematological patients. *Mediterr J Hematol Infect Dis.* 2011; **3:** e2011012.
488. Petrikkos G, Skiada A, Drogari-Apiranthitou M. Epidemiology of mucormycosis in Europe. *Clin Microbiol Infect.* 2014; **20 (Suppl. 6):** 67-73.
489. Petrikkos G, Skiada A, Lortholary O, *et al.* Epidemiology and clinical manifestations of mucormycosis. *Clin Infect Dis.* 2012; **54:** S23-34.
490. Phulpin-Weibel A, Rivier A, Leblanc T, *et al.* Focus on invasive mucormycosis in paediatric haematology oncology patients: A series of 11 cases. *Mycoses.* 2013; **56:** 236-40.
491. Poirier P, Nourrisson C, Gibold L, *et al.* Three cases of cutaneous mucormycosis with *Lichtheimia* spp. (ex *Absidia/Mycocladus*) in ICU. Possible cross-transmission in an intensive care unit between 2 cases. *J Mycol Med.* 2013; **23:** 265-9.
492. Pongas GN, Lewis RE, Samonis G, *et al.* Voriconazole-associated zygomycosis: A significant consequence of evolving antifungal prophylaxis and immunosuppression practices? *Clin Microbiol Infect.* 2009; **15 (Suppl. 5):** 93-7.
493. Pourahmad M, Sepidkar A, Farokhnia MH, *et al.* Mucormycosis after scorpion sting: Case report. *Mycoses.* 2013; **56:** 589-91.
494. Prakash H, Ghosh AK, Rudramurthy SM, *et al.* The environmental source of emerging *Apophysomyces variabilis* infection in India. *Med Mycol.* 2016; **54:** 567-75.
495. Prasad K, Lalitha RM, Reddy EK, *et al.* Role of early diagnosis and multimodal treatment in rhinocerebral mucormycosis: Experience of 4 cases. *J Oral Maxillofac Surg.* 2012; **70:** 354-62.
496. Prasad PA, Vaughan AM, Zaoutis TE. Trends in zygomycosis in children. *Mycoses.* 2012; **55:** 352-6.

497. Prasanna Kumar S, Ravikumar A, Somu L. Fungal necrotizing fasciitis of the head and neck in 3 patients with uncontrolled diabetes. *Ear Nose Throat J.* 2014; **93:** e18-21.
498. Pyrgos V, Shoham S, Walsh TJ. Pulmonary zygomycosis. *Semin Respir Crit Care Med.* 2008; **29:** 111-20.
499. Qu LF, Yang J, Wu XW, *et al.* Endobronchial mucormycosis showing as a bronchial tumor. *Chin Med J (Engl).* 2013; **126:** 3620.
500. Rabie NB, Althaqafi AO. *Rhizopus*-associated soft tissue infection in an immunocompetent air-conditioning technician after a road traffic accident: A case report and review of the literature. *J Infect Public Health.* 2012; **5:** 109-11.
501. Radhakrishnan N, Yadav SP, Oberoi J, *et al.* Intestinal mucormycosis: A rare entity in pediatric oncology. *Pediatr Hematol Oncol.* 2013; **30:** 178-83.
502. Radowsky JS, Strawn AA, Sherwood J, *et al.* Invasive mucormycosis and aspergillosis in a healthy 22-year-old battle casualty: Case report. *Surg Infect (Larchmt).* 2011; **12:** 397-400.
503. Rahman A, Akter K, Hossain S, *et al.* Rhino-orbital mucourmycosis in a non-immunocompromised patient. *BMJ Case Rep.* 2013; pii: bcr2012007863. PMID: 23391952.
504. Rajagopal K, Watkins AC, Gibber M, *et al.* Re-operative lung transplantation for donor-derived pulmonary mucormycosis. *Ann Thorac Surg.* 2014; **98:** 327-9.
505. Ram R, Swarnalatha G, Naidu GD, *et al.* Multiple ring enhancing lesions in brain due to disseminated zygomycosis in a renal transplant recipient. *Nephrology (Carlton).* 2013; **18:** 479-80.
506. Ramachandran L, Dewan S, Kumar V, *et al.* Mucormycosis causing pulmonary artery aneurysm. *Respir Med Case Rep.* 2015; **16:** 71-3.
507. Ramaswami A, Pisharam JK, Aung H, *et al.* Co-incidental *Plasmodium knowlesi* and mucormycosis infections presenting with acute kidney injury and lower gastrointestinal bleeding. *Am J Case Rep.* 2013; **14:** 103-5.
508. Rammaert B, Angebault C, Scemla A, *et al.* *Mucor irregularis*-associated cutaneous mucormycosis: Case report and review. *Med Mycol Case Rep.* 2014; **6:** 62-5.
509. Rammaert B, Lanternier F, Poiree S, *et al.* Diabetes and mucormycosis: A complex interplay. *Diabetes Metab.* 2012; **38:** 193-204.
510. Rammaert B, Lanternier F, Zahar JR, *et al.* Healthcare-associated mucormycosis. *Clin Infect Dis.* 2012; **54:** S44-54.
511. Raveenthiran V. Gastrointestinal mucormycosis mimicking necrotizing enterocolitis of newborn. *J Neonatal Surg.* 2013; **2:** 41.
512. Reddy IS, Rao NR, Shankar Reddy VM, *et al.* Primary cutaneous mucormycosis (zygomycosis) caused by *Apophysomyces elegans. Indian J Dermatol Venereol Leprol.* 2008; **74:** 367-70.
513. Reddy SS, Rakesh N, Chauhan P, *et al.* Rhinocerebral mucormycosis among diabetic patients: An emerging trend. *Mycopathologia.* 2015; **180:** 389-96.
514. Reinbold C, Derder M, Hivelin M, *et al.* Using free flaps for reconstruction during infections by mucormycosis: A case report and a structured review of the literature. *Ann Chir Plast Esthet.* 2015. pii: S0294-1260(15)00089-8.
515. Relloso S, Romano V, Landaburu MF, *et al.* *Saksenaea erythrospora* infection following a serious sailing accident. *J Med Microbiol.* 2014; **63:** 317-21.
516. Requena L, Sitthinamsuwan P, Santonja C, *et al.* Cutaneous and mucosal mucormycosis mimicking pancreatic panniculitis and gouty panniculitis. *J Am Acad Dermatol.* 2012; **66:** 975-84.
517. Ribeiro EF, dos Santos VM, Paixao GT, *et al.* Mucormycosis in a patient with acute myeloid leukemia successfully treated with liposomal amphotericin B associated with deferasirox and hyperbaric oxygen. *Mycopathologia.* 2013; **175:** 295-300.
518. Ribeiro LC, Wanke B, da Silva M, *et al.* Mucormycosis in Mato Grosso, Brazil: A case reports, caused by *Rhizopus microsporus* var. *oligosporus* and *Rhizopus microsporus* var. *rhizopodiformis. Mycopathologia.* 2012; **173:** 187-92.
519. Ribes JA, Vanover-Sams CL, Baker DJ. Zygomycetes in human disease. *Clin Microbiol Rev.* 2000; **13:** 236-301.
520. Richardson M. The ecology of the zygomycetes and its impact on environmental exposure. *Clin Microbiol Infect.* 2009; 15 **(Suppl. 5):** 2-9.
521. Riley TT, Muzny CA, Swiatlo E, et al. Breaking the Mold: A review of mucormycosis and current pharmacological treatment options. *Ann Pharmacother.* 2016; **50:** 747-57.
522. Robin C, Alanio A, Cordonnier C. Mucormycosis: A new concern in the transplant ward? *Curr Opin Hematol.* 2014; **21:** 482-90.
523. Rodrigues S, Santos L, Nuak J, *et al.* Colonic mucormycosis. *Endoscopy.* 2013; **45 (Suppl. 2)**: e20.
524. Rodriguez-Gutierrez G, Carrillo-Casas EM, Arenas R, *et al.* Mucormycosis in a non-Hodgkin lymphoma patient caused by *Syncephalastrum racemosum:* Case report and review of literature. *Mycopathologia.* 2015; **180:** 89-93.
525. Rodriguez JY, Rodriguez GJ, Morales-Lopez SE, *et al.* *Saksenaea erythrospora* infection after medical tourism for esthetic breast augmentation surgery. *Int J Infect Dis.* 2016; **49:** 107-10.
526. Rodriguez-Lobato E, Ramirez-Hobak L, Aquino-Matus JE, *et al.* Primary cutaneous mucormycosis caused by *Rhizopus oryzae:* A case report and review of literature. *Mycopathologia.* 2017; **182:** 387-92.
527. Rogers TR. Treatment of zygomycosis: Current and new options. *J Antimicrob Chemother.* 2008; **61 (Suppl. 1):** i35-40.
528. Roilides E, Antachopoulos C, Simitsopoulou M. Pathogenesis and host defence against Mucorales: The role of cytokines and interaction with antifungal drugs. *Mycoses.* 2014; **57 (Suppl. 3):** 40-7.
529. Roilides E, Kontoyiannis DP, Walsh TJ. Host defenses against zygomycetes. *Clin Infect Dis.* 2012; **54:** S61-6.
530. Rojek NW, Norton SA. Violaceous necrotic plaques on the leg of an immunosuppressed patient. *JAMA Dermatol.* 2014; **150:** 674.
531. Roussy JF, Allard C, St-Germain G, *et al.* Gastrointestinal mucormycosis following a *Streptococcus pyogenes* toxic shock syndrome in a previously healthy patient: A case report. *Case Rep Infect Dis.* 2012; 476719. PMID: 22900216.

532. Royer M, Cervera P, Kahan A, *et al.* Mucormycosis cerebral arteritis mimicking a flare in ANCA-associated vasculitis. *Lancet Infect Dis.* 2013; **13:** 182.
533. Royer M, Puechal X. Mucormycosis in systemic autoimmune diseases. *Joint Bone Spine.* 2014; **81:** 303-7.
534. Rudler M, Barret M, Poynard T, *et al.* Gastric mucormycosis: A rare cause of gastrointestinal bleeding in cirrhosis. *Clin Res Hepatol Gastroenterol.* 2012; **3:** e32-3.
535. Ryan LJ, Ferrieri P, Powell RD Jr, *et al.* Fatal *Cokeromyces recurvatus* pneumonia: Report of a case highlighting the potential for histopathologic misdiagnosis as *Coccidioides. Int J Surg Pathol.* 2011; **19:** 373-6.
536. Ryan O, Frohlich S, Crotty TB, *et al. Rhizopus microsporus* infection in an immunocompetent host: A case of immunoparalysis? *Anaesth Intensive Care.* 2012; **40:** 367-8.
537. Sachdeva K. Rhino-oculo-cerebral mucormycosis with multiple cranial nerve palsy in diabetic patient: Review of six cases. *Indian J Otolaryngol Head Neck Surg.* 2013; **65:** 375-9.
538. Saegeman V, Maertens J, Ectors N, *et al.* Epidemiology of mucormycosis: Review of 18 cases in a tertiary care hospital. *Med Mycol.* 2010; **48:** 245-54.
539. Saegeman V, Maertens J, Meersseman W, *et al.* Increasing incidence of mucormycosis in University Hospital, Belgium. *Emerg Infect Dis.* 2010; **16:** 1456-8.
540. Sagar S, Kumar S, Gupta A, *et al.* Successful management of a case of primary cutaneous mucormycosis in an immunocompetent host. *Wounds.* 2013; **25:** 136-40.
541. Salas V, Pastor FJ, Calvo E, *et al.* Experimental murine model of disseminated infection by *Saksenaea vasiformis:* Successful treatment with posaconazole. *Med Mycol.* 2012; **50:** 710-5.
542. Salas V, Pastor FJ, Calvo E, Sutton DA, Chander J, Mayayo E, Alvarez E, Guarro J. Efficacy of posaconazole in a murine model of disseminated infection caused by *Apophysomyces variabilis. J Antimicrob Chemother.* 2012; **67:** 1712-5.
543. Saltmarsh G, Plurad D, Bricker S, *et al.* Gastric necrosis and recurrent small bowel perforation resulting from gastrointestinal mucormycosis. *Am Surg.* 2014; **80:** e293-4.
544. Samanta DR, Senapati SN, Sharma PK, *et al.* Hard palate perforation in acute lymphoblastic leukemia due to mucormycosis - A case report. *Indian J Hematol Blood Transfus.* 2009; **25:** 36-9.
545. Samarei R, Gharebaghi N, Zayer S. Evaluation of 30 cases of mucormycosis at a university hospital in Iran. *Mycoses.* 2017; **60:** 426-32.
546. Sanavi S, Afshar R, Afshin-Majd S. Rhino-orbitocerebral mucormycosis in a patient with idiopathic crescentic glomerulonephritis. *Saudi J Kidney Dis Transpl.* 2013; **24:** 768-72.
547. Sankar J, Arun S, Sankar MJ, *et al.* 'Primary cutaneous mucormycosis during induction chemotherapy in a child with acute lymphoblastic leukemia.' *Indian J Pediatr.* 2009; 76: 1161-3.
548. Saraiya HA. Successful management of cutaneous mucormycosis by delaying debridement. *Ann Plast Surg.* 2012; **69:** 301-6.
549. Saravia-Flores M, Guaran DM, Argueta V. Invasive cutaneous infection caused by *Apophysomyces elegans* associated with a spider bite. *Mycoses.* 2010; **53:** 259-61.
550. Sarin YK. Intestinal mucormycosis in a neonate: A case report and review. *J Indian Assoc Pediatr Surg.* 2010; **15:** 98-100.
551. Sarkar S, Jash D, Maji A, *et al.* Solitary pulmonary nodule: A rare presentation of pulmonary mucormycosis in an immunocompetent adult. *Lung India.* 2014; **31:** 70-2.
552. Sarrami AH, Setareh M, Izadinejad M, *et al.* Fatal disseminated mucormycosis in an immunocompetent patient: A case report and literature review. *Int J Prev Med.* 2013; **4:** 1468-71.
553. Sasannejad P, Ghabeli-Juibary A, Aminzadeh S, *et al.* Cerebellar infarction and aneurysmal subarachnoid hemorrhage: An unusual presentation and rare complications of rhinocerebral mucormycosis. *Iran J Neurol.* 2015; **14:** 222-4.
554. Sathe KP, Mehta KP. Irreversible fatal renal failure resulting from isolated renal mucormycosis. *Saudi J Kidney Dis Transpl.* 2014; **25:** 1312-4.
555. Sathe PA, Ghodke RK, Kandalkar BM. A survivor of neonatal intestinal mucormycosis. *J Clin Diagn Res.* 2015; **9:** ED24-5.
556. Schell WA, O'Donnell K, Alspaugh JA. Heterothallic mating in *Mucor irregularis* and first isolate of the species outside of Asia. *Med Mycol.* 2011; **49:** 714-23.
557. Schmidt S, Tramsen L, Schneider A, *et al.* Immunotherapeutic strategies against mucormycosis in haematopoietic stem cell transplantation. *Mycoses.* 2014; **57 (Suppl. 3):** 8-12.
558. Schneidawind D, Nann D, Vogel W, *et al.* Allogeneic hematopoietic cell transplantation in patients with acute myeloid leukemia and pulmonary mucormycosis. *Transpl Infect Dis.* 2012; **14:** e166-72.
559. Schofield C, Stern A, Jevtic A. Disseminated zygomycosis due to *Mycocladus corymbifera* with cutaneous and cerebral involvement. *Australas J Dermatol.* 2013; **54:** e8-11.
560. Schrodl W, Heydel T, Schwartze VU, *et al.* Direct analysis and identification of pathogenic *Lichtheimia* species by matrix-assisted laser desorption ionization-time of flight analyzer-mediated mass spectrometry. *J Clin Microbiol.* 2012; **50:** 419-27.
561. Schwartze VU, Hoffmann K, Nyilasi I, *et al. Lichtheimia* species exhibit differences in virulence potential. *PLoS One.* 2012; **7:** e40908.
562. Schwartze VU, Jacobsen ID. Mucormycoses caused by *Lichtheimia* species. *Mycoses.* 2014; **57 (Suppl. 3):** 73-8.
563. Scully MA, Yeaney GA, Compton ML, *et al.* SWAN MRI revealing multiple microhemorrhages secondary to septic emboli from mucormycosis. *Neurology.* 2012; **79:** 1932-3.
564. Scully MA. SWAN MRI revealing multiple microhemorrhages secondary to septic emboli from mucormycosis. Author response. *Neurology.* 2013; **81:** 200.
565. Serio B, Rosamilio R, Giudice V, *et al.* Successful management of pulmonary mucormycosis with liposomal amphotericin B and surgery treatment: A case report. *Infez Med.* 2012; **20 (Suppl. 2):** 43-7.

566. Sethi PK, Khanna L, Batra A, *et al.* Central nervous system fungal infections: observations from a large tertiary hospital in northern India. *Clin Neurol Neurosurg.* 2012; **114:** 1232-7.
567. Shah RJ, Katyayan MK, Katyayan PA, *et al.* Prosthetic rehabilitation of acquired maxillary defects secondary to mucormycosis: Clinical cases. *J Contemp Dent Pract.* 2014; **15:** 242-9.
568. Shah S, Suresh PV, Maheshwari S, *et al.* Cardiac mucormycosis with T-cell immunodeficiency. *Indian Pediatr.* 2009; **46:** 257-9.
569. Shakoor S, Jabeen K, Idrees R, *et al.* Necrotising fasciitis due to *Absidia corymbifera* in wounds dressed with non-sterile bandages. *Int Wound J.* 2011; **8:** 651-5.
570. Sharma N, Batra H, Mehta M, Chander J. Maxillary osteomyelitis caused by *Apophysomyces variabilis* - Emerging trends. *Maced J Med Sci.* 2014; **7:** 305-10.
571. Sharma SK, Agarwal N, Mukherjee A, *et al.* Co-existing pulmonary tuberculosis and mucormycosis in a patient with aplastic anemia post allogenic stem cell transplantation. *Mediterr J Hematol Infect Dis.* 2011; **3:** e2011036.
572. Shatriah I, Mohd-Amin N, Tuan-Jaafar TN, *et al.* Rhino-orbito-cerebral mucormycosis in an immunocompetent patient: Case report and review of literature. *Middle East Afr J Ophthalmol.* 2012; **19:** 258-61.
573. Shekar V, Sikander J, Rangdhol V, *et al.* Facial nerve paralysis: A case report of rare complication in uncontrolled diabetic patient with mucormycosis. *J Nat Sci Biol Med.* 2015; **6:** 226-8.
574. Shetty S, Mambatta AK, Penmatsa KR, *et al.* Ileal mucormycosis: A rare cause of lower gastrointestinal bleeding. *Ann Gastroenterol.* 2014; **27:** 169.
575. Sheybani F, Naderi HR, Sarvghad M, *et al.* How should we manage a patient with invasive mucormycosis who develops life-threatening reaction to amphotericin B? Report of two cases and literature review. *Med Mycol Case Rep.* 2015; **8:** 29-31.
576. Shinde RV, Karande GS, Mohite ST, Patil SR. Rhino-orbital mucormycosis in diabetes mellitus. *J Clin Diagn Res.* 2013; **7:** 1145-7.
577. Shiraishi K, Sasaki S, Sadamoto Y. Cutaneous mucormycosis in a patient with acute lymphocytic leukemia. *Eur J Dermatol.* 2014; **24:** 116-7.
578. Shiva Prasad BN, Shenoy A, Nataraj KS, *et al.* Primary gastrointestinal mucormycosis in an immunocompetent person. *J Postgrad Med.* 2008; **54:** 211-3.
579. Shivananda P, Mahabala C, Kausalya S, *et al.* Cutaneous mucormycosis with necrotising fasciitis in a young immunocompetent individual. *Trop Doct.* 2011; **41:** 183-4.
580. Silberstein E, Krieger Y, Rosenberg N, *et al.* Facial reconstruction of a mucormycosis survivor by free rectus abdominis muscle flap, tissue expansion and ocular prosthesis. *Ophthal Plast Reconstr Surg.* 2016; **32:** e131-2.
581. Simbli M, Hakim F, Koudieh M, *et al.* Nosocomial post-traumatic cutaneous mucormycosis: A systematic review. *Scand J Infect Dis.* 2008; **40:** 577-82.
582. Singh JI, Singhal MK. Zygomycosis presenting as acute bilateral renal artery thrombosis. *J Assoc Physicians India.* 2016; **64:** 104.
583. Singh P, Paul S, Shivaprakash MR, *et al.* Stress response in medically important Mucorales. *Mycoses.* 2016; **59:** 628-35.
584. Singh SK, Wadhwa P, Sakhuja V. Isolated bilateral renal mucormycosis. *Urology.* 2004; **63:** 979-80.
585. Singh V, Sharma B, Sen R, *et al.* Rhinocerebral mucormycosis: A diagnostic challenge and therapeutic dilemma in immunocompetent host. *J Oral Maxillofac Surg.* 2012; **70:** 1369-75.
586. Singh V, Singh M, Joshi C, Sangwan J. Rhinocerebral mucormycosis in a patient with type 1 diabetes presenting as toothache: A case report from Himalayan region of India. *BMJ Case Rep.* 2013; pii: bcr2013200811.
587. Sirignano S, Blake P, Turrentine JE, *et al.* Primary cutaneous zygomycosis secondary to minor trauma in an immunocompromised pediatric patient: A case report. *Dermatol Online J.* 2014; **20.** pii: 13030/qt4r2455h9.
588. Skiada A, Lanternier F, Groll AH, *et al.* Diagnosis and treatment of mucormycosis in patients with hematological malignancies: Guidelines from the 3rd European Conference on Infections in Leukemia (ECIL 3). *Haematologica.* 2013; **98:** 492-504.
589. Skiada A, Petrikkos G. Cutaneous mucormycosis. *Skinmed.* 2013; **11:** 155-9; Quiz 159-60.
590. Skiada A, Rigopoulos D, Larios G, *et al.* Global epidemiology of cutaneous zygomycosis. *Clin Dermatol.* 2012; **30:** 628-32.
591. Skiada A, Vrana L, Polychronopoulou H, *et al.* Disseminated zygomycosis with involvement of the central nervous system. *Clin Microbiol Infect.* 2009; **15 (Suppl. 5):** 46-9.
592. Snyder AB, Churey JJ, Worobo RW. Characterization and control of *Mucor circinelloides* spoilage in yogurt. *Int J Food Microbiol.* 2016; **228:** 14-21.
593. Soman R, Gupta N, Shetty A, *et al.* Deferasirox in mucormycosis: Hopefully, not defeated. *J Antimicrob Chemother.* 2012; **67:** 783-4.
594. Soman R, Madan S. Mucormycosis - A formidable challenge. *J Assoc Physicians India.* 2013; **61:** 303.
595. Song Y, Qiao J, Giovanni G, *et al.* Mucormycosis in renal transplant recipients: Review of 174 reported cases. *BMC Infect Dis.* 2017; **17:** 283. PMID: 28420334.
596. Spellberg B, Edwards J Jr, Ibrahim A. Novel perspectives on mucormycosis: pathophysiology, presentation and management. *Clin Microbiol Rev.* 2005; **18:** 556-69.
597. Spellberg B, Ibrahim A, Roilides E, *et al.* Combination therapy for mucormycosis: why, what, and how? *Clin Infect Dis.* 2012; **54:** S73-8.
598. Spellberg B, Ibrahim AS. Recent advances in the treatment of mucormycosis. *Curr Infect Dis Rep.* 2010; **12:** 423-9.
599. Spellberg B, Kontoyiannis DP, Fredricks D, *et al.* Risk factors for mortality in patients with mucormycosis. *Med Mycol.* 2012; **50:** 611-8.
600. Spellberg B, Walsh TJ, Kontoyiannis DP, *et al.* Recent advances in the management of mucormycosis: from bench to bedside. *Clin Infect Dis.* 2009; **48:** 1743-51.

601. Spellberg B. Gastrointestinal mucormycosis: An evolving disease. *Gastroenterol Hepatol (NY).* 2012; **8:** 140-2.
602. Springer J, Lackner M, Ensinger C, *et al.* Clinical evaluation of Mucorales-specific real-time PCR assay in tissue and serum samples. *J Med Microbiol.* 2016; **65:** 1414-21.
603. Sravani T, Uppin SG, Uppin MS, Sundaram C. Rhinocerebral mucormycosis: Pathology revisited with emphasis on perineural spread. *Neurol India.* 2014; **62:** 383-6.
604. Sreenath G, Prakash AR, Kanth MR, *et al.* Rhinomaxillary mucormycosis with palatal perforation: A case report. *J Clin Diagn Res.* 2014; **8:** ZD01-3.
605. Srivastava N, Bansal V, Kantoor P. Palatal mucormycosis in an infant. *J Dent Child (Chic).* 2015; **82:** 153-6.
606. Stanistreet B, Bell D. Burn wound mucormycosis: A case study on poor wound healing. *J Burn Care Res.* 2017; **38:** e582-4.
607. Stelzmueller I, Lass-Floerl C, Geltner C, *et al.* Zygomycosis and other rare filamentous fungal infections in solid organ transplant recipients. *Transpl Int.* 2008; **21:** 534-46.
608. Stephen S, Subashini B, Thomas R, *et al.* Skull base osteomyelitis caused by an elegant fungus. *J Assoc Physicians India.* 2016; **64:** 70-1.
609. Stewart JI, D'Alonzo GE, Ciccolella DE, *et al.* Reverse halo sign on chest imaging in a renal transplant recipient. *Transpl Infect Dis.* 2014; **16:** 115-8.
610. Strasfeld L, Espinosa-Aguilar L, Gajewski JL, *et al.* Emergence of *Cunninghamella* as a pathogenic invasive mold infection in allogeneic transplant recipients. *Clin Lymphoma Myeloma Leuk.* 2013; **13:** 622-8.
611. Su H, Thompson GR 3rd, Cohen SH. Hepatic mucormycosis with abscess formation. *Diagn Microbiol Infect Dis.* 2012; **73:** 192-4.
612. Su YY, Chang TY, Wang CJ, *et al.* Disseminated *Cunninghamella bertholletiae* infection during induction chemotherapy in a girl with high-risk acute lymphoblastic leukemia. *Pediatr Neonatol.* 2016; **57:** 531-4.
613. Sugui JA, Christensen JA, Bennett JE, *et al.* Hematogenously disseminated skin disease caused by *Mucor velutinosus* in a patient with acute myeloid leukemia. *J Clin Microbiol.* 2011; **49:** 2728-32.
614. Sun M, Hou X, Wang X, *et al.* Gastrointestinal mucormycosis of the jejunum in an immunocompetent patient: A case report. *Medicine (Baltimore).* 2017; **96:** e6360.
615. Sun HY, Singh N. Mucormycosis: Its contemporary face and management strategies. *Lancet Infect Dis.* 2011; **11:** 301-11.
616. Sunagawa K, Ishige T, Kusumi Y, *et al.* Renal abscess involving mucormycosis by immunohistochemical detection in a patient with acute lymphocytic leukemia: A case report and literature review. *Jpn J Infect Dis.* 2013; **66:** 345-7.
617. Sutton DA, Stchigel AM, Chander J, Guarro J, Cano-Lira JF. *Saksenaea loutrophoriformis,* sp. nov. *Fungal Planet.* 2017; 558-624.
618. Suwan Y, Punyawattanaporn A, Preechawai P. Rhino-orbital fungal infection: Two cases report. *J Med Assoc Thai.* 2012; **95:** 739-42.
619. Symeonidis AS. The role of iron and iron chelators in zygomycosis. *Clin Microbiol Infect.* 2009; **15 (Suppl. 5):** 26-32.
620. Tacke D, Koehler P, Markiefka B, *et al.* Our 2014 approach to mucormycosis. *Mycoses.* 2014; **57:** 519-24.
621. Taj-Aldeen SJ, Falamarzi A, AlMuzrkchi A, *et al.* Rare pediatric rhino-orbital infection caused by *Saksenaea vasiformis. Infection.* 2012; **40:** 703-7.
622. Taj-Aldeen SJ, Gamaletsou MN, Rammaert B, *et al.* Bone and joint infections caused by mucormycetes: A challenging osteoarticular mycosis of the twenty-first century. *Med Mycol.* 2017; pii: myw136. PMID: 28053147.
623. Tajima S, Yamazaki H, Waki M, *et al.* Timing of sputum cultivation in cases of pulmonary mucormycosis. *Intern Med.* 2015; **54:** 1159-60.
624. Talebi-Taher M, Alavi Niakou SN, Javad-Mousavi SA, *et al.* Pulmonary mucormycosis in a patient with chronic rejection of kidney transplant: A case report. *Tanaffos.* 2015; **14:** 149-52.
625. Talwalkar PG. Rhino-occulo-cerebral mucormycosis in patients with Type 2 diabetes mellitus. *J Assoc Physicians India.* 2009; **57:** 407-9.
626. Tan J, Manickam R, Pisharam J, *et al.* Mucormycosis - A possible trigger pathogen for encapsulating peritoneal sclerosis. *Perit Dial Int.* 2012; **32:** 479-81.
627. Tan M, Gibney EM. Lung mass in a kidney transplant recipient. *Am J Transplant.* 2015; **15:** 281-2; Quiz 283.
628. Tan S, Aronowitz P. A common fungus, an unusual (and deadly) infection. *Am J Med.* 2011; **124:** 1023-4.
629. Tapish S, Taha M, Naresh G, *et al.* Primary mucormycosis of abdominal wall: A rare fungal infection in a immunocompetent patient. *Indian J Surg.* 2010; **72 (Suppl. 1):** 306-8.
630. Tathe SP, Dani AA, Chawhan SM, *et al.* Gastric mucormycosis: Diagnosis by imprint cytology. *Diagn Cytopathol.* 2016; **44:** 820-2.
631. Teixeira CA, Medeiros PB, Leushner P, *et al.* Rhinocerebral mucormycosis: Literature review apropos of a rare entity. *BMJ Case Rep.* 2013; pii: bcr2012008552.
632. Terry AR, Kahle KT, Larvie M, *et al.* A 43-year-old man with altered mental status and a history of alcohol use. *N Engl J Med.* 2016; **374:** 671-80.
633. Thami GP, Kaur S, Bawa AS, Chander J, *et al.* Post-surgical zygomycotic necrotizing subcutaneous infection caused by *Absidia corymbifera. Clin Exp Dermatol.* 2003; **28:** 251-3.
634. Thawani R, Singh S, Sharma S, Sharma N. Infective gangrene in extremity trauma - Are we targeting the right organisms? *Indian J Surg.* 2015; **77 (Suppl. 1):** 10-2.
635. Thomas AJ, Shah S, Mathews MS, *et al. Apophysomyces elegans*: Renal mucormycosis in a healthy host - A case report from south India. *Indian J Med Microbiol.* 2008; **26:** 269-71.
636. Thomas S, Singh VD, Vaithilingam Y, *et al.* Rhinocerebral mucormycosis - A case report. *Oral Maxillofac Surg.* 2012; **16:** 233-6.
637. Tilak R, Raina P, Gupta SK, *et al.* Cutaneous zygomycosis: A possible post-operative complication in immunocompetent individuals. *Indian J Dermatol Venereol Leprol.* 2009; **75:** 596-9.
638. Tomita H, Muroi E, Takenaka M, *et al. Rhizomucor variabilis* infection in human cutaneous mucormycosis. *Clin Exp Dermatol.* 2011; **36:** 312-4.

639. Toumi A, Larbi Ammari F, Loussaief C, *et al.* Rhino-orbito-cerebral mucormycosis: Five cases. *Med Mal Infect.* 2012; **42:** 591-8.
640. Trasmonte MV, Jimenez JD, Santiago MA, *et al.* Association of topical amphotericin B lipid complex treatment to standard therapy for rhinomaxillary mucormycosis after liver transplantation: A case report. *Transplant Proc.* 2012; **44:** 2120-3.
641. Tribble DR, Warkentien T, Rodriguez C; *et al.* Mucormycosis after a tornado in Joplin, Missouri. *N Engl J Med.* 2013; **368:** 1067.
642. Trief D, Gray ST, Jakobiec FA, *et al.* Invasive fungal disease of the sinus and orbit: A comparison between mucormycosis and *Aspergillus*. *Br J Ophthalmol.* 2016; **100:** 184-8.
643. Trzaska WJ, Correia JN, Villegas MT, *et al.* pH manipulation as a novel strategy for treating mucormycosis. *Antimicrob Agents Chemother.* 2015; **59:** 6968-74.
644. Tully CC, Romanelli AM, Sutton DA, *et al.* Fatal *Actinomucor elegans* var. *kuwaitiensis* infection following combat trauma. *J Clin Microbiol.* 2009; **47:** 3394-9.
645. Turan MN, Tatar E, Yaprak M, *et al.* A mucormycosis case presented with orbital apex syndrome and hemiplegia in a renal transplant patient. *Int Urol Nephrol.* 2013; **45:** 1815-9.
646. Tuysuz G, Ozdemir N, Senyuz OF, *et al.* Successful management of hepatic mucormycosis in an acute lymphoblastic leukaemia patient: A case report and review of the literature. *Mycoses.* 2014; **57:** 513-8.
647. Tyll T, Lyskova P, Hubka V, *et al.* Early diagnosis of cutaneous mucormycosis due to *Lichtheimia corymbifera* after a traffic accident. *Mycopathologia.* 2016; **181:** 119-24.
648. Uchida Y, Tsukino M, Shigemori W, *et al.* Diagnosis of pulmonary mucormycosis aiding the diagnosis of small cell lung cancer. *J Med Microbiol.* 2012; **61:** 1610-3.
649. Udupa K, Ganesan P, Majhi U, *et al.* Favorable outcome in a patient with vulvar mucormycosis during acute myeloid leukemia induction with medical management alone. *Hematol Oncol Stem Cell Ther.* 2012; **5:** 125-6.
650. Upton A, Gabriel R, la Fougere C, *et al.* A patient with cutaneous zygomycosis due to *Saksenaea vasiformis*. *Infect Dis Clin Pract.* 2002; **11:** 137-9.
651. Uy EM, Rustagi T, Khera S. Cerebral mucormycosis in a diabetic man. *Conn Med.* 2011; **75:** 273-9.
652. Vaezi A, Moazeni M, Rahimi MT, *et al.* Mucormycosis in Iran: A systematic review. *Mycoses.* 2016; **59:** 402-15.
653. Vallabhaneni S, Mody RK. Gastrointestinal mucormycosis in neonates: A review. *Curr Fungal Infect Rep.* 2015; **9:** 269-74.
654. Vallabhaneni S, Walker TA, Lockhart SR, *et al.* Fatal gastrointestinal mucormycosis in a premature infant associated with a contaminated dietary supplement - Connecticut, 2014. *MMWR.* 2015; **64:** 155-6.
655. Varghese A, Thomas S. Orbital apex syndrome secondary to mucormycosis after a tooth extraction in an immunocompetent patient. *Ear Nose Throat J.* 2010; **89:** e24-6.
656. Vashi N, Avedian R, Brown J, *et al.* Successful surgical and medical treatment of *Rhizopus* osteomyelitis following hematopoietic cell transplantation. *Orthopedics.* 2012; **35:** e1556-61.
657. Vehreschild JJ, Birtel A, Vehreschild MJ, *et al.* Mucormycosis treated with posaconazole: Review of 96 case reports. *Crit Rev Microbiol.* 2013; **39:** 310-24.
658. Vercillo MS, Liptay MJ, Seder CW. Early pneumonectomy for pulmonary mucormycosis. *Ann Thorac Surg.* 2015; **99:** e67-8.
659. Verma A, Singh V, Jindal N, *et al.* Necrosis of maxilla, nasal and frontal bone secondary to extensive rhino-cerebral mucormycosis. *Natl J Maxillofac Surg.* 2013; **4:** 249-51.
660. Verma R, Nair V, Vasudevan B, *et al.* Rare case of primary cutaneous mucormycosis of the hand caused by *Rhizopus microsporus* in an immunocompetent patient. *Int J Dermatol.* 2014; **53:** 66-9.
661. Verma R, Vij M, Agrawal V, Jain M. Renal mucormycosis in immunocompetent patients: Report of three cases. *Basic Appl Pathol.* 2011; **4:** 66-70.
662. Vidovic A, Arsic-Arsenijevic V, Tomin D, *et al.* Proven invasive pulmonary mucormycosis successfully treated with amphotericin B and surgery in patient with acute myeloblastic leukemia: A case report. *J Med Case Rep.* 2013; **7:** 263.
663. Vijayabala GS, Annigeri RG, Sudarshan R. Mucormycosis in a diabetic ketoacidosis patient. *Asian Pac J Trop Biomed.* 2013; **3:** 830-3.
664. Ville S, Talarmin JP, Gaultier-Lintia A, *et al.* Disseminated mucormycosis with cerebral involvement owing to *Rhizopus microsporus* in a kidney recipient treated with combined liposomal amphotericin B and posaconazole therapy. *Exp Clin Transplant.* 2016; **14:** 96-9.
665. Vinay K, Chandrasegaran A, Kanwar AJ, *et al.* Primary cutaneous mucormycosis presenting as a giant plaque: Uncommon presentation of a rare mycosis. *Mycopathologia.* 2014; **178:** 97-101.
666. Vinh DC, Freeman AF, Shea YR, *et al.* Mucormycosis in chronic granulomatous disease: Association with iatrogenic immunosuppression. *J Allergy Clin Immunol.* 2009; **123:** 1411-3.
667. Vironneau P, Kania R, Morizot G, *et al.* Local control of rhino-orbito-cerebral mucormycosis dramatically impacts survival. *Clin Microbiol Infect.* 2014; **20:** 336-9.
668. Vondran FW, Knitsch W, Krech T, *et al.* Intestinal mucormycosis with *Rhizopus microsporus* after liver transplantation - successful treatment of a rare but life-threatening complication. *Transplantation.* 2014; **97:** e11-3.
669. Wali U, Balkhair A, Al-Mujaini A. Cerebro-rhino orbital mucormycosis: An update. *J Infect Public Health.* 2012; **5:** 116-26.
670. Walsh TJ, Bloom BE, Kontoyiannis DP. Meeting the challenges of an emerging pathogen: the Henry Schueler 41 & 9 Foundation International Forum on Mucormycosis. *Clin Infect Dis.* 2012; **54:** S1-4.
671. Walsh TJ, Gamaletsou MN, McGinnis MR, *et al.* Early clinical and laboratory diagnosis of invasive pulmonary, extra-pulmonary and disseminated mucormycosis (zygomycosis). *Clin Infect Dis.* 2012; **54:** S55-60.
672. Walsh TJ, Kontoyiannis DP. Editorial commentary: What is the role of combination therapy in management of zygomycosis? *Clin Infect Dis.* 2008; **47:** 372-4.

673. Walsh TJ, Skiada A, Cornely OA, *et al.* Development of new strategies for early diagnosis of mucormycosis from bench to bedside. *Mycoses.* 2014; **57 (Suppl. 3):** 2-7.
674. Waness A, Dawsari GA, Al Jahdali H. The rise of an opportunistic infection called "Invasive Zygomycosis". *J Glob Infect Dis.* 2009; **1:** 131-8.
675. Weddle G, Gandy K, Bratcher D, et al. *Apophysomyces trapeziformis* infection associated with a tornado-related injury. *Pediatr Infect Dis J.* 2012; **31:** 640-2.
676. Weddle G, Pahud B, Jackson MA. Mucormycosis after a tornado in Joplin, Missouri. *N Engl J Med.* 2013; **368:** 1066-7.
677. Williams KE, Parish JM, Lyng PJ, *et al.* Pseudomembranous tracheobronchitis caused by *Rhizopus* sp. after allogeneic stem cell transplantation. *J Bronchology Interv Pulmonol.* 2014; **21:** 166-9.
678. Wolkow N, Jakobiec FA, Stagner AM, *et al.* Chronic orbital and calvarial fungal infection with *Apophysomyces variabilis* in an immunocompetent patient. *Surv Ophthalmol.* 2017; **62:** 70-82.
679. Woo PC, Lau SK, Ngan AH, *et al. Lichtheimia hongkongensis* sp. nov., a novel *Lichtheimia* spp. associated with rhinocerebral, gastrointestinal and cutaneous mucormycosis. *Diagn Microbiol Infect Dis.* 2010; **66:** 274-84.
680. Woo PC, Leung SY, Ngan AH, *et al.* A significant number of reported *Absidia corymbifera (Lichtheimia corymbifera)* infections are caused by *Lichtheimia ramosa* (Syn. *Lichtheimia hongkongensis*): An emerging cause of mucormycosis. *Emerg Microbes Infect.* 2012; **1:** e15.
681. Woo PC, Leung SY, To KK, *et al.* Internal transcribed spacer region sequence heterogeneity in *Rhizopus microsporus:* Implications for molecular diagnosis in clinical microbiology laboratories. *J Clin Microbiol.* 2010; **48:** 208-14.
682. Xess I, Mohapatra S, Shivaprakash MR, *et al.* Evidence implicating *Thamnostylum lucknowense* as an etiological agent of rhino-orbital mucormycosis. *J Clin Microbiol.* 2012; **50:** 1491-4.
683. Xia XJ, Shen H, Liu ZH. Primary cutaneous mucormycosis caused by *Mucor irregularis. Clin Exp Dermatol.* 2015; **40:** 875-8.
684. Xia ZK, Wang WL, Yang RY. Slowly progressive cutaneous, rhinofacial and pulmonary mucormycosis caused by *Mucor irregularis* in an immunocompetent woman. *Clin Infect Dis.* 2013; **56:** 993-5.
685. Xu S, Nambudiri VE, Tahan S, *et al.* Violaceous necrotic plaques on the leg of an immunosuppressed patient. Cutaneous mucormycosis. *JAMA Dermatol.* 2014; **150:** 79-81.
686. Yamaguchi S, Okubo Y, Katano A, *et al.* Primary cutaneous mucormycosis caused by *Mucor irregularis* in an elderly person. *J Dermatol.* 2015; **42:** 210-4.
687. Yang C, Friess SH, Dehner LP. Hepatic mucormycosis mimicking veno-occlusive disease: Report of a case and review of the literature. *Pediatr Dev Pathol.* 2016; **19:** 150-3.
688. Yano S, Minami J, Nishiwaki K, *et al.* Rapid progression and unusual premortal diagnosis of mucormycosis in patients with hematologic malignancies: Analysis of eight patients. *Int J Hematol.* 2011; **93:** 344-50.
689. Ye B, Yu D, Zhang X, *et al.* Disseminated *Rhizopus microsporus* infection following allogeneic hematopoietic stem cell transplantation in a child with severe aplastic anemia. *Transpl Infect Dis.* 2013; **15:** e216-23.
690. Ye W, Wang Y, Wen Y, *et al.* Dramatic remission of nephrotic syndrome after unusual complication of mucormycosis in idiopathic membranous nephropathy. *Int Urol Nephrol.* 2014; **46:** 1247-51.
691. Yu J, Walther G, van Diepeningen AD, *et al.* DNA barcoding of clinically relevant *Cunninghamella* species. *Med Mycol.* 2015; **53:** 99-106.
692. Yuda J, Kato K, Kikushige Y, *et al.* Successful treatment of invasive zygomycosis based on a prompt diagnosis using molecular methods in a patient with acute myelogenous leukemia. *Intern Med.* 2014; **53:** 1087-91.
693. Zaki SM, Elkholy IM, Elkady NA, *et al.* Mucormycosis in Cairo, Egypt: Review of 10 reported cases. *Med Mycol.* 2014; **52:** 73-80.
694. Zaman K, Kaur H, Rudramurthy SM, *et al.* Cutaneous mucormycosis of scalp and eyelids in a child with Type I diabetes mellitus. *Indian J Dermatol Venereol Leprol.* 2015; **81:** 275-8.
695. Zanation A, Fleischman D, Chavala SH. Ptosis, erythema and rapidly decreasing vision. *JAMA.* 2013; **309:** 2382-3.
696. Zaoutis TE, Roilides E, Chiou CC, *et al.* Zygomycosis in children: A systematic review and analysis of reported cases. *Pediatr Infect Dis J.* 2007; **26:** 723-7.
697. Zhang L, Tian X, Wang P, *et al.* Recurrent pulmonary mucormycosis after lobectomy in a non-smoking patient without predisposing risk factors. *Braz J Infect Dis.* 2012; **16:** 590-3.
698. Zhao L, Wang CX, Zhang L, *et al.* Mucormycosis extending from the surgical wound to the transplanted kidney: Case report and literature review. *Exp Clin Transplant.* 2012; **10:** 403-5.
699. Zhao Y, Zhang Q, Li L, *et al.* Primary cutaneous mucormycosis caused by *Rhizomucor variabilis* in an immunocompetent patient. *Mycopathologia.* 2009; **168:** 243-7.
700. Zhu X, Liu H, Wang W, *et al.* Two cases of transplant renal artery thrombosis and spontaneous rupture caused by mucormycosis. *Transpl Infect Dis.* 2015; **17:** 442-8.
701. Ziakas PD, Adraktas P, Ntountas J, *et al.* Isolated orbital mucormycosis: A rare presentation ending in a fatal outcome. *Int J Infect Dis.* 2009; **13:** 112-4.
702. Zilberberg MD, Shorr AF, Huang H, *et al.* Hospital days, hospitalization costs and inpatient mortality among patients with mucormycosis: A retrospective analysis of US hospital discharge data. *BMC Infect Dis.* 2014; **14:** 310.

CHAPTER

27

Entomophthoramycosis

Entomophthoramycosis is a chronic subcutaneous infection caused by fungal agents belonging to sub-phylum Entomophthoromycotina under phylum Glomeromycota. Some investigators prefer to call mycoses caused by members belonging to sub-phylum Mucoromycotina as mucormycosis and to sub-phylum Entomophthoromycotina as entomophthoromycosis, which is further subdivided into two types *i.e.* **conidiobolomycosis** and **basidiobolomycosis**.

Both these mycoses are usually described under section Subcutaneous Mycoses but being described in this Chapter under the Section of Opportunistic Mycoses due to their increasing involvement of immunocompromised patients in recent times. Rather there are drastic changes in the epidemiology of these infections ranging from involvement of subcutaneous tissue to the visceral organs of the body, particularly gastrointestinal basidiobolomycosis. The anatomical areas earlier were mainly the domain of mucormycosis and not entomophthoramycosis. Hence, it is described in a separate chapter under this section, after the chapter on mucormycosis.

In the tissues, they are seen as broad, non-septate hyphae, without any predilection to invade blood vessels as seen in mucormycosis, with an additional finding as Splendore-Hoeppli phenomenon. Since morphology of these organisms is sometimes indistinguishable in histopathological sections hence culture is essential for correct identification of the causative fungal species. The course and outcome of disease differs according to anatomical site affected as well as nature of the underlying fungal species involved in a particular patient.

Historical Perspective

The first human infection caused by *Basidiobolus* was reported in 1956 in Indonesia. However, the first recognition of the disease in Africa came from Uganda in 1961 where isolated cases were reported from all its four provinces. Subsequently eight cases were reported from Indonesia and other tropical areas of Southeast Asia, East as well as West Africa. Rather Dr D P Burkitt was actively involved in studies on non-Hodgkin lymphoma among African children (Burkitt's Lymphoma) and during the same period he came across patients of subcutaneous swellings which turned out to be basidiobolomycosis. Consequently in 1964, Burkitt *et al,* reported 31 cases from Uganda, giving the detailed description of this disease. It is also prevalent in Uganda, Sudan, Kenya, Cameroon Ghana as well as Nigeria and presents as a single, painless, hard, circumscribed subcutaneous nodule. Recently Atadokpede *et al,* reported seven cases of basidiobolomycosis from Benin in 2017 with subcutaneous involvement.

The initial cases of conidiobolomycosis were described in literature during early 1960s, particularly as nasal granulomatous disease in horse in Texas in 1961. Subsequently its first human case was reported in Jamaica in 1965.

Mycology

In the order Entomophthorales, spores are forcibly ejected in contrast to the Orders Mucorales, wherein spores arise by cleavage of sporangial plasma and are passively liberated. The fungal species of order Entomophthorales are essentially arthropod parasites. They cause uncommon clinical entity called as **entomophthoramycosis**, which is a chronic granulomatous type of subcutaneous, mucocutaneous and very rarely visceral infections. The name of Order itself is derived from Greek word ***'Entomon'*** meaning ***'insect'*** as these fungi are primarily insect pathogens, which is the largest group of arthropods. These fungi are found in soil, decaying vegetation, insects and gastrointestinal tracts of amphibians. The literal meaning of ***'Entomophthora'*** is ***'insect destroyer'*** from where nomenclature of the Order has been derived. Entomophthoromycosis may also be written as entomophthoramycosis by replacing alphabet **'o'** with **'a'** and both are correct rather interchangeably used in medical practice.

Entomophthorales cause chronic subcutaneous, mucocutaneous and visceral infections, mainly in immunocompetent individuals residing in tropical and subtropical regions in contrast to members of the Order Mucorales, which cause acute, angioinvasive infections among immunocompromised patients with high level of mortality. There are two clinical entities under the heading of entomophthoromycosis: (i) conidiobolomycosis and (ii) basidiobolomycosis, which are uncommon diseases without any known predisposing factors such as diabetic acidosis or immunodeficiency.

Therefore, these infections are invariably found in otherwise healthy immunocompetent individuals. Earlier there used to be only subcutaneous infections but recently, some of visceral infections of both the clinical entities have been reported among immunocompetent as well as immunocompromised patients. Therefore, based on the latest reports both diseases may be clinically sub-divided under three headings namely (i) **subcutaneous** (ii) **mucocutaneous** and (iii) **visceral** types.

These clinical entities are caused by species belonging to **two** genera of order Entomophthorales *i.e.* ***Conidiobolus*** and ***Basidiobolus***. These fungi should be distinguished from Mucorales by production of **forcibly** discharging sporangioles borne on tubular sporophores as well as markedly compact and glabrous mycelial morphology. When tissue sections are stained with H&E, members of Entomophthorales demonstrate typical hyphal encasement by pink-colored eosinophilic material, which is popularly known as **Splendore-Hoeppli** phenomenon. Sometimes this may be the first indication that patient has infection with either *Conidiobolus* or *Basidiobolus* species in contrast to members of Mucorales, which do not demonstrate such phenomenon rather are angioinvasive in nature. However, similar feature of Splendore-Hoeppli may also be seen in some other fungal infections of chronic nature.

Within order Entomophthorales, morphological differences of zygospores are important identification features so that both principal causative genera can be differentiated from each other. In genus *Conidiobolus,* zygospores are formed within larger of two gametangia, whereas in *Basidiobolus,* zygospores have two closely appraised **beak-like** appendages. Some of the differences in these clinical entities are summarized at a glance in **Table 27.1**.

The disease entity, entomophthoromycosis is found prevalent in tropical countries **(Fig. 27.1)**. The fungi are widely distributed in these countries as saprotrophic and there are **two** medically significant genera having **four** fungal species, which are responsible for the following **two** clinical entities:

1. Conidiobolomycosis—*Conidiobolus coronatus, C.incongruus* and *C.lamprauges.*
2. Basidiobolomycosis—*Basidiobolus ranarum.*

Table 27.1. Comparative Clinical Features of Conidiobolomycosis and Basidiobolomycosis.

Features	*Conidiobolomycosis*	*Basidiobolomycosis*
Anatomical sites	Face, nasal area	Trunk, thigh, buttocks
Lesions	Lesions are firmly attached to underlying tissue	Lesions are not firmly attached to the underlying tissue
Age	Adults	Children
Visceral Involvement	Not usually seen	Gastrointestinal
Causative Agents	*Conidiobolus coronatus, C.incongruus* and *C.lamprauges*	*Basidiobolus ranarum*
Colony Morphology	Flat, waxy colonies, becoming powdery, with a short aerial mycelium with age; petri dish lid becomes covered with conidia ejected from older cultures; colonies are white, buff, tan or brown	Flat furrowed colonies with a waxy texture; yellowish to gray
Hyphal Morphology	Conidia are round to pyriform, have prominent papillae on the wall, which give rise to secondary conidia.	The primary asexual spores are globose, forcibly discharged phototropic and produced singly at tip of conidiophore
Sexual Spores	Rarely produced but within larger of two gametangia	The zygospores have two closely appraised beak-like appendages

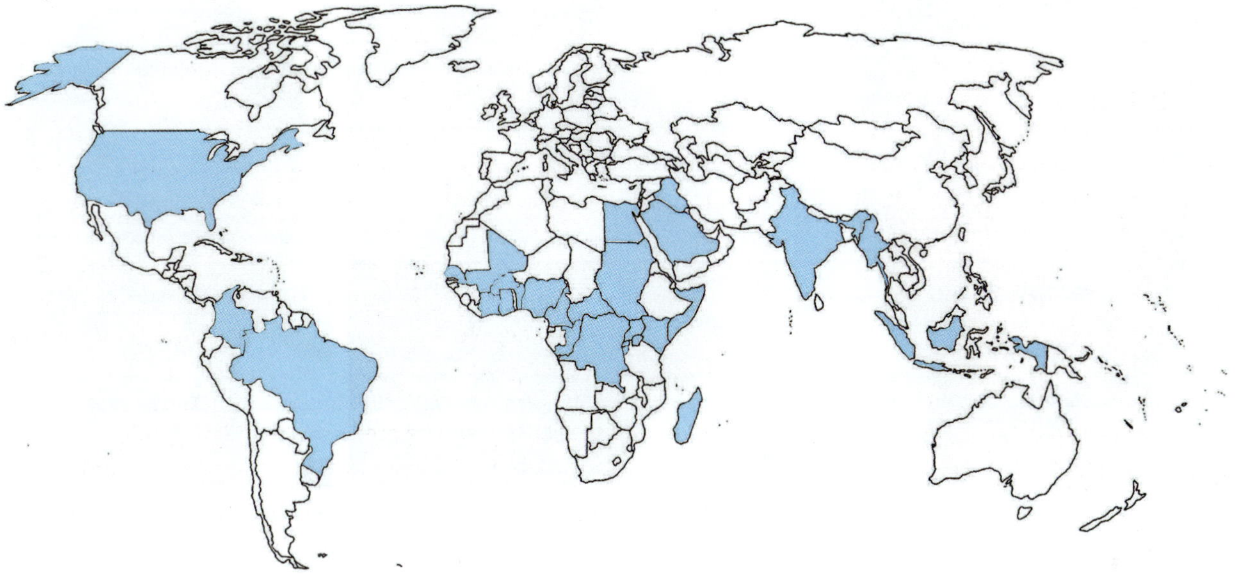

Fig. 27.1. Map showing worldwide distribution of entomophthoramycosis cases, mainly involving the tropical areas.

I. CONIDIOBOLOMYCOSIS

Conidiobolomycosis is a chronic subcutaneous fungal infection caused by ***Conidiobolus coronatus***, *C.incongruus* and *C.lamprauges*, majority of them are caused by *C.coronatus*. This was initially designated as *Entomophthora coronatus*, linking this organism to the insect source from where it was originally identified.

Epidemiology

These fungi are found ubiquitously in the soil and environment. In the past, these used to be traditionally considered pathogenic to arthropods and non-pathogenic to man as well as animals. Now, understanding about the taxonomy of Entomophthorales has entirely changed as these are **emerging** pathogenic organisms invariably entailing chronic course of disease without any obvious underlying predisposing factor. It is in contrast to cases of mucormycosis, wherein underlying factor(s) is usually present like diabetes mellitus. However, cases are now being reported in immunocompromised patients.

Conidiobolus coronatus is found in soil where it is pathogenic to termites, spiders and other insects. The distribution of infection is nevertheless more restricted as the causal fungus being an insect pathogen, which is found on decaying vegetation in most warm climates. This may also be isolated from soil or leaf detritus. The initial cases of conidiobolomycosis were described in literature during early 1960s, particularly as nasal granulomatous disease in horses in Texas in 1961. Subsequently the first human case was reported in Jamaica in 1965.

The infection is mainly prevalent in **adult** males, predominately among those living or working in tropical rain forests of Western Africa, South and Central America as well as Southeast Asia. The disease occurs in tropical Africa, India, Puerto Rico, Colombia and Brazil and presents as tumor-like masses in nasal skin, nasal obstruction or sinusitis. The pathogenesis of infection is unclear but rarity of disease indicates its low intrinsic pathogenicity. However, in this clinical setting this disease may be kept as one of the differential diagnosis even if the patient is immunocompromised.

Clinical Features

The commonest presentation is rhinoentomophthoromycosis, which is characterized by chronic, indolent and localized swelling of nose, perinasal tissues, paranasal sinuses, cheeks and upper lips **(Figs. 27.2 to 27.4)**. The swelling is firm and painless and may progress slowly to involve nasal bridge, upper face including orbit as well as lower face. The lesions are **firmly attached** to underlying tissue, however, bones are spared. These painless subcutaneous nodular masses can be palpated through intact skin. Obstruction of nasal passages, rhinorrhea and

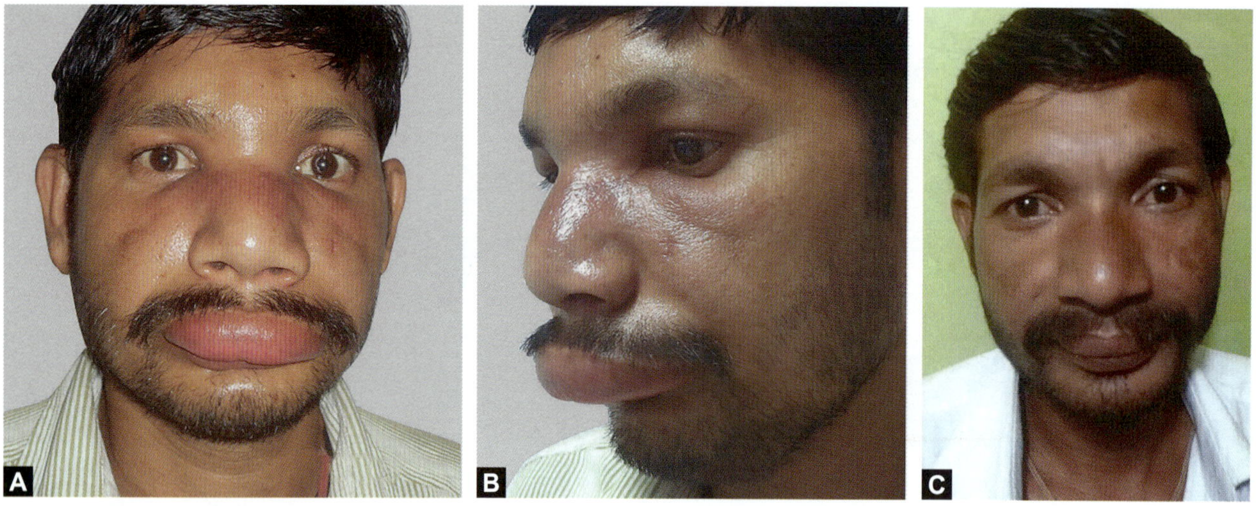

Figs. 27.2A to C. Facial swelling in conidiobolomycosis patient before (A and B) and after (C) the treatment with KI for six weeks.

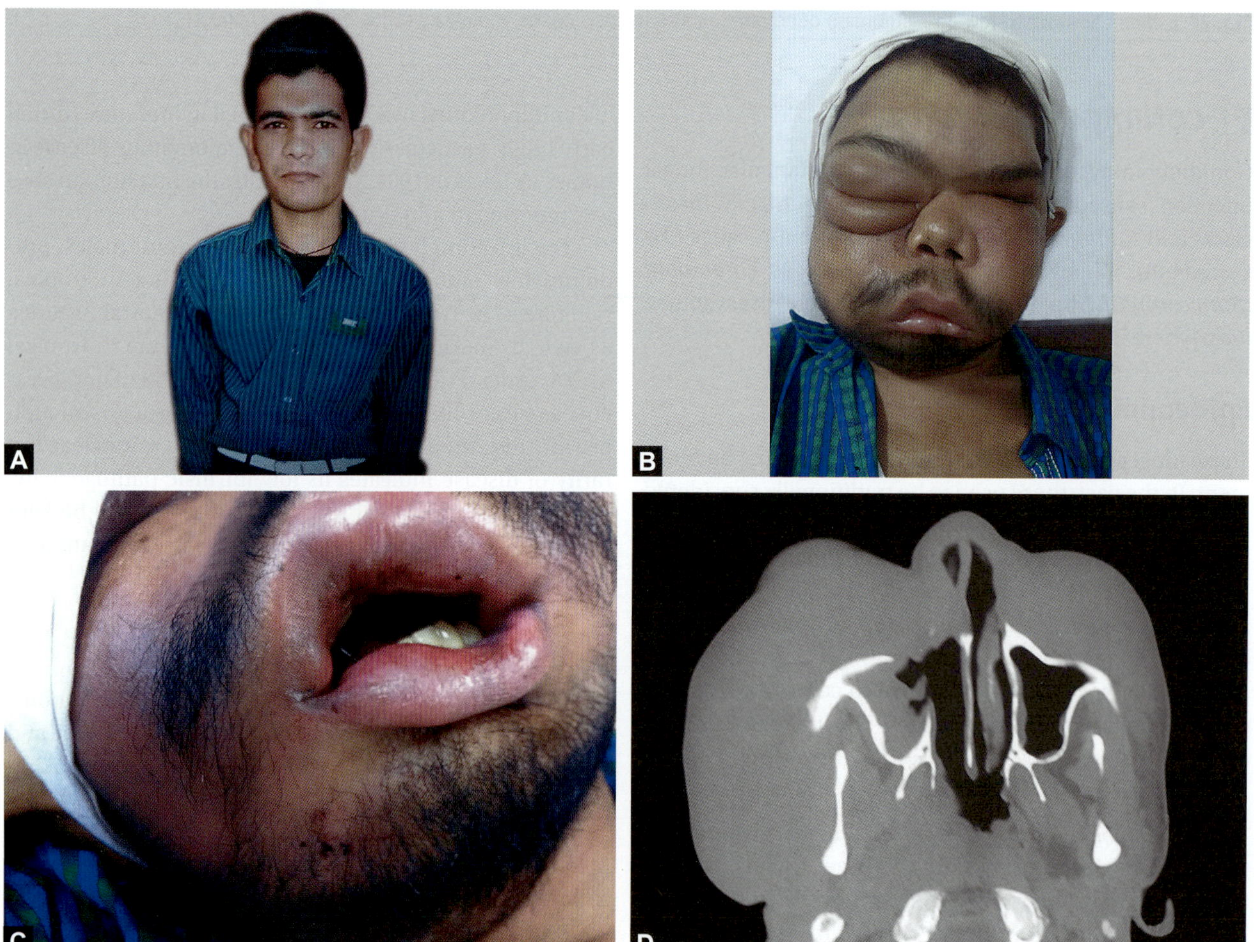

Figs. 27.3A to D. Facial swelling in conidiobolomycosis at different stages of development including involvement of right maxillary sinus.

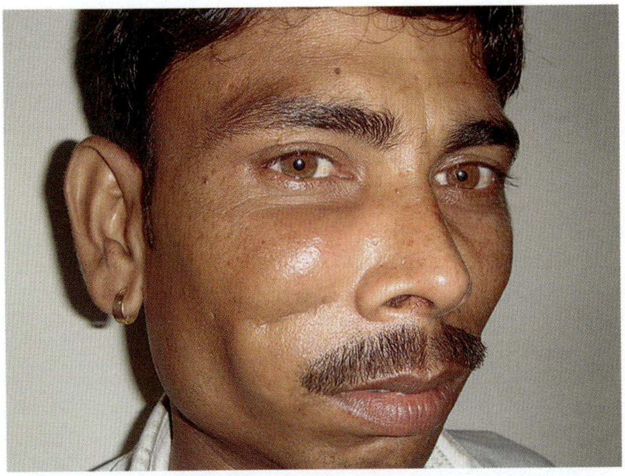

Fig. 27.4. Right-sided facial swelling in a patient of conidiobolomycosis without any involvement of underlying tissue.

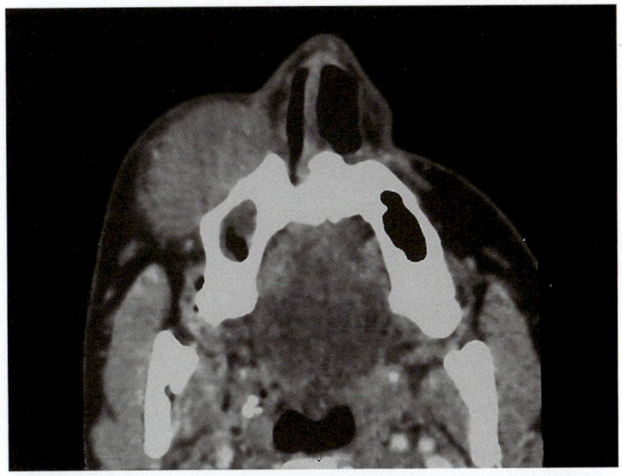

Fig. 27.5. CT scan of face showing swelling of conidiobolomycosis on the right side without any involvement of underlying tissue.

nosebleeds are common symptoms. The maxillary sinus may also be involved as the disease progresses. The generalized facial swelling occurs as disease further progresses entailing inability to open the eyes. There may occasionally be severe disfigurement of the involved anatomical site. The systemic symptoms and signs of the disease are generally absent among these patients.

The portal of entry of spores is believed to be by inhalation, which then invades tissues of the nasal cavity, paranasal sinuses and facial soft tissues. The infection probably originates in inferior nasal turbinate and spreads to adjacent subcutaneous tissue of face causing these painless swelling, discomfort and disfigurement. Among these patients vascular invasion and thrombosis as well as tissue infarction are notably absent, which are the typical features of the infection caused by Mucorales.

In recent times, number of cases have been reported where the infection develops among patients with underlying pathologies such as neutropenia, Burkitt's lymphoma or any other similar disease process. Pulmonary, pericardial and other systemic infections are also occasionally encountered but due to lack of angioinvasion, intracranial extension or dissemination is rare. However, these deeply invasive infections are rarely seen as fatal nature. Therefore, *Conidiobolus coronatus* may now be considered as an emerging opportunist fungus.

Radiodiagnosis

The patient of conidiobolomycosis may undergo radiological investigation to know the extent of involvement. Most of times it is subcutaneous swelling as seen in **Figure 27.4** without any involvement of underlying tissue or bones as seen in **Figure 27.5** in CT scan of facial area. However, rarely it may invade structures surrounding its vicinity as seen in **Figure 27.3D**, wherein right-sided maxillary sinus is also involved.

Differential Diagnosis

The subcutaneous disease in entomophthoramycosis can mimic soft-tissue tumors. Hence, a high index of suspicion is essential to avoid any misdiagnosis thereby inappropriate treatment. A newly described ***'doughnut-lifting sign'***, wherein the swelling grows like a discoid mass, easily lifted from deeper tissues, may be helpful in establishing the clinical diagnosis. On histopathology, it may be confused with mucormycosis due to broad, ribbon-like, nonseptate hyphae with wide-angle branching **(Fig. 27.6)**. However, demonstration of Splendore-Hoeppli phenomenon is useful finding for their differentiation **(Fig. 27.7)**.

Laboratory Diagnosis

The clinical specimen like biopsy taken from the affected site, is gently teased apart in a solution containing 500 mg of streptomycin and 1 mega unit of penicillin. On microscopy in KOH wet mount, the coenocytic hyphae are short, 6-15 μm wide, with cross-walls separating empty hyphal fragments from actively growing parts. The hyphal walls are doubly refractile with some branching where granular inclusions are easily seen **(Fig. 27.6)**. Histopathologically, diagnosis of conidiobolomycosis is made by H&E

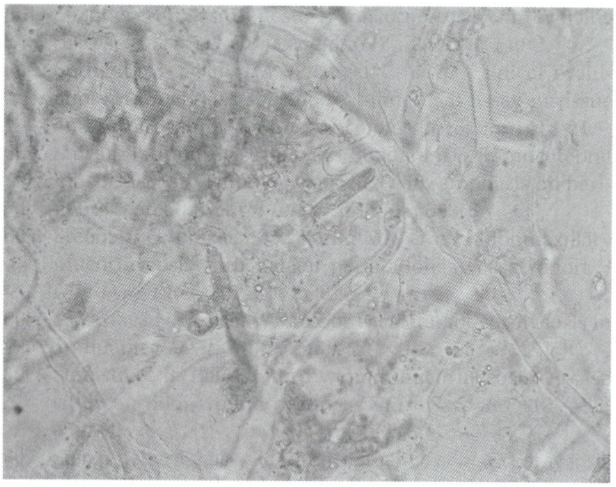

Fig. 27.6. Non-septate hyphae of *Conidiobolus coronatus* with right-angle branching (KOH × 400).

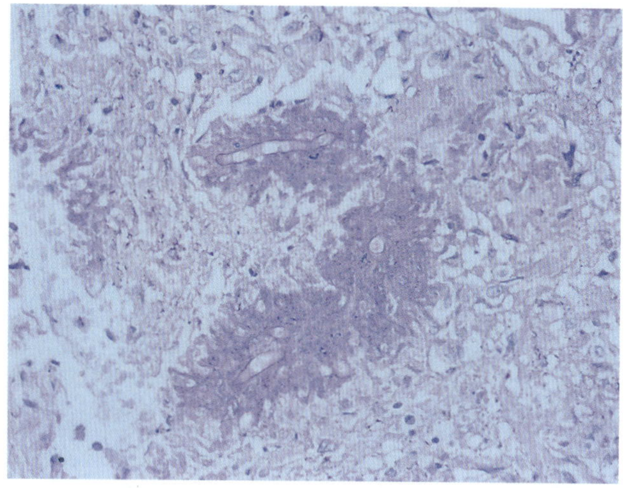

Fig. 27.7. Broad non-septate hyphae seen embedded in eosinophilic material of Splendore-Hoeppli phenomenon (H&E × 400).

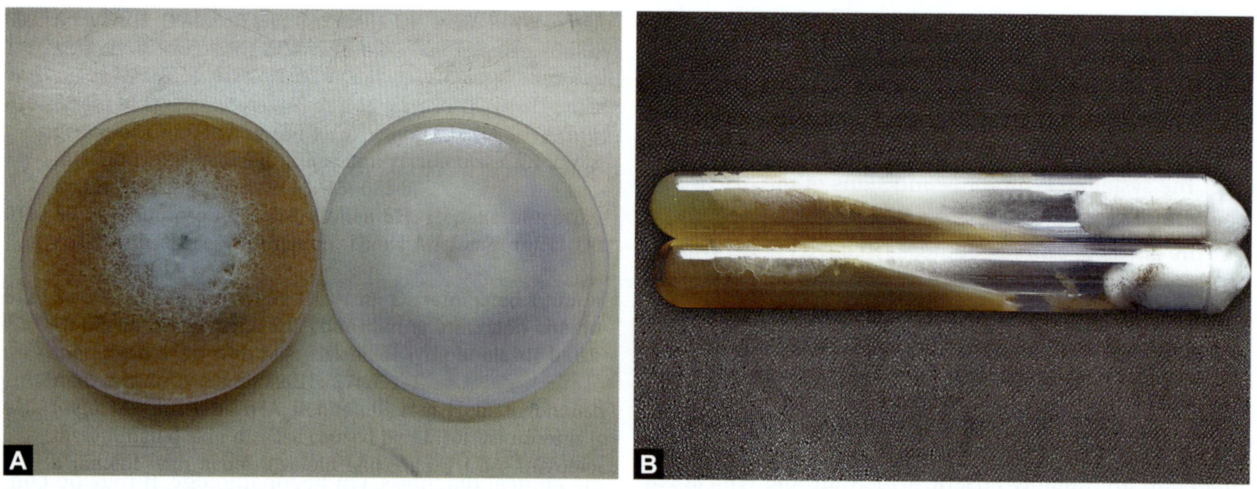

Figs. 27.8A and B. Culture of *Conidiobolus coronatus* on SDA plate along with its lid and opposite wall of tubes showing growth on its undersurface.

staining by demonstrating distinctive non-septate hyphae with surrounding eosinophilic sleeve *i.e.* Splendore-Hoeppli phenomenon in the tissue sections **(Fig. 27.7)**.

The cultivation of *C.coronatus* is done on SDA, PDA without cycloheximide at 25-35°C, which shows rapidly growing, creamy, glabrous colonies adherent to surface with pale reverse. Subsequently furrows and folds appear particularly when organism is grown at 37°C. In due course, waxy colonies become powdery from center due to production of short, white aerial mycelia. There are satellite colonies arising from germination of ejected sporangioles. The conidia are forcibly discharged and stick to the inner side of the culture container, like tube or under-surface of petri dish lid, completely clouding the view into the culture with passage of time. The older colonies become covered with short, aerial mycelia and conidiophores **(Figs. 27.8A and B)**.

The primary conidia are globose and approximately 40 µm in diameter. They are produced singly, on 50-70 µm long phototrophic conidiophores from which they are forcibly discharged. This results in a visible film on culture medium. They have a characteristic protruding **papilla** on one side **(Figs. 27.9 and 27.10)**. Many papillae and buddings on original spore give appearance of a corona

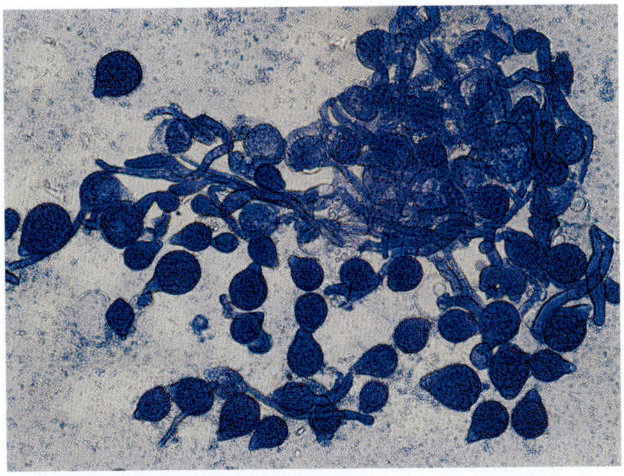

Fig. 27.9. Papillated conidia of *Conidiobolus coronatus* (LCB × 400).

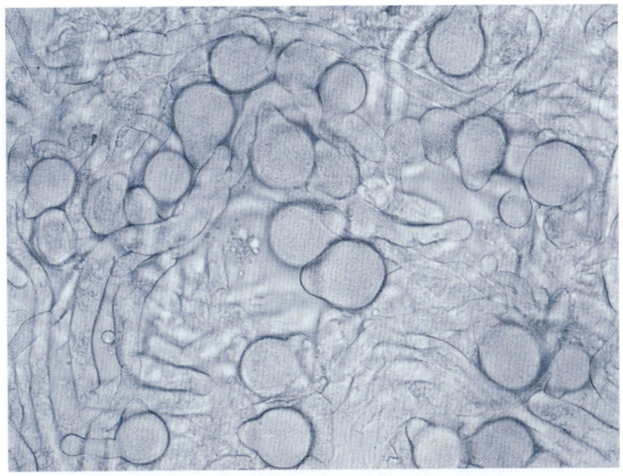

Fig. 27.10. Papillated conidia with non-septate, broad hyphae of *Conidiobolus coronatus* (Phase Contrast × 400).

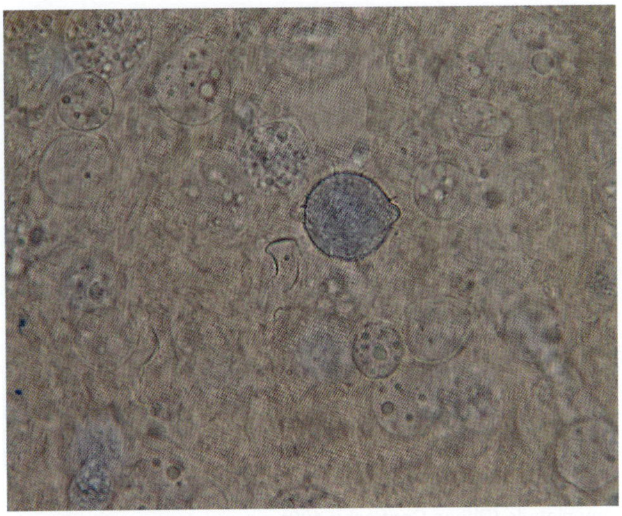

Fig. 27.11. Papillated and villose conidia of *Conidiobolus coronatus* (LCB × 400).

of secondary spores. Due to this corona-like appearance, the species name is given as '***coronatus***'. A few conidia also show multiple fine short hair-like appendages on surface called **villae**. Several smaller conidia may also arise on discharged primary conidia. The species, *C.coronatus* is characterized by fact that its resting spores are covered with long soft hairs **(Figs. 27.11 to 13)**.

Capilliconidia are also present, which are elongated cells often with a terminal adhesive knob, borne by a thin supporting hypha. The speciation of Conidiobolus genus is feasible depending on the sexual reproduction, whether is *C.coronatus, C.incongruus* or *C.lamprauges* **(Figs. 27.14 and 27.15; Table 27.2)**.

The colonies of *C.incongruus* are also similar to *C.coronatus* but are dry becoming aerial with increased humidity. Growth is optimal at 37°C, producing little or no growth at room temperature. The vegetative mycelia are wide *i.e.* 10-22 µm; conidiophores are 30-130 µm tall and inflate sub-apically.

The experimental studies of *C.coronatus* have been successfully conducted in mice **(Table 27.2)**.

II. BASIDIOBOLOMYCOSIS

Basidiobolomycosis is also predominantly a subcutaneous mycotic infection of trunk and limbs in immunocompetent hosts, which is caused by ***Basidiobolus ranarum***, which has two synonym species *i.e. B.haptosporus* and *B.meristosporus*. These fungi were previously given independent species status but subsequent taxonomic studies, based on antigenic analysis, isoenzyme banding and restriction enzyme analysis of rDNA, indicate that all these human pathogens, with different names, are *B.ranarum*. However, recently in 2017, Sitterle *et al*, have reported *B.meristosporus* as new fungal pathogen based on ultra deep sequencing (UDS).

This disease used to be known with variety of names in the past such as entomophthoromycosis basidiobolae, subcutaneous phycomycosis or subcutaneous zygomycosis. Its presentation is not very common hence it is invariably misdiagnosed thereby not managed or treated on fungal etiological lines.

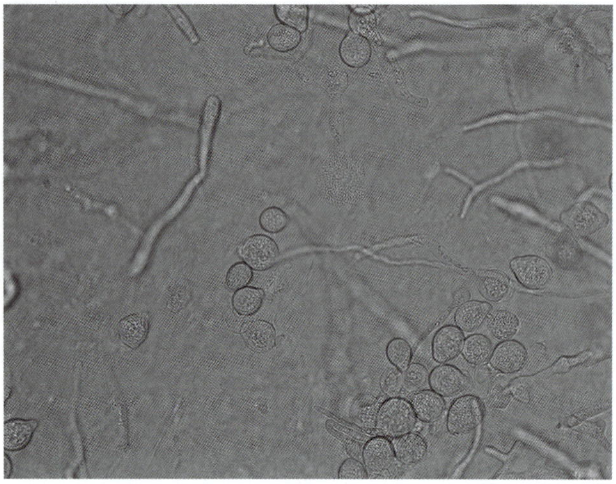

Fig. 27.12. Papillated and villose conidia of *Conidiobolus coronatus* in another view (Phase Contrast × 400).

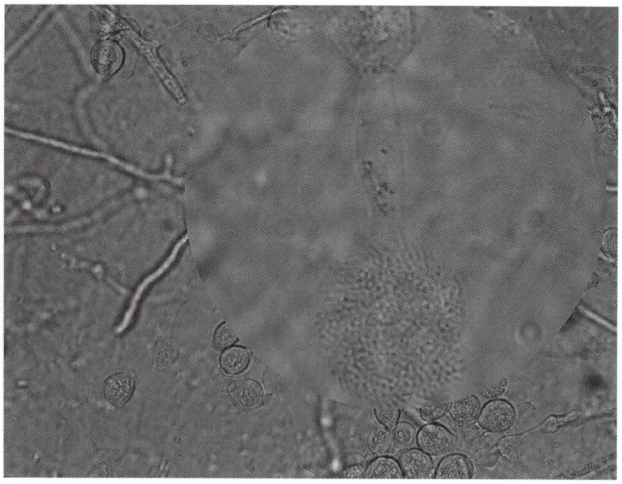

Fig. 27.13. Papillated and villose conidia of *C.coronatus* as an enlarged view in the center *i.e.* inset (Phase Contrast × 400).

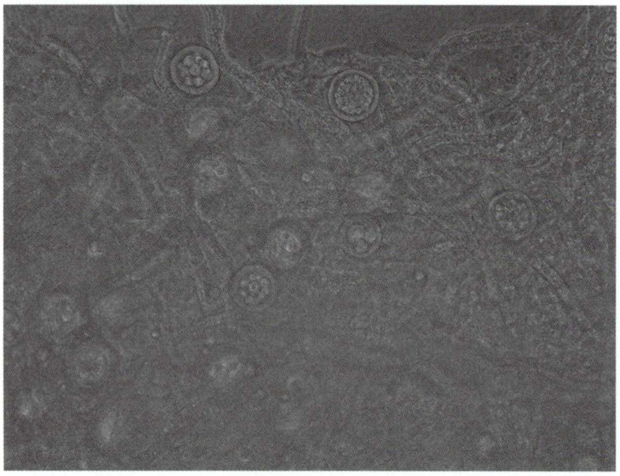

Fig. 27.14. Sexual spores of *Conidiobolus incongruus* with numerous globules (Phase Contrast × 400).

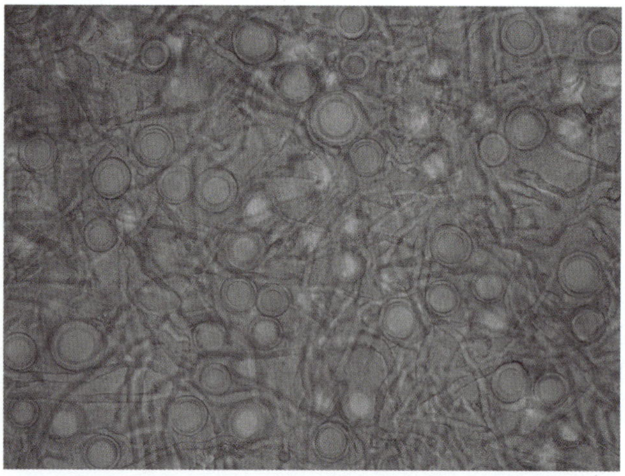

Fig. 27.15. Sexual spores of *Conidiobolus lamprauges* without any globule (Phase Contrast × 400).

Table 27.2. Comparison of various Clinical Significant Entomophthorales species (*Conidiobolus* and *Basidiobolus*).

Features	*C.coronatus*	*C.incongruus*	*C.lamprauges*	*B.ranarum*
Sexual conidia (Zygospores)	Not produced	Larger in size 15-40 µm with numerous globules	Smaller in size 12-18 µm with single large globule	With beak-like appendages
Asexual conidia	Globose 25-45 µm	Globose 16-34 µm	Globose to obovoid 15-22 µm	Globose
Replicate conidia	Present	Present	Present	-
Secondary conidia	Present	Present	Present	-
Villose conidia	Present	Absent	Absent	-
Multi-replicative conidia	Present	Present in a few strains	Absent	-

Epidemiology

Basidiobolus ranarum is commonly found in leaf detritus, on decaying plant material in tropical and subtropical regions. It is also found in gastrointestinal tract of amphibians such as frogs, toads and small reptiles like lizards. The majority of clinical reports concerning this fungus have been from tropical and subtropical regions despite the fact that this environmental saprotroph has worldwide distribution.

There is a significant change in the epidemiology of this clinical entity. Earlier only cases of subcutaneous disease were reported, however, now gastrointestinal involvement is very significant. It is mainly limited to Middle-East but cases have been reported from other areas also. This is a total transformation in fungal epidemiology. A disease remaining confined to the subcutaneous tissue of lower portion of the body for years, is now involving the gastrointestinal tract. That is the reason that clinicians being unfamiliar are not able to contemplate this disease or if at all suspected, it is considered as gastrointestinal mucormycosis, wherein treatment modalities are entirely different.

Clinical Features

The clinical presentations of basidiobolomycosis are two types, involving the subcutaneous tissue of lower part of the body and the gastrointestinal tract. It is in contrast to conidiobolomycosis, wherein upper part of the body is affected mainly over the facial areas without any gastrointestinal involvement, however, dissemination may be reported. Both these subtypes of basidiobolomycosis are briefly described below:

(i) Subcutaneous Basidiobolomycosis

The disease is usually characterized by an insidious onset of substantial induration of subcutaneous tissue involving limbs, trunk or buttocks. Infections occur mainly in **children** and young adults, usually males below the age of twenty years. This is generally presumed that patients become infected through abrasions of skin but it is not clear why this fungus predominantly affects children. The mode of infection is presumed to be through subcutaneous implantation of fungus with thorn prick or an insect bite. The invasive retroperitoneal infections have also been reported, independently as well as associated with subcutaneous infections.

The disease develops as a well-circumscribed, firm but painless subcutaneous nodule, which slowly progresses to form large mass that is attached to skin but moves freely over underlying muscles. The spread of infection is very slow, developing into a large mass, which is attached to skin but not to underlying tissue, unlike that of conidiobolomycosis. The development of ulceration or involvement of underlying tissue is rare.

The lesions are usually found over gluteal region, thighs although any part of body may be affected. They are in the form of development of extensive and swollen areas. The most common presentation is by involvement of thighs and buttocks as a **'bathing suit'** distribution. There is no functional impairment as long as the dimensions of swelling do not block the joints.

(ii) Gastrointestinal Basidiobolomycosis

The gastrointestinal basidiobolomycosis (GIB) is an emerging disease in southwestern Saudi Arabia and should be considered in a patient with an abdominal mass and eosinophilia coming from this region. Persistent elevation of leukocytes and eosinophils after surgical resection of the affected tissue could be used as a predictor of the fungal dissemination.

In recent times, more than seventy cases of systemic involvement by *B.ranarum* have been reported, particularly involving gastrointestinal tract and retroperitoneal area. The gastrointestinal lesions are associated with mural thickening, nodular masses and ulcerations of the intestine that resemble Crohn's disease. The mural thickening is responsible for the subacute obstruction of the intestinal lumen thereby causing acute presentation of such patients in surgical emergencies. It can involve the stomach, small intestine, colon or rectum. However, further research is needed for a better understanding of the patients of gastrointestinal basidiobolomycosis.

There are rare instances of involvement of respiratory system or invasive fungal sinusitis caused by this fungus.

Differential Diagnosis

Basidiobolomycosis should be differentiated from onchocerciasis, elephantiasis, mycetoma, sporotrichosis and *Mycobacterium ulcerans* infections when there is subcutaneous involvement. When there is gastrointestinal involvement it may be confused with intestinal obstruction or Crohn's disease. The lesions may occasionally be mistaken clinically for malignancy, eosinophilic granuloma,

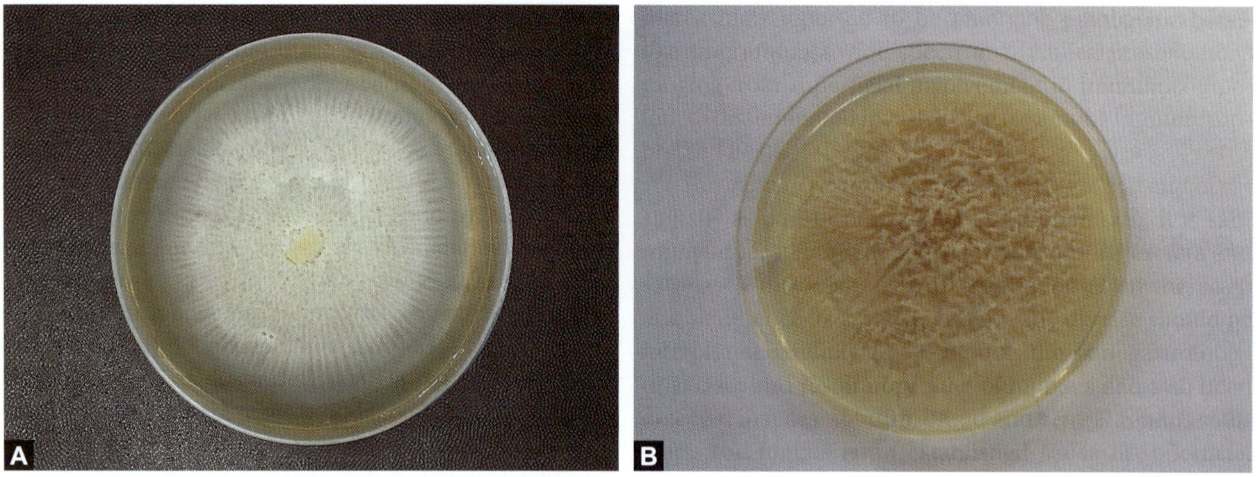

Figs. 27.16A and B. Culture of *Basidiobolus ranarum* showing wrinkled and folded growth on SDA plate after one and two weeks.

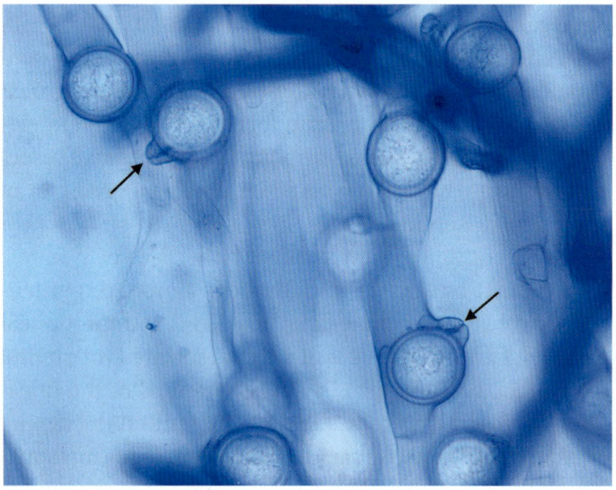

Fig. 27.17. Sexual spores of *Basidiobolus ranarum* with beak-like appearance (LCB × 400).

angiolymphoid hyperplasia with eosinophilia and Kimura disease. The characteristic histopathology of lesions reveal fungal elements showing short, broad fragments and pauci-septation with right-angle irregular non-dichotomous branching, broad ribbon-like hyphae with Splendore-Hoeppli phenomenon. This disease should not be confused with Fournier's gangrene.

Laboratory Diagnosis

The direction examination of biopsy specimen is done in KOH wet mount, which shows non-septate, broad hyphae. The histopathological examination of this disease is very similar to that of conidiobolomycosis. Staining with H&E shows, broad non-septate hyphae encircled by eosinophilic mass of Splendore-Hoeplli phenomenon. Immunodiffusion test can be done to establish the diagnosis. Moreover, previous difficulties with identification based on morphological characteristics can be overcome by precise and quick identification of *Conidiobolus* and *Basidiobolus* using PCR with taxon-specific primers and adopting other molecular based techniques.

The organism grows well on conventional fungal culture media like SDA, CMA and PDA from 25° to 37°C. The moderately fast growing colonies of *B.ranarum* at 30°C are flat, yellowish to gray and later on becoming radiated, folded, furrowed and waxy textured with pale reverse. There is cerebriform centre without aerial mycelia **(Figs. 27.16A and B)**. The growth is less rapid at body temperature *i.e.* 37°C. There is brownish pigmentation within less than a week time. There are satellite colonies, which may be seen due to germination of ejected sporangioles. Microscopically, there are coenocytic hyphae, 8-20 μm wide and may fragment into short hyphal bodies.

The primary asexual spores are globose, produced singly at tip of conidiophore and forcibly discharged. The secondary conidia are obclavate, wider at base and have an adhesive tip and these are not discharged. Sporangiospores may undergo cleavage to produce **meristospores**.

The distinctive feature of genus *Basidiobolus* is production of typical smooth-walled **zygospores** with two closely appressed **beak-shaped** gametangial remnants of copulation tubes, the feature that differentiates it from *Conidiobolus* species **(Fig. 27.17)**. The conjugation beaks represent the remnants of copulatory tubes. The zygospores are 20-50 μm in diameter having smooth to undulated thick-wall.

Entomophthoramycosis should be considered in the differential diagnosis of slow growing subcutaneous mass. This should not be mistaken for soft tissue tumors thereby avoid mismanagement. In this disease now increasing incidence of multidrug resistant strains and clinical fatalities are a source of great concern.

Treatment and Prophylaxis

The infections caused by genera of order Entomophthorales are not life-threatening, unlike in case of the Mucorales. Oral potassium iodide as saturated solution (SSKI) is given as 20 mg/kg/day, up to maximum dose of 2 gm/day. It is drug of choice for both types of entomophthoromycosis and should be continued for one month even after lesions have disappeared.

A combination of itraconazole and SSKI is an effective treatment modality for rhinofacial conidiobolomycosis (RFC) with relatively faster onset of action, low relapse rates and minimal adverse effects. It can be considered as first-line treatment in patients with RFC.

The disease usually takes a favorable course when timely treatment is provided. Hyperbaric oxygen is believed to accelerate healing process by enhancing vasodilation thereby improving perfusion of infected tissue and augmenting immune response. Conidiobolomycosis often requires surgical intervention along with KI treatment. KI is not effective in mucormycosis, caused by Mucorales, which are characterized by vascular invasion and tissue necrosis.

As there is occasionally involvement of visceral organs when such types of infections are present, amphotericin B may be given intravenously similar to as it is indicated in other systemic fungal diseases. Emergence of multi-drug resistant infection is also an issue of concern.

Further Reading

1. Agrawal S, Meshram P, Qazi MS. Rhinoentomophthoromycosis: A rare case report. *Indian J Med Microbiol.* 2013; **31:** 401-3.
2. Al-Asmi MM, Faqeehi HY, Alshahrani DA, *et al.* A case of pediatric gastrointestinal basidiobolomycosis mimicking Crohn's disease. A review of pediatric literature. *Saudi Med J.* 2013; **34:** 1068-72.
3. Al-Baradi BA, Babiker AM, Al-Qahtani HS. Successful treatment of gastrointestinal basidiobolomycosis with voriconazole without surgical intervention. *J Trop Pediatr.* 2014; **60:** 476-9.
4. Al-Hajjar S, Perfect J, Hashem F, *et al.* Orbitofascial conidiobolomycosis in a child. *Pediatr Infect Dis J.* 1996; **15:** 1130-2.
5. Al-Huraiji A, Alqaraawi A, Alaraj A, *et al.* Chronic abdominal pain and intestinal obstruction in a 24-year-old woman. *Clin Infect Dis.* 2014; **58:** 990 & 1035-6.
6. Al-Jarie A, Al Azraki T, Al Mohsen I, *et al.* Basidiobolomycosis: Case series. *J Mycol Med.* 2011; **21:** 37-45.
7. Al-Jarie A, Al-Mohsen I, Al Jumaah S, *et al.* Pediatric gastrointestinal basidiobolomycosis. *Pediatr Infect Dis J.* 2003; **22:** 1007-14.
8. Al-Maani AS, Paul G, Jardani A, *et al.* Gastrointestinal basidiobolomycosis: First case report from Oman and literature review. *Sultan Qaboos Univ Med J.* 2014; **14:** e241-4.
9. Al-Naemi AQ, Khan LA, Al-Naemi I, *et al.* A case report of gastrointestinal basidiobolomycosis treated with voriconazole: A rare emerging entity. *Medicine (Baltimore).* 2015; **94:** e1430.
10. Al-Saleem K, Al-Mehaidib A, Banemai M, *et al.* Gastrointestinal basidiobolomycosis: Mimicking Crohn's disease - Case report and review of the literature. *Ann Saudi Med.* 2013; **33:** 500-4.
11. Al-Shanafey S, Al Robean F, Bin Hussain I. Surgical management of gastrointestinal basidiobolomycosis in pediatric patients. *J Pediatr Surg.* 2012; **47:** 949-51.
12. Anand M, Deshmukh SD, Pande DP, *et al.* Subcutaneous zygomycosis due to *Basidiobolus ranarum:* A case report from Maharashtra, India. *J Trop Med.* 2010; 950390. PMID: 21209703.
13. Anaparthy UR, Deepika G. A case of subcutaneous zygomycosis. *Indian Dermatol Online J.* 2014; **5:** 51-4.
14. Arjmand R, Karimi A, Sanaei Dashti A, *et al.* A child with intestinal basidiobolomycosis. *Iran J Med Sci.* 2012; **37:** 134-6.
15. Arora N, Bhargava EK, Rai V. Nasal conidiobolomycosis-The unknown threat. *J Clin Diagn Res.* 2016; **10:** MJ01.
16. Arora P, Sardana K, Bansal S, *et al.* Entomophthoromycosis (basidiobolomycosis) presenting with "saxophone" penis and responding to potassium iodide. *Indian J Dermatol Venereol Leprol.* 2015; **81:** 616-8.
17. Arya AA, Naik C, Desmukh S, *et al.* Mastoid infection caused by Entomophthorales: A rare fungal disease. *J Laryngol Otol.* 2011; **125:** 630-2.
18. Atadokpede F, Gnossike J, Adegbidi H, *et al.* Cutaneous basidiobolomycosis: Seven cases in southern Benin. *Ann Dermatol Venereol.* 2017; **144:** 250-4.
19. Auxiliadora-Martins M, Alkmim-Teixeira GC, Machado-Viana J, *et al.* Meningoencephalitis caused by a zygomycete fungus (*Basidiobolus*) associated with septic shock in an immunocompetent patient: 1-year follow-up after treatment. *Braz J Med Biol Res.* 2010; **43:** 794-8.
20. Balraj A, Anandi V, Raman R. Nasofacial conidiobolomycosis. *Ear Nose Throat J.* 1991; **70:** 737-9.
21. Bamba S, Konsegre V, Zida A, *et al.* A case of rhinofacial entomophthoromycosis in Soudano-Sahelian tropical climate in Burkina Faso. *J Mycol Med.* 2017 Feb 14. pii: S1156-5233(16)30296-7.

22. Bento DP, Tavares R, Martins Mda L, *et al.* Atypical presentation of entomophthoromycosis caused by *Conidiobolus coronatus. Med Mycol.* 2010; **48:** 1099-104.
23. Bhaskaran CS, Rao G, Rao PV. Subcutaneous phycomycosis. *Indian J Pathol Microbiol.* 1985; **28:** 171-4.
24. Bhat RM, Srinivas M, Sequeira F, *et al.* Rhinoentomophthoromycosis with mutilation. *Clin Exp Dermatol.* 2010; **35:** 501-4.
25. Bhatia N, Singh C, Kawatra R, *et al.* Rhinoentomophthoromycosis: Rarity revisited. *Ear Nose Throat J.* 2010; **89:** 268-71.
26. Bigliazzi C, Poletti V, Dell'Amore D, *et al.* Disseminated basidiobolomycosis in an immunocompetent woman. *J Clin Microbiol.* 2004; **42:** 1367-9.
27. Bittencourt AL, Arruda SM, de Andrade JA, *et al.* Basidiobolomycosis: A case report. *Pediatr Dermatol.* 1991; **8:** 325-8.
28. Bittencourt AL, Marback R, Nossa LM. Mucocutaneous entomophthoramycosis acquired by conjunctival inoculation of the fungus. *Am J Trop Med Hyg.* 2006; **75:** 936-8.
29. Blumentrath CG, Grobusch MP, Matsiegui PB, *et al.* Classification of rhinoentomophthoromycosis into atypical, early, intermediate and late disease: A proposal. *PLoS Negl Trop Dis.* 2015; **9:** e0003984. PMID: 26426120.
30. Boonsarngsuk V, Suankratay C, Wilde H. Presumably entomophthoramycosis in an HIV-infected patient: The first in Thailand. *J Med Assoc Thai.* 2001; **84:** 1635-40.
31. Burkitt DP, Wilson AMM, Jelliffe DB. Subcutaneous phycomycosis: A review of 31 cases seen in Uganda. *Br Med J.* 1964; **1:** 1669-72.
32. Busapakum R, Youngchaiyud U, Sriumpai S, *et al.* Disseminated infection with *Conidiobolus incongruus. Sabouraudia.* 1983; **21:** 323-30.
33. Cherian LM, Varghese L, Panchatcharam BS, *et al.* Nasal conidiobolomycosis: A successful treatment option for localized disease. *J Postgrad Med.* 2015; **61:** 143-4.
34. Chetambath R, Deepa Sarma MS, Suraj KP, *et al. Basidiobolus:* An unusual cause of lung abscess. *Lung India.* 2010; **27:** 89-92.
35. Chiewchanvit S, Khamwan C, Pruksachatkunakorn C, *et al.* Entomophthoromycosis in Maharaj Nakorn Chiang Mai Hospital. *J Med Assoc Thai.* 2002; **85:** 1089-94.
36. Chokka KM, Reddy SR, Rayi NR, *et al.* Entomophthoromycosis of face. *Indian J Pathol Microbiol.* 2010; **53:** 564-5.
37. Choon SE, Kang J, Neafie RC, *et al.* Conidiobolomycosis in a young Malaysian woman showing chronic localized fibrosing leukocytoclastic vasculitis: A case report and meta-analysis focusing on clinicopathologic and therapeutic correlations with outcome. *Am J Dermatopathol.* 2012; **34:** 511-22.
38. Choonhakarn C, Inthraburan K. Concurrent subcutaneous and visceral basidiobolomycosis in a renal transplant patient. *Clin Exp Dermatol.* 2004; **29:** 369-72.
39. Chowdhary A, Randhawa HS, Khan ZU, *et al.* Rhinoentomophthoromycosis due to *Conidiobolus coronatus:* A case report and an overview of the disease in India. *Med Mycol.* 2010; **48:** 870-9.
40. Costa AR, Porto E, Pegas JR, *et al.* Rhinofacial zygomycosis caused by *Conidiobolus coronatus.* A case report. *Mycopathologia.* 1991; **115:** 1-8.
41. Dasgupta LR, Agarwal SC, Varma RA, *et al.* Subcutaneous phycomycosis from Pondicherry, south India. *Sabouraudia.* 1976; **14:** 123-7.
42. Das P, Vijay MK, Joshi P, *et al.* Histological identification of entomophthoromycosis in biopsy samples is required. *Indian J Pathol Microbiol.* 2014; **57:** 514-6.
43. de Godoy I, de Campos CG, Pescador CA, *et al.* Experimental infection in gerbils by *Conidiobolus lamprauges. Microb Pathog.* 2017; **105:** 251-4.
44. de Ment SH, Davis MS, Morris GL, *et al.* Nasofacial zygomycosis: A case report of *Conidiobolus coronatus* in a Hispanic immigrant in South Carolina. *J S C Med Assoc.* 2005; **101:** 93-6.
45. Drouhet E, Ravisse P. Entomophthoromycosis. *Curr Top Med Mycol.* 1993; **5:** 215-45.
46. Dutta S, Sarkar S, Linka U, *et al.* Conidiobolomycosis: A case report of rare fungal infection from the eastern India. *Indian Dermatol Online J.* 2015; **6:** 393-5.
47. Ejtehadi F, Anushiravani A, Bananzadeh A, *et al.* Gastrointestinal basidiobolomycosis accompanied by liver involvement: A case report. *Iran Red Crescent Med J.* 2014; **16:** e14109. PMID: 25593711.
48. El-Shabrawi MH, Arnaout H, Madkour L, *et al.* Entomophthoromycosis: A challenging emerging disease. *Mycoses.* 2014; **57 (Suppl. 3):** 132-7.
49. El-Shabrawi MH, Kamal NM, Jouini R, *et al.* Gastrointestinal basidiobolomycosis: An emerging fungal infection causing bowel perforation in a child. *J Med Microbiol.* 2011; **60:** 1395-402.
50. El-Shabrawi MH, Kamal NM, Kaerger K, *et al.* Diagnosis of gastrointestinal basidiobolomycosis: A mini-review. *Mycoses.* 2014; **57 (Suppl. 3):** 138-43.
51. El-Shabrawi MH, Kamal NM. Gastrointestinal basidiobolomycosis in children: An overlooked emerging infection? *J Med Microbiol.* 2011; **60:** 871-80.
52. Fahimzad A, Karimi A, Tabatabaei SR. Gastrointestinal basidiobolomycosis as a rare etiology of bowel obstruction. *Turk J Med Sci.* 2006; **36:** 239-41.
53. Feng MG, Xu Q. A simple method for routine maintenance and preservation of entomophthoraceous cultures. *J Invertebr Pathol.* 2001; **77:** 141-3.
54. Fischer N, Ruef CH, Ebnother C, *et al.* Rhinofacial *Conidiobolus coronatus* infection presenting with nasal enlargement. *Infection.* 2008; **36:** 594-6.
55. Flicek KT, Vikram HR, De Petris GD, *et al.* Abdominal imaging findings in gastrointestinal basidiobolomycosis. *Abdom Imaging.* 2015; **40:** 246-50.
56. Geramizadeh B, Foroughi R, Keshtkar-Jahromi M, *et al.* Gastrointestinal basidiobolomycosis, an emerging infection in the immunocompetent host: A report of 14 patients. *J Med Microbiol.* 2012; **61:** 1770-4.
57. Geramizadeh B, Heidari M, Shekarkhar G. Gastrointestinal basidiobolomycosis, a rare and under-diagnosed fungal infection in immunocompetent hosts: A review article. *Iran J Med Sci.* 2015; **40:** 90-7.

58. Geramizadeh B, Modjalal M, Nabai S, *et al.* Gastrointestinal zygomycosis: A report of three cases. *Mycopathologia.* 2007; **164:** 35-8.
59. Geramizadeh B, Sanai Dashti A, Kadivar MR, *et al.* Isolated hepatic basidiobolomycosis in a 2-year-old girl: The first case report. *Hepat Mon.* 2015; **15:** e30117. PMID: 26322112.
60. Ghorpade A, Sarma PS, Iqbal SM. Elephantine nose due to rhinoentomophthoromycosis. *Eur J Dermatol.* 2006; **16:** 87-9.
61. Girish M, Arora A, Bhalla L, *et al.* Elephantine but not elephantiasis: Subcutaneous zygomycosis. *Indian J Pediatr.* 2011; **78:** 1142-4.
62. Gomez-Munoz MT, Fernandez-Barredo S, Martinez-Diaz RA, *et al.* Development of a specific polymerase chain reaction assay for the detection of *Basidiobolus. Mycologia.* 2012; **104:** 585-91.
63. Gordon CL, Whiting S, Haran G, *et al.* Entomophthoromycosis caused by *Basidiobolus ranarum* in tropical northern Australia. *Pathology.* 2012; **44:** 375-9.
64. Goyal A, Gupta N, Das S, *et al.* Basidiobolomycosis of the nose and face: A case report and a mini-review of unusual cases of basidiobolomycosis. *Mycopathologia.* 2010; **170:** 165-8.
65. Gryganskyi AP, Humber RA, Smith ME, *et al.* Molecular phylogeny of the Entomophthoromycota. *Mol Phylogenet Evol.* 2012; **65:** 682-94.
66. Gryganskyi AP, Humber RA, Smith ME, *et al.* Phylogenetic lineages in Entomophthoromycota. *Persoonia.* 2013; **30:** 94-105.
67. Guarro J, Aguillar C, Pujol I. *In vitro* antifungal susceptibilities of *Basidiobolus* and *Conidiobolus* spp. strains. *J Antimicrob Chemotherapy.* 1999; **44**: 557-60.
68. Gugnani HC. A review of zygomycosis due to *Basidiobolus ranarum. Eur J Epidemiol.* 1999; **15:** 923-9.
69. Gugnani HC. Entomophthoromycosis due to *Conidiobolus. Eur J Epidemiol.* 1992; **8:** 391-6.
70. Gupta M, Narang T, Kaur RJ, *et al.* A prospective case series evaluating efficacy and safety of combination of itraconazole and potassium iodide in rhinofacial conidiobolomycosis. *Int J Dermatol.* 2016; **55:** 208-14.
71. Hamid ME, Joseph MR, Al-Qahtani AS. Chronic rhinofacial basidiobolomycosis caused by *Basidiobolus ranarum:* Report of a case from Aseer Region, Kingdom of Saudi Arabia. *J Mycol Med.* 2015; **25:** 306-9.
72. Hassan HA, Majid RA, Rashid NG, *et al.* Eosinophilic granulomatous gastrointestinal and hepatic abscesses attributable to basidiobolomycosis and fasciolias: A simultaneous emergence in Iraqi Kurdistan. *BMC Infect Dis.* 2013; **13:** 91.
73. Hemmat M, Singh S. Rare case of *Conidiobolus* lung infection. *Am J Respir Crit Care Med.* 2015; **191:** A1862.
74. Hernandez MJ, Landaeta W, Salazar BN, *et al.* Subcutaneous zygomycosis due to *Conidiobolus incongruus. Int J Infect Dis.* 2007; **11:** 468-9.
75. Hoogendijk CF, Pretorius E, Marx J, *et al.* Detection of villous conidia of *Conidiobolus coronatus* in a blood sample by scanning electron microscopy investigation. *Ultrastruct Pathol.* 2006; **30:** 53-8.
76. Hoogendijk CF, Pretorius E. Cavernous sinus anatomy as a basis for interpretation of the clinical picture and radiological investigations in a case of Entomophthorales infection. *Clin Anat.* 2006; **19:** 535-9.
77. Hoogendijk CF, van Heerden WF, Pretorius E, *et al.* Rhino-orbito-cerebral entomophthoramycosis. *Int J Oral Maxillofac Surg.* 2006; **35:** 277-80.
78. Humber RA. Entomophthoromycota: A new phylum and reclassification for entomophthoroid fungi. *Mycotaxon.* 2012; **120:** 477-92.
79. Hussein MR, Musalam AO, Assiry MH, *et al.* Histological and ultrastructural features of gastrointestinal basidiobolomycosis. *Mycol Res.* 2007; **111:** 926-30.
80. Ilyas MI, Jordan SA, Nfonsam V. Fungal inflammatory masses masquerading as colorectal cancer: A case report. *BMC Res Notes.* 2015; **8:** 32.
81. Imwidthaya P, Srimuang S. Immunodiffusion test for diagnosing basidiobolomycosis. *Mycopathologia.* 1992; **118:** 127-31.
82. Isa-Isa R, Arenas R, Fernandez RF, *et al.* Rhinofacial conidiobolomycosis (entomophthoramycosis). *Clin Dermatol.* 2012; **30:** 409-12.
83. Jaffey PB, Haque AK, el-Zaatari M, *et al.* Disseminated *Conidiobolus* infection with endocarditis in a cocaine abuser. *Arch Pathol Lab Med.* 1990; **114:** 1276-8.
84. Jayanth ST, Gaikwad P, Promila M, *et al.* The sinus that breeds fungus: Subcutaneous zygomycosis caused by *Basidiobolus ranarum* at the injection site. *Case Rep Infect Dis.* 2013; 534192. PMID: 24349806.
85. Jayawickrama MM, Gamage HU, Perera PD. *Conidiobolus coranatus* infection of the para-nasal sinuses masquerading as a sino-nasal tumour. *Ceylon Med J.* 2012; **57:** 87-8.
86. Jensen AB, Dromph KM. The causal agents of 'entomophthoramycosis' belong to two different orders: A suggestion for modification of the clinical nomenclature. *Clin Microbiol Infect.* 2005; **11:** 249-50.
87. John D, Irodi A, Michael JS. Concurrent infections of *Conidiobolus coronatus* with disseminated tuberculosis presenting as bilateral orbital cellulitis. *J Clin Diagn Res.* 2016; **10:** ND01-2.
88. Joob B, Wiwanitkit V. Visceral basidiobolomycosis. *Afr J Paediatr Surg.* 2016 ; **13:** 158.
89. Kabra SK, Jain Y, Sudhin T, *et al.* Successful treatment of entomophthoromycosis with itraconazole. *Indian Pediatrics.* 1998; **35:** 163-6.
90. Kamalam A, Sentamilselvi G, Thambiah AS. Entomophthoromycosis. *Proc R Col Phys (Edin).* 1992; **22:** 52-9.
91. Kamalam A, Thambiah AS. Basidiobolomycosis following injection injury. *Mykosen.* 1982; **25:** 512-6.
92. Kamalam A, Thambiah AS. Basidiobolomycosis with lymph node involvement. *Sabouraudia.* 1975; **13:** 44-8.
93. Kamalam A, Thambiah AS. Entomophthoromycosis basidiobolae - successfully treated with KI. *Mykosen.* 1979; **22:** 82-4.
94. Kamalam A, Thambiah AS. Lymph node invasion by *Conidiobolus coronatus* and its spore formation *in vivo. Sabouraudia.* 1978; **16:** 175-84.

95. Kamalam A, Thambiah AS. Lymphedema and elephantiasis in basidiobolomycosis. *Mykosen.* 1982; **25:** 508-11.
96. Kamalam A, Thambiah AS. Muscle invasion by *Basidiobolus haptosporus. Sabouraudia.* 1984; **22:** 273-7.
97. Kamalam A, Yesudian P, Thambiah AS. Basidiobolomycosis - A case report. *Australas J Dermatol.* 1973; **14:** 136-9.
98. Karuna T, Asati DP, Biswas D, *et al.* Subcutaneous entomophthoromycoses. *Indian Dermatol Online J.* 2015; **6:** 410-2.
99. Kaufman L, Mendoza L, Standard PG. Immunodiffusion test for serodiagnosing subcutaneous zygomycosis. *J Clin Microbiol.* 1990; **28:** 1887-90.
100. Khalil M, Ekanem IO, Gugnani HC, *et al.* Some deep mycoses diagnosed by histopathology in south-eastern Nigeria. *Rev Iberoam Micol.* 1999; **16:** 221-4.
101. Khan ZU, Khoursheed M, Makar R, *et al. Basidiobolus ranarum* as an etiologic agent of gastrointestinal zygomycosis. *J Clin Microbiol.* 2001; **39:** 2360-3.
102. Khan ZU, Prakash B, Kapoor MM, *et al.* Basidiobolomycosis of the rectum masquerading as Crohn's disease: Case report and review. *Clin Infect Dis.* 1998; **26:** 521-3.
103. Khatami A, Outhred AC, Britton PN, *et al.* Mediastinal mass in a healthy adolescent at the Children's Hospital at Westmead, Australia. *Thorax.* 2015; **70:** 194-7.
104. Kimura M, Yaguchi T, Sutton DA, *et al.* Disseminated human conidiobolomycosis due to *Conidiobolus lamprauges. J Clin Microbiol.* 2011; **49:** 752-6.
105. Kindo AJ, Giri S, Rao S, *et al.* Abscesses that did not respond to just incision drainage and antibiotics. *Int J Low Extrem Wounds.* 2010; **9:** 160-2.
106. Koshi G, Kurien T, Sudarsanam D, *et al.* Subcutaneous phycomycosis caused by *Basidiobolus*: A report of three cases. *Sabouraudia.* 1972; **10:** 237-43.
107. Krishnan SG, Sentamilselvi G, Kamalam A, *et al.* Entomophthoromycosis in India - A 4-year study. *Mycoses.* 1998; **41:** 55-8.
108. Krishnamurthy S, Singh R, Chandrasekaran V, *et al.* Basidiobolomycosis complicated by hydronephrosis and a perinephric abscess presenting as a hypertensive emergency in a 7-year-old boy. *Paediatr Int Child Health.* 2017; 1-4. PMID: 28112037.
109. Kumar A, Viswam V, Regi S, *et al.* Rhinoentomophthoromycosis caused by *Conidiobolus coronatus* in a diabetic patient: The importance of species identification. *Ann Trop Med Public Health.* 2013; **6:** 331-4.
110. Kumar Verma R, Shivaprakash MR, Shanker A, *et al.* Subcutaneous zygomycosis of the cervicotemporal region: Due to *Basidiobolus ranarum. Med Mycol Case Rep.* 2012; **1:** 59-62.
111. Kumaravel S, Bharath K, Rajesh NG, *et al.* Delay and misdiagnosis of basidiobolomycosis in tropical south India: Case series and review of the literature. *Paediatr Int Child Health.* 2016; **36:** 52-7.
112. Leopairut J, Larbcharoensub N, Cheewaruangroj W, *et al.* Rhinofacial entomophthoramycosis; A case series and review of the literature. *Southeast Asian J Trop Med Public Health.* 2010; **41:** 928-35.
113. Li Y, Fang X, Zhou X, *et al.* Pathogenicity of *Conidiobolus coronatus* and *Fusarium solani* in mouse models. *Mycoses.* 2017; **60:** 394-401.
114. Lithander J, Louon A, Worthing E, *et al.* Orbital entomophthoramycosis in an infant: Recovery following surgical debridement, combination antifungal therapy and use of hyperbaric oxygen. *Br J Ophthalmol.* 2001; **85:** 374-5.
115. Lyon GM, Smilack JD, Komatsu KK, *et al.* Gastrointestinal basidiobolomycosis in Arizona: Clinical and epidemiological characteristics and review of the literature. *Clin Infect Dis.* 2001; **32:** 1448-55.
116. Mackey PE, Cappe KG, Mani R, *et al.* Disseminated *Conidiobolus incongruus* in a dog: A case report and literature review. *Med Mycol Case Rep.* 2015; **8:** 24-8.
117. Mahamaytakit N, Singalavanija S, Limpongsanurak W. Subcutaneous zygomycosis in children: 2 case reports. *J Med Assoc Thai.* 2014; **97:** S248-53.
118. Maiti PK, Bose R, Bandyopadhyay S, Bhattacharya S, Dey JB, Ray A. Entomophthoromycosis in south Bengal (Eastern India): A 9 years study. *Indian J Pathol Microbiol.* 2004; **47:** 295-7.
119. Mandhan P, Hassan KO, Samaan SM, *et al.* Visceral basidiobolomycosis: An overlooked infection in immunocompetent children. *Afr J Paediatr Surg.* 2015; **12:** 193-6.
120. Mathew R, Kumaravel S, Kuruvilla S, *et al.* Successful treatment of extensive basidiobolomycosis with oral itraconazole in a child. *Int J Dermatol.* 2005; **44:** 572-5.
121. Mendiratta V, Karmakar S, Jain A, *et al.* Severe cutaneous zygomycosis due to *Basidiobolus ranarum* in a young infant. *Pediatr Dermatol.* 2012; **29:** 121-3.
122. Mendoza L, Vilela R, Voelz K, *et al.* Human fungal pathogens of Mucorales and Entomophthorales. *Cold Spring Harb Perspect Med.* 2014; **5:** pii: a019562.
123. Michael RC, Michael JS, Mathews MS, *et al.* Unusual presentation of entomophthoromycosis. *Indian J Med Microbiol.* 2009; **27:** 156-8.
124. Moncada DC, Montes M, Molina V, *et al.* Orofacial infection by *Conidiobolus coronatus. Biomedica.* 2016; **36:** 15-22.
125. Mondal AK, Saha A, Seth J, *et al.* Subcutaneous zygomycosis: A report of one case responding excellently to potassium iodide. *Indian J Dermatol.* 2015; **60:** 500-2.
126. Moore MH, Murphy D, James C. Nasofacial phycomycosis: Persistence pays. *J Craniofac Surg.* 2007; **18:** 448-50.
127. Mukhopadhyay D, Ghosh LM, Thammayya A, *et al.* Entomophthoromycosis caused by *Conidiobolus coronatus:* Clinicomycological study of a case. *Auris Nasus Larynx.* 1995; **22:** 139-42.
128. Nayak DR, Pillai S, Rao L. Rhinofacial zygomycosis caused by *Conidiobolus coronatus. Indian J Otolaryngol Head Neck Surg.* 2004; **56:** 225-7.
129. Nemenqani D, Yaqoob N, Khoja H, *et al.* Gastrointestinal basidiobolomycosis: An unusual fungal infection mimicking colon cancer. *Arch Pathol Lab Med.* 2009; **133:** 1938-42.
130. Nguyen BD. CT features of basidiobolomycosis with gastrointestinal and urinary involvement. *AJR Am J Roentgenol.* 2000; **174:** 878-9.

131. Okafor BC, Gugnani HC. Nasal entomophthoromycosis in Nigerian Igbos. *Trop Geogr Med.* 1983; **35:** 53-7.
132. Okafor JI, Gugnani HC, Testratke D, *et al.* Extracellular enzyme activities by *Basidiobolus* and *Conidiobolus* isolates on solid media. *Mykosen.* 1987; **30:** 404-7.
133. Onuigbo WI, Gugnani HC, Okafor BC, *et al.* Nasal entomophthorosis in an Igbo from Nigeria. *J Laryngol Otol.* 1975; **89:** 657-61.
134. Pandit V, Rhee P, Aziz H, *et al.* Perforated appendicitis with gastrointestinal basidiobolomycosis: A rare finding. *Surg Infect (Larchmt).* 2014; **15:** 339-42.
135. Pasha TM, Leighton JA, Smilack JD, *et al.* Basidiobolomycosis: An unusual fungal infection mimicking inflammatory bowel disease. *Gastroenterology.* 1997; **112:** 250-4.
136. Pornpanich K, Uiprasertkul M, Luemsamran P, *et al.* Entomophthoramycosis: A rare fungal orbital infection presenting with dacryocystitis. *Orbit.* 2011; **30:** 21-3.
137. Prabhu RM, Patel R. Mucormycosis and entomophthoramycosis: A review of the clinical manifestations, diagnosis and treatment. *Clin Microbiol Infect.* 2004; **10 (Suppl. 1):** 31-47.
138. Pujary P, Pujary K, Mahesh SG, *et al.* *Conidiobolus* coronata granuloma of left inferior turbinate: A rare presentation. *J Laryngol Otol.* 2008; **122:** e15.
139. Rabie ME, El Hakeem I, Al-Shraim M, *et al.* Basidiobolomycosis of the colon masquerading as stenotic colon cancer. *Case Rep Surg.* 2011; 685460. PMID: 22606589.
140. Radhakrishnan N, Sachdeva A, Oberoi J, *et al.* Conidiobolomycosis in relapsed acute lymphoblastic leukemia. *Pediatr Blood Cancer.* 2009; **53:** 1321-3.
141. Radjou AN, Rajesh NG. Intestinal obstruction due to *Basidiobolus ranarum:* An unusual case. *Indian J Med Microbiol.* 2011; **29:** 186-8.
142. Rajan RJ, Mohanraj P, Rose W. Subcutaneous basidiobolomycosis resembling Fournier's gangrene. *J Trop Pediatr.* 2017; **63:** 217-220.
143. Ramesh V, Ramam M, Capoor MR, *et al.* Subcutaneous zygomycosis: Report of 10 cases from two institutions in north India. *J Eur Acad Dermatol Venereol.* 2010; **24:** 1220-5.
144. Ramirez J, Maguina P. Invasive conidiobolomycosis can be successfully treated on burn survivors. *J Burn Care Res.* 2017; **38:** e460-3.
145. Rane SR, Jayaraman A, Puranik SC, *et al.* Entomopthoromycosis—Report of four cases. *Indian J Dermatol Venereol Leprol.* 2002; **68:** 296-7.
146. Raveenthiran V, Mangayarkarasi V, Kousalya M, *et al.* Subcutaneous entomophthoromycosis mimicking soft-tissue sarcoma in children. *J Pediatr Surg.* 2015; **50:** 1150-5.
147. Rermluk N, Laolerd W, Chantharit P, *et al.* Mammary subcutaneous basidiobolomycosis in a male. *Breast J.* 2017; doi: 10.1111/tbj.12817. PMID: 28397341.
148. Ribes JA, Vanover-Sams CL, Baker DJ. Zygomycetes in human disease. *Clin Microbiol Rev.* 2000; **13:** 236-301.
149. Rodrigues O, Commey JO. Basidiobolomycosis in Ghanaian children. *Trop Doct.* 1994; **24:** 170-1.
150. Rose SR, Lindsley MD, Hurst SF, *et al.* Gastrointestinal basidiobolomycosis treated with posaconazole. *Med Mycol Case Rep.* 2013; **2:** 11-4.
151. Saadah OI, Farouq MF, Daajani NA, *et al.* Gastrointestinal basidiobolomycosis in a child: An unusual fungal infection mimicking fistulising Crohn's disease. *J Crohn's Colitis.* 2012; **6:** 368-72.
152. Saeed MA, Al Khuwaitir TS, Attia TH. Gastrointestinal basidiobolomycosis with hepatic dissemination: A case report. *JMM Case Rep.* 2014; **1:** doi: 10.1099/jmmcr.0.003269.
153. Sanaei Dashti A, Nasimfar A, Hosseini Khorami H, *et al.* Gastrointestinal basidiobolomycosis in a 2-year-old boy: Dramatic response to potassium iodide. *Paediatr Int Child Health.* 2017; 1-4. PMID: 27376878.
154. Schmidt JH, Howard RJ, Chen JL, *et al.* First culture-proven gastrointestinal entomophthoromycosis in the United States: A case report and review of the literature. *Mycopathologia.* 1986; **95:** 101-4.
155. Segura JJ, Gonzalez K, Berrocal J, *et al.* Rhinoentomophthoromycosis: Report of the first two cases observed in Costa Rica (Central America) and review of the literature. *Am J Trop Med Hyg.* 1981; **30:** 1078-84.
156. Sethi P, Balakrishnan D, Surendran S, *et al.* Fulminant zygomycosis of graft liver following liver transplantation. *BMJ Case Rep.* 2016; Pii bcr2015214097.
157. Seyedmousavi S, Guillot J, Tolooe A, *et al.* Neglected fungal zoonoses: Hidden threats to man and animals. *Clin Microbiol Infect.* 2015; **21:** 416-25.
158. Shah PA, Pell JK. Entomopathogenic fungi as biological control agents. *Appl Microbiol Biotechnol.* 2003; **61:** 413-23.
159. Shaikh N, Hussain KA, Petraitiene R, *et al.* Entomophthoramycosis: A neglected tropical mycosis. *Clin Microbiol Infect.* 2016; **22:** 688-94.
160. Sharma NL, Mahajan VK, Singh P. Orofacial conidiobolomycosis due to *Conidiobolus incongruus. Mycoses.* 2003; **46:** 137-40.
161. Shreef K, Saleem M, Saeedd MA, *et al.* Gastrointestinal basidiobolomycosis: An emerging and a confusing disease in children (A multicenter experience). *Eur J Pediatr Surg.* 2017; doi: 10.1055/s-0037-1598104. PMID: 28166589.
162. Singh D, Kochhar RC, Seth HN. Rhinoentomophthoromycosis. *J Laryngol Otol.* 1976; **90:** 871-5.
163. Singh R, Xess I, Ramavat AS, *et al.* Basidiobolomycosis: a rare case report. *Indian J Med Microbiol.* 2008; **26:** 265-7.
164. Sitterle E, Rodriguez C, Mounier R, *et al.* Contribution of ultra deep sequencing in the clinical diagnosis of a new fungal pathogen species: *Basidiobolus meristosporus. Front Microbiol.* 2017; **8:** 334.
165. Smilack JD. Gastrointestinal basidiobolomycosis. *Clin Infect Dis.* 1998; **27:** 663-4.
166. Sood S, Sethi S, Banerjee U. Entomophthoromycosis due to *Basidiobolus haptosporus. Mycoses.* 1997; **40:** 345-6.
167. Subramanian C, Sobel JD. A case of *Conidiobolus coronatus* in the vagina. *Med Mycol.* 2011; **49:** 427-9.
168. Sujatha S, Sheeladevi C, Khyriem AB, *et al.* Subcutaneous zygomycosis caused by *Basidiobolus ranarum* - A case report. *Indian J Med Micro.* 2003; **21:** 205-6.
169. Talwar P, Chander J, Radotra BD, *et al.* Entomophthoromycosis: Clinicomycological study of a case. *Indian J Pathol Microbiol.* 1990; **33:** 81-4.

170. Tan DC, Hsu LY, Koh LP, *et al.* Severe conidiobolomycosis complicating induction chemotherapy in a patient with acute lymphoblastic leukaemia. *Br J Haematol.* 2005; **129:** 447.
171. Temple ME, Brady MT, Koranyi KI, *et al.* Periorbital cellulitis secondary to *Conidiobolus incongruus. Pharmacotherapy.* 2001; **21:** 351-4.
172. Thammayya A. Zygomycosis due to *Conidiobolus coronatus* in West Bengal. *Indian J Chest Dis Allied Sci.* 2000; **42:** 305-9.
173. Thomas MM, Bai SM, Jayaprakash C, *et al.* Rhinoentomophthoromycosis. *Indian J Dermatol Venereol Leprol.* 2006; **72:** 296-9.
174. Thotan SP, Kumar V, Gupta A, *et al.* Subcutaneous phycomycosis - Fungal infection mimicking a soft tissue tumor: A case report and review of literature. *J Trop Pediatr.* 2010; **56:** 65-6.
175. Tondolo JS, de Loreto ES, Dutra V, *et al. In vitro* susceptibility of *Conidiobolus lamprauges* recovered from sheep to antifungal agents. *Vet Microbiol.* 2013; **166:** 690-3.
176. Valle AC, Wanke B, Lazera MS, *et al.* Entomophthoramycosis by *Conidiobolus coronatus.* Report of a case successfully treated with the combination of itraconazole and fluconazole. *Rev Inst Med Trop Sao Paulo.* 2001; **43:** 233-6.
177. van den Berk GE, Noorduyn LA, van Ketel RJ, *et al.* A fatal pseudo-tumour: Disseminated basidiobolomycosis. *BMC Infect Dis.* 2006; **6:** 140.
178. Varshney S, Gupta P, Bist SS, *et al. Conidiobolus coronatus* granuloma of the right inferior turbinate: A rare presentation. *Ear Nose Throat J.* 2015; **94:** e32-5.
179. Verma RK, Patro SK, Shivaprakash MR, *et al.* Invasive fungal infections of face - A diagnostic challenge: Case series and review. *Egyptian Journal of Ear, Nose, Throat and Allied Sciences.* 2015; **16:** 231-5.
180. Vianna LM, de Lacerda MV, de Moraes MA. Case report of subcutaneous entomophthoromycosis with retroperitoneal invasion. *Rev Soc Bras Med Trop.* 2005; **38:** 348-50.
181. Vikram HR, Smilack JD, Leighton JA, *et al.* Emergence of gastrointestinal basidiobolomycosis in the United States, with a review of worldwide cases. *Clin Infect Dis.* 2012; **54:** 1685-91.
182. Vilela R, Silva SM, Riet-Correa F, *et al.* Morphologic and phylogenetic characterization of *Conidiobolus lamprauges* recovered from infected sheep. *J Clin Microbiol.* 2010; **48:** 427-32.
183. Waingankar VM, Singh SK, Srinivasan MC. A new thermophilic species of *Conidiobolus* from India. *Mycopathologia.* 2008; **165:** 173-7.
184. Walker SD, Clark RV, King CT, *et al.* Fatal disseminated *Conidiobolus coronatus* infection in a renal transplant patient. *Am J Clin Pathol.* 1992; **98:** 559-64.
185. Walsh TJ, Renshaw G, Andrews J, *et al.* Invasive zygomycosis due to *Conidiobolus incongruus. Clin Infect Dis.* 1994; **19:** 423-30.
186. Wasim Yusuf N, Assaf HM, Rotowa NA. Invasive gastrointestinal *Basidiobolus ranarum* infection in an immunocompetent child. *Pediatr Infect Dis J.* 2003; **22:** 281-2.
187. Wuppenhorst N, Lee MK, Rappold E, *et al.* Rhino-orbitocerebral zygomycosis caused by *Conidiobolus incongruus* in an immunocompromised patient in Germany. *J Clin Microbiol.* 2010; **48:** 4322-5.
188. Yang X, Li Y, Zhou X, *et al.* Rhinofacial conidiobolomycosis caused by *Conidiobolus coronatus* in a Chinese rice farmer. *Mycoses.* 2010; **53:** 369-73.
189. Yousef OM, Smilack JD, Kerr DM, *et al.* Gastrointestinal basidiobolomycosis: Morphologic findings in a cluster of six cases. *Am J Clin Pathol.* 1999; **112:** 610-6.
190. Yu WN, Chen CJ, Chen CM. Pharyngeal entomophthoromycosis. *J Chin Med Assoc.* 2014; **77:** 337-40.
191. Zabolinejad N, Naseri A, Davoudi Y, *et al.* Colonic basidiobolomycosis in a child: Report of a culture-proven case. *Int J Infect Dis.* 2014; **22:** 41-3.
192. Zahir ST, Sharahjin NS, Kargar S. Basidiobolomycosis - A mysterious fungal infection mimic small intestinal and colonic tumour with renal insufficiency and ominous outcome. *BMJ Case Rep.* 2013; pii: bcr2013200244.
193. Zavasky DM, Samowitz W, Loftus T, *et al.* Gastrointestinal zygomycotic infection caused by *Basidiobolus ranarum:* Case report and review. *Clin Infect Dis.* 1999; **28:** 1244-8.
194. Zubaidah H, Irfan M, Amirozi A, *et al.* Rhinoentomophthoromycosis: A case report. *Int Med J.* 2014; **21:** 316-8.

CHAPTER

28

Miscellaneous Opportunistic Mycoses

In the era of modern advancements in technology in medical sciences, there is prolonged survival of patients who have even life-threatening underlying diseases along with impaired immune status. In such circumstances the invasive fungal diseases have emerged as major cause of morbidity and mortality in the hospitalized patients. With the exception of cryptococcal meningitis and perhaps catheter-associated candidemia, these infections are difficult to diagnose, particularly at their early stages. Despite limitations of diagnostic methods and overwhelming nature of invasive fungal infections in the immunocompromised patients there is necessity for initiation of antifungal therapy on an empirical basis. Moreover, due to high mortality associated with invasive fungal diseases and inability to recognize active cases reliably, prophylaxis is considered as reasonable therapeutic approach.

As such invasive fungal diseases contribute substantially to death and illness associated with the prolonged, profound neutropenia resulting from intensive chemotherapy for hematologic cancers and from myeloablation for allogeneic hematopoietic stem-cell transplantation. In addition, the development of graft-versus-host disease (GVHD) puts transplant recipients at higher risk for infection for several weeks or months after engraftment.

The fungi are otherwise taken as the opportunistic agents except for few ones, which can infect the immunocompetent individuals. The fungal infections caused by *Histoplasma, Blastomyces, Coccidioides* and *Paracoccidioides* species, although true pathogens, are also now frequently encountered in immunocompromised patients. The opportunistic infections caused by *Candida, Cryptococcus, Aspergillus* and Mucorales species are described in the separate Chapters of this Section of the Textbook. But some of the recently established fungal agents, which also cause substantial number of infections in the immunocompromised patients with various underlying factors, are discussed in this Chapter. This Chapter is also dealing with the healthcare-associated infections (nosocomial) caused by various fungi, which are also becoming significant clinical entity in any of the hospital settings. In addition, some of the emerging fungal infections are also briefly described.

Invasive fungal infections (IFIs) are a major cause of morbidity and mortality in immunocompromised patients. Host factors such as severe and prolonged neutropenia, allogeneic stem cell transplantation, prolonged use of corticosteroids, prolonged hospitalization at an intensive care unit (ICU) and receipts of recognized T cell immunosuppressants may predispose patients for developing such infections.

A. MISCELLANEOUS MYCOSES

Many individuals with compromised defenses are susceptible to infections by various microorganisms including fungi. Some of ubiquitous saprotrophic fungi cause rare infections. Such mycoses occur much less frequently than the established opportunistic infections like candidiasis, aspergillosis and mucormycosis because these are rarely known etiological agents. They have low pathogenetic potential due to their weak virulence and high susceptibility to the host defenses. The exposure may only be limited to environmental factors.

Man and animals are constantly exposed to the hundreds of species of yeasts and molds that thrive in nature. The pulmonary and systemic *Talaromyces marneffei* infection has been reported sporadically and it is endemic in northern Thailand and southern China. It is the third commonest opportunistic infection in the patients of AIDS in Thailand after tuberculosis and cryptococcosis. Similar type of agents, which were previously thought to be insignificant and have now acquired an significant place in medical sciences in general and in the field of Medical Mycology, in particular.

Species of another common saprotroph, *Fusarium*, have been isolated from burn patients. The appearance of *Aspergillus, Fusarium, Scedosporium, Lomentospora, Pseudallescheria, Paecilomyces* and other molds in tissue are indistinguishable. Infections with any of these molds are characterized by presence of hyaline, septate hyphae in the tissue and are termed hyalohyphomycosis. Hence fungal culture must be done to establish definitive etiological diagnosis.

In contrast, phaeohyphomycosis refers to opportunistic infection by the phaeoid fungi such as *Alternaria alternata* that may be diagnosed in AIDS patients and have been described separately as independent Chapter. The agents of fungal peritonitis in patients undergoing peritoneal dialysis include environmental saprotrophs such as *Bipolaris spicifera* and *Rhodotorula rubra* as well as *Candida* species. The yeasts such as *Hansenula, Rhodotorula* and *Saccharomyces* species, occasionally cause systemic infections.

In patients with severely compromised defenses, almost any fungus may become opportunistic pathogen. There are hundreds of species, which have been documented as agents of human mycoses. Both the number of opportunistic mycoses and the causative fungi continue to increase since last three decades.

The opportunistic fungal infections such as cryptococcosis and mucosal form of candidiasis have emerged as major problems in AIDS patients. The classical infections such as histoplasmosis and dermatophytosis have appeared in altered forms among these individuals. The following opportunistic fungal infections are of immense clinical significance:

1. Trichosporonosis
2. Geotrichosis
3. *Magnusiomyces* Infections
4. *Rhodotorula* Infections
5. *Saccharomyces* Infections

The rest of clinically significant opportunistic infections caused by *Malassezia* species, *Pichia anomala* (*Hansenula anomala*), phaeohyphomycosis, hyalohyphomycosis and talaromycosis, etc. are discussed in their respective Chapters.

1. Trichosporonosis

The increase in severely immunocompromised patients has been accompanied by an upsurge not only in the frequency of opportunistic infections but also in variety of fungal species involved. *Trichosporon* species are being increasingly reported to be emerging pathogenic yeast-like fungi among a wide-spectrum of clinical presentations ranging from superficial involvement in immunocompetent to disseminated invasive disease among immunosuppressed hosts.

The genus *Trichosporon* is inhabitant of soil and occasionally as part of normal flora of human skin. It is saprophytic yeast found in air, soil, water, animal excreta and sewage. It may also exist as a commensal on the scalp and skin hence its association with any disease should be interpreted carefully. They colonize the skin, gastrointestinal tract and respiratory tract of humans and also been detected in feces, sputum, blood and central venous catheters. The fungus is an established cause of superficial infection known as **White Piedra**.

The other such presentations include purpuric papules and nodules with central necrosis or ulceration, indurated erythema and tender lesion mimicking cellulitis. Trichosporonosis is uncommon disease but its incidence is increasing in the immunocompromised patients. It is caused by *Trichosporon beigelii* (formerly *T.capitatum*). This can cause fatal condition in the patients with hematological malignancies, granulocytopenia and those receiving corticosteroids for a prolonged period of time.

In 1992, Gueho, *et al* classified human pathogenic *Trichosporon beigelii* group into six species *(T.ovoides, T.inkin, T.mucoides, T.asteroides, T.asahii* and *T.cutaneum).* Two more species *T.loubieri* and *T.pullulans* were added subsequently. Presently, there are 51 species of *Trichosporon*, 16 of which are able to infect human hosts.

In the literature, name of *Trichosporon beigelii* is still used for the etiological representative of systemic trichosporonosis. However, this name is historically associated with capital white piedra **(*See* Chapter 9)**. It is now recognized that these clinical entities are caused by two separate species. The proper name for predominant species involved in systemic mycosis is *T.asahii* and for agent of capital white piedra is *T.ovoides.*

White piedra is one such superficial mycosis caused by this genus. *Trichosporon* species most commonly associated with white piedra are *T.ovoides* and *T.inkin*. Both species have a strict distribution on the body. *T.ovoides* mostly causes white piedra of the scalp, while *T.inkin* is restricted to the groin.

On the other hand, two species of this genus involved with deep fungal infections are *T.asahii* and *T.mucoides. T.asahii* infections are often present with hematogenous dissemination, while *T.mucoides* has been recovered from patients with central nervous system disease. The other three species, *T.cutaneum*, *T.inkin* and *T.ovoides* are more difficult to distinguish and have been casually grouped as the *T.cutaneum* or *T.beigelii*-complex.

(a) Clinical Features

Trichosporon species can cause both superficial and disseminated infections. The early clinical manifestations of trichosporonosis include fever, pneumonia and skin lesions. The patients may have pulmonary infiltrate also. The dissemination may occur in the other body sites like spleen, liver, kidney, bone marrow, brain and eye. Recently, *Trichosporon* species have been implicated in systemic infections in both normal and immunocompromised host. Such infections have included endocarditis, pneumonia, cellulitis, glomerulonephritis, endophthalmitis, brain abscess, otomycosis, onychomycosis, hypersensitivity pneumonitis and life-threatening systemic dissemination. *Trichosporon* species are also found to be associated with lymphadenopathy in Job's Syndrome.

(b) Laboratory Diagnosis

The diagnosis of *Trichosporon* species is usually based on morphological and biochemical tests, though inconsistency in identification by conventional methods had been reported. The fungus may be demonstrated in the healthy skin, other body sites, wood, air and fomites. The direct examination is done depending on the clinical material received in the laboratory with the help of KOH wet mount and Giemsa staining **(Figs. 28.1A to D)**. The other stainings like H&E, PAS **(Figs. 28.2A and B)** and GMS are also done. The organism may be encountered in the blood cultures also.

On Sabouraud dextrose agar there are yeast-like colonies, which are cream colored and highly wrinkled or radially folded **(Figs. 28.3A to D; *See* Fig. 9.2)**. Microscopically, there are pleomorphic yeast-like cells, 3-8 µm diameter, septate hyaline hyphae with arthrospores and blastospores **(Figs. 28.4A and B; *See* Fig. 9.3)**. The arthrospores can be demonstrated on cornmeal agar also **(Fig. 28.5)**. However, no annelloconidia are seen as found in *Magnusiomyces capitatus*. The diagnosis of *T.beigelii* is slightly difficult because of its close resemblance to *Geotrichum*, *Magnusiomyces* and *Candida* species in both clinical features and histopathological findings and some of the features are given in the **Figure 20.8**. It is being one of the basidiomycete hence **urease** test is **positive**, which is negative in non-basidiomycetous fungi like *Magnusiomyces capitatus* and *Geotrichum* species. Hence *Trichosporon* species are differentiated from *Magnusiomyces* on basis of absence of arthroconidia, urease negative test, presence of annelloconidia and growth positive in cycloheximide containing media at 25°C. *Magnusiomyces* species do not assimilate sugars other than dextrose and galactose while *Trichosporon* are able to do so namely dextrose, lactose, xylose by all but maltose, sucrose, galactose, mellibiose, cellobiose, inositol, raffinose, trehalose and dulcitol with strain variation.

The characteristic assimilation patterns of some of the clinically significant *Trichosporon* species are given in **Table 28.1**. New diagnostic techniques for infections with *Trichosporon* species are needed so that specific antifungal therapy can be instituted at the earliest. Cultures from myelosuppressed patients with invasive fungal infections are often sterile and serological testing helps in early establishment of diagnosis. Serologically, the cell wall antigen of *Trichosporon* cross-reacts with the capsular polysaccharide of *Cryptococcus neoformans* on latex agglutination test, which is done for establishing diagnosis of cryptococcosis. Rather this cross-reactivity may be utilized to screen the patients of trichosporonosis.

Recently *Galleria mellonella* and murine infection models for the study of *Trichosporon* infections have been developed. The utility of these experimental models is demonstrated through the assessment of virulence and antifungal efficacy for clinical isolates of *Trichosporon asahii*, *T.asteroides* and *T.inkin*.

(c) Treatment and Prophylaxis

Amphotericin B, ketoconazole and miconazole are the drugs that are commonly used in the patients of invasive trichosporonosis but the patients invariably do not survive. Flucytosine in combination with amphotericin B has also been tried. *Trichosporon* isolates are proved to be almost resistant to flucytosine. Voriconazole may offer a clinical solution in trichosporonosis when the other drugs fail. Caspofungin is not found effective for its treatment, however, when combined with amphotericin B or azoles, the results are stated to be satisfactory. Some of the emerging species of this genus like *T.asahii* are invariably found resistant to amphotericin B hence very difficult to treat.

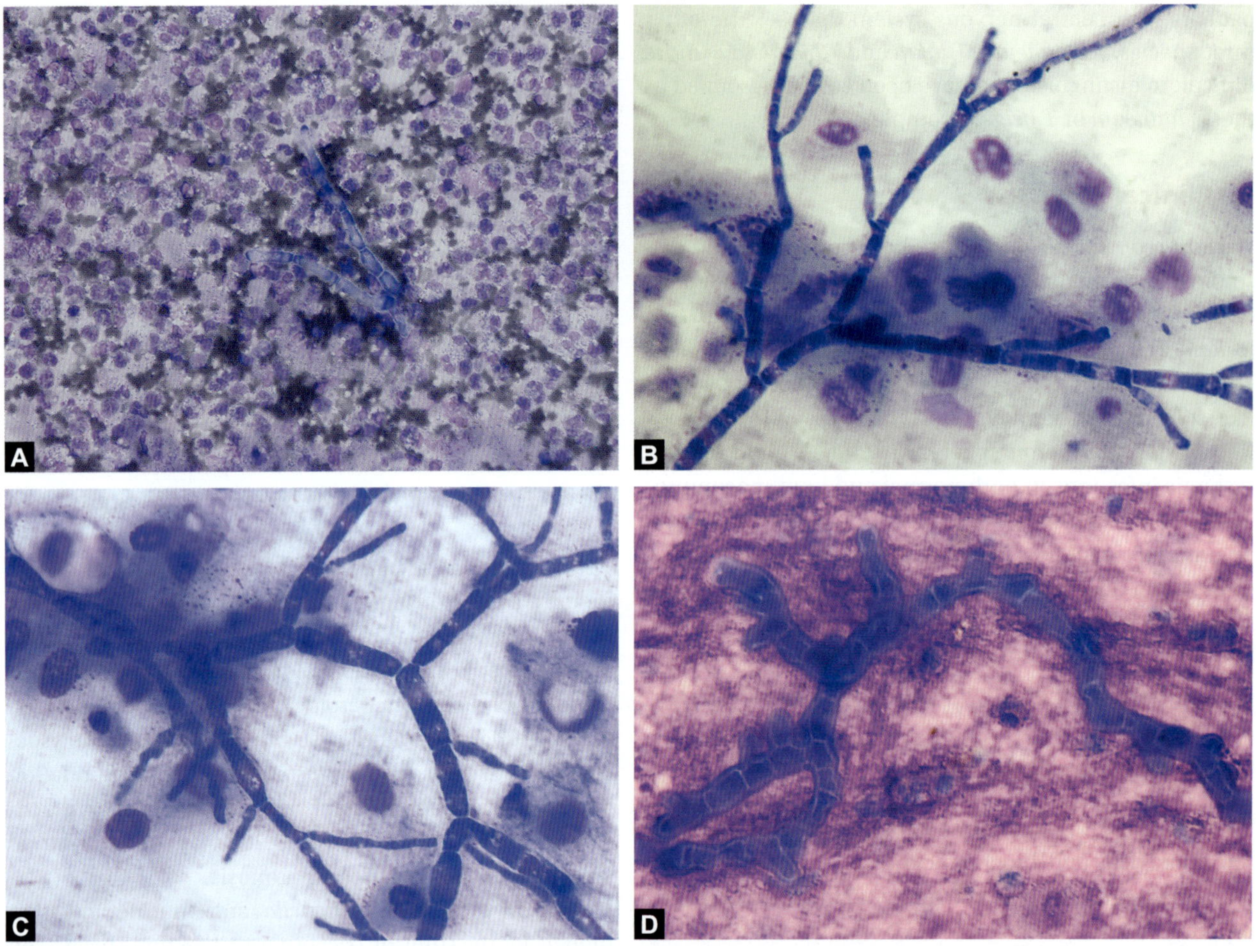

Figs. 28.1A to D. Giemsa stain showing yeast-like cells and pseudohyphae of *Trichosporon* species in smears (MGG × 200, 400, 1000).

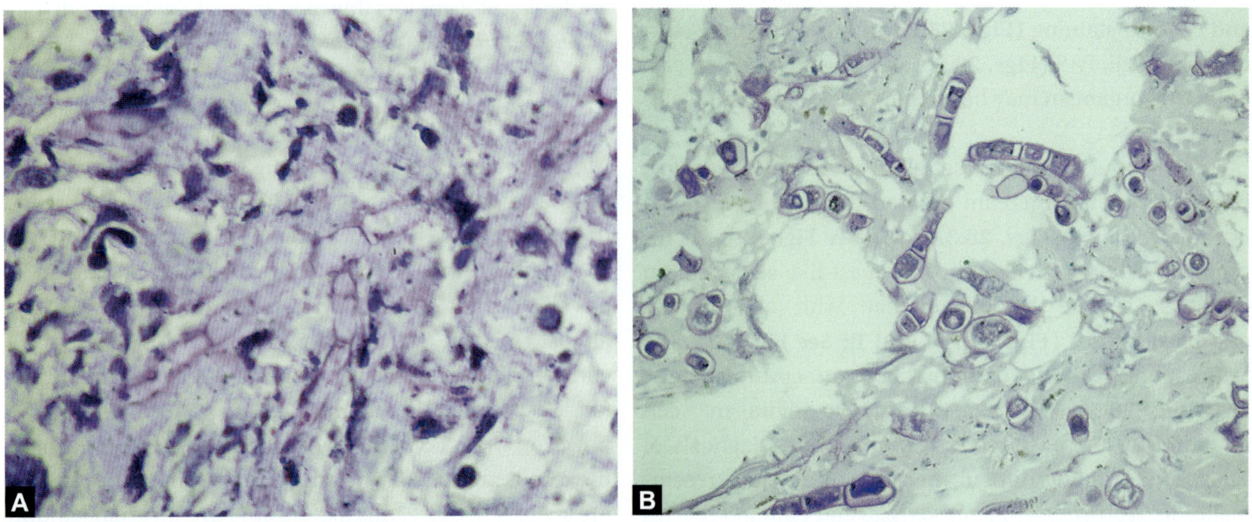

Figs. 28.2A and B. Histopathological sections showing yeast-like cells and pseudohyphae of *Trichosporon* species (PAS × 400).

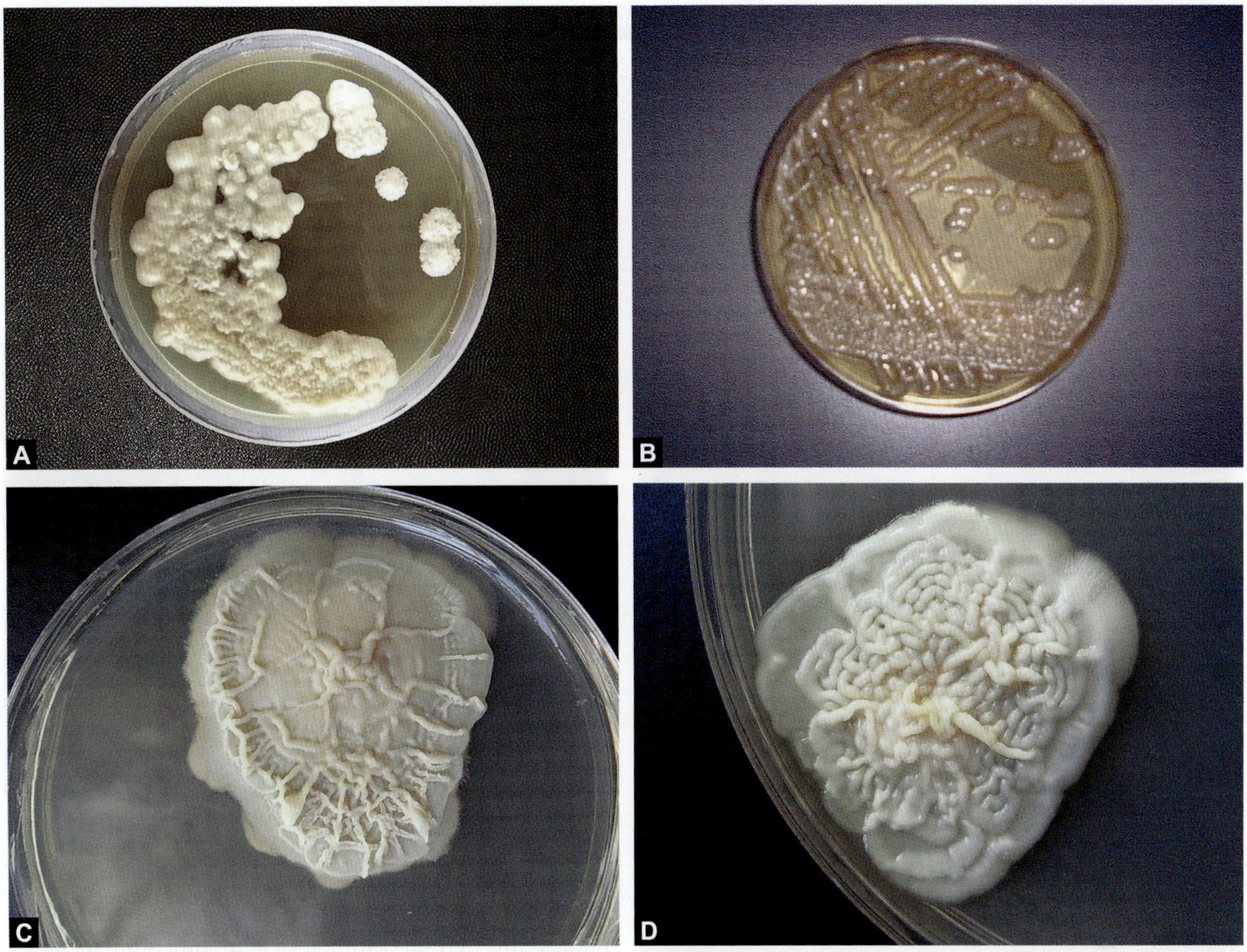

Figs. 28.3A to D. Yeast-like cream-colored, wrinkled and folded colonies of *T.mucoides* (A & B) and other *Trichosporon* species (C & D).

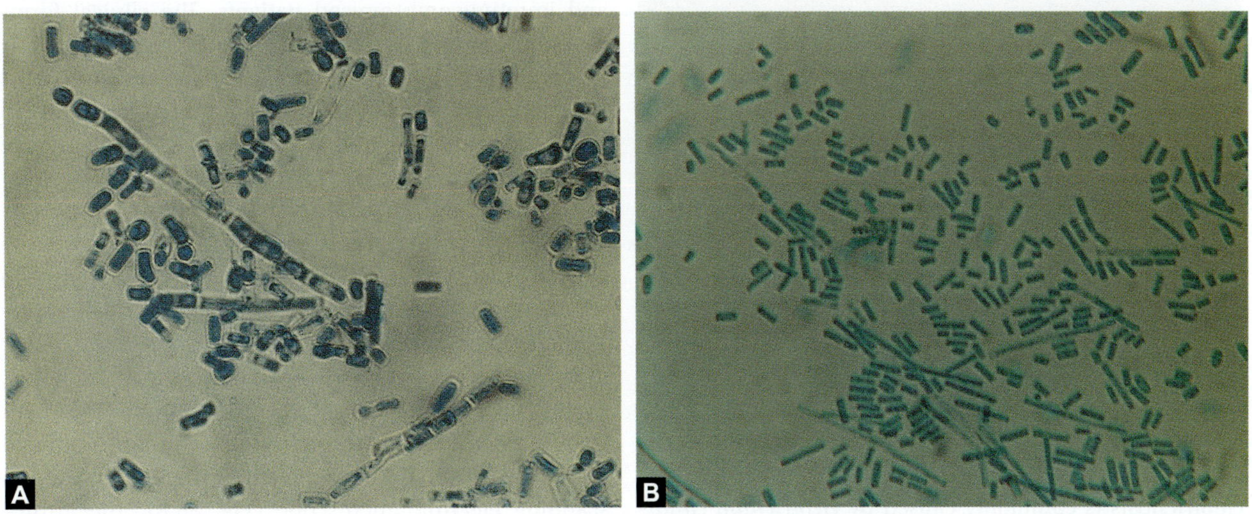

Figs. 28.4A and B. Hyphae and rectangular arthrospores of *Trichosporon* species (LCB x 400).

Table 28.1. Assimilation Patterns of Clinically Significant *Trichosporon* Species.

Trichosporon Species	*Growth at 37°C*	*L-rham*	*Mel*	*Raff*	*Rib*	*Xyl*	*L-arab*	*Gala*	*Sorb*	*Ino*
Trichosporon asahii	+	+	–	–	*v*	*v*	+	–	–	–
Trichosporon cutaneum	–	+	+	+	+	+	+	–	+	+
Trichosporon inkin	+	–	–	–	–	–	–	–	–	+
Trichosporon mucoides	+	+	+	+	+	+	+	+	+	+
Trichosporon ovoides	+	+	–	*v*	–	*v*	–	–	–	–

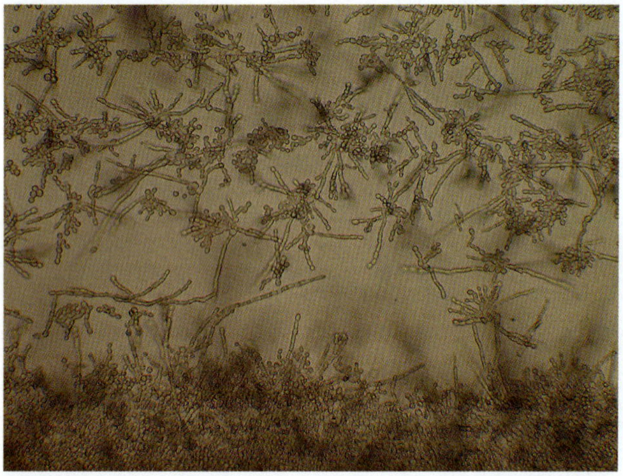

Fig. 28.5. Hyphae and rectangular arthrospores of *Trichosporon inkin* (CMA x 100).

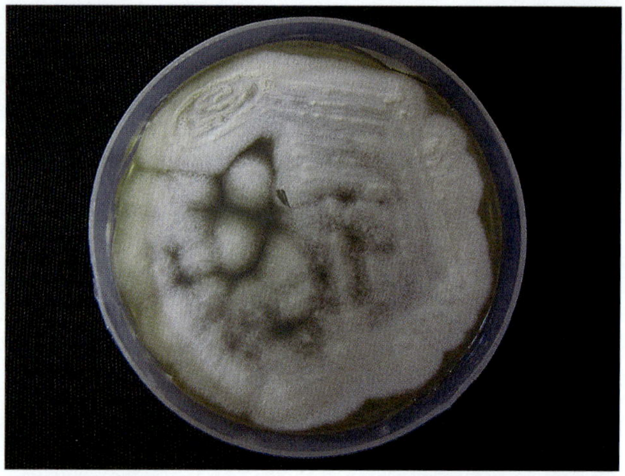

Fig. 28.6. Dry, white to cream-colored colonies of *Geotrichum candidum* on SDA.

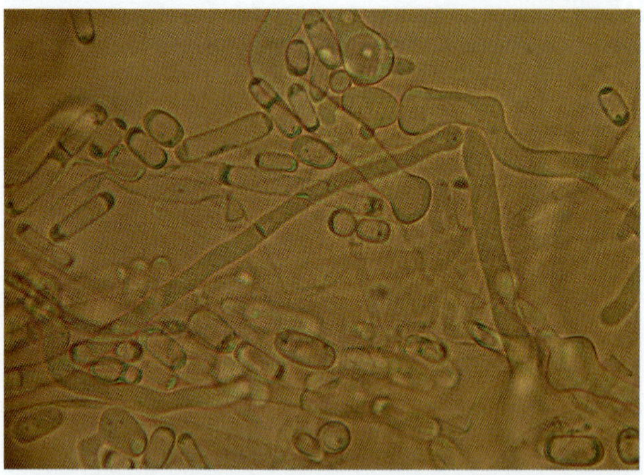

Fig. 28.7. Yeast and hyphae of *Geotrichum candidum* (LCB x 400).

2. Geotrichosis

Geotrichosis is commonly caused by *Geotrichum candidum*, the fungus found prevalent as saprotrophic in nature. This may be commensal in mouth, bronchi, lungs, gastrointestinal and genitourinary tract. It is usually considered as contaminant, except in extremely debilitated and immunocompromised patients. The clinical forms of the infections are pulmonary geotrichosis that mimic tuberculosis in its clinical manifestations and superficial infections of the skin, oral mucosa and gastrointestinal tract. The tissue reaction is non-specific showing acute inflammatory process.

The laboratory diagnosis of geotrichosis is made by cultural identification and biochemical reactions of the clinical isolates. At 25°C *Geotrichum* grows as dry, white to cream, smooth or hairy colonies and at 37°C growth is very slow and subsurface however mature colonies may appear more mold-like **(Fig. 28.6)**. On microscopic examination, hyaline hyphae are seen, which fragment into rectangular arthroconidia, measuring 4-8 µm and have rounded ends **(Fig. 28.7)**. The morphology is more delineated in scanning electron microscope **(Fig. 28.8)**. The arthroconidia may produce hyphal extension from one corner,

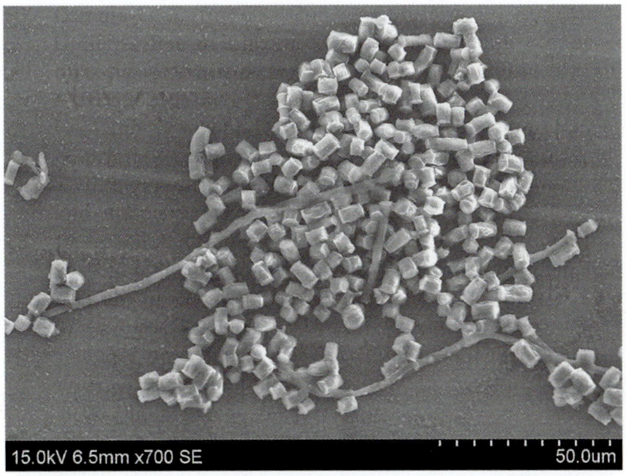

Fig. 28.8. Hyphae of *Geotrichum candidum* under scanning electron microscope.

producing *"hockey-stick"* appearance. There are no blastoconidia produced and **urease** test is **negative** in contrast to the *Trichosporon* species. The comparative features of other fungi are given in **Table 28.2**. *Geotrichum candidum* can only assimilate glucose, galactose, lactose and xylose but not saccharose and maltose. The assimilation pattern is same as seen in *B.capitatus* except xylose utilization. The treatment includes aerosolized nystatin, iodides, gentian violet and amphotericin B.

3. *Magnusiomyces* Infections

Magnusiomyces capitatus previously used to be known with various names such as ***Blastoschizomyces capitatus*** or *Trichosporon capitatum* or *Geotrichum capitatum.* Its anamorph state is *Saprochaete capitata* and is also a rare cause of disseminated infection. The teleomorph state of this fungus is *Dipodascus capitatus,* a heterothallic ascomycete that produces asci usually with two ascospores.

This can also be isolated from the environment and may be microflora of the skin and mucosa of digestive or respiratory tracts. It is one of the rare causes of invasive fungal diseases in immunocompromised hosts. It is seen in neutropenic patients, leukemia and bone marrow transplant (BMT) recipients for aplastic anemia and also endocarditis. It is an opportunistic mycotic pathogen and can cause an infection, especially in neutropenic hemato-oncology patients in Italy, Spain and France of Europe, the Mediterranean region. However, the cases have been observed from other countries also which are not typical for this infection including Nepal, Kuwait and Switzerland.

Table 28.2. Comparative Features of *Trichosporon*, *Magnusiomyces capitatus* and *Geotrichum* species.

Species	*Blastoconidia*	*Annelloconidia*	*Urease*
Trichosporon Species	+	–	+
M.capitatus Species	+	+	–
Geotrichum Species	–	–	–

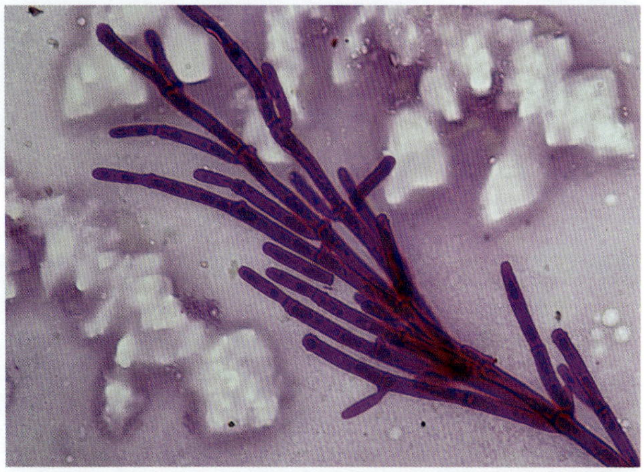

Fig. 28.9. Hyphae and pseudohyphae of *Magnusiomyces capitatus* (PAS x 400).

It is frequently isolated from sputum and is widely distributed in nature and considered to be minor part of the normal skin flora. It has also been reported recently as cause of onychomycosis also. The disseminated type of disease is usually associated with immunosuppressive conditions.

In the direct microscopic examination in KOH wet mount, long hyphae, pseudohyphae of *Magnusiomyces capitatus* can be seen and may be confused with *Trichosporon* or *Geotrichum* species. In the special stainings like PAS **(Fig. 28.9)** or GMS, the same findings are seen. The colonies of *Magnusiomyces capitatus* on Sabouraud dextrose agar are glabrous with radiating edges, white to cream-colored **(Fig. 28.10)**. Microscopically the isolates produce true hyphae, pseudohyphae and **annelloconidia** resembling arthroconidia **(Figs. 28.11A and B)**. This can be distinguished from other yeast-like species such as *Geotrichum candidum* and *Trichosporon* species on the basis of its formation of annelloconidia at the tip of proliferating conidiogenous cell **(Fig. 28.12)**, its inability to utilize urea, its resistance to cycloheximide and growth at higher temperature of 45°C.

Magnusiomyces capitatus assimilates only glucose and galactose, in contrast to *Trichosporon* species, which

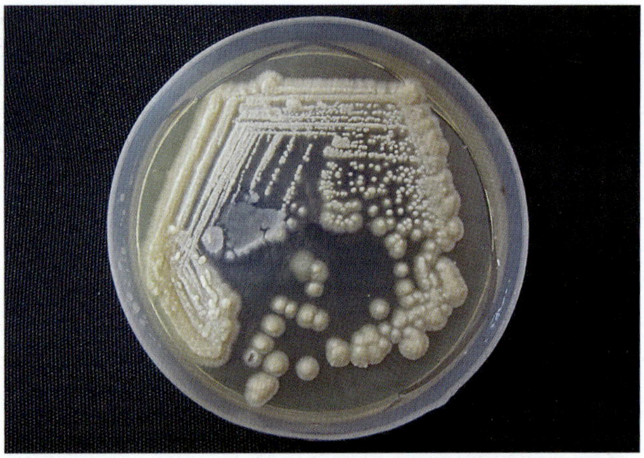

Fig. 28.10. Yeast-like colonies of *Magnusiomyces capitatus* on SDA.

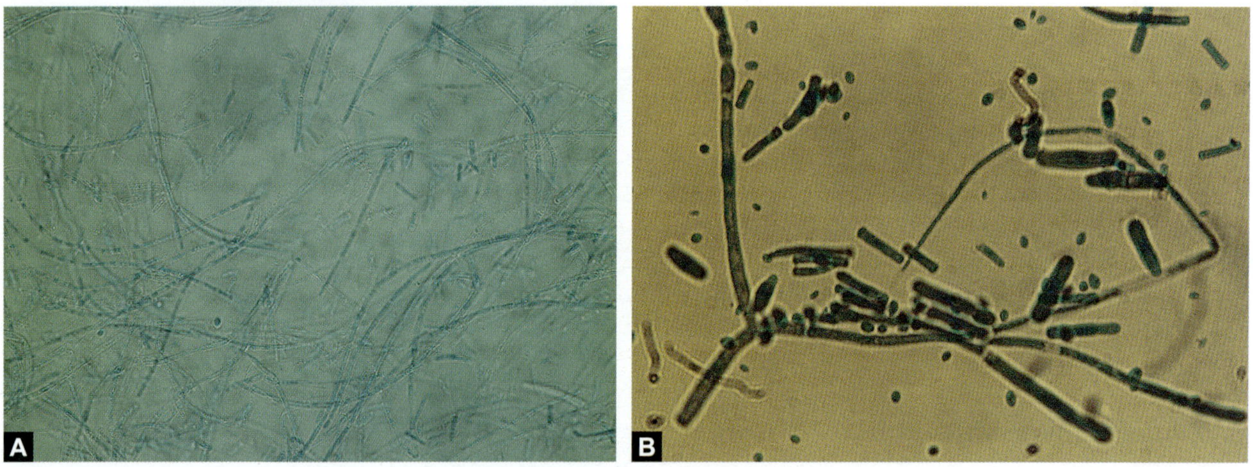

Figs. 28.11A and B. Hyphae and annelloconidia of *Magnusiomyces capitatus* (LCB x 200 / 400).

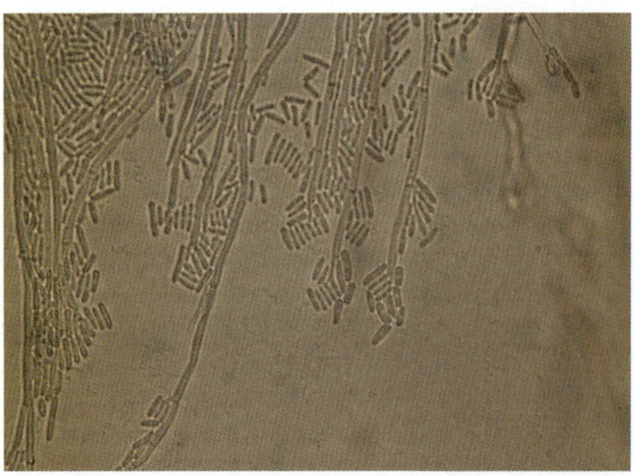

Fig. 28.12. Hyphae and annelloconidia of *Magnusiomyces capitatus* (CMA × 400).

assimilates most carbohydrates, providing easy way to identify and differentiation between these two fungal genera. This is reported to be susceptible *in vitro* to wide range of antifungal drugs like flucytosine, amphotericin B and various azole derivatives but results *in vivo* are not very successful.

4. *Rhodotorula* Infections

This species although a rare cause of infections in humans but cases of systemic infections are being increasingly observed. *Rhodotorula* belongs to family Sporidiobolaceae in phylum Basidiomycota. *R.mucilaginosa*, *R.glutinis* are clinically significant species. It has affinity for foreign material hence commonly found among patients with VP shunts, etc.

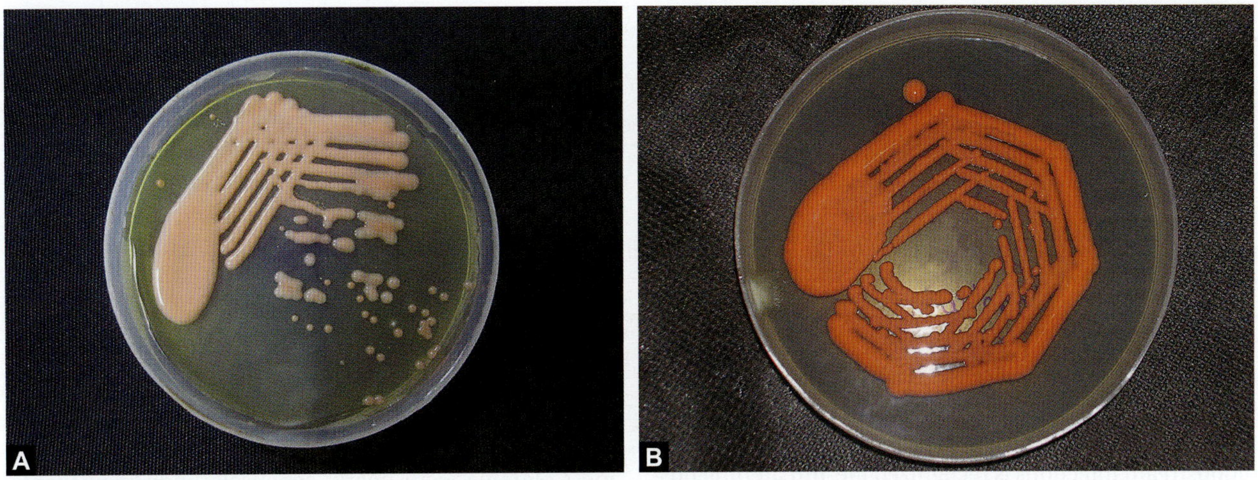

Figs. 28.13A and B. Mucoid, yeast-like colonies of *Rhodotorula* species seen on Sabouraud's dextrose agar.

This organism is commonly encountered in culture from vaginal smears, abscesses, feces, etc. The colonies are glistening, creamy-mucoid, showing various shades of pink, red or orange color, which is its pathognomonic feature **(Figs. 28.13A and B)**. Microscopically, organism is seen as budding cells similar to *Cryptococcus* species except for the presence of carotinoid pigment.

Rhodotorula species are separated from other genera of Cryptococcaceae by their lack of inositol fermentation, urease and pigment production. There is clear distinction between two basidiomycetous fungi, in one pigment and in other capsule, still it is usually confused with *Cryptococcus* species hence identification should be made carefully. Ascospores are not produced and it cannot assimilate inositol. This basidiomycete is one of the potentially significant emerging yeasts and is associated with number of hematogenous infections, fungemia, endocarditis, meningitis and peritonitis. Some of the species are involved in keratitis and endophthalmitis and are discussed in **Chapter 29**.

Exocellular mannan from *Rhodotorula glutinis* has been used as immunoreactive antigen for detection of anti-*Leptospira* antibodies in the diagnosis of leptospirosis. The treatment is similar to as given in other systemic fungal infections.

5. *Saccharomyces* Infections

The commonest species of this genus is *Saccharomyces cerevisiae*. This fungus is also known by its more common names like **Baker's** or **Brewer's yeast**. It is commercially used in fermentation of beer, wine and mash from that distilled liquor is made. It is usually grown on hops or another substrate similar to the plant utilized in the beer-making industry, after which it is harvested and killed. The final product is generally half composed of protein, as well as large amount of vitamin B and minerals and depending on the technology, diverse number of other healthy compounds. Typically, Brewer's yeast is used as protein supplement, energy booster, immune enhancer or other vehicle where some compounds can be inserted to create commercialized health product. It is found prevailing in nature on ripening fruits, in soil and sewage and is frequently isolated as contaminants from skin. The same species of this fungus is used in both processes *i.e.* baking and fermentation, however, strains may be different. The bread making strain is genetically selected to produce more CO_2 and much less alcohol while the opposite is true of the alcohol-making strains. Another species of this genus *i.e. Saccharomyces boulardii* has recently been used as probiotic in common use.

The role of *Saccharomyces cerevisiae* as clinically significant pathogen has been unclear. It has rarely been reported to cause thrush, vulvovaginitis and fungemia. It can colonize in human respiratory, gastrointestinal and urinary tracts in setting of chronic underlying diseases. It is also considered that fungemia may be attributed to usage of *S.boulardii* as probiotics. Microscopically, the budding yeast cells are round to oval and occasionally short rudimentary pseudohyphae may be formed. This is multiplying asexually, however, when sexual reproduction takes place, there are four haploid spores seen **(*See* Fig. 36.15)**.

The treatment of this infection is similar to that is given in other systemic fungal infections.

B. HEALTHCARE-ASSOCIATED MYCOSES

The healthcare-associated infections are defined as the infections those are neither present nor are in the incubation period when patient is admitted to the hospital *i.e.* purely acquired in the hospital setting. These are also called as nosocomial infections and have been recognized as significant ground for adverse outcome of patient and major public health problem. Nosocomial bloodstream infections are particularly serious and there is evidence to suggest that gradually these are becoming more and more common. Nosocomial fungemia, subset of all nosocomial bloodstream infections is recognized life-threatening hospital infection.

Healthcare-associated fungal infections have gained more importance especially in association with or as consequence of the extraordinary progress in the management of seriously ill patients during the past three decades. However, despite this increase, there has been comparatively little progress in elucidating the pathogenesis of nosocomial fungal infections or in their prevention, diagnosis and treatment.

Numerous systemic, opportunistic and dermatophytic fungal pathogens have been reported to cause infection in the hospitalized patients. There is need for the classification and definition of terms in order to better understand these infections. These have been divided into the following types:

Type I (Hospital-acquired)*:* A superficial, locally invasive or disseminated mycosis acquired from the hospital environment.

Type II (Hospital-associated)*:* These infections are developing during the course of patients' hospitalization from fungus previously colonizing or latently infecting patient. This type of fungal infections are further classified as infection developing from endogenous colonizing fungi *(Subtype IIA)* or infections developing from reactivation of latently infecting fungi *(Subtype IIB)*.

The most commonly encountered infections are caused by *Candida* species, *Pichia anomala* and *Aspergillus* species and these are invariably resistant to most commonly used antifungal agents.

C. EMERGING FUNGAL INFECTIONS

With the continuous expanding spectrum of fungi causing invasive disease, it is important to consider the clinical relevance of every fungus cultured from samples of immunocompromised patients. Coordination between the microbiologist and clinician is therefore indispensable. The case reports illustrate the difficulty in identifying the etiological agents of an invasive fungal infection. According to generally accepted consensus criteria both *H.aspergillata* and *Aspergillus* species could be considered as probable causes of fungal pneumonia. However, this uncertainty may have significant implications for choosing the most appropriate antifungal therapy.

"The use of whole-genome sequence typing (WGST) to investigate fungal outbreaks has become integral to epidemiologic investigations," Dr. Engelthaler said. "Our WGST analysis demonstrated that the patient isolates from Chile and Colombia were nearly genetically indistinguishable from those recovered from the unopened medication vials, indicating the likely presence of a single-source infection."

D. CHRONIC GRANULOMATOUS DISEASE

The invasive fungal infections are a major threat for chronic granulomatous disease (CGD) patients. Now, various studies provide comprehensive overview of published invasive fungal infections in the CGD host through an extensive review of epidemiological, clinical, diagnostic and therapeutic data. In addition to the often mild clinical presentation, the currently used diagnostics for invasive aspergillosis have low sensitivity in CGD patients and cannot be easily translated to the non-neutropenic host.

Aspergillus fumigatus and *A.nidulans* are the most commonly isolated species. *A.nidulans* infections are seldom reported in other immunocompromised patients, indicating a unique interaction between this fungus and the CGD host. The occurrence of mucormycosis is mainly noted in the setting of treatment of inflammatory complications with immunosuppressive drugs. *Candida* infections are infrequently seen and do not cause mucocutaneous disease but do show an age-dependent clinical presentation. The CGD patient is susceptible to a wide range of fungal pathogens, indicating the need to determine the causative fungus, often by invasive diagnostics, to guide optimal and rational treatment. The current understanding of invasive fungal infections in patients with CGD serves as a starting point to guide optimal treatment strategies and to direct further research aimed at improving outcomes.

Treatment and Prophylaxis

The miscellaneous opportunistic mycoses may pose life-threatening condition hence specific diagnosis is to be established well in time thereby to avoid delay in institution of proper antifungal therapy. The first and foremost point is rapid correction of underlying predisposing factor of the host.

Further Reading

1. Agarwal R, Kalita J, Marak RS, *et al.* Spectrum of fungal infection in a neurology tertiary care center in India. *Neurol Sci.* 2012; **33:** 1305-10.
2. Agarwal V, Joyce M. Microbiologically documented fungal infections in pediatric patients with acute myeloid leukemia: 12.5-year experience at a single institution. *J Pediatr Hematol Oncol.* 2014; **36:** e528-32.
3. Agrawal S, Jones B, Barnes R, *et al.* A practical critique of antifungal treatment guidelines for haemato-oncologists. *Crit Rev Microbiol.* 2012; **38:** 203-16.
4. Akan H, Antia VP, Kouba M, *et al.* Preventing invasive fungal disease in patients with haematological malignancies and the recipients of haematopoietic stem cell transplantation: Practical aspects. *J Antimicrob Chemother.* 2013; **68 (Suppl. 3):** iii5-16.
5. Alangaden GJ. Nosocomial fungal infections: Epidemiology, infection control and prevention. *Infect Dis Clin North Am.* 2011; **25:** 201-25.
6. Allou N, Allyn J, Montravers P. When and how to cover for fungal infections in patients with severe sepsis and septic shock. *Curr Infect Dis Rep.* 2011; **13:** 426-32.
7. Almeida-Silva F, Damasceno LS, Serna MJ, *et al.* Multiple opportunistic fungal infections in an individual with severe HIV disease: A case report. *Rev Iberoam Micol.* 2016. **33:** 118-21.
8. Anesi JA, Baddley JW. Approach to the solid organ transplant patient with suspected fungal infection. *Infect Dis Clin North Am.* 2016; **30:** 277-96.
9. Angarone M. Fungal infections in cancer patients. *Cancer Treat Res.* 2014; **161:** 129-55.
10. Ansari Sh, Shirzadi E, Elahi M. The prevalence of fungal infections in children with hematologic malignancy in Ali-Asghar Children Hospital between 2005 and 2010. *Iran J Ped Hematol Oncol.* 2015; **5:** 1-10.
11. Arendrup MC, Boekhout T, Akova M, *et al.* ESCMID and ECMM joint clinical guidelines for the diagnosis and management of rare invasive yeast infections. *Clin Microbiol Infect.* 2014; **20 (Suppl. 3):** 76-98.
12. Arendrup MC, Fisher BT, Zaoutis TE. Invasive fungal infections in the paediatric and neonatal population: Diagnostics and management issues. *Clin Microbiol Infect.* 2009; **15:** 613-24.
13. Aridogan IA, Ilkit M, Izol V, *et al.* Glans penis and prepuce colonisation of yeast fungi in a paediatric population: Pre- and post-circumcision results. *Mycoses.* 2009; **52:** 49-52.
14. Ariza-Heredia EJ, Kontoyiannis DP. Our recommendations for avoiding exposure to fungi outside the hospital for patients with haematological cancers. *Mycoses.* 2014; **57:** 336-41.
15. Asimakopoulos P, Supriya M, Kealey S, *et al.* A case-based discussion on a patient with non-otogenic fungal skull base osteomyelitis: Pitfalls in diagnosis. *J Laryngol Otol.* 2013; **127:** 817-21.
16. Atalla A, Garnica M, Maiolino A, *et al.* Risk factors for invasive mold diseases in allogeneic hematopoietic cell transplant recipients. *Transpl Infect Dis.* 2015; **17:** 7-13.
17. Aysel K, Alev Y, Nur C, *et al.* Acute renal failure caused by fungus balls in renal pelvises. *Pediatr Int.* 2009; **51:** 836-8.
18. Azoulay E, Dupont H, Tabah A, *et al.* Systemic antifungal therapy in critically ill patients without invasive fungal infection. *Crit Care Med.* 2012; **40:** 813-22.
19. Baddley JW. Pinpointing prognostic indicators of fungal infections in transplant patients. *Future Microbiol.* 2009; **4:** 77-84.
20. Badiee P, Alborzi A, Joukar M. Molecular assay to detect nosocomial fungal infections in intensive care units. *Eur J Intern Med.* 2011; **22:** 611-5.
21. Badiee P, Alborzi A. Invasive fungal infections in renal transplant recipients. *Exp Clin Transplant.* 2011; **9:** 355-62.
22. Badiee P, Hashemizadeh Z. Opportunistic invasive fungal infections: Diagnosis & clinical management. *Indian J Med Res.* 2014; **139:** 195-204.
23. Badiee P, Kordbacheh P, Alborzi A, *et al.* Study on invasive fungal infections in immunocompromised patients to present a suitable early diagnostic procedure. *Int J Infect Dis.* 2009; **13:** 97-102.
24. Bajwa S, Kulshrestha A. Fungal infections in intensive care unit: Challenges in diagnosis and management. *Ann Med Health Sci Res.* 2013; **3:** 238-44.
25. Ballot DE, Bosman N, Nana T, *et al.* Background changing patterns of neonatal fungal sepsis in a developing country. *J Trop Pediatr.* 2013; **59:** 460-4.
26. Bandh SA, Kamili AN, Ganai BA, *et al.* Opportunistic fungi in lake water and fungal infections in associated human population in Dal Lake, Kashmir. *Microb Pathog.* 2016; **93:** 105-10.
27. Baradkar VP, Kumar S. Meningitis caused by *Rhodotorula mucilaginosa* in human immunodeficiency virus seropositive patient. *Ann Indian Acad Neurol.* 2008; **11:** 245-7.
28. Barnes RA, Stocking K, Bowden S, *et al.* Prevention and diagnosis of invasive fungal disease in high-risk patients within an integrative care pathway. *J Infect.* 2013; **67:** 206-14.
29. Basiri K, Meidani M, Rezaie F, *et al.* A rare case of *Trichosporon* brain abscess, successfully treated with surgical excision and antifungal agents. *Neurol Neurochir Pol.* 2012; **46:** 92-5.
30. Bassetti M, Bouza E. Invasive mould infections in the ICU setting: Complexities and solutions. *J Antimicrob Chemother.* 2017; **72 (Suppl. 1):** i39-47.
31. Bassetti M, Garnacho-Montero J, Calandra T, *et al.* Intensive care medicine research agenda on invasive fungal infection in critically ill patients. *Intensive Care Med.* 2017; doi: 10.1007/s00134-017-4731-2. PMID: 28255613.
32. Basu S, Tilak R, Kumar A. Multidrug-resistant *Trichosporon*: An unusual fungal sepsis in preterm neonates. *Pathog Glob Health.* 2015; **109:** 202-6.

33. Ben-Ami R, Halaburda K, Klyasova G, *et al.* A multidisciplinary team approach to the management of patients with suspected or diagnosed invasive fungal disease. *J Antimicrob Chemother.* 2013; **68 (Suppl. 3):** iii25-33.
34. Ben-Ami R, Lewis RE, Kontoyiannis DP. Invasive mould infections in the setting of hematopoietic cell transplantation: Current trends and new challenges. *Curr Opin Infect Dis.* 2009; **22:** 376-84.
35. Benedict K, Chiller TM, Mody RK. Invasive fungal infections acquired from contaminated food or nutritional supplements: A review of the literature. *Foodborne Pathog Dis.* 2016; **13:** 343-9.
36. Benedict K, Park BJ. Invasive fungal infections after natural disasters. *Emerg Infect Dis.* 2014; **20:** 349-55.
37. Bentubo HD, Mantovani A, Yamashita JT, *et al.* Yeasts of the genital region of patients attending the dermatology service at Hospital Sao Paulo, Brazil. *Rev Iberoam Micol.* 2015; **32:** 229-34.
38. Bhari N, Xess I, Pandey M, *et al.* Primary cutaneous trichosporonosis responsive to voriconazole. *JAMA Dermatol.* 2015; **151:** 1139-41.
39. Bhaskaran A, Hosseini-Moghaddam SM, Rotstein C, *et al.* Mold infections in lung transplant recipients. *Semin Respir Crit Care Med.* 2013; **34:** 371-9.
40. Bhowmik D, Dinda AK, Xess I, *et al.* Fungal panniculitis in renal transplant recipients. *Transpl Infect Dis.* 2008; **10:** 286-9.
41. Birrenbach T, Bertschy S, Aebersold F, *et al.* Emergence of *Blastoschizomyces capitatus* yeast infections, Central Europe. *Emerg Infect Dis.* 2012; **18:** 98-101.
42. Biswas D, Agarwal S, Sindhwani G, *et al.* Fungal colonization in patients with chronic respiratory diseases from Himalayan region of India. *Ann Clin Microbiol Antimicrob.* 2010; **9:** 28.
43. Bitar D, Lortholary O, Le Strat Y, *et al.* Population-based analysis of invasive fungal infections, France, 2001-2010. *Emerg Infect Dis.* 2014; **20:** 1149-55.
44. Blot S, Charles PE. Fungal sepsis in the ICU: Are we doing better? Trends in incidence, diagnosis and outcome. *Minerva Anestesiol.* 2013; **79:** 1396-405.
45. Blyth CC, Gilroy NM, Guy SD, *et al.* Consensus guidelines for the treatment of invasive mould infections in haematological malignancy and haemopoietic stem cell transplantation, 2014. *Intern Med J.* 2014; **44:** 1333-49.
46. Blyth DM, Chung KK, Cancio LC, *et al.* Clinical utility of fungal screening assays in adults with severe burns. *Burns.* 2013; **39:** 413-9.
47. Bodro M, Sabe N, Gomila A, *et al.* Risk factors, clinical characteristics and outcomes of invasive fungal infections in solid organ transplant recipients. *Transplant Proc.* 2012; **44:** 2682-5.
48. Bonifaz A, Vazquez-Gonzalez D, Macias B, *et al.* Oral geotrichosis: Report of 12 cases. *J Oral Sci.* 2010; **52:** 477-83.
49. Bonnal C, Leleu C, Brugiere O, *et al.* Relationship between fungal colonisation of the respiratory tract in lung transplant recipients and fungal contamination of the hospital environment. *PLoS One.* 2015; **10:** e0144044.
50. Bow EJ. Invasive fungal infection in haematopoietic stem cell transplant recipients: epidemiology from the transplant physician's viewpoint. *Mycopathologia.* 2009; **168:** 283-97.
51. Brissaud O, Guichoux J, Harambat J, *et al.* Invasive fungal disease in PICU: Epidemiology and risk factors. *Ann Intensive Care.* 2012; **2:** 6.
52. Brown GD, Meintjes G, Kolls JK, *et al.* AIDS-related mycoses: The way forward. *Trends Microbiol.* 2014; **22:** 107-9.
53. Bryant PA, Baddley JW. Opportunistic infections in biological therapy, risk and prevention. *Rheum Dis Clin North Am.* 2017; **43:** 27-41.
54. Caira M, Trecarichi EM, Mancinelli M, *et al.* Uncommon mold infections in hematological patients: epidemiology, diagnosis and treatment. *Expert Rev Anti Infect Ther.* 2011; **9:** 881-92.
55. Calley JL, Warris A. Recognition and diagnosis of invasive fungal infections in neonates. *J Infect.* 2017; **74:** S108-13.
56. Capoor MR, Aggarwal S, Raghvan C, *et al.* Clinical and microbiological characteristics of *Rhodotorula mucilaginosa* infections in a tertiary-care facility. *Indian J Med Microbiol.* 2014; **32:** 304-9.
57. Capoor MR, Gupta DK, Verma PK, *et al.* Rare yeasts causing fungemia in immunocompromised and haematology patients: Case series from Delhi. *Indian J Med Microbiol.* 2015; **33:** 576-9.
58. Capoor MR, Gupta S, Sarabahi S, *et al.* Epidemiological and clinico-mycological profile of fungal wound infection from largest burn centre in Asia. *Mycoses.* 2012; **55:** 181-8.
59. Capoor MR, Sarabahi S, Tiwari VK, *et al.* Fungal infections in burns: Diagnosis and management. *Indian J Plast Surg.* 2010; **43:** S37-42.
60. Caroti L, Zanazzi M, Rogasi P, *et al.* Subcutaneous nodules and infectious complications in renal allograft recipients. *Transplant Proc.* 2010; **42:** 1146-7.
61. Castagnola E, Faraci M, Fioredda F, *et al.* Invasive mould infections in newborns and children. *Early Hum Dev.* 2011; **87:** S67-9.
62. Caston-Osorio JJ, Rivero A, Torre-Cisneros J. Epidemiology of invasive fungal infection. *Int J Antimicrob Agents.* 2008; **32:** S103-9.
63. Chagas-Neto TC, Chaves GM, Melo AS, *et al.* Bloodstream infections due to *Trichosporon* spp.: Species distribution, *Trichosporon asahii* genotypes determined on the basis of ribosomal DNA intergenic spacer 1 sequencing and antifungal susceptibility testing. *J Clin Microbiol.* 2009; **47:** 1074-81.
64. Chai LY, Naesens R, Khoo AL, *et al.* Invasive fungal infection in an elderly patient with defective inflammatory macrophage function. *Clin Microbiol Infect.* 2011; **17:** 1546-9.
65. Chaitanya V, Lakshmi BS, Kumar AC, *et al.* Disseminated *Trichosporon* infection in a renal transplant recipient. *Transpl Infect Dis.* 2015; **17:** 605-9.
66. Chakrabarti A, Chatterjee SS, Shivaprakash MR. Overview of opportunistic fungal infections in India. *Jpn J Med Mycol.* 2008; **49:** 165-72.

67. Chakrabarti A, Singh R. The emerging epidemiology of mould infections in developing countries. *Curr Opin Infect Dis.* 2011; **24:** 521-6.
68. Chan GF, Sinniah S, Idris TI, Puad MS, Abd Rahman AZ. Multiple rare opportunistic and pathogenic fungi in persistent foot skin infection. *Pak J Biol Sci.* 2013; **16:** 208-18.
69. Chandran SK, Lyons KM, Divi V, *et al.* Fungal laryngitis. *Ear Nose Throat J.* 2009; **88:**1026-7.
70. Chang CC, Ananda-Rajah M, Belcastro A, *et al.* Consensus guidelines for implementation of quality processes to prevent invasive fungal disease and enhanced surveillance measures during hospital building works, 2014. *Intern Med J.* 2014; **44:** 1389-97.
71. Chaudhary SC, Sawlani KK, Arora R, *et al.* Native aortic valve fungal endocarditis. *BMJ Case Rep.* 2013; pii: bcr2012007144. PMID: 23429013.
72. Chauhan S, Tiwari S, Singh P, *et al.* Pulmonary infection by *Blastoschizomyces capitatus* in an old tuberculosis patient. *J Clin Diagn Res.* 2015; **9:** DD01-2.
73. Chayakulkeeree M, Vongwiwatana A. Invasive mold infection in kidney transplant recipients: Observation of early-onset mucormycosis. *Transplant Proc.* 2014; **46:** 595-7.
74. Chellapandian D, Lehrnbecher T, Phillips B, *et al.* Bronchoalveolar lavage and lung biopsy in patients with cancer and hematopoietic stem-cell transplantation recipients: a systematic review and meta-analysis. *J Clin Oncol.* 2015; **33:** 501-9.
75. Chen K, Wang Q, Pleasants RA, *et al.* Empiric treatment against invasive fungal diseases in febrile neutropenic patients: A systematic review and network meta-analysis. *BMC Infect Dis.* 2017; **17:** 159. PMID: 28219330.
76. Chen SC, Blyth CC, Sorrell TC, *et al.* Pneumonia and lung infections due to emerging and unusual fungal pathogens. *Semin Respir Crit Care Med.* 2011; **32:** 703-16.
77. Chen SC, Sorrell TC, Chang CC, *et al.* Consensus guidelines for the treatment of yeast infections in the haematology, oncology and intensive care setting, 2014. *Intern Med J.* 2014; **44:** 1315-32.
78. Chen YT, Yang WC, Chen TW, *et al. Trichosporon mucoides* peritonitis in a continuous ambulatory peritoneal dialysis patient. *Perit Dial Int.* 2013; **33:** 341-2.
79. Chitasombat MN, Kofteridis DP, Jiang Y, *et al.* Rare opportunistic (non-*Candida*, non-*Cryptococcus*) yeast bloodstream infections in patients with cancer. *J Infect.* 2012; **64:** 68-75.
80. Cho O, Matsukura M, Sugita T. Molecular evidence that the opportunistic fungal pathogen *Trichosporon asahii* is part of the normal fungal microbiota of the human gut based on rRNA genotyping. *Int J Infect Dis.* 2015; **39:** 87-8.
81. Chou N, Burbridge R, Karram S. Pancreatic fungal ball presenting as pseudomass. *ACG Case Rep J.* 2017; **4:** e55. PMID: 28459079.
82. Chotirmall SH, McElvaney NG. Fungi in the cystic fibrosis lung: Bystanders or pathogens? *Int J Biochem Cell Biol.* 2014; **52:** 161-73.
83. Chretien ML, Legouge C, Pages PB, *et al.* Emergency and elective pulmonary surgical resection in haematological patients with invasive fungal infections: A report of 50 cases in a single centre. *Clin Microbiol Infect.* 2016. **22:** 782-7.
84. Cindolo L, Berardinelli F, Castellan P, *et al.* A fatal mycotic sepsis after retrograde intrarenal surgery: A case report and literature review. *Urologia.* 2017; **84:** 106-8.
85. Colombo AL, Padovan AC, Chaves GM. Current knowledge of *Trichosporon* spp. and trichosporonosis. *Clin Microbiol Rev.* 2011; **24:** 682-700.
86. Cornely OA, Bohme A, Reichert D, *et al.* Risk factors for breakthrough invasive fungal infection during secondary prophylaxis. *J Antimicrob Chemother.* 2008; **61:** 939-46.
87. Cornely OA. *Aspergillus* to zygomycetes: Causes, risk factors, prevention and treatment of invasive fungal infections. *Infection.* 2008; **36:** 296-313.
88. Cortegiani, A. *et al.* The paradox of the evidence about invasive fungal infection prevention. *Critical Care.* 2016; **20:** 114.
89. Corzo-Leon DE, Satlin MJ, Soave R, *et al.* Epidemiology and outcomes of invasive fungal infections in allogeneic haematopoietic stem cell transplant recipients in the era of antifungal prophylaxis: A single-centre study with focus on emerging pathogens. *Mycoses.* 2015; **58:** 325-36.
90. d'Enfert C. Hidden killers: Persistence of opportunistic fungal pathogens in the human host. *Curr Opin Microbiol.* 2009; **12:** 358-64.
91. Dabritz J, Schneider M, Just-Nuebling G, *et al.* Mini-review: Invasive fungal infection complicating acute *Plasmodium falciparum* malaria. *Mycoses.* 2011; **54:** 311-7.
92. David C, Martin DB, Deng A, *et al.* Disseminated *Trichosporon inkin* and *Histoplasma capsulatum* in a patient with newly diagnosed AIDS. *J Am Acad Dermatol.* 2008; **59:** S13-5.
93. Davies GE, Thornton CR. Differentiation of the emerging human pathogens *Trichosporon asahii* and *Trichosporon asteroides* from other pathogenic yeasts and moulds by using species-specific monoclonal antibodies. *PLoS One.* 2014; **9:** e84789. PMID: 24392156.
94. Dayal D, Jain P, Kumar R, *et al.* Clinical spectrum and outcome of invasive filamentous fungal infections in children with Type 1 diabetes: North Indian experience. *Clin Pediatr Endocrinol.* 2015; **24:** 51-7.
95. de Almeida GM, Costa SF, Melhem M, *et al. Rhodotorula* spp. isolated from blood cultures: Clinical and microbiological aspects. *Med Mycol.* 2008; **46:** 547-56.
96. de Almeida JN Junior, Favero Gimenes VM, Francisco EC, et al. Evaluating and improving Vitek MS™ for identification of clinically relevant species of *Trichosporon* and the close-related genera *Cutaneotrichosporon* and *Apiotrichum. J Clin Microbiol.* 2017; pii: JCM.00461-17. PMID: 28539340.
97. de Almeida Junior JN, Figueiredo DS, Toubas D, *et al.* Usefulness of matrix-assisted laser desorption ionisation-time-of-flight mass spectrometry for identifying clinical *Trichosporon* isolates. *Clin Microbiol Infect.* 2014; **20:** 784-90.

98. de Almeida Junior JN, Hennequin C. Invasive *Trichosporon* infection: A systematic review on a re-emerging fungal pathogen. *Front Microbiol.* 2016; **7:** 1629. PMID: 27799926.
99. de Hoog GS, Smith MTH, Gueho E. A revision of the genus *Geotrichum* and its teleomorphs. *Stud Mycol.* 1986; **29:** 1-131.
100. de Oliveira RB, Atobe JH, Souza SA, *et al.* Epidemiology of invasive fungal infections in patients with acquired immunodeficiency syndrome at a reference hospital for infectious diseases in Brazil. *Mycopathologia.* 2014; **178:** 71-8.
101. de Pascale G, Tumbarello M. Fungal infections in the ICU: Advances in treatment and diagnosis. *Curr Opin Crit Care.* 2015; **21:** 421-9.
102. Debourgogne A, Dorin J, Machouart M. Emerging infections due to filamentous fungi in humans and animals: Only the tip of the iceberg? *Environ Microbiol Rep.* 2016; **8:** 332-42.
103. Deepa A, Nair BJ, Sivakumar T, Joseph AP. Uncommon opportunistic fungal infections of oral cavity: A review. *J Oral Maxillofac Pathol.* 2014; **18:** 235-43.
104. Denis J, Sabou M, Degot T, *et al.* First case of *Arthrographis kalrae* fungemia in a patient with cystic fibrosis. *Med Mycol Case Rep.* 2016; **14:** 8-11.
105. des Champs-Bro B, Leroy-Cotteau A, Mazingue F, *et al.* Invasive fungal infections: Epidemiology and analysis of antifungal prescriptions in onco-haematology. *J Clin Pharm Ther.* 2011; **36:** 152-60.
106. Diekema D, Arbefeville S, Boyken L, *et al.* The changing epidemiology of healthcare-associated candidemia over three decades. *Diagn Microbiol Infect Dis.* 2012; **73:** 45-8.
107. Dinleyici EC. Pediatric invasive fungal infections: Realities, challenges, concerns, myths and hopes. *Expert Rev Anti Infect Ther.* 2011; **9:** 273-4.
108. Dodds Ashley E, Drew R, Johnson M, *et al.* Cost of invasive fungal infections in the era of new diagnostics and expanded treatment options. *Pharmacotherapy.* 2012; **32:** 890-901.
109. Dogen A, Kaplan E, Oksuz Z, *et al.* Dishwashers are a major source of human opportunistic yeast-like fungi in indoor environments in Mersin, Turkey. *Med Mycol.* 2013; **51:** 493-8.
110. Doligalski CT, Benedict K, Cleveland AA, *et al.* Epidemiology of invasive mold infections in lung transplant recipients. *Am J Transplant.* 2014; **14:** 1328-33.
111. Donnelly JP. A multidisciplinary approach to managing invasive fungal disease. Introduction and Aims. *J Antimicrob Chemother.* 2013; **68 (Suppl. 3):** iii3-4.
112. Donnelly JP. Care pathways for managing invasive mould diseases. *Int J Antimicrob Agents.* 2013; **42:** 5-9.
113. Douglas AP, Thursky KA, Slavin MA. Invasive fungal infections in ALL: A new 'growth' area. *Leuk Lymphoma.* 2016; **57:** 1985-7.
114. Drgona L, Khachatryan A, Stephens J, *et al.* Clinical and economic burden of invasive fungal diseases in Europe: Focus on pre-emptive and empirical treatment of *Aspergillus* and *Candida* species. *Eur J Clin Microbiol Infect Dis.* 2014; **33:** 7-21.
115. Duarte-Oliveira C, Rodrigues F, Goncalves SM, *et al.* The cell biology of the *Trichosporon*-host interaction. *Front Cell Infect Microbiol.* 2017; **7:** 118. PMID: 28439501.
116. Ducassou S, Rivaud D, Auvrignon A, *et al.* Invasive fungal infections in pediatric acute myelogenous leukemia. *Pediatr Infect Dis J.* 2015; **34:** 1262-4.
117. Duggal R, Rajwanshi A, Gupta N, *et al.* Polymicrobial lung infection in post-renal transplant recipient diagnosed by fine-needle aspiration cytology. *Diagn Cytopathol.* 2010; **38:** 294-6.
118. Duggal S, Jain H, Tyagi A, *et al.* *Rhodotorula* fungemia: Two cases and a brief review. *Med Mycol.* 2011; **49:** 879-82.
119. Durand CM, Alonso CD, Subhawong AP, *et al.* Rapidly progressive cutaneous *Rhizopus microsporus* infection presenting as Fournier's gangrene in a patient with acute myelogenous leukemia. *Transpl Infect Dis.* 2011; **13:** 392-6.
120. Echenique IA, Angarone MP, Gordon RA, *et al.* Invasive fungal infection after heart transplantation: A seven-year, single-center experience. *Transpl Infect Dis.* 2017; **19:** doi: 10.1111/tid.12650. PMID: 27943503.
121. Einollahi B, Lessan-Pezeshki M, Aslani J, *et al.* Two decades of experience in mucormycosis after kidney transplantation. *Ann Transplant.* 2011; **16:** 44-8.
122. Einollahi B, Lessan-Pezeshki M, Pourfarziani V, *et al.* Invasive fungal infections following renal transplantation: A review of 2410 recipients. *Ann Transplant.* 2008; **13:** 55-8.
123. El-Cheikh J, Venton G, Crocchiolo R, *et al.* Efficacy and safety of micafungin for prophylaxis of invasive fungal infections in patients undergoing haplo-identical hematopoietic SCT. *Bone Marrow Transplant.* 2013; **48:** 1472-7.
124. Emori TG. An overview of nosocomial infections, including the role of the microbiology laboratory. *Clin Microbiol Rev.* 1993, **6:** 428-42.
125. Enoch DA, Yang H, Aliyu SH, *et al.* The changing epidemiology of invasive fungal infections. *Methods Mol Biol.* 2017; **1508:** 17-65.
126. Eves S, Sayeed R, Potter M. A case of rib fungal osteomyelitis. *J Plast Reconstr Aesthet Surg.* 2014; **67:** e81-3.
127. Ezzatzadegan S, Chen S, Chapman JR. Invasive fungal infections after renal transplantation. *Int J Organ Transplant Med.* 2012; **3:** 18-25.
128. Faksri K, Kaewkes W, Chaicumpar K, *et al.* Epidemiology an identification of potential fungal pathogens causing invasive fungal infection in a tertiary care hospital in northeast Thailand. *Med Mycol.* 2014; **52:** 810-8.
129. Falcone EL, Holland SM. Invasive fungal infection in chronic granulomatous disease: Insights into pathogenesis and management. *Curr Opin Infect Dis.* 2012; **25:** 658-69.
130. Falcone M, Concia E, Iori I, *et al.* Identification and management of invasive mycoses in internal medicine: A roadmap for physicians. *Intern Emerg Med.* 2014; **9:** 501-11.
131. Fan YC, Li WG, Zheng MH, *et al.* Invasive fungal infection in patients with systemic lupus erythematosus: Experience from a single institute of northern China. *Gene.* 2012; **506:** 184-7.
132. Fan YM, Huang WM, Yang YP, *et al.* Primary cutaneous trichosporonosis caused by *Trichosporon dermatis* in an immunocompetent man. *J Am Acad Dermatol.* 2011; **65:** 434-6.

133. Fanfair RN, Heslop O, Etienne K, *et al. Trichosporon asahii* among intensive care unit patients at a medical center in Jamaica. *Infect Control Hosp Epidemiol.* 2013; **34:** 638-41.
134. Farah CS, Lynch N, McCullough MJ. Oral fungal infections: An update for the general practitioner. *Aust Dent J.* 2010; **55 (Suppl. 1):** 48-54.
135. Farmakiotis D, Kontoyiannis DP. Emerging issues with diagnosis and management of fungal infections in solid organ transplant recipients. *Am J Transplant.* 2015; **15:** 1141-7.
136. Fernandez-Ruiz M, Guinea J, Puig-Asensio M, *et al.* Fungemia due to rare opportunistic yeasts: Data from a population-based surveillance in Spain. *Med Mycol.* 2017; **55:** 125-36.
137. Fleming S, Yannakou CK, Haeusler GM, *et al.* Consensus guidelines for antifungal prophylaxis in haematological malignancy and haemopoietic stem cell transplantation, 2014. *Intern Med J.* 2014; **44:** 1283-97.
138. Florescu DF, Sandkovsky U. Fungal infections in intestinal and multivisceral transplant recipients. *Curr Opin Organ Transplant.* 2015; **20:** 295-302.
139. Fortun J, Ruiz I, Martin-Davila P, *et al.* Fungal infection in solid organ recipients. *Enferm Infecc Microbiol Clin.* 2012; **30 (Suppl. 2):** 49-56.
140. Foster CE, Edwards MS, Brackett J, *et al.* Trichosporonosis in pediatric patients with a hematologic disorder. *J Pediatric Infect Dis Soc.* 2017; doi: 10.1093/jpids/pix031. PMID: 28510690.
141. Fridkin SK, Jarvis WR. Epidemiology of nosocomial fungal infections. *Clin Microbiol Rev.* 1996; **9:** 499-511.
142. Gabriel F, Noel T, Accoceberry I. Fatal invasive trichosporonosis due to *Trichosporon loubieri* in a patient with T-lymphoblastic lymphoma. *Med Mycol.* 2011; **49:** 306-10.
143. Galimberti R, Torre AC, Baztan MC, *et al.* Emerging systemic fungal infections. *Clin Dermatol.* 2012; **30:** 633-50.
144. Garbee DD, Pierce SS, Manning J. Opportunistic fungal infections in critical care units. *Crit Care Nurs Clin North Am.* 2017; **29:** 67-79.
145. Garcia-Vidal C, Viasus D, Carratala J. Pathogenesis of invasive fungal infections. *Curr Opin Infect Dis.* 2013; **26:** 270-6.
146. Gardenier JC, Chopra VK, Filicori F, *et al.* Angioinvasive mold in the surgical and burn intensive care unit: A case series and review of the literature. *Surg Infect Case Reports.* 2016; **1:** 72-8.
147. Garnacho-Montero J, Olaechea P, Alvarez-Lerma F, *et al.* Epidemiology, diagnosis and treatment of fungal respiratory infections in the critically ill patient. *Rev Esp Quimioter.* 2013; **26:** 173-88.
148. Gautam G, Singh AK, Kumar R, *et al.* Beware! Fungal urosepsis may follow endoscopic intervention for prolonged indwelling ureteral stent. *J Endourol.* 2006; **20:** 522-4.
149. Gavalda J, Meije Y, Fortun J, et al. Invasive fungal infections in solid organ transplant recipients. *Clin Microbiol Infect.* 2014; **20 (Suppl. 7):** 27-48.
150. Gayet-Ageron A, Iten A, van Delden C, *et al.* In-hospital transfer is a risk factor for invasive filamentous fungal infection among hospitalized patients with hematological malignancies: A matched case-control study. *Infect Control Hosp Epidemiol.* 2015; **36:** 320-8.
151. Gazzoni FF, Severo LC, Marchiori E, *et al.* Fungal diseases mimicking primary lung cancer: Radiologic-pathologic correlation. *Mycoses.* 2014; **57:** 197-208.
152. Gefen A, Zaidman I, Shachor-Meyouhas Y, *et al.* Serum galactomannan screening for diagnosis of invasive pulmonary aspergillosis in children after stem cell transplantation or with high-risk leukemia. *Pediatr Hematol Oncol.* 2015; **32:** 146-52.
153. Gill PK, Gill JS. *Blastoschizomyces capitatus* pneumonia in an immunocompetent female. *Indian J Tuberc.* 2011; **58:** 88-9. PMID: 21644397.
154. Girmenia C, Ferretti A, Barberi W. Epidemiology and risk factors for invasive fungal diseases in hematopoietic stem cell transplantation. *Curr Opin Hematol.* 2014; **21:** 459-65.
155. Gomez CA, Singh N. Donor-derived filamentous fungal infections in solid organ transplant recipients. *Curr Opin Infect Dis.* 2013; **26:** 309-16.
156. Graf C, Gavazzi G. *Saccharomyces cerevisiae* fungemia in an immunocompromised patient not treated with *Saccharomyces boulardii* preparation. *J Infect.* 2007; **54:** 310-1.
157. Groll AH, Castagnola E, Cesaro S, *et al.* Fourth European Conference on Infections in Leukaemia (ECIL-4): Guidelines for diagnosis, prevention and treatment of invasive fungal diseases in paediatric patients with cancer or allogeneic haemopoietic stem-cell transplantation. *Lancet Oncol.* 2014; **15:** e327-40.
158. Groll AH, Lumb J. New developments in invasive fungal disease. *Future Microbiol.* 2012; **7:** 179-84.
159. Guimaraes MD, Marchiori E, de Souza Portes Meirelles G, *et al.* Fungal infection mimicking pulmonary malignancy: Clinical and radiological characteristics. *Lung.* 2013; **191:** 655-62.
160. Gullo A. Invasive fungal infections: The challenge continues. *Drugs.* 2009; **69 (Suppl. 1):** 65-73.
161. Guo LN, Xiao M, Kong F, *et al.* Three-locus identification, genotyping and antifungal susceptibilities of medically important *Trichosporon* species from China. *J Clin Microbiol.* 2011; **49:** 3805-11.
162. Gurgui M, Sanchez F, March F, *et al.* Nosocomial outbreak of *Blastoschizomyces capitatus* associated with contaminated milk in a haematological unit. *J Hosp Infect.* 2011; **78:** 274-8.
163. Haidar G, Zerbe CS, Cheng M, *et al. Phellinus* species: An emerging cause of refractory fungal infections in patients with X-linked chronic granulomatous disease. *Mycoses.* 2017; **60:** 155-60.
164. Hajdu S, Obradovic A, Presterl E, *et al.* Invasive mycoses following trauma. *Injury.* 2009; **40:** 548-54.
165. Hamoud S, Keidar Z, Hayek T. Recurrent *Saccharomyces cerevisiae* fungemia in an otherwise healthy patient. *Isr Med Assoc J.* 2011; **13:** 575-6.
166. Harris M, Ananth Narayan S, Orchard EA. Fungal myocarditis in a preterm neonate. *BMJ Case Rep.* 2012; pii: bcr-2012-007174. PMID: 23166173.

167. Harrison N, Mitterbauer M, Tobudic S, *et al.* Incidence and characteristics of invasive fungal diseases in allogeneic hematopoietic stem cell transplant recipients: A retrospective cohort study. *BMC Infect Dis.* 2015; **15:** 584.
168. Hashino S, Takahashi S, Morita R, *et al.* Fungemia due to *Trichosporon dermatis* in a patient with refractory Burkitt's leukemia. *Blood Res.* 2013; **48:** 154-6.
169. Hazen KC. New and emerging yeast pathogens. *Clin Microbiol Rev.* 1995; **8:** 462-78.
170. He SY, Makhzoumi ZH, Singer JP, *et al.* Practice variation in *Aspergillus* prophylaxis and treatment among lung transplant centers: A national survey. *Transpl Infect Dis.* 2015; **17:** 14-20.
171. Henriet S, Verweij PE, Holland SM, *et al.* Invasive fungal infections in patients with chronic granulomatous disease. *Adv Exp Med Biol.* 2013; **764:** 27-55.
172. Herb A, Sabou M, Delhorme JB, *et al. Pseudozyma aphidis* fungemia after abdominal surgery: First adult case. *Med Mycol Case Rep.* 2015; **8:** 37-9.
173. Heslop OD, Nyi Nyi MP, Abbott SP, *et al.* Disseminated trichosporonosis in a burn patient: Meningitis and cerebral abscess due to *Trichosporon asahii. J Clin Microbiol.* 2011; **49:** 4405-8.
174. Hirai Y, Asahata S, Ainoda Y, *et al.* Nosocomial *Candida parapsilosis* candidaemia: Risk factors, antifungal susceptibility and outcome. *J Hosp Infect.* 2014; **87:** 54-8.
175. Hoenigl M, Seeber K, Koidl C, *et al.* Sensitivity of galactomannan enzyme immunoassay for diagnosing breakthrough invasive aspergillosis under antifungal prophylaxis and empirical therapy. *Mycoses.* 2013; **56:** 471-6.
176. Hof H. IFI = Invasive Fungal Infections. What is that? A misnomer because a non-invasive fungal infection does not exist! *Int J Infect Dis.* 2010; **14:** e458-9.
177. Hogen R, Dhanireddy K. Invasive fungal infections following liver transplantation. *Curr Opin Organ Transplant.* 2017. doi: 10.1097/MOT.0000000000000431. PMID: 28548995.
178. Hol JA, Wolfs TF, Bierings MB, *et al.* Predictors of invasive fungal infection in pediatric allogeneic hematopoietic SCT recipients. *Bone Marrow Transplant.* 2014; **49:** 95-101.
179. Hope W, Natarajan P, Goodwin L. Invasive fungal infections. *Clin Med (Lond).* 2013; **13:** 507-10.
180. Horowitz A, Spendel D, Kraut R, *et al.* Cavernous sinus thrombosis as a result of a fungal infection: A case report. *J Oral Maxillofac Surg.* 2013; **71:** 1899.e1-5.
181. Hosokawa K, Yamazaki H, Mochizuki K, *et al.* Successful treatment of *Trichosporon* fungemia in a patient with refractory acute myeloid leukemia using voriconazole combined with liposomal amphotericin B. *Transpl Infect Dis.* 2012; **14:** 184-7.
182. Hosseini-Moghaddam SM, Husain S. Fungi and molds following lung transplantation. *Semin Respir Crit Care Med.* 2010; **31:** 222-33.
183. Hsu LY, Lee DG, Yeh SP, *et al.* Epidemiology of invasive fungal diseases among patients with haematological disorders in the Asia-Pacific: A prospective observational study. *Clin Microbiol Infect.* 2015; **21:** 594. e7-11.
184. Hsu LY, Ng ES, Koh LP. Common and emerging fungal pulmonary infections. *Infect Dis Clin North Am.* 2010; **24:** 557-77.
185. Hundalani S, Pammi M. Invasive fungal infections in newborns and current management strategies. *Expert Rev Anti Infect Ther.* 2013; **11:** 709-21.
186. Huprikar S, Shoham S, *et al.* Emerging fungal infections in solid organ transplantation. *Am J Transplant.* 2013; **13 (Suppl. 4):** 262-71.
187. Iatta R, Napoli C, Borghi E, *et al.* Rare mycoses of the oral cavity: A literature epidemiologic review. *Oral Surg Oral Med Oral Pathol Oral Radiol Endod.* 2009; **108:** 647-55.
188. Ichikawa T, Yoshiyama N, Ohgane Y, *et al.* Switching of colony morphology and adhesion activity of *Trichosporon asahii* clinical isolates. *Med Mycol.* 2016; **54:** 189-96.
189. Iken M, Belkouch A, Bellarj B, et al. A rare case of urinary tract infection due to *Trichosporon asahii* in a diabetic patient. *Pan Afr Med J.* 2015; **20:** 127.
190. Imko-Walczuk BB, Predota A, Okuniewska A, *et al.* Superficial fungal infections in renal transplant recipients. *Transplant Proc.* 2014; **46:** 2738-42.
191. Ingham A, Gilbert JD, Byard RW. Disseminated fungal infection with renal infarction simulating homicide. *Am J Forensic Med Pathol.* 2010; **31:** 390-2.
192. Islam A, Mody CH. Management of fungal lung disease in the immunocompromised. *Ther Adv Respir Dis.* 2011; **5:** 305-24.
193. Issarachaikul R, Lertwannawit O, Detporntewan P, *et al.* Catheter-related bloodstream infections caused by *Trichosporon* species. *Southeast Asian J Trop Med Public Health.* 2014; **45:** 421-9.
194. Iturrieta-Gonzalez IA, Padovan AC, Bizerra FC, *et al.* Multiple species of *Trichosporon* produce biofilms highly resistant to triazoles and amphotericin B. *PLoS One.* 2014; **9:** e109553.
195. Jain A, Jain S, Rawat S. Emerging fungal infections among children: A review on its clinical manifestations, diagnosis and prevention. *J Pharm Bioallied Sci.* 2010; **2:** 314-20.
196. Jakobs O, Schoof B, Klatte TO, *et al.* Fungal periprosthetic joint infection in total knee arthroplasty: A systematic review. *Orthop Rev (Pavia).* 2015; **7:** 5623.
197. Joo H, Choi YG, Cho SY, *et al. Pseudozyma aphidis* fungaemia with invasive fungal pneumonia in a patient with acute myeloid leukaemia: Case report and literature review. *Mycoses.* 2016; **59:** 56-61.
198. Kalawat U, Sharma KK. *Trichosporon* peritonitis following duodenal perforation. *Saudi J Gastroenterol.* 2010; **16:** 43-5.
199. Kanamori H, Rutala WA, Sickbert-Bennett EE, *et al.* Review of fungal outbreaks and infection prevention in healthcare settings during construction and renovation. *Clin Infect Dis.* 2015; **61:** 433-44.
200. Kapur S, Louie BE. Hemoptysis and thoracic fungal infections. *Surg Clin North Am.* 2010; **90:** 985-1001.
201. Karabinos IK, Kokladi M, Katritsis D. Fungal endocarditis of the superior *vena cava:* The role of transesophageal echocardiography. *Hellenic J Cardiol.* 2010; **51:** 538-9.

202. Karapinar DY, Karadas N, Yazici P, *et al.* *Trichosporon asahii*, sepsis and secondary hemophagocytic lymphohistiocytosis in children with hematologic malignancy. *Pediatr Hematol Oncol.* 2014; **31:** 282-4.
203. Katragkou A, Roilides E. Best practice in treating infants and children with proven, probable or suspected invasive fungal infections. *Curr Opin Infect Dis.* 2011; **24:** 225-9.
204. Katz T, Wasiak J, Cleland H, *et al.* Incidence of non-candidal fungal infections in severe burn injury: an Australian perspective. *Burns.* 2014; **40:** 881-6.
205. Kauffman CA, Freifeld AG, Andes DR, *et al.* Endemic fungal infections in solid organ and hematopoietic cell transplant recipients enrolled in the Transplant-Associated Infection Surveillance Network (TRANSNET). *Transpl Infect Dis.* 2014; **16:** 213-24.
206. Kaur R, Mehra B, Dhakad MS, *et al.* Fungal opportunistic pneumonias in HIV/AIDS patients: An Indian tertiary care experience. *J Clin Diagn Res.* 2017; **11:** DC14-9.
207. Kazancioglu R, Kirikci G, Albaz M, *et al.* Fungal peritonitis among the peritoneal dialysis patients of four Turkish centres. *J Ren Care.* 2010; **36:** 186-90.
208. Kenyon C, Bonorchis K, Corcoran C, *et al.* A dimorphic fungus causing disseminated infection in South Africa. *N Engl J Med.* 2013; **369:** 1416-24.
209. Khan ID, Sahni AK, Basu A, *et al.* *Trichosporon asahii* urinary tract infection in immunocompetent patients. *Med J Armed Forces India.* 2015; **71:** 373-6.
210. Kim HA, Hyun M, Ryu S-Y. Catheter-associated *Rhodotorula mucilaginosa* fungemia in an immunocompetent host. *Infect Chemother.* 2013; **45:** 339-42.
211. Kim SH, Ha YE, Youn JC, *et al.* Fatal scedosporiosis in multiple solid organ allografts transmitted from a nearly-drowned donor. *Am J Transplant.* 2015; **15:** 833-40.
212. Kim SH, Kim DH, Joo SI, *et al.* Chronic cutaneous disseminated *Trichosporon asahii* infection in a non-immunocompromised patient. *J Am Acad Dermatol.* 2008; **59:**S37-9.
213. Klingspor L, Saaedi B, Ljungman P, *et al.* Epidemiology and outcomes of patients with invasive mould infections: A retrospective observational study from a single centre (2005-2009). *Mycoses.* 2015; **58:** 470-7.
214. Kobayashi R, Kaneda M, Sato T, *et al.* The clinical feature of invasive fungal infection in pediatric patients with hematologic and malignant diseases: A 10-year analysis at a single institution at Japan. *J Pediatr Hematol Oncol.* 2008; **30:** 886-90.
215. Kobayashi S, Murayama S, Takanashi S, *et al.* Clinical features and prognoses of 23 patients with chronic granulomatous disease followed for 21 years by a single hospital in Japan. *Eur J Pediatr.* 2008; **167:** 1389-94.
216. Kontoyiannis DP, Marr KA, Park BJ, *et al.* Prospective surveillance for invasive fungal infections in hematopoietic stem cell transplant recipients, 2001-2006: Overview of the Transplant-Associated Infection Surveillance Network (TRANSNET) Database. *Clin Infect Dis.* 2010; **50:** 1091-100.
217. Kornberger A, Walter V, Jaeger F, *et al.* Necrotizing pulmonary aspergillosis and ventricular assist device infection: Case report and review of literature. *Transpl Infect Dis.* 2015; **17:** 737-43.
218. Kriengkauykiat J, Ito JI, Dadwal SS. Epidemiology and treatment approaches in management of invasive fungal infections. *Clin Epidemiol.* 2011; **3:** 175-91.
219. Krishnan PA. Fungal infections of the oral mucosa. *Indian J Dent Res.* 2012; **23:** 650-9.
220. Kroner C, Kappler M, Grimmelt AC, *et al.* The basidiomycetous yeast *Trichosporon* may cause severe lung exacerbation in cystic fibrosis patients - Clinical analysis of *Trichosporon* positive patients in a Munich cohort. *BMC Pulm Med.* 2013; **13:** 61.
221. Kumar A, Udayakumaran S, Babu R, *et al.* *Trichosporon asahii* infection presenting as chronic meningo-ventriculitis and intra-ventricular fungal ball: A case report and literature review. *Mycoses.* 2015; **58:** 99-103.
222. Kumar DA, Muralidhar S, Banerjee U, *et al.* Diversity and antifungal resistance patterns of prevalent opportunistic pathogenic yeasts colonizing the oral cavities of asymptomatic human immunodeficiency virus-infected individuals and their relation to CD4(+) counts. *Indian J Sex Transm Dis.* 2015; **36:** 53-8.
223. Kurosawa M, Yonezumi M, Hashino S, *et al.* Epidemiology and treatment outcome of invasive fungal infections in patients with hematological malignancies. *Int J Hematol.* 2012; **96:** 748-57.
224. Kushima H, Ishii H, Komiya K, *et al.* Prognostic significance of serum β-D-glucan levels in 78 patients with *Trichosporon* fungemia. *Int J Infect Dis.* 2013; **17:** e134-5.
225. Kwong CA, Puloski SK, Hildebrand KA. Fungal periprosthetic joint infection following total elbow arthroplasty: A case report and review of the literature. *J Med Case Rep.* 2017; **11:** 20.
226. Lacasse A, Cleveland KO. *Trichosporon mucoides* fungemia in a liver transplant recipient: Case report and review. *Transpl Infect Dis.* 2009; **11:** 155-9.
227. Lalla RV, Latortue MC, Hong CH, *et al.* A systematic review of oral fungal infections in patients receiving cancer therapy. Support Care Cancer. 2010; **18:** 985-92.
228. Lamoth F, Calandra T. Early diagnosis of invasive mould infections and disease. *J Antimicrob Chemother.* 2017; **72 (Suppl. 1):** i19-28.
229. Lamoth F, Chung SJ, Damonti L, *et al.* Changing epidemiology of invasive mold infections in patients receiving azole prophylaxis. *Clin Infect Dis.* 2017; **64:** 1619-21.
230. Lanternier F, Cypowyj S, Picard C, *et al.* Primary immunodeficiencies underlying fungal infections. *Curr Opin Pediatr.* 2013; **25:** 736-47.
231. Lass-Florl C. Invasive fungal infections in pediatric patients: A review focusing on antifungal therapy. *Expert Rev Anti Infect Ther.* 2010; **8:** 127-35.
232. Lass-Florl C. The changing face of epidemiology of invasive fungal disease in Europe. *Mycoses.* 2009; **52:** 197-205.
233. Lass-Florl C, Cuenca-Estrella M. Changes in the epidemiological landscape of invasive mould infections and disease. J Antimicrob Chemother. 2017; **72 (Suppl. 1):** i5-11.
234. Lehrnbecher T, Groll AH. Invasive fungal infections in the pediatric population. *Expert Rev Anti Infect Ther.* 2011; **9:** 275-8.

235. Leon M, Alave J, Bustamante B, *et al.* A probable association between HTLV-1 and endemic mycosis in Latin America. *J Infect Dev Ctries.* 2012; **6:** 301.
236. Lepak A, Andes D. Fungal sepsis: Optimizing antifungal therapy in the critical care setting. *Crit Care Clin.* 2011; **27:** 123-47.
237. Lerolle N, Raffoux E, Socie G, *et al.* Breakthrough invasive fungal disease in patients receiving posaconazole primary prophylaxis: a 4-year study. *Clin Microbiol Infect.* 2014; **20:** O952-9.
238. Leroux S, Ullmann AJ. Management and diagnostic guidelines for fungal diseases in infectious diseases and clinical microbiology: Critical appraisal. *Clin Microbiol Infect.* 2013; **19:** 1115-21.
239. Leventakos K, Lewis RE, Kontoyiannis DP. Fungal infections in leukemia patients: How do we prevent and treat them? *Clin Infect Dis.* 2010; **50:** 405-15.
240. Lewandowski L, Purcell R, Fleming M, *et al.* The use of dilute Dakin's solution for the treatment of angioinvasive fungal infection in the combat wounded: A case series. *Mil Med.* 2013; **178:** e503-7.
241. Li XS, Zhu HX, Fan HX, *et al.* Pulmonary fungal infections after bone marrow transplantation: The value of high-resolution computed tomography in predicting their etiology. *Chin Med J (Engl).* 2011; **124:** 3249-54.
242. Liao Y, Chen M, Hartmann T, *et al.* Epidemiology of opportunistic invasive fungal infections in China: Review of literature. *Chin Med J (Engl).* 2013; **126:** 361-8.
243. Liao Y, Hartmann T, Ao JH, *et al.* Serum glucuronoxylomannan may be more appropriate for the diagnosis and therapeutic monitoring of *Trichosporon* fungemia than serum β-D-glucan. *Int J Infect Dis.* 2012; **16:** e638.
244. Lim DS, Tan PL, Jureen R, *et al.* Cutaneous emboli of invasive basidiomycosis in a child with aplastic anemia. *Am J Dermatopathol.* 2017; **39:** 204-7.
245. Limper AH. Clinical approach and management for selected fungal infections in pulmonary and critical care patients. *Chest.* 2014; **146:** 1658-66.
246. Lin CY, Liu WL, Chang CC, *et al.* Invasive fungal tracheobronchitis in mechanically ventilated critically ill patients: Underlying conditions, diagnosis and outcomes. *Ann Intensive Care.* 2017; **7:** 9.
247. Lin SS, Pranikoff T, Smith SF, *et al.* Central venous catheter infection associated with *Pseudozyma aphidis* in a child with short gut syndrome. *J Med Microbiol.* 2008; **57:** 516-8. Ustilago?
248. Lin YP, Yang YP, Huang WM, *et al.* Experimental pathogenicity of a clinical isolate of *Trichosporon dermatis* in a murine model. *Mycopathologia.* 2011; **172:** 381-7.
249. Liu J, You C, Tang J, *et al.* Fungal pituitary abscess: Case report and review of the literature. *Neurol India.* 2013; **61:** 210-2.
250. Liu X, Ling Z, Li L, Ruan B. Invasive fungal infections in liver transplantation. *Int J Infect Dis.* 2011; **15:** e298-304.
251. Liu YC, Chien SH, Fan NW, *et al.* Incidence and risk factors of probable and proven invasive fungal infection in adult patients receiving allogeneic hematopoietic stem cell transplantation. *J Microbiol Immunol Infect.* 2016; **49:** 567-74.
252. Lortholary O, Charlier C, Lebeaux D, *et al.* Fungal infections in immunocompromised travelers. *Clin Infect Dis.* 2013; **56:** 861-9.
253. Machouart M, Garcia-Hermoso D, Rivier A, *et al.* Emergence of disseminated infections due to *Geosmithia argillacea* in patients with chronic granulomatous disease receiving long-term azole antifungal prophylaxis. *J Clin Microbiol.* 2011; **49:** 1681-3.
254. Maertens J, Lyon S. Meeting the challenge of invasive fungal infections: Part 1. *Future Microbiol.* 2013; **8:** 1063-5.
255. Maertens J, Lyon S. Meeting the challenge of invasive fungal infections: Part 2. *Future Microbiol.* 2013; **8:** 1245-7.
256. Maertens J, Marchetti O, Herbrecht R, *et al.* European guidelines for antifungal management in leukemia and hematopoietic stem cell transplant recipients: summary of the ECIL 3 - 2009 update. *Bone Marrow Transplant.* 2011; **46:** 709-18.
257. Magee K, Hata DJ, Meza D. *Trichosporon asahii* infection in a patient with metastatic prostate cancer as an example of an emerging fungal pathogen. *Lab Med.* 2015; **46:** e74-8.
258. Mahajan VA, Mahajan AL, Harte BH, *et al.* Optimal environment for free flaps can also be a haven for fungi! *J Plast Reconstr Aesthet Surg.* 2009; **62:** e661-2.
259. Maly J, Szarszoi O, Netuka I, *et al.* Fungal infections associated with long-term mechanical circulatory support-diagnosis and management. *J Card Surg.* 2014; **29:** 95-100.
260. Mandhaniya S, Iqbal S, Sharawat SK, *et al.* Diagnosis of invasive fungal infections using real-time PCR assay in paediatric acute leukaemia induction. *Mycoses.* 2012; **55:** 372-9.
261. Marchetti O, Lamoth F, Mikulska M, *et al.* ECIL recommendations for the use of biological markers for the diagnosis of invasive fungal diseases in leukemic patients and hematopoietic SCT recipients. *Bone Marrow Transplant.* 2012; **47:** 846-54.
262. Marom EM, Kontoyiannis DP. Imaging studies for diagnosing invasive fungal pneumonia in immunocompromised patients. *Curr Opin Infect Dis.* 2011; **24:** 309-14.
263. Maron GM, Hayden RT, Rodriguez A, *et al.* Voriconazole prophylaxis in children with cancer: Changing outcomes and epidemiology of fungal infections. *Pediatr Infect Dis J.* 2013; **32:** e451-5.
264. Marques DS, Pinho Vaz C, Branca R, *et al.* *Rhizomucor* and *Scedosporium* infection post-hematopoietic stem-cell transplant. *Case Rep Med.* 2011; 830769.
265. Marques SA. Fungal infections of the mucous membrane. *Dermatol Ther.* 2010; **23:** 243-50.
266. Marr KA. Fungal infections in hematopoietic stem cell transplant recipients. *Med Mycol.* 2008; **46:** 293-302.
267. Marr KA. Fungal infections in oncology patients: update on epidemiology, prevention, and treatment. *Curr Opin Oncol.* 2010; **22:** 138-42.
268. Martin IW, Tonner R, Trivedi J, *et al.* *Saccharomyces boulardii* probiotic-associated fungemia: Questioning the

safety of this preventive probiotic's use. *Diagn Microbiol Infect Dis.* 2017; **87:** 286-8.
269. Martin J, Welch KC. Wegener granulomatosis (WG) with superimposed fungal infection. *Arch Otolaryngol Head Neck Surg.* 2012; **138:** 981-3.
270. Marukutira T, Huprikar S, Azie N, *et al.* Clinical characteristics and outcomes in 303 HIV-infected patients with invasive fungal infections: Data from the Prospective Antifungal Therapy Alliance registry, a multicenter, observational study. *HIV AIDS (Auckl).* 2014; **6:** 39-47.
271. Marzaban R, Salah M, Mukhtar AM, *et al.* Fungal infections in liver transplant patients admitted to the intensive care unit. *Ann Transplant.* 2014; **19:** 667-73.
272. Maschmeyer G, Beinert T, Buchheidt D, *et al.* Diagnosis and antimicrobial therapy of lung infiltrates in febrile neutropenic patients: Guidelines of the infectious diseases working party of the German Society of Haematology and Oncology. *Eur J Cancer.* 2009; **45:** 2462-72.
273. Maschmeyer G. Invasive fungal disease: better survival through early diagnosis and therapeutic intervention. *Expert Rev Anti Infect Ther.* 2011; **9:** 279-81.
274. Maschmeyer G. The changing face of febrile neutropenia - from monotherapy to moulds to mucositis. Prevention of mould infections. *J Antimicrob Chemother.* 2009; **63 (Suppl. 1):** i27-30.
275. Matthaiou DK, Christodoulopoulou T, Dimopoulos G. How to treat fungal infections in ICU patients. *BMC Infect Dis.* 2015; **15:** 205.
276. Matuszkiewicz-Rowinska J. Update on fungal peritonitis and its treatment. *Perit Dial Int.* 2009; **29:** S161-5.
277. Maxfield L, Matthews JJ, Ambrosetti DR, *et al. Trichosporon* fungemia in a pediatric patient with acute lymphoblastic leukemia. *ID Cases.* 2015; **2:** 106-8.
278. Mekha N, Sugita T, Ikeda R, *et al.* Genotyping and antifungal drug susceptibility of the pathogenic yeast *Trichosporon asahii* isolated from Thai patients. *Mycopathologia.* 2010; **169:** 67-70.
279. Menegueti MG, Ferreira LR, Silva MF, *et al.* Assessment of microbiological air quality in hemato-oncology units and its relationship with the occurrence of invasive fungal infections: An integrative review. *Rev Soc Bras Med Trop.* 2013; **46:** 391-6.
280. Menon S, Gupta HR, Sequeira R, *et al. Rhodotorula glutinis* meningitis: A case report and review of literature. *Mycoses.* 2014; **57:** 447-51.
281. Mercier T, Maertens J. Clinical considerations in the early treatment of invasive mould infections and disease. *J Antimicrob Chemother.* 2017; **72 (Suppl. 1):** i29-38.
282. Messias Silvestre AJ, Alexandre Bandeira Rampazzo Miranda M, Pires de Camargo Z. *Trichosporon* species isolated from the perigenital region, urine and catheters of a Brazilian population. *Braz J Microbiol.* 2010; **41:** 628-34.
283. Miceli MH, Churay T, Braun T, *et al.* Risk factors and outcomes of invasive fungal infections in allogeneic hematopoietic cell transplant recipients. *Mycopathologia.* 2017; **182:** 495-504.
284. Miceli MH, Diaz JA, Lee SA. Emerging opportunistic yeast infections. *Lancet Infect Dis.* 2011; **11:** 142-51.
285. Miceli MH, Lee SA. Emerging moulds: Epidemiological trends and antifungal resistance. *Mycoses.* 2011; **54:** e666-78.
286. Michallet M, Sobh M, Morisset S, *et al.* Risk factors for invasive aspergillosis in acute myeloid leukemia patients prophylactically treated with posaconazole. *Med Mycol.* 2011; **49:** 681-7.
287. Middleton PG, Chen SC, Meyer W. Fungal infections and treatment in cystic fibrosis. *Curr Opin Pulm Med.* 2013; **19:** 670-5.
288. Mihon C, Alexandre T, Pereira A. Clinical experience in invasive fungal infections: Multiple fungal infections as the first presentation of HIV. *Clin Drug Investig.* 2013; **33:** S37-40.
289. Miller R, Assi M; AST Infectious Diseases Community of Practice. Endemic fungal infections in solid organ transplantation. *Am J Transplant.* 2013; **13 (Suppl. 4):** 250-61.
290. Mohd Nor F, Tan LH, Na SL, *et al.* Meningitis caused by *Rhodotorula mucilaginosa* in HIV-infected patient: A case report and review of the literature. *Mycopathologia.* 2015; **180:** 95-8.
291. Montagna MT, Caggiano G, Lovero G, *et al.* Epidemiology of invasive fungal infections in the intensive care unit: Results of a multicenter Italian survey (AURORA Project). *Infection.* 2013; **41:** 645-53.
292. Montagna MT, De Giglio O, Napoli C, *et al.* Invasive fungal infections in patients with hematologic malignancies (AURORA Project): Lights and shadows during 18-months surveillance. *Int J Mol Sci.* 2012; **13:** 774-87.
293. Montagna MT, Lovero G, Coretti C, *et al.* SIMIFF study: Italian fungal registry of mold infections in hematological and non-hematological patients. *Infection.* 2014; **42:** 141-51.
294. Montoya AM, Sanchez Gonzalez A, Palma-Nicolas JP, *et al.* Genotyping, extracellular compounds and antifungal susceptibility testing of *Trichosporon asahii* isolated from Mexican patients. *Med Mycol.* 2015; **53:** 505-11.
295. Morace G, Borghi E. Fungal infections in ICU patients: Epidemiology and the role of diagnostics. *Minerva Anestesiol.* 2010; **76:** 950-6.
296. Moreno-Coutino G, Aquino MA, Vega-Memije M, *et al.* Necrotic ulcer caused by *Trichosporon asahii* in an immunocompetent adolescent. *Mycoses.* 2012; **55:** 93-4.
297. Mori T, Nakamura Y, Kato J, *et al.* Fungemia due to *Rhodotorula mucilaginosa* after allogeneic hematopoietic stem cell transplantation. *Transpl Infect Dis.* 2012; **14:** 91-4.
298. Morrissey CO, Gilroy NM, Macesic N, *et al.* Consensus guidelines for the use of empiric and diagnostic-driven antifungal treatment strategies in haematological malignancy, 2014. *Intern Med J.* 2014; **44:** 1298-314.
299. Muller FM, Seidler M. Characteristics of pathogenic fungi and antifungal therapy in cystic fibrosis. *Expert Rev Anti Infect Ther.* 2010; **8:** 957-64.
300. Muskett H, Shahin J, Eyres G, *et al.* Risk factors for invasive fungal disease in critically ill adult patients: A systematic review. *Crit Care.* 2011; **15:** R287.

301. Muzyka BC, Epifanio RN. Update on oral fungal infections. *Dent Clin North Am.* 2013; **57:** 561-81.
302. Nadeem M, Zafar H, Effendi MS. Localised fungal infection in a prosthetic mesh treated conservatively. *BMJ Case Rep.* 2011; pii: bcr0720103177.
303. Nagao M, Fujimoto Y, Yamamoto M, *et al.* Epidemiology of invasive fungal infections after liver transplantation and the risk factors of late-onset invasive aspergillosis. *J Infect Chemother.* 2016; **22:** 84-9.
304. Negi V, Sharma M, Juyal D, *et al.* Disseminated trichosporonosis due to *Trichosporon asahii* in a diabetic patient. *Indian J Pathol Microbiol.* 2015; **58:** 246-8.
305. Neofytos D, Fishman JA, Horn D, *et al.* Epidemiology and outcome of invasive fungal infections in solid organ transplant recipients. *Transpl Infect Dis.* 2010; **12:** 220-9.
306. Neofytos D, Horn D, Anaissie E, *et al.* Epidemiology and outcome of invasive fungal infection in adult hematopoietic stem cell transplant recipients: analysis of Multicenter Prospective Antifungal Therapy (PATH) Alliance registry. *Clin Infect Dis.* 2009; **48:** 265-73.
307. Neofytos D, Lu K, Hatfield-Seung A, *et al.* Epidemiology, outcomes and risk factors of invasive fungal infections in adult patients with acute myelogenous leukemia after induction chemotherapy. *Diagn Microbiol Infect Dis.* 2013; **75:** 144-9.
308. Neofytos D, Treadway S, Ostrander D, *et al.* Epidemiology, outcomes and mortality predictors of invasive mold infections among transplant recipients: A 10-year, single-center experience. *Transpl Infect Dis.* 2013; **15:** 233-42.
309. Neumann U, Langrehr J, Kaisers U, *et al.* Simultaneous splenectomy increases risk for opportunistic pneumonia in patients after liver transplantation. *Transpl Int.* 2002; **15:** 226-32.
310. Nigo M, Vial MR, Munita JM, *et al.* Fungal empyema thoracis in cancer patients. *J Infect.* 2016; **72:** 615-21.
311. Nobrega de Almeida Junior J, Buccheri de Oliveira R, Duarte A, *et al. Trichosporon inkin* as an emergent pathogen in patients with severe pemphigus. *JAMA Dermatol.* 2015; **151:** 642-5.
312. Noman EA, Al-Gheethi AA, Rahman NN, *et al.* Assessment of relevant fungal species in clinical solid wastes. *Environ Sci Pollut Res Int.* 2016; **23:** 19806-24.
313. Nourbakhsh E, Goodman S, Nash S, *et al.* Generalized lymphadenopathy secondary to an invasive fungal infection in an apparently healthy patient. *Am J Med Sci.* 2010; **340:** 84-8.
314. Nucci M, Queiroz-Telles F, Tobon AM, *et al.* Epidemiology of opportunistic fungal infections in Latin America. *Clin Infect Dis.* 2010; **51:** 561-70.
315. Nunes JM, Bizerra FC, Ferreira RC, *et al.* Molecular identification, antifungal susceptibility profile and biofilm formation of clinical and environmental *Rhodotorula* species isolates. *Antimicrob Agents Chemother.* 2013; **57:** 382-9.
316. Obana Y, Sano M, Jike T, *et al.* Differential diagnosis of trichosporonosis using conventional histopathological stains and electron microscopy. *Histopathology.* 2010; **56:** 372-83.
317. Odom DC, Reno HEL. Fungal infections in the hospitalized patient. *Hosp Med Clin.* 2015; **4:** 313-27.
318. Oh SB, Word AP, Dominguez AR. Multiple erythematous papules in a child with neutropenia. *Pediatr Dermatol.* 2015; **32:** 729-30.
319. Ohkubo T, Sugawara Y, Takayama T, *et al.* The risk factors of fungal infection in living-donor liver transplantations. *J Hepatobiliary Pancreat Sci.* 2012; **19:** 382-8.
320. Ok Atılgan A, Ozdemir BH, Kırnap M, *et al.* Invasive fungal infections in liver transplant recipients. *Exp Clin Transplant.* 2014; **12 (Suppl. 1):** 110-6.
321. Oliveira dos Santos C, Zijlstra JG, Porte RJ, *et al.* Emerging pan-resistance in *Trichosporon* species: A case report. *BMC Infect Dis.* 2016; **16:** 148 & 279.
322. Ozkan VK, Sulun Y. Microfungal contaminants on mobile phones of health services vocational school students in Marmaris, Turkey. *Mycopathologia.* 2014; **177:** 59-64.
323. Ozkaya-Parlakay A, Karadag-Oncel E, Cengiz AB, *et al. Trichosporon asahii* sepsis in a patient with pediatric malignancy. *J Microbiol Immunol Infect.* 2016; **49:** 146-9.
324. Pacheco P, Ventura AS, Branco T, *et al.* Clinical experience in invasive fungal infections. *Clin Drug Investig.* 2013; **33:** S23-6.
325. Pacholczyk M, Lagiewska B, Lisik W, *et al.* Invasive fungal infections following liver transplantation - risk factors, incidence and outcome. *Ann Transplant.* 2011; **16:** 14-6.
326. Padhi S, Dash M, Pattanaik S, *et al.* Fungemia due to *Trichosporon mucoides* in a diabetes mellitus patient: A rare case report. *Indian J Med Microbiol.* 2014; **32:** 72-4.
327. Padhye AA, Verghese S, Ravichandran P, *et al. Trichosporon loubieri* infection in a patient with adult polycystic kidney disease. *J Clin Microbiol.* 2003; **41:** 479-82.
328. Pagano L, Lumb J. Update on invasive fungal disease. *Future Microbiol.* 2011; **6:** 985-9.
329. Pagano L, Stamouli M, Tumbarello M, *et al.* Risk of invasive fungal infection in patients affected by acute promyelocytic leukaemia. A report by the SEIFEM-D registry. *Br J Haematol.* 2015; **170:** 434-9.
330. Pahlavan S, Bhatia NN. Fungal spine. *Semin Spine Surg.* 2016; **28:** 163-72.
331. Pana ZD, Kougia V, Roilides E. Therapeutic strategies for invasive fungal infections in neonatal and pediatric patients: An update. *Expert Opin Pharmacother.* 2015; **16:** 693-710.
332. Pang KA, Godet C, Fekkar A, *et al.* Breakthrough invasive mould infections in patients treated with caspofungin. *J Infect.* 2012; **64:** 424-9.
333. Pappas PG, Alexander BD, Andes DR, *et al.* Invasive fungal infections among organ transplant recipients: Results of the Transplant-Associated Infection Surveillance Network (TRANSNET). *Clin Infect Dis.* 2010; **50:** 1101-11.
334. Parahym AM, da Silva CM, Domingos Ide F, *et al.* Pulmonary infection due to *Pseudozyma aphidis* in a patient with Burkitt lymphoma: First case report. *Diagn Microbiol Infect Dis.* 2013; **75:** 104-6.
335. Paramythiotou E, Frantzeskaki F, Flevari A, *et al.* Invasive fungal infections in the ICU: How to approach, how to treat. *Molecules.* 2014; **19:** 1085-119.

336. Parize P, Angebault C, de Fonbrune FS, *et al.* Naturotherapy as a potential source of mould infections in patients with haematological malignancies. *J Hosp Infect.* 2013; **85:** 163-4.
337. Park BJ, Pappas PG, Wannemuehler KA, *et al.* Invasive non-*Aspergillus* mold infections in transplant recipients, United States, 2001-2006. *Emerg Infect Dis.* 2011; **17:** 1855-64.
338. Parsonage MJ, Stafford ND, Lillie P, *et al.* Oral voriconazole for invasive fungal skull base infection. *J Laryngol Otol.* 2010; **124:** 1010-3.
339. Patel MH, Patel RD, Vanikar AV, *et al.* Invasive fungal infections in renal transplant patients: A single center study. *Ren Fail.* 2017; **39:** 294-8.
340. Patel MS, Wright AJ, Kohn R, *et al.* Successful long-term management of invasive cerebral fungal infection following liver transplantation. *Mycoses.* 2015; **58:** 181-6.
341. Patra S, Vij M, Chirla DK, et al. Unsuspected mixed invasive intestinal mycosis in an infant with metabolic disorder. *Trop Gastroenterol.* 2013; **34:** 283-5.
342. Pechlivanoglou P, Le HH, Daenen S, *et al.* Mixed treatment comparison of prophylaxis against invasive fungal infections in neutropenic patients receiving therapy for haematological malignancies: a systematic review. *J Antimicrob Chemother.* 2014; **69:** 1-11.
343. Peghin M, Monforte V, Martin-Gomez MT, *et al.* Epidemiology of invasive respiratory disease caused by emerging non-*Aspergillus* molds in lung transplant recipients. *Transpl Infect Dis.* 2016; **18:** 70-8.
344. Perard B, Rougeron A, Favre S, *et al. Trichosporon faecale* invasive infection in a patient with severe aplastic anemia: Efficacy of voriconazole and liposomal amphotericin B before neutrophil recovery. *Med Mycol Case Rep.* 2015; **9:** 12-4.
345. Pereira C, Ribeiro S, Lopes V, *et al. Rhodotorula mucilaginosa* fungemia and pleural tuberculosis in an immunocompetent patient: An uncommon association. *Mycopathologia.* 2016; **181:** 145-9.
346. Perfect JR, Hachem R, Wingard JR. Update on epidemiology of and preventive strategies for invasive fungal infections in cancer patients. *Clin Infect Dis.* 2014; **59:** S352-5.
347. Perusquia-Ortiz AM, Vázquez-Gonzalez D, Bonifaz A. Opportunistic filamentous mycoses: Aspergillosis, mucormycosis, phaeohyphomycosis and hyalohyphomycosis. *J Dtsch Dermatol Ges.* 2012; **10:** 611-22.
348. Pesut D, Adzic T, Tosic T, *et al.* Bilateral pulmonary shadows and incapsulated mediastinal effusion in a patient with respiratory blastoschizomycosis. *Pneumologia.* 2008; **57:** 88 & 90-1.
349. Pfaller MA, Diekema DJ. Epidemiology of invasive mycoses in North America. *Crit Rev Microbiol.* 2010; **36:** 1-53.
350. Philips BJ. What should we be doing about fungal infections in intensive care? *Br J Anaesth.* 2011; **107:** 299-302.
351. Pignataro L, Torretta S, Capaccio P, *et al.* Unusual otolaryngological manifestations of certain systemic bacterial and fungal infections in children. *Int J Pediatr Otorhinolaryngol.* 2009; **73:** S33-7.
352. Pinney MF, Rosenberg AF, Hampp C, *et al.* Invasive fungal infections in lung transplant recipients not receiving routine systemic antifungal prophylaxis: 12-year experience at a university lung transplant center. *Pharmacotherapy.* 2011; **31:** 537-45.
353. Prakash A, Wankhede S, Singh PK, *et al.* First neonatal case of fungaemia due to *Pseudozyma aphidis* and a global literature review. *Mycoses.* 2014; **57:** 64-8.
354. Quan C, Spellberg B. Mucormycosis, pseudallescheriasis and other uncommon mold infections. *Proc Am Thorac Soc.* 2010; **7:** 210-5.
355. Queiroz-Telles F, Nucci M, Colombo AL, *et al.* Mycoses of implantation in Latin America: An overview of epidemiology, clinical manifestations, diagnosis and treatment. *Med Mycol.* 2011; **49:** 225-36.
356. Raggam RB, Fischbach LM, Prattes J, *et al.* Detection of (1→3)-β-D-glucan in same-day urine and serum samples obtained from patients with haematological malignancies. *Mycoses.* 2015; **58:** 394-8.
357. Ramesha KN, Kate MP, Kesavadas C, *et al.* Fungal infections of the central nervous system in HIV-negative patients: Experience from a tertiary referral center of South India. *Ann Indian Acad Neurol.* 2010; **13:** 112-6.
358. Ramos-E-Silva M, Lima CM, Schechtman RC, *et al.* Superficial mycoses in immunodepressed patients (AIDS). *Clin Dermatol.* 2010; **28:** 217-25.
359. Ramos-e-Silva M, Lima CM, Schechtman RC, *et al.* Systemic mycoses in immunodepressed patients (AIDS). *Clin Dermatol.* 2012; **30:** 616-27.
360. Ranjan R, Chowdhary P, Pandey A, *et al. Trichosporon asahii* as a cause of urinary tract infection: A rare human pathogen. *Indian J Pathol Microbiol.* 2015; **58:** 413-4.
361. Rastogi V, Honnavar P, Rudramurthy SM, *et al.* Molecular characterisation and antifungal susceptibility of clinical *Trichosporon* isolates in India. *Mycoses.* 2016; **59:** 528-34.
362. Rastogi V, Nirwan PS. Invasive trichosporonosis due to *Trichosporon asahii* in a non-immunocompromised host: A rare case report. *Indian J Med Microbiol.* 2007; **25:** 59-61.
363. Rastogi V, Rudramurthy SM, Maheshwari M, *et al.* Non-healing ulcer due to *Trichosporon loubieri* in an immunocompetent host and review of published reports. *Mycopathologia.* 2013; **176:** 107-11.
364. Rath PC, Purohit BV, Agrawal B, *et al.* Pacemaker lead endocarditis due to *Trichosporon* species. *J Assoc Physicians India.* 2015; **63:** 66-8.
365. Rebolledo M, Sarria JC. Intra-abdominal fungal infections. *Curr Opin Infect Dis.* 2013; **26:** 441-6.
366. Repetto EC, Giacomazzi CG, Castelli F. Hospital-related outbreaks due to rare fungal pathogens: a review of the literature from 1990 to June 2011. *Eur J Clin Microbiol Infect Dis.* 2012; **31:** 2897-904.
367. Riches ML, Trifilio S, Chen M, *et al.* Risk factors and impact of non-*Aspergillus* mold infections following allogeneic HCT: A CIBMTR infection and immune reconstitution analysis. *Bone Marrow Transplant.* 2016; **51:** 277-82.
368. Rodriguez CJ, Weintrob AC, Shah J, *et al.* Risk factors associated with invasive fungal infections in combat trauma. *Surg Infect (Larchmt).* 2014; **15:** 521-6.

369. Roilides E, Sigler L, Bibashi E, *et al.* Disseminated infection due to *Chrysosporium zonatum* in a patient with chronic granulomatous disease and review of non-*Aspergillus* fungal infections in patients with this disease. *J Clin Microbiol.* 1999; **37:** 18-25.
370. Roilides E. Emerging fungi causing human infection: New or better identified? *Clin Microbiol Infect.* 2016; **22:** 660-1.
371. Roman AD, Salvana EM, Guzman-Penamora MA, *et al.* Invasive trichosporonosis in an AIDS patient: Case report and review of the literature. *Int J STD AIDS.* 2014; **25:** 70-5.
372. Roy U, Jessani LG, Rudramurthy SM, *et al.* Seven cases of *Saccharomyces* fungaemia related to use of probiotics. *Mycoses.* 2017; **60:** 375-80.
373. Ruan SY, Chien JY, Hsueh PR. Invasive trichosporonosis caused by *Trichosporon asahii* and other unusual *Trichosporon* species at a medical center in Taiwan. *Clin Infect Dis.* 2009; **49:** e11-7.
374. Rubic Z, Novak A, Tomic Z, *et al.* Prompt diagnosis and effective treatment of *Trichosporon asahii* catheter-related infection in non-immunocompromised neurosurgical patient. *Mycopathologia.* 2015; **179:** 125-8.
375. Ruhnke M, Schwartz S. Recent developments in the management of invasive fungal infections in patients with oncohematological diseases. *Ther Adv Hematol.* 2016; **7:** 345-59.
376. Ruping MJ, Vehreschild JJ, Cornely OA. Patients at high risk of invasive fungal infections: When and how to treat. *Drugs.* 2008; **68:** 1941-62.
377. Sabharwal ER. Successful management of *Trichosporon asahii* urinary tract infection with fluconazole in a diabetic patient. *Indian J Pathol Microbiol.* 2010; **53:** 387-8.
378. Sadique Z, Grieve R, Harrison DA, *et al.* An integrated approach to evaluating alternative risk prediction strategies: A case study comparing alternative approaches for preventing invasive fungal disease. *Value Health.* 2013; **16:** 1111-22.
379. Saliba F. Emerging fungal infections. *Expert Rev Anti Infect Ther.* 2012; **10:** 419-21.
380. Santino I, Alari A, Bono S, *et al.* *Saccharomyces cerevisiae* fungemia, a possible consequence of the treatment of *Clostridium difficile* colitis with a probioticum. *Int J Immunopathol Pharmacol.* 2014; **27:** 143-6.
381. Santos T, Aguiar B, Santos L, *et al.* Invasive fungal infections after kidney transplantation: A single-center experience. *Transplant Proc.* 2015; **47:** 971-5.
382. Sarabahi S, Tiwari VK, Arora S, *et al.* Changing pattern of fungal infection in burn patients. Burns. 2012; **38:** 520-8.
383. Schaal JV, Leclerc T, Pasquier P, *et al.* Epidemiology of fungal infection in burns: Therapeutic implications. *Burns.* 2012; **38:** 942-4.
384. Schaal JV, Leclerc T, Soler C, *et al.* Epidemiology of filamentous fungal infections in burned patients: A French retrospective study. *Burns.* 2015; **41:** 853-63.
385. Schaenman JM, Rosso F, Austin JM, *et al.* Trends in invasive disease due to *Candida* species following heart and lung transplantation. *Transpl Infect Dis.* 2009; **11:** 112-21.
386. Scheffel RS, Dora JM, Weinert LS, *et al.* Invasive fungal infections in endogenous Cushing's syndrome. *Infect Dis Rep.* 2010; **2:** e4.
387. Schoof B, Jakobs O, Schmidl S, *et al.* Fungal periprosthetic joint infection of the hip: A systematic review. *Orthop Rev (Pavia).* 2015; **7:** 5748.
388. Scully EP, Baden LR, Katz JT. Fungal brain infections. *Curr Opin Neurol.* 2008; **21:** 347-52.
389. Serio S, Burgess B, Voigt D. Fungal periapical abscess and the burn patient: A report of two cases of an unreported source for systemic infection. *Burns.* 2017. pii: S0305-4179(17)30291-7. PMID: 28602586.
390. Sganga G, Bianco G, Frongillo F, *et al.* Fungal infections after liver transplantation: Incidence and outcome. *Transplant Proc.* 2014; **46:** 2314-8.
391. Shah AV, McColley SA, Weil D, *et al.* *Trichosporon mycotoxinivorans* infection in patients with cystic fibrosis. *J Clin Microbiol.* 2014; **52:** 2242-4.
392. Shahapur PR, Bidri RC. Recent trends in the spectrum of opportunistic infections in human immunodeficiency virus infected individuals on antiretroviral therapy in South India. *J Nat Sci Biol Med.* 2014; **5:** 392-6.
393. Sharifkashani B, Farshidpour M, Droudinia A, *et al.* Cutaneous alternariosis with *Trichosporon* infection in a heart transplant recipient: A case report. *Exp Clin Transplant.* 2013; **11:** 464-6.
394. Sharma S, Bajaj D, Sharma P. Fungal infection in thermal burns: A prospective study in a tertiary care centre. *J Clin Diagn Res.* 2016; **10:** PC05-7.
395. Sharma SK, Kumar S, Singh AK, *et al.* Feasibility and outcome of CT-guided lung biopsy in patients with hematological diseases and suspected fungal pneumonia. *J Infect Dev Ctries.* 2013; **7:** 748-52.
396. Shinde RS, Mantur BG, Patil G, *et al.* Meningitis due to *Rhodotorula glutinis* in an HIV infected patient. *Indian J Med Microbiol.* 2008; **26:** 375-7.
397. Shoham S, Hinestrosa F, Moore J Jr, *et al.* Invasive filamentous fungal infections associated with renal transplant tourism. *Transpl Infect Dis.* 2010; **12:** 371-4.
398. Shoham S, Marr KA. Invasive fungal infections in solid organ transplant recipients. *Future Microbiol.* 2012; **7:** 639-55.
399. Shoham S, Marwaha S. Invasive fungal infections in the ICU. *J Intensive Care Med.* 2010; **25:** 78-92.
400. Siddiqui W, Ahmed Y, Albrecht H, *et al.* *Pseudozyma* spp. catheter-associated blood stream infection, an emerging pathogen and brief literature review. *BMJ Case Rep.* 2014; pii: bcr2014206369.
401. Silva MF, Ferriani MP, Terreri MT, *et al.* A multicenter study of invasive fungal infections in patients with childhood-onset systemic lupus erythematosus. *J Rheumatol.* 2015; **42:** 2296-303.
402. Silveira FP, Husain S. Fungal infections in lung transplant recipients. *Curr Opin Pulm Med.* 2008; **14:** 211-8.
403. Silveira FP, Kwak EJ, Paterson DL, *et al.* Post-transplant colonization with non-*Aspergillus* molds and risk of development of invasive fungal disease in lung transplant recipients. *J Heart Lung Transplant.* 2008; **27:** 850-5.

404. Singh GK, Capoor MR, Nair D, *et al.* Spectrum of fungal infection in head and neck cancer patients on chemoradiotherapy. *J Egypt Natl Canc Inst.* 2017; **29:** 33-37.

405. Singh N, Huprikar S, Burdette SD, *et al.* Donor-derived fungal infections in organ transplant recipients: Guidelines of the American Society of Transplantation, Infectious Diseases Community of Practice. *Am J Transplant.* 2012; **12:** 2414-28.

406. Sinko J, Bryan J. Managing invasive fungal infection in hematopoietic stem-cell transplantation. *Future Microbiol.* 2010; **5:** 1309-14.

407. Sipsas NV, Kontoyiannis DP. Invasive fungal infections in patients with cancer in the Intensive Care Unit. *Int J Antimicrob Agents.* 2012; **39:** 464-71.

408. Slavin MA, Chakrabarti A. Opportunistic fungal infections in the Asia-Pacific region. *Med Mycol.* 2012; **50:** 18-25.

409. Smith JA, Kauffman CA. Recognition and prevention of nosocomial invasive fungal infections in the intensive care unit. *Crit Care Med.* 2010; **38:** S380-7.

410. Sole A, Salavert M. Fungal infections after lung transplantation. *Curr Opin Pulm Med.* 2009; **15:** 243-53.

411. Spiliopoulou A, Anastassiou ED, Christofidou M. *Rhodotorula* fungemia of an intensive care unit patient and review of published cases. *Mycopathologia.* 2012; **174:** 301-9.

412. Sreeja S, Banashankari GS, Bhavana MV, *et al. Blastoschizomyces capitatus* pneumonia: A rare case. *Indian J Pathol Microbiol.* 2011; **54:** 846-7.

413. Stewart E, Waldman S, Sutton DA, *et al. Ustilago echinata:* Infection in a mixed martial artist following an open fracture. *Mycopathologia.* 2016; **181:** 311-4.

414. Struck MF, Gille J. Fungal infections in burns: A comprehensive review. *Ann Burns Fire Disasters.* 2013; **26:** 147-53.

415. Suarez F, Olivier G, Garcia-Hermoso D, *et al.* Breakthrough *Hormographiella aspergillata* infections arising in neutropenic patients treated empirically with caspofungin. *J Clin Microbiol.* 2011; **49:** 461-5.

416. Subramanya Supram H, Gokhale S, Chakrabarti A, *et al.* Emergence of *Magnusiomyces capitatus* infections in western Nepal. *Med Mycol.* 2016; **54:** 103-10.

417. Sun W, Su J, Xu S, *et al. Trichosporon asahii* causing nosocomial urinary tract infections in intensive care unit patients: Genotypes, virulence factors and antifungal susceptibility testing. *J Med Microbiol.* 2012; **61 (Pt. 12):** 1750-7.

418. Suzuki K, Nakase K, Kyo T, *et al.* Fatal *Trichosporon* fungemia in patients with hematologic malignancies. *Eur J Haematol.* 2010; **84:** 441-7.

419. Swiss T, Cervantes SS, Hinni M, *et al.* Invasive fungal laryngopharyngitis resulting in laryngeal destruction with complete laryngotracheal separation: Report of a case. *Ear Nose Throat J.* 2017; **96:** e33-6.

420. Sychev D, Maya ID, Allon M. Clinical outcomes of dialysis catheter-related candidemia in hemodialysis patients. *Clin J Am Soc Nephrol.* 2009; **4:** 1102-5.

421. Tacke D, Koehler P, Cornely OA. Fungal endocarditis. *Curr Opin Infect Dis.* 2013; **26:** 501-7.

422. Taj-Aldeen SJ, AbdulWahab A, Kolecka A, *et al.* Uncommon opportunistic yeast bloodstream infections from Qatar. *Med Mycol.* 2014; **52:** 552-6.

423. Taj-Aldeen SJ, Al-Ansari N, El Shafei S, *et al.* Molecular identification and susceptibility of *Trichosporon* species isolated from clinical specimens in Qatar: Isolation of *Trichosporon dohaense* Taj-Aldeen, Meis & Boekhout sp. nov. *J Clin Microbiol.* 2009; **47:** 1791-9.

424. Tanuskova D, Horakova J, Svec P, *et al.* First case of invasive *Magnusiomyces capitatus* infection in Slovakia. *Med Mycol Case Rep.* 2017; **16:** 12-5.

425. Tap RM, Sabaratnam P, Ramli NY, *et al.* Subcutaneous infection associated with *Trichosporon ovoides:* A case report and review of literature. *Mycopathologia.* 2016; **181:** 285-90.

426. Tattevin P, Revest M, Lefort A, *et al.* Fungal endocarditis: Current challenges. *Int J Antimicrob Agents.* 2014; **44:** 290-4.

427. Taverna CG, Cordoba S, Murisengo OA, *et al.* Molecular identification, genotyping and antifungal susceptibility testing of clinically relevant *Trichosporon* species from Argentina. *Med Mycol.* 2014; **52:** 356-66.

428. Tessari G, Cagalli A, Girolomoni G. Opportunistic deep cutaneous mycoses in solid organ transplant recipients. *G Ital Dermatol Venereol.* 2014; **149:** 417-22.

429. Thion LA, Coutard A, Eloy O, *et al. Trichosporon inkin* disseminated infection. *Intensive Care Med.* 2017; doi: 10.1007/s00134-017-4862-5. PMID: 28612088.

430. Thygesen JB, Glerup H, Tarp B. *Saccharomyces boulardii* fungemia caused by treatment with a probioticum. *BMJ Case Rep.* 2012; pii: bcr0620114412.

431. Tortorano AM, Dho G, Prigitano A, *et al.* Invasive fungal infections in the intensive care unit: A multicentre, prospective, observational study in Italy (2006-2008). *Mycoses.* 2012; **55:** 73-9.

432. Trabelsi H, Neji S, Gargouri L, *et al. Geotrichum capitatum* septicemia: Case report and review of the literature. *Mycopathologia.* 2015; **179:** 465-9.

433. Trabelsi H, Neji S, Sellami, *et al.* Invasive fungal infections in renal transplant recipients: About 11 cases. *J Mycol Med.* 2013; **23:** 255-60.

434. Tragiannidis A, Kyriakidis I, Zundorf I, *et al.* Invasive fungal infections in pediatric patients treated with tumor necrosis alpha (TNF-α) inhibitors. *Mycoses.* 2017; 60: 222-9.

435. Trevino M, Garcia-Riestra C, Areses P, *et al.* Emerging *Trichosporon asahii* in elderly patients: Epidemiological and molecular analysis by the Diversi Lab system. *Eur J Clin Microbiol Infect Dis.* 2014; **33:** 1497-503.

436. Tribble DR, Rodriguez CJ, Weintrob AC, *et al.* Environmental factors related to fungal wound contamination after combat trauma in Afghanistan, 2009-2011. *Emerg Infect Dis.* 2015; **21:** 1759-69.

437. Tribble DR, Rodriguez CJ. Combat-related invasive fungal wound infections. *Curr Fungal Infect Rep.* 2014; **8:** 277-286.

438. Tsai MJ, Chang WA, Tsai KB, *et al.* Probable invasive pulmonary trichosporonosis in a diabetic patient. *Am J Respir Crit Care Med.* 2011; **184:** 982.

439. Tsai MS, Yang YL, Wang AH, *et al.* Susceptibilities to amphotericin B, fluconazole and voriconazole of *Trichosporon* clinical isolates. *Mycopathologia.* 2012; **174:** 121-30.
440. Tuon FF, Costa SF. *Rhodotorula* infection. A systematic review of 128 cases from literature. *Rev Iberoam Micol.* 2008; **25:** 135-40.
441. Ullah K, Raza S, Ahmed P, *et al.* Post-transplant infections: single center experience from the developing world. *Int J Infect Dis.* 2008; **12:** 203-14.
442. Valerio M, Vena A, Bouza E, *et al.* How much European prescribing physicians know about invasive fungal infections management? *BMC Infect Dis.* 2015; **15:** 80.
443. Vallejo C, Vazquez L, Cabrera Martín JR, *et al.* Treatment of invasive fungal infections in high-risk haematological patients: What have we learnt in the past 10 years? *Rev Esp Quimioter.* 2013; **26:** 378-86.
444. Varkey JB, Perfect JR. Rare and emerging fungal pulmonary infections. *Semin Respir Crit Care Med.* 2008; **29:** 121-31.
445. Vazquez JA. Invasive fungal infections in the intensive care unit. *Semin Respir Crit Care Med.* 2010; **31:** 79-86.
446. Vazquez R, Vazquez-Guillamet MC, Suarez J, *et al.* Invasive mold infections in lung and heart-lung transplant recipients: Stanford University experience. *Transpl Infect Dis.* 2015; **17:** 259-66.
447. Vazquez-Gonzalez D, Perusquia-Ortiz AM, Hundeiker M, *et al.* Opportunistic yeast infections: Candidiasis, cryptococcosis, trichosporonosis and geotrichosis. *J Dtsch Dermatol Ges.* 2013; **11:** 381-93; Quiz 394.
448. Vehreschild JJ, Ruping MJ, Steinbach A, *et al.* Diagnosis and treatment of fungal infections in allogeneic stem cell and solid organ transplant recipients. *Expert Opin Pharmacother.* 2010; **11:** 95-113.
449. Verduyn Lunel FM, Meis JF, Voss A. Nosocomial fungal infections: Candidemia. *Diagn Microbiol Infect Dis.* 1999; **34:** 213-20.
450. Vidya KM, Rao UK, Nittayananta W, *et al.* Oral mycoses and other opportunistic infections in HIV: Therapy and emerging problems - A workshop report. *Oral Dis.* 2016; **22 (Suppl. 1):** 158-65.
451. Vishwanath M, Ilanthodi S, Bhandary R, *et al.* Invasive fungal infection in acute myeloid leukemia associated with myeloid sarcoma of the sinonasal cavity: A rare case. *Indian J Otolaryngol Head Neck Surg.* 2017; **69:** 266-8.
452. Wadhwa A, Kaur R, Bhalla P. *Saccharomyces cerevisiae* as a cause of oral thrush & diarrhoea in an HIV/AIDS patient. *Trop Gastroenterol.* 2010; **31:** 227-8.
453. Wan J, Piette EW, Rosenbach M. Purpuric and cream-colored plaques in an immunocompromised person: A case of disseminated trichosporonosis. *JAAD Case Rep.* 2016; **2:** 275-7.
454. Wang LR, Barber CE, Johnson AS, *et al.* Invasive fungal disease in systemic lupus erythematosus: A systematic review of disease characteristics, risk factors and prognosis. *Semin Arthritis Rheum.* 2014; **44:** 325-30.
455. Wang ES. Common fungal infections in patients with leukemia. *Clin Adv Hematol Oncol.* 2017; **15:** 352-4.
456. Wang Y, Zhao Z, Lu H, *et al.* Fungal infection involvement in primary biliary cirrhosis: A review of 2 cases. *Exp Ther Med.* 2017; **13:** 489-94.
457. Warkentien T, Rodriguez C, Lloyd B, *et al.* Invasive mold infections following combat-related injuries. *Clin Infect Dis.* 2012; **55:** 1441-9.
458. Watanabe N, Matsumoto K, Kojima S, *et al.* Invasive fungal infections in pediatric patients with hematologic malignancies receiving oral amphotericin B solution and early intravenous administration of fluconazole. *J Pediatr Hematol Oncol.* 2011; **33:** 270-5.
459. Wattier RL, Dvorak CC, Hoffman JA, *et al.* A Prospective, international cohort study of invasive mold infections in children. *J Pediatric Infect Dis Soc.* 2015; **4:** 313-22.
460. Weinhold B. Rare fungal illness follows tornado. *Environ Health Perspect.* 2013; **121:** A116.
461. Weng TF, Ho MW, Lin HC, *et al.* Successful treatment of disseminated mixed invasive fungal infection after hematopoietic stem cell transplantation for severe aplastic anemia. *Pediatr Transplant.* 2012; **16:** e35-8.
462. Wheat J. Endemic mycoses in AIDS: a clinical review. *Clin Microbiol Rev.* 1995; **8:** 146-59.
463. Wirth F, Goldani LZ. Epidemiology of *Rhodotorula*: An emerging pathogen. *Interdiscip Perspect Infect Dis.* 2012; 465717. PMID: 23091485.
464. Xiao M, Guo LN, Kong F, *et al.* Practical identification of eight medically important *Trichosporon* species by reverse line blot hybridization (RLB) assay and rolling circle amplification (RCA). *Med Mycol.* 2013; **51:** 300-8.
465. Yang CH, He XS, Chen J,*et al.* Fungal infection in patients after liver transplantation in years 2003 to 2012. *Ann Transplant.* 2012; **17:** 59-63.
466. Yilmaz Karapinar D, Karadas N, Onder Sivis Z, *et al.* Rare severe mycotic infections in children receiving empirical caspofungin treatment for febrile neutropenia. *Braz J Infect Dis.* 2015; **19:** 549-52.
467. Yong AMY, Yang SS, Tan KB, *et al.* Disseminated cutaneous trichosporonosis in an adult bone marrow transplant patient. *Indian Dermatol Online J.* 2017; **8:** 192-4.
468. Zaoutis TE, Webber S, Naftel DC, *et al.* Invasive fungal infections in pediatric heart transplant recipients: Incidence, risk factors and outcomes. *Pediatr Transplant.* 2011; **15:** 465-9.
469. Zekavat OR, Amanati A, Saleh F, *et al.* Current strategies against invasive fungal infections in patients with aplastic anemia, strong power and weak weapon - A case report and review of literature. *Med Mycol Case Rep.* 2016; **11:** 16-20.
470. Zhang E, Sugita T, Tsuboi R, *et al.* The opportunistic yeast pathogen *Trichosporon asahii* colonizes the skin of healthy individuals: Analysis of 380 healthy individuals by age and gender using a nested polymerase chain reaction assay. *Microbiol Immunol.* 2011; **55:** 483-8.
471. Zicker M, Colombo AL, Ferraz-Neto BH, *et al.* Epidemiology of fungal infections in liver transplant recipients: A six-year study of a large Brazilian liver transplantation centre. *Mem Inst Oswaldo Cruz.* 2011; **106:** 339-45.
472. Zilberberg MD, Shorr AF. Fungal infections in the ICU. *Infect Dis Clin North Am.* 2009; **23:** 625-42.

SECTION VI

Miscellaneous Mycoses

Chapters

This Section consists of diverse topics covering miscellaneous fungal infections and a few related issues to Medical Mycology. These could not be sorted out under any of the previously described Sections hence they are here clubbed together. As these clinical entities are causing significant disorders of fungal origin thus their comprehensive understanding is essential. The fungal infections of eye and its adnexa are now detected more often due to increased awareness among medical personnel and availability of newer advances in diagnostic modalities. Previously, keratomycosis was the only established fungal infection of eye but involvement of internal structures *i.e.* endophthalmitis is now becoming common due to exogenous and/or endogenous invasion by the fungi. Moreover, adnexa of the eye are also substantially infected by fungi. Similarly, fungal infections of ear, particularly otomycosis, is very common during the rainy season and is described in a separate Chapter. The infection may extend to the middle ear as well as mastoid also leading to irreparable damage.

Hyalohyphomycosis, a disease entity caused by a variety of hyaline fungi, on an analogous pattern of phaeohyphomycosis, has been described including some of the emerging infections. Adiaspiromycosis, an unusual zoonotic fungal disease caused by *Emmonsia crescens* and *Ea.parva* but incidentally encountered among humans also. The spherule formed in this entity may mimic other endosporulating organisms, signifying it from differential diagnosis point of view. The other species previously categorized under the genus *Emmonsia* or *Emmonsia*-like, have now been assigned new genus name as *Emergomyces* with its three clinically significant species *i.e. Es.pasteurianus, Es.africanus* and *Es.orientalis.* These species came to limelight in recent times due to their strong association with AIDS. The disease name, Emergomycosis, is being used for the first time in medical literature through this Textbook.

The allergic diseases of fungal origin constitute a substantial number of patients and have been briefly described. During recent times lot of cases of fungal rhinosinusitis have been encountered, therefore, a separate Chapter is devoted on this emerging entity dealing with its clinical and mycological aspects. Mycotoxicosis and mycetismus are responsible for a large number of disorders in human as well as animals and sometimes may be found in epidemic form. Therefore, their clinical and public health importance is comprehensively described in this Section. The Chapter on Mycology Laboratory Contaminants, which was withdrawn in previous edition but again added keeping in view of frequent contamination in diagnostic microbiology laboratory despite the fact that very same fungi may be causing significant disease in one or the other form.

CHAPTER 29

Oculomycosis

The fungi are significant pathogens causing ocular infections due to their frequent involvement and difficulty in establishing definitive diagnosis. Moreover, there are devastating consequences if these infective processes are not accurately diagnosed at an early stage of their onset. **Oculomycosis** encompasses fungal disease of eye denoting manifestations related to external and internal structures of the eyeball. However, to be more precise, **keratomycosis** is defined as fungal infection of cornea and **endophthalmitis** when posterior chamber is involved.

The clinical descriptions of fungal infections of eye and adnexa are divided into **three** broad categories:

A. Keratomycosis
B. Fungal Endophthalmitis
C. Fungal Infections of Ocular Adnexa

A. KERATOMYCOSIS

This clinical entity, **keratomycosis** is defined as invasive infection of corneal stroma caused by variety of fungal species. This is also called **mycotic keratitis** or simply **fungal keratitis** or **fungal corneal ulcer**. It is exclusively infection of the cornea only and not the keratinised layers of skin as its name sounds like that hence should not be mistaken.

The majority of cases occur among agricultural workers following corneal trauma with vegetative matter contaminated by the fungi. These are opportunistic organisms and colonize when natural defenses of eye are abrogated by corneal trauma, use of topical corticosteroids or any other predisposing factors. Essentially, most of them are saprotrophic fungi and are occasionally associated with true infections among healthy individuals.

Historical Perspective

Leber described fungal keratitis caused by *Aspergillus* species for the first time in 1879 in Germany. It was considered as rare entity in the past because up to 1951 only 63 cases were reported. As a result of awareness among medical personnel about use of corticosteroids predisposed to keratomycosis as well as use of calcofluor white stain in mid-1980s as diagnostic tool in this clinical entity, have improved early detection of these cases. In recent times, confocal microscopy has been very useful in delineating the nature and extent of keratomycosis. Earlier, phaeoid (dematiaceous) fungi were not considered to be significant but now these are also important cause of keratomycosis.

Fungal keratitis has been a neglected tropical disease. The WHO is taking an initiative to document this disease so that it may be categorized under the neglected tropical disease.

Epidemiology

The causative fungi of keratomycosis are ubiquitous organisms and are responsible for 6-53% of all corneal infections worldwide. This entity is frequently encountered, resulting in visual disability, particularly in tropical countries. This scenario is truer where trachoma and other infectious agents of ocular diseases are commonly prevalent. Corneal opacification following various infections is second major cause of blindness after cataract.

Due to large agrarian population and environmental factors, keratomycosis is common in India. The condition frequently develops after corneal trauma by vegetative matter. Rarely, it may occur following surgery like keratoplasty. It may also be found in contact lens wearers like that of *Acanthamoeba* infection. Injudicious use of topical corticosteroids and antibacterial agents for external ocular diseases enhance risk of keratomycosis. Its management really poses a challenge to the microbiologists for the diagnosis and ophthalmologists for the treatment. However, in recent times, increasing awareness about this entity due to addition of various newer techniques for fungal isolation and identification, has essentially contributed to its early diagnosis entailing institution of appropriate antifungal therapy.

Table 29.1. Common Agents of Keratomycosis.

A. Hyaline Hyphomycetes
Aspergillus flavus, A.fumigatus
Fusarium solani, F.oxysporum
Acremonium species
Paecilomyces species
Purpureocillium lilacinum
Penicillium species
Scedosporium apiospermum
B. Phaeoid Hyphomycetes
Curvularia species
Bipolaris species
Alternaria species
Exserohilum species
Fonsecaea pedrosoi
Aureobasidium pullulans
Colletotrichum dematium
C. Yeast-like Fungi
Candida albicans
Candida tropicalis
Candida krusei

The males are more commonly affected as compared to females due to their higher outdoor exposure. It may occur at any age but the highest incidence coincides with period of maximal activity in the middle age. The peak months of disease correspond to harvesting season when chances of injury with infected vegetative matter are very high. In northern India *Aspergillus* is the commonest cause whereas in South India, it is *Fusarium*, which is most common. In India fungal keratitis cases are 113 per 100000 of population.

Mycology

The list of fungal causative agents isolated from patients of keratomycosis is long and varied. There is substantial fungal flora in normal eye and even greater in the diseased eye. Some of the fungi are found as commensals in normal conjunctival sac and are usually present bilaterally. The infection is substantiated by demonstration of fungal elements in debrided materials as well as culture. The causative fungi belong to at least 70 species from 40 genera. These may be **hyaline** or **phaeoid** hyphomycetes and **yeast-like** fungi. The majority of cases of keratomycosis have been caused by ***Aspergillus, Fusarium*** and phaeoid (dematiaceous) fungal species.

The filamentous fungi are responsible for up to one-third of all cases of infectious keratitis in certain parts of the world. *Aspergillus* genus is listed as the most frequent cause of keratomycosis. *Aspergillus* and *Fusarium* are responsible for one-third of all traumatic infectious keratitis. Other fungal genera may be: *Curvularia, Candida, Acremonium, Paecilomyces, Penicillium, Alternaria, Fonsecaea, Scedosporium, Bipolaris* and *Aureobasidium*. Out of aspergilli causing keratomycosis, *Aspergillus flavus, A.fumigatus, A.niger* and *A.glaucus* are in order of their frequency. The phaeoid fungi are now considered as significant cause of fungal keratitis, which were not frequently encountered previously. This is now the third major group of fungi after *Aspergillus* and *Fusarium* species.

In India, *Aspergillus* genus is the commonest fungal agent followed by other genera. *Fusarium* is the most common agent in southeastern USA and *Candida* and *Aspergillus* species, in northern USA. In one of Indian studies, *Curvularia* species has been reported second commonest agent of mycotic keratitis after *Aspergillus* species.

Keratomycosis caused by dimorphic fungi is rarely encountered in their respective areas of endemicity. Occasionally mucormycetes such as *Lichtheimia* or *Rhizopus* species may be implicated in keratitis. The commonly encountered fungal agents in keratomycosis are listed in **Table 29.1**.

Pathogenesis and Pathology

The keratomycosis occurs due to interaction between different agent factors such as invasiveness and toxigenicity of fungal strains, host factors, hypersensitivity reaction and underlying defect of neutrophils. The predisposing factors, which render immune system very weak are trauma, administration of topical corticosteroids, antibacterial antibiotics and indigenous remedies.

The fungi are unable to penetrate intact corneal epithelium hence any trauma particularly, organic matter facilitate penetration of fungal inoculum into corneal stroma. The fungal hyphae invade from corneal ulcer bed to stroma. Coagulation necrosis associated with loss of keratocytes and edematous changes of collagen fibres occur. Satellite lesions are formed around main site of involvement. Late in the course of disease process, hyphae may be seen to penetrate Descemet's membrane, encased in dense neutrophilic exudate of **hypopyon**. Leukocytic infiltrate with feathery borders in corneal stroma is characteristic of fungal keratitis **(Fig. 29.1)**.

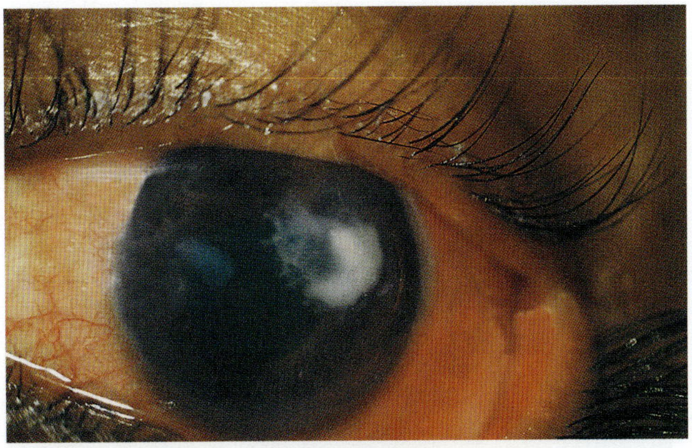

Fig. 29.1. Fungal corneal ulcer with feathery hyphate edges.

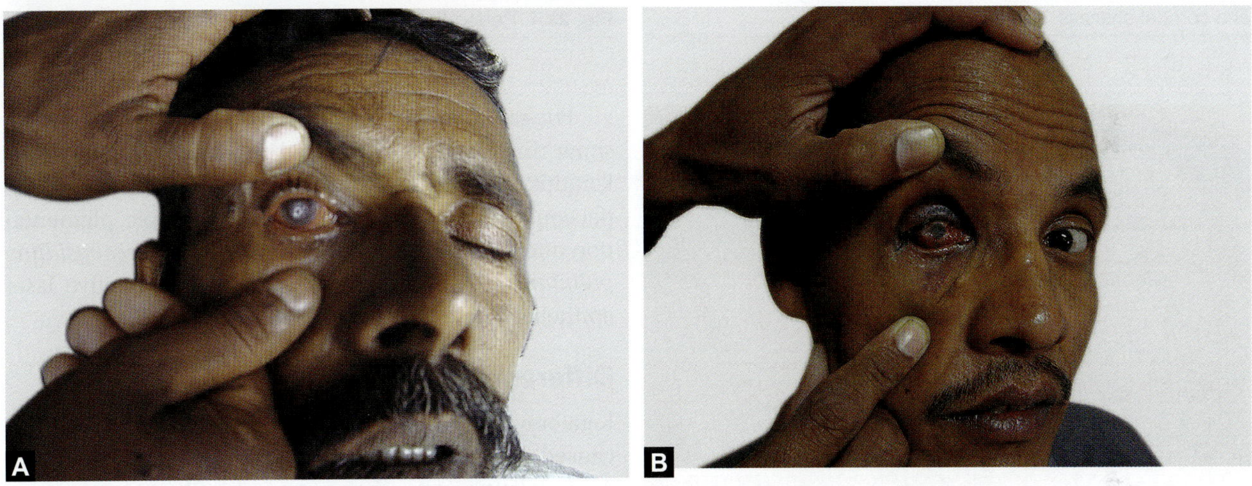

Figs. 29.2A and B. Fungal corneal ulcers with hypopyon in the right eye of both patients.

Fusarium is more aggressive and less responsive to treatment than *Aspergillus* hence poses therapeutic challenges. The phaeoid fungi are of low virulence but produce protracted infections. Stromal keratitis caused by *Candida* species may be more localized in contrast to filamentous fungi and have ***'collar-button'*** configuration often with small ulceration and expanding infiltrate.

Clinical Features

The patient of keratomycosis feels gritty, foreign-body sensation in eye, burning and stinging along with ocular discomfort. After few days, pain becomes more intense and is associated with photophobia. Usually it is presented unilaterally and eye becomes red and vision is blurred. The slit-lamp examination reveals breach in corneal epithelium and in Bowman's membrane, which becomes grossly evident as disease progress. The classical signs are central, shaggy-edged ulcer with satellite lesions, marked hypopyon and endothelial plaque. The surrounding corneal stroma are also edematous **(Figs. 29.2A and B)**.

Stromal keratitis caused by yeasts, such as *Candida albicans* is more localized, there is a small ulceration with expanding but discrete stromal infiltration. Fungal keratitis associated with contact lens wearers is invariably due to *Candida* species although *Cryptococcus laurentii* has also been reported in one case. Keratitis due to *Curvularia* species is usually mild and not associated with much corneal sloughing and may heal by medical therapy along with debridement. The presence of **pigmented infiltrate** may be an important diagnostic clue for suspecting a phaeoid fungus in keratitis.

The clinical manifestations of keratitis caused by filamentous fungi also include severe ocular reaction and folds in Descemet's membrane; hyphate margins that extend

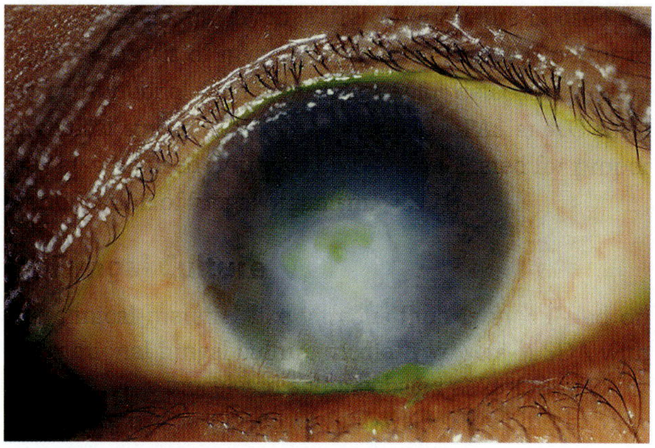

Fig. 29.3. Fungal corneal ulcer with immune ring (Wessely's ring) and corneal abscess.

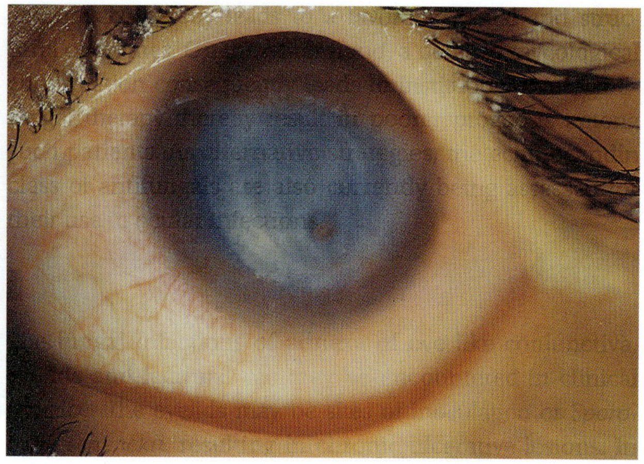

Fig. 29.4. Perforated fungal corneal ulcer.

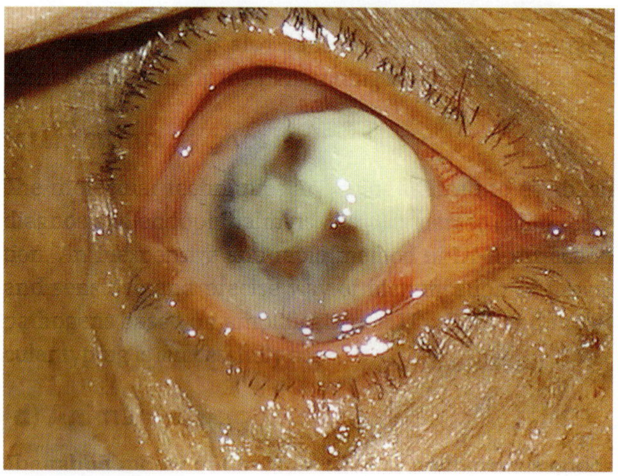

Fig. 29.5. Perforated fungal corneal ulcer of patient after penetrating keratoplasty.

beyond edge of ulcer, which is in contrast to margins of bacterial ulcers and precede appearance of satellite lesions; elevated slough that is usually firm in consistency; endothelial plaques of whitish material and presence of an immune ring *i.e.* Wessely's ring (**Fig. 29.3**) along with corneal abscess. The satellite lesions around central wound are called ***'ring-abscesses'*** which may be visible with passage of time. Sometimes, endophthalmitis may result in later stage of infection due to perforation of corneal layers (**Fig. 29.4**). There may be mixed infection particularly due to protozoal organisms like *Acanthamoeba* species as well as bacterial agents. Sometimes corneal ulcer may develop after doing penetrating keratoplasty also (**Fig. 29.5**).

Fungal keratitis caused by *Fusarium* species neither show ring infiltrate nor raised lesion; *Aspergillus flavus* keratitis shows ring infiltrate, feathery borders and dematiaceous fungal keratitis shows macroscopic pigmentation and raised corneal lesion. Moreover, *Aureobasidium pullulans* keratitis is reported following refractive laser epithelial keratomileusis (LASEK).

Differential Diagnosis

Keratomycosis may be difficult to distinguish from keratitis caused by viruses, bacteria and *Acanthamoeba* species. An indolent course, dense stromal infiltrate, early commencement of hypopyon, satellite lesions suggest keratomycosis, however, laboratory confirmation is essential.

The corneal involvement in pythiosis insidiosi should also be differentiated from keratomycosis, which shows keratitis and orbital cellulitis. There is progressive periorbital edema with chemosis, erythema leading to periorbital cellulitis. It should be one of diagnostic considerations in a patient with corneal ulcer unresponsive to routine antibiotic therapy, particularly when etiologic agent is not readily identified.

Laboratory Diagnosis

The diagnosis of fungal keratitis is usually difficult and thorough clinical examination of the affected eye should be performed using slit-lamp biomicroscope. The basic difficulty is often failure to suspect such infection. The clinical suspicion by ophthalmologists is unequivocally, key element in making diagnosis of fungal infections of cornea. The involvement of fungi may be delineated with

Fig. 29.6. Fungal corneal ulcer seen in confocal microscope.

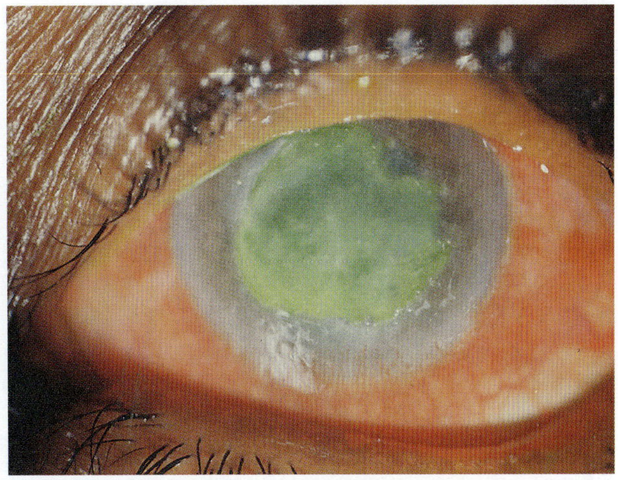

Fig. 29.7. Fungal corneal ulcer stained with fluorescein dye.

the help of confocal microscope also **(Fig. 29.6)**. The fluorescein dye may also be used to demarcate the margins and extent of corneal ulcer **(Fig. 29.7)**. The conventional techniques and newer modalities are employed for laboratory identification of etiological fungal agents.

The diagnostic material is harvested by an experienced ophthalmologist, by taking scrapings of cornea with no. 15 Bard-Parker surgical blade from base and margins of ulcer aseptically using local anesthetics (4% xylocaine) drops. Kimura's platinum spatula is an alternative particularly to get material from base and borders of the ulcer. The base should be thoroughly scraped to remove all necrotic tissue. Biopsy from lesion with 4 mm or larger trephine can be taken. The superficial swabs from lesion may yield causative fungal agent, which mainly depends on methodology of collecting clinical specimens from affected cornea and is the key to establish diagnosis.

(a) Direct Examination

The direct examination of corneal scrapings is performed in 10% KOH wet mount; calcofluor white stain or Gram stain for demonstration of yeast cells, pseudohyphae and/or hyphae. Septate hyphae are easily visible in KOH preparation as shown in **Figures 29.8A to C**. The Gram stain is especially more useful when etiological agent is a yeast although hyphal elements are also stained. Non-specific fluorescent stains like calcofluor white, blankophor and Uvitex 2B are used in tissue sections and cytopathologic preparations for rapid diagnosis of mycotic infections. The calcofluor white stained smears are examined under fluorescent microscope and barrier filter (300 to 412 nm), which is superior to conventional staining techniques for detection of fungi in clinical specimens particularly where number of fungal agents is scanty **(Fig. 29.9)**. Acridine orange staining may be useful in early diagnosis of keratomycosis. Special stains like periodic acid-Schiff **(Figs. 29.10A to D)** and GMS **(Fig. 29.11)** may also be used for the delineation of fungi in the corneal tissue.

(b) Fungal Culture

For fungal cultures, two sets of SDA with antibacterial antibiotics but without actidione are inoculated and incubated at 25°C and 37°C, separately over a period of four weeks. Similarly, one set of brain-heart infusion agar containing antibiotics and cycloheximide is incubated at 37°C for the same period. Specimens are also inoculated on a set of blood agar as well.

The corneal scrapings are inoculated on agar plates in a special manner to differentiate growth over inoculum from fungal contaminants found prevalent in the diagnostic laboratory. A simple technique is to make **'C'** or **'S'** -shaped streak on the plates. Fungal growth in form of streak ensures that growth is from inoculum/specimen rather than a laboratory contaminant.

All cultures are checked every day during first week and twice a week during next three weeks. The positive cultures are more convincing when growth is obtained on more than one occasion. The mycelial isolates are identified by their colony characteristic and microscopic morphology in LCB stained mount and finally by slide cultures.

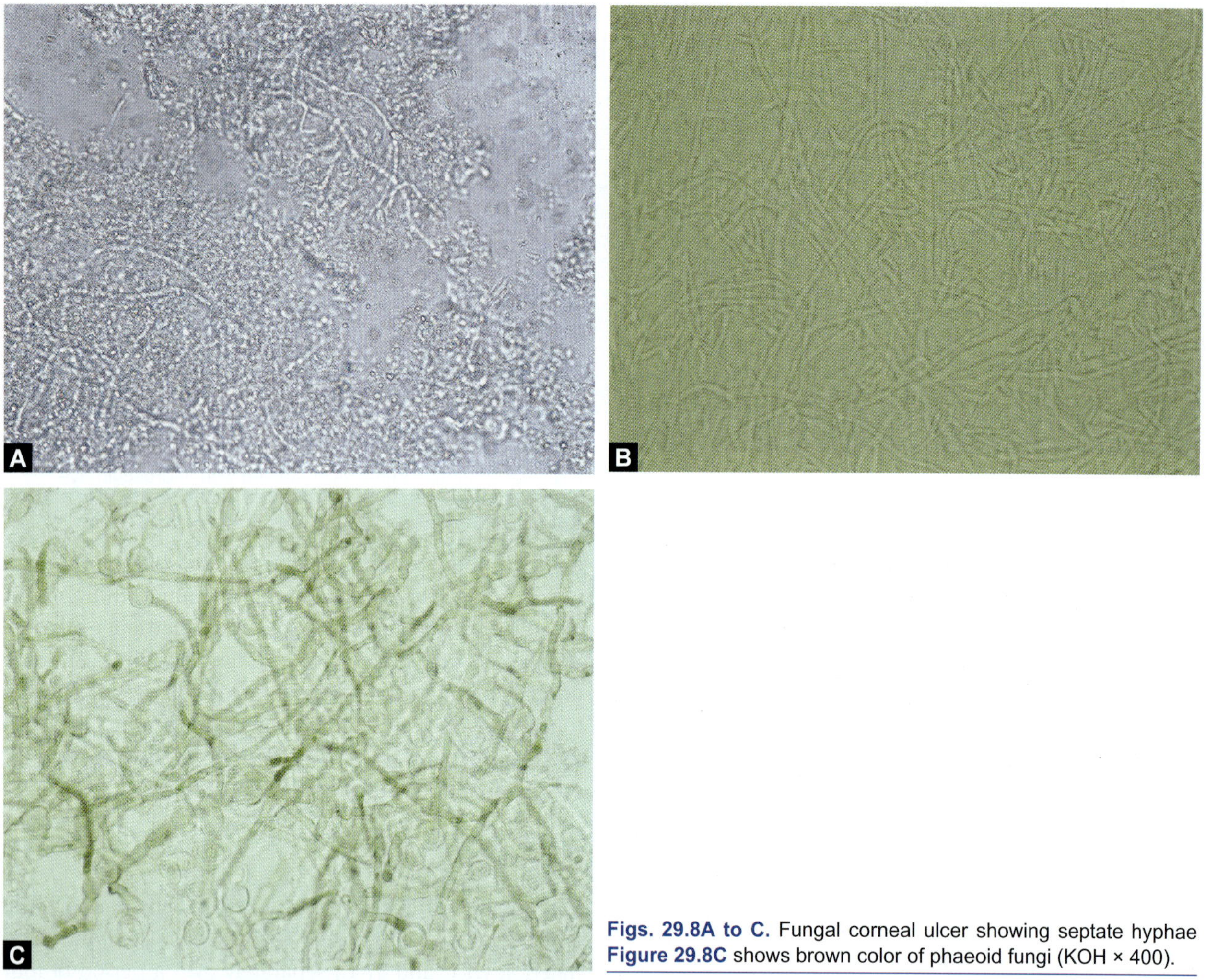

Figs. 29.8A to C. Fungal corneal ulcer showing septate hyphae **Figure 29.8C** shows brown color of phaeoid fungi (KOH × 400).

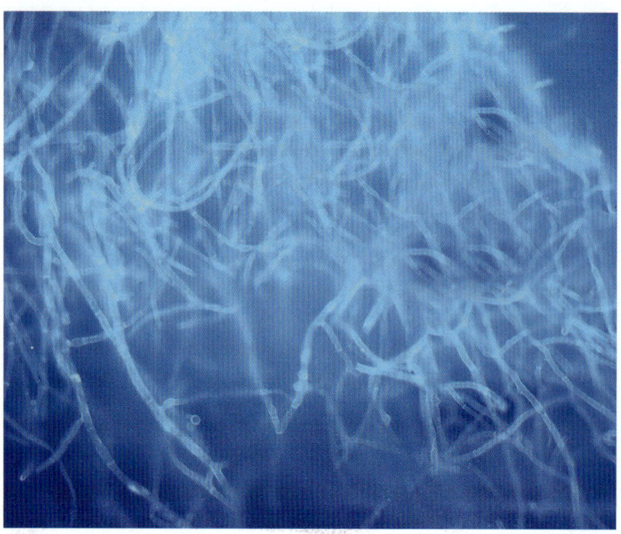

Fig. 29.9. Fungal corneal ulcer showing septate hyphae in wet mount with fluorescent brightener (CFW × 400).

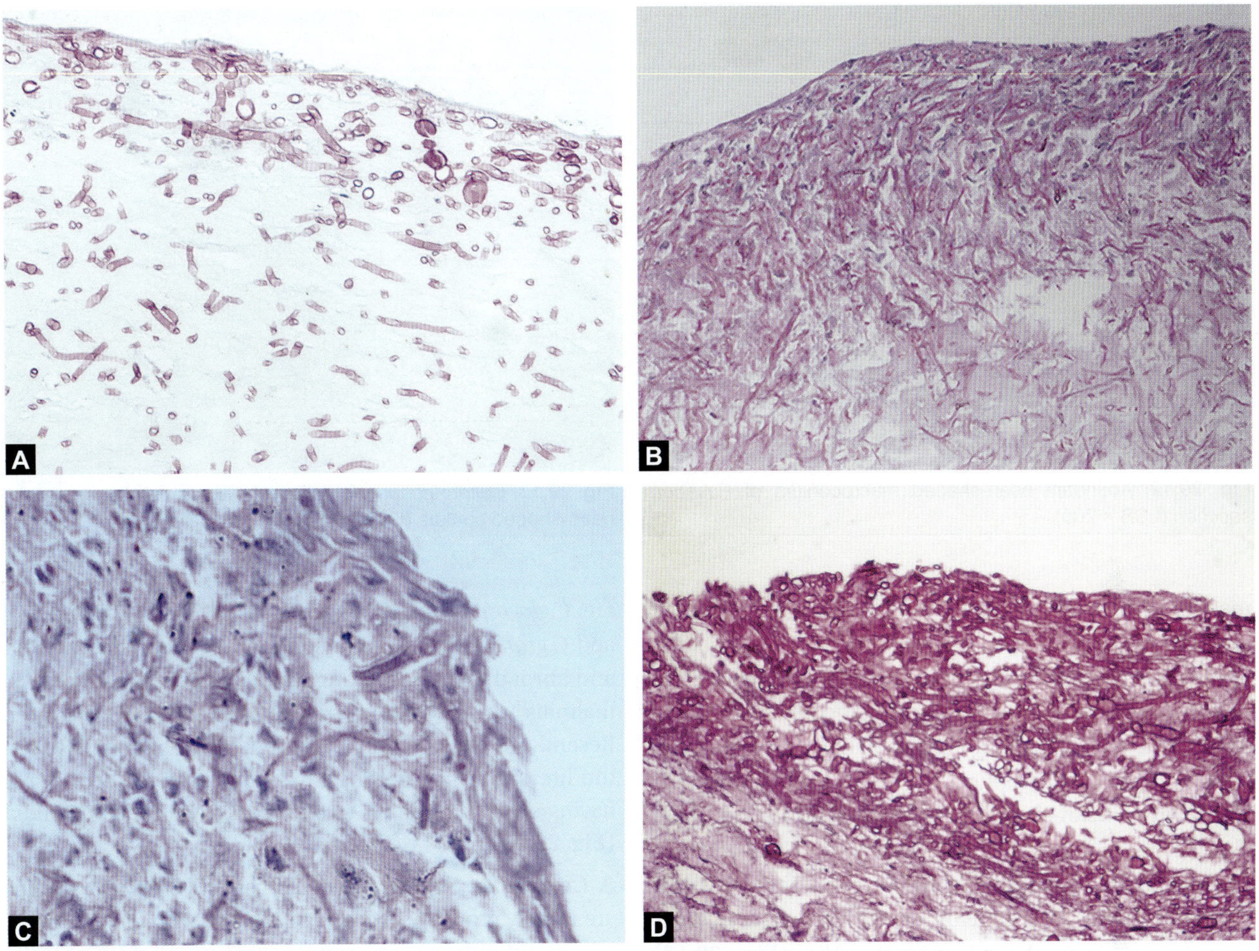

Figs. 29.10A to D. Fungal corneal ulcer showing septate hyphae in histopathology sections of cornea (PAS × 400).

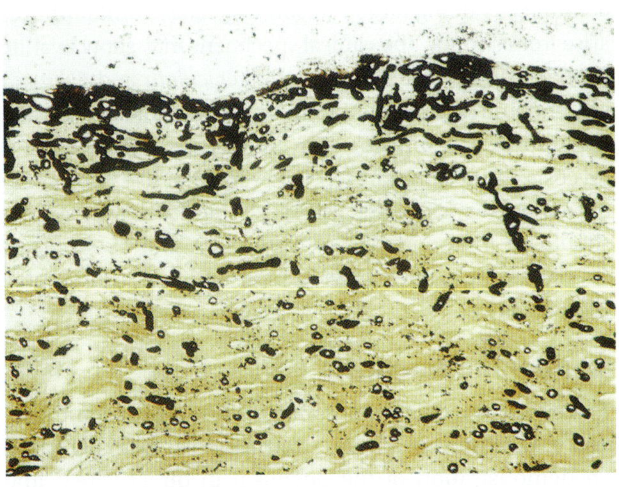

Fig. 29.11. Histopathological section of fungal corneal ulcer showing black-colored branching septate hyphae (GMS × 400).

The yeast isolates are identified by standard tests like germ tube, reduction of tetrazolium medium, chlamydospores production on cornmeal agar or rice starch agar, urease test, sugar fermentation, assimilation, etc.

Investigations of the donor and recipient cornea should also be done regularly for fungal agents to run the corneal transplantation services smoothly.

The descriptions of fungi involved in fungal keratitis have already been given with respect to a particular Chapter. As *Aspergillus* is the commonest genus and its description is dealt in details in **Chapter 25**. However, some of the fungi involved are being described here as below:

***1. Fusarium* Species:** The colonies of *Fusarium* are pluffy to cottony owing to extensive mycelium. Sometimes, diffusible pigment is produced on reverse. The conidiophores are single or grouped into **sporodochia**, a compact

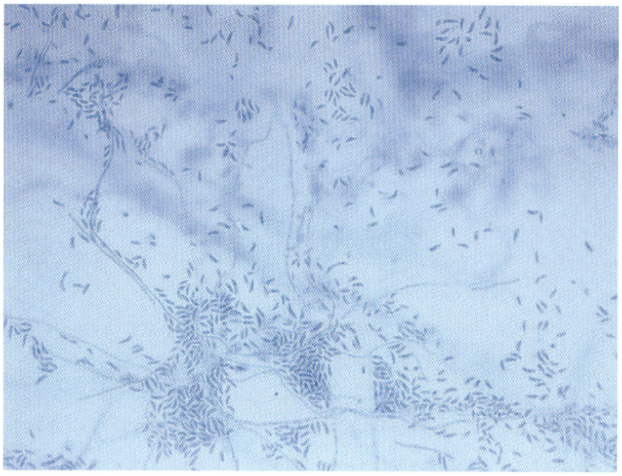

Fig. 29.12. Abundant bean-shaped macroconidia of *Fusarium sacchari* (LCB × 200).

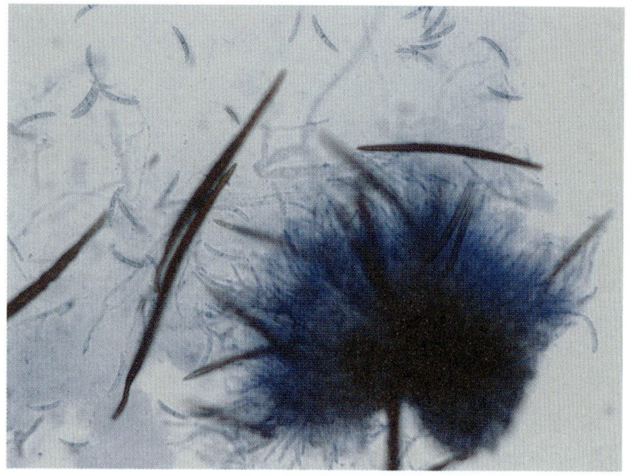

Fig. 29.13. *Colletotrichum dematium* showing sclerotia, setae and bean-shaped conidia (LCB × 200).

mass of interwoven conidiophores. This may be present in colors of red, pink, purple, green and various other shades. Conidia are produced singly or in conidial balls, hyaline and unicellular or transversely septate. Microconidia are single-celled and often found in chains or balls. Macroconidia are oblongate and cylindrical but more often **crescent** or **falciform** or **bean-shaped** **(Fig. 29.12 and *See* Fig. 31.2)**. This genus is invariably confused with *Colletotrichum* species, which also cause fungal keratitis and has bean-shaped conidia. But *Colletotrichum* has setae in addition to the bean-shaped macroconidia **(Fig. 29.13)**. The other one is *Cylindrocarpon*, which has chlamydospores without foot cells. The differentiating features between genus *Fusarium* and *Cylindrocarpon* are given in **Chapter 36** along with **Figure 36.2**.

Colletotrichum* Species:* The genus *Colletotrichum* has gained importance in the recent past. There are more than hundred species of this genus belonging to phylum Ascomycota. These are mainly causative agents of variety of plant infections. This is included in the coelomycetous order Melanconiales and characterized by the formation of acervular conidiomata when the fungus invade the plant tissue. The species concepts in this genus are based on morphology of the fungi on natural substrate. It is recognized as an asexual genus of fungi, with a number of species linked to sexual morphs assigned to the genus *Glomerella*. Various species of *Colletotrichum* are established pathogens but now the human infections are also increasingly being reported. The commonly involved species among human are: *C.gloeosporioide*, C.*dematium, C.coccodes, C.truncatum* and *C.graminicola*. There are many case reports from India and abroad focused on fungal keratitis as well as endophthalmitis caused by various species of *Colletotrichum*. Recently Pote *et al*, in 2017 have reported case and reviewed the literature highlighting the importance of this fungus having setae in addition to the bean-shaped macroconidia **(Fig. 29.13)**.

***3. Curvularia* Species:** The colonies of *Curvularia* species are rapidly growing, floccose and brown with black reverse. The conidiophores are simple, bearing conidia apically from new sympodial growth. The conidia are transversely septate and cylindrical or slightly curved with one of the central cells being larger. In some species, conidiophores arise from stroma. The clinical specimens may show hyaline hyphae. Most of times *Curvularia geniculata* and *C.lunata* are encountered as contaminant or pathogenic causing keratomycosis, sinusitis, onychomycosis, phaeohyphomycosis, eumycetoma, etc. However, other less commonly encountered species are also seen. The *Curvularia* species should be differentiated from *Bipolaris* and *Exserohilum* species, which have false septae and not true septae **(Fig. 29.14)**.

(c) Immunodiagnosis

There is no specific immunological test for the confirmation of keratomycosis. Moreover, diagnosis is finally based on demonstration of the organism in direct smear and culture confirmation, which is sufficient to the extent that serological tests are not really required.

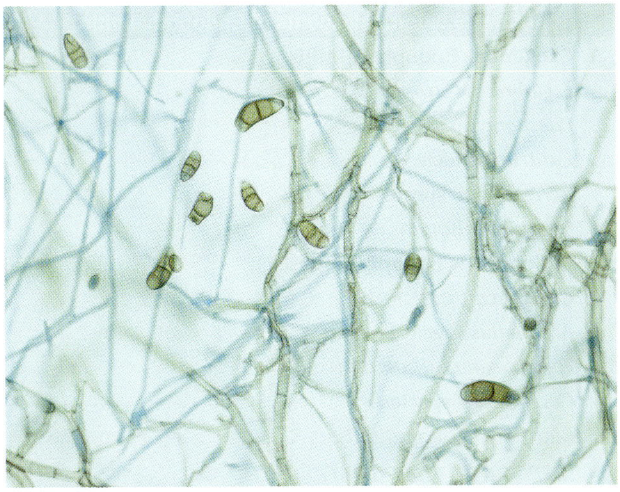

Fig. 29.14. *Curvularia lunata* showing three septae in the conidia (LCB × 200).

Table 29.2. Antifungal Regimen Used in Keratomycosis.
A. Filamentous Fungi
First Choice
Natamycin 5% Ointment
Second Choice
Amphotericin B 0.5% drops + Flucytosine
Itraconazole (1% cream + 200-400 mg PO)
Silver Sulfadiazine Cream 1%
B. Yeast-like Fungi
First Choice
Amphotericin B 0.15% drops plus
Flucytosine (1% drops + 150 mg/kg PO)
Second Choice
Fluconazole (0.5% drops + 200 mg PO)
Miconazole 1% drops

(d) Animal Pathogenicity

In experimental animals, corneal ulcer may be produced by interlamellar injection of fungal spore suspension in eyes of previously immunosuppressed albino and black wild rabbits. Intraocular penetration of ketoconazole in rabbits has been tried as a therapeutic modality.

Treatment and Prophylaxis

The topical antifungal preparations are mainstay of therapy. Pimaricin (natamycin) as 5% suspension is used for treatment of keratomycosis. Amphotericin B, flucytosine and nystatin lavage can also be tried. The antifungal agents, dosage their routes of administration employed in patients of keratomycosis are given in **Table 29.2**.

Successful antifungal therapy for keratomycosis requires frequent drug administration for a prolonged period *i.e.* eye drops are applied hourly, when patient is awake. Twelve weeks therapy is probably minimum requirement in most of cases of keratitis caused by filamentous fungi and for all patients in whom infection has produced a clinical infiltrate in posterior half of corneal stroma.

Oral fluconazole, when added to topical natamycin, has been reported as an effective modality in case of fungal keratitis caused by filamentous fungi. Topical corticosteroids therapy for any of eye ailment is contraindicated in keratomycosis because even low dose steroids may lead to bad prognosis.

As a surgical measure, extensive debridement is essential in keratomycosis. Therefore, therapeutic surgery *i.e.* **penetrating keratoplasty** or **lamellar keratectomy** is done for management of patients who are not responding to medical treatment or impending perforation with presence of a descemetocele. Although prognosis after therapeutic penetrating keratoplasty (PK) is good but Garg *et al*, have found a poor outcome in their patients. If this clinical condition is not recognized early or treated in time, evisceration may be required.

In a recent study intrastromal corneal injections of amphotericin B have shown a less invasive, alternative to repeated penetrating keratoplasty for treating fungal keratitis.

B. FUNGAL ENDOPHTHALMITIS

The fungal endophthalmitis is suppurative inflammation of inner ocular coats and their adjacent structures with the involvement of vitreous fluid. It is caused by variety of fungal agents either from external **(exogenous)** or internal source through hematogenous route **(endogenous)**.

Historical Perspective

Fungal endophthalmitis used to be one of rare conditions during first half of the 20th century. In 1914, Dimmer reported choroidoscleritis presumably caused by *Aspergillus fumigatus*. In the same year, Churchil *et al,* found blastomycosis from vitreous of patient having systemic infection. The first histologically proven case of endophthalmitis resulting from systemic candidiasis was reported by Milae in 1943. The first case of endogenous cryptococcal endophthalmitis was reported by Weiss *et al,* in the year 1948. In 1958, Van

Buren reported first clinically recognized case of endogenous candidial endophthalmitis. Levitt was first to report case of coccidioidal endophthalmitis. Lieberman *et al,* described case of endogenous endophthalmitis due to *Fusarium solani* in a healthy, immunocompetent individual who did not have any history of intraocular infection or manipulation.

Epidemiology

Fungal endophthalmitis accounts for about 4-11% of all cases of culture-proven endophthalmitis and is usually acquired from an endogenous source via hematogenous spread. However, it may be secondary to intraocular surgery, corneal ulceration or even trauma.

Candida species are most frequently encountered cause of both types of fungal endophthalmitis. The other fungi may be *Aspergillus* species, *Histoplasma capsulatum, Coccidioides immitis, Blastomyces dermatitidis, Cryptococcus* species, *Sporothrix schenckii* and *Fusarium* species. Some of the important and frequently encountered species are listed in **Table 29.3**.

Pathogenesis and Pathology

This clinical entity arises out of an interaction between different agent factors such as invasiveness and toxigenic nature of causative fungal strains, host factors and other predisposing conditions. Sustained fungemia with even saprotrophic fungi may lead to endophthalmitis. The patients with exogenous endophthalmitis are usually healthy and immunocompetent whereas in endogenous type of endophthalmitis they are invariably found to be immunocompromised.

Clinical Features

Fungal endophthalmitis may have classical appearance with progressive granulomatous uveitis, diffuse retinitis and deep vitreous abscesses. The time to make diagnosis from onset of symptoms has been reported to be varying from three days to four months during which bilateral ocular disease may cause severe morbidity, especially in patients who are already debilitated. *Candida* species are most common cause of fungal endophthalmitis and may present in a similar manner to chronic bacterial endophthalmitis. Fungal endophthalmitis is clinically of **two types** depending upon mode of infection:

1. Exogenous Endophthalmitis
2. Endogenous Endophthalmitis

Table 29.3. Fungal Agents of Endophthalmitis.
A. Exogenous Endophthalmitis
C.glabrata, C.tropicalis, C.parapsilosis
Paecilomyces species
Fusarium species
Acremonium species
Curvularia lunata
B. Endogenous Endophthalmitis
Candida albicans
Aspergillus species
Histoplasma capsulatum
Coccidioides species
Blastomyces dermatitidis
Cryptococcus neoformans
Sporothrix schenckii
Fusarium species

1. Exogenous Endophthalmitis

Exogenous fungal endophthalmitis occurs by introduction of organisms into eye from external source. The patients are otherwise healthy hence do not have any underlying immunodeficiency. This clinical entity develops after ocular surgery, trauma by intraocular spread from fungal keratitis or therapeutic keratoplasty for keratitis. It is also found in patients taking contaminated intravenous fluids. It may be as a result of pre-existing scleritis or keratitis. Mucormycosis in the surrounding soft tissue may also lead to exogenous endophthalmitis. Cataract removal followed by placement of prosthetic lens and corneal transplantation are surgical procedures most often associated with postoperative exogenous endophthalmitis. This entity may have latency of weeks to months before clinically detectable disease occurs.

Although *Candida* species are the commonest agent of exogenous endophthalmitis but many other fungal agents are also reported like *Aspergillus, Fusarium, Paecilomyces, Curvularia* and mucormycete species. In one of studies Lily Therese, *et al,* have reported *Penicillium* species in endophthalmitis cases from south India.

2. Endogenous Endophthalmitis

The incidence of endogenous fungal endophthalmitis has increased in second half of outgoing 20th century because of advent of antibiotics and indwelling catheters. The disease process can produce highly suggestive, ocular sepsis, which assists clinician in making diagnosis. Granulomatous intraocular inflammation in patients with one or more of risk factors should raise suspicion of fungal endophthalmitis. The patients are generally immunocompromised and have underlying predisposing condition. *Aspergillus* is seen more in organ transplant recipients with granulocytopenia or in cardiac surgery patients as compared to *Candida* endophthalmitis.

The ocular manifestations of histoplasmosis are rarely described at the time of initial infection. Late sequelae of subclinical *Histoplasma capsulatum* infection called Presumed Ocular Histoplasmosis Syndrome **(POHS)**, is common and characterized by bilateral, multifocal chorioretinal scarring. Small, indolent inflammatory choroidal foci may persist for many years and can lead to subfoveal neo-vascularization and loss of central vision, often decades later. The punched out chorioretinal scars, ***'histo spots'*** seen in POHS are thought to represent areas of past granulomatous inflammation from choroidal dissemination of *Histoplasma capsulatum.*

In addition to *Candida* species, some of dimorphic fungi are involved in clinical entity which are not frequently seen in India as their prevalence is either very less or not at all. The other dimorphic fungal species mentioned in **Table 29.3** are also rarely encountered as case of endogenous endophthalmitis. A case of *Ochroconis gallopava* endophthalmitis in fludarabine treated chronic lymphatic leukaemia has also been reported.

Chorioretinitis: Intraocular infection with *Cryptococcus* species in patients with AIDS is very uncommon compared with meningeal and optic nerve involvement. When it does occur it is mostly in the posterior segment. The cryptococcal chorioretinitis is a recognized cause of permanent visual impairment. It is almost always associated either with severe disseminated disease like meningitis or with local inoculation. The characteristic primary choroidal involvement helps to distinguish cryptococcal chorioretinitis from that due to other opportunistic pathogens, such as cytomegalovirus and *Toxoplasma gondii*, which primarily affect the retina rather than the choroid.

In addition, cryptococcal intraocular infection is a rare disease and usually associated with generalized systemic disease in immunocompromised patients. The diagnosis may be difficult to arrive because of the rarity of this disease and its similarities to other uveitic entities. Cryptococcal iridociliary granuloma has been diagnosed by anterior chamber tap and fine-needle aspiration cytology in an immunocompetent woman with acute granulomatous uveitis.

The histopathological examination of the choroid can reveal granulomatous inflammation with microabscesses and giant cells centered around the capsule. As ocular involvement can precede CNS infection, the diagnosis of cryptococcal choroiditis should lead to appropriate evaluation and treatment for systemic dissemination. The choroidal infections with *Candida* species, *Histoplasma capsulatum* and *Aspergillus* species have also been reported.

Differential Diagnosis

Fungal endophthalmitis should be differentiated from infections caused by bacterial, viral or parasitic origin which are presenting with similar types of manifestations.

Moreover, Acute Zonal Outer Optic Retinitis (AZOOR) has been linked with some of the *Candida* species like *C.fomata*, wherein patients usually see spots in the outer boundaries of the field of vision, which should be differentiated from endophthalmitis.

Laboratory Diagnosis

A detailed examination of the affected eye should be performed. The primary obstacle is often failure to suspect fungal nature of infection. The confirmation of suspected clinical disease currently presents challenge to clinician with difficulty in making definitive diagnosis consequently delaying treatment. In case of typical history and clinical signs of fungal endophthalmitis, intraocular specimen like aqueous or vitreous fluid, is required. As intraocular samples are often culture negative, cases that respond to antifungal treatment are considered to be fungal in origin. Blood cultures are useful but insensitive for detection of candidaemia and serological testing remains experimental. Isolation of fungi may also be difficult, as coexisting bacteremia may suppress fungemia in culture because of bacteria competing with fungi for nutrients *in vitro.* The conventional techniques along with newer methods are employed for laboratory identification of etiological agents of fungal endophthalmitis.

(a) Direct Examination

This is performed with KOH wet mount; calcofluor white stain or Gram stain for demonstration of yeast cells, pseudohyphae and hyphae. Calcofluor white is used for rapid diagnosis of mycotic infections. This is superior to the conventional staining techniques for detection of fungi in clinical specimens. Special stains like PAS and Giemsa may also be used. Vitreous biopsy is indicated for proper identification of fungal organism.

(b) Fungal Culture

The clinical material like vitreous fluid is inoculated on conventional fungal culture media. Two sets of SDA with antibiotics are inoculated and incubated at 25°C and 37°C, separately over a period of four weeks, keeping in view of the possibility of dimorphic fungi. Similarly, one set of brain-heart infusion agar containing antibiotics and cycloheximide is incubated at 37°C for same period of four weeks. The other procedures for identification of fungi are as per routinely used conventional laboratory techniques.

Fungal culture is also done in case of donor and recipient cornea during the penetrating keratoplasty.

(c) Immunodiagnosis

The molecular techniques are being utilized to establish diagnosis of endophthalmitis like polymerase chain reaction. An inexpensive and rapid method of DNA extraction and sensitive and precise way of identification of fungal pathogens based on amplification of ITS, DNA and molecular typing is successfully tried.

(d) Animal Pathogenicity

The rabbit model of *Candida albicans* endogenous endophthalmitis has been developed.

Treatment and Prophylaxis

There are three major modes of treatment of fungal endophthalmitis: (i) systemic administration of antifungal agents, primarily amphotericin B or an azole derivative; (ii) intravitreal injection of antifungal agent, usually voriconazole or amphotericin B and (iii) vitrectomy. As such vitrectomy removes bulk of infectious burden and inflammatory debris thereby facilitates drug diffusion within eye. Most of the time a combination of treatment modalities is used. Vitrectomy and intravitreal amphotericin B have good prognosis among these patients. Exogenous fungal endophthalmitis resulting from keratomycosis generally has very poor prognosis as far as restoring vision is concerned. Hence a strong clinical suspicion of fungal etiology translates into early vitrectomy plus intravitreal antifungals thereby leads to a favorable visual and structural outcomes. On the other hand, a delay in mycological confirmation consequently leads to institution of late antifungal therapy thereby result in poorer prognosis among such patients. As alternative strategies, the echinocandin class of antifungals are also currently being assessed for their use in ocular infections.

C. MYCOSES OF OCULAR ADNEXA

The fungal infections of adnexa of eye like conjunctiva, eyelids and lacrimal sac may be encountered in clinical practice. The eyelids may be sites of inoculation of *Sporothrix schenckii*, resulting in chronic ulcerative lesions. In addition, *Blastomyces dermatitidis* and *Paracoccidioides brasiliensis* may also cause lesions of eyelids. Palpebral paracoccidioidomycosis has been reported by Burnier in 1997. Eyelashes may be infected by *Microsporum canis*, *Malassezia furfur* and *Candida* species.

The eyelids and surrounding soft tissues of eyes may be swollen or impinged by adjacent tissue involved in mycotic infections. The conventional protocol for diagnosing the fungal infections can establish the diagnosis like KOH wet mounts and histopathological stains like H&E and special stains such as PAS or GMS. In **Figures 29.15A and B**, the histopathological section is clearly showing infection by *Aspergillus* species in H&E staining in low power and PAS staining in high power. The proptosis of eyes is commonly seen in rhino-orbito-cerebral mucormycosis, which may arise from buccal or nasal cavity extension to eyes and the optic nerves and is discussed in details in **Chapter 26**.

Infection and obstruction of lacrimal duct system or dacryocystitis may also be due to fungal infection particularly caused by *Candida* species. Acute dacryocystitis has been reported with *Sporothrix braziliensis*. In addition, chronic dacryocystitis cases are reported due to *Candida* and *Aspergillus* species.

Pneumocystis jirovecii is documented to cause conjunctivitis apart from retinal or choroidal lesions. The fungal agents do not seem to pose any threat to the patients with anophthalmos.

The clinical appearance of coccidioidal infection of the eyelid can mimic bacterial disease or malignant neoplasm. A delay in arriving an appropriate diagnosis thereby its treatment can result in extension of infection in the orbit, loss of the globe or even death. Coccidioidal infection of the eyelid should be suspected in a patient who presents with an atypical lesions, which are unresponsive to routine

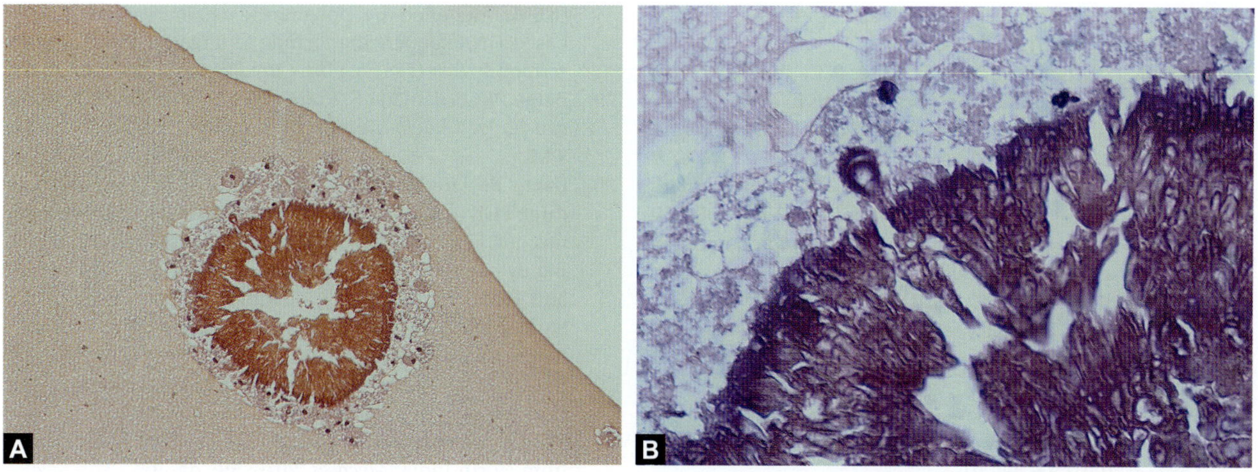

Figs. 29.15A and B. Histopathological section from a cyst of eyelid showing *Aspergillus* species (H&E × 100, PAS × 400).

antibiotic therapy, particularity, when the patient has recently traveled through or lived in an area where *Coccidioides* species are found to be endemic.

The fungal infections in the surrounding vicinity of the eyeball may also extend as oculomycosis. This usually happens when paranasal sinuses are involved as discussed in **Chapter 34** in detail. In addition, Cheon *et al*, have described a case of fungal ball within Onodi cell mucocele causing visual loss for the first time. It was completely removed via transnasal endoscopic approach. The Onodi cell is a pneumatized posterior ethmoid cell located laterally and superiorly to the sphenoid sinus and closely related to the optic nerve. As such mucocele is a benign, expansile, cyst-like lesion of the paranasal sinuses that is filled with mucoid secretion. Therefore, optic neuropathy caused by an infected mucocele in an Onodi cell is not common. Furthermore, fungal infection superimposed on an Onodi cell mucocele is extremely rare and is not reported previously before this report.

The laboratory diagnosis and therapeutic modalities of these conditions are same as seen in any of systemic infections caused by fungal agents.

The infections of cornea and ocular adnexa caused by other fungal and pseudofungal organisms are discussed in their respective Chapter in detail *i.e.* microsporidiosis, pythiosis insidiosi and rhinosporidiosis. The conjunctiva may be infected by pseudofungal organism *i.e. Rhinosporidium seeberi* which is commonly found in southern states of India as well as in Sri Lanka. The conjunctival rhinosporidiosis associated with scleral melting and staphyloma formation has also been reported.

Further Reading

1. Abbey AM, Shah NV, Forster RK, *et al.* Infectious *Pseudomonas* and *Bipolaris* scleritis following history of pterygium surgery. *Indian J Ophthalmol.* 2016; **64:** 674-6.
2. Adam CR, Sigler EJ. Multimodal imaging findings in endogenous *Aspergillus* endophthalmitis. *Retina.* 2014; **34:** 1914-5.
3. Adam MK, Vahedi S, Nichols MM, *et al.* Inpatient ophthalmology consultation for fungemia: Prevalence of ocular involvement and necessity of funduscopic screening. *Am J Ophthalmol.* 2015; **160:** 1078-1083.e2.
4. Agarwal A, Agarwal R. Role of potassium hydroxide preparation in the management of mycotic corneal ulcers. *Indian J Ophthalmol.* 2012; **60:** 336.
5. Aggarwal S, Yamaguchi T, Dana R, *et al. Exophiala phaeomuriformis* fungal keratitis: Case report and *in vivo* confocal microscopy findings. *Eye Contact Lens.* 2017; **43:** e4-6.
6. Al-Badriyeh D, Leung L, Davies GE, *et al.* Successful salvage treatment of *Scedosporium apiospermum* keratitis with topical voriconazole after failure of natamycin. *Ann Pharmacother.* 2009; **43:** 1139-42.
7. Al-Badriyeh D, Neoh CF, Stewart K, *et al.* Clinical utility of voriconazole eye drops in ophthalmic fungal keratitis. *Clin Ophthalmol.* 2010; **4:** 391-405.
8. Al-Hatmi AM, Bonifaz A, de Hoog GS, *et al.* Keratitis by *Fusarium temperatum*, a novel opportunist. *BMC Infect Dis.* 2014; **14:** 588.
9. Al-Hatmi AM, Meletiadis J, Curfs-Breuker I, *et al. In vitro* combinations of natamycin with voriconazole, itraconazole and micafungin against clinical *Fusarium* strains causing keratitis. *J Antimicrob Chemother.* 2016; **71:** 953-5.
10. Alex D, Li D, Calderone R, *et al.* Identification of *Curvularia lunata* by polymerase chain reaction in a case of fungal endophthalmitis. *Med Mycol Case Rep.* 2013; **2:** 137-40.
11. Ali TK, Amescua G, Miller D, *et al.* Contact-lens-associated *Purpureocillium* keratitis: risk factors, microbiologic

characteristics, clinical course and outcomes. *Semin Ophthalmol.* 2017; **32:** 157-62.

12. Alkatan H, Athmanathan S, Canites CC. Incidence and microbiological profile of mycotic keratitis in a tertiary care eye hospital: A retrospective analysis. *Saudi J Ophthalmol.* 2012; **26:** 217-21.
13. Alzahrani YA, Aziz HA, Shrestha NK, *et al.* Cryptococcal iridociliary granuloma. *Surv Ophthalmol.* 2016; **61:** 498-501.
14. Amin RM, Hamdy I, Osman IM. Chronic *Candida* endophthalmitis as a cause of intermediate uveitis. *BMJ Case Rep.* 2015; pii: bcr2014208847. PMID: 25870216.
15. An N, Liu XN, Wang YN, *et al.* The pathogenic spectrum of fungal keratitis in northwestern China. *Int J Ophthalmol.* 2016; **9:** 1846-8.
16. Anandi V, George JA, Thomas R, *et al.* Phaeohyphomycosis of the eye caused by *Exserohilum rostratum* in India. *Mycoses.* 1991; **34:** 489-91.
17. Anita KB, Fernandez V, Rao R. Fungal endophthalmitis caused by *Paecilomyces variotii*, in an immunocompetent patient, following intraocular lens implantation. *Indian J Med Microbiol.* 2010; **28:** 253-4.
18. Ansari Z, Miller D, Galor A. Current thoughts in fungal keratitis: Diagnosis and treatment. *Curr Fungal Infect Rep.* 2013; **7:** 209-18.
19. Anutarapongpan O, Thanathanee O, Suwan-Apichon O. *Penicillium* keratitis in a HIV-infected patient. *BMJ Case Rep.* 2016; pii: bcr2016216139. PMID: 27535731.
20. Arnoldner MA, Kheirkhah A, Jakobiec FA, *et al.* Successful treatment of *Paecilomyces lilacinus* keratitis with oral posaconazole. *Cornea.* 2014; **33:** 747-9.
21. Arora R, Gupta D, Goyal J, Kaur R. Voriconazole versus natamycin as primary treatment in fungal corneal ulcers. *Clin Experiment Ophthalmol.* 2011; **39:** 434-40.
22. Azari AA, Nehls S, Ghoghawala SY, *et al. Rhizopus* keratitis following corneal trauma. *JAMA Ophthalmol.* 2013; **131:** 776.
23. Babu K, Murthy PR, Prakash PY, *et al.* Chronic endophthalmitis due to *Pyrenochaeta romeroi* in an immunocompetent host—A case report from southern India. *Retin Cases Brief Rep.* 2014; **8:** 197-9.
24. Badenoch P, Wetherall B, Woolley M, *et al.* Newer emerging pathogens of ocular non-sporulating molds (NSM) identified by polymerase chain reaction (PCR)-based DNA sequencing technique targeting internal transcribed spacer (ITS) region. *Curr Eye Res.* 2008; **33:** 903-4.
25. Badenoch PR, Halliday CL, Ellis DH, *et al. Ulocladium atrum* keratitis. *J Clin Microbiol.* 2006; **44:** 1190-3.
26. Badiee P, Alborzi A, Nejabat M. Detection of *Aspergillus* keratitis inocular infections by culture and molecular method. *Int Ophthalmol.* 2011; **31:** 291-6.
27. Badiee P, Nejabat M, Alborzi A, *et al.* Comparative study of Gram stain, potassium hydroxide smear, culture and nested PCR in the diagnosis of fungal keratitis. *Ophthalmic Res.* 2010; **44:** 251-6.
28. Bagyalakshmi R, Therese KL, Prasanna S, *et al.* Newer emerging pathogens of ocular non-sporulating molds (NSM) identified by polymerase chain reaction (PCR)-based DNA sequencing technique targeting internal transcribed spacer (ITS) region. *Curr Eye Res.* 2008; **33:** 139-47.
29. Balne PK, Nalamada S, Kodiganti M, *et al.* Fungal keratitis caused by *Chaetomium atrobrunneum. Cornea.* 2012; **31:** 94-5.
30. Balne PK, Reddy AK, Kodiganti M, *et al.* Evaluation of three PCR assays for the detection of fungi in patients with mycotic keratitis. *Br J Ophthalmol.* 2012; **96:** 911-2.
31. Bandyopadhyay S, Das D, Mondal KK, *et al.* Epidemiology and laboratory diagnosis of fungal corneal ulcer in the Sundarban region of West Bengal, eastern India. *Nepal J Ophthalmol.* 2012; **4:** 29-36.
32. Bansal Y, Chander J, Kaistha N, Singla N, Sood S, van Diepeningen AD. *Fusarium sacchari*, a cause of mycotic keratitis among sugarcane farmers—A series of four cases from North India. *Mycoses.* 2016; **59:** 705-9.
33. Baradkar V, Samal B, Mali SA, *et al. Acanthamoeba* on Sabouraud's agar from a patient with keratitis. *Trop Parasitol.* 2011; **1:** 141-2.
34. Baranyi N, Kocsube S, Szekeres A, *et al.* Keratitis caused by *Aspergillus pseudotamarii. Med Mycol Case Rep.* 2013; **2:** 91-4.
35. Barut Selver O, Egrilmez S, Palamar M, *et al.* Therapeutic corneal transplant for fungal keratitis refractory to medical therapy. *Exp Clin Transplant.* 2015; **13:** 355-9.
36. Behera UC, Budhwani M, Das T, *et al.* Role of early vitrectomy in the treatment of fungal endophthalmitis. *Retina.* 2017. doi: 10.1097/IAE.0000000000001727. PMID: 28541964.
37. Berbel RF, Casella AM, de Freitas D, *et al. Curvularia lunata* endophthalmitis. *J Ocul Pharmacol Ther.* 2011; **27:** 535-7.
38. Bhandary SV, VijayaPai H, Rao LG, *et al. Cunninghamella spinosum* fungal corneal ulcer - First case report. *Indian J Ophthalmol.* 2014; **62:** 375-6.
39. Bharathi MJ, Ramakrishnan R, Meenakshi R, *et al.* Analysis of the risk factors predisposing to fungal, bacterial & *Acanthamoeba* keratitis in south India. *Indian J Med Res.* 2009; **130:** 749-57.
40. Bhuta S, Hsu CC, Kwan GN. *Scedosporium apiospermum* endophthalmitis: Diffusion-weighted imaging in detecting subchoroidal abscess. *Clin Ophthalmol.* 2012; **6:** 1921-4.
41. Biju R, Sushil D, Georgy NK. Successful management of presumed *Candida* endogenous endophthalmitis with oral voriconazole. *Indian J Ophthalmol.* 2009; **57:** 306-8.
42. Binnani A, Gupta PS, Gupta A. Epidemio-clinico-microbiological study of mycotic keratitis in north-east region of Rajasthan. *Mycopathologia.* 2016 Dec 29. doi: 10.1007/s11046-016-0102-8. PMID: 28035615.
43. Birnbaum FA, Gupta G. The role of early vitrectomy in the treatment of fungal endogenous endophthalmitis. *Retin Cases Brief Rep.* 2016; **10:** 232-5.
44. Blennow O, Tallstedt L, Hedquist B, *et al.* Duration of treatment for candidemia and risk for late-onset ocular candidiasis. *Infection.* 2013; **41:** 129-34.

45. Braich PS, Chang JS, Albini TA, *et al.* Irido-lenticular abscess as the initial sign of *Candida albicans* endogenous endophthalmitis. *Ophthalmic Surg Lasers Imaging.* 2011; **42:** e107-9.
46. Buchta V, Feuermannova A, Vasa M, *et al.* Outbreak of fungal endophthalmitis due to *Fusarium oxysporum* following cataract surgery. *Mycopathologia.* 2014; **177:** 115-21.
47. Bui AQ, Espana EM, Margo CE. Chromoblastomycosis of the conjunctiva mimicking melanoma of the ciliary body. *Arch Ophthalmol.* 2012; **130:** 1615-7.
48. Bullock JD, Elder BL, Khamis HJ, *et al.* Effects of time, temperature and storage container on the growth of *Fusarium* species: Implications for the worldwide *Fusarium* keratitis epidemic of 2004-2006. *Arch Ophthalmol.* 2011; **129:** 133-6.
49. Bullock JD, Elder BL, Warwar RE, *et al.* Mechanism of drug failure in *Fusarium* keratitis, 2004-2006. *N Engl J Med.* 2014; **370:** 88-9.
50. Bullock JD, Khamis HJ. A retrospective statistical analysis of the *Fusarium* keratitis epidemic of 2004-2006. *Ophthalmic Epidemiol.* 2010; **17:** 179-84.
51. Cagini C, Iannone A, Fiore T, *et al.* Post-operative endophthalmitis caused by *Acremonium falciforme* with orbital and extra-orbital involvement following combined cataract and glaucoma surgery: A case report. *J Med Case Rep.* 2014; **8:** 373.
52. Cakir M, Imamoglu S, Cekic O, *et al.* An outbreak of early-onset endophthalmitis caused by *Fusarium* species following cataract surgery. *Curr Eye Res.* 2009; **34:** 988-95.
53. Calderaro A, Motta F, Larini S, *et al.* Two rare cases of *Acremonium* acute endophthalmitis after cataract surgery in a tertiary care hospital. *New Microbiol.* 2013; **36:** 427-31.
54. Caldwell MC, Perfect JR, Carlson AN, *et al.* *Candida glabrata* endophthalmitis following penetrating keratoplasty. *J Cataract Refract Surg.* 2009; **35:** 598-602.
55. Campbell S. *Aspergillus terreus* recovered from a corneal scraping. *Clin Lab Sci.* 2014; **27:** 67-71.
56. Campos-Herrero MI, Tandon L, Horcajada I, *et al.* Endophthalmitis caused by *Phialophora verrucosa:* A case report and literature review of *Phialophora* ocular infections. *Enferm Infecc Microbiol Clin.* 2012; **30:** 163-5.
57. Cano J, Guarro J, Gene J. Molecular and morphological identification of *Colletotrichum* species of clinical interest. *J Clin Microbiol.* 2004; **42:** 2450-4.
58. Caretti L, Babighian S, Rapizzi E, *et al.* Fungal keratitis following deep lamellar keratoplasty. *Semin Ophthalmol.* 2011; **26:** 33-5.
59. Carrasco MA, Genesoni G. Treatment of severe fungal keratitis with subconjunctival amphotericin B. *Cornea.* 2011; **30:** 608-11.
60. Chaidaroon W, Supalaset S, Tananuvat N, *et al.* Corneal phaeohyphomycosis caused by *Bipolaris hawaiiensis. Case Rep Ophthalmol.* 2016; **7:** 364-71.
61. Chaidaroon W, Tananuvat N, Chavengsaksongkram P, *et al.* Corneal chromoblastomycosis caused by *Fonsecaea pedrosoi. Case Rep Ophthalmol.* 2015; **6:** 82-7.
62. Chakrabarti A, Shivaprakash MR, Singh R, *et al.* Fungal endophthalmitis: Fourteen years' experience from a center in India. *Retina.* 2008; **28:** 1400-7.
63. Chander J, Chakrabarti A, Sharma A, *et al.* Evaluation of calcofluor staining in the diagnosis of fungal corneal ulcer. *Mycoses.* 1993; **36:** 243-5.
64. Chander J, Chakrabarti A, Sharma A, *et al.* Mycotic keratitis: A five-year study from Chandigarh. *Indian J Med Microbiol.* 1993; **11:** 218-22.
65. Chander J, Sharma A. Prevalence of fungal corneal ulcers in northern India. *Infection.* 1994; **22:** 207-9.
66. Chander J, Singla N, Agnihotri N, Arya SK, Deep A. Keratomycosis in and around Chandigarh: A five-year study from a north Indian tertiary care hospital. *Indian J Pathol Microbiol.* 2008; **51:** 304-6.
67. Chander J, Singla N, Gulati N, Sood S. *Fusarium sacchari:* A cause of exogenous fungal endophthalmitis: First case report and review of literature. *Mycopathologia.* 2011; **171:** 431-4.
68. Chandra J, Mukherjee PK. *Candida* Biofilms: Development, architecture and resistance. *Microbiol Spectr.* 2015; **3.** PMID: 26350306.
69. Chang HY, Chodosh J. Diagnostic and therapeutic considerations in fungal keratitis. *Int Ophthalmol Clin.* 2011; **51:** 33-42.
70. Chang YF, Yang CS, Lee FL, *et al.* Voriconazole for *Candida* endophthalmitis. *Ophthalmology.* 2012; **119:** 2414-5.e4.
71. Chatterjee S, Agrawal D. Confocal microscopy. *Ophthalmology.* 2012; **119:** 428-9.
72. Chatterjee S, Agrawal D. Fungal keratitis in lattice dystrophy. *Indian J Ophthalmol.* 2010; **58:** 162-4.
73. Chavan R, Mustafa MZ, Narendran N, *et al.* A case of *Candida albicans* endophthalmitis with no predisposing risk factors and a distant source of infection. *Case Rep Ophthalmol.* 2012; **3:** 277-82.
74. Chawla B, Sharma N, Titiyal JS, Nayak N, Satpathy G. *Aureobasidium pullulans* keratitis following automated lamellar therapeutic keratoplasty. *Ophthalmic Surg Lasers Imaging.* 2010: 1-3. PMID: 20337364.
75. Chawla B, Sharma S, Kashyap S, *et al.* Primary orbital mycosis in immunocompetent infants. *J AAPOS.* 2011; **15:** 211-3.
76. Chee YE, Eliott D. The role of vitrectomy in the management of fungal endophthalmitis. *Semin Ophthalmol.* 2017; **32:** 29-35.
77. Chen J, Chai C, Teoh SC. Cryptococcal-related exudative retinal detachment. *Eye (Lond).* 2011; **25:** 1234-5.
78. Chen KJ, Chen TL. Man with blurred vision. Endogenous *Candida* endophthalmitis caused by urinary tract infection. *Ann Emerg Med.* 2012; **59:** e3-4.
79. Chen KJ, Sun CC, Hsiao CH, *et al.* Treatment for fungal endophthalmitis resulting from keratitis. *Am J Ophthalmol.* 2011; **151:** 185-6.
80. Chen KJ, Wu WC, Sun MH, *et al.* Endogenous fungal endophthalmitis: Causative organisms, management strategies, and visual acuity outcomes. *Am J Ophthalmol.* 2012; **154:** 213-4.
81. Chen S, Aronow ME, Wang C, *et al.* Classical pathology of sympathetic ophthalmia presented in a unique case. *Open Ophthalmol J.* 2014; **8:** 32-8.

82. Chen WL, Tsai YY, Lin JM, *et al.* Unilateral *Candida parapsilosis* interface keratitis after laser *in situ* keratomileusis: Case report and review of the literature. *Cornea.* 2009; **28:** 105-7.
83. Cheng HH, Ding Y, Wu M, *et al.* Endogenous *Aspergillus* endophthalmitis after kidney transplantation. *Int J Ophthalmol.* 2011; **4:** 567-71.
84. Cheng ML, Leibowitz M, Ha E. Coccidioidal endophthalmitis in immunocompetent person, California, USA. *Emerg Infect Dis.* 2012; **18:** 1015-6.
85. Cheng SC, Lin YY, Kuo CN, *et al. Cladosporium keratitis* - A case report and literature review. *BMC Ophthalmol.* 2015; **15:** 106. PMID: 26286482.
86. Cheon YI, Hong SL, Roh HJ, *et al.* Fungal ball within Onodi cell mucocele causing visual loss. *J Craniofac Surg.* 2014; **25:** 512-4.
87. Chew AC, Mehta JS, Li L, *et al.* Fungal endophthalmitis after Descemet stripping automated endothelial keratoplasty - A case report. *Cornea.* 2010; **29:** 346-9.
88. Chew HF, Jungkind DL, Mah DY, *et al.* Post-traumatic fungal keratitis caused by *Carpoligna* sp. *Cornea.* 2010; **29:** 449-52.
89. Chew R, Dorman A, Woods ML. *Purpureocillium lilacinum* keratitis: A case series and review of the literature. *Can J Ophthalmol.* 2016; **51:** 382-5.
90. Chhablani J, Panchal B, Das T, *et al.* Endophthalmitis in Boston keratoprosthesis: Case series and review of literature. *Int Ophthalmol.* 2015; **35:** 673-8.
91. Chhablani J. Fungal endophthalmitis. *Expert Rev Anti Infect Ther.* 2011; **9:** 1191-201.
92. Chiam N, Rose LV, Waters KD, *et al. Scedosporium prolificans* endogenous endophthalmitis. *J AAPOS.* 2013; **17:** 627-9.
93. Cho KJ, Kim MS. Fungal hyphae growing into anterior chamber from cornea. *Can J Ophthalmol.* 2014; **49:** e151-4.
94. Choi KS, Yoon SC, Rim TH, *et al.* Effect of voriconazole and ultraviolet - A combination therapy compared to voriconazole single treatment on *Fusarium solani* fungal keratitis. *J Ocul Pharmacol Ther.* 2014; **30:** 381-6.
95. Choy BN, Tang EW, Lai JS. A case series of contact lens-assocaited *Fusarium* keratitis in Hong Kong. *Cornea.* 2009; **28:** 955.
96. Clamp MF, Jumper JM, Ku CW, *et al.* Chronic exogenous *Exophiala dermatitidis* endophthalmitis. *Retin Cases Brief Rep.* 2014; **8:** 265-8.
97. Comer GM, Stem MS, Saxe SJ. Successful salvage therapy of *Fusarium* endophthalmitis secondary to keratitis: An interventional case series. *Clin Ophthalmol.* 2012; **6:** 721-6.
98. Comez AT, Komur B, Akcali A, *et al.* Ocular aspergillosis: Obtaining a specimen is crucial for diagnosis. A report of three cases. *Med Mycol Case Rep.* 2012; **1:** 39-41.
99. Cui Y, Jia H, He D, *et al.* Characterization of *Edenia gomezpompae* isolated from a patient with keratitis. *Mycopathologia.* 2013; **176:** 75-81.
100. Dalmon C, Porco TC, Lietman TM, *et al.* The clinical differentiation of bacterial and fungal keratitis: A photographic survey. *Invest Ophthalmol Vis Sci.* 2012; **53:** 1787-91.
101. Das M, Murthy SI, Dikshit S, *et al.* Natamycin and voriconazole in fungal keratitis. *Arch Ophthalmol.* 2011; **129:** 814-5.
102. Das S, Samant M, Garg P, *et al.* Role of confocal microscopy in deep fungal keratitis. *Cornea.* 2009; **28:** 11-3.
103. Das S, Sharma S, Kar S, *et al.* Is inclusion of Sabouraud dextrose agar essential for the laboratory diagnosis of fungal keratitis? *Indian J Ophthalmol.* 2010; **58:** 281-6.
104. Das S, Sharma S, Mahapatra S, *et al. Fusarium* keratitis at a tertiary eye care centre in India. *Int Ophthalmol.* 2015; **35:** 387-93.
105. Das S, Sharma S, Priyadarshini O, *et al.* Association between culture results of corneal scrapings and culture and histopathology results of corneal tissues in therapeutic keratoplasty. *Cornea.* 2011; **30:** 1003-6.
106. Dave VP, Majji AB, Suma N, *et al.* A rare case of *Aspergillus terreus* endogenous endophthalmitis in a patient of acute lymphoid leukemia with good clinical outcome. *Eye (Lond).* 2011; **25:** 1094-6.
107. Dave VP, Sharma S, Yogi R, *et al. Apophysomyces elegans:* A novel cause of endogenous endophthalmitis in an immunocompetent individual. *Int Ophthalmol.* 2014; **34:** 1285-9.
108. De Silva SR, Menezo V. Bilateral endogenous endophthalmitis secondary to *Candida albicans. Postgrad Med J.* 2011; **87:** 652.
109. Deng SX, Kamal KM, Hollander DA. The use of voriconazole in the management of post-penetrating keratoplasty *Paecilomyces* keratitis. *J Ocul Pharmacol Ther.* 2009; **25:** 175-7.
110. Deorukhkar S, Katiyar R, Saini S. Epidemiological features and laboratory results of bacterial and fungal keratitis: A five-year study at a rural tertiary-care hospital in western Maharashtra, India. *Singapore Med J.* 2012; **53:** 264-7.
111. Diaz RI, Sigler EJ, Randolph JC, *et al.* Subacute exogenous *Rhizoctonia solani* endophthalmitis following cataract extraction with intraocular lens implantation. *J Cataract Refract Surg.* 2013; **39:** 1609-10.
112. Doan T, Vemulakonda GA, Choi D, *et al.* Retinal neovascularization and endogenous fungal endophthalmitis in intravenous drug users. *Ophthalmology.* 2014; **121:** 1847-8. e2.
113. do Carmo A, Costa E, Marques M, *et al. Fusarium dimerum* species complex (*Fusarium penzigii*) keratitis after corneal trauma. *Mycopathologia.* 2016; **181:** 879-84.
114. Dorin J, Debourgogne A, Zaidi M, *et al.* First unusual case of keratitis in Europe due to the rare fungus *Metarhizium anisopliae. Int J Med Microbiol.* 2015; **305:** 408-12.
115. Dozier CC, Tarantola RM, Jiramongkolchai K, *et al.* Fungal eye disease at a tertiary care center: The utility of routine inpatient consultation. *Ophthalmology.* 2011; **118:** 1671-6.
116. Duggan J, Walls HM. Ocular complications of cryptococcal meningitis in patients with HIV: Report of two cases and review of the literature. *J Int Assoc Physicians AIDS Care (Chic).* 2012; **11:** 283-8.
117. Durkin SR, Henderson T, Raju R, *et al.* Successful treatment of phaeohyphomycotic keratitis caused by *Bipolaris australiensis. Clin Experiment Ophthalmol.* 2008; **36:** 697-9.

118. Farrell S, McElnea E, Moran S, *et al.* Fungal keratitis in the Republic of Ireland. *Eye (Lond).* 2017. doi: 10.1038/eye.2017.82. PMID: 28524886.
119. Echavez MI, Agahan AL, Carino NS. Fungal keratitis caused by *Drechslera* spp. treated with voriconazole: A case report. *Case Rep Ophthalmol Med.* 2013; 626704. PMID: 24151572.
120. Edelstein SL, Akduman L, Durham BH, *et al.* Resistant *Fusarium* keratitis progressing to endophthalmitis. *Eye Contact Lens.* 2012; **38:** 331-5.
121. Eghrari AO, Gibas C, Watkins T, *et al.* First human case of fungal keratitis caused by a putatively novel species of *Lophotrichus. J Clin Microbiol.* 2015; **53:** 3063-7.
122. Eguchi H, Toibana T, Hotta F, *et al.* Severe fungal sclerokeratitis caused by *Metarhizium anisopliae:* A case report and literature review. *Mycoses.* 2015; **58:** 88-92.
123. Ehlers JP, Chavala SH, Woodward JA, *et al.* Delayed recalcitrant fungal endophthalmitis secondary to *Curvularia. Can J Ophthalmol.* 2011; **46:** 199-200.
124. El-Mofty HM, Abdelhakim MA, El-Miligi MF, *et al.* A new combination formula for treatment of fungal keratitis: An experimental study. *J Ophthalmol.* 2014; 173298. PMID: 24872888.
125. Erdem E, Yagmur M, Boral H, *et al. Aspergillus flavus* keratitis: Experience of a tertiary eye clinic in Turkey. *Mycopathologia.* 2017; **182:** 379-85.
126. Esgin H, Bulut E, Orüm C. *Candida pelliculosa* endophthalmitis after cataract surgery: A case report. *BMC Res Notes.* 2014; **7:** 169. PMID: 24656053.
127. Espinosa-Heidmann DG, McMillan BD, Lasala PR, *et al. Candida dubliniensis* endophthalmitis: First case in North America. *Int Ophthalmol.* 2012; **32:** 41-5.
128. Esposito E, Maccio JP, Monti R, *et al. Alternaria* keratitis and hypopyon after clear-cornea phacoemulsification. *J Cataract Refract Surg.* 2014; **40:** 331-4.
129. Fadzillah MT, Ishak SR, Ibrahim M. Refractory *Scedosporium apiospermum* keratitis successfully treated with combination of amphotericin B and voriconazole. *Case Rep Ophthalmol Med.* 2013; 413953. PMID: 23509650.
130. Farooq AV, Patel RM, Lin AY, *et al.* Fungal orbital cellulitis: Presenting features, management and outcomes at a referral center. *Orbit.* 2015; **34:** 152-9.
131. Fernandes M, Vira D, Dey M, *et al.* Comparison between polymicrobial and fungal keratitis: Clinical features, risk factors and outcome. *Am J Ophthalmol.* 2015; **160:** 873-881.e2.
132. Fernandez-Flores A, Lopez-Medrano R, Fuster-Foz C. Histopathological clues in the diagnosis of fungal infection by *Scedosporium* in a case of endophthalmitis starting as conjunctivitis. *J Cutan Pathol.* 2016; **43:** 461-7.
133. Ferrer C, Alio JL. Evaluation of molecular diagnosis in fungal keratitis.Ten years of experience. *J Ophthalmic Inflamm Infect.* 2011; **1:** 15-22.
134. Ferrer C, Perez-Santonja JJ, Rodriguez AE, *et al.* New *Pyrenochaeta* species causing keratitis. *J Clin Microbiol.* 2009; **47:** 1596-8.
135. Fierro JL, Prasad PA, Fisher BT, *et al.* Ocular manifestations of candidemia in children. *Pediatr Infect Dis J.* 2013; **32:** 84-6.
136. Figueira L, Pinheiro D, Moreira R, *et al. Beauveria bassiana* keratitis in bullous keratopathy: Antifungal sensitivity testing and management. *Eur J Ophthalmol.* 2012; **22:** 814-8.
137. Fintelmann RE, Gilmer W, Bloomer MM, *et al.* Recurrent *Lecythophora mutabilis* keratitis and endophthalmitis after deep anterior lamellar keratoplasty. *Arch Ophthalmol.* 2011; **129:** 108-10.
138. FlorCruz NV, Evans JR. Medical interventions for fungal keratitis. *Cochrane Database Syst Rev.* 2015; **4:** CD004241. PMID: 25855311.
139. Fox AR, Houser KH, Morris WR, Walton RC. Dematiaceous fungal endophthalmitis: Report of a case and review of the literature. *J Ophthalmic Inflamm Infect.* 2016; **6:** 43.
140. Funakoshi Y, Yakushijin K, Matsuoka H, *et al.* Fungal endophthalmitis successfully treated with intravitreal voriconazole injection. *Intern Med.* 2011; **50:** 941.
141. Gajjar DU, Pal AK, Ghodadra BK, *et al.* Microscopic evaluation, molecular identification, antifungal susceptibility and clinical outcomes in *Fusarium, Aspergillus* and dematiaceous keratitis. *Biomed Res Int.* 2013; 605308. PMID: 24260740.
142. Gajjar DU, Pal AK, Parmar TJ, *et al.* Fungal scleral keratitis caused by *Phomopsis phoenicicola. J Clin Microbiol.* 2011; **49:** 2365-8.
143. Gajjar DU, Pal AK, Santos JM, *et al.* Severe pigmented keratitis caused by *Cladorrhinum bulbillosum. Indian J Med Microbiol.* 2011; **29:** 434-7.
144. Galperin G, Berra M, Tau J, *et al.* Treatment of fungal keratitis from *Fusarium* infection by corneal cross-linking. *Cornea.* 2012; **31:** 176-80.
145. Ganguly S, Salma KC, Kansakar I, *et al.* Pattern of fungal isolates in cases of corneal ulcer in the western periphery of Nepal. *Nepal J Ophthalmol.* 2011; **3:** 118-22.
146. Gao H, Song P, Echegaray JJ, *et al.* Big bubble deep anterior lamellar keratoplasty for management of deep fungal keratitis. *J Ophthalmol.* 2014; 209759. PMID: 25105019.
147. Garcia-Delpech S, Diaz-Llopis M, Udaondo P, *et al. Fusarium* keratitis 3 weeks after healed corneal cross-linking. *J Refract Surg.* 2010; **26:** 994-5.
148. Garg P, Roy A, Roy S. Update on fungal keratitis. *Curr Opin Ophthalmol.* 2016; **27:** 333-9.
149. Garg P. Fungal, Mycobacterial and *Nocardia* infections and the eye: An update. *Eye (Lond).* 2012; **26:** 245-51.
150. Gaujoux T, Borsali E, Goldschmidt P, *et al.* Fungal keratitis in France. *Acta Ophthalmol.* 2011; **89:** e215-6.
151. Gaujoux T, Chatel MA, Chaumeil C, *et al.* Outbreak of contact lens-related *Fusarium* keratitis in France. *Cornea.* 2008; **27:** 1018-21.
152. Georgala A, Layeux B, Kwan J, *et al.* Inaugural bilateral *Aspergillus* endophthalmitis in a seriously immunocompromised patient. *Mycoses.* 2011; **54:** e631-3.
153. Geraymovych E, Conduff JH, Braich PS, *et al.* Prevalence and factors predictive of intraocular fungal infection in patients with fungemia at an academic urban tertiary care center. *Clin Ophthalmol.* 2015; **9:** 1853-8.
154. Gharamah AA, Moharram AM, Ismail MA, *et al.* Bacterial and fungal keratitis in Upper Egypt: *In vitro*

screening of enzymes, toxins and antifungal activity. *Indian J Ophthalmol.* 2014; **62:** 196-203.

155. Gharamah AA, Moharram AM, Ismail MA, *et al.* Bacterial and fungal endophthalmitis in upper Egypt: Related species and risk factors. *Asian Pac J Trop Biomed.* 2012; **2:** 655-9.
156. Ghodasra DH, Eftekhari K, Shah AR, *et al.* Outcomes, impact on management and costs of fungal eye disease consults in a tertiary care setting. *Ophthalmology.* 2014; **121:** 2334-9.
157. Ghosh AK, Gupta A, Rudramurthy SM, *et al.* Fungal keratitis in north India: Spectrum of agents, risk factors and treatment. *Mycopathologia.* 2016; **181:** 843-50.
158. Gluck S, Headdon WG, Tang D, *et al.* The incidence of ocular candidiasis and evaluation of routine opthalmic examination in critically ill patients with candidaemia. *Anaesth Intensive Care.* 2015; **43:** 693-7.
159. Goldschmidt P, Degorge S, Benallaoua D, *et al.* New strategy for rapid diagnosis and characterization of keratomycosis. *Ophthalmology.* 2012; **119:** 945-50.
160. Goldschmidt P, Degorge S, Che Sarria P, *et al.* New strategy for rapid diagnosis and characterization of fungal infections: The example of corneal scrapings. *PLoS One.* 2012; **7:** e37660. PMID: 22768289.
161. Gong H, Gong X. Combined application of 5% natamycin and 0.2% fluconazole for the treatment of fungal keratitis. *Eye Sci.* 2013; **28:** 84-7.
162. Gonzalez-Granado LI. *Aspergillus* endophthalmitis: Pars plana vitrectomy is an alternative. *Indian J Med Sci.* 2009; **63:** 366-7.
163. Gower EW, Keay LJ, Oechsler RA, *et al.* Trends in fungal keratitis in the United States, 2001 to 2007. *Ophthalmology.* 2010; **117:** 2263-7.
164. Goyal J, Fernandes M, Shah SG. Intracameral voriconazole injection in the treatment of fungal endophthalmitis resulting from keratitis. *Am J Ophthalmol.* 2010; **150:** 939-40.
165. Grancini A, Colosimo M, Mapelli C, *et al.* A case of *Histoplasma capsulatum* endophthalmitis diagnosed in Italy. *Travel Med Infect Dis.* 2013; **11:** 256-8.
166. Gregory ME, Macdonald EC, Lockington D, *et al.* Recurrent fungal keratitis following penetrating keratoplasty: An unusual source of infection. *Arch Ophthalmol.* 2010; **128:** 1490-1.
167. Gregory ME, Weir CR, Roberts F, *et al. Aspergillus* endophthalmitis following orthotopic heart transplant. *Can J Ophthalmol.* 2009; **44:** 607-8.
168. Grieser EJ, Tuli SS, Chabi A, *et al.* Blueberry eye: Acquired total anterior staphyloma after a fungal corneal ulcer. *Cornea.* 2009; **28:** 231-2.
169. Gungel H, Eren MH, Pınarcı EY, *et al.* An outbreak of *Fusarium solani* endophthalmitis after cataract surgery in an eye training and research hospital in Istanbul. *Mycoses.* 2011; **54:** e767-74.
170. Gupta A, Capoor MR, Gupta S, *et al.* Clinico-demographical profile of keratomycosis in Delhi, North India. *Indian J Med Microbiol.* 2014; **32:** 310-4.
171. Gupta G, Feder RS, Lyon AT. Fungal keratitis with intracameral extension following penetrating keratoplasty. *Cornea.* 2009; **28:** 930-2.
172. Gupta MK, Chandra A, Prakash P, *et al.* Fungal keratitis in north India; Spectrum and diagnosis by calcofluor white stain. *Indian J Med Microbiol.* 2015; **33:** 462-3.
173. Gupta N, Samantaray JC, Duggal S, Srivastava V, Dhull CS, Chaudhary U. *Acanthamoeba* keratitis with *Curvularia* co-infection. *Indian J Med Microbiol.* 2010; **28:** 67-71.
174. Gupta R, Hari P, Anandula V. Eyelid fungal infections are rare and often associated with predisposing factors. We present the case of a child with an ulcerated lesion of fungal etiology on the eyelid. *Can J Ophthalmol.* 2017; **52:** e28-9.
175. Gupta S, Loudill C, Tammara A, *et al.* A rare case of bilateral *Aspergillus* endophthalmitis. *J Community Hosp Intern Med Perspect.* 2015; **5:** 28984. PMID: 26653687.
176. Gupta S, Shrivastava RM, Tandon R, *et al.* Role of voriconazole in combined *Acanthamoeba* and fungal corneal ulcer. *Cont Lens Anterior Eye.* 2011; **34:** 287-9.
177. Ha J, Heath CH, Suzanne R, *et al.* Ophthalmic manifestation of *Candida*: Case report and review of the literature. *Eur J Pediatr Surg.* 2009; **19:** 197-9.
178. Haddad RS, El-Mollayess GM. Combination of intracameral and intrastromal voriconazole in the treatment of recalcitrant *Acremonium* fungal keratitis. *Middle East Afr J Ophthalmol.* 2012; **19:** 265-8.
179. Hager JL, Phillips DL, Goins KM, *et al.* Boston type 1 keratoprosthesis for failed keratoplasty. *Int Ophthalmol.* 2016; **36:** 73-8.
180. Hamada Y, Okuma R, Katori Y, *et al.* Bibliographical investigation (domestic and overseas) on the treatment of endogenous *Candida* endophthalmitis over an 11-year period. *Med Mycol J.* 2013; **54:** 53-67.
181. Hariprasad SM, Mieler WF, Lin TK, *et al.* Voriconazole in the treatment of fungal eye infections: a review of current literature. *Br J Ophthalmol.* 2008; **92:** 871-8.
182. Hashemi SB, Shishegar M, Nikeghbalian S, *et al.* Endogenous *Aspergillus* endophthalmitis occurring after liver transplantation: A case report. *Transplant Proc.* 2009; **41:** 2933-5.
183. Hassan A, Poon W, Baker M, *et al.* Confirmed *Candida albicans* endogenous fungal endophthalmitis in a patient with chronic candidiasis. *Med Mycol Case Rep.* 2012; **1:** 42-4.
184. Hassan AS, Al-Hatmi AM, Shobana CS, *et al.* Antifungal susceptibility and phylogeny of opportunistic members of the genus *Fusarium* causing human keratomycosis in South India. *Med Mycol.* 2016; **54:** 287-94.
185. Hassan HM, Papanikolaou T, Mariatos G, *et al. Candida albicans* keratitis in an immunocompromised patient. *Clin Ophthalmol.* 2010; **4:** 1211-5.
186. Hayashi Y, Eguchi H, Toibana T, *et al.* Polymicrobial sclerokeratitis caused by *Scedosporium apiospermum* and *Aspergillus cibarius. Cornea.* 2014; **33:** 875-7.
187. He D, Hao J, Gao S, *et al.* Etiological analysis of fungal keratitis and rapid identification of predominant fungal pathogens. *Mycopathologia.* 2016; **181:** 75-82.

188. He D, Hao J, Zhang B, *et al.* Pathogenic spectrum of fungal keratitis and specific identification of *Fusarium solani*. *Invest Ophthalmol Vis Sci.* 2011; **52:** 2804-8.
189. He H, Liu H, Chen X, *et al.* Diagnosis and treatment of *Pythium insidiosum* corneal ulcer in a Chinese child: A case report and literature review. *Am J Case Rep.* 2016; **17:** 982-8.
190. Hirai Y, Asahata S, Ainoda Y, *et al.* Nosocomial *Candida parapsilosis* candidaemia: Risk factors, antifungal susceptibility and outcome. *J Hosp Infect.* 2014; **87:** 54-8.
191. Hirst LW, Choong K, Playford EG. Non-traumatic *Paecilomyces* anterior segment infection: A pathognomonic clinical appearance. *Cornea.* 2014; **33:** 1031-7.
192. Hodkin MJ, Gustus RC. Fungal keratitis associated with airborne organic debris and soft contacts lenses: Case reports and review of the literature. *Eye Contact Lens.* 2017. doi: 10.1097/ICL.0000000000000395. PMID: 28520596.
193. Ho HC, Liew OH, Teh SS, *et al.* Unilateral rhino-orbital-cerebral mucormycosis with contralateral endogenous fungal endophthalmitis. *Clin Ophthalmol.* 2015; **9:** 553-6.
194. Ho JW, Fernandez MM, Rebong RA, *et al.* Microbiological profiles of fungal keratitis: A 10-year study at a tertiary referral center. *J Ophthalmic Inflamm Infect.* 2016; **6:** 5.
195. Hoenigl M, Krause R. Antifungal therapy of aspergillosis of the central nervous system and *Aspergillus* endophthalmitis. *Curr Pharm Des.* 2013; **19:** 3648-68.
196. Homa M, Shobana CS, Singh YR, *et al.* *Fusarium* keratitis in South India: Causative agents, their antifungal susceptibilities and a rapid identification method for the *Fusarium solani* species complex. *Mycoses.* 2013; **56:** 501-11.
197. Hong KH, Ryoo NH, Chang SD. *Phialemonium obovatum* keratitis after penetration injury of the cornea. *Korean J Ophthalmol.* 2012; **26:** 465-8.
198. Hosseini H, Saki S, Saki N, *et al.* *Aspergillus* endophthalmitis in orthotopic liver transplantation. *Indian J Med Sci.* 2009; **63:** 253-6.
199. Hotta F, Eguchi H, Nishimura K, *et al.* A super-infection in the cornea caused by *Stemphylium, Acremonium* and *α-Streptococcus. Ann Clin Microbiol Antimicrob.* 2017; **16:** 11.
200. Hsu YJ, Huang JS, Tsai JH, *et al.* Early-onset severe donor-related *Candida* keratitis after Descemet stripping automated endothelial keratoplasty. *J Formos Med Assoc.* 2014; **113:** 874-6.
201. Hu J, Ma H, Li T, *et al.* Retina translocation for a patient with endogenous fungal endophthalmitis: Long-term follow-up. *Graefes Arch Clin Exp Ophthalmol.* 2012; **250:** 1237-40.
202. Hua X, Yuan X, Di Pietro A, *et al.* The molecular pathogenicity of *Fusarium* keratitis: A fungal transcriptional regulator promotes hyphal penetration of the cornea. *Cornea.* 2010; **29:** 1440-4.
203. Huang YH, Lin IH, Chang TC, *et al.* Early diagnosis and successful treatment of *Cryptococcus albidus* keratitis: A case report and literature review. *Medicine (Baltimore).* 2015; **94:** e885. PMID: 25984681.
204. Hurtado-Sarrio M, Duch-Samper A, Cisneros-Lanuza A, *et al.* Successful topical application of caspofungin in the treatment of fungal keratitis refractory to voriconazole. *Arch Ophthalmol.* 2010; **128:** 941-2.
205. Huynh TK, Lee LR, Ellis D. Late-onset post-traumatic *Phaeoacremonium parasiticum* endophthalmitis. *Clin Experiment Ophthalmol.* 2007; **35:** 366-8.
206. Ibrahim MM, Vanini R, Ibrahim FM, *et al.* Epidemiologic aspects and clinical outcome of fungal keratitis in southeastern Brazil. *Eur J Ophthalmol.* 2009; **19:** 355-61.
207. Idiga J, Rootman DB, Nagiel A, *et al.* Isolated zygomycetes endophthalmitis: A case report. *Ophthal Plast Reconstr Surg.* 2015; **31:** e145-7.
208. Ito M, Yokoi T, Sugita S, *et al.* Endogenous *Candida* chorioretinitis in a healthy infant. *Jpn J Ophthalmol.* 2010; **54:** 629-31.
209. Jafarinasab MR, Feizi S, Yazdizadeh F, *et al.* *Aspergillus flavus* keratitis after deep anterior lamellar keratoplasty. *J Ophthalmic Vis Res.* 2012; **7:** 167-71.
210. Jain V, Dabir S, Shome D, *et al.* *Aspergillus* iris granuloma: A case report with review of literature. *Surv Ophthalmol.* 2009; **54:** 286-91.
211. Jain V, Mhatre K, Nair AG, *et al.* *Aspergillus* keratitis in vernal shield ulcer - A case report and review. *Int Ophthalmol.* 2010; **30:** 641-4.
212. Jain V, Mhatre K, Shome D, *et al.* Fungal keratitis with the Type 1 Boston keratoprosthesis: Early Indian experience. *Cornea.* 2012; **31:** 841-3.
213. Jain V, Shome D, Sajnani M, *et al.* Fungal keratitis associated with ocular rosacea. *Int Ophthalmol.* 2010; **30:** 239-44.
214. Jalili M, Hashemi H, Jabarvand M, *et al.* *Aspergillus* endophthalmitis in one eye subsequent to bilateral anterior chamber phakic intraocular lens implantation. *J Refract Surg.* 2012; **28:** 363-5.
215. Jaramillo S, Varon CL. *Curvularia lunata* endophthalmitis after penetrating ocular trauma. *Retin Cases Brief Rep.* 2013; **7:** 315-8.
216. Jastaneiah SS, Al-Rajhi AA, Abbott D. Ocular mycosis at a referral center in Saudi Arabia: A 20-year study. *Saudi J Ophthalmol.* 2011; **25:** 231-8.
217. Jhanji V, Yohendran J, Constantinou M, *et al.* *Scedosporium* scleritis or keratitis or both: Case series. *Eye Contact Lens.* 2009; **35:** 312-5.
218. Jiang K, Brownstein S, Baig K, *et al.* Clinicopathologic case reports of *Alternaria* and *Fusarium* keratitis in Canada. *Can J Ophthalmol.* 2013; **48:** e151-4.
219. Jin KW, Jeon HS, Hyon JY, *et al.* A case of fungal keratitis and onychomycosis simultaneously infected by *Trichophyton* species. *BMC Ophthalmol.* 2014; **14:** 90.
220. Joe SG, Lim J, Lee JY, *et al.* Case report of *Acremonium* intraocular infection after cataract extraction. *Korean J Ophthalmol.* 2010; **24:** 119-22.
221. Jorgensen JS, Prause JU, Kiilgaard JF. Bilateral endogenous *Fusarium solani* endophthalmitis in a liver-transplanted patient: A case report. *J Med Case Rep.* 2014; **8:** 101.
222. Joseph NM, Kumar MA, Stephen S, *et al.* Keratomycosis caused by *Exserohilum rostratum. Indian J Pathol Microbiol.* 2012; **55:** 248-9.

223. Jurkunas U, Behlau I, Colby K. Fungal keratitis: Changing pathogens and risk factors. *Cornea.* 2009; **28:** 638-43.
224. Jutley G, Koukkoulli A, Forbes J, *et al.* Unusual case of *Scedosporium apiospermum* keratitis following phacoemulsification in a systemically well patient. *J Cataract Refract Surg.* 2015; **41:** 230-3.
225. Kaburaki T, Takamoto M, Araki F, *et al.* Endogenous *Candida albicans* infection causing subretinal abscess. *Int Ophthalmol.* 2010; **30:** 203-6.
226. Kah TA, Yong KC, Rahman RA. Disseminated fusariosis and endogenous fungal endophthalmitis in acute lymphoblastic leukemia following platelet transfusion possibly due to transfusion-related immunomodulation. *BMC Ophthalmol.* 2011; **11:** 30.
227. Kalaiselvi G, Narayana S, Krishnan T, *et al.* Intrastromal voriconazole for deep recalcitrant fungal keratitis: A case series. *Br J Ophthalmol.* 2015; **99:** 195-8.
228. Kaliamurthy J, Kalavathy CM, Nelson Jesudasan CA, *et al.* Keratitis due to *Chaetomium* sp. *Case Rep Ophthalmol Med.* 2011; 696145. PMID: 22606471.
229. Kaliamurthy J, Thomas PA. Is inclusion of Sabouraud dextrose agar essential for the laboratory diagnosis of fungal keratitis? *Indian J Ophthalmol.* 2011; **59:** 263-4.
230. Kalkanci A, Ozdek S. Ocular fungal infections. *Curr Eye Res.* 2011; **36:** 179-89.
231. Kamada R, Monden Y, Uehara K, *et al.* Rare case of fungal keratitis caused by *Plectosporium tabacinum. Clin Ophthalmol.* 2012; **6:** 1623-7.
232. Kamoshita M, Matsumoto Y, Nishimura K, *et al. Wickerhamomyces anomalus* fungal keratitis responds to topical treatment with antifungal micafungin. *J Infect Chemother.* 2015; **21:** 141-3.
233. Kapoor A, Jain R, Sahu S, *et al.* Fungal keratitis presenting as radial keratoneuritis. *BMJ Case Rep.* 2014; pii: bcr2013202200. PMID: 24717856.
234. Kapp M, Schargus M, Deuchert T, *et al.* Endophthalmitis as primary clinical manifestation of fatal fusariosis in an allogeneic stem cell recipient. *Transpl Infect Dis.* 2011; **13:** 374-9.
235. Kapur R, Aakalu VK, August CZ, *et al.* Mucormycosis infection of the lacrimal sac. *Ophthal Plast Reconstr Surg.* 2009; **25:** 494-6.
236. Keating A, Pineda R 2nd. *Trichosporon asahii* keratitis in a patient with a type I Boston keratoprosthesis and contact lens. *Eye Contact Lens.* 2012; **38:** 130-2.
237. Keay LJ, Gower EW, Iovieno A, *et al.* Clinical and microbiological characteristics of fungal keratitis in the United States, 2001-2007: A multicenter study. *Ophthalmology.* 2011; **118:** 920-6.
238. Kepez Yildiz B, Hasanreisoglu M, Aktas Z, *et al.* Fungal keratitis secondary to *Scedosporium apiospermum* infection and successful treatment with surgical and medical intervention. *Int Ophthalmol.* 2014; **34:** 305-8.
239. Kernt M, Neubauer AS, De Kaspar HM, *et al.* Intravitreal voriconazole: *In vitro* safety-profile for fungal endophthalmitis. *Retina.* 2009; **29:** 362-70.
240. Khan B, Vohra R, Kaur R, *et al.* Excellent outcome of *Aspergillus* endophthalmitis in a case of allergic bronchopulmonary aspergillosis. *Indian J Ophthalmol.* 2014; **62:** 352-4.
241. Khan S, Pillai GS, Vivek V, *et al.* Post-operative endophthalmitis due to *Fusarium dimerum. Southeast Asian J Trop Med Public Health.* 2012; **43:** 1484-8.
242. Khater MM, Shehab NS, El-Badry AS. Comparison of mycotic keratitis with non-mycotic keratitis: An epidemiological study. *J Ophthalmol.* 2014; 254302. PMID: 25548657.
243. Khochtali S, Hriz A, Abid F, *et al. Alternaria* keratitis after uneventful phacoemulsification in an otherwise healthy adult. *J Ophthalmic Inflamm Infect.* 2016; **6:** 4.
244. Kim DY, Moon HI, Joe SG, *et al.* Recent clinical manifestation and prognosis of fungal endophthalmitis: A 7-year experience at a tertiary referral center in Korea. *J Korean Med Sci.* 2015; **30:** 960-4.
245. Kim EL, Patel SR, George MS, *et al. Ochroconis gallopava* endophthalmitis successfully treated with intravitreal voriconazole and amphotericin B. *Retin Cases Brief Rep.* 2016. doi: 10.1097/ICB.0000000000000505. PMID: 28002285.
246. Kim JW, Rha MS, Kim JH, *et al.* Orbital apex syndrome caused by invasive aspergillosis. *J Craniofac Surg.* 2014; **25:** e191-3.
247. Kim KS, Kim HJ. Periorbital pain induced by the concha bullosa fungus ball: A case report. *J Oral Facial Pain Headache.* 2014; **28:** 277-9.
248. Kim SJ, Cho YW, Seo SW, *et al.* Clinical experiences in fungal keratitis caused by *Acremonium. Clin Ophthalmol.* 2014; **8:** 283-7.
249. Kindo AJ, Anita S, Kalpana S. *Nattrassia mangiferae* causing fungal keratitis. *Indian J Med Microbiol.* 2010; **28:** 178-81.
250. Kitzmann AS, Wagoner MD, Syed NA, *et al.* Donor-related *Candida* keratitis after Descemet stripping automated endothelial keratoplasty. *Cornea.* 2009; **28:** 825-8.
251. Kleinschmidt-Demasters BK. Disseminated *Fusarium* infection with brain abscesses in a lung transplant recipient. *Clin Neuropathol.* 2009; **28:** 417-21.
252. Klotz SA, Penn CC, Negvesky GJ, *et al.* Fungal and parasitic infections of the eye. *Clin Microbiol Rev.* 2000; **13:** 662-85.
253. Knape RM, Motamarry SP, Sakhalkar MV, *et al.* Pseudodendritic fungal epithelial keratitis in an extended wear contact lens user. *Eye Contact Lens.* 2011; **37:** 36-8.
254. Koenig SB, Wirostko WJ, Fish RI, *et al. Candida* keratitis after Descemet stripping and automated endothelial keratoplasty. *Cornea.* 2009; **28:** 471-3.
255. Kokuzawa S, Suemori S, Mochizuki K, *et al. Aspergillus tubingenesis* endophthalmitis after cataract surgery with implantation of preloaded intraocular lens. *Semin Ophthalmol.* 2014; **29:** 218-21.
256. Konidaris V, Mersinoglou A, Vyzantiadis TA, *et al.* Corneal transplant infection due to *Alternaria alternata:* A case report. *Case Rep Ophthalmol Med.* 2013; 589620. PMID: 23573439.

257. Kothari SG, Kothari RS. Successful treatment of *Fusarium* keratitis after photo refractive keratectomy. *Indian J Ophthalmol.* 2014; **62:** 661.
258. Kotwal A, Biswas D, Kakati B, *et al.* Non-traumatic keratitis due to *Colletotrichum coccodes:* A case report. *J Clin Diagn Res.* 2015; **9:** DD01-2. PMID: 25859457.
259. Kredics L, Narendran V, Shobana CS, *et al.* Filamentous fungal infections of the cornea: A global overview of epidemiology and drug sensitivity. *Mycoses.* 2015; **58:** 243-60.
260. Kredics L, Varga J, Kocsube S, *et al.* Infectious keratitis caused by *Aspergillus tubingensis. Cornea.* 2009; **28:** 951-4.
261. Krishnan T, Prajna NV, Gronert K, *et al.* Gender differences in re-epithelialisation time in fungal corneal ulcers. *Br J Ophthalmol.* 2012; **96:** 137-8.
262. Krishnan T, Rishi P. Management of a case of *Candida* endogenous endophthalmitis in a neonate. *Ocul Immunol Inflamm.* 2014; **22:** 77-8.
263. Kumar P, Thomas S, Papagiannuli E, *et al.* A case of *Phoma* fungal keratitis in a contact lens user. *JRSM Open.* 2015; **6:** 2054270415577760. PMID: 25893109.
264. Kuo MT, Chang HC, Cheng CK, *et al.* A highly sensitive method for molecular diagnosis of fungal keratitis: A dot hybridization assay. *Ophthalmology.* 2012; **119:** 2434-42.
265. Kurbanyan K, Hoesl LM, Schrems WA, *et al.* Corneal nerve alterations in acute *Acanthamoeba* and fungal keratitis: An *in vivo* confocal microscopy study. *Eye (Lond).* 2012; **26:** 126-32.
266. Kuruba SL, Prabhakaran VC, Nagarajappa AH, *et al.* Orbital *Aspergillus* infection diagnosed by FNAC. *Diagn Cytopathol.* 2011; **39:** 523-6.
267. Labbe A, Gabison E, Cochereau I, *et al.* Diagnosis of fungal keratitis by *in vivo* confocal microscopy: A case report. *Eye (Lond).* 2011; **25:** 956-8.
268. Labiris G, Troeber L, Gatzioufas Z, *et al.* Bilateral *Fusarium oxysporum* keratitis after laser *in situ* keratomileusis. *J Cataract Refract Surg.* 2012; **38:** 2040-4.
269. Lai J, Pandya V, McDonald R, *et al.* Management of *Fusarium* keratitis and its associated fungal iris nodule with intracameral voriconazole and amphotericin B. *Clin Exp Optom.* 2014; **97:** 181-3.
270. Laich FS, Alcoba-Florez J, Perez-Roth E, *et al.* Molecular characterization of *Alternaria alternata* causing ocular infection: Detection of IGS-RFLP intraspecific polymorphism. *Med Mycol.* 2008; **46:** 615-9.
271. Lalitha P, Prajna NV, Manoharan G, *et al.* Trends in bacterial and fungal keratitis in South India, 2002-2012. *Br J Ophthalmol.* 2015; **99:** 192-4.
272. Lalitha P, Prajna NV, Oldenburg CE, *et al.* Organism, minimum inhibitory concentration and outcome in a fungal corneal ulcer clinical trial. *Cornea.* 2012; **31:** 662-7.
273. Lalitha P, Shapiro BL, Loh AR, *et al.* Amphotericin B and natamycin are not synergistic *in vitro* against *Fusarium* and *Aspergillus* spp. isolated from keratitis. *Br J Ophthalmol.* 2011; **95:** 744-5.
274. Lalitha P, Sun CQ, Prajna NV, *et al. In vitro* susceptibility of filamentous fungal isolates from a corneal ulcer clinical trial. *Am J Ophthalmol.* 2014; **157:** 318-26.
275. Lamarca J, Vilaplana F, Nadal J, *et al.* Treatment resistant fungal keratitis caused by *Colletotrichum gloeosporioides. Arch Soc Esp Oftalmol.* 2016; **91:** 97-101.
276. Lari HB, Mirani N, Chu DS. Corneal chromoblastomycosis caused by *Cladophialophora carrionii* after cataract surgery. *J Cataract Refract Surg.* 2011; **37:** 963-6.
277. Lavine JA, Mititelu M. Multimodal imaging of refractory *Candida* chorioretinitis progressing to endogenous endophthalmitis. *J Ophthalmic Inflamm Infect.* 2015; **5:** 54.
278. Le Q, Wu D, Li Y, *et al.* Early-onset *Candida glabrata* interface keratitis after deep anterior lamellar keratoplasty. *Optom Vis Sci.* 2015; **92:** e93-6.
279. Leal AF, Leite MC, Medeiros CS, *et al.* Antifungal activity of a liposomal itraconazole formulation in experimental *Aspergillus flavus* keratitis with endophthalmitis. *Mycopathologia.* 2015; **179:** 225-9.
280. Lebas M, De Lahitte GD, Fardeau C, *et al.* Bilateral fungal endophthalmitis masquerading as multifocal chorioretinitis. *Retin Cases Brief Rep.* 2011; **5:** 279-83.
281. Leck A, Burton M. Distinguishing fungal and bacterial keratitis on clinical signs. *Community Eye Health.* 2015; **28:** 6-7.
282. Lee B, Grossniklaus HE, Edelhauser HF. Concurrent *Acanthamoeba* and *Fusarium* keratitis with silicone hydrogel contact lens use. *Cornea.* 2010; **29:** 210-3.
283. Lee CY, Ho YJ, Sun CC, *et al.* Recurrent fungal keratitis and blepharitis caused by *Aspergillus flavus. Am J Trop Med Hyg.* 2016; **95:** 1216-8.
284. Lee JH, Kim JS, Park YH. Diagnosis and treatment of postpartum *Candida* endophthalmitis. *J Obstet Gynaecol Res.* 2012; **38:** 1220-2.
285. Lee SJ, Lee JJ, Kim SD. Topical and oral voriconazole in the treatment of fungal keratitis. *Korean J Ophthalmol.* 2009; **23:** 46-8.
286. Lee WB, Foster JB, Kozarsky AM, *et al.* Interface fungal keratitis after endothelial keratoplasty: A clinicopathological report. *Ophthalmic Surg Lasers Imaging.* 2011; **42:** e44-8.
287. Leema G, Kaliamurthy J, Geraldine P, *et al.* Keratitis due to *Aspergillus flavus*: Clinical profile, molecular identification of fungal strains and detection of aflatoxin production. *Mol Vis.* 2010; **16:** 843-54.
288. Lekhanont K, Nonpassopon M, Nimvorapun N, *et al.* Treatment with intrastromal and intracameral voriconazole in 2 eyes with *Lasiodiplodia theobromae* keratitis: Case reports. *Medicine (Baltimore).* 2015; **94:** e541.
289. Leung C, Farmer JP, Zoutman DE, *et al. Histoplasma capsulatum* endophthalmitis in southeastern Ontario. *Can J Ophthalmol.* 2010; **45:** 90-1.
290. Leung EH, Moskalewicz R, Parada JP, *et al. Exophiala jeanselmei* keratitis after laser *in situ* keratomileusis. *J Cataract Refract Surg.* 2008; **34:** 1809-11.
291. Li L, Wang Z, Li R, *et al. In vitro* evaluation of combination antifungal activity against *Fusarium* species isolated from ocular tissues of keratomycosis patients. *Am J Ophthalmol.* 2008; **146:** 724-8.
292. Li LM, Zhao LQ, Qu LH, *et al.* Excimer laser phototherapeutic keratectomy for the treatment of clinically presumed fungal keratitis. *J Ophthalmol.* 2014; 963287.

293. Li ST, Yiu EP, Wong AH, *et al.* Successful treatment of *Lasiodiplodia theobromae* keratitis - Assessing the role of voriconazole. *Case Rep Ophthalmol.* 2016; **7:** 179-85.
294. Li Z, Jhanji V, Tao X, *et al.* Riboflavin/ultraviolet light-mediated crosslinking for fungal keratitis. *Br J Ophthalmol.* 2013; **97:** 669-71.
295. Lin CM, Pao SI, Chen YH, *et al.* Fungal endophthalmitis caused by *Trichophyton* spp. after cataract surgery. *Clin Experiment Ophthalmol.* 2014; **42:** 696-7.
296. Lin HC, Lin JL, Lin-Tan DT, *et al.* Early keratectomy in the treatment of moderate *Fusarium* keratitis. *PLoS One.* 2012; **7:** e42126. PMID: 22936982.
297. Lin TT, Wei RH, Yang RB, *et al.* Fungal keratitis associated with viral keratitis. *Chin Med J (Engl).* 2015; **128:** 2823-5.
298. Lingappan A, Wykoff CC, Albini TA, *et al.* Endogenous fungal endophthalmitis: causative organisms, management strategies, and visual acuity outcomes. *Am J Ophthalmol.* 2012; **153:** 162-6.e1.
299. Liu K, Fang F, Li H. Reliability of vitreous histological detection of pathogenic fungi in the diagnosis of fungal endophthalmitis. *Eye (Lond).* 2015; **29:** 424-7.
300. Liu M, Xin X, Li J, *et al.* The first case of endophthalmitis due to *Rhinocladiella basitona* in an immunocompetent patient. *Diagn Microbiol Infect Dis.* 2015; **83:** 49-52.
301. Liu Y, Jia H, Shi X, *et al.* Minimal trephination penetrating keratoplasty for severe fungal keratitis complicated with hypopyon. *Can J Ophthalmol.* 2013; **48:** 529-34.
302. Liu Z, Zhang P, Liu C, *et al.* Split of Descemet's membrane and pre-Descemet's layer in fungal keratitis: New definition of corneal anatomy incorporating new knowledge of fungal infection. *Histopathology.* 2015; **66:** 1046-9.
303. Llamos R, Al-Hatmi AM, Martinez G, *et al.* Non-traumatic keratitis due to *Colletotrichum truncatum. JMM Case Rep.* 2016; **3:** e005047. PMID: 28348770.
304. Loh AR, Hong K, Lee S, *et al.* Practice patterns in the management of fungal corneal ulcers. *Cornea.* 2009; **28:** 856-9.
305. Luong PM, Kalpakian B, Jaeger LJ, *et al. Rhodotorula* endogenous endophthalmitis: A novel harbinger of the injection drug epidemic in the United States. *Case Rep Infect Dis.* 2017; 9686353. PMID: 28480087.
306. Ma SK, So K, Chung PH, *et al.* A multi-country outbreak of fungal keratitis associated with a brand of contact lens solution: The Hong Kong experience. *Int J Infect Dis.* 2009; **13:** 443-8.
307. Mack RJ, Shott S, Schatz S, *et al.* Association between moxifloxacin ophthalmic solution and fungal infection in patients with corneal ulcers and microbial keratitis. *J Ocul Pharmacol Ther.* 2009; **25:** 279-84.
308. Mahdy RA, Nada WM, Wageh MM, *et al.* Assessment safety and efficacy of a combination therapy of topical amphotericin B and subconjunctival fluconazole for the treatment of fungal keratitis. *Cutan Ocul Toxicol.* 2010; **29:** 193-7.
309. Mahdy RA, Nada WM, Wageh MM. Topical amphotericin B and subconjunctival injection of fluconazole (combination therapy) versus topical amphotericin B (monotherapy) in treatment of keratomycosis. *J Ocul Pharmacol Ther.* 2010; **26:** 281-5.
310. Mainville N, Jordan DR. Orbital aspergillosis treated with retrobulbar amphotericin B. *Orbit.* 2012; **31:** 15-7.
311. Malathi J, Sowmiya M, Khetan V, *et al.* Complication of *Salmonella* bacteremia in a case of treated fungal endophthalmitis. *Case Rep Ophthalmol Med.* 2012; 198637. PMID: 22606488.
312. Malhotra S, Khare A, Grover K, *et al.* Design and evaluation of voriconazole eye drops for the treatment of fungal keratitis. *J Pharm (Cairo).* 2014; 490595. PMID: 26556196.
313. Mamandhar A, Bajracharya L. Endogenous *Aspergillus* endophthalmitis in a healthy individual. *Nepal J Ophthalmol.* 2012; **4:** 179-83.
314. Mandell KJ, Colby KA. Penetrating keratoplasty for invasive fungal keratitis resulting from a thorn injury involving *Phomopsis* species. *Cornea.* 2009; **28:** 1167-9.
315. Manikandan P, Varga J, Kocsube S, *et al.* Epidemiology of *Aspergillus* keratitis at a tertiary care eye hospital in South India and antifungal susceptibilities of the causative agents. *Mycoses.* 2013; **56:** 26-33.
316. Manikandan P, Varga J, Kocsube S, *et al.* Keratitis caused by the recently described new species *Aspergillus brasiliensis:* two case reports. *J Med Case Rep.* 2010; **4:** 68.
317. Manikandan P, Varga J, Kocsube S, *et al.* Mycotic keratitis due to *Aspergillus nomius. J Clin Microbiol.* 2009; **47:** 3382-5.
318. Manning RA, Carlson JN, Hein EW, *et al.* Bilateral *Candida* endophthalmitis in a premature infant. *J AAPOS.* 2013; **17:** 206-7.
319. Martone G, Pichierri P, Franceschini R, *et al. In vivo* confocal microscopy and anterior segment optical coherence-tomography in a case of *Alternaria* keratitis. *Cornea.* 2011; **30:** 449-53.
320. Mascarenhas J, Lalitha P, Prajna NV, *et al.* Acanthamoeba, fungal and bacterial keratitis: a comparison of risk factors and clinical features. *Am J Ophthalmol.* 2014; **157:** 56-62.
321. Matoba AY. Fungal keratitis responsive to moxifloxacin monotherapy. *Cornea.* 2012; **31:** 1206-9.
322. Matsumoto Y, Murat D, Kojima T, *et al.* The comparison of solitary topical micafungin or fluconazole application in the treatment of *Candida* fungal keratitis. *Br J Ophthalmol.* 2011; **95:** 1406-9.
323. Matthews BJ, Partridge D, Sheard RM, *et al.* A unique case of phaeohyphomycosis subretinal abscess in a patient with arthropathy and lung pathology. *Indian J Ophthalmol.* 2013; **61:** 763-5.
324. Mayercik VA, Eller AW, Pihlblad MS. Fungal endophthalmitis developing in asthmatic individuals treated with inhaled corticosteroids. *Arch Ophthalmol.* 2011; **129:** 952-3.
325. McDonald EM, Ram FS, Patel DV, *et al.* Effectiveness of topical antifungal drugs in the management of fungal keratitis: A systematic review and meta-analysis of randomized controlled trials. *Asia Pac J Ophthalmol (Phila).* 2014; **3:** 41-7.
326. McGowan G, Lockington D, Imrie F. The firework display of fungal endogenous endophthalmitis. *Postgrad Med J.* 2010; **86:** 509.

327. McLintock CA, Lee GA, Atkinson G. Management of recurrent *Paecilomyces lilacinus* keratitis. *Clin Exp Optom.* 2013; **96:** 343-5.
328. McMillan BD, Miller GJ, Nguyen J. Rare case of exogenous *Candida dubliniensis* endophthalmitis: A case report and brief review of the literature. *J Ophthalmic Inflamm Infect.* 2014; **4:** 11.
329. Mehta H, Mehta HB, Garg P, *et al.* Voriconazole for the treatment of refractory *Aspergillus fumigatus* keratitis. *Indian J Ophthalmol.* 2008; **56:** 243-5.
330. Mehta P, Joganathan V, Oppenheim B, *et al. Aspergillus* infection of supramid orbital implant and hyperostosis of orbital bone: Report of a unique case. *Orbit.* 2010; **29:** 370-2.
331. Mela EK, Anastassiou ED, Gartaganis SP, *et al.* Fungal isolation from disinfectant solutions of contact lens storage cases among asymptomatic users. *Eye Contact Lens.* 2015; **41:** 87-90.
332. Menassa N, Bosshard PP, Kaufmann C, *et al.* Rapid detection of fungal keratitis with DNA-stabilizing FTA filter paper. *Invest Ophthalmol Vis Sci.* 2010; **51:** 1905-10.
333. Merle H, Guyomarch J, Joyaux JC, *et al.* Keratomycosis complicating pterygium excision. *Clin Ophthalmol.* 2011; **5:** 1435-7.
334. Mesa Varona D, Celis Sanchez J, Alfaya Munoz L, *et al.* Keratitis caused by *Absidia corymbifera* in an immunocompetent male with no corneal injuries. *Arch Soc Esp Oftalmol.* 2015; **90:** 139-41.
335. Michal W, Olena W, Wojciech O. Bilateral endogenous fungal endophthalmitis. *Int Ophthalmol.* 2014; **34:** 321-5.
336. Mikosz CA, Smith RM, Kim M, *et al.* Fungal endophthalmitis associated with compounded products. *Emerg Infect Dis.* 2014; **20:** 248-56.
337. Mimouni M, Tam G, Paitan Y, *et al.* Safety and efficacy of intrastromal injection of 5% natamycin in experimental fusarium keratitis. *J Ocul Pharmacol Ther.* 2014; **30:** 543-7.
338. Minckler D, Small KW, Walsh TJ. Clinical and pathologic features of *Bipolaris* endophthalmitis after intravitreal triamcinolone. *JAMA Ophthalmol.* 2014; **132:** 630-2.
339. Mitani A, Shiraishi A, Miyamoto H, *et al.* Fungal keratitis caused by *Beauveria bassiana:* Drug and temperature sensitivity profiles: A case report. *BMC Res Notes.* 2014; **7:** 677.
340. Mitani A, Shiraishi A, Uno T, *et al. In vivo* and *in vitro* investigations of fungal keratitis caused by *Colletotrichum gloeosporioides. J Ocul Pharmacol Ther.* 2009; **25:** 563-5.
341. Mitchell BM, Kanellopoulos AJ, Font RL. Post-intrastromal corneal ring segments insertion complicated by *Candida parapsilosis* keratitis. *Clin Ophthalmol.* 2013; **7:** 443-8.
342. Mithal K, Pathengay A, Bawdekar A, *et al.* Filamentous fungal endophthalmitis: Results of combination therapy with intravitreal amphotericin B and voriconazole. *Clin Ophthalmol.* 2015; **9:** 649-55.
343. Mitra A, Savant V, Aralikatti A, *et al.* The use of voriconazole in the treatment of *Cylindrocarpon* keratomycosis. *Cornea.* 2009; **28:** 217-8.
344. Mittal V, Jain R, Mittal R, *et al.* Post-laser in situ keratomileusis interface fungal keratitis. *Cornea.* 2014; **33:** 1022-30.
345. Mittal V, Mittal R, Sharma PC. Voriconazole-refractory fungal infection of phacoemulsification tunnel. *Indian J Ophthalmol.* 2010; **58:** 434-7.
346. Mittal V, Mittal R. Intracameral and topical voriconazole for fungal corneal endoexudates. *Cornea.* 2012; **31:** 366-70.
347. Mochizuki K, Murase H, Yasuda Y, *et al.* Discrepancy of *in vitro* data and clinical efficacy of micafungin against *Candida tropicalis* endophthalmitis. *J Infect Chemother.* 2012; **18:** 786-9.
348. Mochizuki K, Shiraki I, Murase H, *et al.* Identification and sensitivity of two rare fungal species isolated from two patients with *Fusarium* keratomycosis. *J Infect Chemother.* 2012; **18:** 939-44.
349. Mochizuki K, Suemori S, Udo K, *et al.* Intraocular penetration of micafungin in patient with *Candida albicans* endophthalmitis. *J Ocul Pharmacol Ther.* 2011; **27:** 531-3.
350. Mody KH, Ali MJ, Vemuganti GK, *et al.* Orbital aspergillosis in immunocompetent patients. *Br J Ophthalmol.* 2014; **98:** 1379-84.
351. Mohammadinia M, Amiri MA, Delavari F, *et al.* Antifungal efficacy of soft contact lens disinfecting solutions against *Fusarium solani* and *Candida albicans. Clin Exp Optom.* 2012; **95:** 207-11.
352. Mohd-Tahir F, Norhayati A, Siti-Raihan I, *et al.* A 5-year retrospective review of fungal keratitis at hospital Universiti Sains Malaysia. *Interdiscip Perspect Infect Dis.* 2012; 851563.
353. Moloney TP, Park J. *Pseudallescheria* endophthalmitis: Four cases over 15 years in Queensland, Australia and a review of the literature. *Retina.* 2014; **34:** 1683-701.
354. Mondal KK, Chattopadhyay C, Ray B, *et al.* Corneal ulcer with *Scopulariopsis brevicaulis* and *Staphylococcus aureus* - A rare case report. *J Indian Med Assoc.* 2012; **110:** 253-4.
355. Monden Y, Yamamoto S, Yamakawa R, *et al.* First case of fungal keratitis caused by *Pestalotiopsis clavispora. Clin Ophthalmol.* 2013; **7:** 2261-4.
356. Monno R, Alessio G, Guerriero S, *et al. Alternaria* is an infrequent cause of keratitis: A case report and review of the literature. *Eye Contact Lens.* 2015; **41:** e14-7.
357. Monno R, Alessio G, Guerriero S, *et al. Paecilomyces lilacinus* keratitis in a soft contact lens wearer. *Eye Contact Lens.* 2016 Nov 29. PMID: 27902509.
358. Morrison AS, Lockhart SR, Bromley JG, *et al.* An environmental *Sporothrix* as a cause of corneal ulcer. *Med Mycol Case Rep.* 2013; **2:** 88-90.
359. Moss HB, Fowler WC, Schell WA, *et al.* Dematiaceous fungi mimicking a conjunctival melanoma. *Ophthal Plast Reconstr Surg.* 2013; **29:** e65-6.
360. Motley WW, Melson AT, Mortensen JE. Pediatric *Metarrhizium anisopliae* keratitis. *J AAPOS.* 2011; **15:** 101-3.
361. Motukupally SR, Nanapur VR, Chathoth KN, *et al.* Ocular infections caused by *Candida* species: Type of species, *in vitro* susceptibility and treatment outcome. *Indian J Med Microbiol.* 2015; **33:** 538-46.
362. Mravicic I, Dekaris I, Gabric N, *et al. Trichophyton* spp. fungal keratitis in 22 years old female contact lenses wearer. *Coll Antropol.* 2010; **34 (Suppl. 2):** 271-4.

363. Mukherjee A, Pramanik S, Das D, *et al.* Polymicrobial chronic endophthalmitis diagnosed by culture and molecular technique. *Indian J Med Microbiol.* 2014; **32:** 331-2.
364. Mukherjee B, Chatterjee R, Biswas J. Reverse masquerade syndrome: Fungal adnexal infection mimicking carcinoma in a HIV-positive patient. *Indian J Ophthalmol.* 2013; **61:** 521-3.
365. Mukherjee PK, Chandra J, Yu C, *et al.* Characterization of *Fusarium* keratitis outbreak isolates: Contribution of biofilms to antimicrobial resistance and pathogenesis. *Invest Ophthalmol Vis Sci.* 2012; **53:** 4450-7.
366. Mutlu FM, Yildiran ST, Saracli MA, *et al.* The first case of fungal endophthalmitis caused by *Emericella nidulans* after cataract surgery. *J Mycol Med.* 2016; **26:** 271-6.
367. Muzaliha MN, Adil H, Ibrahim M, *et al. Candida glabrata* endophthalmitis following penetrating keratoplasty in a patient with negative donor rim culture. *BMC Ophthalmol.* 2010; **10:** 18. PMID: 20537193.
368. Myint T, Dykhuizen MJ, McDonald CH, *et al.* Post-operative fungal endophthalmitis due to *Geotrichum candidum. Med Mycol Case Rep.* 2015; **10:** 4-6.
369. Mythili A, Babu Singh YR, Priya R, *et al. In vitro* and comparative study on the extracellular enzyme activity of molds isolated from keratomycosis and soil. *Int J Ophthalmol.* 2014; **7:** 778-84.
370. Nachiappan K, Kadekar A, Gopal L, *et al.* Intravitreal ranibizumab in the treatment of fungal endophthalmitis scar-related choroidal neovascular membrane. *Retin Cases Brief Rep.* 2011; **5:** 175-8.
371. Nagao M, Saito T, Doi S, *et al.* Clinical characteristics and risk factors of ocular candidiasis. *Diagn Microbiol Infect Dis.* 2012; **73:** 149-52.
372. Naik M, Mohd Shahbaaz, Sheth J, *et al. Alternaria* keratitis after deep anterior lamellar keratoplasty. *Middle East Afr J Ophthalmol.* 2014; **21:** 92-4.
373. Narang S, Gupta A, Gupta V, *et al.* Fungal endophthalmitis following cataract surgery: clinical presentation, microbiological spectrum and outcome. *Am J Ophthalmol.* 2001; **132:** 609-17.
374. Narendran N, Balasubramaniam B, Johnson E, *et al.* Five-year retrospective review of guideline-based management of fungal endophthalmitis. *Acta Ophthalmol.* 2008; **86:** 525-32.
375. Narita A, Seguchi J, Shiraga F. *Paecilomyces lilacinus*-induced scleritis following bleb-associated endophthalmitis after trabeculectomy. *Acta Med Okayama.* 2015; **69:** 313-8.
376. Natarajan SV, Rekha NS, Sharda RD, *et al. Colletotrichum keratitis:* A rare but definite clinical entity. *J Clin Diagn Res.* 2013; **7:** 1430-3.
377. Nath R, Baruah S, Saikia L, *et al.* Mycotic corneal ulcers in upper Assam. *Indian J Ophthalmol.* 2011; **59:** 367-71.
378. Nath R, Gogoi RN, Saikia L. Keratomycosis due to *Scedosporium apiospermum. Indian J Med Microbiol.* 2010; **28:** 414-5.
379. Nath R, Saikia L, Baruah S, *et al.* Bilateral mycotic keratitis in a case of hyperthyroidism-induced exophthalmos. *Indian J Med Microbiol.* 2010; **28:** 182.
380. Nayak N, Satpathy G, Prasad S, *et al.* Correlation of proteinase production with amphotericin B resistance in fungi from mycotic keratitis. *Ophthalmic Res.* 2010; **44:** 113-8.
381. Nayak N, Satpathy G, Prasad S, *et al.* Molecular characterization of drug-resistant and drug-sensitive *Aspergillus* isolates causing infectious keratitis. *Indian J Ophthalmol.* 2011; **59:** 373-7.
382. Nayak N. Fungal infections of the eye - Laboratory diagnosis and treatment. *Nepal Med Coll J.* 2008; **10:** 48-63.
383. Neoh CF, Daniell M, Chen SC, *et al.* Clinical utility of caspofungin eye drops in fungal keratitis. *Int J Antimicrob Agents.* 2014; **44:** 96-104.
384. Neoh CF, Leung L, Vajpayee RB, *et al.* Treatment of *Alternaria* keratitis with intrastromal and topical caspofungin in combination with intrastromal, topical and oral voriconazole. *Ann Pharmacother.* 2011; **45:** e24.
385. Ng J, Ho S, Krishnan P, *et al. Aspergillus terreus* endogenous endophthalmitis in a non-immunocompromised patient with a history of bronchiectasis. *Ocul Immunol Inflamm.* 2013; **21:** 231-3.
386. Ng J, Minckler D, Walsh TJ, *et al.* An intractable case of *Prototheca* keratitis and chronic endophthalmitis in Stevens-Johnson syndrome with Boston Type 1 keratoprosthesis. *Cornea.* 2016; **35:** 1257-60.
387. Nielsen E, Heegaard S, Prause JU, *et al.* Fungal keratitis - improving diagnostics by confocal microscopy. *Case Rep Ophthalmol.* 2013; **4:** 303-10.
388. Nielsen SE, Nielsen E, Julian HO, *et al.* Incidence and clinical characteristics of fungal keratitis in a Danish population from 2000 to 2013. *Acta Ophthalmol.* 2015; **93:** 54-8.
389. Niki M, Eguchi H, Hayashi Y, *et al.* Ineffectiveness of intrastromal voriconazole for filamentous fungal keratitis. *Clin Ophthalmol.* 2014; **8:** 1075-9.
390. Nunes TE, Brazil NT, Fuentefria AM, *et al. Acanthamoeba* and *Fusarium* interactions: A possible problem in keratitis. *Acta Trop.* 2016; **157:** 102-107.
391. Odorcic S, Haas W, Gilmore MS, *et al.* Fungal infections after Boston Type 1 keratoprosthesis implantation: Literature review and in vitro antifungal activity of hypochlorous acid. *Cornea.* 2015; **34:** 1599-605.
392. Oechsler RA, Feilmeier MR, Miller D, *et al. Fusarium* keratitis: Genotyping, *in vitro* susceptibility and clinical outcomes. *Cornea.* 2013; **32:** 667-73.
393. Oechsler RA, Yamanaka TM, Bispo PJ, *et al. Fusarium* keratitis in Brazil: Genotyping, *in vitro* susceptibilities and clinical outcomes. *Clin Ophthalmol.* 2013; **7:** 1693-701.
394. Ogawa A, Matsumoto Y, Yaguchi T, *et al.* Successful treatment of *Beauveria bassiana* fungal keratitis with topical voriconazole. *J Infect Chemother.* 2016; **22:** 257-60.
395. Ogawa M, Sugita S, Watanabe K, *et al.* Novel diagnosis of fungal endophthalmitis by broad-range real-time PCR detection of fungal 28S ribosomal DNA. *Graefes Arch Clin Exp Ophthalmol.* 2012; **250:** 1877-83.
396. Oh JY, Lee MJ, Wee WR, *et al.* A case of necrotizing sclerokeratitis and endophthalmitis caused by *Beauveria bassiana. Jpn J Ophthalmol.* 2009; **53:** 551-3.

397. Ohtomo K, Ueta T, Nagahara M. Congenital nasolacrimal duct obstruction with fungal dacryoliths. *Can J Ophthalmol.* 2015; **50:** e7-8.
398. Oldenburg CE, Prajna VN, Prajna L, *et al.* Clinical signs in dematiaceous and hyaline fungal keratitis. *Br J Ophthalmol.* 2011; **95:** 750-1.
399. Opremcak EM. Endogenous fungal endophthalmitis. *Insight.* 2012; **37:** 20. PMID: 22685879.
400. Ortiz-Gomariz A, Higueras-Esteban A, Gutierrez-Ortega AR, *et al.* Late-onset *Candida* keratitis after Descemet stripping automated endothelial keratoplasty: Clinical and confocal microscopic report. *Eur J Ophthalmol.* 2011; **21:** 498-502.
401. Oude Lashof AM, Rothova A, Sobel JD, *et al.* Ocular manifestations of candidemia. *Clin Infect Dis.* 2011; **53:** 262-8.
402. Ozdek S, Urgancioglu B, Ozturk S. Recurrent endogenous *Candida. Ann Ophthalmol (Skokie).* 2009; **41:** 118-20.
403. Pai HV, Jamal E, Yegneswaran PP. Corneal ulcer due to a rare pleosporalean member of the genus *Bipolaris* following cow tail injury to the eye: A case report and review of literature. *Indian J Ophthalmol.* 2017; **65:** 403-5.
404. Pakzad-Vaezi K, Levasseur SD, Schendel S, *et al.* The corneal ulcer one-touch study: A simplified microbiological specimen collection method. *Am J Ophthalmol.* 2015; **159:** 37-43.e1.
405. Palanisamy M, Venkatapathy N, Rajendran V, *et al.* Keratomycosis caused by *Graphium eumorphum* (Graphium State of *Scedosporium apiospermum*). *J Clin Diagn Res.* 2015; **9:** DD03-4.
406. Panigrahi PK, Roy R, Pal SS, *et al. Aspergillus terreus* endogenous endophthalmitis: Report of a case and review of literature. *Indian J Ophthalmol.* 2014; **62:** 887-9.
407. Papakostas TD, Kohanim S, Skondra D, *et al.* Medical management of *Alternaria* keratitis with endophthalmitis in a patient with a corneal graft. *Clin Experiment Ophthalmol.* 2014; **42:** 496-7.
408. Parchand S, Gupta A, Ram J, *et al.* Voriconazole for fungal corneal ulcers. *Ophthalmology.* 2012; **119:** 1083.
409. Pariseau B, Nehls S, Ogawa GS, *et al. Beauveria* keratitis and biopesticides: Case histories and a random amplification of polymorphic DNA comparison. *Cornea.* 2010; **29:** 152-8.
410. Parmar TJ, Gajjar DU, Pal AK, *et al.* Fungal keratitis associated with mite embedded in cornea. *Indian J Pathol Microbiol.* 2011; **54:** 214-5.
411. Pathengay A, Miller DM, Flynn HW Jr, *et al. Curvularia* endophthalmitis following open globe injuries. *Arch Ophthalmol.* 2012; **130:** 652-4.
412. Patil A, Majumdar S. Echinocandins in ocular therapeutics. *J Ocul Pharmacol Ther.* 2017. doi: 10.1089/jop.2016.0186. : 28437176.
413. Payne JF, Keenum DG, Sternberg P Jr, *et al.* Concentrated intravitreal amphotericin B in fungal endophthalmitis. *Arch Ophthalmol.* 2010; **128:** 1546-50.
414. Pelegrin L, Mesquida M, Adan A, *et al. Candida dubliniensis* endophthalmitis in a HIV-infected intravenous drug abuser. *Mycoses.* 2011; **54:** e856-8.
415. Pellegrino F, Carrasco MA. Argon laser phototherapy in the treatment of refractory fungal keratitis. *Cornea.* 2013; **32:** 95-7.
416. Peponis V, Rosenberg P, Chalkiadakis SE, *et al.* Fungal scleral keratitis and endophthalmitis following pterygium excision. *Eur J Ophthalmol.* 2009; **19:** 478-80.
417. Perez-Balbuena AL, Vanzzini-Rosano V, Valadez-Virgen Jde J, *et al. Fusarium* keratitis in Mexico. *Cornea.* 2009; **28:** 626-30.
418. Phan CM, Subbaraman L, Jones L. Contact lenses for antifungal ocular drug delivery: A review. *Expert Opin Drug Deliv.* 2014; **11:** 537-46.
419. Pinna A, Usai D, Zanetti S, *et al. In vitro* efficacy of contact lens solutions against various corneal fungal isolates. *J Ocul Pharmacol Ther.* 2013; **29:** 694-9.
420. Posteraro B, Mattei R, Trivella F, *et al.* Uncommon *Neosartorya udagawae* fungus as a causative agent of severe corneal infection. *J Clin Microbiol.* 2011; **49:** 2357-60.
421. Pote ST, Chakraborty A, Lahiri KK, *et al.* Keratitis by a rare pathogen *Colletotrichum gloeosporioides:* A case report. *J Mycol Med.* 2017. pii: S1156-5233(16)30208-6. doi: 10.1016/j.my med.2017.04.009. PMID: 28501467.
422. Pradhan L, Sharma S, Nalamada S, *et al.* Natamycin in the treatment of keratomycosis: Correlation of treatment outcome and in vitro susceptibility of fungal isolates. *Indian J Ophthalmol.* 2011; **59:** 512-4.
423. Prajna NV, Krishnan T, Mascarenhas J, *et al.* Predictors of outcome in fungal keratitis. *Eye (Lond).* 2012; **26:** 1226-31.
424. Prajna NV, Krishnan T, Mascarenhas J, *et al.* The mycotic ulcer treatment trial: A randomized trial comparing natamycin *vs* voriconazole. *JAMA Ophthalmol.* 2013; **131:** 422-9.
425. Prajna NV, Mascarenhas J, Krishnan T, *et al.* Comparison of natamycin and voriconazole for the treatment of fungal keratitis. *Arch Ophthalmol.* 2010; **128:** 672-8.
426. Prajna NV, O'Brien KS, Acharya NR, *et al.* Voriconazole for fungal keratitis. *Ophthalmology.* 2013; **120:** e62-3.
427. Prajna NV, Prajna L, O'Brien KS, *et al.* Association of pre-treatment with antifungal medication and fungal resistance in the mycotic ulcer treatment trial I. *JAMA Ophthalmol.* 2015; **133:** 1210-1.
428. Prajna NV, Srinivasan M, Lalitha P, *et al.* Differences in clinical outcomes in keratitis due to fungus and bacteria. *JAMA Ophthalmol.* 2013; **131:** 1088-9.
429. Prajna VN, Lalitha PS, Mascarenhas J, *et al.* Natamycin and voriconazole in *Fusarium* and *Aspergillus* keratitis: Subgroup analysis of a randomised controlled trial. *Br J Ophthalmol.* 2012; **96:** 1440-1.
430. Prakash G, Sharma N, Goel M, *et al.* Evaluation of intrastromal injection of voriconazole as a therapeutic adjunctive for the management of deep recalcitrant fungal keratitis. *Am J Ophthalmol.* 2008; **146:** 56-59.
431. Premamalini T, Ambujavalli BT, Vijayakumar R, *et al.* Fungal keratitis caused by *Macrophomina phaseolina* - A case report. *Med Mycol Case Rep.* 2012; **1:** 123-6.
432. Proença-Pina J, Ssi Yan Kai I, Bourcier T, *et al. Fusarium* keratitis and endophthalmitis associated with lens contact wear. *Int Ophthalmol.* 2010; **30:** 103-7.

433. Punia RS, Kundu R, Chander J, Arya SK, Handa U, *et al.* Spectrum of fungal keratitis: Clinicopathologic study of 44 cases. *Int J Ophthalmol.* 2014; **7:** 114-7.
434. Pushker N, Meel R, Kashyap S, *et al.* Invasive aspergillosis of orbit in immunocompetent patients: Treatment and outcome. *Ophthalmology.* 2011; **118:** 1886-91.
435. Qiu S, Zhao GQ, Lin J, *et al.* Natamycin in the treatment of fungal keratitis: A systematic review and meta-analysis. *Int J Ophthalmol.* 2015; **8:** 597-602.
436. Qiu WY, Yao YF. Mycotic keratitis caused by concurrent infections of *Exserohilum mcginnisii* and *Candida parapsilosis. BMC Ophthalmol.* 2013; **13:** 37.
437. Quintero-Estades JA, Walter S, Valenzuela F, *et al.* Delayed-onset post-operative endophthalmitis secondary to *Exophiala. BMJ Case Rep.* 2015; pii: bcr2014208680.
438. Rachitskaya AV, Reddy AK, Miller D, *et al.* Prolonged *Curvularia* endophthalmitis due to organism sequestration. *JAMA Ophthalmol.* 2014; **132:** 1123-6.
439. Rajaraman R, Bhat P, Vaidee V, *et al.* Topical 5% natamycin with oral ketoconazole in filamentous fungal keratitis: A randomized controlled trial. *Asia Pac J Ophthalmol (Phila).* 2015; **4:** 146-50.
440. Rajaraman R. Topical 5% nantamycin with oral ketoconazole in filamentous fungal keratitis: A randomized controlled trial. *Asia Pac J Ophthalmol (Phila).* 2015; **4:** 399.
441. Rajmane VS, Rajmane ST, Ghatole MP. *Rhodotorula* species infection in traumatic keratitis - A case report. *Diagn Microbiol Infect Dis.* 2011; **71:** 428-9.
442. Ramakrishnan R, Bharathi MJ, Shivkumar C, *et al.* Microbiological profile of culture-proven cases of exogenous and endogenous endophthalmitis: A 10-year retrospective study. *Eye.* 2009; **23:** 945-56.
443. Ramakrishnan T, Constantinou M, Jhanji V, *et al.* Factors affecting treatment outcomes with voriconazole in cases with fungal keratitis. *Cornea.* 2013; **32:** 445-9.
444. Ramani R, Chaturvedi V. Evaluations of shorter exposures of contact lens cleaning solutions against *Fusarium oxysporum* species complex and *Fusarium solani* species complex to simulate inappropriate usage. *Antimicrob Agents Chemother.* 2011; **55:** 2265-75.
445. Ramli SR, Francis AL, Yusof Y, *et al.* A severe case of *Arthrographis kalrae* keratomycosis. *Case Rep Infect Dis.* 2013; 851875. PMID: 24377053.
446. Rana M, Fahad B, Abid Q. Embolic *Aspergillus* endophthalmitis in an immunocompetent patient from aortic root *Aspergillus* endocarditis. *Mycoses.* 2008; **51:** 352-3.
447. Rath S, Kar S, Sahu SK, *et al.* Fungal periorbital necrotizing fasciitis in an immunocompetent adult. *Ophthal Plast Reconstr Surg.* 2009; **25:** 334-5.
448. Rathi H, Venugopal A, Rameshkumar G, *et al.* Fungal keratitis caused by *Exserohilum*, An emerging pathogen. *Cornea.* 2016; **35:** 644-6.
449. Rathi HS, Venugopal A, Rengappa R, *et al. Scedosporium* keratitis: An experience from a tertiary eye hospital in South India. *Cornea.* 2016; **35:** 1575-7.
450. Rautaraya B, Sharma S, Kar S, *et al.* Diagnosis and treatment outcome of mycotic keratitis at a tertiary eye care center in eastern India. *BMC Ophthalmol.* 2011; **11:** 39.
451. Ravuri S, Mittal R, Bagga B. Topical 5% natamycin with oral ketoconazole in filamentous fungal keratitis: A randomized controlled trial. *Asia Pac J Ophthalmol (Phila).* 2015; **4:** 399.
452. Rebong RA, Santaella RM, Goldhagen BE, *et al.* Polyhexamethylene biguanide and calcineurin inhibitors as novel antifungal treatments for *Aspergillus keratitis. Invest Ophthalmol Vis Sci.* 2011; **52:** 7309-15.
453. Reddy JC, Murthy SI, Reddy AK, *et al.* Risk factors and clinical outcomes of bacterial and fungal scleritis at a tertiary eye care hospital. *Middle East Afr J Ophthalmol.* 2015; **22:** 203-11.
454. Reddy JC, Rapuano CJ, Eagle RC Jr, *et al.* Late-onset necrotizing scleritis due to pigmented mycetoma (dematiaceous fungi) in 2 cases. *Cornea.* 2013; **32:** 450-3.
455. Reed DC, Shah KH, Hubschman JP. Resolution of *Coccidioides immitis* endophthalmitis with an aggressive surgical and medical therapeutic approach. *Semin Ophthalmol.* 2013; **28:** 251-2.
456. Refojo N, Minervini P, Hevia AI, *et al.* Keratitis caused by moulds in Santa Lucia Ophthalmology Hospital in Buenos Aires, Argentina. *Rev Iberoam Micol.* 2016; **33:** 1-6.
457. Rensch F, Schlichtenbrede FC, Jonas JB. Postoperative mycotic endophthalmitis. *J Cataract Refract Surg.* 2010; **36:** 1233-4.
458. Retuerto MA, Szczotka-Flynn L, Ho D, *et al.* Efficacy of care solutions against contact lens-associated *Fusarium* biofilms. *Optom Vis Sci.* 2012; **89:** 382-91.
459. Riddell J 4th, Comer GM, Kauffman CA. Treatment of endogenous fungal endophthalmitis: Focus on new antifungal agents. *Clin Infect Dis.* 2011; **52:** 648-53.
460. Rodrigues IA, Jackson TL. A high-definition view of endogenous fungal endophthalmitis. *Lancet Infect Dis.* 2014; **14:** 358.
461. Rogers GM, Goins KM, Sutphin JE, *et al.* Outcomes of treatment of fungal keratitis at the University of Iowa Hospitals and Clinics: A 10-year retrospective analysis. *Cornea.* 2013; **32:** 1131-6.
462. Rosenberger E, Youssef DA, Safdar S, *et al.* Third case of *Candida dubliniensis* endogenous endophthalmitis in North America: Case report and review of the literature. *Int Ophthalmol.* 2014; **34:** 945-50.
463. Roy R, Panigrahi PK, Pal SS, *et al.* Post-traumatic endophthalmitis secondary to keratomycosis caused by *Scedosporium apiospermum. Ocul Immunol Inflamm.* 2016; **24:** 107-9.
464. Ruban VV, Kaliamurthy J, Dineshkumar M, *et al.* Keratitis due to the wood saprobic ascomycete, *Auerswaldia lignicola* (Family Botryosphaeriaceae), in a carpenter in India. *Mycopathologia.* 2013; **176:** 463-6.
465. Sachdev N, Gupta P, Singh R, *et al.* Bilateral simultaneous endogenous *Aspergillus* endophthalmitis in an immunocompetent patient. *Retin Cases Brief Rep.* 2010; **4:** 14-7.
466. Sadiq MA, Hassan M, Agarwal A, *et al.* Endogenous endophthalmitis: diagnosis, management and prognosis. *J Ophthalmic Inflamm Infect.* 2015; **5:** 32.

467. Saeedi OJ, Iyer SA, Mohiuddin AZ, *et al. Exophiala jeanselmei* keratitis: Case report and review of literature. *Eye Contact Lens.* 2013; **39:** 410-2.
468. Saffra NA, Desai RU, Seidman CJ, *et al.* Endogenous fungal endophthalmitis after cardiac surgery. *Ophthalmic Surg Lasers Imaging.* 2010; **41:** e1-3.
469. Saha S, Banerjee D, Khetan A, *et al.* Epidemiological profile of fungal keratitis in urban population of West Bengal, India. *Oman J Ophthalmol.* 2009; **2:** 114-8.
470. Saha S, Sengupta J, Banerjee D, *et al. Lasiodiplodia theobromae* keratitis: A case report and review of literature. *Mycopathologia.* 2012; **174:** 335-9.
471. Saha S, Sengupta J, Banerjee D, *et al. Schizophyllum commune*: A new organism in eye infection. *Mycopathologia.* 2013; **175:** 357-60.
472. Saha S, Sengupta J, Chatterjee D, *et al. Rhodotorula mucilaginosa* keratitis: A rare fungus from Eastern India. *Indian J Ophthalmol.* 2014; **62:** 341-4.
473. Sahu SK, Dora J, Hota G. Keratomycosis caused by *Blastoschizomyces capitatus. Indian J Pathol Microbiol.* 2016; **59:** 117-8.
474. Said DG, Otri M, Miri A, *et al.* The challenge of fungal keratitis. *Br J Ophthalmol.* 2011; **95:** 1623-4.
475. Sallam A, Taylor SR, Khan A, *et al.* Factors determining visual outcome in endogenous *Candida* endophthalmitis. *Retina.* 2012; **32:** 1129-34.
476. Sangwan J, Lathwal S, Juyal D, *et al. Fonsecaea pedrosoi:* A rare etiology in fungal keratitis. *J Clin Diagn Res.* 2013; **7:** 2272-3.
477. Scartozzi R, Tom D, Liggett PE, *et al.* Post-operative endophthalmitis caused by *Candida parapsilosis:* clinical features and treatment outcomes. *Ophthalmic Surg Lasers Imaging.* 2011; **42:** e37-9.
478. Schein OD. Evidence-based treatment of fungal keratitis. *JAMA Ophthalmol.* 2016; **134:** 1372-3.
479. Schlaen A, Ingolotti M, Couto C, *et al.* Endogenous *Histoplasma capsulatum* endophthalmitis in an immunocompetent patient. *Eur J Ophthalmol.* 2015; **25:** e53-5.
480. Schlaen A, Ingolotti M, Lorenzon P, *et al. Fusarium solani* subretinal abscess in a patient with acute myeloid leukemia. *Retin Cases Brief Rep.* 2016 Nov 8. PMID: 27828901.
481. Sedaghat MR, Hosseinpoor SS. *Candida albicans* interface infection after deep anterior lamellar keratoplasty. *Indian J Ophthalmol.* 2012; **60:** 328-30.
482. Selvakumar P, Sofia O, Gopal L, *et al.* Recurrent fungal iris granuloma in a 10-year-old child. *Ocul Immunol Inflamm.* 2012; **20:** 221-3.
483. Selvin SS, Korah SM, Michael JS, *et al.* Series of five cases of *Papulaspora equi* keratomycosis. *Cornea.* 2014; **33:** 640-3.
484. Sengupta J, Khetan A, Saha S, *et al. Candida* keratitis: Emerging problem in India. *Cornea.* 2012; **31:** 371-5.
485. Sengupta J, Saha S, Khetan A, *et al. Candida fermentati:* A rare yeast involved in fungal keratitis. *Eye Contact Lens.* 2013; 39: e15-8.
486. Sengupta J, Saha S, Khetan A, *et al.* Effects of lactoferricin B against keratitis-associated fungal biofilms. *J Infect Chemother.* 2012; **18:** 698-703.
487. Sengupta S, Rajan S, Reddy PR, *et al.* Comparative study on the incidence and outcomes of pigmented versus non-pigmented keratomycosis. *Indian J Ophthalmol.* 2011; **59:** 291-6.
488. Shah AR, Ruggiero J. Factors determining visual outcome in endogenous *Candida* endophthalmitis. *Retina.* 2013; **33:** 453-4.
489. Shapiro BL, Lalitha P, Loh AR, *et al.* Susceptibility testing and clinical outcome in fungal keratitis. *Br J Ophthalmol.* 2010; **94:** 384-5.
490. Sharma A, Mohan K, Chander J, Sharma S. Comment on: Is inclusion of Sabouraud dextrose agar essential for the laboratory diagnosis of fungal keratitis? *Indian J Ophthalmol.* 2011; **59:** 260-1.
491. Sharma S. Diagnosis of fungal keratitis: Current options. *Expert Opin Med Diagn.* 2012; **6:** 449-55.
492. Sharma V, Angrup A, Panwar P, *et al.* Keratitis by *Paecilomyces lilacinus:* A case report from Sub-Himalayan region. *Indian J Med Microbiol.* 2015; **33:** 585-7.
493. Sharma Y, Jain S, Jayachandran. Keratomycosis: Etiology, risk factors and differential diagnosis - A mini-review on *Trichophyton* spp. *J Clin Diagn Res.* 2014; **8:** DD01-2.
494. Shekhar H, Gogia V, Vashisht N, *et al.* Endogenous endophthalmitis by *Chrysosporium:* An opportunistic pathogen. *Ocul Immunol Inflamm.* 2014; **22:** 158-60.
495. Sheybani A, Nguyen AT, Harocopos G, *et al.* A masquerader? *Paecilomyces* must be distinguished from *Penicillium* in fungal keratitis: A report of two contrasting cases. *BMJ Case Rep.* 2015; pii: bcr2014207605. PMID: 26286907.
496. Sheyman AT, Cohen BZ, Friedman AH, *et al.* An outbreak of fungal endophthalmitis after intravitreal injection of compounded combined bevacizumab and triamcinolone. *JAMA Ophthalmol.* 2013; **131:** 864-9.
497. Shi W, Wang T, Xie L, *et al.* Risk factors, clinical features, and outcomes of recurrent fungal keratitis after corneal transplantation. *Ophthalmology.* 2010; **117:** 890-6.
498. Shigeyasu C, Yamada M, Nakamura N, *et al.* Keratomycosis caused by *Aspergillus viridinutans:* An *Aspergillus fumigatus*-resembling mold presenting distinct clinical and antifungal susceptibility patterns. *Med Mycol.* 2012; **50:** 525-8.
499. Shiraishi A, Araki-Sasaki K, Mitani A, *et al.* Clinical characteristics of keratitis due to *Colletotrichum gloeosporioides. J Ocul Pharmacol Ther.* 2011; **27:** 487-91.
500. Shivaprakash MR, Appannanavar SB, Dhaliwal M, *et al. Colletotrichum truncatum:* An unusual pathogen causing mycotic keratitis and endophthalmitis. *J Clin Microbiol.* 2011; **49:** 2894-8.
501. Shulman J, de la Cruz EL, Latkany P, *et al.* Cryptococcal chorioretinitis with immune reconstitution inflammatory syndrome. *Ocul Immunol Inflamm.* 2009; **17:** 314-5.
502. Silva RA, Sridhar J, Miller D, *et al.* Exogenous fungal endophthalmitis: An analysis of isolates and susceptibilities to antifungal agents over a 20-year period (1990-2010). *Am J Ophthalmol.* 2015; **159:** 257-64.e1.
503. Silva-Vergara ML, de Camargo ZP, Silva PF, *et al.* Disseminated *Sporothrix brasiliensis* infection with endocardial and ocular involvement in an HIV-infected patient. *Am J Trop Med Hyg.* 2012; **86:** 477-80.

504. Slocumb RW, Elner SG, Hall EF. Chronic postoperative fungal endophthalmitis caused by *Trichosporon asahii*. *Retin Cases Brief Rep.* 2010; **4:** 366-7.
505. Slowik M, Biernat MM, Urbaniak-Kujda D, *et al.* Mycotic Infections of the Eye. *Adv Clin Exp Med.* 2015; **24:** 1113-7.
506. Small KW, Chan CK, Silva-Garcia R, *et al.* Onset of an outbreak of *Bipolaris hawaiiensis* fungal endophthalmitis after intravitreal injections of triamcinolone. *Ophthalmology.* 2014; **121:** 952-8.
507. Smith TC, Benefield RJ, Kim JH. Risk of fungal endophthalmitis associated with cataract surgery: A mini-review. *Mycopathologia.* 2015; **180:** 291-7.
508. Smith WM, Fahle G, Nussenblatt RB, *et al.* A rare case of endogenous *Aspergillus conicus* endophthalmitis in an immunocompromised patient. *J Ophthalmic Inflamm Infect.* 2013; **3:** 37.
509. Spierer O, Dugar J, Miller D, *et al.* Comparative antifungal susceptibility analysis of *Candida albicans* versus non-*albicans Candida* corneal isolates. *Cornea.* 2015; **34:** 576-9.
510. Squissato V, Yucel YH, Richardson SE, *et al. Colletotrichum truncatum* species complex: Treatment considerations and review of the literature for an unusual pathogen causing fungal keratitis and endophthalmitis. *Med Mycol Case Rep.* 2015; **9:** 1-6.
511. Sridhar J, Flynn HW Jr, Kuriyan AE, *et al.* Endogenous fungal endophthalmitis: Risk factors, clinical features and treatment outcomes in mold and yeast infections. *J Ophthalmic Inflamm Infect.* 2013; **3:** 60.
512. Sridhar J, Lam BL, Pasol J, *et al.* Neuro-ophthalmic manifestations of fungal disease associated with post-hurricane environment. *J Neuroophthalmol.* 2012; **32:** 197-201.
513. Sugita S, Kamoi K, Ogawa M, *et al.* Detection of *Candida* and *Aspergillus* species DNA using broad-range real-time PCR for fungal endophthalmitis. *Graefes Arch Clin Exp Ophthalmol.* 2012; **250:** 391-8.
514. Sun CQ, Lalitha P, Prajna NV, *et al.* Association between *in vitro* susceptibility to natamycin and voriconazole and clinical outcomes in fungal keratitis. *Ophthalmology.* 2014; **121:** 1495-500.e1.
515. Sun CQ, Prajna NV, Krishnan T, *et al.* Effect of pre-treatment with antifungal agents on clinical outcomes in fungal keratitis. *Clin Exp Ophthalmol.* 2016; **44:** 763-7.
516. Sun S, Yuan G, Zhao G, *et al.* Endophthalmitis caused by *Phialophora verrucosa* and *Streptococcus intermedius:* A case report. *Med Mycol.* 2010; **48:** 1108-11.
517. Takezawa Y, Shiraishi A, Noda E, *et al.* Effectiveness of *in vivo* confocal microscopy in detecting filamentous fungi during clinical course of fungal keratitis. *Cornea.* 2010; **29:** 1346-52.
518. Talwar P, Sehgal SC. Mycotic infections of the eye in Chandigarh and neighbouring areas. *Indian J Med Res.* 1978; **67:** 929-33.
519. Tanaka H, Ishida K, Yamada W, *et al.* Study of ocular candidiasis during nine-year period. *J Infect Chemother.* 2016; **22:** 149-56.
520. Taylan Sekeroglu H, Erdem E, Yagmur M, *et al.* Successful medical management of recalcitrant *Fusarium solani* keratitis: Molecular identification and susceptibility patterns. *Mycopathologia.* 2012; **174:** 233-7.
521. Thamke DC, Mendiratta DK, Dhabarde A, *et al.* Mycotic keratitis due to *Engyodontium album:* First case report from India. *Indian J Med Microbiol.* 2015; **33:** 303-4.
522. Theoulakis P, Goldblum D, Zimmerli S, *et al.* Keratitis resulting from *Thielavia subthermophila* Mouchacca. *Cornea.* 2009; **28:** 1067-9.
523. Thomas BC, Zimmermann S, Volcker HE, *et al.* Severe *Arthrographis kalrae* keratomycosis in an immunocompetent patient. *Cornea.* 2011; **30:** 364-6.
524. Thomas PA, Jesudasan CA, Geraldine P, *et al.* Adventitious sporulation in *Fusarium* keratitis. *Graefes Arch Clin Exp Ophthalmol.* 2011; **249:** 1429-31.
525. Thomas PA, Kaliamurthy J, Jesudasan CA, *et al.* Use of chlorazol black E mounts of corneal scrapes for diagnosis of filamentous fungal keratitis. *Am J Ophthalmol.* 2008; **145:** 971-6.
526. Thomas PA, Kaliamurthy J. Mycotic keratitis: Epidemiology, diagnosis and management. *Clin Microbiol Infect.* 2013; **19:** 210-20.
527. Thomas PA, Teresa PA, Theodore J, *et al.* PCR for the molecular diagnosis of mycotic keratitis. *Expert Rev Mol Diagn.* 2012; **12:** 703-18.
528. Thomas PA. Current perspectives on ophthalmic mycoses. *Clin Microbiol Rev.* 2003; **16:** 730-797.
529. Tilak R, Singh A, Maurya OP, *et al.* Mycotic keratitis in India: A five-year retrospective study. *J Infect Dev Ctries.* 2010; **4:** 171-4.
530. Tirpack AR, Duker JS, Baumal CR. An outbreak of endogenous fungal endophthalmitis among intravenous drug abusers in New England. *JAMA Ophthalmol.* 2017; **135:** 534-40.
531. Todokoro D, Yamada N, Fukuchi M, Kishi S. Topical voriconazole therapy of *Purpureocillium lilacinum* keratitis that occurred in disposable soft contact lens wearers. *Int Ophthalmol.* 2014; **34:** 1159-63.
532. Toker E, Ziyade N, Atici S, *et al.* Postoperative keratitis due to *Paecilomyces*: A rare pediatric case. *Pan Afr Med J.* 2016; **24:** 317. PMID: 28154672.
533. Toutain-Kidd CM, Porco TC, Kidd EM, *et al.* Evaluation of fungal keratitis using a newly developed computer program, Optscore, for grading digital corneal photographs. *Ophthalmic Epidemiol.* 2014; **21:** 24-32.
534. Trachsler S, Eberhard R, Kocher C, *et al. Paecilomyces lilacinus* endophthalmitis following cataract surgery: A therapeutic challenge. *Klin Monbl Augenheilkd.* 2012; **229:** 441-2.
535. Troke P, Obenga G, Gaujoux T, *et al.* The efficacy of voriconazole in 24 ocular *Fusarium* infections. *Infection.* 2013; **41:** 15-20.
536. Tsai SH, Lin YC, Hsu HC, *et al.* Subconjunctival injection of fluconazole in the treatment of fungal *Alternaria* keratitis. *Ocul Immunol Inflamm.* 2016; **24:** 103-6.
537. Tu EY. *Alternaria* keratitis: Clinical presentation and resolution with topical fluconazole or intrastromal voriconazole and topical caspofungin. *Cornea.* 2009; **28:** 116-9.

538. Tuft SJ, Tullo AB. Fungal keratitis in the United Kingdom 2003-2005. *Eye (Lond).* 2009; **23:** 1308-13.
539. Tuli SS. Fungal keratitis. *Clin Ophthalmol.* 2011; **5:** 275-9.
540. Tupaki-Sreepurna A, Al-Hatmi AM, Kindo AJ, *et al.* Multidrug-resistant *Fusarium* in keratitis: A clinico-mycological study of keratitis infections in Chennai, India. *Mycoses.* 2017; **60:** 230-3.
541. Ugarte M, Nicol DA, Jones NP, *et al.* Chronic intraocular copper foreign body and *Candida*: A unique combination. *Ocul Immunol Inflamm.* 2009; **17:** 356-60.
542. Ursea R, Tavares LA, Feng MT, *et al.* Non-traumatic *Alternaria* keratomycosis in a rigid gas-permeable contact lens patient. *Br J Ophthalmol.* 2010; **94:** 389-90.
543. Usui T, Misawa Y, Honda N, *et al.* Non-traumatic keratomycosis caused by *Alternaria* in a glaucoma patient. *Int Ophthalmol.* 2009; **29:** 529-31.
544. Vaddavalli PK, Garg P, Sharma S, *et al.* Role of confocal microscopy in the diagnosis of fungal and *Acanthamoeba* keratitis. *Ophthalmology.* 2011; **118:** 29-35.
545. Vanzzini Zago V, Alcantara Castro M, Naranjo Tackman R. Support of the laboratory in the diagnosis of fungal ocular infections. *Int J Inflam.* 2012; 643104. PMID: 22518339.
546. Vanzzini-Zago V, Corredor-Casas S, Rodriguez-Reyes A, *et al.* Endophthalmitis of probable endogenous origin caused by *Scedosporium boydii:* A case report. *Rev Iberoam Micol.* 2016; **33:** 122-5.
547. Vasantha Ruban V, Geraldine P, Kaliamurthy J, *et al.* Keratitis due to *Fusarium langsethiae:* Clinical profile, molecular identification and susceptibility to antifungals. *Mycopathologia.* 2015;1 **79:** 453-8.
548. Vasconcelos-Santos DV, Lim JI, Rao NA. Chronic coccidioidomycosis endophthalmitis without concomitant systemic involvement: A clinicopathological case report. *Ophthalmology.* 2010; **117:** 1839-42.
549. Vela JI, Diaz-Cascajosa J, Sanchez F, *et al.* Management of endogenous cryptococcal endophthalmitis with voriconazole. *Can J Ophthalmol.* 2009; **44:** e61-2.
550. Vengayil S, Panda A, Satpathy G, Nayak N, *et al.* Polymerase chain reaction-guided diagnosis of mycotic keratitis: A prospective evaluation of its efficacy and limitations. *Invest Ophthalmol Vis Sci.* 2009; **50:** 152-6.
551. Verghese S. Post traumatic fungal keratitis caused by *Acremonium recifei. Indian J Pathol Microbiol.* 2010; **53:** 587-8.
552. Vergoulidou M, Krause L, Foerster MH, *et al.* Endogenous filamentous fungal endophthalmitis - Single-centre survey in patients with acute leukaemia or postallogeneic stem cell transplantation and review of the literature. *Mycoses.* 2011; **54:** e704-11.
553. Verkley GJ, Gene J, Guarro J, *et al. Pyrenochaeta keratinophila* sp. nov., isolated from an ocular infection in Spain. *Rev Iberoam Micol.* 2010; **27:** 22-4.
554. Verma S, Sharma V, Kanga A, *et al.* Current spectrum of oculomycosis in north India: A 5-year retrospective evaluation of clinical and microbiological profile. *Indian J Med Microbiol.* 2016; **34:** 72-5.
555. Vilela RC, Vilela L, Vilela P, *et al.* Etiological agents of fungal endophthalmitis: diagnosis and management. *Int Ophthalmol.* 2014; **34:** 707-21.
556. Villani E, Baudouin C, Efron N, *et al. In vivo* confocal microscopy of the ocular surface: From bench to bedside. *Curr Eye Res.* 2014; **39:** 213-31.
557. Villarrubia A, Cano-Ortiz A. *Candida* keratitis after Descemet stripping with automated endothelial keratoplasty. *Eur J Ophthalmol.* 2014; **24:** 964-7.
558. Vinekar A, Dogra MR, Avadhani K, *et al.* Management of recurrent post-operative fungal endophthalmitis. *Indian J Ophthalmol.* 2014; **62:** 136-40.
559. Vinod Mootha V, Shahinpoor P, Sutton DA, *et al.* Identification problems with sterile fungi, illustrated by a keratitis due to a non-sporulating *Chaetomium*-like species. *Med Mycol.* 2012; **50:** 361-7.
560. Vyas N, Al-Hashimi S, Munir W. *Microcyclosporella mali:* A novel fungal keratitis in a post-penetrating keratoplasty patient. *BMJ Case Rep.* 2015; pii: bcr2014207416. PMID: 25725025.
561. Vyawahare CR, Misra RN, Gandham NR, *et al. Penicillium* keratitis in an immunocompetent patient from Pune, Maharashtra, India. *J Clin Diagn Res.* 2014; **8:** DD01-2.
562. Walia H, Tucci VT, Greene JN, *et al.* A case of endogenous *Trichosporon* endophthalmitis treated with micafungin and voriconazole. *J Glob Infect Dis.* 2009; **1:** 71-4.
563. Wang A, Cao J, Hu S, *et al.* Anterior and pan-endophthalmitis caused by *Candida albicans* and *Aspergillus fumigatus:* Report of two cases. *Mycopathologia.* 2013; **176:** 451-6.
564. Wang L, Sun S, Jing Y, *et al.* Spectrum of fungal keratitis in central China. *Clin Experiment Ophthalmol.* 2009; **37:** 763-71.
565. Wang L, Wang L, Han L, *et al.* Study of pathogens of fungal keratitis and the sensitivity of pathogenic fungi to therapeutic agents with the disk diffusion method. *Curr Eye Res.* 2015; **40:** 1095-101.
566. Wang L, Al-Hatmi AM, Lai X, *et al. Bipolaris oryzae*, a novel fungal opportunist causing keratitis. *Diagn Microbiol Infect Dis.* 2016; **85:** 61-5.
567. Warner DB, Wright III HE, Rosenbaum ER. *Rhizopus* keratitis associated with poor contact lens hygiene. *Case Rep Ophthalmol.* 2016; **7:** 21-4.
568. Weng CY, Parke DW 3rd, Walter SD, *et al. Candida glabrata* endophthalmitis transmitted from graft to host after Descemet stripping automated endothelial keratoplasty. *JAMA Ophthalmol.* 2014; **132:** 1381-3.
569. Wessel JM, Bachmann BO, Meiller R, *et al.* Fungal interface keratitis by *Candida orthopsilosis* following deep anterior lamellar keratoplasty. *BMJ Case Rep.* 2013; pii: bcr2012008361. PMID: 23349184.
570. Willermain F, Bradstreet C, Kampouridis S, *et al.* Different presentations of ophthalmic aspergillosis. *Eur J Ophthalmol.* 2008; **18:** 827-30.
571. Wiwanitkit S, Wiwanitkit V. The first case of *Candida dubliniensis* endophthalmitis in North America. *Int Ophthalmol.* 2013; **33:** 3.

572. Wiwanitkit V. Sabouraud dextrose agar for the diagnosis of fungal keratitis. Indian J Ophthalmol. 2010; **58:** 549-50.
573. Wong AC, Mak ST, Luk S, *et al.* Successful treatment of *Pseudallescheria boydii* keratitis with topical natamycin as monotherapy. *J Ocul Pharmacol Ther.* 2010; **26:** 519-21.
574. Wood M, Wright D, Witherspoon P. Fungal endophthalmitis: An unusual complication of GI surgery and endoluminal vacuum therapy. *BMJ Case Rep.* 2015; pii: bcr2015209902.
575. Wu L, Tapia-Herrera G. Fungal endophthalmitis. *Curr Fungal Infect Rep.* 2009; **3:** 55-61.
576. Wu PC, Lai CH, Tan HY, *et al.* The successful medical treatment of a case of *Paecilomyces lilacinus* keratitis. *Cornea.* 2010; **29:** 357-8.
577. Xu LJ, Song XS, Zhao J, *et al.* Hypopyon in patients with fungal keratitis. *Chin Med J (Engl).* 2012; **125:** 470-5.
578. Xu Y, Gao C, Li X, *et al. In vitro* antifungal activity of silver nanoparticles against ocular pathogenic filamentous fungi. *J Ocul Pharmacol Ther.* 2013; **29:** 270-4.
579. Xu Y, He Y, Li X, *et al.* Antifungal effect of ophthalmic preservatives phenylmercuric nitrate and benzalkonium chloride on ocular pathogenic filamentous fungi. *Diagn Microbiol Infect Dis.* 2013; **75:** 64-7.
580. Yamada H, Takahashi N, Hori N, *et al.* Rare case of fungal keratitis caused by *Corynespora cassiicola. J Infect Chemother.* 2013; **19:** 1167-9.
581. Yamazoe K, Den S, Yamaguchi T, *et al.* Severe donor-related *Candida* keratitis after Descemet's stripping automated endothelial keratoplasty. *Graefes Arch Clin Exp Ophthalmol.* 2011; **249:** 1579-82.
582. Yegneswaran PP, Pai V, Bairy I, *et al. Colletotrichum graminicola* keratitis: First case report from India. *Indian J Ophthalmol.* 2010; **58:** 415-7.
583. Yildiz EH, Abdalla YF, Elsahn AF, *et al.* Update on fungal keratitis from 1999 to 2008. *Cornea.* 2010; **29:** 1406-11.
584. Yildiz EH, Ailani H, Hammersmith KM, *et al. Alternaria* and *Paecilomyces* keratitis associated with soft contact lens wear. *Cornea.* 2010; **29:** 564-8.
585. Young LH, Bazari H, Durand ML, *et al.* A 22-year-old woman with blurred vision and renal failure. *N Engl J Med.* 2010; **363:** 1749-58.
586. Yuan X, Wilhelmus KR, Matoba AY, *et al.* Pathogenesis and outcome of *Paecilomyces* keratitis. *Am J Ophthalmol.* 2009; **147:** 691-6.
587. Zaigraykina N, Tomkins O, Garzozi HJ, *et al. Fusarium* keratitis acquired during travel to Namibia. *J Travel Med.* 2010; **17:** 209-11.
588. Zeng B, Wang P, Xu LJ, *et al.* Amniotic membrane covering promotes healing of cornea epithelium and improves visual acuity after debridement for fungal keratitis. *Int J Ophthalmol.* 2014; **7:** 785-9.
589. Zhang M, Chelnis J, Mawn LA. Periorbital necrotizing fasciitis secondary to *Candida parapsilosis* and *Streptococcus pyogenes. Ophthal Plast Reconstr Surg.* 2015. PMID: 25902389.
590. Zhang W, Yang H, Jiang L, *et al.* Use of potassium hydroxide, Giemsa and calcofluor white staining techniques in the microscopic evaluation of corneal scrapings for diagnosis of fungal keratitis. *J Int Med Res.* 2010; **38:** 1961-7.
591. Zhang X, Sun X, Wang Z, *et al.* Keratitis-associated fungi form biofilms with reduced antifungal drug susceptibility. *Invest Ophthalmol Vis Sci.* 2012; **53:** 7774-8.
592. Zhang ZY. Corneal cross-linking for the treatment of fungal keratitis. *Cornea.* 2013; **32:** 217-8.
593. Zhao G, Zhai H, Yuan Q, *et al.* Rapid and sensitive diagnosis of fungal keratitis with direct PCR without template DNA extraction. *Clin Microbiol Infect.* 2014; **20:** 776-82.
594. Zhao J, Cheng Y, Song X, *et al.* A Comparative treatment study of intravitreal voriconazole and liposomal amphotericin B in an *Aspergillus fumigatus* endophthalmitis model. *Invest Ophthalmol Vis Sci.* 2015; **56:** 7369-76.
595. Zloty P, Villavicencio O, Belin MW. Aggressive debridement improves outcome of fungal keratitis. *Asia Pac J Ophthalmol (Phila).* 2013; **2:** 217-20.

CHAPTER 30

Otomycosis

Otomycosis is an acute, subacute or chronic superficial fungal infection of external auditory canal and auricle. It is one of the frequently encountered clinical entities of the infective ear. The causative agents are saprotrophic fungi found prevalent in the environment. The true fungi (dimorphic) are found to be very rarely involved. The commonest group of fungi belongs to genus *Aspergillus*, particularly *A.niger*. The fungi are usually secondary invaders of tissue already rendered prone by bacterial infections, physical injury or excessive accumulation of cerumen in the external auditory canal. Sometimes, it may be simply colonization of non-pathogenic fungi in the auditory canal.

The term **'otitis externa'** is preferred as compared to 'otomycosis' because it covers the entire gamut of causative agents including fungi. Moreover, the term 'otomycosis' needs to be re-defined to incorporate fungal infection of external ear, middle ear, open mastoid cavity as against the current definition limited to otitis externa arising from fungal etiology. Otherwise there is no other terminology to cover the deeper fungal infections beyond external auditory canal. In this Chapter all fungal infections of the ear have been described. **Otomycoses**, a less popular term, is pleural of otomycosis. The disease is also called as Swimmer's ear or Singapore ear.

Historical Perspective

In the year 1843, Andral *et al* and Mayer in 1844, first of all described fungal infections of external auditory canal. During the period of description by Mayer, there was a long-lasting debate as to whether fungi were to be considered as primary infectious agent or secondary to bacterial infection. In 1851, Pacini was the first to describe various preparations for the treatment of otomycosis. During that period of time Virchow suggested the term **'otomycosis'** in the year 1856.

In due course of time additional reports were published during the beginning of 20th century. Subsequently Wolf published first extensive review concerning relation of various fungi to otomycosis and listed 53 species of fungi causing this clinical entity. Now, there is enough data, which supports the view that fungi are the primary pathogens of this disease.

Gregson *et al,* in 1961 realized significance of fungal infections in the etiology of otitis externa and labeled it as a neglected disease. Nelson and Damato believed that recurrence was likely if cleansing and drying were adopted as the only management regimens. Damato later on established predominant role of *A.niger* in the 1960's and 1970's and further emphasized the role of tolnaftate in its treatment.

Cleansing and drying are essential aspects of management and provide symptomatic relief but whether additional therapy is necessary is controversial. Stern, *et al,* in 1988 suggested that most cases of otomycosis would resolve with meticulous cleansing and drying. Otomycosis classically has been described as a fungal infection of the external auditory canal but in 1989 Paulose *et al,* suggested that the term should be expanded and redefined to include fungal infections of the middle ear and open mastoid cavities. They also reported that gentian violet, an aniline dye, was clinically effective in treating 81% of otomycoses.

The malignant (invasive) otitis externa, the most severe form of disease, was first described by Chandler in 1968, wherein cause was *P.aeruginosa*. However, subsequently fungal etiology was also established, particularly involvement of *Aspergillus* species in the background of immunocompromised patients. Haruna, *et al* updated the histopathological findings of otomycosis, which were further elaborated by Vennewald and colleagues.

In 2009, *Candida auris* was first found in an ear infection in a patient in Japan. This is now evident that this yeast has emerged as an important nosocomial pathogen with clonal inter- and intra-hospital transmission and it has become widespread across several countries. A total of 77 patients from 7 states in the USA itself have been diagnosed with this highly drug-resistant yeast, maximum from New York. The federal health officials first warned clinicians in June 2016 to be on the lookout for this emerging pathogen, which is spreading around the world. It has been reported in a dozen countries *i.e.* Colombia, India, Israel, Kenya, Kuwait, Panama, Pakistan, South Africa, South Korea, Venezuela, UK and USA.

Epidemiology

The otitis externa is a common condition and manifests as an acute, subacute or chronic form. The acute form affects four in 1,000 persons annually and the chronic form affects 3-5% of the population. Acute disease commonly results from bacterial (90%) or fungal (10%) overgrowth in ear canal subjected to excess moisture or to local trauma.

Fungi have been implicated overall in ~9% of cases of external otitis. In various studies, it is estimated that about 5-25% of total cases of otitis externa are due to otomycosis. The disease is worldwide in distribution. Otomycosis is more prevalent in warm, humid climates, particularly in the rainy season as compared to arid or cold climates. It is more frequent in individuals of lower socio-economic status with poor hygienic conditions. It is most commonly seen between 2nd and 3rd decades of life. Workers in moldy or dusty settings are usually more affected. Its prevalence is higher in malnourished children as compared to the normal children.

There are many predisposing factors of otomycosis such as chronic infection of the ear, use of oils, ear drops, steroids, swimming and evidence of fungal infection elsewhere in the body such as vaginitis or onychomycosis, etc. Local lesions observed in bacterial otitis externa create favorable conditions for the growth of fungi in the external and middle ear, as well as in post-operative cavities, especially in cases of canal wall down mastoidectomy. Persistent wetness of the external auditory canal predisposes to fungal infection.

The physicians need to keep high index of suspicion of otomycosis as a cause of persistent otorrhea, especially following treatment with topical antibiotic drops. Ofloxacin remains a common choice for bacterial otorrhea but it appears to increase the incidence of otomycosis, which might have led an increase in the prevalence of otomycosis in recent years. There has been an increase in the prevalence of otomycosis in recent years possibly linked to the extensive use of antibacterial ear drops.

Some of the fungal species like *A.fumigatus* may cause necrotizing (malignant) external otitis, particularly among the patients with AIDS.

Mycology

Aspergillus and *Candida* species are the most common fungal genera responsible for otomycosis although as many as 61 different species have been identified to cause this disease. In about 75% cases of otomycosis, *Aspergillus* genus alone is the causative agent and *A.niger* is the most common cause, with occasional cases caused by *A.flavus* and *A.fumigatus*. In another study, *Aspergillus* was the predominant genus of fungi isolated from cases of otomycosis comprising 92% with *A.niger* implicated in 71% of the total isolated fungi. Within the black colored *Aspergillus* species *i.e. Aspergillus* section *Nigri*, other species have also been reported like *A.awamori* and *A.tubingensis*, which look very similar to *A.niger* but may be resistant to the commonly used antifungals hence imparting necessity of sequence-based identification.

Isolated cases of otitis externa due to *Trichoderma longibrachiatum* are reported. Unusual invasive mastoid infection by *Lecythophora hoffmannii* has also been reported as chronic infection of the mastoid and middle ear space in an immunologically competent child, a site and host where one would rarely encounter an invasive fungal infection. *Malassezia* is one of the important fungal agents causing otomycosis but it is the most neglected organism because in this disease no regular microscopy is advised and even the culture is also neither suggested nor it is easy to grow in routine. The list of various fungal agents responsible for otomycosis is given in **Table 30.1**.

Clinical Features

Otomycosis usually presents with a history of intractable itching, irritation, discomfort, pain and scanty discharge (otorrhea) from the affected ear canal. Initially there is not much debris however as the disease progresses, moist fungal growth may completely occlude the lumen. There is a feeling of blockage also in the affected ear due to fungal growth and collection of debris material in external auditory canal. Irritation is more marked in fungal as compared to bacterial otitis externa **(Fig. 30.1)**.

Table 30.1. Common Causative Agents of Otomycosis.

• *Aspergillus niger*	• *Scedosporium apiospermum*
• *A.fumigatus*	• *Fusarium solani*
• *A.flavus*	• *Neoscytalidium dimidiatum*
• *A.terreus*	• *Mucor* species
• *A.candidus*	• *Paecilomyces variotii*
• *A.hollandicus*	• *Penicillium* species
• *A.alliaceus*	• *Cryptococcus neoformans*
• *A.janus*	• *Candida parapsilosis*
• *A.versicolor*	• *C.albicans*
• *A.nidulans*	• *C.guilliermondii*
• *Trichophyton rubrum*	• *C.glabrata*
• *T.mentagrophytes*	• *C.auris*
• *Microsporum canis*	• *Rhodotorula rubra*
• *E.floccosum*	• *Malassezia* species

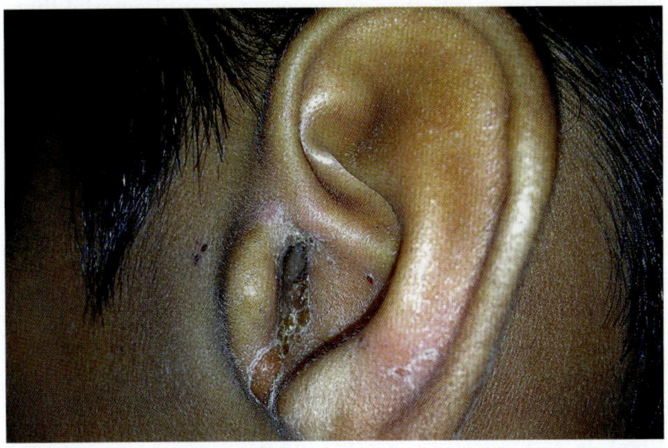

Fig. 30.1. Clinical appearance of the discharge coming out of the external auditory canal of a patient of otomycosis.

Pruritus and discharge are the most common symptoms, with reddened epidermis and lining of the tympanic cavity, being common. These manifestations are usually unilateral but rarely bilateral involvement is also observed either with same fungus or two different species. If there is a concurrent perforation of the tympanic membrane and otalgia is a significant feature, suppurative otitis media caused by *Aspergillus* species or other fungi should be considered. Fungal infection should be suspected in all cases of chronic otitis externa, which do not respond to the conventional topical antibacterial therapy. An established middle ear cleft **cholesteatoma** is associated with keratinous debris, which is likely to be an ideal medium for saprophytic fungal colonization.

The cases of aspergillosis of the middle ear and mastoid have been well-documented and have occurred mostly

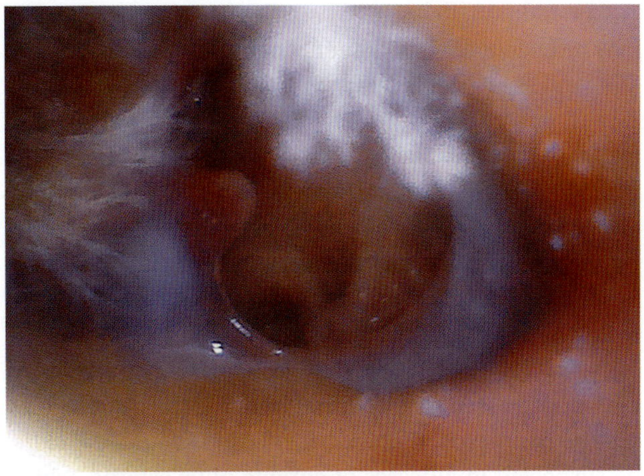

Fig. 30.2. Otoscopic examination showing fungal mass in the external auditory canal of a patient of otomycosis.

in immunosuppressed patients, including those with AIDS. In such cases, deep pain is the most common presentation and may be accompanied by facial palsy from invasion of the seventh cranial nerve as it courses through the medial wall of the middle ear.

Otoscopic examination reveals infection confined to the ear canal. There is typically fluffy, white to off-white discharge but may be black, gray, bluish-green or yellow growth on cerumen or debris resembling wet **'blotting paper'**, which may fill up the entire meatus. Small black or white conidiophores on white hyphae of *Aspergillus* species may be seen **(Fig. 30.2)**. There may be slight conductive deafness also due to mechanical obstruction of the external auditory canal. The local area may be hyperemic and sometimes bleeding may be observed. In immunocompromised patients especially diabetics, *Aspergillus* may invade locally to adjacent anatomical sites like mastoid cavity/bone or even brain. Most of the cases of fungal mastoiditis are immunocompromised, particularly with deficient CMI and caused by *Aspergillus* species.

Aspergillus species may cause invasive external otitis (necrotizing or malignant otitis externa) with local spread to bone and cartilage, which is a severe and potentially life-threatening condition. The rare cases of fungal malignant external otitis are most often due to *A.fumigatus*; although *A.niger*, *Scedosporium apiospermum*, *Pseudallescheria boydii*, *Candida ciferri* and *Malassezia sympodialis* are implicated in a few case reports. This may be associated with underlying immunocompromised situation, diabetes mellitus or patient receiving hemodialysis entailing high mortality. Invasive otitis externa is more frequently caused by *A.fumigatus* than *A.niger*.

In addition, invasive *Aspergillus* tympano-mastoiditis may be encountered in immunocompetent patients as well. The skull base osteomyelitis typically arises as a complication of ear infection in older diabetic patients, involves the temporal bone and has *Pseudomonas aeruginosa as* the usual pathogen. It is now being reported to be caused by *Apophysomyces elegans* also.

In case of *Candida* infection, the clinical appearance is more variable. The external auditory canal may be filled with whitish cheesy material. At times there is no exudate but canal wall is edematous, dull and red. Discrete organisms are normally seen. The appearance of ear canal sometimes may be non-descriptive. Fungal infection should, therefore, be suspected in all cases of chronic otitis externa, which do not respond to conventional topical therapy. *Malassezia sympodialis* has also been found as cause of malignant otitis externa in which the infection may spread to the bone and cartilage, which is serious and potentially life-threatening disease. Other fungi involved in this clinical entity are *Aspergillus*, *Candida*, *Scedosporium* and Mucormycetes. Very rarely dermatophyte infection of ear has been reported. In a report, 14 cases of AIDS-associated otic pneumocystosis have been reported with external auditory canal masses and otorrhea.

Differential Diagnosis

On the basis of history and clinical examination establishing diagnosis of otomycosis is not very difficult. However, it needs to be differentiated from similarly looking clinical conditions such as bacterial otitis externa, seborrheic dermatitis, impetigo, furunculosis and contact dermatitis.

Pathology and Pathogenesis

Hall and Farrior in 1993, classified *Aspergillus* infection in temporal bone into three types:

(i) Non-invasive: Localized and does not invade tissue, responds to conservative removal.

(ii) Invasive: Bony invasion with granulomatous response and fibrosis, occurs in immunocompetent individuals.

(iii) Fulminant: Tissue and angioinvasion with no granulomatous response, occurs in immunocompromised patients.

Moreover, infection can extend through the soft tissue and vascular planes into adjacent anatomical structures. Without any intervention, the disease usually turns out to be fatal. Even after the recommended therapy, mortality and morbidity among patients related to its complications are very common.

Radiodiagnosis

The radiological techniques like CT scan or MRI are done to delineate any involvement of the adjoining anatomical sites. These investigations are clinically very significant particularly in malignant otitis externa, involvement of middle ear or formation of cholesteatoma in delineating the extent of lesions.

Laboratory Diagnosis

The clinical finding based diagnosis should be confirmed by microscopy and culture of debris material taken from the infected site. The confirmation of diagnosis is also helpful to decide whether antimycotic or antibacterial therapy is to be given in a suspected patient.

(a) Direct Examination

The clinical diagnosis of otomycosis should be supplemented by microscopy of debris material taken out from the external auditory canal or mastoid area in case of cholesteatoma. The presence of fungal elements is seen in KOH, calcofluor white or blankophor wet mounts. The microscopic examination shows discrete clumps of hyphae with conidiophores. In otitis externa caused by *A.niger*, septate hyphae, sporulating vesicles and abundant black spores are also seen **(Fig. 30.3)**.

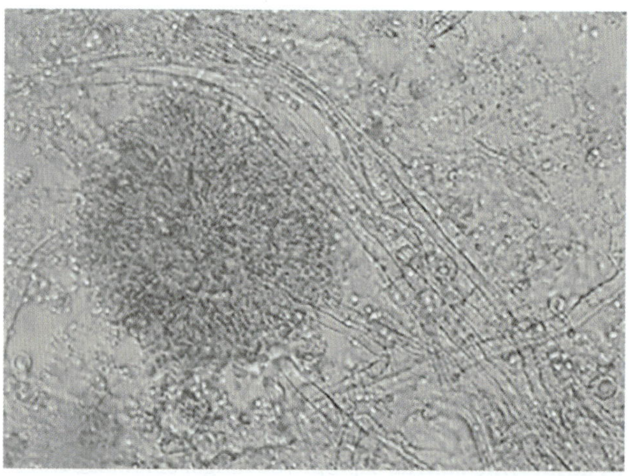

Fig. 30.3. Conidia, vesicles, conidiophores and hyphae of *Aspergillus niger* in material from external auditory canal (KOH × 400).

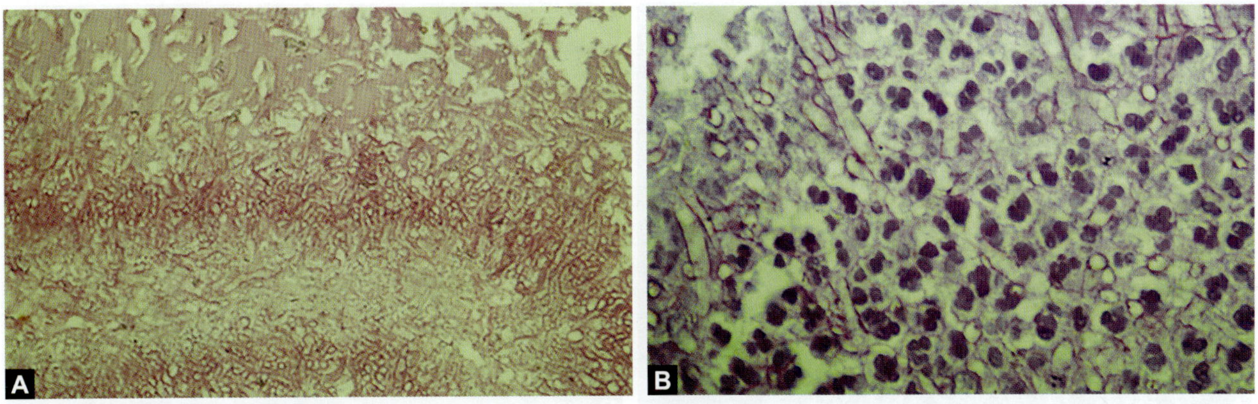

Figs. 30.4A and B. HPE showing septate fungal hyphae with acute angle branching in (A) low and high (B) powers (H&E × 200 / 400).

Fig. 30.5. Mycelial growth of *Aspergillus niger* after two days on Sabouraud dextrose agar plate.

If infection appears to involve the middle ear and mastoid, biopsy should be taken for direct demonstration of invasive fungal infection as well as for fungal culture **(Figs. 30.4A and B)**.

(b) Fungal Culture

The colonies of *A.niger* grown on Sabouraud dextrose agar are black, *A.flavus* are yellow and *A.fumigatus* are smoky-green. *A.niger* growth is woolly, at first white to yellow, then turning dark brown to black due to color of the conidia **(Fig. 30.5)**. The colonies may also be seen under the stereozoom microscope as the conidia are visible **(Figs. 30.6A and B)**. The identification of colonies is done in LCB mount, which shows sporulating vesicles, abundant black spores, conidiophores of variable length, phialides as biseriate *i.e.* arranged in two rows, covering the entire vesicle forming radiate head **(Figs. 30.7A and B)**.

Isolation of non-*albicans* species is having particular interest in therapy of otitis externa because of their reduced susceptibility to antifungals. One of such species, *Candida auris*, isolated for the first time in 2009 from ear created havoc in many countries due to its resistance to commonly used antifungals.

If perforation in tympanic membrane is observed, *A.fumigatus* cultured, local invasion clinically apparent or in immunocompromised patient, then invasive otitis externa should be strongly suspected.

(c) Immunodiagnosis

The immunodiagnosis has not much role in the establishment of the diagnosis of otomycosis. However, immunofluorescence microscopy, using fungal-specific monoclonal antibodies, has been performed as a rapid, accurate and sensitive diagnostic technique.

(d) Animal Pathogenicity

The detailed animal pathogenicity has also not been studied as far as this disease is concerned. However, some studies on animals with experimentally induced fungal infections have shown evidence for the risk of the infections spreading to the inner ear thereby causing serious damage to the organ of Corti, wherein, indirect damage to these structures by mycotoxins cannot be ruled out.

Moreover, it is seen that ototopical medications carry a risk of ototoxicity in patients with perforated ear drums.

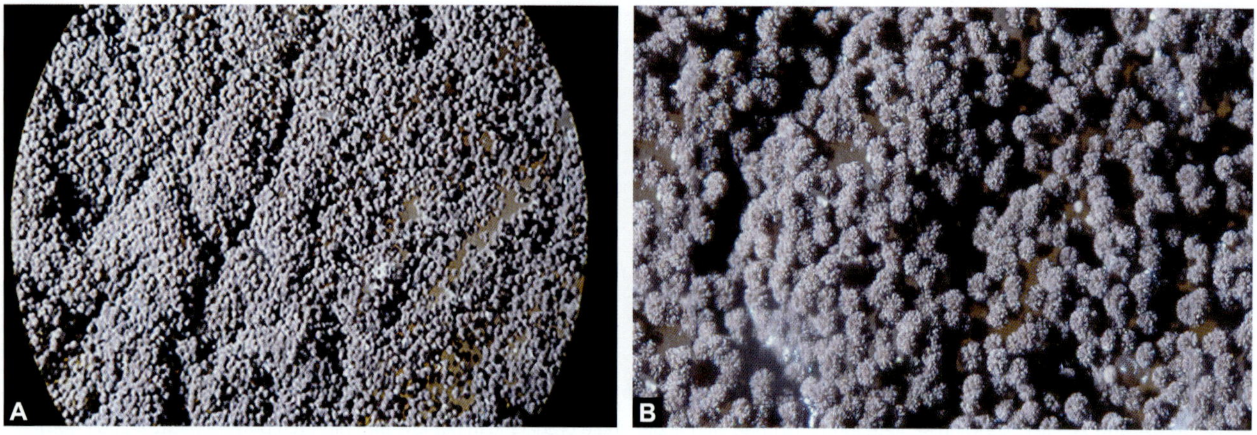

Figs. 30.6A and B. (A) *Aspergillus niger* showing vesicles with conidia under stereo zoom microscope. (B) *Aspergillus niger* showing vesicles with conidia as an enlarged view under stereo zoom microscope.

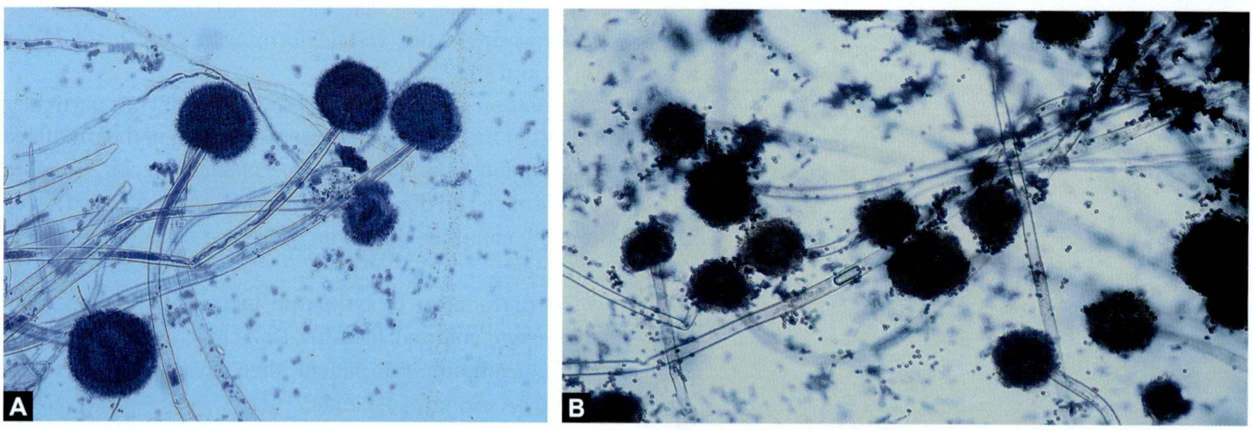

Figs. 30.7A and B. Conidia, vesicles, conidiophores and hyphae of *Aspergillus niger* (LCB × 400).

Hence in a study, it was shown that nystatin, an antifungal that can be ototopically used in treating otomycosis, may cause a decrease in otoacoustic emissions among experimental rats when administered into the middle ear cavities.

Treatment and Prophylaxis

First of all, the predisposing factors have to be taken into consideration. The fungi thrive in the moist atmosphere and in the presence of epithelial debris. Moreover, the topical drugs do not act unless debris material and discharge is removed. Therefore, careful drying and cleaning of external auditory canal is the first step in the treatment of otomycosis, which is done preferably by suction evacuation. Syringing of external auditory canal should be avoided as it may sometimes lead to flare up of the infection to deeper anatomical sites, particularly when the underlying tympanic membrane is perforated and is not visible due to impacted debris material overlying it. If the facilities of suction evacuation are not available, syringing should be done under all aseptic conditions with normal saline mixed with antifungals. The ear should be mopped up absolutely dry following syringing as moisture support continued growth of fungi. Although slightly painful, methylated spirit may be used to dry mop the ear.

(i) Topical Therapy: Antifungal treatment of otomycosis depends upon the nature of disease, especially whether it is uncomplicated or complicated wherein there is breech in the surface of the external auditory canal. The topical therapy with antifungal or other antimicrobial agents is also necessary. Effective choices include amphotericin B

(3%), flucytosine (10%), econazole cream (1%), clotrimazole cream, powder or solution (1%), thiomersal or cresyl acetate solution. Thiomersal (US=thimerosal, trade name=Merthiolate), though very effective but has not been encouraged very much, being a mercury preparation. Other agents like salicylic acid, griseofulvin and ketoconazole are less effective. Econazole (1%) solution is very effective *in vivo* in the treatment of otomycosis within 1-3 weeks. This may be valuable in cases of mixed otitis externa owing to high broad-spectrum antifungal activity as the treatment of choice in otomycosis and can be used safely as otic drops. If the tympanic membrane is perforated, tolnaftate 1% solution (Tinactin) should be used in order to prevent ototoxicity. In such cases even topical Castellani's solution is the best one. Burow's solution or 5% aluminum acetate solution should be used to reduce the swelling and remove the debris in otomycosis.

The use of topical antifungal agents such as nystatin or imidazole derivatives is suggested. Nystatin is particularly effective in the infection by *Candida* species but less active against *Aspergillus* species. Hamycin is also having the same range of efficacy as that of nystatin. Econazole nitrate is broad-spectrum antifungal agent, which is more effective for *Aspergillus* infection. It is useful against some of the Gram-positive bacteria also. The powder is preferred as compared to the ear drops because of following reasons:

(a) Ear drops produce humidity.
(b) Local irritant or pain in underlying inflamed areas.
(c) Concentration of powder is much higher after direct infiltration.
(d) Ear drops may not coat entire involved deeper areas of canal due to effect of gravity.

The other mode of treatment is by applying antifungal ointment to the external auditory canal. The only drawback of powder and ointment is that the patient himself cannot apply it properly hence needs to visit the otologist. An alternative is simply asking the patient to put 4-6 drops and he should be lying in lateral position with affected ear upward for 10-15 minutes. The patient is examined after a gap of one week for follow up.

Mercurochrome, a water-soluble organic mercurial compound, is commonly used as an antibacterial agent and also known to have antifungal properties on topical application. It is often used as 1-2% solution and has been specifically used in cases with humid environment with a reported efficacy range between 96% and 100%. However, FDA no longer approves it due to the fact that it contains mercury.

In rural areas of some developing countries, people are traditionally putting various types of oils in their external auditory canal. However, many studies have now shown that these oils are substantially antifungal in nature, endorsing the therapeutic relevance of such tradition. Keratolytic agents with nonspecific activity such as boric acid are widely used. Administration of 4% boric acid solution in alcohol and frequent suction cleaning of the ear canal might be a cost-effective treatment for otomycosis since 77% of the patients are treated effectively in this way. In another study, 1% clotrimazole solution was found to be more effective than 3% boric acid in 70% alcohol for treating otomycosis. In a randomized controlled trial of the treatment of otitis externa, it was found that 1% silver nitrate gel to be useful in 92% of patients.

Antifungal drops, given three or four times daily for five to seven days, are usually adequate to complete treatment. Because the infection can persist asymptomatically, the patient should be re-evaluated at the end of the course of treatment. At this time any further cleansing can be performed as needed. *Aspergillus* infections may be resistant to clotrimazole and may require the use of oral itraconazole. Malaysian henna leaves have also been successfully tried for the treatment of otomycosis.

(ii) Systemic Therapy: Systemic antifungal therapy is only required if patients fail to respond to the topical therapy or have invasive external otitis. Itraconazole can be used for superficial otitis externa but any invasion, including perforation of the tympanic membrane should be treated with voriconazole. Tympanoplasty may be required if perforation does not heal spontaneously.

Itraconazole resistance in *A.fumigatus* and *A.niger* are described in isolates causing otomycosis as well as more widely in nature. It is not known if such isolates are resistant to econazole but this is likely. Some may be resistant to voriconazole and posaconazole. The treatment should be continued for minimum three weeks as fungal spores in deeper layers of canal may take longer time for eradication.

Amphotericin B may be used, which has similar range of activity as nystatin. Pimaricin can be applied in case of *A.fumigatus* infection. Systemic antifungal therapy for otomycosis is required in immunocompromised patients. Any concurrent bacterial infection or other underlying abnormality should also be treated. A proper aural hygiene and drying after exposure to water are desirable. Recently efficacy of antifungal therapy with voriconazole 200 mg twice a day, in invasive otitis externa caused by *Aspergillus*, has been reviewed.

Although antifungal agents are effective in eradicating infection in the majority of patients but are unable to reverse any cranial nerve palsies. There are recent reports suggesting use of hyperbaric oxygen, which may be capable of reversing cranial nerve damage if started early enough thus reinforcing the need for early identification of the invasive fungal disease.

Further Reading

1. Abdelazeem M, Gamea A, Mubarak H, *et al.* Epidemiology, causative agents and risk factors affecting human otomycosis infections. *Turk J Med Sci.* 2015; **45:** 820-6.
2. Abou-Halawa AS, Khan MA, Alrobaee AA, *et al.* Otomycosis with perforated tympanic membrane: Self medication with topical antifungal solution versus medicated ear wick. *Int J Health Sci (Qassim).* 2012; **6:** 73-7.
3. Adoga AS, Iduh AA. Otomycosis in Jos: Predisposing factors and management. *Afr J Med Sci.* 2014; **43 (Suppl. 1):** 209-13.
4. Agnihotri VP. Aspergilli causing otomycosis. *Die Natur.* 1961; 702.
5. Alcazar-Fuoli L, Mellado E, Alastruey-Izquierdo A, *et al.* Species identification and antifungal susceptibility patterns of species belonging to *Aspergillus* section Nigri. *Antimicrob Agents Chemother.* 2009; **53:** 4514-7.
6. Alnawaiseh S, Almomani O, Alassaf S, *et al.* Treatment of otomycosis: A comparative study using miconazole cream with clotrimazole otic drops. *JRMS.* 2011; **18:** 34-7.
7. Amonoo-Kuofi K, Tostevin P, Knight JR. *Aspergillus* mastoiditis in a patient with systemic lupus erythematosus: A case report. *Skull Base.* 2005; **15:** 109-12.
8. Ananthanarayan R. Otomycosis in South Indians. *Indian J Surg.* 1951; **13:** 344-9.
9. Andral G, Gavarret J. *Ann Chim Phys.* 1843; **88.**
10. Aneja KR, Sharma C, Joshi R. Fungal infection of the ear: A common problem in the north-eastern part of Haryana. *Int J Pediatr Otorhinolaryngol.* 2010; **74:** 604-7.
11. Anwar K, Gohar MS. Otomycosis: Clinical features, predisposing factors and treatment implications. *Pak J Med Sci.* 2014; **30:** 564-7.
12. Arndal E, Glad H, Homoe P. Large discrepancies in otomycosis treatment in private ear, nose and throat clinics in Denmark. *Dan Med J.* 2016; **63:** pii: A5231.
13. Aron M, Manoukian JJ, Al-Saab FA. Deep neck abscess via direct bony erosion of the external ear canal. *J Otolaryngol Head Neck Surg.* 2008; **37:** E26-7.
14. Arya AA, Naik C, Desmukh S, *et al.* Mastoid infection caused by Entomophthorales: A rare fungal disease. *J Laryngol Otol.* 2011; **125:** 630-2.
15. Aswani V, Shukla SK. Two unusual pediatric cases of fungal infections in farming families. *J Agromedicine.* 2011; **16:** 153-7.
16. Badali H, Fakhim H, Zarei F, *et al. In vitro* activities of five antifungal drugs against opportunistic agents of *Aspergillus* Nigri complex. *Mycopathologia.* 2016.**181:** 235-40.
17. Bae WK, Lee KS, Park JW, *et al.* A case of malignant otitis externa caused by *Candida glabrata* in a patient receiving haemodialysis. *Scand J Infect Dis.* 2007; **39:** 370-2.
18. Barati B, Okhovvat SA, Goljanian A, *et al.* Otomycosis in central Iran: A clinical and mycological study. *Iran Red Crescent Med J.* 2011; **13:** 873-6.
19. Bassiouny A, Kamel T, Moawad MK, *et al.* Broad-spectrum antifungal agents in otomycosis. *J Laryngol Otol.* 1986; **100:** 867-73.
20. Baylancicek S, Serin GM, Ciprut A, *et al.* Ototoxic effect of topical ciclopirox as an antimycotic preparation. *Otol Neurotol.* 2008; **29:** 910-3.
21. Beaney GPE, Broughton A. Tropical otomycosis. *J Laryngol Otol.* 1967; **81:** 987-97.
22. Bellini C, Antonini P, Ermanni S, *et al.* Malignant otitis externa due to *Aspergillus niger. Scand J Infect Dis.* 2003; **35:** 284-8.
23. Bennett JE. Fungi as a cause of otitis. *JAMA.* 1994; **271:** 1628.
24. Bezjak V, Arya OP. Otomycosis due to *Aspergillus niger. East Afr Med J.* 1970; **47:** 247-53.
25. Bickley LS, Betts RF, Parkins CW. Atypical invasive external otitis from *Aspergillus. Arch Otolaryngol Head Neck Surg.* 1988; **114:** 1024-8.
26. Block TK, Munson E. A 19-year-old male with a 3-day history of congestion and ear pain. *J Clin Microbiol.* 2012; **50:** Que. 2185, Ans. 2544.
27. Bovo R, Benatti A, Ciorba A, *et al. Pseudomonas* and *Aspergillus* interaction in malignant external otitis: Risk of treatment failure. *Acta Otorhinolaryngol Ital.* 2012; **32:** 416-9.
28. Bryce GE, Phillips P, Lepawsky M, *et al.* Invasive *Aspergillus* tympanomastoiditis in an immunecompetent patient. *J Otolaryngol.* 1997; **26:** 266-9.
29. Buzina W, Lang-Loidolt D, Ginter-Hanselmayer G. *Trichophyton rubrum* in the external auditory meatus. *Mycoses.* 2004; **47:** 85-6.
30. Carfrae MJ, Kesser BW. Malignant otitis externa. *Otolaryngol Clin North Am.* 2008; **41:** 537-49.
31. Caruso VG. When the patient has otitis externa. *Geriatrics.* 1980; **35:** 35-42.
32. Celis-Aguilar E, Macias-Valle L, Coutinho-De Toledo H. Auditory neuropathy secondary to cryptococcal central nervous system infection in two immunocompromised patients. *Otolaryngol Head Neck Surg.* 2012; **147:** 597-8.
33. Chai FC, Auret K, Christiansen K, *et al.* Malignant otitis externa by *Malassezia sympodialis. Head Neck.* 2000; **22:** 87-9.
34. Chander J, Maini S, Subhramanyan S, Handa A. Otomycosis: A clinico-mycological study and efficacy of mercurochrome in its treatment. *Mycopathologia.* 1996; **135:** 9-12.
35. Chander J. *Aspergillus* Otomycosis. 2009. *Aspergillus* & Aspergillosis Website *(http://www.aspergillus.org.uk).*
36. Chandler JR. Malignant external otitis. *Laryngoscope.* 1968; **78:** 1257-94.
37. Chang CY, Schell WA, Perfect JR, *et al.* Novel use of a swimming pool biocide in the treatment of a rare fungal mastoiditis. *Laryngoscope.* 2005; **115:** 1065-9.

38. Cheatle A. Otomycosis. *J Laryngol Rhinol Otol.* 1920; **35:** 3-4.
39. Chen CM, Chiang CW. Allergic fungal otomastoiditis: A case report. *Laryngoscope.* 2013; **123:** 1040-2.
40. Chen D, Lalwani AK, House JW, Choo D. *Aspergillus* mastoiditis in acquired immunodeficiency syndrome. *Am J Otol.* 1999; **20:** 561-7.
41. Cheong CS, Tan LM, Ngo RY. Clinical audit of the microbiology of otorrhea referred to a tertiary hospital in Singapore. *Singapore Med J.* 2012; **53:** 244-8.
42. Choi HI, An J, Hwang JJ, *et al.* Otomastoiditis caused by *Candida auris:* Case report and literature review. *Mycoses.* 2017; doi: 10.1111/myc.12617. PMID: 28378904.
43. Civila ES, Tost JF, Jimenez AL, *et al.* Root-like hyphal structures in aspergillar otomycosis. *Arq Inst Biol (Sao Paulo).* 1972; **39:** 137-9.
44. Clark JH, Lin FR, Salaria SN, *et al.* Malignant otitis externa caused by *Aspergillus fumigatus:* A case report. *Otol Neurotol.* 2011; **32:** e22-3.
45. Cohen SR, Thompson JW. Otitic candidiasis in children: An evaluation of the problem and effectiveness of ketoconazole in 10 patients. *Ann Otol Rhinol Laryngol.* 1990; **99:** 427-31.
46. Cunningham M, Yu VL, Turner J, *et al.* Necrotizing otitis externa due to *Aspergillus* in an immunocompetent patient. *Arch Otolaryngol Head Neck Surg.* 1988; **114:** 554-6.
47. Damato PJ. Therapy with tolnaftate in otomycosis due to *Aspergillus niger. Minerva Med (Italian).* 1973; **64:** 24-6.
48. Damato PJ. Treatment of otomycosis due to *A.niger. The St Lukes Hosp Gaz.* 1966; **1:** 66-88.
49. Deguine C, Pulec JL. Otomycosis with pus. *Ear Nose Throat J.* 2002; **81:** 428.
50. Del Palacio A, Cuetara MS, Lopez-Suso MJ, *et al.* Randomized prospective comparative study: short-term treatment with ciclopiroxolamine (cream and solution) versus boric acid in the treatment of otomycosis. *Mycoses.* 2002; **45:** 317-28.
51. Desai KJ, Malek SS, Italia IK, *et al.* Fungal spectrum in otomycosis at tertiary care hospital. *NJIRM.* 2012; **3:** 58-61.
52. Deshmukh J, Surpam R, Band A. Mycological study of *Aspergillus* infections in otomycosis in eastern part of Maharashtra. *Int J Health Sci Res.* 2014; **4:** 77-82.
53. Dhindsa MK, Naidu J, Singh SM, *et al.* Chronic suppurative otitis media caused by *Paecilomyces variotii. J Med Vet Mycol.* 1995; **33:** 59-61.
54. Effat KG, Madany NM. Mycological study on cholesteatoma keratin obtained during primary mastoid surgery. *J Laryngol Otol.* 2014; **128:** 881-4.
55. English MP. Otomycosis caused by a ringworm fungus. *J Laryngol Otol.* 1957; **71:** 207-8.
56. Fasunla J, Ibekwe T, Onakoya P. Otomycosis in western Nigeria. *Mycoses.* 2008; **51:** 67-70.
57. Fayemiwo SA, Ogunleye VO, Adeosun AA, *et al.* Prevalence of otomycosis in Ibadan: a review of laboratory reports. *Afr J Med Med Sci.* 2010; **39 (Suppl):** 219-22.
58. Finer G, Greenberg D, Leibovitz E, *et al.* Conservative treatment of malignant (invasive) external otitis caused by *Aspergillus flavus* with oral itraconazole solution in a neutropenic patient. *Scand J Infect Dis.* 2002; **34:** 227-9.
59. Galloway DJ. Otomycosis of the Malay Archipelago. *J Laryngol.* 1903; 64-70.
60. Garcia-Agudo L, Aznar-Marin P, Galan-Sanchez F, *et al.* Otomycosis due to filamentous fungi. *Mycopathologia.* 2011; **172:** 307-10.
61. Gharaghani M, Seifi Z, Zarei Mahmoudabadi A. Otomycosis in Iran: A review. *Mycopathologia.* 2015; **179:** 415-24.
62. Glikson E, Sagiv D, Wolf M, *et al.* Necrotizing otitis externa: Diagnosis, treatment and outcome in a case series. *Diagn Microbiol Infect Dis.* 2017; **87:** 74-8.
63. Gokale SK, Suligavi SS, Baragundi M, *et al.* Otomycosis: A clinico mycological study. *Int J Med Health Sci.* 2013; **2:** 218-23.
64. Gopalakrishnan S, Chadha SK, Gopalan G, *et al.* Role of mastoid obliteration in patients with persistent cavity problems following modified radical mastoidectomy. *J Laryngol Otol.* 2001; **115:** 967-72.
65. Gordon G, Giddings NA. Invasive otitis externa due to *Aspergillus* species: Case report and review. *Clin Infect Dis.* 1994; **19:** 866-70.
66. Goyal A, Tyagi I, Syal R, *et al. Apophysomyces elegans* causing acute otogenic cervicofacial zygomycosis involving salivary glands. *Med Mycol.* 2007; **45:** 457-61.
67. Gregson AEW, La Touche CJ. Otomycosis: A neglected disease. *J Laryngol Otol.* 1961a; **75:** 45-69.
68. Gregson AEW, La Touche CJ. The significance of mycotic infections in the etiology of otitis externa. *J Laryngol Otol.* 1961b; **75:** 167-70.
69. Gugnani HC, Okafor BC, Nzelibe F, *et al.* Etiological agents of otomycosis in Nigeria. *Mycoses.* 1989; **32:** 224-9.
70. Gurr PA, Evans K, Dewey FM, *et al.* Otomycosis: The detection of fungi in ears by immunofluorescence microscopy. *Clin Otolaryngol.* 1997; **22:** 275-83.
71. Hall PJ, Farrior JB. *Aspergillus* mastoiditis. *Otolaryngol Head Neck Surg.* 1993; **108:** 167-70.
72. Halsey C, Lumley H, Luckit J. Necrotising external otitis caused by *Aspergillus wentii:* A case report. *Mycoses.* 2011; **54:** e211-3.
73. Harima N, Inoue T, Kubota T, *et al.* A case of otomycosis caused by *Aspergillus sclerotiorum. J Dermatol.* 2004; **31:** 949-50.
74. Harley WB, Dummer JS, Anderson TL, *et al.* Malignant external otitis due to *Aspergillus flavus* with fulminant dissemination to the lungs. *Clin Infect Dis.* 1995; **20:** 1052-4.
75. Haruna S, Haruna Y, Schachern PA, *et al.* Histopathology update: Otomycosis. *Am J Otolaryngol.* 1994; **15:** 74-8.
76. Harvey RP, Pappagianis D, Cochran J, *et al.* Otomycosis due to coccidioidomycosis. *Arch Intern Med.* 1978; **138:** 1434-5.
77. Hazarika P, Zachariah J, Victor J, *et al.* Mucormycosis of the middle ear: A case report with review of literature. *Indian J Otolaryngol Head Neck Surg.* 2012; **64:** 90-4.
78. Hennequin C, Chouaki T, Pichon JC, Strunski V, *et al.* Otitis externa due to *Trichoderma longibrachiatum. Eur J Clin Microbiol Infect Dis.* 2000; **19:** 641-2.

79. Herasym K, Bonaparte JP, Kilty S. A comparison of Locacorten-Vioform and clotrimazole in otomycosis: A systematic review and one-way meta-analysis. *Laryngoscope.* 2016. **126:** 1411-9.
80. Ho HC, Hsiao SH, Lee CY, *et al.* Treatment of refractory *Aspergillus* otomycosis with voriconazole: Case series and review. *J Laryngol Otol.* 2014; **128:** 547-51.
81. Ho T, Vrabec JT, Yoo D, *et al.* Otomycosis: clinical features and treatment implications. *Otolaryngol Head Neck Surg.* 2006; **135:** 787-91.
82. Hoshino T, Matsumoto M. Otomycosis: Subdermal growth in calcified mass. *Eur Arch Otorhinolaryngol.* 2006; **263:** 875-8.
83. Hsu Chao-Lan, Chen Cheng-Wei, Wang Hua-Kung. Iatrogenic invasive otomycosis. *Tzu Chi Med J.* 2011; **23:** 66-8.
84. Hueso Gutierrez P, Jimenez Alvarez S, Gil-Carcedo Sanudo E, *et al.* Presumption diagnosis: Otomycosis. A 451 patients study. *Acta Otorrinolaringol Esp.* 2005; **56:** 181-6.
85. Huguenin A, Noel V, Rogez A, *et al. Scedosporium apiospermum* otitis complicated by a temporomandibular arthritis: A case report and mini-review. *Mycopathologia.* 2015; **180:** 257-64.
86. Hurst WB. Outcome of 22 cases of perforated tympanic membrane caused by otomycosis. *J Laryngol Otol.* 2001; **115:** 879-80.
87. Hutchin M, Chenoweth C, Ma L, *et al.* Auricular erythema with nodules and scale. *Arch Dermatol.* 2007; **143:** 1441-6.
88. Ibekwe AO, al-Shareef Z, Benayam A. Anaerobes and fungi in chronic suppurative otitis media. *Ann Otol Rhinol Laryngol.* 1997; **106:** 649-52.
89. Ismail HK. Otomycosis. *J Laryngol Otol*, 1962; **76:** 713-21.
90. Jackman A, Ward R, April M, *et al.* Topical antibiotic induced otomycosis. *Int J Pediatr Otorhinolaryngol.* 2005; **69:** 857-60.
91. Jadhav VJ, Pal M, Mishra GS. Etiological significance of *Candida albicans* in otitis externa. *Mycopathologia.* 2003; **156:** 313-5.
92. Jain SK, Agrawal SC, Jain PC. *In vitro* evaluation of sulfa drugs against fungi causing otomycosis. *Hindustan Antibiot Bull.* 2002; **44:** 42-6.
93. Jain SK, Agrawal SC. Fungitoxic effect of some organic volatile substances against fungi causing otomycosis. *Mycoses.* 1994; **37:** 299-301.
94. Jia X, Liang Q, Chi F, Cao W. Otomycosis in Shanghai: Aetiology, clinical features and therapy. *Mycoses.* 2012; **55:** 404-9.
95. Jinn TH, Kim PD, Russell PT, *et al.* Determination of ototoxicity of common otic drops using isolated cochlear outer hair cells. *Laryngoscope.* 2001; **111:** 2105-8.
96. Joy MJ, Agarwal MK, Samant HC. Mycological and bacteriological studies in otomycosis. *Indian J Otolaryngol.* 1980; **32:** 72-5.
97. Karaarslan A, Arikan S, Ozcan M, *et al. In vitro* activity of terbinafine and itraconazole against *Aspergillus* species isolated from otomycosis. *Mycoses.* 2004; **47:** 284-7.
98. Kaur R, Mittal N, Kakkar M, *et al.* Otomycosis: A clinico-mycologic study. *Ear Nose Throat J.* 2000; **79:** 606-9.
99. Kaushik V, Malik T, Saeed SR. Interventions for acute otitis externa. *Cochrane Database Syst Rev.* 2010; **1:** CD004740. PMID: 20091565.
100. Kaya AD, Kiraz N. *In vitro* susceptibilities of *Aspergillus* spp. causing otomycosis to amphotericin B, voriconazole and itraconazole. *Mycoses.* 2007; **50:** 447-50.
101. Kazemi A, Majidinia M, Jaafari A, *et al.* Etiologic agents of otomycosis in the North-Western area of Iran. *Jundishapur J Microbiol.* 2015; **8:** e21776.
102. Khan F, Muhammad R, Khan MR, *et al.* Efficacy of topical clotrimazole in treatment of otomycosis. *J Ayub Med Coll Abbottabad.* 2013; **25:** 78-80.
103. Khanna V, Chander J, Nagarkar NM, *et al.* Clinicomicrobiological evaluation of active tubotympanic type chronic suppurative otitis media. *J Otolaryngol.* 2000; **29:** 148-53.
104. Kiakojuri K, Rajabnia R, Jalili B, *et al.* Otomycosis in adolescent patients referred to the therapeutic centers in Babol City, Iran. *Jundishapur J Microbiol.* 2015; **8:** e17138.
105. Kiakojuri K, Roushan MR, Sepidgar SA. Suction clearance and 2% topical miconazole versus the same combination with acidic drops in the treatment of otomycosis. *Southeast Asian J Trop Med Public Health.* 2007; **38:** 749-53.
106. Kikuchi A, Funakubo T, Kohsyu H. A study of fungal infections in otorhinolaryngology. *Acta Otolaryngol (Stockh).* 1994; **51:** 224-7.
107. Kintzel PE, Trausch DE, Copfer AL. Otic administration of amphotericin B 0.25% in sterile water. *Ann Pharmacother.* 1994; **28:** 333-5.
108. Kirschner R, Sun PL, Huang SL, *et al.* A case of bilateral otomycosis associated with *Aspergillus flavus* and *A.terreus* in Taiwan. *J Mycol Med.* 2017; pii: S1156-5233(16)30273-6. PMID: 28501466.
109. Kley E. A contribution to the treatment of otomycosis. *Laryngol Rhinol Otol.* 1976; **55:** 765-7.
110. Koch S, Rudel B, Tietz HJ. Lethal otogenic *Candida* meningitis. *Mycoses.* 2004; **47:** 450-3.
111. Kolling WM, McPherson TB, McJunkin JL. The use of extemporaneously compounded 1% tetracaine to improve adherence with clotrimazole 1% topical solution in the treatment of ear infection: A case report. *Am J Otolaryngol.* 2013; **34:** 757-8.
112. Kumar A. Fungal spectrum in otomycosis patients. 2005; *JK Sci.* **7:** 152-5.
113. Kurnatowski P, Filipiak A. Otomycosis: prevalence, clinical symptoms, therapeutic procedure. *Mycoses.* 2001; **44:** 472-9.
114. Kwok P, Hawke M. Clotrimazole powder in the treatment of otomycosis. *J Otolaryngol.* 1987; **16:** 98.
115. Lakshmipati G, Murti RB. Otomycosis. *J Indian Med Assoc.* 1960; **34:** 439-41.
116. Latha R, Sasikala R, Muruganandam N. Chronic otomycosis due to *Malassezia* spp. *J Glob Infect Dis.* 2010; **2:** 189-90.
117. Lawrence TL, Ayers LW, Saunders WH. Drug therapy in otomycosis: an *in vitro* study. *Laryngoscope.* 1978; **88:** 1755-60.

118. Lee A, Tysome JR, Saeed SR. Interventions for otomycosis. *Cochrane Database of Syst Rev.* 2011; **9:** CD009289.
119. Lee H, Kim do Y, Lee KH, *et al.* Deformity of the earlobe caused by fixed cutaneous sporotrichosis in a pediatric patient. *Int J Dermatol.* 2015; **54:** e187-9.
120. Ling SS, Sader C. Fungal malignant otitis externa treated with hyperbaric oxygen. *Int J Infect Dis.* 2008; **12:** 550-2.
121. Lohoue-Petmy J, Bengono-Toure G, Founda-Onana A. A study of otomycoses in Yaounde. *Rev Laryngol Otol Rhinol.* 1996; **117:** 119-21.
122. Lopez L, Evens RP. Drug therapy of *Aspergillus* otitis externa. *Otolaryngol Head Neck Surg.* 1980; **88:** 649-51.
123. Lucente FE. Fungal infections of the external ear. *Otolaryngol Clin North Am.* 1993; **26:** 995-1006.
124. Mahlakwane MS, Ramdial PK, Sing Y, *et al.* Otic pneumocystosis in acquired immune deficiency syndrome. *Am J Surg Pathol.* 2008; **32:** 1038-43.
125. Makary CA, Roberts TD, Wetmore SJ. Disseminated blastomycosis presenting as mastoiditis and epidural abscess. *W V Med J.* 2014; **110:** 46-7.
126. Marsh RR, Tom LW. Ototoxicity of antimycotics. *Otolaryngol Head Neck Surg.* 1989; **100:** 134-6.
127. Martin TJ, Kerschner JE, Flanary VA. Fungal causes of otitis externa and tympanostomy tube otorrhea. *Int J Pediatr Otorhinolaryngol.* 2005; **69:** 1503-8.
128. Mayer AFJK. Beobachtung von Cysten mit Fadenpilzen aus dem aussern Gehorgange eines Madchens. *Arch Anat Physiol. Wissench Med.* 1844; **11:** 404-8.
129. McGonigle JJ, Jillson OF. Otomycosis: An entity. *Arch Dermatol.* 1967; **95:** 45-6.
130. McLaren O, Potter C. *Scedosporium apiospermum*: A rare cause of malignant otitis externa. *BMJ Case Rep.* 2016; pii: bcr2016217015. PMID: 27613266.
131. Menachof M R, Jackler RK. Otogenic skull base osteomyelitis caused by invasive fungal infection. *Otolaryngol Head Neck Surg.* 1990; **102:** 285-9.
132. Meunier A, Blassin R, Meunier F. Remarquable efficacite due to naftate sur les otomycoses (30 observations). *Annales D'otolaryngol (Paris).* 1975; **92:** 699-702.
133. Mgbor N, Gugnani HC. Otomycosis in Nigeria: treatment with mercurochrome. *Mycoses.* 2001; **44:** 395-7.
134. Mishra GS, Mehta N, Pal M. Chronic bilateral otomycosis caused by *Aspergillus niger*. *Mycoses.* 2004; **47:** 82-4.
135. Moberly AC, Naumann IC, Cordes SR. Cryptococcal meningitis with isolated otologic symptoms. *Am J Otolaryngol.* 2010; **31:** 49-53.
136. Mofatteh MR, Naseripour Yazdi Z, Yousefi M, *et al.* Comparison of the recovery rate of otomycosis using betadine and clotrimazole topical treatment. *Braz J Otorhinolaryngol.* 2017; pii: S1808-8694(17)30056-3. PMID: 28549873.
137. Morinaka S. Dermatophytosis of the ear. *Otolaryngol Head Neck Surg.* 2005; **133:** 113-5.
138. Mugliston T, O'Donoghue G. Otomycosis - A continuing problem. *J Laryngol Otol.* 1985; **99:** 327-33.
139. Munguia R, Daniel SJ. Ototopical antifungals and otomycosis: A review. *Int J Pediatr Otorhinolaryngol.* 2008; **72:** 453-9.
140. Munoz A, Martinez-Chamorro E. Necrotizing external otitis caused by *Aspergillus fumigatus*: computed tomography and high resolution magnetic resonance imaging in an AIDS patient. *J Laryngol Otol.* 1998; **112:** 98-102.
141. Murai H, Kira J, Kobayashi T, *et al.* Hypertrophic cranial pachymeningitis due to *Aspergillus flavus*. *Clin Neurol Neurosurg.* 1992; **94:** 247-50.
142. Murakami A, Tutumi T, Watanabe K. Middle ear effusion and fungi. *Ann Otol Rhinol Laryngol.* 2012; **121:** 609-14.
143. Narozny W, Kuczkowski J, Stankiewicz C, *et al.* Value of hyperbaric oxygen in bacterial and fungal malignant external otitis treatment. *Eur Arch Otorhinolaryngol.* 2006; **263:** 680-4.
144. Navaneethan N, Yaadhava Krishnan RP. Type of antifungals: Does it matter in empirical treatment of otomycosis? *Indian J Otolaryngol Head Neck Surg.* 2015; **67:** 64-7.
145. Neji S, Ines H, Houaida T, *et al.* Externa otitis caused by the *Graphium* stage of *Pseudallescheria apiosperma*. *Med Mycol Case Rep.* 2013; **2:** 113-5.
146. Nemati S, Hassanzadeh R, Khajeh Jahromi S, *et al.* Otomycosis in the north of Iran: Common pathogens and resistance to antifungal agents. *Eur Arch Otorhinolaryngol.* 2014; **271:** 953-7.
147. Nguyen JT, Challapalli M, McElheny K, *et al.* Blastomycosis presenting as isolated otitis and otomastoiditis. *Pediatr Infect Dis J.* 2013; **32:** 301-2.
148. Nowrozi H, Arabi FD, Mehraban HG, *et al.* Mycological and clinical study of otomycosis in Tehran, Iran. *Bull Environ Pharmacol Life Sci.* 2014; **3:** 29-31.
149. Ozcan M, Ozcan KM, Karaarslan A, *et al.* Concomitant otomycosis and dermatomycoses: A clinical and microbiological study. *Eur Arch Otorhinolaryngol.* 2003; **260:** 24-7.
150. Pahwa VK, Chamiyal PC, Suri PN. Mycological study in otomycosis. *Indian J Med Res.* 1983; **77:** 334-8.
151. Pak MW, Soo G, van Hasselt CA. Flourishing otomycosis. *Ear Nose Throat J.* 1997; **76:** 10.
152. Palmer KK, Vaghela DB, Manjusha R. A clinical study on the effect of Arka Taila in the management of Karnasrava (Otomycosis). *Ayu.* 2011; **32:** 349-52.
153. Panchal P, Pethani J, Patel D, *et al.* Analysis of various fungal agents in clinically suspected cases of otomycosis. *Indian J Basic Appl Med Res.* 2013; **2:** 865-9.
154. Parize P, Chandesris MO, Lanternier F, *et al.* Antifungal therapy of *Aspergillus* invasive otitis externa: Efficacy of voriconazole and review. *Antimicrob Agents Chemother.* 2009; **53:** 1048-53.
155. Paulose KO, Al Khalifa S, Shenoy P, *et al.* Mycotic infection of the ear (otomycosis): A prospective study. *J Laryngol Otol.* 1989; **103:** 30-5.
156. Petrak N, Pottage JC, Levin S. Invasive external otitis caused by *Aspergillus fumigatus* in an immunocompromised patient. *J Infect Dis.* 1985; **151:** 196.
157. Philip A, Thomas R, Job A, *et al.* Effectiveness of 7.5 percent povidone iodine in comparison to 1 percent clotrimazole with lignocaine in the treatment of otomycosis. *ISRN Otolaryngol.* 2013; 239730. PMID: 23984100.

158. Philips P, Bryce G, Shepherd J, *et al.* Invasive external otitis caused by *Aspergillus. Rev Infect Dis.* 1990; **12:** 277-81.
159. Pontes ZB, Silva AD, Lima Ede O, *et al.* Otomycosis: A retrospective study. *Braz J Otorhinolaryngol.* 2009; **75:** 367-70.
160. Pradhan B, Tuladhar NR, Amatya RM. Prevalence of otomycosis in outpatient department of otolaryngology in Tribhuvan University Teaching Hospital, Kathmandu, Nepal. *Ann Otol Rhinol Laryngol.* 2003; **112:** 384-7.
161. Prasad SC, Kotigadde S, Shekhar M, *et al.* Primary otomycosis in the Indian subcontinent: Predisposing factors, microbiology and classification. *Int J Microbiol.* 2014; 636493. PMID: 24949016.
162. Pulec JL, Deguine C. Otomycosis. *Ear Nose Throat J.* 2002; **81:** 370.
163. Rainsbury P, Mitchell-Innes A, Wilson H, *et al.* Aspergilloma of the middle ear mimicking necrotising otitis externa: Case report. *J Laryngol Otol.* 2010; **124:** 1209-11.
164. Reiss P, Hadderingh R, Schot LJ, Danners SA. Invasive external otitis caused by *Aspergillus fumigatus* in two patients with AIDS. *AIDS.* 1991; **5:** 605-6.
165. Rijal AS, Joshi RR, Regmi S, *et al.* Ear diseases in children presenting at Nepal Medical College Teaching Hospital. *Nepal Med Coll J.* 2011; **13:** 164-8.
166. Romsaithong S, Tomanakan K, Tangsawad W, *et al.* Effectiveness of 3 per cent boric acid in 70 per cent alcohol versus 1 per cent clotrimazole solution in otomycosis patients: A randomised, controlled trial. *J Laryngol Otol.* 2016; **130:** 811-5.
167. Rotoli M, Sascaro G, Cavalieri S. *Aspergillus versicolor* infection of the external auditory canal successfully treated with terbinafine. *Dermatology.* 2001; **202:** 143.
168. Rubin Grandis J, Branstetter BFT, Yu VL. The changing face of malignant (necrotising) external otitis: Clinical, radiological and anatomic correlations. *Lancet Infect Dis.* 2004; **4:** 34-9.
169. Rutt AL, Sataloff RT. *Aspergillus* otomycosis in an immunocompromised patient. *Ear Nose Throat J.* 2008; **87:** 622-3.
170. Saki N, Rafiei A, Nikakhlagh S, *et al.* Prevalence of otomycosis in Khouzestan Province, south-west Iran. *J Laryngol Otol.* 2013; **127:** 25-7.
171. Salamat AA, Archer C, Basarab A, *et al. Scedosporium apiospermum* causing otomycosis in an immunocompetent child with tympanostomy tubes: Management of this rare entity. *Int J Pediatr Otorhinolaryngol.* 2015; **79:** 1785-7.
172. Sarvan RR, Kikani KM, Mehta SJ, *et al.* Clinico-mycological study of otomycosis. *Int J Biol Med Res.* 2012; **3:** 2469-70.
173. Satish HS, Viswanatha B, Manjuladevi M. A clinical study of otomycosis. *IOSR J Dent Med Sci.* 2013; **5:** 57-62.
174. Satoh K, Makimura K, Hasumi Y, *et al. Candida auris* sp. nov., a novel ascomycetous yeast isolated from the external ear canal of an inpatient in a Japanese hospital. *Microbiol Immunol.* 2009; **53:** 41-4.
175. Saunders JE, Raju RP, Boone JL, *et al.* Antibiotic resistance and otomycosis in the draining ear: culture results by diagnosis. *Am J Otolaryngol.* 2011; **32:** 470-6.
176. Schneider ML. Merthiolate in treatment of otomycosis. *Laryngoscope.* 1981; **91:** 1194-5.
177. Schonebeck J, Zakrisson JE. Topical 5-fluorocytosine therapy in otomycosis. *J Laryngol Otol.* 1974; **88:** 227-31.
178. Schrader N, Isaacson G. Fungal otitis externa - its association with fluoroquinolone ear drops. *Pediatrics.* 2003; **111:** 1123.
179. Sengupta RP, Kacker SK. Otomycosis. *Indian J Med Sci.* 1978; **32:** 5-7.
180. Sheikh MS, Quzi By, Rammen B. Otomycosis in Khozistan. *Indian J Otolaryngol Head Neck Surg.* 1993; **2:** 73-7.
181. Song JE, Haberkamp TJ, Patel R, *et al.* Fungal otitis externa as a cause of tympanic membrane perforation: A case series. *Ear Nose Throat J.* 2014; **93:** 332-6.
182. Sood VP, Sinha A, Mohapatra LN. Otomycosis: A clinical entity - clinical and experimental study. *J Laryngol Otol.* 1967; **81:** 999-1004.
183. Stephen S, Subashini B, Thomas R, *et al.* Skull base osteomyelitis caused by an elegant fungus. *J Assoc Physicians India.* 2016; **64:** 70-1.
184. Stern JC, Lucente FE. Otomycosis. *Ear Nose Throat J.* 1988; **67:** 804-10.
185. Stern JC, Shah MK, Lucente FE. *In vitro* effectiveness of 13 agents in otomycosis and review of the literature. *Laryngoscope.* 1988; **98:** 1173-7.
186. Strauss M, Fine E. *Aspergillus* otomastoiditis in acquired immunodeficiency syndrome. *Am J Otol.* 1991; **12:** 49-53.
187. Supiyaphun P, Sampatanukul P, Vongfu S, *et al.* Aural aspergilloma: An unusual delayed complication of tympanoplasty. *J Med Assoc Thai.* 2003; **86:** S358-61.
188. Szigeti G, Kocsube S, Doczi I, *et al.* Molecular identification and antifungal susceptibilities of black *Aspergillus* isolates from otomycosis cases in Hungary. *Mycopathologia.* 2012; **174:** 143-7.
189. Szigeti G, Sedaghati E, Mahmoudabadi AZ, *et al.* Species assignment and antifungal susceptibilities of black aspergilli recovered from otomycosis cases in Iran. *Mycoses.* 2012; **55:** 333-8.
190. Talwar P, Chakrabarti A, Kaur P, *et al.* Fungal infection of ear with special reference to chronic suppurative otitis media. *Mycopathologia.* 1988; **104:** 47-50.
191. Than KM, Naing KS, Min M. Otomycosis in Burma and its treatment. *Am J Trop Med Hyg.* 1980; **29:** 620-3.
192. Tisner J, Millan J, Rivas P, *et al.* Otomycosis and topical application of thimerosal: study of 152 cases. *Acta Otorinolaringol Esp.* 1995; **46:** 85-9.
193. Tiwari S, Singh SM, Jain S. Chronic bilateral suppurative otitis media caused by *Aspergillus terreus. Mycoses.* 1995; **38:** 297-300.
194. Tom LW. Ototoxicity of common topical antimycotic preparations. *Laryngoscope.* 2000; **110:** 509-16.
195. Tuzcu A, Bahceci M, Celen MK, *et al.* Necrotizing (malignant) otitis externa: an unusual localization of mucormycosis. *Indian J Med Microbiol.* 2006; **24:** 289-91.
196. van Hasselt P, Gudde H. Randomized controlled trial on the treatment of otitis externa with one percent silver nitrate gel. *J Laryngol Otol.* 2004; **118:** 93-6.

197. van Tol A, van Rijswijk J. *Aspergillus* mastoiditis, presenting with unexplained progressive otalgia, in an immunocompetent (older) patient. *Eur Arch Otorhinolaryngol.* 2009; **266:** 1655-7.
198. Varghese L, Chacko R, Varghese GM, *et al.* Septic arthritis of the temporomandibular joint caused by *Aspergillus flavus* infection as a complication of otitis externa. *Ear Nose Throat J.* 2015; **94:** e24-6.
199. Varghese R, Nair RM, Kavalakkat FJ. Fungal otomastoiditis: A case series in immunocompetent adults. *Indian J Otolaryngol Head Neck Surg.* 2014; **66:** 110-3.
200. Velesrubio F, Walters D, Fulcher J, *et al.* Auricular cryptococcosis as the initial presentation of AIDS. *Otolaryngol Head Neck Surg.* 2007; **136;** 493-4.
201. Vennewald I, Klemm E. Otomycosis: Diagnosis and treatment. *Clin Dermatol.* 2010; **28:** 202-11.
202. Vennewald I, Schonlebe J, Klemm E. Histologic studies on otomycosis. *Mycoses.* 2002; **45 (Suppl. 1):** 47-52.
203. Vennewald I, Schonlebe J, Klemm E. Mycological and histological investigations in humans with middle ear infections. *Mycoses.* 2003; **46:** 12-8.
204. Viswanatha B, Naseeruddin K. Fungal infections of the ear in immunocompromised host: A review. *Mediterr J Hematol Infect Dis.* 2011; **3:** e2011003.
205. Viswanatha B, Sumatha D, Vijayashree MS. Otomycosis in immunocompetent and immunocompromised patients: Comparative study and literature review. *Ear Nose Throat J.* 2012; **91:** 114-21.
206. Vural A, Sahin MI, Aydın M, *et al.* The effect of nystatin solution on otoacoustic emissions in rats. *J Int Adv Otol.* 2017; **13:** 105-9.
207. Wall SJ, Lee KH, Alvarez JD, *et al.* Cutaneous mucormycosis of the external ear. *Arch Otolaryngol Head Neck Surg.* 2000; **126:** 236-9.
208. Wang MC, Liu CY, Shiao AS, *et al.* Ear problems in swimmers. *J Chin Med Assoc.* 2005; **68:** 347-52.
209. Wolf FT. Relation of various fungi to otomycosis. *Arch Otolaryngol.* 1947; **46:** 361-74.
210. Yaganeh MA, Asadi M, Dehghani R, *et al.* Evaluating the effects of mixture of alcohol and acetic acid for otomycosis therapy. *Jundishapur J Microbiol.* 2010; **3:** 66-70.
211. Yamashita K. Fungus diseases in otolaryngology, with special reference to otomycosis. *Iryo.* 1969; **23:** 1233-40.
212. Yaroko AA, Salim R, Mohamed, Z, *et al.* Antifungal activity of Malaysian henna leaves extracts on pathogenic fungi of otomycosis. *Int Med J.* 2015; **22:** 389-91.
213. Yassin A, Maher A, Moawad MK. Otomycosis: A survey in the eastern province of Saudi Arabia. *J Laryngol Otol.* 1978; **92:** 869-76.
214. Yassin A, Mostofa MA, Moawad MK. Fungus infection of the ear. *J Laryngol Otol.* 1964; **78:** 591-602.
215. Yates PD, Upile T, Axon PR, de Carpentier J. *Aspergillus* mastoiditis in a patient with acquired immunodeficiency syndrome. *J Laryngol Otol.* 1997; **111:** 560-1.
216. Yehia MM, Al-Habib HM, Shehab NM. Otomycosis: A common problem in north Iraq. *J Laryngol Otol.* 1990; **104:** 387-9.
217. Yenisehirli G, Bulut Y, Guven M, *et al. In vitro* activities of fluconazole, itraconazole and voriconazole against otomycotic fungal pathogens. *J Laryngol Otol.* 2009; **123:** 978-81.
218. Youssef M, Bassim M, Shabb N, *et al. Aspergillus* mastoiditis in an immunocompetent patient: A case report and review of the literature. *Scand J Infect Dis.* 2014; **46:** 325-30.
219. Youssef YA, Abdou MH. Studies on fungus infection of the external ear. I. Mycological and clinical observations. *J Laryngol Otol.* 1967a; **81:** 401-12.
220. Youssef YA, Abdou MH. Studies on fungus infection of the external ear. II. On the chemotherapy of otomycosis. *J Laryngol Otol.* 1967b; **81:** 1005-12.

CHAPTER

31

Hyalohyphomycosis

Hyalohyphomycosis is fungal infection in which the basic tissue form of fungus is hyaline *i.e.* without any pigment in its cell wall. The disease may present as cutaneous, subcutaneous or systemic infection among immunocompromised as well as immunocompetent individuals. Many hyaline fungi are responsible for causing this clinical entity and number is increasing day-by-day as newly recognized fungi belonging to this group are being reported in the medical literature.

This is a general term used to group together all infections caused by unusual and uncommon hyaline fungal pathogens. These are not the agents of otherwise-named common hyaline fungal infections such as aspergillosis. Therefore, this term is not actually intended to replace the well-established disease names *i.e.* aspergillosis, fusariosis, scedosporiosis/pseudallescheriasis or penicilliosis. This allows hyaline hyphomycetes which cause infection to be categorized until it is established that a particular hyaline fungus causes infections with regularity. However, it has been seen that the terminology, hyalohyphomycosis, is being commonly used interchangeably in parallel with other diseases like fusariosis, scedosporiosis or penicilliosis.

Historical Perspective

In the year 1974, Ajello, *et al* proposed term **hyalohyphomycosis** (Gk. ***hyalos*** - glass) because the fungi involved have basic tissue form as hyaline, septate mycelial elements that may be branched and are occasionally toruloid. This terminology came into existence in contrast to **phaeohyphomycosis**, wherein the phaeoid or dematiaceous fungi with brown to blackish pigmented cell wall, are involved.

As Ajello originally created term hyalohyphomycosis and subsequently Ajello and McGinnis after ten years in 1984 further reviewed it to accommodate more infections caused by molds that lacked melanin in their cell wall. As seen in phaeohyphomycosis, in this disease also spectrum and taxonomic divergence of non-phaeoid hyphomycetes reported as agents of mycosis and varied clinical presentations have significantly increased.

Both these terms are very successful in eliminating need to create unnecessary disease names each time that an infection or new etiological agent is described. The basic intention was to provide stable nomenclature to the medical community that could accommodate current and future knowledge regarding infections and their non-melanized septate, filamentous fungi.

Therefore, just for the sake of convenience, diseases caused by phaeoid and non-phaeoid fungi are assigned two major categories, phaeohyphomycosis and hyalohyphomycosis, respectively. The division is based on one readily evident differentiating characteristic *i.e.* presence or absence of brown pigment in mycelia in tissue or culture. Masson-Fontana stain is negative in hyalohyphomycosis due to absence of melanin. This is an artificial system of nomenclature because taxonomically unrelated organisms fall into same category but it is believed that grouping is useful to categorize case reports.

Thus disease names phaeohyphomycosis and hyalohyphomycosis were introduced not to disrupt the already existing nomenclature of well-established disease names but to deal with this issue in a rational manner with seemingly ever-growing number of mycelial fungal genera and species which cause superficial, cutaneous, subcutaneous, systemic or even opportunistic infections.

Mycology

There are 74 species of fungi belonging to 20 genera that meet requirements and qualify as agents of hyalohyphomycosis and are proved to cause infection in humans and animals. This number is gradually increasing as many newer hyaline hyphomyces are being added from time to time. Spectrum of disease caused by these hyaline hyphomycetes is diverse and disease state is largely determined by local as well as general immunological and physiological state of host, which may be symptomatic or as such remain asymptomatic.

The important human pathogens of this clinical entity are *Scedosporium, Fusarium, Scopulariopsis* and *Penicillium* species. The other fungi, which can cause opportunistic hyalohyphomycosis, include *Acremonium, Paecilomyces* and *Chrysosporium* species. The emerging basidiomycetes pathogens are *Schizophyllum commune, Hormographiella aspergillata* including its sexual state *(Coprinus cinereus)* and *Tilletiopsis minor*, which are found, associated with various hyalohyphomycosis infections.

Moreover, few scientists think that *Scedosporium* and *Scopulariopsis* species should also be included in phaeohyphomycosis rather than hyalohyphomycosis. Their argument is that some element of melanin is produced by these fungi during their life cycle. There are case reports also in the literature to this effect. Patel *et al,* have described a case of phaeohyphomycosis caused by *Scopulariopsis brumptii* in a liver transplant recipient in 1994. However, others are contesting it and traditionally these fungi are being described as hyalohyphomycosis and not phaeohyphomycosis. In this Textbook such fungi of gray area are described in this very Chapter on Hyalohyphomycosis.

The epidemiology, clinical features, laboratory diagnosis, treatment and prophylaxis of fungal agents causing hyalohyphomycosis are discussed under following headings:

1. *Scedosporium* Species
2. *Fusarium* Species
3. *Penicillium* Species
4. *Purpureocillium/Paecilomyces* Species
5. *Scopulariopsis* Species
6. *Acremonium/Sarocladium* Species
7. *Trichoderma* Species
8. *Chrysosporium* Species
9. *Beauveria* Species
10. *Hormographiella* Species
11. *Schizophyllum commune*

1. *Scedosporium* Species

There are lots of taxonomic changes in this genus. Now the preferred name for the holomorph is *Scedosporium*, which replaces *Pseudallescheria*. However, *Pseudallescheria boydii* remains the perfect state of *Scedosporium* and *Graphium* species. It is an ascomycete and belongs to family Microascaceae of order Microascales in sub-phylum Pezizomycotina of phylum Ascomycota. *Scedosporium-like* organism *i.e. Lomentospora* is also discussed under this very heading of hyalohyphomycosis.

Scedosporium is characterized by production of globose, brown to black cleistothecia in culture and pale yellow to golden brown, lemon-shaped ascospores. Strains without ascomata can be recognized by their two typical anamorphs *i.e. Scedosporium apiospermum*, with long, cylindrical conidiogenous cells producing slimy heads of subhyaline to brown, sub-spherical to elongated conidia and *Graphium* species, with erect synnemata.

The genus *Scedosporium* currently comprises six species, *Scedosporium apiospermum, Scedosporium boydii, Pseudallescheria angusta, Scedosporium minutisporum, Scedosporium dehoogii,* and *Scedosporium aurantiacum,* most of which can be distinguished with the primary fungal DNA barcode, the ITS1/2 region of the rDNA gene cluster.

However, there is controversy regarding the inclusion of certain species as dematiaceous, particularly *Scedosporium (Lomentospora) prolificans.* Some experts do not feel that this species is truly melanized to the same extent as other fungi labeled as dematiaceous. However, others do consider it dematiaceous, as it is pigmented in culture and does react with the Masson-Fontana stain, which is specific for the melanin.

Pseudallescheria boydii was previously known as *Allescheria boydii* and *Petriellidium boydii* and its asexual form, *Scedosporium apiospermum* was known as *Monosporium apiospermum.* Recently, more species have been proposed in the *P.boydii*-complex like *P.minutispora, S.aurantiacum* and *S.dehoogii.*

Scedosporium boydii is frequently found as soil saprotrophic and in many other natural substrates. This species is now emerging as an opportunistic fungus causing severe systemic infections among the immunocompromised patients. This is an uncommon but highly infectious organism in granulocytopenic patients that produce pattern of infections very similar to that of *Aspergillus* species, with acute-angle branching, invasion of blood vessels, respiratory tract, etc. The disease caused by *Pseudallescheria (Scedosporium) boydii* was called pseudallescheriasis, now being replaced with scedosporiosis.

Lomentospora prolificans (formerly *Scedosporium prolificans* or *Scedosporium inflatum)* causes numerous important infections, which have been reported in recent years. It is especially significant in immunocompromised patients due to tropism to CNS. Its virulence is variable in healthy animals. This is also an opportunistic fungal pathogen that is normally resistant to the commonly used antifungals and has occasionally been involved in symptomless colonization, localized infection and disseminated infections in immunocompromised hosts. It has also been implicated in nosocomial infections.

This was formerly named as, *Scedosporium inflatum,* owing to its ballooned conidiogenous cells. It is very similar to *S.apiospermum* but conidiogenous cells are inflated at the base and are frequently in brush-like arrangement. It has been considered pathogenic since 1984 and can produce localized or disseminated infections mainly in patients, who are neutropenic and fatal infections in undergoing organ transplants and with leukemia. The disease spectrum, diagnosis and treatment of infections caused by *L.prolificans* resemble those of pseudallescheriasis. Unlike *P.boydii,* however, *L.prolificans* appears to be **resistant** to most of antifungals. It is anamorphic species without teleomorph connections and closely related to order Microascales and hence described with *Scedosporium* and *P.boydii.*

Scedosporium (Lomentospora) prolificans or *Scedosporium inflatum* is newly recognized pathogen. It has been isolated from soil of potted plants and is most likely to be soil saprotrophic. The clinical spectrum of infections caused by this fungus can be expanded to include disseminated infection in granulocytopenic patients and that such infection can present with triad of fever, rash and myalgia, previously thought to be specific for disseminated candidiasis.

Epidemiology

Scedosporium apiospermum and *S.boydii* are saprotrophic fungi which are isolated from soil, manure, decaying vegetation, polluted water and other natural sites. The infections caused by this group of fungi are mostly reported after near-drowning. The fungus enters into body via respiratory route or by penetrating injury.

For long time it escaped attention as significant human pathogen because it was known by number of different names. This nomenclatural confusion was caused by its polymorphism, different strains being potentially capable of producing *Scedosporium*- and *Graphium*-like anamorphs in addition to *Pseudallescheria* cleistothecia.

Clinical Features

The main causative fungi of scedosporiosis *i.e. Scedosporium apiospermum* and *Scedosporium boydii* may cause deep-seated infections such as sinusitis, endophthalmitis and pneumonia with an inclination to disseminate and involve central nervous system. Among emerging fungal agents, *Scedosporium* are relatively uncommon but well-documented potential CNS pathogen. The cutaneous infection may spread locally with subsequent bone, joint and bursal involvement. The other infections, which are produced by *Scedosporium,* are white grain mycetoma, disseminated mycosis in lungs with leukemia or cystic fibrosis, otitis externa and keratomycosis.

Scedosporiosis often resembles and may be mistaken for aspergillosis, particularly in the ear, paranasal sinus, lung and eye. *Scedosporium* species can grow in cerumen and desquamated debris of external auditory canal causing otomycosis. The fungus is histopathologically analogous to *Aspergillus* species with typical septate hyphae with acute angle-branching. Its HPE report is usually despatched as, *"consistent with aspergillosis."* Hence culture of the causative fungus in a clinical specimen is pertinent to reach the correct diagnosis thereby proper treatment.

The colonization in a poorly draining paranasal sinus may lead to formation of fungus ball that later may extend into orbit or meninges especially in an immunosuppressed patient. The colonization of fungus in cysts, cavities and ectatic bronchi of patients with prior tuberculosis or other chronic lung diseases ultimately result in productive cough or hemoptysis. A fungus ball may sometimes be detected radiologically. A patient with chronic lung disease has also been reported to develop pleural pseudallescheriasis.

In subcutaneous tissue, fungus may grow exclusively, in grossly visible clumps, called grains. This chronic, suppurative infection resulting from cutaneous inoculation is white-grain eumycetoma. However, *Scedosporium* species infections, which clinically do not resemble mycetoma, are called scedosporiosis, name that encompasses variety of clinical conditions.

Scedosporium apiospermum is ubiquitous fungus that has been isolated from soil, polluted water and sewage. It is the most frequent cause of mycetoma in temperate

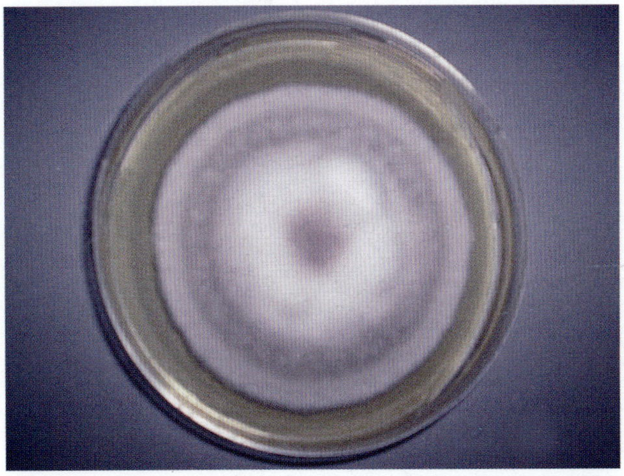

Fig. 31.1. Grayish mycelial colonies of *Scedosporium apiospermum* on Sabouraud's dextrose agar.

regions. This type of localized chronic infection is typically found in immunocompetent patients and is characterized by induration associated with draining sinuses. Although organism has shown low inherent virulence, an increasing number of invasive infections caused by this organism have been reported in past few years, mainly in patients with hematological malignancies or in those who have undergone solid organ transplantation. Over the recent years this pathogen has emerged as an increasingly common cause of disease in immunocompromised patients, with potentially devastating consequences.

Ocular scedosporiosis may result from corneal trauma *i.e.* keratomycosis or complication of cataract extraction. Hematogenous endophthalmitis from an unknown portal is rarely seen in immunosuppressed patient. Meningitis is usually outcome of the damage to blood brain barrier or trauma. Similarly, arthritis and osteomyelitis can occur. Endocarditis due to scedosporiosis has rarely been reported and seems to involve patients with heart valve prosthesis.

The most severe cases of scedosporiosis have been reported in patients who are receiving immunosuppressive therapy for organ transplantation, leukemia, lymphoma, systemic lupus erythematosus or Crohn's disease.

Laboratory Diagnosis

All relevant details are taken and an appropriate clinical specimen in sufficient amount should be collected from an active site of lesion keeping aseptic precautions. On histopathological examination, septate hyphae, 2–7 µm in diameter, are found within abscesses, accompanied by granulomatous inflammation. As there is an acute angle branching, it is not feasible to differentiate from aspergillosis unless the causative agent is cultured.

Scedosporium is cottony fluffy white mold that rapidly turns brownish-gray (mousy) with age **(Fig. 31.1)**. The reverse of colony is black. The fungus grows well at 37°C on routine culture media like SDA. Microscopically, septate hyphae with oval or club-shaped, pale brown conidia are seen. It produces elliptical, sperm-shaped, single-celled conidia borne singly from tips of long or short conidiophores. The ascocarp and ascospores are produced on PDA or CMA. *Scedosporium* gives *Graphium*-like appearance hence it should be meticulously examined **(Figs. 31.2 to 31.4)**. The type of conidiation in various fungal species causing hyalohyphomycosis is given in **Table 31.1**.

Fungal culture is only reliable means of diagnosis but histopathological examination of lung or sinus lesions is also necessary to establish invasion. The fungus may be found as matted clump of hyphae but not invading wall of pulmonary cavities, ectatic bronchi or paranasal sinuses. Invasive infection also occurs and is characterized by necrosis and infiltration of lesion with neutrophils. Hyphae are seen scattered throughout abscess. The vascular invasion by hyphae, which is characteristic of aspergillosis and mucormycosis, may also occur in *Scedosporium* infections. The hyphae tend to be 2-7 µm in diameter but may be larger or show areas of bulbous swelling. Masses of hyphae in ventilated alveolar spaces of lung or paranasal sinus may show conidia or clumped groups of hyphae in parallel array, termed coremium or synnema. These structures strongly suggest diagnosis of pseudallescheriasis. The patient with non-invasive pulmonary disease may have precipitating antibody to *Scedosporium* antigen.

Scedosporium is usually recognized by its morphological features but to facilitate its identification, several serological techniques have also been developed which can recognize pathogen even in tissues. The physiological properties of *Scedosporium* have also been explored for use as diagnostic tool to differentiate it from similar species such as *Lomentospora (Scedosporium) prolificans* **(Fig. 31.5)**. These tests include cycloheximide tolerance and assimilation of glucose, fructose, sucrose, raffinose, inulin, biotin, thiamine, nitrates, higher alcohol and sugar acids.

Treatment and Prophylaxis

The treatment of pseudallescheriasis is almost similar to that of aspergillosis. However, *P.boydii* may be found to

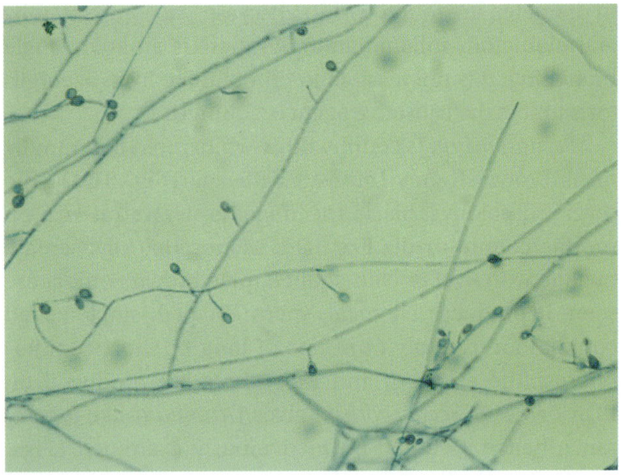

Fig. 31.2. Microscopic appearance of *Scedosporium apiospermum* (LCB × 400).

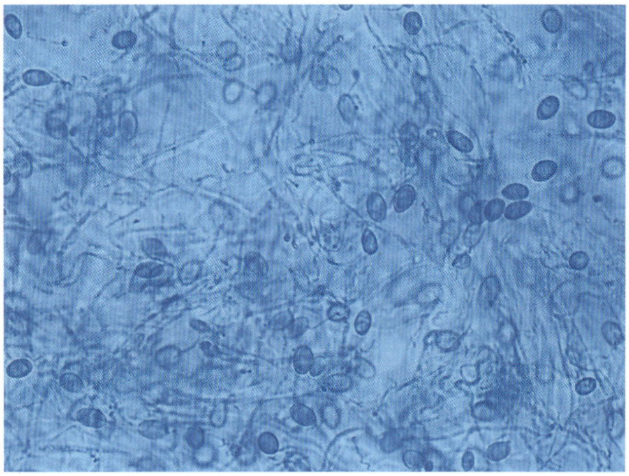

Fig. 31.3. Microscopic appearance of *Scedosporium apiospermum* (LCB × 400).

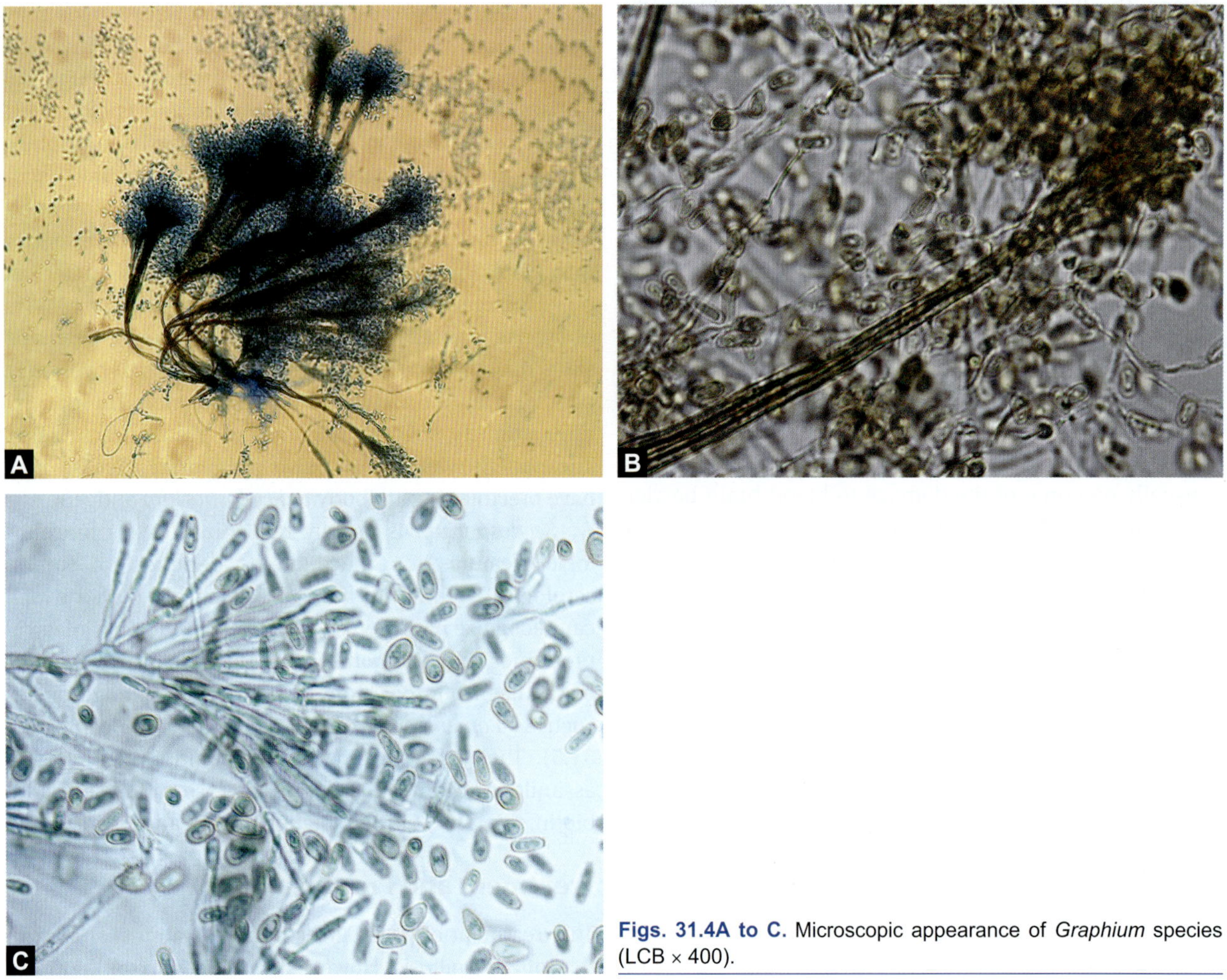

Figs. 31.4A to C. Microscopic appearance of *Graphium* species (LCB × 400).

Table 31.1. Type of Conidiation in Hyalohyphomycetes.

Category	*Hyalohyphomycetes*
Conidia Singly	*Scedosporium, Lomentospora, Pseudallescheria boydii*
Conidia in Clusters	*Acremonium* species
Conidia in Chains	*Aspergillus, Penicillium, Paecilomyces, Purpureocillium, Scopulariopsis* species

have resistance to amphotericin B as well as flucytosine. It is susceptible to triazoles like itraconazole, which is the drug of choice. Therefore, cultural substantiation of fungal isolate in diagnosis is pertinent to have an effective treatment. Surgical resection of localized lesion remains a key to successful outcome. In setting of persistent immunosuppression, results of therapy are invariably poor. The response of disseminated infection to combination therapy with interferon-γ and antifungal agents is encouraging.

2. *Fusarium* Species

Fusarium species are very important fungi in agriculture as plant pathogens. These are responsible for loss of crops of millions of rupees. Now, these are being increasingly reported from various types of clinical presentations among humans. The disease caused is known as fusariosis, which is invariably diagnosed as aspergillosis due to its similarities on histopathological examination of biopsy specimens. Hence fungal culture becomes imperative to reach its final diagnosis as seen in scedosporiosis.

Epidemiology

Various species of genus *Fusarium* are involved to cause hyalohyphomycosis, which is also called fusariosis. Several species are newly emerging fungal pathogens, which are causing morbidity and mortality among patients with some immunocompromised background like those who have received bone marrow transplantation, malignancies, etc. The disseminated infections may occur, mostly in patients with hematologic malignancies and occasionally in patients with extensive burns. The most common species involved in human infections are *Fusarium solani, F.oxysporum* and *F.verticillioides* (formerly, *F.moniliforme*). The number of species became enormous and it is being known by many corresponding species complex *e.g. Fusarium solani* species complex (FSSC).

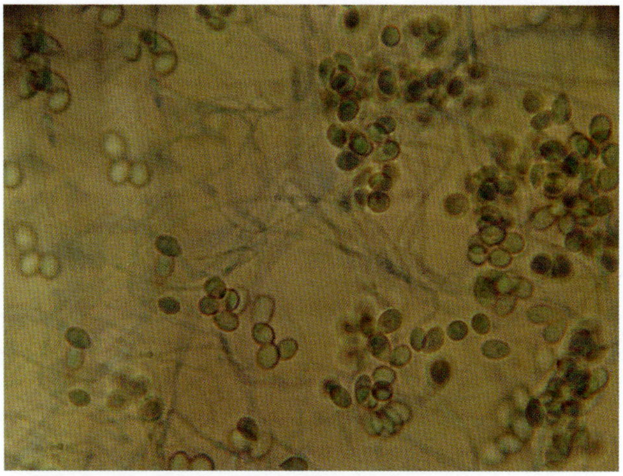

Fig. 31.5. Microscopic appearance of *Lomentospora prolificans* (LCB × 400).

Fusarium is found in soil of regions with wide range of climate. The infection can be acquired through inhalation of aerosolized conidia or through breaks in integumentary barriers, particularly in patients with onychomycosis caused by *Fusarium* species. Fungal pneumonia may be caused by *Fusarium* species, particularly by *F.verticillioides*. Infection by *F.oxysporum* has been reported in diabetic patients. Newer species of *Fusarium* are being reported in human infections. Thus, some of these are considered as emerging pathogens *e.g. Fusarium sacchari* has been isolated from a series of fungal keratitis and/or endophthalmitis cases among sugarcane farmers. Some of the *Fusarium* species infecting banana are also found among human infections contemplating that it may be either a carrier or source of infection.

Clinical Features

There are three types of clinico-pathological features, which are described: (i) locally invasive infection; (ii) disseminated infections and (iii) toxicosis from mycotoxins. Locally invasive infections such as wound, burn or ulcerative lesions, septic arthritis and endophthalmitis are uncommon but are described in both healthy and immunocompromised hosts.

The cutaneous lesions occur more frequently in patients with fusarial infections than in those with aspergillosis. Various cutaneous lesions may be observed: ecthyma-like lesions similar to those observed in patients with disseminated aspergillosis; target lesions consisting of ecthyma-like lesions surrounded by thin rim of erythema and multiple subcutaneous nodules. Cellulitis with

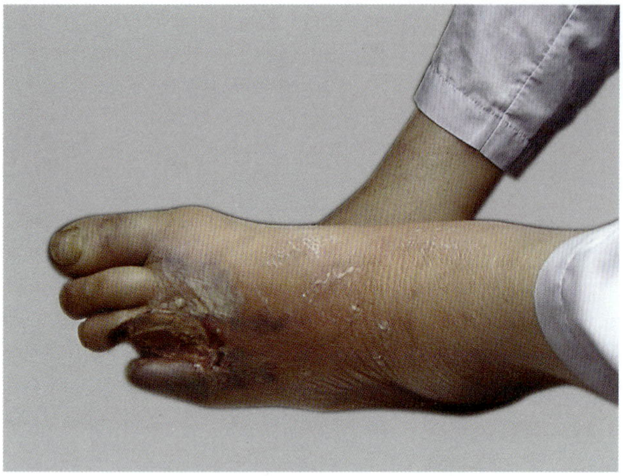

Fig. 31.6. Toe infection caused by *Fusarium* species in a diabetic patient and 4th toe of left foot had to be amputated.

or without fasciitis may also occur. Members of this genus are important cause of non-dermatophytic dermomycosis as well as onychomycosis, particularly *F. oxysporum*.

The disseminated infections usually present as persistent fever in profoundly neutropenic patients with cancer. Sinusitis and/or rhinocerebral infection, endophthalmitis or pyomyositis may at times be the presenting feature. There are many clues about the presence of fusarial infection, which include growth of mold from bloodstream, preceding or concomitant cellulitis of toe or finger and presence of cutaneous or subcutaneous lesions **(Fig. 31.6)**.

Fusarium species are regularly recovered from blood of infected patients in contrast to rare isolation of *Aspergillus* species from patients with similar clinical manifestations. If a mold grows in blood culture bottle, it is usually considered as a contaminant but if *Fusarium* species grows, it is taken as significant and patient's thorough workup is to be done for any underlying immunocompromising factor. As stated above the pathogenesis resembles aspergillosis histopathologically but can be differentiated microbiologically on cultural identification and up to some extent clinically, particularly by the pattern of skin involvement.

The recovery from myelo-suppression may lead to resolution of infection or establishment of chronic focus such as lungs, eyes or soft tissue, with potentiation for dissemination upon reinstitution of therapy with cytotoxic agents. The mortality from disseminated fusarial infection is high among patients with persistent immunosuppression.

Interest in toxins of this genus arises chiefly as a result of its role as plant pathogen and as major cause of range of animal and human disorder resulting from ingestion of mycotoxins produced in contaminated grains. Various *Fusarium* species are responsible for diverse clinical manifestations of mycotoxicosis caused by fumonisins, trichothecenes and zearalenones.

Laboratory Diagnosis

The diagnosis depends on keeping high index of suspicion concerning neutropenic patient whose fever persists despite antibacterial and antifungal therapy, especially with sinopulmonary infection, cellulitis at site of previous toenail infection, presence of skin lesions described above, endophthalmitis, myositis or blood cultures positive for molds. The final diagnosis relies on culture of organism from bloodstream or from tissues obtained as biopsy since histopathological examination cannot differentiate between fusariosis and aspergillosis, scedosporiosis and/or penicilliosis, etc. All have septate hyphae with acute-angle dichotomous branching with vascular invasion. The culture shows septate hyphae with bean-shaped macroconidia **(Figs. 31.7 to 31.10)**.

The fungal cultivation for *Fusarium* is very similar to that of *Aspergillus* species. However, species like *Colletotrichum* and *Cylindrocarpon* may sometimes be confused with *Fusarium* on culture. But *Colletotrichum* has setae in addition to the bean-shaped macroconidia and *Cylindrocarpon* has chlamydospores without foot cells. The differentiating features between genus *Fusarium* and *Cylindrocarpon* are given in **Chapter 36** along with **Figure 36.2**. *Fusarium* infections should be strongly considered in the absence of *Aspergillus* isolation among patients at high risk of mold infections with positive galactomannan assay test results. The identification of *Fusarium* species requires molecular methods *i.e.* PCR also as morphologically it is easily feasible up to genus level.

Treatment and Prophylaxis

The successful outcome of *Fusarium* infections is related to control of underlying disease process and recovery from neutropenia. Surgical resection of infected tissue may be useful if infection is localized. One of the most important clinical aspects about *Fusarium* is its relative lack of responsiveness to conventional amphotericin B. Recently, increasing numbers of patients have responded well to

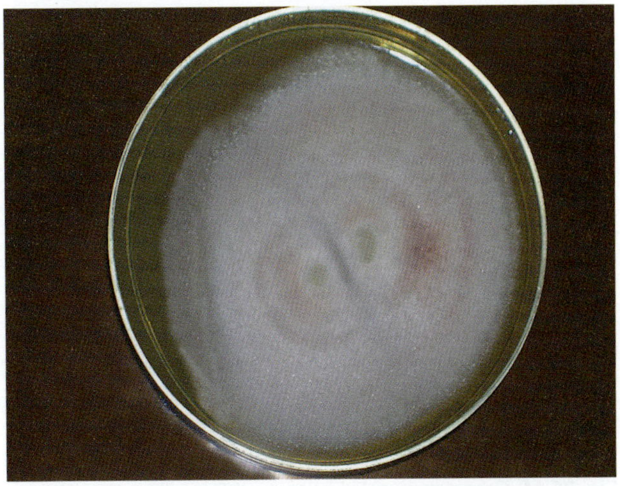

Fig. 31.7. Greyish mycelial colonies of *Fusarium* Species.

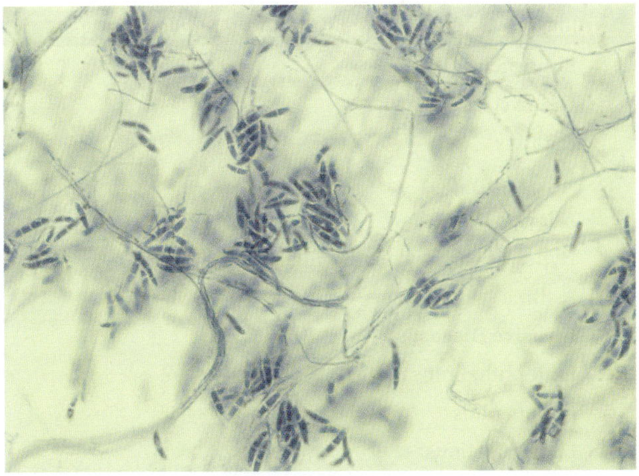

Fig. 31.8. Microscopic appearance of *Fusarium* Species showing sickle-shaped macroconidia (LCB × 400).

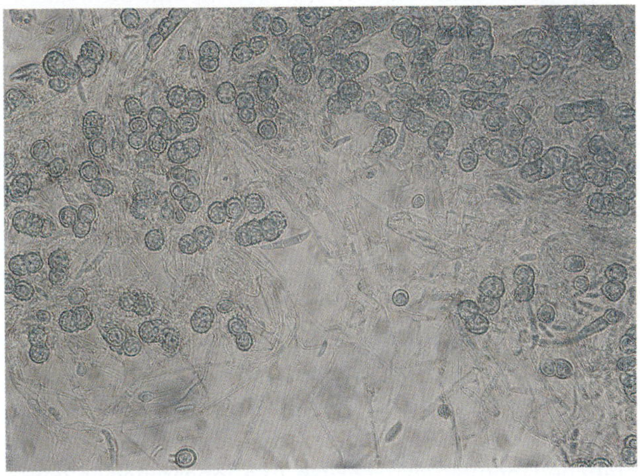

Fig. 31.9. *Fusarium chlamydosporum* with sickle-shaped macroconidia and numerous chlamydospores (LCB × 400).

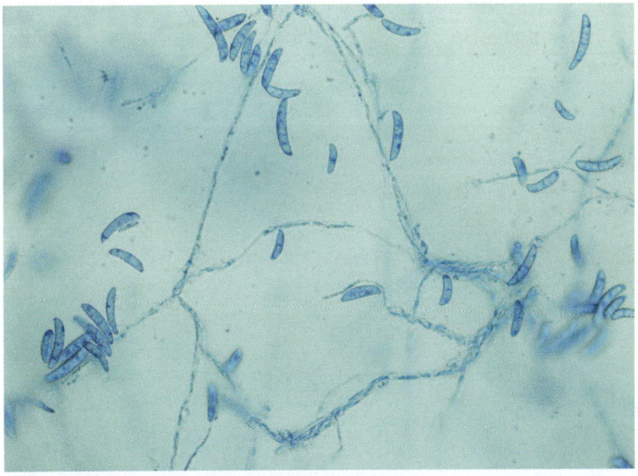

Fig. 31.10. Microscopic appearance of *Fusarium* Species showing sickle-shaped macroconidia (LCB × 400).

this drug when it is given at higher doses. This is successfully achieved by liposomal formulations of amphotericin B, which produce fewer side effects. Azole derivatives have also been given successfully like voriconazole either alone or in combination with terbinafine, particularly in disseminated infections.

The effective preventive measures are urgently needed to curtail this infections. Moreover, infection control policies aimed at reducing number of airborne fungal spores should be properly carried out. Prophylaxis with ideal antifungal agents that have substantial activity against *Fusarium* species may also prove useful for preventing these overwhelming infections.

3. *Penicillium* Species

There are more than 300 species of genus *Penicillium* found in the environment, which are usually considered as common mycology laboratory contaminants. Their main role is found in the production of antibiotics. Hence some species are routinely used in manufacturing antimicrobial agents like *Penicillium chrysogenum* (former *P.notatum*) for producing penicillin, an antibacterial antibiotic and *Penicillium griseofulvum* for griseofulvin, another effective antifungal antibiotic, used for treating various forms of dermatophytosis.

Penicillium species are involved in allergic fungal diseases, fungal keratitis and mycotoxicosis as well due

Table 31.2. Diseases Spectrum by *Penicillium* species.

1.	**Superficial Diseases**
	Otomycosis
	Keratitis
	Onychomycosis
2.	**Allergic Diseases**
	Hypersensitivity pneumonitis
	Asthma
3.	**Invasive Diseases**
	Endophthalmitis
	Endocarditis
	Lung infections
	Intracranial infections
	Peritonitis
	Kidney infections
	Pericarditis
	Esophagitis

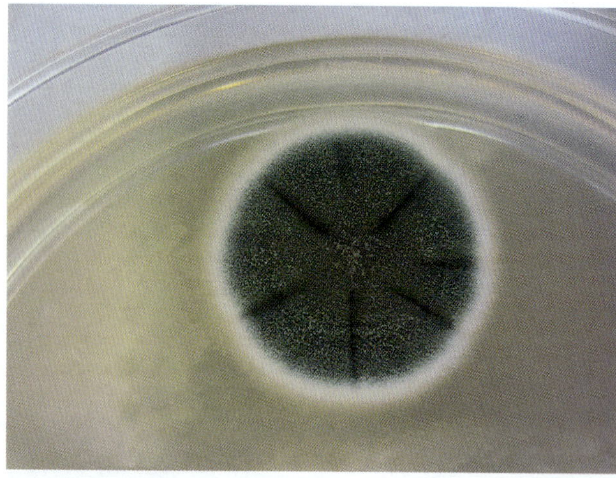

Fig. 31.11. Greenish mycelial colonies of *Penicillium* Species.

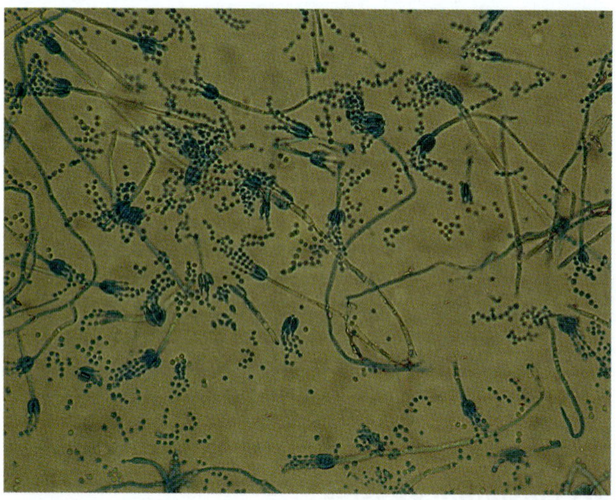

Fig. 31.12. Paintbrush like appearance of *Penicillium* species along with chains of conidia (LCB × 400).

to their ability to produce potential mycotoxins. Therefore, most of the species of genus *Penicillium* are usually encountered as contaminants in diagnostic microbiology laboratory and very rarely as significant pathogenic agents.

The species of this genus are accidentally seen as causative agent of hyalohyphomycosis once in a while. They have also been isolated among patients with underlying predisposing conditions. These species of genus *Penicillium* causing hyalohyphomycosis are *P.brevicompactum, P.chrysogenum, P.citrinum, P.commune, P.crustaceum, P.decumbens, P.dupontii, P.funiculosum, P.glaucum* and *P.purpurogenum. Penicillium marneffei* has been re-named as *Talaromyces marneffei* because of its shift to genus *Talaromyces* and is described as separate **Chapter 24** (Talaromycosis) in this Textbook. The spectrum of fungal diseases caused by genus *Penicillium* is depicted at glance in **Table 31.2**.

Whenever species of *Penicillium* is isolated from clinical specimens, it is essential to demonstrate its tissue reaction in histopathological sections of the infected part for having correlation to establish its clinical significance.

The growth of *Penicillium* species is rapid to develop the colony **(Fig. 31.11)**. The conidiophores arise in various forms producing phialides singly or in groups or from branched metulae giving **brush-like** appearance *i.e.* dense brush-like spore-bearing structures called penicilli (sing. penicillus). The name *Penicillium* comes from the resemblance of the conidiophore of the fungus to a paintbrush - Penicillus is the Latin word for paintbrush.

The conidia are unicellular and found in chain with youngest at base *i.e.* **basipetal (Fig. 31.12)**. It should not be confused with other penicillate fungi such as *Paecilomyces, Purpureocillium* and *Scopulariopsis. Paecilomyces* produces characteristic phialides, which are elongated, tapered and end in several **elongated** or **elliptical** conidia.

4. *Purpureocillium/Paecilomyces* Species

The species of genus *Paecilomyces* are somewhat similar to *Penicillium* hence their culture are invariably confused with it. In the past they were originally classified with *Penicillium* species due to morphological similarities.

But its colonies are not usually green and conidiogenous cells *i.e.* phialides are flask-shaped with strongly tapering tip often pointed at the end and usually aggregated in divergent clusters. Therefore, they were re-classified as member of genus *Paecilomyces* in 1974.

This is also heterogeneous ascomycetous genus whose teleomorphs belong to different families like Trichocomaceae and Thermoascaceae. The species of both families are identified by their morphology, color of colonies, size and shape of conidia. There are approximately 78 species and subspecies described in this genus, however; very few of them are implicated in human infections.

Paecilomyces species are generally encountered as soil saprotrophic and occasionally airborne contaminant in diagnostic microbiology laboratories. These are occasionally implicated as pathogen in human or animal diseases as superficial or invasive infections.

Purpureocillium is a genus recently proposed to accommodate *Paecilomyces lilacinus*, a well-studied species that has biotechnological properties and an ability to cause human infections also. Most of cases are involving *Purpureocillium lilacinum* among healthy, immunocompetent hosts with endophthalmitis following intraocular lens implantation as these lenses are infected by storage in contaminated neutralizing solution. The other ocular infections caused by *Purpureocillium* species are keratitis, infection of corneal transplant and orbital granuloma. The cases of maxillary sinusitis, soft tissue infection after prepatellar bursitis, have also been reported.

In the immunocompromised patients, *P.lilacinum* is being reported to cause cutaneous infections, catheter-related fungemia and disseminated disease secondary to application of contaminated skin lotion. *Paecilomyces* species other than *P.lilacinum,* such as *P.variotii, P.marquandii, P.javanicus,* are also reported to cause some of the opportunistic infections.

About ten cases of *Paecilomyces* infections have been reported in the literature. Half of which are from eye or skin infection due to *P.lilacinum*, while majority of former resulted in poor outcome, later generally responded to various antifungal drugs. Deep infections due to *P.lilacinum, P.variotii* and other *Paecilomyces* species often respond poorly to drug therapy thus leading to serious complications or even fatal consequences among such patients.

The appearance of *Purpureocillium/Paecilomyces* in tissues is commonly confused with that of other molds. These genera, unlike common pathogenic mold *(Aspergillus)*, may sporulate in tissues. The association of budding conidial cells, free conidia and short hyphae may lead to confusion with yeast cells which emphasize considerable importance of fungal cultures in these patients for establishing the correct diagnosis.

Fig. 31.13. Lilac colored mycelial colonies of *Purpureocillium lilacinum*.

Purpureocillium lilacinum grows on SDA, within three days, producing white, flat, powdery colony with red-brown reverse. As culture of *Purpureocillium* matures, colony develops slight lilac appearance **(Fig. 31.13)**, eventually darkening to violet and not green as seen in *Penicillium* species. Microscopically, it produces characteristically phialides which are elongated, tapered end in several elongated or elliptical conidia **(Fig. 31.14)**.

In cases of *Paecilomyces variotii,* the mycelial colonies are yellowish-brown in color and conidiophores bear dense, verticillate branches, each bearing 2–7 phialides, may be up to 150 μm in length and 3.5–6.5 μm wide. The phialides are cylindrical or ellipsoidal, tapering abruptly into long, thin, cylindrical neck **(Fig. 31.15)**.

5. *Scopulariopsis* Species

Scopulariopsis is ubiquitous fungus usually found in soil, dead plants and animal matters. Members of this genus are thus saprotrophic fungi and occasionally pathogenic to man and animals. Being soil saprotrophs, *Scopulariopsis* species are isolated from wide variety of substrates. Out of 22 species of this genus, 9 are known to cause human diseases, which are named alphabetically as *Scopulariopsis acremonium, S.asperula, S.brevicaulis, S.brumptii, S.candida, S.carbonaria, S.flava, S.fusca* and *S.koningii.* However, two most important are *S.brevicaulis* and

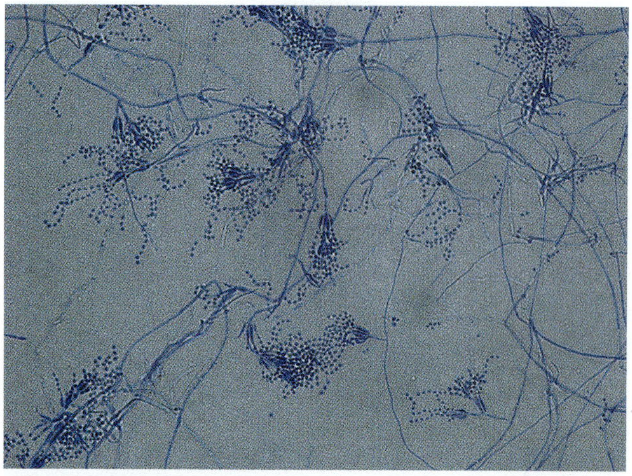

Fig. 31.14. Microscopic appearance of *Purpureocillium lilacinum* showing chains of conidia (LCB × 400).

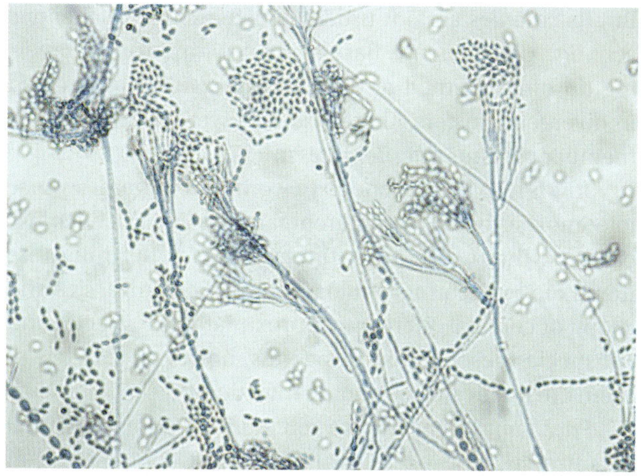

Fig. 31.15. Microscopic appearance of *Paecilomyces variotii* showing chains of conidia (LCB × 400).

Fig. 31.16. Whitish-powdery colonies of *Scopulariopsis brevicaulis* on Sabouraud's dextrose agar.

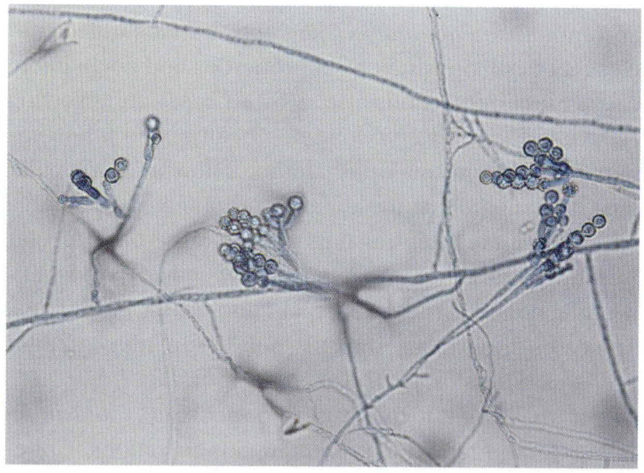

Fig. 31.17. Microscopic appearance of *Scopulariopsis brevicaulis* showing chains of conidia on penicillate phialides (LCB × 400).

S.brumptii, which are found to be regularly involved in various clinical settings.

Scopulariopsis brevicaulis is associated with cases of onychomycosis, fungal balls in preformed cavities, pneumonia in leukemia patients and even cutaneous infections as hyalohyphomycosis. It produces moderately fast growing, whitish powdery colonies that are either tan or beige-colored with reverse as cream-colored to brownish **(Fig. 31.16)**. The truncate conidia are formed in chains on well-developed annellidic conidiogenous cells, which are produced either singly or in a penicillate arrangement **(Figs. 31.17 and 31.18)**. The conidial structures when present as penicillate, it looks somewhat similar to *Penicillium, Paecilomyces* and *Purpureocillium* species. The conidia are thick walled, roughened and arranged in chains with the youngest at the base *i.e.* **basipetal**.

Scopulariopsis brumptii is reported from cases of hypersensitive pneumonitis in intravenous drug abusers. This has also been isolated from pulmonary nodules, hydrothorax in AIDS patient and liver transplantation recipient. Moreover, *S.brumptii* was also reported from onychomycosis by Dr Jaishree Naidu and her colleagues from Jabalpur (Madhya Pradesh). The conidiophores are erect, short, often branched and pale. The conidiogenous cells are ampulliform to cylindrical, elongating with annellated zones on which conidia are produced.

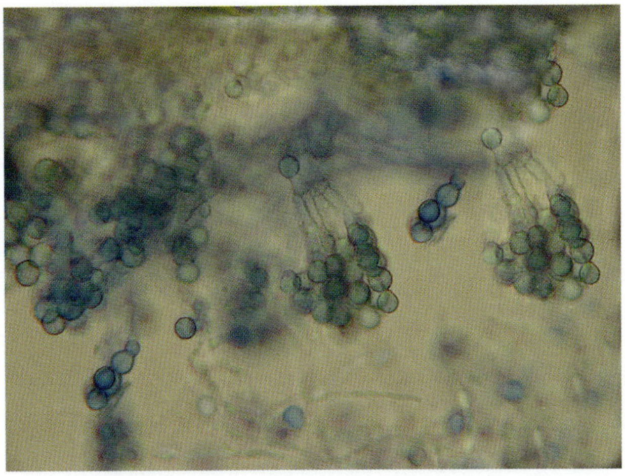

Fig. 31.18. *Scopulariopsis brevicaulis* showing penicillate phialides with chains of conidia (LCB × 400).

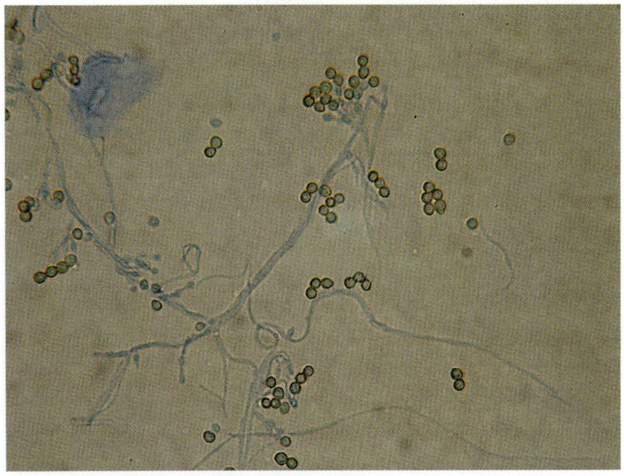

Fig. 31.19. Microscopic appearance of *Scopulariopsis brumptii* showing chains of conidia (LCB × 400).

Fig. 31.20. Grayish-white fluffy colonies of *Sarocladium kiliense.*

Fig. 31.21. Grayish-white membranous colonies of *Sarocladium kiliense.*

The conidia are sub-spherical, obovoidal and broadly clavate with truncate base, smooth-walled, punctate or verrucose, arranged in dry, basipetal chains, hyaline to pale brown **(Fig. 31.19)**.

The treatment of scopulariopsis is very similar to other hyaline infections. However, *Scopulariopsis* species are also found to be resistant to amphotericin B and flucytosine.

6. *Acremonium* Species

The genus *Acremonium* has currently about 100 species, which are environmental contaminants. They are saprotrophic being found in decaying vegetation, foodstuff and soil. Its species are usually slow growing and are initially compact and moist.

The colonies of *Acremonium* species are often moderately fast growing, flat at first becoming fluffy with age occasionally raised at the center, velvety to membranaceous and glabrous, turn whitish-yellow or pinkish to gray **(Figs. 31.20 and 31.21)**. The conidiophores are absent and phialidic conidiogenous cells are erect, gradually tapering towards apex and arising singly from creeping vegetative hyphae. The hyphae are fine and hyaline and produce mostly simple phialides. Their conidia are usually one-celled (ameroconidia), hyaline or pigmented, globose to cylindrical and mostly aggregated in slimy heads at

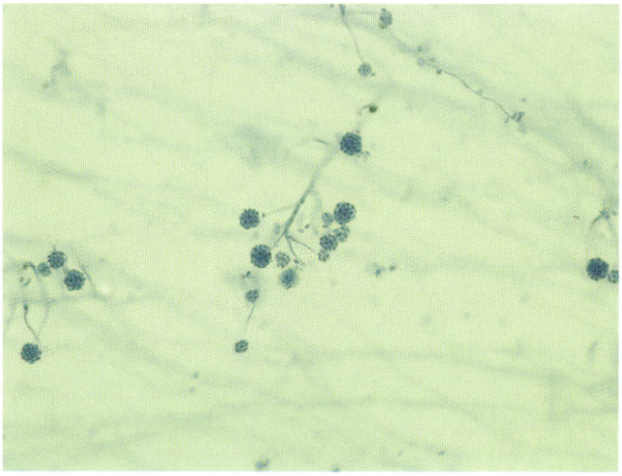

Fig. 31.22. Microscopic appearance of *Sarocladium kiliense* showing conidia in groups (LCB × 400).

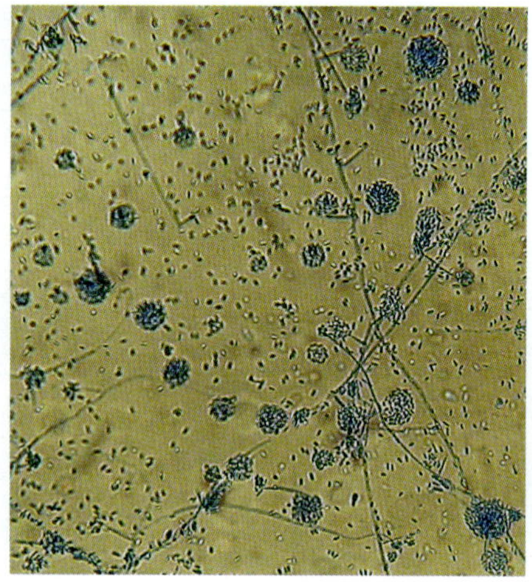

Fig. 31.23. Microscopic appearance of *Sarocladium kiliense* showing conidia in groups (LCB × 400).

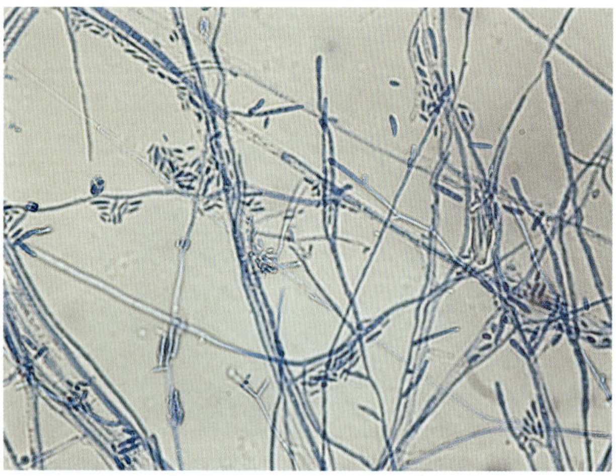

Fig. 31.24. Microscopic appearance of *Sarocladium kiliense* showing conidia in groups (LCB × 400).

apex of each phialide due to mucin matrix. Occasionally, chlamydospores may also be present.

This genus was previously called *Cephalosporium* due to accumulation of conidia as balls or heads, which is obsolete now. These fungi are mainly involved in production of antibiotics like cephalosporins, which was designated after its old name. *Acremonium kiliense*, the commonly encountered one is now known by its new genus name as *Sarocladium kiliense* **(Figs. 31.22 to 31.24)**.

These are not usually implicated as causative agents of human infections except very rarely in mycetoma, keratomycosis, postoperative endophthalmitis and onychomycosis. The clinical manifestations of hyalohyphomycosis caused by *Acremonium* include arthritis, osteomyelitis, peritonitis, endocarditis, pneumonia, cerebritis and subcutaneous infection. In addition to these, there are only few well-documented cases of hyalohyphomycosis caused by *Acremonium* species and *A.falciforme, S.kiliense* are usually encountered among these patients.

7. *Trichoderma* Species

Trichoderma species are soil-borne, filamentous fungi and have been known as harmless organisms to the humans. These are used in biotechnology as source of enzymes and antibiotics. Moreover, these are applied to agricultural crops as plant growth promoters and biofungicides. These saprotrophic ubiquitous fungi are phialosporous deuteromycetes and found growing naturally on wood and soil as well as other substrates. There are nine aggregates of *Trichoderma* species, which are been described on the basis of types of branching systems of the conidiophores, the phialide disposition and the character of the phialoconidia.

However, as recently emerging fungal pathogens, *Trichoderma* species have been isolated from skin, lungs, oral mucosa in necrotizing stomatitis and as causative agents of peritonitis in patients undergoing peritoneal dialysis. These are found to disseminate to the liver, brain and heart of the immunocompromised patients. Histopathologically, these may be confused with aspergillosis. There are six clinically significant species of genus *Trichoderma i.e. T.citrinoviride, T.harzianum, T.koningii, T.longibrachiatum, T.pseudokoningii* and *T.viride*, which are identified as etiological agents of various infections mainly in the immunocompromised hosts.

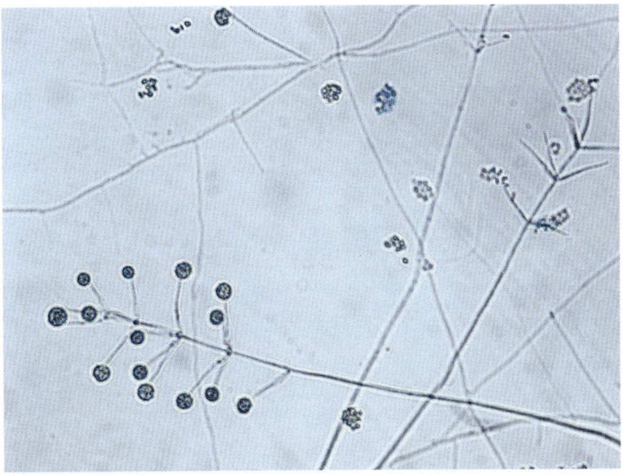

Fig. 31.25. Microscopic appearance of *Trichoderma* showing phialides and conidia (LCB × 400).

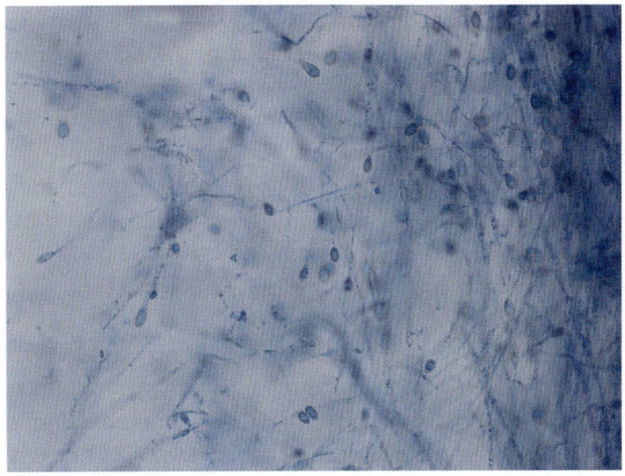

Fig. 31.26. *Chrysosporium* showing pyriform to clavate conidia with truncate bases (LCB × 400).

Out of these, the most common emerging species is *T.longibrachiatum*, which grows very fast. At 25°C, the colonies on SDA after 7 days are white initially then turn light to olivaceous green and woolly with distinct tufts of conidiophores in zonal radial pattern. There is striking lemon yellow diffusible pigment evident throughout agar on the reverse. The LCB mount shows branched conidiophores with pyramidal arrangement. The phialides are mostly single and flask-shaped bearing globose, semi-globose or ellipsoidal, greenish conidia on the tip **(Fig. 31.25)**. The cultures should be differentiated from *Penicillium* and *Paecilomyces* species.

Microscopically, hyphae of *T.longibrachiatum* are smooth, septate and branched and bear conidiophores with long main branches, which in turn produce relatively few, short, often slightly curved side branches at right angles, each terminating in a phialide. The phialides are bottle-shaped, mostly solitary, often somewhat inflated in the middle and bent at the apex and slightly constricted at the base and measured mostly 5–11 × 2–3 μm. The terminal phialides are more elongate, are not constricted at the base and are up to 14 μm long. The phialoconidia are smooth walled, ellipsoidal to cylindrical and green, measured 3.4–6.4 × 2.4–3.0 μm. Occasional smooth, thick-walled, subglobose to ellipsoidal chlamydoconidia that measure from 5 to 10 μm in diameter are also observed.

8. *Chrysosporium* Species

This fungal species is commonly found in the soil. Although it is considered as a contaminant, its presence in cultures and histological sections is evidence of its pathogenicity. Human infection if any exists, is limited to rare, poorly documented nail invasion. Some species are keratinophilic, thermophilic or cellulolytic. They were earlier called *Chrysosporium parvum* var. *parvum or Emmonsia parva* var. *parva* and *Chrysosporium parvum* var. *crescens or Emmonsia parva* var. *crescens*, respectively. Now, these two species are placed in genus *Emmonsia* causing adiaspiromycosis.

This is a moderately fast-growing fungus, producing usually spreading white to pale colony, often powdery or granular. The hyaline, smooth-walled, branched and fertile hyphae have terminal and lateral conidia. One-celled (ameroconidia) are produced directly on vegetative hyphae by non-specialized conidiogenous cells. These are typically pyriform to clavate with truncate bases (6–7 × 3.5–4 μm) and are formed either intercalary (arthroconidia), laterally often on pedicels or terminally **(Fig. 31.26)**. The intercalary conidia as well as chlamydospores occasionally present. This fungus closely resembles dermatophytes, *Histoplasma* and *Blastomyces* species.

Chrysosporium infection mostly affects the lungs but dissemination to the brain, sinus, skin, bone, heart, liver and kidney has been reported especially in immunocompromised patients, such as those undergoing transplantation or patients with diabetes or AIDS. Invasive pulmonary infection caused by *Chrysosporium articulatum* has been reported.

The pulmonary infection secondary to *Chrysosporium zonatum* in an immunocompetent individual and disseminated infection in a patient with chronic granulomatous disease are also seen. Its aleurioconidia are formed at the ends of short stalks that arise at an acute angle and are often slightly curved. The conidia also formed on the sides of the hyphae *i.e.* sessile or less commonly as intercalary (arthroconidia). The aleurioconidia are club-shaped to obovoid are warty, measuring 5–7 × 3–4 µm and arthroconidia as 7–11 × 2–3 µm **(Fig. 31.27)**.

In addition, localized infections, such as keratomycosis or osteomyelitis, may also occur in healthy individuals. It has been reported in a heart transplant patient who presented with a scaly erythematous, multinodular lesion above the right knee with a history of trauma during a camping trip. The prognosis is usually favorable but very poor in the setting of persistent profound immunosuppression. The primary cutaneous infection by this fungus is relatively rare. However, a case of primary cutaneous *Chrysosporium* infection following ear piercing has been reported.

It is important for clinicians to consider this condition in patients with slow-onset skin and soft tissue infection following cutaneous injury, particularly if they do not respond to conventional antibiotic treatment, even in an immunocompetent individuals.

The molecular characterization of reptile pathogens is currently known as members of the *Chrysosporium* anamorph of *Nannizziopsis vriesii* complex and relationship with some human-associated isolates.

9. *Beauveria* Species

Beauveria bassiana is a well-known pathogen in silkworms and insects in addition to being a soil fungus. This species has historical importance in that it was first microorganism to be recognized as disease-producing agent. It has been isolated from bronchial aspirates and lung tissue.

10. *Hormographiella* Species

There are two species of *Hormographiella i.e. H.aspergillata* and *H.verticillate.* Its teleomorphic state, *Coprinus cinereus*, is a basidiomycete with a typical mushroom form normally occurring in compost and sewage. *H.aspergillata* causes pulmonary infections in neutropenic patients. The cases were from the same geographical region *i.e.* Belgium and the Netherlands. Earlier it has rarely been associated with human infections. A case of endocarditis with fatal outcome and another with endophthalmitis have been reported. In other reports, two strains from skin lesions

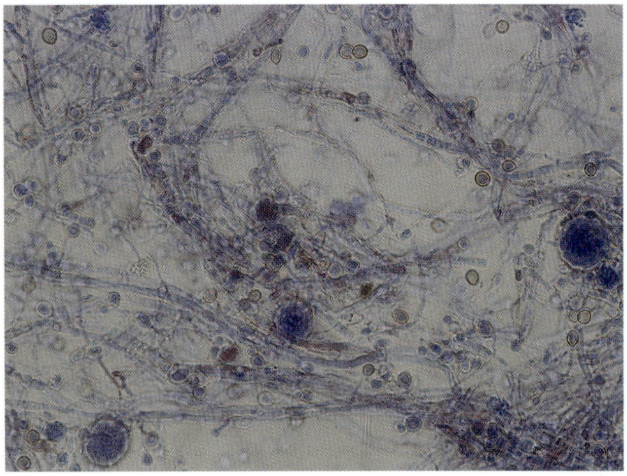

Fig. 31.27. *Chrysosporium zonatum* showing club-shaped to obovoid aleurioconidia (LCB × 400).

and one from a catheter have also been studied however, no causal relationship with the diseases was proven.

The identification of these filamentous basidiomycetes is difficult because key diagnostic features are often lacking and usually only arthroconidia are formed. The other fungi regularly cultured from clinical samples in which arthroconidia predominate, should be differentiated from *H.aspergillata*. These are *Geotrichum candidum*, *Trichosporon* species, *Geomyces pannorum*, *Malbranchea* species, *Arthrographis kalrae* and *Scytalidium* species. The conidiophores are only present in *Arthrographis*, *Geomyces* and *Hormographiella*. The characteristic for *Hormographiella* is that the conidiophore is clearly broader than the conidiogenous hyphae. The availability of molecular techniques may facilitate the identification of these fungal species in the future, which may lead to a better understanding of incidence of invasive infections.

11. *Schizophyllum commune*

Schizophyllum commune is an emerging filamentous holobasidiomycetes, which has recently been reported from a number of clinical entities, mainly sinusitis. It is a ubiquitous white rot fungus with worldwide distribution and usually thrives on dead and decaying matter. This macrofungus, commonly called as "split gill", is two centuries old and is gradually gaining lot of medical importance. Being omnipresent, it is one of the most studied mushrooms on the earth. When seen from above, it is just another small white bracket-like fungus but beneath the cap are radial gill-like folds, each of which is centrally split hence the name"split-gill" or "split leaves" is given to it. The genus name, *Schizophyllum,* has literal meaning as "split leaves" and its species name as *Commun-* means "common" or "in common" referring to the commonness of this species, relating to its omnipresence. The scientific name, *S.commune,* was originally given by Swedish mycologist, Elias Magnus Fries, in 1815.

Schizophyllum commune is classified under phylum Basidiomycota, class Agaricomycetes, order Agaricales and family *Schizophyllaceae.* Like most basidiomycetes, it has dikaryotic (binucleate) or monokaryotic stages in its life cycle. Dikaryotic isolates can be easily identified by the presence of clamp connections and spicules on hyphae and production of basidiocarps (mushroom), which are distinctive features of this organism. However, several strains of this basidiomycete isolated from human specimens fail to exhibit the macroscopic and microscopic morphological characteristics of the dikaryotic phase, when it is in monokaryotic stage. The microscopic examination of the monokaryotic mold shows hyaline, septate, branched hyphae of variable widths. However, for the induction of fruiting bodies in monokaryotic strains, exposure of culture plate to alternate cycles of light and dark, cultivating them on autoclaved decayed wooden bark pieces of *Syzygium cumini* (blackberry tree) and twigs of *Quercus acutissimia* (sawtooth oak) have been successfully tried.

Although it is very old fungus but the first human infection involving nail was reported in 1950 by Kligman. Since then there have been occasionally case reports, however, out of all reports, sinusitis is the commonest clinical presentation. Other entities include brain abscess, fungus ball in the lung, allergic bronchopulmonary mycosis (ABPM), chronic lung disease, meningitis, onychomycosis, mycotic keratitis, etc. This fungus has special predilection for respiratory tract hence it is the site of infection in most of the infections caused by *S.commune* while extrapulmonary infections are rarely reported. The infections by *S.commune* have been reported worldwide with maximum cases reported from Japan while there have been no case reports from African continent as yet.

The mycelial colonies of *S.commune* are spreading, woolly, whitish to pale, grayish-brown, soon forming macroscopically visible fruiting bodies. Some isolates may take up to even 12 weeks to form these fruiting bodies, which are sessile, kidney-shaped, lobed with "split-gills" on the lower side. The basidia bear four basidiospores on erect sterigmata. The basidiospores are hyaline, smooth-walled, elongate with lateral scar at lower end having size as 6–7 × 2–3 µm. On microscopic examination, hyphae are hyaline, septate with variable thickness and show clamp connections and curved lateral pegs (spicules) when they are present in their dikaryotic stage. As most of the clinical isolates are monokaryotic hence do not show either clamp connections or spicules making their identification slightly difficult. In such circumstances, any white, rapidly growing, sterile isolate showing substantial growth at 37°C with tolerance to benomyl, susceptibility to cycloheximide and distinctive bleach like odor, should be suspected as *S.commune.* Hence SDA should be supplemented with 10 µg/ml of benomyl and without cycloheximide. Further, molecular sequencing plays an important role in such a situation to identify this fungus.

Fig. 31.28. Grayish-white mycelial colonies of *Schizophyllum commune* on Sabouraud's dextrose agar.

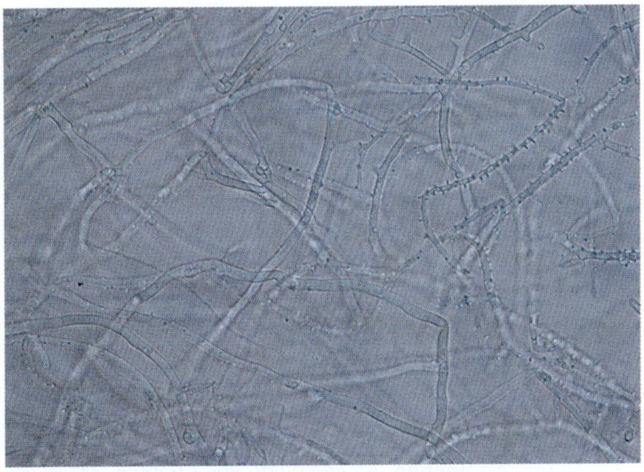

Fig. 31.29. Microscopic appearance of *Schizophyllum commune* showing curved lateral pegs *i.e.* spicules. (LCB × 400).

The morphological identification of *S.commune* as a basidiomycete fungus is not very difficult in dikaryotic isolates but in case of monokaryotic isolates, it becomes challenging due to devoid of clamp connections and/or spicules. A failure to exhibit morphological characteristics, typical of the dikaryotic phase in a clinical isolate, eventually leads to failure of morphological identification in the diagnostic microbiology laboratory resulting in misdiagnosis, incorrect therapy thereby hindering patient's recovery. Moreover, failed identification of this fungus entails its decreased frequency of reporting from clinical specimens. Hence it is one of the underreported fungal species in Medical Mycology and the disease load is much higher than reported **(*See* Figs. 31.28 and 31.29)**.

The other infections, which have been included in hyalohyphomycosis like trichosporosis, geotrichosis, etc. are already discussed in **Chapter 28** as Opportunistic Fungal Infections.

Treatment and Prophylaxis

The hyalohyphomycosis may be treated as the other hyaline fungal diseases.

Further Reading

1. Abarca ML, Martorell J, Castella G, *et al.* Cutaneous hyalohyphomycosis caused by a *Chrysosporium* species related to *Nannizziopsis vriesii* in two green iguanas (*Iguana iguana*). *Med Mycol.* 2008; **46:** 349-54.
2. Ahmad S, Zia S, Sarwari AR. *Scedosporium prolificans* endocarditis: Case report and review of literature. *W V Med J.* 2010; **106:** 24-6.
3. Ajello L. Hyalohyphomycosis and phaeohyphomycosis: Two global disease entities of public health importance. *Eur J Epidemiol.*1986; **2:** 243-51.
4. Akagi T, Kawamura C, Terasawa N, *et al.* Suspected pulmonary infection with *Trichoderma longibrachiatum* after allogeneic stem cell transplantation. *Intern Med.* 2017; **56:** 215-9.
5. Al-Hatmi AM, Hagen F, Menken SB, *et al.* Global molecular epidemiology and genetic diversity of *Fusarium*, a significant emerging group of human opportunists from 1958 to 2015. *Emerg Microbes Infect.* 2016; **5:** e124.
6. Al-Hatmi AM, van Diepeningen AD, Curfs-Breuker I, *et al.* Specific antifungal susceptibility profiles of opportunists in the *Fusarium fujikuroi* complex. *J Antimicrob Chemother.* 2015; **70:** 1068-71.
7. Al-Jehani H, Guiot MC, Torres C, Marcoux J. *Scedosporium* cerebral abscesses after extra-corporeal membrane oxygenation. *Can J Neurol Sci.* 2010; **37:** 671-6.
8. Alanio A, Brethon B, Feuilhade de Chauvin M, *et al.* Invasive pulmonary infection due to *Trichoderma longibrachiatum* mimicking invasive aspergillosis in a neutropenic patient successfully treated with voriconazole combined with caspofungin. *Clin Infect Dis.* 2008; **46:** e116-8.
9. Alanio A, Brethon B, Feuilhade de Chauvin M, *et al.* Invasive pulmonary infection due to *Trichoderma longibrachiatum* mimicking invasive aspergillosis in a neutropenic patient successfully treated with voriconazole combined with caspofungin. *Clin Infect Dis.* 2008; **46:** e116-8.
10. Alastruey-Izquierdo A, Cuenca-Estrella M, Monzon A, *et al.* Prevalence and susceptibility testing of new species of *Pseudallescheria* and *Scedosporium* in a collection of clinical mold isolates. *Antimicrob Agents Chemother.*2007; **51:** 748-51.
11. Alvarez E, Sanhueza C. New record of *Scedosporium dehoogii* from Chile: Phylogeny and susceptibility profiles to classic and novel putative antifungal agents. *Rev Iberoam Micol.* 2016; **33:** 224-9.
12. Amirrajab N, Aliyali M, Mayahi S, *et al.* Co-infection of invasive pulmonary aspergillosis and cutaneous *Fusarium* infection in a patient with pyoderma gangrenosum. *J Res Med Sci.* 2015; **20:** 199-203.
13. Antas PR, Brito MM, Peixoto E, *et al.* Neglected and emerging fungal infections: Review of hyalohyphomycosis by *Paecilomyces lilacinus* focusing in disease burden, *in vitro* antifungal susceptibility and management. *Microbes Infect.* 2012; **14:** 1-8.
14. Atty C, Alagiozian-Angelova VM, Kowal-Vern A. Black plaques and white nodules in a burn patient. *Fusarium* and mucormycosis. *JAMA Dermatol.* 2014; **150:** 1355-6.
15. Aviles-Robles M, Gomez-Ponce C, Resendiz-Sanchez J, *et al.* Disseminated penicilliosis due to *Penicillium chrysogenum* in a pediatric patient with Henoch-Schönlein syndrome. Int J Infect Dis. 2016; **51:** 78-80.
16. Azor M, Gene J, Cano J, *et al.* Less-frequent *Fusarium* species of clinical interest: Correlation between morphological and molecular identification and antifungal susceptibility. *J Clin Microbiol.* 2009; **47:** 1463-8.
17. Bansal Y, Chander J, Kaistha N, Singla N, Sood S, van Diepeningen AD. *Fusarium sacchari*, a cause of mycotic keratitis among sugarcane farmers - A series of four cases from North India. *Mycoses.* 2016; **59:** 705-9.
18. Beier F, Kittan N, Holzmann T, *et al.* Successful treatment of *Scedosporium apiospermum* soft tissue abscess with caspofungin and voriconazole in a severely immunocompromised patient with acute myeloid leukemia. *Transpl Infect Dis.* 2010; **12:** 538-42.
19. Bernal-Martinez L, Buitrago MJ, Castelli MV, *et al.* Detection of invasive infection caused by *Fusarium solani* and non-*Fusarium solani* species using a duplex quantitative PCR-based assay in a murine model of fusariosis. *Med Mycol.* 2012; **50:** 270-5.
20. Bernhardt A, Seibold M, Rickerts V, *et al.* Cluster analysis of *Scedosporium boydii* infections in a single hospital. *Int J Med Microbiol.* 2015; **305:** 724-8.
21. Bhagavatula S, Vale L, Evans J, *et al. Scedosporium prolificans* osteomyelitis following penetrating injury: A case report. *Med Mycol Case Rep.* 2014; **4:** 26-9.

22. Borghi E, Iatta R, Manca A, *et al.* Chronic airway colonization by *Scedosporium apiospermum* with a fatal outcome in a patient with cystic fibrosis. *Med Mycol.* 2010; **48:** S108-13.
23. Boyce Z, Collins N. *Scedosporium apiospermum:* An unreported cause of fungal sporotrichoid-like lymphocutaneous infection in Australia and review of the literature. *Australas J Dermatol.* 2015; **56:** e39-42.
24. Brito MM, Lima Mda S, Morgado FN, *et al.* Characteristics of *Paecilomyces lilacinus* infection comparing immunocompetent with immunosuppressed murine model. *Mycoses.* 2011; **54:** e513-21.
25. Buzina W, Feierl G, Haas D, *et al.* Lethal brain abscess due to the fungus *Scedosporium apiospermum* (teleomorph *Pseudallescheria boydii*) after a near-drowning incident: Case report and review of the literature. *Med Mycol.* 2006; **44:** 473-7.
26. Cabanes FJ, Sutton DA, Guarro J. *Chrysosporium*-related fungi and reptiles: A fatal attraction. *PLoS Pathog.* 2014; **10:** e1004367. PMID: 25329049.
27. Caggiano G, Cantisani P, Rolli M, *et al.* The importance of a proper aetiological diagnosis in the management of patients with invasive mycoses: A case report of a brain abscess by *Scedosporium apiospermum. Mycopathologia.* 2011; **172:** 317-22.
28. Caira M, Girmenia C, Valentini CG, *et al.* Scedosporiosis in patients with acute leukemia: a retrospective multicenter report. *Haematologica.* 2008; **93:** 104-10.
29. Cardoso JC, Serra D, Cardoso R, *et al.* Cutaneous *Pseudallescheria boydii* infection in a renal transplant patient: A case report. *Dermatol Online J.* 2009; **15:** 8.
30. Castelli MV, Alastruey-Izquierdo A, Cuesta I, *et al.* Susceptibility testing and molecular classification of *Paecilomyces* spp. *Antimicrob Agents Chemother.* 2008; **52:** 2926-8.
31. Caston JJ, Linares MJ, Rivero A, *et al.* Clinical differences between invasive pulmonary infection by *Scedosporium apiospermum* and invasive pulmonary aspergillosis. *Mycoses.* 2011; **54:** e468-73.
32. Chan JF, Teng JL, Li IW, *et al.* Fatal empyema thoracis caused by *Schizophyllum commune* with cross-reactive cryptococcal antigenemia. *J Clin Microbiol.* 2014; **52:** 683-7.
33. Chen M, Houbraken J, Pan W, *et al.* Pulmonary fungus ball caused by *Penicillium capsulatum* in a patient with Type 2 diabetes: A case report. *BMC Infect Dis.* 2013; **13:** 496.
34. Chander J, Singla N, Gulati N, Sood S. *Fusarium sacchari:* A cause of exogenous fungal endophthalmitis: First case report and review of literature. *Mycopathologia.* 2011; **171:** 431-4.
35. Chandrakar S, Dias M, Shetty P, *et al.* Isolation and characterization of an unknown *Chrysosporium* sp. producing subcutaneous mycosis in an immunocompromised patient. *Mycopathologia.* 2016; **181:** 115-8.
36. Chang YC, Stewart C, Chu EY, *et al.* Rare angio-invasive fungal infection in association with leukemia cutis. *Cutis.* 2015; **95:** 332-5.
37. Chen M, Houbraken J, Pan W, *et al.* Pulmonary fungus ball caused by *Penicillium capsulatum* in a patient with Type 2 diabetes: A case report. *BMC Infect Dis.* 2013; **13:** 496.
38. Chen M, Zeng J, De Hoog GS, *et al.* The 'species complex' issue in clinically relevant fungi: A case study in *Scedosporium apiospermum. Fungal Biol.* 2016; **120:** 137-46.
39. Chen QX, Li CX, Huang WM, *et al.* Cutaneous hyalohyphomycosis caused by *Fusarium subglutinans. Eur J Dermatol.* 2010; **20:** 526-7.
40. Chowdhary A, Agarwal K, Kathuria S, *et al.* Clinical significance of filamentous basidiomycetes illustrated by isolates of the novel opportunist *Ceriporia lacerata* from the human respiratory tract. *J Clin Microbiol.* 2013; **51:** 585-90.
41. Chowdhary A, Agarwal K, Kathuria S, *et al.* First human case of pulmonary fungal ball due to a *Perenniporia* species (a basidiomycete). *J Clin Microbiol.* 2012; **50:** 3786-91.
42. Chowdhary A, Kathuria S, Agarwal K, *et al.* Recognizing filamentous basidiomycetes as agents of human disease: A review. *Med Mycol.* 2014; **52:** 782-97.
43. Chowdhary A, Kathuria S, Agarwal K, *et al.* Voriconazole-resistant *Penicillium oxalicum:* An emerging pathogen in immunocompromised hosts. *Open Forum Infect Dis.* 2014; **1:** ofu029. PMID: 25734109.
44. Chowdhary A, Randhawa HS, Gaur SN, *et al. Schizophyllum commune* as an emerging fungal pathogen: A review and report of two cases. *Mycoses.* 2013; **56:** 1-10.
45. Cocchi S, Codeluppi M, Venturelli C, *et al. Fusarium verticillioides* fungemia in a liver transplantation patient: Successful treatment with voriconazole. *Diagn Microbiol Infect Dis.* 2011; **71:** 438-41.
46. Collado C, Medina L, Zorraquino A, *et al.* Cutaneous fusariosis by a species of the *Fusarium dimerum* species complex in a patient with acute myeloblastic leukemia. *Rev Iberoam Micol.* 2013; **30:** 119-21.
47. Cortez KJ, Roilides E, Quiroz-Telles F, *et al.* Infections caused by *Scedosporium* spp. *Clin Microbiol Rev.* 2008; **21:** 157-97.
48. Coulibaly O, Marinach-Patrice C, Cassagne C, *et al. Pseudallescheria/Scedosporium* complex species identification by matrix-assisted laser desorption ionization time-of-flight mass spectrometry. *Med Mycol.* 2011; **49:** 621-6.
49. Dalyan Cilo B, Al-Hatmi AM, Seyedmousavi S, *et al.* Emergence of fusarioses in a university hospital in Turkey during a 20-year period. *Eur J Clin Microbiol Infect Dis.* 2015; **34:** 1683-91.
50. Das S, Saha R, Dar SA, Ramachandran VG. *Acremonium* species: A review of the etiological agents of emerging hyalohyphomycosis. *Mycopathologia.* 2010; **170:** 361-75.
51. Day S, Lalitha P, Haug S, *et al.* Activity of antibiotics against *Fusarium* and *Aspergillus. Br J Ophthalmol.* 2009; **93:** 116-9.
52. de Almeida Junior JN, Ibrahim KY, Del Negro GM, *et al. Rhizopus arrhizus* and *Fusarium solani* concomitant infection in an immunocompromised host. *Mycopathologia.* 2016; **181:** 125-9.
53. Debourgogne A, de Hoog S, Lozniewski A, *et al.* Amphotericin B and voriconazole susceptibility profiles for the *Fusarium solani* species complex: Comparison between the E-test and CLSI M38-A2 microdilution methodology. *Eur J Clin Microbiol Infect Dis.* 2012; **31:** 615-8.

54. del Alcazar E, Jaka A, Camino N, *et al.* Fever and skin lesions in an immunocompromised patient. *Clin Exp Dermatol.* 2015; **40:** 219-21.
55. Demitsu T, Nagashima K, Okabe T, *et al.* Subcutaneous hyalohyphomycosis due to *Purpureocillium lilacinum* in an immunocompromised patient after renal transplantation. *J Dermatol.* 2017; **44:** 725-6.
56. Detporntewan P, Chindamporn A, Worasilchai N, *et al.* A case of invasive pulmonary infection caused by novel species of *Perenniporia. Mycoses.* 2014; **57:** 703-5.
57. Donta B, Naik DD, Mali BN, *et al.* "Fruiting body" *Penicillium* species: Papanicolaou stained conventional cervical smear findings. *Diagn Cytopathol.* 2010; **38:** 34-5.
58. Dony A, Perpoint T, Ducastelle S, *et al.* Disseminated fusariosis with immune reconstitution syndrome and cracking mycotic aortic aneurysm in a 55-year-old patient with acute myeloid leukaemia. *BMJ Case Rep.* 2013; pii: bcr2013201050. PMID: 24114601.
59. Douglas AP, Chen SC, Slavin MA. Emerging infections caused by non-*Aspergillus* filamentous fungi. *Clin Microbiol Infect.* 2016; **22:** 670-80.
60. Dutta P, Premkumar A, Chakrabarti A, *et al. Fusarium falciforme* infection of foot in a patient with Type 2 diabetes mellitus: A case report and review of the literature. *Mycopathologia.* 2013; **176:** 225-32.
61. Edupuganti S, Rouphael N, Mehta A, *et al. Fusarium falciforme* vertebral abscess and osteomyelitis: Case report and molecular classification. *J Clin Microbiol.* 2011; **49:** 2350-3.
62. Elad D. Infections caused by fungi of the *Scedosporium / Pseudallescheria* complex in veterinary species. *Vet J.* 2011; **187:** 33-41.
63. Elm MK, Ahmed A, Goksel D, *et al.* Cutaneous and systemic infection with *Scedosporium apiospermum. Cutis.* 2009; **84:** 275-8.
64. Etienne KA, Roe CC, Smith RM, *et al.* Whole-Genome Sequencing to determine origin of multinational outbreak of *Sarocladium kiliense* bloodstream infections. *Emerg Infect Dis.* 2016; **22:** 476-81.
65. Fakharian A, Dorudinia A, Alavi Darazam I, *et al. Acremonium* pneumonia: Case report and literature review. *Tanaffos (Respiration).* 2015; **14:** 156-60.
66. Fan F, Lui G, Ip M, *et al.* Foot ulcer, skin nodules and blurred vision. *Clin Infect Dis.* 2012; **54:** 706 & 736-7.
67. Fanci R, Pini G, Bartolesi AM, *et al.* Refractory disseminated fusariosis by *Fusarium verticillioides* in a patient with acute myeloid leukaemia relapsed after allogeneic hematopoietic stem cell transplantation: A case report and literature review. *Rev Iberoam Micol.* 2013; **30:** 51-3.
68. Feramisco JD, Hsiao JL, Fox LP, *et al.* Angioinvasive *Fusarium* and concomitant *Enterococcus* infection arising in association with leukemia cutis. *J Cutan Pathol.* 2011; **38:** 926-9.
69. Fernandez Guerrero ML, Askari E, Prieto E, *et al.* Emerging infectious endocarditis due to *Scedosporium prolificans:* A model of therapeutic complexity. *Eur J Clin Microbiol Infect Dis.* 2011; **30:** 1321-4.
70. Fernandez-Silva F, Capilla J, Mayayo E, *et al.* Combination therapy in the treatment of experimental invasive fungal infection by *Sarocladium (Acremonium) kiliense. Int J Antimicrob Agents.* 2014; **44:** 136-9.
71. Forghieri F, Rossi G, Potenza L, *et al.* Splenic hyalohyphomycosis, molecularly and immunologically consistent with invasive aspergillosis, in a patient with acute lymphoblastic leukemia. *Am J Hematol.* 2010; **85:** 188-9.
72. Garcia RR, Min Z, Narasimhan S, *et al. Fusarium* brain abscess: Case report and literature review. *Mycoses.* 2015; **58:** 22-6.
73. Garcia-Ruiz JC, Olazabal I, Adan Pedroso RM, *et al.* Disseminated fusariosis and hematologic malignancies, a still devastating association. Report of three new cases. *Rev Iberoam Micol.* 2015; **32:** 190-6.
74. Garnica M, da Cunha MO, Portugal R, *et al.* Risk factors for invasive fusariosis in patients with acute myeloid leukemia and in hematopoietic cell transplant recipients. *Clin Infect Dis.* 2015; **60:** 875-80.
75. Geltner C, Lass-Florl C, Bonatti H, et al. Invasive pulmonary mycosis due to *Penicillium chrysogenum:* A new invasive pathogen. *Transplantation.* 2013; **95:** e21-3.
76. Georgiadou SP, Velegraki A, Arabatzis M, *et al.* Cluster of *Fusarium verticillioides* bloodstream infections among immunocompetent patients in an internal medicine department after reconstruction works in Larissa, Central Greece. *J Hosp Infect.* 2014; **86:** 267-71.
77. Giraldo A, Gene J, Sutton DA, *et al.* Phylogeny of *Sarocladium (Hypocreales). Persoonia.* 2015; **34:** 10-24.
78. Giraud S, Favennec L, Bougnoux ME, *et al. Rasamsonia argillacea* species complex: Taxonomy, pathogenesis and clinical relevance. *Future Microbiol.* 2013; **8:** 967-78.
79. Gopinath M, Cherian A, Baheti NN, *et al.* An elusive diagnosis: *Scedosporium apiospermum* infection after near-drowning. *Ann Indian Acad Neurol.* 2010; **13:** 213-5.
80. Gottesman-Yekutieli T, Shwartz O, Edelman A, *et al. Pseudallescheria boydii* infection of a prosthetic hip joint - An uncommon infection in a rare location. *Am J Med Sci.* 2011; **342:** 250-3.
81. Grenouillet F, Botterel F, Crouzet J, *et al. Scedosporium prolificans:* An emerging pathogen in France? *Med Mycol.* 2009; **47:** 343-50.
82. Guarro J, Mendiratta DK, De Sequeira H, *et al. Acrophialophora fusispora*: an emerging agent of human mycoses. A report of 3 new clinical cases. *Diagn Microbiol Infect Dis.* 2007; **59:** 85-8.
83. Guarro J. Fusariosis, a complex infection caused by a high diversity of fungal species refractory to treatment. *Eur J Clin Microbiol Infect Dis.* 2013; **32:** 1491-500.
84. Gutierrez Paredes EM, Gamez Perez L, Gonzalez Rodriguez AJ, *et al.* Disseminated fusariosis in immunocompromised patients. *Eur J Dermatol.* 2011; **21:** 753-5.
85. Halliday CL, Chen SC, Kidd SE, *et al.* Antifungal susceptibilities of non-*Aspergillus* filamentous fungi causing invasive infection in Australia: Support for current antifungal guideline recommendations. *Int J Antimicrob Agents.* 2016; **48:** 453-8.

86. Harun A, Blyth CC, Gilgado F, *et al.* Development and validation of a multiplex PCR for detection of *Scedosporium* spp. in respiratory tract specimens from patients with cystic fibrosis. *J Clin Microbiol.* 2011; **49:** 1508-12.
87. Harun A, Gilgado F, Chen SC, *et al.* Abundance of *Pseudallescheria/Scedosporium* species in the Australian urban environment suggests a possible source for scedosporiosis including the colonization of airways in cystic fibrosis. *Med Mycol.* 2010; **48:** S70-6.
88. Harun A, Serena C, Gilgado F, *et al.* *Scedosporium aurantiacum* is as virulent as *S.prolificans* and shows strain-specific virulence differences, in a mouse model. *Med Mycol.* 2010; **48:** S45-51.
89. Heath CH, Slavin MA, Sorrell TC, *et al.* Population-based surveillance for scedosporiosis in Australia: Epidemiology, disease manifestations and emergence of *Scedosporium aurantiacum* infection. *Clin Microbiol Infect.* 2009; **15:** 689-93.
90. Hilmioglu S, Metin DY, Tasbakan M, *et al.* Skin infection on both legs caused by *Acremonium strictum* (case report). *Ann Saudi Med.* 2015; **35:** 406-8.
91. Horn DL, Freifeld AG, Schuster MG, *et al.* Treatment and outcomes of invasive fusariosis: Review of 65 cases from the PATH alliance registry. *Mycoses.* 2014; **57:** 652-8.
92. Houbraken J, Giraud S, Meijer M, *et al.* Taxonomy and antifungal susceptibility of clinically important *Rasamsonia* species. *J Clin Microbiol.* 2013; **51:** 22-30.
93. Houbraken J, Verweij PE, Rijs AJ, *et al.* Identification of *Paecilomyces variotii* in clinical samples and settings. *J Clin Microbiol.* 2010; **48:** 2754-61.
94. Huang CY, Sun PL, Tseng HK. Cutaneous hyalohyphomycosis caused by *Paecilomyces lilacinus* successfully treated by oral voriconazole and nystatin packing. *Mycopathologia.* 2011; **172:** 141-5.
95. Hughart R, Katz PJ, Ray TL, *et al.* Paronychia and necrotic nodules. *Arch Dermatol.* 2011; **147:** 1317-22.
96. Ikeda I, Ohno T, Ohno H, *et al.* Case of *Fusarium* paronychia successfully treated with occlusive dressing of antifungal cream. *J Dermatol.* 2014; **41:** 340-2.
97. Inano S, Kimura M, Iida J, *et al.* Combination therapy of voriconazole and terbinafine for disseminated fusariosis: Case report and literature review. *J Infect Chemother.* 2013; **19:** 1173-80.
98. Ioakimidou A, Vyzantiadis TA, Sakellari I, *et al.* An unusual cluster of *Acremonium kiliense* fungaemias in a haematopoietic cell transplantation unit. *Diagn Microbiol Infect Dis.* 2013; **75:** 313-6.
99. Ishiguro T, Takayanagi N, Kagiyama N, *et al.* Clinical characteristics of biopsy-proven allergic bronchopulmonary mycosis: Variety in causative fungi and laboratory findings. *Intern Med.* 2014; **53:** 1407-11.
100. Ishiguro T, Takayanagi N, Tokunaga D, *et al.* Pulmonary *Schizophyllum commune* infection developing mucoid impaction of the bronchi. *Yale J Biol Med.* 2007; **80:** 105-11.
101. Iwen PC, Schutte SD, Florescu DF, *et al.* Invasive *Scopulariopsis brevicaulis* infection in an immunocompromised patient and review of prior cases caused by *Scopulariopsis* and *Microascus* species. *Med Mycol.* 2012; **50:** 561-9.
102. Jagielski T, Sandoval-Denis M, Yu J, *et al.* Molecular taxonomy of *Scopulariopsis*-like fungi with description of new clinical and environmental species. *Fungal Biol.* 2016; **120:** 586-602.
103. Jain D, Oberoi JK, Shahi SK, *et al.* *Scopulariopsis brevicaulis* infection of prosthetic valve resembling aspergilloma on histopathology. *Cardiovasc Pathol.* 2011; **20:** 381-3.
104. Johnson LS, Shields RK, Clancy CJ. Epidemiology, clinical manifestations and outcomes of *Scedosporium* infections among solid organ transplant recipients. *Transpl Infect Dis.* 2014; **16:** 578-87.
105. Junior MC, de Moraes Arantes A, Silva HM, *et al.* *Acremonium kiliense:* Case report and review of published studies. *Mycopathologia.* 2013; **176:** 417-21.
106. Kadri SS, Remy KE, Strich JR, *et al.* Role of granulocyte transfusions in invasive fusariosis: Systematic review and single-center experience. *Transfusion.* 2015; **55:** 2076-85.
107. Kang Y, Li L, Zhu J, *et al.* Identification of *Fusarium* from a patient with fungemia after multiple organ injury. *Mycopathologia.* 2013; **176:** 151-5.
108. Kantarcioglu AS, Guarro J, de Hoog GS. Central nervous system infections by members of the *Pseudallescheria boydii* species complex in healthy and immunocompromised hosts: Epidemiology, clinical characteristics and outcome. *Mycoses.* 2008; **51:** 275-90.
109. Kantarcioglu AS, Summerbell RC, Sutton DA, *et al.* A dark strain in the *Fusarium solani* species complex isolated from primary subcutaneous sporotrichoid lesions associated with traumatic inoculation via a rose bush thorn. *Med Mycol.* 2010; **48:** 103-9.
110. Katz T, Wasiak J, Cleland H, *et al.* Incidence of non-candidal fungal infections in severe burn injury: An Australian perspective. *Burns.* 2014; **40:** 881-6.
111. Kaur R, Maheshwari M. Hyperkeratotic warty skin lesion of foot caused by *Fusarium oxysporum. Indian J Dermatol.* 2013; **58:** 159.
112. Kawashima N, Yoshida N, Matsushita N, *et al.* Intra-articular injection of voriconazole for *Fusarium solani* arthritis after bone marrow transplantation. *J Infect.* 2012; **65:** 366-7.
113. Kebabcı N, van Diepeningen AD, Ener B, *et al.* Fatal breakthrough infection with *Fusarium andiyazi:* New multi-resistant aetiological agent cross-reacting with *Aspergillus* galactomannan enzyme immunoassay. *Mycoses.* 2014; **57:** 249-55.
114. Kelly M, Stevens R, Konecny P. *Lomentospora prolificans* endocarditis - Case report and literature review. *BMC Infect Dis.* 2016; **16:** 36. PMID: 26822980.
115. Keshtkar-Jahromi M, McTighe AH, Segalman KA, *et al.* Unusual case of cutaneous and synovial *Paecilomyces lilacinus* infection of hand successfully treated with voriconazole and review of published literature. *Mycopathologia.* 2012; **174:** 255-8.
116. Keskar VS, Wanjare S, Jamale TE, *et al.* Subcutaneous hyalohyphomycosis caused by *Fusarium* in a kidney transplant recipient. *Ren Fail.* 2014; **36:** 1129-32.

117. Khan Z, Ahmad S, Al-Ghimlas F, *et al.* *Purpureocillium lilacinum* as a cause of cavitary pulmonary disease: A new clinical presentation and observations on atypical morphologic characteristics of the isolate. *J Clin Microbiol.* 2012; **50:** 1800-4.
118. Kilvington S, Lam A, Nikolic M, *et al.* Resistance and growth of *Fusarium* species in contact lens disinfectant solutions. *Optom Vis Sci.* 2013; **90:** 430-8.
119. Kim MS, Lee HM, Sung HS, *et al.* Breakthrough disseminated fusariosis in an immunocompromised patient on voriconazole therapy. *Int J Dermatol.* 2012; **51:** 621-3.
120. Kimura M, Maenishi O, Ito H, *et al.* Unique histological characteristics of *Scedosporium* that could aid in its identification. *Pathol Int.* 2010; **60:** 131-6.
121. Kordalewska M, Jagielski T, Brillowska-Dabrowska A. Rapid assays for specific detection of fungi of *Scopulariopsis* and *Microascus* genera and *Scopulariopsis brevicaulis* species. *Mycopathologia.* 2016; **181:** 465-74.
122. Kravitz JN, Steed LL, Judson MA. Intracavitary voriconazole for the treatment of hemoptysis complicating *Pseudallescheria angusta* pulmonary mycetomas in fibrocystic sarcoidosis. *Med Mycol.* 2011; **49:** 198-201.
123. Kuruvilla TS, Dias M. *Fusarium Solani:* A causative agent of skin and nail infections. *Indian J Dermatol.* 2012; **57:** 308-9.
124. Labois A, Gray C, Lepretre S. Successful treatment of disseminated fusariosis with voriconazole in an acute lymphoblastic leukaemia patient. *Mycoses.* 2011; **54 (Suppl. 4):** 8-11.
125. Lackner M, de Hoog GS. *Scedosporium* species: Emerging agents of systemic disease. 2011; *J Invasive Fungal Infect.* 2011; **5:** 43-7.
126. Lackner M, de Hoog GS. *Parascedosporium* and its relatives: Phylogeny and ecological trends. *IMA Fungus.* 2011; **2:** 39-48.
127. Lackner M, Hoog GS de, Yang L, *et al.* Proposed nomenclature for *Pseudallescheria, Scedosporium* and related genera. *Fungal Divers.* 2014; **67:** 1-10.
128. Lackner M, Klaassen CH, Meis JF, *et al.* Molecular identification tools for sibling species of *Scedosporium* and *Pseudallescheria. Med Mycol.* 2012; **50:** 497-508.
129. Lackner M, Rezusta A, Villuendas MC, *et al.* Infection and colonisation due to *Scedosporium* in northern Spain. An *in vitro* antifungal susceptibility and molecular epidemiology study of 60 isolates. *Mycoses.* 2011; **54 (Suppl. 3):** 12-21.
130. Lam SM, Lau AC, Ma MW, *et al.* *Pseudallescheria boydii or Aspergillus fumigatus* in a lady with an unresolving lung infiltrate and a literature review. *Respirology.* 2008; **13:** 478-80.
131. Lavergne RA, Cassaing S, Nocera T, *et al.* Simultaneous cutaneous infection due to *Paecilomyces lilacinus* and *Alternaria* in a heart transplant patient. *Transpl Infect Dis.* 2012; **14:** e156-60.
132. Le Clech L, Hutin P, Le Gal S, *et al.* Skin nodules in a patient with acute lymphoblastic leukaemia. *BMJ Case Rep.* 2014 pii: bcr2013010481. PMID: 24408938.
133. Lennon PA, Cooper CR Jr, Salkin IF, *et al.* Ribosomal DNA internal transcribed spacer analysis supports synonymy of *Scedosporium inflatum* and *Lomentospora prolificans. J Clin Microbiol.*1994; **32:** 2413-6.
134. Li FG, Yang YP, Li W, *et al.* Spontaneous remission of subcutaneous scedosporiosis caused by *Scedosporium dehoogii* in a psoriatic patient. *Mycopathologia.* 2017; **182:** 561-7.
135. Lin D, Kamili Q, Lai S, *et al.* Cerebral *Scedosporium apiospermum* infection presenting with intestinal manifestations. *Infection.* 2013; **41:** 723-6.
136. Lindsley MD, Guarro J, Khairy RN, *et al.* *Pseudallescheria fusoidea,* A new cause of osteomyelitis. *J Clin Microbiol.* 2008; **46:** 2141-3.
137. Litvinov N, da Silva MT, van der Heijden IM, *et al.* An outbreak of invasive fusariosis in a children's cancer hospital. *Clin Microbiol Infect.* 2015; **21:** 268.e1-7.
138. Liu JY, Chen WT, Ko BS, *et al.* Combination antifungal therapy for disseminated fusariosis in immunocompromised patients: A case report and literature review. *Med Mycol.* 2011; **49:** 872-8.
139. Lu Q, Gerrits van den Ende AH, Bakkers JM, *et al.* Identification of *Pseudallescheria* and *Scedosporium* species by three molecular methods. *J Clin Microbiol.* 2011; **49:** 960-7.
140. Makino K, Fukushima S, Maruo K, *et al.* Cutaneous hyalohyphomycosis by *Scedosporium apiospermum* in an immunocompromised patient. *Mycoses.* 2011; **54:** 259-61.
141. Mapelli ET, Cerri A, Borghi E, *et al.* A rare case of subcutaneous fusariosis in an immunocompetent patient. *G Ital Dermatol Venereol.* 2017; **152:** 178-80.
142. Matsumoto Y, Oh-I T, Nagai A, *et al.* Case of cutaneous *Scedosporium apiospermum* infection successfully treated with voriconazole. *J Dermatol.* 2009; **36:** 98-102.
143. Mazumder SA, Cleveland KO, Norwood J. *Scedosporium apiospermum* infection presenting as a pneumothorax and cavitary lung lesions in a patient with acquired immune deficiency syndrome. *South Med J.* 2010; **103:** 960-2.
144. Mehta A, Bellam N. Disseminated fusariosis during acute myelogenous leukemia induction treatment. *Blood.* 2014; **123:** 3379.
145. Mehta V. Cutaneous hyalohyphomycosis in an immunocompetent host. *Indian J Dermatol.* 2013; **58:** 411.
146. Mijiti J, Pan B, de Hoog S, *et al.* Severe chromoblastomycosis-like cutaneous infection caused by *Chrysosporium keratinophilum. Front Microbiol.* 2017; **8:** 83.
147. Mikulska M, Furfaro E, Del Bono V, *et al.* Galactomannan testing might be useful for early diagnosis of fusariosis. *Diagn Microbiol Infect Dis.* 2012; **72:** 367-9.
148. Miossec C, Morio F, Lepoivre T, *et al.* Fatal invasive infection with fungemia due to *Microascus cirrosus* after heart and lung transplantation in a patient with cystic fibrosis. *J Clin Microbiol.* 2011; **49:** 2743-7.
149. Miraldi F, Anile M, Ruberto F, *et al.* *Scedosporium apiospermum* atrial mycetomas after lung transplantation for cystic fibrosis. *Transpl Infect Dis.* 2012; **14:** 188-91.
150. Morel LN, Cid PM, De Celada RM, *et al.* Disseminated fusariosis in a pediatric population. *Pediatr Dermatol.* 2013; **30:** e255-6.

151. Muhammed M, Anagnostou T, Desalermos A, *et al.* *Fusarium* infection: Report of 26 cases and review of 97 cases from the literature. *Medicine (Baltimore).* 2013; **92:** 305-16.
152. Muhammed M, Coleman JJ, Carneiro HA, *et al.* The challenge of managing fusariosis. *Virulence.* 2011; **2:** 91-6.
153. Muraosa Y, Schreiber AZ, Trabasso P, *et al.* Development of cycling probe-based real-time PCR system to detect *Fusarium* species and *Fusarium solani* species complex (FSSC). *Int J Med Microbiol.* 2014; **304:** 505-11.
154. Naidu J, Singh SM, Pouranik M. Onychomycosis caused by *Scopulariopsis brumptii. Mycopathologia.* 1991; **113:** 159-64.
155. Nakai K, Yoneda K, Imataki O, *et al.* Transepidermal growth in disseminated *Fusarium* infection. *J Dermatol.* 2014; **41:** 770-1.
156. Nelson PE, Dignani MC, Anaissie EJ. Taxonomy, biology and clinical aspects of *Fusarium* species. *Clin Microbiol Rev.* 1994; **7:** 479-504.
157. Nucci F, Nouer SA, Capone D, *et al.* Fusariosis. *Semin Respir Crit Care Med.* 2015; **36:** 706-14.
158. Nucci M, Anaissie E. *Fusarium* infections in immunocompromised patients. *Clin Microbiol Rev.* 2007; **20:** 695-704.
159. Nucci M, Carlesse F, Cappellano P, *et al.* Earlier diagnosis of invasive fusariosis with *Aspergillus* serum galactomannan testing. *PLoS One.* 2014; **9:** e87784. PMID: 24489964.
160. Nucci M, Marr KA, Vehreschild MJ, *et al.* Improvement in the outcome of invasive fusariosis in the last decade. *Clin Microbiol Infect.* 2014; **20:** 580-5.
161. Nucci M, Varon AG, Garnica M, *et al.* Increased incidence of invasive fusariosis with cutaneous portal of entry, Brazil. *Emerg Infect Dis.* 2013; **19:** 1567-72.
162. Nunes Mdo C, Barbosa FB, Gomes GH, *et al.* Fatal right-sided endocarditis caused by *Fusarium* in an immunocompromised patient: A case report. *Mycoses.* 2011; **54:** 460-2.
163. Obana Y, Sano M, Jike T, *et al.* Differential diagnosis of trichosporonosis using conventional histopathological stains and electron microscopy. *Histopathology.* 2010; **56:** 372-83.
164. Ocampo-Garza J, Herz-Ruelas ME, Chavez-Alvarez S, *et al.* Disseminated fusariosis with endophthalmitis after skin trauma in acute lymphoblastic leukaemia. *J Eur Acad Dermatol Venereol.* 2016; **30:** e121-3.
165. Ochi Y, Hiramoto N, Takegawa H, *et al.* Infective endocarditis caused by *Scedosporium prolificans* infection in a patient with acute myeloid leukemia undergoing induction chemotherapy. *Int J Hematol.* 2015; **101:** 620-5.
166. Ohashi R, Kato M, Katsura Y, *et al.* Breakthrough lung *Scedosporium prolificans* infection with multiple cavity lesions in a patient receiving voriconazole for probable invasive aspergillosis associated with monoclonal gammopathy of undetermined significance (MGUS). *Med Mycol J.* 2011; **52:** 33-8.
167. Ong A, Blyth CC, Bency R, *et al.* Fatal mycotic aneurysms due to *Scedosporium* and *Pseudallescheria* infection. *J Clin Microbiol.* 2011; **49:** 2067-71.
168. Ortmann C, Wüllenweber J, Brinkmann B, *et al.* Fatal mycotic aneurysm caused by *Pseudallescheria boydii* after near-drowning. *Int J Legal Med.* 2010; **124:** 243-7.
169. Oshikata C, Tsurikisawa N, Saito A, *et al.* Fatal pneumonia caused by *Penicillium digitatum:* A case report. *BMC Pulm Med.* 2013; **13:** 16. PMID: 23522080.
170. Ounissi M, Abderrahim E, Trabelsi S, *et al.* Hyalohyphomycosis caused by *Paecilomyces lilacinus* after kidney transplantation. *Transplant Proc.* 2009; **41:** 2917-9.
171. Ozgun G, Borsa BA. A rare but diagnostic fungal microorganism in cervical Pap smears. *Indian J Pathol Microbiol.* 2012; **55:** 606-7.
172. Pai R, Boloor R, Shreevidya K, *et al.* *Fusarium solani:* An emerging fungus in chronic diabetic ulcer. *J Lab Physicians.* 2010; **2:** 37-9.
173. Panackal AA, Marr KA. *Scedosporium/Pseudallescheria* infections. *Semin Respir Crit Care Med.* 2004; **25:** 171-81.
174. Paredes K, Capilla J, Mayayo E, Guarro J. Virulence and resistance to antifungal therapies of *Scopulariopsis* species. *Antimicrob Agents Chemother.* 2016; **60:** 2063-8.
175. Pate MJ, Hemmige V, Woc-Colburn L, *et al.* Successful eradication of invasive *Scopulariopsis brumptii* in a liver transplant recipient. *Transpl Infect Dis.* 2016; **18:** 275-9.
176. Pei SN, Lee CH, Liu JW. Hemophagocytic syndrome in a patient with acquired immunodeficiency syndrome and acute disseminated penicilliosis. *Am J Trop Med Hyg.* 2008; **78:** 11-3.
177. Peixoto ML, Santos DO, Souza Ide C, *et al.* Interaction of an opportunistic fungus *Purpureocillium lilacinum* with human macrophages and dendritic cells. *Rev Soc Bras Med Trop.* 2014; **47:** 613-7.
178. Perdomo H, Cano J, Gene J, *et al.* Polyphasic analysis of *Purpureocillium lilacinum* isolates from different origins and proposal of the new species *Purpureocillium lavendulum. Mycologia.* 2013; **105:** 151-61.
179. Pereira GH, de Angelis DA, Brasil RA, *et al.* Disseminated amphotericin-resistant fusariosis in acute leukemia patients: Report of two cases. *Mycopathologia.* 2013; **175:** 107-14.
180. Petit A, Levine E, Epaud R, *et al.* *Scopulariopsis brevicaulis* abscess in a child treated for myeloblastic leukaemia. *Lancet Infect Dis.* 2011; **11:** 416.
181. Priyadharshini G, Varghese RG, Phansalkar M, *et al.* Subcutaneous fungal cyst masquerading as benign lesions - A series of eight cases. *J Clin Diagn Res.* 2015; **9:** EM01-4.
182. Quan C, Spellberg B. Mucormycosis, pseudallescheriasis and other uncommon mold infections. *Proc Am Thorac Soc.* 2010; **7:** 210-5.
183. Rahman FU, Irfan M, Fasih N, *et al.* Pulmonary scedosporiosis mimicking aspergilloma in an immunocompetent host: A case report and review of the literature. *Infection.* 2016; **44:** 127-32.
184. Randhawa PS, Nouraei SR, Howard DJ, Sandhu GS, Petrou MA. *Penicillium* species as a rare isolate in tracheal granulation tissue: A case series. *J Med Case Reports.* 2008; **2:** 84.
185. Rekha A, Kindo AJ, Ravi A. *Fusarium solani* in the post-transplant patient: An unusual fungus. *Int J Low Extrem Wounds.* 2008; **7:** 38-40.

186. Rimawi RH, Carter Y, Ware T, *et al.* Use of voriconazole for the treatment of *Paecilomyces lilacinus* cutaneous infections: Case presentation and review of published literature. *Mycopathologia.* 2013; **175:** 345-9.
187. Roan JN, Hsieh HY, Tsai HW, *et al.* Pulmonary nodules caused by *Schizophyllum commune* after cardiac transplantation. *J Infect.* 2009; **58:** 164-7.
188. Rodriguez MM, Pastor FJ, Salas V, *et al.* Experimental murine scedosporiosis: Histopathology and azole treatment. *Antimicrob Agents Chemother.* 2010; **54:** 3980-4.
189. Rodriguez-Tudela JL, Berenguer J, Guarro J, *et al.* Epidemiology and outcome of Scedosporium prolificans infection, a review of 162 cases. *Med Mycol.* 2009; **47:** 359-70.
190. Rougeron A, Schuliar G, Leto J, *et al.* Human-impacted areas of France are environmental reservoirs of the *Pseudallescheria boydii/Scedosporium apiospermum* species complex. *Environ Microbiol.* 2015; **17:** 1039-48.
191. Saegeman VS, Dupont LJ, Verleden GM, *et al. Paecilomyces lilacinus* and *Alternaria infectoria* cutaneous infections in a sarcoidosis patient after double-lung transplantation. *Acta Clin Belg.* 2012; **67:** 219-21.
192. Saghrouni F, Saidi W, Ben Said Z, *et al.* Cutaneous hyalohyphomycosis caused by *Purpureocillium lilacinum* in an immunocompetent patient: Case report and review. *Med Mycol.* 2013; **51:** 664-8.
193. Salah H, Al-Hatmi AM, Theelen B, *et al.* Phylogenetic diversity of human pathogenic *Fusarium* and emergence of uncommon virulent species. *J Infect.* 2015; **71:** 658-66.
194. Salmon A, Debourgogne A, Vasbien M, *et al.* Disseminated *Scopulariopsis brevicaulis* infection in anallogeneic stem cell recipient: Case report and review of the literature. *Clin Microbiol Infect.* 2010; **16:** 508-12.
195. Sandoval-Denis M, Gene J, Sutton DA, *et al.* Redefining *Microascus, Scopulariopsis* and allied genera. *Persoonia.* 2016; **36:** 1-36.
196. Sandoval-Denis M, Sutton DA, Cano-Lira JF, *et al.* Phylogeny of the clinically relevant species of the emerging fungus *Trichoderma* and their antifungal susceptibilities. *J Clin Microbiol.* 2014; **52:** 2112-25.
197. Sandoval-Denis M, Sutton DA, Fothergill AW, *et al. Scopulariopsis,* a poorly known opportunistic fungus: Spectrum of species in clinical samples and *in vitro* responses to antifungal drugs. *J Clin Microbiol.* 2013; **51:** 3937-43.
198. Saghrouni F, Saidi W, Ben Said Z, *et al.* Cutaneous hyalohyphomycosis caused by *Purpureocillium lilacinum* in an immunocompetent patient: Case report and review. *Med Mycol.* 2013; **51:** 664-8.
199. Santillan Salas CF, Joshi AY, Dhiman N, *et al.* Fatal post-operative *Trichoderma longibrachiatum* mediastinitis and peritonitis in a paediatric patient with complex congenital cardiac disease on peritoneal dialysis. *J Med Microbiol.* 2011; **60 (Pt. 12):** 1869-71.
200. Sarva ST, Manjunath SK, Baldwin HS, *et al.* Lung scedosporiosis in human immunodeficiency virus/acquired immunodeficiency syndrome. *Am J Med Sci.* 2010; **339:** 300-3.
201. Satirapoj B, Ruangkanchanasetr P, Treewatchareekorn S, *et al. Pseudallescheria boydii* brain abscess in a renal transplant recipient: First case report in Southeast Asia. *Transplant Proc.* 2008; **40:** 2425-7.
202. Satyavani M, Viswanathan R, Harun NS, *et al.* Pulmonary scopulariopsis in a chronic tobacco smoker. *Singapore Med J.* 2010; **51:** e137-9.
203. Sayah DM, Schwartz BS, Kukreja J, *et al. Scedosporium prolificans* pericarditis and mycotic aortic aneurysm in a lung transplant recipient receiving voriconazole prophylaxis. *Transpl Infect Dis.* 2013; **15:** e70-4.
204. Schafer K, Di Pietro A, Gow NA, *et al.* Murine model for *Fusarium oxysporum* invasive fusariosis reveals organ-specific structures for dissemination and long-term persistence. *PLoS One.* 2014; **9:** e89920. PMID: 24587124.
205. Schwarz C, Brandt C, Antweiler E, *et al.* Prospective multicenter German study on pulmonary colonization with *Scedosporium/Lomentospora* species in cystic fibrosis: Epidemiology and new association factors. *PLoS One.* 2017; **12:** e0171485. PMID: 28178337.
206. Sedlacek L, Graf B, Schwarz C, *et al.* Prevalence of *Scedosporium* species and *Lomentospora prolificans* in patients with cystic fibrosis in a multicenter trial by use of a selective medium. *J Cyst Fibros.* 2015; **14:** 237-41.
207. Shah PJ, Bergman S, Vegi S, *et al. Fusarium* peritonitis successfully managed with posaconazole and catheter removal. *Perit Dial Int.* 2014; **34:** 566-8.
208. Shankar EM, Vignesh R, Barton RC, *et al.* Hydrothorax in association with *Scopulariopsis brumptii* in an AIDS patient in Chennai, India. *Trans R Soc Trop Med Hyg.* 2007; **101:** 1270-2.
209. Sharifkashani B, Farshidpour M, Droudinia A, *et al.* Cutaneous alternariosis with *Trichosporon* infection in a heart transplant recipient: A case report. *Exp Clin Transplant.* 2013; **11:** 464-6.
210. Sharma A, Hazarika NK, Barua P, *et al. Acremonium strictum:* Report of a rare emerging agent of cutaneous hyalohyphomycosis with review of literatures. *Mycopathologia.* 2013; **176:** 435-41.
211. Shaver CM, Castilho JL, Cohen DN, *et al.* Fatal *Scopulariopsis* infection in a lung transplant recipient: Lessons of organ procurement. *Am J Transplant.* 2014; **14:** 2893-7.
212. Sheu R, Bricker AO, Sahi H, et al, *Pseudallescheria boydii (Scedosporium* species*)* in 3 lung transplant recipients: Computed tomography findings and literature review. *J Comput Assist Tomogr.* 2009; **33:** 247-52.
213. Shinohara MM, George E. *Scedosporium apiospermum:* An emerging opportunistic pathogen that must be distinguished from *Aspergillus* and other hyalohyphomycetes. *J Cutan Pathol.* 2009; **36 (Suppl. 1):** 39-41.
214. Shivaprasad A, Ravi GC, Shivapriya, *et al.* A rare case of nasal septal perforation due to *Purpureocillium lilacinum:* Case report and review. *Indian J Otolaryngol Head Neck Surg.* 2013; **65:** 184-8.
215. Short DP, O'Donnell K, Thrane U, *et al.* Phylogenetic relationships among members of the *Fusarium solani* species

complex in human infections and the descriptions of *F.keratoplasticum* sp. nov. and *F.petroliphilum* stat. nov. *Fungal Genet Biol.* 2013; **53:** 59-70.
216. Siddiqui AS, Zimmerman JL. Pulmonary infection secondary to *Chrysosporium zonatum* in an immunocompetent man. *Ann Am Thorac Soc.* 2016; **13:** 757-8.
217. Sidhu S, Chander J, Singh K. Perinephric abscess caused by *Fusarium chlamydosporum* in an immunocompetent child: case report and identification of the morphologically atypical fungal strain. *Indian J Pathol Microbiol.* 2013; **56:** 312-4.
218. Sigler L, Hambleton S, Pare JA. Molecular characterization of reptile pathogens currently known as members of the *Chrysosporium* anamorph of *Nannizziopsis vriesii* complex and relationship with some human-associated isolates. *J Clin Microbiol.* 2013; **51:** 3338-57.
219. Silva GM, Silveira AR, Betania CA, *et al.* Disseminated fusariosis secondary to neuroblastoma with fatal outcome. *Mycopathologia.* 2013; **176:** 233-6.
220. Singh PK, Kathuria S, Agarwal K, *et al.* Clinical significance and molecular characterization of non-sporulating moulds isolated from the respiratory tract of bronchopulmonary mycoses patients with special reference to basidiomycetes. *J Clin Microbiol.* 2013; **51:** 3331-7.
221. Singhal KV, Saoji V, Saoji SV. *Fusarium* skin infection: A case report. *Dermatol Online J.* 2012; **18:** 6.
222. Siqueira JP, Sutton D, Gene J, *et al. Schizophyllum radiatum*, an emerging fungus from human respiratory tract. *J Clin Microbiol.* 2016; **54:** 2491-7.
223. Sireesha P, Manoj Kumar CH, Setty CR. Thyroid abscess due to *Scedosporium apiospermum. Indian J Med Microbiol.* 2010; **28:** 409-11.
224. Sitterle E, Giraud S, Leto J, *et al.* Matrix-assisted laser desorption ionization-time of flight mass spectrometry for fast and accurate identification of *Pseudallescheria/Scedosporium* species. *Clin Microbiol Infect.* 2014; **20:** 929-35.
225. Slavin M, van Hal S, Sorrell TC, *et al.* Invasive infections due to filamentous fungi other than *Aspergillus*: Epidemiology and determinants of mortality. *Clin Microbiol Infect.* 2015; **21:** 490.
226. Sohn JY, Jang MA, Lee JH, *et al.* Isolation and identification of *Geosmithia argillacea* from a fungal ball in the lung of a tuberculosis patient. *Ann Lab Med.* 2013; **33:** 136-40.
227. Soman R, Mahashur AA, Naphade D, *et al. Pseudallescheria boydii* lung infection in an immunocompetent adult, difficulties in diagnosis and management. *J Assoc Physicians India.* 2010; **58:** 633-4.
228. Spanevello M, Morris KL, Kennedy GA. Pseudoaneurysm formation by *Scedosporium prolificans* infection in acute leukaemia. *Intern Med J.* 2010; **40:** 793.
229. Suankratay C, Dhissayakamol O, Uaprasert N, *et al.* Invasive pulmonary infection caused by *Chrysosporium articulatum*: The first case report. *Mycoses.* 2015; **58:** 1-3.
230. Suchonwanit P, Chaiyabutr C, Vachiramon V. Primary cutaneous *Chrysosporium* infection following ear piercing: A case report. *Case Rep Dermatol.* 2015; **7:** 136-40.
231. Summerbell RC, Gueidan C, Schroers HJ, *et al. Acremonium* phylogenetic overview and revision of *Gliomastix, Sarocladium* and *Trichothecium. Stud Mycol.* 2011; **68:** 139-62.
232. Swick BL, Reddy SC, Friedrichs A, *et al.* Disseminated scopulariopsis - Culture is required to distinguish from other disseminated mould infections. *J Cutan Pathol.* 2010; **37:** 687-91.
233. Taj-Aldeen SJ, Rammaert B, Gamaletsou M, *et al.* Osteoarticular infections caused by non-*Aspergillus* filamentous fungi in adult and pediatric patients: A systematic review. *Medicine (Baltimore).* 2015; **94:** e2078.
234. Takeuchi M, Yoshida C, Ota Y, *et al.* Deep skin infection of *Scedosporium apiospermum* in a patient with refractory idiopathic thrombocytopenic purpura. *Intern Med.* 2011; **50:** 1339-43.
235. Takenaka M. A case of cutaneous fusariosis of the scrotum as a complication of acute myeloid leukemia. *Med Mycol J.* 2016; **57:** J65-70.
236. Tamaki M, Nozaki K, Onishi M, *et al.* Fungal meningitis caused by *Lomentospora prolificans* after allogeneic hematopoietic stem cell transplantation. *Transpl Infect Dis.* 2016; **18:** 601-5.
237. Tammer I, Tintelnot K, Braun-Dullaeus RC, *et al.* Infections due to *Pseudallescheria/Scedosporium* species in patients with advanced HIV disease - A diagnostic and therapeutic challenge. *Int J Infect Dis.* 2011; **15:** e422-9.
238. Tascini C, Cardinali G, Barletta V, *et al.* First case of *Trichoderma longibrachiatum* CIED (Cardiac Implantable Electronic Device)-associated endocarditis in a non-immunocompromised host: Biofilm removal and diagnostic problems in the light of the current literature. *Mycopathologia.* 2016; **181:** 297-303.
239. Terada M, Fujita J, Watanabe S, *et al.* Olecranon bursa with *Fusarium solani* infection in an otherwise healthy patient. *Mycoses.* 2011; **54:** e853-5.
240. Terasaki JM, Shah SK, Schnadig VJ, *et al.* Airway complication contributing to disseminated fusariosis after lung transplantation. *Transpl Infect Dis.* 2014; **16:** 621-4.
241. Tintelnot K, Just-Nubling G, Horre R, *et al.* A review of German *Scedosporium prolificans* cases from 1993 to 2007. *Med Mycol.* 2009; **47:** 351-8.
242. Tortorano AM, Prigitano A, Esposto MC, *et al.* European Confederation of Medical Mycology (ECMM) epidemiological survey on invasive infections due to *Fusarium* species in Europe. *Eur J Clin Microbiol Infect Dis.* 2014; **33:** 1623-30.
243. Tortorano AM, Richardson M, Roilides E, *et al.* ESCMID and ECMM joint guidelines on diagnosis and management of hyalohyphomycosis: *Fusarium* spp., *Scedosporium* spp. and others. *Clin Microbiol Infect.* 2014; **20 (Suppl. 3):** 27-46.
244. Trabelsi S, Hariga D, Khaled S. First case of *Trichoderma longibrachiatum* infection in a renal transplant recipient in Tunisia and review of the literature. *Tunis Med.* 2010; **88:** 52-7.
245. Triest D, Stubbe D, De Cremer K, *et al.* Banana infecting fungus, *Fusarium musae,* is also an opportunistic human

pathogen: Are bananas potential carriers and source of fusariosis? *Mycologia.* 2015; **107:** 46-53.

246. Triest D, Stubbe D, De Cremer K, *et al.* Use of matrix-assisted laser desorption ionization-time of flight mass spectrometry for identification of molds of the *Fusarium* genus. *J Clin Microbiol.* 2015; **53:** 465-76.
247. Trubiano JA, Paratz E, Wolf M, *et al.* Disseminated *Scedosporium prolificans* infection in an 'extensive metaboliser': navigating the minefield of drug interactions and pharmacogenomics. *Mycoses.* 2014; **57:** 572-6.
248. Tsang CC, Chan JFW, Pong WM, *et al.* Cutaneous hyalohyphomycosis due to *Parengyodontium album* gen. et comb. nov. *Med Mycol.* 2016; **54:** 699-713.
249. Tullio V, Mandras N, Banche G, *et al. Schizophyllum commune:* An unusual of agent bronchopneumonia in an immunocompromised patient. *Med Mycol.* 2008; **46:** 735-8.
250. Uno K, Kasahara K, Kutsuna S, *et al.* Infective endocarditis and meningitis due to *Scedosporium prolificans* in a renal transplant recipient. *J Infect Chemother.* 2014; **20:** 131-3.
251. van Diepeningen AD, Brankovics B, Iltes J, *et al.* Diagnosis of *Fusarium* infections: Approaches to identification by the clinical mycology laboratory. *Curr Fungal Infect Rep.* 2015; **9:** 135-43.
252. van Diepeningen AD, Feng P, Ahmed S, *et al.* Spectrum of *Fusarium* infections in tropical dermatology evidenced by multilocus sequencing typing diagnostics. *Mycoses.* 2015; **58:** 48-57.
253. van Schooneveld T, Freifeld A, Lesiak B, *et al. Paecilomyces lilacinus* infection in a liver transplant patient: Case report and review of the literature. *Transpl Infect Dis.* 2008; **10:** 117-22.
254. Varon AG, Nouer SA, Barreiros G, *et al.* Superficial skin lesions positive for *Fusarium* are associated with subsequent development of invasive fusariosis. *J Infect.* 2014; **68:** 85-9.
255. Vasudevan B, Hazra N, Verma R, *et al.* First reported case of subcutaneous hyalohyphomycosis caused by *Paecilomyces variotii. Int J Dermatol.* 2013; **52:** 711-3.
256. Wakabayashi Y, Okugawa S, Tatsuno K, *et al. Scedosporium prolificans* endocarditis: Case report and literature review. *Intern Med.* 2016; **55:** 79-82.
257. Wang H, Wan Z, Li R, *et al.* Molecular identification and susceptibility of clinically relevant *Scedosporium* spp. in China. *Biomed Res Int.* 2015; 109656. PMID: 26550562.
258. Wessolossky M, Haran JP, Bagchi K. *Paecilomyces lilacinus* olecranon bursitis in an immunocompromised host: Case report and review. *Diagn Microbiol Infect Dis.* 2008; **61:** 354-7.
259. Woudenberg JHC, Meijer M, Houbraken J, *et al. Scopulariopsis* and *Scopulariopsis*-like species from indoor environments. *Stud Mycol.* 2017; **88:** 1-35.
260. Wu CH, Lu PL, Hsiao HH, *et al.* Breakthrough *Fusarium solani* infection in a patient with acute myeloid leukemia receiving posaconazole prophylaxis. *Ann Hematol.* 2014; **93:** 1079-81.
261. Yamada Y, Nishioka G, Makimura K, *et al.* Localized cutaneous fusariosis after long-term topical application of corticosteroids in a healthy elderly woman. *Eur J Dermatol.* 2012; **22:** 555-9.
262. Yan X, Yu C, Shi Z, *et al.* Nasal cutaneous infection in a healthy boy caused by *Fusarium moniliforme. Pediatr Dermatol.* 2013; **30:** e43-5.
263. Yang Q, Wei J, Chen Z. Fatal bronchial invasion of *Scopulariopsis brevicaulis* in an acute monocytic leukemia patient. *Diagn Microbiol Infect Dis.* 2012; **73:** 369-71.
264. Zhu A, Htet S, Kalro A, *et al.* Breakthrough fusariosis presenting initially with pyomyositis. *Pathology.* 2017; **49:** 105-7.
265. Zhou X, Kong F, Sorrell TC, *et al.* Practical method for detection and identification of *Candida, Aspergillus,* and *Scedosporium* spp. by use of rolling-circle amplification. *J Clin Microbiol.* 2008; **46:** 2423-7.

CHAPTER 32

Adiaspiromycosis and Emergomycosis

The genus ***Emmonsia*** currently consists of **two** medically significant species associated with human diseases *i.e.* ***Ea.crescens*** and ***Ea.parva***. Both species are known for the last half a century and are responsible for unique granulomatous zoonotic pulmonary disease called as **adiaspiromycosis**. It is very rarely reported among humans also.

On the other hand **emergomycosis** is a systemic fungal infection presenting with cutaneous and disseminated clinical manifestations among immunocompromised patients. This is caused by three dimorphic fungi belonging to a newly described genus ***Emergomyces*** *i.e. Emergomyces pasteurianus* comb. nov., *Emergomyces africanus* sp. nov. and *Emergomyces orientalis* sp. nov., which were formerly classified under genus *Emmonsia* either as *Emmonsia pasteuriana* or *Emmonsia*-like species. As nomenclature of genus is replaced for these unique fungal species from *Emmonsia* to *Emergomyces*, consequently disease name is also changed from emmonsiosis to emergomycosis for the convenience of deliberations.

Emergomyces pasteurianus however known for quite some time but came to limelight recently by causing clinical entity due to its strong association with immunocompromised conditions. *Es.africanus*, another dimorphic fungal species, recently reported particularly among cluster of cases in South Africa and other places. This is very close to *Es.pasteurianus*, not reported till date in any of the animal or environmental niche but has come to the international attention.

However, for convenience of discussion both these diseases (adiaspiromycosis and emergomycosis), caused by various species of genus ***Emmonsia*** *(Ea.crescens* and *Ea.parva)* and ***Emergomyces*** *(Es.pasteurianus* and *Es.africanus* and *Es.orientalis)* are clubbed together in this Chapter and described comprehensively as separate clinical entities.

I. ADIASPIROMYCOSIS

This disease is primarily fungal infection of rodents and lower animals of wildlife. It is prevalent worldwide mainly involving respiratory system of mice, rats, rabbits, squirrels, etc, which live or burrow in the soil. However, man can rarely be an incidental host and get infection by inhaling dust-borne conidia of causative saprobic soil fungi of genus *Emmonsia*. There are two clinically important species of this genus *i.e. Ea.crescens* and *Ea.parva*, which cause **adiaspiromycosis.** This had also been known by various synonym like **haplomycosis** and **adiaspirosis** but the name recommended by ISHAM is popularly used *i.e.* **adiaspiromycosis.** However, when the term emergomycosis is used, it is meant for the disease caused by three opportunistic dimorphic species of genus *Emergomyces - Es.pasteurianus, Es.africanus* and *Es.orientalis* given in detail in later part of this very Chapter.

Historical Perspective

In 1942, Chester Wilson Emmons and Ashburn, during an active search for another fungus *i.e. Coccidioides immitis* in Lower Sonoran Deserts (Arizona), unexpectedly encountered a new fungus in the lungs of rodents. This was forming spherule-like structure in lung tissue, measuring up to 20 μm having thick laminated walls however without any budding or endosporulation. They also observed that large spherule grew from smaller asexual conidia (2-4 μm) thereby classified the fungus as *Haplosporangium parvum* thinking it to be a mucormycete because of its resemblance to *Haplosporangium bisporale*. The fungus was subsequently transferred to a new genus *Emmonsia* as *Ea.parvum*.

Five years later in 1947, Dowding in Alberta, Canada isolated a similar type of organism from rodents but spherules were about six times larger than that of previously discovered fungus in Arizona. A close relationship of both fungal isolates of Arizona and Canada, prompted classification of this newly described fungus also, in the genus *Haplosporangium* and therefore disease was initially named as **haplomycosis**. In 1959, Ciferri *et al*, isolated *Haplosporangium parvum* from soil in Italy. They renamed the isolate and placed it in the genus earlier described *i.e. Emmonsia* as *Emmonsia parvum*.

As *Emmonsia* species differed on the basis of size of mature adiaconidia hence fungal isolate from Arizona was named as *Emmonsia parva* whereas Canadian isolate was designated as *Emmonsia crescens* (Latin = ***'crescens'*** - to grow or come forth) due to its **larger** size observed in the tissues. As the larger adiaconidia were being discovered in material from Canada and northern USA, therefore, in the year 1960, strains that produced spherules with a diameter up to 400-500 µm were all designated under new species as *Emmonsia crescens*.

Subsequently due to taxonomical changes, it became imperative to propose a new name for this infection in **wild animals**, as disease was not known to cause infection in man at that time. Thus haplomycosis gave way to a new name as **adiaspiromycosis** due to production of unique structure (adiaconidia/adiaspore). The name of disease as adiaspiromycosis (***a***- = not or without + ***dia***- = through + ***speirein***- = to scatter or disperse) suggested by Emmons and Jellison, literally translating as mycotic infection of lungs in which there is no multiplication or dissemination of inhaled organism beyond its site of implantation.

The term adiaconidium (formerly **adiaspore**) is derived from typical conidia of causative fungi, which exhibit a unique property of progressive enlargement of its size without replication *in vivo* as well as *in vitro* at an elevated temperature of approximately 37°C or even more. But other workers considered it as a variety of genus *Chrysosporium*. Hence in 1962, *Ea.crescens* was re-named as *Chrysosporium parvum* var. *crescens*.

In 1964 Doby-Dubois *et al*, while demonstrating a solitary pulmonary nodule, reported the first case of human adiaspiromycosis from France. In 1979, Kamalam *et al*, reported two cases of cutaneous infection from India caused by *Emmonsia (Chrysosporium) crescens*. In 1997, a case of disseminated bilateral pulmonary adiaspiromycosis was reported in a two year old Finnish girl. She was rather the first human case of adiaspiromycosis in Scandinavia, the youngest child with this disease reported so far and recovered successfully after treatment with amphotericin B. Later on more cases were reported from other continents as well, substantiating it to be a significant disease not only in wild animals but humans as well.

Mycology

Adiaspiromycosis is caused by two species of genus *Emmonsia i.e.* ***Ea.parva*** and ***Ea.crescens***. Earlier they were called as *Chrysosporium parvum* var. *parvum or Emmonsia parva* var. *parva* and *Chrysosporium parvum* var. *crescens or Emmonsia parva* var. *crescens*, respectively. Both the organisms have recently been given independent **species** status as *parva* and *crescens* instead of designating them with their **variety** names. The genus name, *Emmonsia*, is favored over *Chrysosporium* as correct generic name for agents of adiaspiromycosis.

These species of *Emmonsia* produce non-replicating cells called as adiaconidia, which are globose, thick-walled, mono-or multinucleate cells commonly found in enlarged cyst-like structures in the lungs of rodents and other small terrestrial wild animals. The term adiaconidium/adiaspore was derived from Greek verb - ***'-speirein'*** meaning to scatter or disperse, with ***'adia-'*** being negative and was proposed for spherule which enlarges from inhaled asexual conidium signifying increasing size without its replication under influence of elevated temperature. This feature is seen *in vivo* as well as *in vitro*. The terminology **adiaconidium** is frequently used interchangeably as **adiaspore** also but former one is correct. Adiaspiromycosis is the term used when such development occurs within animal or human body.

These two species of genus *Emmonsia* differ in their growth requirement of tolerable temperature (37°C for *Ea.crescens* versus 40°C for *Ea.parva*), sensitivity to cycloheximide, size of adiaconidia, host range, ability to form a teleomorph as well as their geographical distribution. The optimum temperature required for the growth of *Ea.crescens* is 37°C forming adiaconidia and at 40°C it is inhibited, whereas at 37°C *Ea.parva* grows as mycelial fungus but at 40°C it is converted to adiaconidium. The adiaconidia of *Ea.crescens* are about 5-10 times larger than that of *Ea.parva* (250-400 µm *vs.* 10-25 µm) and wall thickness is about 20-70 µm as compared to 2 µm in case of *Ea.parva*. This remains as **mononucleate** in *Ea.parva* even on attaining its full size but in *Ea.crescens* becomes **multinucleate** when it begins its phenomenal increase in size. The differentiating features between *Ea.crescens* and *Ea.parva vis-a-vis* species of *Emergomyces* are given in **Table 32.1**.

Table 32.1. Differentiating Features among Species of Genera *Emmonsia* and *Emergomyces*.

Characteristics	*Ea.crescens*	*Ea.parva*	*Es.pasteurianus*	*Es.africanus*	*Es.orientalis*
Adiaconidia (Adiaspores)	Produced	Produced	Not produced	Not produced	Not produced
Size of Adiaconidia	50-500 µm (Max. 700)	10-25 µm (Max. 40)	Not applicable	Not applicable	Not applicable
Spherule Wall Thickness	20-70 µm	2 µm	Not applicable	Not applicable	Not applicable
Nucleus	Multinucleate	Mononucleate	Not applicable	Not applicable	Not applicable
Growth at 37°C	Adiaconidia	Mycelial	Yeasts (2-4 µm)	Yeasts (2-4 µm)	Yeasts (4-10 µm)
Growth at 40°C	Inhibited	Adiaconidia	-	-	-
Cycloheximide	Reduced growth	No inhibition	No inhibition	No inhibition	-

The taxonomy pertaining to causative fungi of adiaspiromycosis has been controversial. These fungi were placed in family **Moniliaceae** of class **Hyphomycetes** in phylum **Deuteromycetes**. Traditionally nomenclature of fungus as genus ***Emmonsia*** was designated as a tribute to **Chester Wilson Emmons** (1900-1985) eminent mycologist of 20th century but adiaspiromycosis is currently accepted name for this clinical entity caused by *Ea.crescens* and *Ea.parva* as per the ISHAM recommendations.

The teleomorph state has recently been demonstrated in *Ea.crescens*. It is also a member of *Ajellomyces* clade, which belongs to the family Ajellomycetaceae. The genus *Ajellomyces* contained heterothallic species, *A.dermatitidis, A.capsulatus* and *A.crescens*, anamorphs of which are placed in different genera *i.e. Blastomyces, Histoplasma* and *Emmonsia,* respectively. Now, newly described genus is also added to this family *i.e. Emergomyces.* The taxonomic arrangement of the group is currently in flux, complicated by the fact that the type species, *Ea.parva*, appears more related to *Blastomyces* than to *Ea.crescens.*

The teleomorphic state of *Ea.crescens* has been named *Ajellomyces crescens* in the same genus as most of the dimorphic fungi. This designation was given because *A.crescens* closely resembles *Ajellomyces dermatitidis*, teleomorph state of *Blastomyces dermatitidis* and is causative dimorphic fungus of systemic disease, blastomycosis. The phylogenetic analysis of 18S rDNA sequences of *Ea.crescens* places this pathogen within order **Onygenales** where various **dimorphic fungi** are placed in genus *Ajellomyces* and taxonomically it may also be taken as a dimorphic fungus. There is no known sexual state in *Emmonsia parva* but phylogenetic diversity from one group to other and to *B.dermatitidis* strains suggest diverging groups, may be varieties or sibling species. Therefore, *Emmonsia* genus is also grouped under **dimorphic fungi** as its species are converted from mycelial to adiaconidia at the body temperature.

The species of another genus, *Emmonsiellopsis,* were morphologically found to be related to thermally dimorphic fungi of family Ajellomycetaceae but analysis of ITS region of rDNA and domains D1 and D2 of the 28S rRNA gene sequences confirmed that they were different from all species described in that family. They were accommodated in new genus *Emmonsiellopsis* as *E.coralliformis* sp. nov. and *E.terrestris* sp. nov. and neither related to adiaspiromycosis nor to emergomycosis.

Epidemiology

Adiaspiromycosis is a worldwide, non-infectious, non-arthropod transmitted fungal infection of lower vertebrates, most commonly among the rodents. The humans become an accidental host by inhaling dust-borne aerosolized conidia of the saprophytic soil fungus of genus *Emmonsia.* It is primarily a respiratory infection of burrowing rodents with worldwide geographical and ecological distribution. The disease has been reported in 124 mammalian species and sub-species. Despite its ubiquitous nature occurring in lower wild animals, disease rarely affects man. The mammals become hosts following inhalation of minute globose asexual **conidia** (2-4 µm) that eventually develop into large, thick-walled **adiaconidia** in lung tissue of host and may persist for rest of their lives. The conidia are **aleurioconidia** as these are solitary ones released by lytic disintegration of part of supportive hyphal cell. The adiaconidia in large number can significantly damage pulmonary tissue and respiratory capacity of the infected animals. The disease name represents this long described mycosis mainly infecting rodents and is characterized by presence of adiaconidia.

The rodents may serve as zoonotic reservoir of the disease although fungus causing infection in **rodents**, *Ea.parva***, rarely** infects **humans**. *Emmonsia crescens* is found in nature predominantly in temperate zones and likely mode of human infection is accidental inhalation of asexual fungal conidia aerosolized from contaminated soil. There is development of scattered pulmonary granulomas due to development of conidia into adiaconidia. *Ea.crescens* is prevalent worldwide in temperate zones except in Africa and Australia whereas *Ea.parva* is associated with fewer animal hosts and geographically more restricted as it is only confined to south-western USA, Eastern Europe, Kenya and Australia.

Doby-Dubois *et al*, reported first case of human adiaspiromycosis in 1964 from France in formalin-fixed lung tissue of a patient suffering from aspergillosis with cystic disease. The majority of human infections are caused by *Ea.crescens*. Only two patients with underlying AIDS have been found to be infected with *Ea.parva*. The natural habitat of *Ea.crescens* and *Ea.parva* is soil. It is an under-reported as well as misdiagnosed pulmonary infection among humans. The disease is non-communicable in nature as animal-to-animal transmission has not been documented. The infection is more common in male patients engaged in outdoor activities and/or occupations.

Emmonsia parva and *Ea.crescens* are etiologic agents of adiaspiromycosis. Usually this presents as an asymptomatic pulmonary infection in animals and more recently in immunocompromised patients after the inhalation of the fungal conidia. Lung biopsy may be necessary to diagnose the illness, as the organisms may not be present in sputa or bronchial alveolar lavages. Dissemination of the infection may occur more readily in immunocompromised hosts. *Ea.crescens* is more commonly isolated from humans while *Ea.parva* is isolated more often from animals.

Human adiaspiromycosis is uncommon but there are nearly **seventy** cases reported in the literature from various countries like Israel, India, Czechoslovakia, erstwhile USSR, France, Germany, Spain, Finland, Argentina, Brazil, Venezuela, Guatemala, Honduras and USA. There were reports of this disease among animals in England but in 2009 a human case has also been described by Denson *et al*, who was primarily suspected to be having pulmonary adenocarcinoma. Since the original report of human pulmonary infections, *Ea.parva* has also been isolated from nail, skin, sinuses and bones, where it produces severe osteomyelitis. As it is with pulmonary disease, skin and nail infections occur in immunocompetent individuals. However, disseminated infection has been reported in immunocompromised patients.

Immunity

In adiaspiromycosis, delayed-type of hypersensitivity reaction to antigens of *Ea.crescens* and precipitating antibodies in sera can be demonstrated in human infections. The environmental exposure to *Emmonsia* species was first detected by positive skin test reactions to antigens prepared from culture extracts in a group of school children. The strain-as well as species-specific antigens were detected by immunodiffusion tests using antisera raised against the partially purified cytoplasmic antigens.

Pathogenesis and Pathology

The adiaconidia may infect normal or underlying diseased lung. After inhalation and reaching alveoli, asexual conidia progressively grows in size as adiaconidia without any replication or budding, etc. Moreover, they neither germinate nor multiply but induce a granulomatous type of chronic inflammatory reaction.

The lesions resemble foreign body or tuberculoid granuloma but without any necrosis or caseation. Adiaconidia are surrounded by granulomatous inflammatory reaction composed of epithelioid cells, macrophages, lymphocytes, plasma cells, scattered neutrophils and fibrous tissue whereas eosinophils are uncommon. These are 4-5 μm in size and on maturity attain size of about 50 to 500 μm, representing about hundred-fold increment. The adiaconidia are closely surrounded by foreign body and Langhans' type of giant cells. They remain in alveoli of host and eventually become calcified in due course of time. The pulmonary lesions produced by enlarging conidia consist of discrete or confluent fibrotic granuloma, each containing one or more adiaconidia.

In adiaspiromycosis, granular proteinaceous radiating material peripheral to outer zone of adiaconidia show Splendore-Hoeppli phenomenon. The huge adiaconidia become encased in granuloma with epithelioid cells and multinucleate giant cells. However, collapsed adiaconidia are seen as crescent-shaped and are believed to be non-viable. The disease is unique among fungal infections as conidia apparently do not multiply in host tissues but only increase in size and illness seems to remain usually confined to the lungs. The thick-walled adiaconidia when mature, are large enough to be seen even with unaided eye. The internal contents of adiaconidia are invariably not seen but when visible, their appearance is variable and may mimic cut-sections of internal organs of helminths.

Clinical Features

Adiaspiromycosis is an uncommon but chronic pulmonary fungal disease caused by two species of genus *Emmonsia i.e. Ea.crescens* and *Ea.parva*. This is more common in rodents however man may get infection depending upon amount of initial inoculum of inhaled conidia. The majority of documented cases among humans have been found to be asymptomatic. The natural disease, more or less remains confined to the lungs. The patient may become symptomatic as a result of dense and diffuse nodular involvement of lungs. The clinical manifestations are mainly due to the mechanical damage by adiaconidia to its surrounding tissue.

At present the description of this disease is mainly based on clinical manifestations of various case reports available in the literature. There are three clinical forms of human adiaspiromycosis, which are recognized (1) solitary granuloma (2) localized granulomatous and (3) diffuse-disseminated granulomatous disease, which are briefly described here.

1. Solitary Adiaspiromycosis

This clinical form of adiaspiromycosis is an asymptomatic and self-limiting respiratory infection, which is incidentally diagnosed by pathologist during histopathological examination of lung tissue removed for some other reasons. There are no significant radiological abnormalities attributable to this type of infection.

2. Localized Adiaspiromycosis

The disease remains localized at primary implantation site in case of a limited inoculum and pulmonary nodules may be detected radiographically, sometimes incidentally at autopsy or in surgical specimens of lung. The severity of disease depends upon number of conidia inhaled. In a recent report from Brazil, 17 cases of acute conjunctivitis with episcleritis and anterior uveitis were linked to adiaspiromycosis and freshwater sponges in the northern Amazon region. The demonstration of adiaconidia in scleral biopsy specimen from children with severe disease in this outbreak suggest that conjunctival irritation was most likely followed by constant exposure to conidia of this fungus, perhaps in dust caused in part by dry environmental condition, similar to the exposure of respiratory mucosa.

3. Disseminated Adiaspiromycosis

A diffuse and disseminated type of granulomatous adiaspiromycosis is more common as compared to the solitary one. The patient usually presents with low-grade fever, cough, progressive dyspnea, fatigue and night sweats. There are crepitations on auscultation due to compression as well as displacement of distal airways and alveolar parenchyma by gradually expanding granulomas. There is **no** replication of adiaconidia in the lungs. The other organs like skin, peritoneum and bones may rarely be involved as a result of extrapulmonary dissemination causing extensive osteomyelitis. The bone marrow involvement has been found in patient with AIDS, wherein *Ea.parva* was isolated. Moreover, severe cases with inhalation of many conidia can result in acute respiratory failure as it occurs in other well-known dimorphic fungal infections like histoplasmosis and coccidioidomycosis.

Radiodiagnosis

The chest X-rays of granulomatous pulmonary lesions of adiaspiromycosis show multiple nodular shadows of about 1 to 3 mm diameter and usually mistaken for miliary tuberculosis. The most common pattern among these patients is bilateral reticulonodular infiltrates with areas of honeycombing. The tentative diagnosis is invariably based on radiological findings. The high-resolution computed tomography (HRCT) of thorax may also be very useful to delineate lesions, which are eventually confirmed on histopathological examination. The chest radiography of disseminated pulmonary adiaspiromycosis usually shows bilateral reticulonodular infiltrates mimicking those of miliary tuberculosis. As the culture of causative organism from sputum or bronchoalveolar lavage has rarely been achieved and lung biopsy is usually required for diagnosing pulmonary infection, it is difficult to assess that these lesions are caused by tuberculosis, adiaspiromycosis or emergomycosis.

Differential Diagnosis

The most common diseases, which are to be differentiated histopathologically from adiaspiromycosis, are coccidioidomycosis and rhinosporidiosis due to formation of spherule-like structures in all of them. The spherule of *C.immitis* is smaller as its size on an average about 30-60 μm and has 2 μm thick trilaminar walls and contain endospores of 2-5 μm in diameter. Contrarily the center of adiaconidia does not contain any endospore-like structures hence appear empty. The lack of such findings in adiaspiromycosis actually indicates that no intra-host replication occurs.

The sporangium of *Rhinosporidium seeberi* should be distinguished from this disease, which is 200-300 μm in diameter. Sometimes, adiaconidia may be confused with helminthic nematodes like *Dirofilaria* and *Strongyloides* species in their cut-section. In dirofilariasis, chest x-rays usually show ***'coin-lesion'*** of 2 to 3 cm, much larger than miliary nodules of adiaspiromycosis of size 1 to 2 mm. The degenerated and apparently dead adiaconidia with a flattened and folded aspect may resemble roundworm. The disease should not be confused with miscellaneous structures such as vegetable matter, starch granules of lentil pneumonia or corpora amylacea. The tissue form of *Sporothrix luriei* may sometimes be confused but it has thick-walled horizontal and longitudinal septate and classical eyeglass-shaped cells. The differences in morphological features of spherules and similar structures producing fungal and pseudofungal organisms are given in **Table 5.1 and Table 18.1**.

Another entity, **myospherulosis**, which is a granulomatous type of reaction containing sac-like structures with spherules; associated with lipid medications either being applied to open wounds or being injected leading to solid subcutaneous or dermal nodules. Therefore, it should also not be confused with adiaspiromycosis.

Laboratory Diagnosis

The diagnosis of adiaspiromycosis rests upon radiological and histopathological findings by demonstration of adiaconidia in various stages of their development and/or degeneration. The disease being very rare hence is usually not suspected on clinical grounds because as such there is no pathognomonic feature among these patients. However, it is encountered while examining histopathology specimen for some other disease or accidentally seen during autopsy. In lungs, numerous globose, white to grey nodules can be seen with even naked eye. Moreover, as yet there is no reliable serological test for establishing diagnosis of disease, therefore, significant feature of adiaspiromycosis is finding adiaconidia in biopsy or autopsy tissues.

(a) Direct Examination

The conventional methods of direct examination of fungi are usually utilized to demonstrate organism in fine needle aspiration or biopsy tissue. The respiratory system is invariably involved in adiaspiromycosis but sputum examination and bronchial washing are usually negative for adiaconidia even in a severe case. Therefore, H&E, PAS, GMS, Papanicolaou and Gridley fungal stains are used in biopsy specimen.

The biopsy tissue shows characteristic large thick-walled adiaconidia surrounded by a fibrous granulomatous reaction in lungs **(Figs. 32.1A and B)**. They are symmetrical and round to oval. Neither budding nor endosporulation is seen in any of adiaconidia. Their size range from 50-500 μm and thickness of **trilaminar** wall varying from 20-70 μm. The wall appears **bilaminar** when stained with H&E stain in contrast to other stains like GMS and PAS showing trilaminar wall. The walls of adiaconidia are refractile and when stained with H&E, appear to be comprised of

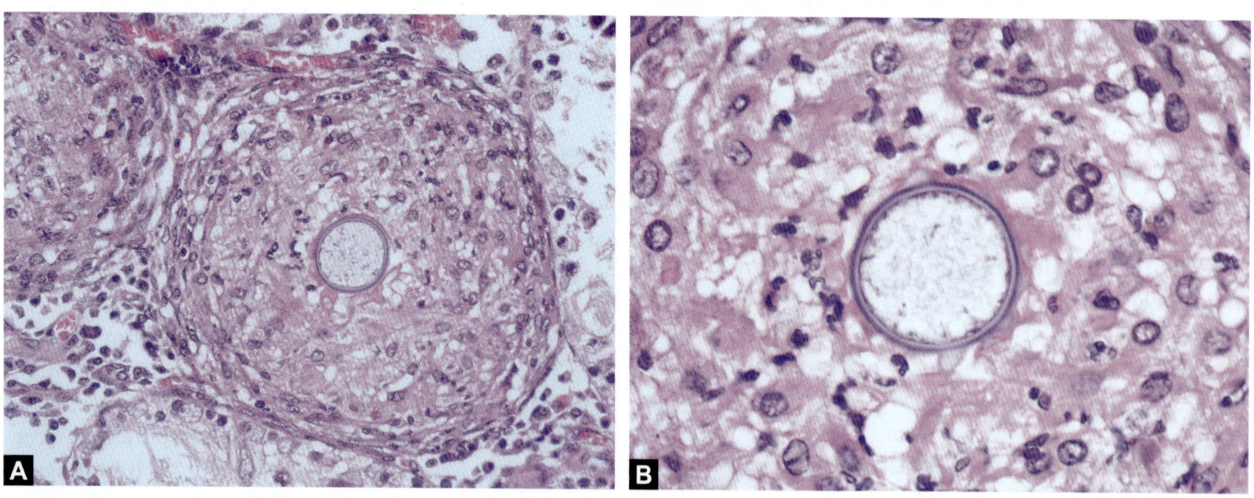

Figs. 32.1A and B. Histopathology shows "honeycomb-like" adiaconidia surrounded by epithelioid macrophages and giant cells of granuloma (H&E × 400 & 1000).

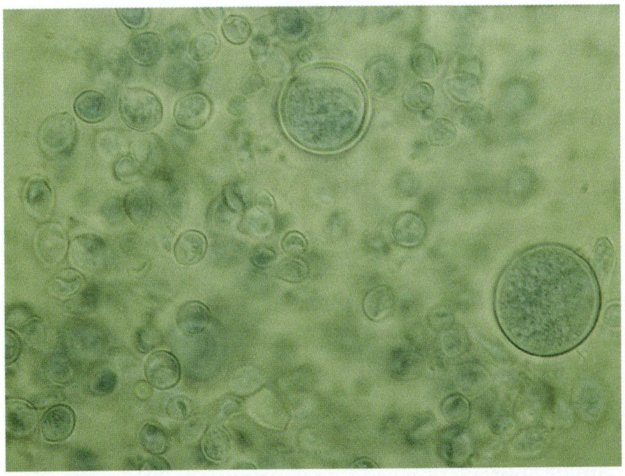

Fig. 32.2. *In vitro* growth of *Emmonsia crescens* showing adiaconidia (LCB × 400).

two layers, a narrow, outer, eosinophilic layer containing periodic fenestrations and a broad, hyaline, inner layer composed predominantly of chitin. Recently fine needle aspiration cytology (FNAC) has also been successfully tried for establishing definitive diagnosis of pulmonary adiaspiromycosis.

The adiaconidia of *Ea.crescens* is about 5-10 times larger than that of *Ea.parva*. Sometimes, especially in *Ea.crescens*, they appear as **empty rings**, which can be seen easily with even naked eye when slide of tissue section is held against source of light as center is always empty. In human lung tissue, adiaconidia are usually empty but may contain small eosinophilic globules along inner surface of walls giving honeycomb-like appearance **(Figs. 32.1A and B)**. The conidial walls stain with GMS, PAS and Gridley fungus stainings but not with Mayer's mucicarmine as in case of rhinosporidiosis or cryptococcosis. The adiaconidia although are empty and round to oval but due to collapse, probably during processing, crescent, sickle or barbell forms can be seen and may also look like a cut-section of helminth.

(b) Fungal Culture

The culture is done on conventional fungal media like SDA and BHI blood agar but **without actidione**. The organism grows moderately rapidly initially as a glabrous to velvety colony that later on produces aerial mycelia. The colonies are white and may have buff to pale brown center, however, on reverse side, the color is cream to pale brown. Microscopically, there are branching septate hyphae with diameter of about 0.5-2 μm. Many conidia, with 2-5 μm diameter, are produced on conidiophores that branch at right angles from vegetative hyphae.

The conidiation in both species is enhanced on potato dextrose agar. For their differentiation, it is mandatory to grow at higher than the ambient temperature *i.e.* 25°C on enriched media. When conidia and hyphae of *Ea.parva* are inoculated on media at **40°C**, mycelium degenerates but **mononucleate** adiaconidia enlarge to a size of 15-25 μm and some of these may develop up to 40 μm or more with a 2 μm thick wall. On the other hand, *Ea.crescens* can grow *in vitro* on **phytone yeast-extract agar** within 10-12 days at **37°C** and not at 40°C. Its thick-walled **multinucleate** adiaconidia may be demonstrated, which attain diameter of 50-500 μm with wall thickness upto 70 μm **(Fig. 32.2)**. Some of the adiaconidia presented with rudimentary hyphae resemble germ tubes.

Urease test found to be positive, which is otherwise commonly seen in basidiomycetes. The exoantigen testing is also helpful to establish an early diagnosis of adiaspiromycosis, which may show non-specific precipitin lines against *B.dermatitidis* and *H.capsulatum* antisera. The PCR amplification and sequencing of rDNA gene fragments of cultured organism is also employed to confirm diagnosis.

(c) Immunodiagnosis

At present there is no reliable serological test available to establish diagnosis of adiaspiromycosis. Haplosporangin skin test has been done in a few case reports in which antigen is prepared from *Ea.parva*. However, there may be cross-reactivity with other dimorphic fungi like *Blastomyces*, *Histoplasma*, *Paracoccidioides* and *Coccidioides* species.

(d) Animal Pathogenicity

Various studies have been carried out experimentally in mice and other rodents. The conidial suspensions of *Ea.crescens* were inoculated through intranasal, intraperitoneal, subcutaneous or intravenous routes. The asexual conidia developed into adiaconidia, which are distributed in almost all soft tissues of body viscera especially lungs, liver and spleen. Experimental animal infected by intraperitoneal inoculation develop adiaconidia within granuloma in liver, spleen, kidneys, peritoneum or other organs.

Treatment and Prophylaxis

For the treatment of adiaspiromycosis, it is yet to establish that antifungal therapy has any significant role or not.

There are claims of imidazole derivatives like ketoconazole successfully tried among these patients with a dose of 200 mg per day for one year. The disappearance of symptoms and clearing of lungs may be direct result of antifungal therapy with ketoconazole. Thiabendazole has been successfully tried in some patients. Itraconazole may also be given 200 mg per day. Surgical intervention with limited resection may be required if condition of patient progresses to deterioration. The role of corticosteroids in its treatment remains more or less unexplored.

II. EMERGOMYCOSIS

Emergomycosis is a systemic fungal infection presenting with cutaneous and disseminated clinical manifestations. This is caused by three dimorphic fungal species belonging to a newly described genus *Emergomyces i.e. Emergomyces pasteurianus* comb. nov. and *Emergomyces africanus* sp. nov. and *Emergomyces orientalis* sp. nov., which were formerly classified under genus *Emmonsia* either as *Emmonsia pasteuriana* or *Emmonsia*-like species. As the name of the genus is replaced from *Emmonsia* to *Emergomyces*, consequently name of the disease has also been changed from emmonsiosis to emergomycosis.

Only a very few cases of emergomycosis caused by *Es.pasteurianus i.e.* five, have been reported in the literature till date. In South Africa, the etiological agent of this disease is another species *i.e. Emergomyces africanus*, till recently was under taxonomical speciation. It is most closely related to *Es.pasteurianus*, which is a very rare cause of disseminated infection among immunocompromised hosts. It may be merely the "tip of an iceberg" and more cases may be reported in the times to come. This disease highlights the emerging nature of this infection *i.e.* other than those caused by classical dimorphic fungal genera like *Histoplasma* and related fungi. As far as the third species is concerned, *Emergomyces orientalis*, its case report is from China.

The members of the sister genus *Emmonsia* are *Ea.parva* and *Ea.crescens*, the agents of adiaspiromycosis, a granulomatous lung disease of rodents and occasionally humans following inhalation of asexual conidia that once lodge in the lungs, transform into large adiaconidia. This has already been described in the earlier part of this Chapter and now emergomycosis is precisely being described. In this section the geographical distribution, clinical presentation, diagnostic findings and management of five cases caused by *Es.pasteurianus,* first 56 patients with disseminated caused by *Emergomyces africanus* reported in South Africa as well as a case of *Emergomyces orientalis*, are summarized to highlight this disease.

Historical Perspective

On June 5, 1999, at the ECMM Congress held in the historical city Dresden (Germany), Edouard Drouhet delivered amazing lecture on *Emmonsia pasteuriana*, the new dimorphic fungal species of genus *Emmonsia* he had identified and named to commemorate Louis Pasteur, one of the greatest scientist of 19th century. Now, this very species is renamed as *Emergomyces pasteurianus* by Dukik *et al* with its abbreviation as *Es.pasteurianus* in 2017.

The first thermally dimorphic *Emergomyces* species to be recognized was *Es.pasteurianus*, isolated in 1994 from a HIV-infected Italian woman with a disseminated mycosis. Since then, 5 cases have been reported in immunocompromised patients from Spain, China and India. *Es.pasteurianus*, is a non-adiaconidia forming species, had been isolated from one patient from Italy, one patient from Spain and two patients from China. The Indian case is the fifth one of disseminated infection in the world and the first case from India.

A new fungus causing disseminated infections in patients with AIDS was identified in 2009. Molecular identification on the basis of ITS1 and ITS2 sequencing showed that all isolates of this new species were tightly clustered and were most similar to *Es.pasteurianus* and *Ea.parva* and slightly more distantly related to *Histoplasma capsulatum*. Now, it has been named as *Emergomyces africanus* by Dukik *et al* in 2017 and is abbreviated as *Es.africanus*.

The third type of fungal strain, reported from Shanxi (China) by Wang *et al,* in 2005 was identified as *Emmonsia* species in 2009 and subsequently established as novel species of this genus *i.e. Emergomyces orientalis* in 2017 and is abbreviated as *Es.orientalis*. The disease caused by all these three species is called as **emergomycosis**.

Mycology

This polyphyletic group of pathogens includes at least seven separate taxa, among which three species are properly named. In addition, there are two other members of the sister genus, *Ea.parva* and *Ea.crescens*, these do not cause disseminated disease but rather cause a distinct clinical syndrome called adiaspiromycosis. This granulomatous lung disease is caused by the host response to adiaconidia, enormous, sessile cells that follow transformation from inhaled conidia. Infection with dimorphic *Emergomyces* species is also presumed to follow inhalation of conidia, however, instead of forming adiaconidia, they transform to yeast-like cells capable of replication, extra-pulmonary dissemination and a full fledged disease.

New species, *Emergomyces africanus*, with phylogenetic and clinical similarities to genera *Blastomyces* and *Histoplasma*, have emerged as causes of systemic human mycoses worldwide. They differ from classical *Emmonsia* species by producing a thermally dependent, yeast-like phase rather than adiaconidia and by causing disseminated infections, predominantly in immunocompromised patients and often with high mortality. Such differences are important for clinicians to consider diagnosis and management of patient as well as for microbiologists who may encounter these fungi with increasing frequency.

At present there are three species of the genus *Emergomyces*, which are known as agents of emergomycosis among the mammals. *Es.pasteurianus* is a recently described species from cases of disseminated cutaneous mycosis in an AIDS patients. This species was morphologically compatible with the genus *Emmonsia* but phylogenetically distinct from *Ea.parva* and *Ea.crescens*. In addition, at 37°C on BHI agar or in host tissue, yeast-like budding cells with a few larger thick-walled cells are formed but adiaconidia are definitely absent. The differentiating features between species of *Emergomyces vis-a-vis Ea.crescens* and *Ea.parva* are given in **Table 32.1**.

At 25°C on light microscopy, preparations show, thin hyaline septate hyphae, 1 to 1.2 µm in diameter, with numerous smooth-walled conidia that are oval to subglobose. The conidia are borne on thin pedicles (0.4 to 0.5 µm wide) that formed perpendicular to an ampulliform vesicle. These vesicles gave rise to single pedicles at first, then four to eight pedicles, each forming a terminal conidium, establishing a **floret** of four to eight conidia grouped together. Conidia are attached to the pedicles by their long axes, appearing flattened on top with slightly rough-ended walls after 21 to 30 days. Scanning electron microscopy can confirm these features.

There are three species grouped in genus *Emergomyces*. One of them was given binomial nomenclature by Drouhet *et al*, in 1999 as *Emmonsia pasteuriana*, which is now re-named as *Emergomyces pasteurianus*. *Emergomyces* is placed in family **Onygenaceae** of order **Onygenales** under class **Plectomycetes** in phylum **Ascomycota**. The mycelial colonies of this fungus have been successfully converted to yeasts in golden hamster by intratesticular inoculation.

Epidemiology

The global epidemiology of thermally dimorphic fungi causing disseminated emergomycosis has recently been reviewed. Very little is known about *Emergomyces* species or the disease caused by this fungus. Emergomycosis is caused by dimorphic fungi, which were very near to the genus *Emmonsia*. Earlier, *Es.pasteurianus* infection was described in a single human case that of an Italian patient with late-stage of AIDS but has not been found in any animal or environmental source.

There are total 56 cases caused by *Emergomyces africanus* till date, with (sub)cutaneous infections have been reported. None of these were identified as *Ea.crescens* or *Ea.parva*. Two disseminated (sub)cutaneous infections due to *Es.pasteurianus*, presented with scattered nodular lesions and some ulcerated with a central necrotic crust. Previously unknown fungus, *Emergomyces africanus* was implicated in disseminated skin infections in a case series of 18 HIV-infected adults in South Africa. Most of the patients concerned had widespread skin lesions, varying from erythematous papules and plaques with or without scales to ulcers and crustation. Only a single patient showed evidence of central umbilication, which is typically observed in dimorphic fungal skin infections such as talaromycosis and histoplasmosis. Similar to the second case of *Es.pasteurianus* infection and case 2 in Heys *et al*'s report, the patient was also an organ transplant recipient who received long-term immunosuppressive therapy. The patient developed respiratory failure soon after admission and CT of the lungs revealed widespread bilateral pulmonary infiltrates with multiple small nodules.

Molecular sequencing has since confirmed that these fungi belong to a cluster of novel *Emergomyces africanus* species. It remains unclear whether this cluster of *Es.africanus* species only emerged recently as human pathogens or whether previous infections were merely underestimated. Support for the latter hypothesis comes from South Africa, where the introduction of molecular identification tools resulted in a dramatic increase in the number of cases of disseminated emergomycosis, commensurate with a decline in the number of cases of confirmed histoplasmosis.

Es.pasteurianus is a thermally dimorphic fungus identified subsequently in very few human cases. There was a recent report of case of 43-year-old male renal transplant patient from China presenting with multiple painful skin eruptions on his head, nose and left thigh, later accompanied by respiratory failure. Histopathology of the biopsy collected from the left thigh upper ulcer and occipital nodule both demonstrated chronic inflammation with granuloma formation and yeast-like elements. *Es.pasteurianus* was cultured from two biopsy specimens and their identity was confirmed by sequencing of the rDNA internal transcribed spacer. The patient in intensive care showed marked clinical improvement with antifungal treatment.

Hence only five human cases of *Es.pasteurianus* have been reported, including an Italian woman with AIDS and a transplant recipient with HIV from Spain. The third case of (sub)cutaneous infection due to *Es.pasteurianus*; a fourth, very similar case from the same city, Guangzhou in China, was found recently in a clinic.

On the other hand, the highest burden of emergomycosis appears to be in South Africa, where a novel *Emergomyces africanus* species was recently recognized as the cause of an AIDS-related mycosis. Between 2008 and 2015, when molecular techniques were introduced to augment phenotypic identification of dimorphic fungi in South Africa, 56 cases were diagnosed, although this likely represents the "tip of the iceberg". All cases have occurred in profoundly immunocompromised individuals. Earlier reports of disease in immunocompetent patients were, in fact, due to another novel species of dimorphic fungus endemic to South Africa that appears to be more closely related to *Blastomyces* species, which eventually turned out to be a novel species, *Blastomyces percursus.*

The true geographic range of these organisms is currently unknown, although cases have been reported sporadically from North America, Europe, and Asia. The understanding of the range within sub-Saharan Africa is hampered, in part, by limited capacity for diagnostic mycology in many countries. The possibility is not excluded that other cases have been missed due to lack of awareness of the disease and misidentification of the etiological agent, as was observed with the *Es.africanus* epidemic in South Africa.

Emergomyces is a cosmopolitan soil saprobe found in all temperate climates and lives of decaying organic matter. It has been isolated from a variety of mammalian species, particularly small rodents. The geographic distribution of cases of disseminated emergomycosis was wider than previously reported and patients were diagnosed in 5 of 9 South African provinces. However, it is presumed that the true burden of disease is underestimated. Public hospitals, which provide care to roughly 85% of the population, contributed only 24% of emergomycosis cases described from outside Western Cape Province. In contrast, 92% of cases diagnosed in the Western Cape were from public hospitals.

The pulmonary route is commonly observed in dimorphic fungal disseminated skin infections. This also holds true for other case, where respiratory failure was probably due to pulmonary dissemination of the emergomycosis although no proven infection in the lungs was achieved. Successful antifungal treatment and increasing levels of β-D-glucan on G-Test strongly imply that a pulmonary infection due to *Es.pasteurianus* was the progressive underlying disease.

Disseminated emergomycosis presented a diagnostic challenge, even to experienced clinicians and a quarter of patients died prior to the diagnosis of a deep fungal infection. This series included only cases with a confirmed mycological diagnosis and thus it is likely that the true proportion of patients who remain undiagnosed or misdiagnosed is even higher. The most common diagnostic dilemma encountered by clinicians was differentiating emergomycosis from tuberculosis, inferred by the fact that three-quarters of patients were treated for later disease.

Pathogenesis and Pathology

Reports of *Emergomyces* have increased in recent years, presenting as disseminated cutaneous mycosis in persons with underlying AIDS infections. Unlike *Ea.parva* and *Ea.crescens*, *Emergomyces* does not produce adiaconidia and therefore does not cause adiaspiromycosis. *Emergomyces* species appear as yeasts in infected tissue.

Clinical Features

The clinical presentation of the patients of emergomycosis is very similar to the progressive disseminated histoplasmosis or talaromycosis among the immunocompromised hosts. The clinical features of infection include fever, loss of weight, anaemia, skin lesions similar to those in disseminated histoplasmosis and chest radiograph similar to that noted in pulmonary tuberculosis. There are mainly two types of presentations *i.e.* cutaneous and disseminated, which are being described here:-

1. Cutaneous Emergomycosis

The cutaneous presentation is most common, occurring in more than 95% of patients at the time of diagnosis. Skin lesions usually include widespread papules and/or plaques, although other morphologies can also occur. The infection frequently becomes apparent as part of an unmasking immune reconstitution inflammatory syndrome (IRIS), with cutaneous lesions appearing weeks or months after initiating antiretroviral therapy (ART).

Cutaneous manifestations of emergomycosis are also commonly misdiagnosed, most often as varicella, Kaposi sarcoma and drug reactions. The protean nature of cutaneous involvement is highlighted by the fact that deep fungal infections were considered unlikely or not considered at all, in the documented differential diagnoses of dermatologists in more than one-quarter of the cases examined by them.

All the patients are found to be anemic and all had widespread skin lesions. Morphologically, these lesions varied from erythematous papules and plaques with or without scales to ulcers and crusted, boggy plaques. The larger plaques and ulcers healed with marked scarring and the smaller lesions healed with residual hyperpigmentation. Mucosal ulceration of the penis was noted in one patient and another had scaly papules on the lips. The lesions in two patients showed evidence of central necrosis and in one there was evidence of central umbilication.

2. Disseminated Emergomycosis

The dissemination emergomycosis may lead to multifocal disease that involve extrapulmonary sites also. In retrospect, ART was usually started, along with empiric antitubercular therapy, in response to constitution and pulmonary symptoms that likely represented protean manifestations of the disease. In addition to cutaneous and lung disease, any of the body system may become involved, including hepatic, reticuloendothelial and gastrointestinal systems. Most of these patients have clinically advanced HIV disease with very low CD4+ T-cell counts.

Differential Diagnosis

The clinical findings of emergomycosis are non-specific and pulmonary disease can easily be mistaken for bacterial or *Pneumocystis* pneumonia, tuberculosis or Kaposi's sarcoma. It should be differentiated from tuberculosis where caseation is the prominent feature. The chest radiography are frequently abnormal, even when respiratory symptoms are not part of the presenting complaint. Over the last four decades, reports have emerged of patients with unusual mycoses: in the laboratory, cultures have isolated molds with asexual reproductive structures that resembled *Emmonsia* but clinical and histopathological pictures were more compatible with blastomycosis or histoplasmosis than adiaspiromycosis. At various stages of development, *Emergomyces* species may resemble other fungi in their phenotypic stages such as *Histoplasma capsulatum, Blastomyces dermatitidis, Paracoccidioides brasiliensis, Talaromyces marneffei* and *Chrysosporium* species.

Laboratory Diagnosis

As such any case of emergomycosis is defined when a patient with a clinically compatible illness has laboratory confirmation of infection with *Emergomyces* species, identified either by fungal culture or nucleic acid detection from the clinical specimens. The skin biopsy is a valuable diagnostic tool, with yeasts seen on histology in majority of the biopsied patients. Skin biopsy for histological examination is also inexpensive, non-invasive technique and has the potential for earlier turnaround of results than culture-based diagnostics.

(a) Direct Examination

The causative fungus is dimorphic in nature hence direct demonstration of yeast form is pertinent to establish the diagnosis by using various conventional and molecular techniques. The biopsy of affected tissue for histopathology is the most expedient way to diagnose a deep-seated fungal infection, although features are identical to those of other dimorphic mycoses including histoplasmosis and sporotrichosis. Therefore, KOH/CFW wet mounts, H&E, PAS and GMS stains are usually used for establishing the diagnosis. The typical yeast cells with narrow-base budding are visualized.

Emergomyces species are usually present in the infected tissue as intrahistiocytic budding yeast cells. Hence this can be easily mistaken for other fungal and/or protozoal infections characterized by the presence of histiocytes with phagocytosed organisms. While doing histopathological examination, it should be differentiated from other intracellular organisms like *H.capsulatum* and *Talaromyces marneffei*, the dimorphic fungi in their yeast phase as well as two protozoa - *Leishmania donovani* and *Toxoplasma gondii. L.donovani* contains kinetoplast, which is not present in the yeasts. The tachyzoites of *T.gondii*, being protozoa are not stained with GMS stain.

(b) Fungal Culture

Culture of skin biopsy tissue or blood on standard fungal media is usually diagnostic, although growth may take

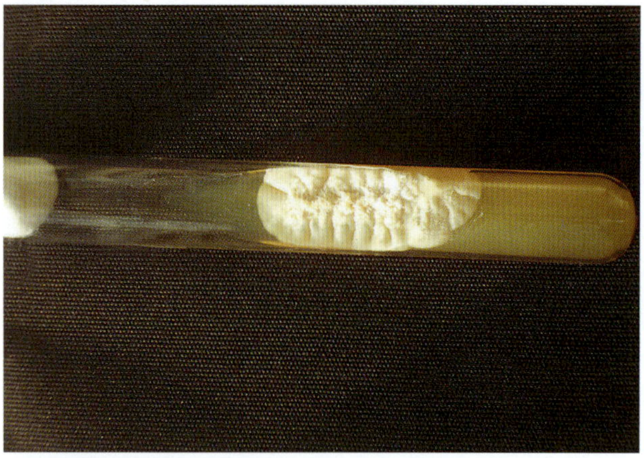

Fig. 32.3. Growth of *Emergomyces pasteurianus* showing yeast-like colonies on SDA at 37°C.

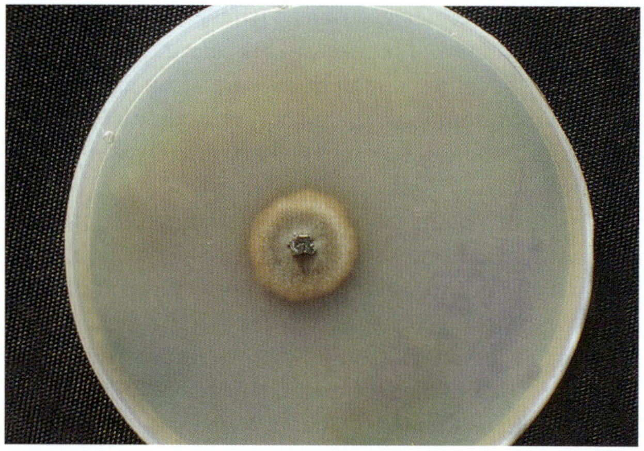

Fig. 32.4. Growth of *Emergomyces pasteurianus* showing mycelial growth on SDA at 25°C.

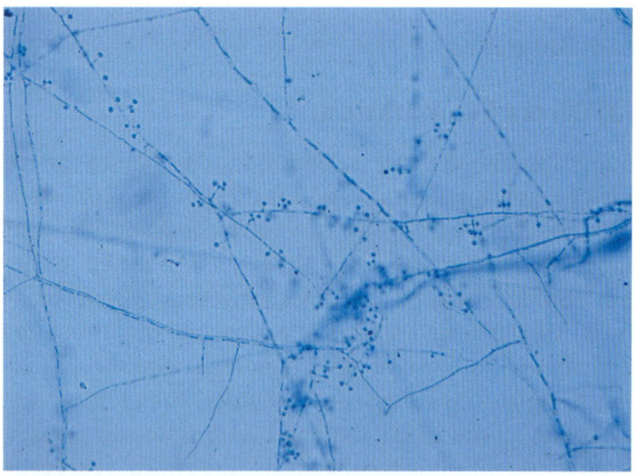

Fig. 32.5. *Emergomyces pasteurianus* showing thin hyaline septate hyphae, perpendicular slender conidiophore and terminal single or multiple conidia (LCB × 400).

up to four weeks of incubation. The sensitivity of blood cultures is enhanced by use of mycobacterial/fungal blood culture bottles. Bone marrow cultures may also yield the diagnosis.

Emergomyces is not inhibited by cycloheximide and therefore can be grown on media containing this selective agent. It exhibits slow to moderate growth at 30°C, with a colony diameter of about 25 mm after 10 days and 60 mm after 21 days. The colony appears velvety to powdery in texture with folded, wrinkled or cerebriform surface contours **(Fig. 32.3)**. The colony spontaneously acquired splits in the surface as it aged. The colony is primarily white in color and remains so and may develop a light brown color as it ages **(Fig. 32.4)**. The reverse is tan or light brown in color. On potato dextrose agar at 25°C, colonies grow moderately rapid, white to tan colored. The colonies produce no diffusible pigment as seen in *Talaromyces marneffei*. Slide cultures of the mycelial growth shows hyaline, septate hyphae. The conidia are one-celled, sessile or on stalks alongside hyphae or on ampulliform conidiophores. Initially a conidium may be found at the end of delicate pedicle, which may then develop further to form four to eight pedicles with conidia, establishing a **'floret'**. The conidia occurring in small bunches, each on its own pedicle, branching off from a central delicate pedicle **(Fig. 32.5)**.

The pure, single-conidium subcultures are obtained from primary cultures grown on Sabouraud dextrose agar, incubated at 25°C in the dark. In general, colonies grow at a slow-to-moderate rate, reaching a diameter of 15 to 25 mm after 7 to 14 days, up to a maximum of 50 to 60 mm in 21 days. They appear glabrous in some early cultures or showed densely white growth, taking on a folded, wrinkled or cerebriform appearance and became light brown with powdery segments over time. Cultures grew well at 30°C on both Sabouraud dextrose agar and brain-heart infusion agar, forming slightly larger colonies as compared with those grown at 25°C. No diffusible pigment is noted and reverse side of colonies is either tan or yellow-brown, with darker brown pigment in the folds or groves of older colonies. None of the strains grow at 40°C and no adiaconidia are produced at this temperature.

Emergomyces is a dimorphic fungus as it exhibits the filamentous form at one temperature and the yeast form at another. The filamentous fungus form can be converted to the yeast form by incubating a freshly inoculated culture at 37°C for 10 to 14 days. The conversion is enhanced by cultivation on the nutritionally richer Brain-Heart Infusion (BHI) agar. Smooth, butyrous colonies appear cream to beige in color, which may darken as they age. Colonies consist of globose to oval yeast colonies, which may show narrow-based budding.

The mycelial cultures are converted to the yeast phase by incubating streaked or single-colony subcultures on brain-heart infusion agar at 37°C after 10 to 14 days. Yeast colonies appeared smooth, with a cream-to-beige color that became light brown with age. Light microscopy reveals small, thin-walled, globose-to-oval yeasts (2-4 µm), with mostly single but occasionally multiple, polar budding yeasts formed from a narrow base attachment. The macroscopic and microscopic features of these isolates are indistinguishable from those of *Es.pasteurianus.*

At 37°C on brain heart infusion agar, a rough and wrinkled yeast-like colony are is formed. Light microscopy reveal budding yeast-like cells, some with larger, thick-walled cells at the tip of short hyphae. There are no adiaconidia are seen at 37°C.

The cultured isolates are confirmed as *Emergomyces* species with a broad-range fungal polymerase chain reaction (PCR) and sequencing of the internal transcribed interspacer region or based on phenotypic characteristics of colonies based on macroscopic and microscopic examination. To confirm the identity of the isolate, molecular verification of mycelium and yeast-like cells are performed. Internal transcribed spacer (ITS) rDNA is amplified using primers ITS 5 and ITS 4.3. ITS sequences of standard strains showing 99.3% similarity with the type strain of *Es.pasteurianus*. Molecular investigation and histopathological observations confirm and expand the mycological diagnosis, leading to a final diagnosis of a disseminated mycosis due to *Es.pasteurianus.*

(c) Immunodiagnosis

Urine *Histoplasma* antigen may prove a useful diagnostic tool, given the apparent cross-reactivity with *Emergomyces* species. Infection, wherein 2 of 3 patients with disseminated emergomycosis give positive tests. It has been noted that some cross-reaction giving false positive results may occur between *Emergomyces* species and *Blastomyces dermatitidis* with both DNA probe and direct immunofluorescent-antigen tests.

(d) Animal Pathogenicity

The yeast tissue phase of *Es.pasteurianus* was demonstrated by inoculating in golden hamster through intratesticular route establishing the dimorphic nature of the newly emerged fungal species.

Treatment and Prophylaxis

There are no significant prospective studies conducted on emergomycosis for evaluating treatment for future strategies. Therefore, in the absence of more definitive data, it is empirically recommended to follow the IDSA guidelines applied to other endemic mycoses in immunocompromised hosts. As such amphotericin B appears to be important followed by suppressive itraconazole therapy until immune reconstitution. An optimal timing of ART initiation among such patients remains unknown and clinicians should be aware of the potential for clinical worsening due to the IRIS.

The disseminated emergomycosis is invariability fatal if remain untreated. The patients are treated with an antifungal regimen that included amphotericin B are more likely to survive than those treated with a triazole alone. Moreover, both the triazoles *i.e.* fluconazole and itraconazole are associated with poor outcomes. However, in most patients treated with fluconazole, which is prescribed incidentally for esophageal candidiasis in doses and for durations considered inadequate for the treatment of severe disseminated disease caused by other endemic mycoses.

Further Reading

1. Anstead GM, Sutton DA, Graybill JR. Adiaspiromycosis causing respiratory failure and a review of human infections due to *Emmonsia* and *Chrysosporium* spp. *J Clin Microbiol.* 2012; **50:** 1346-54.
2. Bambirra EA, Nogueira AMMF. Human pulmonary granulomas caused by *Chrysosporium parvum* var. *crescens (Emmonsia crescens). Am J Trop Med Hyg.* 1983; **32:** 1184-5.
3. Barbas Filho JV, Amato MB, Deheinzelin D, *et al.* Respiratory failure caused by adiaspiromycosis. *Chest.* 1990; **97:** 1171-5.
4. Borman AM, Simpson VR, Palmer MD, *et al.* Adiaspiromycosis due to *Emmonsia crescens* is widespread in native British mammals. *Mycopathologia.* 2009; **168:** 153-63.
5. Bottone EJ. Histologic diagnosis of cutaneous adiaspiromycosis in the absence of successful culture of *Emmonsia crescens* or *Emmonsia parva. Clin Microbiol Newsletter.* 2011; **33:** 131-4.
6. Buyuksirin M, Ozkaya S, Yucel N, *et al.* Pulmonary adiaspiromycosis: The first reported case in Turkey. *Resp Med CME.* 2011; **4:** 166-9.

7. Chandler FW, Kaplan W, Ajello L. Adiaspiromycosis. *In:* Histopathology of Mycotic Diseases. Wolfe Medical Publications, London: 1980; 30-3 & 138-43.
8. Ciferri R, Montemartini A. Taxonomy of *Haplosporangium parvum*. *Mycopathol Mycol Appl.* 1959.**10:** 303-16.
9. Cueva JA, Little MD. *Emmonsia crescens* infection (adiaspiromycosis) in man in Honduras. Report of a case. *Am J Trop Med Hyg.*1971; **20:** 282-7.
10. de Almeida Barbosa A, Moreira Lemos AC, Severo LC. Acute pulmonary adiaspiromycosis - Report of three cases and a review of 16 other cases collected from the literature. *Rev Iberoam Micol.*1997; **14:** 177-80.
11. Denson JL, Keen CE, Froeschle PO, *et al.* Adiaspiromycosis mimicking widespread malignancy in a patient with pulmonary adenocarcinoma. *J Clin Pathol.* 2009; **62:** 837-9.
12. Doby-Dubois M, Chevrel ML, Doby JM, *et al.* 1st human case of adiaspiromycosis, caused by *Emmonsia crescens*, Emmons and Jellison, 1960. *Bull Soc Pathol Exot.* 1964; **57:** 240-4.
13. dos Santos VM, Fatureto MC, Saldanha JC, *et al.* Pulmonary adiaspiromycosis: Report of two cases. *Rev Soc Bras Med Trop.* 2000; **33:** 483-8.
14. Dot JM, Debourgogne A, Champigneulle J, *et al.* Molecular diagnosis of disseminated adiaspiromycosis due to *Emmonsia crescens*. *J Clin Microbiol.* 2009; **47:** 1269-73.
15. Dowding ES. The pulmonary fungus, *Haplosporangium parvum* and its relationship with some human pathogens. *Can J Res Sect E Med Sci.* 1947; **25:** 195-206.
16. Drouhet E, Huerre M. Yeast tissue phase of *Emmonsia pasteuriana* inoculated in golden hamster by intratesticular way. *Mycoses.* 1999; **42 (Suppl. 2):** 11-8.
17. Drouhet E, Gueho E, Gori S, *et al.* Mycological, ultrastructural and experimental aspects of a new dimorphic fungus *Emmonsia pasteuriana* sp. nov. isolated from a cutaneous disseminated mycosis in AIDS. *J Mycol Med.* 1998; **8:** 64-77.
18. Dukik K, Munoz JF, Jiang Y, *et al.* Novel taxa of thermally dimorphic systemic pathogens in the Ajellomycetaceae (Onygenales). *Mycoses.* 2017; **60:** 296-309.
19. Echevarria E, Cano EL, Restrepo A. Disseminated adiaspiromycosis in a patient with AIDS. *J Med Vet Mycol.* 1993; **31:** 91-7.
20. Emmons CW, Ashburn LL. The isolation of *Haplosporangium parvum* n. sp. and *Coccidioides immitis* from wild rodents. *Public Health Rep.* 1942; **57:** 1715-27.
21. Emmons CW, Jellison WL. *Emmonsia crescens* sp. nov. and adiaspiromycosis (haplomycosis) in mammals. *Ann N Y Acad Sci.* 1960; **89:** 91-101.
22. England DM, Hochholzer L. Adiaspiromycosis: An unusual fungal infection of the lung: Report of 11 cases. *Am J Surg Pathol.* 1993; **17:** 876-86.
23. Feng P, Yin S, Zhu G, *et al.* Disseminated infection caused by *Emmonsia pasteuriana* in a renal transplant recipient. *J Dermatol.* 2015. **42:** 1179-82.
24. Filho JVB, Amato MBP, Deheinzelin D, *et al.* Respiratory failure caused by adiaspiromycosis. *Chest.* 1990; **97:** 1171-5.
25. Gori S, Drouhet E. Cutaneous disseminated mycosis in a patient with AIDS due to a new dimorphic fungus. *J Mycol Med.* 1998; **8:** 57-63.
26. Heys I, Taljaard J, Orth H. An *Emmonsia* species causing disseminated infection in South Africa. *N Engl J Med.* 2014; **370:** 283-4.
27. Hubalek Z, Burda H, Scharff A, *et al.* Emmonsiosis of subterranean rodents *(Bathyergidae, Spalacidae)* in Africa and Israel. *Med Mycol.* 2005; **43:** 691-7.
28. Hubalek Z, Nesvadbova J, Halouzka J. Emmonsiosis of rodents in an agro-ecosystem. *Med Mycol.* 1998; **36:** 387-90.
29. Hubalek Z. Emmonsiosis of wild rodents and insectivores in Czechland. *J Wildl Dis.* 1999; **35:** 243-9.
30. Kamalam A, Thambiah AS. Adiaspiromycosis of human skin caused by *Emmonsia crescens*. *Sabouraudia.* 1979; **17:** 377-81.
31. Kappagoda S, Adams JY, Luo R, *et al.* Fatal *Emmonsia* sp. infection and fungemia after orthotopic liver transplantation. *Emerg Infect Dis.* 2017; **23:** 346-9.
32. Kemna ME, Weinberger M, Sigler L, *et al.* A primary oral blastomycosis-like infection in Israel. Abstr. F-75. 94th General Meeting of ASM. Washington, DC; 1994. p.601.
33. Kenyon C, Bonorchis K, Corcoran C, *et al.* A dimorphic fungus causing disseminated infection in South Africa. *N Engl J Med.* 2013; **369:** 1416-24.
34. Kenyon C, Corcoran C, Govender NP. An *Emmonsia* species causing disseminated infection in South Africa. *N Engl J Med.* 2014; **370:** 283-4.
35. Kodousek R, Vortel V, Fingerland A, *et al.* Pulmonary adiaspiromycosis in man caused by *Emmonsia crescens*: Report of a unique case. *Am J Clin Pathol.* 1971; **56:** 394-9.
36. Krivanec K. Adiaspiromycosis in Czechoslovakian mammals. *Sabouraudia.* 1977; **15:** 221-3.
37. Latge JP. Oh, to be new. *N Engl J Med.* 2013; **369:** 1464-6.
38. Lochan H, Naicker P, Maphanga T, *et al.* A case of emmonsiosis in an HIV-infected child. *S Afr J HIV Med.* 2015; **16:** 1-4.
39. Malik R, Capoor MR, Vanidassane I, *et al.* Disseminated *Emmonsia pasteuriana* infection in India: A case report and a review. *Mycoses.* 2016; **59:** 127-32.
40. Maphanga TG, Britz E, Zulu TG, *et al. In vitro* antifungal susceptibility of the yeast- and mould-phases of the dimorphic fungal pathogen, *Emergomyces africanus* (formerly *Emmonsia* species) from HIV-infected South African patients. *J Clin Microbiol.* 2017; **55:** 1812-20.
41. Marin-Felix Y, Stchigel AM, Cano-Lira JF, *et al. Emmonsiellopsis*: A new genus related to the thermally dimorphic fungi of the family Ajellomycetaceae. *Mycoses.* 2015; **58:** 451-60.
42. Matsuda K, Niki H, Yukawa A, *et al.* First detection of adiaspiromycosis in the lungs of a deer. *J Vet Med Sci.* 2015; **77:** 981-3.
43. Mendes MO, Moraes MA, Renoiner EI, *et al.* Acute conjunctivitis with episcleritis and anterior uveitis linked to adiaspiromycosis and freshwater sponges, Amazon region, Brazil, 2005. *Emerg Infect Dis.* 2009; **15:** 633-9.

44. Munoz JF, Gauthier GM, Desjardins CA, *et al.* The dynamic genome and transcriptome of the human fungal pathogen *Blastomyces* and close relative *Emmonsia*. *PLoS Genet.* 2015; **11:** e1005493. PMID: 26439490.
45. Mutyaba A, Sonderup MW, Locketz M, *et al.* Disseminated *Emmonsia* in an HIV-HBV co-infected man. *ID Cases.* 2015; **2:** 35-6.
46. Nuorva K, Pitkanen R, Issakainen J, *et al.* Pulmonary adiaspiromycosis in a two-year old girl. *J Clin Pathol.* 1997; **50:** 82-5.
47. Otcenasek M, Ditrich O. Comparative susceptibility of the agents of adiaspiromycosis to imidazole derivatives *in vitro. Sabouraudia.* 1983; **21:** 239-42.
48. Pelegrin I, Alastruey-Izquierdo A, Ayats J, *et al.* A second look at *Emmonsia* infection can make the difference. *Transpl Infect Dis.* 2014; **16:** 519-20.
49. Pelegrin I, Ayats J, Xiol X, *et al.* Disseminated adiaspiromycosis: Case report of a liver transplant patient with human immunodeficiency infection and literature review. *Transpl Infect Dis.* 2011; **13:** 507-14.
50. Peres LC, Figueiredo F, Peinado M, *et al.* Fulminant disseminated pulmonary adiaspiromycosis in humans. *Am J Trop Med Hyg.* 1992; **46:** 146-50.
51. Peterson SW, Sigler L. Molecular genetic variation in *Emmonsia crescens* and *Emmonsia parva*, etiologic agents of adiaspiromycosis and their phylogenetic relationship to *Blastomyces dermatitidis* (*Ajellomyces dermatitidis*) and other systemic fungal pathogens. *J Clin Microbiol.* 1998; **36:** 2918-25.
52. Pfaller MA, Diekema DJ. Unusual fungal and pseudofungal infections of humans. *J Clin Microbiol.* 2005; **43:** 1495-504.
53. Randhawa HS. Respiratory and systemic mycoses: An overview. *Indian J Chest Dis Allied Sci.* 2000; **42:** 207-19.
54. Roy P, Sahni AK, Sriram R *et al. Sporotrichum pruinosum* causing a rare invasive infection in an immunocompromised patient. *Med J Armed Forces India.* 2015; **71:** S456-9.
55. Sanche S, Wong A, Sigler L, *et al.* Invasive infection caused by a novel *Emmonsia* species in a renal transplant patient. *Focus on Fungal Infections.* Miami; 2005. Abstract 87.
56. Schwartz IS, Govender NP, Corcoran C, *et al.* Clinical characteristics, diagnosis, management and outcomes of disseminated emmonsiosis: A retrospective case series. *Clin Infect Dis.* 2015; **61:** 1004-12.
57. Schwartz IS, Kenyon C, Feng P, *et al.* 50 years of *Emmonsia* disease in humans: The dramatic emergence of a cluster of novel fungal pathogens. *PLoS Pathog.* 2015; **11:** e1005198.
58. Schwartz IS, Kenyon C, Thompson III GR. Endemic mycoses: What's new about old diseases? *Curr Clin Micro Rpt.* 2016; **3:** 71-80.
59. Schwarz J. Adiaspiromycosis. *Pathol Annu.* 1978; **13:** 41-53.
60. Severo LC, Geyer GR, Camargo JJ, *et al.* Adiaspiromycosis treated successfully with ketoconazole. *J Med Vet Mycol.* 1989; **27:** 265-8.
61. Sigler L, Peterson SW. Molecular genetic analysis supports recognition of new species among *Emmonsia* and *Blastomyces* isolates. *Int J Antimicrob Agents.* 2009; **34:** S593.
62. Sigler L. *Ajellomyces crescens* sp. nov. taxonomy of *Emmonsia* spp. and relatedness with *Blastomyces dermatitidis* (teleomorph *Ajellomyces dermatitidis*). *J Med Vet Mycol.* 1996; **34:** 303-14.
63. Stebbins WG, Krishtul A, Bottone EJ, *et al.* Cutaneous adiaspiromycosis: A distinct dermatologic entity associated with *Chrysosporium* species. *J Am Acad Dermatol.* 2004; **51:** S185-9.
64. Sun Y, Bhuiya T, Wasil T, *et al.* Fine needle aspiration of pulmonary adiaspiromycosis: A case report. *Acta Cytol.* 2007; **51:** 217-21.
65. Tang XH, Zhou H, Zhang XQ, *et al.* Cutaneous disseminated emmonsiosis due to *Emmonsia pasteuriana* in a patient with cytomegalovirus enteritis. *JAMA Dermatol.* 2015; **151:** 1263-4.
66. Thammayya A Sanyal M. *Chrysosporium parvum* from eczematous lesions in man. *Bull Calcutta Sch Trop Med.* 1973; **21:** 23-4.
67. Tomssikova A, Hejtmanek M. Antigenic activity of *Emmonsia crescens* mutants. *Mycopathologia.* 1978; **66:** 83-90.
68. Turner D, Burke M, Bashe E, *et al.* Pulmonary adiaspiromycosis in a patient with acquired immunodeficiency syndrome. *Eur J Clin Microbiol Infect Dis.* 1999; **18:** 893-5.
69. van Hougenhouck-Tulleken WG, Papavarnavas NS, Nel JS, *et al.* HIV-associated disseminated emmonsiosis, Johannesburg, South Africa. *Emerg Infect Dis.* 2014; **20:** 2164-6.
70. Wang P, Guo LN, Fan HW, *et al.* A dimorphic fungus *Emmonsia* sp. caused a disseminated infection in a diabetic patient in China. Abstr. ISHAM Satellite Congress, Beijing, 2009; 92.
71. Wang P, Kenyon C, de Hoog S, *et al.* A novel dimorphic pathogen, *Emergomyces orientalis* (Onygenales), agent of disseminated infection. *Mycoses.* 2017; **60:** 310-9.
72. Watts JC, Callaway CS, Chandler FW, *et al.* Human pulmonary adiaspiromycosis. *Arch Pathol.* 1975; **99:** 11-5.
73. Watts JC, Chandler FW. Adiaspiromycosis: An uncommon disease caused by an unusual pathogen. *Chest.* 1990; **97:** 1030-1.
74. Wellinghausen N, Kern WV, Haase G, *et al.* Chronic granulomatous lung infection caused by the dimorphic fungus *Emmonsia* sp. *Int J Med Microbiol.* 2003; **293:** 441-5.

CHAPTER 33

Allergic Fungal Diseases

Many fungi are known to be significant cause of allergy among human beings. The clinical entities produced by these organisms are designated with different names. In Medical Mycology, traditionally disease names are ending with suffix ***'mycosis'*** but allergic diseases are not called mycosis rather known by name of the allergen involved. The fungi can cause hypersensitivity reactions including Types I, III and IV.

The hypersensitivity to inhalation of conidia, fungal products or debris can manifest in various ways. The atopic individuals are more prone to allergic fungal diseases. It may involve normal individuals who are chronically exposed to conidia of a particular fungus during their occupational activities. The protective immunity to *Candida* and other superficial fungal pathogen is based on the induction of cellular immunity, cytokines and effector phagocytic cells. An interleukin 12 (IL-12)-mediated Th1 host response can resist *Candida* infection. Skin recall antigen testing with *Candida* whole cell lysate gives a DTH indicating that CD4+ cells play a crucial role in antifungal defences. Eosinophils and their stored mediators in response to fungal allergen lead to allergic fungal sinusitis.

The allergic manifestations in dermatophytosis and candidiasis *i.e.* **'id' reactions**, have already been discussed in previous **Chapters 10 and 20**, respectively. Some of the Gram-positive filamentous bacteria *i.e.* actinomycetes, are also producing similar type of disease and differentiation of bacteria and fungi, on clinical grounds, becomes difficult, Hence, these are mentioned along with the fungi just to enrich the discussion on this subject matter. A list of few well-defined allergic syndromes of respiratory system caused by different fungi and actinomycetes is given in **Table 33.1**.

1. Farmer's Lung Disease

This is an allergic disease of lungs induced by inhalation of fungal as well as bacterial antigens present in the moldy hay. It is an immunologically mediated inflammatory disease of the lungs involving the terminal airways. Most frequently antigens belong to molds and also to thermophilic actinomycetes. It is often called extrinsic allergic alveolitis or hypersensitivity pneumonitis. The most common fungi and bacteria responsible for Farmer's lung are depicted in **Table 33.2**.

Table 33.1. Allergic Respiratory Diseases caused by Different Fungi and Actinomycetes.

Allergic Disease	*Source of Allergy*	*Etiological Agents*
Farmer's lung	Stored hay	*Faenia rectivirgula, T.vulgaris*
Bagassosis	Sugar cane	*Thermoactinomyces saccharii*
Maple bark stripper's lung	Maple tree bark	*Cryptostroma corticale*
Cheese washer's disease	Cheese	*Penicillium casei*
Maltster's lung	Barley malt	*Aspergillus clavatus*
Sequoiosis	Redwood sawdust	*Aureobasidium pullulans*
Suberosis	Cork	*Penicillium frequentans*
Wood pulp worker's disease	Wood pulp	*Alternaria species*
Humidifier lung	Humidifier air-conditioner	*T.vulgaris, T.candidus*

Table 33.2. Causative Agents of Farmer's Lung.	
A.	**Fungal Agents - Hyaline Hyphomycetes**
	Aspergillus fumigatus, A.flavus, A.niger
	Penicillium simplicissimum, P.rubrum
B.	**Bacterial Agents - Actinomycetes**
	Faenia rectivirgula (Micropolyspora faeni)
	Thermoactinomyces (Micromonospora) candidus
	Thermoactinomyces (Micromonospora) vulgaris

This disease is found more often in temperate than tropical zones particularly in areas of high rainfall where conditions for mold on hay are conducive to optimum growth. It mostly occurs in late winters and early spring, depends on the type of farming practices and the climate. Majority of patients are between third and fourth decade of their life. Males are usually more affected than the females.

The colonization over moist hay by bacteria results in elevation of temperature of hay due to fermentation of substrate. As temperature increases, thermophilic actinomycetes begin to grow faster and supersede organisms that are susceptible to higher degrees. Some thermotolerant fungi also grow well in hay. There is repeated exposure after sensitization, which results in severe disease. In dry places these are easily aerosolized and inhalation by sensitive individual induces an asthmatic reaction. It is an occupational disease of individual who frequently come across with stored hay and similar reservoirs of these fungi and actinomycetes. Apparently Farmer's lung is mediated by these allergens through Type I or Type III hypersensitivity reactions.

There are three main stages of clinical manifestations of Farmer's lung:

(a) Acute Stage: This is found in harvesters and thrashers, who have overwhelming initial exposure to conidia.

(b) Subacute Stage: This stage is seen in harvesters, animal keepers and grain mill workers who have long-spaced exposure periods.

(c) Chronic Stage: The chronic stage is found in mill workers, who have low-level but constant exposure to the allergens.

The initial response is granulomatous reaction associated with an interstitial fibrosis in the lungs. If exposure to moldy hay is prolonged, progressive pulmonary fibrosis and sensitization follows. Sometimes, permanent pulmonary damage may occur even after a single exposure.

There are three serological methods, which have been described to establish diagnosis of Farmer's lung. These are agar gel double diffusion, biotin-avidin linked enzyme immunoassay and dot immunobinding assay. As far as managing this disease, the first line of treatment involves avoidance of exposure to the causative agents. However, acute and/or severe cases may require oral corticosteroids over a variable period of time ranging from weeks to even months.

2. Bronchial Asthma

In this disease there are sprouted conidia and mycelia in bronchial lumen and patients have increased IgG as well as IgE antibodies. These fungal elements are common allergens in triggering bronchial asthma. Of the estimated number of more than one million different fungal species, approximately 80 fungi have been connected with respiratory allergy. The following fungi are usually responsible for this clinical entity *i.e. Alternaria, Bipolaris, Cladosporium, Penicillium* and *Aspergillus* species. There are seasonal variations, which correspond to peak growth of the involved fungi. The colony morphology and typical microscopic appearance of *Alternaria* species having muriform conidia with horizontal and transverse septae, are shown in **Figures 33.1 and 33.2**.

The childhood form of asthma reaches its peak in early adolescence and gradually subsides. The other form of disease begins after forty and may become severe with increasing age. This type of disease may have more rapid and fatal course if patient develops status asthmatics. The reaginic antibodies (IgE) to allergic substances are detected among these patients. There may be an Arthus-type of reaction later on, which is IgG-mediated and may further complicate the disease process. Severe asthma with fungal sensitization (SAFS), fungal-associated (severe) asthma is another clinical presentation associated with this disease.

3. Allergic Bronchopulmonary Aspergillosis

Allergic bronchopulmonary aspergillosis (ABPA) is also called mucomembranous *Aspergillus* bronchitis or (Hinson-Pepys disease). This is a hypersensitivity reaction to the presence of *Aspergillus* species colonizing in the bronchial tree. With increasing awareness among physicians and improved diagnostic modalities, ABPA is no longer an uncommonly identified entity. This form of aspergillosis may develop as an exaggeration of above disease, in chronic or extrinsic asthma, late onset asthma or as some

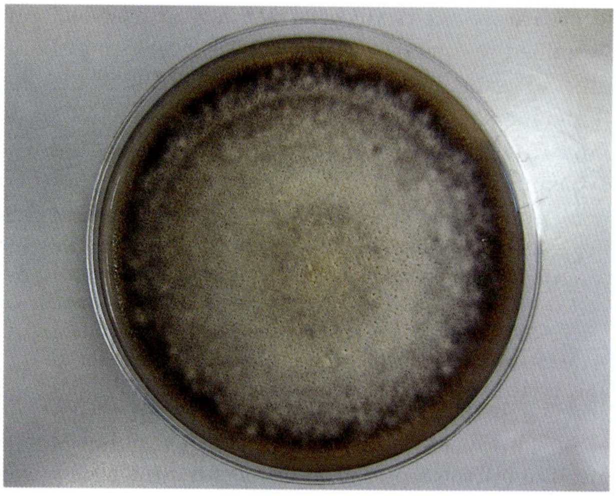

Fig. 33.1. Grayish-black mycelial phaeoid colonies of *Alternaria* species on SDA after three days of incubation.

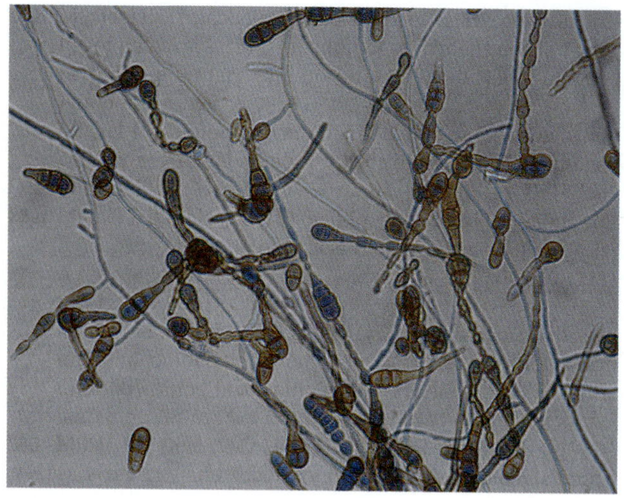

Fig. 33.2. Muriform conidia with horizontal and transverse septae seen in *Alternaria* species (LCB × 400).

other manifestations of atopic diathesis. About 5% patients with asthma and allergic rhinitis have features indicative of ABPA, like raised IgG and IgE against *A.fumigatus*, eosinophilia and characteristic radiological findings.

The clinical features of ABPA are almost similar to those of asthma but are more chronic and severe. Bronchial plugging may also be there. The peripheral shadows are consistent with areas of collapse distal to bronchial occlusion resembling pulmonary infiltrate and eosinophilia (PIE) syndrome. Bronchoscopy reveals fungus growing in patches of various sizes, perpendicularly to bronchi. Fruiting heads and conidia are sometimes produced within the bronchi. The underlying mucous membrane is congested but not invaded. Coughing produces sputum in which fungus may be seen by direct smear examination, which should be cultured on appropriate media. However, positive sputum culture is regular finding and persists during entire length of disease. The Charcot-Leyden crystals may also be seen.

An Immediate as well as Delayed-Type skin reactions are demonstrated among these patients. In most of the cases, Arthus Type III hypersensitivity reaction is elicited. The pathology of disease is thought to be owing to both Immediate and Delayed Types of hypersensitivity. The most frequently isolated organism, *A.fumigatus*, is known to produce endotoxins and C-substance and former may also contribute to the disease. An absolute eosinophilia is present among these patients. Precipitins are demonstrable in sera of almost all patients. Thus both IgG and IgE levels are significantly increased in this disease. The IgE antibodies may be specific for *Aspergillus* species involved. It is considered that ABPA is one of the most common causes of pulmonary eosinophilia.

The other diagnostic criteria are: episodic obstruction, rather than a restriction deficit, one-second forced expiratory volume (FEV_1) less than 70% of capacity and abundant fungi in culture (*See* **Chapter 25**). The later criterion consists of culturing more than two colonies of fungus from one specimen. The sputum produced is gelatinous and sometimes blood tinged hence confusing with tuberculosis.

The radiological picture is that of segmental shadowing and a migrating type of recurrent pneumonitis, usually in upper lobe and especially in apical segments. The other lobes and lingula are seldom involved. The radiological shadows of mucoid impaction are continuous with hilum proximally and distally blunted. The bronchial wall thickening is a common and regular finding. Dilated bronchi sometimes appear opaque and other times lucent. Cylindrical bronchiectasis of upper lobes is seen on bronchograms. A prominent feature of this disease is marked change in X-ray patterns associated with relatively mild clinical symptoms but massive shadowing, however, band shadows and *'gloved fingers'* are infrequent and cavitation is rare **(Fig. 33.3)**.

The features of ABPA are basically characterized by asthma, recurrent roentgenographic infiltrates, peripheral blood and sputum eosinophilia with significant lung damage. The **central bronchiectasis** is one of the complications of ABPA and its presence clinches the diagnosis.

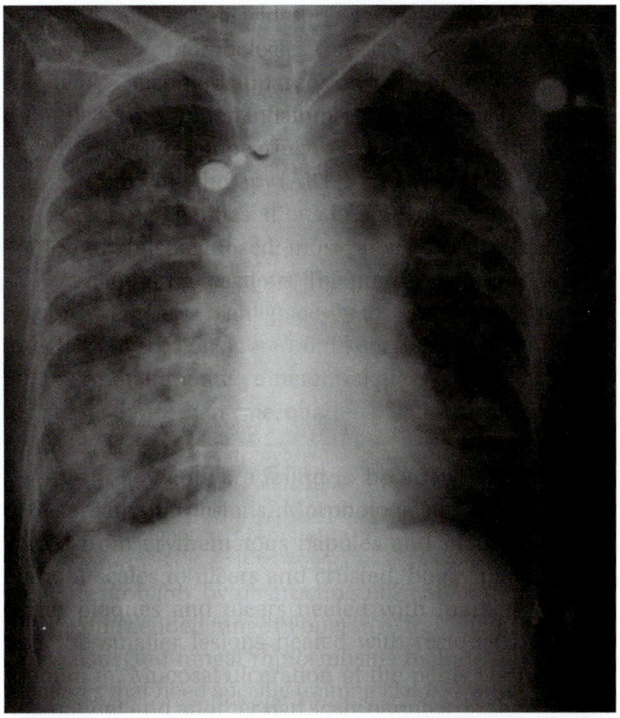

Fig. 33.3. X-rays chest of an ABPA patient showing ring shadows in both lung fields suggestive of bilateral bronchiectasis.

Radiologically **fleeting** pulmonary opacities on serial X-rays are considered to be an important indicator suggestive of this disease.

Although most of the patients with ABPA are adults in whom disease developed as sequelae of asthma, considerable numbers of pediatric cases are also being described. The fungi grow in lumen of airways hence systemic antifungal agents are of no use. Steroids and bronchodilators are effective in its management thereby averting subsequent complications in the clinical condition. In case of patients who are being tapered off steroids, antifungal may be given and itraconazole is the best in the dose of 400 mg per day for a period of 3 months.

Criteria for the diagnosis of ABPA in patients with asthma include: (i) asthma (ii) positive skin test for fungal allergens (iii) elevated IgE levels (iv) *Aspergillus*-specific IgE and (v) proximal bronchiectasis. Similar criteria for ABPM are not established but may include elevated IgE levels, positive skin tests and response to systemic steroids. ABPA is often misdiagnosed as pulmonary tuberculosis by an inexperienced clinician. This is due to lack of awareness, index of suspicion and requisite diagnostic facilities especially serological investigations.

Aspergillus species and other molds are widely distributed in outdoor and indoor environments and are capable of causing variety of hypersensitive respiratory disorders, predominantly extrinsic asthma and allergic rhinitis. In addition, *Aspergillus* and some other fungi are also known to cause fungal sinusitis and ABPA. As prevalence of bronchial asthma and other hypersensitive lung disorders has steadily increased in last three decades, there is renewed interest in study of aero-allergens. In the geographical areas where *A.flavus* is found commonly associated, the prominent cause of ABPA still remains as *A.fumigatus* may be because of the size of its spores, being smaller than *A.flavus*.

Some species of *Penicillium* genus like *P.chrysogenum* are also involved in pathogenesis of a few respiratory diseases, such as bronchitis, bronchial asthma and extrinsic allergic alveolitis (EAA), also known as hypersensitivity pneumonitis, a lung disease occurring in a variety of occupational environment. As there is also involvement of other fungi than *Aspergillus* hence it is now called Allergic Bronchopulmonary Mycoses (ABPM). Various fungi, other than *Aspergillus* species are responsible for the ABPM. This is similar in presentation to allergic bronchopulmonary aspergillosis (ABPA), which is typically seen in patients with asthma or cystic fibrosis. There is suggestion that allergic fungal sinusitis and allergic bronchopulmonary mycosis may actually be a continuum of disease and should be referred to as Sinobronchial Allergic Mycosis (SAM).

Early diagnosis of ABPA is important to prevent irreversible damage of the bronchi and the lungs. The tissue specimen or lavage can be directly examined or may be transferred to relevant media for culture and morphological identification of fungal species. Skin Prick test (SPT) can identify whether patient is sensitive to a particular antigen preparation if the immediately developing wheal diameter is greater than 4 mm. SPT is simple, provides results within minutes and is inexpensive. Serological test with total IgE or specific IgE antibodies, an ImmunoCAP assay can measure specific IgE antibodies to rAsp f1. Substantial cross reactivity among allergens of various fungal species and genera, often, leads to erroneous diagnosis. Currently, about 70 fungal allergens have been approved by the International Allergen Nomenclature Committee. However, not many of these allergens have been analyzed for their diagnostic specificity and sensitivity.

This is also resolved now that *A fumigatus*-specific IgE is sufficient to detect *Aspergillus* hypersensitivity. The assay for *A.niger*-specific IgE done earlier is redundant, less sensitive rather unnecessary if the assay for specific IgE for *A.fumigatus* is performed to establish the diagnosis.

Peterson vs. Ritesh Aggarwal Criteria

ABPA is still underdiagnosed in India despite high prevalence, more so because of misdiagnosis of tuberculosis as chest X-rays findings are often very similar. According to Patterson, ABPA is divided into I-IV stages:

A. Stage I-Acute
B. Stage II-Remission
C. Stage III-Exacerbation
D. Stage IV-End stage.

Criteria for ABPA (Patterson): Asthma, proximal bronchiectasis (dilated bronchi in 2/3rd of CT); immediate reaction to *Aspergillus* antigen, high total serum IgE >417 KIU/L or 1000 μg/ml; high specific IgE or IgG for *A.fumigates* and (any Two of the following):

- Presence of transient or fixed pulmonary infiltrates (fleeting shadows)
- Peripheral blood eosinophilia >500 cells/μl
- Serum precipitating antibodies (or IgG) against *A.fumigatus*
- Central/proximal bronchiectasis with normal tapering of distal bronchi

Criteria for follow up of patients with IgE levels:

(a) Remission-IgE levels fall by >35%. Chest X-rays lesion clear after 6 weeks of steroids.
(b) Relapse or exacerbations-double levels of IgE rises
(c) Complete remission - no exacerbations after 3 months of stopping therapy, monitor with IgE levels at 6 months and later once annually.
(d) End stage ABPA - extensive bronchiectasis or Cor pulmonale or type II respiratory failure

Well-recognized and approved **Patterson** criteria includes presence of more than 6 criteria.

However, **Agarwal** criteria - based on the presence of all obligatory criteria and three of five other criteria have simplified the diagnostic criteria further.

Obligatory criteria: Asthma, elevated total IgE levels (1000 μg/ml) elevated serum *A.fumigatus* specific IgE (0.35 kU_A/L).

Other Criteria: Type I *Aspergillus* skin test positive, presence of serum precipitating antibodies against *A.fumigatus*, radiographic pulmonary opacities (fixed/transient), total eosinophilia (1000 cells/μl), central bronchiectasis on high resolution computed tomography of the chest.

4. Allergic Fungal Rhinosinusitis

Allergic fungal sinusitis is an increasingly recognised cause of refractory chronic sinusitis in young immunocompetent hosts, analogous to allergic bronchopulmonary aspergillosis, a related process in lower respiratory tract. Allergic fungal rhinosinusitis (AFRS) is found in patients with allergic *Aspergillus* sinusitis who have recurrent bouts of severe sinusitis with *A.fumigatus* or other fungi growing *in situ*. The drainage shows mucinous material containing eosinophils, Charcot-Leyden crystals and fungal hyphae. The patients with AFRS tend to have elevated total serum IgE and fungal specific IgG at diagnosis but not fungal-specific IgE or precipitins.

According to Bent & Kuhn the major criteria include a history of type I hypersensitivity by skin testing or *in vitro* testing; nasal polyposis; computed tomography (CT) scan findings; the presence of eosinophilic mucin without invasion; and a positive fungal stain of sinus contents removed at the time of surgery. The minor criteria include a history of asthma, unilateral predominance of disease, radiographic evidence of bone erosion, fungal cultures, presence of Charcot-Leyden crystals in surgical specimens and serum eosinophilia.

The allergic fungal rhinosinusitis may also be caused by phaeoid fungi like *Alternaria alternata, Bipolaris hawaiiensis, Curvularia lunata, Exserohilum rostratum* and *Cladosporium* species.

The histopathologic findings in haematoxylin & eosin (H&E) in AFRS are important for its diagnosis, which shows typical inflammatory infiltrate composed of eosinophils, lymphocytes and plasma cells. The various types of fungal rhinosinusitis are depicted in **Table 34.3** as well as discussed in detail in **Chapter 34**.

5. Maple Bark Stripper's Disease

This disease is mainly caused by inhalation of conidia of *Cryptostroma corticale*, usually in individuals working with maple bark tree and it is also called **Cryptostromosis**. The clinical features include cough, dyspnoea, fever, night sweats and substernal pain and they may resolve uneventfully.

The lung biopsy shows small cream-colored granulomas. These chronic granulomatous changes consist of histiocytes, fibroblasts, giant cells, lymphocytes and a few neutrophils. Sometimes small conidia, 4-5 μm, are present within phagocytic cells. These are brownish-red in color without budding. Initially, it was confused with yeast cells of *Histoplasma capsulatum* but conidia were shown to be from *C.corticale*. The inhaled conidia do not grow in lung and may remain dormant. When lung tissue containing spores is inoculated on Sabouraud dextrose agar and incubated at 25°C, white pleomorphic colonies of fungus develop. The hyphae are septate and branched and conidia are oval 4-5 µm and brownish-red in color. The patients show positive skin tests and precipitins by immunodiffusion test. The allergic pneumonitis caused by *Alternaria* species growing on logs is called **wood pulp worker's disease**.

6. Scopulariopsosis

Allergic scopulariopsosis is caused by *Scopulariopsis brumptii*, however, *S.brevicaulis* is more common in other type of fungal infections. This is almost similar to cryptostromosis, with additional respiratory distress among the patients. On chest X-rays, several nodular shadows are seen in the lower lung fields. Histopathological examination of nodule may reveal granulomatous reaction and presence of many brown bodies of 4-7 µm. The material when cultured on fungal culture media shows numerous greyish white fungal colonies. The organism can be identified as *S.brumptii* by its typical morphology. Antibodies can be demonstrated in sera of these patients. The disease may resolve uneventfully even without treatment. This may also be reproduced in experimental animals.

7. Bagassosis

Bagassosis is caused by inhalation of dust from bagasse that is material left after juice has been expressed from sugar cane. This is used in making wallboard and similar items and not burned as it was done previously. In addition to fiber, silica and quartz, bagasse dust contains numerous fungal spores of various species. The thermophilic actinomycetes are also involved in this disease.

The clinical manifestations of this disease are cough, dyspnoea, fever and blood streaked sputum. The disease should be differentiated from farmer's lung. On chest X-rays, there is mottling in lung fields. Histopathology shows interstitial inflammatory reaction with giant cells. Intra-alveolar foam cells are seen in early stages of disease. The giant cells are observed to contain birefringent material. As disease progresses interstitial fibrosis occurs along with chronic non-specific changes in lungs. A continuous exposure for prolonged duration may result in fatal pulmonary disease. These patients have precipitating antibodies in their sera.

8. Byssinosis

Byssinosis is similar to bagassosis but involves inhalation of dust from cotton processing plants *i.e.* **dusty cotton disease**. It has been observed that pericarp of cottonseed pods is an active material and cause histamine release. There is bronchoconstriction and in chronic exposure, inflammatory, fibrotic and non-specific changes may occur. The fungal conidia are found in dust but their contribution to allergy is not that frequent.

9. Lycoperdonosis

Lycoperdonosis is caused by inhalation of conidia of a fleshy basidiomycete, *Lycoperdon* species, commonly called **puffball**. The disease is also referred to as extrinsic allergic alveolitis or hypersensitivity pneumonitis. This fungus is used to treat epistaxis in folk medicine. Inhalation of large number of conidia results in this allergic pulmonary disease. The clinical features include fever, cough and crepitations. The conidia can be seen in sputum and they sprout when placed on culture media. Chest X-rays reveal fine miliary nodules in lung fields.

Laboratory Diagnosis

The allergic fungal infections are established in the diagnostic mycology laboratory depending up on the disease involved in a particular patient. The diagnostic protocol is conventional based with direct demonstration of fungi in the clinical material followed by cultivation and serological procedures by demonstrating either antigen or antibodies in serum samples and skin testings.

As far as the investigations of environmental factors is concerned, the concentrations of *Aspergillus* species below 1 CFU/mm^3 may be sufficient to cause infection among high-risk patients. However, new group of patients without classical risk factor of neutropenia like chronic pulmonary disease, critically ill and HIV are emerging. A high concentrations of average count of 100 CFU/mm^3 may potentially increase the risk for development of invasive aspergillosis in hospital-attending patients.

Treatment and Prophylaxis

This is essential that all those conditions which produce allergy, should be prevented by avoiding exposure to particular allergen. Hence, exposure to fungi is reduced thereby to decrease the chances of getting allergic fungal disease. Moreover, removing or cleaning mold-laden objects and maintaining low humidity *i.e.* less than 50%, may be beneficial in reducing fungal allergens. Using dehumidifier and/or air-conditioning unit can reduce both mold and filamentous bacteria. Therefore, care must be taken to ensure that the dehumidifiers or air-conditioning units do not get contaminated with mold, which act new source of allergens or non-specific irritants. In tropical and subtropical climates, fungi might grow on walls of house due to water leakage and humidity. Hence to avoid such fungal growth, walls may be tiled or cleaned as and when necessary.

Drug therapy is given by prescribing anti-allergic drugs, as there is no significant role of antifungal agents. Second generation antihistamines are not only effective to control the symptoms of allergic rhinitis but are mostly devoid of side effects. Topical (intranasal) glucocorticosteroids are one of the most effective modes of therapy since they control nasal congestion, one of the most difficult symptom associated with allergic rhinitis to resolve. Intranasal steroids (INS) are the most effective medication in controlling the symptoms of allergic rhinitis. INS may provide significant relief of symptoms of intermittent allergic rhinitis. Apart from isotonic and hypertonic nasal sprays, which are already widely used, alternative methods, such as thermal water applications, endonasal phototherapy or acupuncture are available and have been investigated in clinical trials. Allergen immunotherapy (AIT) is an immune-modulating therapy aiming at restoring normal immunity against allergens.

The general prophylactic measures are adopted to avoid the fungal spores taking entry in the body. The masks like N-95 may be used to decrease the intensity of exposure to various fungi.

Further Reading

1. Agarwal K, Gaur SN, Chowdhary A. The role of fungal sensitisation in clinical presentation in patients with chronic obstructive pulmonary disease. *Mycoses.* 2015; **58:** 531-5.
2. Agarwal R, Aggarwal AN, Dhooria S, *et al.* A randomised trial of glucocorticoids in acute-stage allergic bronchopulmonary aspergillosis complicating asthma. *Eur Respir J.* 2016; **47:** 490-8.
3. Agarwal R, Aggarwal AN, Garg M, *et al.* Allergic bronchopulmonary aspergillosis with aspergilloma: An immunologically severe disease with poor outcome. *Mycopathologia.* 2012; **174:** 193-201.
4. Agarwal R, Aggarwal AN, Garg M, *et al.* Cut-off values of serum IgE (total and *A.fumigatus*-specific) and eosinophil count in differentiating allergic bronchopulmonary aspergillosis from asthma. *Mycoses.* 2014; **57:** 659-63.
5. Agarwal R, Aggarwal AN, Gupta D, *et al.* Case report: A rare cause of miliary nodules-allergic bronchopulmonary aspergillosis. *Br J Radiol.* 2009; **82:** e151-4.
6. Agarwal R, Aggarwal AN, Gupta D, *et al. Aspergillus* hypersensitivity and allergic bronchopulmonary aspergillosis in patients with bronchial asthma: Systematic review and meta-analysis. *Int J Tuberc Lung Dis.* 2009; **13:** 936-44.
7. Agarwal R, Aggarwal AN, Gupta N, *et al.* A rare cause of acute respiratory failure-allergic bronchopulmonary aspergillosis. *Mycoses.* 2011; **54:** e223-7.
8. Agarwal R, Aggarwal AN, Sehgal IS, *et al.* Performance of serum galactomannan in patients with allergic bronchopulmonary aspergillosis. *Mycoses.* 2015; **58:** 408-12.
9. Agarwal R, Aggarwal AN, Sehgal IS, *et al.* Utility of IgE (total and *Aspergillus fumigatus* specific) in monitoring for response and exacerbations in allergic bronchopulmonary aspergillosis. *Mycoses.* 2016; **59:** 1-6.
10. Agarwal R, Chakrabarti A, Shah A, *et al.* Allergic bronchopulmonary aspergillosis: Review of literature and proposal of new diagnostic and classification criteria. *Clin Exp Allergy.* 2013; **43:** 850-73.
11. Agarwal R, Chakrabarti A. Allergic bronchopulmonary aspergillosis in asthma: Epidemiological, clinical and therapeutic issues. *Future Microbiol.* 2013; **8:** 1463-74.
12. Agarwal R, Denning DW, Chakrabarti A. Estimation of the burden of chronic and allergic pulmonary aspergillosis in India. *PLoS One.* 2014; **9:** e114745.
13. Agarwal R, Devi D, Gupta D, *et al.* A questionnaire-based study on the role of environmental factors in allergic bronchopulmonary aspergillosis. *Lung India.* 2014; **31:** 232-6.
14. Agarwal R, Dua D, Choudhary H, *et al.* Role of *Aspergillus fumigatus*-specific IgG in diagnosis and monitoring treatment response in allergic bronchopulmonary aspergillosis. *Mycoses.* 2017; **60:** 33-9.
15. Agarwal R, Garg M, Aggarwal AN, *et al.* Serologic allergic bronchopulmonary aspergillosis (ABPA-S): Long-term outcomes. *Respir Med.* 2012; **106:** 942-7.
16. Agarwal R, Gupta D, Aggarwal AN, *et al.* Clinical significance of decline in serum IgE levels in allergic bronchopulmonary aspergillosis. *Respir Med.* 2010; **104:** 204-10.
17. Agarwal R, Gupta D. Severe asthma and fungi: Current evidence. *Med Mycol.* 2011; **49:** S150-7.
18. Agarwal R, Hazarika B, Gupta D, *et al. Aspergillus* hypersensitivity in patients with chronic obstructive pulmonary disease: COPD as a risk factor for ABPA? *Med Mycol.* 2010; **48:** 988-94.
19. Agarwal R, Khan A, Aggarwal AN, *et al.* Clinical relevance of peripheral blood eosinophil count in allergic bronchopulmonary aspergillosis. *J Infect Public Health.* 2011; **4:** 235-43.
20. Agarwal R, Khan A, Aggarwal AN, *et al.* Link between CFTR mutations and ABPA: A systematic review and meta-analysis. *Mycoses.* 2012; **55:** 357-65.

21. Agarwal R, Khan A, Aggarwal AN, *et al.* Role of inhaled corticosteroids in the management of serological allergic bronchopulmonary aspergillosis (ABPA). *Intern Med.* 2011; **50:** 855-60.
22. Agarwal R, Khan A, Garg M, *et al.* Pictorial essay: Allergic bronchopulmonary aspergillosis. *Indian J Radiol Imaging.* 2011; **21:** 242-52.
23. Agarwal R, Khan A, Gupta D, *et al.* An alternate method of classifying allergic bronchopulmonary aspergillosis based on high-attenuation mucus. *PLoS One.* 2010; **5:** e15346.
24. Agarwal R, Khan A. ABPA or AFAA: That is the question. *Am J Respir Crit Care Med.* 2011; **183:** 1281-2.
25. Agarwal R, Maskey D, Aggarwal AN, *et al.* Diagnostic performance of various tests and criteria employed in allergic bronchopulmonary aspergillosis: A latent class analysis. *PLoS One.* 2013; **8:** e61105.
26. Agarwal R, Nath A, Aggarwal AN, *et al. Aspergillus* hypersensitivity and allergic bronchopulmonary aspergillosis in patients with acute severe asthma in a respiratory intensive care unit in North India. *Mycoses.* 2010; **53:** 138-43.
27. Agarwal R, Sehgal IS, Dhooria S, *et al.* Radiologic criteria for the diagnosis of high-attenuation mucus in allergic bronchopulmonary aspergillosis. *Chest.* 2016; **149:** 1109-10.
28. Agarwal R, Singh N, Gupta D. Pulmonary hypertension as a presenting manifestation of allergic bronchopulmonary aspergillosis. *Indian J Chest Dis Allied Sci.* 2009; **51:** 37-40.
29. Agarwal R, Srinivas R, Agarwal AN, *et al.* Pulmonary masses in allergic bronchopulmonary aspergillosis: Mechanistic explanations. *Respir Care.* 2008; **53:** 1744-8.
30. Agarwal R. Allergic bronchopulmonary aspergillosis: Lessons learnt from genetics. *Indian J Chest Dis Allied Sci.* 2011; **53:** 137-40.
31. Agarwal R. Allergic bronchopulmonary aspergillosis: Lessons for the busy radiologist. *World J Radiol.* 2011; **3:** 178-81.
32. Agarwal R. Allergic bronchopulmonary aspergillosis. *Chest.* 2009; **135:** 805-26.
33. Agarwal R. Burden and distinctive character of allergic bronchopulmonary aspergillosis in India. *Mycopathologia.* 2014; **178:** 447-56.
34. Agarwal R. Severe asthma with fungal sensitization. *Curr Allergy Asthma Rep.* 2011; **11:** 403-13.
35. Agarwal R. What is the current place of azoles in allergic bronchopulmonary aspergillosis and severe asthma with fungal sensitization. *Expert Rev Respir Med.* 2012; **6:** 363-71.
36. Amin MU, Mahmood R. Multiple bronchoceles in a non-asthmatic patient with allergic bronchopulmonary aspergillosis. *J Pak Med Assoc.* 2008; **58:** 514-6.
37. Aneja P, Singh UP, Kaur B, *et al.* Miliary nodules: An unusual presentation of allergic bronchopulmonary aspergillosis. *Lung India.* 2014; **31:** 285-8.
38. Arikoglu T, Batmaz SB, Coskun T, *et al.* The characteristics of indoor and outdoor fungi and their relation with allergic respiratory diseases in the southern region of Turkey. *Environ Monit Assess.* 2016; **188:** 380.
39. Awosika SA, Olajubu FA, Amusa NA. Microbiological assessment of indoor air of a teaching hospital in Nigeria. *Asian Pac J Trop Biomed.* 2012; **2:** 465-8.
40. Azad C, Jat KR, Aggarwal P. Bronchial asthma with ABPA presenting as PTE. *Indian J Crit Care Med.* 2013; **17:** 188-9.
41. Bains SN, Judson MA. Allergic bronchopulmonary aspergillosis. *Clin Chest Med.* 2012; **33:** 265-81.
42. Bakri SJ, Sears JE, Procop GW, *et al.* Excellent visual outcome following *Aspergillus flavus* endogenous endophthalmitis—farmer's lung disease. *Int Ophthalmol.* 2010; **30:** 99-101.
43. Barreiros G, Akiti T, Magalhaes AC, *et al.* Effect of the implosion and demolition of a hospital building on the concentration of fungi in the air. *Mycoses.* 2015; **58:** 707-13.
44. Barton RC, Hobson RP, Denton M, *et al.* Serologic diagnosis of allergic bronchopulmonary aspergillosis in patients with cystic fibrosis through the detection of immunoglobulin G to *Aspergillus fumigatus. Diagn Microbiol Infect Dis.* 2008; **62:** 287-91.
45. Baxi SN, Portnoy JM, Larenas-Linnemann D, *et al.* Exposure and health effects of fungi on humans. *J Allergy Clin Immunol Pract.* 2016; **4:** 396-404.
46. Baxter CG, Denning DW, Jones AM, *et al.* Performance of two *Aspergillus* IgG EIA assays compared with the precipitin test in chronic and allergic aspergillosis. *Clin Microbiol Infect.* 2013; **19:** e197-204.
47. Baxter CG, Dunn G, Jones AM, *et al.* Novel immunologic classification of aspergillosis in adult cystic fibrosis. *J Allergy Clin Immunol.* 2013; **132:** 560-6.e10.
48. Bedi RS. Allergic bronchopulmonary aspergillosis: Indian perspective. *Indian J Chest Dis Allied Sci.* 2009; **51:** 73-4.
49. Bedi RS. Azoles in allergic bronchopulmonary aspergillosis. *Indian J Chest Dis Allied Sci.* 2015; **57:** 5-6.
50. Bellanger AP, Reboux G, Botterel F, *et al.* New evidence of the involvement of *Lichtheimia corymbifera* in farmer's lung disease. *Med Mycol.* 2010; **48:** 981-7.
51. Bent JP, Kuhn FA. Diagnosis of allergic fungal sinusitis. *Otolaryngol Head Neck Surg.* 1994; **111:** 580-8.
52. Bhan U, Newstead MJ, Zeng X, *et al. Stachybotrys chartarum*-induced hypersensitivity pneumonitis is TLR9 dependent. *Am J Pathol.* 2011; **179:** 2779-87.
53. Biswas D, Agarwal S, Sindhwani G, *et al.* Fungal colonization in patients with chronic respiratory diseases from Himalayan region of India. *Ann Clin Microbiol Antimicrob.* 2010; **9:** 28.
54. Bobolea I, Fernandez Rodriguez C, Diaz-Campos R, *et al.* Measuring total IgE is useful in detecting exacerbations in patients with allergic bronchopulmonary aspergillosis receiving omalizumab. *J Allergy Clin Immunol Pract.* 2016; **4:** 361-3.
55. Borchers AT, Chang C, Eric Gershwin M. Mold and human health: A reality check. *Clin Rev Allergy Immunol.* 2017; **52:** 305-22.
56. Bordon Y. Microbiome: Fungal monopoly promotes allergy. *Nat Rev Microbiol.* 2014; **12:** 234-5.
57. Bordon Y. Mucosal immunology: Fungal monopoly promotes allergy. *Nat Rev Immunol.* 2014; **14:** 138-9.

58. Braude S, Bowen D. Allergic bronchopulmonary aspergillosis: A reversible cause of complete lung collapse. *Intern Med J.* 2011; **41:** 754-5.
59. Bumbacea RS, Balea MI, Ghiordanescu IM, *et al.* Allergic bronchopulmonary aspergillosis associated with eosinophilic gastroenteritis. *J Investig Allergol Clin Immunol.* 2015; **25:** 380-1.
60. Cabo Verde S, Almeida SM, Matos J, *et al.* Microbiological assessment of indoor air quality at different hospital sites. *Res Microbiol.* 2015; 166: 557-63.
61. Callejas CA, Douglas RG. Fungal rhinosinusitis: What every allergist should know. *Clin Exp Allergy.* 2013; **43:** 835-49.
62. Challa S, Uppin SG, Hanumanthu S, *et al.* Fungal rhinosinusitis: A clinicopathological study from South India. *Eur Arch Otorhinolaryngol.* 2010; **267:** 1239-45.
63. Chang FY, Lee JH, Yang YH, *et al.* Analysis of the serum levels of fungi-specific immunoglobulin E in patients with allergic diseases. *Int Arch Allergy Immunol.* 2011; **154:** 49-56.
64. Chang GH, Wang WH. Intranasal fungal (*Alternaria*) infection related to nasal steroid spray. *Am J Otolaryngol.* 2013; **34:** 743-5.
65. Chaudhary N, Marr KA. Impact of *Aspergillus fumigatus* in allergic airway diseases. *Clin Transl Allergy.* 2011; **1:** 4.
66. Chen TH, Hollingsworth H. Allergic bronchopulmonary aspergillosis. *N Engl J Med.* 2008; **359:** e7.
67. Chhabra SK, Sahay S, Ramaraju K. Allergic bronchopulmonary aspergillosis complicating childhood asthma. *Indian J Pediatr.* 2009; **76:** 331-2.
68. Chiba S, Okada S, Suzuki Y, *et al.* *Cladosporium* species-related hypersensitivity pneumonitis in household environments. *Intern Med.* 2009; **48:** 363-7.
69. Chishimba L, Langridge P, Powell G, *et al.* Efficacy and safety of nebulised amphotericin B (NAB) in severe asthma with fungal sensitisation (SAFS) and allergic bronchopulmonary aspergillosis (ABPA). *J Asthma.* 2015; **52:** 289-95.
70. Chishimba L, Niven RM, Cooley J, *et al.* Voriconazole and posaconazole improve asthma severity in allergic bronchopulmonary aspergillosis and severe asthma with fungal sensitization. *J Asthma.* 2012; **49:** 423-33.
71. Chowdhary A, Agarwal K, Kathuria S, *et al.* Allergic bronchopulmonary mycosis due to fungi other than *Aspergillus*: A global overview. *Crit Rev Microbiol.* 2014; **40:** 30-48.
72. Chowdhary A, Agarwal K, Randhawa HS, *et al.* A rare case of allergic bronchopulmonary mycosis caused by *Alternaria alternata. Med Mycol.* 2012; **50:** 890-6.
73. Chowdhary A, Kathuria S, Agarwal K, *et al.* Reply to "implications of high antifungal susceptibility on *Schizophyllum commune*-associated allergy in clinical practice". *Antimicrob Agents Chemother.* 2013; **57:** 5784-5.
74. Chowdhary A, Kathuria S, Singh PK, *et al.* Molecular characterization and *in vitro* antifungal susceptibility profile of *Schizophyllum commune*, an emerging basidiomycete in bronchopulmonary mycoses. *Antimicrob Agents Chemother.* 2013; **57:** 2845-8.
75. Chrdle A, Mustakim S, Bright-Thomas RJ, *et al.* *Aspergillus* bronchitis without significant immunocompromise. *Ann N Y Acad Sci.* 2012; **1272:** 73-85.
76. Coop CA. Immunotherapy for mold allergy. *Clin Rev Allergy Immunol.* 2014; **47:** 289-98.
77. Crameri R, Garbani M, Rhyner C, *et al.* Fungi: The neglected allergenic sources. *Allergy.* 2014; **69:** 176-85.
78. Crameri R, Zeller S, Glaser AG, *et al.* Cross-reactivity among fungal allergens: A clinically relevant phenomenon? *Mycoses.* 2009; **52:** 99-106.
79. De Soyza A, Aliberti S. Bronchiectasis and *Aspergillus*: How are they linked? *Med Mycol.* 2017; **55:** 69-81.
80. Delhaes L, Frealle E, Pinel C. Serum markers for allergic bronchopulmonary aspergillosis in cystic fibrosis: State of the art and further challenges. *Med Mycol.* 2010; **48:** S77-87.
81. Denning DW, O'Driscoll BR, Powell G, *et al.* Randomized controlled trial of oral antifungal treatment for severe asthma with fungal sensitization: The Fungal Asthma Sensitization Trial (FAST) study. *Am J Respir Crit Care Med.* 2009; **179:** 11-8.
82. Denning DW, Pashley C, Hartl D, *et al.* Fungal allergy in asthma-state of the art and research needs. *Clin Transl Allergy.* 2014; **4:** 14.
83. Denning DW, Pleuvry A, Cole DC. Global burden of allergic bronchopulmonary aspergillosis with asthma and its complication chronic pulmonary aspergillosis in adults. *Med Mycol.* 2013; **51:** 361-70.
84. Dhooria S, Agarwal R. Diagnosis of allergic bronchopulmonary aspergillosis: A case-based approach. *Future Microbiol.* 2014; **9:** 1195-208.
85. Dhooria S, Kumar P, Saikia B, *et al.* Prevalence of *Aspergillus* sensitisation in pulmonary tuberculosis-related fibrocavitary disease. *Int J Tuberc Lung Dis.* 2014; **18:** 850-5.
86. Dickson SD, Tankersley MS. Fatal hypersensitivity pneumonitis from exposure to *Fusarium vasinfectum* in a home environment: A case Report. *Int Arch Allergy Immunol.* 2015; **166:** 150-3.
87. Dixit R, George J, Nirwan PS. Allergic bronchopulmonary aspergillosis presenting with cavitary lesion and simulating a lung abscess. *Indian J Med Microbiol.* 2010; **28:** 402-4.
88. Dogra V, Sinha AK, Saxena R, *et al.* *Aspergillus* March: From ABPA to aspergilloma to subacute invasive aspergillosis. *Allergy Asthma Clin Immunol.* 2016; **12:** 64.
89. Dournes G, Berger P, Refait J, *et al.* Allergic bronchopulmonary aspergillosis in cystic fibrosis: MR imaging of airway mucus contrasts as a tool for diagnosis. *Radiology.* 2017; 162350. PMID: 28530849.
90. Dyer ZA, Wright RS, Rong IH, *et al.* Back pain associated with endobronchial mucus impaction due to *Bipolaris australiensis* colonization representing atypical allergic bronchopulmonary mycosis. *Med Mycol.* 2008; **46:** 589-94.
91. Eickmeier O, Rieber N, Eckrich J, *et al.* Immune response, diagnosis and treatment of allergic bronchopulmonary aspergillosis in cystic fibrosis lung disease. *Curr Pharm Des.* 2013; **19:** 3669-78.
92. Elphick HE, Southern KW. Antifungal therapies for allergic bronchopulmonary aspergillosis in people with cystic fibrosis. *Cochrane Database Syst Rev.* 2016; **11:** CD002204.

93. Er M, Emri SA, Demir AU, *et al.* Byssinosis and COPD rates among factory workers manufacturing hemp and jute. *Int J Occup Med Environ Health.* 2016; **29:** 55-68.
94. Esch RE, Codina R. Fungal raw materials used to produce allergen extracts. *Ann Allergy Asthma Immunol.* 2017; **118:** 399-405.
95. Faerden K, Lund MB, Mogens Aalokken T, *et al.* Hypersensitivity pneumonitis in a cluster of sawmill workers: A 10-year follow-up of exposure, symptoms and lung function. *Int J Occup Environ Health.* 2014; **20:** 167-73.
96. Farrant J, Brice H, Fowler S, *et al.* Fungal sensitisation in severe asthma is associated with the identification of *Aspergillus fumigatus* in sputum. *J Asthma.* 2016; **53:** 732-5.
97. Fekadu S, Getachewu B. Microbiological assessment of indoor air of teaching hospital wards: A case of Jimma University specialized hospital. *Ethiop J Health Sci.* 2015; **25:** 117-22.
98. Fukahori S, Matsuse H, Tsuchida T, *et al.* Clearance of *Aspergillus fumigatus* is impaired in the airway in allergic inflammation. *Ann Allergy Asthma Immunol.* 2014; **113:** 180-6.
99. Fukutomi Y, Taniguchi M. Sensitization to fungal allergens: Resolved and unresolved issues. *Allergol Int.* 2015; **64:** 321-31.
100. Fukutomi Y, Tanimoto H, Yasueda H, *et al.* Serological diagnosis of allergic bronchopulmonary mycosis: Progress and challenges. *Allergol Int.* 2016; **65:** 30-6.
101. Gabriel MF, Postigo I, Tomaz CT, *et al. Alternaria alternata* allergens: Markers of exposure, phylogeny and risk of fungi-induced respiratory allergy. *Environ Int.* 2016; **89-90:** 71-80.
102. Galiatsatos P, Melia MT, Silhan LL. Human immunodeficiency virus and allergic bronchopulmonary aspergillosis: Case report and review of literature. *Open Forum Infect Dis.* 2016; **3:** ofw116. PMID: 27419184.
103. Garg K, Saini V. Mass-like lesions disappearing with treatment in a young patient. *Indian J Med Res.* 2015; **141:** 251-2.
104. Garg MK, Gupta P, Agarwal R, *et al.* MRI: A new paradigm in imaging evaluation of allergic bronchopulmonary aspergillosis? *Chest.* 2015; **147:** e58-9.
105. Gernez Y, Walters J, Mirkovic B, *et al.* Blood basophil activation is a reliable biomarker of allergic bronchopulmonary aspergillosis in cystic fibrosis. *Eur Respir J.* 2016; **47:** 177-85.
106. Ghosh G, Sharma B, Chauhan A, *et al.* Coexistence of allergic bronchopulmonary aspergillosis and allergic aspergillus sinusitis in a patient without clinical asthma. *BMJ Case Rep.* 2013; pii: bcr2013008683. PMID: 23645638.
107. Ghosh K, Sanders BE. Allergic bronchopulmonary aspergillosis causing total lung collapse. *BMJ Case Rep.* 2012; pii: bcr1220115349. PMID: 22605607.
108. Ghosh S, Hoselton SA, Schuh JM. Allergic inflammation in *Aspergillus fumigatus*-induced fungal asthma. *Curr Allergy Asthma Rep.* 2015; **15:** 59.
109. Ghosh T, Dey A, Biswas D, Chatterjee S, Haldar N, Maiti PK. *Aspergillus* hypersensitivity and allergic bronchopulmonary aspergillosis among asthma patients in eastern India. *J Indian Med Assoc.* 2010; **108:** 863-5.
110. Godet C, Meurice JC, Roblot F, *et al.* Efficacy of nebulised liposomal amphotericin B in the attack and maintenance treatment of ABPA. *Eur Respir J.* 2012; **39:** 1261-3.
111. Goldman DL, Huffnagle GB. Potential contribution of fungal infection and colonization to the development of allergy. *Med Mycol.* 2009; **47:** 445-56.
112. Gonzalez de Olano D, Gonzalez-Mancebo E, Gandolfo Cano M, *et al.* Successful treatment of allergic bronchopulmonary candidiasis with a recombinant anti-immunoglobulin E antibody. *J Investig Allergol Clin Immunol.* 2009; **19:** 416-7.
113. Greenberger PA, Bush RK, Demain JG, *et al.* Allergic bronchopulmonary aspergillosis. *J Allergy Clin Immunol Pract.* 2014; **2:** 703-8.
114. Greenberger PA. When to suspect and work up allergic bronchopulmonary aspergillosis. *Ann Allergy Asthma Immunol.* 2013; **111:** 1-4.
115. Gupta A, McKean M, Haynes S, *et al.* Chronic granulomatous disease presenting as fulminant *Aspergillus* pneumonitis: a lethal combination? *Pediatr Crit Care Med.* 2009; **10:** e43-5.
116. Gupta K, Das A, Joshi K, *et al. Aspergillus* endocarditis in a known case of allergic bronchopulmonary aspergillosis: An autopsy report. *Cardiovasc Pathol.* 2010; **19:** e137-9.
117. Gupta RK, Chandra A, Gautam PB. Allergic bronchopulmonary aspergillosis - A clinical review. *J Assoc Physicians India.* 2012; **60:** 46-51.
118. Guy JP, Raza S, Bondi E, *et al. Cryptococcus* pneumonia presenting in an immunocompetent host with pulmonary asbestosis: A case report. *J Med Case Rep.* 2012; **6:** 170.
119. Hafen GM, Hartl D, Regamey N, *et al.* Allergic bronchopulmonary aspergillosis: The hunt for a diagnostic serological marker in cystic fibrosis patients. *Expert Rev Mol Diagn.* 2009; **9:** 157-64.
120. Hagemeyer O, Bunger J, van Kampen V, *et al.* Occupational allergic respiratory diseases in garbage workers: Relevance of molds and actinomycetes. *Adv Exp Med Biol.* 2013; **788:** 313-20.
121. Harrison E, Singh A, Morris J, *et al.* Mannose-binding lectin genotype and serum levels in patients with chronic and allergic pulmonary aspergillosis. *Int J Immunogenet.* 2012; **39:** 224-32.
122. Hartl D. Immunological mechanisms behind the cystic fibrosis-ABPA link. *Med Mycol.* 2009; **47:** S183-91.
123. Hasnain SM, Akhter T, Waqar MA. Airborne and allergenic fungal spores of the Karachi environment and their correlation with meteorological factors. *J Environ Monit.* 2012; **14:** 1006-13.
124. Hassoun A. Cauliflower mass obstructing the left mainstem bronchus. *Clin Infect Dis.* 2008; **47:** 540-1 & 585-6.
125. Hayashi S, Naitoh K, Matsubara S, *et al.* Pulmonary colonization by *Chrysosporium zonatum* associated with allergic inflammation in an immunocompetent subject. *J Clin Microbiol.* 2002; **40:** 1113-5.

126. Hayes D Jr, Murphy BS, Lynch JE, *et al.* Aerosolized amphotericin for the treatment of allergic bronchopulmonary aspergillosis. *Pediatr Pulmonol.* 2010; **45:** 1145-8.
127. Hogan C, Denning DW. Allergic bronchopulmonary aspergillosis and related allergic syndromes. *Semin Respir Crit Care Med.* 2011; **32:** 682-92.
128. Hope J. A review of the mechanism of injury and treatment approaches for illness resulting from exposure to water-damaged buildings, mold and mycotoxins. *Sci World J.* 2013; 767482. PMID: 23710148.
129. Horner WE, Heibling A, Salvaggio JE, *et al.* Fungal allergens. *Clin Microbiol Rev.* 1995; **8:** 161-79.
130. Howard SJ, Pasqualotto AC, Denning DW. Azole resistance in allergic bronchopulmonary aspergillosis and *Aspergillus* bronchitis. *Clin Microbiol Infect.* 2010; **16:** 683-8.
131. Hox V, Steelant B, Fokkens W, *et al.* Occupational upper airway disease: How work affects the nose. *Allergy.* 2014; **69:** 282-91.
132. Huang WC, Lu YH, Lin ZG, *et al.* Sauna lung: Hypersensitivity pneumonitis due to *Exophiala jeanselmei. Respirology.* 2010; **15:** 573-6.
133. Hurra J, Heinzow B, Aurbach U, *et al.* Medical diagnostics for indoor mold exposure. *Int J Hyg Environ Health.* 2017; **220 (2 Pt B):** 305-28 6.
134. Ishiguro T, Takayanagi N, Kagiyama N, *et al.* Clinical characteristics of biopsy-proven allergic bronchopulmonary mycosis: Variety in causative fungi and laboratory findings. *Intern Med.* 2014; **53:** 1407-11.
135. Ishiguro T, Takayanagi N, Takaku Y, *et al.* Allergic bronchopulmonary aspergillosis with repeated isolation of non-tuberculous mycobacteria. *Intern Med.* 2013; **52:** 1721-6.
136. Ishiguro T, Takayanagi N, Takaku Y, *et al.* Combined allergic bronchopulmonary aspergillosis and eosinophilic granulomatosis with polyangiitis: Three cases and a review of the literature. *Intern Med.* 2016; **55:** 793-7.
137. Jat KR, Walia DK, Khairwa A. Anti-IgE therapy for allergic bronchopulmonary aspergillosis in people with cystic fibrosis. *Cochrane Database Syst Rev.* 2015; **11:** CD010288.
138. Jensen SG, Nielsen TL. Chronic necrotizing and allergic bronchopulmonary aspergillosis in an asthmatic patient - A therapeutic challenge. *Clin Respir J.* 2008; **2:** 245-6.
139. Kasprzyk I. Aeromycology - main research fields of interest during the last 25 years. *Ann Agric Environ Med.* 2008; **15:** 1-7.
140. Kaur H, Rudramurthy SM, Mohindra S, *et al. Ochroconis humicola* coexisting with esthesioneuroblastoma: An incidental coloniser or allergen? *Mycopathologia.* 2014; **178:** 79-83.
141. Kern I, Lopert A. Allergic bronchopulmonary aspergillosis with co-existent aspergilloma: A case report. *J Med Case Rep.* 2010; **4:** 309.
142. Kespohl S, Raulf M. Mould allergens: Where do we stand with molecular allergy diagnostics? *Allergo J Int.* 2014; **23:** 120-5.
143. Khalil KF. Therapeutic bronchoalveolar lavage with conventional treatment in allergic bronchopulmonary aspergillosis. *J Coll Physicians Surg Pak.* 2015; **25:** 359-62.
144. Kin NW, Stefanov EK, Dizon BL, *et al.* Antibodies generated against conserved antigens expressed by bacteria and allergen-bearing fungi suppress airway disease. *J Immunol.* 2012; **189:** 2246-56.
145. Kirschner AN, Kuhlmann E, Kuzniar TJ. Eosinophilic pleural effusion complicating allergic bronchopulmonary aspergillosis. *Respiration.* 2011; **82:** 478-81.
146. Knutsen AP, Bush RK, Demain JG, *et al.* Fungi and allergic lower respiratory tract diseases. *J Allergy Clin Immunol.* 2012; **129:** 280-91; Quiz 292-3.
147. Knutsen AP. Allergic bronchopulmonary aspergillosis in asthma. *Expert Rev Clin Immunol.* 2017; **13:** 11-14.
148. Knutsen AP. Allergic bronchopulmonary aspergillosis. *Clin Exp Allergy.* 2015; **45:** 298-9.
149. Kochar S, Ahlawat M, Dahiya P, *et al.* Assessment of allergenicity to fungal allergens of Rohtak city, Haryana, India. *Allergy Rhinol (Providence).* 2014; **5:** 56-65.
150. Kordalewska M, Brillowska-Dabrowska A, Jagielski T, *et al.* PCR and real-time PCR assays to detect fungi of *Alternaria alternata* species. *Acta Biochim Pol.* 2015; **62:** 707-12.
151. Kuhn DM, Ghannoum MA. Indoor mold, toxigenic fungi and *Stachybotrys chartarum*: Infectious disease perspective. *Clin Microbiol Rev.* 2003; **16:**144-72.
152. Kumar R, Goel N. Allergic bronchopulmonary aspergillosis: A clinico-serological correlation with radiologic profile. *J Asthma.* 2013; **50:** 759-63.
153. Kumar R, Poongadan MN, Singh M. Allergic bronchopulmonary aspergillosis presenting as lobar or total lung collapse. *Pneumonol Alergol Pol.* 2015; **83:** 144-50.
154. Laoudi Y, Paolini JB, Grimfed A, *et al.* Nebulised corticosteroid and amphotericin B: An alternative treatment for ABPA? *Eur Respir J.* 2008; **31:** 908-9.
155. Lebecque P, Pepermans X, Marchand E, *et al.* ABPA in adulthood: A CFTR-related disorder. *Thorax.* 2011; **66:** 540-1.
156. Lee JY, Choi H, Chon GR. Allergic bronchopulmonary aspergillosis mimicking relapsing chronic eosinophilic pneumonia in non-asthma patient. *Allergol Int.* 2014; **63:** 133-5.
157. Lee W, Teo FS, Santosa A, *et al.* Eosinophilic granulomatosis with polyangiitis preceding allergic bronchopulmonary aspergillosis. *Eur Ann Allergy Clin Immunol.* 2015; **47:** 228-31.
158. Levetin E, Horner WE, Scott JA, *et al.* Taxonomy of allergenic fungi. *J Allergy Clin Immunol Pract.* 2016; **4:** 375-85.e1.
159. Lin RY, Sethi S, Bhargave GA. Measured immunoglobulin E in allergic bronchopulmonary aspergillosis treated with omalizumab. *J Asthma.* 2010; **47:** 942-5.
160. Lowes D, Chishimba L, Greaves M, *et al.* Development of chronic pulmonary aspergillosis in adult asthmatics with ABPA. *Respir Med.* 2015; **109:** 1509-15.
161. Madan K, Bal A, Agarwal R. Pleural effusion in a patient with allergic bronchopulmonary aspergillosis. *Respir Care.* 2012; **57:** 1509-13.
162. Madan K, Guleria R. Vanishing lung mass in a patient with asthma. *J Thorac Dis.* 2013; **5:** e45-9.

163. Madani Y, Barlow A, Taher F. Severe asthma with fungal sensitization: A case report and review of literature. *J Asthma.* 2010; **47:** 2-6.
164. Mahdavinia M, Grammer LC. Management of allergic bronchopulmonary aspergillosis: A review and update. *Ther Adv Respir Dis.* 2012; **6:** 173-87.
165. Martinez S, Heyneman LE, McAdams HP, *et al.* Mucoid impactions: Finger-in-glove sign and other CT and radiographic features. *Radiographics.* 2008; **28:** 1369-82.
166. Matsuse H, Tsuchida T, Fukahori S, *et al.* Dissociation between sensitizing and colonizing fungi in patients with allergic bronchopulmonary aspergillosis. *Ann Allergy Asthma Immunol.* 2013; **111:** 190-3.
167. Matsuwaki Y, Uno K, Okushi T, *et al.* Total and antigen-(fungi, mites and staphylococcal enterotoxins) specific IgEs in nasal polyps is related to local eosinophilic inflammation. *Int Arch Allergy Immunol.* 2013; **161 (Suppl. 2):** 147-53.
168. Maturu VN, Agarwal R. Acute invasive pulmonary aspergillosis complicating allergic bronchopulmonary aspergillosis: Case report and systematic review. *Mycopathologia.* 2015; **180:** 209-15.
169. Maturu VN, Agarwal R. Prevalence of *Aspergillus* sensitization and allergic bronchopulmonary aspergillosis in cystic fibrosis: Systematic review and meta-analysis. *Clin Exp Allergy.* 2015; **45:** 1765-78.
170. Meng J, Barnes CS, Rosenwasser LJ; *et al.* Identity of the fungal species present in the homes of asthmatic children. *Clin Exp Allergy.* 2012; **42:** 1448-58.
171. Menzies D, Holmes L, McCumesky G, *et al. Aspergillus* sensitization is associated with airflow limitation and bronchiectasis in severe asthma. *Allergy.* 2011; **66:** 679-85.
172. Min KH, Park SJ, Kim SR, *et al.* Coexistence of allergic bronchopulmonary aspergillosis and active pulmonary tuberculosis. *Am J Respir Crit Care Med.* 2011; **183:** 137-9.
173. Mir E, Shah A. Allergic bronchopulmonary aspergillosis in a patient with chronic obstructive pulmonary disease. *Prim Care Respir J.* 2012; **21:** 111-4.
174. Misra R, Malik A, Singhal S. Comparison of the activities of amphotericin B, itraconazole and voriconazole against clinical and environmental isolates of *Aspergillus* species. *Indian J Pathol Microbiol.* 2011; **54:** 112-6.
175. Mittal R, Gupta P, Chhabra SK. Occupational bronchiolitis induced by cotton dust exposure in a non-smoker. *Indian J Occup Environ Med.* 2016; **20:** 118-20.
176. Montani D, Zendah I, Achouh L, *et al.* Association of pulmonary aspergilloma and allergic bronchopulmonary aspergillosis. *Eur Respir Rev.* 2010; **19:** 349-51.
177. Moreira AS, Silva D, Ferreira AR, *et al.* Antifungal treatment in allergic bronchopulmonary aspergillosis with and without cystic fibrosis: A systematic review. *Clin Exp Allergy.* 2014; **44:** 1210-27.
178. Moreno-Ancillo A, Gil-Adrados AC, Pineda F, *et al.* Fungal allergens in a saxophonist who had never smoked with allergic bronchopulmonary aspergillosis previously diagnosed as COPD. *J Investig Allergol Clin Immunol.* 2016; **26:** 386-8.
179. Moss RB. Allergic bronchopulmonary aspergillosis and *Aspergillus* infection in cystic fibrosis. *Curr Opin Pulm Med.* 2010; **16:** 598-603.
180. Moss RB. Fungi in cystic fibrosis and non-cystic fibrosis bronchiectasis. *Semin Respir Crit Care Med.* 2015; **36:** 207-16.
181. Moss RB. The use of biological agents for the treatment of fungal asthma and allergic bronchopulmonary aspergillosis. *Ann N Y Acad Sci.* 2012; **1272:** 49-57.
182. Moss RB. Treating allergic bronchopulmonary aspergillosis: The way forward. *Eur Respir J.* 2016; **47:** 385-7.
183. Moss RB. Treatment options in severe fungal asthma and allergic bronchopulmonary aspergillosis. *Eur Respir J.* 2014; **43:** 1487-500.
184. Mou Y, Ye L, Ye M, *et al.* A retrospective study of patients with a delayed diagnosis of allergic bronchopulmonary aspergillosis/allergic bronchopulmonary mycosis. *Allergy Asthma Proc.* 2014; **35:** e21-6.
185. Muthu V, Agarwal R. A report of a successfully treated case of ABPA in an HIV-infected individual. *BMJ Case Rep.* 2014; pii: bcr2014206236. PMID: 25385562.
186. Muthu V, Behera D, Agarwal R. Allergic bronchopulmonary aspergillosis or pulmonary tuberculosis: A case of mistaken identity? *Lung India.* 2015; **32:** 529-30.
187. Natarajan S, Rudrawar P. Bronchiectasis in western India: Clinical presentations and socio-economic burden. *J Indian Med Assoc.* 2014; **112:** 89-92.
188. Natarajan S, Subramanian P. Allergic bronchopulmonary aspergillosis: A clinical review of 24 patients: Are we right in frequent serologic monitoring? *Ann Thorac Med.* 2014; **9:** 216-20.
189. Nevalainen A, Taubel M, Hyvarinen A. Indoor fungi: Companions and contaminants. *Indoor Air.* 2015; **25:** 125-56.
190. No authors listed. What is allergic bronchopulmonary aspergillosis (ABPA)? *Am J Respir Crit Care Med.* 2014; **190:** P3-4.
191. O'Driscoll BR, Powell G, Chew F, *et al.* Comparison of skin prick tests with specific serum immunoglobulin E in the diagnosis of fungal sensitization in patients with severe asthma. *Clin Exp Allergy.* 2009; **39:** 1677-83.
192. Ogawa H, Fujimura M, Ohkura N, *et al.* A proposal of guidance for identification of *Schizophyllum commune*-associated sinobronchial allergic mycosis. *Allergol Int.* 2014; **63:** 287-8.
193. Ogawa H, Fujimura M, Ohkura N, *et al.* Fungus-associated asthma: overcoming challenges in diagnosis and treatment. *Expert Rev Clin Immunol.* 2014; **10:** 647-56.
194. Ogawa H, Fujimura M, Ohkura N, *et al.* Implications of high antifungal susceptibility on *Schizophyllum commune*-associated allergy in clinical practice. *Antimicrob Agents Chemother.* 2013; **57:** 5783.
195. Ogawa H, Fujimura M, Takeuchi Y, *et al.* A case of sinobronchial allergic mycosis; Possibility of basidiomycetous fungi as a causative antigen. *Intern Med.* 2011; **50:** 59-62.
196. Ogawa H, Fujimura M, Takeuchi Y, *et al.* Impact of *Schizophyllum* sensitization on decline of lung function in asthma. *J Asthma.* 2013; **50:** 764-8.

197. Ogawa H, Fujimura M, Takeuchi Y, *et al.* Possible roles of 2 basidiomycetous fungi in allergic fungal respiratory disease. *J Allergy Clin Immunol.* 2012; **130:** 279-80.
198. Ogawa H, Fujimura M, Takeuchi Y, *et al.* The definitive diagnostic process and successful treatment for ABPM caused by *Schizophyllum commune*: A report of two cases. *Allergol Int.* 2012; **61:** 163-9.
199. Ogawa H, Fujimura M, Takeuchi Y, et al. The influence of *Schizophyllum commune* on asthma severity. *Lung.* 2011; **189:** 485-92.
200. Ogawa H, Fujimura M, Takeuchi Y, *et al.* Two cases of *Schizophyllum* asthma: Is this a new clinical entity or a precursor of ABPM? *Pulm Pharmacol Ther.* 2011; **24:** 559-62.
201. Oliveira M, Ribeiro H, Delgado JL, *et al.* Aeromycological profile of indoor and outdoor environments. *J Environ Monit.* 2009; **11:** 1360-7.
202. Ong KH, Lewis RD, Dixit A, *et al.* Inactivation of dust mites, dust mite allergen and mold from carpet. *J Occup Environ Hyg.* 2014; **11:** 519-27.
203. Osborne NJ, Thornton CR, Sharpe RA. Indoor fungal exposure and allergic respiratory disease. *Curr Allergy Asthma Rep.* 2015; **15:** 71.
204. Oshikata C, Watanabe M, Saito A, *et al.* Allergic bronchopulmonary mycosis caused by *Penicillium luteum. Med Mycol Case Rep.* 2016; **15:** 9-11.
205. Park JH, Cox-Ganser JM. Mold exposure and respiratory health in damp indoor environments. *Front Biosci (Elite Ed).* 2011; **3:** 757-71.
206. Park MR, Nam YH, Lee SK, *et al.* Uncertain areas in the diagnosis of allergic bronchopulmonary aspergillosis in patients with asthma. *Allergol Immunopathol (Madr).* 2014; **42:** 433-8.
207. Parulekar AD, Diamant Z, Hanania NA. Antifungals in severe asthma. *Curr Opin Pulm Med.* 2015; **21:** 48-54.
208. Pashley CH. Fungal culture and sensitisation in asthma, cystic fibrosis and chronic obstructive pulmonary disorder: What does it tell us? *Mycopathologia.* 2014; **178:** 457-63.
209. Pasqualotto AC, Powell G, Niven R, *et al.* The effects of antifungal therapy on severe asthma with fungal sensitization and allergic bronchopulmonary aspergillosis. *Respirology.* 2009; **14:** 1121-7.
210. Patterson K, Strek ME. Allergic bronchopulmonary aspergillosis. *Proc Am Thorac Soc.* 2010; **7:** 237-44.
211. Pettigrew HD, Selmi CF, Teuber SS, *et al.* Mold and human health: Separating the wheat from the chaff. *Clin Rev Allergy Immunol.* 2010; **38:** 148-55.
212. Phuyal S, Garg MK, Agarwal R, *et al.* High-attenuation mucus impaction in patients with allergic bronchopulmonary aspergillosis: Objective criteria on high-resolution computed tomography and correlation with serologic parameters. *Curr Probl Diagn Radiol.* 2016. **45:** 168-73.
213. Pihet M, Carrere J, Cimon B, *et al.* Occurrence and relevance of filamentous fungi in respiratory secretions of patients with cystic fibrosis - A review. *Med Mycol.* 2009; **47:** 387-97.
214. Pizzimenti S, Bussolino C, Badiu I, *et al.* Itraconazole as 'bridge-therapy' to anti-IgE in a patient with severe asthma with fungal sensitisation. *BMJ Case Rep.* 2013; pii: bcr2012008462. PMID: 23598928.
215. Porter P, Polikepahad S, Qian Y, *et al.* Respiratory tract allergic disease and atopy: Experimental evidence for a fungal infectious etiology. *Med Mycol.* 2011; **49:** S158-63.
216. Porter PC, Roberts L, Fields A, *et al.* Necessary and sufficient role for T helper cells to prevent fungal dissemination in allergic lung disease. *Infect Immun.* 2011; **79:** 4459-71.
217. Prasad R, Garg R, Sanjay, *et al.* Allergic bronchopulmonary aspergillosis: A review of 42 patients from a tertiary care center in India. *Lung India.* 2009; **26:** 38-40.
218. Ram B, Aggarwal AN, Dhooria S, *et al.* A pilot randomized trial of nebulized amphotericin in patients with allergic bronchopulmonary aspergillosis. *J Asthma.* 2016; **53:** 517-24.
219. Ramirez RM, Jacobs RL. Hypersensitivity pneumonitis by *Fusarium vasinfectum* in a home environment. *J Allergy Clin Immunol Pract.* 2014; **2:** 483-4.
220. Randhawa HS, Chowdhary A, Preeti Sinha K, *et al.* Evaluation of peptone glucose fluconazole agar as a selective medium for rapid and enhanced isolation of *Aspergillus fumigatus* from the respiratory tract of bronchopulmonary aspergillosis patients colonized by *Candida albicans. Med Mycol.* 2006; **44:** 343-8.
221. Rankin BT, Jariwala S. Graded challenge protocol for fluconazole hypersensitivity in a patient with cryptococcal pneumonitis. *Ann Allergy Asthma Immunol.* 2012; **108:** 466.
222. Rapaka RR, Kolls JK. Pathogenesis of allergic bronchopulmonary aspergillosis in cystic fibrosis: Current understanding and future directions. *Med Mycol.* 2009; **47:** S331-7.
223. Ren S. Combined Churg-Strauss syndrome and allergic bronchopulmonary aspergillosis—Case report and review of the literature. *Clin Respir J.* 2013; **7:** e6-10.
224. Rick EM, Woolnough K, Pashley CH, *et al.* Allergic fungal airway disease. *J Investig Allergol Clin Immunol.* 2016; **26:** 344-54.
225. Rodrigues J, Caruthers C, Azmeh R, *et al.* The spectrum of allergic fungal diseases of the upper and lower airways. *Expert Rev Clin Immunol.* 2016; **12:** 531-50.
226. Romani L. Parasites and autoimmunity: The case of fungi. *Autoimmun Rev.* 2008; **8:** 129-33.
227. Rosenberg M, Patterson R, Mintzer R, *et al.* Clinical and immunologic criteria for the diagnosis of allergic bronchopulmonary aspergillosis. *Ann Intern Med.* 1977; **86:** 405-14.
228. Roth R, Schatz M. Allergic bronchopulmonary aspergillosis presenting as chronic cough in an elderly woman without previously documented asthma. *Perm J.* 2013; **17:** e103-8.
229. Rudramurthy SM, Singh G, Hallur V, *et al.* High fungal spore burden with predominance of *Aspergillus* in hospital air of a tertiary care hospital in Chandigarh. *Indian J Med Microbiol.* 2016; **34:** 529-32.
230. Santos A, Loureiro G, Faria E, *et al.* Itraconazole, an effective adjunctive treatment for allergic bronchopulmonary aspergillosis. *J Investig Allergol Clin Immunol.* 2009; **19:** 404-8.
231. Sarkar A, Mukherjee A, Ghoshal AG, *et al.* Occurrence of allergic bronchopulmonary mycosis in patients with asthma: An Eastern India experience. *Lung India.* 2010; **27:** 212-6.

232. Sauni R, Verbeek JH, Uitti J, *et al.* Remediating buildings damaged by dampness and mould for preventing or reducing respiratory tract symptoms, infections and asthma. *Cochrane Database Syst Rev.* 2015; **2:** CD007897. PMID: 25715323.
233. Schwerk N, Rochwalsky U, Brinkmann F, *et al.* Don't forget other causes of wheeze. ABPA in a boy with asthma. A case report and review of the literature. *Acta Paediatr.* 2011; **100:** 307-10.
234. Segerer F, Morbach H, Hassold N, *et al.* A 58-year-old man with respiratory insufficiency after a 50-year history of hypersensitivity pneumonitis and pulmonary *Aspergillus* infections. *J Allergy Clin Immunol Pract.* 2013; **1:** 677-80.
235. Sehgal IS, Agarwal R. Specific IgE is better than skin testing for detecting *Aspergillus* sensitization and allergic bronchopulmonary aspergillosis in asthma. *Chest.* 2015; **147:** e194.
236. Sehgal IS, Dhooria S, Bal A, *et al.* Allergic bronchopulmonary aspergillosis in an adult with Kartagener syndrome. *BMJ Case Rep.* 2015; pii: bcr2015211493. PMID: 26250371.
237. Seki M, Ohno H, Gotoh K, *et al.* Allergic bronchopulmonary mycosis due to co-infection with *Aspergillus fumigatus* and *Schizophyllum commune. IDCases.* 2014; **1:** 5-8.
238. Selman M, Lacasse Y, Pardo A, *et al.* Hypersensitivity pneumonitis caused by fungi. *Proc Am Thorac Soc.* 2010; **7:** 229-36.
239. Selvaggi TA, Walco JP, Parikh S, *et al.* Hypersensitivity testing for *Aspergillus fumigatus* IgE is significantly more sensitive than testing for *Aspergillus niger* IgE. *Am J Clin Pathol.* 2012; **137:** 203-6.
240. Shah A, Behera S, Panjabi C. Middle lobe syndrome: A rare presentation of allergic bronchopulmonary aspergillosis. *Eur Ann Allergy Clin Immunol.* 2014; **46:** 147-51.
241. Shah A, Gera K, Panjabi C. Childhood allergic bronchopulmonary aspergillosis presenting as a middle lobe syndrome. *Asia Pac Allergy.* 2016; **6:** 67-9.
242. Shah A, Kala J, Sahay S, *et al.* Frequency of familial occurrence in 164 patients with allergic bronchopulmonary aspergillosis. *Ann Allergy Asthma Immunol.* 2008; **101:** 363-9.
243. Shah A, Panjabi C. Allergic aspergillosis of the respiratory tract. *Eur Respir Rev.* 2014; **23:** 8-29.
244. Shah A, Panjabi C. Allergic bronchopulmonary aspergillosis: A perplexing clinical entity. *Allergy Asthma Immunol Res.* 2016; **8:** 282-97.
245. Shah A. *Aspergillus*-associated hypersensitivity respiratory disorders. *Indian J Chest Dis Allied Sci.* 2008; **50:** 117-28.
246. Shah A. Concurrent allergic bronchopulmonary aspergillosis and aspergilloma: Is it a more severe form of the disease? *Eur Respir Rev.* 2010; **19:** 261-3.
247. Sharif MA, Mahmood A, Khan A, *et al.* Allergic bronchopulmonary aspergillosis presenting as flitting opacities on chest radiograph. *J Coll Physicians Surg Pak.* 2010; **20:** 484-6.
248. Sharma B, Ghosh G, Kamble U, *et al.* Allergic bronchopulmonary aspergillosis complicated by nocardiosis. *Case Rep Pulmonol.* 2012; 758630. PMID: 23320238.
249. Sharma VK, Raj D, Xess I, *et al.* Prevalence and risk factors for allergic bronchopulmonary aspergillosis in Indian children with cystic fibrosis. *Indian Pediatr.* 2014; **51:** 295-7.
250. Sharpe RA, Bearman N, Thornton CR, *et al.* Indoor fungal diversity and asthma: A meta-analysis and systematic review of risk factors. *J Allergy Clin Immunol.* 2015; **135:** 110-22.
251. Shin SY, Ye YM, Eun YG, *et al.* Local IgE-mediated hypersensitivity to *Alternaria* in pediatric adenoid tissue. *Int J Pediatr Otorhinolaryngol.* 2012; **76:** 1423-8.
252. Simon-Nobbe B, Denk U, Poll V, *et al.* The spectrum of fungal allergy. Int Arch Allergy Immunol. 2008; **145:** 58-86.
253. Singh AB, Shahi S. Aeroallergens in clinical practice of allergy in India-ARIA Asia Pacific Workshop report. *Asian Pac J Allergy Immunol.* 2008; **26:** 245-56.
254. Singh B, Denning DW. Allergic bronchopulmonary mycosis due to *Alternaria*: Case report and review. *Med Mycol Case Rep.* 2012; **1:** 20-3.
255. Singh B, Singh S, Asif AR, *et al.* Allergic aspergillosis and the antigens of *Aspergillus fumigatus. Curr Protein Pept Sci.* 2014; **15:** 403-23.
256. Singh M, Das S, Chauhan A, *et al.* The diagnostic criteria for allergic bronchopulmonary aspergillosis in children with poorly controlled asthma need to be re-evaluated. *Acta Paediatr.* 2015; **104:** e206-9.
257. Singh PK, Kathuria S, Agarwal K, *et al.* Clinical significance and molecular characterization of non-sporulating molds isolated from the respiratory tracts of bronchopulmonary mycosis patients with special reference to basidiomycetes. *J Clin Microbiol.* 2013; **51:** 3331-7.
258. Singh Sehgal I, Agarwal R. Pulse methylprednisolone in allergic bronchopulmonary aspergillosis exacerbations. *Eur Respir Rev.* 2014; **23:** 149-52.
259. Soeria-Atmadja D, Onell A, Borga A. IgE sensitization to fungi mirrors fungal phylogenetic systematics. *J Allergy Clin Immunol.* 2010; **125:** 1379-86.e1.
260. Sotiriou A, Koulouris N, Bakakos P. Fever and multilobular mass of the right lung in a young adult with asthma. *Med Mycol Case Rep.* 2015; **10:** 7-10.
261. Speth C, Rambach G, Wurzner R, *et al.* Complement and fungal pathogens: An update. *Mycoses.* 2008; **51:** 477-96.
262. Sproson EL, Thomas KM, Lau LC, *et al.* Common airborne fungi induce species-specific effects on upper airway inflammatory and remodelling responses. *Rhinology.* 2016; **54:** 51-5.
263. Sugita T, Ikeda R, Nishikawa A. Analysis of *Trichosporon* isolates obtained from the houses of patients with summer-type hypersensitivity pneumonitis. *J Clin Microbiol.* 2004; **42:** 5467-71.
264. Sutrave H, Ward A, Smyth AR, *et al.* Pneumomediastinum as a presenting feature of allergic bronchopulmonary aspergillosis in a child with cystic fibrosis. *J R Soc Med.* 2012; **105:** S36-9.
265. Takazono T, Sheppard DC. *Aspergillus* in chronic lung disease: Modeling what goes on in the airways. *Med Mycol.* 2017; **55:** 39-47.

266. Takeuchi Y, Shirai T, Sakurai S, *et al.* Allergic bronchopulmonary aspergillosis presenting with a pulmonary mass mimicking lung cancer. *Respirol Case Rep.* 2013; **1:** 5-7.
267. Templeton SP, Buskirk AD, Green BJ, *et al.* Murine models of airway fungal exposure and allergic sensitization. *Med Mycol.* 2010; **48:** 217-28.
268. Terr AI. Sick Building Syndrome: Is mould the cause? *Med Mycol.* 2009; **47:** S217-22.
269. Tham R, Katelaris CH, Vicendese D, *et al.* The role of outdoor fungi on asthma hospital admissions in children and adolescents: A 5-year time stratified case-crossover analysis. *Environ Res.* 2017; **154:** 42-9.
270. Tripathi SK, Srivastava R, Tg R, *et al.* Allergic bronchopulmonary aspergillosis: Still a mirage or ignorance. *BMJ Case Rep.* 2012; pii: bcr0120125584. PMID: 22787178.
271. Tsukatani T, Ogawa H, Anzawa K, *et al. Schizophyllum commune*-induced allergic fungal rhinosinusitis and sinobronchial mycosis. *Med Mycol Case Rep.* 2015; **8:** 10-3.
272. Upadhyay R, Kant S, Prakash V, *et al.* Sinobronchial allergic aspergillosis with allergic bronchopulmonary aspergillosis: A less common co-existence. *BMJ Case Rep.* 2014; pii: bcr2013201946. PMID: 25371437.
273. Vacher G, Niculita-Hirzel H, Roger T. Immune responses to airborne fungi and non-invasive airway diseases. *Semin Immunopathol.* 2015; **37:** 83-96.
274. Vardhan V, Mulajker DS. Allergic bronchopulmonary candidiasis. *Med J Armed Forces India.* 2012; **68:** 395-7.
275. Vermani M, Vijayan VK, Kausar MA, *et al.* Quantification of airborne *Aspergillus* allergens: Redefining the approach. *J Asthma.* 2010; **47:** 754-61.
276. Vijay HM, Kurup VP. Fungal allergens. *Clin Allergy Immunol.* 2008; **21:** 141-60.
277. Wardhana, Datau EA. A patient with allergic bronchopulmonary mycosis caused by *Aspergillus fumigatus* and *Candida albicans. Acta Med Indones.* 2012; **44:** 317-23.
278. Weber RW, Levetin E. Allergen of the month—*Fusarium. Ann Allergy Asthma Immunol.* 2014; **112:** A11.
279. Weber RW. Allergen of the month—*Mucor. Ann Allergy Asthma Immunol.* 2015; **115:** A15.
280. Weber RW. Allergen of the month—*Stemphylium. Ann Allergy Asthma Immunol.* 2015; **114:** A11.
281. Williams R, Krishnadas R, Patel B, *et al.* Endobronchial valves in the management of bronchial fistulae caused by bronchopulmonary aspergillosis. *BMJ Case Rep.* 2015; pii: bcr2015211955. PMID: 26628306.
282. Wiszniewska M, Walusiak-Skorupa J, Gutarowska B, *et al.* Is the risk of allergic hypersensitivity to fungi increased by indoor exposure to moulds? *Int J Occup Med Environ Health.* 2009; **22:** 343-54.
283. Woodcock AA, Steel N, Moore CB, *et al.* Fungal contamination of bedding. *Allergy.* 2006; **61:** 140-2.
284. Woodfolk JA. Allergy and dermatophytes. *Clin Microbiol Rev.* 2005; **18:** 30-43.
285. Woolnough K, Fairs A, Pashley CH, *et al.* Allergic fungal airway disease: Pathophysiologic and diagnostic considerations. *Curr Opin Pulm Med.* 2015; **21:** 39-47.
286. Woolnough KF, Richardson M, Newby C, *et al.* The relationship between biomarkers of fungal allergy and lung damage in asthma. *Clin Exp Allergy.* 2017; **47:** 48-56.
287. Yamamoto N, Shendell DG, Peccia J. Assessing allergenic fungi in house dust by floor wipe sampling and quantitative PCR. *Indoor Air.* 2011; **21:** 521-30.
288. Yilmaz I, Aydin O, Ozdemir SK, *et al.* Co-existence of allergic bronchopulmonary aspergillosis and atopic dermatitis: Is total IgE level useful to identify relapses of allergic bronchopulmonary aspergillosis? *Allergol Immunopathol (Madr).* 2013; **41:** 281-3.
289. Yoshida Y, Shirai T, Mikamo M, *et al.* Development of allergic bronchopulmonary aspergillosis with central bronchiectasis over a 10-year period: The need to recheck allergen sensitization. *Intern Med.* 2013; **52:** 2135-8.
290. Zhang M, Liu F, Liu H, *et al.* Sensitization and cross-reactions of dermatophyte and *Candida albicans* allergens in patients with chronic urticaria. *Int J Dermatol.* 2016; **55:** 1138-42.
291. Zhou Y, Xu D, Zhang Y, *et al.* Allergic bronchopulmonary aspergillosis in children. *Pediatr Int.* 2015; **57:** e73-6.
292. Zicari AM, Celani C, De Castro G, *et al.* Anti-IgE antibody as treatment of allergic bronchopulmonary aspergillosis in a patient with cystic fibrosis. *Eur Rev Med Pharmacol Sci.* 2014; **18:** 1839-41.

CHAPTER 34

Fungal Rhinosinusitis

Fungal rhinosinusitis is the mycotic infections of nasal and paranasal sinuses. This term coined to define a spectrum of clinical conditions associated with sinonasal inflammation due to fungal etiopathogenesis. Various agents belonging to genus *Aspergillus*, Mucorales, phaeoid fungi and others are usually found involved in this clinical entity. This disease may spread to the surrounding vicinity like orbit or brain hence is taken very seriously. It occurs in both acute and chronic forms. Its incidence has increased worldwide over the past few decades, both in immunocompromised as well as immunocompetent individuals. The geographical distribution of fungal rhinosinusitis (FRS) varies widely with more cases are being reported in hot and dry areas.

Historical Perspective

Fungal rhinosinusitis was first reported way back in the 18th century in 1791 by Plaignaud as a 'fungus tumor' in the maxillary sinus of a young soldier. The next case was of non-invasive fungal sinusitis reported by Mackenzie in 1893. A more specific diagnosis of nasal and paranasal sinus aspergillosis was reported by Schubert in 1885 with the cause first time identified as *Aspergillus* and hence called *Aspergillus* sinusitis. In 1897, the first case of invasive fungal sinusitis was described by Oppe that affected sphenoid sinus and extended into cerebrum through bony erosion, caused by *Aspergillus* species. Wright in 1927, reported two more cases of invasive fungal sinusitis extending to the orbit, which were also caused by *Aspergillus* species. The fungal sinusitis in old reports was hence termed sinus aspergillosis.

The rhinocerebral mucormycosis, another form of invasive fungal sinusitis was first described in 1957 by Baker in a patient with diabetes mellitus with broad hyphae in vessels, lymphatics and in tissues. Later, in 1969, primary paranasal *Aspergillus* granuloma was seen by Milosev *et al* in 17 immunocompetent cases in Sudan with granuloma composed of multinucleated giant cells, bone and artery destruction by the chronic inflammatory infiltrate and fungal elements, present in the vessels.

In 1971, McCarthy and Pepys observed that about 10% of patients with allergic bronchopulmonary aspergillosis (ABPA) complained nasal plugs similar to expectorated from bronchi. Maxillary sinusitis was seen on X-ray films in 41.4% of patients having ABPA. Cultures of the nasal plugs and sinus lavages grew *Aspergillus fumigatus.* In 1976, Safirstein noted a combination of nasal polyposis, crust formation and sinus cultures yielding *Aspergillus* species and clinical similarity of findings of allergic FRS with ABPA. He also reported a significant finding that obstruction produced by nasal casts improved when ABPA patient was given oral corticosteroids.

Later on in 1981, a sinonasal corollary of ABPA in patients with chronic pansinusitis was termed as "allergic aspergillosis of the maxillary sinuses". In 1983 pathophysiologic resemblance of chronic rhinosinusitis (CRS) with ABPA was described as "allergic *Aspergillus* sinusitis" (AAS) in healthy patients with asthma, nasal polyposis and sinusitis diagnosed on the basis of a histologic triad caused due to non-invasive fungal hyphae of *Aspergillus* species and also gave a unique term "allergic mucin".

Later on *Aspergillus* species were considered responsible for about 15% of cases of AFS, subsequent reports also established different fungi besides *Aspergillus*. This led to the renaming of "allergic *Aspergillus* sinusitis" as "allergic fungal sinusitis" (AFS or AFRS) in 1987. In 1991, dematiaceous fungi were isolated among patients with allergic fungal sinusitis.

In the year 1994, Bent and Kuhn published diagnostic criteria focused on histologic, radiographic and immunologic characteristics of the disease. Others have proposed several set of criteria that have served to further the discussion of and investigation into this unique disease.

However, these criteria are largely regarded as the standard for diagnosis.

The International Society for Human and Animal Mycology (ISHAM) has established a working group on 'Fungal Sinusitis' in 2006 in Paris. This group held an international consortium to assess the prevalence of FRS among patients with CRS; categorization; etiological agents; pathogenesis and management by organizing a Workshop in 2008 at Chandigarh.

Categorization of Fungal Rhinosinusitis

The classification and categorization of FRS has been attempted since long. FRS classification originated, evolved and underwent modifications and endless controversies over the years. The present classification of FRS was thereby formulated into invasive and non-invasive diseases based on histopathological evidence of tissue invasion by fungi that can occur in both immunocompetent or immunocompromised individuals.

The FRS is classified into two broad categories *i.e.* invasive and non-invasive disease forms based on clinical and histopathological evidence of tissue invasion by the causative fungi **(Table 34.1)**.

I. Invasive Disease

1. Acute invasive (fulminant) FRS
2. Granulomatous invasive FRS and
3. Chronic invasive FRS.

II. Non-invasive Disease

1. Allergic FRS
2. Eosinophilic FRS
3. Localized Colonization
4. Fungus Ball

Further acute FRS is used for illness with duration of <1 month and mainly includes the acute invasive type of FRS only. Chronic FRS is of duration >3 months and includes the other five types. In addition, a new term subacute FRS is being used when duration of illness is between 1 and 3 months.

Acute invasive (fulminant) FRS is the most common form of invasive FRS. It is a rapidly progressive tissue invasive fungal disease. It is an opportunistic infection affecting almost exclusively the severely immunocompromised individuals. Mortality has been reported to be as high as 80 percent. Identification of the population at risk with high index of suspicion for early diagnosis and aggressive medical and surgical intervention is crucial for survival. On histopathology, there is necrotic tissue invaded by plenty of fungal hyphae with acute neutrophilic infiltrate.

Table 34.1. Classification of Fungal Rhinosinusitis (FRS).

A. Non-invasive
- (i) Allergic Fungal Rhinosinusitis
- (ii) Eosinophilic Fungal Rhinosinusitis
- (iii) Localized Colonization
- (iv) Fungal Ball

B. Invasive
- (i) Acute Invasive (Fulminant)
- (ii) Chronic Invasive
  - Immunocompetent
  - Immunocompromised
- (iii) Granulomatous Invasive

Acute fulminant invasive FRS must be approached and treated as two possible entities. Firstly, in diabetic patients having a higher mortality, morbidity and a higher neurological and visual sequelae of invasion into cerebrum and orbits, respectively. Its incidence is increasing due to high index of suspicion coupled with high incidence of diabetes, specially uncontrolled or undetected at the time of presentation in association with the causative organism as mucormycetes species, which are more virulent pathogens and behaves in a more aggressive manner than *Aspergillus* species. Secondly, in neutropenic patients more commonly having invasive aspergillosis (IA) in which recovery and recurrence are related to fall and rise in absolute neutrophil count (ANC).

Granulomatous invasive FRS is usually seen in young adults and is described by a time course of >12 weeks presenting with an enlarging mass in the cheek, orbit, nose and paranasal sinuses and proptosis in immunocompetent hosts. The disease has primarily been seen in Sudan, India, Pakistan and Saudi Arabia and *Aspergillus flavus* is the most commonly isolated fungal agent. Histopathology shows granulomatous response with considerable fibrosis. Non-caseating granuloma with foreign body or Langerhans-type giant cells may be seen with scanty number of hyphae.

The chronic invasive fungal rhinosinusitis is relatively rare slowly destructive disease with a time frame of >12 weeks. It is more common among the individuals with AIDS, diabetes mellitus or in those on corticosteroid therapy. Most commonly it affects the ethmoid and sphenoid sinuses but may involve any of the paranasal sinuses. Cultures of tissue are positive in >50% of cases and *Aspergillus flavus* is the most common agent isolated in this clinical entity. On histopathology, there is dense accumulation

Table 34.2. Bent and Kuhn Diagnostic Criteria of FRS.

Major	*Minor*
Type I Hypersensitivity	Asthma
Nasal Polyposis	Unilateral Disease
Characteristic CT Findings	Bone Erosion
Eosinophilic Mucin Without Invasion	Fungal Cultures
Positive Fungal Stains	Charcot-Lyden Crystal
	Eosinophilia

of hyphae, occasional presence of vascular invasion and sparse inflammatory reaction in association with involvement of local structures lying in the surrounding vicinity. The abundance and dense accumulation of fungal hyphae in chronic invasive fungal rhinosinusitis is thought to be due to progression of fungal ball becoming invasive in nature, which sometimes loosely termed as **"sinus mycetoma"**.

The saprophytic fungal infestation is superficial sinonasal mycoses presenting as asymptomatic colonization of mucous crusts within the nasal cavity, often in patients who have had previous sinus surgery. Though in itself this is not serious but it is perceived to represent opportunistic nature of fungi and has potential to be nidus for further growth into a more significant fungal rhinosinusitis.

The fungal ball is a extramucosal dense conglomeration/accumulation of matted fungal hyphae in just one sinus cavity, usually maxillary sinus followed by sphenoid sinus. It is more commonly seen in immunocompetent middle-aged and elderly females. The diagnostic criteria include radiological evidence of sinus opacification, mucopurulent cheesy or clay-like material within the sinus, a dense conglomeration of hyphae separate from the sinus mucosa, non-specific chronic inflammation of the mucosa, no predominance of eosinophils or granuloma or allergic mucin and no histopathological evidence of fungal invasion of mucosa. Approximately 70% of fungal balls turn out to be culture negative. *Aspergillus* species can be implied by galactomannan detection in the sinus material. It can become invasive after immunosuppression and can develop allergic mucin surrounding the fungal balls when corticosteroids are tapered.

As per Bent and Kuhn criteria, the patients must meet all the major criteria for diagnosis, while minor criteria serve to support diagnosis and describe individual patients but are not used to make diagnosis. The major criteria include a history of Type I hypersensitivity, skin testing or *in vitro* testing; nasal polyposis; characteristic CT scan findings; the presence of eosinophilic mucin without invasion and a positive fungal stain of sinus contents removed during surgery. The minor criteria include a history of asthma, unilateral predominance of disease, radiographic evidence of bone erosion, fungal cultures, presence of Charcot-Leyden crystals in surgical specimens and eosinophilia **(Table 34.2)**.

The common causes of AFRS are the hyaline molds, such as *Aspergillus, Fusarium, Scedosporium* and dematiaceous fungi like *Curvularia, Bipolaris* species. However, *A.flavus* is isolated in more than 80% of the cases of AFRS from India and in 50% cases from the Middle East. In USA, AFRS is seen in urban population while in Asia young villagers are found to be more involved.

The eosinophilic fungal rhinosinusitis (EFRS) comprises detection of fungi along with eosinophil and eosinophil degraded products as mucus in clusters along with Charcot-Leyden crystals in absence of specific IgE response. High levels of toxic major basic protein (MBP) from eosinophils in the mucus of patients with CRS damages nasal epithelium from luminal side and may permit secondary bacterial infection on the damaged epithelium. So, it is eosinophilic mucin without allergic component.

The eosinophilic mucin rhinosinusitis (EMRS or EFRS-like) arises from confusion in categorization of AFRS when hyphae are sparse and defined when fungus is not visualized microscopically in addition to absence of specific IgE response but presence of eosinophilic mucin. It is uniformly bilateral, combined with a significantly higher frequency of asthma and an increased incidence of aspirin sensitivity and frequently IgG_1 deficiency.

The AFRS-like group is defined because sometimes the mucous is not provided to the pathologists. It may be due to the use of micro-debriders, which suck away removed tissue as well as the mucus. Therefore, even if fungus is not demonstrable, fungal-specific IgE response is demonstrable.

Epidemiology

A number of host and environmental factors are responsible for the development of fungal sinusitis, the most common being any obstruction in the sinuses leading to impaired ventilation.

The environmental factors also play an equally important role as fungi are ubiquitous and elusive in nature. They are found in air, dust, soil, plants and decaying organic matter. They adhere to dust particles, are inhaled and deposited over sinonasal mucosa, proliferate in the ideal warm moist environment of upper respiratory tract in normal individuals and become pathogenic in susceptible individuals.

A significantly higher incidence is reported in the restricted zones of north Sudan, Saudi Arabia and in the south-western states of the USA, which have a warm and dry climate. In India, north India, including western Utter Pradesh, north Rajasthan, Haryana, Punjab and Chandigarh, also having a warm and dry climate during the summer months, have a high incidence. So much so, north India has been identified as the endemic zone for paranasal mycoses possibly due to the dust and frequent sand storms during the summer months. These storms contain large numbers of *Aspergillus* conidia that can easily settle on to the injured mucosa of the sinuses of young men working outdoors, who are regularly exposed to the warm dry climate.

The increasing incidence of FRS is attributed to various predisposing factors related to the host like uncontrolled diabetes mellitus, immunosuppressed states, allergy to dust, allergic rhinitis, bronchial asthma and anatomical risk factors like deviated nasal septum. There is evidence of gradual shift of risk factors towards increasing immunosuppressive states like hematologic malignancies, solid organ transplant, bone marrow transplant, chemotherapy, radiotherapy, chronic corticosteroid use and Acquired Immunodeficiency Syndrome (AIDS). The other risk factors include autoimmune diseases and systemic lupus erythematosus. In addition, the use of broad-spectrum antibiotics, antihistamines and corticosteroids used in the treatment of nasal allergy and sinusitis may instead play a role in the pathogenesis of the disease by suppressing bacterial flora and supporting fungal overgrowth.

Mycology

There are many types of fungi involved in rhinosinusitis. These may be hyaline or dematiaceous hyphomycetes, dimorphic fungi, mucormycetes as well as yeast like fungi. A wide range of fungal species are involved in disease but the most common are those belonging to genus *Aspergillus*. The etiology has been found to be different in different types and subtypes of fungal sinusitis.

Aspergillus fumigatus has always been considered the primary etiological agent for AFRS but now-a-days, dematiaceous fungi belonging to *Bipolaris, Exserohilum, Curvularia* and *Alternaria* species are increasingly being reported especially in western literature. However, *A.flavus* is the predominant agent isolated from cases of AFRS in Indian subcontinent.

In 1998, in USA cases of allergic fungal sinusitis had *Bipolaris spicifera* as the most prevalent fungus (67%) followed by other dematiaceous fungi. Eighty-one percent of the cultured fungal species from cases of allergic AFRS were members of pigmented dematiaceous family of fungi such as *Bipolaris, Drechslera, Exserohilum, Curvularia, Alternaria* and *Cladosporium* species and were concluded as the predominant etiological agents in several subsequent studies.

More recently, other non-*Aspergillus*, non-dematiaceous, hyaline fungi have been reported in a substantial number of cases. These include *Fusarium, Chrysosporium*, members of family mucormycetes like *Rhizomucor* amongst the many more and constantly newly emerging fungal agents. *Aspergillus flavus* is the most frequent isolate in all types of paranasal sinus mycoses in India especially in north India. In Sudan also, *A.flavus* is isolated in most of the cases. However, in the United States, *A.fumigatus* and *Rhizopus arrhizus* have been found in an increasing number of cases. *Bipolaris* is an emerging pathogen in humans, particularly being reported in the southwestern states of the USA, Australia, Pakistan and India, areas having a hot climate. The common fungi involved in the FRS are listed in **Table 34.3**.

Pathogenesis and Pathology

The pathogenesis of fungal rhinosinusitis is poorly understood. It is presumed that the fungi get trapped in the sinuses of individuals who have mucociliary functional or anatomical disorders. In such persons, they stimulate release of IgE, IgA and IgG production. Both Type I and Type III immune mechanisms may be involved leading to mucosal edema and inflammation. It is believed to be secondary to hypersensitivity reaction to the colonizing fungus in the sinus lumen, which is indicated by eosinophilia, raised serum titers of IgE directed against the incriminating fungi and hypersensitivity to mold extracts as demonstrated by cutaneous testing.

The disease appears to be more aggressive in pediatric patients with rhinocranial and rhinorbital complications. The allergic mucin and polyps form a partially calcified expansile mass that obstructs sinus drainage and lead to

Table 34.3. Causative Agents of Fungal Rhinosinusitis.

A. Hyaline Hyphomycetes
Aspergillus flavus, A.fumigatus
Fusarium solani, F.oxysporum
Acremonium species
Paecilomyces species
Penicillium species
Scedosporium apiospermum
Trichoderma species
B. Phaeoid Hyphomycetes
Curvularia species
Bipolaris species
Alternaria species
Exserohilum species
Cladosporium species
Fonsecaea pedrosoi var. cladosporium
Aureobasidium pullulans
Lasiodiplodia theobromae
C. Yeast-like Fungi
Candida albicans
Candida tropicalis
Candida krusei

sinusitis. The growth of such mass may also lead to pressure erosion of bone, rupture of sinus walls and leakage of sinus contents into adjacent orbit or brain.

The mucosal hypertrophy with fungus ball, which is the most common finding in both types, with primary and secondary cases and aspergillosis is the most common, followed by mucormycosis. In secondary cases, the most common concomitant disease is found to be diabetes mellitus.

The recent research in the field of fungal rhinosinusitis has shown that there is involvement of superantigens of *Staphylococcus aureus* especially enterotoxin. The *Aspergillus* facilitates action of the superantigens by breaking the epithelial barrier and initiating a Th2 response.

Clinical Features

The clinical features of FRS are nasal obstruction, post-nasal discharge and/or along with cerebral and orbital symptoms in invasive forms as well as allergic symptoms in allergic forms. The patients with fulminant invasive disease present with rapidly progressive swelling of the face with discoloration of the overlying skin, proptosis and invasion into the brain. Various clinical presentations involving different sinuses are shown in **Figures 34.1A to D**.

On physical examination, symptoms can be localized to the affected sinus. Clinical examination aided by anterior rhinoscopy shows mucosal edema, crusts, purulence and/or polyps. Further nasal endoscopy can be done if no obvious inflammation is seen till the extent rhinoscopy is able to visualize. The simple colonization of nasal or paranasal sinuses without any symptoms has been described as saprotrophic fungal infestation. The colonization occurs over mucous crust often in patients who had a history of previous sinus surgery and is detected upon endoscopic examination. It remains silent until it is detected or presents with foul odor. Further extension of the growth may lead to fungal ball formation.

In immunocompromised patients, symptoms and the pathogen spectrum of sinusitis are frequently atypical. If progressive loss of vision occurs, an infection of the anterior skull base or nasal sinuses should be considered. The possibility of opportunistic infections by saprophytic fungi must be taken into account in immunocompromised patients, as they may endanger both vision and survival. In orbit-associated symptoms, CT-imaging reveal bony defects in sinus borders to orbits or endocranium.

Hence a very high index of clinical suspicion, is required for establishing these risk factors. It must be included in the differential diagnosis of all patients with unexplained, recalcitrant or aggressive disease afflicting the nose and paranasal sinuses especially when not responding to the conventional medical therapy including antibacterial antibiotics.

Radiodiagnosis

The imaging technique of choice for diagnosis is computed tomography (CT) scan of nose and paranasal sinuses (PNS) with air fluid levels in acute FRS and polyps in chronic FRS in addition to X-ray PNS (Water's view) and Magnetic Resonance Imaging (MRI) of brain. The CT findings of heterogenous opacity with central high attenuation and MRI T2-weighted images depicting low signal intensity to signal void in the corresponding areas are well-defined diagnostic features of FRS, so is the opacification of affected sinuses on the seldom used X-rays.

The typical radiological findings in fungal sinusitis include haziness, calcification and bone destruction of the involved sinuses. The CT findings may show mucosal

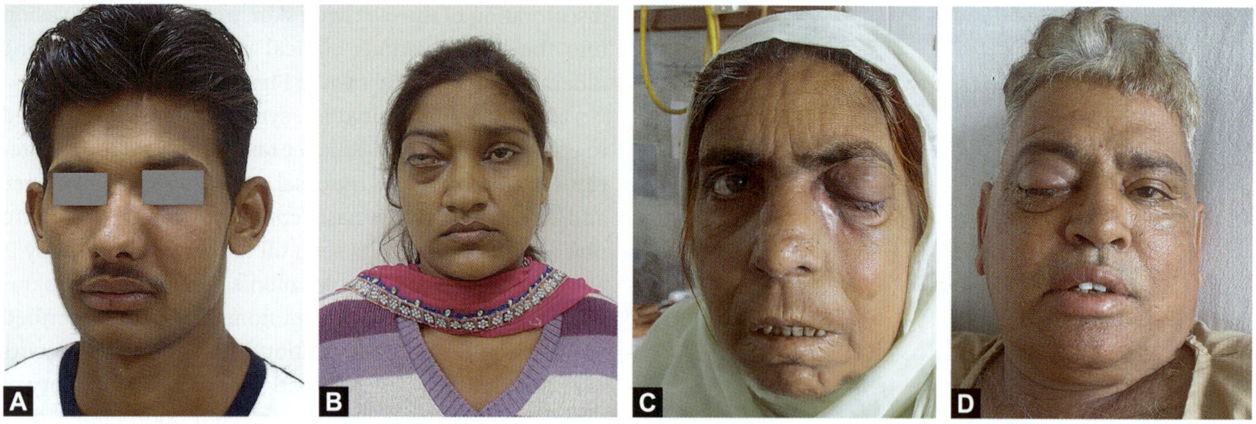

Figs. 34.1A to D. Fungal rhinosinusitis seen in patients with various clinical presentations in (A) left maxillary (B) right ophthalmic (C) left ophthalmic (D) right ophthalmic areas.

changes within the osteomeatal complex (OMC) and/or sinuses. CT scans in about 80% of the primary and 70% of the secondary cases reveal calcific densities in paranasal soft tissue mass.

The bony erosion may be present but is caused by pressure remodeling rather than fungal invasion and destruction. MRI demonstrates iso-intense or slightly hypodense regions on Tl-weighted images and marked hypodensity on T2-weighted images, which correlate to fungus, surrounded by mucosal inflammation with increased density. Various radiological presentations involving different sinuses are shown in **Figures 34.2A to P**.

Laboratory Diagnosis

The establishment of exact diagnosis of fungal rhinosinusitis is pertinent as other similarly presenting diseases, particularly caused by bacteria, are to be differentiated from this clinical entity so that specific treatment can be imparted. Allergic fungal rhinosinusitis, amongst all forms of FRS, has been most controversial regarding its definition, pathogenesis and the diagnostic criteria. It is the most common form of FRS. It is believed to result from fungal allergens induced Type I, IgE-mediated allergic and possibly Type III, immune complex mediated mucosal inflammation in an atopic host. The five diagnostic criteria include (1) Type I hypersensitivity, (2) nasal polyposis, (3) characteristic findings on CT scan, (4) presence of fungi on direct microscopy or culture and (5) allergic mucin containing fungal elements and many eosinophils without tissue invasion. This is established based on conventional and molecular methods. These are as follows:

(a) Direct Examination

The direct mycological examination includes KOH wet mount of specimen from post-operative or endoscopic sinonasal tissue/debris looking for septate or aseptate hyphae **(Figs. 34.3A to F)**. The fungal hyphae are more clearly seen in calcofluor fluorescent brightener staining **(Figs. 34.4A and B)**. The role of CFW is very significant in delineating fungi as sometimes other stains are not able to visualize fungal elements in the tissue and/or debris. Further, for histopathologic correlation, staining of tissue sections with Hematoxylin and Eosin (H&E) **(Figs. 34.5A and B)** and special fungal stains like Periodic acid Schiff (PAS) and Grocott-Gomori's methenamine silver (GMS) staining is done to assess for the tissue reaction: acute neutrophilic, allergic or granulomatous; invasiveness and look for the fungal elements **(Figs. 34.6A to C)**.

Therefore, to diagnose AFRS, the presence of allergic mucin in histopathologic specimens is important in addition to the demonstration of fungal elements. However, allergic mucin is not uniformly distributed throughout sinus content. An inadequate sampling may thus pose difficulties in proper categorization of the cases. Hence a detailed examination of all the fragments is mandatory.

Histologic examination characteristically contains allergic mucin with sheets of eosinophils, necrotic eosinophils and cellular debris within an amorphous stroma. High-power examination reveals Charcot-Leyden crystals with a hexagonal cross-section and a bipyramidal longitudinal section. The pathologist should be cognizant of the presumptive diagnosis of AFS to closely examine the

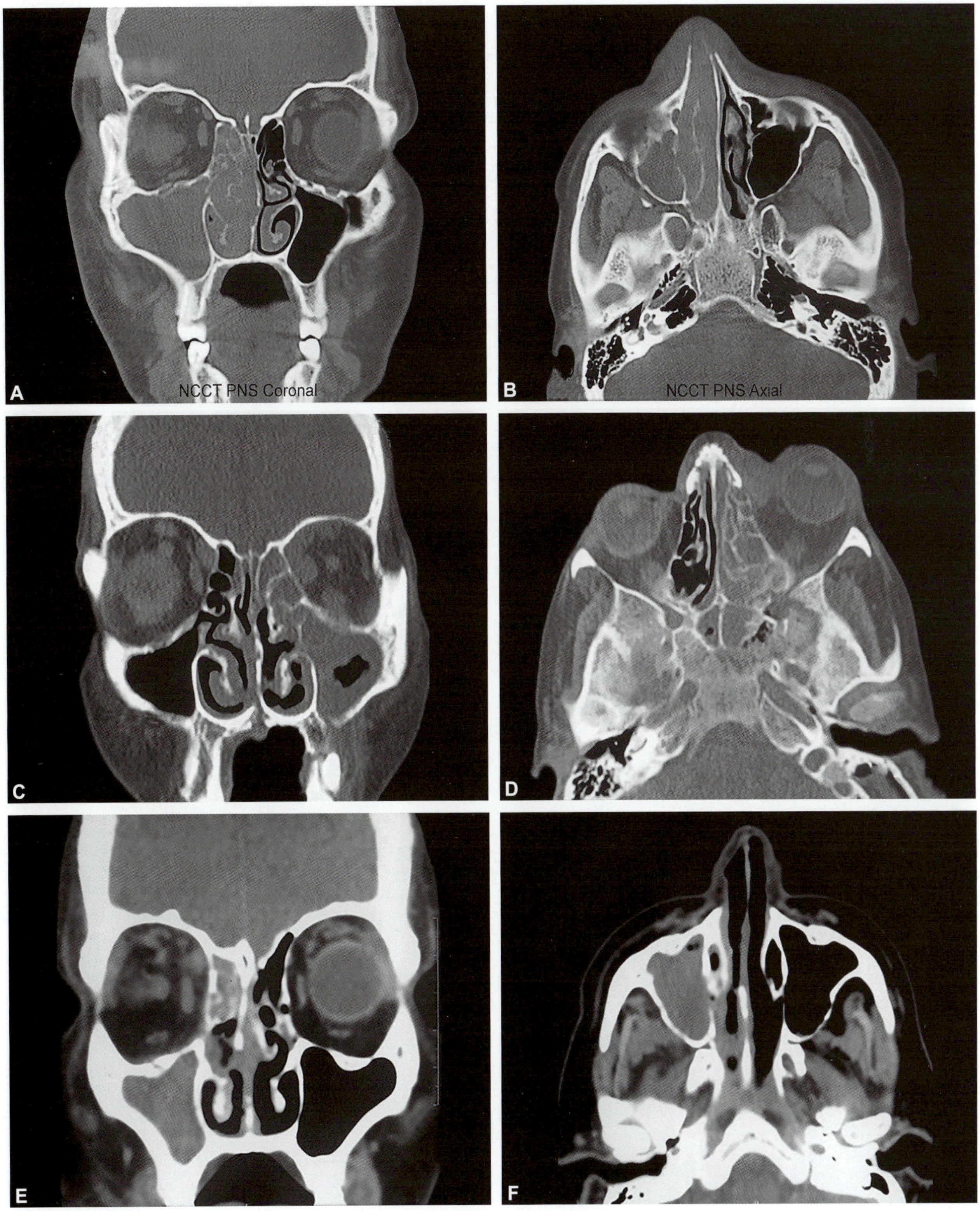

Figs. 34.2A to F

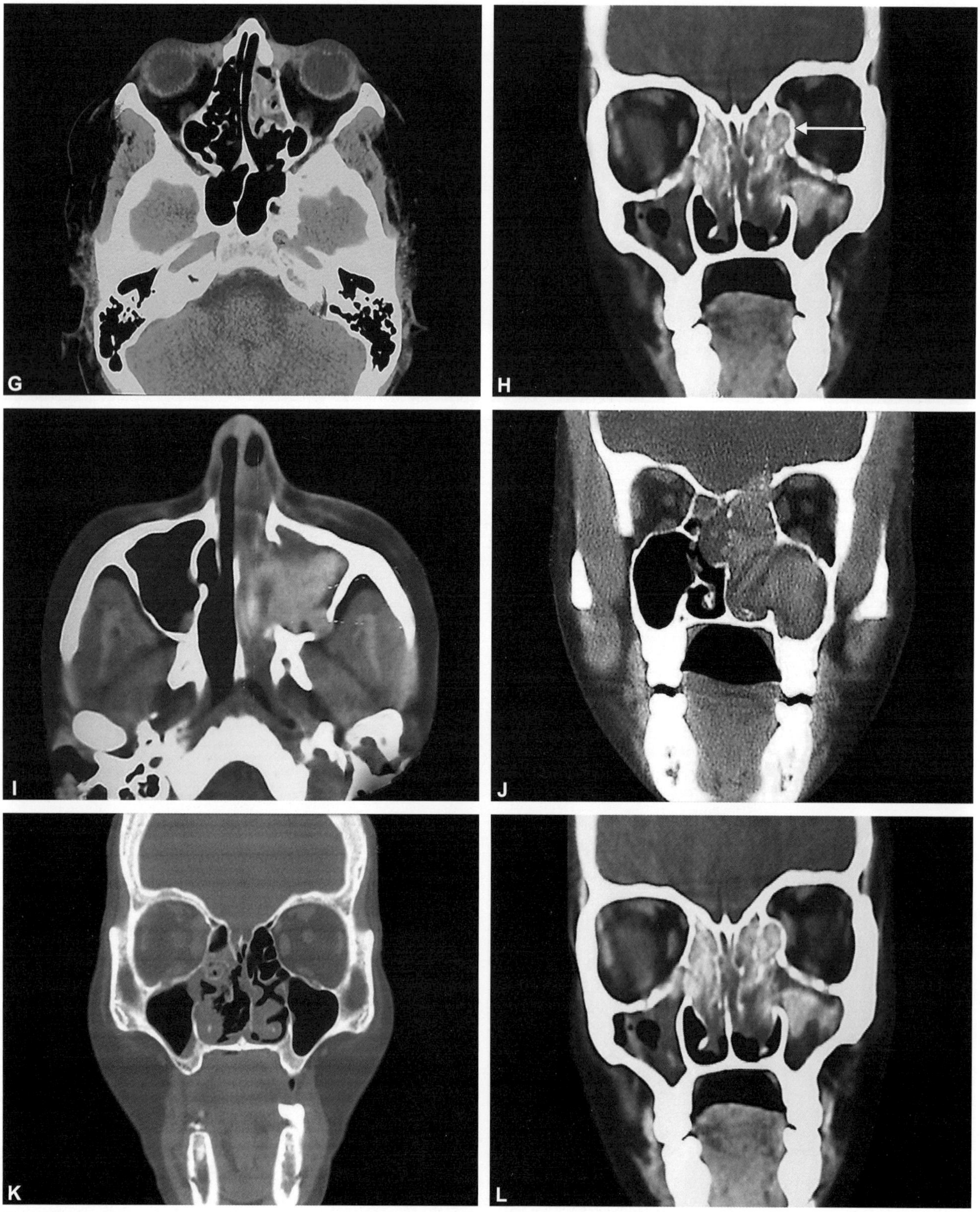

Figs. 34.2G to L

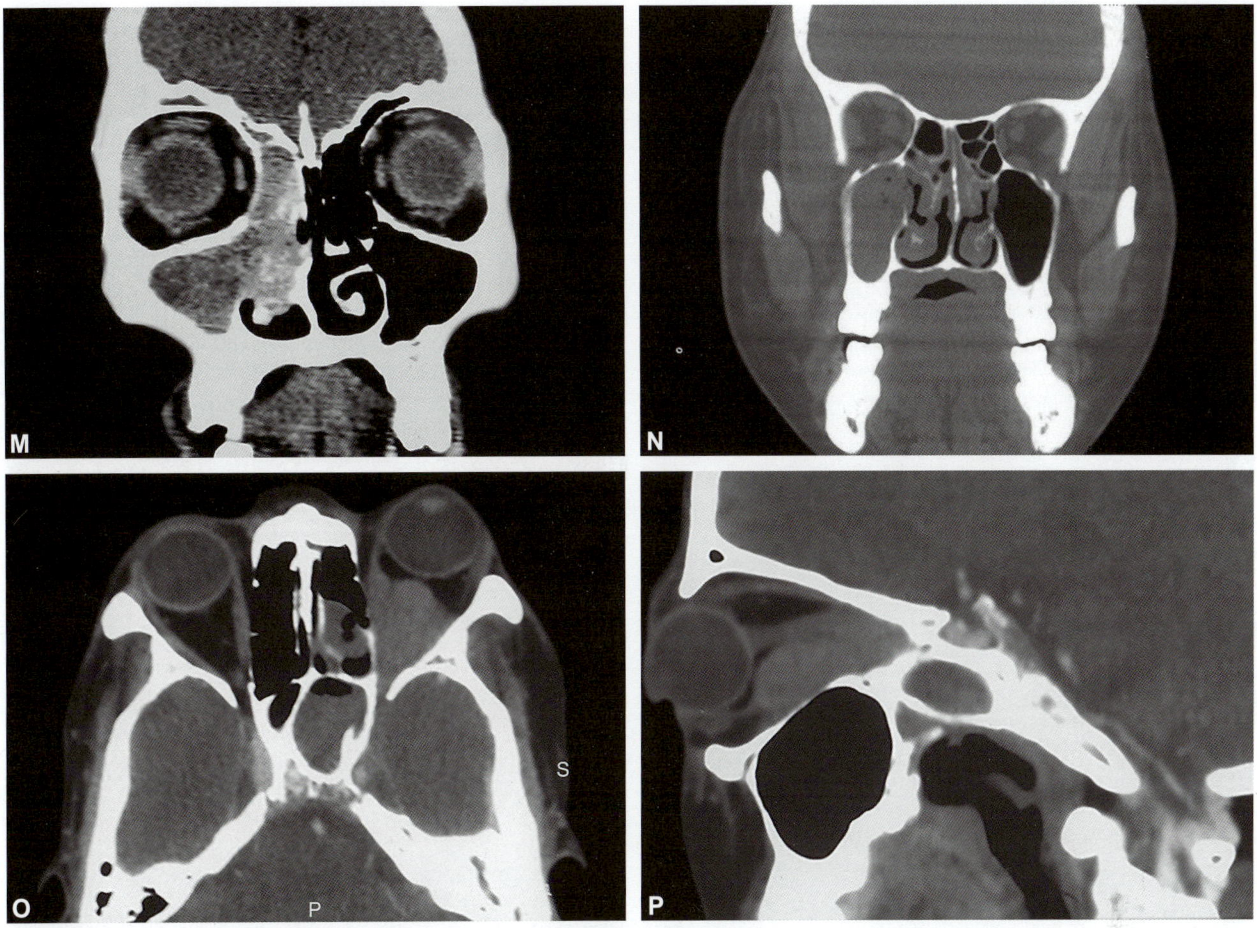

Figs. 34.2A to P. MRI/CT scans of PNS showing involvement of various sinuses among the patients of fungal rhinosinusitis.

specimen for mucosal or bony invasion. Tissue invasion precludes the diagnosis of AFS and favors chronic invasive fungal rhinosinusitis.

In addition, frozen section biopsies may be done, which remain key in obtaining an early diagnosis among patients with a high clinical suspicion for invasive fungal sinusitis. The positivity for invasive fungal pathology given by this modality is invariably consistent with definitive diagnosis made during the regular course of diagnostic services.

(b) Fungal Culture

The clinical samples obtained after endoscopic sinus surgery (ESS) are cultured on Sabouraud dextrose agar. The *Aspergillus* species culture takes around 3-4 days to grow. Dematiaceous fungi may take longer time to grow. Identification of fungal isolate is done by the morphological examination of growth and the microscopic examination of the teased LCB mounts made from the fungal colonies.

For establishing a causal relationship between the fungal culture results and the clinical presentation of AFS can be difficult since many of the fungi isolated are ones that are more commonly considered as contaminants. Sometimes, more than one fungus may be grown in the clinical samples and that may be significant also. It has been observed that *Aspergillus* and Mucormycetes are present in the same patient. Clinical correlation is often necessary for interpretation of direct smears and fungal cultures among such patients.

Moreover, in recent times molecular techniques are ideally suited for fungal identification and assessment of drug resistance mechanisms. The sinus contents can be used for PCR amplification with universal fungal primers for 28S ribosomal DNA and amplicon identification by hybridization with species-specific probes for various

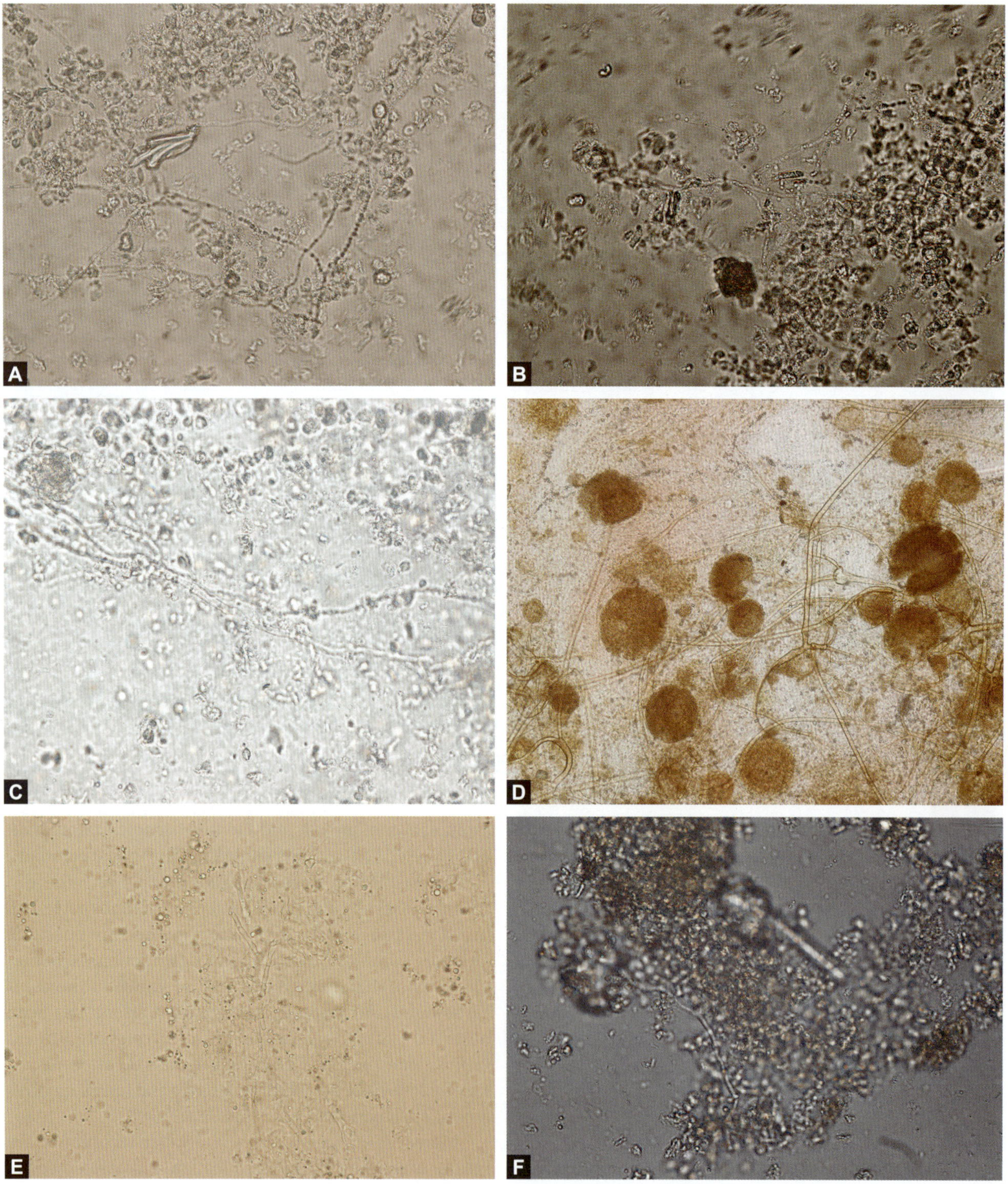

Figs. 34.3A to F. Wet mounts material drained from sinuses (A) septate hyphae with acute-angle branching with debris (B) septate fungal hyphae embedded with debris (C) Long filamentous septate hyphae along with sinus debris (D) septate hyphae with conidiophore and vesicles of *Aspergillus* species (E) Irregular septate hyphae of phaeoid fungi from sinus aspirate (F) septate hyphae with mucous debris (KOH × 400).

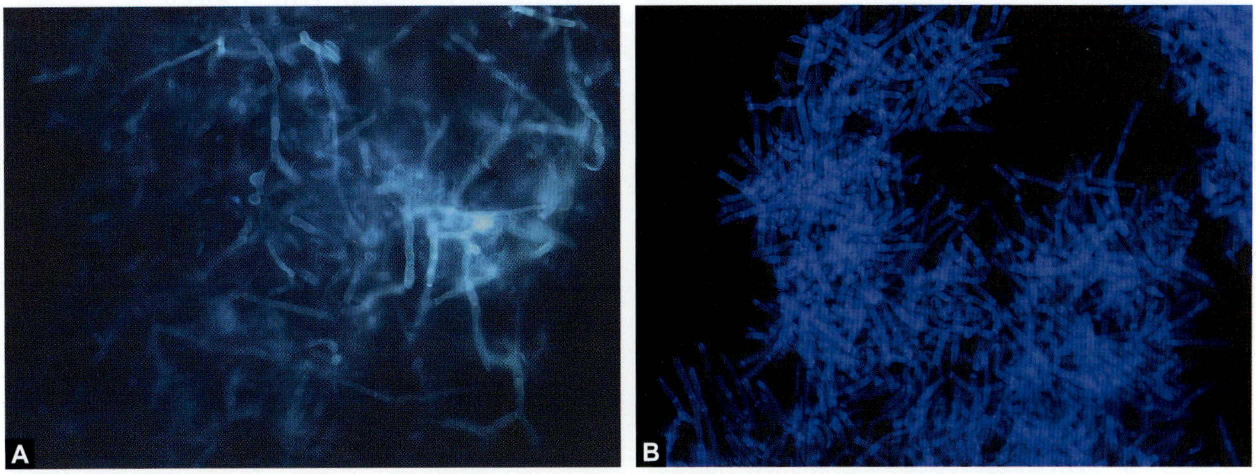

Figs. 34.4A and B. Septate hyphae seen with fluorescent brightener in patients of fungal rhinosinusitis (CFW × 400).

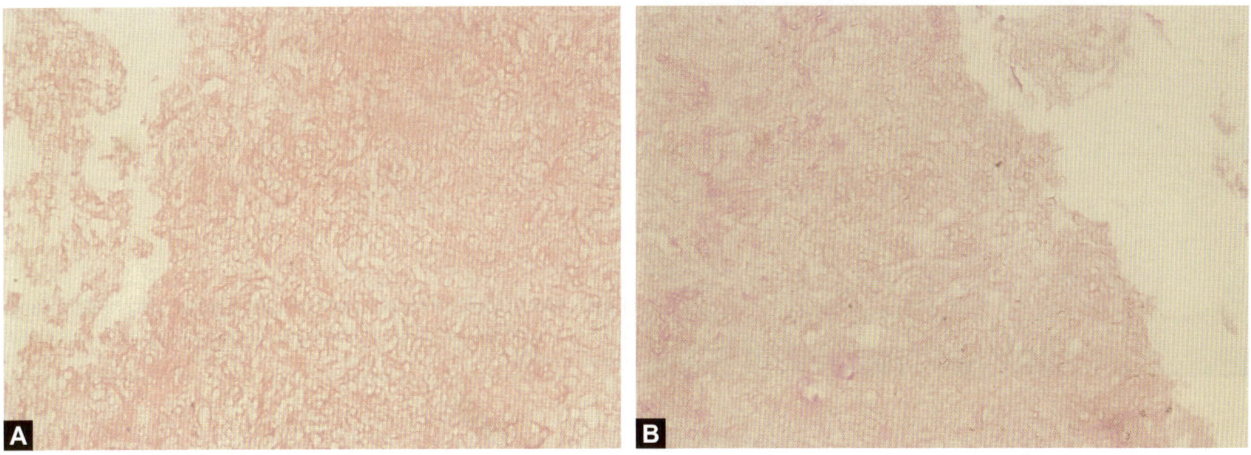

Figs. 34.5A and B. Septate hyphae with acute-angle branching without any inflammatory reaction in a fungal ball (H&E × 400)

species of *Aspergillus* and other fungi. Furthermore, PCR products can be sequenced by an appropriate technique.

The identification of sinus-orbital mucormycosis caused by *Rhizomucor pusillus* can be achieved by sequencing of the internal transcribed spacer (ITS) regions of the rRNA gene and by expression of zygospores in mating. The molecular diagnostic platforms are thus, ideal for rapid detection of fungal pathogens and they provide an opportunity to develop in parallel molecular assays that can evaluate antifungal drug resistance mechanisms.

(c) Immunodiagnosis

Certain hematological parameters like absolute eosinophil count (AEC) and serological investigations for detection of fungal antigens like galactomannan directly in tissue samples or patient's serum are done to establish the diagnosis. The total IgE and/or fungus specific IgE and other antibodies like IgG, may also be quite useful. There is need of Type I hypersensitivity skin testing or demonstration of elevated levels of IgE to label the patient as allergic fungal rhinosinusitis.

(d) Animal Pathogenicity

The animal experiments in the study of fungal sinus disease date back to the last century, when several studies on non-invasive fungal sinusitis were conducted. The models are required to study chronic invasive, antigen stimulation, pressure effects, impact of blockage, etc. in

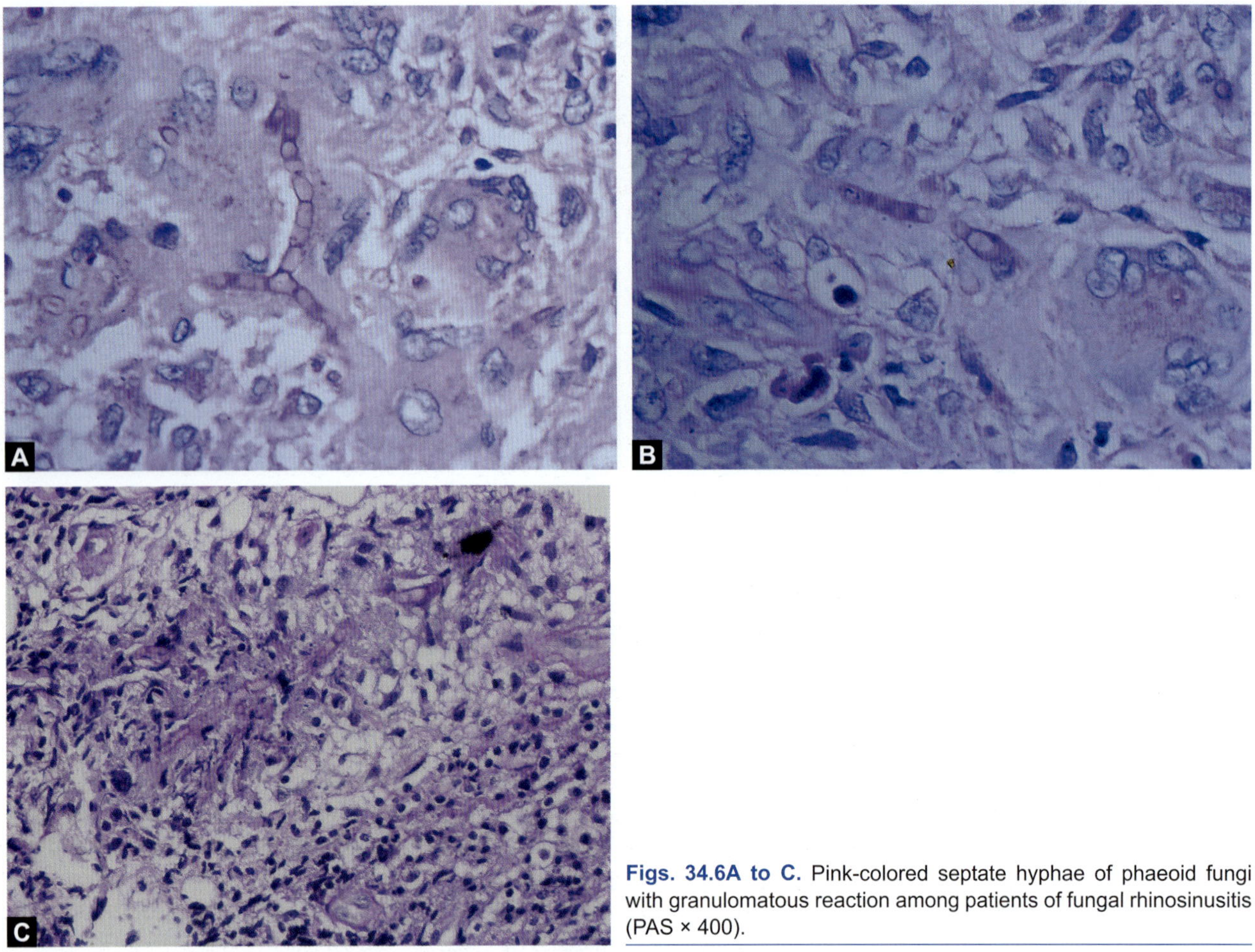

Figs. 34.6A to C. Pink-colored septate hyphae of phaeoid fungi with granulomatous reaction among patients of fungal rhinosinusitis (PAS × 400).

an appropriate animal. The experimental non-invasive fungal sinusitis model was established in rabbits caused by *Aspergillus fumigatus*. Chakrabarti *et al*, have developed a rabbit model for demonstration of paranasal sinus mycosis caused by *Aspergillus* species. Another model was developed in mice of *Aspergillus*- and *Alternaria*-induced non-invasive fungal rhinosinusitis. In addition, a sheep model was reported and it was found that significant fungal biofilm occurred when *Staphylococcus aureus* and fungus were the co-inoculums. Recently a novel invasive fungal rhinosinusitis model in rats has also been developed to meet the research need on this disease.

Treatment and Prophylaxis

The treatment options for fungal rhinosinusitis can be medical or surgical but a combination of both is usually adopted. As far as the medical treatment is concerned, primarily topical or oral decongestants, mucolytics, corticosteroids and antihistamines are prescribed for symptomatic relief in addition to culture guided systemic oral or intravenous antifungals. The various antifungals useful for fungal rhinosinusitis include conventional amphotericin B or the less toxic liposomal amphotericin B preparations that remain the treatment of choice for invasive fungal sinusitis occurring due to the mucormycetes and voriconazole for invasive sinusitis due to *Aspergillus* species. Other azoles are also useful in filamentous fungi except fluconazole. The azole derivatives are however, not effective against mucormycetes, except posaconazole and isavuconazole. The echinocandins are of highly active antifungals used for almost all the fungal agents except mucormycetes.

The patients unresponsive to medical therapy are indicated for sinus surgery. The nasal polyps may occur in the presence of chronic rhinosinusitis and complicate its mana-

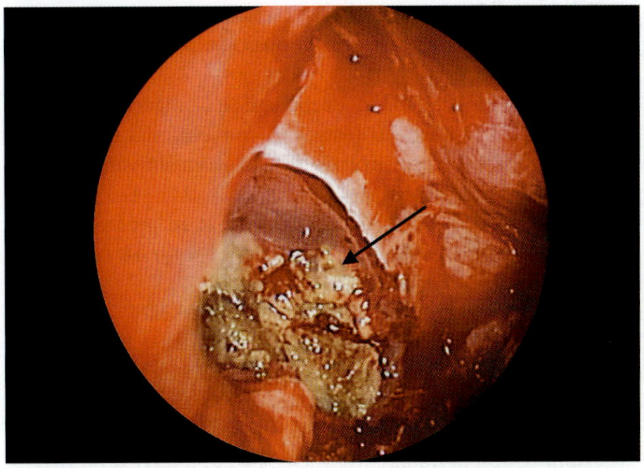

Fig. 34.7. Endoscopic view of right maxillary sinus showing fungal debris.

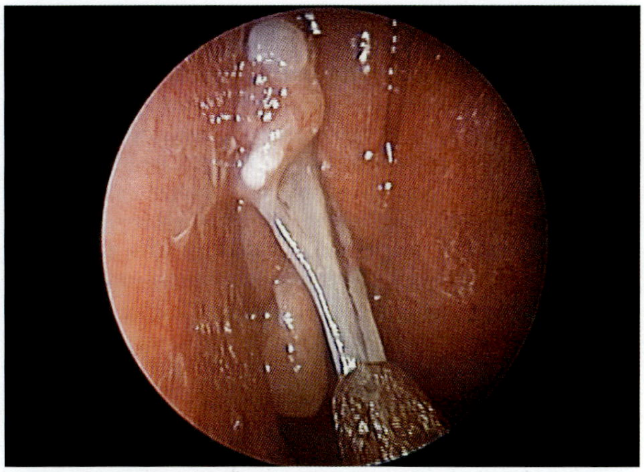

Fig. 34.8. Endoscopic view of nasal cavity showing allergic mucin being sucked.

gement, probably by obstructing normal sinus drainage. The recommended range of operative treatments for clinically suspected fungal rhinosinusitis include endoscopic measures that need specific training like ESS under general anesthesia or FESS under local anesthesia that may include ethmoidectomy and middle meatal antrostomy in addition to drainage of purulent material and aspiration of sinus contents **(Figs. 34 7 and 34.8)**. In addition, it also provides symptomatic relief to the patients and helps to remove obstruction and decreases the fungal load. Besides this, orbital exenteration or neurosurgical intervention may be needed in invasive conditions.

The prognosis of the non-invasive fungal rhinosinusitis is fairly good with rare exacerbations and recurrences. The various complications of fungal rhinosinusitis are mainly in the invasive forms with orbital and intracranial manifestations and have a poor prognosis especially in the immunocompromised or with other debilitating diseases like diabetes mellitus. Under such circumstances it has high tendency to disseminate to other organs thereby further worsening the prognosis. Morbidity is high in many cases though no mortality is as such related directly to any forms of fungal rhinosinusitis.

Treatment of fungal rhinosinusitis usually involves surgical debridement to remove the hypertrophic tissue and mucinous concretions. Nasal and oral corticosteroids are often used to modulate the immune response. In refractory cases, a trial of systemic antifungal therapy may be warranted. An accurate diagnosis of the condition is important as can guide the surgeon on appropriate surgical approach, the extent of resection and also any perioperative adjuvant medical therapy, if needed.

Non-invasive paranasal sinus aspergillosis is most common and successfully treated by endoscopic sinus surgery (ESS) with post-operative sinus irrigation. For the treatment of fungal sinusitis, ESS with or without antifungals and control of predisposing factors for secondary cases are recommended. All patients undergo surgery including biopsy, ESS and Caldwell-Luc's operation. Amphotericin B or itraconazole may be given post-operatively depending on the fungus isolated. There are very high chance of recurrence of this disease during the follow-up period.

The past decade has witnessed a significant increase in the prevalence of resistance to antifungal agents that has significant implications for morbidity, mortality and healthcare costs in hospitals as well as in the community. Antifungal susceptibility testing is performed on causative fungi especially if they belong to a species exhibiting resistance to commonly used antifungals. Although *in vitro* susceptibility testing is often used to select antimicrobial agents that are most likely to be active clinically, perhaps the most important role of such testing is rather in detecting *in vitro* resistance; that is, determining those agents that will not work among generally susceptible population or to detect the development of resistance during the course of therapy to enable favorable therapeutic outcome. Dilution methods are used to establish the MICs of antifungals. Other methods like Etest, API, Vitek and disc diffusion, may also be applied.

Further Reading

1. Adhikary R, Joshi S. *Schizophyllum commune* sinusitis in an immunocompetent host. *Indian J Med Microbiol.* 2012; **30:** 249.

2. Agarwal K, Kathuria S, Sundar G, *et al.* A case of allergic fungal rhinosinusitis due to *Ceratocystis adiposa*. *Diagn Microbiol Infect Dis.* 2014; **78:** 196-8.
3. Agarwal R, Bansal S, Chakrabarti A. Are allergic fungal rhinosinusitis and allergic bronchopulmonary aspergillosis lifelong conditions? *Med Mycol.* 2017; **55:** 87-95.
4. Ahmed MK, Ishino T, Takeno S, *et al.* Bilateral allergic fungal rhinosinusitis caused by *Schizophyllum commune* and *Aspergillus niger*. A case report. *Rhinology.* 2009; **47:** 217-21.
5. Ahmed Y, Delaney S, Markarian A. Successful isavuconazole therapy in a patient with acute invasive fungal rhinosinusitis and acquired immune deficiency syndrome. *Am J Otolaryngol.* 2016; **37:** 152-5.
6. Ahmeti H, Balschun K, Mehdorn HM, *et al.* Unusual coincidence of a pituitary adenoma and an aspergilloma of the sphenoid sinus. *J Neurooncol.* 2014; **119:** 447-8.
7. Ait-Mansour A, Pezzettigotta S, Genty E, *et al.* Evaluation of the prevalence and specificities of asymptomatic paranasal sinus aspergillosis: Retrospective study of 59 cases. *Eur Ann Otorhinolaryngol Head Neck Dis.* 2015; **132:** 19-23.
8. Akhaddar A, Gazzaz M, Albouzidi A, *et al.* Invasive *Aspergillus terreus* sinusitis with orbitocranial extension: Case report. *Surg Neurol.* 2008; **69:** 490-5.
9. Al Anazy FH, Al Dousary SH. Ophthalmic manifestations of paranasal sinus disease: A clinical grading system. *Int Forum Allergy Rhinol.* 2012; **2:** 331-5.
10. Al-Dousary SH. Allergic fungal sinusitis: Radiological and microbiological features of 59 cases. *Ann Saudi Med.* 2008; **28:** 17-21.
11. Al Qahtani A, Alshaikh N, Alzarei A, *et al.* Contralateral sinus involvement of surgically treated unilateral allergic fungal rhinosinusitis. *Eur Arch Otorhinolaryngol.* 2017; doi: 10.1007/s00405-017-4604-x. PMID: 28501959.
12. Al-Radadi AM, Alnoury KI. Optic chiasma involvement secondary to allergic fungal rhinosinusitis. *J Pak Med Assoc.* 2011; **61:** 704-7.
13. Alves MD, Pinheiro L, Manica D, *et al.* *Histoplasma capsulatum* sinusitis: Case report and review. *Mycopathologia.* 2011; **171:** 57-9.
14. Anand V, Santosh S, Aishwarya A. Canine fossa approaches in endoscopic sinus surgery - Our experience. *Indian J Otolaryngol Head Neck Surg.* 2008; **60:** 214-7.
15. Arndt S, Aschendorff A, Echternach M, *et al.* Rhino-orbital-cerebral mucormycosis and aspergillosis: Differential diagnosis and treatment. *Eur Arch Otorhinolaryngol.* 2009; **266:** 71-6.
16. Azar MM, Assi R, Patel N, *et al.* Fungal mycotic aneurysm of the internal carotid artery associated with sphenoid sinusitis in an immunocompromised patient: A case report and review of the literature. *Mycopathologia.* 2016; **181:** 425-33.
17. Bachert C, Zhang L, Gevaert P. Current and future treatment options for adult chronic rhinosinusitis: Focus on nasal polyposis. *J Allergy Clin Immunol.* 2015; **136:** 1431-40.
18. Badali H, Hedayati MT, Bahoosh M, *et al.* *Exophiala oligosperma* involved in a refractory chronic rhinosinusitis. *Eur Rev Med Pharmacol Sci.* 2011; **15:** 319-23.
19. Badiee P, Gandomi B, Sabz G, *et al.* Evaluation of nested PCR in diagnosis of fungal rhinosinusitis. *Iran J Microbiol.* 2015; **7:** 62-6.
20. Badiee P, Moghadami M, Rozbehani H. Comparing immunological and molecular tests with conventional methods in diagnosis of acute invasive fungal rhinosinusitis. *J Infect Dev Ctries.* 2016; **10:** 90-5.
21. Bakhshaee M, Bojdi A, Allahyari A, *et al.* Acute invasive fungal rhinosinusitis: Our experience with 18 cases. *Eur Arch Otorhinolaryngol.* 2016; **273:** 4281-7.
22. Bakhshaee M, Fereidouni M, Mohajer MN, *et al.* The prevalence of allergic fungal rhinosinusitis in sinonasal polyposis. *Eur Arch Otorhinolaryngol.* 2013; **270:** 3095-8.
23. Bakhshizadeh M, Hashemian HR, Najafzadeh MJ, *et al.* First report of rhinosinusitis caused by *Neoscytalidium dimidiatum* in Iran. *J Med Microbiol.* 2014; **63 (Pt. 7):** 1017-9.
24. Bansal S, Grover G, Grover M, *et al.* Isolated sphenoid mucormycosis presenting as visual impairment: Changing trends? *Am J Otolaryngol.* 2010; **31:** 64-6.
25. Banuprakash S, Jolappara M, Kesavadas C, *et al.* Atypical fungal granuloma of the sphenoid wing. *J Neuroradiol.* 2009; **36:** 233-6.
26. Barac A, Pekmezovic M, Spiric VT, *et al.* Chronic rhinosinusitis: Association of recalcitrant nasal polyposis and fungal finding in polyp's single-cell suspension. *Eur Arch Otorhinolaryngol.* 2015; **272:** 3727-34.
27. Barac A, Stevanovic G, Pekmezovic M, *et al.* Study toward resolving the controversy over the definition of allergic fungal rhinosinusitis. *Med Mycol.* 2017; doi: 10.1093/mmy/myx032. PMID: 28482010.
28. Baradkar VP, Mathur M, Kumar S. Invasive fungal sinusitis resulting in orbital apex syndrome in HIV patient. *Indian J Pathol Microbiol.* 2010; **53:** 862-3.
29. Bassiouny A, Ragab A, Attia AF, *et al.* Prevalence of extramucosal fungal elements in sinonasal polyposis: A mycological an pathologic study in an Egyptian population. *Am J Otolaryngol.* 2011; **32:** 308-17.
30. Bathoorn E, Escobar Salazar N, Sepehrkhouy S, *et al.* Involvement of the opportunistic pathogen *Aspergillus tubingensis* in osteomyelitis of the maxillary bone: A case report. *BMC Infect Dis.* 2013; **13:** 59.
31. Beltrame A, Sarmati L, Cudillo L, *et al.* A fatal case of invasive fungal sinusitis by *Scopulariopsis acremonium* in a bone marrow transplant recipient. *Int J Infect Dis.* 2009; **13:** e488-92.
32. Bent JP, Kuhn FA. Diagnosis of allergic fungal sinusitis. *Otolaryngol Head Neck Surg.* 1994; **111:** 580-8.
33. Bernardini E, Karligkiotis A, Fortunato S, *et al.* Surgical and pathogenetic considerations of frontal sinus fungus ball. *Eur Arch Otorhinolaryngol.* 2017; **274:** 2493-7.
34. Bhandarkar AM, Kudva R, Damry K, *et al.* Fungus ball in the nasal cavity mimicking a rhinolith. *BMJ Case Rep.* 2016; pii: bcr2016215490. PMID: 27353176.
35. Biswas SS, Al-Amin Z, Razib FA, *et al.* Acute invasive fungal rhinosinusitis: Our experience in immunocompromised host. *Mymensingh Med J.* 2013; **22:** 814-9.

36. Bosi GR, de Braga GL, de Almeida TS, *et al.* Fungus ball of the paranasal sinuses: Report of two cases and literature review. *Int Arch Otorhinolaryngol.* 2012; **16:** 286-90.
37. Boutarbouch M, Arkha Y, El Ouahabi A, *et al.* Sphenoid sinus aspergillosis simulating pituitary tumor in immunocompetent patient. *J Clin Neurosci.* 2009; **16:** 840-1.
38. Bozeman S, deShazo RD, Stringer S, *et al.* Complications of allergic fungal sinusitis. *Am J Med.* 2011; **124:** 359-68.
39. Brecheteau C, Laccourreye L, Breheret R. A chronic supra-orbital swelling. *Eur Ann Otorhinolaryngol Head Neck Dis.* 2013; **130:** 161-3.
40. Broglie MA, Tinguely M, Holzman D. How to diagnose sinus fungus balls in the paranasal sinus? An analysis of an institution's cases from January 1999 to December 2006. *Rhinology.* 2009; **47:** 379-84.
41. Burnham R, Bridle C. Aspergillosis of the maxillary sinus secondary to a foreign body (amalgam) in the maxillary antrum. *Br J Oral Maxillofac Surg.* 2009; **47:** 313-5.
42. Callejas CA, Douglas RG. Fungal rhinosinusitis: What every allergist should know. *Clin Exp Allergy.* 2013; **43:** 835-49.
43. Cardoso IC, de Mattos Oliveira F, Hochhegger B, *et al.* Sphenoid sinus fungus ball by filaments of actinomycetes and *Aspergillus fumigatus. Mycopathologia.* 2015; **179:** 471-3.
44. Carr TF. Complications of sinusitis. *Am J Rhinol Allergy.* 2016; **30:** 241-5.
45. Cavanna C, Seminari E, Pusateri A, *et al.* Allergic fungal rhinosinusitis due to *Curvularia lunata. New Microbiol.* 2014; **37:** 241-5.
46. Chaaban MR, Bell W, Woodworth BA. Invasive mucormycosis in an immunocompetent patient with allergic fungal rhinosinusitis. *Otolaryngol Head Neck Surg.* 2013; **148:** 174-5.
47. Chaganti J, Marriott D, Steel T, *et al.* Perineural trigeminal nerve abscess secondary to *Mucor* sinusitis: Serial diffusion-weighted MRI and literature review. *Clin Radiol.* 2011; **66:** 1106-9.
48. Chakrabarti A, Das A, Panda NK. Controversies surrounding the categorization of fungal sinusitis. *Med Mycol.* 2009; **47:** S299-308.
49. Chakrabarti A, Das A, Radotra B. Observations on 'Allergic fungal sinusitis: Innocence under suspicion'. *Med Mycol.* 2013; **51:** 223-4.
50. Chakrabarti A, Denning DW, Ferguson BJ, *et al.* Fungal rhinosinusitis: A categorization and definitional schema addressing current controversies. *Laryngoscope.* 2009; **119:** 1809-18.
51. Chakrabarti A, Jatana M, Sharma SC. Rabbit as an animal model of paranasal sinus mycoses. *J Med Vet Mycol.* 1997; **35:** 295-7.
52. Chakrabarti A, Kaur H. Allergic *Aspergillus* rhinosinusitis. *J Fungi.* 2016; **2:** 32.
53. Chakrabarti A, Rudramurthy SM, Panda N, *et al.* Epidemiology of chronic fungal rhinosinusitis in rural India. *Mycoses.* 2015; **58:** 294-302.
54. Chakrabarti A, Sharma SC, Chander J. Epidemiology and pathogenesis of paranasal sinus mycoses. *Otolaryngol Head Neck Surg.* 1992; **107:** 745-50.
55. Challa S, Uppin SG, Hanumanthu S, *et al.* Fungal rhinosinusitis: A clinicopathological study from South India. *Eur Arch Otorhinolaryngol.* 2010; **267:** 1239-45.
56. Champagne JP, Antisdel JL, Woodard TD, *et al.* Epidemiologic factors affect surgical outcomes in allergic fungal sinusitis. *Laryngoscope.* 2010; **120:** 2322-4.
57. Chan KO, Genoway KA, Javer AR. Effectiveness of itraconazole in the management of refractory allergic fungal rhinosinusitis. *J Otolaryngol Head Neck Surg.* 2008; **37:** 870-4.
58. Chan Y, Kuhn FA. An update on the classifications diagnosis, and treatment of rhinosinusitis. *Curr Opin Otolaryngol Head Neck Surg.* 2009; **17:** 204-8.
59. Chang C, Gershwin ME, Thompson GR 3rd. Fungal disease of the nose and sinuses: An updated overview. *Curr Allergy Asthma Rep.* 2013; **13:** 152-61.
60. Chang YT, Fang SY. Tissue-specific immunoglobulin E in maxillary sinus mucosa of allergic fungal sinusitis. *Rhinology.* 2008; **46:** 226-30.
61. Charakorn N, Snidvongs K. Chronic sphenoid rhinosinusitis: Management challenge. *J Asthma Allergy.* 2016; **9:** 199-205.
62. Chatterjee SS, Chakrabarti A. Epidemiology and medical mycology of fungal rhinosinusitis. *Otorhinolaryngol Clin Int J.* 2009; **1:** 1-13.
63. Chavan SS, Bhople KS, Deshmukh SD, *et al.* Chronic invasive fungal granuloma - A diagnostic dilemma in an immunocompetent host. *Iran J Otorhinolaryngol.* 2016; **28:** 83-8.
64. Chen CY, Sheng WH, Cheng A, *et al.* Invasive fungal sinusitis in patients with hematological malignancy: 15 years experience in a single university hospital in Taiwan. *BMC Infect Dis.* 2011; **11:** 250.
65. Chen JC, Ho CY. The significance of computed tomographic findings in the diagnosis of fungus ball in the paranasal sinuses. *Am J Rhinol Allergy.* 2012; **26:** 117-9.
66. Cheung EJ, Scurry WC, Isaacson JE, *et al.* Cavernous sinus thrombosis secondary to allergic fungal sinusitis. *Rhinology.* 2009; **47:** 105-8.
67. Cho HJ, Hong SD, Kim HY, *et al.* Clinical implications of serum galactomannan measurement in patients with acute invasive fungal rhinosinusitis. *Rhinology.* 2016; **54:** 336-41.
68. Cho HJ, Jang MS, Hong SD, *et al.* Prognostic factors for survival in patients with acute invasive fungal rhinosinusitis. *Am J Rhinol Allergy.* 2015; **29:** 48-53.
69. Cho HS, Yang HS, Kim KS. Mucormycosis (*Mucor* fungus ball) of the maxillary sinus. *Ear Nose Throat J.* 2014; **93:** e18-22.
70. Cho SH, Jin BJ, Lee YS, *et al.* Orbital apex syndrome in a patient with sphenoid fungal balls. *Clin Exp Otorhinolaryngol.* 2009; **2:** 52-4.
71. Chopra H, Dua K, Bhatia S, *et al.* Invasive rhino-orbital fungal sinusitis following dental manipulation. *Mycoses.* 2009; **52:** 368-71.
72. Chua JL, Cullen JF. Fungal pan-sinusitis with severe visual loss in uncontrolled diabetes. *Ann Acad Med Singapore.* 2008; **37:** 964-7.
73. Ciger E, Demiray U, Onal K, *et al.* An unusual location for a fungus ball: The concha bullosa. *J Laryngol Otol.* 2012; **126:** 844-6.

74. Clark DW, Wenaas A, Luong A, *et al.* *Staphylococcus aureus* prevalence in allergic fungal rhinosinusitis vs other subsets of chronic rhinosinusitis with nasal polyps. *Int Forum Allergy Rhinol.* 2013; **3:** 89-93.
75. Cleland EJ, Bassiouni A, Boase S, *et al.* The fungal microbiome in chronic rhinosinusitis: Richness, diversity, postoperative changes and patient outcomes. *Int Forum Allergy Rhinol.* 2014; **4:** 259-65.
76. Comacle P, Belaz S, Jegoux F, *et al.* Contribution of molecular tools for the diagnosis and epidemiology of fungal chronic rhinosinusitis. *Med Mycol.* 2016; **54:** 794-800.
77. Costa A, Branca V, Pigatto PD, *et al.* Aspergillosis, maxillary foreign body and mercury amalgam. *Br J Oral Maxillofac Surg.* 2010; **48:** 64-5.
78. Costa F, Polini F, Zerman N, *et al.* Functional endoscopic sinus surgery for the treatment of *Aspergillus* mycetomas of the maxillary sinus. *Minerva Stomatol.* 2008; **57:** 117-25.
79. Cyriac JT, Cherian T, Hadi WA, *et al.* Optic neuropathy due to allergic fungal rhinosinusitis. *J Neurosci Rural Pract.* 2011; **2:** 180-2.
80. D'Anza B, Stokken J, Greene JS, *et al.* Chronic invasive fungal sinusitis: Characterization and shift in management of a rare disease. *Int Forum Allergy Rhinol.* 2016; **6:** 1294-300.
81. Das A, Bal A, Chakrabarti A, *et al.* Spectrum of fungal rhinosinusitis; histopathologist's perspective. *Histopathology.* 2009; **54:** 854-9.
82. Dayal D, Bakshi J, Jain P, *et al.* Outcome of rhino-sinus mucormycosis in children with Type 1 diabetes. *Indian J Pediatr.* 2015; **82:** 651-2.
83. De Young K, Wentzel JL, Schlosser RJ, *et al.* Systematic review of immunotherapy for chronic rhinosinusitis. *Am J Rhinol Allergy.* 2014; **28:** 145-50.
84. DeShazo RD, Chapin K, Swain RE. Fungal sinusitis. *N Eng J Med.* 1997; **337:** 254-9.
85. deShazo RD, Stringer S, Skipworth LB. Possible allergic fungal sinusitis. *Ann Allergy Asthma Immunol.* 2016; **116:** 290-4.
86. deShazo RD. Syndromes of invasive fungal sinusitis. *Med Mycol.* 2009; **47:** S309-14.
87. Desmots F, Gabaudan C, Geffroy Y, *et al.* Allergic fungal sinusitis. Answer to March *e-quid. Diagn Interv Imaging.* 2012; **93:** 413-9.
88. Devars du Mayne M, Gratacap M, *et al.* An uncommon cause of allergic fungal sinusitis: *Rhizopus oryzae. Ear Nose Throat J.* 2015; **94:** e17-20.
89. Dhirawani R, Asrani S, Pathak S, *et al.* Facial translocation approach for management of invasive sinonasal aspergillosis. *J Maxillofac Oral Surg.* 2015; **14 (Suppl. 1):** 482-7.
90. Doellman MS, Dion GR, Weitzel EK, *et al.* Immunotherapy in allergic fungal sinusitis: The controversy continues. A recent review of literature. *Allergy Rhinol (Providence).* 2013; **4:** e32-5.
91. Doni BR, Peerapur BV, Thotappa LH, *et al.* Sequence of oral manifestations in rhino-maxillary mucormycosis. *Indian J Dent Res.* 2011; **22:** 331-5.
92. Donovan ST, Thompson JW, Sandlund JT, *et al.* Imaging of acute invasive fungal rhinosinusitis in a patient with Gorlin syndrome and acute lymphocytic leukemia. *Case Rep Otolaryngol.* 2013; 272314. PMID: 23853729.
93. Duggal P, Wise SK. Invasive fungal rhinosinusitis. *Am J Rhinol Allergy.* 2013; **27:** S28-30.
94. Dunmade AD, Afolabi OA, Alabi BS, *et al.* Intra-antral application of an anti-fungal agent for recurrent maxillary fungal rhinosinusitis: A case report. *J Med Case Rep.* 2012; **6:** 245.
95. Dupret-Bories A, Vergez S, de Bonnecaze G, *et al.* Surgical treatment options for maxillary sinus fungus balls. *B-ENT.* 2013; **9:** 37-43.
96. Durbec M, Bienvenu AL, Picot S, *et al.* Maxillary sinus fungal infection by *Acremonium. Eur Ann Otorhinolaryngol Head Neck Dis.* 2011; **128:** 41-3.
97. Dutre T, Al Dousary S, Zhang N, *et al.* Allergic fungal rhinosinusitis-more than a fungal disease? *J Allergy Clin Immunol.* 2013; **132:** 487-9.
98. Ebbens FA, Georgalas C, Fokkens WJ. Fungus as the cause of chronic rhinosinusitis: the case remains unproven. *Curr Opin Otolaryngol Head Neck Surg.* 2009; **17:** 43-9.
99. Ebbens FA, Georgalas C, Fokkens WJ. The mold conundrum in chronic hyperplastic sinusitis. *Curr Allergy Asthma Rep.* 2009; **9:** 114-20.
100. Effat KG, Madany NM. Microbiological study of role of fungi in primary atrophic rhinitis. *J Laryngol Otol.* 2009; **123:** 631-4.
101. El-Morsy SM, Khafagy YW, El-Naggar MM, *et al.* Allergic fungal rhinosinusitis: Detection of fungal DNA in sinus aspirate using polymerase chain reaction. *J Laryngol Otol.* 2010; **124:** 152-60.
102. Eloy P, Grenier J, Pirlet A, *et al.* Sphenoid sinus fungal ball: A retrospective study over a 10-year period. *Rhinology.* 2013; **51:** 181-8.
103. Ergun O, Tahir E, Kuscu O, *et al.* Acute invasive fungal rhinosinusitis: Presentation of 19 cases, review of the literature and a new classification system. *J Oral Maxillofac Surg.* 2017; **75:** 767.e1-9.
104. Ferguson BJ. Eosinophilic mucin rhinosinusitis: A distinct clinicopathological entity. *Laryngoscope.* 2000; **110:** 799-813.
105. Ferguson BJ. Mucormycosis of the nose and paranasal sinuses. *Otolaryngol Clin N Am.* 2000; **33:** 349-65.
106. Ferguson BJ. Definitions of fungal rhinosinusitis. *Otolaryngol Clin N Am.* 2000; **33:** 227-35.
107. Ferraro RA, Ivanidze J, Margolskee E, *et al.* Invasive granulomatous cryptococcal sinusitis in an adult with multiple myeloma. *Clin Imaging.* 2017; **41:** 65-8.
108. Finkelstein A, Contreras D, Pardo J, *et al.* Paranasal sinuses computed tomography in the initial evaluation of patients with suspected invasive fungal rhinosinusitis. *Eur Arch Otorhinolaryngol.* 2011; **268:** 1157-62.
109. Fokkens WJ, Ebbens F, van Drunen CM. Fungus: A role in pathophysiology of chronic rhinosinusitis, disease modifier, a treatment target or no role at all? *Immunol Allergy Clin North Am.* 2009; **29:** 677-88.

110. Fokkens WJ, van Drunen C, Georgalas C, *et al.* Role of fungi in pathogenesis of chronic rhinosinusitis: The hypothesis rejected. *Curr Opin Otolaryngol Head Neck Surg.* 2012; **20:** 19-23.
111. Foreman A, Boase S, Psaltis A, *et al.* Role of bacterial and fungal biofilms in chronic rhinosinusitis. *Curr Allergy Asthma Rep.* 2012; **12:** 127-35.
112. Foshee J, Luminais C, Casey J, *et al.* An evaluation of invasive fungal sinusitis outcomes with subsite analysis and use of frozen section analysis. *Int Forum Allergy Rhinol.* 2016; **6:** 807-11.
113. Fu KA, Nguyen PL, Sanossian N. Basilar artery territory stroke secondary to invasive fungal sphenoid sinusitis: A case report and review of the literature. *Case Rep Neurol.* 2015; **7:** 51-8.
114. Gamba P, Lombardi C. Eosinophil fungal rhinosinusitis caused by *Fusarium* infection secondary to odontogenic maxillary sinus disease: When collaboration between otolaryngologist and allergologist leads to the correct diagnosis and therapy. *Eur Ann Allergy Clin Immunol.* 2017; **49:** 138-42.
115. Gan EC, Thamboo A, Rudmik L, *et al.* Medical management of allergic fungal rhinosinusitis following endoscopic sinus surgery: An evidence-based review and recommendations. *Int Forum Allergy Rhinol.* 2014; **4:** 702-15.
116. Ganju SA, Bhagra S, Kanga AK, *et al.* A case report of an uncommon phaeoid fungal infection in nasal polyposis and review of literature. *Indian J Med Microbiol.* 2013; **31:** 196-8.
117. Garlapati K, Chavva S, Vaddeswarupu RM, *et al.* Fulminant mucormycosis involving paranasal sinuses: A rare case report. *Case Rep Dent.* 2014; 465919. PMID: 24523970.
118. Gavito-Higuera J, Mullins CB, Ramos-Duran L, *et al.* Sinonasal fungal infections and complications: A pictorial review. *J Clin Imaging Sci.* 2016; **6:** 23.
119. Gelber JT, Cope EK, Goldberg AN, *et al.* Evaluation of *Malassezia* and common fungal pathogens in subtypes of chronic rhinosinusitis. *Int Forum Allergy Rhinol.* 2016; **6:** 950-5.
120. Gerlinger I, Fittler A, Fonai F, *et al.* Post-operative application of amphotericin B nasal spray in chronic rhinosinusitis with nasal polyposis, with a review of the antifungal therapy. *Eur Arch Otorhinolaryngol.* 2009; **266:** 847-55.
121. Giri S, Kindo AJ, Rao S, *et al.* Unusual causes of fungal rhinosinusitis: A study from a tertiary care centre in South India. *Indian J Med Microbiol.* 2013; **31:** 379-84.
122. Glass D. Amedee RG. Allergic fungal rhinosinusitis: A review. *Ochsner J.* 2011; **11:** 271-5.
123. Gode S, Turhal G, Ozturk K, *et al.* Acute invasive fungal rhinosinusitis: Survival analysis and the prognostic indicators. *Am J Rhinol Allergy.* 2015; **29:** 164-9.
124. Gorovoy IR, Kazanjian M, Kersten RC, *et al.* Fungal rhinosinusitis and imaging modalities. *Saudi J Ophthalmol.* 2012; **26:** 419-26.
125. Green KK, Barham HP, Allen GC, *et al.* Prognostic factors in the outcome of invasive fungal sinusitis in a pediatric population. *Pediatr Infect Dis J.* 2016; **35:** 384-6.
126. Greenhaw B, deShazo RD, Arnold J, *et al.* Fungal immunotherapy in patients with allergic fungal sinusitis. *Ann Allergy Asthma Immunol.* 2011; **107:** 432-6.
127. Gu HJ, Kim YJ, Lee HJ, *et al.* Invasive fungal sinusitis by *Lasiodiplodia theobromae* in a patient with aplastic anemia: An extremely rare case report and literature review. *Mycopathologia.* 2016; **181:** 901-8.
128. Gungor AA. On chronic rhinosinusitis and the prevalence of fungal sinus disease: Problems of diagnostic accuracy and a proposed classification of chronic rhinosinusitis. *Am J Otolaryngol.* 2012; **33:** 543-8.
129. Guo C, Ghadersohi S, Kephart GM, *et al.* Improving the detection of fungi in eosinophilic mucin: Seeing what we could not see before. *Otolaryngol Head Neck Surg.* 2012; **147:** 943-9.
130. Gupta AK, Bansal S, Rijuneeta, *et al.* Invasive fungal sinusitis. *Clin Rhinol An Int J.* 2012; **5:** 63-71.
131. Gupta AK, Gupta AK. Postgraduate institute management protocol for invasive *Aspergillus flavus* sinusitis: Is it effective? *Int J Infect Dis.* 2009; **13:** 134-9.
132. Gupta A, Xess I, Sharma SC, *et al.* Invasive rhinosinusitis by *Exserohilum rostratum* in an immunocompetent child. *BMJ Case Rep.* 2014; pii: bcr2013202380. PMID: 24711469.
133. Gupta K, Saggar K. Analysis of computed tomography features of fungal sinusitis and their correlation with nasal endoscopy and histopathology findings. *Ann Afr Med.* 2014; **13:** 119-23.
134. Gupta P, Verma N, Singh AK, *et al.* Molecular detection of *Schizophyllum commune* in a case of allergic fungal rhinosinusitis. *BMJ Case Rep.* 2015; pii: bcr2015209955.
135. Gupta R, Gupta AK. Isolated primary frontal sinus aspergillosis: Role of endonasal endoscopic approach. *J Laryngol Otol.* 2013; **127:** 274-8.
136. Halderman A, Shrestha R, Sindwani R. Chronic granulomatous invasive fungal sinusitis: An evolving approach to management. *Int Forum Allergy Rhinol.* 2014; **4:** 280-3.
137. Hall AG, deShazo RD. Immunotherapy for allergic fungal sinusitis. *Curr Opin Allergy Clin Immunol.* 2012; **12:** 629-34.
138. Hamilos DL. Allergic fungal rhinitis and rhinosinusitis. *Proc Am Thorac Soc.* 2010; **7:** 245-52.
139. Handra-Luca A, Carette MF. Nasal fungal infection in chronic myelodysplastic syndrome-associated acute leukemia occurring in hereditary hemorrhagic telangiectasia. *J Craniofac Surg.* 2014; **25:** 718-20.
140. Hariri A, Choudhury N, Saleh HA. *Scytalidium dimidiatum* associated invasive fungal sinusitis in an immunocompetent patient. *J Laryngol Otol.* 2014; **128:** 1018-21.
141. Hashemia F, Hashemian F, Bakhshaei M. The prevalence of positive fungal cultures in patients with chronic rhinosinusitis in a high altitude region of Iran. *Iran J Otorhinolaryngol.* 2012; **24:** 29-33.
142. Henriques JC, Kreich EM, Rosa RR, *et al.* Non-invasive aspergillosis as a maxillary antrolith: Report of a rare case. *Quintessence Int.* 2012; **43:** 143-6.
143. Hoenigl M, Aspeck E, Valentin T, *et al.* Sinusitis and frontal brain abscess in a diabetic patient caused by the basidiomycete *Schizophyllum commune*: Case report and review of the literature. *Mycoses.* 2013; **56:** 389-93.

144. Holbrook JF, Eastwood JD, Kilani RK. Intracranial abscess as a complication of allergic fungal sinusitis. *J Neuroimaging.* 2014; **24:** 95-8.
145. Hoyt AE, Borish L, Gurrola J, *et al.* Allergic fungal rhinosinusitis. *J Allergy Clin Immunol Pract.* 2016; **4:** 599-604.
146. Hsin LJ, Yang SW. Concomitant inverted papilloma and fungus ball in unilateral maxillary sinus. *B-ENT.* 2013; **9:** 71-5.
147. Hu L, Wang D, Yu H. Isolated sphenoid fungal sinusitis and vision loss: The case for early intervention. *J Laryngol Otol.* 2009; **123:** e8.
148. Huang YD, Yu HW, Xia SW, *et al.* Expression of maspin in invasive fungal rhinosinusitis. *J Laryngol Otol.* 2017; **131:** 150-4.
149. Hunsaker DH, Leid JG. The relationship of biofilms to chronic rhinosinusitis. *Curr Opin Otolaryngol Head Neck Surg.* 2008; **16:** 237-41.
150. Hutcheson PS, Schubert MS, Slavin RG. Distinctions between allergic fungal rhinosinusitis and chronic rhinosinusitis. *Am J Rhinol Allergy.* 2010; **24:** 405-8.
151. Hwang SH, Kang JM, Cho JH, *et al.* What is the relationship between the localization of maxillary fungal balls and intranasal anatomic variations? *Clin Exp Otorhinolaryngol.* 2012; **5:** 213-7.
152. Idris N, Lim LH. Nasal eschar: A warning sign of potentially fatal invasive fungal sinusitis in immunocompromised children. *J Pediatr Hematol Oncol.* 2012; **34:** e134-6.
153. Ikram M, Abbas A, Suhail A, *et al.* Management of allergic fungal sinusitis with postoperative oral and nasal steroids: A controlled study. *Ear Nose Throat J.* 2009; **88:** e8-11.
154. Ilica AT, Mossa-Basha M, Maluf F, *et al.* Clinical and radiologic features of fungal diseases of the paranasal sinuses. *J Comput Assist Tomogr.* 2012; **36:** 570-6.
155. Illing EA, Dunlap Q, Woodworth BA. Outcomes of pressure-induced cranial neuropathies from allergic fungal rhinosinusitis. *Otolaryngol Head Neck Surg.* 2015; **152:** 541-5.
156. Iqbal J, Rashid S, Darira J, *et al.* Diagnostic accuracy of CT scan in diagnosing paranasal fungal infection. *J Coll Physicians Surg Pak.* 2017; **27:** 271-4.
157. Isaacs S, Fakhri S, Luong A, *et al.* A meta-analysis of topical amphotericin B for the treatment of chronic rhinosinusitis. *Int Forum Allergy Rhinol.* 2011; **1:** 250-4.
158. Isaacson G. Surgical treatment of pediatric rhinosinusitis. *Minerva Pediatr.* 2015; **67:** 357-68.
159. Iyer SP, Movva K, Wiebel M, *et al.* Cryptococcal meningitis presenting as sinusitis in a renal transplant recipient. *Transpl Infect Dis.* 2013; **15:** e187-90.
160. Jain R, Singhal SK, Singla N, Punia RS, Chander J. Mycological profile and antifungal susceptibility of fungal isolates from clinically suspected cases of fungal rhinosinusitis in a tertiary care hospital in north India. *Mycopathologia.* 2015; **180:** 51-9.
161. Jain S, Das S, Gupta N, *et al.* Frequency of fungal isolation and antifungal susceptibility pattern of the fungal isolates from nasal polyps of chronic rhinosinusitis patients at a tertiary care centre in north India. *Med Mycol.* 2013; **51:** 164-9.
162. Janagond A, Krishnan KM, Kindo AJ, *et al. Trichosporon inkin,* an unusual agent of fungal sinusitis: A report from south India. *Indian J Med Microbiol.* 2012; **30:** 229-32.
163. Johnson J, Gupta ND, Palacios E, *et al.* Persistent headaches: Sinus mycetoma. *J La State Med Soc.* 2008; **160:** 255-7.
164. Joshi RR, Bhandary S, Khanal B, *et al.* Fungal maxillary sinusitis: A prospective study in a tertiary care hospital of eastern Nepal. *Kathmandu Univ Med J.* 2007; **5:** 195-8.
165. Jung H, Park SK. Indolent mucormycosis of the paranasal sinus in immunocompetent patients: Are antifungal drugs needed? *J Laryngol Otol.* 2013; **127:** 872-5.
166. Jung JH, Cho GS, Chung YS, *et al.* Clinical characteristics and outcome in patients with isolated sphenoid sinus aspergilloma. *Auris Nasus Larynx.* 2013; **40:** 189-93.
167. Kalayci CB, Cevik H. Isolated sphenoid fungal sinusitis and coexisting intrasellar mass lesion presenting with headache as the sole symptom. *Acta Radiol Short Rep.* 2014; **3:** 1-4.
168. Kale P, Rudramurthy SM, Panda NK, *et al.* The inflammatory response of eosinophil-related fungal rhinosinusitis varies with inciting fungi. *Med Mycol.* 2015; **53:** 387-95.
169. Kalin-Hajdu E, Hirabayashi KE, Vagefi MR, *et al.* Invasive fungal sinusitis: Treatment of the orbit. *Curr Opin Ophthalmol.* 2017; doi: 10.1097/ICU.0000000000000394. PMID: 28505035.
170. Kamath PM, Prasad V, Shenoy VS, *et al. Chrysosporium:* An uncommon fungus in chronic rhinosinusitis. *J Clin Diagn Res.* 2015; **9:** MD01-2.
171. Karkas A, Rtail R, Reyt E, *et al.* Sphenoid sinus fungus ball. *Eur Arch Otorhinolaryngol.* 2013; **270:** 893-8.
172. Karthikeyan P, Nirmal Coumare V. Incidence and presentation of fungal sinusitis in patient diagnosed with chronic rhinosinusitis. *Indian J Otolaryngol Head Neck Surg.* 2010; **62:** 381-5.
173. Kasapoglu F, Coskun H, Ozmen OA, *et al.* Acute invasive fungal rhinosinusitis: Evaluation of 26 patients treated with endonasal or open surgical procedures. *Otolaryngol Head Neck Surg.* 2010; **143:** 614-20.
174. Katotomichelakis M, Anastassakis K, Gouveris H, *et al.* Clinical significance of *Alternaria alternata* sensitization in patients with allergic rhinitis. *Am J Otolaryngol.* 2012; **33:** 232-8.
175. Kaur R, Lavanya S, Khurana N, *et al.* Allergic fungal rhinosinusitis: A study in a tertiary care hospital in India. *J Allergy (Cairo).* 2016; 7698173. PMID: 26904136.
176. Kaur R, Wadhwa A, Gulati A, *et al.* An unusual phaeoid fungi: *Ulocladium,* as a cause of chronic allergic fungal sinusitis. *Iran J Microbiol.* 2010; **2:** 95-7.
177. Keane C, Colreavy M, Lynch M, *et al. Aspergillus flavus* sinusitis in ALL. *Am J Hematol.* 2009; **84:** 123.
178. Killeen DE, Sedaghat AR, Cunnane ME, *et al.* Objective radiographic density measurements of sinus opacities are not strong predictors of non-invasive fungal disease. *Am J Rhinol Allergy.* 2014; **28:** 483-6.
179. Kim C, Kacker A, Chee RI, *et al.* Allergic fungal sinusitis causing nasolacrimal duct obstruction. *Orbit.* 2013; **32:** 143-5.

180. Kim DW, Heo ST, Jeon SY, *et al.* Invasive paranasal mucormycosis with peripheral eosinophilia in an immunocompetent patient. *Med Mycol.* 2010; **48:** 406-9.
181. Kim IA, Thompson CF, Kedeshian PA, *et al.* Invasive fungal sinusitis in a healthy athlete due to long-term anabolic steroid use. *Laryngoscope.* 2014; **124:** 1756-9.
182. Kim JH, Kang BC, Lee JH, *et al.* The prognostic value of gadolinium-enhanced magnetic resonance imaging in acute invasive fungal rhinosinusitis. *J Infect.* 2015; **70:** 88-95.
183. Kim JS, So SS, Kwon SH. The increasing incidence of paranasal sinus fungus ball: A retrospective cohort study in two hundred forty-five patients for fifteen years. *Clin Otolaryngol.* 2017; **42:** 175-9.
184. Kim KS, Lee JH. Spontaneous resolution of acute headache by migration of fungus ball. *Headache.* 2011; **51:** 992-4.
185. Kim TH, Jang HU, Jung YY, *et al.* Granulomatous invasive fungal rhinosinusitis extending into the pterygopalatine fossa and orbital floor: A case report. *Med Mycol Case Rep.* 2012; **1:** 107-11.
186. Kindo AJ, Pramod C, Anita S, *et al.* Maxillary sinusitis caused by *Lasiodiplodia theobromae. Indian J Med Microbiol.* 2010; **28:** 167-9.
187. Kini JR, Shetty SS, Kini H. Spectrum of paranasal sinus mycoses in coastal India. *Ear Nose Throat J.* 2012; **91:** e15-8.
188. Knollman P, Naclerio R. Progressive unilateral nasal congestion, facial pain and vision change in an immunocompetent patient. *JAMA Otolaryngol Head Neck Surg.* 2015; **141:** 579-80.
189. Krishnan KU, Agatha D, Selvi R. Fungal rhinosinusitis: A clinicomycological perspective. *Indian J Med Microbiol.* 2015; **33:** 120-4.
190. Kumar Behera S, Patro M, Mishra D, *et al.* Fine needle aspiration in aspergilloma of frontal sinus: A case report. *Acta Cytol.* 2008; **52:** 500-4.
191. Kumar MS, Panella NJ, Magliocca KR, *et al.* Otopolyposis with middle ear allergic mucin in a patient with allergic fungal rhinosinusitis. *Ann Otol Rhinol Laryngol.* 2016; **125:** 862-5.
192. Kurimoto T, Tonari M, Ishizaki N, *et al.* A case of eosinophilic chronic rhinosinusitis associated with optic neuropathy. *Clin Ophthalmol.* 2011; **5:** 853-6.
193. Kwon MO, Kim KS. Headache induced by isolated sphenoid fungal sinusitis: Sinus headache? *J Headache Pain.* 2009; **10:** 473-6.
194. Lai JC, Lee HS, Chen MK, Tsai YL. Patient satisfaction and treatment outcome of fungus ball rhinosinusitis treated by functional endoscopic sinus surgery. *Eur Arch Otorhinolaryngol.* 2011; **268:** 227-30.
195. Lammers MJ, Lo Galbo AM, Buwalda J. Allergic fungal rhinosinusitis (AFRS). *Arch Otolaryngol Head Neck Surg.* 2012; **138:** 426-8.
196. Larson DA, Han JK. Microbiology of sinusitis: Does allergy or endoscopic sinus surgery affect the microbiologic flora? *Curr Opin Otolaryngol Head Neck Surg.* 2011; **19:** 199-203.
197. Laury AM, Wise SK. Allergic fungal rhinosinusitis. *Am J Rhinol Allergy.* 2013; **27:** S26-7.
198. Leboime A, Berthelot JM, Allanore Y, *et al.* Sinus aspergilloma in rheumatoid arthritis before or during tumor necrosis factor-alpha antagonist therapy. *Arthritis Res Ther.* 2009; **11:** R164.
199. Ledderose GJ, Braun T, Betz CS, *et al.* Functional endoscopic surgery of paranasal fungus ball: Clinical outcome, patient benefit and health-related quality of life. *Eur Arch Otorhinolaryngol.* 2012; **269:** 2203-8.
200. Lee AS, Przybyszewski B, Montone K, *et al.* Oral antifungal therapy for *Fusarium*-associated chronic rhinosinusitis. *Int Forum Allergy Rhinol.* 2012; **2:** 45-50.
201. Lee DH, Kim SK, Joo YE, *et al.* Fungus ball within a mucocele of the sphenoid sinus and infratemporal fossa: Case report with radiological findings. *J Laryngol Otol.* 2012; **126:** 210-3.
202. Lee DH, Lim SC. Maxillary fungus ball caused by retained foreign bodies for 25 years. *J Craniofac Surg.* 2012; **23:** e176-7.
203. Lee JH, Jeong HM. A rare occurrence of a fungus ball in the sphenoethmoid recess. *Ear Nose Throat J.* 2014; **93:** e47-8.
204. Lee JH, Jeong HM. Fungal sinusitis with a nasal septal perforation. *Ear Nose Throat J.* 2013; **92:** e24-5.
205. Lee SH, Kim HJ, Lee JW, *et al.* Categorization and clinicopathological features of chronic rhinosinusitis with eosinophilic mucin in a Korean population. *Clin Exp Otorhinolaryngol.* 2015; **8:** 39-45.
206. Lee TJ, Huang SF, Chang PH. Characteristics of isolated sphenoid sinus aspergilloma: Report of twelve cases and literature review. *Ann Otol Rhinol Laryngol.* 2009; **118:** 211-7.
207. Leyngold I, Olivi A, Ishii M, *et al.* Acute chiasmal abscess resulting from perineural extension of invasive sino-orbital aspergillosis in an immunocompetent patient. *World Neurosurg.* 2014; **81:** 203. e1-6.
208. Li Y, Li Y, Li P, *et al.* Diagnosis and endoscopic surgery of chronic invasive fungal rhinosinusitis. *Am J Rhinol Allergy.* 2009; **23:** 622-5.
209. Lin JJ, Wu CT, Hsia SH, *et al.* Pneumocephalus: A rare presentation of *Candida* sphenoid sinusitis. *Pediatr Neurol.* 2009; **40:** 398-400.
210. Loftus PA, Wise SK. Allergic fungal rhinosinusitis: The latest in diagnosis and management. *Adv Otorhinolaryngol.* 2016; **79:** 13-20.
211. London NR Jr, Reh DD. Differential diagnosis of chronic rhinosinusitis with nasal polyps. *Adv Otorhinolaryngol.* 2016; **79:** 1-12.
212. Lop-Gros J, Gras-Cabrerizo JR, Bothe-Gonzalez C, *et al.* Fungus ball of the paranasal sinuses: Analysis of our series of patients. *Acta Otorrinolaringol Esp.* 2016; **67:** 220-5.
213. Lorentz C, Rivier A, Debourgogne A, et al. Ethmoidomaxillary sinusitis caused by the basidiomycetous fungus *Schizophyllum commune. Mycoses.* 2012; **55:** e8-12.
214. Lu X, Huang Y, Zhang M, *et al.* Matrix-assisted laser desorption ionization time-of-flight mass spectrometry for rapid identification of fungal rhinosinusitis pathogens. *J Med Microbiol.* 2017; **66:** 328-33.

215. Lu-Myers Y, Deal AM, Miller JD, *et al.* Comparison of socioeconomic and demographic factors in patients with chronic rhinosinusitis and allergic fungal rhinosinusitis. *Otolaryngol Head Neck Surg.* 2015; **153:** 137-43.
216. Lusardi JJ, Naclerio RM. Radiologic findings in allergic fungal rhinosinusitis. *J Allergy Clin Immunol Pract.* 2016; **4:** 749-50.
217. Macedo DP, Neves RP, Fontan J, *et al.* A case of invasive rhinosinusitis by *Fusarium verticillioides* (Saccardo) Nirenberg in an apparently immunocompetent patient. *Med Mycol.* 2008; **46:** 499-503.
218. Mahdavinia M, Bishehsari F, Hayat W, *et al.* Prevalence of allergic rhinitis and asthma in patients with chronic rhinosinusitis and gastroesophageal reflux disease. *Ann Allergy Asthma Immunol.* 2016; **117:** 158-162.e1.
219. Mahomed S, Basanth S, Mlisana K. The successful use of amphotericin B followed by oral posaconazole in a rare case of invasive fungal sinusitis caused by co-infection with mucormycosis and aspergillus. *IDCases.* 2015; **2:** 116-7.
220. Mandegari E, Fu L, Arambu C, *et al.* Mucormycosis rhinosinusitis at diagnosis of acute lymphoblastic leukemia: Diagnostics and management challenges in a low-middle-income country. *J Pediatr Hematol Oncol.* 2015; **37:** e173-7.
221. Manning SC, Vuitch F, Weinberg AG, *et al.* Allergic aspergillosis: A newly recognized form of sinusitis in the pediatric population. *Laryngoscope.* 1989; **99:** 681-5.
222. Manning SC. Holman M. Further evidence for allergic pathophysiology in allergic fungal sinusitis. *Laryngoscope* 1998; **108:** 1485-96.
223. Marfani MS, Jawaid MA, Shaikh SM, *et al.* Allergic fungal rhinosinusitis with skull base and orbital erosion. *J Laryngol Otol.* 2010; **124:** 161-5.
224. Marglani O. Update in the management of allergic fungal sinusitis. *Saudi Med J.* 2014; **35:** 791-5.
225. Masterson L, Egro FM, Bewick J, *et al.* Quality-of-life outcomes after sinus surgery in allergic fungal rhinosinusitis versus non-fungal chronic rhinosinusitis. *Am J Rhinol Allergy.* 2016; **30:** e30-5.
226. McGill TJ, Simpson G, Healy GB. Fulminant aspergillosis of the nose and paranasal sinuses: A new clinical entity. 1980. *Laryngoscope.* 2015; **125:** 782.
227. Meccariello G, Deganello A, Mannelli G, *et al.* Allergic fungal rhinosinusitis infiltrating anterior skull base and clivus. *Auris Nasus Larynx.* 2013; **40:** 405-8.
228. Melancon CC, Clinger JD. The use of frozen section in the early diagnosis of acute invasive fungal sinusitis. *Otolaryngol Head Neck Surg.* 2017; doi: 10.1177/0194599817697279. PMID: 28349786.
229. Meltzer EO, Hamilos DL. Rhinosinusitis diagnosis and management for the clinician: A synopsis of recent consensus guidelines. *Mayo Clin Proc.* 2011; **86:** 427-43.
230. Melzer JM, Driskill BR, Clenney TL, *et al.* Sublingual immunotherapy for allergic fungal sinusitis. *Ann Otol Rhinol Laryngol.* 2015; **124:** 782-7.
231. Michael RC, Michael JS, Ashbee RH, *et al.* Mycological profile of fungal sinusitis: An audit of specimens over a 7-year period in a tertiary care hospital in Tamil Nadu. *Indian J Pathol Microbiol.* 2008; **51:** 493-6.
232. Michel J, Maubon D, Varoquaux DA, *et al. Schizophyllum commune:* An emergent or misdiagnosed fungal pathogen in rhinology? *Med Mycol.* 2016; **54:** 301-9.
233. Michel J, Radulesco T, Mancini J, *et al.* Maxillary sinus volume: New physiopathological data in fungal ball genesis? A retrospective study. *Clin Otolaryngol.* 2017; doi: 10.1111/coa.12813. PMID: 28004496.
234. Mignogna MD, Fortuna G, Leuci S, *et al.* Mucormycosis in immunocompetent patients: A case-series of patients with maxillary sinus involvement and a critical review of the literature. *Int J Infect Dis.* 2011; **15:** e533-40.
235. Miller JD, Deal AM, McKinney KA, *et al.* Markers of disease severity and socioeconomic factors in allergic fungal rhinosinusitis. *Int Forum Allergy Rhinol.* 2014; **4:** 272-9.
236. Mimouni O, Curto CL, Danvin JB, *et al.* Sinonasal mucormycosis: Case report. *Eur Ann Otorhinolaryngol Head Neck Dis.* 2010; **127:** 27-9.
237. Mistry SG, Kumar BN. The value of antifungal therapy in allergic fungal rhinosinusitis. *Rhinology.* 2014; **52:** 9-18.
238. Mitsimponas KT, Walsh S, Collyer J. Bilateral maxillary sinus fungus ball: Report of a case. *Br J Oral Maxillofac Surg.* 2009; **47:** 242.
239. Mohammadi A, Hashemi SM, Abtahi SH, *et al.* An investigation on non-invasive fungal sinusitis: Molecular identification of etiologic agents. *J Res Med Sci.* 2017; **22:** 67.
240. Mohindra S, Mehta R, Bal A. ABPA Concomitantly occurring with invasive sinus aspergillosis: A short report on two patients. *Indian J Otolaryngol Head Neck Surg.* 2014; **66 (Suppl. 1):** 329-33.
241. Mohindra S, Mohindra S, Bal A. Allergic fungal sinusitis: Innocence under suspicion. *Med Mycol.* 2012; **50:** 179-86.
242. Mohindra S, Mohindra S, Bal A. Reply to Chakrabarti *et al. Med Mycol.* 2013; **51:** 224.
243. Molnar-Gabor E, Doczi I, Hatvani L, *et al.* Isolated sinusitis sphenoidal is caused by *Trichoderma longibrachiatum* in an immunocompetent patient with headache. *J Med Microbiol.* 2013; **62 (Pt 8):** 1249-52.
244. Monroe MM, McLean M, Sautter N, *et al.* Invasive fungal rhinosinusitis: A 15-year experience with 29 patients. *Laryngoscope.* 2013; **123:** 1583-7.
245. Montone KT, Livolsi VA, Feldman MD, *et al.* Fungal rhinosinusitis: A retrospective microbiologic and pathologic review of 400 patients at a single university medical center. *Int J Otolaryngol.* 2012; 684835. PMID: 22518160.
246. Montone KT, LiVolsi VA, Lanza DC, *et al. In situ* hybridization for specific fungal organisms in acute invasive fungal rhinosinusitis. *Am J Clin Pathol.* 2011; **135:** 190-9.
247. Montone KT. Pathology of fungal rhinosinusitis: A review. *Head Neck Pathol.* 2016; **10:** 40-6.
248. Montone KT. Role of fungi in the pathophysiology of chronic rhinosinusitis: An update. *Curr Allergy Asthma Rep.* 2013; **13:** 224-8.

249. Morgand M, Rammaert B, Poiree S, *et al.* Chronic invasive *Aspergillus* sinusitis and otitis with meningeal extension successfully treated with voriconazole. *Antimicrob Agents Chemother.* 2015; **59:** 7857-61.
250. Mossa-Basha M, Ilica AT, Maluf F, *et al.* The many faces of fungal disease of the paranasal sinuses: CT and MRI findings. *Diagn Interv Radiol.* 2013; **19:** 195-200.
251. Mostafa Bel-D, Taha MS, Abdel Hamid T, *et al.* Evaluation of vitamin D levels in allergic fungal sinusitis, chronic rhinosinusitis and chronic rhinosinusitis with polyposis. *Int Forum Allergy Rhinol.* 2016; **6:** 185-90.
252. Murosaki T, Nagashima T, Honne K, *et al.* Invasive sphenoid sinus aspergillosis mimicking giant cell arteritis. *Int J Rheum Dis.* 2014; **17:** 476-8.
253. Murr AH, Goldberg AN, Pletcher SD, *et al.* Some chronic rhinosinusitis patients have elevated populations of fungi in their sinuses. *Laryngoscope.* 2012; **122:** 1438-45.
254. Myles IA, Gada S. Allergen immunotherapy in an HIV+ patient with allergic fungal rhinosinusitis. *Case Reports Immunol.* 2015; 875260. PMID: 25954557.
255. Myoken Y, Sugata T, Katayama Y, *et al.* Prompt and definitive diagnosis of acute invasive *Aspergillus* rhinosinusitis in a patient with acute myeloid leukaemia using in situ hybridization: A case report. *Mycoses.* 2012; **55:** e23-6.
256. Nayak N, Satpathy G, Prasad S, *et al.* Clinical implications of microbial biofilms in chronic rhinosinusitis and orbital cellulitis. *BMC Ophthalmol.* 2016; **16:** 165.
257. Nazeri M, Hashemi SJ, Ardehali M, *et al.* Fungal rhinosinusitis in Tehran, Iran. *Iran J Public Health.* 2015; **44:** 374-9.
258. Negoro E, Morinaga K, Taga M, *et al.* Mycotic aneurysm due to *Aspergillus* sinusitis. *Int J Hematol.* 2013; **98:** 4-5.
259. Ni Mhurchu E, Ospina J, Janjua AS, *et al.* Fungal rhinosinusitis: A radiological review with intraoperative correlation. *Can Assoc Radiol J.* 2017; **68:** 178-86.
260. Nithyanandam S, Correa MA. Rhino-orbital mucormycosis and aspergillosis: Differences in outcome, clinical and imaging characteristics. *Eur Arch Otorhinolaryngol.* 2010; **267:** 161-2.
261. Nomura K, Asaka D, Nakayama T, *et al.* Sinus fungus ball in the Japanese population: Clinical and imaging characteristics of 104 cases. *Int J Otolaryngol.* 2013; 731640.
262. Odessey E, Cohn A, Beaman K, *et al.* Invasive mucormycosis of the maxillary sinus: Extensive destruction with an indolent presentation. *Surg Infect (Larchmt).* 2008; **9:** 91-8.
263. Ogawa Y, Sato M, Tashiro M, *et al.* Rapid development of a mycotic aneurysm of the intracranial artery secondary to *Scedosporium apiospermum* sinusitis. *Med Mycol Case Rep.* 2016; **14:** 30-2.
264. Orhan K, Kocyigit D, Turkoglu K, *et al.* Aspergillosis of maxillary sinus in immunocompromised patient. Case report. *N Y State Dent J.* 2012; **78:** 46-9.
265. Orlandi RR, Kingdom TT, Hwang PH, *et al.* International consensus statement on allergy and rhinology: Rhinosinusitis. *Int Forum Allergy Rhinol.* 2016; **6:** S22-209.
266. Orlandi RR, Marple BF. Fungus and chronic rhinosinusitis: Weighing the evidence. *Otolaryngol Head Neck Surg.* 2010; **143:** 611-3.
267. Orlandi RR, Marple BF. The role of fungus in chronic rhinosinusitis. *Otolaryngol Clin North Am.* 2010; **43:** 531-7.
268. Otto RA. Nasal discharge. Mucormycosis sinusitis. *J Fam Pract.* 2013; **62:** 765.
269. Padhye AA, Ajello L, Wieden MA, *et al.* Phaeohyphomycosis of the nasal sinuses caused by a new species of *Exserohilum. J Clin Microbiol.* 1986; **24:** 245-9.
270. Pagella F, De Bernardi F, Dalla Gasperina D, *et al.* Invasive fungal rhinosinusitis in adult patients: Our experience in diagnosis and management. *J Craniomaxillofac Surg.* 2016; **44:** 512-20.
271. Pagella F, Pusateri A, Matti E, *et al.* Sphenoid sinus fungus ball: Our experience. *Am J Rhinol Allergy.* 2011; **25:** 276-80.
272. Pakdaman MN, Corry DB, Luong A. Fungi linking the pathophysiology of chronic rhinosinusitis with nasal polyps and allergic asthma. *Immunol Invest.* 2011; **40:** 767-85.
273. Pandey A, Bansal V, Asthana AK, *et al.* Maxillary osteomyelitis by mucormycosis: Report of four cases. *Int J Infect Dis.* 2011; **15:** e66-9.
274. Panjabi C, Shah A. Allergic *Aspergillus* sinusitis and its association with allergic bronchopulmonary aspergillosis. *Asia Pac Allergy.* 2011; **1:** 130-7.
275. Pant H, Beroukas D, Kette FE, *et al.* Nasal polyp cell populations and fungal-specific peripheral blood lymphocyte proliferation in allergic fungal sinusitis. *Am J Rhinol Allergy.* 2009; **23:** 453-60.
276. Pant H, Ferguson BJ, Macardle PJ. The role of allergy in rhinosinusitis. *Curr Opin Otolaryngol Head Neck Surg.* 2009; **17:** 232-8.
277. Pant H, Schembri MA, Wormald PJ, *et al.* IgE-mediated fungal allergy in allergic fungal sinusitis. *Laryngoscope.* 2009; **119:** 1046-52.
278. Pao KY, Yakopson V, Flanagan JC, *et al.* Allergic fungal sinusitis involving the lacrimal sac: A case report and review. *Orbit.* 2014; **33:** 311-3.
279. Papagiannopoulos P, Lin DM, Al-Khudari S, *et al.* Utility of intraoperative frozen sections in surgical decision making for acute invasive fungal rhinosinusitis. *Int Forum Allergy Rhinol.* 2017; **7:** 502-7.
280. Park JH, Kreiss K, Cox-Ganser JM. Rhinosinusitis and mold as risk factors for asthma symptoms in occupants of a water-damaged building. *Indoor Air.* 2012; **22:** 396-404.
281. Patadia MO, Welch KC. Role of immunotherapy in allergic fungal rhinosinusitis. *Curr Opin Otolaryngol Head Neck Surg.* 2015; **23:** 21-8.
282. Patro SK, Verma RK, Panda NK, *et al.* Efficacy of preoperative itraconazole in allergic fungal rhinosinusitis. *Am J Rhinol Allergy.* 2015; **29:** 299-304.
283. Patro SK, Verma RK, Panda NK, *et al.* Understanding paediatric allergic fungal sinusitis: Is it more aggressive? *Int J Pediatr Otorhinolaryngol.* 2015; **79:** 1876-80.
284. Payne SJ, Mitzner R, Kunchala S, *et al.* Acute invasive fungal rhinosinusitis: A 15-year experience with 41 patients. *Otolaryngol Head Neck Surg.* 2016; **154:** 759-64.
285. Peral-Cagigal B, Redondo-Gonzalez LM, Verrier-Hernandez A. Invasive maxillary sinus aspergillosis: A case

report successfully treated with voriconazole and surgical debridement. *J Clin Exp Dent.* 2014; **6:** e448-51.

286. Peregud-Pogorzelski J, Wawrykow P, Wozniak S, *et al.* Highly effective unconventional management of aspergillosis of the left maxillary sinus in an 11-year-old girl with rhabdomyosarcoma embryonale of the frontal sinus. *J Med Microbiol.* 2013; **62 (Pt 4):** 652-4.
287. Peric A, Vojvodic D, Zolotarevski L, *et al.* Nasal polyposis and fungal *Schizophyllum commune* infection: A case report. *Acta Medica (Hradec Kralove).* 2011; **54:** 83-6.
288. Permi HS, Sunil KY, Karnaker VK, *et al.* A rare case of fungal maxillary sinusitis due to *Paecilomyces lilacinus* in an immunocompetent host, presenting as a subcutaneous swelling. *J Lab Physicians.* 2011; **3:** 46-8.
289. Philpott CM, Javer AR, Clark A. Allergic fungal rhinosinusitis - A new staging system. *Rhinology.* 2011; **49:** 318-23.
290. Philpott CM, Thamboo A, Lai L, *et al.* Olfactory dysfunction in allergic fungal rhinosinusitis. *Arch Otolaryngol Head Neck Surg.* 2011; **137:** 694-7.
291. Piao YS, Zhang Y, Yang X, *et al.* The use of MUC5B antibody in identifying the fungal type of fungal sinusitis. *Hum Pathol.* 2008; **39:** 650-6.
292. Piromchai P, Thanaviratananich S. Impact of treatment time on the survival of patients suffering from invasive fungal rhinosinusitis. *Clin Med Insights Ear Nose Throat.* 2014; **7:** 31-4.
293. Piromchai P, Thanaviratananich S. Invasive fungal rhinosinusitis versus bacterial rhinosinusitis with orbital complications: A case-control study. *Sci World J.* 2013; 453297. PMID: 24298218.
294. Plonk DP, Luong A. Current understanding of allergic fungal rhinosinusitis and treatment implications. *Curr Opin Otolaryngol Head Neck Surg.* 2014; **22:** 221-6.
295. Ponikau JU, Sherris DA, Kern EB. Immunologic response to fungus is not universally associated with chronic rhinosinusitis. *Otolaryngol Head Neck Surg.* 2010; **143:** 607-10.
296. Popalzai MJ, Kushawaha A, Mobarakai N, *et al.* Chronic fungal sinusitis leading to disastrous cerebral aspergillosis: A case report. *Cases J.* 2009; **2:** 9406. PMID: 20069073.
297. Popko M, Broglie MA, Holzmann D. Isolated fungus ball mimicking mucocele or frontal sinus tumour: A diagnostic pitfall. *J Laryngol Otol.* 2010; **124:** 1111-5.
298. Posteraro B, Scarano E, La Sorda M, *et al.* Eosinophilic fungal rhinosinusitis due to the unusual pathogen *Curvularia inaequalis.* Mycoses. 2010; **53:** 84-8.
299. Prateek S, Banerjee G, Gupta P, *et al.* Fungal rhinosinusitis: A prospective study in a University hospital of Uttar Pradesh. *Indian J Med Microbiol.* 2013; **31:** 266-9.
300. Premamalini T, Ambujavalli BT, Anitha S, *et al. Schizophyllum commune,* a causative agent of fungal sinusitis: A case report. *Case Rep Infect Dis.* 2011; 821259.
301. Rae W, Doffinger R, Shelton F, *et al.* A novel insight into the immunologic basis of chronic granulomatous invasive fungal rhinosinusitis. *Allergy Rhinol (Providence).* 2016; **7:** 102-6.
302. Ragab A, Samaka RM, Salem M. Impact of fungal load on diagnosis and outcome of allergic fungal rhinosinusitis. *Eur Arch Otorhinolaryngol.* 2014; **271:** 93-101.
303. Ragab A, Samaka RM. Immunohistochemical dissimilarity between allergic fungal and non-fungal chronic rhinosinusitis. *Am J Rhinol Allergy.* 2013; **27:** 168-76.
304. Rai S, Tiwari R, Sandhu SV, *et al.* Hyalohyphomycosis of maxillary antrum. *J Oral Maxillofac Pathol.* 2012; **16:** 149-52.
305. Rajmane VS, Rajmane ST, Patil VC, *et al.* Maxillary rhinosinusitis due to *Fusarium* species leading to cavernous sinus thrombosis. *J Mycol Med.* 2013; **23:** 53-6.
306. Rallis G, Gkinis G, Dais P, *et al.* Visual loss due to paranasal sinus invasive aspergillosis in a diabetic patient. *Ann Maxillofac Surg.* 2014; **4:** 247-50.
307. Rank MA, Adolphson CR, Kita H. Antifungal therapy for chronic rhinosinusitis: The controversy persists. *Curr Opin Allergy Clin Immunol.* 2009; **9:** 67-72.
308. Rao GSP, Mann SBS, Talwar P, *et al.* Primary mycotic infection of paranasal sinuses. *Mycopathologia.* 1983-84; **84:** 73-6.
309. Rassi SJ, Melkane AE, Rizk HG, *et al.* Sinonasal mucormycosis in immunocompromised pediatric patients. *J Pediatr Hematol Oncol.* 2009; **31:** 907-10.
310. Reddy CE, Gupta AK, Singh P, *et al.* Imaging of granulomatous and chronic invasive fungal sinusitis: Comparison with allergic fungal sinusitis. *Otolaryngol Head Neck Surg.* 2010; **143:** 294-300.
311. Ren K, Wang W, Ma X, *et al.* Nasopharyngeal carcinoma mimicking *Aspergillosis* rhinosinusitis: An unusual case report and review of the literature. *Int J Clin Exp Pathol.* 2014; **7:** 9050-5.
312. Robey AB, O'Brien EK, Richardson BE, *et al.* The changing face of paranasal sinus fungus balls. *Ann Otol Rhinol Laryngol.* 2009; **118:** 500-5.
313. Rodrigues J, Caruthers C, Azmeh R, *et al.* The spectrum of allergic fungal diseases of the upper and lower airways. *Expert Rev Clin Immunol.* 2016; **12:** 531-50.
314. Roehm CE, Salazar JC, Hagstrom N, *et al. Phoma* and *Acremonium* invasive fungal rhinosinusitis in congenital acute lymphocytic leukemia and literature review. *Int J Pediatr Otorhinolaryngol.* 2012; **76:** 1387-91.
315. Roy P, Chanda S, Gupta SC, Aggarwal AK. Orbital proptosis in a young immunocompetent female patient. *Med J Armed Forces India.* 2015; **71:** 389-92.
316. Rudagi BM, Halli R, Kalburge J, *et al.* Management of maxillary aspergillosis in a patient with diabetic mellitus followed by prosthetic rehabilitation. *J Maxillofac Oral Surg.* 2010; **9:** 297-301.
317. Rupa V, Thomas M. Different types of fungal sinusitis occurring concurrently: Implications for therapy. *Eur Arch Otorhinolaryngol.* 2013; **270:** 603-8.
318. Ryan MW, Clark CM. Allergic fungal rhinosinusitis and the unified airway: The role of antifungal therapy in AFRS. *Curr Allergy Asthma Rep.* 2015; **15:** 75.
319. Ryan MW. Allergic fungal rhinosinusitis. *Otolaryngol Clin North Am.* 2011; **44:** 697-710.
320. Sa HS, Ko KS, Woo KI, et al. A case of sino-orbital infection caused by the *Schizophyllum commune. Diagn Microbiol Infect Dis.* 2012; **73:** 376-7.

321. Sacks PL, Harvey RJ, Rimmer J, *et al.* Antifungal therapy in the treatment of chronic rhinosinusitis: A meta-analysis. *Am J Rhinol Allergy.* 2012; **26:** 141-7.
322. Sacks PL, Harvey RJ, Rimmer J, *et al.* Topical and systemic antifungal therapy for the symptomatic treatment of chronic rhinosinusitis. *Cochrane Database Syst Rev.* 2011; **8:** CD008263. PMID: 21833965.
323. Safirstein BH. Allergic bronchopulmonary aspergillosis with obstruction of the upper respiratory tract. *Chest.* 1976; **70:** 788-90.
324. Sajko T, Gnjidic Z, Sesar N, *et al.* Sphenoid sinus aspergilloma in trans-sphenoidal surgery for pituitary adenomas. *Acta Neurochir (Wien).* 2015; **157:** 1345-51.
325. Sato FR, Sawazaki R, Berretta D, *et al.* Aspergillosis of the maxillary sinus associated with a zygomatic implant. *J Am Dent Assoc.* 2010; **141:** 1231-5.
326. Sattler L, Sabou M, Ganeval-Stoll A, *et al.* Sinusitis caused by *Scopulariopsis brevicaulis:* Case report and review of the literature. *Med Mycol Case Rep.* 2014; **5:** 24-7.
327. Schubert MS. Allergic fungal sinusitis: Pathophysiology, diagnosis and management. *Med Mycol.* 2009; **47 :** S324-30.
328. Selvamani M, Donoghue M, Bharani S, *et al.* Mucormycosis causing maxillary osteomyelitis. *J Nat Sci Biol Med.* 2015; **6:** 456-9.
329. Sharma BN, Panta OB, Lohani B, *et al.* Computed tomography in the evaluation of pathological lesions of paranasal sinuses. *J Nepal Health Res Counc.* 2015; **13:** 116-20.
330. Sharma D, Mahajan N, Rao S, *et al.* Invasive maxillary aspergillosis masquerading as malignancy in two cases: Utility of cytology as a rapid diagnostic tool. *J Cytol.* 2012; **29:** 194-6.
331. Shivaprakash MR, Jain N, Gupta S, *et al.* Allergic fungal rhinosinusitis caused by *Neosartorya hiratsukae* from India. *Med Mycol.* 2009; **47:** 317-20.
332. Shockley R, Roy S. Allergic fungal rhinosinusitis in children. *Ear Nose Throat J.* 2012; **91:** 280-1.
333. Shrestha S, Kafle P, Akhter J, *et al.* Allergic fungal rhinosinusitis in chronic rhinosinusitis. *J Nepal Health Res Counc.* 2011; **9:** 6-9.
334. Silva MP, Baroody FM. Allergic fungal rhinosinusitis. *Ann Allergy Asthma Immunol.* 2013; **110:** 217-22.
335. Singh N, Nadig A, Pal P, *et al.* Aspergillosis sinusitis with cranio-orbital spread in an apparently immunocompetent patient. *J Neurol Surg A Cent Eur Neurosurg.* 2013; **74 (Suppl. 1):** e237-8.
336. Singh V, Sharma B, Sen R, *et al.* Abnormal growth of the maxillary sinus and orbit. *J Oral Maxillofac Surg.* 2011; **69:** 2167-72.
337. Singhal D, Baker L, Wormald PJ, *et al. Aspergillus fumigatus* biofilm on primary human sinonasal epithelial culture. *Am J Rhinol Allergy.* 2011; **25:** 219-25.
338. Singhal N, Raghubanshi G, Handa U, Punia RP, Singhal S. Fine needle aspiration cytology: A useful technique for diagnosis of invasive fungal rhinosinusitis. *Head Neck Pathol.* 2013; **7:** 236-40.
339. Smith A, Thimmappa V, Shepherd B, *et al.* Invasive fungal sinusitis in the pediatric population: Systematic review with quantitative synthesis of the literature. *Int J Pediatr Otorhinolaryngol.* 2016; **90:** 231-5.
340. Soler ZM, Schlosser RJ. The role of fungi in diseases of the nose and sinuses. *Am J Rhinol Allergy.* 2012; **26:** 351-8.
341. Soontrapa P, Larbcharoensub N, Luxameechanporn T, *et al.* Fungal rhinosinusitis: A retrospective analysis of clinicopathologic features and treatment outcomes at Ramathibodi Hospital. *Southeast Asian J Trop Med Public Health.* 2010; **41:** 442-9.
342. Srivastava RM, Rijuneeta, Gupta AK, *et al.* Quality of life, disability scores and distress index in fungal rhinosinusitis. *Med Mycol.* 2014; **52:** 706-14.
343. Stewart TA, Carter C, Seiberling K. Temporal lobe abscess in a patient with isolated fungal sphenoiditis. *Laryngoscope.* 2010; **120:** S247.
344. Sun GH, Elluru RG. Allergic fungal rhinosinusitis (AFRS). *Arch Otolaryngol Head Neck Surg.* 2010; **136:** 307 & 308-9.
345. Suresh S, Arumugam D, Zacharias G, *et al.* Prevalence and clinical profile of fungal rhinosinusitis. *Allergy Rhinol (Providence).* 2016; **7:** 115-20.
346. Suslu AE, Ogretmenoglu O, Suslu N, *et al.* Acute invasive fungal rhinosinusitis: Our experience with 19 patients. *Eur Arch Otorhinolaryngol.* 2009; **266:** 77-82.
347. Suwan Y, Punyawattanaporn A, Preechawai P. Rhino-orbital fungal infection: Two cases report. *J Med Assoc Thai.* 2012; **95:** 739-42.
348. Swain B, Panigrahy R, Panigrahi D. *Schizophyllum commune* sinusitis in an immunocompetent host. *Indian J Med Microbiol.* 2011; **29:** 439-42.
349. Swami T, Pannu S, Kumar M, *et al.* Chronic invasive fungal rhinosinusitis by *Paecilomyces variotii:* A rare case report. *Indian J Med Microbiol.* 2016; **34:** 103-6.
350. Tamgadge AP, Mengi R, Tamgadge S, *et al.* Chronic invasive aspergillosis of paranasal sinuses: A case report with review of literature. *J Oral Maxillofac Pathol.* 2012; **16:** 460-4.
351. Tarkan O, Karagun B, Ozdemir S, *et al.* Endonasal treatment of acute invasive fungal rhinosinusitis in immunocompromised pediatric hematology-oncology patients. *Int J Pediatr Otorhinolaryngol.* 2012; **76:** 1458-64.
352. Taxy JB, El-Zayaty S, Langerman A. Acute fungal sinusitis: Natural history and the role of frozen section. *Am J Clin Pathol.* 2009; **132:** 86-93.
353. Telmesani LM. Prevalence of allergic fungal sinusitis among patients with nasal polyps. *Ann Saudi Med.* 2009; **29:** 212-4.
354. Tendolkar U, van Diepeningen AD, Joshi A, *et al.* Rhinosinusitis caused by *Saksenaea erythrospora* in an immunocompetent patient in India: A first report. *JMM Case Rep.* 2015; **2:** 1-4.
355. Thakar A, Lal P, Dhiwakar M, *et al.* Optic nerve compression in allergic fungal sinusitis. *J Laryngol Otol.* 2011; **125:** 381-5.

356. Thanasumpun T, Batra PS. Oral antifungal therapy for chronic rhinosinusitis and its subtypes: A systematic review. *Int Forum Allergy Rhinol.* 2011; **1:** 382-9.
357. Thery A, Espitalier F, Cassagnau E, *et al.* Clinical features and outcome of sphenoid sinus aspergillosis: A retrospective series of 15 cases. *Eur Ann Otorhinolaryngol Head Neck Dis.* 2012; **129:** 179-84.
358. Thomas J, Munson E, Christianson JC. Unexpected *Blastomyces dermatitidis* etiology of fungal sinusitis and erosive palatal infection in a diabetic patient. *J Clin Microbiol.* 2014; **52:** 3130-3.
359. Thompson GR 3rd, Patterson TF. Fungal disease of the nose and paranasal sinuses. *J Allergy Clin Immunol.* 2012; **129:** 321-6.
360. Thompson LD. Allergic fungal sinusitis. *Ear Nose Throat J.* 2011; **90:** 106-7.
361. Thorp BD, McKinney KA, Rose AS, *et al.* Allergic fungal sinusitis in children. *Otolaryngol Clin North Am.* 2012; **45:** 631-42.
362. Thurtell MJ, Chiu AL, Goold LA, *et al.* Neuro-ophthalmology of invasive fungal sinusitis: 14 consecutive patients and a review of the literature. *Clin Experiment Ophthalmol.* 2013; **41:** 567-76.
363. Toplu Y, Toplu SA, Can S, *et al.* Fungus ball in concha bullosa: An unusual cause of retro-orbital pain. *J Craniofac Surg.* 2014; **25:** e138-40.
364. Toros SZ, Karaca CT, Kulekci S, *et al.* Choanal fungus ball mimicking a tumour. *J Craniomaxillofac Surg.* 2012; **40:** e24-7.
365. Toya T, Shinohara A, Tatsuno K, *et al.* A case of *Schizophyllum commune* sinusitis following unrelated cord blood transplantation for acute lymphoblastic leukemia. *Int J Hematol.* 2013; **98:** 261-3.
366. Trinidade A, Shakeel M, Chapman A, *et al.* Isolated sphenoid aspergilloma: A differential diagnosis for solitary abducens nerve palsy. *West Indian Med J.* 2009; **58:** 285-7.
367. Tsai TL, Lan MY, Ho CY. There is no structural relationship between nasal septal deviation, concha bullosa and paranasal sinus fungus balls. *Sci World J.* 2012; 181246. PMID: 23326212.
368. Tsukatani T, Ogawa H, Anzawa K, *et al.* *Schizophyllum commune*-induced allergic fungal rhinosinusitis and sinobronchial mycosis. *Med Mycol Case Rep.* 2015; **8:** 10-3.
369. Turhan O, Bostanci A, Ozbudak IH, *et al.* Chronic invasive non-granulomatous fungal rhinosinusitis in immunocompetent individuals. *Case Rep Otolaryngol.* 2016; 6854121.
370. Twaruzek M, Soszczynska E, Winiarski P, *et al.* The occurrence of molds in patients with chronic sinusitis. *Eur Arch Otorhinolaryngol.* 2014; **271:** 1143-8.
371. Upadhyay R, Kant S, Prakash V, *et al.* Sinobronchial allergic aspergillosis with allergic bronchopulmonary aspergillosis: A less common co-existence. *BMJ Case Rep.* 2014; pii: bcr2013201946. PMID: 25371437.
372. Uri N, Ronen O, Marshak T, *et al.* Allergic fungal sinusitis and eosinophilic mucin rhinosinusitis: Diagnostic criteria. *J Laryngol Otol.* 2013; **127:** 867-71.
373. Vaidya D, Shah P. Coinfection by *Aspergillus* and zygomycetes species in a case of acute rhinosinusitis. *Case Rep Otolaryngol.* 2011; 382473. PMID: 22937365.
374. Valera FC, do Lago T, Tamashiro E, *et al.* Prognosis of acute invasive fungal rhinosinusitis related to underlying disease. *Int J Infect Dis.* 2011; **15:** e841-4.
375. van de Beek D, Patel R, Campeau NG, *et al.* Insidious sinusitis leading to catastrophic cerebral aspergillosis in transplant recipients. *Neurology.* 2008; **70 (24 Pt 2):** 2411-3.
376. Vascellari M, Carminato A, Danesi P, *et al.* Severe, chronic, pyogranulomatous rhinosinusitis with necrosis and fungal septate hyphae consistent with *Fusarium* spp infection. *J Am Vet Med Assoc.* 2011; **238:** 449-51.
377. Vashishth A. Extensive allergic fungal rhinosinusitis: Ophthalmic and skull base complications. *Indian J Otolaryngol Head Neck Surg.* 2015; **67:** 227-33.
378. Verma RK, Patro SK, Francis AA, *et al.* Role of preoperative *versus* postoperative itraconazole in allergic fungal rhinosinusitis. *Med Mycol.* 2017; pii: myw125. PMID: 27838640.
379. Verma RK, Shankar A, Kumar Arora S, *et al.* *Histoplasma* rhinosinusitis presenting as a non-healing oral ulcer in an immunocompetent patient. *J Otolaryngol Head Neck Surg.* 2011; **40:** e47-50.
380. Verma S, Gupta P, Singh D, Kanga A, Mohindroo NK, Jhobta A. *Schizophyllum commune* causing sinusitis with nasal polyposis in the sub-Himalayan region: First case report and review. *Mycopathologia.* 2014; **177:** 103-10.
381. Vinay K, Khullar G, Yadav S, *et al.* Granulomatous invasive aspergillosis of paranasal sinuses masquerading as actinomycosis and review of published literature. *Mycopathologia.* 2014; **177:** 179-85.
382. Vinciguerra A, Saibene AM, Lozza P, *et al.* Unusual case of bilateral maxillary fungus ball. *BMJ Case Rep.* 2016; pii: bcr2016217930. PMID: 27979848.
383. Vlaardingerbroek H, van der Flier M, Borgstein JA, *et al.* Fatal *Aspergillus* rhinosinusitis during induction chemotherapy in a child with acute lymphoblastic leukemia. *J Pediatr Hematol Oncol.* 2009; **31:** 367-9.
384. Wang T, Su J, Feng Y. The effectiveness topical amphotericin B in the management of chronic rhinosinusitis: A meta-analysis. *Eur Arch Otorhinolaryngol.* 2015; **272:** 1923-9.
385. Wang X, Dong D, Cheng J, *et al.* Relationship between biofilms and clinical features in patients with sinus fungal ball. *Eur Arch Otorhinolaryngol.* 2015; **272:** 2363-9.
386. Watts JC, Chandler FW. Fungal sinusitis. *N Engl J Med.* 1997; **337:** 1633-4.
387. Webb BJ, Vikram HR. Chronic invasive sinus aspergillosis in immunocompetent hosts: A geographic comparison. *Mycopathologia.* 2010; **170:** 403-10.
388. White LC; Jang DW; Yelvertan JC. Bony erosion patterns in patients with allergic fungal sinusitis. *Am J Rhinol Allergy.* 2015, **29:** 243-5.
389. Wise SK, Ahn CN, Lathers DM, *et al.* Antigen-specific IgE in sinus mucosa of allergic fungal rhinosinusitis patients. *Am J Rhinol.* 2008; **22:** 451-6.
390. Wise SK, Rogers GA, Ghegan MD, *et al.* Radiologic staging system for allergic fungal rhinosinusitis (AFRS). *Otolaryngol Head Neck Surg.* 2009; **140:** 735-40.

391. Won EJ, Shin JH, Lim SC, et al. Molecular identification of *Schizophyllum commune* as a cause of allergic fungal sinusitis. *Ann Lab Med.* 2012; **32:** 375-9.
392. Won HR, Park JH, Kim KS. Simultaneous actinomycosis with aspergillosis in maxillary sinus. *Br J Oral Maxillofac Surg.* 2013; **51:** e51-3.
393. Wong G, Nash R, Barai K, *et al. Paecilomyces lilacinus* causing debilitating sinusitis in an immunocompetent patient: A case report. *J Med Case Rep.* 2012; **6:** 86.
394. Woods CM, Hooper DN, Ooi EH, *et al.* Human lysozyme has fungicidal activity against nasal fungi. *Am J Rhinol Allergy.* 2011; **25:** 236-40.
395. Xu R, Ma L, Shi J, *et al.* Endoscopic sinus surgery for fungal ball rhinosinusitis in South China: Long-term results and analysis of prognostic factors. *Acta Otolaryngol.* 2012; **132:** 519-24.
396. Yakirevitch A, Barg AA, Bedrin L, *et al.* Acute invasive fungal rhinosinusitis in children with hematologic malignancies: Outcome of surgical treatment. *Pediatr Hematol Oncol.* 2015; **32:** 568-75.
397. Yan Y, Zhao Z, Wan H, *et al.* A novel fungus concentration-dependent rat model for acute invasive fungal rhinosinusitis: An experimental study. *BMC Infect Dis.* 2014; **14:** 3856. PMID: 25526739.
398. Yin X, Liang Y, Zeng L, *et al.* A case of sinusitis caused by *Schizophyllum commune* and bacteria in acute myelocytic leukemia. *Clin Lab.* 2015; **61:** 1799-801.
399. Younis RT, Ahmed J. Predicting revision sinus surgery in allergic fungal and eosinophilic mucin chronic rhinosinusitis. *Laryngoscope.* 2017; **127:** 59-63.
400. Zakirullah, Nawaz G, Sattar SF. Presentation and diagnosis of allergic fungal sinusitis. *J Ayub Med Coll Abbottabad.* 2010; **22:** 53-7.
401. Zhang F, An Y, Li Z, *et al.* A novel model of invasive fungal rhinosinusitis in rats. *Am J Rhinol Allergy.* 2013; **27:** 361-6.
402. Zhang J, Li Y, Lu X, *et al.* Analysis of fungal ball rhinosinusitis by culturing fungal clumps under endoscopic surgery. *Int J Clin Exp Med.* 2015; **8:** 5925-30.
403. Zhao Z, Li L, Wan Z, *et al.* Simultaneous detection and identification of *Aspergillus* and Mucorales species in tissues collected from patients with fungal rhinosinusitis. *J Clin Microbiol.* 2011; **49:** 1501-7.
404. Zuluaga A, Ospina-Medina J, Castano-Gallego I, *et al.* Frequency of fungal agents identified in sinus samples from patients with clinically suspected rhinosinusitis. *Diagn Microbiol Infect Dis.* 2015; **81:** 208-12.
405. Zuniga MG, Turner JH. Treatment outcomes in acute invasive fungal rhinosinusitis. *Curr Opin Otolaryngol Head Neck Surg.* 2014; **22:** 242-8.

CHAPTER 35

Mycotoxicoses and Mycetismus

A substantial number of fungi are utilized for preparing several products useful for the mankind like antimicrobial antibiotics and food-related products. However, a few of the fungi are found to be occurring naturally in food but produce secondary metabolites, which are known as **mycotoxins**. The human beings are exposed to such substances worldwide although their amount and frequency fluctuate according to the geographical variations.

Toxins as such are the compounds that have harmful effects on other cells, tissues and organisms. The toxins produced by fungi are called **Mycotoxins**. These toxins are secondary metabolites produced by fungi and are as such not essentially required for the fungal growth. They appear to have no function in normal metabolism of fungal cell but may play an important role in eliminating other microorganisms competing in the same environment. Additionally, they also help the pathogenic fungi to invade host tissues. These natural products, which are yielded by fungi, evoke toxic response in the higher vertebrates and animals when introduced through natural routes into the body in a low concentration. Their toxic properties may be carcinogenicity, genotoxicity, immunotoxicity, mutagenicity as well as reproductive and developmental toxicity.

The literal meaning of **Mycotoxin** is a compound that is toxic to fungi but in real sense these are the toxins produced by the fungi. As such mycotoxins are neither contagious nor infectious but can occur in clusters in a population with varying severity. Some are lethal, cause identifiable diseases, weaken the immune system without producing symptoms specific to that toxin or may act as an allergen. However, some toxins are harmless to the human beings, yet others are harmful to farm animals and some produce inhibitory effects on other microorganisms. The **five** most common mycotoxins found worldwide are aflatoxins, zearalenone, deoxynivalenol, fumonisins and ochratoxin A.

As per the definition **Mycotoxins** are naturally occurring low-molecular weight secondary metabolites produced by certain filamentous fungi. These substances are injurious and even lethal in low concentrations. The diverse group of these metabolites are produced by taxonomically wide range of food-borne filamentous ascomycetous fungi, three genera of which are the most important *i.e. Aspergillus, Fusarium* and *Penicillium*, which are responsible for most of these diseases. Ingestion of food material containing mycotoxin may lead to wide range of diseases, which are called **mycotoxicoses** and study of mycotoxins is called **Mycotoxicology**. Mycotoxicosis is an unintentional and rather accidental development. These should be distinguished from **mycoses** *i.e.* diseases produced by direct invasion by the pathogenic fungi.

A somewhat similar type of clinical condition, produced by ingestion of basidiomycetous fungi *i.e.* mushrooms, is known as **mycetismus** or **mycetism,** a synonym to **muscarinism**. The presentation in mycetismus is analogous to infective/invasive-toxic type of bacterial food poisoning. Most of the times, these clinical entities are documented as outbreaks involving large number of human beings or animals since isolated cases remain unreported or rarely reported. Mycetismus is due to intentional intake of the edible mushroom, incidentally that turns out to be poisonous one. The polypore mushroom (*Irpex lacteus*), a wood-decaying agent and other basidiomycetous filamentous fungi such as *Hormographiella aspergillata* and *Schizophyllum commune* are now emerging as potential pathogens among humans and clinically they are not presenting as mushroom poisonings. Hence they are dealt with other fungal diseases as significant pathogenic fungi.

Historical Perspective

The ergot poisoning is being observed since ancient times. This was once confused with witchcraft thereby leading to infamous Salem Witch Trials in Massachusetts, USA, conducted in the year 1692, wherein 24 females and males were convicted. Out of these, 19 were hanged, 4 died waiting for completion of the trial and one was crushed under stones to death. This very instance of convulsive ergotism prompted Robin Cook to write his 1995 novel **Acceptable Risk** after even three centuries of the actual happenings. In addition in 1938, anti-Jewish Nazi propaganda book, titled **Der Giftpilz** (The Poisonous Mushroom), was also widely circulated by Adolf Hitler. Moreover, during the **Cold War** period in 1953, CIA started *"Program of Research in Behavioral Modification"* code-named as **MKUltra** using ergot alkaloid derivative, LSD, for interrogation and torture in order to weaken individuals thereby to force them for confessions through their mind controls.

In the last fifty years, it has become amply clear that commonly occurring fungi may produce mycotoxins. These toxins have caused significant epidemics in both man and animals in recent times. The most important epidemics have been reported in the literature like ergotism, which has killed thousands of people in Europe in the last century. **Stachybotryotoxicosis** killed thousands of horses in the erstwhile USSR in 1930s. **Alimentary toxic aleukia** was responsible for death of at least 1,00,000 Russian people between 1942 and 1948. **Aflatoxicosis** killed about 1,00,000 young turkeys in UK in 1960 and caused disease/death in other animals as well as in man. From 1961 to 1971, the US military sprayed millions of liters of toxic herbicides, mainly **Agent Orange**, over Vietnam to destroy its vegetation used by liberation forces as food and also for their hiding cover.

An outbreak of aflatoxicosis, reported from India in 1974 in the adjoining districts of Gujarat and Rajasthan states, affected about four hundred individuals and killing one-fourth of them. A self-limiting outbreak by trichothecenes was reported from the Kashmir valley in 1987 due to rain-damaged breads. Bhat *et al*, reported an outbreak of fumonisin toxicity in 1994 from south India, which affected thousands of people but no fatality, was reported.

The International Society for Mycotoxicology organized **Mycotoxin Conference** in Beijing on May 19-23, 2014, with the theme, *"Global Prevention and Control of Mycotoxins"*. Moreover, in view of the new data available on the damage caused due to foodborne illnesses, the WHO raised the slogan *"From Farm to Plate, Make Food Safe"* on the World Health Day celebrated on April 7, 2015, thereby highlighting cross-border action required to improve the entire food supply chain. The **2nd International Conference on Mycology & Mushrooms** will be held on September 25-26, 2017at Chicago, Illinois (USA).

Epidemiology

In the past, there were no proper facilities of prolonged storage of perishable foodstuffs so food had to be consumed within a short span of preparation/procurement. As the world is becoming more industrialized and technologically advanced, storage of food has become an important issue. Now, food is commonly stored for longer periods thereby increasing the chances for fungal contamination.

The fungi involved in mycotoxicosis belong to ascomycetous genera like *Aspergillus, Fusarium, Penicillium, Claviceps*, etc., whereas in mycetismus are the frequently encountered basidiomycetes like *Amanita* species. The documentation of excessive levels of aflatoxin in foods and blood samples from people in developing countries, along with synergistic effects of hepatitis, reflect that these toxins are significant cause of death in some parts of Africa and Southeast Asia. Based on data by Pitt, *et al*, it is estimated that number of deaths from liver cancer due to aflatoxin in Indonesia alone exceeds 20,000 per annum. Human fatalities have also been reported in India due to acute aflatoxin poisoning because of un-seasonal rains and scarcity of food, thereby, compelling the poor people to consume heavily contaminated maize and maize-based products, particularly in rural areas.

Such diseases in the past were domains of veterinary importance but now due to increasing involvement of humans; mycotoxicosis has become the field of interest for medical investigators and public health personnel too. The routine diagnostic microbiology laboratory is seldom approached for establishing such a diagnosis. However, once the responsible fungus, which otherwise encountered routinely is suspected, the identification is not very difficult. This is perhaps the major reason that diagnostic and therapeutic aspects of mycotoxicoses and mycetismus are not mentioned in Medical Mycology textbooks rather found in reference books pertaining to the public health importance.

The effects produced by these fungi and/or their metabolites are broadly classified under **two** broad headings: (I) Mycotoxicoses (II) Mycetismus. Therefore, mycotoxicosis and mycetismus are actually intoxications rather than fungal infections and both these topics are briefly discussed in the Textbook of Medical Mycology in this Chapter under two separate headings.

I. MYCOTOXICOSES

The clinical entity, **mycotoxicosis** is defined as an illness of man or domestic animals due to ingestion of pre-formed secondary metabolites - mycotoxins, produced by molds on certain food items. Therefore, consumption of such food, contaminated with mycotoxins, quite often leads to fatal consequences. Maize and groundnut are the most frequently contaminated foodstuffs.

There are about 350 to 400 metabolites that are considered toxigenic and new compounds are regularly being added to this list. These toxins are non-volatile, have low molecular weight with diverse chemical structures. The most significant fungi producing mycotoxicoses belong mainly to toxigenic ascomycetous species of *Aspergillus, Fusarium* and *Penicillium*. There are **four** basic types of mycotoxicosis: **acute, chronic, mutagenic** and **teratogenic**.

Mycotoxicoses in man and animals, due to ingestion of mold-damaged cereals, has been suspected from time to time. Out of all the known mycotoxins causing diseases in both human beings and animals - aflatoxins, *Fusarium* toxins and ochratoxins, are considered to be the most significant. Certain species of *Penicillium* are also associated with spoilage of food and some of them have been found to be potential mycotoxin producers and implicated in causing serious health hazards in man and animals.

These are more commonly seen in males as compared to females. Such diseases are mainly found in rural and tribal populations in the developing countries. Acute mycotoxicoses may cause serious diseases, sometimes with fatal outcomes. This should be suspected when an acute disease affects several individuals and as such there is no evidence of apparent infection with a known etiological agent and there is no improvement in clinical condition following institution of routine therapeutic regimen.

The production of mycotoxin depends upon several factors. Its amount depends on physical (moisture, relative humidity, temperature and mechanical damage), chemical (carbon dioxide, oxygen, composition of substrate, insecticides and fungicides) and biological factors (plant variety, stress, insects, spore load). However, moisture and temperature have major influence on fungal growth rate thereby mycotoxin production.

Most of the reported outbreaks of mycotoxicoses have occurred as a consequence to consumption of food items contaminated with mycotoxins. A brief account of medically significant mycotoxins, which produce various diseases in man and animals, is given in **Table 35.1** and are described below in **alphabetical** order:

1. Aflatoxins
2. Cyclopiazonic Acid
3. Ergot Poisoning
4. Fumonisins
5. Ochratoxins
6. Patulin
7. Trichothecenes
8. Zearalenone

Table 35.1. Important Mycotoxins producing Fungi, Source of Exposure and Clinical Conditions.

Mycotoxins	*Producing Fungal Species*	*Source of Exposure*	*Clinical Conditions*
Aflatoxins	*Aspergillus flavus, A.parasiticus, A.nomius, Penicillium puberulum*	Nuts, Maize	Aflatoxicosis, Reye's Syndrome, Hepatitis, Hepatoma
Ergot Alkaloid Poisoning	*Claviceps purpurea, C.fusiformis*	Rye Flour	St. Anthony's Fire
Cyclopiazonic Acid	*Aspergillus flavus, A.versicolor A.oryzae, Penicillium cyclopium*	Groundnut, Corn, Meat	Co-contaminant, Kodua Poisoning
Fumonisins	*Fusarium verticillioides*	Maize	ELEM, PPE
Ochratoxins	*Aspergillus ochraceus, A.niger, Penicillium verrucosum*	Cereals, Coffee-Beans, Bread	Nephropathies *i.e.* BEN and MPN
Trichothecenes	*Fusarium graminearum, F.sporotrichioides*	Maize, Sorghum	Human Toxicosis, ATA, Biological Warfare
Zearalenones	*Fusarium graminearum*	Wheat, Maize, Barley, Sorghum	Genital disorders in animals *i.e.* Pigs

1. Aflatoxins

Aflatoxins were discovered as a cause of strange manifestations of mysterious disease termed as **Turkey-x-disease** during early 1960s in England, which had killed approximately 1,00,000 birds and ruined turkey industry as a whole. It was later realized that this disease resulted from the consumption of animal feed prepared using contaminated Brazilian groundnut. After this outbreak, general awareness and interest about mycotoxins arose among medical community as well as public health personnel.

The contamination of crops occurs predominantly in sub-tropical and tropical climates where conditions are favorable for toxin production by the toxigenic fungi. In November 1974, an outbreak of curious liver disease occurred in the adjoining districts of Gujarat (Panchamahal) and Rajasthan (Banswada and Dungarpur), the states of western India. These areas experienced unseasoned rains during previous month and as a result, standing maize crop was heavily drenched and got contaminated with *A.flavus*, containing aflatoxin level up to 15 mg/kg in food samples. The maize was harvested immediately after rains and stored in damp places having moisture as per the traditional practices adopted in those areas. More than 400 individuals were affected and about 100 lives were lost due to aflatoxin toxicity. The aflatoxin toxicity was also observed in dogs of the affected areas during the corresponding period.

Aflatoxins are secondary metabolites produced by two species of *Aspergillus*, a fungus that is especially found in areas with hot and humid climates. Most common aflatoxins are B_1 and B_2 which are produced by *A.flavus*. In addition to these, *A.parasiticus* produces G_1 and G_2 metabolites. These toxins are also produced by *Penicillium puberulum*.

Aflatoxins can occur in raw agricultural products, such as ground nuts, corn, maize, rice, figs as well as other dried foods, spices, crude vegetable oils and cocoa beans, as a result of fungal contamination before and after harvest. Most of these substrates are used as major staple food by some of the African tribes. Corn, wheat and rice normally contain more aflatoxins than sorghum, peanut and soybean. The products like peanut butter, beer, stored-grains, nut meal and animal feed should be monitored during their process of production in such a way to avoid intoxication by aflatoxins. Since they are known to be genotoxic and carcinogenic, their exposure through food should be minimized.

It is observed that as a result of large-scale contamination of cereals, millets and oilseeds by fungi, there is an accompanying discoloration of the kernel. Bhat *et al*, revealed that 21% samples of groundnut exceeded the Indian tolerance limit for aflatoxin (30 μg/kg) at that time. A similar analysis of maize samples by same group of workers showed that 26% of the samples exceeded the tolerance levels. Indian tolerance limit for aflatoxins is now 15 μg/kg for raw agricultural products like cereal, pulses, nuts, oilseeds, spices, whereas it is 10 μg/kg for figs, ready to eat nuts and oilseeds. Indian tolerance has been reduced from 30 μg/kg to 15 μg/kg and even 10 μg/kg keeping in view of internationally strict limits recently adopted.

The acronym of **aflatoxin** has been derived from the first alphabet of ***A**spergillus **fla**vus* **toxin**. These are series of compounds with similar activity and are designated as aflatoxin - B_1, B_2, G_1 and G_2 on the basis of their metabolites, which exhibit blue (B) or green (G) fluorescence when irradiated under ultraviolet (UV) light on thin layer chromatography plates. There are three main types of aflatoxin mycotoxins: Aflatoxins B: This group includes aflatoxin B_1 and B_2. Aflatoxin B_1 is the commonest as well as the most toxic and carcinogenic. Aflatoxins G: this group includes aflatoxin G_1 and aflatoxin G_2. Aflatoxins M: this group includes aflatoxins M_1 and M_2. These aflatoxins are metabolic products, which are found in the urine and milk produced by animals, which have been given feed with aflatoxins B_1 and B_2 in it. The acute toxicities produced by aflatoxins M_1 and M_2 are of lesser severity than those of B_1 and B_2. They are now abbreviated as AFM_1 and AFM_2 also.

The exposure to aflatoxin has also been implicated in diseases like liver cirrhosis, kwashiorkor and Reye's syndrome. Aflatoxin B_1 is biologically the most potent mycotoxin because of its common occurrence in food items and is highly toxic as well as carcinogenic. The International Agency for Research on Cancer (IARC) concluded that there was sufficient evidence of carcinogenicity in humans by aflatoxin B_1 and classified it as Group 1 human carcinogen. Chemically it is related to coumarins. The effect of aflatoxin varies considerably from species to species for a particular fungal metabolite *i.e.* it can be progressive as well as cumulative. Most of the clinical manifestations have been studied in animals. The ducklings and trout are more sensitive to aflatoxins than rats and mice. The human cell lines are not very sensitive to the effects of these toxins. At levels greater than 50 ppb, aflatoxins can cause signs and symptoms such as feed refusal, fever, weight loss, sluggishness, liver damage, jaundice, bloody diarrhea, kidney damage, birth defects, tumors and suppressed immune function.

The importance of aflatoxins in human disease is still being evaluated. Incidental exposure of adults and especially children, may lead to demonstrable abnormalities and even neoplasia. **Aflatoxicosis** is a sort of poisoning that results from ingesting aflatoxins. Two clinical forms of the disease have been identified, an acute severe intoxication, which results in direct liver damage and subsequent illness or death and chronic sub-symptomatic aflatoxicosis. The signs and symptoms are lethargy, anorexia and muscle weakness followed by spasm. The other clinical manifestations may be fever, jaundice, hepatomegaly, ascites, portal hypertension, pedal edema and eventually death of the patient.

Reye's syndrome is also thought to be an acute aflatoxicosis wherein patient presents with signs and symptoms of encephalopathy and fatty degeneration of viscera. This is an endemic disease of children in developing countries. Aflatoxins have also been implicated in the outbreaks of hepatitis in India where maize forms the major part of staple diet of the population at large. It has been suggested that Indian childhood cirrhosis (ICC) is also caused by aflatoxins. This disease is prevalent in some parts of India and is characterized by enlargement and fatty degeneration of liver.

2. Cyclopiazonic Acid

The cyclopiazonic acid (CPA) is another toxic fungal metabolite. Chemically it is indole tetramic acid, derived from tryptophan and is related to ergoline alkaloids. It was originally produced by plant pathogens like *Penicillium aurantiogriseum* (formerly *P.cyclopium*) and subsequently from other fungi - *P.griseofulvum, P.camemberti, P.commune, Aspergillus flavus, A.tamarii, A.oryzae* and *A.versicolor.* This appears to be toxic only in high concentrations and biologically, it is a specific inhibitor of ATPase in the intracellular calcium storage sites. It occurs naturally in agriculture products such as groundnut and corn, mostly as co-contaminant with aflatoxin. It has been experimentally studied that neither of the toxins augments action of the other.

The clinical symptoms of cyclopiazonic acid intoxication includes loss of appetite and weight, weakness, vomiting, diarrhea, dehydration, depression, opisthotonus, convulsions and death. This has been suspected to be causing symptoms of **Kodua poisoning**, which is toxic syndrome in cattle and man often encountered in areas where kodo millet seeds of rice grass (*Paspalum scrobiculatum* - Tamil: Varagu) are consumed as staple food by poor people in several regions of India. Eating of this grain results in giddiness, sleepiness and tremors, with recovery after 1 to 3 days.

3. Ergot Alkaloids

The poisoning by ergot alkaloids is the oldest known mycotoxicosis and is caused by fungal species of the genus *Claviceps*. The toxic effects in humans result from consumption of bread made of rye, millet or other grain flours contaminated with *Claviceps purpurea*. A purple, hard, curved body thus formed called **Sclerotium**, replaces almost the whole grain.

Ergot is common name of sclerotia of fungal species within the genus *Claviceps*, which produce ergot alkaloids. The toxins involved in this type of poisonings are otherwise very useful in medicine. The natural ergot alkaloids isolated from sclerotia of *Claviceps* are the tetracyclic indole containing compounds, which are considered as derivatives of lysergic acid.

Ergotism exists in two forms: gangrenous and convulsive. Gangrenous ergotism was common from the 9th to the 14th century in central Europe. Symptoms and signs included paraesthesia, swelling, inflammation and vasoconstriction of the extremities, which sometimes resulted in limb loss. Convulsive ergotism that involved the nervous system occurred from late 16th to late 19th century in Europe and USA. Even the 1692 Salem Witch Trials in USA were conducted under confusion, whereas these were actually cases of ergotism. Such cases are extremely rare these days, primarily because normal grain cleaning and milling processes remove most of the ergots thereby only very low levels of alkaloids remain in the resultant flours. In addition, alkaloids are relatively labile and usually destroyed during baking and cooking. Some of the drugs also occasionally produce similar types of alkaloids effects.

The alkaloids are now produced in culture by strains of *Claviceps fusiformis* and *C.paspalii*. The ergot alkaloids are divided into two groups:-

(a) Amine Alkaloid—Ergometrine

(b) Aminoacid Alkaloid—Ergotamine

(a) Ergometrine: This amine derivative is a water-soluble group and has either lysergic acid or its isomer, which is linked to an amino-alcohol. Ergometrine or its semisynthetic analogue, methylergometrine is often used clinically for its oxytocic effects.

(b) Ergotamine: This alkaloid is water insoluble polypeptide in which either lysergic acid or isolysergic acid is linked to an aminoacid. It has α-adrenergic blocking and vasoconstrictor property hence is widely used in the treatment of migraine.

The animals feeding on grass contaminated with this fungus may suffer from such type of poisoning. There is inflammation of affected tissues, followed by necrosis and gangrene. These ergots produce α-adrenergic blockade that inhibits responses to epinephrine and 5-hydroxytryptamine. If progressing inflammatory condition is not corrected in time, it may lead to marked peripheral vasoconstriction and restricts blood circulation thereby resulting in necrosis and gangrene. Due to constriction of arteries and veins, extremities become cold, followed by burning sensation due to ischemia. There may be epileptic fits, vomiting and insanity. The symptoms may be coupled with a feeling as if the sufferer has visited St. Anthony's shrine in a hope of getting relief from intense burning sensation, which led to disease state being named as **St. Anthony's Fire** (***'ignis sacer'*** = Sacred Fire).

An outbreak of ergot poisoning was reported in 1953 in France. In 1975, another type of ergotism was seen in India, which was convulsive form related to intoxication with clavine alkaloids from *Claviceps fusiformis*. This time the infections last seen was affecting around 78 persons without any fatal outcome. This was characterized by gastrointestinal symptoms like nausea, vomiting and giddiness followed by involvement of central nervous system like drowsiness, prolonged sleepiness, twitching, convulsions, blindness and paralysis.

4. Fumonisins

The fumonisins are group of mycotoxins produced primarily by *Fusarium verticillioides* (formerly, *F.moniliforme*) and *F.proliferatum*, although few other *Fusarium* species are also involved. There are at least 28 different forms of fumonisins, most designated as A-series, B-series, C-series and P-series. The B series of the fumonisin analogues consists mainly of fumonisin B_1 (FB_1), fumonisin B_2 (FB_2) and fumonisin B_3 (FB_3). Fumonisin B_1 is the most common and economically important form, followed by B_2 and B_3. These toxins are believed to cause toxicity by blocking ceramide synthetase, a key enzyme in sphingolipid biochemistry, which converts sphingosine and acetyl-CoA to ceramide. Maize is the most commonly contaminated crop although these toxins can occur in a few other crops as well. These have similar toxicological profiles, although FB_1 is the most prevalent analogue in the naturally contaminated corn, however, *F.verticillioides* contaminated-corn in both culture and under natural conditions produces these secondary metabolites and may occur even in the corn that is not visibly moldy. Thus, fumonisins are frequently found in corn and corn-products at the levels ranging from 0.5 to 5 ppm and can be present in much higher amounts (up to 300 to 400 ppm) in moldy corn, which can be observed by corn screening.

Fumonisins are carcinogenic to the laboratory animals and in humans as consumption of fumonisin-contaminated maize is associated with higher rates of esophageal cancer and neural tube birth defects. Fumonisins were first characterized in the late 1980s, as a result of studies, which looked into the causes of esophageal cancer in endemic regions of South Africa, particularly Transkei and also some parts of China, coupled with outbreaks of equine leukoencephalomalacia and porcine pulmonary edema in the US. An IARC working group has classified the mycotoxins from *F.verticillioides* as class 2B human carcinogens as it is believed to play a significant role in esophageal cancer in man.

The primary health concerns associated with fumonisins are acute (single exposure) toxic effects in horses and swine with carcinogenic properties. Horses that are poisoned with fumonisins may develop a fatal disease known as equine leukoencephalomalacia. Symptoms of this disease include drowsiness, blindness, staggering and liquefaction of brain tissue. Swines that are poisoned with fumonisins may experience reduced feed intake and weight gain, liver damage and may develop a fatal pulmonary edema.

In southern India, Bhat *et al*, reported an outbreak in 1997 that affected 1325 individuals in 27 villages by consumption of maize and sorghum. All patients recovered fully when the exposure ceased and there were no fatalities. There is widespread prevalence of fumonisin B_1 in maize and poultry feeds in different districts of Haryana state. In 1999, an outbreak was reported in poultry due to consumption of fumonisin-contaminated feed containing rain-damaged maize.

5. Ochratoxins

The **ochratoxin A** is a naturally occurring mycotoxin, which was discovered in 1965. This toxin was characterized and defined after it was originally isolated in the pure form from ***Aspergillus ochraceus*** hence designated ochratoxin based on its species name. Subsequently, it was found to be produced by other fungi also, including species of *Penicillium i.e. P.verrucosum*. Earlier it was wrongly thought to be produced by *P.viridicatum*. Recently, studies have shown that strains of *Aspergillus niger* are also capable of producing ochratoxin.

Ochratoxins are phenylalanine-containing dihydroisocoumarins and it has been suggested that they act by disrupting phenylalanine metabolism. Ochratoxin A also possesses chlorinated phenol moiety, a pro-carcinogenic group that causes electrophilic attack on DNA.

Out of the many ochratoxins produced by fungal species, only ochratoxin A is medically significant. These are fat-soluble toxins and do not get readily excreted. Hence these get accumulated as depot in the fat of affected animals and while eating pork they are ingested by humans. They accumulate primarily in the kidneys as a result of high vascular flow and tubular re-uptake, therefore they are involved in various renal disorders. It is responsible for many systemic toxicities as well as cause nephropathies, a chronic kidney disease in humans when it is digested incombination with mycotoxin citrinin. Further ochratoxin A was found in several studies to be carcinogenic which was subsequently confirmed. The IARC has also classified ochratoxin A as a compound possibly carcinogenic to humans group 2B and is one of the most common contaminant in foodstuff.

Almost four years after ochratoxins were first detected, they were found to occur naturally in cereals and meat products. This was the beginning of several researches for the natural occurrences of ochratoxins. Subsequently, toxins were found in commodities and products such as cereal grains (barley, wheat, oats and rye), green coffee beans, peas, peanuts, bread, sorghum, olives, beer, mixed feeds, meats, cheese, milk powder and hay. More recently, ochratoxin A has been found in raisins, currants, sultanas, wine and grape juice.

This toxin has nephrotoxic, immunosuppressive, teratogenic, mutagenic and carcinogenic properties in all experimental animals tested so far. Human exposure to ochratoxins can take place either directly by consumption of contaminated plant food or indirectly by consumption of animal tissues previously exposed to contaminated material. This has caused cases of fatal renal disease called **endemic nephropathy** in rural populations of Croatia, Bosnia and Herzegovina, Yugoslavia, Bulgaria and Romania, where around 20,000 people were suspected to be either suffering from or exposed to this toxin. Due to its presence in these Balkan states of southeast Europe, this is also called **Balkan endemic nephropathy** (BEN). This is an unusual kidney disease found in middle-aged rural population and commonly prevalent in the vicinity of big rivers. In the Balkans, it is encountered almost exclusively along the tributary system of river Danube. Clinically, the disease starts without any acute episode and progresses slowly with appearance of non-specific symptoms and signs like fatigue, headache, loss of body weight, pale skin and typical signs of kidney impairment.

The importance of ochratoxin A has also increased after it was determined to be a major cause of disease in Denmark, called **mycotoxic porcine nephropathy (MPN)**. This disease has also been reported from USA and urinary tract tumors are associated with it. Experimental feeding on ochratoxin diets has produced adverse effects on a number of animal species.

6. Patulin

Patulin was first derived in 1943 from ***Penicillium patulum*** (now *P.griseofulvum*). It was initially thought to be an antiviral antibiotic, which could relieve symptoms of the common cold. However, subsequently it was realized to be a mycotoxin. Patulin is isolated from several fungal species belonging to genera *Penicillium, Aspergillus, Paecilomyces* and *Byssochlamys*. It is also produced by *Penicillium expansum*. It is the most common mycotoxin found in apples and apple-derived products such as juice, cider, compotes and other food intended for young children. It has also been detected in other fruits including cherries, blueberries, plums, bananas, strawberries and grapes. Exposure to patulin is associated with immunological, neurological and gastrointestinal outcomes. Assessment of health risks due to its consumption by humans has led many countries to regulate its quantity in food.

7. Trichothecenes

The trichothecenes is a group of cyclic compounds, often classified into four categories, of which type A trichothecenes, (including T-2 toxin, deoxynivalenol - DON or vomitoxin), nivalenol (NIV) and diacetoxyscirpenol (DAS), are the most significant substances. They were discovered in 1960s and 1970s and are mainly produced by *Fusarium* species. A fungal pathogen from gramineous plants especially wheat, *F.graminearum* (sexual stage: *Gibberella zeae*), is known to produce trichothecenes. Other fungal genera like *Trichoderma, Trichothecium, Myrothecium* and *Acremonium* may also produce these compounds. *F.sporotrichioides* produces primarily T-2 toxin and *F.graminearum* and *F.culmorum* produce deoxynivalenol, nivalenol and other related trichothecenes.

The effects of these toxins are weight loss, vomiting, feed refusal, dermatitis, diarrhea, hemorrhages and necrosis of epithelium of stomach and intestine and immunosuppression of lymphoid cells in thymus, bone marrow, spleen, testis and ovary. Vomitoxin is responsible for emesis and also refusal of feed in pigs when they eat maize infected with *F.graminearum* containing approximately 8 mg/kg of feed of trichothecenes.

The depression of immune system due to trichothecenes is probably the most significant effect in both humans as well as animals. They impair cellular immunity, increase skin graft rejection time and host's susceptibility to various opportunistic fungal infections such as candidiasis and cryptococcosis. T-2 toxin is highly toxic, inhibiting eukaryotic protein synthesis and inducing DNA fragmentation, a characteristic of apoptosis, possesses cytostatic potential to inhibit platelet aggregation, increase prothrombin time and cause cardiomyopathy. They do not appear to be carcinogenic but may synergize the effect of aflatoxins.

Alimentary toxic aleukia (ATA) is caused by cereals infected with *F.graminearum,* wherein there are necrotic lesions of oral cavity, esophagus and stomach with marked leukopenia. This was once common in 1930s and 1940s in the erstwhile USSR but has not been seen now for the last many decades. At the time of World War-II, circumstances necessitated storage of grains for human consumption in open fields during winter season thereby leading to contamination by trichothecenes. These toxins were also implicated as biological warfare agents, the putative incident of **Yellow rain** in Laos is one of the classical examples.

During June to September 1987, an outbreak of gastrointestinal disorder occurred in the Kashmir valley, especially in Srinagar and its surrounding areas. The outbreak was identified due to trichothecenes present in bread made from rain-damaged moldy wheat and was self-limiting. Human toxicosis by these toxins has also been reported from China, Japan and Korea.

8. Zearalenone

The zearalenone (ZEN/ZEA/ZON, previous name F-2) is a potent estrogenic toxin and it is produced by *Fusarium* and *Gibberella* species particularly, *F.graminearum* and its closely related species. This is found in wide variety of cereals like maize, barley, wheat grains and sorghum infected with this fungus and cause genital disorders in the farm animals like pigs, both in males as well as females. It has low order of acute toxicity in mice, rats and dogs. In female animal species it causes infertility, vulval edema, vaginal prolapse and mammary hypertrophy. In males there are feminization signs like atrophy of testes, decreased libido and enlargement of mammary glands. In human beings, this is reported to have possible involvement in cervical cancer and precocious pubertal changes in children. The recent evidence also suggests that it has potential to stimulate growth of human breast cancer cells.

In addition to the mycotoxins described above, there are other toxins like citrinin, nitropropionic acid, etc, which are produced by many genera of fungi and are known to affect man and animals. *Fusarium* species also produce emerging mycotoxins like beauvericin (BEA) enniatins (ENN), fusaproliferin (FUS) and moniliformin (MON) in addition to the most common ones described earlier such as trichothecenes, fumonisins and zearalenone.

II. MYCETISMUS

The basidiomycetous fungi (mushrooms) are recognised as important components of human diet with versatile medicinal properties. Some of them are very popular worldwide and are of commercial use for their nutritional and therapeutic properties. The edible mushroom is the most tempting food item with splendid flavor. The basidiomycetes have the potential to provide us with the next generation of antibiotics, to reduce environmental pollution and even produce fuel. A unique Taiwanese mushroom *i.e. Antrodia camphorata*, has been used as a traditional medicine for protection of diverse health-related conditions like cancer. Similarly another Asian fungus *i.e. Ganoderma lucidum* has anti-cancer, antiviral, hepato/cardioprotective and hypoglycemic effects. Hence mushrooms currently represent one of the great untapped resources of nature.

However, after ingestion some of these closely related fungi may produce dramatic symptoms of food poisoning called **mycetism/mycetismus** or simply **mushroom poisoning**. This is due to intentional consumption of fleshy fungi *i.e.* **mycophagy**, which is practiced by man since antiquity. There are more than five thousand known species of mushrooms, however, only about one hundred are known to be poisonous.

As such **fungus** is essentially a Latin word that means **mushroom**. The very origin of the word **mycology** came into existence from Greek word ***Mykes***, mushroom/fungus (basidiomycetes). The branch of biological sciences, which deals with study of fungi, is known as **Mycology** as described in the introductory chapter of this textbook (Myco + logy = Fungus + study; Gr. *Mykes*, mushroom/fungus, *logos*, discourse/study). The term ***toxicum*** refer to **toxin** making it to **mycotoxin**.

The most popular edible basidiomycete-"button mushroom" - ***Agaricus bisporus***, is one of the commonly grown species throughout the world. When immature it is found in either white or brown in color. The white variety one is known by number of synonyms *i.e.* "white mushroom", "gilled mushroom", "commercial mushroom" "table mushroom", "champignon mushroom" and "Paris mushroom". This is routinely eaten by millions of people worldwide daily as vegetable and/or diet. A collection of these **edible fungi** *i.e. Agaricus bisporus* is shown in **Figure 35.1**. During this very process of food intake, the poisonous ones are mistakenly eaten thereby leading to **mycetismus.**

Fig. 35.1. Popular button mushroom *(Agaricus bisporus)* eaten worldwide as routine vegetable and/or diet.

As such, there is no way to detect which mushroom are edible ones and which ones are poisonous except a careful identification by an experienced mycologist. The patients usually present in medical emergency services with this disease. The clinical manifestations of mushroom poisoning range from minor gastrointestinal disturbances, hallucinations, delirium and cramps to fatal consequences depending upon the underlying fungal species involved. Mushroom toxidromes are classified according to the toxins produced and the consequent clinical presentations. A brief account of medically significant fungi causing mycetismus that can cause distinctive diseases in man and animals is given in **Table 35.2**. These clinical entities are described below in **alphabetical** order:

1. Coprine Poisoning
2. Cyclopeptides
3. Ibotenic Acid, Muscimol, Muscazone and Muscarine Poisoning
4. Orellanus Syndrome

1. Coprine Poisoning

The mushroom, *Coprine atrementarius,* when taken as cream butter sauce, followed by alcohol intake, leads to **Coprinus syndrome**, an antabuse-like reaction. The symptoms include unpleasant, hot flushed sensation due to dilatation of blood vessels. There is also metallic taste, palpitation, chest pain, headache, vertigo, nausea, vomiting, weakness and confusion. The recovery of the patient is spontaneous,

Table 35.2. Medically significant Mycetismus, Fungal species, Source of Exposure and Clinical Conditions.

Mushroom Poisonings	*Causative Fungal Species*	*Source of Exposure*	*Clinical Conditions*
Coprine Poisoning	*Coprine atrementarius*	Cream Butter Sauce	Antabuse-like reaction
Cyclopeptides	*Amanita phalloides, A.verna, A.virosa*	Toadstools	Hepatocellular failure 'Green Death Cap'
Ibotenic Acid, Muscimol and Muscazone	*Amanita pantherina, A.muscaria*	Edible Mushroom	Abdominal pain, vomiting, diarrhea, etc.
Muscarine Poisoning	*Inocybe fastigiata, I.napipes*	Food Materials	Cholinergic effect
Orellanus Syndrome	*Cortinarius rubellus, C.orellanus, C.orellanosus, C.orellanoides,*	Lethal Webcap	Irreversible renal failure

approximately within five days after eating the mushrooms, if no more alcohol is consumed. The severity of symptoms is proportional to the amount of alcohol consumed by the deceased.

Coprine is metabolized into glutamic acid and 1-aminocyclopropanol, the biologically active substance that inhibits the enzyme acetaldehyde dehydrogenase. This enzyme is involved in the metabolism of alcohol. When it is blocked, acetaldehyde accumulates causing the adverse reactions of this poisoning.

2. Cyclopeptides

Cyclopeptide toxins represent family of nine cyclic octapeptides and produce cellular necrosis through inhibition of mRNA synthesis by blocking specific enzyme RNA polymerase II thereby interrupting protein synthesis and consequently causing death of human beings consuming such toxins. Ingestion of wild mushrooms of genus *Amanita phalloides* containing these cyclopeptides, may account for more than 90% fatalities worldwide, which is attributable to mushroom ingestion. It is also called **Green Death Cap**. *Amanita* toxin has two important groups of toxic peptides *i.e.* amatoxins and phallotoxins. Of these amatoxins is by far more potent poison of *'green death cap'*. Amatoxin contain α, β and γ-**amanitin,** whereas phallotoxins include **phalloidin** and **phalloin.** *Amanita phalloides* derives its poisonous properties mainly from amanitin that is intensely cytotoxic. The phallotoxins are cyclic heptapeptides and not absorbed through gastrointestinal tract hence these are active only if injected parenterally in experimental animals and their role in mycetismus after ingestion is doubtful. Liver is the main target organ for both the families of these toxins.

Phalloides syndrome is also a type of mushroom poisoning characterized by gastrointestinal symptoms such as abdominal pain, vomiting and cholera-like diarrhea followed by late-onset hepato-renal dysfunction. The toxicity in phalloides syndrome results mainly from α-amanitin, one of the amatoxins. In January 1997, such type of mushroom poisoning was observed in northern California, USA, where nine patients had required hospitalization.

Out of these two groups, amatoxins are the most potent. Amatoxin poisoning caused by *Amanita virosa* (destroying angel) has also been reported from South Korea. This type of intoxication is a true medical emergency, as an early diagnosis and prompt treatment is required for a successful management.

3. Ibotenic Acid, Muscimol, Muscazone and Muscarine Poisoning

Amanita poisoning is widespread as this mushroom is often consumed in the form of food. ***Amanita pantherina***, the false blusher and ***Amanita muscaria*** (fly agaric mushroom), the dry agaric, are two of the common poisonous basidiomycetous fungi, which are mistakenly eaten for edible mushrooms and owe their poisonous effect due to the presence of muscarine.

In sublethal doses, *Amanita muscaria* may produce abdominal pain, vomiting and severe diarrhea. In addition, there may be sweating, twitching, miosis, muscular in-coordination, visual distortions, delirium convulsions and coma. The symptoms appear rapidly, within few minutes to six hours of consumption. This is perhaps one of the oldest known hallucinogens. Recent studies suggest that this mushroom was the mysterious God-narcotic **Divine Soma**, the narcotic consumed by Gods of ancient India. Thousands years ago, Aryan conquerors, who swept across India, worshiped soma, drinking it during religious ceremonies. Many hymns in the Indian Rig-Veda are devoted to soma and describe the mushroom as well as its effects.

The primary active chemicals known in *Amanita muscaria* and *A.pantherina* are: muscimol, ibotenic acid, muscazone and muscarine. Muscimol is considered the principal psychoactive in these mushrooms. Muscimol's primary action is at GABA receptor sites as a potent GABA-A agonist. Ibotenic acid is also active orally but at doses 5-8 times higher than those of muscimol. The current view is that some ibotenic acid does cross the blood brain barrier but some is partially metabolized into muscimol and the rest excreted. Ibotenic acid, however, has been shown to be a potent neurotoxin when injected directly in the brains of mice and rats. It is structurally similar to glutamate and activates NMDA receptors. The other two chemicals *i.e.* muscarine and muscazone are only present in trace amounts in these mushrooms, well below their active levels in humans.

The muscarine is also a natural product found in certain other mushrooms of the genus *Inocybe* and *Clitocybe*. Among the *Inocybe* species, particularly *Inocybe fastigiata* and *I.napipes,* contain large quantities of muscarine. Muscarine is known to affect acetylcholine levels and acts at muscarinic receptors and its effects are due to its cholinergic properties. The major symptoms are perspiration, lacrimation and salivation, which are unique feature of this type of mushroom poisoning. There may be cramps, diarrhea, headache and blurred vision, sometimes followed by constriction of pupils, hypotension and slow pulse.

4. Orellanus Syndrome

Isolated renal injury in any of the mushroom poisoning is very rare. However, in some of the cases severe and irreversible renal failure may occur. **Orellanus syndrome** is caused by the ingestion of nephrotoxic toxin, **Orellanine**, produced by some lethal Webcap mushrooms species of genus ***Cortinarius*** *i.e. C.rubellus* - formerly *C.speciosissimus* (Deadly Webcap), *C.orellanus* (Fool's Webcap), *C.orellanosus* or *C.orellanoides*, which are found in Europe and North America. The nephrotoxicity by these toxins was first recognized in the 1950s when this mushroom poisoning was reported in Poland.

This is the classical example of organic renal damage related to consumption of mushrooms. It is characterized by progressive clinical phases with a predominant kidney involvement, finally requiring renal transplantation in a substantial number of patients. The damage in kidneys is often late during the course of illness and manifests as interstitial nephritis. The initial symptoms are usually delayed by 3-20 days due to long latency period from the time of ingestion until the onset of flu-like symptoms, headache and vomiting. Some patients suffer mild gastrointestinal discomfort in the latency period before developing signs of renal impairment due to severe interstitial nephritis, acute focal tubular damage and interstitial fibrosis, entailing irreversible renal failure. Its diagnosis is broadly clinical and as such no specific therapy is found to be effective in preventing as well as treating the kidney damage. There is no specific antidote and the mainstay of treatment is the prevention of secondary complications of kidney failure, adequate dialysis and in case of partial recovery, management of chronic renal insufficiency.

Amanita smithiana is a mushroom (North America), which also has nephrotoxic effects that can be easily mistaken for the edible and highly prized **Pine mushroom** *(Tricholoma magnivelare)*. Initially, it causes gastrointestinal symptoms, followed by acute reversible renal failure. The syndrome induced by *Amanita smithiana/ proxima* consists of acute tubulopathy that appears earlier but does not have the same poor prognosis as the orellanine-induced syndrome characterized by a delayed onset of severe and often irreversible renal failure. It has been described since 1992 in the US and Canada in association with *A.smithiana*; in France, Spain and Italy with *A.proxima* (Mediterranean area) and in Japan with *A.pseudoporphyria. Amanita smithiana* mushroom poisoning presents within 6 hours of ingestion with GI toxicity and leads to a delayed onset renal insufficiency over the first 1 to 4 days. The chemical structure of the *A.smithiana* toxin is unknown but it can be identified as a characteristic spot on thin-layer chromatography. Treatment includes dialysis and supportive care until renal functions of patient improve.

Detection of Mycotoxins

In the developed countries, stringent sorting and clean up procedures are regularly practiced to reduce aflatoxins to lower levels in foods with perceived risk. For peanuts, where fungal growth is usually accompanied by discoloration of the kernel, use of sophisticated color sorting equipments has been practiced. The diagnosis of mycotoxicosis is based on estimation of toxin by *in vitro* and *in vivo* testing. The mycotoxins can be detected in urine, feces, milk, bile, feeds and food-stuffs by various methods. The physiological methods like thin layer chromatography (TLC), gas liquid chromatography (GLC), high pressure liquid chromatography (HPLC) and spectrophotometry as well as ELISA are used for detection of mycotoxins. Now-a-days, HPLC-MS/MS analysis is used for quantification and confirmation, as HPLC alone may yield false positive results.

The biological assays are commonly performed in duckling, rat, rabbit and trout. The cytological assays are also done, which can be performed by producing cytotoxic effects in the tissue cultures. Primary fetal bovine kidney (PFBK) cells are used for studying effects of mycotoxins. The cytotoxic effects in the cultured cells can also be analyzed by colorimetric bioassays using MTT (Methylthiazyl tetrazolium). This dye is a yellow salt, which is reduced to dark purple formazan crystals by mitochondrial enzymes in living cells and it is now frequently used in cytotoxic studies of mycotoxins. Microbial assays are also performed involving certain bacteria belonging to genus, *Bacillus*. Newer techniques have been used which include patterns of isoenzyme electrophoresis, DNA-DNA homology, restriction fragment length polymorphism (RFLP), ubiquinone systems, etc.

A number of biosensors and imaging systems for detection of mycotoxins have also been developed. The biomarkers have been developed for assessing risk of exposures for aflatoxin B_1 and measured by ELISA with detection limit of 15.6 pmol aflatoxin B_1-N7-guanine.

The ELISA technique is simpler, rapid and sensitive to use than conventional analytical methods. Several commercial ELISA-based kits using monoclonal antibodies are also available for aflatoxin analysis. Similarly, fumonisin B_1 is measured using HPLC with a detection limit of 8 ng/ml of urine. Radioimmunoassay may also be used for detection of mycotoxins. A bio-wipe, faecal material collection kit, is also used to detect the presence of FB_1, which is considered as a rapid method to determine human exposure to FB_1.

Treatment

The treatment of patients suffering from mycotoxicosis and mycetismus is mostly symptomatic and supportive as specific antidotes are not available. The mainstay of treatment is vigorous intravenous fluid replacement and correction of electrolyte imbalance and coagulopathy. Hemodialysis and hemoperfusion may be effective in removing toxin if initiated within 24 hours of intake. Sometimes, the ailment may itself be self-limiting. The other methods like combined intensive hemodialysis, charcoal hemoperfusion, penicillin and intravenous silibinin may be utilized as used in the treatment of amatoxin poisonings. Silibinin is a derivative of milk thistle and protects liver by preventing the uptake of amatoxin by the hepatocytes. Once the amatoxins are eliminated from the body, the liver can quickly recover.

Gastric lavage is very useful in the treatment at an early stage of poisoning. The treatment of *Amanita* poisoning in previous reports was disappointing with mortality rates ranging between 30 to 90 percent. Some of the cases of treatment with plasma exchange showed better survival rates with mortality rates of 10 to 12 percent. Atropine is effective for symptomatic treatment of mycetismus. Since the para-sympathomimetic nervous system is involved in case of muscarine poisoning, atropine counters the cholinergic effects of the toxic mushrooms. The corticosteroids may also prove useful and such patients should be offered a trial.

Prevention and Control Measures

As both clinical entities may lead to fatal consequences, therefore, preventive measures should be taken at public health level. Since the mycotoxins cause substantial economic loss hence several countries have instituted regulatory guidelines controlling permissible aflatoxin levels in foodstuffs for dairy cattle. The aflatoxin exposure in human population has been assessed using either analysis of foods and total diet for aflatoxin or measurement of aflatoxin biomarkers in human biological fluids like urine and breast milk. High levels of aflatoxin have been reported in commodities, particularly in the regions of Africa, Southeast Asia and Southern China. The aquaculture industry is also now looking for such types of issues. The **Fantastic Fungi**, a film about mushrooms by Louie Schwartzberg is informative on this topic.

In the past there were no International Regulatory Guidelines for mycotoxins in food , however, the Food and Agriculture Organization (FAO) and World Health Organization (WHO) advised countries in developing their own national guidelines. There has been a global increase in the number of countries coming up with regulatory guidelines for mycotoxins in foods. As the available technology cannot prevent mycotoxins from getting produced thus it is necessary to monitor foods from contamination.

In India, under the Prevention of Food Adulteration (PFA) Act, tolerance limit for aflatoxins in all foodstuffs meant for human consumption were fixed as 30 μg/kg based on the recommendations of Protein Advisory Group of the United Nations System of 1967. However, now under the Food Safety and Standard Act, 2006 and regulations 2011 no article of food shall contain the crop contaminant in excess of quantity as specified under the law and tolerance limit for aflatoxins in food articles meant for human consumption has been re-fixed as mentioned earlier. Maximum acceptance level of contamination is set for food articles which are ready to eat or destined for further processing so as consumers' health is protected.

The Codex Alimentarius Commission has been recognized as an international standardizing body of 165 member countries administered jointly by FAO and WHO. It is committed to protect the health of consumers; instituting fair practices in food trade and to facilitate international trade in food, trade barriers and disputes. The Codex General Principles of Food Hygiene recommended Hazard Analysis Critical Control Point (HACCP)-based approach as the means to enhance food safety and has advised how to implement the principles.

Mushrooms are now being cultivated on commercial scale in India and the annual production is about 10,000 tons. The mushroom cultivation and processing units are being encouraged to implement the HACCP program to improve the safety and quality of mushrooms. The preventive and control measures related to mycotoxicosis and mycetismus may be in the following ways.

1. Prevention

Prevention using appropriate technology (a) preventing food and feed contamination at the source *e.g.* by reducing environmental pollution (b) applying appropriate technology control measures in food and feed production, manufacturing, processing, preparation, packing and transport *e.g.* mushrooms.

2. Control Measures

The control measures about mycotoxicosis and mycetismus are undertaken as the following headings: (a) Decontamination (b) Detoxification and (c) Legislation.

(a) Decontamination: A large number of approaches have been investigated for removal of mycotoxins from feeds. There have been attempts to study physical decontamination procedures that are simple, practical and efficient. Hand picking and density segregation with water or sodium chloride are found to be most practical approaches for groundnut, maize, wheat and rice reducing about 70-90% amount of the mycotoxins.

(b) Detoxification: The detoxification occurs with the conversion of aflatoxin B_1 to less active hydroxylated metabolites, such as aflatoxin M_1, formation of glutathione conjugate and glucuronides. There are many detoxification methods based on physical, chemical and microbiological basis. But they have inherent difficulty of practical application because of residual toxicity. The use of chemical treatments to detoxify mycotoxin is of limited success. Formaldehyde and ammonium hydroxide treatments reduced zearalenone levels in spiked corn grits but not in whole kernel corn. In the physical methods, use of heat in combination with pressure during processing resulted in substantial decrease in levels of mycotoxins in corn. Presently, the bio-technological methods particularly microbial detoxification of moldy grains has attained more significance.

In the human food and animal feed industry, it has become common practice now to add toxin-binding agents, which reverse the adverse effects of mycotoxins. The characteristics features of these agents are low effective inclusion rate, stability over a wide pH range, capacity to adsorb high concentration of mycotoxins, being non-toxic and environment-friendly. Since not all mycotoxins can be bound to such agents, the latest approach to mycotoxin control is deactivating them to non-toxic metabolites using enzymes (esterase, epoxidase) yeasts (*Trichosporon mycotoxinivorans*) or bacterial strains (*Eubacteria* BBSH-797).

(c) Legislation: With the latest notification of Food Safety and Standards Authority of India (FSSAI), the maximum permissible levels of mycotoxins in foods have been revised to protect the consumers' health. These levels are in close proximity to the international Codex Alimentarius standards adopted for international food trade. Substantial differences exist between the limits set by different countries that underscore the need for international harmonization of mycotoxin regulation.

Further Reading

1. Abrar M, Anjum FM, Butt MS, *et al.* Aflatoxins: Biosynthesis, occurrence, toxicity and remedies. *Crit Rev Food Sci Nutr.* 2013; **53:** 862-74.
2. Adhikari M, Negi B, Kaushik N, *et al.* T-2 mycotoxin: Toxicological effects and decontamination strategies. *Oncotarget.* 2017; **8:** 33933-52.
3. Afsah-Hejri L, Jinap S, Hajeb P, *et al.* A review on mycotoxins in food and feed: Malaysia case study. *Compr Rev Food Sci Food Saf.* 2013; **12:** 629-51.
4. Ahishali E, Boynuegri B, Ozpolat E, *et al.* Approach to mushroom intoxication and treatment: Can we decrease mortality? *Clin Res Hepatol Gastroenterol.* 2012; **36:** 139-45.
5. Al-Gabr HM, Zheng T, Yu X. Fungi contamination of drinking water. *Rev Environ Contam Toxicol.* 2014; **228:** 121-39.
6. Ali S, Malik A, Shahid M, *et al.* Pulmonary aspergillosis and aflatoxins in chronic lung diseases. *Mycopathologia.* 2013; **176:** 287-94.
7. Alonso VA, Pereyra CM, Keller LA, *et al.* Fungi and mycotoxins in silage: An overview. *J Appl Microbiol.* 2013; **115:** 637-43.
8. Alm T. The witch trials of Finnmark, northern Norway, during the 17th century: Evidence for ergotism as a contributing factor. *Economic Botany.* 2003; **57:** 403-16.
9. Ampere A, Delhaes L, Soots J, *et al.* Hypersensitivity pneumonitis induced by Shiitake mushroom spores. *Med Mycol.* 2012; **50:** 654-7.
10. Anater A, Manyes L, Meca G, *et al.* Mycotoxins and their consequences in aquaculture: A review. *Aquaculture.* 2016; **451:** 1-10.
11. Antonissen G, Martel A, Pasmans F, *et al.* The impact of *Fusarium* mycotoxins on human and animal host susceptibility to infectious diseases. *Toxins (Basel).* 2014; **6:** 430-52.
12. Apperley S, Kroeger P, Kirchmair M, *et al.* Laboratory confirmation of *Amanita smithiana* mushroom poisoning. *Clin Toxicol (Phila).* 2013; **51:** 249-51.
13. Arroyo-Manzanares N, Huertas-Perez JF, Gamiz-Gracia L, *et al.* Simple and efficient methodology to determine mycotoxins in cereal syrups. *Food Chem.* 2015; **177:** 274-9.

14. Ashiq S, Hussain M, Ahmad B. Natural occurrence of mycotoxins in medicinal plants: A review. *Fungal Genet Biol.* 2014; **66:** 1-10.
15. Ashiq S. Natural occurrence of mycotoxins in food and feed: Pakistan perspective. *Compr Rev Food Sci Food Saf.* 2015; **14:** 159-75.
16. Awad WA, Ghareeb K, Bohm J, *et al.* Decontamination and detoxification strategies for the *Fusarium* mycotoxin deoxynivalenol in animal feed and the effectiveness of microbial biodegradation. *Food Addit Contam. Part A. Chem Anal Control Expo Risk Assess.* 2010; **27:** 510-20.
17. Baniasad N, Oghabian Z, Mehrpour O. Hepatotoxicity due to mushroom poisoning: A case report. *Int J Med Tox Foren Med.* 2014; **4:** 68-73.
18. Belser-Ehrlich S, Harper A, Hussey J, *et al.* Human and cattle ergotism since 1900: Symptoms, outbreaks and regulations. *Toxicol Ind Health.* 2013; **29:** 307-16.
19. Bennett JW, Klich M. Mycotoxins. *Clin Microbiol Rev.* 2003; **16:** 497-516.
20. Bergis D, Friedrich-Rust M, Zeuzem S, *et al.* Treatment of *Amanita phalloides* intoxication by fractionated plasma separation and adsorption. *J Gastrointestin Liver Dis.* 2012; **21:** 171-6.
21. Berthiller F, Crews C, Dall'Asta C, *et al.* Masked mycotoxins: A review. *Mol Nutr Food Res.* 2013; **57:** 165-86.
22. Bhat R, Rai RV, Karim AA. Mycotoxins in food and feed: Present status and future concerns. *Compr Rev Food Sci Food Saf.* 2010; **9:** 57-81.
23. Bhat RV, Beedu SR, Ramakrishna Y, *et al.* Outbreak of trichothecene mycotoxicosis associated with consumption of mould-damaged wheat products in Kashmir Valley, India. *Lancet.* 1989; **i:** 35-7.
24. Bhat RV, Krishnamachari KAVR. Follow-up study of aflatoxic hepatitis in parts of western India. *Indian J Med Res.* 1977; **66:** 55-8.
25. Bhat RV, Moy GG. Monitoring and assessment of dietary exposure to chemical contaminants. *World Health Stat Q.* 1997; **50:** 132-49.
26. Bhat RV, Shetty PH, Amruth RP, *et al.* A food-borne disease outbreak due to the consumption of moldy sorghum and maize containing fumonisin mycotoxins. *J Toxicol Clin Toxicol.* 1997; **35:** 249-55.
27. Bhat RV. Aflatoxin and liver injury. *Trop Gastroenterol.* 1996; **17:** 153-60.
28. Bhatnagar D, Yu J, Ehrlich KC. Toxins of filamentous fungi. *Chem Immunol.* 2002; **81:** 167-206.
29. Bloom E, Nyman E, Must A, *et al.* Molds and mycotoxins in indoor environments - A survey in water-damaged buildings. *J Occup Environ Hyg.* 2009; **6:** 671-8.
30. Boevre MD, Graniczkowska K, Saeger SD. Metabolism of modified mycotoxins studied through *in vitro* and *in vivo* models: An overview. *Toxicol Lett.* 2015; **233:** 24-8.
31. Brewer JH, Thrasher JD, Hooper D. Chronic illness associated with mold and mycotoxins: Is naso-sinus fungal biofilm the culprit? *Toxins (Basel).* 2013; **6:** 66-80.
32. Broekaert N, Devreese M, De Baere S, *et al.* Modified *Fusarium* mycotoxins unmasked: From occurrence in cereals to animal and human excretion. *Food Chem Toxicol.* 2015; **80:** 17-31.
33. Brown DW, Baker SE. Mycotoxins: A fungal genomics perspective. *Methods Mol Biol.* 2017; **1542:** 367-79.
34. Buzina W, Lass-Florl C, Kropshofer G, *et al.* The polypore mushroom *Irpex lacteus:* A new causative agent of fungal infections. *J Clin Microbiol.* 2005; **43:** 2009-11.
35. Cai Q, Cui YY, Yang ZL. Lethal *Amanita* species in China. *Mycologia.* 2016; **108:** 993-1009.
36. Campbell AW. Molds and mycotoxins: A brief review. *Altern Ther Health Med.* 2016; **22:** 8-11.
37. Chan CK, Lam HC, Chiu SW, *et al.* Mushroom poisoning in Hong Kong: A ten-year review. *Hong Kong Med J.* 2016; **22:** 124-30.
38. Chaudhary S, Chaurasia RK, Patel S, *et al.* Clinical profile and outcome of patients presenting with mushroom poisoning in a tertiary care center of eastern Nepal. *J Nepal Med Assoc.* 2013; **52:** 543-8.
39. Chauhan R, Singh J, Sachdev T, *et al.* Recent advances in mycotoxins detection. *Biosens Bioelectron.* 2016; **81:** 532-45.
40. Cheli F, Campagnoli A, Dell'Orto V. Fungal populations and mycotoxins in silages: From occurrence to analysis. *Anim Feed Sci Technol.* 2013; **183:** 1-16.
41. Chen WC, Kassi M, Saeed U, *et al.* A rare case of amatoxin poisoning in the state of Texas. *Case Rep Gastroenterol.* 2012; **6:** 350-7.
42. Chilaka CA, De Boevre M, Atanda OO, *et al.* The status of *Fusarium* mycotoxins in Sub-Saharan Africa: A review of emerging trends and post-harvest mitigation strategies towards food control. *Toxins (Basel).* 2017; 9. pii: e19.
43. Chulze SN. Strategies to reduce mycotoxin levels in maize during storage: A review. *Food Addit Contam Part A Chem Anal Control Expo Risk Assess.* 2010; **27:** 651-7.
44. Coilly A. Keep watch for the appearing mushroom! *Clin Res Hepatol Gastroenterol.* 2012; **36:** 95-6.
45. Comelli I, Lippi G, De Blasio A, *et al.* Accidental mushroom poisoning mimicking stroke: A case report and literature review. *Acta Biomed.* 2014; **84:** 229-33.
46. Copetti MV, Iamanaka BT, Pitt JI, *et al.* Fungi and mycotoxins in cocoa: From farm to chocolate. *Int J Food Microbiol.* 2014; **178:** 13-20.
47. Crews C. Analysis of Ergot alkaloids. *Toxins (Basel).* 2015; **7:** 2024-50.
48. de Mattos-Shipley KM, Ford KL, Alberti F, *et al.* The good, the bad and the tasty: The many roles of mushrooms. *Stud Mycol.* 2016; **85:** 125-57.
49. Deutsch CJ, Swallow D. Deliberate ingestion of "magic" mushrooms may also cause renal failure. *BMJ.* 2012; **345:** e6388.
50. Dhatt PS, Parida NK, Das Chaudhury PK, *et al.* Aflatoxins and Indian childhood cirrhosis. *Indian J Pediatr.* 1982; **19:** 407-8.
51. Diaz JH. Evolving global epidemiology, syndromic classification, general management and prevention of unknown mushroom poisonings. *Crit Care Med.* 2005; **33:** 419-26.

52. Diaz JH. Syndromic diagnosis and management of confirmed mushroom poisonings. *Crit Care Med.* 2005; **33:** 427-36.
53. Dinis-Oliveira RJ, Soares M, Rocha-Pereira C, *et al.* Human and experimental toxicology of orellanine. *Hum Exp Toxicol.* 2016; **35:** 1016-29.
54. Duarte SC, Lino CM, Pena A. Mycotoxin food and feed regulation and the specific case of ochratoxin A: A review of the worldwide status. *Food Addit Contam. Part A. Chem Anal Control Expo Risk Assess.* 2010; **27:** 1440-50.
55. Dutta TK, Das P. Isolation of aflatoxigenic strains of *Aspergillus* and detection of aflatoxin B_1 from feeds in India. *Mycopathologia.* 2001; **151:** 29-33.
56. Erden A, Esmeray K, Karagoz H, *et al.* Acute liver failure caused by mushroom poisoning: A case report and review of the literature. *Int Med Case Rep J.* 2013; **6:** 85-90.
57. Escriva L, Font G, Manyes L. *In vivo* toxicity studies of *Fusarium* mycotoxins in the last decade: A review. *Food Chem Toxicol.* 2015; **78:** 185-206.
58. Esposito P, La Porta E, Calatroni M, *et al.* Renal involvement in mushroom poisoning: The case of Orellanus syndrome. *Hemodial Int.* 2015; **19:** E1-5.
59. Evans N, Hamilton A, Bello-Villalba MJ, *et al.* Irreversible renal damage from accidental mushroom poisoning. *BMJ.* 2012; **345:** e5262.
60. Frank H, Zilker T, Kirchmair M, *et al.* Acute renal failure by ingestion of *Cortinarius* species confounded with psychoactive mushrooms: A case series and literature survey. *Clin Nephrol.* 2009; **71:** 557-62.
61. Garcia de la Fuente I, McLin VA, Rimensberger PC, *et al.* *Amanita* poisoning and liver transplantation: Do we have the right decision criteria? *J Pediatr Gastroenterol Nutr.* 2011; **53:** 459-62.
62. Garcia J, Costa VM, Carvalho A, *et al.* *Amanita phalloides* poisoning: Mechanisms of toxicity and treatment. *Food Chem Toxicol.* 2015; **86:** 41-55.
63. Garcia J, Costa VM, Carvalho AT, *et al.* A breakthrough on *Amanita phalloides* poisoning: An effective antidotal effect by polymyxin B. *Arch Toxicol.* 2015; **89:** 2305-23.
64. Gatselis NK, Liamis G, Makaritsis KP, *et al.* Metabolic acidosis during treatment of mushroom poisoning: A diagnostic pitfall. *Intern Med.* 2012; **51:** 1077-80.
65. Geethangili M, Tzeng YM. Review of pharmacological effects of *Antrodia camphorata* and its bioactive compounds. *Evid Based Complement Alternat Med.* 2011; 212641. PMID: 19687189.
66. George P, Hegde N. Muscarinic toxicity among family members after consumption of mushrooms. *Toxicol Int.* 2013; **20:** 113-5.
67. Grabhorn E, Nielsen D, Hillebrand G, *et al.* Successful outcome of severe *Amanita phalloides* poisoning in children. *Pediatr Transplant.* 2013; **17:** 550-5.
68. Graeme KA. Mycetism: A review of the recent literature. *J Med Toxicol.* 2014; **10:** 173-89.
69. Grenier B, Oswald IP. Mycotoxin co-contamination of food and feed: Meta-analysis of publications describing toxicological interactions, *World Mycotoxin J.* 2011; **4:** 285-313.
70. Gruber-Dorninger C, Novak B, Nagl V, *et al.* Emerging mycotoxins: Beyond traditionally determined food contaminants. *J Agric Food Chem.* 2017; doi: 10.1021/acs.jafc.6b03413. PMID: 27599910.
71. Hof H. Mycotoxins: Pathogenicity factors or virulence factors? *Mycoses.* 2008; **51:** 93-4.
72. Hori K, Fukui H, Fujimori T. Hemorrhagic enteritis due to poisonous mushroom. *Int J Surg Pathol.* 2008; **16:** 62.
73. Hymery N, Vasseur V, Coton M, *et al.* Filamentous fungi and mycotoxins in cheese: A review. *Compr Rev Food Sci Food Saf.* 2014; **13:** 437-56.
74. Jan MA, Siddiqui TS, Ahmed N, *et al.* Mushroom poisoning in children: Clinical presentation and outcome. *J Ayub Med Coll Abbottabad.* 2008; **20:** 99-101.
75. Jang J, Kim CH, Yoo JJ, *et al.* An elderly man with fatal respiratory failure after eating a poisonous mushroom *Podostroma cornu-damae. Tuberc Respir Dis (Seoul).* 2013; **75:** 264-8.
76. Jard G, Liboz T, Mathieu F, *et al.* Review of mycotoxin reduction in food and feed: From prevention in the field to detoxification by adsorption or transformation. *Food Addit Contam. Part A. Chem Anal Control Expo Risk Assess.* 2011; **28:** 1590-609.
77. Jindal N, Mahipal SK, Rottinghaus GE. Occurrence of fumonisin B_1 in maize and poultry feeds in Haryana, India. *Mycopathologia.* 1999; **148:** 37-40.
78. Jo WS, Hossain MA, Park SC. Toxicological profiles of poisonous, edible and medicinal mushrooms. *Mycobiology.* 2014; **42:** 215-20.
79. Joshi A, Awale P, Shrestha A, *et al.* Acute mushroom poisoning: A report of 41 cases. *J Nepal Med Assoc.* 2007; **46:** 7-12.
80. Judge BS, Ammirati JF, Lincoff GH, *et al.* Ingestion of a newly described North American mushroom species from Michigan resulting in chronic renal failure: *Cortinarius orellanosus. Clin Toxicol (Phila).* 2010; **48:** 545-9.
81. Kew MC. Aflatoxins as a cause of hepatocellular carcinoma. *J Gastrointestin Liver Dis.* 2013; **22:** 305-10.
82. Kim HN, Do HH, Seo JS, *et al.* Two cases of incidental *Podostroma cornu-damae* poisoning. *Clin Exp Emerg Med.* 2016; **3:** 186-9.
83. Kirchmair M, Carrilho P, Pfab R, *et al.* *Amanita* poisonings resulting in acute, reversible renal failure: New cases, new toxic *Amanita* mushrooms. *Nephrol Dial Transplant.* 2012; **27:** 1380-6.
84. Klich MA, Tang S, Denning DW. Aflatoxin and ochratoxin production by *Aspergillus* species under *ex-vivo* conditions. *Mycopathologia.* 2009; **168:** 185-91.
85. Kowalczyk M, Sekuła A, Mleczko P, *et al.* Practical aspects of genetic identification of hallucinogenic and other poisonous mushrooms for clinical and forensic purposes. *Croat Med J.* 2015; **56:** 32-40.
86. Krishnamachari KAVR, Bhat RV, Nagarajan V, *et al.* Hepatitis due to aflatoxicosis: An outbreak in western India. *Lancet.* 1975; **1:** 1061-3.
87. Krishnamachari KAVR, Bhat RV, Nagarajan V, *et al.* Investigations into an outbreak of hepatitis in parts of western India. *Indian J Med Res.* 1975; **63:** 1036-48.

88. Krishnamachari KAVR, Bhat RV. Poisoning of ergoty *bajra* (pearl millet) in man. *Indian J Med Res.* 1976; **64:** 1624-8.
89. Kumar P, Mahato DK, Kamle M, *et al.* Aflatoxins: A global concern for food safety, human health and their management. *Front Microbiol.* 2017; **7:** 2170. PMID: 28144235.
90. Lalitha Rao B, Husain A. Presence of cyclopiazonic acid in kodo millet *(Paspalum scrobiculatum)* causing 'kodua poisoning' in man and its production by associated fungi. *Mycopathologia.* 1985; **89:** 177-80.
91. Lawton LD, Bhraonain SN. Accidental poisoning by death cap mushrooms: Be careful what you eat. *Wilderness Environ Med.* 2013; **24:** 168-70.
92. Lee HJ, Ryu D. Advances in mycotoxin research: Public health perspectives. *J Food Sci.* 2015; **80:** T2970-83.
93. Levine M, Ruha AM, Graeme K, *et al.* Toxicology in the ICU: Part 3: Natural toxins. *Chest.* 2011; **140:** 1357-70.
94. Lieberman SM, Jacobs JB, Lebowitz RA, *et al.* Measurement of mycotoxins in patients with chronic rhinosinusitis. *Otolaryngol Head Neck Surg.* 2011; **145:** 327-9.
95. Lima AD, Costa Fortes R, Carvalho Garbi Novaes MR, *et al.* Poisonous mushrooms: A review of the most common intoxications. *Nutr Hosp.* 2012; **27:** 402-8.
96. Liu Y. Perspectives on global prevention and control of mycotoxins. *Food Addit Contam. Part A. Chem Anal Control Expo Risk Assess.* 2015; **32:** 439.
97. Liu Y, Wu F. Global burden of aflatoxin-induced hepatocellular carcinoma: A risk assessment. *Environ Health Perspect.* 2010; **118:** 818-24.
98. Madhok M, Scalzo AJ, Blume CM, *et al. Amanita bisporigera* ingestion: Mistaken identity, dose-related toxicity and improvement despite severe hepatotoxicity. *Pediatr Emerg Care.* 2006; **22:** 177-80.
99. Magnussen A, Parsi MA. Aflatoxins, hepatocellular carcinoma and public health. *World J Gastroenterol.* 2013; **19:** 1508-12.
100. Malik A, Ali S, Shahid M, *et al.* Occupational exposure to *Aspergillus* and aflatoxins among food-grain workers in India. *Int J Occup Environ Health.* 2014; **20:** 189-93.
101. Marin S, Ramos AJ, Cano-Sancho G, *et al.* Mycotoxins: Occurrence, toxicology and exposure assessment. *Food Chem Toxicol.* 2013; **60:** 218-37.
102. McCormick SP. Microbial detoxification of mycotoxins. *J Chem Ecol.* 2013; **39:** 907-18.
103. Meggs WJ. Epidemics of mold poisoning past and present. *Toxicol Ind Health.* 2009; **25:** 571-6.
104. Mendonca CN, Silva PM, Avelleira JC, *et al.* Shiitake dermatitis. *An Bras Dermatol.* 2015; **90:** 276-8.
105. Merlet A, Dauchy FA, Dupon M. Hyperprocalcitonemia due to mushroom poisoning. *Clin Infect Dis.* 2012; **54:** 307-8.
106. Michelot D, Melendez-Howell LM. *Amanita muscaria*: Chemistry, biology, toxicology and ethnomycology. *Mycol Res.* 2003; **107:** 131-46.
107. Mollison LC. Mushroom poisoning: A personal vignette. *Med J Aust.* 2011; **195:** 720-1.
108. Moretti A, Logrieco AF, Susca A. Mycotoxins: An underhand food problem. *Methods Mol Biol.* 2017; **1542:** 3-12.
109. Mudili V, Siddaih CN, Nagesh M, *et al.* Mould incidence and mycotoxin contamination in freshly harvested maize kernels originated from India. *J Sci Food Agric.* 2014; **94:** 2674-83.
110. Mullins ME, Schwarz ES, Dribben WH, *et al.* Silibinin: Where is the ethical conundrum? *Chest.* 2015; **147:** e105.
111. Munkvold GP. *Fusarium* species and their associated mycotoxins. *Methods Mol Biol.* 2017; **1542:** 51-106.
112. Nazari F, Sulyok M, Kobarfard F, *et al.* Evaluation of emerging *Fusarium* mycotoxins beauvericin, enniatins, fusaproliferin and moniliformin in domestic rice in Iran. *Iran J Pharm Res.* 2015; **14:** 505-12.
113. Nesic K, Ivanovic S, Nesic V. Fusarial toxins: Secondary metabolites of *Fusarium* fungi. *Rev Environ Contam Toxicol.* 2014; **228:** 101-20.
114. Nici A, Kim S. *Amanita bisporigera*-induced hepatic failure: A fatal case of mushroom ingestion. *Case Reports Hepatol.* 2011; 936867. PMID: 25954548.
115. No authors listed. Fungi producing significant mycotoxins. *IARC Sci Publ.* 2012; **158:** 1-30.
116. Nogueira L, Foerster C, Groopman J, *et al.* Association of aflatoxin with gallbladder cancer in Chile. *JAMA.* 2015; **313:** 2075-7.
117. Oldenburg E, Hoppner F, Ellner F, *et al. Fusarium* diseases of maize associated with mycotoxin contamination of agricultural products intended to be used for food and feed. *Mycotoxin Res.* 2017; doi: 10.1007/s12550-017-0277-y. PMID: 28455556.
118. Paterson RR, Lima N. Thermophilic fungi to dominate aflatoxigenic/mycotoxigenic fungi on food under global warming. *Int J Environ Res Public Health.* 2017; **14.** pii: e199.
119. Patowary BS. Mushroom Poisoning - An overview. *J Col Med Sci Nepal.* 2010; **6:** 56-61.
120. Pfliegler WP, Pusztahelyi T, Pocsi I. Mycotoxins - Prevention and decontamination by yeasts. *J Basic Microbiol.* 2015; **55:** 805-18.
121. Phoku JZ, Dutton MF, Barnard TG, *et al.* Use of a bio-wipe kit to detect fumonisin B_1 in faecal materials. *Food Addit Contam. Part A. Chem Anal Control Expo Risk Assess.* 2014; **31:** 1760-8.
122. Piacentini KC, Rocha LO, Fontes LC, *et al.* Mycotoxin analysis of industrial beers from Brazil: The influence of fumonisin B_1 and deoxynivalenol in beer quality. *Food Chem.* 2017; **218:** 64-9.
123. Pitt JI. Toxigenic fungi and mycotoxins. *Br Med Bull.* 2000; **56:** 184-92.
124. Pitt JI, Miller JD. A concise history of mycotoxin research. *J Agric Food Chem.* 2017; doi: 10.1021/acs.jafc.6b04494. PMID: 27960261.
125. Rahmani F, Ebrahimi Bakhtavar H, Ghavidel A. Acute hepatorenal failure in a patient following consumption of mushrooms: A case report. *Iran Red Crescent Med J.* 2015; **17:** e17973.
126. Rai M, Jogee PS, Ingle AP. Emerging nanotechnology for detection of mycotoxins in food and feed. *Int J Food Sci Nutr.* 2015; **66:** 363-70.

127. Raiola A, Tenore GC, Manyes L, *et al.* Risk analysis of main mycotoxins occurring in food for children: An overview. *Food Chem Toxicol.* 2015; **84:** 169-80.
128. Ramakrishna Y, Bhat RV, Ravindranath V. Production of deoxynivalenol by *Fusarium* isolates from samples of wheat associated with a human mycotoxicosis outbreak and from sorghum cultivars. *Appl Environ Microbiol.* 1989; **55:** 2619-20.
129. Rashmi R, Ramana MV, Shylaja R, *et al.* Evaluation of a multiplex PCR assay for concurrent detection of four major mycotoxigenic fungi from foods. *J Appl Microbiol.* 2013; **114:** 819-27.
130. Reddy KR, Farhana NI, Salleh B, *et al.* Microbiological control of mycotoxins: Present status and future concerns. *Current Research, Technology and Education Topics in Applied Microbiology and Microbial Biotechnology.* 2010; **2:** 1078-86.
131. Reddy KR, Reddy CS, Muralidharan K. Detection of *Aspergillus* spp. and aflatoxin B_1 in rice in India. *Food Microbiol.* 2009; **26:** 27-31.
132. Reddy UM, Rani PC. Effect of processing on detoxification of aflatoxins in maize. *Indian J Nut Diet.* 2000; **37:** 59-63.
133. Rippon JW. Mycetismus: 'mushroom poisoning'. *Mycopathologia.* 1992; **118:** 1-2.
134. Rodrigues I, Naehrer K. Prevalence of mycotoxins in feedstuffs and feed surveyed worldwide in 2009 and 2010. *Phytopathol Mediterr.* 2012; **51:** 175-92.
135. Rosenthal P. Auxiliary liver transplantation for *Amanita phalloides* poisoning. *J Pediatr Gastroenterol Nutr.* 2012; **54:** 438.
136. Samson RA. Mycotoxins: a mycologist's perspective. *J Med Vet Mycol.* 1992; **30:** S9-18.
137. Santi L, Maggioli C, Mastroroberto M, *et al.* Acute liver failure caused by *Amanita phalloides* poisoning. *Int J Hepatol.* 2012; 487480.
138. Sarma UP, Bhetaria PJ, Devi P, *et al.* Aflatoxins: Implications on health. *Indian J Clin Biochem.* 2017; **32:** 124-33.
139. Schenk-Jaeger KM, Rauber-Luthy C, Bodmer M, *et al.* Mushroom poisoning: A study on circumstances of exposure and patterns of toxicity. *Eur J Intern Med.* 2012; **23:** e85-91.
140. Shao D, Tang S, Healy RA, *et al.* A novel orellanine containing mushroom *Cortinarius armillatus. Toxicon.* 2016; **114:** 65-74.
141. Shotar AM, Alzyoud SA, Samara O, *et al.* Mushroom poisoning: A case report from Jordan. *Pak J Biol Sci.* 2012; **15:** 208-11.
142. Sinha KK. Testing methods for aflatoxins in foods. *Food Nut Bull.* 2000; **21:** 458-64.
143. Siruguri V, Kumar PU, Raghu P, *et al.* Aflatoxin contamination in stored rice variety PAU 201 collected from Punjab, India. *Indian J Med Res.* 2012; **136:** 89-97.
144. Smith MR, Davis RL. Mycetismus: A review. *Gastroenterol Rep (Oxf).* 2016; **4:** 107-12.
145. Stein CM, Wu PE, Scott JA, *et al.* Fulminant hepatic failure following ingestion of wild mushrooms. *CMAJ.* 2015; **187:** 822-4.
146. Stiborova M, Arlt VM, Schmeiser HH. Balkan endemic nephropathy: An update on its aetiology. *Arch Toxicol.* 2016; **90:** 2595-2615.
147. Stone R. Epidemiology. Will a midsummer's nightmare return? *Science.* 2010; **329:** 132-4.
148. Stone R. Toxicology. Heart-stopping revelation about how Chinese mushroom kills. *Science.* 2012; **335:** 1293.
149. Streit E, Schatzmayr G, Tassis P, *et al.* Current situation of mycotoxin contamination and co-occurrence in animal feed - Focus on Europe. *Toxins (Basel).* 2012; **4:** 788-809.
150. Sun XD, Su P, Shan H. Mycotoxin contamination of rice in China. *J Food Sci.* 2017; **82:** 573-84.
151. Syed VA, Lakshmi R, Karki P, *et al.* Scenario of mushroom poisoning in a tertiary care hospital. *J Nepal Med Assoc.* 2009; **48:** 296-300.
152. Tandon BN, Krishnamurthy L, Koshy A, *et al.* Study of an epidemic of jaundice, presumably due to toxic hepatitis, in northwest India. *Gasteroenterology.* 1977; **72:** 488-94.
153. Tandon HD. Mycotoxins: A review with experience in India. *Prog Clin Med.* 1978; **20:** 304-26.
154. Tomkova J, Ondra P, Valka I. Simultaneous determination of mushroom toxins α-amanitin, β-amanitin and muscarine in human urine by solid-phase extraction and ultra-high-performance liquid chromatography coupled with ultra-high-resolution TOF mass spectrometry. *Forensic Sci Int.* 2015; **251:** 209-13.
155. Tonon KM, Reiter MGR, Scussel VM. Mycotoxins levels in human milk: A menace to infants and children health. *Curr Food Sci.* 2013; **9:** 33-42.
156. Trabulus S, Altiparmak MR. Clinical features and outcome of patients with amatoxin-containing mushroom poisoning. *Clin Toxicol (Phila).* 2011; **49:** 303-10.
157. Tuddenham EG. "Magic" mushrooms don't cause renal failure. *BMJ.* 2012; **345:** e6893.
158. Tulpule PG, Bhat RV. Food toxins and their implication in human health. *Indian J Med Res.* 1978; **68:** S99-108.
159. Umesha S, Manukumar HM, Chandrasekhar B, *et al.* Aflatoxins and food pathogens: Impact of biologically active aflatoxins and their control strategies. *J Sci Food Agric.* 2017; **97:** 1698-707.
160. Urusov AE, Zherdev AV, Petrakova AV, *et al.* Rapid multiple immunoenzyme assay of mycotoxins. *Toxins (Basel).* 2015; **7:** 238-54.
161. Vanooteghem S, Arts J, Decock S, *et al.* Four patients with *Amanita phalloides* poisoning. *Acta Gastroenterol Belg.* 2014; **77:** 353-6.
162. Varma A, Gaur KJ, Bhatia P. Mushrooms and poisoning. *J Indian Med Assoc.* 2011; **109:** 826-8.
163. Varvenne D, Retornaz K, Metge P, *et al.* Amatoxin-containing mushroom *(Lepiota brunneoincarnata)* familial poisoning. *Pediatr Emerg Care.* 2015; **31:** 277-8.
164. Vasanthi S, Bhat RV. Mycotoxins in foods: Occurrence, health and economic significance and food control measures. *Indian J Med Res.* 1998; **108:** 212-4.
165. Vekiru E, Hametner C, Mitterbauer R, *et al.* Cleavage of zearalenone by *Trichosporon mycotoxinivorans* to a novel nonestrogenic metabolite. *Appl Environ Microbiol.* 2010; **76:** 2353-9.

166. Vendramin A, Brvar M. *Amanita muscaria* and *Amanita pantherina* poisoning: Two syndromes. *Toxicon.* 2014; **90:** 269-72.
167. Venkataramana M, Rashmi R, Uppalapati SR, *et al.* Development of sandwich dot-ELISA for specific detection of Ochratoxin A and its application on to contaminated cereal grains originating from India. *Front Microbiol.* 2015; **6:** 511.
168. Verma N, Bhalla A, Kumar S, *et al.* Wild mushroom poisoning in north India: Case series with review of literature. *J Clin Exp Hepatol.* 2014; **4:** 361-5.
169. Wagacha JM, Muthomi JW. Mycotoxin problem in Africa: Current status, implications to food safety and health and possible management strategies. *Int J Food Microbiol.* 2008; **124:** 1-12.
170. Wagner C. Critical practicalities in sampling for mycotoxins in feed. *J AOAC Int.* 2015; **98:** 301-8.
171. Wambacq E, Vanhoutte I, Audenaert K, *et al.* Occurrence, prevention and remediation of toxigenic fungi and mycotoxins in silage: A review. *J Sci Food Agric.* 2016; **96:** 2284-302.
172. Ward J, Kapadia K, Brush E, *et al.* Amatoxin poisoning: Case reports and review of current therapies. *J Emerg Med.* 2013; **44:** 116-21.
173. Watson SA, Mirocha CJ, Hayes AW. Analysis for trichothecenes in samples from southeast Asia associated with 'yellow rain'. *Fund Appl Toxicol.* 1984; **4:** 700-17.
174. West PL, Lindgren J, Horowitz BZ. *Amanita smithiana* mushroom ingestion: A case of delayed renal failure and literature review. *J Med Toxicol.* 2009; **5:** 32-8.
175. White M. Disease facts: Mycotoxicosis. *Livestock.* 2012; **17:** 42.
176. Woolf A. Witchcraft or mycotoxin? The Salem witch trials. *J Toxicol Clin Toxicol.* 2000; **38:** 457-60.
177. Xu L, Zhang Z, Zhang Q, *et al.* Mycotoxin determination in foods using advanced sensors based on antibodies or aptamers. *Toxins (Basel).* 2016; **8:** pii: e239.
178. Yildiz BD, Abbasoglu O, Saglam A, *et al.* Urgent liver transplantation for *Amanita phalloides* poisoning. *Pediatr Transplant.* 2008; **12:** 105-8.
179. Yilmaz I, Kaya E, Sinirlioglu ZA, *et al.* Clinical importance of toxin concentration in *Amanita verna* mushroom. *Toxicon.* 2014; **87:** 68-75.

CHAPTER 36

Mycology Laboratory Contaminants

In a diagnostic mycology laboratory sometimes it becomes very difficult to draw a line between pathogenic and contaminating fungus, as many agents may be clinically significant in different circumstances. An isolate without corroborating direct examination findings is usually taken as either insignificant or a contaminant unless it is a true fungal pathogen. Thus, it becomes imperative to demonstrate the causative fungus either in the KOH wet mount or on histopathology to make out the clinical significance of that particular isolate on culture, otherwise the very same strain is taken as a laboratory contaminant.

On the other hand, an isolate alleged to be a contaminant if repeatedly grown in clinical specimen may be considered as significant. The mycology laboratory encounters more contaminants than the real pathogenic fungi. Therefore, the exact identification of any isolate even if it is considered as least important becomes more relevant keeping in view of its circumstantial background.

Most of these fungal agents have already been described either in the full-fledged Chapters or as passing reference to morphological description of the pathogenic fungi. These may be taken as common laboratory contaminants unless proved otherwise, depending upon the circumstantial factors of a particular patient. Therefore, it is pertinent to summarize such fungi as a ready reference in a separate Chapter.

The fungi included here are considered in the following broad categories for convenience of discussion. The description of colony morphology is based on the growth observed on the commonly used culture media like Sabouraud dextrose agar or on special media such as PDA or MEA. One or two lines are mentioned about the pathogenic potential of the fungus under discussion as a reference, if at all the very same is involved in significant human disease. Moreover, some of these laboratory contaminants have useful properties also, which is also mentioned as passing reference. The common fungal laboratory contaminants are categorized below along with allied issues:

A. Mucormycetes Fungi
B. Hyaline Hyphomycetes
C. Phaeoid (Dematiaceous) Fungi
D. Yeasts and Yeast-like Fungi
E. Look-alike Fungi
F. Miscellaneous Contaminants
G. General Safety Guidelines

The mucormycetes fungi have broad, non-septate or sparsely septate hyphae. The hyaline hyphomycetes have septate hyphae without distinct melanin pigment in their cell wall neither in culture nor in tissue section. The phaeoid/dematiaceous hyphomycetes are those that generally have melanin in their cell wall when grown in culture as well as in tissue section involved in the infective process. The yeasts and yeast-like fungi are round, elliptical or irregular shaped depending on the fungal species involved. The gross colonial morphology and microscopic descriptions are described in their respective chapters also but here again narrated at the cost of somewhat repetition. The individual fungal agents, usually encountered as contaminants, are described in the following sections, in an alphabetical order under various headings.

A. Mucormycetes Fungi

The fungi of this group are aseptate or sparsely septate mycelial organisms. Although some of these fungi are found as ***"bread molds"*** or laboratory contaminants but they may also cause various clinical forms of life-threatening mucormycosis. The description of these fungi is given below:

1. *Cunninghamella* Species

The colonies of *Cunninghamella* genus are rapidly growing and grey in color. The sporangiophores are erect with a whorl of short lateral branches at the apical region, each branch ending in a swollen vesicle. One-spored sporangioles cover the entire surface and the sporangiospores are oval to tear-shaped **(*See* Figs. 26.17A and B)**. *Cunninghamella bertholletiae* is the commonly encountered species in certain infections.

2. *Lichtheimia (Absidia)* Species

Lichtheimia (Absidia) species produce many branched, arching stolons that are similar to those of *Rhizopus* species. They form rhizoids at nodes that penetrate the substrate. The sporangiophores are **internodal** and occur at the curve of stolons, in contrast to those of *Rhizopus* species that occur at nodes. The sporangiophores gradually expand, forming an apophysis topped by a columella. The prominent apophysis helps to differentiate from *Rhizopus, Rhizomucor* and *Mucor*. The sporangiophore is also branched and has a septation so that walls do not collapse which helps in differentiation from other mucormycetes like *Apophysomyces* species. The columellae are hemispherical or elongate, sometimes topped by a long spine **(*See* Figs. 26.18A to C)**.

3. *Mucor* Species

This is a rapidly growing mucormycete, light to dull grey or brownish, with abundant aerial and substrate mycelia with rhizoids or stolons being absent. The hyphae are broad with thin walls, mostly non-septate, sporangiophores are branched, bearing spherical sporangia at the tip of each branch; the sporangia are thin-walled with columellae becoming dark as the sporangiospores mature. When sporangia rupture, fragments of wall remain at base of columellae. The sporangiospores are globose, oval or cylindrical with smooth walls **(*See* Fig. 26.2)**.

4. *Rhizopus* Species

The colonies of *Rhizopus* are fast spreading white to light grey in the beginning but become darker as sporangia mature. The hyphae are broad non-septate with thin walls submerged or aerial stolon with brown root-like short branches *i.e.* rhizoids, under the points where sporangiospore develops. The sporangiophores are hyaline to brown, erect, single or in groups, producing sporangia with columellae at the tip. The sporangia are globose, hyaline at first, turning dark brown, grey or black as the sporangiospores mature **(*See* Fig. 26.3 and Figs. 36.1A and B)**.

5. *Syncephalastrum* Species

The colony of *Syncephalastrum* species is cottony with aerial mycelia that quickly cover the agar surface. The aerial mycelia are white at first that later turns grey to black against a white background. The reverse is white. Microscopically, chains of spores are formed within long, finger-like tubular sporangia **(merosporangia)** that project from gently swollen tip of short branches of the aerial hyphae called sporangiophores **(*See* Figs. 26.21A to F)**. They lack phialides, which is helpful to differentiate this genus from morphologically mimicking structure of *Aspergillus* genus.

B. Hyaline Hyphomycetes

The agents involved in this group are hyaline or colorless mycelial fungi which are septate usually encountered as agents of hyalohyphomycosis. However, when there is no evidence of tissue involvement such isolates are considered as laboratory contaminants. Some of the hyaline mycelial agents are described below but *Chrysosporium*, *Malbranchea* and *Sepedonium* are described under the heading of look-alike fungi:

1. *Acremonium* Species

The colonies of *Acremonium* species are often flat at first becoming fluffy with age and turn whitish-yellow or pinkish to grey. The conidiophores are absent and phialidic conidiogenous cells are erect, gradually tapering towards the apex and arising singly from creeping vegetative hyphae. The unicellular conidia are globose, ovoid, cylindrical or slightly curved, produced as balls or fragile chains at the tip of phialides. Occasionally, chlamydospores may also be present. It used to be earlier known as *Cephalosporium* species due to the accumulation of conidia as balls or heads. *Acremonium kiliense*, the commonly encountered one is now known by its new genus name as *Sarocladium kiliense* **(*See* Figs. 31.20 to 31.24)**.

2. *Aspergillus* Species

Various species of the genus *Aspergillus* are responsible for large number of fungal diseases including allergy and mycotoxicosis. But simultaneously these species are encountered as laboratory contaminants as well. The septate hyphae of *Aspergillus* grow rapidly and non-septate conidiophores arise from a specialized hyphal cell called **foot cell** and terminate in a bullous head, a **vesicle**. The

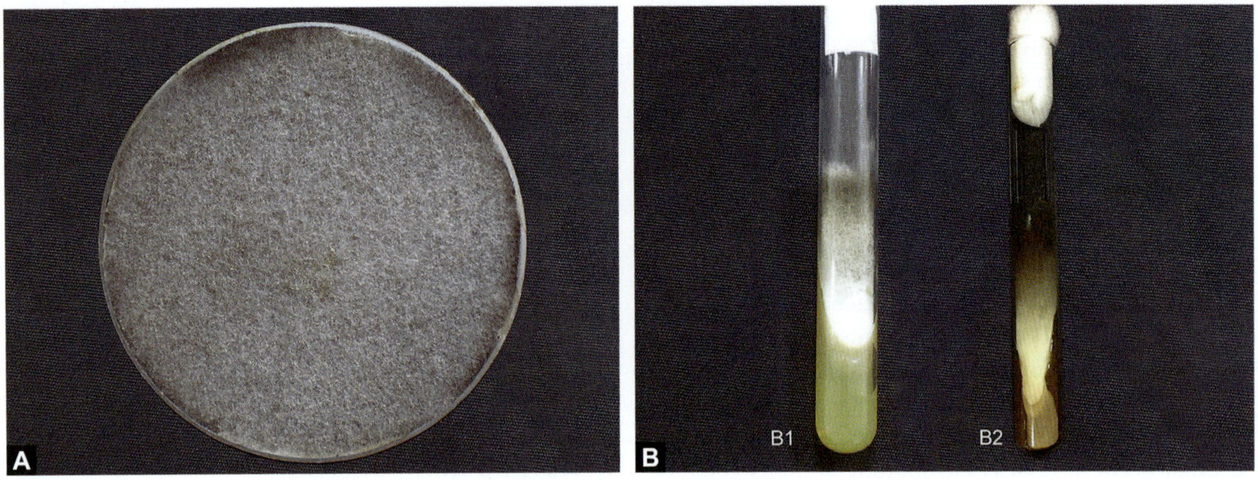

Figs. 36.1A and B. Greyish mycelial growth of *Rhizopus arrhizus* on SDA plate and tube (B1) and *Rhizopus microsporus* (B2).

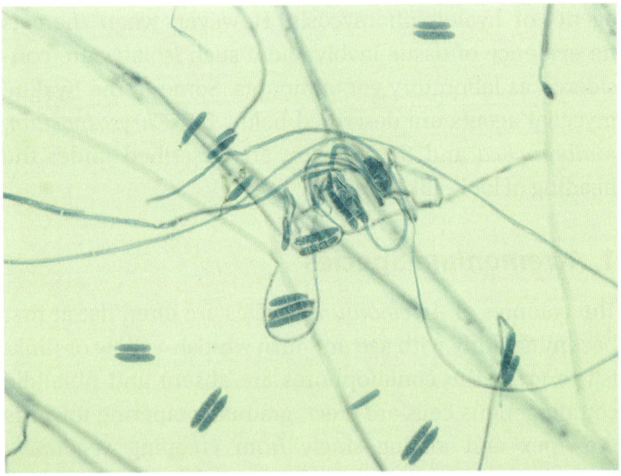

Fig. 36.2. Hyaline straight conidia of *Cylindrocarpon* species with rounded tip (LCB × 200).

vesicle produces phialides/metulae in one or two series giving rise to uniseriate or biseriate arrangements. The unicellular conidia are arranged in a way that the youngest conidium is at the base *i.e.* basipetal **(*See* Figs. 25.11A to D)**.

The conidia are colorless to begin with but become darkly pigmented in colors of yellow, green, blue-green, grey, black, tan, brown etc., giving unique color to a particular species. This is the basis of their nomenclature also like *A.flavus* (yellowish-green), *A.fumigatus* (fumes like smoky-green) or *A.niger* (black). The cell wall of the genus remain hyaline hence it should not be mistaken for phaeoid fungi due to apparent black color of conidia in some species.

3. *Cylindrocarpon* Species

This saprophytic fungus inhabit woody and herbaceous plants and commonly found in the soi1. Besides plant pathogenicity, in humans its species include *Cylindrocarpon lichenicola, C.destructans, C.cyanescens* and *C.vaginae.* The conidia are straight but sometimes slightly curved, hyaline, smooth-walled, cylindrical to fusiform with rounded tip and distinctly truncate base. There are usually three septae but occasionally up to five septae, 18-40 × 5-7 µm may be present. The microconidia are absent **(Fig. 36.2)**.

In the mycology laboratory, it may be encountered as contaminant. These are causing fungal keratitis, eumycetoma, dermatomycosis, peritonitis in case of CAPD, disseminated infection in neutropenic, AML and other immunocompromised patients.

This fungal species resembles *Fusarium* species morphologically and taxonomically. It differs by (i) forming macroconidia that are mainly straight rather than curved; (ii) have rounded apical cells rather than tapering; (iii) have basal cells with truncate and offset rather than attenuated foot cell; (iv) lacking microconidia (v) having pigmented chlamydoconidia and by the formation of a brown rather than cream colored colony. Moreover, macroconidia lack the distinctive foot cell, which is characteristic of *Fusarium* genus.

4. *Penicillium* Species

There are more than 300 known species of the genus ***Penicillium***, which are widely distributed in nature. They

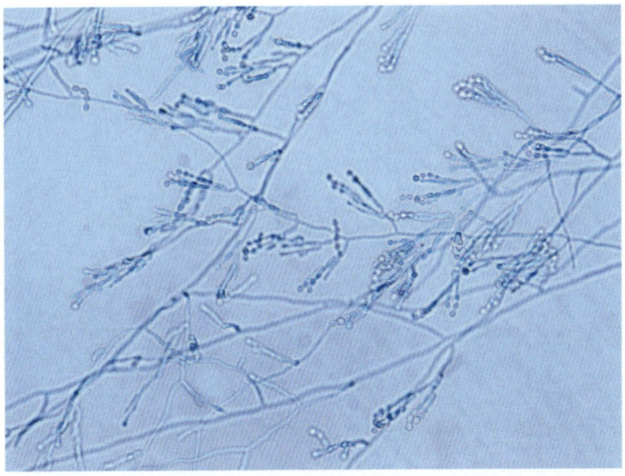

Fig. 36.3. Microscopic appearance of *Scopulariopsis brevicaulis* showing penicillate phialides with chains of conidia (LCB × 200).

are found in soil as well as decomposed organic debris. The genus *Penicillium* is characterized by rapidly growing colonies having conidial structures resembling brushes and commonly producing strong musty odor. Most of these species are common contaminants found on various substances and cause food spoilage, colonize leather articles and are considered as indicator organisms for indoor dampness. The human population may be adversely affected due to environmental contamination by the genus *Penicillium.*

The hyphae are hyaline, septate and multi-branched. The growth is rapid to develop colony and conidiophores arise in various forms producing phialides singly or in groups or from branched metulae giving **brush-like** appearance. The conidia are unicellular and in chain with youngest at base *i.e.* **basipetal (*See* Figs. 31.11 and 31.12)**. They are colorless and may have dark-green or blue-green pigmentation. Sometimes, conidiophores are aggregated to stalk forming synnemata which are called as coremia. Morphologically, the colonies of *Penicillium* are usually green and the conidiogenous cells are cylindrical to bottle-shaped and aggregated in compact penicillia as compared to *Paecilomyces,* where the conidiogenous cells are flask-shaped.

5. *Purpureocillium/Paecilomyces* Species

Purpureocillium is a genus recently proposed to accommodate *Paecilomyces lilacinus, Paecilomyces* species other than *P.lilacinum,* such as *P.variotii, P.marquandii, P.javanicus.* These are also reported to cause some of the opportunistic infections. The colonies of *Purpureocillium/Paecilomyces* species are golden, greenish-lilac or tan colored but never blue or green as seen in *Penicillium* species. The conidiogenous cells *i.e.* phialides are single or verticillate found on well-developed conidiophores or arising directly from vegetative hyphae. The flask-shaped phialides end in a long tapered tip which bear long chains of lemon-shaped or nearly cylindrical conidia **(*See* Figs. 31.4A to C)**.

6. *Scopulariopsis* Species

This is a large genus comprising mainly soil species, which are frequently isolated from food, paper and other material and also occur as laboratory contaminants. In *Scopulariopsis* species the turf is white at first, thin or floccose and then become grey to tan in color, often resembling *Microsporum gypseum.* The conidiophore is well-developed and conidia are produced by annellides singly or in groups directly from the mycelium. They produce dry, hydrophobic conidia in chains rather than slimy conidia in heads. The conidial structures are penicillate like *Penicillium, Paecilomyces* and *Purpureocillium* species. The conidia are thick walled, roughened and arranged in chains with the youngest at the base *i.e.* **basipetal (Fig. 36.3)**.

There are nine pathogenic species, which differ from each other mainly due to color of colonies and size of the conidiogenous cells. However, two species are regularly found associated with human diseases. *Scopulariopsis brevicaulis* is a clinically significant species that is responsible for onychomycosis, pulmonary mycosis and invasive infections as hyalohyphomycosis. The other species, *S.brumptii* is found associated with fungal allergic manifestations and as an opportunistic pathogen among immunocompromised patients.

7. *Trichoderma* Species

Trichoderma species are saprophytic filamentous fungi with worldwide distribution in the soil, plant material, decaying vegetation and wood. But actually, *Trichoderma* shows increasing medical importance as an opportunistic human pathogen particularly in immunocompromised patients. *Trichoderma* species can cause localized infections, such as pulmonary mycosis, peritonitis, sinusitis, otitis or brain abscess and fatal disseminated disease. These infections are characterized by the presence of fine septate hyphae in tissue sections, entities called hyalohyphomycosis, for which differential diagnosis with invasive aspergillosis is difficult. *Trichoderma longibrachiatum*

Fig. 36.4. Whitish mycelial colonies of *Trichoderma* species on Sabauraud's dextrose agar.

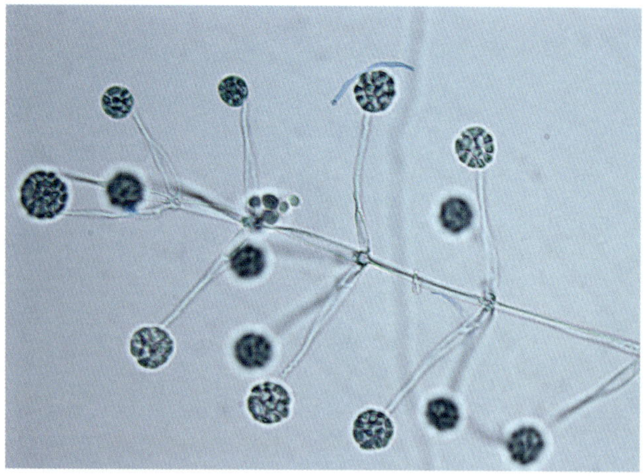

Fig. 36.5. Microscopic appearance of *Trichoderma* showing phialides and conidia (LCB × 400).

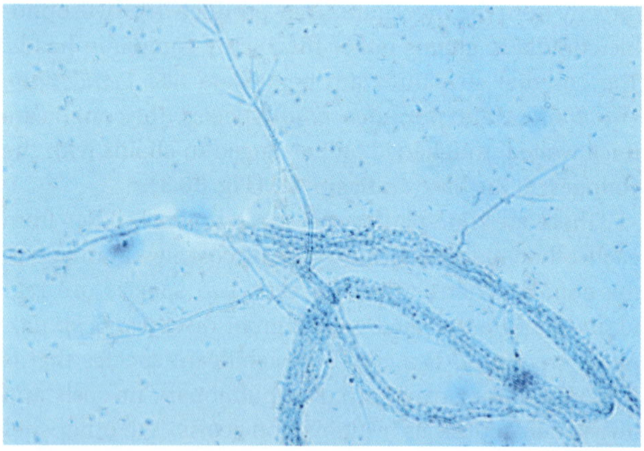

Fig. 36.6. *Verticillium* species showing pointed conidiophore arising from septate hyphae (LCB × 400).

had been identified as the causal agent in the majority of reported *Trichoderma* mycoses, particularly after solid-organ transplant *e.g.* renal transplant patients.

The colony of *Trichoderma longibrachiatum* is at first thin, cottony, white that soon becomes dense woolly, matted with shades of green, yellowish-green or white. They may grossly resemble *Penicillium* species. Microscopically, short irregular branching conidiophores develop from septate hyphae. The terminal conidiophores are flask-shaped and produce rounded clusters of single-celled, colorless to green conidia at tips **(Figs. 36.4 and 36.5)**. *Trichoderma* species are used in biotechnology as sources of enzymes and antibiotics. Moreover, they are applied to agricultural crops as plant growth promoters and biofungicides.

8. *Trichothecium* Species

The colonies of *Trichothecium* grow moderately fast, producing white, thin, later becoming rosy, pink, fleshy or orange-pink. The conidiophores are erect, unbranched, producing conidia at the apex in groups. The conidia are terminal, single or two-celled, produced one after another forming a head or chain. They are pyriform, two-celled; with the apical cell larger.

9. *Verticillium* Species

This hyaline fungal species, widely found in the environment, is a plant pathogen. It develops at first a white colony with a thin radial surface growth. In due course of time, it becomes powdery because of heavy sporulation and may have shades of yellow, green or red.

There are elongate conidiophores with pointed apex originated from septate hyphae. The conidiophores are arranged in **whorls**. There are oval or elliptical conidia, which are single-celled and appear either singly or in clusters at the end of conidiophores. The conidia are borne on phialides, which are specialized hyphae produced in a whorl around each conidiophore. It is named for the "verticillate" (=whorled) arrangement of phialides on the conidiophore **(Fig. 36.6)**.

C. Phaeoid/Dematiaceous Fungi

This group consists of mycelial or yeast-like fungi in which the cell wall is either brown or black in color. Most of the

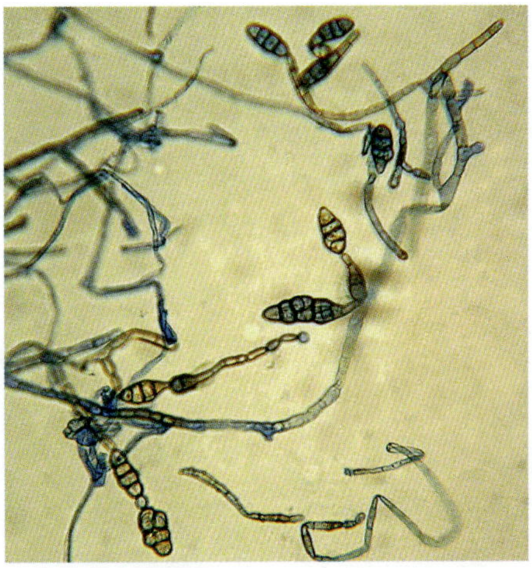

Fig. 36.7. Microscopic appearance of *Alternaria* species showing muriform conidia (LCB × 400).

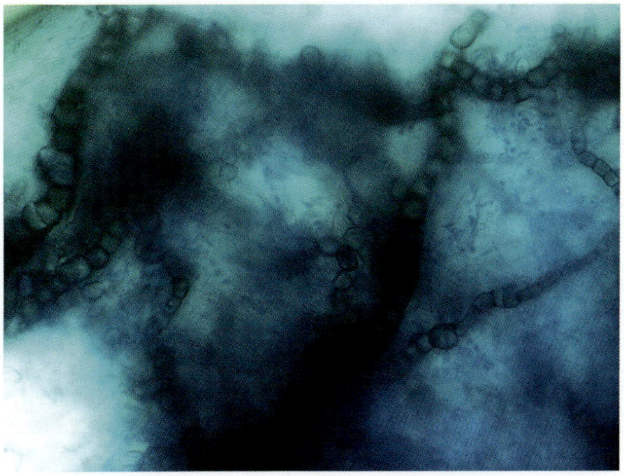

Fig. 36.8. Microscopic appearance of *Aureobasidium pullulans* showing thick and thin phaeoid septate hyphae (LCB × 400).

members of this group are causing mycetoma, chromoblastomycosis, phaeohyphomycosis, etc., but some of them or even same are encountered as laboratory contaminants, which are given below:

1. *Alternaria* Species

The colonies are floccose with phaeoid/dematiaceous hyphae on obverse and reverse is black. Conidiophores are simple, sometimes branched short or elongate, bearing chains of conidia. The conidia are **muriform** cells with septation in both planes *i.e.* transverse and longitudinal. The conidiophore elongate in chain with the youngest at top *i.e.* **acropetal**. Conidiophores extend sympodially and often form zig-zag structures. Out of its 5-6 species, *Alternaria alternata* is commonly encountered **(Fig. 36.7)**. Moreover, microscopically *Alternaria* species should be clearly differentiated from *Ulocladium*, *Stemphylium* and *Epicoccum* species.

2. *Aureobasidium* species

The initial growth of *Aureobasidium* is often looking white to pink yeast, turning black, wrinkled with a fringe of submerged mycelia. The hyphae are not extensive, becoming multi-septate with thick walls containing pigment. Little wart-like pegs are formed and give rise to hyaline ameroconidia. These are formed from several places immediately on the conidiogenous cells. The conidia then may bud to form a chain that later on become dark. Several conidia may be seen aggregating in a mass around conidiogenous cells. The hyalinized hyphae also produce hyaline single-celled synchronous conidia. Other hyaline hyphae give rise to conidia from a common point, which are non-synchronous **(Fig. 36.8)**. All these hyaline conidia may then bud and grow as non-pigmented yeasts resembling *Candida* species. This **black yeast** is most commonly isolated in clinical laboratories and morphology is extremely variable. *A.pullulans* is found on decaying vegetation and is a plant pathogen and also involved in some of the fungal infections.

3. *Bipolaris* Species

The colonies of *Bipolaris* species are fluffy to velvety and grey to brown or black in color. The hyphae are septate and darkly pigmented. The conidiophores are twisted where the large multi-septate elongate conidia with distoseptae are sympodially attached. The production of conidia on the alternate sides of conidiophore is manifested as a zigzag appearance where growth continues until a conidium is formed on alternate side. The conidia germinate by production of germination tubes by polar cells hence called *Bipolaris*. Its two species are frequently encountered in phaeohyphomycosis and keratomycosis *i.e.* *Bipolaris hawaiiensis* and *Bipolaris spicifera* **(Figs. 36.9A to C)**.

4. *Chaetomium* Species

Chaetomium is the only genus of ascomycetes, which produces black ascocarps and is commonly encountered in foods. The ascocarps of *Chaetomium* are perithecia, similar in appearance to cleistothecia but having an opening

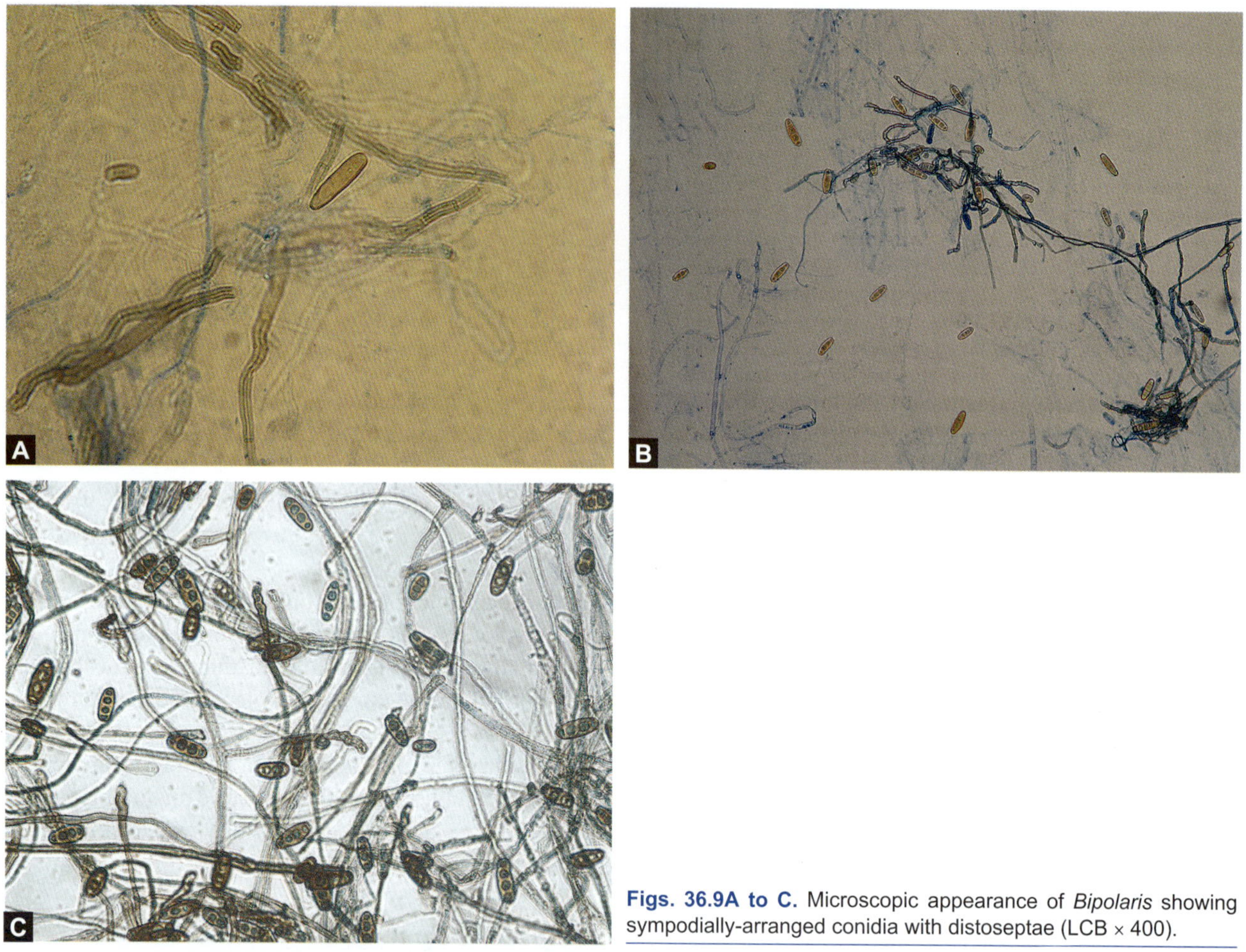

Figs. 36.9A to C. Microscopic appearance of *Bipolaris* showing sympodially-arranged conidia with distoseptae (LCB × 400).

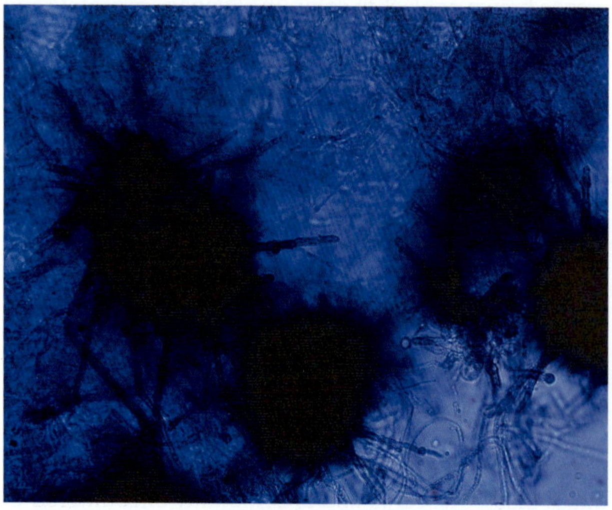

Fig. 36.10. *Chaetomium* species with perithecia, conidia and setae covering ascomata (LCB × 400).

through which the ascospores are released. Stout black hyphae visible in culture under a low power microscope to perithecium walls are the characteristics.

The genus *Chaetomium* is one of the largest genera of saprophytic ascomycetes, which includes more than 180 species. The main characteristic of nearly all *Chaetomium* species is the presence of hairs or setae covering ascomata **(Fig. 36.10)**. These species are responsible for cutaneous infections and occasionally encountered in systemic phaeohyphomycosis. Several species of *Chaetomium* like *Ch.globosum*, *Ch.strumarium* and *Ch.atrobrunneum* are reported as causative agents of opportunistic infections in humans. *Chaetomium atrobrunneum* has been isolated from brain abscess. *Chaetomium* species produce dark, subglobose to vasiform perithecia covered with lateral and terminal septate hairs. In many species, terminal hairs are coiled while lateral hairs are flexuous. Asci are prototunicate, usually club-shaped and contain eight ascospores.

5. *Cladosporium* Species

In *Cladosporium* there are floccose dematiaceous hyphae on obverse and black on reverse. Many common species are velvety or suede-like in the primary cultures. Conidiophores are tall branched variously near apex. The branching and conidiation change axis and often have swellings. They continue to extend producing irregular chain with the youngest cell at top *i.e.* **acropetal**. Attachment scars are prominent often with evident hilum (*See* **Fig. 13.7**).

6. *Curvularia* Species

The colonies of *Curvularia* species are rapidly growing, floccose and brown with black reverse. The conidiophores are simple, bearing conidia apically from new sympodial growth. The conidia are transversely septate and cylindrical or slightly curved with one of the central cells being larger. In some species, conidiophores arise from stroma. The clinical specimens may show hyaline hyphae. Most commonly *Curvularia geniculata* and *C.lunata* are encountered as contaminant or pathogenic causing keratomycosis, sinusitis, onychomycosis, phaeohyphomycosis, eumycetoma, etc. However, other less commonly encountered species are also seen. The *Curvularia* species should be differentiated from *Bipolaris* and *Exserohilum* species, which have false septae and not true septae (*See* **Figs. 2.14C, 14.13A and B** as well as **Fig. 29.14**).

7. *Drechslera* Species

The spores of *Drechslera* species occur individually and have very thick walls. In addition the septa between cells appear to be incomplete and are called pseudosepta. The shape of the spores varies as they can be straight, slightly curved, club-shaped, cylindrical or even elliptical. The texture on the surface of the spores is usually smooth but occasionally may be warty. Several genera of very similar fungi including the genera *Drechslera, Bipolaris, Helminthosporium* and *Exserohilum* have very similar spores and this spore type is called *Drechslera*-type spores. These fungi are either plant pathogens or saprobes in the natural environment. *Drechslera* spores represent one of the more common spore types found prevalent in many atmospheric conditions. These are known to be allergenic and can also cause fungal rhinosinusitis.

It is pertinent to differentiate *Drechslera* species from all other graminicolous helminthosporoid genera by its ability to develop a germ tube from any area of the cells in the distoseptate conidia. The hilum morphology can also be used to differentiate *Bipolaris* and *Drechslera*. In *Drechslera* a flat scar exists within the lowest part of the basal cell, whereas in *Bipolaris* it is inconspicuous or very slightly protuberant. The sexual morphology of *Drechslera* is linked to *Pyrenophora* whereas the sexual morphology of *Bipolaris* are regarded as *Cochliobolus*. However, *Exserohilum* can be differentiated from other graminicolous helminthosporoid genera by a truncate, strongly protruding hilum, often with an enveloping bubble. The sexual morphology of *Exserohilum* is placed in genus *Setosphaeria*.

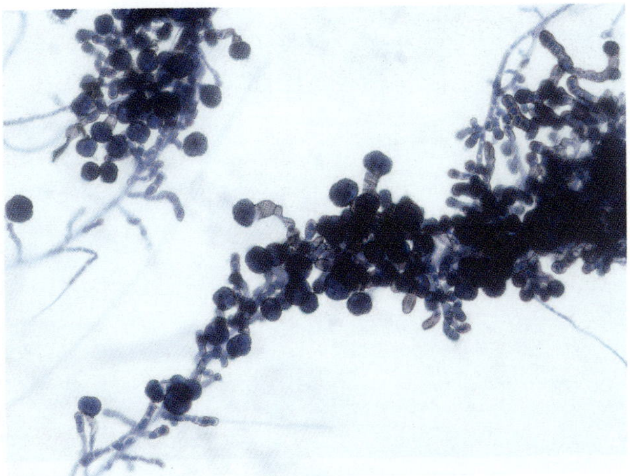

Fig. 36.11. Microscopic appearance of *Epicoccum* species showing darkly colored, warted, spherical conidia (LCB × 400).

8. *Epicoccum* Species

Epicoccum is known for its potential to produce diverse classes of biologically active secondary metabolites. This fungal species is ubiquitous and cosmopolitan. It usually grows well on cellulose surfaces. It is often found growing in soil, plant litter, decaying plants and/or damaged plant tissue. It is often found growing on a variety of building materials. As such it is linked to any human or animal infection, however, some patients may experience hay fever and/or asthma. The antibiotics flavipin and epicorazine are produced from this type of mold.

The fungus grows as felty colonies in bright shades of yellow, orange and red, often with brown or black throughout. The colonies grow quickly, reaching about 6 cm in diameter in two days at room temperature. *Epicoccum nigrum* forms blastoconidia that are darkly colored, warted and spherical, reaching 15 to 25 μm in diameter. Conidia grow on a sporodochium, formed by warty and fibrous hyphae (**Fig. 36.11**). The sporets are found to contain up to 15 cells. It may be sometimes misidentified as *Ulocladium, Stemphylium* or *Alternaria* species.

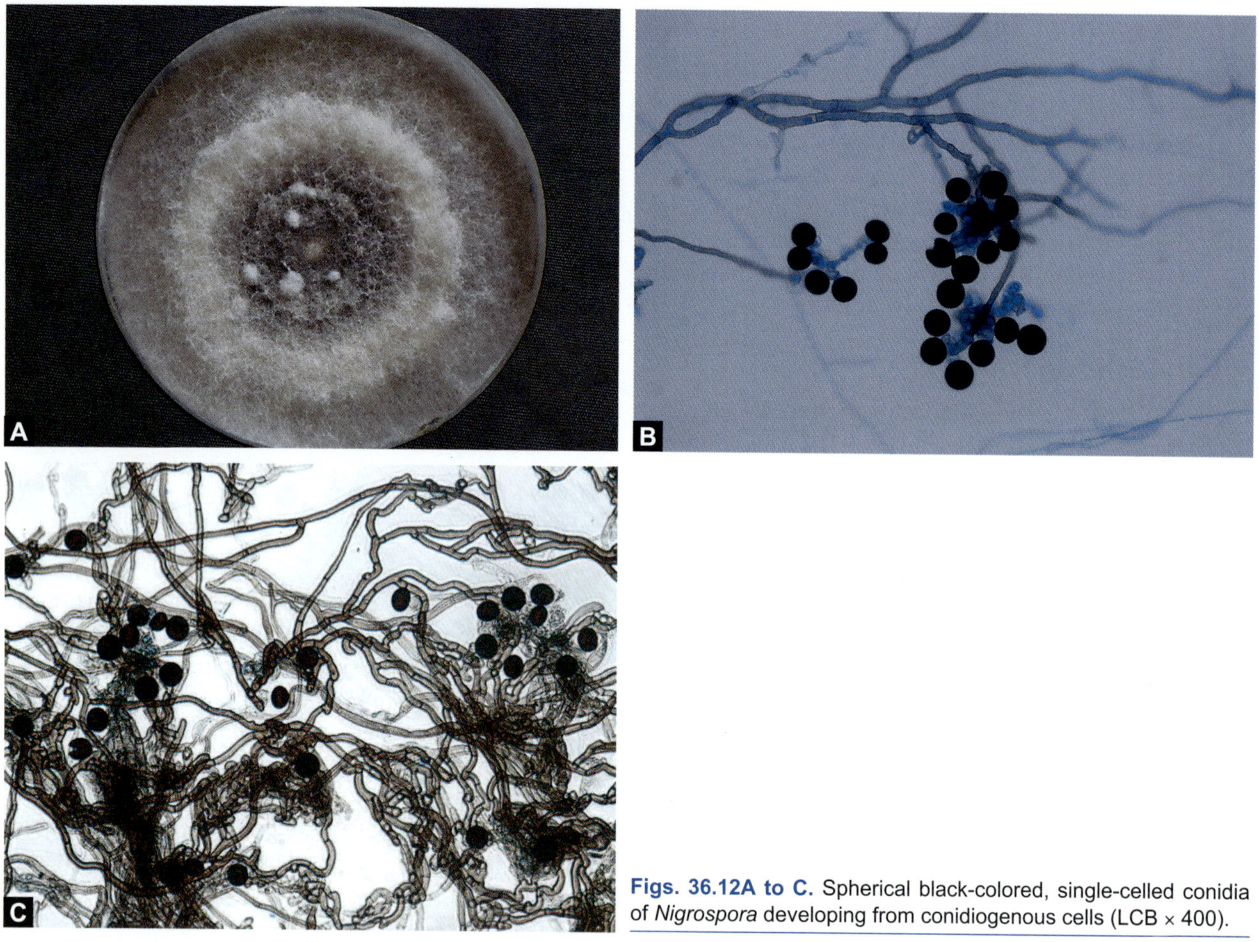

Figs. 36.12A to C. Spherical black-colored, single-celled conidia of *Nigrospora* developing from conidiogenous cells (LCB × 400).

9. *Helminthosporium* Species

The colony of *Helminthosporium* is at first greyish brown or greenish in color, often forms a black, matted centre with raised greyish periphery on obverse and reverse is black. The conidiophores are erect, simple, determinate and found often in clusters. The conidia are produced from pores on conidiophore and are brown, elongate and contain four or more cells. The pores and poroconidia often appear in whorls. This fungus should be differentiated from other similar-looking fungi *i.e. Bipolaris, Drechslera* and *Exserohilum* species.

10. *Nigrospora* Species

This phaeoid fungus is usually encountered as plant pathogen and *N.sphaerica* is the commonest species. Its colonies grow rapidly appearing compact, hairy or woolly. The conidiophores are short and clustered surfacing from mycelium. They appear translucent in color and have an average range of 8-11 μm in diameter. The conidiophores are often straight stalks or slightly curved. The conidia grow from the tips of the translucent conidiophores, which are brownish-black, oblate spheroid and single-celled. On an average they range from 16-18 μm in diameter. The initial white translucent looking colony turns brown to black due to mass sporulation of conidia from the conidiophores and the reserve is black. It is very rarely reported from non-dermatophytic onychomycosis and fungal keratitis cases **(Figs. 36.12A to C)**.

11. *Phoma* Species

Phoma species are phaeoid fungi that are commonly isolated from soil. They are usually encountered as phytopathogen in the nature. The growth on SDA shows turf of light grey to soft brown broadly spreading colonies. The mycelia are abundant in some species and usually

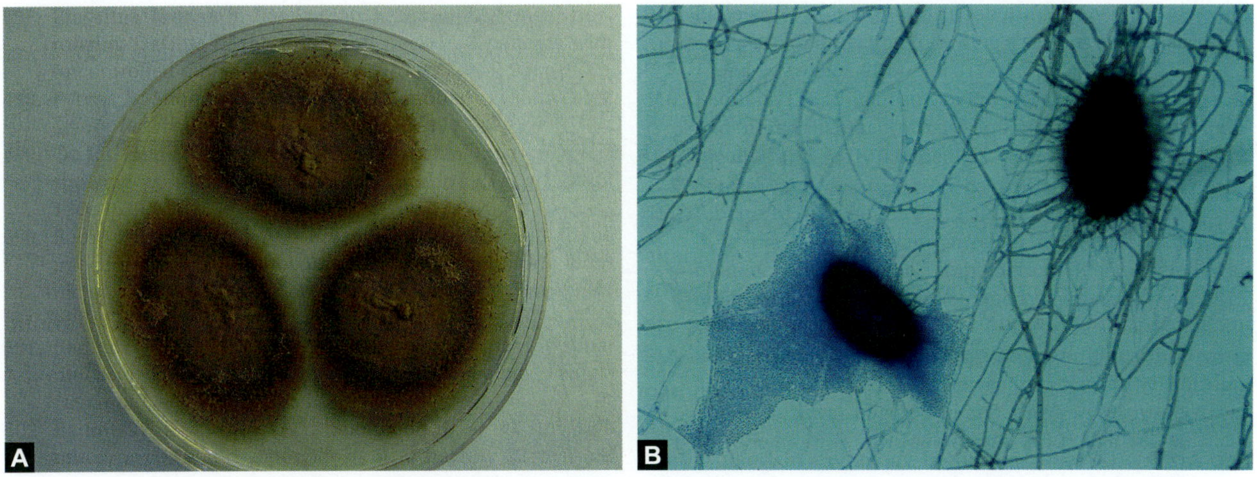

Figs. 36.13A and B. Spherical black-colored, single-celled conidia of *Phoma* developing from conidiogenous cells (LCB × 400).

submerged in others. The ovoid fruiting body is a **pycnidium** that may be 600 μm across which is a tough leathery to membranous structure that is globose to lens-shaped, with a small papilla at apex. The conidia are produced in abundance within the pycnidia on thread-like conidiophores. At maturity conidia ooze out of volcano-shaped pycnidia **(Figs. 36.13A and B)**. Although it is a common plant pathogen but rarely reported in human diseases causing subcutaneous phaeohyphomycosis. *Phoma hibernica, P.minutella* and *P.sorghina* are the common species, which are reported among human cases.

12. *Stemphylium* Species

Stemphylium species are plant pathogens in the phylum Ascomycota and *S.solani* is the commonly encountered fungus from this genus. Its teleomorph is *Pleospora* though there are no naturally known occurrences of sexual reproduction. It is the causal pathogen for grey leaf spot in tomatoes, alliums and cotton. The diagnosis of specific species is determined based on distinct conidia and conidiophore morphology though substantial overlap in characteristics makes identification difficult. Hence DNA studies are currently the most conclusive method of correctly distinguishing its species. This fungus is easily mistaken for *Alternaria, Ulocladium* and *Epicoccum* species.

13. *Ulocladium* Species

Ulocladium species are usually recovered from soil. However, when grown from clinical specimens, they are normally considered as contaminants; however, there are few reports of their involvement in cases of human phaeohyphomycosis. The growth is rapid, which matures in 4 to 5 days on SDA at 25°C. The colonies are cottony, sometimes fluffy, dark brown to olive-brown and the reverse is black. The hyphae are dark and septate. The conidiophores are simple or branched and bent at the point where the conidia originate, giving a characteristic zig-zag appearance (sympodial geniculate conidiogenesis). The conidia (poroconidia) are round or oval (10-12 × 10-15 μm), brown or olive-grey, with both transverse and longitudinal septations.

On microscopic morphology, isolates of genus *Ulocladium* are to be differentiated from those of *Alternaria* by having obovate conidia with tapered and narrow bases and with no or very short apical beaks (false beak). Additionally, conidia of *Ulocladium* are in short chains or non-catenate **(Fig. 36.14)**. *U.chartarum* differs from *A.alternata*, species by having conidia that are coarsely verrucose and have short, spindle-shaped apex. Therefore, *Ulocladium* species should not be confused with *Alternaria, Stemphylium* and *Epicoccum* species.

D. Yeasts and Yeast-like Fungi

The yeasts or yeast-like fungi which may be significantly pathogenic but sometimes may be encountered in the diagnostic mycology laboratory as contaminant are described below:

1. *Geotrichum* Species

This produces yeast-like colonies with waxy moist surface and thin disk of radiating grey or white moist mycelia.

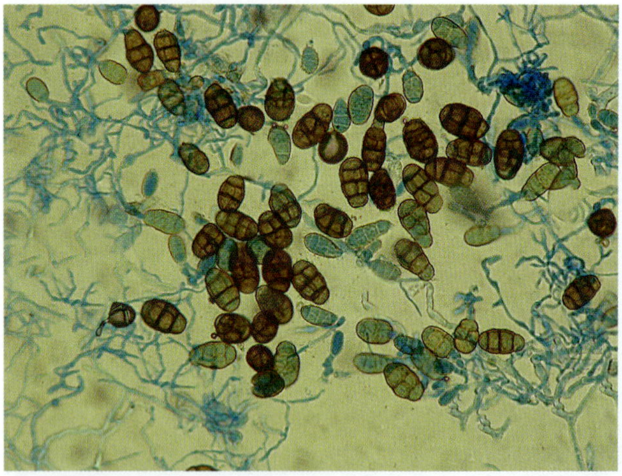

Fig. 36.14. *Ulocladium* species showing sympodial geniculate conidiophore with poroconidia (LCB × 400).

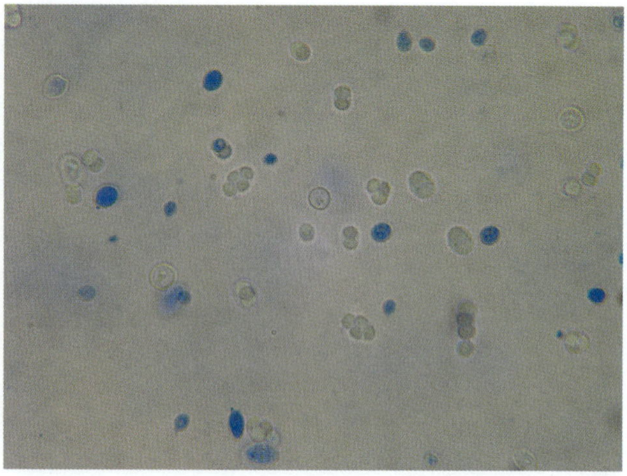

Fig. 36.15. *Saccharomyces cerevisiae* showing round to ovoid yeast cells and four haploid spores in some of them (LCB × 400).

Microscopically, it shows true mycelium that disarticulate into thin-walled rectangular, barrel-shaped cells *i.e.* arthrospores. Neither blastospores nor ascospores are produced **(*See* Figs. 28.6 to 28.8)**. *Geotrichum* is encountered frequently in specimens from mouth or anogenital region. *G.candidum* is commonly found in sour milk, milk-products and the sewage.

2. *Rhodotorula* Species

Rhodotorula has been an infrequent cause of infection in humans but there are case reports about systemic infection. It occurs commonly in culture from vaginal smears, abscesses and feces. The colonies are glistening, creamy to mucoid, showing various shades of pink, red or orange colors **(*See* Figs. 28.13A and B)**. Microscopically, organisms are seen with budding cells similar to *Cryptococcus* species except for the presence of a carotinoid pigment.

3. *Saccharomyces* Species

Saccharomyces cerevisiae is a commercial yeast used in fermentation of beer, wine and mash from that distilled liquor is made. This is also called **Brewers'** or **Bakers'** yeast. It is normally prevailing in nature on ripening fruits, soil, sewage and frequently isolated as contaminants from skin. It has been reported occasionally to cause thrush, vulvovaginitis and even fungemia. There are multilateral budding yeast cells, which are round to oval and rarely short rudimentary pseudohyphae may be formed. This is multiplying asexually however when sexual reproduction takes place, there are four haploid spores seen **(Fig. 36.15)**.

4. *Trichosporon* Species

The young yeast-like colonies are cream-colored, smooth and shiny, later on becoming dry and membranous. It produces a grayish-white or a yellow, heaped up, rugose, waxy colony with a mycelial border and is easily mistaken for *Geotrichum* and any of *Candida* species. Microscopic examination reveals true mycelia breaking into arthrospores which in turn bud, producing many blastospores **(*See* Figs. 9.2 and 9.3; Figs. 28.3 to 28.5)**. *Trichosporon* species are responsible for white piedra and other opportunistic systemic infections. It is a common organism encountered in sputum specimens and skin scrapings.

5. *Ustilago* Species

Ustilago is phytopathogen responsible for corn smut, a disease of corn characterized by black smutty leaves and marked hypertrophy of the infected seeds. On the other hand, *U.maydis* is used as traditional Mexican food eaten in many parts of country. This is frequently encountered organism in mycology laboratory and is invariably mistaken for various yeasts-like fungi. Initially it produces moist, white colonies, which later on become membranous, tan-colored velvety to powdery. Microscopically, cells are irregular, yeast-like or elongate, spindle-shaped and may resemble arthroconidia. The individual spindle-shaped cells reproduce by budding to form black spores that rarely become attached to form a false mycelium. The Gram's stain and LCB mount show irregular cells as diagrammatically shown in **Figure 36.16**.

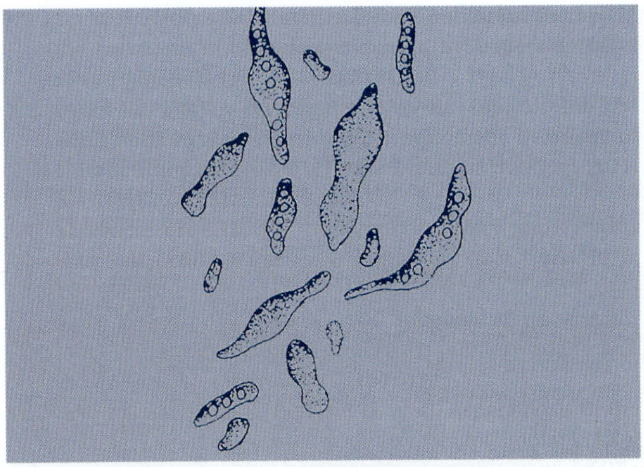

Fig. 36.16. Diagrammatic representation of *Ustilago* species showing irregular cells resembling yeasts.

E. Look-alike Fungi

This is heterogenous group, not related to each other and invariably mistaken for clinically significant pathogens, particularly the dimorphic fungi. There are chances if an isolate is not meticulously examined and wrong report is despatched due to inexperience and over-enthusiasm. These fungal species are commonly mistaken for their morphological counterparts like *Histoplasma* and *Coccidioides* species in their mycelial phase. Sometimes, it becomes very difficult to differentiate from pathogenic fungal agents due to their morphological similarities. Sequences of partial 18S rDNA gene can demonstrate close relationship between *Coccidioides* species and non-pathogenic *Malbranchea* due to production of alternate, barrel-shaped arthroconidia.

In addition, other members of family Gymnoascaceae under order Onygenales of class Plectomycetes in phylum Ascomycota like *Auxarthron*, *Uncinocarpus* and *Arachniotus*, which are non-pathogenic, may also confuse with *Coccidioides* species. Due to production of arthroconidia, *Coccidioides* species may mimic *Geotrichum candidum*. Other fungi like *Chrysosporium* species look similar to *Blastomyces dermatitidis* and *Paracoccidioides brasiliensis; Acrodontium* and *Acremonium* species are similar to *Sporothrix schenckii* at room temperature. Some of the look-like fungi are described below and also listed in **Table 36.1**.

Table 36.1. Commonly Encountered Look-alike Fungi.

Dimorphic Fungi	*Look-alike Fungi*
H.capsulatum	*Sepedonium, Renispora* species
B.dermatitidis	*S.apiospermum, Chrysosporium* spp.
P.brasiliensis	*Chrysosporium* species
Coccidioides species	*Malbranchea* species
Sporothrix schenckii	*Acrodontium, Acremonium* species

1. *Chrysosporium* Species

The fungal species belonging to genus *Chrysosporium* are keratinophilic and found in the soil. They are usually considered as laboratory contaminants and rarely found associated with fungal infection as hyalohyphomycosis. It is a moderately fast-growing fungus, producing usually spreading white to pale colony, often powdery or granular. The hyphae are hyaline, smooth-walled, branched and the fertile hyphae have terminal and lateral conidia. The conidia are sessile or borne on short protrusions or side branches, subspherical, truncate, clavate, pyriform or obovoid, one-celled, rarely two-celled. The intercalary conidia as well as chlamydospores are occasionally present. This fungus closely resembles dermatophytes, *Histoplasma* and *Blastomyces* species **(Fig. 36.17)**.

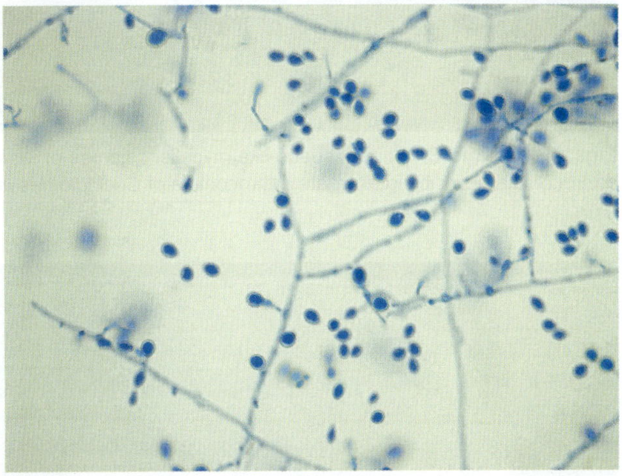

Fig. 36.17. *Chrysosporium* species showing pyriform to clavate one-celled conidia with truncate bases (LCB × 400).

2. *Malbranchea* Species

Malbranchea species are soil fungi of worldwide distribution which microscopically resemble *Coccidioides immitis* and *C.posadasii*. This hyphomycete is producing alternate arthroconidia with disjunctor cells. The colonies are white to sulfur-yellow to ochre-brown in color, suede-like in texture, with a reddish-brown reverse and often a reddish diffusible pigment **(Fig. 36.18A)**.

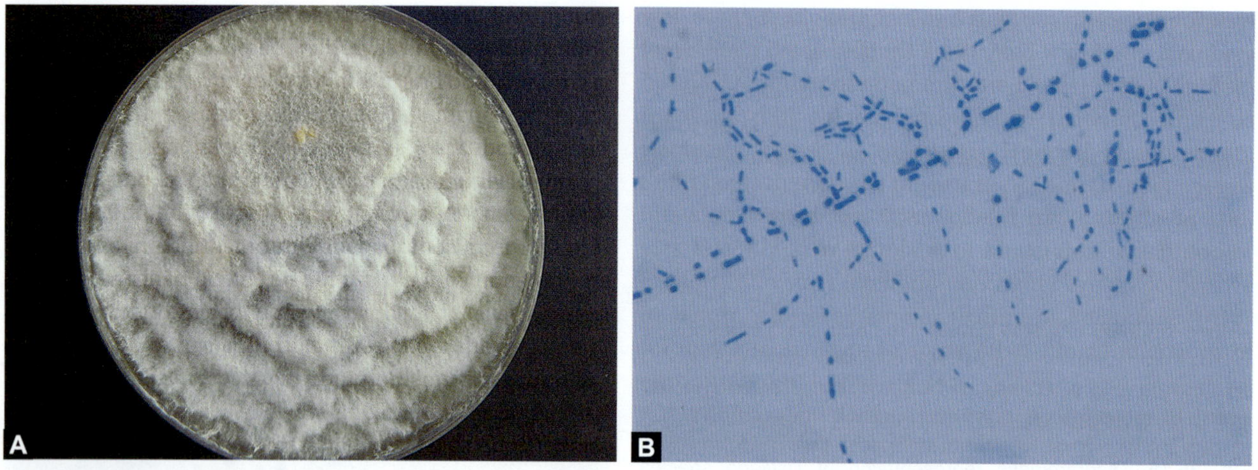

Figs. 36.18A and B. Whitish, suede-like mycelial colonies of *Malbranchea* species showing hyaline, one-celled, cylindrical, truncated arthroconidia resembling alternate arthroconidia of *Coccidioides immitis* grown at ambient temperature (LCB × 400).

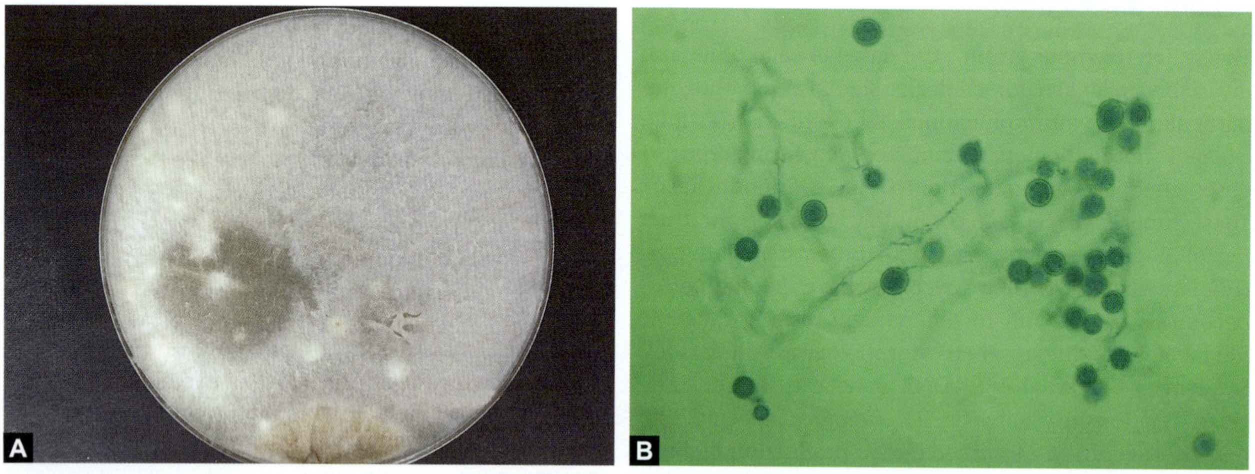

Figs. 36.19A and B. White to golden yellow, suede-like to downy fluffy mycelial colony of *Sepedonium* species showing macroconidia resembling tuberculate macroconidia of *Histoplasma* species grown at ambient temperature (LCB × 400).

The microscopic morphology shows typical hyaline, one-celled, cylindrical, truncate, alternate arthroconidia produced in tightly coiled, terminal fertile branches of the hyphae **(Fig. 36.18B)**. The arthroconidia are released by lysis of the disjunctor cells. These arthroconidia may be perceived as a yellow dust when released at maturity. The identification by exoantigen test or ITS sequencing remain the method of choice for its differentiation from *C.immitis* and *C.posadasii.* A substantial number of isolates of *Malbranchea* are wrongly identified as *C.immitis* in different parts of the world. Hence adequate awareness about similarities between *C.immitis* and *Malbranchea* is of paramount importance to avoid chances of misdiagnosis.

3. *Sepedonium* Species

The moderately fast-growing colonies of *Sepedonium* species are usually white to golden yellow, suede-like to downy, becoming fluffy with age **(Fig. 36.19A)**. The hyphae are multi-branched and conidiogenous cells are non-specialized, resembling short branches of vegetative hyphae. Large warty-covered macroconidia and small, thin-walled, egg-shaped microconidia are produced and both of them are aleuric. Microconidia are also produced from phialides. The microscopic morphology resembles that of *Histoplasma capsulatum* grown at 25°C, which was originally included in this genus **(Fig. 36.19B)**. Recently

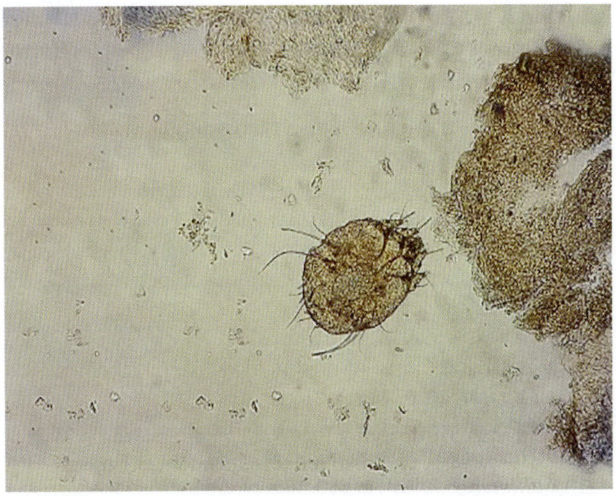

Fig. 36.20. Microscopic appearance of mite, a media contaminant showing four pairs of legs being Arachnid (KOH × 100).

there are reports due to this fungus as intraabdominal infections and fatal pneumonia in patient undergoing stem cell transplantation.

F. Miscellaneous Contaminants

Some of the contaminants in a diagnostic mycology laboratory may not be fungal in nature but they create significant spoilage of either culture media or fungi grown over there. The Arachnida ectoparasite, particularly, mites play an important role as contaminant in this regard. The identification of such parasites based on the fact that they have four pairs of legs, in contrast to three pairs in Insecta and five pairs in Crustacea **(Fig. 36.20)**.

Culture Contamination by Mites

Mostly it is seen that mites feed on to mycelial fungi and contaminate fungal culture by wandering from one culture plate or cotton plugged tube to the other. Hence the mites can be a real challenge to get rid of fungal culture media in the diagnostic mycology laboratory. The dermatophyte fungal cultures, stock cultures, etc., usually become infested with these mites. The difficulty with transferring hyphae from a contaminated sample is the potential of transferring mite eggs. In the event that it is decided to try a transfer to clean agar it would be wise to select hyphal tips under the microscope. Seal with paraffin film and store in with moth balls. It would be the best to discard the contaminated cultures and decontaminate the laboratory.

Control of Mites: The decontamination of mites in fungal cultures is very important issue In such circumstances fumigation is to be done to destroy mites. Addition of Lindane powder (0.1 gm/l) to medium will eliminate mites without affecting fungal growth. An alternative measure for prevention of mites is usage of polythene bags which can be sealed. The same bags can be re-used hence it is the most suitable and economical method. The use of naphthalene balls in the incubator is also applied. In addition, the culture infested with mites should be kept at -20°C overnight so that mites and their eggs lose their viability at this much low temperature and the next day subcultures are done on fresh medium.

When decontaminating, it is important that the entire laboratory should be covered not just the working bench area. Place moth balls to discourage mites but it would be good to find where the mites came from and put in measures to minimize their return once you have decontaminated the laboratory. The mites can also be exterminated by incubating stock cultures with 1,4 dichlorobenzene crystals in a large-mouthed bottle for four hours and repeating this process in a week time interval.

G. General Safety Guidelines

There are defined do's and don'ts, essentially followed in the diagnostic medical mycology laboratory to keep general safety measures thereby avoid fungal contamination. The precise and abridged safety guidelines in general are as follows:

- Don't eat, drink or smoke within mycology laboratory premises.
- Don't lick the adhesive labels.
- Don't put fingers, pencils or any other object in mouth.
- Wash hands thoroughly with soap and water before leaving laboratory.
- Always put on laboratory apron before venturing into routine work.
- Keep all the samples always upright in the racks.
- Requisition form should never be wrapped around sample, which is common practice in use.
- Discard sample if found to be soiled or leaking.
- Open specimen container slowly to avoid spurting and aerosol formation.
- Don't overcrowd and keep the bench tidy.
- Carry out all procedures in biosafety cabinet.
- Do not put up slide culture for dimorphic fungi.
- Use rubber teat or automatic suction device. Don't suck with mouth.

- Keep disinfectant solution, discard jar and dust bin near working bench.
- Pour disinfectant immediately in case of spillage of fungal culture. Keep it at least for ten minutes.
- Wipe the laboratory area daily at the end of day and discard the unwanted material.
- Place contaminated material in separate box and autoclave.
- Report all hazards to the in-charge of the laboratory.
- Don't wear loose sleeves, shorts and sandals.
- Know locations of exit, safety shower, eye wash, fire blanket, fire extinguisher and emergency telephone.
- Extinguish burners when away from the desk.
- Disinfect the whole laboratory at least weekly.

If these general type of safety guidelines are followed properly in letter and spirit in a diagnostic mycology laboratory, the chances of getting infection to the staff working over there will be minimized thereby over all fungal contamination of the laboratory culture media will decrease to a great extent. The laboratory will be able to make out clear cut difference between a fungal laboratory contaminants and the real pathogens.

Further Reading

1. Alangaden GJ. Miscellaneous fungi. *Curr Treat Option Infect Dis.* 2000; **2:** 499-506.
2. Arellano-Galindo J, Eugenio VM, Elva JH, *et al.* A saprophytic fungus (*Sepedonium*) associated with fatal pneumonia in a patient undergoing stem cell transplantation. *J Int Med Res.* 2017; 300060517708103. PMID: 28606018.
3. Benda TJ Jr, Corey JP. *Malbranchea pulchella* fungal sinusitis. *Otolaryngol Head Neck Surg.* 1994; **110:** 501-4.
4. Champa H, Sreeshma P, Yegneswaran Prakash P, *et al.* Cutaneous infection with *Cylindrocarpon lichenicola. Med Mycol Case Rep.* 2013; **2:** 55-8.
5. de Almondes AI, de Araujo JO, Mendes de Sirqueira Amaral L, *et al.* Fungal contamination and disinfection of dental chairs, Teresina, Piaui, Brazil. *Acta Odontol Latinoam.* 2016; **29:** 225-9.
6. de Hoog GS, Guarro J, Gene H, Figueras MJ. *Atlas of Clinical Fungi.* 2nd edn. Centraalbureau voor Schimmelcultures, Utrecht, The Netherlands: 2000.
7. de Hoog GS, Guarro J, Gene J, Figueras MJ. *Atlas of Clinical Fungi: The Ultimate Benchtool for Diagnostics.* Electronic Version 4.1.4. Westerdijk Fungal Biodiversity Institute (Former CBS-KNAW), Utrecht, The Netherlands and Universitat Rovira i Virgili, Reus, Spain. Interactive CD-ROM is based on book with same title; 2017.
8. Diongue K, Diallo MA, Seck MC, *et al.* Tinea pedis due to *Cylindrocarpon lichenicola* beginning onycholysis. *Med Mycol Case Rep.* 2016; **11:** 13-5.
9. Duran MT, Del Pozo J, Yebra MT, *et al.* Cutaneous infection caused by *Ulocladium chartarum* in a heart transplant recipient: Case report and review. *Acta Derm Venereol.* 2003; **83:** 218-21.
10. Elston DM. Demodex mites: Facts and controversies. *Clin Dermatol.* 2010; **28:** 502-4.
11. Evans EGV and Richardson MD. *Medical Mycology: A Practical Approach.* IRL Press; 1989: 171-85.
12. Fan YM, Huang WM, Li W, *et al.* Onychomycosis caused by *Nigrospora sphaerica* in an immunocompetent man. *Arch Dermatol.* 2009; **145:** 611-2.
13. Fatima N, Ismail T, Muhammad SA, *et al. Epicoccum* sp., an emerging source of unique bioactive metabolites. *Acta Pol Pharm.* 2016; **73:** 13-21.
14. Fincher R-ME, Fisher JF, Lovell RD, *et al.* Infection due to the fungus *Acremonium (Cephalosporium). Medicine (Baltimore).* 1991; **70:** 398-409.
15. Gaujoux T, Borsali E, Gavrilov JC, *et al.* Fungal keratitis caused by *Cylindrocarpon lichenicola. J Fr Ophtalmol.* 2012; **35:** 356.e1-5.
16. Gokhale S, Punia H, Ohri VC, Thammayya A. *Malbranchea*, an imitator of *Coccidioides immitis*: A case report. *Indian J Med Microbiol.* 1995, **13:** 48-50.
17. Hofman V, Butori C, Long E, *et al. Aureobasidium pullulans* contamination in bronchial aspirates mimicking cryptococcosis: A rare diagnostic pitfall. *Pathology.* 2008; **40:** 729-32.
18. Kwon-Chung KJ, Bennett JE. *Medical Mycology.* 1st edn. Lea & Febiger; 1992: 797-815.
19. Larone DH. *Medically Important Fungi: A Guide to Identification.* 5th edn. ASM Press, Washington DC: 2011.
20. Manamgoda DS, Rossman AY, Castlebury LA, *et al.* The genus *Bipolaris. Stud Mycol.* 2014; **79:** 221-88.
21. Mulkey DA, Vollmer J. Mite contamination in Sabouraud's agar. *Lab Med.* 1976; **7:** 25.
22. Nevalainen A, Taubel M, Hyvarinen A. Indoor fungi: Companions and contaminants. *Indoor Air.* 2015; **25:** 125-56.
23. Pastor FJ, Guarro J. *Alternaria* infections: Laboratory diagnosis and relevant clinical features. *Clin Microbiol Infect.* 2008; **14:** 734-46.
24. Rai MK. *Phoma sorghina* infection in human being. *Mycopathologia.* 1989; **105**: 167-70.
25. Rippon JW. *Medical Mycology*, 3rd edn. WB Saunders; 1988: 747-78.
26. Stewart E, Waldman S, Sutton DA, *et al. Ustilago echinata:* Infection in a mixed martial artist following an open fracture. *Mycopathologia.* 2016; **181:** 311-4.
27. Yogo N, Shapiro L, Erlandson KM. *Sepedonium* intra-abdominal infection: A case report and review of an emerging fungal infection. *J Antimicrob Chemother.* 2014; **69:** 2583-5.
28. Wang XW, Lombard L, Groenewald JZ, *et al.* Phylogenetic reassessment of the *Chaetomium globosum* species complex. *Persoonia.* 2016; **36:** 83-133.

SECTION VII

Pseudofungal Infections

Chapters

This section consists of diverse topics, covering organisms other than Fungi because of their increasing medical importance. These infections could not be sorted out under any of the previously described Sections of this Textbook as their causative organisms are non-fungal in nature hence clubbed together. As these entities are medically relevant and their presentation being very similar to fungal diseases thus a comprehensive understanding is essential to establish exact diagnosis or rule out the fungal etiology. In the beginning of their recognition, they were thought to be fungal infections but in due course of time there was substantial progress in molecular taxonomy hence their non-fungal nature was established. Consequent to these advanced taxonomic developments, two organisms, earlier thought as protozoa are now established as fungi and incorporated recently in the kingdom Fungi (*Pneumocystis jirovecii* and *Microsporidium*) as described under Opportunistic Fungal Infections and three controversial organisms are decisively removed from kingdom Fungi and are now being described in this Section of Pseudofungal Infections *i.e. Prototheca, Rhinosporidium* and *Pythium* species.

The infections caused by pseudofungal agents are algal diseases (Protothecosis and Chlorellosis). Another infection usually encountered via subcutaneous route (Rhinosporidiosis) and is caused by a protist, *Rhinosporidium seeberi*, which is hydrophilic in nature due to its affinity to water. It has been disputed since long to be a fungus, then in between even prokaryote (*Cyanobacterium*) but its taxonomical status has now resolved to a great extent as a Mesomycetozoea and is dealt in detail. It is also being considered as pseudofungal infection hence described in this Section and not with subcutaneous fungal infections. This is mainly found prevalent in southern India and Sri Lanka. The other infections caused by Oomycetes (Pythiosis and Lagenidiosis) are comprehensively described in a separate Chapter. Recently number of *Pythium* keratitis cases has increased substantially. Moreover, recognition of such infections is pertinent to plan out correct diagnostic as well as appropriate therapeutic modalities since traditional antifungals are not effective for their treatments. Hence this section deals comprehensively with three groups of diseases caused by pseudofungi - *Prototheca, Rhinosporidium, Pythium* and its allied genera.

CHAPTER 37

Protothecosis and Chlorellosis

The creamy yeast-like algae—*Prototheca* and *Chlorella;* Mesomycetozoea—*Rhinosporidium seeberi* and mycelial Oomycetes—*Pythium* and *Lagenidium* species are **not** classified in the kingdom **Fungi** due to their taxonomical features. The diseases caused by these organisms are very similar in presentation to the fungal infections hence are called **pseudofungi**. Therefore, due to apparent fungus-like morphology as well as their clinical resemblances, these organisms are being described in this Textbook of Medical Mycology despite the fact that these are not fungal agents.

The algae are placed in the kingdom **Protista**. In general, these are considered non-infectious agents to mammals, however, a few of these are pathogenic. One among them belongs to genus ***Prototheca***, which contains unicellular, achlorophyllous, aerobic organism with its life cycle and morphology very similar to those of **green-algae** belonging to the genus ***Chlorella***. Both these algal genera are commonly prevalent among domestic animals but infection in humans is very rare. By now more than hundred cases of human **protothecosis** have been reported. Majority of them are caused by ***Prototheca wickerhamii*** although ***P.zopfii*** and some of the newly discovered species have also been isolated from few patients. On the other hand, there are very few cases of human **chlorellosis**, which have been reported till date. A brief description of both these human algal diseases are dealt in this Chapter.

I. PROTOTHECOSIS

Protothecosis is an uncommon but emerging disease of animals and man caused by achlorophyllous aerobic algae-like organisms belonging to genus ***Prototheca***. In animals its species cause bovine mastitis and visceral diseases, whereas, in man, infection usually remains confined to skin, subcutaneous tissues and regional lymph-nodes with very occasional systemic dissemination.

Out of **seven** established species of genus *Prototheca*, only **two** namely—***P.wickerhamii*** and ***P.zopfii*** are mainly pathogenic to man and domestic animals. The human protothecosis is reported from widely scattered geographical areas. In recent times, this infection has acquired immense importance because of an increase in the number of cases with systemic involvement getting reported not only among immunocompromised patients but in immunocompetent individuals as well.

Historical Perspective

Wilhelm Kruger in 1894 originally referred genus *Prototheca* to group of organisms sharing very similar morphological and physiological characteristics, after he isolated one from slime flux of trees. Because of yeast-like morphology on culture, *Prototheca* was then classified as a fungus. But in 1913, Chodat classified it as an algae and in 1916, West noted sporulation identical to the algae and re-classified it as an achlorophyllic **mutant** of green algae.

The first case of bovine mastitis was reported by Lerche from Germany in 1952. A few years later in 1964, Davies *et al,* reported first case of human protothecosis in a rice farmer from Sierra Leone in western Africa, who was having localized, chronic skin lesion over the foot. *Prototheca zopfii* was established as the causative organism. However, in subsequent reports in the literature *P.wickerhamii* was found to be the predominant causative agent of human protothecosis. In majority of the cases, skin or olecranon bursa was involved.

The first case of multi-organ systemic protothecal infection was described in 1974 by Cox *et al,* in a 29-year-old man who presented with involvement of liver, peritoneum, lymph nodes, skin and blood. In India, Pal *et al,* reported association of *Prototheca* species with bovine mastitis in 1997. Dhar *et al,* have reported this disease from Rajasthan state of India in 1999. In 2006, Roesler *et al,* proposed two genotypes of *P.zopfii i.e.* genotype 1 (GT1) and genotype 2 (GT2), wherein GT2 is mainly responsible for severe forms of animal infections. However, molecular characterization of *Prototheca* strains isolated in China revealed the first cases of protothecosis associated with *Prototheca zopfii* genotype 1 as reported by Hirose *et al,* in 2017.

In the recent times, a Working Group-**Medical Phycology: Protothecosis and Chlorellosis**, was also approved on May 4, 2014 under the aegis of International Society for Human and Animal Mycology (ISHAM) to support all the scientific aspects and advancements dealing with such algal diseases.

Organism (*Prototheca* Species)

Prototheca species are aerobic algae-like organisms that **lack chlorophyll** but are closely related to the **green-algae** —*Chlorella,* as they exist as saprobes in nature and have very similar life cycle. It lost the ability to photosynthesize and are widely distributed in nature throughout the world in sewage, soil, lakes and marshes. These are spherical unicellular ubiquitous organisms, 3-30 µm in diameter reproducing by asexual internal septation resulting in **morula-like** morphology, resembling **mulberry** or **spoke-wheel** or **daisy-like** appearance. After further divisions of the daughter cells within parent cells, sporangia, containing as many as 50 sporangiospores are formed as morula-like structure. The sporangiospores are passively released upon rupture of the wall of mother cell as a result of pressure exerted by enlarging sizes of daughter cells. Thereafter, each spore, when liberated, passes through the same life cycle. The resting cell stage between an aplanospore and formation of new sporangium is called as **dauer cell**, which can be recognized by thick cell wall and no cell division.

As these are algae-like, aerobic, unicellular eukaryotes that lack chlorophyll thus being unable to produce energy by photosynthesis. They are heterotrophic, requiring external sources of carbon, nitrogen and thiamine, which they obtain from detritus and the waste products of other organisms. The genus *Prototheca* belongs to the family *Chlorellaceae*. It is a unicellular organism that reproduces asexually by internal septation and cleavage, to form sporangia (6 to 26 µm in diameter) containing sporangiospores. The size of the organisms is variable depending upon the different developmental stages of their maturity.

There has been controversy over taxonomic classification of *Prototheca* since it shares some of morphological features with both algae and fungi. The absence of cell wall components like glucosamine and muramic acid distinguish algae from fungi and bacteria, respectively. *Prototheca* can be differentiated from fungi on the basis of cell wall composition, mitotic apparatus, synthetic pathway of amino-acid and nucleic acid sequences. *Chlorella* and *Prototheca* species can be differentiated from bacteria and fungi by their size, shape and mode of reproduction. *Prototheca* contain cellulose in their cell wall and divide mitotically involving centerioles, which is lacking in fungi. It synthesizes lysine through diaminopimelic acid (DAP) pathway, whereas fungi use aminoadipic acid (AAA) pathway. This achlorophyllous algae, grow readily on most of conventional culture media. On electron microscopy, *Prototheca* also differs from *Chlorella* by the presence of **two** rather than **three**-layered cell wall, without chloroplasts and prominent cytoplasmic granules; the outer layer is thinner as compared to the inner one.

The taxonomic status of *Prototheca* is still uncertain and is considered to be a possible mutation of the green algae of genus *Chlorella* belonging to the kingdom **Protista** as they sporulate in similar manner. Initially, they were considered as yeasts, on the basis of morphology hence were included in the kingdom Fungi. But now the genus *Prototheca* belongs to family **Chlorellaceae**, order **Chlorellales** in class **Trebouxiophyceae** of phylum **Chlorophyta**, which include unicellular organisms that reproduce asexually by internal septation and cleavage. There are **seven** valid species in the genus, which are currently recognized, namely, *P.wickerhamii, P.zopfii, P.blaschkeae, P.cutis, P.stagnora, P.ulmea* and *P.miyajii*. Others like *P.moriformis* was found to be synonym of *P.zopfii* and *P.filamenta* is now found identical to *Trichosporon inkin*. Out of seven species, only **two**, *P.wickerhamii* and *P.zopfii,* predominantly cause human and animal diseases, respectively. Further *P.zopfii* is classified into two genotypes (genotype 1 and 2) based on its biochemical, serological and genetic assays.

Structurally, *Prototheca* species are found either as unicellular endospores or as multicellular clusters called sporangia. The endospores reproduce by internal septation and evolve into sporangia. The sporangia of *Prototheca* are distinctive and are often referred to as **morula**. A morula has a central endospore that is surrounded by 5 endospores, which are molded in single layer, peripherally forming soccer ball-like configuration. The sporangia of *P.wickerhamii* (7-13 µm) are generally smaller than those of *P.zopfii* (14-16 µm) but this may be influenced by environmental factors. The cells of *P.wickerhamii* are round in comparison to oval or cylindrical shape of most strains of *P.zopfii*.

On electron microscopy *Prototheca* species may be distinguished from *Chlorella* by absence of chloroplasts and presence of a double layered cell wall in contrast to triple-layered wall of *Chlorella* species.

Epidemiology

Prototheca species are ubiquitous organisms that live saprophytically on decaying organic matter. They are isolated from a wide variety of sources like sewage, soil and many water sources, including both fresh and salty water. It has been recovered from fresh and marine water, fish tanks, swimming pools, waste stabilization ponds, vegetables and meat products such as beef, pork, crabs and clams. The organisms are found in a wide range of environmental sites like tree slime flux, sewage, fresh or marine water, soil and foodstuff with high humidity and decaying organic matter. They have been isolated from flowing water, manure, drinking water sources and wet soil. These heterotrophic algae have adapted to natural environment as slime flux, which contain carbohydrates, alcohol and fatty acids for their growth. As *P.zopfii* and *P.wickerhamii* are ubiquitous inhabitants of raw and treated sewage of rural and urban areas, they may contaminate aquatic systems and are subsequently taken by man or animals and then excreted thereby completing the cycle. *Prototheca* species have also been normally isolated from the skin, feces and sputum of healthy individuals. *P. zopfii* is classified into genotypes 1 and 2; genotype 1 is predominantly found in bovine and swine feces and genotype 2, isolated from bovine mastitis cases.

In the last few decades, comprehensive studies focusing on the genotyping of *Prototheca* species have enabled advancements in the epidemiology of protothecosis among domestic animals, particularly in relation to the geographic distribution of these microalgae. The genotypic characterization using multiplex PCR confirmed that most of the significant isolates are *P.zopfii* and further identified as genotype 2.

Protothecosis in man and animals usually follows traumatic inoculation in the exposed parts of the body, which may be too trivial to be remembered by the patient. In this disease person-to-person transmission does not occur. As such it is non-communicable disease and the incubation period is considered to be about two weeks. The highest prevalence of *P.zopfii* is reported in bovine mastitis and the organism can be isolated from various dairy farm habitats such as stalls, bedding and feces. The contamination of milk and dairy products with *Prototheca* species and resistance to thermal treatment represents public health risk of transmission of this algal zoonosis. There is no age or racial predilection for protothecosis. The condition is more commonly seen in adult males as compared to the females.

Human protothecosis is caused by either of these two common species but mainly it is *P.wickerhamii*, as *P.zopfii* is predominantly an animal pathogen. Although *P.zopfii* was isolated from the index case (from Sierra Leone) but subsequently *P.wickerhamii* outnumbered it in humans as observed by Krcmery's review of 108 cases over a period of twenty-five years, wherein hardly four cases were caused by *P.zopfii*. When *Prototheca* is involved to cause disease in animals, it is almost always *P.zopfii* and its genotype 2 is the main pathogenic type involved in bovine mastitis globally. So in humans, protothecosis is mainly caused by *P.wickerhamii*. Protothecosis is an uncommon infection but it is seen worldwide, with cases predominately reported from North America and Asia. Since the first report of human protothecosis, 150 cases have been documented upto 2011 as per the report of Todd *et al.* However, about 50 cases have been subsequently reported making it to around 200 cases during the entire period of last five decades of this disease.

Immunity

Immunological changes associated with protothecosis have not been adequately studied. However, it is observed that almost half of patients have some underlying immunodeficient condition. It is still not known whether antibodies can be protective to *Prototheca* or not. Intracellular killing of *P.wickerhamii* by human polymorphonuclear neutrophils *in vitro* is enhanced in the presence of humoral factors, which stimulate oxidative metabolism and degranulation. The use of corticosteroids may make the individual vulnerable to opportunistic protothecosis by suppressing cell-mediated immunity. In addition, cytotoxic agents and specific immunosuppressants are also common contributory factors.

Pathogenesis and Pathology

The pathogenesis of human protothecosis is not very well understood. Histopathologically, skin lesions often show hyperkeratosis, mild acanthosis and dense dermal granulomatous infiltrate consisting of predominantly histiocytes, multinucleated giant cells and occasional lymphocytes, macrophages with focal areas of necrosis. The olecranon bursal lesions have oval and stellate areas of fibrinoid necrosis, surrounded by granulomatous response. In soft tissue lesions, microabscesses are very common. The tissue reaction in protothecosis is almost similar to that seen in algosis. There is production of biofilm, which is species-dependent and *P. zopfii* genotype 2 strains are considered as strong biofilm producers. This genotype is reported in severe forms of disseminated protothecosis in animals.

There are multinodular granulomatous or pyogranulomatous chronic lesion with intralesional algal *Prototheca* organisms. The macrophages and giant cells contain most of the algal cells intracytoplasmic. These algal cells are large single or binucleated cells as sporangium as well as septated cells as endosporulation, about 8-15 μm in diameter. In addition, similar or larger morula-like cells, 10-30 μm diameter, containing 4, 8 or more spores within, may be seen admixed. In sites with a strong inflammatory response and necrosis is common to see empty shell-like and refractile structures. There is occasionally a very minimal inflammatory response seen in this disease.

Clinical Features

The human protothecosis although is a sporadic disease but is seen worldwide affecting both immunocompromised as well as immunocompetent individuals. The species involved is invariably always *P.wickerhamii* except a very few human cases caused by *P.zopfii.* As the organisms are ubiquitous in nature hence usually enter the body following minor trauma or surgical incision. The disease may present with protean manifestations however majority of infections are localized to the skin and subcutaneous tissue. The clinical manifestations can be grouped into following three categories, based on the cases reported so far in the literature:

1. Cutaneous-Subcutaneous Protothecosis
2. Protothecal Olecranon Bursitis
3. Systemic-Disseminated Protothecosis

1. Cutaneous-Subcutaneous Protothecosis

This is the most common type of presentation in protothecosis, accounting for almost half of the total cases. It is characterized by the development of single or multiple painless, slowly progressive, well-circumscribed plaques and papulonodular lesion of skin or underlying subcutaneous tissue, which may also have an **apple-jelly** hue. There may be ulcer formation too. The infection may occur following trivial trauma over exposed parts of body. Onychoprotothecosis involving nails due to *P.wickerhamii* has also been reported. The erythematous plaques of this disease are usually confused with atopic dermatitis. Cutaneous protothecosis presenting with erythema and induration following excision of an un-diagnosed case of olecranon bursitis is seen in **Figure 37.1**.

2. Protothecal Olecranon Bursitis

There is another subset of patients who present with olecranon bursal infections. Together they constitute about one-third of total cases of protothecosis. In this clinical entity both the species, *P.wickerhamii* and *P.zopfii*, are involved. Protothecal olecranon bursitis is usually seen in immunocompetent individuals following non-penetrating trauma. The patients have painful swelling of elbow, mild erythema and tenderness at local site due to articular inflammation. The dissemination of infection to regional lymphnodes or internal organs has not been reported.

3. Systemic-Disseminated Protothecosis

About one-third of cases of protothecosis present as systemic infections. However, disseminated disease is more frequently encountered in animals as compared to humans.

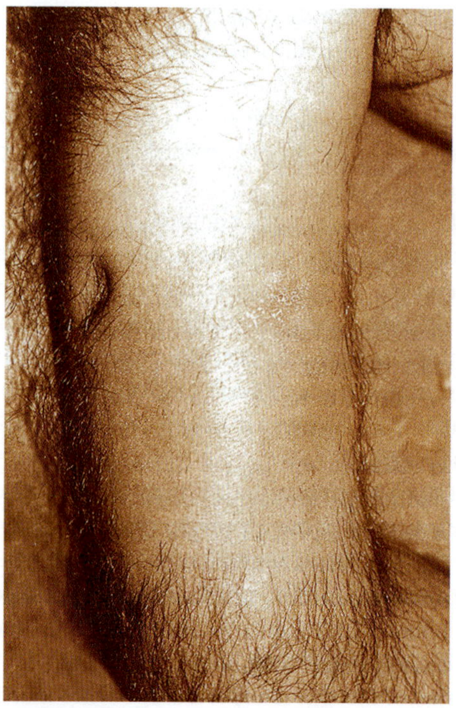

Fig. 37.1. Subcutaneous protothecosis of arm with erythema and induration following protothecal olecranon bursitis.

It may also present as an opportunistic infection as observed in increasing number of human cases. The opportunistic wound infections are probably the second most common type of human protothecosis and infection in immunosuppressed patients tends to be more florid. There are some case reports that this disease is found among AIDS patients also. Meningitis and granulomatous encephalitis may also be caused by *Prototheca* species. The catheter-related disease, disseminated abdominal and respiratory infections have occasionally been reported. *Prototheca* species are now implicated as an intestinal pathogen causing tropical sprue in patients with chronic mucocutaneous candidiasis as the organism has been isolated from stool culture. Moreover, peritonitis is also reported among patients undergoing continuous ambulatory peritoneal dialysis (CAPD).

The ocular manifestations, which were previously reported in animals, are now found in human protothecosis also. Bilateral choroiditis has been reported in case with *P.wickerhamii* algaemia. Moreover, algaemia caused by *Prototheca* species has also been reported in patients of myasthenia gravis. There is a report of an elderly man having developed disseminated protothecosis and later died during treatment with infliximab for refractory graft-versus-host disease. Protothecosis has been reported in an immunocompromised host on long-term prednisolone therapy for the treatment of systemic lupus erythematosus (SLE). There are case reports of urinary tract infections. The disseminated infection with *P.zopfii*, following stem cell transplantation for leukemia, is reported. The cancerous conditions are found to be underlying predisposing factors for protothecosis. Disseminated *P.wickerhamii* infection with arthritis and tenosynovitis is also reported. This organism is also found to be complicating prolonged endotracheal intubation.

Protothecosis in Animals

The genus *Prototheca* is prevalent in dairy environment, particularly in damp areas contaminated with manure. Hence protothecosis involves a wide variety of animals both as individual case or herd infection. Most cases have been reported in dogs and cattle. The first case of bovine mastitis was reported in 1952. The canine form tends to be more severe with systemic manifestations involving multiple internal organs entailing fatal consequences. However, in bovine infections, only udder is usually involved and protothecal bovine mastitis is currently prevalent worldwide. There may be severe local and systemic infections in other domestic animals, especially in dogs. The animal infections are considered to be relatively more opportunistic as compared to humans and subclinical bovine mastitis has been found to adversely affect the milk production among cattle. *Prototheca* can transiently colonize alimentary canal of both animals and humans without causing disease. When there is protothecosis in animals, it is almost always caused by *P.zopfii* and that too by genotype 2.

There is no serological test available for screening of dairy herds infected with *P.zopfii*. An ELISA-based technique for detection of *P.zopfii*-specific IgA and IgG_1 antibodies in whey and IgG antibodies in serum, have been tried. The test provides rapid, reliable and inexpensive screening methods for *P.zopfii* compared to plate culturing. An early detection of subclinical animal disease reduces risk of fresh infection and contamination of surrounding environment.

Differential Diagnosis

This clinical entity is very rare hence it is usually not suspected at first instance. Protothecosis should be differentiated from eczematous dermatitis, neurodermatitis, dermatophytosis, sporotrichosis, chromoblastomycosis and African histoplasmosis. Several bovine infections caused by green algae *i.e. Chlorella,* have been misdiagnosed as protothecosis. The visceral involvement is related to hepatic and biliary tree hence protothecosis should be included as one of the differential diagnoses of chronic inflammatory process involving gastrointestinal or biliary tract, particularly in animals **(Table 37.1)**.

The appearance of protothecal sporangia is diagnostic and described as moruloid, daisy-like, spoke-like and frambesiform. The differentiation is based on structural differences, with morula configuration being characteristic for *P.wickerhamii.* Therefore, *Prototheca* species must be differentiated from other endosporulating organisms particularly those which undergo internal septation *e.g. Coccidioides* species. In tissues, the organism is easily distinguished from *Coccidioides*, which produce sporangia upto 10 to 100 times larger in size, whereas the endospores are smaller (2-4 μm). The spores are 1.3-16.1 μm in diameter and have internal septation of cart-wheel-like appearance. Similarly, in rhinosporidiosis sporangia are larger and contain more endospores and are somewhat different from the morula-like appearance seen in protothecosis. These comparative features are described in **Table 18.1** for a ready reference. Histopathological differential diagnoses include other systemic and deep fungal infections such as cryptococcosis, blastomycosis and paracoccidioidomycosis.

Table 37.1. Differences between Various *Prototheca* and *Chlorella* Species.

Sr. No.	Features	Prototheca Species	Chlorella Species
1.	Gross lesions (Color)	Multinodular granulomatous reaction colorless or non-specific yellowish discoloration	Same multinodular granulomatous reaction but there is characteristic greenish and bright discoloration in fresh tissues. The greenish color persists even after formalin fixation
2.	Histopathology (H&E)	Multinodular granulomatous or pyogranulomatous chronic lesion with intralesional algal organisms.	Almost same as that of *Prototheca* species
3.	Histochemistry, PAS and GMS stains	*Prototheca* species have cell wall that reacts positive. Intracellular components are poorly stained	*Chlorella* species have cell wall positive as well as distinctive and strongly positive intracytoplasmic granules that sometimes occlude the cell completely. Empty shells appear thicker and also strongly positive
4.	Chloroplasts and starch bodies (TEM)	Absent	Present

Laboratory Diagnosis

In the laboratory a definitive diagnosis of human protothecosis is established by direct demonstration of organisms in the clinical material followed by culture on various media and/or specific immunohistochemical reactivity *in situ*. The other non-specific laboratory parameters include hypergammaglobulinemia, elevated ESR and eosinophilia, etc.

(a) Direct Examination

Protothecosis is diagnosed on smear examination in KOH wet mount and biopsy by histopathology. The inflammatory response shows both microabscess and granuloma formation with multinucleate giant cells. The organisms can be demonstrated in clinical specimens by various staining methods. Routine Gram's stain shows Gram-positive yeast-like cells. On H&E stain, organism may appear poorly stained as clear spaces with or without visible cell wall. The cells usually range from 8-20 μm in diameter and are better demonstrated by PAS **(Fig. 37.2)** or GMS staining The wall of sporangium and sporangiospores is uniformly stained. The morphological form "morula" is uncommon in infections caused by *P.zopfii*. Moreover, endosporulating cells of *P.zopfii* are oval and their larger nuclei are more conspicuous than those of *P.wickerhamii*. It is important to recognize spores and sporangia, containing as many as 50 spores lying clustered together as morula-like structures, in tissue sections on staining with PAS or GMS.

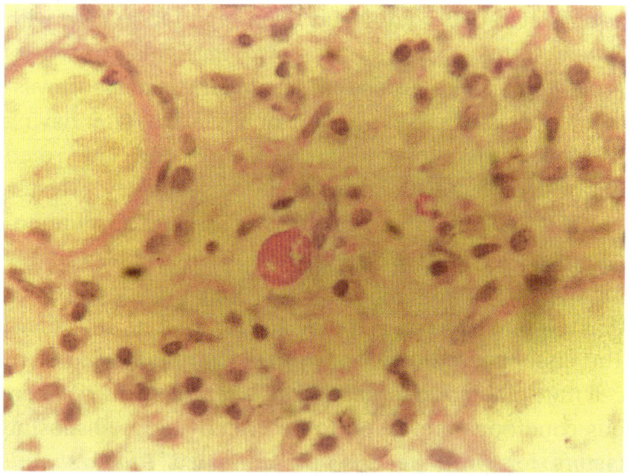

Fig. 37.2. Thick-walled sporangia with sporangiospores of *Prototheca* species in an inflammatory infiltrate (PAS × 400).

An immunofluorescence stain has been developed for identification of *Prototheca* in tissues and it may be used for species identification. They often contain 2-8 tightly packed sporangiospores in each cell or **sporangium**. The sporangia are usually smaller than those of *Coccidioides* species and *Rhinosporidium seeberi,* whereas sporangiospores of *P.zopfii,* seen in tissues, are larger than endospores of *Coccidioides* species and *R.seeberi*. Indirect fluorescent antibody test is also very useful for identification of the disease. Immunofluorescence tests are developed for identification of *Prototheca* species and for their differentiation from morphologically similar fungi of various genera in formalin-fixed tissues. Histopathology shows numerous PAS-positive organisms with distinctive mulberry-like endosporulation in the dermis.

An organism can be recognized as *Prototheca* species if the characteristic endosporulating cells are visible; some of which may resemble small *Coccidioides*-like spherules. However, if these structures are not visible in tissue sections, the non-sporulating cells may resemble a number of other fungi, *e.g. Blastomyces dermatitidis, Cryptococcus* species, *Paracoccidioides brasiliensis* and some stages of *Coccidioides*. Several bovine infections due to green algae *i.e. Chlorella* species, have been mis-diagnosed as protothecosis. Therefore, to establish an accurate diagnosis of protothecosis, it is essential to cultivate the organism and document its nature *in situ i.e.* demonstrate absence of chloroplasts, which are seen in the infections caused by *Chlorella* species.

Fourier Transform Infrared (FT-IR) spectroscopy has been reported to be suitable, efficient method for distinguishing and characterizing human-pathogenic yeasts and animal-pathogenic algae and is also suitable for diagnostic differentiation between *P.zopfii* and *P.wickerhamii*.

(b) Culture

The clinical material from patients is collected in the form of pus swab, biopsy or blood sample for culture. *Prototheca* grows within 24-72 hours on conventional fungal culture media including SDA **(without cycloheximide)**, beef-infusion broth, blood agar and BHI agar with 5% sheep blood. Like *Cryptococcus* species, *Prototheca* species are also inhibited by cycloheximide *in vitro*. Hence selective media containing cycloheximide should not be used for inoculating the clinical sample suspected to be having *Prototheca* species.

The colonies of *Prototheca* species on SDA are yeast-like, creamy-white and dull or moist to mucoid in appearance at both the temperature of incubation *i.e.* 30°C and 37°C. In the upper portion the colonies of *P.zopfii* and lower one *P.wickerhamii* after 3 days of incubation **(Fig. 37.3)**. On CHROMagar Candida the colonies are frequently confused with yeasts of *Candida* and *Cryptococcus* species. *Prototheca* species also produce cream-colored colonies after 3 days of incubation **(Fig. 37.4)**, similar to those of *C.parapsilosis* but *Prototheca* colonies are slightly smaller in size as compared to that of *C.parapsilosis*. In the upper portion the colonies of *P.zopfii* and lower one *P.wickerhamii*.

The LCB/PHOL mounts show typical morula/mulberry-like multi-cellular *Prototheca* containing slightly smaller sporangiospores of *P.wickerhamii* **(Fig. 37.5)** and bigger ones *P.zopfii* **(Fig. 37.6)**. Electron microscopy shows sporangium with five endospores with single-layered wall. The isolates may be identified by PCR genotyping and MALDI-TOF mass spectrometry proteomic phenotyping, particularly for epidemiological analysis of world-wide strains from bovine, canine and human infections. The PAS staining done from the culture of *P.zopfii* shows slightly bigger sporangiospores **(Fig. 37.7)**.

The diagnosis of protothecosis is generally made by direct identification of the organism in histologic sections or by culture from infected tissue. The cells of *P.wickerhamii*

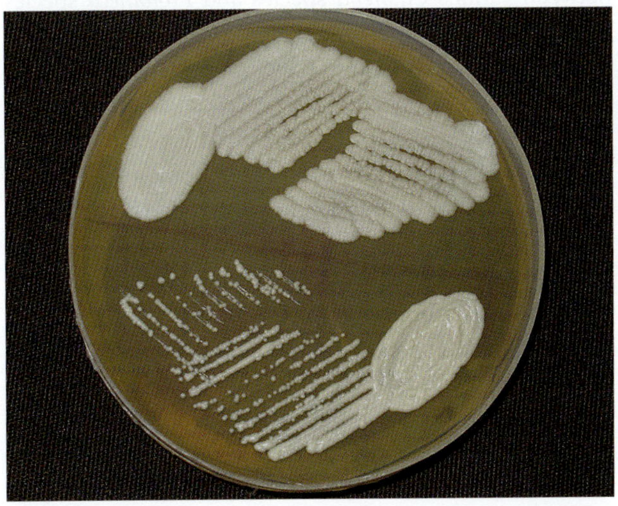

Fig. 37.3. Yeast-like colonies of *Prototheca* species (upper half *P.zopfii* and lower half *P.wickerhamii*) after three days on SDA.

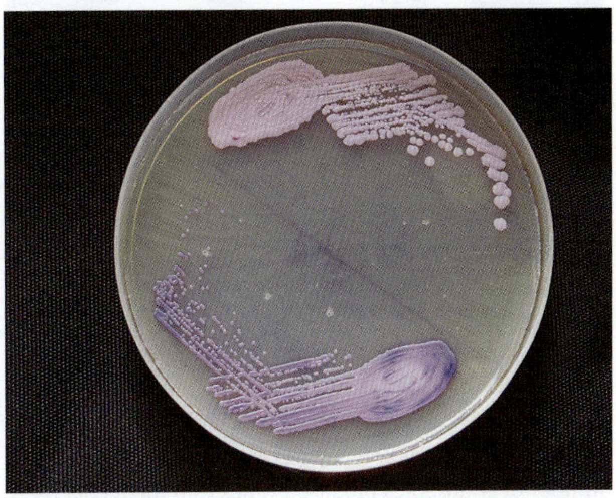

Fig. 37.4. Yeast-like colonies of *P.zopfii* in upper and *P.wickerhamii* in lower half after three days on CHROMagar Candida plate.

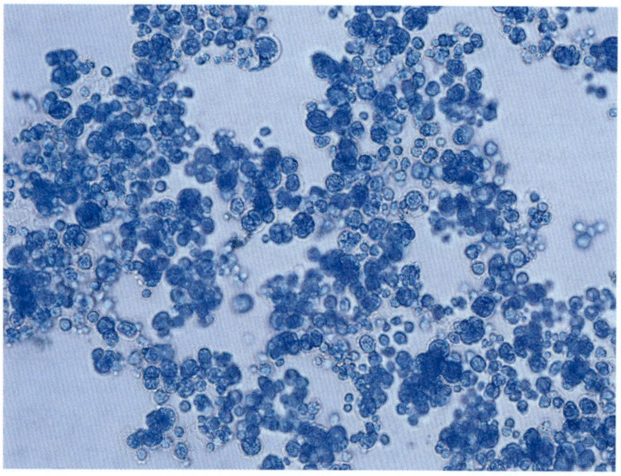

Fig. 37.5. Sporangia and sporangiospores of *Prototheca wickerhamii* forming smaller morula-like patterns (LCB × 400).

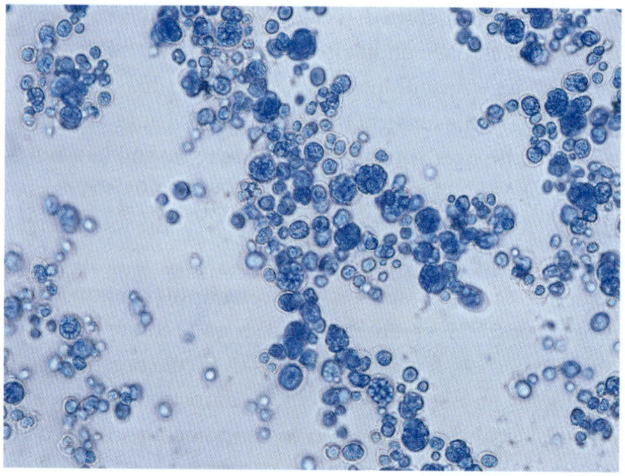

Fig. 37.6. Sporangia and sporangiospores of *Prototheca zopfii* forming relatively bigger morula-like patterns (LCB × 400).

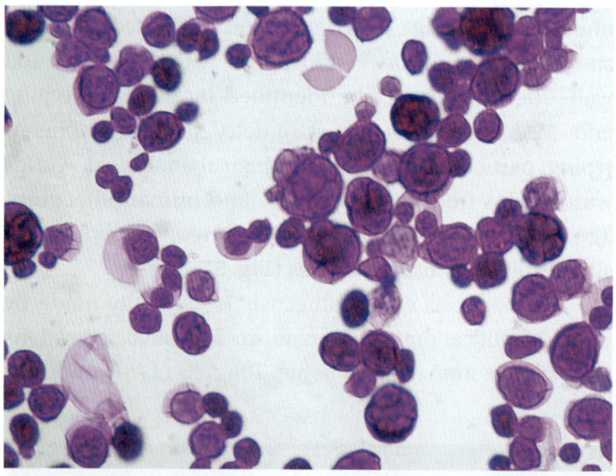

Fig. 37.7. Sporangia and sporangiospores of *Prototheca zopfii* forming morula-like patterns (PAS × 400).

are rounded in comparison to oval or cylindrical cells of *P.zopfii.* Both species reproduce by internal septation, resulting in the formation of sporangia. Each sporangium usually contains up to 20 endospores that are released on rupture.

The sporangia of *P.wickerhamii* are smaller, round with symmetrical morula formation and diameter of 7-13 μm than those of *P.zopfii*, which are oval with relatively asymmetric random septation having diameter of 14-16 μm. The number of endospores produced varies with the media used. These thick-walled structures range from 1.3 to 13.4 μm in diameter depending upon the species. The morula form, containing 2 to 20 endospores in spoke-like pattern, is the characteristic feature of *P.wickerhamii.*

The organisms that morphologically mimic *Prototheca* include those capable of forming endosporulating spherules, such as *Coccidioides* species, *Rhinosporidium seeberi* and *Chlorella* species. *Coccidioides* species and *R.seeberi* form sporangia 10 to 100 times larger than protothecal sporangia. Lesions produced by *Chlorella* species are **greenish** on gross examination of specimens. Moreover, they contain greater amount of PAS-positive, diastase-sensitive granules than do *Prototheca* species, owing to starch present within the cytoplasm. The differences between various *Chlorella* species and *Prototheca* species are given in. The non-spherulating organisms mimicking *Prototheca* species include *Cryptococcus* species, *Blastomyces dermatitidis* and *Paracoccidioides brasiliensis. Cryptococcus* is encapsulated budding yeast, whereas *B.dermatitidis* produces broad-based budding yeast cells and *Paracoccidioides* forms multiple buds on single/ mother yeast cell.

Biochemical tests may be done to identify *Prototheca* species. They assimilate galactose, trehalose, glycerol and glucose but no other carbohydrates. Moreover, *P.zopfii* or *P.stagnora* can be differentiated from *P.wickerhamii*, as both do not assimilate galactose or trehalose. Hence **trehalose** assimilation remains the key test to differentiate *P.wickerhamii* from *P.zopfii.* Susceptibility to neomycin is helpful in differentiating *P.wickerhamii* and *P.zopfii* from *P.stagnora*.

***Prototheca* isolation medium** (PIM) has been used as selective medium for routine detection and isolation of *Prototheca* species from natural sources, where lot of contamination is expected. In this medium, phthalate inhibits bacteria and fluorocytosine inhibits yeasts. This can be modified by adding Rose Bengal for further suppression of bacterial and fungal growth. Sheep blood agar can also be used for enhancing growth of *Prototheca*. The colonies are yeast-like and usually mistaken for similar fungi. The daughter cells, **without budding** can be observed within the theca *i.e.* **sporangiospores**. The reproduction is by formation of variable number of sporangiospores within sporangium.

The species differ from each other in the number of sporangiospores produced and size of the organism. The laboratory differentiation of these species may be difficult, which is usually based on macroscopic and microscopic studies, temperature requirement for growth and sugar assimilation pattern. *P.wickerhamii* can be differentiated from sporangiospore diameter when grown on PIM, which is larger in *P.zopfii* with average 6.5 μm as compared to *P.wickerhamii* with average size 3.2 μm and corresponding sporangia measuring about 25 μm and 9.4 μm, respectively. Moreover, trehalose assimilation is **positive** in *P.wickerhamii* and negative in rest of the species.

It has also been isolated from the blood cultures of patients with systemic protothecosis having algaemia. *Prototheca* species may morphologically be confused with yeasts such as *Candida* and *Cryptococcus* on culture and even on Gram staining. PHOL staining, a technique similar to LCB mount, may also be used to examine morphological details of *Prototheca* species.

On CHROMagar Candida medium, *Prototheca* produces cream-colored colonies after 48-72 hours of incubation at 30°C similar to those of *Candida parapsilosis* but are slightly smaller in size. However, the CHROMagar cannot be used to differentiate among *Prototheca* species since all species *i.e. P.wickerhamii, P.zopfii, P.stagnora* produce almost similar colored growth. The colonies are sub-cultured on cornmeal agar and incubated at 30°C. Growth on Sabouraud dextrose agar is obtained at both temperatures, which distinguishes *P.wickerhamii* and *P.zopfii* from *P.stagnora* which grows only at 30°C.

The conventional auxanographic method of testing for the assimilation of carbohydrates and alcohols by various species of *Prototheca* requires at least two weeks of incubation at 25 to 30°C before definitive results are obtained. The commercially available ready-to-use micro-method like API 20C, API 20C AUX and Vitek 2 yeast identification systems, are found useful for the definitive identification of *P.stagnora*, *P.wickerhamii* and *P.zopfii*. The characteristic carbohydrate assimilation pattern of *P.zopfii* is glucose and glycerol positive; galactose and **trehalose negative** at 37°C of incubation. Moreover, in case *P.zopfii*, genotyping is important as genotype 2 is significantly pathogenic as compared to genotype 1.

On sub-cultures, *P.stagnora* produces typical bright, moist, mucilaginous, cream-to-brownish colored colonies. The India ink preparation shows capsule, surrounding the cells of *P.stagnora*, which resembles that of *Cryptococcus*. However, its inability to produce urease or to grow at 37°C, differentiates it from *Cryptococcus* species. The molecular based techniques are used to identify *Prototheca* species in clinical material, including ribosomal internal transcribed spacer (ITS) sequencing and 26S rRNA gene sequencing. Moreover, successful identification with mass spectrometry has also been reported.

(c) Immunodiagnosis

The isolation of *Prototheca* species from stools with positive serology should be considered as strong presumptive evidence of visceral protothecosis. Both, isolation and the serological studies are relatively easier tasks for any clinical laboratory and should yield negative results among healthy individuals.

(d) Animal Pathogenicity

This is very difficult to produce infection with *Prototheca* species in the experimental animals. Subcutaneous and intraperitoneal injections of *P.zopfii* are able to produce local and metastatic lesions in mice, rats and monkeys. There is a weak inflammatory response but the chronic infection of tubule and capsule of kidney is significant manifestation. *P.wickerhamii* infection could not be established in healthy Swiss albino mice when inoculated with one million cells of the organism by intravenous route. This finding suggests that *Prototheca* species are able to produce infection in immunocompromised host.

The differences in morphological features of **spherules** and similar structures producing fungal as well as pseudofungal organisms are already given at a glance in **Table 18.1**.

Treatment and Prophylaxis

The disease has little tendency for self-healing, with the typical course being chronic and slowly progressive with low-grade inflammation. Most cases have been successfully treated by excising the lesion or with antifungals. Amphotericin B, both conventional as well as liposomal, may be recommended for the treatment since *Prototheca* cell membrane has approximately 4% ergosterol, which is the target site of its action. If cutaneous lesion is localized, surgical excision followed by intravenous amphotericin B may be successful. Simple bursectomy cures the lesions of olecranon bursitis.

The efficacy of azoles is unclear, with variable clinical response and no standardized susceptibility testing method being available. Treatment with oral ketoconazole has documented variable response and *Prototheca* is resistant to flucytosine. Both species, *P.wickerhamii* and *P.zopfii* were found to be resistant to fluconazole and itraconazole by Etest in one of the studies but in others this combination has been found to be successful. However, in another report, voriconazole was found to be effective in the treatment of protothecosis. Isolated case reports showing variable responses to ketoconazole, fluconazole, itraconazole, tetracycline and gentamicin have been described. Moreover, adjunctive thermal therapy to the isolated lesions may be effective. The cutaneous protothecosis caused by *P.zopfii* has been successfully treated with amikacin combined with tetracyclines. Some of the newer agents like miltefosine may be promising therapies in the times to come. *P.wickerhamii* is generally considered as susceptible to antifungal agents and is easily treated while *P.zopfii* is resistant to treatment.

In a recent study of 2017 by Alves *et al, P.zopfii* (genotype 2) strains of animal origin were found to be *in vitro* susceptible to guanidine with minimal algaecide concentrations ranging from 0.001 to 0.035%. This is known to have a high microbicidal effect and is considered to be a new generation microbicidal compound. It is not toxic to human mucous membranes and conjunctivas at low concentrations and has been used as a disinfectant, in swimming pools and as an antiseptic for human wounds. The algicidal action of guanidine at low concentrations indicates that it could be an alternative disinfectant or antiseptic for cleaning of the dairy environment and milking equipment, in pre- and post-dipping solutions, in the chemical dry therapy of bovine teats and even in the intramammary therapy of *P.zopfii* infections.

II. CHLORELLOSIS

Chlorellosis is very similar to protothecosis and is caused by species of unicellular green algae belonging to genus *Chlorella*. There are only a very few reports of the human cases till date. The organism contains chloroplasts in their cells that give typical **greenish** color to its pathological lesions.

Chlorella species are unicellular, ovoid, spherical or polygonal organisms with size of about 4 to 5 μm. They also reproduce asexually by internal septation and cytoplasmic cleavage *i.e.* endosporulation, producing upto 20 sporangiospores within sporangium. All the *Chlorella* species have three-layered wall except *Auxenochlorella protothecoides* (Chlorophyceae), which has double-layering, which is one of the most distinctive features. On transmission electron microscopy, the organisms contain numerous green chloroplasts, which appear as cytoplasmic granules and stain intensely with PAS, GMS and Gridley fungal stains. On maturation, outer wall of sporangium ruptures, releasing sporangiospores, each of which further produces a sporangium.

The clinical features of chlorellosis in domestic and wild animals range from cutaneous and subcutaneous lesions to deeper involvement of lymph nodes and visceral organs presumably related to exposure to water. On physical examination, fresh lesions of chlorellosis are seen as **green-colored**, a feature clearly distinguishing it from protothecosis on clinical grounds.

The diagnosis of chlorellosis is made by direct demonstration of organisms followed by culture of material taken from the infected tissue. The wet mounts of wound exudates and/or impression smear of infected tissue show ovoid, endosporulating cells with typical **greenish** refractile cytoplasmic granules, representing chloroplasts. The organisms are intensely stained with PAS and GMS as compared to the H&E stain.

The *Chlorella* species are easily distinguished from other algae as they contain abundant PAS-positive cytoplasmic granules, which represent starch and chloroplasts, whereas *Prototheca* species are achlorophyllous in nature. The organisms grow readily on conventional fungal culture media producing characteristic bright green colonies. The differences between various *Chlorella* species and *Prototheca* species are given in **Table 37.1**.

The treatment of chlorellosis is surgical debridement and irrigation of the local wound. However, medical treatment with amphotericin B and tetracycline may also be useful, as it is effective in some cases of protothecosis.

Further Reading

1. Ahn A, Choe YJ, Chang J, *et al.* Chronic eosinophilic meningoencephalitis by *Prototheca wickerhamii* in an immunocompetent boy. *Pediatr Infect Dis J.* 2017; **36:** 687-9.

2. Ahrholdt J, Murugaiyan J, Straubinger RK, *et al.* Epidemiological analysis of worldwide bovine, canine and human clinical *Prototheca* isolates by PCR genotyping and MALDI-TOF mass spectrometry proteomic phenotyping. *Med Mycol.* 2012; **50:** 234-43.
3. Alves AC, Capra E, Morandi S, *et al. In vitro* algicidal effect of guanidine on *Prototheca zopfii* genotype 2 strains isolated from clinical and subclinical bovine mastitis. *Lett Appl Microbiol.* 2017; **64:** 419-23.
4. Bandaranayake TD, Paniz Mondolfi A, Peaper DR, *et al. Prototheca wickerhamii* algaemia: An emerging infection in solid organ transplant recipients. *Transpl Infect Dis.* 2015; **17:** 599-604.
5. Blaschke-Hellmessen R. Fluconazole and itraconazole susceptibility testing with clinical yeast isolates and algae of the genus *Prototheca* by means of the Etest. *Mycoses.* 1996; **39:** S39-43.
6. Boyce F, Longcrier DL, Young JW, *et al. Prototheca* as a pathogen. *Clin Microbiol Newsl.* 1979; **1:** 4-5.
7. Boyd AS, Langley M, King LE Jr. Cutaneous manifestations of *Prototheca* infections. *J Am Acad Dermatol.* 1995; **32:** 758-64.
8. Bozzo G, Bonerba E, Di Pinto A, *et al.* Occurrence of *Prototheca* spp. in cow milk samples. *New Microbiol.* 2014; **37:** 459-64.
9. Bussian AH, Diwan AH, Haley HR, *et al.* Cutaneous protothecosis in a patient with myasthenia gravis. *Am J Dermatopathol.* 2002; **24:** 156-60.
10. Buzzini P, Turchetti B, Branda E, *et al.* Large-scale screening of the *in vitro* susceptibility of *Prototheca zopfii* towards polyene antibiotics. *Med Mycol.* 2008; **46:** 511-4.
11. Camboim EK, Garino FJ, Dantas AF, *et al.* Protothecosis by *Prototheca wickerhamii* in goats. *Mycoses.* 2011; **54:** e196-200.
12. Carey WP, Kaykova Y, Bandres JC, *et al.* Cutaneous protothecosis in a patient with AIDS and a severe functional neutrophil defect: Successful therapy with amphotericin B. *Clin Infect Dis.* 1997; **25:** 1265-6.
13. Carfora V, Noris G, Caprioli A, *et al.* Evidence of a *Prototheca zopfii* genotype 2 disseminated infection in a dog with cutaneous lesions. *Mycopathologia.* 2017; **182:** 603-8.
14. Casal M, Linares MJ, Solis F, *et al.* Appearance of colonies of *Prototheca* on CHROMagar Candida medium. *Mycopathologia.* 1997; **137:** 79-82.
15. Cox GE, Wilson JD, Brown P. Protothecosis: A case of disseminated algal infection. *Lancet.* 1974; **2:** 379-82.
16. Cunliffe CH, DiPersio JR. Localised cutaneous protothecosis in a patient with acquired immunodeficiency syndrome. *Clin Microbiol Newsletter.* 2000; **22:** 133-5.
17. Dalmau J, Pimentel CL, Alegre M, *et al.* Treatment of protothecosis with voriconazole. *J Am Acad Dermatol.* 2006; **55:** S122-3.
18. Davies RR, Spencer H, Wakelin PO. A case of human protothecosis. *Trans R Soc Trop Med Hyg.* 1964; **58:** 448-51.
19. Davies RR. Protothecosis and opportunist fungal infections. *Trans St Johns Hosp Dermatol Soc.* 1972; **58:** 38-42.
20. Dhar S, Dhar S. Protothecosis. *Indian J Dermatol Venereol Leprol.* 1999; **65:** 104-5.
21. DiPersio JR. *Prototheca* and protothecosis. *Clin Microbiol Newslett.* 2001; **23:** 115-21.
22. Endo S, Sekiguchi M, Kishimoto Y, *et al.* The first case of feline *Prototheca wickerhamii* infection in Japan. *J Vet Med Sci.* 2010; **72:** 1351-3.
23. Figueroa CJ, Camp BJ, Varghese GI, *et al.* A case of protothecosis in a patient with multiple myeloma. *J Cutan Pathol.* 2014; **41:** 409-13.
24. Follador I, Bittencourt A, Duran F, *et al.* Cutaneous protothecosis: Report of the second Brazilian case. *Rev Inst Med Trop Sao Paulo.* 2001; **43:** 287-90.
25. Fong K, Tee SI, Ho MS, *et al.* Cutaneous protothecosis in a patient with previously undiagnosed HIV infection. *Australas J Dermatol.* 2015; **56:** e71-3.
26. Gandham NR, Vyawahare CR, Chaudhaury N, *et al.* Onychoprotothecosis: An uncommon presentation of protothecosis. *Indian J Med Microbiol.* 2015; **33:** 435-7.
27. Gaur S, Marrin C, Barnes RA. Disseminated protothecosis following traumatic Hickman line removal in a patient with leukaemia. *Med Mycol.* 2010; **48:** 410-2.
28. Grzesiak B, Glowacka A, Krukowski H, *et al.* The *in vitro* efficacy of essential oils and antifungal drugs against *Prototheca zopfii. Mycopathologia.* 2016; **181:** 609-15.
29. Gupta A, Gumber S, Bauer RW, *et al.* What is your diagnosis? Cerebrospinal fluid from a dog. Eosinophilic pleocytosis due to protothecosis. *Vet Clin Pathol.* 2011; **40:** 105-6.
30. Hariprasad SM, Prasad A, Smith M, *et al.* Bilateral choroiditis from *Prototheca wickerhamii* algaemia. *Arch Ophthalmol.* 2005; **123:** 1138-41.
31. Hart J, Mooney L, Arthur I, *et al.* First case of *Chlorella* wound infection in a human in Australia. *New Microbes New Infect.* 2014; **2:** 132-3.
32. Hench J, Roschanski N, Hewer E, *et al.* Granulomatous encephalitis: Protothecosis excluded? *Histopathology.* 2016; **69:** 1082-4
33. Hightower KD, Messina JL. Cutaneous protothecosis: A case report and review of the literature. *Cutis.* 2007; **80:** 129-31.
34. Hillesheim PB, Bahrami S. Cutaneous protothecosis. *Arch Pathol Lab Med.* 2011; **135:** 941-4.
35. Hirose N, Hua Z, Kato Y, *et al.* Molecular characterization of *Prototheca* strains isolated in China revealed the first cases of protothecosis associated with *Prototheca zopfii* genotype 1. *Med Mycol.* 2017. doi: 10.1093/mmy/myx039. PMID: 28525645.
36. Hirose N, Nishimura K, Inoue-Sakamoto M, *et al.* Ribosomal internal transcribed spacer of *Prototheca wickerhamii* has characteristic structure useful for identification and genotyping, *PLoS One.* 2013; **8:** e81223.
37. Humphrey S, Martinka M, Lui H. Cutaneous protothecosis following a tape-stripping injury. *J Cutan Med Surg.* 2009; **13:** 273-5.
38. Iacoviello VR, DeGirolami PC, Lucarini J, *et al.* Protothecosis complicating prolonged endotracheal intubation: Case report and literature review. *Clin Infect Dis.* 1992; **15:** 959-67.

39. Jagielski T, Buzzini P, Lassa H, *et al.* Multicentre Etest evaluation of *in vitro* activity of conventional antifungal drugs against European bovine mastitis *Prototheca* spp. isolates. *J Antimicrob Chemother.* 2012; **67:** 1945-7.
40. Jagielski T, Lagneau PE. Protothecosis. A pseudofungal infection. *J Med Mycol.* 2007; **17:** 261-70.
41. Jagielski T, Lassa H, Ahrholdt J, *et al.* Genotyping of bovine *Prototheca* mastitis isolates from Poland. *Vet Microbiol.* 2011; **149:** 283-7.
42. Jagielski T, Lassa H, Ahrholdt J, *et al.* Molecular characterization of Polish *Prototheca zopfii* mastitis isolates and first isolation of *Prototheca blaschkeae* in Poland. *Pol J Vet Sci.* 2010; **13:** 725-9.
43. Jeunon T, Fantin-Ribeiro A, Jeunon-Sousa MA, *et al.* Erythematous plaque and shallow ulcers on right arm and forearm. *Int J Dermatol.* 2009; **48:** 1171-3.
44. Jones JW, McFadden HW, Chandler FW, *et al.* Green algal infection in a human. *Am J Clin Pathol.* 1983; **80:** 102-7.
45. Joshi KR, Gavin JB, Wheeler EF. The ultrastructure of *Prototheca wickerhamii. Mycopathologia.* 1975; **56:** 9-13.
46. Kaminski ZC, Kapila R, Sharer LR, *et al.* Meningitis due to *Prototheca wickerhamii* in a patient with AIDS. *Clin Infect Dis.* 1992; **15:** 704-6.
47. Kano R, Matsumoto T. ISHAM-WG Symposium: Introduction of Medical Phycology, retrospective overview and prospective aspect. *Med Mycol J.* 2015; **56:** J93-7.
48. Kano R, Sobukawa H, Suzuki M, *et al.* Immunohistopathology of *Prototheca wickerhamii* in cutaneous lesions of protothecosis. *Med Mycol J.* 2014; **55:** E29-32.
49. Kantrow SM, Boyd AS. Protothecosis. *Dermatol Clin.* 2003; **21:** 249-55.
50. Kapica L. First case of human protothecosis in Canada: Laboratory aspects. *Mycopathologia.* 1981; **73:** 43-8.
51. Khoury JA, Dubberke ER, Devine SM. Fatal case of protothecosis in a hematopoietic stem cell transplant recipient after infliximab treatment for graft-versus-host disease. *Blood.* 2004; **104:** 3414-5.
52. Kim ST, Suh KS, Chae YS, *et al.* Successful treatment with fluconazole of protothecosis developing at the site of an intralesional corticosteroid injection. *Br J Dermatol.*1996; **135:** 803-6.
53. Kovalyshyn I, Brankov N, Fernandez AP. Erythematous and edematous plaques on the bilateral extremities in an immunocompromised patient. *Int J Dermatol.* 2016; **55:** e59-61.
54. Krcmery V Jr. Systemic chlorellosis, an emerging infection in humans caused by algae. *Int J Antimicrob Agents.* 2000; **15:** 235-7.
55. Kwong JC, Ward PB, Johnson PD. Cutaneous protothecosis in a patient with hypogammaglobulinemia. *Med Mycol Case Rep.* 2013; **2:** 132-3.
56. Laeng RH, Egger C, Schaffner T, *et al.* Protothecosis in an HIV-positive patient. *Am J Surg Pathol.* 1994; **18:** 1261-4.
57. Lass-Florl C, Fille M, Gunsilius E, *et al.* Disseminated infection with *Prototheca zopfii* after unrelated stem cell transplantation for leukemia. *J Clin Microbiol.* 2004; **42:** 4907-8.
58. Lass-Florl C, Mayr A. Human protothecosis. *Clin Microbiol Rev.* 2007; **20:** 230-42.
59. Le Roux A, Gumber S, Bauer RW, *et al.* Algal meningoencephalitis due to *Prototheca* spp. in a dog. *Case Reports in Veterinary Medicine.* 2013; ID 474731: 1-5.
60. Lee JS, Moon GH, Lee NY, *et al.* Case report: protothecal tenosynovitis. *Clin Orthop Relat Res.* 2008; **466:** 3143-6.
61. Leimann BC, Monteiro PC, Lazera M, *et al.* Protothecosis. *Med Mycol.* 2004; **42:** 95-106.
62. Lu S, Xi L, Qin W, *et al.* Cutaneous protothecosis: Two new cases in China and literature review. *Int J Dermatol.* 2012; **51:** 328-31.
63. Macesic N, Fleming S, Kidd S, *et al.* Protothecosis in hematopoietic stem cell transplantation: Case report and review of previous cases. *Transpl Infect Dis.* 2014; **16:** 490-5. Comment in: Protothecosis: A lethal infection due to an achlorophyllous algae. *Clin Infect Dis.* 2014; **59:** iv.
64. Marques S, Huss VA, Pfisterer K, *et al.* Internal transcribed spacer sequence-based rapid molecular identification of *Prototheca zopfii* and *Prototheca blaschkeae* directly from milk of infected cows. *J Dairy Sci.* 2015; **98:** 3001-9.
65. Marr KA, Hirschmann JV, Thorning D, *et al.* Protothecosis. *Clin Infect Dis.*1998; **26:** 575 & 756-7.
66. Masuda M, Hirose N, Ishikawa T, *et al.* *Prototheca miyajii* sp. nov., isolated from a patient with systemic protothecosis. *Int J Syst Evol Microbiol.* 2016; **66:** 1510-20.
67. Mathew LG, Pulimood S, Thomas M, *et al.* Disseminated protothecosis. *Indian J Pediatr.* 2010; **77:** 198-9.
68. Matsuda T, Matsumata T. Protothecosis: A report of two cases in Japan and a review of the literature. *Eur J Epidemiol.* 1992; **8:** 397-406.
69. Matsumoto Y, Shibata M, Adachi A, *et al.* Two cases of protothecosis in Nagoya, Japan. *Australas J Dermatol.* 1996; **37:** S42-3.
70. Mayorga J, Barba-Gomez JF, Verduzco-Martinez AP, *et al.* Protothecosis. *Clin Dermatol.* 2012; **30:** 432-6.
71. McAnally T, Parry EL. Cutaneous protothecosis presenting as recurrent chromomycosis. *Arch Dermatol.* 1985; **121:** 1066-9.
72. McMullan B, Muthiah K, Stark D, *et al.* *Prototheca wickerhamii* mimicking yeast: A cautionary tale. *J Clin Microbiol.* 2011; **49:** 3078-81.
73. McMullan B, Pollett S, Biswas C, *et al.* Successful treatment of cutaneous protothecosis with liposomal amphotericin and oral itraconazole. *Med Mycol Case Rep.* 2016; **12:** 21-3.
74. Mejia-Otero C, Singh S, Arias Urdaneta L, *et al.* A rare case of *Prototheca* algaemia in a patient with systemic lupus erythematosus and recent belimumab infusion. *Case Reports Immunol.* 2012; 754901. PMID: 25374733.
75. Melville PA, Watanabe ET, Benites NR, *et al.* Evaluation of the susceptibility of *Prototheca zopfii* to milk pasteurization. *Mycopathologia.* 1999; **146:** 79-82.
76. Mendez CM, Silva-Lizama E, Logemann H. Human cutaneous protothecosis. *Int J Dermatol.* 1995; **34:** 554-5.
77. Min Z, Moser SA, Pappas PG. *Prototheca wickerhamii* algaemia presenting as cholestatic hepatitis in a patient with systemic lupus erythematosus: A case report and literature review. *Med Mycol Case Rep.* 2012; **2:** 19-22.
78. Modly CE, Burnett JW. Cutaneous algal infections: Protothecosis and chlorellosis. *Cutis.* 1989; **44:** 23-4.

79. Mohabeer AJ, Kaplan PJ, Southern PM Jr, *et al.* Algaemia due to *Prototheca wickerhamii* in a patient with myasthenia gravis. *J Clin Microbiol.* 1997; **35:** 3305-7.
80. Mohd Tap R, Sabaratnam P, Mohd Salleh MA, *et al.* Characterization of *Prototheca wickerhamii* isolated from disseminated algaemia of kidney transplant patient from Malaysia. *Mycopathologia.* 2012; **173:** 173-8.
81. Morandi S, Cremonesi P, Capra E, *et al.* Molecular typing and differences in biofilm formation and antibiotic susceptibilities among *Prototheca* strains isolated in Italy and Brazil. *J Dairy Sci.* 2016; **99:** 6436-45.
82. Murata M, Ito T, Nagae K, *et al.* Disseminated protothecosis manifesting with multiple, rapidly-progressing skin ulcers: Successful treatment with amphotericin B. *Eur J Dermatol.* 2015; **25:** 208-9.
83. Murugaiyan J, Eravci M, Weise C, *et al.* Label-free quantitative proteomic analysis of harmless and pathogenic strains of infectious microalgae, *Prototheca* spp. *Int J Mol Sci.* 2016; 18. pii: e59. PMID: 28036087.
84. Nabai H, Mehregan AH. Cutaneous protothecosis. Report of a case from Iran. *J Cutan Pathol.* 1974; **1:** 180-5.
85. Narita M, Muder RR, Cacciarelli TV, *et al.* Protothecosis after liver transplantation. *Liver Transpl.* 2008; **14:** 1211-5.
86. Nelson AM, Neafie RC, Connor DH. Cutaneous protothecosis and chlorellosis, extraordinary 'aquatic-borne' algal infections. *Clin Dermatol.* 1987; **5:** 76-87.
87. Nguyen Q, Rosen T. Cutaneous protothecosis in a patient with chronic lymphocytic leukemia: A case report and literature review. *J Fungi.* 2015; **1:** 4-12.
88. Nosanchuk JS, Greenberg RD. Protothecosis of the olecranon bursa caused by achloric algae. *Am J Clin Pathol.* 1973; **59:** 567-73.
89. Okazaki C, Wakusawa C, Chikama R, *et al.* A case of cutaneous protothecosis in a polyarteritis nodosa patient and review of cases reported in Japan. *Dermatol Online J.* 2011; **17:** 2.
90. Ottuso P. Aquatic dermatology: Encounters with the denizens of the deep (and not so deep) - a review. Part II: The vertebrates, single-celled organisms and aquatic biotoxins. *Int J Dermatol.* 2013; **52:** 268-78; Quiz 277-8.
91. Padhye AA, Baker JG, D'Amato RF. Rapid identification of *Prototheca* species by the API 20C system. *J Clin Microbiol.* 1979; **10:** 579-82.
92. Pal M, Hasegawa A, Ono K, Lee CW. A new staining solution for the morphological studies of fungi and *Prototheca.* *Jpn J Vet Sci.* 1990; **52:** 527-31.
93. Pal M, Kenubih A, Mammo B. Chlorellosis: An emerging algal disease of animals and humans. *Indian J Comp Microbiol Immunol Infect Dis.* 2015; **36:** 1-6.
94. Pal M, Lee CW. Association of *Prototheca* with bovine mastitis. *Korean J Vet Clin Med.* 1997; **14:** 349-51.
95. Panchabhai TS, Bandyopadhyay D, Piliang M, *et al.* Fungus or parasite or both: A diagnostic challenge. *J Glob Infect Dis.* 2015; **7:** 40-2.
96. Pascual JS, Balos LL, Baer AN. Disseminated *Prototheca wickerhamii* infection with arthritis and tenosynovitis. *J Rheumatol.* 2004; **31:** 1861-5.
97. Pednekar M, Chandra PA, Margulis Y, *et al.* Protothecal olecranon bursitis: An unusual algal infection. *Am J Med Sci.* 2011; **342:** 424.
98. Pimentel CL, Alegre M, Dalmau J, *et al.* Erythematous papules on the leg: Protothecosis. *Arch Dermatol.* 2006; **142:** 921-6.
99. Piyophirapong S, Linpiyawan R, Mahaisavariya P, *et al.* Cutaneous protothecosis in an AIDS patient. *Br J Dermatol.* 2002; **146:** 713-5.
100. Polk P, Sanders DY. Cutaneous protothecosis in association with the acquired immunodeficiency syndrome. *South Med J.* 1997; **90:** 831-2.
101. Pore RS. *Prototheca* taxonomy. *Mycopathologia.* 1985; **90:** 129-39.
102. Quigley RR, Knowles KE, Johnson GC. Disseminated chlorellosis in a dog. *Vet Pathol.* 2009; **46:** 439-43.
103. Qureshi HS, Dunne WM, Nakhleh RE. Unremitting ulcer in a scuba diver: Subcutaneous protothecosis. *Arch Pathol Lab Med.* 2001; **125:** 450-2.
104. Ramirez I, Nieto-Rios JF, Ocampo-Kohn C, *et al.* Protothecal bursitis after simultaneous kidney/liver transplantation: A case report and review. *Transpl Infect Dis.* 2016; **18:** 266-74.
105. Ramirez-Romero R, Rodriguez-Tovar LE, Nevarez-Garza AM, *et al.* *Chlorella* infection in a sheep in Mexico and minireview of published reports from humans and domestic animals. *Mycopathologia.* 2010; **169:** 461-6.
106. Raz R, Rottem M, Bisharat N, *et al.* Intestinal protothecosis in a patient with chronic mucocutaneous candidiasis. *Clin Infect Dis.* 1998; **27:** 399-400.
107. Ricchi M, Cammi G, Garbarino CA, *et al.* A rapid real-time PCR/DNA resolution melting method to identify *Prototheca* species. *J Appl Microbiol.* 2011; **110:** 27-34.
108. Roesler U, Moller A, Hensel A, *et al.* Diversity within the current algal species *Prototheca zopfii:* A proposal for two *Prototheca zopfii* genotypes and description of a novel species, *Prototheca blaschkeae* sp. nov. *Int J Syst Evol Microbiol.* 2006; **56:** 1419-25.
109. Roesler U, Scholz H, Hensel A. Immunodiagnostic identification of dairy cows infected with *Prototheca zopfii* at various clinical stages and discrimination between infected and uninfected cows. *J Clin Microbiol.* 2001; **39:** 539-43.
110. Satoh K, Ooe K, Nagayama H, *et al.* *Prototheca cutis* sp. nov., A newly discovered pathogen of protothecosis isolated from inflamed human skin. *Int J Syst Evol Microbiol.* 2010; **60:** 1236-40.
111. Schumann K, Hollandsworth K, Ormsby A. Non-healing leg ulceration: Protothecosis. *Arch Dermatol.* 2000; **136:** 1263 & 68.
112. Segal E, Padhye AA, Ajello L. Susceptibility of *Prototheca* species to antifungal agents. *Antimicrob Agents Chemother.* 1976; **10:** 75-9.
113. Seok JY, Lee Y, Lee H, *et al.* Human cutaneous protothecosis: Report of a case and literature review. *Korean J Pathol.* 2013; **47:** 575-8.
114. Shaoxi W, Guixia L, Yongnian S, *et al.* The first case of protothecosis zopfii in China. *Chin Med Sci J.* 2000; **15:** 121-3.

115. Sheikh-Ahmad M, Goldstein S, Potasman I. *Prototheca wickerhamii* hand infection successfully treated by itraconazole and voriconazole. *J Travel Med.* 2006; **13:** 321-3.
116. Silva PC, Costa e Silva SB, Lima RB, *et al.* Cutaneous protothecosis - Case report. *An Bras Dermatol.* 2013; **88 (6 Suppl. 1):** 183-5.
117. Sirikulchayanonta V, Visuthikosol V, Tanphaichitra D, *et al.* Protothecosis following hand injury: A case report. *J Hand Surg.* 1989; **14:** 88-90.
118. Skedros JG, Keenan KE, Trachtenberg JD. *Candida glabrata* olecranon bursitis treated with bursectomy and intravenous caspofungin. *J Surg Orthop Adv.* 2013; **22:** 179-82.
119. Small LN, Ross JJ. Suppurative tenosynovitis and septic bursitis. *Infect Dis Clin North Am.* 2005; **19:** 991-1005.
120. Solky AC, Laver NM, Williams J, *et al. Prototheca wickerhamii* infection of a corneal graft. *Cornea.* 2011; **30:** 1173-5.
121. Srisuttiyakorn C, Sindhuphak W. Cutaneous protothecosis: A case report from Thailand. *Int J Dermatol.* 2012; **51:** 1340-2.
122. Stenner VJ, Mackay B, King T, *et al.* Protothecosis in 17 Australian dogs and a review of the canine literature. *Med Mycol.* 2007; **45:** 249-66.
123. Sudman MS, Kaplan W. Identification of the *Prototheca* species by immunofluorescence. *Appl Microbiol.* 1973; **25:** 981-90.
124. Sudman MS. Protothecosis: A critical review. *Am J Clin Pathol.* 1974; **61:** 10-9.
125. Sykora T, Horakova J, Buzzasyova D, *et al.* Protothecal peritonitis in child after bone marrow transplantation: Case report and literature review of paediatric cases. *New Microbes New Infect.* 2014; **2:** 156-60.
126. Takano M, Hoshi S, Nagai K, *et al.* The first case of human protothecosis caused by *Prototheca zopfii* in Japan. *J Infect Chemother.* 2014; **20:** 647-9.
127. Tan M, Pan JY, Chia HY, *et al.* Cutaneous protothecosis on the bilateral wrists of a food handler. *Clin Exp Dermatol.* 2017; **42:** 72-4.
128. Tan RM, Aw MM, Quak SH, *et al.* Pulmonary protothecosis in a pediatric liver transplant patient. *J Pediatric Infect Dis Soc.* 2014; **3:** e31-4.
129. Tang WY, Lo KK, Lam WY, *et al.* Cutaneous protothecosis: Report of a case in Hong Kong. *Br J Dermatol.* 1995; **133:** 479-82.
130. Tello-Zavala MC, Mercado-Lara A, Gomez-Hernandez N, *et al.* Disseminated protothecosis in a Mexican child. *Pediatr Infect Dis J.* 2013; **32:** e476-7.
131. Thiele D, Bergmann A. Protothecosis in human medicine. *Int J Hyg Environ Health.* 2002; **204:** 297-302.
132. Todd JR, King JW, Oberle A, *et al.* Protothecosis: Report of a case with 20-year follow-up and review of previously published cases. *Med Mycol.* 2012; **50:** 673-89.
133. Torres HA, Bodey GP, Tarrand JJ, *et al.* Protothecosis in patients with cancer: Case series and literature. *Clin Microbiol Infect.* 2003; **9:** 786-92.
134. Tortorano AM, Prigitano A, Dho G, et al. *In vitro* activity of conventional antifungal drugs and natural essences against the yeast-like alga *Prototheca. J Antimicrob Chemother.* 2008; **61:** 1312-4.
135. Tuney I; Sukatar A. Association of *Prototheca* species and protothecosis in man and animals. *Rev Med Micro.* 2006; **17:** 77-82.
136. van Bezooijen BP, Newling DW. Protothecosis of the urinary tract. *J Urol.* 2002; **167:** 252.
137. Van den Bossche D, De Bel A, Hendrickx M, *et al.* Galactomannan enzymatic immunoassay cross-reactivity caused by *Prototheca* species. *J Clin Microbiol.* 2012; **50:** 3371-3.
138. Van den Bossche D, de Haan R, Van der Werff Ten Bosch J, *et al.* Case report: Infrapatellar bursitis caused by *Prototheca wickerhamii. Med Mycol Case Rep.* 2012; **1:** 13-6.
139. von Bergen M, Eidner A, Schmidt F, *et al.* Identification of harmless and pathogenic algae of the genus *Prototheca* by MALDI-MS. *Proteomics Clin Appl.* 2009; **3:** 774-84.
140. Walsh SV, Johnson RA, Tahan SR. Protothecosis: An unusual cause of chronic subcutaneous and soft tissue infection. *Am J Dermatopathol.* 1998; **20:** 379-82.
141. Woo JY, Suhng EA, Byun JY, *et al.* A case of cutaneous protothecosis in an immunocompetent patient. *Ann Dermatol.* 2016; **28:** 273-4.
142. Woolrich A, Koestenblatt E, Don P, *et al.* Cutaneous protothecosis and AIDS. *J Am Acad Dermatol.* 1994; **31:** 920-4.
143. Yamada N, Yoshida Y, Ohsawa T, *et al.* A case of cutaneous protothecosis successfully treated with local thermal therapy as an adjunct to itraconazole therapy in an immunocompromised host. *Med Mycol.* 2010; **48:** 643-6.
144. Young M, Bush W, Sanchez M, *et al.* Serial MRI and CSF analysis in a dog treated with intrathecal amphotericin B for protothecosis. *J Am Anim Hosp Assoc.* 2012; **48:** 125-31.
145. Yun CH, Jeong JH, Ryu HR, *et al.* Cutaneous protothecosis responds rapidly to voriconazole. *Int J Dermatol.* 2016; **55:** 1373-7.
146. Zaitz C, Godoy AM, Colucci FM, *et al.* Cutaneous protothecosis: Report of a third Brazilian case. *Int J Dermatol.* 2006; **45:** 124-6.
147. Zaitz C, Godoy AM, de Sousa VM, *et al.* Onychoprotothecosis: Report of the first case in Brazil. *Int J Dermatol.* 2006; **45:** 1071-3.
148. Zak I, Jagielski T, Kwiatkowski S, *et al. Prototheca wickerhamii* as a cause of neuro-infection in a child with congenital hydrocephalus. First case of human protothecosis in Poland. *Diagn Microbiol Infect Dis.* 2012; **74:** 186-9.
149. Zhang Q, Li L, Yuli K, *et al.* A case of cutaneous protothecosis mimics eczema. *Mycopathologia.* 2015; **179:** 163-6.
150. Zhang Q, Weng X, Li L, *et al.* An unusual case of granulomatous lymphadenitis due to *Prototheca zopfii* var. *portoricensis* in an immunocompetent man in China. *Int J Infect Dis.* 2010; **14 (Suppl. 3):** e32-5.
151. Zhang QQ, Li L, Zhu LP, *et al.* Cutaneous protothecosis in patient with diabetes mellitus and review of published case reports. *Mycopathologia.* 2012; **173:** 163-71.

CHAPTER 38

Rhinosporidiosis

Rhinosporidiosis is a chronic granulomatous disease of man and animals, characterized by polyposis of nasal cavity, conjunctiva and other mucosal as well as non-mucosal body sites. In recent past, giant lesions and disseminated type of disease, has also been reported in some of the clinical settings. It is caused by ***Rhinosporidium seeberi***, which is a taxonomically debated endosporulating aquatic eukaryotic organism since long. This disease is hyperendemic in the Indian subcontinent, particularly in Sri Lanka and southern India. However, sporadic cases have occasionally been reported from other continents like East Africa, South America, North America and even Europe.

Rhinosporidium seeberi is yet to be successfully grown on artificial culture media. It was thought to be fungus for a long time but on the basis of advanced molecular studies, it is now considered as a **hydrophilic protistan** parasite hence called **pseudofungal** organism. Taxonomically, it is now included with some of the fish and amphibian pathogens, in a new clade *i.e.* **Mesomycetozoea**.

Historical Perspective

The clinical entity, rhinosporidiosis, although endemic in the Indian subcontinent, was first reported and described from Argentina. In 1892, Carlos Malbran first observed this disease from Buenos Aires. It was not until 1900 that **Guillermo Seeber** meticulously reported the condition in a 19-year-old agricultural worker in Argentina whose breathing was impaired due to a nasal mass. He published the first two cases describing this organism as protozoa allied to the coccidian group in his doctorate dissertation in 1900. Both these Argentinean scientists from Buenos Aires considered this organism to be a sporozoan. However, later on Roberto Wernicke designated it as *Coccidium seeberia*, adopting the genus name from protozoal sub-class Coccidia and the species name from his pupil, Guillermo Seeber. As happened with *Coccidioides immitis*, this organism was also classified as a protozoan at that time.

Later on, unaware of the earlier works, Major F. O'Kinealy in 1903, presented a paper on this subject in a meeting of London Laryngological Society, which earned another name to the organism *i.e. Rhinosporidium kinealyi.* David in his paper of 1974 reviewed that O'Kinealy had considered it to be a ***'Localized Psorospermiosis'*** of mucous membrane of nasal septum. He had described a case, material of which was obtained nine years ago in 1894 from Gaya, the territory under Calcutta province (now Kolkata) during days of the British Empire. In 1905, Minchin and Fantham, who were unaware of Seeber's findings, introduced a new genus *Rhinosporidium* and named this organism *Rhinosporidium kinealyi.* Therefore, in old reports published in early part of 20th century, particularly in the *Lancet* and other leading medical journals, the organism was mentioned with the nomenclature as *Rhinosporidium kinealyi.*

Ashworth in the year 1923, showed that the organism was not a sporozoan but a lower order fungus belonging to class **Phycomycetes**. He studied its life cycle in detail and coined the term ***Rhinosporidium seeberi***, changing its species name in honor of Guillermo Seeber, who had already described the first two cases. The reasons put forward by Ashworth for classifying this organism under kingdom **Fungi** were presence of cellulose in sporangial wall, a mucoid substance between spores, absence of residual cytoplasm and occurrence of repeated synchronous mitosis before the cytoplasmic division.

Chester Emmons thought it to be a *Synchytrium* species, essentially a parasite of plants. Most of the mycologists during those days contemplated that it was related to genus *Chytridium* of order **Chytridiales** and class **Chytridiomycetes**. These organisms have parasitic phase in algae and higher plants, closely resembling the sporangia (spherules) of *R.seeberi* in man.

The major contribution about rhinosporidiosis was by Prof. William Arthur Edward Karunaratne, who reviewed the literature on this disease covering almost all aspects and published an excellent monograph in 1964. Kennedy *et al,* gave an alternative nomenclature proposing new terminology for ontogenetic stages of endospores and sporangia in 1995. The taxonomy of the organism remained under question almost all through the 20th century, which has recently resolved by designating it into Mesomycetozoea clade, however, its nomenclature remained stable as such till date without any change. The century old disease is yet to be described in the books of Protozoology (Parasitology) hence traditionally it is given as a separate Chapter in the Textbook of Medical Mycology despite the fact that it is not a fungus.

Organism *(Rhinosporidium seeberi)*

Rhinosporidium seeberi was the centre of taxonomic debate throughout 20th century although its nomenclature relatively remained stable for the last eight decades or so. Since then, numerous investigators have considered *R.seeberi* to be a Protozoa, a Fungus, more recently *Cyanobacterium* and even carbohydrate waste product. It was believed to be fungus-like organism, though few postulated it as protozoan belonging to **Sporozoa** hence origin of its name (***'rhino'*** - nose + ***'sporidiosis'*** - sporozoan) implying literally sporozoan infection of nose. However, terminology of ontogenetic stages of *R.seeberi* are proposed to be under revision.

The failure to cultivate *R.seeberi* prevented it from being phylogenetically classified. Levy *et al,* cultured it for the first time from nasal polyp in a canine, employing monolayers of an epithelioid human rectal tumor (HRT) cell line, incubated at 34°C in humidified air with 5% CO_2. In the cell line all stages of the entire life cycle, including the trophic form and resting states as well as sporangium-endospore formation, were produced. The spores induced the epithelial cells to form polyps around them. The endospore viability has been assessed by MTT-reduction test, which has been used to define its sensitivity to biocides such as povidone-iodine, 70% ethanol, 1% formalin and chlorhexidine.

The wall of sporangium is approximately 5 µm thick during the young developing stage but becomes thinner at maturity thereby pore is formed. The wall of mature sporangium shows two distinct layers *i.e.* **outer chitinous** and **inner cellulose** layer, except at the pore, where cellulose layer is thinned out. This is also a carminophilic organism like *Cryptococcus,* which takes up pink color. Therefore, Mayer's mucicarmine stains the cellulose-like material in the inner layer of sporangium and the outer layer of newly released endospores, as pink, suggesting acidic polysaccharides.

This organism was previously classified as fungus on the basis of morphological and histochemical characteristics. Its morphology resembles that of *Coccidioides* species, since both these organisms have mature stages that consist of large, thick-walled, spherical structure containing daughter cells (endospores). The mature sporangia are much larger and thicker than spherule of *Coccidioides* species. In addition, *R.seeberi* is visualized with special stains such as GMS, PAS as well as Mayer's mucicarmine. A large number of endospores are formed within sporangia, which on maturity move towards pore and are released out passing through it. These spores measure about 6-9 µm at the time of release and each may develop to form new sporangium as shown in the diagrammatic **Figure 38.1A**.

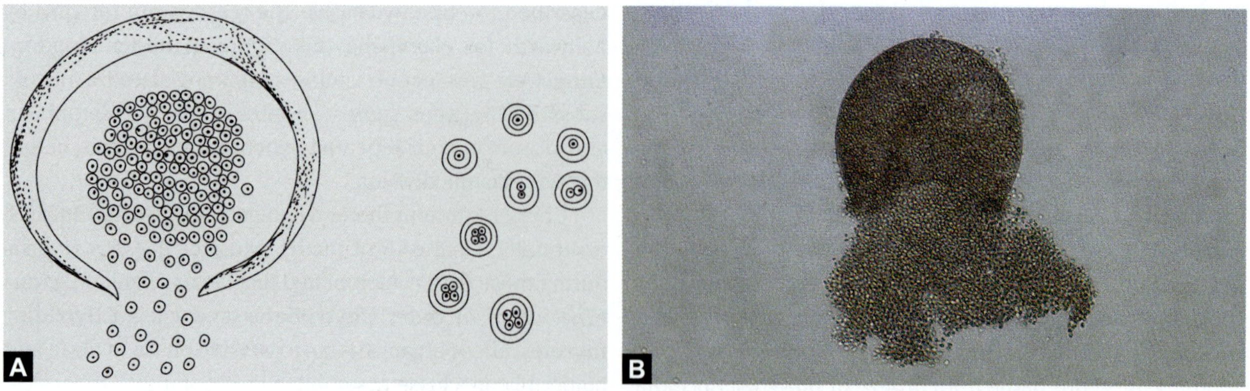

Figs. 38.1A and B. (A) Diagrammatic representation of ruptured sporangium of *Rhinosporidium seeberi* showing endospores at various stages of their development. (B) Photomicrograph of the same findings in wet mount taken from a patient (KOH × 200).

as well as photomicrograph taken from KOH wet mount of a patient's smear **(Fig. 38.1B)**.

The analysis of aligned sequence and inference of phylogenetic relationships showed that *R.seeberi* is a protistan from novel clade of parasites that infect fish and amphibians. The fluorescence *in situ* hybridization (FISH) and *R.seeberi*-specific PCR showed that unique 18S rRNA gene was also present in other tissues infected with *R.seeberi*. The data supports that *R.seeberi* phylogeny is same as suggested by another group of workers (Ragan, *et al*) that the organism is not a fungus rather first known human pathogen from group of fish and amphibian parasites called **DRIP** clade, a novel group of aquatic protistan species. The name DRIP is an acronym for protozoa identified as members of group: ***D**ermocystidium*, the **R**osette agent (now *Sphaerothecum destruens*), ***I**chthyophonus* and ***P**sorospermium*.

Most studies conducted by Karvita Ahluwalia *et al*, labeling this organism as a prokaryote *i.e. Cyanobacterium*, blue-green algae-*Microcystis aeruginosa*, are not reproducible. The version of Herr *et al*, as well as another independent molecular study by Fredricks *et al*, showed that the organism is linked to DRIP clade of protistan parasites (Ichthyosporea) and proposed to rename this monophyletic clade as **Mesomycetozoea**. This name signifies its phylogenetic location at the animal-fungal divergence in the evolutionary tree (between fungi and animals). The Mesomycetozoea or DRIP clade is a small group of protist, mostly parasites of fish and other aquatic animals. At present *R.seeberi* is a eukaryotic organism, classified within the protistan Mesomycetozoea clade.

Therefore, phylogenetic analysis of *R.seeberi* 18S rRNA gene sequencing suggests that this culture-resistant organism is not a member of eumycetes rather it is the first known human pathogen from a novel clade of aquatic protistan parasites that have branched in the evolutionary tree near animal-fungal divergence.

Life Cycle *(Rhinosporidium seeberi)*

There are three principal stages involved in the life cycle of *Rhinosporidium seeberi*. These are (i) **trophocyte** maturation (ii) **sporangia** development and (iii) **endospores** formation. The organism starts its life cycle as trophic stage in tissue as a rounded or oval structure, which is about the size of a red blood cell, measuring 6-8 µm with a cell membrane, clear cytoplasm, vesicular nucleus and nucleolus. This is covered by chitin, lies freely in connective tissue and undergoes developmental changes by mitotic division along with condensation of cytoplasm. It gradually increases in size and its wall thickens. On attaining a size of approximately 45 µm diameter, nucleus begins to divide. As time passes by around 14th division, the trophocyte becomes 140 µm in diameter and contains young endospores, each measuring about 3 µm.

The organism develops into sporangium containing approximately 12,000 to 16,000 endospores. The sporangium has a thick outer **chitinous** and a thin inner **cellulose** layer with a germinal pore situated eccentrically. The maturation of endospores occurs with the formation of mucoid envelope with a chitinous film, enlarging up to 7-9 µm with appearance of refractive sporangium. The mature sporangium is >450 µm in diameter, which eventually ruptures to release endospores. The endospores escape from the sporangium in two ways, if internal pressure is not high enough, spores escape from the pore, one by one, without breaking the membrane. Alternatively, if internal pressure is high, sporangium ruptures at the weakest point. The spores now enter in surrounding connective tissue, where they further develop into trophic stages and may be carried by lymphatic channels also.

Epidemiology

Rhinosporidiosis in humans is known since 1892 and its cases, both sporadic and two reports as outbreaks, have been reported from over 90 countries. The disease is generally found prevalent in Southeast Asian countries like Sri Lanka, southern India mainly, where it is hyperendemic and the South American continent particularly Argentina and Brazil. Recently two cases of ocular rhinosporidiosis, both acquired in Canada, have been reported. A few cases have been reported from African countries like Burundi, Egypt, Rwanda, Malawi, Kenya, Tanzania, Burkina Faso, Nigeria, Chad and Zaire. In Nepal, this disease is predominately prevalent in three adjoining districts of Saptari, Siraha and Dhanusa, located in the southeastern areas.

The maximum incidence is seen in endemic areas of the Indian subcontinent having warm climates and nearly 90% of the cases of rhinosporidiosis are from two countries *i.e.* India and Sri Lanka. It is rare in Europe, North America and other continents and is not yet reported from Australia. It has been observed that majority (>90%) of patients are from Asia, while South America accounts for less than 5%. The awareness regarding this disease is gradually increasing among clinicians both in endemic as well as non-endemic areas. The only outbreak of human rhinosporidiosis has been reported from Serbia in which infection got widespread because of mass level bathing in a lake.

Despite high endemicity in India, distribution of cases throughout the country is not uniform. It has been extensively reported from southern states of Tamil Nadu, Kerala, Pondicherry, Andhra Pradesh, West Bengal and Chhattisgarh. In a report published in 2007, Sudarshan *et al,* have reported 462 cases of this disease from Raipur, Chhattisgarh. The sporadic cases of rhinosporidiosis have been reported from non-endemic areas also. Isolated cases have been reported from other states either among patients who have migrated from endemic areas or as original cases. The states of Tamil Nadu and Kerala are hyperendemic areas of this disease. The cases have been reported from Eastern and Northern parts but rarely from Central India. From Bankura 116 cases have been reported over a period of five years. Prabhakar and Bhatnagar have reported 36 cases from Amritsar. Moses *et al,* have reported 19 and 112 cases in 1987 and 1988, respectively from Kanyakumari district of Tamil Nadu.

There are two outbreaks, which have been described, ocular and nasal disease in humans and ocular and cutaneous disease in swans. The human outbreak reported in 1992-93 from Balkans wherein patients of northern Serbia (erstwhile Yugoslavia) were involved after taking bath in stagnant water of Silver Lake. In an avian outbreak, the first of its kind, over a period of three years, 41 captive swans (*Cygnus olor* and *C.atratus*), dwelling on a lake in Central Florida city, developed conjunctival and cutaneous polypoid lesions of this disease.

There are various modes of transmission of rhinosporidiosis, which have been postulated but none of them has yet been proven. It is not a contagious disease and exogenous source of infection is not exactly known. The mode of transmission from animal-to-man has never been proved and direct transmission in man from another man is also unknown. However, trauma can explain direct spread of infection from one anatomical site to adjacent tissues as well as its dissemination to distant body parts. The incidence is more common among people who are habitual of taking bath in ponds, tanks and lakes along with the domestic animals.

Very little is known about the habitat of this organism, as its exact ecological niche is still unknown. But it is believed that fresh or stagnant water, which come frequently in contact of man and animals, may act as a reservoir of infection hence *R.seeberi* is believed to be a **hydrophilic** organism. It is a very common practice among people residing in endemic areas to have a dip in holy ponds especially during festival seasons, which are celebrated almost throughout the year. This practice may be a major risk factor for transmission of disease possibly from one person to other. An association between starchy tapioca plant *(Manihot esculenta)* and the disease has also been studied from south India.

Rhinosporidium seeberi is not only found in aquatic environment but has also been reported in the Middle East deserts where there is scarcity of water and also in dry regions of Argentina, north India and Sri Lanka. It has been observed in a study that in dry and dusty areas most of the cases have ocular rather than nasal rhinosporidiosis. This finding strongly supports the notion that this organism possesses resistant spores, which may be airborne and are transmitted in the desert areas, especially during sand storms.

The disease is especially prevalent in rural environment, particularly among individuals working in stagnant water or sand. Although no age is exempt, the cases are most commonly seen between 2nd and 3rd decade of life. The disease affects predominantly the male population, with a male-female ratio of 3:1 to 4:1, possibly due to their more number of outdoor activities. In some studies, it is observed that the incidence is either equal or slightly higher in females, particularly in cases of oculosporidiosis. It is mainly prevalent among poor people of developing countries and is not a life-threatening condition. Hence it could not warrant significant international attention during the last one century of its existence. There is no barrier for any race as far as this disease is concerned. This has also been reported by Stricker *et al,* in a patient of AIDS as well as by Sahoo *et al,* in HIV-positive patient.

Rhinosporidiosis in Animals

This disease is also found in cattle, horses, mules, dogs and wood duck. In a study from south India, it has been reported most often in bullocks and cows. Recently, it has also been reported in captive swans of genus *Cygnus,* which further supports the contention that an aquatic environmental niche acts as the reservoir for *R.seeberi.*

Immunity

The mechanisms of immunity in rhinosporidiosis are not fully documented. Moreover, this is not a successfully cultivable organism, which limits its detailed studies. There is no significant underlying condition or nutritional deficiency implicated in this disease. Although role of cell-mediated immunity is regarded as an important defense mechanism against this infection but information is inadequate. The cell-mediated immune response has been investigated by injection of endospores into the skin, which shows non-specific response in the form of local, transient erythematous reactions in the normal population that is also present in the patients.

Recently, the inner sporangial layer has been proven to possess antigenic properties. It was found that antibodies, present in individuals with rhinosporidiosis, detected mucoid layer beneath cell wall of the sporangium. As this layer has immunogenic properties hence it was suggested that it might play a significant role in immunology. Based on Herr and colleague's observations it has been speculated that perhaps this electrolucent layer is important in the immunology of this disease.

Pathogenesis and Pathology

The histopathological examination shows a polyp, with epithelial lining of the region of origin. The characteristic sporangia are present in large numbers, at various stages of development. The liberated endospores lodge in the host tissue and mature into endosporulating sporangia, repeating their *in vivo* cycle. In early stages, trophocyte, sporangia and spores are present in the tissues along with chronic inflammatory infiltrate, consisting of polymorphs and eosinophils, particularly in relation to ruptured sporangia. In the later stage, few viable forms of organism are seen but numerous collapsed chitinous shells may provoke a foreign body giant cell reaction. The stroma shows a chronic non-specific cellular inflammatory exudate, with a predominance of lymphocytes, plasma cells and granulocytes.

The use of MTT-reduction assay and Evan's blue-staining tests, have also been described for assessment of viability and morphological integrity of endospores after exposure to sera from patients with high titres of anti-rhinosporidial antibody. The salt (MTT) is reduced by mitochondrial dehydrogenases in viable mammalian and fungal cells, to a purple-colored, **formazan** that is quantifiable spectrophotometrically, after extraction from cells or is detected microscopically in an intact microbial cell.

Clinical Features

The clinical presentation depends on the affected site, size and extent of the lesions. The organism infects mucocutaneous surfaces of humans and animals. The incubation period in rhinosporidiosis is variable as duration of symptoms ranges from one to ten months. The disease is a localized condition, which usually presents as polypoid growth in nasal cavity. It is a slowly growing, tumor like, fleshy polypoidal mass in the nasal cavity with white studded sporangia giving it a strawberry-appearance. The general health of patients with *R.seeberi* infection is not affected, as polypoidal swelling usually causes no discomfort unless it is large enough to create mechanical obstruction and bleeding at the local site. The ocular manifestations of rhinosporidiosis, which should preferably be called as **oculosporidiosis**, are less common as compared to the nasal involvement.

The clinical manifestations of disease may involve different organs of body and present in following forms:

1. Nasal Rhinosporidiosis
2. Ocular Rhinosporidiosis
3. Cutaneous Rhinosporidiosis
4. Disseminated Rhinosporidiosis
5. Miscellaneous Manifestations

1. Nasal Rhinosporidiosis

The nasal cavity is the commonest site of infection caused by *R.seeberi*, where it forms friable polypoidal lesions, which are highly vascular and bleed easily. The papillary projections and lobules give characteristic raspberry, strawberry or cauliflower-like appearance to the lesions. Anterior nasal sites, particularly along the septum are more commonly affected but polypoidal mass may also be seen in nasopharynx and that may sometimes protrude down into oropharynx as seen in **Figures 38.2 to 38.4**. The presence of tiny whitish spots underneath the epithelium

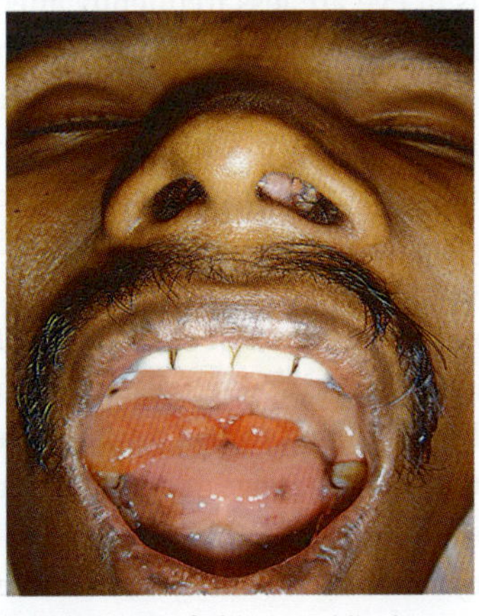

Fig. 38.2. Polypoid mass of rhinosporidiosis seen in left nasal cavity and oropharynx.

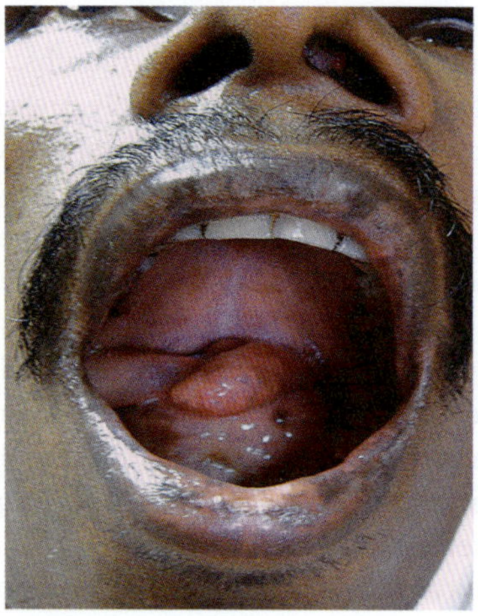

Fig. 38.3. Polypoid mass of rhinosporidiosis seen in left nasal cavity and prominent in oropharynx.

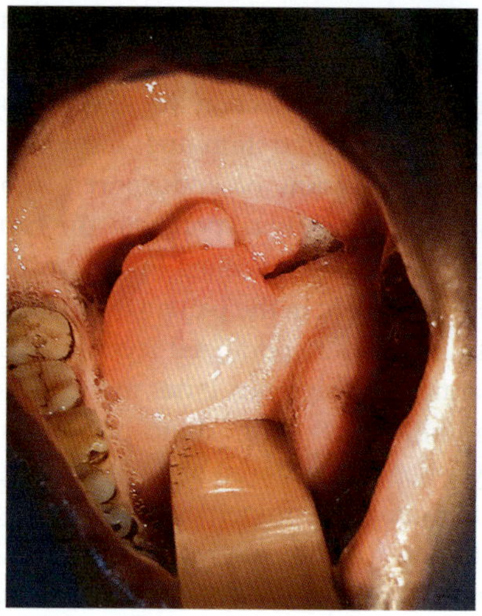

Fig. 38.4. Polypoid mass of rhinosporidiosis seen in oropharynx.

represent sporangia, which is a diagnostic feature of this disease. A thin mucoid nasal discharge, at times blood-stained, is often noted that contains spores as well as sporangia of *R.seeberi*. The other common presenting symptoms are epistaxis, unilateral nasal obstruction and foreign body sensation inside nose. There is no pain unless it is associated with secondary infection or sinusitis and rarely itching may also be observed.

2. Ocular Rhinosporidiosis

The eye is the second most common site involved after nose and accounts for about 15% cases of *R.seeberi* infections. The lesions are seen mainly in bulbar and palpebral conjunctiva. It is also called as **oculosporidiosis**. The bulbar conjunctiva, corneal limbus, lacrimal sac, caruncle and inner canthus of eye are less frequently affected. There is typically monocular involvement, although cases with binocular infection have also been observed. The lesion appears as a fleshy polyp with grayish-white granular spots scattered all over its surface representing superficial sporangia. It has a pedicular attachment to bulbar, fornix or palpebral conjunctiva.

The lesions are more often seen than felt and foreign body sensation or blood stained discharge may be noted. The polypoid masses tend to bleed with slightest touch or trauma due to rich and congested vasculature. The large-sized lesions may obstruct blinking of eyelids, causing exposure conjunctivitis. The conjunctival rhinosporidiosis can also cause scleral staphyloma in young, healthy, adult men. The polypoidal mass may grow rapidly within a very short period. This entity should be included in differential diagnosis of well-vascularized polypoid conjunctival lesions. There are also reports of peripheral keratitis being caused by this organism. The lacrimal sac is also frequently affected by extension of disease from nasal mucosa. Very rarely oculosporidiosis of lacrimal sac may also present as idiopathic orbital inflammatory disease (IOID).

3. Cutaneous Rhinosporidiosis

The cutaneous manifestations in rhinosporidiosis are rare. There may be wart-like, papillomatous or sessile masses in areas adjoining the nose and face. In addition to that there are subcutaneous scattered nodules, which ulcerate and are seen fungating over the skin. The cutaneous lesions are rarely pedunculated as seen over the pinna in **Figure 38.5**. The mode of infection may be direct inoculation to the local site or it may be through hematogenous route as extra-cutaneous lesions have also been reported in some patients. The dissemination results in painless, firm to hard, subcutaneous nodules that remain un-attached to skin. The subcutaneous rhinosporidiosis resembling a tumor is described by various adjectives and nomenclature such as 'giant cutaneous', 'tumoral disseminated' and 'generalized' one; however, 'rhinosporidioma' suggested by Arvindan *et al*, seems to be the most appropriate term.

4. Disseminated Rhinosporidiosis

In addition to local infiltration and hematogenous dissemination, lymphatic spread has also recently been described.

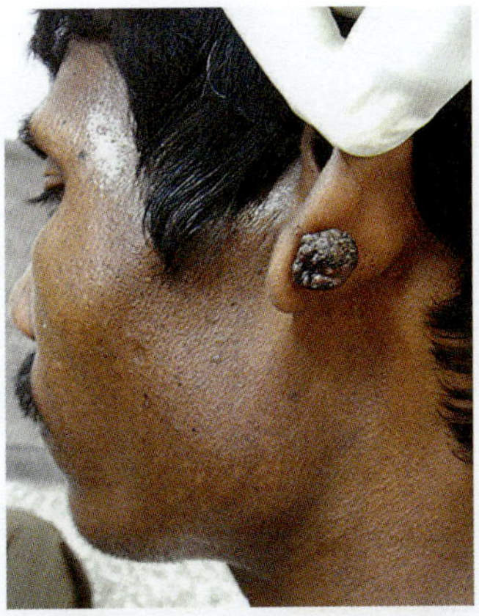

Fig. 38.5. Pedunculated mass of rhinosporidiosis seen over the posterior side of left pinna.

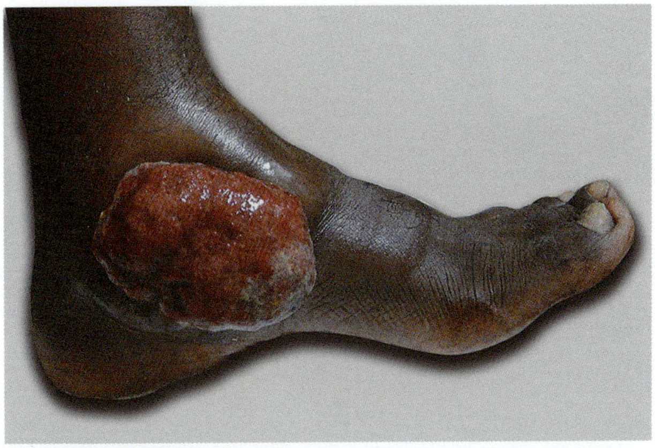

Fig. 38.6. Pedunculated giant mass of rhinosporidiosis seen over medial aspect of left ankle.

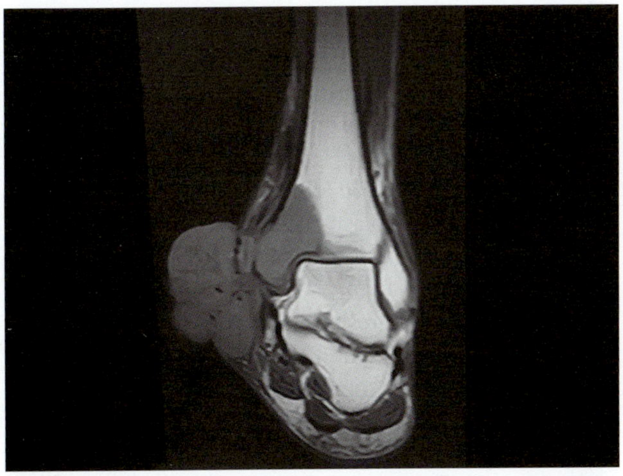

Fig. 38.7. Bony involvement of rhinosporidiosis seen over medial aspect of left ankle in the lower end of tibia.

There are many reports of disseminated type of infections. Most of the times the lesions are manifesting over one part of the body and in due course of time they spread to distant anatomical sites and are multiple in nature. The dissemination may manifest over skin or even visceral organs. On autopsy, lesions are found involving the bones, liver, lungs, spleen and brain. Now, there are reports of even nail involvement also due to this organism.

5. Miscellaneous Manifestations

In rhinosporidiosis, papillomatous lesions in buccal cavity, vagina, vulva, urethra and over penis have been reported. The lesions over genitourinary sites may suggest possibility of sexual route of transmission. But probability of its being sexually transmitted disease is doubtful because the spouses of these patients with such lesions are free from *R.seeberi* infection. Therefore, a possible source of infection could be taking bath in a contaminated pond water and not actually sexual route.

Urethral involvement is manifested by a friable, pink, discrete painless polyp protruding from urethral meatus, however, multiple lesions have also been described. Hematuria, intermittent bleeding, discharge and polypoid growth from external urethral meatus are usual modes of presentation. The differential diagnosis includes condyloma acuminata, transitional cell carcinoma and adenocarcinoma of the urethra.

The tumor-like masses of bone, clinically and radiologically, resemble chondrosarcoma. The patients presenting with such lesions show extensive bony destruction. A large pedunculated mass of rhinosporidiosis and its corresponding bony involvement is seen in **Figures 38.6 and 38.7**. The pedunculated mass was initially diagnosed as some sort of malignancy by the local physician. After amputation of the lower portion of the leg, the cut section of the same lesion is showing its involvement of lower potion of tibia **(Fig. 38.8)**. The disease has also been reported from scalp, tracheobronchial tree, larynx, lips, palate and rectum. The anal lesions resemble condyloma, rectal polyps or hemorrhoids. The multiple abscesses with sinuses have also been reported. In some of the cases, swelling over parotid gland may masquerade many diseases that eventually turns out to be rhinosporidiosis of the parotid duct.

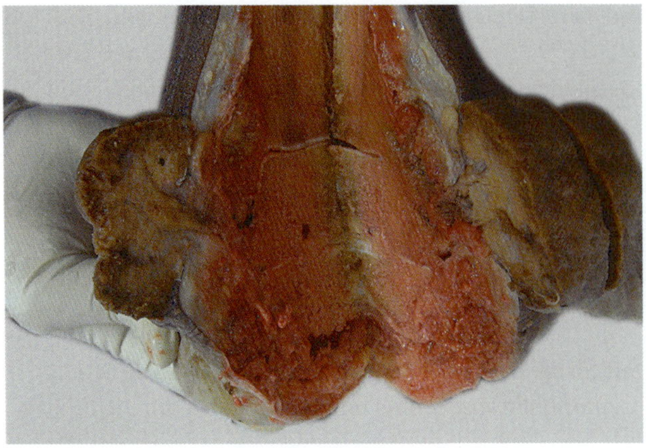

Fig. 38.8. After amputation of lower portion of leg, cut section of same lesion is clearly demarcated along with its inner involvement.

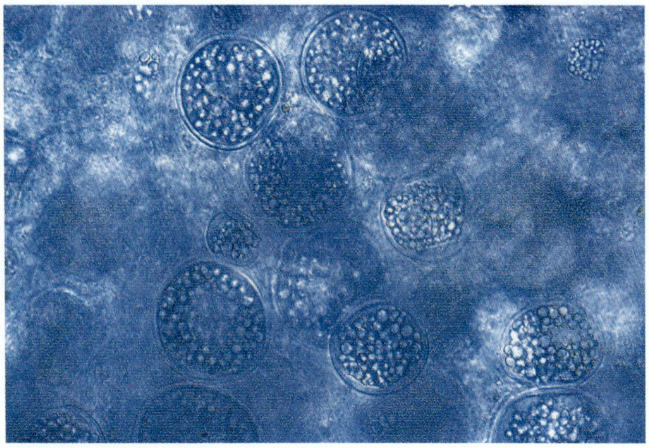

Fig. 38.9. KOH wet mount of sporangia of *Rhinosporidium seeberi* showing endospores at various stages (KOH × 200).

Differential Diagnosis

Rhinosporidiosis is one of the enigmatic diagnosis hence usually mistaken for many conditions, particularly neoplastic disorders. It should be included as one of the differential diagnosis in patients from the Indian subcontinent who present with nasal polypoid mass. The characteristic appearance of the lesions is very helpful for establishing diagnosis but may resemble inflammatory granulomas such as syphilitic, tubercular, leishmanoid and lepromatous; rhinoscleroma, neoplasia such as polyposis, hemangioma, angiofibroma, sarcoma or epithelioma. *Cryptococcus neoformans* is the only fungus known to cause polypoidal tumors.

On histopathological examination, the most common diseases to be differentiated from rhinosporidiosis are coccidioidomycosis and adiaspiromycosis due to formation of similar type of large rounded structures, which may confuse with the endosporulating sporangium. However, spherules of *Coccidioides* species are smaller with an average size being 30-60 μm, have about 2 μm thick walls and contain endospores of 2-5 μm in diameter. In contrast sporangium of *R.seeberi* is much larger and has thicker walls than the spherule of *Coccidioides* species. Moreover, size and number of endospores in sporangium of *R.seeberi* are greater than those of *Coccidioides* species. The large trophocyte stage of *Rhinosporidium* resembles tissue phase of adiaspiromycosis caused by *Emmonsia* species (*Ea.crescens* and *Ea.parva*). The spherules in these infections do not multiply and the wall thickness is more as compared to sporangium of *R.seeberi*. The differences in morphological features of spherules and similar structures produced by fungal and pseudofungal organisms are already given at a glance in **Table 18.1**.

The cutaneous lesions of rhinosporidiosis mimic verruca vulgaris. The ocular manifestations of disease may be confused with mucocele of lacrimal sac. Moreover, oculosporidiosis should also be differentiated from chalazion, dacryocystitis and neoplastic conditions. The penile lesions of rhinosporidiosis mimic malignancy. The disease should also be differentiated from prolapsed internal hemorrhoids. The bony involvement resembles chondrosarcoma. Therefore, single unilateral papilloma-like lesions should be clinically suspected as rhinosporidiosis.

Laboratory Diagnosis

While establishing diagnosis of rhinosporidiosis, a detailed study of clinical manifestations, appearance of lesions, geographical residence of patients, history of recurrence and demonstration of the organism (*R.seeberi*) in the tissue, should be included. It is essentially based on histopathological demonstration of pathognomonic endosporulating sporangia of *R.seeberi* in various stages of development, as it cannot be grown on artificial culture media.

(a) Direct Examination

The diagnosis of rhinosporidiosis mainly depends on direct microscopy by histopathological examination, scrape cytology and fine needle aspiration cytology (FNAC). A smear from excision biopsy of lesion or nasal washing may be observed in KOH wet mount as seen in **Figure 38.9** or stained with H&E, which shows giant cells and lymphocytic reaction around mature sporangium **(Fig. 38.10)**. The other techniques like GMS, PAS **(Fig. 38.11)** and

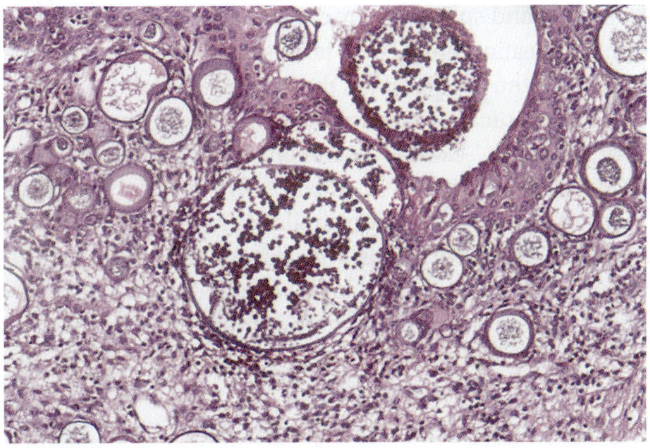

Fig. 38.10. Sporangia of *Rhinosporidium seeberi* showing endospores at various stages of their development (H&E × 400).

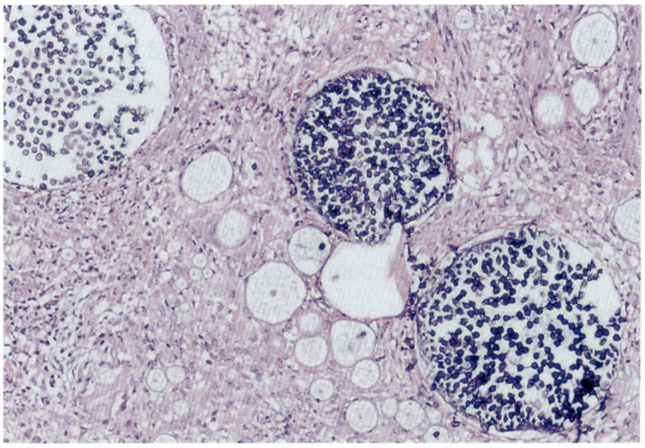

Fig. 38.11. Sporangia of *Rhinosporidium seeberi* showing endospores at various stages of their development (PAS × 400).

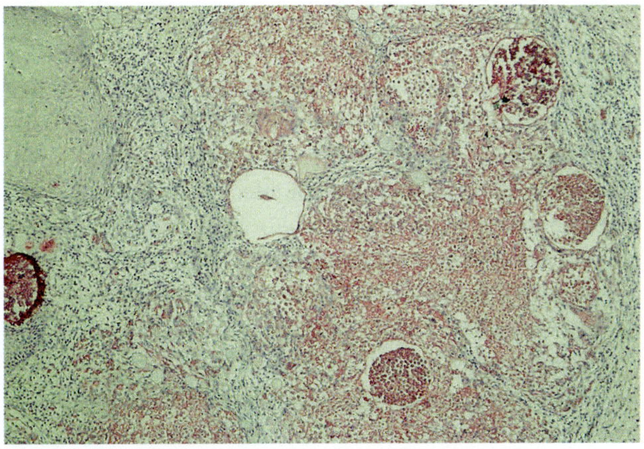

Fig. 38.12. Sporangia of *Rhinosporidium seeberi showing* endospores at various stages (Mayer's Mucicarmine × 200).

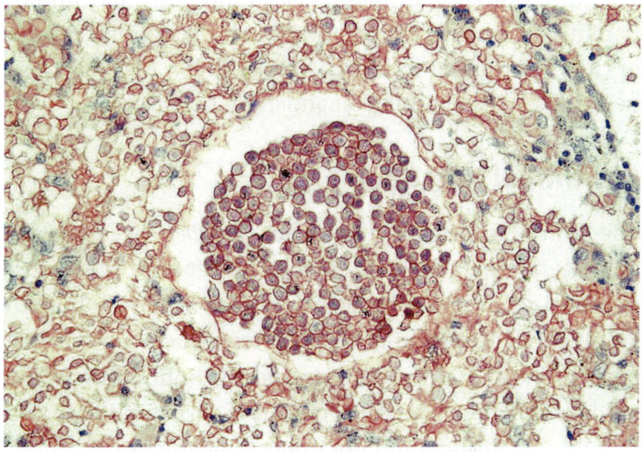

Fig. 38.13. Sporangium of *Rhinosporidium seeberi* showing endospores at various stages (Mayer's Mucicarmine × 400).

May-Grunwald-Giemsa (MGG) stains are performed. MGG shows dark blue thick chitinous walls of sporangium. Mayer's mucicarmine stains *Cryptococcus* species, is also used for demonstration of *R.seeberi* as seen in **Figures 38.12 and 38.13**. The Papanicolaou technique is also used to stain ethanol fixed smears of clinical specimen. The cytological techniques are particularly useful in providing quick diagnosis. Such techniques are very helpful for screening in endemic areas.

In stained sections, epidermis is usually seen as hyperplastic with edematous connective tissue. This contains **sporangia**, having hyaline wall and filled with **endospores** (sporangiospores). The sporangia, which are only known phenotypic structures produced by this pathogen, measure from 10 to > 450 μm in diameter. Occasional foreign body type giant cells may also be seen, usually in association with ruptured or degenerating sporangia.

(b) Culture

So far *R.seeberi* has not been successfully grown on artificial culture media. However, it has been shown to grow *in vivo* in epithelial carcinoma cell lines but permanent cultivation has not yet been achieved. Percoll, a sterile preparation of colloidal silica, which is used to isolate blood and tissue cells, subcellular particles, viruses and bacteria, has also been used to isolate spores of *R.seeberi*. It has successfully been used to separate different ontogenic stages of this organism in homogenates of rhinosporidial tissues. MTT reduction, an indicator of viability of endospores, is used to demonstrate efficacy of several biocides used in laboratories and wards, after exposure of endospores to these agents for periods ranging from 7 minutes to 24 hours—povidone iodine, 70% ethanol, chloroxylenol, chlorhexidine, formaldehyde are especially effective. The organism has been reported to be isolated and cultured *in vitro* successfully in 2005 at Kerala University (India) by Dr. V. Thankamani. However, in subsequent studies also, she made similar types of claims but these were not found to be tenable on scientific grounds.

(c) Immunodiagnosis

As very little is known about immunology or serology of rhinosporidiosis therefore no serological or skin test is useful for establishing its diagnosis. Even with the use of sensitive methods like counterimmunoelectrophoresis and immunodiffusion, no antibodies could be demonstrated in sera of rhinosporidiosis patients. Dot blot ELISA assay has also been tried for establishing diagnosis of this disease.

(d) Animal Pathogenicity

The experimental infections with *R.seeberi* in laboratory animals have not been successful. Injections of saline suspensions of infectious tissue into rabbits and even human volunteers have failed not only to reproduce symptoms of disease but also to demonstrate specific antibodies. In the study conducted by Arseculeratne *et al,* also it was observed that this organism could not infect congenitally immunodeficient, SCID and nude mice. No animal model of experimental rhinosporidiosis has been established till date but two promising approaches to study pathogenicity of *R.seeberi* include direct hemagglutination and adherence to Hep2 cells *in vitro*.

Treatment and Prophylaxis

The treatment of rhinosporidiosis is a real therapeutic challenge. It is treated by radical surgery, either by excision or electrocautery of the polyp base. The snaring alone may be associated with excessive amount of bleeding during removal hence it is to be avoided. There is a high rate of recurrence of disease after removal. Therefore, surgical resection of the lesion is done *in toto* by diathermy knife to minimize chances of recurrence. The electrocautery is preferred over excision to prevent undue bleeding and implantation of spores on the surrounding surfaces cut by scalpel and snare thereby curtailing recurrence. In addition, application of KTP-532 laser shows numerous advantages over the conventional therapeutic modality.

The medical treatment has not been found to be very effective except diaminodiphenylsulphone (DDS or dapsone), which is a widely used drug in the treatment of leprosy. There are similarities in tissue response between rhinosporidiosis and leprosy as there is failure to cultivate *in vitro* both causative organisms *i.e. R.seeberi* and *Mycobacterium leprae*. The antimicrobial effect of dapsone in this disease had also prompted trial of the drug. It has been tried successfully in cutaneous manifestations and for reducing recurrence of nasal lesions. Therefore, dapsone is required post-operatively to prevent recurrence caused by sub-epithelial and subcutaneous tissue spread. It is given orally 100 mg once daily for one year. This drug appears to arrest maturation of sporangia and promote fibrosis in stroma, when used as an adjunct to surgery.

Intravenous pentavalent antimonial compounds like neostibasan have also been tried. Although rhinosporidiosis is a benign disease but rarely fatalities are known to occur in disseminated type of infection. Airway obstruction, profuse bleeding and multi-organ failure often get precipitated during the disease process.

Despite all medical and surgical measures, recurrence is very common in rhinosporidiosis and many a times repeat excision is required. Now, coblation system can be a promising new tool in the surgical resection to avoid recurrence. In this technique, radiofrequency energy penetrates the surrounding tissues to a depth of only 100 μm. This technique enables to obtain better resection of lesions due to constant suctioning, with minimum lateral thermal damage. The chances of recurrence are less due to very little contamination of the surrounding areas by preventing autoinoculation. This technique also helps to resect the difficult to reach areas of the lesions. Moreover, coblation has also been used to treat other ENT conditions like obstructive sleep apnea, tonsillectomy, turbinate reduction and benign tumors of head and neck. As the temperature does not exceed 60°C, there is less thermal damage to surrounding tissues in comparison to the cautery and laser treatment. Moreover, due to simultaneous irrigation there is constant cooling.

The preventive measures are in general for any of the infectious disease. As the mode of infection is considered aquatic in nature hence contaminated water should be avoided. As such there is no vaccine available for the prophylaxis of this disease.

Further Reading

1. Adiga BK, Singh N, Arora VK, *et al.* Rhinosporidiosis: report of a case with an unusual presentation with bony involvement. *Acta Cytol.* 1997; **41:** 889-91.
2. Agirdir BV, Derin AT, Ozbilim G, Akman A. Cutaneous rhinosporidiosis presents with recurrent nasal philtrum mass in southern Turkey. *Int J Dermatol.*2008; **47:** 700-3.
3. Ahluwalia KB, Dhaulakhandi DB, Garg LC. Sequence analysis of 16S rRNA gene in *Rhinosporidium seeberi* shows similarity to plant chloroplast DNA. *Bioinformation.* 2010; **5:** 89-96.
4. Ahluwalia KB, Maheshwari N, Deka RC. Rhinosporidiosis: A study that resolves etiologic controversies. *Am J Rhinol.* 1997; **11:** 479-83.
5. Ahluwalia KB, Sharma N, Kacker SK, *et al.* Association of *Tapioca* and chronic inflammation with rhinosporidiosis. *Indian J Otolaryngol Head Neck Surg.* 1994; **3:** 25-7.
6. Ahluwalia KB. Causative agent of rhinosporidiosis. *J Clin Microbiol.* 2001; **39:** 413-5.
7. Ahluwalia KB. Culture of the organism that causes rhinosporidiosis. *J Laryngol Otol.* 1999; **113:** 523-8.
8. Al-Hilli F. Rhinosporidiosis in Bahrain, Arabian Gulf. *Mycopathologia.* 1985; **89:** 155-9.
9. Ali A, Flieder D, Guiter G, *et al.* Rhinosporidiosis: An unusual affliction. *Arch Pathol Lab Med.* 2001; **125:** 1392-3.
10. Aliyu D, Sahabi SM. Nasal rhinosporidiosis in a Nigerian Child: A rare clinical entity. *Int J Otolaryngol Head Neck Surg.* 2014; **3:** 51-3.
11. Almeida FA, Feitoza Lde M, Pinho JD, *et al.* Rhinosporidiosis: The largest case series in Brazil. *Rev Soc Bras Med Trop.* 2016; **49:** 473-6.
12. Alonso MJ, Gomez De Tejada R, Diaz P, *et al.* Description of a case of rhinosporidiosis in Spain. *Mycopathologia.* 1987; **97:** 9-16.
13. Amritanand R, Nithyananth M, Cherian VM, *et al.* Disseminated rhinosporidiosis destroying the talus: A case report. *J Orthop Surg.* 2008; **16:** 99-101.
14. Angunawela P, De Tissera A, Dissanaike AS. Rhinosporidiosis presenting with two soft tissue tumors followed by dissemination. *Pathology.* 1999; **31:** 57-8.
15. Anuradha S, Bharathi K, Khalique A. Oculosporidial polyp infected secondarily by *Enterobius vermicularis. Adv Biomed Res.* 2014; **3:** 195.
16. Aquino TA, Costa MC, Costa IM, *et al.* Nasal rhinosporidiosis: A report of two cases. *Int J Dermatol.* 2016; **55:** 1275-7.
17. Aravindan KP, Viswanathan MK, Jose L. Rhinosporidioma of bone: A case report. *Indian J Pathol Microbiol.* 1989; **32:** 312-3.
18. Arjun P, Sasikumar S. Tracheal rhinosporidiosis. *J Assoc Physicians India.* 2010; **58:** 692.
19. Arjun R, Khan F, Aravind B. Disseminated rhinosporidiosis masquerading as sarcoma. *J Assoc Physicians India.* 2016; **64:** 86-7.
20. Arora R, Gupta R, Dinda AK. Rhinosporidiosis of trachea: A clinical cause for concern. *J Laryngol Otol.* 2008; **122:** e13.
21. Arora R, Ramachandran V, Raina U, *et al.* Oculosporidiosis in Northern India. *Indian Pediatr.* 2001; **38:** 540-3.
22. Arseculeratne SN, Atapattu DN, Balasooriya P, *et al.* The effects of biocides (antiseptics and disinfectants) on the endospores of *Rhinosporidium seeberi. Indian J Med Microbiol.* 2006; **24:** 85-91.
23. Arseculeratne SN, Atapattu DN, Eriyagama NB. Human anti-rhinosporidial antibody does not cause metabolic inactivation or morphological damage in endospores of *Rhinosporidium seeberi, in vitro. Indian J Med Microbiol.* 2005; **23:** 14-9.
24. Arseculeratne SN, Atapattu DN. The assessment of the viability of the endospores of *Rhinosporidium seeberi* with MTT (3-[4, 5-dimethyl-2-thiazolyl]-2, 5-diphenyl-2H-tetrazolium bromide). *Mycol Res.* 2004; **108:** 1423-30.
25. Arseculeratne SN, Hussein FN, Atapattu DN, *et al.* Failure to infect congenitally immunodeficient SCID and nude mice with *Rhinosporidium seeberi. Med Mycol.* 2000; **38:** 393-5.
26. Arseculeratne SN, Kumarasiri PV, Rajapakse RP, *et al.* Anti-rhinosporidial antibody levels in patients with rhinosporidiosis and in asymptomatic persons, in Sri Lanka. *Mycopathologia.* 2004; **158:** 157-64.
27. Arseculeratne SN, Panabokke RG, Atapattu DN. Lymphadenitis, transepidermal elimination and unusual histopathology in human rhinosporidiosis. *Mycopathologia.* 2002; **153:** 57-69.
28. Arseculeratne SN, Sumathipala S, Eriyagama NB. Patterns of rhinosporidiosis in Sri Lanka: Comparison with international data. *Southeast Asian J Trop Med Public Health.* 2010; **41:** 175-91.
29. Arseculeratne SN. Recent advances in rhinosporidiosis and *Rhinosporidium seeberi. Indian J Med Microbiol.* 2002; **20:** 119-31.
30. Arseculeratne SN. *Microcystis aeruginosa* as the causative organism of rhinosporidiosis. *Mycopathologia.* 2001; **151:** 3-4.
31. Arseculeratne SN. Rhinosporidiosis: What is the cause? *Curr Opin Infect Dis.* 2005; **18:** 113-8.
32. Ashique KT, Sajid M, Anjit U. Strawberry-shaped lesion on the chest: Cutaneous rhinosporidiosis. *Indian Dermatol Online J.* 2014; **5:** S125-7.
33. Ashworth JH. On *Rhinosporidium seeberi* (Wernicke, 1903) with special reference to its sporulation and affinities. *Trans R Soc Edin.* 1923; **53:** 46.
34. Attapattu DN, Arseculeratne SN, Rajapake RPVJ, *et al.* Purification of the endospores and sporangia of *Rhinosporidium seeberi* on Percoll columns. *Mycopathologia.* 1999; **145:** 113-9.
35. Ayub-ur-Rehman, Muhammad MN, Moallam FA. Rhinosporidiosis. *J Coll Physicians Surg Pak.* 2012; **22:** 671-2.
36. Azad NS, Ahmad Z, Kayani N. Rhinosporidiosis presenting as an urethral polyp. *J Coll Physicians Surg Pak.* 2008; **18:** 314-5.
37. Azizi MH. Rhinosporidiosis. *Arch Iran Med.* 2006; **9:** 81-2 & 85.

38. Bakshi SS, Rajalakshmi P. Rhinosporidiosis. *Acta Clin Belg.* 2017; **72:** 217-8.
39. Bakshi SS. Rhinosporidiosis. *Med Clin (Barc).* 2016; **146:** e73.
40. Balachandran C, Muthiah V, Moses JS. Incidence and clinicopathological studies on rhinosporidiosis in Tamil Nadu. *J Indian Med Assoc.* 1990; **88:** 274-5.
41. Bandyopadhyay SN, Das S, Majhi TK, *et al.* Disseminated rhinosporidiosis. *J Laryngol Otol.* 2013; **127:** 1020-4.
42. Banerjee SB, Sarkar A, Mukherjee S, *et al.* Laryngeal rhinosporidiosis. *J Indian Med Assoc.* 1996; **94:** 148-50.
43. Banjara H, Panda RK, Daharwal AV, *et al.* Bronchial rhinosporidiosis: An unusual presentation. *Lung India.* 2012; **29:** 173-5.
44. Basu SK, Bain J, Maity K, *et al.* Rhinosporidiosis of lacrimal sac: An interesting case of orbital swelling. *J Nat Sci Biol Med.* 2016; **7:** 98-101.
45. Behera SK, Nayak S, Mishra D. Diagnosis by fine needle aspiration of disseminated cutaneous rhinosporidiosis presenting as cutaneous and subcutaneous tumor-like nodules. *Acta Cytol.* 2008; **52:** 635-6.
46. Belliveau MJ, Strube YN, Dexter DF, *et al.* Bloody tears from lacrimal sac rhinosporidiosis. *Can J Ophthalmol.* 2012; **47:** e23-4.
47. Bhanu TS. New drug regimen for rhinosporidiosis. *Indian J Otolaryngol.* 1980; **32:** 96-7.
48. Bhat V. Comments on 'novel multidrug therapy for disseminated rhinosporidiosis, refractory to dapsone - case report'. *Trop Doct.* 2014; **44:** 59-60.
49. Bhomaj S, Das JC, Chaudhuri Z, *et al.* Rhinosporidiosis and peripheral keratitis. *Ophthalmic Surg Lasers.* 2001; **32:** 338-40.
50. Bothra N, Mittal R, Tripathy D. Scleromalacia perforans from long-standing bulbar conjunctival rhinosporidiosis. *Ophthalmology.* 2016; **123:** 1493.
51. Branscomb R. Rhinosporidiosis update. *Lab Med.* 2002; **33:** 631-3.
52. Capoor MR, Khanna G, Rajni, *et al.* Rhinosporidiosis in Delhi, north India: Case series from a non-endemic area and mini-review. *Mycopathologia.* 2009; **168:** 89-94.
53. Castelino AM, Rao SK, Biswas J, *et al.* Conjunctival rhinosporidiosis associated with scleral melting and staphyloma formation: Diagnosis and management. *Cornea.* 2000; **19:** 30-3.
54. Chakraborty S, Chakrabarti A, Mukhopadhyay S, *et al.* Orbital rhinosporidiosis. *J Indian Med Assoc.* 2005; **103:** 383-4.
55. Chandra Shekharyga SH, Suligavi SS, Chogule S, *et al.* Nasal rhinosporidiosis revisited. *Clinical Rhinology: An International Journal.* 2010; **3:** 735-75.
56. Chao SS, Loh KS. Rhinosporidiosis: An unusual cause of nasal masses gains prominence. *Singapore Med J.* 2004; **45:** 224-6.
57. Chatterjee K, Banka S, Patra AC, *et al.* A curious ulcer on the pinna: Rhinosporidiosis at an unusual place. *Int J Dermatol.* 2015; **54:** e277-9.
58. Chatterjee PK, Khatua CR, Chatterjee SN, *et al.* Recurrent multiple rhinosporidiosis with osteolytic lesions in hand and foot: A case report. *J Laryngol Otol.* 1977; **91:** 729-34.
59. Chaudhary SK, Joshi KR. Diagnosis of rhinosporidiosis by nasal smear examination. *J Indian Med Assoc.* 1986; **84:** 274-6.
60. Chen L, Buonocore D, Wang B, *et al.* Delayed recurrence of sinonasal rhinosporidiosis. *Am J Otolaryngol.* 2015; **36:** 778-80.
61. Chitravel V, Subramanian S, Sundaram BM, *et al.* Rhinosporidiosis in a south Indian village. *Sabouraudia.* 1980; **18:** 241-7.
62. Chitravel V, Sundaram BM, Subramanian S, *et al.* Rhinosporidiosis in man: Case reports. *Mycopathologia.* 1990; **109:** 11-2.
63. Choudhury M. *Rhinosporidium seeberi* in nasal smears. *Diagn Cytopathol.* 2011; **39:** 593-4.
64. Chowdhury RK, Behera S, Bhuyan D, *et al.* Oculosporidiosis in a tertiary care hospital of western Orissa, India: A case series. *Indian J Ophthalmol.* 2007; **55:** 299-301.
65. Cohen J, Carvalho RC, Guimaraes R, *et al.* Oculosporidiosis. *Arch Ophthalmol.* 1997; **115:** 1340-1.
66. Daharwal A, Banjara H, Singh D, *et al.* A rare case of laryngeal rhinosporidiosis. *J Laryngol Voice.* 2011; **1:** 30-2.
67. Darbari BS, Gupta RL, Shukla IM, *et al.* Rhinosporidiosis in Raipur: A clinicopathological study of 348 cases. *Indian J Path Bacteriol.* 1972; **15:** 103-7.
68. Das S, Kashyap B, Barua M, *et al.* Nasal rhinosporidiosis in humans: New interpretations and a review of the literature of this enigmatic disease. *Med Mycol.* 2011; **49:** 311-5.
69. Dash A, Satpathy S, Devi K, *et al.* Cytological diagnosis of rhinosporidiosis with skeletal involvement - a case report. *Indian J Pathol Microbiol.* 2005; **48:** 215-7.
70. Date A, Ramakrishna B, Lee VN, *et al.* Tumoral rhinosporidiosis. *Histopathology.* 1995; **27:** 288-90.
71. deDoncker RM, de Keizer RJ, Oosterhuis JA, *et al.* Scleral melting in a patient with conjunctival rhinosporidiosis. *Br J Ophthalmol.*1990; **74:** 635-7.
72. de Silva NR, Huegel H, Atapattu DN, *et al.* Cell-mediated immune responses (CMIR) in human rhinosporidiosis. *Mycopathologia.* 2001; **152:** 59-68.
73. Delfino D, Mendoza L, Vilela R. *Rhinosporidium seeberi* nuclear cycle activities using confocal microscopy. *J Parasitol.* 2016; **102:** 60-8.
74. Deshpande AH, Agarwal S, Kelkar AA. Primary cutaneous rhinosporidiosis diagnosed on FNAC: A case report with review of literature. *Diagn Cytopathol.* 2009; **37:** 125-7.
75. Dey AK, Sharma R, Mittal K, *et al.* Rhinosporidiosis: A rare cause of proptosis and an imaging dilemma for sinonasal masses. *Case Rep Otolaryngol.* 2016; 3573512. PMID: 28042489.
76. Dhaulakhandi DB, Ahluwalia KB, Ravi AK, *et al.* Detection of 16S rRNA gene in round bodies isolated from polyps of rhinosporidiosis. *Infect Genet Evol.* 2006; **6:** 331-6.
77. Dutta S, Haldar D, Dutta M, *et al.* Socio-demographic correlates of rhinosporidiosis: A hospital-based epidemiologic study in Purulia, India. *Indian J Otolaryngol Head Neck Surg.* 2017; **69:** 108-12.

78. Echejoh GO, Manasseh AN, Tanko MN, *et al.* Nasal rhinosporidiosis. *J Natl Med Assoc.* 2008; **100:** 713-5.
79. Elston DM. Fungal foes: *Rhinosporidium seeberi. Cutis.* 2009; **84:** 131-2.
80. Fernandez-Flores A, Saeb-Lima M, Arenas-Guzman R. Morphological findings of deep cutaneous fungal infections. *Am J Dermatopathol.* 2014; **36:** 531-53; Quiz 554-6.
81. Fredricks DN, Jolley JA, Lepp PW, *et al. Rhinosporidium seeberi*: A human pathogen from a novel group of aquatic protistan parasites. *Emerg Infect Dis.* 2000; **6:** 273-82.
82. Ganne P, Babu R, Mohan Girija G. Conjunctival rhinosporidiosis. *JAMA Ophthalmol.* 2015; **133:** 39.
83. George L, Dincy P, Chopra M, *et al.* Novel multidrug therapy for disseminated rhinosporidiosis, refractory to dapsone-case report. *Trop Doct.* 2013; **43:** 110-2.
84. Ghorpade A, Gurumurthy J, Banerjee PK, *et al.* Oculosporidiosis presenting as an under-eye swelling. *Indian J Dermatol Venereol Leprol.* 2007; **73:** 196-7.
85. Ghorpade A, Ramanan C. Verrucoid cutaneous rhinosporidiosis. *J Eur Acad Dermatol Venereol.* 1998; **10:** 269-70.
86. Ghorpade A. Giant cutaneous rhinosporidiosis. *J Eur Acad Dermatol Venereol.* 2006; **20:** 88-9.
87. Ghorpade A. Polymorphic cutaneous rhinosporidiosis. *Eur J Dermatol.* 2006; **16:** 190-2.
88. Ghorpade A. Polymorphic ecthymatoid dermosporidiosis. *Indian J Dermatol Venereol Leprol.* 2008; **74:** 298.
89. Ghorpade A. Rhinosporidiosis: Gigantic cells with engulfed sporangia of *Rhinosporidium seeberi* in the case of dermosporidiosis. *Int J Dermatol.* 2008; **47:** 694-5.
90. Ghosh A, Saha S, Srivastava A, *et al.* Rhinosporidiosis - Unusual presentations. *Indian J Otolaryngol Head Neck Surg.* 2008; **60:** 159-62.
91. Ghosh AK, De Sarkar A, Bhaduri G, *et al.* Ocular rhinosporidiosis. *J Indian Med Assoc.* 2004; **102:** 732, 764.
92. Gichuhi S, Onyuma T, Macharia E, *et al.* Ocular rhinosporidiosis mimicking conjunctival squamous papilloma in Kenya - A case report. *BMC Ophthalmol.* 2014; **14:** 45.
93. Godwins O Echejoh, A Gabus N, *et al.* Nasal rhinosporidiosis. *J Natl Med Assoc.* 2008; **100:** 713-5.
94. Gokhale S, Ohri VC, Subramanya H, *et al.* Subcutaneous and osteolytic rhinosporidiosis. *Indian J Pathol Microbiol.* 1997; **40:** 95-8.
95. Gori S, Scasso A. Cytologic and differential diagnosis of rhinosporidiosis. *Acta Cytol.* 1994; **38:** 361-6.
96. Goswami R. Rhinosporidiosis: Bizarre clinical presentation and mode of dissemination. *Int J Med Res Rev.* 2014; **2:** 58-62.
97. Grover S. *Rhinosporidium seeberi*: A preliminary study of the morphology and life cycle. *Sabouraudia.* 1970; **7:** 249-51.
98. Gupta K, Gupta AK, Jain SB. Rhinosporidiosis of conjunctiva: A case report from Patiala. *J Postgrad Med.* 1984; **30:** 119-20.
99. Gupta RL, Darbari BS, Dwivedi MP, *et al.* An epidemiological study of rhinosporidiosis in and around Raipur. *Indian J Med Res.* 1976; **64:** 1293-9.
100. Gupta SB, Ghai HS. Pictorial CME: Rhinosporidiosis. *J Assoc Physicians India.* 2000; **48:** 224.
101. Hamersley ER, Perez AJ, Gallagher TQ. Nasal dyspnea and intermittent epistaxis. *JAMA Otolaryngol Head Neck Surg.* 2014; **140:** 179-80.
102. Hanchanale VS, Rao AR, Joseph G. A case of urethral rhinosporidiosis. *Urology.* 2005; **66:** 1106-7.
103. Harissi-Dagher M, Robillard N, Corriveau C, *et al.* Histopathologically confirmed ocular rhinosporidiosis in two Canadians. *Can J Ophthalmol.* 2006; **41:** 226-9.
104. Herr RA, Ajello L, Taylor JW, *et al.* Phylogenetic analysis of *Rhinosporidium seeberi's* 18S small-subunit ribosomal DNA groups this pathogen among members of the protoctistan Mesomycetozoa clade. *J Clin Microbiol.* 1999; **37:** 2750-4.
105. Herr RA, Mendoza L, Arseculeratne SN, *et al.* Immunolocalization of an endogenous antigenic material of *Rhinosporidium seeberi* expressed only during mature sporangial development. *FEMS Immunol Med Microbiol.* 1999; **23:** 205-12.
106. Houreih MA, Eyden B, Lucas SB, *et al. Rhinosporidium seeberi*: A case clinically mimicking a soft tissue sarcoma. *Histopathology.* 2006; **49:** 208-10.
107. Hussein MR, Rashad UM. Rhinosporidiosis in Egypt: A case report and review of literature. *Mycopathologia.* 2005; **159:** 205-7.
108. Iqbal MS, Dani SK. Clinicoepidemiological profile of rhinosporidiosis. *Indian J Otolaryngol Head Neck Surg.* 1993; **2:** 66-9.
109. Jacob P, Rose JS, Hoshing A, *et al.* Tectonic corneal graft for conjunctival rhinosporidiosis with scleral melt. *Indian J Ophthalmol.* 2011; **59:** 251-3.
110. Jahan S, Haque MA, Nessa F, *et al.* Vaginal rhinosporidiosis: A case report. *Mymensingh Med J.* 2014; **23:** 572-4.
111. Jain MR, Dhruva AK, Mogra N, *et al.* Rhinosporidiosis of the conjunctiva. *J Indian Med Assoc.* 1984; **82:** 140-1.
112. Jaiswal V, Kumar M, Gupta S, *et al.* Cytodiagnosis of rhinosporidiosis. *J Trop Med Hyg.*1992; **95:** 71-2.
113. James J, Thulaseedharan NK. Rhinosporidiosis. *QJM.* 2017; **110:** 391.
114. Jamison A, Crofts K, Roberts F, *et al.* Educational report: A case of lacrimal sac rhinosporidiosis. *Orbit.* 2016; **35:** 254-7.
115. Janardhanan J, Patole S, Varghese L, *et al.* Elusive treatment for human rhinosporidiosis. *Int J Infect Dis.* 2016; **48:** 3-4.
116. Job A, Venkateswaran S, Mathan M, *et al.* Medical therapy of rhinosporidiosis with dapsone. *J Laryngol Otol.* 1993; **107:** 809-12.
117. John D, Selvin SST, Irodi A, *et al.* Disseminated rhinosporidiosis with conjunctival involvement in an immunocompromised patient. *Middle East Afr J Ophthalmol.* 2017; **24:** 51-3.
118. John SS, Mohandas SG. Conjunctival oculosporidiosis with scleral thinning and staphyloma formation. *Indian J Ophthalmol.* 2005; **53:** 272-4.
119. Justice JM, Solyar AY, Davis KM, *et al.* Progressive left nasal obstruction and intermittent epistaxis. *JAMA Otolaryngol Head Neck Surg.* 2013; **139:** 955-6.

120. Kaluarachchi K, Sumathipala S, Eriyagama N, *et al.* The identification of the natural habitat of *Rhinosporidium seeberi* with *R.seeberi*-specific *in situ* hybridization probes. *J Infect Dis Antimicrob Agents.* 2008; **25:** 25-32.
121. Kamal MM, Luley AS, Mundhada SG, *et al.* Rhinosporidiosis: Diagnosis by scrape cytology. *Acta Cytol.* 1995; **39:** 931-5.
122. Kannan-Kutty M, Teh EC. *Rhinosporidium seeberi*: An ultrastructural study of its endosporulation phase and trophocyte phase. *Arch Pathol.* 1975; **99:** 51-4.
123. Karthikeyan P, Vijayasundaram S, Pulimoottil DT. A Retrospective epidemiological study of rhinosporidiosis in a rural tertiary care centre in Pondicherry. *J Clin Diagn Res.* 2016; **10:** MC04-8.
124. Karunaratne WAE. The pathology of rhinosporidiosis. *J Pathol Bacteriol.* 1936; **42:** 193-202.
125. Kaushal S, Mathur SR, Mallick SR, *et al.* Disseminated cutaneous, laryngeal, nasopharyngeal and recurrent obstructive nasal rhinosporidiosis in an immunocompetent adult: A case report and review of literature. *Int J Dermatol.* 2011; **50:** 340-2.
126. Kavishwar VS, Naik LP, Vora IM. Fine needle aspiration diagnosis of subcutaneous and osteolytic rhinosporidiosis. *Cytopathology.*1998; **9:** 215-7.
127. Kennedy FA, Buggage RR, Ajello L. Rhinosporidiosis: A description of an unprecedented outbreak in captive swans (*Cygnus* spp.) and a proposal for revision of the ontogenic nomenclature of *Rhinosporidium seeberi. J Med Vet Mycol.* 1995; **33:** 157-65.
128. Khan I, Gogia S, Agarwal A, *et al.* Recurrent rhinosporidiosis: Coblation assisted surgical resection - A novel approach in management. *Case Rep Otolaryngol.* 2014; 609784. PMID: 25574413.
129. Khoo JJ, Kumar KS. Rhinosporidiosis presenting as recurrent nasal polyps. *Med J Malaysia.* 2003; **58:** 282-5.
130. Kini G, Gopalkrishna, Thomas D. Anaesthesia for unusual presentation of rhinosporidiosis: A case report. *Sri Lankan J Anaesthesiol.* 2010; **18:** 39-42.
131. Kini U, Amirtham U, Shetty SC, *et al.* Rhinosporidiosis of the parotid duct cyst: Cytomorphological diagnosis of an unusual extra-nasal presentation. *Diagn Cytopathol.* 2001; **25:** 244-7.
132. Kishan Prasad HL, Rao C, Girisha BS, *et al.* Subcutaneous rhinosporidiosis masquerading as soft tissue tumor: diagnosed by fine-needle aspiration cytology. *Indian J Dermatol.* 2015; **60:** 215.
133. Krishnamoorthy S, Sreedharan VP, Koshy P, *et al.* Culture of *Rhinosporidium seeberi*: Preliminary report. *J Laryngol Otol.* 1989; **103:** 178-80.
134. Kumar BV, Osmani M, Mudhar HS. Rhinosporidiosis: An unusual presentation. *Int Ophthalmol.* 2005; **26:** 243-5.
135. Kumar S, Mathew J, Cherian V, *et al.* Laryngeal rhinosporidiosis: Report of a rare case. *Ear Nose Throat J.* 2004; **83:** 568 & 570.
136. Kumari R, Laxmisha C, Thappa DM. Disseminated cutaneous rhinosporidiosis. *Dermatol Online J.* 2005; **11:** 19.
137. Kumari R, Nath AK, Rajalakshmi R, *et al.* Disseminated cutaneous rhinosporidiosis: Varied morphological appearances on the skin. *Indian J Dermatol Venerol Leprol.* 2009; **75:** 68-71.
138. Kundu AK, Phuljhele S, Jain M, *et al.* Osseous involvement in rhinosporidiosis. *Indian J Orthop.* 2013; **47:** 523-5.
139. Lavaju P, Arya SK, Kumar B, *et al.* Conjunctival rhinosporidiosis presenting as a cystic mass - An unusual presentation. *Nepal J Ophthalmol.* 2010; **2:** 157-9.
140. Laxmisha C, Thappa DM. Subcutaneous swellings due to rhinosporidiosis. *J Dermatol.* 2005; **32:** 150-2.
141. Levy MG, Meuten DJ, Breitschwerdt EB. Cultivation of *Rhinosporidium seeberi in vitro:* Interaction with epithelial cells. *Science.* 1986; **234:** 474-6.
142. Loh KS, Chong SM, Pang YT, *et al.* Rhinosporidiosis: Differential diagnosis of a large nasal mass. *Otolaryngol Head Neck Surg.* 2001; **124:** 121-2.
143. Madana J, Yolmo D, Gopalakrishnan S, *et al.* Rhinosporidiosis of the upper airways and trachea. *J Laryngol Otol.* 2010; **124:** 1139-41.
144. Madana J, Yolmo D, Gopalakrishnan S, Saxena SK. Laryngotracheal rhinosporidiosis. *Ear Nose Throat J.* 2013; **92:** E27-30.
145. Madhavan M, Ratnakar C, Mehndiratta KS. Rhinosporidial infection of the forehead: report of a case. *J Postgrad Med.* 1978; **24:** 235-6.
146. Madke B, Mahajan S, Kharkar V, *et al.* Disseminated cutaneous with nasopharyngeal rhinosporidiosis: Light microscopy changes following dapsone therapy. *Australas J Dermatol.* 2011; **52:** e4-6.
147. Mahanta J, Deka NC, Das TK. Generalised rhinosporidiosis: A case report. *Indian J Med Microbiol.* 1992; **10:** 135-8.
148. Mahapatra S, Tripathy S, Rath G, *et al.* Rhinosporidiosis of parotid duct: A rare case report. *Indian J Pathol Microbiol.* 2007; **50:** 320-2.
149. Makannavar JH, Chavan SS. Rhinosporidiosis: A clinicopathological study of 34 cases. *Indian J Pathol Microbiol.* 2001; **44:** 17-21.
150. Mallick AA, Majhi TK, Pal DK. Rhinosporidiosis affecting multiple parts of the body. *Trop Doct.* 2012; **42:** 174-5.
151. Mandal SK, Bhakta A, Mandal A, *et al.* Giant mass conjunctival rhinosporidiosis causing severe mechanical ectropion. *J Indian Med Assoc.* 2012; **110:** 328-9.
152. Mathew JS, Padhy S, Lata S, *et al.* Case report: Telelaryngoscopy-guided flexible fiberoptic intubation for laryngeal rhinosporidiosis. *Anesth Analg.* 2010; **110:** 1066-8.
153. Mathur DR, Vyas MC. Disseminated cutaneous rhinosporidiosis. *Indian J Pathol Microbiol.* 1995; **38:** 223.
154. Matusik J, Hira PR, Saad HT. Rhinosporidiosis in Kuwait. *Trop Geogr Med.* 1986; **38:** 190-2.
155. Mears TP, Amerasinghe C. View from beneath - Pathology in focus: Rhinosporidiosis. *J Laryngol Otol.* 1992; **106:** 468.
156. Mendoza L, Herr RA, Ajello L. Causative agent of rhinosporidiosis. *J Clin Microbiol.* 2001; **39:** 413-5.
157. Mendoza L, Herr RA, Arseculeratne SN, *et al. In vitro* studies on the mechanisms of endospore release by *Rhinosporidium seeberi. Mycopathologia.* 1999; **148:** 9-15.
158. Mendoza L, Taylor JW, Ajello L. The class Mesomycetozoea: A heterogeneous group of microorganisms at the animal-fungal boundary. *Annu Rev Microbiol.* 2002; **56:** 315-44.

159. Mendoza L, Vilela R, Rosa PS, *et al.* *Lacazia loboi* and *Rhinosporidium seeberi:* A genomic perspective. *Rev Iberoam Micol.* 2005; **22:** 213-6.
160. Mendoza L, Vilela R. Presumptive synchronized nuclear divisions without cytokinesis in the *Rhinosporidium seeberi* parasitic life cycle. *Microbiology.* 2013; **159 (Pt. 8):** 1545-51.
161. Misra RC. Rhinosporidiosis in north-western Orissa. *J Commun Dis.* 1989; **21:** 153-5.
162. Mithal C, Agarwal P, Mithal N. Ocular and adnexal rhinosporidiosis : The clinical profile and treatment outcomes in a tertiary eye care centre. *Nepal J Ophthalmol.* 2012; **4:** 45-8.
163. Mitra K, Maity PK. Cutaneous rhinosporidiosis. *J Indian Med Assoc.* 1996; **94:** 84.
164. Mohan H, Chander J, Dhir R, Singhal U. Rhinosporidiosis in India: A case report and review of literature. *Mycoses.* 1995; **38:** 223-5.
165. Mohanty RC, Khalkho J, Sahu A. Diagnosis of disseminated rhinosporidiosis by fine needle aspiration cytology. *Acta Cytol.* 2006; **50:** 234.
166. Mohapatra M, Banushree CS. Two rare cases of rhinosporidiosis of parotid duct: Case reports and review of literature. *Ann Maxillofac Surg.* 2014; **4:** 234-6.
167. Mondal S, Chowdhury A. Rhinosporidiosis of the left wrist joint: A case report. *J Orthop Surg (Hong Kong).* 2013; **21:** 245-8.
168. Morelli L, Polce M, Piscioli F, *et al.* Human nasal rhinosporidiosis: An Italian case report. *Diagn Pathol.* 2006; **1:** 25.
169. Moses JS, Balachandran C, Sandhanam S, *et al.* Ocular rhinosporidiosis in Tamil Nadu, India. *Mycopathologia.* 1990; **111:** 5-8.
170. Moses JS, Shanmugham A, Kingsly N, *et al.* Epidemiological survey of rhinosporidiosis in Kanyakumari district of Tamil Nadu. *Mycopathologia.* 1988; **101:** 177-9.
171. Muhammed K, Abdul Latheef K. Subcutaneous rhinosporidiosis. *Indian J Dermatol Venereol Leprol.* 1997; **63:** 320-3.
172. Mukherjee B, Mohan A, Sumathi V, *et al.* Infestation of the lacrimal sac by *Rhinosporidium seeberi:* A clinicopathological case report. *Indian J Ophthalmol.* 2013; **61:** 588-90.
173. Mukhopadhyay S, Datta H, Sen D. Intracorneal rhinosporidiosis managed with deep anterior lamellar keratoplasty. *Middle East Afr J Ophthalmol.* 2014; **21:** 361-2.
174. Mukhopadhyay S, Shome S, Bar PK, *et al.* Ocular rhinosporidiosis presenting as recurrent chalazion. *Int Ophthalmol.* 2015; **35:** 705-7.
175. Nair AG, Ali MJ, Kaliki S, *et al.* Rhinosporidiosis of the tarsal conjunctiva. *Indian J Ophthalmol.* 2015; **63:** 462-3.
176. Nambiar SS, Radhakrishnan S, Vijayan A. Rhinosporidiosis: Report of an extra-ductal facial lesion. *IDCases.* 2016; **7:** 40-3.
177. Nanda BK, Naik UP, Panda GK, *et al.* Rhinosporidiosis in western Orissa. *J Indian Med Assoc.* 1969; **53:** 489-92.
178. Nath AK, Madana J, Yolmo D, *et al.* Disseminated rhinosporidiosis with unusual involvement of the nail apparatus. *Clin Exp Dermatol.* 2009; **34:** e886-8.
179. Nayak S, Acharjya B, Devi B, *et al.* Disseminated cutaneous rhinosporidiosis. *Indian J Dermatol Venereol Leprol.* 2007; **73:** 185-7.
180. Neelakandan RS, Krithika C, Devadoss P, *et al.* Rhinosporidiosis of the parotid duct. *J Oral Maxillofac Surg.* 2012; **70:** e39-42.
181. Nerurkar NK, Bradoo RA, Joshi AA, *et al.* Lacrimal sac rhinosporidiosis: A case report. *Am J Otolaryngol.* 2004; **25:** 423-5.
182. Ngamdu YB, Ngadda HA, Kodiya AM, *et al.* Nasal rhinosporidiosis: A case report and review of literature. *J Case Rep.* 2014; **4:** 26-8.
183. Nuruddin M, Mudhar HS, Osmani M, *et al.* Lacrimal sac rhinosporidiosis: Clinical profile and surgical management by modified dacryocystorhinostomy. *Orbit.* 2014; **33:** 29-32.
184. Nwana EJ, da Lilly-Tariah OB, Manasseh A. Rhinosporidiosis in the nasal cavity. *Trop Doct.* 1998; **28:** 109-10.
185. Padmavathi L, Lakshmana L, Selvam Siva S, *et al.* Disseminated cutaneous rhinosporidiosis in a HIV seropositive patient. *Indian J Dermatol Venerol Leprol.* 2001; **67**: 332-3.
186. Pai SA, Naresh KN, Shinde SR, *et al.* Rhinosporidioma of bone: Diagnosis by fine needle aspiration. *Acta Cytol.* 1996; **40:** 45-6.
187. Pal DK, Bhattacharyya N, Bag AK, *et al.* Recurrent rhinosporidiosis of male urethra. *Indian J Surg.* 2006; **68:** 97-8.
188. Pal DK, Mallick AA, Majhi TK, *et al.* Rhinosporidiosis in southwest Bengal. *Trop Doct.* 2012; **42:** 150-3.
189. Pal DK, Moulik D, Chowdhury MK Genitourinary rhinosporidiosis. *Indian J Urol.* 2008; **24:** 419-21.
190. Pal S, Chakrabarti S, Biswas BK, *et al.* Cytodiagnosis of Extra-nasal rhinosporidiosis: A study of 16 cases from endemic area. *J Lab Physicians.* 2014; **6:** 80-3.
191. Panda S, Lenka S, Padhiary SK, *et al.* Rhinosporidiosis in the parotid duct: A rare case report. *J Investig Clin Dent.* 2013; **4:** 271-4.
192. Pandey N, Chandrakar AK, Garg ML, *et al.* Oculosporidiosis. *Indian J Ophthalmol.* 2008; **56:** 81-2.
193. Pandey P, Shukla S, Kundu A, *et al.* Urethral polyp: An uncommon presentation of rhinosporidiosis. *Indian J Dermatol.* 2014; **59:** 95-6.
194. Parthasarathi R, Sen S, Khan M. Conjunctival rhinosporidiosis. *J Indian Med Assoc.* 2003; **101:** 667-8 & 670.
195. Pathak D, Neelaiah S. Disseminated cutaneous rhinosporidiosis: Diagnosis by fine needle aspiration cytology. *Acta Cytol.* 2006; **50:** 111-2.
196. Patnaik K, Vasal PC. Rhinosporidiosis presenting as urethral polyp. *Indian J Pathol Microbiol.* 1994; **37:** 339-42.
197. Paul J, Khan AR, Gupta JP. Rhinosporidiosis in Jammu. *Indian J Pathol Microbiol.* 1978; **21:** 73-7.
198. Pe'er J, Gnessin H, Levinger S, *et al.* Conjunctival oculosporidiosis in east Africa caused by *Rhinosporidium seeberi.* *Arch Pathol Lab Med.* 1996; **120:** 854-8.
199. Prabhakar BR, Bhatnagar GS. Rhinosporidiosis in Amritsar (Punjab). *Indian J Pathol Microbiol.* 1985; **28:** 239-43.
200. Prabhu SM, Irodi A, Khiangte HL, *et al.* Imaging features of rhinosporidiosis on contrast CT. *Indian J Radiol Imaging.* 2013; **23:** 212-8.
201. Prakash M, Carlton JJ. Nasal rhinosporidiosis with an atypical presentation. *J Pharm Bioallied Sci.* 2015; **7:** S72-3.

202. Prakash M, Johnny JC. Rhinosporidiosis and the pond. *J Pharm Bioallied Sci.* 2015; **7:** S59-62.
203. Prakash PY. Observations regarding the clinical manifestations and management of extranasal rhinosporidiosis in southern India. *Clin Microbiol Newsl.* 2009; **31:** 30-1.
204. Prasad K, Veena S, Permi HS, *et al.* Disseminated cutaneous rhinosporidiosis. *J Lab Physicians.* 2010; **2:** 44-6.
205. Prasad V, Shenoy VS, Rao RA, *et al.* Rhinosporidiosis: A chronic tropical disease in lateral pharyngeal wall. *J Clin Diagn Res.* 2015; **9:** MD01-2.
206. Priyadarshini GR, Srinivasa KS, Parijatham BO, *et al.* A rare case of disseminated cutaneous rhinosporidiosis. *J Clin Diagn Res.* 2015; **9:** EL01-2.
207. Pushker N, Kashyap S, Bajaj MS, *et al.* Primary lacrimal sac rhinosporidiosis with grossly dilated sac and nasolacrimal duct. *Ophthal Plast Reconstr Surg.* 2009; **25:** 234-5.
208. Qureshi F, Abdulmannan MS, Eltayeb AA, *et al.* Rhinosporidiosis of the male urethra. *Ann Saudi Med.* 2004; **24:** 299-300.
209. Ragan MA, Goggin CL, Cawthorn RJ, *et al.* A novel clade of protistan parasites near the animal-fungal divergence. *Proc Natl Acad Sci.* 1996; **93:** 11907-12.
210. Rajakannu M, Sri Vengadesh G, Pai D, *et al.* Disseminated rhinosporidiosis - An unusual presentation with pulmonary involvement. *Int J Dermatol.* 2006; **45:** 297-8.
211. Raju GC, Jamalabadi MH. Rhinosporidiosis in Trinidad. *Trop Geogr Med.* 1983; **35:** 257-8.
212. Ramanan C, Ghorpade A. Giant cutaneous rhinosporidiosis. *Int J Dermatol.*1996; **35:** 441-2.
213. Rameshkumar A, Gnanaselvi UP, Dineshkumar T, *et al.* Rhinosporidiosis presenting as a facial swelling: a case report. *J Int Oral Health.* 2015; **7:** 58-60.
214. Rath R, Baig SA, Debata T. Rhinosporidiosis presenting as an oropharyngeal mass: A clinical predicament? *J Nat Sci Biol Med.* 2015; **6:** 241-5.
215. Ratnakar C, Madhavan M, Sankaran V, *et al.* Rhinosporidiosis in Pondicherry. *J Trop Med Hyg.* 1992; **95:** 280-3.
216. Raveenthiran V. Metastatic rhinosporidioma in a child. *J Pediatr Surg.* 2006; **41:** 853-5.
217. Ravi R, Mallikarjuna VS, Chaturvedi HK. Rhinosporidiosis mimicking penile malignancy. *Urol Int.* 1992; **49:** 224-6.
218. Reidy JJ, Sudesh S, Klafter AB, *et al.* Infection of the conjunctiva by *Rhinosporidium seeberi. Surv Ophthalmol.* 1997; **41:** 409-13.
219. Rekha P, Thomas B, Pappachan JM, *et al.* Tracheal rhinosporidiosis. *J Thorac Cardiovasc Surg.* 2006; **132:** 718-9.
220. Rogers S, Waring D, Martin P. Recurrent lacrimal sac rhinosporidiosis involving the periocular subcutaneous tissues, nasolacrimal duct and nasopharynx. *Orbit.* 2012; **31:** 358-60.
221. Rv S, Ag P. A rare case of systemic rhinosporidiosis. *J Assoc Physicians India.* 2016; **64:** 155.
222. Sah SP, Singh RK, Ashok Raj G. Urethral rhinosporidiosis: Report of a case presenting with a penile growth and urethral fistula. *Trans R Soc Trop Med Hyg.* 1999; **93:** 298-9.
223. Saha KL, Siddiquee BH, Uddin M, *et al.* Tracheal rhinosporidiosis: An uncommon life threatening benign condition. *J Coll Physicians Surg Pak.* 2014; **24:** 526-7.
224. Sahoo GC, Raghavan R. Rhinosporidiosis in HIV: A case report. *Indian J Otolaryngol Head Neck Surg.* 2000; **52:** 281-2.
225. Sahu B, Sahu P, Puhan MR. Disseminated cutaneous rhinosporidiosis: A polymorphic presentation in an immunocompetent patient. *Indian J Dermatol.* 2015; **60:** 218.
226. Salazar Campos MC, Surka J, Garcia Jardon M, *et al.* Ocular rhinosporidiosis. *S Afr Med J.* 2005; **95:** 950 & 952.
227. Salim T, Komu F. Varied presentations of cutaneous rhinosporidiosis: A report of three cases. *Indian J Dermatol.* 2016; **61:** 209-12.
228. Samaddar RR, Sen MK. Rhinosporidiosis in Bankura. *Indian J Pathol Microbiol.* 1990; **33:** 129-36.
229. Sarkar S, Panja S, Bandyopadhyay A, *et al.* Rhinosporidiosis of parotid duct presenting as consecutive bilateral facial swelling: A rare case report and literature review. *J Clin Diagn Res.* 2016; **10:** PD14-6.
230. Sasidharan K, Subramonian P, Moni VN, *et al.* Urethral rhinosporidiosis: Analysis of 27 cases. *Br J Urol.* 1987; **59:** 66-9.
231. Senaratne T, Edussuriya K, Dhanapala M, *et al.* Ocular rhinosporidiosis with staphyloma formation: A case with unusual features. *Eye Brain.* 2011; **3:** 1-4.
232. Seth J, Saha A, Mondal AK, *et al.* Penile rhinosporidiosis: An unusual presentation and review of literature. *Indian J Dermatol Venereol Leprol.* 2017. doi: 10.4103/0378-6323.199583. PMID: 28195081.
233. Shah S, Lavaju P, Bharati P, *et al.* A case report of an unusual presentation of ocular rhinosporidiosis as a conjunctival cystic mass. *Orbit.* 2017; **36:** 55-7.
234. Shenoy MM, Girisha BS, Bhandari SK, *et al.* Cutaneous rhinosporidiosis. *Indian J Dermatol Venereol Leprol.* 2007; **73:** 179-81.
235. Shrestha SP, Hennig A, Parija SC. Prevalence of rhinosporidiosis of the eye and its adnexa in Nepal. *Am J Trop Med Hyg.* 1998; **59:** 231-4.
236. Silva V, Pereira CN, Ajello L, *et al.* Molecular evidence for multiple host-specific strains in the genus *Rhinosporidium. J Clin Microbiol.* 2005; **43:** 1865-8.
237. Singh RK, Christopher DJ, Isaac TJ, *et al.* Endobronchial rhinosporidiosis. *J Bronchology Interv Pulmonol.* 2013; **20:** 164-6.
238. Sinha A, Phukan JP, Bandyopadhyay G, *et al.* Clinicopathological study of rhinosporidiosis with special reference to cytodiagnosis. *J Cytol.*2012; **29:** 246-9.
239. Sivapathasundharam B, Saraswathi TR, Manjunath K, *et al.* Rhinosporidiosis of parotid duct. *Indian J Dent Res.* 2009; **20:** 388-9.
240. Sood A, Badhu B. Oculosporidiosis in a tertiary care hospital of Western Orissa, India. *Indian J Ophthalmol.* 2008; **56:** 165.
241. Sood N, Agarwal MC, Gugnani HC. Ocular rhinosporidiosis: A case report from Delhi. *J Infect Dev Ctries.* 2012; **6:** 825-7.
242. Stricker JB, Hurley DL, Slovacek KJ, *et al.* An unusual infection in a human immunodeficiency virus-positive man: Pathologic diagnosis - Rhinosporidiosis. *Arch Pathol Lab Med.* 1999; **123:** 1121-2.

243. Sudarshan V, Gahine R, Daharwal A, *et al.* Rhinosporidiosis of the parotid duct presenting as a parotid duct cyst - A report of three cases. *Indian J Med Microbiol.* 2012; **30:** 108-11.
244. Sudarshan V, Goel NK, Gahine R, *et al.* Rhinosporidiosis in Raipur, Chhattisgarh: A report of 462 cases. *Indian J Pathol Microbiol.* 2007; **50:** 718-21.
245. Sudasinghe T, Rajapakse RP, Perera NA, *et al.* The regional sero-epidemiology of rhinosporidiosis in Sri Lankan humans and animals. *Acta Trop.* 2011; **120:** 72-81.
246. Suh LH, Barron J, Dubovy SR, *et al.* Ocular rhinosporidiosis presenting as chronic follicular conjunctivitis in a contact lens wearer. *Arch Ophthalmol.* 2009; **127:** 1076-7.
247. Sunderraj P. Scleral melting and conjunctival rhinosporidiosis. *Br J Ophthalmol.* 1992; **76:** 191.
248. Suryawanshi PV, Rekhi B, Desai S, *et al.* Rhinosporidiosis isolated to the distal clavicle: A rare presentation clinicoradiologically mimicking a bone tumor. *Skeletal Radiol.* 2011; **40:** 225-8.
249. Thakur SK, Sah SP, Badhu BP. Oculosporidiosis in eastern Nepal: A report of five cases. *Southeast Asian J Trop Med Public Health.* 2002; **33:** 362-4.
250. Thankamani V, LipinDev MS. *Rhinosporidium seeberi* proven as a fungus for the first time after a century since its discovery. *Res Biotechnol.* 2012; **3:** 41-6.
251. Thankamani V. *Rhinosporidium seeberi* grown in culture - A ray of hope for the thousands of patients with rhinosporidiosis! *South Asian J Socio-Political Studies (SAJOSPS).* 2005; **5:** 118-23.
252. Thappa DM, Venkatesan S, Sirka CS, *et al.* Disseminated cutaneous rhinosporidiosis. *J Dermatol.* 1998; **25:** 527-32.
253. Thianprasit M, Thagerngpol K. Rhinosporidiosis. *Curr Topic Med Mycol.* 1989; **3:** 64-85.
254. Thompson LD. Rhinosporidiosis. *Ear Nose Throat J.* 2016; **95:** 101.
255. Tiwari R, Karthik K, Dhama K, *et al.* Rhinosporidiosis: A riddled disease of man and animals. *Adv Anim Vet Sci.* 2015; **3:** 54-63.
256. Tolat SN, Gokhale NR, Belgaumkar VA, *et al.* Disseminated cutaneous rhinosporidiomas in an immunocompetent male. *Indian J Dermatol Venereol Leprol.* 2007; **73:** 343-5.
257. Tubachi P, Naalla R, Koduru V, *et al.* Primary cutaneous rhinosporidiosis: An unusual lesion with an unusual presentation. *BMJ Case Rep.* 2015; pii: bcr2015209596.
258. Uledi S, Fauzia A. Human nasal rhinosporidiosis: A case report from Malawi. *Pan Afr Med J.* 2011; **9:** 27.
259. Vallarelli AF, Rosa SP, Souza EM. Rhinosporidiosis: Cutaneous manifestation. *An Bras Dermatol.* 2011; **86:** 795-6.
260. Van der Coer JM, Marres HAM, Wielinga EW, *et al.* Rhinosporidiosis in Europe. *J Laryngol Otol.* 1992; **106:** 440-3.
261. Venkatachalam VP, Anand N, Bhooshan O. Rhinosporidiosis: Its varied presentations. *Indian J Otolaryngol Head Neck Surg.* 2007; **59:** 142-4.
262. Verma PL, Kohli GS. Rhinosporidiosis in Haryana. *Indian J Otolaryngol.* 1970; **22:** 220.
263. Vijaikumar M, Thappa DM, Karthikeyan K, *et al.* A verrucous lesion of the palm. *Postgrad Med J.* 2002; **78:** 302 & 305-6.
264. Vilela R, Mendoza L. The taxonomy and phylogenetics of the human and animal pathogen *Rhinosporidium seeberi*: A critical review. *Rev Iberoam Micol.* 2012; **29:** 185-99.
265. Viswanatha B. Rhinosporidiosis: An unusual presentation. *Ear Nose Throat J.* 2014; **93:** E22-4.
266. Von Haacke NP, Mugliston TA. Rhinosporidiosis. *J Laryngol Otol.* 1982; **96:** 743-50.
267. Vukovic Z, Bobic-Radovanovic A, Latkovic Z, *et al.* An epidemiological investigation of the first outbreak of rhinosporidiosis in Europe. *J Trop Med Hyg.* 1995; **98:** 333-7.
268. Yesudian P. Cutaneous rhinosporidiosis mimicking verruca vulgaris. *Int J Dermatol.* 1988; **27:** 47-8.

CHAPTER 39

Pythiosis and Lagenidiosis

On the basis of ecology and natural history, pathogenic fungi and related organisms were traditionally divided into **three** categories: **anthropophilic**, **zoophilic** and **geophilic**. However, **fourth** group, called **hydrophilic** category, has also emerged that includes free-living pathogens found prevalent in aquatic environment. This hydrophilic group consists of pathogenic fungi *(Lacazia loboi)*, Mesomycetozoea *(Rhinosporidium seeberi)*, already discussed in previous Chapters and pseudofungal organisms *(Pythium insidiosum* and *Lagenidium* species*)*, which are being described in this Chapter.

I. PYTHIOSIS

The mycelial and unicellular organisms that cannot be classified under the kingdom Fungi but cause infections having clinical and histopathological characteristics resembling the fungal diseases are called **parafungal** or **pseudofungal** organisms *e.g. Pythium* and *Prototheca* species. The incidence and prevalence of these infections is increasing almost parallel to the true fungal diseases. The infected individual may be immunocompetent or immunocompromised with underlying debilitating illnesses such as malignancies (leukemia or lymphoma), on immunosuppressive therapy or those suffering from AIDS are at risk. There is an association of such diseases with thalassemia also among the Asian patients.

Among the emerging pseudofungal organisms of aquatic origin, there are pathogenic oomycetes that belong to genus ***Pythium***, with ***P.insidiosum*** being its representative species. Earlier, only one species of the genus *i.e. P.insidiosum* was clinically significant but recently another species, *P.aphanidermatum*, has also emerged as a significant pathogen. These are flagellated **water-molds**, which have fungus-like filamentous morphology causing, **pythiosis**, an uncommon emerging life-threatening infection prevalent among animals and rarely encountered in man.

Pythiosis is one of the under-diagnosed infection. It is invariably mistaken for fungal etiology because of its clinical and diagnostic resemblance, particularly with mucormycosis. This confusion may lead to improper and/or delayed diagnosis entailing late initiation of treatment thereby implying poor prognosis or even fatal consequences. In addition, cases of infection with a member of another genus *i.e. Lagenidium*, have also been recently reported. Therefore, description of these two emerging infections *i.e.* **Pythiosis** and **Lagenidiosis**, becomes imperative in the Textbook of Medical Mycology despite their non-fungal nature. The Chapter focuses mainly on pythiosis insidiosi, caused by *Pythium insidiosum,* followed by brief discussion of other similar oomycotic infections.

Historical Perspective

Pythiosis insidiosi as a clinical entity was first described more than a century ago among horses in tropical, subtropical and temperate countries including India, Indonesia and USA. A British veterinarian, Smith, while working in India first reported this disease in 1884 as cutaneous granuloma in horses popularly known as ***'Bursatee'*** based on its Indian vernacular name denoting ***'burus'*** for rain sores, encountered mainly in the rainy season. He also observed that the disease could be transmitted to the healthy animals by inoculation of infected tissues.

In 1901, de Haan and Hoogkamer, two Dutch investigators, isolated the etiological agent of pythiosis insidiosi for the **first** time from cutaneous granuloma of horses in Indonesia. Since no spores could be demonstrated from the culture isolates at that time hence they could not assign any name to the organism. However, the disease was designated as ***hyphomycosis destruens***, implying it as a 'fungal' infection of destructive nature. As the disease was prevalent among horses, its name was subsequently modified during the following year (1902) as ***hyphomycosis***

destruens-equi by adding suffix ***'equi'*** thereby signifying involvement of horses.

In 1925, another Dutch investigator, Witkamp, while working in Indonesia, isolated the same organism. This was **second** instance of its isolation but this time also it did not show sporulation. He also developed several serological tests for diagnosis of pythiosis insidiosi and published his comprehensive findings. With the passage of time, it became evident that the disease was not only affecting horses but other mammalian species as well, including humans.

In due course of time, Bridges and Emmons in 1961, isolated a non-sporulating filamentous organism from equines in Texas and Florida, which was similar to the organism previously studied and described earlier by de Haan and Hoogkamer in the beginning of the 20th century. Therefore, based on the disease name as earlier designated *i.e.* **hyphomycosis destruens**; they called the organism as *Hyphomyces destruens*. They considered it to be a fungus *i.e.* phycomycetes (mucormycetes), perhaps a species of the genus *Mortierella*. But neither asexual nor sexual spores could be induced to support their contention. This was **third** instance when it was isolated as 'sterile' filamentous organism.

In 1974, Austwick and Copland reported that an organism isolated from equine cutaneous granulomas in New Guinea produced biflagellated zoospores in water and identified it as *Pythium* species. Ichitani and Amemiya in Japan reported cases with similar findings among equines having granular dermatitis. They identified the isolate as *Pythium gracile*. However, in subsequent studies by de Cock *et al*, it was found that all *Pythium* isolates recovered from humans and animals, including the Japanese isolate were in fact single species *i.e. P.insidiosum*. In 1980, Chandler *et al*, labeled clinical and pathological manifestations produced by genus *Pythium* as **pythiosis** but species name still remained unresolved.

In the year 1987, de Cock *et al*, from Central American country-Costa Rica, proposed a name for the unnamed species of genus *Pythium* as ***insidiosum*** keeping in view the **insidious** onset of infection thus completing binomial nomenclature of the organism with Latin description. The first case of human *P.insidiosum* keratitis was reported in the year 1988. Mendoza and Marin studied another isolate in 1989, which was indistinguishable from the strains isolated from previous cases. Hence they recognized the new strain as synonym of *P.insidiosum*. In 1990, Chaiprasert described the method to induce zoospore in this organism and later on, in 1993, Mendoza *et al*, explained its detailed life cycle. Sharma *et al,* carried out detailed studies in relation to the ophthalmic involvement including the zoospore induction in Indian context in the year 2015.

The taxonomical history of organism started with its consideration as member of kingdom **Fungi** but later on shifted to Protista. In 1994, Corliss proposed two subdivisions of Protista *i.e.* Protozoa and Chromista and subsequently placed this organism under Chromista. At present, *P.insidiosum* is classified in the kingdom **Stramenopila** on the basis of taxonomical and molecular developments. As members of this kingdom inhabit wet environment such as ponds, swamps and develop hypha-like elements so they used to be classified as ***'aquatic fungi'*** in the past. However, recent phylogenetic analysis has demonstrated that they are protistan microbes close to the algae and plants and distant from kingdom Fungi.

Organism *(Pythium insidiosum)*

Pythium species were originally considered to be oomycetous members of kingdom **Fungi** due to their coenocytic filamentous nature and feeding on decaying matter. The literal meaning of 'Oomycota' is **egg fungi**, which refer to large round **Oogonium**, the structure containing female gametes. They remain diploid throughout their life cycle with meiosis occurring in the gametangia just before fertilization.

The organism has several peculiar characteristics, very different from fungi: (i) biflagellate zoospores (asexual reproduction); (ii) oogamic sexual reproduction; (iii) cell wall composed of β-glucan, cellulose and hydroxyproline; (iv) diploid thallus; (v) tubular cristae mitochondria and (vi) molecular and biochemical characteristics such as an alternative route for lysine synthesis.

There are more than 500 species in Oomycota, which include **'water-molds'** and downy mildews. These filamentous protists absorb their food material from surrounding water or soil. They may invade body of another organism in search of food. Their nomenclature, classification and taxonomy have undergone several revisions over the years and up to some extent it still continues to be controversial. The taxonomy of genus *Pythium* is mainly based on morphological descriptions supplemented with molecular studies.

Earlier it was classified in the kingdom **Fungi**, therefore, called as **'water-mold'** due to its physical resemblance. But because of certain basic morphological differences, now it is placed under the kingdom **Protista**. The cell wall of *Pythium* species is composed of cellulose and β-glucan

Table 39.1. Important Morphological and Functional Differences between Oomycetes and Fungi.

Features	*Oomycetes*	*Fungi*
Kingdom	Stramenopila – Protista	Fungi
Hyphae	Coenocytic hyphae with few septae	Variable but often with septae except mucormycetes
Cell walls	Cellulose, polymerized β-1, 3-glucan and hydroxyproline	Chitin (β-1, 4-N-acetyl-glucosamine) is major component
Sterol type	If synthesized: Fucosterol, Cycloartenol, Cholesterol, Desmosterol, Lanosterol	Ergosterol
Nuclear condition	Diploid dominant	Haploid dominant (with rare derived exceptions)
Nuclear division	Intranuclear with centrioles	Extranuclear with centrioles
Mitochondrial type	Tubular cristae	Plate-like cristae
Mitochondrial gene content	Additional genes present (esp. RP genes, ATPase alpha) but no introns (at least in *Phytophthora infestans* and *megasperma*)	Low gene content (Similar to animals) but loaded with group I and II introns
Golgi bodies	Well-defined	Not well-defined
Acyclic polyols (sugar alcohols)	Absent	Common (particularly Mannitol)
Storage carbohydrate	Mycolaminarins (β 1-3 glucan)	Glycogen
Lysine synthesis	Diaminopimelic acid pathway (also in bacteria and higher plants)	Aminoadipic acid pathway
Steroid pheromones	Antheridiol and Oogonial	Pheromones may be present but not steroids
Flagella	Two anterior or lateral (one whiplash, one tinsel)	If present, single posterior whiplash (Chytridiomycota)
Sexual spores	Sexual spores are oospore (2N)	Sexual spores (N) - product of meiosis
Gametangia	Oogonia and antheridia common	If present, rarely Oogamous
Meiosis	Meiosis is gametangial	Meiosis is not gametangial (some fungi have gametangia)

rather than chitin. The plasma membrane does not have ergosterol that is susceptible to action of majority of antifungal drugs. Moreover, in the vegetative state, the organism has diploid nuclei, whereas fungi have haploid nuclei. *P.insidiosum*, being an oomycete is not only physiologically distinct from members of kingdom Fungi but phylogenetically also. It is differentiated from true fungi on the basis of 18s rRNA and other sequences as well as on various biological grounds, which are summarized in **Table 39.1**.

The members of genus *Pythium* (*Gr.* ***'Pythein'***-*'to make rot'*) have long been recognized as destructive pathogens of plants and fish. There are more than 200 species in the genus, distributed worldwide with *P.insidiosum* and *P.aphanidermatum* being implicated in infections of various mammalian hosts *i.e.* animals as well as man. The knowledge of intra-species variability in *P.insidiosum* might be important for the management of pythiosis in mammals.

The **sexual** reproduction is by **oogamy (Fig. 39.1)** in which **sexual** spores are produced. These are thick-walled oospores resulting by the union of antheridia and oogonia

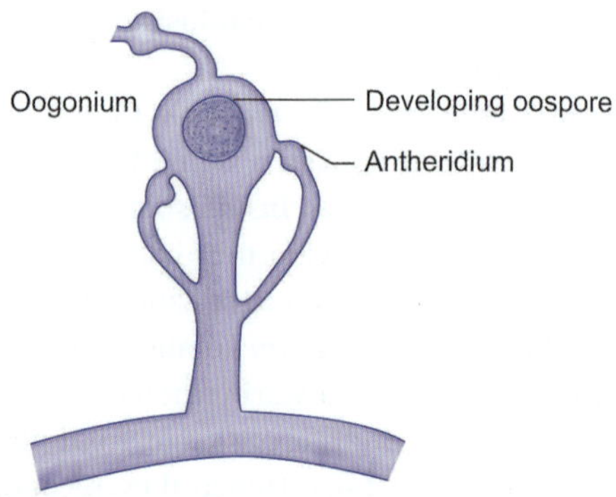

Fig. 39.1. Diagrammatic representation of Oospore of *Pythium* species of Oomycetes.

i.e. male and female sexual structures, respectively and are the non-motile gametes. Oogonium produces the oosphere *i.e.* female gamete and a fertilized oosphere is called an oospore. The oospores are significant survival structures as well as source of genetic-recombination. On the other hand, **asexual** reproduction takes place by formation of **zoospores** in filamentous or spherical types of sporangia. Most of the medically important fungi also have both asexual and sexual stages in their life cycles.

The oomycetes are eukaryotic organisms, which produce motile spores called as **zoospores** with **two** dissimilar flagella originating from common point but are of unequal length. The posterior smooth one is called as **whiplash** while anterior one where the main stem has fine hair-like projection is called as **tinsel** flagellum. The biflagellate zoospores are formed inside vesicle through progressive cleavage and the vegetative hyphae and have diploid nuclei. Because of such features, the organism was once labeled in the past as **'zoosporic fungus'**.

The detection of zoosporogenesis and oosporogenesis in genus *Pythium* with subsequent clarifications resulted in its re-classification to kingdom **Protista**. This kingdom was further sub-divided into two kingdoms: **Protozoa** and **Chromista.** *P.insidiosum* was placed under phylum Pseudofungi of kingdom Chromista. The current molecular and biochemical analysis suggests that these organisms are closer to algae and higher plants and are now classified amongst **Stramenopiles**, one of the five major eukaryotic kingdoms, which include organisms characterized by formation of motile spores. Therefore, taxonomically, *P.insidiosum* is at present placed in family **Pythiaceae** of order **Pythiales** in class **Oomycetes** of phylum **Oomycota** in the kingdom **Stramenopila** and **not** in the kingdom Fungi.

Epidemiology

In the environment, the oomycetes play an important role in decomposition and recycling of decaying organic matter. Some of the species are found to be parasitic to crops and fish. *Pythium* species are amongst the most destructive phytopathogen, which cause major economic loss because of their ability to cause infection in crops, horses and cattle. *P.insidiosum* is a cosmopolitan organism found in soil and water bodies with an ecological preference for swampy environment. This may be the reason of increase in cases during rainy season. Therefore, the disease is popularly called as **'swamp cancer'**, which is a well-known veterinary condition.

Pythium species prefer wet environment but can live and survive in soil for considerably long time. It is contemplated that since this oomycete resides in wet habitat hence it requires a plant to complete its life cycle. The studies show that it may involve more than one species of aquatic plants to complete its life cycle in nature. Hence several plants like water lilies present in aquatic environment keep the organism viable and support their life cycle.

The asexual propagules of *P.insidiosum* appear to be chemotactically attracted to plant leaves, certain grasses or human and animal hairs, which serve as potential infectious particles. Its asexual motile zoospores are usually found in freshwater environment associated with plants. However, there are no case reports documenting transmission of infection from animal-to-animal or animal-to-man. The reservoir for *P.insidiosum* remains unknown despite preponderance of its cases among horses and dogs in tropical and subtropical countries known to have visited marshy areas or having history of swimming.

The common route of infection of pythiosis insidiosi is via entry of zoospores through small cuts in body cavities of individuals who either spend time in water or drink it from stagnant swampy areas as many fungal species, including other *Pythium* species, thrive in such environments. The motile biflagellate **zoospore** is the **infective stage** of organism and has a predilection for the traumatized tissue and requires water to move towards the host. However, the exact incubation period of pythiosis insidiosi is not yet known. Upon reaching the host they lose their flagella, attach to tissue using sticky substance, get encysted and then germinate by producing germ tubes and later on abundant hyphal elements, which eventually invade the host tissue. Individuals with thalassemia, hemoglobin E disease and farmers are predisposed to acquire this infection.

The well-documented cases of human pythiosis were published for the first time in 1987. However, the pseudo-fungal nature of this disease has only been recognized recently, in last three decades or so. Since then sporadic cases of this disease have been found in several countries, majority of them are reported from **Thailand**. In addition to man, disease is found prevalent among horses, dogs, cats and cattle residing in tropical, subtropical and temperate regions.

The clinical cases have been reported from tropical and subtropical areas of Asian subcontinent and Far East (Thailand, Indonesia, India, Japan, New Guinea, North Korea, Australia and New Zealand), Central America and Caribbean islands (Costa Rica, Guatemala, Panama, Nicaragua and Haiti), South America (Argentina, Brazil, Colombia and Venezuela) and North America (USA), particularly states bordering Gulf of Mexico like Florida, Texas, Louisiana and Mississippi. In recent times, numerous cases have been reported from the Midwest and Northeast USA and as far north in Wisconsin and Washington States. Although the early cases of this disease were reported from India, still the disease is not very common here and/or it may be under-diagnosed.

Most of the cases observed in Thailand are commonly associated with α- and β-thalassemia, which is prevalent in Southeast Asia. More than 150 cases of human pythiosis have been reported and nearly 80% of the total global cases are from Thailand with over 90% of these individuals being farmers. It has also been reported in a dog from Bamako, capital of Mali, a country in north-western Africa, the continent that till recent times was thought to be free from this disease. It is not yet observed in the European continent but imported cases are reported from non-endemic countries, among the individuals who have visited the endemic areas.

Previously, it was believed that only fungi possessed chitin synthase genes. But it is now established that these genes are present in this pathogen as well. However, phylogenetic analysis shows that the chitin synthase present in Oomycetes is different from fungi. This suggests that its molecules were passed from lower protist ancestor to Oomycetes and Fungi. Recently, in phylogenetic analysis, using ITS1, 5.8S and ITS2 sequences from *P.insidiosum*, it was found that these isolates were grouped according to their geographical locations. This finding was in contrast to serological studies, which indicated that all isolates, irrespective of their geographical locations, were antigenically identical.

Immunity

The development of an immune response to *P.insidiosum* antigens has been comprehensively studied in equine infections. The cellular and humoral immune responses are active in early stages of the disease. There is production of cellular immune response in skin tests and antibodies can also be demonstrated in active cases of pythiosis insidiosi. There are broadly two antigenic groups involved in immune response to *P.insidiosum*. One is **culture filtrate antigens** (CFA) and other group is **sonicated hyphal antigen**. On the basis of immunological process, two vaccines have been successfully used in the management of disease in animals and also in clinical trials in humans. However, animal becomes anergic to vaccine when the disease reaches chronicity *i.e.* beyond two months.

Pathogenesis and Pathology

The oomycetes have evolved their ability to infect plants and animals independently of other eukaryotic microbes and have probably developed unique mechanism of pathogenicity. As pythiosis insidiosi is primarily a zoonotic disease hence most of the studies are based on clinical and experimental findings in animals. The **infective stage** is **zoospore**, however, other stages of organism like hyphae, oogonia or resting spores may also be infective as evident from a case who had no prior history of contact with stagnant water. The zoospores of *P.insidiosum* have an affinity towards animal and plant tissues. Moreover, zoospores are known to secrete a sticky substance, which helps the organism in maintaining a close contact with host tissue during early stages of invasion.

When immune response is fairly weak, eosinophils try to phagocytose the organism and degranulated eosinophils are deposited over the hyphae. The new eosinophils fuse around already degranulated eosinophils and subsequently they are also degranulated. The eosinophilic precipitate around hyphae form the basis of **Splendore-Hoeppli phenomenon**, very similar to that seen in entomophthoramycosis. This organism is also angioinvasive like some of fungi *e.g.* mucormycetes and *Aspergillus* species hence this organism is also responsible for arteritis. In addition, histopathologically, sequestered eosinophilic granular, yellow-white, coral-like, stony masses of necrotic tissue composed of degranulated eosinophils interlaced with viable hyphae of *P.insidiosum* are seen in horses, called as **'kunkers'.** However, the 'kunkers' are not seen in human cases of pythiosis insidiosi.

In case of vascular involvement, patients usually have a cutaneous wound at the lower extremities. The susceptible patients become infected after contact with zoospores in contaminated water through a skin wound. The organism invades the artery and spreads proximally along the arterial walls. In this way all layers of arterial wall are usually involved. In due course, the invasion of the arterial wall leads to thrombosis and aneurysm formation. Due to this underlying pathology, the patients mostly present with signs and symptoms of acute or chronic arterial occlusion, including non-healing type of ulcer or even toe gangrene. However, in the terminal stage of the disease, iliac or abdominal aortic aneurysm may also be observed.

Clinical Features

Pythiosis insidiosi is primarily a zoonotic disease. Human beings acquire infection by coming in contact with aquatic habitats harboring infectious asexual propagules of this pseudofungus (*P.insidiosum*), leading to subcutaneous, ocular, vascular and disseminated infections. The vascular form has been reported mainly from Thailand, particularly among thalassemic patients. The rest of the clinical presentations are observed both among Thai as well as in patients from other countries. Thus thalassemic hemoglobinopathy appears to be an important predisposing factor for pythiosis insidiosi. Although human infection is very rare but if not treated, disease advances rapidly entailing fatal consequences. The morbidity and mortality levels associated with pythiosis insidiosi are very high. Recently, pythiosis caused by another species, *P.aphanidermatum*, is also being reported. Although it is almost exclusively a plant pathogen but cases of infection in human beings have been reported among soldiers injured during combat operations.

Approximately, fifty percent of published cases of pythiosis insidiosi are classified as arterial, twenty five percent as ocular and remainder as cutaneous or cardiopulmonary types. The arterial type is usually characterized by erosive and occlusive disease of major arteries, with occasional systemic spread. Therefore, on the basis of common presenting clinical features, human pythiosis may be categorized into following **four** types:

1. Cutaneous-Subcutaneous Pythiosis
2. Ocular Pythiosis
3. Vascular-Systemic Pythiosis
4. Disseminated Pythiosis

1. Cutaneous-Subcutaneous Pythiosis

Pythiosis insidiosi predominantly involving cutaneous and subcutaneous tissues affects mainly facial area or limbs. This type of disease is found mostly in thalassemic patients often leading to granulomatous tissue reaction, diffuse infiltration and edema. The acute type of lesions ulcerate and rapidly enlarge as subcutaneous extension with necrosis, granulomatous reaction followed by formation of fistulous tracts. There is formation of yellowish necrotic lesions, which contain masses of organisms. The chronic lesions may sometimes be confused with either types of entomophthoramycosis *i.e.* conidiobolomycosis or basidiobolomycosis.

2. Ocular Pythiosis

In case of ocular involvement in human pythiosis, patients present with keratitis and orbital cellulitis. The clinical picture of keratitis is characterized by tentacle-like or dot-like infiltrates surrounding the main infiltrate, hyphate edges, and reticular subepithelial as well as superficial infiltration. The features are indistinguishable from keratomycosis caused by fungi. Therefore, pythiosis insidiosi should be considered as one of the differential diagnoses among patients with corneal ulcer, who are refractory to various antibiotic therapies, particularly when etiologic agent is not readily identified. In addition to that there is progressive periorbital edema with chemosis and erythema due to cellulitis. Individuals using contact lens, corneal trauma and exposure to water are more prone to such infections.

The patients with orbital cellulitis, shows swelling, erythema and tenderness of the periorbital area. CT of the face reveals preseptal cellulitis with maxillary opacification. Ocular infection may or may not be associated with thalassemia. The condition usually deteriorates, as the diagnosis is not properly established in time thereby enucleation or evisceration may have to be resorted upon in such patients.

3. Vascular-Systemic Pythiosis

The human pythiosis may present as systemic disease involving vascular system in immunodeficient individuals. This is rather a new clinical entity and is found preferentially among patients having underlying hematologic disorders such as thalassemia, aplastic anemia or paroxysmal nocturnal hemoglobinuria (PNH) in Thailand, especially those who either live or work in wet or swampy areas. This is also called as **pythiosis arteritis**. There is ascending type of progressive arteritis with thrombosis due to *P.insidiosum* involving major arteries especially those of lower extremities. There is inflammation and necrosis of blood vessels resulting in ischemia, thereby leading to gangrene or formation of aneurysm. This may lead to amputation of affected extremities or fatal blood loss from the involved arteries. Necrotizing arteritis and thrombosis of major arteries of lower limbs present with marked inflammation in outer layers of involved vessel wall. The prognosis is poor if main arteries, such as common iliac, aorta or renal, are involved and patient usually dies of systemic pythiosis insidiosi within a period of weeks. The highest mortality rate has been observed in infection involving proximal part of superficial femoral artery. A case report of pleuropericarditis complicating pneumonia in a child with leukemia is also documented.

The radiological investigations like magnetic resonance angiography or computed tomographic angiography are useful in identifying arterial lesions. Among all clinical forms of human pythiosis, arterial form has the worst prognosis. The vascular pythiosis insidiosi is rare type of systemic infection, which does not respond to antifungal agents. The only treatment is surgical resection and most patients die within 1 to 2 years. Since antifungal drugs are not effective against this pathogen hence immunotherapy is to be resorted upon in the management of infections among humans and animals.

4. Disseminated Pythiosis

Pythiosis may sometimes leads to disseminated infection, which usually follows any of the above-mentioned clinical entities. Invariably it occurs after vascular involvement. By the time it is established that the disease is pythiosis, a lot of damage to various organs of the body with irreversible changes has already commenced eventually leading to death. Therefore, the disease process in pythiosis is to be taken very seriously as this may lead to disseminated infection with fatal consequences.

Pythiosis Insidiosi in Animals

Pythiosis insidiosi is primarily a disease of equines, bovines and canines. Horses are most commonly affected with cutaneous and subcutaneous lesions followed by dogs, rarely cats and other animals such as captive camels. The disease presents as granulomatous and necrotic lesions usually seen on head and lower portions of legs, the anatomical parts first to come in contact with swampy water. The lesions enlarge by subcutaneous extension and may subsequently ulcerate. The intestinal lesions are very common in animals, particularly in dogs, in addition to the cutaneous involvement and there may also be bony involvement. The granulomas contain firm, yellowish coral-like stony masses of necrotic tissue known as ***'kunkers'***, which can be removed intact. The 'kunkers' are basically foci of coagulative necrosis in vessels that have sequestered from surrounding tissue and consist of broad, branching non-septate hyphae of 3-10 μm diameter.

The animal pythiosis has also been known by various colloquial terminologies such as **'swamp cancer'**, 'Florida horse leeches' under the impression that it was an infection caused by nematode larva, 'equine phycomycosis', 'granular dermatitis' and 'hyphomycosis destruens-equi'. The increase in the prevalence of this disease is considered to be associated with the periods of extreme rainfall and subsequent flooding. However, it may also be due to greater awareness and more frequent testing by the veterinary community.

Lagenidiosis, different clinical entity, caused by another oomycete, *Lagenidium giganteum,* a facultative parasite of mosquito larvae also presents with similar features as that of pythiosis insidiosi affecting mainly dogs and sometimes human beings too and is separately described in the trailing part of this Chapter.

Differential Diagnosis

The infection caused by *P.insidiosum* is well known for its development of eosinophilic granuloma, also seen in diseases like entomophthoramycosis caused by *Conidiobolus* and *Basidiobolus*, which may look alike both clinically, histopathologically and initially on culture from pythiosis insidiosi. Other differential diagnosis include focal nodular dermatosis and neoplasia. Peripheral eosinophilia of unexplained etiology is also useful but a non-specific diagnostic feature of pythiosis insidiosi. Fungal keratitis particularly caused by sewage-contaminating agents like *Scedosporium apiospermum,* may resemble with this disease. The other clinical entities like mucormycosis and aspergillosis also may present similarly with pythiosis insidiosi due to the vascular involvement. However, in aspergillosis there are septate hyphae sized between 3 to 6 μm in diameter showing acute-angle dichotomous branching. The index of suspicion should be high for making the diagnosis of this disease. The comparison of various differential diagnoses is given in **Table 39.2**.

Another clinical entity, lagenidiosis an emerging disease affecting cutaneous/subcutaneous tissue and multiple organs has recently been reported from southern United States affecting dogs. It is also caused by water-mold and clinically is very similar to cutaneous form of pythiosis insidiosi. However, multi-system involvement is more frequently seen in lagenidiosis.

Laboratory Diagnosis

An early diagnosis of pythiosis insidiosi is advantageous for the specific management of patients. However, most of the laboratories are unfamiliar with this disease and/or organism because of its morphological similarities with mucormycetes. Thus, this disease is usually misdiagnosed and wrongly treated on the lines of fungal infections.

Table 39.2. Comparative Clinical and Related Features of Pythiosis, Mucormycosis and Entomophthoramycosis.

Features	*Pythiosis*	*Mucormycosis*	*Entomophthoramycosis*
Prevalence	Tropical, subtropical and temperate areas	Ubiquitous	Tropical and subtropical
Causative Agents	*Pythium insidiosum*	*Rhizopus, Mucor, Absidia, Apophysomyces, Saksenaea* species	*Conidiobolus, Basidiobolus* species
Taxonomic Status	Stramenopila (Protista)	Mucorales (Fungi)	Entomophthorales (Fungi)
Status of Host	Apparently normal, thalassemic patients are more prone	Compromised, particularly underlying diabetes	Apparently normal
Duration of Illness	Protracted course	Rapidly progressive	Indolent progressive
Inflammatory Response	Granulomatous, marked eosinophilia, Splendore-Hoeppli phenomenon, *kunkers* formation in horses	Suppurative, marked necrosis, neutrophils	Granulomatous, marked eosinophilia, Splendore-Hoeppli phenomenon
Angioinvasion	Common	Very Common	Unusual
Size of Hyphae in the Infected Tissue	Approx. 3-10 µm diameter, uniform in size, non-septate, if septa visible, they are double walled	Approx. 10-15 µm diameter, non-septate, ribbon-like	Approx. 7-20 µm diameter, non-septate, ribbon-like

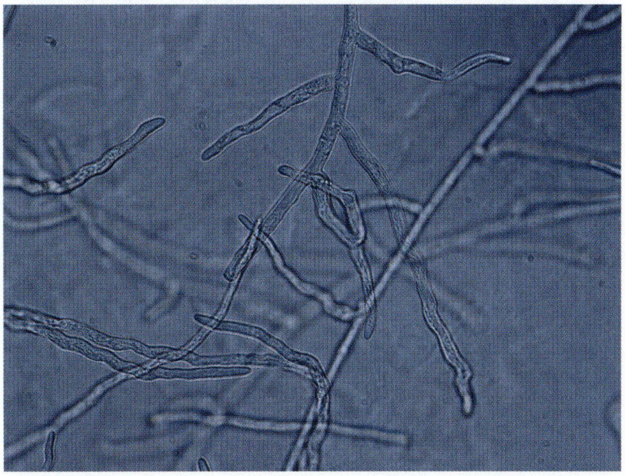

Fig. 39.2. Wet mount of *Pythium insidiosum* showing non-septate hyphae resembling those of mucormycetes (KOH × 400).

The diagnosis of this pseudofungal infection is established by using conventional mycological techniques despite the fact that it is not a fungal infection. The clinical specimen taken from the lesions should be processed at the earliest since delay may adversely affect viability of the organisms.

(a) Direct Examination

The clinical specimens such as pus, exudate and biopsy material are examined in KOH and/or CFW wet mount, which may reveal non-septate or sparsely septate hyphae of 3-10 µm in diameter and tend to have wide-angle branching pattern of approximately 90 degrees as seen in mucormycetes, which has relatively broad non-septate hyphae of 10-15 µm diameter **(Fig. 39.2)**.

The histopathology reveals yellowish masses representing foci of coagulative necrosis containing numerous neutrophils, eosinophils and occasionally giant cells along with masses of branching non-septate or sparsely septate hyphae. The histopathological examination is often difficult since *P.insidiosum* frequently seen only as clear spaces outlining hyphae or unevenly stained hyphae seen in H&E stain. Therefore, GMS and PAS stainings are recommended for properly delineating the hyphal structure in tissues. Its hyphae are better stained with PAS than H&E stain, which is in contrast to the pattern seen in mucormycosis and is very useful clue for differentiation.

Pythiosis insidiosi should also be differentiated from acute-angle dichotomous branching pattern of septate hyphae typically seen in various fungal species of genus *Aspergillus, Fusarium* or *Scedosporium*. **Splendore-Hoeppli** phenomenon of eosinophilic material is often found around hyphae, as is also seen in entomophthoramycosis caused by *Conidiobolus coronatus* or *Basidiobolus ranarum*. The immunofluorescence studies by specific fluorescent antibody test of histopathological section are also recommended for the identification of *P.insidiosum*.

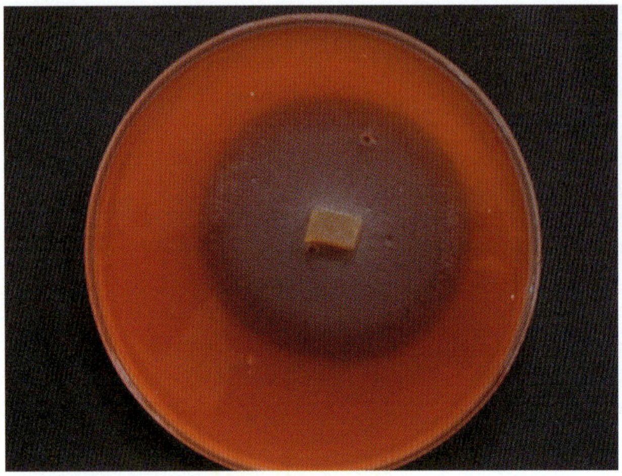

Fig. 39.3. Blood agar plate showing growth of *Pythium insidiosum* after 18 hrs of subculture at 37°C.

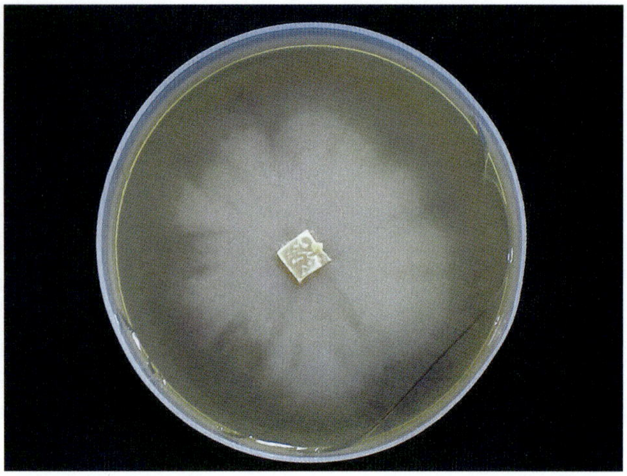

Fig. 39.4. Sabouraud dextrose agar plate showing growth of *Pythium insidiosum* after 48 hrs of subculture at 37°C.

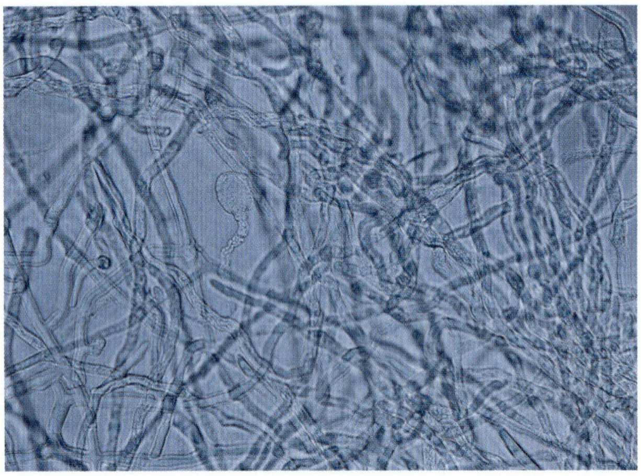

Fig. 39.5. Culture mount of *Pythium insidiosum* showing early zoosporangium in the center (KOH × 400).

(b) Culture

The biopsy specimens and/or corneal scrapings from lesions are inoculated on conventional fungal culture media like Blood/chocolate Agar, Sabouraud Dextrose Agar and Brain-Heart Infusion Agar and incubated at 37°C and 25°C. After 24 to 48 hours of incubation, colonies are flat, feathery, glabrous, yellowish-white at 37°C **(Figs. 39.3 and 39.4)**, which later on turn to cream or light brown-colored. This organism does not grow at 25°C. Although it is neither a fungus nor dimorphic in nature still incubation is done at both the temperatures as it is customary in a diagnostic mycology laboratory, keeping in view of likely possibility of a dimorphic fungus in the clinical specimens. The LCB mount of the culture isolate reveals broad non-septate or sparsely septate hyaline hyphae of 3-10 μm in diameter with granular cytoplasm. The hyphae are wide and extensively branched, with some terminal swellings, very similar to the mucormycetes.

The isolate usually fails to sporulate using conventional laboratory techniques on solid culture media. Therefore, definitive identification is based on production of zoosporangia and motile zoospores in an induction liquid medium. A few boiled grass leaves are placed on *P.insidiosum* culture and after 24 hours of incubation at 37°C, zoosporangium containing zoospores may be observed at the edge of leaves **(Fig. 39.5)**. The parasitized grass leaves are then placed in a beaker containing dilute salt solution for 2 to 3 hours at 37°C, which are then transferred on to the slides and microscopically motile biflagellate zoospores are observed which are diagnostic for this disease **(*See* Appendix C)**.

Some of Mucorales like *Apophysomyces* species and *Saksenaea* species also invariably do not sporulate hence should not be confused with *P.insidiosum*. However, infections caused by these fungi do not trigger any eosinophilic inflammatory response as seen in case of *P.insidiosum*. Hence simultaneously histopathological correlation is imperative in diagnosing this infection.

The strains of *P.insidiosum* usually perish after primary isolation if not sub-cultured properly. Most of the clinicians and microbiologists are not familiar with characteristics of *Pythium* species hence there are chances of failing to arrive at a definitive diagnosis. The task of zoospore induction may be bypassed by employing immunodiagnosis techniques such as detection of **exoantigen** by reverse immunodiffusion using *P.insidiosum* antiglobulins.

(c) Immunodiagnosis

The serological tests are useful both for diagnosis as well as prognosis by monitoring course of the treatment among patients of pythiosis insidiosi. The immunodiffusion test is done using **culture filtrate antigens** (CFA) of *P.insidiosum* reacting to patients' and reference rabbit antisera, which shows 1 to 6 precipitin bands and line of identity also. This test is specific but less sensitive. Therefore, other tests like CFT, ELISA, Immunohistochemical Assays and Western Blotting are also utilized to establish the diagnosis of pythiosis insidiosi.

ELISA with soluble antigen from broken hyphae (SABH) of *P.insidiosum* appears to be more sensitive than immunodiffusion and is equally specific for detection of the disease in areas where it is found to be endemic, regardless of types of hosts. This test also appears to be useful for monitoring response to therapy in affected individuals. In addition, this test may be useful for detecting pythiosis insidiosi in horses and other animals in cases where immunodiffusion has failed. Due to its sensitivity, ELISA may be valuable for diagnosing early stages of diseases and also for conducting epidemiological surveys.

An agglutination assay, for detecting antibodies in human and other mammalian sera, is also described. This uses latex particles and extracellular antigens of *P.insidiosum*, which bind to serum antibodies leading to agglutination. Such type of assay is also important for detecting early stages of pythiosis insidiosi. Many strains of *P.insidiosum* do not sporulate despite the best efforts, so they can only be identified by exoantigen testing, indirect immunoperoxidase techniques for *P.insidiosum* antigen, immunodiffusion or specific fluorescent antibody test.

The CFT is less sensitive and less specific than immunodiffusion. The skin test is also carried out in horses by using culture filtrate antigens. Pythiosis keratitis may be confirmed by DNA sequence analysis. Recently, nested PCR/RT-PCR assays have been developed for identifying *P.insidiosum* in animals and this technique has also been used to detect an isolate from human periorbital lesion. In addition, immunohistochemical (IHC) techniques using polyclonal antibodies may also be done for establishing the diagnosis of pythiosis.

(d) Animal Pathogenicity

At present, *P.insidiosum* is true and established pathogen and not just merely an opportunistic organism. All previous experimental attempts to produce pythiosis insidiosi in dogs, horses and mice have failed. The disease has been experimentally produced in healthy and immunosuppressed rabbits (*Oryctolagus cuniculus*) through various routes leading to typical histopathological findings. The subcutaneous injection of infective zoospores produces progressive granulomatous eosinophilic abscesses whereas intraperitoneal injections cause severe necrotizing hepatitis, granulomatous colitis with omental adhesions. The intravenous route leads to various embolic lesions, especially in the kidneys.

Treatment and Prophylaxis

There are no recorded cases of spontaneous remission in human pythiosis. Therefore, if left untreated, infection is invariably life-threatening in due course of time. The treatment modalities include antimicrobial agents, immunotherapy and surgical debridement. *P.insidiosum,* being an oomycete, does not utilize ergosterol as the main sterol in its cytoplasmic membrane. Therefore, currently used antifungal agents are **not** effective since most of them act by interfering with ergosterol synthesis. Imidazoles, triazoles and polyenes have been tried but without much success and give conflicting results. It seems that these drugs do appear to decrease the growth rate of *P.insidiosum* in previously documented cases but are unable to kill the organism. Itraconazole combined with terbinafine for a period of about one year has yielded fair results in some of patients with arteritis.

The patients of subcutaneous pythiosis insidiosi have also been treated with saturated solution of potassium iodide (SSKI) in a manner similar to that used for lymphocutaneous sporotrichosis as well as entomophthoramycosis. But such treatment cannot be applicable to patients with keratitis or systemic vasculitis. Recently in 2017, intraocular minocycline for the treatment of ocular pythiosis has been successfully tried.

Therefore, only treatment option available for potential cure is surgery and the goal is complete resection of the affected tissue. In case of localized lesions, radical excision of infected tissue is usually employed as the last resort. In patients of pythiosis keratitis, keratoplasty, evisceration and enucleation may be required depending on deteriorating clinical status of patient. However, timely recognition of the disease and treatment are essential for successful management. The other factors that influence prognosis include size and site of lesion as well as duration of infection.

Immunotherapy with antigens derived from *P.insidiosum* is shown to have curative properties in equine pythiosis insidiosi hence vaccination appears to be an alternative therapeutic choice. There are **two** *Pythium insidiosum* vaccines **(PIV)**, which are successfully used in horses. Mendoza's soluble concentrated antigen vaccine **(SCAV)** is prepared from culture filtrate antigens and is used in Costa Rica. On the other hand, Miller's preparation, **cell-mass vaccine (CMV)**, consisting of sonicated hyphal antigens is used in Australia. The modified Mendoza's vaccine is successfully used as a therapeutic vaccine in life-threatening arteritic infection among Thai patients.

The evaluation studies of these vaccines have been carried out in infected horses and it has been observed that Mendoza's vaccine is more stable, atleast for one year after preparation with longer shelf-life and less side effects. However, both vaccines are ineffective in chronic infection of more than two months' duration. In 2003, Mendoza *et al*, have introduced new immunotherapeutic vaccine to treat pythiosis insidiosi of human and animal origin. This formulation contains exo- and endoproteins extracted from culture of *P.insidiosum* and is able to cure acute as well as chronic cases of this disease. These findings have generated new perspective for research, which is essential to investigate mechanisms by which therapeutic vaccines elicit an immunological response that kills pathogens in the infected tissues.

II. LAGENIDIOSIS

The term, **Lagenidiosis**, was coined by Grooters *et al*, in 2003 recognizing one of the newly established infections as a separate clinical entity, from pythiosis despite many similarities between both the diseases. During the last century, *P.insidiosum* was the only pseudofungal species, from the phylum Oomycota, which used to be known to cause significant infections among the mammals. However, a novel group of pathogenic Oomycota was isolated in south-eastern USA in 2003 from dogs having invasive cutaneous infections, which resembled pythiosis but strains had morphological and molecular affinities with the genus *Lagenidium*.

The representative species, *L.giganteum* is a facultative parasite of mosquito larvae that infect by production of biflagellate zoospores that selectively recognize and attach to the larval cuticle. After penetration of the cuticle, organism proliferates within the host, killing it within 2-3 days.

Like *P.insidiosum*, *Lagenidium* species also cause infections ranging from involvement of superficial, cutaneous to subcutaneous tissues besides arterial involvement. Systemic infection is although very rare but if established it is even more aggressive than pythiosis. In humans and animals the sites of infection are cornea, gastrointestinal tract and extremities.

Lagenidium species is also an oomycete like *P.insidiosum* that causes infections in mammals but rarely in humans. Members of the genus *Lagenidium* also cause infections in lower animals including crabs, nematodes and mosquito larvae. Among mammals the infection presents with involvement of the skin and subsequently involvement of the blood vessels.

Organism *(Lagenidium giganteum)*

The genus *Lagenidium* was first introduced by Schenk in 1857 and comprises numerous saprotrophic species, as well as several members, which are pathogenic to algae, phytoplankton, pollen, barnacles, blue crabs, mosquito larvae, shrimp and mammals. Subsequently, other strains similar to the previously studied *Lagenidium* species, were found to be associated with infections in dogs, cats and humans. These organisms are currently classified within the kingdom Stramenopila, phylum Heterokontophyta, class Oomycota, order Lagenidiales and family Lagenidiaceae. The genus *Lagenidium,* in the class Oomycetes includes more than 50 species, majority of them occur as parasites of algae, fungi, nematodes, crustaceans or insect larvae. *Lagenidium* also completes its life cycle in aquatic environments, possibly using plants or lower animal hosts. It was once thought to be cryptic isolate of genus *P.insidiosum.*

This was first recognized as a medically important genus in 1999, when a previously un-described species of *Lagenidium* was identified as a cause of subcutaneous and systemic disease in dogs. More recently, second pathogenic *Lagenidium* species was isolated from tissues obtained from four dogs and one human being with chronic subcutaneous infections. Recent molecular phylogenetic analysis of *Lagenidium* strains recovered from subcutaneous lesions in cats, dogs and a human case with lagenidiosis resolved into four clades; one of them being *Lagenidium giganteum* while the rest three are new ones. Hence *Lagenidium ajelloi, L.albertoi* and *L.vilelae* are now designated as three novel species.

Epidemiology

Most of the cases of lagenidiosis in mammals presently are reported in USA almost overlapping in the same areas as pythiosis insidiosi *i.e.* states bordering Gulf of Mexico, Arkansas, Georgia, Illinois, Indiana, Maryland, both North and South Carolina, Tennessee as well as Virginia. In addition to the regions noted, case of human keratitis caused by *Lagenidium* has been reported from Thailand. The disease has also been reported in a dog from Australia.

This is considered that infection is initiated when zoospores present in a contaminated environment, gain entry through open skin injuries. It does not appear to be transmitted from one infected host to the other and attempts to establish infection in mice have been unsuccessful. The discovery of second mammalian pathogen in the genus *Lagenidium* and its identification for the first time infecting humans suggest that the oomycetes represent an emerging and possibly under-diagnosed group of medically important pseudofungi.

Clinical Features

This is mainly an oomycotic infection of dogs characterized by progressive, multifocal cutaneous and subcutaneous lesions, most often affecting the extremities, mammary region, perineum or trunk. Clinical findings in both human and canine are characterized by progressive deep cellulitis usually with many draining tracts involving one or more extremities. Cutaneous lesions are firm dermal or subcutaneous nodules or as ulcerated, thickened, edematous areas of deep cellulitis with regions of necrosis and numerous draining tracts. In dogs with aggressive *Lagenidium* species, involvement of distant sites is often seen in contrast to the clinical course of cutaneous pythiosis. Thoracic and abdominal lymph nodes, lungs and especially larger blood vessels may be affected. Animals with big vessel or sublumbar lymph node involvement typically have cutaneous or subcutaneous lesions on the hind limbs and often develop edema. The great vessel aneurysms may acutely rupture, resulting in hemoabdomen and sudden death. Overall the prognosis in this disease remain as such poor to grave.

Laboratory Diagnosis

The diagnosis of lagenidiosis in the laboratory may be made by using direct microscopy and culture of the clinical material taken from the site of infection. On microscopic examination of specimens stained with Giemsa, *Lagenidium* species appear as broad branched hyphae. Culture on SDA at 37°C, shows whitish-yellow flat glabrous colonies, submerged in the agar after 24-48 hours. On microscopy the hyphae are non-septate and broad (9-18 μm), with large spherical structures connected by short segments of hyphae. The observation of spherical structures connected by small tubules is helpful to differentiate *Lagenidium* species from *P.insidiosum.* Detection of antibodies using ELISA and Western blot assays are used for both diagnosis and monitoring of response to therapy. A strong cross-reaction with *P.insidiosum* antigens are observed with ELISA testing.

On histopathological examination, lesions are characterized by granulomatous inflammation centered around broad, infrequently septate hyphae. Eosinophilic inflammation, necrosis and suppuration are invariably present. Immunoblot analysis of the serological response of affected dogs to soluble mycelial extracts of *L.giganteum* and of other pathogenic *Lagenidium* species are performed and indicate that each dog's serum recognized a large number of antigens of both *Lagenidium* species.

In contrast to other pathogens covered in this chapter, *Lagenidium* species grow readily on routine fungal isolation media. On agar based media these organisms grow readily at 37°C as white to yellow submerged colonies without aerial mycelia. Similar to *P.insidiosum*, *Lagenidium* species produces 9-18 μm ribbon-like hyphae with spherical structures 20-45 μm in diameter. In liquid media vesicles may be seen at the tips of the undifferentiated hyphae. The presence of sexual structures (oogonia) has not yet been described. The two species mentioned in the literature thus far, *L.caninum* and *L.karlingii*, have not undergone a formal taxonomic description.

The culture of affected tissues on peptone-yeast-glucose (PYG) agar at 30°C for 3 days yields growth of a colorless to white usually submerged colony. The human isolate produce small, restricted, heaped, white to cream-colored colonies approximately 3 to 5 mm in diameter after 10 days incubation at 25°C on potato flakes agar. Short hyphal fragments of varying widths (10-30 μm) are present; however large, globose to oval-shaped cells are the prominent feature. Molecular identification of the isolates, is performed by amplifying and sequencing the 18s and ITS regions of the ribosomal RNA gene. The 18s rRNA region shared 97.3% identity with *L.giganteum* and previously described pathogenic *Lagenidium* species.

Treatment and Prophylaxis

Similar to pythiosis, an early and extensive surgical resection followed by long-term itraconazole and terbinafine, are recommended as the treatments of choice, which are found to be effective among dogs with multiple recurrent cutaneous lesions. In the human patient, chronic cellulitis involving a single distal lower extremity failed to respond to voriconazole but showed marked improvement after administration of posaconazole. Being oomycetes, *Lagenidium* species lack ergosterol in cytoplasmic membrane thereby making antifungals as the ineffective agents, which usually target sterol pathway. Despite this fact, many antifungals are still being used both clinically and *in vitro*, with mixed results. In this disease, general preventive measures are similar to that followed for pythiosis insidiosi.

Further Reading

1. Argenta JS, Alves SH, Silveira F, *et al. In vitro* and *in vivo* susceptibility of two-drug and three-drug combinations of terbinafine, itraconazole, caspofungin, ibuprofen and fluvastatin against *Pythium insidiosum. Vet Microbiol.* 2012; **157:** 137-42.
2. Azevedo MI, Botton SA, Pereira DI, *et al.* Phylogenetic relationships of Brazilian isolates of *Pythium insidiosum* based on ITS rDNA and cytochrome oxidase II gene sequences. *Vet Microbiol.* 2012; **159:** 141-8.
3. Badenoch PR, Mills RA, Chang JH, *et al. Pythium insidiosum keratitis* in an Australian child. *Clin Experiment Ophthalmol.* 2009; **37:** 806-9.
4. Barequet IS, Lavinsky F, Rosner M. Long-term follow-up after successful treatment of *Pythium insidiosum* keratitis in Israel. *Semin Ophthalmol.* 2013; **28:** 247-50.
5. Bosco Sde M, Bagagli E, Araujo JP Jr, *et al.* Human pythiosis, Brazil. *Emerg Infect Dis.* 2005; **11:** 715-8.
6. Bridges CH, Emmons JW. A phycomycosis of horses caused by *Hyphomycosis destruens. J Am Vet Assoc.* 1961; **138:** 579-89.
7. Brown TA, Grooters AM, Hosgood GL. *In vitro* susceptibility of *Pythium insidiosum* and a *Lagenidium* sp to itraconazole, posaconazole, voriconazole, terbinafine, caspofungin and mefenoxam. *Am J Vet Res.* 2008; **69:** 1463-8.
8. Calvano TP, Blatz PJ, Vento TJ, *et al. Pythium aphanidermatum* infection following combat trauma. *J Clin Microbiol.* 2011; **49:** 3710-3.
9. Cavalheiro AS, Maboni G, de Azevedo MI, *et al. In vitro* activity of terbinafine combined with caspofungin and azoles against *Pythium insidiosum. Antimicrob Agents Chemother.* 2009; **53:** 2136-8.
10. Chaiprasert A, Samerpitak K, Wanachiwanawin W, *et al.* Induction of zoospore formation in Thai isolates of *Pythium insidiosum. Mycoses.* 1990; **33:** 317-23.
11. Chareonsirisuthigul T, Khositnithikul R, Intaramat A, *et al.* Performance comparison of immunodiffusion, enzyme-linked immunosorbent assay, immunochromatography and hemagglutination for serodiagnosis of human pythiosis. *Diagn Microbiol Infect Dis.* 2013; **76:** 42-5.
12. Chetchotisakd P, Pairojkul C, Porntaveevudhi O, *et al.* Human pythiosis in Srinagarind Hospital: one year's experience. *J Med Assoc Thai.* 1992; **75:** 248-54.
13. de Cock AW, Mendoza L, Padhye AA, *et al. Pythium insidiosum* sp. nov., the etiologic agent of pythiosis. *J Clin Microbiol.* 1987; **25:** 344-9.
14. Farmer AR, Murray CK, Driscoll IR, *et al.* Combat-related *Pythium aphanidermatum* invasive wound infection: Case report and discussion of utility of molecular diagnostics. *J Clin Microbiol.* 2015; **53:** 1968-75.
15. Fernandes CP, Giordani C, Grecco FB, *et al.* Gastric pythiosis in a dog. *Rev Iberoam Micol.* 2012; **29:** 235-7.
16. Fonseca AO, BottonSde A, Nogueira CE, *et al. In vitro* reproduction of the life cycle of *Pythium insidiosum* from kunkers' equine and their role in the epidemiology of pythiosis. *Mycopathologia.* 2014; **177:** 123-7.
17. Franco DM, Aronson JF, Hawkins HK, *et al.* Systemic *Pythium insidiosum* in a pediatric burn patient. *Burns.* 2010; **36:** e68-71.
18. Gaastra W, Lipman LJ, de Cock AW, *et al. Pythium insidiosum*: An overview. *Vet Microbiol.* 2010; **146:** 1-16.
19. Grooters AM, Proia LA, Sutton D, *et al.* Characterization of a previously un-described *Lagenidium* pathogen associated with soft tissue infection: Initial description of a new human oomycosis [Abstract]. *Focus on Fungal Infections.* 14.2004 Mar 24-26; New Orleans, Louisiana, USA. p. 174.
20. Grooters AM. Pythiosis, lagenidiosis and zygomycosis in small animals. *Vet Clin North Am Small Anim Pract.* 2003; **33:** 695-720.
21. Hahtapornsawan S, Wongwanit C, Chinsakchai K, *et al.* Suprainguinal vascular pythiosis: effective long-term outcome of aggressive surgical eradication. *Ann Vasc Surg.* 2014; **28:** 1797. e1-6.
22. Hartfield JN, Grooters AM, Waite KJ. Development and evaluation of an ELISA for the quantitation of anti-*Lagenidium giganteum* forma *caninum* antibodies in dogs. *J Vet Intern Med.* 2014; **28:** 1479-84.
23. Heath JA, Kiehn TE, Brown AE, *et al. Pythium insidiosum* pleuropericarditis complicating pneumonia in a child with leukemia. *Clin Infect Dis.* 2002; **35:** e60-4.
24. Hilton RE, Tepedino K, Glenn CJ, Merkel KL. Swamp cancer: A case of human pythiosis and review of the literature. *Br J Dermatol.* 2016; **175:** 394-7.
25. Hoffman MA, Cornish NE, Simonsen KA. A painful thigh lesion in an immunocompromised 11-year-old boy. *Pediatr Infect Dis J.* 2011; **30:** 1011 & 1017-8.
26. Holloway SA. Editorial: The Oomycetes - Fungi with teeth and flippers? *J Vet Intern Med.* 2003; **17:** 607-8.
27. Hummel J, Grooters A, Davidson G, *et al.* Successful management of gastrointestinal pythiosis in a dog using itraconazole, terbinafine and mefenoxam. *Med Mycol.* 2011; **49:** 539-42.
28. Hung C, Leddin D. Keratitis caused by *Pythium insidiosum* in an immunosuppressed patient with Crohn's disease. *Clin Gastroenterol Hepatol.* 2014; **12:** A21-2.
29. Imwidthaya P. Human pythiosis in Thailand. *Postgrad Med J.* 1994; **70:** 558-60.

30. Jesus FP, Ferreiro L, Loreto ES, *et al.* *In vitro* synergism observed with azithromycin, clarithromycin, minocycline or tigecycline in association with antifungal agents against *Pythium insidiosum. Antimicrob Agents Chemother.* 2014; **58:** 5621-5.
31. Jesus FP, Loreto ES, Ferreiro L, *et al.* *In vitro* and *in vivo* antimicrobial activities of minocycline in combination with azithromycin, clarithromycin or tigecycline against *Pythium insidiosum. Antimicrob Agents Chemother.* 2015; **60:** 87-91.
32. Jindayok T, Piromsontikorn S, Srimuang S, *et al.* Hemagglutination test for rapid serodiagnosis of human pythiosis. *Clin Vaccine Immunol.* 2009; **16:** 1047-51.
33. Keeratijarut A, Karnsombut P, Aroonroch R, *et al.* Evaluation of an in-house immunoperoxidase staining assay for histodiagnosis of human pythiosis. *Southeast Asian J Trop Med Public Health.* 2009; **40:** 1298-305.
34. Keeratijarut A, Lohnoo T, Yingyong W, *et al.* A peptide ELISA to detect antibodies against *Pythium insidiosum* based on predicted antigenic determinants of exo-1,3-beta-glucanase. *Southeast Asian J Trop Med Public Health.* 2013; **44:** 672-80.
35. Keoprasom N, Chularojanamontri L, Chayakulkeeree M, *et al.* Vascular pythiosis in a thalassemic patient presenting as bilateral leg ulcers. *Med Mycol Case Rep.* 2012; **2:** 25-8.
36. Kerwin JL. Oomycetes: *Lagenidium giganteum. J Am Mosq Control Assoc.* 2007; **23 (Suppl. 2):** 50-7.
37. Khunkhet S, Rattanakaemakorn P, Rajatanavin N. Pythiosis presenting with digital gangrene and subcutaneous nodules mimicking medium vessel vasculitis. *JAAD Case Rep.* 2015; **1:** 399-402.
38. Kirzhner M, Arnold SR, Lyle C, *et al.* *Pythium insidiosum:* A rare necrotizing orbital and facial infection. *J Pediatric Infect Dis Soc.* 2015; **4:** e10-3.
39. Kosrirukvongs P, Chaiprasert A, Uiprasertkul M, *et al.* Evaluation of nested PCR technique for detection of *Pythium insidiosum* in pathological specimens from patients with suspected fungal keratitis. *Southeast Asian J Trop Med Public Health.* 2014; **45:** 167-73.
40. Krajaejun T, Chongtrakool P, Angkananukul K, *et al.* Effect of temperature on growth of the pathogenic oomycete *Pythium insidiosum. Southeast Asian J Trop Med Public Health.* 2010; **41:** 1462-6.
41. Krajaejun T, Imkhieo S, Intaramat A, Ratanabanangkoon K. Development of an immunochromatographic test for rapid serodiagnosis of human pythiosis. *Clin Vaccine Immunol.* 2009; **16:** 506-9.
42. Krajaejun T, Kunakorn M, Niemhom S, *et al.* Development and evaluation of an in-house enzyme-linked immunosorbent assay for early diagnosis and monitoring of human pythiosis. *Clin Diagn Lab Immunol.* 2002; **9:** 378-82.
43. Krajaejun T, Kunakorn M, Pracharktam R, *et al.* Identification of a novel 74-kilodalton immunodominant antigen of *Pythium insidiosum* recognized by sera from human patients with pythiosis. *J Clin Microbiol.* 2006; **44:** 1674-80.
44. Krajaejun T, Pracharktam R, Wongwaisayawan S, *et al.* Ocular pythiosis: Is it under-diagnosed? *Am J Ophthalmol.* 2004; **137:** 370-2.
45. Krajaejun T, Sathapatayavongs B, Pracharktam R, *et al.* Clinical and epidemiological analyses of human pythiosis in Thailand. *Clin Infect Dis.* 2006; **43:** 569-76.
46. Laohapensang K, Rerkasem K, Supabandhu J, *et al.* Necrotizing arteritis due to emerging *Pythium insidiosum* infection in patients with thalassemia: Rapid diagnosis with PCR and serological tests - case reports. *Int J Angiol.* 2005; **14:** 123-8.
47. Laohapensang K, Rutherford RB, Supabandhu J, *et al.* Vascular pythiosis in a thalassemic patient. *Vascular.* 2009; **17:** 234-8.
48. Leal AT, Santurio JM, Leal ABM, *et al.* Characterization of the specificity of the humoral response to *Pythium insidiosum* antigens. *J Mycol Med.* 2005; **15:** 63-8.
49. Lekhanont K, Chuckpaiwong V, Chongtrakool P, *et al.* *Pythium insidiosum* keratitis in contact lens wear: A case report. *Cornea.* 2009; **28:** 1173-7.
50. Lelievre L, Borderie V, Garcia-Hermoso D, *et al.* Imported *Pythium insidiosum* keratitis after a swim in Thailand by a contact lens-wearing traveler. *Am J Trop Med Hyg.* 2015; **92:** 270-3.
51. Levesque CA, de Cock AW. Molecular phylogeny and taxonomy of the genus *Pythium. Mycol Research.* 2004; **108:** 1363-83.
52. Loreto ES, Mario DA, Denardi LB, *et al.* *In vitro* susceptibility of *Pythium insidiosum* to macrolides and tetracycline antibiotics. *Antimicrob Agents Chemother.* 2011; **55:** 3588-90.
53. Loreto ES, Tondolo JS, Pilotto MB, *et al.* New insights into the *in vitro* susceptibility of *Pythium insidiosum. Antimicrob Agents Chemother.* 2014; **58:** 7534-7.
54. Mahl DL, Jesus FP, Loreto ES, *et al.* *In vitro* susceptibility of *Pythium insidiosum* isolates to aminoglycoside antibiotics and tigecycline. *Antimicrob Agents Chemother.* 2012; **56:** 4021-3.
55. Marques SA, Bagagli E, Bosco SMG, *et al.* *Pythium insidiosum*: Report of the first case of human infection in Brazil. *An Bras Dermatol.* 2006; **81:** 483-5.
56. Mendoza L, Hernandez F, Ajello L. Life cycle of the human and animal oomycete pathogen *Pythium insidiosum. J Clin Microbiol.* 1993; **31:** 2967-73.
57. Mendoza L, Kaufman L, Mandy W, *et al.* Serodiagnosis of human and animal pythiosis using an enzyme-linked immunosorbent assay. *Clin Diagn Lab Immunol.* 1997; **4:** 715-8.
58. Mendoza L, Kaufman L, Standard P. Antigenic relationship between the animal and human pathogen *Pythium insidiosum* and non-pathogenic *Pythium* species. *J Clin Microbiol.* 1987; **25:** 2159-62.
59. Mendoza L, Mandy W, Glass R. An improved *Pythium insidiosum*-vaccine formulation with enhanced immunotherapeutic properties in horses and dogs with pythiosis. *Vaccine.* 2003; **21:** 2797-804.

60. Mendoza L, Newton JC. Immunology and immunotherapy of the infections caused by *Pythium insidiosum. Med Mycol.* 2005; **43:** 477-86.
61. Mendoza L, Prasla SH, Ajello L. Orbital pythiosis: A non-fungal disease mimicking orbital mycotic infections, with a retrospective review of the literature. *Mycoses.* 2004; **47:** 14-23.
62. Mendoza L, Schurko A, Newton JC. Are strains identified as *Lagenidium* sp. from dogs actually cryptic isolates of *Pythium insidiosum*? *Am J Vet Res.* 2009; **70:** 163.
63. Mendoza L, Taylor JW, Walker ED, *et al.* Description of three novel *Lagenidium* (Oomycota) species causing infection in mammals. *Rev Iberoam Micol.* 2016; **33:** 83-91.
64. Mendoza L, Vilela R. *Pythium insidiosum* or *Lagenidium* sp.? *Vet Clin Pathol.* 2009; **38:** 273-4.
65. Murdoch D, Parr D. *Pythium insidiosum* keratitis. *Aust N Z J Ophthalmol.* 1997; **25:** 177-9.
66. Pal M, Mahendra R. Pythiosis: An emerging oomycetic disease of humans and animals. *Int J Livest Res.* 2014; **4:** 1-9.
67. Pan JH, Kerkar SP, Siegenthaler MP, *et al.* A complicated case of vascular *Pythium insidiosum* infection treated with limb-sparing surgery. *Int J Surg Case Rep.* 2014; **5:** 677-80.
68. Pannanusorn S, Chaiprasert A, Prariyachatigul C, *et al.* Random amplified polymorphic DNA typing and phylogeny of *Pythium insidiosum* clinical isolates in Thailand. *Southeast Asian J Trop Med Public Health.* 2007; **38:** 383-91.
69. Pereira DI, Botton SA, Azevedo MI, *et al.* Canine gastrointestinal pythiosis treatment by combined antifungal and immunotherapy and review of published studies. *Mycopathologia.* 2013; **176:** 309-15.
70. Permpalung N, Worasilchai N, Plongla R, *et al.* Treatment outcomes of surgery, antifungal therapy and immunotherapy in ocular and vascular human pythiosis: A retrospective study of 18 patients. *J Antimicrob Chemother.* 2015; **70:** 1885-92.
71. Pier AC, Cabanes FJ, Chermette R, *et al.* Prominent animal mycoses from various regions of the world. *Med Mycol.* 2000; **38 (Suppl. 1):** 47-58.
72. Pracharktam R, Changtrakool P, Sathapatayavongs B, *et al.* Immunodiffusion test for diagnosis and monitoring of human pythiosis insidiosi. *J Clin Microbiol.* 1991; **29:** 2661-2.
73. Prasertwitayakij N, Louthrenoo W, Kasitanon N, *et al.* Human pythiosis, a rare cause of arteritis: Case report and literature review. *Semin Arthritis Rheum.* 2003; **33:** 204-14.
74. Pupaibool J, Chindamporn A, Patrakul K, *et al.* Human pythiosis. *Emerg Infect Dis.* 2006; **12:** 517-8.
75. Rajeev S, Ilha MR, Harrison JM, *et al.* Pathology in Practice. *Lagenidium* infection. *J Am Vet Med Assoc.* 2012; **241:** 447-9.
76. Ramappa M, Nagpal R, Sharma S, *et al.* Successful medical management of presumptive *Pythium insidiosum* keratitis. *Cornea.* 2017; **36:** 511-4.
77. Reanpang T, Orrapin S, Orrapin S, *et al.* Vascular pythiosis of the lower extremity in northern Thailand: Ten years' experience. *Int J Low Extrem Wounds.* 2015; **14:** 245-50.
78. Reinprayoon U, Permpalung N, Kasetsuwan N, *et al. Lagenidium* sp. ocular infection mimicking ocular pythiosis. *J Clin Microbiol.* 2013; **51:** 2778-80.
79. Riefler JF. Expanded access for a little boy. *Front Public Health.* 2013; **1:** 54.
80. Rinaldi MG, Seidenfeld SM, Fotherbell AM, *et al. Pythium insidiosum* causes severe disease in a healthy boy. *Mycol Observ.* 1989; **9:** 7-8.
81. Rivierre C, Laprie C, Guiard-Marigny O, *et al.* Pythiosis in Africa. *Emerg Infect Dis.* 2005: **11:** 479-81.
82. Ros Castellar F, Jimenez CS, Zarzuelo ADH, *et al.* Intraocular minocycline for the treatment of ocular pythiosis. *Am J Health Syst Pharm.* 2017; **74:** 821-5.
83. Salipante SJ, Hoogestraat DR, Sengupta DJ, *et al.* Molecular diagnosis of subcutaneous *Pythium insidiosum* infection by use of PCR screening and DNA sequencing. *J Clin Microbiol.* 2012; **50:** 1480-3.
84. Santos CE, Marques LC, Zanette RA, *et al.* Does immunotherapy protect equines from reinfection by the oomycete *Pythium insidiosum*? *Clin Vaccine Immunol.* 2011; **18:** 1397-9.
85. Santurio JM, Leal AT, Leal AB, *et al.* Three types of immunotherapies against pythiosis insidiosi developed and evaluated. *Vaccine.* 2003; **21:** 2535-40.
86. Schurko AM, Mendoza L, de Cock AW, *et al.* Development of a species-specific probe for *Pythium insidiosum* and the diagnosis of pythiosis. *J Clin Microbiol.* 2004; **42:** 2411-8.
87. Schurko AM, Mendoza L, de Cock AW, *et al.* Evidence for geographic clusters: molecular genetic differences among strains of *Pythium insidiosum* from Asia, Australia and the Americas are explored. *Mycologia.* 2003; **95:** 200-8.
88. Schurko AM, Mendoza L, Levesque CA, *et al.* A molecular phylogeny of *Pythium insidiosum. Mycol Res.* 2003; **107:** 537-44.
89. Sekhon AS, Padhye AA, Garg AK. *In vitro* sensitivity of *Penicillium marneffei* and *Pythium insidiosum* to various antifungal agents. *Eur J Epidemiol.*1992; **8:** 427-32.
90. Sermsathanasawadi N, Praditsuktavorn B, Hongku K, *et al.* Outcomes and factors influencing prognosis in patients with vascular pythiosis. *J Vasc Surg.* 2016; **64:** 411-7.
91. Sharma S, Balne PK, Motukupally SR, *et al. Pythium insidiosum* keratitis: Clinical profile and role of DNA sequencing and zoospore formation in diagnosis. *Cornea.* 2015; **34:** 438-42.
92. Shenep JL, English BK, Kaufman L, *et al.* Successful medical therapy for deeply invasive facial infection due to *Pythium insidiosum* in a child. *Clin Infect Dis.* 1998; **27:** 1388-93.
93. Smith F. The pathology of bursattee. *Vet J.* 1884; **19:** 16-7.
94. Spies CF, Grooters AM, Levesque CA, *et al.* Molecular phylogeny and taxonomy of *Lagenidium*-like oomycetes pathogenic to mammals. *Fungal Biol.* 2016; **120:** 931-47.
95. Sudjaritruk T, Sirisanthana V. Successful treatment of a child with vascular pythiosis. *BMC Infect Dis.* 2011; **11:** 33.
96. Supabandhu J, Vanittanakom P, Laohapensang K, *et al.* Application of immunoblot assay for rapid diagnosis of human pythiosis. *J Med Assoc Thai.* 2009; **92:** 1063-71.

97. Tanhehco TY, Stacy RC, Mendoza L, *et al. Pythium insidiosum* keratitis in Israel. *Eye Contact Lens.* 2011; **37:** 96-8.
98. Thanathanee O, Enkvetchakul O, Rangsin R, *et al.* Outbreak of *Pythium keratitis* during rainy season: A case series. *Cornea.* 2013; **32:** 199-204.
99. Thianprasit M, Chaiprasert A, Imwidthaya P. Human pythiosis. *Curr Top Med Mycol.* 1996; **7:** 43-54.
100. Thianprasit M. Human pythiosis. *Trop Dermatol.* 1990; **4:** 1-4.
101. Thitithanyanont A, Mendoza L, Chuansumrit A, *et al.* Use of an immunotherapeutic vaccine to treat a life-threatening human arteritic infection caused by *Pythium insidiosum. Clin Infect Dis.* 1998; **27:** 1394-400.
102. Thongsri Y, Wonglakorn L, Chaiprasert A, *et al.* Evaluation for the clinical diagnosis of *Pythium insidiosum* using a single-tube nested PCR. *Mycopathologia.* 2013; **176:** 369-76.
103. Triscott JA, Weedon D, Cabana E. Human subcutaneous pythiosis. *J Cutan Pathol.* 1993; **20:** 267-71.
104. Uzuhashi S, Tojo M, Kakishima M. Phylogeny of the genus *Pythium* and description of new genera. *Mycoscience.* 2010; **51:** 337-65.
105. Vanittanakom N, Supabandhu J, Khamwan C, *et al.* Identification of emerging human-pathogenic *Pythium insidiosum* by serological and molecular assay-based methods. *J Clin Microbiol.* 2004; **42:** 3970-4.
106. Vento S, Cainelli F, Cesario F. Infections and thalassaemia. *Lancet Infect Dis.* 2006; **6:** 226-33.
107. Videla R, van Amstel S, O'neill SH, *et al.* Vulvar pythiosis in two captive camels *(Camelus dromedarius). Med Mycol.* 2012; **50:** 219-24.
108. Vilela R, Taylor JW, Walker ED, Mendoza L. *Lagenidium giganteum* pathogenicity in mammals. *Emerg Infect Dis.* 2015; **21:** 290-7.
109. Vilela R, Viswanathan P, Mendoza LA. A biochemical screening approach to putatively differentiate mammalian pathogenic Oomycota species in the clinical laboratory. *J Med Microbiol.* 2015; **64:** 862-8.
110. Virgile R, Perry HD, Pardanani B, *et al.* Human infectious corneal ulcer caused by *Pythium insidiosum. Cornea.* 1993; **12:** 81-3.
111. Wanachiwanawin W, Mendoza L, Visuthisakchai S, *et al.* Efficacy of immunotherapy using antigens of *Pythium insidiosum* in the treatment of vascular pythiosis in humans. *Vaccine.* 2004; **22:** 3613-21.
112. Wanachiwanawin W. Infections in E-beta thalassemia. *J Pediatr Hematol Oncol.* 2000; **22:** 581-7.
113. Zanette RA, Jesus FP, Pilotto MB, *et al.* Micafungin alone and in combination therapy with deferasirox against *Pythium insidiosum. J Mycol Med.* 2015; **25:** 91-4.

SECTION VIII

Appendices

Appendices

This Section deals with practical aspects in Medical Mycology referred in the text of almost each Chapter. The dealing in details was not feasible in that context. Hence these technical issues are being described in the end of this Textbook as Appendices. These issues being more or less technical hence usually are described comprehensively in various Manuals and Practical Reference Books on Medical Mycology and briefly given in this Section as well.

The first Appendix deals with Fungal Culture Media that are routinely used in diagnostic medical mycology laboratory in any of the clinical setting. The requirement of media, their appropriate incubation at specific temperature after inoculation, are pertinent to isolate causative fungus from the clinical material. If a suspected fungus is fastidious in nature, special media may be required for its primary isolation, which are to be incubated for a prolonged period of time. Similarly, suitable Fungal Reagents and Stains are important in delineating fungal agents either in wet mounts or histopathological sections thereby to correlate direct findings later on with culture. An experienced medical mycologist should be able to detect organism in direct examination so that cultural isolate can supplement importance of such findings thereby establishing significance of a particular isolate to be pathogenic or merely a contaminant.

The diagnostic techniques, which are essentially used in medical mycology laboratory, are being dealt in separate Appendix as Conventional Mycological Techniques. The emphasis is given on the routine procedures, which can be adopted in any of the simplest type of available infrastructure. However, for the molecular based diagnostic techniques, reference books and manuals may be consulted, as this Textbook may not be adequate except for providing the basic outlines only.

The Antifungal Susceptibility Testing is an important component of management of fungal diseases. In the previous editions, these aspects were given in the chapter on antifungals. But keeping in view the relevance of technicalities involved in the treatment of any fungal infection, a separate Appendix is given on AFST. The Quality Controls, Preservation and Culture Collections of Fungi are essential components for proper functioning of the diagnostic mycology laboratory. These issues have been dealt at length in one of the Appendices.

Some of the reference books may be required for acquiring thorough knowledge on particular aspect of Medical Mycology. Therefore, comprehensive list of selected Bibliography is given in one of the Appendices. These books are not referred in respective Chapters hence these are collectively listed out in this Appendix. Medical Mycology being relatively a new and upcoming subject of biological sciences and particularly for beginners in this field, therefore, Glossary of commonly used terms is also given as separate Appendix to oversimplify text of the book. This will be very useful for readers to grasp subject matter in a lucid manner with an in-depth understanding.

The References and/or Bibliography pertaining to the content of these Appendices are also given under the heading of Further Reading as a ready reference instead of citing in the corresponding Chapter.

Appendix A: Fungal Culture Media

The selection of proper medium for isolation, identification and maintenance of fungal strains is one of the most important tasks of personnel working in diagnostic medical mycology laboratory. There are various types of fungal culture media, which are required for this purpose. The selection of media to be used is based upon the experience of the mycologist and the fungi that are usually recovered in that particular geographical region. The following culture media are commonly used in the mycology laboratory.

I. Basal Media

The basal media are used for isolation, growth and maintenance of common fungal pathogens, which are described as follows:

1. Sabouraud Dextrose Agar (SDA)

The Sabouraud dextrose agar (SDA) is name recommended for present day versions of medium originally designed in the year 1892 by French dermatologist and mycologist, Raymond Jacques Adrien Sabouraud (1864-1938). This is now used by almost every diagnostic mycology laboratory for primary isolation of fungi from clinical specimens as well as for subcultures and short-term maintenance of fungal isolates. It is peptone containing medium supplemented with 4% dextrose and has an acidic pH, to support growth of fungi and is inhibitory to contaminating bacteria present in clinical specimens. The colony morphology, microscopic characteristics and other physiological properties of fungi cultivated on SDA may vary considerably according to type of substrates used in the medium. This medium is ideal for studying colony morphology of dermatophytes. The colony morphology of fungi is best studied on malt extract agar but still SDA is the most popular medium used in the diagnostic mycology laboratory for primary isolation.

This culture medium is also called as Sabouraud glucose agar (SGA) but its nomenclature as Sabouraud dextrose agar, is more popular. The mycologists universally accept title of medium as Sabouraud dextrose agar instead of Sabouraud's dextrose agar *i.e.* without apostrophe 's'. At some places it is also known as SAB or Sabdex medium. The peptones are source of nitrogenous growth factors and dextrose being carbon source provides energy to the growing fungi. This formulation serves as an excellent basal medium to which antibiotics and other inhibitory substances may be added for the selective cultivation of various fungal pathogens. The ingredients of the traditional SDA are as follows:

Peptone	10 gm
Dextrose	40 gm
Agar	20 gm
Distilled water	1000 ml

After mixing the above ingredients, adjust pH at 5.6. Sterilize (autoclave) the medium at 121°C at 15 PSI for 15 min. After cooling, the medium is poured into sterile test tubes. These tubes are now cooled at an angle to form an agar slant. When the medium has solidified and cooled to room temperature, it is ready for inoculation and stored in the refrigerator for future use.

The test tubes (25 × 125 mm) with screw caps and Petri dishes (100 mm) are usually recommended for pouring this medium. The test tubes remain ideal containers as these have adequate surface area, minimum aerosol production on opening up and can be incubated for longer period for weeks due to very slow drying of medium, which is not feasible with the Petri dishes. The drawbacks with this medium are: (i) saprotrophic fungi may grow faster obscuring the true fungal pathogen (ii) acidic pH may also inhibit some of the significant pathogenic fungal species.

Instead of cycloheximide, SDA may be supplemented with 10 µg/ml of benomyl to make it selective for the growth of basidiomycetes like *Schizophyllum commune.*

2. Neutral SDA (Emmons' Modification)

Chester W. Emmons modified the original formulation of SDA by adjusting the pH closer to neutral range and by lowering dextrose concentration from 40 to 20 g/L for increasing isolation rate of fungi. Hence **Emmons' modification** of this medium differs from original Sabouraud's formulation with lower content of glucose and pH being almost towards neutral. It contains 2% dextrose instead of 4% and neopeptone in place of peptone with final pH of 6.8 ± 7.0 hence it is also called as **neutral** Sabouraud dextrose agar.

The modified medium is generally used in diagnostic mycology laboratory for primary isolation of opportunistic and dimorphic fungi from the clinical specimens. Emmons' modification with less glucose is preferred, as isolation medium because some fungi, like *Blastomyces dermatitidis* may not be recovered using the traditional Sabouraud formulation. This is also the standard medium for maintenance of a wide variety of fungi commonly isolated in the clinical laboratory. The ingredients of this medium are as follows:

Neopeptone	10 gm
Dextrose	20 gm
Agar	20 gm
Distilled water	1000 ml

The procedure of preparing this medium will also remains the same as in case of traditional SDA except that adjustment of final pH to 6.8±7.0.

3. SDA with Antibiotics

A selective medium is required for isolation of pathogenic fungi and simultaneously avoiding bacterial and saprotrophic fungal contaminants. Add the following antibiotics into SDA while it is cooling after autoclaving at 121°C to make it selective for isolation of fungal pathogens:

Cycloheximide (Actidione)	500 mg
Chloramphenicol	50 mg
Gentamicin	20 mg

Therefore, antibiotics i.e. cycloheximide is used at 0.5 gm/liter and chloramphenicol and gentamicin at 50 mg/liter. Dissolve cycloheximide in 2 ml acetone, add it to boiling medium and mix properly. Similarly, dissolve chloramphenicol/gentamicin in 10 ml of 95% ethanol and add to boiling medium. The antibiotics should only be added after media has been autoclaved and then cooled to ~45 to 50°C. Then remove from heating and mix well. The rest of the procedure will remain the same as in case of Emmons' modification. Keep all plates at ~4°C until they are used, regardless of whether they contain antibiotics.

Whenever a medium containing antimicrobial agents is used, a second medium without these agents is also necessary for parallel inoculation because presence of cycloheximide is inhibitory to most of the saprophytic fungi and upto some extent to the growth of many clinically significant molds and yeasts *i.e. Aspergillus, Fusarium, Scopulariopsis,* mucormycetes, dematiaceous fungi, *Cryptococcus* and some of *Candida* species whereas chloramphenicol and gentamicin suppress bacterial contamination.

4. Sterile Bread (as Culture Medium)

The recovery of mucormycetes from clinical specimens is sometimes not easy despite the fact that this group of fungi is not fastidious. In such circumstances, inoculation on Sterile Bread may be tried as an additional method, which may reveal good results. A bread is taken without adding any sort of preservatives and sterilized in humidified Petri dish. The clinical specimen is directly inoculated on the bread, when it is taken from the sterile site. However, if the specimen is taken from the contaminated site, it should be treated first with antibacterials before inoculation. Mucormycetes grow rapidly usually filling the entire Petri dish within a few days hence labeled with adjectives like "lid-lifters" bread molds.

II. Nutritionally Deficient Media

The nutritionally deficient culture media are low in nutrients hence suppress vegetative growth and stimulate sporulation in fungi. They are also used for preservation of fungi as stock cultures.

1. Corn Meal Agar

Benham in 1931 described use of corn meal agar (CMA) for stimulation of chlamydospore formation. Since then, many modifications have been made to increase its efficiency. The **Dalmau plate culture** on corn meal agar is commonly used to observe chlamydospore production in yeasts and it has following ingredients:

Corn Meal	8 gm
Agar	4 gm
Distilled water	200 ml
Tween 80 (1%)	2 ml

Heat corn meal and water at 60°C for 1 hour and pass it through the filter-paper. Add distilled water to make it

200 ml and then add agar. Autoclave at 121°C for 15 minutes and pour in plates. Zein can also be used in place of corn meal, which is a protein extracted from corn meal (Zein 40 gm/100 ml). **Corn meal-Tween agar** can also be prepared by adding Tween 80 (Polysorbate) before autoclaving. The role of Tween 80 is to decrease surface tension thereby stimulating production of chlamydospores.

Yeasts are grown on corn meal agar to examine their morphological details. A heavy inoculum of yeast is streaked across plate containing medium **(ploughing)** and coverslip is placed over it to create slight anaerobic condition. The streak should project beyond cover slip. Incubate the plates for 24 to 48 hours at 25°C. Examine under **low power** microscopy over the edge of cover slip. It is a sort of junction of aerobic and relatively anaerobic conditions. The chlamydospores are induced in maximum number in this area.

There is formation of large, highly refractive, thick-walled, terminal chlamydospores, which are observed in most of the clinical isolates of *C.albicans* and *C.dubliniensis*. Moreover, trypan blue, a vital dye, can also be added to the medium, which is selectively absorbed by the chlamydospores making their morphology clearer. The chlamydospores are seldom produced at 37°C. It is also pertinent to note that morphological examination of yeasts on corn meal agar is essential to avoid confusion between organisms with almost identical biochemical profiles.

2. Rice Starch Agar

The rice starch agar (RSA) is also used to produce chlamydospore of *C.albicans* and *C.dubliniensis* faster as compared to corn meal agar. The ingredients of this medium are:

Cream of rice	2 gm
Tween 80	2 ml
Agar	4 gm
Distilled water	200 ml

Add cream of rice to boiling water and continue boiling for 30 seconds. Filter through cotton tampon and add water to make up one liter. Add agar and Tween 80 and dissolve while heating. These ingredients are autoclaved at 121°C for period of 15 min. Pour into cylinder and leave overnight in water bath at 60°C. Decant clear portion and re-filter through cotton and again autoclave at 121°C for 15 min. Adjust pH at 6.2 and pour in Petri plates.

III. Enriched and Selective Media

In some of the culture media extra nutritional substances are added to enrich them for easy and luxuriant growth of fungi during primary isolation. If some contamination of specimen is apprehended, practicality when it is from the non-sterile site, chemicals may be added to make the medium as selective. These are given below:

1. Brain-Heart Infusion Agar

The brain-heart infusion agar (BHIA) is used to grow fastidious and true pathogenic fungi such as *Histoplasma capsulatum* and *Blastomyces dermatitidis*. This is used for primary isolation as well as *in vitro* conversion of mycelial to yeast form of dimorphic fungi and/or *vice versa* (M↔Y). The ingredients of this medium are:

Brain-heart infusion	37 gm
Glucose	20 gm
L-Cysteine hydrochloride	1 gm
Agar	20 gm
Distilled water	900 gm

Dissolve ingredients by boiling and dispense into screw-cap tubes and autoclave at 121°C for 15 min. The antibiotics *i.e.* cycloheximide, chloramphenicol and gentamicin are added to this medium in similar manner as they are added to the Sabouraud dextrose agar to inhibit undesired organisms. The pH is adjusted to 6.7. Cool in slanted position with one inch butt. Store in refrigerator.

2. Cysteine-Heart and Hemoglobin Agar

Cysteine-heart and hemoglobin agar (CHHA) may also be used in place of brain-heart infusion agar. The ingredients of this medium are:

Dehydrated cysteine-heart agar	10.2 gm
Hemoglobin	2.0 gm
Distilled water	100 ml

Antibiotics are then added. This medium can be used instead of BHIA for primary isolation of the dimorphic fungi from the clinical samples as well as *in vitro* conversion of their mycelial to yeast form and/or *vice versa* (M↔Y) with good results.

3. Biphasic Medium

This medium is used for recovery of fungi from blood. It contains brain-heart infusion agar plus broth. This technique is called Castaneda's method of growing microorganisms from blood or sterile fluids. The BHIA is sterilized first and bottles are kept at slanting position. Then BHI broth is added to it aseptically to cover nearly 40% of agar in standing position. This medium can be used with or without antibiotics.

4. Bird Seed Agar (Niger Seed Agar)

Bird seed agar or niger seed (*Guizotia abyssinica*) agar or **Staib Agar** with antibiotics is an important medium used in primary isolation of *Cryptococcus* species.

In the 1964, Staib proposed a culture medium containing *Guizotia abyssinica* seed extract, which allowed for the presumptive identification of this genus by its brownish pigmentation.

The organisms of this genus utilize creatinine as sole source of nitrogen. The colonies are brown in color because phenoloxidase enzyme produced by organism oxidizes caffeic acid (3, 4-dihydroxy-transcinnamic acid) into melanin or a melanin-like pigment in niger seed extract. It is also known as **phenoloxidase test**. Other yeasts produce little or no enzyme hence there is no color change.

Limitation: A Sabouraud Dextrose Agar control should be inoculated simultaneously to ensure that the colonies do not naturally produce a dark pigment. *Aureobasidium, Sporothrix, Exophiala* and *Phialophora* may grow as yeast form in early cultures and develop dark brown colonies but the pigment will be evident on all media and not be a result of an enzymatic reaction.

The BSA contains:

Niger seed extract	200 ml
Glucose	1 gm
Diphenyl solution	10 ml
Agar	20 gm
Distilled water	800 ml

Autoclave at 121°C for 15 min and dispense into tubes. Diphenyl inhibits saprotrophic fungi. After inoculation of clinical specimen it is kept at 37°C for fortnight and periodically examined for brown-colored, mucoid colonies, which later on turn black or brown in case of genus *Cryptococcus.*

Caffeic Acid Agar: This medium is used for tentative identification of *Cr.neoformans*. Its principle is almost similar to that of niger seed *(Guizotia abyssinica)* agar. Caffeic acid agar must be protected from light to prevent the medium from turning dark.

A modification of Dichloran Rose-Bengal Chloramphenicol Medium (DRBCm) is also being used to detect colonies of *Cr.neoformans*, particularly from food or environment. Nandhakumar *et al,* have described use of mustard seed agar for the differentiation of *Cryptococcus neoformans*. In addition, Tendolkar *et al,* used tobacco agar, new medium for the pigment production in *Cryptococcus neoformans*. Tobacco agar has also been used for differentiation of *C.dubliniensis* from *C.albicans* by Khan *et al,* like Staib agar and caffeic acid-ferric citrate agar

5. Sunflower Seed Agar

The sunflower (*Helianthus annuus*) seed agar (SSA) with antibiotics can also be used for the primary isolation of *Cryptococcus* species, similar to the birdseed agar. This is called **Pal' medium**, after name of discoverer, Dr. Mahendra Pal. All known strains of *Cryptococcus* species produce light to dark brown colored colonies. This medium contains:

Pulverized sunflower seed	45 gm
Chloramphenicol	100 mg
Agar	20 gm
Distilled water	1000 ml

Autoclave at 121°C for 15 min and dispense into plates. The pH of this medium is to be adjusted at 5.6 and is incubated at 30°C after inoculation of specimens.

Moreover, Khan *et al,* have used the simplified sunflower seed agar for differentiation of *C.dubliniensis* from *C.albicans.*

6. DRBC Agar

Rose bengal is xanthene dye, that is extensively used in dyeing, printing industries and also as an insecticide. This has been used in some of the fungal culture media also. The overgrowth by spreading molds such as *Rhizopus* and *Mucor* species is a problem with fungal enumeration media used for the examination of food samples. Dichloran Rose Bengal Chloramphenicol (DRBC) agar is a modification of Rose Bengal Chloramphenicol Agar which additionally contains dichloran. The DRBC Agar is formulated as described by King *et al* and is recommended for selective isolation of yeasts and moulds especially in food samples.

Suspend 31.6 grams in 1000 ml distilled water. Heat to boiling to dissolve the medium completely. Sterilize by autoclaving at 15 lbs pressure (121°C) for 15 min. Mix well and pour into sterile Petri plates.

This medium has also been used to differentiate *Candida dubliniensis* from other yeast species.

In addition, an improved medium for isolating *Cryptococcus* species from environmental samples, DRBC can be modified medium (DRBCm), which is prepared with an infusion of 50 gm of *Guizotia abyssinica* seeds in 1000 ml of distilled water containing 2 gm of pure creatinine compound and 15 gm of DRBC commercially formulated agar. The performance of the DRBCm is considered better than the classic *Guizotia abyssinica* agar (BSA) for recovering *Cr.neoformans* colonies.

7. Blood Agar

The blood agar is routinely used in bacteriology laboratory and it can be used to grow fungal agents. It contains following ingredients:

Agar base	40 gm
Sheep blood	50 ml
Distilled water	1000 ml

This is used for isolation of *Histoplasma capsulatum* and *Blastomyces dermatitidis* from the clinical specimens.

8. Lowenstein-Jensen Medium

This medium is primarily used for isolation and cultivation of *Mycobacterium tuberculosis* and it can also be used to isolate *Madurella* and *Nocardia* species. It consists of coagulated hen's egg, mineral salt solution, asparagine and malachite green. Malachite green acts as selective agent inhibiting undesired bacteria. *Nocardia asteroides* can even grow at higher temperature of 44°C due to its thermotolerance.

9. Dermatophyte Test Medium

The dermatophyte test medium (DTM) is used for the presumptive identification of dermatophytes in hair, nail and skin specimens, which may be heavily contaminated with bacteria. Most of the dermatophytes species produce red color due to liberation of alkaline metabolites, which increase pH entailing change in color of phenol red indicator from yellow to red. This medium contains:

Phyton	10 gm
Dextrose	10 gm
Phenol red solution	40 ml
8N HCl	6 ml
Actidione	500 mg
Gentamicin	100 mg
Agar	20 gm
Distilled water	1000 ml

Final pH is 5.5 +/- 0.4

The phenol red solution is 0.5 gm in 15 ml of 1N NaOH made up to 100 ml with distilled water. The medium is incubated at 25°C and there is change in color within 3 to 6 days. If there is no change upto two weeks, the medium is discarded, as there are other fungi, which may give rise to false positive results due to incubation beyond this period. However, there are some disadvantages of this medium, which are as follows:

- It does not inhibit saprotrophic fungi, which may grow inducing color change.
- The true pathogens like *H.capsulatum* and *B.dermatitidis* are also not inhibited leading to change in color.

10. Dermatophyte Identification Medium

The Dermatophyte Identification Medium (DIM) is like DTM and it is also used for presumptive identification of dermatophytes. It contains dextrose (20 mg/ml), neopeptone (10 mg/ml), cycloheximide (4 mg/ml), penicillin (20 U/ml), streptomycin (40 U/ml) and bromocresol purple (1.6 mg/ml). Unlike DTM it is incubated at 37°C and if there is growth of dermatophytes, color of medium changes from greenish blue to purple within 24 to 48 hours. It has shown less false-positive results with fewer fungi than DTM. Most of fungi giving false-positive reactions are morphologically dissimilar to dermatophytes. In addition, the DIM may also be used for primary isolation of dimorphic fungi.

11. Czapek-Dox Agar

This media is synthetic and includes sodium nitrate as its sole source of nitrogen. Fungi and bacteria capable of using inorganic nitrogen will grow on the media. The media is recommended as a subculture media for filamentous fungi and is useful for identification of *Aspergillus* and *Penicillium* species. It contains following ingredients:

$NaNO_3$	3.0 gm
K_2HPO_4	1.0 gm
KCl	0.5 gm
$MgSO_4.7H_2O$	0.5 gm
$FeSO_4.7H_2O$	0.01 gm
Glucose	30 gm
Agar	15 gm
Distilled water	1000 ml

Add all dry ingredients, except glucose, to water and boil to dissolve. Add glucose and sterilize by autoclaving.

12. PYG Agar Medium

The peptone-yeast extract glucose agar/broth medium contains following ingredients per liter:

Bactopeptone	5 gm
Yeast extract	5 gm
Glucose	20 gm

Adjust pH at 7.2

13. Malt Extract Agar

This medium can be used in place of SDA because there is an early growth of fungus during primary isolation and sporulation is much better than SDA. It is also useful for maintenance of fungi in mycology laboratory. It can also be acidified with hydrochloric acid to a pH <4 for purifying bacterially contaminated yeast cultures. It contains the following ingredients:

Malt extract	20.0 gm
Peptone	1.0 gm
Glucose	20.0 gm
Agar	15 gm
Distilled water	1000 ml

Mix reagents and bring to boil. Autoclave at 121°C for 15 min. The shelf life of this medium is 30 days in the tubes and 15 days in Petri dishes.

14. Oatmeal Agar

This medium enhances sporulation in many of phaeoid fungi. It contains following ingredients:

Tomato paste	10.0 gm
Oatmeal	10.0 gm
Magnesium sulphate	1.0 gm
Potassium phosphate	1.0 gm
Sodium nitrate	1.0 gm
Agar	15.0 gm
Distilled water	1000 ml

Adjust the pH at 5.6.

Autoclave at 121°C for 20 min. Dispense in tubes or Petri dishes.

15. Potato Dextrose Agar

This plant based medium is used for the stimulation of sporulation, maintaining stock cultures and differentiation of typical varieties of dermatophytes on the basis of pigment production. Potato infusion and dextrose promote luxuriant fungal growth. Therefore, it may also be used for primary isolation of fungi from the clinical specimens. Most of the times it is used for secondary inoculation to have better description of the morphology of fungal culture. The morphological description of majority of fungal isolates is given on this medium. In this medium heating after adjusting pH should be avoided as it may hydrolyze the agar, which can render it unable to solidify. One liter of this medium contains the following ingredients:

Potatoes	250.0 gm
Dextrose	20.0 gm
Agar	20.0 gm

Final pH is 5.6±0.2

For preparing the medium, scrub, slice and boil potatoes in 100 ml of distilled water for one hour. Filter infusion through gauze and add agar and boil till it dissolves completely. Add dextrose and make upto one liter by adding distilled water. Sterilize by autoclaving at 15 lbs pressure at 121ºC for 15 min. Cool to 50ºC and dispense the medium in tubes or Petri dishes. Alternatively, PDA can also be prepared from commercially available dehydrated medium according to instructions of the manufacturer.

16. Modified Dixon Agar

The modified version of Dixon agar is used for the growth of *Malassezia* species. It contains malt extract (3.6%), mycological peptone (0.6%), ox bile (2%), Tween 40 (1%), oleic acid (0.2%), glycerol (0.2%) and agar (1.2%) with addition of chloramphenicol (0.5%) as well as cycloheximide (1 gm/l) and incubation at 30-32°C.

17. Leeming and Notman Agar

Leeming and Notman medium gives better results *i.e.* higher recovery rate of *Malassezia* species, than variety of oil overlays and has longer shelf-life than poured plates. The modified Leeming and Notman agar composed of (per 100 gm) peptone (1.0 gm), glucose (1.0 gm), yeast extract (0.2 gm), ox bile (0.8 gm), glycerol (1.0 ml), glycerol monostearate (0.05 gm), Tween 60 (0.5 ml), olive oil (2.0 ml), agar (1.5 gm) and sterilized by autoclaving. However, modified Dixon medium permits an easier colony counting and better appreciation of colony morphology hence is usually favored instead of Leeming and Notman formula.

A new medium for rapid diagnosis and determination of antifungal susceptibility testing against *Malassezia* species has been devised and designated as Dutta and Dikshit Medium (DDM).

IV. Sporulation Induction Media

Some of the culture media are used for the induction of spores, which is essential component of the phenotypic identification of fungal isolates. These may be sexual or asexual spores. The media are briefly described here:

1. Alphacel-Yeast Extract Agar

This is used for mating and formation of cleistothecia in *Histoplasma capsulatum* var. *capsulatum* and *H.capsulatum* var. *duboisii*. Alphacel is α-cellulose. It contains following ingredients:

Alphacel	20 gm
$MgSO_4$	1.0 gm
KH_2PO_4	1.5 gm
Yeast extract	0.5 gm
Agar	20 gm
Distilled water	1000 ml

The pH is adjusted to 5.7 with 1N NaOH.

2. Soil Extract Agar

This is used for mating of *Histoplasma capsulatum* and *Blastomyces dermatitidis* for development of cleistothecia. It is also used as maintenance medium for these fungi. Moreover, it is also used for sporulating mucormycetes like *Apophysomyces* and *Saksenaea* species, which otherwise do not sporulate easily on the routine media. It contains following ingredients:

Garden soil	500 gm
Tap water	1200 ml

Autoclave the soil at 121°C for 3 hours, filter through Whatman No. 2 filter paper. Take filtrate and make it up to 1000 ml and then add following:

Yeast extract	1 gm
Dextrose	2 gm
Agar	15 gm
KH_2PO_4	0.5 gm

Adjust pH at 7.0 before autoclaving at 121°C for 20 min.

V. Media used for Biochemical Tests

In a diagnostic mycology laboratory, some of the culture media are required for performing biochemical reactions, for the differentiation of various fungi for their identification. These are as follows:

1. Tetrazolium Reduction Medium

The tetrazolium reduction medium (TRM) is used to differentiate various *Candida* species. Tetrazolium is reduced in different gradients by various species of *Candida* (*See* **Table 20.8**).

Peptone	1 gm
Glucose	4 gm
Beef extract	0.1 gm
Tetrazolium	20 mg
Neomycin	50 mg
Distilled water	100 ml

Adjust pH 5.6-6.2

Tetrazolium and neomycin are added after autoclaving. The following colors are seen in various species of genus *Candida*:

***Candida* species**	**Color on TRM**
Candida albicans	Pale Pink
C.tropicalis	Orange Pink
C.pseudotropicalis	Salmon Pink
C.parapsilosis	Rose Pink
C.guilliermondii	Pink and Pasty
C.krusei	Pink and Dry
C.glabrata	Pale Pink

2. Carbohydrate Fermentation Media

This medium contains peptone 1%, NaCl 0.5% and Andrade's indicator 0.005%. Different sugars like dextrose, lactose, maltose, sucrose, etc. at a concentration of 2% are added to it. In bacteriology, the sugar concentration is 1% instead of 2%. Inverted Durham's tube is used for detection of gas. Examine the tubes every 24-48 hours intervals for the production of acid and gas (in Durham's tube). Autoclave at 10 lbs pressure. Incubation is done at 37°C.

3. Carbohydrate Assimilation Media

Yeasts and yeast-like fungi utilize specific carbohydrate substrates aerobically for energy and growth. The organisms are inoculated on carbohydrate-free medium (**Auxanographic** method). Carbohydrate containing filter paper disks are added and a positive assimilation/ utilization of carbohydrate is determined by presence of growth around disk. Glucose is assimilated by all species of yeasts and is included in the test as an internal quality control. The inability to assimilate a particular carbohydrate is the result of either due to substrates inability to enter the yeast cell or the absence of specific enzyme. The technique consists of sugar disk with 10% concentration, 2% extra pure agar, KNO_3 disk 4%, Yeast Carbon Base (YCB) 1.17% for nitrogen assimilation and Yeast Nitrogen Base (YNB) 6.7% for carbohydrate assimilation.

4. Urease Medium

This medium detects ability of various yeasts to produce urease enzyme and hydrolyze urea. In presence of suitable substrates, urease splits urea producing ammonia, which raises pH and color changes from amber to pinkish red due to phenol red as the indicator. The test helps in differentiating of basidiomycetous urease positive fungi like *Cryptococcus, Trichosporon* and *Rhodotorula* species from *Geotrichum* and *Candida* species, which are urease negative.

It is also used in differentiation of dermatophytes such as *T.mentagrophytes* species, which hydrolyzes urea and color of medium becomes deep red while *T.rubrum* is negative. The Christensen's urea agar contains:

Peptic digest	1.0 gm
Dextrose	1.5 gm
NaCl	5.0 gm
Disodium phosphate	1.2 gm
Monopotassium phosphate	0.8 gm
Phenol red	0.012 gm
Agar	15.0 gm
Distilled water	1000 ml

Final pH is adjusted to 6.8 ± 0.2.

Mix properly all the ingredients in 1000 ml distilled water and autoclave at 121°C for 15 min. After cooling make slant. The indicator used in this medium is phenol red.

5. Rapid Urease Test

This medium gives result about urease producing organisms within minutes hence it is called as rapid test. Take sterile swab stick, sweep it across yeast colony and dip in Christensen' urea agar base. Place swab stick in test tube containing 3 drops of 1% benzalkonium chloride (pH 4.86) and press cotton tip firmly against tube. Plug tube with cotton. Incubate at 45°C and examine after 10, 15, 20 and 30 min for color change from yellow to purple, which indicates urease production by *Cryptococcus* species.

6. CHROMagar Candida Medium

The CHROMagar Candida is selective and differential type of chromogenic medium, which is useful for identification of various *Candida* species and other yeast-like organisms like *Prototheca*. Due to the chromogenic substrate in the medium, the colony morphology and color have been well defined when it is used to isolate the yeasts directly from clinical material, including stool, urine, respiratory, vaginal, oropharyngeal and esophageal sources.

This medium is also formulated to provide presumptive identification of medically significant yeast-like organisms based on colony morphology, particularly their color. It is based on direct detection of specific enzymatic activities by adding multiple chemical dyes *i.e.* substrates of fluorochromes. This facilitates presumptive differentiation of clinically significant yeast-like organisms. It is one such medium, which can be used for simultaneous isolation as well as presumptive identification of various *Candida* species like *C.albicans, C.krusei, C.tropicalis, C.glabrata, C.parapsilosis* and *C.dubliniensis* in addition to *Prototheca* and *Trichosporon* species **(Figs. 5.10 and 20. 2)**.

The benefits of using this medium are that it shortens time for presumptive identification of organisms and it also allows for easier detection of multiple yeast species present in the clinical specimen with mixed infection. The CHROMagar Candida shows following colors of colonies after incubation at 30°C for a period of 48 to 72 hrs:

C.albicans	Light green
C.dubliniensis	Dark green
C.glabrata	Pink to Purple
C.krusei	Pink
C.parapsilosis	Cream to pale pink
C.tropicalis	Blue with pink halo

Prototheca species also produce cream-colored colonies, similar to those of *C.parapsilosis* but *Prototheca* colonies are slightly smaller in size as compared to that of *C.parapsilosis*. Similarly, confusion may arise between *C.krusei* and *C.glabrata* due to their similarity in pink color of the colonies. Therefore, this medium should be prepared from a standard and reliable source instead of local brand companies, which may provide cheap quality products.

CHROMagar Malassezia*:* On this differential chromogenic medium, *Malassezia furfur* colonies are large, pale to pink and wrinkled. However, the other *Malassezia* species, including *M.globosa* and *M.restricta,* are mostly pink to purple in color.

7. CGB Agar

The canavanine glycine bromothymol blue (CGB) agar is significant tool for biotyping of various species of *Cryptococcus*. It is used to differentiate *Cryptococcus gattii* from other species of this genus. The principle of this simple biotype test is based on the ability of *Cr.gattii* isolates to grow in the presence of L-canavanine and to assimilate glycine as sole source of carbon. The medium has two different solutions and their composition is given further:

Solution A

Glycine	10 gm
KH_2PO_4	1 gm
$MgSO_4$	1 gm
L-Canavanine Sulphate	30 mg
Thiamine hydrochloride	1 mg
Distilled water	100 ml

Solution B

Sodium Bromothymol Blue	0.4 gm
NaOH (0.01 M)	64 ml
Distilled water	36 ml

To prepare the CGB medium, two solutions are prepared: for the glycine-L-canavanine solution **(Solution A)**, 10 gm of glycine, 1 gm of KH_2PO_4, 1 gm of $MgSO_4$, 1 mg of thiamine-HCl and 30 gm of L-canavanine sulphate (Sigma) are added to 100 ml of distilled water and the pH is adjusted to 5.6 and the solution is millipore sterilized (0.45 m).

Solution B - Add 2 ml of Solution B to 88 ml of distilled water. Add 2 gm of agar, autoclave at 121°C for 15 minutes and cool to temperature of about 55°C. Add 10 ml of Solution A, mix well and pour in small tubes for making slants.

For the testing, *Cryptococcus* species, to be identified, is inoculated over medium and incubated at 37°C for 1 to 5 days. *Cr.gattii* (Serotypes B and C) will turn the medium cobalt blue while other *Cryptococcus* species will not be able to change the color, which remains as such *i.e.* yellow to light green.

Further Reading

1. Al Mosaid A, Sullivan D, Salkin IF, *et al.* Differentiation of *Candida dubliniensis* from *Candida albicans* on Staib agar and caffeic acid-ferric citrate agar. *J Clin Microbiol.* 2001; **39:** 323-7.
2. Al Mosaid A, Sullivan DJ, Coleman DC. Differentiation of *Candida dubliniensis* from *Candida albicans* on Pal's agar. *J Clin Microbiol.* 2003; **41:** 4787-9.
3. Akgul O, Cerikcioglu N. Hypertonic Sabouraud dextrose agar as a substrate for differentiation of *Candida dubliniensis. Mycopathologia.* 2009; **167:** 357-9.
4. Arora DR, Saini S, Aparna GN. Evaluation of germ tube test in various media. *Indian J Pathol Microbiol.* 2003; **46:** 124-6.
5. Ayyar S, Tendolkar U, Deodhar L. A comparison of three media for isolation of *Nocardia* species from clinical specimens. J Postgrad Med. 1992; **38:** 70-2.
6. Bosshard PP. Incubation of fungal cultures: How long is long enough? *Mycoses.* 2011; **54:** e539-45.
7. Carmichael JW. Lactofuchsin: A new medium for mounting fungi. *Mycologia.* 1955; **47**: 611.
8. Casal M, Linares MJ, Solis F, *et al.* Appearance of colonies of *Prototheca* on CHROMagar Candida medium. *Mycopathologia.* 1997; **137:** 79-82.
9. Castro e Silva Dde M, Santos DC, Pukinskas SR, *et al.* A new culture medium for recovering the agents of cryptococcosis from environmental sources. *Braz J Microbiol.* 2015; **46:** 355-8.
10. Chowdhary A, Randhawa HS, Kowshik T, *et al.* Application of hypertonic Sabouraud glucose agar for differentiation of *Candida dubliniensis* from *Candida albicans. Diagn Microbiol Infect Dis.* 2011; **69:** 440-2.
11. Khan ZU, Ahmad S, Mokaddas E, *et al.* Simplified sunflower *(Helianthus annuus)* seed agar for differentiation of *Candida dubliniensis* from *Candida albicans. Clin Microbiol Infect.* 2004; **10:** 590-2.
12. Khan ZU, Ahmad S, Mokaddas E, *et al.* Sunflower seed husk agar: A new medium for the differentiation of *Candida dubliniensis* from *Candida albicans. Indian J Med Microbiol.* 2005; **23:** 182-5.
13. Khan ZU, Ahmad S, Mokaddas E, *et al.* Tobacco agar, a new medium for differentiating *Candida dubliniensis* from *Candida albicans. J Clin Microbiol.* 2004; **42:** 4796-8.
14. Kindo AJ, Tupaki-Sreepurna A, Yuvaraj M. Banana peel culture as an indigenous medium for easy identification of late-sporulation human fungal pathogens. *Indian J Med Microbiol.* 2016; **34:** 457-61.
15. King DA Jr, Hocking AD, Pitt JI. Dichloran rose bengal medium for enumeration and isolation of molds from foods. *Appl Environ Microbiol.* 1979; **37:** 959-64.
16. Nadeem SG, Hakim ST, Kazmi SU. Use of CHROMagar Candida for the presumptive identification of *Candida* species directly from clinical specimens in resource-limited settings. *Libyan J Med.* 2010; 5. PMID: 21483597.
17. Nandhakumar B, Kumar CP, Prabu D, *et al.* Mustard seed agar, a new medium for differentiation of *Cryptococcus neoformans. J Clin Microbiol.* 2006; **44:** 674.
18. Odds FC. Sabouraud's agar. *J Med Vet Mycol.* 1991; **29:** 355-9.
19. Pal M, Hasegawa A, Ono K, Lee CW. A new staining solution for the morphological studies of fungi and *Prototheca. Jpn J Vet Sci.* 1990; **52:** 527-31.
20. Pal M. Use of Pal's sunflower medium for an early diagnosis of cryptococcosis. *Antiseptic.* 1997; **95:** 175.
21. Scheid LA, Nunes Mario DA, Soares IH, *et al.* DRBC agar: A new tool for *Candida dubliniensis* identification. *Folia Microbiol (Praha).* 2010; **55:** 155-8.
22. Tan GL, Peterson EM. CHROMagar Candida medium for direct susceptibility testing of yeast from blood cultures. *J Clin Microbiol.* 2005; **43:** 1727-31.
23. Tendolkar U, Tainwala S, Jog S, *et al.* Use of a new medium - Tobacco agar, for pigment production of *Cryptococcus neoformans. Indian J Med Microbiol.* 2003; **21:** 277-9.
24. Yazdanpanah A, Khaithir TM. Issues in identifying germ tube positive yeasts by conventional methods. *J Clin Lab Anal.* 2014; **28:** 1-9.

Appendix B: Fungal Reagents and Stains

In a diagnostic mycology laboratory various reagents and stainings are used while performing procedures to establish the diagnosis of fungal diseases. These may be either wet mounts or differential stained preparations, adopted depending on the nature of clinical specimen received. These are briefly described below:

I. Wet Preparations

The wet preparations are used either for diagnosis of mycotic diseases in the clinical specimen for direct demonstration of the fungus or to study morphological features of a fungal isolates after the growth is obtained on culture. These are as follows:

1. Potassium Hydroxide Wet Mount

This is primary screening tool used in diagnostic microbiology laboratory. The aqueous potassium hydroxide **(KOH)** softens and digests protein debris as well as dissolves cement substance, which holds keratinized cells together. The fungal elements become quite clear and are easy to visualize in clinical specimens. It is prepared from the following ingredients:

Potassium hydroxide	10 gm
Glycerol	10 ml
Distilled water (DW)	80 ml

In a solution of KOH (10%), glycerol (10%) is added to prevent drying, as it is hygroscopic in nature. Mix these ingredients properly and store the solution at room temperature. The concentration of KOH may be increased depending upon the nature of the clinical material, as solid specimens require more than 10% concentration, may be upto 40%. Dimethyl sulfoxide (DMSO), a colorless organosulfur compound, can also be added to KOH (20%) solution, which acts as useful cleansing agent of fungi (KOH - 20 gm, DMSO - 40 ml, DW - 60 ml). The wet mount with DMSO can be examined immediately, without waiting for dissolution. The wet mounts of KOH are prepared by the following **two** methods:

Slide KOH

Place epidermal scales, skin scrapings, nail clippings or homogenized biopsy tissue on clean glass slide. In case of suspected mucormycosis, tissue may be placed as such without homogenization to retain the cellular morphology intact. Pour drop of KOH (10%) on specimen and place coverslip over it. Press it slightly to get monolayer and excessive fluid is properly mopped up, which may otherwise damage microscope due to corrosive nature being a strong alkali.

The slide is gently warmed over flame or slide warmer to speed up clearing process. Too much of warming may lead to crystallization of KOH hence should be avoided. In addition to clearing, it also helps in getting rid of air bubbles. However, while using DMSO, heating is not required. Now, prepared slide is kept for 5-30 min. at room temperature and then examine under microscope. If specimen is not properly dissolved, it may be kept for some more time in a wet petri dish and examined later on. Moreover, a mixture of KOH (20%) solution and equal part of Parker ink may be used for better visualization/delineation of fungal elements, particularly *Malassezia* species. A drop of acridine orange (0.01%) may also be added to KOH slide and is observed under fluorescent microscope.

Tube KOH

The tube KOH is prepared mainly for those biopsy specimens, which take longer time for dissolution. The homogenized biopsy/piece of tissue is dissolved in KOH (10%) and kept in an incubator at 37°C. After an overnight incubation, it is examined by preparing a slide mount. This procedure is also adopted in case of nail clippings and skin biopsies, which also dissolve with difficulty hence KOH concentration may be increased upto 40% while processing such specimens.

The wet preparations are examined under 10X, then 20X and finally 40X powers of microscope. While examining of either slide or tube KOH, sufficient time should be spent to screen the entire smear in a **ploughing** manner searching for fungal elements like hyphae, yeast cells, pseudohyphae, spherule or sclerotic bodies. Sometimes clinical material is very scanty, particularly in corneal scrapings where direct microscopy may play a decisive role in managing fungal corneal ulcer.

If none of the organic fungal structure is found in entire smear, report is dispatched as ***"No fungal element seen"***. However, care should be taken not to report an artifact considering as the fungus. This difficulty can be overcome by experience and also performing simultaneously different wet mounts and differential stains. The KOH may be used along with fluorescent brightener such as calcofluor white (CFW) for increasing sensitivity of this technique and obtaining best results.

In KOH wet mount, experienced person can make out type of fungus to the extent whether it is yeast or mycelial (septate/non-septate), dermatophyte or non-dermatophyte and/or phaeoid. These findings should be matched with corresponding fungal growth obtained on culture. If KOH mount is negative and there is growth in one or both culture tubes, repeat the culture. If same fungal growth is obtained, it shows the significance of organism. The report is despatched as *"Culture shows growth of.......... (Name of Fungus).............Please repeat sample, if feasible."*

This is the simplest and cheapest key procedure, which provide basic, reliable and very significant information for the course of management of fungal diseases. It should be compulsorily done by all the diagnostic microbiology laboratories. The other alkali like sodium hydroxide (NaOH) may also be used in place of potassium hydroxide however, it forms the crystals very fast.

2. India Ink Stain

This is one of the negative staining techniques used for visualizing encapsulated microorganisms like *Cryptococcus* species, *Streptococcus pneumoniae* and *Klebsiella pneumoniae*. The stain contains following ingredients:

India ink	150 ml
Merthiolate (1 : 1000)	3 ml
Tween 80 (1 : 10,000)	0.1 ml

Mix, filter and keep in the dropping bottle. This stain should be filtered periodically to remove bacterial contamination and carbon particles.

A drop of clinical material, such as CSF, is mixed with an equal volume of India ink on the glass side. The coverslip is put up and gently pressed so that extra amount of ink is expelled, then mopped and examined under microscope. If properly made, one should be able to barely read newsprint through the preparation. Alternatively, clinical material is placed on the slide followed by coverslip and India ink is added on the side of coverslip to allow it to diffuse underneath. The cell outline of *Cryptococcus* ranging from 5-20 µm in size with distinct halo is observed. The clear halo around budding yeasts against a black background shows that polysaccharide capsule has repelled opaque medium, which may be twice the size of yeast cells.

The demonstration of encapsulated organism on an India ink preparation is pathognomonic for *Cryptococcus* species. However, observer must distinguish yeast cells from erythrocytes, leukocytes, air bubbles talc granules from gloves. Sometimes, macrophages may produce similar appearance, therefore, slide should be meticulously examined for budding yeast cells. The India ink staining is positive in about 75% of cases.

A modification of India ink staining technique may also be used by employing 2% chromium mercury. A drop of CSF is placed on clean glass slide. Then, drop of 2% chromium mercury is mixed with CSF on the slide. Immediately a small drop of India ink is added and coverslip is placed and preparation is observed with bright-field microscope. This helps in visualizing external and internal structures of organism, layer-wise from outer capsule to internal ones, which used to be distinguished only by electron microscopy.

In addition to the wet mount, dry India ink preparation may also be made by mounting with DPX solution to keep it permanently for teaching purposes.

3. Nigrosin Stain

This is also one of the negative stainings and used as an alternative to India ink stain. It contains following reagents:

Nigrosin granules	10 gm
Formalin (10%)	100 ml

Add nigrosin to formalin in 25 ml, boil in water bath for 30 min. Add more formalin to make volume to 100 ml. Filter through double Whatman No. 1. This is better than India ink stain as it can last for more than one year and no carbon particles appear in stain solution. Moreover, India ink does not contain any of antimicrobial agents and there is risk of getting infection while handling the stain, which is not possible with nigrosin as it contains formalin. Rest of the technique, description and interpretation are same as that of India ink wet mount preparation.

4. Calcofluor White Stain

The fluorescent dyes or optical brighteners (Calcofluor white, Blankophor or Uvitex 2B) that specifically stains chitin contained in cell wall, coupled with a clearing and dissociating agent, such as KOH, provides better sensitivity of detecting fungi in the clinical specimen. Calcofluor white **(CFW)** is a water-soluble, colorless dye and fluorescent whitener used in textile and paper industry. Calcofluor binds non-specifically to beta-linked polysaccharides found in the cell walls of various organisms such as chitin and cellulose, which is visualized when exposed to long-wavelength ultraviolet and short-wavelength visible light. It is used as non-toxic biological marker to demonstrate cellulose in microorganisms. It fluoresces light-blue in color when exposed to UV light of fluorescent microscope fitted with barrier filter (300 to 412 nm). It is very popular technique because of its higher sensitivity as compared to other conventional methods like KOH wet mounts. In addition, mixed infection by various types of fungi is also better visualized by this technique.

Calcofluor white, 10 µg/ml is prepared in 10% KOH and 300 mg% Evans blue. Potassium hydroxide acts as clearing agent.

CFW M2R	100 mg
Evans blue	50 mg
Distilled water	100 ml

Mix well and store in refrigerator. The calcofluor solution is prepared by dissolving calcofluor white powder in distilled water with final concentration of 0.1%. Evans blue is also added (0.1%) in order to reduce non-specific background fluorescence thereby clearly revealing fungi in the surrounding tissue. Evans blue counterstain also produces contrasting orange to ruby-red background, thereby, enhancing detection of fungi. This stain gives apple green fluorescence of fungi. Sometimes, artifacts may be confusing hence it should be freshly prepared or filtered to avoid the same. When used with Evans blue, it is called as **calcofluor-blue** for the diagnosis of *P.jirovecii* pneumonia in BAL fluids and may be extended to the other fungi as well.

Ultraviolet light with excitation filter that transmits wavelengths between 300 and 412 nm (maximum absorbance of CFW is 347 nm) and barrier filter that removes UV while transmitting visible blue light and longer wavelength is used. This requires light source *i.e.* mercury vapor lamp; quartz halogen bulbs are usually suitable, as energy output is very low.

By conventional KOH wet mounts, it is difficult to differentiate hyphae from collagen fibers and other artifacts. Hence CFW stain is far superior to these staining techniques for detection of fungi in clinical specimens. It is technically simple, quick and highly reliable to detect fungi even if observer is relatively inexperienced. It requires use of fluorescent microscope, which is usually available in teaching medical institutions and universities only.

Blankophor (1%) may also be used instead of calcofluor white staining which gives less artifacts and does not get precipitated while mixing with KOH.

Therefore, these fluorescent brighteners can be utilized for diagnosing most of the fungal infections including *Pneumocystis* and *Microsporidium* species. These may also be routinely applied to non-fungal agents such as free-living amoebae *(Acanthamoeba, Naegleria* and *Balamuthia)* and larva of *Dirofilaria*.

5. Lactophenol Cotton Blue Stain

The Lactophenol Cotton Blue (LCB/LPCB) mount is used to study morphological features of fungal isolates after the strain is grown on culture medium. This is of two types:

Plain LCB

It contains following ingredients:

Melted phenol	20 ml
Lactic acid	20 ml
Glycerol	40 ml
Cotton blue	0.05 gm
Distilled water	20 ml

Mix all reagents properly and dissolve 0.05 gm of aniline dye (cotton blue) in distilled water before mixing with the remaining reagents. The phenol acts as disinfectant, lactic acid preserves fungal morphology and glycerol is hygroscopic agent thereby prevents drying. The cotton blue stains outer wall of the fungus.

For preparing a mount, take out a small fragment of fungal culture on a glass slide with round loop if yeasts and with L-shaped loop in case of mold. Add a drop of LCB stain and tease the culture properly with the help of two teasing needles. Put up the coverslip and examine under the microscope. The edges of coverslip can be sealed with nail polish to keep it for a pretty long time.

An alternative to the coverslip is that take a transparent tape and touch its sticky surface to the fungal colony grown in the culture plate and place on the slide where a drop of LCB is already added. Then examine under the microscope. The arrangement of conidia remain as such like that seen in the slide culture, very helpful to speciate fungi despite the fact that the vision is slightly hazy.

LCB with Polyvinyl Alcohol

It contains following ingredients:

Staining Reagents:

Polyvinyl alcohol (PVA) powder	15 gm
Distilled water	100 ml

Mix powder at 80°C in a beaker placed in water bath. A thick viscous solution is formed and if this mixture is lumpy, it is filtered through double-layered cloth.

Staining Solution:

PVA stock solution	56 ml
Melted phenol	22 ml
Lactic acid	22 ml
Cotton blue	0.05 gm

Add lactic acid to stock polyvinyl alcohol, mix well, then phenol is added, mix and warm gently to dissolve. The LCB preparation can be permanently preserved if PVA is used. Moreover, instead of cotton blue, safranine can also be used.

The LCB mount serves as the best method to study morphological details of fungal isolates. All fungal strains, encountered in diagnostic mycology laboratory, should be routed through this technique. Sometimes, an isolate is presumed as merely contaminant and insignificant on gross morphology but turns out to be very significant on examination of the LCB mount.

6. PHOL Stain

This is used similar to the LCB stain for examination of fungal isolates and *Prototheca* species. The acronym PHOL is derived from initials of surnames of four scientists *i.e.* Pal, Hasegawa, Ono, Lee. It contains formalin instead of phenol and methylene blue in place of cotton blue contained in the Lactophenol Cotton Blue stain.

Like LCB and PHOL stainings, Narayan Stain may also be used to study morphology of fungal isolates, which contains glycerine (4.0 ml), methylene blue (0.5 ml of 3%) and dimethyl sulfoxide (7.0 ml).

7. Neutral Red Stain

Neutral red staining is useful and easily applicable method for evaluation of viability of fungal elements. It is water-soluble dye that is capable of passing through intact plasma membrane and is stored in lysosome of viable cells. When cell membrane and lysosomes are damaged, dye uptake ceases. Therefore, it is used as vital stain mainly in dermatophytes and *Candida* species. Its properties have led to development of procedure for measuring effect of test substances on normal human epidermal keratinocytes *in vitro*.

8. Diazonium Blue B Stain

The heterogeneous genera, basidiomycetous *Trichosporon* and ascomycetous *Geotrichum* share number of morphological and physiological characteristics hence confused with each other. Diazonium Blue B (DBB) reagent is used for differentiation of species of these genera. Alkali-ethanol DBB staining reaction is most reliable out of different types of procedures utilized for performing this color test showing red-stained colonies in most of the basidiomycetes.

The positive reaction is indicated if dark-red or violet red color develops within two min. *Trichosporon* isolates form pink to violet pigments following reaction with DBB that reliably differentiates them from *Geotrichum* species which are DBB-negative. The inoculum for performing reaction can be grown in static culture without diminishing color intensity. This modification facilitates use of DBB reaction in clinical laboratories. The molecular basis of this reaction is not known. The test represents reproducible and empirical method of differentiating yeasts with basidiomycetous affinity from those with ascomycetous affinity as colonies of ascomycetes remain unstained. Therefore, *Malassezia*, *Cryptococcus* and *Rhodotorula* species are DBB-positive. Hence DBB reaction is also of paramount importance for identification of the yeast-like fungi.

II. Differential Stains

There are many stains, which are utilized to examine fungal elements in clinical specimens. Some of them are routinely done and rests are specifically done as special stains when particular specimen is suspected to be having fungal etiology. These are described as follows:

1. Gram's Stain

The Gram's staining method is named after the Danish bacteriologist Hans Christian Gram who devised it in 1882, which was published in 1884. This staining method differentiates various bacterial species into Gram-positive and Gram-negative bacteria based on the chemical and physical properties of their cell walls. Gram's stain is effective for detecting some of fungal pathogens too. The fungi are Gram-positive and seen as violet-colored in stained smear.

For routine **Gram's** staining, fix smear by passing it over flame. Place 0.5% aqueous crystal violet solution on slide for 20 seconds. Wash the smear gently under tap water. Apply Gram's iodine solution over slide for 20 seconds. Wash again with tap water. Decolorize quickly with solution of equal parts of acetone and 95% ethanol and wash immediately in running tap water. Counterstain with 0.5% aqueous safranin for 10 seconds. Again wash with tap water, air or blot dry and examine under the microscope.

Weigert's modification of the Gram's stain is particularly useful for staining tissue sections. The primary stain carbol gentian violet is prepared using saturate alcoholic solution of gentian violet and 5% phenol solution. Gram's iodine is used as a mordant and aniline-xylol is used as a decolorizer. The counterstain carmalum (carminic acid and potassium alum in water), however is used ahead of primary stain. This method may be used to stain *Pneumocystis* cysts.

Brown and Brenn modification of the Gram's stain is used for *Nocardia*, *Actinomyces* and *Microsporidium* species in tissues. In general, procedure is more suited to sections than to smears. The yeast cells usually show well-stained morphology but filamentous fungi in smears become desiccated and their morphological characteristics are invariably lost. Although finding fungi in clinical specimens by Gram staining is not routinely advocated but it is a very good method, which help in suspecting presence of fungal pathogens in tissues.

Procedure (Brown-Brenn):

- Deparaffinise and hydrate to distilled water.
- Drop crystal violet stain on to tissue section for 1 min.
- Wash in tap water.
- Put Lugol's iodine for 1 min.
- Wash in tap water.
- Blot sections dry, breath on section then quickly pour acetone until no color runs off.
- Wash in tap water.
- Place slides on staining rack, drop Basic fuchsin on tissue sections, stain for 3 min.
- Wash in tap water, blot gently but not completely dry.
- Dip quickly into acetone, 2 dips.
- Dip directly into picric acid-acetone mixture until a 'salmon' color.
- Dip quickly into two changes of acetone.
- Air dry, dip into xylene and put coverslip and see under microscope.

Observance:

Gram-positive organisms:	Blue
Gram-negative organisms	Red
Nuclei	Red
Background	Yellow

2. Modified Acid-Fast Stain

The aerobic actinomycetes are important cause of actinomycetoma and difficult to be differentiated because all are filamentous Gram-positive **bacteria**. Only *Nocardia* species are positive for modified acid-fast stain (with 1% H_2SO_4 instead of 25% used for *Mycobacterium tuberculosis*). Therefore, smear should be stained with modified Ziehl-Neelsen or with **cold** acid-fast staining **(Kinyoun's method)** because *Nocardia* species are weakly **acid-fast** giving pink or red color.

In modified Ziehl-Neelsen staining, make smear and fix it by passing 2-3 times over flame. Flood the slide with carbol-fuchsin (1%) with intermittent heating and then wash it with water. Decolorize with aqueous sulfuric acid (1%) and again wash with tap water. Counterstain with methylene blue (0.3%) for 1 min. Rinse with water, dry and examine under oil immersion microscopy. A positive and negative control should be put up every time to check for any procedural lapse or lacunae.

In the cold acid-fast staining **(Kinyoun's method)**, concentration of phenol/carbolic acid is increased (from 4.5 to 6.4%) thereby allowing the stain to penetrate through lipid part of cell wall without the need of heating.

For the fungi, hat-shaped ascospores of *Pichia anomala* are also seen pink colored after staining with modified Ziehl-Neelsen stain (1% H_2SO_4). The ascospores of *Saccharomyces cerevisiae* are also seen as pink-colored. The spores of *Microsporidium* species are also stained by this method. In addition, cold AFB also stains the coccidian protozoa *i.e.*

cysts of *Cryptosporidium hominis*, *Cyclospora cayetanensis* and *Cystoisospora belli*. Moreover, Ziehl-Neelsen as well as Wade-Fite stainings are also used to demonstrate Medlar bodies of chromoblastomycosis.

3. Hematoxylin and Eosin Stain

Hematoxylin and Eosin (H&E) staining is one of the basic stainings used in any of the diagnostic pathology setting. It is used to stain nuclei in shade of blue by oxidized Hematoxylin (hematein) through mordant such as aluminium, iron, etc., followed by counter-staining by xanthene dye - Eosin, which colors in varying shades of red and pink with different tissue fibers and cytoplasm. A general background tissue reaction is seen hence it serves as the basic diagnostic staining technique. When there is any doubt that fungal elements are observed in the tissue based on H&E findings, special stains like PAS and GMS are supplemented for confirmation.

For nuclear stain it is necessary to oxidize hematoxylin -weakly anionic purple dye. This oxidation or ripening may be accomplished by natural process of exposure to light and air, over long period or by addition of oxidizing agents. Hematein is anionic having no particular affinity for nucleic acids of cell nuclei. It is necessary to combine metallic salt or mordant with hematoxylin to confer net positive charge to the dye-mordant compound. The cationic dye-mordant/metal complex will bind to anionic nuclear chromatin. The choice of hematoxylin is subjective decision, as all of them are capable of giving good results like Ehrlich's hematoxylin, Mayer's hematoxylin, Harris's hematoxylin, Cole's hematoxylin, Carazzi's hematoxylin and Gill's hematoxylin.

The differentiation of hematoxylin is done by using 1% acid alcohol. There are various forms of eosin available but most commonly used is water soluble (ws) Eosin Y (yellowish). Others are Eosin B (bluish), Ethyl Eosin, an alcohol-soluble form.

Staining Solution:

- **Hematoxylin Stain**

  Cole's Hematoxylin

Hematoxylin	3 gm
Ethyl alcohol	125 ml
Potassium or ammonium alum	160 gm
Iodine	1 gm
Distilled water	2000 ml

- **Eosin 1% aqueous**
- **Differentiator**

  HCl (1%) in 70% alcohol
- **Bluing agent**

  Scott's Tap Water

Sodium bicarbonate	2 gm
Magnesium sulfate	20 gm
Water	2000 ml

Staining Technique:

- Bring sections to water.
- Stain with hematoxylin solution for requisite time.
- Wash briefly in water and differentiate in acid alcohol (one rapid dip).
- Wash well in water and blue for 3-5 min.
- Check microscopically - nuclei should be deep blue with vesicular nuclei showing well-marked chromatin pattern.
- Wash well in running water and counterstain with eosin solution for 30 seconds to 1 min.
- Wash quickly in water, dehydrate in alcohol. Clear and mount with DPX.

Observance:

- Keratohyalin, nuclei, cytoplasmic RNA, some calcium salts, bacteria, fungi = Blue to purple
- Muscle, keratin, coarse elastic fibers, fibrin, fibrinoid = Deep pink/Bright red
- Collagen, reticulin, myelinated nerve fibers, amyloid = Pink
- Red blood cells = Reddish orange

4. May-Grunwald Giemsa Stain

Giemsa is a compound stain formed by interaction of methylene blue and eosin. On exposure to acids, alkalis and ultraviolet light, number of oxidation products (methylene azures) are formed from methylene blue and give contrast staining. It is commonly used to demonstrate intracellular yeasts cells of *Histoplasma capsulatum* and other fungi. The modified method *i.e.* May-Grunwald Giemsa **(MGG)** technique is now routinely used.

Staining Solutions:

May-Grunwald stock solution

May-Grunwald dye	1 gm
Methanol	300 ml

Dissolve the dye in methanol and heat up to 60°C for one hour.

Working dilution

Working dilution 1 : 1 (to be prepared fresh)

Giemsa stock solution

Giemsa dye	1 gm
Methanol	85 ml
Glycerol	54 ml

Mix the ingredients.

Working dilution

Working dilution 1 : 9 (to be prepared fresh)

Staining Procedure:

- Fix smears in methanol for 5 to 10 min.
- Stain in diluted MG solution for 10 min.
- Without washing transfer the slides to working Giemsa stain
- Stain in Giemsa solution for 15 min.
- Wash and differentiate by keeping the water on slides for 1-2 min. Air dry, mount and examine under microscope.

Observance:

Nuclei	Purple
Cell cytoplasm	Blue to mauve
Red blood cells	Pink

5. Diff-Quik Stain

This is differential stain consists of three reagents: Fixative, Stain Solution 1 and Stain Solution 2. This is 'dip' stain and should be dispensed in coplin jars.

Fix smear by immersing slide in Diff-Quik Fixative for 10 seconds. Blot edges of slide. Stain smear by immersing fixed slide in Diff-Quik Solution 1 for 20 seconds. Blot edges of slide. Counterstain by immersing slide in Diff-Quik Solution 2 for 20 seconds. Blot the edges of slide. Rinse slide gently in distilled water. Blot and allow to dry.

It rapidly stain and differentiate variety of smears, commonly blood and vaginal smears. It is based on modification of Wright stain pioneered by Bernard Witlin in 1970. It has advantages over older Wright Giemsa staining technique, as it reduces 4 min. process into simplified 15 second operation and allows for selective increased eosinophilic or basophilic staining depending upon time smear is left in staining solutions.

Solution Required:

- Fixative (Fast green in methanol) - pale green color
- Stain solution 1 (Eosin G in phosphate buffer) - red color
- Stain solution 2 (Thiazine dye in phosphate buffer)- blue color

Staining Procedure:

- Allow smears to dry
- Dip slide or tape-strip five times, for one second each, into Fixative. Allow excess to drain after each dip.
- Dip slide or tape-strip five times, for one second each, into Stain 1. Allow excess to drain after each dip.
- Dip slide or tape-strip five times, for one second each, into Stain 2. Allow excess to drain after each dip.
- Rinse slide or tape-strip in distilled water or Weise's buffer, pH 7.2.
- Stick tape strip to slide (sticky side down) and remove excess tape.
- Blot or allow to dry in air.
- Examine at low power to identify structures and then under oil immersion.

Observance:

Yeasts, bacteria and squames generally stain blue.
Inflammatory cells stain as for Romanowsky stains.

6. Periodic Acid-Schiff Stain

The principle of PAS stain is based on **Feulgen reaction**, wherein hydrolysis with hydrochloric acid liberates aldehydes which recolor Schiff's reagent. The polysaccharides of fungi and bacteria are oxidized by periodic acid to form aldehyde groups that yield magenta colored compounds with Schiff's fuchsin sulfite. The protein and nucleic acids remain unstained. The nuclei stain blue, fungi magenta or deep pink.

This stain is very useful for demonstrating fungi in tissue that are usually stained darker than surrounding tissues. The disadvantages are that many components of tissue that are carbohydrate in composition, are also stained by PAS stain (***See* Table C.2**). Moreover, actinomycetes such as *Nocardia* species are not stained by PAS method but are stained by methenamine stains.

Solution Required:

- Periodic acid (1% aq).
- Schiff's reagent:

Basic fuchsin	1 gm
Potassium/sodium metabisulfite	2 gm
Concentrated hydrochloric acid	2 ml
Activated charcoal	2 gm
Distilled water	200 ml

Heat distilled water, remove flame to avoid excessive effervescence and add basic fuchsin. Mix, cool to 50°C and add metabisulfite. Mix, cool to room temperature then add hydrochloric acid and charcoal. Leave overnight in darkness at room temperature. Filter and store at 4°C in dark container. The filtrate should be colorless and should give deep pink color when mixed with formalin which will check efficacy of reagent. In use, it will change from colorless to pale yellow to pink due to loss of SO_2 causing restoration of basic fuchsin color.

Staining Procedure:

- Bring sections to water.
- Oxidize with periodic acid solution for 5 min.
- Rinse well in distilled water.
- Treat with Schiff's reagent for 15 min.
- Wash in running tap water for 5 to 10 min. This intensifies color reaction.
- Stain nuclei with hematoxylin solution.
- Dehydrate in alcohol, clear in xylene and mount in synthetic resin medium.

Observance:

PAS-positive substance	Magenta/deep pink
Nuclei	Blue

7. Gridley's Fungal Stain

In Gridley's fungal stain mycelia, yeasts, elastic tissue and mucin as purplish red and background as yellow. This is like PAS staining but chromic acid is used as oxidizing agent. The aldehydes that are produced recolor Schiff's reagent, giving **fungi purplish red color**. The filamentous bacteria like actinomycetes are not stained by this technique.

The technique is combination of Bauer, Schiff and aldehyde fuchsin techniques. Some workers feel that Gridley technique reveals more of internal structure of fungi than other methods. The types of fungus demonstrated are paralleled by much simpler PAS technique. Another point to be noted is that elastic fibers and some connective tissue mucin stain purple, making fungus demonstration more difficult in tissues such as skin.

Staining Solutions:

Chromium trioxide (chromic acid - 4% aq)
Schiff's reagent
Aldehyde fuchsin solution
Saturated tartrazine in Cellosolve.
Sulfurous acid rinse:

Sodium metabisulfite (10% aq)	6 ml
Molar hydrochloric acid	5 ml
Distilled water	100 ml

Staining Procedure:

- Bring sections to water.
- Treat with chromic acid solution for 1 hour. Wash in tap water, then in distilled water.
- Treat with Schiff's reagent for 15 min.
- Treat with sulfurous acid solution for three changes over 6 min.
- Wash in water for 10 min.
- Treat with aldehyde fuchsin solution for 20-30 min.
- Wash in 50% alcohol, then in water.
- Wash in absolute alcohol and stain with tartrazine solution for half to one min.
- Wash in Cellosolve, dehydrate, clear and mount in DPX-type mountant.

Observance:

Fungi, elastin, some mucins	Purple
Background	Yellow

8. Gomori's Methenamine Silver Stain

This stain works on principle of liberation of aldehyde groups and their subsequent identification by reduced silver method, popularly known as **GMS** stain. It is used for the demonstration of polysaccharide content of fungus in tissue sections. The aldehydes reduce methenamine silver nitrate complex, resulting in brown-black, staining fungal cell wall due to deposition of reduced silver wherever aldehydes are located. Gomori prescribed an alcoholic fixative for glycogen however, Grocott in 1955 prescribed 10% formalin. Now-a-days **Grocott's** modification of GMS stain is usually used. Tendolkar *et al*, devised simplified Grocott's silver staining by the use of running tap water for washing, avoiding use of alcohol, xylene and expensive gold chloride which did not affect staining character of the fungi.

The fungi and bacteria are stained **black**, mucopolysaccharide dark grey and cytoplasm old rose and tissue pale green. GMS stain is better than other stainings as:

- This stains both live and dead fungi as compared to PAS which stain only live fungi.
- It also stains filamentous bacteria *i.e.* Actinomycetes (*Nocardia, Actinomyces, Streptomyces* and *Actinomadura*) which are not stained by other fungal stainings.

The only disadvantage of GMS is that the tissue reaction is not clearly visible. Therefore, sometimes, Gomori's methenamine silver stain may be combined with H&E to simultaneously visualize fungi along with the tissue reaction.

This is the method of choice for demonstration of fungi including *Pneumocystis* species. Chromic acid form aldehyde from certain structures, reduces hexamine silver mixture at alkaline pH which are selectively blackened. Over impregnation results in blackening of connective tissues, making fungal identification more difficult. Terminate treatment with hexamine-silver when fungi appear as dark brown color and background is still clear *i.e.* aim at slightly under impregnating. The inclusion of borate gives final pH of approximately 8.0.

Solutions Required:

Aqueous Chromium Trioxide:

Chromic acid (4%) = chromic acid (CRO_3) 4.0 g in 100 ml distilled water

Methenamine (3%) = Hexamethylene tetramine USP ($(CH_2)_6 N_4$ = 3 gm in 100 ml distilled water

Methenamine Silver Nitrate Stock Solution:

Silver nitrate solution 5% - 5 ml

Methenamine solution 3% - 100 ml

Keep stock solution refrigerated (life period is 1 to 2 months).

Working Solution:

Borax 5% solution 2 ml, distilled water 25 ml

Mix and add 25 ml of stock solution

Ferric chloride (0.1%)

Sodium metabisulfite (1% aq)

Sodium thiosulfate (5% aq)

Stock Light-Green

Light green SF (yellow)	0.2 gm
Distilled water	100 ml
Glacial acetic acid	0.2 ml

Working Light-Green

Light green stock solution	10 ml
Distilled water	50 ml

Staining Procedure:

- Bring test and control sections to water, oxidize in 5% chromic acid solution for one hour. Wash in water.
- Bleach in metabisulfite solution for 1 min.
- Wash in tap water for 5 to 10 min. then in four changes of distilled water.
- The working hexamine solution should have been pre-heated to 56°C for one hour using Coplin jar in water bath. Treat section in this solution at 56°C and examine after 10-20 min. and after that at three min. intervals until fungi are blackened but background is clear.
- Wash in six changes of distilled water and tone in 0.1% aqueous ferric chloride for 3 min.
- Wash in water, fix in sodium thiosulfate for 5 min. Wash again.
- Counterstain in light green for 1-2 min. Wash in water. Dehydrate, clear and mount as desired.

Observance:

Cellulose, chitin, fungi including *Pneumocystis jirovecii*, amoebae, some mucins, melanin, glycogen and starch stain **black** in a green background.

9. Mayer's Mucicarmine Stain

This is used for staining of *Cryptococcus* and *Rhinosporidium* species. *Cryptococcus* stains deep rose red, nuclei black, tissue yellow. In case of rhinosporidiosis, sporangium and endospores are stained by mucicarmine stain.

Being polysaccharide stain, it is used in case of *Cryptococcus* infection but for mucin demonstration it has little value. It uses complex solution, has lengthy staining time and is imprecise in staining. It has been shown that neutral mucin stain weakly or not at all. The strongly sulfated mucin is variable in their reaction, whilst other acidic mucin particularly hyaluronic acid stain strongly. The fact shows that like Alcian blue it also seems to be specific for mucin.

The solution incorporates aluminium salts and rationale of staining is one of positively charged carmine-mordant complex bonding with negatively charged acid mucin. The nuclei are well-stained to avoid subsequent masking by mucicarmine. Ehrlich's hematoxylin is not ousted as it stains some mucin.

Staining Solution:

Grind 1 gm carmine and place in large conical flask. Add 100 ml of 50% alcohol and mix. Add 1 gm aluminium hydroxide, mix and add 0.5 gm anhydrous aluminium chloride. Mix and boil gently for 2.5 min. Cool and filter. Store at 4°C where it can be kept for a period of 6 months or so.

Staining Procedure:

- Bring sections to water.
- Stain nuclei with an alum hematoxylin solution. Differentiate well in acid-alcohol and blue appropriately.
- Stain with mucicarmine solution for 20 min.
- Wash in water, dehydrate, clear and mount as desired.

Observance:

Mucins	Red
Nuclei	Blue

10. Masson-Fontana Stain

The Masson-Fontana Silver Staining (MFSS) is used to identify melanin containing phaeoid (dematiaceous) fungi in tissue or smear and popularly known as **MF stain**. Melanin possesses the ability to bind silver from silver solutions and reduce it directly to metallic silver without the use of an external reducer. The histopathological examination of tissue is one of the most reliable means of documenting invasive fungal infections. However, despite these advantages, histopathological stains are non-specific and they do not provide identification of fungal pathogens, which remain as problematic.

The phaeoid fungi are now among a group of emerging significant pathogens. These organisms are classified on the basis that they have **melanin** in their cell wall. In some cases, pigmentation of fungi in tissue may be weak and require staining with melanin-specific techniques such as Masson-Fontana stain to bring out this character prominently. The hyphae typically appear black, more fragmented in tissue as compared to hyaline fungi like *Aspergillus*, with irregular, septation and beaded, yeast-like forms. This is also pertinent to note that GMS stains all the fungal elements as black with greenish background, whereas MF stain only take up the phaeoid fungi as black in color with brownish background.

This is widely held that fungi, which are stained positive with MF stain and are classified as phaeoid organisms. Kimura and McGinnis have disputed this concept because MF process may stain many non-phaeoid fungi as well. These findings indicate that although MF stain may be useful for detecting phaeoid fungi but it should be used carefully and on selective basis, keeping in view that some non-phaeoid fungi may also be stained positively with this stain.

Masson's technique using Fontana's silver solution does not employ reducing bath and only demonstrates substances capable of reducing silver salt to metallic silver. This is a simple technique and its success depends on preparation of ammoniacal silver solution. A positive control section should be put up along with the test. A longer room temperature treatment is preferred for more precise results. This may also be written by its synonym *i.e.* Fontana-Masson stain and both the nomenclatures are correct.

Staining Solution:

- Silver solution. Add strong ammonia drop by drop to 20 ml of 10% aqueous silver nitrate, mixing well in between each drop, until formed precipitate almost re-dissolves. A faint opalescence will be evident when end point is reached. Finally, add 20 ml of distilled water, mix and filter. Store in dark container. The solution may be used for one month although fresh solution gives better results.
- Aqueous sodium thiosulfate (0.5%).
- Aqueous neutral red (1%).

Staining Procedure:

- Bring sections to distilled water.
- Treat with ammoniacal silver solution in dark container *e.g.* Coplin jar painted black, for 20-40 min at 56°C or overnight at room temperature.
- Wash well in several changes of distilled water.
- Treat with 0.5% sodium thiosulfate for 2 min. A stronger solution tends to bleach reduced silver.
- Wash, counterstain in neutral red solution for 3-5 min.
- Wash, dehydrate, clear and mount.

Observance:

Melanin, argentaffin granules/cells, chromaffin and some lipofuscin pigment stain **black** and nuclei as red.

11. Schmorl's Melanin Stain

This is one of the melanin-based stainings like Masson-Fontana. It is done to test for the presence of reducing substances in tissues like melanin, argentaffin granules, thyroid colloid, keratin, keratohyalin and lipofuscin pigments. The ferricyanide is converted into ferrocyanide, which is further converted to insoluble Prussian blue in the presence of ferric ions. The phaeoid (dematiaceous) fungi contain melanin in their cell wall hence this stain may also be used like MF stain for examining biopsy tissue or smear to ascertain the diagnosis of phaeohyphomycosis.

The commonly used Masson-Fontana method relies upon the melanin granules to reduce ammoniacal silver nitrate but argentaffin, chromaffin and some lipochrome pigments also get stained black. But Schmorl's method uses the reducing properties of melanin to stain granules as bluish-green. The Schmorl's solution is from Lillie's modification of Schmorl's ferricyanide reduction method used for demonstrating tissue reducing substances.

Staining Solution:

- Ferric chloride (1% aq) 37.5 ml
- Potassium ferricyanide (1% aq) 5.0 ml
- Distilled water 7.5 ml
- Make up fresh

Staining Procedure:

- Sections to water.
- Immerse sections in solution in a Coplin jar, examine microscopically at 30 sec intervals - up to 5 min.
- Wash well in running tap water.
- Treat with 1% acetic acid - 5 min.
- Counterstain with nuclear fast red - 5 min.
- Wash in water.
- Dehydrate, clear and mount sections in DPX.

Observance:

Melanin (phaeoid/dematiaceous fungi), other reducing substances like lipofuscins, liver bile, hematoxylin argentaffin cells, chromaffin cells and sulfydryl groups take varying shades of **blue** color. The nuclei are stained as pinkish-red and cytoplasm as pale-pink.

12. Toluidine Blue O Stain

This stain is primarily used for rapid detection of ***Pneumocystis jirovecii*** in lung biopsy, imprint smears and BAL fluid. It stains the cysts as reddish blue or dark purple against a light blue background. The cysts are usually clumped and may be punched in, appearing crescent shaped. The trophozoites are not discernible. It is easy and rapid method yielding reliable results.

Staining Procedure:

Place the air dried slide in sulfation reagent (45 ml of glacial acetic acid mixed with 15 ml of concentrated sulfuric acid) for 10 min. Then rinse in cold water for 5 min, drain and place in Toluidine blue O (0.3 gm of dye in 60 ml of H_2O) for 3 min. Rinse in 95% ethanol, followed by absolute ethanol and then xylene. Dry slide and observe under oil immersion.

Observance:

Pneumocystis jirovecii cysts	Purple
Tissue remnants, debris	Blue
Trophozoites	Not discernible

Toluidine Blue O stain reveals only the cyst form of the organism and staining of background material in smears often makes identification difficult hence GMS, monoclonal antibody and calcofluor white stains are usually preferred.

13. Acridine Orange Stain

The acridine dye has an affinity for nucleic acids usually used in bacteriology and oncology. When fungi are stained with this dye and viewed under ultraviolet light, RNA component of cells fluoresces with shades of orange and red, while DNA component gives green shades.

The fluorescent staining of exfoliative material particularly that of female reproductive tract, has gained popularity owing to clarity with which fluorescence of RNA is seen. Rapidly growing cells *e.g.* malignant cells have an increased cytoplasmic RNA content and this has afforded mean by which malignant cells can be detected. The early popularity of technique has waned due to impermanence of preparations and interpretation, particularly of squamous carcinoma cells still requires expertise and experience on a level with other techniques. The technique requires pH of 6.0 for differential staining of RNA and DNA. Formalin fixed material does not stain satisfactorily, neither does material fixed in Bouin's solution. Alcohol is fixative of choice.

Staining Solution:

(i) Acridine orange solution

Aqueous acridine orange (0.1%). Before use dilute one part stain with 10 parts of pH 6.0 phosphate buffer to make 0.01% solution.

(ii) Buffer

Phosphate buffer (pH 6).

(iii) Differentiator

Calcium chloride (0.1 M).

Staining Procedure:

- Alcohol fixed smears to distilled water.
- Rinse in 1% acetic acid for few seconds then two changes of distilled water over 1 min.
- Stain in diluted acridine orange solution for 3 min.
- Rinse in pH 6.0 buffer for 1 min.
- Differentiate in 0.1 M $CaCl_2$ solution for half a min.
- Wash in phosphate buffer and mount in same and examine under fluorescence microscope.

Observance:

DNA	Yellow-green
RNA, mucins	Red

14. Fluorescent-Antibody Stain

This staining may be used to detect fungal antigen in clinical material such as pus, blood, CSF, tissue impression smears and in paraffin sections of formalin fixed tissue where number of organisms is scanty. However, it is less satisfactory for sputum specimens. Like antibody-coated bacteria, antibody-coated fungi can also be demonstrated in urine in similar way *e.g. Candida* cells coated with antibody in urine. However, its use is limited by restricted availability of specific antisera. The Division of Mycotic Diseases, CDC, Atlanta and other specialized laboratories, have developed battery of sensitive and specific immunofluorescence reagents for detection and identification of pathogenic fungi.

Further Reading

1. Bancroft JD, Gamble M. *Theory and Practice of Histological Techniques.* 6th ed. Churchill Livingstone, Oxford: 2008.
2. Bhavasar RS, Goje SK, Takalkar AA, *et al.* Detection of *Candida* by calcofluor white. *Acta Cytol.* 2010; **54:** 679-84.

3. Chick EW. Acridine orange fluorescent stain for fungi. *Arch Dermatol.* 1961; **83:** 305-9.
4. Dasgupta T, Sahu J. Origins of the KOH technique. *Clin Dermatol.* 2012; **30:** 238-41.
5. Desoubeaux G, Franck-Martel C, Caille A, *et al.* Use of calcofluor-blue brightener for the diagnosis of *Pneumocystis jirovecii* pneumonia in bronchial-alveolar lavage fluids: A single-center prospective study. *Med Mycol.* 2017; **55:** 295-301.
6. Grocott RG. A stain for fungi in tissue sections and smears using Gomori's methenamine-silver nitrate technic. *Am J Clin Pathol.* 1955; **25:** 975-9.
7. Hagler AN, Ahearn DG. Rapid Diazonium Blue B test to detect basidiomycetous yeasts. *Int J Syst Evol Microbiol.* 1981; **31:** 204-8.
8. Hamer EC, Moore CB, Denning DW. Comparison of two fluorescent whiteners, Calcofluor and Blankophor, for the detection of fungal elements in clinical specimens in the diagnostic laboratory. *Clin Microbiol Infect.* 2006; **12:** 181-4.
9. Harrington BJ, Hageage GJ. Calcofluor white: A review of its uses and applications in clinical mycology and parasitology. *Lab Med.* 2003; **34:** 361-7.
10. Hoch HC, Galvani CD, Szarowski DH, *et al.* Two new fluorescent dyes applicable for visualization of fungal cell walls. *Mycologia.* 2005; **97:** 580-8.
11. Kumar RS, Ganvir S, Hazarey V. *Candida* and calcofluor white: Study in precancer and cancer. *J Oral Maxillofac Pathol.* 2009; **13:** 2-8.
12. Lillie RD. *Histopathologic Technique and Practical Histochemistry.* 2nd edn. Blakiston, New York: 1954.
13. Lokuhetty MD, Alahakoon VS, Kularatne BD, *et al.* Ziehl Neelsen and Wade-Fite stains to demonstrate Medlar bodies of chromoblastomycosis. *J Cutan Pathol.* 2007; **34:** 71-2.
14. Naka W, Hanyaku H, Tajima S, *et al.* Application of neutral red staining for evaluation of the viability of dermatophytes and *Candida* in human skin scales. *J Med Vet Mycol.* 1994; **32:** 31-5.
15. Summerbell RC. The staining of filamentous fungi with Diazonium Blue B. *Mycologia.* 1985: **77:** 587-93.
16. Talwar P, Chakrabarti A, Sapra RK, *et al.* Calcofluor white stain for detection of fungi in clinical specimens. *Bull PGI.* 1991; **26:** 139-44.
17. Tendolkar UM, Gogate AS. Simplified Grocott's methenamine silver nitrate method for staining fungi. *Indian J Med Microbiol.* 1997; **15:** 127-9.
18. West KL, Proia AD, Puri PK. Fontana-Masson stain in fungal infections. *J Am Acad Dermatol.* 2017 April 6. doi: 10.1016/j.jaad.2017.02.052. PMID: 28392288.
19. Zerpa R, Huicho L, Guillen A. Modified India ink preparation for *Cryptococcus neoformans* in cerebrospinal fluid specimens. *J Clin Microbiol.* 1996; **34:** 2290-1.

Appendix C: Conventional Mycological Techniques

The procedures for laboratory investigation of mycotic infections are as such not very difficult to master. The methods summarized here are those, which are used in routine diagnostic laboratories. They have been proved to be efficient in experienced hands.

The molecular techniques, which are used for establishing diagnosis and taxonomic purposes, are not dealt in this Appendix and reference books may be consulted from as listed in Select Bibliography given as **Appendix F** of this Textbook.

The following techniques are adopted in clinical diagnostic mycology laboratory to identify fungal or pseudofungal isolates and establish their diagnosis of suspected infections.

1. Germ Tube Test

A number of simple and rapid tests have been devised for the identification of some of clinically significant fungal pathogens. The observation of germ tube production as a quick method for presumptive identification of *Candida albicans* isolates from clinical samples has been used successfully. This test is also a reliable method and is performed in less than three hours. This is also known as **Reynolds-Braude** phenomenon.

The culture of *Candida* species is treated with 0.5 ml mammalian serum from fetal bovine, horse, sheep or normal human being and incubated at 37°C for 2-4 hrs. A drop of suspension is taken on slide and examined under the microscope. The germ tubes are seen as long tube-like projections extending from the yeast cells. They grow at distal end. There is no constriction at the point of attachment to yeast cell. The germ tubes should not be confused with pseudohyphae **(*See* Fig. 20.7A and B and Table 20.5)**.

The germ tubes are formed within two hours of incubation in *C.albicans* and *C.dubliniensis*. The preparation should not be incubated longer than 4 hrs because other hyphae producing yeasts might begin to germinate beyond this time. It is also pertinent that positive and negative controls should be put up while performing this test. The production of germ tube is only a presumptive identification of *C.albicans* and *C.dubliniensis*. Moreover, expertise is required to recognize the germ tubes.

This is also observed that more than 5% *C.albicans* strain may be germ tube negative, while false positive results can occur with other species of this genus. As a consequence, in order to reach an unequivocal identification of *C.albicans,* additional morphological and physiological tests may also be put up simultaneously. It is the cheapest method as the several recently developed alternative diagnostic tests, such as chromogenic agar media or commercially available rapid identification tests, are considered to be too expensive for routine use and/or often characterized by inconsistent results.

A new milk medium has been used for the germ tube and chlamydospores production in *C.albicans*. Similarly, amino acid synthetic medium also exhibits germ tubes and mycelial development at 25°C.

2. Hair Perforation Test

This test is done to differentiate between dermatophytes, *T.mentagrophytes* and *T.rubrum* as well as *M.canis* and *M.equinum*.

Place a sterile filter paper strip into bottom of sterile petri dish and cover surface of strip with sterile distilled water. Add small portion of sterilized prepubertal or infant hair into water. Add 5-6 drops of 10% yeast extract to speed up the reaction. Inoculate colony directly on hair and incubate at 25°C for four weeks. Observe hair by making wet smear or LCB mount for presence of conical perforations of hair shaft **(*See* Fig. 10.12)**. This test is positive in *T.mentagrophytes* and *M.canis* and negative in *T.rubrum* and *M.equinum*.

3. Hair Bait Technique

This technique is used to isolate keratinophilic fungi from soil. It can identify those fungi, which utilize keratinous substance. The following procedure is adopted:

- Place soil sample in sterile petri dish.
- Spread on the top of soil about 30 pieces of 1-inch long, preferably horse's, hair after sterilization by autoclaving.
- Wet surface of soil with sterile, distilled water. Use enough water to wet entire soil area without flooding plate.
- Add more water and incubate at 30°C.
- When white to brown fungal growth appears around hair, transfer piece of hair along with fungal growth to surface of SDA plate. Scatter about 3-4 pieces on each agar plate. Bury one end of hair under agar surface.
- Take hair with growth for microscopic observation under LCB mount.
- When fungal colonies grow on SDA plate, make LCB mount and identify the fungus, accordingly. The fungi which appear on hair piece should be transferred to separate SDA plates for further identification.

4. Lactophenol Cotton Blue Mount

The LCB mounts are prepared to examine detail morphological features of fungi grown on culture medium. The mounts are prepared with plain LCB or with polyvinyl alcohol. A 'teased preparation' is the most common method used in mycology laboratory. A small portion of colony is picked and suspended in a drop of LCB solution on a slide. The mycelial mat is teased apart with dissecting or teasing needles, covered with coverslip and viewed under low and high powers of microscope for types of hyphae as well as conidia and manner these are attached to conidiophore or hyphae, which will often yield identification. Semi-permanent mount of the prepared plain LCB slide can be made by sealing edge of cover slip with nail polish. It can be kept for a fairly long time.

The PHOL stain is also used in place of LCB stain. PHOL (Pal, Hasegawa, Ono, Lee) stain (Pal *et al.*, 1990). This stain contains methylene blue 0.3 ml (3% aqueous solution), glycerol 3.0 ml and formalin (4%).

Lactofuchsin stain may serve the same purpose. The disadvantage of teased LCB mount is that arrangement of conidia, conidiophore and hyphae is dislodged making identification difficult upto species level. A safe and low distortion tape touch method for fungal slide mount is also used. This is simple, safe and easily executed mounting method using tape resulting in high-quality tape mount for microscopic study. Any of the mounting medium like LCB stain, PHOL stain, lactofuchsin or aniline blue, is used in this method.

5. Adhesive Tape Technique

A modification of Borelli's adhesive tape technique is used to make direct smear from site of superficial lesions among patients. A piece of scotch tape, 6-8 cm long × 1-2 cm wide, is applied on the area to be studied and immediately adhered by its glue side containing infected material, to a clean slide. Then, raising one of the edges, a small drop of KOH is placed between tape and glass. Then, press gently with cotton or gauze to uniformly distribute the dye and eliminate any excess. The whole procedure hardly takes one to two minutes. It is used in case of superficial fungal infections caused by *Malassezia* species or dermatophytes, subcutaneous fungal infections like chromoblastomycosis, lacaziosis and even for cutaneous manifestations of systemic infections like paracoccidioidomycosis.

The same procedure may be used for making LCB mounts from cultural isolates for fungi by taking material from colony itself. It permits observation of different biological structures and their normal distribution without distortions produced by other procedures like swabbing, scraping and teasing, which make the procedure almost equivalent to fungal slide culture when used from the colonies.

6. Fungal Slide Culture

For the fastidious fungal pathogens, much laboratory time is spent in attempting to induce sporulation on various culture media so that their typical morphological structures upto genus level can be studied. The mycelial fungi usually grow best on nutritionally enriched media, such as Sabouraud dextrose agar but overproduction of mycelia often results in loss of sporulation. If mycelial isolate fails to produce spores or other recognizable structures after two weeks, it should be subcultured to less rich medium to encourage sporulation. Media such as cornmeal, oatmeal, potato dextrose or malt extract agars may be used for this purpose.

The slide culture in Mycology is used to study undisturbed morphological details of fungi, particularly relationship between reproductive structures like conidial arrangement, conidiophore and hyphae. It is indicated

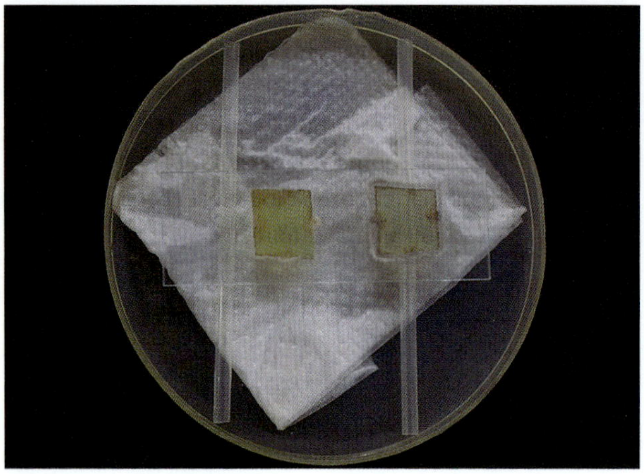

Fig. C.1. Two slide cultures put up for a mycelial fungal strain placed on a glass slide under the coverslips.

when teased mount of LCB is inconclusive in particular fungal isolate.

Place sterile microscopic slide on bent glass rod at the bottom of petri dish. A piece of one square centimeter block of cornmeal agar or potato dextrose agar or malt extract agar is put up on the slide. Inoculate the fungal strain under identification at four sides of agar block. Cover the inoculated block with sterile coverslip and incubate at room temperature (25°C) in the BOD incubator. Add a little sterile distilled water on filter paper placed at the bottom of petri dish to provide moisture to the growing fungus. Two blocks may also be placed instead of one to have wider space for inoculation **(Fig. C.1)**.

After 5-7 days when growth appears, put drop of LCB on new slide and place coverslip from block. Likewise a drop of stain is placed on glass slide of this culture after removing the agar block and put fresh coverslip for examining details under microscope to identify the fungus. The mycelia which adhere to glass surface usually show characteristic microscopic appearance, which may be lost if teasing needle are used, as it happens in routine LCB mounts. The slide culture may be directly examined by putting under low power of microscope.

The fungal slide culture is the most revealing technique in mycology. All non-sporulating mycelial fungi should be routed through it even if isolate is considered to be a contaminant. As the final diagnosis is usually established on the morphological features hence this technique in reality is the key of mycology and should never be compromised.

Now-a-days a new microculture technique is very popular. In this slide culture is directly inoculated on the deficient media plate (CMA, MEA, PDA, OMA). Cut small pieces of agar blocks from the corner of the plate and place on the same plate. Inoculate on the same way as in slide culture on four corners of the block and put coverslip on the block, incubate for 5-7 days at 25°C. In this technique 2-4 blocks can be used as per requirement.

The cellophane tape preparation has come into greater use to overcome obstacles of time consumption and requirement of extra equipment to prepare slide culture. A piece of tape is gently laid over portion of fungal colony and slowly lifted to remove an area of LCB on microscopic slide and covered with coverslip. This preparation becomes an instant slide culture, revealing relationship of various fungal structures.

It is also emphasized not to set-up slide culture on suspected *Coccidioides, Blastomyces, Histoplasma, Talaromyces marneffei* or any isolate mimicking these dimorphic fungi, as they grow well on the routine culture media and induction of sporulation is seldom required. Moreover, these can cause severe infection to the handlers while performing these techniques thereby pose a serious threat. Therefore, handling of slide culture and cellophane tape preparations should be done in a biosafety cabinet, particularly meant for BSL Class III pathogens. However, use of DNA based identification methods may replace procedure of slide culture in future.

7. Induction of *P.insidiosum* Zoospores

Pythiosis insidiosi is one of the pseudofungal infection with cutaneous, ophthalmic and vascular involvement. During the direct microscopy and histopathology, it is invariably confused with mucormycosis due to its hyphal similarities. Therefore, demonstration of its motile zoospores becomes imperative for establishing the final diagnosis. This techniques was originally devised by Mendoza *et al*, in the year 1988.

Induction of Zoospores: For the induction of zoospores, the leaves of the different grasses are cut into small pieces (10 by 4 mm) and placed in tubes of 24 hrs cultures on corn meal agar slants of *P.insidiosum* strains and incubated for 24 hrs at 37°C. The leaves are then transferred to 30 ml induction medium and incubated for 1-2 hrs at 37°C. The development of zoosporangia as well as releasing zoospores are examined under the light microscope.

Preparation of Induction Medium: The induction medium is prepared by adding two solutions; **Solution I** consists of 87.09 gm of K_2HPO_4, 68.05 gm of KH_2PO_4, 66.04 gm of $(NH_4)_2HPO_4$ in 500 ml of distilled water. **Solution II** is prepared by mixing 25.42 gm of $MgCl_2.6H_2O$ and 118.38 gm of $CaCl_2$ in 250 ml of distilled water. Then 0.5 ml of solution I is mixed with 0.1 ml of solution II in a 2-litre flask containing 1000 ml distilled water. The pH of resulting induction solution is ~7.1. The solution is stored at 4°C until used. The taxis behavior of zoospores can be monitored with light microscopy.

8. Serological Tests

The serological tests are useful for systemic fungal infections and have both diagnostic and prognostic values. Most of the immune responses that develop in an individual as a result of antigenic stimulus are highly specific. Based on this principle of specificity, several immunological tests have been developed which are performed *in vivo* as well as *in vitro* for diagnosis of fungal diseases. Most of the commercially produced kits detect antibody and can be difficult to interpret in opportunistic infections.

Generally serodiagnostic tests are referred to specialized or reference laboratories. These tests are used to demonstrate either antigens or antibodies to a particular fungal pathogen. The cryptococcal antigen tests may be useful in tertiary care hospitals where many AIDS patients are being treated. Similarly, antigens of other fungi in suspected clinical specimen may be detected. In the diagnostic laboratory agglutination, immunodiffusion or ELISA can demonstrate antibodies against *Candida* species.

9. Permanent Immunodiffusion Mount

After performing the immunodiffusion tests, wash precipitation bands in normal saline 3 to 4 times after every 8 to 10 hrs to remove nonspecific bands and wash it again with distilled water. The precipitation bands are kept in amido black staining solution for a few minutes. Wash out excess stain in saline followed by differentiation in 2% acetic acid. Wash gel again with water and place it on a glass slide and allow it to dry at 37°C so that it becomes a permanent mount as shown in **Figure 5.11**.

10. Exoantigen Tests

The exoantigens are defined as cell free antigens or soluble immunogenic macromolecules produced by fungi during their growth. These are readily detected in supernatant of culture broths or aqueous extracts of slant cultures. The exoantigens test depends on interaction between antigen produced by mycelial phase of fungi in culture and homologous antibodies, which are specifically generated to precipitate them. These can be detected even with contaminated or non-viable fungi. The technique reduces potential for exposure to bio-hazardous fungi by eliminating need for *in vitro* conversion or *in vivo* cultural inoculations/ manipulations used in experimental animals.

Table C.1. Exoantigens Specific for Pathogenic Fungi.

Systemic Dimorphic Fungi	*Specific Antigens*
Histoplasma capsulatum	H, M
Blastomyces dermatitidis	A
Coccidioides immitis	HS, HL (F)
Paracoccidioides brasiliensis	1, 2, 3

The exoantigens were introduced into mycology by Manych and Sourek in 1966 and later on developed by Kaufman and Standard in 1987 as tools of making identification of fungi causing systemic mycosis thereby resolving various taxonomic issues as well.

The principle of exoantigen test is that antibodies developed against particular mycelial antigens react specifically in gel immunodiffusion precipitin test (**Ouchterlony's** technique). The mold form of dimorphic fungi can be identified easily within a weeks' time, negating need of conversion to yeast phase. Monoclonal antibodies are exceptionally useful for characterizing species-specific exoantigens and for elucidating intraspecies antigenic relationship. Some of the species-specific exoantigens for identification of systemic dimorphic fungal pathogens like *Histoplasma capsulatum, Blastomyces dermatitidis, Coccidioides immitis* and *Paracoccidioides brasiliensis* are shown in **Table C.1**.

The exoantigen tests have also been found useful in the identification of culture of *Pseudallescheria boydii, Talaromyces marneffei, Sporothrix schenckii, Aspergillus fumigatus, Exophiala jeanselmei, Exophiala dermatitidis, Cladosporium* species and dermatophytes.

11. Uses of Animals in Medical Mycology

The experimental animals are frequently used in Medical Mycology laboratory for diagnostic purposes and research studies for a particular fungal disease. The use of animal models in training of students is not being recommended by international authorities due to pressure of social organizations for safety of the animals. There are international standard guidelines regarding handling of experimental animals. The carcasses of animals after use are to be disposed in plastic bags and incinerated. There should be an ethical clearance before use of experimental animals in particular study. The commonly used laboratory animals are mice, guinea pigs, rabbits and hamsters as referenced in respective Chapters. In addition, invertebrate hosts are also used, especially nematode *Caenorhabditis elegans* and model insects *Drosophila melanogaster* as well as *Galleria mellonella* to achieve the study goals in medical mycology.

12. Animal Inoculation Techniques

The inoculation of laboratory animal is of value in establishing pathogenicity of fungi as well as for observation of tissue phase of dimorphic fungi. In some cases where infected material fails to grow in culture, same may produce disease in animals.

Animals can be inoculated intracerebrally, intravenously, intraperitoneally or intratesticular with saline suspensions of organisms using syringe and needle of proper size. All instruments should be sterilized prior to use and care should be taken to see that required amount of suspension is injected into the animal. Disinfect instruments that are used during inoculation of animals. The following routes and quantities of saline suspension of fungi are proper for injection into mice:

(i) Intravenous: An inoculum of 0.2 ml or less depending on concentration with tuberculin syringe with 26 gauge needle. It is injected into one of four veins of tail of mouse.

(ii) Intracerebral: An amount of 0.02 ml suspension of fungal organisms is injected. The needle is inserted just above midpoint joining ear and outer canthus of eye mainly used in *Cryptococcus* species.

(iii) Intratesticular: An inoculum of 0.02 ml suspension in tuberculin syringe is injected. It is necessary in diagnosis of sporotrichosis infection as an animal pathogenicity test. Animals are sacrificed periodically to observe course of infection.

(iv) Intraperitoneal: An amount of 0.5 ml suspension of organisms with needle of 24 gauge. The needle is inserted through skin and peritoneum but not deep into abdominal cavity. Care should be taken that no injury is inflicted to viscera like liver, intestine and urinary bladder of animal.

In addition, intraperitoneal inoculation may also be done with gastric mucin, which is usually required to provide reproducible fungal infections in mice. A 5% hog gastric mucin may enhance virulence of fungi when suspension is inoculated into animal. Without mucin, infections would require the large numbers of organism provided by cultures that are undiluted or only marginally diluted.

Put 5 gm of gastric mucin in 95 ml of distilled water, emulsify with Waring Blender for 5 minutes. If necessary, centrifuge to remove particulate material. Autoclave for 15 minutes at 121°C and cool. Adjust pH to 7.3 with sterile NaOH. Check for sterility and store in refrigerator. Use equal parts of gastric mucin and fungus suspension and inject mixture (1 ml) intraperitoneally into laboratory animal using 23 or 24 gauge needle.

13. Isolation of Pathogenic Fungi from Soil

Soil screening through laboratory animals, especially mice, has been one of the most useful methods for increasing knowledge concerning natural habitats of pathogenic fungi. A number of modified procedures have been used including variations in ratio of soil sample to physiological saline solution, variations in antibiotics used and in length of time to hold mice prior to sacrifice and culture of fungi from internal organs.

A sample of soil is taken by pressing open mouth of sterile 4 oz screw-cap bottle or large test tubes against surface of superficial layer of soil and scooping up sample. Make up soil and physiological suspension in ratio of 1 : 5 without antibiotics. After thoroughly stirring with sterile glass rod or by shaking soil suspension manually after stoppering test tube, sample is allowed to settle for 1 hr.

A portion of supernatant (8 ml) is removed and combined with 2 ml of an antibiotic solution (2 mg streptomycin and 5 mg penicillin per ml of water).

One ml suspension is injected intraperitoneally into total of five mice. Sacrifice mice in one month period. Isolate portions of infected spleen, omentum, liver or other organs aseptically. Mince tissue with glass tissue grinder or scissor and spread same with inoculating needle on surface of medium. Incubate tubes at 25°C for 4 weeks before declaring it sterile thereby discarding.

14. Artifacts

There are large number of structures which may mimic fungal elements in clinical specimens and are called artifacts (UK = Artefact). For example, lipid layer between keratinized cells may give false appearance of pseudohyphae with irregular walls. Potassium hydroxide may form crystals in wet mount preparation of clinical material. Similarly pollen granules, white blood cells, powder from gloves and threads of fabric from patients' clothing may

Table C.2. Artifacts Resembling Fungal Agents.	
A.	**PAS-Positive Stained Materials**
	Corpora amylacea - common in CNS
	Russell bodies
	Ceroid bodies
	Starch granules
B.	**Silver-Positive Stained Materials**
	Calcified bodies-only seen with intense staining
	Elastic fibers - seen with intense staining
	Wood particles
C.	**Other Artifacts**
	Carbon particles in lung
	Suture material
	Talc
	Asteroid and Schaumann bodies in sarcoidosis

create confusion with hyphae. Moreover, calcium deposits, ceroid bodies and silver-stained elastin may be seen in stained preparations. Many tissue components, which are carbohydrate in nature, are stained with PAS staining. The commonly encountered artifacts are listed in **Table C.2**.

There are number of clues, which are helpful in distinguishing artifacts from fungi. The artifacts don't have well-defined cell walls or cross-walls. There is usually no definite internal structure in an artifact and it is not possible to visualize nuclei or cytoplasmic inclusions. Budding is also not seen in suspected yeast cell. The use of fluorescent stains can minimize artifact problems and increase fungal detection. Calcofluor white stain or Blankophor, which preferentially stain cellulose, one of components of cell wall, give fluorescence of bright yellowish-green color of fungal elements. The artifacts don't have morphological consistency when stained with various techniques as they are stained with one fungal stain and often don't take up an alternative stain.

When there is doubtful artifact on CFW the very same field is examined under the KOH hence by toggling between CFW and KOH more clearer view of the suspected fungus can be delineated.

Further Reading

1. Capilla J, Clemons KV, Stevens DA. Animal models: An important tool in mycology. *Med Mycol.* 2007; **45:** 657-84.
2. Kaufman L, Standard PG. Specific and rapid identification of medically important fungi by exoantigen detection. *Annu Rev Microbiol.* 1987; **41:** 209-25.
3. Manych J, Sourek J. Diagnostic possibilities of utilizing precipitation in agar for the identification of *Histoplasma capsulatum, Coccidioides immitis, Blastomyces dermatitidis* and *Paracoccidioides brasiliensis. J Hyg Epidemiol Microbiol Immunol.* 1966; **10:** 74-84.
4. Mendoza L, Prendas J. A method to obtain rapid zoosporogenesis of *Pythium insidiosum. Mycopathologia.* 1988; **104:** 59-62.
5. Sharma S, Balne PK, Motukupally SR, *et al. Pythium insidiosum* keratitis: Clinical profile and role of DNA sequencing and zoospore formation in diagnosis. *Cornea.* 2015; **34:** 438-42.
6. Zimmer BL, Roberts GD. Rapid selective urease test for presumptive identification of *Cryptococcus neoformans. J Clin Microbiol.* 1979; **10:** 380-1.

Appendix D: Antifungal Susceptibility Testing

The need for a reproducible, clinically relevant antifungal susceptibility testing has been prompted by increasing number of invasive fungal infections, expanding use of established and newer antifungals as well as recognition of resistance as a clinical challenge. The collaborative efforts of numerous investigators on Antifungal Susceptibility Testing (AFST) have generated consensus documents describing standardized methods for broth and agar-based testing. As a result, *in vitro* antifungal susceptibility testing is performed, which plays an important role in guiding therapeutic decision making, as an aid in drug development studies and as a means of tracking the development of antifungal resistance in epidemiological studies.

The antifungal susceptibility testing is performed to provide information that allows clinician to select an appropriate drug suitable for treating a particular fungal infection. The changing epidemiology of fungal agents, increasing drug resistance, availability of more and newer antifungals, increasing practice of prophylactic use have made it necessary to go for AFST for isolates obtained from clinical material. The agents may appear resistant *in vitro* and still have clinical efficacy. Sometimes, there has been no correlation between clinical response and susceptibility within a fungal species. This testing for fungi has not progressed to that level as far as tests used for deciding susceptibility of bacteria to antimicrobials.

There are many groups working on this issue of antifungal susceptibility testing and also have their networks in different continents. As the Clinical and Laboratory Standards Institute (CLSI) methods were being developed in the USA, which is re-designated name of National Committee for Clinical and Laboratory Standards (NCCLS) since January 1, 2005. Almost parallel, the subcommittee of the Europe began work on standard methods as well. The European Committee on Antimicrobial Susceptibility Testing (EUCAST) is the standing committee of the European Society of Clinical Microbiology and Infectious Diseases (ESCMID) to decide on all such issues.

CLSI and EUCAST Documents

Both the organizations, CLSI as well as EUCAST release official documents on various aspects of antifungal susceptibility testing from time to time. These documents highlight the procedure involved, break points for a particular drug *vis-a-vis* the fungal species and epidemiological cut offs. Initially, the CLSI documents are labeled as proposed (P), then replaced by approved (A) document after the concurrence from the concerned committee/authority, followed by informational supplement (S) in case some modifications have taken place in due course to time.

The EUCAST also releases its rationale documents on methods, ECOFFs and breakpoints from time to time. The current definitive documents on AFST methods is EDef 7.3.1 (Yeast) and EDef 9.3.1 (Mold), valid since January 15, 2017 replacing all its previously existing documents. In addition, from time to time EUCAST also publishes **technical note** in journals like *Clin Microbiol Infect* or *Mycoses*, on any of the antifungals, as need arises.

This is also pertinent to note that all the EUCAST documents are free of cost and easily available on its website *(http://www.eucast.org/ast_of_fungi)*. In contrast, the CLSI is purely a commercial organization and for every document payment has to be made to procure it *(http://www.clsi.org)*. A comprehensive list of documents released by CLSI as well as EUCAST is given in **Table D.1**.

At the outset it is pertinent to note that lots of difficulties are associated with AFST, which are already referred in **Chapter 6** and are also reproduced in **Table D.2**. Despite of all the difficulties, this testing is important for selection of an appropriate antifungal and as a method to detect development of resistance in certain organisms during the course of therapy. A simple, rapid susceptibility test is needed to find out possible resistance of fungi to commonly used antifungals.

Definitions Used in AFST

There is need to know about certain basic issues/ definitions before undergoing the methodology used for antifungal susceptibility testing:

(a) Susceptible (S): This is a category causative fungus, which implies that isolate is inhibited by usual achievable concentration of antifungal under normal recommended dosage.

(b) Intermediate (I): This category includes isolates which have MIC that reaches near attainable blood and tissue levels of drug but response rate is lower than susceptible isolates. It also includes strains for which available data do not permit them to be clearly recognized as susceptible or resistant.

(c) Susceptible Dose-dependent (SDD): This indicates that a given drug may be effective in patients that can be treated with higher than normal doses and where maximum blood or tissue concentrations has been achieved. The susceptible dose-dependent category relates to only the yeast testing and is not interchangeable with the intermediate category.

(d) Resistant (R): This is a category which implies that the isolate will not be inhibited by the usual achievable concentration of the antifungal under normal recommended dosage.

(e) Minimum Effective Concentration (MEC): This is defined as the lowest concentration of drug producing a macroscopically visible morphological change (aberrant growth). Mostly such terminology is used in cases of echinocandins. Aberrant growth typically continues through the higher concentrations.

(f) Clinical Breakpoints (CBP): The clinical interpretive breakpoints indicate that isolates are likely to respond to treatment with a given antifungal administered using approved dosing regimen for that agent. In medical practice, breakpoint values should predict ideally success or failure of an antifungal drug.

(g) "90-60 Rule": The "90-60 Rule" was proposed by Rex and Pfaller in 2002, which states that infections caused by isolates with MICs in susceptible range respond favorably to the appropriate therapy ~90% of the time, whereas infections caused by isolates with MICs considered to be as resistant can also respond favorably ~60% of the time.

(h) Phenotypes Associated with Azole Resistance: The trailing effect is seen with fluconazole and other azoles, echinocandins when there is difficulty in determining the endpoints or breakpoints. Isolates appear "susceptible" while taking reading at 24 hrs and "resistant" at 48 hrs. However, two independent investigations in murine models of candidiasis demonstrated that isolates should be characterized as "susceptible". Trailing occurs because turbidity continuously decreases as drug concentration increases but the suspension fail to become optically clear due to partial inhibition of growth. However, trailing can be minimized by taking reading at 24 hrs or adding methylene blue to it.

Table D.1. List of Documents of Released by CLSI and EUCAST Pertaining to Antifungal Susceptibility Testing.

1. CLSI Documents

M27-A3	Yeast broth dilution
M27-S4	QC and BPs
M38-A2	Mold broth dilution
M44-A2	Yeast disk diffusion
M44-S3	QC and BPs
M51-A	Mould disk diffusion
M51-S1	QC and ECVs
M57-Ed1	Principles & Procedures of ECVs
M59-Ed1	ECVs for AFST

2. EUCAST Documents

- Methods EDef 7.3.1 (Yeast) and EDef 9.3.1 (Mould)
- Rationale Documents (ECOFFs & breakpoints)

Table D.2. Difficulties Faced During Performing of Antifungal Susceptibility Testing.

1. Problems in relation to the fungal organism
   - *Slow growth rates in relation to bacteria*
   - *Dimorphism in fungal growth as yeast-mold*
2. Problems in relation to the antifungal agent
   - *Solubility in aqueous media*
   - *Stability of the antifungal agent*
   - *Partial inhibition of growth*
   - *Mechanism of action*
3. Test conditions that affect the MIC of antifungal agent
   - *Composition and pH of medium*
   - *Inoculum preparation and size*
   - *Incubation temperature and time*
   - *Endpoint criteria*
4. Lack of correlation between results of antifungal susceptibility testing and the clinical outcome

Some of the antifungal susceptibility testing methods are briefly described below:

I. CONVENTIONAL METHODS

In the diagnostic mycology laboratory the conventional methods are followed for performing the antifungal susceptibility testing. These are briefly described below:

(a) Broth Dilution Method

The nomenclature of the US National Committee for Clinical Laboratory Standards (NCCLS) was re-designated as Clinical and Laboratory Standards Institute (CLSI) at Wayne, Pennsylvania (USA) on January 1, 2005, which deals with susceptibility testing of various microorganisms including fungi.

(i) Yeast: The CLSI M27 document was proposed in 1992 after two multi-centric studies which provided data on critical facts affecting reproducibility of testing from one laboratory to another, including endpoint definition, inoculum preparation and size, incubation time, temperature and media. Presently used approved version of CLSI document (M27-A3) on standardized broth macrodilution and microdilution methods for antifungal susceptibility testing of yeasts was originally released in April, 2008.

The EUCAST method, although almost similar, incorporated some revisions to the CLSI method to include addition of a higher concentration of glucose to the RPMI-1640. This addition facilitates the rate of fungal growth allowing the MIC to be determined at 24 hrs as opposed to the original M27-A mandated 48 hrs. Studies have shown that the two methods are equivalent despite these differences and that a given set of isolates can expect the same categorical placement regardless of the method utilized. The equivalent of CLSI M27A is EUCAST Edef 7.2. The comparison between two methods is depicted in **Table D.3**.

(ii) Mycelial Fungi: The determination of MICs for filamentous fungi can be facilitated with a method that overcomes observer's bias and quantifies hyphal growth of molds. The CLSI had also proposed antifungal susceptibility testing of conidia forming filamentous fungi in 1998, which can be now performed as per M38-A2 document on a similar pattern as that of yeasts.

In dimorphic fungi, as the yeasts or spherules are the state present within the mammalian host, antimicrobial susceptibilities need to be performed with these pathogenic-phase cells, not the mycelia which may lead to erroneous conclusions. Hence testing of the pathogenic-phase *in vitro* should follow recently optimized procedures. Since yeast cells of the dimorphic fungi reside within host phagocytes, it is also advisable for *in vitro* tests to be followed with tests on drug effectiveness on intracellular yeasts, at least during initial drug development stages.

(iii) Fungal Strains Used as Control in AFST: While performing the antifungal susceptibility, various controls strains are required for a comparison. *Candida* species are used as representative genus for yeasts and *Aspergillus* for mycelial fungi. Most of these strains have the ATCC number as an identification. A list of strains both yeasts as well as molds is given in **Table D.4**.

Table D.3. Comparison of CLSI vs. EUCAST Methodologies for Antifungal Susceptibility Testing.

	CLSI M27-A3	*EUCAST Edef 7.2*
Format	Microdilution	Microdilution
Shape of Well (Bottom)	Round	Flat
Medium	RMPI 1640	RPMI 1640
Glucose content	0.2%	2%
Inoculum Size	$0.5\text{-}2.5 \times 10^3$	$0.5\text{-}2.5 \times 10^5$
Incubation Temp	35ºC	35ºC
Incubation time		
AMB, FCZ, Candins	24 hrs	24 hrs
Azoles	48 hrs	24 hrs
Cryptococcus	72 hrs	48 hrs
Reading	Visual	Spectrophotometric
Endpoint		
AMB	100% Inhibition	90% Inhibition
Azoles/Candins	50% Inhibition	50% Inhibition

Table D.4. Various ATCC Standard Fungal Strains used as Control in Performing Antifungal Susceptibility Testing.

1. *Candida* Species
- *Candida albicans* ATCC 90028
- *Candida albicans* ATCC 24433
- *Candida krusei* ATCC 6258
- *Candida tropicalis* ATCC 750
- *Candida parapsilosis* ATCC 90018
- *Candida parapsilosis* ATCC 22019

2. *Aspergillus* Species
- *Aspergillus flavus* ATCC 204304
- *Aspergillus fumigatus* ATCC 204305

In this Textbook, mainly the protocol of the CLSI is being followed. Although the **broth macrodilution** method was firstly proposed, the studies subsequently identified suitable quality control isolates which showed that there was good agreement between broth macrodilution and **microdilution** methods. Now-a-days, most of the laboratories use broth microdilution method only as it is more practical, less cumbersome with more consistent MIC results. The inter-laboratory agreement of the microdilution MICs has also been found to be much better than broth macrodilution methods. The procedure below is as per the CLSI Reference Method for Broth Dilution Antifungal Susceptibility Testing of Filamentous Fungi; Approved Standard - Second Edition (M38-A2) and for Yeasts; Approved Standard - Third Edition (M27-A3).

Procedure for Broth Microdilution Method

(a) Reagents Required: Antifungal drugs, RPMI-1640, Dimethyl sulfoxide (DMSO), sterile distilled water (DW), MOPS buffer, 0.85% normal saline, magnet stirrer, spectrophotometer, sterilized pipette tips, pipettes in ranges of 10-1000 μl, multi-channel pipette (optional for convenience), membrane filters, 96 well microtiter plates with lid, 1N NaOH solution for pH adjustments, micro-centrifuge tubes, beakers, bottles, flasks and test tubes.

(b) Antifungal Drugs: The reference drug powders should be obtained directly from the manufacturers to ensure it's a purity. The commercial clinical preparations or pharmacy stocks should not be used. The generic name, potency and expiration date of the drug should be noted. Powders should be stored properly as per the recommendations.

(i) Media Preparation

RPMI 1640 medium is used with glutamine and phenol red, without bicarbonate. It is supplemented with 0.2% glucose (1000 ml) buffered with 0.165 mol/L MOPS buffer.

The powdered RPMI 1640 medium (10.4 gm) is dissolved in 800 ml of distilled water and 34.53 gm of MOPS buffer is added, followed by stirring with a magnet stirrer without heating till it completely dissolves. Then, adjust the pH to 7.0 at 25°C by adding 1 mol/L NaOH (1N NaOH), prepared by dissolving 4 gm NaOH pellets in 100 ml of distilled water. Add distilled water to make up the total volume 1 litre. It is finally filter sterilized using a membrane filter of pore size 0.22 μm and stored at 4°C in bottles wrapped with aluminium foil.

Table D.5. List of Water Soluble and Insoluble Drugs.

Water Soluble Drugs	*Water Insoluble Drugs*
Caspofungin	Amphotericin B
Fluconazole	Anidulafungin
Flucytosine	Griseofulvin
Micafungin	Itraconazole
	Ketoconazole
	Posaconazole
	Terbinafine
	Voriconazole

(ii) Preparing Stock Solution of Antifungals

(a) Drug Weight Calculation: The drug powder is weighed as per calculation following formula:

$$W\,(mg) = \frac{V\ (ml) \times C\ (\mu g/ml)}{P\ (\mu g/mg)}$$

Where full form of:

W = Weight of drug powder required (μg/mg)

V = Volume of stock solution to be made (ml)

C = Concentration of stock solution (μg/ml)

P = Potency of antifungal powder (μg/mg)

Note: If potency of drug powder is mentioned as 98%, then it can be converted to μg/mg by multiplying it by 10 *i.e.* 98 x 10 = 980 μg/mg.

For example to make a 10 ml stock solution of amphotericin B having a concentration of 1600 μg/ml from a commercially available amphotericin B powder having a potency of 1000 μg/mg (100% potency), the weight of the powder required is = 10 x 1600 / 1000 = 16 mg.

Thus, 16 mg of amphotericin B powder having a potency of 1000 μg/mg when mixed with 10 ml of DMSO will give a solution of drug having a concentration of 1600 μg/ml.

Antifungal agents can be water soluble or insoluble and the concentration as well as preparation varies depending upon the water solubility as shown in the **Table D.5**.

The concentrations tested should include the breakpoint concentration and the drug levels achieved in the patient.

(b) Preparing Stock Solutions: Preparation of stock solutions is done as follows:

For Water Insoluble Drugs: Antifungal stock solutions for water insoluble drugs should be prepared at least

100 times the highest concentration to be tested. The measured amount of antifungal powder is dissolved in 3 ml DMSO in a test tube and labeled as "stock solution".

For Water Soluble Drugs: Antifungal stock solutions for these water soluble drugs should be prepared at least 10 times the highest concentration to be tested. The measured amount of antifungal powder is mixed first in 1 ml distilled water in a test tube. Then add 7 ml of RPMI 1640 prepared medium to it and label it as "T1" as shown in **Table D.6**.

(c) Preparing Dilution Series for Water Insoluble: Preparation of the dilution series of water insoluble drugs is done as follows as shown in **Table D.7**.

(d) Preparing Dilution Series for Water Soluble: Preparation of the dilution series of water soluble drugs is done as follows as shown in **Table D.8**.

(iii) Preparing Final Concentration of Drug Solutions

(a) For Water Soluble Drugs: Further dilution by 1/5 is done for T1 to T10 tubes by adding 4 ml of RPMI + 1 ml of drug dilution from each tube *i.e.* T1 to T10 into a new set of tubes labeled T1′ to T10′.

(b) For Water Insoluble Drugs: Further dilution by 1/50 is done for T1 to T10 tubes by adding 4.9 ml of RPMI + 100 µl of drug dilution from each tube *i.e.* T1 to T10 into a new set of tubes labeled TI′ to T10′

(c) For Final Concentration: For final desired concentration in each well requires further dilution by 1/2 achieved by adding 100 µl of each drug dilution and 100 µl of inoculum.

(iv) Inoculum Preparation

(a) Filamentous Fungi: Non-germinated conidial suspensions are prepared by a spectrophotometric procedure. A conidial test inocula in a range of approximately 0.4×10^4 to 5×10^4 give reproducible MIC.

Procedure: Culture on potato dextrose agar (PDA) for 48 hrs incubation or more (up to 7 days) induces conidia

Table D.6. Serial Dilutions of Water Soluble and Water Insoluble Antifungal Drugs.

Serial dilution for water insoluble drugs		Serial dilution for water soluble drugs	
Stock solution = 1600 µg/ml.		Stock = 5120 µg/ml	
T1	1600 µg/ml (Stock solution)	T1	1 ml stock + 7 ml RPMI-1640 (640 µg/ml)
T2	0.5 ml T1 + 0.5 ml DMSO (800 µg/ml)	T2	1 ml T1 + 1 ml RPMI (320 µg/ml)
T3	0.5 ml T1 + 1.5 ml DMSO (400 µg/ml)	T3	1 ml T1 + 3 ml RPMI (160 µg/ml)
T4	0.5 ml T1 + 3.5 ml DMSO (200 µg/ml)	T4	1 ml T3 + 1 ml RPMI (80 µg/ml)
T5	0.5 ml T4 + 0.5 ml DMSO (100 µg/ml)	T5	0.5 ml T3 + 1.5 ml RPMI (40 µg/ml)
T6	0.5 ml T4 + 1.5 ml DMSO (50 µg/ml)	T6	0. 5 ml T3 + 3.5 ml RPMI (20 µg/ml)
T7	0.5 ml T4 + 3.5 ml DMSO (25 µg/ml)	T7	1 ml T6 + 1 ml RPMI (10 µg/ml)
T8	0.5 ml T7 + 0.5 ml DMSO (12.5 µg/ml)	T8	0.5 ml T6 + 1.5 ml RPMI (5 µg/ml)
T9	0.5 ml T7 + 1.5 ml DMSO (6.25 µg/ml)	T9	0.5 ml T6 + 3.5 ml RPMI (2.5 µg/ml)
T10	0.5 ml T7 + 3.5 ml DMSO (3.125 µg/ml)	T10	1 ml T9 + 1 ml RPMI (1.25 µg/ml)

Table D.7. Preparation of the Dilution Series of Water Insoluble Antifungal Drugs.

		Antimicrobial solution				
Step	*Concentration (µg/ml)*	*Source*	*Volume (ml)*	*Solvent (ml) (DMSO)*	*Intermediate Conc. (µg/ml)*	*Final Conc. at 1:100 (µg/ml)*
1	1600	Stock			1600 µg/ml	16
2	1600	Stock	0.5	0.5	800	8
3	1600	Stock	0.5	1.5	400	4
4	1600	Stock	0.5	3.5	200	2
5	200	Step 4	0.5	0.5	100	1
6	200	Step 4	0.5	1.5	50	0.5
7	200	Step 4	0.5	3.5	25	0.25
8	25	Step 7	0.5	0.5	12.5	0.125
9	25	Step 7	0.5	1.5	6.25	0.0625
10	25	Step 7	0.5	3.5	3.13	0.0313

Table D.8. Preparation of Dilution Series of Water Soluble Drugs.

	Antimicrobial solution					
Step	*Concentration (µg/ml)*	*Source*	*Volume (ml)*	*Medium (ml)*	*Intermediate Conc.(µg/ml)*	*Final Conc. at 1:10 (µg/ml)*
1	5120	Stock	1 ml	7 ml	640 µg/ml	64
2	640	Step 1	1.0	1.0	320	32
3	640	Step 1	1.0	3.0	160	16
4	160	Step 3	1.0	1.0	80	8
5	160	Step 3	0.5	1.5	40	4
6	7160	Step 3	0.5	3.5	20	2
7	20	Step 6	1.0	1.0	10	1
8	20	Step 6	0.5	1.5	5	0.5
9	20	Step 6	0.5	3.5	2.5	0.25
10	2.5	Step 9	1.0	1.0	1.25	0.125

of various filamentous fungi producing growth with high spore concentrations. Take the screw cap test tubes containing around 5 ml of 0.85% normal saline and add a loopful growth of test organism. Now, add 10 µl of Tween 60 to the suspension (only in case of filamentous fungi and never for yeasts).

- Vortex these test tubes so that the spores get suspended and hyphal mass remains as a clump.
- Let the tubes stand for 5-10 minutes for the hyphal fragments to settle down.
- Transfer the upper homogeneous suspension into sterile screw cap tubes and vortex it for 5 seconds.
- Read and adjust the densities of conidial suspensions to an optical densities at 530 nm. OD value of a fungus depends upon the size of the conidia and ranges from 0.09-0.13 for *Aspergillus* and *Sporothrix* species; 0.15-0.17 for *Fusarium, Rhizopus* and other mucormycetes and *Pseudallescheria* species; 0.25-0.3 for *Histoplasma, Alternaria* and *Bipolaris* species.
- First add water to the cuvette and take the wavelength at 530 nm, set it at 0 and then add the prepared inoculum to the cuvette and take the OD reading at 530 nm. If it comes out to be less than desired OD then add measured amount of prepared inoculum and adjust it to the desired OD and if it comes out to be more than desired OD then add the measured amount of water and adjust it to the desired OD.
- Dilute these suspensions 1:50 in the standard medium (4.9 ml of RPMI and 100 µl of final prepared inoculum).
- Further dilution to 1:100 occurs when 100 µl of test inoculum and 100 µl of drug dilution are added to the wells of microtiter plate.

(b) Yeasts: A fresh 24 hrs yeast growth from Sabouraud dextrose agar at 35°C is taken for inoculum preparation. The yeast should be subcultured twice to ensure purity and viability. Pick up 5 colonies, each at least 1 mm in diameter from 24 hrs culture in case of *Candida* and from 48 hrs culture in case of *Cryptococcus*. Use of yeast nitrogen base for *Cr.neoformans* may enhance the growth and improve the clinical relevance of antifungal MICs.

Procedure: The optical density of yeast suspension in sterile 0.85% normal saline is either set by spectrophotometer to optical density 0.08 to 0.1 at 530 nm or matched against the 0.5 McFarland $BaSO_4$ turbidity standard. It produces a cell suspension of 1×10^6 to 5×10^6 cfu/ml. This suspension is diluted to 1:20 by adding 1.9 ml RPMI to 100 µl yeast suspension. This is further diluted by 1:50 by adding 4.9 ml RPMI to 100 µl of the diluted suspension.

(v) MIC Testing

- Mark the 96 well 'U' shaped microtiter sterile plate vertically with the test strains from A to H and horizontally from T1 to T12.
- Now add 100 µl of each prepared drug dilution vertically in all wells in one column for each dilution.
- Do the same for all dilutions from T1 to T10 in respective columns.
- Then add 100 µl of prepared inoculum in the wells marked T1 to T10 horizontally in one row for one fungal strain.
- Do the same for all the respective fungal strains from A to H.

Table D.9. Dilutions from T1 to T10 in Respective Columns and Controls in T11 and T12.

	T1	T2	T3	T4	T5	T6	T7	T8	T9	T10	T11	T12
A												
B												
C												
D												
E												
F												
G												
H												

- To one well T11 add only inoculum without drug as a growth control and to another well T12 add only RPMI medium without drug as medium control, as shown in **Table D.9**.
- After inoculation, it is advisable to mix the trays thoroughly to ensure smooth even suspension of the inocula.

Cover the microtiter plate with lid and incubate the plate at 35°C for 48 hrs. Incubation should be done in a moist chamber, especially not sealed. Take the reading at 24 hrs and again at 48 hrs. Read the results visually with the help of reading mirror. Reading in each well should be compared with that of growth control well.

All microdilution trays are incubated at 35°C and MICs are determined as described in the M38-A2 document and as recommended for some uncommon molds: (i) after 21 to 26 hrs for *Rhizopus* spp; (ii) after 46 to 50 hrs for most other filamentous fungi such as *Aspergillus* species, *Exophiala* species, *Fusarium* species, *Paecilomyces* species, as well as *Sporothrix schenckii* and (iii) after 70 to 74 hrs for *Scedosporium* species. Testing of the dimorphic fungi may require 5 to 7 days of incubation.

The numerical score, which ranges from 0-4, is marked using scale as given below:

0 - Optically clear
1 - Slightly hazy
2 - Prominent decrease in turbidity
3 - Slight reduction in turbidity
4 - No reduction of turbidity

The minimum inhibitory concentration for amphotericin B is defined as the lowest concentration in which the above score of zero is observed whereas for flucytosine and azoles as the lowest concentration in which score of two is observed and results are interpreted, accordingly. The MIC results of test isolates against the antifungal agent amphotericin B are shown in **Figure D.1**.

Amphotericin B MICs determined by the M38-A2 microdilution method are ≤ 1.0 µg/ml for most mold species. High amphotericin B MICs (> 2 µg/ml) have been reported for *Purpureocillium lilacinum,* most *Scedosporium apiospermum* isolates, *Lomentospora prolificans* and some isolates of *Alternaria* species, *Aspergillus* species (especially *A.terreus*), *Fusarium* species, *Talaromyces marneffei*, *S.schenckii* and *Phialophora* species. The MICS of itraconazole are also usually ≤ 2.0 µg/ml, the exceptions being high MICs for *Aspergillus ustus/calidoustus* (some isolates), *Fusarium solani, Fusarium oxysporum, Purpureocillium lilacinum, S.schenckii, Lomentospora prolificans* and some of the isolates of *Rhizopus arrhizus.*

When testing the echinocandins, most *Aspergillus* isolates show trailing growth and conventional MIC determination could categorize these trailing isolates as resistant to caspofungin. A more careful examination of the microdilution wells reveals the presence of compact, round micro-colonies. The hyphae grow abnormally as short, highly branched filaments with swollen germ tubes. The concentration of drug producing these morphologic changes is defined as the minimum effective concentration (MEC) to distinguish it from conventional MICs.

For Dermatophytes: Broth microdilution method for dermatophytes incorporates certain modifications as the use of oatmeal agar for inoculum preparation when testing *Trichophyton rubrum* in order to induce conidium formation and 4 to 5 days of incubation at 35°C for MIC determination (80% growth inhibition endpoints).

For Yeasts: Break points for the yeast are placed into categories susceptible, intermediate, susceptible dose-dependent and resistant. Originally, the MIC was determined at either 48 hrs for *Candida* species or 72 hrs for *Cr.neoformans.* It was established that incubation at 35°C for 48 and 72 hrs provided most consistent results for

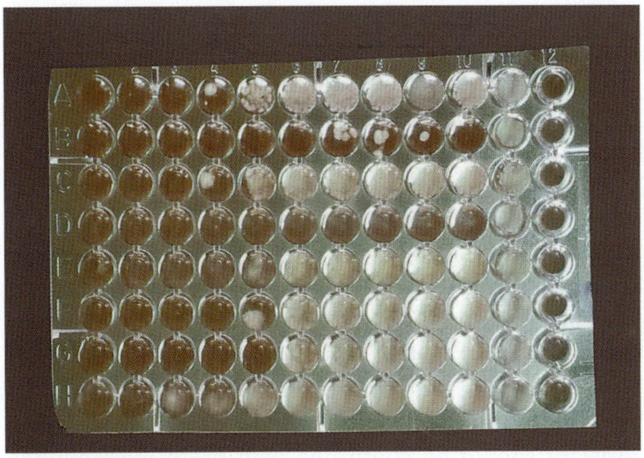

Fig. D.1. Microtiter plate showing MIC results of fungal isolates against the antifungal drug amphotericin B.

Candida species and *Cryptococcus* species, respectively and demonstrated that 48 hrs microdilution MIC was in close agreement with macrodilution MIC. While the recommendation for the candins permits only 24 hrs time point for determining the MIC, both amphotericin B and fluconazole may be read at either 24 or 48 hrs for *Candida* species. However, the M27-S4 document has been released in 2012 which provides updated guidelines after M27-A3 for broth dilution antifungal susceptibility testing of yeasts.

Therefore, based on the evidence available, the document has validated the 24 hrs reading time for broth microdilution of azoles and has also established the clinical breakpoints of five common species of *Candida* namely: *C.albicans, C.parapsilosis, C.glabrata, C.tropicalis, C.krusei* and *C.auris* for echinocandins, fluconazole and voriconazole. These new breakpoints are both drug and species specific, whereas the previous breakpoints were not. For most species, with the exception of *C.parapsilosis* and *C.guilliermondii* against the echinocandins, the breakpoints have been lowered, so that previously susceptible MICs are now classified as resistant. In addition, *C.glabrata* isolates are no longer considered susceptible to fluconazole but are rather classified only as either susceptible dose-dependent or resistant. Additionally, a goal of harmonizing the antifungal breakpoints with those set by EUCAST has been sought. Epidemiological cut off values (ECVs) for individual fungal species have been used to optimize the detection of non-wild type strains thus acquisition of resistance mechanisms and to prevent the breakpoints from dividing wild-type populations.

(b) Disk Diffusion Method

This method has widespread use for antibacterial drug testing. But this has limited application in antifungal drug susceptibility testing. Although useful for *in vitro* testing of flucytosine, early attempts to develop a standardized method for fluconazole were complicated by lack of reference method for comparison. The disk diffusion has potential to provide a simple means of performing *in vitro* tests with various antifungal agents. The method can be easily incorporated in the laboratories with limited resources where broth microdilution tests cannot be performed in routine. Sending the strains to reference laboratory will take substantial time, leading to delay in reports thereby treatment in such cases. In this method a disk is containing antifungal agent as routinely done in antibacterial susceptibility testing, which diffuses in the surrounding medium, inhibiting growth of fungi and measurements of zone of inhibition are taken, accordingly. The CLSI reference document for performing this method is M44-A.

***Procedure*:** The following media and other materials are required in this method:

Mueller-Hinton agar with 2% dextrose and methylene blue: CLSI M44-A2 method uses Mueller-Hinton agar supplemented with 2% glucose and 0.5 μg methylene blue/ml.

The increased glucose helps in improving yeast growth and the methylene blue supplementation provides sharper delineating zones surrounding the disks especially azoles. Medium has the advantage of being easily available because of its common use in antibacterial susceptibility testing in each laboratory, low cost and shows good batch to batch reproducibility.

1. Mueller-Hinton agar can be prepared from a commercially available dehydrated Mueller-Hinton agar base according to the manufacturer's instructions.
2. Dissolve 0.1 gm of methylene blue dye in 20 ml of distilled water and warm gently to dissolve. Do not overheat. Add 100 μl of this solution per liter of agar suspension.
3. Add 20 gm of glucose per liter of agar suspension.
4. Autoclave as directed by manufacturer's instructions.
5. Immediately after autoclaving, allow the agar solution to cool in a 45 to 50°C water bath.
6. Pour the freshly prepared and cooled medium into plastic, flat-bottomed petri dishes on a level, horizontal surface to give a uniform depth of approximately

4 mm. This corresponds to 67 to 70 ml of medium for plates with diameters of 150 mm and 28 to 30 ml for plates with a diameter of 100 mm.

7. The agar medium should be allowed to cool to room temperature and unless the plate is used on the same day of preparation, stored at refrigerator temperature (2 to 8°C). The agar medium should have a pH between 7.2 and 7.4 at room temperature. Plates should be used within seven days after preparation unless adequate precautions, such as wrapping in plastic, have been taken to minimize drying of the agar.

Inoculum Preparation: This is prepared like broth dilution method but the cell concentration is maintained as 10^6 cells/ml each in both, test as well control strains. The swabs soaked in inoculum are inoculated on plates, which are unable to produce confluent growth.

Antifungal Disks: The antifungal drugs, which are water insoluble like ketoconazole, itraconazole and amphotericin B are to be dissolved in dimethyl sulphoxide and rests are dissolved in distilled water like fluconazole and flucytosine. Subsequently 10 µl of each dissolved drug is poured to each disk to make 10 µg of antifungal drug on each disk. Ready made disks are also commercially available.

Susceptibility Testing Procedure: Optimally, within 15 minutes after adjusting the turbidity of the inoculum suspension, a sterile cotton swab is dipped into the suspension. The swab should be rotated several times and pressed firmly against the inside wall of the tube above the fluid level. This will remove excess fluid from the swab. The dried surface of a sterile Mueller-Hinton + GMB agar plate is inoculated by evenly streaking the swab over the entire agar surface. This procedure is repeated by streaking two more times, rotating the plate approximately 60° each time to ensure an even distribution of inoculum. As a final step, the rim of the agar is swabbed. With the help of forceps antifungal disks are placed onto the surface of the inoculated agar plate. Each disk must be pressed down to ensure its complete contact with the agar surface. Ordinarily, no more than 12 disks should be placed on a 150 mm plate or more than five disks on a 100 mm plate. Because the drug diffuses almost instantaneously, a disk should not be moved once it has come into contact with the agar surface.

The plates are incubated at 35°C for 24 hrs and measurements of inhibition zone are taken. Read at 48 hrs only when insufficient growth is observed after 24 hrs incubation.

Results and Interpretation: The zone diameters are measured to the nearest point at which there is prominent reduction in growth. Pinpoint colonies at the zone edge or larger colonies within a zone are frequently encountered and should be ignored. After measurement of zone of inhibition, results of antifungal susceptibility testing are interpreted according to following criteria:

Sensitive. The zone diameter of test strain is more than 80% of control strain.

Intermediate. The zone diameter of test strain is less than 80% of control strain.

Resistant. There is no zone of inhibition.

It is recommended that all the strains that appear resistant should be confirmed with broth microdilution method.

The CLSI M51-A document was released in 2010 as a reference for antifungal disk diffusion susceptibility testing for non-dermatophytic filament fungi that cause invasive infections. The standard disk diffusion method to test antifungal drugs for non-dermatophytic filamentous fungi isolates (M51-A and Supplement M51-S1) provides qualitative results in 8-24 hrs when caspofungin, triazoles and amphotericin B are used, faster than the CLSI reference microdilution method. Among *Aspergillus* species, a lower agreement of results produced by disk diffusion susceptibility tests was reported for *A.flavus* and amphotericin B or voriconazole. Amphotericin B to test *A.fumigatus* susceptibility also showed a lower agreement when compared to the reference method. Amphotericin B disks usually show the lowest correlation between MICs and inhibition zone diameters for filamentous fungi.

Epidemiological Cut Offs (ECVs or ECoffs)

Although clinical breakpoints help in guiding therapy, they do not distinguish between isolates with or without resistance mechanisms, nor do they always allow for their early detection. A wild-type strain is defined by EUCAST, as one in which there is an absence of acquired and mutational resistance mechanisms to the antimicrobial agent in question.

An ECV, also referred as 'ECOFF,' is MIC threshold value that allows the discrimination of wild-type (WT) strains, those without mutational or acquired resistance mechanisms, from non-WT strains, those having mutational or acquired resistance mechanisms. Therefore, ECVs represent the most sensitive measure of the emergence of strains with decreased susceptibility to a given fungal agent.

Non-wild type isolates often have high MICs or MECs that are above the ECV for the species/agent being studied so indicates a decreased susceptibility to the agent evaluated or potential phenotypic resistance. One advantage of defining normal ranges of MIC values is that non-wild-type organisms with acquired resistance mechanisms can be identified. These non-wild type organisms have higher MICs compared with the upper limit of the wild type distribution, which is defined as the epidemiological cut off value (ECV) MIC. The MIC value of the ECV generally encompasses at least 95% of the isolates in the wild type distribution. Because it is species specific, the ECV will not be altered by changing circumstances.

In recent times, CLSI has released two documents, which are important to highlight the principles, procedures and criteria of ECVs. These are as follows:

CLSI M57-Ed1: *"Principles and Procedures for the Development of Epidemiological Cutoff Values for Antifungal Susceptibility Testing, 1st Edition"* This guideline includes the criteria for developing and using epidemiological cut-off values for guiding clinical decisions when testing fungal species and antifungal agent combinations for which there are no breakpoints.

CLSI M59-Ed1: *"Epidemiological Cutoff Values for Antifungal Susceptibility Testing, 1st Edition"* This document includes the epidemiological cutoff value and quality control tables developed according to criteria provided in the Clinical and Laboratory Standards Institute guideline M57.

ECVs play an important role in tracking the emergence of resistance but ECVs do not correlate with clinical outcomes. However, the laboratory can warn the clinician that a resistance mechanism for this drug might be present. *In vitro*, ECVs should be both drug specific and species specific and should be able to identify isolates less likely to respond to a drug due to acquired drug resistance mechanisms when clinical breakpoints are unavailable. CLSI postulates that epidemiological cut off values may aid physicians in managing diseases caused by fungal species where breakpoints are not available.

The CLSI M59 is the document which gives criteria for epidemiological cut offs for antifungal susceptibility testing. The ECVs are normally established on the basis of the MIC distribution data (phenotype) created from testing isolates derived from geographically diverse laboratory surveys. They are determined on the basis of the pooled MIC distribution for each species/agent and incubation time combinations that comprise >100 MIC results that were obtained in at least three independent laboratories. ECVs have been given for *Candida* and *Aspergillus* species.

II. OTHER TESTING METHODS

In addition to the conventional methods, there are other methods also which can be utilized for performing antifungal susceptibility testing. These are briefly given below:

(a) Etest (Epsilometer Test)

The Etest is a commercial method for determination of MIC. It is set up in a similar manner as disk diffusion test except disk is replaced by a calibrated plastic strip impregnated with a concentration gradient of antifungal compound. Comparisons of MIC obtained by Etest and CLSI method have shown promising results for *Candida* species but marked disparity between Etest and broth dilution MIC for fluconazole in testing *Cryptococcus* species have been reported. RPMI 1640 agar supplemented with 0.2% glucose is recommended for use as medium for testing.

(b) Neo-Sensitabs

In this method both modified RPMI-1640 agar supplemented with 0.2% glucose and Mueller-Hinton agar with 2% glucose and 0.5 µg/ml methylene blue, have been used. Biomic plate reader is used to electronically read and interpret zones sizes. A large number of antifungals are available in tablets form *i.e.* amphotericin B (10 µg), ketoconazole (15 µg), itraconazole (8 µg), fluconazole (25 µg), voriconazole (1.0 µg), fluorocytosine (1 and 10 µg), clotrimazole (10 µg), miconazole (10 µg), econazole (10 µg), natamycin (10 µg) and nystatin (50 µg). However, individual disk zone sizes are often not able to differentiate between Susceptible (S) and Susceptible Dose-Dependent (SDD) isolates and the correlation between zone size and MIC is more variable. The Neo-Sensitabs is manufactured by Rosco Laboratory, Denmark *(www.rosco.dk)*.

(c) Colorimetric Methods

This is an alternative to the CLSI reference procedures in which growth of isolates is measured in cultures containing just one or two antimicrobial drug concentrations that distinguish resistant from susceptible strains.

(a) Fungitest: The Fungitest is a commercial micro-plate panel for break point testing of yeasts like *Candida* and

Cryptococcus species against various antifungals *i.e.* amphotericin B, flucytosine, fluconazole, itraconazole, ketoconazole and miconazole. The drugs are provided in dried form and are reconstituted by adding inoculum suspended in RPMI medium. A colorimetric redox indicator is incorporated in medium to facilitate end point interpretation. The growth assessment is based on reduction of the colored indicator, which turns the medium from blue to pink. When growth is inhibited by the antifungal agent, the medium remains blue.

(b) YeastOne Sensititre Assay: A commercial system that includes a colorimetric response is marketed as Sensititre YeastOne antifungal panel. It is based on broth microdilution methodology and is very easy to interpret. In this test, a chromogenic substrate indicator resazurin (Alamar Blue) redox marker facilitates endpoint interpretation by visual and by fluorescence reading. It changes from blue to pink as a result of growth and allows for the visual and spectrophotometric determination of MIC values.

The yeast growth in the antifungal solution is evident as a change in the colorimetric growth indicator from blue (negative) to red (positive). However, some yeasts may not change the indicator completely to red but display more of a purpling of the indicator. In addition, some organisms may show a slight purpling in posaconazole, voriconazole, fluconazole, itraconazole and ketoconazole.

Recently, a new antifungal susceptibility assay has been developed based on a broth medium which allows growth as well as reliable and convenient antifungal testing of all *Malassezia* species. This is compatible with fluorometric indicator resazurin thus enabling rapid and effective determination of MIC values by visual and by fluorescence reading.

(d) Spectrophotometric Methods

There are efforts made to determine MIC end points more objectively by reading broth microdilution plates with spectrophotometer. However, determination of spectrophotometric MICs requires selection of a level of inhibition and different studies have employed different end point definitions. Tetrazolium salts can penetrate rapidly into intact cells and directly into subcellular membrane with dehydrogenase activity, where they are converted to colored formazan derivative. Therefore, they are used as indicators of reducing system *i.e.* colorimetric assays.

One of these methods is an assay using dye 3-(4, 5-dimethyl-2-thiazyl)-2, 5-diphenyl-2H tetrazolium bromide (MTT). This yellow tetrazolium salt is cleaved by dehydrogenases inside mitochondria or in other cellular locations possessing dehydrogenase activity to form its purple formazan derivative that can be measured spectrophotometrically at 550 nm. MTT is cleaved by all living, metabolically active fungi independent of proliferation and irrespective of unicellular or multicellular growth and thus is a measure of metabolic activity.

Hence spectrophotometric determination of MICs could be a useful alternative to the visual endpoint reading if the inoculum is homogeneous. Recently spectrophotometry has been successfully used in EUCAST method for determining MIC of azoles against *Aspergillus fumigatus*, which may be tried for other molds also in due course of time.

(e) Flowcytometry

During last three decades, flowcytometry has been developed as a powerful tool in many diagnostic and research laboratories. A rapid assay of antifungal activity has been developed by utilizing flowcytometry to detect accumulation of a vital dye in drug damaged fungal cells. Various membrane potential sensitive or DNA binding vital dyes, such as FUN-1, propidium iodide, 3, 3'-Dipentyloxacarbocyanine iodide or acridine orange, are used.

The flow cytofluorometric method allows the detection of antifungal resistance and estimates the impact of the molecule tested on the viability and cell damage of the tested fungus. The biggest advantage is that the results can be derived in a short span of time of 4-6 hrs. It is suggested that flowcytometry may provide an improved, rapid method for determining and comparing antifungal activities of compounds with different modes of action. It is usually done via Fluorescence-activated cell sorting (FACS).

(f) VITEK 2 Yeast Susceptibility Test

Antifungal susceptibility testing for yeasts can be performed by automated methods such as VITEK 2 (bioMerieux, Hazelwood, USA). The procedure is performed as per manufacturer's documentation supplied with the equipment. It is fully automated commercial antifungal susceptibility testing system which has been found to be a reliable, reproducible and standardized procedure. By this method, identification of infecting species of *Candida* along with antifungal susceptibility profile is obtained within

24 hrs. Other automated system available is BacT/Alert automated blood culture system for antifungal susceptibility testing.

(g) Bioluminescence Assay

The ATP bioluminescence assay is based on the capacity to measure adenosine triphosphate (ATP) produced by the fungi. As such ATP is the chemical form of energy of all living cells hence it is present in more or less a constant amount in a cell. Therefore, its quantification is used to estimate the microbial population in a sample.

D-luciferin in the presence of the ATP undergoes conversion by luciferase to oxyluciferin that generates light. The quantity of the emitted light is measured by a luminometer and expressed as relative light unit (RLU) which can be converted into RLU/mole of ATP. Thus, there is a linear relationship between cell viability and luminescence measured. Its rapidity is the major advantage that provides quantitative results. Hence this method for determining antifungal susceptibility has been introduced where concentration of intracellular ATP is measured to determine drug action.

(h) Ergosterol Quantitation Method

The method is based on measurement of cellular ergosterol content rather than growth inhibition *i.e.* sterol quantitation method (SQM). Ergosterol is isolated from whole yeast cells by saponification and is finally identified by its spectrophotometric absorbance profile between 240 and 300 nm. The method has been investigated for assessment of activity of fluconazole and itraconazole for *Candida*, particularly heavy trailer *Candida* isolates.

Moreover, in contrast to broth microdilution technique, the sterol quantitation method determines trailing isolates to be susceptible rather than resistant, indicating that it may predict clinical outcome more accurately. Hence this may provide means to enhance current methods of fluconazole susceptibility testing and better correlation of *in vitro* with *in vivo* results, particularly for isolates with trailing end points.

Further Reading

1. Alastruey-Izquierdo A, Melhem MS, Bonfietti LX, *et al.* Susceptibility test for fungi: Clinical and laboratorial correlations in medical mycology. *Rev Inst Med Trop Sao Paulo.* 2015; **57 (Suppl. 19):** 57-64.
2. Albataineh MT, Sutton DA, Fothergill AW, *et al.* Update from the laboratory: Clinical identification and susceptibility testing of fungi and trends in antifungal resistance. *Infect Dis Clin North Am.* 2016; **30:** 13-35.
3. Arendrup MC, Park S, Brown S, *et al.* Evaluation of CLSI M44-A2 disk diffusion and associated breakpoint testing of caspofungin and micafungin using a well-characterized panel of wild-type and *fks* hot spot mutant *Candida* isolates. *Antimicrob Agents Chemother.* 2011; **55:** 1891-5.
4. Arendrup MC, Cuenca-Estrella M, Lass-Florl C, *et al.* EUCAST technical note on the EUCAST definitive document EDef 7.2: Method for the determination of broth dilution minimum inhibitory concentrations of antifungal agents for yeasts EDef 7.2 (EUCAST-AFST). *Clin Microbiol Infect.* 2012; **18:** e246-7.
5. Arendrup MC, Meletiadis J, Mouton JW, *et al.* EUCAST technical note on isavuconazole breakpoints for *Aspergillus*, itraconazole breakpoints for *Candida* and updates for the antifungal susceptibility testing method documents. *Clin Microbiol Infect.* 2016; **22:** 571.e1-4.
6. Arendrup MC, Meletiadis J, Mouton JW, *et al.* Technical note on the EUCAST definitive document EDef 9.3.1: Method for the determination of broth dilution minimum inhibitory concentrations of antifungal agents for conidia forming moulds. EDef 9.3.1 (EUCAST-AFST). Jan. 15, 2017.
7. Arthington-Skaggs BA, Jradi H, Desai T, *et al.* Quantitation of ergosterol content: Novel method for determination of fluconazole susceptibility of *Candida albicans. J Clin Microbiol.* 1999; **37:** 3332-7.
8. Bhatia VK, Sharma PC. Determination of minimum inhibitory concentrations of itraconazole, terbinafine and ketoconazole against dermatophyte species by broth microdilution method. *Indian J Med Microbiol.* 2015; **33:** 533-7.
9. Canton E, Espinel-Ingroff A, Peman J. Trends in antifungal susceptibility testing using CLSI reference and commercial methods. *Expert Rev Anti Infect Ther.* 2009; **7:** 107-19.
10. Clinical and Laboratory Standards Institute/National Committee for Clinical Laboratory Standards. Method for antifungal disk diffusion susceptibility testing of yeasts. Approved Guideline. Document M44-A. CLSI, Wayne, PA: 2004.
11. Clinical and Laboratory Standards Institute. Zone diameter interpretive standards and corresponding minimal inhibitory concentration (MIC) interpretive breakpoints. Supplement M44-S1. CLSI, Wayne, PA: 2006.
12. Clinical and Laboratory Standards Institute. Reference method for broth dilution susceptibility testing of yeasts: Approved Standard - 3rd edn. M27-A3. CLSI, Wayne, PA: 2008.
13. Clinical and Laboratory Standards Institute. Reference method for broth dilution antifungal susceptibility testing of filamentous fungi; Approved Standard - 2nd edn. M38-A2. CLSI, Wayne, PA: 2008.
14. Clinical and Laboratory Standards Institute. Methods for Antifungal Disk Diffusion Susceptibility Testing of Yeasts; Approved Guideline: Approved Standard - 2nd edn. M44-A2. CLSI, Wayne, PA: 2009.
15. Clinical and Laboratory Standards Institute. Performance Standards for Antifungal Disk Diffusion Susceptibility

Testing of Non-dermatophyte Fungi; Informational Supplement. M51-S2. CLSI, Wayne, PA: 2010.
16. Clinical and Laboratory Standards Institute. Reference method for broth dilution antifungal susceptibility testing of yeasts; fourth informational supplement (Document M27-S4). CLSI, Wayne, PA: 2012.
17. Cuenca-Estrella M, Rodriguez-Tudela JL. The current role of the reference procedures by CLSI and EUCAST in the detection of resistance to antifungal agents *in vitro*. *Expert Rev Anti Infect Ther.* 2010; **8:** 267-76.
18. de Vasconcelos AA Jr, Menezes EA, Cunha FA. Chromogenic medium for direct susceptibility testing of *Candida* spp. isolated from urine. *Mycopathologia.* 2011; **172:** 125-30.
19. Eschenauer GA, Carver PL. The evolving role of antifungal susceptibility testing. *Pharmacotherapy.* 2013; **33:** 465-75.
20. Espinel-Ingroff A, Aller AI, Canton E, *et al. Cryptococcus neoformans-Cryptococcus gattii* species complex: An international study of wild-type susceptibility endpoint distributions and epidemiological cutoff values for fluconazole, itraconazole, posaconazole and voriconazole. *Antimicrob Agents Chemother.* 2012; **56:** 5898-906.
21. Espinel-Ingroff A, Arendrup MC, Pfaller MA, *et al.* Using CLSI and EUCAST methods: Should the clinical laboratory be testing this agent? *Antimicrob Agents Chemother.* 2013; **57:** 5836-42.
22. Espinel-Ingroff A, Canton E. Comparison of Neo-Sensitabs tablet diffusion assay with CLSI broth microdilution M38-A and disk diffusion methods for testing susceptibility of filamentous fungi with amphotericin B, caspofungin, itraconazole, posaconazole and voriconazole. *J Clin Microbiol.* 2008; **46:** 1793-803.
23. Espinel-Ingroff A, Chowdhary A, Gonzalez GM, *et al.* Multicenter study of isavuconazole MIC distributions and epidemiological cutoff values for *Aspergillus* spp. for the CLSI M38-A2 broth microdilution method. *Antimicrob Agents Chemother.* 2013; **57:** 3823-8.
24. Espinel-Ingroff A, Chakrabarti A, Chowdhary A, *et al.* Multicenter evaluation of MIC distributions for epidemiologic cutoff value definition to detect amphotericin B, posaconazole and itraconazole resistance among the most clinically relevant species of Mucorales. *Antimicrob Agents Chemother.* 2015; **59:** 1745-50.
25. Espinel-Ingroff A, Turnidge J. The role of epidemiological cutoff values (ECVs/ECOFFs) in antifungal susceptibility testing and interpretation for uncommon yeasts and moulds. *Rev Iberoam Micol.* 2016; **33:** 63-75.
26. Fothergill AW, Sutton DA, McCarthy DI, *et al.* Impact of new antifungal breakpoints on antifungal resistance in *Candida* species. *J Clin Microbiol.* 2014; **52:** 994-7.
27. Hope WW, Cuenca-Estrella M, Lass-Florl C, *et al.* EUCAST technical note on voriconazole and *Aspergillus* spp. *Clin Microbiol Infect.* 2013; **19:** E278-80.
28. Iqbal NJ, Boey A, Park BJ, *et al.* Determination of *in vitro* susceptibility of ocular *Fusarium* spp. isolates from keratitis cases and comparison of Clinical and Laboratory Standards Institute M38-A2 and E-test methods. *Diagn Microbiol Infect Dis.* 2008; **62:** 348-50.
29. Kirkpatrick WR, Zimmerman JD, Haikal FP, *et al.* Screening for drug-resistant *Candida* yeasts with chromogenic agar. *Med Mycol.* 2010; **48:** 807-16.
30. Koga-Ito CY, Lyon JP, Resende MA. Comparison between E-test and CLSI broth microdilution method for antifungal susceptibility testing of *Candida albicans* oral isolates. *Rev Inst Med Trop Sao Paulo.* 2008; **50:** 7-10.
31. Kuper KM, Coyle EA, Wanger A. Antifungal susceptibility testing: A primer for clinicians. *Pharmacotherapy.* 2012; **32:** 1112-22.
32. Lamoth F, Alexander BD. Comparing Etest and broth microdilution for antifungal susceptibility testing of the most-relevant pathogenic molds. *J Clin Microbiol.* 2015; **53:** 3176-81.
33. Lamoth F, Damonti L, Alexander BD. Role of antifungal susceptibility testing in non-*Aspergillus* invasive mold infections. *J Clin Microbiol.* 2016; **54:** 1638-40.
34. Lass-Florl C. *In vitro* susceptibility testing in *Aspergillus* species: An update. *Future Microbiol.* 2010; **5:** 789-99.
35. Lockhart SR, Ghannoum MA, Alexander BD. Establishment and use of epidemiological cutoff values for molds and yeasts using the Clinical and Laboratory Standards Institute M57 standard. *J Clin Microbiol.* 2017; **55:** 1262-8.
36. Lyon GM, Karatela S, Sunay S, *et al.* Antifungal susceptibility testing of *Candida* isolates from the *Candida* surveillance study. *J Clin Microbiol.* 2010; **48:** 1270-5.
37. Meletiadis J, Leth Mortensen K, Verweij PE, *et al.* Spectrophotometric reading of EUCAST antifungal susceptibility testing of *Aspergillus fumigatus. Clin Microbiol Infect.* 2017; **23:** 98-103.
38. Morales BP, Trilles L, Bertho AL, *et al. In vitro* susceptibility testing of amphotericin B for *Cryptococcus neoformans* variety *grubii* AFLP1/VNI and *Cryptococcus gattii* AFLP6/VGII by CLSI and flow cytometry. *Mycoses.* 2015; **58:** 273-9.
39. Morales-Lopez SE, Taverna CG, Bosco-Borgeat ME, *et al. Candida glabrata* species complex prevalence and antifungal susceptibility testing in a culture collection: First description of *Candida nivariensis* in Argentina. *Mycopathologia.* 2016; **181:** 871-8.
40. Ochiuzzi ME, Santiso GM, Arechavala AI. Correlation of Etest and Neo-Sensitabs diffusion assays on Mueller-Hinton-methylene blue agar with broth microdilution reference method (CLSI-M27-A2) for testing susceptibilities of *Cryptococcus neoformans* to amphotericin B and fluconazole. *Med Mycol.* 2010; **48:** 893-6.
41. Oz Y, Dag I, Kiraz N. Broth microdilution and time-kill testing of caspofungin, voriconazole, amphotericin B and their combinations against clinical isolates of *Candida krusei. Mycopathologia.* 2012; **173:** 27-34.
42. Pappalardo MC, Szeszs MW, Martins MA, *et al.* Susceptibility of clinical isolates of *Cryptococcus neoformans* to amphotericin B using time-kill methodology. *Diagn Microbiol Infect Dis.* 2009; **64:** 146-51.
43. Pfaller M, Boyken L, Hollis R, *et al.* Use of epidemiological cutoff values to examine 9-year trends in susceptibility of *Aspergillus* species to the triazoles. *J Clin Microbiol.* 2011; **49:** 586-90.

44. Pfaller MA, Andes D, Diekema DJ, *et al.* Wild-type MIC distributions, epidemiological cutoff values and species-specific clinical breakpoints for fluconazole and *Candida*: time for harmonization of CLSI and EUCAST broth microdilution methods. *Drug Resist Updat.* 2010; **13:** 180-95.
45. Pfaller MA, Diekema DJ, Ghannoum MA, *et al.* Wild-type MIC distribution and epidemiological cutoff values for *Aspergillus fumigatus* and three triazoles as determined by the Clinical and Laboratory Standards Institute broth microdilution methods. *J Clin Microbiol.* 2009; **47:** 3142-6.
46. Pfaller MA, Diekema DL. Progress in antifungal susceptibility testing of *Candida* spp. by use of Clinical and Laboratory Standards Institute broth microdilution methods, 2010 to 2012. *J Clin Microbiol.* 2012: **50:** 2846-56.
47. Pfaller MA, Chaturvedi V, Diekema DJ, *et al.* Comparison of the Sensititre YeastOne colorimetric antifungal panel with CLSI microdilution for antifungal susceptibility testing of the echinocandins against *Candida* spp., using new clinical breakpoints and epidemiological cut off values. *Diagn Microbiol Infect Dis.* 2012; **73:** 365-8.
48. Pina-Vaz C, Rodrigues AG. Evaluation of antifungal susceptibility using flowcytometry. *Methods Mol Biol.* 2010; **638:** 281-9.
49. Posteraro B, Sanguinetti M. The future of fungal susceptibility testing. *Future Microbiol.* 2014; **9:** 947-67.
50. Posteraro B, Spanu T, Fiori B, *et al.* Antifungal susceptibility profiles of bloodstream yeast isolates by Sensititre YeastOne over nine years at a large Italian teaching hospital. *Antimicrob Agents Chemother.* 2015; **59:** 3944-55.
51. Rex JH, Pfaller MA, Rinaldi MG, *et al.* Antifungal susceptibility testing. *Clin Microbiol Rev.* 1993; **6:** 367-81.
52. Rex JH, Pfaller MA, Walsh TJ. Antifungal susceptibility testing: Practical aspects and current challenges. *Clin Microbiol Rev.* 2001; **14:** 643-58.
53. Rodriguez-Tudela JL, Arendrup MC, Barchiesi F, *et al.* EUCAST definitive document EDef 7.1: Method for the determination of broth dilution MICs of antifungal agents for fermentative yeasts. *Clin Microbiol Infect.* 2008; **14:** 398-405.
54. Sanguinetti M, Posteraro B. New approaches for antifungal susceptibility testing. *Clin Microbiol Infect.* 2017 April1. doi: 10.1016/j.cmi.2017.03.025. PMID: 28377311.
55. Santos ER, Dal Forno CF, Hernandez MG, *et al.* Susceptibility of *Candida* spp. isolated from blood cultures as evaluated using the M27-A3 and new M27-S4 approved breakpoints. *Rev Inst Med Trop Sao Paulo.* 2014; **56:** 477-82.
56. Song YB, Suh MK, Ha GY, *et al.* Antifungal susceptibility testing with Etest for *Candida* species isolated from patients with oral candidiasis. *Ann Dermatol.* 2015; **27:** 715-20.
57. Tan TY, Hsu LY, Alejandria MM, *et al.* Antifungal susceptibility of invasive *Candida* bloodstream isolates from the Asia-Pacific region. *Med Mycol.* 2016; **54:** 471-7.
58. Thatai P, Sapra B. Critical review on retrospective and prospective changes in antifungal susceptibility testing for dermatophytes. *Mycoses.* 2016; **59:** 615-27.
59. Tewari A, Behera B, Mathur P, *et al.* Comparative analysis of the Vitek 2 antifungal susceptibility system and E-test with the CLSI M27-A3 broth microdilution method for susceptibility testing of Indian clinical isolates of *Cryptococcus neoformans. Mycopathologia.* 2012; **173:** 427-33.
60. Wanjare S, Gupta R, Mehta P. Caspofungin MIC distribution amongst commonly isolated *Candida* species in a tertiary care centre - An Indian experience. *J Clin Diagn Res.* 2016; **10:** DC11-3.
61. Xiao M, Fan X, Chen SC, *et al.* Antifungal susceptibilities of *Candida glabrata* species complex, *Candida krusei*, *Candida parapsilosis* species complex and *Candida tropicalis* causing invasive candidiasis in China: 3 year national surveillance. *J Antimicrob Chemother.* 2015; **70:** 802-10.

Appendix E: Quality Controls, Preservation and Culture Collections of Fungi

In a diagnostic medical mycology laboratory, quality control, preservation and culture collection of fungi is immensely pertinent. The adoption of these measures brings efficiency as well as accuracy of the mycology laboratory. These may be followed in the laboratory at its own level and/or in collaboration with any of the national or international reference center. A brief account of all such practices is given in this **Appendix E**.

I. Quality Controls in Medical Mycology

The quality control regulates the accuracy of the functioning of a diagnostic laboratory. It is a regular and periodic exercise done in almost all sections of the microbiology laboratory. It is part and parcel of good laboratory practices. Rather it acts as a deterrent on the wrong habits and practices being followed unintentionally in medical mycology laboratory.

The quality control in any laboratory is of two types: internal and external. The Internal Quality Control (IQC) along with External Quality Assessment Scheme (EQAS) is a very important tool for maintaining optimal quality services of any diagnostic laboratory. Hence for any clinical laboratory aspiring to improve its quality standards and provide the best possible results, participation in an EQAS is an important activity towards achieving this goal. The EQAS helps the laboratories to assess their accuracy as well as bias of their results and stability of methods over a longer period of time.

These serve as educational tools by identifying staff training needs and helps to monitor and improve the performance of the laboratory by providing early warning for systematic problems associated with the respective methods. They also provide assurance to consumers both physician as well as patients that the laboratory results are reliable by establishing objective evidence of testing quality. The participation in EQAS program is a pre-condition, rather mandatory for accreditation as they complement each other dynamically to ensure high quality standards. Eventually, a satisfactory EQAS results ensure that all the requirements of the accreditation standards are met with by the diagnostic mycology laboratory.

In India, EQAS program in the field of Medical Mycology was started by Prof H C Gugnani from the Department of Medical Mycology, VPCI, University of Delhi in mid-1990s, which continued for substantial number of years. However, at present, it is being run by the Mycology Division, Department of Medical Microbiology, PGIMER, Chandigarh, wherein five samples are sent to all the participating laboratories, twice a year *i.e.* January and July. These fungal strains are to be identified within two months up to genus and species level. Additionally, one of these fungus (invariably yeast) is meant for the antifungal susceptibility testing also. The individual results of the participating laboratories are kept confidential however, cumulative results are reported just for the introspection of the program.

II. Preservation of Fungi

In a diagnostic medical mycology laboratory, isolation of fungi is a constant process. In due course of time a large number of fungal strains are collected. Therefore, it is very pertinent to keep all these strains alive for further antifungal susceptibility testing, molecular identification and/or epidemiological studies for short-term as well as long-term. Moreover, besides viability of fungal strain, another principal requirement for successful maintenance is the ability to preserve their features remain unchanged *i.e.* growth, morphology, production of metabolites, etc.

The storage of fungi shows some difficulties like contamination by other fungi, bacteria and/or mites as well as pleomorphic changes in the fungal strain itself. The commonest method for storing fungi in routine diagnostic laboratories is periodic serial sub-culturing whereas reference laboratories use other methods like covering with mineral oil, storage in distilled water, storage in glycerol (10%) at -20°C and -80°C, lyophilization and storage under liquid nitrogen. Although sub-culturing is the simplest method but culture is to be transferred to new medium periodically hence it is cumbersome and time consuming. If there is some irregularity in this process, viability and sporulation of fungal strain may be lost in due course of time.

The stock cultures of medically important fungi are essential source material of quality control organisms, for reference cultures to be used in comparative identification, for production of known metabolites and also for teaching collections. When fungi are stored for long period, it is ideal to preserve in duplicate using more than one method so that chances of losing strain are minimized.

In routine laboratory, fungi are grown on Sabouraud dextrose agar and repeatedly subcultured to keep them alive after gap of a few weeks. But it is not an ideal medium for maintaining cultures for long period as it is nutritionally rich medium. For long-term maintenance, media should have less nutrition *i.e.* PDA, Oatmeal Agar, MEA, CZ, PCA, etc., so that heavy sporulation takes place because fungi produce spores in unfavorable circumstances and remain alive for longer period of time.

The primary methods of culture preservation are basically three types (i) continuous growth (ii) drying and (iii) freezing. The continuous growth methods are simple and inexpensive because as such no specialized equipment is required. The drying based methods are the most useful for preservation that produce spores or other resting structures. The cultures can be dried *i.e.* desiccated by removal of water and by prevention of rehydration using soil, silica gel, sand, glass beads, paper strips, grains and various other materials as carriers. Moreover, drying methods are also technically simple and do not require expensive equipment. These methods together with storage under mineral or paraffin oil, in sterile water or cool storage in a standard refrigerator at 4-8°C may be suitable in the absence of alternatives. The freezing based methods are versatile and widely applicable. Most fungi can be preserved, with cryo-protectants, in liquid nitrogen or in standard -20°C and -80°C freezers. In the freeze-drying (lyophilization), the fungal cultures are frozen and subsequently dried under vacuum. The method is highly successful with cultures that mainly produce spores.

Many methods of preserving fungal cultures are used. These methods vary according to organism represented, size of the culture collection and economic considerations. The use of sterile distilled water at room temperature show good results in case of yeasts and sporulated mucormycetes. For yeasts 3-4 days growth is the best for preservation. The preservation of *Malassezia pachydermatis* can be done using skimmed milk as the cryoprotective agent.

However, some of the dermatophytes like *Epidermophyton* species and many mucormycetes may not survive in freezing. Sabouraud agar without dextrose can also be used as medium for maintaining stock cultures. In the later instance, cultures are grown on potato-dextrose agar until mature for several days to two weeks and spores are scraped or washed off from surface of slants by using sterile distilled water and pipette. The resulting fungal suspensions can be sealed in sterile vials to prevent evaporation and stored at room temperature. A less commonly used technique is to cover an agar slant culture of fungus with sterile mineral oil.

For long-term preservation of fungi, preparations of lyophilized cultures and storage at -70°C or in liquid nitrogen at ultra low temperature are well-established procedures, which require cryo-preservatives, expensive equipment such as lyophilizers or liquid nitrogen freezers. However, cultures stored by these techniques may be kept for many years without losing viability, biochemical or morphological characteristics. Lyophilization is considered to be ideal method but not for the fungi having very large propagules. Moreover, procedure is complicated and infrastructure is not available in routine diagnostic mycological laboratories.

The **Castellani's method** is suitable for small collections and is very economical also. It is very useful for preservation of variety of species of pathogenic fungi for humans and animals. In this method, pieces of fungal cultures are introduced in screw cap tube with distilled water and left at room temperature at 25–28°C. Therefore, Castellani's method is very easy, simple and cost-effective to preserve fungal viability including their pathogenicity of the isolated strains. But temperature should be maintained at 25–28°C.

The most suitable method currently used in main Culture Collection Centers (CBS, ATCC and NCPF) are freeze-drying (lyophilization) along with preservation in liquid nitrogen, under mineral oil and/or periodic sub-culturing.

A living culture collection of fungi that states their genetic stability, pathogenicity, purity and viability should be an important part of mycological laboratory, as reference for diagnosis of mycoses which affect man and animals, in comparative and taxonomical studies of isolates and in training of students and associated personnel in Medical Mycology.

In addition preservation of fungi can be done with broth media and 10–15% glycerol, wherein the fungal strains can be kept for a longer period. The other way is to store in PDA with 1% agar. A filter paper method is also very effective for transport of fungi from one place to another.

As such there is no 'perfect' procedure of preservation, which is appropriate for all laboratories and simultaneously applicable to all fungal organisms. However, cost and difficulty of storage are major considerations, sterile water storage method is an excellent choice for simple storage of many commonly encountered fungi in a diagnostic laboratory. The commonly used methods are: Castellani's Sterile Distilled Water Method, Periodic Serial Subcultures, Sterile Mineral Oil Overlay, Sterile Paper Technique, Soil/Sand Method, Freeze Drying (Lyophilization), Freezing (-20 and -80°C), Liquid Nitrogen Method and Cryopreservation on poros beads. Some of these preservation methods are briefly described below:

1. Subculturing

Subculturing or serial transfer is the most common method to preserve fungi in culture collections. It is continuous subculturing on agar slants. This simple and inexpensive method is typically used for short-term storage. The cultures stored at room or higher temperatures must be transferred to the fresh media relatively often. Although this method has a lot of disadvantages, it is still used in almost all small-scale laboratories.

However, unfortunately it does not prevent damage to the cultures consisting in an unpredictable appearance or selection of diverse phenotypic variants *i.e.* pleomorphic growth and genetic mutants. These changes might result in cultures sometimes completely different from the fungal populations originally isolated or acquired. Therefore, this routine subculturing is not very practical for storing of large numbers of fungal cultures. On the other side, this technique is applicable universally and is at present used in a majority of low-budget culture collections or as a second method of maintenance. In many studies, subculturing is compared with other methods of maintenance.

2. Preservation under Mineral Oil

Preservation under mineral or paraffin oil is cheap and easy and is used in laboratories with limited resources. The fungal cultures can be kept for several years at room temperature or at 15–20°C. This method is appropriate for mycelial or non-sporulating cultures that are not amenable to freezing or freeze-drying. In addition, oil also reduces mite infestations. Although many fungi can be maintained this way, however, the growth rates of the cultures may decrease as storage time increase. This method is for limited number of stock culutres.

3. Storage in Sterile Water

The storage of fungi in sterile water is another inexpensive method for maintaining cultures. It was originally described by Castellani, who stored pathogenic fungi in humans. Ellis obtained good viabilities of up to 20 months of isolates that did not sporulate well or gave poor results with lyophilization, including Entomophthorales, Pyrenomycetes, Hymenomycetes and Hyphomycetes. There should be maintenance of temperature with a periodic check.

4. Drying of Fungal Culture

Drying of cultures enables suspending of fungal metabolism to a level at which metabolism is either reduced or preferably stopped but from which the strain can be revived. This method of preservation is useful for cultures that produce spores or other resting structures. Its disadvantages are possible culture degeneration or strain drift. Drying methods are suitable for laboratories with limited funds because they are technically simple and also do not require expensive equipment. Silica gel, glass beads and soil are substrates commonly used in drying. Fungi have been stored successfully on silica gel for up to 11 years.

5. Freeze-drying (Lyophilization)

A more sophisticated and efficient technique is freeze-drying (lyophilization) *i.e.* quick drying of frozen fungal suspensions by sublimation in vacuum. The term implies the common features of this process with cryopreservation. Dehydration of cells, necessary to avoid intracellular crystallization during freezing, is regulated through the cooling rate and depends on the size of the cell and the thickness of the cell-wall.

6. Cryopreservation

The preservation of basidiomycetes presents a specific problem because most of them do not form or at least not in a sufficient amount of asexual spores, their dominating form being water-rich mycelium, which is rather sensitive to environmental conditions and also mycorrhizal relationships play often a certain role. Therefore, the most widely used method of culture maintenance in lyophilized state is rather limited in basidiomycetes, because many of them (especially non-sporulating and mycorrhizal ones) do not survive the process. Storage in liquid nitrogen has been considered the best and most widely applicable preservation technique available for filamentous fungi including basidiomycetes. It is a safe and reliable method for especially long-term maintenance of fungal cultures.

III. Fungal Culture Collections

Whenever new or unfamiliar fungus is isolated in routine Diagnostic Mycology Laboratory, it should be sent to a national or international Reference Mycology Laboratory for final identification and/or confirmation. The Pro forma/ Deposit Form of institute where the strain is being sent is to be properly filled up and signed giving all relevant information. The strains should be sent as per protocol of the Centers for Diseases Control and Prevention through registered parcel. If strain is kept for several days or months, to assure viability and purity, fresh subculture should be done for sending purpose.

Use screw cap tubes/bottles and seal cap with masking tape. Label name of isolate, date of subculture and medium used for transportation. Put on metallic container, packing all sides with cotton and place this container into another container. Finally pack in cardboard box that should contain copy of letter of correspondence with full correct addresses of destination as well as of sender including land-line and/or mobile phone numbers. While transporting fungal material to any other laboratory, procedures to be followed as per *WHO Laboratory Manual* (3rd edn), 2004. While sending the strains, the ICMR, New Delhi guidelines are to be followed in India.

For the purposes of patent procedure, Budapest Treaty provisions are to be followed on the international recognition of the deposit of microorganisms, which was done at Budapest on April 28, 1977 and subsequently amended on September 26, 1980.

There is another option to facilitate the process of preservation is to implement a Quality Management System (QMS) based on the ISO 9001 standard. Its aim is to provide guidelines for the certification of other Culture Collections so that they can also enhance the search and choice of the most consistent, reliable and effective operating methods with assured procedures and validation of preservation and ensure reliable relations with all the stakeholders involved in this process.

1. International Culture Collections

The addresses of main Culture Collections that maintain pathogenic fungi at international level are given below:

1. **American Type Culture Collection (ATCC)**
   12301 Parklawn Drive,
   Rockville, MD, 20852, USA.
2. **Fungus Reference Laboratory**
   Bldg. 5, B-37, G-11,
   Division of Bacterial and Mycotic Diseases,
   National Center for Infectious Diseases,
   Centers for Disease Control and Prevention (CDC),
   Atlanta, Georgia - 30333, USA.
3. **Microfungus Collection and Herbarium**
   Devonian Botanic Garden,
   University of Alberta,
   Edmonton, Alberta - T6G 2EI, Canada.
4. **Westerdijk Fungal Biodiversity Institute**
   (Former CBS-KNAW), PO Box 85167,
   3508 AD, Utrecht, The Netherlands.
5. **Mycology Reference Laboratory**
   Division of Microbiology,
   University of Leeds,
   Leeds, LS2 9JT. UK.
6. **PHLS Mycological Reference Laboratory and National Collection of Pathogenic Fungi (NCPF),**
   Myrtle Road, Kingsdown,
   Bristol, BS2 8EL, UK.
7. **International Mycological Institute (IMI)**
   Bakeham Lane, Egham
   Surrey, TW20 9TY, UK.
8. **National Collection of Yeasts Cultures (NCYC)**
   Institute of Food Research, Norwich Research Park,
   Colney, Norwich. NR4 7UA, UK.
9. **Collection Nationale Cultures de Microorganismes**
   Institut Pasteur, 25-28 Rue du Dr Roux,
   75724, Paris, Cedex 15, France.
10. **Australian National Ref Lab in Medical Mycology**
    The Royal North Shore Hospital of Sydney,
    St. Leonards, NSW, 2065, Australia.
11. **National Collection of Type Cultures**
    Central Public Health Laboratory,
    London, UK.

12. **Mycology Unit, Medical School,**
Rovira i Virgili University,
Sant Llorenc 21, 43201-Reus, Spain.
13. **Regional Mycology Laboratory,**
Education and Research Center,
University Hospital of South Manchester,
Wythenshawe Hospital, Manchester,
M23 9LT, UK.
14. **The Fungus Testing Laboratory,**
Department of Pathology, Mail Code 7750,
The University of Texas Health Science Center,
San Antonio, Texas. 78229-3900 (USA).

Note: *Addresses of a particular Culture Collection may also be obtained from World Directory of Collections of Cultures of Microorganisms.*

2. National Culture Collections

At national level also, addresses of main culture collections/reference laboratory which hold clinically significant fungi and a few are advanced centers where research work on fungal disease is actively going on. These are given below:

1. **Human, Fungal Culture Collection**
Department of Medical Mycology
Vallabhbhai Patel Chest Institute,
University of Delhi, Delhi-110007.
2. **Microbial Type Culture Collection (MTCC)***
Institute of Microbial Technology
Sector 39 A, Chandigarh-160014.
3. **National Culture Collection of Pathogenic Fungi**,
Mycology Division, Center for Advanced Research in Medical Mycology, and WHO Collaborating Center,
Department of Medical Microbiology,
Postgraduate Institute of Medical Education and Research (PGIMER), Chandigarh-160012.
4. **Mycology Section**
Department of Microbiology,
All India Institute of Medical Sciences (AIIMS),
Ansari Nagar, New Delhi-110029.
5. **Department of Medical Mycology**
Calcutta School of Tropical Medicine and Hygiene,
Kolkata (West Bengal) - 700073.
6. **National Fungal Culture Collection of India (NFCCI)***
Agharkar Research Institute, G. G. Agharkar Road,
Pune, Maharashtra-411004.

****Note:*** *The MTCC and NFCCI are meant only for non-pathogenic fungi, as the Institute of Microbial Technology, Chandigarh and Agharkar Research Institute, Pune do not generally deal with human fungal pathogens.*

National Culture Collection of Pathogenic Fungi

The National Culture Collection of Pathogenic Fungi (NCCPF) is the national facility sponsored by the Indian Council of Medical Research (ICMR), New Delhi, for the deposition, maintenance, identification and supply of all pathogenic fungi. It is situated at the Department of Medical Microbiology at the Postgraduate Institute of Medical Education and Research (PGIMER), Chandigarh, It was established with funding from the ICMR in 2010.

The collection now maintains over 2500 strains of fungi, both filamentous and yeasts, representing a large percentage of the pathogenic species in the fungal kingdom, which have been cultured to date. The main aim of the center is to set up a facility of international standard for preservation and to handle deposits of medically important fungi and to supply authentic strains to investigators in India. The scientists and researchers have access to database of preserved medically important fungi. The center also initiated fungal taxonomy research at the facility and impart training on taxonomy to interested researchers.

In diversity of species, it is unchallenged as a reference center for mycological research in India. The preservation of the organisms while maintaining their original characters is an enormous challenge. Due to high level of contamination along with change in characteristics of fungi preserved on agar, there is an increasing demand of reliable long-term preservation, which is the main aim of NCCPF. The center also focuses on the identification of fungi, making use of phenotypic as well as molecular and genomics approaches.

The significance of a collection depends not only on the quality of the strains maintained but also on the data affixed to them. The available databases allow information recovery and access to strain data. The NCCPF is a center of expertise, advising on mycological problems of health-related and scientific nature.

Whenever, the unknown fungal strain is to be submitted for identification and deposition to the NCCPF, the application form is downloaded from its website *(http://www.nccpf.com)*. The pure form of the fungal strain along with the filled up form is to be sent by speed post or courier keeping all the standard biosafety precautions.

Further Reading

1. Belkacemi L, Barton RC, Evans EGV. Cryopreservation of *Aspergillus fumigatus* stock cultures with a commercial bead system. *Mycoses.* 1997; **40:** 103-4.
2. Bhat V, Chavan P, Naresh C, *et al.* The External Quality Assessment Scheme (EQAS): Experiences of a medium-sized accredited laboratory. *Clin Chim Acta.* 2015; **446:** 61-3.
3. Borba C, Mendes da Silva, AM, Oliveira PC. Long-time survival and morphological stability of preserved *Sporothrix schenckii* strains. *Mycoses.* 1992; **35:** 185-8.
4. Borba C, Rodrigues KF. Viability and sporulating capability of Coelomycetes preserved under a range of different storage regimes. *Rev Iber Micol.* 2000; **17:** 142-5.
5. Borman AM, Szekely A, Campbell CK, Johnson EM. Evaluation of the viability of pathogenic filamentous fungi after prolonged storage in sterile water and review of recent published studies on storage methods. *Mycopathologia.* 2006; **161:** 361-8.
6. Brilhante RS, Silva NF, Lima RA, *et al.* Easy storage strategies for *Sporothrix* spp. strains. *Biopreserv Biobank.* 2015; **13:** 131-4.
7. Butterfield W, Jong SC, Alexander MT. Preservation of living fungi pathogenic for man and animals. *Can J Microbiol.* 1974; **20:** 1665-73.
8. Castellani A. Maintenance and cultivation of the common pathogenic fungi of man in sterile distilled water. Further re-searches. *J Trop Med Hyg.* 1967; **70:** 181-4.
9. Chandler D. Cryopreservation of fungal spores using porous beads. *Mycol Res.* 1994; **98:** 525-6.
10. Dahmen H, Staub T, Schwinn FT. Technique for long-term preservation of phytopathogenic fungi in liquid nitrogen. *Phytopathology.* 1983; **73:** 241-6.
11. Forti T, Souto Ada S, do Nascimento CR, *et al.* Evaluation of a fungal collection as certified reference material producer and as a biological resource center. *Braz J Microbiol.* 2016; **47:** 403-9.
12. Gentles JC, Scott E. The preservation of medically important fungi. *Sabouraud.* 1979; **17:** 415-8.
13. Gleason FH, Mozley-Standridge SE, Porter D, *et al.* Preservation of Chytridiomycota in culture collections. *Mycol Res.* 2007; **111:** 129-36.
14. Guimaraes LC, Fernandes AP, Chalfoun SM, *et al.* Methods to preserve potentially toxigenic fungi. *Braz J Microbiol.* 2014; **45:** 43-7.
15. Homolka L, Lisa L, Eichlerovaa I, *et al.* Effect of long-term preservation of different basidiomycetes on perlite in liquid nitrogen on their growth, morphological, enzymatic and genetic characteristics. *Fungal Biol.* 2010; **114:** 929-35.
16. Homolka L, Lisa L. Long-term maintenance of fungal cultures on perlite in cryovials - An alternative for agar slants. *Folia Microbiol (Praha).* 2008; **53:** 534-6.
17. Homolka L, Lisa L, Nerud F. Basidiomycete cryopreservation on perlite, evaluation of a new method. *Cryobiology.* 2006; **52:** 446-53.
18. Homolka L. Methods of Cryopreservation in Fungi. *In:* Laboratory Protocols in Fungal Biology: Current Methods in Fungal Biology. Gupta VK, *et al*, eds. Springer, New York: 2013; p. 9-16.
19. Homolka L. Preservation of live cultures of basidiomycetes - Recent methods. *Fungal Biol.* 2014; **118:** 107-25.
20. Hu X, Webster G, Xie L, *et al.* A new method for the preservation of axenic fungal cultures. *J Microbiol Methods.* 2014; **99:** 81-3.
21. Juarros E, Tortajada C, Garcia MD, *et al.* Storage of stock cultures of filamentous fungi at -80 degrees C: Effects of different freezing-thawing methods. *Microbiologia.* 1993; **9:** 28-33.
22. Kitamoto Y, Suzuki A, Shimada S, *et al.* A new method for the preservation of fungus stock cultures by deep freezing. *Mycoscience.* 2002; **43:** 143-9.
23. Lelieveld HL, Bachmayer H, Boon B, *et al.* Transport of infectious and biological materials. *Appl Microbiol Biotechnol.* 1997; **48:** 135-40.
24. Libeer JC. Role of external quality assurance schemes in assessing and improving quality in medical laboratories. *Clin Chim Acta.* 2001; **309:** 173-7.
25. Lima RF, Borba CM. Viability, morphological characteristics and dimorphic ability of fungi preserved by different methods. *Rev Iber Micol.* 2001; **18:** 191-6.
26. Marx DH, Daniel WJ. Maintaining cultures of ectomycorrhizal and plant pathogenic fungi in sterile water cold storage. *Can J Microbiol.* 1976; **22:** 338-41.
27. McGinnis MR, Padhye AA, Ajello L. Storage of stock cultures of filamentous fungi, yeasts and some aerobic actinomycetes in sterile distilled water. *Appl Microbiol.* 1974; **28:** 218-22.
28. Pasarell L, McGinnis MR. Viability of fungal cultures maintained at -70 degrees C. *J Clin Microbiol.* 1992; **30:** 1000-4.
29. Qiangqiang Z, Jiajun W, Li L. Storage of fungi using sterile distilled water or lyophilization: Comparison after 12 years. *Mycoses.* 1998; **41:** 255-7.
30. Richter DL. Revival of saprotrophic and mycorrhizal basidiomycete cultures after 20 years in cold storage in sterile water. *Can J Microbiol.* 2008; **54:** 595-9.
31. Ryan MJ, Smith D. Cryopreservation and freeze-drying of fungi employing centrifugal and shelf freeze-drying. *Meth Mol Biol.* 2007; **368:** 127-40.
32. Sharma B, Smith D. Recovery of fungi after storage for over a quarter of a century. *World J Microbiol Biotechnol.* 1999; **15:** 517-9.
33. Silva Sergent FA, Acosta-Hernandez B, Raduan Jaber J, *et al.* Preservation of *Malassezia pachydermatis* using skimmed milk as cryoprotective agent. *J Microbiol Methods.* 2017; **132:** 166-7.
34. Simoes MF, Dias N, Santos C, *et al.* Establishment of a quality management system based on ISO 9001 standard in a public service fungal culture collection. *Microorganisms.* 2016; 4. pii: e21. PMID: 27681915.
35. Singh SK, Upadhyay RC, Kamal S, *et al.* Mushroom cryopreservation and its effect on survival, yield and genetic stability. *CryoLetters.* 2004; **25:** 23-32.

Appendix F: Bibliography

The readers may require consulting reference books, manuals, monographs, software packages or online information services in medicine for an in-depth study on a particular aspect of Medical Mycology. The present time has entirely changed from previous one and medical science has already entered into an **electronic-era**. The developments in information technology are so rapid that whole world has virtually become a **global village**. As such there is no dearth of information as abundant literature is available from several sources. A large number of reference articles are accessible from electronic media almost free of cost or on nominal charges. Some of old biomedical journals are even available as microfilms. But information about some of the reference books in Medical Mycology and allied aspects of Microbiology is still difficult to obtain. Hence list of such books under different headings is given in this Appendix as **Bibliography**.

At completion of every Chapter of this Textbook, relevant literature has already been cited in the form of primary scientific papers and review articles published in biomedical journals. The Chapters are updated on the basis of information available in the references, mainly dealing with latest developments on various aspects of Medical Mycology during the period of eight and half years since the publication of its last edition (3rd edn) in October 2008. Therefore, all the important references available in the field of medical mycology published between 01.01.2009 to 31.05.2017 have been appended in the respective Chapter of this Textbook.

A group of editors of general medical journals met informally in Vancouver (Canada) in 1978 to establish guidelines for format of manuscripts submitted to their journals. In due course of time, this became popular as Vancouver Group. Its requirements for manuscripts, including formats for bibliographic references developed by National Library of Medicine (NLM), were first published in 1979. The Vancouver Group expanded and evolved into International Committee of Medical Journal Editors (ICMJE), which meets annually and gradually it has broadened its concerns. The official version of Uniform Requirements for Manuscripts Submitted to Biomedical Journals may be retrieved from website - ***http://www.icmje.org***. The references in this Textbook are cited following **Vancouver** system with minor modifications and standard format complying with instructions of *'Uniform Requirements for Manuscripts Submitted to Biomedical Journals: Writing and Editing for Biomedical Publication'* which were updated in October 2007. The titles of journals also conform to abbreviations used in *Index Medicus*. However, now first six names of authors are recommended but in this textbook only first three are mentioned in the cited references under the heading as **Further Reading** appended in every Chapter of Textbook due to shortage of space. The final format is like - Author 1, Author 2, Author 3, *et al.* Title of Article. *Abbreviated Name of Journal.* Year of Publication; **Volume** (Without Issue Number)**:** Pages 00-0.

The reference books mentioned in this Appendix are both old as well as new ones, to refer developments concerning different aspects of Medical Mycology. In the Central Library of medical institute they may be traced in rack, with series coded as **616.961**. These books are not appended in **Further Reading** of the Chapters to avoid repetition. Hence a comprehensive list of Select Bibliography is separately given in this Appendix of the Textbook.

The following reference books and other available informative resources are recommended for general outlines, historical interest or details of particular aspect of Medical Mycology. These are categorized under the following nine headings:

- Reference Books in Medical Mycology
- Books in Mycology Practical Aspects

- Books in Microbiology and Allied Subjects
- Biomedical Journals Published in Mycology
- Supplementary Issues in Medical Mycology
- Manuals of Mycology Seminars and Workshops
- Online Information Services in Medicine
- Electronic Versions of Books and Journals
- Software Packages in Medical Mycology

Reference Books in Medical Mycology

A large number of Reference as well as Textbooks have been written covering most of theoretical aspects of Medical Mycology or focusing only particular disease or even part of it. These books are listed alphabetically below:

1. Ahmad I, Owais M, Shahid M, Aqil F. *Combating Fungal Infections: Problems and Remedy*. Springer, London: 2010.
2. Ainsworth GC. *Introduction to the History of Medical and Veterinary Mycology*. 1st edn. Cambridge University Press, Cambridge: 1986 (Paperback edn. 2002).
3. Ajello L, Chick EW, Furcolow ML. *Histoplasmosis*. Charles C Thomas, Springfield: 1971.
4. Ajello L. *Coccidioidomycosis: Current Clinical and Diagnostic Status*. Symposia Specialists Medical Books, Miami: 1977.
5. Ajello L. *Proceedings of Coccidioidomycosis*. (Second Coccidioidomycosis Symposium). University of Arizona Press, Tucson: 1967.
6. Al-Doory Y, DiSalvo AF. *Blastomycosis*. Plenum Medical Books, New York: 1992.
7. Al-Doory Y, Domson JF. *Mould Allergy*. Lea & Febiger, Philadelphia: 1984.
8. Al-Doory Y, Wagner GE. *Aspergillosis*. Charles C Thomas, Springfield: 1985.
9. Al-Doory Y. *Chromomycosis*. Mountain Press Publishers, Missoula, Montana: 1972.
10. Al-Doory Y. *The Epidemiology of Human Mycotic Diseases*. Charles C Thomas, Springfield: 1975.
11. Alexopoulos CJ, Mims CW, Blackwell M. *Introductory Mycology*. 4th edn. John Wiley, New York: 1996.
12. Anaissie EJ, McGinnis MR, Pfaller MA. *Clinical Mycology*. 2nd edn. Elsevier, London: 2009.
13. Arora David. *Mushrooms Demystified*. 2nd edn. Ten Speed Press, Berkeley: 1986.
14. Arora DK, Ajello L, Mukerji KG. *Handbook of Applied Mycology: Humans, Animals and Insects - Fungi Pathogenic to Humans*. Marcel Dekker, New York: 1991.
15. Arora DK, Bridge PD, Bhatnagar D. *Handbook of Fungal Biotechnology*. 2nd edn. Marcel Dekker, New York: 2004.
16. Arseculeratne SN, Atapattu DN. *Rhinosporidiosis in Humans and Animals & Rhinosporidium seeberi*. Faculty of Medicine, University of Peradeniya, Sri Lanka: 2011.
17. Ashbee HR, Bignell EM. *Pathogenic Yeasts: The Yeast Handbook*. Springer-Verlag, Berlin, Heidelberg: 2010.
18. Baker RD. *Human Infection with Fungi, Actinomycetes and Algae*. Springer-Verlag, New York: 1971.
19. Barns RA, Warnock DW. *Fungal Infection in the Intensive Care Unit*. Kluwer Academic Publishers, Norwell: 2002.
20. Barug D, Bhatnagar D, Van Egmond HP, Van Der Kamp JW, Van Osenbruggen WA, Visconti A. *The Mycotoxin Factbook - Food and Feed Topics*. 1st edn, Wageningen Academic Publishers: 2006.
21. Beneke ES, Rogers AL. *Medical Mycology and Human Mycoses*. Star Publishing, Belmont: 1996.
22. Bennett JE, Hay RJ, Peterson PK. *New Strategies in Fungal Diseases*. Churchill Livingstone, Edinburgh: 1992.
23. Berger S. *Miscellaneous Invasive Fungi: Global Status*. GIDEON Informatics, Los Angeles: 2017.
24. Bhat RV, Nagarajan V, Tulpule PG. *Health Hazards of Mycotoxins in India*. ICMR Publication, New Delhi: 1986.
25. Bilgrami KS, Sinha KK. *Mycotoxins: Incidence and Effects*. CBS Publications, New Delhi: 1992.
26. Billore OP. *Rhinosporidiosis*. 1st edn. All India Travellers Book Sales (AITBS) Publishers & Distributors, Delhi: 1996.
27. Bodey GP. *Candidiasis: Pathogenesis, Diagnosis and Treatment*. 2nd edn. Raven Press, New York: 1993.
28. Bodey GP. *Fungal Infections in Cancer Patients: An Overview*. Pfizer International, New York: 1990.
29. Boekhout T, Gueho-Kellermann E, Mayser P, Velegraki A. *Malassezia and the Skin - Science and Clinical Practice*. 1st edn. Springer, Heidelberg, 2010.
30. Booth C. *The Genus Fusarium*. CMI, Kew: 1971.
31. Borgers M, Hay RJ, Rinaldi MG. *Current Topics in Medical Mycology*. Vol. 4, Springer-Verlag, New York: 1992.
32. Bossche HV Mackenzie DWR, Cauwenbergh G. *Aspergillus and Aspergillosis*. Plenum Press, New York: 1988.
33. Brakhage AA, Jahn B, Schmidt A. *Aspergillus fumigatus: Biology, Clinical Aspects and Molecular Approaches to Pathogenicity*. Karger, Basal: 1999.
34. Brand AC, MacCallum DM. *Host-Fungus Interactions: Methods and Protocols (Methods in Molecular Biology : Series - 845)*. Humana Press, New York: 2012.
35. Bray GA, Ryan DH. *Mycotoxins, Cancer and Health*. Louisiana State University Press, Baton Rouge: 1990.
36. Breitenbach M, Crameri R, Lehrer SB. *Fungal Allergy and Pathogenicity*. Karger Publishers, Farmington: 2002.
37. Brown DW, Proctor RH. *Fusarium: Genomics, Molecular and Cellular Biology*. Caister Academic Press, Norfolk: 2013.
38. Brown GD, Netea MG. *Immunology of Fungal Infections*. Springer, Dordrecht; 2007.
39. Bulmer GS. *Fungus Diseases in the Orient*. 3rd edn. Rex Publishing, Manila: 1995.
40. Bulmer GS. *Introduction to Medical Mycology*. Year Book Medical Publishers, Chicago: 1979.
41. Bulmer GS. *Medical Mycology in Chinese and English*. Shanghai Scientific and Technical Publishers, 2005.
42. Calderone RA, Cihlar RL. *Fungal Pathogenesis: Principles and Clinical Applications*. Marcel Dekker, New York: 2002.
43. Calderone RA, Clancy CJ. *Candida and Candidiasis*. 2nd edn. ASM Press, Washington, DC: 2011.
44. Cann C. *Medical Mycology*. Lea & Febiger, Philadelphia: 1992.
45. Carlile MJ, Watkinson SC. *The Fungi*. Academic Press, London: 1997.

46. Casadevall A, Mitchell AP, Berman J, Kwon-Chung KJ, Perfect JR, Heitman J. *Human Fungal Pathogens.* Cold Spring Harbor Press, New York: 2014.
47. Casadevall A, Perfect JR. *Cryptococcus neoformans.* 1st edn. ASM Press, Washington, DC: 1998.
48. Chakrabarti A. *Fungal Infections in Asia - The Eastern Frontier of Mycology.* ECAB Clinical Update, Elsevier, New Delhi: 2013.
49. Cihlar RL, Calderone RA. *Candida albicans: Methods and Protocols (Methods in Molecular Biology : Series - 499).* Humana Press, New York: 2009.
50. Chelkowski J. *Fusarium: Mycotoxins, Taxonomy and Pathogenicity.* Elsevier, Amsterdam: 1989.
51. Cole RJ, Cox RH. *Handbook of Toxic Fungal Metabolites.* Academic Press, New York: 1981.
52. Cole RJ, Jarvis BB, Schweikert MA. *Handbook of Secondary Fungal Metabolites.* Academic Press, London: 2003.
53. Cooke RC, Rayner ADM. *Ecology of Saprotrophic Fungi.* Longman, London: 1984.
54. Coste AT, Vandeputte P. *Antifungals: From Genomics to Resistance and the Development of Novel Agents.* Caister Academic Press, Norfolk: 2015.
55. Cox RA. *Immunology of the Fungal Diseases.* CRC Press, Boca Raton: 1989.
56. Creppy EE, Castegnaro M, Dirheimer, G. *Human Ochratoxicosis and its Pathologies.* John Libbey Eurotext Ltd., Montrouge, France: 1993.
57. Daniel TM, Baum GL. *Drama and Discovery: The Story of Histoplasmosis.* Greenwood Press, Westport: 2002.
58. Deacon JW. *Modern Mycology.* 3rd edn. Blackwell Science, Oxford: 1997.
59. Deacon JW. *Fungal Biology.* 4th edn. Blackwell Publishing, Oxford: 2006.
60. de Hoog GS, Smith MT, Weijman ACM. *The Expanding Realm of Yeast-like Fungi.* Elsevier, Amsterdam: 1987.
61. Deshmukh SK, Rai MK. *Biodiversity of Fungi - Their Role in Human Life.* Oxford & IBH, New Delhi: 2005.
62. de Vries RP, Gelber IB, Andersen MR. *Aspergillus and Penicillium in the Post-genomic Era.* Caister Academic Press, Norfolk: 2016.
63. De Vries JW, Trucksess MW, Jackson LS. *Mycotoxins and Food Safety.* Kluwer Academic/Plenum Publishers, New York: 2002.
64. Dey NC, Grueber HLE, Dey TK. *Medical Mycology.* 1st edn. 1957, 1st Central Edn. New Central Book Agency, Kolkata: 2006.
65. DiConsiglio J. *There is a Fungus Among Us! True Stories of Killer Molds.* Franklin Watts, Scholastic Inc., New York: 2008.
66. DiSalvo AF. *Occupational Mycoses.* Lea & Febiger, Philadelphia: 1983.
67. Dismukes WE, Pappas PG, Sobel JD. *Clinical Mycology.* 1st edn. Oxford University Press, New York: 2003.
68. Domer JE, Kobayashi GS. *The Mycota, XII, Human Fungal Pathogens.* Springer-Verlag, Heidelberg: 2003.
69. Eaton DL, Groupman JD. *The Toxicology of Aflatoxins: Human Health and Agricultural Significance.* Academic Press, San Diego: 1994.
70. Einstein HE, Catanzaro A. *Coccidioidomycosis.* National Foundation for Infectious Diseases, Washington, DC: 1996.
71. Elewski BE. *Cutaneous Fungal Infections.* Igaku-Shoin, New York: 1992.
72. Ellis DH. *Clinical Mycology: The Human Opportunistic Mycoses.* Gillington Printers, Australia: 1995.
73. Ellis MB. *Dematiaceous Hyphomycetes.* CMI, CAB Press, Kew: 1990.
74. Emmons CW, Binford CH, Utz JP, Kwon-Chung KJ. *Medical Mycology.* 3rd edn. Lea & Febiger, Philadelphia: 1977. (Indian edn: KM Verghese Company, Mumbai).
75. English MP. *Medical Mycology.* Arnold Publishers, London: 1980.
76. Ernst EJ, Rogers PD. *Antifungal Agents: Methods and Protocols.* Humana Press, New Jersey: 2005.
77. Ernst JF, Schmidt A. *Dimorphism in Human Pathogenic and Apathogenic Yeasts.* Karger, Basel: 2000.
78. Espinel-Ingroff A. *Medical Mycology in the United States: A Historical Analysis (1894-1996).* Kluwer Academic Publishers, New York: 2003.
79. Evans EGV, Gentles JC. *Essentials of Medical Mycology.* 1st edn. Churchill Livingstone, Edinburgh: 1985.
80. Fahal AH. Mycetoma - Clinicopathological Monograph. Khartoum University Press, Khartoum: 2006.
81. Fetter BF, Klintworth GK, Hendry WS. *Mycoses of the Central Nervous System.* Williams & Wilkins, Baltimore: 1967.
82. Fidel P, Huffnagle G. *Fungal Immunology: From an Organ Perspective.* Springer, New York: 2005.
83. Fiese MJ. *Coccidioidomycosis.* Charles C Thomas, Springfield: 1958.
84. Franco MF, Lacaz CS, Restrepo-Moreno A, del Negro G. *Paracoccidioidomycosis.* CRC Press, Boca Raton: 1994.
85. Frisvad JC, Bridge PD, Arora DK. *Chemical Fungal Taxonomy.* Marcel Dekker, New York: 1998.
86. Fromtling RA. *Recent Trends in the Discovery, Development and Evaluation of Antifungal Agents.* Prous Publishers, Barcelona: 1987.
87. Gatti F, De Vroey C, Parsi A. *Human Mycoses in Tropical Countries.* 2nd edn. OCSI, Bologna: 1991.
88. Gehlot S. *Gehlots' Review of Mycology.* 1st edn. 2013.
89. Ghannoum MA, Perfect JR. *Antifungal Therapy.* Informa Healthcare, New York: 2010.
90. Gherbawy Y, Voigt K. *Molecular Identification of Fungi.* Springer, New York: 2010.
91. Goldman GH, Osmani SA. *The Aspergilli: Genomics, Medical Aspects, Biotechnology and Research Methods.* 1st edn. CRC Press, Boca Raton: 2007.
92. Griffin DH. *Fungal Physiology.* 2nd edn. Wiley-Liss, New York: 1994.
93. Grigoriu D, Delacretaz J, Borelli D. *Medical Mycology.* (Translated from French by Stephenson DQ). Hans Huber Publishers, Toronto; Lewiston, NY; Bern, Stuttgart: 1987.
94. Hall GS. *Interactions of Yeasts, Moulds and Antifungal Agents: How to Detect Resistance.* Humana Press, New York: 2012.
95. Hawksworth DL, Hall GS. *International Mycological Directory.* 2nd edn. CAB International, Wallingford: 1990.

96. Hawksworth DL. *Frontiers in Mycology*. CAB International, Wallingford: 1990.
97. Hay RJ. *Fungi and Skin Diseases*. Mosby-Wolfe, London: 1993.
98. Hay RJ. *Tropical Fungal Infections. In:* Hay RJ. *Bailliere's Clinical Tropical Medicine and Communicable Diseases*. Vol. 4. Bailliere Tindall, London: 1989.
99. Heitman J, Filler SG, Edwards JE Jr, Mitchell AP. *Molecular Principles of Fungal Pathogenesis*. ASM Press, Washington, DC: 2006.
100. Heitman J, Kozel TR, Kwon-Chung KJ, Perfect JR, Casadevall A. *Cryptococcus: From Human Pathogen to Model Yeast*. ASM Press, Washington, DC, 2011.
101. Hildick-Smith G, Blank H, Sarkany I. *Fungal Diseases and Their Treatment*. 1st edn. J & A Churchill Ltd, London: 1964.
102. Hocking AD, Pitt JI, Samson RA, Thrane U. *Advances in Food Mycology*. Springer, New York: 2006.
103. Holmes C, Rautemaa-Richardson R, Novak-Frazer L, Moore CB, Richardson MD. *Illustrative Cases in Medical Mycology*. Free iBook. University of Manchester and Mycology Reference Centre, Manchester, UK: 2016.
104. Homei A, Worboys M. *Fungal Disease in Britain and the United States 1850-2000: Mycoses and Modernity (Science, Technology and Medicine in Modern History)*. Palgrave Macmillan. Basingstoke, UK: 2013.
105. Hopkin JM. *Pneumocystis carinii*. Oxford University Press, Oxford: 1991.
106. Hospenthal DR, Rinaldi MG. *Diagnosis and Treatment of Human Mycoses*. Humana Press, New York: 2008.
107. Hospenthal DR, Rinaldi MG. *Diagnosis and Treatment of Fungal Infections*. 2nd edn. Springer, Switzerland: 2015.
108. Howard DH, Miller JD. *The Mycota VI. Human and Animal Relationships*. Springer-Verlag, Berlin: 1996.
109. Howard DH. *Pathogenic Fungi in Humans and Animals*. 2nd edn. Marcel Dekker, New York: 2002.
110. Hughes WT. *Pneumocystis carinii Pneumonia*. 1st edn. Vol. I & II. CRC Press, Boca Raton: 1987.
111. Hydler GW. *Magical Mushrooms, Mischievous Molds*. Princeton University Press, Princeton: 1998.
112. *Index of Fungi*. International Mycological Institute, Egham, UK (Published twice a year).
113. Isaac S, Frankland JC, Watling R, Whalley AJS. *Aspects of Tropical Mycology*. Cambridge University Press, UK: 1993.
114. Iwata K. *Recent Advances in Medical and Veterinary Mycology*. University of Tokyo Press, Tokyo: 1986.
115. Jackson LS, Devries JW, Bullerman LB. *Fumonisins in Food*. Plenum Press, New York: 1996.
116. Jacobs PH, Nall L. *Antifungal Drug Therapy: A Complete Guide for the Practitioner*. Marcel Dekker, New York: 1990.
117. Jacobs PH, Nall L. *Fungal Diseases: Biology, Immunology and Diagnosis*. Marcel Dekker, New York: 1997.
118. Jellison WL. *Adiaspiromycosis*. Mountain Press Publishers, Missoula, Montana: 1969.
119. Joffe AZ. *Fusarium Species: Their Biology and Toxicology*. John Wiley, New York: 1986.
120. Joshi KR, Solanki A, Joshi YR. *Mycetoma*. Agrobios Publishers, Jodhpur: 2000.
121. Jucker E. *Antifungal Agents: Advances and Problems*. Birkhauser, Basel: 2003.
122. Jungerman PF, Schwartzman RM. *Veterinary Medical Mycology*. Lea & Febiger, Philadelphia: 1972.
123. Kalkum M, Semis M. *Vaccines for Invasive Fungal Infections: Methods and Protocols (Methods in Molecular Biology : Series 1625)*. Springer, Science+Business Media, LLC: 2017.
124. Karunaratne WAE. *Rhinosporidiosis in Man*. University of London, Athlone Press, London: 1964.
125. Kaufmann CA, Pappas PG, Sobel JD, Dismukes WE. *Essential of Clinical Mycology*. 2nd edn. Springer, New York: 2011.
126. Kavanagh K. *Medical Mycology: Cellular and Molecular Techniques*. 1st edn. John Wiley & Sons, Chichester: 2007.
127. Kelly SL, Quail MA, *et al. New Approaches for Antifungal Drugs*. 1st edn. Birkhauser, Boston: 1992.
128. Kendrick B. *The Fifth Kingdom*. 3rd edn. Mycologue Publications, British Columbia, Canada: 2001.
129. Kibbler CC, Mackenzie DWR, Odds FC. *Principles and Practice of Clinical Mycology*. John Wiley, Chichester: 1996.
130. Kinghorn JR, Martinelli SD. *Aspergillus: 50 Years on*. Elsevier Science, 1994.
131. Kirk PM, Cannon PF, Minter DW, Stalpers JA. *Ainsworth and Bisby's Dictionary of the Fungi*. 10th edn. CAB International, Wallingford, UK: 2008.
132. Kirsch DR, Kelly R, Kurtz MB. The *Genetics of Candida*. CRC Press, Boca Raton: 1990.
133. Klich MA. *Identification of Common Aspergillus Species*. ASM Press, Washington, DC: 2002.
134. Kostrzewa M, Schubert S. *MALDI-TOF Mass Spectrometry in Microbiology*. Caister Academic Press, Norfolk: 2016.
135. Krogh P. *Mycotoxins in Food*. Academic Press, New York: 1987.
136. Kurata H, Ueno Y. *Toxigenic Fungi: Their Toxin and Health Hazards*. Elsevier, Amsterdam: 1984.
137. Kurstak E. *Immunology of Fungal Diseases*. Marcel Dekker, New York: 1989.
138. Kurtzman CP, Fell JW, Boekhout T. *The Yeasts: A Taxonomic Study*. 5th edn. Elsevier, Amsterdam: 2011.
139. Kushwaha RKS, Guarro J. *Biology of the Dermatophytes and Other Keratinophilic Fungi. Revista Iberoamericana de Micologia:* Bilbao, Pais Vasco, Spain: 2000.
140. Kushwaha RKS. *Fungi in Human and Animal Health*. Scientific Publishers, Jodhpur: 2004.
141. Kwon-Chung KJ, Bennett JE. *Medical Mycology*. 1st edn. Lea & Febiger, Philadelphia: 1992.
142. Latge JP, Boucais D. *Fungal Cell Wall and Immune Response*. Springer-Verlag, Berlin: 1991.
143. Latge JP, Steinbach WJ. *Aspergillus fumigatus and Aspergillosis*. ASM Press, Washington, DC: 2009.
144. Levine HB. *Ketoconazole in the Management of Fungal Disease*. ADIS Press, New York: 1982.
145. Lewis GM, Hopper ME, Wilson JW, Plunkett OA. *An Introduction to Medical Mycology*. Year Book Publications, Chicago: 1958.
146. Lion T. *Human Fungal Pathogen Identification: Methods and Protocols (Methods in Molecular Biology : Series - 1508)*. Humana Press, New York: 2017.

147. Littman ML, Zimmerman LE. *Cryptococcosis*. Grune and Stratton, New York: 1956.
148. Liu D. *Molecular Detection of Human Fungal Pathogens.* CRC Press, Taylor & Francis Group, Boca Raton: 2011.
149. Lyr H. *Modern Selective Fungicides: Properties, Applications and Mechanisms of Action.* Gustav Fisher Verlag, Jena: 1995.
150. Machida M, Gomi K. *Aspergillus: Molecular Biology and Genomics.* Caister Academic Press. Norfolk: 2010.
151. Maertens JA & Marr KA. Diagnosis of Fungal Infections. Informa Healthcare, New York: 2007.
152. Magan N, Olsen M. *Mycotoxins in Food: Detection and Control.* Woodhead Publishing, Cambridge: 2004.
153. Maheshwari R. *Fungi: Experimental Methods in Biology.* CRC Press, Taylor and Francis, Boca Raton: 2005.
154. Mahgoub ES, Murray I. *Mycetoma.* William Heinemann Medical Books, London: 1973.
155. Mahgoub ES. *Tropical Mycoses.* Janssen Research Council, Beerse, Belgium: 1989.
156. Makun HA. *Mycotoxin and Food Safety in Developing Countries.* InTech, Rijeka, Croatia: 2013.
157. Mankekar G. *Invasive Fungal Rhinosinusitis.* 1st Edn. Springer, 2014.
158. Maresca B, Kobayashi GS, Yamaguchi H. *Molecular Biology and Application to Medical Mycology.* Springer, Berlin: 1993.
159. Maresca B, Kobayashi GS. *Molecular Biology of Pathogenic Fungi.* 2nd edn. Telos Press, New York: 1994.
160. Margulis L, Schwartz KV. *Five Kingdoms - An Illustrated Guide to the Phyla of Life on Earth.* 2nd edn. WH Freeman & Co., New York: 1988.
161. Margulis L. *Diversity of Life: The Five Kingdoms.* Enslow Publishers, New Jersey: 1992.
162. McGowen VF, Skerman VBD. *World Dictionary of Collections of Cultures of Microorganisms.* 2nd edn. World Data Centre, The University of Queensland, Brisbane: 1982.
163. Mehrotra RS, Aneja KR. *An Introduction to Mycology.* Wiley Eastern, New Delhi: 1990.
164. Meinhof W. *Oral Therapy in Dermatophytes: A Step Forward.* Medicine Publishing Foundation, Oxford: 1985.
165. Merz WG, Hay RJ. *Medical Mycology. In:* Collier L, Balows A, Sussman M. *Topley and Wilson's Microbiology and Microbial Infections.* 10th edn (8 Vol. Set). Vol. 4. Wiley-Blackwell, New Jersey: 2007.
166. Meunier F. *Invasive Fungal Infection in Cancer Patients.* Vol. 2. No. 1. *Bailliere's Clinical Infectious Diseases.* Bailliere Tindall, London: 1995.
167. Miller JD, Trenholm HL. *Mycotoxins in Grain.* Eagan Press, St. Paul: 1994.
168. Mitchell TG, Cutler JE, Deepe GS, Hazen KG. *Principles of Medical Mycology.* ASM Press, Washington, DC: 1997.
169. Moore-Landecher E. *Fundamentals of the Fungi.* 4th edn. Prentice-Hall International, London: 1996.
170. Moretti A, Susca A. *Mycotoxigenic Fungi: Methods and Protocols (Methods in Molecular Biology : Series - 1542).* Humana Press, New York: 2017.
171. Moss M, Smith JE. *The Applied Mycology of Fusarium.* Cambridge University Press, Cambridge: 1984.
172. Mukerji KG, Srivastava AK, Singh KP, Garg KL. *Advances in Medical Mycology.* Aditya Books, New Delhi: 1992.
173. Murphy JW, Friedman H, Bendinelli M. *Fungal Infections and Immune Responses.* Plenum Press, New York: 1993.
174. Nolting S, Fegeler K. *Medical Mycology.* Springer, New York: 1987.
175. O'Connor L, Glynn B. *Fungal Diagnostics: Methods and Protocols.* Human Press. New York: 2013.
176. Odds FC. *Candida and Candidosis.* 2nd edn. Bailliere Tindall, London: 1988.
177. Olsen LA, Choffnes ER, Relman DA, Pray L. *Fungal Diseases: An Emerging Threat to Human, Animal and Plant Health: Workshop Summary.* The National Academies Press, Washington, DC: 2011.
178. Osiewacz HD. Molecular Biology of Fungal Development. Marcel Decker, New York: 2001.
179. Pal M. *Veterinary and Medical Mycology.* 1st edn. Indian Council of Agricultural Research, New Delhi: 2007.
180. Pasqualotto AC. *Aspergillosis: From Diagnosis to Prevention.* Springer, Dordrecht: 2010.
181. Petersen JH. *The Kingdom of Fungi.* Princeton University Press, New Jersey: 2013.
182. Pitt JI, Hocking AD. *Fungi and Food Spoilage.* 3rd edn. Academic Press, Sydney: 2009.
183. Pitt JI. *A Laboratory Guide to Common Penicillium Species.* 3rd edn. Food Science, North Ryde, Australia: 2000.
184. Prasad R. *Candida albicans: Cellular and Molecular Biology.* Springer-Verlag, New York: 1991.
185. Preusser HJ. *Medical Mycology: Proceedings of Mycological Symposia of the XII International Congress of Microbiology.* Gustav Fischer Verlag, Stuttgart: 1980.
186. Purchase IFH. *Mycotoxins in Human Health.* MacMillan, Edinburgh: 1971.
187. Rai M, Bridge PD. *Applied Mycology. CAB International,* Wallingford: 2009.
188. Raper KB, Fennell DI. *The Genus Aspergillus.* Williams & Wilkins, Baltimore: 1977.
189. Rayner ADM, Brasier CM, Moore D. *Evolutionary Biology of the Fungi.* Cambridge University Press, New York: 1987.
190. Razzaghi-Abyaneh M, Shams-Ghahfarokhi M, Rai M. *Medical Mycology: Current Trends and Future Prospects.* Productivity Press (CRC Press), Boca Raton: 2015.
191. Reiss E, Shadomy HJ, Lyon GM. *Fundamental Medical Mycology,* Wiley-Blackwell, New Jersey: 2012.
192. Reynolds DR, Taylor JW. *The Fungal Holomorph: Mitotic, Meiotic and Pleomorphic Speciation in Fungal Systematics.* CAB International, Wallingford: 1993.
193. Richardson M, Jones BL, Rautemaa R. *Therapeutic Guidelines in Systemic Fungal Infections.* 4th edn. Remedica, London: 2007.
194. Richardson MD, Warnock DW. *Fungal Infection: Diagnosis and Management.* 4th edn. Blackwell, Oxford: 2012.
195. Rippon JW, Fromtling RA. *Cutaneous Antifungal Agents: Selected Compounds in Clinical Practice and Development.* Marcel Dekker, New York: 1993.

196. Rippon JW. *Medical Mycology: The Pathogenic Fungi and the Pathogenic Actinomycetes*. 2nd & 3rd eds. WB Saunders, Philadelphia: 1982 & 1988.
197. Roberts DT, Evans EGV, Allen BR. *Fungal Infection of Nail*. 2nd edn. Mosby-Wolfe, London: 1998.
198. Roberts SOB, Hay RJ, Mackenzie DWR. *A Clinician's Guide to Fungal Disease*. Marcel Dekker, New York: 1984.
199. Ruiz-Herrera J. *Dimorphic Fungi: Their importance as Models for Differentiation and Fungal Pathogenesis*. Bentham Science Publishers, Sharjah, UAE: 2012.
200. Ruiz-Herrera J. *Fungal Cell Wall: Structure, Synthesis and Assembly*. CRC Press, Boca Raton: 1991.
201. Ryley JF. *Chemotherapy of Fungal Diseases*. Springer-Verlag, Berlin: 1990.
202. Saltarelli CG. *Candida albicans: The Pathogenic Fungus*. Hemisphere Publishing, New York: 1989.
203. Samaranayake LP, MacFarlane TW. *Oral Candidosis*. Butterworths Scientific, London: 1990.
204. Samson RA, *et al. Health Implications of Fungi in Indoor Environments*. Elsevier, Amsterdam: 1994.
205. Samson RA, Hocking AD, Pitt JI, King AD. *Modern Methods in Food Mycology*. Elsevier, Amsterdam: 1992.
206. Samson RA, Hoekstra ES, Frisvad JC. *Introduction to Food and Airborne Fungi*. 7th edn. CBS, Utrecht: 2004.
207. Samson RA, Pitt JI. *Advances in Penicillium and Aspergillus Systematics*. Plenum Press, New York: 1985.
208. Samson RA, Pitt JI. *Modern Concepts in Penicillium and Aspergillus Classification*. Plenum Press, New York: 1990.
209. Samson RA, Pitt JI. *Integration of Modern Taxonomic Methods for Penicillium and Aspergillus Classification*. Harwood Academic Publishers, Australia: 2000.
210. San-Blas G, Calderone RA. *Pathogenic Fungi: Structural Biology and Taxonomy*. Vol. 1: Caister Academic Press, Norfolk: 2004.
211. San-Blas G, Calderone RA. *Pathogenic Fungi: Host Interactions and Emerging Strategies for Control*. Vol. 2: Caister Academic Press, Norfolk: 2004.
212. San-Blas G, Calderone RA. *Pathogenic Fungi: Insights in Molecular Biology*. Caister Academic Press, Norfolk: 2008.
213. Sarosi GA, Davies SF. *Fungal Diseases of the Lung*. 3rd edn. Lippincott Williams & Wilkins, Philadelphia: 2000.
214. Sattler FR, Walzer PD. *Pneumocystis carinii*. 1st edn. Bailliere Tindall, London: 1995.
215. Schubler A, Walker C. *The Glomeromycota: A Species List With New Families and New Genera*. Gloucester: 2010.
216. Schwarz J. *Histoplasmosis*. Praeger Publishers, New York: 1981.
217. Segal E, Baum GL. *Pathogenic Yeasts and Yeast Infections*. CRC Press, Boca Raton: 1994.
218. Sharma RP, Salunkhe DK. *Mycotoxins and Phytoalexins*. CRC Press, Boca Raton: 1991.
219. Singh J, Singh J. *Building Mycology: Management of Decay and Health in Buildings*. 1st edn. Taylor & Francis, London: 1994.
220. Singh SM, Naidu J. *Current Advances in Medical Mycology*. Ravi Publishers, Jabalpur: 1998.
221. Sinha KK, Bhatnagar D. *Mycotoxins in Agricultural and Food Safety*. Marcel Dekker, New York: 1998.
222. Smith JE, Henderson RS. *Mycotoxins and Animal Foods*. CRC Press, London: 1991.
223. Smith JE, Moss MO. *Mycotoxins: Formation, Analysis and Significance*. John Wiley, Chichester: 1985.
224. Smith JE. *Aspergillus (Biotechnology Handbooks. Vol. 7)*. Plenum Press, New York: 1994.
225. Smith JMB. *Opportunistic Mycoses of Man and Other Animals*. CAB International, Wallingford: 1989.
226. Sobel JD, Vazquez JA. *Contemporary Diagnosis and Management of Fungal Infections*. Handbooks in Health Care Co., Newtown, Pennsylvania: 2003.
227. Sobel JD. *Clinical Perspectives - Terconazole: An Advance in Vulvovaginal Candidiasis Therapy*. McGraw-Hill, New York: 1998.
228. Speller DCE. *Antifungal Chemotherapy*. John Wiley, Chichester: 1980.
229. Spencer JFT, Spencer DM. *Yeasts in Natural and Artificial Habitats*. Springer, Berlin: 1997.
230. Spoerke DG, Rumack BH. *Handbook of Mushroom Poisoning: Diagnosis and Management*. CRC Press, Boca Raton: 1994.
231. Srivastava OP, Srivastava AK, Shukla PK. *Advances in Medical Mycology*. Evoker Research Perfecting, Lucknow: 1997.
232. Staib F, Huhn D. *Fungal Infections in Defence-Impaired Patients*. Springer-Verlag, Berlin: 1991.
233. Stevens DA. *Coccidioidomycosis: A Text*. Plenum, New York: 1980.
234. Suhonen RE, Dawber RPR, Ellis D. *Fungal Infections of the Skin, Hair and Nails*. Martin Dunitz, London: 1999.
235. Sullivan DJ, Moran GP. *Human Pathogenic Fungi: Molecular Biology and Pathogenic Mechanisms*. Caister Academic Press, Norfolk: 2014.
236. Sutton B. *A Century of Mycology*. Cambridge University Press, Port Chester: 1996.
237. Sutton DA, Fothergill AW, Rinaldi MG. *Guide to Clinically Significant Fungi*. 1st edn. Williams & Wilkins, Baltimore: 1998.
238. Sweany HC. *Histoplasmosis*. Charles C Thomas, Springfield: 1960.
239. Szaniszlo PJ, Harris JL. *Fungal Dimorphism: With Emphasis on Fungi Pathogenic for Humans*. Plenum Press, New York: 1985.
240. Trucksess MW, Pohland AE. *Mycotoxin Protocols*. Humana Press, Totowa: 2001.
241. Tumbay E, Seeliger HPR, Ano O. *Candida and Candidamycosis*. Plenum Press, New York: 1991.
242. Ueno Y. *Trichothecenes: Chemical, Biological and Toxicological Aspects*. Elsevier, Amsterdam: 1983.
243. Usha Sarma P, Sharma GL, Das RH. *Recent Trends in Mycoses*. Centre for Biochemical Technology, Delhi: 1998.
244. Van Egmond HP. *Mycotoxin in Dairy Products*. Elsevier Applied Science, London: 1989.
245. Vanbreuseghem R, De Vroey CH. *Sexuality and Pathogenicity of Fungi*. Masson Publishing, New York: 1981.

246. Vanden Bossche H, Mackenzie DWR, Cauwenbergh G, Van Cutsem J, Drouhet E, Dupont B. *Mycoses in AIDS Patients*. Plenum Press, New York: 1990.
247. Vanden Bossche H, Mackenzie DWR, Cauwenbergh G. *Aspergillus and Aspergillosis*. Plenum Press, New York: 1988.
248. Vanden Bossche H, Odds FC, Kerridge D. *Dimorphic Fungi in Biology and Medicine*. Plenum Press, New York: 1993.
249. Vandepitte J. *Clinical Aspects of Cryptococcosis in Patients with AIDS*. Plenum Press, New York: 1990.
250. Vashisth P. *Veterinary and Medical Mycology*. 1st edn. Oxford Book Company, Jaipur: 2014.
251. Venugopal PV, Venugopal TV. *Mycoses in the New Millennium*. Udayam Offset, Chennai: 2000.
252. Verbou JL. *Superficial Fungal Infections: New Clinical Applications in Dermatology*. MTP Press, Lancaster: 1986.
253. Walzer PD. *Pneumocystis carinii Pneumonia*. 2nd edn. Marcel Dekker, New York: 1994.
254. Warnock DW, Richardson MD. *Fungal Infection in the Compromised Patient*. 2nd edn. John Wiley, Chichester: 1991.
255. Watkinson S, Boddy L, Money N. *The Fungi*. 3rd edn. Academic Press, London: 2016.
256. Webster J, Weber R. *Introduction to Fungi*. 3rd edn. Cambridge University Press, Cambridge: 2007.
257. Weidenborner M. *Encyclopaedia of Food Mycotoxins*. 1st edn. Springer-Verlag, Berlin, Heidelberg: 2001.
258. Weiss LM, Becnel JJ. *Microsporidia: Pathogens of Opportunity*. 1st edn. John Wiley & Sons, New York: 2014.
259. Wheat LJ. *Fungal Diseases of the Lung*. 2nd edn. Raven Press, New York: 1993.
260. Wilson JW, Plunkett OA. *The Fungous Diseases of Man*. University of California Press, Berkeley: 1965.
261. Wingard JR, Anaissie EJ. *Fungal Infections in the Immunocompromised Patient* (*Infectious Disease and Therapy*. Vol. **34**). Taylor & Francis, Boca Raton: 2005.
262. Wyllie TD, Morehouse LG, *Mycotoxic Fungi, Mycotoxins, Mycotoxicoses: An Encyclopedic Handbook*. Marcel Decker, New York: 1978.
263. Yamaguchi H, Kobayashi GS, Takahashi H. *Recent Progress in Antifungal Chemotherapy*. Marcel Dekker, New York: 1992.
264. Young LS. *Pneumocystis carinii Pneumonia: Pathogenesis, Diagnosis and Treatment*. 1st edn. Marcel Dekker, New York: 1984.
265. Zaias N. *The Nail in Health and Disease*. 2nd edn. Appleton & Lange, Norwalk: 1990.
266. Zaini F, Mehbod ASA, Emami M. *Comprehensive Medical Mycology*. 3rd edn. University of Tehran Press, Tehran: 2013.

Books on Mycology Practical Aspects

In dealing with practical aspects in Medical Mycology also, large number of Books, Manuals and Monographs have been written covering most of technical aspects of handling and processing of specimens as well as identification of medically important fungi. These are listed below:

1. Al-Doory Y. *Laboratory Medical Mycology*. Lea & Febiger, Philadelphia: 1980.
2. Bancroft JD, Gamble M. *Theory and Practice of Histological Techniques*. 6th ed. Churchill Livingstone, Oxford: 2008.
3. Barnett JA, Payne RW, Yarrow D. *Yeasts: Characteristics and Identification*, 3rd edn. Cambridge University Press, Cambridge: 2000.
4. Beneke ES, Rogers AL. *Medical Mycology Manual with Human Mycoses Monograph*. 4th edn. Burgess Publishing, Minneapolis: 1980.
5. Beneke ES. *Scope Monograph on Human Mycoses* (Ed.-Thomas BA), The Upjohn Co., Kalamazoo, Michigan: 1970.
6. Bennett JW, Lasure LL. *More Gene Manipulations in Fungi*. Academic Press, San Diego: 1991.
7. Booth C. *Fusarium: Laboratory Guide to the Identification of Major Species*. CMI, Kew: 1977.
8. Bridge P, Couteaudier Y, Clarkson J. *Molecular Variability of Fungal Pathogens*. CAB International, New York: 1998.
9. Bridge PD, Arora DK, Reddy CA, Elander RP. *Applications of PCR in Mycology*, CAB International, Wallingford: 1998.
10. Campbell CK, Johnson EM, Warnock DW. *Identification of Pathogenic Fungi*. 2nd edn. Health Protection Agency, Wiley Blackwell, Chichester: 2013.
11. Campbell MC, Stewart JL. *The Medical Mycology Handbook*. John Wiley, New York: 1980.
12. Chandler FW, Kaplan W, Ajello L. *A Colour Atlas and Textbook of the Histopathology of Mycotic Diseases*. Wolfe Medical Publications, London: 1980.
13. Chandler FW, Watts JC. *Pathologic Diagnosis of Fungal Infections*. ASCP Press, Chicago: 1987.
14. Chandler FW, Watts JC. *Pathologic Diagnosis of Fungal, Actinomycetal and Algae Infections*. Raven Press, New York: 1988.
15. Clayton YM, Midgley G. *Medical Mycology: Pocket Picture Guide Series*. Gower Medical Publishing, London: 1985.
16. Coleman D, Haynes K. *Methods in Molecular Medicine Series: Medical Mycology Protocols*. Humana Press, Totowa: 1999.
17. Conant NF, Smith DT, Baker RD, Callaway JL. *Manual of Clinical Mycology*. 3rd edn. WB Saunders, Philadelphia: 1971.
18. Crissey JT, Lang H, Parish LC. *Manual of Medical Mycology*. Blackwell Science, Cambridge: 1995.
19. de Hoog GS, Guarro J, Gene J, Figueras MJ. *Atlas of Clinical Fungi*. 1st edn. Centraalbureau voor Schimmelcultures, Baarn and Delft, The Netherlands and Universitat Rovira i Virgili, Reus, Spain: 1995.
20. de Hoog GS, Guarro J, Gene J, Figueras MJ. *Atlas of Clinical Fungi: The Ultimate Benchtool for Diagnostics*. 2nd edn/ Electronic Version 4.1. Westerdijk Fungal Biodiversity Institute (Former CBS-KNAW), Utrecht, The Netherlands and Universitat Rovira i Virgili, Reus, Spain: 2000/2017.
21. Delacretaz J, Grigoriu D, Ducel G. *Colour Atlas of Medical Mycology*. Hans Huber Publishers, Bern: 1976.
22. Diamond RD, Mandell GL. *Atlas of Infectious Diseases (Vol. XII): Fungal Infections*. 1st edn. Current Medicine, Philadelphia: 2000.

23. Domsch KH, Gams W, Anderson TH. *Compendium of Soil Fungi*. Vol. 1. Academic Press, London: 1980.
24. Dugan FM. *The Identification of Fungi: An Illustrated Introduction with Keys, Glossary and Guide to Literature*. American Phytopathological Society, St. Paul, Minnesota: 2006 (Digitalized 2010).
25. Elewski BE, Rinaldi MG, Weitzman I. *Diagnosis and Treatment of Onychomycosis: A Clinician's Handbook*. Gardiner-Caldwell SynerMed, California: 1995.
26. Ellis D, Davis S, Alexiou H, Handke R, Bartley R. *Descriptions of Medical Fungi*. 2nd edn. Nexus Print Solutions, Adelaide, South Australia: 2007.
27. Evans EGV, Richardson MD. *Medical Mycology: A Practical Approach*. IRL Press, Oxford: 1989.
28. Fisher F, Cook NB. *Fundamentals of Diagnostic Mycology*. 1st edn. WB Saunders, Philadelphia: 1998.
29. Frey D, Oldfield RJ, Bridger RC. *A Colour Atlas of Pathogenic Fungi*. Wolfe Medical Publications, Sydney: 1979.
30. Gams W, Hoekstra ES, Aptroot A. *CBS Course of Mycology*. 4th edn. CBS, Baarn: 1998.
31. Gupta VK, Tuohy MG, Ayyachamy M, Turner KM, O'Donovan A. *Laboratory Protocols in Fungal Biology: Current Methods in Fungal Biology*. 1st edn. Springer, New York: 2013.
32. Hanlin R, Ulloa M. *Atlas of Introductory Mycology*. 2nd edn. Hunter Textbooks, Inc., Winston-Salem: 1988.
33. Hazen EL. *Laboratory Identification of Pathogenic Fungi Simplified*. 3rd edn. Charles C Thomas, Springfield: 1973.
34. Holmberg K, Meyer RD. *Diagnosis and Therapy of Systemic Fungal Infections*. Raven Press, New York: 1989.
35. Jong SC, Birmingham JM, Ma G. *Stedman's ATCC Fungus Names*. Williams & Wilkins, Baltimore: 1993.
36. Jong SC, Dugan F, Edwards MLS. *ATCC Filamentous Fungi*. ATCC, Rockville: 1996.
37. Joshi KR. *Opportunistic Mycoses: Practical Manual for Diagnosis and Infections*. Scientific Publishers, Jodhpur: 1995.
38. Kane J, Summerbell RC, Sigler L, Krajden S, Land G. *Laboratory Handbook of Dermatophytes: A Clinical Guide and Laboratory Manual of Dermatophytes and other Filamentous Fungi from Skin, Hair and Nails*. Star Publishing, Belmont: 1997.
39. Kauffman CA, Mandell GL, eds. *Atlas of Fungal Infections*. 2nd edn. Springer, Philadelphia: 2007.
40. Kauffman CA. *Essentials of Clinical Mycology*, Springer, Philadelphia: 2011
41. Kern ME, Blevins KS. *Medical Mycology: A Self-Instructional Text*. 2nd edn. FA Davis Press, Philadelphia: 1997.
42. Kidd S, Halliday C, Alexiou H, Ellis D. *Descriptions of Medical Fungi*. 3rd edn. Newstyle Printing, Adelaide, South Australia: 2016.
43. Kirk PM, Ansell AE. *Authors of Fungal Names*. CAB International, Wallingford: 1992.
44. Koneman EW, Roberts GD. *Practical Laboratory Mycology*. 3rd edn. Williams & Wilkins, Baltimore: 1985.
45. Larone DH. *Medically Important Fungi: A Guide to Identification*. 5th edn. ASM Press, Washington, DC: 2011.
46. Mackenzie DWR, Philpot CM. *Isolation and Identification of Ringworm Fungi*. PHLS Monograph No. 15, HMSO, London: 1981.
47. Mackenzie DWR, Philpot CMB, Proctor AGJ. *Basic Serodiagnostic Methods for Diseases caused by Fungi and Actinomycetes*. PHLS Monograph No. 12, HMSO, London: 1980.
48. Maresca B, Kobayashi GS. *Molecular Biology of Pathogenic Fungi: A Laboratory Manual*. 2nd edn. Telos Press, New York: 1994.
49. McGinnis MR, D'Amato RF, Land GA. *Pictorial Handbook of Medically Important Fungi and Aerobic Actinomycetes*. Praeger, New York: 1982.
50. McGinnis MR. *Laboratory Handbook of Medical Mycology*. Academic Press, New York: 1980.
51. Midgley G, Clayton YM, Hay RJ. *Diagnosis in Color: Medical Mycology*. Mosby-Wolfe, Chicago: 1997.
52. Miyaji M. *Animal Models in Medical Mycology*. 1st edn. CRC Press, Boca Raton: 1987.
53. Moore GS, Jaciow DM. *Mycology for the Clinical Laboratory*. Prentice-Hall, Reston: 1979.
54. Moss ES, McQuown AL. *Atlas of Medical Mycology*. 3rd edn. Williams & Wilkins, Baltimore: 1969.
55. Ngan Antonio HY. *Atlas of Fungi*. 1st edn. June Just Printing Company Ltd., Hong Kong, China: 2016.
56. Palmer DF, Kaufman L, Kaplan W, Cavallaro JJ. *Serodiagnosis of Mycotic Diseases*. Charles C Thomas, Springfield: 1977.
57. Rebell G, Taplin D. *Dermatophytes: Their Recognition and Identification*. University of Miami Press, Miami: 1979.
58. Richardson MD, Johnson EM. *The Pocket Guide to Fungal Infection*. 2nd edn. Blackwell Publishing, Oxford: 2006.
59. Richardson MD, Warnock DW, Campbell CK. *Slide Atlas of Fungal Infections: The Set*. Blackwell Science, Oxford: 1995.
60. Roberts GD. *Mycology: Laboratory Procedure Manual*. Mayo Foundation, Rochester: 1987.
61. Salfelder K, de Liscano TR, Sauerteig E. *Atlas of Fungal Pathology*. *Current Histopathology* (Series Vol. 17). Kluwer Academic Publisher, Dordrecht, Boston, London: 1990.
62. Sciortino C. *Atlas of Clinically Important Fungi*. Wiley-Blackwell, New Jersey: 2017.
63. Smith D. *Catalogue of the Culture Collection of the CAB International Mycological Institute*. 9th edn. CAB International, Wallingford: 1988.
64. St-Germain G, Summerbell R. *Identifying Fungi: A Clinical Laboratory Handbook*. Star Publishing, Belmont: 2011.
65. Sugar AA, Lyman CA. *A Practical Guide to Medically Important Fungi and the Diseases They Cause*. Lippincott Williams & Wilkins, Philadelphia: 1997.
66. Tewari RP, Kohler RB, Wheat LJ. *Fungal Antigens, Isolation, Purification and Detection*. Plenum Press, New York: 1989.
67. Vanbreuseghem R, De Vroey CH, Takashio M. *Practical Guide to Medical and Veterinary Mycology*. Masson Publishing, New York: 1978.
68. Wainwright M. *An Introduction to Fungal Biotechnology*. John Wiley, Chichester: 1992.

69. Wang CJK, Zabel RA. *Identification Manual for Fungi from Utility Poles in the Eastern United States*. ATCC, Rockville: 1990.
70. Wentworth BB. *Diagnostic Procedures for Mycotic and Parasitic Infections*. 7th edn. American Public Health Association, Washington, DC: 1988.

Books on Microbiology and Allied Subject

There are many books of a particular speciality covering mycological aspects along with their own subject like Dermatology, Infectious Diseases, etc. They have devoted substantial number of pages or chapters in the voluminous books and are very informative for the readers to go through them. These books are listed below:

1. Aly R, Maibach HI. *Atlas of Infections of the Skin*. Churchill Livingstone. New York: 1999.
2. An Z. *Handbook of Industrial Mycology*. Marcel Dekker, New York: 2005.
3. Anonymous. *Categorization of Pathogens According to Hazard and Categories of Containment*. 2nd edn. HSMO, London: 1990.
4. Armstrong D, Cohen J. *Infectious Diseases*, Mosby, St. Louis: 1999.
5. Bancroft JD, Gamble M. *Theory and Practice of Histopathological Techniques*. Churchill Livingstone/ Elsevier, 6th edn. London: 2008.
6. Baron EJ, Chang RS, Howard DH, Miller JN, Turner JA. *Medical Microbiology: A Short Course*. John Wiley, New York: 1994.
7. Baron S. *Medical Microbiology*. 3rd edn. Churchill Livingstone, Edinburgh: 1991.
8. Bennett IE, Dolin R, Blaser MJ. *Mandell, Douglas and Bennett's Principles and Practice of Infectious Diseases*. 8th edn (2 Vol. Set), Saunders/Elsevier: 2015.
9. Bhatnagar D, Lillehoj EB, Arora DK. *Handbook of Applied Mycology*. Vol. 5. *Mycotoxins in Ecological Systems*. Marcel Dekker; New York: 1992.
10. Blackwell McGinnis MR, Tilton RC. *Fungi and Actinomycetes*. *In:* Howard BJ, Keiser JF, Weissfeld AS, Smith TF, Tilton RC. *Clinical and Pathological Microbiology*. 2nd edn. Mosby, London: 1994.
11. Bowker RR. *Ulrich's International Periodicals Directory*. 35th edn. Vol. 1-5. The Bowker International Serials Database, New Providence: 1997.
12. Braude AI, Davis CE, Fiere J. *Infectious Diseases and Medical Microbiology*. 2nd edn. WB Saunders, Philadelphia: 1986.
13. Braunwald E, Fauci AS, Kasper DL, Hauser SL, Longo DL, Jameson JL. *Harrison's Principles of Internal Medicine*. 15th edn. Vol. 1. McGraw-Hill, New York: 2001: 1168-85.
14. Brown EM, Simione FP. *ATCC Guide to Packaging and Shipping of Biological Materials*. American Type Culture Collection, Rockville: 1994.
15. Carroll KC, Morse SA, Mietzner T, Miller S. *Jawetz, Melnick & Adelberg's Medical Microbiology*. 27th edn. McGraw-Hill, New York: 2015.
16. Cary JW, Linz JE, Bhatnagar D. *Microbial Food-Borne Diseases: Mechanisms of Pathogenesis and Toxin Synthesis*. Technomic Publishing Co., Lancaster: 2000.
17. Collee JG, Fraser AG, Marmion BP, Simmons A. *Mackie and McCartney Practical Medical Microbiology*. 14th edn. Churchill Livingstone, London: 1996.
18. Cook G. *Manson's Tropical Diseases*. 20th edn. ELBS-WB Saunders, London: 1996.
19. Cox CS, Wathes CM. *Bioaerosols Handbook*. CRC Press, Boca Raton: 1995.
20. Drutz DJ. *Fungal Diseases*. *In*: Stites DP, Terr AI. *Basic and Clinical Immunology*. 7th edn. Prentice-Hall International, London: 1990: 657-71.
21. Forbes BA, Sahm DF, Weissfeld AS. *Bailey & Scott's Diagnostic Microbiology*. 12th edn. Mosby, London: 2007.
22. Gorbach SL, Bartlett JG, Blacklow NR. *Infectious Diseases*. WB Saunders, Philadelphia: 1998: 2313-92.
23. Greenwood D, Slack RCB, Peutherer JF. *Medical Microbiology: A Guide to Microbial Infections*. 16th edn. ELST-Churchill Livingstone, Edinburgh: 2006.
24. Guerrant RL, Walker DH, Weller PF. *Tropical Infectious Diseases: Principles, Pathogens and Practice*. 2nd edn (2 Vol.), Churchill Livingstone, New York: 2005.
25. Hardman JG, Limbird LE, Molinoff PB, Ruddon RW, Goodman AG. *Goodman and Gilman's the Pharmaceutical Basis of Therapeutics*. 9th edn. McGraw-Hill, New York: 1996.
26. Hay RJ, Ashbee HR. *Mycology*. *In:* Burns DA, Breathnach SM, Cox N, Griffiths C. *Rook's Textbook of Dermatology*. 8th edn. Vol. 2. Blackwell Science, Oxford: 2010.
27. Hay RJ. *The Mycoses*. *In:* Strickland GT. *Hunter's Tropical Medicine and Emerging Diseases*. 8th edn. WB Saunders, Philadelphia: 2000: 529-573.
28. Hearn VM, Cole GT, Susuki S. *Fungal Antigens*. *In:* Van Regenmortel. *Structure of Antigens*. CRC Press, Boca Raton: 1993: 211-260.
29. Helferich W, Winter CK. *Food Toxicology*. CRC Press, Boca Raton: 2000.
30. Ichhpujani RL, Bhatia R. *Medical Parasitology*. Jaypee Brothers, New Delhi: 1994: 155-57.
31. Innis MA, Gelfand DH, Sninsky JJ, White TJ. *PCR Protocols: A Guide to the Methods and Applications*. Academic Press, San Diego: 1990.
32. Jeffrey C. *Biological Nomenclature*. 3rd edn. Arnold, London: 1989.
33. Joklik WK, Willett HP, Amos DB, Wilfert CM. *Zinsser's Microbiology*. 20th edn. Appleton & Lange, Norwalk: 1992.
34. Kanwar AJ, Mamta, Chander J. *Superficial Fungal Infections*. *In:* Valia RG, Valia AR. *IADVL Textbook and Atlas of Dermatology*. 2nd edn. Bhalani Publishing House, Mumbai: 2001: 215-58.
35. Kirsop BE, Doyle A. *Maintenance of Microorganisms and Cultured Cells: A Manual of Laboratory Methods*. 2nd edn. New York: 1991.
36. Logan J, Edwards K, Saunders N. Real-Time PCR: *Current Technology and Applications*. Caister Academic Press, Norfolk: 2009.

37. Miller JM. *A Guide to Specimen Management in Clinical Microbiology*. ASM Press, Washington, DC: 1996.
38. Moschella SL, Hurley HJ. *Dermatology*. 3rd edn. Vol. 1. WB Saunders, Philadelphia: 1992.
39. Murray PR, Rosenthal KS, Pfaller MA. *Medical Microbiology*. 8th edn. Elsevier, London: 2016.
40. Nahmias AJ, O'Reilly RJ. *Immunology of Human Infections*. Part I. *Bacteria, Mycoplasma, Chlamydia and Fungi*. Plenum Medical Books, New York: 1981.
41. National Academy of Sciences. *Guide for the Care and Use of Laboratory Animals*. National Academy Press, Washington, DC: 1996.
42. Nuovo CJ. *PCR in situ Hybridisation: Protocols and Applications*. 3rd edn. Lippincott-Raven, Philadelphia: 1997.
43. Paddock SW. *Confocal Microscopy: Methods and Protocols*. Humana Press, Totowa: 1998.
44. Parfitt K. *Martindale - The Complete Drug Reference*. 37th edn. Pharmaceutical Press, London: 1999.
45. Parija SC. *Textbook of Medical Parasitology: Protozoology & Helminthology*. 4th edn. All India Publishers & Distributors, Chennai: 2013.
46. Persing DH, Smith TF, Tenover FC, White TJ. *Diagnostic Molecular Microbiology: Principles and Applications*. ASM Press, Washington, DC: 1993.
47. Pohland AE, Dowell VR, Richard JC. *Microbial Toxins in Foods and Feeds: Cellular and Molecular Modes of Action*. Plenum Press, New York: 1990.
48. Procop GW, Koneman EW. *Koneman's Color Atlas and Textbook of Diagnostic Microbiology*. 7th edn. Wolters Kluwer, Philadelphia: 2016.
49. Reiss E. *Molecular Immunology of Mycotic and Actinomycotic Infections*. Elsevier, New York: 1986.
50. Remington JS, Swartz M. *Current Topics of Clinical and Infectious Diseases*. Blackwell Science, Boston: 1997.
51. Rose NR, Hamilton RG, Detrick B. *Manual of Clinical Laboratory Immunology*. 6th edn. ASM Press, Washington, DC: 2002.
52. Scher RK, Daniel CR III. *Nails: Diagnosis, Treatment and Surgery*. 2nd edn. WB Saunders, Philadelphia: 1997.
53. Sebastian A. *A Dictionary of the History of Medicine*. Parthenon, New York: 1999
54. Sell S. *Immunology, Immunopathology and Immunity*. Elsevier, New York: 1987.
55. Shanson DC. *Microbiology in Clinical Practice*. 2nd edn. Wright: 1989.
56. Sherris JC. *Medical Microbiology: An Introduction to Infectious Diseases*. 2nd edn. Elsevier, Hanover: 1990.
57. Spicer WJ. *Clinical Bacteriology, Mycology and Parasitology: An Illustrated Color Text*. Harcourt Publishers Limited, London: 2000.
58. Venugopal PV, Venugopal TV. *Deep Fungal Infections*. *In:* Valia RG, Valia AR. *IADVL Textbook and Atlas of Dermatology*. 2nd edn. Bhalani Publishing House, Mumbai: 2001: 259-84.
59. Walzer PD, Kim CK, Cushion MT. *Pneumocystis carinii*. *In:* Walzer PD, Genta RM. *Parasitic Infections in the Compromised Host*. 1st edn. Marcel Dekker, New York: 1989: 83-178.
60. Warnock DW. *Mycology*. *In:* Versalovic J, Carroll KC, Funke G, Jorgensen JH, Landry ML, Warnock DW. *Manual of Clinical Microbiology*. 10th edn (Vol. 2). ASM Press, Washington, DC: 2011.
61. Warnock DW (Vol. ed), Brandt ME, Johnson EM (Section eds). *Mycology*. *In:* Jorgensen JH, Pfaller MA, Carroll KC, Funke G, Landry ML, Richter SS, Warnock DW. *Manual of Clinical Microbiology*. 11th edn (Vol. 2). ASM Press, Washington, DC: 2015. 1932-2281.
62. Winn WC, Jr. *Mycotic Diseases*. *In:* Henry JB. *Clinical Diagnosis and Management by Laboratory Methods*. 20th edn. WB Saunders, Philadelphia: 2001. 1210-51.

Biomedical Journals Published in Mycology

In addition to the reference books mentioned above in the list of Bibliography and general type of periodicals in biomedical sciences, there are specific journals that are exclusively publishing Original Research articles and Reviews on different domains of Medical Mycology. In addition to hard copies, electronic versions of these journals are also available. Their home pages may be visited and desired reference article or full text in any of the format, on specific aspect of Medical Mycology, may be downloaded. Most of them have very good Impact Factor. However, as far as the Case Reports are concerned, most of them have started publishing their additional journals, during the last 5 to 6 years' period, exclusively for this purpose.

A large number of journals are published in the field of Mycology in general and Medical Mycology in particular. Some of them are Open Access journals hence available free of cost but rest are chargeable and articles can be downloaded only on payment. However, some of them become free of cost after being chargeable during the initial period of publication like six months to a year or so *e.g.* ASM journals. These are listed below in an alphabetical order along with their current websites. However, website may be changed depending on the renewal of contract of a particular journal with its publishing agency:

1. *Abstracts of Mycology (http://scientific.thomson.com)*
2. *Current Fungal Infection Reports (http://www.springer.com)*
3. *Current Medical Mycology (http://cmm.mazums.ac.ir)*
4. *Current Topics in Medical Mycology (http://journalseek.net)*
5. *Experimental Mycology (http://www.sciencedirect.com)*
6. *Fungal Planet (http://www.fungalplanet.org)*
7. *Frontier in Fungi and Their Interactions (http://www.frontiersin.org)*
8. *IMA Fungus (http://www.imafungus.org)*
9. *Japanese Journal of Medical Mycology (http://www.jsmm.org)*
10. *Journal de Mycologie Medicale (http://www.sciencedirect.com)*

11. *Journal of Fungi (http://www.mdpi.com/journal/jof)*
12. *Journal of Invasive Fungal Infections (http://www.invasive-fungalinfections.com)*
13. *Journal of Mycology (http://about.hindawi.com)*
14. *Medical Mycology (Formerly Sabouraudia/Journal of Medical and Veterinary Mycology) (http://mmy.oxfordjournals.org)*
15. *Medical Mycology Case Reports (http://www.elsevier.com/locate/mmcr)*
16. *Micologia (http://www.RevIberoamMicol.com)*
17. *Mycobiology (mycobiology.or.kr)*
18. *Mycologia (http://www.mycologia.org)*
19. *Mycological Research (http://www.elsevier.com)*
20. *Mycologist- Fungal Biodiversity Reviews (http://www.elsevier.com)*
21. *Mycopathologia (Former Mycopathologia et Mycologia Applicata) (http://www.iums.org/index.php/publications)*
22. *Mycoscience (http://www.springer.com)*
23. *Mycoses/Mykoses (Former Mykosen till 1988) (http://www.blackwell-Synergy.com)*
24. *Mycotaxon (http://www.mycotaxon.com)*
25. *Mycotoxin Research (http://www.springer.com)*
26. *Persoonia (http://www.persoonia.org)*
27. *Review of Medical and Veterinary Mycology (http://www.cabi.org)*
28. *Studies in Mycology (http://www.studiesinmycology.org)*
29. *Transactions of the British Mycological Society (http://www.cybertruffle.org.uk)*
30. *Transactions of the Mycological Society of Japan (http://journalseek.net)*
31. *World Mycotoxin Journal (http://wageningen academic.metapress.com)*
32. *Yeast (http://www.wiley.com)*

The *ISHAM Mycoses Newsletter* is also informative source for day-to-day developments and events in Medical Mycology and is being regularly published by International Society for Human and Animal Mycology (ISHAM). The ISHAM News is an online update and can be subscribed through email as well *(website: www.isham.org)*.

On the similar pattern, Society for Indian Human and Animal Mycologists (SIHAM) is also publishing its Newsletter since November 2002. The periodic updates can be seen on its website *(www.siham.in)*.

There are many organizations which have come up in the recent past in India. The Fungal Infection Study Forum (FISF), is one holding time to time meetings on various issues pertaining to fungal infections. This trust was instituted on March 3, 2012 in Mumbai and is an independent working group consisting of clinicians and mycologists *(website: www.fisf.org)*.

There are many biomedical journals dealing with infectious diseases in various specialties besides above-mentioned list of journals. They are also regularly publishing articles on clinical and mycological aspects of fungal diseases from time to time. These biomedical journals of mycological importance are enumerated as follows:

1. *Antonie van Leeuwenhoek (http://www.springer.com)*
2. *British Journal of Dermatology (http://www.blackwellpublishing.com)*
3. *Clinical Infectious Diseases (http://www.journals.uchicago.edu) (Former Reviews of Infectious Diseases)*
4. *Clinical Microbiology and Infection (http://www.blackwell-publishing.com)*
5. *Clinical Microbiology Reviews (http://cmr.asm.org)*
6. *Current Medical Mycology (http://cmm.mazums.ac.ir)*
7. *Dermatology (Since 1992, Former Dermatologica) (http://www.karger.com)*
8. *Emerging Infectious Diseases (http://www.cdc.gov/eid)*
9. *Emerging Microbes & Infections (http://www.nature.com/emi)*
10. *Expert Review of Molecular Diagnostics (http://www.tandfonline.com/toc)*
11. *IDCases (https://www.journals.elsevier.com/idcases)*
12. *Infection (http://www.springer.com)*
13. *Infection and Immunity (http://iai.asm.org)*
14. *International Journal of Dermatology (http://www.blackwellpublishing.com)*
15. *Journal of American Academy of Dermatology (http://www.eblue.org)*
16. *Journal of Clinical Microbiology (http://jcm.asm.org)*
17. *Journal of Infection (http://www.elsevier.com)*
18. *Journal of Infectious Diseases (http://www.journals.uchicago.edu)*
19. *Journal of Medical Microbiology (http://jmm.sgmjournals.org)*
20. *mBio (http://mbio.asm.org)*
21. *Microbes and Infection (http://www.elsevier.com)*
22. *mSphere (http://msphere.asm.org)*
23. *Reviews in Medical Microbiology (http://www.revmedmicrobiol.com)*
24. *Tropical Diseases Bulletin (http://www.cabi.org)*

Supplementary Issues on Medical Mycology

In recent years following journals have been devoted to various aspects of Medical Mycology in the form of either Supplementary or Special Issues. These are referred under authorship of Guest Editors and are as follows:

1. Ampel NM. Symposium on coccidioidomycosis. *Semin Respir Infect.* 2001; **16:** 229-85.
2. Anaissie EJ, Galgiani JN, Rinaldi MG. Focus on Fungal Infections-3. *Clin Infect Dis.* 1994; **19 (Suppl. 1):** S1-58.
3. Anaissie EJ. Focus on Fungal Infections-5. *Clin Infect Dis.* 1996; **22 (Suppl. 2):** S71-184.
4. Behrens-Baumann W. *Mycosis of the Eye and its Adnexa. Dev Ophthalmol.* 1999; **32:** 1-202.

5. Bennett JE. Fluconazole: A Novel Advance in Therapy for Systemic Fungal Infections. *Rev Infect Dis.* 1990; **12 (Suppl. 3):** S263-89.
6. Bernhardt H, Muller J. Fluconazole. *Mycoses.* 1996; **36 (Suppl. 2):** S1-65.
7. Bouchara JP, Mignon B. Dermatophytes and Dermatophytoses. *Mycopathologia.* 2008: **166:** 235-352.
8. Bouchara JP, Mignon B, Chaturvedi V. Dermatophytes and dermatophytoses: A thematic overview of state of the art and the directions for future research and developments (Editorial of Special Issue). *Mycopathologia.* 2017; **182:** 1-4.
9. Chiu AM, Fink JN. Fungal Allergy and Pathogenicity. *Chem Immunol.* 2002; **81:** 1-295.
10. Cornely O. A review of caspofungin: Translating efficacy into effectiveness in the management of invasive fungal infections. *Mycoses.* 2008; **51 (Suppl. 1):** 1-67.
11. Cornely O. Optimizing the use of posaconazole in prophy laxis and therapy of fungal infections. *Mycoses.* 2008; **51 (Suppl. 2):** 1-61.
12. Cornely O. Present options and future trends in invasive fungal infection. *Mycoses.* 2007; **50 (Suppl. 1):** 1-55.
13. Cramer RA, Sheppard DC, Clemons KV. 7th Advances against aspergillosis: Basic, diagnostic, clinical and therapeutic studies. *Med Mycol.* 2017; **55:** 1-3.
14. Cushion MT. *Pneumocystis carinii.* Proceedings of 6th International Workshop on Opportunistic Protists, Raleigh, North Carolina, USA; 1999. *J Eukaryot Microbiol.* 1999; **46 (Suppl):** S85-154.
15. de Hoog S, Ibrahim AS, Voigt K. Preface. Zygomycetes: An emerging problem in the clinical laboratory. *Mycoses.* 2014; **57 (Suppl. 3):** 1-143.
16. Dei-Cas E, Cailliez JC. *Pneumocystis* and Pneumocystosis: Advances in *Pneumocystis* Research. *FEMS Immunol Med Microbiol.* 1998; **22 (Suppl. 1-2):** S1-189.
17. Dinleyici EC. Pediatric invasive fungal infections: Realities, challenges, concerns, myths and hopes. *Expert Rev Anti Infect Ther.* 2011; **9:** 273-367.
18. Drake L. Onychomycosis. *Int J Dermatol.* 1999; **38 (Suppl. 2):** S1-69.
19. Ecology and Evolution of Black Yeasts and their Relations. *Stud Mycol.* 1999; **43:** S208.
20. Elewski BE, Hay RJ. Novel Treatment Strategies for Superficial Mycosis. *J Am Acad Dermatol.* 1999; **40 (Suppl):** S1-42.
21. Elewski BE, Ogawa H. Antifungal Therapy Beyond the Year 2000. *Br J Dermatol.* 1999; **141 (Suppl. 56):** S1-40.
22. Elewski BE. Once-Weekly Fluconazole in the Treatment of Onychomycosis. *J Am Acad Dermatol.* 1998; **38:** S73-116.
23. Elgart ML. Cutaneous Mycology. *Dermatol Clin.* 1996; **14:** 1-169.
24. Evans G, Togami H. Topical Lamisil: New Indications and Dosage Forms. *Dermatology.* 1997; **194 (Suppl. 1):** S1-24.
25. Ferguson BJ. Fungal Rhinosinusitis - A Spectrum of Disease. *Otolaryngol Clin N Am.* 2000; **33:** 227-454.
26. Filler SG. Host–microbe interactions: Fungi - Editorial overview. *Curr Opin Microbiol.* 2011; **14:** 373-4.
27. Fisher JF. *Candida* urinary tract infections - Epidemiology, pathogenesis, diagnosis and treatment: Executive summary. *Clin Infect Dis.* 2011; **52 (Suppl. 6):** S429-32.
28. Guarro J, Summerbell RC, Samson RA. Onygenales: The dermatophytes, dimorphics and keratin degraders in their evolutionary context, *Stud Mycol.* 2002; **47:** 1-220.
29. Gugnani HC, Sandhu RS. Recent Advances in Respiratory Mycoses. *Indian J Chest Dis Allied Sci.* 2000; **42:** 199-368.
30. Gupta AK. Ciclopirox Nail Lacquer: The First Prescription - Topical Therapy for Onychomycosis. *J Am Acad Dermatol.* 2000; **43 (Suppl):** S55-102.
31. Hay RJ, Dupont B, Graybill JR. First International Symposium on Itraconazole. *Rev Infect Dis.* 1987; **9 (Suppl. 1):** S1-152.
32. Hay RJ. Amorolfine, a breakthrough in topical antimycotic therapy. Satellite Symposium, ISHAM. *Dermatology.* 1992; **184 (Suppl. 1):** S1-30.
33. Hipler U. Current topics in dermatomycoses. *Mycoses.* 2007; **50 (Suppl. 2):** 1-35.
34. Hospenthal DR. Pulmonary fungal infections. *Semin Respir Crit Care Med.* 2008; **29:** 101-219.
35. Idnurm A, Lin X. *Cryptococcus*: Model basidiomycetes and deadly pathogens. *Fungal Genet Biol.* 2015; **78:** 1-108.
36. Joseph WS. Special Topics on Onychomycosis. *J Am Podiatr Med Assoc.* Part 1 & 2. 1997; **87:** 497-539 & 545-574.
37. Kauffman CA. Issue Theme: Pulmonary and invasive fungal infections. *Semin Respir Crit Care Med.* 2015; **36:** 639-40.
38. Kauffman CA. Pulmonary fungal infections. *Semin Respir Crit Care Med.* 2004; **25:** 109-239.
39. Kauffman CA. The changing landscape of invasive fungal infections: Epidemiology, diagnosis and pharmacologic options. *Clin Infect Dis.* 2006; **43 (Suppl. 1):** 1-39.
40. Knox KS, Sarosi GA. Updates on individual pathogens as well as the field of fungal diagnostics and therapeutics. *Clin Chest Med.* 2009; **30:** 203-408.
41. Kurup VP, Apter AJ. Allergic Bronchopulmonary aspergillosis. *Immun Allergy Clin N Am.* 1998; **18:** S715.
42. Limper AH. *Pneumocystis carinii. Semin Respir Infect.* 1998; **13:** 277-352.
43. Marra CM. Central Nervous System Infections. *Neurol Clin.* 1999: **17:** 761-81.
44. Miller WT. Fungus Disease of the Chest. *Semin Roentgenol.* 1996; **31:** 4-87.
45. No authors listed. *Chemotherapy.* 1999; **45 (Suppl. 1):** 1-76.
46. No authors listed. Onychomycosis. *Clin Podiatr Med Surg.* 2004; **21:** 533-696.
47. No authors listed. Proceedings of the 1st International Dermatophytosis Workshop. Short Hills, New Jersey, USA. June 11, 1999. *Cutis.* 2001; 67 **(Suppl. 5):** 1-47.
48. No authors listed. Proceedings of the International Conference on the Toxicology of Fumonisin. June 28-30, 1999, Arlington, Virginia. *Environ Health Perspect.* 2001; **109 (Suppl. 2):** 237-342.
49. No authors listed. The IX International Meeting on Paracoccidioidomycosis, Aguas de Lindoia, SP, Brazil, October 2-5, 2005. *Rev Inst Med Trop Sao Paulo.* 2005; **47 (Suppl. 14):** 2-64.

50. Nosanchuk JD. *Histoplasma capsulatum*. pp. 409-26. *In:* Brown GD, Netea MG (eds). *Immunology of Fungal Infections*. Springer, Dordrecht; 2007.
51. Okay TS. Tribute to Prof. Carlos da Silva Lacaz. *Rev Inst Med Trop Sao Paulo*. 2015; **57 (Suppl. 19):** 1-64.
52. Odom RB. Update on Topical Therapy for Superficial Fungal Infections: Focus on Butenafine. *J Am Acad Dermatol*. 1997; **36 (Suppl):** S1-24.
53. Ostrosky-Zeichner L, Sobel JD. Fungal Infections. *Infect Dis Clin North Am*. 2016; **30:** 1-312.
54. Pariser D. The rationale for renewed attention to onychomycosis. *Semin Cutan Med Surg*. 2013; **32:** S1-14.
55. Patterson TF, Moellering RC Jr. Fungal Infections. *Infect Dis Clin North Am*. WB Saunders Company, Philadelphia: 2006.
56. Plestina R. Mycotoxins and Balkan Endemic Nephropathy. *Food Chem Toxic*. 1992; **30:** 177-224.
57. Riley RT, Richard JL. Fumonisins: A Current Perspective and View to the Future. *Mycopathologia*. 1992; **117:** 1-124.
58. Romani L, Kaufmann SHE. Immunity to Fungi. 73rd Forum in Immunology. *Res Immunol*. 1998; **149:** 273-528.
59. Rosen T. Cutaneous Fungal Diseases. *J Fungi*. 2015; **1:** 1-216.
60. Sarosi GA. Blastomycosis. *Semin Respir Infect*. 1997; **12:** 187-267.
61. Scher RK, Nishikawa. Oral Lamisil: 1996 Worldwide Experience Update. *Dermatology*. 1997; **194:** S25-49.
62. Severo CB, Severo LC, Fischer GB. Mini-symposium: Fungi and The Paediatric Lung. *Paediatr Respir Rev*. 2009; **10:** 159-98.
63. Steinbach WJ. Special Issue "*Aspergillus fumigatus:* From diagnosis to therapy" *J Fungi*. 2016; **2:** 31.
64. Ullmann AJ, Johnson A. A multidisciplinary approach to managing invasive fungal disease. Preface. *J Antimicrob Chemother*. 2013; **68 (Suppl. 3):** i1.
65. Walsh TJ, Bloom BE, Kontoyiannis DP. Meeting the challenges of an emerging pathogen: the Henry Schueler 41 & 9 Foundation International Forum on Mucormycosis. *Clin Infect Dis*. 2012; **54 (Suppl. 1):** S1-85.
66. Stevens DA. Managing Fungal Infections in the 21st Century: Focus of itraconazole. *Drugs*. 2001; **61:** 1-56.
67. Vitale RG, de Queiroz Telles F, de Hoog S. The dark world of black fungi - A major area of concern. Preface. *Med Mycol*. 2009; **47:** 1.
68. Welsh O, Arenas R. Superficial mycoses. *Clin Dermatol*. 2010; **28:** 123-236.
69. Welsh O, Arenas R. Systemic mycoses. *Clin Dermatol*. 2012; **30:** 563-661.
70. Wheat LJ. Update on histoplasmosis. *Semin Respir Infect*. 2001; **16:** 89-148.
71. WHO. Paracoccidioidomycosis. Proceedings of the First Pan American Symposium in Medellin (Colombia), Scientific Publication No. 254, Washington, DC: 1972.
72. Wolf JE Jr. An Update on Dermatophytosis. *J Am Acad Dermatol*. 2000; **43:** S103-42.

Manuals of Mycology Seminars/Workshops

The following Manuals or Abstracts Books have been published, covering proceedings of seminars and workshops held in the field of Medical Mycology during recent years at various institutions in India and/or abroad. These are listed in names of Organizing Secretary and associated colleagues of Organizing Committee of that scientific event and are given below:

1. Calderone R. The 7th Conference on *Candida* and Candidiasis, Austin, Texas. *FEMS Yeast Res*. 2004; **4:** 885-6.
2. Chakrabarti A, Shivaprakash MR, *et al. Medical Mycology Laboratory Procedures*. Centre for Advanced Research in Medical Mycology (ICMR), PGIMER, Chandigarh, November 2005 & June 2008.
3. Chakrabarti A, Singh K, Jatana M. *Laboratory Procedures for the International Workshop on Medical Mycology*. Mycology Division, PGIMER, Chandigarh, November 16-21, 1998.
4. Chapman RL, Faix RG. Invasive neonatal candidiasis: An overview. *Semin Perinatol*. 2003; **27:** 352-424.
5. Clemons KV, Laniado-Laborin R, Stevens DA. Sixth International Symposium on Coccidioidomycosis. *Ann N Y Acad Sci*. 2007; **1111:** 1-462.
6. Clemons KV, Steinbach WJ, Denning DW, *et al.* Advances against aspergillosis (September 9-11, 2004, San Francisco, California). *Mycopathologia*. 2005; **159:** 315-7.
7. Fishman JA. Fungal infections in the transplant patient. *Transpl Infect Dis*. 2002; **4 (Suppl. 3):** 3-68.
8. Gugnani HC, *et al. National Symposium on Recent Advances in Respiratory Mycoses*. Department of Medical Mycology, Vallabhbhai Patel Chest Institute (VPCI), University of Delhi, October 29-30, 1998.
9. Gupta P. *SOP Mycology*. KGMC, King George's Medical University, Lucknow (Utter Pradesh): 2013.
10. International Ethical Guidelines for Biomedical Research Involving Human Subjects. Prepared by the Council for International Organizations of Medical Sciences (CIOMS) in collaboration with WHO, Geneva: 2002.
11. Jahagirdar VL, Kothadia SN, Sheikh NK. *Symposium and Workshop on Medical Mycology*. Dept. of Microbiology, Dr V M Medical College, Solapur: 1997.
12. Joshi KR, *et al. National Symposium on Mycetoma*. 2nd Conference of SIHAM, Department of Microbiology, Dr. Sampurnanand Medical College, Jodhpur, February 26-28, 1998.
13. Klis FM, De Groot PW. Trends and highlights in *Candida* research. The 8th ASM Conference about *Candida* and Candidiasis, Denver, Colorado. *FEMS Yeast Res*. 2006; **6:** 667-8.
14. Lakshmi V, *et al. Fungal Infections: An Update*. Department of Microbiology, Nizam's Institute of Medical Sciences, Hyderabad, January 28, 1996.

15. Moran GP, Jabra-Rizk MA. 8th ASM conference on *Candida* and candidiasis: Molecular tools provide insights into host-pathogen interactions. *Mycopathologia*. 2006; **162:** 17-24.
16. Murthy JMK. Central Nervous System Mycoses. *Neurology India*. 2007; **55:** 187-314.
17. No authors listed. A centennial discovery of *Paracoccidioides brasiliensis. Mycopathologia*. 2008; **165:** 179-352.
18. No authors listed. Allergic bronchopulmonary aspergillosis. *Front Biosci*. 2003; **8:** S92-1198.
19. No authors listed. Manual for Second Workshop on Medical Mycology. NICD, Delhi. November, 1979.
20. No authors listed. Proceedings of the 6th International Conference on *Cryptococcus* and Cryptococcosis. June 24-28, 2005. Boston, Massachusetts, USA. *FEMS Yeast Res*. 2006; **6:** 461-6.
21. No authors listed. The changing clinical spectrum of fungal infections and the role of liposomal amphotericin B (AmBisome). *Clin Microbiol Infect*. 2008; **14:** 1-83.
22. Petrikkos G, *et al.* International Forum on Zygomycosis. May 30-June 1, 2008, Athens, Greece.
23. Richardson M. Deep fungal infections. *Acta Biomed*. 2006; 77 (Suppl. 4).
24. Singh SM, Naidu J. *International Conference on Current Status and Challenges in Medical Mycological Research.* 1st Conference of SIHAM (Foundation Conference), Dept. of Biological Sciences, Rani Durgavati University, Jabalpur (Madhya Pradesh), February 26-28, 1996.
25. Soll DR. A review of the 7th ASM conference on *Candida* and Candidiasis, Austin, Texas. *FEMS Yeast Res*. 2004; **4:** 881-3.
26. Steinbach WJ, Latge JP, Stevens DA. Advances against aspergillosis. Proceedings of the Advances Against Aspergillosis Conference. September 9-11, 2004. San Francisco, California. *Med Mycol*. 2005; **43 (Suppl. 1):** S1-319.
27. Sturtevant J. Meeting report: *Candida* and Candidiasis. *Mycopathologia*. 2004; **158:** 141-6.
28. Sullivan D. The 6th ASM *Candida* and Candidiasis Conference. *FEMS Yeast Res.* 2002; **2:** 249-50.
29. Talwar P, Chakrabarti A, Roy P, Nayak N. *Manual for Workshop on Laboratory Diagnosis of Medically Important Yeasts and Opportunistic Fungi.* PGIMER, Chandigarh: 1990.
30. Tendolkar U, *et al. Mycology Laboratory Manual for Pre-Conference Workshop on Laboratory Diagnosis of Common Fungal Infections - Mycology-2000.* Dept. of Microbiology, LTM Medical College, October 12-16, 2000, Sion, Mumbai.
31. US Department of Health and Human Services. *Biosafety in Microbiological and Biomedical Laboratories.* 5th edn. US Government Printing Office, Washington, DC: 2007.
32. Usha Sarma P, Sharma GL, Das RH. *National Symposium on Mycoses.* Centre for Biochemical Technology, May 5-7, 1997, Delhi.
33. Venugopal PV, *et al. Mycoses in the New Millennium - Challenges and Strategies.* 3rd Conference of SIHAM, Sri Ramachandra Medical College and Research Institute, February 9-11, 2000, Chennai.
34. Venugopal PV, Prabhakar K, *et al. International Symposium on Mycotic Infections in the 21st Century.* 4th Conference of SIHAM, Rajah Muthiah Medical College, January 18-20, 2002, Annamalai Nagar, Tamil Nadu.
35. Warris A. European Paediatric Mycology Network (EPMyN). Towards a Better Understanding and Management of Fungal Infections in Children. *Curr Fungal Infect Rep.* 2016; **10:** 7-9.
36. WHO. *Laboratory Biosafety Manual.* 3rd edn. Geneva: 2004.

Online Information Services in Medicine

The pursuit of medical knowledge has always been lifelong process for the medical personnel. The physicians have accepted ongoing roles of both student and teacher, striving to seek out latest knowledge, perpetuating wisdom of experience to each question. In the last quarter of 20th century, however, growth of biomedical information has exceeded ability of any individual physician to keep abreast of advances in Medical Mycology. A possible solution of this issue is familiarity with access and retrieval of data from information systems that greatly surpass human memory. There are online services which are essential to retrieve published research works as well as review articles from biomedical journals. The searches for various studies and review articles in medical mycology were updated in this Textbook following databases up to March 31, 2017 *i.e.* the Cochrane, BioMed Central (1966-2017), MEDLINE (1966-2017), EMBASE (1947-2017), LILACS (1982-2017), SciELO (1997-2017) and CINAHL (1982-2017). The most popular ones are Scopus, PubMed, Medline, Embase/Excerpta Medica, Index Medicus and Index Copernicus. A brief description about these databases is given below:

(i) Internet

The Internet is worldwide collection of networks tied together by common protocol that allows uses of any of the networks to access services provided on other networks. One of the most successful parts of Internet is World Wide Web (*www*) which is multimedia hyperdynamic system distributed throughout world. Accessibility to World Wide Web makes it possible to visit many sites in world, wide rang of resources that include images as well as text. It was invented by Tim Berners-Lee from thousands of interconnected threads of Internet, which proved to be mass medium for the 21st century.

Most of the libraries in teaching medical institutions now-a-days have Internet facility. It provides wide-ranging access to large number of references including peer-reviewed literature. Many search engines provide advanced search techniques, allowing experienced individuals to reduce sheer quantity of sites. By using World Wide Web interface and view via pioneering Adobe Acrobat technology, anyone can access text articles, images and any sort of requisite information. Therefore, by retrieving web pages, information in Medical Mycology is available within site of an academic medical centre or its collections.

The information is easily accessible from MEDLINE search via an excellent search engine, which is one of the most important online databases of Medical Literature Analysis Retrieval System (MEDLARS). A menu driven bibliographic retrieval software, BRS/SEARCH, is used to search this database. The *Index Medicus* is available as MEDLINE database that has been publicly accessible since 1971 and *Abstracts of Mycology*, as a part of BIOSIS. It is one of world's largest computer based bibliographic systems comprising over 28 databases. The *Mycoses* and *Medical Mycology* are included in ADONIS services and copies of individual articles may be printed from CD-ROM on demand or subscribed online. The journal *Medical Mycology* is now available on website *http://**mmy.oxford-journals.org.*** Most frequently used retrieval services are MEDLINE (PubMed), which can be retrieved on website ***http://www.ncbi.nlm.nih.gov/jbrowser.html***. The full text journals are also available on Internet through various websites. Most of the publications now assign unique PubMed ID **(PMID)**, PubMed Central ID **(PMCID)** and/ or Digital Object Identifier **(DOI)** number to every online article, which are keys to retrieve the same. The DOI reference of articles pertaining to medical mycology starts with figure 10.xxxx, followed by rest of the complete citation number. Now, most of the medical journals have started providing the DOI number in the Reference section of published articles.

Therefore, for retrieving references on any of issues of biomedical sciences, website ***http://www.ncbi.nlm.nih.gov/*** should be visited. The online access to full text articles on life sciences research is available at this website and it can be downloaded as per requirements. The individual articles may be assessed through their publishers' websites, which is cost-effective as compared to hard copy of original biomedical journals.

The mycological information may be gathered about different pathogenic fungi from list of the following websites, which are given in an alphabetical order:

1. *htpp://www.abdn.ac.uk/mmfi/*
2. *http://botit.botany.wisc.edu*
3. *http://hstelearning.mit.edu*
4. *http://labmed.ucsf.edu*
5. *http://microbiology.mtsinai.on.ca/mig*
6. *http://ncbi.nlm.nih.gov*
7. *http://timm.main.teikyo-u.ac.jp*
8. *http://www.allfungi.com*
9. *http://www.aspergillus.man.org.uk*
10. *http://www.botany.utoronto.ca*
11. *http://www.cdc.gov*
12. *http://www.cher.ubc.ca.cryptococcus/new/default.htm*
13. *http://www.clinical-mycology.com*
14. *http://www.clsi.org*
15. *http://www.dehs.umn.edu*
16. *http://www.dermnz.org*
17. *http://www.doctorfungus.org*
18. *http://www.eucast.org/ast_of_fungi*
19. *http://www.fisf.org*
20. *http://www.fungalbiodiversitycentre.com*
21. *http://www.fungalforum.com*
22. *http://www.henryschueler.org*
23. *http://www.hpa.org.uk*
24. *http://www.idimages.org*
25. *http://www.invasivefungalinfections.com*
26. *http://www.ipfn.org*
27. *http://www.isham.org*
28. *http://www.life-worldwide.org.*
29. *http://www.med.sc.edu*
30. *http://www.medicalmycology-asia.com*
31. *http://www.microfungi.net*
32. *http://www.mold-help.org*
33. *http://www.mold-survivor.com*
34. *http://www.mycetoma.edu.sd*
35. *http://www.mycology.adelaide.edu.au*
36. *http://www.mycology.net*
37. *http://www.mycologymanchester.org*
38. *http://www.osha.gov*
39. *http://www.pangloss.ucsfmedicalcenter.org*
40. *http://www.pfdb.net*
41. *http://www.siham.in*
42. *http://www.tigr.org*
43. *http://www.TomVolkFungi.net*
44. *http://www.wadsworth.org*

Besides these websites, in recent times many online applications (aps) are available on internet, which provide general type of information but in the same aps medical mycology information may also be sought ***e.g.*** *www.upto-date.com.*

(ii) Medline

MEDLINE is compiled by the US National Library of Medicine (NLM) and published on the World Wide Web. It is the world's most comprehensive source of life sciences and biomedical bibliographic information. It contains nearly

11 million records from more than 7,300 different publications dating from 1965-present. It is updated weekly.

MEDLINE is the primary component of PubMed (*http://pubmed.gov*). A link to PubMed is found on the NLM home page at *http://www.nlm.nih.gov*. The result of a MEDLINE/PubMed search is a list of citations (including authors, title, source and often an abstract) to journal articles and an indication of free electronic full-text availability. Searching is free of charge and does not require registration. MEDLINE in PubMed may also be searched using the NLM Gateway, a single Web interface that searches multiple NLM retrieval systems.

(iii) Pre-Medline

Another database has also been added within PubMed to overcome any delay in updating and transferring data to MEDLINE as 'pre-MEDLINE'. This is symbolic partnership between National Library of Medicine and publishers of biomedical literature who submit data to pre-MEDLINE database electronically. The pre-MEDLINE database is updated daily and contains references well in advance of appearance in MeSH coded MEDLINE.

(iv) MedlinePlus

MedlinePlus is another service offered by the NLM, which provides consumer-oriented health information. They are encouraged to discuss search results with their health care provider.

(v) PubMed

PubMed comprises more than 19 million citations for biomedical articles from MEDLINE and life science journals. Citations may include links to full-text articles from PubMed Central or publisher websites. PubMed is an online, web-based version of MEDLINE containing references in English language dating back to 1966. It is accessible free of charge over World Wide Web and contains whole of MEDLINE citations with abstracts from more than 5000 biomedical journals and over 8.6 millions records. Each reference, before it is added to MEDLINE, is manually coded with keywords using Medical Subject Headings (MeSH).

(vi) PubMed Central

The National Institute of Health, Bethesda (USA) has also established mechanism of dissemination of scientific information in which free global repository for primary literature in life sciences is available for use by research workers. It is called PubMed Central and is accessible on World Wide Web just as PubMed MEDLINE services are available. Therefore, writing, reading as well as processing of scientific papers by authors, referees and editors has taken entirely different form due to such advanced information systems.

(vii) E-Biomed

The Director of National Institute of Health proposed more radical solution to problem of accessibility in the form of 'E-Biomed', website where all biomedical research are present and accessible free of cost through powerful search engine.

(viii) Embase

Embase (Excerpta Medica Database) is a biomedical and pharmacological database produced by Elsevier B.V. This is very similar to PubMed in abstracting and indexing database covering medical journal articles. However, it has relatively more coverage of European journals beginning from 1974. It provides unparalleled coverage of the biomedical literature, with 30 million+ records from 8,500+ currently published journals. Moreover, it includes six million+ records and 2,900+ journals that are not covered by MEDLINE. The indexing is more for the drugs and pharmaceuticals.

(ix) Scopus

This bibliographic database, owned by Elsevier, contains abstracts and citations for academic journal articles. It covers nearly 22,000 titles from over 5,000 publishers, of which 20,000 are peer-reviewed journals in the scientific, technical, medical and social sciences including arts and humanities. However, it is available online by subscription and retrieving also incorporates search of patent databases.

(x) LILACS

The Latin American and Caribbean Health Sciences Literature is a cooperative database built by the institutions which integrate the Latin American and Caribbean of Health Sciences Information System. This has been published since 1982 and registers the health scientific-technique literature published by Latin American and

Caribbean authors. The main objectives of this database are the bibliographic control and the dissemination of health scientific-technique literature from Latin American and Caribbean countries, the absentee from the international databases.

(xi) SciELO

The Scientific Electronic Library Online (SciELO) is a bibliographic database, digital library as well as co-operative electronic publishing model of open access journals. This was created in the year 1997 to meet the scientific communication needs of developing countries and provides an efficient way to increase visibility and access to scientific literature. It was originally established in Brazil, however, now there are fourteen countries, mainly Spanish speaking, in this network. Its journal collections is in Argentina, Bolivia, Brazil, Chile, Colombia, Costa Rica, Cuba, Mexico, Peru, Portugal, South Africa, Spain, Uruguay and Venezuela. Paraguay is developing a journal collection.

(xii) Online Services in Medical Mycology

Large-scale host information systems, generally known as online systems, have been subject of an enormous attention in recent past. The computerized information services are available as databases for online search e.g. *Review of Medical and Veterinary Mycology,* is available as sub file (OL) of CAB ABSTRACTS database on various hosts, including DIALOG, CAN/OLE, ESA-IRS and DIMDI. For detail website ***http://www.cabweb.org*** may be visited. Similarly, free accessibility is available for full text articles of ASM's online primary research journals after few months of publication of that issue. They may be accessed at website ***http://www.journals.asm.org***.

Websites on various infectious diseases are operated by many organizations like WHO, therefore, particular website related to specific fungal infection can be visited. A lot of information regarding fungal infections can be retrieved from website ***www.doctorfungus.org*** and ***www.adelaidemycology.edu***. Some of laboratories have their own websites like **Histoplasmosis Reference Laboratory** which may be visited to access requisite information on *Histoplasma capsulatum* at website ***http://www.iupui.edu/~histodgn***. Similarly, if there is query regarding any other fungus *e.g.* ***Aspergillus*** genus or fungal disease *e.g.* **Coccidioidomycosis**, it can be inquired online from website ***http://www.aspergillus.org.uk*** and ***http://www.arl.arizona.edu/vfce***, respectively. The details of available sources, description of technologies, participants and general information regarding ***Pneumocystis jirovecii*** can be accessed through internet at website: ***http://www.uky.edu/projects/pneumocystis***.

In India, MEDLARS is available through National Informatics Centre (NIC), New Delhi from National Library of Medicine (NLM) of National Institutes of Health. The NIC has designated as Indian MEDLARS Centre (IMC). The Indian Biomedical Research of Indian MEDLARS Centre has designed and developed database entitled IndMED fulfilling international standards. The database is accessible on Internet at website ***http://www.basis.delhi.nic.in/nd.***

The Indian Medlars Centre (medIND), the Biomedical Journals from India is One point resource of peer reviewed Indian biomedical literature covering full text of IndMED journals. It has been designed to provide quick and easy access through searching or browsing.

Electronic Versions of Books and Journals

In Medical Mycology, besides the printed textbooks and references, now-a-days most of these books are available in the form of their electronic versions on CD-ROM. These electronic versions facilitate information retrieval and cross-referencing on any of the topics in Medical Mycology. *Topley and Wilson's Microbiology and Microbial Infections* - ***Medical Mycology (Vol. 4)***, is available as electronic version since 2010.

Most of journals have started publishing articles in HTML and PDF formats in their electronic editions *e.g. Clinical Infectious Diseases* is available at website ***http://www.journal/uchicago.edu/cid/***. These scientific journals have created websites, providing sections from their paper journals (hard copy) in electronic form. Some of them are publishing only electronic version of journal without its hard copy, keeping in view of the cost-effectiveness. Therefore, to avoid any confusion, their pages are marked with prefix ***'e'*** before page number. The full texts of journals are also available on these websites which may be retrieved after subscribing the said journal.

Software Packages in Medical Mycology

There are software packages available on various aspects of Medical Mycology so that handling and retrieving of data in Diagnostic Mycology Laboratory become easier. Some of these software packages are given below:

1. Anaissie EJ, McGinnis MR, Pfaller MA. *Clinical Mycology*. 2nd edn. Elsevier, London: 2009.

2. Berger S. *Miscellaneous Invasive Fungi: Global Status* (eBook Series). GIDEON Informatics, Los Angeles: 2017.
3. de Hoog GS, Guarro J, Gene J, Figueras MJ. *Atlas of Clinical Fungi:* The Ultimate Benchtool for Diagnostics. 2nd edn/ Electronic Version 4.1.4. Westerdijk Fungal Biodiversity Institute (Former CBS-KNAW), Utrecht, The Netherlands and Universitat Rovira i Virgili, Reus, Spain: 2000. Interactive CD-ROM is based on the book with the same title; 2017.
4. Dismukes WE, Pappas PG, Sobel JD. *Clinical Mycology.* 1st edn. Oxford University Press, New York: 2003.
5. Hilger AE, Schell WA, Burford HJ. *Medical Mycology.* Warlock Productions, Hanover: 1993.
6. Holmes C, Rautemaa-Richardson R, Novak-Frazer L, Moore CB, Richardson MD. *Illustrative Cases in Medical Mycology.* Free iBook. University of Manchester and Mycology Reference Centre, Manchester, UK: 2016.
7. Yeast Identification PC Program-Version 5 (May 2000) that is based on the book (Barnett JA, Payne RW, Yarrow D. *Yeasts: Characteristics and Identification,* 3rd edn. Cambridge University Press, Cambridge: 2000). For more detail, e-mail request may be sent to the author at <*j.barnett@uea.ac.uk*>.

Appendix G: Glossary

Note: *Words in parentheses ( ) are plurals, in curly bracket { } description is given and in square brackets [ ] words with cross reference for clarification are being referred.*

A-, an-: Terms are used as prefix to mean without.

Acervulus (acervuli): An open cup or saucer-shaped structure on which conidia are formed as seen in Coelomycetes [Cf. Pycnidium].

Achloric: Without chlorophyll.

Acicular: Needle-shaped or slender, gradually becoming narrower towards apex.

Acid-fast: Retaining primary stain *i.e.* concentrated carbol-fuchsin after decolorization with mineral acids or acid-alcohol.

Acrogenous: Conidia, developing at the apex, borne at the tip of conidiophore {Gr., *Acros* = highest, outermost}.

Acropetal: Conidia developing from base towards apex wherein **youngest** is at **top** of chain as seen in genera *Alternaria*, *Cladosporium* and *Curvularia* [Cf. Basipetal].

Actidione: Trade name of antibiotic, cycloheximide produced by *Streptomyces griseus* and used in culture media to selectively inhibit the contaminating fungi.

Aculeate: Prickly; with narrow spines.

Adamson's fringe: That portion of lower hair shaft at the junction of the living *i.e.* mitotically active, hair bulb and the keratinized, non-living hair shaft. Dermatophytes can only infect hair down to the level of Adamson's fringe.

Adiaconidium: Large, globose, thick-walled structure develops after lodgement of asexual conidia into host tissue *in vivo* or incubated at elevated temperature of 37-40°C *in vitro*. This increases enormously in its size without reproduction, budding, replication or endosporulation eventually developing as **spherule**, which is seen in adiaspiromycosis caused by **two** species of genus *Emmonsia i.e. Ea.crescens* and *Ea.parva*. It is also interchangeably used with former term **adiaspore** {Gr. *a* = not, *dia* = across}.

Aerial hyphae: Mycelia produced above the colony agar interface.

Aflatoxin: Mycotoxin produced by *Aspergillus flavus*. The acronym ***'aflatoxin'*** is derived from ***A****spergillus* ***fla****vus* **toxin**.

Aleurioconidium/aleuriospore: Thick-walled terminal or lateral, single-cell conidium attached by wide base and released by lysis or fracture of supporting cell. These are borne directly on hyphae, short pedicels or on undifferentiated conidiophore as seen in *Microsporum* and *Trichophyton*. Now, the term aleurioconidium is not recommended, instead conidium is preferred to avoid confusion.

Allochthonous: Organism that originated at a distance from its present position [Cf. Autochthonous].

Amero-: Single-cell.

Ameroconidium: Single-celled asexual spore as seen in *Acremonium* species {group} and *Cladosporium cladosporioides* {chain}.

Anamorph: An asexual or imperfect state of fungus and its identification is based on mitotic or asexual reproductive structure *i.e.* ***ana***tomic ***morph***ology, identical to parent [Cf. Teleomorph].

Aneuploid: A nucleus with chromosome number, which is not a multiple of the haploid set {*2n+1, 2n-1*}.

Annellation: Formation of ring-like structures at the conidiogenous end of a conidiophore.

Annellide: Conidiogenous cell where first conidium is formed from all layers of an elongating apex of conidiophore and subsequently from inner cell wall layer in a basipetal manner. It is morphologically distinctive because of apical rings composed of outer cell wall remains at each released conidium [Cf. Phialide].

Annelloconidium: Conidia produced in succession by conidiogenous cell *i.e.* an annellide, usually characterized by rounded apex and flat base *e.g. Exophiala* species.

Annular frill: Skirt or ring-like portion of cell wall material at base of conidium, which is left during its separation from conidiophore.

Antheridium: Motile, smaller, sperm-producing gametangial structure in teleomorph state of some fungi *i.e.* male copulating cell.

Anthropophilic: Preferring man as host as in dermatophytes or ecologically adapted to man.

Antler hypha: Hypha terminating in broad irregular branches resembling antlers.

Apophysis: Bell or Funnel-shaped swelling or widening at the end of sporangiophore just below columella (Gr., *apo* = off + *phyein* = to grow) as seen in some of mucormycetes like *Lichtheimia* and *Apophysomyces* species.

Apothecium (apothecia): A wide, open cup- or saucer-shaped sexual reproductive structures *i.e.* ascocarp, formed by certain ascomycetous fungi.

Appressorium (appressoria): The swollen tip of hypha or germ tube that facilitates attachment and penetration of host by fungus.

Artefact: Something extraneous or artificially created [US = Artifact].

Arthroconidium: Formerly referred to as arthrospores, which are resulted from fragmentation of densely septate hypha into separate cells. The resulting cell may be rectangular or barrel-shaped as seen in dermatophytes, *Trichosporon*, *Geotrichum* and *Coccidioides* spp.

Ascocarp: Fruiting body which bears asci, ascospores and ascoma as seen in Ascomycota [Cf. Basidiocarp].

Ascogenous: Ascus-bearing.

Ascospore: Sexual spore produced within sac-like structure *i.e.* ascus. This is haploid in nature and produced after nuclear fusion {karyogamy} and meiosis.

Ascostroma (ascostromata): Ascocarp characterized by cavities resulting from compression of surrounding tissues in fruiting body prior to ascus formation.

Ascus (asci): Sac-like cell in which sexual spores *i.e.* ascospores are produced as seen in Ascomycota.

Aseptate: Hyphae lacking regular cross-walls as seen in mucormycetes and oomycetes. This term is misnomer as these organisms are in fact pauci-septate or sparsely septate rather than aseptate, having very few septae at irregular intervals [Cf. Septate].

Asexual: Reproduction without involving union of two nuclei. This phase of fungi is also known as imperfect state.

Asporogenous: Non-spore forming.

Assimilation: An aerobic process wherein carbon and nitrogen compounds are utilized with oxygen as final electron acceptor.

Asteroid body: Star-like eosinophilic radiating structure surrounding fungal cell in tissue wherein antibodies are precipitated {IgG and IgM} as seen on HPE of lesions caused by *S.schenckii* complex.

Autochthonous: Organism originating in same area wherein it is found [Cf. Allochthonous].

Autospore: Daughter cells formed within covering of *Prototheca* species.

Autotrophic: Capable of growing without organic substance for source of energy as seen in algae and plants.

Bacilliform: Rod-shaped.

Ballistoconidium: Conidium formed on spine-like outgrowth of cell, forcibly discharged by droplet mechanism.

Basidiocarp: Fruiting body seen in Basidiomycota, which produces basidia [Cf. Ascocarp].

Basidiospore: Sexual spore formed on basidium as seen in basidiomycetes *i.e. Cryptococcus* species. This is also haploid in nature like ascospore and is produced after nuclear fusion {karyogamy} and meiosis.

Basidium (basidia): Structure, usually single club-shaped cell, which bears basidiospores, as seen in basidiomycetes.

Basipetal: Youngest conidium is at the **base** of a chain *i.e.* older conidia are displaced by newer ones, as seen in phialidic *Aspergillus* or annellidic *Scopulariopsis* species. ***'Basi-'*** literally means base [Cf. Acropetal].

Biseriate: Phialides supported by metulae in two series as seen in *Aspergillus flavus*.

Blastic: Conidia formation caused by blowing out of fertile hyphae before being delimited by septum.

Blastoconidium/blastospore: Conidial formation through budding process *i.e.* blown out from part of its parent cell as seen in yeasts.

Break point: In medical practice, breakpoint values should predict ideally success or failure of an antifungal drug.

Budding: Asexual formation of small, rounded outgrowths from parent cell as seen in yeasts.

Campanulate: Shaped like campana or bell.

Capilliconidia: These are elongate cells, often with a terminal adhesive knob and borne by a thin supporting hypha as seen in *Conidiobolus* species.

Capsule: Hyaline, mucopolysaccharide sheath around wall of cell or spore as seen in basidiomycetous *Cryptococcus* and *Rhodotorula* species.

Catenate: End to end arrangement seen in chain of more than three spores.

Catenulate: End to end arrangement seen in short chain of two or three spores.

Chlamydospore: Large, thick-walled usually spherical resting conidium produced in unfavorable conditions from an intercalary or terminal somatic hyphal cell containing stored food and able to function as spore *i.e.* as a survival propagule *e.g. Candida albicans.*

Chloroplast: A chlorophyll-containing plastid found in algal and green plant cells like *Chlorella* and absent in *Prototheca* species.

Circinate: Coiled into partial or complete ring.

Cladistics: It provides simplified method of understanding phylogenetic trees.

Cladogram: Graph showing where different groups diverged from common ancestral line.

Clamp-connection: Hook-like bridging outgrowth on many dikaryotic basidiomycetous hyphae, providing a bypass for one of the nuclei produced during synchronous division of the dikaryon, insuring their equal distribution between the new cells.

Class: Taxonomic rank above Order and below Phylum, which is written with suffix ***-mycetes***.

Clavate: Club-shaped.

Cleistothecium (cleistothecia): Large, fairly round, closed and multicellular fruiting body in which randomly dispersed asci and ascospores are formed and held until the structure bursts as seen in *Aspergillus* species [Cf. Perithecium].

Coenocytic: Ribbon-like hyaline multinucleate vegetative hyphae with few or no septa (lacking cross walls) as seen in mucormycetes and oomycetes [Cf. Septate].

Collarette: Cup- or funnel-shaped remnant of cell wall present at tip of phialide through which conidia are produced.

Columella (columellae): Enlarged, dome-shaped tip of sporangiophore extending into sporangium. When sporangium bursts, it leaves columella bare.

Conidiogenous: A cell that produces conidia.

Conidiophore: Specialized and differentiated hyphal structure that serves as stalk wherein conidia are formed singly or in clusters. The shape and arrangement of conidiophores as well as conidia are identifying feature of a particular genus.

Conidium (conidia): Asexual reproductive structures produced either from transformation of vegetative yeast/hyphal cell or from a specialized conidiogenous cell, which may be simple to complex and elaborate. Conidia may be formed on specialized hyphae, termed conidiophore. Microconidia are relatively smaller in size whereas macroconidia are larger and multicellular.

Coprophilous: Dung-loving.

Coremium (coremia): Group of hyphae or sporophores, which are generally upright and sometimes joined together, producing spores at their tips.

Corymbiform: Arrangement of convex or flat-topped clusters of fruiting branches with branchlets arising from different points on the axis.

Crozier: Hook on ascogenous hyphae formed before formation of asci.

Cryptic species: A group of fungal species which cannot interbreed and that contain individuals that are morphologically identical to each other but belong to different species, identified on molecular sequencing.

Dauer cell: A resting cell stage between an aplanospore and new sporangium formation in *Prototheca* and can be recognized by a thick cell wall and no cell divisions.

Dematiaceous: Pigmented fungi that have brown or black colored cell walls in their conidia, mycelia and sclerotic bodies and form brown to black colonies in culture. The pigment appears to be a form of melanin. This term is now replaced by broader term *i.e.* phaeoid, as all such fungi are not covered under family Dematiaceae.

Dendrogram: It is tree-like diagram used to illustrate arrangement of clusters produced by clustering algorithm {Gk: ***'dendron'*** - tree, ***'-gramma'*** - drawing}, often used in computational biology to illustrate clustering of genes.

Denticle: Small tooth-like projection on which conidium develops as seen in *Sporothrix schenckii* or *Schizophyllum commune.*

Determinate: Adjective applied to conidiophore or conidiogenous cell that ceases extension growth at or before onset of conidium production.

Deuteromycetes: These are fungi without known sexual state and are included in broad artificial group which is also called as *'Fungi Imperfecti'* or mitosporic fungi.

Dichotomous: Type of branching in which two branches are approximately equal in size and stem from which they originated as seen in *Aspergillus* species.

Dikaryon: A fungal stage in which all cells contain two nuclei; it may be homokaryotic or heterokaryotic.

Dimorphic: A fungal growth existing in two distinct morphological forms of certain pathogenic fungi due to difference in temperatures *e.g.* yeast/spherule or parasitic form at body temperature and hyphal or saprotrophic form at room temperature.

Diploid: A fungal stage in which the nuclei contain two genomes {*2n*}. It may be homozygous or heterozygous.

Disjunctor: Structure between conidia to promote their disarticulation as seen in *Coccidioides* and *Malbranchea* species.

Disjunctor cell: An empty cell that fragments and/or undergoes lysis to release conidium as seen in *Coccidioides* and *Malbranchea* species.

Distosepta: False septa, where conidia are subdivided by inner wall layer only as seen in dematiaceous fungi like *Curvularia, Bipolaris,* etc.

Dolipores: Complex barrel-shaped structures formed in the septa of basidiomycetes which are covered on both sides by membrane called parenthesome.

Downy: Having a short and dense mycelial texture.

Dysgonic: Slow growing variant of an organism.

Echinulate: Having small pointed processes or spines.

Ectothrix: Forming sheath of arthroconidia outside hair shaft. Cuticle of hair remains intact [Cf. Endothrix].

Emergomycosis*:* An opportunistic fungal disease caused by three species of a newly described dimorphic genus - *Emergomyces (Es.pasteurianus, Es.africanus* and *Es.orientalis).*

Endemic: Disease found prevalent in limited geographical areas with no major fluctuations in frequency and its causative organism co-exists with its host population. This is in contrast to 'epidemic' where disease tends to occur episodically at sudden high rates.

Endogenous: From within.

Endophyte: Fungus that lives within plant tissues for at least part of its life cycle without causing apparent disease.

Endospore: Spore formed internally in spherule by cleavage process following karyogamy and mitosis.

Endosporulation: Formation of endospores within sporangium as seen in *Coccidioides* species, *Rhinosporidium seeberi* and *Prototheca* species.

Endothrix: Arthroconidia formed inside hair shaft. Cuticle of hair is usually destroyed [Cf. Ectothrix].

Enteroblastic: Formed from inner cell wall layers of conidiogenesis cell.

Enzootic: An outbreak of disease affecting animals in certain areas and its analogous term in humans is called as endemic.

Epithet. Adjective or descriptive phrase used to indicate fungal species, genus or disease name.

Epizootic: An outbreak of disease that spreads quickly among animals, usually in one geographic area. In humans, such an analogous occurrence is known as an epidemic.

Etymology: Study pertaining to history of words, when they got incorporated into a language, from what source and how their form and meaning have changed in due course of time.

Exogenous: From outside.

Favic chandeliers: Hyphae terminating in broad irregular branches resembling antlers typically seen in *T.schoenleinii.*

Faviform: Waxy, restricted honeycomb-like growth of certain dermatophytes.

Favus: Tinea favosa in which cup-like crusts *i.e.* scutula formation around infected hair follicles caused by *T.schoenleinii.*

Fermentation: An enzymatic oxidation-reduction process in which organic substrate serves both as electron donor and acceptor. It is an anaerobic type of metabolism in which organic compounds are degraded in absence of external electron acceptors. The organism produces mixed oxidized and reduced compounds.

Fission: Method of conidium separation from parent cell in which natural break occurs at double septum between walls of conidium and conidiogenous cells. It is an asexual mode of reproduction.

Fission yeast: A yeast multiplying by planate division {not by budding) as observed in *Schizosaccharomyces* species *e.g. S.pombe.* This feature is also observed in the yeast {parasitic} phase of *Talaromyces marneffei.*

Floccose: Cottony texture of hyphae [Cf. Lanose].

Foot Cell: Hyphal cell at the base of conidiophore seen in genus *Aspergillus.* However, in *Fusarium,* it is sharply-angled end of macroconidium where it attaches to conidiophore.

Fruiting body: Large spore-bearing reproductive structure of fungi seen in Ascomycota and Basidiomycota.

Fungicide: Chemical used to kill fungal growth.

Fungistatic: Chemical used to inhibit fungal growth.

Fungus: Eukaryotic, unicellular to filamentous, achlorophyllous organism having an absorptive nutrition, which reproduces by sexual, asexual and/or both means.

Furfuraceous: Fine and loose; scales resembling bran or dandruff.

Fuseaux: Multicellular macroconidia of dermatophytes.

Fusiform: Spindle-shaped cells, swollen near middle and tapering towards both ends.

Geniculate: With repeated knee-like bends.

Genus: (**genera**) Taxonomic category ranking below family and above species. Biologists use genus either alone or followed by Latin adjective to give name to a species.

Geophilic: Soil-inhabiting *e.g. M.gypseum.*

Germ tube: Initial hyphae from sprouting yeast, conidia or spore without constriction at the point of origin.

Giant cell: It is a mass formed by union of several distinct cells like macrophages and arise in response to either infection or foreign body.

Glabrous: Smooth or waxy without aerial mycelia.

Globose: Round.

Granular: Coarse, particulate.

Granule: Compact agglomeration of fungi or actinomycetes formed *in vivo* as seen in mycetoma.

Graphium: Anamorph of Ascomycota that is characterized by having as its fruiting structure a synnema.

Gymnothecium (gymnothecia): Fruiting structure {ascoma} composed of loosely intertwined hyphae.

Halophilic: Able to tolerate higher concentration of salts.

Haploid: A fungal stage in which the nuclei contain one genome {*n*}.

Herbarium: It is a collection of preserved plant specimens. In mycology its analogous is Fungal Culture Collection.

Heterokaryon: Genetically dissimilar nuclei in common cytoplasm.

Heterothallic: Self-sterile and requiring two compatible thalli for sexual reproduction [Cf. Homothallic].

Hilum (Hila): Scar at the base of conidium.

Holomorph: Collective term for teleomorph, anamorph and synanamorph *i.e.* whole fungus.

Holothallic: Involving all cell wall layers of conidiogenous or sporogenous cell in thallic development.

Homothallic: Self-fertile, where sexual reproduction can take place within one thallus [Cf. Heterothallic].

Hulle cells: Thick-walled vegetative cells with characteristic thin-walled pores usually associated with cleistothecia as seen in *Aspergillus* species {*A.nidulans* and *A.versicolor*}.

Hyaline: Colorless or transparent.

Hyalo-: Prefix denoting light-colored or colorless or hyaline.

Hyperendemic: The organisms causing disease exist in its host population at very high rate in particular geographical area *e.g.* coccidioidomycosis, sporotrichosis, rhinosporidiosis.

Hypha (hyphae): Tubular, branching filament, 2-10 µm in width, septate or aseptate, constituting basic fungal growth unit that makes vegetative or aerial mycelia.

Hypopyon: Pus in anterior chamber of eye often seen in keratomycosis.

Id: Allergic manifestations seen in dermatophytosis and candidiasis *i.e.* dermatophytid and candid reactions.

Imperfect Fungi: Reproduced only by an anamorph and lack sexual reproduction [Cf. Perfect Fungi].

Indeterminate: Descriptive of conidiogenous cell that continues extension growth during or between formation of successive conidia, as an annellide.

Indurated: Hard.

Intercalary: Formed within hyphal unit or between two undifferentiated cells; refers to such things as non-apical chlamydospores in many fungi [Cf. Terminal].

Internodal: Portion of hypha that is between two nodes.

Karyogamy: Fusion of two nuclei during sexual process.

Keratitis: Inflammation of cornea.

Keratomycosis: Fungal disease of cornea.

Kerion: An intensely inflammatory dermatophyte lesion of scalp.

Lanose: Woolly texture with abundant loose, regular aerial mycelia [Cf. Floccose].

Lyophilization: Process of freeze-drying.

Macroconidium: The larger of two types of conidia produced in the same manner by a fungus, which is usually multicellular.

Melanized: Blackish brown-pigmented, synonymously used for phaeoid or dematiaceous fungi.

Meristematic: There is continuous increase in fungal biomass in all directions and concordant septum formation.

Meristospore: Sporangiospore with cleavage formation as seen in *Basidiobolus ranarum.*

Merosporangium: Tubular sac producing single row of sporangiospores as seen in *Syncephalastrum* species.

Metula (metulae): Special cells at the apex of a conidiophore which support phialides as seen in *Aspergillus* and *Penicillium* species.

Microconidium: Smaller and unicellular conidia, where two different sizes are produced in the same manner by a fungus.

Microcycle Conidiation*:* This type of conidiation takes place when fungal species are subjected to stress conditions, exhibit an extremely simplified asexual life cycle wherein conidia that germinate directly generate further conidia, without forming mycelia.

Mitospore: Asexual spore.

Mitosporic: Fungi in which only asexual phase has been identified, division by mitosis, also known as Deuteromycetes or *'Fungi Imperfecti'*.

Mnemonic: This term is helpful set of words or sentence to remember order of something important, such as planets or months of the year, etc. and is pronounced *'noo-mon-ick'*. The first letter of each word matches up with the first letter of sequence being tried to remember. It helps to remember hierarchy of Linnaeus' Taxonomy as - **K**ing **P**hilip **C**an **O**nly **F**ind **G**reen **S**ocks, being used for seven groups *i.e.* **K**ingdom, **P**hylum, **C**lass, **O**rder, **F**amily, **G**enus and **S**pecies, respectively (Cf. Abbreviations/Acronym).

Mold: Any hyphal or mycelial colony or form of fungal growth. It is derived from French *'molde'* referring to shape or form while **mould** is from Norse *'mowlde'* used for fuzzy. Both are interchangeably used.

Moniliform: Hyphae with swellings at regular intervals like a string of beads.

Monokaryon: A fungal stage in which all cells contain one single nucleus.

Monomorphic: Single form of propagation.

Monophyletic: Descended from common ancestor or stem cell. Monophyletic groups are also called **clades** and are generally considered as the only 'natural' kind of group. They are very important in phylogenetic classification.

Muriform body: Thick-walled, multicellular bodies with cross and longitudinal septa due to septation in more than one plane, as seen in tissue form of agents causing chromoblastomycosis.

Mycelium (Mycelia): Mass of intertwined hyphae *i.e.* woven into network.

Mycetoma (Mycetomata): Chronic subcutaneous fungal and/or bacterial disease with triad of tumefaction, multiple discharging sinuses and granules formation.

Mycobiome: The fungal microbiota within a host environment.

Mycobiota: A group of all the fungi present in a particular area or geographic region {Only plural noun, no singular}

Mycology: Branch of biology that deals with study of fungi.

Mycophagy: Consumption of fungi usually refers to fleshy fungi *i.e.* mushrooms.

Mycorrhiza: A symbiotic association between fungus and the roots of vascular plant.

Mycosis (mycoses): Fungal disease of warm-blooded animals.

Myospherulosis: A granulomatous reaction containing sac-like structures with spherules; associated with lipid medications applied to open wounds or injected, entailing dermal or subcutaneous nodules.

Nodular organ: Knot of hyphae, which is produced by some species of dermatophytes.

Nomen: Latin word meaning name.

Nosology: Branch of medical science dealing with the classification of diseases.

Obclavate: Club-shaped in reverse *i.e.* distal portion is smaller as in secondary conidia of *Basidiobolus ranarum*.

Oblate: Circular in frontal view and broadly ellipsoidal in lateral view *i.e.* ascospores seen in some of Onygenales.

Oblong: Rectangle-like but with rounded ends.

Obovate: Egg-shaped in reverse.

Oidium (oidia): Asexually produced spore that is presumed not to constitute the main reproductive preoccupation of fungus at that time. The hypha breaks up into component cells and develops into spores. It cannot survive in unfavorable conditions, seen in *Mucor racemosus*.

Olivaceous: Olive-gray in color.

Ontogeny: Complete developmental history of an individual organism.

Onychomadesis: This is a periodic idiopathic shedding of nails commencing at proximal end and perhaps caused by temporary arrest of function of nail matrix.

Oogamy: Fusion of motile sperm and non-motile egg as seen in some Chytridiomycetes and not in fungi.

Oogonium: Large rounded structure in oomycetes containing female gametes.

Oospore: The sexual spore in oomycetes, which are produced in the oogonium after gametangial contact and it also functions as survival structures in soil.

Order: Taxonomic rank above Family and below Class, which is written with suffix ***-ales***.

Ostiolate: Having an ostiole.

Ostiole: An opening or pore in an ascocarp or pycnidium through which spores or conidia escape.

Papilla: Nipple-shaped elevation or small rounded process.

Parafungal: Mycelial and unicellular organisms that are not classified in the kingdom Fungi like *Pythium, Rhinosporidium* and *Prototheca* species but cause infections with clinical and histopathological features resembling those produced by fungi. These are also called as **pseudofungal** organisms.

Paraphyletic: In phylogenetics, group of organisms contain most recent common ancestor {Gr., *para* = near, *phyle* = race} but does not contain all descendants of that ancestor.

Parasexual: A nuclear cycle that results in recombinant haploid genotypes without meiosis, involving heterokaryosis, karyogamy to produce heterozygous diploid nuclei and mitotic crossing over and/or nondisjunction and aneuploidy.

Paronychia: Infection of nail fold, not included in tinea unguium or onychomycosis.

Pathogenicity: Qualitative term which means simply the ability to cause disease [Cf. Virulence].

Pectinate: Like teeth of comb.

Penicillate: Brush-like pattern of conidiophores and metulae of *Penicillium, Paecilomyces* and *Scopulariopsis* spp.

Percurrent: Conidiogenous cell growth where new axis grows through previous apex as seen in *Malassezia* species.

Perfect fungi: Capable of sexual reproduction *i.e.* part of life cycle in which spores are formed after nuclear fusion, synonym to teleomorph [Cf. Imperfect fungi].

Peridium: Outside covering or wall of an asexual or sexual fruiting body.

Perithecium (perithecia): Large, round to pear-shaped fruiting body with apical opening in which asci and ascospores are borne *e.g. Chaetomium* species. The small and round opening distinguishes it from cleistothecium [Cf. Cleistothecium].

Perosporomycetes: The pseudofungal organisms belonging to Class Oomycetes under the Phylum Stramenopila like *Pythium* species.

Phaeoid: Darkly pigmented, formerly dematiaceous.

Phialide: Type of conidiogenous cell, usually flask-shaped, which produces basipetal succession of enteroblastic conidia from an apical opening, without resulting change in length of cell [Cf. Annellide].

Phialoconidium: Conidium that is produced by flask-shaped conidiogenous cell termed phialide as seen in *Aspergillus* or *Phialophora* species.

Phore: Suffix used for a bearing structure.

Phragmoconidia: Conidia with three or more transverse septa as seen in *Curvularia* species.

Phylogenetic systematic: Evolutionary tree.

Phylogenetic: In biological sciences, phylogenetics is the study of evolutionary relatedness among various groups of organisms *e.g.* species, populations {Gk., *phyle* = tribe, race and *genetikos* = related to birth, from genesis = birth}. This is also called as phylogenetic systematics or cladistics. Phylogenetics treats each species as group of lineage-connected individuals.

Phylogeny: Evolutionary history of an organism or group.

Phylum (phyla): A taxon above Class and below Kingdom. In plants, is equivalent is **Division**.

Pleomorphic: Fungal species, which has an ability to form more than one asexual reproductive morphology in culture as seen in *Fonsecaea pedrosoi.*

Polymorphic: Fungal species use more than one type of conidiation process.

Polyphyletic: Descended or derived from more than one ancestral stock or source.

Poroconidia: Conidia formed apically through pore as seen in *Curvularia* species.

Probe: Specific nucleic acid sequence {known} used to detect complimentary sequence in an unknown fungus.

Propagule: A reproductive unit or any part of an organism that is capable of propagation.

Proteinaceous: Protein-like.

Pseudo-: Prefix meaning false indicating deceptive resemblance to the following element of the word.

Pseudofungal: Mycelial and unicellular organisms that are not classified in the kingdom Fungi like *Pythium, Rhinosporidium* and *Prototheca* species but cause infections with clinical and histopathological features resembling those produced by fungi. These are also called as **parafungal** organisms.

Pseudohypha: Fragile chain of cells usually yeasts which have arisen by budding and have elongated without detaching from adjacent cells with morphological characteristics intermediate between chain of yeast cells and hyphae.

Pseudomycelium: Mass of pseudohyphae.

Pusillus: Tiny

Pycnidiospore: Spore/conidia formed within pycnidium.

Pycnidium (pycnidia): Spherical to flask-shaped enclosed structure with an apical opening {ostiole} inside which conidia are formed *i.e.* Coelomycetes [Cf. Acervulus].

Pyriform: Pear- or teardrop-shaped.

Rachis (rachids): Apical elongate part of a cell bearing conidia.

Racquet/racket hyphae: Hyphae having series of cells that are swollen at one end.

Rhexolytic cleavage: Conidial release wherein an intermediate cell is disrupted and lysed leaving behind wall remnants *e.g.* arthroconidium of *Coccidioides* species.

Rhizoid: Root-like structure of branched hyphae, which grows into substrate, anchoring the mycelium as seen in Mucorales {Gk., *rhiza* = root, *oeides* = like}.

Ringworm: Layman term used for dermatophyte skin infections due to ancient belief that these were caused by worm-like organisms. Moreover, derived from the circinate or circular form of the lesions.

Rugose: Wrinkled or folded.

Saprobe: An organism that derives its nutrients by absorption of soluble organic compounds from non-living and decaying organic material, plant or animal.

Sapronosis: An infection acquired from organisms of environment *i.e.* soil, where agents live and reproduce.

Saprophyte: An organism that feeds on dead organic matter especially fungus or bacterium.

Saprotrophic: An organism that feeds on dead organic matter. This term is coined to replace 'saprophyte' in which suffix *'phyte'* gives connotation of plant and now fungi have their own Kingdom and are not being considered as plants.

Sclerotic body: Histopathological appearance of phaeoid fungi seen in chromoblastomycosis consists of dark brown, subglobose, multicellular structures that divide by splitting along septal planes by transverse septa and these internal septa are in more than one plane. These structures are also called Medlar bodies, copper-penny or muriform cells.

Sclerotium (sclerotia): Resistant mass of hyphae with protective, hard covering that remains dormant for long period during unfavorable conditions as seen in ergot alkaloid *i.e. Claviceps purpurea*.

Scutulum: Compact mass of hyphae, keratinous material and hair as seen in favus.

Sensu lato: This is a Latin phrase, used after a binomial species name, often abbreviated as *s.l.*, to indicate a species complex represented by that species. Its literal meaning is *'in the broad sense'*.

Sensu stricto: This Latin phrase is added after a binomial species name to mean the taxon is being used in the sense of the original author or without taxa, which may otherwise be associated with it. Its literal meaning is *'in the narrow sense'* and is abbreviated as *s.s.*

Septum (septa): Hyphal cross-wall, typically perforated.

Seta (setae): Hairs or whiskers in spore-bearing structures of certain fungi {as seen in *Pyrenochaeta* species}, which are sterile.

Sexual reproduction: Reproductive process in which haploid cells of compatible strains of fungal species mate through process of plasmogamy, karyogamy and meiosis.

Shield cell: Shield-shaped cells seen in *Cladosporium* species with one disjunctor {attachment point} at one end and two disjunctors at other end.

Species complex: A group of organisms, which are morphologically indistinguishable to each other hence require molecular techniques, if precise species identification is required.

Spherule: Large, round structure containing endospores as seen in *Coccidioides* species and *R.seeberi.*

Spiral hyphae: Vegetative hyphae with coiled or corkscrew-like turns.

Splendore–Hoeppli reaction: Eosinophilic, pseudomycotic structures formed in infected tissues composed of necrotic debris and immunoglobulins as seen in a variety of fungi like *Conidiobolus, Basidiobolus, Sporothrix,* etc.

Sporangiole: Asexual spore that may be single- or multicelled produced on swollen vesicle that form at the end of aerial sporangiophores called sterigmata as seen in two genera of mucormycetes *i.e. Cunninghamella* and *Cokeromyces.*

Sporangiophore: Specialized hyphal branch or stalk bearing sporangium.

Sporangiospore: Asexual structure, characteristics of mucormycetes, which are mitotic, spores *i.e.* mitospores, produced within an enclosed sporangium borne on specialized aerial hyphae or sporangiophore.

Sporangium (sporangia): Closed sac-like structure, which produces asexual spores *i.e.* sporangiospores, internally by series of 'progressive cleavages' [Cf. Ascospore].

Spore: General term used for reproductive propagule in fungi as a result of either sexual or asexual reproduction. In higher fungi it is taken as result of meiotic division in contrast to the term conidium, however, most of the times both are interchangeably used as seen in adiaspore *vs.* adiaconidia.

Sporodochium (sporodochia): Cushion-shaped stroma or mat of hyphae covered with conidiophores as seen in *Fusarium* and *Epicoccum* species.

Sporophore: Spore-bearing hypha.

Sporulation: Production of spores.

Stellate: Star-shaped; arranged in rosettes.

Sterigma (sterigmata): Narrow pointed specialized structure arising from a cell and supporting a spore as seen in basidiomycetes. Phialides are seen in ascomycetes that were previously referred to as sterigmata.

Stolon: Horizontal aerial hypha, which grows along surface of medium, often bearing rhizoids that penetrate into agar and sporangiophores ascend above surface.

Sub: A prefix meaning below, under, somewhat slightly or approximately.

Subglobose: Almost round.

Submerged: Below the surface of culture medium.

Susceptible dose-dependent: The SDD category of AFST implies that susceptibility of an isolate is dependent on the dosing regimen that is used in a patient.

Suspensor: Hyphal element supporting zygospore.

Symbiont: An organism living in state of symbiosis.

Symbiosis (Symbioses): Partnership formed between two living organisms.

Sympodial: Conidial arrangement characterized by continued growth after main axis has produced terminal spore, by development of succession of apices each of which originates below and to one side of previous apex. Fungi with sympodial conidiation have multicellular conidia, either with transverse septa-phragmoconidia or with both transverse and vertical septa -dictyoconidia.

Synanamorph: Any part of life cycle of fungus, which consists of distinctive asexual form {anamorph} when it has multiple anamorphs, each is referred to as synanamorph because each is component of one fungus. *Nattrassia mangiferae* is synanamorph of *Scytalidium dimidiatum.*

Syngamy: Fusion of gametes in fertilization.

Synnema (synnemata): Large, erect reproductive structure borne by some fungi, bearing compact conidiophores, which fuse together to form a strand resembling a stalk of wheat, with conidia at the end or on the edges [Cf. Coremia].

Synonym: Another name for species or taxonomic group.

Talaromycosis: Opportunistic infection caused by an endemic dimorphic fungus - *Talaromyces marneffei.*

Taxon (taxa): One of hierarchical category into which organisms are classified *e.g.* species, order, class, etc.

Taxonomy: Science of systematic classification of organisms.

Teleomorph: Perfect or sexual reproductive stage of fungus [Cf. Anamorph].

Thallus: Vegetative growth of a fungus, which is composed of mycelia.

Tinea: General term used for fungal skin diseases *i.e.* dermatophytosis or ringworm, pityriasis, nigra, etc, literally derived from 'moth' {L., *Tineidae* = clothes moths}; used as prefix to designate various clinical types coinciding with anatomical sites [Cf. *Taenia* species].

Titan cells: Enlarged polyploid yeasts {11-100 μm}, originally seen in cryptococcal mice {lung} infection; shown to reduce phagocytosis thereby enhance virulence.

Trichomycosis: Any disease of the hair caused by a fungus.

Truncate: Conidium with scar flattened off the point of attachment.

Tubercle: Wart-like protruberance.

Tuberculate: Covered with finger-like outgrowths as seen in macroconidia of *Histoplasma species.*

Umbonate: Colony with central button-shaped elevation.

Uncinate: With bent tip to form hook.

Uniseriate: Vesicle bearing single layer of phialides as seen in *Aspergillus fumigatus* [Cf. Biseriate].

Vasiform: Vase-shaped sporangium {*Saksenaea* spp}.

Velvety: Low aerial mycelia with smooth nap as seen in velvet clothes.

Verrucose: Furrowed or convoluted colony.

Verticil: Group of conidiogenous cells radiating out from single point like spokes in wheel.

Verticillate: Whorled branches of sporing structure or sporophore as seen in *Verticillium* species.

Vesicle: Inflated end of conidiophore bearing conidiogenous cells *e.g. Aspergillus* species.

Villose: Covered with slack hair.

Virulence: Quantitative term which means amount/extent of disease caused. It is degree of disease producing capacity of an organism [Cf. Pathogenicity].

Wood's lamp: Apparatus producing filtered ultraviolet rays to demarcate infected sites caused by the fungi.

Woronin bodies: The highly refractive, electron-dense, spherical, membrane-bound proteinaceous structure found in hyphae of Ascomycota and some Deuteromycota. These are present especially in the cytoplasm on either side of septal pores.

Yeast-like: Soft, pasty, smooth colony usually without filamentous growth.

Yeast: Unicellular, spherical to ellipsoidal fungal cell, 3-15 μm which replicates by blastoconidia formation or fission and without pseudohyphae.

Zonate: Having concentric bands or rings of growth or color.

Zoophilic: Found primarily among animals and occasionally infect humans.

Zoospore: Motile {flagellated} asexual spore as seen in some of the pseudofungal organism like *Pythium* species.

Zygospore: Thick-walled resting sexual spore results from fusion of two similar gametangia, meiosis as seen in mucormycetes.

Index

Note: Page numbers followed by *f* refer to Figure and *t* refer to Table.

B

C

G

N

O

P

Q

R

S

U

V

W

X

Y

Z